INORGANIC CHEMISTRY

FOURTH EDITION

D1352505

CATHERINE E. HOUSECROFT
AND ALAN G. SHARPE

PEARSON

Harlow, England • London • New York • Boston • San Francisco • Toronto • Sydney • Auckland • Singapore • Hong Kong
Tokyo • Seoul • Taipei • New Delhi • Cape Town • São Paulo • Mexico City • Madrid • Amsterdam • Munich • Paris • Milan

Pearson Education Limited
Edinburgh Gate
Harlow
Essex CM20 2JE
England

and Associated Companies throughout the world

Visit us on the World Wide Web at:
www.pearson.com/uk

First published 2001
Second edition 2005
Third edition 2008
Fourth edition published 2012

© Pearson Education Limited 2001, 2012

ISBN: 978-0-273-74275-3

British Library Cataloguing-in-Publication Data
A catalogue record for this book is available from the British Library

Library of Congress Cataloging-in-Publication data
Housecroft, Catherine E., 1955-
  Inorganic chemistry / Catherine E. Housecroft and Alan G. Sharpe. - - 4th ed.
      p. cm.
  Includes index.
  ISBN 978-0-273-74275-3 - - ISBN 978-0-273-74276-0 (solution manual) 1.
Chemistry, Inorganic--Textbooks.   I. Sharpe, A. G. II. Title.
  QD151.3.H685 2012
  546- -dc23

                                                          2012001442

10 9 8 7 6 5 4 3 2 1
15 14 13 12

Typeset in 10 /12.5 pt Times by 73
Printed and bound by Grafos S.A., Arte sobre papel, Barcelona, Spain

Summary
of contents

Contents

Supporting resources

Visit **www.pearsoned.co.uk/housecroft** to find valuable online resources

Companion Website for students

- Multiple choice questions to help test your learning
- Rotatable three-dimensional structures taken from the book
- Interactive periodic table

For instructors

- Downloadable Instructor's Manual
- PowerPoint slides of figures and tables from the book
- Rotatable three-dimensional structures taken from the book

Also: The Companion Website provides the following features:

- Search tool to help locate specific items of content
- E-mail results and profile tools to send results of quizzes to instructors
- Online help and support to assist with website usage and troubleshooting

For more information please contact your local Pearson Education sales representative or visit **www.pearsoned.co.uk/housecroft**

Guided tour

Key definitions are highlighted.

Icons indicate a 3D rotatable graphic of the molecule is available on the companion website (see p. ii)

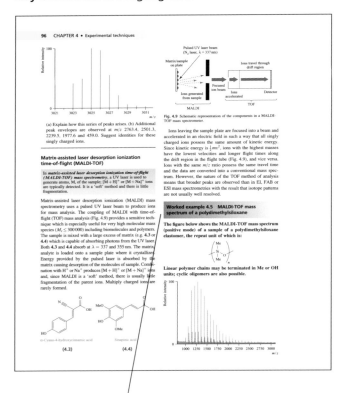

Worked examples are given throughout the text.

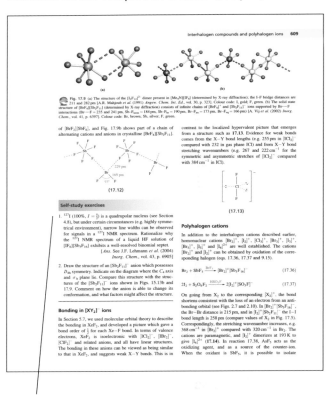

Self-study exercises allow students to test their understanding

Illustrated Topic boxes provide in-depth theoretical background for students

Illustrated Topic boxes reveal how inorganic chemistry is applied to real-life situations

Illustrated Topic boxes relate inorganic chemistry to real-life in the areas of the **Environment**

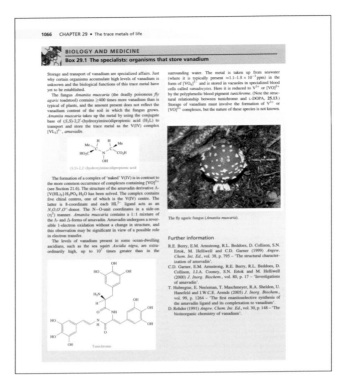

Illustrated Topic boxes relate inorganic chemistry to real-life in the areas of **Biology and Medicine**

End-of-chapter problems, including a set of overview problems, which test the full range of material from each chapter

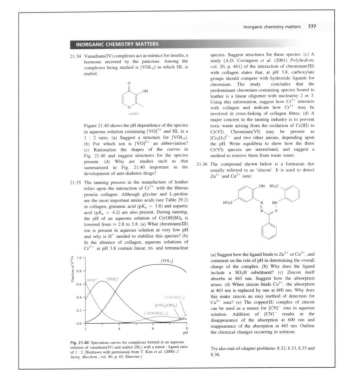

Inorganic Chemistry Matters problems, which are set in a contemporary real-world context

Preface to the fourth edition

Like previous editions of this popular, internationally recognized text, the fourth edition of *Inorganic Chemistry* provides a sound foundation for undergraduate and graduate students in physical inorganic principles, descriptive inorganic chemistry, bioinorganic chemistry and applications, including catalysis, industrial processes and inorganic materials. Maintaining students' attention during discussions of the descriptive chemistry of the elements requires effort on the part of the lecturer. Towards this end, *Inorganic Chemistry* makes extensive use of highly illustrated boxed material to emphasize the roles of inorganic elements and compounds in our everyday lives and in biology, medicine, the environment and industry. The inclusion of up to date literature references allows readers to readily explore the topics further. The eye-catching topic boxes achieve their aim of bringing inorganic chemistry alive. Just as important is boosting the intellectual confidence of students. *Inorganic Chemistry* achieves this through large numbers of worked examples, self-study exercises and end-of-chapter problems. The latter are organized in three sections: problems that focus on specific aspects of a given chapter, overview problems and a set of problems ('inorganic chemistry matters') that link inorganic chemistry to applications and topical research issues. These last problem sets are new to the fourth edition and aim to test students' knowledge in a manner that links the theme of a chapter to the real world.

A major change from previous editions of *Inorganic Chemistry* is the removal of the detailed discussion of nuclear chemistry. This decision was not made lightly, but came after consideration of comments from the review panel set up by the publisher and from discussions with a number of colleagues. A proportion of the material still appears in the text. For example, an introduction to decay chains is now described with the actinoid metals in Chapter 27.

Chapter 4 is new to the fourth edition and pulls together the experimental techniques that previously were scattered through the book in themed boxes. The inclusion of a large number of worked examples, self-study exercises and end-of-chapter problems in this chapter benefits students and teachers alike, and also ensures that the text can support inorganic practical classes in addition to lecture courses. The techniques covered in Chapter 4 include vibrational, electronic, NMR, EPR, Mössbauer and photoelectron spectroscopies and mass spectrometry in addition to purification methods, elemental analysis, thermogravimetric analysis, diffraction methods and computational methods. The practical issues of IR spectroscopy detailed in Chapter 4 complement the group theory approach in Chapter 3.

I am mindful of the ever-changing nomenclature guidelines of the IUPAC. Changes were made on going from the second to the third edition as a result of the 2005 recommendations, and this new edition of *Inorganic Chemistry* incorporates further revisions (e.g. oxido and chlorido in place of oxo and chloro ligands).

Three-dimensional molecular structures in *Inorganic Chemistry* have been drawn using atomic coordinates accessed from the Cambridge Crystallographic Data Base and implemented through the ETH in Zürich, or from the Protein Data Bank (http://www/rcsb.org/pdb).

Accompanying this text is a *Solutions Manual* written by Catherine E. Housecroft, and there is an accompanying Website with multiple choice questions and rotatable structures based on the graphics in the hard-copy text. The site can be accessed from www.pearsoned.co.uk/housecroft.

It is always a pleasure to receive emails from those who read and use *Inorganic Chemistry*. It is their feedback that helps to shape the next edition. On going from the third to fourth editions, I would particularly like to acknowledge the following colleagues for their suggestions: Professor Enzo Alessio, Professor Gareth Eaton, Dr Evan Bieske, Dr Mark Foreman and Dr Jenny Burnham. I am very grateful for the time that my colleagues have spent reading and commenting upon specific sections of text: Dr Henk Bolink (solid-state devices), Dr Cornelia Palivan (EPR spectroscopy), Dr Markus Neuburger (diffraction methods), Professor Helmut Sigel (equilibria and stability constants) and Professor Jan Reedijk (IUPAC nomenclature). The publishing team at Pearson ensure that I keep to schedule and for the fourth edition, particular thanks go to Kevin Ancient, Wendy Baskett, Sarah Beanland, Melanie Beard, Patrick Bond, Rufus Curnow, Mary Lince, Darren Prentice and Ros Woodward.

Working on the fourth edition has been a very different experience for me compared to the previous editions of the book. Dr Alan Sharpe passed away a few months after the publication of the third edition. Although in his eighties when we worked on the last edition, Alan's enthusiasm for inorganic chemistry remained undiminished. He was not one for computers, and his contributions and corrections came to me in longhand by regular mail from the UK to Switzerland. I have missed his thought-provoking letters and comments, and I dedicate the new edition of the book to his memory.

No writing project is complete without input from my husband, Edwin Constable. His critical evaluation of text and problems is never ending, but always invaluable. The writing team in our home is completed by Rocco and Rya whose boundless energy and feline mischief seem to outweigh their eagerness to learn any chemistry ... perhaps with time.

Catherine E. Housecroft
Basel
July 2011

In the 3-dimensional structures, unless otherwise stated, the following colour coding is used: C, grey; H, white; O, red; N, blue; F and Cl, green; S, yellow; P, orange; B, blue.

Acknowledgements

We are grateful to the following for permission to reproduce copyright material:

Figures
Figure 4.13 reprinted with permission from Infrared Spectra of Metallic Complexes. IV. Comparison of the Infrared Spectra of Unidentate and Bidentate Metallic Complexes, *Journal of the American Chemical Society*, volume 79, pp. 4904–8 (Nakamoto, K. *et al.* 1957). Copyright 1957 American Chemical Society; Figure 4.34 reprinted from *Encyclopedia of Spectroscopy and Spectrometry*, Hargittai, I. *et al.*, Electron diffraction theory and methods, pp. 461–465. Copyright 2009, with permission from Elsevier; Figure 21.40 reprinted from *Journal of Inorganic Biochemistry*, volume 80, Kiss, T. *et al.*, Speciation of insulin-mimetic (VO(IV)-containing drugs in blood serum, pp. 65–73. Copyright 2000, with permission from Elsevier; Figure 22.16 reprinted with permission from Where Is the Limit of Highly Fluorinated High-Oxidation-State Osmium Species?, *Inorganic Chemistry*, volume 45, 10497, (Riedel, S. and Kaupp, M. 2006). Copyright 2006 American Chemical Society; Figure 26.1 adapted from Mechanistic Studies of Metal Aqua Ions: A Semi-Historical Perspective, *Helvetica Chimica Acta*, volume 88 (Lincoln, S.F. 2005), Figure 1. Copyright © 2005 Verlag Helvetica Chimica Acta AG, Zurich, Switzerland, with permission from John Wiley & Sons; Figure 29.8 reprinted from *Free Radical Biology and Medicine*, volume 36, Kim-Shapiro, D.B., Hemoglobin–nitric acid cooperativity: is NO the third respiratory ligand?, p. 402. Copyright 2004, with permission from Elsevier.

Photographs
(Key: b-bottom; c-centre; l-left; r-right; t-top)

5 Science Photo Library Ltd: Dept of Physics, Imperial College; 30 Science Photo Library Ltd: Dept of Physics, Imperial College; 89 © Edwin Constable; 99 Alamy Images: Mikael Karlsson; 102 Science Photo Library Ltd; 103 Science Photo Library Ltd; 115 Pearson Education Ltd: Corbis; 121 C.E. Housecroft; 123 M; Neuburger; 126 Science Photo Library Ltd; 179 Science Photo Library Ltd; 181 Alamy Images: Richard Handley; 182 Corbis: Ted Soqui; 186 Science Photo Library Ltd: Maximilian Stock Ltd; 242 © Edwin Constable; 244 Alamy Images: Andrew Lambert / LGPL; 250 Emma L; Dunphy; 254 Alamy Images: Ashley Cooper pics; 284 Science Photo Library Ltd: NASA; 286 © Edwin Constable; 293 Science Photo Library Ltd: Maximilian Stock Ltd; 302 Science Photo Library Ltd: David A; Hardy; 304 Alamy Images: Caro; 318 Alamy Images: Iain Masterton; 326 © Edwin Constable; 331 DK Images: Dr Donald Sullivan / National Institute of Standards and Technology; 332 Alamy Images: Nigel Reed QED Images; 337

Science Photo Library Ltd: James Holmes, Hays Chemicals; 348 © Edwin Constable; 349 Science Photo Library Ltd; 353 Alamy Images: Paul White – UK Industries; 355 Science Photo Library Ltd: Astrid 358 Corbis: Vincon / Klein / plainpicture; 360 Alamy Images: Jeff Morgan 07; 363 Corbis: Ted Spiegel; 375 NASA: Greatest Images of NASA (NASA-HQ-GRIN); 396 Science Photo Library Ltd: David Parker; 428 Science Photo Library Ltd: Paul Rapson; 430 © Edwin Constable; 431 Photolibrary;com: Phil Carrick; 455 Science Photo Library Ltd: NASA; 465 Science Photo Library Ltd: Steve Gschmeissner; 466 Science Photo Library Ltd: Pascal Goetgheluck; 467 Science Photo Library Ltd: Power and Syred; 469 Science Photo Library Ltd: Getmapping plc; 475 Corbis: Nancy Kaszerman/ZUMA; 478 C.E. Housecroft; 485 © Edwin Constable; 497 Alamy Images: AGStockUSA; 504 DK Images: Kyocera Corporation; 519 Corbis: Reuters; 523 Science Photo Library Ltd: Charles D; Winters; 531 DK Images: Bill Tarpenning / U;S; Department of Agriculture; 548 Science Photo Library Ltd: C;S Langlois, Publiphoto Diffusion; 563 Alamy Images: Larry Lilac; 574 Corbis: Paul Almasy; 576 USGS: Cascades Volcano Observatory/J; Vallance; 579 Alamy Images: Bon Appetit; 591 © Edwin Constable; 595 Science Photo Library Ltd: Scott Camazine; 598 Corbis: Ricki Rosen; 626 Science Photo Library Ltd: Philippe Plailly/Eurelios; 627 Science Photo Library Ltd: David A; Hardy, Futures: 50 Years in Space; 698 Science Photo Library Ltd: Hank Morgan; 719 © Edwin Constable; 720 Science Photo Library Ltd; 723 Science Photo Library Ltd; 739 DK Images: Tom Bochsler/PH College; 751 Corbis: Micro Discovery; 755 Alamy Images: INTERFOTO; 758 DK Images: Judith Miller/ Sloan's; 780 Corbis: Document General Motors/Reuter R; 781 Corbis: Laszlo Balogh/ Reuters; 782 Science Photo Library Ltd: Scott Camazine; 795 Alamy Images: Pictorium; 808 Science Photo Library Ltd: David Parker; 835 Science Photo Library Ltd: Eye of Science; 860 Alamy Images: Stock Connection; 925 Science Photo Library Ltd: Hattie Young; 929 Science Photo Library Ltd: Harris Barnes Jr/Agstockusa; 960 Image originally created by IBM Corporation; 976 © Edwin Constable; 1005 Science Photo Library Ltd: Lawrence Livermore National Laboratory; 1038 Peter Visser, OLED-LIGHT-ING;com; 1041 Alamy Images: Phil Degginger; 1045 Science Photo Library Ltd: Adam Hart-Davis; 1053 Corbis: ULI DECK / epa; 1055 Science Photo Library Ltd: Rosenfeld Images Ltd; 1057 Science Photo Library Ltd: James King-Holmes; 1059 Dr Amina Wirth-Heller; 1060 Science Photo Library Ltd: Delft University of Technology; 1066 © Edwin Constable; 1080 Science Photo Library Ltd; Applications topic box header image © Edwin Constable.

Cover images: *Front:* Gary Thompson

In some instances we have been unable to trace the owners of copyright material, and we would appreciate any information that would enable us to do so;

1 Basic concepts: atoms

1.1 Introduction

Inorganic chemistry: it is not an isolated branch of chemistry

If organic chemistry is considered to be the 'chemistry of carbon', then inorganic chemistry is the chemistry of all elements except carbon. In its broadest sense, this is true, but of course there are overlaps between branches of chemistry. A topical example is the chemistry of the *fullerenes* (see Section 14.4) including C_{60} (see Fig. 14.5) and C_{70}; this was the subject of the award of the 1996 Nobel Prize in Chemistry to Professors Sir Harry Kroto, Richard Smalley and Robert Curl. An understanding of such molecules, carbon nanotubes and graphene sheets (see Sections 28.8 and 28.9) involves studies by organic, inorganic and physical chemists, physicists and materials scientists.

Inorganic chemistry is not simply the study of elements and compounds; it is also the study of physical principles. For example, in order to understand why some compounds are soluble in a given solvent and others are not, we apply laws of thermodynamics. If our aim is to propose details of a reaction mechanism, then a knowledge of reaction kinetics is needed. Overlap between physical and inorganic chemistry is also significant in the study of molecular structure. In the solid state, X-ray diffraction methods are routinely used to obtain pictures of the spatial arrangements of atoms in a molecule or molecular ion. To interpret the behaviour of molecules in solution, we use physical techniques such as nuclear magnetic resonance (NMR) spectroscopy; the equivalence or not of particular nuclei on a spectroscopic timescale may indicate whether a molecule is static or undergoing a dynamic process. The application

of a wide range of physical techniques in inorganic chemistry is the topic of Chapter 4.

The aims of Chapters 1 and 2

In Chapters 1 and 2, we outline some concepts fundamental to an understanding of inorganic chemistry. We have assumed that readers are to some extent familiar with most of these concepts and our aim is to give a point of reference for review purposes.

1.2 Fundamental particles of an atom

An ***atom*** is the smallest unit quantity of an element that is capable of existence, either alone or in chemical combination with other atoms of the same or another element. The fundamental particles of which atoms are composed are the ***proton***, ***electron*** and ***neutron***.

A neutron and a proton have approximately the same mass and, relative to these, an electron has negligible mass (Table 1.1). The charge on a proton is positive and of equal magnitude, but opposite sign, to that on a negatively charged electron. A neutron has no charge. In an atom of any element, there are equal numbers of protons and electrons and so an atom is neutral. The *nucleus* of an atom consists of protons and (with the exception of *protium*, see Section 10.3) neutrons, and is positively charged; the nucleus of protium consists of a single proton. The electrons occupy a region of space around the nucleus. Nearly all the mass of an atom is concentrated in the nucleus, but the volume of the nucleus is only a tiny fraction of that of the atom; the radius of the nucleus is

Table 1.1 Properties of the proton, electron and neutron.

	Proton	Electron	Neutron
Charge / C	$+1.602 \times 10^{-19}$	-1.602×10^{-19}	0
Charge number (relative charge)	1	-1	0
Rest mass / kg	1.673×10^{-27}	9.109×10^{-31}	1.675×10^{-27}
Relative mass	1837	1	1839

about 10^{-15} m while the atom itself is about 10^{5} times larger than this. It follows that the density of the nucleus is enormous, more than 10^{12} times that of the metal Pb.

Although chemists tend to consider the electron, proton and neutron as the fundamental (or elementary) particles of an atom, particle physicists deal with yet smaller particles.

1.3 Atomic number, mass number and isotopes

Nuclides, atomic number and mass number

A *nuclide* is a particular type of atom and possesses a characteristic *atomic number*, Z, which is equal to the number of protons in the nucleus. Because the atom is electrically neutral, Z also equals the number of electrons. The *mass number*, A, of a nuclide is the number of protons *and* neutrons in the nucleus. A shorthand method of showing the atomic number and mass number of a nuclide along with its symbol, E, is:

Mass number $\longrightarrow \ ^{A}_{Z}E \longleftarrow$ Element symbol   e.g. $^{20}_{10}Ne$
Atomic number $\longrightarrow$

Atomic number $= Z =$ number of protons in the nucleus $=$ number of electrons

Mass number $= A =$ number of protons $+$ number of neutrons

Number of neutrons $= A - Z$

Relative atomic mass

Since the electrons are of minute mass, the mass of an atom essentially depends upon the number of protons and neutrons in the nucleus. As Table 1.1 shows, the mass of a single atom is a very small, non-integral number, and for convenience a system of *relative atomic masses* is adopted. The atomic mass unit is defined as 1/12th of the mass of a $^{12}_{6}C$ atom so that it has the value 1.660×10^{-27} kg. *Relative atomic masses* (A_r) are therefore all stated relative to

$^{12}_{6}C = 12.0000$. The masses of the proton and neutron can be considered to be ≈ 1 u where u is the *atomic mass unit* ($1\,u \approx 1.660 \times 10^{-27}$ kg).

Isotopes

Nuclides of the same element possess the same number of protons and electrons but may have different mass numbers. The number of protons and electrons defines the element but the number of neutrons may vary. Nuclides of a particular element that differ in the number of neutrons and, therefore, their mass number, are called *isotopes* (see Appendix 5). Isotopes of some elements occur naturally while others may be produced artificially.

Elements that occur naturally with only one nuclide are *monotopic* and include phosphorus, $^{31}_{15}P$, and fluorine, $^{19}_{9}F$. Elements that exist as mixtures of isotopes include C ($^{12}_{6}C$ and $^{13}_{6}C$) and O ($^{16}_{8}O$, $^{17}_{8}O$ and $^{18}_{8}O$). Since the atomic number is constant for a given element, isotopes are often distinguished only by stating the atomic masses, e.g. ^{12}C and ^{13}C.

Worked example 1.1   Relative atomic mass

Calculate the value of A_r for naturally occurring chlorine if the distribution of isotopes is 75.77% $^{35}_{17}Cl$ and 24.23% $^{37}_{17}Cl$. Accurate masses for ^{35}Cl and ^{37}Cl are 34.97 and 36.97.

The relative atomic mass of chlorine is the weighted mean of the mass numbers of the two isotopes:

Relative atomic mass,

$$A_r = \left(\frac{75.77}{100} \times 34.97 \right) + \left(\frac{24.23}{100} \times 36.97 \right) = 35.45$$

Self-study exercises

1. If A_r for Cl is 35.45, what is the ratio of $^{35}Cl : ^{37}Cl$ present in a sample of Cl atoms containing naturally occurring Cl?

[*Ans.* 3.17 : 1]

THEORY

Box 1.1 Isotopes and allotropes

Do not confuse *isotope* and *allotrope*! Sulfur exhibits both isotopes and allotropes. Isotopes of sulfur (with percentage naturally occurring abundances) are $^{32}_{16}S$ (95.02%), $^{33}_{16}S$ (0.75%), $^{34}_{16}S$ (4.21%), $^{36}_{16}S$ (0.02%).

Allotropes of an element are different structural modifications of that element. Allotropes of sulfur include cyclic structures, e.g. S_6 (see below) and S_8 (Fig. 1.1c), and S_x-chains of various lengths (poly*catena*sulfur).

Further examples of isotopes and allotropes appear throughout the book.

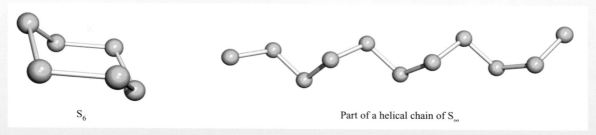

S_6

Part of a helical chain of S_∞

2. Calculate the value of A_r for naturally occurring Cu if the distribution of isotopes is 69.2% ^{63}Cu and 30.8% ^{65}Cu; accurate masses are 62.93 and 64.93. [*Ans.* 63.5]

3. Why in question 2 is it adequate to write ^{63}Cu rather than $^{63}_{29}Cu$?

4. Calculate A_r for naturally occurring Mg if the isotope distribution is 78.99% ^{24}Mg, 10.00% ^{25}Mg and 11.01% ^{26}Mg; accurate masses are 23.99, 24.99 and 25.98. [*Ans.* 24.31]

Isotopes can be separated by *mass spectrometry* and Fig. 1.1a shows the isotopic distribution in naturally occurring Ru. Compare this plot (in which the most abundant isotope is set to 100) with the values listed in Appendix 5. Figure 1.1b shows a mass spectrometric trace for molecular S_8, the structure of which is shown in Fig. 1.1c; five peaks are observed due to combinations of the isotopes of sulfur. (See end-of-chapter problem 1.5.)

Isotopes of an element have the same atomic number, Z, but different atomic masses.

1.4 Successes in early quantum theory

We saw in Section 1.2 that electrons in an atom occupy a region of space around the nucleus. The importance of electrons in determining the properties of atoms, ions and

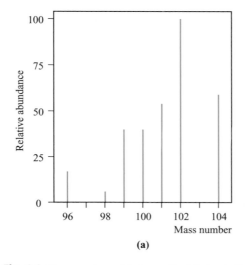

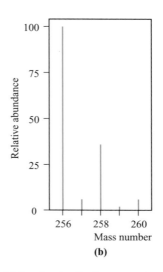

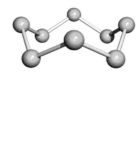

(a)   (b)   (c)

Fig. 1.1 Mass spectrometric traces for (a) atomic Ru and (b) molecular S_8; the mass : charge ratio is m/z and in these traces $z = 1$. (c) The molecular structure of S_8.

molecules, including the bonding between or within them, means that we must have an understanding of the electronic structures of each species. No adequate discussion of electronic structure is possible without reference to *quantum theory* and *wave mechanics*. In this and the next few sections, we review some crucial concepts. The treatment is mainly qualitative, and for greater detail and more rigorous derivations of mathematical relationships, the references at the end of Chapter 1 should be consulted.

The development of quantum theory took place in two stages. In older theories (1900–1925), the electron was treated as a particle, and the achievements of greatest significance to inorganic chemistry were the interpretation of atomic spectra and assignment of electronic configurations. In more recent models, the electron is treated as a wave (hence the name *wave mechanics*) and the main successes in chemistry are the elucidation of the basis of stereochemistry and methods for calculating the properties of molecules (exact *only* for species involving light atoms).

Since all the results obtained by using the older quantum theory may also be obtained from wave mechanics, it may seem unnecessary to refer to the former; indeed, sophisticated treatments of theoretical chemistry seldom do. However, most chemists often find it easier and more convenient to consider the electron as a particle rather than a wave.

Some important successes of classical quantum theory

Historical discussions of the developments of quantum theory are dealt with adequately elsewhere, and so we focus only on some key points of *classical* quantum theory (in which the electron is considered to be a particle).

At low temperatures, the radiation emitted by a hot body is mainly of low energy and occurs in the infrared, but as the temperature increases, the radiation becomes successively dull red, bright red and white. Attempts to account for this observation failed until, in 1901, Planck suggested that energy could be absorbed or emitted only in *quanta* of magnitude ΔE related to the frequency of the radiation, ν, by eq. 1.1. The proportionality constant is h, the Planck constant ($h = 6.626 \times 10^{-34}$ J s).

$$\Delta E = h\nu \qquad \text{Units: } E \text{ in J; } \nu \text{ in s}^{-1} \text{ or Hz} \qquad (1.1)$$

$$c = \lambda\nu \qquad \text{Units: } \lambda \text{ in m; } \nu \text{ in s}^{-1} \text{ or Hz} \qquad (1.2)$$

The hertz, Hz, is the SI unit of frequency.

The frequency of radiation is related to the wavelength, λ, by eq. 1.2, in which c is the speed of light in a vacuum ($c = 2.998 \times 10^8$ m s^{-1}). Therefore, eq. 1.1 can be rewritten in the form of eq. 1.3. This relates the energy of radiation to its wavelength.

$$\Delta E = \frac{hc}{\lambda} \qquad (1.3)$$

On the basis of this relationship, Planck derived a relative intensity/wavelength/temperature relationship which was in good agreement with experimental data. This derivation is not straightforward and we shall not reproduce it here.

When energy is provided (e.g. as heat or light) to an atom or other species, one or more electrons may be promoted from a ground state level to a higher energy state. This excited state is transient and the electron falls back to the ground state. This produces an *emission spectrum*.

One of the most important applications of early quantum theory was the interpretation of the atomic spectrum of hydrogen on the basis of the Rutherford–Bohr model of the atom. When an electric discharge is passed through a sample of dihydrogen, the H_2 molecules dissociate into atoms, and the electron in a particular *excited* H atom may be *promoted* to one of many high energy levels. These states are transient and the electron falls back to a lower energy state, emitting energy as it does so. The consequence is the observation of *spectral lines* in the emission spectrum of hydrogen. The spectrum (a part of which is shown in Fig. 1.2) consists of groups of discrete lines corresponding to electronic transitions, each of *discrete energy*. In 1885, Balmer pointed out that the wavelengths of the spectral lines observed in the visible region of the atomic spectrum of hydrogen obeyed eq. 1.4, in which R is the Rydberg constant for hydrogen, $\bar{\nu}$ is the wavenumber in cm^{-1}, and n is an integer 3, 4, 5... This series of spectral lines is known as the *Balmer series*.

Wavenumber is the reciprocal of wavelength; convenient (non-SI) units are 'reciprocal centimetres', cm^{-1}

$$\bar{\nu} = \frac{1}{\lambda} = R\left(\frac{1}{2^2} - \frac{1}{n^2}\right) \qquad (1.4)$$

$$R = \text{Rydberg constant for hydrogen}$$
$$= 1.097 \times 10^7 \text{ m}^{-1} = 1.097 \times 10^5 \text{ cm}^{-1}$$

Other series of spectral lines occur in the ultraviolet (Lyman series) and infrared (Paschen, Brackett and Pfund series). All lines in all the series obey the general expression given in eq. 1.5 where $n' > n$. For the Lyman series, $n = 1$, for the Balmer series, $n = 2$, and for the Paschen, Brackett and Pfund series, $n = 3$, 4 and 5 respectively. Figure 1.3 shows some of the allowed transitions of the Lyman and Balmer series in the emission spectrum of atomic H. Note the use of the word *allowed*; the transitions must obey *selection rules*, to which we return in Section 20.7.

$$\bar{\nu} = \frac{1}{\lambda} = R\left(\frac{1}{n^2} - \frac{1}{n'^2}\right) \qquad (1.5)$$

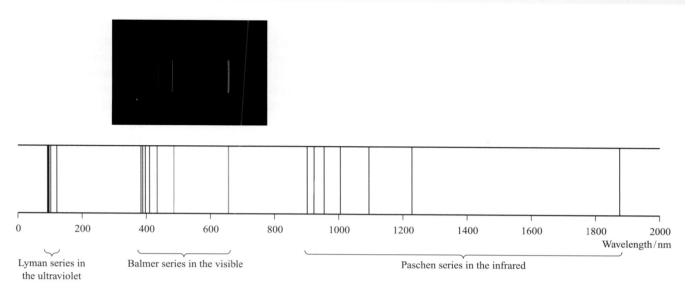

Fig. 1.2 A schematic representation of part of the emission spectrum of hydrogen showing the Lyman, Balmer and Paschen series of emission lines. The photograph shows the predominant lines in the observed, visible part of the spectrum of hydrogen which appear at 656.3 (red), 486.1 (cyan) and 434.0 nm (blue). Other fainter lines are not visible in this photograph.

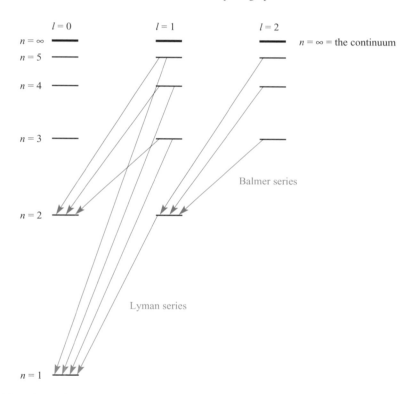

Fig. 1.3 Some of the transitions that make up the Lyman and Balmer series in the emission spectrum of atomic hydrogen.

Bohr's theory of the atomic spectrum of hydrogen

In 1913, Niels Bohr combined elements of quantum theory and classical physics in a treatment of the hydrogen atom. He stated two postulates for an electron in an atom:

- *Stationary states* exist in which the energy of the electron is constant; such states are characterized by *circular orbits* about the nucleus in which the electron has an angular momentum mvr given by eq. 1.6. The integer, n, is the *principal quantum number*.

$$mvr = n\left(\frac{h}{2\pi}\right) \quad (1.6)$$

where m = mass of electron; v = velocity of electron; r = radius of the orbit; h = the Planck constant; $h/2\pi$ may be written as $\hbar$.

• Energy is absorbed or emitted only when an electron moves from one stationary state to another and the energy change is given by eq. 1.7 where n_1 and n_2 are the principal quantum numbers referring to the energy levels E_{n_1} and E_{n_2} respectively.

$$\Delta E = E_{n_2} - E_{n_1} = h\nu \qquad (1.7)$$

If we apply the Bohr model to the H atom, the radius of each allowed circular orbit can be determined from eq. 1.8. The origin of this expression lies in the centrifugal force acting on the electron as it moves in its circular orbit. For the orbit to be maintained, the centrifugal force must equal the force of attraction between the negatively charged electron and the positively charged nucleus.

$$r_n = \frac{\varepsilon_0 h^2 n^2}{\pi m_e e^2} \qquad (1.8)$$

where ε_0 = permittivity of a vacuum
$$= 8.854 \times 10^{-12}\,\mathrm{F\,m^{-1}}$$
$$h = \text{Planck constant} = 6.626 \times 10^{-34}\,\mathrm{J\,s}$$
$$n = 1, 2, 3 \ldots \text{describing a given orbit}$$
$$m_e = \text{electron rest mass} = 9.109 \times 10^{-31}\,\mathrm{kg}$$
$$e = \text{charge on an electron (elementary charge)}$$
$$= 1.602 \times 10^{-19}\,\mathrm{C}$$

From eq. 1.8, substitution of $n = 1$ gives a radius for the first orbit of the H atom of 5.293×10^{-11} m, or 52.93 pm. This value is called the *Bohr radius* of the H atom and is given the symbol a_0.

An increase in the principal quantum number from $n = 1$ to $n = \infty$ has a special significance. It corresponds to the ionization of the atom (eq. 1.9) and the ionization energy, *IE*, can be determined by combining eqs. 1.5 and 1.7, as shown in eq. 1.10. Values of *IE*s are quoted *per mole of atoms*.

One mole of a substance contains the **Avogadro number**, *L*, of particles:

$$L = 6.022 \times 10^{23}\,\mathrm{mol^{-1}}$$

$$\mathrm{H(g)} \longrightarrow \mathrm{H^+(g)} + \mathrm{e^-} \qquad (1.9)$$

$$IE = E_\infty - E_1 = \frac{hc}{\lambda} = hcR\left(\frac{1}{1^2} - \frac{1}{\infty^2}\right) \qquad (1.10)$$

$$= 2.179 \times 10^{-18}\,\mathrm{J}$$
$$= 2.179 \times 10^{-18} \times 6.022 \times 10^{23}\,\mathrm{J\,mol^{-1}}$$
$$= 1.312 \times 10^{6}\,\mathrm{J\,mol^{-1}}$$
$$= 1312\,\mathrm{kJ\,mol^{-1}}$$

Although the SI unit of energy is the joule, ionization energies are often expressed in electron volts (eV)

($1\,\mathrm{eV} = 96.4853 \approx 96.5\,\mathrm{kJ\,mol^{-1}}$). Therefore, the ionization energy of hydrogen can also be given as 13.60 eV.

Impressive as the success of the Bohr model was when applied to the H atom, extensive modifications were required to cope with species containing more than one electron. We shall not pursue this topic further here.

1.5 An introduction to wave mechanics

The wave-nature of electrons

The quantum theory of radiation introduced by Max Planck and Albert Einstein implies a particle theory of light, in addition to the wave theory of light required by the phenomena of interference and diffraction. In 1924, Louis de Broglie argued that if light were composed of particles and yet showed wave-like properties, the same should be true of electrons and other particles. This phenomenon is referred to as *wave–particle duality*. The de Broglie relationship (eq. 1.11) combines the concepts of classical mechanics with the idea of wave-like properties by showing that a particle with momentum mv (m = mass and v = velocity of the particle) possesses an associated wave of wavelength λ.

$$\lambda = \frac{h}{mv} \qquad \text{where } h \text{ is the Planck constant} \qquad (1.11)$$

An important physical observation which is a consequence of the de Broglie relationship is that electrons accelerated to a velocity of $6 \times 10^6\,\mathrm{m\,s^{-1}}$ (by a potential of 100 V) have an associated wavelength of ≈ 120 pm and such electrons are diffracted as they pass through a crystal. This phenomenon is the basis of electron diffraction techniques used to determine structures of chemical compounds (see Section 4.10).

The uncertainty principle

If an electron has wave-like properties, there is an important and difficult consequence: it becomes impossible to know exactly both the momentum and position of the electron *at the same instant in time*. This is a statement of Heisenberg's *uncertainty principle*. In order to get around this problem, rather than trying to define its exact position and momentum, we use the *probability of finding the electron* in a given volume of space. The probability of finding an electron at a given point in space is determined from the function ψ^2 where ψ is a mathematical function called the *wavefunction* which describes the behaviour of an electron-wave.

The probability of finding an electron at a given point in space is determined from the function ψ^2 where ψ is the **wavefunction**.

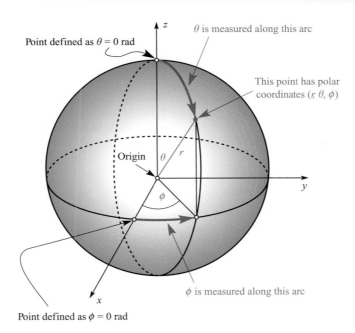

Point defined as $\theta = 0$ rad

θ is measured along this arc

This point has polar coordinates (r, θ, ϕ)

Origin

θ is measured along this arc

ϕ is measured along this arc

Point defined as $\phi = 0$ rad

Fig. 1.4 Definition of the polar coordinates (r, θ, ϕ) for a point shown here in pink; r is the radial coordinate and θ and ϕ are angular coordinates. θ and ϕ are measured in radians (rad). Cartesian axes (x, y and z) are also shown.

The Schrödinger wave equation

Information about the wavefunction is obtained from the Schrödinger wave equation, which can be set up and solved either exactly or approximately. The Schrödinger equation can be solved *exactly only* for a species containing a nucleus and *only one* electron (e.g. ^1H, $^4_2\text{He}^+$), i.e. a *hydrogen-like* system.

A *hydrogen-like atom* or *hydrogen-like ion* contains a nucleus and only one electron.

The Schrödinger wave equation may be represented in several forms and in Box 1.2 we examine its application to the motion of a particle in a 1-dimensional box. Equation 1.12 gives the form of the Schrödinger wave equation that is appropriate for motion in the x direction.

$$\frac{d^2\psi}{dx^2} + \frac{8\pi^2 m}{h^2}(E - V)\psi = 0 \qquad (1.12)$$

where m = mass

E = total energy

V = potential energy of the particle

Of course, in reality, electrons move in 3-dimensional space and an appropriate form of the Schrödinger wave equation is given in eq. 1.13.

$$\frac{\partial^2\psi}{\partial x^2} + \frac{\partial^2\psi}{\partial y^2} + \frac{\partial^2\psi}{\partial z^2} + \frac{8\pi^2 m}{h^2}(E - V)\psi = 0 \qquad (1.13)$$

Solving this equation will not concern us, although it is useful to note that it is advantageous to work in spherical polar coordinates (Fig. 1.4). When we look at the results obtained from the Schrödinger wave equation, we talk in terms of the *radial and angular parts of the wavefunction*, and this is represented in eq. 1.14 where $R(r)$ and $A(\theta, \phi)$ are radial and angular wavefunctions respectively.[†]

$$\psi_{\text{Cartesian}}(x, y, z) \equiv \psi_{\text{radial}}(r)\psi_{\text{angular}}(\theta, \phi) = R(r)A(\theta, \phi)$$
$$(1.14)$$

Having solved the wave equation, what are the results?

- The wavefunction ψ is a solution of the Schrödinger equation and describes the behaviour of an electron in a region of space called the *atomic orbital*.
- We can find energy values that are associated with particular wavefunctions.
- The quantization of energy levels arises naturally from the Schrödinger equation (see Box 1.2).

A *wavefunction* ψ is a mathematical function that contains detailed information about the behaviour of an electron. An atomic wavefunction ψ consists of a *radial component*, $R(r)$, and an *angular component*, $A(\theta, \phi)$. The region of space defined by a wavefunction is called an *atomic orbital*.

[†] The radial component in eq. 1.14 depends on the quantum numbers n and l, whereas the angular component depends on l and m_l, and the components should really be written as $R_{n,l}(r)$ and $A_{l,m_l}(\theta, \phi)$.

THEORY

Box 1.2 Particle in a box

The following discussion illustrates the so-called *particle in a 1-dimensional box* and illustrates quantization arising from the Schrödinger wave equation.

The *Schrödinger wave equation* for the motion of a particle in one dimension is given by:

$$\frac{d^2\psi}{dx^2} + \frac{8\pi^2 m}{h^2}(E - V)\psi = 0$$

where m is the mass, E is the total energy and V is the potential energy of the particle. The derivation of this equation is considered in the set of exercises following this box. For a given system for which V and m are known, Schrödinger's equation is used to obtain values of E (the *allowed energies of the particle*) and ψ (the *wavefunction*). The wavefunction itself has no physical meaning, but ψ^2 is a probability (see main text) and for this to be the case, ψ must have certain properties:

- ψ must be finite for all values of x;
- ψ can only have one value for any value of x;
- ψ and $\dfrac{d\psi}{dx}$ must vary continuously as x varies.

Now, consider a particle that is undergoing simple-harmonic wave-like motion in one dimension, i.e. we can fix the direction of wave propagation to be along the x axis (the choice of x is arbitrary). Let the motion be further constrained such that the particle cannot go outside the fixed, vertical walls of a box of width a. There is no force acting on the particle *within* the box and so the potential energy, V, is zero. If we take $V = 0$, we are placing limits on x such that $0 \leq x \leq a$, i.e. the particle cannot move outside the box. The only restriction that we place on the total energy E is that it must be positive and cannot be infinite. There is one further restriction that we shall simply state: the *boundary condition* for the particle in the box is that ψ must be zero when $x = 0$ and $x = a$.

Now rewrite the Schrödinger equation for the specific case of the particle in the 1-dimensional box where $V = 0$:

$$\frac{d^2\psi}{dx^2} = -\frac{8\pi^2 mE}{h^2}\psi$$

which may be written in the simpler form:

$$\frac{d^2\psi}{dx^2} = -k^2\psi \quad \text{where} \quad k^2 = \frac{8\pi^2 mE}{h^2}$$

The solution to this (a known general equation) is:

$$\psi = A\sin kx + B\cos kx$$

where A and B are integration constants. When $x = 0$, $\sin kx = 0$ and $\cos kx = 1$; hence, $\psi = B$ when $x = 0$. However, the boundary condition above stated that $\psi = 0$ when $x = 0$, and this is only true if $B = 0$. Also from the boundary condition, we see that $\psi = 0$ when $x = a$, and hence we can rewrite the above equation in the form:

$$\psi = A\sin ka = 0$$

Since the probability, ψ^2, that the particle will be at points between $x = 0$ and $x = a$ cannot be zero (i.e. the particle must be somewhere inside the box), A cannot be zero and the last equation is only valid if:

$$ka = n\pi$$

where $n = 1, 2, 3 \ldots$; n cannot be zero as this would make the probability, ψ^2, zero meaning that the particle would no longer be in the box.

Combining the last two equations gives:

$$\psi = A\sin\frac{n\pi x}{a}$$

and, from earlier:

$$E = \frac{k^2 h^2}{8\pi^2 m} = \frac{n^2 h^2}{8ma^2}$$

where $n = 1, 2, 3 \ldots$; n is the *quantum number* determining the energy of a particle of mass m confined within a 1-dimensional box of width a. So, the limitations placed on the value of ψ have led to *quantized energy levels*, the spacing of which is determined by m and a.

The resultant motion of the particle is described by a series of standing sine waves, three of which are illustrated below. The wavefunction ψ_2 has a wavelength of a, while wavefunctions ψ_1 and ψ_3 possess wavelengths of $\dfrac{a}{2}$ and $\dfrac{3a}{2}$ respectively. Each of the waves in the diagram has an amplitude of zero at the origin (i.e. at the point $a = 0$); points at which $\psi = 0$ are called *nodes*. For a given particle of mass m, the separations of the energy levels vary according to n^2, i.e. the spacings are not equal.

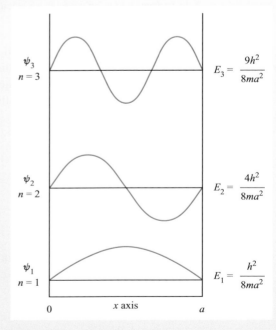

Consider a particle that is undergoing simple-harmonic wave-like motion in one dimension, with the wave propagation along the x axis. The general equation for the wave is:

$$\psi = A \sin \frac{2\pi x}{\lambda}$$

where A is the amplitude of the wave.

1. If $\psi = A \sin \frac{2\pi x}{\lambda}$, find $\frac{d\psi}{dx}$ and hence show that

$$\frac{d^2\psi}{dx^2} = -\frac{4\pi^2}{\lambda^2}\psi$$

2. If the particle in the box is of mass m and moves with velocity v, what is its kinetic energy, KE? Using the de Broglie equation (1.11), write an expression for KE in terms of m, h and λ.

3. The equation you derived in part (2) applies only to a particle moving in a space in which the potential energy, V, is constant, and the particle can be regarded as possessing only kinetic energy, KE. If the potential energy of the particle does vary, the total energy, $E = KE + V$. Using this information and your answers to parts (1) and (2), derive the Schrödinger equation (stated on p. 8) for a particle in a 1-dimensional box.

1.6 Atomic orbitals

The quantum numbers n, l and m_l

An atomic orbital is usually described in terms of three integral *quantum numbers*. We have already encountered the *principal quantum number*, n, in the Bohr model of the hydrogen atom. The principal quantum number is a positive integer with values lying between the limits $1 \leq n \leq \infty$. Allowed values of n arise when the radial part of the wavefunction is solved.

Two more quantum numbers, l and m_l, appear when the angular part of the wavefunction is solved. The quantum number l is called the *orbital quantum number* and has allowed values of $0, 1, 2 \ldots (n - 1)$. The value of l determines the shape of the atomic orbital, and the *orbital angular momentum* of the electron. The value of the *magnetic quantum number*, m_l, gives information about the directionality of an atomic orbital and has integral values between $+l$ and $-l$.

Each atomic orbital may be uniquely labelled by a set of three *quantum numbers*: n, l and m_l.

Given that the principal quantum number, n, is 2, write down the allowed values of l and m_l, and determine the number of atomic orbitals possible for $n = 3$.

For a given value of n, the allowed values of l are $0, 1, 2 \ldots (n - 1)$, and those of m_l are $-l \ldots 0 \ldots +l$.

For $n = 2$, allowed values of $l = 0$ or 1.

For $l = 0$, the allowed value of $m_l = 0$.

For $l = 1$, allowed values of $m_l = -1, 0, +1$

Each set of three quantum numbers defines a particular atomic orbital, and, therefore, for $n = 2$, there are four atomic orbitals with the sets of quantum numbers:

$n = 2$, $l = 0$, $m_l = 0$
$n = 2$, $l = 1$, $m_l = -1$
$n = 2$, $l = 1$, $m_l = 0$
$n = 2$, $l = 1$, $m_l = +1$

1. If m_l has values of $-1, 0, +1$, write down the corresponding value of l.   [*Ans. $l = 1$*]

2. If l has values 0, 1, 2 and 3, deduce the corresponding value of n.   [*Ans. $n = 4$*]

3. For $n = 1$, what are the allowed values of l and m_l?   [*Ans. $l = 0$; $m_l = 0$*]

4. Complete the following sets of quantum numbers: (a) $n = 4, l = 0, m_l = \ldots$; (b) $n = 3, l = 1, m_l = \ldots$ [*Ans. (a) 0; (b) $-1, 0, +1$*]

The distinction among the *types* of atomic orbital arises from their *shapes* and *symmetries*. The four types of atomic orbital most commonly encountered are the s, p, d and f orbitals, and the corresponding values of l are 0, 1, 2 and 3 respectively. Each atomic orbital is labelled with values of n and l, and hence we speak of $1s$, $2s$, $2p$, $3s$, $3p$, $3d$, $4s$, $4p$, $4d$, $4f$ etc. orbitals.

For an s orbital, $l = 0$.   For a p orbital, $l = 1$.
For a d orbital, $l = 2$.   For an f orbital, $l = 3$.

Worked example 1.3   Quantum numbers: types of orbital

Using the rules that govern the values of the quantum numbers *n* and *l*, write down the possible types of atomic orbital for *n* = 1, 2 and 3.

The allowed values of *l* are integers between 0 and $(n-1)$.
For $n = 1$, $l = 0$.
The only atomic orbital for $n = 1$ is the 1*s* orbital.

For $n = 2$, $l = 0$ or 1.
The allowed atomic orbitals for $n = 2$ are the 2*s* and 2*p* orbitals.

For $n = 3$, $l = 0$, 1 or 2.
The allowed atomic orbitals for $n = 3$ are the 3*s*, 3*p* and 3*d* orbitals.

Self-study exercises

1. Write down the possible types of atomic orbital for $n = 4$.   [*Ans.* 4*s*, 4*p*, 4*d*, 4*f*]

2. Which atomic orbital has values of $n = 4$ and $l = 2$?   [*Ans.* 4*d*]

3. Give the three quantum numbers that describe a 2*s* atomic orbital.   [*Ans.* $n = 2$, $l = 0$, $m_l = 0$]

4. Which quantum number distinguishes the 3*s* and 5*s* atomic orbitals?   [*Ans.* *n*]

Degenerate orbitals possess the same energy.

Now consider the consequence on these orbital types of the quantum number m_l. For an *s* orbital, $l = 0$ and m_l can only equal 0. This means that for any value of *n*, there is only one *s* orbital; it is said to be singly degenerate. For a *p* orbital, $l = 1$, and there are three possible m_l values: $+1$, 0, -1. This means that there are three *p* orbitals for a given value of *n* when $n \geq 2$; the set of *p* orbitals is said to be triply or three-fold degenerate. For a *d* orbital, $l = 2$, and there are five possible values of m_l: $+2$, $+1$, 0, -1, -2, meaning that for a given value of *n* $(n \geq 3)$, there are five *d* orbitals; the set is said to be five-fold degenerate. As an exercise, you should show that there are seven *f* orbitals in a degenerate set for a given value of *n* $(n \geq 4)$.

For a given value of *n* $(n \geq 1)$ there is one *s* atomic orbital.
For a given value of *n* $(n \geq 2)$ there are three *p* atomic orbitals.
For a given value of *n* $(n \geq 3)$ there are five *d* atomic orbitals.
For a given value of *n* $(n \geq 4)$ there are seven *f* atomic orbitals.

The radial part of the wavefunction, *R*(*r*)

The mathematical forms of some of the wavefunctions for the H atom are listed in Table 1.2. Figure 1.5 shows plots of the radial parts of the wavefunction, $R(r)$, against distance, *r*, from the nucleus for the 1*s* and 2*s* atomic orbitals of the hydrogen atom, and Fig. 1.6 shows plots of $R(r)$ against *r* for the 2*p*, 3*p*, 4*p* and 3*d* atomic orbitals; the nucleus is at $r = 0$.

From Table 1.2, we see that the radial parts of the wavefunctions decay exponentially as *r* increases, but the decay is slower for $n = 2$ than for $n = 1$. This means that the likelihood of the electron being further from the

Table 1.2   Solutions of the Schrödinger equation for the hydrogen atom which define the 1*s*, 2*s* and 2*p* atomic orbitals. For these forms of the solutions, the distance *r* from the nucleus is measured in atomic units.

Atomic orbital	*n*	*l*	m_l	Radial part of the wavefunction, $R(r)^\dagger$	Angular part of wavefunction, $A(\theta, \phi)$
1*s*	1	0	0	$2e^{-r}$	$\dfrac{1}{2\sqrt{\pi}}$
2*s*	2	0	0	$\dfrac{1}{2\sqrt{2}}(2-r)\,e^{-r/2}$	$\dfrac{1}{2\sqrt{\pi}}$
2*p*$_x$	2	1	+1	$\dfrac{1}{2\sqrt{6}}r\,e^{-r/2}$	$\dfrac{\sqrt{3}(\sin\theta\cos\phi)}{2\sqrt{\pi}}$
2*p*$_z$	2	1	0	$\dfrac{1}{2\sqrt{6}}r\,e^{-r/2}$	$\dfrac{\sqrt{3}(\cos\theta)}{2\sqrt{\pi}}$
2*p*$_y$	2	1	−1	$\dfrac{1}{2\sqrt{6}}r\,e^{-r/2}$	$\dfrac{\sqrt{3}(\sin\theta\sin\phi)}{2\sqrt{\pi}}$

† For the 1*s* atomic orbital, the formula for $R(r)$ is actually $2\left(\frac{Z}{a_0}\right)^{\frac{3}{2}}e^{-Zr/a_0}$ but for the hydrogen atom, $Z = 1$ and $a_0 = 1$ atomic unit. Other functions are similarly simplified.

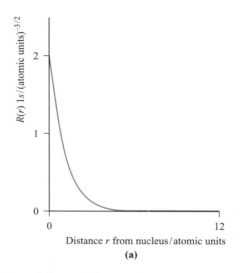

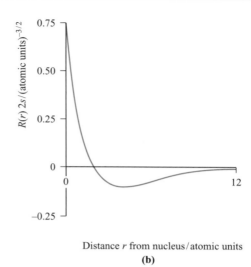

Fig. 1.5 Plots of the radial parts of the wavefunction, $R(r)$, against distance, r, from the nucleus for (a) the $1s$ and (b) the $2s$ atomic orbitals of the hydrogen atom; the nucleus is at $r = 0$. The vertical scales for the two plots are different but the horizontal scales are the same.

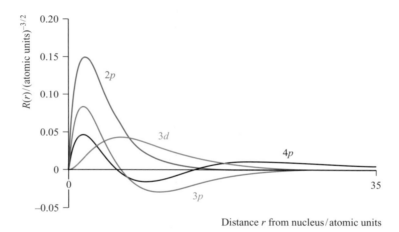

Fig. 1.6 Plots of radial parts of the wavefunction $R(r)$ against r for the $2p$, $3p$, $4p$ and $3d$ atomic orbitals; the nucleus is at $r = 0$.

nucleus increases as n increases. This pattern continues for higher values of n. The exponential decay can be seen clearly in Fig. 1.5a. Several points should be noted from the plots of the radial parts of wavefunctions in Figs. 1.5 and 1.6:

- s atomic orbitals have a finite value of $R(r)$ at the nucleus;
- for all orbitals other than s, $R(r) = 0$ at the nucleus;
- for the $1s$ orbital, $R(r)$ is always positive; for the first orbital of other types (i.e. $2p$, $3d$, $4f$), $R(r)$ is positive everywhere except at the origin;
- for the second orbital of a given type (i.e. $2s$, $3p$, $4d$, $5f$), $R(r)$ may be positive or negative but the wavefunction has only one sign change; the point at which $R(r) = 0$ (not including the origin) is called a radial node;
- for the third orbital of a given type (i.e. $3s$, $4p$, $5d$, $6f$), $R(r)$ has two sign changes, i.e. it possesses two radial nodes.

Radial nodes:

ns orbitals have $(n - 1)$ radial nodes.
np orbitals have $(n - 2)$ radial nodes.
nd orbitals have $(n - 3)$ radial nodes.
nf orbitals have $(n - 4)$ radial nodes.

The radial distribution function, $4\pi r^2 R(r)^2$

Let us now consider how we might represent atomic orbitals in 3-dimensional space. We said earlier that a useful description of an electron in an atom is the *probability of finding the electron* in a given volume of space. The function ψ^2 (see Box 1.3) is proportional to the *probability density* of the electron at a point in space. By considering values of ψ^2 at points around the nucleus, we can define a *surface boundary* which encloses the volume of space in which the electron will spend, say, 95% of its time. This effectively gives us a physical representation of the atomic orbital,

THEORY

Box 1.3 Notation for ψ^2 and its normalization

Although we use ψ^2 in the text, it should strictly be written as $\psi\psi^*$ where ψ^* is the complex conjugate of ψ. In the x-direction, the probability of finding the electron between the limits x and $(x + dx)$ is proportional to $\psi(x)\psi^*(x)\,dx$. In 3-dimensional space this is expressed as $\psi\psi^*\,d\tau$ in which we are considering the probability of finding the electron in a volume element $d\tau$. For just the radial part of the wavefunction, the function is $R(r)R^*(r)$.

In all of our mathematical manipulations, we must ensure that the result shows that the electron is *somewhere* (i.e. it has not

vanished!) and this is done by *normalizing* the wavefunction to unity. This means that the probability of finding the electron somewhere in space is taken to be 1. Mathematically, the normalization is represented as follows:

$$\int \psi^2\,d\tau = 1 \qquad \text{or more correctly} \qquad \int \psi\psi^*\,d\tau = 1$$

and this effectively states that the integral ($\int$) is over all space ($d\tau$) and that the total integral of ψ^2 (or $\psi\psi^*$) must be unity.

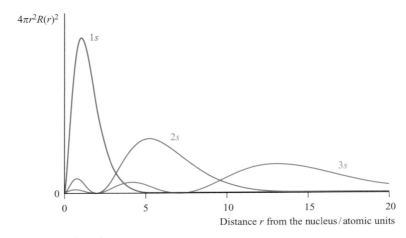

Fig. 1.7 Radial distribution functions, $4\pi r^2 R(r)^2$, for the $1s$, $2s$ and $3s$ atomic orbitals of the hydrogen atom.

since ψ^2 may be described in terms of the radial and angular components $R(r)^2$ and $A(\theta, \phi)^2$.

First consider the radial components. A useful way of depicting the probability density is to plot a *radial distribution function* (eq. 1.15) and this allows us to envisage the region in space in which the electron is found.

Radial distribution function $= 4\pi r^2 R(r)^2$ (1.15)

The radial distribution functions for the $1s$, $2s$ and $3s$ atomic orbitals of hydrogen are shown in Fig. 1.7, and Fig. 1.8 shows those of the $3s$, $3p$ and $3d$ orbitals. *Each* function is zero at the nucleus, following from the r^2 term and the fact that at the nucleus $r = 0$. Since the function depends on $R(r)^2$, it is always positive in contrast to $R(r)$, plots for which are shown in Figs. 1.5 and 1.6. Each plot of $4\pi r^2 R(r)^2$ shows at least one maximum value for the function, corresponding to a distance from the nucleus at which the electron has the highest probability of being found. Points at which $4\pi r^2 R(r)^2 = 0$ (ignoring $r = 0$) correspond to radial nodes where $R(r) = 0$.

The angular part of the wavefunction, $A(\theta, \phi)$

Now let us consider the angular parts of the wavefunctions, $A(\theta, \phi)$, for different types of atomic orbitals. These are *independent* of the principal quantum number as Table 1.2 illustrates for $n = 1$ and 2. Moreover, for s orbitals, $A(\theta, \phi)$ is independent of the angles θ and ϕ and is of a constant value. Thus, an s orbital is spherically symmetric about the nucleus. We noted above that a set of p orbitals is triply degenerate; by convention the three orbitals that make up the degenerate set are given the labels p_x, p_y and p_z. From Table 1.2, we see that the angular part of the p_z wavefunction is independent of ϕ. The orbital can be represented as two spheres (touching at the origin)[†], the centres of which lie on the z axis. For the p_x and p_y orbitals, $A(\theta, \phi)$ depends on both the angles θ and

---

[†] In order to emphasize that ϕ is a continuous function we have extended boundary surfaces in representations of orbitals to the nucleus, but for p orbitals, this is strictly not true if we are considering $\approx 95\%$ of the electronic charge.

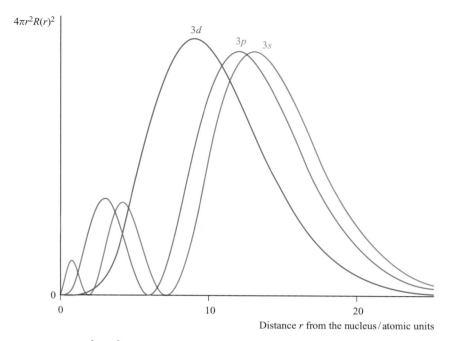

Fig. 1.8 Radial distribution functions, $4\pi r^2 R(r)^2$, for the 3s, 3p and 3d atomic orbitals of the hydrogen atom.

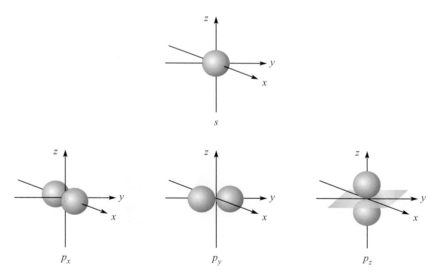

Fig. 1.9 Boundary surfaces for the angular parts of the 1s and 2p atomic orbitals of the hydrogen atom. The nodal plane shown in grey for the $2p_z$ atomic orbital lies in the xy plane.

ϕ; these orbitals are similar to p_z but are oriented along the x and y axes.

Although we must not lose sight of the fact that wavefunctions are mathematical in origin, most chemists find such functions hard to visualize and prefer pictorial representations of orbitals. The boundary surfaces of the s and three p atomic orbitals are shown in Fig. 1.9. The different colours of the *lobes* are significant. The boundary surface of an s orbital has a constant *phase*, i.e. the amplitude of the wavefunction associated with the boundary surface of the s orbital has a constant sign. For a p orbital, there is *one* phase change with respect to the boundary surface and this occurs at a *nodal plane* as is shown for the p_z orbital in Fig. 1.9. The amplitude of a wavefunction may be positive or negative; this is shown using + and − signs, or by shading the lobes in different colours as in Fig. 1.9.

Just as the function $4\pi r^2 R(r)^2$ represents the probability of finding an electron at a distance r from the nucleus, we use a function dependent upon $A(\theta, \phi)^2$ to represent the probability in terms of θ and ϕ. For an s orbital, squaring $A(\theta, \phi)$ causes no change in the spherical symmetry, and the surface boundary for the s atomic orbital shown in

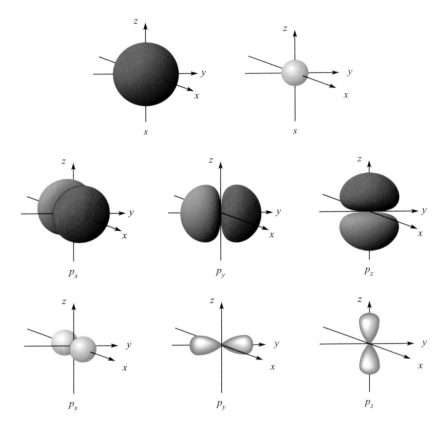

Fig. 1.10 Representations of an s and a set of three degenerate p atomic orbitals. The lobes of the p_x orbital are elongated like those of the p_y and p_z but are directed along the axis that passes through the plane of the paper. The figure shows 'cartoon' diagrams of the orbitals alongside more realistic representations generated using the program *Orbital Viewer* (David Manthey, www.orbitals.com/orb/index.html).

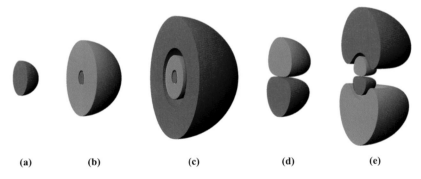

Fig. 1.11 Cross-sections through the (a) $1s$ (no radial nodes), (b) $2s$ (one radial node), (c) $3s$ (two radial nodes), (d) $2p$ (no radial nodes) and (e) $3p$ (one radial node) atomic orbitals of hydrogen. The orbitals have been generated using the program *Orbital Viewer* (David Manthey, www.orbitals.com/orb/index.html).

Fig. 1.10 looks similar to that in Fig. 1.9. For the p orbitals, however, going from $A(\theta, \phi)$ to $A(\theta, \phi)^2$ has the effect of altering the shapes of the lobes as illustrated in Fig. 1.10. Squaring $A(\theta, \phi)$ necessarily means that the signs ($+$ or $-$) disappear, but in practice chemists often indicate the amplitude by a sign or by shading (as in Fig. 1.10) because of the importance of the signs of the wavefunctions with respect to their overlap during bond formation (see Section 2.3). The consequence of the radial nodes that were introduced in Figs. 1.7 and 1.8

can be seen by looking at cross-sections through the atomic orbitals (Fig. 1.11).

Figure 1.12 shows the boundary surfaces for five hydrogen-like d orbitals. We shall not consider the mathematical forms of these wavefunctions, but merely represent the orbitals in the conventional manner. Each d orbital possesses *two* nodal planes and as an exercise you should recognize where these planes lie for each orbital. We consider d orbitals in more detail in Chapters 19 and 20, and f orbitals in Chapter 27.

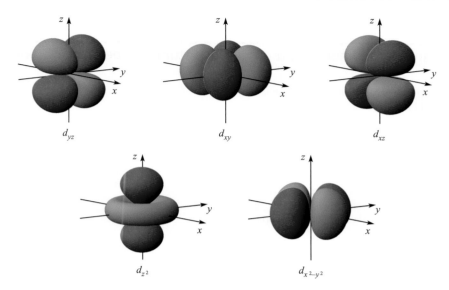

Fig. 1.12 Representations of a set of five degenerate d atomic orbitals. The orbitals have been generated using the program *Orbital Viewer* (David Manthey, www.orbitals.com/orb/index.html).

Orbital energies in a hydrogen-like species

Besides providing information about the wavefunctions, solutions of the Schrödinger equation give orbital energies, E (energy levels), and eq. 1.16 shows the dependence of E on the principal quantum number for *hydrogen-like species* where Z is the atomic number. For the hydrogen atom, $Z = 1$, but for the hydrogen-like He$^+$ ion, $Z = 2$. The dependence of E on Z^2 therefore leads to a significant lowering of the orbitals on going from H to He$^+$.

$$E = -\frac{kZ^2}{n^2} \qquad k = \text{a constant} = 1.312 \times 10^3 \text{ kJ mol}^{-1}$$

$$(1.16)$$

By comparing eq. 1.16 with eq. 1.10, we can see that the constant k in eq. 1.16 is equal to the ionization energy of the H atom, i.e. $k = hcR$ where h, c and R are the Planck constant, the speed of light and the Rydberg constant, respectively.

For each value of n there is only one energy solution and for *hydrogen-like species*, all atomic orbitals with the same principal quantum number (e.g. $3s$, $3p$ and $3d$) are degenerate. It follows from eq. 1.16 that the orbital energy levels get closer together as the value of n increases. This result is a general one for all other atoms.

$1 \text{ eV} = 96.485 \text{ kJ mol}^{-1}$

1. Show that the energy of both the $2s$ and $2p$ orbitals for a hydrogen atom is -328 kJ mol^{-1}.

2. For a hydrogen atom, confirm that the energy of the $3s$ orbital is -1.51 eV.

3. The energy of a hydrogen ns orbital is -13.6 eV. Show that $n = 1$.

4. Determine the energy (in kJ mol^{-1}) of the $1s$ orbital of an He$^+$ ion and compare it with that of the $1s$ orbital of an H atom.
   [*Ans.* $-5248 \text{ kJ mol}^{-1}$ for He$^+$; $-1312 \text{ kJ mol}^{-1}$ for H]

Size of orbitals

For a given atom, a series of orbitals with different values of n but the same values of l and m_l (e.g. $1s$, $2s$, $3s$, $4s$, ...) differ in their relative size (spatial extent). The larger the value of n, the larger the orbital, although this relationship is not linear. The relative spatial extents of the $1s$, $2s$ and $3s$ orbitals, and of the $2p$ and $3p$ orbitals, are shown in Fig. 1.11. An increase in size also corresponds to an orbital being more *diffuse*.

The spin quantum number and the magnetic spin quantum number

Before we place electrons into atomic orbitals, we must define two more quantum numbers. In a classical model, an electron is considered to spin about an axis passing through it and to have *spin angular momentum* in addition to orbital angular momentum (see Box 1.4). The *spin quantum number*, s, determines the magnitude of the spin angular momentum of an electron and has a value of $\frac{1}{2}$. Since angular momentum is a vector quantity, it must have direction, and this is determined by the *magnetic spin quantum number*, m_s, which has a value of $+\frac{1}{2}$ or $-\frac{1}{2}$.

Whereas an atomic orbital is defined by a unique set of *three* quantum numbers, an electron in an atomic orbital is defined by a unique set of *four* quantum numbers: n, l, m_l

THEORY

Box 1.4 Angular momentum, the inner quantum number, j, and spin–orbit coupling

The value of l determines not only the shape of an orbital but also the amount of orbital angular momentum associated with an electron in the orbital:

Orbital angular momentum $= \left[\sqrt{l(l+1)}\right]\dfrac{h}{2\pi}$

The axis through the nucleus about which the electron (considered classically) can be thought to rotate defines the direction of the orbital angular momentum. The latter gives rise to a magnetic moment the direction of which is in the same sense as the angular vector and the magnitude of which is proportional to the magnitude of the vector.

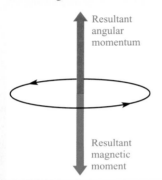

In a classical model, an electron moving in a circular orbit has an angular momentum defined by the resultant vector shown in red.

Because the electron is a charged particle, there is an associated magnetic moment, the direction of which is represented in the diagram by the blue arrow.

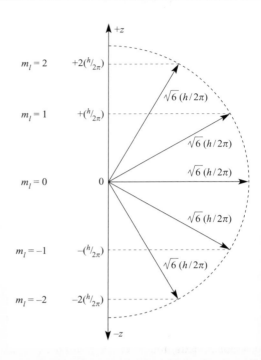

An electron in an s orbital ($l = 0$) has no orbital angular momentum, an electron in a p orbital ($l = 1$) has angular momentum $\sqrt{2}(h/2\pi)$, and so on. The orbital angular momentum vector has $(2l + 1)$ possible directions in space corresponding to the $(2l + 1)$ possible values of m_l for a given value of l. Consider the component of the angular momentum vector along the z axis; this has a different value for each of the possible orientations that this vector can take up. The actual magnitude of the z component is given by $m_l(h/2\pi)$. Thus, for an electron in a d orbital ($l = 2$), the orbital angular momentum is $\sqrt{6}(h/2\pi)$, and the z component of this may have values of $+2(h/2\pi)$, $+(h/2\pi)$, 0, $-(h/2\pi)$ or $-2(h/2\pi)$ as shown in the figure opposite.

The orbitals in a sub-shell of given n and l, are degenerate. If, however, the atom is placed in a magnetic field, this degeneracy is removed. If we arbitrarily define the direction of the magnetic field as the z axis, electrons in the various d orbitals will interact to different extents with the magnetic field because of their different values of the z components of their angular momentum vectors (and, hence, orbital magnetic moment vectors).

An electron also has spin angular momentum which can be regarded as originating in the rotation of the electron about its own axis. The magnitude of this is given by:

Spin angular momentum $= \left[\sqrt{s(s+1)}\right]\dfrac{h}{2\pi}$

where $s =$ spin quantum number. The axis defines the direction of the spin angular momentum vector, but again it is the orientation of this vector with respect to the z direction in which we are interested. The z component is given by $m_s(h/2\pi)$; since m_s can only equal $+\frac{1}{2}$ or $-\frac{1}{2}$, there are only two possible orientations of the spin angular momentum vector, and these give rise to z components of magnitude $+\frac{1}{2}(h/2\pi)$ and $-\frac{1}{2}(h/2\pi)$.

For an electron having both orbital and spin angular momentum, the total angular momentum vector is given by:

Total angular momentum $= \left[\sqrt{j(j+1)}\right]\dfrac{h}{2\pi}$

where j is the *inner quantum number*; j may take values of $|l + s|$ or $|l - s|$, i.e. $\left|l + \frac{1}{2}\right|$ or $\left|l - \frac{1}{2}\right|$. The symbol '| |' is a modulus sign and signifies that the quantities $(l + s)$ and $(l - s)$ must have *positive* values. Thus, when $l = 0$, j can only be $\frac{1}{2}$ because $\left|0 + \frac{1}{2}\right| = \left|0 - \frac{1}{2}\right| = \frac{1}{2}$. (When $l = 0$ and the electron has no orbital angular momentum, the total angular momentum is $\left[\sqrt{s(s+1)}\right]\dfrac{h}{2\pi}$ because $j = s$.) The z component of the total angular momentum vector is now $j(h/2\pi)$ and there are $(2j + 1)$ possible orientations in space.

For an electron in an ns orbital ($l = 0$), j can only be $\frac{1}{2}$. When the electron is promoted to an np orbital, j may be $\frac{3}{2}$ or $\frac{1}{2}$, and the energies corresponding to the different j values are not equal. In the emission spectrum of sodium, for example, transitions from the $3p_{3/2}$ and $3p_{1/2}$ levels to the $3s_{1/2}$ level correspond to slightly different amounts of energy, and this *spin–orbit coupling* is the origin of the doublet structure of the strong yellow line in the spectrum of atomic sodium (see Fig. 1.18, p. 30). The fine structure of many other spectral lines arises in analogous ways, though the number of lines observed depends on the difference in energy between states differing only in j value and on the resolving power of the spectrometer. The difference in energy between levels for which $\Delta j = 1$ (the spin–orbit coupling constant, λ) increases with the atomic number of the element; e.g. that between the $np_{3/2}$ and $np_{1/2}$ levels for Li, Na and Cs is 0.23, 11.4 and 370 cm^{-1} respectively.

and m_s. As there are only two values of m_s, an orbital can accommodate only two electrons.

> An *orbital is fully occupied* when it contains two *spin-paired electrons*; one electron has $m_s = +\frac{1}{2}$ and the other has $m_s = -\frac{1}{2}$.

Self-study exercises

1. What do you understand by the term 'the orbital angular momentum of an electron in an orbital'?

2. Explain why it is incorrect to write that 'the quantum number $s = \pm\frac{1}{2}$'.

3. For an *s* orbital, $l = 0$. Explain why this leads to a classical picture of an electron in an *s* orbital *not* moving around the nucleus.

4. By considering a $2p$ orbital with values of $m_l = +1, 0$ and -1, explain the physical significance of the quantum number m_l.

5. Show that for an electron in a $2s$ orbital, the quantum number j can only take the value $\frac{1}{2}$.

6. Show that for an electron in a $2p$ orbital, the quantum number j can take the value of $\frac{3}{2}$ or $\frac{1}{2}$.

7. For a *p* electron circulating clockwise or counter-clockwise about an axis, the value of m_l is $+1$ or -1. What can you say about a *p* electron for which $m_l = 0$?

Worked example 1.4   Quantum numbers: an electron in an atomic orbital

Write down two possible sets of quantum numbers that describe an electron in a $2s$ atomic orbital. What is the physical significance of these unique sets?

The $2s$ atomic orbital is defined by the set of quantum numbers $n = 2, l = 0, m_l = 0$.

An electron in a $2s$ atomic orbital may have one of two sets of four quantum numbers:

$$n = 2, \quad l = 0, \quad m_l = 0, \quad m_s = +\tfrac{1}{2}$$

or

$$n = 2, \quad l = 0, \quad m_l = 0, \quad m_s = -\tfrac{1}{2}$$

If the orbital were fully occupied with two electrons, one electron would have $m_s = +\frac{1}{2}$, and the other electron would have $m_s = -\frac{1}{2}$, i.e. the two electrons would be spin-paired.

Self-study exercises

1. Write down two possible sets of quantum numbers to describe an electron in a $3s$ atomic orbital.
   [*Ans.* $n = 3, l = 0, m_l = 0, m_s = +\frac{1}{2}; n = 3, l = 0, m_l = 0, m_s = -\frac{1}{2}$]

2. If an electron has the quantum numbers $n = 2, l = 1, m_l = -1$ and $m_s = +\frac{1}{2}$ which type of atomic orbital is it occupying?
   [*Ans.* $2p$]

3. An electron has the quantum numbers $n = 4, l = 1, m_l = 0$ and $m_s = +\frac{1}{2}$. Is the electron in a $4s$, $4p$ or $4d$ atomic orbital?
   [*Ans.* $4p$]

4. Write down a set of quantum numbers that describes an electron in a $5s$ atomic orbital. How does this set of quantum numbers differ if you are describing the second electron in the same orbital?
   [*Ans.* $n = 5, l = 0, m_l = 0, m_s = +\frac{1}{2}$ or $-\frac{1}{2}$]

The ground state of the hydrogen atom

So far we have focused on the atomic orbitals of hydrogen and have talked about the probability of finding an electron in different atomic orbitals. The most energetically favourable (stable) state of the H atom is its *ground state* in which the single electron occupies the $1s$ (lowest energy) atomic orbital. The electron can be promoted to higher energy orbitals (see Section 1.4) to give *excited states*.

> The notation for the *ground state electronic configuration* of an H atom is $1s^1$, signifying that one electron occupies the $1s$ atomic orbital.

1.7 Many-electron atoms

The helium atom: two electrons

The preceding sections have been devoted mainly to hydrogen-like species containing one electron, the energy of which depends on n and Z (eq. 1.16). The atomic spectra of such species contain only a few lines associated with changes in the value of n (Fig. 1.3). It is *only* for such species that the Schrödinger equation has been solved exactly.

The next simplest atom is He ($Z = 2$), and for its two electrons, three electrostatic interactions must be considered:

- attraction between electron (1) and the nucleus;
- attraction between electron (2) and the nucleus;
- repulsion between electrons (1) and (2).

The *net* interaction determines the energy of the system.

In the ground state of the He atom, two electrons with $m_s = +\frac{1}{2}$ and $-\frac{1}{2}$ occupy the $1s$ atomic orbital, i.e. the electronic configuration is $1s^2$. For all atoms except hydrogen-like species, orbitals of the same principal quantum number but differing l are *not* degenerate. If one of the $1s^2$ electrons is promoted to an orbital with $n = 2$, the energy of the system depends upon whether the electron goes into a $2s$ or $2p$ atomic orbital, because each situation gives rise to different electrostatic interactions involving the two electrons and the nucleus. However, there is no energy distinction among the three different $2p$ atomic orbitals. If promotion is to an orbital with $n = 3$, different amounts of energy are needed depending upon whether $3s$, $3p$ or $3d$ orbitals are involved, although there is no energy difference among the three $3p$ atomic orbitals, or among the five $3d$ atomic orbitals. The emission spectrum of He arises as the electrons fall back to lower energy states or to the ground state and it follows that the spectrum contains more lines than that of atomic H.

In terms of obtaining wavefunctions and energies for the atomic orbitals of He, it has not been possible to solve the Schrödinger equation exactly and only approximate solutions are available. For atoms containing more than two electrons, it is even more difficult to obtain accurate solutions to the wave equation.

In a ***multi-electron atom***, orbitals with the same value of n but different values of l are *not* degenerate.

Ground state electronic configurations: experimental data

Now consider the ground state electronic configurations of isolated atoms of all the elements (Table 1.3). These are experimental data, and are nearly always obtained by analysing atomic spectra. Most atomic spectra are too complex for discussion here and we take their interpretation on trust.

We have already seen that the ground state electronic configurations of H and He are $1s^1$ and $1s^2$ respectively. The $1s$ atomic orbital is fully occupied in He and its configuration is often written as [He]. In the next two elements, Li and Be, the electrons go into the $2s$ orbital, and then from B to Ne, the $2p$ orbitals are occupied to give the electronic configurations [He]$2s^2 2p^m$ ($m = 1$–6). When $m = 6$, the energy level (or *shell*) with $n = 2$ is fully occupied, and the configuration for Ne can be written as [Ne]. The filling of the $3s$ and $3p$ atomic orbitals takes place in an analogous sequence from Na to Ar, the last element in the series having the electronic configuration [Ne]$3s^2 3p^6$ or [Ar].

With K and Ca, successive electrons go into the $4s$ orbital, and Ca has the electronic configuration [Ar]$4s^2$. At this point, the pattern changes. To a first approximation, the 10 electrons for the next 10 elements (Sc to Zn) enter the $3d$ orbitals, giving Zn the electronic configuration $4s^2 3d^{10}$. There are some irregularities (see Table 1.3) to which we return later. From Ga to Kr, the $4p$ orbitals are filled, and the electronic configuration for Kr is [Ar]$4s^2 3d^{10} 4p^6$ or [Kr].

From Rb to Xe, the general sequence of filling orbitals is the same as that from K to Kr although there are once again irregularities in the distribution of electrons between s and d atomic orbitals (see Table 1.3).

From Cs to Rn, electrons enter f orbitals for the first time; Cs, Ba and La have configurations analogous to those of Rb, Sr and Y, but after that the configurations change as we begin the sequence of the *lanthanoid* elements (see Chapter 27).[†] Cerium has the configuration [Xe]$4f^1 6s^2 5d^1$ and the filling of the seven $4f$ orbitals follows until an electronic configuration of [Xe]$4f^{14} 6s^2 5d^1$ is reached for Lu. Table 1.3 shows that the $5d$ orbital is not usually occupied for a lanthanoid element. After Lu, successive electrons occupy the remaining $5d$ orbitals (Hf to Hg) and then the $6p$ orbitals to Rn which has the configuration [Xe]$4f^{14} 6s^2 5d^{10} 6p^6$ or [Rn]. Table 1.3 shows some irregularities along the series of d-block elements.

For the remaining elements in Table 1.3 beginning at francium (Fr), filling of the orbitals follows a similar sequence as that from Cs but the sequence is incomplete and some of the heaviest elements are too unstable for detailed investigations to be possible. The metals from Th to Lr are the *actinoid* elements, and in discussing their chemistry, Ac is generally considered with the actinoids (see Chapter 27).

A detailed inspection of Table 1.3 makes it obvious that there is no one sequence that represents accurately the occupation of different sets of orbitals with increasing atomic number. The following sequence is *approximately* true for the relative energies (lowest energy first) of orbitals in *neutral atoms*:

$$1s < 2s < 2p < 3s < 3p < 4s < 3d < 4p < 5s < 4d < 5p$$
$$< 6s < 5d \approx 4f < 6p < 7s < 6d \approx 5f$$

The energies of different orbitals are close together for high values of n and their relative energies can change significantly on forming an ion (see Section 19.2).

Penetration and shielding

Although it is not possible to calculate the dependence of the energies of orbitals on atomic number with the degree of accuracy that is required to obtain agreement with all the electronic configurations listed in Table 1.3, some useful information can be gained by considering the different *screening effects* that electrons in different atomic orbitals have on one another. Figure 1.13 shows the radial distribution functions for the $1s$, $2s$ and $2p$ atomic orbitals of the H atom. (It is a common approximation to assume

---

[†] The IUPAC recommends the names lanthanoid and actinoid in preference to lanthanide and actinide; the ending '-ide' usually implies a negatively charged ion. However, lanthanide and actinide are still widely used.

Table 1.3 Ground state electronic configurations of the elements up to $Z = 103$.

Atomic number	Element	Ground state electronic configuration	Atomic number	Element	Ground state electronic configuration
1	H	$1s^1$	53	I	$[Kr]5s^24d^{10}5p^5$
2	He	$1s^2 = [He]$	54	Xe	$[Kr]5s^24d^{10}5p^6 = [Xe]$
3	Li	$[He]2s^1$	55	Cs	$[Xe]6s^1$
4	Be	$[He]2s^2$	56	Ba	$[Xe]6s^2$
5	B	$[He]2s^22p^1$	57	La	$[Xe]6s^25d^1$
6	C	$[He]2s^22p^2$	58	Ce	$[Xe]4f^16s^25d^1$
7	N	$[He]2s^22p^3$	59	Pr	$[Xe]4f^36s^2$
8	O	$[He]2s^22p^4$	60	Nd	$[Xe]4f^46s^2$
9	F	$[He]2s^22p^5$	61	Pm	$[Xe]4f^56s^2$
10	Ne	$[He]2s^22p^6 = [Ne]$	62	Sm	$[Xe]4f^66s^2$
11	Na	$[Ne]3s^1$	63	Eu	$[Xe]4f^76s^2$
12	Mg	$[Ne]3s^2$	64	Gd	$[Xe]4f^76s^25d^1$
13	Al	$[Ne]3s^23p^1$	65	Tb	$[Xe]4f^96s^2$
14	Si	$[Ne]3s^23p^2$	66	Dy	$[Xe]4f^{10}6s^2$
15	P	$[Ne]3s^23p^3$	67	Ho	$[Xe]4f^{11}6s^2$
16	S	$[Ne]3s^23p^4$	68	Er	$[Xe]4f^{12}6s^2$
17	Cl	$[Ne]3s^23p^5$	69	Tm	$[Xe]4f^{13}6s^2$
18	Ar	$[Ne]3s^23p^6 = [Ar]$	70	Yb	$[Xe]4f^{14}6s^2$
19	K	$[Ar]4s^1$	71	Lu	$[Xe]4f^{14}6s^25d^1$
20	Ca	$[Ar]4s^2$	72	Hf	$[Xe]4f^{14}6s^25d^2$
21	Sc	$[Ar]4s^23d^1$	73	Ta	$[Xe]4f^{14}6s^25d^3$
22	Ti	$[Ar]4s^23d^2$	74	W	$[Xe]4f^{14}6s^25d^4$
23	V	$[Ar]4s^23d^3$	75	Re	$[Xe]4f^{14}6s^25d^5$
24	Cr	$[Ar]4s^13d^5$	76	Os	$[Xe]4f^{14}6s^25d^6$
25	Mn	$[Ar]4s^23d^5$	77	Ir	$[Xe]4f^{14}6s^25d^7$
26	Fe	$[Ar]4s^23d^6$	78	Pt	$[Xe]4f^{14}6s^15d^9$
27	Co	$[Ar]4s^23d^7$	79	Au	$[Xe]4f^{14}6s^15d^{10}$
28	Ni	$[Ar]4s^23d^8$	80	Hg	$[Xe]4f^{14}6s^25d^{10}$
29	Cu	$[Ar]4s^13d^{10}$	81	Tl	$[Xe]4f^{14}6s^25d^{10}6p^1$
30	Zn	$[Ar]4s^23d^{10}$	82	Pb	$[Xe]4f^{14}6s^25d^{10}6p^2$
31	Ga	$[Ar]4s^23d^{10}4p^1$	83	Bi	$[Xe]4f^{14}6s^25d^{10}6p^3$
32	Ge	$[Ar]4s^23d^{10}4p^2$	84	Po	$[Xe]4f^{14}6s^25d^{10}6p^4$
33	As	$[Ar]4s^23d^{10}4p^3$	85	At	$[Xe]4f^{14}6s^25d^{10}6p^5$
34	Se	$[Ar]4s^23d^{10}4p^4$	86	Rn	$[Xe]4f^{14}6s^25d^{10}6p^6 = [Rn]$
35	Br	$[Ar]4s^23d^{10}4p^5$	87	Fr	$[Rn]7s^1$
36	Kr	$[Ar]4s^23d^{10}4p^6 = [Kr]$	88	Ra	$[Rn]7s^2$
37	Rb	$[Kr]5s^1$	89	Ac	$[Rn]6d^17s^2$
38	Sr	$[Kr]5s^2$	90	Th	$[Rn]6d^27s^2$
39	Y	$[Kr]5s^24d^1$	91	Pa	$[Rn]5f^27s^26d^1$
40	Zr	$[Kr]5s^24d^2$	92	U	$[Rn]5f^37s^26d^1$
41	Nb	$[Kr]5s^14d^4$	93	Np	$[Rn]5f^47s^26d^1$
42	Mo	$[Kr]5s^14d^5$	94	Pu	$[Rn]5f^67s^2$
43	Tc	$[Kr]5s^24d^5$	95	Am	$[Rn]5f^77s^2$
44	Ru	$[Kr]5s^14d^7$	96	Cm	$[Rn]5f^77s^26d^1$
45	Rh	$[Kr]5s^14d^8$	97	Bk	$[Rn]5f^97s^2$
46	Pd	$[Kr]5s^04d^{10}$	98	Cf	$[Rn]5f^{10}7s^2$
47	Ag	$[Kr]5s^14d^{10}$	99	Es	$[Rn]5f^{11}7s^2$
48	Cd	$[Kr]5s^24d^{10}$	100	Fm	$[Rn]5f^{12}7s^2$
49	In	$[Kr]5s^24d^{10}5p^1$	101	Md	$[Rn]5f^{13}7s^2$
50	Sn	$[Kr]5s^24d^{10}5p^2$	102	No	$[Rn]5f^{14}7s^2$
51	Sb	$[Kr]5s^24d^{10}5p^3$	103	Lr	$[Rn]5f^{14}7s^26d^1$
52	Te	$[Kr]5s^24d^{10}5p^4$			

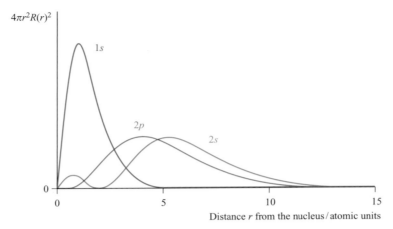

Fig. 1.13 Radial distribution functions, $4\pi r^2 R(r)^2$, for the $1s$, $2s$ and $2p$ atomic orbitals of the hydrogen atom.

hydrogen-like wavefunctions for multi-electron atoms.) Although values of $4\pi r^2 R(r)^2$ for the $1s$ orbital are much greater than those of the $2s$ and $2p$ orbitals at distances relatively close to the nucleus, the values for the $2s$ and $2p$ orbitals are still significant. We say that the $2s$ and $2p$ atomic orbitals *penetrate* the $1s$ atomic orbital. Calculations show that the $2s$ atomic orbital is more penetrating than the $2p$ orbital.

Now let us consider the arrangement of the electrons in Li ($Z = 3$). In the ground state, the $1s$ atomic orbital is fully occupied and the third electron could occupy either a $2s$ or $2p$ orbital. Which arrangement will possess the lower energy? An electron in a $2s$ or $2p$ atomic orbital experiences the *effective nuclear charge*, Z_{eff}, of a nucleus partly *shielded* by the $1s$ electrons. Since the $2p$ orbital penetrates the $1s$ orbital less than a $2s$ orbital does (Fig. 1.13), a $2p$ electron is shielded more than a $2s$ electron. Thus, occupation of the $2s$ rather than the $2p$ atomic orbital gives a lower energy system. Although we should consider the energies of the *electrons* in atomic orbitals, it is common practice to think in terms of the orbital energies themselves: $E(2s) < E(2p)$. Similar arguments lead to the sequence $E(3s) < E(3p) < E(3d)$ and $E(4s) < E(4p) < E(4d) < E(4f)$. As we move to atoms of elements of higher atomic number, the energy differences between orbitals with the same value of n become smaller, the validity of assuming hydrogen-like wavefunctions becomes more doubtful, and predictions of ground states become less reliable. The treatment above also ignores electron–electron interactions within the same principal quantum shell.

A set of empirical rules (Slater's rules) for estimating the effective nuclear charges experienced by electrons in different atomic orbitals is described in Box 1.5.

1.8 The periodic table

In 1869 and 1870 respectively, Dmitri Mendeléev and Lothar Meyer stated that the *properties of the elements can be represented as periodic functions of their atomic weights*, and set out their ideas in the form of a *periodic table*. As new elements have been discovered, the original form of the periodic table has been extensively modified, and it is now recognized that *periodicity* is a consequence of the variation in ground state electronic configurations. A modern periodic table (Fig. 1.14) emphasizes the blocks of 2, 6, 10 and 14 elements which result from the filling of the s, p, d and f atomic orbitals respectively. An exception is He, which, for reasons of its chemistry, is placed in a *group* with Ne, Ar, Kr, Xe and Rn. A more detailed periodic table is given inside the front cover of the book.

The IUPAC (International Union of Pure and Applied Chemistry) has produced guidelines[†] for naming blocks and groups of elements in the periodic table. In summary,

- blocks of elements may be designated by use of the letters s, p, d or f (Fig. 1.14);
- elements (except H) in groups 1, 2 and 13–18 are called *main group elements*;
- with the exception of group 18, the first two elements of each main group are called *typical elements*;
- elements in groups 3–12 (the d-block elements) are also commonly called the *transition elements*, although elements in group 12 are not always included;
- the f-block elements are sometimes called the *inner transition elements*.

Collective names for some of the groups of elements in the periodic table are given in Table 1.4.

In mid-2011, the number of elements in the periodic table stood at 117 (see Section 27.5).[‡]

[†] IUPAC: *Nomenclature of Inorganic Chemistry* (*Recommendations 2005*), senior eds N.G. Connelly and T. Damhus, RSC Publishing, Cambridge.
[‡] P.J. Karol, H. Nakahara, B.W. Petley and E. Vogt (2003) *Pure Appl. Chem.*, vol. 75, p. 1601 – 'On the claims for the discovery of elements 110, 111, 112, 114, 116 and 118'; K. Tatsumi and J. Corish (2010) *Pure Appl. Chem.*, vol. 82, p. 753 – 'Name and symbol of the element with atomic number 112'.

THEORY

Box 1.5 Effective nuclear charge and Slater's rules

Slater's rules

Effective nuclear charges, Z_{eff}, experienced by electrons in different atomic orbitals may be estimated using *Slater's rules*. These rules are based on experimental data for electron promotion and ionization energies, and Z_{eff} is determined from the equation:

$$Z_{eff} = Z - S$$

where Z = nuclear charge, Z_{eff} = effective nuclear charge, S = screening (or shielding) constant.

   Values of S may be estimated as follows:

1. Write out the electronic configuration of the element in the following order and groupings: $(1s)$, $(2s, 2p)$, $(3s, 3p)$, $(3d)$, $(4s, 4p)$, $(4d)$, $(4f)$, $(5s, 5p)$ etc.
2. Electrons in any group higher in this sequence than the electron under consideration contribute nothing to S.
3. Consider a particular electron in an ns or np orbital:
   (i) Each of the other electrons in the (ns, np) group contributes $S = 0.35$.
   (ii) Each of the electrons in the $(n-1)$ shell contributes $S = 0.85$.
   (iii) Each of the electrons in the $(n-2)$ or lower shells contributes $S = 1.00$.
4. Consider a particular electron in an nd or nf orbital:
   (i) Each of the other electrons in the (nd, nf) group contributes $S = 0.35$.
   (ii) Each of the electrons in a lower group than the one being considered contributes $S = 1.00$.

An example of how to apply Slater's rules

Question: Confirm that the experimentally observed electronic configuration of K, $1s^2 2s^2 2p^6 3s^2 3p^6 4s^1$, is energetically more stable than the configuration $1s^2 2s^2 2p^6 3s^2 3p^6 3d^1$.

   For K, $Z = 19$.

   Applying Slater's rules, the effective nuclear charge experienced by the $4s$ electron for the configuration $1s^2 2s^2 2p^6 3s^2 3p^6 4s^1$ is:

$$Z_{eff} = Z - S$$

The nuclear charge, $Z = 19$

The screening constant, $S = (8 \times 0.85) + (10 \times 1.00)$
$$= 16.8$$

$Z_{eff} = 19 - 16.8 = 2.2$

The effective nuclear charge experienced by the $3d$ electron for the configuration $1s^2 2s^2 2p^6 3s^2 3p^6 3d^1$ is:

$$Z_{eff} = Z - S$$

The nuclear charge, $Z = 19$

The screening constant, $S = (18 \times 1.00)$
$$= 18.0$$

$Z_{eff} = 19 - 18.0 = 1.0$

Thus, an electron in the $4s$ (rather than the $3d$) atomic orbital is under the influence of a greater effective nuclear charge and in the ground state of potassium, it is the $4s$ atomic orbital that is occupied.

Slater versus Clementi and Raimondi values of Z_{eff}

Slater's rules have been used to estimate ionization energies, ionic radii and electronegativities. More accurate effective nuclear charges have been calculated by Clementi and Raimondi by using *self-consistent field* (SCF) methods, and indicate much higher Z_{eff} values for the d electrons. However, the simplicity of Slater's approach makes this an attractive method for 'back-of-the-envelope' estimations of Z_{eff}.

Self-study exercises

1. Show that Slater's rules give a value of $Z_{eff} = 1.95$ for a $2s$ electron in a Be atom.

2. Show that Slater's rules give a value of $Z_{eff} = 5.20$ for a $2p$ electron of F.

3. Use Slater's rules to estimate values of Z_{eff} for (a) a $4s$ and (b) a $3d$ electron in a V atom.
   [*Ans.* (a) 3.30; (b) 4.30]

4. Using your answer to question 3, explain why the valence configuration of the ground state of a V^+ ion is likely to be $3d^3 4s^1$ rather than $3d^2 4s^2$.

Further reading

J.L. Reed (1999) *J. Chem. Educ.*, vol. 76, p. 802 – 'The genius of Slater's rules'.

D. Tudela (1993) *J. Chem. Educ.*, vol. 70, p. 956 – 'Slater's rules and electronic configurations'.

G. Wulfsberg (2000) *Inorganic Chemistry*, University Science Books, Sausalito, CA – Contains a fuller treatment of Slater's rules and illustrates their application, particularly to the assessment of electronegativity.

s-block elements   **d-block elements**   **p-block elements**

Group 1	Group 2	Group 3	Group 4	Group 5	Group 6	Group 7	Group 8	Group 9	Group 10	Group 11	Group 12	Group 13	Group 14	Group 15	Group 16	Group 17	Group 18
1 H																	2 He
3 Li	4 Be											5 B	6 C	7 N	8 O	9 F	10 Ne
11 Na	12 Mg											13 Al	14 Si	15 P	16 S	17 Cl	18 Ar
19 K	20 Ca	21 Sc	22 Ti	23 V	24 Cr	25 Mn	26 Fe	27 Co	28 Ni	29 Cu	30 Zn	31 Ga	32 Ge	33 As	34 Se	35 Br	36 Kr
37 Rb	38 Sr	39 Y	40 Zr	41 Nb	42 Mo	43 Tc	44 Ru	45 Rh	46 Pd	47 Ag	48 Cd	49 In	50 Sn	51 Sb	52 Te	53 I	54 Xe
55 Cs	56 Ba	57–71 La–Lu	72 Hf	73 Ta	74 W	75 Re	76 Os	77 Ir	78 Pt	79 Au	80 Hg	81 Tl	82 Pb	83 Bi	84 Po	85 At	86 Rn
87 Fr	88 Ra	89–103 Ac–Lr	104 Rf	105 Db	106 Sg	107 Bh	108 Hs	109 Mt	110 Ds	111 Rg	112 Cn	113 Uut	114 Uuq	115 Uup	116 Uuh		118 Uuo

f-block elements

Lanthanoids	58 Ce	59 Pr	60 Nd	61 Pm	62 Sm	63 Eu	64 Gd	65 Tb	66 Dy	67 Ho	68 Er	69 Tm	70 Yb	71 Lu
Actinoids	90 Th	91 Pa	92 U	93 Np	94 Pu	95 Am	96 Cm	97 Bk	98 Cf	99 Es	100 Fm	101 Md	102 No	103 Lr

Fig. 1.14 The modern periodic table in which the elements are arranged in numerical order according to the number of protons (and electrons) they possess. The division into *groups* places elements with the same number of valence electrons into vertical columns within the table. Under IUPAC recommendations, the groups are labelled from 1 to 18 (Arabic numbers). The vertical groups of three *d*-block elements are called *triads*. Rows in the periodic table are called *periods*. The first period contains H and He, but the row from Li to Ne is usually referred to as the first period. Strictly, the lanthanoids include the 14 elements Ce–Lu, and the actinoids include Th–Lr; however, common usage places La with the lanthanoids, and Ac with the actinoids (see Chapter 27).

Table 1.4 IUPAC recommended names for groups of elements in the periodic table.

Group number	Recommended name
1	Alkali metals
2	Alkaline earth metals
15	Pnictogens
16	Chalcogens
17	Halogens
18	Noble gases

1.9 The *aufbau* principle

Ground state electronic configurations

In the previous two sections, we have considered experimental electronic configurations and have seen that the organization of the elements in the periodic table depends on the number, and arrangement, of electrons that each element possesses. Establishing the ground state electronic configuration of an atom is the key to understanding its chemistry, and we now discuss the *aufbau* principle (*aufbau* means 'building up' in German) which is used in conjunction with Hund's rules and the Pauli exclusion principle to determine electronic ground state configurations:

- Orbitals are filled in order of energy, the lowest energy orbitals being filled first.
- Hund's first rule (often referred to simply as Hund's rule): in a set of degenerate orbitals, electrons may not be spin-paired in an orbital until *each* orbital in the set contains one electron; electrons singly occupying orbitals in a degenerate set have parallel spins, i.e. they have the same values of m_s.
- Pauli exclusion principle: no two electrons in the same atom may have the same set of n, l, m_l and m_s quantum numbers; it follows that each orbital can accommodate a maximum of two electrons with different m_s values (different spins = spin-paired).

Worked example 1.5   Using the *aufbau* principle

Determine (with reasoning) the ground state electronic configurations of (a) Be ($Z = 4$) and (b) P ($Z = 15$).

The value of Z gives the number of electrons to be accommodated in atomic orbitals in the ground state of the atom.

Assume an order of atomic orbitals (lowest energy first) as follows: $1s < 2s < 2p < 3s < 3p$

(a) Be $Z = 4$

Two electrons (spin-paired) are accommodated in the lowest energy $1s$ atomic orbital.

The next two electrons (spin-paired) are accommodated in the $2s$ atomic orbital.

The ground state electronic configuration of Be is therefore $1s^2 2s^2$.

(b) P $Z = 15$

Two electrons (spin-paired) are accommodated in the lowest energy $1s$ atomic orbital.

The next two electrons (spin-paired) are accommodated in the $2s$ atomic orbital.

The next six electrons are accommodated in the three degenerate $2p$ atomic orbitals, two spin-paired electrons per orbital.

The next two electrons (spin-paired) are accommodated in the $3s$ atomic orbital.

Three electrons remain and, applying Hund's rule, these singly occupy each of the three degenerate $3p$ atomic orbitals.

The ground state electronic configuration of P is therefore $1s^2 2s^2 2p^6 3s^2 3p^3$.

Self-study exercises

1. Where, in the worked example above, is the Pauli exclusion principle applied?

2. Will the three electrons in the P $3p$ atomic orbitals possess the same or different values of the spin quantum number?   [*Ans.* Same; parallel spins]

3. Show, with reasoning, that the ground state electronic configuration of O ($Z = 8$) is $1s^2 2s^2 2p^4$.

4. Determine (with reasoning) how many unpaired electrons are present in a ground state Al atom ($Z = 13$).   [*Ans.* 1]

Worked example 1.6   The ground state electronic configurations of the noble gases

The atomic numbers of He, Ne, Ar and Kr are 2, 10, 18 and 36 respectively. Write down the ground state electronic configurations of these elements and comment upon their similarities or differences.

Apply the *aufbau* principle using the atomic orbital energy sequence:

$1s < 2s < 2p < 3s < 3p < 4s < 3d < 4p$

The ground state electronic configurations are:

He	$Z = 2$	$1s^2$
Ne	$Z = 10$	$1s^2 2s^2 2p^6$
Ar	$Z = 18$	$1s^2 2s^2 2p^6 3s^2 3p^6$
Kr	$Z = 36$	$1s^2 2s^2 2p^6 3s^2 3p^6 4s^2 3d^{10} 4p^6$

Each element Ne, Ar and Kr has a ground state electronic configuration $\ldots ns^2 np^6$. Helium is the odd one out, but still possesses a filled quantum level; this is a characteristic property of a noble gas.

Self-study exercises

1. Values of Z for Li, Na, K and Rb are 3, 11, 19 and 37 respectively. Write down their ground state configurations and comment on the result.
   [*Ans.* All are of the form [X]ns^1 where X is a noble gas]

2. How are the ground state electronic configurations of O, S and Se ($Z = 8$, 16, 34 respectively) alike? Give another element related in the same way.
   [*Ans.* All are of the form [X]$ns^2 np^4$ where X is a noble gas; Te or Po]

3. State two elements that have ground state electronic configurations of the general type [X]$ns^2 np^1$.
   [*Ans.* Any two elements from group 13]

Valence and core electrons

Core electrons occupy lower energy quantum levels than *valence electrons*. The *valence electrons* of an element determine its chemical properties.

The configuration of the outer or *valence electrons* is of particular significance. These electrons determine the chemical properties of an element. Electrons that occupy lower energy quantum levels are called *core electrons*. The core electrons shield the valence electrons from the nuclear charge, resulting in the valence electrons experiencing only the effective nuclear charge, Z_{eff}. For an element of low atomic number, the core and valence electrons are readily recognized by looking at the ground state electronic configuration. That of oxygen is $1s^2 2s^2 2p^4$. The core electrons of oxygen are those in the $1s$ atomic orbital; the six electrons with $n = 2$ are the valence electrons.

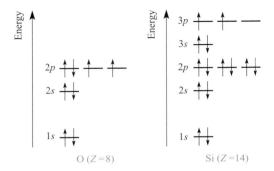

Fig. 1.15 Diagrammatic representations of the ground state electronic configurations of O and Si. The complete configurations are shown here, but it is common to simply indicate the valence electrons. For O, this consists of the $2s$ and $2p$ levels, and for Si, the $3s$ and $3p$ levels.

Diagrammatic representations of electronic configurations

The notation we have used to represent electronic configurations is convenient and is commonly adopted, but sometimes it is also useful to indicate the relative energies of the electrons. In this case, the electrons are represented by arrows ↑ or ↓ with the direction of the arrow corresponding to $m_s = +\frac{1}{2}$ or $-\frac{1}{2}$. Figure 1.15 gives qualitative energy level diagrams which describe the ground state electronic configurations of O and Si.

Worked example 1.7   Quantum numbers for electrons

Confirm that the ground state configuration shown for oxygen in Fig. 1.15 is consistent with each electron possessing a unique set of four quantum numbers.

Each atomic orbital is designated by a unique set of three quantum numbers:

$1s$	$n = 1$	$l = 0$	$m_l = 0$
$2s$	$n = 2$	$l = 0$	$m_l = 0$
$2p$	$n = 2$	$l = 1$	$m_l = -1$
	$n = 2$	$l = 1$	$m_l = 0$
	$n = 2$	$l = 1$	$m_l = +1$

If an atomic orbital contains two electrons, they must have opposite spins so that the sets of quantum numbers for the two electrons are different: e.g. in the $1s$ atomic orbital:

one electron has $n = 1$ $l = 0$ $m_l = 0$ $m_s = +\frac{1}{2}$

the other electron has $n = 1$ $l = 0$ $m_l = 0$ $m_s = -\frac{1}{2}$

[This discussion is extended in Section 20.6.]

Self-study exercises

1. Show that the ground state electronic configuration $1s^2 2s^2 2p^1$ for B corresponds to each electron having a unique set of four quantum numbers.

2. The ground state of N is $1s^2 2s^2 2p^3$. Show that each electron in the $2p$ level possesses a unique set of four quantum numbers.

3. Explain why it is *not* possible for C to possess a ground state electronic configuration of $1s^2 2s^2 2p^2$ with the $2p$ electrons having paired spins.

1.10 Ionization energies and electron affinities

Ionization energies

The ionization energy of hydrogen (eqs. 1.9 and 1.10) was discussed in Section 1.4. Since the H atom has only one electron, no additional ionization processes can occur. For multi-electron atoms, successive ionizations are possible.

The first ionization energy, IE_1, of an atom is the internal energy change at 0 K, $\Delta U(0\,\text{K})$, associated with the removal of the first valence electron (eq. 1.17). The energy change is defined for a *gas-phase* process. The units are kJ mol^{-1} or electron volts (eV).[†]

$$X(g) \longrightarrow X^+(g) + e^- \qquad (1.17)$$

It is often necessary to incorporate ionization energies into thermochemical calculations (e.g. Born–Haber or Hess cycles) and it is convenient to define an associated *enthalpy change*, $\Delta H(298\,\text{K})$. Since the difference between $\Delta H(298\,\text{K})$ and $\Delta U(0\,\text{K})$ is very small (see Box 1.6), values of IE can be used in thermochemical cycles so long as extremely accurate answers are not required.

The ***first ionization energy*** (IE_1) of a gaseous atom is the internal energy change, ΔU, at 0 K associated with the removal of the first valence electron:

$$X(g) \longrightarrow X^+(g) + e^-$$

For thermochemical cycles, an associated change in enthalpy, ΔH, at 298 K is used:

$$\Delta H(298\,\text{K}) \approx \Delta U(0\,\text{K})$$

The second ionization energy, IE_2, of an atom refers to step 1.18; this is equivalent to the first ionization of the ion X^+. Equation 1.19 describes the step corresponding

---

[†] An electron volt is a non-SI unit with a value of $\approx 1.60218 \times 10^{-19}$ J; to compare eV and kJ mol^{-1} units, it is necessary to multiply by the Avogadro number. $1\,\text{eV} = 96.4853 \approx 96.5\,\text{kJ mol}^{-1}$.

THEORY

Box 1.6 The relationship between ΔU and ΔH

The relationship between the change in internal energy, ΔU, and change in enthalpy, ΔH, of the system for a reaction at a given temperature is given by the equation:

$$\Delta U = \Delta H - P\Delta V$$

where P is the pressure and ΔV is the change in volume. The $P\Delta V$ term corresponds to the work done, e.g. in expanding the system against the surroundings as a gas is liberated during a reaction. Often in a chemical reaction, the pressure P corresponds to atmospheric pressure (1 atm = 101 300 Pa, or 1 bar = 10^5 Pa).

In general, the work done by or on the system is much smaller than the enthalpy change, making the $P\Delta V$ term negligible with respect to the values of ΔU and ΔH. Thus:

$$\Delta U(T\,\text{K}) \approx \Delta H(T\,\text{K})$$

However, in Section 1.10, we are considering two different temperatures and state that:

$$\Delta U(0\,\text{K}) \approx \Delta H(298\,\text{K})$$

In order to assess the variation in ΔH with temperature, we apply Kirchhoff's equation where C_P = molar heat capacity at constant pressure:

$$\Delta C_P = \left(\frac{\partial \Delta H}{\partial T}\right)_P$$

the integrated form of which (integrating between the limits of the temperatures 0 and 298 K) is:

$$\int_0^{298} \text{d}(\Delta H) = \int_0^{298} \Delta C_P \, \text{d}T$$

Integrating the left-hand side gives:

$$\Delta H(298\,\text{K}) - \Delta H(0\,\text{K}) = \int_0^{298} \Delta C_P \, \text{d}T$$

Consider the ionization of an atom X:

$$X(g) \longrightarrow X^+(g) + e^-(g)$$

If X, X^+ and e^- are all ideal monatomic gases, then the value of C_P for each is $\frac{5}{2}R$ (where R is the molar gas constant = 8.314×10^{-3} kJ K^{-1} mol^{-1}), giving for the reaction a value of ΔC_P of $\frac{5}{2}R$. Therefore:

$$\Delta H(298\,\text{K}) - \Delta H(0\,\text{K}) = \int_0^{298} \tfrac{5}{2}R \, \text{d}T$$

$$= \left(\frac{5 \times 8.314 \times 10^{-3}}{2}\right)[T]_0^{298}$$

$$= 6.2\,\text{kJ mol}^{-1}$$

Inspection of typical values of ionization energies in Appendix 8 shows that a correction of this magnitude is relatively insignificant because values of *IE* are so large.

---

to the third ionization energy, IE_3, of X, and successive ionizations are similarly defined, all for gas phase processes.

$$X^+(g) \longrightarrow X^{2+}(g) + e^- \tag{1.18}$$

$$X^{2+}(g) \longrightarrow X^{3+}(g) + e^- \tag{1.19}$$

Values of ionization energies for the elements are listed in Appendix 8. Figure 1.16 shows the variation in the values

of IE_1 as a function of Z. Several repeating patterns are apparent and some features to note are:

- the high values of IE_1 associated with the noble gases;
- the very low values of IE_1 associated with the group 1 elements;
- the *general* increase in values of IE_1 as a given period is crossed;

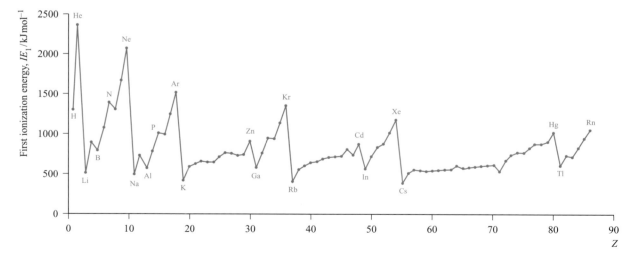

Fig. 1.16 The values of the first ionization energies of the elements up to Rn.

THEORY

Box 1.7 Exchange energies

Filled and half-filled shells are often referred to as possessing a 'special stability'. However, this is misleading, and we should really consider the *exchange energy* of a given configuration. This can only be justified by an advanced quantum mechanical treatment but we can summarize the idea as follows. Consider two electrons in *different* orbitals. The repulsion between the electrons if they have anti-parallel spins is greater than if they have parallel spins, e.g. for a p^2 configuration:

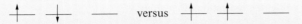

The difference in energy between these two configurations is the *exchange energy*, K, i.e. this is the extra stability that the right-hand configuration has with respect to the left-hand one.

The total exchange energy is expressed in terms of K (the actual value of K depends on the atom or ion):

$$\text{Exchange energy} = \sum \frac{N(N-1)}{2} K$$

where N = number of electrons with parallel spins.

Further reading

A.B. Blake (1981) *J. Chem. Educ.*, vol. 58, p. 393.
B.J. Duke (1978) *Educ. Chem.*, vol. 15, p. 186.
D.M.P. Mingos (1998) *Essential Trends in Inorganic Chemistry*, Oxford University Press, Oxford, p. 14.

- the discontinuity in values of IE_1 on going from an element in group 15 to its neighbour in group 16;
- the decrease in values of IE_1 on going from an element in group 2 or 12 to its neighbour in group 13;
- the rather similar values of IE_1 for a given row of *d*-block elements.

Each of these trends can be rationalized in terms of ground state electronic configurations. The noble gases (except for He) possess ns^2np^6 configurations which are particularly stable (see Box 1.7) and removal of an electron requires a great deal of energy. The ionization of a group 1 element involves loss of an electron from a singly occupied ns orbital with the resultant X^+ ion possessing a noble gas configuration. The *general* increase in IE_1 across a given period is a consequence of an increase in Z_{eff}. A group 15 element has a ground state electronic configuration ns^2np^3 and the np level is *half-occupied*. A certain stability (see Box 1.7) is associated with such configurations and it is more difficult to ionize a group 15 element than its group 16 neighbour. In going from Be (group 2) to B (group 13), there is a marked decrease in IE_1 and this may be attributed to the relative stability of the filled shell $2s^2$ configuration compared with the $2s^22p^1$ arrangement; similarly, in going from Zn (group 12) to Ga (group 13), we need to consider the difference between $4s^23d^{10}$ and $4s^23d^{10}4p^1$ configurations. Trends among IE values for *d*-block metals are discussed in Section 19.3.

Electron affinities

The first electron affinity (EA_1) is *minus* the internal energy change (eq. 1.20) for the gain of an electron by a *gaseous* atom (eq. 1.21). The second electron affinity of atom Y is defined for process 1.22. Each reaction occurs in the gas phase.

$$EA = -\Delta U(0\,\text{K}) \tag{1.20}$$

$$Y(g) + e^- \longrightarrow Y^-(g) \tag{1.21}$$

$$Y^-(g) + e^- \longrightarrow Y^{2-}(g) \tag{1.22}$$

As we saw for ionization energies, it is convenient to define an enthalpy change, $\Delta_{EA}H$, associated with each of the reactions 1.21 and 1.22. We approximate $\Delta_{EA}H(298\,\text{K})$ to $\Delta_{EA}U(0\,\text{K})$. Selected values of these enthalpy changes are given in Table 1.5.

Table 1.5 Approximate *enthalpy changes* $\Delta_{EA}H(298\,\text{K})$ associated with the attachment of an electron to an atom or anion.[†]

Process	$\approx \Delta_{EA}H/\text{kJ mol}^{-1}$
$H(g) + e^- \longrightarrow H^-(g)$	-73
$Li(g) + e^- \longrightarrow Li^-(g)$	-60
$Na(g) + e^- \longrightarrow Na^-(g)$	-53
$K(g) + e^- \longrightarrow K^-(g)$	-48
$N(g) + e^- \longrightarrow N^-(g)$	≈ 0
$P(g) + e^- \longrightarrow P^-(g)$	-72
$O(g) + e^- \longrightarrow O^-(g)$	-141
$O^-(g) + e^- \longrightarrow O^{2-}(g)$	$+798$
$S(g) + e^- \longrightarrow S^-(g)$	-201
$S^-(g) + e^- \longrightarrow S^{2-}(g)$	$+640$
$F(g) + e^- \longrightarrow F^-(g)$	-328
$Cl(g) + e^- \longrightarrow Cl^-(g)$	-349
$Br(g) + e^- \longrightarrow Br^-(g)$	-325
$I(g) + e^- \longrightarrow I^-(g)$	-295

[†] Tables of data differ in whether they list values of EA or $\Delta_{EA}H$ and it is essential to note which is being used.

The *first electron affinity*, EA_1, of an atom is *minus* the internal energy change at 0 K associated with the gain of one electron by a gaseous atom:

$$Y(g) + e^- \longrightarrow Y^-(g)$$

For thermochemical cycles, an associated enthalpy change is used:

$$\Delta_{EA}H(298\,K) \approx \Delta_{EA}U(0\,K) = -EA$$

The attachment of an electron to an *atom* is usually exothermic. Two electrostatic forces oppose one another: the repulsion between the valence shell electrons and the additional electron, and the attraction between the nucleus and the incoming electron. In contrast, *repulsive* interactions are dominant when an electron is added to an *anion* and the process is endothermic (Table 1.5).

KEY TERMS

The following terms were introduced in this chapter. Do you know what they mean?

- ❑ atom
- ❑ proton
- ❑ electron
- ❑ neutron
- ❑ nucleus
- ❑ protium
- ❑ nuclide
- ❑ atomic number
- ❑ mass number
- ❑ relative atomic mass
- ❑ isotope
- ❑ allotrope
- ❑ emission spectrum
- ❑ ground state
- ❑ excited state
- ❑ quanta

- ❑ wavenumber
- ❑ Avogadro number
- ❑ wavefunction, ψ
- ❑ hydrogen-like species
- ❑ principal quantum number, n
- ❑ orbital quantum number, l
- ❑ magnetic quantum number, m_l
- ❑ magnetic spin quantum number, m_s
- ❑ degenerate
- ❑ radial distribution function
- ❑ radial part of a wavefunction
- ❑ angular part of a wavefunction
- ❑ atomic orbital
- ❑ nodal plane
- ❑ radial node

- ❑ ground state electronic configuration
- ❑ effective nuclear charge
- ❑ screening effects of electrons
- ❑ penetration
- ❑ shielding
- ❑ Slater's rules
- ❑ periodic table
- ❑ *aufbau* principle
- ❑ Hund's rules
- ❑ Pauli exclusion principle
- ❑ valence electrons
- ❑ core electrons
- ❑ ionization energy
- ❑ electron affinity

FURTHER READING

First-year chemistry: basic principles

C.E. Housecroft and E.C. Constable (2010) *Chemistry*, 4th edn, Prentice Hall, Harlow – A readable text covering fundamental aspects of inorganic, organic and physical chemistry which gives detailed background of all material that is taken as assumed knowledge in this book. An accompanying multiple-choice test bank and Solutions Manual can be found through www. pearsoned.co.uk/housecroft.

P. Atkins and J. de Paula (2005) *The Elements of Physical Chemistry*, 4th edn, Oxford University Press, Oxford – An excellent introductory text which covers important areas of physical chemistry.

Basic quantum mechanics

P. Atkins and J. de Paula (2010) *Atkins' Physical Chemistry*, 9th edn, Oxford University Press, Oxford – This text

gives a solid and well-tested background in physical chemistry.

D.O. Hayward (2002) *Quantum Mechanics for Chemists*, RSC Publishing, Cambridge – An undergraduate student text that covers the basic principles of quantum mechanics.

Ionization energies and electron affinities

P.F. Lang and B.C. Smith (2003) *J. Chem. Educ.*, vol. 80, p. 938 – 'Ionization energies of atoms and atomic ions'.

D.M.P. Mingos (1998) *Essential Trends in Inorganic Chemistry*, Oxford University Press, Oxford – This text includes detailed discussions of trends in ionization energies and electron attachment enthalpies within the periodic table.

J.C. Wheeler (1997) *J. Chem. Educ.*, vol. 74, p. 123 – 'Electron affinities of the alkaline earth metals and the sign convention for electron affinity'.

PROBLEMS

1.1 Chromium has four isotopes, $^{50}_{24}Cr$, $^{52}_{24}Cr$, $^{53}_{24}Cr$ and $^{54}_{24}Cr$. How many electrons, protons and neutrons does each isotope possess?

1.2 'Arsenic is monotopic.' What does this statement mean? Using Appendix 5, write down three other elements that are monotopic.

1.3 Using the list of naturally occurring isotopes in Appendix 5, determine the number of electrons, protons and neutrons present in an atom of each isotope of (a) Al, (b) Br and (c) Fe, and give appropriate notation to show these data for each isotope.

1.4 Hydrogen possesses three isotopes, but tritium (^{3}H), which is radioactive, occurs as less than 1 in 10^{17} atoms in a sample of natural hydrogen. If the value of A_r for hydrogen is 1.008, estimate the percentage abundance of protium, ^{1}H, and deuterium, ^{2}H (or D) present in a sample of natural hydrogen. Point out any assumptions that you make. Explain why your answers are not the same as those quoted in Appendix 5.

1.5 (a) By using the data in Appendix 5, account for the isotopic distribution shown in Fig. 1.1b. (b) The mass spectrum of S_8 shows other peaks at lower values of m/z. By considering the structure of S_8 shown in Fig. 1.1c, suggest the origin of these lower-mass peaks.

1.6 Calculate the corresponding wavelengths of electromagnetic radiation with frequencies of (a) 3.0×10^{12} Hz, (b) 1.0×10^{18} Hz and (c) 5.0×10^{14} Hz. By referring to Appendix 4, assign each wavelength or frequency to a particular type of radiation (e.g. microwave).

1.7 State which of the following $n' \longrightarrow n$ transitions in the emission spectrum of atomic hydrogen belong to the Balmer, Lyman or Paschen series: (a) $3 \longrightarrow 1$; (b) $3 \longrightarrow 2$; (c) $4 \longrightarrow 3$; (d) $4 \longrightarrow 2$; (e) $5 \longrightarrow 1$.

1.8 Calculate the energy (in kJ per mole of photons) of a spectroscopic transition, the corresponding wavelength of which is 450 nm.

1.9 Four of the lines in the Balmer series are at 656.28, 486.13, 434.05 and 410.17 nm. Show that these wavelengths are consistent with eq. 1.4.

1.10 Using the Bohr model, determine the values of the radii of the second and third orbits of the hydrogen atom.

1.11 How is the (a) energy and (b) size of an ns atomic orbital affected by an increase in n?

1.12 Write down a set of quantum numbers that uniquely defines each of the following atomic orbitals: (a) $6s$, (b) each of the five $4d$ orbitals.

1.13 Do the three $4p$ atomic orbitals possess the same or different values of (a) principal quantum number, (b) the orbital quantum number and (c) the magnetic quantum number? Write down a set of quantum numbers for each $4p$ atomic orbital to illustrate your answer.

1.14 How many radial nodes does each of the following orbitals possess: (a) $2s$; (b) $4s$; (c) $3p$; (d) $5d$; (e) $1s$; (f) $4p$?

1.15 Comment on differences between plots of $R(r)$ against r, and $4\pi r^2 R(r)^2$ against r for each of the following atomic orbitals of an H atom: (a) $1s$; (b) $4s$; (c) $3p$.

1.16 Write down the sets of quantum numbers that define the (a) $1s$, (b) $4s$, (c) $5s$ atomic orbitals.

1.17 Write down the three sets of quantum numbers that define the three $3p$ atomic orbitals.

1.18 How many atomic orbitals make up the set with $n = 4$ and $l = 3$? What label is given to this set of orbitals? Write down a set of quantum numbers that defines each orbital in the set.

1.19 Which of the following species are hydrogen-like: (a) H^+; (b) He^+; (c) He^-; (d) Li^+; (e) Li^{2+}?

1.20 (a) Will a plot of $R(r)$ for the $1s$ atomic orbital of He^+ be identical to that of the H atom (Fig. 1.5a)? [**Hint:** look at Table 1.2.] (b) On the *same axis set*, sketch approximate representations of the function $4\pi r^2 R(r)^2$ for H and He^+.

1.21 Calculate the energy of the $3s$ atomic orbital of an H atom. [**Hint:** see eq. 1.16.] Is the energy of the hydrogen $3p$ atomic orbital the same as or different from that of the $3s$ orbital?

1.22 Using eq. 1.16, determine the energies of atomic orbitals of hydrogen with $n = 1, 2, 3, 4$ and 5. What can you say about the relative spacings of the energy levels?

1.23 Write down the six sets of quantum numbers that describe the electrons in a degenerate set of $5p$ atomic orbitals. Which pairs of sets of quantum numbers refer to spin-paired electrons?

1.24 For a neutral atom, X, arrange the following atomic orbitals in an approximate order of their relative energies (not all orbitals are listed): $2s$, $3s$, $6s$, $4p$, $3p$, $3d$, $6p$, $1s$.

1.25 Using the concepts of shielding and penetration, explain why a ground state configuration of $1s^2 2s^1$ for an Li atom is energetically preferred over $1s^2 2p^1$.

1.26 For each of the following atoms, write down a ground state electronic configuration and indicate which electrons are core and which are valence: (a) Na, (b) F, (c) N, (d) Sc.

1.27 Draw energy level diagrams (see Fig. 1.15) to represent the ground state electronic configurations of the atoms in problem 1.26.

1.28 Write down the ground state electronic configuration of boron, and give a set of quantum numbers that uniquely defines each electron.

1.29 Write down (with reasoning) the ground state electronic configurations of (a) Li, (b) O, (c) S, (d) Ca, (e) Ti, (f) Al.

1.30 Draw energy level diagrams to show the ground state electronic configurations of only the *valence* electrons in an atom of (a) F, (b) Al and (c) Mg.

1.31 The ground state electronic configuration of a group 16 element is of the type $[X]ns^2np^4$ where X is a group 18 element. How are the outer four electrons arranged, and what rules are you using to work out this arrangement?

1.32 (a) Write down an equation that defines the process to which the value of IE_4 of Sn refers. Is this process exothermic or endothermic? (b) To what overall process does a value of $(IE_1 + IE_2 + IE_3)$ for Al refer?

1.33 The first four ionization energies of an atom X are 403, 2633, 3900 and 5080 kJ mol^{-1}. Suggest to what periodic group X belongs and give reasons for your choice.

1.34 In Fig. 1.16, identify the trends in the first ionization energies of the elements in (a) descending group 1, (b) descending group 13, (c) crossing the first row of the *d*-block, (d) crossing the row of elements from B to Ne, (e) going from Xe to Cs, and (f) going from P to S. Rationalize each of the trends you have described.

1.35 Figure 1.17 shows the values of IE_1 for the first 10 elements. (a) Label each point with the symbol of the appropriate element. (b) Give detailed reasons for the observed trend in values.

1.36 (a) Using the data in Table 1.5, determine a value for ΔH for the process:

$$O(g) + 2e^- \longrightarrow O^{2-}(g)$$

(b) Comment on the relevance of the sign and magnitude of your answer to part (a) in the light of the fact that many metal oxides with ionic lattices are thermodynamically stable.

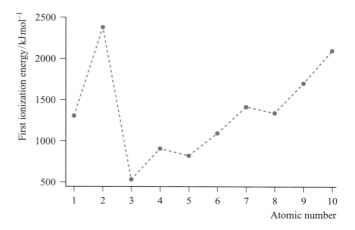

Fig. 1.17 Graph for problem 1.35.

OVERVIEW PROBLEMS

1.37 How do you account for the fact that, although potassium is placed after argon in the periodic table, it has a lower relative atomic mass?

1.38 What is the evidence that the *aufbau* principle is only approximately true?

1.39 The first list in the table opposite contains symbols or phrases, each of which has a 'partner' in the second list. Match the 'partners'; there is only one match for each pair of symbols or phrases.

1.40 Suggest explanations for the following.

(a) High values of ionization energies are associated with the noble gases.

(b) The enthalpy changes associated with the attachment of the first and second electrons to an O atom are exothermic and endothermic, respectively.

(c) In an Li atom in its ground state, the outer electron occupies a *2s* rather than a *2p* orbital.

List 1	List 2
S_6 and S_8	electron
^{19}F and ^{31}P	proton
isotope of hydrogen	pnictogens
^{12}C and ^{13}C	*d*-block elements
hydrogen ion	protium
group 1 elements	fundamental particles
same energy	$m_s = \pm\frac{1}{2}$
negatively charged particle	allotropes
spin-paired electrons	degenerate
electron, proton and neutron	monotopic elements
group 15 elements	alkali metals
Cr, Mn, Fe	isotopes of an element

1.41   Using data from Appendix 8, construct a graph to show the trend in the third ionization energies of the elements from Li to Kr. Compare the graph with that shown in Fig. 1.16, and rationalize what you observe.

1.42   The sign convention for electron affinity can often cause confusion for students. In this textbook, why have we referred to 'an enthalpy change for the attachment of an electron' rather than to an 'electron affinity'?

1.43   (a) How would Fig. 1.9 have to be modified to show boundary surfaces for the $2s$ and the $3p$ wavefunctions of a one-electron species?

(b) 'The probability of finding the electron of a ground-state hydrogen atom at a distance r from the proton is at a maximum when $r = 52.9 \, pm$.' Why is this statement compatible with the maximum in the value of $R(r)$ at $r = 0$?

INORGANIC CHEMISTRY MATTERS

1.44   Ruthenium, osmium, rhodium, iridium, palladium and platinum (Fig. 1.14) are called the platinum group metals. Most of the world's reserves of these metals are in mineral deposits in Russia, Canada and South Africa. The platinum group metals are important as catalysts for air pollution control (e.g. in catalytic converters) and in the manufacture of organic and inorganic chemicals, and they have applications in the electronics industry. Thus, countries such as the US depend upon importing the metals and upon their recycling. The table below gives import data for the US for 2008:

Metal	Ru	Os	Rh	Ir	Pd	Pt
Imported amount/kg	49 800	11	12 600	2550	120 000	150 000

Plot bar charts to illustrate these data, first using mass on the vertical axis, and then using a logarithmic scale. Comment on the advantages or disadvantages of the two plots.

1.45   Figure 1.18 shows the emission spectrum of sodium. Low-pressure sodium street lamps depend upon this bright yellow emission from sodium atoms excited by an electrical discharge. Figure 1.18 shows a

Fig. 1.18 The emission spectrum of sodium. The apparent single line consists of two very close emissions at 589.0 and 589.6 nm.

simple spectrum (see figure caption), but the National Institute of Standards and Technology (NIST) atomic spectra database lists 5888 lines in the emission spectrum of sodium. Suggest *three* reasons why no other lines are visible in Fig. 1.18. (b) The wavelengths of the yellow lines in Fig. 1.18 are close to 589 nm. To what frequency does this correspond? (c) Give a *general* explanation of how a series of spectral lines such as those in Fig. 1.18 arises.

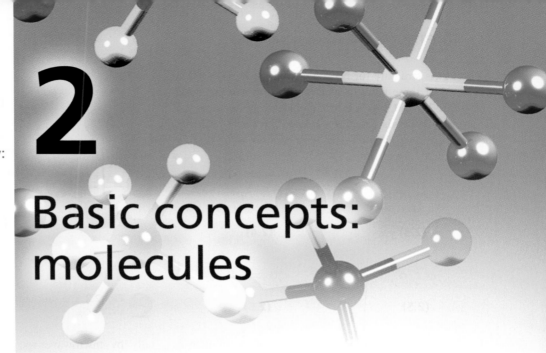

Topics

Lewis structures
Valence bond theory
Molecular orbital theory:
   diatomics
Octet rule
Isoelectronic species
Electronegativity
Dipole moments
VSEPR model
Stereoisomerism

2

Basic concepts: molecules

2.1 Bonding models: an introduction

In Sections 2.1–2.3 we summarize valence bond (VB) and molecular orbital (MO) theories of homonuclear bond formation, and include practice in generating *Lewis structures*.

A historical overview

The foundations of modern chemical bonding theory were laid in 1916–1920 by G.N. Lewis and I. Langmuir, who suggested that ionic species were formed by electron transfer while electron sharing was important in covalent molecules. In some cases, it was suggested that the shared electrons in a bond were provided by one of the atoms but that once the bond (sometimes called a *coordinate bond*) is formed, it *is indistinguishable from a 'normal' covalent bond*.

In a ***covalent*** species, electrons are shared between atoms. In an ***ionic*** species, one or more electrons are transferred between atoms to form ions.

Modern views of atomic structure are, as we have seen, based largely on the applications of wave mechanics to atomic systems. Modern views of *molecular structure* are based on applying wave mechanics to molecules; such studies provide answers as to how and why atoms combine. The Schrödinger equation can be written to describe the behaviour of electrons in molecules, but it can be solved only approximately. Two methods of doing this are the valence bond approach, developed by Heitler and Pauling, and the molecular orbital approach associated with Hund and Mulliken:

- *Valence bond (VB) theory* treats the formation of a molecule as arising from the bringing together of complete atoms which, when they interact, to a large extent retain their original character.
- *Molecular orbital (MO) theory* allocates electrons to molecular orbitals formed by the overlap (interaction) of atomic orbitals.

Although familiarity with both VB and MO concepts is necessary, it is often the case that a given situation is more conveniently approached by using one or other of these models. We begin with the conceptually simple approach of Lewis for representing the bonding in covalent molecules.

Lewis structures

Lewis presented a simple, but useful, method of describing the arrangement of valence electrons in molecules. The approach uses dots (or dots and crosses) to represent the number of *valence electrons*, and the nuclei are indicated by appropriate elemental symbols. A basic premise of the theory is that electrons in a molecule should be paired; the presence of a single (odd) electron indicates that the species is a *radical*.

Diagram **2.1** shows the Lewis structure for H_2O with the O$-$H bonds designated by pairs of dots (electrons). An alternative representation is given in structure **2.2** where each line stands for *one pair* of electrons, i.e. a *single covalent bond*. Pairs of valence electrons which are not involved in bonding are *lone pairs*.

$$
\begin{array}{cc}
\text{H} & \text{H} \\
\overset{\cdot\cdot}{\underset{\cdot\cdot}{:\text{O}:}}\ \text{H} & \overset{|}{\underset{\cdot\cdot}{:\text{O}}}\!\!-\!\!\text{H} \\
(2.1) & (2.2)
\end{array}
$$

The Lewis structure for N_2 shows that the N−N bond is composed of three pairs of electrons and is a *triple bond* (structures **2.3** and **2.4**). Each N atom has one lone pair of electrons. The Lewis structures **2.5** and **2.6** for O_2 indicate the presence of a *double bond*, with each O atom bearing two lone pairs of electrons.

:N : N :

(2.3)

:N≡N :

(2.4)

:O : O:

(2.5)

:O=O:

(2.6)

Lewis structures give the connectivity of an atom in a molecule, the bond order and the number of lone pairs, and these may be used to derive structures using the valence-shell electron-pair repulsion model (see Section 2.8).

2.2 Homonuclear diatomic molecules: valence bond (VB) theory

Uses of the term *homonuclear*

The word *homonuclear* is used in two ways:

- A *homonuclear covalent bond* is one formed between two atoms of the same element, e.g. the H−H bond in H_2, the O=O bond in O_2, and the O−O bond in H_2O_2 (Fig. 2.1).
- A *homonuclear molecule* contains one type of element. Homonuclear diatomic molecules include H_2, N_2 and F_2, homonuclear triatomics include O_3 (ozone) and examples of larger homonuclear molecules are P_4, S_8 (Fig. 2.2) and C_{60}.

Covalent bond distance, covalent radius and van der Waals radius

Three important definitions are needed before we discuss covalent bonding.

The length of a covalent bond (*bond distance*), d, is the *internuclear separation* and may be determined experimen-

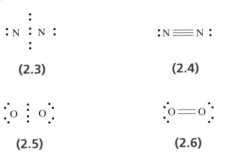

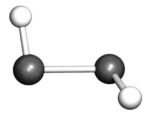

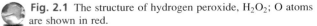

Fig. 2.1 The structure of hydrogen peroxide, H_2O_2; O atoms are shown in red.

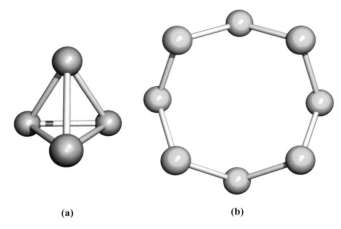

(a)   (b)

Fig. 2.2 The structures of the homonuclear molecules (a) P_4 and (b) S_8.

tally by microwave spectroscopy or diffraction methods (see Chapter 4). It is convenient to define the covalent radius, r_{cov}, of an atom: for an atom X, r_{cov} is half of the covalent bond length of a homonuclear X−X *single* bond. Thus, $r_{cov}(S)$ can be determined from the solid state structure of S_8 (Fig. 2.2) determined by X-ray diffraction methods or, better still, by averaging the values of the bond distances of S−S single bonds found for all the allotropes of sulfur.

> For an atom X, the value of the single bond *covalent radius*, r_{cov}, is half of the internuclear separation in a homonuclear X−X single bond.

The α- and β-forms of sulfur (orthorhombic and monoclinic sulfur, respectively) both crystallize with S_8 molecules stacked in a regular arrangement. The packing in the α-form (Fig. 2.3, density = $2.07 \, g \, cm^{-3}$) is more efficient than that in the β-form (density = $1.94 \, g \, cm^{-3}$). Van der Waals forces operate between the molecules, and half of the distance of closest approach of two sulfur atoms belonging to *different* S_8 rings is defined as the van der Waals radius, r_v, of sulfur. The weakness of the bonding is evidenced by the fact that S_8 vaporizes, retaining the ring

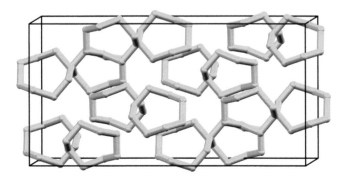

Fig. 2.3 The packing of S_8 rings in the α-allotrope of sulfur. The black box defines the unit cell (see Section 6.2). [Data: S. J. Rettig *et al.* (1987) *Acta Crystallogr., Sect. C*, vol. 43, p. 2260.]

structure, without absorbing much energy. The van der Waals radius of an element is necessarily larger than its covalent radius, e.g. r_v and r_{cov} for S are 185 and 103 pm, respectively. Van der Waals forces include dispersion and dipole–dipole interactions; dispersion forces are discussed in the latter part of Section 6.13 and dipole moments in Section 2.6. Because van der Waals forces operate *between* molecules, they are crucial in controlling the way in which molecules pack in the solid state. Values of r_v and r_{cov} are listed in Appendix 6.

The **van der Waals radius**, r_v, of an atom X is half of the distance of closest approach of two non-bonded atoms of X.

The valence bond (VB) model of bonding in H_2

Valence bond theory considers the interactions between separate atoms as they are brought together to form molecules. We begin by considering the formation of H_2 from two H atoms, the nuclei of which are labelled H_A and H_B, and the electrons of which are 1 and 2, respectively. When the atoms are so far apart that there is no interaction between them, electron 1 is exclusively associated with H_A, while electron 2 resides with nucleus H_B. Let this state be described by a wavefunction ψ_1.

When the H atoms are close together, we cannot tell which electron is associated with which nucleus since, although *we* gave them labels, the two nuclei are actually indistinguishable, as are the two electrons. Thus, electron 2 could be with H_A and electron 1 with H_B. Let this be described by the wavefunction ψ_2.

Equation 2.1 gives an overall description of the covalently bonded H_2 molecule; $\psi_{covalent}$ is a *linear combination* of wavefunctions ψ_1 and ψ_2. The equation contains a *normalization factor*, N (see Box 1.3). In the general case where:

$$\psi_{covalent} = c_1\psi_1 + c_2\psi_2 + c_3\psi_3 + \cdots$$

$$N = \frac{1}{\sqrt{c_1{}^2 + c_2{}^2 + c_3{}^2 + \cdots}}$$

$$\psi_{covalent} = \psi_+ = N(\psi_1 + \psi_2) \tag{2.1}$$

Another linear combination of ψ_1 and ψ_2 can be written as shown in eq. 2.2.

$$\psi_- = N(\psi_1 - \psi_2) \tag{2.2}$$

In terms of the spins of electrons 1 and 2, ψ_+ corresponds to spin-pairing, and ψ_- corresponds to parallel spins (non-spin-paired). Calculations of the energies associated with these states as a function of the internuclear separation of H_A and H_B show that, while ψ_- represents a repulsive state (high energy), the energy curve for ψ_+ reaches a

minimum value when the internuclear separation, d, is 87 pm and this corresponds to an H–H bond dissociation energy, ΔU, of 303 kJ mol^{-1}. While these are near enough to the experimental values of $d = 74$ pm and $\Delta U = 458$ kJ mol^{-1} to suggest that the model has some validity, they are far enough away from them to indicate that the expression for ψ_+ needs refining.

The **bond dissociation energy** (ΔU) and **enthalpy** (ΔH) values for H_2 are defined for the process:

$$H_2(g) \longrightarrow 2H(g)$$

Improvements to eq. 2.1 can be made by:

- allowing for the fact that each electron screens the other from the nuclei to some extent;
- taking into account the possibility that *both* electrons 1 and 2 may be associated with either H_A *or* H_B, i.e. allowing for the transfer of one electron from one nuclear centre to the other to form a pair of ions $H_A{}^+H_B{}^-$ or $H_A{}^-H_B{}^+$.

The latter modification is dealt with by writing two additional wavefunctions, ψ_3 and ψ_4 (one for each ionic form), and so eq. 2.1 can be rewritten in the form of eq. 2.3. The coefficient c indicates the relative contributions made by the two sets of wavefunctions. For a *homonuclear diatomic* such as H_2, the situations described by ψ_1 and ψ_2 are equally probable, as are those described by ψ_3 and ψ_4.

$$\psi_+ = N[(\psi_1 + \psi_2) + c(\psi_3 + \psi_4)] \tag{2.3}$$

Since the wavefunctions ψ_1 and ψ_2 arise from an internuclear interaction involving the *sharing* of electrons among nuclei, and ψ_3 and ψ_4 arise from electron *transfer*, we can simplify eq. 2.3 to 2.4, in which the overall wavefunction, $\psi_{molecule}$, is composed of covalent and ionic terms.

$$\psi_{molecule} = N[\psi_{covalent} + (c \times \psi_{ionic})] \tag{2.4}$$

Based on this model of H_2, calculations with $c \approx 0.25$ give values of 75 pm for d(H–H) and 398 kJ mol^{-1} for the bond dissociation energy. Modifying eq. 2.4 still further leads to a value of ΔU very close to the experimental value, but details of the procedure are beyond the scope of this book.[†]

Now consider the physical significance of eqs. 2.3 and 2.4. The wavefunctions ψ_1 and ψ_2 represent the structures shown in **2.7** and **2.8**, while ψ_3 and ψ_4 represent the ionic forms **2.9** and **2.10**. The notation $H_A(1)$ stands for 'nucleus H_A with electron (1)', and so on.

$H_A(1)H_B(2)$	$H_A(2)H_B(1)$	$[H_A(1)(2)]^- \ H_B{}^+$	$H_A{}^+ \ [H_B(1)(2)]^-$
(2.7)	**(2.8)**	**(2.9)**	**(2.10)**

[†] For detailed discussion, see: R. McWeeny (1979) *Coulson's Valence*, 3rd edn, Oxford University Press, Oxford.

Dihydrogen is described as a *resonance hybrid* of these contributing *resonance* or *canonical structures*. In the case of H_2, an example of a homonuclear diatomic molecule which is necessarily symmetrical, we simplify the picture to **2.11**. Each of structures **2.11a**, **2.11b** and **2.11c** is a *resonance structure* and the double-headed arrows indicate the *resonance* between them. The contributions made by **2.11b** and **2.11c** are equal. The term 'resonance hybrid' is somewhat unfortunate but is too firmly established to be eradicated.

$$H\text{—}H \longleftrightarrow H^+ \ H^- \longleftrightarrow H^- \ H^+$$

$$\textbf{(2.11a)} \qquad\qquad \textbf{(2.11b)} \qquad\qquad \textbf{(2.11c)}$$

The bonding in a molecule is described in terms of contributing *resonance structures*. The *resonance* between these contributing structures results in a *resonance stabilization*. The relationship between resonance structures is indicated by using a double-headed arrow.

A crucial point about resonance structures is that they *do not exist as separate species*. Rather, they indicate extreme bonding pictures, the combination of which gives a description of the molecule overall. In the case of H_2, the contribution made by resonance structure **2.11a** is significantly greater than that of **2.11b** or **2.11c**.

Notice that **2.11a** describes the bonding in H_2 in terms of a *localized 2-centre 2-electron*, 2c-2e, covalent bond. A particular resonance structure will always indicate a localized bonding picture, although the combination of several resonance structures may result in the description of the bonding in the species as a whole being delocalized (see Section 5.3).

The valence bond (VB) model applied to F_2, O_2 and N_2

Consider the formation of F_2. The ground state electronic configuration of F is $[He]2s^2 2p^5$, and the presence of the single unpaired electron indicates the formation of an F—F single bond. We can write down resonance structures **2.12** to describe the bonding in F_2, with the expectation that the covalent contribution will predominate.

$$F\text{—}F \longleftrightarrow F^+ \ F^- \longleftrightarrow F^- \ F^+$$

$$\textbf{(2.12)}$$

The formation of O_2 involves the combination of two O atoms with ground state electronic configurations of $1s^2 2s^2 2p^4$. Each O atom has two unpaired electrons and so VB theory predicts the formation of an O=O double bond. Since VB theory works on the premise that electrons are paired wherever possible, the model predicts that O_2 is diamagnetic. One of the notable failures of VB theory is its inability to predict the observed *paramagnetism* of O_2. As we shall see,

molecular orbital theory is fully consistent with O_2 being a diradical. When two N atoms ($[He]2s^2 2p^3$) combine to give N_2, an N≡N triple bond results. Of the possible resonance structures, the predominant form is covalent and this gives a satisfactory picture of the bonding in N_2.

In a *diamagnetic* species, all electrons are spin-paired; a diamagnetic substance is repelled by a magnetic field. A *paramagnetic* species contains one or more unpaired electrons; a paramagnetic substance is attracted by a magnetic field.

Self-study exercises

1. Within VB theory, the wavefunction that describes the bonding region between two H atoms in H_2 can be written in the form:

$$\psi_{\text{molecule}} = N\,[\psi_{\text{covalent}} + (c \times \psi_{\text{ionic}})]$$

Explain the meaning of this equation, including the reason why the factor N is included.

2. It is *incorrect* to draw an equilibrium symbol between two resonance structures. The correct notation is a double-headed arrow. Explain why the distinction between these notations is so important.

3. Although O_2 is paramagnetic, VB theory results in a prediction that it is diamagnetic. Explain why this is the case.

2.3 Homonuclear diatomic molecules: molecular orbital (MO) theory

An overview of the MO model

In molecular orbital (MO) theory, we begin by placing the nuclei of a given molecule in their equilibrium positions and then calculate the *molecular orbitals* (i.e. regions of space spread over the entire molecule) that a single electron might occupy. Each MO arises from interactions between orbitals of atomic centres in the molecule, and such interactions are:

- allowed if the *symmetries* of the atomic orbitals are compatible with one another;
- efficient if the region of *overlap* between the two atomic orbitals is significant;
- efficient if the atomic orbitals are relatively close in energy.

An important ground-rule of MO theory is that *the number of MOs that can be formed must equal the number of atomic orbitals of the constituent atoms*.

Each MO has an associated energy and, to derive the electronic ground state of a molecule, the available electrons are placed, according to the *aufbau* principle, in MOs beginning with that of lowest energy. The sum of the individual energies of the electrons in the orbitals (after correction for electron–electron interactions) gives the total energy of the molecule.

Molecular orbital theory applied to the bonding in H_2

An approximate description of the MOs in H_2 can be obtained by considering them as *linear combinations of atomic orbitals* (LCAOs). Each of the H atoms has one $1s$ atomic orbital; let the two associated wavefunctions be ψ_1 and ψ_2. In Section 1.6, we mentioned the importance of the *signs of the wavefunctions* with respect to their overlap during bond formation. The sign of the wavefunction associated with the $1s$ atomic orbital may be either $+$ or $-$. Just as transverse waves interfere in a constructive (in-phase) or destructive (out-of-phase) manner, so too do orbitals. Mathematically, we represent the possible combinations of the two $1s$ atomic orbitals by eqs. 2.5 and 2.6, where N and N^* are the normalization factors. Whereas ψ_{MO} is an in-phase (*bonding*) interaction, ψ_{MO}^* is an out-of-phase (*antibonding*) interaction.

$$\psi_{MO(in-phase)} = \psi_{MO} = N[\psi_1 + \psi_2] \qquad (2.5)$$

$$\psi_{MO(out-of-phase)} = \psi_{MO}^* = N^*[\psi_1 - \psi_2] \qquad (2.6)$$

The values of N and N^* are determined using eqs. 2.7 and 2.8 where S is the *overlap integral*. This is a measure of the extent to which the regions of space described by the two wavefunctions ψ_1 and ψ_2 coincide. Although we mentioned earlier that orbital interaction is efficient if the region of overlap between the two atomic orbitals is significant, the numerical value of S is still much less than unity and is often neglected giving the approximate results shown in eqs. 2.7 and 2.8.

$$N = \frac{1}{\sqrt{2(1+S)}} \approx \frac{1}{\sqrt{2}} \qquad (2.7)$$

$$N^* = \frac{1}{\sqrt{2(1-S)}} \approx \frac{1}{\sqrt{2}} \qquad (2.8)$$

The interaction between the H $1s$ atomic orbitals on forming H_2 may be represented by the energy level diagram in Fig. 2.4. The bonding MO, ψ_{MO}, is stabilized with respect to the $1s$ atomic orbitals, while the antibonding MO, ψ_{MO}^*, is destabilized.[†] Each H atom contributes one electron and, by the *aufbau* principle, the two electrons occupy the lower of the two MOs in the H_2 molecule and are spin-paired (Fig. 2.4). It is important to remember that in MO theory

[†] The difference between the energies of the $1s$ atomic orbitals and ψ_{MO}^* is slightly greater than between those of the $1s$ atomic orbitals and ψ_{MO}, i.e. an antibonding MO is slightly more antibonding than the corresponding bonding MO is bonding; the origin of this effect is beyond the scope of this book.

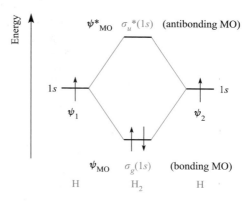

Fig. 2.4 An orbital interaction diagram for the formation of H_2 from two hydrogen atoms. By the *aufbau* principle, the two electrons occupy the lowest energy (bonding) molecular orbital.

we construct the orbital interaction diagram first and then put in the electrons according to the aufbau principle.

The bonding and antibonding MOs in H_2 are given the symmetry labels σ and σ^* ('*sigma*' and '*sigma-star*') or, more fully, $\sigma_g(1s)$ and $\sigma_u^*(1s)$ to indicate their atomic orbital origins and the *parity* of the MOs (see Box 2.1). In order to define these labels, consider the pictorial representations of the two MOs. Figure 2.5a shows that when the $1s$ atomic orbitals interact in phase, the two wavefunctions reinforce each other, especially in the region of space between the nuclei. The two electrons occupying this MO will be found predominantly between the two nuclei, and the build-up of electron density reduces internuclear repulsion. Figure 2.5b illustrates that the out-of-phase interaction results in a *nodal plane between the two H nuclei*. If the antibonding orbital were to be occupied, there would be a zero probability of finding the electrons at any point on the nodal plane. This lack of electron density raises the internuclear repulsion and, as a result, destabilizes the MO.

Now let us return to the σ and σ^* labels. An MO has σ-symmetry if it is symmetrical with respect to a line joining the two nuclei; i.e. if you rotate the orbital about the internuclear axis (the axis joining the two nuclear centres marked in Figs. 2.5a and 2.5b), there is no phase change. A σ^*-orbital must exhibit two properties:

- the σ label means that rotation of the orbital about the internuclear axis generates no phase change, *and*
- the * label means that there is a nodal plane *between* the nuclei, and this plane is orthogonal to the internuclear axis.

The ground state electronic configuration of H_2 may be written using the notation $\sigma_g(1s)^2$, indicating that two electrons occupy the $\sigma_g(1s)$ MO.

The orbital interaction diagram shown in Fig. 2.4 can be used to predict several properties of the H_2 molecule. Firstly, the electrons are paired and so we expect H_2 to be diamagnetic as is found experimentally. Secondly, the formal bond order can be found using eq. 2.9. For H_2 this

THEORY

Box 2.1 The parity of MOs for a molecule that possesses a centre of inversion

We consider symmetry in Chapter 3, but it is useful at this point to consider the labels that are commonly used to describe the *parity of a molecular orbital*. A *homonuclear* diatomic molecule (e.g. H_2, Cl_2) possesses a centre of inversion (centre of symmetry), and the parity of an MO describes the way in which the orbital behaves with respect to this centre of inversion.

First find the centre of inversion in the molecule; this is the point through which you can draw an infinite number of straight lines such that each line passes through a pair of similar points, one on each side of the centre of symmetry and at equal distances from it:

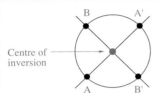

Point A is related to A' by passing through the centre of inversion. Similarly, B is related to B'.

Now ask the question: 'Does the wavefunction have the same *sign* at the same distance but in opposite directions from the centre of symmetry?'

If the answer is 'yes', then the orbital is labelled *g* (from the word *gerade*, German for 'even'). If the answer is 'no',

then the orbital is labelled *u* (from the word *ungerade*, German for 'odd'). For example, the σ-bonding MO in H_2 (Fig. 2.5a) is labelled σ_g, while the antibonding MO (Fig. 2.5b) is σ_u^*.

Parity labels *only* apply to MOs in molecules that possess a centre of inversion (*centrosymmetric* molecules), e.g. homonuclear X_2, octahedral EX_6 and square planar EX_4 molecules. Heteronuclear XY, or tetrahedral EX_4 molecules, for example, do not possess a centre of inversion and are called *non-centrosymmetric* species.

Self-study exercises

Look at Fig. 2.7 which may be applied to the MOs in the homonuclear diatomic O_2.

1. Why does a σ-MO formed by the overlap of two $2p_z$ orbitals (Fig. 2.7a) have the label σ_g?

2. Why does a π-MO formed by the overlap of two $2p_x$ orbitals (Fig. 2.7c) have the label π_u?

3. The antibonding MOs shown at the right-hand sides of Figs. 2.7b and 2.7d carry the labels σ_u^* and π_g^*, respectively. Explain the difference in the parity labels.

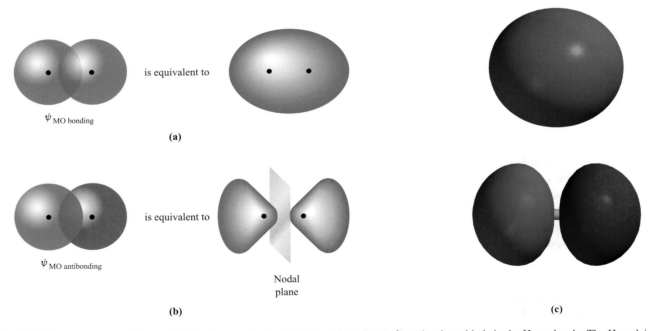

Fig. 2.5 Schematic representations of (a) the bonding (σ_g) and (b) the antibonding (σ_u^*) molecular orbitals in the H_2 molecule. The H nuclei are represented by black dots. The red orbital lobes could equally well be marked with a + sign, and the blue lobes with a − sign (or vice versa) to indicate the sign of the wavefunction. (c) More realistic representations of the molecular orbitals of H_2, generated computationally using Spartan '04, ©Wavefunction Inc. 2003.

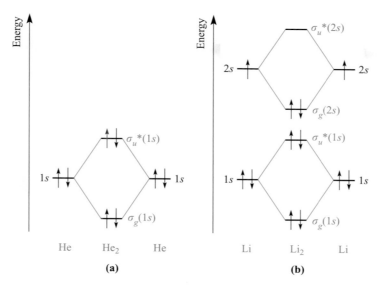

Fig. 2.6 Orbital interaction diagrams for the formation of (a) He_2 from two He atoms and (b) Li_2 from two Li atoms.

gives a bond order of 1.

$$\text{Bond order} = \tfrac{1}{2}[(\text{Number of bonding electrons})$$
$$- (\text{Number of antibonding electrons})]$$

$$(2.9)$$

We cannot measure the bond order experimentally but we can make some useful correlations between bond order and the experimentally measurable bond distances and bond dissociation energies or enthalpies. Along a series of species related by electron gain (reduction) or loss (oxidation), inspection of the corresponding MO diagram shows how the bond order may change (assuming that there are no major changes to the energy levels of the orbitals). For example, the oxidation of H_2 to $[H_2]^+$ (a change brought about by the action of an electric discharge on H_2 at low pressures) can be considered in terms of the removal of one electron from the bonding MO shown in Fig. 2.4. The bond order of $[H_2]^+$ is (eq. 2.9) 0.5, and we expect the H—H bond to be weaker than in H_2. Experimentally, the bond dissociation energy, ΔU, for H_2 is $458\,kJ\,mol^{-1}$ and for $[H_2]^+$ is $269\,kJ\,mol^{-1}$. Similar correlations can be made between bond order and bond length: the lower the bond order, the larger the internuclear separation; the experimentally determined bond lengths of H_2 and $[H_2]^+$ are 74 and 105 pm. While such correlations are useful, they must be treated with caution[†] and *only* used in series of closely related species.

The bonding in He_2, Li_2 and Be_2

Molecular orbital theory can be applied to any homonuclear diatomic molecule, but as more valence atomic orbitals

---

[†] See for example: M. Kaupp and S. Riedel (2004) *Inorg. Chim. Acta*, vol. 357, p. 1865 – 'On the lack of correlation between bond lengths, dissociation energies and force constants: the fluorine-substituted ethane homologues'.

become available, the MO diagram becomes more complex. Treatments of the bonding in He_2, Li_2 and Be_2 are similar to that for H_2. In practice, He does not form He_2, and the construction of an MO diagram for He_2 is a useful exercise because it rationalizes this observation. Figure 2.6a shows that when the two $1s$ atomic orbitals of two He atoms interact, σ and σ^* MOs are formed as in H_2. However, each He atom contributes two electrons, meaning that in He_2, both the bonding *and* antibonding MOs are fully occupied. The bond order (eq. 2.9) is zero and so the MO picture of He_2 is consistent with its non-existence. Using the same notation as for H_2, the ground state electronic configuration of He_2 is $\sigma_g(1s)^2\sigma_u^*(1s)^2$.

The ground state electronic configuration of Li $(Z = 3)$ is $1s^2 2s^1$ and when two Li atoms combine, orbital overlap occurs efficiently between the $1s$ atomic orbitals and between the $2s$ atomic orbitals. To a first approximation we can ignore $1s$–$2s$ overlap since the $1s$ and $2s$ orbital energies are poorly matched. An approximate orbital interaction diagram for the formation of Li_2 is given in Fig. 2.6b. Each Li atom provides three electrons, and the six electrons in Li_2 occupy the lowest energy MOs to give a ground state electronic configuration of $\sigma_g(1s)^2\sigma_u^*(1s)^2\sigma_g(2s)^2$. Effectively, we could ignore the interaction between the core $1s$ atomic orbitals since the net bonding is determined by the interaction between the valence atomic orbitals, and a simpler, but informative, electronic ground state is $\sigma_g(2s)^2$. Figure 2.6b also shows that Li_2 is predicted to be diamagnetic in keeping with experimental data. By applying eq. 2.9, we see that MO theory gives a bond order in Li_2 of one. Note that the terminology 'core and valence orbitals' is equivalent to that for 'core and valence electrons' (see Section 1.9).

Like Li, Be has available $1s$ and $2s$ atomic orbitals for bonding; these atomic orbitals constitute the *basis set of*

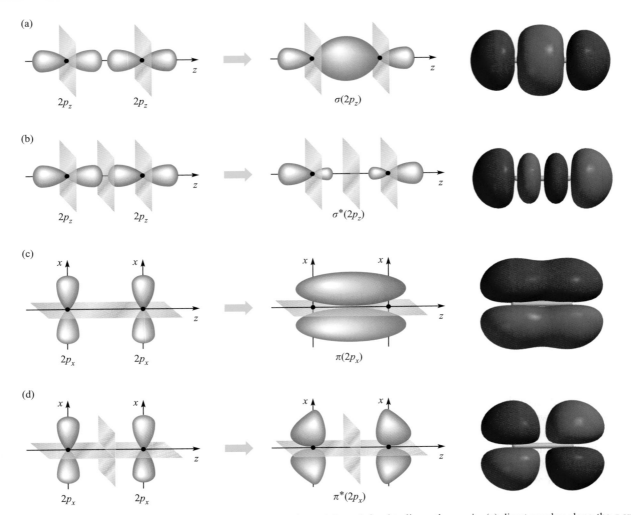

Fig. 2.7 The overlap of two $2p$ atomic orbitals for which the atomic nuclei are defined to lie on the z axis: (a) direct overlap along the z axis gives a $\sigma_g(2p_z)$ MO (bonding); (b) the formation of the $\sigma_u^*(2p_z)$ MO (antibonding); (c) sideways overlap of two $2p_x$ atomic orbitals gives a $\pi_u(2p_x)$ MO (bonding); (d) the formation of $\pi_g^*(2p_x)$ MO (antibonding). Atomic nuclei are marked in black and nodal planes in grey. The diagrams on the right-hand side are more realistic representations of the MOs and have been generated computationally using Spartan '04, ©Wavefunction Inc. 2003.

orbitals. An orbital interaction diagram similar to that for Li_2 (Fig. 2.6b) is appropriate. The difference between Li_2 and Be_2 is that Be_2 has two more electrons than Li_2 and these occupy the $\sigma^*(2s)$ MO. The predicted bond order in Be_2 is thus zero. In practice, this prediction is essentially fulfilled, although there is evidence for an extremely unstable Be_2 species with bond length 245 pm and bond energy 10 kJ mol^{-1}.

A ***basis set of orbitals*** is composed of those which are available for orbital interactions.

In each of Li_2 and Be_2, it is unnecessary to include the core ($1s$) atomic orbitals in order to obtain a useful bonding picture. This is true more generally, and throughout this book, MO treatments of bonding focus only on the interactions between the valence orbitals of the atoms concerned.

The bonding in F₂ and O₂

The valence shell of an F atom contains $2s$ and $2p$ atomic orbitals, and the formation of an F_2 molecule involves $2s$–$2s$ and $2p$–$2p$ orbital interactions. Before we can construct an MO diagram for the formation of F_2, we must consider what types of interactions are possible between p atomic orbitals.

By convention, each p atomic orbital is directed along one of the three Cartesian axes (Fig. 1.10), and, in considering the formation of a diatomic X_2, it is convenient to fix the positions of the X nuclei on one of the axes. In diagram **2.13**, the nuclei are placed on the z axis, but this choice of axis is arbitrary. Defining these positions also defines the relative orientations of the two sets of p orbitals (Fig. 2.7).

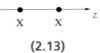

(2.13)

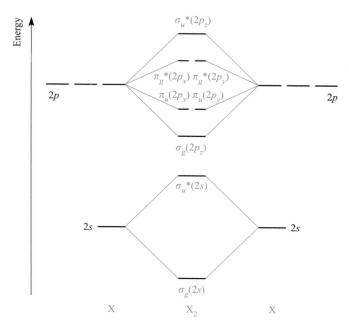

Fig. 2.8 A general orbital interaction diagram for the formation of X_2 in which the valence orbitals of atom X are the $2s$ and $2p$. In constructing this diagram we assume that the s–p separation is sufficiently large that no orbital mixing occurs. The X nuclei lie on the z axis.

Figures 2.7a and 2.7b show the in-phase and out-of-phase combinations of two $2p_z$ atomic orbitals. In terms of the region between the nuclei, the p_z–p_z interaction is similar to that of two s atomic orbitals (Fig. 2.5) and the symmetries of the resultant MOs are consistent with the σ_g and σ_u^* labels. Thus, the direct interaction of two p atomic orbitals (i.e. when the orbitals lie along a common axis) leads to $\sigma_g(2p)$ and $\sigma_u^*(2p)$ MOs. The p_x orbitals of the two atoms X can overlap only in a sideways manner, an interaction which has a smaller overlap integral than the direct overlap of the p_z atomic orbitals. The in-phase and out-of-phase combinations of two $2p_x$ atomic orbitals are shown in Figs. 2.7c and 2.7d. The bonding MO is called a π-orbital ('*pi-orbital*'), and its antibonding counterpart is a π^*-orbital ('*pi-star-orbital*'). Note the positions of the nodal planes in each MO. A π molecular orbital is asymmetrical with respect to rotation about the internuclear axis, i.e. if you rotate the orbital about the internuclear axis (the z axis in Fig. 2.7), there is a phase change. A π^*-orbital must exhibit two properties:

- the π label means that rotation of the orbital about the internuclear axis generates a phase change, *and*
- the * label means that there must be a nodal plane *between* the nuclei.

The parity (see Box 2.1) of a π-orbital is u, and that of a π^*-orbital is g. These labels are the reverse of those for σ and σ^*-orbitals, respectively (Fig. 2.7). The overlap between two p_y atomic orbitals generates an MO which has the same symmetry properties as that derived from the combination of the two p_x atomic orbitals, but the $\pi_u(p_y)$ MO lies in a plane perpendicular to that of the $\pi_u(p_x)$ MO. The $\pi_u(p_x)$ and $\pi_u(p_y)$ MOs lie at the same energy: they are *degenerate*. The $\pi_g^*(p_y)$ and $\pi_g^*(p_x)$ MOs are similarly related.

Now let us return to the formation of F_2. The valence orbitals of F are the $2s$ and $2p$, and Fig. 2.8 shows a general orbital interaction diagram for the overlap of these orbitals. We may assume to a first approximation that the energy separation of the fluorine $2s$ and $2p$ atomic orbitals (the s–p separation) is sufficiently great that only $2s$–$2s$ and $2p$–$2p$ orbital interactions occur. Notice that the stabilization of the $\pi_u(2p_x)$ and $\pi_u(2p_y)$ MOs relative to the $2p$ atomic orbitals is less than that of the $\sigma_g(2p_z)$ MO, consistent with the relative efficiencies of orbital overlap discussed above. In F_2 there are 14 electrons to be accommodated and, according to the *aufbau* principle, this gives a ground state electronic configuration of $\sigma_g(2s)^2\sigma_u^*(2s)^2\sigma_g(2p_z)^2\pi_u(2p_x)^2\pi_u(2p_y)^2\pi_g^*(2p_x)^2\pi_g^*(2p_y)^2$. The MO picture for F_2 is consistent with its observed diamagnetism. The predicted bond order is 1, in keeping with the result of the VB treatment (see Section 2.2).

Figure 2.8 can also be used to describe the bonding in O_2. Each O atom has six valence electrons $(2s^22p^4)$ and the total of 12 electrons in O_2 gives an electronic ground state of $\sigma_g(2s)^2\,\sigma_u^*(2s)^2\,\sigma_g(2p_z)^2\,\pi_u(2p_x)^2\pi_u(2p_y)^2\pi_g^*(2p_x)^1\pi_g^*(2p_y)^1$. This result is one of the triumphs of early MO theory: the model correctly predicts that O_2 possesses two unpaired electrons and is paramagnetic. From eq. 2.9, the bond order in O_2 is 2.

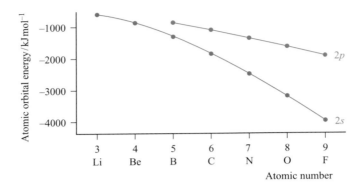

Fig. 2.9 In crossing the period from Li to F, the energies of the $2s$ and $2p$ atomic orbitals decrease owing to the increased effective nuclear charge.

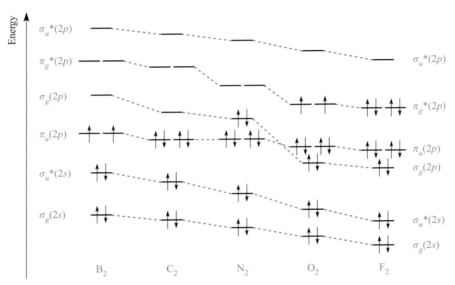

Fig. 2.10 Changes in the energy levels of the MOs and the ground state electronic configurations of homonuclear diatomic molecules involving first-row p-block elements.

What happens if the s–p separation is small?

A comparison of theoretical with experimental data for F_2 and O_2 indicates that the approximations we have made above are appropriate. However, this is not the case if the s–p energy difference is relatively small. In going from Li to F, the effective nuclear charge experienced by an electron in a $2s$ or $2p$ atomic orbital increases and the orbital energy decreases. This is shown in Fig. 2.9: the trend is non-linear and the s–p separation increases significantly from B to F. The relatively small s–p separation observed for B and C means that the approximation made when constructing the orbital interaction diagram in Fig. 2.8 is no longer valid when we construct similar diagrams for the formation of B_2 and C_2. Here, *orbital mixing* may occur[†] between orbitals of similar

---

[†] This effect is dealt with in detail but at a relatively simple level in Chapter 4 of C.E. Housecroft and E.C. Constable (2010) *Chemistry*, 4th edn, Prentice Hall, Harlow.

symmetry and energy, with the result that the ordering of the MOs in B_2, C_2 and N_2 differs from that in F_2 and O_2. Figure 2.10 compares the energy levels of the MOs and the ground state electronic configurations of the diatomics X_2 for X = B, C, N, O and F. Notice the so-called σ–π *crossover* that occurs between N_2 and O_2.

Since the MO approach is a theoretical model, what experimental evidence is there for this σ–π crossover? The actual electronic configurations of molecules are nearly always determined spectroscopically, particularly by *photoelectron spectroscopy*, a technique in which electrons in different orbitals are distinguished by their ionization energies (see Section 4.11). Experimental data support the orbital orderings shown in Fig. 2.10. Table 2.1 lists experimental bond distances and bond dissociation enthalpies for diatomics of the second period including Li_2 and Be_2, and also gives their bond orders calculated from MO theory. Since the nuclear charges change along the series, we should not expect all bonds

Table 2.1 Experimental data and bond orders for homonuclear diatomic molecules X_2 in which X is an atom in the period Li to F.

Diatomic	Bond distance / pm	Bond dissociation enthalpy / kJ mol^{-1}	Bond order	Magnetic properties
Li_2	267	110	1	Diamagnetic
Be_2†	–	–	0	–
B_2	159	297	1	Paramagnetic
C_2	124	607	2	Diamagnetic
N_2	110	945	3	Diamagnetic
O_2	121	498	2	Paramagnetic
F_2	141	159	1	Diamagnetic

† See text on p. 38.

of order 1 to have the same bond dissociation enthalpy. However, the general relationship between the bond order, dissociation enthalpy and distance is unmistakable. Table 2.1 also states whether a given molecule is diamagnetic or paramagnetic. We have already seen that MO theory correctly predicts (as does VB theory) that Li_2 is diamagnetic. Similarly, both the MO and VB models are consistent with the diamagnetism of C_2, N_2 and F_2. The paramagnetism of O_2 is predicted by MO theory as we have already seen, and this result is independent of whether the crossover of the $\sigma_g(2p)$ and $\pi_u(2p)$ occurs or not (Fig. 2.10). However, the MO model is only consistent with B_2 being paramagnetic *if* the $\pi_u(2p)$ level is at a lower energy than the $\sigma_g(2p)$. Consider in Fig. 2.10 what would happen if the relative orbital energies of the $\sigma_g(2p)$ and $\pi_u(2p)$ were reversed.

Worked example 2.1   Molecular orbital theory: properties of diatomics

The bond dissociation enthalpies for the nitrogen–nitrogen bond in N_2 and $[N_2]^-$ are 945 and 765 kJ mol^{-1} respectively. Account for this difference in terms of MO theory, and state whether $[N_2]^-$ is expected to be diamagnetic or paramagnetic.

Each N atom has the ground state configuration of $[He]2s^2 2p^3$.

An MO diagram for N_2, assuming only 2s–2s and 2p–2p orbital interactions, can be constructed, the result being as shown in Fig. 2.10. From this diagram, the bond order in N_2 is 3.0.

The change from N_2 to $[N_2]^-$ is a one-electron reduction and, assuming that Fig. 2.10 is still applicable, an electron is added to a $\pi_g^*(2p)$ orbital. The calculated bond order in $[N_2]^-$ is therefore 2.5.

The lower bond order of $[N_2]^-$ compared with N_2 is consistent with a lower bond dissociation enthalpy.

The electron in the $\pi_g^*(2p)$ orbital is unpaired and $[N_2]^-$ is expected to be paramagnetic.

Self-study exercises

1. Using Fig. 2.10 as a basis, account for the fact that $[N_2]^+$ is paramagnetic.

2. Using MO theory, rationalize why the N–N bond distance in $[N_2]^+$ is greater (112 pm) than in N_2 (109 pm).
   [*Ans.* Loss of electron from $\sigma_g(2p)$ MO]

3. Use Fig. 2.10 to rationalize why the bond orders in $[N_2]^+$ and $[N_2]^-$ are *both* 2.5.

4. Classify the changes from (a) N_2 to $[N_2]^+$, (b) from $[N_2]^-$ to N_2 and (c) from $[N_2]^+$ to $[N_2]^-$ as 1- or 2-electron, oxidation or reduction steps.
   [*Ans.* (a) 1e oxidation; (b) 1e oxidation; (c) 2e reduction]

2.4 The octet rule and isoelectronic species

The octet rule: first row *p*-block elements

The ground state electronic configurations in Table 1.3 reveal a pattern illustrating that filled quantum levels provide 'building blocks' within the electronic configurations of the heavier elements. Worked example 1.6 emphasized that each noble gas is characterized by having a filled quantum level. With the exception of He, this configuration is of the form $ns^2 np^6$, and this gives rise to the concept of the *octet rule*.

An atom obeys the **octet rule** when it gains, loses or shares electrons to give an **outer** shell containing eight electrons (an octet) with a configuration $ns^2 np^6$.

Ions such as Na^+ ($2s^22p^6$), Mg^{2+} ($2s^22p^6$), F^- ($2s^22p^6$), Cl^- ($3s^23p^6$) and O^{2-} ($2s^22p^6$) do in fact obey the octet rule, and they typically exist in environments in which electrostatic interaction energies compensate for the energies needed to form the ions from atoms (see Chapter 6). In general, the octet rule is most usefully applied in covalently bonded compounds involving *p*-block elements.

In structures **2.14–2.16**, Lewis structures are used to illustrate how the octet rule is obeyed by elements from the first row of the *p*-block. A carbon atom has four valence electrons ($2s^22p^2$) and if it forms four covalent single bonds, it achieves an octet of electrons in its valence shell (structure **2.14**). A boron atom has three valence electrons ($2s^22p^1$) and the formation of three single bonds generates a sextet (six electrons). The BH_3 molecule deals with this problem by dimerizing as we discuss in Section 5.7. In $[BH_4]^-$, the negative charge can formally be assigned to the B centre. By forming four single bonds, the B atom achieves an octet of electrons as shown in structure **2.15**. Nitrogen is in group 15 and an N atom has five valence electrons ($2s^22p^3$). In $[NH_4]^+$, if we formally assign the positive charge to the N atom, this centre then has four valence electrons and the formation of four single bonds provides the N atom with an octet of electrons (structure **2.16**).

H
··
H : C : H
··
H

CH$_4$

(2.14)

H
··
H : B$^-$: H
··
H

[BH$_4$]$^-$

(2.15)

H
··
H : N$^+$: H
··
H

[NH$_4$]$^+$

(2.16)

In these examples, only bonding electrons contribute to the octet of electrons. Lone pairs of electrons may also contribute as illustrated in H_2O (**2.17**) and HF (**2.18**).

··
H : O :
··
H

H$_2$O

(2.17)

··
H : F :
··

HF

(2.18)

Self-study exercises

1. Show that *each* atom in each of the following molecules obeys the octet rule: NF_3, CF_4, OF_2, F_2. Draw Lewis structures for the molecules and confirm that the bond order of each bond is one.

2. Show that *each* atom in the $[BF_4]^-$ ion obeys the octet rule.

3. Show that *each* atom in each of the following molecules obeys the octet rule: CO_2, OCF_2, ONF. Draw Lewis structures for the molecules and state the bond order of each bond.

Isoelectronic species

The series of molecular species shown in structures **2.14–2.16** illustrates the important concept of *isoelectronic species*.

Two species are **isoelectronic** if they possess the same *total* number of electrons.

Boron, carbon and nitrogen are adjacent in the periodic table, and atoms of B, C and N contain three, four and five valence electrons, respectively. It follows that each of B^-, C and N^+ possesses four valence electrons, and $[BH_4]^-$, CH_4 and $[NH_4]^+$ are therefore *isoelectronic*. The word *isoelectronic* is often used in the context of meaning 'same number of valence electrons', but strictly such usage should always be qualified. For example, HF, HCl and HBr are isoelectronic *with respect to their valence electrons*.

The isoelectronic principle is simple but important. Often, species that are isoelectronic possess the same structure, i.e. they are *isostructural*, e.g. $[BH_4]^-$, CH_4 and $[NH_4]^+$.

If two species are **isostructural,** they possess the same structure.

Worked example 2.2   Isoelectronic molecules and ions

Show that N_2 and $[NO]^+$ are isoelectronic.

N is in group 15 and has five valence electrons.

O is in group 16 and has six valence electrons.

O^+ has five valence electrons.

Therefore, each of N_2 and $[NO]^+$ possesses 10 valence electrons and the species are isoelectronic.

Self-study exercises

1. Show that $[SiF_6]^{2-}$ and PF_6 are isoelectronic.

2. Confirm that $[CN]^-$ and $[NO]^+$ are isoelectronic.

3. Are I_2 and F_2 isoelectronic?

4. In terms only of valence electrons, which of the following species is not isoelectronic with the remaining three: NH_3, $[H_3O]^+$, BH_3 and AsH_3?   [*Ans.* BH_3]

The octet rule: heavier *p*-block elements

As one descends a given group in the *p*-block, there is a tendency towards increased coordination numbers. Thus for example, a coordination number of 6 is found in SF_6, $[PF_6]^-$ and $[SiF_6]^{2-}$, but is not found in simple molecular species for the analogous first row elements O, N and C. Similarly, the heavier group 17 elements form compounds such as ClF_3, BrF_5 and IF_7 in which F is always a terminal atom and forms only one single bond. The Lewis structures **2.19** and **2.20** for ClF_3 imply that the Cl atom is surrounded by 10 valence electrons, i.e. it has 'expanded its octet'. Such species are referred to as being *hypervalent*.

(2.19)   **(2.20)**

It is, however, not necessary to exceed a valence octet if we make use of *charge-separated* species as contributing resonance structures. In order to maintain the octet of electrons around the Cl centre in ClF_3, we have to follow a similar strategy to that adopted in $[NH_4]^+$ described above (**2.16**). Whereas a Cl atom $(3s^2 3p^5)$ can form only one bond while obeying the octet rule, a Cl^+ centre can form two bonds:

Thus, we can write a Lewis structure for ClF_3 in terms of the charge-separated species **2.21**.

(2.21)

There is, however, a problem: structure **2.21** implies that one Cl–F interaction is ionic, while the other two are covalent. This problem is readily overcome by drawing a set of three resonance structures:

We look again at the bonding in hypervalent species in Sections 5.2, 5.7 and 15.3.

Self-study exercises

1. Show that As in AsF_3 obeys the octet rule.

2. Show that Se in H_2Se obeys the octet rule.

3. In which of the following molecules is it necessary to invoke charge-separated resonance structures in order that the central atom obeys the octet rule: (a) H_2S; (b) HCN; (c) SO_2; (d) AsF_5; (e) $[BF_4]^-$; (f) CO_2; (g) BrF_3.   [*Ans.* (c); (d); (g)]

4. Draw Lewis structures for the following ions, ensuring that all atoms obey the octet rule: (a) $[NO]^+$; (b) $[CN]^-$; (c) $[AlH_4]^-$; (d) $[NO_2]^-$.

2.5 Electronegativity values

In a homonuclear diatomic molecule X_2, the electron density in the region between the nuclei is symmetrical; each X nucleus has the same effective nuclear charge. On the other hand, the disposition of electron density in the region between the two nuclei of a *heteronuclear* diatomic molecule X–Y may be asymmetrical. If the effective nuclear charge of Y is greater than that of X, the pair of electrons in the X–Y covalent bond will be drawn towards Y and away from X.

Pauling electronegativity values, χ^P

Electronegativity, χ^P, was defined by Pauling as 'the power of an atom in a molecule to attract electrons to itself'.

In the early 1930s, Linus Pauling established the concept of *electronegativity*. The symbol for electronegativity is χ but we distinguish between different electronegativity scales by use of a superscript, e.g. χ^P for Pauling. Pauling first developed the idea in response to the observation that experimentally determined bond dissociation enthalpy values for heteronuclear bonds often did not agree with those obtained by simple additivity rules. Equation 2.10 shows the relationship between the bond dissociation enthalpy, D, of the gas phase homonuclear diatomic X_2 and the enthalpy change of atomization, $\Delta_a H^\circ$, of X(g). Effectively, this partitions bond enthalpy into a contribution made by each atom and, in the case of X_2, the contribution made by each atom is the same.

$$\Delta_a H^\circ(X, g) = \tfrac{1}{2} \times D(X–X) \qquad (2.10)$$

In eq. 2.11, the same type of additivity is applied to the bond in the heteronuclear diatomic XY. Estimates obtained for $D(X–Y)$ using this method sometimes agree quite well

Table 2.2   Pauling electronegativity (χ^P) values for the *s*- and *p*-block elements.

Group 1	Group 2		Group 13	Group 14	Group 15	Group 16	Group 17
H 2.2							
Li 1.0	Be 1.6		B 2.0	C 2.6	N 3.0	O 3.4	F 4.0
Na 0.9	Mg 1.3		Al(III) 1.6	Si 1.9	P 2.2	S 2.6	Cl 3.2
K 0.8	Ca 1.0		Ga(III) 1.8	Ge(IV) 2.0	As(III) 2.2	Se 2.6	Br 3.0
Rb 0.8	Sr 0.9	(*d*-block elements)	In(III) 1.8	Sn(II) 1.8 Sn(IV) 2.0	Sb 2.1	Te 2.1	I 2.7
Cs 0.8	Ba 0.9		Tl(I) 1.6 Tl(III) 2.0	Pb(II) 1.9 Pb(IV) 2.3	Bi 2.0	Po 2.0	At 2.2

with experimental data (e.g. ClBr and ClI), but may differ significantly (e.g. HF and HCl) as worked example 2.3 shows.

$$D(X-Y) = \tfrac{1}{2} \times [D(X-X) + D(Y-Y)] \qquad (2.11)$$

Worked example 2.3   Bond enthalpy additivity

Given that $D(H-H)$ and $D(F-F)$ in H_2 and F_2 are 436 and 158 kJ mol^{-1}, estimate the bond dissociation enthalpy of HF using a simple additivity rule. Compare the answer with the experimental value of 570 kJ mol^{-1}.

Assume that you can transfer the contribution made to $D(H-H)$ by an H atom to $D(H-F)$, and similarly for F.

$$D(H-F) = \tfrac{1}{2} \times [D(H-H) + D(F-F)]$$
$$= \tfrac{1}{2} \times [436 + 158]$$
$$= 297 \text{ kJ mol}^{-1}$$

Clearly, this model is unsatisfactory since it grossly underestimates the value of $D(H-F)$ which, experimentally, is found to be 570 kJ mol^{-1}.

Self-study exercises

1. Given that $D(H-H)$, $D(Cl-Cl)$, $D(Br-Br)$ and $D(I-I)$ in H_2, Cl_2, Br_2 and I_2 are 436, 242, 193 and 151 kJ mol^{-1} respectively, estimate (by the above method) values of $D(H-X)$ in HCl, HBr and HI.

[*Ans.* 339; 315; 294 kJ mol^{-1}]

2. Compare your answers to question 1 with experimental values of 432, 366 and 298 kJ mol^{-1} for $D(H-X)$ in HCl, HBr and HI.

Within the framework of the VB approach, Pauling suggested that the difference, ΔD, between an experimental value of $D(X-Y)$ and that obtained using eq. 2.11 could be attributed to the ionic contribution to the bond (eq. 2.4). The greater the *difference* in electron attracting powers (the *electronegativities*) of atoms X and Y, the greater the contribution made by X^+Y^- (or X^-Y^+), and the greater the value of ΔD. Pauling determined an approximately self-consistent scale of electronegativities, χ^P, as follows. He first converted ΔD values (obtained from $D_{\text{experimental}} - D_{\text{calculated}}$, the calculated value coming from eq. 2.11) from units of kJ mol^{-1} to eV in order to obtain a numerically small value of ΔD. He then arbitrarily related $\sqrt{\Delta D}$ to the difference in electronegativity values between atoms X and Y (eq. 2.12).

$$\Delta\chi = \chi^P(Y) - \chi^P(X) = \sqrt{\Delta D} \qquad \text{units of } \Delta D = \text{eV}$$
$$(2.12)$$

Over the years, the availability of more accurate thermochemical data has allowed Pauling's initial values of χ^P to be more finely tuned. Values listed in Table 2.2 are those in current use. Some intuition is required in deciding whether X or Y has the higher electronegativity value and in order to avoid giving an element a negative value of

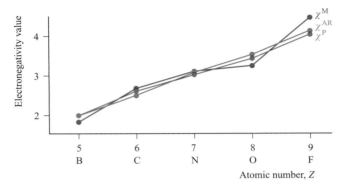

Fig. 2.11 Although electronegativity values for a given element from different scales cannot be expected to be the same, *trends* in values along a series of elements are comparable. This is illustrated with scaled values of χ^P (Pauling; red), χ^M (Mulliken; green) and χ^{AR} (Allred–Rochow; blue) for first row elements from the *p*-block.

χ^P, $\chi^P(H)$ has been taken as 2.2. Although eq. 2.12 implies that the units of χ^P are $eV^{\frac{1}{2}}$, it is not customary to give units to electronegativity values. By virtue of their different definitions, values of χ on different electronegativity scales (see below) possess different units.

In Table 2.2, more than one value of χ^P is listed for some elements. This follows from the fact that the electron withdrawing power of an element varies with its oxidation state (see Section 8.1); remember that the Pauling definition of χ^P refers to an atom *in a compound*. Electronegativity values also vary with bond order. Thus for C, χ^P has the values of 2.5 for a C−C bond, 2.75 for a C=C bond and 3.3 for a C≡C bond. For most purposes, the value of $\chi^P(C) = 2.6$ suffices, although the variation underlines the fact that such values must be used with caution.

Following from the original concept of electronegativity, various scales based upon different ground rules have been devised. We focus on two of the more commonly used scales, those of Mulliken and of Allred and Rochow; χ values from these scales are *not directly comparable* with Pauling values, although *trends* in the values should be similar (Fig. 2.11). Scales may be adjusted so as to be comparable with the Pauling scale.

Mulliken electronegativity values, χ^M

In one of the simplest approaches to electronegativity, Mulliken took the value of χ^M for an atom to be the mean of the values of the first ionization energy, IE_1, and the first electron affinity, EA_1 (eq. 2.13).

$$\chi^M = \frac{IE_1 + EA_1}{2} \qquad \text{where } IE_1 \text{ and } EA_1 \text{ are in eV}$$

$$(2.13)$$

Allred–Rochow electronegativity values, χ^{AR}

Allred and Rochow chose as a measure of electronegativity of an atom the electrostatic force exerted by the effective nuclear charge Z_{eff} (estimated from Slater's rules, see Box 1.5) on the valence electrons. The latter are assumed to reside at a distance from the nucleus equal to the covalent radius, r_{cov}, of the atom. Equation 2.14 gives the method of calculating values of the Allred–Rochow electronegativity, χ^{AR}.

$$\chi^{AR} = \left(3590 \times \frac{Z_{eff}}{r_{cov}^2} \right) + 0.744 \qquad \text{where } r_{cov} \text{ is in pm}$$

$$(2.14)$$

Since, however, Slater's rules are partly empirical and covalent radii are unavailable for some elements, the Allred–Rochow scale is no more rigid or complete than the Pauling one.

Electronegativity: final remarks

Despite the somewhat dubious scientific basis of the three methods described above, the trends in electronegativities obtained by them are roughly in agreement, as Fig. 2.11 exemplifies. The most useful of the scales for application in inorganic chemistry is probably the Pauling scale, which, being based empirically on thermochemical data, can reasonably be used to predict similar data. For example, if the electronegativities of two elements X and Y have been derived from the single covalent bond enthalpies of HX, HY, X_2, Y_2 and H_2, we can estimate the bond dissociation enthalpy of the bond in XY with a fair degree of reliability.

Worked example 2.4   Estimation of a bond dissociation enthalpy from χ^P values

Using the following data, estimate a value for $D(Br–F)$:
$D(F–F) = 158 \text{ kJ mol}^{-1}$ $D(Br–Br) = 224 \text{ kJ mol}^{-1}$
$\chi^P(F) = 4.0$ $\chi^P(Br) = 3.0$

First, use the values of χ^P to find ΔD:

$$\sqrt{\Delta D} = \chi^P(F) - \chi^P(Br) = 1.0$$

$$\Delta D = 1.0^2 = 1.0$$

This gives the value in eV; convert to kJ mol^{-1}:

$$1.0 \text{ eV} \approx 96.5 \text{ kJ mol}^{-1}$$

ΔD is defined as follows:

$$\Delta D = [D(Br–F)_{experimental}] - \{\tfrac{1}{2} \times [D(Br–Br) + D(F–F)]\}$$

So an estimate of $D(Br-F)$ is given by:

$$D(Br-F) = \Delta D + \{\tfrac{1}{2} \times [D(Br-Br) + D(F-F)]\}$$
$$= 96.5 + \{\tfrac{1}{2} \times [224 + 158]\}$$
$$= 287.5\,kJ\,mol^{-1}$$

[This compares with an experimental value of $250.2\,kJ\,mol^{-1}$.]

Self-study exercises

1. Use the following data to estimate the bond dissociation enthalpy of BrCl: $D(Br-Br) = 224\,kJ\,mol^{-1}$; $D(Cl-Cl) = 242\,kJ\,mol^{-1}$; $\chi^P(Br) = 3.0$; $\chi^P(Cl) = 3.2$.
   [*Ans.* $\approx 237\,kJ\,mol^{-1}$; actual experimental value $= 218\,kJ\,mol^{-1}$]

2. Use the following data to estimate the bond dissociation enthalpy of HF: $D(H-H) = 436\,kJ\,mol^{-1}$; $D(F-F) = 158\,kJ\,mol^{-1}$; $\chi^P(H) = 2.2$; $\chi^P(F) = 4.0$.
   [*Ans.* $\approx 610\,kJ\,mol^{-1}$; actual experimental value $= 570\,kJ\,mol^{-1}$]

3. Estimate the bond dissociation enthalpy of ICl given that $\chi^P(I) = 2.7$, $\chi^P(Cl) = 3.2$, and $D(I-I)$ and $D(Cl-Cl) = 151$ and $242\,kJ\,mol^{-1}$ respectively.
   [*Ans.* $221\,kJ\,mol^{-1}$]

In this book we avoid the use of the concept of electronegativity as far as possible and base the systemization of descriptive inorganic chemistry on rigidly defined and independently measured thermochemical quantities such as ionization energies, electron affinities, bond dissociation enthalpies, lattice energies and hydration enthalpies. However, some mention of electronegativity values is unavoidable.

2.6 Dipole moments

Polar diatomic molecules

The symmetrical electron distribution in the bond of a homonuclear diatomic renders the bond *non-polar*. In a heteronuclear diatomic, the electron withdrawing powers of the two atoms may be different, and the bonding electrons are drawn closer towards the more electronegative atom. The bond is *polar* and possesses an *electric dipole moment* (μ). Be careful to distinguish between *electric* and *magnetic* dipole moments (see Section 20.10).

The dipole moment of a diatomic XY is given by eq. 2.15 where d is the distance between the point electronic charges (i.e. the internuclear separation), e is the charge on the electron (1.602×10^{-19} C) and q is point charge. The SI unit of μ is the coulomb metre (C m) but for convenience, μ tends to be given in units of debyes (D) where $1\,D = 3.336 \times 10^{-30}\,C\,m$.

$$\mu = q \times e \times d \tag{2.15}$$

Worked example 2.5   Dipole moments

The dipole moment of a gas phase HBr molecule is 0.827 D. Determine the charge distribution in this diatomic if the bond distance is 141.5 pm. ($1\,D = 3.336 \times 10^{-30}\,C\,m$)

To find the charge distribution, you need to find q using the expression:

$$\mu = qed$$

Units must be consistent:

$d = 141.5 \times 10^{-12}\,m$

$\mu = 0.827 \times 3.336 \times 10^{-30} = 2.76 \times 10^{-30}\,C\,m$ (to 3 sig. fig.)

$$q = \frac{\mu}{ed}$$

$$= \frac{2.76 \times 10^{-30}}{1.602 \times 10^{-19} \times 141.5 \times 10^{-12}}$$

$$= 0.12 \text{ (no units)}$$

The charge distribution can be written as $\overset{+0.12}{H}\text{——}\overset{-0.12}{Br}$ since Br is more electronegative than H.

Self-study exercises

1. The bond length in HF is 92 pm, and the dipole moment is 1.83 D. Determine the charge distribution in the molecule.   [*Ans.* $\overset{+0.41}{H}\text{——}\overset{-0.41}{F}$]

2. The bond length in ClF is 163 pm. If the charge distribution is $\overset{+0.11}{Cl}\text{——}\overset{-0.11}{F}$, show that the molecular dipole moment is 0.86 D.

In worked example 2.5, the result indicates that the electron distribution in HBr is such that effectively 0.12 electrons have been transferred from H to Br. The partial charge separation in a polar diatomic molecule can be represented by use of the symbols δ^+ and δ^- assigned to the appropriate nuclear centres, and an arrow represents the direction in which the dipole moment acts. By SI convention, the arrow points from the δ^- end of the bond to the δ^+ end, which is contrary to long-established chemical practice. This is shown for HF in structure **2.22**. Keep in mind that a dipole moment is a vector quantity.

$$\overset{\delta^+}{H} \text{——} \overset{\delta^-}{F}$$

$\longleftarrow$

(2.22)

A word of caution: attempts to calculate the degree of ionic character of the bonds in heteronuclear diatomics from their observed dipole moments and the moments calculated on the basis of charge separation neglect the effects of any lone pairs of electrons and are therefore of doubtful validity. The significant effects of lone pairs are illustrated below in Example 3.

Molecular dipole moments

Polarity is a *molecular property*. For polyatomic species, the net molecular dipole moment depends upon the magnitudes and relative directions of all the bond dipole moments in the molecule. In addition, lone pairs of electrons may contribute significantly to the overall value of μ. We consider three examples below, using the Pauling electronegativity values of the atoms involved to give an indication of individual bond polarities. This practice is useful but must be treated with caution as it can lead to spurious results, e.g. when the bond multiplicity is not taken into account when assigning a value of χ^P. Experimental values of molecular electric dipole moments are determined by microwave spectroscopy or other spectroscopic methods.

Example 1: CF$_4$

(2.23)

The values of $\chi^P(C)$ and $\chi^P(F)$ are 2.6 and 4.0, respectively, indicating that each C–F bond is polar in the sense $C^{\delta+}-F^{\delta-}$. The CF$_4$ molecule (**2.23**) is tetrahedral and the four bond moments (each a vector of equivalent magnitude) oppose and cancel one another. The effects of the F lone pairs also cancel out, and the net result is that CF$_4$ is non-polar.

Example 2: H$_2$O

(2.24)

For O and H, $\chi^P = 3.4$ and 2.2, respectively, showing that each O–H bond is polar in the sense $O^{\delta-}-H^{\delta+}$. Since the H$_2$O molecule is non-linear, resolution of the two bond vectors gives a resultant dipole moment which acts in the direction shown in structure **2.24.** In addition, the O atom has two lone pairs of electrons which will reinforce the overall moment. The experimental value of μ for H$_2$O in the gas phase is 1.85 D.

Example 3: NH$_3$ and NF$_3$

X = H or F

(2.25)

The molecules NH$_3$ and NF$_3$ have trigonal pyramidal structures (**2.25**), and have dipole moments of 1.47 and 0.24 D respectively. This significant difference may be rationalized by considering the bond dipole moments and the effects of the N lone pair. The values of $\chi^P(N)$ and $\chi^P(H)$ are 3.0 and 2.2, so each bond is polar in the sense $N^{\delta-}-H^{\delta+}$. The resultant dipole moment acts in a direction that is reinforced by the lone pair. Ammonia is a polar molecule with N carrying a partial negative charge. In NF$_3$, each N–F bond is polar in the sense $N^{\delta+}-F^{\delta-}$ since F is more electronegative ($\chi^P(F) = 4.0$) than N. The resultant dipole moment *opposes* the effects of the lone pair, rendering the NF$_3$ molecule far less polar than NH$_3$.

Clearly, molecular shape is an important factor in determining whether a molecule is polar or not and the examples below and in end-of-chapter problem 2.19 consider this further.

Worked example 2.6   Molecular dipole moments

Use electronegativity values in Table 2.2 to work out whether or not the following molecule is polar and, if so, in what direction the dipole acts.

First, look up values of χ^P from Table 2.2: $\chi^P(H) = 2.2$, $\chi^P(C) = 2.6$, $\chi^P(F) = 4.0$. The molecule is therefore polar with F atoms δ^-, and the molecular dipole moment acts as shown below:

Self-study exercises

1. Use electronegativity values in Table 2.2 to confirm that each of the following molecules is polar. Draw diagrams to show the directions of the molecular dipole moments.

Br —— F

H — S — H

H — C (with Cl, Cl)

2. Explain why each of the following molecules is non-polar.

Br — B (Br, Br)

Cl — Si (Cl, Cl, Cl)

S＝C＝S

2.7 MO theory: heteronuclear diatomic molecules

In this section, we return to MO theory and apply it to heteronuclear diatomic molecules. In each of the orbital interaction diagrams constructed in Section 2.3 for *homonuclear* diatomics, the resultant MOs contained *equal* contributions from each atomic orbital involved. This is represented in eq. 2.5 for the bonding MO in H_2 by the fact that each of the wavefunctions ψ_1 and ψ_2 contributes equally to ψ_{MO}, and the representations of the MOs in H_2 (Fig. 2.5) depict *symmetrical* orbitals. Now we look at representative examples of diatomics in which the MOs may contain *different* atomic orbital contributions, a scenario that is typical for heteronuclear diatomics.

First, we must consider likely restrictions when we are faced with the possibility of combining different types of atomic orbitals.

Which orbital interactions should be considered?

At the beginning of Section 2.3 we stated some general requirements that should be met for orbital interactions to take place efficiently. We stated that orbital interactions are

allowed if the *symmetries* of the atomic orbitals are compatible with one another. In our approach to the bonding in a diatomic, we made the assumption that only the interactions between *like* atomic orbitals, e.g. $2s$–$2s$, $2p_z$–$2p_z$, need be considered. Such interactions are *symmetry-allowed*, and in addition, in a *homonuclear* diatomic the energies of like atomic orbitals on the two atoms are exactly matched.

In a heteronuclear diatomic, we often encounter two atoms that have different basis sets of atomic orbitals, or have sets of similar atomic orbitals lying at different energies. For example, in CO, although both C and O possess valence $2s$ and $2p$ atomic orbitals, the greater effective nuclear charge of O means that its atomic orbitals lie at a lower energy than those of C. Before we look more closely at some examples of heteronuclear diatomics, let us briefly consider some symmetry-allowed and -disallowed orbital interactions. It is important to remember that we are looking at these symmetry properties *with respect to the internuclear axis*. In our earlier discussion of homonuclear diatomics (e.g. Fig. 2.8), we ignored the possibility of overlap between the p_x and p_y orbitals. Such an interaction between orthogonal p atomic orbitals (Fig. 2.12a) would give a zero overlap integral. Similarly, for nuclei lying on the z axis, interaction between p_x and p_z, or p_y and p_z, orbitals gives zero overlap. An interaction between an s and a p atomic orbital *may* occur depending upon the orientation of the p orbital. In Fig. 2.12b, overlap would be partly bonding and partly antibonding and the net effect is a *non-bonding* interaction. On the other hand, Fig. 2.12c shows an s–p interaction that *is* allowed by symmetry. Whether or not this leads to effective overlap depends upon the relative energies of the two atomic orbitals. This is illustrated in Fig. 2.13 for a diatomic XY. Let the interaction between ψ_X and ψ_Y be symmetry-allowed; the orbital energies are not the same but are close enough that overlap between the orbitals is efficient. The orbital interaction diagram shows that the energy of the bonding MO is closer to $E(\psi_Y)$ than to $E(\psi_X)$ and the consequence of this is that the bonding orbital possesses *greater Y than X character*. This is expressed in eq. 2.16 in which $c_2 > c_1$. For the antibonding MO, the situation is reversed, and ψ_X contributes more than ψ_Y; in eq. 2.17, $c_3 > c_4$.

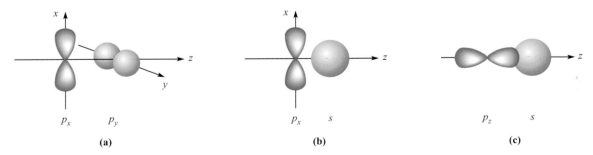

Fig. 2.12 Overlap between atomic orbitals is not always allowed by symmetry. Combinations (a) and (b) lead to non-bonding situations but (c) is symmetry-allowed and gives rise to a bonding interaction.

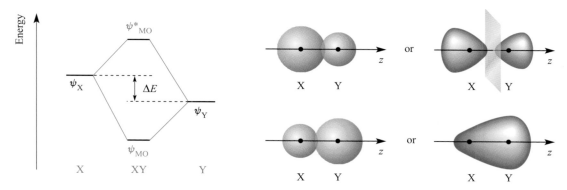

Fig. 2.13 The relative energies of atomic orbitals of X and Y will dictate whether an interaction (formally allowed by symmetry) will lead to efficient overlap or not. Here, an interaction occurs but the contribution made by ψ_Y to ψ_{MO} is greater than that made by ψ_X, while ψ_X contributes more than ψ_Y to the antibonding MO. The diagrams on the right give pictorial representations of the bonding and antibonding MOs.

$$\psi_{MO} = N[(c_1 \times \psi_X) + (c_2 \times \psi_Y)] \qquad (2.16)$$

$$\psi_{MO}^* = N^*[(c_3 \times \psi_X) + (c_4 \times \psi_Y)] \qquad (2.17)$$

The energy separation ΔE in Fig. 2.13 is critical. If it is large, interaction between ψ_X and ψ_Y will be poor (the overlap integral is very small). In the extreme case, there is no interaction at all and both ψ_X and ψ_Y appear in the XY molecule as unperturbed *non-bonding* atomic orbitals. This is exemplified below.

Hydrogen fluoride

The ground state configurations of H and F are $1s^1$ and $[\text{He}]2s^2 2p^5$ respectively. Since $Z_{eff}(F) > Z_{eff}(H)$, the F $2s$ and $2p$ atomic orbital energies are significantly lowered with respect to the H $1s$ atomic orbital (Fig. 2.14).

We now have to consider which atomic orbital interactions are symmetry-allowed and then ask whether the atomic orbitals are sufficiently well energy-matched. First, define the axis set for the orbitals; let the nuclei lie on the

z axis. Overlap between the H $1s$ and F $2s$ orbitals is allowed by symmetry, but the energy separation is very large (note the break on the energy axis in Fig. 2.14). Overlap between the H $1s$ and F $2p_z$ atomic orbitals is also symmetry-allowed and there is a reasonable orbital energy match. As Fig. 2.14 shows, an interaction occurs leading to σ and σ^* MOs; the σ-orbital has greater F than H character. Notice that, because HF is *non-centrosymmetric* (see Box 2.1), the symmetry labels of the orbitals for HF do *not* involve g and u labels. The two F $2p_x$ and $2p_y$ atomic orbitals become non-bonding orbitals in HF since no net bonding interaction with the H $1s$ atomic orbital is possible. Once the orbital interaction diagram has been constructed, the eight valence electrons are accommodated as shown in Fig. 2.14, giving a bond order of 1 in HF. The MO picture of HF indicates that the electron density is greater around the F than H nucleus; the model is consistent with a polar H–F bond in the sense $H^{\delta+}$–$F^{\delta-}$.

Self-study exercise

Sketch pictorial representations of the σ and σ^* MOs in HF.

Carbon monoxide

In Chapter 24 we discuss the chemistry of compounds containing metal–carbon bonds (*organometallic compounds*) of which *metal carbonyls* of the type $M_x(CO)_y$ are one group. In order to investigate the way in which CO bonds to metals, we must appreciate the electronic structure of the carbon monoxide molecule.

Before constructing an orbital interaction diagram for CO, we note the following:

- $Z_{eff}(O) > Z_{eff}(C)$;
- the energy of the O $2s$ atomic orbital is lower than that of the C $2s$ atomic orbital;

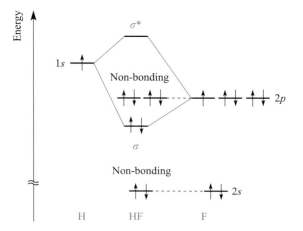

Fig. 2.14 An orbital interaction diagram for the formation of HF. Only the valence atomic orbitals and electrons are shown. The break in the vertical (energy) axis indicates that the energy of the F $2s$ atomic orbital is much lower than is actually shown.

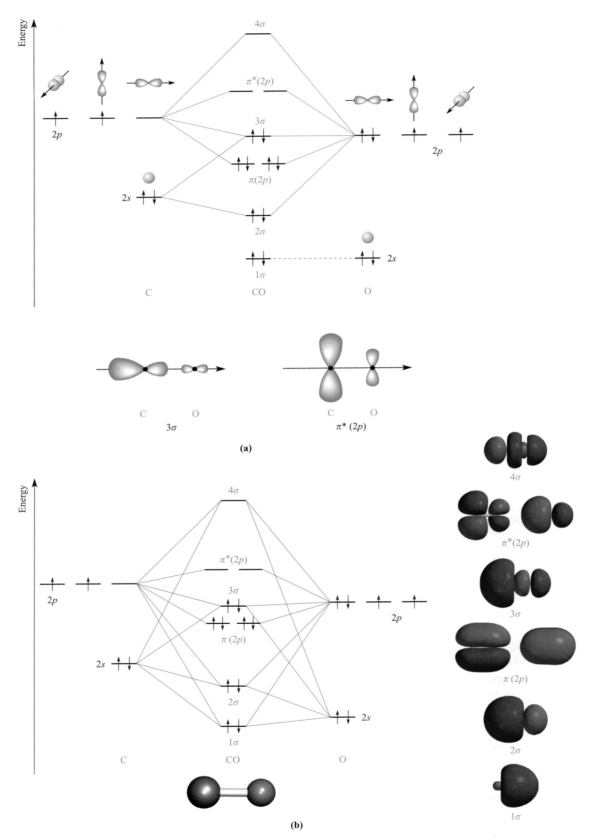

Fig. 2.15 (a) A simplified orbital interaction diagram for CO which allows for the effects of some orbital mixing. The labels 1σ, 2σ ... rather than $\sigma(2s)$ ... are used because some orbitals contain both s and p character. (b) A more rigorous (but still qualitative) orbital interaction diagram for CO. The diagrams on the right-hand side show representations of the MOs and have been generated computationally using Spartan '04, ©Wavefunction Inc. 2003. These diagrams illustrate that the 1σ MO has mainly oxygen character, while the 2σ, 3σ and $\pi^*(2p)$ MOs have more carbon than oxygen character.

- the $2p$ level in O is at lower energy than that in C;
- the $2s$–$2p$ energy separation in O is greater than that in C (Fig. 2.9).

We could generate an approximate orbital interaction diagram by assuming that only $2s$–$2s$ and $2p$–$2p$ overlap occurs, but, as a consequence of the relative atomic orbital energies, such a picture is too simplistic. Figure 2.15a gives a more accurate MO picture of the electronic structure of CO obtained computationally, although even this is over-simplified. Figure 2.15b illustrates more fully the extent of orbital mixing, but for our discussion, the simplified picture presented in Fig. 2.15a suffices. Two of the more important features to notice are:

- The highest occupied MO (HOMO) is σ-bonding and possesses predominantly *carbon* character; occupation of this MO effectively creates an outward-pointing lone pair centred on C.
- A degenerate pair of $\pi^*(2p)$ MOs make up the lowest unoccupied MOs (LUMOs); each MO possesses more C than O character.

Pictorial representations of the HOMO and one of the LUMOs are given in Fig. 2.15; refer to end-of-chapter problem 2.21.

HOMO = highest occupied molecular orbital.
LUMO = lowest unoccupied molecular orbital.

2.8 Molecular shape and the VSEPR model

Valence-shell electron-pair repulsion model

The *valence-shell electron-pair repulsion (VSEPR)* model is used to rationalize or predict the shapes of molecular species. It is based on the assumption that electron pairs adopt arrangements that minimize repulsions between them.

The shapes of molecules containing a central p-block atom tend to be controlled by the number of electrons in the valence shell of the central atom. The *valence-shell electron-pair repulsion* (VSEPR) model provides a simple model for predicting the shapes of such species. The model combines original ideas of Sidgwick and Powell with extensions developed by Nyholm and Gillespie, and may be summarized as follows:

- Each valence shell electron pair of the central atom E in a molecule EX_n containing E–X single bonds is stereochemically significant, and repulsions between them determine the molecular shape.
- Electron–electron repulsions decrease in the sequence:
  lone pair–lone pair > lone pair–bonding pair > bonding pair–bonding pair.
- Where the central atom E is involved in multiple bond formation to atoms X, electron–electron repulsions decrease in the order:

triple bond–single bond > double bond–single bond > single bond–single bond.

- Repulsions between the bonding pairs in EX_n depend on the difference between the electronegativities of E and X; electron–electron repulsions are less the more the E—X bonding electron density is drawn away from the central atom E.

The VSEPR model works best for simple halides of the p-block elements, but may also be applied to species with other substituents. However, the model does *not* take *steric factors* (i.e. the relative sizes of substituents) into account.

In a molecule EX_n, there is a minimum energy arrangement for a given number of electron pairs. In $BeCl_2$ (Be, group 2), repulsions between the two pairs of electrons in the valence shell of Be are minimized if the Cl—Be—Cl unit is linear. In BCl_3 (B, group 13), electron–electron repulsions are minimized if a trigonal planar arrangement of electron pairs (and thus Cl atoms) is adopted. The structures in the left-hand column of Fig. 2.16 represent the minimum energy structures for EX_n molecules for $n = 2$–8 and in which there are no lone pairs of electrons associated with E. Table 2.3 gives further representations of these structures, along with their *ideal* bond angles. Ideal bond angles may be expected when all the X substituents are identical, but in, for example, BF_2Cl (**2.26**) some distortion occurs because Cl is larger than F, and the shape is only *approximately* trigonal planar.

$\angle$ F–B–F = $118°$

(2.26)   **(2.27)**

The presence of lone pairs is taken into account using the guidelines above and the 'parent structures' in Fig. 2.16. In H_2O (**2.27**), repulsions between the two bonding pairs and two lone pairs of electrons lead to a tetrahedral arrangement, but owing to the inequalities between the lone pair–lone pair, lone pair–bonding pair and bonding pair–bonding pair interactions, distortion from an ideal arrangement arises and this is consistent with the observed H—O—H bond angle of $104.5°$.

Worked example 2.7   The VSEPR model

Predict the structures of (a) XeF_2 and (b) $[XeF_5]^-$.

Xe is in group 18 and possesses eight electrons in its valence shell. F is in group 17, has seven valence electrons and forms one covalent single bond. Before applying the VSEPR model, decide which is the central atom in the molecule. In each of XeF_2 and $[XeF_5]^-$, Xe is the central atom.

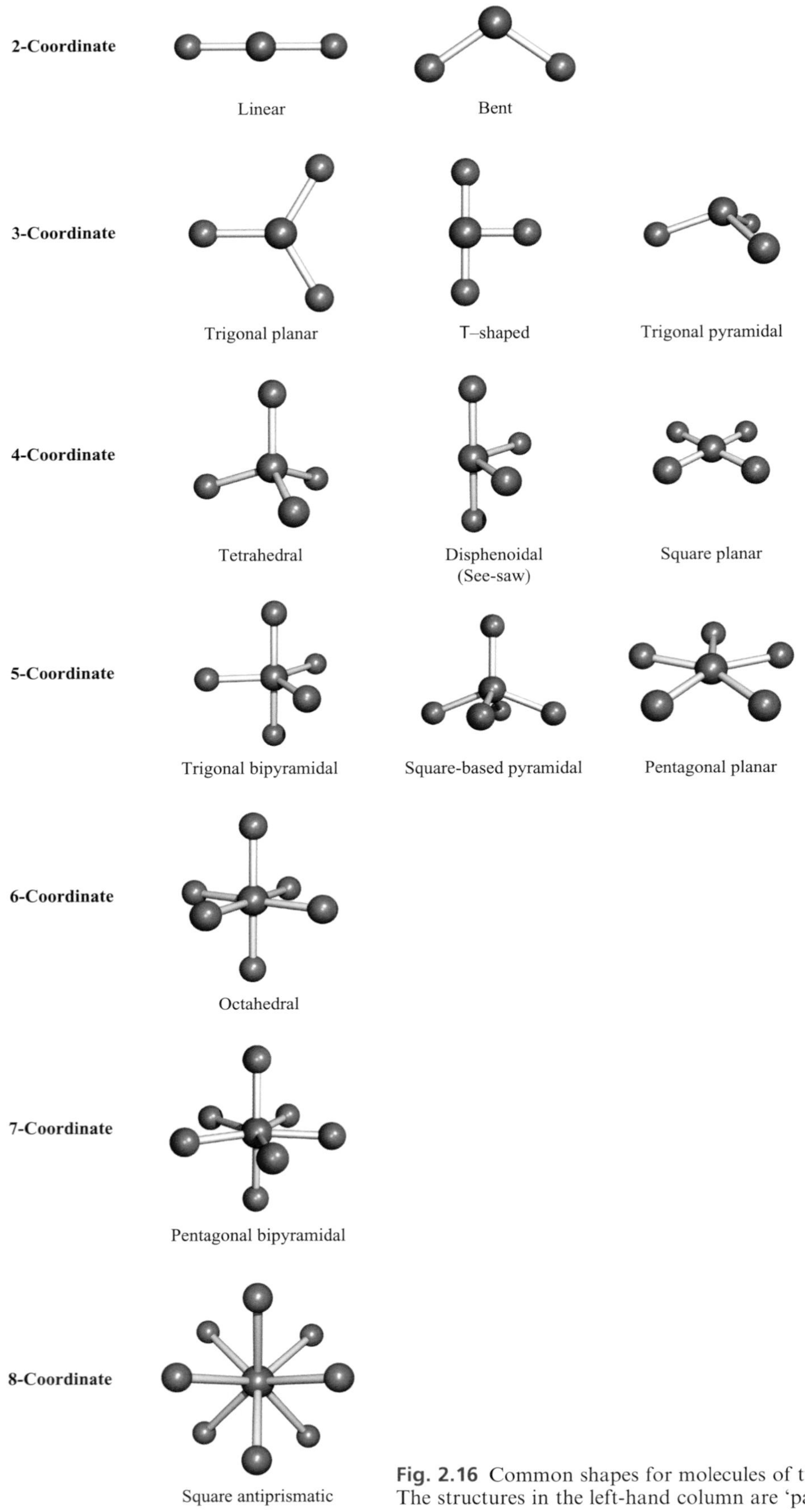

2-Coordinate

Linear

Bent

3-Coordinate

Trigonal planar

T–shaped

Trigonal pyramidal

4-Coordinate

Tetrahedral

Disphenoidal
(See-saw)

Square planar

5-Coordinate

Trigonal bipyramidal

Square-based pyramidal

Pentagonal planar

6-Coordinate

Octahedral

7-Coordinate

Pentagonal bipyramidal

8-Coordinate

Square antiprismatic

Fig. 2.16 Common shapes for molecules of type EX_n or ions of type $[EX_n]^{m+/-}$. The structures in the left-hand column are 'parent' shapes used in the **VSEPR** model.

Table 2.3 'Parent' shapes for EX_n molecules ($n = 2$–8).

Formula EX_n	Coordination number of atom E	Shape	Spatial representation	Ideal bond angles ($\angle$X–E–X) / degrees
EX_2	2	Linear	X——E——X	180
EX_3	3	Trigonal planar		120
EX_4	4	Tetrahedral		109.5
EX_5	5	Trigonal bipyramidal		$\angle X_{ax}\text{–}E\text{–}X_{eq} = 90$ $\angle X_{eq}\text{–}E\text{–}X_{eq} = 120$
EX_6	6	Octahedral		$\angle X_1\text{–}E\text{–}X_2 = 90$
EX_7	7	Pentagonal bipyramidal		$\angle X_{ax}\text{–}E\text{–}X_{eq} = 90$ $\angle X_{eq}\text{–}E\text{–}X_{eq} = 72$
EX_8	8	Square antiprismatic		$\angle X_1\text{–}E\text{–}X_2 = 78$ $\angle X_1\text{–}E\text{–}X_3 = 73$

(a) XeF_2. Two of the eight valence electrons of the Xe atom are used for bonding (two Xe–F single bonds), and so around the Xe centre there are two bonding pairs of electrons and three lone pairs.

The parent shape is a trigonal bipyramid (Fig. 2.16) with the three lone pairs in the equatorial plane to minimize lone pair–lone pair repulsions. The XeF_2 molecule is therefore linear:

(b) $[XeF_5]^-$. The electron from the negative charge is conveniently included within the valence shell of the central atom. Five of the nine valence electrons are used for bonding and around the Xe centre there are five bonding pairs and two lone pairs of electrons.

The parent shape is a pentagonal bipyramid (Fig. 2.16) with the two lone pairs opposite to each other to minimize lone pair–lone pair repulsions. The $[XeF_5]^-$ anion is therefore pentagonal planar:

When structures are determined by diffraction methods, *atom* positions are effectively located. Thus, in terms of a molecular structure, XeF_2 is linear and $[XeF_5]^-$ is pentagonal planar. In the diagrams above, two representations of each species are shown, one with the lone pairs to emphasize the origin of the prediction from the VSEPR model.

Self-study exercise

Show that the VSEPR model is in agreement with the following molecular shapes:

BF_3	trigonal planar
$[IF_5]^{2-}$	pentagonal planar
$[NH_4]^+$	tetrahedral
SF_6	octahedral
XeF_4	square planar
AsF_5	trigonal bipyramidal
$[AlCl_4]^-$	tetrahedral

Worked example 2.8   VSEPR: molecules with double bonds

Is the VSEPR model consistent with a linear or bent structure for $[NO_2]^+$?

N is in group 15 and has five valence electrons. Allow the positive charge to be localized on the nitrogen centre; an N^+ centre has four valence electrons. O is in group 16 and has six valence electrons; an atom of O requires two electrons to complete its octet. All four electrons in the valence shell of the N^+ centre are involved in bonding, forming two double bonds in $[NO]^+$. Since there are no lone pairs on the N atom, the VSEPR model is consistent with a linear structure:

$$\left[O = N = O \right]^+$$

Self-study exercises

1. Show that the VSEPR model is consistent with a trigonal planar structure for SO_3.

2. Using the VSEPR model, rationalize why a CO_2 molecule is linear whereas an $[NO_2]^-$ ion is bent.

3. The sulfite ion, $[SO_3]^{2-}$, has the following structure:

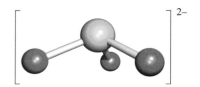

Show that the VSEPR model is consistent with this structure.

Structures derived from a trigonal bipyramid

In this section, we consider the structures of species such as ClF_3 and SF_4 which have five electron pairs in the valence shell of the central atom. The experimentally determined structure of ClF_3 is shown in Fig. 2.17, and the VSEPR model can be used to rationalize this T-shaped arrangement. The valence shell of the Cl atom contains three bonding pairs and two lone pairs of electrons. If both lone pairs occupy equatorial sites (see Table 2.3), then a T-shaped ClF_3 molecule results. The choice of locations for the bonding and lone pairs arises from a consideration of the difference between the X_{ax}–E–X_{eq} and X_{eq}–E–X_{eq} bond angles (Table 2.3), coupled with the relative magnitudes of lone pair–lone pair, bonding pair–lone pair and bonding pair–bonding pair repulsions. It follows that the chlorine lone pairs in ClF_3 preferentially occupy the equatorial sites where there is greatest space. The small departure of the F–Cl–F bond angle from the ideal value of 90° (Table 2.3) may be attributed to lone pair–bonding pair repulsion. Figure 2.17 also shows that there is a significant difference between the axial and equatorial Cl–F bond lengths, and this is a trend that is seen in a range of structures of molecules derived from a trigonal bipyramidal arrangement. In PF_5, the axial (ax) and equatorial (eq) bond distances are 158 and 153 pm respectively, in SF_4 (**2.28**), they are 165 and 155 pm, and in BrF_3, they are 181 and 172 pm.[†] Bond distance variation is, however, not restricted to species derived from a trigonal bipyramid. For example, in BrF_5 (**2.29**), the Br atom lies a little below the plane containing the basal F atoms ($\angle F_{ax}$–Br–F_{bas} = 84.5°) and the Br–F_{ax} and Br–F_{bas} bond distances are 168 and 178 pm respectively.

(2.28)   (2.29)

Limitations of the VSEPR model

The generalizations of the VSEPR model are useful, but there are limitations to its use. In this section, we give examples that illustrate some problems.

The isoelectronic species IF_7 and $[TeF_7]^-$ are predicted by the VSEPR model to be pentagonal bipyramidal and this is observed. However, electron diffraction data for IF_7 and X-ray diffraction data for $[Me_4N][TeF_7]$ reveal that the equatorial F atoms are not coplanar, a result that cannot be predicted by the VSEPR model. Moreover, in IF_7, the I–F_{ax} and I–F_{eq} distances are 179 and 186 pm respectively, and in

---

[†] For further discussion of this topic, see: R.J. Gillespie and P.L.A. Popelier (2001) *Chemical Bonding and Molecular Geometry*, Oxford University Press, Oxford, Chapter 4.

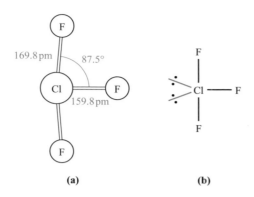

Fig. 2.17 (a) The experimentally determined structure of ClF_3 and (b) the rationalization of this structure using the VSEPR model.

$[TeF_7]^-$, the $Te-F_{ax}$ bond distance is 179 pm and the $Te-F_{eq}$ distances lie in the range 183 to 190 pm.

Among species in which the VSEPR model appears to fail are $[SeCl_6]^{2-}$, $[TeCl_6]^{2-}$ and $[BrF_6]^-$ (see also Section 16.7). When characterized as alkali metal salts, these anions are found to possess *regular octahedral* structures *in the solid state*, whereas the VSEPR model suggests shapes based on there being seven electron pairs around the central atom. Although these structures cannot readily be predicted, we can rationalize them in terms of having a *stereochemically inactive* pair of electrons. Stereochemically inactive lone pairs are usually observed for the heaviest members of a periodic group, and the tendency for valence shell s electrons to adopt a non-bonding role in a molecule is called the *stereochemical inert pair effect*. Similarly, $[SbCl_6]^-$ and $[SbCl_6]^{3-}$ *both* possess regular octahedral structures. Finally, consider $[XeF_8]^{2-}$, $[IF_8]^-$ and $[TeF_8]^{2-}$. As expected from the VSEPR model, $[IF_8]^-$ and $[TeF_8]^{2-}$ are square antiprismatic; this structure is related to the cube but with one face of the cube rotated through $45°$. However, $[XeF_8]^{2-}$ also adopts this structure, indicating that the lone pair of electrons is stereochemically inactive.

It is important to note that whereas the VSEPR model may be applicable to p-block species, it is *not* appropriate to apply it to d-electron configurations of transition metal compounds.

If the presence of a lone pair of electrons influences the shape of a molecule or ion, the lone pair is **stereochemically active**. If it has no effect, the lone pair is **stereochemically inactive**. The tendency for the pair of **valence** s electrons to adopt a non-bonding role in a molecule or ion is termed the **stereochemical inert pair effect**.

2.9 Molecular shape: stereoisomerism

An **isomer** is one of several species that have the same atomic composition (molecular formula), but have different constitutional formulae (atom connectivities) or different stereochemical formulae. Isomers exhibit different physical and/or chemical properties.

In this section we discuss *stereoisomerism*. Examples are taken from both p- and d-block chemistry. Other types of isomerism are described in Section 19.8.

If two species have the same molecular formula and the same atom connectivity, but differ in the spatial arrangement of different atoms or groups about a central atom or a double bond, then the compounds are **stereoisomers**.

Stereoisomers fall into two categories, *diastereoisomers* and *enantiomers*.

Diastereoisomers are stereoisomers that *are not* mirror-images of one another. **Enantiomers** are stereoisomers that *are* mirror-images of one another.

In this section, we shall only be concerned with diastereoisomers. We return to enantiomers in Sections 3.8 and 19.8.

Square planar species

In a square planar species such as $[ICl_4]^-$ or $[PtCl_4]^{2-}$ (**2.30**), the four Cl atoms are equivalent. Similarly, in $[PtCl_3(PMe_3)]^-$ (**2.31**), there is only one possible arrangement of the groups around the square planar Pt(II) centre. (The use of arrows or lines to depict bonds in coordination compounds is discussed in Section 7.11.)

(2.30)   **(2.31)**

The introduction of two PMe_3 groups to give $[PtCl_2(PMe_3)_2]$ leads to the possibility of two *stereoisomers*, i.e. two possible spatial arrangements of the groups around the square planar Pt(II) centre. These are shown in structures **2.32** and **2.33** and the names *cis* and *trans* refer to the positioning of the Cl (or PMe_3) groups, adjacent to or opposite one another.

cis-isomer   *trans*-isomer

(2.32)   **(2.33)**

Square planar species of the general form EX_2Y_2 or EX_2YZ may possess **cis-** and **trans-**isomers.

Octahedral species

There are two types of stereoisomerism associated with octahedral species. In EX_2Y_4, the X groups may be mutually *cis* or *trans* as shown for $[SnF_4Me_2]^{2-}$ (**2.34** and **2.35**). In the solid state structure of $[NH_4]_2[SnF_4Me_2]$, the anion is present as the *trans*-isomer.

$$\left[\begin{array}{c} F \\ F_{\prime\prime\prime\prime}\,Sn\,^{\backslash\backslash\backslash}F \\ F \quad Me \\ Me \end{array}\right]^{2-} \qquad \left[\begin{array}{c} Me \\ F_{\prime\prime\prime\prime}\,Sn\,^{\backslash\backslash\backslash}F \\ F \quad F \\ Me \end{array}\right]^{2-}$$

cis-isomer   *trans*-isomer

(2.34)   **(2.35)**

If an octahedral species has the general formula EX_3Y_3, then the X groups (and also the Y groups) may be arranged so as to define one face of the octahedron or may lie in a plane that also contains the central atom E (Fig. 2.18). These stereoisomers are labelled *fac* (facial) and *mer* (meridional) respectively. In $[PCl_4][PCl_3F_3]$, the $[PCl_3F_3]^-$ anion exists as both *fac*- and *mer*-isomers (**2.36** and **2.37**).

$$\left[\begin{array}{c} F \\ F_{\prime\prime\prime\prime}\,P\,^{\backslash\backslash\backslash}Cl \\ F \quad Cl \\ Cl \end{array}\right]^{-} \qquad \left[\begin{array}{c} Cl \\ F_{\prime\prime\prime\prime}\,P\,^{\backslash\backslash\backslash}F \\ F \quad Cl \\ Cl \end{array}\right]^{-}$$

fac-isomer   *mer*-isomer

(2.36)   **(2.37)**

> An *octahedral* species containing two identical groups (e.g. of type EX_2Y_4) may possess *cis*- and *trans*-arrangements of these groups. An octahedral species containing three identical groups (e.g. of type EX_3Y_3) may possess *fac*- and *mer*-isomers.

Trigonal bipyramidal species

In trigonal bipyramidal EX_5, there are two types of X atom: axial and equatorial. This leads to the possibility of stereoisomerism when more than one type of substituent is attached to the central atom. Iron pentacarbonyl, $Fe(CO)_5$,

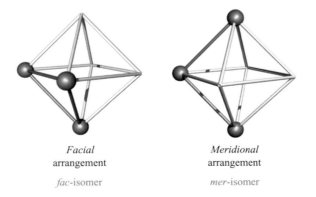

Facial   *Meridional*
arrangement   arrangement

fac-isomer   *mer*-isomer

Fig. 2.18 The origin of the names *fac*- and *mer*-isomers. For clarity, the central atom is not shown.

is trigonal bipyramidal and when one CO is exchanged for PPh_3, two stereoisomers are possible depending on whether the PPh_3 ligand is axially (**2.38**) or equatorially (**2.39**) sited.

$$OC\,—\,\underset{PPh_3}{\overset{CO}{Fe}}\!\!\!\!\!\!\overset{\textstyle{}}{\underset{\textstyle{}}{{}^{\prime\prime\prime\prime}CO}}\!\!\!\searrow CO \qquad Ph_3P\,—\,\underset{CO}{\overset{CO}{Fe}}\!\!\!\!\!\!\overset{\textstyle{}}{\underset{\textstyle{}}{{}^{\prime\prime\prime\prime}CO}}\!\!\!\searrow CO$$

(2.38)   **(2.39)**

For trigonal bipyramidal EX_2Y_3, three stereoisomers (**2.40** to **2.42**) are possible depending on the relative positions of the X atoms. Steric factors may dictate which isomer is preferred for a given species; e.g. in the static structure of PCl_3F_2, the F atoms occupy the two axial sites, and the larger Cl atoms reside in the equatorial plane.

$$Y\,—\,\underset{X}{\overset{X}{E}}\!\!\!\!\!\!\overset{}{\underset{}{{}^{\prime\prime\prime\prime}Y}}\!\!\!\searrow Y \qquad X\,—\,\underset{Y}{\overset{Y}{E}}\!\!\!\!\!\!\overset{}{\underset{}{{}^{\prime\prime\prime\prime}Y}}\!\!\!\searrow X \qquad X\,—\,\underset{X}{\overset{Y}{E}}\!\!\!\!\!\!\overset{}{\underset{}{{}^{\prime\prime\prime\prime}Y}}\!\!\!\searrow Y$$

(2.40)   **(2.41)**   **(2.42)**

> In a *trigonal bipyramidal* species, stereoisomerism arises because of the presence of *axial* and *equatorial* sites.

High coordination numbers

The presence of axial and equatorial sites in a pentagonal bipyramidal molecule leads to stereoisomerism in a similar manner to that in a trigonal bipyramidal species. In a square antiprismatic molecule EX_8, each X atom is identical (Fig. 2.16). Once two or more different atoms or groups are present, e.g. EX_6Y_2, stereoisomers are possible. As an exercise, draw out the four possibilities for square antiprismatic EX_6Y_2.

Double bonds

In contrast to a single (σ) bond where free rotation is generally assumed, rotation about a double bond is not a low energy process. The presence of a double bond may therefore lead to stereoisomerism as is observed for N_2F_2. Each N atom carries a lone pair as well as forming one N—F single bond and an N=N double bond. Structures **2.43** and **2.44** show the *trans*- and *cis*-isomers[†] respectively of N_2F_2.

$$\underset{F}{\overset{\textstyle F}{N=N}} \qquad \underset{F}{\overset{}{N=N}}\underset{F}{\overset{}{}}$$

(2.43)   **(2.44)**

[†] In organic chemistry, IUPAC nomenclature uses the prefix (E)- for a *trans*-arrangement of groups and (Z)- for a *cis*-arrangement, but for inorganic compounds, the terms *trans*- and *cis*- remain in use.

Self-study exercises

1. Draw the structures of the two isomers of octahedral $[Cr(OH_2)_4Cl_2]^+$ and give labels that distinguish the isomers.

2. $[PtCl_2(PEt_3)_2]$ possesses two stereoisomers. Is the complex square planar or tetrahedral? Rationalize your answer.

3. Draw the structures of *mer*- and *fac*-$[RhCl_3(OH_2)_3]$. What is the coordination geometry at the metal centre?

4. Tetrahydrofuran (THF) has the following structure and coordinates to metal ions through the oxygen atom. Draw the structures of the three possible isomers of trigonal bipyramidal $[MnI_2(THF)_3]$.

THF

FURTHER READING

P. Atkins and J. de Paula (2010) *Atkins' Physical Chemistry*, 9th edn, Oxford University Press, Oxford – This text gives a solid and well-tested background in physical chemistry.

J. Barrett (2002) *Structure and Bonding*, RSC Publishing, Cambridge – An introductory text that includes valence bond, molecular orbital and VSEPR theories.

R.J. Gillespie (2008) *Coord. Chem. Rev.*, vol. 252, p. 1315 – 'Fifty years of the VSEPR model'.

R.J. Gillespie and E.A. Robinson (2005) *Chem. Soc. Rev.*, vol. 34, p. 396 – A 'tutorial review' 'Models of molecular geometry' that considers the VSEPR model and the more recently developed ligand close-packing (LCP) model.

D.O. Hayward (2002) *Quantum Mechanics for Chemists*, RSC Publishing, Cambridge – An undergraduate student text that covers the basic principles of quantum mechanics.

C.E. Housecroft and E.C. Constable (2010) *Chemistry*, 4th edn, Prentice Hall, Harlow – This text provides clear discussion of the fundamental principles of bonding in molecules at an introductory level.

R. McWeeny (1979) *Coulson's Valence*, 3rd edn, Oxford University Press, Oxford – A classic book containing a general treatment of chemical bonding with a detailed mathematical approach.

D.W. Smith (2004) *J. Chem. Educ.*, vol. 81, p. 886 – A useful article entitled 'Effects of exchange energy and spin-orbit coupling on bond energies'.

M.J. Winter (1994) *Chemical Bonding*, Oxford University Press, Oxford – This 'primer' for first year undergraduates approaches chemical bonding non-mathematically.

PROBLEMS

2.1 Draw Lewis structures to describe the bonding in the following molecules: (a) F_2; (b) BF_3; (c) NH_3; (d) H_2Se; (e) H_2O_2; (f) $BeCl_2$; (g) SiH_4; (h) PF_5.

2.2 Use the Lewis structure model to deduce the type of nitrogen–nitrogen bond present in (a) N_2H_4, (b) N_2F_4, (c) N_2F_2 and (d) $[N_2H_5]^+$.

2.3 Draw the resonance structures for the O_3 molecule. What can you conclude about the net bonding picture?

2.4 Draw Lewis structures for (a) CO_2, (b) SO_2, (c) OF_2 and (d) H_2CO.

2.5 Each of the following is a radical. For which does a Lewis structure correctly confirm this property: (a) NO, (b) O_2, (c) NF_2?

2.6 (a) Use VB theory to describe the bonding in the diatomic molecules Li_2, B_2 and C_2. (b) Experimental data show that Li_2 and C_2 are diamagnetic whereas B_2 is paramagnetic. Is the VB model consistent with these facts?

2.7 Using VB theory and the Lewis structure model, determine the bond order in (a) H_2, (b) Na_2, (c) S_2, (d) N_2 and (e) Cl_2. Is there any ambiguity with finding the bond orders by this method?

2.8 Does VB theory indicate that the diatomic molecule He_2 is a viable species? Rationalize your answer.

2.9 (a) Use MO theory to determine the bond order in each of $[He_2]^+$ and $[He_2]^{2+}$. (b) Does the MO picture of the bonding in these ions suggest that they are viable species?

2.10 (a) Construct an MO diagram for the formation of O_2; use only the valence orbitals of the oxygen atoms. (b) Use the diagram to rationalize the following trend in O–O bond distances: O_2, 121 pm; $[O_2]^+$, 112 pm; $[O_2]^-$, 134 pm; $[O_2]^{2-}$, 149 pm. (c) Which of these species are paramagnetic?

2.11 Confirm that the octet rule is obeyed by *each* of the atoms in the following molecules: (a) CF_4, (b) O_2, (c) $AsBr_3$, (d) SF_2.

2.12 Draw charge-separated resonance structures to give a representation of the bonding in PF_5 such that the octet rule is strictly obeyed.

2.13 One member of each of the following sets of compounds is not isoelectronic with the others. Which one in each set is the odd one out?

(a) $[NO_2]^+$, CO_2, $[NO_2]^-$ and $[N_3]^-$
(b) $[CN]^-$, N_2, CO, $[NO]^+$ and $[O_2]^{2-}$
(c) $[SiF_6]^{2-}$, $[PF_6]^-$, $[AlF_6]^{3-}$ and $[BrF_6]^-$

2.14 In the following table, match a species in list 1 with an isoelectronic partner in list 2. Some species may have more than one partner. Qualify how you have interpreted the term *isoelectronic*.

List 1	List 2
F_2	$[H_3O]^+$
NH_3	$[GaCl_4]^-$
$[GaBr_4]^-$	Cl_2
$[SH]^-$	$[NH_4]^+$
$[BH_4]^-$	$[OH]^-$
$[AsF_6]^-$	$[O_2]^{2-}$
$[PBr_4]^+$	SeF_6
HF	$SiBr_4$

2.15 Using the data in Table 2.2, determine which of the following covalent single bonds is polar and (if appropriate) in which direction the dipole moment acts. (a) N−H; (b) F−Br; (c) C−H; (d) P−Cl; (e) N−Br.

2.16 Pick out *pairs* of isoelectronic species from the following list; not all species have a 'partner': HF; CO_2; SO_2; NH_3; PF_3; SF_4; SiF_4; $SiCl_4$; $[H_3O]^+$; $[NO_2]^+$; $[OH]^-$; $[AlCl_4]^-$.

2.17 Use the VSEPR model to predict the structures of (a) H_2Se, (b) $[BH_4]^-$, (c) NF_3, (d) SbF_5, (e) $[H_3O]^+$, (f) IF_7, (g) $[I_3]^-$, (h) $[I_3]^+$, (i) SO_3.

2.18 Use the VSEPR model to rationalize the structure of SOF_4 shown in Fig. 2.19. What are the bond orders of (a) each S−F bond and (b) the S−O bond?

2.19 Determine the shapes of each of the following molecules and then, using the data in Table 2.2, state whether each is expected to be polar or not: (a) H_2S; (b) CO_2; (c) SO_2; (d) BF_3; (e) PF_5; (f) *cis*-N_2F_2; (g) *trans*-N_2F_2; (h) HCN.

2.20 State whether you expect the following species to possess stereoisomers and, if so, draw their structures and give them distinguishing labels: (a) BF_2Cl; (b) $POCl_3$; (c) $MePF_4$; (d) $[PF_2Cl_4]^-$.

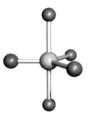

Fig. 2.19 The structure of SOF_4.

OVERVIEW PROBLEMS

2.21 (a) Draw resonance structures for CO, choosing only those that you think contribute significantly to the bonding.
(b) Figure 2.15a shows an MO diagram for CO. Two MOs are illustrated by schematic representations. Draw similar diagrams for the remaining six MOs.

2.22 (a) On steric grounds, should *cis*- or *trans*-$[PtCl_2(PPh_3)_2]$ be favoured?
(b) Use the VSEPR model to rationalize why SNF_3 is tetrahedral but SF_4 has a see-saw geometry.
(c) Suggest why KrF_2 is a linear rather than bent molecule.

2.23 Account for each of the following observations.

(a) IF_5 is a polar molecule.
(b) The first ionization energy of K is lower than that of Li.
(c) BI_3 is trigonal planar while PI_3 is trigonal pyramidal in shape.

2.24 Suggest reasons for the following observations.

(a) The second ionization energy of He is higher than the first despite the fact that both electrons are removed from the $1s$ atomic orbital.
(b) Heating N_2F_2 at 373 K results in a change from a non-polar to polar molecule.
(c) S_2 is paramagnetic.

2.25 Account for each of the following observations.

(a) The mass spectrum of molecular bromine shows three lines for the parent ion $[Br_2]^+$.
(b) In the structure of solid bromine, each Br atom has one nearest neighbour at a distance of 227 pm, and several other next nearest neighbours at 331 pm.
(c) In the salt formed from the reaction of Br_2 and SbF_5, the Br–Br distance in the $[Br_2]^+$ ion (215 pm) is shorter than in Br_2.

2.26 (a) Draw possible stereoisomers for the trigonal bipyramidal $[SiF_3Me_2]^-$ anion (Me = CH_3). An X-ray diffraction study of a salt of $[SiF_3Me_2]^-$ shows that two F atoms occupy axial sites. Suggest why this stereoisomer is preferred over the other possible structures that you have drawn.

(b) Account for the fact that members of the series of complexes $[PtCl_4]^{2-}$, $[PtCl_3(PMe_3)]^-$, $[PtCl_2(PMe_3)_2]$ and $[PtCl(PMe_3)_3]^+$ do not possess the same number of stereoisomers.

2.27 (a) Write down the ions that are present in the compound $[PCl_4][PCl_3F_3]$. What shape do you expect each ion to adopt? In theory, does either ion possess stereoisomers?

(b) Use the VSEPR model to rationalize why BCl_3 and NCl_3 do not adopt similar structures. Is either molecule expected to be polar? Rationalize your answer.

2.28 Assuming that the VSEPR model can be applied successfully to each of the following species, determine how many different fluorine environments are present in each molecule or ion: (a) $[SiF_6]^{2-}$, (b) XeF_4, (c) $[NF_4]^+$, (d) $[PHF_5]^-$, (e) $[SbF_5]^{2-}$.

2.29 Critically compare the VB and MO treatments of the bonding in O_2, paying particular attention to the properties of O_2 that the resulting bonding models imply.

INORGANIC CHEMISTRY MATTERS

2.30 The table below gives the average composition of the Earth's atmosphere (ppm = parts per million). Water vapour is also present in small and variable amounts.

Gas	Average amount/ppm	Gas	Average amount/ppm
He	5.2	CH_4	1.72
Ne	18	CO	0.12
Ar	9340	CO_2	355
Kr	1.1	N_2O	0.31
Xe	0.09	NO	<0.01
H_2	0.58	O_3	0.1–0.01
N_2	780840	SO_2	$<10^{-4}$
O_2	209460	NH_3	$<10^{-3}$

(a) Draw a Lewis structure for N_2O, ensuring that each atom obeys the octet rule. (b) Use the VSEPR model to predict the molecular shapes of SO_2, NH_3, N_2O, CH_4 and CO_2. (c) Which of the gases in the table are radicals? For each of the gases you have chosen, explain how the radical nature arises. (d) O_3 (ozone) is only present in <0.1 ppm. Nonetheless, it is a vital component of the Earth's atmosphere. Why? (e) Draw an MO diagram for the formation of N_2 from two N atoms, using only the valence orbitals. Use the diagram to rationalize why N_2 is chemically very inert. (f) What is the relationship between the monoatomic gases in the Earth's atmosphere?

2.31 Carbon monoxide is a toxic pollutant which arises from the partial combustion of carbon-based fuels. Complete combustion produces CO_2. The toxicity of CO is a result of its competition for the O_2-binding sites in blood, i.e. the iron present in haemoglobin (see Chapter 29). When CO binds to the iron, it prevents O_2 from being carried in the bloodstream. The following are resonance structures for CO:

(a) Comment on these structures in terms of the octet rule. (b) How is the right-hand resonance structure related to a Lewis structure for N_2? (c) A primary interaction between CO and iron in haemoglobin involves a lone pair of electrons on the carbon atom. Using MO theory, explain how this lone pair arises. (d) Without treatment, severe CO poisoning is fatal. Explain why a hyperbaric chamber containing pure O_2 at a pressure of 1.4 bar is used to treat a patient with severe CO poisoning. Normal air contains 21% O_2.

2.32 Volcanoes and deep sea hydrothermal vents are both associated with sulfur-rich environments. Mount Etna is classed as a continuously degassing volcano and emissions of SO_2 and H_2S are around 1.5 Tg y^{-1} and 100 Gg y^{-1}, respectively (Tg = 10^{12} g; Gg = 10^9 g). (a) Draw Lewis structures for H_2S and SO_2, ensuring that the octet rule is obeyed by the S and O atoms. Your answer must be consistent with the fact that in SO_2, the two sulfur–oxygen bonds are the same length. (b) Are H_2S and SO_2 polar or non-polar molecules? If polar, draw a diagram to show the direction of the molecular dipole moment. (c) In the troposphere, SO_2 reacts with $HO^{\bullet}$ radicals. Construct an MO diagram for $HO^{\bullet}$ from H and O atoms, and deduce what you can about the bonding in $HO^{\bullet}$.

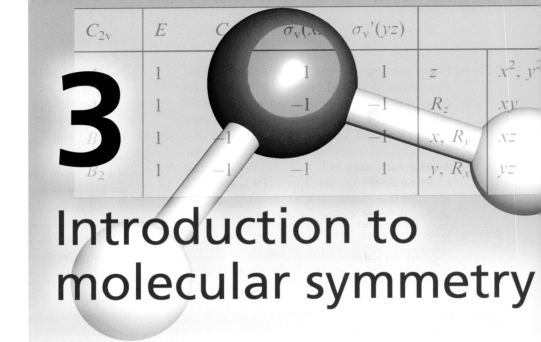

3

Introduction to molecular symmetry

3.1 Introduction

In chemistry, symmetry is important both at a molecular level and within crystalline systems. An understanding of symmetry is essential in discussions of molecular spectroscopy and calculations of molecular properties. A discussion of *crystal symmetry* is not included in this book, and we introduce only *molecular symmetry*. For qualitative purposes, it is sufficient to refer to the shape of a molecule using terms such as tetrahedral, octahedral or square planar. However, the common use of these descriptors is not always precise, e.g. consider the structures of BF_3, **3.1**, and BF_2H, **3.2**, both of which are planar. A molecule of BF_3 is correctly described as being trigonal planar, since its symmetry properties are fully consistent with this description; all the F−B−F bond angles are $120°$ and the B−F bond distances are all identical (131 pm). It is correct to say that the boron centre in BF_2H, **3.2**, is in a *pseudo-trigonal planar* environment but the molecular symmetry properties are not the same as those of BF_3. The F−B−F bond angle in BF_2H is smaller than the two H−B−F angles, and the B−H bond is shorter (119 pm) than the B−F bonds (131 pm).

F
120°   120°
B
F   F
120°

(3.1)

H
121°   121°
B
F   F
118°

(3.2)

The descriptor *symmetrical* implies that a species possesses a number of indistinguishable configurations. When structure **3.1** is rotated in the plane of the paper through $120°$, the resulting structure is indistinguishable from the first; another $120°$ rotation results in a third indistinguishable molecular orientation (Fig. 3.1). This is *not* true if we carry out the same rotational operations on BF_2H.

Group theory is the mathematical treatment of symmetry. In this chapter, we introduce the fundamental language of group theory (*symmetry operator*, *symmetry element*, *point group* and *character table*). The chapter does not set out to give a comprehensive survey of molecular symmetry, but rather to introduce some common terminology and its meaning. We include an introduction to the vibrational spectra of simple inorganic molecules and show how to use this technique to distinguish between possible structures for XY_2, XY_3 and XY_4 molecules. Complete normal coordinate analysis of such species is beyond the remit of this book.

3.2 Symmetry operations and symmetry elements

In Fig. 3.1, we applied $120°$ rotations to BF_3 and saw that each rotation generated a representation of the molecule that was indistinguishable from the first. Each rotation is an example of a *symmetry operation*.

> A **symmetry operation** is an operation performed on an object which leaves it in a configuration that is indistinguishable from, and superimposable on, the original configuration.

The rotations described in Fig. 3.1 are performed about an axis perpendicular to the plane of the paper and passing through the boron atom; the axis is an example of a *symmetry element*.

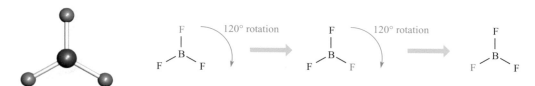

Fig. 3.1 Rotation of the trigonal planar BF$_3$ molecule through $120°$ generates a representation of the structure that is indistinguishable from the first; one F atom is marked in red simply as a label. A second $120°$ rotation gives another indistinguishable structural representation.

> A symmetry operation is carried out with respect to points, lines or planes, the latter being the *symmetry elements*.

Rotation about an *n*-fold axis of symmetry

The symmetry operation of rotation about an *n*-fold axis (the symmetry element) is denoted by the symbol C_n, in which the angle of rotation is $\dfrac{360°}{n}$; *n* is an integer, e.g. 2, 3 or 4. Applying this notation to the BF$_3$ molecule in Fig. 3.1 gives a value of $n = 3$ (eq. 3.1), and therefore we say that the BF$_3$ molecule contains a C_3 *rotation axis*. In this case, the axis lies perpendicular to the plane containing the molecule.

$$\text{Angle of rotation} = 120° = \frac{360°}{n} \qquad (3.1)$$

In addition, BF$_3$ also contains three 2-fold (C_2) rotation axes, each coincident with a B—F bond as shown in Fig. 3.2.

If a molecule possesses more than one type of *n*-axis, the axis of highest value of *n* is called the *principal axis*; it is the axis of *highest molecular symmetry*. For example, in BF$_3$, the C_3 axis is the principal axis.

In some molecules, rotation axes of lower orders than the principal axis may be coincident with the principal axis. For example, in square planar XeF$_4$, the principal axis is a C_4 axis but this also coincides with a C_2 axis (see Fig. 3.4).

Where a molecule contains more than one type of C_n axis with the same value of *n*, they are distinguished by using prime marks, e.g. C_2, C_2' and C_2''. We return to this in the discussion of XeF$_4$ (see Fig. 3.4).

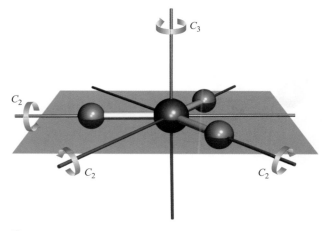

Fig. 3.2 The 3-fold (C_3) and three 2-fold (C_2) axes of symmetry possessed by the trigonal planar BF$_3$ molecule.

Self-study exercises

1. Each of the following contains a 6-membered ring: benzene, borazine (see Fig. 13.23), pyridine and S$_6$ (see Box 1.1). Explain why only benzene contains a 6-fold principal rotation axis.

2. Among the following, why does only XeF$_4$ contain a 4-fold principal rotation axis: CF$_4$, SF$_4$, [BF$_4$]$^-$ and XeF$_4$?

3. Draw the structure of [XeF$_5$]$^-$. On the diagram, mark the C_5 axis. The ion contains five C_2 axes. Where are these axes?   [*Ans.* for structure, see worked example 2.7]

4. Look at the structure of B$_5$H$_9$ in Fig. 13.29a. Where is the C_4 axis in this molecule?

Reflection through a plane of symmetry (mirror plane)

If reflection of all parts of a molecule through a plane produces an indistinguishable configuration, the plane is a *plane of symmetry*; the symmetry operation is one of reflection and the symmetry element is the mirror plane (denoted by σ). For BF$_3$, the plane containing the molecular framework (the brown plane shown in Fig. 3.2) is a *mirror plane*. In this case, the plane lies perpendicular to the vertical principal axis and is denoted by the symbol σ_h.

The framework of atoms in a linear, bent or planar molecule can always be drawn in a plane, but this plane can be labelled σ_h *only* if the molecule possesses a C_n axis *perpendicular* to the plane. If the plane *contains* the principal axis, it is labelled σ_v. Consider the H$_2$O molecule. This possesses a C_2 axis (Fig. 3.3) but it also contains *two* mirror planes, one containing the H$_2$O framework, and one perpendicular to it. Each plane contains the principal axis of rotation and so may be denoted as σ_v but in order to distinguish between them, we use the notations σ_v and σ_v'. The σ_v label refers to the plane that bisects the H—O—H bond angle and the σ_v' label refers to the plane in which the molecule lies.

A special type of σ plane which contains the principal rotation axis, but which bisects the angle between two adjacent 2-fold axes, is labelled σ_d. A square planar

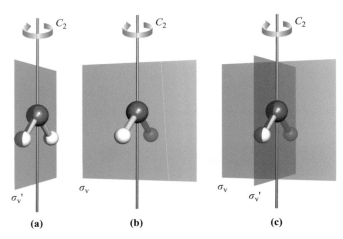

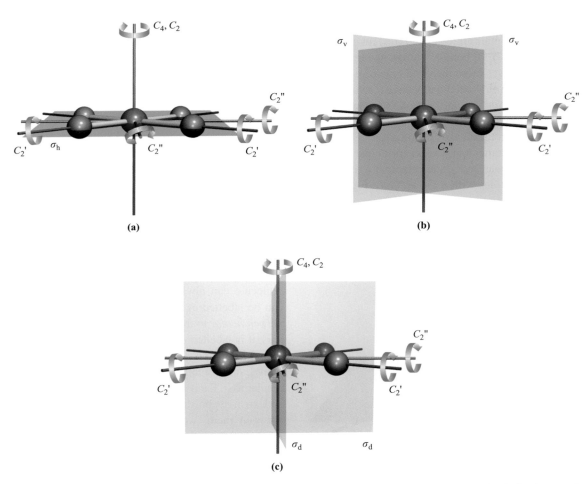

(a)   **(b)**   **(c)**

Fig. 3.3 The H_2O molecule possesses one C_2 axis and two mirror planes. (a) The C_2 axis and the plane of symmetry that contains the H_2O molecule. (b) The C_2 axis and the plane of symmetry that is perpendicular to the plane of the H_2O molecule. (c) Planes of symmetry in a molecule are often shown together on one diagram; this representation for H_2O combines diagrams (a) and (b).

molecule such as XeF_4 provides an example. Figure 3.4a shows that XeF_4 contains a C_4 axis (the principal axis) and perpendicular to this is the σ_h plane in which the molecule lies. Coincident with the C_4 axis is a C_2 axis. Within the plane of the molecule, there are two sets of C_2 axes. One type (the C_2' axis) coincides with F–Xe–F bonds, while the second type (the C_2'' axis) bisects the F–Xe–F 90° angle (Fig. 3.4). We can now define two sets of mirror planes: one type (σ_v) contains the principal axis and a C_2' axis (Fig. 3.4b), while the second type (σ_d) contains the principal axis and a C_2'' axis (Fig. 3.4c). Each σ_d plane bisects the angle between two C_2' axes.

In the notation for planes of symmetry, σ, the subscripts h, v and d stand for horizontal, vertical and dihedral respectively.

Self-study exercises

1. N_2O_4 is planar (Fig. 15.15). Show that it possesses three planes of symmetry.

Fig. 3.4 The square planar molecule XeF_4. (a) One C_2 axis coincides with the principal (C_4) axis; the molecule lies in a σ_h plane which contains two C_2' and two C_2'' axes. (b) Each of the two σ_v planes contains the C_4 axis and one C_2' axis. (c) Each of the two σ_d planes contains the C_4 axis and one C_2'' axis.

2. B_2Br_4 has the following staggered structure:

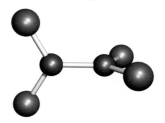

Show that B_2Br_4 has one less plane of symmetry than B_2F_4 which is planar.

3. Ga_2H_6 has the following structure in the gas phase:

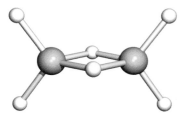

Show that it possesses three planes of symmetry.

4. Show that the planes of symmetry in benzene are one σ_h, three σ_v and three σ_d.

Reflection through a centre of symmetry (inversion centre)

If reflection of *all* parts of a molecule through the centre of the molecule produces an indistinguishable configuration, the centre is a *centre of symmetry*. It is also called a *centre of inversion* (see also Box 2.1) and is designated by the symbol *i*. Each of the molecules CO_2 (**3.3**), *trans*-N_2F_2 (see worked example 3.1), SF_6 (**3.4**) and benzene (**3.5**) possesses a centre of symmetry, but H_2S (**3.6**), *cis*-N_2F_2 (**3.7**) and SiH_4 (**3.8**) do not.

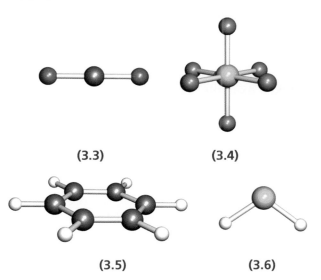

(3.3)   **(3.4)**

(3.5)   **(3.6)**

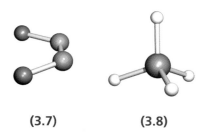

(3.7)   **(3.8)**

Self-study exercises

1. Draw the structures of each of the following species and confirm that each possesses a centre of symmetry: CS_2, $[PF_6]^-$, XeF_4, I_2, $[ICl_2]^-$.

2. $[PtCl_4]^{2-}$ has a centre of symmetry, but $[CoCl_4]^{2-}$ does not. One is square planar and the other is tetrahedral. Which is which?

3. Why does CO_2 possess an inversion centre, but NO_2 does not?

4. CS_2 and HCN are both linear. Explain why CS_2 possesses a centre of symmetry whereas HCN does not.

Rotation about an axis, followed by reflection through a plane perpendicular to this axis

If rotation through $\dfrac{360°}{n}$ about an axis, followed by reflection through a plane perpendicular to that axis, yields an indistinguishable configuration, the axis is an *n-fold rotation–reflection axis*, also called an *n-fold improper rotation axis*. It is denoted by the symbol S_n. Tetrahedral species of the type XY_4 (all Y groups must be equivalent) possess three S_4 axes, and the operation of one S_4 rotation–reflection in the CH_4 molecule is illustrated in Fig. 3.5.

Self-study exercises

1. Explain why BF_3 possesses an S_3 axis, but NF_3 does not.

2. C_2H_6 in a staggered conformation possesses an S_6 axis. Show that this axis lies along the C–C bond.

3. Figure 3.5 shows one of the S_4 axes in CH_4. On going from CH_4 to CH_2Cl_2, are the S_4 axes retained?

Identity operator

All objects can be operated upon by the identity operator *E*. This is the simplest operator (although it may not be easy to appreciate why we identify such an operator!) and effectively identifies the molecular configuration. The operator *E* leaves the molecule unchanged.

Axis bisects the
H–C–H bond
angle

Rotate
through
90°

Reflect
through a
plane that is
perpendicular
to the original
rotation axis

🌐 **Fig. 3.5** An improper rotation (or rotation–reflection), S_n, involves rotation about $\dfrac{360°}{n}$ followed by reflection through a plane that is perpendicular to the rotation axis. The diagram illustrates the operation about one of the S_4 axes in CH_4; three S_4 operations are possible for the CH_4 molecule. [Exercise: where are the three rotation axes for the three S_4 operations in CH_4?]

Worked example 3.1   Symmetry properties of *cis-* and *trans-*N_2F_2

How do the rotation axes and planes of symmetry in *cis-* and *trans-*N_2F_2 differ?

First draw the structures of *cis-* and *trans-*N_2F_2; both are planar molecules.

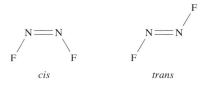

cis   *trans*

1. The identity operator E applies to each isomer.
2. Each isomer possesses a plane of symmetry which contains the molecular framework. However, their labels differ (see point 5 below).
3. The *cis-*isomer contains a C_2 axis which lies in the plane of the molecule, but the *trans-*isomer contains a C_2 axis which bisects the N–N bond and is perpendicular to the plane of the molecule.

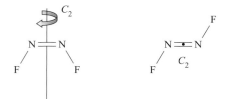

4. The *cis-* (but not the *trans-*) isomer contains a mirror plane, σ_v, lying perpendicular to the plane of the molecule and bisecting the N–N bond:

5. The consequence of the different types of C_2 axes, and the presence of the σ_v plane in the *cis-*isomer, is that the symmetry planes containing the *cis-* and *trans-*N_2F_2 molecular frameworks are labelled σ_v' and σ_h respectively.

Self-study exercises

1. How do the rotation axes and planes of symmetry in Z- and E-CFH=CFH differ?

2. How many planes of symmetry do (a) $F_2C=O$, (b) ClFC=O and (c) $[HCO_2]^-$ possess?
   [*Ans.* (a) 2; (b) 1; (c) 2]

Worked example 3.2   Symmetry elements in NH_3

The symmetry elements for NH_3 are E, C_3 and $3\sigma_v$. (a) Draw the structure of NH_3. (b) What is the meaning of the E operator? (c) Draw a diagram to show the symmetry elements.

(a) The molecule is trigonal pyramidal.

(b) The E operator is the identity operator and it leaves the molecule unchanged.

(c) The C_3 axis passes through the N atom, perpendicular to a plane containing the three H atoms. Each σ_v plane contains one N−H bond and bisects the opposite H−N−H bond angle.

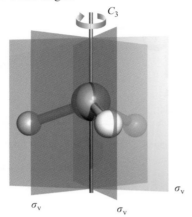

Self-study exercises

1. What symmetry elements are lost in going from NH_3 to NH_2Cl? [*Ans.* C_3; two σ_v]

2. Compare the symmetry elements possessed by NH_3, NH_2Cl, $NHCl_2$ and NCl_3.

3. Draw a diagram to show the symmetry elements of $NClF_2$. [*Ans.* Show one σ plane; only other operator is E]

Worked example 3.3   Trigonal planar BCl_3 *versus* trigonal pyramidal PCl_3

What symmetry elements do BCl_3 and PCl_3 (a) have in common and (b) not have in common?

PCl_3 is trigonal pyramidal (use the VSEPR model) and so possesses the same symmetry elements as NH_3 in worked example 3.2. These are E, C_3 and $3\sigma_v$.

BCl_3 is trigonal planar (use VSEPR) and possesses all the above symmetry elements:

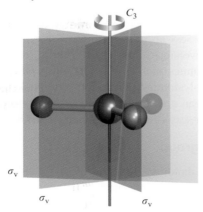

In addition, BCl_3 contains a σ_h plane and three C_2 axes (see Fig. 3.2).

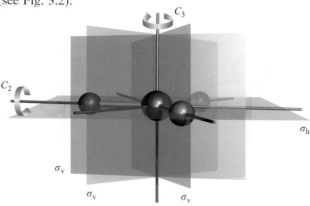

Rotation through 120° about the C_3 axis, followed by reflection through the plane perpendicular to this axis (the σ_h plane), generates a molecular configuration indistinguishable from the first – this is an improper rotation S_3.

Conclusion

The symmetry elements that BCl_3 and PCl_3 have in common are E, C_3 and $3\sigma_v$. The symmetry elements possessed by BCl_3 but not by PCl_3 are σ_h, $3C_2$ and S_3.

Self-study exercises

1. Show that BF_3 and $F_2C{=}O$ have the following symmetry elements in common: E, two mirror planes, one C_2.

2. How do the symmetry elements of ClF_3 and BF_3 differ? [*Ans:* BF_3, as for BCl_3 above; ClF_3, E, σ_v', σ_v, C_2]

3.3 Successive operations

As we have seen in Section 3.2, a particular symbol is used to denote a specific symmetry element. To say that NH_3 possesses a C_3 axis tells us that we can rotate the molecule through 120° and end up with a molecular configuration that is indistinguishable from the first. However, it takes three such operations to give a configuration of the NH_3 molecule that *exactly* coincides with the first. The three separate 120° rotations are identified by using the notation in Fig. 3.6. We cannot *actually* distinguish between the three H atoms, but for clarity they are labelled H(1), H(2) and H(3) in the figure. Since the third rotation, C_3^3, returns the NH_3 molecule to its initial configuration, we can write eq. 3.2, or, in general, eq. 3.3.

$$C_3^3 = E \tag{3.2}$$

$$C_n^n = E \tag{3.3}$$

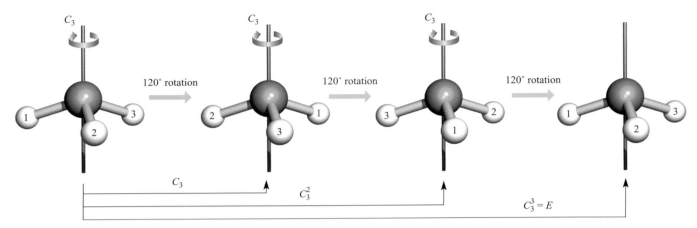

Fig. 3.6 Successive C_3 rotations in NH_3 are distinguished using the notation C_3 (or C_3^1), C_3^2 and C_3^3. The effect of the last operation is the same as that of the identity operator acting on NH_3 in the initial configuration.

Similar statements can be written to show the combined effects of successive operations. For example, in planar BCl_3, the S_3 improper axis of rotation corresponds to rotation about the C_3 axis followed by reflection through the σ_h plane. This can be written in the form of eq. 3.4.

$$S_3 = C_3 \times \sigma_h \qquad (3.4)$$

Self-study exercises

1. $[PtCl_4]^{2-}$ is square planar; to what rotational operation is C_4^2 equivalent?

2. Draw a diagram to illustrate what the notation C_6^4 means with respect to rotational operations in benzene.

3.4 Point groups

The number and nature of the symmetry elements of a given molecule are conveniently denoted by its *point group*, and give rise to labels such as C_2, C_{3v}, D_{3h}, D_{2d}, T_d, O_h or I_h. These point groups belong to the classes of C groups, D groups and special groups, the latter containing groups that possess special symmetries, i.e. tetrahedral, octahedral and icosahedral.

To describe the symmetry of a molecule in terms of one symmetry element (e.g. a rotation axis) provides information only about this property. Each of BF_3 and NH_3 possesses a 3-fold axis of symmetry, but their structures and overall symmetries are different: BF_3 is trigonal planar and NH_3 is trigonal pyramidal. On the other hand, if we describe the symmetries of these molecules in terms of their respective point groups (D_{3h} and C_{3v}), we are providing information about *all* their symmetry elements.

Before we look at some representative point groups, we emphasize that it is not essential to memorize the symmetry elements of a particular point group. These are listed in *character tables* (see Sections 3.5, 5.4 and 5.5, and Appendix 3) which are widely available.

Table 3.1 summarizes the most important classes of point group and gives their characteristic types of symmetry elements; E is, of course, common to every group. Some particular features of significance are given below.

C_1 point group

Molecules that appear to have no symmetry at all, e.g. **3.9**, must possess the symmetry element E and effectively possess at least one C_1 axis of rotation. They therefore belong to the C_1 point group, although since $C_1 = E$, the rotational symmetry operation is ignored when we list the symmetry elements of this point group.

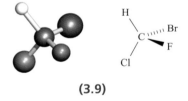

(3.9)

$C_{\infty v}$ point group

C_∞ signifies the presence of an ∞-fold axis of rotation, i.e. that possessed by a linear molecule (Fig. 3.7); for the molecular species to belong to the $C_{\infty v}$ point group, it must also possess an infinite number of σ_v planes but *no* σ_h plane or inversion centre. These criteria are met by asymmetrical diatomics such as HF, CO and $[CN]^-$ (Fig. 3.7a), and linear polyatomics (throughout this book, polyatomic is used to mean a species containing three or more atoms) that do not possess a centre of symmetry, e.g. OCS and HCN.

$D_{\infty h}$ point group

Symmetrical diatomics (e.g. H_2, $[O_2]^{2-}$) and linear polyatomics that contain a centre of symmetry (e.g. $[N_3]^-$, CO_2, HC≡CH) possess a σ_h plane in addition to a C_∞

Table 3.1 Characteristic symmetry elements of some important classes of point groups. The characteristic symmetry elements of the T_d, O_h and I_h are omitted because the point groups are readily identified (see Figs. 3.8 and 3.9). No distinction is made in this table between σ_v and σ_d planes of symmetry. For complete lists of symmetry elements, character tables (Appendix 3) should be consulted.

Point group	Characteristic symmetry elements	Comments
C_s	E, one σ plane	
C_i	E, inversion centre	
C_n	E, one (principal) n-fold axis	
C_{nv}	E, one (principal) n-fold axis, n σ_v planes	
C_{nh}	E, one (principal) n-fold axis, one σ_h plane, one S_n-fold axis which is coincident with the C_n axis	The S_n axis necessarily follows from the C_n axis and σ_h plane. For $n = 2$, 4 or 6, there is also an inversion centre.
D_{nh}	E, one (principal) n-fold axis, n C_2 axes, one σ_h plane, n σ_v planes, one S_n-fold axis	The S_n axis necessarily follows from the C_n axis and σ_h plane. For $n = 2$, 4 or 6, there is also an inversion centre.
D_{nd}	E, one (principal) n-fold axis, n C_2 axes, n σ_v planes, one S_{2n}-fold axis	For $n = 3$ or 5, there is also an inversion centre.
T_d		Tetrahedral
O_h		Octahedral
I_h		Icosahedral

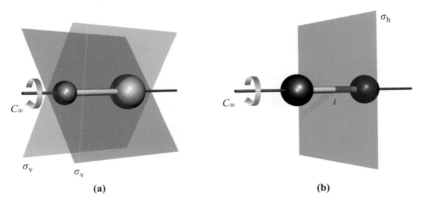

Fig. 3.7 Linear molecular species can be classified according to whether they possess a centre of symmetry (inversion centre) or not. All linear species possess a C_∞ axis of rotation and an infinite number of σ_v planes; in (a), two such planes are shown and these planes are omitted from (b) for clarity. Diagram (a) shows an asymmetrical diatomic belonging to the point group $C_{\infty v}$, and (b) shows a symmetrical diatomic belonging to the point group $D_{\infty h}$.

axis and an infinite number of σ_v planes (Fig. 3.7). These species belong to the $D_{\infty h}$ point group.

T_d, O_h or I_h point groups

Molecular species that belong to the T_d, O_h or I_h point groups (Fig. 3.8) possess many symmetry elements, although it is seldom necessary to identify them all before the appropriate point group can be assigned. Species with tetrahedral symmetry include SiF_4, $[ClO_4]^-$, $[CoCl_4]^{2-}$, $[NH_4]^+$, P_4 (Fig. 3.9a) and B_4Cl_4 (Fig. 3.9b). Those with octahedral symmetry include SF_6, $[PF_6]^-$, $W(CO)_6$

(Fig. 3.9c) and $[Fe(CN)_6]^{3-}$. There is no centre of symmetry in a tetrahedron but there is one in an octahedron, and this distinction has consequences with regard to the observed electronic spectra of tetrahedral and octahedral metal complexes (see Section 20.7). Members of the icosahedral point group are uncommon, e.g. $[B_{12}H_{12}]^{2-}$ (Fig. 3.9d).

Determining the point group of a molecule or molecular ion

The application of a *systematic* approach to the assignment of a point group is essential, otherwise there is the risk that

Tetrahedron   Octahedron   Icosahedron

Fig. 3.8 The tetrahedron (T_d symmetry), octahedron (O_h symmetry) and icosahedron (I_h symmetry) possess 4, 6 and 12 vertices respectively, and 4, 8 and 20 equilateral-triangular faces respectively.

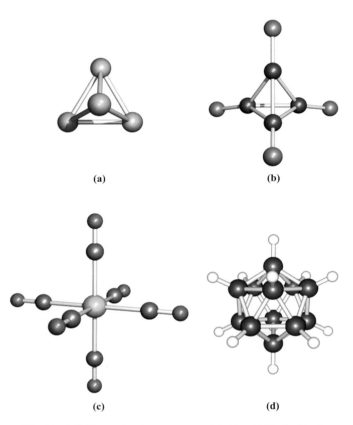

(a)   (b)

(c)   (d)

Fig. 3.9 The molecular structures of (a) P_4, (b) B_4Cl_4 (the B atoms are shown in blue), (c) $[W(CO)_6]$ (the W atom is shown in yellow and the C atoms in grey) and (d) $[B_{12}H_{12}]^{2-}$ (the B atoms are shown in blue).

symmetry elements will be missed with the consequence that an incorrect assignment is made. Figure 3.10 shows a procedure that may be adopted; some of the less common point groups (e.g. S_n, T, O) are omitted from the scheme. Notice that it is *not* necessary to find all the symmetry elements (e.g. improper axes) in order to determine the point group.

We illustrate the application of Fig. 3.10 in four worked examples, with an additional example in Section 3.8. Before assigning a point group to a molecule, its structure must be determined by, for example, microwave

spectroscopy, or X-ray, electron or neutron diffraction methods (see Section 4.10).

Worked example 3.4   Point group assignments: 1

Determine the point group of *trans*-N_2F_2.

First draw the structure.

Apply the strategy shown in Fig. 3.10:

START $\Longrightarrow$

Is the molecule linear?	No
Does *trans*-N_2F_2 have T_d, O_h or I_h symmetry?	No
Is there a C_n axis?	Yes; a C_2 axis perpendicular to the plane of the paper and passing through the midpoint of the N–N bond
Are there two C_2 axes perpendicular to the principal axis?	No
Is there a σ_h plane (perpendicular to the principal axis)?	Yes

$\Longrightarrow$ STOP

The point group is C_{2h}.

Self-study exercises

1. Show that the point group of *cis*-N_2F_2 is C_{2v}.

2. Show that the point group of *E*-CHCl=CHCl is C_{2h}.

Worked example 3.5   Point group assignments: 2

Determine the point group of PF_5.

First, draw the structure.

                        Axial

                        Equatorial

                        Axial

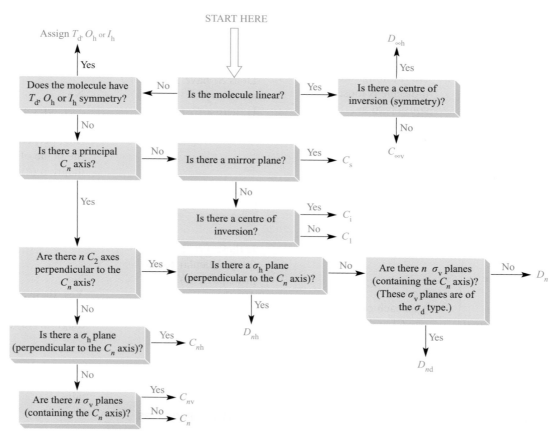

Fig. 3.10 Scheme for assigning point groups of molecules and molecular ions. Apart from the cases of $n = 1$ or ∞, n most commonly has values of 2, 3, 4, 5 or 6.

In the trigonal bipyramidal arrangement, the three equatorial F atoms are equivalent, and the two axial F atoms are equivalent.

Apply the strategy shown in Fig. 3.10:

START ⟹

Is the molecule linear?	No
Does PF$_5$ have T_d, O_h or I_h symmetry?	No
Is there a C_n axis?	Yes; a C_3 axis containing the P and two axial F atoms
Are there three C_2 axes perpendicular to the principal axis?	Yes; each lies along an equatorial P−F bond
Is there a σ_h plane (perpendicular to the principal axis)?	Yes; it contains the P and three equatorial F atoms.

⟹ **STOP**

The point group is D_{3h}.

Self-study exercises

1. Show that BF$_3$ belongs to the D_{3h} point group.

2. Show that OF$_2$ belongs to the C_{2v} point group.

3. Show that BF$_2$Br belongs to the C_{2v} point group.

Worked example 3.6   Point group assignments: 3

To what point group does POCl$_3$ belong?

The structure of POCl$_3$ is:

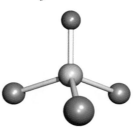

Apply the strategy shown in Fig. 3.10:

START ⟹

Is the molecule linear?	No
Does POCl$_3$ have T_d, O_h or I_h symmetry?	No (although this molecule is loosely considered as being tetrahedral in shape, it does *not* possess T_d symmetry)
Is there a C_n axis?	Yes; a C_3 axis running along the O−P bond
Are there three C_2 axes perpendicular to the principal axis?	No
Is there a σ_h plane (perpendicular to the principal axis)?	No
Are there n σ_v planes (containing the principal axis)?	Yes; each contains the one Cl and the O and P atoms

⟹ STOP

The point group is C_{3v}.

Self-study exercises

1. Show that CHCl$_3$ possesses C_{3v} symmetry, but that CCl$_4$ belongs to the T_d point group.

2. Assign point groups to (a) [NH$_4$]$^+$ and (b) NH$_3$.
[*Ans.* (a) T_d; (b) C_{3v}]

Worked example 3.7   Point group assignments: 4

Three projections of the cyclic structure of S$_8$ are shown below, and the structure can also be viewed on the accompanying website. All S−S bond distances are equivalent, as are all S−S−S bond angles. To what point group does S$_8$ belong?

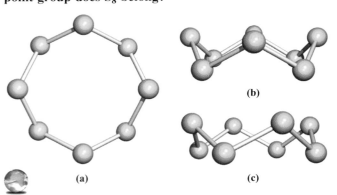

(a)   (c)

(b)

Follow the scheme in Fig. 3.10:

START ⟹

Is the molecule linear?	No
Does S$_8$ have T_d, O_h or I_h symmetry?	No
Is there a C_n axis?	Yes; a C_4 axis running through the centre of the ring; perpendicular to the plane of the paper in diagram (a)
Are there four C_2 axes perpendicular to the principal axis?	Yes; these are most easily seen from diagram (c)
Is there a σ_h plane (perpendicular to the principal axis)?	No
Are there n σ_d planes (containing the principal axis)?	Yes; these are most easily seen from diagrams (a) and (c)

⟹ STOP

The point group is D_{4d}.

Self-study exercises

1. Why does the S$_8$ ring not contain a C_8 axis?

2. Copy diagram (a) above. Show on the figure where the C_4 axis and the four C_2 axes lie.

3. S$_6$ has the chair conformation shown in Box 1.1. Confirm that this molecule contains a centre of inversion.

Earlier, we noted that it is not necessary to find all the symmetry elements of a molecule or ion to determine its point group. However, if one needs to identify all the operations in a point group, the following check of the total number can be carried out:[†]

- assign 1 for C or S, 2 for D, 12 for T, 24 for O or 60 for I;
- multiply by n for a numerical subscript;
- multiply by 2 for a letter subscript (s, v, d, h, i).

For example, the C_{3v} point group has $1 \times 3 \times 2 = 6$ operations, and D_{2d} has $2 \times 2 \times 2 = 8$ operations.

3.5 Character tables: an introduction

While Fig. 3.10 provides a point group assignment using certain diagnostic symmetry elements, it may be necessary to establish whether any additional symmetry elements are exhibited by a molecule in a given point group.

[†] See O.J. Curnow (2007) *J. Chem. Educ.*, vol. 84, p. 1430.

Table 3.2 The character table for the C_{2v} point group. For more character tables, see Appendix 3.

C_{2v}	E	C_2	$\sigma_v(xz)$	$\sigma_v'(yz)$		
A_1	1	1	1	1	z	x^2, y^2, z^2
A_2	1	1	-1	-1	R_z	xy
B_1	1	-1	1	-1	x, R_y	xz
B_2	1	-1	-1	1	y, R_x	yz

Each point group has an associated *character table*, and that for the C_{2v} point group is shown in Table 3.2. The point group is indicated at the top left-hand corner and the symmetry elements possessed by a member of the point group are given across the top row of the character table. The H_2O molecule has C_{2v} symmetry and when we looked at the symmetry elements of H_2O in Fig. 3.3, we labelled the two perpendicular planes. In the character table, taking the z axis as coincident with the principal axis, the σ_v and σ_v' planes are defined as lying in the xz and yz planes, respectively. Placing the molecular framework in a convenient orientation with respect to a Cartesian set of axes has many advantages, one of which is that the atomic orbitals on the central atom point in convenient directions. We return to this in Chapter 5.

Table 3.3 shows the character table for the C_{3v} point group. The NH_3 molecule possesses C_{3v} symmetry, and worked example 3.2 illustrated the principal axis of rotation and planes of symmetry in NH_3. In the character table, the presence of three σ_v planes in NH_3 is represented by the notation '$3\sigma_v$' in the top line of the table. The notation '$2C_3$' summarizes the two operations C_3^1 and C_3^2 (Fig. 3.6). The operation C_3^3 is equivalent to the identity operator, E, and so is not specified again.

Figure 3.4 showed the proper axes of rotation and planes of symmetry in the square planar molecule XeF_4. This has D_{4h} symmetry. The D_{4h} character table is given in Appendix 3, and the top row of the character table that summarizes the symmetry operations for this point group is as follows:

D_{4h}	E	$2C_4$	C_2	$2C_2'$	$2C_2''$	i	$2S_4$	σ_h	$2\sigma_v$	$2\sigma_d$

In Fig. 3.4 we showed that a C_2 axis is coincident with the C_4 axis in XeF_4. The C_2 operation is equivalent to C_4^2. The character table summarizes this information by stating '$2C_4$ C_2', referring to C_4^1 and C_4^3, and $C_4^2 = C_2$. The operation C_4^4 is

Table 3.3 The character table for the C_{3v} point group. For more character tables, see Appendix 3.

C_{3v}	E	$2C_3$	$3\sigma_v$		
A_1	1	1	1	z	$x^2 + y^2, z^2$
A_2	1	1	-1	R_z	
E	2	-1	0	$(x, y)\ (R_x, R_y)$	$(x^2 - y^2, xy)\ (xz, yz)$

taken care of in the identity operator E. The two sets of C_2 axes that we showed in Fig. 3.4 and labelled as C_2' and C_2'' are apparent in the character table, as are the σ_h, two σ_v and two σ_d planes of symmetry. The symmetry operations that we did not show in Fig. 3.4 but that are included in the character table are the centre of symmetry, i, (which is located on the Xe atom in XeF_4), and the S_4 axes. Each S_4 operation can be represented as $(C_4 \times \sigma_h)$.

The left-hand column in a character table gives a list of *symmetry labels*. These are used in conjunction with the numbers, or *characters*, from the main part of the table to label the symmetry properties of, for example, molecular orbitals or modes of molecular vibrations. As we shall see in Chapter 5, although the symmetry labels in the character tables are upper case (e.g. A_1, E, T_{2g}), the corresponding symmetry labels for orbitals are lower case (e.g. a_1, e, t_{2g}). Symmetry labels give us information about degeneracies as follows:

- A and B (or a and b) indicate non-degenerate;
- E (or e) refers to doubly degenerate;
- T (or t) means triply degenerate.

In Chapter 5, we use character tables to label the symmetries of orbitals, and to understand what orbital symmetries are allowed for a molecule possessing a particular symmetry.

Appendix 3 gives character tables for the most commonly encountered point groups, and each table has the same format as those in Tables 3.2 and 3.3.

3.6 Why do we need to recognize symmetry elements?

So far in this chapter, we have described the possible symmetry elements that a molecule might possess and, on the basis of these symmetry properties, we have illustrated how a molecular species can be assigned to a particular point group. Now we address some of the reasons why the recognition of symmetry elements in a molecule is important to the inorganic chemist.

Most of the applications of symmetry fall into one of the following categories:

- constructing molecular and hybrid orbitals (see Chapter 5);
- interpreting spectroscopic (e.g. vibrational and electronic) properties;
- determining whether a molecular species is chiral.

The next two sections introduce the consequences of symmetry on observed bands in infrared spectra and with the relationship between molecular symmetry and chirality. In Chapter 22, we consider the electronic spectra of octahedral and tetrahedral d-block metal complexes and discuss the effects that molecular symmetry has on electronic spectroscopic properties.

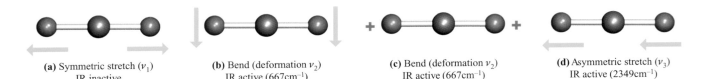

(a) Symmetric stretch (ν_1)
IR inactive

(b) Bend (deformation ν_2)
IR active (667cm^{-1})

(c) Bend (deformation ν_2)
IR active (667cm^{-1})

(d) Asymmetric stretch (ν_3)
IR active (2349cm^{-1})

Fig. 3.11 The vibrational modes of CO_2 ($D_{\infty h}$). In the vibrational mode shown in (a), the carbon atom remains stationary. Vibrations (a) and (d) are stretching modes. Bending mode (b) occurs in the plane of the paper, while bend (c) occurs in a plane perpendicular to that of the paper; the + signs designate motion towards the reader. The two bending modes require the same amount of energy and are therefore *degenerate*.

3.7 Vibrational spectroscopy

Infrared (IR) and Raman spectroscopies (see also Section 4.7) are branches of *vibrational spectroscopy* and the former technique is much the more widely available of the two in student teaching laboratories. The discussion that follows is necessarily selective and is pitched at a relatively simplistic level. We derive the number of vibrational modes for some simple molecules, and determine whether these modes are infrared (IR) and/or Raman active (i.e. whether absorptions corresponding to the vibrational modes are observed in the IR and/or Raman spectra). We also relate the vibrational modes of a molecule to its symmetry by using the character table of the relevant point group. A rigorous group theory approach to the normal modes of vibration of a molecule is beyond the scope of this book. The reading list at the end of the chapter gives sources of more detailed discussions.

How many vibrational modes are there for a given molecular species?

Vibrational spectroscopy is concerned with the observation of the *degrees of vibrational freedom*, the number of which can be determined as follows. The motion of a molecule containing n atoms can conveniently be described in terms of the three Cartesian axes. The molecule has *3n degrees of freedom* which together describe the *translational, vibrational* and *rotational* motions of the molecule.

The translational motion of a molecule (i.e. movement through space) can be described in terms of three degrees of freedom relating to the three Cartesian axes. If there are $3n$ degrees of freedom in total and three degrees of freedom for translational motion, it follows that there must be $(3n - 3)$ degrees of freedom for rotational and vibrational motion. For a *non-linear molecule* there are three degrees of rotational freedom, but for a *linear molecule*, there are only two degrees of rotational freedom. Having taken account of translational and rotational motion, the number of degrees of vibrational freedom can be determined (eqs. 3.5 and 3.6).[†]

[†] For further detail, see: P. Atkins and J. de Paula (2010) *Atkins' Physical Chemistry*, 9th edn, Oxford University Press, Oxford, p. 471.

Number of degrees of vibrational freedom for a *non-linear* molecule = $3n - 6$ \hfill (3.5)

Number of degrees of vibrational freedom for a *linear* molecule = $3n - 5$ \hfill (3.6)

For example, from eq. 3.6, the linear CO_2 molecule has four *normal modes of vibration* and these are shown in Fig. 3.11. Two of the modes are *degenerate*; i.e. they possess the same energy and could be represented in a single diagram with the understanding that one vibration occurs in the plane of the paper and another, identical in energy, takes place in a plane perpendicular to the first.

Self-study exercises

1. Using the **VSEPR** model to help you, draw the structures of CF_4, XeF_4 and SF_4. Assign a point group to each molecule. Show that the number of degrees of vibrational freedom is independent of the molecular symmetry.
   [*Ans.* T_d; D_{4h}; C_{2v}]

2. Why do CO_2 and SO_2 have a different number of degrees of vibrational freedom?

3. How many degrees of vibrational freedom do each of the following possess: $SiCl_4$, BrF_3, $POCl_3$? [*Ans.* 9; 6; 9]

Selection rules for an infrared or Raman active mode of vibration

One of the important consequences of precisely denoting molecular symmetry is seen in IR and Raman spectroscopy. For example, an IR spectrum records the frequency of a molecular vibration, i.e. bond stretching and molecular deformation (e.g. bending) modes. However, not all modes of vibration of a particular molecule give rise to observable absorption bands in the IR spectrum. This is because the following *selection rule* must be obeyed:

For a mode of vibration to be *infrared (IR) active,* it must give rise to a change in the molecular electric dipole moment.

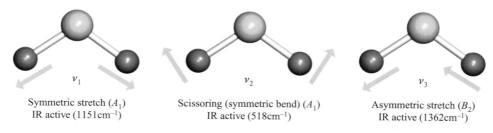

v_1

Symmetric stretch (A_1)
IR active (1151cm^{-1})

v_2

Scissoring (symmetric bend) (A_1)
IR active (518cm^{-1})

v_3

Asymmetric stretch (B_2)
IR active (1362cm^{-1})

Fig. 3.12 The vibrational modes of SO_2 (C_{2v}). In each mode, the S atom is also displaced from the position shown.

A different selection rule applies to Raman spectroscopy:

For a mode of vibration to be ***Raman active,*** it must give rise to a change in the polarizability of the molecule. ***Polarizability*** is the ease with which the electron cloud associated with the molecule is distorted.

In addition to these two selection rules, molecules with a centre of symmetry (e.g. linear CO_2, and octahedral SF_6) are subject to the *rule of mutual exclusion*:

For centrosymmetric molecules, the ***rule of mutual exclusion*** states that vibrations that are IR active are Raman inactive, and vice versa.

Application of this rule means that the presence of a centre of symmetry in a molecule is readily determined by comparing its IR and Raman spectra. Although Raman spectroscopy is now a routine technique, it is IR spectroscopy that remains the more accessible of the two for everyday compound characterization. Hence, we restrict most of the following discussion to IR spectroscopic absorptions. Furthermore, we are concerned only with *fundamental* absorptions, these being the dominant features of IR spectra.

The transition from the vibrational ground state to the first excited state is the ***fundamental*** transition.

Linear ($D_{\infty h}$ or $C_{\infty v}$) and bent (C_{2v}) triatomic molecules

We can readily illustrate the effect of molecular symmetry on molecular dipole moments, and thus on IR active modes of vibration, by considering the linear molecule CO_2. The two C−O bond distances are equal (116 pm) and the molecule is described as being 'symmetrical'. Strictly, CO_2 possesses $D_{\infty h}$ symmetry. As a consequence of its symmetry, CO_2 is non-polar. The number of degrees of vibrational freedom is determined from eq. 3.6:

Number of degrees of vibrational freedom for CO_2
$$= 3n - 5 = 9 - 5 = 4$$

The four fundamental modes of vibration are shown in Fig. 3.11. Although both the asymmetric stretch and the bend (Fig. 3.11) give rise to a change in dipole moment (generated transiently as the vibration occurs), the symmetric stretch does not. Thus, only two fundamental absorptions are observed in the IR spectrum of CO_2.

Now consider SO_2 which is a bent molecule (C_{2v}). The number of degrees of vibrational freedom for a non-linear molecule is determined from eq. 3.5:

Number of degrees of vibrational freedom for SO_2
$$= 3n - 6 = 9 - 6 = 3$$

The three fundamental modes of vibration are shown in Fig. 3.12. In the case of a triatomic molecule, it is simple to deduce that the three modes of vibration are composed of two stretching modes (symmetric and asymmetric) and a bending mode. However, for larger molecules it is not so easy to visualize the modes of vibration. We return to this problem in the next section. The three normal modes of vibration of SO_2 *all* give rise to a change in molecular dipole moment and are therefore IR active. A comparison of these results for CO_2 and SO_2 illustrates that vibrational spectroscopy can be used to determine whether an X_3 or XY_2 species is linear or bent.

Linear molecules of the general type XYZ (e.g. OCS or HCN) possess $C_{\infty v}$ symmetry and their IR spectra are expected to show three absorptions: the symmetric stretching, asymmetric stretching and bending modes are all IR active. In a linear molecule XYZ, provided that the atomic masses of X and Z are significantly different, the absorptions observed in the IR spectrum can be assigned to the X–Y stretch, the Y–Z stretch and the XYZ bend. The reason that the stretching modes can be assigned to individual bond vibrations rather than to a vibration involving the whole molecule is that each of the symmetric and asymmetric stretches is dominated by the stretching of one of the two bonds. For example, absorptions at 3311, 2097 and 712 cm^{-1} in the IR spectrum of HCN are assigned to the H–C stretch, the C≡N stretch and the HCN bend, respectively.

A ***stretching mode*** is designated by the symbol v, while a ***deformation*** (bending) is denoted by δ. For example, v_{CO} stands for the stretch of a C−O bond.

C_{2v}	E	C_2	$\sigma_v(xz)$	$\sigma_v'(yz)$		
A_1	1	1	1	1	z	x^2, y^2, z^2
A_2	1	1	-1	-1	R_z	xy
B_1	1	-1	1	-1	x, R_y	xz
B_2	1	-1	-1	1	y, R_x	yz

Worked example 3.8   IR spectra of triatomic molecules

The IR spectrum of $SnCl_2$ exhibits absorptions at 352, 334 and 120 cm^{-1}. What shape do these data suggest for the molecule, and is this result consistent with the VSEPR model?

For linear $SnCl_2$, $D_{\infty h}$, the asymmetric stretch and the bend are IR active, but the symmetric stretch is IR inactive (no change in molecular dipole moment).

For bent $SnCl_2$, C_{2v}, the symmetric stretching, asymmetric stretching and scissoring modes are all IR active.

The data therefore suggest that $SnCl_2$ is bent, and this is consistent with the VSEPR model since there is a lone pair in addition to two bonding pairs of electrons:

Self-study exercises

1. The vibrational modes of XeF_2 are at 555, 515 and 213 cm^{-1} but only two are IR active. Explain why this is consistent with XeF_2 having a linear structure.

2. How many IR active vibrational modes does CS_2 possess, and why? [*Hint:* CS_2 is isostructural with CO_2.]

3. The IR spectrum of SF_2 has absorptions at 838, 813 and 357 cm^{-1}. Explain why these data are consistent with SF_2 belonging to the C_{2v} rather than $D_{\infty h}$ point group.

4. To what point group does F_2O belong? Explain why the vibrational modes at 928, 831 and 461 cm^{-1} are all IR active.　　　　　　　　　　　　　　　　[*Ans.* C_{2v}]

Bent molecules XY$_2$: using the C$_{2v}$ character table

The SO_2 molecule belongs to the C_{2v} point group, and in this section we look again at the three normal modes of vibration of SO_2, but this time use the C_{2v} character table to determine:

- whether the modes of vibration involve stretching or bending;
- the symmetry labels of the vibrational modes;
- which modes of vibration are IR and/or Raman active.

The C_{2v} character table is shown here, along with a diagram that relates the SO_2 molecule to its C_2 axis and two mirror planes. The z axis coincides with the C_2 axis, and the molecule lies in the yz plane.

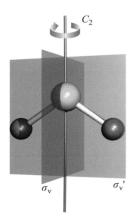

In a molecule, stretching or bending modes can be described in terms of changes made to the bond vectors or bond angles, respectively. Let us first consider vibrations involving bond stretching in SO_2. (Since a triatomic molecule is a simple case, it is all too easy to wonder why we need the following exercise; however, it serves as an instructive example before we consider larger polyatomics.) *Without* thinking about the relative directions in which the bonds may be stretched, consider the effect of each symmetry operation of the C_{2v} point group on the bonds in SO_2. Now ask the question: how many bonds are left *unchanged* by each symmetry operation? The E operator leaves both S–O bonds unchanged, as does reflection through the $\sigma_v'(yz)$ plane. However, rotation about the C_2 axis affects both bonds, and so does reflection through the $\sigma_v(xz)$ plane. These results can be summarized in the row of characters shown below, where '2' stands for 'two bonds unchanged', and '0' stands for 'no bonds unchanged':

E	C_2	$\sigma_v(xz)$	$\sigma_v'(yz)$
2	0	0	2

This is known as a *reducible representation* and can be rewritten as the sum of rows of characters from the C_{2v} character table. Inspection of the character table reveals that summing the two rows of characters for the A_1 and B_2 representations gives us the result we require, i.e.:

A_1	1	1	1	1
B_2	1	−1	−1	1
Sum of rows	2	0	0	2

This result tells us that there are two non-degenerate stretching modes, one of A_1 symmetry and one of B_2 symmetry. For a bent XY_2 molecule, it is a straightforward matter to relate these labels to schematic representations of the stretching modes, since there can be only two options: bond stretching in-phase or out-of-phase. However, for the sake of completeness, we now work through the assignments using the C_{2v} character table.

The modes of vibration of SO_2 are defined by vectors which are illustrated by yellow arrows in Fig. 3.12. In order to assign a symmetry label to each vibrational mode, we must consider the effect of each symmetry operation of the C_{2v} point group on these vectors. For the symmetric stretch (ν_1) of the SO_2 molecule, the vectors are left unchanged by the E operator and by rotation about the C_2 axis. There is also no change to the vectors when the molecule is reflected through either of the $\sigma_v(xz)$ or $\sigma_v'(yz)$ planes. If we use the notation that a '1' means 'no change', then the results can be summarized as follows:

E	C_2	$\sigma_v(xz)$	$\sigma_v'(yz)$
1	1	1	1

Now compare this row of characters with the rows in the C_{2v} character table. There is a match with the row for symmetry type A_1, and therefore the symmetric stretch is given the A_1 symmetry label. Now consider the asymmetric stretching mode (ν_3) of the SO_2 molecule (Fig. 3.12). The vectors are unchanged by the E and $\sigma_v'(yz)$ operations, but their directions are altered by rotation about the C_2 axis and by reflection through the $\sigma_v(xz)$ plane. Using the notation that a '1' means 'no change', and a '−1' means 'a reversal of the direction of the vector', we can summarize the results as follows:

E	C_2	$\sigma_v(xz)$	$\sigma_v'(yz)$
1	−1	−1	1

This corresponds to symmetry type B_2 in the C_{2v} character table, and so the asymmetric stretching mode is labelled B_2.

Now recall that SO_2 has a total of $(3n-6) = 3$ degrees of vibrational freedom. Having assigned two of these to stretching modes, the third must arise from a bending (or scissoring) mode. The bending mode (ν_2) can be defined in terms of changes in the O–S–O bond angle. To assign a symmetry label to this mode of vibration, we consider the effect of each symmetry operation of the C_{2v} point group on the

bond angle. Each of the E, C_2, $\sigma_v(xz)$ and $\sigma_v'(yz)$ operations leaves the angle unchanged and, therefore, we can write:

E	C_2	$\sigma_v(xz)$	$\sigma_v'(yz)$
1	1	1	1

The scissoring mode is therefore assigned A_1 symmetry.

Finally, how can we use a character table to determine whether a particular mode of vibration is IR or Raman active? At the right-hand side of a character table, there are two columns containing functions x, y and/or z, or products of these functions (e.g. x^2, xy, yz, $(x^2 - y^2)$, etc.). We will not detail the origins of these terms, but will focus only on the information that they provide:

> If the symmetry label (e.g. A_1, B_1, E) of a normal mode of vibration is associated with x, y or z in the character table, then the mode is ***IR active***.
>
> If the symmetry label (e.g. A_1, B_1, E) of a normal mode of vibration is associated with a product term (e.g. x^2, xy) in the character table, then the mode is ***Raman active***.

The SO_2 molecule has A_1 and B_2 normal modes of vibration. In the C_{2v} character table, the right-hand columns for the A_1 representation contain z and also x^2, y^2 and z^2 functions. Hence, the A_1 modes are *both* IR and Raman active. Similarly, the right-hand columns for the B_2 representation contain y and yz functions, and the asymmetric stretch of SO_2 is both IR and Raman active.

The most common bent triatomic molecule that you encounter daily is H_2O. Like SO_2, H_2O belongs to the C_{2v} point group and possesses three modes of vibration, all of which are IR and Raman active. These are illustrated in Fig. 3.13a which shows a calculated IR spectrum of gaseous H_2O. (An experimental spectrum would also show rotational fine structure.) In contrast, the IR spectrum of liquid water shown in Fig. 3.13b is broad and the two absorptions around 3700 cm^{-1} are not resolved. The broadening arises from the presence of hydrogen bonding between water molecules (see Section 10.6). In addition, the vibrational wavenumbers in the liquid and gas phase spectra are shifted with respect to one another.

Self-study exercises

1. In the vibrational spectrum of H_2O vapour, there are absorptions at 3756 and 3657 cm^{-1} corresponding to the B_2 and A_1 stretching modes, respectively. Draw diagrams to show these vibrational modes.

2. The symmetric bending of the non-linear NO_2 molecule gives rise to an absorption at 752 cm^{-1}. To what point group does NO_2 belong? Explain why the symmetric bending mode is IR active. Why is it assigned an A_1 symmetry label?

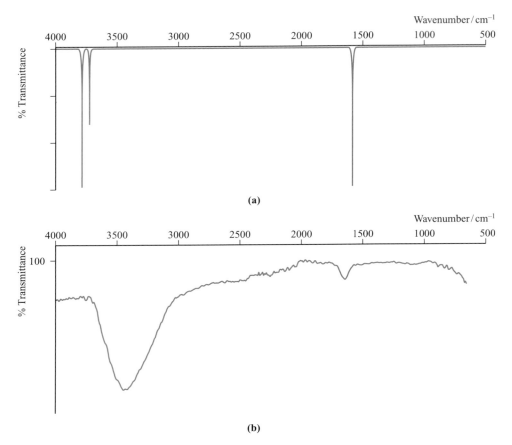

Fig. 3.13 (a) Calculated IR spectrum of gaseous H_2O (Spartan '04, ©Wavefunction Inc. 2003) showing the three fundamental absorptions. Experimental values are 3756, 3657 and 1595 cm^{-1}. (b) IR spectrum of liquid H_2O.

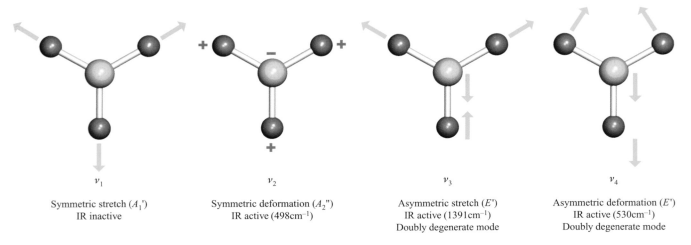

ν_1	ν_2	ν_3	ν_4
Symmetric stretch (A_1') IR inactive	Symmetric deformation (A_2'') IR active (498cm^{-1})	Asymmetric stretch (E') IR active (1391cm^{-1}) Doubly degenerate mode	Asymmetric deformation (E') IR active (530cm^{-1}) Doubly degenerate mode

Fig. 3.14 The vibrational modes of SO_3 (D_{3h}); only three are IR active. The + and − notation is used to show the 'up' and 'down' motion of the atoms during the mode of vibration. Two of the modes are doubly degenerate, giving a total of six normal modes of vibration.

XY_3 molecules with D_{3h} symmetry

An XY_3 molecule, irrespective of shape, possesses $(3 \times 4) - 6 = 6$ degrees of vibrational freedom. Let us first consider planar XY_3 molecules belonging to the D_{3h} point group.

Examples are SO_3, BF_3 and $AlCl_3$, and the six normal modes of vibration of SO_3 are shown in Fig. 3.14. The symmetries of the stretching modes stated in the figure are deduced by considering how many bonds are left *unchanged*

Table 3.4 The character table for the D_{3h} point group.

D_{3h}	E	$2C_3$	$3C_2$	σ_h	$2S_3$	$3\sigma_v$		
A_1'	1	1	1	1	1	1		$x^2+y^2,\ z^2$
A_2'	1	1	−1	1	1	−1	R_z	
E'	2	−1	0	2	−1	0	(x, y)	$(x^2-y^2,\ xy)$
A_1''	1	1	1	−1	−1	−1		
A_2''	1	1	−1	−1	−1	1	z	
E''	2	−1	0	−2	1	0	(R_x, R_y)	(xz, yz)

by each symmetry operation of the D_{3h} point group (refer to Fig. 3.2, worked example 3.3 and Table 3.4). The E and σ_h operators leave all three bonds unchanged. Each C_2 axis coincides with one X–Y bond and therefore rotation about a C_2 axis leaves one bond unchanged; similarly for reflection through a σ_v plane. Rotation about the C_3 axis affects all three bonds. The results can be summarized in the following row of characters:

E	C_3	C_2	σ_h	S_3	σ_v
3	0	1	3	0	1

If we rewrite this reducible representation as the sum of rows of characters from the D_{3h} character table, we can determine the symmetries of the vibrational modes of the planar XY$_3$ molecule:

A_1'	1	1	1	1	1	1
E'	2	−1	0	2	−1	0
Sum of rows	3	0	1	3	0	1

Inspection of Fig. 3.14 reveals that the symmetric stretch (the A_1' mode) does *not* lead to a change in molecular dipole moment and is therefore IR inactive. This can be verified by looking at the D_{3h} character table (Table 3.4) where the entries in the two right-hand columns show that the A_1' mode is IR inactive, but Raman active. The asymmetric stretch (E') of a D_{3h} XY$_3$ molecule is doubly degenerate, and Fig. 3.14 shows one of these modes. The vibration is accompanied by a change in molecular dipole moment, and so is IR active. In Table 3.4, the entries in the right-hand columns for the E' representation show that the mode is both IR and Raman active.

The symmetries of the deformation modes of D_{3h} XY$_3$ (Fig. 3.14) are E' and A_2'' (see end-of-chapter problem 3.25). From the D_{3h} character table we can deduce that the A_2'' mode is IR active, while the E' mode is both IR and Raman active. We can also deduce that both deformations are IR active by showing that each deformation in Fig. 3.14 leads to a change in molecular dipole moment.

Molecules with D_{3h} symmetry (e.g. SO$_3$, BF$_3$ and AlCl$_3$) therefore exhibit three absorptions in their IR spectra: one band arises from a stretching mode and two from deformations. The IR spectra of anions such as $[NO_3]^-$ and $[CO_3]^{2-}$ may also be recorded, but the counter-ion may also give rise to IR spectroscopic bands. Therefore, simple salts such as those of the alkali metals are chosen because they give spectra in which the bands can be assigned to the anion (see Table 4.2 on page 101).

XY$_3$ molecules with C$_{3v}$ symmetry

An XY$_3$ molecule belonging to the C_{3v} point group has six degrees of vibrational freedom. Examples of C_{3v} molecules are NH$_3$, PCl$_3$ and AsF$_3$. The normal modes of vibration of NH$_3$ are shown in Fig. 3.15; two modes are doubly degenerate. The symmetry labels can be verified by using the C_{3v} character table (Table 3.3 on p. 71). For example, each of the E, C_3 and σ_v operations leaves the vectors that define the symmetric vibration unchanged and, therefore, we can write:

E	C_3	σ_v
1	1	1

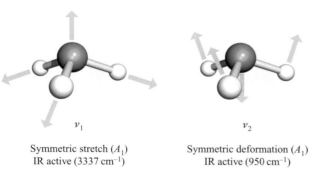

ν_1

Symmetric stretch (A_1)
IR active (3337 cm^{-1})

ν_2

Symmetric deformation (A_1)
IR active (950 cm^{-1})

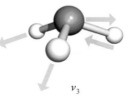

ν_3

Asymmetric stretch (E)
IR active (3414 cm^{-1})
(Doubly degenerate mode)

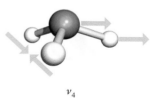

ν_4

Asymmetric deformation (E)
IR active (1627 cm^{-1})
(Doubly degenerate mode)

Fig. 3.15 The vibrational modes of NH$_3$ (C_{3v}), all of which are IR active.

This corresponds to the A_1 representation in the C_{3v} character table, and therefore the symmetric stretch has A_1 symmetry. Each of the vibrational modes shown in Fig. 3.15 has either A_1 or E symmetry, and the functions listed in the right-hand columns of Table 3.3 reveal that each of the vibrational modes is both IR and Raman active. We therefore expect to observe four absorptions in the IR spectrum of species such as gaseous NH_3, NF_3, PCl_3 and AsF_3.

Differences in the number of bands in the IR spectra of XY_3 molecules possessing C_{3v} or D_{3h} symmetry is a method of distinguishing between these structures. Further, XY_3 molecules with T-shaped structures (e.g. ClF_3) belong to the C_{2v} point group, and vibrational spectroscopy may be used to distinguish their structures from those of C_{3v} or D_{3h} XY_3 species.

$$
\begin{array}{c}
F_{ax} \\
| \\
Cl \!-\!-\! F_{eq} \\
| \\
F_{ax}
\end{array}
$$

See also Fig. 2.17

(3.10)

For the C_{2v} molecules ClF_3 (**3.10**) or BrF_3, there are six normal modes of vibration, approximately described as equatorial stretch, symmetric axial stretch, asymmetric axial stretch and three deformation modes. All six modes are IR active.

Self-study exercises

1. The IR spectrum of BF_3 shows absorptions at 480, 691 and 1449 cm^{-1}. Use these data to decide whether BF_3 has C_{3v} or D_{3h} symmetry. [*Ans.* D_{3h}]

2. In the IR spectrum of NF_3, there are four absorptions. Why is this consistent with NF_3 belonging to the C_{3v} rather than D_{3h} point group?

3. The IR spectrum of BrF_3 in an argon matrix shows six absorptions. Explain why this observation confirms that BrF_3 cannot have C_{3v} symmetry.

4. Use the C_{3v} character table to confirm that the symmetric deformation mode of NH_3 (Fig. 3.15) has A_1 symmetry.

XY_4 molecules with T_d or D_{4h} symmetry

An XY_4 molecule with T_d symmetry has nine normal modes of vibration (Fig. 3.16). In the T_d character table (see Appendix 3), the T_2 representation has an (x,y,z) function, and therefore the two T_2 vibrational modes are IR active. The character table also shows that the T_2 modes are Raman active. The A_1 and E modes are IR inactive, but Raman active. The IR spectra of species such as CCl_4, $TiCl_4$, OsO_4, $[ClO_4]^-$ and $[SO_4]^{2-}$ exhibit *two* absorptions (see Table 4.2).

There are nine normal modes of vibration for a square planar (D_{4h}) XY_4 molecule. These are illustrated for $[PtCl_4]^{2-}$ in Fig. 3.17, along with their appropriate symmetry labels. In the D_{4h} character table (see Appendix 3), the A_{2u} and E_u representations contain z and (x,y) functions, respectively. Therefore, of the vibrational modes shown in Fig. 3.17, only the A_{2u} and E_u modes are IR active. Since $[PtCl_4]^{2-}$ contains an inversion centre, the rule of mutual exclusion applies, and the A_{2u} and E_u modes are Raman inactive. Similarly, the A_{1g}, B_{1g} and B_{2g} modes that are Raman active, are IR inactive. Among compounds of the p-block elements, D_{4h} XY_4 structures are rare; the observation of absorptions at 586, 291 and 161 cm^{-1} in the IR spectrum of XeF_4 is consistent with the structure predicted by the VSEPR model.

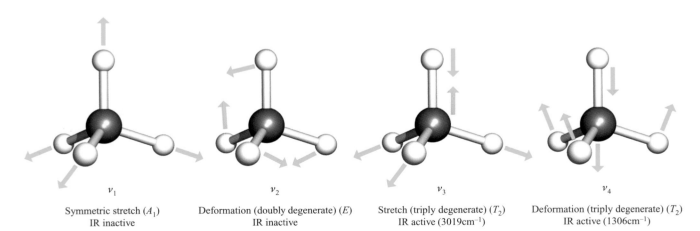

ν_1	ν_2	ν_3	ν_4
Symmetric stretch (A_1) IR inactive	Deformation (doubly degenerate) (E) IR inactive	Stretch (triply degenerate) (T_2) IR active (3019cm^{-1})	Deformation (triply degenerate) (T_2) IR active (1306cm^{-1})

Fig. 3.16 The vibrational modes of CH_4 (T_d), only two of which are IR active.

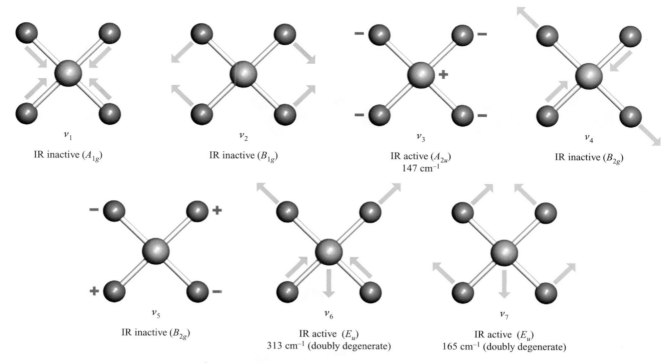

Fig. 3.17 The vibrational modes of $[PtCl_4]^{2-}$ (D_{4h}); only the three modes (two of which are degenerate) ν_3, ν_6 and ν_7 are IR active. The $+$ and $-$ notation is used to show the 'up' and 'down' motion of the atoms during the mode of vibration.

Self-study exercises

1. Use the D_{4h} character table in Appendix 3 to confirm that the A_{1g}, B_{1g} and B_{2g} modes of $[PtCl_4]^{2-}$ are IR inactive, but Raman active. Why does this illustrate the rule of mutual exclusion?

2. The IR spectrum of gaseous ZrI_4 shows absorptions at 55 and 254 cm^{-1}. Explain why this observation is consistent with molecules of ZrI_4 having T_d symmetry.

3. The $[PdCl_4]^{2-}$ ion gives rise to three absorptions in its IR spectrum (150, 321 and 161 cm^{-1}). Rationalize why this provides evidence for a D_{4h} rather than T_d structure.

4. SiH_2Cl_2 is described as having a tetrahedral structure; SiH_2Cl_2 has eight IR active vibrations. Comment on these statements.

XY$_6$ molecules with O_h symmetry

An XY_6 molecule belonging to the O_h point group has $(3 \times 7) - 6 = 15$ degrees of vibrational freedom. Figure 3.18 shows the modes of vibration of SF_6 along with their symmetry labels. Only the T_{1u} modes are IR active; this can be confirmed from the O_h character table in Appendix 3. Since the S atom in SF_6 lies on an inversion centre, the T_{1u} modes are Raman inactive (by the rule of mutual exclusion). Of the T_{1u} modes shown in Fig. 3.18, one can

be classified as a stretching mode (939 cm^{-1} for SF_6) and one a deformation (614 cm^{-1} for SF_6).

Metal carbonyl complexes, M(CO)$_n$

Infrared spectroscopy is especially useful for the characterization of metal carbonyl complexes $M(CO)_n$ since the absorptions arising from C–O bond stretching modes (ν_{CO}) are strong and easily observed in an IR spectrum. These modes typically give rise to absorptions close to 2000 cm^{-1} (see Section 24.2) and these bands are usually well separated from those arising from M–C stretches, M–C–O deformations and C–M–C deformations. The ν_{CO} modes can therefore be considered separately from the remaining vibrational modes. For example, $Mo(CO)_6$ belongs to the O_h point group. It has $(3 \times 13) - 6 = 33$ modes of vibrational freedom, of which 12 comprise four T_{1u} (i.e. IR active) modes: ν_{CO} 2000 cm^{-1}, δ_{MoCO} 596 cm^{-1}, ν_{MoC} 367 cm^{-1} and δ_{CMoC} 82 cm^{-1}. The other 21 modes are all IR inactive. A routine laboratory IR spectrometer covers a range from $\simeq 400$ to 4000 cm^{-1} and, therefore, only the ν_{CO} and δ_{MoC} modes are typically observed. We can confirm why an O_h $M(CO)_6$ species exhibits only one absorption in the C–O stretching region by comparing it with SF_6 (Fig. 3.18). The set of six C–O bonds in $M(CO)_6$ can be considered analogous to the set of six S–F bonds in SF_6. Therefore, an O_h $M(CO)_6$ molecule possesses A_{1g}, E_g and T_{1u} carbonyl stretching modes, but only the T_{1u} mode is IR active.

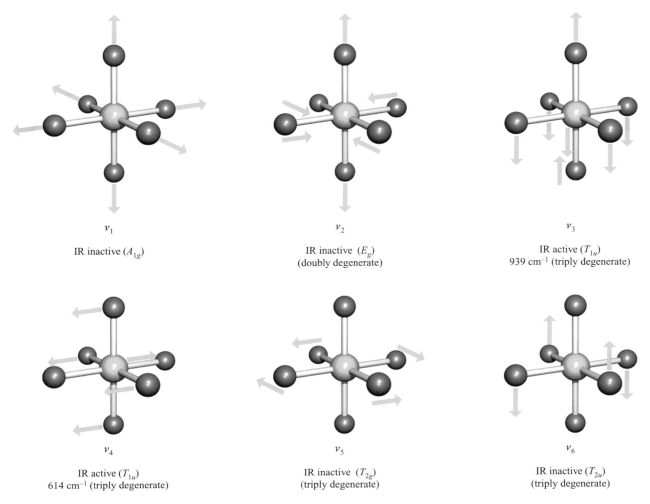

ν_1

IR inactive (A_{1g})

ν_2

IR inactive (E_g)
(doubly degenerate)

ν_3

IR active (T_{1u})
939 cm^{-1} (triply degenerate)

ν_4

IR active (T_{1u})
614 cm^{-1} (triply degenerate)

ν_5

IR inactive (T_{2g})
(triply degenerate)

ν_6

IR inactive (T_{2u})
(triply degenerate)

Fig. 3.18 The vibrational modes of SF_6 (O_h). Only the T_{1u} modes are IR active.

Self-study exercises

1. By considering only the six CO groups in $Cr(CO)_6$ (O_h), sketch diagrams to represent the A_{1g}, E_g and T_{1u} stretching modes. Use the O_h character table to deduce which modes are IR active.

[*Ans.* See Fig. 3.18; each C—O acts in the same way as an S—F bond]

2. In its IR spectrum, $W(CO)_6$ exhibits an absorption at 1998 cm^{-1}. Sketch a diagram to show the mode of vibration that corresponds to this absorption.

[*Ans.* Analogous to the IR active T_{1u} mode in Fig. 3.18]

Metal carbonyl complexes M(CO)$_{6-n}$X$_n$

In this section, we illustrate the relationship between the numbers of IR active ν_{CO} modes and the symmetries of $M(CO)_{6-n}X_n$ complexes. The metal carbonyls $M(CO)_6$, $M(CO)_5X$, *trans*-$M(CO)_4X_2$ and *cis*-$M(CO)_4X_2$ are all described as being 'octahedral' but only $M(CO)_6$ belongs to the O_h point group (Fig. 3.19). We saw above that an O_h $M(CO)_6$ complex exhibits one absorption in the CO stretching region of its IR spectrum. In contrast, C_{4v} $M(CO)_5X$ shows three absorptions, e.g. in the IR spectrum of $Mn(CO)_5Br$, bands are observed at 2138, 2052 and 2007 cm^{-1}. The origins of these bands can be understood by using group theory. Consider how many C—O bonds in the $M(CO)_5X$ molecule (Fig. 3.19) are left *unchanged* by each symmetry operation (E, C_4, C_2, σ_v and σ_d) of the C_{4v} point group (the C_{4v} character table is given in Appendix 3). The diagram below shows the C_4 and C_2 axes and the σ_v planes of symmetry. The σ_d planes bisect the σ_v planes (look at Fig. 3.4). The E operator leaves all five C—O bonds unchanged, while rotation around each axis and reflection through a σ_d plane leaves one C—O

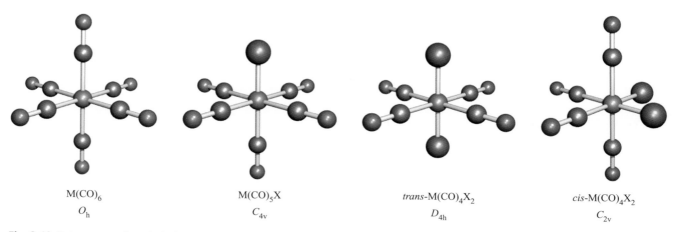

Fig. 3.19 Point groups of octahedral metal carbonyl complexes $M(CO)_6$, $M(CO)_5X$, *trans*-$M(CO)_4X_2$ and *cis*-$M(CO)_4X_2$. Colour code: metal M, green; C, grey; O, red; group X, brown.

bond unchanged. Reflection through a σ_v plane leaves three C–O bonds unchanged.

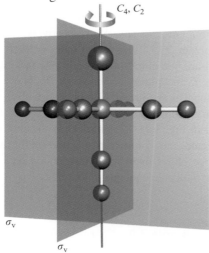

The results can be summarized in the following row of characters:

E	C_4	$2C_2$	σ_v	σ_d
5	1	1	3	1

This representation can be reduced to rows of characters from the C_{4v} character table:

A_1	1	1	1	1	1
A_1	1	1	1	1	1
B_1	1	−1	1	1	−1
E	2	0	−2	0	0
Sum of rows	5	1	1	3	1

The vibrational modes of $M(CO)_5X$ therefore have A_1, B_1 and E symmetries and the C_{4v} character table shows that only the two A_1 and the E modes are IR active, consistent with the observation of three absorptions in the IR spectrum.

A similar strategy can be used to determine the number of IR active modes of vibration for *cis*- and *trans*-$M(CO)_4X_2$, as well as for other complexes. Table 3.5 gives representative examples.

Self-study exercises

1. Draw a diagram to show the structure of *fac*-$M(CO)_3X_3$. Mark on the C_3 axis and one of the σ_v planes.

2. Using the C_{3v} character table (Appendix 3), confirm that the CO stretching modes of *fac*-$M(CO)_3X_3$ have A_1 and E symmetries. Confirm that both are IR active.

3. Rationalize why the IR spectrum of *fac*-$[Fe(CO)_3(CN)_3]^-$ has two strong absorptions at 2121 and 2096 cm^{-1}, as well as two weaker bands at 2162 and 2140 cm^{-1}.
   [*Ans.* See: J. Jiang *et al.* (2002) *Inorg. Chem.*, vol. 41, p. 158.]

Observing IR spectroscopic absorptions

In this section we have focused on how to establish the number of vibrational degrees of freedom for a simple molecule with n atoms, how to deduce the total number of normal modes of vibration for the molecule and how to determine the number of absorptions in its IR spectrum. We expand the discussion at a practical level in Section 4.5.

Table 3.5 Carbonyl stretching modes (ν_{CO}) for some families of mononuclear metal carbonyl complexes; X is a general group other than CO.

Complex	Point group	Symmetries of CO stretching modes	IR active modes	Number of absorptions observed in the IR spectrum
$M(CO)_6$	O_h	A_{1g}, E_g, T_{1u}	T_{1u}	1
$M(CO)_5X$	C_{4v}	A_1, A_1, B_1, E	A_1, A_1, E	3
trans-$M(CO)_4X_2$	D_{4h}	A_{1g}, B_{1g}, E_u	E_u	1
cis-$M(CO)_4X_2$	C_{2v}	A_1, A_1, B_1, B_2	A_1, A_1, B_1, B_2	4
fac-$M(CO)_3X_3$	C_{3v}	A_1, E	A_1, E	2
mer-$M(CO)_3X_3$	C_{2v}	A_1, A_1, B_1	A_1, A_1, B_1	3

3.8 Chiral molecules

A molecule is ***chiral*** if it is non-superposable on its mirror image.[†]

Helical chains such as Se_∞ (Fig. 3.20a) may be right- or left-handed and are chiral. 6-Coordinate complexes such as $[Cr(acac)_3]$ ($[acac]^-$, see Table 7.7) in which there are three bidentate chelating ligands also possess non-superposable mirror images (Fig. 3.20b). Chiral molecules can rotate the plane of plane-polarized light. This property is known as *optical activity* and the two mirror images are known as *optical isomers* or *enantiomers*. We return to this in Chapter 19.

The importance of chirality is clearly seen in, for example, dramatic differences in the activities of different enantiomers of chiral drugs.[‡]

A helical chain such as Se_∞ is easy to recognize, but it is not always such a facile task to identify a chiral compound by attempting to convince oneself that it is, or is not, non-superposable on its mirror image. Symmetry considerations come to our aid:

A ***chiral*** molecule lacks an improper (S_n) axis of symmetry.

Another commonly used criterion for identifying a chiral species is the lack of an inversion centre, i, and plane of

symmetry, σ. However, both of these properties are compatible with the criterion given above, since we can rewrite the symmetry operations i and σ in terms of the improper rotations S_2 and S_1 respectively. (See end-of-chapter problem 3.35.) A word of caution: there are a few species that are non-chiral (achiral) despite lacking an inversion centre, i, and plane of symmetry, σ. These 'problem' species belong to an S_n point group in which n is an even number. An example is the tetrafluoro derivative of spiropentane shown in Fig. 3.21. This molecule does not contain an inversion centre, nor a mirror plane, and might therefore be thought to be chiral. However, this conclusion is incorrect because the molecule contains an S_4 axis.

Worked example 3.9 Chiral species

The oxalate ligand, $[C_2O_4]^{2-}$, is a bidentate ligand and the structure of the complex ion $[Cr(ox)_3]^{3-}$ is shown below. The view in the right-hand diagram is along one O−Cr−O axis. Confirm that the point group to which the ion belongs is D_3 and that members of this point group are chiral.

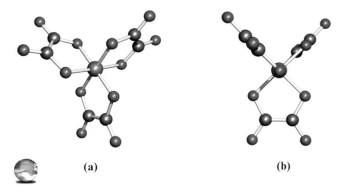

(a)   (b)

[†] This definition is taken from 'Basic terminology of stereochemistry: IUPAC recommendations 1996' (1996) *Pure Appl. Chem.*, vol. 68, p. 2193.

[‡] Relevant articles are: E. Thall (1996) *J. Chem. Educ.*, vol. 73, p. 481 – 'When drug molecules look in the mirror'; H. Caner *et al.* (2004) *Drug Discovery Today*, vol. 9, p. 105 – 'Trends in the development of chiral drugs'; M.C. Núñez *et al.* (2009) *Curr. Med. Chem.*, vol. 16, p. 2064 – 'Homochiral drugs: a demanding tendency of the pharmaceutical industry'.

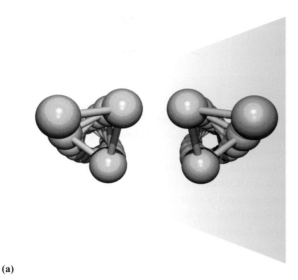

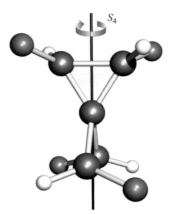

Fig. 3.21 A tetrafluoro derivative of spiropentane which belongs to the S_4 point group. This is an example of a molecule that contains no inversion centre and no mirror plane but is, nonetheless, achiral.

Using the scheme in Fig. 3.10:

START $\Longrightarrow$

Is the molecular ion linear?	No
Does it have T_d, O_h or I_h symmetry?	No
Is there a C_n axis?	Yes; a C_3 axis; perpendicular to the plane of the paper in diagram (a)
Are there three C_2 axes perpendicular to the principal axis?	Yes; one runs vertically, in the plane of the paper through the Fe centre in diagram (b)
Is there a σ_h plane (perpendicular to the principal axis)?	No
Are there n σ_d planes (containing the principal axis)?	No

$\Longrightarrow$ STOP

The point group is D_3.

No centre of symmetry or planes of symmetry have been identified and this confirms that molecular species in the D_3 point group are chiral.

Self-study exercise

By referring to the character table (Appendix 3) for the D_3 point group, confirm that the symmetry elements of the D_3 point group do not include i, σ or S_n axis.

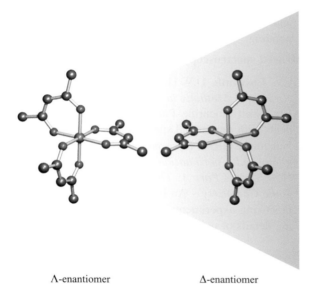

Λ-enantiomer   Δ-enantiomer

(b)

Fig. 3.20 A pair of enantiomers consists of two molecular species which are mirror images of each other and are non-superposable. (a) Helical Se_∞ has either right- or left-handedness. (b) The 6-coordinate complex $[Cr(acac)_3]$ contains three identical bidentate, chelating ligands; the labels Λ and Δ describe the absolute configuration of the molecule (see Box 19.3).

(a)

KEY TERMS

The following terms have been introduced in this chapter. Do you know what they mean?

- ❏ symmetry element
- ❏ symmetry operator
- ❏ identity operator (E)
- ❏ rotation axis (C_n)

- ❏ plane of reflection (σ_h, σ_v or σ_d)
- ❏ centre of symmetry or inversion centre (i)
- ❏ improper rotation axis (S_n)

- ❏ point group
- ❏ translational degrees of freedom
- ❏ rotational degrees of freedom
- ❏ vibrational degrees of freedom

❏ normal mode of vibration
❏ degenerate modes of vibration
❏ selection rule for an IR active mode

❏ selection rule for a Raman active mode
❏ rule of mutual exclusion

❏ fundamental absorption
❏ chiral
❏ enantiomer (optical isomer)

FURTHER READING

Symmetry and group theory

P.W. Atkins, M.S. Child and C.S.G. Phillips (1970) *Tables for Group Theory*, Oxford University Press, Oxford – A set of character tables with useful additional notes and symmetry diagrams.

R.L. Carter (1998) *Molecular Symmetry and Group Theory*, Wiley, New York – An introduction to molecular symmetry and group theory as applied to chemical problems including vibrational spectroscopy.

M.E. Cass, H.S. Rzepa, D.R. Rzepa and C.K. Williams (2005) *J. Chem. Educ.*, vol. 82, p. 1736 – 'The use of the free, open-source program Jmol to generate an interactive web site to teach molecular symmetry'.

F.A. Cotton (1990) *Chemical Applications of Group Theory*, 3rd edn, Wiley, New York – A more mathematical treatment of symmetry and its importance in chemistry.

G. Davidson (1991) *Group Theory for Chemists*, Macmillan, London – An excellent introduction to group theory with examples and exercises.

S.F.A. Kettle (2007) *Symmetry and Structure*, 3rd edn, Wiley, Chichester – A detailed, but readable, account of symmetry and group theory.

J.S. Ogden (2001) *Introduction to Molecular Symmetry*, Oxford University Press, Oxford – A concise introduction to group theory and its applications.

A. Rodger and P.M. Rodger (1995) *Molecular Geometry*, Butterworth-Heinemann, Oxford – A useful, clear text for student use.

R.B. Shirts (2007) *J. Chem. Educ.*, vol. 84, p. 1882 – A note entitled: 'Correcting two long-standing errors in point group symmetry character tables'.

A.F. Wells (1984) *Structural Inorganic Chemistry*, 5th edn, Oxford University Press, Oxford – A definitive work on structural inorganic chemistry; Chapter 2 gives a concise introduction to crystal symmetry.

D. Willock (2009) *Molecular Symmetry*, Wiley, Chichester – A student text introducing symmetry and group theory and their applications to vibrational spectroscopy and bonding.

Infrared spectroscopy

E.A.V. Ebsworth, D.W.H. Rankin and S. Cradock (1991) *Structural Methods in Inorganic Chemistry*, 2nd edn, Blackwell Scientific Publications, Oxford – Chapter 5 deals with vibrational spectroscopy in detail.

K. Nakamoto (1997) *Infrared and Raman Spectra of Inorganic and Coordination Compounds*, 5th edn, Wiley, New York – *Part A*: *Theory and Applications in Inorganic Chemistry* – An invaluable reference book for all practising experimental inorganic chemists, and including details of normal coordinate analysis.

PROBLEMS

Some of these problems require the use of Fig. 3.10.

3.1 Give the structures of the following molecules: (a) BCl_3; (b) SO_2; (c) PBr_3; (d) CS_2; (e) CHF_3. Which molecules are polar?

3.2 In group theory, what is meant by the symbols (a) E, (b) σ, (c) C_n and (d) S_n? What is the distinction between planes labelled σ_h, σ_v, σ_v' and σ_d?

3.3 For each of the following 2-dimensional shapes, determine the highest order rotation axis of symmetry.

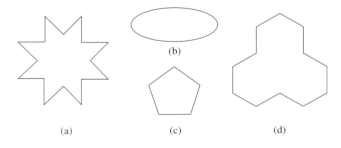

(a)   (c)   (d)
(b)

3.4 Draw the structure of SO_2 and identify its symmetry properties.

3.5 The structure of H_2O_2 was shown in Fig. 2.1. Apart from the operator E, H_2O_2 possesses only one other symmetry operator. What is it?

3.6 By drawing appropriate diagrams, illustrate the fact that BF_3 possesses a 3-fold axis, three 2-fold axes, and four planes of symmetry. Give appropriate labels to these symmetry elements.

3.7 Using the answer to problem 3.6 to help you, deduce which symmetry elements are lost on going from (a) BF_3 to $BClF_2$ and (b) $BClF_2$ to $BBrClF$. (c) Which symmetry element (apart from E) is common to all three molecules?

3.8 Which of the following species or ions contain (a) a C_3 axis but no σ_h plane, and (b) a C_3 axis *and* a σ_h plane: NH_3; SO_3; PBr_3; $AlCl_3$; $[SO_4]^{2-}$; $[NO_3]^-$?

3.9   Which of the following species contain a C_4 axis *and* a σ_h plane: CCl_4; $[ICl_4]^-$; $[SO_4]^{2-}$; SiF_4; XeF_4?

3.10   How many mirror planes do each of the following molecules contain: (a) SF_4; (b) H_2S; (c) SF_6; (d) SOF_4; (e) SO_2; (f) SO_3?

3.11   (a) What structure would you expect Si_2H_6 to possess? (b) Draw the structure of the conformer most favoured in terms of steric energy. (c) Does this conformer possess an inversion centre? (d) Draw the structure of the conformer least favoured in terms of steric energy. (e) Does this conformer possess an inversion centre?

3.12   Which of the following species contain inversion centres? (a) BF_3; (b) SiF_4; (c) XeF_4; (d) PF_5; (e) $[XeF_5]^-$; (f) SF_6; (g) C_2F_4; (h) $H_2C=C=CH_2$.

3.13   Explain what is meant by an ∞-fold axis of rotation.

3.14   To which point group does NF_3 belong?

3.15   The point group of $[AuCl_2]^-$ is $D_{\infty h}$. What shape is this ion?

3.16   Determine the point group of SF_5Cl.

3.17   The point group of BrF_3 is C_{2v}. Draw the structure of BrF_3 and compare your answer with the predictions of the VSEPR model.

3.18   In worked example 2.7, the structure of the $[XeF_5]^-$ ion was predicted. Confirm that this structure is consistent with D_{5h} symmetry.

3.19   Assign a point group to each member in the series (a) CCl_4, (b) CCl_3F, (c) CCl_2F_2, (d) $CClF_3$ and (e) CF_4.

3.20   (a) Deduce the point group of SF_4. (b) Is SOF_4 in the same point group?

3.21   Which of the following point groups possesses the highest number of symmetry elements: (a) O_h; (b) T_d; (c) I_h?

3.22   Determine the number of degrees of vibrational freedom for each of the following: (a) SO_2; (b) SiH_4; (c) HCN; (d) H_2O; (e) BF_3.

3.23   How many normal modes of vibration are IR active for (a) H_2O, (b) SiF_4, (c) PCl_3, (d) $AlCl_3$, (e) CS_2 and (f) HCN?

3.24   Use the C_{2v} character table to confirm that D_2O ('heavy water') has three IR active modes of vibration.

3.25   By considering the effect of each symmetry operation of the D_{3h} point group on the symmetric deformation mode shown in Fig. 3.14, confirm that this mode has A_2'' symmetry.

3.26   To what point group does CBr_4 belong? Using the appropriate character table, construct a reducible representation for the stretching modes of vibration. Show that this reduces to $A_1 + T_2$.

3.27   Six of the nine vibrational degrees of freedom of SiF_4 are IR active. Why are IR absorptions observed only at 389 and 1030 cm^{-1} for this compound?

3.28 Al_2Cl_6 belongs to the D_{2h} point group:

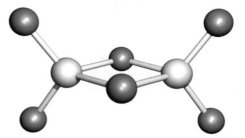

(a) How many degrees of vibrational freedom does Al_2Cl_6 possess?
(b) Use the D_{2h} character table in Appendix 3 to determine the symmetries of the IR active stretching modes of vibration.

3.29   The IR spectra of salts of $[AlF_6]^{3-}$ (O_h) exhibit absorptions around 540 and 570 cm^{-1}. Using a group theory approach, confirm that only one of these absorptions arises from a stretching mode.

3.30   Determine how many CO stretching modes are possible for *trans*-$M(CO)_4X_2$. What are their symmetries, and how many are IR active?

3.31   In 1993, the $[Pt(CO)_4]^{2+}$ ion was reported for the first time [G. Hwang *et al.* (1993) *Inorg. Chem.*, vol. 32, p. 4667]. One strong absorption at 2235 cm^{-1} in the IR spectrum was assigned to ν_{CO}, and this was absent in the Raman spectrum. In the Raman spectrum, two absorptions (ν_{CO}) at 2257 and 2281 cm^{-1} (absent in the IR spectrum) were observed. Show that these data are consistent with $[Pt(CO)_4]^{2+}$ having D_{4h} symmetry.

3.32   Explain how you could distinguish between *cis*-$M(CO)_2X_2$ and *trans*-$M(CO)_2X_2$ by using information from the CO stretching region of IR spectra. Include in your answer a derivation of the number of ν_{CO} modes for each molecule.

3.33   (a) To which point group does a trigonal bipyramidal XY_5 belong? Determine the number and symmetries of the stretching modes of vibration for this molecule. (b) The IR spectrum of gaseous PF_5 exhibits absorptions at 1026 and 944 cm^{-1}. Show that this observation is consistent with your answer to part (a). How many absorptions would you expect to observe in the Raman spectrum of gaseous PF_5 arising from stretching modes?

3.34   Explain what is meant by the terms (a) chiral; (b) enantiomer; (c) helical chain.

3.35 Confirm that the symmetry operation of (a) inversion is equivalent to an S_2 improper rotation, and (b) reflection through a plane is equivalent to an S_1 improper rotation.

WEB-BASED PROBLEMS

These problems are designed to introduce you to the website that accompanies this book. Visit the website: *www.pearsoned.co.uk/housecroft* and then navigate to the Student Resources site for Chapter 3 of the 4th edition of *Inorganic Chemistry* by Housecroft and Sharpe.

3.36 Open the structure file for problem 3.36: this is the structure of PF_5. (a) Orientate the structure so that you are looking down the C_3 axis. Where is the σ_h plane with respect to this axis? (b) Locate three C_2 axes in PF_5. (c) Locate three σ_v planes in PF_5. (d) To what point group does PF_5 belong?

3.37 Open the structure file for problem 3.37 which shows the structure of NH_2Cl. (a) How many planes of symmetry does NH_2Cl possess? (b) Does NH_2Cl possess any axes of rotation? (c) Confirm that NH_2Cl belongs to the C_s point group. (d) Detail what is meant by the statement: 'On going from NH_3 to NH_2Cl, the symmetry is lowered'.

3.38 Open the structure file for problem 3.38: this shows the structure of OsO_4, which has T_d symmetry. (a) Orientate the molecule so that you are looking down an O–Os bond, O atom towards you. What rotation axis runs along this bond? (b) The character table for the T_d point group shows the notation '$8C_3$'. What does this mean? By manipulating the structure, perform the corresponding symmetry operations on OsO_4.

3.39 Open the structure file for problem 3.39: this shows the structure of $[Co(en)_3]^{3+}$ where en stands for the bidentate ligand $H_2NCH_2CH_2NH_2$; the H atoms are omitted from the structure. The complex $[Co(en)_3]^{3+}$ is generally described as being octahedral. Look at the character table for the O_h point group. Why does $[Co(en)_3]^{3+}$ not possess O_h symmetry? What does this tell you about the use of the word 'octahedral' when used in a description of a complex such as $[Co(en)_3]^{3+}$?

3.40 Open the structure file for problem 3.40: this shows the structure of C_2Cl_6 in the preferred staggered conformation. (a) Orientate the structure so you are looking along the C–C bond. You should be able to see six Cl atoms forming an apparent hexagon around two superimposed C atoms. Why is the principal axis a C_3 axis and not a C_6 axis? (b) Explain why an S_6 axis is coincident with the C_3 axis. (c) By referring to the appropriate character table in Appendix 3, confirm that C_2Cl_6 has D_{3d} symmetry.

3.41 Open the structure file for problem 3.41: this shows the structure of α-P_4S_3. (a) Orientate the structure so that the unique P atom is closest to you and the P_3 triangle coincides with the plane of the screen. You are looking down the principal axis of α-P_4S_3. What type of axis is it? (b) Show that the molecule does not have any other axes of rotation. (c) How many planes of symmetry does the molecule possess? Are they σ_v, σ_h or σ_d planes? (d) Confirm that α-P_4S_3 belongs to the C_{3v} point group.

INORGANIC CHEMISTRY MATTERS

3.42 Carbon monoxide is a controlled emission from vehicle exhausts. Catalytic converters catalyse the conversion of CO to CO_2. Emissions of CO can be quantified using IR spectroscopy with detection limits of $<0.5\,mg\,m^{-3}$. (a) The IR absorption monitored during analysis for CO is at $2143\,cm^{-1}$. Draw a diagram of the vibrational mode that gives rise to this absorption. (b) Why might the presence of CO_2 interfere with detection of CO? (c) 78% of the Earth's atmosphere is N_2, and the fundamental stretching mode gives rise to an absorption at $2359\,cm^{-1}$. Why does this absorption not interfere with CO analysis by IR spectroscopy? (d) By referrring to Chapter 29, comment on why CO is toxic. (e) Calculate the number of molecules of CO present in $1\,dm^3$ of air containing $0.20\,mg\,m^{-3}$ CO.

3.43 Haemoglobin is the iron-containing metalloprotein responsible for transporting O_2 in the bloodstream of mammals. When O_2 binds to the Fe centre, it does so in an 'end-on' manner and gives rise to a band in the IR spectrum at $1107\,cm^{-1}$. No absorption is present in the IR spectrum of gaseous O_2. Rationalize these observations. (It is not necessary to discuss the bonding in O_2 or oxyhaemoglobin.)

Topics

4 Experimental techniques

4.1 Introduction

Analysing chemical compounds and confirming their identities is a crucial part of practical chemistry. In addition to applications in research laboratories, analysis is a daily part of scientists' work in areas such as the pharmaceutical industry, food and drink quality control, environmental monitoring and forensics. Modern laboratories offer a wide range of analytical techniques, and in this chapter, the theory and applications of some of these methods are detailed. We focus on the techniques used to determine the identity and spectroscopic properties of a compound, and will not be concerned with determining the amounts of trace elements or compounds in, for example, water and air samples.[†]

Analytical methods can be grouped roughly according to their use in:

- compositional analysis and formula determination;
- investigating bonding, connectivity of atoms and oxidation states of elements in a compound;
- determining molecular structure.

Introductions to several techniques (cyclic voltammetry, transmission electron microscopy, magnetism, emission spectroscopy) have been left until later in the book to facilitate association with appropriate theoretical details or specific applications.

> The component being analysed in a system is called the **analyte**.

[†] For an introduction to methods of solving analytical problems in environmental, forensic, pharmaceutical and food sciences, see Chapter 36 in C.E. Housecroft and E.C. Constable (2010) *Chemistry*, 4th edn, Prentice Hall, Harlow.

4.2 Separation and purification techniques

Before you begin to analyse a compound, it is crucial to ensure its purity. This section introduces routine methods of separation and purification of chemical compounds.

Gas chromatography (GC)

> **Gas chromatography (GC)** is a separation technique in which the mobile phase is a gas; the stationary phase is packed inside a capillary or microbore column.

Gas chromatography (GC) is used to separate *volatile* components of a mixture and depends upon the different interactions of the components in a *mobile phase* (the carrier gas) with a *stationary phase* (e.g. alumina or silica) contained in a chromatography column. The sample is injected into a flow of dried carrier gas which is heated so that the components of the sample are in the vapour phase. The sample is carried into a capillary or microbore chromatography column (Fig. 4.1), the temperature of which is controlled to maintain the vapour state for the samples. The times taken for the components in the mixture to be eluted from the chromatography column are characteristic *retention times* and the data are output as a chromatogram (a plot of relative intensity against time). Components in the sample that interact least well with the stationary phase have the shortest retention times, while more strongly adsorbed components move more slowly through the column. While GC is applied routinely to separate volatile compounds, liquid chromatography (LC) has wider applications in inorganic chemistry.

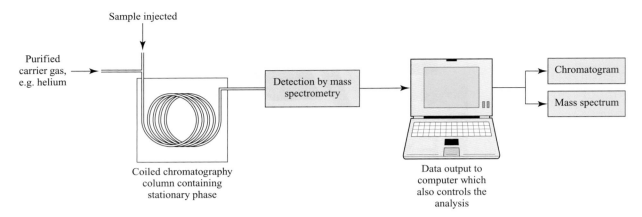

Fig. 4.1 Schematic diagram showing the components of a computer-controlled gas chromatography-mass spectrometer (GC-MS) instrument.

Liquid chromatography (LC)

Liquid chromatography (LC) is a separation technique in which the mobile phase is a liquid; the stationary phase is either packed inside a column or adhered to a glass plate.

For the purification of a product after synthesis, column liquid chromatography is often used. The crude material is adsorbed onto a *stationary phase* (e.g. alumina or silica with particle diameter $\approx 20\,\mu m$). The column is eluted with a solvent or solvent mixture (the *mobile phase*) chosen so as to separate the components based on their different solubilities and interactions with the stationary phase. Test separations are first carried out using thin-layer chromatography (TLC). The ratio of the distance travelled by the analyte to that of the solvent front is called the retention

factor (R_f value). An equilibrium is set up between surface-bound and solution species, and the preference of a given species for the stationary or mobile phase is given by the equilibrium constant K, where:

$$K = \frac{a_{\text{stationary}}}{a_{\text{mobile}}}$$

At low concentrations, activities, a, can be approximated to concentrations. As the column is eluted with solvent, some components are removed from the stationary phase more readily than others, and separation of the mixture is achieved. Figure 4.2 shows an open-column set-up. For coloured compounds, separation can be monitored by eye but instrumental methods of detection (e.g. UV absorption spectroscopy or mass spectrometry) can be integrated into the system (Fig. 4.3). In column chromatography, fractions are eluted under gravity flow or under pressure (flash chromatography).

High-performance liquid chromatography (HPLC)

High-performance liquid chromatography (HPLC) is a form of liquid chromatography in which the mobile phase is introduced under pressure and the stationary phase consists of very small particles (diameter 3–$10\,\mu m$).

The technique of high-performance liquid chromatography (HPLC) is summarized in Fig. 4.4. All parts of the separation process are computer controlled. Solvents are degassed before use, and the rate of flow of each solvent is controlled so that the solvent mixture entering the pump has a pre-determined composition. The pump operates at pressures up to $\approx 40\,MPa$, and the flow-rate delivered by the pump is varied as required. After injection of the sample, the mobile phase enters the chromatography column. The eluted fractions are monitored using a detector (e.g. UV-VIS, fluorescence, IR or circular dichroism spectroscopies or mass spectrometry) and the data are recorded in terms of, for example, absorbance against retention time.

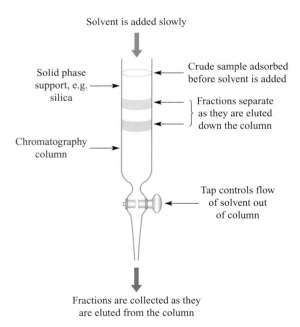

Fig. 4.2 Schematic representation of an open column chromatography set-up.

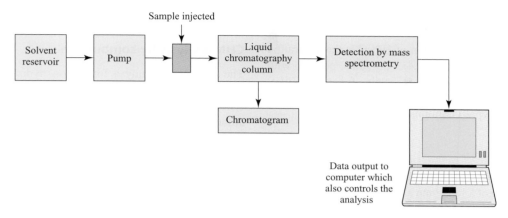

Fig. 4.3 Schematic diagram showing the components of a computer-controlled liquid chromatography-mass spectrometer (LC-MS) instrument.

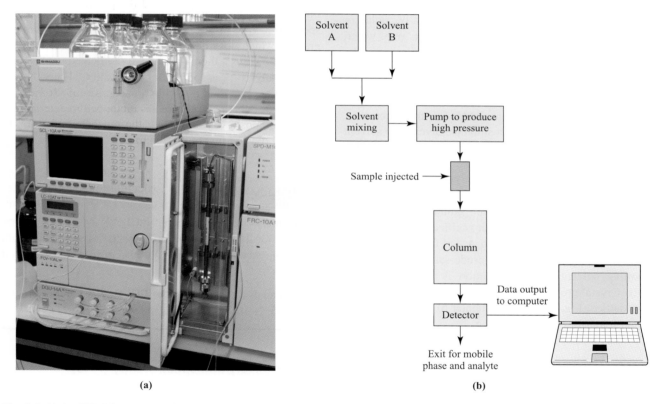

(a)   (b)

Fig. 4.4 (a) An HPLC instrument. The solvents are contained in the bottles at the top. The column is in the chamber with the open door. (b) Schematic diagram illustrating the components of a computer-controlled HPLC instrument.

Analytical HPLC uses columns of length 3–25 cm and width 2–4 mm, and stationary phase particle diameter of 3–10 μm. Preparative HPLC may use up to 1 g of sample, with ≈25 cm long columns of diameter 10–150 mm. In *normal-phase* HPLC, the stationary phase is a polar adsorbent such as silica (uniform particle diameter ≈10 μm) packed into a column. The solvents are usually non-polar

or of low polarity. Fractions are eluted from the column in order of their polarity, with non-polar components eluting first. The steric properties of substituents in a compound have a significant influence on the rate of elution. In *reversed-phase* HPLC, the surface of the stationary phase is modified so that it is hydrophobic. This is combined with polar solvents, usually water mixed with MeOH,

MeCN or THF, and the fractions are eluted from the column in order of decreasing polarity.

An example of the application of HPLC is the separation of fullerenes (see Section 14.4). Columns packed with stationary phases designed specifically for preparative-scale fullerene separation are commercially available (e.g. Cosmosil™ Buckyprep columns).

Recrystallization

> *Recrystallization* is a purification step involving the dissolution and crystallization of a solid from a solvent or solvent mixture.

After synthesis and chromatographic separation of a solid compound, recrystallization is usually used as the final purification step. Suppose you wish to separate a compound **A** from *minor* impurities. First, it is necessary to find a solvent in which **A** is insoluble at low temperatures, but is very soluble at higher temperatures. Typically, polar or ionic compounds are soluble in polar solvents (e.g. water, acetonitrile, dichloromethane, methanol), while non-polar compounds are soluble in non-polar solvents (e.g. hexanes, toluene). The impurities may also be soluble in the same solvent or solvent mixture but, ideally, a solvent is found in which the impurities are insoluble. For the latter case, after preliminary tests to find a suitable solvent, the solvent is added to crude **A** and the mixture is heated. The insoluble impurities are removed by filtering the hot solution, and the filtrate is allowed to cool either at or below room temperature. As the solution cools, the solubility of **A** decreases and, once the solution has become saturated, **A** begins to crystallize from the solution. If crystallization is relatively rapid, microcrystals form. If both **A** and the impurities are soluble in the same solvent, the crude sample is dissolved in boiling solvent and the solution is allowed to cool slowly. Crystal growth involves the assembly of an ordered crystal lattice (Fig. 4.5 and see Chapter 6) and the aim is to exclude impurities from the lattice.

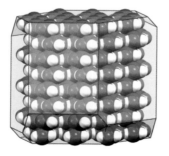 

Fig. 4.5 When boric acid, $B(OH)_3$, crystallizes, the molecules pack in layers and this produces crystals of well-defined shape (*morphology*).

4.3 Elemental analysis

CHN analysis by combustion

The quantitative analysis for carbon, hydrogen and/or nitrogen in a compound is carried out simultaneously using a *CHN analyser*. The compound is fully combusted and the amounts of products are measured. An accurately known mass of sample (2–5 mg) is sealed in an aluminium or tin capsule. After the sample has been introduced into the analyser, its combustion and product analysis are fully automated. The sample passes into a pyrolysis tube and is heated at $\approx 900\,°C$ in an atmosphere of pure O_2. Carbon, hydrogen and nitrogen are oxidized to CO_2, H_2O and nitrogen oxides. A stream of helium carries the gaseous products into a chamber containing heated copper where the oxides of nitrogen are reduced to N_2, and excess O_2 is removed. The carrier gas then transports the mixture of CO_2, H_2O and N_2 into the analysing chamber where they are separated using a specialized type of gas chromatography (frontal chromatography) which does not require a mobile phase. Detection of the gases uses a thermal conductivity detector. The automated analysis takes about 5 minutes, and C, H and N percentages are recorded to within an accuracy of <0.3%.

Modern CHN analysers are also equipped to determine S and O compositions. Sulfur is determined by oxidation to SO_2, and oxygen by conversion to CO, then to CO_2.

Self-study exercises

1. CH analysis of a compound $[PtCl_2(PR_3)_2]$ gives C 54.65, H 3.83%. Is R = Ph or Bu? Are the experimental data within acceptable limits compared to the calculated values?

2. CHN analysis for a complex $[TiCl_n(py)_{6-n}]$ gives C 46.03, H 3.85, N 10.72%. What is the value of n?

Atomic absorption spectroscopy (AAS)

> The quantitative determination of a metal can be carried out using *atomic absorption spectroscopy (AAS)* by observing the diagnostic absorption spectrum of gaseous atoms of the metal.

The emission spectrum of atomic hydrogen (Section 1.4) consists of a series of sharp lines, each line corresponding to a discrete electronic transition from higher to lower energy levels (Fig. 1.3). Conversely, if atomic hydrogen is irradiated, it will give rise to an *absorption spectrum*. Every element has a characteristic atomic emission and absorption spectrum, and the most common analytical method for quantitative determination of a given metal is *atomic absorption spectroscopy (AAS)*. Usually, radiation

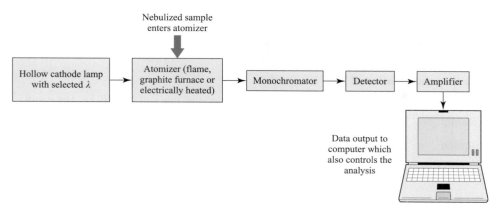

Fig. 4.6 Schematic diagram showing the components of an atomic absorption spectrometer.

using a wavelength corresponding to one specific transition is selected for the analysis (e.g. 217.0 or 283.3 nm for Pb).

The metal being analysed is not usually in its elemental form, and so the first step is *digestion* (decomposition) of the sample and of a series of standards. The standards contain known concentrations of the metal being analysed and are used to construct a calibration curve (see below). The atomic absorption spectrometer (Fig. 4.6) contains either a flame atomizer, a graphite furnace or an electrically heated atomizer. Each standard for the calibration is initially *aspirated*, i.e. air is sucked through the liquid sample as it passes into a *nebulizing chamber*. The fine spray of liquid that forms is then atomized. Radiation of the appropriate wavelength for the metal being detected is generated in a hollow cathode lamp and passed through the atomized standard or sample. The amount of radiation absorbed is recorded. Modern AAS instruments are computer controlled, and data are automatically recorded and processed. A linear relationship between absorbance and concentration follows from the Beer–Lambert law (see Section 4.7). However, at absorbances greater than ≈0.5, deviations from a linear plot occur, and it is therefore necessary to work with suitably dilute solutions. Starting from a stock solution of accurately known concentration, a series of five standard solutions is prepared and used to obtain the calibration curve. The latter must include a blank (i.e. solvent) to account for background effects. Once the calibration curve has been drawn, the process is repeated with the sample of unknown composition, and the measured absorbance is used to determine the concentration of metal in the sample. The AAS technique is extremely sensitive and the limit of detection is of the order of $\mu g\,dm^{-3}$.

Worked example 4.1 Sodium ions trapped in an inorganic complex

The cation, $[X]^{2+}$, drawn below was prepared as the salt $[X][PF_6]_2$. Elemental analysis showed the compound to be $[X][PF_6]_2 \cdot xNa[PF_6]$. The complex is soluble in acetonitrile (MeCN), but not in water. The sodium content of the bulk sample was determined by AAS.

A stock solution ($1.0 \times 10^{-4}\,mol\,dm^{-3}$) of NaCl was prepared. Standards were made by adding 2.0, 3.0, 4.0, 6.0 or $10.0\,cm^3$ of the stock solution to a $100\,cm^3$ volumetric flask; $5\,cm^3$ of MeCN were added to each flask which was then filled to the $100\,cm^3$ mark with deionized water. A blank was also prepared. The sample for analysis was prepared by dissolving 8.90 mg of $[X][PF_6]_2 \cdot xNa[PF_6]$ in $2.5\,cm^3$ of MeCN in a $50\,cm^3$ volumetric flask. This was filled to the mark with deionized water. This solution was diluted 10-fold with a 5% (by volume) MeCN/water solution. Using a sodium hollow cathode lamp ($\lambda = 589$ nm), AAS was used to determine the absorbance of each standard. Each absorbance reading was corrected for the absorbance of the blank and the data are tabulated below:

$[Na^+]/\mu mol\,dm^{-3}$	2.0	3.0	4.0	6.0	10.0
Corrected absorbance, A	0.0223	0.0340	0.0481	0.0650	0.1144

[Data: P. Rösel, Ph.D. Thesis, University of Basel, 2009]

(a) Confirm the concentrations of the standards given in the table. (b) How is the blank prepared? (c) Plot a calibration curve and determine the relationship between A and $[Na^+]$. (d) The sample of $[X][PF_6]_2 \cdot xNa[PF_6]$ gave an absorbance of 0.0507. Determine the concentration of Na^+ ions in the sample. Hence determine x in the formula.

(a) For the stock solution: $[Na^+] = 1.0 \times 10^{-4} \, mol \, dm^{-3}$

In $2.0 \, cm^3$ of this solution:

$$\text{Number of moles of } Na^+ = \frac{2.0 \times 1.0 \times 10^{-4}}{1000}$$
$$= 2.0 \times 10^{-7} \text{ moles}$$

After making up to $100 \, cm^3$:

$$[Na^+] = \frac{2.0 \times 10^{-7} \times 10^3}{100} = 2.0 \times 10^{-6} \, mol \, dm^{-3}$$
$$= 2.0 \, \mu mol \, dm^{-3}$$

The other concentrations can be similarly confirmed.

(b) The blank must contain the same solvents as the sample, so should be a mixture of deionized water and 5% MeCN by volume.

(c) Using the data given, construct a plot of A against $[Na^+]$. The linear least squares fit gives a line going through the origin (a result of the correction made using the blank).

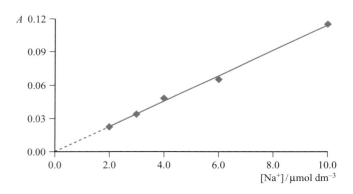

The equation for the line is:

$$A = 0.0114 \times [Na^+]$$

(d) From the above equation, you can determine $[Na^+]$ for the sample using the value of $A = 0.0507$:

$$[Na^+] = \frac{0.0507}{0.0114} = 4.45 \, \mu mol \, dm^{-3}$$

Therefore the concentration of $Na[PF_6]$ is $4.45 \, \mu mol \, dm^{-3} = 4.45 \times 10^{-6} \, mol \, dm^{-3}$

The solution of the sample used in the analysis comprises $8.90 \, mg \, [X][PF_6]_2 \cdot xNa[PF_6]$ in $50 \, cm^3$, diluted 10-fold, i.e. $8.90 \, mg$ in $500 \, cm^3$.

Moles of $Na[PF_6]$ in $500 \, cm^3$

$$= \frac{500 \times 4.45 \times 10^{-6}}{1000} = 2.225 \times 10^{-6}$$
$$= 2.23 \times 10^{-6} \text{ (to 3 sig. fig.)}$$

Molecular mass of $Na[PF_6] = 168 \, g \, mol^{-1}$

Mass of $Na[PF_6]$ in the sample

$$= (2.23 \times 10^{-6}) \times 168$$
$$= 3.75 \times 10^{-4} \, g$$
$$= 0.375 \, mg$$

Therefore, $8.90 \, mg$ of $[X][PF_6]_2 \cdot xNa[PF_6]$ contains $0.375 \, mg \, Na[PF_6]$.

x can be determined by finding the ratio of the number of moles of $[X][PF_6]_2 : Na[PF_6]$.

$8.90 \, mg$ of $[X][PF_6]_2 \cdot xNa[PF_6]$ is composed of $0.375 \, mg \, Na[PF_6]$ and $8.53 \, mg \, [X][PF_6]_2$.

Amount of $Na[PF_6]$ in $0.375 \, mg = 2.23 \, \mu mol$ (see above).

The molecular mass of $[X][PF_6]_2 = 1362 \, g \, mol^{-1}$

$$\text{Amount of } [X][PF_6]_2 \text{ in } 8.53 \, mg = \frac{8.53}{1362} = 6.26 \, \mu mol$$
$$\text{(to 3 sig. fig.)}$$

Ratio of moles of $[X][PF_6]_2 : Na[PF_6] = 6.26 : 2.23$
$$= 1 : 0.36$$

Therefore, $x = 0.36$.

Self-study exercises

1. By referring to Section 7.12, comment on the structural features of $[X]^{2+}$ (drawn on p. 91) that lead to Na^+ ions being trapped in the complex.

2. What is the physical significance of a value of $x = 0.36$ on the solid state structure of the complex?

4.4 Compositional analysis: thermogravimetry (TG)

In **_thermogravimetric analysis (TGA)_**, the change in mass of a sample is monitored as the sample is heated.

When a compound crystallizes from solution, the crystals may contain solvent of crystallization. The compound is then called a *solvate*, and if the solvent is water, the compound is a *hydrate*. The presence of the solvent is revealed by elemental analysis, and thermogravimetric analysis (TGA) shows whether the solvent molecules are loosely or strongly bound in the crystal lattice. More generally, TGA is used to investigate the thermal degradation of inorganic compounds or polymers or the gas uptake of a solid (e.g. H_2 uptake by WO_3). A TGA instrument is able to simultaneously heat (at a constant rate) and record the mass of a sample. Samples are usually heated in air or N_2, or in an atmosphere of a reactive gas (e.g. H_2) for studying the uptake of a particular gas. The output data are presented as described in the example below.

Worked example 4.2   Three-stage decomposition of $CaC_2O_4 \cdot H_2O$

TGA data for hydrated calcium oxalate, $CaC_2O_4 \cdot H_2O$, are shown below:

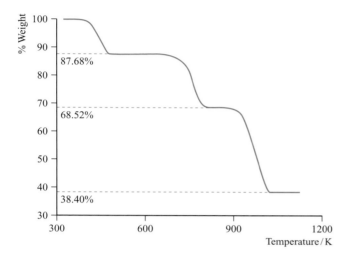

Explain the shape of the curve, and give a series of equations to summarize the thermal decomposition.

The curve shows three steps in the decomposition of $CaC_2O_4 \cdot H_2O$.

In the graph, the % weight losses are with respect to the original mass of $CaC_2O_4 \cdot H_2O$. Determine the % weight loss corresponding *to each step*.

Weight loss in step 1 = $100 - 87.7 = 12.3\%$

Weight loss in step 2 = $87.7 - 68.5 = 19.2\%$

Weight loss in step 3 = $68.5 - 38.4 = 30.1\%$

$M_r(CaC_2O_4 \cdot H_2O) = 40.08 + 2(12.01) + 5(16.00)$
$$+ 2(1.01) = 146.12 \text{ g mol}^{-1}$$

The weight loss of 12.3% corresponds to a mass of
$$\left(\frac{12.3}{100}\right) \times 146.12 = 18.0 \text{ g}$$

$M_r(H_2O) = 18.02 \text{ g mol}^{-1}$

Therefore, the lowest temperature thermal event is loss of H_2O.

The next weight loss of 19.2% corresponds to a mass of
$$\left(\frac{19.2}{100}\right) \times 146.12 = 28.1 \text{ g}$$

$M_r(CO) = 28.01 \text{ g mol}^{-1}$

Therefore, CO is lost in step 2.

The next weight loss of 30.1% corresponds to a mass of
$$\left(\frac{30.1}{100}\right) \times 146.12 = 44.0 \text{ g}$$

This corresponds to loss of CO_2 ($M_r = 44.01 \text{ g mol}^{-1}$). The three steps are therefore:

$CaC_2O_4 \cdot H_2O \longrightarrow CaC_2O_4 + H_2O$

$CaC_2O_4 \longrightarrow CaCO_3 + CO$

$CaCO_3 \longrightarrow CaO + CO_2$

Self-study exercises

1. 100.09 mg of $CaCO_3$ loses 43.97% of its weight when heated from 298 to 1100 K. Confirm that this corresponds to the formation of CaO.

2. When 2.50 g of $CuSO_4 \cdot 5H_2O$ are heated from 298 to 573 K, three decomposition steps are observed. The % weight losses for the consecutive steps with respect to the original mass are 14.42, 14.42 and 7.21%. Rationalize these data.

4.5 Mass spectrometry

Mass spectrometry is the separation of ions (atomic or molecular) according to their mass-to-charge (m/z) ratio.

A variety of mass spectrometric techniques is available and we focus on electron ionization (or electron impact), fast atom bombardment, matrix-assisted laser desorption ionization time-of-flight and electrospray ionization methods, all of which are routinely available.

Electron ionization (EI)

In *electron ionization (EI) mass spectrometry*, ions are produced by bombarding gaseous molecules with high-energy electrons. It is a 'hard' technique and causes fragmentation of the parent molecule.

A schematic representation of an electron ionization (EI) mass spectrometer is shown in Fig. 4.7. The compound (solid, liquid or gas) to be analysed is introduced into the instrument and is vaporized by heating (unless the sample is gaseous at 298 K). The vapour is subjected to a stream of high-energy (≈ 70 eV) electrons which ionizes the sample (eq. 4.1). The $[M]^+$ ion is a radical cation and is strictly written as $[M]^{\cdot +}$. The energy of the electrons used for the bombardment is much greater than covalent bond energies and causes fragmentation of the parent molecule, M. Thus, in addition to the molecular (or parent) ion,

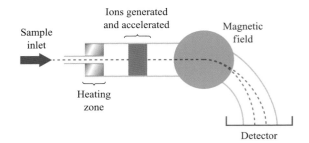

Fig. 4.7 Schematic diagram of an electron ionization (or electron impact, EI) mass spectrometer. During operation, the interior of the instrument is evacuated.

$[M]^+$, fragment ions are also formed. Electron ionization mass spectrometry is therefore classed as a 'hard' technique.

$$M(g) + \underset{\text{high energy}}{e^-} \longrightarrow [M]^+(g) + \underset{\text{low energy}}{2e^-} \qquad (4.1)$$

The positive ions pass through a magnetic field where their paths are deflected. The amount of deflection depends on the mass of the ion: the larger the m/z value, the larger the radius of the path that the ion follows (Fig. 4.7). The output from the mass spectrometer (a *mass spectrum*) is a plot of signal intensity against mass-to-charge (m/z) ratio. For an ion with $z = 1$, the m/z value is the same as the molecular mass of the ion. If $z = 2$, the recorded mass is half the actual mass of the ion, and so on. In EI mass spectrometry, most ions have $z = 1$. In a mass spectrum, the signal intensity is plotted in terms of relative values, with the most intense signal (the *base peak*) being arbitrarily assigned a value of 100%.

In addition to fragmentation patterns, the appearance of a mass spectrum depends upon the naturally occurring isotopes of the elements (see Appendix 5). The presence of several isotopes of an element leads to the observation of *peak envelopes*.

Worked example 4.3   EI mass spectrum of POCl₃

Assign the peaks in the EI mass spectrum of POCl₃ shown below.

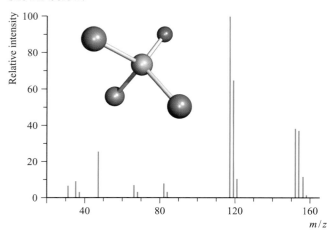

Look in Appendix 5 for the isotopes of P, O and Cl:

Cl   75.77% ^{35}Cl, 24.23% ^{37}Cl
P   100% ^{31}P
O   99.76% ^{16}O, 0.04% ^{17}O, 0.20% ^{18}O
         (i.e. close to 100% ^{16}O)

Note that, because Cl has two isotopes both of which are present in significant abundance, the mass spectrum exhibits groups of peaks (*peak envelopes*).

Determine the m/z value for the peak that you expect for the parent molecular ion, $[M]^+$, using the most abundant isotopes:

$$[M]^+ = [(^{31}P)(^{16}O)(^{35}Cl)_3]^+ \qquad m/z \; 152 \quad \text{for } z = 1$$

This matches the left-hand peak in the group of highest mass peaks in the mass spectrum. The peaks at m/z 154, 156 and 158 arise from $[(^{31}P)(^{16}O)(^{35}Cl)_2(^{37}Cl)]^+$, $[(^{31}P)(^{16}O)(^{35}Cl)(^{37}Cl)_2]^+$ and $[(^{31}P)(^{16}O)(^{37}Cl)_3]^+$ and the relative intensities of these peaks correspond to the probabilities of finding the particular combinations of isotopes.

Now consider the possible fragmentations: cleavage of P–Cl or P–O bonds. The peaks at m/z 117, 119 and 121 are assigned to $[POCl_2]^+$ with two ^{35}Cl, one ^{35}Cl and one ^{37}Cl, and two ^{37}Cl, respectively. This ion can also be written as $[M-Cl]^+$.

The pairs of peaks at m/z 82 and 84, and m/z 66 and 68 have the same isotope pattern and this is diagnostic of one chlorine atom. The assignments are $[M-2Cl]^+$ and $[M-2Cl-O]^+$.

The single peak at m/z must arise from an ion which does not contain Cl, i.e. to $[M-3Cl]^+$. The remaining peaks at m/z 35 and 37, and at m/z 31 are due to the atomic ions $[Cl]^+$ and $[P]^+$.

Although EI mass spectrometry is widely used (especially for organic compounds), it has limitations. It is suitable for the mass determination of relatively low mass compounds ($M_r < 1500$). If not a gas at 298 K, the compound must be volatile when heated and must be stable at the temperature required for vaporization. EI mass spectrometry cannot be used for ionic compounds.

Fast atom bombardment (FAB)

In *fast atom bombardment (FAB) mass spectrometry*, ions are produced by bombarding the sample (neutral molecules or ionic salts) with high energy xenon or argon atoms. It is a 'soft' technique and usually causes little fragmentation.

Fast atom bombardment (FAB) mass spectrometry was developed in the 1980s and has a number of advantages over EI: molecular masses up to $\approx 10\,000$ can be determined, the instrument can be run in both positive and negative modes permitting both positive and negative ions to be detected, and both ionic salts and neutral compounds can be investigated. Unlike EI, FAB mass spectrometry is a 'soft' technique and fragmentation of the parent ion is usually minimal.

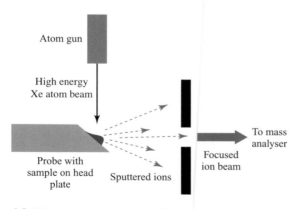

Glycerol
m/z 92

3-Nitrobenzyl alcohol (NOBA)
m/z 153

(4.1)   **(4.2)**

The sample to be analysed is mixed with a high boiling point liquid matrix such as glycerol (**4.1**) or 3-nitrobenzyl alcohol (**4.2**). The mixture is applied to the probe of the instrument and is bombarded with fast Xe or Ar atoms (Fig. 4.8). These are generated in an atom gun by ionizing Xe to Xe^+ (or Ar to Ar^+), and then accelerating the ions in an environment where electron capture occurs to give a stream of atoms with energies $\leq 10\,keV$. The beam of atoms bombards the matrix/sample obliquely (Fig. 4.8) and sample ions are produced with energies close to 1 eV or $96.5\,kJ\,mol^{-1}$. When the instrument is run in positive mode, atom bombardment of the sample leads to:

- direct release of parent $[M]^+$ ions from an ionic salt M^+X^-, or $[MX]^+$ from an ionic salt $M^{2+}(X^-)_2$;
- combination of a neutral molecule with a cation (usually H^+ or Na^+) to give $[M+H]^+$ or $[M+Na]^+$;
- ions such as $[M-H]^+$;
- aggregation to form ions such as $[2M+H]^+$, $[2M+Na]^+$ or $[M+matrix+H]^+$.

With the instrument in negative mode, atom bombardment results in:

- direct release of $[X]^-$ from an ionic salt containing X^- anions;
- combination of a neutral molecule with an anion (usually H^- or Cl^-) to give $[M+H]^-$ or $[M+Cl]^-$;
- proton loss to give $[M-H]^-$;
- aggregation to form ions such as $[2M-H]^-$.

The origin of Na^+ and Cl^- ions may be impurities in the sample, but NaCl may be added to the analyte. The ions leaving the probe are said to be *sputtered*, and the beam of ions is focused for entry into the mass analyser (Fig. 4.8).

Worked example 4.4   Analysing a multinuclear metal cluster compound by FAB mass spectrometry

When Ph_3PAuCl reacts with $[HRe_3(CO)_{12}]^{2-}$, and the product is treated with $K[PF_6]$, a yellow crystalline solid is isolated. The FAB (positive mode) mass spectrum exhibits peak envelopes at *m/z* 2135.1, 1845.1 and 1582.8. Show that these data are consistent with the product being $[(Ph_3PAu)_4Re(CO)_4][PF_6]$, and assign all the peaks. The structure of the $\{(PAu)_4Re(CO)_4\}$ core of the $[(Ph_3PAu)_4Re(CO)_4]^+$ cation is shown below:

Colour code: Au, blue; P, orange; Re, silver; C, grey; O, red.
[Data: G. Pivoriunas *et al.* (2005) *Inorg. Chim. Acta*, vol. 358, p. 4301]

The compound is ionic. In positive mode FAB MS, direct release of $[(Ph_3PAu)_4Re(CO)_4]^+$ leads to an ion with calculated *m/z* 2135.2. This is consistent with the observed value for the highest mass peak.

Fragmentation is expected to involve loss of CO or PPh_3 ligands. The peaks at *m/z* 1845.1 and 1582.8 are assigned to $[M-PPh_3-CO]^+$ and $[M-2PPh_3-CO]^+$, respectively.

Self-study exercises

1. In the FAB mass spectra of gold(I) triphenylphosphane complexes, a peak envelope at *m/z* 721.2 is often observed. Assign this peak.

2. The highest mass peak envelope in the FAB (positive mode) mass spectrum of the compound $[(Ph_3PAu)_6Re(CO)_3][PF_6]$ appears as follows:

Fig. 4.8 Schematic representation of the components in a FAB mass spectrometer.

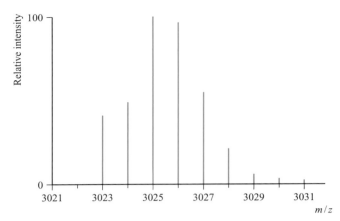

(a) Explain how this series of peaks arises. (b) Additional peak envelopes are observed at m/z 2763.4, 2501.3, 2239.5, 1977.6 and 459.0. Suggest identities for these singly charged ions.

Matrix-assisted laser desorption ionization time-of-flight (MALDI-TOF)

> In *matrix-assisted laser desorption ionization time-of-flight (MALDI-TOF) mass spectrometry*, a UV laser is used to generate atoms, M, of the sample; $[M + H]^+$ or $[M + Na]^+$ ions are typically detected. It is a 'soft' method and there is little fragmentation.

Matrix-assisted laser desorption ionization (MALDI) mass spectrometry uses a pulsed UV laser beam to produce ions for mass analysis. The coupling of MALDI with time-of-flight (TOF) mass analysis (Fig. 4.9) provides a sensitive technique which is especially useful for very high molecular mass species ($M_r \leq 300\,000$) including biomolecules and polymers. The sample is mixed with a large excess of matrix (e.g. **4.3** or **4.4**) which is capable of absorbing photons from the UV laser. Both **4.3** and **4.4** absorb at $\lambda = 337$ and 355 nm. The matrix/ analyte is loaded onto a sample plate where it crystallizes. Energy provided by the pulsed laser is absorbed by the matrix causing desorption of the molecules of sample. Combination with H^+ or Na^+ produces $[M + H]^+$ or $[M + Na]^+$ ions and, since MALDI is a 'soft' method, there is usually little fragmentation of the parent ions. Multiply charged ions are rarely formed.

α-Cyano-4-hydroxycinnamic acid

(4.3)

Sinapinic acid

(4.4)

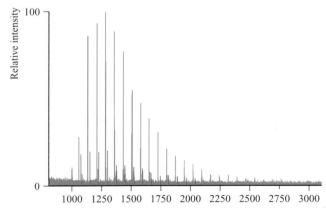

Fig. 4.9 Schematic representation of the components in a MALDI-TOF mass spectrometer.

Ions leaving the sample plate are focused into a beam and accelerated in an electric field in such a way that all singly charged ions possess the same amount of kinetic energy. Since kinetic energy is $\frac{1}{2}mv^2$, ions with the highest masses have the lowest velocities and longer flight times along the drift region in the flight tube (Fig. 4.9), and vice versa. Ions with the same m/z ratio possess the same travel time and the data are converted into a conventional mass spectrum. However, the nature of the TOF method of analysis means that broader peaks are observed than in EI, FAB or ESI mass spectrometries with the result that isotope patterns are not usually well resolved.

Worked example 4.5   MALDI-TOF mass spectrum of a polydimethylsiloxane

The figure below shows the MALDI-TOF mass spectrum (positive mode) of a sample of a polydimethylsiloxane elastomer, the repeat unit of which is:

Linear polymer chains may be terminated in Me or OH units; cyclic oligomers are also possible.

The main spectrum (in blue) is accompanied by a sub-spectrum (in red), and in each spectrum, the spacings between consecutive lines is m/z 74. Rationalize the appearance of the mass spectrum.

[Data: S.M. Hunt *et al.* (2000) *Polym. Int.*, vol. 49, p. 633]

The value of m/z 74 corresponds to an Me_2SiO monomer unit. Therefore the series of blue peaks arises from polymer chains of different lengths. Similarly for the set of red peaks. The observation of two series indicates that both cyclic and linear polymer chains are present. A cyclic oligomer has the formula $(Me_2SiO)_n$. For a particular chain length, a linear oligomer has a higher molecular mass than the corresponding cyclic oligomer because of the presence of the end groups (Me or OH).

The relative intensities of the two spectra reveal that cyclic oligomers dominate over linear chains.

Self-study exercises

These questions refer to the data in the worked example.

1. Polydimethylsiloxanes are neutral species. How do the positive ions observed in the MALDI-TOF experiment arise?
2. Give a formula for the positive ion observed at m/z 985.
3. Use Appendix 5 to explain why each of the peaks shown in the figure above consists of an envelope of peaks.

Electrospray ionization (ESI)

In *electrospray ionization (ESI) mass spectrometry*, ions are formed from a fine spray of solution under an applied electrical potential; it is a 'soft' technique used for both neutral molecules or ionic salts. Singly and multiply charged ions are observed.

The 'soft' technique of electrospray ionization (ESI) mass spectrometry has widespread applications in chemical and biochemical analysis of high-molecular mass compounds ($M_r \leq 200\,000$). In contrast to the methods described above, both singly and multiply charged ions are observed, making the ESI technique valuable for analysis of ionic compounds.

The sample is dissolved in a volatile solvent (e.g. MeCN or MeOH) and the solution is converted into a fine spray (nebulized) at atmospheric pressure by the application of a high electrical potential (Fig. 4.10). The ESI mass spectrometer can be operated in positive-ion mode which gives positively charged droplets of sample, or in negative-ion mode which produces negatively charged droplets. An electrical potential is applied between the needle through which the sample is injected and a counter electrode. The charged droplets travel towards the counter electrode, during which time the solvent evaporates. The gas-phase ions so-formed pass into a mass analyser. Zoom scans of peaks in the mass spectrum reveal the isotope patterns and allow the peak separations to be determined. If the peaks in an envelope are one mass unit apart, the ion is singly charged. If they

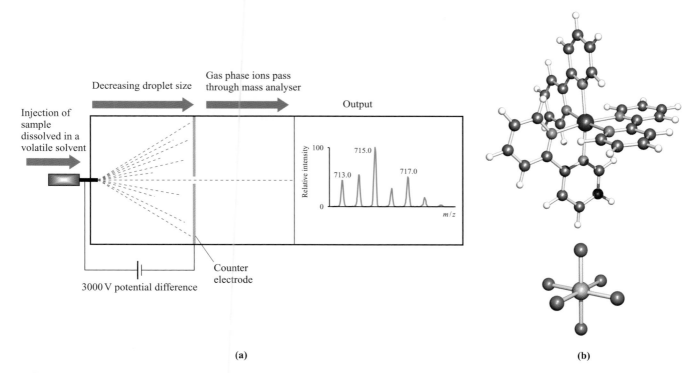

(a)   **(b)**

Fig. 4.10 (a) Schematic representation of the components in an ESI mass spectrometer. The *zoom scan* shown on the right corresponds to the highest mass peak envelope in the ESI mass spectrum of the complex $[Ru(bpy)_3][PF_6]_2$ (bpy = 2,2'-bipyridine) and is assigned to the $[M - PF_6]^+$ ion. (b) The structures of the $[Ru(bpy)_3]^{2+}$ and $[PF_6]^-$ ions. Colour code: Ru, red; N, blue; C, grey; H, white; P, orange; F, green.

are half a mass unit apart, the ion is doubly charged, and so on. The zoom scan shown in Fig. 4.10 shows the highest mass peak envelope in the ESI mass spectrum of the coordination compound $[Ru(bpy)_3][PF_6]_2$ (bpy = 2,2'-bipyridine, **4.5**). The pattern of peaks is largely controlled by the fact that Ru possesses seven naturally occurring isotopes (see Appendix 5), and the peak envelope is assigned to the $[M - PF_6]^+$ ion.

2,2'-bipyridine
(bpy)

2,2'-bipyridine coordinated
to an M^{n+} ion

(4.5)

As in FAB and MALDI-TOF mass spectrometries, neutral molecules are converted to positive ions by combination with H^+ or Na^+ ions. Aggregation may also occur to produce ions of the type $[2M + Na]^+$, and combination with the solvent molecules gives ions such as $[M + MeCN + H]^+$.

Self-study exercises

1.

H_2L

The ligand, H_2L, reacts with copper(II) acetate to give a neutral complex. In the ESI mass spectrum of a CH_2Cl_2/MeOH solution of the product, the base peak is at m/z 494.1 and exhibits an isotope pattern characteristic of one Cu atom. Assign the peak and, assuming no fragmentation occurs, suggest a formula for the complex.

2. Elemental analysis for an ionic cobalt(II) compound is consistent with a formula of $C_{30}H_{22}N_6CoP_2F_{12}$. (a) In the ESI mass spectrum of an MeCN solution of the compound, the highest group of peaks appears at m/z 670.1 (100%), 671.1 (35%) and 672.1 (6%). By referring to Appendix 5, explain how this group of peaks arises.

(b) In a second group of peaks, the most intense peak has m/z 262.6 and the peaks in the envelope are at half-mass separations. Rationalize these data.

4.6 Infrared and Raman spectroscopies

Infrared (IR) and Raman spectroscopies are concerned with transitions between vibrational energy levels. For a vibrational mode to be *IR active*, it must give rise to a *change in dipole moment*. For a vibrational mode to be *Raman active*, it must give rise to a *change in polarizability*.

Infrared (IR) and Raman spectroscopies are types of *vibrational spectroscopy*. In Chapter 3, we showed how to determine the number of degrees of vibrational freedom of a molecule, and stated the selection rules for an IR or Raman active mode of vibration (summarized above). We detailed the use of character tables to determine the types of vibrational modes and their symmetry labels.

Energies and wavenumbers of molecular vibrations

The *zero point energy* of a molecule corresponds to the energy of its lowest vibrational level (vibrational ground state).

When a molecule absorbs infrared radiation, it undergoes vibrational motions (Section 3.7). Stretching a covalent bond is like stretching a spring, and molecules undergo *anharmonic oscillations*. The vibrational energies are quantized and are given by eq. 4.2. The energy level with $v = 0$ is the *zero point energy* of the molecule. If, as is usually the case, we are concerned only with a transition from the vibrational ground state to the first excited state, then the motion is approximately that of a *simple harmonic oscillator* (eq. 4.3).[†]

$$E_v = (v + \tfrac{1}{2})h\nu - (v + \tfrac{1}{2})^2 h\nu x_e \qquad (E_v \text{ in J}) \qquad (4.2)$$

where: v is the vibrational quantum number

h = Planck constant

ν = frequency of vibration

x_e = anharmonicity constant

$$E_v = (v + \tfrac{1}{2})h\nu \qquad (4.3)$$

Consider a diatomic molecule, XY. The vibrational frequency of the bond depends on the masses of atoms X and Y, and on the *force constant*, k, of the bond. The force constant is a property of the bond and is related to its strength. If we return to stretching a spring, then the force constant is a measure of the stiffness of the spring. If atoms X and Y are of similar mass, they contribute almost equally to the molecular vibration. However, if the masses

[†]For more basic detail, see Chapter 12 in: C.E. Housecroft and E.C. Constable (2010) *Chemistry*, 4th edn, Prentice Hall, Harlow.

are very different, the lighter atom moves more than the heavier one. It is therefore necessary to define a quantity that describes the mass of the oscillator in a way that reflects the relative masses of X and Y. This is the *reduced mass*, μ, and for X and Y with masses of m_X and m_Y, μ is given by eq. 4.4.

$$\frac{1}{\mu} = \frac{1}{m_X} + \frac{1}{m_Y} \quad \text{or} \quad \mu = \frac{m_X m_Y}{m_X + m_Y} \qquad (4.4)$$

For a diatomic molecule, the transition from the vibrational ground state to the first excited state gives rise to a *fundamental absorption* in the vibrational spectrum. Equation 4.5 gives the frequency of this absorption.

$$\nu = \frac{1}{2\pi}\sqrt{\frac{k}{\mu}} \qquad (4.5)$$

where: ν = fundamental vibrational frequency (Hz)

$\quad k$ = force constant ($N\,m^{-1}$)

$\quad \mu$ = reduced mass (kg)

Absorptions in IR spectra are usually quoted in wavenumbers (wavenumber = 1/wavelength, $\bar{\nu} = 1/\lambda$) rather than frequencies, and eq. 4.5 then becomes eq. 4.6.

$$\bar{\nu} = \frac{1}{2\pi c}\sqrt{\frac{k}{\mu}} \qquad (4.6)$$

where: $\bar{\nu}$ = wavenumber (cm^{-1})

$\quad c$ = speed of light = $3.00 \times 10^{10}\,cm\,s^{-1}$

The Fourier transform infrared (FT-IR) spectrometer and sample preparation

The IR region is from $20\,cm^{-1}$ (far IR) to $14\,000\,cm^{-1}$ (near IR), but most laboratory IR spectrometers probe only the mid-IR (400 to $4000\,cm^{-1}$). In addition to being invaluable as a routine analytical tool in the laboratory, IR spectroscopy has applications in fields as far apart as forensic sciences (Fig. 4.11) and astronomy. The output data may be in the

Fig. 4.11 Fourier transform (FT) IR spectrometer in a forensics crime laboratory.

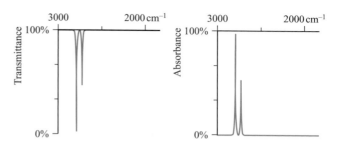

Fig. 4.12 An IR spectrum can be recorded in terms of transmittance or absorbance, and provides information about the energy and intensity of an absorption.

form of an absorption or a transmission spectrum (Fig. 4.12): 100% transmission corresponds to 0% absorption.

Infrared spectra can be recorded using gaseous, liquid or solid samples. Spectra of gases usually exhibit rotational structure in addition to bands arising from vibrational modes. Solid samples have traditionally been prepared in the form of a mull in an organic oil (e.g. nujol) or as a pressed disc in which the sample is ground with an alkali metal halide (e.g. KBr). The disadvantage of a mull is the appearance of absorptions arising from the matrix, while the use of an alkali metal halide restricts the window of observation: KBr is optically transparent from 4000 to $450\,cm^{-1}$, whereas the window for NaCl is 4000 to $650\,cm^{-1}$. The use of a modern diamond attenuated total reflectance (ATR) accessory allows the IR spectrum of a solid to be measured directly, and avoids the need for mulls or pressed discs.

Diagnostic absorptions

Infrared spectra of inorganic and organic compounds can generally be separated into two regions. Absorptions above $1500\,cm^{-1}$ can typically be assigned to specific groups (Table 4.1), while bands below $1500\,cm^{-1}$ tend to arise from single bond stretching modes, deformations, and vibrational modes of the molecular framework. This latter region is called the *fingerprint region* and the absorptions within it are usually taken together to provide a diagnostic signature of the compound. However, the distinction between the two regions is not definitive. For example, the T_{1u} vibrational mode (see Fig. 3.18) of the $[PF_6]^-$ ion gives rise to a strong and easily recognizable absorption at $865\,cm^{-1}$. Table 4.2 lists IR active bands for some common inorganic anions with D_{3h}, C_{3v}, T_d, D_{4h} or O_h symmetries; the vibrational modes are illustrated in Figs. 3.14–3.18. Table 4.2 shows that the vibrational wavenumbers for the D_{4h} $[PdCl_4]^{2-}$ and $[PtCl_4]^{2-}$ ions are below $400\,cm^{-1}$, and the associated IR absorptions, like those of most other metal halides, are not observed using a typical laboratory mid-IR spectrometer.

Ions such as $[NO_3]^-$, $[CO_3]^{2-}$, $[SO_4]^{2-}$ and $[ClO_4]^-$ can coordinate to metal centres and, as a result, the symmetry of the anion is lowered. Structures **4.6–4.8** illustrate common

Table 4.1 Typical ranges of fundamental stretching wavenumbers of selected groups (M = metal atom).

Functional group	ν/cm^{-1}	Functional group	ν/cm^{-1}
O–H	3700–3500	CO ligand (terminal MCO)[†]	2200–1900
O–H (hydrogen bonded)	3500–3000	CO ligand (bridging M_2CO)[†]	1900–1700
N–H	3500–3200	NO (terminal, linear MNO)	1900–1650
C–H	3000–2850	NO (terminal, bent MNO)	1690–1525
B–H (terminal)	2650–2350	CN ligand (terminal MCN)	2200–2000
B–H (bridge)	2100–1600	B–Cl	1000–600
S–H	2700–2550	C–Cl	1000–600
P–H	2450–2275	Si–Cl	750–600
Si–H	2250–2100	N–Cl	800–600
Al–H	1800–1700	P–Cl	600–450
C=O (organic)	1750–1650	O–Cl	1200–700
C=N (organic)	1690–1630	S–Cl	750–400

[†]See Fig. 24.2 and accompanying discussion.

coordination modes for the sulfato ligand; charges are ignored in the diagrams. The effects of coordination can be seen in Fig. 4.13 which shows part of the IR spectra of $[Co(NH_3)_6][SO_4]_3 \cdot 5H_2O$ (which contains non-coordinated $[SO_4]^{2-}$ ions) and $[Co(NH_3)_5(OSO_3)]Br$ (which contains a metal-bound sulfato ligand in coordination mode **4.6**).

(4.6)   **(4.7)**   **(4.8)**

Section 3.7 gives examples of the fundamental stretching wavenumbers of simple inorganic molecules and anions, as well as exercises to help you to relate an observed spectrum to the symmetry of a species.

Deuterium/hydrogen exchange

When the hydrogen atom in an X–H bond is exchanged by deuterium D,[†] the reduced mass of the pair of bonded atoms changes and shifts the position of the absorption in the IR spectrum due to the X–H stretching mode. Shifts of this kind can be used to confirm assignments in IR spectra.

For example, N–H, O–H and C–H bonds all absorb around 3000–3600 cm^{-1} (Table 4.1). However, if a compound containing C–H and N–H or O–H bonds is shaken with D_2O, usually, only the OH and NH groups undergo rapid *deuterium exchange reactions* (eq. 4.7). An H atom attached directly to C exchanges extremely slowly except in cases where it is acidic (e.g. a terminal alkyne).

$$R-OH + D_2O \rightleftharpoons R-OD + HOD \qquad (4.7)$$

By observing which IR spectroscopic bands shift (and by how much), it is possible to confirm the assignment of an N–H, O–H or C–H absorption.

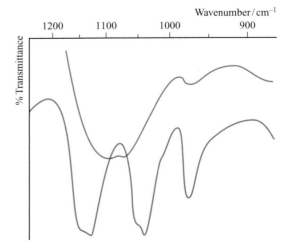

Fig. 4.13 IR spectra of $[Co(NH_3)_6]_2[SO_4]_3 \cdot 5H_2O$ (in red) and $[Co(NH_3)_5(OSO_3)]Br$ (in green). [Redrawn with permission from: K. Nakamoto *et al.* (1957) *J. Am. Chem. Soc.*, vol. 79, p. 4904.]

[†]Rather than use the full notation for the $_1^2H$ isotope, we use the less rigorous, but still unambiguous, notation with only the mass number, e.g. ^2H. In addition, the label D for deuterium is introduced.

Table 4.2 Vibrational modes and wavenumbers of selected anions.[†] See Figs. 3.14–3.18 for schematic representations of the vibrational modes.

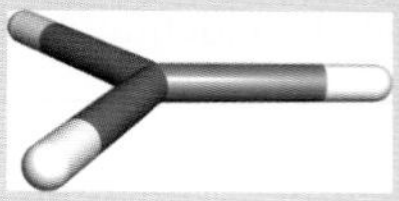

D_{3h}	$\nu_2\ (A_2'')$	$\nu_3\ (E')$		$\nu_4\ (E')$
$[CO_3]^{2-}$ (solid $CaCO_3$)	879	1429–1492 broad		706
$[NO_3]^-$ (solid $NaNO_3$)	831	1405		692
$[NO_3]^-$ (solid KNO_3)	828	1370		695

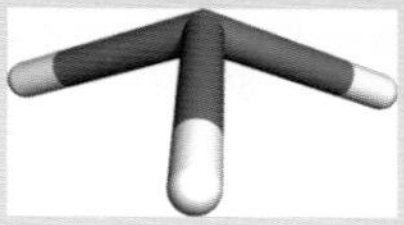

C_{3v}	$\nu_1\ (A_1)$	$\nu_2\ (A_1)$	$\nu_3\ (E)$	$\nu_4\ (E)$
$[SO_3]^{2-}$	967	620	933	469
$[ClO_3]^-$	933	608	977	477

D_{4h}	$\nu_3\ (A_{2u})$	$\nu_6\ (E_u)$	$\nu_7\ (E_u)$
$[PdCl_4]^{2-}$	150	321	161
$[PtCl_4]^{2-}$	147	313	165

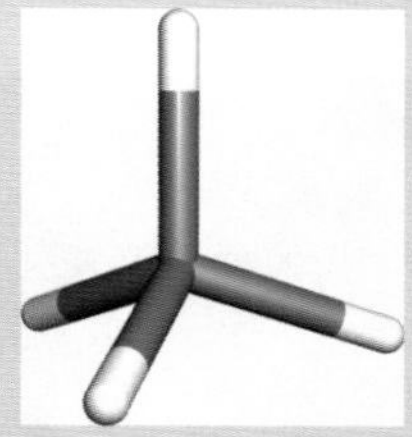

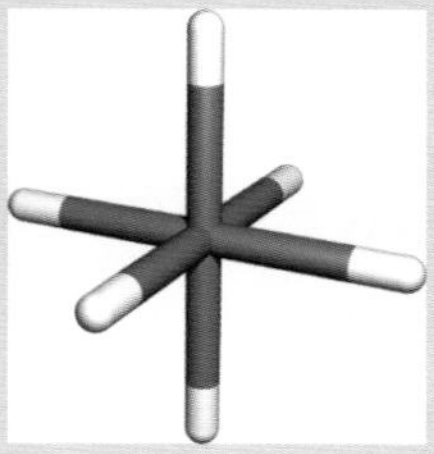

T_d	$\nu_3\ (T_2)$	$\nu_4\ (T_2)$	O_h	$\nu_3\ (T_{1u})$	$\nu_4\ (T_{1u})$
$[BF_4]^-$	1070	533	$[SiF_6]^{2-}$	741	483
$[BCl_4]^-$	722	278	$[PF_6]^-$	865	559
$[AlCl_4]^-$	498	182	$[PCl_6]^-$	444	285
$[SiO_4]^{4-}$	956	527	$[AsF_6]^-$	700	385
$[PO_4]^{3-}$	1017	567	$[AsCl_6]^-$	333	220
$[SO_4]^{2-}$	1105	611	$[SbF_6]^-$	669	350
$[ClO_4]^-$	1119	625	$[SbCl_6]^-$	353	180

[†]Data: K. Nakamoto (1997) *Infrared and Raman Spectra of Inorganic and Coordination Compounds*, Part A, 5th edn, Wiley, New York.

Worked example 4.6   The effects of deuteration on $\bar{\nu}_{O-H}$ in an IR spectrum

An absorption at $3650 \, \text{cm}^{-1}$ in the IR spectrum of a compound X has been assigned to an O–H stretching mode. To what wavenumber is this band expected to shift upon deuteration? What assumption have you made in the calculation?

The vibrational wavenumber, $\bar{\nu}$, is given by:

$$\bar{\nu} = \frac{1}{2\pi c} \sqrt{\frac{k}{\mu}}$$

If we assume that the force constants of O–H and O–D bonds are the same, then the only variables in the equation are $\bar{\nu}$ and μ. Thus, the O–H vibrational wavenumber, $\bar{\nu}$, is related to the reduced mass, μ, by the equation:

$$\bar{\nu}_{O-H} \propto \frac{1}{\sqrt{\mu_{O-H}}}$$

For comparison of the O–H and O–D stretching frequencies, we can write:

$$\frac{\bar{\nu}_{O-D}}{\bar{\nu}_{O-H}} = \frac{\sqrt{\mu_{O-H}}}{\sqrt{\mu_{O-D}}}$$

and since we are now dealing with a *ratio*, it is not necessary to convert the atomic masses to kg (see eq. 4.5). The relative atomic masses of O, H and D are 16.00, 1.01 and 2.01, respectively. The reduced masses of O–H and O–D bonds are found as follows:

$$\frac{1}{\mu_{O-H}} = \frac{1}{m_1} + \frac{1}{m_2} = \frac{1}{16.00} + \frac{1}{1.01} = 1.0526 \quad \mu_{O-H} = 0.9500$$

$$\frac{1}{\mu_{O-D}} = \frac{1}{m_1} + \frac{1}{m_2} = \frac{1}{16.00} + \frac{1}{2.01} = 0.5600 \quad \mu_{O-D} = 1.7857$$

The vibrational wavenumber of the O–D bond is therefore:

$$\bar{\nu}_{O-D} = \bar{\nu}_{O-H} \times \sqrt{\frac{\mu_{O-H}}{\mu_{O-D}}} = 3650 \times \sqrt{\frac{0.9500}{1.7857}} = 2662 \, \text{cm}^{-1}$$

$$= 2660 \, \text{cm}^{-1} \text{ (to 3 sig. fig.)}$$

The calculation makes the assumption that the force constants of O–H and O–D bonds are the same.

Self-study exercises

Data: atomic masses $H = 1.01$; $D = 2.01$; $N = 14.01$; $C = 12.01$; $O = 16.00$

1. Show that upon converting NH_3 to ND_3, an absorption in the vibrational spectrum at $3337 \, \text{cm}^{-1}$ shifts to $2440 \, \text{cm}^{-1}$.

2. An absorption at $3161 \, \text{cm}^{-1}$ in an IR spectrum is assigned to a C–H stretching mode. Show that upon deuteration, the band appears at $2330 \, \text{cm}^{-1}$.

3. An absorption in the IR spectrum of a compound containing an X–H bond shifts from 3657 to $2661 \, \text{cm}^{-1}$ upon deuteration. Show that X is likely to be O rather than C. What assumption have you made in the calculation?

Spectroscopic studies of isotopically substituted molecules often involve special syntheses which must be designed so as to make the best possible use of the isotope to be incorporated. For example, deuterated ammonia, ND_3, would not be prepared by exchange between NH_3 and D_2O since a large proportion of the deuterium is wasted in conversion to HOD. A better method is to react D_2O with Mg_3N_2 (eq. 4.8).

$$Mg_3N_2 + 6D_2O \longrightarrow 2ND_3 + 3Mg(OD)_2 \qquad (4.8)$$

Raman spectroscopy

The nature of the selection rules for IR and Raman spectroscopies results in these techniques being complementary. Chandrasekhara V. Raman was awarded the 1930 Nobel Prize in Physics 'for his work on the scattering of light and for the discovery of the effect named after him'. When radiation (usually from a laser, Fig. 4.14) of a frequency, ν_0, falls on a vibrating molecule, most of the radiation is scattered without a change in frequency. This is called *Rayleigh scattering*. A small amount of the scattered radiation has frequencies of $\nu_0 \pm \nu$, where ν is the fundamental frequency of a vibrating mode of the molecule. This is *Raman scattering*. For recording the Raman spectra of inorganic compounds, the radiation source is usually a visible noble gas laser (e.g. a red krypton laser, $\lambda = 647 \, \text{nm}$). One of the advantages of Raman spectroscopy is that it extends to lower wavenumbers than routine laboratory IR spectroscopy, thereby permitting the observation of,

Fig. 4.14 Laser Raman spectroscopy being used to measure ambient flame pressure at the Combustion Research Facility and Sandia National Laboratory, California. The research is aimed at creating heat engines and systems that burn fuel more efficently while creating less pollution.

APPLICATIONS

Box 4.1 Analysing paints and pigments

Analysis of paints is a routine part of forensic investigations of hit-and-run accidents or collisions involving vehicles. Paints have complex formulations. Automotive paints are applied in layers and the sequence and colours of the layers provide information on the origins of the vehicle. By matching IR and Raman spectroscopic data from a vehicle paint sample to those in databases (e.g. the European Collection of Automotive Paints and Paint Data Query), it is possible to determine the vehicle's manufacturer and year of production. The main components of a paint fall into four categories:

- binders (resins)
- pigments
- solvents
- additives.

Binders are typically organic resins, while pigments include inorganic compounds. The table below gives IR spectroscopic absorptions that can be used to identify some of the most common inorganic pigments. For the identification of a wider range of pigments, Raman spectroscopy must be used.

Pigment	Dominant IR absorptions/cm^{-1}
TiO_2 (rutile)	600
$BaCrO_4$	860
$Al_2Si_2O_5(OH)_4$ (kaolinite)	3700, 3620, 1030, 1010
Fe_2O_3	560
$BaSO_4$	1180, 1080, 630

Raman spectroscopy plays an important role in the identification of pigments in artwork. In *Raman microscopy*, a sample is irradiated with a laser beam and the scattered light is detected by a combination of an optical microscope and a Raman spectrometer. This relatively new technique is ideally suited to the analysis of pigments in manuscripts and artwork, because analysis takes place *in situ* without the need for scraping samples from the source material. In addition to identifying pigments, art historians gain insight into methods of conservation (e.g. do pigments degrade with time?) and are aided in authenticating an artwork. Over the centuries, a wide variety of artistic inorganic pigments has been used. Of those listed below, all occur naturally except for Prussian blue which was first manufactured in 1704, and viridian (Guignet's green) which has been manufactured since the mid-1800s.

Pigment	Chemical formula	Colour
Azurite	$2CuCO_3 \cdot Cu(OH)_2$	Blue
Lazurite (lapis lazuli)	$Na_8[Al_6Si_6O_{24}][S_n]$	Blue
Prussian blue	$Fe_4[Fe(CN)_6]_3 \cdot nH_2O$	Blue
Ochre	$Fe_2O_3 \cdot H_2O$	Brown/orange
Malachite	$CuCO_3 \cdot Cu(OH)_2$	Green
Realgar	As_4S_4	Red
Orpiment	As_2S_3	Yellow
Barytes	$BaSO_4$	White
Cadmium yellow	CdS	Yellow
Viridian	$Cr_2O_3 \cdot 2H_2O$	Green
Litharge	PbO	Red
Vermilion (Cinnabar)	HgS	Red

Further reading

P. Buzzini and W. Stoecklein (2005) in *Encyclopedia of Analytical Science*, 2nd edn, eds. P. Worsfold, A. Townshend and C. Poole, Elsevier, Oxford, p. 453 – 'Paints, varnishes and lacquers'.

R.J.H. Clark (1995) *Chem. Soc. Rev.*, vol. 24, p. 187 – 'Raman microscopy: application to the identification of pigments in medieval manuscripts'.

R.J.H. Clark (2002) *C. R. Chimie*, vol. 5, p. 7 – 'Pigment identification by spectroscopic means: an arts/science interface'.

R.J.H. Clark (2007) *Appl. Phys. A*, vol. 89, p. 833 – 'The scientific investigation of artwork and archaeological artefacts: Raman microscopy as a structural, analytical and forensic tool'.

Illustrations in medieval manuscripts used natural pigments from plants, animals and minerals.

for example, metal–ligand vibrational modes. A disadvantage of the Raman effect is its insensitivity because only a tiny percentage of the scattered radiation undergoes Raman scattering. This can be overcome by using the Fourier transform (FT) technique. For coloured compounds, sensitivity can be enhanced by using *resonance Raman spectroscopy*. This relies on using laser excitation wavelengths that coincide with wavelengths of absorptions in the electronic spectrum of a compound. This leads to resonance enhancement and an increase in the intensities of lines in the Raman spectrum.

An early success of Raman spectroscopy was in 1934 when Woodward reported the spectrum of mercury(I) nitrate. After the assignment of lines to the $[NO_3]^-$ ion, a line at $169\,cm^{-1}$ remained which he assigned to the stretching mode of the Hg–Hg bond in $[Hg_2]^{2+}$. This was one of the first pieces of evidence for the dimeric nature of the 'mercury(I) ion'. Resonance Raman spectroscopy is now used extensively for the investigation of coloured *d*-block metal complexes and for probing the active metal sites in metalloproteins. Application of Raman spectroscopy in pigment analysis is described in Box 4.1.

Self-study exercises

1. The fundamental stretching vibration for O_2 is at $1580\,cm^{-1}$. Why would you use Raman, and not IR, spectroscopy to observe this absorption?

2. Each of the Raman and IR spectra of O_3 exhibits three bands at 1135, 1089 and $716\,cm^{-1}$. Explain why this provides evidence that O_3 is a non-linear molecule.

3. Why is the fundamental stretching vibration of NO both Raman and IR active?

4.7 Electronic spectroscopy

Electronic spectroscopy is concerned with transitions of electrons between energy levels and covers both *absorption and emission spectroscopies*.

Electronic spectra arise from transitions of electrons between energy levels. Transitions from lower to higher energy levels produce absorption spectra, while those from higher to lower energy levels give rise to emission spectra. *Atomic* absorption and emission spectra were discussed in Sections 1.4 and 4.3. In this section, we describe the electronic spectra of *molecular species*.

UV-VIS absorption spectroscopy

Molecular orbitals may be bonding (σ or π), non-bonding (*n*) or antibonding (σ^* or π^*) in character. If a molecule

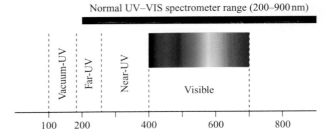

Fig. 4.15 The ultraviolet (UV) and visible regions of the electromagnetic spectrum. Normal laboratory UV-VIS spectrophotometers operate between 200 and 900 nm.

absorbs an appropriate amount of energy, an electron in an occupied orbital is excited to an unoccupied or partially occupied orbital. The HOMO-LUMO separation (HOMO = highest occupied MO, LUMO = lowest occupied MO) is such that absorption spectra are usually observed in the ultraviolet (UV) or visible region of the electromagnetic spectrum (Fig. 4.15). The energies of MOs are quantized, and so an electronic transition is associated with a specific amount of energy, ΔE. You might expect, therefore, that transitions from one electronic level to another in a molecular species would give rise to sharp spectral lines just as they do in atomic absorption spectra. However, molecular electronic absorption spectra usually consist of broad bands. Unlike atoms, molecules undergo vibrational and rotational motions and these are much slower than the absorption of a photon ($\approx 10^{-18}\,s$): the *Franck–Condon approximation* states that electronic transitions are very much faster than nuclear motion (because nuclear mass is far greater than electronic mass). Thus, an electronic transition is a 'snapshot' of the molecule in a particular vibrational and rotational state. Molecular orbital energies depend on the molecular geometry, and so a range of ΔE values (corresponding to different vibrational and rotational states as the molecular geometry changes) is observed. The result is that broad bands are observed in electronic absorption spectra. Sharper spectra may be seen at low temperatures because less energy is available for molecular motion. A spectrum is also sharp if electronic transitions are localized on a single atomic centre, e.g. for compounds of the *f*-block metals (see Chapter 27).

Types of absorption

The *notation for electronic transitions* shows the higher energy level first.

Emission: (high energy level) $\longrightarrow$ (low energy level), e.g. $\pi^* \longrightarrow \pi$.

Absorption: (high energy level) $\longleftarrow$ (low energy level), e.g. $\pi^* \longleftarrow \pi$.

Electronic transitions giving rise to absorption spectra usually fall into the following categories:

- $\sigma^* \leftarrow \sigma$
- $\sigma^* \leftarrow n$
- $\pi^* \leftarrow \pi$
- $\pi^* \leftarrow n$
- charge transfer
- 'd–d'

The energies of most $\sigma^* \leftarrow \sigma$ electronic transitions are too high (i.e. the wavelength is too short and occurs in the vacuum UV) to be observed using a normal UV-VIS spectrophotometer. The most intense absorptions in the 200–800 nm region (Fig. 4.15) typically arise from $\pi^* \leftarrow \pi$, $\sigma^* \leftarrow n$, $\pi^* \leftarrow n$ or charge transfer transitions. Examples of the latter include metal-to-ligand and ligand-to-metal charge transfer in which electronic charge is transferred between metal and ligand orbitals (see Section 20.7).

Absorbance and the Beer–Lambert law

Absorption bands in *molecular electronic spectra* are often broad, and may be described in terms of wavelength, λ_{max} (nm), and the molar extinction (or absorption) coefficient, ε_{max} ($dm^3 mol^{-1} cm^{-1}$):

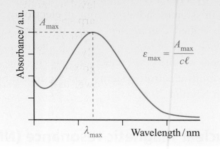

$$\varepsilon_{max} = \frac{A_{max}}{c\ell}$$

A UV-VIS spectrophotometer typically records absorption spectra in the 200–800 nm range, and Fig. 14.16 shows a simplified diagram of the instrument. The light source consists of a deuterium lamp and a tungsten or halogen lamp which cover the UV and visible regions, respectively. The solution sample and a solvent reference are held in quartz cuvettes, usually with a 1 cm *path length*; the path length is the distance that

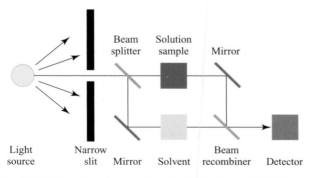

Fig. 4.16 Schematic representation of a double beam UV-VIS spectrophotometer.

the radiation travels through the cuvette. The two cuvettes are placed between the source of radiation and a detector and are simultaneously irradiated. The detector monitors the radiation transmitted through the sample. The *transmittance, T*, of the sample is equal to the ratio of the intensity of the transmitted radiation (I) to that of the incident radiation (I_0), and eq. 4.9 shows how the absorbance, A, of the sample is determined. This is done automatically by the spectrophotometer and the data are output as an absorption spectrum. The whole range of wavelengths is detected simultaneously using a diode array detector (DAD), and the double beam set up allows the absorption spectrum of the solvent to be subtracted from that of the sample solution.

$$\text{Absorbance, } A = -\log T = -\log \frac{I}{I_0} \quad (4.9)$$

It can be seen from eq. 4.9 that absorbance is dimensionless. The absorbance is related to the concentration of the solution by the Beer–Lambert law (eq. 4.10) where ε is the *molar extinction* (or *absorption*) *coefficient* of the dissolved sample. The extinction coefficient is a property of the compound and is independent of concentration for dilute solutions. Solution concentrations of the order of 10^{-5} mol dm^{-3} are typical, giving absorbance values ≤ 3.

$$\text{Absorbance, } A = \varepsilon \times c \times \ell \quad (4.10)$$

Since the concentration, c, is measured in mol dm^{-3} and the cell path length, ℓ, in cm, the units of ε are $dm^3 mol^{-1} cm^{-1}$. Values of ε_{max} range from close to zero (a very weak absorption) to >10 000 $dm^3 mol^{-1} cm^{-1}$ (an intense absorption).

Electronic absorption spectra may be presented as plots of absorbance against wavenumber or wavelength, or as plots of ε_{max} against wavenumber or wavelength (Fig. 4.17).

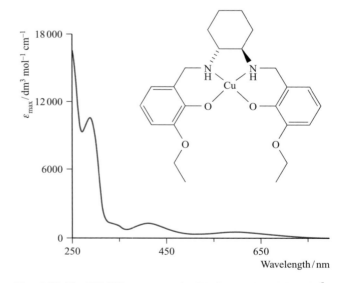

Fig. 4.17 The UV-VIS spectrum of a CH_2Cl_2 solution (8.8×10^{-5} mol dm^{-3}) of the copper(II) complex shown in the figure. Absorption maxima are at 252, 292, 347 (shoulder), 415 and 612 nm. [G. Zhang is thanked for recording the spectrum.]

The advantage of plotting ε_{max} rather than absorbance on the vertical axis is that spectra of different compounds can be readily compared with one another. Remember that ε_{max} is the characteristic of a given compound, whereas absorbance depends on both ε_{max} and concentration. The effect of plotting wavenumber, $\bar{\nu}$ (in cm^{-1}), rather than wavelength, λ (in nm), is to invert the spectrum; from eq. 4.11, it follows that 400 nm = 25 000 cm^{-1} and 200 nm = 50 000 cm^{-1}.

$$\bar{\nu} = \frac{1}{\lambda} \qquad (4.11)$$

When comparing spectra, we are often interested in whether bands shift to lower or higher energy. A *bathochromic effect* or *red shift* is a shift in an absorption towards the red end of the spectrum (longer wavelengths, lower energy). A *hypsochromic effect* or *blue shift* is a shift in a band towards shorter wavelength (i.e. towards the blue end of the spectrum, higher energy).

Worked example 4.7 The UV-VIS spectrum of a copper(II) complex

Figure 4.17 shows the UV-VIS spectrum of a CH_2Cl_2 solution of a copper(II) complex. A 1 cm cuvette was used for the measurement. Solutions of the complex are brown. The intense bands in the spectrum arise from ligand-based $\pi^* \longleftarrow \pi$ transitions. (a) Suggest how the $\pi^* \longleftarrow \pi$ transitions arise. (b) Which absorption or absorptions give rise to the observed colour of the complex in solution? (c) Calculate the absorbance that corresponds to the absorption at 292 nm.

(a) A $\pi^* \longleftarrow \pi$ transition can occur when an electron in a high-lying π-orbital is promoted to a low-lying, vacant π^*-orbital. The question states that the transitions are 'ligand-based'. Therefore, the π- and π^*-orbitals must be associated with the arene rings.

(b) The visible region is 400–700 nm. The absorption at 612 nm is in the visible region, and that at 415 nm is partly in the visible region.

(c) Use the Beer–Lambert law:

Absorbance, $A = \varepsilon \times c \times \ell$

ℓ is the path length = 1 cm

From Fig. 4.17, for the 292 nm absorption, $\ell \approx 10\,000$ dm^3 mol^{-1} cm^{-1} and $c = 8.8 \times 10^{-5}$ mol dm^{-3} (see the figure caption). Therefore:

$A = (10\,000 \text{ dm}^3 \text{ mol}^{-1} \text{ cm}^{-1}) \times (8.8 \times 10^{-5} \text{ mol dm}^{-3}) \times (1 \text{ cm})$
$= 0.88$ (dimensionless)

Self-study exercises

1. Show that 625 nm = 16 000 cm^{-1}.

2. The data in Fig. 4.17 are redrawn as a plot of ε_{max} against wavenumber. Sketch the appearance of this absorption spectrum, and include the scales.

3. Show that for the absorption at 415 nm in Fig. 4.17, $A \approx 0.1$.

Further details and applications of electronic absorption spectra and the assignment of absorptions are given in Sections 17.4 (charge transfer complexes of the halogens), 19.5 (colours of *d*-block metal complexes), 20.7 (electronic spectra of *d*-block metal complexes) and 27.4 (electronic spectra of *f*-block metal complexes).

Emission spectroscopy

The energy of the absorbed radiation corresponds to the energy of a transition from ground to an excited state. Decay of an excited state back to the ground state may take place by a radiative or non-radiative process. The spontaneous emission of radiation from an electronically excited species is called *luminescence* and this term covers both *fluorescence* and *phosphorescence*. A discussion of these phenomena requires an understanding of the electronic states of multi-electron systems, and we return to emission spectra in Section 20.8.

4.8 Nuclear magnetic resonance (NMR) spectroscopy

Nuclear magnetic resonance (NMR) spectroscopy is a resonance technique involving absorption of radiofrequency energy. The magnetic environment of a nucleus affects its resonance frequency and allows structural information to be deduced.

In this section we introduce one of the most important routine analytical tools of the synthetic chemist: nuclear magnetic resonance (NMR) spectroscopy. It is used to determine the relative numbers and environments of NMR active nuclei, and to investigate (usually in solution) the dynamic behaviour of molecular species. Although ^{1}H and ^{13}C NMR spectra are most commonly recorded, many more nuclei can be studied by this technique.

NMR active nuclei and isotope abundance

Many nuclei possess a property described as spin. The nuclear spin (nuclear angular momentum) is quantized and is described by the spin quantum number I which can have values of $0, \frac{1}{2}, 1, \frac{3}{2}, 2, \frac{5}{2}$ etc. If the value of I for a nucleus is zero, the nucleus is *NMR inactive*, e.g. ^{12}C. For both ^1H and ^{13}C, $I = \frac{1}{2}$ and these nuclei are *NMR active*. Later, we introduce other NMR active nuclei with different (non-zero) values of I. In the absence of an applied magnetic field, the different nuclear spin states of a nucleus are degenerate. However, when a magnetic field is applied, they are split (become non-degenerate) and this allows nuclear spin transitions to occur when radiofrequency (RF) radiation is absorbed.

When a ^1H NMR spectrum of a hydrogen-containing compound is recorded, virtually all the H atoms in the sample contribute to the observed spectrum; in a naturally occurring hydrogen sample, the abundance of ^1H is 99.985%. The fact that only 1% of naturally occurring carbon is ^{13}C means that if a ^{13}C NMR spectrum of a carbon-containing compound is recorded, only 1% of the carbon atoms present are observed. This has important consequences with respect to ^1H–^{13}C coupling as we see below.

Which nuclei are suitable for NMR spectroscopic studies?

A wide range of nuclei may be observed by NMR spectroscopy, but the inherent properties of some nuclei (e.g. a large quadrupole moment) may make their observation difficult. The main criterion is that the nucleus possesses a value of the nuclear spin quantum number $I \geq \frac{1}{2}$ (Table 4.3). Secondly, it is advantageous (but not essential) for the nucleus to occur in significant abundance. Carbon-13 is an example of a low-abundance isotope which is, nevertheless, extensively used for NMR spectroscopy. Isotopic enrichment may be used to improve signal:noise ratios. A third requirement is that the nucleus possesses a relatively short *spin-relaxation time* (τ_1). This property depends not only on the nucleus itself but also on its molecular environment. Some elements exhibit more than one NMR active nucleus and the choice for experimental observation may depend upon the relative inherent values of τ_1. For example, ^6Li and ^7Li are NMR active, but whereas τ_1 values for ^7Li

Table 4.3 Properties of selected NMR active nuclei.

Nucleus	Natural abundance / %	I	Frequency of observation / MHz (referred to ^1H at 100 MHz)[†]	Chemical shift reference (δ 0 ppm)[‡]
^1H	>99.9	$\frac{1}{2}$	100	$SiMe_4$
^2H	0.015	1	15.35	$SiMe_4$
^7Li	92.5	$\frac{3}{2}$	38.9	LiCl (1 M in H_2O)
^{11}B	80.1	$\frac{3}{2}$	32.1	$F_3B \cdot OEt_2$
^{13}C	1.1	$\frac{1}{2}$	25.1	$SiMe_4$
^{17}O	0.04	$\frac{5}{2}$	13.5	H_2O
^{19}F	100	$\frac{1}{2}$	94.0	$CFCl_3$
^{23}Na	100	$\frac{3}{2}$	26.45	NaCl (1 M in H_2O)
^{27}Al	100	$\frac{5}{2}$	26.1	$[Al(OH_2)_6]^{3+}$
^{29}Si	4.67	$\frac{1}{2}$	19.9	$SiMe_4$
^{31}P	100	$\frac{1}{2}$	40.5	H_3PO_4 (85%, aq)
^{77}Se	7.6	$\frac{1}{2}$	19.1	$SeMe_2$
^{103}Rh	100	$\frac{1}{2}$	3.2	Rh (metal)
^{117}Sn	7.68	$\frac{1}{2}$	35.6	$SnMe_4$
^{119}Sn	8.58	$\frac{1}{2}$	37.3	$SnMe_4$
^{129}Xe	26.4	$\frac{1}{2}$	27.7	$XeOF_4$
^{183}W	14.3	$\frac{1}{2}$	4.2	Na_2WO_4 (in D_2O)
^{195}Pt	33.8	$\frac{1}{2}$	21.5	$Na_2[PtCl_6]$
^{199}Hg	16.84	$\frac{1}{2}$	17.9	$HgMe_2$

[†] The operating frequency of an instrument is defined by the field of the magnet and is designated by the frequency at which the ^1H nuclei of $SiMe_4$ resonate.
[‡] It is important to quote the reference when reporting NMR spectra since alternative references may be used.

are typically <3 s, those for ^{6}Li lie in the range $\approx 10\text{--}80$ s. ^{7}Li is thus more appropriate for NMR spectroscopic observation and this choice is also favoured by the fact that ^{7}Li is more abundant (92.5%) than ^{6}Li. Another nuclear property that may militate against easy observation is the *quadrupole moment* arising from a non-spherical charge distribution of the nucleus and which is associated with values of $I > \frac{1}{2}$. Although the possession of a quadrupole moment leads to short values of τ_1, it generally causes the signals in the NMR spectrum to be broad (e.g. ^{11}B). Signal broadening is also seen in the spectra of nuclei *attached* to nuclei with quadrupole moments, e.g. the ^{1}H NMR spectrum of protons attached to ^{11}B.

Resonance frequencies and chemical shifts

A particular nucleus (e.g. ^{1}H, ^{13}C, ^{31}P) absorbs characteristic radiofrequencies, i.e. it *resonates* at a characteristic frequency. If an NMR spectrometer is tuned to a particular resonance frequency, *only* a selected NMR active nucleus is observed. For example, only ^{1}H nuclei are observed if a 400 MHz spectrometer is tuned to 400 MHz, but if the same spectrometer is retuned to 162 MHz, only ^{31}P nuclei are observed. This is analogous to tuning a radio and receiving only one station at a time.

In a ^{1}H NMR experiment, protons in different chemical environments resonate at different frequencies. The same is true of, for example, non-equivalent ^{13}C nuclei in a ^{13}C NMR experiment, or non-equivalent ^{19}F nuclei in a ^{19}F NMR spectroscopic experiment, and so on. Each signal in an NMR spectrum is denoted by a *chemical shift value*, δ, a value that is given relative to the signal observed for a specified reference compound (see below).

The parameter δ is independent of the applied magnetic field strength and is defined as follows. The frequency difference $(\Delta\nu)$, in Hz, between the signal of interest and some defined reference frequency (ν_0) is divided by the absolute frequency of the reference signal (eq. 4.12):

$$\delta = \frac{(\nu - \nu_0)}{\nu_0} = \frac{\Delta\nu}{\nu_0} \tag{4.12}$$

Typically, this leads to a very small number. In order to obtain a more convenient number for δ, it is usual to multiply the ratio in eq. 4.12 by 10^6. This gives δ in units of parts per million, ppm. The IUPAC[†] defines δ according to eq. 4.12, but eq. 4.13 gives a method of calculating δ in ppm.

$$\delta \text{ in ppm} = \frac{(\nu - \nu_0) \text{ in Hz}}{\nu_0 \text{ in MHz}} \tag{4.13}$$

It follows that you use eq. 4.14 to work out the frequency difference between two spectroscopic peaks in Hz when

the chemical shifts have been measured in ppm.

$$\Delta\nu \text{ (in Hz)} = \text{(spectrometer frequency in MHz)} \\ \times \Delta\delta \text{ (in ppm)} \tag{4.14}$$

The standard reference (for which δ is defined as 0 ppm) for both ^{1}H and ^{13}C NMR spectroscopies is tetramethylsilane, SiMe$_4$ (TMS). When the NMR spectrum of a compound is recorded, signals due to particular nuclei are said to be *shifted* with respect to the standard reference signal. A shift to more positive δ is 'shifted to higher frequency' and a shift to negative (or to less positive) δ is 'shifted to lower frequency'. Older terminology which may still be encountered relates a positive δ value to a 'downfield shift' and a negative δ value to an 'upfield shift'.

Chemical shift ranges

The range of chemical shifts over which NMR spectroscopic signals appear is dependent on the nucleus. The most commonly observed nucleus is ^{1}H and, in organic compounds, a *spectral window* from δ +15 to 0 ppm usually encompasses most signals. In inorganic compounds, the window may have to be widened if, for example, ^{1}H nuclei attached to metal centres are to be observed, or if signals are *paramagnetically shifted* (see Box 4.2). The chemical shift range for ^{13}C NMR spectra is typically δ +250 to -50 ppm, for ^{31}P NMR spectra, $\approx\delta$ +300 to -300 ppm, and for ^{77}Se NMR spectra $\approx\delta$ +2000 to -1000 ppm. Figure 4.18 illustrates the change in chemical shift for the ^{31}P nucleus on going from triphenylphosphane to the corresponding oxide. Such a shift to higher frequency is typical when a tertiary phosphane (R$_3$P) is oxidized, and also tends to occur when a phosphane ligand coordinates to a *d*-block metal centre.

Solvents for solution studies

Samples for solution NMR spectroscopy are generally prepared using *deuterated solvents*. One reason for this is that, were non-deuterated solvents to be used (e.g. CH$_3$Cl in place of CD$_3$Cl) for a ^{1}H NMR spectroscopic experiment, the signals due to the solvent would 'swamp' those due to the sample. Deuterated solvents are commercially available, typically with $>99.5\%$ ^{2}H label incorporated. The remaining unlabelled compound provides a useful *internal reference* signal in the ^{1}H NMR spectrum of the sample under study.

Integration of signals and signal broadening

Under normal conditions of measuring ^{1}H NMR spectra, the ratio of the peak areas (*integrals*) of the signals in the spectrum is proportional to the number of nuclei giving rise to the signals. For example, in a ^{1}H NMR spectrum of HC≡CCH$_3$, two signals with relative integrals 1:3 are observed. However, the integration of signals must be treated

---

[†] R.K. Harris, E.D. Becker, S.M. Cabral de Menezes, R. Goodfellow and P. Granger (2001) *Pure and Applied Chemistry*, vol. 73, p. 1795 – 'NMR nomenclature. Nuclear spin properties and conventions for chemical shifts (IUPAC recommendations 2001)'.

THEORY

Box 4.2 Paramagnetically shifted ¹H NMR spectra

The presence of a paramagnetic *centre* (i.e. a centre with one or more unpaired electrons) in a compound has significant consequences on the ¹H NMR spectrum of the compound. Firstly, the *local magnetic field* at each ¹H nucleus is affected. The energy difference between nuclear spin states – a consequence of applying an external magnetic field in an NMR experiment – arises from the interaction of the magnetic fields of the spinning nuclei with the applied field. However, the local field experienced by the nuclei is not the same as the applied field because electron pairs in the vicinity of the ¹H nucleus generate small local magnetic fields. The local magnetic field is the sum of the applied and all the smaller fields. The latter depend on the chemical environment of the ¹H nucleus. Typically, the differences in local magnetic fields for protons in different environments are small and, as a consequence, the chemical shift range over which the ¹H NMR signals occur is not large. In a paramagnetic compound, there is an additional factor: a large, local magnetic field arising from the unpaired

electron or electrons on the paramagnetic centre. This contributes to the energy difference between nuclear spin states, and as a consequence, the chemical shift range for the ¹H NMR signals is much larger than in a diamagnetic compound. The second effect that is observed in ¹H NMR spectra of paramagnetic compounds is a broadening of the signals. This effect has its origins in a significant shortening of the excited state lifetime, i.e. the relaxation time is very short. In some cases, the broadening is so great that no well-resolved signals are observed.

An example of a paramagnetic centre is a Co^{2+} ion which, in an octahedral complex, has one or three unpaired electrons (see Chapter 20). The figure below shows the ¹H NMR spectrum of the Co^{2+} complex $[Co(phen)_3]^{2+}$ (phen = 1,10-phenanthroline), the structure of which is shown below. There are four different aromatic proton environments in the complex, and the chemical shifts of the signals assigned to these ¹H nuclei fall in the range δ +110 to +15 ppm.

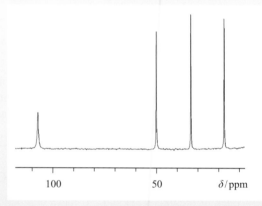

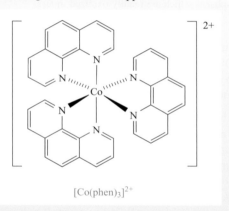

$[Co(phen)_3]^{2+}$

[Barbara Brisig is acknowledged for recording the spectrum shown above.]

Further reading

I. Bertini and C. Luchinat (1996) *Coord. Chem. Rev.*, vol. 150 – 'NMR of paramagnetic substances'.

---

with caution since the peak integral is dependent upon the *relaxation time* of the nucleus in question, i.e. the time taken for the nucleus to relax from an excited to ground state during the NMR spectroscopic experiment. (Further details of this phenomenon may be found in references cited at the end of this chapter.) One particular problem is the relative integrals of signals in a ¹³C NMR spectrum.

In some cases, signals may be broadened and this can affect the measurement of the relative integrals of signals. For example, signals arising from protons attached to N are broadened due to *quadrupolar relaxation* by ¹⁴N ($I = 1$). Exchange with solvent protons also causes broadening, e.g.:

$$CH_3CH_2OH + HOH \rightleftharpoons CH_3CH_2OH + HOH$$

Homonuclear spin–spin coupling: ¹H–¹H

A ¹H nucleus ($I = \frac{1}{2}$) may be in one of two spin states ($m_I = +\frac{1}{2}$, $m_I = -\frac{1}{2}$) and the energy difference between the spin states depends on the applied magnetic field of the NMR spectrometer. Consider a system in which there are two magnetically non-equivalent ¹H nuclei, H_A and H_B. There are two possible situations:

- The local magnetic field generated by the spin of H_A is *not* detected by H_B; the ¹H NMR spectrum consists of two resonances, each a *singlet* because there is *no coupling* between the two ¹H nuclei.
- H_A *is* affected by the magnetic fields associated with H_B; the ¹H NMR signal for H_A is *split into two equal lines*

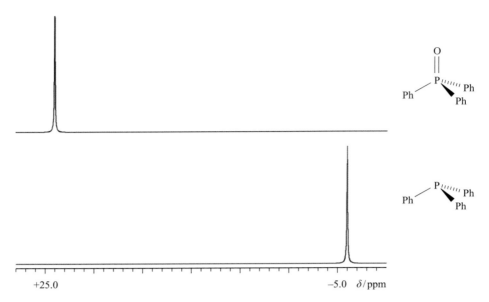

Fig. 4.18 The 162 MHz ^{31}P NMR spectra of PPh$_3$ and O=PPh$_3$. A shift to more positive δ (higher frequency) generally accompanies the oxidation of a tertiary phosphane and recording the ^{31}P NMR spectrum of a phosphane before use in the laboratory is an easy way of checking the purity of phosphanes which are readily oxidized in air.

depending on which of the two spin states of H$_B$ (equal probabilities) it 'sees'. Similarly, the signal for H$_B$ is composed of two equal lines. Protons H$_A$ and H$_B$ *couple* with each other and the spectrum consists of two *doublets*.

The separation between the two lines in each of the doublets described above must be equal, and this splitting is called the *coupling constant*, *J*, and is measured in hertz (Hz). In general, coupling to one proton gives a doublet, to two equivalent protons gives a triplet, to three equivalent protons gives a quartet, and so on. The relative intensities of the lines in the *multiplet* are given by a binomial distribution, readily determined using a Pascal's triangle:

$$
\begin{array}{ccccccccc}
 & & & & 1 & & & & \leftarrow \text{singlet} \\
 & & & 1 & & 1 & & & \leftarrow \text{doublet} \\
 & & 1 & & 2 & & 1 & & \leftarrow \text{triplet} \\
 & 1 & & 3 & & 3 & & 1 & \leftarrow \text{quartet} \\
1 & & 4 & & 6 & & 4 & & 1 \leftarrow \text{quintet} \\
\end{array}
$$

The number and spins of the *attached nuclei* determine the *multiplicity* (number of lines) and pattern of the NMR spectroscopic signal of the observed nucleus. The coupling constant between nuclei X and Y is denoted as J_{XY} and is measured in Hz.

In general the multiplicity of an NMR spectroscopic signal can be determined using eq. 4.15 where the nucleus being observed is coupling to n equivalent nuclei with quantum number *I*.

Multiplicity (number of lines) $= 2nI + 1$ (4.15)

Self-study exercise

The 100 MHz ^{1}H NMR spectrum of butanone is shown in Fig. 4.19 and consists of a quartet, a singlet and a triplet. The coupling constants *J* for the triplet and quartet are equal. Account for the observed spectrum.

Heteronuclear spin–spin coupling: ^{13}C–^{1}H

Each of the nuclei ^{1}H and ^{13}C has a magnetic spin quantum number $I = \frac{1}{2}$, and when ^{13}C and ^{1}H nuclei are in close proximity, they can couple. However, in molecules containing a natural isotopic distribution of carbon atoms, only 1% are ^{13}C nuclei. From a statistical consideration, it follows that in a ^{1}H NMR spectrum of, for example, acetone, ^{13}C–^{1}H coupling is *not* observed, although it *is* observed in the ^{13}C NMR spectrum of the *same* sample. The ^{13}C NMR spectrum of acetone exhibits a singlet due to the C=O carbon atom, and a quartet due to the two equivalent methyl ^{13}C nuclei.

Self-study exercise

Why do you not observe ^{13}C–^{13}C coupling in the ^{13}C NMR spectrum of acetone?

The following case studies illustrate the use of NMR spectroscopy to study compounds containing spin-active nuclei other than ^{1}H and ^{13}C.

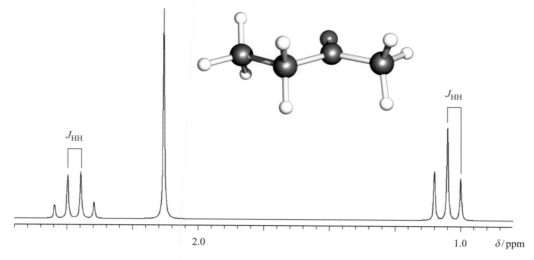

 Fig. 4.19 The 100 MHz NMR spectrum of butanone. In the structure, the colour code is: C, grey; H, white; O, red.

Case studies

Case study 1: ^{31}P NMR spectrum of [PF$_6$]$^-$

The ^{31}P NMR spectrum of a salt containing the octa-hedral [PF$_6$]$^-$ ion exhibits a binomial septet (Fig. 4.20) consistent with six equivalent ^{19}F nuclei ($I = \frac{1}{2}$) attached to the central ^{31}P centre. The large value of J_{PF} 708 Hz is typical of ^{31}P–^{19}F coupling constants for *directly attached* nuclei; the magnitudes of coupling constants usually diminish with nuclear separation, but a consequence of large values for directly attached nuclei is that *long-range couplings* may be observed (see Case study 2).

Case study 2: ^{31}P NMR spectrum of Ph$_2$PCH$_2$CH$_2$P(Ph)CH$_2$CH$_2$PPh$_2$

(4.9)

Structure **4.9** shows that Ph$_2$PCH$_2$CH$_2$P(Ph)CH$_2$CH$_2$PPh$_2$ contains two phosphorus environments, labelled a and b.

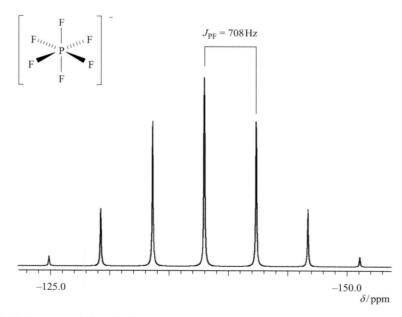

Fig. 4.20 The 162 MHz ^{31}P NMR spectrum of a salt of [PF$_6$]$^-$ consists of a binomial septet. The value of J_{PF} can be measured between any pair of adjacent lines in the signal.

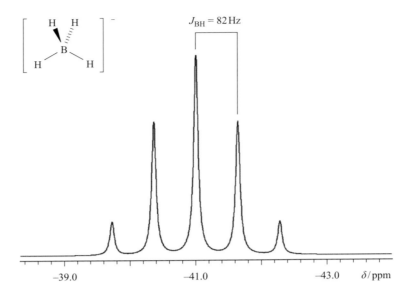

Fig. 4.21 The 128 MHz ^{11}B NMR spectrum of a solution of NaBH$_4$ in CD$_3$C(O)CD$_3$. The value of J_{BH} can be measured between any pair of adjacent lines in the signal.

The ^{31}P NMR spectrum exhibits two signals with an integral ratio of 1:2. For directly attached inequivalent phosphorus atoms, values of J_{PP} are typically 450–600 Hz; in compound **4.9**, *long-range coupling* between non-equivalent ^{31}P nuclei is observed. The signals due to atoms P$_b$ and P$_a$ are a triplet and doublet respectively; values of J_{PP} (29 Hz) measured from the two signals are necessarily equal. Additionally, coupling between the ^{31}P and closest ^{1}H nuclei may be observed. Two types of heteronuclear NMR spectra are routinely recorded: one in which coupling to protons is observed and one in which protons are instrumentally *decoupled* from the observed nucleus.

> The notation ^{31}P{^{1}H} means ***proton-decoupled*** ^{31}P;
> corresponding notations are used for other proton-decoupling.

Case study 3: ^{11}B NMR spectrum of [BH$_4$]$^-$

The ^{11}B NMR spectrum of Na[BH$_4$] is shown in Fig. 4.21. The 1:4:6:4:1 pattern of signal integrals corresponds to the binomial quintet expected for four equivalent ^{1}H nuclei coupling to ^{11}B. Although $I = \frac{3}{2}$ for ^{11}B, it is the $I = \frac{1}{2}$ of the attached protons that determines the nature of the signal in the ^{11}B NMR spectrum of [BH$_4$]$^-$.

Case study 4: ^{31}P{^{1}H} NMR spectrum of PhMe$_2$P·BH$_3$

Figure 4.22 shows the structure of the adduct PhMe$_2$P·BH$_3$ and its ^{31}P{^{1}H} NMR spectrum. The signal is a four-line multiplet (but *not* a binomial quartet) and arises primarily from coupling between ^{31}P and ^{11}B nuclei. For ^{11}B, $I = \frac{3}{2}$;

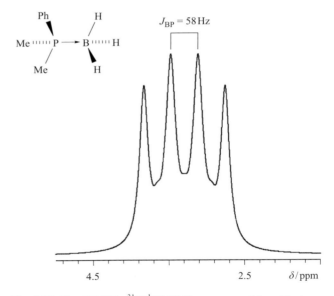

Fig. 4.22 The 162 MHz ^{31}P{^{1}H} NMR spectrum of the adduct PhMe$_2$P·BH$_3$. The four-line pattern is *not* a binomial quartet but an approximate 1:1:1:1 multiplet.

this means there are four spin states with values $+\frac{3}{2}$, $+\frac{1}{2}$, $-\frac{1}{2}$ and $-\frac{3}{2}$. There is an *equal probability* that the ^{31}P nucleus will 'see' the ^{11}B nucleus in each of the four spin states, and this gives rise to the ^{31}P signal being split into four equal intensity lines: a 1:1:1:1 multiplet. The observed signal is complicated by the fact that ^{11}B has an 80% abundance and the second isotope, ^{10}B, is also NMR active ($I = 3$). It too couples to the ^{31}P nucleus, giving a seven-line multiplet (1:1:1:1:1:1:1), but the value of $J_{^{31}P^{10}B}$ is smaller than $J_{^{31}P^{11}B}$. The result is two overlapping signals, but the dominant feature is the 1:1:1:1 multiplet,

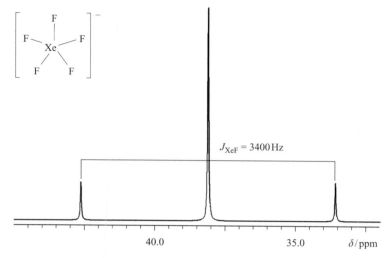

Fig. 4.23 The 376 MHz ^{19}F NMR spectrum of $[XeF_5]^-$, simulated using literature parameters. The isotopic abundance of ^{129}Xe is 26.4%; the centre of the doublet coincides with the position of the singlet. (K.O. Christe *et al.* (1991) *J. Am. Chem. Soc.*, vol. 113, p. 3351.)

the signal shape of which is affected by both the underlying seven-line multiplet and relaxation effects.

Case study 5: ^{19}F NMR spectrum of $[XeF_5]^-$

The planar $[XeF_5]^-$ ion contains five equivalent F atoms (see worked example 2.7). Both the ^{19}F and ^{129}Xe nuclei are NMR active: ^{19}F, $I = \frac{1}{2}$, 100% abundance; ^{129}Xe, $I = \frac{1}{2}$, 26.4%. The ^{19}F NMR spectrum of $[XeF_5]^-$ is shown in Fig. 4.23. The chemical equivalence of the ^{19}F nuclei gives rise to one signal. However, 26.4% of the F centres are attached to ^{129}Xe, while the remainder are bonded to other Xe nuclei. The spectrum can be interpreted in terms of a singlet (the central line) due to 73.6% of the ^{19}F nuclei, plus an overlapping doublet due to the 26.4% of the ^{19}F nuclei that couple to ^{129}Xe. The centre of the doublet coincides with the position of the singlet because *all* the ^{19}F nuclei resonate at the same frequency. The two side peaks in Fig. 4.23 are called *satellite peaks*.

Stereochemically non-rigid species

The NMR spectroscopic examples discussed so far have assumed that, with the exception of free rotation about single bonds, the molecule or ion is static in solution. For the majority of organic and inorganic species, this assumption is valid, but the possibility of *stereochemical non-rigidity* (*fluxionality*) *on the NMR spectroscopic timescale* must be considered. Five-coordinate species such as Fe(CO)$_5$, **4.10**, PF$_5$, **4.11**, and BrF$_5$, **4.12**, constitute one group of compounds for which the activation barrier for dynamic behaviour in solution is relatively low, and exchange of substituent positions is facile.

(4.10)   **(4.11)**   **(4.12)**

The inclusion of the qualifier 'on the NMR spectroscopic timescale' is important. The timescale[†] of the NMR spectroscopic technique (10^{-1} to 10^{-5} s, depending on the observed nucleus) is relatively long, and is significantly longer than that of IR spectroscopy; Fe(CO)$_5$ appears static on the IR spectroscopic timescale, but dynamic within the timescale of a ^{13}C NMR spectroscopic experiment. Lowering the temperature slows down the dynamic behaviour, and *may* make it slower than the spectroscopic timescale. However, some fluxional processes have very low energy barriers. Even at 103 K, the axial and equatorial CO groups in Fe(CO)$_5$ exchange positions and the ^{13}C NMR spectrum consists of one signal corresponding to the average ^{13}C environment. On the other hand, the room temperature solution ^{19}F NMR spectrum of BrF$_5$ exhibits a doublet and a binomial quintet (due to ^{19}F–^{19}F coupling) with relative integrals of 4:1, and this is consistent with structure **4.12**. Above 450 K, one signal is observed, indicating that the five F atoms are equivalent on the NMR timescale, i.e. the BrF$_5$ molecule is fluxional. On going from the low to high temperature limit, the two signals *coalesce* to give a single resonance.

[†] See: A.B.P. Lever (2003) in *Comprehensive Coordination Chemistry II*, eds J.A. McCleverty and T.J. Meyer, Elsevier, Oxford, vol. 2, p. 435 – 'Notes on time frames'.

BIOLOGY AND MEDICINE
Box 4.3 Magnetic resonance imaging (MRI)

Magnetic resonance imaging (MRI) is a rapidly expanding clinical technique to obtain an image of, for example, a human organ or tumour. In 2003, approximately 10 000 MRI units performed ≈75 million scans worldwide and since then these numbers have risen dramatically. The impact of this non-invasive technique on clinical medicine was recognized by the award of the 2003 Nobel Prize in Medicine to Paul Lauterbur and Peter Mansfield. The MRI image is generated from information obtained from the ^{1}H NMR spectroscopic signals of water. The signal intensity depends upon the proton relaxation times and the concentration of water. The relaxation times can be altered, and the image enhanced, by using *MRI contrast agents*. Coordination complexes containing paramagnetic Gd^{3+}, Fe^{3+} or Mn^{2+} are potentially suitable as contrast agents and, of these, complexes containing the Gd^{3+} ion have so far proved to be especially useful. As a free ion, Gd^{3+} is extremely toxic and, therefore, to minimize side-effects in patients, Gd^{3+} must be

introduced in the form of a complex that will not dissociate in the body. Chelating ligands are particularly suitable (see Chapter 7 for a discussion of stability constants). Ligands such as $[DTPA]^{5-}$ (the conjugate base of H_5DTPA, drawn below) possess both *O*- and *N*-donor atoms, and form gadolinium(III) complexes in which the Gd^{3+} ion exhibits a high coordination number (see Chapter 19). For example, in $[Gd(DTPA)(OH_2)]^{2-}$, the Gd^{3+} centre is 9-coordinate. The complex $[Gd(DTPA)(OH_2)]^{2-}$ was approved in 1988 for medical use as an MRI contrast agent and is used under the brand name of Magnevist. Two other approved contrast agents are $[Gd(DTPA-BMA)(OH_2)]$ (trade name Omniscan) and $[Gd(HP-DO3A)(OH_2)]$ (ProHance). The solid state structure of $[Gd(DTPA-BMA)(OH_2)]$ is shown below, and confirms a 9-coordinate metal centre. Magnevist, Omniscan and ProHance are classed as *extra-cellular* contrast agents, meaning that, once injected into a patient, they are distributed non-specifically

H₅DTPA

H₃DTPA-BMA

H₅BOPTA

H₃HP-DO3A

The molecular structure of [Gd(DTPA-BMA)(OH₂)] (Omniscan) determined by X-ray diffraction [A. Aukrust *et al.* (2001) *Org. Proc. Res. Develop.*, vol. 5, p. 361]. Colour code: Gd, green; N, blue; O, red; C, grey; H, white.

throughout the plasma and extra-cellular fluids in the body. Clearance through the kidneys occurs rapidly, the elimination half-life being ≈ 90 minutes.

Two other types of MRI contrast agents are *hepatobiliary* and *blood pool* agents. Hepatocytes are the main cell types present in the liver, and a hepatobiliary contrast agent is designed to target the liver, and then be excreted through the bile ducts, gall bladder and intestines. The gadolinium(III) complex [Gd(BOPTA)(OH$_2$)]$^{2-}$ (brand name Multihance) is an approved hepatobiliary contrast agent. [BOPTA]$^{5-}$ is the conjugate base of H$_5$BOPTA, the structure of which is shown opposite. [BOPTA]$^{5-}$ is structurally similar to [DTPA]$^{5-}$, differing only in the presence of the pendant hydrophobic group which is responsible for making Multihance cell-specific. Blood pool contrast agents remain intravascular for a significant period of time. In 2005, MS-325 (trade name Vasovist) received approval by the European Commission and US Food and Drug Administration for medical use. The chelating ligand in Vasovist is structurally related to [DTPA]$^{5-}$, but carries a phosphate-containing group. The structure of the ligand, L^{6-}, in Vasovist is shown below. Vasovist itself is the sodium salt Na$_3$[GdL(OH$_2$)], and the ligand binds Gd^{3+} in the same environment as in Omniscan (structure opposite).

The introduction of the diphenylcyclohexylphosphate group increases the hydrophilic character of the contrast agent and this allows Vasovist to bind reversibly to human serum albumin. This results in enhanced images of vascular structures. This particular type of MR imaging is known as *magnetic resonance angiography* (MRA) and is a significant development in imaging techniques. Prior to the availability of non-invasive MRA, vascular structures could only be imaged by conventional angiography, an invasive procedure that involves injection into the blood of a substance that absorbs X-rays. An angiogram is then obtained by exposing the patient to X-rays.

A second class of MRI contrast agents consists of iron oxide nanoparticles (magnetite, Fe$_3$O$_4$, and maghemite, γ-Fe$_2$O$_3$). These paramagnetic materials are able to shorten proton relaxation times. They are categorized according to particle size: superparamagnetic iron oxide (SPIO) nanoparticles have diameters of 50 nm to hundreds of nanometres, whereas ultrasmall superparamagnetic iron oxide (USPIO) nanoparticles are <50 nm in diameter. The nanoparticles can be prepared by a number of routes, the simplest of which involves precipitation from aqueous solution under carefully controlled pH conditions starting with a mixture of iron(II) and iron(III) salts (e.g. chlorides):

$$2Fe^{3+} + Fe^{2+} + 8[OH]^- \longrightarrow Fe_3O_4 + 4H_2O$$

$$Fe_3O_4 + 2H^+ \longrightarrow \gamma\text{-}Fe_2O_3 + Fe^{2+} + H_2O$$

The iron oxide nanoparticles are coated in polysaccharides (dextran or carboxydextran) to make them biocompatible and are administered to patients as aqueous colloids. Clinical MRI contrast agents of this type include Feridex, Resovist and Combidex which are selectively taken up by the liver, spleen and bone marrow. The synthesis and potential applications in MRI of related iron oxide nanoparticles is currently an active area of research.

Doctors with a patient entering an MRI scanner.

Dependence upon the observation of proton signals in some organs (e.g. lungs) presents problems with respect to MRI. The use of ^{129}Xe magnetic imaging has been tested as a means of overcoming some of the difficulties associated with proton observation. Under the right conditions, gaseous ^{129}Xe taken into mouse lungs allows excellent images to be observed.

Further reading

A. Accardo, D. Tesauro, L. Aloj, C. Pedone and G. Morelli (2009) *Chem. Rev.*, vol. 253, p. 2193 – 'Supramolecular aggregates containing lipophilic Gd(III) complexes as contrast agents in MRI'.

P. Hermann, J. Kotek, V. Kubíček and I. Lukeš (2008) *Dalton Trans.*, p. 3027 – 'Gadolinium(III) complexes as MRI contrast agents: ligand design and properties of the complexes'.

S. Laurent, D. Forge, M. Port, A. Roch, C. Robic, L. Vander Elst and R.N. Muller (2008) *Chem. Rev.*, vol. 108, p. 2064 – 'Magnetic iron oxide nanoparticles: synthesis, stabilization, vectorization, physicochemical characterizations and biological applications'.

H.B. Na and T. Hyeon (2009) *J. Mater. Chem.*, vol. 19, p. 6267 – 'Nanostructured T1 MRI contrast agents'.

H.B. Na, I.C. Song and T. Hyeon (2009) *Adv. Mater.*, vol. 21, p. 2133 – 'Inorganic nanoparticles for MRI contrast agents'.

E.L. Que and C.J. Chang (2010) *Chem. Soc. Rev.*, vol. 39, p. 51 – 'Responsive magnetic resonance imaging contrast agents as chemical sensors for metals in biology and medicine'.

O. Taratula and I.J. Dmochowski (2010) *Curr. Opin. Chem. Biol.*, vol. 14, p. 97 – 'Functionalized ^{129}Xe contrast agents for magnetic resonance imaging'.

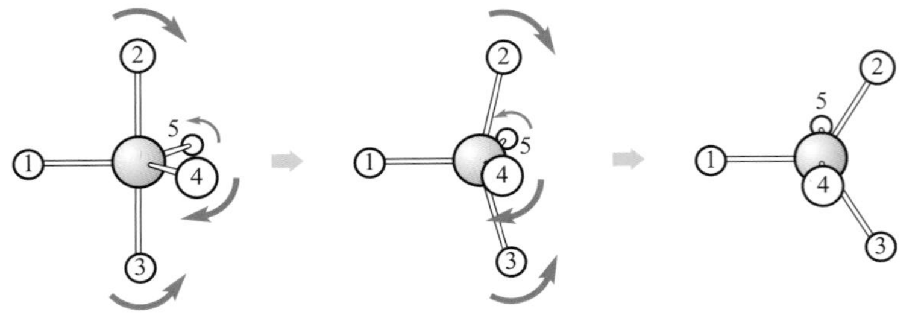

Fig. 4.24 Berry pseudo-rotation interconverts one trigonal bipyramidal structure into another via a square-based pyramidal transition state. The numbering scheme illustrates that axial and equatorial sites in the trigonal bipyramid are interchanged.

The usual dynamic process in which 5-coordinate species are involved in solution is *Berry pseudo-rotation.*[†] Although ligand–ligand repulsions are minimized in a trigonal bipyramidal arrangement, only a small amount of energy is needed to convert it into a square-based pyramid. The interconversion involves small perturbations of the bond angles subtended at the central atom, and continued repetition of the process results in each substituent 'visiting' both equatorial and axial sites in the trigonal bipyramidal structure (Fig. 4.24).

Exchange processes in solution

A number of hydrated cations in aqueous solution undergo exchange with the solvent at rates slow enough to be observed on the NMR spectroscopic timescale by using ^{17}O isotopic labelling; ^{17}O has $I = \frac{5}{2}$, while both ^{16}O and ^{18}O are NMR inactive. Different chemical shifts are observed for the ^{17}O nuclei in bulk and coordinated water, and from the signal intensity ratios, hydration numbers can be obtained. For example, Al^{3+} has been shown to be present as $[Al(OH_2)_6]^{3+}$.

Reactions such as that in eq. 4.16 are known as *redistribution reactions.*

$$PCl_3 + P(OEt)_3 \rightleftharpoons PCl_2(OEt) + PCl(OEt)_2 \qquad (4.16)$$

A **redistribution reaction** is one in which substituents exchange between species but the types and numbers of each type of bond remain the same.

The position of equilibrium can be followed by using ^{31}P NMR spectroscopy, since each of the four species has a characteristic chemical shift. Rate data are obtained by following the variation in relative signal integrals with time, and equilibrium constants (and hence values of ΔG° since $\Delta G^{\circ} = -RT \ln K$) can be found from the relative signal integrals when no further change takes place (i.e. equilibrium has been established). By determining ΔG° at different temperatures, values of ΔH° and ΔS° can be found using eqs. 4.17 and 4.18.

$$\Delta G^{\circ} = \Delta H^{\circ} - T\Delta S^{\circ} \qquad (4.17)$$

$$\frac{d \ln K}{dT} = \frac{\Delta H^{\circ}}{RT^2} \qquad (4.18)$$

Values of ΔH° for these types of reactions are almost zero, the redistribution of the groups being driven by an increase in the entropy of the system.

4.9 Electron paramagnetic resonance (EPR) spectroscopy

What is EPR spectroscopy?

Electron paramagnetic resonance (EPR) spectroscopy is a resonance technique involving microwave-induced transitions between magnetic energy levels of electrons which possess a net spin and orbital angular momentum. An EPR spectrum provides information about *paramagnetic species*.

Electron paramagnetic resonance (EPR) spectroscopy (also called electron spin resonance (ESR) spectroscopy), is used to study paramagnetic species with one or more unpaired electrons, e.g. free radicals, diradicals, metal complexes containing paramagnetic metal centres, defects in semiconductors and irradiation effects in solids. While diamagnetic materials are EPR silent, paramagnetic species always exhibit an EPR spectrum. This consists of one or more lines, depending on the interactions between the unpaired electron (which acts as a 'probe') and the molecular framework in which it is located. Analysis of the shape of the EPR spectrum (the number and positions of EPR lines, their intensities and line widths) provides information

---

[†] A discussion that goes beyond Berry pseudo-rotation and considers the 'lever mechanism' in SF_4 (based on a trigonal bipyramidal structure with an equatorial site occupied by a lone pair of electrons) and related species is: M. Mauksch and P. von R. Schleyer (2001) *Inorg. Chem.*, vol. 40, p. 1756.

about the paramagnetic species, e.g. the structure of a free radical, characterization of the coordination sphere around the metal centre in a coordination complex, or the presence of multiple paramagnetic species.

EPR spectroscopic measurements can be performed at high, room or low (≥ 4 K) temperature. Samples may be solid (single crystal or powder) or liquid (fluid or frozen solution, a 'glass'). In this introduction to EPR spectroscopy, we shall be concerned only with *magnetically dilute systems* in which the unpaired electrons are involved in intramolecular (not intermolecular) interactions. We shall focus attention on the application of the technique to mononuclear, metal-containing systems.

The Zeeman electronic effect

For a paramagnetic metal ion such as Ti^{3+} (d^1), V^{4+} (d^1) or Cu^{2+} (d^9) with a single unpaired electron, the total spin quantum number $S = \frac{1}{2}$. There are two possible spin states: $M_S = +\frac{1}{2}$ and $M_S = -\frac{1}{2}$ (see Box 1.4 for one-electron systems and Section 20.6 for quantum numbers for multi-electron systems). In the absence of a magnetic field, these states are degenerate. Consider a one-electron case. By applying a magnetic field, B_0, the interaction between the unpaired electron and the magnetic field leads to a splitting of the energy levels (Fig. 4.25). This is called the Zeeman electronic effect and the energy difference, ΔE, is given by eq. 4.19.

$$\Delta E = g\mu_B B_0 \qquad (4.19)$$

where: g = Landé g-factor ('g-value')

B_0 = applied magnetic field (in tesla, T)

μ_B = Bohr magneton
   ($1\mu_B = eh/4\pi m_e = 9.2740 \times 10^{-24}$ J T^{-1})

The g-value is given by the ratio $2\mu_e/\mu_B$ where μ_e is the electron magnetic moment (9.2848×10^{-24} J T^{-1}); g is dimensionless. For a free electron, $g = 2.0023$. For a metal ion, spin–orbit coupling (see Chapter 20) leads to g-values that are significantly different from that of a free electron. The energy separation between the α and β states (Fig. 4.25) corresponds to the microwave region of the elec-

tromagnetic spectrum. Thus, by supplying appropriate microwave radiation to the sample, electron spin transitions between the two energy states occur. The system is then *in resonance*, and the recording of these transitions represents the EPR spectrum. (Compare this with the nuclear spin transitions resulting from radiofrequency radiation in NMR spectroscopy described in Section 4.8.) Usually, an EPR spectrometer operates at a constant microwave frequency (measured in gigahertz, GHz) and the magnetic field (measured in gauss or tesla, 1 G $= 10^{-4}$ T) is varied until the energy separation of the two spin states coincides with the microwave radiation energy. Standard EPR spectrometers operate at 9–10 GHz (so-called 'X-band'), but there are also domains of lower and higher microwave frequencies: 1–2 GHz (L-band), 2–4 GHz (S-band), 35 GHz (Q-band) and 95 GHz (W-band). Recently developed FT-EPR spectrometers (as opposed to continuous wave instruments) produce increased spectral resolution and their use has widened the scope of systems that can be investigated (e.g. the second coordination sphere around a paramagnetic metal centre in a metalloprotein).

EPR spectra

The form in which an EPR spectrum is recorded is the first derivative of an absorption peak (Fig. 4.26), because in this form the detection is more sensitive and the signal : noise ratio is improved due to intrinsic electronic properties resulting from modulation of the magnetic field. The point at which the derivative curve is zero (i.e. crosses the baseline) corresponds to the absorption maximum (the hashed line in Fig. 4.26), and the magnetic field, B_{sample}, at this

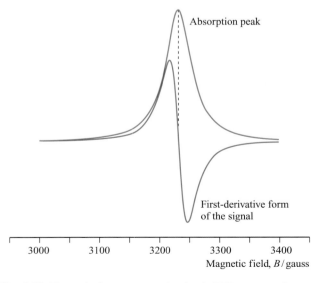

Absorption peak

First-derivative form of the signal

Magnetic field, B / gauss

Fig. 4.26 The typical appearance of a simple EPR spectrum is shown by the blue line. This is the first derivative of the absorption peak (shown in red).

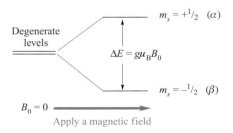

$m_s = +\frac{1}{2}$ (α)

Degenerate levels

$\Delta E = g\mu_B B_0$

$m_s = -\frac{1}{2}$ (β)

$B_0 = 0$

Apply a magnetic field

Fig. 4.25 Under an applied magnetic field, B_0, the interaction between an unpaired electron and the magnetic field results in a splitting of the energy levels (the Zeeman electronic effect).

point is recorded. The g-value for the sample is found by substituting the value of B_{sample} into eq. 4.19 (eq. 4.20).

$$\Delta E = h\nu = g_{\text{sample}} \times \mu_B \times B_{\text{sample}} \qquad (4.20)$$

The experimental g-value can be found directly since the frequency, ν, of a modern spectrometer is known accurately, $h = $ Planck constant, and $\mu_B = $ Bohr magneton (a constant). For old spectrometers, or where a calibration is required, g_{sample} can be found by comparing the value of B_{sample} with that of an internal reference material for which g is known (e.g. for the reference DPPH, **4.13**, $g = 2.0036$). Equation 4.21 follows from $\Delta E = g\mu_B B_0$ because μ_B is a constant.

$$g_{\text{sample}} \times B_{\text{sample}} = g_{\text{reference}} \times B_{\text{reference}} \qquad (4.21)$$

2,2-Diphenyl-1-picrylhydrazyl radical
(DPPH)

(4.13)

The g-value obtained from an EPR experiment provides diagnostic information about the system being investigated. For a paramagnetic metal centre, the g-value is characteristic of the oxidation state (i.e. the number of unpaired electrons), the coordination environment and the molecular symmetry. However, unless a system has cubic symmetry (i.e. it belongs to the T_h, O_h or I_h point groups and is *isotropic*), the g-value depends on the orientation of the molecular principal axis with respect to the magnetic field. Such systems are said to be *anisotropic*. By rotation of the sample placed in the magnetic field in three orthogonal planes, three g-values are therefore obtained. Each g-value is associated with one of the three orthogonal axes. Three cases must now be considered:

- For an *isotropic* system (e.g. an MX_6 species with O_h symmetry), the three g-values are equal to one another ($g_{xx} = g_{yy} = g_{zz} = g_{\text{iso}}$).
- A system that is *anisotropic*, but has *axial symmetry*, has two axes (x and y) that are equivalent but are different from the principal axis, z. This gives rise to two g-values labelled $g_\parallel$ and $g_\perp$ ($g_{xx} = g_{yy} = g_\parallel$ and $g_{zz} = g_\perp$) depending on whether the molecular principal axis is aligned parallel to or perpendicular to the magnetic field.
- An *anisotropic* system in which each of the x, y and z axes is unique gives rise to three g-values (g_{xx}, g_{yy} and g_{zz}).

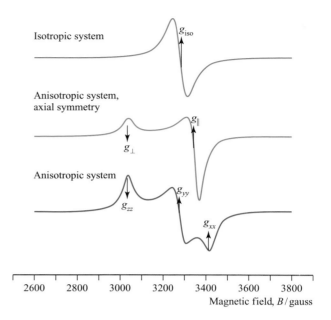

Fig. 4.27 Typical line shapes for EPR spectra of an isotropic system (blue line), an anisotropic system with axial symmetry (red line), and an anisotropic system (green line). [Spectra simulated by Dr C. Palivan, University of Basel.]

These three cases are illustrated in Fig. 4.27. In addition to the information available from g-values, we can obtain information about nuclei with nuclear spin quantum number $I \neq 0$ which are close to the paramagnetic centre. The spins of such nuclei interact magnetically with the unpaired electron and give rise to a *hyperfine interaction*. There is a direct analogy here with coupling of nuclear spins in NMR spectroscopy. The hyperfine interaction is added to the Zeeman electronic interaction, leading to a further splitting of the energy levels (eq. 4.22).

$$\Delta E = g\mu_B B_0 + SAI \qquad (4.22)$$

where: $S = $ total electron spin quantum number

$A = $ hyperfine coupling constant (in MHz)

$I = $ nuclear spin quantum number

The additional term SAI results in an EPR spectrum that is more complicated than those illustrated in Fig. 4.27. The EPR spectrum of one line (as for a scalar gyromagnetic factor) is split and the number of lines in the hyperfine pattern is given by $2nI + 1$, where n is the number of equivalent nuclei with spin quantum number I (compare this with eq. 4.15). For example, cobalt possesses one isotope, ^{59}Co, with $I = \frac{7}{2}$. An unpaired electron on a Co^{2+} centre couples to the ^{59}Co nucleus giving rise to an 8-line splitting pattern (Fig. 4.28a). Many elements possess more than one isotope (see Appendix 5). For example, naturally occurring Cu consists of ^{63}Cu (69.2%, $I = \frac{3}{2}$) and ^{65}Cu (30.8%, $I = \frac{3}{2}$). An unpaired electron on a Cu^{2+} ion couples to ^{63}Cu and to ^{65}Cu, giving rise to two, superimposed 4-line hyperfine patterns (in the case of a scalar gyromagnetic factor). As

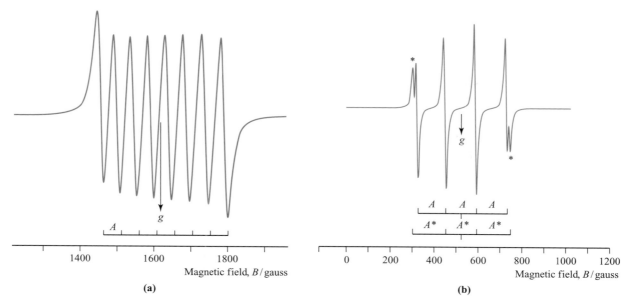

Fig. 4.28 (a) Coupling between an unpaired electron on a Co^{2+} ion with the ^{59}Co nucleus (100%, $I = \frac{7}{2}$) gives rise to an 8-line hyperfine splitting pattern for the EPR signal (microwave frequency = 9.785 GHz). (b) Coupling between an unpaired electron on a Cu^{2+} ion with the ^{63}Cu and ^{65}Cu nuclei (69.2% and 30.8%, respectively, both $I = \frac{3}{2}$) produces two superimposed 4-line splitting patterns. A and A^* are the hyperfine coupling constants. [Spectra simulated by Dr C. Palivan, University of Basel.]

$A(^{65}Cu) = 1.07 \times A(^{63}Cu)$, a well-resolved spectrum is still observed (Fig. 4.28b, A and A^* are the hyperfine coupling constants for ^{63}Cu and ^{65}Cu, respectively). Hyperfine interactions will also arise if there is delocalization of the unpaired electron from the paramagnetic metal centre onto ligands in the first coordination sphere, atoms in which possess $I > 0$ (e.g. ^{19}F, 100%, $I = \frac{1}{2}$). In this case, the spectrum is more complex because of the presence of the transitions arising from this so called, 'superhyperfine' interaction.

As in the case of g-values, hyperfine coupling constants, A, can be either isotropic or anisotropic depending on the symmetry of the system, and the shape of the EPR spectra reflects this. As a result, g and A-values can be used to give detailed information about the coordination sphere of a paramagnetic metal centre (e.g. geometry, symmetry, nature of adjacent nuclei having $I > 0$). Depending on the paramagnetic metal centre, further insight into their structure can be gained by considering other interactions (e.g. zero field interactions, quadrupolar interaction, Zeeman nuclear interactions).

EPR spectroscopy has a wide range of applications, including its use in bioinorganic systems, for example, blue copper proteins (see Section 29.4).

Worked example 4.8   EPR spectrum of [VO(acac)$_2$]

The complex [VO(acac)$_2$] has the structure shown below (Hacac = pentane-2,4-dione or acetylacetone). Vanadium has two isotopes (^{50}V, 0.25%, $I = 6$; ^{51}V, 99.75%, $I = \frac{7}{2}$).

The solution EPR spectrum of [VO(acac)$_2$] at 298 K shows an 8-line signal ($A = 120$ G, $g = 1.971$). When the sample is frozen in liquid N$_2$ (77 K), the EPR spectrum consists of two overlapping 8-line patterns for which the g-values are 1.985 and 1.942, respectively. Rationalize these data.

[Data: A. Serianz *et al.* (1976) *J. Chem. Educ.*, vol. 53, p. 394.]

Vanadium is in group 5 and in [VO(acac)$_2$], it is in oxidation state +4. Therefore the V centre has one unpaired electron.

Vanadium is almost monotopic. The pattern in the EPR spectrum is dominated by hyperfine coupling between the unpaired electron and the ^{51}V nucleus (99.75% abundant). The $I = \frac{7}{2}$ spin gives rise to the 8-line pattern:

$$2nI + 1 = 2(1)(\tfrac{7}{2}) + 1 = 8$$

Look at the structure of the [VO(acac)$_2$] molecule. It is not spherically symmetrical, but instead has axial symmetry. The principal axis contains the V=O unit. At 298 K, the EPR spectrum corresponds to an isotropic system and this is because the molecules are tumbling rapidly in solution and an average orientation is observed. On cooling to

77 K, the spectrum splits into two superimposed 8-line patterns. This corresponds to the principal axis of the molecule aligning parallel or perpendicular to the applied magnetic field. The two g values of 1.985 and 1.942 correspond to $g_\parallel$ and $g_\perp$. The value at 298 K ($g = 1.971$) is in between the values of $g_\parallel$ and $g_\perp$.

Self-study exercises

1. Why is copper(I) EPR silent?

2. Manganese is monotopic (^{55}Mn). The EPR spectrum of an aqueous solution of $[Mn(OH_2)_6]^{2+}$ shows a 6-line pattern. Why is $[Mn(OH_2)_6]^{2+}$ isotropic? What is the value of I for ^{55}Mn?

3. Show that the spectrum in Fig. 4.28a is consistent with a g-value of 4.32.

4.10 Mössbauer spectroscopy

The technique of Mössbauer spectroscopy

The *Mössbauer effect* is the emission and resonant absorption of nuclear γ-rays studied under conditions such that the nuclei have negligible recoil velocities when γ-rays are emitted or absorbed. This is only achieved by working with *solid samples* in which the nuclei are held rigidly in a crystal lattice. The energy, and thus the frequency of the γ-radiation involved, corresponds to the transition between the ground state and the short-lived excited state of the nuclide concerned. Table 4.4 lists properties of several nuclei which can be observed using Mössbauer spectroscopy.

We illustrate the study of the Mössbauer effect by reference to ^{57}Fe spectroscopy. The basic apparatus includes a radioactive source, a solid absorber with the ^{57}Fe-containing sample and a γ-ray detector. For ^{57}Fe samples, the radioactive source is ^{57}Co and is incorporated into stainless steel; the ^{57}Co source decays by capture of an extra-nuclear electron to give the excited state of ^{57}Fe which emits γ-radiation as it decays to its ground state. If ^{57}Fe is present in the

same form in both source and absorber, resonant absorption occurs and no radiation is transmitted. However, if the ^{57}Fe in the source and absorber is present in two different forms, absorption does *not* occur and γ-radiation reaches the detector. Moving the source at different velocities towards or away from the ^{57}Fe absorber has the effect of varying the energy of the γ-radiation (i.e. by the Doppler effect). The velocity of movement required to bring about maximum absorption relative to stainless steel (defined as an arbitrary zero for iron) is called the *isomer shift* of ^{57}Fe in the sample, with units of mm s^{-1} (see Fig. 20.31).

What can isomer shift data tell us?

The isomer shift gives a measure of the electron density on the ^{57}Fe centre, and isomer shift values can be used to determine the oxidation state of the Fe atom. Similarly, in ^{197}Au Mössbauer spectroscopy, isomer shifts can be used to distinguish between Au(I) and Au(III). Three specific examples are chosen here from iron chemistry.

The cation $[Fe(NH_3)_5(NO)]^{2+}$ has presented chemists with an ambiguity of the description of the bonding which has, in some instances, been described in terms of an $[NO]^+$ unit bound to an Fe(I) centre. Results of ^{57}Fe Mössbauer spectroscopy have revealed that the correct description is that of an $[NO]^-$ ligand bound to an Fe(III) centre.

The formal oxidation states of the iron centres in $[Fe(CN)_6]^{4-}$ and $[Fe(CN)_6]^{3-}$ are +2 and +3. However, the closeness of the isomer shifts for these species suggests that the actual oxidation states are similar and this may be interpreted in terms of the extra electron in $[Fe(CN)_6]^{4-}$ being delocalized on the cyanido ligands rather than the iron centre.

Differences in isomer shifts can be used to distinguish different iron environments in the same molecule: the existence of two signals in the Mössbauer spectrum of $Fe_3(CO)_{12}$ provided the first evidence for the presence of two types of iron atom in the solid state structure (Fig. 4.29), a fact that has been confirmed by X-ray diffraction methods.

Table 4.4 Properties of selected nuclei observed by Mössbauer spectroscopy. The radioisotope source provides the γ-radiation required for the Mössbauer effect.

Nucleus observed	Natural abundance / %	Ground spin state	Excited spin state	Radioisotope source[†]
^{57}Fe	2.2	$\frac{1}{2}$	$\frac{3}{2}$	^{57}Co
^{119}Sn	8.6	$\frac{1}{2}$	$\frac{3}{2}$	^{119m}Sn
^{99}Ru	12.7	$\frac{3}{2}$	$\frac{5}{2}$	^{99}Rh
^{197}Au	100	$\frac{3}{2}$	$\frac{1}{2}$	^{197m}Pt

[†] m = metastable.

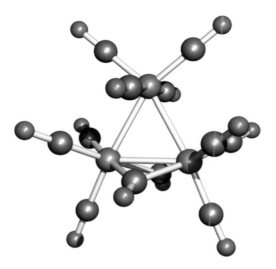

Fig. 4.29 The solid state structure of $Fe_3(CO)_{12}$ as determined by X-ray diffraction methods. The molecule contains two Fe environments by virtue of the arrangement of the CO groups. Colour code: Fe, green; C, grey; O, red.

The use of Mössbauer spectroscopy to investigate different electronic spin states of iron(II) is exemplified in Fig. 20.31 and the accompanying discussion.

4.11 Structure determination: diffraction methods

Chemists rely on diffraction methods for the structural determination of molecular solids (i.e. solids composed of discrete molecules), non-molecular solids (e.g. ionic materials) and, to a lesser extent, gaseous molecules. As the technique has been developed, its range of applications has expanded to include polymers, proteins and other macromolecules. The most commonly applied techniques are single crystal and powder X-ray diffraction. Electron diffraction is important for the structural elucidation of molecules in the gas phase and for the study of solid surfaces. Neutron diffraction is used for the accurate location of light atoms (e.g. H, D or Li), or if one needs to distinguish between atoms of similar atomic numbers, e.g. C and N, or Ni and Cu.

X-ray diffraction (XRD)

In **X-ray diffraction (XRD)**, X-rays are diffracted by electrons surrounding the nuclei in atoms in a crystalline or polycrystalline solid.

The wavelength of X-rays ($\approx 10^{-10}$ m, i.e. ≈ 100 pm) is of the same order of magnitude as the internuclear distances in molecules or non-molecular solids. As a consequence of this, diffraction is observed when X-rays interact with the electrons in an array of atoms in the crystalline solid.

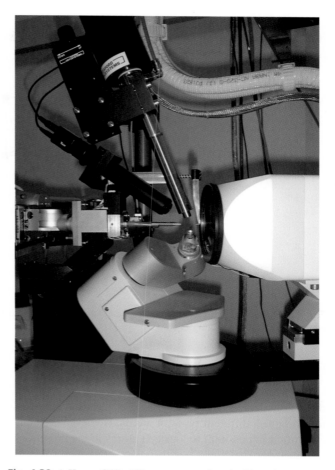

Fig. 4.30 A Kappa-CCD diffractometer equipped with a nitrogen gas, low-temperature cryostat (upper, centre in the photograph). The X-ray source and the detector are on the left- and right-hand sides of the photograph, respectively. The crystal is mounted on the goniometer head (centre). The black 'tube' shown at the upper left is a microscope.

This permits atomic resolution to be achieved when a structure is determined from the X-ray diffraction data which are collected. An X-ray diffractometer (Fig. 4.30) typically consists of an X-ray source, a mounting for the crystal, turntables which allow the operator to alter the orientation of the crystal with respect to the incident X-ray beam, and an X-ray detector. The source provides *monochromatic radiation*, i.e. X-rays of a single wavelength. The detector records X-rays that are scattered by the crystal. Modern diffractometers incorporate *imaging plate detectors* or *charge-coupled device (CCD) area detectors* which make the process of data collection much faster. For both these detectors, radiation must be converted to light before it can be recorded. Currently, *pixel detectors* are being developed which detect radiation directly, and their use will avoid the step of transforming radiation to light.

X-rays are scattered by the electrons surrounding the nuclei in atoms in the solid. Because the *scattering power*

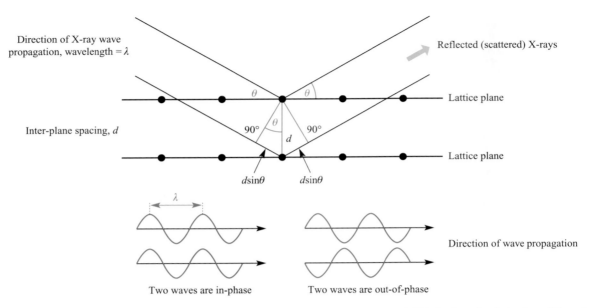

Fig. 4.31 Schematic representation of the interaction of X-rays with layers of atoms in a crystal. This leads to the derivation of Bragg's equation (eq. 4.23).

of an atom depends on the number of electrons, it is possible to distinguish different types of atoms. However, it is difficult (often impossible) to locate H atoms in the presence of heavy atoms.

Figure 4.31 shows an ordered array of atoms with the latter represented as black dots. The atoms are arranged in layers or *lattice planes*. Consider the case in which the two waves of incident radiation are in-phase. Let one wave be reflected from an atom in the first lattice plane, and the second wave reflected from an atom in the second lattice plane (Fig. 4.31). The two scattered (reflected) waves will be in-phase only if the additional distance travelled by the second wave is equal to a multiple of the wavelength, i.e. $n\lambda$. If the *lattice spacing* (the distance between the planes of atoms in the crystal) is d, then by trigonometry it follows that the additional distance travelled by the second wave is $2d \times \sin\theta$. For the two waves (originally in-phase) to stay in-phase as they are scattered, eq. 4.23 must hold. This relationship between the wavelength, λ, of the incident X-ray radiation and the lattice spacings, d, of the crystal is Bragg's equation and is the basis of the techniques of X-ray and neutron diffraction.

$$2d \sin\theta = n\lambda \qquad \text{Bragg's equation} \qquad (4.23)$$

The angle θ in eq. 4.23 is half of the diffraction angle (Fig. 4.31), and diffraction data are often referred to in terms of an angle 2θ. Scattering data are collected over a wide range of θ (or 2θ) values and for a range of crystal orientations. Each setting results in a different *diffraction pattern* as discussed below.

Single crystal X-ray diffraction

Analysis by *single crystal X-ray diffraction* leads to the full determination of the structure of a compound.

Single crystals suitable for X-ray diffraction can be grown by a number of methods,[†] all of which lower the solubility of the compound in solution:

- evaporation of solvent from a solution of the compound;
- cooling a solution of the compound;
- diffusing a solvent in which the compound is insoluble or poorly soluble into a solution of the compound in a second solvent;
- allowing the vapour of a volatile solvent to diffuse into a solution of the compound in a second solvent;

In addition, it may be possible to grow crystals by sublimation or from a melt. Problems sometimes occur with the intergrowth of two crystals (*crystal twinning*). Large crystals are not necessary for use with modern diffractometers, and ideal crystal dimensions are 0.1 to 0.3 mm.

A single crystal gives rise to a diffraction pattern consisting of well-defined spots. Figure 4.32a shows the diffraction pattern in one *frame* recorded during a data collection of a single crystal of 4'-azido-2,2':6':2''-terpyridine (**4.14**). This is an example of an organic ligand used in metal coordination chemistry (see Sections 7.11 and 19.7).

[†]See: W. Clegg (2004) in *Comprehensive Coordination Chemistry II*, eds. J.A. McCleverty and T.J. Meyer, Elsevier, Oxford, vol. 1, p. 579.

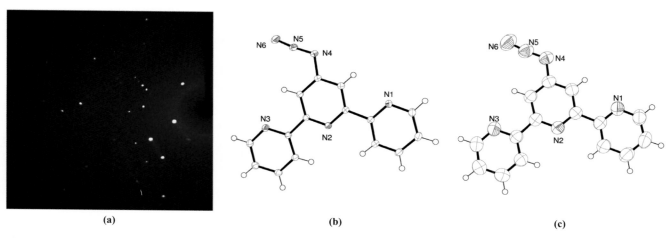

Fig. 4.32 Solving the structure of the ligand 4'-azido-2,2':6',2''-terpyridine. (a) One photographic frame from the single crystal X-ray data collection showing diffraction spots. ORTEP drawings of the molecular structure of 4'-azido-2,2':6',2''-terpyridine from data collected at (b) 123 K and (c) 293 K with ellipsoids plotted at a 50% probability level. The H atoms were refined isotropically (see text). Colour code for atoms: C, grey; N, blue; H, white. [Data: M. Neuburger and N. Hostettler, University of Basel; R. Al-Fallahpour *et al.* (1999) *Synthesis*, p. 1051.]

(4.14)

In order to obtain a complete data set (i.e. enough reflection data to solve the structure of the compound under investigation), many hundreds or thousands of frames are recorded. Each frame contains diffraction data from different reflections, and at the end of the experiment, the data are extracted and assembled to produce a numerical datafile. The methods of solving a crystal structure from the reflection data are beyond the scope of this book, and further details can be found in the texts listed at the end of the chapter.

Compounds such as **4.14** consist of discrete molecules, and the results of a structural determination are usually discussed in terms of molecular structure (atomic coordinates, bond distances, bond angles and torsion angles) and in terms of the packing of the molecules in the lattice. Packing involves intermolecular interactions (e.g. hydrogen bonding, van der Waals forces, π-stacking of aromatic rings). Molecular structures are routinely represented as ORTEP diagrams (Figs. 4.32b and c),[†] in which the atoms are drawn as ellipsoids. Each ellipsoid delineates the volume

in space in which there is a probability of finding a particular atom and indicates the *thermal motion* of the atom. Since this motion depends upon the amount of energy that the molecule possesses, the temperature of the X-ray data collection is important. Accurate bond distances and angles can only be obtained if thermal motions are minimized. A comparison of Figs. 4.32b and c reveals the differences in thermal motion of the atoms in compound **4.14** at 123 and 293 K. Low-temperature data collection is now a routine part of single crystal X-ray structure determination. Note that the H atoms in Figs. 4.32b and c are drawn as circles (*isotropic*) rather than as ellipsoids (*anisotropic*). This is because the H atoms are not directly located. Instead, their positions are fixed in chemically sensible sites with respect to the C atoms which have been directly located using the X-ray diffraction data. It is not uncommon for solid state structures to suffer from *disorder* (see Box 15.5).

In Chapter 6, we discuss solid state structures of metals and ionic compounds, and detail the *unit cells* of a number of prototype stuctures. The unit cell is the smallest repeating unit in a crystal lattice, and its dimensions are characteristic of a particular *polymorph* of a compound. A unit cell is characterized by three cell edge lengths (*a*, *b* and *c*) and three angles (α, β and γ). Distances are often given in the non-SI unit of the ångström (Å), because $1\,\text{Å} = 10^{-10}\,\text{m}$ and bond distances typically lie in the range 1–3 Å; in this book we use SI units of the picometre ($1\,\text{pm} = 10^{-12}\,\text{m}$).

The smallest repeating unit in a solid state lattice is the ***unit cell***.

Polymorphs are different phases of the same chemical compound with different crystal structures.

[†]ORTEP software: www.chem.gla.ac.uk/~louis/software; L.J. Farrugia (1997) *J. Appl. Cryst.*, vol. 30, p. 565.

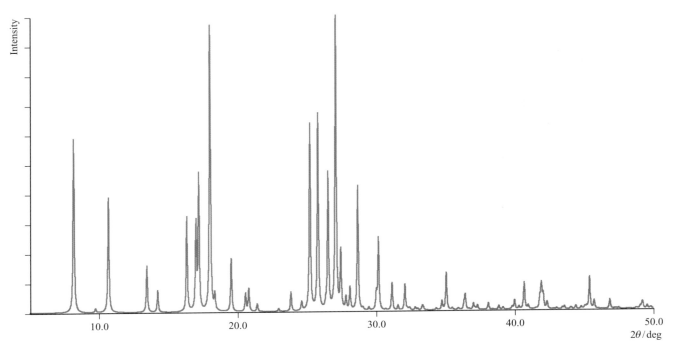

Fig. 4.33 Calculated powder pattern for 4'-azido-2,2':6',2''-terpyridine (compound **4.14**) using single crystal X-ray diffraction data. The pattern is a fingerprint of the bulk powder material.

Powder X-ray diffraction

A powder is a polycrystalline sample; *powder X-ray diffraction* data are routinely used for identifying a bulk sample of a material, and for screening different phases of a compound.

Powders (as distinct from amorphous materials) consist of huge numbers of microcrystals and are classed as polycrystalline materials. When X-rays interact with powders, they are scattered in all directions because the microcrystals lie in random orientations. This simultaneously provides all the reflection data that are produced in a single crystal experiment by making changes to the orientation of the crystal with respect to the incident X-ray beam. As a consequence, and in contrast to the single crystal diffraction experiment, all information about the relative orientations of the scattering vectors in space is lost, and the data consist of scattered intensities as a function of diffraction angle (2θ). The output from a powder diffractometer is in the form of a powder pattern (Fig. 4.33) and this is a diagnostic fingerprint of the polycrystalline compound. Different phases of the same compound exhibit different powder patterns and powder X-ray diffraction is a powerful tool for distinguishing between phases. Under certain conditions, it is possible to solve a molecular structure using powder data. This can be achieved for relatively

simple molecules,[†] but it is not the aim of most powder diffraction experiments. On the other hand, if a single crystal structure is known, a powder pattern can be calculated and Fig. 4.33 shows the calculated powder pattern of compound **4.14**. Having this pattern to hand allows you to match it to powder patterns obtained for new bulk samples of the compound that are synthesized, giving a means of rapidly screening bulk materials. This is particularly important in the pharmaceutical industry where different phases of a drug exhibit different pharmacological properties.

Single crystal neutron diffraction

In *neutron diffraction*, Bragg scattering of neutrons occurs when neutrons interact with the nuclei of the atoms in a single crystal; both light and heavy atoms can be directly detected.

Neutron diffraction is less commonly used than X-ray diffraction because of the limitations and costs of sources of neutrons. However, because the technique relies upon the diffraction of neutrons by atomic nuclei, neutron diffraction

[†]See: K.D.M. Harris, R.L. Johnston, E.Y. Cheung, G.W. Turner, S. Habershon, D. Albesa-Jové, E. Tedesco and B.M. Kariuki (2002) *CrystEngComm*, vol. 4, p. 356; W.I.F. David, K. Shankland, J. van de Streek, E. Pidcock, W.D.S. Motherwell and J.C. Cole (2006) *J. Appl. Cryst.*, vol. 39, p. 910.]

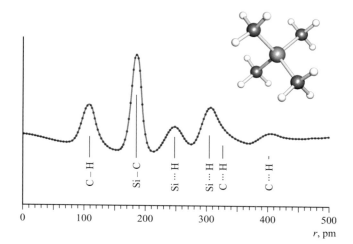

Fig. 4.34 Experimental (dots) and calculated (solid line) radial distributions for SiMe$_4$. The vertical scale is arbitrary. The shortest distances (C–H and Si–C) correspond to bonded contacts, and the remaining ones to non-bonded contacts. [Redrawn with permission from I. Hargittai *et al.* (2009) *Encyclopedia of Spectroscopy and Spectrometry*, Elsevier, p. 461.]

has the advantage that it can accurately locate light atoms such as H. Before a beam of neutrons from a nuclear reactor can be used for a neutron diffraction experiment, its energy must be reduced. Neutrons released from the fission of $^{235}_{92}$U nuclei lose most of their kinetic energy by passage through a moderator such as heavy water (D_2O). It follows from the deBroglie relationship (eq. 1.11) that the energy of the thermal neutrons so-produced corresponds to a wavelength of 100–500 pm (1–5 Å), i.e. appropriate for diffraction bya molecular or ionic solid. The scattering of neutrons follows Bragg's equation, and the diffraction experiment is similar to that for X-ray diffraction.

Electron diffraction

> In *gas phase electron diffraction*, electrons are scattered by the electric fields of atomic nuclei in gas phase molecules; intramolecular bond parameters are determined.

Electrons that have been accelerated through a potential difference of 50 kV possess a wavelength of 5.5 pm (from the de Broglie relationship, eq. 1.11). A monochromated beam of 50 kV electrons is therefore suitable for diffraction by molecules in the gas phase. An electron diffraction apparatus is maintained under high vacuum and the electron beam interacts with a gas stream emerging from a nozzle. Unlike X-rays and neutrons, electrons are charged, and electron scattering is mainly caused by the interactions of the electrons with the electric fields of the atomic nuclei in a sample. Because the sample molecules are in the gas phase, they are continually in motion and are, therefore, in random orientations and well separated from one another.

Thus, the diffraction data mainly provide information about *intra*molecular bond parameters.

The initial electron diffraction data relate scattering angle of the electron beam to intensity. Corrections must be made for atomic, inelastic and extraneous scatterings, and after this, *molecular scattering data* are obtained. Fourier transformation of these data produces radial distribution data which reveal the interatomic distances between all pairs of bonded and non-bonded atoms in the gaseous molecule. Converting these data into a 3-dimensional molecular structure is not trivial, particularly for large molecules. Experimental radial distribution curves are compared to those calculated for modelled structures. This is shown for SiMe$_4$ in Fig. 4.34, where the best match between experimental and calculated radial distributions is found for a molecule with T_d symmetry, Si–C and C–H bond distances of 187.7 and 111.0 pm, and Si–C–H bond angles of 111.0°.

Self-study exercise

Results of electron diffraction data for BCl$_3$ give bonded B–Cl distances of 174 pm (all bonds of equal length) and non-bonded Cl...Cl distances of 301 pm (three equal distances). Show that these data are consistent with BCl$_3$ being trigonal planar rather than trigonal pyramidal.

Low-energy electron diffraction (LEED)

Electron diffraction is not confined to the study of gases. Low-energy electrons (10–200 eV) are diffracted from the surface of a solid and the diffraction pattern so obtained provides information about the arrangement of atoms on the surface of the solid.

Structural databases

Structural data are stored in, and can be accessed from, three primary databases which are continually updated. Data for organic and metal–organic compounds and boranes are compiled by the Cambridge Structural Database (www.ccdc. cam.ac.uk), while crystallographic data for purely inorganic compounds (e.g. metal and non-metal halides and oxides) are deposited in the Inorganic Crystal Structure Database (www.fiz-karlsruhe.de). Structural data for biological macromolecules are collected in the Protein Data Bank (www.rcsb.org/pdb).

4.12 Photoelectron spectroscopy (PES, UPS, XPS, ESCA)

> *Photoelectron spectroscopy (PES)* is a technique used to study the energies of occupied atomic or molecular orbitals.

The technique of *photoelectron spectroscopy* (PES, also called *photoemission spectroscopy*), was developed in the 1960s independently by Turner (in Oxford), Spicer (in Stanford), Vilesov (in Leningrad, now St. Petersburg) and Siegbahn (in Uppsala). In a PES experiment, atoms or molecules are excited with monochromated electromagnetic radiation of energy, E_{ex}, causing electrons to be ejected, i.e. photoionization occurs (eq. 4.24).

$$X \xrightarrow{E_{ex} = h\nu} X^+ + e^- \tag{4.24}$$

The atom or molecule X is in its ground state, and X^+ is in either its ground or an excited state. The ejected electrons are called *photoelectrons*. Each electron in an atom or molecule possesses a characteristic *binding energy* and must absorb an amount of energy equal to, or in excess of, the binding energy to be ionized. The kinetic energy, KE, of the ejected photoelectron is the energy in excess of the ionization (or binding) energy, IE (eq. 4.25).

$$KE = E_{ex} - IE \tag{4.25}$$

Since the excess energy can be measured and E_{ex} is known, the binding energy of the electron can be determined. A photoelectron spectrum records the number of photoelectrons with a particular kinetic energy against binding energy. The quantization of energy states leads to a photoelectron spectrum having a discrete band structure. *Koopmans' theorem* relates ionization energy to the energy of the atomic or molecular orbital in which the electron resides before ejection. Thus, binding energies measured from PES give a measure of atomic or molecular orbital energies.

As an example, consider the photoelectron spectrum of gaseous N_2, one of the first PES results reported by Turner in 1963. A helium(I) lamp ($E_{ex} = 21.2\,eV$) is a suitable source of photons for this experiment since 21.2 eV exceeds the binding energies of interest. The photoelectron spectrum of N_2 consists of three peaks corresponding to binding energies of 15.57, 16.72 and 18.72 eV. These three ionizations arise from the ejection of an electron from the $\sigma_g(2p)$, $\pi_u(2p)$ or $\sigma_u^*(2s)$ MOs of N_2 (Fig. 2.10), respectively.

The applications of PES can be diversified by using different irradiation sources. In the example above, a helium(I) emission photon source with energy of 21.2 eV was used; the 'He(I) emission' corresponds to a transition from an excited state of configuration $1s^1 2p^1$ to the ground state $(1s^2)$ of He. The ionization of more tightly bound electrons can be achieved by using higher energy sources. For example, a 'helium(II) source' corresponds to radiation from He^+, and a helium(II) lamp provides an excitation energy of 40.8 eV. Both helium(I) and helium(II) radiation are in the vacuum UV region of the electromagnetic spectrum and therefore PES using these excitation energies is known as UPS (*UV photoelectron spectroscopy*). The ionization of core electrons in molecules can be achieved using excitation sources in the X-ray region. The Mg and Al X-ray emissions (Mg $K\alpha$ with $E_{ex} = 1254\,eV$, and Al $K\alpha$ with $E_{ex} = 1487\,eV$) are typical examples. With an X-ray source, the technique is referred to as XPS (*X-ray photoelectron spectroscopy*) and Fig. 4.35 shows a modern X-ray photoelectron spectrometer. XPS is also known as *electron spectroscopy for chemical analysis*, ESCA, and is a valuable analytical tool because core ionization energies have characteristic values for a given element. Thus, the technique can be applied for the detection of any element except for hydrogen, and can also be used to differentiate between oxidation states of an element. XPS is widely used for the analysis of surfaces, and applications include that in the semiconductor industry (e.g. to distinguish between Si and SiO_2) and the study of surface corrosion.

4.13 Computational methods

Computational methods are now used extensively by experimental chemists. Information that can be calculated includes the equilibrium geometry of a molecule, transition state geometries, heats of formation, composition of molecular orbitals, vibrational frequencies, electronic spectra, reaction mechanisms and (from molecular mechanics calculations) strain energies. The last 20 years have witnessed a huge increase in the use of computational methods in chemistry. Two factors have revolutionized the ways in which computational chemistry may be applied. The first is that calculations can now be performed on small computers (including laptops) or small clusters of computers, instead of on a mainframe computer. The second is the development of the computational methods themselves. The importance of

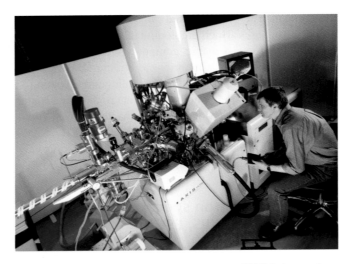

Fig. 4.35 An X-ray photoelectron spectrometer (XPS) being used to analyse the surface of a material.

the latter was recognized by the award of the 1998 Nobel Prize in Chemistry jointly to John Pople 'for his development of computational methods in quantum chemistry' and to Walter Kohn 'for his development of the density-functional theory'. Many of the computational packages available to chemists fall into the following categories: *ab initio* methods, self-consistent field (SCF) MO methods, semi-empirical methods, density functional methods and molecular mechanics.

Hartree–Fock theory

The Schrödinger equation (Section 1.5) can be solved exactly only for one-electron species, i.e. hydrogen-like systems. This is very restrictive and quantum chemists have invested a great deal of effort into finding ways to obtain approximate solutions to the Schrödinger equation for many-electron systems. Towards this end, the work of Hartree, Fock and Slater in the 1930s led to the development of Hartree–Fock theory. Equations in Hartree–Fock theory are solved by an iterative process and the calculation *converges to self-consistency*, hence the term 'self-consistent'. Various levels of theory deal differently with the approximations made when solving the Schrödinger equation, in particular with regard to electron correlation (i.e. taking into account interactions between electrons). The higher the level of calculation, the closer the result should come to experimental observation. A number of semi-empirical methods, which are parameterized and consider only valence electrons, have been developed. These include CNDO (*complete neglect of differential overlap*), INDO (*intermediate neglect of differential overlap*), MNDO (*modified neglect of diatomic overlap*), AM1 (*Austin model 1*) and PM3 (*parametric method 3*). While these methods reduce the time required for computation, they may not always produce reliable results for complex systems. Hence, they should be used with a degree of caution.

Density functional theory

In contrast to other methods, density functional theory (DFT) focuses on the electron density distribution in a system rather than on many-electron wavefunctions. Within DFT, there are several levels of calculation, two common ones being BLYP (after Becke, Lee, Yang and Parr) and B3LYP. The great advantage of DFT is that it can be applied to a wide variety of systems, ranging from transition metal complexes to solids, surfaces and metalloproteins. The computation time is not excessive, and the results are generally reliable. Conventional DFT cannot be used to investigate systems in which van der Waals (dispersion) forces are a dominant feature (e.g. biomolecules).

Development of the theory to allow these intermolecular interactions to be accurately computed is currently (as of 2011) ongoing.[†]

Hückel MO theory

At a simple level, Hückel MO theory (proposed in the 1930s by Erich Hückel) works well for dealing with the π-systems of unsaturated organic molecules. By extending the basis set and including overlap and all interactions (σ and π), Roald Hoffmann showed that *extended Hückel theory* could be applied to most hydrocarbons. This theory has since been developed further and continues to be a useful method to determine the relative energies of different conformers of an organic molecule.

Molecular mechanics (MM)

Before attempting a new chemical synthesis, you might wish to compute the structure of the target molecule in order to investigate, for example, steric crowding of substituents. For this purpose, *molecular mechanics* (MM) has become a routine tool. Pure MM does not have a quantum mechanical basis. Instead, it calculates a *strain energy* which is the sum of energy terms involving bond deformation, angle deformation, torsion deformation and non-bonded interactions. The equation for the strain energy together with a number of input parameters that describe the atoms and bonds are known as a *force field*. When an MM calculation is running, the conformation of the molecule changes until it reaches an optimized structure for which the strain energy is minimized. As well as minimizing a ground state structure, it is also possible to investigate time-dependent processes by using *molecular dynamics* (MD). Force fields for such modelling take into account the cleavage and formation of bonds, so that MD simulations can explore potential energy surfaces associated with the dynamic systems. Examples of MD force fields are AMBER (*assisted model building and energy refinement*) and CHARMM (*chemistry at Harvard macromolecular mechanics*). Molecular mechanics and dynamics can be applied to both small (discrete molecules) and large (e.g. nucleic acids and proteins) systems. Force field parameters for metal ions bound in the active sites of metalloproteins have been, and continue to be, developed, permitting the application of molecular dynamics simulations to these systems.

[†]See: E.R. Johnson, I.D. Mackie and G.A. DiLabio (2009) *J. Phys. Org. Chem.*, vol. 22, p. 1127; F.O. Kannemann and A.D. Becke (2010) *J. Chem. Theory Comput.*, vol. 6, p. 1081; V.R. Cooper (2010) *Phys. Rev. B*, vol. 81, p. 161104.

KEY TERMS

The following terms were introduced in this chapter. Do you know what they mean and what the techniques are used for?

- gas chromatography
- column liquid chromatography
- plate liquid chromatography
- high-performance liquid chromatography
- mobile and stationary phases
- recrystallization
- atomic absorption spectroscopy
- aspirate
- nebulize
- thermogravimetric analysis
- electron ionization mass spectrometry
- fast atom bombardment mass spectrometry
- matrix-assisted laser desorption ionization mass spectrometry
- time-of-flight
- electrospray mass spectrometry
- infrared spectroscopy
- force constant
- reduced mass

- zero point energy
- fingerprint region (in an IR spectrum)
- Raman spectroscopy
- resonance Raman spectroscopy
- electronic spectroscopy
- absorption spectroscopy
- Beer–Lambert law
- Franck–Condon approximation
- molar extinction coefficient
- bathochromic effect (red shift)
- hypsochromic effect (blue shift)
- emission spectroscopy
- nuclear magnetic resonance spectroscopy
- spin-active nucleus
- chemical shift (in NMR spectroscopy)
- spin–spin coupling (in NMR spectroscopy)
- proton decoupled NMR spectrum
- stereochemical non-rigidity

- electron paramagnetic resonance spectroscopy
- *g*-value (in EPR spectroscopy)
- isotropic and anisotropic
- Mössbauer spectroscopy
- isomer shift (in Mössbauer spectroscopy)
- X-ray diffraction
- solvate
- hydrate
- monochromatic radiation
- scattering power of an atom (in XRD)
- Bragg's equation
- unit cell
- polymorph
- neutron diffraction
- electron diffraction
- low-energy electron diffraction
- photoelectron spectroscopy
- Koopmans' theorem
- computational methods

IMPORTANT ACRONYMS: WHAT DO THEY STAND FOR?

- GC
- LC
- TLC
- HPLC
- AAS
- TGA
- EI
- FAB
- MALDI-TOF

- ESI
- IR
- UV-VIS
- NMR
- MRI
- EPR
- XRD
- LEED
- PES

- UPS
- XPS
- ESCA
- CNDO
- MNDO
- AM1
- PM3
- DFT
- MM

FURTHER READING

General

A.K. Brisdon (1998) *Inorganic Spectroscopic Methods*, Oxford University Press, Oxford – An OUP Primer covering basic vibrational and electronic spectroscopies and mass spectrometry.

J.R. Dean, A.M. Jones, D. Holmes, R. Reed, J. Weyers and A. Jones (2002) *Practical Skills in Chemistry*, Pearson, Harlow – An excellent introduction to a wide range of experimental techniques with many examples.

S. Duckett and B. Gilbert (2000) *Foundations of Spectroscopy*, Oxford University Press, Oxford – An OUP Primer that introduces mass spectrometry, IR, UV-VIS and NMR spectroscopies and X-ray diffraction.

R.P. Wayne (1994) *Chemical Instrumentation*, Oxford University Press, Oxford – An introductory text focusing on the operations of instruments.

Mass spectrometry

W. Henderson and J.S. McIndoe (2005) *Mass Spectrometry of Inorganic and Organometallic Compounds*, Wiley, Chichester – A useful textbook that focuses on applications in inorganic chemistry.

J.T. Watson and D. Sparkman (2007) *Introduction to Mass Spectrometry: Instrumentation, Applications, and Strategies for Data Interpretation*, 4th edn, Wiley, Chichester – A detailed introduction to the topic with many examples.

Vibrational spectroscopy

J.A. McCleverty and T.J. Meyer, eds (2004) *Comprehensive Coordination Chemistry II*, Elsevier, Oxford – Volume 2 contains three articles covering Raman, FT-Raman and resonance Raman spectroscopies including applications in bioinorganic chemistry.

K. Nakamoto (1997) *Infrared and Raman Spectra of Inorganic and Coordination Compounds*, 5th edn, Wiley, New York – Two volumes comprising an invaluable reference source for inorganic chemists.

NMR spectroscopy

C. Brevard and P. Granger (1981) *Handbook of High Resolution Multinuclear NMR*, Wiley-Interscience, New York – A reference book listing nuclear properties, standard references, typical chemical shift ranges and coupling constants.

R. Freeman (2003) *Magnetic Resonance in Chemistry and Medicine*, Oxford University Press, Oxford – An up-to-date treatment of high-resolution NMR spectroscopy, illustrating applications from discrete molecular to human body levels.

C.E. Housecroft (1994) *Boranes and Metallaboranes: Structure, Bonding and Reactivity*, 2nd edn, Ellis Horwood, Hemel Hempstead – Chapter 2 includes an account of the interpretation of ^{11}B and ^{1}H NMR spectra of boranes and their derivatives.

B.K. Hunter and J.K.M. Sanders (1993) *Modern NMR Spectroscopy: A Guide for Chemists*, 2nd edn, Oxford University Press, Oxford – An excellent, detailed and readable text.

J.A. Iggo (1999) *NMR Spectroscopy in Inorganic Chemistry*, Oxford University Press, Oxford – A primer that introduces the theory of NMR spectroscopic techniques as well as their use in structure determination.

Mössbauer spectroscopy

G.J. Long and F. Grandjean (2004) in *Comprehensive Coordination Chemistry II*, eds J.A. McCleverty and T.J. Meyer, Elsevier, Oxford, vol. 2, p. 269 – An introduction to Mössbauer spectroscopy with references to pertinent literature examples.

A.G. Maddock (1997) *Mössbauer Spectroscopy: Principles and Applications*, Horwood Publishing, Chichester – A comprehensive account of the technique and its uses.

Diffraction methods

P. Atkins and J. de Paula (2010) *Atkins' Physical Chemistry*, 9th edn, Oxford University Press, Oxford – Chapter 19 covers crystal lattices and diffraction methods.

W. Clegg (1998) *Crystal Structure Determination*, Oxford University Press, Oxford – An excellent introductory text.

W. Clegg (2004) in *Comprehensive Coordination Chemistry II*, eds J.A. McCleverty and T.J. Meyer, Elsevier, Oxford, vol. 2, p. 57 – A short review of X-ray diffraction methods.

C. Hammond (2001) *The Basics of Crystallography and Diffraction*, 2nd edn, Oxford University Press, Oxford – A detailed treatment of crystal symmetry and diffraction methods.

M.F.C. Ladd and R.A. Palmer (2003) *Structure Determination by X-ray Crystallography*, 4th edn, Kluwer/Plenum, New York – A detailed introduction to crystal symmetry, lattices and diffraction methods.

A.G. Orpen (2002) *Acta Crystallogr., Sect. B*, vol. 58, p. 398 – A discussion of the applications of the Cambridge Structural Database to inorganic compounds.

Computational methods

L. Banci (2003) *Curr. Opin. Chem. Biol.*, vol. 7, p. 143 – A short review focusing on molecular dynamics simulations in metalloproteins (e.g. zinc enzymes, haem proteins and copper proteins).

G.H. Grant and W.G. Richards (1995) *Computational Chemistry*, Oxford University Press, Oxford – An OUP Primer covering the basics of computational methods in chemistry.

J.A. McCleverty and T.J. Meyer, eds (2004) *Comprehensive Coordination Chemistry II*, Elsevier, Oxford – Volume 2 contains a section 'Theoretical models, computational methods, and simulation' consisting of a series of articles covering computational methods including molecular mechanics, semi-empirical SCF MO methods, and density functional theory.

PROBLEMS

[Additional problems on IR and electronic spectroscopies can be found in Chapters 3 and 20, respectively.]

4.1 The conversion of solar energy into chemical energy using artificial photosynthesis involves the photocatalytic conversion of H^+ to H_2. Why is GC suitable for the detection and quantification of H_2?

4.2 A test TLC plate (silica) using a 1 : 3 mixture of $CH_3CN : H_2O$ as eluent shows that two compounds have R_f values of 0.52 and 0.15:

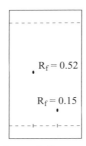

You plan to use column LC to separate a mixture of the two compounds. How can you attempt to ensure that their behaviour on the column will closely mimic that on the TLC plate?

4.3 What are the near UV and visible ranges in nm? In its UV-VIS spectrum, $Ru_3(CO)_{12}$ absorbs at 392 nm. Explain why, in a column chromatographic separation of this compound, visual detection is possible.

4.4 Why can a CHN analysis of a compound not distinguish between a monomer and dimer of the species? What technique would you use to confirm that a dimer was present?

4.5 The reaction of $NbCl_4(THF)_2$ with pyridine in the presence of a reducing agent gives $NbCl_x(py)_y$ which contains 50.02% C, 4.20% H and 11.67% N. Determine the values of x and y.

4.6 During purification of 2,2'-bipyridine (see structure **4.5**), the compound was accidentally exposed to a mineral acid. Elemental analysis gave the following results: C 62.35, H 4.71, N 14.54%. Suggest the identity of the isolated compound.

4.7 On being heated, the fullerene solvate $C_{60} \cdot x\text{CHBr}_3$ loses solvent in a two-step process. The final weight loss is 41%. Account for these data and determine x.

4.8 When gypsum ($CaSO_4 \cdot 2H_2O$) is heated to 433 K, it converts to the hemihydrate $CaSO_4 \cdot \frac{1}{2}H_2O$, and at 463 K, it forms γ-$CaSO_4$. Calculate the % weight changes at 433 and 463 K, and sketch what you expect to see in a TGA curve.

4.9 Birnessite, $[\text{Na,K}][\text{Mn}^{IV}\text{Mn}^{III}]O_4 \cdot xH_2O$, is a mineral with a layered structure of the same type as CdI_2 (see Fig. 6.23) comprising octahedral MnO_6 units. Na^+ and K^+ ions and H_2O molecules are sited between the layers. The Na/K composition is variable. Analysis for Na is carried out using AAS. A 20 mg sample of birnessite is dissolved in concentrated HCl/HNO_3 and the solution made up to $200 \, cm^3$ in a volumetric flask. (a) Why is birnessite treated with acid? (b) Detail how you would proceed with the sodium analysis. (c) How would you determine the water content in the sample of birnessite?

4.10 The EI mass spectrum and structure of $Cr(CO)_6$ is shown in Fig. 4.36. Rationalize the peaks in the spectrum. Why is the EI technique suitable for recording the mass spectrum of $Cr(CO)_6$?

4.11 In the FAB mass spectrum of $[Pd(PPh_3)_4]$ with NOBA matrix, the base peak appears at m/z 279.1. The isotope pattern shows that Pd is absent from the ion. Suggest an identity for the ion. Why can a peak at m/z 154.0 be ignored?

4.12 The EI mass spectrum of lead(II) acetate shows four peak envelopes, each with an isotope pattern characteristic of Pb. The most intense peak in each envelope appears at m/z 326.0, 267.0, 224.0 and 208.0, respectively. (a) By using Appendix 5, sketch the pattern of each peak envelope. (b) Assign the peaks.

4.13 Four MeCN solutions were made up containing $AgNO_3$ and PPh_3 in molar ratios of 1 : 1, 1 : 2, 1 : 3 and 1 : 4, respectively. ESI mass spectra (positive mode) of the solutions were recorded, and the data are tabulated below. Account for the data, including

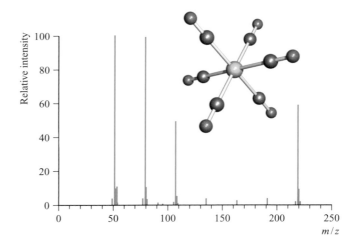

Fig. 4.36 The EI mass spectrum and structure of $Cr(CO)_6$. Colour code: Cr, yellow; C, grey; O, red. [Dr. P. Rösel and H. Nadig are acknowledged for the data.]

the differences between the spectra. Comment on the isotope patterns that you expect to see for each peak (see Appendix 5).

	Relative intensities (%) for different ratios AgNO$_3$: PPh$_3$			
m/z	1 : 1	1 : 2	1 : 3	1 : 4
369	42	32	0	0
410	100	89	0	0
633	20	100	100	100
802	11	2	0	0
893	1	3	22	22
1064	2	4	0	0

[Data: L.S. Bonnington *et al.* (1999) *Inorg. Chim. Acta*, vol. 290, p. 213.]

4.14 The ESI mass spectrum (positive mode) of the ligand shown below exhibits two peaks at m/z 299.2 (base peak) and 321.1. (a) What is a 'base peak'? (b) Suggest how the observed peaks arise. [Data: C.J. Sumby *et al.* (2009) *Tetrahedron*, vol. 65, p. 4681.]

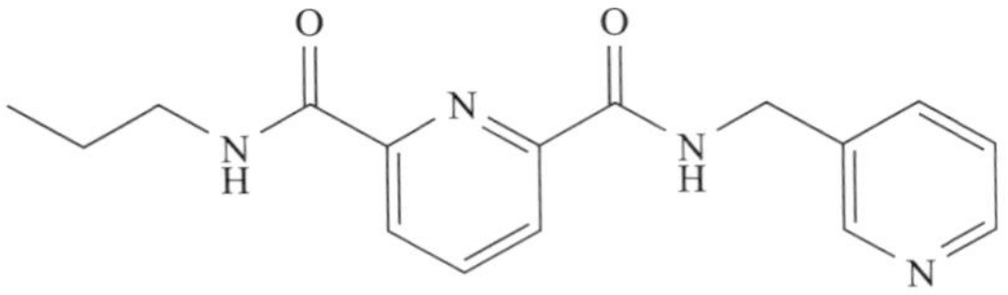

4.15 The ESI mass spectrum (positive mode) of the complex shown below contained a peak envelope with m/z 527.9 (100%), 528.9 (15%), 529.9 (46%), 530.9 (7%), 531.9 (0.5%). A group of peaks of low intensity and with spacings of $m/z = 1$ was also observed around m/z 994. (a) What is the oxidation state of Cu in the complex? (b) Assign the major peak and account for the isotope pattern. (c) Suggest how the minor peak arises. (Isotopes: see Appendix 5.)

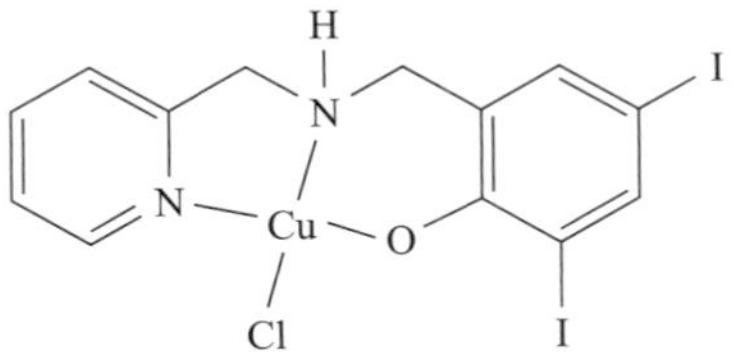

[Data: S.S. Hindo *et al.* (2009) *Eur. J. Med. Chem.*, vol. 44, p. 4353.]

4.16 Both positive and negative-ion ESI mass spectra of [Me$_4$Sb][Ph$_2$SbCl$_4$] were recorded. In one spectrum, peaks at m/z 181 (100%), 182 (4.5%), 183 (74.6%) and 184 (3.4%) were observed. The other mass spectrum revealed peaks at m/z 415 (48.8%), 416 (6.4%), 417 (100%), 418 (13.1%), 419 (78.6%), 420 (10.2%), 421 (30.1%), 422 (3.9%) and 423 (5.7%). (a) Account for the results, stating which mass spectrum was recorded in positive and which in negative mode. (b) Draw the structures of the ions present in [Me$_4$Sb][Ph$_2$SbCl$_4$]. Does either of the ions possess isomers? If so, do the mass spectrometric data provide information about which isomer or isomers are present? (Isotopes: see Appendix 5.) [Data: H.J. Breunig *et al.* (2010) *J. Organomet. Chem.*, vol. 695, p. 1307.]

4.17 1.0 mmol of the ligand, L, shown below was reacted with 0.50 mmol of PtCl$_2$. The positive mode MALDI-TOF mass spectrum of the purified product was run in α-cyano-4-hydroxycinnamic acid matrix. The most intense peaks in the peak envelopes in the mass spectrum were m/z 891.1, 869.1, 833.2 and 302.1. Assign these peaks and suggest a formula for the product that is consistent with the data.

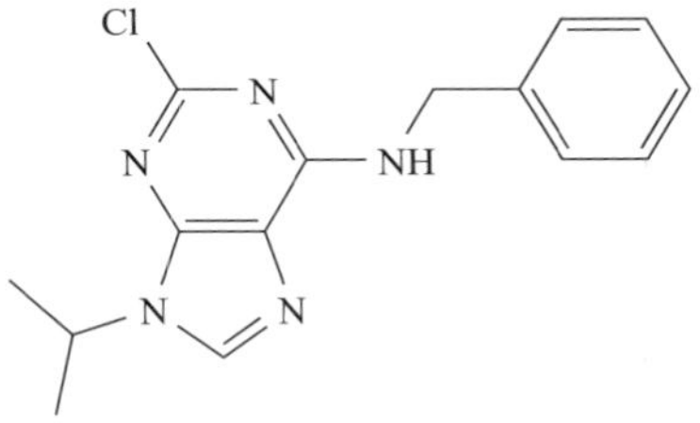

[Data: L. Szücová *et al.* (2008) *Polyhedron*, vol. 27, p. 2710.]

4.18 In the MALDI-TOF mass spectrum of the macrocyclic ligand shown below in 1,8,9-trihydroxyanthracene matrix, the dominant peaks are at m/z 615.7 (base peak), 637.7. Assign the peaks.

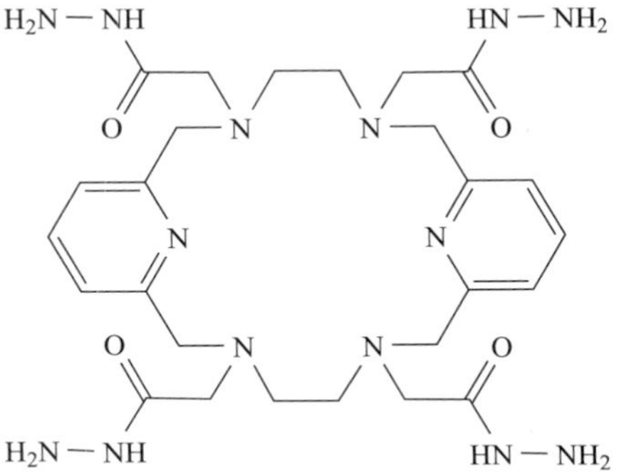

[Data: C. Nuñez *et al.* (2009) *Inorg. Chim. Acta*, vol. 362, p. 3454.]

4.19 Reaction of H_3L (drawn below) with $Cu(O_2CMe)_2 \cdot H_2O$ in MeOH with addition of pyridine (py) yields $[Cu_4L_2(O_2CMe)_2(py)_4(MeOH)_2]$. Show that a MALDI-TOF mass spectrum with peak envelopes at m/z 977 and 611 is consistent with this formulation.

H_3L

[Data: J.D. Crane *et al.* (2004) *Inorg. Chem. Comm.*, vol. 7, p. 499.]

4.20 What is the rule of mutual exclusion? Give two examples of molecular species to which this rule applies.

4.21 The ν_3 vibrational wavenumber for $[BF_4]^-$ comes at $1070\,cm^{-1}$, whereas the corresponding band for $[BCl_4]^-$, $[BBr_4]^-$ and $[BI_4]^-$ comes at 722, 620 and $533\,cm^{-1}$, respectively. Rationalize this trend.

4.22 Vibrational wavenumbers for $K[N_3]$ are 2041, 1344 and $645\,cm^{-1}$. Draw the structure of the $[N_3]^-$ ion and sketch the three vibrational modes. Which are IR active?

4.23 Resonance structures for urea are represented below:

The IR spectrum of free urea has absorptions at 3500 and 3350 ($\nu(NH_2)$), 1683 ($\nu(CO)$) and $1471\,cm^{-1}$ ($\nu(CN)$). Urea can bond to metal ions through either an *N*- or *O*-donor atom. When urea bonds through the O atom, the contribution from resonance form **A** decreases. In the IR spectrum of $[Pt(urea)_6]Cl_2$, bands at 3390, 3290, 3130, 3030, 1725 and $1395\,cm^{-1}$ are assigned to the vibrational modes of metal-bound urea. Suggest why these data suggest the formation of Pt–N rather than Pt–O bonds.

4.24 The IR spectrum of $Li_3[PO_4]$ shows absorptions at 1034 and $591\,cm^{-1}$. There are no bands below the $400\,cm^{-1}$ cutoff of the IR spectrometer. Why are these data consistent with the $[PO_4]^{3-}$ ion being tetrahedral rather than square planar?

4.25 The UV-VIS spectrum of a CH_3CN solution $(2.0 \times 10^{-5}\,mol\,dm^{-3})$ of an iron(II) complex is: $\lambda_{max}(\varepsilon) = 245$ (48 200), 276 (74 100), 284 (81 700), 324 (45 100), 569 nm $(25\,000\,dm^3\,mol^{-1}\,cm^{-1})$. A quartz cuvette with path length 1 cm was used for the measurement. (a) Explain why the compound is coloured. What colour do you expect the compound to be? (b) Which is the lowest energy absorption? Rationalize your answer. (c) The initial data were recorded as a plot of absorbance against wavelength. What was the value of A_{max} for the band at 245 nm?

4.26 The UV-VIS spectrum of a CH_2Cl_2 solution of the gold(I) compound shown below with R = Ph is: $\lambda_{max}(\varepsilon) = 239$ (92 500), 269 (67 000), 286 (72 000), 303 (28 000), 315 nm $(21\,000\,dm^3\,mol^{-1}\,cm^{-1})$.

[Data: E.C. Constable *et al.* (2009) *Eur. J. Inorg. Chem.*, p. 4710.]

(a) $\pi^* \longleftarrow \pi$ transitions contribute to the observed spectrum. How do these arise? (b) Is the compound coloured? (c) You are asked to compare the UV-VIS spectra of a series of these compounds with different R substituents. Why should you compare plots of ε against λ rather than A against λ?

4.27 Two isomers, **A** and **B**, of a complex can be distinguished because the UV-VIS spectrum of **B** is blue shifted with respect to that of **A**. Explain what this means. What is another term used for a blue shift?

In problems 4.28 to 4.51, refer to Table 4.3 for isotopic abundances where needed.

4.28 Why is a coupling constant measured in Hz and is not recorded as a chemical shift difference?

4.29 Long-range couplings are often observed between ^{31}P and ^{19}F nuclei, between ^{31}P and ^1H nuclei, but not between remote non-equivalent ^1H nuclei. What does this tell you about the relative magnitudes of values of J_{PF}, J_{PH} and J_{HH} for the respective pairs of nuclei when they are directly attached?

4.30 Rationalize the fact that the ^{13}C NMR spectrum of CF_3CO_2H consists of two binomial quartets with coupling constants of 44 and 284 Hz respectively.

4.31 How might you use ^{31}P NMR spectroscopy to distinguish between Ph_2PH and Ph_3P?

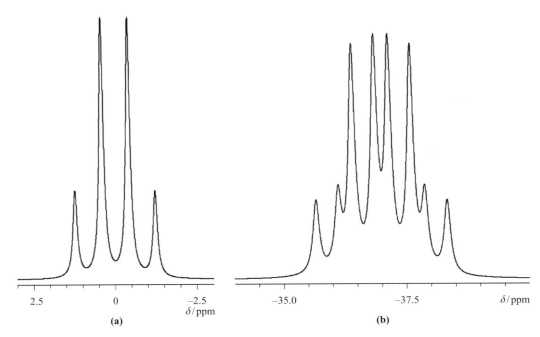

Fig. 4.37 Figure for problem 4.34.

4.32 The ^{31}P NMR spectrum of PMe$_3$ consists of a binomial decet (J 2.7 Hz). (a) Account for this observation. (b) Predict the nature of the ^{1}H NMR spectrum of PMe$_3$.

4.33 The ^{29}Si NMR spectrum of compound **4.15** shows a triplet with a coupling constant of 194 Hz. (a) Rationalize these data and (b) predict the nature of the signal in the ^{1}H NMR spectrum of **4.15** that is assigned to the silicon-bound protons. [^{29}Si: 4.7% abundant; $I = \frac{1}{2}$]

(4.15)

4.34 Figure 4.37 shows the ^{11}B NMR spectra of (a) THF·BH$_3$ (**4.16**) and (b) PhMe$_2$P·BH$_3$. Interpret the observed coupling patterns and mark on the figure where you would measure relevant coupling constants.

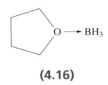

(4.16)

4.35 (a) Predict the structure of SF$_4$ using the VSEPR model. (b) Account for the fact that at 298 K and in solution the ^{19}F NMR spectrum of SF$_4$ exhibits a singlet but that at 175 K, two equal-intensity triplets are observed.

4.36 The ^{19}F NMR spectrum of each of the following molecules exhibits one signal. For which species is this observation consistent with a static molecular structure as predicted by the VSEPR model: (a) SiF$_4$; (b) PF$_5$; (c) SF$_6$; (d) SOF$_2$; (e) CF$_4$?

4.37 Outline the mechanism of Berry pseudo-rotation, giving two examples of molecules that undergo this process.

4.38 Is it correct to interpret the phrase 'static solution structure' as meaning necessarily rigid? Use the following molecules to exemplify your answer: PMe$_3$; OPMe$_3$; PPh$_3$; SiMe$_4$.

4.39 Account for the fact that the ^{29}Si NMR spectrum of a mixture of SiCl$_4$ and SiBr$_4$ that has been standing for 40 h contains five singlets which include those assigned to SiCl$_4$ (δ −19 ppm) and SiBr$_4$ (δ −90 ppm).

4.40 The structure of [P$_5$Br$_2$]$^+$ is shown in diagram **4.17**. Account for the fact that the ^{31}P NMR spectrum of this cation at 203 K consists of a doublet of triplets (J 321 Hz, 149 Hz), a triplet of triplets (J 321 Hz, 26 Hz) and a triplet of doublets (J 149 Hz, 26 Hz).

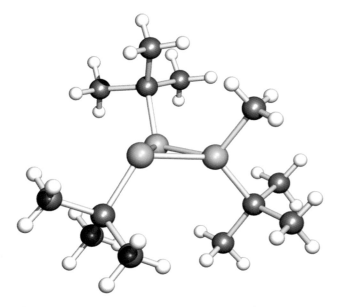

(4.17)

4.41 Tungsten hexacarbonyl (**4.18**) contains six equivalent CO ligands. With reference to Table 4.3, suggest what you would expect to observe in the ^{13}C NMR spectrum of a ^{13}C-enriched sample of $W(CO)_6$.

(4.18)

4.42 The compounds Se_nS_{8-n} with $n = 1$–5 are structurally similar to S_8. Structure **4.19** shows a representation of the S_8 ring (it is actually non-planar) and the atom numbering scheme; all the S atoms are equivalent. Using this as a guide, draw the structures of SeS_7, 1,2-Se_2S_6, 1,3-Se_2S_6, 1,2,3-Se_3S_5, 1,2,4-Se_3S_5, 1,2,5-Se_3S_5 and 1,2,3,4-Se_4S_4. How many signals would you expect to observe in the ^{77}Se ($I = \frac{1}{2}$, 7.6%) NMR spectrum of each compound?

(4.19)

4.43 Explain why the ^{19}F NMR spectrum of $BFCl_2$ consists of a 1:1:1:1 quartet. What would you expect to observe in the ^{19}F NMR spectrum of BF_2Cl?

4.44 Rationalize the fact that at 173 K, ^1H NMR spectroscopy shows that $SbMe_5$ possesses only one type of Me group.

4.45 MeCN solutions of $NbCl_5$ and HF contain a mixture of octahedral $[NbF_6]^-$, $[NbF_5Cl]^-$, $[NbF_4Cl_2]^-$, $[NbF_3Cl_3]^-$ and $[NbF_2Cl_4]^-$. Predict the number and coupling patterns of the signals in the ^{19}F NMR spectrum of each separate component in this mixture, taking into account possible isomers. (Assume static structures and no coupling to ^{193}Nb.)

4.46 (a) Explain why the ^{19}F NMR spectrum of $[PF_6]^-$ appears as a doublet.

(b) The $^{31}P\{^1H\}$ NMR spectrum of *trans*-$[PtI_2(PEt_3)_2]$ (**4.20**) shows a three-line pattern, the lines in which have relative integrals of $\approx$1 : 4 : 1. What is the origin of this pattern?

(4.20)

4.47 (a) In the ^1H NMR spectrum of compound **4.21**, there is a triplet at δ 3.60 ppm (J 10.4 Hz). Assign the signal and explain the origin of the coupling. What would you observe in the $^{31}P\{^1H\}$ NMR spectrum of compound **4.21**?

(4.21)

(b) Figure 4.38 shows the solid state structure of a phosphorus-containing cation. The ^{31}P NMR spectrum of a $CDCl_3$ solution of the $[CF_3SO_3]^-$ salt was recorded. How many signals (ignore spin–spin

Fig. 4.38 The structure of the $[(PCMe_3)_3Me]^+$ cation in the salt $[(PCMe_3)_3Me][CF_3SO_3]$ determined by X-ray diffraction [N. Burford *et al.* (2005) *Angew. Chem. Int. Ed.*, vol. 44, p. 6196]. Colour code: P, orange; C, grey; H, white.

coupling) would you expect to see in the spectrum, assuming that the solid state structure is retained in solution?

4.48 The ^{19}F NMR spectrum of the octahedral ion $[PF_5Me]^-$ shows two signals (δ −45.8 and −57.6 ppm). Why are *two* signals observed? From these signals, three coupling constants can be measured: $J_{PF} = 829$ Hz, $J_{PF} = 680$ Hz and $J_{FF} = 35$ Hz. Explain the origins of these coupling constants.

4.49 The ^{31}P{^{1}H} NMR spectrum of a CDCl$_3$ solution of the square planar rhodium(I) complex **4.22** exhibits a doublet of doublets (J 38 Hz, 145 Hz) and a doublet of triplets (J 38 Hz, 190 Hz). Rationalize these data. [*Hint:* look at Table 4.3.]

PPh$_3$
|
Ph$_3$P — Rh — PPh$_3$
|
Cl

(4.22)

4.50 NaBH$_4$ contains the tetrahedral $[BH_4]^-$ ion. Although NaBH$_4$ hydrolyses slowly in water, it is possible to obtain a clean ^{1}H NMR spectrum of the compound in D$_2$O. Naturally occurring boron consists of two isotopes: ^{11}B, 80.1%, $I = \frac{3}{2}$, and ^{10}B, 19.9%, $I = 3$. Assuming that a sharp, well-resolved spectrum is obtained, sketch the expected 400 MHz ^{1}H NMR spectrum (including a scale) if the signal for the protons occurs at δ −0.2 ppm, and the values of $J_{^{11}B^1H} = 80.5$ Hz and $J_{^{10}B^1H} = 27.1$ Hz. How would the spectrum differ if it were recorded at 100 MHz?

4.51 (a) Predict what you would expect to see in the ^{15}N NMR spectrum of the isotopically labelled compound *cis*-[Pt(^{15}NH$_3$)$_2$Cl$_2$]. (b) The observed coupling constants for this compound are $J_{^{15}N^1H} = 74$ Hz and $J_{^{15}N^{195}Pt} = 303$ Hz. Using your predicted spectrum as a starting point, explain why the observed spectrum is an *apparent* octet.

4.52 (a) If Na has the ground state electronic configuration of $[Ne]3s^1$, why is NaCl EPR silent? (b) Sketch an EPR spectrum for an isotropic system in which an electron interacts with a ^{14}N ($I = 1$) nucleus.

4.53 Vanadium has two isotopes (^{50}V, 0.25%; ^{51}V, 99.75%). The EPR spectrum of an aqueous solution of $[VO(OH_2)_5]^{2+}$ shows an 8-line pattern, and g-values of $g_{zz} = 1.932$, $g_{xx} = 1.979$ and $g_{yy} = 1.979$ were determined. What can you deduce from the data?

4.54 Figure 4.39 shows the single crystal EPR spectrum arising from Cu^{2+} doped into CaCd(O$_2$CMe)$_4 \cdot$6H$_2$O for one orientation of the crystal relative to the external magnetic field. The spectrum illustrates the

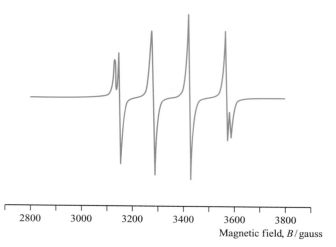

Fig. 4.39 The single crystal EPR spectrum of Cu^{2+} doped-CaCd(O$_2$CMe)$_4 \cdot$6H$_2$O. [Simulated by Dr. C. Palivan, University of Basel.]

presence of both isotopes of Cu (^{63}Cu and ^{65}Cu, natural abundances 69.2% and 30.8%, respectively, both $I = \frac{3}{2}$). (a) How many peaks do you expect to observe for each isotope? (b) Explain why the intensities of the central peaks in Fig. 4.39 are higher than those of the outer peaks. (c) The hyperfine coupling constant $A(^{65}$Cu) is 1.07 times higher than $A(^{63}$Cu). Which signals in Fig. 4.39 arise from each isotope? Calculate the value of A (in G) for each isotope. (d) The EPR spectrum in Fig. 4.39 was measured as 9.75 GHz. Calculate the gyromagnetic factor of the copper(II) paramagnetic species.

4.55 Figure 4.40 shows the EPR spectra for two isotropic systems in which the unpaired electron interacts with two ^{14}N nuclei ($I = 1$). (a) Use Fig. 4.40a to calculate the gyromagnetic factor of the paramagnetic species if the spectrum was measured at 9.75 GHz. (b) Which EPR spectrum in Fig. 4.40 indicates the interaction of the unpaired electron with two equivalent nitrogen nuclei? (c) Calculate the values of the hyperfine coupling constants for both cases in Fig. 4.40.

4.56 Predict what you expect to observe in the EPR spectrum of a species in which an unpaired electron interacts with one ^{14}N nucleus ($I = 1$) *and* one ^{1}H nucleus ($I = \frac{1}{2}$) if the hyperfine coupling constants are (a) $A(^{14}$N) = $A(^1$H) = 30 G; (b) $A(^{14}$N) = 30 G, $A(^1$H) = 10 G.

4.57 Figure 4.41 shows the 9.214 GHz EPR spectrum arising from the two sites of Co^{2+} (^{59}Co, 100% abundant, $I = \frac{7}{2}$) doped into magnesium acetate. The sites differ slightly in their effective hyperfine coupling constants, and more significantly in their g-values. (a) Calculate the g-values for the two sites.

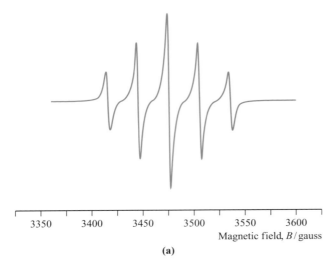

(a)

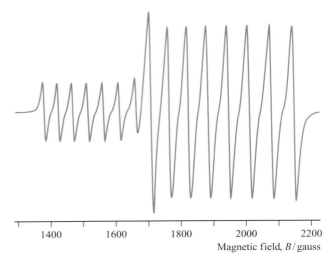

Fig. 4.41 EPR spectrum for problem 4.57. [Simulated by Dr. C. Palivan, University of Basel.]

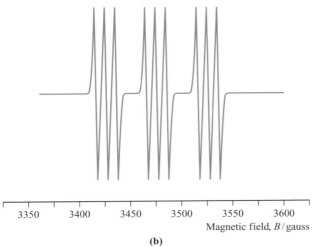

(b)

Fig. 4.40 EPR spectra for problem 4.55. [Simulated by Dr. C. Palivan, University of Basel.]

(b) Explain why the central peak is more intense than the peaks appearing at higher magnetic field. (c) What would happen if the differences in *g*-values of the two sites is smaller than is observed in Fig. 4.41?

4.58 Suggest a suitable technique for investigating the presence of Fe^{2+} and/or Fe^{3+} ions in clays, and for distinguishing between tetrahedrally and octahedrally sited Fe centres. Give reasons for your choice.

4.59 (a) Explain why the location of H atoms by X-ray diffraction is difficult. (b) What technique (and why) is used for the accurate determination of H atom positions in a structure? Comment on any limitations of this method.

4.60 What are the features that distinguish X-ray quality single crystals, a polycrystalline solid and an amorphous solid?

4.61 By referring to Section 10.6, suggest how the following solid state structures (i.e. the organization of the molecules in the solid state) may be affected by hydrogen bonding: (a) $PhB(OH)_2$, (b) H_2SO_4, (c) $MeCO_2H$.

4.62 Comment on ways in which you can ensure that the X-ray crystal structure of a compound that you have prepared is representative of the bulk sample of the compound.

4.63 Molecular structures determined by X-ray diffraction are often represented in the form of ORTEP diagrams showing thermal ellipsoids for non-hydrogen atoms. (a) What do you understand by a *thermal ellipsoid*? (b) Why are H atoms usually represented isotropically?

OVERVIEW PROBLEMS

4.64 Suggest a suitable experimental technique for each of the following problems.

(a) Single crystal X-ray diffraction data show that a compound crystallizes as a dihydrate. How could you show that the bulk sample was anhydrous?

(b) The single crystal structure of a compound has been determined, and you wish to prepare the same polymorph in bulk in a series of syntheses over a period of several weeks. How can you ensure that the products fulfil this aim?

(c) Changing the conditions under which a compound **X** crystallizes leads to two batches of crystals with melting points of 97 and 99 °C, respectively. How can you show that the compounds are polymorphs?

4.65 (a) The IR spectrum of naturally occurring CO shows an absorption at $2170 \, cm^{-1}$ assigned to the molecular vibrational mode. If the sample is enriched in ^{13}C, what change do you expect to see in the IR spectrum?

(b) The IR spectra of salts of $[Fe(CO)_6]^{2+}$ and $[Fe(CO)_4]^{2-}$, respectively show absorptions at 2204 and $1788 \, cm^{-1}$ assigned to carbonyl vibrational modes. What do these tell you about the strengths of the carbon–oxygen bonds in the complexes compared to that in free CO?

(c) $[Fe(CO)_4]^{2-}$ and $Fe(CO)_5$ have T_d and D_{3h} symmetries respectively. What do you expect to observe in the room temperature ^{13}C NMR spectra of $Na_2[Fe(CO)_4]$ and $Fe(CO)_5$? Rationalize your answer.

4.66 You have prepared the complex $[Ru(py)_6][BF_4]_2$ (py = pyridine). (a) What information can you obtain from the elemental analysis? (b) How would you confirm the presence of the $[BF_4]^-$ ion? (c) How would you confirm that all the pyridine ligands were in the same environment in solution? (d) How would you confirm the presence of an octahedral $[Ru(py)_6]^{2+}$ ion in the solid state? (e) What would you expect to see in the ESI mass spectrum of $[Ru(py)_6][BF_4]_2$?

4.67 The question of whether you have synthesized a monomeric (**X**) or dimeric (**X**$_2$) species can be solved in a number of ways. (a) Will elemental analysis help you? (b) The compound contains an organic unit that can be probed by ^1H NMR spectroscopy. Will this technique distinguish between monomer and dimer? (c) What information could you gain from mass spectrometry? (d) A single crystal X-ray diffraction study at 223 K shows the presence of the dimer **X**$_2$? Does this mean that monomeric **X** never forms?

4.68 NMR spectroscopy is considered to be one of the most powerful, routine analytical tools for the characterization of new compounds. Comment on the validity of this statement.

4.69 A student has prepared a sample of $[Zn(en)_3]Cl_2$ (en = $H_2NCH_2CH_2NH_2$) but is worried that the complex appears blue when $[Zn(en)_3]Cl_2$ should be colourless. The student wonders if she picked up a bottle of nickel(II) chloride instead of zinc(II) chloride. The experimental CHN analysis for the complex is C 23.00, H 7.71, N 26.92%. (a) Do the elemental analytical data distinguish between $[Zn(en)_3]Cl_2$ and $[Ni(en)_3]Cl_2$? Comment on your answer. (b) How would mass spectrometry help you to distinguish between the two compounds? (c) By referring to Chapter 20, suggest why ^1H NMR spectroscopy might be useful in distingushing between $[Zn(en)_3]Cl_2$ and $[Ni(en)_3]Cl_2$. (d) A single crystal X-ray diffraction study was carried out and confirms the presence of $[M(en)_3]Cl_2$. Can this technique unambiguously assign M to Zn or Ni? (e) Explain why AAS could be used to confirm the identity of the metal.

INORGANIC CHEMISTRY MATTERS

4.70 Portable mass spectrometers are now available for monitoring gaseous emissions from volcanoes. Analyses after the eruption of the Turrialba Volcano in Costa Rica in January 2010 showed the presence of $[M]^+$ ions (i.e. atomic ions and non-fragmented molecular ions) at m/z 64, 44, 40, 34, 32, 28 and 18. Mass spectra recorded before the eruption confirmed that the peak at m/z 64 was missing, and the peak at m/z 34 was substantially diminished. (a) To what do you assign the peaks at m/z 64 and 34? (b) Bearing in mind the composition of the Earth's atmosphere, what molecular ion interferes with the detection of the volcanic emission observed at m/z 34? (c) Volcanic gases mix with the surrounding air. Suggest what atmospheric gases give rise to the ions at m/z 40, 32 and 28. (d) Sampling of the volcanic plume peak before and after the eruption reveals that the molecular ion with m/z 18 is present on both occasions, but in greater amounts after the eruption. Identify this ion. (e) Helium forms under the Earth's crust and is released during volcanic activity. Mass spectrometric ground monitoring of vents (fumaroles) in the Turrialba crater showed 20 ppm levels of He being released. What m/z value characterizes helium emission and what is the formula of the corresponding ion?

[Data: J.A. Diaz *et al.* (2010) *Int. J. Mass. Spec.*, vol. 295, p. 105.]

4.71 Ultramarines are bright blue pigments based on a zeolite (sodalite) that hosts the *colour centres* $[S_2]^-$ and $[S_3]^-$. Ultramarines include lapis lazuli and have been in use for over 5500 years. The UV-VIS spectrum of $[S_2]^-$ exhibits a broad band centred at 370 nm, while $[S_3]^-$ absorbs at 595 nm. (a) Draw Lewis structures for $[S_2]^-$ and $[S_3]^-$. What shape is the $[S_3]^-$ ion? (b) Why can EPR spectroscopy be used to study these ions? Suggest a reason why both ions behave isotropically even at low temperatures. (c) Explain why an ultramarine containing only $[S_3]^-$ colour centres appears violet-blue in colour. (d) Why does $[S_2]^-$ contribute to the colour of ultramarine pigments even though λ_{max} is 370 nm, i.e. in the UV region? (e) In synthetic ultramarines, the ratio of $[S_3]^- : [S_2]^-$ can be altered to produce pigments ranging from violet-blue through blues to green. Account for this.

Topics

Hybridization of atomic
  orbitals
Molecular orbital theory:
  polyatomics
Application of character
  tables
Delocalized bonding

5

Bonding in polyatomic molecules

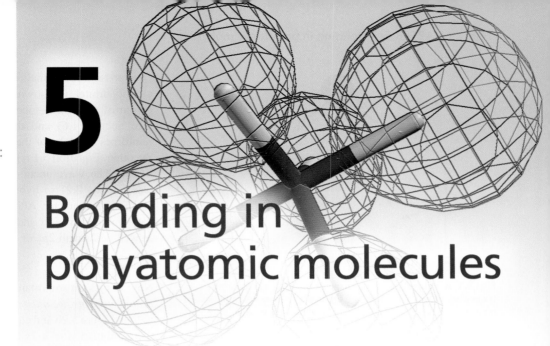

5.1 Introduction

In Chapter 2, we considered three approaches to the bonding in diatomic molecules:

- Lewis structures;
- valence bond (VB) theory;
- molecular orbital (MO) theory.

In this chapter we extend the discussion to polyatomic molecules (i.e. those containing three or more atoms). Within the valence bond model, treatment of a molecule XY_n ($n \geq 2$) raises the question of compatibility (or not) between the positions of the Y atoms and the directionalities of the atomic orbitals on the central atom X. Although an s atomic orbital is spherically symmetric, other atomic orbitals possess directional properties (see Section 1.6). Consider H_2O: Fig. 5.1 illustrates that, if the atoms of the H_2O molecule lie in (for example) the yz plane, the directionalities of the $2p_y$ and $2p_z$ atomic orbital of oxygen are not compatible with the directionalities of the two O−H bonds. Although we could define the z axis to coincide with one O−H bond, the y axis could not (at the same time) coincide with the other O−H bond. Hence, there is a problem in trying to derive a localized bonding scheme in terms of an atomic orbital basis set (see Section 2.3). The next section describes a bonding model within valence bond (VB) theory that overcomes this problem. After we have considered how VB theory views the bonding in a range of XY_n species, we move on to the problems of applying molecular orbital theory to polyatomic species.

A **polyatomic species** contains three or more atoms.

5.2 Valence bond theory: hybridization of atomic orbitals

What is orbital hybridization?

The word 'hybridization' means 'mixing' and when used in the context of atomic orbitals, it describes a way of deriving *spatially directed orbitals* which may be used within VB theory. Like all bonding theories, *orbital hybridization is a model*, and should *not* be taken to be a real phenomenon.

Hybrid orbitals may be formed by mixing the characters of atomic orbitals that are close in energy. The character of a hybrid orbital depends on the atomic orbitals involved and their percentage contributions. The labels given to hybrid orbitals reflect the contributing atomic orbitals, e.g. an sp hybrid possesses equal amounts of s and p orbital character.

Hybrid orbitals are generated by mixing the characters of atomic orbitals.

The reason for creating a set of hybrid orbitals is to produce a convenient bonding scheme for a particular molecular species. An individual hybrid orbital points along a given internuclear axis within the framework of the molecule under consideration, and use of a set of hybrid orbitals provides a bonding picture in terms of *localized σ-bonds*. In working through the rest of this section, notice that each hybridization scheme for an atom X in a molecule XY_n is appropriate only for a particular shape, the shape being defined by the number of attached groups and any lone pairs.

A set of hybrid orbitals provides a bonding picture for a molecule in terms of **localized σ-bonds**.

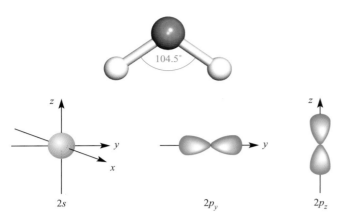

Fig. 5.1 A comparison of the shape of the H_2O molecule (the framework of which is taken as lying in the yz plane) with the spatial properties of the $2s$, $2p_y$ and $2p_z$ atomic orbitals of oxygen.

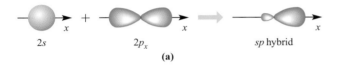

(a)

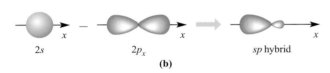

(b)

Fig. 5.2 The formation of two sp hybrid orbitals from one $2s$ atomic orbital and one $2p$ atomic orbital.

sp Hybridization: a scheme for linear species

The notation **sp** means that one s atomic orbital and one p atomic orbital mix to form a set of two hybrid orbitals with different directional properties.

One possible combination of a $2s$ atomic orbital and $2p_x$ atomic orbital is shown in Fig. 5.2a. In the figure, the colour of the orbital lobe corresponds to a particular phase (see Section 1.6) and the addition of the $2s$ component reinforces one lobe of the $2p_x$ atomic orbital but diminishes the other. Equation 5.1 represents the combination mathematically. The wavefunction $\psi_{sp\,hybrid}$ describes a normalized (see Section 2.2) sp hybrid orbital which possesses 50% s and 50% p character. Although eq. 5.1 and Fig. 5.2a refer to the combination of $2s$ and $2p_x$ atomic orbitals, this could just as well be $2s$ with $2p_y$ or $2p_z$, or $3s$ with $3p_x$, and so on.

$$\psi_{sp\,hybrid} = \frac{1}{\sqrt{2}}(\psi_{2s} + \psi_{2p_x}) \tag{5.1}$$

Now comes an important general rule: *if we begin with n atomic orbitals, we must end up with n orbitals after hybridization.* Figure 5.2b and eq. 5.2 show the second possibility for the combination of a $2s$ and a $2p_x$ atomic orbital. The sign change for the combination changes the phase of the $2p_x$ orbital and so the resultant hybrid points in the opposite direction to the one shown in Fig. 5.2a. (Remember that p atomic orbitals have vector properties.)

$$\psi_{sp\,hybrid} = \frac{1}{\sqrt{2}}(\psi_{2s} - \psi_{2p_x}) \tag{5.2}$$

Equations 5.1 and 5.2 represent two wavefunctions which are equivalent in every respect *except for their directionalities* with respect to the x axis. Although the orbital energies of the initial $2s$ and $2p_x$ atomic orbitals were different, mixing leads to two hybrid orbitals of equal energy.

The model of sp hybridization can be used to describe the σ-bonding in a linear molecule such as $BeCl_2$ in which the Be—Cl bonds are of equal length. The ground state electronic configuration of Be is $[He]2s^2$ and the valence shell contains the $2s$ atomic orbital and three $2p$ atomic orbitals (Fig. 5.3). If we use two of these atomic orbitals,

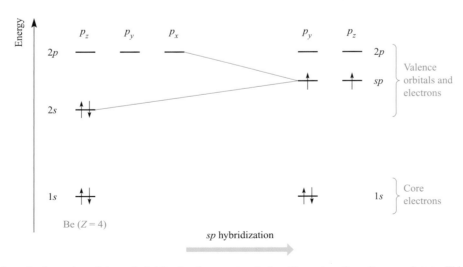

Fig. 5.3 Scheme to show the formation of the sp hybridized valence state of a beryllium atom from its ground state. This is a formalism and is not a 'real' observation, e.g. the valence state *cannot* be observed by spectroscopic techniques. The choice of using the $2p_x$ orbital for hybridization is arbitrary.

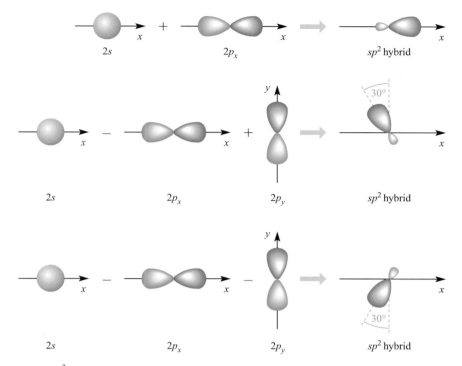

Fig. 5.4 The formation of three sp^2 hybrid orbitals from one $2s$ atomic orbital and two $2p$ atomic orbitals. The choice of p_x and p_y is arbitrary. (If we started with $2p_x$ and $2p_z$ atomic orbitals, the hybrids would lie in the xz plane; using the $2p_y$ and $2p_z$ atomic orbitals gives hybrid orbitals in the yz plane.) The directionalities of the hybrid orbitals follow from the relative contributions of the atomic orbitals (see eqs. 5.3–5.5).

treating them separately, to form two localized Be−Cl bonds, we cannot rationalize the bond equivalence. However, if we take the $2s$ atomic orbital and one $2p$ atomic orbital, mix their characters to form sp hybrids, and use one hybrid orbital to form one Be−Cl interaction and the other hybrid orbital for the second interaction, then the equivalence of the Be−Cl interactions is a natural consequence of the bonding picture. Effectively, we are representing the valence state of Be in a linear molecule as consisting of two degenerate sp hybrids, each containing one electron; this is represented by the notation $(sp)^2$. Figure 5.3 represents the change from the ground state electronic configuration of Be to an sp valence state. This is a *theoretical state* which can be used to describe σ-bonding in a linear molecule.

sp^2 Hybridization: a scheme for trigonal planar species

The notation sp^2 means that one s and two p atomic orbitals mix to form a set of three hybrid orbitals with different directional properties.

Let us consider the combination of $2s$, $2p_x$ and $2p_y$ atomic orbitals. The final hybrid orbitals must be equivalent in every way except for their directional properties; sp^2 hybrids must contain the same amount of s character as each other and the same amount of p character as one another. We begin by giving one-third of the $2s$ character to each sp^2 hybrid orbital. The remaining two-thirds of

each hybrid orbital consists of $2p$ character, and the normalized wavefunctions are given in eqs. 5.3 to 5.5.

$$\psi_{sp^2\ \text{hybrid}} = \frac{1}{\sqrt{3}}\psi_{2s} + \sqrt{\frac{2}{3}}\psi_{2p_x} \qquad (5.3)$$

$$\psi_{sp^2\ \text{hybrid}} = \frac{1}{\sqrt{3}}\psi_{2s} - \frac{1}{\sqrt{6}}\psi_{2p_x} + \frac{1}{\sqrt{2}}\psi_{2p_y} \qquad (5.4)$$

$$\psi_{sp^2\ \text{hybrid}} = \frac{1}{\sqrt{3}}\psi_{2s} - \frac{1}{\sqrt{6}}\psi_{2p_x} - \frac{1}{\sqrt{2}}\psi_{2p_y} \qquad (5.5)$$

Figure 5.4 gives a pictorial representation of the way in which the three sp^2 hybrid orbitals are constructed. Remember that a change in sign for the atomic wavefunction means a change in phase. The resultant directions of the lower two hybrid orbitals in Fig. 5.4 are determined by resolving the vectors associated with the $2p_x$ and $2p_y$ atomic orbitals.

The model of sp^2 hybridization can be used to describe the σ-bonding in trigonal planar molecules such as BH_3. The valence state of the B atom is $(sp^2)^3$ (i.e. three sp^2 hybrid orbitals, each with one electron) and the equivalence of the B−H interactions follows by considering that each interaction is formed by the overlap of one B sp^2 hybrid orbital with the $1s$ atomic orbital of an H atom (Fig. 5.5). Each H atom contributes one electron to the bonding scheme and, so, each B−H σ-bond is a localized 2c-2e interaction (see Section 2.2). A diagram similar to that shown in Fig. 5.3 can be constructed to show the formation of a valence state for the trigonal planar B atom.

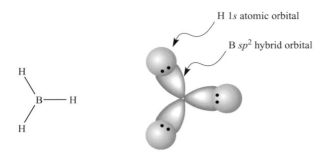

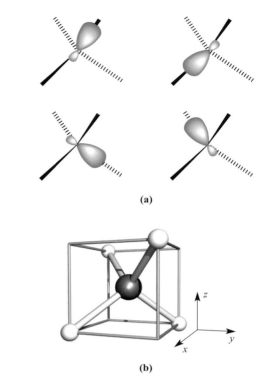

Fig. 5.5 The bonding in trigonal planar BH_3 can be conveniently described in terms of the interactions between a set of sp^2 hybrid orbitals centred on the B atom and three H $1s$ atomic orbitals. Three pairs of electrons are available (three electrons from B and one from each H) to give three 2c-2e σ-bonds.

(a)

(b)

 Fig. 5.6 (a) The directions of the orbitals that make up a set of four sp^3 hybrid orbitals correspond to a tetrahedral array. (b) The relationship between a tetrahedron and a cube; in CH_4, the four H atoms occupy alternate corners of a cube, and the cube is easily related to a Cartesian axis set.

sp^3 Hybridization: a scheme for tetrahedral and related species

The notation sp^3 means that one s and three p atomic orbitals mix to form a set of four hybrid orbitals with different directional properties.

A similar scheme to those described above can be derived to generate four sp^3 hybrid orbitals from one $2s$ and three $2p$ atomic orbitals. The sp^3 hybrid orbitals are described by the normalized wavefunctions in eqs. 5.6–5.9 and are shown pictorially in Fig. 5.6a. Each sp^3 hybrid orbital possesses 25% s character and 75% p character, and the set of four equivalent orbitals defines a tetrahedral framework.

$$\psi_{sp^3 \text{ hybrid}} = \frac{1}{2}(\psi_{2s} + \psi_{2p_x} + \psi_{2p_y} + \psi_{2p_z}) \tag{5.6}$$

$$\psi_{sp^3 \text{ hybrid}} = \frac{1}{2}(\psi_{2s} + \psi_{2p_x} - \psi_{2p_y} - \psi_{2p_z}) \tag{5.7}$$

$$\psi_{sp^3 \text{ hybrid}} = \frac{1}{2}(\psi_{2s} - \psi_{2p_x} + \psi_{2p_y} - \psi_{2p_z}) \tag{5.8}$$

$$\psi_{sp^3 \text{ hybrid}} = \frac{1}{2}(\psi_{2s} - \psi_{2p_x} - \psi_{2p_y} + \psi_{2p_z}) \tag{5.9}$$

Figure 5.6b illustrates how the tetrahedral structure of CH_4 relates to a cubic framework. This relationship is important because it allows you to describe a tetrahedron in terms of a Cartesian axis set. Within valence bond theory, the bonding in CH_4 can conveniently be described in terms of an sp^3 valence state for C, i.e. four degenerate orbitals, each containing one electron. Each hybrid orbital overlaps with the $1s$ atomic orbital of one H atom to generate one of four equivalent, localized 2c-2e C–H σ-interactions.

Worked example 5.1   Hybridization scheme for the nitrogen atom in NH₃

Use the VSEPR model to account for the structure of NH_3, and suggest an appropriate hybridization scheme for the N atom.

The ground state electronic configuration of N is $[\text{He}]2s^2 2p^3$.

Three of the five valence electrons are used to form three N–H single bonds, leaving one lone pair.

The structure is trigonal pyramidal, derived from a tetrahedral arrangement of electron pairs:

The N atom has four valence atomic orbitals: $2s$, $2p_x$, $2p_y$ and $2p_z$. An sp^3 hybridization scheme gives a tetrahedral arrangement of hybrid orbitals, appropriate for accommodating the four pairs of electrons:

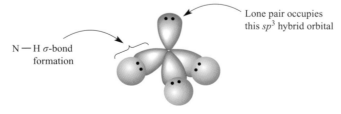

N–H σ-bond formation

Lone pair occupies this sp^3 hybrid orbital

Self-study exercises

1. Use the VSEPR model to account for the tetrahedral structure of $[NH_4]^+$.

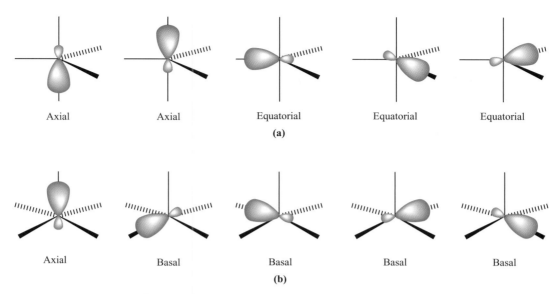

Fig. 5.7 A schematic representation of sp^3d hybridization. (a) A combination of s, p_x, p_y, p_z and d_{z^2} atomic orbitals gives a set of five sp^3d hybrid orbitals corresponding to a trigonal bipyramidal arrangement; the axial sp^3d hybrid orbitals are directed along the z axis. (b) A combination of s, p_x, p_y, p_z and $d_{x^2-y^2}$ atomic orbitals gives a set of five sp^3d hybrid orbitals corresponding to a square-based pyramidal arrangement; the axial sp^3d hybrid orbital is directed along the z axis.

2. Rationalize why H_2O is bent but XeF_2 is linear. Explain why an sp^3 hybridization scheme can be applied to H_2O, but not to XeF_2.

3. Give a suitable hybridization scheme for the central atom in each of the following: (a) $[NH_4]^+$; (b) H_2S; (c) BBr_3; (d) NF_3; (e) $[H_3O]^+$.

   [*Ans.* (a) sp^3; (b) sp^3; (c) sp^2; (d) sp^3; (e) sp^3]

Other hybridization schemes

For molecular species with other than linear, trigonal planar or tetrahedral-based structures, it is usual to involve d orbitals within valence bond theory. We shall see later that this is not necessarily the case within molecular orbital theory. We shall also see in Chapters 15 and 16 that the bonding in so-called *hypervalent compounds* such as PF_5 and SF_6, can be described without invoking the use of d-orbitals. One should therefore be cautious about using sp^nd^m hybridization schemes in compounds of p-block elements with apparently expanded octets around the central atom. *Real* molecules do not have to conform to simple theories of valence, nor must they conform to the sp^nd^m schemes that we consider in this book. Nevertheless, it is convenient to visualize the bonding in molecules in terms of a range of simple hybridization schemes.

The mixing of s, p_x, p_y, p_z and d_{z^2} atomic orbitals gives a set of five sp^3d hybrid orbitals, the mutual orientations of which correspond to a trigonal bipyramidal arrangement (Fig. 5.7a). The five sp^3d hybrid orbitals are *not* equivalent and divide into sets of two axial and three equatorial orbitals; the axial orbital lobes lie along

the z axis.[†] The model of sp^3d hybridization can be used to describe the σ-bonding in 5-coordinate species such as $[Ni(CN)_5]^{3-}$ (see Section 21.11).

The σ-bonding framework in a square-pyramidal species may also be described in terms of an sp^3d hybridization scheme. The change in spatial disposition of the five hybrid orbitals from trigonal bipyramidal to square-based pyramidal is a consequence of the participation of a different d orbital. Hybridization of s, p_x, p_y, p_z and $d_{x^2-y^2}$ atomic orbitals generates a set of five sp^3d hybrid orbitals (Fig. 5.7b).

Hybridization of s, p_x, p_y, p_z, d_{z^2} and $d_{x^2-y^2}$ atomic orbitals gives six sp^3d^2 hybrid orbitals corresponding to an octahedral arrangement. The bonding in MoF_6 can be described in terms of sp^3d^2 hybridization of the central atom. If we remove the z-components from this set (i.e. p_z and d_{z^2}) and hybridize only the s, p_x, p_y and $d_{x^2-y^2}$ atomic orbitals, the resultant set of four sp^2d hybrid orbitals corresponds to a square planar arrangement, e.g. $[PtCl_4]^{2-}$.

Each set of hybrid orbitals is associated with a particular shape, although this may not coincide with the molecular shape if lone pairs also have to be accommodated:

- sp linear
- sp^2 trigonal planar
- sp^3 tetrahedral
- sp^3d (d_{z^2})   trigonal bipyramidal
- sp^3d ($d_{x^2-y^2}$) square-based pyramidal
- sp^3d^2 octahedral
- sp^2d square planar

[†]Choice of coincidence between the z axis and the axial lobes is convenient and tends to be conventional.

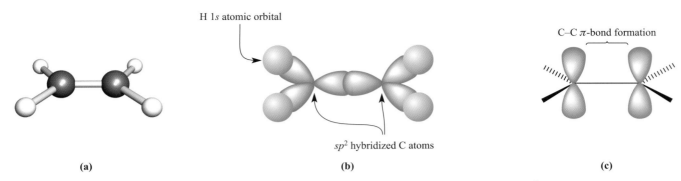

Fig. 5.8 (a) Ethene is a planar molecule with H−C−H and C−C−H bond angles close to 120°. (b) An sp^2 hybridization scheme is appropriate to describe the σ-bonding framework. (c) This leaves a $2p$ atomic orbital on each C atom; overlap between them gives a C−C π-interaction.

5.3 Valence bond theory: multiple bonding in polyatomic molecules

In the previous section, we emphasized that hybridization of some or all of the valence atomic orbitals of the central atom in an XY_n species provided a scheme for describing the X−Y σ-bonding. In, for example, the formation of sp, sp^2 and sp^3d hybrid orbitals, some p or d atomic orbitals remain unhybridized and, if appropriate, may participate in the formation of π-bonds. In this section we use the examples of C_2H_4, HCN and BF_3 to illustrate how multiple bonds in polyatomic molecules are treated within VB theory. Before considering the bonding in any molecule, the ground state electronic configurations of the atoms involved should be noted.

C_2H_4

C $[He]2s^22p^2$
H $1s^1$

Ethene, C_2H_4, is a planar molecule (Fig. 5.8a) with C−C−H and H−C−H bond angles of 121.3° and 117.4° respectively. Thus, each C centre is approximately trigonal planar and the σ-bonding framework within C_2H_4 can be described in terms of an sp^2 hybridization scheme (Fig. 5.8b). The three

σ-interactions per C atom use three of the four valence electrons, leaving one electron occupying the unhybridized $2p$ atomic orbital. The interaction between the two $2p$ atomic orbitals (Fig. 5.8c) and the pairing of the two electrons in these atomic orbitals generates a C−C π-interaction. The bond order of the C−C bond in C_2H_4 is therefore 2, in keeping with Lewis structure **5.1**. The π-component of the overall carbon–carbon bond is weaker than the σ-component and hence a C=C double bond, though stronger than a C−C single bond, is not twice as strong. The C−C bond enthalpy terms in C_2H_4 and C_2H_6 are 598 and 346 kJ mol^{-1} respectively.

(structure of ethene showing C=C with H atoms)

(5.1)

HCN

C $[He]2s^22p^2$
N $[He]2s^22p^3$
H $1s^1$

Figure 5.9a shows the linear HCN molecule, a Lewis structure (**5.2**) for which indicates the presence of an H−C single

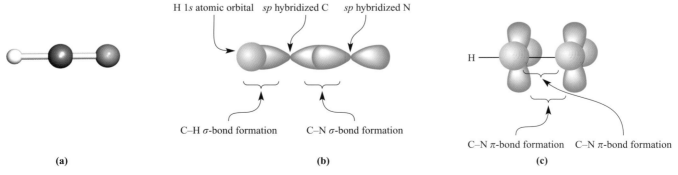

Fig. 5.9 (a) The linear structure of HCN; colour code: C, grey; N, blue; H, white. (b) An sp hybridization scheme for C and N can be used to describe the σ-bonding in HCN. (c) The π-character in the C−N bond arises from $2p$–$2p$ overlap.

$$H\!\!-\!\!C\!\!\equiv\!\!N\!:$$

(5.2)

bond, a C≡N triple bond, and a lone pair of electrons on N. An sp hybridization scheme is appropriate for both C and N; it is consistent with the linear arrangement of atoms around C and with the placement of the lone pair on N as far away as possible from the bonding electrons. Figure 5.9b shows the σ-bonding framework in HCN (each region of orbital overlap is occupied by a pair of electrons) and the out-

ward-pointing sp hybrid on N that accommodates the lone pair. If we arbitrarily define the HCN axis as the z axis, then after the formation of the σ-interactions, a $2p_x$ and a $2p_y$ atomic orbital remain on each of the C and N atoms. Each atomic orbital contains one electron. Overlap between the two $2p_x$ and between the two $2p_y$ orbitals leads to two π-interactions (Fig. 5.9c). The overall C–N bond order is 3, consistent with Lewis structure **5.2**.

BF₃

B $[\text{He}]2s^2 2p^1$
F $[\text{He}]2s^2 2p^5$

Boron trifluoride (Fig. 5.10a) is trigonal planar (D_{3h}), and sp^2 hybridization is appropriate for the B atom. Each of the three B–F σ-interactions arises by overlap of an sp^2 hybrid on the B atom with, for example, an sp^2 orbital on the F atom. After the formation of the σ-bonding framework, the B atom is left with an *unoccupied* $2p$ atomic orbital lying perpendicular to the plane containing the BF₃ molecule. As Fig. 5.10b shows, this is ideally set up for interaction with a *filled* $2p$ atomic orbital on one of the F atoms to give a localized B–F π-interaction. Notice that the two electrons occupying this π-bonding orbital both originate from the F atom. This picture of the bonding in BF₃ is analogous to one of the resonance forms shown in pink in Fig. 5.10c. All three resonance forms (see Section 2.2) are needed to account for the experimental observation that the three B–F bonds are of equal length (131 pm).

Worked example 5.2   Valence bond treatment of the bonding in [NO₃]⁻

(a) The [NO₃]⁻ ion has D_{3h} symmetry. What does this tell you about its structure? (b) Draw a set of resonance structures (focusing only on those that contribute significantly) for the nitrate ion. (c) Use an appropriate hybridization scheme to describe the bonding in [NO₃]⁻.

(a) If [NO₃]⁻ has D_{3h} symmetry, it must be planar, possess O–N–O bond angles of 120°, and have equal N–O bond distances.
(b) First, write down the electronic configurations for N ($Z = 7$) and O ($Z = 8$).

N $[\text{He}]2s^2 2p^3$ O $[\text{He}]2s^2 2p^4$

There is an additional electron from the negative charge giving a total of 24 valence electrons.

Both N and O obey the octet rule and so the most important resonance forms are expected to be:

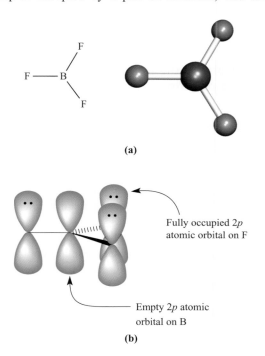

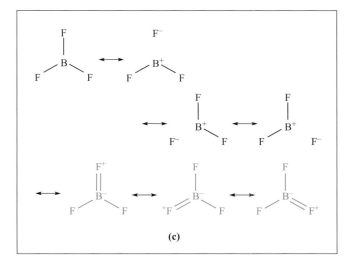

Fig. 5.10 (a) BF₃ possesses a trigonal planar structure. (b) $2p$–$2p$ overlap between B and F leads to the formation of a π-interaction. (c) Boron–fluorine double bond character is also deduced by considering the resonance structures for BF₃; only those forms that contribute significantly are shown.

(c) Using a hybridization scheme, we should end up with a bonding picture that corresponds to that depicted by the resonance structures.

An sp^2 hybridized nitrogen centre is consistent with the trigonal planar shape of $[NO_3]^-$. Allow the hybrid orbitals to overlap with suitable orbitals from oxygen. A choice of sp^2 hybridization on the O atom provides suitable orbitals to accommodate the oxygen lone pairs. Occupation of each bonding orbital by a pair of electrons gives three equivalent N−O σ-bonds:

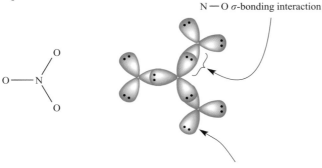

N−O σ-bonding interaction

sp^2 hybrid orbital on oxygen, occupied by a lone pair of electrons

Of the 24 valence electrons, 18 are accommodated either in σ-bonds or as oxygen lone pairs.

The next step is to consider multiple bonding character. Each N and O atom has an unused $2p$ atomic orbital lying perpendicular to the plane of the molecule. Six valence electrons are available. Overlap between the $2p$ atomic orbital on nitrogen with one of those on an oxygen atom gives rise to *one* localized π-bond. The six valence electrons are allocated as follows with the N centre treated as N^+:

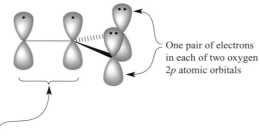

One pair of electrons in each of two oxygen $2p$ atomic orbitals

One pair of electrons for N−O π-bond formation

The combination of the σ- and π-bonding pictures gives one nitrogen–oxygen double bond and two single bonds. Three such schemes must be drawn (with the π-character in one of each of the N−O bonds) in order that the overall scheme is in keeping with the observed D_{3h} symmetry of $[NO_3]^-$.

Self-study exercises

1. Why are resonance structures containing two N=O double bonds not included in the set shown above for $[NO_3]^-$?

2. Use an appropriate hybridization scheme to describe the bonding in $[BO_3]^{3-}$.

5.4 Molecular orbital theory: the ligand group orbital approach and application to triatomic molecules

Despite its successes, the application of valence bond theory to the bonding in polyatomic molecules leads to conceptual difficulties. The method dictates that bonds are localized and, as a consequence, sets of resonance structures and bonding pictures involving hybridization schemes become rather tedious to establish, even for relatively small molecules (e.g. see Fig. 5.10c). We therefore turn our attention to molecular orbital (MO) theory.

Molecular orbital diagrams: moving from a diatomic to polyatomic species

As part of our treatment of the bonding in diatomics in Section 2.3, we constructed MO diagrams such as Figs. 2.8, 2.14 and 2.15. In each diagram, the atomic orbitals of the two atoms were represented on the right- and left-hand sides of the diagram with the MOs in the middle. Correlation lines connecting the atomic and molecular orbitals were constructed to produce a readily interpretable diagram.

Now consider the situation for a triatomic molecule such as CO_2. The molecular orbitals contain contributions from the atomic orbitals of three atoms, and we are presented with a problem of trying to draw an MO diagram involving four sets of orbitals (three sets of atomic orbitals and one of molecular orbitals). A description of the bonding in CF_4 involves five sets of atomic orbitals and one set of molecular orbitals, i.e. a six-component problem. Similarly, SF_6 is an eight-component problem. It is obvious that such MO diagrams are complicated and, probably, difficult both to construct and to interpret. In order to overcome this difficulty, it is common to resolve the MO description of a polyatomic molecule into a three-component problem, a method known as the *ligand group orbital (LGO) approach*.

MO approach to bonding in linear XH₂: symmetry matching by inspection

Initially, we illustrate the ligand group orbital approach by considering the bonding in a linear triatomic XH₂ in which the valence orbitals of X are the $2s$ and $2p$ atomic orbitals. Let us orient the H−X−H framework so that it coincides with the z axis as shown in Fig. 5.11. Consider the two $1s$ atomic orbitals of the two H atoms. Each $1s$ atomic orbital has two possible phases and, when the *two 1s orbitals are taken as a group*, there are two possible phase combinations. These are called *ligand group orbitals*

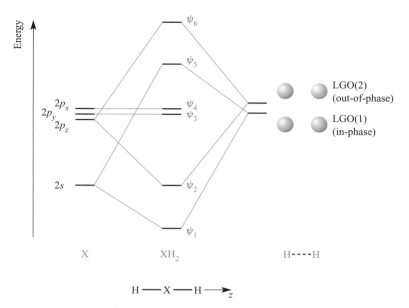

Fig. 5.11 Application of the ligand group orbital (LGO) approach to construct a qualitative MO diagram for the formation of a linear XH_2 molecule from the interactions of the valence orbitals of X ($2s$ and $2p$ atomic orbitals) and an H---H fragment. For clarity, the lines marking the $2p$ orbital energies are drawn apart, although these atomic orbitals are degenerate.

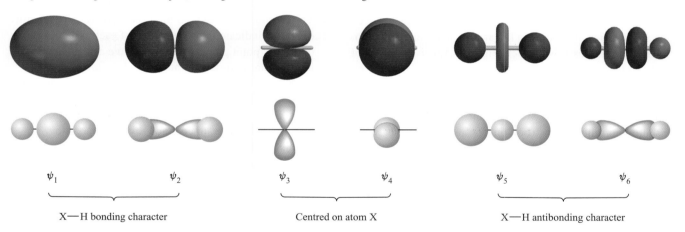

Fig. 5.12 The lower diagrams are schematic representations of the MOs in linear XH_2. The wavefunction labels correspond to those in Fig. 5.11. The upper diagrams are more realistic representations of the MOs and have been generated computationally using Spartan '04, © Wavefunction Inc. 2003.

(LGOs) and are shown at the right-hand side of Fig. 5.11.[†] Effectively, we are *transforming* the description of the bonding in XH_2 from one in which the basis sets are the atomic orbitals of atoms X and H, into one in which the basis sets are the atomic orbitals of atom X and the ligand group orbitals of an H---H fragment. This is a valuable approach for polyatomic molecules.

> The number of *ligand group orbitals* formed = the number of atomic orbitals used.

[†] In Fig. 5.11, the energies of the two ligand group orbitals are close together because the H nuclei are far apart; compare this with the situation in the H_2 molecule (Fig. 2.5). Similarly, in Fig. 5.17, the LGOs for the H_3 fragment form two sets (all in-phase, and the degenerate pair of orbitals) but their respective energies are close because of the large H---H separations.

In constructing an MO diagram for XH_2 (Fig. 5.11), we consider the interactions of the valence atomic orbitals of X with the ligand group orbitals of the H---H fragment. Ligand group orbital LGO(1) has the correct symmetry to interact with the $2s$ atomic orbital of X, giving an MO with H−X−H σ-bonding character. The symmetry of LGO(2) is matched to that of the $2p_z$ atomic orbital of X. The resultant bonding MOs (ψ_1 and ψ_2) and their antibonding counterparts (ψ_5 and ψ_6) are shown in Fig. 5.12, and the MO diagram in Fig. 5.11 shows the corresponding orbital interactions. The $2p_x$ and $2p_y$ atomic orbitals of X become non-bonding orbitals in XH_2. The final step in the construction of the MO diagram is to place the available electrons in the MOs according to the *aufbau* principle (see Section 1.9). An important result of the MO treatment of the bonding in

XH$_2$ is that the σ-bonding character in orbitals ψ_1 and ψ_2 is spread over all three atoms, indicating that the bonding character is *delocalized* over the H−X−H framework. Delocalized bonding is a general result within MO theory.

MO approach to bonding in linear XH$_2$: working from molecular symmetry

The method shown above for generating a bonding description for linear XH$_2$ cannot easily be extended to larger molecules. A more rigorous method is to start by identifying the point group of linear XH$_2$ as $D_{\infty h}$ (Fig. 5.13a). The $D_{\infty h}$ character table is used to assign symmetries to the orbitals on atom X, and to the ligand group orbitals. The MO diagram is then constructed by allowing interactions between orbitals of the same symmetry. *Only ligand group orbitals that can be classified within the point group of the whole molecule are allowed.*

Unfortunately, although a linear XH$_2$ molecule is structurally simple, the $D_{\infty h}$ character table is not. This, therefore, makes a poor first example of the use of group theory in orbital analysis. We can, however, draw an analogy between the symmetries of orbitals in linear XH$_2$ and those in homonuclear diatomics (also $D_{\infty h}$). Figure 5.13b is a repeat of Fig. 5.11, but this time the symmetries of the orbitals on atom X and the two ligand group orbitals are given. Compare these symmetry labels with those in Figs. 2.6 and 2.7. The construction of the MO diagram in Fig. 5.13b follows by allowing interactions (bonding or antibonding) between orbitals on atom X and ligand group orbitals with the same symmetry labels.

A bent triatomic: H$_2$O

The H$_2$O molecule has C_{2v} symmetry (Fig. 5.14) and we now show how to use this information to develop an MO picture of the bonding in H$_2$O. Part of the C_{2v} character table is shown below:

C_{2v}	E	C_2	$\sigma_v(xz)$	$\sigma_v{'}(yz)$
A_1	1	1	1	1
A_2	1	1	−1	−1
B_1	1	−1	1	−1
B_2	1	−1	−1	1

The inclusion of the xz and yz terms in the last two columns of the character table specifies that the H$_2$O molecule is taken to lie in the yz plane, i.e. the z axis coincides with the principal axis (Fig. 5.14). The character table has several important features.

- The labels in the first column (under the point group symbol) tell us the symmetry types of orbitals that are permitted within the specified point group.
- The numbers in the column headed E (the identity operator) indicate the degeneracy of each type of orbital; in the C_{2v} point group, all orbitals have a degeneracy of 1, i.e. they are non-degenerate.
- Each row of numbers following a given symmetry label indicates how a particular orbital behaves when operated upon by each symmetry operation. A number 1 means that the orbital is unchanged by the operation, a −1 means the orbital changes sign, and a 0 means that the orbital changes in some other way.

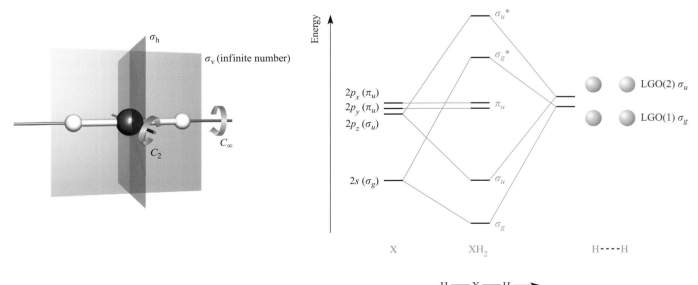

(a)

(b)

 Fig. 5.13 (a) A linear XH$_2$ molecule belongs to the $D_{\infty h}$ point group. Some of the symmetry elements are shown; the X atom lies on a centre of symmetry (inversion centre). (b) A qualitative MO diagram for the formation of linear XH$_2$ from atom X and two H atoms.

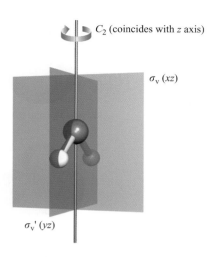

C_2 (coincides with z axis)

$\sigma_{\mathrm{v}}(xz)$

$\sigma_{\mathrm{v}}'(yz)$

 Fig. 5.14 The H_2O molecule possesses a C_2 axis and two σ_{v} planes and belongs to the $C_{2\mathrm{v}}$ point group.

To illustrate its use, let us consider the $2s$ atomic orbital of the O atom *in water*:

2s atomic orbital of oxygen

Apply each symmetry operation of the $C_{2\mathrm{v}}$ point group in turn. Applying the E operator leaves the $2s$ atomic orbital unchanged. Rotation about the C_2 axis leaves the atomic orbital unchanged. Reflections through the σ_{v} and σ_{v}' planes leave the $2s$ atomic orbital unchanged. These results correspond to the following row of characters:

E	C_2	$\sigma_{\mathrm{v}}(xz)$	$\sigma_{\mathrm{v}}'(yz)$
1	1	1	1

and this matches those for the symmetry type A_1 in the $C_{2\mathrm{v}}$ character table. We therefore label the $2s$ atomic orbital on the oxygen atom *in water* as an a_1 orbital. (Lower case letters are used for the orbital label, but upper case for the symmetry type in the character table.) The same test is now carried out on each atomic orbital of the O atom. The oxygen $2p_x$ orbital is left unchanged by the E operator and by reflection through the $\sigma_{\mathrm{v}}(xz)$ plane. Each of rotation about the C_2 axis and reflection through the $\sigma_{\mathrm{v}}'(yz)$ plane inverts the phase of the $2p_x$ orbital. This is summarized as follows:

E	C_2	$\sigma_{\mathrm{v}}(xz)$	$\sigma_{\mathrm{v}}'(yz)$
1	-1	1	-1

This matches the row of characters for symmetry type B_1 in the $C_{2\mathrm{v}}$ character table, and the $2p_x$ orbital therefore possesses b_1 symmetry. The $2p_y$ orbital is left unchanged by the E operator and by reflection through the $\sigma_{\mathrm{v}}'(yz)$ plane, but rotation about the C_2 axis and reflection through the

$\sigma_{\mathrm{v}}(xz)$ plane each inverts the phase of the orbital. This is summarized by the row of characters:

E	C_2	$\sigma_{\mathrm{v}}(xz)$	$\sigma_{\mathrm{v}}'(yz)$
1	-1	-1	1

This corresponds to symmetry type B_2 in the $C_{2\mathrm{v}}$ character table, and the $2p_y$ orbital is labelled b_2. The $2p_z$ orbital is left unchanged by the E operator, by reflection through either of the $\sigma_{\mathrm{v}}(xz)$ and $\sigma_{\mathrm{v}}'(yz)$ planes, and by rotation about the C_2 axis. Like the $2s$ orbital, the $2p_z$ orbital therefore has a_1 symmetry.

The next step is to work out the nature of the H---H ligand group orbitals that are allowed within the $C_{2\mathrm{v}}$ point group. Since we start with *two* H $1s$ orbitals, only *two* LGOs can be constructed. The symmetries of these LGOs are deduced as follows. By looking at Fig. 5.14, you can see what happens to each of the two H $1s$ orbitals when each symmetry operation is performed: both $1s$ orbitals are left unchanged by the E operator and by reflection through the $\sigma_{\mathrm{v}}'(yz)$ plane, but both are affected by rotation about the C_2 axis and by reflection through the $\sigma_{\mathrm{v}}(xz)$ plane. This information is summarized in the following row of characters:

E	C_2	$\sigma_{\mathrm{v}}(xz)$	$\sigma_{\mathrm{v}}'(yz)$
2	0	0	2

in which a '2' shows that 'two orbitals are unchanged by the operation', and a '0' means that 'no orbitals are unchanged by the operation'. Next, we note two facts: (i) we can construct only *two* ligand group orbitals, and (ii) the symmetry of each LGO must correspond to one of the symmetry types in the character table. We now compare the row of characters above with the *sums of two rows of characters* in the $C_{2\mathrm{v}}$ character table. A match is found with the sum of the characters for the A_1 and B_2 representations. As a result, we can deduce that the two LGOs must possess a_1 and b_2 symmetries, respectively. In this case, it is relatively straightforward to use the a_1 and b_2 symmetry labels to sketch the LGOs shown in Fig. 5.15, i.e. the a_1 orbital corresponds to an in-phase combination of H $1s$ orbitals, while the b_2 orbital is the out-of-phase combination of H $1s$ orbitals. However, once their symmetries are known, the rigorous method of determining the nature of the orbitals is as follows.

Let the $1s$ orbitals of the two H atoms shown in Fig. 5.14 be designated as ψ_1 and ψ_2. We now look at the effect of each symmetry operation of the $C_{2\mathrm{v}}$ point group on ψ_1. The E operator and reflection through the $\sigma_{\mathrm{v}}'(yz)$ plane (Fig. 5.14) leave ψ_1 unchanged, but a C_2 rotation and reflection through the $\sigma_{\mathrm{v}}(xz)$ plane each

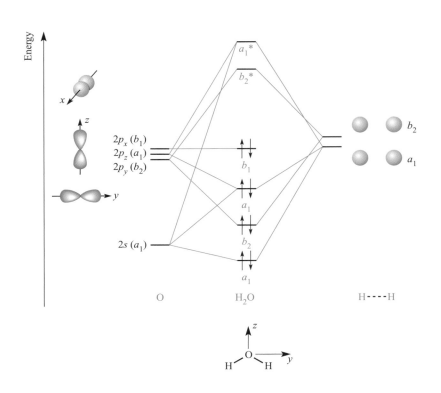

Representation of the b_1 MO

Representation of the higher energy a_1 MO

Representation of the b_2 MO

Representation of the lower energy a_1 MO

Fig. 5.15 A qualitative MO diagram for the formation of H_2O using the ligand group orbital approach. The two H atoms in the H_2 fragment are out of bonding range with each other, their positions being analogous to those in H_2O. For clarity, the lines marking the oxygen $2p$ orbital energies are drawn apart, despite their being degenerate. Representations of the occupied MOs are shown at the right-hand side of the figure (generated with Spartan '04, ©Wavefunction Inc. 2003). For the a_1 and b_2 MOs, the H_2O molecule is in the plane of the paper; for the b_1 MO, the plane containing the molecule is perpendicular to the plane of the paper.

transforms ψ_1 into ψ_2. The results are written down as a row of characters:

E	C_2	$\sigma_v(xz)$	$\sigma_v'(yz)$
ψ_1	ψ_2	ψ_2	ψ_1

To determine the composition of the a_1 LGO of the H---H fragment in H_2O, we multiply each character in the above row by the corresponding character for the A_1 representation in the C_{2v} character table, i.e.

C_{2v}	E	C_2	$\sigma_v(xz)$	$\sigma_v'(yz)$
A_1	1	1	1	1

The result of the multiplication is shown in eq. 5.10 and gives the unnormalized wavefunction for the a_1 orbital.

$$\psi(a_1) = (1 \times \psi_1) + (1 \times \psi_2) + (1 \times \psi_2) + (1 \times \psi_1)$$
$$= 2\psi_1 + 2\psi_2 \tag{5.10}$$

This can be simplified by dividing by 2 and, after normalization (see Section 2.2), gives the final equation for the

wavefunction (eq. 5.11).

$$\psi(a_1) = \frac{1}{\sqrt{2}}(\psi_1 + \psi_2) \quad \textit{in-phase combination} \tag{5.11}$$

Similarly, by using the B_2 representation in the C_{2v} character table, we can write down eq. 5.12. Equation 5.13 gives the equation for the normalized wavefunction.

$$\psi(b_2) = (1 \times \psi_1) - (1 \times \psi_2) - (1 \times \psi_2) + (1 \times \psi_1)$$
$$= 2\psi_1 - 2\psi_2 \tag{5.12}$$

$$\psi(b_2) = \frac{1}{\sqrt{2}}(\psi_1 - \psi_2) \quad \textit{out-of-phase combination} \tag{5.13}$$

The MO diagram shown in Fig. 5.15 is constructed as follows. Each of the $2s$ and $2p_z$ orbitals of the O atom possesses the correct symmetry (a_1) to interact with the a_1 orbital of the H---H fragment. These orbital interactions must lead to *three* MOs: two bonding MOs with a_1 symmetry and one antibonding ($a_1{}^*$) MO. On symmetry grounds, the lower energy a_1 MO could also include $2p_z$ character, but $2s$ character dominates because of the energy separation of the $2s$ and $2p_z$ atomic orbitals. The interaction between the

$2p_y$ atomic orbital and the LGO with b_2 symmetry leads to two MOs which possess H−O−H bonding and antibonding character respectively. The oxygen $2p_x$ orbital has b_1 symmetry and there is no symmetry match with a ligand group orbital. Thus, the oxygen $2p_x$ orbital is non-bonding in H_2O.

The eight valence electrons in H_2O occupy the MOs according to the *aufbau* principle, and this gives rise to two occupied H−O−H bonding MOs and two occupied MOs with mainly oxygen character. (To appreciate this fully, see end-of-chapter problem 5.12.) Although this bonding model for H_2O is approximate, it is *qualitatively* adequate for most descriptive purposes.

5.5 Molecular orbital theory applied to the polyatomic molecules BH₃, NH₃ and CH₄

We begin this section by considering the bonding in BH_3 and NH_3. The bonding in both molecules involves σ-interactions, but whereas BH_3 has D_{3h} symmetry, NH_3 belongs to the C_{3v} point group.

BH₃

The existence of BH_3 in the gas phase has been established even though the molecule readily dimerizes; the bonding in B_2H_6 is described in Section 5.7. The BH_3 molecule belongs to the D_{3h} point group. By considering the orbital interactions between the atomic orbitals of the B atom and the LGOs of an appropriate H_3 fragment, we can establish a molecular bonding scheme. We begin by choosing an appropriate axis set. The z axis coincides with the C_3 axis of BH_3 and all of the atoms lie in the xy plane. Part of the D_{3h} character table is shown in Table 5.1. By using the same approach as we did for the orbitals of the O atom in H_2O, we can assign symmetry labels to the orbitals of the B atom in BH_3:

- the $2s$ orbital has $a_1{}'$ symmetry;
- the $2p_z$ orbital has $a_2{}''$ symmetry;
- the $2p_x$ and $2p_y$ orbitals are degenerate and the orbital set has e' symmetry.

Table 5.1 Part of the D_{3h} character table; the complete table is given in Appendix 3.

D_{3h}	E	$2C_3$	$3C_2$	σ_h	$2S_3$	$3\sigma_v$
$A_1{}'$	1	1	1	1	1	1
$A_2{}'$	1	1	−1	1	1	−1
E'	2	−1	0	2	−1	0
$A_1{}''$	1	1	1	−1	−1	−1
$A_2{}''$	1	1	−1	−1	−1	1
E''	2	−1	0	−2	1	0

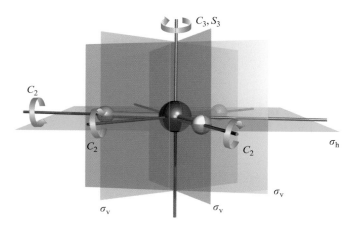

Fig. 5.16 The BH_3 molecule has D_{3h} symmetry.

We now consider the nature of the three ligand group orbitals that are formed from linear combinations of the three H $1s$ orbitals. By referring to the H_3-fragment in BH_3, we work out how many H $1s$ orbitals are left *unchanged* by each symmetry operation in the D_{3h} point group (Fig. 5.16). The result is represented by the following row of characters:

E	C_3	C_2	σ_h	S_3	σ_v
3	0	1	3	0	1

This same row of characters can be obtained by summing the rows of characters for the $A_1{}'$ and E' representations in the D_{3h} character table. Thus, the three LGOs have $a_1{}'$ and e' symmetries; recall that the e label designates a doubly degenerate set of orbitals. We must now determine the wavefunction for each LGO. Let the three H $1s$ orbitals in the H_3 fragment in BH_3 be ψ_1, ψ_2 and ψ_3. The next step is to see how ψ_1 is affected by each symmetry operation of the D_{3h} point group (Fig. 5.16). For example, the C_3 operation transforms ψ_1 into ψ_2, the C_3^2 operation transforms ψ_1 into ψ_3, and the three C_2 operations, respectively, leave ψ_1 unchanged, transform ψ_1 into ψ_2, and transform ψ_1 into ψ_3. The following row of characters gives the complete result:

E	C_3	C_3^2	$C_2(1)$	$C_2(2)$	$C_2(3)$	σ_h	S_3
ψ_1	ψ_2	ψ_3	ψ_1	ψ_3	ψ_2	ψ_1	ψ_2

S_3^2	$\sigma_v(1)$	$\sigma_v(2)$	$\sigma_v(3)$
ψ_3	ψ_1	ψ_3	ψ_2

The unnormalized wavefunction (eq. 5.14) for the $a_1{}'$ ligand group orbital is found by multiplying each character in the above row by the corresponding character for the $A_1{}'$ representation in the D_{3h} character table. After

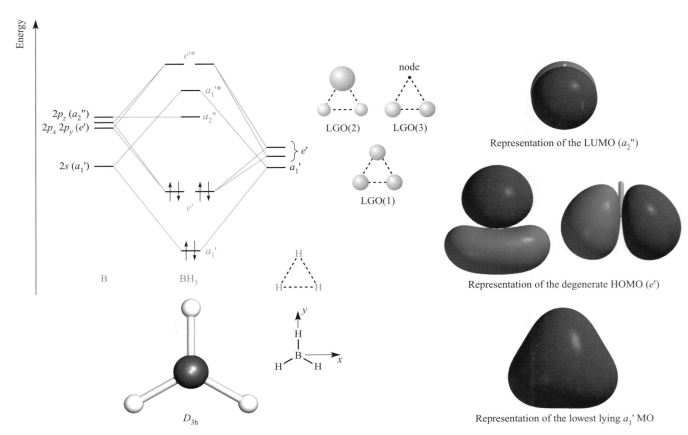

Fig. 5.17 A qualitative MO diagram for the formation of BH_3 using the ligand group orbital approach. The three H atoms in the H_3 fragment are out of bonding range with each other, their positions being analogous to those in the BH_3 molecule. Orbitals LGO(2) and LGO(3) form a degenerate pair (e' symmetry) although, for clarity, the lines marking their orbital energies are drawn apart; similarly for the three $2p$ atomic orbitals of boron. [Exercise: where do the nodal planes lie in LGO(2) and LGO(3)?] The diagrams at the right-hand side show representations of the three occupied MOs and the LUMO (generated with Spartan '04, ©Wavefunction Inc. 2003).

simplification (dividing by 4) and normalizing, the wavefunction can be written as eq. 5.15, and can be described schematically as the in-phase combination of $1s$ orbitals shown as LGO(1) in Fig. 5.17.

$$\psi(a_1') = \psi_1 + \psi_2 + \psi_3 + \psi_1 + \psi_3 + \psi_2 + \psi_1 + \psi_2 + \psi_3$$
$$\qquad + \psi_1 + \psi_3 + \psi_2$$
$$\qquad = 4\psi_1 + 4\psi_2 + 4\psi_3 \qquad\qquad (5.14)$$

$$\psi(a_1') = \frac{1}{\sqrt{3}}(\psi_1 + \psi_2 + \psi_3) \qquad\qquad (5.15)$$

A similar procedure can be used to deduce that eq. 5.16 describes the normalized wavefunction for one of the degenerate e' orbitals. Schematically, this is represented as LGO(2) in Fig. 5.17; the orbital contains one nodal plane.

$$\psi(e')_1 = \frac{1}{\sqrt{6}}(2\psi_1 - \psi_2 - \psi_3) \qquad\qquad (5.16)$$

Each e' orbital must contain a nodal plane, and the planes in the two orbitals are orthogonal to one another. Thus, we can write eq. 5.17 to describe the second e' orbital; the nodal plane passes through atom H(1) and the $1s$ orbital on this

atom makes *no contribution* to the LGO. This is represented as LGO(3) in Fig. 5.17.

$$\psi(e')_2 = \frac{1}{\sqrt{2}}(\psi_2 - \psi_3) \qquad\qquad (5.17)$$

The MO diagram for BH_3 can now be constructed by allowing orbitals of the same symmetry to interact. The $2p_z$ orbital on the B atom has a_2'' symmetry and no symmetry match can be found with an LGO of the H_3 fragment. Thus, the $2p_z$ orbital is non-bonding in BH_3. The MO approach describes the bonding in BH_3 in terms of three MOs of a_1' and e' symmetries. The a_1' orbital possesses σ-bonding character which is *delocalized over all four atoms*. The e' orbitals also exhibit delocalized character, and the bonding in BH_3 is described by considering a *combination of all three bonding MOs*.

NH_3

The NH_3 molecule has C_{3v} symmetry (Fig. 5.18) and a bonding scheme can be derived by considering the interaction between the atomic orbitals of the N atom and the ligand group orbitals of an appropriate H_3 fragment. An

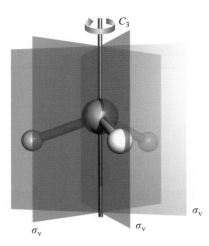

Fig. 5.18 The NH₃ molecule has C_{3v} symmetry.

appropriate axis set has the z axis coincident with the C_3 axis of NH₃ (see worked example 3.2); the x and y axes are directed as shown in Fig. 5.19. Table 5.2 shows part of the C_{3v} character table. By seeing how each symmetry operation affects each orbital of the N atom in NH₃, the orbital symmetries are assigned as follows:

- each of the $2s$ and $2p_z$ orbitals has a_1 symmetry;
- the $2p_x$ and $2p_y$ orbitals are degenerate and the orbital set has e symmetry.

Table 5.2 Part of the C_{3v} character table; the complete table is given in Appendix 3.

C_{3v}	E	$2C_3$	$3\sigma_v$
A_1	1	1	1
A_2	1	1	−1
E	2	−1	0

To determine the nature of the ligand group orbitals, we consider how many H $1s$ orbitals are left *unchanged* by each symmetry operation in the C_{3v} point group (Fig. 5.18). The result is represented by the row of characters:

E	C_3	σ_v
3	0	1

It follows that the three ligand group orbitals have a_1 and e symmetries. Although the symmetry labels of the LGOs of the H₃ fragments in NH₃ and BH₃ differ because the molecules belong to different point groups, the normalized wavefunctions for the LGOs are the same (eqs. 5.15–5.17). Schematic representations of the LGOs are shown in Fig. 5.19.

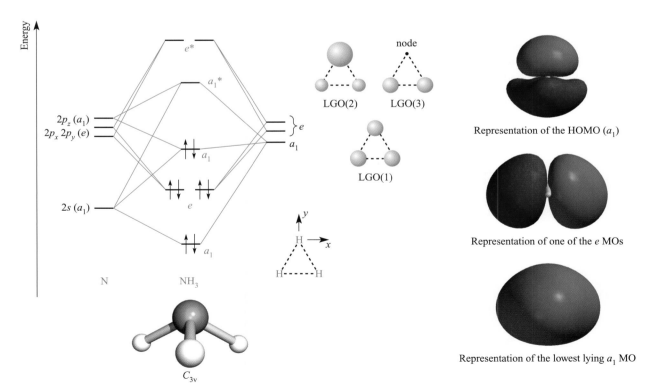

Fig. 5.19 A qualitative MO diagram for the formation of NH₃ using the ligand group orbital approach. For clarity, the lines marking degenerate orbital energies are drawn apart. The diagrams on the right-hand side (generated with Spartan '04, ©Wavefunction Inc. 2003) show representations of three of the occupied MOs; the orientation of the NH₃ molecule in each diagram is the same as in the structure at the bottom of the figure.

1. Give a full explanation of how one derives the symmetries of the LGOs of the H_3 fragment in NH_3.

2. By following the same procedure as we did for BH_3, derive equations for the normalized wavefunctions that describe the LGOs shown schematically in Fig. 5.19.

The qualitative MO diagram shown in Fig. 5.19 is constructed by allowing interactions between orbitals of the same symmetries. Because the nitrogen $2s$ and $2p_z$ orbitals have a_1 symmetry, they can both interact with the a_1 LGO. This leads to three a_1 MOs. On symmetry grounds, the lowest-lying a_1 MO could also contain N $2p_z$ character, but the energy separation of the $2s$ and $2p$ atomic orbitals is such that $2s$ character predominates. This is analogous to the case for H_2O described earlier. After constructing the MO diagram, the eight valence electrons are placed in the MOs according to the *aufbau* principle. The characters of three of the occupied orbitals are shown at the right-hand side of Fig. 5.19. The lowest energy orbital (a_1) has delocalized N–H bonding character. The highest occupied MO (HOMO) has some N–H bonding character, but retains an outward-pointing orbital lobe; this a_1 MO is essentially the nitrogen lone pair.

List differences between the MO diagrams for BH_3 and NH_3 shown in Figs. 5.17 and 5.19. Explain why these differences occur. In particular, explain why the $2p_z$ orbital on the central atom is non-bonding in BH_3, but can interact with the LGOs of the H_3 fragment in NH_3.

CH₄

The CH_4 molecule has T_d symmetry. The relationship between a tetrahedron and cube (illustrated in Fig. 5.6) is seen formally by the fact that the T_d point group belongs to the *cubic point group* family. This family includes the T_d and O_h point groups. Table 5.3 shows part of the T_d character table. The C_3 axes in CH_4 coincide with the C–H bonds, and the C_2 and S_4 axes coincide with the x, y and z axes defined in Fig. 5.6. Under T_d symmetry, the orbitals of the C atom in CH_4 (Fig. 5.20a) are classified as follows:

- the $2s$ orbital has a_1 symmetry;
- the $2p_x$, $2p_y$ and $2p_z$ orbitals are degenerate and the orbital set has t_2 symmetry.

In order to construct the LGOs of the H_4 fragment in CH_4, we begin by working out the number of H $1s$ orbitals left

Table 5.3   Part of the T_d character table; the complete table is given in Appendix 3.

T_d	E	$8C_3$	$3C_2$	$6S_4$	$6\sigma_d$
A_1	1	1	1	1	1
A_2	1	1	1	-1	-1
E	2	-1	2	0	0
T_1	3	0	-1	1	-1
T_2	3	0	-1	-1	1

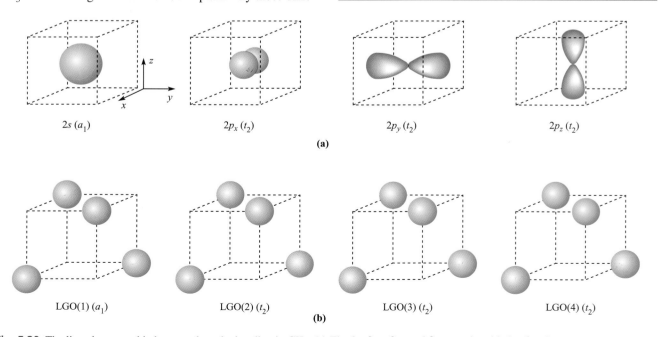

$2s\ (a_1)$ $2p_x\ (t_2)$ $2p_y\ (t_2)$ $2p_z\ (t_2)$

(a)

$LGO(1)\ (a_1)$ $LGO(2)\ (t_2)$ $LGO(3)\ (t_2)$ $LGO(4)\ (t_2)$

(b)

Fig. 5.20 The ligand group orbital approach to the bonding in CH_4. (a) The $2s$, $2p_x$, $2p_y$ and $2p_z$ atomic orbitals of carbon. (b) The four hydrogen $1s$ atomic orbitals combine to generate four ligand group orbitals (LGOs).

unchanged by each symmetry operation of the T_d point group. The result is summarized in the row of characters:

E	C_3	C_2	S_4	σ_d
4	1	0	0	2

This same row of characters results by summing the rows of characters for the A_1 and T_2 representations in the T_d character table (Table 5.3). The four ligand group orbitals therefore have a_1 and t_2 symmetries; the t label designates a triply degenerate set of orbitals. Normalized wavefunctions for these LGOs are given by eqs. 5.18–5.21.

$$\psi(a_1) = \tfrac{1}{2}(\psi_1 + \psi_2 + \psi_3 + \psi_4) \tag{5.18}$$

$$\psi(t_2)_1 = \tfrac{1}{2}(\psi_1 - \psi_2 + \psi_3 - \psi_4) \tag{5.19}$$

$$\psi(t_2)_2 = \tfrac{1}{2}(\psi_1 + \psi_2 - \psi_3 - \psi_4) \tag{5.20}$$

$$\psi(t_2)_3 = \tfrac{1}{2}(\psi_1 - \psi_2 - \psi_3 + \psi_4) \tag{5.21}$$

These four LGOs are shown schematically in Fig. 5.20b. By comparing Figs. 5.20a and 5.20b, the symmetries of the four ligand group orbitals can be readily matched to those of the $2s$, $2p_x$, $2p_y$ and $2p_z$ atomic orbitals of the C atom. This allows us to construct a qualitative MO diagram (Fig. 5.21) in which the interactions between the carbon atomic orbitals

and the ligand group orbitals of the H₄ fragment lead to four MOs with delocalized σ-bonding character and four anti-bonding MOs. Representations of the four bonding MOs are shown on the right-hand side of Fig. 5.21.

A comparison of the MO and VB bonding models

When we considered how valence bond theory can be used to describe the bonding in BH₃, CH₄ and NH₃, we used appropriate hybridization schemes so that bonds known to be structurally equivalent would be equivalent in the bonding scheme. One hybrid orbital contributed to each *localized* X–H (X = B, C or N) bond. On the other hand, the results of MO theory indicate that the bonding character is *delocalized*. Moreover, in each of BH₃, NH₃ and CH₄, there are two different *types* of bonding MO: a unique MO involving the $2s$ atomic orbital of the central atom, and a degenerate set of two (in BH₃ and NH₃) or three (in CH₄) MOs involving the $2p$ atomic orbitals of the central atom. Evidence for these orderings of MOs comes from photoelectron spectroscopy (see Section 4.12). How can the results of MO theory account for the experimentally observed equivalence of the X–H bonds in a given molecule?

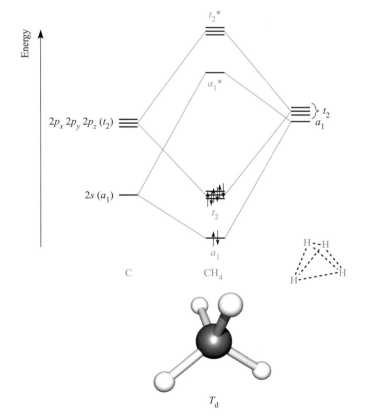

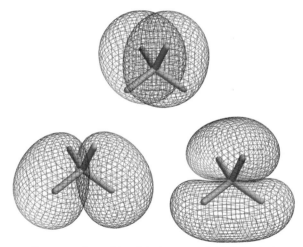

Representations of the triply degenerate (t_2) bonding MOs

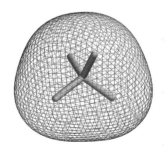

Representation of the a_1 bonding MO

Fig. 5.21 A qualitative MO diagram for the formation of CH₄ from the orbital basis set shown in Fig. 5.20. The diagrams on the right-hand side (generated with Spartan '04, ©Wavefunction Inc. 2003) illustrate the four bonding MOs. The orbitals are drawn in a 'mesh' representation so that the molecular framework is visible. Each of the t_2 MOs contains a nodal plane.

As we have already mentioned, it is essential to understand that, in MO theory, the bonding in a molecule is described by combining the characters of *all* the occupied MOs with bonding character. Take CH_4 as an example. The a_1 orbital (Fig. 5.21) is spherically symmetric and provides equal bonding character in all four C−H interactions. The t_2 orbitals must be considered as a set and not as individual orbitals. Taken together, this set of orbitals provides a picture of four equivalent C−H bonding interactions and, therefore, the overall picture is one of C−H bond equivalence.

5.6 Molecular orbital theory: bonding analyses soon become complicated

In this section, we consider the bonding in BF_3 using the ligand group orbital approach. Although BF_3 is a simple molecule, the following discussion demonstrates the complexity of the treatment when the atomic orbital basis set of each atom contains both s and p orbitals. The BF_3 molecule has D_{3h} symmetry. The z axis is defined to coincide with the C_3 axis and the BF_3 molecule lies in the xy plane (Fig. 5.22). Just as in BH_3, the atomic orbitals of the B atom in BF_3 are assigned the following symmetries:

- the $2s$ orbital has a_1' symmetry;
- the $2p_z$ orbital has a_2'' symmetry;
- the $2p_x$ and $2p_y$ orbitals are degenerate and the orbital set has e' symmetry.

Ligand group orbitals involving the F $2s$ orbitals in BF_3 and having a_1' and e' symmetries can be derived in the same way as those for the H_3 fragment in BH_3. These are shown as LGO(1)–LGO(3) in Fig. 5.22. The p orbitals on the F atoms can be partitioned into two types: those lying in the plane of the molecule ($2p_x$ and $2p_y$) and those perpendicular to the plane ($2p_z$). Ligand group orbitals can be formed from combinations of $2p_z$ orbitals, and from combinations of the in-plane $2p$ orbitals. Let us first consider the $2p_z$ orbitals. The procedure for deriving the wavefunctions that describe the LGOs allowed within the D_{3h} point group is the same as we have used before, but there is one important difference: when we consider how a $2p_z$ orbital is changed by a symmetry operation, we must look not only for the orbital being transformed to another position, but also for a change in phase. For example, if a p_z orbital is perpendicular to a σ_h plane, reflection through the plane will change its phase, but its position remains the same. This is exemplified when we work out how many F $2p_z$ orbitals are unchanged by each symmetry operation in the D_{3h} point group. The following row of characters summarizes the result; a negative sign means that the orbital is unmoved, but its phase has changed:

E	C_3	C_2	σ_h	S_3	σ_v
3	0	−1	−3	0	1

This row of characters is also produced by summing the rows of characters for the A_2'' and E'' representations in the D_{3h} character table (Table 5.1), and therefore the LGOs are of a_2'' and e'' symmetries. By considering the effects of every operation on one of the F $2p_z$ orbitals in the F_3 fragment, we can (as before) arrive at an equation for the unnormalized wavefunction of each LGO. Let the three F $2p_z$ orbitals be ψ_1, ψ_2 and ψ_3. We now generate

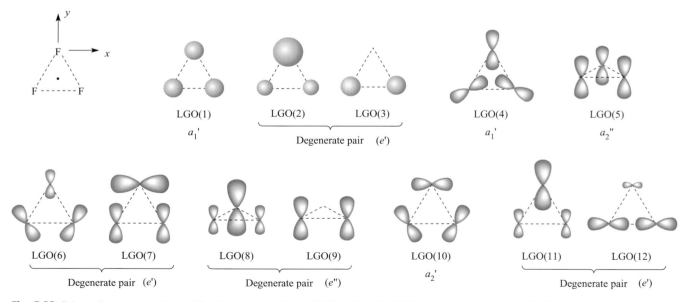

Fig. 5.22 Schematic representations of the ligand group orbitals (LGOs) for a D_{3h} F_3 fragment, the geometry of which is analogous to that in BF_3 (the position of the B atom is marked by the dot in the top left-hand diagram); the F_3 triangle lies in the xy plane. Orbitals LGO(5), LGO(8) and LGO(9) contain contributions from the $2p_z$ atomic orbitals, directed perpendicular to the F_3 triangle. The relative sizes of the lobes in each diagram *approximately* represent the relative contributions made by the fluorine atomic orbitals to each ligand group orbital.

the following row of characters, including a negative sign whenever the operation produces a change of orbital phase:

E	C_3	C_3^2	$C_2(1)$	$C_2(2)$	$C_2(3)$	σ_h	S_3
ψ_1	ψ_2	ψ_3	$-\psi_1$	$-\psi_3$	$-\psi_2$	$-\psi_1$	$-\psi_2$

S_3^2	$\sigma_v(1)$	$\sigma_v(2)$	$\sigma_v(3)$
$-\psi_3$	ψ_1	ψ_3	ψ_2

Multiplying each character in this row by the corresponding character in the row for the A_2'' representation in the D_{3h} character table (Table 5.1) gives the unnormalized form of the wavefunction for the a_2'' LGO (eq. 5.22). Simplification and normalization gives eq. 5.23. The a_2'' LGO can thus be described as an in-phase combination of $2p_z$ orbitals and is shown schematically in Fig. 5.22 as LGO(5).

$$\psi(a_2'') = \psi_1 + \psi_2 + \psi_3 + \psi_1 + \psi_3 + \psi_2 + \psi_1$$
$$+ \psi_2 + \psi_3 + \psi_1 + \psi_3 + \psi_2$$
$$= 4\psi_1 + 4\psi_2 + 4\psi_3 \qquad (5.22)$$

$$\psi(a_2'') = \frac{1}{\sqrt{3}}(\psi_1 + \psi_2 + \psi_3) \qquad (5.23)$$

Similarly, eqs. 5.24 and 5.25 can be derived for the e'' orbitals; these are represented in Fig. 5.22 as LGO(8) and LGO(9).

$$\psi(e'')_1 = \frac{1}{\sqrt{6}}(2\psi_1 - \psi_2 - \psi_3) \qquad (5.24)$$

$$\psi(e'')_2 = \frac{1}{\sqrt{2}}(\psi_2 - \psi_3) \qquad (5.25)$$

The same procedure can be used to derive the fact that the in-plane F $2p$ orbitals combine to give two LGOs with a_1' and a_2' symmetries respectively, and two sets of e' LGOs. These are shown schematically in Fig. 5.22 as LGOs (4), (6), (7), (10), (11) and (12).

We are now in a position to construct a qualitative MO diagram to describe the bonding in BF_3. The symmetries of the B orbitals under D_{3h} symmetry are given at the left side of Fig. 5.23, and those of the LGOs are shown in Fig. 5.22. The problem is best tackled in three steps:

- look for orbital interactions that give rise to σ-MOs;
- look for orbital interactions that give rise to π-orbitals;
- look for any orbital that has a symmetry that precludes orbital interactions between fragments.

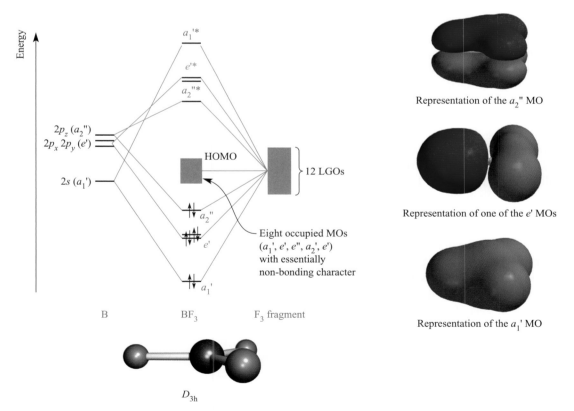

Fig. 5.23 A qualitative MO diagram for the formation of BF_3; the ligand group orbitals (LGOs) are shown in Fig. 5.22. The light grey rectangle in the stack of MOs in BF_3 represents a group of eight non-bonding MOs. The diagram is an over-simplification of the bonding in BF_3, but is sufficiently detailed to account for the B–F bonds possessing partial π-character. The characters of three of the occupied B–F bonding MOs are shown at the right-hand side of the figure (generated with Spartan '04, ©Wavefunction Inc. 2003); the orientation of the BF_3 molecule in each diagram is the same as in the structure at the bottom of the figure.

The σ-bonding in BF_3 evolves from interactions involving the fragment a_1' and e' orbitals. Inspection of Fig. 5.22 reveals that there are two F_3-fragment LGOs with a_1' symmetry, and three sets of e' orbitals. The extent of mixing between fragment orbitals of the same symmetry depends on their relative energies, and is impossible to predict with any degree of reliability. At the simplest level, we can assume a σ-bonding picture that mimics that in BH_3 (Fig. 5.17). This picture involves LGO(1) in the formation of the a_1' and $a_1'^*$ MOs labelled in Fig. 5.23, but leaves LGO(4) as a non-bonding orbital. This model can be fine-tuned by allowing some of the character of LGO(4) to be mixed into the a_1' and $a_1'^*$ MOs with B−F bonding or antibonding character. In order to 'balance the books', some character from LGO(1) must then end up in a non-bonding a_1' orbital. Similarly, we could allow contributions from the fragment e' MOs containing F $2p_x$ and $2p_y$ character to mix into the e' and e'^* MOs with B−F bonding or antibonding character. In the simplest bonding picture, these MOs contain F $2s$ character, and LGOs (6), (7), (11) and (12) become non-bonding MOs in BF_3. Assessing the extent of orbital mixing is difficult, if not impossible, at a qualitative level. It is best attempted by computational programs (many of which are available for use on desktop or laptop computers) which run at a variety of levels of sophistication (see Section 4.13).

The a_2'' symmetry of the B $2p_z$ orbital matches that of LGO(5) and an in-phase orbital interaction gives rise to an MO that has π-bonding character delocalized over all three B−F interactions.

The only orbitals on the F_3 fragment for which there is no symmetry match on the B atom are those with e'' and a_2' symmetries (i.e. LGOs 8, 9 and 10). These orbitals are carried across into BF_3 as non-bonding MOs.

The overall bonding picture for BF_3 is summarized in Fig. 5.23. There are four bonding MOs, four antibonding MOs and eight non-bonding MOs. The B atom provides three electrons and each F atom provides seven electrons, giving a total of 12 electron pairs to occupy the 12 bonding and non-bonding MOs shown in Fig. 5.23. This is a simple picture of the bonding which does not allow for orbital mixing. However, it provides a description that includes partial π-character in each B−F bond, and is therefore consistent with the VB treatment that we discussed in Section 5.3.

1. Based on symmetry arguments, why does the $2p_z$ orbital on boron remain non-bonding in BH_3 but is involved in a bonding interaction in BF_3?

2. Explain why LGO(4) in Fig. 5.22 can become involved in B−F bonding in BF_3, but is treated as a non-bonding MO in Fig. 5.23.

3. Explain why the a_2' orbital of the F_3 fragment is non-bonding in BF_3.

5.7 Molecular orbital theory: learning to use the theory objectively

The aim of this section is not to establish complete bonding pictures for molecules using MO theory, but rather to develop an objective way of using the MO model to rationalize particular features about a molecule. This often involves drawing a *partial MO diagram* for the molecule in question. In each example below, you should consider the implications of this partial treatment: it can be dangerous because bonding features, other than those upon which you are focusing, are ignored. However, with care and practice, the use of partial MO treatments is extremely valuable as a method of understanding structural and chemical properties in terms of bonding and we shall make use of it later in the book.

π-Bonding in CO_2

In this section, we develop an MO description of the π-bonding in CO_2. Before beginning, we must consider what valence orbitals are unused after σ-bonding. The CO_2 molecule belongs to the $D_{\infty h}$ point group; the z axis is defined to coincide with the C_∞ axis (structure **5.3**). The σ-bonding in an XH_2 molecule was described in Fig. 5.13. A similar picture can be developed for the σ-bonding in CO_2, with the difference that the H $1s$ orbitals in XH_2 are replaced by O $2s$ and $2p_z$ orbitals in CO_2. Their overlap with the C $2s$ and $2p_z$ orbitals leads to the formation of six MOs with σ_g or σ_u symmetry, four occupied and two unoccupied.

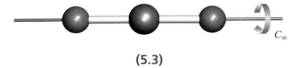

(5.3)

After the formation of C−O σ-interactions, the orbitals remaining are the C and O $2p_x$ and $2p_y$ orbitals. We now use the ligand group orbital approach to describe the π-bonding in terms of the interactions between the C $2p_x$ and $2p_y$ orbitals and the LGOs (derived from O $2p_x$ and $2p_y$ orbitals) of an O---O fragment. The LGOs are shown in Fig. 5.24. An in-phase combination of $2p$ orbitals is non-centrosymmetric and has π_u symmetry, while an out-of-phase combination is centrosymmetric and has π_g symmetry. Only the π_u LGOs have the correct symmetry to interact with the C $2p_x$ and $2p_y$ orbitals, leaving the π_g LGOs as non-bonding MOs in CO_2. After filling the lower-lying σ-bonding MOs, there are eight electrons left.

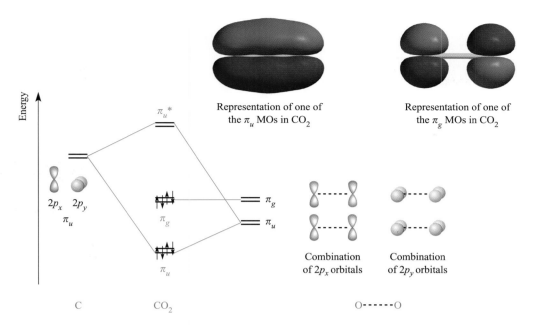

Fig. 5.24 A partial MO diagram that illustrates the formation of delocalized C−O π-bonds using the ligand group orbital approach. The CO_2 molecule is defined as lying on the z axis. The characters of the π_g and π_u MOs are shown in the diagrams at the top of the figure (generated with Spartan '04, ©Wavefunction Inc. 2003).

These occupy the π_u and π_g MOs (Fig. 5.24). The characters of one π_u MO and one π_g MO are shown at the top of Fig. 5.24; for each degenerate set of MOs, the character of the second π_u MO is the same as the first but is orthogonal to it. Each π_u MO has delocalized O−C−O π-bonding character, and the net result of having both π_u orbitals occupied is a π-bond order of 1 per C−O interaction.

Self-study exercise

Work out a qualitative MO description for the σ-bonding in CO_2 and show that this picture is consistent with leaving eight electrons to occupy the π-type MOs shown in Fig. 5.24.

[NO₃]⁻

In worked example 5.2, we considered the bonding in $[NO_3]^-$ using a VB approach. Three resonance structures (one of which is **5.4**) are needed to account for the equivalence of the N−O bonds, in which the net bond order per N−O bond is 1.33. Molecular orbital theory allows us to represent the N−O π-system in terms of delocalized interactions.

(5.4)

The $[NO_3]^-$ ion has D_{3h} symmetry and the z axis is defined to coincide with the C_3 axis. The valence orbitals of each N and O atom are $2s$ and $2p$ orbitals. The π-bonding in $[NO_3]^-$ can be described in terms of the interactions of the N $2p_z$ orbital with appropriate LGOs of the O_3 fragment. Under D_{3h} symmetry, the N $2p_z$ orbital has a_2'' symmetry (see Table 5.1). The LGOs that can be constructed from O $2p_z$ orbitals are shown in Fig. 5.25 along with their symmetries. The method of derivation is identical to that for the corresponding LGOs for the F_3 fragment in BF_3 (eqs. 5.23–5.25). The partial MO diagram shown in Fig. 5.25 can be constructed by symmetry-matching of the orbitals. The MOs that result have π-bonding (a_2''), non-bonding (e'') and π-antibonding ($a_2''^*$) character; the a_2'' and $a_2''^*$ MOs are illustrated at the right-hand side of Fig. 5.25. Six electrons occupy the a_2'' and e'' MOs. This number of electrons can be deduced by considering that of the 24 valence electrons in $[NO_3]^-$, six occupy σ-bonding MOs and 12 occupy oxygen-centred MOs with essentially non-bonding character, leaving six electrons for the π-type MOs (see end-of-chapter problem 5.18).

Molecular orbital theory therefore gives a picture of $[NO_3]^-$ in which there is *one* occupied MO with π-character and this is delocalized over all four atoms giving an N−O π-*bond order* of $\frac{1}{3}$. This is in agreement with the valence bond picture, but it is easier to visualize the delocalized bonding scheme than the resonance between three contributing forms of the type of structure **5.4**. The bonding in the isoelectronic species $[CO_3]^{2-}$ and $[BO_3]^{3-}$ (both D_{3h}) can be treated in a similar manner.

SF₆

Sulfur hexafluoride (**5.5**) provides an example of a so-called *hypervalent* molecule, i.e. one in which the central atom

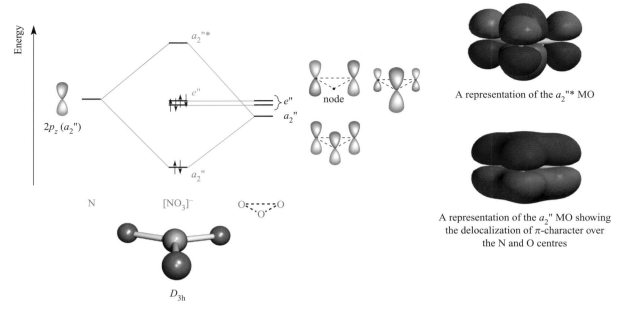

Fig. 5.25 A qualitative, partial MO diagram to illustrate the formation of a delocalized π-system in $[NO_3]^-$; a ligand group orbital approach is used. The characters of the a_2'' and $a_2''^*$ MOs are shown in the diagrams at the right-hand side of the figure (generated with Spartan '04, ©Wavefunction Inc. 2003).

appears to expand its octet of valence electrons. However, a valence bond picture of the bonding in SF_6 involving resonance structures such as **5.6** allows the S atom to obey the octet rule. A set of resonance structures is needed to rationalize the observed equivalence of the six S–F bonds. Other examples of 'hypervalent' species of the p-block elements are PF_5, $POCl_3$, AsF_5 and $[SeCl_6]^{2-}$. The bonding in each compound can be described within VB theory by a set of resonance structures in which the octet rule is obeyed for each atom (see Sections 15.3 and 16.3).

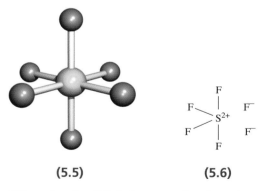

(5.5)   **(5.6)**

The SF_6 molecule, **5.5**, belongs to the O_h point group, which is one of the cubic point groups. The relationship between the octahedron and cube is shown in Fig. 5.26a; the x, y and z axes for the octahedron are defined as being parallel to the edges of the cube. In an octahedral molecule such as SF_6, this means that the x, y and z axes coincide with the S–F bonds. Table 5.4 gives part of the O_h character table, and the positions of the rotation axes are shown in Fig. 5.26b. The SF_6 molecule is centrosymmetric, the S

atom being on an inversion centre. Using the O_h character table, the valence orbitals of the S atom in SF_6 can be classified as follows:

- the $3s$ orbital has a_{1g} symmetry;
- the $3p_x$, $3p_y$ and $3p_z$ orbitals are degenerate and the orbital set has t_{1u} symmetry.

Ligand group orbitals for the F_6 fragment in SF_6 can be constructed from the F $2s$ and $2p$ orbitals. For a qualitative picture of the bonding, we can assume that the $s-p$ separation for fluorine is relatively large (see Section 2.3) and, as a consequence, there is negligible $s-p$ mixing. Separate sets of LGOs can therefore be formed from the F $2s$ orbitals and from the F $2p$ orbitals. Furthermore, the $2p$ orbitals fall into two classes: those that point towards the S atom (radial orbitals, diagram **5.7**) and those that are tangential to the octahedron (diagram **5.8**).

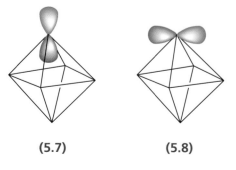

(5.7)   **(5.8)**

The S–F σ-bonds involve the radial $2p$ orbitals, and therefore the partial MO diagram that we construct for SF_6 focuses only on these fluorine orbitals. The wavefunctions

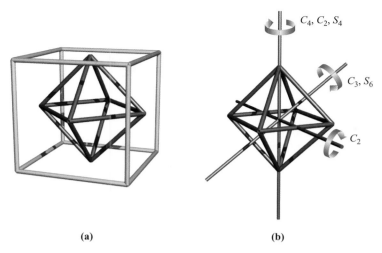

(a)   **(b)**

Fig. 5.26 (a) An octahedron can be inscribed in a cube; each vertex of the octahedron lies in the middle of a face of the cube. (b) The diagram shows one of each type of rotation axis of an octahedron. An inversion centre lies at the centre of the octahedron. [Exercise: Work out where the σ_h and σ_d planes lie; see Table 5.4.]

Table 5.4 Part of the O_h character table; the complete table is given in Appendix 3.

O_h	E	$8C_3$	$6C_2$	$6C_4$	$3C_2$ $(=C_4^2)$	i	$6S_4$	$8S_6$	$3\sigma_\mathrm{h}$	$6\sigma_\mathrm{d}$
A_{1g}	1	1	1	1	1	1	1	1	1	1
A_{2g}	1	1	−1	−1	1	1	−1	1	1	−1
E_g	2	−1	0	0	2	2	0	−1	2	0
T_{1g}	3	0	−1	1	−1	3	1	0	−1	−1
T_{2g}	3	0	1	−1	−1	3	−1	0	−1	1
A_{1u}	1	1	1	1	1	−1	−1	−1	−1	−1
A_{2u}	1	1	−1	−1	1	−1	1	−1	−1	1
E_u	2	−1	0	0	2	−2	0	1	−2	0
T_{1u}	3	0	−1	1	−1	−3	−1	0	1	1
T_{2u}	3	0	1	−1	−1	−3	1	0	1	−1

that describe the LGOs for the F_6 fragment in SF_6 are derived as follows. We first work out how many of the six radial $2p$ orbitals are unchanged under each O_h symmetry operation. The following row of characters gives the result:

E	$8C_3$	$6C_2$	$6C_4$	$3C_2$ $(=C_4^2)$	i	$6S_4$	$8S_6$	$3\sigma_\mathrm{h}$	$6\sigma_\mathrm{d}$
6	0	0	2	2	0	0	0	4	2

This same row of characters can be obtained by summing the characters for the A_{1g}, T_{1u} and E_g representations in the O_h character table (Table 5.4). Therefore, the LGOs have a_{1g}, t_{1u} and e_g symmetries.

It is now helpful to introduce the concept of a *local axis set*. When the LGOs for a Y_n group in an XY_n molecule involve orbitals other than spherically symmetric s orbitals, it is often useful to define the axis set on each Y atom so that the z axis points towards X. Diagram **5.9** illustrates this for the F_6 fragment.

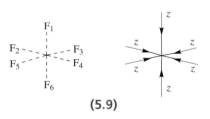

(5.9)

Thus, the six radial $3p$ orbitals that constitute the basis set for the LGOs of the F_6 fragment in SF_6 can be taken to be six $3p_z$ orbitals. Let these be labelled ψ_1–ψ_6 (numbering as in **5.9**). By using the same method as in previous examples in this chapter, we can derive the wavefunctions for the a_{1g}, t_{1u} and e_g LGOs (eqs. 5.26–5.31). These LGOs are represented schematically in Fig. 5.27.

$$\psi(a_{1g}) = \frac{1}{\sqrt{6}}(\psi_1 + \psi_2 + \psi_3 + \psi_4 + \psi_5 + \psi_6) \tag{5.26}$$

$$\psi(t_{1u})_1 = \frac{1}{\sqrt{2}}(\psi_1 - \psi_6) \tag{5.27}$$

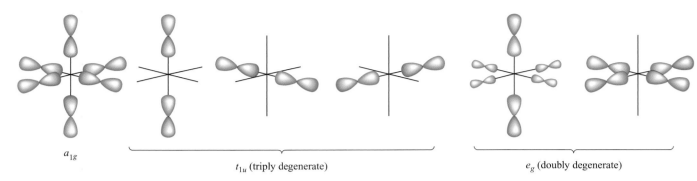

a_{1g} t_{1u} (triply degenerate) e_g (doubly degenerate)

Fig. 5.27 Ligand group orbitals for the F_6 fragment in SF_6 (O_h). These orbitals only include contributions from the radial $2p$ orbitals on fluorine (see text).

$$\psi(t_{1u})_2 = \frac{1}{\sqrt{2}}(\psi_2 - \psi_4) \qquad (5.28)$$

$$\psi(t_{1u})_3 = \frac{1}{\sqrt{2}}(\psi_3 - \psi_5) \qquad (5.29)$$

$$\psi(e_g)_1 = \frac{1}{\sqrt{12}}(2\psi_1 - \psi_2 - \psi_3 - \psi_4 - \psi_5 + 2\psi_6) \qquad (5.30)$$

$$\psi(e_g)_2 = \frac{1}{2}(\psi_2 - \psi_3 + \psi_4 - \psi_5) \qquad (5.31)$$

The partial MO diagram in Fig. 5.28 is constructed by matching the symmetries of the S valence orbitals and the LGOs of the F_6 fragment. Orbital interactions occur between the a_{1g} orbitals and between the t_{1u} orbitals, but the e_g set on the F_6 fragment is non-bonding in SF_6.

There are 48 valence electrons in SF_6. These occupy the a_{1g}, t_{1u} and e_g MOs shown in Fig. 5.28, in addition to

18 MOs that possess mainly fluorine character. The qualitative MO picture of the bonding in SF_6 that we have developed is therefore consistent with six equivalent S—F bonds. Based on Fig. 5.28, the S—F bond order is $\frac{2}{3}$ because there are four bonding pairs of electrons for six S—F interactions.

Three-centre two-electron interactions

We have already described several examples of bonding pictures that involve the delocalization of electrons. In cases such as BF_3 and SF_6, this leads to fractional bond orders. We now consider two linear XY_2 species in which there is only one occupied MO with Y—X—Y bonding character. This leads to the formation of a three-centre two-electron (3c-2e) bonding interaction.

In a **3c-2e bonding interaction,** two electrons occupy a bonding MO which is delocalized over three atomic centres.

The $[HF_2]^-$ ion (see Fig. 10.9) has $D_{\infty h}$ symmetry and the z axis coincides with the C_∞ axis. The bonding in $[HF_2]^-$ can be described in terms of the interactions of the H $1s$ orbital (σ_g symmetry) with the LGOs of an F---F fragment. If we assume a relatively large s–p separation for fluorine, then sets of LGOs can be constructed as follows:

- LGOs formed by combinations of the F $2s$ orbitals;
- LGOs formed by combinations of the F $2p_z$ orbitals;
- LGOs formed by combinations of the F $2p_x$ and $2p_y$ orbitals.

The method of deriving the wavefunctions that describe these LGOs is as before, and the results are summarized schematically at the right-hand side of Fig. 5.29. Although the H $1s$ orbital is of the correct symmetry to interact with either of the F---F σ_g LGOs, there is a poor energy match between the H $1s$ orbital and F---F $2s$–$2s$ combination. Thus, the qualitative MO diagram in Fig. 5.29 shows the H $1s$ orbital interacting only with the higher-lying σ_g LGO, giving rise to σ_g and σ_g^* MOs, the character of which is shown in the diagrams at the top of Fig. 5.29. All

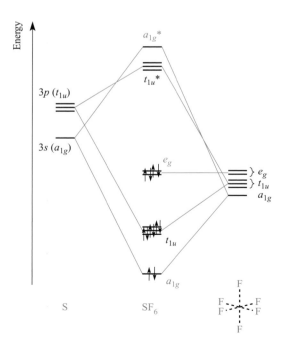

Fig. 5.28 Qualitative, partial MO diagram for the formation of SF_6 using the ligand group orbital approach with a basis set for sulfur that is composed of the $3s$ and $3p$ atomic orbitals.

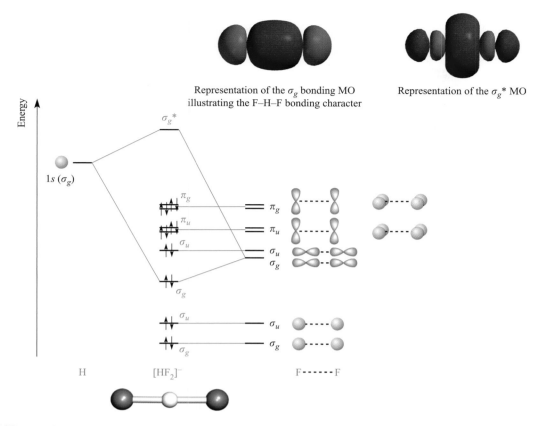

Fig. 5.29 A qualitative MO diagram for the formation of $[HF_2]^-$ using a ligand group orbital approach. The characters of the σ_g and $\sigma_g{}^*$ MOs are shown at the top of the figure (generated with Spartan '04, ©Wavefunction Inc. 2003).

other MOs have non-bonding character. Of the nine MOs, eight are fully occupied. Since there is only one MO that has H−F bonding character, the bonding in $[HF_2]^-$ can be described in terms of a 3c-2e interaction. The formal bond order for each H−F 'bond' is $\frac{1}{2}$.

Self-study exercise

How many nodal planes does each of the σ_g and $\sigma_g{}^*$ MOs shown at the top of Fig. 5.29 possess? Where do these lie in relation to the H and F nuclei? From your answers, confirm that the σ_g MO contains delocalized F−H−F bonding character, and that the $\sigma_g{}^*$ MO has H−F antibonding character.

The second example of a linear triatomic with a 3c-2e bonding interaction is XeF_2 ($D_{\infty h}$). The bonding is commonly described in terms of the partial MO diagram shown in Fig. 5.30. The Xe $5p_z$ orbital (σ_u symmetry) interacts with the combination of F $2p_z$ orbitals that has σ_u symmetry, giving rise to σ_u and $\sigma_u{}^*$ MOs. The combination of F

$2p_z$ orbitals with σ_g symmetry becomes a non-bonding MO in XeF_2. There are 22 valence electrons in XeF_2 and all MOs except one (the $\sigma_u{}^*$ MO) are occupied. The partial MO diagram in Fig. 5.30 shows only those MOs derived from p_z orbitals on Xe and F. There is only one MO that has

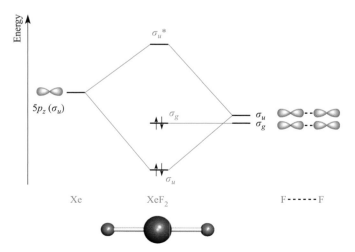

Fig. 5.30 A qualitative MO diagram for the formation of XeF_2 using a ligand group orbital approach and illustrating the 3c-2e bonding interaction.

Bridging H atom
(H$_{bridge}$)

Terminal H atom
(H$_{term}$)

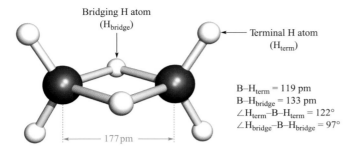

B–H$_{term}$ = 119 pm
B–H$_{bridge}$ = 133 pm
∠H$_{term}$–B–H$_{term}$ = 122°
∠H$_{bridge}$–B–H$_{bridge}$ = 97°

177 pm

 Fig. 5.31 The structure of B_2H_6 determined by electron diffraction.

Xe−F bonding character and therefore the bonding in XeF_2 can be described in terms of a 3c-2e interaction.[†]

Three-centre two-electron interactions are not restricted to triatomic molecules, as we illustrate in the next section with a bonding analysis of B_2H_6.

A more advanced problem: B_2H_6

Two common features of boron hydrides (see Sections 13.5 and 13.11) are that the B atoms are usually attached to more than three atoms and that *bridging* H atoms are often present. Although a valence bond model has been developed by Lipscomb to deal with the problems of generating localized bonding schemes in boron hydrides,[‡] the bonding in these compounds is not readily described in terms of VB theory. The structure of B_2H_6 (D_{2h} symmetry) is shown in Fig. 5.31. Features of particular interest are that:

- despite having only one valence electron, each *bridging* H atom is attached to *two* B atoms;
- despite having only three valence electrons, each B atom is attached to four H atoms;
- the B−H bond distances are not all the same and suggest two types of B−H bonding interaction.

Often, B_2H_6 is described as being *electron deficient*; it is a dimer of BH_3 and possesses 12 valence electrons. The formation of the B−H−B bridges can be envisaged as in diagram **5.10**. Whereas each terminal B−H interaction is taken to be a localized 2c-2e bond, each bridging unit is considered as a 3c-2e bonding interaction. Each *half* of the 3c-2e interaction is expected to be weaker than a terminal 2c-2e bond and this is consistent with the observed bond distances in Fig. 5.31. Bonding pictures for B_2H_6 which

assume either sp^3 or sp^2 hybridized B centres are frequently adopted, but this approach is not entirely satisfactory.

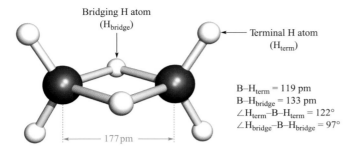

(5.10)

Although the molecular orbital treatment given below is an oversimplification, it still provides valuable insight into the distribution of electron density in B_2H_6. Using the ligand group orbital approach, we can consider the interactions between the *pair* of bridging H atoms and the residual B_2H_4 fragment (Fig. 5.32a).

The B_2H_6 molecule has D_{2h} symmetry, and the D_{2h} character table is given in Table 5.5. The x, y and z axes are defined in Fig. 5.32a. The molecule is centrosymmetric, with the centre of symmetry lying midway between the two B atoms. In order to describe the bonding in terms of the interactions of the orbitals of the B_2H_4 and H---H fragments (Fig. 5.32a), we must determine the symmetries of the allowed LGOs. First, we consider the H---H fragment and work out how many H $1s$ orbitals are left unchanged by each symmetry operation in the D_{2h} point group. The result is as follows:

E	$C_2(z)$	$C_2(y)$	$C_2(x)$	i	$\sigma(xy)$	$\sigma(xz)$	$\sigma(yz)$
2	0	0	2	0	2	2	0

This row of characters is produced by adding the rows of characters for the A_g and B_{3u} representations in the D_{2h} character table. Therefore, the LGOs for the H---H fragment have a_g and b_{3u} symmetries. Now let the two H $1s$ orbitals be labelled ψ_1 and ψ_2. The wavefunctions for these LGOs are found by considering how ψ_1 is affected by each symmetry operation of the D_{2h} point group. The following row of characters gives the result:

E	$C_2(z)$	$C_2(y)$	$C_2(x)$	i	$\sigma(xy)$	$\sigma(xz)$	$\sigma(yz)$
ψ_1	ψ_2	ψ_2	ψ_1	ψ_2	ψ_1	ψ_1	ψ_2

Multiplying each character in the row by the corresponding character in the A_g or B_{3u} representations in the D_{2h} character table gives the unnormalized wavefunctions for the LGOs. The normalized wavefunctions are represented by

---

[†] In the chemical literature, the bonding in XeF_2 is sometimes referred to as a 3c-4e interaction. Since two of the electrons occupy a non-bonding MO, we consider that a 3c-2e interaction description is more meaningful.

[‡] For detailed discussion of the VB model (called *styx* rules) see: W.N. Lipscomb (1963) *Boron Hydrides*, Benjamin, New York. A summary of *styx* rules and further discussion of the use of MO theory for boron hydrides are given in: C.E. Housecroft (1994) *Boranes and Metallaboranes: Structure, Bonding and Reactivity*, 2nd edn, Ellis Horwood, Hemel Hempstead.

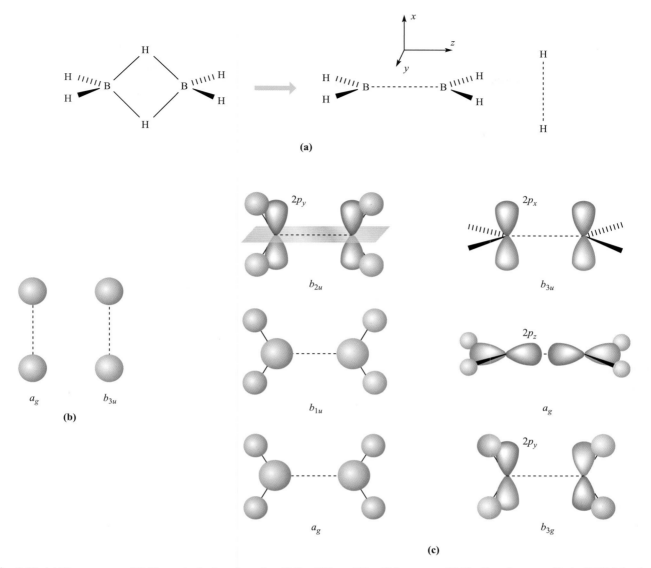

Fig. 5.32 (a) The structure of B_2H_6 can be broken down into H_2B---BH_2 and H---H fragments. (b) The ligand group orbitals (LGOs) for the H---H fragment. (c) The six lowest energy LGOs for the B_2H_4 unit; the nodal plane in the b_{2u} orbital is shown.

Table 5.5 Part of the D_{2h} character table; the complete table is given in Appendix 3.

D_{2h}	E	$C_2(z)$	$C_2(y)$	$C_2(x)$	i	$\sigma(xy)$	$\sigma(xz)$	$\sigma(yz)$
A_g	1	1	1	1	1	1	1	1
B_{1g}	1	1	-1	-1	1	1	-1	-1
B_{2g}	1	-1	1	-1	1	-1	1	-1
B_{3g}	1	-1	-1	1	1	-1	-1	1
A_u	1	1	1	1	-1	-1	-1	-1
B_{1u}	1	1	-1	-1	-1	-1	1	1
B_{2u}	1	-1	1	-1	-1	1	-1	1
B_{3u}	1	-1	-1	1	-1	1	1	-1

eqs. 5.32 and 5.33, and the LGOs are drawn schematically in Fig. 5.32b.

$$\psi(a_g) = \frac{1}{\sqrt{2}}(\psi_1 + \psi_2) \tag{5.32}$$

$$\psi(b_{3u}) = \frac{1}{\sqrt{2}}(\psi_1 - \psi_2) \tag{5.33}$$

The same procedure can be used to determine the LGOs of the B_2H_4 fragment. Since the basis set comprises four orbitals per B atom and one orbital per H atom, there are

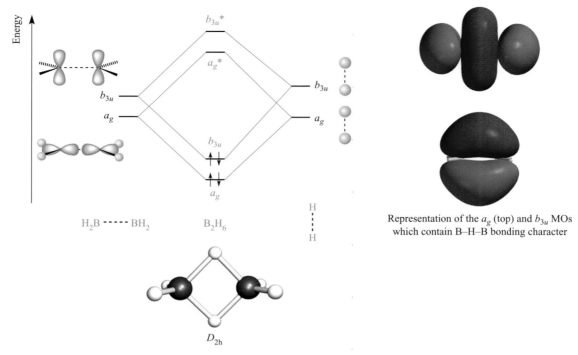

Fig. 5.33 A qualitative, partial MO diagram showing the formation of the B−H−B bridging interactions. The B−H and B−H−B bonding character of the a_g MO, and the B−H−B bonding character of the b_{3u} MO are shown in the diagrams on the right-hand side (generated with Spartan '04, ©Wavefunction Inc. 2003); the orientation of the molecule is the same as in the structure at the bottom of the figure.

Representation of the a_g (top) and b_{3u} MOs which contain B–H–B bonding character

12 LGOs in total. Figure 5.32c shows representations of the six lowest energy LGOs. The higher energy orbitals possess antibonding B−H or B---B character. Of those LGOs drawn in Fig. 5.32c, three have symmetries that match those of the LGOs of the H---H fragment. In addition to symmetry-matching, we must also look for a good energy match. Of the two a_g LGOs shown in Fig. 5.32c, the one with the lower energy is composed of B $2s$ and H $1s$ character. Although difficult to assess with certainty at a qualitative level, it is reasonable to assume that the energy of this a_g LGO is not well matched to that of the H---H fragment.

We now have the necessary information to construct a qualitative, partial MO diagram for B_2H_6. The diagram in Fig. 5.33 focuses on the orbital interactions that lead to the formation of B−H−B bridging interactions.

Consideration of the number of valence electrons available leads us to deduce that both the bonding MOs will be occupied. An important conclusion of the MO model is that the boron–hydrogen bridge character is delocalized over all *four* atoms of the bridging unit in B_2H_6. Since there are two such bonding MOs containing four electrons, this result is consistent with the 3c-2e B−H−B model that we described earlier.

KEY TERMS

The following terms were introduced in this chapter. Do you know what they mean?

- ❏ orbital hybridization
- ❏ sp, sp^2, sp^3, sp^3d, sp^2d and sp^3d^2 hybridization
- ❏ ligand group orbital (LGO) approach
- ❏ basis set of orbitals
- ❏ delocalized bonding interaction
- ❏ symmetry matching of orbitals
- ❏ energy matching of orbitals
- ❏ 3c-2e bonding interaction

FURTHER READING

J. Barrett (1991) *Understanding Inorganic Chemistry: The Underlying Physical Principles*, Ellis Horwood (Simon & Schuster), New York – Chapters 2 and 4 give a readable introduction to group theory and bonding in polyatomic molecules.

J.K. Burdett (1997) *Chemical Bonds, A Dialog*, Wiley, New York – An original résumé of modern valence theory presented in the form of a 19th century style dialogue between teacher and pupil.

M.E. Cass and W.E. Hollingsworth (2004) *J. Chem. Educ.*, vol. 81, p. 997 – A clearly written article, invaluable to both teachers and students: 'Moving beyond the single center – ways to reinforce molecular orbital theory in an inorganic course'.

F.A. Cotton (1990) *Chemical Applications of Group Theory*, 3rd edn, Wiley, New York – An excellent text that includes the applications of group theory in bonding analyses.

G. Davidson (1991) *Group Theory for Chemists*, Macmillan, London – Chapter 10 provides a useful discussion and also illustrates the use of group theory.

R.L. DeKock and H.B. Gray (1989) *Chemical Structure and Bonding*, University Science Books, California – A readable text, treating VB and MO theories and giving examples of the relationship between photoelectron spectra and MO energy levels.

H.B. Gray (1994) *Chemical Bonds*, University Science Books, California – An introduction to atomic and molecular structure with numerous illustrations.

S.F.A. Kettle (2007) *Symmetry and Structure*, 3rd edn, Wiley, Chichester – An advanced discussion which includes carefully explained applications of group theory.

L. Pauling (1960) *The Nature of the Chemical Bond*, 3rd edn, Cornell University Press, Ithaca, NY – A classic book dealing with covalent, metallic and hydrogen bonding from the viewpoint of VB theory.

M.J. Winter (1994) *Chemical Bonding*, Oxford University Press, Oxford – Chapters 5 and 6 give a basic introduction to hybridization and MO theory in polyatomics.

PROBLEMS

5.1   (a) State what is meant by the *hybridization of atomic orbitals*. (b) Why does VB theory sometimes use hybrid orbital rather than atomic orbital basis sets? (c) Show that eqs. 5.1 and 5.2 correspond to normalized wavefunctions.

5.2   Figure 5.4 shows the formation of three sp^2 hybrid orbitals (see eqs. 5.3–5.5). (a) Confirm that the directionalities of the three hybrids are as specified in the figure. (b) Show that eqs. 5.3 and 5.5 correspond to normalized wavefunctions.

5.3   Use the information given in Fig. 5.6b and eqs. 5.6 to 5.9 to reproduce the directionalities of the four sp^3 hybrid orbitals shown in Fig. 5.6a.

5.4   (a) Derive a set of diagrams similar to those in Figs. 5.2 and 5.4 to describe the formation of sp^2d hybrid orbitals. (b) What is the percentage character of each sp^2d hybrid orbital in terms of the constituent atomic orbitals?

5.5   Suggest an appropriate hybridization scheme for the central atom in each of the following species: (a) SiF_4; (b) $[PdCl_4]^{2-}$; (c) NF_3; (d) F_2O; (e) $[CoH_5]^{4-}$; (f) $[FeH_6]^{4-}$; (g) CS_2; (h) BF_3.

5.6   (a) The structures of *cis*- and *trans*-N_2F_2 were shown in worked example 3.1. Give an appropriate hybridization scheme for the N atoms in each isomer. (b) What hybridization scheme is appropriate for the O atoms in H_2O_2 (Fig. 2.1)?

5.7   (a) PF_5 has D_{3h} symmetry. What is its structure? (b) Suggest an appropriate bonding scheme for PF_5 within VB theory, giving appropriate resonance structures.

5.8   (a) Draw the structure of $[CO_3]^{2-}$. (b) If all the C–O bond distances are equal, write a set of resonance structures to describe the bonding in $[CO_3]^{2-}$. (c) Describe the bonding in $[CO_3]^{2-}$ in terms of a hybridization scheme and compare the result with that obtained in part (b).

5.9   (a) Is CO_2 linear or bent? (b) What hybridization is appropriate for the C atom? (c) Outline a bonding scheme for CO_2 using the hybridization scheme you have suggested. (d) What C–O bond order does your scheme imply? (e) Draw a Lewis structure for CO_2. Is this structure consistent with the results you obtained in parts (c) and (d)?

Table 5.6   Results of a self-consistent field quantum chemical calculation for H_2O using an orbital basis set of the atomic orbitals of the O atom and the ligand group orbitals of an H---H fragment. The axis set is defined in Fig. 5.15.

Atomic orbital or ligand group orbital	Percentage character of MOs with the sign of the eigenvector given in parentheses					
	ψ_1	ψ_2	ψ_3	ψ_4	ψ_5	ψ_6
O $2s$	71 (+)	0	7 (−)	0	0	22 (−)
O $2p_x$	0	0	0	100 (+)	0	0
O $2p_y$	0	59 (+)	0	0	41 (−)	0
O $2p_z$	0	0	85 (−)	0	0	15 (+)
H---H LGO(1)	29 (+)	0	8 (+)	0	0	63 (+)
H---H LGO(2)	0	41 (−)	0	0	59 (−)	0

5.10   What is meant by a *ligand group orbital*?

5.11   VB and MO approaches to the bonding in linear XH_2 (X has $2s$ and $2p$ valence atomic orbitals) give pictures in which the X−H bonding is localized and delocalized respectively. Explain how this difference arises.

5.12   Table 5.6 gives the results of a self-consistent field (SCF) quantum chemical calculation for H_2O using an orbital basis set of the atomic orbitals of O and the LGOs of an H---H fragment. The axis set is as defined in Fig. 5.15. (a) Use the data to construct pictorial representations of the MOs of H_2O and confirm that Fig. 5.15 is consistent with the results of the calculation. (b) How does MO theory account for the presence of lone pairs in H_2O?

5.13   Refer to Fig. 5.17 and the accompanying discussion. (a) Why does the B $2p_z$ atomic orbital become a non-bonding MO in BH_3? (b) Draw schematic representations of each bonding and antibonding MO in BH_3.

5.14   The diagrams at the right-hand side of Fig. 5.19 show three of the MOs in NH_3. Sketch representations of the other four MOs.

5.15   Use a ligand group orbital approach to describe the bonding in $[NH_4]^+$. Draw schematic representations of each of the bonding MOs.

5.16   The I−I bond distance in I_2 (gas phase) is 267 pm, in the $[I_3]^+$ ion is 268 pm, and in $[I_3]^-$ is 290 pm (for the

$[AsPh_4]^+$ salt). (a) Draw Lewis structures for these species. Do these representations account for the variation in bond distance? (b) Use MO theory to describe the bonding and deduce the I−I bond order in each species. Are your results consistent with the structural data?

5.17   (a) BCl_3 has D_{3h} symmetry. Draw the structure of BCl_3 and give values for the bond angles. NCl_3 has C_{3v} symmetry. Is it possible to state the bond angles from this information? (b) Derive the symmetry labels for the atomic orbitals on B in BCl_3 and on N in NCl_3.

5.18   Using Figs. 5.22, 5.23 and 5.25 to help you, compare the MO pictures of the bonding in BF_3 and $[NO_3]^-$. What approximations have you made in your bonding analyses?

5.19   By considering the structures of the following molecules, confirm that the point group assignments are correct: (a) BH_3, D_{3h}; (b) NH_3, C_{3v}; (c) B_2H_6, D_{2h}.

5.20   In the description of the bonding of B_2H_6, we draw the conclusion that the two bonding MOs in Fig. 5.33 have B−H bonding character delocalized over the four bridge atoms. (a) What other character do these MOs possess? (b) Does your answer to (a) alter the conclusion that this approximate MO description is consistent with the valence bond idea of there being two 3c-2e bridge bonds?

5.21   In $[B_2H_7]^-$ (**5.11**), each B atom is *approximately* tetrahedral. (a) How many valence electrons are present in the anion? (b) Assume that each B atom is sp^3 hybridized. After localization of the three terminal B–H bonds per B, what B-centred orbital remains for use in the bridging interaction? (c) Following from your answer to part (b), construct an approximate orbital diagram to show the formation of $[B_2H_7]^-$ from two BH_3 units and H^-. What does this approach tell you about the nature of the B–H–B bridge?

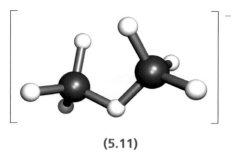

(5.11)

OVERVIEW PROBLEMS

5.22 (a) What hybridization scheme would be appropriate for the Si atom in SiH_4?
(b) To which point group does SiH_4 belong?
(c) Sketch a qualitative MO diagram for the formation of SiH_4 from Si and an H_4-fragment. Label all orbitals with appropriate symmetry labels.

5.23 Cyclobutadiene, C_4H_4, is unstable but can be stabilized in complexes such as $(C_4H_4)Fe(CO)_3$. In such complexes, C_4H_4 is planar and has equal C–C bond lengths:

(a) After the formation of C–H and C–C σ-bonds in C_4H_4, what orbitals are available for π-bonding?
(b) Assuming D_{4h} symmetry for C_4H_4, derive the symmetries of the four π-MOs. Derive equations for

the normalized wavefunctions that describe these MOs, and sketch representations of the four orbitals.

5.24 (a) Draw a set of resonance structures for the hypothetical molecule PH_5, ensuring that P obeys the octet rule in each structure. Assume a structure analogous to that of PF_5.
(b) To what point group does PH_5 belong?
(c) Using PH_5 as a model compound, use a ligand group orbital approach to describe the bonding in PH_5. Show clearly how you derive the symmetries of both the P atomic orbitals, and the LGOs of the H_5 fragment.

5.25 What hybridization scheme would be appropriate for the C atom in $[CO_3]^{2-}$? Draw resonance structures to describe the bonding in $[CO_3]^{2-}$. Figure 5.34 shows representations of three MOs of $[CO_3]^{2-}$. The MOs in diagrams (a) and (b) in Fig. 5.34 are occupied; the MO in diagram (c) is unoccupied. Comment on the characters of these MOs and assign a symmetry label to each orbital.

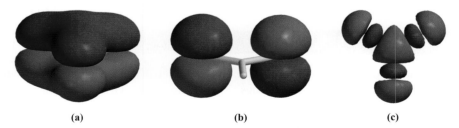

(a)   (b)   (c)

Fig. 5.34 Figure for problem 5.25.

5.26 The hydrido complex $[FeH_6]^{4-}$ has O_h symmetry. The bonding in $[FeH_6]^{4-}$ can be described in terms of the interactions between the atomic orbitals of Fe and the LGOs of the H_6-fragment.
(a) Derive the six LGOs of the H_6 fragment, showing clearly how you determine their symmetries.
(b) The basis set for the Fe atom consists of valence $3d$ (see Fig. 1.12), $4s$ and $4p$ orbitals. Determine the symmetries of these orbitals under O_h symmetry.
(c) Construct an MO diagram for the formation of $[FeH_6]^{4-}$ from Fe and the H_6-fragment, showing which MOs are occupied. Comment on the characters of the MOs. How does this bonding picture differ from that described for SF_6 in Fig. 5.28?

5.27 (a) The lists below show wrongly paired molecules or ions and point groups. Assign the correct point group to each species.

Molecule or ion	Point group
$[H_3O]^+$	D_{3h}
C_2H_4	$D_{\infty h}$
CH_2Cl_2	T_d
SO_3	C_{3v}
CBr_4	$C_{\infty v}$
$[ICl_4]^-$	D_{2h}
HCN	D_{4h}
Br_2	C_{2v}

(b) A molecule X_2H_6 belongs to the D_{3d} point group. Does the molecule have an eclipsed or a staggered conformation?
(c) Figure 5.35 shows the lowest energy a_{1g} MO in ethane. If the molecule had an eclipsed rather than a staggered conformation, what would be the symmetry label of the corresponding MO?

5.28 The structures below show (on the left) an octahedral and (on the right) a trigonal prismatic XY_6 molecule.

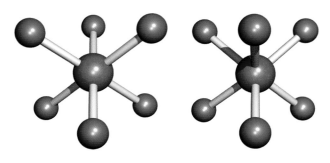

(a) To what point groups do these molecules belong?
(b) The bonding MOs in octahedral XY_6 have a_{1g}, e_g and t_{1u} symmetries. Confirm that these symmetries are consistent with the point group that you have assigned.
(c) Can there be a triply degenerate set of bonding orbitals for the trigonal prismatic XY_6 molecule? Rationalize your answer.

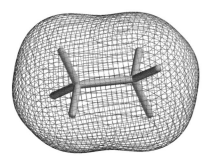

Fig. 5.35 A representation of the lowest occupied bonding MO (a_{1g}) in C_2H_6.

5.29 The industrial manufacture of NH_3 from N_2 and H_2 is carried out on a huge scale using heterogeneous catalysis, i.e. the reaction between gaseous N_2 and H_2 is carried out over a solid catalyst. (a) Construct an MO diagram for N_2 and use the diagram to explain why N_2 is a chemically inert species. (b) To what point group does NH_3 belong? (c) Using the appropriate character table in Appendix 3, construct a set of ligand group orbitals for a triangular H_3-fragment. Give symmetry labels to the LGOs. (d) Construct an MO diagram for NH_3 showing the interactions between the N atomic orbitals and the LGOs of the H_3-fragment. Use the MO diagram to determine the N–H bond order and to confirm that NH_3 is diamagnetic.

5.30 The compound $H_3N \cdot BH_3$ is an adduct of NH_3 and BH_3. It is currently being investigated as a possible hydrogen storage material. (a) What is the hydrogen storage capacity (percentage by weight) of $H_3N \cdot BH_3$? (b) Using the VSEPR model, draw structures for BH_3, NH_3 and $H_3N \cdot BH_3$. (c) Figure 5.36 shows how the orbital energies of a D_{3h} XY_3 molecule alter as the molecular geometry changes. This type of diagram is called a Walsh diagram. Why are the orbital symmetry labels different at the two sides of the diagram? (d) Which of the orbitals in Fig. 5.36 represent the HOMO and LUMO of BH_3 and NH_3? (e) Sketch representations of the MOs to which the orbital symmetry labels in Fig. 5.36 refer. (f) Suggest how BH_3 and NH_3 interact to form $H_3N \cdot BH_3$. What type of N–B bond is present?

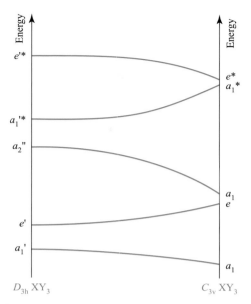

Fig. 5.36 Changes in orbital energies on going from D_{3h} to C_{3v} XY_3 as the Y atoms are moved out of the plane into a pyramidal arrangement.

Topics

Packing of spheres
Polymorphism
Alloys and intermetallic
   compounds
Band theory
Semiconductors
Sizes of ions
Prototype structures
Lattice energy
Born–Haber cycle
Defects in solid state
   structures

6

Structures and energetics of metallic and ionic solids

6.1 Introduction

In the solid state, both metallic and ionic compounds possess ordered arrays of atoms or ions and form crystalline materials with *lattice* structures. Studies of their structures may conveniently be considered as related topics because both are concerned with the packing of *spherical* atoms or ions. However, differences in *bonding* result in quite distinct properties for metallic and ionic solids. In metals, the bonding is essentially covalent. The bonding electrons are delocalized over the whole crystal, giving rise to the high electrical conductivity that is characteristic of metals. Ionic bonding in the solid state arises from electrostatic interactions between charged species (ions), e.g. Na^+ and Cl^- in rock salt. Ionic solids are *insulators*.

An *anion* is a negatively charged ion and a *cation* is a positively charged ion.

Although metallic and ionic solids have 3-dimensional structures, it does *not* follow that 3-dimensional structures are necessarily metallic or ionic. Diamond, for example, is a non-metal (see Sections 6.11 and 6.12). In Sections 2.2 and 2.5, we considered the inclusion of ionic contributions to 'covalent' bonding pictures. Later in this chapter we discuss how including some covalent character in a predominantly ionic model comes closer to reality for some so-called 'ionic' compounds.

6.2 Packing of spheres

Many readers will be familiar with descriptions of metal lattices based upon the packing of spherical atoms, and in this section we provide a résumé of common types of packing.

Cubic and hexagonal close-packing

Let us place a number of equal-sized spheres in a rectangular box, with the restriction that there must be a *regular arrangement* of spheres. Figure 6.1 shows the most efficient way in which to cover the floor of the box. Such an arrangement is *close-packed*, and spheres that are not on the edges of the assembly are in contact with six other spheres within the layer. A motif of hexagons is produced within the assembly. Figure 6.2a shows part of the same close-packed arrangement of spheres; hollows lie between the spheres and we can build a second layer of spheres upon the first by placing spheres in these hollows. However, if we arrange the spheres in the second layer so that close-packing is again achieved, it is possible to occupy only every other hollow. This is shown on going from Fig. 6.2a to 6.2b.

Now consider the hollows that are visible in layer B in Fig. 6.2b. There are *two distinct types of hollows*. Of the four hollows between the grey spheres in layer B, one lies over a red sphere in layer A, and three lie over hollows in layer A. The consequence of this is that when a third layer of spheres is constructed, two different close-packed arrangements are possible as shown in Figs. 6.2c and 6.2d. The arrangements shown can, of course, be extended sideways, and the sequences of layers can be repeated such

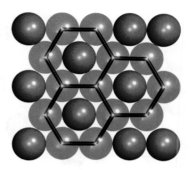

Fig. 6.1 Part of one layer of a close-packed arrangement of equal-sized spheres. It contains hexagonal motifs.

The ABABAB... and ABCABC... packing arrangements are called *hexagonal close-packing* (hcp) and *cubic close-packing* (ccp), respectively. In each structure, any given sphere is surrounded by (and touches) 12 other spheres and is said to have 12 *nearest neighbours*, to have a *coordination number* of 12, or to be *12-coordinate*. Figure 6.3 shows representations of the ABABAB... and ABCABC... arrangements which illustrate how this coordination number arises. In these diagrams, 'ball-and-stick' representations of the lattice are used to allow the connectivities to be seen. This type of representation is commonly used *but does not imply* that the spheres do not touch one another.

that the fourth layer of spheres is equivalent to the first, and so on. The two close-packed arrangements are distinguished in that one contains *two repeating layers*, ABABAB..., while the second contains *three repeating layers*, ABCABC... (Figs. 6.2d and 6.2c respectively).

> *Close-packing of spheres* results in the most efficient use of the space available; 74% of the space is occupied by the spheres.

The unit cell: hexagonal and cubic close-packing

A *unit cell* is a fundamental concept in solid state chemistry (see Section 4.11). It is the smallest repeating unit of the structure which carries *all* the information necessary to construct *unambiguously* an infinite lattice. The unit cells in Fig. 6.4 characterize cubic (ccp) and hexagonal

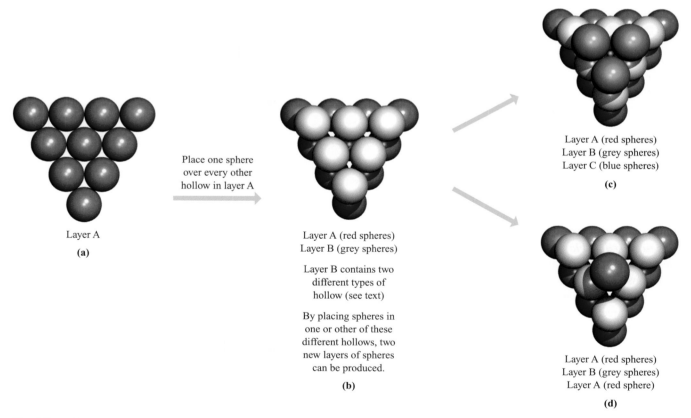

Layer A

(a)

Place one sphere over every other hollow in layer A

Layer A (red spheres)
Layer B (grey spheres)

Layer B contains two different types of hollow (see text)

By placing spheres in one or other of these different hollows, two new layers of spheres can be produced.

(b)

Layer A (red spheres)
Layer B (grey spheres)
Layer C (blue spheres)

(c)

Layer A (red spheres)
Layer B (grey spheres)
Layer A (red sphere)

(d)

Fig. 6.2 (a) One layer (layer A) of close-packed spheres contains hollows that exhibit a regular pattern. (b) A second layer (layer B) of close-packed spheres can be formed by occupying every other hollow in layer A. In layer B, there are two types of hollow; one lies over a sphere in layer A, and three lie over hollows in layer A. By stacking spheres over these different types of hollow, two different third layers of spheres can be produced. The blue spheres in diagram (c) form a new layer C; this gives an ABC sequence of layers. Diagram (d) shows that the second possible third layer replicates layer A; this gives an ABA sequence.

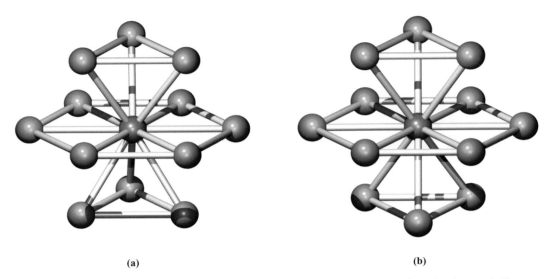

(a)　　　　　　　　　　　　　　　　　　**(b)**

 Fig. 6.3 In both the (a) ABA and (b) ABC close-packed arrangements, the coordination number of each atom is 12.

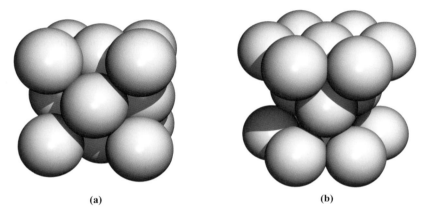

(a)　　　　　　　　　　　　　　　　　**(b)**

 Fig. 6.4 Unit cells of (a) a cubic close-packed (face-centred cubic) lattice and (b) a hexagonal close-packed lattice.

close-packing (hcp). Whereas these respective descriptors are not obviously associated with the packing sequences shown in Figs. 6.2 and 6.3, their origins are clear in the unit cell diagrams. Cubic close-packing is also called *face-centred cubic* (fcc) packing, and this name clearly reflects the nature of the unit cell shown in Fig. 6.4a. The relationship between the ABABAB... sequence and the hcp unit cell is easily recognized; the latter consists of parts of three ABA layers. However, it is harder to see the ABCABC... sequence within the ccp unit cell since the close-packed layers are not parallel to the base of the unit cell but instead lie along the body-diagonal of the cube.

Interstitial holes: hexagonal and cubic close-packing

Close-packed structures contain *octahedral* and *tetrahedral holes* (or *sites*). Figure 6.5 shows representations of two layers of close-packed spheres: Fig. 6.5a is a 'space-filling'

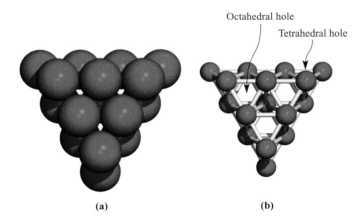

Octahedral hole

Tetrahedral hole

(a)　　　　　　　　　**(b)**

Fig. 6.5 Two layers of close-packed atoms shown (a) with the spheres touching, and (b) with the sizes of the spheres reduced so that connectivity lines are visible. In (b), the tetrahedral and octahedral holes are indicated.

representation, while in Fig. 6.5b, the sizes of the spheres have been reduced so that connectivity lines can be shown (a 'ball-and-stick' diagram). This illustrates that the spheres lie at the corners of either tetrahedra or octahedra. Conversely, the spheres pack such that there are octahedral and tetrahedral holes between them. There is one octahedral hole per sphere, and there are twice as many tetrahedral as octahedral holes in a close-packed array. The octahedral holes are larger than the tetrahedral holes. Whereas a tetrahedral hole can accommodate a sphere of radius ≤ 0.23 times that of the close-packed spheres, a sphere of radius 0.41 times that of the close-packed spheres fits into an octahedral hole.

Non-close-packing: simple cubic and body-centred cubic arrays

Spheres are not always packed as efficiently as in close-packed arrangements. Ordered arrays can be constructed in which the space occupied by the spheres is less than the 74% found for a close-packed arrangement.

If spheres are placed so as to define a network of cubic frameworks, the unit cell is a simple cube (Fig. 6.6a). In the extended lattice, each sphere has a coordination number of 6. The hole within each cubic unit is not large enough to accommodate a sphere equal in size to those in the array, but if the eight spheres in the cubic cell are pulled apart slightly, another sphere is able to fit inside the hole. The result is the *body-centred cubic* (bcc) arrangement (Fig. 6.6b). The coordination number of each sphere in a bcc lattice is 8.

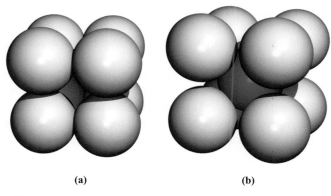

 Fig. 6.6 Unit cells of (a) a simple cubic lattice and (b) a body-centred cubic lattice.

Worked example 6.1   Packing efficiency

Show that in a simple cubic lattice, (a) there is one sphere per unit cell, and (b) approximately 52% of the volume of the unit cell is occupied.

(a) The diagram on the left-hand side below is a space-filling representation of the unit cell of a simple cubic lattice, while the right-hand side diagram shows a ball-and-stick representation.

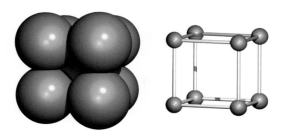

In the complete lattice, unit cells are packed side by side in three dimensions. Therefore, each sphere is shared between eight unit cells.

Number of spheres per unit cell $= (8 \times {}^{1}/_{8}) = 1$

(b) Let the radius of each sphere $= r$. From the diagrams above, it follows that:

Unit cell edge length $= 2r$
   Unit cell volume $= 8r^3$

The unit cell contains one sphere, and the volume of a sphere $= {}^{4}/_{3}\,\pi r^3$. Therefore:

Volume of the unit cell occupied $= {}^{4}/_{3}\,\pi r^3$

$$\approx 4.19r^3$$

Percentage occupancy $= \dfrac{4.19r^3}{8r^3} \times 100 \approx 52\%$

Self-study exercises

1. Show that a face-centred cubic unit cell contains four complete spheres.

2. If the radius of each sphere in an fcc arrangement is r, show that the unit cell edge length is $\sqrt{8}r$.

3. Using the answers to questions (1) and (2), show that the packing efficiency of a cubic close-packed arrangement is 74%.

4. Show that in a bcc arrangement of spheres, a unit cell contains two complete spheres.

5. Confirm that the packing efficiency of a bcc arrangement is 68%.

6.3 The packing-of-spheres model applied to the structures of elements

In Section 6.2, we considered some of the ways in which *hard spheres* may pack together to give ordered arrays. Although the idea of hard, spherical atoms is at odds with modern quantum theory, the packing-of-spheres model is extremely useful for depicting many solid state structures. The model is applicable to the group 18 elements because

Table 6.1 Selected physical data for the group 18 elements.

Element	Melting point / K	$\Delta_{fus}H$(mp) / kJ mol^{-1}	Boiling point / K	$\Delta_{vap}H$(bp) / kJ mol^{-1}	van der Waals radius (r_v) / pm
Helium	‡	–	4.2	0.08	99
Neon	24.5	0.34	27	1.71	160
Argon	84	1.12	87	6.43	191
Krypton	116	1.37	120	9.08	197
Xenon	161	1.81	165	12.62	214
Radon	202	–	211	18	–

‡Helium cannot be solidified under atmospheric pressure, the pressure condition for which all other phase changes in the table are considered.

they are monatomic, to metals, and to H_2 and F_2 because these diatomic molecules are freely rotating in the solid state and so can be regarded as spherical entities.

Group 18 elements in the solid state

The group 18 elements are the 'noble gases' (see Chapter 18), and Table 6.1 lists selected physical data for these elements. Each element (with the exception of helium, see footnote in Table 6.1) solidifies only at low temperatures. The enthalpy changes accompanying the fusion processes are very small, consistent with the fact that only weak van der Waals forces operate between the atoms in the solid state. In the crystalline solid, ccp structures are adopted by each of solid Ne, Ar, Kr and Xe.

H$_2$ and F$_2$ in the solid state

The liquefaction of gaseous H_2 occurs at 20.4 K[†] and solidification at 14.0 K. However, even in the solid state, H_2 molecules have sufficient energy to rotate about a fixed lattice point and consequently the space occupied by each diatomic can be represented by a sphere. In the solid state, these spheres adopt an hcp arrangement.

Difluorine solidifies at 53 K, and on cooling to 45 K, a phase change occurs to give a distorted close-packed structure. This description is applicable because, like H_2, each F_2 molecule rotates freely about a fixed lattice point. (The second phase above 45 K has a more complicated structure.)

The application of the packing-of-spheres model to the crystalline structures of H_2 and F_2 is *only* valid because they contain freely rotating molecules. Other diatomics such as the heavier halogens do not behave in this manner (see Section 17.4).

[†] All phase changes mentioned in this chapter are at atmospheric pressure, unless otherwise stated.

Metallic elements in the solid state

With the exception of Hg, all metals are solid at 298 K. The statement 'solid at room temperature' is ambiguous because the low melting points of Cs (301 K) and Ga (303 K) mean that in some hot climates, these metals are liquids. Table 6.2 shows that most metals crystallize with ccp, hcp or bcc lattices. However, many metals are *polymorphic* and exhibit more than one structure depending upon the conditions of temperature and/or pressure; we return to this later.

On the basis of the hard sphere model, close-packing represents the most efficient use of space with a common packing efficiency of 74%. The bcc structure is not much less efficient in packing terms, for although there are only eight nearest neighbours, each at a distance x (compared with 12 in the close-packed lattices), there are six more neighbours at distances of $1.15x$, leading to a packing efficiency of 68%.

Among the few metals that adopt structures other than ccp, hcp or bcc lattices are those in group 12. The structures of Zn and Cd are based upon hcp lattices but distortion leads to each atom having only six nearest neighbours (within the same layer of atoms) and six others at a greater distance. Mercury adopts a distorted simple cubic lattice, with the distortion leading to a coordination number of 6. Manganese stands out among the *d*-block metals as having an unusual structure. The atoms are arranged in a complex cubic lattice such that there are four environments with coordination numbers of 12, 13 or 16. Atypical structures are also exhibited by most of the *p*-block metals. In group 13, Al and Tl adopt ccp and hcp lattices respectively, but Ga (the α-form) and In adopt quite different structures. Atoms of Ga are organized so that there is only one nearest neighbour (at 249 pm), with six next-nearest neighbours lying at distances within the range 270 and 279 pm, i.e. there is a tendency for the atoms to pair together. Indium forms a distorted ccp lattice, and the twelve near neighbours separate into two groups, four at 325 pm and eight at

Table 6.2 Structures (at 298 K), melting points (K) and values of the standard enthalpies of atomization of the metallic elements.
◆ = hcp; ● = ccp (fcc); ● = bcc

Be
◆
1560
324
112

← Metal lattice type
← Melting point (K)
← Standard enthalpy of atomization (kJ mol^{-1})
← Metallic radius for 12-coordinate atom (pm)

1	2	3	4	5	6	7	8	9	10	11	12	13	14	15
Li ● 454 161 157	Be ◆ 1560 324 112													
Na ● 371 108 191	Mg ◆ 923 146 160											Al ● 933 330 143		
K ● 337 90 235	Ca ◆ 1115 178 197	Sc ◆ 1814 378 164	Ti ◆ 1941 470 147	V ● 2183 514 135	Cr ● 2180 397 129	Mn see text 1519 283 137	Fe ● 1811 418 126	Co ◆ 1768 428 125	Ni ● 1728 430 125	Cu ● 1358 338 128	Zn see text 693 130 137	Ga see text 303 277 153		
Rb ● 312 82 250	Sr ◆ 1040 164 215	Y ◆ 1799 423 182	Zr ◆ 2128 609 160	Nb ● 2750 721 147	Mo ● 2896 658 140	Tc ◆ 2430 677 135	Ru ◆ 2607 651 134	Rh ● 2237 556 134	Pd ● 1828 377 137	Ag ● 1235 285 144	Cd see text 594 112 152	In see text 430 243 167	Sn see text 505 302 158	
Cs ● 301 78 272	Ba ● 1000 178 224	La ◆ 1193 423 188	Hf ◆ 2506 619 159	Ta ● 3290 782 147	W ● 3695 850 141	Re ◆ 3459 774 137	Os ◆ 3306 787 135	Ir ● 2719 669 136	Pt ● 2041 566 139	Au ● 1337 368 144	Hg see text 234 61 155	Tl ◆ 577 182 171	Pb ● 600 195 175	Bi † 544 210 182

†See Fig. 15.3c and associated text.

338 pm.† In group 14, Pb adopts a ccp structure, but in white Sn (the stable allotrope at 298 K), each atom possesses a coordination number of only 6 (for the structure of grey Sn, see Section 6.4). Metals with coordination numbers of less than 8 are among those that are the most volatile.

6.4 Polymorphism in metals

Polymorphism: phase changes in the solid state

It is generally convenient to consider the structures of metals in terms of the observed structure type at 298 K and

atmospheric pressure,‡ but these data do not tell the whole story. When subjected to changes in temperature and/or pressure, the structure of a metal may change; each form of the metal is a particular *polymorph* (see Section 4.11). For example, scandium undergoes a reversible transition from an hcp lattice (α-Sc) to a bcc lattice (β-Sc) at 1610 K. Some metals undergo more than one change: at atmospheric pressure, Mn undergoes transitions from the α- to β-form at 983 K, from the β- to γ-form at 1352 K, and from γ- to σ-Mn at 1416 K. Although α-Mn adopts a complex lattice (see above), the β-polymorph has a somewhat simpler structure containing two 12-coordinate Mn environments, the γ-form possesses a distorted ccp

---

†For more detailed discussions of the origin of the distorted ccp structure of indium and an overall view of the structures of the group 13 metals, see: U. Häussermann *et al.* (1999) *Angew. Chem. Int. Ed.*, vol. 38, p. 2017; U. Häussermann *et al.* (2000) *Angew. Chem. Int. Ed.*, vol. 39, p. 1246.

‡ Although we often refer to 'atmospheric pressure', a pressure of 1 bar (1.00 × 10^5 Pa) has been defined by the IUPAC as the *standard pressure*. Until 1982, the standard pressure was 1 atmosphere (1 atm = 101 300 Pa) and this pressure remains in use in some tables of physical data.

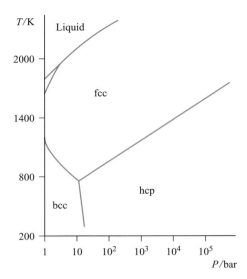

Fig. 6.7 A pressure–temperature phase diagram for iron.

structure, and the σ-polymorph adopts a bcc lattice. Phases that form at high temperatures may be *quenched* to lower temperatures (i.e. rapidly cooled with retention of structure), allowing the structure to be determined at ambient temperatures. Thermochemical data show that there is usually very little difference in energy between different polymorphs of an element.

An interesting example of polymorphism is observed for tin. At 298 K and 1 bar pressure, β-Sn (white tin) is the thermodynamically stable polymorph but lowering the temperature to 286 K results in a slow transition to α-Sn (grey tin). The β ⟶ α transition is accompanied by a change in coordination number from 6 to 4, and α-Sn adopts a diamond-type lattice (see Fig. 6.20). The density of Sn *decreases* from 7.31 to 5.75 g cm^{-3} during the β ⟶ α transition, whereas it is more usual for there to be an increase in density in going from a higher to lower temperature polymorph.

Phase diagrams

In order to appreciate the effects on an element of changing the temperature and pressure, a *phase diagram* must be consulted. Figure 6.7 shows the phase diagram for Fe; each line on the diagram is a *phase boundary* and crossing a boundary (i.e. changing the phase of the metal) requires a change of temperature and/or pressure. For example, at 298 K and 1 bar pressure, Fe has a bcc structure (α-Fe). Raising the temperature to 1185 K (still at 1 bar) results in a transition to γ-Fe with an fcc structure. A transition from α- to γ-Fe also occurs by increasing the pressure on Fe maintained at, e.g., 800 K.

Self-study exercises

1. Use Fig. 6.7 to describe what happens to the structure of iron if the pressure is raised from 1 bar while maintaining the temperature at 900 K.

2. In general, the bcc structure is the high-temperature form of a metal which is close-packed at a lower temperature. What happens to the density of the metal during this phase change?

3. Give two examples of metals that have a bcc structure at 298 K. [*Ans*: see Table 6.2]

6.5 Metallic radii

The *metallic radius*, r_{metal}, is defined as half of the distance between the nearest-neighbour atoms in a solid state metallic lattice. However, structural data for different polymorphs of the same metal indicate that r_{metal} varies with the coordination number. For example, the ratio of the interatomic distances (and, therefore, of r_{metal}) in a bcc polymorph to those in close-packed forms of the same metal is 0.97 : 1.00, corresponding to a change in coordination number from 8 to 12. If the coordination number decreases further, r_{metal} also decreases:

Coordination number	12	8	6	4
Relative radius	1.00	0.97	0.96	0.88

> The *metallic radius* is half of the distance between the *nearest-neighbour* atoms in a solid state metal lattice, and is dependent upon coordination number.

The values of r_{metal} listed in Table 6.2 refer to 12-coordinate metal centres. Since not all metals actually adopt structures with 12-coordinate atoms, some values of r_{metal} have been estimated. The need for a *consistent* set of data is obvious if one is to make meaningful comparisons within a periodic sequence of elements. Values of r_{metal} (Table 6.2) increase down each of groups 1, 2, 13 and 14. In each of the triads of the *d*-block elements, r_{metal} generally increases on going from the first to second row element, but there is little change on going from the second to third row metal. This latter observation is due to the presence of a filled 4*f* level, and the so-called *lanthanoid contraction* (see Sections 22.3 and 27.3).

Worked example 6.2   Metallic radii

Use values of r_{metal} in Table 6.2 to deduce an appropriate value for the metallic radius (a) r_K in metallic K at 298 K and 1 bar pressure, and (b) r_{Sn} in α-Sn. Is the answer for part (b) consistent with the observed interatomic distance in α-Sn of 280 pm?

The values of r_{metal} in Table 6.2 refer to 12-coordinate metal atoms, and values of K and Sn are 235 and 158 pm respectively.

(a) The structure of K at 298 K and 1 bar pressure is bcc, and the coordination number of each K atom is 8. From the relative radii listed in the text:

$$\frac{r_{12\text{-coordinate}}}{r_{8\text{-coordinate}}} = \frac{1}{0.97}$$

The appropriate radius for a K atom in a bcc lattice is:

$$r_{8\text{-coordinate}} = 0.97 \times (r_{12\text{-coordinate}}) = 0.97 \times 235 = 228 \, \text{pm}$$

(b) In α-Sn, each Sn atom is 4-coordinate. From the relative radii listed in the text:

$$\frac{r_{12\text{-coordinate}}}{r_{4\text{-coordinate}}} = \frac{1}{0.88}$$

The radius for a Sn atom in α-Sn is estimated from:

$$r_{4\text{-coordinate}} = 0.88 \times (r_{12\text{-coordinate}}) = 0.88 \times 158 = 139 \, \text{pm}$$

The interatomic distance is twice the value of r_{metal}, and so the calculated value of the Sn−Sn distance of 278 pm is in good agreement with the observed value of 280 pm.

Self-study exercises

Use data in Table 6.2.
1. Estimate a value for the metallic radius, r_{Na}, in metallic Na (298 K, 1 bar). [*Ans.* 185 pm]

2. The internuclear separation of two Na atoms in the metal (298 K, 1 bar) is 372 pm. Estimate a value of r_{metal} appropriate for 12-coordination. [*Ans.* 192 pm]

6.6 Melting points and standard enthalpies of atomization of metals

The melting points of the metallic elements are given in Table 6.2 and periodic trends are easily observed. The metals with the lowest melting points are in groups 1, 12, 13 (with the exception of Al), 14 and 15. These metals are, in general, those that do *not* adopt close-packed structures in the solid state. The particularly low melting points of the alkali metals (and correspondingly low values of the standard enthalpies of fusion which range from $3.0 \, \text{kJ mol}^{-1}$ for Li to $2.1 \, \text{kJ mol}^{-1}$ for Cs) often give rise to interesting practical observations. For example, when a piece of potassium is dropped on to water, exothermic reaction 6.1 occurs, providing enough heat energy to melt the unreacted metal; the molten potassium continues to react vigorously (Fig. 6.8).

$$2K + 2H_2O \longrightarrow 2KOH + H_2 \tag{6.1}$$

Values of the standard enthalpies of atomization, $\Delta_a H^\circ(298 \, \text{K})$, (or sublimation) in Table 6.2 refer to the

Fig. 6.8 Potassium reacting with water (eq. 6.1). The exothermic reaction causes the metal to melt.

processes defined in eq. 6.2, and correspond to the destruction of the metallic lattice. Mercury is an exception, since at 298 K it is a liquid.

$$\frac{1}{n} M_n(\text{standard state}) \longrightarrow M(g) \tag{6.2}$$

Those metals with the lowest values of $\Delta_a H^\circ(298 \, \text{K})$ are again those with non-close-packed structures. Since $\Delta_a H^\circ$ appears in thermochemical cycles such as the Born–Haber cycle (see Section 6.14), it is clear that $\Delta_a H^\circ$ is an important factor in accounting for the reactivity patterns of these metals.

In general, there appears to be a rough correlation between values of $\Delta_a H^\circ(298 \, \text{K})$ and the number of unpaired electrons. In any long period (K to Ga, Rb to Sn, and Cs to Bi in Table 6.2), the maximum values are reached in the middle of the d-block (with the exception of Mn which has the atypical structure described in Section 6.3).

6.7 Alloys and intermetallic compounds

The physical properties of many metals render them unsuitable for fabrication and engineering purposes. By combining two or more metals, or metals with non-metals, one can form *alloys* with enhanced properties such as strength, malleability, ductility, hardness or resistance to corrosion. For example, copper is alloyed with zinc (2–40% by weight) to produce different types of brasses which are stronger than copper but retain good fabrication properties.

An *alloy* is an intimate mixture or, in some cases, a compound of two or more metals, or metals and non-metals; alloying changes the physical properties and resistance to corrosion, heat etc. of the material.

Alloys are manufactured by combining the component elements in the molten state followed by cooling. If the

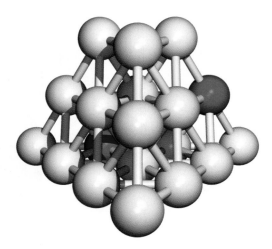

Fig. 6.9 In a substitutional alloy, some of the atom sites in the host lattice (shown in grey) are occupied by solute atoms (shown in red).

melt is quenched (cooled rapidly), the distribution of the two types of metal atoms in the *solid solution* will be random; the element in excess is termed the solvent, and the minor component is the solute. Slow cooling may result in a more ordered distribution of the solute atoms. The subject of alloys is not simple, and we shall introduce it only by considering the classes of substitutional and interstitial alloys, and intermetallic compounds.

Substitutional alloys

In a substitutional alloy, atoms of the solute occupy sites in the lattice of the solvent metal (Fig. 6.9). To maintain the original structure of the host metal, atoms of both components should be of a similar size. The solute atoms must also tolerate the same coordination environment as atoms in the host lattice. An example of a substitutional alloy is sterling silver (used for silver cutlery and jewellery) which contains 92.5% Ag and 7.5% Cu; elemental Ag and Cu both adopt ccp lattices and $r_{metal}(Ag) \approx r_{metal}(Cu)$ (Table 6.2).

Interstitial alloys

A close-packed lattice contains tetrahedral *and* octahedral interstitial holes (see Fig. 6.5). Assuming a hard-sphere model for the atomic lattice,[†] one can calculate that an atom of radius 0.41 times that of the atoms in the close-packed array can occupy an octahedral hole, while significantly smaller atoms may be accommodated in tetrahedral holes.

We illustrate interstitial alloys by discussing *carbon steels* in which C atoms occupy a small proportion of the octahedral holes in an Fe lattice. α-Iron possesses a bcc

structure at 298 K (1 bar pressure), and a transition to γ-Fe (ccp) occurs at 1185 K; over the range 1674 to 1803 K, α-Fe is again observed (Fig. 6.7). Carbon steels are extremely important industrially (see Box 6.1), and there are three basic types designated by their carbon content. *Low-carbon steel* contains between 0.03 and 0.25% carbon and is used for steel sheeting, e.g. in the motor vehicle industry and in the manufacture of steel containers. *Medium-carbon steel* contains 0.25–0.70% C, and is suited for uses such as bolts, screws, machine parts, connecting rods and railings. The strongest of the carbon steels, *high-carbon steel*, contains 0.8–1.5% C and finds applications in cutting and drilling tools. The corrosion of carbon steels is a disadvantage of the material, but coatings can be applied to inhibit such action. *Galvanized steel* possesses a Zn coating; Zn has a low mechanical strength but a high resistance to corrosion and, combined with the high mechanical strength of the steel, galvanized steel meets the demands of many industrial applications. If the Zn coating is scratched revealing the Fe beneath, it is the Zn that oxidizes in preference to the Fe (see Box 8.4).

An alternative method of enhancing the properties of steel is to alloy it with another metal, M. This combines both interstitial and substitutional alloy structures, with C occupying holes in the Fe lattice, and M occupying lattice sites. *Stainless steel* is an example of an *alloy steel* and is discussed in Box 6.2. For high-wear resistance (e.g. in rail and tram tracks), Mn is alloyed with steel. Other alloy steels contain Ti, V, Co or W, and each solute metal confers specific properties on the finished product. Specific steels are described in Sections 21.2 and 22.2.

Intermetallic compounds

When melts of some metal mixtures solidify, the alloy formed may possess a definite structure type that is different from those of the pure metals. Such systems are classified as *intermetallic compounds*, e.g. β-brass, CuZn. At 298 K, Cu has a ccp lattice and Zn has a structure related to an hcp array, but β-brass adopts a bcc structure. The relative proportions of the two metals are crucial to the alloy being described as an intermetallic compound. Alloys labelled 'brass' may have variable compositions, and the α-phase is a substitutional alloy possessing the ccp structure of Cu with Zn functioning as the solute. β-Brass exists with Cu : Zn stoichiometries around 1 : 1, but increasing the percentage of Zn leads to a phase transition to γ-brass (sometimes written as Cu_5Zn_8, although the composition is not fixed), followed by a transition to ε-brass which has an approximate stoichiometry of 1 : 3.[‡]

[†] It is important not to lose sight of the fact that the hard-sphere model is approximate and conflicts with the wave-mechanical view of the atom.

[‡] The variation of phases with temperature and Cu : Zn stoichiometry is more complex than this description implies; see N.N. Greenwood and A. Earnshaw (1997) *Chemistry of the Elements*, 2nd edn, Butterworth-Heinemann, Oxford, p. 1178.

ENVIRONMENT

Box 6.1 Iron and steel production and recycling

The major raw materials for the commercial production of Fe are haematite (Fe_2O_3), magnetite (Fe_3O_4) and siderite ($FeCO_3$) (see also Section 21.2). The extraction of iron is carried out on an enormous scale to meet the consumer demands for both iron and steel. In 2010, China and Japan led the world in the production of crude steel.

The industrial manufacturing processes for iron and steel can be summarized as follows. Iron ore is mixed with limestone ($CaCO_3$) and coke in a blast furnace in which temperatures vary from $\approx$750 to 2250 K. Carbon is converted to CO in the highest temperature zone, but both C and CO may reduce the iron ore:

$$2C + O_2 \longrightarrow 2CO$$

$$Fe_2O_3 + 3C \longrightarrow 2Fe + 3CO$$

$$Fe_2O_3 + 3CO \longrightarrow 2Fe + 3CO_2$$

The function of the limestone is to remove impurities and the product of these reactions is *slag*, which contains, for example, calcium silicate. Molten Fe from the furnace is collected and cooled in salt-moulds as *pig iron*, which contains 2–4% C plus small amounts of P, Si, S and Mn. After remelting and moulding, the product is *cast iron*; this is brittle and its exact nature depends upon the relative amounts of secondary elements. A high Si content results in the C being in the form of graphite, and the cast iron so formed is called *grey cast iron*. On the other hand, *white cast iron* forms when the Si content is low and carbon is present within the iron–carbon phase *cementite*, Fe_3C.

The *puddling process* is used to convert cast iron to wrought iron. During this process, C, S and other impurities are oxidized, leaving wrought iron with <0.2% C content. Unlike cast iron, wrought iron is tough and malleable and is readily worked; its applications, in wrought iron railings and window and door grills, are widespread.

Iron can be converted into steel by the Bessemer, Siemens electric arc or basic oxygen processes. The Bessemer process was the first to be patented, but the electric arc (see below) and basic oxygen processes are used in modern steel production. In the basic oxygen process, O_2 oxidizes the carbon in pig iron, reducing its content to the levels required for commercial steel (see main text). The manufacture of steel involves the addition of ferrosilicon (an alloy of iron and silicon, made by the high-temperature carbon reduction of a mixture of SiO_2 and iron ore or scrap iron). Ferrosilicon serves two purposes: it deoxidizes steel by conversion of Si to SiO_2, and Si is also used as an alloying element (see above).

Impact on the environment: recycling of steel

In the description above, we focused on steel production in a blast furnace or by the basic oxygen process using iron ore, limestone and coke as raw materials. In contrast, the *electric*

arc furnace (see photograph) relies entirely on scrap steel as its 'raw' material. The furnace is first charged with scrap, which is then melted by being electrically heated to about 1500 K by an arc that passes between graphite electrodes and the furnace walls. After refining, the recycled steel can be cast as required. Steel can be recycled over and over again, and as recycling becomes more important, so the use of the electric arc furnace increases. In the US, the Steel Recycling Institute encourages recycling of steel products from cans to domestic appliances, automobiles and construction materials. For example, in 2008, 14.5 million vehicles were recycled, accounting for about 15 million tonnes of scrap steel. The percentage of steel cans recycled in the US has risen from 15% in 1988 to 65% in 2008, while 90% of steel domestic appliances were recycled in 2008.

Use of the electric arc furnace not only reduces the need for raw iron ore, limestone and coke. It also reduces the emissions of CO_2, one of the 'greenhouse gases' targeted by the 1997 Kyoto Protocol, through which most industrialized countries have agreed to limit their output of CO_2, CH_4, N_2O, SF_6, hydrofluorocarbons (HFCs) and perfluorocarbons (PFCs). The 1997 Kyoto Protocol, which runs until the end of 2012, aims to reduce emissions of these gases so that in the period 2008–2012, levels are 5% lower than those in 1990; this corresponds to a 29% reduction compared with emission levels projected for 2010 if the protocol were not in place. Regulations to succeed the Kyoto protocol are not as yet (2011) in place.

Further information: www.worldsteel.org

An electric arc furnace receiving a charge of scrap metal.

APPLICATIONS

Box 6.2 Stainless steel: corrosion resistance by adding chromium

Stainless steels (so-called because they do not become stained with rust) are examples of *alloy steels*, i.e. ones that contain a *d*-block metal in addition to carbon. Stainless steels have a significant content of the alloy metal and are of high commercial value because of their resistance to corrosion. All contain at least 10.5% (by mass) of chromium, the minimum that renders the steel corrosion-resistant under normal aqueous conditions (i.e. in the absence of acid, alkali or pollutants such as chloride ions). The resistance to corrosion arises from the formation of a thin layer of Cr_2O_3 ($\approx 13\,nm$ thick) over the surface of the steel. The oxide layer passivates (see Section 10.4) the steel and is self-repairing, i.e. if some of the oxide coating is scratched off, further oxidation of the chromium in the steel necessarily repairs the 'wound'.

There are four main classes of stainless steel (austenitic, ferritic, ferritic-austenitic (duplex) and martensitic), and within these, a variety of different grades. The names ferritic and austenitic follow from their structures: ferrite (β-Fe) and austenite (γ-Fe) structures hosting the alloying elements. The presence of Cr promotes the formation of the ferrite structure, while the austenite structure forms when Ni is introduced. While ferritic and martensitic stainless steels are magnetic, austenitic stainless steel is non-magnetic. Further additives to some stainless steels are molybdenum (which improves corrosion resistance) and nitrogen (which adds strength and improves corrosion resistance).

Ferritic stainless steels commonly contain 17% Cr and $\leq 0.12\%$ C. Such steels are used in household appliances (e.g. washing machines and dishwashers) and in vehicle trim. Increasing the carbon content of ferritic stainless steels results in the formation of martensitic stainless steels (which usually contain 11–13% Cr). These steels are strong, hard and can be sharpened, and are used to make knives and other blades. Austenitic stainless steels contain $\geq 7\%$ nickel (the most common grade contains 18% Cr, 9% Ni and $\leq 0.08\%$ C) and are ductile, making them suitable for use in the manufacture of forks and spoons. The toughness and ease of welding of austenitic stainless steels lead to their widespread use in the manufacturing industry. In the home, austenitic stainless steels are used in food processors and kitchen sinks. A combination of ferritic and austenitic stainless steels leads to duplex stainless steels (22% Cr, 5% Ni, 3% Mo, 0.15% N, $\leq 0.03\%$ C) with properties that make them suitable for use in, for example, hot-water tanks. Further modifications to the main classes of stainless steel lead to additional grades for specialized applications. Alloying with Cu, Mo and Ni produces stainless steels which are resistant to corrosion by organic and sulfuric acids. A high Mo content gives resistance to phosphoric acid, while increasing the Cr content of the alloy leads to a nitric acid-resistant material. Corrosion is typically observed by pitting of the surface of the steel, and the resistance of a particular grade of stainless steel is empirically estimated from its pitting resistance equivalence number (PREN) where:

$$PREN = \%Cr + (3.3 \times \%Mo) + (16 \times \%N)$$

An alternative empirical relationship is the measure for alloying for resistance to corrosion (MARC) where:

$$\begin{aligned} MARC = \%Cr &+ (3.3 \times \%Mo) + (20 \times \%N) \\ &- (0.5 \times \%Mn) + (20 \times \%C) - (0.25 \times \%Ni) \end{aligned}$$

There is a linear correlation between the MARC value and the critical pitting temperature, i.e. the lowest temperature at which pitting corrosion is first observed under specified conditions. The equation for MARC takes into account the fact that alloying with Cr, Mo, N and C improves corrosion resistance, whereas the presence of Mn and Ni (added to enhance other properties of the material) has the opposite effect.

Stainless steels appear in every facet of our lives, from consumer goods (especially in the kitchen, where cleanliness and corrosion-resistance are essential) to industrial storage tanks, chemical plant components, vehicle parts, including exhaust pipes and catalytic converters (see Section 26.7), and a wide range of industrial corrosion-resistant components. A further property that makes stainless steels commercially important is that they can be polished to satin or mirror finishes and this is easily appreciated in the ranges of stainless steel cutlery available to the consumer. Building projects also make wide use of stainless steels, both in construction and in external decorative parts.

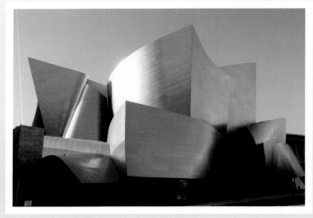

The stainless steel walls of the Walt Disney concert hall in Los Angeles.

Further reading

Web-based site: www.worldstainless.org
Related information: Box 21.1 Chromium: resources and recycling.

6.8 Bonding in metals and semiconductors

If we consider the various structure types adopted by metals and then try to provide a model for localized metal–metal bonding, we run into a problem: there are not enough valence shell orbitals or electrons for each metal atom to form 2-centre 2-electron bonds with all its neighbours. For example, an alkali metal atom in the bulk metal has eight near neighbours (Table 6.2), but only one valence electron. We must therefore use a bonding model with multi-centre orbitals (see Sections 5.4–5.7). Further, the fact that metals are good electrical conductors means that the multi-centre orbitals must spread over the whole metal crystal so that we can account for the electron mobility. Several bonding theories have been described, and *band theory* is the most general. Before discussing band theory, we review *electrical conductivity* and *resistivity*.

Electrical conductivity and resistivity

An ***electrical conductor*** offers a low resistance (measured in ohms, Ω) to the flow of an electrical current (measured in amperes, A).

The electrical resistivity of a substance measures its resistance to an electrical current (eq. 6.3). For a wire of uniform cross-section, the resistivity (ρ) is given in units of ohm metre ($\Omega\,m$).

$$\text{Resistance (in } \Omega) = \frac{\text{resistivity (in } \Omega\,m) \times \text{length of wire (in m)}}{\text{cross-sectional area of wire (in m}^2)}$$

$$R = \frac{\rho \times l}{a} \qquad (6.3)$$

Figure 6.10 shows the variation in resistivity of three metals with temperature. In each case, ρ increases with temperature, and the electrical conductivity (which is the inverse of the resistance) decreases as the temperature is raised. This property distinguishes a metal from a *semiconductor*, which is a material in which the electrical conductivity increases as the temperature increases (Fig. 6.11).

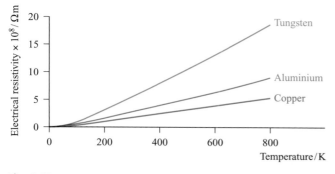

Fig. 6.10 A metal is characterized by the fact that its *electrical resistivity increases* as the temperature increases, i.e. its *electrical conductivity decreases* as the temperature increases.

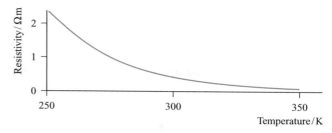

Fig. 6.11 A semiconductor, such as germanium, is characterized by the fact that its electrical resistivity *decreases* as the temperature increases. Its electrical conductivity *increases* as the temperature increases.

The ***electrical conductivity*** of a metal decreases with temperature; that of a semiconductor increases with temperature.

Band theory of metals and insulators

The fundamental concept of band theory is to consider the energies of the molecular orbitals in an assembly of metal atoms. An MO diagram describing the bonding in a metallic solid is characterized by having groups of MOs (i.e. *bands*) which are very close in energy. We can readily see how bands arise by constructing an approximate MO diagram for lithium metal, Li_n.

The valence orbital of an Li atom is the $2s$ atomic orbital, and Fig. 6.12 shows schematic MO diagrams for the formation of species incorporating different numbers of Li atoms (see Section 2.3). If two Li atoms combine, the overlap of the two $2s$ atomic orbitals leads to the formation of two MOs. If three Li atoms combine, three MOs are

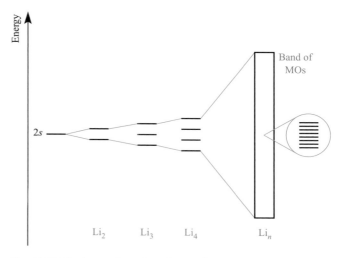

Fig. 6.12 The interaction of two $2s$ atomic orbitals in Li_2 leads to the formation of two MOs. With three Li atoms, three MOs are formed, and so on. For Li_n, there are n molecular orbitals, but because the $2s$ atomic orbitals are all of the same energy, the energies of the MOs are very close together and constitute a *band* of orbitals.

formed, and so on. For *n* Li atoms, there are *n* MOs, but because the 2*s* atomic orbitals possess the same energy, the energies of the resultant MOs are very close together and so are termed a *band* of orbitals. Now, let us apply the *aufbau* principle and consider the occupation of the MOs in Fig. 6.12. Each Li atom contributes one electron. In Li_2, this leads to the lowest MO being filled, and in Li_3, the lowest MO is fully occupied and the next MO is half-filled. In Li_n, the band must be half-occupied. Since the band of MOs in Li_n contains contributions from all the Li atoms, the model provides a delocalized picture of the bonding in the metal. Moreover, because the energies of the MOs within the band are very close together and not all the MOs are populated in the ground state, electrons can move into vacant MOs *within the band* under the influence of an electric field. Because of the delocalization, we can readily rationalize the movement of electrons from one Li atom to another, and understand why electrical conductivity results. This model indicates that electrical conductivity is a characteristic property of *partially filled bands* of MOs. In theory, no resistance should oppose the flow of a current if the nuclei are arranged at the points of a perfectly ordered lattice, and the increased *thermal population* of higher energy levels within the band at higher temperatures might be expected to lead to an increase in the electrical conductivity. In practice, however, thermal vibrations of the nuclei produce electrical resistance and this effect is sufficiently enhanced at higher temperatures so as to result in a *decrease* in the conductivity of the metal as the temperature increases.

A ***band*** is a group of MOs, the energy differences between which are so small that the system behaves as if a continuous, non-quantized variation of energy within the band is possible.

The model just described for Li is oversimplified. Bands are also formed by the overlap of higher energy (unoccupied) atomic orbitals, and the 2*p* band overlaps with the 2*s* band to some extent since the *s*–*p* separation in atomic Li is relatively small. This is also true for Be and, of course, this is of great significance since the ground state electronic configuration of Be is $[He]2s^2$. Were the energy separation of the 2*s* and 2*p* bands in Be large, the 2*s* band would be fully occupied and Be would be an insulator. In reality, the 2*s* and 2*p* bands overlap, and generate, in effect, a single, partially occupied band, thereby giving Be its metallic character. Figure 6.13a–c illustrates that:

• a fully occupied band separated from the next (empty) band by a large energy separation (the *band gap*) leads to the material being an insulator;
• a partially occupied band leads to the material being metallic;
• metallic character is also consistent with the overlap of an occupied and a vacant band.

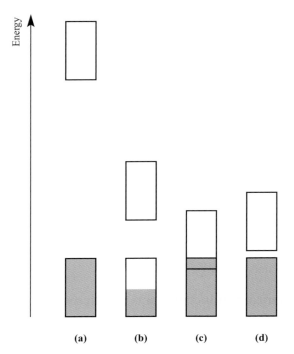

Fig. 6.13 The relative energies of occupied and empty bands in (a) an insulator, (b) a metal in which the lower band is only *partially* occupied, (c) a metal in which the occupied and empty bands overlap, and (d) a semiconductor.

A ***band gap*** occurs when there is a significant energy difference between two bands. The magnitude of a band gap is typically given in electron volts (eV); $1 \text{ eV} = 96.485 \text{ kJ mol}^{-1}$.

The Fermi level

The energy level of the highest occupied orbital in a metal at absolute zero is called the *Fermi level*. At this temperature, the electronic configuration predicted by the *aufbau* principle appertains and so, in Li for example, the Fermi level lies exactly at the centre of the half-filled band. For other metals, the Fermi level lies at or near the centre of the band. At temperatures above 0 K, electrons thermally populate MOs just above the Fermi level, and some energy levels just below it remain unoccupied. In the case of a metal, the thermal populations of different energy states cannot be described in terms of a Boltzmann distribution, but are instead given by the *Fermi–Dirac distribution*.[†]

Band theory of semiconductors

Figure 6.13d illustrates a situation in which a fully occupied band is separated from an unoccupied band by a *small band gap*. This property characterizes a *semiconductor*. In this case, electrical conductivity depends upon there being sufficient energy available for thermal population of the upper

---

[†] For a mathematical treatment of Fermi–Dirac statistics, see Appendix 17 in M. Ladd (1994) *Chemical Bonding in Solids and Fluids*, Ellis Horwood, Chichester.

band, and it follows that the conductivity increases as the temperature is raised. In the next section, we look more closely at the types and properties of semiconductors.

6.9 Semiconductors

Intrinsic semiconductors

If a material behaves as a semiconductor without the addition of dopants (see below), it is an *intrinsic semiconductor*.

In the macromolecular structures of diamond, silicon, germanium and α-tin, each atom is tetrahedrally sited (see Fig. 6.20). An atom of each element provides four valence orbitals and four valence electrons, and, in the bulk element, this leads to the formation of a fully occupied band and an unoccupied band lying at higher energy. The corresponding band gap can be measured spectroscopically since it is equal to the energy needed to promote an electron across the energy gap. For C, Si, Ge and α-Sn, the room temperature band gaps are 5.39, 1.10, 0.66 and 0.08 eV respectively. The variation down group 14 leads to C being an insulator, while for α-Sn, the band structure approaches that of a single, partially occupied band and this allotrope of Sn tends towards being metallic.

Each of Si, Ge and α-Sn is classed as an *intrinsic semiconductor*, the extent of occupation of the upper band increasing with increasing temperature. Electrons present in the upper *conduction* band act as charge carriers and result in the semiconductor being able to conduct electricity. Additionally, the removal of electrons from the lower *valence* band creates *positive holes* into which electrons can move, again leading to the ability to conduct charge.

A *charge carrier* in a *semiconductor* is either a positive hole or an electron that is able to conduct electricity.

Extrinsic (n- and p-type) semiconductors

The semiconducting properties of Si and Ge can be enhanced by *doping* these elements with atoms of a group 13 or group 15 element. Doping involves the introduction of a minutely small proportion of dopant atoms, less than 1 in 10^6, and extremely pure Si or Ge must first be produced. The reduction of SiO_2 in an electric furnace gives Si, and the Czochralski process (see Box 6.3) is used to draw single crystals of Si from the melt. We describe how dopants are introduced into semiconductors in Section 28.6.

Extrinsic semiconductors contain dopants; a *dopant* is an impurity introduced into a semiconductor in minute amounts to enhance its electrical conductivity.

In Ga-doped Si, the substitution of a Ga (group 13) for a Si (group 14) atom in the bulk solid produces an electron-deficient site. This introduces a discrete, unoccupied level

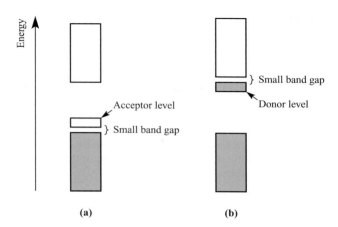

Fig. 6.14 (a) In a p-type semiconductor (e.g. Ga-doped Si), electrical conductivity arises from thermal population of an acceptor level which leaves vacancies (positive holes) in the lower band. (b) In an n-type semiconductor (e.g. As-doped Si), a donor level is close in energy to the conduction band.

into the band structure (Fig. 6.14a). The band gap that separates this level from the lower-lying occupied band is small (≈0.10 eV) and thermal population of the *acceptor level* is possible. The acceptor levels remain discrete if the concentration of Ga atoms is low, and in these circumstances the electrons in them do not contribute directly to the electrical conductance of the semiconductor. However, the positive holes left behind in the valence band act as charge carriers. One can think either in terms of an electron moving into the hole, thereby leaving another hole into which another electron can move and so on, or in terms of the movement of positive holes (in the opposite direction to the electron migration). This gives rise to a *p-type* (*p* stands for positive) semiconductor. Other group 13 dopants for Si are B and Al.

In As-doped Si, replacing an Si (group 14) by an As (group 15) atom introduces an electron-rich site. The extra electrons occupy a discrete level below the conduction band (Fig. 6.14b), and, because of the small band gap (≈0.10 eV), electrons from the *donor level* can thermally populate the conduction band where they are free to move. Electrical conduction can be described in terms of the movement of negatively charged electrons and this generates an *n-type* (*n* stands for negative) semiconductor. Phosphorus atoms can similarly be used as dopants in silicon.

The n- and p-type semiconductors are *extrinsic semiconductors*, and their precise properties are controlled by the choice and concentration of dopant. Semiconductors are discussed further in Section 28.6.

6.10 Sizes of ions

Before we discuss the structures of ionic solids, we must say something about the sizes of ions, and define the term *ionic radius*. The process of ionization (e.g. eq. 6.4) results in a

APPLICATIONS

Box 6.3 The production of pure silicon for semiconductors

Semiconductors demand the use of silicon of extreme purity. The native element does not occur naturally and silica (SiO_2) and silicate minerals are its principal sources. Silicon can be extracted from silica by reduction with carbon in an electric furnace, but the product is far too impure for the semiconductor industry. A number of purification methods are used but, of these, two are important for producing single crystals of Si.

Zone melting

Beginning with a polycrystalline Si rod, a small zone (which lies perpendicular to the direction of the rod) is melted. The focus-point of the zone is gradually moved along the length of the rod. Under carefully controlled conditions, cooling, which takes place behind the melt-zone, produces single crystals while impurities migrate along the rod with the molten material. Since the first experiments in the 1950s to develop this technique, the method has been adapted commercially and involves many passes of the melt-zone along the silicon rod before crystals suitable for use in semiconductors are obtained.

The Czochralski process

The Czochralski process is the most widely used method of producing large single crystals of Si. Pure Si is first produced by chemical vapour deposition (see Section 28.6). This is usually referred to as polysilicon (a contraction of polycrystalline silicon), indicating that it is crystalline and not amorphous. Small pieces of polysilicon are added to a quartz crucible which is then heated under vacuum to just above the melting point of Si (1687 K). Once a melt is achieved, a seed crystal of silicon is added and then slowly withdrawn while being rotated. At the same time, the crucible is rotated in the opposite direction and the temperature of the system is maintained at 1687 K. These conditions result in the formation of a cylindrical, single crystal of silicon which grows below the seed crystal as the latter is pulled upwards from the melt. Typical dimensions of the silicon ingot are 20–40 cm in diameter and 1–2 m in length as shown in the photograph. The rotation of the crystal and counter-rotation of the crucible also helps to produce a uniform distribution of any remaining impurities.

In contrast to crystals grown by the zone melting method, those produced by the Czochralski process contain close to 10^{18} atoms of O per cm^3 of Si. The impurity originates from the reaction of Si with the hot SiO_2 crucible:

$$Si + SiO_2 \longrightarrow 2SiO$$

A flow of argon is used to remove most of the gaseous SiO. Provided that the oxygen content does not exceed 10^{18} atoms cm^{-3}, its presence is not problematical. In fact, the mechanical strength of the end product benefits from the oxygen contamination. The second contaminant is carbon which originates from the oxidation of the graphite casing which surrounds the crucible:

$$SiO + C \longrightarrow CO + Si$$

Most of the gaseous CO is flushed out of the system by the argon flow.

Thin silicon wafers are cut from the single silicon ingots and then etched to produce integrated circuits. The typical electrical resistivity of a wafer produced by the Czochralski process is $\approx 150 \ \Omega \, cm$, whereas the resistivity of a wafer manufactured by zone melting may be as high as $20\,000 \ \Omega \, cm$.

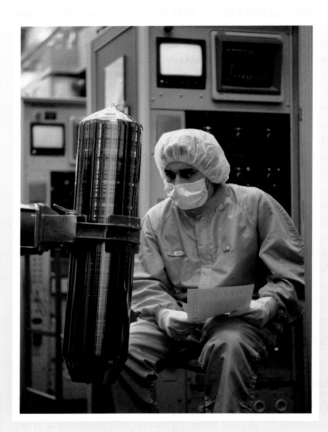

A silicon crystal ingot grown in a clean room facility in the semiconductor industry.

Further reading

J. Evers, P. Klüfers, R. Staudigl and P. Stallhofer (2003) *Angew. Chem. Int. Ed.*, vol. 42, p. 5684 – 'Czochralski's creative mistake: a milestone on the way to the gigabit era'.

K.A. Jackson and W. Schröter, eds (2000) *Handbook of Semiconductor Technology*, Wiley-VCH, Weinheim.

See also: Section 14.6 (hydrides of group 14 elements) and Section 28.6 (chemical vapour deposition).

contraction of the species owing to an increase in the effective nuclear charge. Similarly, when an atom gains an electron (e.g. eq. 6.5), the imbalance between the number of protons and electrons causes the anion to be larger than the original atom.

$$Na(g) \longrightarrow Na^+(g) + e^- \qquad (6.4)$$

$$F(g) + e^- \longrightarrow F^-(g) \qquad (6.5)$$

Ionic radii

Although from a wave-mechanical viewpoint, the radius of an individual ion has no precise physical significance, for purposes of descriptive crystallography, it is convenient to have a compilation of values obtained by partitioning measured interatomic distances in 'ionic' compounds. Values of the *ionic radius* (r_{ion}) may be derived from X-ray diffraction data. However, experimental data only give the *internuclear distance* and we generally take this to be the sum of the ionic radii of the cation and anion (eq. 6.6).

Internuclear distance between
a cation and the closest anion
in a lattice $\qquad = r_{cation} + r_{anion} \qquad (6.6)$

Equation 6.6 assumes a hard sphere model for the ions, with ions of opposite charge touching one another in the crystal lattice. Use of such an approximation means that the assignment of individual radii is somewhat arbitrary. Among many approaches to this problem we mention three.

Landé assumed that in the solid-state structures of the lithium halides, LiX, the anions were in contact with one another (see diagram **6.1** and Fig. 6.16a with the accompanying discussion). Landé took half of each anion–anion distance to be the radius of that anion, and then obtained r_{Li^+} by substituting into eq. 6.6 values of r_{X^-} and the measured internuclear Li–X distances.

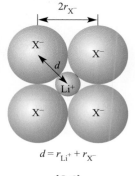

$$d = r_{Li^+} + r_{X^-}$$

(6.1)

Pauling considered a series of alkali metal halides, each member of which contained isoelectronic ions (NaF, KCl, RbBr, CsI). In order to partition the ionic radii, he assumed that the radius of each ion was inversely proportional to its actual nuclear charge less an amount due to screening effects. The latter were estimated using Slater's rules (see Box 1.5).

Goldschmidt and, more recently, Shannon and Prewitt, concentrated on the analysis of experimental data (mostly fluorides and oxides) with the aim of obtaining a set of ionic radii which, when combined in pairs (eq. 6.6), reproduced the observed internuclear distances. In view of the approximate nature of the concept of the ionic radius, no great importance should be attached to small differences in quoted values so long as self-consistency is maintained in any one set of data. Some dependence of ionic size on coordination number is expected if we consider the different electrostatic interactions that a particular ion experiences in differing environments in an ionic crystal. The value r_{ion} for a given ion increases slightly with an increase in coordination number. For example, the Shannon values of r_{ion} for Zn^{2+} are 60, 68 and 74 pm for coordination numbers of 4, 5 and 6, respectively.

Values of ionic radii for selected ions are listed in Appendix 6. Ionic radii are sometimes quoted for species such as Si^{4+} and Cl^{7+}, but such data are highly artificial. The sums of the appropriate ionization energies of Si and Cl (9950 and 39 500 kJ mol^{-1} respectively) make it inconceivable that such ions exist in stable species. Nonetheless, a value for the radius of 'Cl^{7+}' can be calculated by subtracting $r_{O^{2-}}$ from the Cl–O internuclear distance in $[ClO_4]^-$.

In the few cases in which the variation in electron density in a crystal has been accurately determined (e.g. NaCl), the minimum electron density does not in fact occur at distances from the nuclei indicated by the ionic radii in general use. For example, in LiF and NaCl, the minima are found at 92 and 118 pm from the nucleus of the cation, whereas tabulated values of r_{Li^+} and r_{Na^+} are 76 and 102 pm, respectively. Such data make it clear that discussing the structures of ionic solids in terms of the ratio of the ionic radii is, at best, only a rough guide. For this reason, we restrict our discussion of *radius ratio rules* to Box 6.4.

Self-study exercises

1. Explain why NaF and KCl each contains isoelectronic ions.

2. Comment on the following data. For Na: $r_{metal} = 191$ pm, $r_{ion} = 102$ pm; for Al: $r_{metal} = 143$ pm; $r_{ion} = 54$ pm; for O: $r_{cov} = 73$ pm; $r_{ion} = 140$ pm.

Periodic trends in ionic radii

Figure 6.15 illustrates trends in ionic radii on descending representative groups and on crossing the first row of the *d*-block. In each case, r_{ion} corresponds to that of a 6-coordinate ion. The cation size increases on descending groups 1 and 2, as does the anion size on descending group 17. Figure 6.15 also allows comparisons of the

THEORY

Box 6.4 Radius ratio rules

The structures of many ionic crystals can be rationalized *to a first approximation* by considering the relative sizes and relative numbers of the ions present. For monatomic ions, cations are *usually* smaller than anions (see Appendix 6), although examples such as KF and CsF show that this is not always true. The *radius ratio* $\frac{r_+}{r_-}$ can be used to make a first prediction of the likely coordination number and geometry around the cation using a set of simple rules:

Value of $\frac{r_+}{r_-}$	Predicted coordination number of cation	Predicted coordination geometry of cation
<0.15	2	Linear
0.15–0.22	3	Trigonal planar
0.22–0.41	4	Tetrahedral
0.41–0.73	6	Octahedral
>0.73	8	Cubic

For a given compound stoichiometry, predictions about the coordination type of the cation necessarily make predictions about the coordination type of the anion. Use of radius ratios meets with some success, but there are *many* limitations. We can exemplify this by looking at the group 1 halides. The ionic radii are as follows:

Cation	Li$^+$	Na$^+$	K$^+$	Rb$^+$	Cs$^+$
r_+ / pm	76	102	138	149	170

Anion	F$^-$	Cl$^-$	Br$^-$	I$^-$
r_- / pm	133	181	196	220

For LiF, the radius ratio is 0.57 and so an octahedral coordination around the Li$^+$ cation is predicted. This corresponds to an NaCl-type structure (Fig. 6.16), in agreement with that observed. Each of the group 1 halides (except CsCl, CsBr and CsI) at 298 K and 1 bar pressure adopts the NaCl-type structure; CsCl, CsBr and CsI adopt the CsCl-type structure (Fig. 6.17). Radius ratio rules predict the correct structures in only some cases. They predict tetrahedral coordination for the cations in LiBr and LiI, octahedral coordination in LiF, LiCl, NaCl, NaBr, NaI, KBr and KI, and cubic coordination in NaF, KF, KCl, RbF, RbCl, RbBr, CsF, CsCl, CsBr and CsI. Radius ratio rules give only one prediction for any one ionic crystal, and some compounds undergo phase changes under the influence of temperature and pressure, e.g. when CsCl is sublimed onto an amorphous surface, it crystallizes with the NaCl structure and, under high-pressure conditions, RbCl adopts a CsCl-type structure.

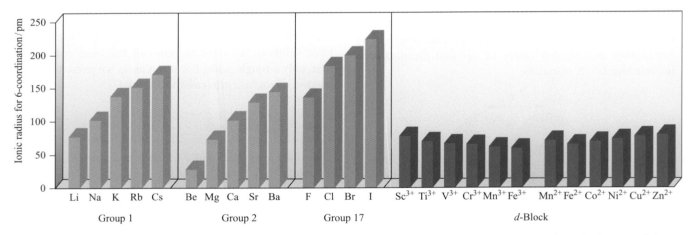

Fig. 6.15 Trends in ionic radii, r_{ion}, within the metal ions of groups 1 and 2, the anions of group 17, and metal ions from the first row of the *d*-block.

relative sizes of cations and anions in alkali metal and alkaline earth metal halide salts (see Section 6.11).

The right-hand side of Fig. 6.15 illustrates the small variation in size for M^{3+} and M^{2+} ions of the *d*-block metals. As expected, the decrease in nuclear charge in going from Fe^{3+} to Fe^{2+}, and from Mn^{3+} to Mn^{2+}, causes an increase in r_{ion}.

6.11 Ionic lattices

In this section we describe some common structure types adopted by ionic compounds of general formulae MX, MX$_2$ or M$_2$X, as well as that of the mineral *perovskite*, CaTiO$_3$. Such structures are usually determined by X-ray diffraction methods (see Section 4.11). Different ions scatter

X-rays to differing extents depending on the total number of electrons in the ion and, consequently, different types of ions can generally be distinguished from one another. Use of X-ray diffraction methods does have some limitations. Firstly, the location of light atoms (e.g. H) in the presence of much heavier atoms is difficult and, sometimes, impossible (see Section 4.11). Secondly, X-ray diffraction is seldom able to identify the state of ionization of the species present; only for a few substances (e.g. NaCl) has the electron density distribution been determined with sufficient accuracy for this purpose.

Throughout our discussion, we refer to 'ionic' lattices, suggesting the presence of discrete ions. Although a *spherical ion model* is used to describe the structures, we shall see in Section 6.13 that this picture is unsatisfactory for some compounds in which covalent contributions to the bonding are significant. Useful as the hard sphere model is in describing common crystal structure types, it must be understood that it is at odds with modern quantum theory. As we saw in Chapter 1, the wavefunction of an electron does not suddenly drop to zero with increasing distance from the nucleus, and in a close-packed or any other crystal, there is a finite electron density everywhere. Thus *all treatments of the solid state based upon the hard sphere model are approximations.*

Each structure type is designated by the name of one of the compounds crystallizing with that structure, and phrases such as 'CaO adopts an NaCl structure' are commonly found in the chemical literature.

The rock salt (NaCl) structure type

In salts of formula MX, the coordination numbers of M and X must be *equal*.

Rock salt (or halite, NaCl) occurs naturally as cubic crystals, which, when pure, are colourless or white. Figure 6.16 shows two representations of the unit cell (see Section 6.2) of NaCl. Figure 6.16a illustrates the way in which the ions occupy the space available. The larger Cl^- ions ($r_{Cl^-} = 181$ pm) define an fcc arrangement with the Na^+ ions ($r_{Na^+} = 102$ pm) occupying the octahedral holes. This description relates the structure of the ionic lattice to the close-packing-of-spheres model. Such a description is often employed, but is not satisfactory for salts such as KF. While KF adopts an NaCl lattice, the K^+ and F^- ions are almost the same size ($r_{K^+} = 138$, $r_{F^-} = 133$ pm) (see Box 6.4). Although Fig. 6.16a is relatively realistic, it hides most of the structural details of the unit cell and is difficult to reproduce when drawing the unit cell. The more open representation shown in Fig. 6.16b tends to be more useful.

The complete NaCl structure is built up by placing unit cells next to one another so that ions residing in the corner, edge or face sites (Fig. 6.16b) are *shared* between adjacent unit cells. Bearing this in mind, Fig. 6.16b shows that *each* Na^+ and Cl^- ion is 6-coordinate in the crystal lattice, while within a single unit cell, the octahedral environment is defined completely only for the central Na^+ ion.

Figure 6.16b is not a unique representation of a unit cell of the NaCl structure. It is equally valid to draw a unit cell with Na^+ ions in the corner sites; such a cell has a Cl^- ion in the unique central site. This shows that the Na^+ ions are also in an fcc arrangement, and the NaCl structure could therefore be described in terms of two interpenetrating fcc lattices, one consisting of Na^+ ions and one of Cl^- ions.

Among the many compounds that crystallize with the NaCl structure type are NaF, NaBr, NaI, NaH, halides of

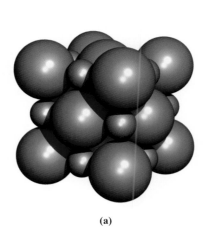

(a)

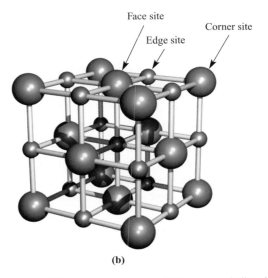

(b)

Fig. 6.16 Two representations of the unit cell of NaCl: (a) shows a space-filling representation, and (b) shows a 'ball-and-stick' representation which reveals the coordination environments of the ions. The Cl^- ions are shown in green and the Na^+ ions in purple; since both types of ion are in equivalent environments, a unit cell with Na^+ ions in the corner sites is also valid. There are four types of site in the unit cell: central (not labelled), face, edge and corner positions.

Li, K and Rb, CsF, AgF, AgCl, AgBr, MgO, CaO, SrO, BaO, MnO, CoO, NiO, MgS, CaS, SrS and BaS.

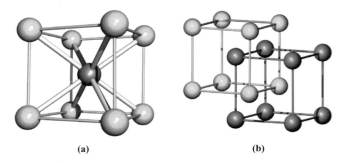

(a)   (b)

🌐 **Fig. 6.17** (a) The unit cell of CsCl; Cs^+ ions are shown in yellow and Cl^- in green, but the unit cell could also be drawn with the Cs^+ ion in the central site. The unit cell is defined by the yellow lines. (b) One way to describe the CsCl structure is in terms of interpenetrating cubic units of Cs^+ and Cl^- ions.

Worked example 6.3   Compound stoichiometry from a unit cell

Show that the structure of the unit cell for sodium chloride (Fig. 6.16b) is consistent with the formula NaCl.

In Fig. 6.16b, 14 Cl^- ions and 13 Na^+ ions are shown. However, all but one of the ions are shared between two or more unit cells.

There are four types of site:

- unique central position (the ion belongs entirely to the unit cell shown);
- face site (the ion is shared between two unit cells);
- edge sites (the ion is shared between four unit cells);
- corner site (the ion is shared between eight unit cells).

The total number of Na^+ and Cl^- ions belonging to the unit cell is calculated as follows:

Site	Number of Na^+	Number of Cl^-
Central	1	0
Face	0	$(6 \times \frac{1}{2}) = 3$
Edge	$(12 \times \frac{1}{4}) = 3$	0
Corner	0	$(8 \times \frac{1}{8}) = 1$
TOTAL	4	4

The ratio of $Na^+ : Cl^-$ ions is $4:4 = 1:1$
This ratio is consistent with the formula NaCl.

Self-study exercises

1. Show that the structure of the unit cell for caesium chloride (Fig. 6.17) is consistent with the formula CsCl.

2. MgO adopts an NaCl structure. How many Mg^{2+} and O^{2-} ions are present per unit cell?   [*Ans.* 4 of each]

3. The unit cell of AgCl (NaCl structure type) can be drawn with Ag^+ ions at the corners of the cell, or Cl^- at the corners. Confirm that the number of Ag^+ and Cl^- ions per unit cell remains the same whichever arrangement is considered.

The caesium chloride (CsCl) structure type

In the CsCl structure, each ion is surrounded by eight others of opposite charge. A single unit cell (Fig. 6.17a) makes the connectivity obvious only for the central ion.

However, by extending the lattice, one sees that it is constructed of interpenetrating cubes (Fig. 6.17b). The coordination number of each ion is 8. Because the Cs^+ and Cl^- ions are in the same environments, it is valid to draw a unit cell with either Cs^+ or Cl^- at the corners of the cube. Note the relationship between the structure of the unit cell and bcc packing.

The CsCl structure is relatively uncommon but is also adopted by CsBr, CsI, TlCl and TlBr. At 298 K, NH_4Cl and NH_4Br possess CsCl structures; $[NH_4]^+$ is treated as a spherical ion (Fig. 6.18), an approximation that can be made for a number of simple ions in the solid state due to their rotating or lying in random orientations about a fixed point. Above 457 and 411 K respectively, NH_4Cl and NH_4Br adopt NaCl structures.

The fluorite (CaF_2) structure type

> In salts of formula MX_2, the coordination number of X must be *half* that of M.

Calcium fluoride occurs naturally as the mineral *fluorite* (fluorspar). Figure 6.19a shows a unit cell of CaF_2. Each cation is 8-coordinate and each anion 4-coordinate; six of

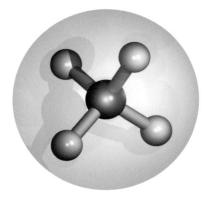

🌐 **Fig. 6.18** The $[NH_4]^+$ ion can be treated as a sphere in descriptions of solid state lattices; some other ions (e.g. $[BF_4]^-$, $[PF_6]^-$) can be treated similarly.

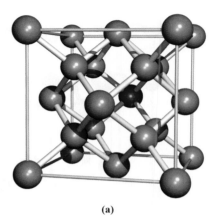 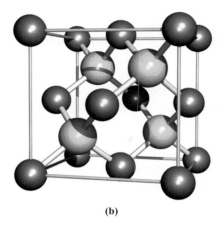

(a) **(b)**

Fig. 6.19 (a) The unit cell of CaF_2; the Ca^{2+} ions are shown in red and the F^- ions in green. (b) The unit cell of zinc blende (ZnS); the zinc centres are shown in grey and the sulfur centres in yellow. Both sites are equivalent, and the unit cell could be drawn with the S^{2-} ions in the grey sites.

the Ca^{2+} ions are shared between two unit cells and the 8-coordinate environment can be appreciated by envisaging two adjacent unit cells. [*Exercise:* How does the coordination number of 8 for the remaining Ca^{2+} ions arise?] Other compounds that adopt this structure type include group 2 metal fluorides, $BaCl_2$, and dioxides of the *f*-block metals including CeO_2, ThO_2, PaO_2, UO_2, PrO_2, AmO_2 and NpO_2.

The antifluorite structure type

If the cation and anion sites in Fig. 6.19a are exchanged, the coordination number of the anion becomes *twice* that of the cation, and it follows that the compound formula is M_2X. This arrangement corresponds to the antifluorite structure, and is adopted by the group 1 metal oxides and sulfides of type M_2O and M_2S; Cs_2O is an exception and instead adopts an anti-$CdCl_2$ structure.

The zinc blende (ZnS) structure type: a diamond-type network

Figure 6.19b shows the structure of zinc blende (ZnS). A comparison of this with Fig. 6.19a reveals a relationship between the structures of zinc blende and CaF_2. In going from Fig. 6.19a to 6.19b, half of the anions are removed and the ratio of cation : anion changes from 1 : 2 to 1 : 1.

An alternative description is that of a *diamond-type network*. Figure 6.20a gives a representation of the structure of diamond. Each C atom is tetrahedrally sited and the structure is very rigid. This structure type is also adopted by Si, Ge and α-Sn (grey tin). Figure 6.20b (with atom

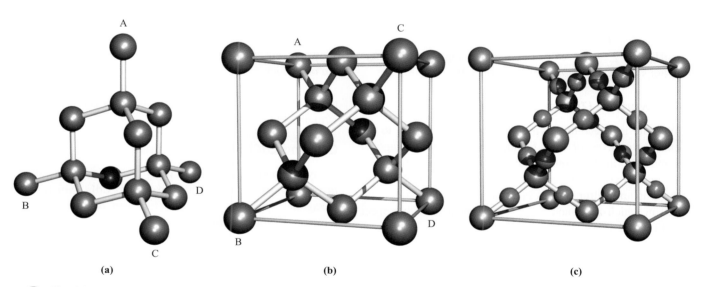

(a) **(b)** **(c)**

Fig. 6.20 (a) A typical representation of the diamond structure. (b) Reorientation of the network shown in (a) provides a representation that can be compared with the unit cell of zinc blende (Fig. 6.19b); the atom labels correspond to those in diagram (a). This structure type is also adopted by Si, Ge and α-Sn. (c) The unit cell of β-cristobalite, SiO_2; colour code: Si, purple; O, red.

labels that relate it to Fig. 6.20a) shows a view of the diamond network that is comparable with the unit cell of zinc blende in Fig. 6.19b. In zinc blende, every other site in the diamond-type array is occupied by either a zinc or a sulfur centre. The fact that we are comparing the structure of an apparently ionic compound (ZnS) with that of a covalently bonded species should not cause concern. As we have already mentioned, the hard sphere ionic model is a convenient approximation but does not allow for the fact that the bonding in many compounds such as ZnS is neither wholly ionic nor wholly covalent.

At 1296 K, zinc blende undergoes a transition to wurtzite, the structure of which we consider later; zinc blende and wurtzite are *polymorphs* (see Section 6.4). Zinc(II) sulfide occurs naturally as both zinc blende (also called *sphalerite*) and wurtzite, although the former is more abundant and is the major ore for Zn production. Although zinc blende is thermodynamically favoured at 298 K by 13 kJ mol^{-1}, the transition from wurtzite to zinc blende is *extremely* slow, allowing both minerals to exist in nature. This scenario resembles that of the diamond $\longrightarrow$ graphite transition (see Chapter 14 and Box 14.4), graphite being thermodynamically favoured at 298 K. If the latter transition were *not* infinitesimally slow, diamonds would lose their place in the world gemstone market!

The β-cristobalite (SiO₂) structure type

Before discussing the structure of wurtzite, we consider β-cristobalite, the structure of which is related to that of the diamond-type network. β-Cristobalite is one of several forms of SiO_2 (see Fig. 14.20). Figure 6.20c shows the unit cell of β-cristobalite. Comparison with Fig. 6.20b shows that it is related to the structure of Si by placing an O atom between adjacent Si atoms. The idealized structure shown in Fig. 6.20c has an Si$-$O$-$Si bond angle of $180°$ whereas in practice this angle is $147°$ (almost the same as in $(SiH_3)_2O$, $\angle Si-O-Si = 144°$), indicating that the interactions in SiO_2 are *not* purely electrostatic.

The wurtzite (ZnS) structure type

Wurtzite is a second polymorph of ZnS. In contrast to the cubic symmetry of zinc blende, wurtzite has hexagonal symmetry. In the three unit cells shown in Fig. 6.21, the 12 ions in corner sites define a hexagonal prism. Each of the zinc and sulfur centres is tetrahedrally sited, and a unit cell in which Zn^{2+} and S^{2-} are interchanged with respect to Fig. 6.21 is equally valid.

The rutile (TiO₂) structure type

The mineral rutile (a white crystalline material) occurs in granite rocks and is an important industrial source of TiO_2 (see Box 21.3). Figure 6.22 shows the unit cell of rutile.

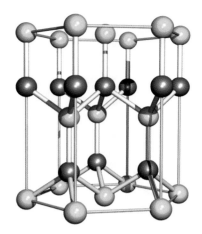

Fig. 6.21 Three unit cells of wurtzite (a second polymorph of ZnS) define a hexagonal prism; the Zn^{2+} ions are shown in grey and the S^{2-} ions in yellow. Both ions are tetrahedrally sited and an alternative unit cell could be drawn by interchanging the ion positions.

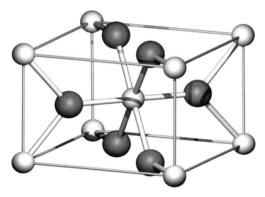

Fig. 6.22 The unit cell of rutile (one polymorph of TiO_2). Colour code: Ti, silver; O, red.

The coordination numbers of titanium and oxygen are 6 (octahedral) and 3 (trigonal planar) respectively, consistent with the 1:2 stoichiometry of rutile. Two of the O^{2-} ions shown in Fig. 6.22 reside fully within the unit cell, while the other four are in face-sharing positions.

The rutile structure type is adopted by SnO_2 (*cassiterite*, the most important tin-bearing mineral), β-MnO_2 (*pyrolusite*) and PbO_2.

CdI₂ and CdCl₂: layer structures

Many compounds of formula MX_2 crystallize in so-called *layer structures*, a typical one being CdI_2 which has hexagonal symmetry. This structure can be described in terms of I^- ions arranged in an hcp array with Cd^{2+} ions occupying the octahedral holes in every other layer (Fig. 6.23, in which the hcp array is denoted by the ABAB layers). Extending the lattice infinitely gives a structure which can be described in terms of 'stacked sandwiches', each 'sandwich' consisting of a layer of I^- ions, a parallel layer of Cd^{2+} ions, and another parallel

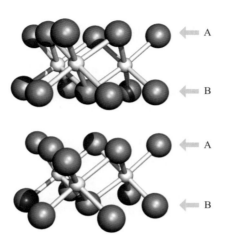

Fig. 6.23 Parts of two layers of the CdI_2 lattice; Cd^{2+} ions are shown in pale grey and I^- ions in gold. The I^- ions are arranged in an hcp array.

layer of I^- ions; each 'sandwich' is electrically neutral. Only weak van der Waals forces operate between the 'sandwiches' (the central gap between the layers in Fig. 6.23) and this leads to CdI_2 crystals exhibiting pronounced cleavage planes parallel to the layers.

> If a crystal breaks along a plane related to the lattice structure, the plane is called a *cleavage plane*.

Other compounds crystallizing with a CdI_2 structure type include $MgBr_2$, MgI_2, CaI_2, iodides of many *d*-block metals, and many metal hydroxides including $Mg(OH)_2$ (the mineral *brucite*) in which the $[OH]^-$ ions are treated as spheres for the purposes of structural description.

The structure of $CdCl_2$ is related to the CdI_2 layer structure but with the Cl^- ions in a *cubic* close-packed arrangement. Examples of compounds adopting this structure are $FeCl_2$ and $CoCl_2$. Other layer structures include *talc* and *mica* (see Section 14.9).

The perovskite ($CaTiO_3$) structure type: a double oxide

Perovskite is an example of a *double oxide*. It does not, as the formula might imply, contain $[TiO_3]^{2-}$ ions, but is a mixed Ca(II) and Ti(IV) oxide. Figure 6.24a shows one representation of a unit cell of perovskite (see end-of-chapter problem 6.13). The cell is cubic, with Ti(IV) centres at the corners of the cube, and O^{2-} ions in the 12 edge sites. The 12-coordinate Ca^{2+} ion lies at the centre of the unit cell. Each Ti(IV) centre is 6-coordinate, and this can be appreciated by considering the assembly of adjacent unit cells in the crystal lattice.

Many double oxides or fluorides such as $BaTiO_3$, $SrFeO_3$, $NaNbO_3$, $KMgF_3$ and $KZnF_3$ crystallize with a perovskite structure type. Deformations of the lattice may be caused as a consequence of the relative sizes of the ions, e.g. in $BaTiO_3$, the Ba^{2+} ion is relatively large ($r_{Ba^{2+}} = 142$ pm compared with $r_{Ca^{2+}} = 100$ pm) and causes a displacement of each Ti(IV) centre such that there is one short Ti–O contact. This leads to $BaTiO_3$ possessing *ferroelectric* properties (see Section 28.6).

The structures of some high-temperature superconductors are also related to that of perovskite (see Figs. 28.12 and 28.13). Another mixed oxide structure type is that of *spinel*, $MgAl_2O_4$ (see Box 13.7).

6.12 Crystal structures of semiconductors

Because of the importance of semiconductors in our everyday lives, it is important to draw attention to the structural similarities that exist within this group of materials. The majority of semiconductors exhibit either a diamond (Fig. 6.20) or zinc blende (Fig. 6.19) structure type, with the wurtzite structure (Fig. 6.21) being less common. In each of these prototype structures, atoms are in tetrahedral environments. Diamond-type structures are adopted by Si and Ge, and the addition of dopants (see Section 6.9)

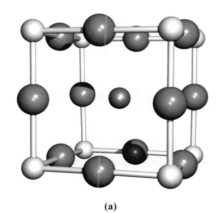

(a)

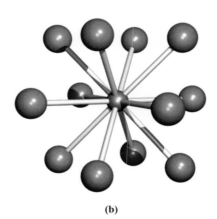

(b)

Fig. 6.24 (a) One representation of a unit cell of perovskite ($CaTiO_3$); (b) the Ca^{2+} ion is 12-coordinate with respect to the O^{2-} ions. Colour code: Ca, purple; O, red; Ti, pale grey.

occurs without structural change. Replacing alternate group 14 atoms in the crystal lattice by group 13 or 15 atoms leads to the family of semiconductors which include GaP, GaAs, InP and InAs (so-called III–V[†] semiconductors). These and the II–VI[†] semiconductors ZnSe, ZnTe, CdS, CdTe, HgS, HgSe and HgTe adopt the zinc blende structure type which is, as described earlier, a diamond-type network. The semiconducting materials AlN, GaN and CdSe crystallize with a wurtzite structure type.

6.13 Lattice energy: estimates from an electrostatic model

The *lattice energy*, $\Delta U(0\,\text{K})$, of an ionic compound is the change in internal energy that accompanies the formation of one mole of the solid from its constituent gas-phase ions at $0\,\text{K}$.[‡]

For a salt MX_n, eq. 6.7 defines the reaction, the energy change for which corresponds to the lattice energy.

$$M^{n+}(g) + nX^-(g) \longrightarrow MX_n(s) \tag{6.7}$$

The lattice energy can be *estimated* by assuming an electrostatic model for the solid state ionic lattice. The ions are considered to be point charges. Later in this chapter, we consider to what extent this approximation is valid.

Coulombic attraction within an isolated ion-pair

We begin by considering the equation for the change in internal energy when two oppositely charged ions M^{z+} and X^{z-} are brought together from infinite separation to form the *isolated ion-pair*, MX (eq. 6.8).

$$M^{z+}(g) + X^{z-}(g) \longrightarrow MX(g) \tag{6.8}$$

Let the ions carry charges of z_+e and z_-e where e is the electronic charge and z_+ and z_- are integers. The ions attract each other, and energy is released as the ion-pair is formed. The change in internal energy can be estimated from eq. 6.9 by considering the Coulombic attraction between the ions. For an isolated ion-pair:

$$\Delta U = -\left(\frac{|z_+||z_-|e^2}{4\pi\varepsilon_0 r}\right) \tag{6.9}$$

where ΔU = change in internal energy (unit = joules); $|z_+|$ = modulus[*] of the positive charge (for K^+, $|z_+| = 1$; for Mg^{2+}, $|z_+| = 2$); $|z_-|$ = modulus[*] of the negative charge (for F^-, $|z_-| = 1$; for O^{2-}, $|z_-| = 2$); e = charge on the electron = 1.602×10^{-19} C; ε_0 = permittivity of a

vacuum = $8.854 \times 10^{-12}\,\text{F m}^{-1}$; and r = internuclear distance between the ions (units = m).

Coulombic interactions in an ionic lattice

Now consider a salt MX which has an NaCl structure. A study of the coordination geometry in Fig. 6.16 (remembering that the lattice extends indefinitely) shows that each M^{z+} ion is surrounded by:

- 6 X^{z-} ions, each at a distance r
- 12 M^{z+} ions, each at a distance $\sqrt{2}r$
- 8 X^{z-} ions, each at a distance $\sqrt{3}r$
- 6 M^{z+} ions, each at a distance $2r$

and so on.

The change in Coulombic energy when an M^{z+} ion is brought from infinity to its position in the lattice is given by eq. 6.10.

$$\Delta U = -\frac{e^2}{4\pi\varepsilon_0}\left[\left(\frac{6}{r}|z_+||z_-|\right) - \left(\frac{12}{\sqrt{2}r}|z_+|^2\right)\right.$$
$$\left. + \left(\frac{8}{\sqrt{3}r}|z_+||z_-|\right) - \left(\frac{6}{2r}|z_+|^2\right)\cdots\right]$$
$$= -\frac{|z_+||z_-|e^2}{4\pi\varepsilon_0 r}\left[6 - \left(\frac{12|z_+|}{\sqrt{2}|z_-|}\right) + \left(\frac{8}{\sqrt{3}}\right)\right.$$
$$\left. - \left(3\frac{|z_+|}{|z_-|}\right)\cdots\right] \tag{6.10}$$

The ratio of the charges on the ions, $\frac{|z_+|}{|z_-|}$, is constant for a given type of structure (e.g. 1 for NaCl) and so the series in square brackets in eq. 6.10 (which slowly converges and may be summed algebraically) is a function only of the crystal geometry. Similar series can be written for other crystal lattices, but for a particular structure type, the series is independent of $|z_+|$, $|z_-|$ and r. Erwin Madelung first evaluated such series in 1918, and the values appropriate for various structure types are *Madelung constants, A* (see Table 6.4)[**]. Equation 6.10 can therefore be written in the more simple form of eq. 6.11, in which the lattice energy is estimated in joules *per mole* of compound.

$$\Delta U = -\frac{LA|z_+||z_-|e^2}{4\pi\varepsilon_0 r} \tag{6.11}$$

where L = Avogadro number = $6.022 \times 10^{23}\,\text{mol}^{-1}$

A = Madelung constant (no units)

Although we have derived this expression by considering the ions that surround M^{z+}, the same equation results by starting from a central X^{z-} ion.

[†] The names III–V and II–VI arise from the old names for groups 13 (IIIA), 15 (VA), 12 (IIB) and 16 (VIA).
[‡] You should note that some textbooks define lattice energy for the reverse process, i.e. the energy needed to convert an ionic solid into its constituent gaseous ions.
[*] The modulus of a real number is its *positive* value, e.g. $|z_+|$ and $|z_-|$ are both positive.

[**] See also: J.G. McCaffrey (2009) *J. Chem. Educ.*, vol. 86, p. 1450 – 'A simple spreadsheet program for the calculation of lattice-site distributions'.

Table 6.3 Values of the Born exponent, n, given for an ionic compound MX in terms of the electronic configuration of the ions [M^+][X^-]. The value of n for an ionic compound is determined by averaging the component values, e.g. for MgO, $n = 7$; for LiCl, $n = \dfrac{5 + 9}{2} = 7$.

Electronic configuration of the ions in an ionic compound MX	Examples of ions	n (no units)
[He][He]	H^-, Li^+	5
[Ne][Ne]	F^-, O^{2-}, Na^+, Mg^{2+}	7
[Ar][Ar], or [$3d^{10}$][Ar]	Cl^-, S^{2-}, K^+, Ca^{2+}, Cu^+	9
[Kr][Kr] or [$4d^{10}$][Kr]	Br^-, Rb^+, Sr^{2+}, Ag^+	10
[Xe][Xe] or [$5d^{10}$][Xe]	I^-, Cs^+, Ba^{2+}, Au^+	12

Self-study exercise

In Fig. 16.16b, the central Na^+ (purple) ion is surrounded by six Cl^- (green) ions, each at a distance r. Confirm that the next-nearest neighbours are (i) 12 Na^+ ions at a distance $\sqrt{2}r$, (ii) 8 Cl^- ions at a distance $\sqrt{3}r$, and (iii) 6 Na^+ ions at a distance $2r$.

Born forces

Coulombic interactions are not the only forces operating in a real ionic lattice. The ions have finite size, and electron–electron and nucleus–nucleus repulsions also arise. These are *Born forces*. Equation 6.12 gives the simplest expression for the increase in repulsive energy upon assembling the lattice from gaseous ions.

$$\Delta U = \frac{LB}{r^n} \tag{6.12}$$

where B = repulsion coefficient

n = Born exponent

Values of the Born exponent (Table 6.3) can be evaluated from compressibility data and depend on the electronic configurations of the ions involved. Effectively, this says that n shows a dependence on the sizes of the ions.

Worked example 6.4   Born exponents

Using the values given in Table 6.3, determine an appropriate Born exponent for BaO.

Ba^{2+} is isoelectronic with Xe, and so $n = 12$
O^{2-} is isoelectronic with Ne, and $n = 7$
The value of n for BaO $= \dfrac{12 + 7}{2} = 9.5$

Self-study exercises

Use data in Table 6.3.

1. Calculate an appropriate Born exponent for NaF.
[*Ans.* 7]

2. Calculate an appropriate Born exponent for AgF.
[*Ans.* 8.5]

3. What is the change in the Born exponent in going from BaO to SrO?
[*Ans.* −1]

The Born–Landé equation

In order to write an expression for the lattice energy that takes into account both the Coulombic and Born interactions in an ionic lattice, we combine eqs. 6.11 and 6.12 to give eq. 6.13.

$$\Delta U(0\,K) = -\frac{LA|z_+||z_-|e^2}{4\pi\varepsilon_0 r} + \frac{LB}{r^n} \tag{6.13}$$

We evaluate B in terms of the other components of the equation by making use of the fact that at the equilibrium separation where $r = r_0$, the differential $\dfrac{d\Delta U}{dr} = 0$. Differentiating with respect to r gives eq. 6.14, and rearrangement gives an expression for B (eq. 6.15).

$$0 = \frac{LA|z_+||z_-|e^2}{4\pi\varepsilon_0 r_0{}^2} - \frac{nLB}{r_0{}^{n+1}} \tag{6.14}$$

$$B = \frac{A|z_+||z_-|e^2 r_0{}^{n-1}}{4\pi\varepsilon_0 n} \tag{6.15}$$

Combining eqs. 6.13 and 6.15 gives an expression for the lattice energy that is based on an electrostatic model and takes into account Coulombic attractions, Coulombic repulsions and Born repulsions between ions in the infinite crystal lattice. Equation 6.16 is the *Born–Landé equation*.

$$\Delta U(0\,K) = -\frac{LA|z_+||z_-|e^2}{4\pi\varepsilon_0 r_0}\left(1 - \frac{1}{n}\right) \tag{6.16}$$

Because of its simplicity, the Born–Landé expression is the one that chemists tend to use. Many chemical problems involve the use of estimated lattice energies, e.g. for hypothetical compounds. Often lattice energies are incorporated into thermochemical cycles, and so an associated *enthalpy* change is needed (see Section 6.14).

Madelung constants

Values of Madelung constants for selected lattices are given in Table 6.4. Remembering that these values are derived by considering the coordination environments (near and far neighbours) of ions in the crystal lattice, it may seem

Table 6.4 Madelung constants, A, for selected structure types. Values of A are numerical and have no units.

Structure type	A
Sodium chloride (NaCl)	1.7476
Caesium chloride (CsCl)	1.7627
Wurtzite (α-ZnS)	1.6413
Zinc blende (β-ZnS)	1.6381
Fluorite (CaF$_2$)	2.5194
Rutile (TiO$_2$)	2.408[†]
Cadmium iodide (CdI$_2$)	2.355[†]

[†] For these structures, the value depends slightly on the lattice parameters for the unit cell.

surprising that, for example, the values for the NaCl and CsCl structures (Figs. 6.16 and 6.17) are similar. This is simply a consequence of the infinite nature of the structures: although the first (attractive) term in the algebraic series for A is greater by a factor of $\frac{8}{6}$ for the CsCl structure, the second (repulsive) term is also greater, and so on.

Table 6.4 shows that Madelung constants for MX$_2$ structures are ≈50% higher than those for MX lattices. We return to this difference in Section 6.16.

Worked example 6.5   Use of the Born–Landé equation

Sodium fluoride adopts the NaCl structure type. Estimate the lattice energy of NaF using an electrostatic model.

Data required:

$L = 6.022 \times 10^{23}$ mol^{-1},
$A = 1.7476$, $e = 1.602 \times 10^{-19}$ C,
$\varepsilon_0 = 8.854 \times 10^{-12}$ F m^{-1},   **Born exponent for NaF = 7,**
internuclear Na−F distance = 231 pm

The change in internal energy (the lattice energy) is given by the Born–Landé equation:

$$\Delta U(0\,\text{K}) = -\frac{LA|z_+||z_-|e^2}{4\pi\varepsilon_0 r_0}\left(1 - \frac{1}{n}\right)$$

r must be in m: 231 pm = 2.31×10^{-10} m

$$\Delta U_0 = -\left(\frac{\begin{array}{c}6.022 \times 10^{23} \times 1.7476 \times 1 \\ \times 1 \times (1.602 \times 10^{-19})^2\end{array}}{4 \times 3.142 \times 8.854 \times 10^{-12} \times 2.31 \times 10^{-10}}\right)$$
$$\times \left(1 - \frac{1}{7}\right)$$

$$= -900\,624\,\text{J mol}^{-1}$$

$$\approx -901\,\text{kJ mol}^{-1}$$

Self-study exercises

1. Show that the worked example above is dimensionally correct given that C, F and J in SI base units are:
   C = A s; F = m^{-2} kg^{-1} s^4 A^2; J = kg m^2 s^{-2}.

2. Estimate the lattice energy of KF (NaCl structure) using an electrostatic model; the K−F internuclear separation is 266 pm.   [*Ans.* -798 kJ mol^{-1}]

3. By assuming an electrostatic model, estimate the lattice energy of MgO (NaCl structure); values of r_{ion} are listed in Appendix 6.   [*Ans.* -3926 kJ mol^{-1}]

Refinements to the Born–Landé equation

Lattice energies obtained from the Born–Landé equation are *approximate*, and for more accurate evaluations, several improvements to the equation can be made. The most important of these arises by replacing the $\frac{1}{r^n}$ term in eq. 6.12 by $e^{-\frac{r}{\rho}}$, a change reflecting the fact that wavefunctions show an exponential dependence on r; ρ is a constant that can be expressed in terms of the compressibility of the crystal. This refinement results in the lattice energy being given by the *Born–Mayer equation* (eq. 6.17).

$$\Delta U(0\,\text{K}) = -\frac{LA|z_+||z_-|e^2}{4\pi\varepsilon_0 r_0}\left(1 - \frac{\rho}{r_0}\right) \qquad (6.17)$$

The constant ρ has a value of 35 pm for all alkali metal halides. Note that r_0 appears in the Born repulsive term (compare eqs. 6.16 and 6.17).

Further refinements in lattice energy calculations include the introduction of terms for the *dispersion energy* and the *zero point energy* (see Section 4.6 and Fig. 10.2). Dispersion forces[†] arise from momentary fluctuations in electron density which produce temporary dipole moments that, in turn, induce dipole moments in neighbouring species. Dispersion forces are also referred to as *induced-dipole–induced-dipole interactions*. They are non-directional and give rise to a dispersion energy that is related to the internuclear separation, r, and the *polarizability*, α, of the atom (or molecule) according to eq. 6.18.

$$\text{Dispersion energy} \propto \frac{\alpha}{r^6} \qquad (6.18)$$

The polarizability of a species is a measure of the degree to which it may be distorted, e.g. by the electric field due to an adjacent atom or ion. In the hard sphere model of ions in lattices, we assume that there is no polarization of the ions. This is a gross approximation. The polarizability increases rapidly with an increase in atomic size, and large ions

[†] Dispersion forces are also known as London dispersion forces.

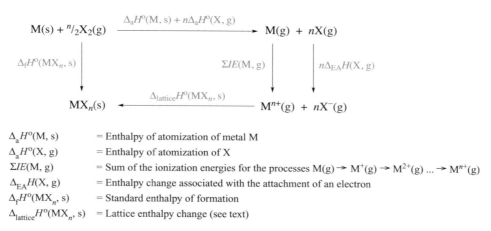

$\Delta_a H^{\circ}(M, s)$ = Enthalpy of atomization of metal M

$\Delta_a H^{\circ}(X, g)$ = Enthalpy of atomization of X

$\Sigma IE(M, g)$ = Sum of the ionization energies for the processes $M(g) \rightarrow M^{+}(g) \rightarrow M^{2+}(g) \ldots \rightarrow M^{n+}(g)$

$\Delta_{EA} H(X, g)$ = Enthalpy change associated with the attachment of an electron

$\Delta_f H^{\circ}(MX_n, s)$ = Standard enthalpy of formation

$\Delta_{lattice} H^{\circ}(MX_n, s)$ = Lattice enthalpy change (see text)

Fig. 6.25 A Born–Haber thermochemical cycle for the formation of a salt MX_n. This gives an *enthalpy change* associated with the formation of the ionic lattice MX_n.

(or atoms or molecules) give rise to relatively large induced dipoles and, thus, significant dispersion forces. Values of α can be obtained from measurements of the relative permittivity (*dielectric constant*, see Section 9.2) or the refractive index of the substance in question.

In NaCl, the contributions to the total lattice energy ($-766\,kJ\,mol^{-1}$) made by electrostatic attractions, electrostatic and Born repulsions, dispersion energy and zero point energy are -860, $+99$, -12 and $+7\,kJ\,mol^{-1}$ respectively. The error introduced by neglecting the last two terms (which always tend to compensate each other) is very small.

Overview

Lattice energies derived using the electrostatic model are often referred to as 'calculated' values to distinguish them from values obtained using thermochemical cycles. It should, however, be appreciated that values of r_0 obtained from X-ray diffraction studies are *experimental* quantities and may conceal departures from ideal ionic behaviour. In addition, the actual charges on ions may well be less than their formal charges. Nevertheless, the concept of lattice energy is of immense importance in inorganic chemistry.

6.14 Lattice energy: the Born–Haber cycle

By considering the *definition* of lattice energy, it is easy to see why these quantities are not measured directly. However, an associated *lattice enthalpy* of a salt can be related to several other quantities by a thermochemical cycle called the *Born–Haber cycle*. If the anion in the salt is a halide, then all the other quantities in the cycle have been determined independently. The reason for this statement will become clearer when we look at applications of lattice energies in Section 6.16.

Let us consider a general metal halide MX_n. Figure 6.25 shows a thermochemical cycle describing the formation of

crystalline MX_n from its constituent elements in their standard states. The quantity $\Delta_{lattice} H^{\circ}(298\,K)$ is the enthalpy change that accompanies the formation of the crystalline salt from the gaseous ions under standard conditions. The same approximation is made as for ionization energies and electron affinities (see Section 1.10), i.e. $\Delta U(0\,K) \approx \Delta H(298\,K)$. Usually, relatively little error is introduced by using this approximation.[†] A value of $\Delta_{lattice} H^{\circ}$ can be determined using eq. 6.19 (by application of Hess's law of constant heat summation) and represents an *experimental value* since it is derived from experimentally determined data.

$$\Delta_f H^{\circ}(MX_n, s) = \Delta_a H^{\circ}(M, s) + n\Delta_a H^{\circ}(X, g)$$
$$+ \Sigma IE(M, g) + n\Delta_{EA} H(X, g)$$
$$+ \Delta_{lattice} H^{\circ}(MX_n, s) \qquad (6.19)$$

Rearranging this expression and introducing the approximation that the lattice energy $\Delta U(0\,K) \approx \Delta_{lattice} H(298\,K)$ gives equation 6.20. All the quantities on the right-hand side of the equation are obtained from standard tables of data. (Enthalpies of atomization: see Appendix 10; ionization energies: see Appendix 8; electron affinities: see Appendix 9.)

$$\Delta U(0\,K) \approx \Delta_f H^{\circ}(MX_n, s) - \Delta_a H^{\circ}(M, s) - n\Delta_a H^{\circ}(X, g)$$
$$- \Sigma IE(M, g) - n\Delta_{EA} H(X, g) \qquad (6.20)$$

Worked example 6.6 Application of the Born–Haber cycle

Given that the standard enthalpy of formation at 298 K of CaF_2 is $-1228\,kJ\,mol^{-1}$, determine the lattice energy for CaF_2 using appropriate data from the Appendices.

[†] In some cases, significant error can be incurred. For an in-depth discussion of the relationship between lattice energy and lattice enthalpy, see: H.D.B. Jenkins (2005) *J. Chem. Educ.*, vol. 82, p. 950.

First, construct an appropriate thermochemical cycle:

$$Ca(s) + F_2(g) \xrightarrow{\Delta_a H^\circ(Ca, s) + 2\Delta_a H^\circ(F, g)} Ca(g) + 2F(g)$$

$$\downarrow \Delta_f H^\circ(CaF_2, s) \qquad IE_1 + IE_2(Ca, g) \qquad \downarrow 2\Delta_{EA}H(F, g)$$

$$CaF_2(s) \xleftarrow{\Delta_{lattice}H^\circ(CaF_2, s) \sim \Delta U(0\,K)} Ca^{2+}(g) + 2F^-(g)$$

Values that need to be found in the Appendices are:

Appendix 10: $\Delta_a H^\circ(Ca, s) = 178\,\text{kJ mol}^{-1}$
$\Delta_a H^\circ(F, g) = 79\,\text{kJ mol}^{-1}$
Appendix 8: $IE_1(Ca, g) = 590\,\text{kJ mol}^{-1}$
$IE_2(Ca, g) = 1145\,\text{kJ mol}^{-1}$
Appendix 9: $\Delta_{EA}H(F, g) = -328\,\text{kJ mol}^{-1}$

Use of Hess's law gives:

$$\Delta U(0\,K) \approx \Delta_f H^\circ(CaF_2, s) - \Delta_a H^\circ(Ca, s)$$
$$- 2\Delta_a H^\circ(F, g) - \Sigma IE(Ca, g) - 2\Delta_{EA}H(F, g)$$
$$\approx -1228 - 178 - 2(79) - 590 - 1145 + 2(328)$$

$$\Delta U(0\,K) \approx -2643\,\text{kJ mol}^{-1}$$

Self-study exercises

Use data from the Appendices.

1. If $\Delta_f H^\circ(298\,K)$ for $CaCl_2 = -795\,\text{kJ mol}^{-1}$, determine its lattice energy. [*Ans.* $-2252\,\text{kJ mol}^{-1}$]

2. If the lattice energy of CsF is $-744\,\text{kJ mol}^{-1}$, determine $\Delta_f H^\circ(298\,K)$ for the compound. [*Ans.* $-539\,\text{kJ mol}^{-1}$]

3. If $\Delta_f H^\circ(298\,K)$ for $MgCl_2 = -641\,\text{kJ mol}^{-1}$, calculate the lattice energy of $MgCl_2$. [*Ans.* $-2520\,\text{kJ mol}^{-1}$]

4. Comment on any approximations made in the calculations in questions 1–3.

6.15 Lattice energy: 'calculated' versus 'experimental' values

If we take NaCl as a typical example, $\Delta U(0\,K)$ determined by using a Born–Haber cycle is approximately $-783\,\text{kJ mol}^{-1}$. The value calculated (using an experimental value of r_0 from X-ray diffraction data) from the Born–Mayer equation is $-761\,\text{kJ mol}^{-1}$. A more refined calculation, the basis of which was outlined in Section 6.13, gives $-768\,\text{kJ mol}^{-1}$. This level of agreement is observed for all the alkali metal halides (including those of Li), and for the group 2 metal fluorides. While this is not rigid proof that all these compounds are wholly ionic, the close agreement does support our use of the electrostatic model as a basis for discussing the thermochemistry of these compounds.

For compounds with layer structures, the situation is different. There is a significant difference between the calculated ($-1986\,\text{kJ mol}^{-1}$) and experimental ($-2435\,\text{kJ mol}^{-1}$) values of $\Delta U(0\,K)$ for CdI_2, indicating that the electrostatic model is unsatisfactory. We noted earlier that in the CdI_2 lattice (Fig. 6.23), van der Waals forces operate between layers of adjacent I^- centres. The electrostatic model is also unsatisfactory for Cu(I) halides (zinc blende lattice) and for AgI (wurtzite lattice). For the Ag(I) halides, the discrepancy between $\Delta U(0\,K)_{calculated}$ and $\Delta U(0\,K)_{experimental}$ follows the sequence AgF < AgCl < AgBr < AgI. Contributions due to covalent character in the lattice are significant for the larger halides, and are the origin of the decreasing solubility of the Ag(I) halides in water on going from AgF to AgI (see Section 7.9).

6.16 Applications of lattice energies

We now consider some typical applications of lattice energies; further examples are given in later chapters.

Estimation of electron affinities

The availability of laser photodetachment techniques has permitted more accurate experimental determinations of electron affinities. Even so, tables of electron affinities list some calculated values, in particular for the formation of multiply charged ions. One method of estimation uses the Born–Haber cycle, with a value for the lattice energy derived using an electrostatic model. Compounds for which this is valid are limited (see Section 6.15).

Consider the estimation of $\Sigma\{\Delta_{EA}H^\circ(298\,K)\}$ for process 6.21.

$$O(g) + 2e^- \longrightarrow O^{2-}(g) \qquad (6.21)$$

We can apply the Born–Haber cycle to a metal oxide having a structure type of known Madelung constant, and for which an electrostatic model is a reasonably valid approximation. Magnesium(II) oxide fits these criteria: it has an NaCl structure type, r_0 has been accurately determined by X-ray diffraction methods, and compressibility data are available; an electrostatic model gives $\Delta U(0\,K) = -3975\,\text{kJ mol}^{-1}$. All other quantities in the appropriate Born–Haber cycle are independently measurable and a value for $\Sigma\{\Delta_{EA}H^\circ (298\,K)\}$ for reaction 6.21 can be evaluated. A series of similar values for $\Sigma\{\Delta_{EA}H^\circ(298\,K)\}$ for reaction 6.21 can be obtained using different group 2 metal oxides.

The attachment of two electrons to an O atom can be considered in terms of the consecutive processes in scheme 6.22, and accepted values for the associated enthalpy changes for the two steps are -141 and $+798\,\text{kJ mol}^{-1}$.

$$\left.\begin{array}{l} O(g) + e^- \longrightarrow O^-(g) \\ O^-(g) + e^- \longrightarrow O^{2-}(g) \end{array}\right\} \qquad (6.22)$$

The second step is highly *endothermic*. It appears that the only reason the O^{2-} ion exists is because of the high lattice energies of oxide salts, e.g. $\Delta U(0\,K)$ for Na_2O, K_2O, MgO and CaO are -2481, -2238, -3795 and $-3414\,kJ\,mol^{-1}$.

Fluoride affinities

Fluoride acceptors such as BF_3, AsF_5 and SbF_5 readily form the anions $[BF_4]^-$, $[AsF_6]^-$ and $[SbF_6]^-$ respectively, and the F^- affinity for each acceptor can be determined using a thermochemical cycle such as that in scheme 6.23.

$$KBF_4(s) \xrightarrow{\;\;\Delta H^\circ_1\;\;} KF(s) + BF_3(g)$$
$$\Delta_{lattice}H^\circ(KBF_4,\,s) \uparrow \qquad\qquad \uparrow \Delta_{lattice}H^\circ(KF,\,s)$$
$$K^+(g) + [BF_4]^-(g) \xleftarrow{\;\;\Delta H^\circ_2\;\;} K^+(g) + F^-(g) + BF_3(g)$$

$$(6.23)$$

The high-temperature form of KBF_4 crystallizes with a CsCl structure and we can estimate the lattice energy using an electrostatic model, assuming that the $[BF_4]^-$ ion can be treated as a sphere (see Fig. 6.18). The lattice energy of KF is known, and ΔH°_1 can be determined from the temperature variation of the dissociation pressure of solid KBF_4. Use of Hess's law allows ΔH°_2 to be determined; this value $(-360\,kJ\,mol^{-1})$ corresponds to the enthalpy change associated with the attachment of F^- to BF_3.

Estimation of standard enthalpies of formation and disproportionation

For well-established ionic compounds, it is seldom the case that the lattice energy is known while the standard enthalpy of formation is not. However, in theoretical studies of hypothetical compounds, one may wish to estimate a value of $\Delta_f H^\circ(298\,K)$ using a Born–Haber cycle incorporating a calculated value of the lattice energy. The earliest example of this method addressed the question of whether it was conceivable that neon might form a salt Ne^+Cl^-. On the basis that the size of the Ne^+ ion would be similar to that of Na^+, and that NeCl should possess an NaCl structure, the lattice energy of NeCl was estimated to be $\approx -840\,kJ\,mol^{-1}$. This leads to a value of $\Delta_f H^\circ(NeCl,\,s) \approx +1010\,kJ\,mol^{-1}$, the very high first ionization energy of Ne $(2081\,kJ\,mol^{-1})$ being responsible for making the process so highly endothermic and unlikely to occur in practice.

Much later, lattice energy considerations pointed towards the feasibility of preparing the first compound of a noble gas. The first ionization energies of Xe and O_2 are similar, and the discovery that O_2 reacted with PtF_6 to give $[O_2]^+[PtF_6]^-$ led to the suggestion that Xe (see Chapter 18) might also react with PtF_6. The trend in first ionization energies on descending group 18 is shown in Fig. 6.26.

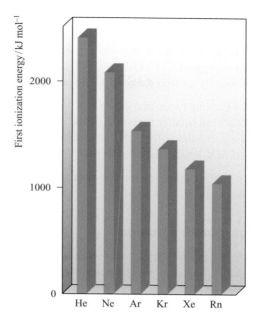

Fig. 6.26 The trend in the values of the first ionization energies of the noble gases (group 18).

Although radon is the easiest to ionize, it is highly radioactive and xenon is more readily handled in the laboratory. The reaction between Xe and PtF_6 was successful, although the exact nature of the product '$Xe[PtF_6]$' remains uncertain, even though the reaction was first studied by Neil Bartlett as long ago as 1962.

A further example considers the possible formation of CaF (in contrast to the more usual CaF_2). Here, a simple Born–Haber cycle is not helpful since CaF is not thermodynamically unstable with respect to decomposition into its *constituent elements*, but is unstable with respect to *disproportionation* (eq. 6.24).

$$2CaF(s) \longrightarrow Ca(s) + CaF_2(s) \qquad (6.24)$$

> A species ***disproportionates*** if it undergoes simultaneous oxidation and reduction.

The thermochemical cycle to be considered is given in eq. 6.25, in which the values of $\Delta_a H^\circ(Ca,\,s)$ $(178\,kJ\,mol^{-1})$ and the difference between IE_1 and IE_2 for Ca $(-555\,kJ\,mol^{-1})$ are significantly smaller in magnitude than the lattice energy of CaF_2 $(-2610\,kJ\,mol^{-1})$.

$$2CaF(s) \xrightarrow{\;\;\Delta H^\circ\;\;} Ca(s) + CaF_2(s)$$
$$2\Delta_{lattice}H^\circ(CaF,\,s) \uparrow \qquad\qquad \uparrow \begin{array}{l}\Delta_{lattice}H^\circ(CaF_2,\,s)\\ -\Delta_a H^\circ(Ca,\,s)\end{array}$$
$$2Ca^+(g) + 2F^-(g) \xleftarrow[IE_1(Ca,\,g)\,-\,IE_2(Ca,\,g)]{} Ca(g) + Ca^{2+}(g) + 2F^-(g)$$

$$(6.25)$$

The magnitude and sign of the enthalpy change, ΔH°, for the disproportionation reaction therefore depend largely on the balance between the lattice energy of CaF_2 and twice the lattice energy of CaF. The value of $\Delta U(0\,K)$ for CaF_2 will significantly exceed that of CaF because:

- $|z_+|$ for Ca^{2+} is twice that of Ca^+;
- r_0 for Ca^{2+} is smaller than that of Ca^+;
- Madelung constants for MX_2 structures are ≈ 1.5 times those of MX lattices (see Table 6.4).

The net result is that ΔH° for the disproportionation reaction shown in eq. 6.25 is negative.

The Kapustinskii equation

A problem in estimating the lattice energy of a hypothetical compound is deciding what ionic structure type to assume. Attempts have been made to use the fact that Madelung constants for MX and MX_2 structure types (Table 6.4) are in an approximate ratio of $2:3$. In 1956, Kapustinskii derived what has become the best known *general* expression for estimating lattice energies, and one form of this is given in eq. 6.26.

$$\Delta U(0\,K) = -\frac{(1.07 \times 10^5)v|z_+||z_-|}{r_+ + r_-}$$ (6.26)

where v = number of ions in the formula of the salt (e.g. 2 for NaCl, 3 for CaF_2); r_+ and r_- = radius for 6-coordinate cation and anion, respectively, in pm.

This expression has its origins in the Born–Landé equation, with a value of 9 for the Born exponent (the value for NaCl) and half the value of the Madelung constant for NaCl. The inclusion of the factor v shows why *half* of A is included. Although the Kapustinskii equation is useful, it is a gross *approximation* and values obtained in this way must be treated with caution.

6.17 Defects in solid state lattices

So far in this chapter, we have assumed implicitly that all the pure substances considered have ideal lattices in which every site is occupied by the correct type of atom or ion. This state appertains only at $0\,K$, and above this temperature, *lattice defects* are always present. The energy required to create a defect is more than compensated for by the resulting increase in entropy of the structure. In this section, we introduce some common types of defect. Electrical conductivity that arises as a result of defects in ionic solids is detailed in Section 28.2.

Intrinsic defects occur in lattices of pure compounds.
Extrinsic defects result from the addition of dopants.

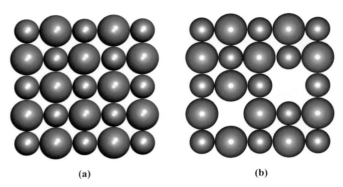

Fig. 6.27 (a) Part of one face of an ideal NaCl structure; compare this with Fig. 6.16. (b) A Schottky defect involves vacant cation and anion sites; equal numbers of cations and anions must be absent to maintain electrical neutrality. Colour code: Na, purple; Cl, green.

Schottky defect

A *Schottky defect* is an example of a *point defect* in a crystal lattice and arises from vacant lattice sites. The stoichiometry and electrical neutrality of the compound are retained. Defects that fall in this category include:

- a vacant atom site in a metal lattice;
- a vacant cation and a vacant anion site in an MX lattice;
- two vacant cations and a vacant anion site in an M_2X lattice;
- a vacant cation and two vacant anion sites in an MX_2 lattice.

Figure 6.27 illustrates a Schottky defect in an NaCl lattice; holes are present (Fig. 6.27b) where ions are expected on the basis of the ideal lattice (Fig. 6.27a).

Frenkel defect

In ionic lattices in which there is a significant difference in size between the cation and anion, the smaller ion may occupy a site that is vacant in the ideal lattice. This is a *Frenkel defect* (a *point defect*) and does not affect the stoichiometry or electrical neutrality of the compound. Figure 6.28 illustrates a Frenkel defect in AgBr which adopts an NaCl structure. The radii of the Ag^+ and Br^- ions are 115[†] and 196 pm. The central Ag^+ ion in Fig. 6.28a is in an octahedral hole with respect to the fcc arrangement of Br^- ions. Migration of the Ag^+ ion to a vacant tetrahedral hole (Fig. 6.28b) creates a Frenkel defect in the lattice.

Experimental observation of Schottky and Frenkel defects

There are several methods that may be used to study the occurrence of Schottky and Frenkel defects in stoichiometric crystals, but the simplest, in principle, is to measure

---

[†] 115 pm refers to a 6-coordinate Ag^+ ion; r_{ion} for a 4-coordinate ion is 100 pm.

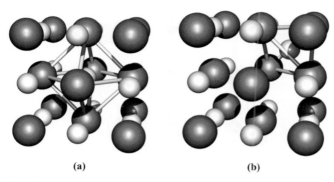

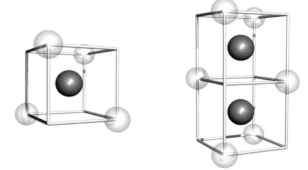

Fe^{3+} shown in green; vacancies are the transparent spheres

(6.2)   **(6.3)**

Fig. 6.28 Silver bromide adopts an NaCl structure. (a) An ideal lattice can be described in terms of Ag^+ ions occupying octahedral holes in a cubic close-packed array of bromide ions. (b) A Frenkel defect in AgBr involves the migration of Ag^+ ions into tetrahedral holes; in the diagram, one Ag^+ ion occupies a tetrahedral hole which was originally vacant in (a), leaving the central octahedral hole empty. Colour code: Ag, pale grey; Br, gold.

the density of the crystal extremely accurately. Low concentrations of Schottky defects lead to the observed density of a crystal being lower than that calculated from X-ray diffraction and data based on the size and structure of the unit cell (see end-of-chapter problems 6.21 and 6.22). On the other hand, since the Frenkel defect does not involve a change in the number of atoms or ions present, no such density differences will be observed.

Non-stoichiometric compounds

Many defects result in a compound being *non-stoichiometric*. Such defects often occur in crystalline solids of *d*-block metal compounds when the metal exhibits variable oxidation states. Metal oxides and sulfides are particularly prone to non-stoichiometric defects and these often lead to dramatic changes in physical properties even at low levels of crystal imperfection. For example, treatment of TiO_2 with H_2 at 770–970 K results in an O deficiency such that the stoichiometry is $TiO_{1.993}$. The change from TiO_2 to $TiO_{1.993}$ causes a decrease in electrical resistivity by more than five orders of magnitude. Some crystalline metal oxides are extremely difficult to prepare in a stoichiometric form. For example, although TiO exists with a NaCl-type structure exhibiting Schottky defects at room temperature (see Section 21.5), non-stoichiometric compounds in the range $TiO_{0.82}$–$TiO_{1.23}$ are also formed. FeO is always Fe-deficient and is formulated as $Fe_{1-x}O$ ($0.04 < x < 0.11$), occurring naturally in this form in meteorites and oceanic basalt. Thus, in $Fe_{1-x}O$, some Fe^{3+} ions are present to counter what would otherwise be a charge imbalance caused by Fe^{2+} vacancies. The Fe^{3+} centres tend to occupy interstitial sites between vacancies giving well-defined clusters such as **6.2** and **6.3**. The clusters can be described as zones of an Fe_3O_4-type structure.

Metal oxides having a CaF_2 structure are subject to various non-stoichiometric defects. Uranium(IV) oxide has the stoichiometry UO_{2+x}, i.e. it has an *anion-excess structure* in which excess O^{2-} ions are accommodated in interstitial sites. The addition of dopants to crystalline solids produces defects which can be of commercial significance. Dopant cations must be of a similar size to those in the parent lattice. Adding CaO to ZrO_2 stabilizes the cubic form of zirconia (see Section 22.5) and prevents a phase change from cubic to monoclinic that would otherwise occur on cooling below 1143 K. The introduction of Ca^{2+} into the ZrO_2 structure results in replacement of Zr^{4+} by Ca^{2+} and the creation of an O^{2-} vacancy to counter the charge imbalance. The doped zirconia, $Ca_xZr_{1-x}O_{2-x}$, is *anion-deficient*.

The introduction of a dopant may result in a change in the oxidation state of metal sites in the parent lattice. A well-cited example is the doping of NiO with Li_2O in the presence of air/O_2. When an Ni^{2+} ion is replaced by Li^+, electrical neutrality is retained by the oxidation of another Ni^{2+} to Ni^{3+} (Fig. 6.29).

Self-study exercises

1. What is meant by a *non-stoichiometric metal oxide*?

2. How is electrical neutrality maintained in $Fe_{1-x}O$?

3. Tungsten bronzes, Na_xWO_3 ($0 < x < 1$), possess defect perovskite structures. How is electrical neutrality maintained in these compounds? What important properties do they exhibit?   [*Ans.* See Section 22.7]

Colour centres (F-centres)

Defects that result from the presence of trapped electrons in a crystal lattice cause colour changes. If NaCl is heated in Na vapour, Na atoms enter the NaCl lattice, and Na is oxidized to Na^+ (eq. 6.26).

$$Na \longrightarrow Na^+ + e^-$$ (6.26)

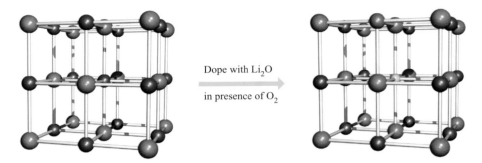

Fig. 6.29 NiO possesses an NaCl structure. Doping with Li_2O in the presence of air/O_2 results in the replacement of an Ni^{2+} centre (blue) with an Li^+ ion (yellow), and the oxidation of one Ni^{2+} to an Ni^{3+} centre (green). Oxide ions are shown in red.

The electron produced in oxidation step 6.26 remains trapped in the crystal lattice and occupies a lattice site, leaving a Cl^- site vacant. Excitation and subsequent relaxation of the electron results in the emission of radiation in the visible region. The electron centre is known as an *F-centre* (from the German *Farbe* for colour). The origins of F-centres are varied, but their presence has some dramatic consequences. For example, some variants of transparent minerals are coloured owing to the presence of F-centres, e.g. Blue John is a rare blue-purple form of fluorspar and is much prized in jewellery and decorative ornaments.

Thermodynamic effects of crystal defects

The presence of defects means that almost all crystals are imperfect. The creation of a defect from an ideal crystal is an endothermic process, but is entropically favoured since a degree of disorder is introduced into the otherwise perfectly ordered lattice. The balance between the ΔH and $T\Delta S$ terms (eq. 6.27) is therefore important.

$$\Delta G = \Delta H - T\Delta S \qquad (6.27)$$

At temperatures above 0 K, the thermodynamic balance favours the presence of defects, and the minimum value of ΔG is attained at a given equilibrium concentration of defects, the concentration being temperature-dependent. For example, bulk copper metal has around one vacancy for every 10^{14} atoms at 298 K, while at 1300 K one site in every 10^3 may be vacant.

KEY TERMS

The following terms were introduced in this chapter. Do you know what they mean?

- ❏ close-packing (of spheres or atoms)
- ❏ cubic close-packed (ccp) lattice
- ❏ hexagonal close-packed (hcp) lattice
- ❏ face-centred cubic (fcc) lattice
- ❏ simple cubic lattice
- ❏ body-centred cubic (bcc) lattice
- ❏ coordination number (in a lattice)
- ❏ unit cell
- ❏ interstitial hole
- ❏ phase diagram
- ❏ metallic radius
- ❏ alloy
- ❏ electrical resistivity
- ❏ band theory

- ❏ band gap
- ❏ insulator
- ❏ semiconductor
- ❏ intrinsic and extrinsic semiconductors
- ❏ n- and p-type semiconductors
- ❏ doping (a semiconductor)
- ❏ ionic radius
- ❏ NaCl structure type
- ❏ CsCl structure type
- ❏ CaF_2 (fluorite) structure type
- ❏ antifluorite structure type
- ❏ zinc blende structure type
- ❏ diamond network
- ❏ wurtzite structure type

- ❏ β-cristobalite structure type
- ❏ TiO_2 (rutile) structure type
- ❏ CdI_2 and $CdCl_2$ (layer) structures
- ❏ perovskite structure type
- ❏ lattice energy
- ❏ Born–Landé equation
- ❏ Madelung constant
- ❏ Born exponent
- ❏ Born–Haber cycle
- ❏ disproportionation
- ❏ Kapustinskii equation
- ❏ Schottky defect
- ❏ Frenkel defect
- ❏ non-stoichiometric solid
- ❏ colour centre

FURTHER READING

Packing of spheres and structures of ionic lattices

C.E. Housecroft and E.C. Constable (2010) *Chemistry*, 4th edn, Prentice Hall, Harlow – Chapters 8 and 9 give detailed accounts at an introductory level.

A.F. Wells (1984) *Structural Inorganic Chemistry*, 5th edn, Clarendon Press, Oxford – Chapters 4 and 6 present careful descriptions, ranging from basic to more advanced material.

Dictionary of Inorganic Compounds (1992), Chapman and Hall, London – The introduction to Vol. 4 gives a useful summary of structure types.

Alloys

A.F. Wells (1984) *Structural Inorganic Chemistry*, 5th edn, Clarendon Press, Oxford – Chapter 29 provides excellent coverage of metal and alloy lattice types.

Semiconductors

M. Hammonds (1998) *Chem. & Ind.*, p. 219 – 'Getting power from the sun' illustrates the application of the semiconducting properties of Si.
J. Wolfe (1998) *Chem. & Ind.*, p. 224 – 'Capitalising on the sun' describes the applications of Si and other materials in solar cells.

Solid state: for more general information

A.K. Cheetham and P. Day (1992) *Solid State Chemistry*, Clarendon Press, Oxford.
M. Ladd (1994) *Chemical Bonding in Solids and Fluids*, Ellis Horwood, Chichester.
M. Ladd (1999) *Crystal Structures: Lattices and Solids in Stereoview*, Ellis Horwood, Chichester.
L. Smart and E. Moore (1992) *Solid State Chemistry: An Introduction*, Chapman and Hall, London.
A.R. West (1999) *Basic Solid State Chemistry*, 2nd edn, Wiley-VCH, Weinheim.

PROBLEMS

6.1 Outline the similarities and differences between cubic and hexagonal close-packed arrangements of spheres, paying particular attention to (a) coordination numbers, (b) interstitial holes and (c) unit cells.

6.2 State the coordination number of a sphere in each of the following arrangements: (a) ccp; (b) hcp; (c) bcc; (d) fcc; (e) simple cubic.

6.3 (a) Lithium metal undergoes a phase change at 80 K (1 bar pressure) from the α- to β-form; one form is bcc and the other is a close-packed lattice. Suggest, with reasons, which form is which. What name is given to this type of structural change? (b) Suggest why tin buttons on nineteenth-century military uniforms crumbled in exceptionally cold winters.

6.4 Refer to Table 6.2. (a) Write an equation for the process for which the standard enthalpy of atomization of cobalt is defined. (b) Suggest reasons for the trend in standard enthalpies of atomization on descending group 1. (c) Outline possible reasons for the trend in values of $\Delta_a H^\circ$ on going from Cs to Bi.

6.5 'Titanium dissolves nitrogen to give a solid solution of composition $TiN_{0.2}$; the metal lattice defines an hcp arrangement.' Explain what is meant by this statement, and suggest whether, on the basis of this evidence, $TiN_{0.2}$ is likely to be an interstitial or substitutional alloy. Relevant data may be found in Appendix 6 and Table 6.2.

6.6 What do you understand by the 'band theory of metals'?

6.7 (a) Draw a representation of the structure of diamond and give a description of the bonding. (b) Is the same picture of the bonding appropriate for silicon, which is isostructural with diamond? If not, suggest an alternative picture of the bonding.

6.8 (a) Give a definition of electrical resistivity and state how it is related to electrical conductivity. (b) At 273–290 K, the electrical resistivities of diamond, Si, Ge and α-Sn are approximately 1×10^{11}, 1×10^{-3}, 0.46 and 11×10^{-8} Ω m. Rationalize this trend in values. (c) How does the change in electrical resistivity with temperature vary for a typical metal and for a semiconductor?

6.9 Distinguish between an intrinsic and extrinsic semiconductor, giving examples of materials that fall into these classes, and further classify the types of extrinsic semiconductors. What structural features do many semiconductors have in common?

6.10 The metallic, covalent and ionic radii of Al are 143, 130 and 54 pm respectively; the value of r_{ion} is for a 6-coordinate ion. (a) How is each of these quantities defined? (b) Suggest reasons for the trend in values.

6.11 With reference to the NaCl, CsCl and TiO_2 structure types, explain what is meant by (a) coordination number, (b) unit cell, (c) ion sharing between unit cells, and (d) determination of the formula of an ionic salt from the unit cell.

6.12 Determine the number of formula units of (a) CaF_2 in a unit cell of fluorite, and (b) TiO_2 in a unit cell of rutile.

6.13 (a) Confirm that the unit cell for perovskite shown in Fig. 6.24a is consistent with the stoichiometry $CaTiO_3$. (b) A second unit cell can be drawn for perovskite. This has Ti(IV) at the centre of a cubic cell; Ti(IV) is in an octahedral environment with respect to the O^{2-} ions. In what sites must the Ca^{2+} ions lie in order that the unit cell depicts the correct compound stoichiometry? Draw a diagram to illustrate this unit cell.

6.14 (a) Give a definition of lattice energy. Does your definition mean that the associated enthalpy of reaction will be positive or negative? (b) Use the Born–Landé equation to calculate a value for the lattice energy of KBr, for which $r_0 = 328$ pm. KBr adopts an NaCl structure; other data may be found in Tables 6.3 and 6.4.

6.15 Using data from the Appendices and the fact that $\Delta_f H^\circ(298\,K) = -859\,kJ\,mol^{-1}$, calculate a value for the lattice energy of $BaCl_2$. Outline any assumptions that you have made.

6.16 (a) Given that $\Delta U(0\,K)$ and $\Delta_f H^\circ(298\,K)$ for MgO are -3795 and $-602\,kJ\,mol^{-1}$ respectively, derive a value for $\Delta_{EA} H^\circ(298\,K)$ for the reaction:

$$O(g) + 2e^- \longrightarrow O^{2-}(g)$$

Other data: see Appendices. (b) Compare the calculated value with that obtained using electron affinity data from Appendix 9, and suggest reasons for any differences.

6.17 Discuss the interpretation of the following:

(a) $\Delta_f H^\circ(298\,K)$ becomes less negative along the series LiF, NaF, KF, RbF, CsF, but more negative along the series LiI, NaI, KI, RbI, CsI.

(b) The thermal stability of the isomorphous sulfates of Ca, Sr and Ba with respect to decomposition into the metal oxide (MO) and SO_3 increases in the sequence $CaSO_4 < SrSO_4 < BaSO_4$.

6.18 Data from Tables 6.3 and 6.4 are needed for this problem. (a) Estimate the lattice energy of CsCl if the Cs−Cl internuclear distance is 356.6 pm. (b) Now consider a polymorph of CsCl that crystallizes with an NaCl structure. Estimate its lattice energy given that the Cs−Cl distance is 347.4 pm. (c) What conclusions can you draw from your answers to parts (a) and (b)?

6.19 Which of the following processes are expected to be exothermic? Give reasons for your answers.

(a) $Na^+(g) + Br^-(g) \longrightarrow NaBr(s)$
(b) $Mg(g) \longrightarrow Mg^{2+}(g) + 2e^-$
(c) $MgCl_2(s) \longrightarrow Mg(s) + Cl_2(g)$
(d) $O(g) + 2e^- \longrightarrow O^{2-}(g)$
(e) $Cu(l) \longrightarrow Cu(s)$
(f) $Cu(s) \longrightarrow Cu(g)$
(g) $KF(s) \longrightarrow K^+(g) + F^-(g)$

6.20 Explain what is meant by Frenkel and Schottky defects in an NaCl structure type.

6.21 (a) How many ion-pairs are present in a unit cell of NaCl? (b) A unit cell length of 564 pm for NaCl has been determined by X-ray diffraction studies. Determine the volume of a unit cell of NaCl. (c) Using the data from part (b), determine the density of NaCl. (d) By comparing your answer to (c) with an observed density of 2.17 g cm^{-3}, confirm that the structure of NaCl is free of defects.

6.22 (a) VO, TiO and NiO all have defect rock salt structures. Explain what this statement means. (b) In NiO, the Ni–O internuclear separation is 209 pm. Determine the volume of a unit cell of NiO, and its density, assuming a non-defect structure. Given that the *observed* density of NiO is 6.67 g cm^{-3}, determine the extent of vacancies in the NiO lattice. Express your answer as a percentage.

6.23 Explain what is meant by (a) a Schottky defect in $CaCl_2$, and (b) a Frenkel defect in AgBr. (c) Suggest what effect doping crystals of AgCl with $CdCl_2$ might have on the AgCl lattice structure.

6.24 Why are *d*-block metal oxides much more frequently non-stoichiometric than are non-*d*-block metal oxides?

6.25 Suggest why doping NiO with Li_2O in air (or the presence of O_2) leads to an increase in electrical conductivity, and comment on the dependence of this increase on the amount of lithium dopant.

6.26 When nickel(II) oxide is heated in O_2, some of the cations are oxidized and vacant cation sites are formed according to the equation:

$$4Ni^{2+}(s) + O_2(g) \rightleftharpoons 4Ni^{3+}(s) + 2\square_+ + 2O^{2-}(s)$$

where $\square_+$ denotes a vacant cation site and (s) denotes an ion in the solid. Account for the fact that the electrical conductivity of the product is, for small deviations from stoichiometry, proportional to the sixth root of the pressure of O_2.

6.27 Comment on each of the following: (a) the difference between extrinsic and intrinsic defects; (b) why CaO is added to ZrO_2 used in refractory materials; (c) the formation of solid solutions of Al_2O_3 and Cr_2O_3.

6.28 Comment on the structural and compositional implications of (a) the Fe-deficiency of iron(II) oxide, and (b) the anion-excess nature of uranium(IV) oxide.

OVERVIEW PROBLEMS

6.29 Give explanations for the following observations.

(a) Raising the temperature of a sample of α-Fe from 298 K to 1200 K (at 1 bar pressure) results in a change of coordination number of each Fe atom from 8 to 12.

(b) Although a non-metal, graphite is often used as an electrode material.

(c) The semiconducting properties of silicon are improved by adding minute amounts of boron.

6.30 ReO_3 is a structure-prototype. Each Re(VI) centre is octahedrally sited with respect to the O^{2-} centres. The unit cell can be described in terms of a cubic array of Re(VI) centres, with each O^{2-} centre at the centre of each edge of the unit cell. Draw a representation of the unit cell and use your diagram to confirm the stoichiometry of the compound.

6.31 Suggest an explanation for each of the following observations.

(a) The Cr and Ni content of stainless steels used to make knife blades is different from that used in the manufacture of spoons.

(b) There is a poor match between experimental and calculated (Born–Landé) values of the lattice energy for AgI, but a good match for NaI.

(c) ThI_2 has been formulated as the Th(IV) compound $Th^{4+}(I^-)_2(e^-)_2$. Comment on why this is consistent with the observation of ThI_2 having a low electrical resistivity.

6.32 The first list below contains words or phrases, each of which has a 'partner' in the second list, e.g. 'sodium' in the first list can be matched with 'metal' in the second list. Match the 'partners'; there is only one match for each pair of words or phrases.

List 1	List 2
Sodium	Antifluorite structure
Cadmium iodide	Extrinsic semiconductor
Octahedral site	Double oxide
Gallium-doped silicon	Polymorphs
Sodium sulfide	Fluorite structure
Perovskite	Metal
Calcium fluoride	Intrinsic semiconductor
Gallium arsenide	Layered structure
Wurtzite and zinc blende	6-Coordinate
Tin(IV) oxide	Cassiterite

INORGANIC CHEMISTRY MATTERS

6.33 Despite the worldwide dependence on Si as a semiconductor, by far the major uses of Si are in the metallurgy industries. By reference to sections in this book, comment on the importance of silicon to the iron and steel industry, and to the Mg manufacturing industry.

6.34 Almost all natural diamonds are impure and contain substitutional nitrogen. Clusters of N atoms are present in Type 1a diamonds (the most common defect structure), while Type 1b diamonds contain well-separated N atoms. (a) What is meant by 'substitutional nitrogen'? (b) In what type of environment will an N atom reside in natural diamond? (c) Sketch a diagram to illustrate the band structure of diamond, assuming no defects. Explain why diamond is an insulator. (d) The donor-level associated with the nitrogen lies about 1.7 eV below the conduction band. How does this donor level arise? Why do natural diamonds not behave as semiconductors?

6.35 TiO_2 occurs naturally as three polymorphs: rutile (Fig. 6.22), anatase (Fig. 6.30) and brookite. (The unit cells of rutile and anatase can be viewed in 3D by going to www.pearsoned.co.uk/housecroft and following the links to the figures.) Rutile is commercially important as a white pigment in paints, paper and plastics, and in the last stages of

its manufacture, seed crystals of rutile are added. Anatase acts as a photocatalyst and has a vital role in, for example, dye-sensitized solar cells. (a) What analytical method is routinely used to screen polymorphs of a compound? Explain briefly how the technique can be used to distinguish anatase and rutile. (b) What are the coordination environments of each Ti(IV) centre and O^{2-} ion in anatase? (c) Use the unit cell in Fig. 6.30 to confirm the stoichiometry of the anatase phase. (d) Anatase preferentially crystallizes under ambient conditions and transforms to rutile at higher temperatures. At the structural level, how do you envisage the phase change taking place?

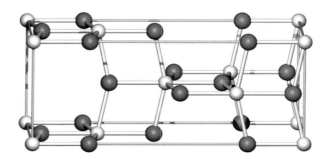

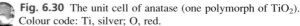

Fig. 6.30 The unit cell of anatase (one polymorph of TiO_2). Colour code: Ti, silver; O, red.

6.36 The mineral brucite, $Mg(OH)_2$, adopts a CdI_2-type structure, and substitution of Al^{3+} for Mg^{2+} results in a series of materials of general formula $[Mg_{1-x}Al_x(OH)_2](X^{n-})_{x/n} \cdot mH_2O$ where $x \approx 0.10$–0.35. (a) With the aid of a diagram, describe the CdI_2 prototype structure. Detail the interactions that operate within and between the layers. (b) Brucite crystals are easily cleaved in one direction. Why is this? (c) What is the effect of going from $Mg(OH)_2$ to $[Mg_{1-x}Al_x(OH)_2](X^{n-})_{x/n} \cdot mH_2O$ on the charge of the sheets in the crystal lattice? (d) Suggest where the X^{n-} anions and water molecules reside in $[Mg_{1-x}Al_x(OH)_2](X^{n-})_{x/n} \cdot mH_2O$, and comment on intermolecular forces that operate within the crystal lattice. (e) $[Mg_{1-x}Al_x(OH)_2](X^{n-})_{x/n} \cdot mH_2O$ materials may be used to remove toxic anions (e.g. arensite, $[AsO_2]^-$) from water. Suggest why they are suitable for this application and how the anion extraction works.

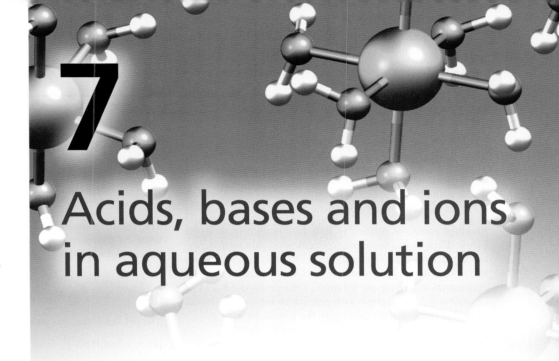

7

Acids, bases and ions in aqueous solution

7.1 Introduction

The importance of water as a medium for inorganic reactions stems not only from the fact that it is far more readily available than any other solvent, but also because of the abundance of accurate physicochemical data for aqueous solutions compared with the relative scarcity of such data for solutions in non-aqueous solvents. This chapter is concerned mainly with *equilibria* and in Section 7.2 and Box 7.1, we review calculations involving acid–base equilibrium constants.

Liquid water is approximately 55 molar H_2O, a fact commonly overlooked in the study of classical physical chemistry where, by convention, we take the *activity* (see Section 7.3) (and hence, the approximate concentration) of water to be unity.[†]

Worked example 7.1   Molarity of water

Show that pure water is approximately 55 molar.

Density of water $= 1\,\text{g}\,\text{cm}^{-3}$
Thus, $1000\,\text{cm}^3$ (or $1\,\text{dm}^3$) has a mass of $1000\,\text{g}$
For H_2O, $M_r = 18$

   Number of moles in $1000\,\text{g} = \dfrac{1000}{18} = 55.5 =$ number of moles per dm^3

   Therefore, the concentration of pure water $\approx 55\,\text{mol}\,\text{dm}^{-3}$.

[†] The use of [ ] for concentration should not be confused with the use of [ ] to show the presence of an ion. For example, $[OH]^-$ means 'hydroxide ion', but $[OH^-]$ means 'the concentration of hydroxide ions'.

7.2 Properties of water

Structure and hydrogen bonding

At atmospheric pressure, solid H_2O can adopt one of two polymorphs, depending upon the conditions of crystallization. We describe only the normal form of ice, the structure of which has been accurately determined using neutron diffraction (see Section 4.11). Ice possesses an infinite 3-dimensional structure. The key to making the structure rigid is *intermolecular hydrogen bonding* (see Section 10.6). There are 13 crystalline polymorphs of ice which crystallize under different conditions of temperature and pressure. Under atmospheric pressure, ordinary ice crystallizes with the structure shown in Fig. 7.1. The hydrogen-bonded network may be described in terms of a wurtzite structure type (see Fig. 6.21) in which the O atoms occupy the sites of *both* the Zn and S centres. This places each O atom in a tetrahedral environment with respect to other O atoms. Each O atom is involved in four hydrogen bonds through the use of two lone pairs and two H atoms (Fig. 7.1). The hydrogen bonds are asymmetrical (O–H distances $= 101\,\text{pm}$ and $175\,\text{pm}$) and non-linear; each H atom lies slightly off the O····O line, so that the intramolecular H–O–H bond angle is $105°$. The wurtzite structure is very

THEORY

Box 7.1 The equilibrium constants K_a, K_b and K_w

In dealing with acid–base equilibria in aqueous solution, three equilibrium constants are of special significance:

- K_a is the acid dissociation constant.
- K_b is the base dissociation constant.
- K_w is the self-ionization constant of water.

Essential equations relating to acid–base equilibria are listed below. Expressions involving concentrations are approximations, since we should strictly be using activities (see main text). Moreover, for a weak acid, HA, we *assume* that the concentration in aqueous solution of the dissociated acid *at equilibrium* is negligible with respect to the concentration of acid present initially; similarly for a weak base.

For a general weak acid HA in aqueous solution:

$$HA(aq) + H_2O(l) \rightleftharpoons [H_3O]^+(aq) + A^-(aq)$$

$$K_a = \frac{[H_3O^+][A^-]}{[HA][H_2O]} = \frac{[H_3O^+][A^-]}{[HA]}$$

By convention, $[H_2O] = 1$; strictly, the *activity* of the solvent H_2O is 1 (see Section 7.3).

For a general weak base B in aqueous solution:

$$B(aq) + H_2O(l) \rightleftharpoons BH^+(aq) + [OH]^-(aq)$$

$$K_b = \frac{[BH^+][OH^-]}{[B][H_2O]} = \frac{[BH^+][OH^-]}{[B]}$$

$$pK_a = -\log K_a \qquad K_a = 10^{-pK_a}$$

$$pK_b = -\log K_b \qquad K_b = 10^{-pK_b}$$

$$K_w = [H_3O^+][OH^-] = 1.00 \times 10^{-14}$$

$$pK_w = -\log K_w = 14.00$$

$$K_w = K_a \times K_b$$

$$pH = -\log[H_3O^+]$$

Review example 1: Calculate the pH of aqueous 0.020 M acetic acid ($K_a = 1.7 \times 10^{-5}$)

The equilibrium in aqueous solution is:

$$MeCO_2H(aq) + H_2O(l) \rightleftharpoons [MeCO_2]^-(aq) + [H_3O]^+(aq)$$

and K_a is given by:

$$K_a = \frac{[MeCO_2^-][H_3O^+]}{[MeCO_2H][H_2O]} = \frac{[MeCO_2^-][H_3O^+]}{[MeCO_2H]}$$

since $[H_2O]$ is taken to be unity where we are dealing with *equilibrium concentrations*.

Since $[MeCO_2^-] = [H_3O^+]$

$$K_a = \frac{[H_3O^+]^2}{[MeCO_2H]}$$

$$[H_3O^+] = \sqrt{K_a \times [MeCO_2H]}$$

The *initial* concentration of $MeCO_2H$ is $0.020 \, mol \, dm^{-3}$, and since the degree of dissociation is very small, the *equilibrium* concentration of $MeCO_2H \approx 0.020 \, mol \, dm^{-3}$.

$$[H_3O^+] = \sqrt{1.7 \times 10^{-5} \times 0.020}$$

$$[H_3O^+] = 5.8 \times 10^{-4} \, mol \, dm^{-3}$$

The pH value can now be determined:

$$pH = -\log[H_3O^+]$$

$$= -\log(5.8 \times 10^{-4})$$

$$= 3.2$$

To check that the assumption that $[MeCO_2H]_{equilm} \approx [MeCO_2H]_{initial}$ is valid:

initial concentration of acid was $0.020 \, mol \, dm^{-3}$, and the equilibrium concentration of $[H_3O]^+$ was found to be $5.8 \times 10^{-4} \, mol \, dm^{-3}$.

$$\text{Degree of dissociation} = \frac{5.8 \times 10^{-4}}{0.020} \times 100\% = 2.9\%$$

This confirms that the degree of dissociation is very small and the assumption made was valid.

Review example 2: Find the concentration of $[OH]^-$ present in a $5.00 \times 10^{-5} \, mol \, dm^{-3}$ solution of $Ca(OH)_2$

At a concentration of $5.00 \times 10^{-5} \, mol \, dm^{-3}$, $Ca(OH)_2$ is *fully* ionized, with two moles of $[OH]^-$ provided by each mole of $Ca(OH)_2$.

$$[OH^-] = 2 \times 5.00 \times 10^{-5} = 1.00 \times 10^{-4} \, mol \, dm^{-3}$$

To find the pH, we need to find $[H_3O^+]$:

$$K_w = [H_3O^+][OH^-] = 1.00 \times 10^{-14} \text{ (at 298 K)}$$

$$[H_3O^+] = \frac{1.00 \times 10^{-14}}{1.00 \times 10^{-4}} = 1.00 \times 10^{-10} \, mol \, dm^{-3}$$

$$pH = -\log[H_3O^+] = 10.0$$

Review example 3: The value of K_a for HCN is 4.0×10^{-10}. What is the value of pK_b for $[CN]^-$?

K_a for HCN and K_b for $[CN]^-$ are related by the expression:

$$K_a \times K_b = K_w = 1.00 \times 10^{-14} \text{ (at 298 K)}$$

$$K_b = \frac{K_w}{K_a} = \frac{1.00 \times 10^{-14}}{4.0 \times 10^{-10}} = 2.5 \times 10^{-5}$$

$$pK_b = -\log K_b = 4.6$$

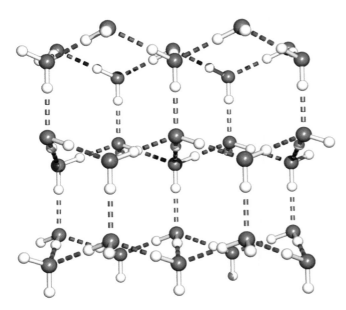

Fig. 7.1 Part of the structure of ordinary ice; it consists of a 3-dimensional network of hydrogen-bonded H_2O molecules.

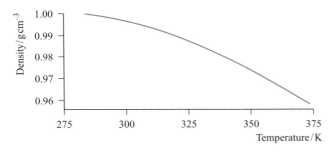

Fig. 7.2 The variation in the value of the density of water between 283 and 373 K.

open and, as a result, ice has a relatively low density ($0.92\,g\,cm^{-3}$). On melting (273 K), the lattice partly collapses, allowing some of the lattice cavities to be occupied by H_2O molecules. Consequently, the density increases, reaching a maximum at 277 K. Between 277 and 373 K, thermal expansion is the dominant effect, causing the density to decrease (Fig. 7.2). Even at the boiling point (373 K), much of the hydrogen bonding remains and is responsible for water having high values of the enthalpy and entropy of vaporization (Table 7.1 and see Section 10.6). The strength of a hydrogen bond in ice or water is $\approx 25\,kJ\,mol^{-1}$, and within the bulk liquid, intermolecular bonds are continually being formed and broken (thus transferring a proton between species) and the lifetime of a given H_2O molecule is only $\approx 10^{-12}\,s$. Water clusters such as $(H_2O)_{10}$ with ice-like arrangements of H_2O molecules have been structurally characterized in some compounds in the solid state.[†]

[†] See: L. Infantes and S. Motherwell (2002) *CrystEngComm.*, vol. 4, p. 454; A.D. Buckingham, J.E. Del Bene and S.A.C. McDowell (2008) *Chem. Phys. Lett.*, vol. 463, p. 1.

Table 7.1 Selected physical properties of water.

Property	Value
Melting point / K	273.00
Boiling point / K	373.00
Enthalpy of fusion, $\Delta_{fus}H^{\circ}(273\,K)/kJ\,mol^{-1}$	6.01
Enthalpy of vaporization, $\Delta_{vap}H^{\circ}(373\,K)/kJ\,mol^{-1}$	40.65
Entropy of vaporization, $\Delta_{vap}S^{\circ}(373\,K)/JK^{-1}\,mol^{-1}$	109
Relative permittivity (at 298 K)	78.39
Dipole moment, μ/debye	1.84

When water acts as a solvent, hydrogen bonds between water molecules are destroyed as water–solute interactions form. The latter may be ion–dipole interactions (e.g. when NaCl dissolves) or new hydrogen bonds (e.g. when H_2O and MeOH mix).

The self-ionization of water

Water itself is ionized to a very small extent (eq. 7.1) and the value of the self-ionization constant, K_w (eq. 7.2), shows that the equilibrium lies well to the left-hand side. The self-ionization in eq. 7.1 is also called autoprotolysis.

$$2H_2O(l) \rightleftharpoons [H_3O]^+(aq) + [OH]^-(aq) \qquad (7.1)$$

Water   Oxonium ion   Hydroxide ion

$$K_w = [H_3O^+][OH^-] = 1.00 \times 10^{-14} \quad (at\ 298\,K) \qquad (7.2)$$

Although we use concentrations in eq. 7.2, this is an approximation (see Section 7.3).

In aqueous solution, protons are solvated and so it is more correct to write $[H_3O]^+(aq)$ than $H^+(aq)$. Even this is oversimplified because the oxonium ion is further hydrated and species such as $[H_5O_2]^+$ (see Fig. 10.1), $[H_7O_3]^+$ and $[H_9O_4]^+$ are also present.

If a pure liquid partially dissociates into ions, it is ***self-ionizing***.

Water as a Brønsted acid or base

A ***Brønsted acid*** can act as a proton donor, and a ***Brønsted base*** can function as a proton acceptor.

Equilibrium 7.1 illustrates that water can function as both a Brønsted acid and a Brønsted base. In the presence of other Brønsted acids or bases, the role of water depends on the relative strengths of the various species in solution. When HCl is bubbled into water, the gas dissolves and equilibrium 7.3 is established.

$$HCl(aq) + H_2O(l) \rightleftharpoons [H_3O]^+(aq) + Cl^-(aq) \qquad (7.3)$$

Hydrogen chloride is a much stronger acid than water. This means that HCl will donate a proton to H_2O and equilibrium 7.3 lies well over to the right-hand side, so much so that hydrochloric acid is regarded as being fully dissociated, i.e. it is a *strong acid*. Water accepts a proton to form $[H_3O]^+$, and thus behaves as a Brønsted base. In the reverse direction, $[H_3O]^+$ acts as a *weak* acid and Cl^- as a *weak* base. The ions $[H_3O]^+$ and Cl^- are, respectively, the *conjugate acid* and *conjugate base* of H_2O and HCl.

In an aqueous solution of NH_3, water behaves as a Brønsted acid, donating H^+ (eq. 7.4). In eq. 7.4, $[NH_4]^+$ is the conjugate acid of NH_3, while H_2O is the conjugate acid of $[OH]^-$. Conversely, NH_3 is the conjugate base of $[NH_4]^+$, and $[OH]^-$ is the conjugate base of H_2O.

$$NH_3(aq) + H_2O(l) \rightleftharpoons [NH_4]^+(aq) + [OH]^-(aq) \quad (7.4)$$

Equation 7.5 gives the value of K for equilibrium 7.4 and shows that NH_3 acts as a *weak base* in aqueous solution. This is explored further in worked example 7.2.

$$K = \frac{[NH_4^+][OH^-]}{[NH_3]} = 1.8 \times 10^{-5} \quad \text{(at 298 K)} \quad (7.5)$$

Conjugate acids and bases are related as follows:

$$
\begin{array}{ccccccc}
HA(aq) & + & H_2O(l) & \rightleftharpoons & A^-(aq) & + & [H_3O]^+(aq) \\
\text{conjugate} & & \text{conjugate} & & \text{conjugate} & & \text{conjugate} \\
\text{acid 1} & & \text{base 2} & & \text{base 1} & & \text{acid 2}
\end{array}
$$

conjugate | acid–base pair

conjugate acid–base pair

Worked example 7.2 Manipulating equilibrium constant data

Using the values K_a for $[NH_4]^+ = 5.6 \times 10^{-10}$ and $K_w = 1.00 \times 10^{-14}$, determine a value of K for equilibrium 7.4.

First, write down the equilibria to which the data in the question refer:

$$[NH_4]^+(aq) + H_2O(l) \rightleftharpoons NH_3(aq) + [H_3O]^+(aq)$$
$$K_a = 5.6 \times 10^{-10}$$

$$H_2O(l) + H_2O(l) \rightleftharpoons [H_3O]^+(aq) + [OH]^-(aq)$$
$$K_w = 1.00 \times 10^{-14}$$

$$NH_3(aq) + H_2O(l) \rightleftharpoons [NH_4]^+(aq) + [OH]^-(aq) \quad K = ?$$

Now write down expressions for each K:

$$K_a = 5.6 \times 10^{-10} = \frac{[NH_3][H_3O^+]}{[NH_4^+]} \quad (1)$$

$$K_w = 1.00 \times 10^{-14} = [H_3O^+][OH^-] \quad (2)$$

$$K = \frac{[NH_4^+][OH^-]}{[NH_3]} \quad (3)$$

The right-hand side of eq. (3) can be written in terms of the right-hand sides of eqs. (1) and (2):

$$\frac{[NH_4^+][OH^-]}{[NH_3]} = \frac{[H_3O^+][OH^-]}{\left(\dfrac{[NH_3][H_3O^+]}{[NH_4^+]}\right)}$$

Substituting in the values of K_a and K_w gives:

$$\frac{[NH_4^+][OH^-]}{[NH_3]} = \frac{1.00 \times 10^{-14}}{5.6 \times 10^{-10}} = 1.8 \times 10^{-5}$$

This value agrees with that quoted in the text (eq. 7.5).

Self-study exercises

These exercises all refer to the equilibria in the worked example.

1. Confirm that $[NH_4]^+$ is a stronger acid in aqueous solution than H_2O.

2. Confirm that NH_3 acts as a base in aqueous solution.

3. For each equilibrium, write down the conjugate acid–base pairs.

7.3 Definitions and units in aqueous solution

In this section, we discuss the conventions and units generally used in the study of aqueous solutions. In some respects, these are *not* the same as those used in many other branches of chemistry. At the level of working within this text and, often, in the practical laboratory, certain approximations can be made, but it is crucial to understand their limitations.

Molarity and molality

A one molar aqueous solution (1 M or $1 \, mol \, dm^{-3}$) contains one mole of solute dissolved in a sufficient volume of water to give $1 \, dm^3$ *of solution*. In contrast, if one mole of solute is dissolved in $1 \, kg$ *of water*, the solution is said to be one molal ($1 \, mol \, kg^{-1}$).

Standard state

We are already used to the concept of standard state in respect of pure solids, liquids and gases. The standard state of a liquid or solid substance, whether pure or in a mixture, or for a solvent is taken as the state of the pure substance at 298 K and 1 bar pressure (1 bar = 1.00×10^5 Pa). The standard state of a gas is that of the pure gas at 298 K, 1 bar pressure and exhibiting ideal gas behaviour.

For a *solute in a solution*, the definition of its standard state is referred to a situation of *infinite dilution*: it is the state (a hypothetical one) at standard molality (m^{o}), 1 bar pressure, and exhibiting infinitely dilute solution behaviour. In the standard state, interactions between solute molecules or ions are negligible.

Activity

When the concentration of a solute is greater than about $0.1\ mol\,dm^{-3}$, interactions between the solute molecules or ions are significant, and the *effective* and real concentrations are no longer equal. It becomes necessary to define a new quantity called the *activity*, which is a measure of concentration but takes into account the interactions between the solution species. The *relative activity*, a_i, of a component i is dimensionless and is defined by eq. 7.6 where μ_i is the chemical potential of component i, μ_i^{o} is the standard chemical potential of i, R is the molar gas constant, and T is the temperature in kelvin.[†]

$$\mu_i = \mu_i^{o} + RT \ln a_i \qquad (7.6)$$

> The *activity* of any pure substance in its standard state is defined to be unity.

The relative activity of a solute is related to its *molality* by eq. 7.7 where γ_i is the activity coefficient of the solute, and m_i and m_i^{o} are the molality and standard state molality, respectively. Since the latter is defined as being unity, eq. 7.7 reduces to eq. 7.8.

$$a_i = \frac{\gamma_i m_i}{m_i^{o}} \qquad (7.7)$$

$$a_i = \gamma_i m_i \qquad (7.8)$$

Although all thermodynamic expressions dealing with aqueous solutions should strictly be expressed in terms of activities, inorganic chemists, and students in particular, may be dealing with situations in which two criteria are true:

- problem solving involving very dilute solutions ($\leq 1 \times 10^{-3}\ mol\,dm^{-3}$);
- very accurate answers are not required.

If these criteria hold, you can approximate the activity of the solute to its concentration, the latter being measured, most often, in *molarity*. We use this approximation throughout the book, but it is crucial to keep in mind the limitations of this approach.

7.4 Some Brønsted acids and bases

> The larger the value of K_a, the stronger the acid.
> The smaller the value of pK_a, the stronger the acid.
> The larger the value of K_b, the stronger the base.
> The smaller the value of pK_b, the stronger the base.

Carboxylic acids: examples of mono-, di- and polybasic acids

In organic compounds, acidity is quite often associated with the presence of a carboxylic acid group (CO_2H) and it is relatively easy to determine the number of *ionizable hydrogen atoms* in the system. Acetic acid,[‡] **7.1**, is a *monobasic acid* since it can donate only one proton. Ethanedioic acid (oxalic acid), **7.2**, can donate two protons and so is a *dibasic acid*. The *tetrabasic acid*, **7.3**, and the anions derived from it are commonly encountered in coordination chemistry; the trivial name for this acid is *N,N,N',N'*-ethylenediaminetetraacetic acid (see Table 7.7) and is generally abbreviated to H$_4$EDTA.

(7.1)   **(7.2)**

H$_4$EDTA

(7.3)

Equilibrium 7.9 describes the acid dissociation of $MeCO_2H$ in aqueous solution. It is a weak acid with $K_a = 1.75 \times 10^{-5}$ at 298 K.

$$MeCO_2H(aq) + H_2O(l) \rightleftharpoons [H_3O]^+(aq) + [MeCO_2]^-(aq)$$

Acetic acid   Acetate ion   (7.9)

(Ethanoic acid)   (Ethanoate ion)

Acids **7.2** and **7.3** undergo *stepwise dissociation* in aqueous solution, and eqs. 7.10 and 7.11 describe the steps for oxalic acid.

$$K_a(1) = 5.90 \times 10^{-2} \quad (298\ K) \qquad (7.10)$$

[†] For further discussion, see: P. Atkins and J. de Paula (2010) *Atkins' Physical Chemistry*, 9th edn, OUP, Oxford, p. 190.

[‡] The systematic name for $MeCO_2H$ is ethanoic acid, but acetic acid is the IUPAC-accepted trivial name.

THEORY

Box 7.2 Systematic oxoacid nomenclature

In 2005, the IUPAC published a set of new guidelines for the systematic naming of inorganic acids and their derivatives. Many inorganic oxoacids possess non-systematic (trivial) names that are in everyday use, and the IUPAC recognizes that it is unrealistic to abandon names such as sulfuric acid, nitric acid, phosphoric acid, boric acid and perchloric acid. However, these names provide no information about composition and structure.

The method of giving a systematic name to an inorganic oxoacid uses an *additive name*. This shows the connectivity of the central atom, as well as the groups attached to that central atom. The structure of a molecule of sulfuric acid is shown below:

The formula is usually written as H_2SO_4, but $SO_2(OH)_2$ gives more information. This way of writing the formula tells you immediately that the central S atom is connected to two OH groups and two O atoms. The systematic additive name is similarly constructed: dihydroxidodioxidosulfur (dihydroxido = 2 OH, dioxido = 2 O).

Although the formula of phosphinic acid is typically written as H_3PO_2, it has the structure shown below:

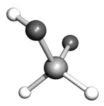

A more representative formula is $PH_2O(OH)$ and the additive name is dihydridohydroxidooxidophosphorus (dihydrido = 2 H, hydroxido = OH, oxido = O).

The table below lists some inorganic oxoacids with both common and systematic names. The last example (hypochlorous acid) illustrates how to name a compound in which the central atom is oxygen.

Formula	Accepted common name	Systematic additive name
$H_3BO_3 = B(OH)_3$	Boric acid	Trihydroxidoboron
$H_2CO_3 = CO(OH)_2$	Carbonic acid	Dihydroxidooxidocarbon
$H_4SiO_4 = Si(OH)_4$	Silicic acid	Tetrahydroxidosilicon
$HNO_3 = NO_2(OH)$	Nitric acid	Hydroxidodioxidonitrogen
$HNO_2 = NO(OH)$	Nitrous acid	Hydroxidooxidonitrogen
$H_3PO_4 = PO(OH)_3$	Phosphoric acid	Trihydroxidooxidophosphorus
$H_2SO_4 = SO_2(OH)_2$	Sulfuric acid	Dihydroxidodioxidosulfur
$H_2SO_3 = SO(OH)_2$	Sulfurous acid	Dihydroxidooxidosulfur
$HClO_4 = ClO_3(OH)$	Perchloric acid	Hydroxidotrioxidochlorine
$HClO_2 = ClO(OH)$	Chlorous acid	Hydroxidooxidochlorine
$HClO = O(H)Cl$	Hypochlorous acid	Chloridohydridooxygen

For the conjugate acids or bases of oxoacids, additive rules are again applied. In addition, the overall charge is shown in the name. For example, $[H_3SO_4]^+$ has the following structure:

The systematic name is trihydroxidooxidosulfur(1+). The conjugate base of sulfuric acid is $[HSO_4]^-$ and this is commonly called the hydrogensulfate ion. The formula may also be written as $[SO_3(OH)]^-$, and the additive name is hydroxidotrioxidosulfate(1−). The table opposite lists examples of conjugate bases of some oxoacids. Notice that, using systematic nomenclature, the ending '-ate' is used for all the anions. This contrasts with the old methods of distinguishing between, for example, $[NO_2]^-$ and $[NO_3]^-$ by using nitr*ite* and nitr*ate*.

Formula	Accepted common name	Systematic additive name
$[BO_3]^{3-}$	Borate	Trioxidoborate(3–)
$[HCO_3]^- = [CO_2(OH)]^-$	Hydrogencarbonate	Hydroxidodioxidocarbonate(1–)
$[CO_3]^{2-}$	Carbonate	Trioxidocarbonate(2–)
$[NO_3]^-$	Nitrate	Trioxidonitrate(1–)
$[NO_2]^-$	Nitrite	Dioxidonitrate(1–)
$[H_2PO_4]^- = [PO_2(OH)_2]^-$	Dihydrogenphosphate	Dihydroxidodioxidophosphate(1–)
$[HPO_4]^{2-} = [PO_3(OH)]^{2-}$	Hydrogenphosphate	Hydroxidotrioxidophosphate(2–)
$[PO_4]^{3-}$	Phosphate	Tetraoxidophosphate(3–)
$[HSO_4]^- = [SO_3(OH)]^-$	Hydrogensulfate	Hydroxidotrioxidosulfate(1–)
$[SO_4]^{2-}$	Sulfate	Tetraoxidosulfate(2–)
$[OCl]^-$	Hypochlorite	Chloridooxygenate(1–)

For the most part in this book, we use common names for inorganic acids. For complete details of systematic nomenclature, including 'hydrogen names' (an alternative nomenclature for hydrogen-containing compounds and ions) and how to deal with ring and chain structures, refer to:

Nomenclature of Inorganic Chemistry (IUPAC 2005 Recommendations), senior eds N.G. Connelly and T. Damhus, RSC Publishing, Cambridge, p. 124.

$$\begin{array}{ccc} CO_2^- \\ | \quad (aq) + H_2O(l) \; \rightleftharpoons \; [H_3O]^+(aq) + \\ CO_2H \end{array} \quad \begin{array}{c} CO_2^- \\ | \quad (aq) \\ CO_2^- \end{array}$$

$$K_a(2) = 6.40 \times 10^{-5} \quad (298\,K) \tag{7.11}$$

Each dissociation step has an associated equilibrium constant (acid dissociation constant), and for polybasic acids, it is general that $K_a(1) > K_a(2)$, and so on; it is more difficult to remove H^+ from an anion than from a neutral species. Values of equilibrium constants may be temperature-dependent, and the inclusion of the temperature to which the stated value applies is important. In general, quoted values usually refer to 293 or 298 K. In this book, unless otherwise stated, values of K_a refer to 298 K.

Inorganic acids

In inorganic chemistry, *hydrogen halides* and *oxoacids* are of particular significance in terms of acidic behaviour in aqueous solution. Each of the hydrogen halides is monobasic (eq. 7.12) and for X = Cl, Br and I, the equilibrium lies far to the right-hand side, making these strong acids. In each case, $K_a > 1$. Note that this means that the pK_a values are negative (pK_a HCl ≈ -7; HBr ≈ -9; HI ≈ -11) since $pK_a = -\log K_a$. In many instances, eq. 7.12 for X = Cl, Br or I is written showing only the forward reaction, thereby emphasizing strong acid behaviour. Hydrogen fluoride, on the other hand, is a weak acid ($pK_a = 3.45$).

$$HX(aq) + H_2O(l) \rightleftharpoons [H_3O]^+(aq) + X^-(aq) \tag{7.12}$$

The IUPAC definition of an **oxoacid** is 'a compound which contains oxygen, at least one other element, at least one hydrogen bound to oxygen, and which produces a conjugate base by proton loss.'

Examples of oxoacids include hypochlorous acid (HOCl), perchloric acid ($HClO_4$), nitric acid (HNO_3), sulfuric acid (H_2SO_4) and phosphoric acid (H_3PO_4). Many well-recognized common names exist for oxoacids, and the IUPAC has recommended that such names be retained. In this book, we follow this recommendation, although in Box 7.2 we introduce systematic nomenclature.

A wide variety of oxoacids exists and later chapters introduce many of them. Note that:

- oxoacids may be mono-, di- or polybasic;
- not all the hydrogen atoms in an oxoacid are necessarily ionizable.

Nitric acid, nitrous acid and hypochlorous acid are examples of monobasic acids. HNO_3 is essentially fully ionized in aqueous solution (eq. 7.13), but HNO_2 and HOCl behave as weak acids (eqs. 7.14 and 7.15).

$$\underset{\text{Nitric acid}}{HNO_3(aq)} + H_2O(l) \rightleftharpoons [H_3O]^+(aq) + \underset{\text{Nitrate ion}}{[NO_3]^-(aq)}$$

$$pK_a = -1.64 \tag{7.13}$$

$$\underset{\text{Nitrous acid}}{HNO_2(aq)} + H_2O(l) \rightleftharpoons [H_3O]^+(aq) + \underset{\text{Nitrite ion}}{[NO_2]^-(aq)}$$

$$pK_a = 3.37 \; (285\,K) \tag{7.14}$$

$$\underset{\text{Hypochlorous acid}}{HOCl(aq)} + H_2O(l) \rightleftharpoons [H_3O]^+(aq) + \underset{\text{Hypochlorite ion}}{[OCl]^-(aq)}$$

$$pK_a = 4.53 \tag{7.15}$$

Sulfuric acid is dibasic. In aqueous solution, the first dissociation step lies well over to the right-hand side (eq. 7.16), but $[HSO_4]^-$ is a weaker acid (eq. 7.17). Two series of salts can be isolated, e.g. sodium hydrogensulfate(1−) ($NaHSO_4$) and sodium sulfate (Na_2SO_4).

$$H_2SO_4(aq) + H_2O(l) \rightleftharpoons [H_3O]^+(aq) + [HSO_4]^-(aq)$$

Sulfuric
acid

Hydrogensulfate(1−) ion

$$pK_a \approx -2.0 \qquad (7.16)$$

$$[HSO_4]^-(aq) + H_2O(l) \rightleftharpoons [H_3O]^+(aq) + [SO_4]^{2-}(aq)$$

Sulfate ion

$$pK_a = 1.92 \qquad (7.17)$$

Tables of data and the existence of crystalline salts can sometimes be misleading, as is the case for 'sulfurous acid'. It is *not* possible to isolate pure H_2SO_3, even though we often refer to 'sulfurous acid' and values of acid dissociation constants are available (eqs. 7.18 and 7.19).

$$H_2SO_3(aq) + H_2O(l) \rightleftharpoons [H_3O]^+(aq) + [HSO_3]^-(aq)$$

Sulfurous
acid

Hydrogensulfite(1−) ion

$$pK_a = 1.82 \qquad (7.18)$$

$$[HSO_3]^-(aq) + H_2O(l) \rightleftharpoons [H_3O]^+(aq) + [SO_3]^{2-}(aq)$$

Sulfite ion

$$pK_a = 6.92 \qquad (7.19)$$

An aqueous solution of 'sulfurous acid' can be prepared by dissolving SO_2 in water (eq. 7.20), but the equilibrium constant indicates that such solutions contain mainly dissolved SO_2. A similar situation arises for 'carbonic acid', H_2CO_3 (see Section 14.9).

$$SO_2(aq) + H_2O(l) \rightleftharpoons H_2SO_3(aq) \qquad K < 10^{-9} \qquad (7.20)$$

In the oxoacids above, *each* hydrogen atom is attached to oxygen in the free acid, and the number of H atoms corresponds to the basicity of the acid. However, this is not always the case: e.g. although phosphinic acid has the formula H_3PO_2, there is only one O−H bond (structure **7.4**) and H_3PO_2 is *monobasic* (eq. 7.21). Further examples of this type are given in Section 15.11.

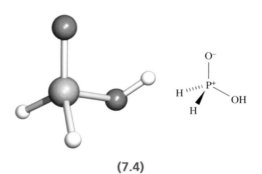

(7.4)

$$H_3PO_2(aq) + H_2O(l) \rightleftharpoons [H_3O]^+(aq) + [H_2PO_2]^-(aq)$$

Phosphinic acid

$$(7.21)$$

Inorganic bases: hydroxides

Many inorganic bases are hydroxides, and the term *alkali* is commonly used. The group 1 hydroxides NaOH, KOH, RbOH and CsOH are strong bases, being essentially fully ionized in aqueous solution; LiOH is weaker ($pK_b = 0.2$).

Inorganic bases: nitrogen bases

The term 'nitrogen bases' tends to suggest ammonia and organic amines (RNH_2), but there are a number of important inorganic nitrogen bases related to NH_3. Ammonia dissolves in water, and functions as a weak base, accepting H^+ to form the ammonium ion (eq. 7.4). Although solutions of NH_3 in water are often referred to as ammonium hydroxide, it is not possible to isolate solid samples of 'NH_4OH'. Commercially available 'NH_4OH' comprises aqueous solutions of NH_3.

Confusion may arise from tables of data for the dissociation constants for bases. Some tables quote K_b or pK_b, while others list values of K_a or pK_a. For the relationship between K_a and K_b, see Box 7.1. Thus, a value of pK_a for 'ammonia' of 9.25 is really that of the ammonium ion and refers to equilibrium 7.22, while a value of pK_b of 4.75 refers to equilibrium 7.4.

$$[NH_4]^+(aq) + H_2O(l) \rightleftharpoons [H_3O]^+(aq) + NH_3(aq)$$

$$pK_a = 9.25 \qquad (7.22)$$

Worked example 7.3   Relationship between pK_a and pK_b for a weak base

The degree of dissociation of NH_3 in aqueous solution can be described in terms of a value of either K_a or K_b. Deduce a relationship between the values of pK_a and pK_b.

K_b refers to the equilibrium:

$$NH_3(aq) + H_2O(l) \rightleftharpoons [NH_4]^+(aq) + [OH]^-(aq)$$

$$K_b = \frac{[NH_4^+][OH^-]}{[NH_3]}$$

K_a refers to the equilibrium:

$$[NH_4]^+(aq) + H_2O(l) \rightleftharpoons [H_3O]^+(aq) + NH_3(aq)$$

$$K_a = \frac{[NH_3][H_3O^+]}{[NH_4^+]}$$

Combining the two expressions gives:

$$\frac{[NH_4^+]}{[NH_3]} = \frac{K_b}{[OH^-]} = \frac{[H_3O^+]}{K_a}$$

$$K_b \times K_a = [H_3O^+][OH^-]$$

The right-hand side product is equal to the self-dissociation constant for water, K_w:

$$K_b \times K_a = K_w = 1.00 \times 10^{-14}$$

and so:

$$pK_b + pK_a = pK_w = 14.00$$

Self-study exercises

1. If pK_a for the conjugate acid of $PhNH_2$ is 4.63, what is pK_b for $PhNH_2$? To what equilibria do K_a and K_b refer?
   [*Ans.* 9.37]

2. For N_2H_4, $pK_b = 6.05$. What is K_b?
   [*Ans.* 8.91×10^{-7}]

3. pK_a for the pyridinium ion is 5.25. Calculate the K_b value of pyridine.

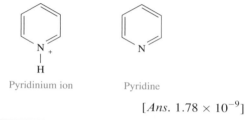

Pyridinium ion   Pyridine

[*Ans.* 1.78×10^{-9}]

Hydrazine, N_2H_4, **7.5**, is a weak Brønsted base ($pK_b = 6.05$), weaker than NH_3. It reacts with strong acids to give hydrazinium salts (eq. 7.23).

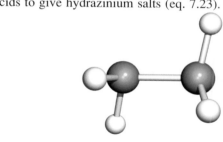

(7.5)

$$N_2H_4(aq) + HCl(aq) \longrightarrow [N_2H_5]Cl(aq) \qquad (7.23)$$

The value of pK_b for hydroxylamine, NH_2OH, is 8.04, showing it to be a weaker base than either NH_3 or N_2H_4.

7.5 The energetics of acid dissociation in aqueous solution

Hydrogen halides

The strengths of different acids in aqueous solutions tend to be discussed in elementary textbooks on a qualitative basis. In the case of the hydrogen halides, an exact treatment in terms of independently measurable thermodynamic quantities is *almost* possible. Consider the dissociation of HX

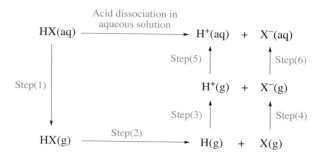

Fig. 7.3 The energetics of the dissociation of a hydrogen halide, HX (X = F, Cl, Br or I), in aqueous solution can be considered in terms of a cycle of steps. The significance of each step is discussed in the text.

(X is F, Cl, Br or I) in aqueous solution (equilibrium 7.24 or 7.25):

$$HX(aq) + H_2O(l) \rightleftharpoons [H_3O]^+(aq) + X^-(aq) \qquad (7.24)$$

$$HX(aq) \rightleftharpoons H^+(aq) + X^-(aq) \qquad (7.25)$$

The factors that influence the degree of dissociation are summarized in Fig. 7.3. Equation 7.26 relates K_a for the dissociation of HX in aqueous solution to ΔG°, and the latter depends on changes in both enthalpy and entropy (eq. 7.27).

$$\Delta G^\circ = -RT \ln K \qquad (7.26)$$

$$\Delta G^\circ = \Delta H^\circ - T\Delta S^\circ \qquad (7.27)$$

A Hess cycle relates ΔH° for each of steps (1) to (6) in Fig. 7.3 to that of the solution dissociation step. In Fig. 7.3, step (2) is the cleavage of the H–X bond for the gas-phase molecule. Steps (3) and (5) are the ionization of the gaseous H atom and the hydration of the gaseous H^+ ion, respectively. These two steps are common to all four hydrogen halides. Step (4) is the attachment of an electron to the gaseous X atom, and the associated enthalpy change is $\Delta_{EA}H$ (see Appendix 9). Step (6) is the hydration of gaseous X^-.

Step (1) causes some experimental difficulty. It is the reverse of the dissolution of gaseous HX in water to form solvated *undissociated* HX. Since HCl, HBr and HI are essentially fully dissociated in aqueous solution, measurement of enthalpy or entropy changes for step (1) must be estimated from somewhat unsatisfactory comparisons with noble gases and methyl halides. For HF, which is a weak acid in dilute aqueous solution, it might appear that values of ΔH° and ΔS° for step (1) could be obtained directly. However, IR spectroscopic data indicate that the species present in solution is the strongly hydrogen-bonded ion-pair $F^- \cdots HOH_2^+$.

We shall focus mainly on the conclusions drawn from calculations using the cycle in Fig. 7.3.[†] Firstly, consider

---

[†] For a fuller discussion, see: W.E. Dasent (1984) *Inorganic Energetics*, 2nd edn, Cambridge University Press, Chapter 5.

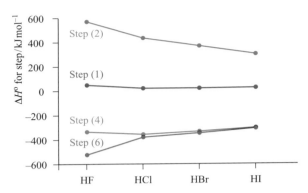

Fig. 7.4 Trends in the values of ΔH° for steps (1), (2), (4) and (6) defined in Fig. 7.3. [Data: W.E. Dasent (1984) *Inorganic Energetics*, 2nd edn, Cambridge University Press, and references cited therein.]

Table 7.2 Thermodynamic data and calculated values of pK_a for the dissociation of the hydrogen halides in aqueous solution. The values of ΔH°, $T\Delta S^\circ$, ΔG° and pK_a refer to the acid dissociation process shown in Fig. 7.3. For steps (3) and (5) in Fig. 7.3, the values of ΔH° are 1312 and $-1091\,kJ\,mol^{-1}$ respectively.

	HF	**HCl**	**HBr**	**HI**
$\Delta H^\circ / kJ\,mol^{-1}$	-22	-63	-71	-68
$T\Delta S^\circ / kJ\,mol^{-1}$	-30	-10	-4	$+3$
$\Delta G^\circ / kJ\,mol^{-1}$	$+8$	-53	-67	-71
Calculated pK_a	1.4	-9.3	-11.7	-12.4

the change in enthalpy for the dissociation of HX(aq). Since values of ΔH° for each of steps (3) and (5) are independent of the halide, it is the sum of the values of ΔH° for steps (1), (2), (4) and (6) that determines the trend in the values of ΔH° for reaction 7.25. Figure 7.4 summarizes the data and illustrates why there is, in fact, rather little difference between the values of the overall enthalpy change for reaction 7.25 for each of the hydrogen halides. Each reaction is exothermic, with ΔH° values in the order HF < HCl < HBr ≈ HI. If we now consider the $T\Delta S^\circ$ term for reaction 7.25 for each halide, the effect of its inclusion is rather dramatic, and leads to ΔG° for reaction 7.25 for X = F being positive while values of ΔG° for HCl, HBr and HI are negative (Table 7.2). Calculated values of pK_a can now be obtained using eq. 7.26 and are listed in Table 7.2. For comparison, the *experimental* value of pK_a for HF is 3.45. Of great significance is that pK_a for HF is positive compared with negative values for HCl, HBr and HI. The enthalpy of dissolution of HF ($-\Delta H^\circ$ for step(1)) is larger than those for the other hydrogen halides: $-48\,kJ\,mol^{-1}$ for HF compared with -18, -21 and $-23\,kJ\,mol^{-1}$ for HCl, HBr and HI, respectively. This, along with the much stronger bond in HF, outweighs the more negative enthalpy of hydration of F$^-$, making ΔH° for the dissociation process much less negative for HF than any of the other halides (Table 7.2). Entropy effects, although smaller, contribute in the same direction. It is easy to see that an explanation of the relative acid strengths of the hydrogen halides is not a trivial exercise. Moreover, electronegativity does *not* enter into the discussion: one must exercise care because it is all too easy to conclude from electronegativity values (see Table 2.2) that HF is expected to be the strongest acid in the series.

H$_2$S, H$_2$Se and H$_2$Te

Similar cycles to that in Fig. 7.3 can be constructed for H$_2$S, H$_2$Se and H$_2$Te, allowing values of K_a to be

estimated. Equations 7.28 to 7.30 give the first acid dissociation steps.

$$H_2S(aq) + H_2O(l) \rightleftharpoons [H_3O]^+(aq) + [HS]^-(aq)$$
$$pK_a(1) = 7.04 \qquad (7.28)$$

$$H_2Se(aq) + H_2O(l) \rightleftharpoons [H_3O]^+(aq) + [HSe]^-(aq)$$
$$pK_a(1) = 3.9 \qquad (7.29)$$

$$H_2Te(aq) + H_2O(l) \rightleftharpoons [H_3O]^+(aq) + [HTe]^-(aq)$$
$$pK_a(1) = 2.6 \qquad (7.30)$$

Although the explanation of the trend in values is not simple, and some data must be estimated (rather than being experimentally determined), it is apparent that the decrease in the X–H bond strength with the increasing atomic number of X plays an important role in accounting for what is often thought to be a puzzling observation: as group 16 is descended and X becomes more metallic, its hydride becomes more acidic.

7.6 Trends within a series of oxoacids EO$_n$(OH)$_m$

For some elements with varying oxidation states, series of oxoacids with different numbers of oxygen atoms may exist (Table 7.3). There is no adequate thermodynamic treatment for rationalizing the observed trends within a series, but there are certain empirical methods for estimating K_a. The best known of these is Bell's rule (eq. 7.31) which relates the first acid dissociation constant to the number of 'hydrogen-free' O atoms in an acid of formula EO$_n$(OH)$_m$.

$$pK_a \approx 8 - 5n \qquad (7.31)$$

Table 7.3 illustrates some comparisons between experimentally determined values of pK_a and those estimated from Bell's rule. Of course, this empirical approach does not take into account the effects of changing element E.

Table 7.3 Examples of series of oxoacids $EO_n(OH)_m$ for an element E; not all experimentally determined values of pK_a are known to the same degree of accuracy.

Formula of acid	$EO_n(OH)_m$ notation	Oxidation state of E	$pK_a(1)$	$pK_a(1)$ estimated by using Bell's rule
HNO_2	$N(O)(OH)$	+3	3.37	3
HNO_3	$N(O)_2(OH)$	+5	−1.64	−2
H_2SO_3	$S(O)(OH)_2$	+4	1.82	3
H_2SO_4	$S(O)_2(OH)_2$	+6	≈ -3	−2
$HOCl$	$Cl(OH)$	+1	7.53	8
$HClO_2$	$Cl(O)(OH)$	+3	2.0	3
$HClO_3$	$Cl(O)_2(OH)$	+5	−1.0	−2
$HClO_4$	$Cl(O)_3(OH)$	+7	≈ -8	−7

It is often the case (experimentally) that successive values of pK_a for members of a series $EO_n(OH)_m$ (e.g. HOCl, $HClO_2$, $HClO_3$ and $HClO_4$) differ by about 4 or 5. The increase in acid strength with increase in the number of O atoms attached to atom E is generally attributed to the greater possibility in the conjugate base of delocalization of negative charge onto the O atoms.

7.7 Aquated cations: formation and acidic properties

Water as a Lewis base

Although in this chapter we are mainly concerned with *Brønsted* acids and bases, it is important not to lose sight of the definition of *Lewis* acids and bases. Relevant to this chapter is the fact that water functions as a Lewis base when it acts as a solvent.

A *Lewis acid* is an electron acceptor, and a *Lewis base* is an electron donor.

When a metal salt dissolves in water, the cation and anion are hydrated. We discuss the energetics of this process in Section 7.9, but for now we consider the interactions between the individual ions (freed from their ionic lattice on dissolution) and the solvent molecules. Consider the dissolution of NaCl. Figure 7.5a shows a schematic representation of the formation of the inner hydration shell around Na^+. The $O \cdots Na$ interaction can be described in terms of an *ion–dipole interaction*, while the solvation of the anion can be described in terms of the formation of hydrogen bonds between Cl^- and H atoms of surrounding H_2O molecules.

Hydration is the specific case of solvation when the solvent is water.

Figure 7.5b shows another representation of a hexaaqua ion. Each O atom donates a pair of electrons to the metal M^{n+} ion, and each H_2O molecule acts as a Lewis base while the metal ion functions as a Lewis acid. We are implying that the M−O interaction is essentially covalent, in contrast to the case for Na^+ in Fig. 7.5a. In practice, the character of the metal $\cdots$ oxygen interaction varies with

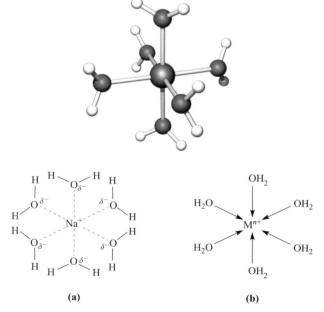

(a)   **(b)**

Fig. 7.5 Octahedral hexaaquametal ions. (a) The first hydration shell of an Na^+ ion; ion–dipole interactions operate between the Na^+ ion and the H_2O molecules. (b) If the metal–oxygen bond possesses significant covalent character, the first hydration shell can be reasonably represented showing oxygen-to-metal ion coordinate bonds; however, there is also an ionic contribution to the bonding interaction.

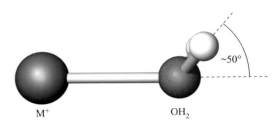

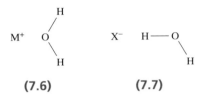

Fig. 7.6 If the plane of each water molecule in $[M(OH_2)_6]^+$ makes an angle of $\approx 50°$ with the $M^+ \cdots O$ axis, it suggests that the metal–oxygen interaction involves the use of an oxygen lone pair.

the nature of the metal ion and relevant to this is the electro-neutrality principle (see Section 19.6).

$$
\begin{array}{cc}
\text{(7.6)} & \text{(7.7)} \\
\end{array}
$$

The configurations **7.6** and **7.7** have been established in the first hydration shell for *dilute* solutions of LiCl and NaCl by neutron diffraction studies. In concentrated solutions, the plane of the water molecule in **7.6** makes an angle of up to $50°$ with the $M^+ \cdots O$ axis (Fig. 7.6) implying interaction of the cation with a lone pair of electrons rather than an ion–dipole interaction.

For both the cation and anion in NaCl, there are six H_2O molecules in the primary hydration shell (Fig. 7.5). Spectroscopic studies suggest that the hydration of other halide ions is similar to that of Cl^-. Experimental techniques that are used to investigate the hydration shells around metal ions include X-ray and neutron diffraction[†], extended X-ray absorption fine structure (EXAFS) spectroscopy and NMR (particularly ^{17}O NMR) spectroscopy. Modern computational methods (e.g. molecular dynamics) are also invaluable.[‡]

Self-study exercises

1. Suggest why the coordination number of Li^+ in its first hydration sphere in aqueous solution is 4, whereas that for Na^+ is 6.

2. EXAFS data reveal that the first hydration shell of Zn^{2+} contains six H_2O molecules ($Zn–O = 200 \, pm$) and in the

second hydration shell, there are ≈ 12 H_2O molecules ($Zn–O = 410 \, pm$). What interactions do you think are responsible for the formation of the second hydration shell?

Aquated cations as Brønsted acids

In the aqueous chemistry of cations, **hydrolysis** refers to the reversible loss of H^+ from an aqua species. The term hydrolysis is, however, also used in a wider context, e.g. the reaction:

$$PCl_3 + 3H_2O \longrightarrow H_3PO_3 + 3HCl$$

is a hydrolysis process.

Aquated cations can act as Brønsted acids by loss of H^+ from a coordinated water molecule (eq. 7.32).

$$[M(OH_2)_6]^{n+}(aq) + H_2O(l)$$
$$\rightleftharpoons [H_3O]^+(aq) + [M(OH_2)_5(OH)]^{(n-1)+}(aq) \qquad (7.32)$$

The position of the equilibrium (and thus, the strength of the acid) depends on the degree to which the $O–H$ bonds are *polarized*, and this is affected by the charge density of the cation (eq. 7.33).

$$\text{Charge density of an ion} = \frac{\text{charge on the ion}}{\text{surface area of the ion}} \qquad (7.33)$$

Surface area of sphere $= 4\pi r^2$

When H_2O coordinates to M^{n+}, charge is withdrawn towards the metal centre, leaving the H atoms more δ^+ (structure **7.8**) than in bulk water. Small cations such as Li^+, Mg^{2+}, Al^{3+}, Fe^{3+} and Ti^{3+} possess high charge densities, and in the corresponding hydrated ions, the H atoms carry significant positive charge. The pK_a values for $[Al(OH_2)_6]^{3+}$ and $[Ti(OH_2)_6]^{3+}$ (eqs. 7.34 and 7.35) illustrate the effect when the charge on the ion is high.

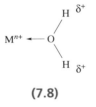

(7.8)

$$[Al(OH_2)_6]^{3+}(aq) + H_2O(l)$$
$$\rightleftharpoons [Al(OH_2)_5(OH)]^{2+}(aq) + [H_3O]^+(aq) \qquad pK_a = 5.0 \qquad (7.34)$$

$$[Ti(OH_2)_6]^{3+}(aq) + H_2O(l)$$
$$\rightleftharpoons [Ti(OH_2)_5(OH)]^{2+}(aq) + [H_3O]^+(aq) \qquad pK_a = 3.9 \qquad (7.35)$$

[†]See: G.W. Neilson *et al.* (2001) *Phil. Trans. R. Soc. Lond. A*, vol. 359, p. 1575 – 'Neutron and X-ray scattering studies of hydration in aqueous solution'.
[‡]See: S. Varma *et al.* (2006) *Biophys. Chem.*, vol. 124, p. 192 – 'Coordination numbers of alkali metal ions in aqueous solution'; J.S. Rao *et al.* (2008) *J. Phys. Chem. A*, vol. 112, p. 12944 – 'Comprehensive study on the solvation of mono- and divalent metal ions'; L.H.V. Lim *et al.* (2009) *J. Phys. Chem. B*, vol. 113, p. 4372 – 'The hydration structure of Sn(II)'.

It is instructive to compare acid strengths of hexaaqua ions with other acids. The pK_a values of $MeCO_2H$ (eq. 7.9) and $HOCl$ (eq. 7.15) are similar to that of $[Al(OH_2)_6]^{3+}$, while pK_a for $[Ti(OH_2)_6]^{3+}$ is close to that of HNO_2 (eq. 7.14).

The characteristic colour of the $[Fe(OH_2)_6]^{3+}$ ion is purple, but aqueous solutions appear yellow due to the formation of the hydroxido species $[Fe(OH_2)_5(OH)]^{2+}$ and $[Fe(OH_2)_4(OH)_2]^+$ (eqs. 7.36 and 7.37); see also structure 21.34 in Chapter 21 and accompanying discussion.

$$[Fe(OH_2)_6]^{3+}(aq) + H_2O(l)$$
$$\rightleftharpoons [Fe(OH_2)_5(OH)]^{2+}(aq) + [H_3O]^+(aq) \qquad pK_a = 2.0$$
$$(7.36)$$

$$[Fe(OH_2)_5(OH)]^{2+}(aq) + H_2O(l)$$
$$\rightleftharpoons [Fe(OH_2)_4(OH)_2]^+(aq) + [H_3O]^+(aq) \qquad pK_a = 3.3$$
$$(7.37)$$

The facile acid dissociation of $[Fe(OH_2)_6]^{3+}$ means that its aqueous solutions must be stabilized by the addition of acid, which (by Le Chatelier's principle) drives equilibrium 7.36 to the left-hand side.

Proton loss is, in some cases, accompanied by the formation of dinuclear or polynuclear species in aqueous solution. For example, after the dissociation of H^+ from $[Cr(OH_2)_6]^{3+}$, the product undergoes an intermolecular condensation (eq. 7.38). The resulting dichromium species (Fig. 7.7) contains *bridging*[†] hydroxy groups.

$$2[Cr(OH_2)_5(OH)]^{2+}(aq)$$
$$\rightleftharpoons [(H_2O)_4Cr(\mu\text{-}OH)_2Cr(OH_2)_4]^{4+}(aq) + 2H_2O(l)$$
$$(7.38)$$

A similar reaction occurs in the corresponding V(III) system. On going from V(III) to V(IV), the charge density on the vanadium centre increases. As a result, the dissociation of two protons from *one* coordinated H_2O occurs, and the blue oxidovanadium(IV) ion, **7.9**, is formed. It is common for this cation to be written simply as $[VO]^{2+}$, even though this is not a 'naked' vanadium oxido species.

(7.9)

[†] The prefix μ means that the specified group is in a *bridging* position; μ_3 means a bridge between three atoms, etc.

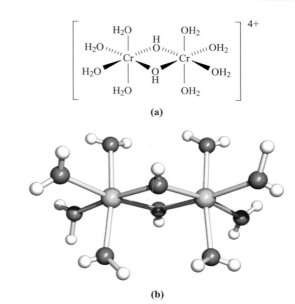

Fig. 7.7 (a) A schematic representation of the structure of the dinuclear cation $[Cr_2(\mu\text{-}OH)_2(OH_2)_8]^{4+}$. (b) The structure (X-ray diffraction) of this cation as determined for the salt $[Cr_2(\mu\text{-}OH)_2(OH_2)_8][2,4,6\text{-}Me_3C_6H_2SO_3]_4\cdot4H_2O$ [L. Spiccia *et al.* (1987) *Inorg. Chem.*, vol. 26, p. 474]. Colour code: Cr, yellow; O, red; H, white.

Self-study exercises

1. Suggest why Be^{2+} has a higher affinity for H_2O molecules than Mg^{2+}.

2. Explain why $[Ga(OH_2)_6]^{3+}$ is acidic ($pK_a = 2.6$).

7.8 Amphoteric oxides and hydroxides

Amphoteric behaviour

If an oxide or hydroxide is able to act as either an acid or a base, it is said to be *amphoteric*.

Some oxides and hydroxides are able to react with both acids and bases, thereby functioning as both bases and acids, respectively. Water is probably the most common example, but in this section we consider the *amphoteric* nature of metal oxides and hydroxides. The γ-form of aluminium oxide, γ-Al_2O_3, reacts with acids (eq. 7.39) and with hydroxide ions (eq. 7.40).[†]

$$\gamma\text{-}Al_2O_3(s) + 3H_2O(l) + 6[H_3O]^+(aq) \longrightarrow 2[Al(OH_2)_6]^{3+}(aq)$$
$$(7.39)$$

$$\gamma\text{-}Al_2O_3(s) + 3H_2O(l) + 2[OH]^-(aq) \longrightarrow 2[Al(OH)_4]^-(aq)$$
$$(7.40)$$

[†] The α-form of aluminium oxide is resistant to attack by acids (see Section 13.7).

The hexaaqua ion, **7.10**, may be isolated as, for example, the sulfate salt after reaction with H_2SO_4. The ion $[Al(OH)_4]^-$, **7.11**, can be isolated as, for example, the Na^+ salt if the source of hydroxide is NaOH.

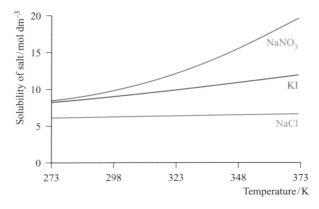

(7.10)   **(7.11)**

Similarly, aluminium hydroxide is amphoteric (eqs. 7.41 and 7.42).

$$Al(OH)_3(s) + KOH(aq) \longrightarrow K[Al(OH)_4](aq) \qquad (7.41)$$

$$Al(OH)_3(s) + 3HNO_3(aq) \longrightarrow Al(NO_3)_3(aq) + 3H_2O(l) \qquad (7.42)$$

Periodic trends in amphoteric properties

As we discuss in later chapters, the character of the oxides of the elements across a row of the periodic table (*s*- and *p*-blocks) changes from basic to acidic, consistent with a change from metallic to non-metallic character of the element. Elements that lie close to the so-called 'diagonal line' (Fig. 7.8) possess amphoteric oxides and hydroxides. In group 2, $Be(OH)_2$ and BeO are amphoteric, but $M(OH)_2$ and MO (M = Mg, Ca, Sr or Ba) are basic. Among the oxides of the *p*-block, Al_2O_3, Ga_2O_3, In_2O_3, GeO, GeO_2, SnO, SnO_2, PbO, PbO_2, As_2O_3, Sb_2O_3 and Bi_2O_3 are amphoteric. Within group 13, Ga_2O_3 is more acidic than Al_2O_3, whereas In_2O_3 is more *basic* than either Al_2O_3 or Ga_2O_3. For most of its chemistry, In_2O_3 can be regarded as having a basic rather than amphoteric

nature. In group 14, both the metal(II) and metal(IV) oxides of Ge, Sn and Pb are amphoteric. In group 15, only the lower oxidation state oxides exhibit amphoteric behaviour, with the M_2O_5 oxides being acidic. For the oxides M_2O_3, basic character predominates as the group is descended: $As_2O_3 < Sb_2O_3 < Bi_2O_3$.

7.9 Solubilities of ionic salts

Solubility and saturated solutions

When an ionic solid, MX, is added to water, equilibrium 7.43 is established (if the ions formed are singly charged). When equilibrium is reached, the solution is *saturated*.

$$MX(s) \rightleftharpoons M^+(aq) + X^-(aq) \qquad (7.43)$$

The *solubility* of a solid *at a specified temperature* is the amount of solid (*solute*) that dissolves in a specified amount of solvent when equilibrium is reached in the presence of excess solid. The solubility may be expressed in several ways, for example:

- mass of solute in a given mass of solvent (g of solute per 100 g of water);
- moles of solute in a given mass of solvent;
- concentration (mol dm^{-3});
- molality (mol kg^{-1});
- mole fraction.

It is crucial to state the temperature, since solubility may depend significantly on temperature as illustrated in Fig. 7.9 for KI and $NaNO_3$. In contrast, Fig. 7.9 shows that between 273 and 373 K, the solubility of NaCl is essentially constant.

> Tabulated values of *solubilities of ionic salts* refer to the maximum amount of solid that will dissolve in a given mass of water to give a saturated solution. Solubilities may also be expressed in concentrations, molalities or mole fractions.

Group 1	Group 2		Group 13	Group 14	Group 15	Group 16	Group 17	Group 18
Li	Be		B	C	N	O	F	Ne
Na	Mg		Al	Si	P	S	Cl	Ar
K	Ca	*d*-block	Ga	Ge	As	Se	Br	Kr
Rb	Sr		In	Sn	Sb	Te	I	Xe
Cs	Ba		Tl	Pb	Bi	Po	At	Rn

■ = Non-metallic elements   ☐ = Metallic elements

Fig. 7.8 The so-called 'diagonal line' divides metals from non-metals, although some elements that lie next to the line (e.g. Si) are semi-metals.

Fig. 7.9 The temperature-dependence of the solubilities in water of KI and $NaNO_3$. The solubility of NaCl is essentially temperature independent in the range 273–373 K.

For very dilute solutions at 298 K, the numerical value of a concentration in $mol\,kg^{-1}$ is equal to that in $mol\,dm^{-3}$, and the solubilities of sparingly soluble salts (see below) are generally expressed in $mol\,dm^{-3}$.

Sparingly soluble salts and solubility products

If the solubility of an ionic salt is extremely small (i.e. a saturated solution contains very few ions), the salt is said to be *sparingly soluble*. Such salts may include some that we might loosely refer to as being 'insoluble', for example AgCl and $BaSO_4$. Equation 7.44 shows the equilibrium that is established in aqueous solution when CaF_2 dissolves.

$$CaF_2(s) \rightleftharpoons Ca^{2+}(aq) + 2F^-(aq) \qquad (7.44)$$

An expression for the equilibrium constant should strictly be given in terms of the activities (see Section 7.3) of the species involved, but since we are dealing with very dilute solutions, we may express K in terms of concentrations (eq. 7.45).

$$K = \frac{[Ca^{2+}][F^-]^2}{[CaF_2]} \qquad (7.45)$$

The activity of any solid is, *by convention*, unity. The equilibrium constant is thereby given in terms of the *equilibrium concentrations* of the dissolved ions and is referred to as the *solubility product*, or *solubility constant*, K_{sp} (eq. 7.46).

$$K_{sp} = [Ca^{2+}][F^-]^2 \qquad (7.46)$$

Values of K_{sp} for a range of sparingly soluble salts are listed in Table 7.4.

Worked example 7.4   Solubility product

The solubility product for PbI_2 is 8.49×10^{-9} (298 K). Calculate the solubility of PbI_2.

The equilibrium for the dissolution of lead(II) iodide is:

$$PbI_2(s) \rightleftharpoons Pb^{2+}(aq) + 2I^-(aq)$$

$$K_{sp} = [Pb^{2+}][I^-]^2$$

One mole of PbI_2 dissolves to give one mole of Pb^{2+} and two moles of I^-, and the solubility of PbI_2 (in $mol\,dm^{-3}$) equals the concentration of aqueous Pb^{2+}. Since $[I^-] = 2[Pb^{2+}]$, we can rewrite the expression for K_{sp}, and thus find $[Pb^{2+}]$:

$$K_{sp} = 4[Pb^{2+}]^3$$

$$8.49 \times 10^{-9} = 4[Pb^{2+}]^3$$

Table 7.4   Values of K_{sp}(298 K) for selected sparingly soluble salts.

Compound	Formula	K_{sp}(298 K)
Barium sulfate	$BaSO_4$	1.07×10^{-10}
Calcium carbonate	$CaCO_3$	4.96×10^{-9}
Calcium hydroxide	$Ca(OH)_2$	4.68×10^{-6}
Calcium phosphate	$Ca_3(PO_4)_2$	2.07×10^{-33}
Iron(II) hydroxide	$Fe(OH)_2$	4.87×10^{-17}
Iron(II) sulfide	FeS	6.00×10^{-19}
Iron(III) hydroxide	$Fe(OH)_3$	2.64×10^{-39}
Lead(II) iodide	PbI_2	8.49×10^{-9}
Lead(II) sulfide	PbS	3.00×10^{-28}
Magnesium carbonate	$MgCO_3$	6.82×10^{-6}
Magnesium hydroxide	$Mg(OH)_2$	5.61×10^{-12}
Silver(I) chloride	$AgCl$	1.77×10^{-10}
Silver(I) bromide	$AgBr$	5.35×10^{-13}
Silver(I) iodide	AgI	8.51×10^{-17}
Silver(I) chromate	Ag_2CrO_4	1.12×10^{-12}
Silver(I) sulfate	Ag_2SO_4	1.20×10^{-5}

$$[Pb^{2+}] = \sqrt[3]{2.12 \times 10^{-9}} = 1.28 \times 10^{-3}\,mol\,dm^{-3}$$

The solubility of PbI_2 is thus $1.28 \times 10^{-3}\,mol\,dm^{-3}$ at 298 K.

Self-study exercises

1. The solubility product for Ag_2SO_4 is 1.20×10^{-5} (298 K). What is the solubility of Ag_2SO_4 in (a) $mol\,dm^{-3}$, and (b) g per 100 g of water?
   [*Ans.* (a) $1.44 \times 10^{-2}\,mol\,dm^{-3}$; (b) 0.45 g per 100 g]

2. If the solubility of AgI is $2.17 \times 10^{-6}\,g\,dm^{-3}$, calculate K_{sp}.   [*Ans.* 8.50×10^{-17}]

3. The value of K_{sp} for lithium carbonate is 8.15×10^{-4} (298 K). Calculate the solubility of Li_2CO_3 in (a) $mol\,dm^{-3}$ and (b) g per 100 g of water.
   [*Ans.* (a) $5.88 \times 10^{-2}\,mol\,dm^{-3}$; (b) 0.434 g per 100 g]

4. The solubility of iron(II) hydroxide in water is $2.30 \times 10^{-6}\,mol\,dm^{-3}$ at 298 K. Determine the equilibrium constant for the process:

   $$Fe(OH)_2(s) \rightleftharpoons Fe^{2+}(aq) + 2[OH]^-(aq)$$
   [*Ans.* 4.87×10^{-17}]

The energetics of the dissolution of an ionic salt: $\Delta_{sol}G^o$

We can consider the equilibrium between a solid salt MX and its ions in saturated aqueous solution in terms of the thermodynamic cycle in eq. 7.47.

$$
\begin{array}{c}
\text{MX(s)} \xrightarrow{-\Delta_{\text{lattice}}G^o} \text{M}^+(\text{g}) \;+\; \text{X}^-(\text{g}) \\[2mm]
\Delta_{\text{sol}}G^o \searrow \qquad \swarrow \Delta_{\text{hyd}}G^o \\[2mm]
\text{M}^+(\text{aq}) \;+\; \text{X}^-(\text{aq})
\end{array}
\qquad (7.47)
$$

where: $\Delta_{\text{lattice}}G^o$ = standard Gibbs energy change accompanying the formation of the ionic lattice from gaseous ions; $\Delta_{\text{hyd}}G^o$ = standard Gibbs energy change accompanying the hydration of the gaseous ions; and $\Delta_{\text{sol}}G^o$ = standard Gibbs energy change accompanying the dissolution of the ionic salt.

In this cycle, $\Delta_{\text{sol}}G^o$ is related by eq. 7.48 to the equilibrium constant, K, for the dissolution process; for a sparingly soluble salt, the equilibrium constant is K_{sp}.

$$\Delta_{\text{sol}}G^o = -RT \ln K \qquad (7.48)$$

In principle, it is possible to use Gibbs energy data to calculate values of K and this is particularly valuable for accessing values of K_{sp}. However, there are two difficulties with determining values of $\Delta_{\text{sol}}G^o$ using cycle 7.47. First, $\Delta_{\text{sol}}G^o$ is a small difference between two much larger quantities (eq. 7.49), neither of which is usually accurately known. The situation is made worse by the exponential relationship between K and $\Delta_{\text{sol}}G^o$. Second, hydration energies are not very accessible quantities, as we shall discuss later on.

$$\Delta_{\text{sol}}G^o = \Delta_{\text{hyd}}G^o - \Delta_{\text{lattice}}G^o \qquad (7.49)$$

An alternative method of accessing values of $\Delta_{\text{sol}}G^o$ is by using eq. 7.50, which relates the energies of formation for the species involved to the energy change for the dissolution of MX(s) (reaction 7.43).

$$\Delta_{\text{sol}}G^o = \Delta_{\text{f}}G^o(\text{M}^+, \text{ aq}) + \Delta_{\text{f}}G^o(\text{X}^-, \text{ aq}) - \Delta_{\text{f}}G^o(\text{MX, s}) \qquad (7.50)$$

Values of $\Delta_{\text{f}}G^o(\text{M}^+, \text{ aq})$ and $\Delta_{\text{f}}G^o(\text{X}^-, \text{ aq})$ can often be determined from standard reduction potentials (see Appendix 11) using eq. 7.51, and tables giving values of $\Delta_{\text{f}}G^o(\text{MX, s})$ for a wide range of salts are readily available. Equation 7.51 and its uses are discussed in detail in Chapter 8, and worked example 8.9 is especially relevant.

$$\Delta G^o = -zFE^o \qquad (7.51)$$

where F = Faraday constant = $96\,485\,\text{C}\,\text{mol}^{-1}$

The magnitude of $\Delta_{\text{sol}}G^o$ depends upon the balance between the corresponding $T\Delta_{\text{sol}}S^o$ and $\Delta_{\text{sol}}H^o$ terms (eq. 7.52).

$$\Delta_{\text{sol}}G^o = \Delta_{\text{sol}}H^o - T\Delta_{\text{sol}}S^o \qquad (7.52)$$

Thermochemical experiments (i.e. measuring the heat evolved or taken in during dissolution of an ionic salt) provide a method of determining values of the enthalpy change, $\Delta_{\text{sol}}H^o$. If $\Delta_{\text{sol}}G^o$ has been determined, then $\Delta_{\text{sol}}S^o$ can be derived using eq. 7.52. Observed trends in the values of these thermodynamic parameters are not easily discussed, since a wide variety of factors contribute to the signs and magnitudes of $\Delta_{\text{sol}}S^o$ and $\Delta_{\text{sol}}H^o$, and hence to $\Delta_{\text{sol}}G^o$ and the actual solubility of a given salt. Table 7.5 lists relevant data for sodium and silver halides. The increase in solubility on going from NaF to NaBr corresponds to a progressively more negative value for $\Delta_{\text{sol}}G^o$, and the $\Delta_{\text{sol}}H^o$ and $T\Delta_{\text{sol}}S^o$ terms *both* contribute to this trend. In contrast, the silver halides show the opposite behaviour, with the solubility in aqueous solution following the sequence AgF > AgCl > AgBr > AgI. While the values of the $T\Delta_{\text{sol}}S^o$ term become more positive on going from AgF to AgI (i.e. the same trend as for the sodium halides), the $\Delta_{\text{sol}}H^o$ term also becomes more positive. Combined in eq. 7.52, these lead to values of $\Delta_{\text{sol}}G^o$ for AgF, AgCl, AgBr and AgI that become increasingly positive (Table 7.5). The origin of this result lies in the non-electrostatic contribution to the lattice energy, which progressively stabilizes the solid with respect to aqueous ions on going from AgF to AgI (see Section 6.15). Even from a consideration of only two sets of metal halides, it is clear that providing general explanations for the observed trends in the solubilities of ionic salts is not possible.

The energetics of the dissolution of an ionic salt: hydration of ions

We have already seen (eq. 7.47) that the energy change accompanying the hydration of an ionic salt contributes towards the solubility of the salt, and we have also mentioned that values of $\Delta_{\text{hyd}}G^o$ and the corresponding enthalpy and entropy changes are not readily accessible quantities. In this section, we look more closely at $\Delta_{\text{hyd}}G^o$, $\Delta_{\text{hyd}}H^o$ and $\Delta_{\text{hyd}}S^o$; eq. 7.53 gives the general hydration processes to which these quantities refer.

$$
\left.
\begin{array}{l}
\text{M}^+(\text{g}) \longrightarrow \text{M}^+(\text{aq}) \\
\text{X}^-(\text{g}) \longrightarrow \text{X}^-(\text{aq})
\end{array}
\right\}
\qquad (7.53)
$$

The primary problem is that individual ions cannot be studied in isolation, and experimental measurements of $\Delta_{\text{hyd}}H^o$ are restricted to those involving pairs of ions that do not interact. Even then, the problem is non-trivial.

Table 7.5 Solubilities and values of the changes in Gibbs energy, enthalpy and entropy of solution at 298 K for the halides of sodium and silver; the entropy change is given in the form of a $T\Delta_{sol}S^{\circ}$ term ($T = 298$ K). Hydrate formation by solid NaBr, NaI and AgF has been neglected in the calculation of $\Delta_{sol}G^{\circ}$ for these compounds.

Compound	Solubility / g per 100 g of water at 298 K	Solubility / mol dm^{-3} at 298 K	$\Delta_{sol}G^{\circ}$ / kJ mol^{-1}	$\Delta_{sol}H^{\circ}$ / kJ mol^{-1}	$T\Delta_{sol}S^{\circ}$ / kJ mol^{-1} (for $T = 298$ K)
NaF	4.2	1.0	+7.9	+0.9	−7.0
NaCl	36	6.2	−8.6	+3.9	+12.5
NaBr	91	8.8	−17.7	−0.6	+17.1
NaI	184	12.3	−31.1	−7.6	+23.5
AgF	182	14.3	−14.4	−20.3	−5.9
AgCl	1.91×10^{-4}	1.33×10^{-5}	+55.6	+65.4	+9.8
AgBr	1.37×10^{-5}	7.31×10^{-7}	+70.2	+84.4	+14.2
AgI	2.16×10^{-7}	9.22×10^{-9}	+91.7	+112.3	+20.6

In principle, the value of $\Delta_{hyd}G^{\circ}$ (in J mol^{-1}) for an ion of charge ze and radius r_{ion} (in m) can be calculated on the basis of electrostatics using eq. 7.54.

$$\Delta_{hyd}G^{\circ} = -\frac{Lz^2e^2}{8\pi\varepsilon_0 r_{ion}}\left(1 - \frac{1}{\varepsilon_r}\right) \qquad (7.54)$$

where: L = Avogadro number = 6.022×10^{23} mol^{-1};
e = charge on the electron = 1.602×10^{-19} C;
ε_0 = permittivity of a vacuum = 8.854×10^{-12} F m^{-1};
and ε_r = relative permittivity of the water (dielectric constant) = 78.7

In practice, this expression gives unsatisfactory results since the relative permittivity (see Section 9.2) of bulk water is not valid close to the ion, and available values of r_{ion} refer to ionic lattices rather than hydrated ions.

The simplest way of obtaining thermodynamic functions of hydration for individual ions rests on the assumption that very large ions such as [Ph$_4$As]$^+$ and [BPh$_4$]$^-$ have the same values of $\Delta_{hyd}G^{\circ}$ etc. From data for salts containing appropriate cation–anion pairs (e.g. [Ph$_4$As][BPh$_4$], [Ph$_4$As]Cl and K[BPh$_4$]), data for individual ions can be derived (e.g. K$^+$ and Cl$^-$). However, direct *experimental* measurements involving [Ph$_4$As][BPh$_4$] are not feasible because of the low solubility of this salt in water. Hence, data for this compound come from theory.

An alternative method for obtaining thermodynamic functions of hydration is based upon an arbitrary assignment of a value of $\Delta_{hyd}H^{\circ}(H^+, g) = 0$. From this starting point, and using values of $\Delta_{hyd}H^{\circ}$ for a range of ionic *salts* and the hydrogen halides, a self-consistent set of *relative* hydration enthalpies can be obtained. More sophisticated methods are based upon the estimation of $\Delta_{hyd}H^{\circ}(H^+, g) =$

-1091 kJ mol^{-1}, and Table 7.6 lists corresponding absolute values of $\Delta_{hyd}H^{\circ}$ for a range of ions.

Values of hydration entropies, $\Delta_{hyd}S^{\circ}$, can be derived by assigning (by convention) a value of zero for the absolute entropy, S°, of gaseous H$^+$. Table 7.6 lists values of $\Delta_{hyd}S^{\circ}$ for selected ions, and the corresponding values of $\Delta_{hyd}G^{\circ}$ are obtained by substitution of $\Delta_{hyd}S^{\circ}$ and $\Delta_{hyd}H^{\circ}$ into eq. 7.52 ($T = 298$ K). Inspection of Table 7.6 reveals several points of interest:

- Highly charged ions have more negative values of $\Delta_{hyd}H^{\circ}$ and $\Delta_{hyd}S^{\circ}$ than singly charged ions. The more negative enthalpy term is rationalized in terms of simple electrostatic attraction, and the more negative $\Delta_{hyd}S^{\circ}$ values can be considered in terms of highly charged ions imposing more order on H$_2$O molecules close to the ion.
- For ions of a given charge, $\Delta_{hyd}H^{\circ}$ and $\Delta_{hyd}S^{\circ}$ show some dependence on ion size (i.e. r_{ion}); smaller ions possess more negative values of both $\Delta_{hyd}H^{\circ}$ and $\Delta_{hyd}S^{\circ}$.
- The variation in $\Delta_{hyd}H^{\circ}$ outweighs that in $T\Delta_{hyd}S^{\circ}$, and as a result, the most negative values of $\Delta_{hyd}G^{\circ}$ arise for small ions (comparing those with a constant charge), and for highly charged ions (comparing those of similar size).
- For monatomic ions of about the same size (e.g. K$^+$ and F$^-$), anions are more strongly hydrated than cations (more negative $\Delta_{hyd}G^{\circ}$).

Solubilities: some concluding remarks

Let us now return to eq. 7.47, and relate the observed solubility of a salt to the magnitude of the difference between $\Delta_{lattice}G^{\circ}$ and $\Delta_{hyd}G^{\circ}$ (eq. 7.49), and in particular to the sizes of the ions involved.

Table 7.6 Absolute values of $\Delta_{hyd}H^o$, $\Delta_{hyd}S^o$, $\Delta_{hyd}G^o$ (at 298 K), and ionic radii for selected ions.

Ion	$\Delta_{hyd}H^o$ / kJ mol^{-1}	$\Delta_{hyd}S^o$ / J K^{-1} mol^{-1}	$T\Delta_{hyd}S^o$ / kJ mol^{-1} (for $T = 298$ K)	$\Delta_{hyd}G^o$ / kJ mol^{-1}	r_{ion} / pm[†]
H$^+$	−1091	−130	−39	−1052	−
Li$^+$	−519	−140	−42	−477	76
Na$^+$	−404	−110	−33	−371	102
K$^+$	−321	−70	−21	−300	138
Rb$^+$	−296	−70	−21	−275	149
Cs$^+$	−271	−60	−18	−253	170
Mg^{2+}	−1931	−320	−95	−1836	72
Ca^{2+}	−1586	−230	−69	−1517	100
Sr^{2+}	−1456	−220	−66	−1390	126
Ba^{2+}	−1316	−200	−60	−1256	142
Al^{3+}	−4691	−530	−158	−4533	54
La^{3+}	−3291	−430	−128	−3163	105
F$^-$	−504	−150	−45	−459	133
Cl$^-$	−361	−90	−27	−334	181
Br$^-$	−330	−70	−21	−309	196
I$^-$	−285	−50	−15	−270	220

[†]Values of r_{ion} refer to a coordination number of 6 in the solid state.

First, we reiterate that $\Delta_{sol}G^o$ is generally a *relatively* small value, being the difference between two much larger values ($\Delta_{lattice}G^o$ and $\Delta_{hyd}G^o$). Moreover, as Table 7.5 illustrates, $\Delta_{sol}G^o$ can be either positive or negative, whereas $\Delta_{lattice}G^o$ and $\Delta_{hyd}G^o$ are always negative values (provided they are defined as in eq. 7.47).

As Table 7.6 shows, of the two terms $\Delta_{hyd}H^o$ and $T\Delta_{hyd}S^o$, the dominant factor in determining the magnitude of $\Delta_{hyd}G^o$ is $\Delta_{hyd}H^o$. Similarly, for $\Delta_{lattice}G^o$, the dominant factor is $\Delta_{lattice}H^o$. Thus, in considering the relationship between the solubility of a salt and the sizes of the component ions, we turn our attention to the relationships between r_{ion}, $\Delta_{hyd}H^o$ and $\Delta_{lattice}H^o$ given in eq. 7.55 and 7.56. The actual *values* of $\Delta_{hyd}H^o$ and $\Delta_{lattice}H^o$ (defined for the processes given in eq. 7.47) are always negative.

$$\Delta_{lattice}H^o \propto \frac{1}{r_+ + r_-} \tag{7.55}$$

$$\Delta_{hyd}H^o \propto \frac{1}{r_+} + \frac{1}{r_-} \tag{7.56}$$

where r_+ = radius of cation; r_- = radius of anion

Now consider the application of these two expressions to a series of salts of similar lattice type. For a series of MX salts

where X$^-$ is constant and M$^+$ varies, if $r_- \gg r_+$, eq. 7.55 shows that there will be little variation in $\Delta_{lattice}H^o$. However, upon dissolution, if $r_- \gg r_+$, $\Delta_{hyd}H^o$(cation) will be much more negative than $\Delta_{hyd}H^o$(anion) for all values of r_+. Thus, $\Delta_{hyd}H^o$(MX) will be roughly proportional to $\frac{1}{r_+}$. Thus, along a series of related salts with increasing r_+, but with $r_- \gg r_+$, $\Delta_{lattice}H^o$ will remain nearly constant while $\Delta_{hyd}H^o$ becomes *less* negative. Hence, $\Delta_{sol}H^o$ (and thus $\Delta_{sol}G^o$) will become less negative (eq. 7.57) and solubility will decrease.

$$\Delta_{sol}H^o = \Delta_{hyd}H^o - \Delta_{lattice}H^o \tag{7.57}$$

Such a series is exemplified by the alkali metal hexachloridoplatinates. The hydrated sodium salt has a very high solubility, while the solubilities of K$_2$[PtCl$_6$], Rb$_2$[PtCl$_6$] and Cs$_2$[PtCl$_6$] are 2.30×10^{-2}, 2.44×10^{-3} and 1.04×10^{-3} mol dm^{-3} (at 293 K). A similar trend is observed for alkali metal hexafluoridophosphates (MPF$_6$).

Although the above, and similar, arguments are qualitative, they provide a helpful means of assessing the pattern in solubilities for series of *ionic* salts. We stress 'ionic' because eqs. 7.55 and 7.56 assume an electrostatic model.

Our discussions in Section 6.15 and earlier in this section indicated how partial covalent character in silver halides affects solubility trends.

Self-study exercise

At room temperature, the solubilities in water of $LiIO_3$, $NaIO_3$ and KIO_3 are 4.4, 0.45 and 0.22 $mol\,dm^{-3}$, respectively. Rationalize this trend.

7.10 Common-ion effect

So far, we have focused on aqueous solutions containing a single, dissolved ionic salt, MX. Now we consider the effect of adding a second salt which has one of its ions in common with the first salt.

If a salt MX is added to an aqueous solution containing the solute MY (the ion M^{n+} is common to both salts), the presence of the dissolved M^{n+} ions suppresses the dissolution of MX compared with that in pure water; this is the *common-ion effect.*

The origin of the common-ion effect is seen by applying Le Chatelier's principle. In eq. 7.58, the presence of Cl^- in solution (from a soluble salt such as KCl) will suppress the dissolution of AgCl, i.e. additional Cl^- ions will shift the equilibrium to the left-hand side.

$$AgCl(s) \rightleftharpoons Ag^+(aq) + Cl^-(aq) \qquad (7.58)$$

The effect is analogous to that of mixing a weak acid with the salt of that acid (e.g. acetic acid and sodium acetate) to form a buffer solution.

Worked example 7.5   The common-ion effect

The value of K_{sp} for AgCl is 1.77×10^{-10} (at 298 K). Compare the solubility of AgCl in water and in $0.0100\,mol\,dm^{-3}$ hydrochloric acid.

First, determine the solubility of AgCl in water.

$$AgCl(s) \rightleftharpoons Ag^+(aq) + Cl^-(aq)$$

$$K_{sp} = [Ag^+][Cl^-] = 1.77 \times 10^{-10}$$

Since the concentrations of $[Ag^+]$ and $[Cl^-]$ in aqueous solution are equal, we can write:

$$[Ag^+]^2 = 1.77 \times 10^{-10}$$

$$[Ag^+] = 1.33 \times 10^{-5}\,mol\,dm^{-3}$$

The solubility of AgCl is therefore $1.33 \times 10^{-5}\,mol\,dm^{-3}$.

Now consider the solubility of AgCl in $0.0100\,mol\,dm^{-3}$ aqueous HCl solution. HCl is essentially fully dissociated and thus, $[Cl^-] = 0.0100\,mol\,dm^{-3}$.

$$AgCl(s) \rightleftharpoons Ag^+(aq) + Cl^-(aq)$$

Initial aqueous ion concentrations/ $mol\,dm^{-3}$:   0   0.0100

Equilibrium concentrations/ $mol\,dm^{-3}$: x $(0.0100 + x)$

$$K_{sp} = 1.77 \times 10^{-10} = [Ag^+][Cl^-]$$

$$1.77 \times 10^{-10} = x(0.0100 + x)$$

Since x is obviously much less than 0.0100, we can make the approximation that $0.0100 + x \approx 0.0100$.

$$1.77 \times 10^{-10} \approx 0.0100x$$

$$x \approx 1.77 \times 10^{-8}\,mol\,dm^{-3}$$

The solubility of AgCl is therefore $1.77 \times 10^{-8}\,mol\,dm^{-3}$.

Conclusion: the solubility of AgCl is ≈ 1000 times less in $0.0100\,mol\,dm^{-3}$ aqueous HCl solution than in water.

Self-study exercises

K_{sp} data: AgCl, 1.77×10^{-10}; $BaSO_4$, 1.07×10^{-10} (298 K).

1. How much more soluble is AgCl in water than in $5.00 \times 10^{-3}\,mol\,dm^{-3}$ aqueous HCl at 298 K?
   [*Ans.* ≈ 375 times]

2. What is the solubility of AgCl in $0.0200\,mol\,dm^{-3}$ aqueous KCl?   [*Ans.* $8.85 \times 10^{-9}\,mol\,dm^{-3}$]

3. What is the solubility of $BaSO_4$ (at 298 K) in (a) water and (b) in $0.0150\,mol\,dm^{-3}$ aqueous Na_2SO_4.
   [*Ans.* (a) $1.03 \times 10^{-5}\,mol\,dm^{-3}$; (b) $7.13 \times 10^{-9}\,mol\,dm^{-3}$]

Worked example 7.5 illustrates the use of the common-ion effect in gravimetric analysis. AgCl is always precipitated from a solution containing a slight excess of a common ion, Cl^- or Ag^+, in the determination of silver or chloride respectively.

Gravimetric analysis is a quantitative technique in which the material under study is isolated as a precipitate.

7.11 Coordination complexes: an introduction

Definitions and terminology

In this section we introduce some general principles concerning the coordination of *ligands* to ions in aqueous solution. These definitions and principles will be used again when we discuss complex formation in detail later in the book. The word *ligand* is derived from the Latin verb 'ligare' meaning 'to bind'.

> In a ***coordination complex***, a central atom or ion is coordinated by one or more molecules or ions (***ligands***) which act as Lewis bases, forming ***coordinate bonds*** with the central atom or ion; the latter acts as a Lewis acid. Atoms in the ligands that are directly bonded to the central atom or ion are ***donor atoms***.

Examples of coordination complexes include those involving *d*-block metal ions (e.g. $[Co(NH_3)_6]^{2+}$, **7.12**) and species with a central *p*-block element (e.g. $[BF_4]^-$, **7.13**, and $H_3B\cdot THF$, **7.14**) (THF = tetrahydrofuran), although **7.14** is unstable with respect to hydrolysis in aqueous solution. Equations 7.59–7.61 show the formation of these coordination complexes.

> In a complex:
> • a *line* is strictly used to denote the interaction between an *anionic* ligand and the acceptor;
> • an *arrow* is strictly used to show the donation of an electron pair from a *neutral* ligand to an acceptor.

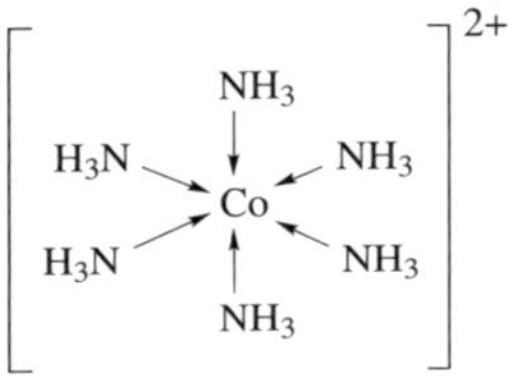

(7.12) 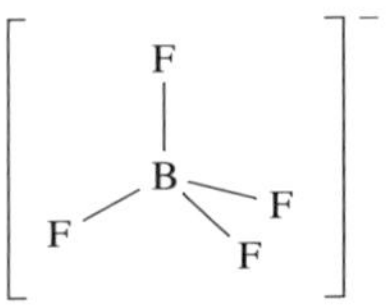   **(7.13)**

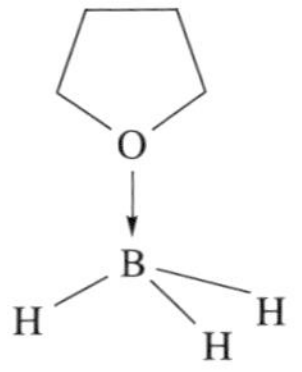

(7.14)

$$[Co(OH_2)_6]^{2+} + 6NH_3 \rightleftharpoons [Co(NH_3)_6]^{2+} + 6H_2O \quad (7.59)$$

$$BF_3 + F^- \rightleftharpoons [BF_4]^- \quad (7.60)$$

$$\tfrac{1}{2}B_2H_6 + THF \rightleftharpoons H_3B\cdot THF \quad (7.61)$$

> When a Lewis base donates a pair of electrons to a Lewis acid, a ***coordinate bond*** is formed and the resulting species is an ***adduct***. The centred dot in, for example, $H_3B\cdot THF$ indicates the formation of an adduct.

In $[BF_4]^-$, the B—F bond formed in reaction 7.60 *is identical* to the other three B—F bonds; all are 2c-2e covalent bonds. In structures **7.12–7.14**, the coordinate bond between the central atom or ion and a *neutral ligand* is denoted by an *arrow*, but if the ligand is *anionic*, the coordinate bond is indicated by a line. However, this convention is often ignored, for example, when the stereochemistry of the coordination complex is illustrated: compare **7.12** with **7.15**:

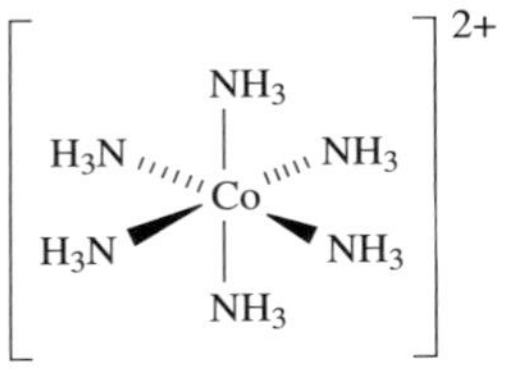

(7.15)

Investigating coordination complex formation

The formation of complexes in aqueous solution may be studied by a number of methods. Physical methods (e.g. electronic and vibrational spectroscopic, solubility or conductivity measurements) usually provide reliable information and, in some cases, allow the determination of equilibrium constants for complex formation. It is also possible to test for modifications of chemical properties, but this has to be carried out with caution. *All* reactions are equilibria, and chemical tests are often only investigations of *relative* values of equilibrium constants. For example, in an aqueous solution of an Ag^+ salt saturated with NH_3, nearly all the Ag^+ is present as the complex $[Ag(NH_3)_2]^+$ (eq. 7.62).

$$Ag^+(aq) + 2NH_3(aq) \rightleftharpoons [Ag(NH_3)_2]^+(aq) \quad (7.62)$$

On adding a chloride-containing solution, *no* AgCl precipitate is observed. However, the addition of an iodide-containing solution results in the precipitation of AgI. These observations can be rationalized as follows: AgI ($K_{sp} = 8.51 \times 10^{-17}$) is much less soluble in aqueous solution than AgCl ($K_{sp} = 1.77 \times 10^{-10}$). The fact that no AgCl is precipitated means that the equilibrium constant for reaction 7.62 is sufficiently large that the AgCl formed is soluble in the solution (i.e. very little uncomplexed Ag^+ is available for combination with Cl^-). On the other hand, the solubility of AgI is so low that even the formation of a small amount produces a precipitate.

Neutral complexes are usually only sparingly soluble in water, but are often readily soluble in organic solvents. For example, the red complex $[Fe(acac)_3]$ (Fig. 7.10) can be extracted from aqueous solution into benzene or chloroform, and the formation of $[Fe(acac)_3]$ is used as a means of

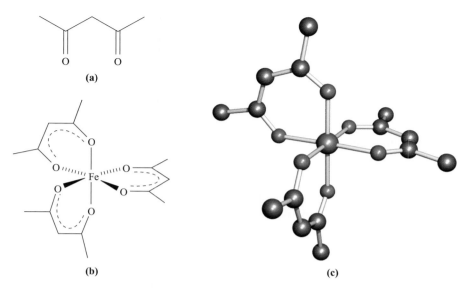

Fig. 7.10 (a) The structure of pentane-2,4-dione (acetylacetone), Hacac (see Table 7.7); (b) Fe(III) forms an octahedral complex with [acac]⁻; (c) the structure of the coordination complex [Fe(acac)₃], determined by X-ray diffraction [J. Iball *et al.* (1967) *Acta Crystallogr.*, vol. 23, p. 239]; colour code: Fe, green; C, grey; O, red.

extracting Fe(III) from aqueous solution. Pentane-2,4-dione (Hacac) is a β-*diketone* and deprotonation gives [acac]⁻, a β-*diketonate* (eq. 7.63). The formation of [Fe(acac)₃] in aqueous solution involves equilibria 7.63 and 7.64.

$$K_a = 1 \times 10^{-9} \qquad (7.63)$$

$$Fe^{3+}(aq) + 3[acac]^-(aq) \rightleftharpoons [Fe(acac)_3](aq)$$

$$K = 1 \times 10^{26} \qquad (7.64)$$

The amount of complex formed depends on the pH of the solution. If the pH is too low, H^+ ions compete with Fe^{3+} ions for the ligand (i.e. the back reaction 7.63 competes with the forward reaction 7.64). If the pH is too high, Fe(III) is precipitated as $Fe(OH)_3$ for which $K_{sp} = 2.64 \times 10^{-39}$. Thus, there is an optimum pH for the extraction of Fe(III) from aqueous media using Hacac and a given organic solvent (e.g. $CHCl_3$). Although we have defined ligands as being *Lewis* bases, most are also Brønsted bases, and accurate pH control is of great importance in studies of complex formation. Solvent extraction is important in the analytical and industrial separation of many metals (see Box 7.3).

Solvent extraction involves the extraction of a substance using a suitable solvent; in a two-phase solvent system, the solute is extracted from one solvent into another, the extracting solvent being chosen so that impurities remain in the original solvent.

7.12 Stability constants of coordination complexes

As we saw earlier, metal ions in aqueous solution are hydrated. The aqua species may be denoted as $M^{z+}(aq)$ where this often represents the hexaaqua ion $[M(OH_2)_6]^{z+}$. Now consider the addition of a neutral ligand L to the solution, and the formation of a series of complexes $[M(OH_2)_5L]^{z+}$, $[M(OH_2)_4L_2]^{z+}$, ... , $[ML_6]^{z+}$. Equilibria 7.65–7.70 show the stepwise displacements of coordinated H_2O by L.

$$[M(OH_2)_6]^{z+}(aq) + L(aq)$$
$$\rightleftharpoons [M(OH_2)_5L]^{z+}(aq) + H_2O(l) \qquad (7.65)$$

$$[M(OH_2)_5L]^{z+}(aq) + L(aq)$$
$$\rightleftharpoons [M(OH_2)_4L_2]^{z+}(aq) + H_2O(l) \qquad (7.66)$$

$$[M(OH_2)_4L_2]^{z+}(aq) + L(aq)$$
$$\rightleftharpoons [M(OH_2)_3L_3]^{z+}(aq) + H_2O(l) \qquad (7.67)$$

$$[M(OH_2)_3L_3]^{z+}(aq) + L(aq)$$
$$\rightleftharpoons [M(OH_2)_2L_4]^{z+}(aq) + H_2O(l) \qquad (7.68)$$

$$[M(OH_2)_2L_4]^{z+}(aq) + L(aq)$$
$$\rightleftharpoons [M(OH_2)L_5]^{z+}(aq) + H_2O(l) \qquad (7.69)$$

$$[M(OH_2)L_5]^{z+}(aq) + L(aq)$$
$$\rightleftharpoons [ML_6]^{z+}(aq) + H_2O(l) \qquad (7.70)$$

APPLICATIONS

Box 7.3 The use of solvent extraction in nuclear reprocessing

Nuclear fission can be successfully harnessed to produce nuclear energy. An advantage of nuclear energy is that it is not associated with emissions into the atmosphere of CO_2, SO_2 and NO_x (see Boxes 14.9, 16.5 and 15.7). Disadvantages include the problems of disposing of radioactive isotopes generated as fission products, and the risks involved if a nuclear reactor 'goes critical'.

The generation of nuclear energy must be a controlled process. Neutrons produced from the fission of $^{235}_{92}U$ lose most of their kinetic energy by passage through a moderator (graphite or D_2O) before they undergo the following nuclear reactions. The first is neutron capture by $^{235}_{92}U$ leading to further fission. The second is neutron capture by $^{238}_{92}U$ (*breeding*):

$$^{238}_{92}U + {}^{1}_{0}n \longrightarrow {}^{239}_{92}U + \gamma\text{-energy}$$

$$^{239}_{92}U \xrightarrow{-\beta^-} {}^{239}_{93}Np \xrightarrow{-\beta^-} {}^{239}_{94}Pu$$

Catastrophic branching chain reactions are prevented by inserting boron-containing steel or boron carbide rods which control the number of neutrons.

Eventually, the $^{235}_{92}U$ fuel is spent, and requires reprocessing. This recovers uranium and also separates $^{235}_{92}U$ from the fission products. First, the spent fuel is kept in pond storage to allow short-lived radioactive products to decay. The uranium is then converted into the soluble salt $[UO_2][NO_3]_2$ and finally into UF_6:

$$[UO_2][NO_3]_2 \xrightarrow{570\,K} UO_3 + NO + NO_2 + O_2$$

$$UO_3 + H_2 \xrightarrow{970\,K} UO_2 + H_2O$$

$$UO_2 + 4HF \longrightarrow UF_4 + 2H_2O$$

$$UF_4 + F_2 \xrightarrow{720\,K} UF_6$$

After separation of $^{235}_{92}UF_6$ and $^{238}_{92}UF_6$ in a centrifuge (application of Graham's law of effusion), $^{235}_{92}U$-enriched UF_6 is produced which is converted back to uranium-235 metal for reuse in the nuclear reactor.

One of the complicating factors in the process described above is that the $[UO_2][NO_3]_2$ contains plutonium and fission products in addition to uranium, and their separation involves the use of two solvent extraction processes.

Stage 1: separation of the fission products from plutonium and uranium nitrates

The mixture to be separated contains $[UO_2]^{2+}$ and Pu(IV) nitrates, as well as metal ions such as $^{90}_{38}Sr^{2+}$. Kerosene (a mixture of hydrocarbons, mainly dodecane) is added to the aqueous solution of metal salts, giving a *two-phase* system (i.e. these solvents are immiscible). Tributyl phosphate (TBP, a phosphate ester) is added to form complexes with the uranium-containing

and plutonium ions, extracting them into the kerosene layer. The fission products remain in the aqueous solution, and separation of the solvent layers thus achieves separation of the fission products from Pu- and U-containing species. Repeated extraction from the aqueous layer by the same process increases the efficiency of the separation.

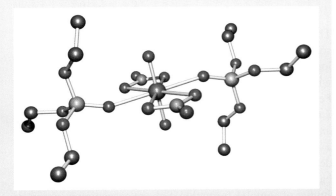

Tributyl phosphate (TBP)

Stage 2: separation of plutonium and uranium nitrates

The kerosene fraction is now subjected to a second solvent extraction. Addition of iron(II) sulfamate, $Fe(NH_2SO_3)_2$, and shaking of the kerosene fraction with water, results in the formation of plutonium(III) nitrate which is partitioned into the aqueous layer. $[UO_2][NO_3]_2$ resists reduction, is complexed by TBP and remains in the organic layer. Separation of the two solvent fractions thus separates the uranium and plutonium salts; repeated extractions result in a highly efficient separation. The extraction of $[UO_2][NO_3]_2$ from kerosene back into an aqueous phase can be achieved by adding nitric acid; under these conditions, the uranium–TBP complex dissociates and $[UO_2][NO_3]_2$ returns to the aqueous layer.

The triethyl phosphate ligand is related to TBP, and the figure above shows the structure (X-ray diffraction) of the complex $[UO_2(NO_3)_2\{OP(OEt)_3\}_2]$. This is a model complex for species present in the extraction process described above. [Data: B. Kanellakopulos *et al.* (1993) *Z. Anorg. Allg. Chem.*, vol. 619, p. 593.] Colour code: U, green; O, red; N, blue; P, orange; C, grey.

The equilibrium constant, K_1, for reaction 7.65 is given by eq. 7.71; [H$_2$O] (strictly, the *activity* of H$_2$O) is unity (see Section 7.3) and does not appear in the expression for K.

$$K_1 = \frac{[M(OH_2)_5L^{z+}]}{[M(OH_2)_6{}^{z+}][L]} \qquad (7.71)$$

> In the formation of a complex $[ML_6]^{z+}$ from $[M(OH_2)_6]^{z+}$, each displacement of a coordinated water molecule by ligand L has a characteristic *stepwise stability constant*, K_1, K_2, K_3, K_4, K_5 or K_6.

Alternatively, we may consider the overall formation of $[ML_6]^{z+}$ (eq. 7.72). In order to distinguish stepwise and overall stability constants, the symbol β is generally used for the latter. Equation 7.73 gives an expression for β_6 for $[ML_6]^{z+}$. We must refer to β_6 and not just β, because overall stability constants for the products of each of reactions 7.65–7.70 can also be defined (see end-of-chapter problem 7.25).

$$[M(OH_2)_6]^{z+}(aq) + 6L(aq) \rightleftharpoons [ML_6]^{z+}(aq) + 6H_2O(l) \qquad (7.72)$$

$$\beta_6 = \frac{[ML_6{}^{z+}]}{[M(OH_2)_6{}^{z+}][L]^6} \qquad (7.73)$$

Values of K and β are related. For equilibrium 7.72, β_6 can be expressed in terms of the six stepwise stability constants according to eq. 7.74.

$$\left.\begin{array}{l} \beta_6 = K_1 \times K_2 \times K_3 \times K_4 \times K_5 \times K_6 \\[4pt] \text{or} \\[4pt] \log \beta_6 = \log K_1 + \log K_2 + \log K_3 \\[2pt] \qquad\qquad + \log K_4 + \log K_5 + \log K_6 \end{array}\right\} \qquad (7.74)$$

Self-study exercise

Write expressions for each of K_1, K_2, K_3, K_4, K_5 and K_6 for equilibria 7.65–7.70, and then show that $\beta_6 = K_1 \times K_2 \times K_3 \times K_4 \times K_5 \times K_6$.

> For the formation of a complex $[ML_n]^{z+}$ from $[M(OH_2)_m]^{z+}$ and ligand L, the *overall stability constant β_n* is given by the expression:
>
> $$\beta_n = \frac{[ML_n{}^{z+}]}{[M(OH_2)_m{}^{z+}][L]^n}$$

Worked example 7.6   Formation of $[Ni(OH_2)_{6-x}(NH_3)_x]^{2+}$

Results of a pH study using a glass electrode (in 2 M NH$_4$NO$_3$ aqueous solution) give values of the stepwise stability constants (at 303 K) of $[Ni(OH_2)_{6-x}(NH_3)_x]^{2+}$ ($x = 1$–6) as: $\log K_1 = 2.79$; $\log K_2 = 2.26$; $\log K_3 = 1.69$; $\log K_4 = 1.25$; $\log K_5 = 0.74$; $\log K_6 = 0.03$. Calculate (a) β_6 for $[Ni(NH_3)_6]^{2+}$ and (b) $\Delta G^\circ{}_1(303\,K)$. (c) If the value of $\Delta H^\circ{}_1(303\,K) = -16.8\,kJ\,mol^{-1}$, calculate $\Delta S^\circ{}_1(303\,K)$. ($R = 8.314\,J\,K^{-1}\,mol^{-1}$)

(a) $\quad \beta_6 = K_1 \times K_2 \times K_3 \times K_4 \times K_5 \times K_6$

$$\log \beta_6 = \log K_1 + \log K_2 + \log K_3$$
$$\qquad\qquad + \log K_4 + \log K_5 + \log K_6$$
$$\log \beta_6 = 2.79 + 2.26 + 1.69 + 1.25 + 0.74 + 0.03$$
$$= 8.76$$
$$\beta_6 = 5.75 \times 10^8$$

(b) $\Delta G^\circ{}_1(303\,K)$ refers to the stepwise formation of $[Ni(OH_2)_5(NH_3)]^{2+}$.

$$\Delta G^\circ{}_1(303\,K) = -RT \ln K_1$$
$$= -(8.314 \times 10^{-3} \times 303) \ln 10^{2.79}$$
$$= -16.2\,kJ\,mol^{-1}$$

(c) $\qquad \Delta G^\circ{}_1 = \Delta H^\circ{}_1 - T\Delta S^\circ{}_1$

$$\Delta S^\circ{}_1 = \frac{\Delta H^\circ{}_1 - \Delta G^\circ{}_1}{T}$$

$$\Delta S^\circ{}_1(303\,K) = \frac{-16.8 - (-16.2)}{303}$$

$$= -1.98 \times 10^{-3}\,kJ\,K^{-1}\,mol^{-1}$$

$$= -1.98\,J\,K^{-1}\,mol^{-1}$$

Self-study exercises

These questions refer to $[Ni(OH_2)_{6-x}(NH_3)_x]^{2+}$ ($x = 1$–6), with data quoted at 303 K.

1. Determine $\Delta G^\circ{}_2(303\,K)$ if $\log K_2 = 2.26$.

    [*Ans.* $-13.1\,kJ\,mol^{-1}$]

2. If $\Delta S^\circ{}_1(303\,K) = -1.98\,J\,K^{-1}\,mol^{-1}$, confirm that $\Delta H^\circ{}_1(303\,K) = -16.8\,kJ\,mol^{-1}$, given that $\log K_1 = 2.79$.

3. Given the values $\log K_1 = 2.79$, $\log K_2 = 2.26$ and $\log K_3 = 1.69$, use the appropriate value to determine $\Delta G^\circ (303\,K)$ for the equilibrium:

$$[Ni(OH_2)_4(NH_3)_2]^{2+} + NH_3 \rightleftharpoons [Ni(OH_2)_3(NH_3)_3]^{2+} + H_2O$$

$$[Ans.\ -9.80\,kJ\,mol^{-1}]$$

Determination of stability constants

For a given aqueous solution containing known concentrations of a metal ion M^{z+} and ligand L, it may have been found that only *one* coordination complex of known formula is present in solution. If this is the case, then the stability constant for this complex can be obtained directly from a determination of the concentration of uncomplexed M^{z+}, L or complexed M^{z+} in that solution. Such determinations can be made by polarographic or potentiometric measurements (if a suitable reversible electrode exists), by pH measurements (if the ligand is the conjugate base of a weak acid), or by ion-exchange, spectrophotometric (i.e. observation of electronic spectra and use of the Beer–Lambert law), NMR spectroscopic or distribution methods.

In the past, the use of empirical methods to predict stability constants for metal ion complexes has had only limited application. The use of DFT theory (see Section 4.13) to calculate values of ΔG for the gas-phase equilibrium:

$$[M(OH_2)_6]^{n+}(g) + NH_3(g) \rightleftharpoons [M(OH_2)_5(NH_3)]^{n+}(g) + H_2O(g)$$

for various metal ions M^{n+} has been assessed. Despite the fact that this gas-phase study fails to take into account the effects of solvation, the results of the DFT calculations provide ΔG values that correlate quite well with experimental data. This suggests that the DFT method may be valuable in estimating thermodynamic data for systems for which experimental data are inaccessible.[†]

Trends in stepwise stability constants

Figure 7.11 shows that for the formation of the complex ions $[Al(OH_2)_{6-x}F_x]^{(3-x)+}$ $(x = 1-6)$, the stepwise stability constants become smaller as more F^- ligands are introduced. A similar trend is also observed in the formation of $[Ni(OH_2)_{6-x}(NH_3)_x]^{2+}$ $(x = 1-6)$ in worked example 7.6. This decrease in values of K is typical of many systems. However, the trend is not always as smooth as in Fig. 7.11 (see Section 20.11).

Thermodynamic considerations of complex formation: an introduction

A detailed discussion of the thermodynamics of complex formation in aqueous solution lies beyond the scope of

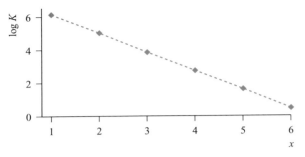

Fig. 7.11 Stepwise stability constants for the formation of $[Al(OH_2)_{6-x}F_x]^{(3-x)+}$ $(x = 1-6)$.

this book, but we discuss briefly entropy changes that accompany the formation of coordination compounds in solution, and the so-called *chelate effect*. In Chapter 20, we look further at the thermodynamics of complex formation.

We saw in Section 7.9 that highly charged ions have more negative values of $\Delta_{hyd}S^\circ$ than singly charged ions, and this can be viewed in terms of the highly charged ions imposing more order on H_2O molecules in the environment of the ion. When complex formation occurs between highly charged cations and anions, with a resulting partial or total cancellation of charges, the changes in enthalpy for these processes are significantly *negative*. However, the accompanying changes in entropy are significantly *positive* because less order is imposed on the H_2O molecules around the complex ion than around the uncomplexed, metal cations and anionic ligands. The corresponding values of ΔG° are, therefore, substantially negative, indicating that very stable complexes are formed. For example, $\Delta S^\circ (298\,K)$ for reaction 7.75 is $+117\,J\,K^{-1}\,mol^{-1}$ and $\Delta G^\circ (298\,K)$ is $-60.5\,kJ\,mol^{-1}$. The ligand in eq. 7.75 is $[EDTA]^{4-}$.[†]

$$Ca^{2+}(aq) \quad + \quad [EDTA]^{4-} \quad \longrightarrow$$

$$(7.75)$$

Another source of increase in entropy is important: when we are dealing with *comparable* uncharged ligands (e.g. NH_3 and $H_2NCH_2CH_2NH_2$), *polydentate* ligands form more stable complexes than *monodentate* ones.

[†] For full details, see: R.D. Hancock and L.J. Bartolotti (2005) *Inorg. Chem.*, vol. 44, p. 7175; R.D. Hancock, L.J. Bartolotti and N. Kaltsoyannis (2006) *Inorg. Chem.*, vol. 45, p. 10780.

[†] In the *solid state*, the complex formed between Ca^{2+} and $[EDTA]^{4-}$ is cation-dependent and is 7- or 8-coordinate; the additional coordination sites are occupied by H_2O, and similarly in $[Mg(EDTA)(OH_2)]^{2-}$.

1,3-propanediamine (pn, **7.18**).

1,2-ethanediamine
(en)

1,3-propanediamine
(pn)

(7.17)   **(7.18)**

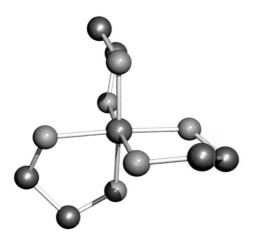

Fig. 7.12 This modelled structure of a complex $[M(en)_3]^{n+}$ illustrates that the ligand en coordinates to give a puckered chelate ring. Colour code: M, green; N, blue; C, grey.

The number of donor atoms through which a ligand coordinates to a metal ion is defined as the *denticity* of the ligand. A monodentate ligand possesses one donor atom (e.g. NH_3), a bidentate ligand two (e.g. $[acac]^-$) and so on. In general, a ligand with more than one donor atom is termed polydentate.

Coordination of a polydentate ligand to an ion leads to the formation of a *chelate ring*, and five such rings can be seen in $[Ca(EDTA)]^{2-}$ in eq. 7.75. The word *chelate* is derived from the Greek for a crab's claw. Table 7.7 lists some common ligands. The ligands en, $[ox]^{2-}$ and bpy form 5-membered chelate rings on coordination to a metal ion (Fig. 7.12), whereas coordination of $[acac]^-$ gives a 6-membered ring (Fig. 7.10). Both 5- and 6-membered chelate rings are common in metal complexes. Each ring is characterized by a *bite angle*, i.e. the X–M–Y angle where X and Y are the two donor atoms of the chelating ligand (structure **7.16**). Ring-strain causes the formation of 3- and 4-membered rings to be relatively unfavourable.

X — Bite angle

M^{n+}

Y

(7.16)

The 6-membered ring formed when $[acac]^-$ chelates to a metal ion (Fig. 7.10) is planar and is stabilized by delocalized π-bonding. Ligands such as bpy and $[ox]^{2-}$ also produce planar chelate rings upon interaction with a metal centre. A saturated diamine such as en (**7.17**) is more flexible and adopts a puckered ring as is shown in Fig. 7.12 for a general $[M(en)_3]^{n+}$ complex. Adding one more carbon atom to the backbone of the ligand en gives

For flexible, saturated *N*-donor ligands of this type, experimental data reveal that small metal ions favour ligands that form 6-membered chelate rings, whereas larger metal ions favour ligands that give 5-membered chelate rings. A general conclusion that '5-membered rings are more stable than 6-membered chelate rings' is often cited in textbooks. However, this statement needs to be qualified, taking into account the size of the metal ion. The enhanced complex stability observed when a small metal ion resides within a 6-membered rather than a 5-membered chelate ring (the ligand being a saturated one such as a diamine) has been explained in terms of a model in which the metal ion replaces an sp^3 hybridized C atom in cyclohexane. For this replacement to be optimized, the bite angle (**7.16**) should be close to 109.5° (i.e. the angle for a tetrahedral C atom), and the M–N bond length should be 160 pm. When diamines coordinate to larger metal ions (e.g. Pb^{2+}, Fe^{2+}, Co^{2+}), the most stable complexes tend to be those involving ligands that form 5-membered chelate rings. The ideal parameters are a bite angle of 69° and an M–N bond length of 250 pm.[†]

We now compare the stability of complexes formed between a given metal ion and related monodentate and bidentate ligands, and address the so-called *chelate effect*. In order to make meaningful comparisons, it is important to choose appropriate ligands. An NH_3 molecule is an approximate (but not perfect) model for half of the ligand en. Equations 7.76–7.78 show equilibria for the displacement of pairs of NH_3 ligands in $[Ni(OH_2)_{6-2n}(NH_3)_{2n}]^{2+}$ ($n = 1$, 2 or 3) by en ligands. The log K and ΔG° values refer to the equilibria at 298 K.

$$[Ni(OH_2)_4(NH_3)_2]^{2+}(aq) + en(aq)$$
$$\rightleftharpoons [Ni(OH_2)_4(en)]^{2+}(aq) + 2NH_3(aq)$$
$$\log K = 2.41 \qquad \Delta G^\circ = -13.7 \, kJ \, mol^{-1} \qquad (7.76)$$

$$[Ni(OH_2)_2(NH_3)_4]^{2+}(aq) + 2en(aq)$$
$$\rightleftharpoons [Ni(OH_2)_2(en)_2]^{2+}(aq) + 4NH_3(aq)$$
$$\log K = 5.72 \qquad \Delta G^\circ = -32.6 \, kJ \, mol^{-1} \qquad (7.77)$$

$$[Ni(NH_3)_6]^{2+}(aq) + 3en(aq) \rightleftharpoons [Ni(en)_3]^{2+}(aq) + 6NH_3(aq)$$
$$\log K = 9.27 \qquad \Delta G^\circ = -52.9 \, kJ \, mol^{-1} \qquad (7.78)$$

[†] For more detailed discussion, see: R.D. Hancock (1992) *J. Chem. Educ.*, vol. 69, p. 615 – 'Chelate ring size and metal ion selection'.

Table 7.7 Names and structures of selected ligands.

Name of ligand	Abbreviation (if any)	Denticity	Structure with donor atoms marked in red
Water		Monodentate	
Ammonia		Monodentate	
Tetrahydrofuran	THF	Monodentate	
Pyridine	py	Monodentate	
1,2-Ethanediamine[†]	en	Bidentate	
Dimethylsulfoxide	DMSO	Monodentate	
Acetylacetonate ion	[acac]$^-$	Bidentate	
Oxalate or ethanedioate ion	[ox]$^{2-}$	Bidentate	
2,2'-Bipyridine	bpy or bipy	Bidentate	
1,10-Phenanthroline	phen	Bidentate	
1,4,7-Triazaheptane[†]	dien	Tridentate	
1,4,7,10-Tetraazadecane[†]	trien	Tetradentate	
N,N,N',N'-Ethylenediaminetetraacetate ion[‡]	[EDTA]$^{4-}$	Hexadentate	See eq. 7.75

[†] The older names (still in use) for 1,2-ethanediamine, 1,4,7-triazaheptane and 1,4,7,10-tetraazadecane are ethylenediamine, diethylenetriamine and triethylenetetramine.
[‡] Although not systematic by the IUPAC rules, this is the commonly accepted name for this anion.

For each ligand displacement, ΔG° is negative and these data (or the values of log K) illustrate that the formation of each chelated complex is thermodynamically more favourable than the formation of the corresponding ammine complex. This phenomenon is called the *chelate effect* and is a general observation.

> For a given metal ion, the thermodynamic stability of a chelated complex involving bidentate or polydentate ligands is greater than that of a complex containing a corresponding number of comparable monodentate ligands. This is called the **chelate effect**.

The value of ΔG° for a reaction such as 7.78 gives a measure of the chelate effect and from the equation:

$$\Delta G^\circ = \Delta H^\circ - T\Delta S^\circ$$

we can see that the relative signs and magnitudes of the contributing ΔH° and $T\Delta S^\circ$ terms are critical.[†] For reaction 7.78 at 298 K, $\Delta H^\circ = -16.8 \, \text{kJ mol}^{-1}$ and $\Delta S^\circ = +121 \, \text{J K}^{-1} \, \text{mol}^{-1}$; the $T\Delta S^\circ$ term is $+36.1 \, \text{kJ mol}^{-1}$. Thus, *both* the negative ΔH° and positive $T\Delta S^\circ$ terms contribute to the overall negative value of ΔG°. In this particular case, the $T\Delta S^\circ$ term is larger than the ΔH° term. However, the mutual reinforcement of these two terms is *not* a general observation as the following examples illustrate. For reaction 7.79, $\Delta G^\circ(298 \, \text{K}) = -8.2 \, \text{kJ mol}^{-1}$. This favourable energy term arises from entropy and enthalpy contributions of $T\Delta S^\circ = -8.8 \, \text{kJ mol}^{-1}$ and $\Delta H^\circ = -17.0 \, \text{kJ mol}^{-1}$, i.e. a favourable enthalpy term that more than compensates for the unfavourable entropy term.

$$\text{Na}^+(\text{aq}) + \text{L}(\text{aq}) \rightleftharpoons [\text{NaL}]^+(\text{aq}) \tag{7.79}$$

where

$$\text{L} = \underset{\text{Me} - \text{O}}{} \diagdown \diagup \text{O} \diagdown \diagup \text{O} \diagdown \diagup \text{O} \diagdown \diagup \text{O} \diagdown \diagup \underset{\text{O} - \text{Me}}{}$$

In reaction 7.80, the enthalpy term is unfavourable, but is outweighed by a very favourable entropy term: at 298 K, $\Delta H^\circ = +13.8 \, \text{kJ mol}^{-1}$, $\Delta S^\circ = +218 \, \text{J K}^{-1} \, \text{mol}^{-1}$, $T\Delta S^\circ = +65.0 \, \text{kJ mol}^{-1}$ and $\Delta G = -51.2 \, \text{kJ mol}^{-1}$.

$$\text{Mg}^{2+}(\text{aq}) + [\text{EDTA}]^{4-} \rightleftharpoons [\text{Mg(EDTA)}]^{2-}(\text{aq}) \tag{7.80}$$

[†] For more in-depth discussions of the chelate and macrocyclic effects, see: M. Gerloch and E.C. Constable (1994) *Transition Metal Chemistry: The Valence Shell in d-Block Chemistry*, VCH, Weinheim (Chapter 8); J. Burgess (1999) *Ions in Solution: Basic Principles of Chemical Interaction*, 2nd edn, Horwood Publishing, Westergate; L.F. Lindoy (1989) *The Chemistry of Macrocyclic Ligand Complexes*, Cambridge University Press, Cambridge (Chapter 6); A.E. Martell, R.D. Hancock and R.J. Motekaitis (1994) *Coord. Chem. Rev.*, vol. 133, p. 39.

In order to examine the origins of the enthalpy and entropy contributions, we again consider reaction 7.78. It has been suggested that the enthalpy contribution to the chelate effect arises from several effects:

- a reduction in the electrostatic repulsion between the δ^- donor atoms (or negatively charged donor atoms in the case of some ligands) on going from two monodentate ligands to one bidentate ligand;
- desolvation effects involving the disruption of ligand–H_2O hydrogen-bonded interactions upon complex formation – such hydrogen-bonded interactions will be greater for, for example, NH_3 than for en;
- an inductive effect of the CH_2CH_2 bridges in bidentate or polydentate ligands which increases the donor strength of the ligand with respect to a corresponding monodentate ligand, e.g. en versus NH_3.

The entropy contribution to the chelate effect is easier to visualize. In eqs. 7.81 and 7.82, two comparable reactions are shown.

$$\underbrace{[\text{Ni(OH}_2)_6]^{2+}(\text{aq}) + 6\text{NH}_3(\text{aq})}_{\text{7 complex ions/molecules}}$$

$$\rightleftharpoons \underbrace{[\text{Ni(NH}_3)_6]^{2+}(\text{aq}) + 6\text{H}_2\text{O(l)}}_{\text{7 complex ions/molecules}} \tag{7.81}$$

$$\underbrace{[\text{Ni(NH}_3)_6]^{2+}(\text{aq}) + 3\text{en(aq)}}_{\text{4 complex ions/molecules}}$$

$$\rightleftharpoons \underbrace{[\text{Ni(en)}_3]^{2+}(\text{aq}) + 6\text{NH}_3(\text{aq})}_{\text{7 complex ions/molecules}} \tag{7.82}$$

In reaction 7.81, monodentate ligands are involved on both sides of the equation, and there is no change in the number of molecules or complex ions on going from reactants to products. However, in reaction 7.82 which involves bidentate ligands replacing monodentate ligands, the number of species in solution increases on going from reactants to products and there is a corresponding increase in entropy (ΔS is positive). Another way of looking at the entropy effect is illustrated in diagram **7.19**. In forming a chelate ring, the probability of the metal ion attaching to the second donor atom is high because the ligand is already anchored to the metal centre. In contrast, the probability of the metal ion associating with a second monodentate ligand is much lower.

(7.19)

Entropy effects associated with desolvation of the ligands prior to complex formation also play a role.

So far, we have considered only the coordination of monodentate or acyclic polydentate ligands. A wealth of coordination chemistry involves *macrocyclic ligands* which include the family of crown ethers (for example, 18-crown-6, **7.20**, and benzo-12-crown-4, **7.21**), and the encapsulating *cryptand ligands* (see Fig. 11.8).

(7.20)   **(7.21)**

Complex stability is enhanced when a macrocyclic ligand replaces a comparable acyclic (open-chain) ligand. For example, values of $\log K_1$ for complexes **7.22** and **7.23** are 23.9 and 28.0 respectively, revealing the thermodynamic stability of the macrocyclic complex.

(7.22)   **(7.23)**

It is not easy to generalize about the origins of the macrocyclic effect. In considering comparable open- and closed-chain complexes such as **7.22** and **7.23**, entropy factors tend, in most cases, to favour the formation of the macrocyclic complex. However, the enthalpy term does not always favour the macrocyclic complex, although the value of ΔG° (i.e. the ultimate arbiter) always favours the formation of the macrocycle. We consider the formation of macrocyclic compounds further in Chapter 11.

7.13 Factors affecting the stabilities of complexes containing only monodentate ligands

Although there is no single generalization relating values of stability constants of complexes of *different cations* with the *same ligand*, a number of useful correlations exist, and in this section we explore some of the most important of them.

Ionic size and charge

The stabilities of complexes of the non-*d*-block metal ions of a given charge normally decrease with increasing cation size (the 'size' of the ion is in a crystallographic sense). Thus, for a complex with a given ligand, L, the order of stability is $Ca^{2+} > Sr^{2+} > Ba^{2+}$. Similar behaviour is found for the lanthanoid M^{3+} ions.

> For ions of similar size, the stability of a complex with a specified ligand increases substantially as the ionic charge increases, e.g. $Li^+ < Mg^{2+} < Al^{3+}$.

For a metal with two (or more) oxidation states, the more highly charged ion is the smaller. The effects of size and charge reinforce each other, leading to greater stability for complexes involving the higher oxidation state metal ion.

Hard and soft metal centres and ligands

When we consider the acceptor properties of metal ions towards ligands (i.e. Lewis acid–Lewis base interactions), two classes of metal ion can be identified, although the distinction between them is not clear-cut. Consider equilibria 7.83 and 7.84.

$$Fe^{3+}(aq) + X^-(aq) \rightleftharpoons [FeX]^{2+}(aq) \qquad (7.83)$$

$$Hg^{2+}(aq) + X^-(aq) \rightleftharpoons [HgX]^+(aq) \qquad (7.84)$$

Table 7.8 gives stability constants for the complexes $[FeX]^{2+}$ and $[HgX]^+$ for different halide ions; while the stabilities of the Fe^{3+} complexes *decrease* in the order $F^- > Cl^- > Br^-$, those of the Hg^{2+} complexes *increase* in the order $F^- < Cl^- < Br^- < I^-$. More generally, in examinations of stability constants by Ahrland, Chatt and Davies, and by Schwarzenbach, the same sequence as for Fe^{3+} was observed for the lighter *s*- and *p*-block cations, other early *d*-block metal cations, and lanthanoid and actinoid metal cations. These cations were collectively termed *class (a) cations*. The same sequence as for Hg^{2+} complexes was observed for halide complexes of the later *d*-block metal ions, tellurium, polonium and thallium. These ions were

Table 7.8 Stability constants for the formation of Fe(III) and Hg(II) halides $[FeX]^{2+}(aq)$ and $[HgX]^+(aq)$; see eqs. 7.83 and 7.84.

Metal ion	$\log K_1$			
	X = F	X = Cl	X = Br	X = I
$Fe^{3+}(aq)$	6.0	1.4	0.5	–
$Hg^{2+}(aq)$	1.0	6.7	8.9	12.9

Table 7.9 Selected hard and soft metal centres (Lewis acids) and ligands (Lewis bases) and those that exhibit intermediate behaviour. Ligand abbreviations are defined in Table 7.7; R = alkyl and Ar = aryl.

	Ligands (Lewis bases)	Metal centres (Lewis acids)
Hard; class (a)	F^-, Cl^-, H_2O, ROH, R_2O, $[OH]^-$, $[RO]^-$, $[RCO_2]^-$, $[CO_3]^{2-}$, $[NO_3]^-$, $[PO_4]^{3-}$, $[SO_4]^{2-}$, $[ClO_4]^-$, $[ox]^{2-}$, NH_3, RNH_2	Li^+, Na^+, K^+, Rb^+, Be^{2+}, Mg^{2+}, Ca^{2+}, Sr^{2+}, Sn^{2+}, Mn^{2+}, Zn^{2+}, Al^{3+}, Ga^{3+}, In^{3+}, Sc^{3+}, Cr^{3+}, Fe^{3+}, Co^{3+}, Y^{3+}, Th^{4+}, Pu^{4+}, Ti^{4+}, Zr^{4+}, $[VO]^{2+}$, $[VO_2]^+$
Soft; class (b)	I^-, H^-, R^-, $[CN]^-$ (C-bound), CO (C-bound), RNC, RSH, R_2S, $[RS]^-$, $[SCN]^-$ (S-bound), R_3P, R_3As, R_3Sb, alkenes, arenes	Zero oxidation state metal centres, Tl^+, Cu^+, Ag^+, Au^+, $[Hg_2]^{2+}$, Hg^{2+}, Cd^{2+}, Pd^{2+}, Pt^{2+}, Tl^{3+}
Intermediate	Br^-, $[N_3]^-$, py, $[SCN]^-$ (N-bound), $ArNH_2$, $[NO_2]^-$, $[SO_3]^{2-}$	Pb^{2+}, Fe^{2+}, Co^{2+}, Ni^{2+}, Cu^{2+}, Os^{2+}, Ru^{3+}, Rh^{3+}, Ir^{3+}

collectively called *class (b) cations*. Similar patterns were found for other donor atoms: ligands with *O*- and *N*-donors form more stable complexes with class (a) cations, while those with *S*- and *P*-donors form more stable complexes with class (b) cations.

In an important development of these generalizations by Pearson, cations (Lewis acids) and ligands (Lewis bases) were classed as being either 'hard' or 'soft'. The *principle of hard and soft acids and bases* (HSAB) is used to rationalize observed patterns in complex stability. In aqueous solution, complexes formed between *class (a), or hard, metal ions* and ligands containing particular donor atoms exhibit trends in stabilities as follows:

$$F > Cl > Br > I$$

$$O \gg S > Se > Te$$

$$N \gg P > As > Sb$$

In contrast, trends in stabilities for complexes formed between *class (b), or soft, metal ions* and ligands containing these donor atoms are:

$$F < Cl < Br < I$$

$$O \ll S > Se \approx Te$$

$$N \ll P > As > Sb$$

Table 7.8 illustrated these trends for halide ions with Fe^{3+} (a hard metal ion) and Hg^{2+} (a soft metal ion):

F^- Cl^- Br^- I^-

Hard   Soft

Similarly, ligands with hard *N*- or *O*-donor atoms form more stable complexes with light *s*- and *p*-block metal cations (e.g. Na^+, Mg^{2+}, Al^{3+}), early *d*-block metal cations (e.g. Sc^{3+}, Cr^{3+}, Fe^{3+}) and *f*-block metal ions (e.g. Ce^{3+}, Th^{4+}). On the other hand, ligands with soft *P*- or *S*-donors show a preference for heavier *p*-block metal ions (e.g. Tl^+) and later *d*-block metal ions (e.g. Cu^+, Ag^+, Hg^{2+}).

Pearson's classification of hard and soft acids comes from a consideration of a series of donor atoms placed in order of electronegativity:

$$F > O > N > Cl > Br > C \approx I \approx S > Se > P > As > Sb$$

A hard acid is one that forms the most stable complexes with ligands containing donor atoms from the left-hand end of the series. The reverse is true for a soft acid. This classification gives rise to the hard and soft acids listed in Table 7.9. A number of metal ions are classed as 'borderline' because they do not show preferences for ligands with particular donor atoms.

The terms 'hard' and 'soft' acids arise from a description of the polarizabilities (see Section 6.13) of the metal ions. Hard acids (Table 7.9) are typically either small monocations with a relatively high charge density or are highly charged, again with a high charge density. These ions are not very polarizable and show a preference for donor atoms that are also not very polarizable, e.g. F^-. Such ligands are called *hard bases*. Soft acids tend to be large monocations with a low charge density, e.g. Ag^+, and are very polarizable. The solution properties of Au^+ indicate that it is the softest metal ion and, therefore, the metal–ligand bonds in complexes of Au(I) exhibit a high degree of covalent character. The group 11 metal(I) ions (Cu^+, Ag^+ and Au^+) are all soft. The transuranium element Rg lies under Au in group 11 but isotopes are not readily accessible for experimental studies of its complexation reactions. The results of DFT calculations conclude that Rg^+ is even softer than Au^+.[†] Soft metal ions prefer to form coordinate bonds with donor atoms that are also highly polarizable, e.g. I^-. Such ligands are called *soft bases*. Table 7.9 lists a range of hard and soft ligands. Note the relationships between the classifications of the ligands and the relative electronegativities of the donor atoms in the series above.

Hard acids (hard metal cations) form more stable complexes with ***hard bases*** (hard ligands), while ***soft acids*** (soft metal cations) show a preference for ***soft bases*** (soft ligands).

[†] See: R.D. Hancock, L.J. Bartolotti and N. Kaltsoyannis (2006) *Inorg. Chem.*, vol. 45, p. 10780.

The HSAB principle is qualitatively useful. Pearson has pointed out that the hard–hard or soft–soft matching of acid and base represents a stabilization that is *additional* to other factors that contribute to the strength of the bonds between donor and acceptor. These factors include the sizes of the cation and donor atom, their charges, their electronegativities and the orbital overlap between them. There is another problem. Complex formation usually involves ligand substitution. In aqueous solution, for example, ligands displace H_2O and this is a *competitive* rather than simple combination reaction (equilibrium 7.85).

$$[M(OH_2)_6]^{2+}(aq) + 6L(aq) \rightleftharpoons [ML_6]^{2+}(aq) + 6H_2O(l)$$
$$(7.85)$$

Suppose M^{2+} is a hard acid. It is already associated with hard H_2O ligands, i.e. there is a favourable hard–hard interaction. If L is a soft base, ligand substitution will not be favourable. If L is a hard base, there are several competing interactions to consider:

- aquated L possesses hard–hard $L-OH_2$ interactions;
- aquated M^{2+} possesses hard–hard $M^{2+}-OH_2$ interactions;
- the product complex will possess hard–hard $M^{2+}-L$ interactions.

Overall, it is observed that such reactions lead to only moderately stable complexes, and values of ΔH° for complex formation are close to zero.

Now consider the case where M^{2+} in eq. 7.85 is a soft acid and L is a soft base. The competing interactions will be:

- aquated L possesses soft–hard $L-OH_2$ interactions;
- aquated M^{2+} possesses soft–hard $M^{2+}-OH_2$ interactions;
- the product complex will possess soft–soft $M^{2+}-L$ interactions.

In this case, experimental data indicate that stable complexes are formed with values of ΔH° for complex formation being large and negative.

Although successful, the HSAB principle initially lacked a satisfactory quantitative basis. However, it is now possible to use DFT theory to derive electronic chemical potential values (electronic chemical potential, μ, is the negative of absolute electronegativity) and chemical hardness values, η.[†] These results complement the use of DFT theory to predict stability constants as described earlier in this chapter.

[†] See: P.W. Ayers, R.G. Parr and R.G. Pearson (2006) *J. Chem. Phys.*, vol. 124, p. 194107; R.G. Pearson (2008) *Int. J. Quantum Chem.*, vol. 108, p. 821.

KEY TERMS

The following terms were introduced in this chapter. Do you know what they mean?

- ❏ self-ionization
- ❏ self-ionization constant of water, K_w
- ❏ Brønsted acid
- ❏ Brønsted base
- ❏ conjugate acid and base pair
- ❏ molality (as distinct from molarity)
- ❏ standard state of a solute in solution
- ❏ activity
- ❏ acid dissociation constant, K_a
- ❏ base dissociation constant, K_b
- ❏ mono-, di- and polybasic acids
- ❏ stepwise dissociation (of an acid or base)
- ❏ Bell's rule
- ❏ Lewis base
- ❏ Lewis acid
- ❏ ion–dipole interaction
- ❏ hydration shell (of an ion)

- ❏ hexaaqua ion
- ❏ hydrolysis (of a hydrated cation)
- ❏ use of the prefix μ, μ_3 ...
- ❏ polarization of a bond
- ❏ charge density of an ion
- ❏ amphoteric
- ❏ 'diagonal line' in the periodic table
- ❏ saturated solution
- ❏ solubility (of an ionic solid)
- ❏ sparingly soluble
- ❏ solubility product
- ❏ standard enthalpy (or Gibbs energy, or entropy) of hydration
- ❏ standard enthalpy (or Gibbs energy, or entropy) of solution
- ❏ common-ion effect
- ❏ gravimetric analysis
- ❏ solvent extraction
- ❏ stepwise stability constant (of a complex)

- ❏ overall stability constant (of a complex)
- ❏ ligand
- ❏ denticity (of a ligand)
- ❏ chelate
- ❏ chelate effect
- ❏ macrocyclic effect
- ❏ hard and soft cations (acids) and ligands (bases)

You should be able to give equations to relate the following quantities:

- ❏ pH and $[H_3O^+]$
- ❏ K_a and pK_a
- ❏ pK_a and pK_b
- ❏ K_a and K_b
- ❏ ΔG° and K
- ❏ ΔG°, ΔH° and ΔS°

FURTHER READING

H₂O: structure

A.F. Goncharov, V.V. Struzhkin, M.S. Somayazulu, R.J. Hemley and H.K. Mao (1996) *Science*, vol. 273, p. 218 – An article entitled 'Compression of ice at 210 gigapascals: Infrared evidence for a symmetric hydrogen-bonded phase'.

R. Ludwig (2001) *Angew. Chem. Int. Ed.*, vol. 40, p. 1808 – A review of recent work on the structures of ice and water.

G. Malenkov (2009) *J. Phys.: Condens. Matter*, vol. 21, p. 283101 – 'Liquid water and ices: understanding the structure and physical properties'.

A.F. Wells (1984) *Structural Inorganic Chemistry*, 5th edn, Clarendon Press, Oxford – Chapter 15 includes a description of the various polymorphs of ice and illustrates the phase diagram of H₂O.

Acid–base equilibria: review material

C.E. Housecroft and E.C. Constable (2010) *Chemistry*, 4th edn, Prentice Hall, Harlow – Chapter 16 includes acid–base equilibria in aqueous solutions, and reviews calculations involving pH, pK_a and pK_b.

Ions in aqueous solution

J. Burgess (1999) *Ions in Solution: Basic Principles of Chemical Interaction*, 2nd edn, Horwood Publishing, Westergate – A very readable introduction to the chemistry of ions in aqueous solution.

W.E. Dasent (1984) *Inorganic Energetics*, 2nd edn, Cambridge University Press, Cambridge – Chapter 5 discusses in detail the energetics of salt dissolution in aqueous solution.

S.F. Lincoln, D.T. Richens and A.G. Sykes (2004) in *Comprehensive Coordination Chemistry II*, eds J.A. McCleverty and T.J. Meyer, Elsevier, Oxford, vol. 1, p. 515 – 'Metal aqua ions' covers aqua ions of elements from groups 1 to 16, and the lanthanoids.

Y. Marcus (1985) *Ion Solvation*, Wiley, New York – A detailed and thorough account of this subject.

H. Ohtaki and T. Radnal (1993) *Chem. Rev.*, vol. 93, p. 1157 – A review dealing with the structure and dynamics of hydrated ions.

A.G. Sharpe (1990) *J. Chem. Educ.*, vol. 67, p. 309 – A short review of the solvation of halide ions and its chemical significance.

E.B. Smith (1982) *Basic Chemical Thermodynamics*, 3rd edn, Clarendon Press, Oxford – Chapter 7 introduces the concept of activity in a very understandable fashion.

Stability constants

A.E. Martell and R.J. Motekaitis (1988) *Determination and Use of Stability Constants*, VCH, New York – A detailed account of the experimental methods for the determination of stability constants, and an overview of their applications.

The IUPAC Stability Constants Database (SC-Database) provides an electronic source of stability constants; the database is kept up to date through regular upgrades (http://www.acadsoft.co.uk/index.html).

Hardness and softness

R.G. Pearson (1997) *Chemical Hardness*, Wiley-VCH, Weinheim – By the originator of the theory of chemical hardness, this book provides an account of its applications in chemistry.

R.D. Hancock and A.E. Martell (1995) *Adv. Inorg. Chem.*, vol. 42, p. 89 – A discussion of the implications of HSAB for metal ions in biology.

PROBLEMS

7.1 The values of pK_a(1) and pK_a(2) for chromic acid (H₂CrO₄) are 0.74 and 6.49 respectively. (a) Determine values of K_a for each dissociation step. (b) Write equations to represent the dissociation steps of chromic acid in aqueous solution.

7.2 Four pK_a values (1.0, 2.0, 7.0, 9.0) are tabulated for the acid H₄P₂O₇. Write equations to show the dissociation steps in aqueous solution and assign, with reasoning, a pK_a value to each step.

7.3 The values of pK_a for CH₃CO₂H and CF₃CO₂H are 4.75 and 0.23, both of which are very nearly independent of temperature. Suggest reasons for this difference.

7.4 (a) To what equilibria do the values of pK_a(1) = 10.71 and pK_a(2) = 7.56 for the conjugate acid of H₂NCH₂CH₂NH₂ refer? (b) Calculate the corresponding values of pK_b and write equations to show the equilibria to which these values refer.

7.5 (a) Write equations to show how you expect compounds **7.24** to **7.28** to dissociate in aqueous solution. (b) Suggest how compound **7.29** will react with NaOH in aqueous solution. What salts would it be possible to isolate? (c) While it is convenient to draw the structures of compounds **7.24** to **7.28** as shown below, these suggest that the P and S atoms violate the octet rule. Redraw structures **7.24** and **7.25** so that the P atoms obey the octet rule.

(7.24)   **(7.25)**   **(7.26)**

(7.27)   **(7.28)**

(7.29)

7.6   In aqueous solution, boric acid behaves as a weak acid ($pK_a = 9.1$) and the following equilibrium is established:

$$B(OH)_3(aq) + 2H_2O(l)$$
$$\rightleftharpoons [B(OH)_4]^-(aq) + [H_3O]^+(aq)$$

(a) Draw the structures of $B(OH)_3$ and $[B(OH)_4]^-$. (b) How would you classify the acidic behaviour of $B(OH)_3$? (c) The formula of boric acid may also be written as H_3BO_3; compare the acidic behaviour of this acid with that of H_3PO_3.

7.7   When NaCN dissolves in water, the resulting solution is basic. Account for this observation given that pK_a for HCN is 9.31.

7.8   Write equations to illustrate the amphoteric behaviour of $[HCO_3]^-$ in aqueous solution.

7.9   Which of the following oxides are likely to be acidic, basic or amphoteric in aqueous solution: (a) MgO; (b) SnO; (c) CO_2; (d) P_2O_5; (e) Sb_2O_3; (f) SO_2; (g) Al_2O_3; (h) BeO?

7.10   Explain what is meant by the terms (a) saturated solution; (b) solubility; (c) sparingly soluble salt; (d) solubility product (solubility constant).

7.11   Write down expressions for K_{sp} for the following ionic salts: (a) AgCl; (b) $CaCO_3$; (c) CaF_2.

7.12   Using your answers to problem 7.11, write down expressions for the solubility (in $mol\,dm^{-3}$) of (a) AgCl, (b) $CaCO_3$ and (c) CaF_2 in terms of K_{sp}.

7.13   Calculate the solubility of $BaSO_4$ at 298 K in g per 100 g of water given that $K_{sp} = 1.07 \times 10^{-10}$.

7.14   Outline the changes that occur (a) to the salt, and (b) to the water molecules, when solid NaF dissolves in water. How do these changes affect (qualitatively) the entropy of the system?

7.15   The values of log K for the following two equilibria are 7.23 and 12.27, respectively:

$$Ag^+(aq) + 2NH_3(aq) \rightleftharpoons [Ag(NH_3)_2]^+(aq)$$
$$Ag^+(aq) + Br^-(aq) \rightleftharpoons AgBr(s)$$

Determine (a) K_{sp} for AgBr, and (b) K for the reaction:

$$[Ag(NH_3)_2]^+(aq) + Br^-(aq) \rightleftharpoons AgBr(s) + 2NH_3(aq)$$

7.16   (a) What are the conjugate bases of the acids HF, $[HSO_4]^-$, $[Fe(OH_2)_6]^{3+}$ and $[NH_4]^+$? (b) What are the conjugate acids of the bases $[HSO_4]^-$, PH_3, $[NH_2]^-$ and $[OBr]^-$? (c) What is the conjugate acid of $[VO(OH)]^+$? (d) $[Ti(OH_2)_6]^{3+}$ has a pK_a value of 2.5. Comment on the fact that when $TiCl_3$ dissolves in dilute hydrochloric acid, the main solution species is $[Ti(OH_2)_6]^{3+}$.

7.17   (a) Discuss the factors that contribute towards KCl being a readily soluble salt (35 g per 100 g H_2O at 298 K). (b) Develop your answer to part (a) by using the following data: $\Delta_{hyd}H^o(K^+, g) = -330\,kJ\,mol^{-1}$; $\Delta_{hyd}H^o(Cl^-, g) = -370\,kJ\,mol^{-1}$; $\Delta_{lattice}H^o(KCl, s) = -715\,kJ\,mol^{-1}$.

7.18   Potassium chromate is used as an indicator in titrations for the determination of chloride ion. At the end-point of a titration of an aqueous solution of a metal chloride salt (e.g. NaCl) against silver nitrate solution in the presence of potassium chromate, red Ag_2CrO_4 precipitates. Give equations for the pertinent reactions occurring during the titration, and, using relevant data from Table 7.4, explain how the indicator works.

7.19   The formation of a buffer solution is an example of the common-ion effect. Explain how a buffer works with reference to a solution containing acetic acid and sodium acetate.

7.20 Calculate the solubility of AgBr ($K_{sp} = 5.35 \times 10^{-13}$) (a) in aqueous solution and (b) in 0.5 M KBr solution.

7.21 Discuss the interpretation of the observation that magnesium oxide is more soluble in aqueous magnesium chloride than in pure water.

7.22 Soda-water is made by saturating H_2O with CO_2. If you titrate soda-water with alkali using phenolphthalein as indicator, you obtain a fading end-point. What does this suggest?

7.23 What explanation can you give for the decrease in solubility of the alkaline earth metal sulfates in the sequence $CaSO_4 > SrSO_4 > BaSO_4$?

7.24 Construct a thermochemical cycle for the decomposition of the phosphonium halides according to the equation:

$$PH_4X(s) \rightleftharpoons PH_3(g) + HX(g)$$

and use it to account for the fact that the most stable phosphonium halide is the iodide.

7.25 (a) Give expressions to define the stepwise stability constants for equilibria 7.66 and 7.68. (b) For each of the complex ions formed in steps 7.66 and 7.68, gives expressions to define the overall stability constants, β_2 and β_4.

7.26 A pH study using a glass electrode at 303 K for complex formation between Al^{3+} ions and $[acac]^-$ (Table 7.7) in aqueous solution gives values of $\log K_1$, $\log K_2$ and $\log K_3$ as 8.6, 7.9 and 5.8. (a) To what equilibria do these values refer? (b) Determine values for $\Delta G^\circ_1(303\,K)$, $\Delta G^\circ_2(303\,K)$ and $\Delta G^\circ_3(303\,K)$ and comment on the relative ease with which successive ligand displacement reactions occur.

7.27 How many chelate rings are present in each of the following complexes? Assume that all the donor atoms are involved in coordination. (a) $[Cu(trien)]^{2+}$; (b) $[Fe(ox)_3]^{3-}$; (c) $[Ru(bpy)_3]^{2+}$; (d) $[Co(dien)_2]^{3+}$; (e) $[K(18\text{-}crown\text{-}6)]^+$.

OVERVIEW PROBLEMS

7.28 Comment on the following observations.

(a) In its complexes, Co(III) forms strong bonds to O- and N-donor ligands, moderately strong bonds to P-donor ligands, but only weak bonds to As-donor ligands.

(b) The values of $\log K$ for the reaction:

$$Zn^{2+}(aq) + X^- \rightleftharpoons [ZnX]^+(aq)$$

are 0.7 for X = F, −0.2 for X = Cl, −0.6 for X = Br, and −1.3 for X = I.

(c) Phosphine adducts of Cr(III) halides can be prepared, but crystallographic studies reveal very long Cr−P bonds (e.g. 247 pm).

7.29 Suggest reasons for the following observations.

(a) Although Pd(II) complexes with monodentate O-donor ligands are not as plentiful as those with P-, S- and As-donor ligands, Pd(II) forms many stable complexes with bidentate O,O'-donor ligands.

(b) $[EDTA]^{4-}$ forms very stable complexes with first row d-block metal ions M^{2+} (e.g. $\log K = 18.62$ for the complex with Ni^{2+}); where the M^{3+} ion is accessible, complexes between M^{3+} and $[EDTA]^{4-}$ are more stable than between the corresponding M^{2+} and $[EDTA]^{4-}$ (e.g. $\log K$ for the complex with Cr^{2+} is 13.6, and for Cr^{3+} is 23.4).

7.30 (a) Explain why water is described as being *amphoteric*.

(b) Draw the structures of the conjugate acid of each of the following:

(c) The value of $K_{sp}(298\,K)$ for Ag_2CrO_4 is 1.12×10^{-12}. What mass of Ag_2CrO_4 dissolves in 100 g of water?

7.31 (a) Comment on the fact that, of the group 1 cations, Li^+ is the most strongly solvated in aqueous solution, even though the first coordination shell only contains four H_2O molecules compared with six for each of the later members of the group.

(b) Suggest how ligand **7.30** coordinates to Ru^{2+} in the 6-coordinate complex $[Ru(\mathbf{7.30})_2]^{2+}$. How many chelate rings are formed in the complex?

(7.30)

(c) For $[Au(CN)_2]^-$, the stability constant $K \approx 10^{39}$ at 298 K. Write an equation that describes the process to which this constant refers, and calculate $\Delta G^\circ(298\,K)$ for the process. Comment on the magnitude of the value you obtain. This cyanide complex is used in the extraction of gold from its ore using the reactions:

$$4Au + 8[CN]^- + O_2 + 2H_2O$$
$$\longrightarrow 4[Au(CN)_2]^- + 4[OH]^-$$
$$2[Au(CN)_2]^- + Zn \longrightarrow [Zn(CN)_4]^{2-} + 2Au$$

What processes take place in this extraction process?

7.32 The structure of H_5DTPA (see Box 4.3) is shown below:

(a) Write equilibria to show the stepwise acid dissociation of H_5DTPA. Which step do you expect to have the largest value of K_a?
(b) In the complex $[Gd(DTPA)(OH_2)]^{2-}$, the Gd^{3+} ion is 9-coordinate. Draw a diagram that illustrates how the $DTPA^{5-}$ ion binds to the metal centre in this complex. How many chelate rings are formed?
(c) Values of $\log K$ for the formation of $[M(DTPA)]^{n+}$ complexes in aqueous media are as follows: Gd^{3+}, 22.5; Fe^{3+}, 27.3; Ag^+, 8.7. Comment on these data.

7.33 (a) For $[Pd(CN)_4]^{2-}$, a value of $\log \beta_4$ of 62.3 (at 298 K in aqueous medium) has been determined. To what equilibrium process does this value refer?
(b) For the equilibrium:
$$Pd(CN)_2(s) + 2CN^-(aq) \rightleftharpoons [Pd(CN)_4]^{2-}$$
the value of $\log K$ is 20.8. Use this value and the data in part (a) to determine K_{sp} for $Pd(CN)_2$.

7.34 (a) Aqueous solutions of copper(II) sulfate contain the $[Cu(OH_2)_6]^{2+}$ ion. The pH of a $0.10\,mol\,dm^{-3}$ aqueous $CuSO_4$ solution is 4.17. Explain the reason why the solution is acidic, and determine K_a for the $[Cu(OH_2)_6]^{2+}$ ion.
(b) When NH_3 is added to aqueous $CuSO_4$, the complex $[Cu(OH_2)_2(NH_3)_4]^{2+}$ is ultimately formed. Initially, however, addition of NH_3 results in the formation of a precipitate of $Cu(OH)_2$ ($K_{sp} = 2.20 \times 10^{-20}$). What is the origin of $[OH]^-$ ions in this solution, and why does $Cu(OH)_2$ form? [Other data: for NH_3, $K_b = 1.8 \times 10^{-5}$]

INORGANIC CHEMISTRY MATTERS

7.35 *Iron overload* is a medical condition where the body cannot cope with abnormally high levels of iron in the system. *Chelation therapy* by administering desferrioxamine, **7.31**, is used to treat the problem. Suggest the origin of the name *chelation therapy*. What form should the iron be in for the therapy to be most effective? Suggest how the therapy works using compound **7.31**; donor sites in the ligand are marked with red arrows and the OH groups can be deprotonated.

(7.31)

7.36 Among the naturally occurring minerals of Al are diaspore (α-AlO(OH)), boehmite (γ-AlO(OH)) and gibbsite (γ-Al(OH)$_3$). At low pH, reactions with H^+ give rise to water-soluble $[Al(OH_2)_6]^{3+}$, abbreviated to Al^{3+}(aq). Thus, Al becomes mobile in rivers and other water courses, but the species present are pH dependent and also depend on other inorganic ions in solution. Using known stability constants, the speciation has been modelled over a pH range of 4.0 to 6.5 and with sulfate and fluoride ions present (Fig. 7.13). (a) What can you deduce about values of K_{sp} for diaspore, boehmite and gibbsite? (b) Draw the structures of the species present in solution at pH 4.5. (c) In the absence of $[SO_4]^{2-}$ and F^-, Al does not enter solution until the pH is lowered to ≈ 4.8. Suggest likely forms in which the Al is present. (d) Write equilibria that describe the relationships between the curves drawn in Fig. 7.13 for Al^{3+} and $[AlOH]^{2+}$, and for $[AlF]^{2+}$ and $[AlF_2]^+$. (e) For the reaction of Al^{3+}(aq) with F^-,

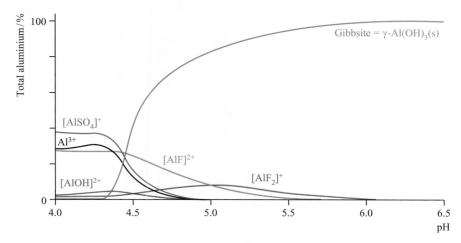

Fig. 7.13 Modelled distribution of Al as a function of pH for an aqueous solution initially containing 50 μmol dm^{-3} total Al. The solution also contains CaSO$_4$ (1 mmol dm^{-3}) and F$^-$ ions (15 μmol dm^{-3}). [Adapted from *Encyclopedia of Soils in the Environment*, D. R. Parker, Aluminium Speciation, pp. 50–56, © 2004 with permission from Elsevier.]

log $\beta_1 = 7.0$, and log $\beta_2 = 12.7$. To what equilibria do these values refer? Determine values of K_1 and K_2 for appropriate stepwise reactions of Al^{3+}(aq) with F$^-$. (f) Give a qualitative explanation for the shapes of the curves in Fig. 7.13.

7.37 The extraction of metals from primary (naturally occurring ores) and secondary (recycled materials) sources is of huge industrial importance. The manipulation of equilibria in, for example, solvent extraction processes is critically important. The family of bidentate ligands shown below is used to extract metal ions including Cu^{2+}:

The extraction can be represented by the equilibrium:

$$2HL(org) + Cu^{2+} \rightleftharpoons [CuL_2](org) + 2H^+$$

in which 'org' refers to extraction into an organic phase. (a) Why is the extraction process pH dependent? (b) In [CuL$_2$], the Cu^{2+} ion is in a square planar environment and the complex can be described as a pseudo-macrocyclic species because of the formation of N–OH⋯O hydrogen bonds. Draw a structure for [CuL$_2$] that is consistent with these observations. (c) The free ligand, HL, is said to be *preorganized* towards the formation of [CuL$_2$]. Suggest what this statement means, and comment on relevant equilibria relating to HL in solution.

7.38 The natural sulfur cycle involves many sulfur-reducing and sulfur-oxidizing bacteria. For example, S^{2-} is oxidized to elemental sulfur and to [SO$_4$]$^{2-}$, and the reverse processes convert mobile [SO$_4$]$^{2-}$ to immobilized S^{2-}. (a) At the end of the sulfur-reduction sequence, HS$^-$ may be produced instead of S^{2-}. What influences this outcome, and what other product is possible? (b) Sulfur-reducing bacteria can be applied to the removal of mobile heavy metals (e.g. Pb, Cd, Hg) from the environment. Explain why the metals are termed 'mobile', and describe the chemical processes and equilibria that lead to the immobilization of the metals.

Topics

Redox reactions
Oxidation states
Reduction potentials
Nernst equation
Disproportionation
Potential diagrams
Frost–Ebsworth
   diagrams
Ellingham diagrams

8

Reduction and oxidation

8.1 Introduction

This chapter is concerned with equilibria involving oxidation and reduction processes. Firstly, we review concepts that will be familiar to most readers: definitions of oxidation and reduction, and the use of oxidation states (oxidation numbers).

Oxidation and reduction

The terms oxidation and reduction are applied in a number of different ways, and you must be prepared to be versatile in their uses.

> ***Oxidation*** refers to gaining oxygen, losing hydrogen or losing one or more electrons. ***Reduction*** refers to losing oxygen, gaining hydrogen or gaining one or more electrons.

Oxidation and reduction steps complement one another, e.g. in reaction 8.1, magnesium is oxidized, while oxygen is reduced. Magnesium acts as the *reducing agent* or *reductant*, while O_2 acts as the *oxidizing agent* or *oxidant*.

$$2Mg + O_2 \longrightarrow 2MgO \qquad (8.1)$$

oxidation
reduction

This reaction could be written in terms of the two half-reactions 8.2 and 8.3, but it is important to remember that neither reaction occurs in isolation.

$$Mg \longrightarrow Mg^{2+} + 2e^- \qquad \textit{oxidation} \qquad (8.2)$$

$$O_2 + 4e^- \longrightarrow 2O^{2-} \qquad \textit{reduction} \qquad (8.3)$$

> ***Redox*** is an abbreviation for reduction–oxidation.

In an *electrolytic cell,* the passage of an electrical current initiates a redox reaction, e.g. in the Downs process (see Section 9.12 and Fig. 11.2) for the manufacture of Na and Cl_2 (eq. 8.4).

$$\left. \begin{array}{l} Na^+ + e^- \longrightarrow Na \\ Cl^- \longrightarrow \tfrac{1}{2}Cl_2 + e^- \end{array} \right\} \qquad (8.4)$$

In a *galvanic cell*, a spontaneous redox reaction occurs and generates an electrical current (see Section 8.2).

> In an ***electrolytic cell***, the passage of an electrical current through an electrolyte causes a chemical reaction to occur. In a ***galvanic cell***, a spontaneous redox reaction occurs and generates an electrical current.

Self-study exercises

1. Which species is being oxidized and which reduced in each of the following reactions or half-reactions?

   $$Fe^{2+} \longrightarrow Fe^{3+} + e^-$$

   $$2H_2 + O_2 \longrightarrow 2H_2O$$

   $$2Ag^+ + Zn \longrightarrow 2Ag + Zn^{2+}$$

   $$Br_2 + 2e^- \longrightarrow 2Br^-$$

   $$2CO + O_2 \longrightarrow 2CO_2$$

   $$Mg + 2H^+ \longrightarrow Mg^{2+} + H_2$$

2. 'A half-reaction refers only to a reduction *or* oxidation process, whereas a balanced redox equation consists of both reduction *and* oxidation processes.' Use your answers to the exercises above to illustrate this statement.

Oxidation states

Many reactions are more complicated than those shown above, and interpreting them in terms of oxidation and reduction steps requires care. The assignment of oxidation states (or oxidation numbers) facilitates this process.

Oxidation states can be assigned to each atom of an element in a compound but *are a formalism*. We assume that readers of this book are already familiar with this concept, but practice is given in end-of-chapter problems 8.1 and 8.2. The oxidation state of an *element* is taken to be zero, irrespective of whether the element exists as atoms (e.g. Ne), molecules (e.g. O_2, P_4) or an infinite lattice (e.g. Si). In addition, in the assignment of oxidation states to elements in a compound, any *homonuclear bond* is ignored. For example, in H_2O_2, **8.1**, the oxidation state of each O atom is -1.

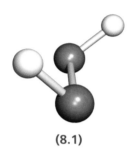

(8.1)

Self-study exercises

What is the oxidation state of each element in the following compounds? Use Pauling electronegativity values in Appendix 7 to help you. Check that in each compound, the positive and negative oxidation numbers balance.
(a) NaCl; (b) CaF_2; (c) BaO; (d) H_2S; (e) BF_3; (f) SiO_2;
(g) PCl_3; (h) $POCl_3$; (i) KI; (j) Cs_2O; (k) N_2O; (l) NO_2;
(m) P_4O_{10}.

An *oxidation* process is accompanied by an increase in the oxidation state of the element involved; conversely, a decrease in the oxidation state corresponds to a *reduction* step.

In reaction 8.5, the oxidation state of Cl in HCl is -1, and in Cl_2 is 0; the change indicates an *oxidation* step. In $KMnO_4$, the oxidation state of Mn is $+7$, while in $MnCl_2$ it is $+2$, i.e. $[MnO_4]^-$ is *reduced* to Mn^{2+}.

$$\text{16HCl(conc)} + 2\text{KMnO}_4(\text{s})$$

$$\longrightarrow 5\text{Cl}_2(\text{g}) + 2\text{KCl(aq)} + 2\text{MnCl}_2(\text{aq}) + 8\text{H}_2\text{O(l)} \tag{8.5}$$

reduction

oxidation

The *net change* in oxidation states involved in the oxidation and reduction steps in a given reaction *must balance*. In reaction 8.5:

- the net *change* in oxidation state for Mn $= 2 \times (-5) = -10$;
- the net *change* in oxidation state for Cl $= 10 \times (+1) = +10$.

Although in some formulae, fractional oxidation states might be suggested, the IUPAC[†] recommends that such usage be avoided. For example, in $[O_2]^-$, it is preferable to consider the group as a whole rather than to assign an oxidation state of $-\frac{1}{2}$ to each O atom.

The *net change in oxidation states* for the oxidation and reduction steps in a given reaction *must balance*.

Stock nomenclature

In MgO, Mg is in oxidation state $+2$, consistent with the presence at an Mg^{2+} ion. However, when we write the oxidation state of Mn in $[MnO_4]^-$ as $+7$, this does not imply the presence of an Mn^{7+} ion. On electrostatic grounds, this would be extremely unlikely. *Stock nomenclature* uses Roman numerals to indicate oxidation state, e.g.:

$[MnO_4]^-$	tetraoxidomanganate(VII)
$[Co(OH_2)_6]^{2+}$	hexaaquacobalt(II)
$[Co(NH_3)_6]^{3+}$	hexaamminecobalt(III)

This gives the oxidation state of the central atom without implying the presence of discrete, highly charged ions.

Self-study exercises

Classify each of the following changes as a reduction process, an oxidation process, a redox reaction or a change that does not involve reduction or oxidation.

1. $2H_2O_2 \longrightarrow 2H_2O + O_2$

2. $[MnO_4]^- + 8H^+ + 5e^- \longrightarrow Mn^{2+} + 4H_2O$

3. $C + O_2 \longrightarrow CO_2$

4. $CaCO_3 \longrightarrow CaO + CO_2$

5. $2I^- \longrightarrow I_2 + 2e^-$

6. $H_2O + Cl_2 \longrightarrow HCl + HOCl$

7. $Cu^{2+} + 2e^- \longrightarrow Cu$

8. $Mg + 2HNO_3 \longrightarrow Mg(NO_3)_2 + H_2$

[†] *IUPAC: Nomenclature of Inorganic Chemistry (Recommendations 2005)*, senior eds N. G. Connelly and T. Damhus, RSC Publishing, Cambridge, p. 66.

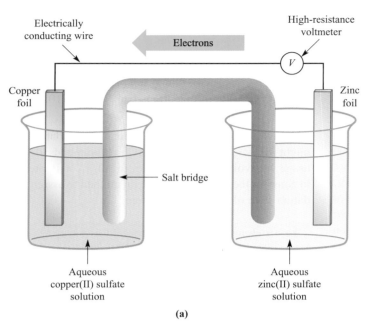

Fig. 8.1 (a) A representation of the Daniell cell. In the left-hand cell, Cu^{2+} ions are reduced to copper metal, and in the right-hand cell, zinc metal is oxidized to Zn^{2+} ions. The cell diagram is written as: $Zn(s)|Zn^{2+}(aq){:}Cu^{2+}(aq)|Cu(s)$. (b) A strip of Zn metal half-submerged in aqueous $CuSO_4$. A spontaneous reaction occurs in which Zn^{2+} ions displace Cu^{2+} in solution and Cu metal is deposited. Eventually, the blue solution of $CuSO_4$ is replaced by a colourless solution of $ZnSO_4$.

8.2 Standard reduction potentials, E^o, and relationships between E^o, ΔG^o and K

Half-cells and galvanic cells

One type of simple electrochemical *half-cell* consists of a metal strip dipping into a solution of its ions, e.g. a Cu strip immersed in an aqueous solution of a Cu(II) salt. No chemical reaction occurs in such a half-cell, although an equation describing the half-cell refers (by convention) to the appropriate *reduction* process (eq. 8.6). The reaction is written as an *equilibrium*.

$$Cu^{2+}(aq) + 2e^- \rightleftharpoons Cu(s) \qquad (8.6)$$

When two such half-cells are combined in an electrical circuit, a redox reaction occurs *if* there is a potential difference between the half-cells. This is illustrated in Fig. 8.1a by the Daniell cell, in which a Cu^{2+}/Cu half-cell (eq. 8.6) is combined with a Zn^{2+}/Zn half-cell (eq. 8.7).

$$Zn^{2+}(aq) + 2e^- \rightleftharpoons Zn(s) \qquad (8.7)$$

The two solutions in the Daniell cell are connected by a *salt-bridge* (e.g. gelatine containing aqueous KCl or KNO_3), which allows the passage of ions between the half-cells without allowing the Cu(II) and Zn(II) solutions to mix too quickly. When the Daniell cell is assembled, redox reaction 8.8 occurs *spontaneously*. The spontaneous nature of the reaction can be confirmed by placing a strip of Zn metal into an aqueous $CuSO_4$ solution (Fig. 8.1b). If a strip of

Cu metal is placed into an aqueous solution of $ZnSO_4$, no reaction occurs.

$$Zn(s) + Cu^{2+}(aq) \longrightarrow Zn^{2+}(aq) + Cu(s) \qquad (8.8)$$

The Daniell cell is an example of a *galvanic cell*. In this type of electrochemical cell, *electrical work* is done *by the system*. The potential difference, E_{cell}, between the two half-cells can be measured (in volts, V) on a voltmeter in the circuit (Fig. 8.1a) and the value of E_{cell} is related to the change in Gibbs energy for the cell reaction. Equation 8.9 gives this relationship under standard conditions, where E^o_{cell} is the *standard cell potential*.

$$\Delta G^o = -zFE^o_{cell} \qquad (8.9)$$

where F = Faraday constant = $96\,485\,C\,mol^{-1}$; z = number of moles of electrons transferred *per mole of reaction*; ΔG^o is in $J\,mol^{-1}$; E^o_{cell} is in volts

Standard conditions for an electrochemical cell are defined as follows:

- unit activity for *each* component in the cell (for *dilute* solutions, activity is approximated to concentration, see Section 7.3);
- the pressure of any gaseous component is 1 bar (10^5 Pa);[†]
- a solid component is in its standard state;
- the temperature is 298 K.

[†] The standard pressure is given in some tables of data as 1 atm ($101\,300\,Pa$), but at the level of accuracy of most tables, this makes no difference to the values of E^o.

For biological electron-transfer processes, the pH of the system is around 7.0, and a biological standard electrode potential, E', is defined instead of E°. We discuss this further in Section 29.4 when we consider the mitochondrial electron-transfer chain.

The equilibrium constant, K, for the cell reaction is related to ΔG° by eq. 8.10, and to E°_{cell} by eq. 8.11.

$$\Delta G^\circ = -RT \ln K \qquad (8.10)$$

$$\ln K = \frac{zFE^\circ_{cell}}{RT} \qquad (8.11)$$

where $R = 8.314\,\mathrm{J\,K^{-1}\,mol^{-1}}$

For $z = 1$, a value of $E^\circ_{cell} = 0.6\,\mathrm{V}$ corresponds to a value of $\Delta G^\circ \approx -60\,\mathrm{kJ\,mol^{-1}}$ and $K \approx 10^{10}$ at 298 K, i.e. this indicates a thermodynamically favourable cell reaction, one that will tend towards completion.

For a ***thermodynamically favourable cell reaction***:

- E°_{cell} is positive;
- ΔG° is negative;
- $K > 1$.

Worked example 8.1   The Daniell cell

The standard cell potential (at 298 K) for the Daniell cell is 1.10 V. Calculate the corresponding values of ΔG° and K and comment on the thermodynamic viability of the cell reaction:

$$\mathbf{Zn(s) + Cu^{2+}(aq) \longrightarrow Zn^{2+}(aq) + Cu(s)}$$

$$\mathbf{(F = 96\,485\,C\,mol^{-1};\ R = 8.314 \times 10^{-3}\,kJ\,K^{-1}\,mol^{-1})}$$

The equation needed is:

$$\Delta G^\circ = -zFE^\circ_{cell}$$

and z is 2 for the cell reaction:

$$Zn(s) + Cu^{2+}(aq) \longrightarrow Zn^{2+}(aq) + Cu(s)$$

$$\Delta G^\circ = -zFE^\circ_{cell}$$

$$= -2 \times 96\,485 \times 1.10$$

$$= -212\,267\,\mathrm{J\ per\ mole\ of\ reaction}$$

$$\approx -212\,\mathrm{kJ\ per\ mole\ of\ reaction}$$

$$\ln K = -\frac{\Delta G^\circ}{RT} = -\frac{-212}{8.314 \times 10^{-3} \times 298}$$

$$\ln K = 85.6$$

$$K = 1.50 \times 10^{37}$$

The large negative value of ΔG° and a value of K which is $\gg 1$ correspond to a thermodynamically favourable reaction, one which virtually goes to completion.

Self-study exercises

1. For the Daniell cell, log $K = 37.2$. Calculate ΔG° for the cell.   [*Ans:* $-212\,\mathrm{kJ\,mol^{-1}}$]

2. The value of ΔG° for the Daniell cell is $-212\,\mathrm{kJ\,mol^{-1}}$. Calculate E°_{cell}.   [*Ans:* 1.10 V]

3. At 298 K, E°_{cell} for the Daniell cell is 1.10 V. Determine the equilibrium ratio $[Cu^{2+}]/[Zn^{2+}]$.   [*Ans:* 6.90×10^{-38}]

4. Show that eqs. 8.10 and 8.11 are dimensionally correct. (*Hint:* you will need SI base units.)

It is possible to obtain values for E°_{cell} *experimentally*, although it is usual in the laboratory to work with solutions of concentrations $<1\,\mathrm{mol\,dm^{-3}}$, and thus measure values of E_{cell} (rather than *standard* cell potentials). Such values are dependent on solution concentration (strictly, activity), and E_{cell} and E°_{cell} are related by the Nernst equation (see eq. 8.21).[†]

It is also possible to *calculate* E°_{cell} (and the corresponding value of ΔG°) using values of *standard reduction potentials* for half-cells, and this is the more routine method of evaluating the thermodynamic viability of redox reactions.

Defining and using standard reduction potentials, E°

Tabulated values of *standard reduction potentials*, E°, refer to single electrodes. For example, for the half-cell reaction 8.6, the value of $E^\circ_{Cu^{2+}/Cu} = +0.34\,\mathrm{V}$. However, it is impossible to measure the potential of an individual electrode and the universal practice is to express all such potentials relative to that of the *standard hydrogen electrode*. The latter consists of a platinum wire immersed in a solution of H^+ ions at a concentration of $1\,\mathrm{mol\,dm^{-3}}$ (strictly, unit activity) in equilibrium with H_2 at 1 bar pressure (eq. 8.12). This electrode is taken to have a standard reduction potential $E^\circ = 0\,\mathrm{V}$ at all temperatures.

$$2H^+(aq,\ 1\,\mathrm{mol\,dm^{-3}}) + 2e^- \rightleftharpoons H_2(g,\ 1\,\mathrm{bar}) \qquad (8.12)$$

Having defined this half-cell, it is now possible to combine it with another half-cell, measure E°_{cell}, and, thus, to find E° for the second half-cell. In order to obtain the correct sign (by convention) for the half-cell, eq. 8.13 must be applied.

$$E^\circ_{cell} = [E^\circ_{reduction\ process}] - [E^\circ_{oxidation\ process}] \qquad (8.13)$$

[†] For an introduction to galvanic cells and the Nernst equation, see: C.E. Housecroft and E.C. Constable (2010) *Chemistry*, 4th edn, Prentice Hall, Harlow, Chapter 18. For a more detailed discussion, see: P. Atkins and J. de Paula (2010) *Atkins' Physical Chemistry*, 9th edn, OUP, Oxford, Chapter 6.

For example, if Zn metal is placed into dilute acid, H_2 is evolved. Thus, when the standard hydrogen electrode is connected in a galvanic cell with a Zn^{2+}/Zn electrode, reaction 8.14 is the spontaneous cell process.

$$Zn(s) + 2H^+(aq) \longrightarrow Zn^{2+}(aq) + H_2(g) \qquad (8.14)$$

The oxidation process is Zn going to Zn^{2+}, and the reduction process involves H^+ ions being converted to H_2. For this cell, the *measured* value of E^o_{cell} is 0.76 V, and, thus, $E^o_{Zn^{2+}/Zn} = -0.76$ V (eq. 8.15). Note that no sign need be included with E_{cell} because it is always positive for the spontaneous reaction.

$$E^o_{cell} = E^o_{2H^+/H_2} - E^o_{Zn^{2+}/Zn}$$

$$0.76 = 0 - E^o_{Zn^{2+}/Zn}$$

$$E^o_{Zn^{2+}/Zn} = -0.76 \text{ V} \qquad (8.15)$$

Selected values of standard reduction potentials are listed in Table 8.1 (see also Appendix 11). Most of these values have been obtained directly from potential difference measurements, but a few values have been calculated from data obtained by calorimetric methods. This latter technique is for systems that cannot be investigated in aqueous media because of solvent decomposition (e.g. $F_2/2F^-$) or for which equilibrium is established only very slowly, such that the electrode is non-reversible (e.g. O_2, $4H^+/2H_2O$). Table 8.1 is organized such that the half-cell with the most positive E^o is at the bottom of the table. The most powerful *oxidizing agent* among the oxidized species in Table 8.1 is F_2, i.e. F_2 is readily reduced to F^- ions. Conversely, at the top of the table, Li is the most powerful *reducing agent*, i.e. Li is readily oxidized to Li^+.

The calculated value $E^o = +1.23$ V for the O_2, $4H^+/2H_2O$ electrode implies that electrolysis of water using this applied potential difference at pH 0 should be possible. Even with a platinum electrode, however, no O_2 is produced. The minimum potential for O_2 evolution to occur is about 1.8 V. The excess potential required (≈ 0.6 V) is the *overpotential* of O_2 on platinum. For electrolytic production of H_2 at a Pt electrode, there is no overpotential. For other metals as electrodes, overpotentials are observed, e.g. 0.8 V for Hg. In general, the overpotential depends on the gas evolved, the electrode material and the current density. It may be thought of as the activation energy for conversion of the species discharged at the electrode into that liberated from the electrolytic cell, and an example is given in worked example 17.3. Some metals do not liberate H_2 from water or acids because of the overpotential of H_2 on them.

Worked example 8.2   Using standard reduction potentials to calculate E^o_{cell}

The following two half-reactions correspond to two half-cells that are combined to form an electrochemical cell:

$$[MnO_4]^-(aq) + 8H^+(aq) + 5e^- \rightleftharpoons Mn^{2+}(aq) + 4H_2O(l)$$

$$Fe^{3+}(aq) + e^- \rightleftharpoons Fe^{2+}(aq)$$

(a) What is the spontaneous cell reaction? (b) Calculate E^o_{cell}.

(a) First, look up values of E^o for the half-reactions.

$$Fe^{3+}(aq) + e^- \rightleftharpoons Fe^{2+}(aq) \qquad E^o = +0.77 \text{ V}$$

$$[MnO_4]^-(aq) + 8H^+(aq) + 5e^- \rightleftharpoons Mn^{2+}(aq) + 4H_2O(l)$$
$$E^o = +1.51 \text{ V}$$

The relative values show that, in aqueous solution under standard conditions, $[MnO_4]^-$ is a more powerful oxidizing agent than Fe^{3+}. The spontaneous cell reaction is therefore:

$$[MnO_4]^-(aq) + 8H^+(aq) + 5Fe^{2+}(aq)$$
$$\longrightarrow Mn^{2+}(aq) + 4H_2O(l) + 5Fe^{3+}(aq)$$

(b) The cell potential difference is the difference between the standard reduction potentials of the two half-cells:

$$E^o_{cell} = [E^o_{reduction process}] - [E^o_{oxidation process}]$$
$$= (+1.51) - (+0.77)$$
$$= 0.74 \text{ V}$$

Self-study exercises

For these exercises, refer to Appendix 11 for data.

1. The following two half-cells are combined:

   $$Zn^{2+}(aq) + 2e^- \rightleftharpoons Zn(s)$$
   $$Ag^+(aq) + e^- \rightleftharpoons Ag(s)$$

   Calculate E^o_{cell}, and state whether the spontaneous reaction reduces Ag^+ or oxidizes Ag. [*Ans.* 1.56 V]

2. For the cell reaction

   $$2[S_2O_3]^{2-} + I_2 \longrightarrow [S_4O_6]^{2-} + 2I^-$$

   write down the two half-cells, and hence determine E^o_{cell}. [*Ans.* 0.46 V]

3. What is the spontaneous reaction if the following two half-cells are combined?

   $$I_2(aq) + 2e^- \rightleftharpoons 2I^-(aq)$$
   $$[MnO_4]^-(aq) + 8H^+(aq) + 5e^- \rightleftharpoons Mn^{2+}(aq) + 4H_2O(l)$$

   Determine a value of E^o_{cell} for the overall reaction. [*Ans.* 0.97 V]

4. Write down the two half-cell reactions that combine to give the following overall reaction:

   $$Mg(s) + 2H^+(aq) \longrightarrow Mg^{2+}(aq) + H_2(g)$$

   Calculate a value of E^o_{cell} for this reaction. [*Ans.* 2.37 V]

Table 8.1 Selected standard reduction potentials (at 298 K); further data are listed in Appendix 11. The concentration of each substance in aqueous solution is 1 mol dm^{-3} and the pressure of a gaseous component is 1 bar (10^5 Pa). Note that where the half-cell contains [OH]$^-$, the value of E° refers to [OH]$^-$ = 1 mol dm^{-3}, and the notation $E^\circ_{[OH^-]=1}$ should be used (see Box 8.1).

Reduction half-equation	E° or $E^\circ_{[OH^-]=1}$ / V
$Li^+(aq) + e^- \rightleftharpoons Li(s)$	-3.04
$K^+(aq) + e^- \rightleftharpoons K(s)$	-2.93
$Ca^{2+}(aq) + 2e^- \rightleftharpoons Ca(s)$	-2.87
$Na^+(aq) + e^- \rightleftharpoons Na(s)$	-2.71
$Mg^{2+}(aq) + 2e^- \rightleftharpoons Mg(s)$	-2.37
$Al^{3+}(aq) + 3e^- \rightleftharpoons Al(s)$	-1.66
$Mn^{2+}(aq) + 2e^- \rightleftharpoons Mn(s)$	-1.19
$Zn^{2+}(aq) + 2e^- \rightleftharpoons Zn(s)$	-0.76
$Fe^{2+}(aq) + 2e^- \rightleftharpoons Fe(s)$	-0.44
$Cr^{3+}(aq) + e^- \rightleftharpoons Cr^{2+}(aq)$	-0.41
$Fe^{3+}(aq) + 3e^- \rightleftharpoons Fe(s)$	-0.04
$2H^+(aq, 1\ mol\ dm^{-3}) + 2e^- \rightleftharpoons H_2(g, 1\ bar)$	0
$Cu^{2+}(aq) + e^- \rightleftharpoons Cu^+(aq)$	$+0.15$
$AgCl(s) + e^- \rightleftharpoons Ag(s) + Cl^-(aq)$	$+0.22$
$Cu^{2+}(aq) + 2e^- \rightleftharpoons Cu(s)$	$+0.34$
$[Fe(CN)_6]^{3-}(aq) + e^- \rightleftharpoons [Fe(CN)_6]^{4-}(aq)$	$+0.36$
$O_2(g) + 2H_2O(l) + 4e^- \rightleftharpoons 4[OH]^-(aq)$	$+0.40$
$I_2(aq) + 2e^- \rightleftharpoons 2I^-(aq)$	$+0.54$
$Fe^{3+}(aq) + e^- \rightleftharpoons Fe^{2+}(aq)$	$+0.77$
$Ag^+(aq) + e^- \rightleftharpoons Ag(s)$	$+0.80$
$[Fe(bpy)_3]^{3+}(aq) + e^- \rightleftharpoons [Fe(bpy)_3]^{2+}(aq)$ †	$+1.03$
$Br_2(aq) + 2e^- \rightleftharpoons 2Br^-(aq)$	$+1.09$
$[Fe(phen)_3]^{3+}(aq) + e^- \rightleftharpoons [Fe(phen)_3]^{2+}(aq)$ †	$+1.12$
$O_2(g) + 4H^+(aq) + 4e^- \rightleftharpoons 2H_2O(l)$	$+1.23$
$[Cr_2O_7]^{2-}(aq) + 14H^+(aq) + 6e^- \rightleftharpoons 2Cr^{3+}(aq) + 7H_2O(l)$	$+1.33$
$Cl_2(aq) + 2e^- \rightleftharpoons 2Cl^-(aq)$	$+1.36$
$[MnO_4]^-(aq) + 8H^+(aq) + 5e^- \rightleftharpoons Mn^{2+}(aq) + 4H_2O(l)$	$+1.51$
$Co^{3+}(aq) + e^- \rightleftharpoons Co^{2+}(aq)$	$+1.92$
$[S_2O_8]^{2-}(aq) + 2e^- \rightleftharpoons 2[SO_4]^{2-}(aq)$	$+2.01$
$F_2(aq) + 2e^- \rightleftharpoons 2F^-(aq)$	$+2.87$

†bpy = 2,2'-bipyridine; phen = 1,10-phenanthroline (see Table 7.7).

THEORY

Box 8.1 Notation for standard reduction potentials

In an electrochemical cell under standard conditions, the concentration of each substance in aqueous solution is 1 mol dm^{-3}. Thus, in Table 8.1, each half-cell listed contains the specified solution species at a concentration of 1 mol dm^{-3}. This leads to the reduction of O_2 being represented by two half-reactions, depending upon the cell conditions:

$$O_2(g) + 4H^+(aq) + 4e^- \rightleftharpoons 2H_2O(l)$$

$$E^{\circ} = +1.23 \text{ V when } [H^+] = 1 \text{ mol dm}^{-3}, \text{ i.e. pH} = 0$$

$$O_2(g) + 2H_2O(l) + 4e^- \rightleftharpoons 4[OH]^-(aq)$$

$$E^{\circ} = +0.40 \text{ V when } [OH^-] = 1 \text{ mol dm}^{-3}, \text{ i.e. pH} = 14$$

Similar situations arise for other species in which the value of the electrode potential is pH-dependent. For clarity, therefore, we have adopted the following notation: for half-cells for which the electrode potential is pH-dependent, E° refers to $[H^+] = 1 \text{ mol dm}^{-3}$ (pH = 0). For other pH values, the concentration of $[H]^+$ or $[OH]^-$ is specifically stated, for example, $E_{[H^+]=0.1}$ or $E_{[OH^-]=0.05}$. For the case of $[OH^-] = 1 \text{ mol dm}^{-3}$, this refers to standard conditions, and the notation used is $E^{\circ}_{[OH^-]=1}$.

Although a positive value of E°_{cell} indicates a spontaneous process, it is more revealing to consider the corresponding value of ΔG° (eq. 8.9). The latter takes into account the number of electrons transferred during the reaction, as well as the magnitude and sign of the cell potential. For example, to investigate the reaction between Fe and aqueous Cl_2, we consider redox couples 8.16–8.18.

$$Fe^{2+}(aq) + 2e^- \rightleftharpoons Fe(s) \qquad E^{\circ} = -0.44 \text{ V} \qquad (8.16)$$

$$Fe^{3+}(aq) + 3e^- \rightleftharpoons Fe(s) \qquad E^{\circ} = -0.04 \text{ V} \qquad (8.17)$$

$$Cl_2(aq) + 2e^- \rightleftharpoons 2Cl^-(aq) \qquad E^{\circ} = +1.36 \text{ V} \qquad (8.18)$$

These data indicate that either reaction 8.19 or 8.20 may occur.

$$Fe(s) + Cl_2(aq) \rightleftharpoons Fe^{2+}(aq) + 2Cl^-(aq)$$

$$E^{\circ}_{\text{cell}} = 1.80 \text{ V} \qquad (8.19)$$

$$2Fe(s) + 3Cl_2(aq) \rightleftharpoons 2Fe^{3+}(aq) + 6Cl^-(aq)$$

$$E^{\circ}_{\text{cell}} = 1.40 \text{ V} \qquad (8.20)$$

The value of E°_{cell} is positive for both reactions, and from their relative magnitudes, it might be thought that reaction 8.19 is favoured over reaction 8.20. *Caution is needed:* the true state of affairs is evident only by comparing values of ΔG°. For reaction 8.19 (where $z = 2$), $\Delta G^{\circ} = -347 \text{ kJ}$ per mole of reaction, while for reaction 8.20 ($z = 6$), $\Delta G^{\circ} = -810 \text{ kJ}$ per mole of reaction. *Per mole of Fe*, the values of ΔG° are -347 and -405 kJ, revealing that reaction 8.20 is thermodynamically favoured over reaction 8.19. This example shows how important it is to consider changes in Gibbs energy, rather than simply the cell potentials.

Dependence of reduction potentials on cell conditions

The discussion above centred on *standard* reduction potentials (see Box 8.1). However, laboratory experiments seldom occur under standard cell conditions, and a change

in conditions can cause a significant change in the ability of a reagent to act as a reducing or oxidizing agent.

Consider a Zn^{2+}/Zn half-cell (at 298 K) in which $[Zn^{2+}] = 0.10 \text{ mol dm}^{-3}$, i.e. *non-standard* conditions. The Nernst equation (eq. 8.21) shows how the reduction potential varies with the concentrations of the species present.

$$E = E^{\circ} - \left\{ \frac{RT}{zF} \times \left(\ln \frac{[\text{reduced form}]}{[\text{oxidized form}]} \right) \right\} \qquad (8.21)^{\dagger}$$

Nernst equation

where R = molar gas constant = $8.314 \text{ J K}^{-1} \text{ mol}^{-1}$

T = temperature in K

F = Faraday constant = $96\,485 \text{ C mol}^{-1}$

z = number of electrons transferred

Application of the Nernst equation to the Zn^{2+}/Zn half-cell ($E^{\circ} = -0.76 \text{ V}$) gives $E = -0.79 \text{ V}$ for $[Zn^{2+}] = 0.10 \text{ mol dm}^{-3}$ (eq. 8.22); the concentration (strictly, activity) of Zn metal is taken to be unity. The more negative value of E, corresponding to a more positive value of ΔG, signifies that it is more difficult to reduce Zn^{2+} at the lower concentration.

$$E = E^{\circ} - \left\{ \frac{RT}{zF} \times \left(\ln \frac{[Zn]}{[Zn^{2+}]} \right) \right\}$$

$$= -0.76 - \left\{ \frac{8.314 \times 298}{2 \times 96\,485} \times \left(\ln \frac{1}{0.10} \right) \right\}$$

$$= -0.79 \text{ V} \qquad (8.22)$$

Now consider the effect of pH (pH = $-\log[H^+]$) on the oxidizing ability of $[MnO_4]^-$ in aqueous solution at 298 K. The crucial factor is that half-reaction 8.23 contains H^+ ions.

$\dagger$ The Nernst equation can also be written in the form:

$$E = E^{\circ} - \left\{ \frac{RT}{zF} \times \ln Q \right\}$$

where Q (the quotient in eq. 8.21) is the *reaction quotient*.

$$[MnO_4]^-(aq) + 8H^+(aq) + 5e^- \rightleftharpoons Mn^{2+}(aq) + 4H_2O(l)$$
$$E^o = +1.51 \text{ V} \qquad (8.23)$$

By applying the Nernst equation, we write eq. 8.24, remembering that the concentration (strictly, activity) of H_2O is, by convention, unity.

$$E = 1.51 - \left\{ \frac{8.314 \times 298}{5 \times 96\,485} \times \left(\ln \frac{[Mn^{2+}]}{[MnO_4^-][H^+]^8} \right) \right\} \qquad (8.24)$$

In eq. 8.24, $E = E^o$ when $[H^+] = 1 \text{ mol dm}^{-3}$ *and* $[Mn^{2+}] = [MnO_4]^- = 1 \text{ mol dm}^{-3}$. As $[H^+]$ increases (i.e. the pH of the solution is lowered), the value of E becomes more positive. The fact that the oxidizing power of $[MnO_4]^-$ is lower in dilute acid than in concentrated acid explains why, for example, $[MnO_4]^-$ will not oxidize Cl^- in neutral solution, but liberates Cl_2 from concentrated HCl.

Worked example 8.3   pH dependence of a reduction potential

Given that E^o for:

$$[MnO_4]^-(aq) + 8H^+(aq) + 5e^- \rightleftharpoons Mn^{2+}(aq) + 4H_2O(l)$$

is +1.51 V, calculate the reduction potential, E, in a solution of pH 2.5 and in which the ratio $[Mn^{2+}]$: $[MnO_4]^- = 1 : 100$.

First, determine $[H^+]$ in a solution of pH 2.5:

$$pH = -\log[H^+]$$

$$[H^+] = 10^{-pH} = 10^{-2.5} = 3.2 \times 10^{-3} \text{ mol dm}^{-3}$$

Now apply the Nernst equation:

$$E = E^o - \left\{ \frac{RT}{zF} \times \left(\ln \frac{[Mn^{2+}]}{[MnO_4^-][H^+]^8} \right) \right\}$$

$$= +1.51 - \left\{ \frac{8.314 \times 298}{5 \times 96\,485} \times \left(\ln \frac{1}{100 \times (3.2 \times 10^{-3})^8} \right) \right\}$$

$$= +1.30 \text{ V}$$

Self-study exercises

These questions all refer to the redox couple in the worked example.

1. Show that $E = +1.25 \text{ V}$ when $pH = 3.0$ and the ratio $[Mn^{2+}] : [MnO_4]^- = 1 : 100$.

2. For a ratio $[Mn^{2+}] : [MnO_4]^- = 1000 : 1$, what must the pH of the solution be to give a value of $E = +1.45 \text{ V}$?
   [*Ans.* 0.26]

3. For a ratio $[Mn^{2+}] : [MnO_4]^- = 1 : 100$, determine E in a solution of pH 1.8.   [*Ans.* 1.36 V]

The potentials for the reduction of water ($[H^+] = 10^{-7} \text{ mol dm}^{-3}$) to H_2, and for the reduction of O_2 to H_2O (the reverse of the oxidation of H_2O to O_2) are of particular significance in aqueous solution chemistry. They provide general guidance (subject to the limitations of thermodynamic versus kinetic control) concerning the nature of chemical species that can exist under aqueous conditions. For reduction process 8.25, $E^o = 0 \text{ V}$ (by definition).

$$2H^+(aq, 1 \text{ mol dm}^{-3}) + 2e^- \rightleftharpoons H_2(g, 1 \text{ bar}) \qquad (8.25)$$

If the pressure of H_2 is maintained at 1 bar, application of the Nernst equation (eq. 8.21) allows us to calculate E over a range of values of $[H^+]$. For neutral water (pH 7), $E_{[H^+]=10^{-7}} = -0.41 \text{ V}$, and at pH 14, $E^o_{[OH^-]=1} = -0.83 \text{ V}$. Whether or not the water (pH 7) or molar aqueous alkali (pH 14) is reduced by a species present in solution depends upon the reduction potential of that species relative to that of the $2H^+/H_2$ couple. Bear in mind that we might be considering the reduction of H_2O to H_2 as a *competitive* process which could occur in preference to the desired reduction. The potential of -0.83 V for the $2H^+/H_2$ electrode in molar alkali is of limited importance in isolation. Many M^{z+}/M systems that should reduce water under these conditions are prevented from doing so by the formation of a coating of hydroxide or hydrated oxide. Others, which are less powerfully reducing, bring about reduction because they are modified by complex formation. An example is the formation of $[Zn(OH)_4]^{2-}$ in alkaline solution (eq. 8.26). The value of $E^o = -0.76 \text{ V}$ for the Zn^{2+}/Zn half-cell (Table 8.1) applies *only* to *hydrated* Zn^{2+} ions. When they are in the form of the stable hydroxido complex $[Zn(OH)_4]^{2-}$, $E^o_{[OH^-]=1} = -1.20 \text{ V}$ (eq. 8.27).

$$Zn^{2+}(aq) + 4[OH]^-(aq) \longrightarrow [Zn(OH)_4]^{2-}(aq) \qquad (8.26)$$

$$[Zn(OH)_4]^{2-}(aq) + 2e^- \rightleftharpoons Zn(s) + 4[OH]^-(aq)$$
$$E^o_{[OH^-]=1} = -1.20 \text{ V} \qquad (8.27)$$

Now consider the reduction of O_2 to H_2O, or the oxidation of H_2O to O_2, by a species present in the cell. Equation 8.28 gives the relevant half-reaction.

$$O_2(g) + 4H^+(aq) + 4e^- \rightleftharpoons 2H_2O(l) \quad E^o = +1.23 \text{ V} \qquad (8.28)$$

For a 1 bar pressure of O_2 at 298 K, applying the Nernst equation shows that the half-cell potential becomes $+0.82 \text{ V}$ in neutral water, and $+0.40 \text{ V}$ in molar aqueous alkali. So, from a thermodynamic standpoint, O_2 in the presence of

THEORY

Box 8.2 Cyclic voltammetry

A large number of inorganic species are able to undergo electrochemical processes. For example, coordination compounds containing *d*-block metal centres may exhibit metal- and/or ligand-centred redox processes. Some electron transfer processes are *reversible*, e.g. metal-centred, one-electron reduction and oxidation in an iron(III)/(II) complex, the potential for which depends on the ligand:

$$[Fe(CN)_6]^{3-} + e^- \rightleftharpoons [Fe(CN)_6]^{4-}$$

$$[Fe(en)_3]^{3+} + e^- \rightleftharpoons [Fe(en)_3]^{2+}$$

Reversible means that the rate of electron transfer at the electrode surface is fast and the concentrations of the oxidized and reduced species at the electrode surface are described by the Nernst equation (eq. 8.21). On the other hand, the transfer of electron(s) to or from a compound may be slow or may be followed by a chemical change, i.e. it is *electrochemically* or *chemically irreversible*.

A number of experimental electrochemical methods are available. These include voltammetry under transient conditions (e.g. cyclic voltammetry) or under steady state conditions (e.g. rotating disc electrode), and spectroelectrochemistry (e.g. using UV-VIS spectroscopy to monitor an electrochemical process). Cyclic voltammetry is a readily available technique and information that can be gained includes:

- whether a species is redox active;
- whether, and under what conditions, a species undergoes reversible or irreversible electron transfer;
- how many electrons are involved in an electron transfer process;
- the kinetics of electron transfer (for systems that are not fully reversible or irreversible).

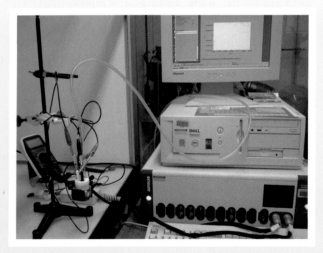

A typical laboratory set-up for CV measurements. The electrochemical cell is on the left-hand side of the photograph. This is connected to a potentiostat with computer interface which displays the CV as it is recorded.

Cyclic voltammetry can be applied to a wide range of inorganic compounds, including bioinorganic systems. Note that the results do not allow one to state the identity or location of the electroactive species.

A typical cyclic voltammetric experiment involves three electrodes: a working electrode (often platinum, gold or glassy carbon), a counter electrode (typically platinum) and a reference electrode (e.g. an AgCl/Ag electrode, see Box 8.3). These electrodes dip into a solution consisting of the sample (analyte) in a suitable solvent. The choice of solvent for an electrochemical experiment is important. The solutes under investigation must be more easily oxidized or reduced than the solvent. The potential window (or electrochemical window) of the solvent is determined by the range of potentials over which the solvent does not undergo any electrochemical processes. For organic solvents, this is less easy to define than for water for which the limits of the potential window correspond to oxidation to O_2 and H^+, and reduction to H_2 and $[OH]^-$. Investigations may be carried out in aqueous or non-aqueous (e.g. acetonitrile or dichloromethane) solvents. In the latter case, special care needs to be taken with reference electrodes. Normal aqueous-based electrodes are not suitable as primary references. The concentration of the analyte is usually low (typically $\leq$ millimolar) and an inert, supporting electrolyte such as $NaClO_4$ or $[^nBu_4N][BF_4]$ is required to ensure a high enough solution conductivity. The electrochemical process being investigated occurs at the working electrode. A potential is applied to the working electrode, and is measured with respect to the reference electrode which remains at a fixed potential. The current response (see later) is passed from the working electrode to the counter electrode.

In a cyclic voltammetry experiment, a potential that changes linearly with time as illustrated in the graph below is applied to the working electrode. The results of the cyclic voltammetry experiment are recorded as a plot of the current response (I) as a function of applied potential (V).

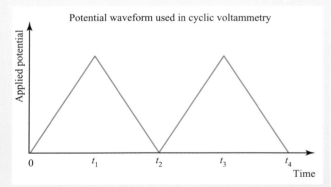

At time $t = 0$, the potential is low and no current flows. The range of potentials swept between $t = 0$ and $t = t_1$ covers the potential at which the electrochemical process being investigated occurs (or potentials, if there is more than one electrochemical process). For example, let the analyte be M^{2+}. As the applied potential approaches $E_{M^{3+}/M^{2+}}$, M^{2+} is oxidized:

$$M^{2+} \longrightarrow M^{3+} + e^-$$

and the current response is recorded. We are only concerned with processes occurring at the surface of the working electrode. Hence, the current increases for a while, but then decreases as the amount of M^{2+} close to the surface of the working electrode is replaced by M^{3+}. Further oxidation is *diffusion controlled*, i.e. no further oxidation of M^{2+} at the working electrode can occur unless M^{2+} ions diffuse to the electrode from the surrounding solution. Reversal of the potential sweep (from time t_1 to t_2) now takes place, and M^{3+} (formed at the working electrode during the initial sweep) is reduced to M^{2+}. The current response reaches a peak and then diminishes. The complete cyclic voltammetric experiment is made up of a series of applied potential sweeps in alternating directions.

The figure below shows the single sweep cyclic voltammogram (CV) recorded for ferrocene (Cp_2Fe, see Section 24.13) in acetonitrile with $NaClO_4$ as the supporting electrolyte. A glassy carbon electrode, a platinum counter electrode, and a silver wire as pseudo-reference electrode were used, and the scan rate for the CV was $200\,mV\,s^{-1}$.

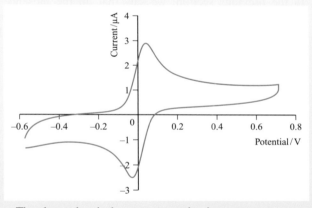

The electrochemical process occurring is:

$$[Cp_2Fe]^+ + e^- \rightleftharpoons Cp_2Fe$$

which can also be written as:

$$Fc^+ + e^- \rightleftharpoons Fc$$

This redox process is used as a convenient internal, secondary reference electrode and, therefore, E^o is defined as $0\,V$ for reference purposes. (Relative to the standard hydrogen electrode, $E^o_{Fc^+/Fc} = +0.40\,V$.)

The symmetrical appearance of the CV shown opposite is typical of a fully reversible redox process. The forward sweep (left to right on the CV above) records a maximum value of current (I_p^{ox}) associated with the oxidation process, while the reverse wave records a minimum value of current (I_p^{red}) associated with reduction. The corresponding values of the potential are E_p^{ox} and E_p^{red} (in V). The separation of these potentials, ΔE, is given by:

$$\Delta E = E_p^{ox} - E_p^{red} \approx \frac{0.059\,V}{z}$$

where z is the number of electrons involved in the chemically reversible, electron transfer process. The reduction potential for the observed electrochemical process is given by:

$$E = \frac{E_p^{ox} + E_p^{red}}{2}$$

The value of E is stated with respect to the reference electrode used in the electrochemical cell.

Cyclic voltammetry may be used to study species not accessible by other means, e.g. the observation of a transient Tl(II) species, $[Tl_2]^{4+}$ (see Section 13.9), and the reversible generation of fulleride anions and of $[C_{60}]^{2+}$ and $[C_{60}]^{3+}$ (see Section 14.4).

Further reading

For an introduction to the subject:
A.M. Bond in *Comprehensive Coordination Chemistry II* (2004), eds J.A. McCleverty and T.J. Meyer, Elsevier, Oxford, vol. 2, p. 197 – 'Electrochemistry: general introduction'.
For an advanced treatment of the topic:
A.J. Bard and L.R. Faulkner (2000) *Electrochemical Methods: Fundamentals and Applications*, 2nd edn, Wiley, New York.

---

water should oxidize any system with a reduction potential less positive than $+1.23\,V$ at pH 0 (i.e. $[H]^+ = 1\,mol\,dm^{-3}$), $+0.82\,V$ at pH 7 and $+0.40\,V$ at pH 14. Conversely, any system with a half-cell potential more positive than $+1.23\,V$ should, at pH 0, oxidize water to O_2, and so on.

We cannot emphasize enough that care has to be taken when considering such processes. Just as the half-cell potentials of the reduction processes considered above vary with experimental conditions, so too do the reduction potentials of other electrodes. It is essential to bear this in mind when using tables of E^o values which are *only* appropriate under *standard conditions*.

Worked example 8.4   Oxidation of Cr^{2+} ions in O_2-free, acidic, aqueous solution

Explain why an acidic, aqueous solution of Cr^{2+} ions liberates H_2 from solution (assume standard conditions). What will be the effect of raising the pH of the solution?

First, write down the half-reactions that are relevant to the question:

$$Cr^{3+}(aq) + e^- \rightleftharpoons Cr^{2+}(aq) \qquad\qquad E^o = -0.41\,V$$

$$2H^+(aq) + 2e^- \rightleftharpoons H_2(g) \qquad\qquad E^o = 0\,V$$

The following redox reaction will occur:

$$2Cr^{2+}(aq) + 2H^+(aq) \longrightarrow 2Cr^{3+}(aq) + H_2(g)$$

In order to check its thermodynamic feasibility, calculate ΔG°.

$$E^\circ{}_{cell} = 0 - (-0.41) = 0.41\,V$$

At 298 K:

$$\Delta G^\circ = -zFE^\circ{}_{cell}$$
$$= -(2 \times 96\,485 \times 0.41)$$
$$= -79.1 \times 10^3\,J \text{ per mole of reaction}$$
$$= -79.1\,kJ \text{ per mole of reaction}$$

Thus, the reaction is thermodynamically favourable, indicating that aqueous Cr^{2+} ions are not stable in acidic (1 M), aqueous solution. (In fact, this reaction is affected by kinetic factors and is quite slow.)

Raising the pH of the solution lowers the concentration of H^+ ions. Let us (arbitrarily) consider a value of pH 3.0 with the ratio $[Cr^{3+}]:[Cr^{2+}]$ remaining equal to 1. The $2H^+/H_2$ electrode now has a new reduction potential.

$$E = E^\circ - \left\{ \frac{RT}{zF} \times \left(\ln \frac{1}{[H^+]^2} \right) \right\}$$
$$= 0 - \left\{ \frac{8.314 \times 298}{2 \times 96\,485} \times \left(\ln \frac{1}{(1 \times 10^{-3})^2} \right) \right\}$$
$$= -0.18\,V$$

Now we must consider the following combination of half-cells, taking Cr^{3+}/Cr^{2+} still to be under standard conditions:

$Cr^{3+}(aq) + e^- \rightleftharpoons Cr^{2+}(aq)$	$E^\circ = -0.41\,V$
$2H^+(aq) + 2e^- \rightleftharpoons H_2(g)$	$E = -0.18\,V$

$$E_{cell} = (-0.18) - (-0.41) = 0.23\,V$$

At 298 K:

$$\Delta G = -zFE_{cell}$$
$$= -(2 \times 96\,485 \times 0.23)$$
$$= -44.4 \times 10^3\,J \text{ per mole of reaction}$$
$$= -44.4\,kJ \text{ per mole of reaction}$$

Thus, although the reaction still has a negative value of ΔG, the increase in pH has made the oxidation of Cr^{2+} less thermodynamically favourable.

[Note: pH plays another important role: at pH values only a few units above zero, precipitation of hydroxides (particularly of Cr^{3+}) will occur.]

1. Calculate E for the reduction of H^+ to H_2 at pH 2.0. Why is this not E°? [*Ans.* −0.12 V]

2. For the half-cell: $O_2 + 4H^+ + 4e^- \rightleftharpoons 2H_2O$
   $E^\circ = +1.23\,V$. Derive a relationship to show how E depends on pH at 298 K and $P(O_2) = 1$ bar. Hence show that at pH 14, $E = +0.40\,V$.

3. Calculate $\Delta G(298\,K)$ for the reaction

   $$2Cr^{2+}(aq) + 2H^+(aq) \rightleftharpoons 2Cr^{3+}(aq) + H_2(g)$$

   in a solution at pH 2.5 and in which $[Cr^{2+}] = [Cr^{3+}] = 1\,mol\,dm^{-3}$. ($E^\circ{}_{Cr^{3+}/Cr^{2+}} = -0.41\,V$.)
   [*Ans.* −50.2 kJ mol^{-1}]

8.3 The effect of complex formation or precipitation on M^{z+}/M reduction potentials

In the previous section, we saw that, in the presence of $[OH]^-$, the potential for the reduction of Zn^{2+} to Zn is significantly different from that of hydrated Zn^{2+}. In this section, we extend this discussion, and discuss how metal ions can be stabilized with respect to reduction by the formation of a precipitate or coordination complex.

Half-cells involving silver halides

Under standard conditions, Ag^+ ions are reduced to Ag (eq. 8.29), but if the concentration of Ag^+ is lowered, application of the Nernst equation shows that the reduction potential becomes less positive (i.e. ΔG is less negative). Consequently, reduction of Ag^+ to Ag becomes less easy. In other words, Ag^+ has been stabilized with respect to reduction (see end-of-chapter problem 8.10.)

$$Ag^+(aq) + e^- \rightleftharpoons Ag(s) \quad E^\circ = +0.80\,V \qquad (8.29)$$
$$\Delta G^\circ = -77.2\,kJ \text{ per mole of Ag}$$

In practice, a lower concentration of Ag^+ ions can be achieved by dilution of the aqueous solution, but it can also be brought about by removal of Ag^+ ions from solution by the formation of a stable complex or by precipitation of a sparingly soluble salt (see Section 7.9). Consider the formation of AgCl (eq. 8.30) for which $K_{sp} = 1.77 \times 10^{-10}$. ΔG° can be found using eq. 8.10.

$$AgCl(s) \rightleftharpoons Ag^+(aq) + Cl^-(aq) \qquad (8.30)$$
$$\Delta G^\circ = +55.6\,kJ \text{ per mole of AgCl}$$

THEORY

Box 8.3 Reference electrodes

Equation 8.31 shows the reduction reaction that occurs in the *silver chloride/silver electrode*, which is written in the form $Cl^-(aq)|AgCl|Ag$ (each vertical bar denotes a phase boundary). This is an example of a half-cell which is constructed by coating a wire of metal M with a solid salt (MX) and immersing this electrode in an aqueous solution containing X^- ions; $[X^-]$ at unit activity $\approx 1\ mol\ dm^{-3}$ for the standard electrode.

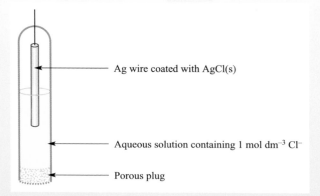

Ag wire coated with AgCl(s)

Aqueous solution containing 1 mol dm^{-3} Cl$^-$

Porous plug

This electrode ($E^o = +0.222\ V$) is used as a reference electrode, being much more convenient to handle in the laboratory than the standard hydrogen electrode. An electrode that requires a cylinder of H_2 at 1 bar pressure is not ideal for routine experimental work! Other reduction potentials may be

quoted *with respect to the silver chloride/silver electrode*, and this effectively gives a scale of relative values on which the standard reduction potential for the reference electrode is set to 0 V.

Another reference electrode which is constructed in a similar manner is the *calomel electrode*, $2Cl^-(aq)|Hg_2Cl_2|2Hg$. The half-cell reaction is:

$$Hg_2Cl_2(s) + 2e^- \rightleftharpoons 2Hg(l) + 2Cl^-(aq) \quad E^o = +0.268\ V$$

The E^o value refers to standard conditions. If the calomel electrode is constructed using 1 M KCl solution, the cell potential, E, is +0.273 V at 298 K. In a *saturated calomel electrode (SCE)*, the Hg_2Cl_2/Hg couple is in contact with a saturated aqueous solution of KCl and for this cell at 298 K, $E = +0.242\ V$. Reduction potentials that are measured 'with respect to SCE $= 0\ V$' are therefore on a relative scale with this reference electrode set to 0 V. Values can be corrected so as to be with respect to the standard hydrogen electrode by adding 0.242 V. For example $E^o_{Ag^+/Ag} = +0.558\ V$ with respect to the SCE, or $E^o_{Ag^+/Ag} = +0.800\ V$ with respect to the standard hydrogen electrode. Clearly, the design of the saturated calomel electrode is not as straightforward as that of the $Cl^-(aq)|AgCl|Ag$ electrode. Mercury is a liquid at 298 K, and contact into an electrical circuit is made by means of a Pt wire which dips into the liquid Hg, itself surrounded by a coating of Hg(I) chloride (calomel). To ensure that the aqueous KCl solution remains saturated, excess KCl crystals are present. A disadvantage of the calomel electrode is the toxicity of Hg.

Reduction of Ag(I) when it is in the form of solid AgCl occurs according to reaction 8.31, and the relationship between equilibria 8.29–8.31 allows us to find, by difference, ΔG^o for reaction 8.31. This leads to a value of $E^o = +0.22\ V$ for this half-cell (see Box 8.3).

$$AgCl(s) + e^- \rightleftharpoons Ag(s) + Cl^-(aq) \quad (8.31)$$

$$\Delta G^o = -21.6\ kJ\ per\ mole\ of\ AgCl$$

The difference in values of E^o for half-reactions 8.29 and 8.31 indicates that it is less easy to reduce Ag(I) in the form of solid AgCl than as hydrated Ag^+.

Silver iodide ($K_{sp} = 8.51 \times 10^{-17}$) is less soluble than AgCl in aqueous solution, and so reduction of Ag(I) in the form of solid AgI is thermodynamically less favourable than reduction of AgCl (see end-of-chapter problem 8.11). However, AgI is much more soluble in aqueous KI than AgCl is in aqueous KCl solution. The species present in the iodide solution is the complex $[AgI_3]^{2-}$, the overall stability constant (see Section 7.12) for which is $\approx 10^{14}$ (eq. 8.32). Following a similar procedure to that detailed above, we can use this value to determine that the half-cell corresponding to reduction process 8.33 has a value of $E^o = -0.03\ V$.

$$Ag^+(aq) + 3I^-(aq) \rightleftharpoons [AgI_3]^{2-}(aq) \quad \beta_3 \approx 10^{14} \quad (8.32)$$

$$[AgI_3]^{2-}(aq) + e^- \rightleftharpoons Ag(s) + 3I^-(aq)$$

$$E^o = -0.03\ V \quad (8.33)$$

Again, Ag(I) has been stabilized with respect to reduction, but this time to a greater extent: the value of E^o indicates that Ag in the presence of $[AgI_3]^{2-}$ and I^- (both $1\ mol\ dm^{-3}$) is as powerful a reducing agent as H_2 in the presence of H^+ (under standard conditions).

Modifying the relative stabilities of different oxidation states of a metal

Just as we can 'tune' the reducing power of Ag by manipulation of the solution species or precipitates present, we can also alter the relative stabilities of two oxidation states of a metal, both of which are subject to removal by precipitation or complexation. As an example, consider the Mn^{3+}/Mn^{2+} couple, for which eq. 8.34 is appropriate for aqua species.

$$Mn^{3+}(aq) + e^- \rightleftharpoons Mn^{2+}(aq) \quad E^o = +1.54\ V \quad (8.34)$$

APPLICATIONS

Box 8.4 Undersea steel structures: sacrificial anodes and cathodic protection

The structural and manufacturing aspects of steel were described in Chapter 6. The use of steel in marine vessels, undersea pipelines, oil rigs and offshore wind turbines carries with it the problem of corrosion. Seawater contains dissolved ionic salts and is able to act as an electrolyte. Typically, it is equivalent to a 3.5% aqueous solution of NaCl and also contains ions such as K^+, Mg^{2+}, Ca^{2+}, Br^-, $[SO_4]^{2-}$ and $[HCO_3]^-$. The concentration of dissolved O_2 in the world's oceans and seas varies with depth. The combination of water, O_2 and an electrolyte leads to oxidation of the iron in submerged steel or steel that is in the splash zone of the waves. At pH 7, the relevant half-equations and reduction potentials are:

$$Fe^{2+} + 2e^- \rightleftharpoons Fe \qquad\qquad E = -0.44\,V$$

$$O_2 + 2H_2O + 4e^- \rightleftharpoons 4OH^- \qquad E_{[OH^-]=10^{-7}} = +0.80\,V$$

In the absence of any protection, the iron spontaneously corrodes:

$$2Fe + O_2 + 2H_2O \longrightarrow 2Fe(OH)_2$$

The iron(II) hydroxide precipitates, since $K_{sp} = 4.87 \times 10^{-17}$. If sufficient O_2 is present, further oxidation occurs to produce red-brown ('rust coloured') $Fe_2O_3{\cdot}H_2O$ (see Section 21.9). The reduction potentials given above refer to specific conditions (pH 7, 298 K and 1 bar pressure of O_2) and will change as pH, temperature and pressure vary, for example, at different depths in the sea. So long as E_{cell} for the conversion of Fe to $Fe(OH)_2$ remains positive, ΔG is negative and corrosion is thermodynamically favourable.

Coatings such as a zinc silicate primer covered with a layer of an epoxy-based polymer are routinely applied to steel structures to protect them against corrosion. However, cracks or flaws in the coating expose Fe which then undergoes oxidation in an anodic process. To prevent this, a second protection system is put in place: cathodic protection. By placing a block of a more electropositive metal on the surface, this second metal is preferentially oxidized. This is the same principle as the use of zinc in galvanized steel (see Section 6.7). From Table 8.1, you can see why Zn, Al and Mg (or alloys of these metals) are typically chosen as *sacrificial anodes*. The most electropositive metals (Li, Na, K and Ca) are unsuitable because they react with cold water. The relevant half-equations (at pH 7) are now:

$$Mg^{2+} + 2e^- \rightleftharpoons Mg \qquad\qquad E = -2.37\,V$$

$$Al^{3+} + 3e^- \rightleftharpoons Al \qquad\qquad E = -1.66\,V$$

$$Zn^{2+} + 2e^- \rightleftharpoons Zn \qquad\qquad E = -0.76\,V$$

$$O_2 + 2H_2O + 4e^- \rightleftharpoons 4OH^- \qquad E_{[OH^-]=10^{-7}} = +0.80\,V$$

Wind turbines in the Solway Firth on the west coast of the UK. The submerged steel and steel in the water's splash zone must be protected to prevent corrosion.

In each case, the metal hydroxide that forms is sparingly soluble ($K_{sp} = 7 \times 10^{-17}$ for $Zn(OH)_2$, 1.9×10^{-33} for $Al(OH)_3$, and 5.61×10^{-12} for $Mg(OH)_2$). While the sacrificial anode block remains in place, the iron is unable to function as an anode and will not corrode. The shape and size of the anode blocks determines their electrical resistance and, therefore, the current that flows from them as oxidation occurs. In turn, this governs the lifetime of the sacrificial anodes which must be replaced on a regular basis.

Zinc and aluminium anodes are typically used for marine applications, whereas magnesium anodes are used widely for protecting steel submerged in freshwater. This preference arises from the different resistivities of freshwater and seawater. The former has a higher resistivity (see Section 6.8) than the latter, and a larger value of E_{cell} is needed to provide adequate current. Ohm's law gives the relationship between potential difference (V in V), current (I in A) and resistance (R in Ω):

$$V = IR \qquad \textit{Ohm's law}$$

The performance of Al anodes is enhanced by alloying with small amounts of Zn and In, Zn and Mg, or Zn and Sn.

There are environmental issues concerning contamination of water with toxic metals, and zinc is among these (see Rousseau *et al.*, below). At pH 7, almost all of the $Zn(OH)_2$ produced by oxidation of the Zn anode is precipitated out of solution. However, an increase in the concentration of $[OH]^-$ ions leads to the formation of $[Zn(OH)_4]^{2-}$, resulting in the metal becoming mobile. Eventually, the zinc accumulates in living organisms.

Further reading

C. Rousseau, F. Baraud, L. Leleyter and O. Gil (2009) *J. Hazard. Mater.*, vol. 167, p. 953 – 'Cathodic protection by zinc sacrificial anodes: impact on marine sediment metallic contamination'.

In alkaline solution, both metal ions are precipitated, but Mn(III) much more completely than Mn(II) since values of K_{sp} for Mn(OH)$_3$ and Mn(OH)$_2$ are $\approx 10^{-36}$ and $\approx 2 \times 10^{-13}$, respectively. Precipitation has the effect of significantly changing the half-cell potential for the reduction of Mn(III). In solutions in which $[OH^-] = 1$ mol dm^{-3}, Mn(III) is stabilized with respect to reduction to Mn(II) as the value of $E^{\circ}_{[OH^-]=1}$ for eq. 8.35 illustrates. Compare this with eq. 8.34.

$$Mn(OH)_3(s) + e^- \rightleftharpoons Mn(OH)_2(s) + [OH]^-(aq)$$

$$E^{\circ}_{[OH^-]=1} = +0.15\,V \qquad (8.35)$$

Worked example 8.5   Oxidation of Mn(II) to Mn(III)

Using data from eqs. 8.34 and 8.35, and from Table 8.1, explain why Mn(II) is not oxidized by O$_2$ in solutions at pH 0, but is oxidized by O$_2$ in solutions in which [OH$^-$] is 1 mol dm^{-3}.

First, find the half-equations that are relevant to the question; note that pH 0 corresponds to standard conditions in which $[H^+] = 1$ mol dm^{-3}.

$$Mn(OH)_3(s) + e^- \rightleftharpoons Mn(OH)_2(s) + [OH]^-(aq)$$
$$E^{\circ}_{[OH^-]=1} = +0.15\,V$$

$$O_2(g) + 2H_2O(l) + 4e^- \rightleftharpoons 4[OH]^-(aq)$$
$$E^{\circ}_{[OH^-]=1} = +0.40\,V$$

$$O_2(g) + 4H^+(aq) + 4e^- \rightleftharpoons 2H_2O(l) \qquad E^{\circ} = +1.23\,V$$

$$Mn^{3+}(aq) + e^- \rightleftharpoons Mn^{2+}(aq) \qquad E^{\circ} = +1.54\,V$$

From this table of reduction potentials (arranged with the most positive value at the bottom of the table), we can see that Mn^{3+}(aq) is the most powerful oxidizing agent of the species listed. Thus, under acidic conditions (pH 0), O$_2$ cannot oxidize Mn^{2+}(aq).

In alkaline medium with $[OH^-] = 1$ mol dm^{-3}, O$_2$ is able to oxidize Mn(OH)$_2$:

$$O_2(g) + 2H_2O(l) + 4Mn(OH)_2(s) \rightleftharpoons 4Mn(OH)_3(s)$$

$$E^{\circ}_{cell} = 0.40 - 0.15$$

$$= 0.25\,V$$

$$\Delta G^{\circ} = -zFE^{\circ}_{cell}$$

$$= -(4 \times 96\,485 \times 0.25)$$

$$= -96\,485\,J \text{ per mole of reaction}$$

$$\approx -96\,kJ \text{ per mole of reaction}$$

or

$$\approx -24\,kJ \text{ per mole of } Mn(OH)_2$$

The large negative value of ΔG° indicates that the oxidation of Mn(OH)$_2$ is thermodynamically favoured.

Self-study exercises

1. Why is the notation $E^{\circ}_{[OH^-]=1}$ used rather than E° for the first two equilibria in the list above?
   [*Ans.* See Box 8.1]

2. For the reaction:

   $$O_2(g) + 2H_2O(l) + 4Mn(OH)_2(s) \rightleftharpoons 4Mn(OH)_3(s)$$

   with $[OH^-] = 1$ mol dm^{-3}, $\Delta G^{\circ} = -24.1$ kJ per mole of Mn(OH)$_2$. Find E°_{cell} for the reaction shown in the equation.
   [*Ans.* 0.25 V]

3. Calculate $\Delta G^{\circ}(298\,K)$ per mole of Mn^{3+} for the reaction:

   $$4Mn^{3+}(aq) + 2H_2O(l) \longrightarrow 4Mn^{2+}(aq) + O_2(g) + 4H^+(aq)$$

   [*Ans.* -30 kJ mol^{-1}]

4. Using the data from the worked example, comment briefly on the pH dependence of the stability of Mn(II) in aqueous solution.

Most d-block metals resemble Mn in that higher oxidation states are more stable (with respect to reduction) in alkaline rather than acidic solutions. This follows from the fact that the hydroxide of the metal in its higher oxidation state is much less soluble than the hydroxide of the metal in its lower oxidation state.

Analogous principles apply when metal ions in different oxidation states form complexes with the same ligand. Usually, the metal ion in the higher oxidation state is stabilized to a greater extent than that in the lower oxidation state. Equations 8.36 and 8.37 show the reduction of hexaaqua and hexaammine complexes of Co(III); remember that M^{z+}(aq) represents $[M(OH_2)_n]^{z+}$(aq) (see Section 7.12).

$$Co^{3+}(aq) + e^- \rightleftharpoons Co^{2+}(aq) \qquad E^{\circ} = +1.92\,V \qquad (8.36)$$

$$[Co(NH_3)_6]^{3+}(aq) + e^- \rightleftharpoons [Co(NH_3)_6]^{2+}(aq)$$
$$E^{\circ} = +0.11\,V \qquad (8.37)$$

It follows from these data that the overall formation constant for $[Co(NH_3)_6]^{3+}$ is $\approx 10^{30}$ times greater than that for $[Co(NH_3)_6]^{2+}$ as is shown below:

$$[Co(OH_2)_6]^{3+}(aq) + 6NH_3(aq) \xrightarrow{\Delta G^{\circ}_1} [Co(NH_3)_6]^{3+}(aq) + 6H_2O(l)$$

$$\Delta G^{\circ}_3 \downarrow \qquad\qquad\qquad\qquad \downarrow \Delta G^{\circ}_4$$

$$[Co(OH_2)_6]^{2+}(aq) + 6NH_3(aq) \xrightarrow{\Delta G^{\circ}_2} [Co(NH_3)_6]^{2+}(aq) + 6H_2O(l)$$

Let β_6 be the formation constant for $[Co(NH_3)_6]^{3+}$ and β_6' be the formation constant for $[Co(NH_3)_6]^{2+}$. A thermochemical cycle can be set up to relate $[Co(NH_3)_6]^{2+}$, $[Co(NH_3)_6]^{3+}$, $[Co(OH_2)_6]^{2+}$ and $[Co(OH_2)_6]^{3+}$, where ΔG^o_1 and ΔG^o_2 refer to complex formation, and ΔG^o_3 and ΔG^o_4 refer to redox reactions.

From the reduction potentials given in eqs. 8.36 and 8.37:

$$\Delta G^o_3 = -zFE^o$$
$$= -(1 \times 96\,485 \times 1.92 \times 10^{-3})$$
$$= -185\,\text{kJ mol}^{-1}$$

$$\Delta G^o_4 = -zFE^o$$
$$= -(1 \times 96\,485 \times 0.11 \times 10^{-3})$$
$$= -11\,\text{kJ mol}^{-1}$$

By Hess's law:

$$\Delta G^o_1 + \Delta G^o_4 = \Delta G^o_2 + \Delta G^o_3$$
$$\Delta G^o_1 - 11 = \Delta G^o_2 - 185$$
$$\Delta G^o_1 - \Delta G^o_2 = -174\,\text{kJ mol}^{-1}$$
$$-RT \ln \beta_6 - (-RT \ln \beta_6') = -174$$
$$-\ln \beta_6 + \ln \beta_6' = -\frac{174}{RT}$$
$$-\ln \frac{\beta_6}{\beta_6'} = -\frac{174}{RT} = -\frac{174}{8.314 \times 10^{-3} \times 298}$$
$$= -70.2$$
$$\ln \frac{\beta_6}{\beta_6'} = 70.2$$
$$\frac{\beta_6}{\beta_6'} = e^{70.2} = 3.1 \times 10^{30}$$

A similar comparison can be made for the reduction of the hexaaqua ion of Fe^{3+} and the cyanido complex (eqs. 8.38 and 8.39), and leads to the conclusion that the overall formation constant for $[Fe(CN)_6]^{3-}$ is $\approx 10^7$ times greater than that of $[Fe(CN)_6]^{4-}$ (see end-of-chapter problem 8.13).

$$Fe^{3+}(aq) + e^- \rightleftharpoons Fe^{2+}(aq) \qquad E^o = +0.77\,\text{V} \qquad (8.38)$$

$$[Fe(CN)_6]^{3-}(aq) + e^- \rightleftharpoons [Fe(CN)_6]^{4-}(aq)$$
$$E^o = +0.36\,\text{V} \qquad (8.39)$$

Some organic ligands, notably 1,10-phenanthroline and 2,2'-bipyridine (Table 7.7), stabilize the *lower* of two oxidation states of a metal. This is apparent from the values of E^o for the appropriate half-reactions in Table 8.1. The observation is associated with the ability of the phen and bpy ligands to accept electrons.[†] Iron(II) complexes of bpy and phen are used as indicators in redox reactions.

For example, in a redox titration of Fe^{2+} with powerful oxidizing agents, all $Fe^{2+}(aq)$ species are oxidized before $[Fe(bpy)_3]^{2+}$ or $[Fe(phen)_3]^{2+}$. The associated colour changes are red to pale blue for $[Fe(bpy)_3]^{2+}$ to $[Fe(bpy)_3]^{3+}$, and orange-red to blue for $[Fe(phen)_3]^{2+}$ to $[Fe(phen)_3]^{3+}$.

8.4 Disproportionation reactions

Disproportionation

Some redox reactions involve *disproportionation* (see Section 6.16), e.g. reactions 8.40 and 8.41.

$$2Cu^+(aq) \rightleftharpoons Cu^{2+}(aq) + Cu(s)$$

oxidation

reduction

$$(8.40)$$

$$3[MnO_4]^{2-}(aq) + 4H^+(aq)$$
$$\rightleftharpoons 2[MnO_4]^-(aq) + MnO_2(s) + 2H_2O(l) \qquad (8.41)$$

oxidation

reduction

Reaction 8.40 takes place when Cu_2SO_4 (prepared by reacting Cu_2O and dimethyl sulfate) is added to water, while reaction 8.41 occurs when acid is added to a solution of K_2MnO_4. Equilibrium constants for such disproportionation reactions can be calculated from reduction potentials as in worked example 8.6.

Worked example 8.6 Disproportionation of copper(I)

Using appropriate data from Table 8.1, determine K (at 298 K) for the equilibrium:

$2Cu^+(aq) \rightleftharpoons Cu^{2+}(aq) + Cu(s)$

Three redox couples in Table 8.1 involve Cu(I), Cu(II) and Cu metal:

(1) $Cu^{2+}(aq) + e^- \rightleftharpoons Cu^+(aq)$ $E^o = +0.15\,\text{V}$

(2) $Cu^{2+}(aq) + 2e^- \rightleftharpoons Cu(s)$ $E^o = +0.34\,\text{V}$

(3) $Cu^+(aq) + e^- \rightleftharpoons Cu(s)$ $E^o = +0.52\,\text{V}$

The disproportionation of Cu(I) is the result of combining half-reactions (1) and (3). Thus:

$$E^o_{\text{cell}} = 0.52 - 0.15$$
$$= 0.37\,\text{V}$$

[†] For a full discussion, see: M. Gerloch and E.C. Constable (1994) *Transition Metal Chemistry: The Valence Shell in d-Block Chemistry*, VCH, Weinheim, p. 176–178.

$$\Delta G^\circ = -zFE^\circ_{cell}$$

$$= -(1 \times 96\,485 \times 0.37 \times 10^{-3})$$

$$= -35.7\,kJ \text{ per mole of reaction}$$

$$\ln K = -\frac{\Delta G^\circ}{RT}$$

$$= \frac{35.7}{8.314 \times 10^{-3} \times 298}$$

$$K = 1.81 \times 10^6$$

The value indicates that disproportionation is thermodynamically favourable.

Self-study exercises

1. For the disproportionation of Cu(I) to Cu and Cu(II), $K(298\,K) = 1.81 \times 10^6$. Calculate ΔG° for the reaction, per mole of Cu(I). [*Ans.* $-17.8\,kJ\,mol^{-1}$]

2. By considering redox couples in Appendix 11 which contain Cr^{2+}, Cr^{3+} and Cr metal, confirm that Cr^{2+} will *not* disproportionate into Cr and Cr^{3+}.

3. Using data from Appendix 11, show that H_2O_2 is unstable with respect to disproportionation into O_2 and H_2O. Calculate $\Delta G^\circ(298\,K)$ for the disproportionation of 1 mole of H_2O_2. [*Ans.* $-104\,kJ\,mol^{-1}$]

Stabilizing species against disproportionation

Species that are unstable with respect to disproportionation, such as Cu^+ in aqueous solution, may be stabilized under appropriate conditions. For example, Cu^+ can be stabilized by precipitation as a sparingly soluble salt such as CuCl ($K_{sp} = 1.72 \times 10^{-7}$; see end-of-chapter problem 8.15) or by the formation in solution of a complex ion such as $[Cu(CN)_4]^{3-}$. In the case of $[MnO_4]^{2-}$ (eq. 8.41), all that is necessary is to make the solution alkaline so as to remove the H^+ ions involved in bringing about the disproportionation.

8.5 Potential diagrams

For an element exhibiting several different oxidation states in aqueous solution, we must consider a number of different half-reactions in order to obtain a clear picture of its solution chemistry. Consider manganese as an example. Aqueous solution species may contain manganese in oxidation states ranging from Mn(II) to Mn(VII), and eqs. 8.42–8.46 give half-reactions for which standard reduction potentials can be determined experimentally.

$$Mn^{2+}(aq) + 2e^- \rightleftharpoons Mn(s) \qquad E^\circ = -1.19\,V \qquad (8.42)$$

$$[MnO_4]^-(aq) + e^- \rightleftharpoons [MnO_4]^{2-}(aq)$$
$$E^\circ = +0.56\,V \qquad (8.43)$$

$$MnO_2(s) + 4H^+(aq) + 2e^- \rightleftharpoons Mn^{2+}(aq) + 2H_2O(l)$$
$$E^\circ = +1.23\,V \qquad (8.44)$$

$$[MnO_4]^-(aq) + 8H^+(aq) + 5e^- \rightleftharpoons Mn^{2+}(aq) + 4H_2O(l)$$
$$E^\circ = +1.51\,V \qquad (8.45)$$

$$Mn^{3+}(aq) + e^- \rightleftharpoons Mn^{2+}(aq) \qquad E^\circ = +1.54\,V \qquad (8.46)$$

These potentials may be used to derive values of E° for other half-reactions such as 8.47. Care must be taken to remember that different numbers of electrons are involved in different reduction steps and, thus, one must calculate E° by first finding the corresponding value of ΔG°.

$$[MnO_4]^-(aq) + 4H^+(aq) + 3e^- \rightleftharpoons MnO_2(s) + 2H_2O(l)$$
$$E^\circ = +1.69\,V \qquad (8.47)$$

Self-study exercise

Confirm that the value of E° for half-equation 8.47 can be obtained from E° values for half-reactions 8.44 and 8.45, but that the method of working must involve determination of ΔG° values for the reactions.

Standard reduction potentials are often tabulated as in Appendix 11, but it is also useful to present data in the form of a *potential diagram* (also known as Latimer diagrams) or a Frost–Ebsworth diagram (see Section 8.6).

Figure 8.2 gives potential diagrams for Mn under conditions of $[H^+] = 1\,mol\,dm^{-3}$ (pH 0) and $[OH^-] = 1\,mol\,dm^{-3}$ (pH 14). Reading from left to right, species are arranged in order of decreasing oxidation state of Mn. The $[MnO_4]^-$ ion (usually in the form of $KMnO_4$) is a common oxidizing agent, and equations 8.45 or 8.47 are the half-reactions that one would usually consider appropriate in acidic solution. The potential diagram (acidic solution) shows an intermediate Mn(VI) species between $[MnO_4]^-$ and MnO_2. However, values of E° show that the $[HMnO_4]^-/MnO_2$ couple is a more powerful oxidant (more negative ΔG°) than the $[MnO_4]^-/[HMnO_4]^-$ couple. This means that $[HMnO_4]^-$ will not accumulate during the reduction of $[MnO_4]^-$ to MnO_2. An alternative way of considering the instability of $[HMnO_4]^-$ in aqueous solution at pH 0 is to note from the potential diagram that $[HMnO_4]^-$ is unstable with respect to disproportionation (eq. 8.48).

$$3[HMnO_4]^-(aq) + H^+(aq)$$
$$\rightleftharpoons MnO_2(s) + 2[MnO_4]^-(aq) + 2H_2O(l) \qquad (8.48)$$

This conclusion can be reached as follows. Extract from the complete potential diagram in Fig. 8.2 the parts

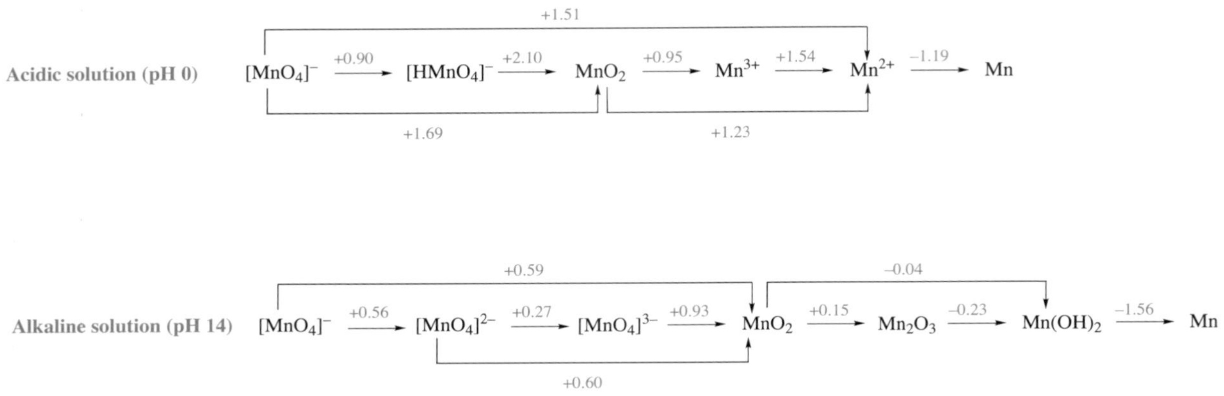

Fig. 8.2 Potential diagrams (Latimer diagrams) for manganese in aqueous solution at pH 0 (i.e. $[H^+] = 1\,mol\,dm^{-3}$), and in aqueous solution at pH 14 (i.e. $[OH^-] = 1\,mol\,dm^{-3}$). For such diagrams, it is essential to specify the pH, and the reason is obvious from comparing the two diagrams. Reduction potentials are given in V.

relevant to reduction and oxidation of $[HMnO_4]^-$ in acidic solution:

$$[MnO_4]^- \xrightarrow{+0.90} [HMnO_4]^- \xrightarrow{+2.10} MnO_2$$

This diagram corresponds to the two half-reactions:

$[MnO_4]^-(aq) + H^+(aq) + e^- \rightleftharpoons [HMnO_4]^-(aq)$

$$E^o = +0.90\,V$$

$[HMnO_4]^-(aq) + 3H^+(aq) + 2e^- \rightleftharpoons MnO_2(s) + 2H_2O(l)$

$$E^o = +2.10\,V$$

Combining these two half-cells gives reaction 8.48 for which $E^o_{cell} = 1.20\,V$ and $\Delta G^o(298\,K) = -231\,kJ\,mol^{-1}$. This indicates that reaction 8.48 is spontaneous. Similarly, at pH 0, Mn^{3+} is unstable with respect to disproportionation to MnO_2 and Mn^{2+} (eq. 8.49; see end-of-chapter problem 8.29).

$$2Mn^{3+}(aq) + 2H_2O(l) \rightleftharpoons Mn^{2+}(aq) + MnO_2(s) + 4H^+(aq) \tag{8.49}$$

We saw in Section 8.2 that the value of the reduction potential for a half-reaction depends on cell conditions, and where the half-reaction involves H^+ or $[OH]^-$ ions, the reduction potential varies with pH. Moreover, the extent of variation depends on the number of moles of H^+ or $[OH]^-$ per mole of reaction. It follows that the potential diagrams in Fig. 8.2 are appropriate *only* at the stated pH values. A new potential diagram is needed for every value of pH, and, therefore, *caution is needed* when using these diagrams.

In using potential diagrams, it is essential to remember that the reduction potential for one step may *not* be derived simply by summation of reduction potentials for steps which contribute to the desired redox half-reaction. For example, in Fig. 8.2, for the reduction of $[MnO_4]^{2-}$ in alkaline solution to MnO_2, $E^o = +0.60\,V$, and this is *not* the sum of the standard reduction potentials for the reduction of $[MnO_4]^{2-}$ to $[MnO_4]^{3-}$ followed by reduction of $[MnO_4]^{3-}$

to MnO_2. Account must be taken of the number of electrons transferred in each step. The most foolproof way of doing this is to determine the corresponding values of ΔG^o for each step as is illustrated below.

Worked example 8.7 Potential diagrams

The following potential diagram summarizes some of the redox chemistry of iron in aqueous solution. Calculate the value of E^o for the reduction of $Fe^{3+}(aq)$ to iron metal.

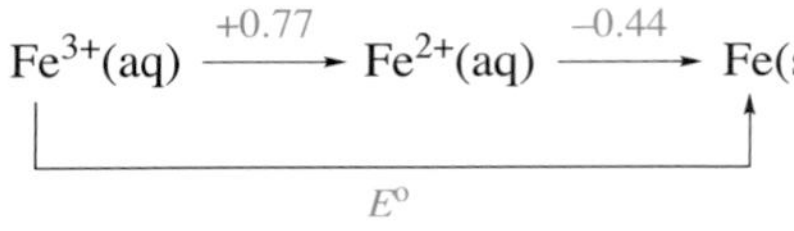

$$Fe^{3+}(aq) \xrightarrow{+0.77} Fe^{2+}(aq) \xrightarrow{-0.44} Fe(s)$$
$$E^o$$

Although there are short cuts to this problem, the most rigorous method is to determine $\Delta G^o(298\,K)$ for each step. Fe^{3+} to Fe^{2+} is a one-electron reduction.

$\Delta G^o_1 = -zFE^o$

$= -[1 \times 96\,485 \times 10^{-3} \times 0.77]$

$= -74.3\,kJ$ per mole of Fe^{3+}

Fe^{2+} to Fe is a two-electron reduction.

$\Delta G^o_2 = -zFE^o$

$= -[2 \times 96\,485 \times 10^{-3} \times (-0.44)]$

$= +84.9\,kJ$ per mole of Fe^{2+}

Next, find ΔG^o for the reduction of Fe^{3+} to Fe:

$\Delta G^o = \Delta G^o_1 + \Delta G^o_2$

$= -74.3 + 84.9$

$= +10.6\,kJ$ per mole of Fe^{3+}

Fe^{3+} to Fe is a 3-electron reduction; the standard reduction potential for the process is found from the corresponding value of ΔG°:

$$E^{\circ} = -\frac{\Delta G^{\circ}}{zF}$$

$$= -\frac{10.6}{3 \times 96\,485 \times 10^{-3}}$$

$$= -0.04\,V$$

Self-study exercises

1. Although the method given here is the most rigorous way to perform the calculation, substitution of a value for the Faraday constant may in fact be excluded. Why?

2. Construct a potential diagram for the reduction of aqueous Cr^{3+} to Cr^{2+}, followed by reduction to Cr. Values of E° for the Cr^{3+}/Cr^{2+} and Cr^{2+}/Cr couples are -0.41 and $-0.91\,V$, respectively. Calculate E° for the Cr^{3+}/Cr couple.   [*Ans.* $-0.74\,V$]

3. Construct a potential diagram (at pH 0) for the reduction of aqueous HNO_2 to NO and then to N_2O given that E° for the HNO_2/NO and NO/N_2O couples are $+0.98$ and $+1.59$ V respectively. Calculate E° for the following half-reaction:

$$2HNO_2(aq) + 4H^{+}(aq) + 4e^{-} \rightleftharpoons N_2O(g) + 3H_2O(l)$$

[*Ans.* $+1.29\,V$]

8.6 Frost–Ebsworth diagrams

Frost–Ebsworth diagrams and their relationship to potential diagrams

Frost–Ebsworth diagrams[†] represent the commonest graphical method of summarizing redox relationships for species containing a given element in different oxidation states. In a Frost–Ebsworth diagram, values of $-\Delta G^{\circ}$ or, more commonly, $-\Delta G^{\circ}/F$ for the formation of M(N) from M(0), where N is the oxidation state, are plotted against increasing N. From the relationship:

$$\Delta G^{\circ} = -zFE^{\circ}$$

it follows that $-\Delta G^{\circ}/F = zE^{\circ}$ and, therefore, a Frost–Ebsworth diagram can equally well be represented as a plot of zE° against oxidation state. Figure 8.3a shows the Frost–Ebsworth diagram for manganese in aqueous solution with $[H^{+}] = 1\,mol\,dm^{-3}$. This diagram can be constructed

---

[†] A.A. Frost (1951) *J. Am. Chem. Soc.*, vol. 73, p. 2680; E.A.V. Ebsworth (1964) *Educ. Chem.*, vol. 1, p. 123.

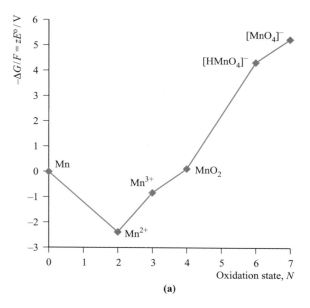

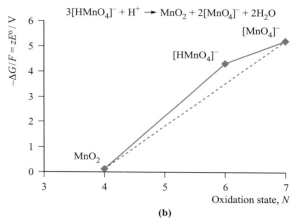

Fig. 8.3 The Frost–Ebsworth diagram for manganese in aqueous solution at pH 0, i.e. $[H^{+}] = 1\,mol\,dm^{-3}$.

from the corresponding potential diagram in Fig. 8.2 as follows.

- For Mn in its standard state, $\Delta G^{\circ} = 0$.
- For Mn(II), the relevant species is $Mn^{2+}(aq)$. E° for the Mn^{2+}/Mn couple is $-1.19\,V$. For the reduction of $Mn^{2+}(aq)$ to Mn(s):

$$\Delta G^{\circ} = -zFE^{\circ} = -2 \times F \times (-1.19) = +2.38F$$

$$-\frac{\Delta G^{\circ}}{F} = -2.38\,V$$

- For Mn(III), the relevant species is $Mn^{3+}(aq)$. E° for the Mn^{3+}/Mn^{2+} couple is $+1.54\,V$. For the reduction of $Mn^{3+}(aq)$ to $Mn^{2+}(aq)$:

$$\Delta G^{\circ} = -zFE^{\circ} = -1 \times F \times 1.54 = -1.54F$$

For $Mn^{3+}(aq)$, relative to Mn(0):

$$-\frac{\Delta G^{\circ}}{F} = -(-1.54 + 2.38) = -0.84\,V$$

• For Mn(IV), the relevant species is $MnO_2(s)$. E^o for the MnO_2/Mn^{3+} couple is $+0.95$ V. For the reduction of $MnO_2(s)$ to $Mn^{3+}(aq)$:

$$\Delta G^o = -zFE^o = -1 \times F \times 0.95 = -0.95F$$

For $MnO_2(s)$, relative to $Mn(0)$:

$$-\frac{\Delta G^o}{F} = -(-0.95 - 1.54 + 2.38) = +0.11 \text{ V}$$

Similarly, values of $-\Delta G^o/F$ for $[HMnO_4]^-$ and $[MnO_4]^-$ can be shown to be $+4.31$ and $+5.21$ V, respectively.

When negative oxidation states are involved, care must be taken in plotting appropriate values of $-\Delta G^o/F$. All points on a Frost–Ebsworth diagram refer to stability with respect to $-\Delta G^o/F = 0$ for the zero oxidation state of the element. Thus, for example, starting from $E^o = +1.09$ V for the $\frac{1}{2}Br_2/Br^-$ couple, a value of $-\Delta G^o/F = +1.09$ V is calculated for the reduction of $\frac{1}{2}Br_2$ to Br^-. For a Frost–Ebsworth diagram, we require a value of $-\Delta G^o/F$ that corresponds to the process $Br^- \longrightarrow \frac{1}{2}Br_2 + e^-$ and therefore the appropriate value of $-\Delta G^o/F$ is -1.09 V. This concept is further explored in end-of-chapter problem 8.24.

Interpretation of Frost–Ebsworth diagrams

Before looking at Fig. 8.3a in detail, we must note some general points about Frost–Ebsworth diagrams. Firstly, Fig. 8.3a and similar diagrams in this book *specifically refer to aqueous solution at pH 0*. For other conditions such as alkaline solution, a new diagram must be constructed for *each* pH value using relevant reduction potentials. Secondly, in Frost–Ebsworth diagrams in this text, the oxidation states are arranged in *increasing* order from left to right. However, some textbooks plot Frost–Ebsworth diagrams in the opposite direction and you should exercise caution when comparing diagrams from a range of data sources. Thirdly, it is usual to connect neighbouring points so that the Frost–Ebsworth diagram appears as a plot made up of linear sections. However, each point represents a chemical species and one can consider the relationship between *any pair* of points, not just neighbouring species. Finally, a Frost–Ebsworth plot provides information about the relative *thermodynamic* stabilities of various species; it says nothing about their kinetic stability.

Now let us use Fig. 8.3a to investigate the relative thermodynamic stabilities of different manganese-containing species in aqueous solution with $[H^+] = 1 \text{ mol dm}^{-3}$.

• The lowest point in Fig. 8.3a represents the most stable oxidation state of Mn in aqueous solution at pH 0, i.e. Mn(II).
• A move *downwards* on the plot represents a thermodynamically favoured process, e.g. at pH 0, $[MnO_4]^-$ is thermodynamically unstable with respect to all other species in Fig. 8.3a.

• A species towards the top-right of the diagram is oxidizing, e.g. $[MnO_4]^-$ is a strong oxidizing agent, stronger than $[HMnO_4]^-$.
• From the gradient of any line drawn between two points on the plot, E^o for the corresponding redox couple can be found. For example, the line between the points for Mn^{2+} and $Mn(0)$ corresponds to the reduction process:

$$Mn^{2+}(aq) + 2e^- \rightleftharpoons Mn(s)$$

and E^o for this half-reaction is found as follows:

$$E^o = \frac{\text{Gradient of line}}{\text{Number of electrons transferred}} = \frac{-2.38}{2}$$
$$= -1.19 \text{ V}$$

A *positive gradient* between two points indicates that E^o for the corresponding reduction process is positive, and a *negative gradient* indicates that E^o for the reduction process is negative.

• Any state represented on a 'convex' point is thermodynamically unstable with respect to *disproportionation*. This is illustrated in Fig. 8.3b where we focus on $[HMnO_4]^-$. It lies *above* a line drawn between two species with higher and lower oxidation states, namely $[MnO_4]^-$ and MnO_2 respectively. $[HMnO_4]^-$ is unstable with respect to these species, as the reaction in Fig. 8.3b shows. In Fig. 8.3a, Mn^{3+} also lies on a 'convex' point and is unstable with respect to Mn(IV) and Mn(II) (eq. 8.49).
• Any state represented on a 'concave' point is thermodynamically stable with respect to disproportionation, e.g. MnO_2 does not disproportionate.

Figure 8.4a shows a Frost diagram for chromium in aqueous solution at pH 0. Inspection of the diagram leads to the following conclusions about chromium species *under these solution conditions*:

• $E^o_{[Cr_2O_7]^{2-}/Cr^{3+}}$ has a positive value, while $E^o_{Cr^{3+}/Cr^{2+}}$ and $E^o_{Cr^{3+}/Cr}$ are both negative;
• $[Cr_2O_7]^{2-}$ is a powerful oxidizing agent and is reduced to Cr^{3+};
• Cr^{3+} is the most thermodynamically stable state;
• no species in the diagram shows a tendency towards disproportionation;
• Cr^{2+} is reducing and is oxidized to Cr^{3+}.

Figures 8.4b and 8.4c show potential diagrams for phosphorus and nitrogen in aqueous solution with $[H^+] = 1 \text{ mol dm}^{-3}$, and these diagrams are the subject of worked example 8.8. We shall make more use of potential (Latimer) diagrams than Frost–Ebsworth diagrams in later chapters in this book, but the latter can readily be constructed from data given in a potential diagram (see end-of-chapter problem 8.24).

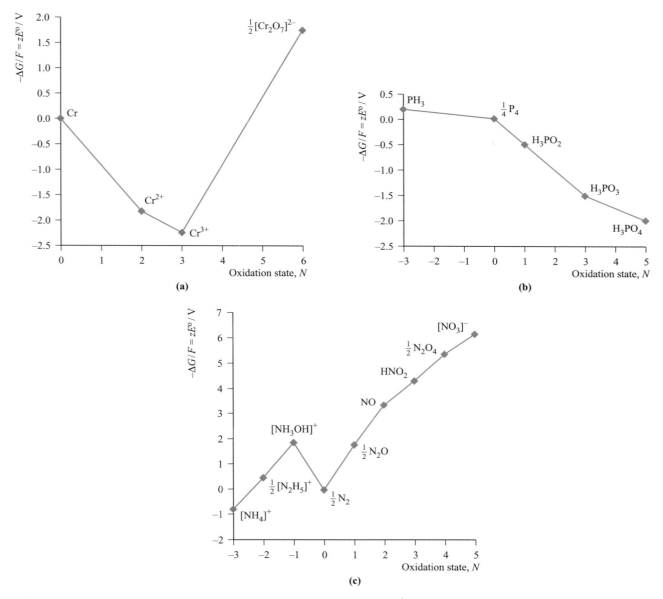

Fig. 8.4 Frost–Ebsworth diagrams in aqueous solution at pH 0, i.e. $[H^+] = 1 \, mol \, dm^{-3}$, for (a) chromium, (b) phosphorus and (c) nitrogen.

Worked example 8.8   Using Frost–Ebsworth diagrams

Use Fig. 8.4b to say something about the relative stabilities of the different oxidation states of phosphorus in aqueous media at pH 0.

Initial analysis of the diagram leads to the following conclusions:

- the most thermodynamically stable state is H_3PO_4 containing P(V);

- PH_3, i.e. P(−III), is the least thermodynamically stable state;

- in aqueous solution at pH 0, P_4 will disproportionate to PH_3 and H_3PO_2 (but see below);

- H_3PO_3 is stable with respect to disproportionation.

By drawing lines between the points for PH_3 and H_3PO_3, and between PH_3 and H_3PO_4, you can see that H_3PO_2 is unstable with respect to disproportionation, either to PH_3 and H_3PO_3, or to PH_3 and H_3PO_4. This illustrates the fact that you should look beyond the lines that are already represented in a given Frost–Ebsworth diagram.

Use Figs. 8.4b and 8.4c to answer these questions; both diagrams refer to the same aqueous solution conditions.

1. On going from N to P, how does the thermodynamic stability of the +5 oxidation state alter?

2. What do the diagrams tell you about the thermodynamic stability of N_2 and of P_4 with respect to other N- or P-containing species?

3. HNO_3 is a strong acid; the pK_a for HNO_2 is 3.37. Explain why the labels on Fig. 8.4c state HNO_2 and $[NO_3]^-$, rather than HNO_2 and HNO_3.

4. Estimate values for $E^o_{N_2/[NH_3OH]^+}$ and $E^o_{[NH_3OH]^+/[N_2H_5]^+}$ and comment on the thermodynamic stability of $[NH_3OH]^+$ in aqueous solution at pH 0.
   [*Ans.* ≈ −1.8 and +1.4 V respectively]

5. Which of the following species will tend to disproportionate: N_2O, NO, N_2, HNO_2?
   [*Ans.* N_2O, NO, HNO_2]

6. Using Fig. 8.4c, state whether you expect HNO_2 to disproportionate at pH 0? If so, what products might form?
   [*Ans.* see Section 15.9.]

7. From Fig. 8.4c, estimate ΔG^o(298 K) for the reduction process:

$$2HNO_2(aq) + 4H^+(aq) + 4e^- \longrightarrow N_2O(g) + 3H_2O(l)$$
   [*Ans.* ≈ −480 ± 10 kJ mol^{-1}]

8.7 The relationships between standard reduction potentials and some other quantities

Factors influencing the magnitudes of standard reduction potentials

In this section, we first consider factors that influence the magnitude of E^o for the Na^+/Na and Ag^+/Ag couples, by correlating these values with those of other, independently determined thermodynamic quantities. This comparison allows us to investigate the reasons why, in aqueous media, Na is so much more reactive than Ag, and gives an example that can be extended to other pairs or families of species.

Whereas the standard reduction potential for half-reaction 8.51 is readily measurable in aqueous solution (see Section 8.2), that for half-reaction 8.50 must be determined by a rather elaborate set of experiments involving Na amalgam electrodes (amalgams, see Box 22.3).

$$Na^+(aq) + e^- \rightleftharpoons Na(s) \qquad E^o = -2.71\,V \qquad (8.50)$$

$$Ag^+(aq) + e^- \rightleftharpoons Ag(s) \qquad E^o = +0.80\,V \qquad (8.51)$$

We can represent the general half-equation for M^+ reduction as taking place in steps as shown in Fig. 8.5. Since all standard reduction potentials are measured with respect to the standard hydrogen electrode (for which, by convention, ΔH^o, ΔG^o and ΔS^o are all zero), we must also consider the second thermodynamic cycle (involving absolute values) in Fig. 8.5. Table 8.2 lists values of ΔH^o for steps in the cycles

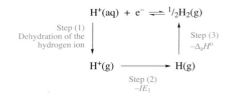

Fig. 8.5 The half-reaction for the reduction of M^{z+} ions to M, or H^+ to $\frac{1}{2}H_2$, can be considered in terms of three contributing steps for which thermodynamic data may be determined independently.

Table 8.2 Factors determining the magnitude of the standard reduction potentials for the Na^+/Na and Ag^+/Ag couples in aqueous solution (pH 0). Steps (1), (2) and (3) are defined in Fig. 8.5.

Redox couple	ΔH^o for step (1) /kJ mol^{-1}	ΔH^o for step (2) /kJ mol^{-1}	ΔH^o for step (3) /kJ mol^{-1}	Overall ΔH^o /kJ mol^{-1}	Calculated E^o /V †
Na^+/Na	404	−496	−108	−200	−2.48
$H^+/\frac{1}{2}H_2$	1091	−1312	−218	−439	0
Ag^+/Ag	480	−731	−285	−536	+1.01

†Values of E^o are estimated by dividing by $-zF$ ($z = 1$), and scaling to give $E^o(H^+/\frac{1}{2}H_2) = 0\,V$.

Table 8.3 Factors determining the magnitude of the standard reduction potentials for the Cu^{2+}/Cu and Zn^{2+}/Zn couples in aqueous solution (pH 0). Steps (1), (2) and (3) are defined in Fig. 8.5.

Redox couple	ΔH^o for step (1) / kJ mol^{-1}	ΔH^o for step (2) / kJ mol^{-1}	ΔH^o for step (3) / kJ mol^{-1}	Overall ΔH^o / kJ mol^{-1}	Calculated E^o / V [†]
Zn^{2+}/Zn	2047	-2639	-130	-722	-0.81
$H^+/\frac{1}{2}H_2$	1091	-1312	-218	-439	0
Cu^{2+}/Cu	2099	-2704	-338	-943	$+0.34$

[†]Values of E^o are estimated by dividing by $-zF$, and scaling to give $E^o(H^+/\frac{1}{2}H_2) = 0\,V$.

defined in Fig. 8.5. In an exact treatment, we ought to consider values of ΔG^o, but to a first approximation, we can ignore entropy changes (which largely cancel one another out in this case). From the thermodynamic data, we derive calculated values of E^o and these are given in the right-hand column of Table 8.2. There is good agreement between these values and the experimental ones for half-reactions 8.50 and 8.51. The enthalpy changes for steps (2) and (3) are both negative, and this is a general result for all elements. The *sign* of E^o is determined by the extent to which $\Delta H^o(1)$ offsets $[\Delta H^o(2) + \Delta H^o(3)]$.

Similar analyses for other metals can be carried out. For example, Cu and Zn are adjacent *d*-block metals, and it is interesting to investigate factors that contribute to the difference between E^o values for the Cu^{2+}/Cu and Zn^{2+}/Zn redox couples, and thus reveal how a balance of thermodynamic factors governs the spontaneous reaction that occurs in the Daniell cell (reaction 8.8). Table 8.3 lists relevant thermodynamic data. It is apparent that the crucial factor in making $E^o_{Cu^{2+}/Cu}$ significantly more positive than $E^o_{Zn^{2+}/Zn}$ is the greater enthalpy of atomization of Cu compared with that of Zn. Thus, what is often regarded as a purely 'physical' property plays a very important role in influencing chemical behaviour. Finally, if we were to consider factors influencing values of E^o for half-reaction 8.52, we would find that the variation in hydration enthalpies plays an important part (oxidizing power of halogens, see Section 17.4).

$$X_2 + 2e^- \rightleftharpoons 2X^- \qquad (X = F, Cl, Br, I) \qquad (8.52)$$

Values of $\Delta_f G^o$ for aqueous ions

In Section 7.9, we saw that the standard Gibbs energies of formation of aqueous ions can often be determined from E^o values. Worked example 8.9 provides an illustration of the use of reduction potential data in a calculation of a standard Gibbs energy of solution of an ionic salt.

Worked example 8.9   Determination of $\Delta_{sol}G^o$ for an ionic salt

Calculate the value of $\Delta_{sol}G^o$(298 K) for NaBr given that $\Delta_f G^o$(NaBr, s) is $-349.0\,kJ\,mol^{-1}$. ($F = 96\,485\,C\,mol^{-1}$)

The process to be considered is:

$$NaBr(s) \rightleftharpoons Na^+(aq) + Br^-(aq)$$

and the equation needed is:

$$\Delta_{sol}G^o = \Delta_f G^o(Na^+, aq) + \Delta_f G^o(Br^-, aq) - \Delta_f G^o(NaBr, s)$$

To find $\Delta_f G^o(Na^+, aq)$ and $\Delta_f G^o(Br^-, aq)$, we need (from Appendix 11) the standard reduction potentials for the processes:

$$Na^+(aq) + e^- \rightleftharpoons Na(s) \qquad E^o = -2.71\,V$$

$$\tfrac{1}{2}Br_2(l) + e^- \rightleftharpoons Br^-(aq) \qquad E^o = +1.09\,V$$

Now determine $\Delta_f G^o$ for each aqueous ion. For Na^+, the standard reduction potential refers to the reverse of the formation of Na^+(aq) and so a change of sign is required:

$$\Delta G^o = -zFE^o$$

$$-\Delta_f G^o(Na^+, aq) = -\frac{96\,485 \times (-2.71)}{1000} = 261.5\,kJ\,mol^{-1}$$

$$\Delta_f G^o(Br^-, aq) = -\frac{96\,485 \times 1.09}{1000} = -105.2\,kJ\,mol^{-1}$$

$$\Delta_{sol}G^o = \Delta_f G^o(Na^+, aq) + \Delta_f G^o(Br^-, aq) - \Delta_f G^o(NaBr, s)$$

$$= -261.5 + (-105.2) - (-349.0)$$

$$= -17.7\,kJ\,mol^{-1}$$

See Appendix 11 for values of E^o.

1. Calculate the value of $\Delta_{sol}G^o(298\,\text{K})$ for NaCl given that $\Delta_f G^o(\text{NaCl, s})$ is $-384.0\,\text{kJ}\,\text{mol}^{-1}$.

[*Ans.* $-8.7\,\text{kJ}\,\text{mol}^{-1}$]

2. $\Delta_{sol}G^o(298\,\text{K})$ for NaF $= +7.9\,\text{kJ}\,\text{mol}^{-1}$. Determine $\Delta_f G^o(\text{NaF, s})$ at 298 K. [*Ans.* $-546.3\,\text{kJ}\,\text{mol}^{-1}$]

3. Given that $\Delta_{sol}G^o(298\,\text{K})$ for KI is $-9.9\,\text{kJ}\,\text{mol}^{-1}$, calculate $\Delta_f G^o(\text{KI, s})$ at 298 K. [*Ans.* $-324.9\,\text{kJ}\,\text{mol}^{-1}$]

8.8 Applications of redox reactions to the extraction of elements from their ores

The Earth's environment is an oxidizing one and, in nature, many elements occur as oxides, sulfides or other compounds in which the element is in an oxidized form, e.g. tin occurs as *cassiterite* (SnO_2), and lead as *galena* (PbS). The extraction of these elements from their ores depends on redox chemistry. Heating cassiterite with carbon reduces Sn(IV) to Sn(0) (eq. 8.53), and Pb is extracted from galena by reaction sequence 8.54.

$$SnO_2 + C \xrightarrow{\Delta} Sn + CO_2 \qquad (8.53)$$

$$PbS \xrightarrow{O_2,\,\Delta} PbO \xrightarrow{C\,\text{or}\,CO,\,\Delta} Pb \qquad (8.54)$$

Examples of this type are numerous, and similar extraction processes are described in Box 6.1 and Chapters 21 and 22.

Ellingham diagrams

The choice of a reducing agent and the conditions for a particular extraction process can be assessed by using an *Ellingham diagram* such as that in Fig. 8.6. This illustrates how $\Delta_f G^o$ for a range of metal oxides and CO varies with temperature. In order that values are mutually comparable, $\Delta_f G^o$ refers to the Gibbs energy of formation *per half-mole of O_2.*[†] Thus for SrO, $\Delta_f G^o$ refers to reaction 8.55, and for Al_2O_3 it corresponds to reaction 8.56.

$$Sr + \tfrac{1}{2}O_2 \longrightarrow SrO \qquad (8.55)$$

$$\tfrac{2}{3}Al + \tfrac{1}{2}O_2 \longrightarrow \tfrac{1}{3}Al_2O_3 \qquad (8.56)$$

[†] Other data could have been plotted, e.g. values of $\Delta_f G^o$ per mole of O_2. *Consistency* is the keyword!

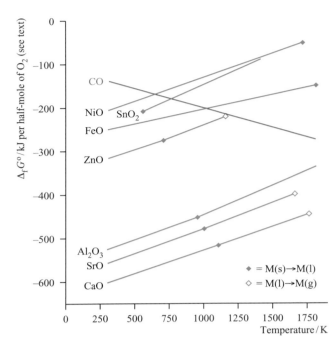

Fig. 8.6 An Ellingham diagram showing how the standard free energies of formation, $\Delta_f G^o$, of several metal oxides and carbon monoxide (the red line) vary with temperature. Values of $\Delta_f G^o$ refer to formation reactions involving a half-mole of O_2: $M + \tfrac{1}{2}O_2 \longrightarrow MO$, $\tfrac{1}{2}M + \tfrac{1}{2}O_2 \longrightarrow \tfrac{1}{2}MO_2$, or $\tfrac{2}{3}M + \tfrac{1}{2}O_2 \longrightarrow \tfrac{1}{3}M_2O_3$. The points marked ◆ and ◇ are the melting and boiling points, respectively, of the elemental metal.

In Fig. 8.6, each plot is either linear (e.g. NiO) or has two linear sections (e.g. ZnO). For the latter, there is a change in gradient at the melting point of the metal.

Three general results arise from Fig. 8.6:

- as the temperature increases, each *metal* oxide becomes *less* thermodynamically stable (less negative $\Delta_f G^o$);
- CO becomes *more* thermodynamically stable at higher temperatures (more negative $\Delta_f G^o$);
- the *relative* stabilities of the oxides at any given temperature can be seen directly from an Ellingham diagram.

The third point indicates how an Ellingham diagram can be applied. For example, at 1000 K, CO is more thermodynamically stable than SnO_2, and carbon can be used at 1000 K to reduce SnO_2 (eq. 8.53). On the other hand, reduction of FeO by carbon occurs at $T > 1000$ K.

The second point has a very important consequence: among the metal oxides in Fig. 8.6, the extraction of *any* of the metals from their respective oxides could involve carbon as the reducing agent. In fact at $T > 1800$ K, a greater range of metal oxides than in Fig. 8.6 may be reduced by carbon. However, on an industrial scale, this method of obtaining a metal from its oxide is often not commercially viable. Alternative methods of extracting metals from their ores are described in later chapters in the book.

KEY TERMS

The following terms were introduced in this chapter. Do you know what they mean?

- ❑ oxidation
- ❑ reduction
- ❑ oxidation state (oxidation number)
- ❑ half-reaction (half-equation)
- ❑ electrolytic cell
- ❑ galvanic cell

- ❑ standard conditions for a half-cell
- ❑ standard hydrogen electrode
- ❑ standard reduction potential, E^o
- ❑ standard cell potential, E^o_{cell}
- ❑ overpotential
- ❑ Nernst equation

- ❑ potential diagram (Latimer diagram)
- ❑ Frost–Ebsworth diagram
- ❑ Ellingham diagram

IMPORTANT THERMODYNAMIC EQUATIONS

$$E^o_{cell} = [E^o_{reduction\ process}] - [E^o_{oxidation\ process}]$$

$$\Delta G^o = -zFE^o_{cell}$$

$$\Delta G^o = -RT \ln K$$

$$E = E^o - \left\{ \frac{RT}{zF} \times \left(\ln \frac{[\text{reduced form}]}{[\text{oxidized form}]} \right) \right\}$$

(Nernst equation)

FURTHER READING

A.J. Bard, R. Parsons and J. Jordan (1985) *Standard Potentials in Aqueous Solution*, Marcel Dekker, New York – A critical compilation of values, the successor to Latimer's famous treatment of this subject.

A.M. Bond (2004) in *Comprehensive Coordination Chemistry II*, eds J.A. McCleverty and T.J. Meyer, Elsevier, Oxford, vol. 2, p. 197 – 'Electrochemistry: general introduction' focuses on principles and methods applicable to the coordination chemist.

J. Burgess (1978) *Metal Ions in Solution*, Ellis Horwood, Chichester and Halsted Press, New York – A thorough treatment of most aspects of metal ions in both aqueous and non-aqueous solutions.

J. Burgess (1999) *Ions in Solution: Basic Principles of Chemical Interaction*, 2nd edn, Horwood Publishing,

Westergate – An excellent introduction to the properties of ions in aqueous solutions, including treatment of the thermodynamics of redox reactions.

R.G. Compton and G.H.W. Sanders (1996) *Electrode Potentials*, Oxford University Press, Oxford – A useful introduction to electrochemical equilibria and electrochemical principles.

D.A. Johnson (1982) *Some Thermodynamic Aspects of Inorganic Chemistry*, 2nd edn, Cambridge University Press, Cambridge – Contains a very useful discussion of solubility and redox potentials.

W.L. Jolly (1991) *Modern Inorganic Chemistry*, 2nd edn, McGraw-Hill, New York – Contains a treatment of redox potentials which complements that given in this chapter by discussing some systems involving non-metals.

PROBLEMS

8.1 Give the oxidation state of each element in the following compounds and ions; Pauling electronegativity values in Appendix 7 may be useful: (a) CaO; (b) H_2O; (c) HF; (d) $FeCl_2$; (e) XeF_6; (f) OsO_4; (g) Na_2SO_4; (h) $[PO_4]^{3-}$; (i) $[PdCl_4]^{2-}$; (j) $[ClO_4]^-$; (k) $[Cr(OH_2)_6]^{3+}$.

8.2 What oxidation state change does each *metal* undergo in the following reactions or half-reactions?
(a) $[Cr_2O_7]^{2-} + 14H^+ + 6e^- \longrightarrow 2Cr^{3+} + 7H_2O$
(b) $2K + 2H_2O \longrightarrow 2KOH + H_2$
(c) $Fe_2O_3 + 2Al \xrightarrow{\Delta} 2Fe + Al_2O_3$
(d) $[MnO_4]^- + 2H_2O + 3e^- \longrightarrow MnO_2 + 4[OH]^-$

8.3 Which of the following reactions are redox reactions? In those that are, identify the oxidation and reduction processes.
(a) $N_2 + 3Mg \xrightarrow{\Delta} Mg_3N_2$
(b) $N_2 + O_2 \longrightarrow 2NO$
(c) $2NO_2 \longrightarrow N_2O_4$
(d) $SbF_3 + F_2 \longrightarrow SbF_5$
(e) $6HCl + As_2O_3 \longrightarrow 2AsCl_3 + 3H_2O$
(f) $2CO + O_2 \longrightarrow 2CO_2$
(g) $MnO_2 + 4HCl \longrightarrow MnCl_2 + Cl_2 + 2H_2O$
(h) $[Cr_2O_7]^{2-} + 2[OH]^- \rightleftharpoons 2[CrO_4]^{2-} + H_2O$

8.4 In each redox reaction in problem 8.3, confirm that the net increases and decreases in oxidation states balance each other.

8.5 Using data from Table 8.1, write down the spontaneous cell process, and calculate E°_{cell} and ΔG° for the following combinations of half-cells:

(a) $Ag^+(aq) + e^- \rightleftharpoons Ag(s)$
$\qquad$ with $Zn^{2+}(aq) + 2e^- \rightleftharpoons Zn(s)$

(b) $Br_2(aq) + 2e^- \rightleftharpoons 2Br^-(aq)$
$\qquad$ with $Cl_2(aq) + 2e^- \rightleftharpoons 2Cl^-(aq)$

(c) $[Cr_2O_7]^{2-}(aq) + 14H^+(aq) + 6e^-$
$\qquad \rightleftharpoons 2Cr^{3+}(aq) + 7H_2O(l)$
$\qquad$ with $Fe^{3+}(aq) + e^- \rightleftharpoons Fe^{2+}(aq)$

8.6 Use the data in Appendix 11 to rationalize *quantitatively* why:

(a) Mg liberates H_2 from dilute HCl, but Cu does not;
(b) Br_2 liberates I_2 from aqueous KI solution, but does not liberate Cl_2 from aqueous KCl solution;
(c) the role of Fe^{3+} ions as an oxidizing agent is influenced by the presence of certain ligands in solution;
(d) a method of growing Ag crystals is to immerse a zinc foil in an aqueous solution of $AgNO_3$.

8.7 Consider the half-reaction:

$$[MnO_4]^-(aq) + 8H^+(aq) + 5e^- \rightleftharpoons Mn^{2+}(aq) + 4H_2O(l)$$
$$E^{\circ} = +1.51 \text{ V}$$

If the ratio of concentrations of $[MnO_4]^- : Mn^{2+}$ is 100:1, determine E at pH values of (a) 0.5; (b) 2.0; and (c) 3.5 ($T = 298$ K). Over this pH range, how does the ability of permanganate(VII) (when being reduced to Mn^{2+}) to oxidize aqueous chloride, bromide or iodide ions change?

8.8 (a) Using appropriate data from Appendix 11, determine E°_{cell} for the disproportionation of H_2O_2. (b) Calculate ΔG° for this process. (c) Comment on the fact that H_2O_2 can be stored without significant decomposition, unless, for example, traces of MnO_2, $[OH]^-$ or iron metal are added.

8.9 Use the following experimental data to determine $E^{\circ}_{Cu^{2+}/Cu}$, and comment on the need (or not) to make use of *all* the data given.

$[Cu^{2+}]/\text{mol dm}^{-3}$	0.001	0.005	0.010	0.050
E/V	0.252	0.272	0.281	0.302

8.10 (a) Calculate $E_{Ag^+/Ag}$ for a half-cell in which the concentration of silver(I) ions is 0.1 mol dm^{-3} ($T = 298$ K). (b) Are silver(I) ions more or less easily reduced by zinc in this solution than under standard conditions? Quantify your answer in thermodynamic terms.

8.11 Given that K_{sp} for AgI is 8.51×10^{-17}, and $E^{\circ}_{Ag^+/Ag} = +0.80$ V, calculate E° for the reduction step:

$$AgI(s) + e^- \rightleftharpoons Ag(s) + I^-(aq)$$

Hence confirm the statement in Section 8.3 that reduction of silver(I) when in the form of solid AgI is thermodynamically less favourable than reduction of AgCl.

8.12 Using data from Table 8.1 and from Section 8.3, explain why H_2 is evolved when powdered Ag is heated with a concentrated solution of HI.

8.13 Calculate the overall formation constant for $[Fe(CN)_6]^{3-}$, given that the overall formation constant for $[Fe(CN)_6]^{4-}$ is $\approx 10^{32}$, and that:

$Fe^{3+}(aq) + e^- \rightleftharpoons Fe^{2+}(aq)$ $\qquad E^{\circ} = +0.77$ V
$[Fe(CN)_6]^{3-}(aq) + e^- \rightleftharpoons [Fe(CN)_6]^{4-}(aq)$ $E^{\circ} = +0.36$ V

8.14 Using data in Appendix 11, determine which of the following species is thermodynamically unstable with respect to disproportionation (and under what conditions) in aqueous solution: (a) Fe^{2+}; (b) Sn^{2+}; (c) $[ClO_3]^-$.

8.15 Determine $\Delta G^{\circ}(298$ K) for the reaction:

$$2CuCl(s) \rightleftharpoons Cu^{2+}(aq) + 2Cl^-(aq) + Cu(s)$$

given the following data:

$2Cu^+(aq) \rightleftharpoons Cu^{2+}(aq) + Cu(s)$ $\quad K = 1.81 \times 10^6$
$CuCl(s) \rightleftharpoons Cu^+(aq) + Cl^-(aq)$ $\quad K_{sp} = 1.72 \times 10^{-7}$

What does the value of ΔG° tell you about the tendency of precipitated CuCl to disproportionate?

8.16 Using appropriate data from eqs. 8.42 to 8.46, confirm the value of E° given for eq. 8.47.

8.17 Write balanced half-equations corresponding to the steps shown in the potential diagrams in Fig. 8.2.

8.18 (a) Use data from Appendix 11 to construct a potential diagram showing the redox chemistry of vanadium in aqueous solution at pH 0. (b) Use your diagram to establish whether any vanadium species is unstable with respect to disproportionation.

8.19 The following potential diagram summarizes the results of electrochemical studies of the aqueous solution (pH 0) chemistry of uranium:

$$[UO_2]^{2+} \xrightarrow{+0.06} [UO_2]^{2+} \xrightarrow{+0.61} U^{4+} \xrightarrow{-0.61} U^{3+} \xrightarrow{-1.80} U$$

with a lower connecting line labelled $+0.33$

Use the information to deduce as much as possible about the chemistry of uranium under these conditions.

$$[ClO_4]^- \xrightarrow{+1.19} [ClO_3]^- \xrightarrow{+1.15} ClO_2 \xrightarrow{+1.28} HClO_2 \xrightarrow{+1.65} HClO \xrightarrow{+1.61} Cl_2 \xrightarrow{+1.36} Cl^-$$

Fig. 8.7 Potential diagram (Latimer diagram) for chlorine in aqueous solution at pH 0, i.e. $[H^+] = 1 \, mol \, dm^{-3}$.

8.20 The following potential diagram is part of that illustrating the redox chemistry of chlorine in aqueous solution at pH 0. (a) Calculate the value of E^o for the reduction of $[ClO_3]^-$ to $HClO_2$. (b) Justify why, *in this case*, the value of E^o can simply be taken to be the mean of $+1.15$ and $+1.28 \, V$.

8.21 By constructing thermodynamic cycles analogous to those shown in Fig. 8.5, discuss the factors that contribute to the trend in values of E^o for the group 1 metals Li to Cs. [$\Delta_{hyd}H^o$: see Table 7.6. *IE* and $\Delta_{atom}H^o$: see Appendices 8 and 10.]

8.22 (a) Using standard reduction potentials from Appendix 11, determine values of $\Delta_f G^o(K^+, \, aq)$ and $\Delta_f G^o(F^-, \, aq)$. (b) Hence, find $\Delta_{sol}G^o(KF, \, s)$ at 298 K, if $\Delta_f G^o(KF, \, s) = -537.8 \, kJ \, mol^{-1}$. (c) What does the value for $\Delta_{sol}G^o(KF, \, s)$ imply about the solubility of KF in water?

8.23 Using data from Appendix 11, and the value for the standard Gibbs energy of formation for PbS of $-99 \, kJ \, mol^{-1}$, determine a value for K_{sp} for this salt.

8.24 Use the data in the potential diagram shown in Fig. 8.7 to construct a Frost–Ebsworth diagram for chlorine. Hence show that Cl^- is the most thermodynamically favoured species of those in the diagram. Which species in the diagram is (a) the best oxidizing agent and (b) the best reducing agent?

OVERVIEW PROBLEMS

8.25 Use the data in Appendix 11 to rationalize the following observations in a *quantitative* manner. What assumption(s) have you made in answering this question?

(a) The dithionate ion, $[S_2O_6]^{2-}$, can be prepared by controlled oxidation of $[SO_3]^{2-}$ using MnO_2.
(b) In the presence of acid, KI and KIO_3 react to form I_2.
(c) Mn^{2+} is instantly oxidized to $[MnO_4]^-$ by aqueous solutions of H_4XeO_6.

8.26 (a) Using the potential diagram below (at pH 14), calculate $E^o{}_{O_3^-/O_2}$.

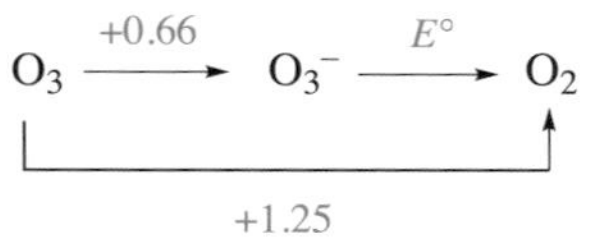

$$O_3 \xrightarrow{+0.66} O_3^- \xrightarrow{E^o} O_2$$
$$\underset{+1.25}{\underline{\hspace{3cm}}}$$

(b) Comment on the following data:

$$Cd^{2+}(aq) + 2e^- \rightleftharpoons Cd(s) \qquad E^o = -0.40 \, V$$
$$[Cd(CN)_4]^{2-}(aq) + 2e^- \rightleftharpoons Cd(s) + 4[CN]^-$$
$$E^o = -1.03 \, V$$

(c) How valid is Fig. 8.4a for aqueous solutions at pH 2?

8.27 In hydrochloric acid, HOI reacts to give $[ICl_2]^-$. Use the potential diagrams below to explain why HOI disproportionates in aqueous acidic solution, but does not when the acid is aqueous HCl.

$$[IO_3]^- \xrightarrow{+1.14} HOI \xrightarrow{+1.44} I_2$$
$$[IO_3]^- \xrightarrow{+1.23} [ICl_2]^- \xrightarrow{+1.06} I_2$$

8.28 Additional data needed for this question can be found in Appendix 11.

(a) Determine $E_{Zn^{2+}/Zn}$ (at 298 K) for a half-cell in which $[Zn^{2+}] = 0.25 \, mol \, dm^{-3}$.
(b) Calculate the reduction potential for the half-reaction:

$$[VO]^{2+}(aq) + 2H^+(aq) + e^- \rightleftharpoons V^{3+}(aq) + H_2O(l)$$

if the ratio of the concentrations of $[VO]^{2+} : V^{3+}$ is 1:2 and the pH of the solution is 2.2.

8.29 (a) In aqueous solution at pH 0, Mn^{3+} disproportionates to MnO_2 and Mn^{2+}. Write equations for the two half-reactions involved in this process. (b) Use Fig. 8.2 to obtain values of E^o for the half-equations in part (a). (c) Determine $E^o{}_{cell}$ and a value of $\Delta G^o(298 \, K)$ for the disproportionation of $Mn^{3+}(aq)$ at pH 0. Write an equation to which this value of $\Delta G^o(298 \, K)$ refers.

8.30 (a) Use appropriate data from Appendix 11 to determine the ratio of the overall stability constants of the complexes $[Fe(phen)_3]^{2+}$ and $[Fe(phen)_3]^{3+}$ at 298 K.

(b) Use the data in Fig. 8.2 to construct a Frost–Ebsworth diagram for manganese in aqueous solution at pH 14. Use your diagram to comment on the stability of $[MnO_4]^{3-}$ under these conditions.

8.31 In each of the following reactions, relate starting materials and products by the processes of *reduction, oxidation, disproportionation* or *no redox change*. In some reactions, more than one process is taking place.

(a) $[HCO_3]^- + [OH]^- \longrightarrow [CO_3]^{2-} + H_2O$
(b) $Au + HNO_3 + 4HCl \longrightarrow HAuCl_4 + NO + 2H_2O$
(c) $2VOCl_2 \longrightarrow VOCl_3 + VOCl$
(d) $SO_2 + 4H^+ + 4Fe^{2+} \longrightarrow S + 4Fe^{3+} + 2H_2O$
(e) $2CrO_2Cl_2 + 3H_2O \longrightarrow [Cr_2O_7]^{2-} + 4Cl^- + 6H^+$
(f) $[IO_4]^- + 2I^- + H_2O \longrightarrow [IO_3]^- + I_2 + 2[OH]^-$
(g) $2KCl + SnCl_4 \longrightarrow K_2[SnCl_6]$
(h) $2NO_2 + H_2O \longrightarrow HNO_2 + HNO_3$

INORGANIC CHEMISTRY MATTERS

8.32 Provide explanations for the following observations.

(a) In moist air, corrosion of iron is spontaneous. However, under anaerobic (O_2 free), wet conditions, corrosion of iron is only marginally favoured.

(b) If iron is exposed to dry, gaseous O_2, surface oxidation occurs but afterwards, the bulk iron is protected against further corrosion.

8.33 A plumber directly connects a galvanized steel pipe to a copper pipe in a system that carries running water. Suggest what will happen over a period of time.

8.34 Consider the following two scenarios: (i) aluminium rivets used to connect two steel plates, and (ii) steel rivets used to connect two aluminium plates. Discuss whether these choices would be sensible.

8.35 Zinc/silver oxide button batteries are used in calculators and watches. Although the silver component makes them expensive, this is outweighed by their high performance. The battery uses KOH as the electrolyte, and the overall cell reaction is:

$$Ag_2O + H_2O + Zn \longrightarrow 2Ag + Zn(OH)_2$$

(a) What are the two half-reactions that make up the cell?

(b) If Ag_2O acts as the cathode, and E^o values for the two half-cells are -1.25 and $+0.34$ V, assign each reduction potential to the correct half-cell in part (a).

(c) What is $E^o{}_{cell}$? Calculate ΔG^o(298 K) for the cell reaction written above.

8.36 The commercial purification of copper metal is carried out in electrolytic cells. The anode is composed of impure ('blister') copper, and the electrolyte is a mixture of aqueous $CuSO_4$ and H_2SO_4. During purification, copper is effectively transferred from the anode to the cathode, and pure copper is thereby produced. (a) How does an electrolytic cell differ from a galvanic cell? (b) Write half equations for the cathode and anode reactions. (c) Is the overall cell reaction spontaneous? If not, how does it occur?

8.37 A tarnished silver knife is placed in a beaker containing hot aqueous $NaHCO_3$. A piece of Al foil is placed in the solution so that it touches the knife. The deposit of Ag_2S disappears and the clean knife is recovered from the solution. (a) Using Appendix 11 and the data given below, write an equation for the electrochemical process that cleans the knife, and determine $E^o{}_{cell}$ and $\Delta G^o{}_{cell}$ (298 K). (b) Al_2S_3 is decomposed by water. Write an equation to show what happens. (c) Write an equation for the equilibria set up when $NaHCO_3$ is added to H_2O. Suggest *two* reasons why aqueous $NaHCO_3$ rather than distilled water is used in the cleaning process.

Data: $Ag_2S + 2e^- \rightleftharpoons 2Ag + S^{2-}$ $\qquad E^o = -0.69$ V

K_{sp} for $Al(OH)_3 = 1.9 \times 10^{-33}$

[Based on: M.M. Ivey *et al.* (2008) *J. Chem. Educ.*, vol. 85, p. 68.]

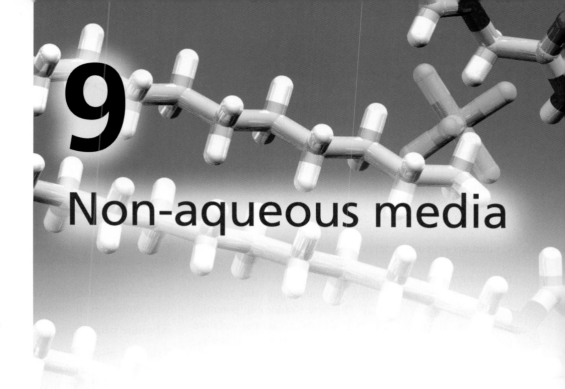

Topics

9

Non-aqueous media

9.1 Introduction

Although many inorganic reactions take place in aqueous solution, it is not always suitable to use water as a solvent. Some reagents may react with H_2O (e.g. the alkali metals) and non-polar molecules are insoluble in water. This chapter discusses *non-aqueous solvents*. The use of solvents other than water is commonplace, and common organic solvents include dichloromethane, hexane, toluene and ethers such as diethyl ether, **9.1**, tetrahydrofuran, **9.2**, and diglyme, **9.3**.

diethyl ether　　　　tetrahydrofuran (THF)

(9.1)　　　　　　　**(9.2)**

diglyme

(9.3)

These solvents are of significant use to the inorganic chemist, but more exotic solvents are also available. These include liquid NH_3, liquid SO_2, H_2SO_4, BrF_3 and ionic liquids such as [pyBu][AlCl$_4$], **9.4** (see Section 9.12).

N-butylpyridinium ion　　　tetrachloroaluminate ion

(9.4)

We can conveniently place non-aqueous solvents into the following categories:

- protic solvents (e.g. HF, H_2SO_4, MeOH);
- aprotic solvents (e.g. N_2O_4, BrF_3);
- coordinating solvents (e.g. MeCN, Et_2O, Me_2CO).

> A *protic solvent* undergoes *self-ionization* (see Section 7.2) to provide protons which are solvated. *If* it undergoes self-ionization, an *aprotic solvent* does so without the formation of protons.

As we discuss the properties and uses of some non-aqueous solvents, we must keep in mind that the extent to which non-aqueous solvents can be used is limited by the fact that many are highly reactive.

Quantitative data are scarce for non-aqueous media, and, in solvents of relative permittivity lower than that of water, data are difficult to interpret because of ion-association. Much of the discussion therefore centres on the properties and uses of selected solvents.

A number of non-aqueous solvents (e.g. NH_3, EtOH, H_2SO_4) exhibit *hydrogen bonding*. An $X-H\cdots Y$ interaction is called a *hydrogen bond* if it constitutes a local bond, and if $X-H$ acts as a proton donor to Y. Whether or not solvent molecules form intermolecular hydrogen bonds affects properties such as boiling point, enthalpy of vaporization and viscosity, as well as the ability of the solvent to solvate particular ions or molecules. If you are unfamiliar with the concept of hydrogen bonding, you may wish to read Section 10.6 before studying Chapter 9.

9.2 Relative permittivity

Before beginning a discussion of non-aqueous solvents, we must define the *relative permittivity*, also referred to as the *dielectric constant*, of a substance. In a vacuum, the Coulombic potential energy of a system of two unit electronic charges is given by eq. 9.1 where ε_0 is the (absolute) permittivity of a vacuum $(8.854 \times 10^{-12} \, \text{F m}^{-1})$, e is the charge on the electron $(1.602 \times 10^{-19} \, \text{C})$ and r is the separation (in metres) between the point charges.

$$\text{Coulombic potential energy} = \frac{e^2}{4\pi\varepsilon_0 r} \qquad (9.1)$$

If a material is placed between the charges, the force is reduced by an amount that depends upon the *relative permittivity* of the material. The new Coulombic potential energy is given by eq. 9.2 where ε_r is the relative permittivity of the material. Since it is a *relative* quantity, ε_r is dimensionless.

$$\text{Coulombic potential energy} = \frac{e^2}{4\pi\varepsilon_0 \varepsilon_r r} \qquad (9.2)$$

For example, at 298 K, ε_r of water (the dielectric constant) is 78.7 but, as Fig. 9.1 shows, ε_r varies with temperature. A value of 78.7 can be considered to be a 'high' value and from eq. 9.2, we see that in aqueous solution, the force between two point charges (or two ions) is considerably reduced compared with that in a vacuum. Thus we can consider a dilute aqueous solution of a salt to contain well-separated, non-interacting ions.

Table 9.1 lists dielectric constants for water and a range of common organic solvents. The *absolute* permittivity of

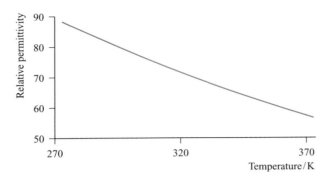

Fig. 9.1 Variation in the relative permittivity (dielectric constant) of water as a function of temperature.

a solvent is found using eq. 9.3, but it is usual to discuss solvent properties in terms of the relative values.

$$\text{Absolute permittivity of a material} = \varepsilon_0 \varepsilon_r \qquad (9.3)$$

Table 9.1 also gives the dipole moment (μ) of each solvent. In general, the trend in values of dipole moments follows that in values of the relative permittivities for solvents having related structures. Ion–solvent interactions are favoured (e.g. to facilitate the dissolution of an ionic salt) by using a solvent with a large dipole moment, but for maximum effect, the solvent molecule should also be small, and both ends of it should be able to interact with the ions in the same way that water interacts with cations through the oxygen atoms (see Fig. 7.5) and with anions through the hydrogen atoms. Thus, ammonia $(\varepsilon_r = 25.0, \mu = 1.47 \, \text{D})$ is a better solvent (see Section 9.6) for ionic salts than dimethylsulfoxide or nitromethane, even though these have ε_r values of 46.7 and 35.9, and dipole moments of 3.96 and 3.46 D, respectively.

Table 9.1 Relative permittivity (dielectric constant) values at 298 K (unless otherwise stated) for water and selected organic solvents.

Solvent	Formula[†]	Relative permittivity, ε_r	Dipole moment, μ / debye
Formamide	$HC(O)NH_2$	109 (293 K)	3.73
Water	H_2O	78.7	1.85
Acetonitrile	MeCN	37.5 (293 K)	3.92
N,N-Dimethylformamide (DMF)	$HC(O)NMe_2$	36.7	3.86
Nitromethane	$MeNO_2$	35.9 (303 K)	3.46
Methanol	MeOH	32.7	1.70
Ethanol	EtOH	24.3	1.69
Dichloromethane	CH_2Cl_2	9.1 (293 K)	1.60
Tetrahydrofuran	C_4H_8O (structure **9.2**)	7.6	1.75
Diethyl ether	Et_2O	4.3 (293 K)	1.15
Benzene	C_6H_6	2.3	0

[†] Me = methyl; Et = ethyl.

In the text, we state that *in general, the trend in dipole moments follows that in values of the relative permittivities for solvents having related structures.* Carry out the following two exercises, and then critically assess this statement.

1. Produce a scatter plot (i.e. no line or curve fitted to the points) for values of μ against ε_r for all the solvents listed in Table 9.1.

2. Produce a scatter plot for values of μ against ε_r for H_2O, MeOH, EtOH and Et_2O using data listed in Table 9.1.

9.3 Energetics of ionic salt transfer from water to an organic solvent

In this section, we consider the changes in enthalpy and Gibbs energy that accompany the transfer of simple ions from water to some organic solvents of high relative permittivity. These data provide us with an idea of the *relative* abilities of water and these organic liquids to act as solvents with regard to the ions considered. Since most organic liquids are soluble in water to some extent, or are completely miscible with water, thermodynamic data for the dissolution of salts are usually obtained by considering the two solvents separately. Data for the transfer of ions ($\Delta_{transfer}G^\circ$ and $\Delta_{transfer}H^\circ$) can be derived from the differences between the values corresponding to the dissolution processes in the two solvents. Our discussion centres on four organic solvents:

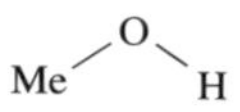

methanol (**9.5**), formamide (**9.6**), N,N-dimethylformamide (DMF, **9.7**) and acetonitrile (**9.8**), relative permittivities and dipole moments for which are listed in Table 9.1.

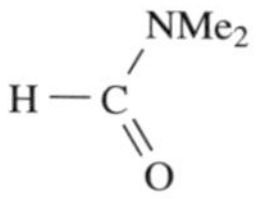

methanol	formamide	DMF	acetonitrile
(**9.5**)	(**9.6**)	(**9.7**)	(**9.8**)

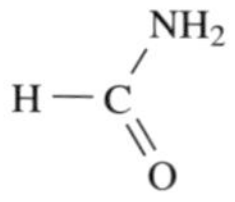

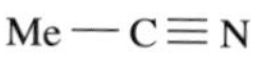

In an analogous approach to that discussed in Section 7.9, we can make the assumption that very large ions such as $[Ph_4As]^+$ and $[BPh_4]^-$ have the same values of $\Delta_{transfer}G^\circ$ and the same values of $\Delta_{transfer}H^\circ$. By considering a series of $[Ph_4As]X$ and $M[BPh_4]$ salts (in conjunction with $[Ph_4As][BPh_4]$), it is possible to obtain the thermodynamic data given in Table 9.2, where $\Delta_{transfer}H^\circ$ and $\Delta_{transfer}G^\circ$ refer to the transfer of the specified ion from water to the organic solvent. A positive value of $\Delta_{transfer}G^\circ$ indicates an unfavourable transfer, while a negative value corresponds to a favourable process.

The data in Table 9.2 show that the large, non-polar $[Ph_4As]^+$ and $[BPh_4]^-$ ions are more solvated in each organic solvent than in water; enthalpy and entropy effects both contribute in the same direction. Alkali metal ions exhibit no simple pattern of behaviour, although in each solvent, values of $\Delta_{transfer}H^\circ$ and $\Delta_{transfer}G^\circ$ are less positive for the alkali metal ions than for the halide ions. For the halide ions, transfer from water to the organic media is thermodynamically unfavourable, but we can go further than this generalization. Methanol and formamide are

Table 9.2 Values of $\Delta_{transfer}H^\circ$ and $\Delta_{transfer}G^\circ$ for the transfer of ions from water to an organic solvent at 298 K.

Ion	Methanol		Formamide		N,N-Dimethylformamide		Acetonitrile	
	$\Delta_{transfer}H^\circ$ / kJ mol^{-1}	$\Delta_{transfer}G^\circ$ / kJ mol^{-1}	$\Delta_{transfer}H^\circ$ / kJ mol^{-1}	$\Delta_{transfer}G^\circ$ / kJ mol^{-1}	$\Delta_{transfer}H^\circ$ / kJ mol^{-1}	$\Delta_{transfer}G^\circ$ / kJ mol^{-1}	$\Delta_{transfer}H^\circ$ / kJ mol^{-1}	$\Delta_{transfer}G^\circ$ / kJ mol^{-1}
F^-	12	20	20	25	–	$\approx$60	–	71
Cl^-	8	13	4	14	18	48	19	42
Br^-	4	11	−1	11	1	36	8	31
I^-	−2	7	−7	7	−15	20	−8	17
Li^+	−22	4	−6	−10	−25	−10	–	25
Na^+	−20	8	−16	−8	−32	−10	−13	15
K^+	−19	10	−18	−4	−36	−10	−23	8
$[Ph_4As]^+, [BPh_4]^-$	−2	−23	−1	−24	−17	−38	−10	−33

capable of forming hydrogen bonds (see structures **9.9**) between the H atoms of the OH or NH_2 groups and the halide ions in solution; MeCN and DMF do not possess this capability.

$\cdots\cdots\cdots$ = hydrogen bond

(9.9)

Not only are the values of $\Delta_{transfer}G^{\circ}$ for the halide ion significantly more positive for MeCN and DMF than for MeOH and formamide, but the variation in values among the halide ions is much greater. We may conclude that halide ions (and F^- and Cl^- in particular) are much less strongly solvated in solvents in which hydrogen bonding is not possible than in those in which hydrogen-bonded interactions can form (these, of course, include water). This difference is the origin of the solvent dependence of reactions involving halide ions. An example is the bimolecular reaction 9.4, for which the rate increases from X = F to I in aqueous solution, but decreases in DMF.

$$CH_3Br + X^- \longrightarrow CH_3X + Br^- \quad (X = F, Cl \text{ or } I) \quad (9.4)$$

Fluoride ion in solvents with which it is not able to form hydrogen bonds is sometimes described as 'naked', but this term is misleading. In DMF, it still has a Gibbs energy of solvation of about $-400 \, kJ \, mol^{-1}$ ($\approx 60 \, kJ \, mol^{-1}$ less negative than in water) and so is still very much less reactive than in the gas phase.

Self-study exercises

You may need to refer to the first part of Section 10.6 before attempting the following exercises.

1. Why can EtOH form hydrogen bonds with chloride ions, whereas Et_2O cannot?

2. Which of the following solvents are capable of forming hydrogen bonds with Br^-: MeOH; THF; DMF; $MeNO_2$; H_2O? [*Ans.* MeOH, H_2O]

3. Figure 7.5 shows H_2O molecules coordinating to Na^+. Explain how THF and MeCN may act as coordinating solvents.

9.4 Acid–base behaviour in non-aqueous solvents

Strengths of acids and bases

When we dealt with acid–base behaviour in aqueous solution in Chapter 7, we saw that the strength of an acid HX (eq. 9.5) depended upon the relative proton donor abilities of HX and $[H_3O]^+$.

$$HX(aq) + H_2O(l) \rightleftharpoons [H_3O]^+(aq) + X^-(aq) \quad (9.5)$$

Similarly, the strength of a base, B, in aqueous solution depends upon the relative proton accepting abilities of B and $[OH]^-$ (eq. 9.6).

$$B(aq) + H_2O(l) \rightleftharpoons [BH]^+(aq) + [OH]^-(aq) \quad (9.6)$$

Tabulated values of K_a (or K_b) generally refer to the ionizations of acids in *aqueous solution*, and in stating that 'HCl is a strong acid', we assume an aqueous medium. However, if HCl is dissolved in acetic acid, the extent of ionization is far less than in water and HCl behaves as a weak acid.

Levelling and differentiating effects

Non-aqueous solvents that are good proton acceptors (e.g. NH_3) encourage acids to ionize in them. Thus, in a *basic solvent*, all acids are strong. The solvent is said to exhibit a *levelling effect* on the acid, since the strength of the dissolved acid cannot exceed that of the protonated solvent. For example, in aqueous solution, no acidic species can exist that is a stronger acid than $[H_3O]^+$. In an acidic solvent (e.g. $MeCO_2H$, H_2SO_4), ionization of bases is facilitated. Most acids are relatively weak under these conditions, and some even ionize as bases.

We noted above that HCl, when dissolved in acetic acid, behaves as a weak acid. Hydrogen bromide and hydrogen iodide behave similarly but the *extent of ionization* of the three hydrogen halides varies along the series: HI > HBr > HCl. This contrasts with the fact that all three compounds are classed as strong acids (i.e. fully ionized) in aqueous solution. Thus, acetic acid exerts a *differentiating effect* on the acidic behaviour of HCl, HBr and HI, whereas water does not.

'Acids' in acidic solvents

The effects of dissolving 'acids' in acidic non-aqueous solvents can be dramatic. When dissolved in H_2SO_4, $HClO_4$ (for which pK_a in aqueous solution is -8) is practically non-ionized and HNO_3 ionizes according to eq. 9.7.

$$HNO_3 + 2H_2SO_4 \rightleftharpoons [NO_2]^+ + [H_3O]^+ + 2[HSO_4]^- \quad (9.7)$$

Reaction 9.7 can be regarded as the summation of equilibria 9.8–9.10, and it is the presence of $[NO_2]^+$ that is responsible for the use of an HNO_3/H_2SO_4 mixture in the nitration of aromatic compounds.

$$HNO_3 + H_2SO_4 \rightleftharpoons [H_2NO_3]^+ + [HSO_4]^- \tag{9.8}$$

$$[H_2NO_3]^+ \rightleftharpoons [NO_2]^+ + H_2O \tag{9.9}$$

$$H_2O + H_2SO_4 \rightleftharpoons [H_3O]^+ + [HSO_4]^- \tag{9.10}$$

These examples signify caution: *just because we name a compound an 'acid', it may not behave as one in non-aqueous media.* Later we consider superacid media in which even hydrocarbons may be protonated (see Section 9.9).

Acids and bases: a solvent-oriented definition

A Brønsted acid is a proton donor, and a Brønsted base accepts protons. In aqueous solution, $[H_3O]^+$ is formed and in bulk water, self-ionization corresponds to the transfer of a proton from one solvent molecule to another (eq. 9.11) illustrating amphoteric behaviour (see Section 7.8).

$$2H_2O \rightleftharpoons [H_3O]^+ + [OH]^- \tag{9.11}$$

In liquid NH_3 (see Section 9.6), proton transfer leads to the formation of $[NH_4]^+$ (eq. 9.12), and, in a liquid ammonia solution, an acid may be described as a substance that produces $[NH_4]^+$ ions, while a base produces $[NH_2]^-$ ions.

$$2NH_3 \rightleftharpoons \underset{\text{ammonium ion}}{[NH_4]^+} + \underset{\text{amide ion}}{[NH_2]^-} \tag{9.12}$$

This solvent-oriented definition can be widened to include behaviour in any solvent which undergoes self-ionization.

> In a **self-ionizing solvent**, an acid is a substance that produces the cation characteristic of the solvent, and a base is a substance that produces the anion characteristic of the solvent.

Liquid dinitrogen tetraoxide, N_2O_4, undergoes the self-ionization shown in eq. 9.13. In this medium, nitrosyl salts such as $[NO][ClO_4]$ behave as acids, and metal nitrates (e.g. $NaNO_3$) behave as bases.

$$N_2O_4 \rightleftharpoons [NO]^+ + [NO_3]^- \tag{9.13}$$

In some ways, this acid–base terminology is unfortunate, since there are other, more common descriptors (e.g. Brønsted, Lewis, hard and soft). However, the terminology has been helpful in suggesting lines of research for the study of non-aqueous systems, and its use will probably continue.

Self-study exercises

1. Why does KOH behave as a base in aqueous solution?

2. Why does NH_4Cl behave as an acid in liquid ammonia?

3. CH_3CO_2H behaves as a weak acid in aqueous solution, but is levelled to a strong acid in liquid ammonia. What does this tell you about the extent of ionization of CH_3CO_2H in each medium?

4. Explain why $NaNH_2$ behaves as a base in liquid NH_3.

Proton-containing and aprotic solvents

In Sections 9.5–9.11, we consider selected inorganic non-aqueous solvents. Most of these solvents are self-ionizing. They can be divided into two categories:

- proton containing (NH_3, HF, H_2SO_4, $HOSO_2F$);
- aprotic (SO_2, BrF_3, N_2O_4).

9.5 Liquid sulfur dioxide

Liquid SO_2 is an important non-aqueous solvent. It is polar and aprotic, and readily solvates anions.[†] Selected properties of SO_2 are listed in Table 9.3, and its liquid range is compared with other solvents in Fig. 9.2.

Table 9.3 Selected physical properties of sulfur dioxide, SO_2.

Property / units	Value
Melting point / K	197.5
Boiling point / K	263.0
Density of liquid / $g\,cm^{-3}$	1.43
Dipole moment / D	1.63
Relative permittivity	17.6 (at boiling point)

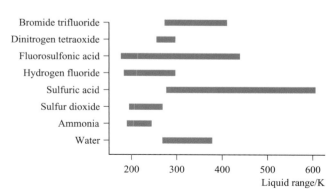

Fig. 9.2 Liquid ranges for water and selected non-aqueous solvents.

[†]See: W. Eisfeld and M. Regitz (1996) *J. Am. Chem. Soc.*, vol. 118, p. 11918 – 'Energetic and structural aspects of the solvation of anions in liquid SO_2'.

Initially, it was proposed that SO_2 underwent self-ionization according to eq. 9.14. However, this equilibrium requires the separation of doubly charged ions, in contrast to the singly charged ions involved in other self-ionization equilibria described in Sections 9.6–9.11. Two observables suggest that no self-ionization occurs: (i) conductance data are not consistent with the presence of ions in liquid SO_2; (ii) when labelled $SOCl_2$ is dissolved in liquid SO_2, neither ^{35}S nor ^{18}O exchanges with the solvent.

$$2SO_2 \rightleftharpoons [SO]^{2+} + [SO_3]^{2-} \qquad (9.14)$$

Liquid SO_2 is an effective, inert solvent for both organic compounds (e.g. amines, alcohols, carboxylic acids, esters) and covalent inorganic substances (e.g. Br_2, CS_2, PCl_3, $SOCl_2$, $POCl_3$) and is quite a good ionizing medium for such compounds as Ph_3CCl (giving $[Ph_3C]^+$). It is also used for the syntheses of some group 16 and 17 cationic species. For example, $[I_3]^+$ and $[I_5]^+$ (eq. 9.15) have been isolated as the $[AsF_6]^-$ salts from the reactions of AsF_5 and I_2 in liquid SO_2, the product depending on the molar ratio of the reactants. Reactions of selenium with AsF_5 (at 350 K) or SbF_5 (at 250 K) in liquid SO_2 have yielded the salts $[Se_4][AsF_6]_2$ and $[Se_8][SbF_6]_2$ respectively.

$$3AsF_5 + 5I_2 \xrightarrow{\text{liquid } SO_2} 2[I_5][AsF_6] + AsF_3 \qquad (9.15)$$

9.6 Liquid ammonia

Liquid ammonia has been widely studied, and in this section we discuss its properties and the types of reactions that occur in it, making comparisons between liquid ammonia and water.

Physical properties

Selected properties of NH_3 are listed in Table 9.4 and are compared with those of water. The liquid range of NH_3 is less than that of H_2O (Fig. 9.2). The lower boiling point of NH_3 compared of H_2O suggests that hydrogen bonding (see Section 10.6) in liquid NH_3 is less extensive than in liquid H_2O, and this is further illustrated by the values of $\Delta_{vap}H^o$ (23.3 and 40.7 kJ mol^{-1} for NH_3 and H_2O respectively). This is in line with what is expected from the fact that in NH_3, there is one lone pair to accept a hydrogen bond, whereas in H_2O, there are two. In fact, one has to be cautious in rationalizing the degree of hydrogen bonding simply in terms of the number of lone pairs. Neutron diffraction studies using hydrogen/deuterium isotopic substitution have shown the presence of two hydrogen bonds per N atom in liquid NH_3. Structural data from X-ray and neutron diffraction and computational methods lead to the conclusion that the hydrogen bonding is relatively weak and, unlike liquid H_2O, liquid NH_3 does not exhibit an extended hydrogen-bonded network.

The relative permittivity of NH_3 is considerably less than that of H_2O and, as a consequence, the ability of liquid NH_3 to dissolve ionic compounds is generally less than that of water. Exceptions include $[NH_4]^+$ salts, iodides and nitrates which are usually readily soluble. For example, AgI, which is sparingly soluble in water, dissolves easily in liquid NH_3 (solubility $= 206.8$ g per 100 g of NH_3), a fact that indicates that both the Ag^+ and I^- ions interact strongly with the solvent; Ag^+ forms an ammine complex (see Section 22.12). Changes in solubility patterns in going from water to liquid NH_3 lead to some interesting precipitation reactions in NH_3. Whereas in aqueous solution, $BaCl_2$ reacts with $AgNO_3$ to precipitate AgCl, in liquid NH_3, AgCl and $Ba(NO_3)_2$ react to precipitate $BaCl_2$. The solubility of AgCl is 0.29 g per 100 g of liquid NH_3 compared with 1.91×10^{-4} g per 100 g of H_2O. Molecular organic compounds are generally more soluble in NH_3 than in H_2O.

Self-ionization

Liquid NH_3 undergoes self-ionization (eq. 9.12), and the small value of K_{self} (Table 9.4) indicates that the equilibrium lies far over to the left-hand side. The $[NH_4]^+$ and $[NH_2]^-$ ions have ionic mobilities approximately equal to those of alkali metal and halide ions. This contrasts with the situation in water, in which $[H_3O]^+$ and $[OH]^-$ are much more mobile than other singly charged ions.

Reactions in liquid NH_3

We described above some precipitations that differ in liquid NH_3 and H_2O. Equation 9.16 shows a further example; the solubility of KCl is 0.04 g per 100 g NH_3, compared with 34.4 g per 100 g H_2O.

$$KNO_3 + AgCl \xrightarrow{} KCl + AgNO_3 \qquad \textit{in liquid } NH_3$$
$$\text{ppt}$$
$$(9.16)$$

Table 9.4 Selected physical properties of NH_3 and H_2O.

Property / units	NH₃	H₂O
Melting point / K	195.3	273.0
Boiling point / K	239.6	373.0
Density of liquid / g cm^{-3}	0.77	1.00
Dipole moment / D	1.47	1.85
Relative permittivity	25.0 (at melting point)	78.7 (at 298 K)
Self-ionization constant	5.1×10^{-27}	1.0×10^{-14}

In water, neutralization reactions follow the general reaction 9.17. The solvent-oriented definition of acids and bases allows us to write an analogous reaction (eq. 9.18) for a neutralization process in liquid NH_3.

$$Acid + Base \longrightarrow Salt + Water \qquad \textit{in aqueous solution}$$
$$(9.17)$$

$$Acid + Base \longrightarrow Salt + Ammonia \qquad \textit{in liquid } NH_3$$
$$(9.18)$$

Thus, in liquid NH_3, reaction 9.19 is a neutralization process which may be followed by conductivity or potentiometry, or by the use of an indicator such as phenolphthalein, **9.10**. This indicator is colourless but is deprotonated by a strong base such as $[NH_2]^-$ to give a red anion just as it is by $[OH]^-$ in aqueous solution.

(9.10)

$$NH_4Br + KNH_2 \longrightarrow KBr + 2NH_3 \qquad (9.19)$$

Liquid NH_3 is an ideal solvent for reactions requiring a strong base, since the amide ion is strongly basic.

As discussed in Section 9.4, the behaviour of 'acids' is solvent-dependent. In aqueous solution, sulfamic acid, H_2NSO_2OH, **9.11**, behaves as a monobasic acid according to eq. 9.20, but in liquid NH_3 it can function as a dibasic acid (eq. 9.21).

(9.11)

$$H_2NSO_2OH(aq) + H_2O(l)$$
$$\rightleftharpoons [H_3O]^+(aq) + [H_2NSO_2O]^-(aq)$$
$$K_a = 1.01 \times 10^{-1} \qquad (9.20)$$

$$H_2NSO_2OH + 2KNH_2 \longrightarrow K_2[HNSO_2O] + 2NH_3 \qquad (9.21)$$

The levelling effect of liquid NH_3 means that the strongest acid possible in this medium is $[NH_4]^+$. Solutions of ammonium halides in NH_3 may be used as acids, for example in the preparation of silane or arsane (eqs. 9.22

and 9.23). Germane, GeH_4, can be prepared from Mg_2Ge in a reaction analogous to the preparation of SiH_4.

$$Mg_2Si + 4NH_4Br \longrightarrow SiH_4 + 2MgBr_2 + 4NH_3 \qquad (9.22)$$

$$Na_3As + 3NH_4Br \longrightarrow AsH_3 + 3NaBr + 3NH_3 \qquad (9.23)$$

A saturated solution of NH_4NO_3 in liquid NH_3 (which has a vapour pressure of less than 1 bar even at 298 K) dissolves many metal oxides and even some metals. Nitrate to nitrite reduction often accompanies the dissolution of metals. Metals that form insoluble hydroxides under aqueous conditions, form insoluble amides in liquid NH_3, e.g. $Zn(NH_2)_2$. Just as $Zn(OH)_2$ dissolves in the presence of excess hydroxide ion (eq. 9.24), $Zn(NH_2)_2$ reacts with amide ion to form soluble salts containing anion **9.12** (eq. 9.25).

$$Zn^{2+} + 2[OH]^- \longrightarrow Zn(OH)_2 \xrightarrow{\text{excess } [OH]^-} [Zn(OH)_4]^{2-}$$
$$(9.24)$$

$$Zn^{2+} + 2[NH_2]^- \longrightarrow Zn(NH_2)_2 \xrightarrow{\text{excess } [NH_2]^-} [Zn(NH_2)_4]^{2-}$$
$$(9.25)$$

(9.12)

Parallels can be drawn between the behaviours of metal nitrides in liquid NH_3 and of metal oxides in aqueous media. Many similar analogies can be drawn.

Complex formation between Mg^{2+} and NH_3 leads to $[Mg(NH_3)_6]^{2+}$, isolated as $[Mg(NH_3)_6]Cl_2$. Similarly, in liquid NH_3, $CaCl_2$ forms $[Ca(NH_3)_6]Cl_2$ and this is the reason that anhydrous $CaCl_2$ (which readily absorbs water, see Section 12.5) cannot be used to dry NH_3. Ammine complexes such as $[Ni(NH_3)_6]^{2+}$ can be prepared in aqueous solution by the displacement of aqua ligands by NH_3. Not all hexaammine complexes are, however, directly accessible by this method. Two examples are $[V(NH_3)_6]^{2+}$ and $[Cu(NH_3)_6]^{2+}$. The ion $[V(OH_2)_6]^{2+}$ is readily oxidized in aqueous solution, making the preparation of V(II) complexes in aqueous conditions difficult. In liquid NH_3, dissolution of VI_2 gives $[V(NH_3)_6]I_2$ containing the octahedral $[V(NH_3)_6]^{2+}$ ion. The $[Cu(NH_3)_6]^{2+}$ ion is not accessible in aqueous solution (see Fig. 20.37) but can be formed in liquid NH_3.

Self-study exercises

1. In eq. 9.10, which reactant acts as a base?

2. In eq. 9.22, why does NH_4Br behave as an acid?

3. Explain why organic amines and amides can function as acids in liquid ammonia.

4. Suggest products for the following reactions:

$$R_2NH + KNH_2 \longrightarrow$$

$$RC(O)NH_2 + KNH_2 \longrightarrow$$

How is KNH_2 functioning in these reactions?

Solutions of *s*-block metals in liquid NH₃

All of the group 1 metals and the group 2 metals Ca, Sr and Ba dissolve in liquid NH_3 to give metastable solutions from which the group 1 metals can be recovered unchanged. The group 2 metals are recoverable as solids of composition $[M(NH_3)_6]$. Yellow $[Li(NH_3)_4]$ and blue $[Na(NH_3)_4]$ may also be isolated at low temperatures.

Dilute solutions of the metals are bright blue, the colour arising from the short wavelength tail of a broad and intense absorption band in the infrared region of the spectrum. The electronic spectra in the visible region of solutions of all the *s*-block metals are the same, indicating the presence of a species common to all the solutions: this is the solvated electron (eq. 9.26).

$$M \xrightarrow{\text{dissolve in liquid NH}_3} M^+(\text{solv}) + e^-(\text{solv}) \qquad (9.26)$$

Each dilute solution of metal in liquid NH_3 occupies a volume greater than the sum of the volumes of the metal plus solvent. These data suggest that the electrons occupy cavities of radius 300–400 pm. Very dilute solutions of the metals are paramagnetic, and the magnetic susceptibility corresponds to that calculated for the presence of one free electron per metal atom.

As the concentration of a solution of an *s*-block metal in liquid NH_3 increases, the molar conductivity initially decreases, reaching a minimum at $\approx 0.05 \text{ mol dm}^{-3}$. Thereafter, the molar conductivity increases, and in saturated solutions, it is comparable with that of the metal itself. Such saturated solutions are no longer blue and paramagnetic, but are bronze and diamagnetic. They are essentially 'metal-like' and have been described as *expanded metals*. The conductivity data can be described in terms of:

- reaction 9.26 at low concentrations;
- association of $M^+(\text{solv})$ and $e^-(\text{solv})$ at concentrations around 0.05 mol dm^{-3};
- metal-like behaviour at higher concentrations.

However, in order to explain why the magnetic susceptibilities of solutions *decrease* as the concentration increases, it is necessary to invoke equilibria 9.27 at higher concentrations.

$$\left. \begin{array}{l} 2M^+(\text{solv}) + 2e^-(\text{solv}) \rightleftharpoons M_2(\text{solv}) \\ M(\text{solv}) + e^-(\text{solv}) \rightleftharpoons M^-(\text{solv}) \end{array} \right\} \qquad (9.27)$$

Hydrogen/deuterium isotopic substitution coupled with neutron diffraction studies confirm that the addition of an alkali metal to liquid NH_3 disrupts the hydrogen bonding present in the solvent. In a saturated lithium–ammonia solution (21 mole percent metal), no hydrogen bonding remains between NH_3 molecules. Saturated Li–NH_3 solutions contain tetrahedrally coordinated Li, whereas saturated K–NH_3 solutions contain octahedrally coordinated K.

The blue solutions of alkali metals in liquid NH_3 decompose very slowly, liberating H_2 (eq. 9.28) as the solvent is reduced.

$$2NH_3 + 2e^- \longrightarrow 2[NH_2]^- + H_2 \qquad (9.28)$$

Although reaction 9.28 is thermodynamically favoured, there is a significant kinetic barrier. Decomposition is catalysed by many *d*-block metal ions. Ammonium salts (which are strong acids in liquid NH_3) decompose immediately (eq. 9.29).

$$2[NH_4]^+ + 2e^- \longrightarrow 2NH_3 + H_2 \qquad (9.29)$$

There are numerous applications of sodium in liquid ammonia in organic synthesis,[†] including the well-established Birch reduction. Dilute solutions of alkali metals in liquid NH_3 have many applications as reducing agents in inorganic syntheses. Reactions 9.30 to 9.34 (in which e^- represents the electron generated in reaction 9.26) provide examples and others are mentioned later in the book. In each of reactions 9.30–9.34, the anion shown is isolated as an alkali metal salt, the cation being provided from the alkali metal dissolved in the liquid NH_3.

$$2GeH_4 + 2e^- \longrightarrow 2[GeH_3]^- + H_2 \qquad (9.30)$$

$$O_2 + e^- \longrightarrow \underset{\text{superoxide ion}}{[O_2]^-} \qquad (9.31)$$

$$O_2 + 2e^- \longrightarrow \underset{\text{peroxide ion}}{[O_2]^{2-}} \qquad (9.32)$$

$$[MnO_4]^- + e^- \longrightarrow [MnO_4]^{2-} \qquad (9.33)$$

$$[Fe(CO)_5] + 2e^- \longrightarrow [Fe(CO)_4]^{2-} + CO \qquad (9.34)$$

Early synthetic routes to *Zintl ions* (see Section 14.7) involved reduction of Ge, Sn or Pb in solutions of Na in liquid NH_3. The method has been developed with the addition of the macrocyclic ligand cryptand-222 (crypt-222) (see Section 11.8) which encapsulates the Na^+ ion and allows the isolation of salts of the type $[Na(\text{crypt-}222)]_2[Sn_5]$ (eq. 9.35). Zintl ions produced in this way include $[Sn_5]^{2-}$ (Fig. 9.3), $[Pb_5]^{2-}$, $[Pb_2Sb_2]^{2-}$, $[Bi_2Sn_2]^{2-}$, $[Ge_9]^{2-}$, $[Ge_9]^{4-}$ and $[Sn_9Tl]^{3-}$.

$$Sn \xrightarrow{\text{Na in liquid NH}_3} \underset{\text{Zintl phase}}{NaSn_{1.0-1.7}}$$

$$\xrightarrow[\text{in 1,2-ethanediamine}]{2,2,2\text{-crypt}} [Na(\text{crypt-}222)]_2[Sn_5] \qquad (9.35)$$

[†]For a summary, see: S.B. Raikar (2007) *Synlett*, p. 341.

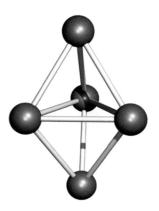

Fig. 9.3 The Zintl ion $[Sn_5]^{2-}$ has a trigonal bipyramidal cluster structure.

A further development in the synthesis of Zintl ions has been to use the reactions of an excess of Sn or Pb in solutions of Li in liquid NH_3. These reactions give $[Li(NH_3)_4]^+$ salts of $[Sn_9]^{4-}$ and $[Pb_9]^{4-}$, and we discuss these Zintl ions further in Section 14.7.

The group 2 metals Ca, Sr and Ba dissolve in liquid NH_3 to give bronze-coloured $[M(NH_3)_x]$ species, and for M = Ca, neutron diffraction data confirm the presence of octahedral $[Ca(ND_3)_6]$. Although pale blue solutions are obtained when Mg is added to NH_3, complete dissolution is not observed and no ammine adducts of Mg have been isolated from these solutions. However, combining an Hg/Mg (22:1 ratio) alloy with liquid NH_3 produces crystals of $[Mg(NH_3)_6Hg_{22}]$ which contain octahedral $[Mg(NH_3)_6]$ units, hosted within an Hg lattice. This material is superconducting (see Section 28.4) with a critical temperature, T_c, of 3.6 K.

Redox reactions in liquid NH_3

Reduction potentials for the reversible reduction of metal ions to the corresponding metal in aqueous solution and in liquid NH_3 are listed in Table 9.5. The values follow the same general trend, but the oxidizing ability of each metal ion is solvent-dependent. Reduction potentials for oxidizing systems cannot be obtained in liquid NH_3 owing to the ease with which the solvent is oxidized.

Information deduced from reduction potentials, and from lattice energies and solubilities, indicates that H^+ and d-block M^{n+} ions have more negative absolute standard Gibbs energies of solvation in NH_3 than in H_2O; for alkali metal ions, values of $\Delta_{solv}G^o$ are about the same in the two solvents. These data are consistent with the observation that the addition of NH_3 to aqueous solutions of d-block M^{n+} ions results in the formation of ammine complexes such as $[M(NH_3)_6]^{n+}$ whereas alkali metal ions are not complexed by NH_3 under aqueous conditions.

Table 9.5 Selected standard reduction potentials (298 K) in aqueous and liquid ammonia media; the concentration of each solution is 1 mol dm^{-3}. The value of $E^o = 0.00$ V for the H^+/H_2 couple is defined by convention.

Reduction half-equation	E^o / V in aqueous solution	E^o / V in liquid ammonia
$Li^+ + e^- \rightleftharpoons Li$	−3.04	−2.24
$K^+ + e^- \rightleftharpoons K$	−2.93	−1.98
$Na^+ + e^- \rightleftharpoons Na$	−2.71	−1.85
$Zn^{2+} + 2e^- \rightleftharpoons Zn$	−0.76	−0.53
$2H^+ + 2e^- \rightleftharpoons H_2$ (g, 1 bar)	0.00	0.00
$Cu^{2+} + 2e^- \rightleftharpoons Cu$	+0.34	+0.43
$Ag^+ + e^- \rightleftharpoons Ag$	+0.80	+0.83

9.7 Liquid hydrogen fluoride

Physical properties

Hydrogen fluoride attacks silica glass (eq. 9.36) thereby corroding glass reaction vessels, and it is only relatively recently that HF has found applications as a non-aqueous solvent. It can be handled in polytetrafluoroethene (PTFE) containers, or, if absolutely free of water, in Cu or Monel metal (a nickel alloy) equipment.

$$4HF + SiO_2 \longrightarrow SiF_4 + 2H_2O \tag{9.36}$$

Hydrogen fluoride has a liquid range from 190 to 292.5 K (Fig. 9.2). The relative permittivity is 84 at 273 K, rising to 175 at 200 K. Liquid HF undergoes self-ionization (equilibrium 9.37), for which $K_{self} \approx 2 \times 10^{-12}$ at 273 K.

$$3HF \rightleftharpoons \underset{\substack{\text{dihydridofluorine}(1+) \\ \text{ion}}}{[H_2F]^+} + \underset{\substack{\text{difluoridohydrogenate}(1-) \\ \text{ion}}}{[HF_2]^-} \tag{9.37}$$

The difference in electronegativities of H ($\chi^P = 2.2$) and F ($\chi^P = 4.0$) results in the presence of extensive intermolecular hydrogen bonding in the liquid. High-energy X-ray and neutron diffraction studies[†] have been used to show that liquid HF (at 296 K) contains chains of hydrogen-bonded molecules (on average, seven molecules per chain). Interchain hydrogen bonding also exists. In the vapour phase, hydrogen fluoride consists of cyclic $(HF)_x$ species as well as clusters.

Acid–base behaviour in liquid HF

Using the solvent-oriented definition (Section 9.4), a species that produces $[H_2F]^+$ ions in liquid HF is an acid, and one that produces $[HF_2]^-$ is a base.

[†] See: S.E. McLain, C.J. Benmore, J.E. Siewenie, J. Urquidi and J.F.C. Turner (2004) *Angew. Chem. Int. Ed.*, vol. 43, p. 1951 – 'The structure of liquid hydrogen fluoride'.

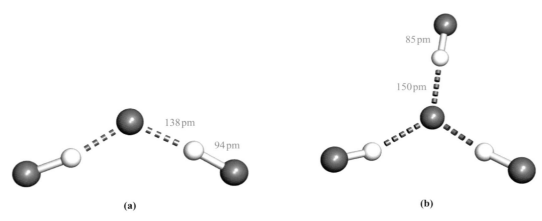

Fig. 9.4 The structures of the anions (a) $[H_2F_3]^-$ and (b) $[H_3F_4]^-$, determined by low-temperature X-ray diffraction for the $[Me_4N]^+$ salts. The distances given are the average values for like internuclear separations; the experimental error on each distance is $\pm 3\text{–}6\,pm$ [D. Mootz *et al.* (1987) *Z. Anorg. Allg. Chem.*, vol. 544, p. 159]. Colour code: F, green; H, white.

Many organic compounds are soluble in liquid HF, and in the cases of, for example, amines and carboxylic acids, protonation of the organic species accompanies dissolution (eq. 9.38). Proteins react immediately with liquid HF, and it produces very serious skin burns.

$$MeCO_2H + 2HF \longrightarrow [MeC(OH)_2]^+ + [HF_2]^- \qquad (9.38)$$

Most inorganic salts are converted to the corresponding fluorides when dissolved in liquid HF, but only a few of these are soluble. Fluorides of the *s*-block metals, silver and thallium(I) dissolve to give salts such as $K[HF_2]$ and $K[H_2F_3]$, and thus exhibit basic character. The single crystal structures of a number of salts containing $[HF_2]^-$ or $[DF_2]^-$ (i.e. deuterated species) have been determined by X-ray or neutron diffraction techniques; these include $[NH_4][HF_2]$, $Na[HF_2]$, $K[HF_2]$, $Rb[HF_2]$, $Cs[HF_2]$ and $Tl[HF_2]$.

The $[HF_2]^-$ anion is linear (structure **9.13**), and its formation is a consequence of the H and F atoms being involved in strong hydrogen bonding. In the solid state structures reported, the F---F distance is $\approx 228\,pm$. This value is greater than twice the H–F bond length in HF ($2 \times 92\,pm$), but an $H\cdots F$ hydrogen bond will always be weaker and longer than a 2-centre covalent H–F bond. Comparison of the values gives some indication of the strength of the hydrogen bonding in $[HF_2]^-$. (See also Figs. 5.29 and 9.4.)

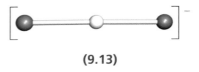

(9.13)

Ammonium fluoride is basic in liquid HF. Studies of the Me_4NF–HF system over a range of compositions and

temperatures reveal the formation of the compounds of composition $Me_4NF \cdot nHF$ ($n = 2, 3, 5$ or 7). X-ray diffraction studies for compounds with $n = 2, 3$ or 5 have confirmed the structures of $[H_2F_3]^-$ (Fig. 9.4a), $[H_3F_4]^-$ (Fig. 9.4b) and $[H_5F_6]^-$, in which strong hydrogen bonding is an important feature (see Section 10.6).

The molecular fluorides CF_4 and SiF_4 are insoluble in liquid HF, but F^- acceptors such as AsF_5 and SbF_5 dissolve according to eq. 9.39 to give very strongly acidic solutions. Less potent fluoride acceptors such as BF_3 function as weak acids in liquid HF (eq. 9.40). PF_5 behaves as a *very* weak acid (eq. 9.41). In contrast, ClF_3 and BrF_3 act as F^- donors (eq. 9.42) and behave as bases.

$$EF_5 + 2HF \rightleftharpoons [H_2F]^+ + [EF_6]^- \quad E = As\ or\ Sb \qquad (9.39)$$

$$BF_3 + 2HF \rightleftharpoons [H_2F]^+ + [BF_4]^- \qquad (9.40)$$

$$PF_5 + 2HF \rightleftharpoons [H_2F]^+ + [PF_6]^- \qquad (9.41)$$

$$BrF_3 + HF \rightleftharpoons [BrF_2]^+ + [HF_2]^- \qquad (9.42)$$

Few protic acids are able to exhibit acidic behaviour in liquid HF, on account of the competition between HF and the solute as H^+ donors. Perchloric acid and fluorosulfonic acid (eq. 9.43) do act as acids.

$$HOSO_2F + HF \rightleftharpoons [H_2F]^+ + [SO_3F]^- \qquad (9.43)$$

With SbF_5, HF forms a *superacid* (eq. 9.44) which is capable of protonating *very* weak bases including hydrocarbons (see Section 9.9).

$$2HF + SbF_5 \rightleftharpoons [H_2F]^+ + [SbF_6]^- \qquad (9.44)$$

Self-study exercises

1. Write equations to show how ClF_3 and AsF_5 behave in liquid HF.

2. Use the VSEPR model to rationalize why $[H_2F]^+$ and $[BrF_2]^+$ are non-linear.

3. Using Fig. 5.29 to help you, give a bonding description for the $[HF_2]^-$ ion.

Electrolysis in liquid HF

Electrolysis in liquid HF is an important preparative route to both inorganic and organic fluorine-containing compounds, many of which are difficult to access by other routes. Anodic oxidation in liquid HF involves half-reaction 9.45 and with NH_4F as substrate, the products of the subsequent fluorination are NFH_2, NF_2H and NF_3.

$$2F^- \rightleftharpoons F_2 + 2e^- \qquad (9.45)$$

In liquid HF, anodic oxidation of water gives OF_2, of SCl_2 yields SF_6, of acetic acid yields CF_3CO_2H and of Me_3N produces $(CF_3)_3N$.

9.8 Sulfuric acid and fluorosulfonic acid

Physical properties of sulfuric acid

Selected physical properties of H_2SO_4 are given in Table 9.6. It is a liquid at 298 K, and the long liquid range (Fig. 9.2) contributes towards making this a widely used non-aqueous solvent. Disadvantages of liquid H_2SO_4 are its high viscosity (27 times that of water at 298 K) and high value of $\Delta_{vap}H^\circ$. Both these properties arise from extensive intermolecular hydrogen bonding, and make it difficult to remove the solvent by evaporation from reaction mixtures. Dissolution of a solute in H_2SO_4 is favourable only if new interactions can be established to compensate for the loss of the extensive hydrogen bonding. Generally, this is possible only if the solute is ionic.

Table 9.6 Selected physical properties of sulfuric acid, H_2SO_4.

Property / units	Value
Melting point / K	283.4
Boiling point / K	≈603
Density of liquid / g cm^{-3}	1.84
Relative permittivity	110 (at 292 K)
Self-ionization constant	2.7×10^{-4} (at 298 K)

The value of the equilibrium constant for the self-ionization process 9.46 is relatively large. Other equilibria such as 9.47 are involved to a lesser extent.

$$2H_2SO_4 \rightleftharpoons [H_3SO_4]^+ + [HSO_4]^-$$
$$\text{hydrogensulfate ion}$$
$$K_{self} = 2.7 \times 10^{-4} \qquad (9.46)$$

$$2H_2SO_4 \rightleftharpoons [H_3O]^+ + [HS_2O_7]^- \quad K_{self} = 5.1 \times 10^{-5} \qquad (9.47)$$

9.14

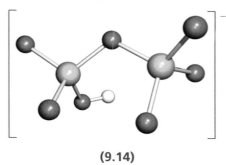

(9.14)

Acid–base behaviour in liquid H_2SO_4

Sulfuric acid is a highly acidic solvent and most other 'acids' are neutral or behave as bases in it. We have already noted the basic behaviour of HNO_3. Initial proton transfer (eq. 9.8) leads to the formation of the 'protonated acid' $[H_2NO_3]^+$ which then eliminates H_2O (eq. 9.9). Overall, reaction 9.48 occurs. Similarly, acetic acid behaves as a base in sulfuric acid (eq. 9.49). A solution of boric acid in H_2SO_4 produces the $[B(HSO_4)_4]^-$ ion (eq. 9.50 and structure **9.15**).

$$HNO_3 + 2H_2SO_4 \longrightarrow [NO_2]^+ + [H_3O]^+ + 2[HSO_4]^- \qquad (9.48)$$
$$MeCO_2H + H_2SO_4 \longrightarrow [MeC(OH)_2]^+ + [HSO_4]^- \qquad (9.49)$$
$$H_3BO_3 + 6H_2SO_4 \longrightarrow [B(HSO_4)_4]^- + 3[H_3O]^+ + 2[HSO_4]^- \qquad (9.50)$$

For the $[B(HSO_4)_4]^-$ ion to be formed, $H[B(HSO_4)_4]$ must act as a strong acid in H_2SO_4 solution. $H[B(HSO_4)_4]$ is a stronger acid even than HSO_3F (see below). The ionization constants (in H_2SO_4) for HSO_3F and $H[B(HSO_4)_4]$ are 3×10^{-3} and 0.4, respectively.

(9.15)

The species '$H[B(HSO_4)_4]$' has not been isolated as a pure compound, but a solution of this acid can be prepared by dissolving boric acid in *oleum* (eq. 9.51) (see Section 16.9) and can be titrated conductometrically against a solution of a strong base such as $KHSO_4$ (eq. 9.52).

$$H_3BO_3 + 2H_2SO_4 + 3SO_3 \longrightarrow [H_3SO_4]^+ + [B(HSO_4)_4]^-$$
(9.51)

$$H[B(HSO_4)_4] + KHSO_4 \longrightarrow K[B(HSO_4)_4] + H_2SO_4 \quad (9.52)$$

In a ***conductometric titration***, the end-point is found by monitoring changes in the electrical conductivity of the solution.[†]

Few species function as strong acids in H_2SO_4 medium. Perchloric acid (a potent acid in aqueous solution) is essentially non-ionized in H_2SO_4 and behaves only as a very weak acid.

In some cases (in contrast to eq. 9.49), the cations formed from carboxylic acids are unstable, e.g. HCO_2H and $H_2C_2O_4$ (eq. 9.53) decompose with loss of CO.

$$\begin{array}{l} CO_2H \\ | \\ CO_2H \end{array} + H_2SO_4 \longrightarrow CO + CO_2 + [H_3O]^+ + [HSO_4]^-$$
(9.53)

The ionic mobilities[‡] of $[H_3SO_4]^+$ and $[HSO_4]^-$ are very high, and the conductivity in H_2SO_4 is almost entirely due to the presence of $[H_3SO_4]^+$ and/or $[HSO_4]^-$. These ions carry the electrical current by proton-switching mechanisms, thus avoiding the need for migration through the viscous solvent.

Self-study exercises

1. The self-ionization constant for H_2SO_4 is 2.7×10^{-4}. To what equilibrium does this refer?

2. When EtOH dissolves in liquid H_2SO_4, $EtOSO_3H$ forms. Write a balanced equation for this reaction, bearing in mind that H_2O acts as a base in H_2SO_4.

3. Explain why $[P(OH)_4]^+$ forms when phosphoric acid dissolves in H_2SO_4.

[†] For an introduction to conductometric titrations, see: C.E. Housecroft and E.C. Constable (2010) *Chemistry*, 4th edn, Prentice Hall, Harlow, Chapter 19.
[‡] For discussions of ion transport see: P. Atkins and J. de Paula (2010) *Atkins' Physical Chemistry*, 9th edn, Oxford University Press, Oxford, Chapter 20; J. Burgess (1999) *Ions in Solution: Basic Principles of Chemical Interactions*, 2nd edn, Horwood Publishing, Westergate, Chapter 2.

Table 9.7   Selected physical properties of fluorosulfonic acid, HSO_3F.

Property / units	Value
Melting point / K	185.7
Boiling point / K	438.5
Density of liquid / g cm^{-3}	1.74
Relative permittivity	120 (at 298 K)
Self-ionization constant	4.0×10^{-8} (at 298 K)

Physical properties of fluorosulfonic acid

Table 9.7 lists some of the physical properties of fluorosulfonic acid,[*] HSO_3F, **9.16**. It has a relatively long liquid range (Fig. 9.2) and a high dielectric constant. It is far less viscous than H_2SO_4 (by a factor of ≈16) and, like H_2SO_4 but unlike HF, can be handled in glass apparatus.

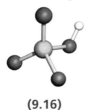

(9.16)

Equation 9.54 shows the self-ionization of HSO_3F for which $K_{self} = 4.0 \times 10^{-8}$.

$$2HSO_3F \rightleftharpoons [H_2SO_3F]^+ + [SO_3F]^- \quad (9.54)$$

9.9 Superacids

A ***superacid*** is a stronger acid than anhydrous H_2SO_4. Superacids are made by dissolving a strong Lewis acid (e.g. SbF_5) in either of the Brønsted acids HF or HSO_3F.

Extremely potent acids, capable of protonating even hydrocarbons, are termed *superacids* and include mixtures of HF and SbF_5 (eq. 9.44) and HSO_3F and SbF_5 (eq. 9.55). The latter mixture is called *magic acid* (one of the strongest acids known) and is available commercially under this name. Antimony(V) fluoride is a strong Lewis acid and forms an adduct with F^- (from HF) or $[SO_3F]^-$ (from HSO_3F). Figure 9.5 shows the crystallographically determined structure of the related adduct $SbF_5OSO(OH)CF_3$.

$$2HSO_3F + SbF_5 \rightleftharpoons [H_2SO_3F]^+ + [F_5SbOSO_2F]^- \quad (9.55)$$

9.17

(9.17)

---

[*] Fluorosulfonic acid is sometimes called fluorosulfuric acid.

Fig. 9.5 The solid state structure (X-ray diffraction) of SbF$_5$OSO(OH)CF$_3$ [D. Mootz *et al.* (1991) *Z. Naturforsch., Teil B*, vol. 46, p. 1659]. Colour code: Sb, brown; F, green; S, yellow; O, red; C, grey; H, white.

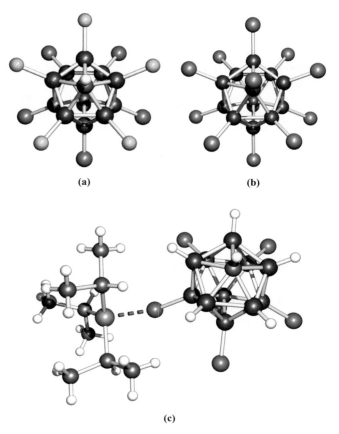

(a)   **(b)**

(c)

Fig. 9.6 (a) The [CHB$_{11}$R$_5$X$_6$]$^-$ (R = H, Me, Cl and X = Cl, Br, I) anions and (b) the [B$_{12}$X$_{12}$]$^{2-}$ (X = Cl, Br) anions that are the conjugate bases of the family of boron cluster superacids. Colour code: C, grey; B, blue; H, white; R, yellow; X, green. (c) The structure of ^{i}Pr$_3$Si(CHB$_{11}$H$_5$Cl$_6$) (X-ray diffraction) showing the long Si–Cl 'bond' which suggests that the structure approaches an ion-pair [Z. Xie *et al.* (1996) *J. Am. Chem. Soc.*, vol. 118, p. 2922]. Colour code: C, grey; B, blue; Cl, green; Si, pink; H, white.

Equilibrium 9.55 is an over-simplification of the SbF$_5$–HSO$_3$F system, but represents it sufficiently for most purposes. The species present depend on the ratio of SbF$_5$: HSO$_3$F, and at higher concentrations of SbF$_5$, species including [SbF$_6$]$^-$, [Sb$_2$F$_{11}$]$^{2-}$, HS$_2$O$_6$F and HS$_3$O$_9$F may exist.

In superacidic media, hydrocarbons act as bases, and this is an important route to the formation of carbenium ions,[†] e.g. eq. 9.56.

$$Me_3CH + [H_2SO_3F]^+ \longrightarrow [Me_3C]^+ + H_2 + HSO_3F \quad (9.56)$$
generated in
magic acid

Superacids have a wide range of applications and have been used to access species such as [HPX$_3$]$^+$ (X = halide), [C(OH)$_3$]$^+$ (by protonation of carbonic acid), [H$_3$S]$^+$ (see Section 16.5), [Xe$_2$]$^+$ (see Section 18.1) and metal carbonyl cations (see Sections 22.9 and 24.4). However, the conjugate bases of traditional superacids are typically strongly oxidizing and strongly nucleophilic, and are not necessarily innocent bystanders in a reaction mixture. A recently developed class of superacids involves icosahedral CB$_{11}$ (*carbaboranes*) or B$_{12}$ cluster cages (see Section 13.11). An advantage of these acids is that they possess chemically inert, extremely weak conjugate bases, and this contrasts with the properties of more traditional superacids. Figures 9.6a and 9.6b show the carbaborane anions [CHB$_{11}$R$_5$X$_6$]$^-$ (R = H, Me, Cl and X = Cl, Br, I) and [B$_{12}$X$_{12}$]$^{2-}$ (X = Cl, Br) which are the conjugate bases of members of this family of superacids. While the anions [CHB$_{11}$R$_5$X$_6$]$^-$ and [B$_{12}$X$_{12}$]$^{2-}$ are chemically unreactive, their conjugate acids are extremely potent Brønsted acids, stronger than fluorosulfonic acid. The superacids HCHB$_{11}$R$_5$X$_6$ and H$_2$B$_{12}$X$_{12}$ are prepared in two steps (eqs. 9.57 and 9.58, or 9.59 and 9.60).

$$R_3SiH + [Ph_3C]^+[CHB_{11}R_5X_6]^-$$
<small>(e.g. R = Et, ^{i}Pr)</small> $\longrightarrow R_3Si(CHB_{11}R_5X_6) + Ph_3CH \quad (9.57)$

$$R_3Si(CHB_{11}R_5X_6) \xrightarrow{\text{anhydrous liquid HCl}} HCHB_{11}R_5X_6 + R_3SiCl$$
<small>(e.g. R = Et, ^{i}Pr)</small> superacid (9.58)

[†] A *carbenium* ion is also called a *carbocation*; the older name of *carbonium* ion is also in use.

$$2Et_3SiH + [Ph_3C]^+{}_2[B_{12}X_{12}]^{2-}$$
$$\longrightarrow (Et_3Si)_2(B_{12}X_{12}) + 2Ph_3CH \quad (9.59)$$

$$(Et_3Si)_2(B_{12}X_{12}) \xrightarrow{\text{anhydrous liquid HCl}} H_2B_{12}X_{12} + 2Et_3SiCl$$
superacid (9.60)

Reactions 9.58 and 9.60 show the formation of intermediate silylium species which are of interest in their own right (see Section 23.5). Figure 9.6c shows the structure of ^{i}Pr$_3$Si(CHB$_{11}$H$_5$Cl$_6$). The Si–Cl distance is 232 pm, significantly greater than the sum of the covalent radii (217 pm). This indicates that ^{i}Pr$_3$Si(CHB$_{11}$H$_5$Cl$_6$) is tending towards an [^{i}Pr$_3$Si]$^+$[CHB$_{11}$H$_5$Cl$_6$]$^-$ ion-pair, although well-separated ions are clearly not present. Similarly, the solid state structure of (Et$_3$Si)$_2$(B$_{12}$Cl$_{12}$) reveals Et$_3$Si units associated with the B$_2$Cl$_{12}$ cage through long Si–Cl bonds (231 pm). The Et$_3$Si units are flattened and the sum of the C–Si–C bond angles is 348° which approaches the 360° expected for an isolated [Et$_3$Si]$^+$ ion. Treatment of the intermediates with anhydrous liquid HCl (eqs. 9.58 and 9.60) yields anhydrous $HCHB_{11}R_5X_6$ and $HB_{12}X_{12}$. These superacids protonate most solvents. They can be handled in liquid SO_2, although complete ionization appears to take place, probably with the formation of $[H(SO_2)_2]^+$ (eqs. 9.61 and 9.62).

$$HCHB_{11}R_5X_6 \xrightarrow{\text{liquid } SO_2} [H(SO_2)_2]^+ + [CHB_{11}R_5X_6]^- \quad (9.61)$$

$$H_2B_{12}X_{12} \xrightarrow{\text{liquid } SO_2} 2[H(SO_2)_2]^+ + [B_{12}X_{12}]^{2-} \quad (9.62)$$

The exceptional acid strength of these superacids is illustrated by their ability to protonate arenes (e.g. C_6H_6, C_6H_5Me, C_6Me_6). The resultant salts are remarkably thermally stable, e.g. $[C_6H_7]^+[CHB_{11}Me_5Br_6]^-$ is stable to 423 K.

9.10 Bromine trifluoride

Physical properties

Bromine trifluoride is a pale yellow liquid at 298 K and is an *aprotic* non-aqueous solvent. Selected physical properties are given in Table 9.8 and the compound is discussed further in Section 17.7. Bromine trifluoride is an extremely powerful fluorinating agent and fluorinates essentially every species that dissolves in it. However, massive quartz is kinetically stable towards BrF_3 and the solvent can be handled in quartz vessels. Apparatus made from Cu, Ni or Monel metal (68% Ni and 32% Cu) can also be used; the metal surface becomes protected by a thin layer of metal fluoride.

The self-ionization of BrF_3 (eq. 9.63) has been demonstrated by the isolation and characterization of salts containing the $[BrF_2]^+$ and $[BrF_4]^-$ ions, and by conductometric titrations of them (see below). Using the solvent-based acid–base definitions, an acid in BrF_3 is a species that produces $[BrF_2]^+$ (**9.18**), and a base is one that gives $[BrF_4]^-$ (**9.19**).

$$2BrF_3 \rightleftharpoons [BrF_2]^+ + [BrF_4]^- \quad (9.63)$$

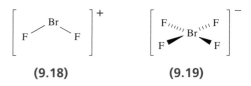

(9.18) **(9.19)**

Table 9.8 Selected physical properties of bromine trifluoride, BrF_3.

Property / units	Value
Melting point / K	281.8
Boiling point / K	408
Density of liquid / g cm^{-3}	2.49
Relative permittivity	107
Self-ionization constant	8.0×10^{-3} (at 281.8 K)

Behaviour of fluoride salts and molecular fluorides in BrF_3

Bromine trifluoride acts as a Lewis acid, readily accepting F^-. When dissolved in BrF_3, alkali metal fluorides, BaF_2 and AgF combine with the solvent to give salts containing the $[BrF_4]^-$ anion, e.g. $K[BrF_4]$ (eq. 9.64), $Ba[BrF_4]_2$ and $Ag[BrF_4]$. On the other hand, if the fluoride solute is a more powerful F^- acceptor than BrF_3, salts containing $[BrF_2]^+$ are formed, e.g. eqs. 9.65–9.67.

$$KF + BrF_3 \longrightarrow K^+ + [BrF_4]^- \quad (9.64)$$

$$SbF_5 + BrF_3 \longrightarrow [BrF_2]^+ + [SbF_6]^- \quad (9.65)$$

$$SnF_4 + 2BrF_3 \longrightarrow 2[BrF_2]^+ + [SnF_6]^{2-} \quad (9.66)$$

$$AuF_3 + BrF_3 \longrightarrow [BrF_2]^+ + [AuF_4]^- \quad (9.67)$$

A conductometric titration of $[BrF_2][SbF_6]$ against $Ag[BrF_4]$ exhibits a minimum at a 1:1 ratio supporting neutralization reaction 9.68.

$$\underset{\text{acid}}{[BrF_2][SbF_6]} + \underset{\text{base}}{Ag[BrF_4]} \longrightarrow Ag[SbF_6] + 2BrF_3 \quad (9.68)$$

Reactions in BrF_3

Much of the chemistry studied in BrF_3 media involves fluorination reactions, and the preparation of highly fluorinated species. For example, the salt $Ag[SbF_6]$ can be prepared in liquid BrF_3 from elemental Ag and Sb in a 1:1 molar ratio (eq. 9.69), and $K_2[SnF_6]$ is produced when KCl and Sn are combined in a 2:1 molar ratio in liquid BrF_3 (eq. 9.70). Further examples are given in eqs. 9.71–9.73.

$$Ag + Sb \xrightarrow{\text{in } BrF_3} Ag[SbF_6] \quad (9.69)$$

$$2KCl + Sn \xrightarrow{\text{in } BrF_3} K_2[SnF_6] \quad (9.70)$$

$$Ag + Au \xrightarrow{\text{in } BrF_3} Ag[AuF_4] \quad (9.71)$$

$$KCl + VCl_4 \xrightarrow{\text{in } BrF_3} K[VF_6] \quad (9.72)$$

$$Ru + KCl \xrightarrow{\text{in } BrF_3} K[RuF_6] \quad (9.73)$$

In contrast to the situation for H_2SO_4, where we noted that it is difficult to separate reaction products from the solvent by evaporation, BrF_3 can be removed *in vacuo* ($\Delta_{vap}H^o = 47.8\,\text{kJ mol}^{-1}$). Some of the compounds prepared in liquid BrF_3 can also be made using F_2 as the fluorinating agent, but use of F_2 generally requires higher reaction temperatures and the reactions are not always product-specific.

Non-aqueous solvents that behave similarly to BrF_3 in that they are good oxidizing and fluorinating agents include ClF_3, BrF_5 and IF_5.

Self-study exercises

1. Suggest how AgF behaves when dissolved in BrF_3.
   [*Ans.* Analogous to reaction 9.64]

2. When AgF and SbF_5 are dissolved in BrF_3, the product is $Ag[SbF_6]$. Comment on the role of the solvent in this reaction.

3. When NaCl and $RhCl_3$ are dissolved in liquid IF_5, a rhodium(V) salt is obtained. Suggest a likely identity for this product.
   [*Ans.* $Na[RhF_6]$]

4. When tungsten metal dissolves in BrF_3, a W(VI) compound is formed. (a) What is this product? (b) The oxidation must be accompanied by a reduction process. Suggest the likely reduction product.
   [*Ans.* (a) WF_6; (b) Br_2]

9.11 Dinitrogen tetraoxide

Physical properties

The data in Table 9.9 and Fig. 9.2 emphasize the very short liquid range of N_2O_4. Despite this and the low relative permittivity (which makes it a poor solvent for most inorganic compounds), N_2O_4 has some important preparative uses.

The low electrical conductivity ($1.3 \times 10^{-13}\ \Omega^{-1}\ cm^{-1}$ at 273 K) of liquid N_2O_4 is inconsistent with the presence of ions in the neat liquid. However, when $[Et_4N][NO_3]$ (which has a very low lattice energy) dissolves in liquid N_2O_4, $[NO_3]^-$ ions rapidly exchange between the solute and solvent. This confirms that $[NO_3]^-$ ions can form in liquid N_2O_4. In terms of a solvent-oriented acid–base definition, basic behaviour in N_2O_4 is characterized by the formation of $[NO_3]^-$, and acidic behaviour by the production of $[NO]^+$. Thus, equilibrium 9.74 is useful for understanding reactions in liquid N_2O_4, even though self-ionization does not occur in the pure liquid.

$$N_2O_4 \rightleftharpoons [NO]^+ + [NO_3]^- \tag{9.74}$$

Table 9.9 Selected physical properties of dinitrogen tetraoxide, N_2O_4.

Property / units	Value
Melting point / K	261.8
Boiling point / K	294.2
Density of liquid / $g\,cm^{-3}$	1.49 (at 273 K)
Relative permittivity	2.42 (at 291 K)

Reactions in N_2O_4

Reactions carried out in liquid N_2O_4 reflect the fact that N_2O_4 is a good oxidizing (see Box 9.1) and nitrating agent. Electropositive metals such as Li and Na react in liquid N_2O_4 liberating NO (eq. 9.75). In this reaction, Li is oxidized to Li^+, and $[NO]^+$ is reduced to NO.

$$Li + N_2O_4 \longrightarrow LiNO_3 + NO \tag{9.75}$$

Less reactive metals may react rapidly if ClNO, $[Et_4N][NO_3]$ or an organic donor such as MeCN is present. These observations can be explained as follows.

- ClNO behaves as a very weak acid in liquid N_2O_4, ionizing to produce $[NO]^+$. This encourages oxidation of metals as $[NO]^+$ is reduced to NO (eq. 9.76).

$$Sn + 2ClNO \xrightarrow{\text{in liquid } N_2O_4} SnCl_2 + 2NO \tag{9.76}$$

- $[Et_4N][NO_3]$ acts as a base in liquid N_2O_4 and its action on metals such as Zn and Al produces nitrato complexes (eq. 9.77). This is analogous to the formation of hydroxido complexes in an aqueous system. Figure 9.7 shows the structure of $[Zn(NO_3)_4]^{2-}$. Again, the oxidation of the metal is accompanied by reduction of $[NO]^+$ to NO.

$$Zn + 2[Et_4N][NO_3] + 2N_2O_4$$
$$\longrightarrow [Et_4N]_2[Zn(NO_3)_4] + 2NO \tag{9.77}$$

- Organic donor molecules appear to facilitate reactions with metals by increasing the degree of self-ionization of the solvent as a result of adduct formation with the $[NO]^+$ cation; e.g. Cu dissolves in liquid N_2O_4/MeCN (eq. 9.78), and Fe behaves similarly, dissolving to give $[NO][Fe(NO_3)_4]$.

$$Cu + 3N_2O_4 \xrightarrow{\text{in presence of MeCN}} [NO][Cu(NO_3)_3] + 2NO \tag{9.78}$$

The presence of $[NO]^+$ cations in compounds such as $[NO][Cu(NO_3)_3]$, $[NO][Fe(NO_3)_4]$, $[NO]_2[Zn(NO_3)_4]$ and $[NO]_2[Mn(NO_3)_4]$ is confirmed by the appearance of a characteristic absorption (ν_{NO}) at $\approx 2300\ cm^{-1}$ in the infrared spectra of the complexes.

Just as hydrolysis of a compound may occur in water (see Section 7.7), solvolysis such as reaction 9.79 can take place in liquid N_2O_4. Such reactions are of synthetic importance as routes to *anhydrous* metal nitrates.

$$ZnCl_2 + 2N_2O_4 \longrightarrow Zn(NO_3)_2 + 2ClNO \tag{9.79}$$

In many of the reactions carried out in liquid N_2O_4, the products are solvates, for example $[Fe(NO_3)_3]\cdot1.5N_2O_4$, $[Cu(NO_3)_2]\cdot N_2O_4$, $[Sc(NO_3)_2]\cdot2N_2O_4$ and $[Y(NO_3)_3]\cdot 2N_2O_4$. Such formulations may, in some cases, be correct, with molecules of N_2O_4 present, analogous to water molecules of crystallization in crystals isolated from an aqueous system. However, the results of X-ray diffraction

APPLICATIONS

Box 9.1 Liquid N_2O_4 as a fuel in the Apollo missions

During the Apollo Moon missions, a propulsion system was required that could be used to alter the velocity of the spacecraft and change its orbit during landing on and take-off from the Moon's surface. The fuel chosen was a mixture of liquid N_2O_4 and derivatives of hydrazine (N_2H_4). Dinitrogen tetraoxide is a powerful oxidizing agent and contact with, for example, $MeNHNH_2$ leads to immediate oxidation of the latter:

$$5N_2O_4 + 4MeNHNH_2 \longrightarrow 9N_2 + 12H_2O + 4CO_2$$

The reaction is highly exothermic, and at the operating temperatures, all products are gases.

Safety is of utmost importance; the fuels clearly must not contact each other before the required moment of landing or lift-off. Further, $MeNHNH_2$ is extremely toxic.

The command module of *Apollo 17* in orbit above the lunar surface.

studies on some solvated compounds illustrate the presence, not of N_2O_4 molecules, but of $[NO]^+$ and $[NO_3]^-$ ions. Two examples are $[Sc(NO_3)_3]\cdot2N_2O_4$ and $[Y(NO_3)_3]\cdot2N_2O_4$, for which the formulations $[NO]_2[Sc(NO_3)_5]$ and $[NO]_2[Y(NO_3)_5]$ have been crystallo-

graphically confirmed. In the $[Y(NO_3)_5]^{2-}$ anion, the Y(III) centre is 10-coordinate with bidentate nitrato ligands, while in $[Sc(NO_3)_5]^{2-}$, the Sc(III) centre is 9-coordinate with one $[NO_3]^-$ ligand being monodentate.

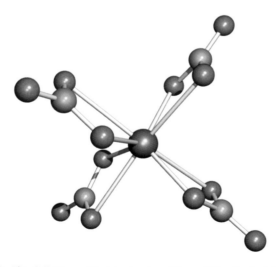

Fig. 9.7 The solid state structure (X-ray diffraction) of the $[Zn(NO_3)_4]^{2-}$ anion in the salt $[Ph_4As]_2[Zn(NO_3)_4]$. Each $[NO_3]^-$ ligand is coordinated to the Zn(II) centre through two O atoms, with one short (average 206 pm) and one long (average 258 pm) Zn–O interaction [C. Bellitto *et al.* (1976) *J. Chem. Soc., Dalton Trans.*, p. 989]. Colour code: Zn, brown; N, blue; O, red.

Self-study exercises

1. In reaction 9.76, why is ClNO considered to act as a weak acid?

2. Reaction of uranium metal with N_2O_4 in $N_2O_4/MeNO_2$ solvent yields $[UO_2(NO_3)_3]^-$ in which the U centre is 8-coordinate. Suggest (a) a structure for $[UO_2(NO_3)_3]^-$, and (b) the identity of the counterion.
   [*Ans.* See M.-J. Crawford *et al.* (2005) *Inorg. Chem.*, vol. 44, p. 8481]

3. Write an equation for the reaction of Na metal in liquid N_2O_4.   [*Ans.* Analogous to reaction 9.75]

4. Why is the following reaction classified as a neutralization process?

$$AgNO_3 + NOCl \xrightarrow{\text{in liquid } N_2O_4} AgCl + N_2O_4$$

9.12 Ionic liquids

The use of *ionic liquids* (also called *molten salts*) as reaction media is a relatively new area, although molten conditions have been well established in industrial processes (e.g. the Downs process, Fig. 11.2) for many years. While some 'molten salts' are hot as the term suggests, others operate at ambient temperatures and the term 'ionic liquids' is more appropriate. Although the terms 'ionic liquids' and 'molten salts' are sometimes used interchangeably, we make a clear distinction between them, using 'ionic liquid' only for a salt with a melting point ≤ 373 K.

Molten salt solvent systems

When an ionic salt such as NaCl melts, the ionic lattice (see Fig. 6.16) collapses, but some order is still retained. Evidence for this comes from X-ray diffraction patterns, from which *radial distribution functions* reveal that the average coordination number (with respect to cation–anion interactions) of each ion in liquid NaCl is ≈ 4, compared with 6 in the crystalline lattice. For cation–cation or anion–anion interactions, the coordination number is higher, although, as in the solid state, the internuclear distances are larger than for cation–anion separations. The solid-to-liquid transition is accompanied by an increase in volume of ≈ 10–15%.

The term *eutectic* is commonly encountered in molten salt systems. The reason for forming a eutectic mixture is to provide a molten system at a convenient working temperature. For example, the melting point of NaCl is 1073 K, but is lowered if $CaCl_2$ is added as in the Downs process.

> A *eutectic* is a mixture of two substances and is characterized by a sharp melting point lower than that of either of the components; a eutectic behaves as though it were a single substance.

Other alkali metal halides behave in a similar manner to NaCl, but metal halides in which the bonding has a significant covalent contribution (e.g. Hg(II) halides) form melts in which equilibria such as 9.80 are established. In the solid state, $HgCl_2$ forms a molecular lattice, and layer structures are adopted by $HgBr_2$ (distorted CdI_2 lattice) and HgI_2.

$$2HgBr_2 \rightleftharpoons [HgBr]^+ + [HgBr_3]^- \qquad (9.80)$$

In terms of the solvent-oriented description of acid–base chemistry in a non-aqueous solvent, eq. 9.80 illustrates that, in molten $HgBr_2$, species producing $[HgBr]^+$ ions may be considered to act as acids, and those providing $[HgBr_3]^-$ ions function as bases. In most molten salts, however, the application of this type of acid–base definition is not appropriate.

An important group of molten salts with more convenient operating temperatures contain the $[AlCl_4]^-$ ion. An example is an $NaCl$–Al_2Cl_6 mixture. The melting point of Al_2Cl_6 is 463 K (at 2.5 bar), and its addition to NaCl (melting point, 1073 K) results in a 1:1 medium with a melting point of 446 K. In this and other Al_2Cl_6–alkali metal chloride melts, equilibria 9.81 and 9.82 are established, with the additional formation of $[Al_3Cl_{10}]^-$ (see Section 13.6).

$$Al_2Cl_6 + 2Cl^- \rightleftharpoons 2[AlCl_4]^- \qquad (9.81)$$

$$2[AlCl_4]^- \rightleftharpoons [Al_2Cl_7]^- + Cl^- \qquad (9.82)$$

Manufacturing processes in which metals are extracted from molten metal salts are important examples of the uses of molten salts and include the Downs process, and the production of Li by electrolysis of molten LiCl, and of Be and Ca from $BeCl_2$ and $CaCl_2$, respectively.

Some unusual cations have been isolated as products from reactions in molten salt media. For example, the reaction of Bi and $BiCl_3$ in KCl–$BiCl_3$ solvent at ≈ 570 K yields $[Bi_9]_2[BiCl_5]_4[Bi_2Cl_8]$ which contains $[Bi_9]^{5+}$, $[BiCl_5]^{2-}$ and $[Bi_2Cl_8]^{2-}$. In a melt containing $AlCl_3$ and MCl (M = Na or K) at ≈ 530 K, Bi and $BiCl_3$ react to form $[Bi_5]^{3+}$ (a trigonal bipyramidal species like $[Sn_5]^{2-}$, Fig. 9.3) and $[Bi_8]^{2+}$, which are isolated as the $[AlCl_4]^-$ salts.

Ionic liquids at ambient temperatures

> An *ionic liquid* is an ionic salt that is a liquid below 373 K.

Interest in the development and applications of ionic liquids continues to grow rapidly, one of the main reasons being that these solvents have implications for green chemistry (see Box 9.2). The first generation of ionic liquids were salts containing alkylpyridinium or dialkylimidazolium cations (Fig. 9.8) with $[AlCl_4]^-$ (**9.20**) or $[Al_2Cl_7]^-$ (**9.21**) anions.

(9.20)   **(9.21)**

For example, reaction between Al_2Cl_6 and *n*-butylpyridinium chloride, [pyBu]Cl, produces [pyBu][AlCl$_4$] (eq. 9.83). This family of ionic liquids suffers from the fact that the $[AlCl_4]^-$ ion readily hydrolyses. Replacement of $[AlCl_4]^-$ by $[BF_4]^-$ or $[PF_6]^-$ increases the stability of the ionic liquids to air and water.

$$Al_2Cl_6 + 2[pyBu]Cl \longrightarrow 2[pyBu][AlCl_4] \qquad (9.83)$$

ENVIRONMENT

Box 9.2 Green chemistry

With the constant drive to protect our environment, 'green chemistry' is now at the forefront of research and is starting to be applied in industry. In its *Green Chemistry Program*, the US Environmental Protection Agency (EPA) defines green chemistry as 'chemistry for pollution prevention, and the design of chemical products and chemical processes that reduce or eliminate the use of hazardous substances.' The European Chemical Industry Council (CEFIC) works through its programme *Sustech* to develop sustainable technologies. Some of the goals of green chemistry are the use of renewable feedstocks, the use of less hazardous chemicals in industry, the use of new solvents to replace, for example, chlorinated and volatile organic solvents, the reduction in the energy consumption of commercial processes, and the minimizing of waste chemicals in industrial processes.

Anastas and Warner (see Further reading) have developed 12 principles of green chemistry. These clearly illustrate the challenges ahead for research and industrial chemists:

- It is better to prevent waste than to treat or clean up waste after it is formed.
- Synthetic methods should be designed to maximize the incorporation of all materials used in the process into the final product.
- Wherever practicable, synthetic methodologies should be designed to use and generate substances that possess little or no toxicity to human health and the environment.
- Chemical products should be designed to preserve efficacy of function while reducing toxicity.
- The use of auxiliary substances (e.g. solvents, separation agents) should be made unnecessary whenever possible and innocuous when used.
- Energy requirements should be recognized for their environmental and economic impacts and should be minimized. Synthetic methods should be conducted at ambient temperature and pressure.
- A raw material feedstock should be renewable rather than depleting whenever technically and economically practical.
- Unnecessary derivatization (e.g. protection/deprotection steps) should be avoided whenever possible.
- Catalytic reagents (as selective as possible) are superior to stoichiometric reagents.
- Chemical products should be designed so that at the end of their function they do not persist in the environment, but break down into innocuous degradation products.
- Analytical methodologies need to be further developed to allow for real-time in-process monitoring and control prior to the formation of hazardous substances.
- Substances and the form of a substance used in a chemical process should be chosen so as to minimize the potential for chemical accidents, including releases, explosions and fires.

At the beginning of the 21st century, green chemistry represents a move towards a sustainable future. The journal *Green Chemistry* (published by the Royal Society of Chemistry since 1999) is a forum for key developments in the area. The American Chemical Society Green Chemistry Institute (ACS Green Chemistry Institute®) has been established with its goal being 'to enable and catalyse the implementation of green chemistry and green engineering into all aspects of the global chemical enterprise'. In the US, the Presidential Green Chemistry Challenge Awards were initiated in 1995 to encourage the development of green technologies, at both academic and commercial levels (see Box 15.1).

A bee-eater in the Camargue, France: our wildlife depends on a clean environment.

Further reading

P.T. Anastas and J.C. Warner (1998) *Green Chemistry Theory and Practice*, Oxford University Press, Oxford.

J.H. Clark and S.J. Tavener (2007) *Org. Process Res. Develop.*, vol. 11, p. 149 – 'Alternative solvents: shades of green'.

I. T. Horváth and P. T. Anastas (eds.) (2007) *Chem. Rev.*, vol. 107, issue 6 presents a series of articles on the theme of green chemistry.

R.A. Sheldon (2005) *Green Chem.*, vol. 7, p. 267 – A review: 'Green solvents for sustainable organic synthesis: state of the art'.

http://www.epa.gov/gcc

http://www.greenchemistrynetwork.org/

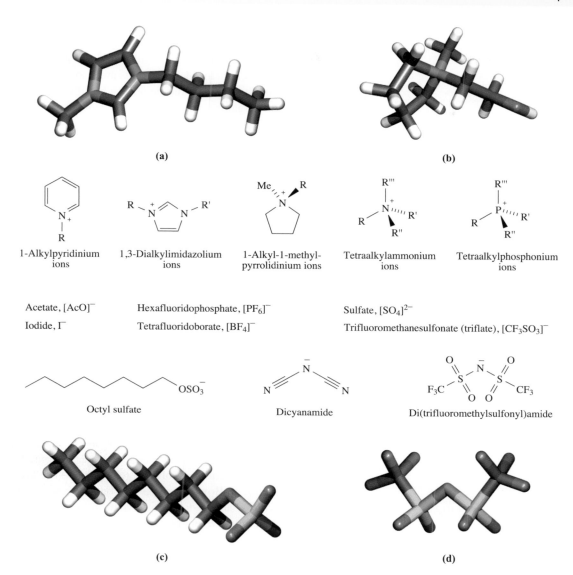

(a)

(b)

1-Alkylpyridinium ions

1,3-Dialkylimidazolium ions

1-Alkyl-1-methyl-pyrrolidinium ions

Tetraalkylammonium ions

Tetraalkylphosphonium ions

Acetate, [AcO]$^-$

Iodide, I$^-$

Hexafluoridophosphate, [PF$_6$]$^-$

Tetrafluoridoborate, [BF$_4$]$^-$

Sulfate, [SO$_4$]$^{2-}$

Trifluoromethanesulfonate (triflate), [CF$_3$SO$_3$]$^-$

Octyl sulfate

Dicyanamide

Di(trifluoromethylsulfonyl)amide

(c)

(d)

Fig. 9.8 Some of the most commonly used cations and anions in ionic liquids. In the cations, the organic groups, R, R' etc., may be unsubstituted alkyl groups or may be functionalized, e.g. (a) 1-butyl-3-methylimidazolium cation and (b) 1-(2-hydroxyethyl)-1-methylpyrrolidinium cation. Structures (c) and (d) show the octyl sulfate and di(trifluoromethylsulfonyl)amide anions. Colour code: C, grey; H, white; N, blue; O, red; S, yellow; F, green.

Currently favoured ionic liquids comprise combinations of cations and anions such as those in Fig. 9.8. By varying (i) the substituents in the cations and (ii) the combinations of anions and cations, a huge number of ionic liquids with differing physical properties (e.g. melting point, viscosity, mobility of ions in solution) can be prepared. A third group of ionic liquids has been designed for use in biocatalysis. These comprise biodegradable, low-toxicity cations and/or anions, for example the choline cation (**9.22**) with alkyl sulfates, alkyl phosphates or anions derived from sugars or amino acids.[†]

(9.22)

Some ionic liquids can be formed by the direct reaction of the N- or P-containing base with an alkylating agent. The latter may provide the desired counter-ion directly (e.g. eq. 9.84), or alkylation may be followed by anion exchange (scheme 9.85).

[†]See: J. Gorke, F. Srienc and R. Kazlauskas (2010) *Biotechnol. Bioprocess Eng.*, vol. 15, p. 40 – 'Toward advanced ionic liquids. Polar, enzyme-friendly solvents for biocatalysis'.

$$+ \; CF_3SO_3Me \longrightarrow \quad [CF_3SO_3]^- \qquad (9.84)$$

$$(9.85)$$

Self-study exercises

1. The following are routes to 1,3-dialkylimidazolium-based ionic liquids. What is the product in each case? Comment on the roles of the reagents.

2. Ionic liquid **B** can be made in two steps:

Identify **A** and **B**. Which step is a metathesis reaction?

The reason that ionic liquids possess such low melting points is that they contain large, unsymmetrical cations usually combined with polyatomic anions (Fig. 9.8). Such ions pack less efficiently in the solid state than symmetrical ions (e.g. $[NH_4]^+$ with Br^-). The relatively poor packing results in low lattice energies (Table 9.10) which, in turn, lead to low melting compounds. The melting point can be tuned by altering the substituents in a given cation. For example, in a series of salts containing different 1-alkyl-3-methylimidazolium cations and a common anion, the melting point decreases as the n-alkyl chain lengthens. However, once the n-alkyl substituent contains more than eight C atoms, the melting point starts to increase because it is energetically favourable for the longer alkyl chains to be aligned parallel to one another with van der Waals forces operating between them. Melting points are also affected by hydrogen bonding. If the ionic liquid is a hexafluoridophosphate salt, the organic cation and $[PF_6]^-$ ion can engage in weak $C–H\cdots F$ non-classical hydrogen bonds (see Section 10.6) in the solid state. This increases the melting point (although not as significantly as would be the case for $O–H\cdots O$ or $N–H\cdots O$ hydrogen bonds). In contrast, the melting points and viscosities of ionic liquids containing $[N(SO_2CF_3)_2]^-$ are low because the negative charge is delocalized over the two SO_2 units, thereby preventing directionalized cation$\cdots$anion interactions.

Ionic liquids are sometimes termed 'task specific', meaning that the large pool of cations and anions available permits the design of a solvent with quite specific properties. The low volatility of ionic liquids gives them a 'green' advantage over volatile organic solvents. Ionic liquids have long liquid ranges, and many are non-flammable and thermally stable. However, it is important to assess both the synthetic route to the ionic liquid as well as the properties of the material itself before labelling the compound environmentally friendly.[†]

Some of the first applications of room temperature ionic liquids (mp $<298\,K$) were in solution electrochemistry, for electrodeposition (electroplating) of metals and as electrolytes in batteries. What advantages do ionic liquids have over conventional solvents (e.g. MeCN) for cyclic voltammetry? In Box 8.2, we pointed out that a supporting electrolyte is added to the solvent in a cyclic voltammetry experiment to ensure a high enough electrical conductivity. Since an ionic liquid is composed of ions, no supporting electrolyte is needed. However, the mobility (not just the number) of ions affects conductivity. The large ions typically present in an ionic liquid exhibit low mobilities and this offsets the advantage of having a solvent composed entirely of ions. Many ionic liquids are too viscous to be used for solution electrochemical studies. Compared with traditional solvents, ionic liquids composed of imidazolium cations with $[AlCl_4]^-$, $[BF_4]^-$, $[CF_3CO_2]^-$, $[CF_3SO_3]^-$ or $[N(SO_2CF_3)_2]^-$ and which exhibit wide potential windows (see Box 8.2) of $\leq 6\,V$, represent possible alternative media for cyclic voltammetry and similar electrochemical measurements.

Commercial electroplating of metals (e.g. Ni, Cu, Zn, Ag, Au) involves the deposition of the metal at the cathode from an aqueous electrolyte in a galvanic cell. Electrodeposition of Al is an exception; during electrolysis, H_2O is reduced at

---

[†] See: M. Deetlefs and K.R. Seddon (2010) *Green Chem.*, vol. 12, p. 17 – 'Assessing the greenness of some typical laboratory ionic liquid preparations'; M. Smiglak *et al.* (2006) *Chem. Commun.*, p. 2554 – 'Combustible ionic liquids by design: is laboratory safety another ionic liquid myth?'.

Table 9.10 Calculated values of lattice enthalpies ($\Delta_{lattice}H^\circ$) for selected ionic liquids. Values refer to the process: gaseous ions $\longrightarrow$ ionic solid. (See Chapter 6 for definitions of lattice energy and lattice enthalpy.)

Ionic liquid	$\Delta_{lattice}H^\circ / kJ\,mol^{-1}$	Ionic liquid	$\Delta_{lattice}H^\circ / kJ\,mol^{-1}$
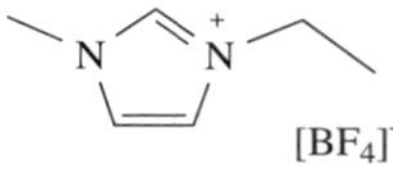 [BF$_4$]$^-$	−492	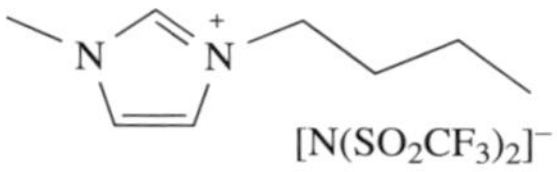 [N(SO$_2$CF$_3$)$_2$]$^-$	−420
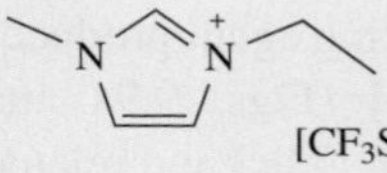 [CF$_3$SO$_3$]$^-$	−464	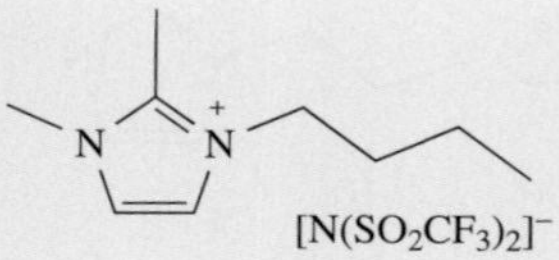 [N(SO$_2$CF$_3$)$_2$]$^-$	−412
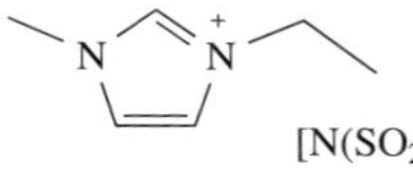 [N(SO$_2$CF$_3$)$_2$]$^-$	−425	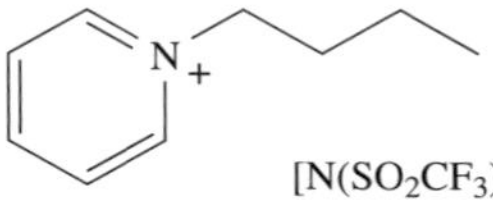 [N(SO$_2$CF$_3$)$_2$]$^-$	−420
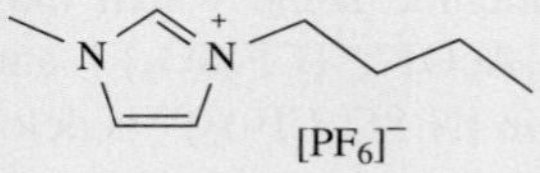 [PF$_6$]$^-$	−457	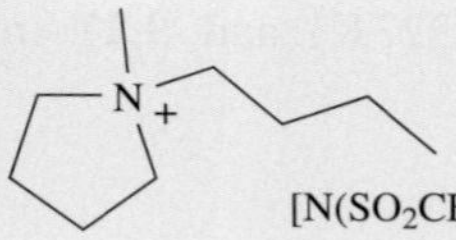 [N(SO$_2$CF$_3$)$_2$]$^-$	−414
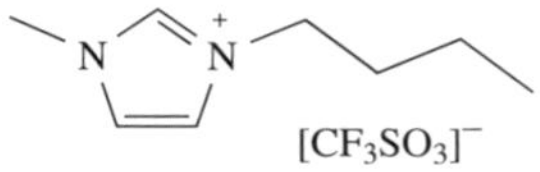 [CF$_3$SO$_3$]$^-$	−449	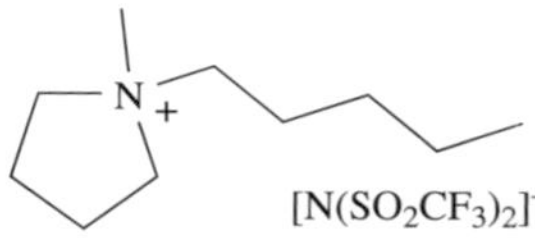 [N(SO$_2$CF$_3$)$_2$]$^-$	−411

Data: I. Krossing *et al.* (2006) *J. Am. Chem. Soc.*, vol. 128, p. 13427; I. Krossing *et al.* (2007) *J. Am. Chem. Soc.*, vol. 129, p. 11296.

the cathode before Al^{3+} is reduced. Thus, commercial electrodeposition of Al is carried out using Et$_3$Al (which is *pyrophoric*) in toluene. A safer alternative is electrodeposition of Al from AlCl$_3$ or related precursor in an ionic liquid. Possible solvents are **9.23** and **9.24**,[†] and further research development should lead to commercial application in the near future. Other metals that could be electroplated using ionic liquids include Mg, Ti, Mo and W.

A *pyrophoric* material is one that spontaneously ignites in air.

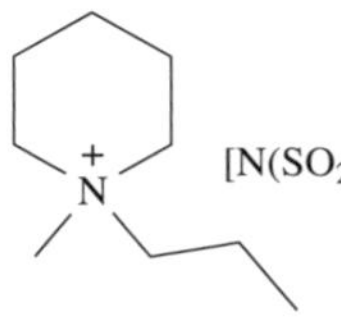

[N(SO$_2$CF$_3$)$_2$]$^-$

[N(SO$_2$CF$_3$)$_2$]$^-$

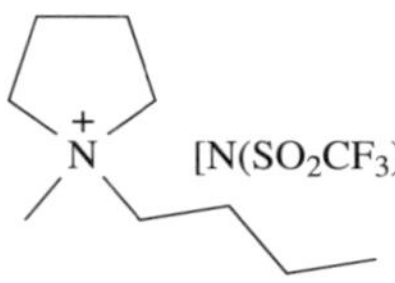

(9.23)   **(9.24)**

Rechargeable lithium-ion batteries have widespread commercial applications in portable electronic devices. However, safety issues associated with the use of flammable organic electrolytes have been a barrier to the development of large lithium-ion batteries of the current composition for use in electric or hybrid electric vehicles (see Box 11.3). The electrolyte in commercial batteries is typically LiPF$_6$ in an alkyl carbonate. The physical properties of ionic liquids suggest that they could be applied in lithium-ion batteries, and among those tested, compound **9.23** functions well. Cell performance is enhanced by adding Li[N(SO$_2$CF$_3$)$_2$] (0.3 mole per kg of **9.23**), and further improved if diethyl carbonate (**9.25**) is added as a co-solvent. The amount of diethyl carbonate must be limited to maintain an electrolyte of low flammability. Although promising,[†] the application of ionic liquids in lithium-ion batteries has not yet (as of 2010) been commercialized.

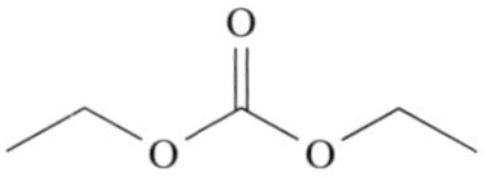

(9.25)

[†]See: T. Rodopoulos, L. Smith, M. D. Horne and T. Rüther (2010) *Chem. Eur. J.*, vol. 16, p. 3815.

[†]See: A. Lewandowski and A. Świderska-Mocek (2009) *J. Power Sources*, vol. 194, p. 601; H.F. Xiang, B. Yin, H. Wang, H.W. Lin, X.W. Ge, S. Xie and C.H. Chen (2010) *Electrochim. Acta*, vol. 55, p. 5204.

Ionic liquids are now used in place of organic solvents in a wide range of organic transformations including Diels–Alder reactions, Friedel–Crafts alkylations and acylations, C–C bond-forming Heck reactions, and syntheses of heterocyclic systems. However, imidazolium-based ionic liquids tend to react with strong bases (e.g. Grignard reagents or KO^{t}Bu), being deprotonated at the C2 position to give a carbene:

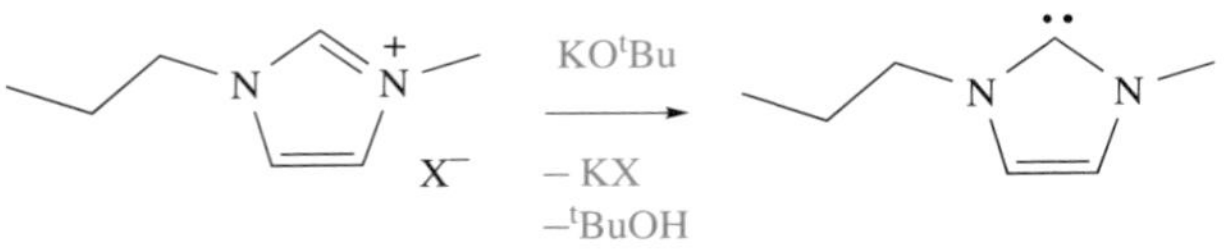

They should therefore not be used under such conditions. Nonetheless, many ionic liquids are good solvents for a wide range of organometallic compounds, thus enabling them to be used in homogeneous catalysis. Asymmetric syntheses may be carried out using chiral catalysts in achiral ionic liquids. Enantiomerically pure ionic liquids containing chiral cations such as **9.26** (mp 327 K) and **9.27** (mp <255 K) are also available.

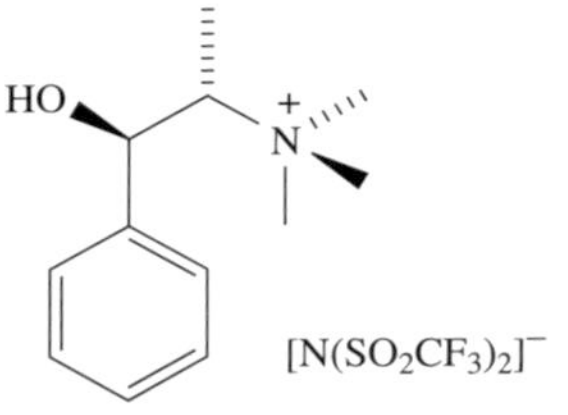

(9.26) 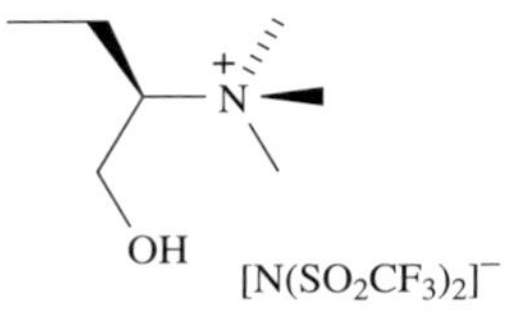   **(9.27)**

The first industrial process to involve ionic liquids was introduced by BASF in 2002. Called BASIL^{TM}, this technology addresses the problem of HCl formed as a byproduct in the synthesis of alkoxyphenylphosphanes (eq. 9.86). Before the advent of ionic liquids, HCl was scavenged by a tertiary amine, forming a solid [R$_3$NH]Cl salt which was separated by filtration. On a commercial scale, this operation is expensive. BASF now uses 1-methylimidazole to scavenge HCl (eq. 9.87). 1-Methylimidazolium chloride is an ionic liquid (mp 348 K) which is immiscible with the reaction solvent. Separation of the two liquid phases is achieved at a lower cost than the filtration previously needed. Deprotonation of the 1-methylimidazolium cation regenerates imidazole which is recycled.

$$\hspace{10cm} (9.86)$$

$$\hspace{10cm} (9.87)$$

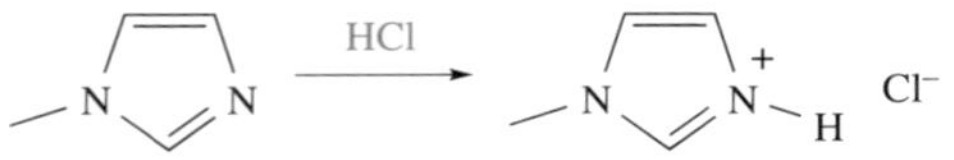

Applications of ionic liquids in inorganic chemistry range from the isolation of molecular species to uses in material chemistry. The choice of an ionic liquid allows access to species that are insoluble in common organic solvents or are unstable in aqueous media. The organic cations in the solvent are often incorporated into the products and can act as templates to facilitate the assembly of 3-dimensional frameworks.

Ionic liquids containing inorganic halide anions may act as a source of halide, thereby functioning as a reagent as well as a solvent. For example, reactions of Bi$_2$O$_3$ or V$_2$O$_5$ in [Bupy][AlCl$_4$] ([Bupy]$^+$ = N-butylpyridinium ion) at 373 K with Bi or In metal as reducing agent produces [Bupy]$_4$[Bi$_4$Cl$_{16}$] or [Bupy]$_4$[V$_4$O$_4$Cl$_{12}$] (Figs. 9.9a and 9.9b). The cluster ion [Zr$_6$MnCl$_{18}$]$^{5-}$ (Fig. 9.9c) and related anions may be 'cut out' from solid state precursors that possess extended structures. Solid Li$_2$Zr$_6$MnCl$_{15}$ comprises octahedral Zr$_6$Mn-units connected by bridging Cl atoms. Heating Li$_2$Zr$_6$MnCl$_{15}$ in [EMim][AlCl$_4$] ([EMim]$^+$ = 1-ethyl-3-methylimidazolium ion) yields the [EMim]$_5$[Zr$_6$MnCl$_{18}$].

Anions used in ionic liquids include some which may coordinate to metal ions, e.g. [AcO]$^-$, [CF$_3$SO$_3$]$^-$ and [N(SO$_2$CF$_3$)$_2$]$^-$. Since the charge in [N(SO$_2$CF$_3$)$_2$]$^-$ is delocalized over the ion, it is typically considered to be only weakly coordinating. However, in an appropriate ionic liquid (i.e. in the absence of competing ligands such as H$_2$O or solvent), [N(SO$_2$CF$_3$)$_2$]$^-$ binds to a range of metal ions and its coordination chemistry is now well developed for s- and f-block metal ions. Reactions 9.88–9.91 give examples ([MPpyr]$^+$ = N-methyl-N-propylpyrrolidinium ion, [BMpyr]$^+$ = N-butyl-N-methylpyrrolidinium ion). The [N(SO$_2$CF$_3$)$_2$]$^-$ ion coordinates through one or two of its oxygen atoms. In [Ca{N(SO$_2$CF$_3$)$_2$}$_4$]$^{2-}$ (Fig. 9.10a) and [Yb{N(SO$_2$CF$_3$)$_2$}$_4$]$^{2-}$, the metal ion is in a square antiprismatic environment (8-coordinate) with four bidentate ligands. In [Pr{N(SO$_2$CF$_3$)$_2$}$_5$]$^{2-}$ and [Nd{N(SO$_2$CF$_3$)$_2$}$_5$]$^{2-}$ (Fig. 9.10b), the metal ion is 9-coordinate with one monodentate and four bidentate ligands.

$$CaI_2 + \text{excess } [MPpyr][N(SO_2CF_3)_2] \xrightarrow{393\,K,\,under\,vacuum}$$
$$[MPpyr]_2[Ca\{N(SO_2CF_3)_2\}_4]$$
$$(9.88)$$

$$YbI_2 + \text{excess } [MPpyr][N(SO_2CF_3)_2] \xrightarrow{393\,K,\,under\,vacuum}$$
$$[MPpyr]_2[Yb\{N(SO_2CF_3)_2\}_4]$$
$$(9.89)$$

$$Pr\{N(SO_2CF_3)_2\}_3 + \text{excess } [BMpyr][N(SO_2CF_3)_2]$$
$$\xrightarrow{393\,K,\,under\,vacuum} [BMpyr]_2[Pr\{N(SO_2CF_3)_2\}_5]$$
$$(9.90)$$

$$Nd\{N(SO_2CF_3)_2\}_3 + \text{excess } [BMpyr][N(SO_2CF_3)_2]$$
$$\xrightarrow{393\,K,\,under\,vacuum} [BMpyr]_2[Pr\{N(SO_2CF_3)_2\}_5]$$
$$(9.91)$$

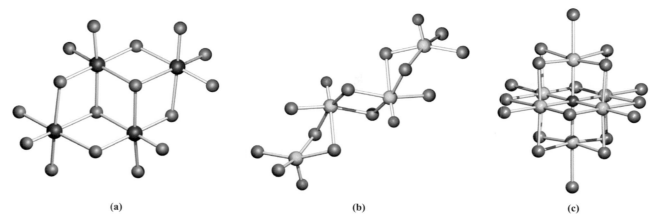

Fig. 9.9 The structures (X-ray diffraction) of (a) the $[Bi_4Cl_{16}]^{4-}$ ion in $[Bupy]_4[Bi_4Cl_{16}]$ (colour code: Bi, blue; Cl, green), (b) the $[V_4O_4Cl_{12}]^{4-}$ ion in $[Bupy]_4[V_4O_4Cl_{12}]$ (colour code: V, yellow; O, red; Cl, green) [P. Mahjoor *et al.* (2009) *Cryst. Growth Des.*, vol. 9, p. 1385], and (c) the $[Zr_6MnCl_{18}]^{5-}$ anion from the 1-ethyl-3-methylimidazolium salt [D. Sun *et al.* (2000) *Inorg. Chem.*, vol. 39, p. 1964]. Colour code: Zr, yellow; Mn, red; Cl, green.

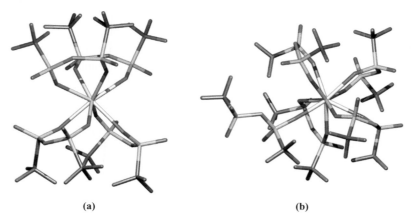

Fig. 9.10 Stick-representations of the structures (X-ray diffraction) of (a) the $[Ca\{N(SO_2CF_3)_2\}_4]^{2-}$ anion in $[MPpyr]_2[Ca\{N(SO_2CF_3)_2\}_4]$ [A. Babai *et al.* (2006) *Inorg. Chem.*, vol. 45, p. 1828], and (b) the $[Nd\{N(SO_2CF_3)_2\}_5]^{2-}$ ion in $[BMpyr]_2[Nd\{N(SO_2CF_3)_2\}_5]$ [A. Babai *et al.* (2006) *Dalton Trans.*, p. 1828.]. Colour code: Ca or Nd, silver; O, red; N, blue; S, yellow; C, grey; F, green.

In reactions 9.90 and 9.91, the precursor is the di(trifluoromethylsulfonyl)amido salt of the metal rather than a halide as in eqs. 9.88 and 9.89. If PrI_3 is reacted with $[BMpyr][N(SO_2CF_3)_2]$ (compare with eq. 9.90), the product is $[BMpyr]_4[PrI_6][N(SO_2CF_3)_2]$ containing the octahedral $[PrI_6]^{3-}$ ion and a non-coordinated $[N(SO_2CF_3)_2]^-$ ion. Separation of products from ionic liquid solvents is facile if the product is insoluble (separation is by filtration or removal of the solvent by cannula). However, if the product is soluble, separation may be difficult since the low volatility of ionic liquids prevents their ready evaporation. If the product is volatile, it can be separated by distillation because the ionic liquid has such a low vapour pressure.

In materials chemistry, ionic liquids are used in a variety of ways, e.g. routes to nanoparticles (some of which have catalytic applications) and to metal-organic frameworks (MOFs). Zeolites represent an important class of crystalline materials (see Section 14.9) with many applications in catalysis and gas/liquid absorption. Their synthesis under hydrothermal conditions (i.e. high temperatures and pressure in aqueous media) from Al_2O_3, SiO_2 and Na_2O or similar precursors requires an organic template to direct the assembly of the 3-dimensional array with template ions (e.g. alkyl ammonium cations) occupying the pores in the open network structure. Using ionic liquids in place of hydrothermal conditions has several advantages: (i) the low vapour pressure of an ionic liquid means that the synthesis takes place at ambient pressures, (ii) the organic template is provided by the ionic liquid, resulting in there being no competition between solvent and template during crystallization, and (iii) new structure types may be assembled. This is an area that is actively being developed, although industrial manufacture of synthetic zeolites remains (as of 2010) reliant on hydrothermal methods.[†]

[†]See: R.E. Morris (2009) *Chem. Commun.*, p. 2990; Z. Ma, J. Yu and S. Dai (2010) *Adv. Mater.*, vol. 22, p. 261.

9.13 Supercritical fluids

Properties of supercritical fluids and their uses as solvents

> Above its critical temperature and critical pressure, an element or compound is in its **supercritical fluid** state.

Since the 1990s, the chemical literature has seen a huge increase in the publication of papers describing the properties and applications of *supercritical fluids*, in particular, supercritical carbon dioxide and water. One of the driving forces for this interest is the search for green solvents to replace volatile organics (see Box 9.2). The meaning of the term *supercritical* is explained in Fig. 9.11 which shows a pressure–temperature phase diagram for a one-component system. The solid blue lines represent the boundaries between the phases. The hashed line illustrates the distinction between a vapour and a gas:

> A *vapour* can be liquefied by increasing the pressure, but a *gas* cannot.

Above the critical temperature, $T_{critical}$, the gas can no longer be liquefied, no matter how high the pressure is increased. If a sample is observed as the critical point is reached, the meniscus at the liquid–gas interface disappears, signifying that there is no longer a distinction between the two phases. At temperatures and pressures above the critical temperature and pressure (i.e. above the critical point), a substance becomes a supercritical fluid.

A supercritical fluid possesses solvent properties that resemble those of a liquid, but also exhibits gas-like transport properties. Thus, not only can a supercritical fluid dissolve solutes, but it is also miscible with ordinary gases and can penetrate pores in solids. Supercritical fluids exhibit lower viscosities and higher diffusion coefficients than liquids. The density of a supercritical fluid increases as the pressure increases, and as the density increases, the solubility of a solute in the supercritical fluid increases dramatically. The fact that the properties can be tuned by varying the pressure and temperature is advantageous in terms of the applications of these fluids as extraction agents. Using a supercritical fluid for the extraction of a given material from a feedstock involves the partitioning of the material into the supercritical liquid, followed by a change in temperature and pressure that results in isolation of the pure solute by vaporization of the solvent, e.g. CO_2. Finally, the supercritical fluid can be recycled by reversing the change in temperature and pressure conditions (see the figure in Box 9.3).

Table 9.11 lists the critical temperatures and pressures of selected elements and compounds that are used as supercritical fluids. Combined with its easy accessibility, low cost, non-toxicity, chemical inertness and non-inflammability, the critical temperature and pressure of CO_2 are convenient enough to make supercritical CO_2 ($scCO_2$) of great value as a solvent, and Box 9.3 gives examples of its commercial applications.

Although $scCO_2$ is a 'clean' alternative to organic solvents for a range of extraction processes, it is non-polar.

Table 9.11 Critical temperatures and pressures of selected elements and compounds with applications as supercritical fluids.

Compound or element	Critical temperature / K	Critical pressure / MPa[†]
Xenon	289.8	5.12
Carbon dioxide	304.2	7.38
Ethane	305.4	4.88
Propane	369.8	4.25
Ammonia	405.6	11.28
Pentane	469.7	3.37
Ethanol	516.2	6.38
Toluene	591.8	4.11
1,2-Ethanediamine	593.0	6.27
Water	647.3	22.05

[†] To convert to bar, multiply by 10.

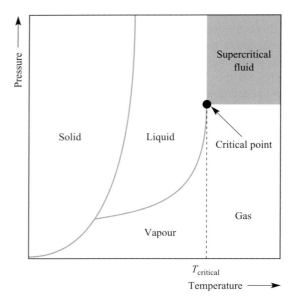

Fig. 9.11 A simple pressure–temperature phase diagram for a one-component system.

APPLICATIONS
Box 9.3 Clean technology with supercritical CO$_2$

Some of the areas in which supercritical CO$_2$ (scCO$_2$) is commercially important are summarized in Fig. 9.12. Extraction processes in the food, tobacco (nicotine extraction) and pharmaceutical industries dominate. Supercritical CO$_2$ is a selective extracting agent for caffeine, and its use in the decaffeination of coffee and tea was the first commercial application of a supercritical fluid, followed by the extraction of hops in the brewing industry. Solvent extractions can be carried out by batch processes, or by a continuous process in which the CO$_2$ is recycled as shown schematically below:

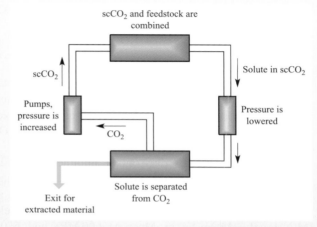

Examples of industrial autoclaves containing scCO$_2$ are illustrated in the photograph opposite. Such autoclaves are used to decaffeinate tea and coffee and to manufacture spice extracts.

Cholesterol (high levels of which in the blood are associated with heart complaints) is soluble in scCO$_2$, and this medium has been used to extract cholesterol from egg yolk, meat and milk. There is potential for wider application of scCO$_2$ in the production of foodstuffs with reduced cholesterol levels.

The extraction of pesticides from rice is carried out commercially using scCO$_2$. Many studies have been carried out to investigate the ability of scCO$_2$ to extract flavours and fragrances from plants, e.g. from ginger root, camomile leaf, vanilla pod, mint leaf, lavender flower and lemon peel. Commercial applications within the food industry include cleaning of rice grains, the extraction of flavours and spices, and the extraction of colouring agents, e.g. from red peppers. Supercritical CO$_2$ can be used to extract compounds from natural products. One example is the anti-cancer drug taxol which may be extracted from the bark of the Pacific yew tree (although the drug can also be synthesized in a multi-step process).

The technique of supercritical fluid chromatography (SFC) is similar to HPLC (see Section 4.2) but has major advantages over the latter: separation is more rapid and the use of organic solvents is minimized. The pharmaceutical industry applies SFC to the separation of chiral and natural products.

The development of new technologies for the manufacture of high-purity polymers using scCO$_2$ in place of organic solvents is an active area of research, and the reduction of large amounts of toxic waste during polymer production is a prime target for the polymer industry. In 2002, DuPont introduced the first commercial Teflon resins manufactured using scCO$_2$ technology.

Commercial applications of scCO$_2$ in dry cleaning clothes, cleaning of optical and electronics components and of heavy-duty valves, tanks and pipes are growing steadily. Tetrachloroethene is still used as the solvent in textile dry cleaning, but alternative methodologies based on scCO$_2$ overcome the negative impact that volatile and chlorinated organics have on the environment. In 2006, the Fred Butler® cleaning company was launched in Europe and has won awards for its innovative scCO$_2$-based textile cleaning process. Hangers® cleaners are responsible for the growing use of scCO$_2$ in dry cleaning facilities in the US.

Supercritical CO$_2$ has found applications within materials processing. The *rapid expansion of supercritical solutions* (RESS) involves saturating the supercritical fluid with a given solute followed by rapid expansion (by reduction in pressure) through a nozzle. The result is the nucleation of the solute (e.g. a polymer such as PVC) and the production of a powder, thin film or fibre as required. Union Carbide has developed a process (UNICARB®) in which scCO$_2$ is used in place of organic solvents for spraying paint onto a range of substrates including vehicles.

The textile industry uses huge amounts of water (>100 dm^3 per kg of fibre), and dyeing of fibres is one of the processes during fabric production that depends on water. Under 25 MPa pressure and at 390 K, scCO$_2$ penetrates textile fibres and can be used as a medium for dyeing. This technology entered the textile industry in 2010: DyeCoo™ Textile Systems (Netherlands) manufacture scCO$_2$-based industrial dyeing equipment and the Yeh Group (Thailand) has begun production of DryDye™ fabrics using the scCO$_2$ technology. Recycling of scCO$_2$ replaces the output of waste water from traditional dyeing processes.

In the examples given above, supercritical CO$_2$ is used in what is termed 'clean technology' with drastic reductions in the use of organic solvents, and the 21st century should see an increase in the use of supercritical fluids in commercial processes.

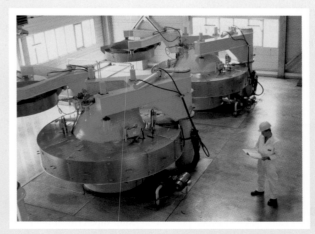

Industrial autoclaves using scCO$_2$ in a food processing plant.

While the behaviour of scCO$_2$ does not parallel a typical non-polar organic solvent, its ability to extract polar compounds is still relatively poor. The dissolution of polar compounds can be aided by introducing a subcritical co-solvent (a modifier) to scCO$_2$, and two common choices are H$_2$O and MeOH. The use of surfactants that possess a water-soluble head and CO$_2$-compatible tail permits water 'pockets' to be dispersed within scCO$_2$. As a result, aqueous chemistry can be carried out in what is essentially a non-aqueous environment. An advantage of this system is that reagents not normally soluble in water, but soluble in scCO$_2$, can be brought into intimate contact with water-soluble reagents.

Two other well-studied solvents are supercritical NH$_3$ and H$_2$O. The critical temperature and pressure of supercritical NH$_3$ are accessible (Table 9.11), but the solvent is chemically very reactive and is relatively hazardous for large-scale applications. Supercritical H$_2$O has a relatively high critical temperature and pressure (Table 9.11) which limit its uses. Even so, it has important applications as a solvent. At its critical point, the density of water is 0.32 g cm^{-3}. The density of the supercritical phase can be controlled by varying the temperature and pressure. Unlike subcritical H$_2$O, supercritical H$_2$O behaves like a *non-polar* solvent. Thus, it is a poor solvent for inorganic salts, but dissolves non-polar organic compounds. This is the basis for its use in *supercritical water oxidation* (or *hydrothermal oxidation*) of toxic and hazardous organic wastes. In the presence of a suitable oxidizing agent, liquid organic waste in scH$_2$O is converted to CO$_2$, H$_2$O, N$_2$ and other gaseous products with efficiencies approaching 100%. The operating temperatures are low enough to prevent the formation of environmentally undesirable products such as oxides of nitrogen and sulfur. In the waste-water industry, sludge disposal can be effected using supercritical water oxidation, and, in 2001, the first commercial plant designed for this purpose commenced operation in Texas, USA.

Initial commercial applications of supercritical fluids were coffee decaffeination (in 1978) and hops extraction (in 1982). Together, these uses accounted for over half of the world's supercritical fluid production processes in 2001 (Fig. 9.12).

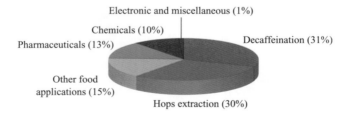

Fig. 9.12 Percentage contributions to the 2001 global US$960 million value of commercial production using supercritical fluid processing [data: Kline & Company, Inc., www.klinegroup.com].

Supercritical fluids as media for inorganic chemistry

In this section, we describe selected examples of inorganic reactions that are carried out in scCO$_2$, supercritical water (scH$_2$O) and ammonia (scNH$_3$), critical temperatures and pressures of which are listed in Table 9.11. An important application of scH$_2$O is in the hydrothermal generation of metal oxides from metal salts (or supercritical hydrothermal crystallization). Equations 9.92 and 9.93 summarize the proposed steps for conversion of metal nitrates to oxides where, for example, M = Fe(III), Co(II) or Ni(II).

$$M(NO_3)_{2x} + 2xH_2O \xrightarrow{\text{scH}_2\text{O}} M(OH)_{2x}(s) + 2xHNO_3$$

Hydrolysis (9.92)

$$M(OH)_{2x}(s) \xrightarrow{\text{scH}_2\text{O}} MO_x(s) + xH_2O$$

Dehydration (9.93)

By altering the precursor, different oxides of a given metal can be obtained. By adjusting the temperature and pressure of the scH$_2$O medium, it is possible to control particle size. Such control is important for the production of optical TiO$_2$ coatings (see Box 21.3).

In Section 9.6, we described metal ammine and amido complex formation in liquid NH$_3$. In scNH$_3$, FeCl$_2$ and FeBr$_2$ form the complexes [Fe(NH$_3$)$_6$]X$_2$ (X = Cl, Br) at 670 K, while reactions of Fe or Mn and I$_2$ in scNH$_3$ yield [M(NH$_3$)$_6$]I$_2$ (M = Fe or Mn). At 600 MPa and 670–870 K, the reaction of Mn with scNH$_3$ gives the manganese nitride, Mn$_3$N$_2$. Single crystals of this compound can be grown by adding I$_2$, K or Rb to the reaction mixture, resulting in the formation of [Mn(NH$_3$)$_6$]I$_2$, K$_2$[Mn(NH$_2$)$_4$] or Rb$_2$[Mn(NH$_2$)$_4$] prior to Mn$_3$N$_2$. Similarly, γ-Fe$_4$N is obtained from [Fe(NH$_3$)$_6$]I$_2$ in scNH$_3$ at 600–800 MPa and 730–850 K. The reaction of CrI$_2$ in scNH$_3$ at 773 K and 600 MPa yields [Cr$_2$(NH$_3$)$_6$(μ-NH$_2$)$_3$]I$_3$ which contains cation **9.28**.

$$
\left[
\begin{array}{c}
\text{H}_3\text{N} \quad \underset{\text{N}}{\overset{\text{H}_2}{\text{N}}} \quad \text{NH}_3 \\
\text{H}_3\text{N} \cdots \text{Cr} \cdots \text{Cr} \cdots \text{NH}_3 \\
\underset{\text{H}_2}{\text{N}} \\
\text{H}_3\text{N} \quad \underset{\text{N}}{\overset{\text{H}_2}{}} \quad \text{NH}_3
\end{array}
\right]^{3+}
$$

(9.28)

Supercritical amines have been found to be useful solvents for the assembly of complex metal sulfides, including K$_2$Ag$_6$S$_4$ (reaction 9.94), KAgSbS$_4$, Rb$_2$AgSbS$_4$, KAg$_2$SbS$_4$, KAg$_2$AsS$_4$ and RbAg$_2$SbS$_4$. Use of scNH$_3$ allows these solid state compounds to be prepared at lower temperatures than more traditional routes used to synthesize related compounds such as $SrCu_2SnS_4$.

$$K_2S_4 + 6Ag \xrightarrow{scNH_3} K_2Ag_6S_4 \qquad (9.94)$$

If the K^+ or Rb^+ ions in this type of compound are replaced by Fe^{2+} (eq. 9.95), Mn^{2+}, Ni^{2+}, La^{3+} (eq. 9.96) or Yb^{3+} (eq. 9.97), the products contain $[M(NH_3)_n]^{2+}$ or $[M(NH_3)_n]^{3+}$ ions. For La^{3+} and Yb^{3+}, these represent the first examples of homoleptic lanthanoid ammine complexes.

> A **homoleptic complex** is of the type $[ML_x]^{n+}$ where all the ligands are identical. In a **heteroleptic complex**, the ligands attached to the metal ion are not all identical.

$$16Fe + 128Cu + 24Sb_2S_3 + 17S_8 \xrightarrow{scNH_3}$$
$$16[Fe(NH_3)_6][Cu_8Sb_3S_{13}] \qquad (9.95)$$

$$La + Cu + S_8 \xrightarrow{scNH_3} [La(NH_3)_8][Cu(S_4)_2] \qquad (9.96)$$

$$Yb + Ag + S_8 \xrightarrow{scNH_3} [Yb(NH_3)_9][Ag(S_4)_2] \qquad (9.97)$$

Extractions of organic compounds dominate the applications of scCO2 (see Box 9.3). The extraction of metal ions (especially toxic heavy metal ions) from contaminated sources currently uses chelating ligands and conventional solvents. In order that scCO2 may be used as the extracting medium, ligands must be designed that are both efficient at binding metal ions and are highly soluble in scCO2. For example, 2,2'-bipyridine (bpy, see Table 7.7) forms many metal complexes with high stability constants, but is not readily soluble in scCO2. Extraction of metal ions using bpy is achieved if the ligand is functionalized with long, fluoroalkyl chains. In general, the presence of perfluorinated alkyl 'tails' on the periphery of a ligand significantly increases the solubility of the ligand and its metal complexes in scCO2.

KEY TERMS

The following terms were introduced in this chapter. Do you know what they mean?

- ❑ non-aqueous solvent
- ❑ relative permittivity
- ❑ coordinating solvent
- ❑ protic solvent
- ❑ aprotic solvent

- ❑ solvent-oriented acid and base
- ❑ levelling effect
- ❑ differentiating effect
- ❑ conductiometric titration
- ❑ superacid

- ❑ molten salt
- ❑ ionic liquid
- ❑ eutectic
- ❑ pyrophoric
- ❑ supercritical fluid

FURTHER READING

General: non-aqueous solvents

J.R. Chipperfield (1999) *Non-aqueous Solvents*, Oxford University Press, Oxford – A book in the OUP 'Primer' series which gives a good introduction to the topic.

R.J. Gillespie and J. Passmore (1971) *Acc. Chem. Res.*, vol. 4, p. 413 – An article that highlights the uses of non-aqueous solvents (HF, SO_2 and HSO_3F) in the preparation of polycations.

K.M. Mackay, R.A. Mackay and W. Henderson (2002) *Modern Inorganic Chemistry*, 6th edn, Blackie, London – Chapter 6 includes a general introduction to non-aqueous solvents.

G. Mamantov and A.I. Popov, eds (1994) *Chemistry of Non-aqueous Solutions: Recent Advances*, VCH, New York – A collection of reviews covering topics in the field of non-aqueous solvents.

T.A. O'Donnell (2001) *Eur. J. Inorg. Chem.*, p. 21 – A review illustrating the generality of inorganic solute speciation in different ionizing solvents.

Superacids

R.J. Gillespie (1968) *Acc. Chem. Res.,* vol. 1, p. 202.

G.A. Olah, G.K.S. Prakash, J. Sommer and A. Molnar (2009) *Superacid Chemistry*, 2nd edn, Wiley-VCH, Weinheim.

C.A. Reed (2010) *Acc. Chem. Res.*, vol 43, p. 121.

Ionic liquids

M.C. Buzzeo, R.G. Evans and R.G. Compton (2004) *ChemPhysChem*, vol. 5, p. 1106.

E.W. Castner, Jr. and J.F. Wishart (2010) *J. Chem. Phys.*, vol. 132, p. 120901.

P.M. Dean, J.M. Pringle and D.R. MacFarlane (2010) *Phys. Chem. Chem. Phys.*, vol. 12, p. 9144.

X. Han and D.W. Armstrong (2007) *Acc. Chem. Res.*, vol. 40, p. 1079.

Z. Ma, J. Yu and S. Dai (2010) *Adv. Mater.*, vol. 22, p. 261.

R.E. Morris (2009) *Chem. Commun.*, p. 2990.

N.V. Plechkova and K.R. Seddon (2008) *Chem. Soc. Rev.*, vol. 37, p. 123.

P. Wasserscheid and T. Welton (eds) (2008) *Ionic Liquids in Synthesis*, 2nd edn, Wiley-VCH, Weinheim (2 volumes).

Supercritical fluids

J.A. Darr and M. Poliakoff (1999) *Chem. Rev.*, vol. 99, p. 495; other articles in this same issue of *Chem. Rev.* deal with various aspects of supercritical fluids.

P. Licence and M. Poliakoff (2008) *NATO Science Series II*, Springer, Düsseldorf, vol. 246, p. 171.

P. Raveendran, Y. Ikushima and S.L. Wallen (2005) *Acc. Chem. Res.*, vol. 38, p. 478.

H. Weingärtner and E.U. Franck (2005) *Angew. Chem. Int. Ed.*, vol. 44, p. 2672.

PROBLEMS

9.1 (a) Give four examples of non-aqueous solvents commonly used in *organic* chemistry, and give one example of a reaction that is carried out in each solvent. (b) Assess the relative importance of the use of aqueous and non-aqueous media in organic and inorganic *general* synthesis.

9.2 Explain what is meant by the relative permittivity of a solvent. What information does this property provide in terms of assisting you to choose a solvent for a given reaction?

9.3 Which of the following solvents are polar: (a) acetonitrile; (b) water; (c) acetic acid; (d) fluorosulfonic acid; (e) dichloromethane; (f) bromine trifluoride; (g) hexane; (h) THF; (i) DMF; (j) liquid sulfur dioxide; (k) benzene?

9.4 Suggest likely products for the following reactions (which are balanced on the left-hand sides) in liquid NH_3.

(a) $ZnI_2 + 2KNH_2 \longrightarrow$

(b) Zinc-containing product of (a) with an excess of KNH_2

(c) $Mg_2Ge + 4NH_4Br \longrightarrow$

(d) $MeCO_2H + NH_3 \longrightarrow$

(e) $O_2 \xrightarrow{\text{Na in liquid NH}_3}$

(f) $HC{\equiv}CH + KNH_2 \longrightarrow$

How does reaction (d) differ from the behaviour of $MeCO_2H$ in aqueous solution?

9.5 Discuss the following observations:

(a) Zinc dissolves in a solution of sodium amide in liquid NH_3 with liberation of H_2; careful addition of ammonium iodide to the resulting solution produces a white precipitate which dissolves if an excess of ammonium iodide is added.

(b) Addition of K to H_2O results in a vigorous reaction; addition of K to liquid NH_3 gives a bright blue solution, which over a period of time liberates H_2.

9.6 Early in the study of chemical reactions in liquid NH_3, it was noted that nitrogen compounds behave in liquid NH_3 in a manner similar to analogous oxygen-containing species in water. For example, $K[NH_2]$ has an analogue in $K[OH]$, and $[NH_4]Cl$ is analogous to $[H_3O]Cl$. What would be the corresponding compounds in the nitrogen system to the following from the oxygen system: (a) H_2O_2; (b) HgO; (c) HNO_3; (d) MeOH; (e) H_2CO_3; (f) $[Cr(OH_2)_6]Cl_3$?

9.7 Give an explanation for the following observations: AlF_3 has only a low solubility in liquid HF, but a combination of NaF and AlF_3 leads to dissolution of the reagents; when BF_3 is added to the solution, a precipitate forms.

9.8 Write equations to show what happens when each of the following dissolves in liquid HF: (a) ClF_3; (b) MeOH; (c) Et_2O; (d) CsF; (e) SrF_2; (f) $HClO_4$.

9.9 $H_2S_2O_7$ functions as a monobasic acid in H_2SO_4. (a) Write an equation to show what happens when $H_2S_2O_7$ dissolves in H_2SO_4. (b) Assess the strength of $H_2S_2O_7$ as an acid given that the ionization constant is 1.4×10^{-2}.

9.10 Give equations to show how the following compounds behave in H_2SO_4: (a) H_2O; (b) NH_3; (c) HCO_2H given that it decomposes; (d) H_3PO_4 given that only one H^+ is transferred between one molecule of H_3PO_4 and one molecule of H_2SO_4; (e) HCl given that HCl acts as an acid and one molecule of HCl reacts with two molecules of H_2SO_4.

9.11 Compare the behaviour of nitric acid in aqueous and sulfuric acid solutions, giving examples from both inorganic and organic chemistries of the uses of HNO_3 in these two media.

9.12 Discuss the following observations:

(a) The alkene $Ph_2C{=}CH_2$ forms a conducting solution in liquid HCl; when such a solution is titrated conductometrically with a solution of

BCl$_3$ in liquid HCl, a sharp end-point is reached when the molar ratio of Ph$_2$C=CH$_2$: BCl$_3$ is 1 : 1.

(b) The following reaction can be described as a neutralization. Explain why this is, identify compound **A**, and determine x.

$$[BrF_2]_2[SnF_6] + 2K[BrF_4] \longrightarrow A + xBrF_3$$

9.13 Confirm that the structures of $[BrF_2]^+$ and $[BrF_4]^-$ (**9.18** and **9.19**) are consistent with the VSEPR model.

9.14 How would you attempt to demonstrate that AsCl$_3$ ionizes slightly according to the equation:

$$2AsCl_3 \rightleftharpoons [AsCl_2]^+ + [AsCl_4]^-$$

and that there exist acids and bases in the AsCl$_3$ system?

9.15 (a) Describe the bonding in the $[Al_2Cl_7]^-$ anion (**9.21**).

(b) Equilibria 9.81 and 9.82 describe part of the NaCl–Al$_2$Cl$_6$ system; additionally $[Al_3Cl_{10}]^-$ is present. Write an equation to show how $[Al_3Cl_{10}]^-$ may be formed, and suggest a structure for this anion.

9.16 Suggest structures for the $[BiCl_5]^{2-}$ and $[Bi_2Cl_8]^{2-}$ anions, the formation of which was described in Section 9.12.

9.17 (a) Give three examples of commonly used ionic liquids. What general properties make ionic liquids attractive in 'green chemistry'? Are the properties of the liquid itself all that determines whether the ionic liquid is environmentally friendly?

(b) The compound shown below was treated with one equivalent of NaOH in methanol to give the ionic liquid **X**:

The positive mode electrospray mass spectrum of **X** exhibits a peak at m/z 126. Compound **X** reacts with chloroauric acid, HAuCl$_4$, to give a salt, **Y**, which contains two square planar Au(III) environments. The anion in **Y** has D_{4h} symmetry. The ^{1}H NMR spectrum of a DMSO-d_6 solution of **Y** exhibits the following signals: δ/ppm 9.07 (s, 1H), 7.92 (br, 2H), 7.74 (overlapping signals, 2H), 4.38 (t, 2H), 3.86 (s, 3H), 3.34 (m, 2H). The electrospray mass spectrum of **Y** shows peak envelopes at m/z 428 and 126 (positive mode), and 339 (negative mode). The peak envelope at m/z 339 consists of four peaks, and that at m/z 428 exhibits three dominant peaks at m/z 428, 429 and 432. Identify **X** and **Y**, and interpret the data. Rationalize the appearance of the mass spectra.

[Data: B. Ballarin et al. (2010) Inorg. Chim. Acta, vol. 363, p. 2055]

9.18 (a) An ionic liquid can be formed by adding ZnCl$_2$ to (2-chloroethyl)trimethylammonium chloride, XCl. When the ratio of ZnCl$_2$: XCl = 2 : 1, fast atom bombardment mass spectrometry shows the presence of $[Zn_xCl_y]^{z-}$ ions with $m/z = 171$, 307 and 443. Suggest identities for these ions and write a series of equilibria to account for their formation.

[Data: A. P. Abbott et al. (2004) Inorg. Chem., vol. 43, p. 3447]

9.19 (a) With the aid of a phase diagram, explain what is meant by a supercritical fluid. Give examples of commercial processes that involve the use of supercritical fluids.

(b) Even though CO$_2$ is classified as a 'greenhouse gas' (see Box 14.7), why is the use of supercritical CO$_2$ regarded as being environmentally friendly?

OVERVIEW PROBLEMS

9.20 (a) Which of the following compounds behave as acids in liquid HF: ClF$_3$, BF$_3$, SbF$_5$, SiF$_4$? Write equations to explain this behaviour.

(b) The salt $[S_8][AsF_6]_2$ can be isolated from the following reaction:

$$S_8 + 3AsF_5 \xrightarrow{\text{liquid HF}} [S_8][AsF_6]_2 + AsF_3$$

What roles does AsF$_5$ play in this reaction?

(c) By first considering its reaction in H$_2$O, suggest how Na might react in liquid N$_2$O$_4$.

9.21 When gallium is dissolved in a solution of KOH in liquid NH$_3$, a salt K[**I**] is formed which is an amido complex of Ga(III). Heating one equivalent of K[**I**] at 570 K under vacuum liberates two equivalents of NH$_3$, and produces a Ga(III) imido complex K[**II**]. Partial neutralization of K[**I**] with NH$_4$Cl yields Ga(NH$_2$)$_3$. Suggest identities for the salts K[**I**] and K[**II**], and write equations for the thermal decomposition and partial neutralization reactions of K[**I**]. *Hint:* an *imido* complex formally contains NH^{2-}.

9.22 (a) SbCl$_3$ may be used as a non-aqueous solvent above its melting point. Suggest a possible self-ionization process for this solvent.

(b) Explain why the reaction of NOCl with AgNO$_3$ in liquid N$_2$O$_4$ can be classed as a neutralization process. Write an equation for the reaction and compare it with that of HCl with Ca(OH)$_2$ in aqueous solution.

(c) In water, Cr^{3+} precipitates as Cr(OH)$_3$ at pH 7, forms [Cr(OH$_2$)$_6$]$^{3+}$ in strongly acidic solution (e.g. HClO$_4$), and [Cr(OH)$_4$]$^-$ in basic solution. Suggest what Cr(III) species are present in liquid NH$_3$ as the pH is varied.

9.23 Suggest explanations for the following observations.

(a) In aqueous solution, AgNO$_3$ and KCl react to give a precipitate of AgCl, whereas in liquid NH$_3$, KNO$_3$ and AgCl react to produce a precipitate of KCl.

(b) Mg dissolves in a concentrated solution of NH$_4$I in liquid NH$_3$.

(c) Most common 'acids' behave as bases in liquid H$_2$SO$_4$.

(d) HClO$_4$ is fully ionized in water and is strongly dissociated in pure (glacial) acetic acid; in liquid HSO$_3$F, the following reaction occurs:

$$KClO_4 + HSO_3F \longrightarrow KSO_3F + HClO_4$$

INORGANIC CHEMISTRY MATTERS

9.24 Ionic liquids may have future applications in lithium-ion batteries. A combination of Li[N(SO$_2$CF$_3$)$_2$] and [EMIm][N(SO$_2$CF$_3$)$_2$] (EMim = 1-ethyl-3-methylimidazolium ion) has been used as a model for a room temperature ionic liquid electrolyte. From this mixture, crystals of Li$_2$[EMIm][N(SO$_2$CF$_3$)$_2$]$_3$ have been isolated. The structure (X-ray diffraction) shows each Li$^+$ ion to be in a 5-coordinate environment, bound by [N(SO$_2$CF$_3$)$_2$]$^-$ ions. The latter adopt either bi- or monodentate modes and coordinate through O-donors. (a) When adopting a bidentate mode, why is it more favourable for a [N(SO$_2$CF$_3$)$_2$]$^-$ ligand (Fig. 9.8d) to coordinate through two O atoms attached to different S atoms rather than through the two O atoms of one CF$_3$SO$_2$ group? (b) What coordination geometries are usually associated with a coordination number of 5? Comment on the energy difference(s) between them. (c) The diagrams below show the two possible arrangements of CF$_3$ groups when [N(SO$_2$CF$_3$)$_2$]$^-$ acts as an O,O'-donor to metal ion M^{n+}. Explain how these arise.

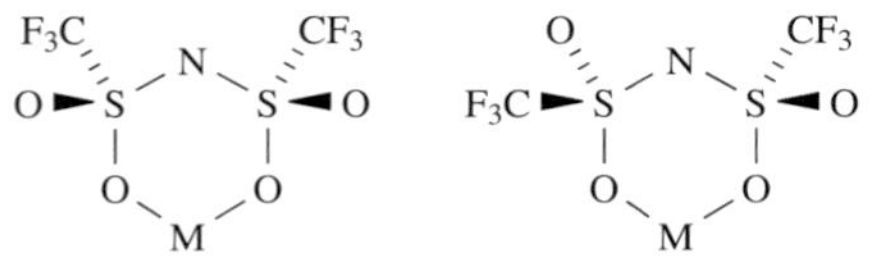

[Data: K. Matsumoto *et al.* (2006) *Solid State Sci.*, vol. 8, p. 1103]

9.25 In the pharmaceutical industry, the active ingredients in drugs are usually manufactured as crystalline solids, and a recurring problem is that of polymorphism. It has been suggested that producing the active pharma ingredients as room temperature ionic liquids could be advantageous. (a) What is polymorphism? (b) Why is polymorphism problematical to the pharmaceutical industry? (c) Distinguish between an 'ionic liquid' and a 'room temperature ionic liquid'. Discuss some of the advantages that room temperature ionic liquids might have over crystalline solids in active pharma ingredients. (d) Hydrochloride salts, RNH$_3$Cl, or quaternary ammonium salts, R$_4$NCl, are common among active pharma ingredients. How would these be converted into ionic liquids? Comment on whether inorganic anions such as [AlCl$_4$]$^-$ or [BF$_4$]$^-$ would be suitable in these ionic liquids.

Topics

H$^+$ and H$^-$ ions
Isotopes of hydrogen
Dihydrogen
Polar and non-polar
   E—H bonds
Hydrogen bonding
Binary hydrides

10

Hydrogen

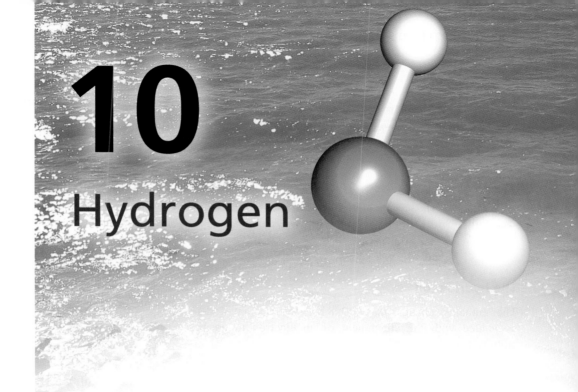

1	2		13	14	15	16	17	18
H								He
Li	Be		B	C	N	O	F	Ne
Na	Mg		Al	Si	P	S	Cl	Ar
K	Ca	*d*-block	Ga	Ge	As	Se	Br	Kr
Rb	Sr		In	Sn	Sb	Te	I	Xe
Cs	Ba		Tl	Pb	Bi	Po	At	Rn
Fr	Ra							

10.1 Hydrogen: the simplest atom

An atom of hydrogen consists of one proton (constituting the nucleus) and one electron. This simplicity of atomic structure means that H is of great importance in theoretical chemistry, and has been central in the development of atomic and bonding theories (see Chapters 1 and 2). The nuclear properties of the hydrogen atom are essential to the technique of ^{1}H NMR spectroscopy (see Section 4.8).

In this chapter, we extend our discussions of hydrogen, looking at the properties of the H$^+$ and H$^-$ ions, properties and reactivity of H$_2$, and aspects of binary hydrides.

A ***binary compound*** is composed of only two different elements.

10.2 The H$^+$ and H$^-$ ions

The hydrogen ion (proton)

The ionization energy of hydrogen (defined for reaction 10.1) is 1312 kJ mol^{-1}, a value that is high enough to preclude the existence of H$^+$ ions under ordinary conditions.

$$H(g) \longrightarrow H^+(g) + e^- \tag{10.1}$$

However, as we discussed in Chapter 7, the *hydrated* proton or *oxonium ion*, [H$_3$O]$^+$, is an important species in aqueous solution; $\Delta_{hyd}H^\circ(H^+, g) = -1091$ kJ mol^{-1} (see Section 7.9). The [H$_3$O]$^+$ ion (**10.1**) is a well-defined species which has been crystallographically characterized in various salts. The ions [H$_5$O$_2$]$^+$ (Fig. 10.1) and [H$_9$O$_4$]$^+$ have also been isolated in crystalline acid hydrates. The [H$_5$O$_2$]$^+$ and [H$_9$O$_4$]$^+$ ions are members of the general family of hydrated protons [H(OH$_2$)$_n$]$^+$ ($n = 1$ to ≈ 20) and we return to these ions in Section 10.6.

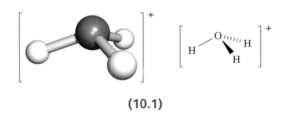

(10.1)

When crystals of a compound are grown from a solvent, they may contain ***solvent of crystallization***; if the solvent is water, the compound is a ***hydrate***. The formula of the solvated compound shows the molar ratio in which the solvent of crystallization is present, e.g. CuSO$_4 \cdot$5H$_2$O, copper(II) sulfate pentahydrate or copper(II) sulfate–water (1/5).

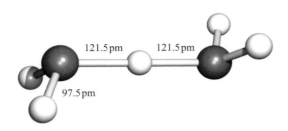

Fig. 10.1 The structure of $[H_5O_2]^+$ determined by neutron diffraction in $[V(OH_2)_6][H_5O_2][CF_3SO_3]_4$. [F.A. Cotton *et al.* (1984) *J. Am. Chem. Soc.*, vol. 106, p. 5319.]

The hydride ion

The enthalpy change $\Delta_{EA}H(298\,\mathrm{K})$ (see Section 1.10) associated with the attachment of an electron to an H atom (reaction 10.2) is $-73\,\mathrm{kJ\,mol^{-1}}$.

$$H(g) + e^- \longrightarrow H^-(g) \qquad (10.2)$$

All alkali metal hydrides (see Sections 10.7 and 11.4) crystallize with the NaCl-type structure. From diffraction data and the ionic radii of the metal ions (Appendix 6), the radius of H^- can be estimated using eq. 10.3. It varies from 130 pm (in LiH) to 154 pm (in CsH), but on average can be considered similar to that of F^- (133 pm).

$$\text{Internuclear distance} = r_{cation} + r_{anion} \qquad (10.3)$$

The large increase in size on going from the H atom ($r_{cov} = 37\,\mathrm{pm}$) to the H^- ion arises from interelectronic repulsion when a second electron enters the 1s atomic orbital. The smaller r_{H^-} in LiH may suggest some degree of covalent bonding, but calculated and experimental values of lattice energies (see Sections 6.13 to 6.16) for *each* of the group 1 metal hydrides are in good agreement, suggesting that an electrostatic model is appropriate for each compound.

Hydrides of the *s*-block metals (excluding Be) can be made by heating the metal with H_2.

$$\tfrac{1}{2}H_2(g) + e^- \longrightarrow H^-(g) \quad \Delta_rH = \tfrac{1}{2}D(H-H) + \Delta_{EA}H$$
$$= \Delta_aH^\circ + \Delta_{EA}H$$
$$= +145\,\mathrm{kJ\,mol^{-1}} \qquad (10.4)$$

When we compare Δ_rH for reaction 10.4 with those for the formations of F^- and Cl^- from F_2 and Cl_2 (-249 and $-228\,\mathrm{kJ\,mol^{-1}}$, respectively), we understand why, since H^- is about the same size as F^-, ionic hydrides are relatively unstable species with respect to dissociation into their constituent elements. Salt-like hydrides of metals in high oxidation states are most unlikely to exist. (There is more about binary hydrides in Section 10.7.)

10.3 Isotopes of hydrogen

Protium and deuterium

Hydrogen possesses three isotopes, *protium*, *deuterium* and *tritium*, selected properties of which are given in Table 10.1. The isotopes of hydrogen exhibit greater differences in physical and chemical properties than isotopes of any other element. The origin of the differences between H and D, or between pairs of compounds such as H_2O and D_2O, lies in the difference in mass, which in turn affects their fundamental vibrational wavenumbers and zero point energies (see Section 4.6 and worked example 4.6). The fundamental vibrations for H_2, HD and D_2 are at 4159, 3630 and $2990\,\mathrm{cm^{-1}}$, respectively. It follows (see exercises below) that the zero point energies of H_2 and D_2 are calculated to be 26.0 and $18.4\,\mathrm{kJ\,mol^{-1}}$, respectively. The total electronic binding energies for these molecules (represented by the overlap of their atomic wavefunctions) are the *same*, and so it follows that their dissociation energies differ by $7.6\,\mathrm{kJ\,mol^{-1}}$, with the D–D bond being stronger than the H–H bond. As Fig. 10.2 shows, it also follows that an X–D bond is stronger than the corresponding X–H bond (where X is any element), and this difference is the basis of the kinetic isotope effect.

Self-study exercises

1. For a simple harmonic oscillator, the vibrational energies, E_v, of a molecule are quantized and the vibrational energy levels are given by:
$$E_v = (v + \tfrac{1}{2})h\nu$$
where v = vibrational quantum number; h = Planck constant; ν = frequency of vibration.
Show that the zero point energies of H_2 and D_2 are 24.9 and $17.9\,\mathrm{kJ\,mol^{-1}}$, respectively.
[Data: $h = 6.626 \times 10^{-34}\,\mathrm{J\,s^{-1}}$; $c = 2.998 \times 10^{10}\,\mathrm{cm\,s^{-1}}$; $L = 6.022 \times 10^{23}\,\mathrm{mol^{-1}}$]

2. Why do the values of the zero point energies of H_2 and D_2 in the above question differ from those given in the text?

Table 10.1 Selected properties of the isotopes of hydrogen.

	Protium	Deuterium	Tritium
Symbols†	^{1}H or H	^{2}H or D	^{3}H or T
Natural abundance	99.985%	0.0156%	<1 in 10^{17} atoms
Isotopic mass / u	1.0078	2.0141	3.0160
Nuclear spin	$\tfrac{1}{2}$	1	$\tfrac{1}{2}$

† Strictly, ^{1}H should be written as $^1_1$H, ^{2}H as $^2_1$H and ^{3}H as $^3_1$H, but the less rigorous symbols are generally used.

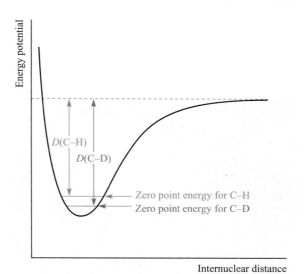

Fig. 10.2 The zero point energy (corresponding to the lowest vibrational state) of a C–D bond is lower than that of a C–H bond and this results in the bond dissociation enthalpy, D, of the C–D bond being greater than that of the C–H bond.

Kinetic isotope effects

Isotopic labelling may be used to probe the mechanism of a reaction. Consider the case where the rate-determining step of a reaction involves breaking a particular C–H bond. Labelling the compound with deuterium (not always a trivial matter experimentally!) at that site will mean that the C–D that replaced the C–H bond is now broken. The bond dissociation energy of a C–D bond is higher than that of a C–H bond because the zero point energy is lowered when the reduced mass, μ, of a bond is increased, i.e. $\mu(\text{C–D}) > \mu(\text{C–H})$ (Fig. 10.2). Since it requires more energy to break a C–D than a C–H bond, the rate-determining step should proceed more slowly for the deuterated compound. This observation is known as the *kinetic isotope effect* and is quantified by comparing the rate constants, k_H and k_D, for the reactions involving the non-deuterated and deuterated compounds respectively. If the value of the ratio $k_H:k_D > 1$, then a kinetic isotope effect has been observed.

Deuterated compounds

A deuterium label in heavy water is indicated by writing $[^2H_2]$water or water-d_2, and similarly for other labelled compounds. The formula for heavy water can be written as ^2H_2O or D_2O.

Compounds in which H atoms have been replaced by D are used for a variety of purposes, e.g. as solvents in ^1H NMR spectroscopy (see Section 4.8). In a fully deuterated material, the D-for-H exchange can have significant effects on the properties of the compound, as is shown in Table 10.2 for H_2O and D_2O. The difference in boiling points indicates that intermolecular hydrogen bonding (see Sections 7.2 and 10.6) is slightly stronger in D_2O than in H_2O. The major industrial use of D_2O is as a moderator in nuclear reactors; D has a much lower cross-section for neutron capture than H, and D_2O is a suitable material for reducing the energies of fast neutrons produced in fission without appreciably diminishing the neutron flux.

Many fully or partially deuterated compounds are available commercially, and the extent of *deuterium labelling* can be determined by mass spectrometry, density measurements (after conversion into water) or IR spectroscopy.

The separation of deuterium from naturally occurring hydrogen is achieved electrolytically with the isotope in the form of D_2O. When an aqueous solution of NaOH (natural isotopic abundances) is electrolysed (eq. 10.5) using an Ni electrode, the separation factor defined in eq. 10.6 is ≈ 6. The choice of electrode is critical for optimizing this value.

$$\left.\begin{array}{ll}\textit{At the anode:} & H_2O \longrightarrow O_2 + 4H^+ + 4e^- \\ \textit{At the cathode:} & 4H_2O + 4e^- \longrightarrow 2H_2 + 4[OH]^-\end{array}\right\} \quad (10.5)$$

$$\text{Separation factor} = \frac{\left(\dfrac{H}{D}\right)_{\text{gas}}}{\left(\dfrac{H}{D}\right)_{\text{solution}}} \quad (10.6)$$

The electrolysis is continued until $\approx 90\%$ of the liquid has been converted into O_2 and H_2. Most of the residual liquid is then neutralized with CO_2, and the water distilled and

Table 10.2 Selected properties of H_2O and D_2O ('heavy water').

Property	H_2O	D_2O
Melting point / K	273.00	276.83
Boiling point / K	373.00	374.42
Temperature of maximum density / K[†]	277.0	284.2
Maximum density / g cm^{-3}	0.999 95	1.105 3
Relative permittivity (at 298 K)	78.39	78.06
K_w (at 298 K)	1×10^{-14}	2×10^{-15}
Symmetric stretch,[‡] $\bar{\nu}_1$ (gaseous molecule) / cm^{-1}	3657	2671

[†] See Fig. 7.2.
[‡] The symmetric stretching mode is illustrated (for SO_2) in Fig. 3.12.

ENVIRONMENT

Box 10.1 Metallic character of hydrogen

The planets Saturn, Jupiter, Uranus and Neptune are referred to as gas giants, in contrast to the terrestrial planets (Mercury, Earth, Venus and Mars) which have solid surfaces. The latter lie closer to the Sun than the gas giants. Through NASA's Pioneer, Voyager, Galileo and Cassini missions (begun in 1972), we now have significant knowledge about the four gaseous planets. Saturn, Jupiter, Uranus and Neptune consist mainly of He and H_2, with small amounts of CH_4 and NH_3. The planets (possibly with the exception of Jupiter) have solid, rocky cores surrounded by ice and then H_2 and He. These elements are subject to extreme conditions of temperature and pressure, and in Saturn and Jupiter (where H_2 and He are the dominant components of the planets) there is a transition from molecular H_2 to metallic hydrogen as the pressure increases towards the centre of the planet. A model based on Saturn's gravitational field and its internal rotation period suggests that the transition occurs about halfway to the centre of the planet. However, establishing the metallic character of hydrogen on Earth is an extremely difficult task.

A report in 1996 from the Livermore Laboratory (US) described how, when a thin layer of *liquid* H_2 is subjected to enormous shock pressure, changes in conductivity are observed that are consistent with the formation of metallic hydrogen. In the experiments, it was observed that at a pressure of 93 GPa (GPa = gigapascal = 10^9 Pa), the resistivity of liquid hydrogen was $\approx 0.01\,\Omega\,m$ (resistivity: see Section 6.8). As the shock compression increased the pressure to 140 GPa, the resistivity of the liquid hydrogen decreased to $5 \times 10^{-6}\,\Omega\,m$ and remained constant up to 180 GPa, the highest pressure tested. A resistivity of $5 \times 10^{-6}\,\Omega\,m$ is typical of a liquid metal; for comparison, that of liquid mercury at 273 K at atmospheric pressure is $9.4 \times 10^{-7}\,\Omega\,m$. At low pressures, liquid hydrogen contains H_2 molecules; the band gap (see Section 6.8) is very large ($\approx 15.0\,eV$) and the element is an electrical insulator. Subjecting liquid H_2 to huge pressures by shock compression results in a drastic reduction in the band gap. The element passes through a semiconducting stage and finally exhibits electrical conductivity typical of a metal when the band gap is $\approx 0.3\,eV$. The tremendous pressure also causes about 10% of the H_2 molecules to dissociate. These results have been applied to update models for the interior of Jupiter. The radius of Jupiter is 71 400 km, and it is proposed that the pressure and temperature conditions are such that the liquid hydrogen is metallic relatively near (7000 km) to the surface of the planet. The magnetic field on the surface of Jupiter is about $10^{-3}\,T$ (T = tesla) compared with a field strength of $5 \times 10^{-5}\,T$ on the Earth's surface. The latter is a consequence of the Earth's magnetic iron core; the former arises from Jupiter's fluid hydrogen core and the high field strength is consistent with the metallic state being achieved relatively close to the planet's surface.

In 1998, experiments were carried out in the US aimed at reproducing theoretical predictions that H_2 would dissociate and form an alkali metal-like lattice under pressures of ≥ 340 GPa. A diamond anvil cell was used to generate a pressure of 342 ± 10 GPa, but there was no experimental evidence for the formation of alkali metal-like hydrogen. In 2002, French physicists performed optical measurements on solid hydrogen up to 320 GPa pressure. No structural changes were observed, but the darkening

The planets in the solar system: the largest planets are Jupiter, Saturn (shown with its characteristic rings), Neptune and Uranus. Pluto (shown at the top left of the illustration) has been reclassified as a dwarf planet.

of the sample is consistent with the band gap becoming smaller. From the data, the band gap is predicted to tend to zero (i.e. the onset of metallic character) around 450 GPa. However, attempts to impart metallic character to *solid* H_2 have so far been unsuccessful. Under extremely high pressures, H_2 (normally a non-polar molecule) undergoes a redistribution of electronic charge such that the ionic contribution to the bonding (represented by the resonance form H^+–H^-) becomes important. This remarkable finding may go some way to helping to explain why attempts to form metallic hydrogen in the solid state have not met with success.

Further reading

J.D. Anderson and G. Schubert (2007) *Science*, vol. 317, p. 1384 – 'Saturn's gravitational field, internal rotation, and interior structure'.

P. Loubeyre, F. Occelli and R. LeToullec (2002) *Nature*, vol. 416, p. 613 – 'Optical studies of solid hydrogen to 320 GPa: Evidence for black hydrogen'.

C. Narayana, H. Luo, J. Orloff and A.L. Ruoff (1998) *Nature*, vol. 393, p. 46 – 'Solid hydrogen at 342 GPa: No evidence for an alkali metal'.

W.J. Nellis (2000) *Scientific American*, May issue, p. 84 – 'Making metallic hydrogen'.

S.T. Weir, A.C. Mitchell and W.J. Nellis (1996) *Phys. Rev. Lett.*, vol. 76, p. 1860 – 'Metallization of fluid molecular hydrogen at 140 GPa (1.4 Mbar)'.

www.llnl.gov/str/Nellis.html

added to the remaining electrolyte. This process is repeated to give ≤99.9% D_2O. In the later stages of the separation, the gas evolved at the cathode is burned to yield partially enriched deuterium oxide that can be electrolysed further. Cheap electrical power is essential for this process to be economically viable.

Tritium

Tritium (Table 10.1) occurs in the upper atmosphere and is formed naturally by reaction 10.7, involving neutrons arriving from outer space. Tritium was first obtained synthetically by the bombardment of deuterium-containing compounds with fast neutrons, but is now prepared from lithium deuteride, LiF or Mg/Li enriched in $^{6}_{3}Li$ (eq. 10.8).

$$^{14}_{7}N + ^{1}_{0}n \longrightarrow ^{12}_{6}C + ^{3}_{1}H \tag{10.7}$$

$$^{6}_{3}Li + ^{1}_{0}n \longrightarrow ^{4}_{2}He + ^{3}_{1}H \tag{10.8}$$

Tritium is radioactive, a weak β-emitter with a half-life of 12.3 yr. It is used extensively as a tracer, in both chemical and biochemical studies. Its weak radioactivity, rapid excretion and failure to concentrate in vulnerable organs make it one of the least toxic radioisotopes. A major use of tritium is in the triggering mechanisms of weapons based on nuclear fusion.

> A *β-particle* (written as β−) is an electron emitted from the nucleus and has a high kinetic energy. The emission of a β-particle results in an increase in the atomic number by 1 and leaves the mass number unchanged.

10.4 Dihydrogen

Occurrence

Hydrogen is the most abundant element in the universe, and, after oxygen and silicon, is the third most abundant element on Earth, where it occurs mainly in the form of water or combined with carbon in organic molecules (hydrocarbons, plant and animal material). In the Earth's atmosphere (see Fig. 15.1b), H_2 occurs to an extent of less than 1 ppm by volume, but those of Jupiter, Neptune, Saturn and Uranus contain large amounts of H_2 (see Box 10.1).

Physical properties

Dihydrogen is a colourless, odourless gas, sparingly soluble in all solvents. At 298 K and 1 bar pressure, it conforms closely to the ideal gas laws. The solid state structure of H_2 can be described in terms of an hcp lattice (see Section 6.3). Values of the melting point, enthalpy of fusion, boiling point and enthalpy of vaporization are all very low (Table 10.3), consistent with there being only weak van der Waals forces between the H_2 molecules. The covalent

Table 10.3 Selected physical properties of H_2.

Physical property	Value
Melting point / K	13.66
Boiling point / K	20.13
Enthalpy of vaporization / kJ mol^{-1}	0.904
Enthalpy of fusion / kJ mol^{-1}	0.117
Density (273 K) / g dm^{-3}	0.090
Bond dissociation enthalpy / kJ mol^{-1}	435.99
Interatomic distance / pm	74.14
Standard entropy (298 K) / J K^{-1} mol^{-1}	130.7

bond in H_2 is unusually strong for a single bond in a diatomic molecule.

Each H nucleus in H_2 could have a nuclear spin of $+\frac{1}{2}$ or $-\frac{1}{2}$. This leads to two forms of H_2 with spin combinations $(+\frac{1}{2}, +\frac{1}{2})$ (equivalent to $(-\frac{1}{2}, -\frac{1}{2})$) or $(+\frac{1}{2}, -\frac{1}{2})$. The former is called *ortho*-dihydrogen and the latter is *para*-dihydrogen. At 0 K, H_2 consists entirely of the lower energy *para*-form. At higher temperatures, an equilibrium is established between *ortho*- and *para*-H_2. At the boiling point of liquid hydrogen (20.1 K), the equilibrium concentration of *ortho*-H_2 is 0.21%, while at room temperature, *normal dihydrogen* consists of 75% *ortho*- and 25% *para*-H_2. This ratio persists above room temperature and therefore, pure *ortho*-H_2 cannot be obtained. The physical properties of *para*- and *ortho*-H_2 are essentially the same, although one significant difference is that the thermal conductivity of *para*-H_2 is 50% greater than that of *ortho*-H_2. The conversion of *ortho*- to *para*-H_2 is exothermic (670 J g^{-1} of normal dihydrogen), leading to vaporization of H_2 during liquefaction. Within the context of research into methods of storage of liquid H_2 (see Box 10.2), there is current interest in accurately controlling and measuring the ratio of *ortho*- to *para*-H_2.[†]

Synthesis and uses

In the laboratory, H_2 may be prepared by electrolysis of water containing an added electrolyte (H_2 is liberated at the cathode), but small quantities of H_2 are most conveniently prepared by reactions between dilute acids and suitable metals (e.g. Fe, Zn, eq. 10.9), by treating metals that form amphoteric hydroxides

---

[†] See: D. Zhou, G.G. Ihas and N.S. Sullivan (2004) *J. Low Temp. Phys.*, vol. 134, p. 401 – 'Determination of the *ortho–para* ratio in gaseous hydrogen mixtures'; R. Muhida, M.M. Rahman, M. David, W.A. Diño, H. Nakanishi and H. Kasai in *Condensed Matter Theories* (2007), vol. 21, eds. H. Akai, A. Hosaka, H. Toki and F. B. Malik, p. 259 – 'Overcoming storage problem of liquid hydrogen: increasing the ortho-para H_2 conversion yield'.

ENVIRONMENT

Box 10.2 Will the fuel cell replace the internal combustion engine?

Refuelling a hydrogen fuel cell vehicle with liquid H_2 in Munich, Germany.

In 1839, William Grove observed that when the current was switched off in an electrolysis cell using Pt electrodes in which water was being electrolysed to give O_2 and H_2, a small current continued to flow, but in the opposite direction to the current that had driven the electrolysis cell. The observation constituted the first *fuel cell*, although this name was not introduced until 1889. Chemical energy produced from the reaction:

$$2H_2 + O_2 \longrightarrow 2H_2O \qquad \textit{catalysed by Pt}$$

is efficiently converted into electrical energy. During the 20th century, there were a number of research efforts to harness the electrical energy from fuel cells. Alkaline fuel cells (containing aqueous KOH electrolyte, carbon electrodes and a Pt catalyst with H_2 as the fuel) and phosphoric acid fuel cells (containing aqueous H_3PO_4 electrolyte, and platinized carbon electrodes, with H_2 fuel) were successfully used to produce electrical energy and provide drinking water for the *Gemini*, *Apollo* and space shuttle missions.

The first decade of the 21st century has seen significant progress in the development and active use of vehicles powered by hydrogen fuel cells, rechargeable batteries (see Box 11.3) and biomass fuels. Combustion of H_2 produces only H_2O and dihydrogen is, therefore, an environmentally clean fuel which is, in principle, ideal for powering the millions of vehicles on our roads. Since 1997, a number of cities worldwide have introduced the Daimler-Benz no-emission bus (the *Nebus*) which contains a fuel cell running on H_2 which is stored in pressurized tanks in the roof-space of the bus. At the end of 2001, Daimler-Chrysler launched the Clean Urban Transport for Europe (CUTE) project under which Amsterdam, Barcelona, Hamburg, London, Luxembourg, Madrid and Reykjavik operate a limited number of fuel cell buses. In Iceland, Shell, Daimler-Chrysler and General Motors are currently making significant advances in hydrogen fuel cell technologies with the result that Iceland is aiming towards a hydrogen economy by 2050. This will work in harmony with Iceland's extensive hydroelectric and geothermal power supplies.

Applying fuel cell technology to the world's transport system as a whole, or even a fraction of it, has huge obstacles. Firstly, to open up a competitive market, any new product from the motor vehicle industry must be at least as efficient as vehicles that rely on the internal combustion engine. Apart from performance, factors to be considered include cost, fuel storage and safety. The public perception of H_2 is that of an explosive gas, and most consumers probably consider H_2 to be more hazardous than hydrocarbon fuels. Secondly, the current infrastructure (e.g. fuel distribution and refuelling) for vehicle transport systems is designed for carbon-based fuels. A change to hydrogen-based fuel would be enormously expensive.

Driven largely by environmental legislation for pollution control, the motor industry is now heavily involved in fuel cell development. Car manufacturers plan to have fuel cell vehicles in the showrooms by 2015, but one of the problems that they must overcome is the form in which hydrogen fuel should be delivered and stored. Hydrogen can be stored in the form of liquid H_2 (at ≤ 1 MPa, 20 K) or compressed gaseous H_2 (at 35–70 MPa, ≈ 295 K). Compression of the gas is necessary because H_2 has such a low density. The high pressures required demand extremely strong tanks for safety, and car manufacturers are opting for lightweight carbon-fibre materials with which to fabricate compressed H_2 fuel tanks. Other methods of H_2 storage include adsorption at low temperatures (cryoadsorption) onto materials that possess high surface areas, e.g. activated carbon. Although porous, zeolites have low storage capacities for H_2. Metal-organic frameworks (MOFs) are also potential storage vessels for H_2 and are the focus of active studies. In all these storage methods, H_2 is physically unchanged. An alternative strategy is to store hydrogen in chemical compounds such as $NaBH_4$ or interstitial hydrides (see Section 10.7). The catalysed hydrolysis of $NaBH_4$ releases H_2 for use as a fuel:

$$NaBH_4 + 4H_2O \longrightarrow 4H_2 + NaOH + H_3BO_3$$

This type of hydrogen storage system has been tested by Chrysler but does not appear suited to widescale application. Interstitial metal hydrides appear to be more promising (see below). In terms of the stored energy *per unit mass*, H_2 supplies 120 MJ kg^{-1}. However, one must consider what this means in terms of the required *volume* of H_2 that has to be stored in a vehicle to permit an acceptable operating distance between refuelling stops. The stored energy capacity of H_2 *per unit volume* is ≈ 2.8 GJ m^{-3} at a pressure of 35 MPa, or ≈ 8.5 GJ m^{-3} for liquid H_2 at 20 K. The chart opposite illustrates that, in terms of stored energy per unit mass, H_2 appears an excellent fuel when compared with a number of carbon-based fuels. However, it compares unfavourably when considered in terms of stored energy per unit volume. High-pressure compression of gaseous H_2 (e.g. 69 MPa in a carbon fibre tank) is the strategy being adopted by many vehicle manufacturers.

The US Department of Energy has proposed that manufacturers should aim for a target of 9 GJ m^{-3} of H_2 in a fuel cell-powered vehicle. The chart opposite shows that interstitial metal hydrides can store around 12 GJ m^{-3} and may be a realistic option for the storage of H_2 in vehicles.

At the end of 2002, Toyota Motor Sales, USA, Inc. announced the delivery of two fuel-cell vehicles to the

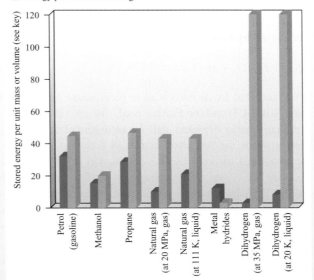

- ■ Energy per unit volume/GJ m^{-3}
- ■ Energy per unit mass/MJ kg^{-1}

[Data: B. McEnaney (2003) *Chem. Brit.*, vol. 39 (Jan. issue), p. 24]

University of California, Irvine, and University of California, Davis, and in a press statement (http://pressroom.toyota.com), the company used this to mark the 'first step in a plan to establish California fuel-cell community partnerships of government, business and higher education that will tackle product, infrastructure and consumer-acceptance challenges.'

Up until 2006, manufacturers including Daimler-Chrysler, Honda, Toyota, BMW and Ford had produced almost 200 prototype hydrogen fuel cell vehicles (FCV). Since 2006, there has been a shift towards hybrid vehicles (FCHV) in which the electrical energy produced by the hydrogen fuel cell is used to drive an electrical motor, and may also recharge a battery (e.g. nickel–metal hydride or lithium-ion) to provide a secondary power supply. These should not be confused with hybrid electric vehicles (HEV) such as the Toyota Prius or Honda Insight which combine electrical motors with a conventional internal combustion engine. HEVs are already established in the automobile market, as are plug-in hybrid electric vehicles which contain batteries that are recharged from an external power supply when the car is stationary. This additional electrical power supply reduces the amount of fossil fuel used by the vehicle.

An alternative to using a direct H$_2$ fuel supply is to refuel a vehicle with a carbon-based fuel such as methanol, and use an on-board fuel processor to transform it into H$_2$. This process has the disadvantage of generating by-products: CO and/or CO$_2$ and N$_2$ or NO$_x$ (see Box 15.7). Thus, the vehicle is classed as reduced-emission rather than zero-emission. An advantage of using an indirect, rather than direct, H$_2$ supply is that there is no longer a need to provide hydrogen-fuel stations. As a consequence, infrastructure costs are reduced.

Finally we come to the fuel cell itself. In addition to the original Grove fuel cell and the alkaline and phosphoric acid fuel cells used in space technology, other types of cell include the molten carbonate fuel cell (with a molten Li$_2$CO$_3$/Na$_2$CO$_3$ electrolyte), the solid oxide fuel cell (containing a solid metal oxide electrolyte) and the polymer electrolyte membrane (PEM) fuel cell. Both the molten carbonate and solid oxide fuel cells

require high operating temperatures ($\approx$900 and 1300 K respectively). In the motor industry, most attention is focused on the PEM fuel cell. The cell contains a proton-conducting polymer membrane, carbon electrodes and a Pt catalyst. The operating temperature of $\approx$350 K is relatively low, and this means that the start-up time is shorter than for the molten carbonate and solid oxide fuel cells. The PEM fuel cell is actually a stack of cells. Each cell is known as a membrane electrode assembly (MEA) and comprises a platinized carbon-fibre paper anode and cathode separated by a proton-conducting membrane. The latter is typically made from Nafion (a perfluorinated polymer with sulfonic acid groups attached along the backbone). The MEA units are connected in series by carbon fibre or polypropylene flow field plates, through which H$_2$ and air can pass (H$_2$ to the anode and O$_2$ to the cathode):

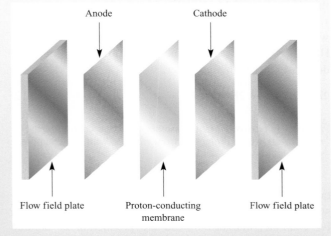

The anode and cathode reactions are, respectively:

$$H_2 \longrightarrow 2H^+ + 2e^-$$

$$O_2 + 4H^+ + 4e^- \longrightarrow 2H_2O$$

The passage of protons across the membrane allows the overall energy-producing cell reaction to take place:

$$2H_2 + O_2 \longrightarrow 2H_2O$$

Each cell generates about 0.7 V, hence the need for the stack of cells to produce sufficient energy for powering an electrical motor.

The high cost of platinum means that the car industry strives to reduce the amount of catalyst required in a PEM fuel cell. Current strategies focus on the use of PtM nanoparticles (M = another *d*-block metal), nanoparticles with a PtM core encased in Pt atoms, and thin Pt films dispersed on nanostructured supports. Iron-based catalysts would be much cheaper, but their performance is usually poor. A promising advance (still at the research stage) is in the application of microporous carbon-supported iron-based catalysts in which the iron cations are thought to be coordinated in {Fe(phen)$_2$}$^{2+}$ sites, the phenanthroline-units being incorporated into graphitic sheets.[†]

More in-depth discussion of fuel cells, and the design and manufacturing problems which have to be overcome to make fuel cell-powered vehicles a viable option for the future, can be found in the references at the end of the chapter.

[†] M. Lefèvre *et al.* (2009) *Science*, vol. 324, p. 71 – 'Iron-based catalysts with improved oxygen reduction activity in polymer electrolyte fuel cells'.

(e.g. Zn, Al) with aqueous alkali (eq. 10.10) or by reacting metal hydrides with water (eq. 10.11).

$$Zn(s) + 2HCl(aq) \longrightarrow ZnCl_2(aq) + H_2(g) \qquad (10.9)$$

$$2Al(s) + 2NaOH(aq) + 6H_2O(l)$$
$$\longrightarrow 2Na[Al(OH)_4](aq) + 3H_2(g) \qquad (10.10)$$

$$CaH_2(s) + 2H_2O(l) \longrightarrow Ca(OH)_2(aq) + 2H_2(g) \qquad (10.11)$$

Group 1 metals liberate H_2 from water (eq. 10.12), but such reactions are not suitable for preparative use because of their extreme vigour (Fig. 6.8). Many other metals that, on thermodynamic grounds, would be expected to react in this way are made *kinetically inert* by the presence of a thin film of insoluble metal oxide. Such metals are *passivated*. Although Be is passivated and does not react with water even on heating, the other group 2 metals react with H_2O to give H_2, reactivity increasing down the group; Mg does not react with *cold* water.

$$2K + 2H_2O \longrightarrow 2KOH + H_2 \qquad (10.12)$$

A metal is ***passivated*** if it possesses a surface coating of the metal oxide which protects it from reaction with, for example, water.

Dihydrogen has industrial applications, the most important being in the Haber–Bosch process (see Sections 15.5 and 25.8), the hydrogenation of unsaturated fats (to produce, for example, margarine), and the production of organic compounds. Among these, methanol is hugely important, with world demand in excess of 35 Mt per year (1 Mt = 10^6 t). It is used as a precursor to formaldehyde, acetic acid, methyl methacrylate and a range of organic solvents, and its application as a fuel is increasing. The first stage in the industrial manufacture of CH_3OH is the production of *synthesis gas (syngas)* by the *steam reforming of methane* (eq. 10.13), and the *water–gas shift reaction* in which some of the CO produced in reaction 10.13 reacts with steam (eq. 10.14). Syngas is converted to CH_3OH using the conditions shown in scheme 10.15. If the available commercial feedstock is coal (e.g. in China) rather than CH_4, the process can be adapted, replacing CH_4 by C in eq. 10.13.

$$CH_4 + H_2O \xrightarrow{\text{Ni catalyst, 1200 k}} CO + 3H_2 \qquad (10.13)$$

$$CO + H_2O \xrightarrow[\text{700 K}]{\text{iron oxide catalyst}} CO_2 + H_2 \qquad (10.14)$$

$$CO + 2H_2 \xrightarrow[473-573\,K,\,50\,bar]{\text{Cu/ZnO catalyst}} CH_3OH$$

or

$$CO_2 + 3H_2 \xrightarrow[473-573\,K,\,50\,bar]{\text{Cu/ZnO catalyst}} CH_3OH + H_2O \qquad (10.15)$$

The very low density and boiling point of H_2 make large-scale transport costs unacceptably high. Thus, for industrial applications such as the Haber–Bosch process, H_2 is produced *in situ* by reactions 10.13 and 10.14. The CO_2 by-product is absorbed in aqueous alkaline K_2CO_3 and can be recovered by heating (eq. 10.16).

$$CO_2 + K_2CO_3 + H_2O \underset{\Delta}{\rightleftharpoons} 2KHCO_3 \qquad (10.16)$$

The facts that fossil fuels lead to greenhouse gas emissions (see Box 14.7) and the Earth's coal, gas and oil reserves are gradually being depleted, call for an urgent change in direction towards the so-called *hydrogen economy*. Energy can be produced directly by the combustion of H_2 (explosive reaction 10.17) or electrochemically in fuel cells (see Box 10.2). Dihydrogen is a clean fuel because its oxidation yields only H_2O, and no regulated emissions are involved. It is also a means of storing energy (each H–H bond stores 436 kJ mol^{-1}) until it is released upon oxidation.

$$2H_2 + O_2 \longrightarrow 2H_2O \qquad (10.17)$$

Although reaction 10.17 appears to be the ideal solution to environmental and economic problems surrounding fossil fuels, H_2 is not naturally abundant in the Earth's atmosphere (see Fig. 15.1b). The production of H_2 from renewable resources is therefore one of the major challenges facing research teams in the 21st century. This is coupled with developing the means to harness clean and sustainable energy sources. Since the Earth receives terawatts of energy per day from the Sun (far more energy than is currently consumed by people worldwide), the conversion of solar energy to electrical energy (see Section 28.3), and the use of solar energy to produce H_2, are two primary goals of current research.

In natural photosynthesis, green plants, green algae and cyanobacteria use solar energy to split H_2O and convert CO_2 to carbohydrates (eq. 10.18). The light-initiated process utilizes the active site in the enzyme Photosystem II (see Section 21.8) to oxidize H_2O, liberating electrons that are passed along a redox pathway to electron acceptors that ultimately reduce CO_2. The protons produced upon water splitting (eq. 10.19) are taken up in the carbohydrate product. Photosynthesis effectively coverts solar energy into stored energy in the carbohydrate. The light is harvested by coloured pigments in chlorophyll (chlorophyll *a*, chlorophyll *b*, xanothophylls and carotenoids) which absorb light across the whole visible spectrum.

$$6H_2O + 6CO_2 \xrightarrow{h\nu} C_6H_{12}O_6 + 6O_2 \qquad (10.18)$$

$$2H_2O \longrightarrow O_2 + 4H^+ + 4e^- \qquad (10.19)$$

The challenge is to devise chemical systems that absorb solar energy which is then harnessed to drive electron transfer processes that result in water splitting, with the electrons

Fig. 10.3 (a) The structure of $[Ru(bpy)_3]^{3+}$ (bpy = 2,2'-bipyridine) determined by X-ray diffraction for the compound $[Ru(bpy)_3][PF_6]_3$ [M. Biner *et al.* (1992) *J. Am. Chem. Soc.*, vol. 114, p. 5197], and (b) a schematic representation of $[Ru(bpy)_3]^{3+}$. Colour code: Ru, red; C, grey; N, blue; H atoms are omitted.

being used to reduce H^+ (e.g. to H_2) or a source of oxidized carbon (e.g. CO_2). In terms of thermodynamics, photolytic water-splitting is an uphill process. Oxidation of H_2O requires a strong oxidizing agent ($E^o > +1.23$ V since the standard reduction potential for the reverse of eq. 10.19 is -1.23 V) and a redox catalyst. Proton reduction to H_2 also requires a redox catalyst. In nature, enzymes called hydrogenases fulfil this role (see Section 29.4). In artificial systems, $[Ru(bpy)_3]^{2+}$ (Fig. 10.3) and related ruthenium(II) complexes are widely used as photosensitizers, and many more coloured, *d*-block metal complexes are currently being investigated for this role. The orange-red complex $[Ru(bpy)_3]^{2+}$ absorbs light of wavelength 452 nm and is converted into an excited state species which is both a better oxidizing and reducing agent than the ground state species. In theory, shuttling between Ru(II) and Ru(III) allows water to be oxidized and protons to be reduced. This is illustrated in the scheme below:

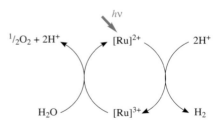

In practice, the system works only in the presence of a quenching agent, $[A]^+$ which accepts an electron from the excited state complex $\{[Ru(bpy)_3]^{2+}\}^*$, and then passes it on to H_2O as A is oxidized back to $[A]^+$. A sacrificial electron donor, D, (often $[EDTA]^{4-}$) is required to reduce the oxidized form of the ruthenium catalyst. The complete

process is summarized below and is discussed fully in Section 22.9 (see Fig. 22.24):

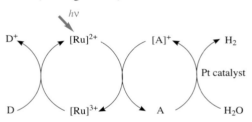

A *photolytic process* (*photolysis*) is initiated by light; in an equation, this is indicated by $h\nu$ over the arrow; the reactants are said to be *photolysed*.

Reactivity

Dihydrogen is not very reactive under ambient conditions, but the lack of reactivity is kinetic rather than thermodynamic in origin, and arises from the strength of the H−H bond (Table 10.3). The branching-chain reaction of H_2 and O_2 is initiated by sparking and the resulting explosion (or 'pop' on a small scale) is well known in the qualitative test for H_2. Part of the reaction scheme is given (in a simplified form) in eqs. 10.20–10.24. Efficient branching results in a rapid, explosive reaction, and is the reason why it is effective in rocket fuels.

$$H_2 \longrightarrow 2H^\bullet \qquad\qquad\qquad \textit{initiation} \qquad (10.20)$$

$$H_2 + O_2 \longrightarrow 2OH^\bullet \qquad\qquad \textit{initiation} \qquad (10.21)$$

$$H^\bullet + O_2 \longrightarrow OH^\bullet + {}^\bullet O^\bullet \qquad \textit{branching} \qquad (10.22)$$

$${}^\bullet O^\bullet + H_2 \longrightarrow OH^\bullet + H^\bullet \qquad \textit{branching} \qquad (10.23)$$

$$OH^\bullet + H_2 \longrightarrow H_2O + H^\bullet \qquad \textit{propagation} \qquad (10.24)$$

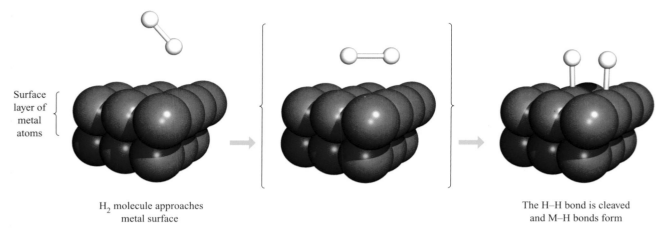

Surface layer of metal atoms

H_2 molecule approaches metal surface

The H–H bond is cleaved and M–H bonds form

Fig. 10.4 A schematic representation of the interaction of an H_2 molecule with a metal surface to give *adsorbed* hydrogen atoms. The scheme does not imply anything about the detailed mechanism of the process. Further details about heterogeneous catalysis are given in Chapter 25.

Halogens react with H_2 (eq. 10.25) with the ease of reaction decreasing down group 17. Even at low temperatures, F_2 reacts explosively with H_2 in a radical chain reaction. In the light-induced reaction of Cl_2 and H_2, the initiation step is the homolytic cleavage of the Cl–Cl bond to give $Cl^{\bullet}$ radicals (eq. 10.26) which react with H_2 to give $H^{\bullet}$ and HCl in one of a series of steps in the radical chain; HCl can be formed in either a propagation or a termination step.

$$H_2 + X_2 \longrightarrow 2HX \qquad X = F, Cl, Br, I \qquad (10.25)$$

$$Cl_2 \xrightarrow{h\nu} 2Cl^{\bullet} \qquad (10.26)$$

Reactions of H_2 with Br_2 or I_2 occur only at higher temperatures and also involve the initial fission of the X_2 molecule. For Br_2 (but not for I_2) the mechanism is a radical chain (equation sequence 10.27).

$$\left.\begin{array}{l} Br_2 \longrightarrow 2Br^{\bullet} \\ Br^{\bullet} + H_2 \longrightarrow HBr + H^{\bullet} \\ H^{\bullet} + Br_2 \longrightarrow HBr + Br^{\bullet} \\ HBr + H^{\bullet} \longrightarrow Br^{\bullet} + H_2 \\ 2Br^{\bullet} \longrightarrow Br_2 \end{array}\right\} \qquad (10.27)$$

Dihydrogen reacts with many metals when heated to give metal hydrides, MH_n, although these are not necessarily stoichiometric (e.g. $TiH_{1.7}$). By the action of an electric discharge, H_2 is partially dissociated into atoms, particularly at low pressures. This provides a reactive source of the element, and facilitates combination with elements (e.g. Sn and As) that do not react directly with H_2.

The reaction between N_2 and H_2 (eq. 10.28) is of major industrial importance. However, the reaction is extremely slow and mixtures of N_2 and H_2 remain indefinitely unchanged. Manipulation of the temperature and pressure and the use of a catalyst are essential. (There is more about

catalysts and their industrial applications in Chapter 25.)

$$3H_2(g) + N_2(g) \rightleftharpoons 2NH_3(g) \qquad (10.28)$$

Interaction between a catalytic surface and H_2 weakens and aids cleavage of the H–H bond (Fig. 10.4). On an industrial scale, the hydrogenation of enormous numbers of unsaturated organic compounds is carried out on surfaces of metals such as Ni, Pd and Pt. The use of homogeneous catalysts is also important, e.g. reaction 10.29 (the *hydroformylation process*) and is detailed in Chapter 25.

$$RHC=CH_2 + H_2 + CO \xrightarrow{Co_2(CO)_8 \text{ catalyst}} RCH_2CH_2CHO \qquad (10.29)$$

10.5 Polar and non-polar E–H bonds

Although we refer to compounds of the type EH_n (E = any element) as *hydrides*, and this tends to suggest the presence of H^- (or at least, $H^{\delta-}$), the difference in electronegativity values between E and H means that the E–H bond may be non-polar, or polar in either of the senses shown in Fig. 10.5. For H, $\chi^P = 2.2$ and a number of E–H bonds in which E is a *p*-block element (e.g. B–H, C–H, Si–H,

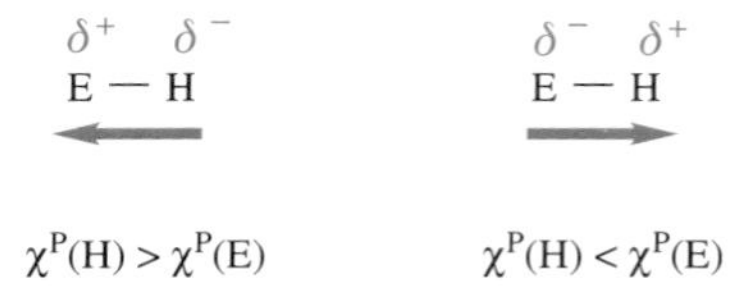

$$\overset{\delta^+ \quad \delta^-}{E — H} \qquad \overset{\delta^- \quad \delta^+}{E — H}$$
$$\longleftarrow \qquad \longleftarrow$$

$$\chi^P(H) > \chi^P(E) \qquad \chi^P(H) < \chi^P(E)$$

Fig. 10.5 The direction of the dipole moment in a polar E–H bond depends upon the relative electronegativity values; Pauling electronegativity values, χ^P, are given in Appendix 7. The direction in which the arrow points ($\delta-$ to $\delta+$) is defined by SI convention.

THEORY

Box 10.3 The $[H_3]^+$ ion

The equilateral triangular $[H_3]^+$ ion may appear to be a theoretical novelty, and it has been the subject of many theoretical studies. However, just as Jupiter and Saturn have provided challenges in regard to metallic hydrogen (see Box 10.1), Jupiter has also proved to be the source of exciting spectroscopic data, analysis of which has confirmed the existence of $[H_3]^+$. The atmosphere of Jupiter consists mainly of dihydrogen, and the formation of $[H_3]^+$ has been explained in terms of the ionization of H_2, brought about by collisions between H_2 molecules and charged particles (with *extremely* high kinetic energies) which originate from Jupiter's magnetosphere:

$$H_2 \longrightarrow [H_2]^+ + e^-$$

It is proposed that further collisions between H_2 and $[H_2]^+$ lead to the formation of $[H_3]^+$.

$$H_2 + [H_2]^+ \longrightarrow [H_3]^+ + H$$

The chemistry of this cation in the atmospheres of Jupiter and Uranus is a subject of continued research.

Further reading

L.M. Grafton, T.R. Geballe, S. Miller, J. Tennyson and G.E. Ballester (1993) *Astrophys. J.*, vol. 405, p. 761 – 'Detection of trihydrogen(1+) ion from Uranus'.
S. Miller and J. Tennyson (1992) *Chem. Soc. Rev.*, vol. 22, p. 281 – '$[H_3]^+$ in space'.
S. Miller *et al.* (2006) *Phil. Trans. R. Soc. A*, vol. 364, p. 3121 – '$[H_3]^+$: the driver of giant planet atmospheres'.
J. Tennyson and S. Miller (2001) *Spectrochim. Acta Part A*, vol. 57, p. 661 – 'Spectroscopy of H_3^+ and its impact on astrophysics'.

P—H) are essentially non-polar. Since metals are electropositive, the H atom in an M—H bond carries a δ^- partial charge. In contrast, N, O and F are more electronegative than H, and in N—H, O—H and F—H bonds, the H atom carries a δ^+ partial charge.

The molecular environment of an E—H bond also influences the magnitude of the bond dipole and properties associated with the bond. This is demonstrated by a comparison of the pK_a values for CH_3CO_2H ($pK_a = 4.75$) and CF_3CO_2H ($pK_a = 0.23$).

10.6 Hydrogen bonding

The hydrogen bond

A **hydrogen bond** is formed between an H atom attached to an electronegative atom, and an electronegative atom that possesses a lone pair of electrons.

Physical and solid state structural data for many compounds provide evidence for the formation of intermolecular hydrogen bonds. Such interactions arise between an H atom attached to an electronegative atom, and an electronegative atom bearing a lone pair of electrons, i.e. X—H$\cdots$Y where atom Y may or may not be the same as X. It is not necessary for the electronegative atom X to be highly electronegative for there to be a meaningful hydrogen-bonded interaction. Thus, in addition to hydrogen bonds of the type F—H$\cdots$F, O—H$\cdots$F, N—H$\cdots$F,

O—H$\cdots$O, N—H$\cdots$O, O—H$\cdots$N and N—H$\cdots$N, it is now well recognized that weaker hydrogen bonds, in particular C—H$\cdots$O interactions, play an important role in the solid state structures of small molecules and biological systems. The wide variety of interactions that are now classed as hydrogen bonds means that the definition of the latter must not be too restrictive. A modern definition of a hydrogen bond which does not rely directly on the concept of electronegativity has been proposed by Steiner:[†]

An X—H$\cdots$Y interaction is called a **hydrogen bond** if it constitutes a local bond, and if X—H acts as a proton donor to Y.

A broad definition has recently been proposed by the IUPAC:[‡]

The **hydrogen bond** is an attractive interaction between the hydrogen from a group X—H and an atom or a group of atoms Y, in the same or different molecule(s), where there is evidence of bond formation.

It is now well recognized that the term 'hydrogen bonding' covers a wide range of interactions with a corresponding variation in strengths of interaction. Table 10.4 lists representative examples.

We have already described the hydrogen-bonded network in ice (see Section 7.2). Here, as in most hydrogen-bonded

---

[†] T. Steiner (2002) *Angew. Chem. Int. Ed.*, vol. 41, p. 48.
[‡] E. Arunan (2007) *Current Science*, vol. 92, p. 17.

Table 10.4 Typical values for the bond dissociation enthalpy of different types of hydrogen bonds. Values are calculated for gas-phase species.[†]

Category of hydrogen bond	Hydrogen bond (····)	Dissociation enthalpy / kJ mol^{-1}
Symmetrical	F····H····F in $[HF_2]^-$ (see eq. 10.26)	163
Symmetrical	O····H····O in $[H_5O_2]^+$ (see structure **10.2**)	138
Symmetrical	N····H····N in $[N_2H_7]^+$ (see structure **10.4**)	100
Symmetrical	O····H····O in $[H_3O_2]^-$ (see structure **10.3**)	96
Asymmetrical	N–H····O in $[NH_4]^+$····OH_2	80
Asymmetrical	O–H····Cl in OH_2····Cl^-	56
Asymmetrical	O–H····O in OH_2····OH_2	20
Asymmetrical	S–H····S in SH_2····SH_2	5
Asymmetrical	C–H····O in $HC{\equiv}CH$····OH_2	9
Asymmetrical	C–H····O in CH_4····OH_2	1 to 3

[†] Data are taken from: T. Steiner (2002) *Angew. Chem. Int. Ed.*, vol. 41, p. 48.

interactions, the H atom is *asymmetrically* positioned with respect to the two atoms with which it interacts. Association in carboxylic acids (see Box 10.4) is a consequence of hydrogen bonding. In a typical X−H····Y interaction, the X−H covalent bond is *slightly* longer and weaker than a comparable bond in the absence of hydrogen bonding. In such cases, the interaction may be considered in terms of an electrostatic interaction between a covalently bonded H with a δ^+ charge, and a lone pair of electrons on the adjacent atom. Some experimental observations cannot be rationalized within a purely electrostatic model, and point towards a covalent contribution, the importance of which increases as the hydrogen bond becomes stronger.

Table 10.4 shows typical values of bond dissociation enthalpies of some hydrogen bonds. The data in the table have been obtained from calculations on isolated species. These enthalpy values are therefore only approximate when applied to hydrogen bonds between molecules in a solid state lattice. Enthalpy values for these interactions cannot be measured directly. An example of how the strengths of hydrogen bonds can be obtained experimentally comes from the dissociation of a carboxylic acid dimer *in the vapour state* (eq. 10.30).

$$R-C \rightleftharpoons 2RCO_2H \quad (10.30)$$

The position of equilibrium 10.30 is temperature-dependent, and ΔH° for the reaction can be obtained from the variation of K_p with temperature:

$$\frac{d(\ln K)}{dT} = \frac{\Delta H^\circ}{RT^2}$$

For formic acid (methanoic acid), ΔH° for the dissociation in eq. 10.30 (R = H) is $+60$ kJ mol^{-1}, or the value can be expressed as $+30$ kJ per mole of hydrogen bonds. This quantity is often referred to as the hydrogen-bond energy, but this is not strictly correct since other bonds change *slightly* when hydrogen bonds are broken (Figs. 10.6a and 10.6b).

In some hydrogen-bonded interactions, the H atom is *symmetrically* positioned, e.g. in $[HF_2]^-$ (see Fig. 10.9) or $[H_5O_2]^+$ (Fig. 10.1). In the formation of $[HF_2]^-$ (eq. 10.31), appreciable stretching of the original covalent H−F bond takes place, to give two equivalent H····F interactions.

$$HF + F^- \longrightarrow [HF_2]^- \quad (10.31)$$

The bonding in symmetrical X····H····X interactions is best considered in terms of a 3c-2e interaction, i.e. as a delocalized interaction such as was described for B_2H_6 in Section 5.7. Each H····F bond is relatively strong (Table 10.4), with the bond dissociation enthalpy being of a similar magnitude to that of the F−F bond in F_2 (158 kJ mol^{-1}). Compare this with the bond dissociation enthalpy of HF (570 kJ mol^{-1}). Strong, symmetrical hydrogen bonds with covalent character usually occur between like atoms (see Table 10.4). Common examples involve interactions between an acid and its conjugate base where there is no distinction between the donor (X) and acceptor (Y) atoms, e.g. eq. 10.31 and structures **10.2–10.5**.

$[H_3O]^+ + H_2O$

(10.2)

$H_2O + [OH]^-$

(10.3)

Figure 7.1 illustrated how hydrogen bonding between H_2O molecules in the solid state produces a rigid network. Hydrogen bonding is responsible for many packing motifs in solid state structures. Among compounds that form dimeric units are carboxylic acids and amides:

Solvents of crystallization may be involved in the packing motifs, e.g. H_2O or MeOH:

Hydrogen bonding between difunctional carboxylic acids can result in the formation of chains:

If three or more carboxylic acid functionalities are present in a molecule, 2-dimensional sheets or 3-dimensional networks may assemble depending upon the spatial arrangement of the CO_2H groups. The same principle can be applied to construct solid state assemblies in which the building blocks are coordination complexes. Consider the following pyridine ligands:

If the ligand coordinates to a metal ion through the *N*-donor, the peripheral carboxylic acid units in the metal complex can associate with one another through hydrogen bonding. An example is *trans*-[PdCl₂L₂] (L = pyridine-3-carboxylic acid) which forms infinite chains:

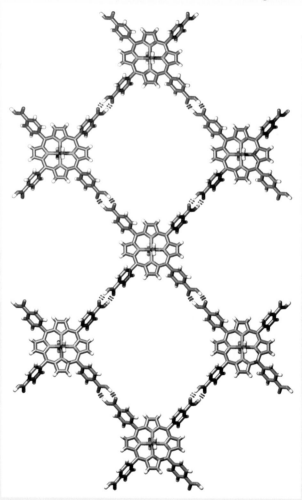

[Data: Z. Qin *et al.* (2002) *Inorg. Chem.*, vol. 41, p. 5174.]

In the zinc(II) porphyrin complex below, the four peripheral CO_2H groups are oriented in such a way that 2-dimensional sheets supported by hydrogen bonds assemble. Each Zn^{2+} ion is octahedrally bound by a porphyrin and two water ligands:

[Data: Y. Diskin-Posner *et al.* (1999) *Chem. Commun.*, p. 1961.]

Further reading

A. M. Beatty (2003) *Coord. Chem. Rev.*, vol. 246, p. 131 – 'Open-framework coordination complexes from hydrogen-bonded networks'.

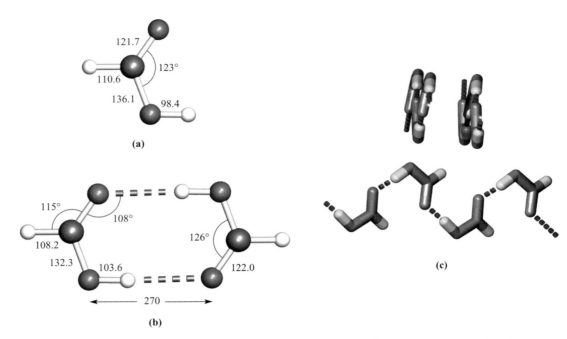

Fig. 10.6 In the vapour state, formic acid exists as both (a) a monomer and (b) a dimer, the structures of which have been determined by electron diffraction. (c) In the solid state, a more complex assembly is formed as revealed in a neutron diffraction study of deuterated formic acid, DCO_2D; the figure shows that molecules assemble into ribbons that lie perpendicular and parallel to one another. [A. Albinati *et al.* (1978) *Acta Crystallogr., Sect. B*, vol. 34, p. 2188.] Distances are in pm. Colour code: C, grey; O, red; H, white; D, yellow.

$[NH_4]^+ + NH_3$

(10.4)

$RCO_2H + [RCO_2]^-$

(10.5)

Neutron diffraction studies have confirmed that adduct **10.6** contains a strong, symmetrical $N\cdots H\cdots O$ hydrogen bond at 90 K ($O-H = N-H = 126$ pm). However, the system is complicated by the observation that the H atom migrates towards the O atom as the temperature is lowered from 200 to 20 K.[†]

(10.6)

The use of the qualitative descriptors 'strong', 'moderate' (or 'normal') and 'weak' for hydrogen bonds is common. For example, strong $O\cdots H\cdots O$ interactions are typified by $O\cdots O$ separations close to 240 pm, while moderate $O-H\cdots O$ interactions are characterized by longer $O\cdots O$ distances, up to ≈ 280 pm. Accurate neutron and X-ray diffraction data[‡] confirm that for $O-H\cdots O$ interactions, shortening of the $O\cdots O$ distance from 280 to 240 pm is accompanied by a change from asymmetrical, electrostatic hydrogen bonds to symmetrical, covalent interactions. Strong hydrogen bonds are usually linear (i.e. the $X-H-Y$ angle is close to $180°$), while in 'moderate' hydrogen bonds, $X-H-Y$ angles may range from $130°$ to $180°$. The transition from 'strong' to 'moderate' hydrogen bonds is not clear-cut. So-called 'weak' hydrogen bonds involve weak electrostatic interactions or dispersion forces, and include $C-H\cdots O$ interactions; we return to these later in the section.

Trends in boiling points, melting points and enthalpies of vaporization for *p*-block binary hydrides

It is generally expected that the melting and boiling points of members of a series of related molecular compounds increase with increasing molecular size, owing to an increase in intermolecular dispersion forces. This is seen, for example, along a homologous series of alkanes. However, a comparison of the melting and boiling points of *p*-block hydrides, EH_n, provides evidence for hydrogen bonding. Figure 10.7 shows that, for E = group 14 element, melting and boiling

---

[†] For details, see: T. Steiner, I. Majerz and C.C. Wilson (2001) *Angew. Chem. Int. Ed.*, vol. 40, p. 2651.

[‡] P. Gilli *et al.* (1994) *J. Am. Chem. Soc.*, vol. 116, p. 909.

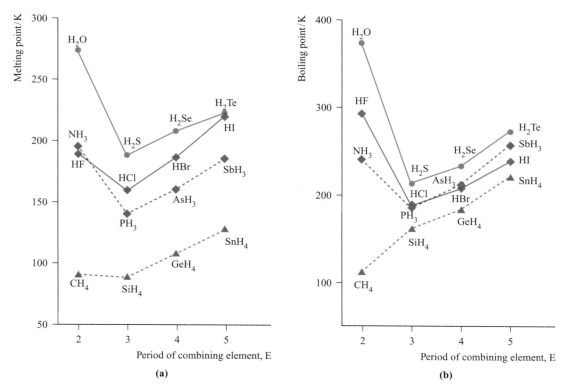

Fig. 10.7 Trends in (a) melting and (b) boiling points for some *p*-block hydrides, EH_n.

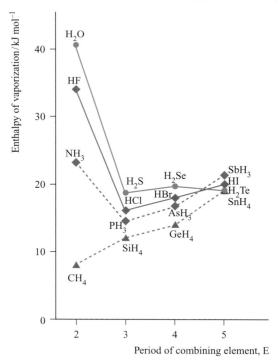

Fig. 10.8 Trends in values of $\Delta_{vap}H$ (measured at the boiling point of the liquid) for some *p*-block hydrides, EH_n.

points follow the expected trends, but for E = group 15, 16 or 17 element, the first member of the group shows anomalous behaviour, i.e. the melting and boiling points of NH_3, H_2O and HF are higher than expected when compared with their heavier congeners. Figure 10.8 illustrates that values of $\Delta_{vap}H$ show a similar pattern. It is tempting to think that Figs. 10.7 and 10.8 indicate that the hydrogen bonding in H_2O is stronger than in HF. Certainly, the values for H_2O appear to be particularly high. However, this is not a sound conclusion. Boiling points and values of $\Delta_{vap}H$ relate to differences between the liquid and gaseous states, and there is independent evidence that while H_2O is hydrogen-bonded in the liquid but not in the vapour state, HF is strongly hydrogen-bonded in both.

Many liquids undergoing a liquid to vapour transition possess similar values of the entropy of vaporization, i.e. the liquids obey Trouton's rule (eq. 10.32). Deviations from Trouton's empirical rule are another way of expressing the data in Figs. 10.7 and 10.8. For HF, H_2O and NH_3, $\Delta_{vap}S = 116$, 109 and 97 J K^{-1} mol^{-1} respectively. Hydrogen bonding in each liquid lowers its entropy, and makes the *change* in the entropy on going from liquid to vapour larger than it would have been had hydrogen bonding not played an important role.

$$\text{For } liquid \rightleftharpoons vapour: \quad \Delta_{vap}S = \frac{\Delta_{vap}H}{\text{bp}}$$

$$\approx 88 \text{ J K}^{-1} \text{ mol}^{-1} \qquad (10.32)$$

Infrared spectroscopy

The IR spectrum of a hydrate, alcohol or carboxylic acid exhibits a characteristic absorption around 3500 cm^{-1} assigned to the $\nu(OH)$ mode (see Fig. 3.13). The typical

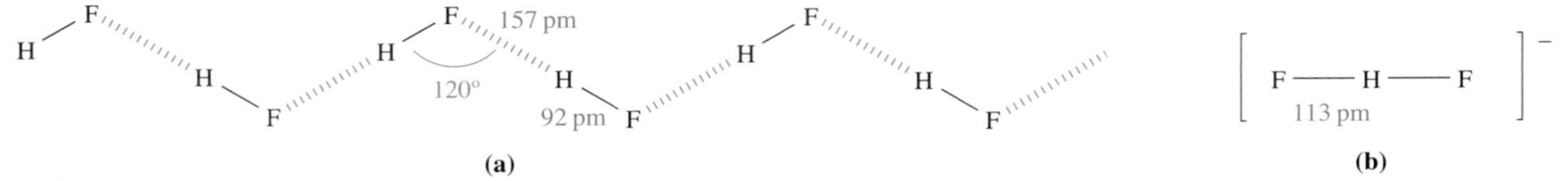

(a)   **(b)**

Fig. 10.9 (a) The solid state structure of HF consists of zigzag chains. (b) The structure of the $[HF_2]^-$ ion, determined by X-ray and neutron diffraction for the K^+ salt.

broadness of this band can be explained in terms of the involvement of the O–H hydrogen atom in hydrogen bonding. In cases where we can compare the stretching frequencies of the same molecule with and without hydrogen-bonded association (e.g. liquid water and water vapour), a shift is observed to higher wavenumber as hydrogen bonding is lost. Similar observations are noted for other hydrogen-bonded systems.

Solid state structures

The presence of hydrogen bonding has important effects on the solid state structures of many compounds, as we have already discussed for ice (Section 7.2) and carboxylic acids (Box 10.4). The *solid* state structures of some simple carboxylic acids are more complex than one might at first imagine. Figure 10.6c shows part of the solid state packing diagram for deuterated formic acid. The orientation of the DCO_2D molecules allows the assembly of a more extensive hydrogen-bonded network than simple dimers. The solid state structure of acetic acid is similarly complex.

The structure of solid HF consists of zigzag chains (Fig. 10.9a), although the positions of the H atoms are not accurately known. Hydrogen-bonded interactions persist in HF in both the liquid and vapour states (see Section 9.7). Structural parameters are available for a number of salts containing $[HF_2]^-$, and include neutron diffraction data for the deuterated species. The anion is linear with the H atom positioned symmetrically between the two F atoms (Fig. 10.9b). The H–F distance is relatively short, consistent with strong hydrogen bonding (see Table 10.4 and earlier discussion).

In describing the $[H_3O]^+$ ion in Section 10.2, we also mentioned $[H_5O_2]^+$ and $[H_9O_4]^+$. These latter species belong to a wider group of ions of general formula $[H(OH_2)_n]^+$. In solution, the formation of these ions is relevant to reactions involving proton transfer. Solid state studies, including neutron diffraction studies in which the positions of the H atoms are accurately determined, have provided structural data for the $[H_5O_2]^+$, $[H_7O_3]^+$, $[H_9O_4]^+$, $[H_{11}O_5]^+$ and $[H_{13}O_6]^+$ ions. In each ion, hydrogen bonding plays a crucial role. Neutron diffraction data for $[H_5O_2]^+$ in $[V(OH_2)_6][H_5O_2][CF_3SO_3]_4$ (Fig. 10.1) reveal a symmetrical $O\cdots H\cdots O$ hydrogen-bonded interaction. A neutron diffraction study of the trihydrate of acid **10.7** shows the presence of $[H_7O_3]^+$ along with the conjugate base of acid **10.7**. Within the $[H_7O_3]^+$ unit,

the $O\cdots O$ distances are 241.4 and 272.1 pm. In this system, the $[H_7O_3]^+$ ion can be described in terms of $[H_5O_2]^+\cdot H_2O$ with one 'strong' hydrogen bond in the $[H_5O_2]^+$ unit and one 'normal' hydrogen-bonded interaction between the $[H_5O_2]^+$ and H_2O units. Crown ethers have been used to stabilize $[H(OH_2)_n]^+$ ions, the stabilizing factor being the formation of hydrogen bonds between the O atoms of the macrocyclic ligand and the H atoms of the $[H(OH_2)_n]^+$ ion. Two examples are shown in Fig. 10.10 and illustrate the encapsulation of an $[H_5O_2]^+$ ion within a single crown ether, and the association of a chain structure involving alternating crown ether and $[H_7O_3]^+$ ions. In the latter, the bond lengths (structure **10.8**) determined by neutron diffraction show two asymmetrical hydrogen bonds and this is consistent with $[H_7O_3]^+$ being considered in terms of $[H_3O]^+\cdot 2H_2O$. No single detailed formulation for a given ion is appropriate in all cases, and the environment and crystal packing of the $[H(OH_2)_n]^+$ ions in a given solid state structure influence the detailed bonding description. The $[H_{14}O_6]^{2+}$ ion, **10.9**, is a rare example of a dicationic $[H_2(OH_2)_n]^{2+}$ species.

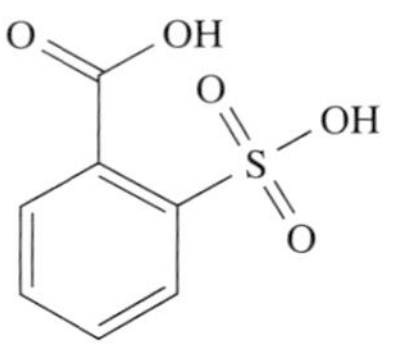

(10.7) 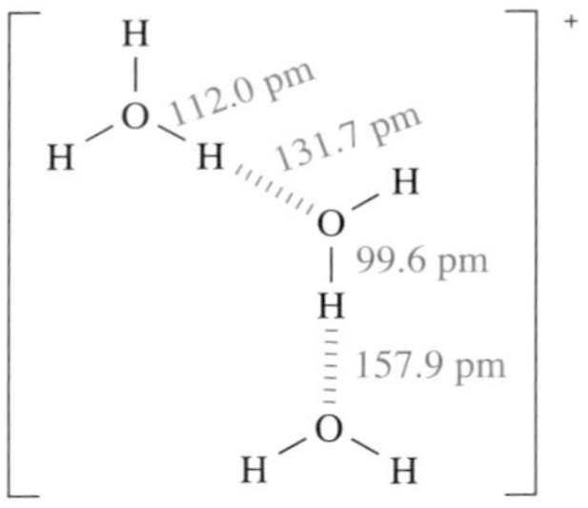   **(10.8)**

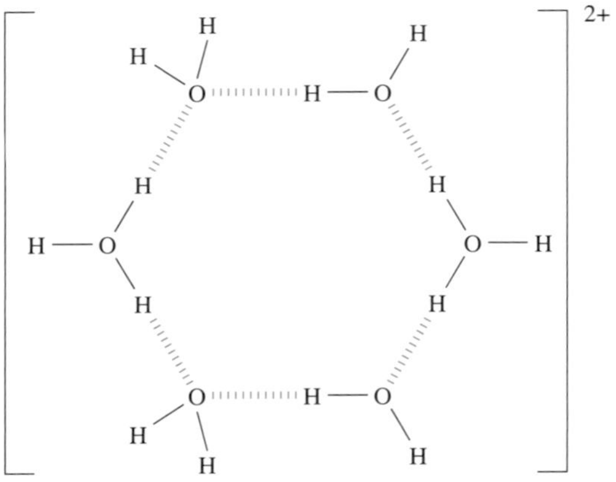

(10.9)

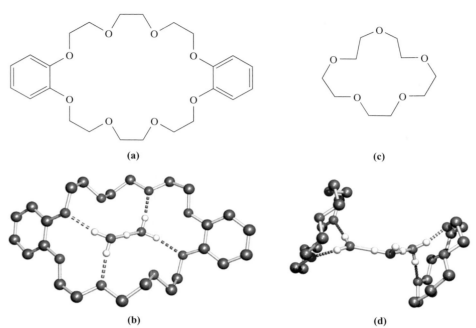

(a)   **(c)**

(b)   **(d)**

Fig. 10.10 The stabilization in the solid state of $[H_5O_2]^+$ and $[H_7O_3]^+$ by hydrogen bonding to crown ethers: (a) the structure of dibenzo-24-crown-8; (b) the structure of $[(H_5O_2)(\text{dibenzo-24-crown-8})]^+$ determined for the $[AuCl_4]^-$ salt by X-ray diffraction [M. Calleja *et al.* (2001) *Inorg. Chem.*, vol. 40, p. 4978]; (c) the stucture of 15-crown-5; and (d) part of the chain structure of $[(H_7O_3)(\text{15-crown-5})]^+$ determined for the $[AuCl_4]^-$ salt by neutron diffraction [M. Calleja *et al.* (2001) *New J. Chem.*, vol. 25, p. 1475]. Hydrogen bonding between the $[H_5O_2]^+$ and $[H_7O_3]^+$ ions and crown ethers is shown by hashed lines; hydrogen atoms in the crown ethers are omitted for clarity. Colour code: C, grey; O, red; H, white.

Although hydrogen bonds commonly involve F, O or N, this, as we have already mentioned, is not an exclusive picture. Examples include the solid state structure of HCN, which exhibits a linear chain with C−H⋯N interactions, the 1:1 complex formed between acetone and chloroform, and the existence of salts containing the $[HCl_2]^-$ anion. Weak (see Table 10.4), asymmetrical C−H⋯O hydrogen bonds play an important role in the assembly of a wide variety of solid state structures ranging from interactions between small molecules to those in biological systems. In the crystal lattice, molecules of Me_2NNO_2 are arranged in chains and, as Fig. 10.11 shows, C−H⋯O hydrogen bonds are responsible for this ordered assembly.

Finally we come to the so-called *dihydrogen bond*. This is a weak electrostatic interaction that may occur between two

hydrogen atoms, one $H^{\delta+}$ and one $H^{\delta-}$. In terms of hydrogen bond classifications, the $H^{\delta+}$ atom behaves as a hydrogen bond donor while $H^{\delta-}$ is a hydrogen bond acceptor. For example, in the solid state structure of the adduct $H_3B\cdot NH_3$, the hydrogen positions have been accurately located by neutron diffraction. The Pauling electronegativity values of B and N are 2.0 and 3.0, respectively, and this leads to polar bonds: $N^{\delta-}-H^{\delta+}$ and $B^{\delta+}-H^{\delta-}$. In crystalline $H_3B\cdot NH_3$, molecules pack as shown in Fig. 10.12 with the shortest $N-H^{\delta+}\cdots H^{\delta-}-B$

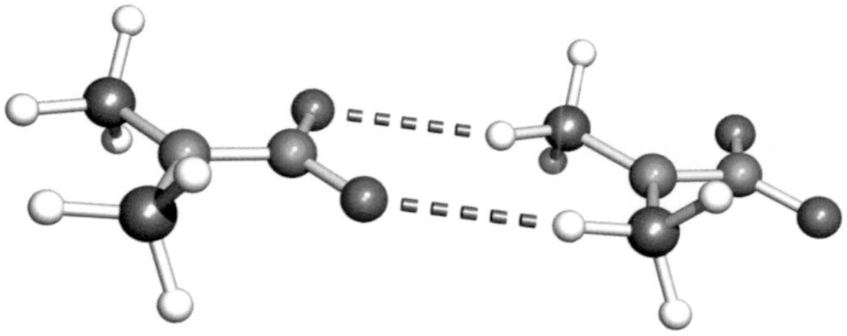

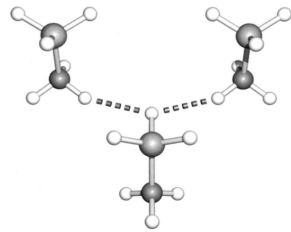

Fig. 10.11 Part of one of the hydrogen-bonded chains in the solid-state structure of Me_2NNO_2 determined by neutron diffraction [A. Filhol *et al.* (1980) *Acta Crystallogr., Sect. B*, vol. 36, p. 575]. Colour code: C, grey; N, blue; O, red; H, white.

Fig. 10.12 In the solid state structure of $H_3B\cdot NH_3$, there are close $N-H^{\delta+}\cdots H^{\delta-}-B$ contacts (202 pm), represented in the figure by grey hashed lines. The structure was determined by neutron diffraction [W.T. Klooster *et al.* (1999) *J. Am. Chem. Soc.*, vol. 121, p. 6337]. Colour code: B, orange; N, blue; H, white.

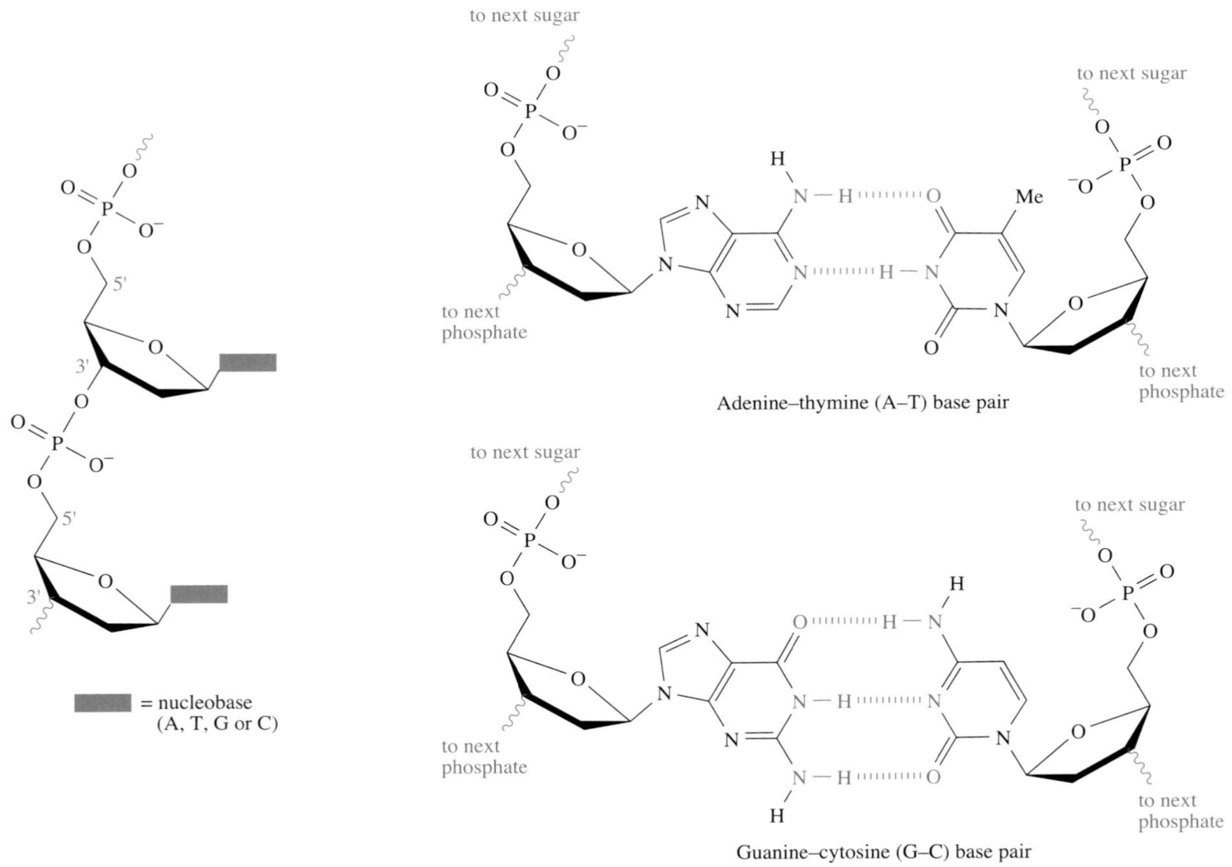

Fig. 10.13 The left-hand diagram shows two units in one strand of DNA; DNA is composed of condensed deoxyribonucleotides and the four possible nucleobases are adenine (A), guanine (G), cytosine (C) and thymine (T). The right-hand diagrams illustrate how complementary base pairs in adjacent strands in DNA interact through hydrogen bonding. (See also Fig. 10.16, p. 323.)

contacts of 202 pm being significantly shorter than the sum of the van der Waals radii of two H atoms (240 pm, see footnote to Appendix 6). Density functional theory (DFT, see Section 4.13) has been used to estimate a value of $\approx 13\ \mathrm{kJ\,mol^{-1}}$ for the $H^{\delta+} \cdots H^{\delta-}$ interaction in solid $H_3B \cdot NH_3$.

Later in the book, we encounter other examples of solid state structures that involve hydrogen bonding. Among these are host–guest systems called *clathrates* in which hydrogen-bonded host molecules form cage structures that encapsulate guest molecules. Examples are given in Fig. 12.9, Section 17.4, Fig. 18.1 and Box 14.5.

Hydrogen bonding in biological systems

Hydrogen bonding is of immense importance in biological systems. One of the best known examples is the formation of the double helical structure of DNA (deoxyribonucleic acid). The structures of adenine and thymine are exactly matched to permit hydrogen bonding between them, and they are referred to as complementary bases. Guanine and cytosine form the second base pair (Fig. 10.13). The hydrogen bonding between these base pairs in the strands of DNA

leads to the assembly of the double helix (see end-of-chapter problem 10.18).[†]

Worked example 10.1 Hydrogen bonding

In which of the following mixtures of solvents will there be intermolecular hydrogen bonding between the different solvent molecules: (a) Et₂O and THF; (b) EtOH and H₂O; (c) EtNH₂ and Et₂O? Give diagrams to show the likely hydrogen-bonded interactions.

In each pair of molecules, look for (i) an electronegative atom in each molecule, and (ii) an H atom attached directly to an electronegative atom in one of the molecules.

(a) Et₂O and THF

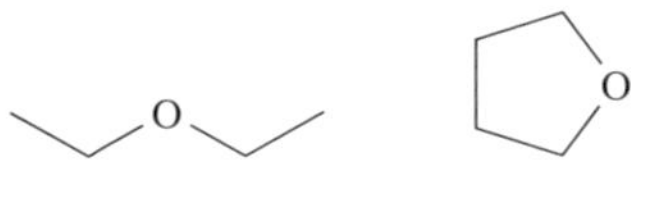

No hydrogen bonding is likely.

[†] For a discussion of DNA, see: C.K. Mathews, K.E. van Holde and K.G. Ahern (2000) *Biochemistry*, 3rd edn, Benjamin/Cummings, New York, Chapter 4.

(b) EtOH and H_2O

Hydrogen bonding is possible:

or

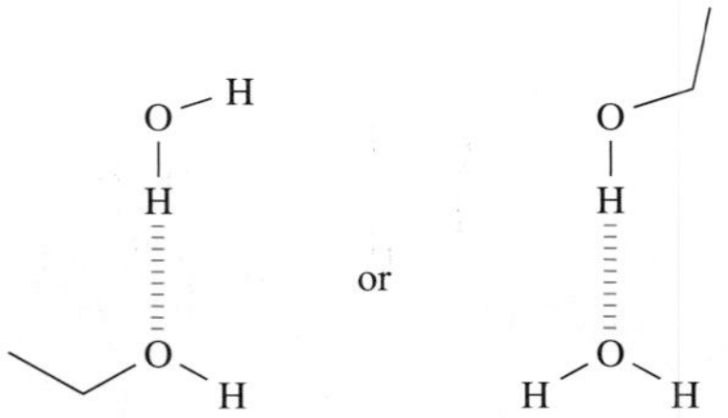

(c) $EtNH_2$ and Et_2O

Hydrogen bonding is possible:

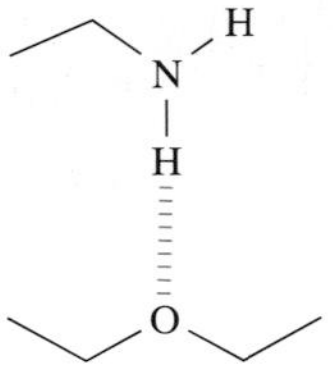

Self-study exercises

1. Suggest why $EtNH_2$ and EtOH are miscible.

2. Suggest how the solid state structure of benzene-1,4-dicarboxylic acid is affected by hydrogen bonding.

[*Ans.* See Box 10.4]

3. Suggest why CH_3CO_2H exists mainly as dimers in hexane, but as monomers in water.

Self-study exercise

The ligand, L:

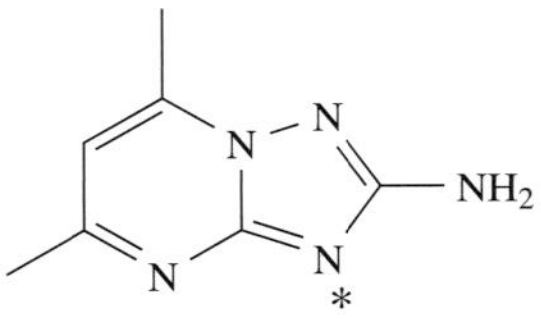

forms an octahedral complex [RuCl$_3$L$_2$(OH$_2$)] in which L coordinates through the N* atom shown above. The chlorido ligands are in a *mer*-arrangement, and the two L ligands are *trans* to one another. The solid state structure of the complex shows an unusual 4-fold intramolecular hydrogen bonding motif. Draw the structure of the complex and suggest how the hydrogen-bonding interactions arise.

[*Ans.* See: A.H. Velders *et al.* (1999) *Eur. J. Inorg. Chem.*, p. 213.]

10.7 Binary hydrides: classification and general properties

Detailed chemistries of most of the hydrides are considered in later chapters.

Classification

The four major classes into which it is convenient to place binary hydrides are:

- metallic;
- saline (salt-like);
- molecular;
- covalent, with extended structures.

A number of hydrides fall into intermediate or borderline categories.

Metallic hydrides

Hydrogen atoms are small enough to occupy the interstitial holes in a metal lattice and the absorption of H_2 by a variety of metals (and also alloys) leads to the formation of metal hydrides in which hydrogen atoms reside in interstitial cavities. In these so-called *metallic* (or *interstitial*) *hydrides*, the metal–hydrogen bonding must compensate for the dissociation of the H–H bond in H_2, and for the expansion of the metal lattice (see below). Non-stoichiometric hydrides $TiH_{1.7}$, $HfH_{1.98}$ and $HfH_{2.10}$ are formed when titanium and hafnium react with H_2. Niobium forms a series of non-stoichiometric hydrides of formula NbH_x ($0 < x \leq 1$) and at low hydrogen content, the bcc structure of Nb metal is retained. An interesting property of these metal hydrides is their ability to release hydrogen upon heating, and this leads to their use as 'hydrogen storage vessels' (see the bar chart in Box 10.2). Nickel–metal hydride batteries rely on hydrogen storage alloys (see Box 10.5).

Palladium is unique in its ability to reversibly absorb large amounts of H_2 or D_2 (but no other gases). The metal can absorb up to 900 times its own volume of H_2 at ambient temperatures. Neutron diffraction studies indicate that the absorbed H occupies octahedral holes in the cubic close-packed Pd lattice. At room temperature, there are two phases of PdH_x. The α-phase contains a low concentration of hydrogen ($x \approx 0.01$) while for the β-phase, $x \approx 0.6$. The different unit cell dimensions for the two phases (389.0 pm for α-PdH_x, and 401.8 pm for β-PdH_x) confirm that the metal lattice expands as the hydrogen content increases. The absorbed hydrogen has a high mobility within the metal. The high selectivity and permeability of palladium-based membranes allow them to be used for separating and purifying H_2, e.g. the ultra-purification of H_2 in the semiconductor industry.[†] Although the use of thin membranes is often favoured, the expansion of the metal lattice that accompanies the α- to β-PdH_x phase transition at 293 K and 20 bar leads to membrane embrittlement.

[†] For a detailed review, see: S.N. Paglieri and J.D. Way (2002) *Separation and Purification Methods*, vol. 31, p. 1 – 'Innovations in palladium membrane research'.

APPLICATIONS

Box 10.5 Nickel–metal hydride batteries

The property of metal hydrides to 'store' hydrogen has been applied to battery technology, and, during the 1980s and 1990s, led to the development of the nickel–metal hydride (NiMH) cell. The NiMH battery uses a metal alloy such as $LaNi_5$ or $M'Ni_5$ where M' is 'misch metal' (typically an alloy of La, Ce, Nd and Pr, see Table 27.1) which can absorb hydrogen and store it as a hydride, e.g. $LaNi_5H_6$. The Ni component of the alloy typically has Co, Al and Mn additives. The metal alloy forms the cathode in an NiMH battery. The anode is made from $Ni(OH)_2$, and the electrolyte is 30% aqueous KOH. The cathode is charged with hydrogen after it is manufactured in its final form. The cell operation can be summarized as follows:

Anode: $$Ni(OH)_2 + [OH]^- \underset{\text{Discharge}}{\overset{\text{Charge}}{\rightleftharpoons}} NiO(OH) + H_2O + e^-$$

Cathode: $$M + H_2O + e^- \underset{\text{Discharge}}{\overset{\text{Charge}}{\rightleftharpoons}} MH + [OH]^-$$

Overall: $$Ni(OH)_2 + M \underset{\text{Discharge}}{\overset{\text{Charge}}{\rightleftharpoons}} NiO(OH) + MH$$

The 'MH' initially formed at the cathode contains hydrogen atoms adsorbed on the surface of the misch metal. Adsorption is followed by surface penetration and diffusion of the hydrogen atoms into the hydrogen storage alloy. The battery recycles hydrogen back and forth between anode and cathode, and can be charged and discharged about 500 times. During charging, hydrogen moves from anode to cathode and is stored in the metal alloy. During discharge, hydrogen is liberated from the alloy, moving from cathode to anode. The designs and discharge characteristics of the NiMH and NiCd batteries (see Section 21.2) are similar, but the newer NiMH batteries are gradually replacing NiCd cells in portable electronic devices such as laptop computers and mobile phones. An NiMH cell has $\approx 40\%$ higher electrical capacity than a NiCd cell operating at the same voltage, and a NiMH battery does not generate hazardous waste, whereas Cd is toxic.

Hybrid electric vehicles (HEV) such as the Toyota Prius, Honda Insight and Honda Civic Hybrid combine a conventional internal combustion engine with a battery-powered motor (see also the discussion at the end of Box 10.2). Nickel–metal hydride batteries are routinely applied in hybrid electric vehicles, although there is increasing competition from

Cut away model of the Toyota Prius hybrid saloon car at the 2009 Frankfurt Motor Show.

lithium-ion batteries, e.g. in the Mercedes-Benz S400 Blue Hybrid, launched in 2010 (see Box 11.3). The Toyota Prius uses a sealed NiMH battery comprising 168×1.2 V cells connected in series to provide a 201.6 V output. A similar array is used in the Honda HEVs. Battery charging is provided by a combination of power from the internal combustion engine and a regenerative braking system. In conventional braking systems, kinetic energy produced during braking is wasted as heat energy. In a regenerative braking system, kinetic energy is converted into electrical energy in the NiMH battery.

Further reading

For discussions of NiMH battery recycling, see:
J.A.S. Tenório and D.C.R. Espinosa (2002) *J. Power Sources*, vol. 108, p. 70.
T. Müller and B. Friedrich (2006) *J. Power Sources*, vol. 158, p. 1498 – 'Development of a recycling process for nickel–metal hydride batteries'.

For a detailed account of hydrogen storage alloys, see:
X. Zhao and L. Ma (2009) *Int. J. Hydrogen Energy*, vol. 34, p. 4788 – 'Recent progress in hydrogen storage alloys for nickel/metal hydride secondary batteries'.

In 1996, a report appeared in *Nature* of experiments in which a 500 nm thick film of yttrium (coated with a 5–20 nm layer of palladium to prevent aerial oxidation) was subjected to 10^5 Pa pressure of H_2 gas at room temperature.[†]

[†] For details of these observations and photographs depicting the mirror to non-reflector transitions, see: J.N. Huiberts, R. Griessen, J.H. Rector, R.J. Wijngaarden, J.P. Dekker, D.G. de Groot and N.J. Koeman (1996) *Nature*, vol. 380, p. 231.

As H_2 diffused through the Pd layer, the latter catalysed the dissociation of H_2 into H atoms which then entered the yttrium lattice. A series of observations followed:

- initially the yttrium film was a reflecting surface, i.e. a mirror;
- a few minutes after H atoms entered the lattice, a partially reflecting surface was observed and this was attributed to the formation of YH_2;

• after more hydrogen had been taken up and a composition of $YH_{2.86}$ had been reached, the surface became yellow and transparent.

These remarkable changes are reversible. The accommodation of the H atoms within the metal lattice is not simple, because the lattice of yttrium atoms undergoes a phase transition from an initially fcc to hcp structure. The fcc lattice is present in the β-YH_2 phase.

Saline hydrides

Saline hydrides are formed when the group 1 or 2 metals (except Be) are heated with H_2. All are white, high melting solids (e.g. LiH, mp = 953 K; NaH, mp = 1073 K with decomposition). The group 1 hydrides crystallize with the NaCl structure, and the presence of the H^- ion (see Section 10.2) is indicated by the good agreement between lattice energies obtained from Born–Haber cycles and from X-ray and compressibility data. Additional evidence comes from the fact that the electrolysis of molten LiH liberates H_2 at the *anode* (eq. 10.33).

$$\left.\begin{array}{ll} 2H^- \longrightarrow H_2 + 2e^- & \textit{at the anode} \\ Li^+ + e^- \longrightarrow Li & \textit{at the cathode} \end{array}\right\} \quad (10.33)$$

The reactivity of the group 1 hydrides increases with an increase in atomic number and ionic size of the metal. In keeping with this, values of $\Delta_f H^\circ$ become less negative, with that of LiH being significantly more negative than those of the other alkali metal hydrides. Table 10.5 lists factors that contribute towards this trend. Since the hydride ion is a common factor in the series, we need to look at the extent to which the value of $\Delta_{lattice} H^\circ$ offsets the sum of $\Delta_a H^\circ$ and IE_1 in order to reconcile the trend in values of $\Delta_f H^\circ$ (scheme 10.34). The H^- ion is similar in size to F^-, and thus the trend parallels that observed for alkali metal fluorides.

$$(10.34)$$

Saline hydrides react immediately with protic solvents such as H_2O (eq. 10.35), NH_3 or EtOH, showing that the H^- ion is an extremely strong base. Widespread use is made of NaH and KH as deprotonating agents (e.g. reaction 10.36).

Table 10.5 Values of the $\Delta_f H^\circ$(298 K) of the alkali metal hydrides, MH, depend upon the relative magnitudes of $\Delta_a H^\circ$(298 K) and IE_1 of the metals, and the lattice energies, $\approx\Delta_{lattice} H^\circ$(298 K), of MH.

Metal	$\Delta_a H^\circ(M)$ / kJ mol^{-1}	$IE_1(M)$ / kJ mol^{-1}	$\Delta_{lattice} H^\circ$ / kJ mol^{-1}	$\Delta_f H^\circ(MH)$ / kJ mol^{-1}
Li	161	521	−920	−90.5
Na	108	492	−808	−56.3
K	90	415	−714	−57.7
Rb	82	405	−685	−52.3
Cs	78	376	−644	−54.2

$$NaH + H_2O \longrightarrow NaOH + H_2 \quad (10.35)$$

$$Ph_2PH + NaH \longrightarrow Na[PPh_2] + H_2 \quad (10.36)$$

Of the saline hydrides, LiH, NaH and KH are the most commonly used, but their moisture sensitivity means that reaction conditions must be water-free. Of particular significance are the reactions between LiH and Al_2Cl_6 to give lithium tetrahydridoaluminate(1−), $Li[AlH_4]$ (also called lithium aluminium hydride or *lithal*), and between NaH and $B(OMe)_3$ or BCl_3 (eqs. 10.37 and 10.38) to give sodium tetrahydridoborate(1−), commonly known as sodium borohydride (see Section 13.5). The compounds $Li[AlH_4]$, $Na[BH_4]$ and NaH are widely used as reducing agents, e.g. reactions 10.39 and 10.40.

$$4NaH + B(OMe)_3 \xrightarrow{520\,K} Na[BH_4] + 3NaOMe \quad (10.37)$$

$$4NaH + BCl_3 \longrightarrow Na[BH_4] + 3NaCl \quad (10.38)$$

$$ECl_4 \xrightarrow{Li[AlH_4]} EH_4 \qquad E = Si,\ Ge\ or\ Sn \quad (10.39)$$

$$[ZnMe_4]^{2-} \xrightarrow{Li[AlH_4]} [ZnH_4]^{2-} \quad (10.40)$$

Molecular hydrides and complexes derived from them

Covalent hydrides with discrete molecular structures are formed by the p-block elements in groups 13 to 17 with the exception of Al (see Section 13.5) and Bi. BiH_3 is thermally unstable, decomposing above 198 K. Hydrides of the halogens, sulfur and nitrogen are prepared by reacting these elements with H_2 under appropriate conditions (e.g. reaction 10.28). The remaining hydrides are formed by treating suitable metal salts with water, aqueous acid or NH_4Br in liquid NH_3, or by use of $[BH_4]^-$ or $[AlH_4]^-$, e.g. reaction 10.39. Specific syntheses are given in later chapters.

Most molecular hydrides are volatile and have simple structures which comply with the VSEPR model (see Section 2.8). However, BH_3, **10.10**, although known in the gas phase, dimerizes to give B_2H_6, **10.11**, and GaH_3 behaves similarly (see Section 13.5).

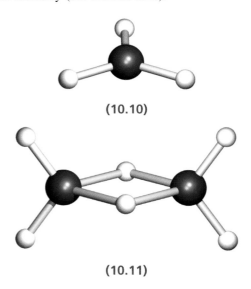

(10.10)

(10.11)

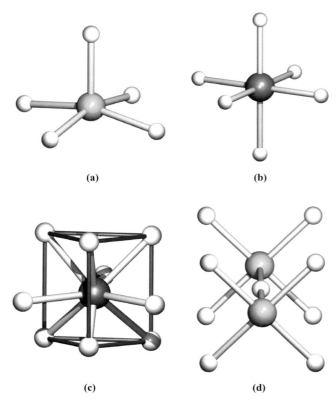

Fig. 10.14 The structures of (a) $[CoH_5]^{4-}$, (b) $[FeH_6]^{4-}$, (c) $[ReH_9]^{2-}$ and (d) $[Pt_2H_9]^{5-}$.

Anionic molecular hydrido complexes of *p*-block elements include tetrahedral $[BH_4]^-$ and $[AlH_4]^-$. Both $LiAlH_4$ and $NaAlH_4$ slowly decompose to give Li_3AlH_6 and Na_3AlH_6, respectively, and Al. Because it is difficult to locate H atoms in the presence of heavy atoms (see Section 4.11), it is common to determine structures of deuterated analogues. Both Li_3AlD_6 and Na_3AlD_6 contain isolated octahedral $[AlD_6]^{3-}$ ions. The solid state reaction of BeD_2 (see Fig. 10.15) with two equivalents of LiH at 833 K and 3 GPa pressure produces Li_2BeD_4. Neutron and X-ray diffraction data confirm the presence of tetrahedral $[BeD_4]^{2-}$ anions.

Molecular hydrido complexes are known for *d*-block metals from groups 7–10 (excluding Mn)[†] and counterions are commonly from group 1 or 2, e.g. K_2ReH_9, Li_4RuH_6, Na_3RhH_6, Mg_2RuH_4, Na_3OsH_7 and Ba_2PtH_6. In the solid state structures of these compounds (the determination of which typically uses deuterated analogues), isolated metal hydrido anions are present with cations occupying the cavities between them. The $[NiH_4]^{4-}$ ion in Mg_2NiH_4 is tetrahedral. X-ray diffraction data have confirmed a square-based pyramidal structure for $[CoH_5]^{4-}$ (Fig. 10.14a), and $[IrH_5]^{4-}$ adopts an analogous structure. These pentahydrido complexes have been isolated as the salts Mg_2CoH_5 and M_2IrH_5 (M = Mg, Ca or Sr). Alkaline earth metal ions

have also been used to stabilize salts containing octahedral $[FeH_6]^{4-}$, $[RuH_6]^{4-}$ and $[OsH_6]^{4-}$ (Fig. 10.14b). Isolated H^- and octahedral $[ReH_6]^{5-}$ ions are present in Mg_3ReH_7. However, in the solid state, Na_3OsH_7 and Na_3RuH_7 contain pentagonal bipyramidal $[OsH_7]^{3-}$ and $[RuH_7]^{3-}$ anions, respectively. The reaction of $Na[ReO_4]$ with Na in EtOH yields Na_2ReH_9, and the K^+ and $[Et_4N]^+$ salts have been prepared by *metathesis* from Na_2ReH_9. The hydrido complex K_2TcH_9 can be made from the reaction of $[TcO_4]^-$ and potassium in EtOH in the presence of 1,2-ethanediamine.

A ***metathesis reaction*** involves an exchange, for example:
$$AgNO_3 + NaCl \longrightarrow AgCl + NaNO_3$$

Neutron diffraction data for $K_2[ReH_9]$ confirm a 9-coordinate Re atom in a tricapped trigonal prismatic environment (Fig. 10.14c) and $[TcH_9]^{2-}$ is assumed to be similar. Despite there being two H environments in $[ReH_9]^{2-}$, only one signal is observed in the solution ^{1}H NMR spectrum, indicating that the dianion is stereochemically non-rigid on the NMR spectroscopic timescale (see Section 4.8). Palladium(II) and platinum(II) form the square planar $[PdH_4]^{2-}$ and $[PtH_4]^{2-}$. The salt $K_2[PtH_4]$ is made by reacting Pt with KH under H_2 (1–10 bar, 580–700 K). 'K_3PtH_5' also

[†] For theoretical insight into $[MnH_9]^{2-}$, see: M. Gupta, R.P. Gupta and D.J. Singh (2009) *Phys. Rev. B*, vol. 80, article 235103.

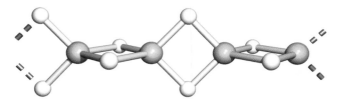

Fig. 10.15 Part of the polymeric chain structure of BeH_2; Be atoms are shown in yellow.

forms in this reaction, but structural data show that this contains $[PtH_4]^{2-}$ and H^- ions. A high pressure of H_2 is also needed to form $Li_5[Pt_2H_9]$ but, once formed, it is stable with respect to H_2 loss; the $[Pt_2H_9]^{5-}$ ion is shown in Fig. 10.14d. The Pt(IV) complex $K_2[PtH_6]$ results if KH

and Pt sponge are heated (775 K) under 1500–1800 bar H_2; neutron diffraction confirms that deuterated $[PtD_6]^{2-}$ is octahedral. The linear $[PdH_2]^{2-}$ ion is present in Na_2PdH_2 and Li_2PdH_2, and contains Pd(0). The reaction of KH with Pd sponge at 620 K yields a compound of formula K_3PdH_3; neutron diffraction data show that this contains isolated H^- and linear $[PdH_2]^{2-}$ ions.

Covalent hydrides with extended structures

Polymeric hydrides (white solids) are formed by Be and Al. In BeH_2 (Fig. 10.15), each Be centre is tetrahedral, giving a chain structure in which multi-centre bonding of the type described for B_2H_6 is present. The structure of AlH_3 consists of an infinite lattice, in which each Al(III) centre is in an AlH_6-octahedral site; H atoms bridge pairs of Al centres.

KEY TERMS

The following terms were introduced in this chapter. Do you know what they mean?

- hydrogen ion (proton)
- oxonium ion
- hydrate
- solvent of crystallization
- hydride ion
- protium
- deuterium
- tritium
- deuterium labelling
- passivate

- synthesis gas (syngas)
- water–gas shift reaction
- heterogeneous catalyst
- homogeneous catalyst
- hydrogen economy
- hydrogen fuel cell
- hydrogen bonding
- asymmetrical hydrogen bond
- symmetrical hydrogen bond

- anomalous properties of HF, H_2O and NH_3
- Trouton's rule
- binary compound
- metallic (interstitial) hydride
- saline (salt-like) hydride
- molecular hydride
- polymeric hydride
- metathesis

FURTHER READING

Hydrogen: the clean fuel

U. Eberle, M. Felderhoff and F. Schüth (2009) *Angew. Chem. Int. Ed.*, vol. 48, p. 6608 – 'Chemical and physical solutions for hydrogen storage'.

J. Graetz (2009) *Chem. Soc. Rev.*, vol. 38, p. 73 – 'New approaches to hydrogen storage' – A review within a themed issue on renewable energy.

L. Hammarström and S. Hammes-Schiffer eds. (2009) *Acc. Chem. Res.*, vol. 42, issue 12 – A themed issue of the journal dealing with artificial photosynthesis and solar cells.

K. Kalyanasundaram and M. Graetzel (2010) *Curr. Opinion Biotech.*, vol. 21, p. 298 – 'Artificial photosynthesis: Biomimetic approaches to solar energy conversion and storage'.

W. Lubitz and B. Tumas eds. (2007) *Chem. Rev.*, vol. 107, issue 10 – An issue of this journal dedicated to reviews on the theme of hydrogen.

C.E.S. Thomas (2009) *Int. J. Hydrogen Energy*, vol. 34, p. 9279 – 'Transportation options in a carbon constrained world: Hybrids, plug-in hybrids, biofuels, fuel cell electric vehicles and battery electric vehicles'.

J. Tollefson (2010) *Nature*, vol. 464, p. 1262 – 'Fuel of the future?'

Hydrogen bonding

A.D. Buckingham, J.E. Del Bene and S.A.C. McDowell (2008) *Chem. Phys. Lett.*, vol. 463, p. 1 – An overview of structural and vibrational spectroscopic properties of hydrogen bonds, and of hydrogen bonding between H_2O molecules.

G. Desiraju and T. Steiner (1999) *The Weak Hydrogen Bond in Structural Chemistry and Biology*, Oxford

University Press, Oxford – A well-illustrated and refer-enced account of modern views of hydrogen bonding.

G.R. Desiraju (2005) *Chem. Commun.*, p. 2995 – 'C–H····O and other weak hydrogen bonds. From crystal engineering to virtual screening'.

P. Gilli, V. Bertolasi, V. Ferretti and G. Gilli (1994) *J. Am. Chem. Soc.*, vol. 116, p. 909 – 'Covalent nature of the strong homonuclear hydrogen bond. Study of the O–H····O system by crystal structure correlation methods'.

G.A. Jeffery (1997) *An Introduction to Hydrogen Bonding*, Oxford University Press, Oxford – A text that introduces modern ideas on hydrogen bonding.

T. Steiner (2002) *Angew. Chem. Int. Ed.*, vol. 41, p. 48 – An excellent review of hydrogen bonding in the solid state.

Metal hydrides

W. Grochala and P.P. Edwards (2004) *Chem. Rev.*, vol. 104, p. 1283 – 'Thermal decomposition of the non-interstitial hydrides for the storage and production of hydrogen'.

I.P. Jain, P. Jain and A. Jain (2010) *J. Alloys Compd.*, vol. 503, p. 303 – A review of lightweight metal hydrides as hydrogen storage materials.

PROBLEMS

10.1 Confirm that the difference in values of $\bar{\nu}(O-H)$ and $\bar{\nu}(O-D)$ given in Table 10.2 is consistent with the isotopic masses of H and D.

10.2 (a) Outline the reasons why it is necessary to use deuterated solvents in ^1H NMR spectroscopy. (b) Draw the structures of THF-d_8 and DMF-d_7.

10.3 For deuterium, $I = 1$. In a fully labelled sample of $CDCl_3$, what is observed in the ^{13}C NMR spectrum?

10.4 In ^1H NMR spectra in which the solvent is acetonitrile-d_3, labelled to an extent of 99.6%, a multiplet is observed at δ 1.94 ppm. How does this multiplet arise, and what is its appearance? [D, $I = 1$; H, $I = \frac{1}{2}$]

10.5 How would you attempt to prepare a sample of pure HD and to establish the purity of the product?

10.6 The IR spectrum of a $0.01 \, mol \, dm^{-3}$ solution of *tert*-butanol in CCl_4 shows a sharp peak at $3610 \, cm^{-1}$; in the IR spectrum of a similar $1.0 \, mol \, dm^{-3}$ solution, this absorption is much diminished in intensity, but a very strong, broad peak at $3330 \, cm^{-1}$ is observed. Rationalize these observations.

10.7 Suggest an explanation for the fact that solid CsCl, but not LiCl, absorbs HCl at low temperatures.

10.8 Suggest a structure for the $[H_9O_4]^+$ ion.

10.9 Write a brief, but critical, account of 'the hydrogen bond'.

10.10 (a) Write equations for the reactions of KH with NH_3 and with ethanol. (b) Identify the conjugate acid–base pairs in each reaction.

10.11 Write equations for the following processes, noting appropriate conditions:

(a) electrolysis of water;
(b) electrolysis of molten LiH;
(c) CaH_2 reacting with water;
(d) Mg treated with dilute nitric acid;
(e) combustion of H_2;
(f) reaction of H_2 with CuO.

10.12 Solutions of H_2O_2 are used as bleaching agents. For the decomposition of H_2O_2 to H_2O and O_2, $\Delta G^o = -116.7 \, kJ \, mol^{-1}$. Why can H_2O_2 be stored for periods of time without significant decomposition?

10.13 Magnesium hydride possesses a rutile lattice. (a) Sketch a unit cell of rutile. (b) What are the coordination numbers and geometries of the Mg and H centres in this structure?

10.14 Confirm the stoichiometry of aluminium hydride as 1:3 from the text description of the infinite structure.

10.15 Discuss the bonding in BeH_2 in terms of a suitable hybridization scheme. Relate this to a bonding description for Ga_2H_6.

10.16 Suggest explanations for the following trends in data.

(a) In gas-phase CH_4, NH_3 and H_2O, $\angle H-C-H = 109.5°$, $\angle H-N-H = 106.7°$ and $\angle H-O-H = 104.5°$.
(b) The dipole moments (in the gas phase) of NH_3 and NH_2OH are 1.47 and 0.59 D.
(c) The ratios of $\Delta_{vap}H$:bp for NH_3, N_2H_4, PH_3, P_2H_4, SiH_4 and Si_2H_6 are, respectively 97.3, 108.2, 78.7, 85.6, 75.2 and $81.9 \, J \, K^{-1} \, mol^{-1}$. However, for HCO_2H, the ratio is $60.7 \, J \, K^{-1} \, mol^{-1}$.

10.17 The structures of [NMe$_4$][HF$_2$] and [NMe$_4$][H$_2$F$_3$] have been determined by X-ray diffraction. The table below shows selected structural data; all F–H–F angles are between 175 and 178°.

Parameter	[NMe$_4$][HF$_2$]	[NMe$_4$][H$_2$F$_3$]
F–H distances	112.9/112.9 pm	89/143 pm
F---F---F angles	–	125.9°

From the data given, draw the structures of the anions in [NMe$_4$][HF$_2$] and [NMe$_4$][H$_2$F$_3$], and say what you can about the bonding in these species.

10.18 Using the information in Figs. 10.13 and 10.16, explain how the two oligonucleotides 5'-CAAAGAAAAG-3' and 5'-CTTTTCTTTG-3' assemble into a double helical structure (see Fig. 10.13 for the 3' and 5' numbering, and definitions of C, A, G and T).

10.19 (a) KMgH$_3$ crystallizes with a CaTiO$_3$-type structure. Draw a diagram to show a unit cell of KMgH$_3$. What is the coordination number of each atom?

(b) Calculate a value of $\Delta_{lattice}H°$ (KMgH$_3$, 298 K) given that the standard enthalpy of formation of KMgH$_3$(s) (298 K) is –278 kJ mol^{-1}.

Fig. 10.16 Two strands of oligonucleotides sequenced 5'-CAAAGAAAAG-3' and 5'-CTTTTCTTTG-3' assemble into a double helix. The structure has been determined by X-ray diffraction [M. L. Kopka *et al.* (1996) *J. Mol. Biol.*, vol. 334, p. 653]. The backbone of each oligonucleotide is depicted as an arrow pointing towards the C3' end of the sequence, and the nucleobases are shown in a 'ladder' representation. The nucleobases are colour coded: G, green; A, red; C, purple; T, turquoise.

OVERVIEW PROBLEMS

10.20 (a) Use data in Appendix 11 to give a quantitative explanation why H$_2$ can be prepared from the reaction of Zn with dilute mineral acid, but not from Cu with a dilute acid.

(b) The ion [H$_{13}$O$_6$]$^+$ can exist in more than one isomeric form. One that has been structurally characterized is described in terms of [(H$_5$O$_2$)(H$_2$O)$_4$]$^+$, in which an [H$_5$O$_2$]$^+$ unit containing a strong hydrogen bond is centrally positioned within the [H$_{13}$O$_6$]$^+$ ion. Draw a schematic representation of this ion and give a description of the bonding within it.

(c) The IR spectrum of gaseous SbH$_3$ shows absorptions at 1894, 1891, 831 and 782 cm^{-1}. Comment on why this provides evidence that SbH$_3$ has C_{3v} rather than D_{3h} symmetry.

10.21 (a) Given that the enthalpy change associated with the addition of H$^+$(g) to H$_2$O(g) is –690 kJ mol^{-1}, and $\Delta_{hyd}H°$(H$^+$, g) = –1091 kJ mol^{-1}, calculate the enthalpy change associated with the solvation of [H$_3$O]$^+$(g) in water.

(b) Outline how the nickel–metal hydride battery works, giving equations for the reactions at each electrode during charging and discharging.

10.22 (a) Sr$_2$RuH$_6$ crystallizes in a lattice that can be described in terms of the CaF$_2$ structure type with octahedral [RuH$_6$]$^{4-}$ ions replacing Ca^{2+} ions, and Sr^{2+} ions replacing F$^-$ ions. Sketch a unit cell of CaF$_2$. Show that in Sr$_2$RuH$_6$, each [RuH$_6$]$^{4-}$ ion is surrounded by eight Sr^{2+} ions in a cubic array.

(b) Suggest products for the following reactions:

$$SiCl_4 + LiAlH_4 \longrightarrow$$
$$Ph_2PH + KH \longrightarrow$$
$$4LiH + AlCl_3 \xrightarrow{Et_2O}$$

10.23 The first list below contains the formula of a hydride. Each has a 'partner' in the second list of phrases. Match the 'partners'; there is only one match for each pair. Structural descriptions refer to the solid state.

List 1	List 2
BeH_2	3D lattice with octahedral metal centres
$[PtH_4]^{2-}$	Non-stoichiometric hydride
NaH	M(0) complex
$[NiH_4]^{4-}$	Polymeric chain
$[PtH_6]^{2-}$	M(IV) complex
$[TcH_9]^{2-}$	Tricapped trigonal prismatic hydrido complex
$HfH_{2.1}$	Square planar complex
AlH_3	Saline hydride

10.24 Suggest explanations for the following observations.

(a) Ammonium fluoride forms solid solutions with ice.

(b) The viscosity decreases along the series of liquids $H_3PO_4 > H_2SO_4 > HClO_4$.

(c) Formic (methanoic) acid has a Trouton constant of $60.7 \, J \, K^{-1} \, mol^{-1}$.

(d) pK_a values for fumaric acid and its geometrical isomer maleic acid are:

	$pK_a(1)$	$pK_a(2)$
Fumaric acid	3.02	4.38
Maleic acid	1.92	6.23

Fumaric acid

INORGANIC CHEMISTRY MATTERS

10.25 In vehicles, the combustion of H_2 rather than a hydrocarbon-based fuel reduces both CO_2 emissions and a dependence on fossil fuels. (a) Using data from Appendix 12, show that the combustion of H_2 releases $120 \, kJ \, g^{-1}$. (b) The chart in Box 10.2 shows that the stored energy per unit mass of compressed (at 35 MPa) H_2 gas and of liquid H_2 are the same, but that liquid H_2 stores more energy per unit volume than compressed H_2. Rationalize these data. (c) 3 kg of gasoline (petrol) is equivalent to 1 kg of H_2 in terms of stored energy. Comment on this fact in terms of the practical application of H_2 as a fuel in a family saloon car.

10.26 $NaAlH_4$ is among lightweight metal hydrides being investigated as a means of storing hydrogen, e.g. for fuel cell applications. Decomposition occurs in three steps upon heating:

$$3NaAlH_4 \longrightarrow Na_3AlH_6 + 2Al + 3H_2 \quad (1)$$

$$Na_3AlH_6 \longrightarrow 3NaH + Al + 1.5H_2 \quad (2)$$

$$3NaH \longrightarrow 3Na + 1.5H_2 \quad (3)$$

(a) Calculate the H content of $NaAlH_4$ as a wt %. (b) Step (3) occurs above 670 K and this limits the practical dehydrogenation steps to (1) and (2). What is the hydrogen storage capacity (in wt %) of $NaAlH_4$ if only steps (1) and (2) are considered? (c) Unfortunately, the kinetics of the dehydrogenation of $NaAlH_4$ militate against practical applications as a hydrogen storage material, but doping the material with Ti improves the kinetics both of dehydrogenation and rehydrogenation. Comment on this statement in terms of the role of the dopant, and the need for both dehydrogenation and rehydrogenation to be viable processes.

10.27 (a) Describe the structure of the ordinary phase of ice. (b) Cow's milk is composed of >85% water. Explain why a carton of milk expands when it is frozen, but returns to its original size when allowed to warm to room temperature.

10.28 Typical compositions of nickel–metal hydride (NiMH) batteries are shown below:

Component	Button cell / % by mass	Cylindrical cell / % by mass
Ni	29–39	36–42
Fe	31–47	22–25
Co	2–3	3–4
La, Ce, Nd, Pr	6–8	8–10
Graphite	2–3	<1
K	1–2	1–2
H / O	8–10	15–17
Plastics	1–2	3–4
Other	2–3	2–3

[Data: T. Müller *et al.* (2006) *J. Power Sources*, vol. 158, p. 1498.]

(a) Write equations to show the processes at the anode and cathode during charging and discharging a NiMH cell. (b) What is the overall cell reaction? Confirm that the changes in oxidation states for reduction and oxidation reactions balance. (c) What role does the mixture of *f*-block elements in the battery play? (d) Why is Fe needed in the battery? (e) During battery recycling, suggest a method of recovering Co and Ni.

Topics

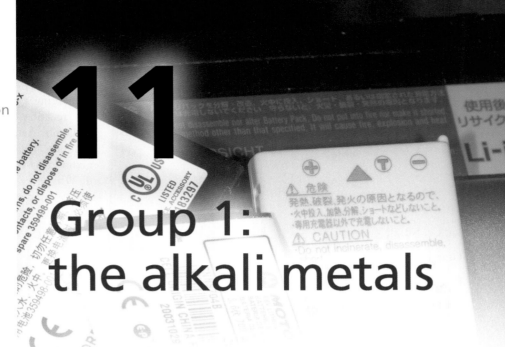

11
Group 1:
the alkali metals

1	2			13	14	15	16	17	18
H									He
Li	Be			B	C	N	O	F	Ne
Na	Mg			Al	Si	P	S	Cl	Ar
K	Ca	*d*-block		Ga	Ge	As	Se	Br	Kr
Rb	Sr			In	Sn	Sb	Te	I	Xe
Cs	Ba			Tl	Pb	Bi	Po	At	Rn
Fr	Ra								

11.1 Introduction

The alkali metals – lithium, sodium, potassium, rubidium, caesium and francium – are members of group 1 of the periodic table, and each has a ground state valence electronic configuration ns^1. Discussions of these metals usually neglect the heaviest member of the group, francium. The isotope ^{233}Fr occurs naturally, but only as the decay product of ^{227}Ac in uranium ores. The half-life of ^{233}Fr is 21.8 min, and it is estimated that, at a given moment, there is ≤ 30 g of francium in the Earth's crust. Isotopes of Fr can be prepared artificially in nuclear reactions, but have no practical applications.

We have already covered several aspects of the chemistry of the alkali metals as follows:

- ionization energies of metals (Section 1.10);
- structures of metal lattices (Section 6.3);

- metallic radii, r_{metal} (Section 6.5);
- melting points and standard enthalpies of atomization of metals (Section 6.6);
- ionic radii, r_{ion} (Section 6.10);
- NaCl and CsCl structure types (Section 6.11);
- energetics of the dissolution of MX (Section 7.9);
- standard reduction potentials, $E^{o}_{M^+/M}$ (Section 8.7);
- energetics of MX transfer from water to organic solvents (Section 9.3);
- alkali metals in liquid NH_3 (Section 9.6);
- saline hydrides, MH (Section 10.7).

11.2 Occurrence, extraction and uses

Occurrence

Sodium and potassium are abundant in the Earth's biosphere (2.6% and 2.4% respectively) but do not occur naturally in the elemental state. The main sources of Na and K are *rock salt* (almost pure NaCl), natural brines and seawater, *sylvite* (KCl), *sylvinite* (KCl/NaCl) and *carnallite* (KCl·MgCl$_2$· 6H$_2$O). The term 'potash' is often used to refer to a range of water-soluble potassium salts, both naturally occurring and manufactured (see Box 11.1). Other Na- and K-containing minerals such as borax (Na$_2$[B$_4$O$_5$(OH)$_4$]·8H$_2$O, see Sections 13.2 and 13.7) and Chile saltpetre (NaNO$_3$, see Section 15.2) are commercially important sources of other elements (e.g. B and N respectively). Unlike many inorganic chemicals, NaCl need not be manufactured since large natural deposits are available (Fig. 11.1). Evaporation of seawater yields a mixture of salts, but since NaCl represents the major component of the mixture, its production in this manner is a viable operation. Evaporation of inland

ENVIRONMENT

Box 11.1 Potassium salts: resources and commercial demand

In statistical tables of mineral production, 'potash' and 'K$_2$O equivalents' are listed. The term 'potash' refers to a variety of water-soluble, potassium-containing salts (KCl, KNO$_3$, NaNO$_3$/KNO$_3$ mixtures, K$_2$SO$_4$ and K$_2$SO$_4$·MgSO$_4$). Historically, the term was used for the water-soluble component of wood ash which consists of K$_2$CO$_3$ and KOH. However, much ambiguity surrounds the word. 'Potash' is used to refer to potassium carbonate and to potassium-containing fertilizers, while caustic potash typically refers to potassium hydroxide. Within agriculture, 'muriate of potash' is a mixture of KCl ($\geq$95%) and NaCl. The potash industry now defines a product's potassium content in terms of 'equivalent percentages of K$_2$O'.

Potash mine in Utah, USA.

Potash is obtained both by underground mining and from the evaporation of natural brines as illustrated in the photograph. World production of potash rose from 0.32 Mt in 1900 to 35 Mt in 2008, with the major producers being Canada, Russia, Belarus and Germany. About 95% of potash produced is destined for use in fertilizers. The potash market collapsed

to 25 Mt in 2009 as a consequence of the world economic crisis, but recovered again in 2010. Potassium is an essential element for the growth of plants, and the chart below illustrates the distribution of potash, phosphate and nitrogen-based fertilizers applied to five of the main crops in the US. Soybeans are legumes and contain nitrogen-fixing bacteria in root nodules and, therefore, fewer nitrogen-based fertilizers are required for this crop:

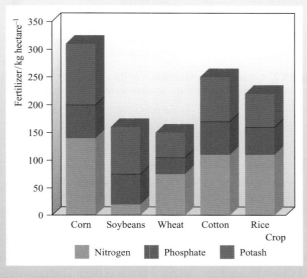

[Data: US Geological Survey Factsheet FS-155-99]

Major agricultural countries such as the US must import large amounts of potash to meet commercial needs. In 2008, imports of 5.8 Mt supplemented the 1.1 Mt of home-produced potash in the US. World reserves of potash are estimated to be approximately 250 billion tonnes.

Fig. 11.1 Salt pans and crude NaCl at Salin-de-Giraud in the Camargue, France. Evaporation of water uses solar energy.

saline seas and lakes is also a commercial source of NaCl. Lake Assal in Djibouti, Africa, is 155 m below sea level. It is fed by saline water from the Red Sea but has no water outlets. Evaporation driven by solar energy yields NaCl at a rate of 4 Mt a year. In contrast to Na and K, natural abundances of Li, Rb and Cs are small (% abundance Rb > Li > Cs). These metals occur as various silicate minerals, e.g. *spodumene* (LiAlSi$_2$O$_6$).

Extraction

Sodium is economically the most important of the alkali metals, and is manufactured by the Downs process in which molten NaCl is electrolysed:

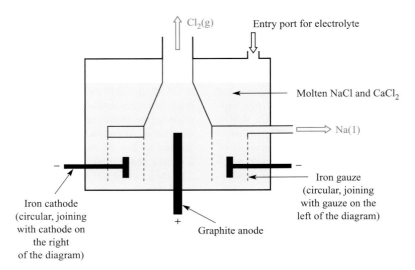

Fig. 11.2 A schematic representation of the electrolysis cell used in the Downs process to produce sodium commercially from NaCl. The products (Na and Cl_2) must be kept separate from each other to prevent recombination to form NaCl.

At the cathode:	$Na^+(l) + e^- \longrightarrow Na(l)$
At the anode:	$2Cl^-(l) \longrightarrow Cl_2(g) + 2e^-$
Overall reaction:	$2Na^+(l) + 2Cl^-(l) \longrightarrow 2Na(l) + Cl_2(g)$

$CaCl_2$ is added to reduce the operating temperature to about 870 K, since pure NaCl melts at 1073 K (see Section 9.12). The design of the electrolysis cell (Fig. 11.2) is critical to prevent reformation of NaCl by recombination of Na and Cl_2. Although the Downs process is the major manufacturing process for Na, the Cl_2 produced contributes only ≈5% of the world's supply. The remaining 95% is produced by the chloralkali process which involves the electrolysis of aqueous NaCl (see Box 11.4).

Lithium is extracted from LiCl in a similar electrolytic process. LiCl is first obtained from the silicate mineral spodumene ($LiAlSi_2O_6$) by heating it with CaO to give LiOH, which is then converted to the chloride. Potassium can be obtained electrolytically from KCl, but a more efficient method of extraction is the action of Na vapour on molten KCl in a counter-current fractionating tower. This yields an Na–K alloy which can be separated into its components by distillation. Similarly, Rb and Cs can be obtained from RbCl and CsCl, small quantities of which are produced as by-products from the extraction of Li from spodumene.

Small amounts of Na, K, Rb and Cs can be obtained by thermal decomposition of their azides (eq. 11.1). An application of NaN_3 is in car airbags (see eq. 15.6). Lithium cannot be obtained from an analogous reaction because the products recombine, yielding the nitride, Li_3N (see eq. 11.6).

$$2NaN_3 \xrightarrow{570\,K} 2Na + 3N_2 \qquad (11.1)$$

Major uses of the alkali metals and their compounds

Lithium has the lowest density ($0.53\,\mathrm{g\,cm^{-3}}$) of all known metals. It is used in the manufacture of alloys, and in certain glasses and ceramics. Lithium carbonate is used in the treatment of bipolar (manic-depressive) disorders, although large amounts of lithium salts damage the central nervous system.

Sodium, potassium and their compounds have many uses of which selected examples are given here. Sodium–potassium alloy is used as a heat-exchange coolant in nuclear reactors. A major use of Na–Pb alloy was in the production of the anti-knock agent $PbEt_4$, but the current demand for unleaded fuels now renders this of minimal importance. The varied applications of compounds of Na include those in the paper, glass, detergent, chemical and metal industries. Figure 11.3 summarizes uses of NaCl and Na_2CO_3. In 2008, the world production of NaCl was 258 Mt. Of this, 47.6 Mt were produced and 60.5 Mt consumed in the US. The major consumption of NaCl is in the manufacture of NaOH, Cl_2 (see Box 11.4) and Na_2CO_3 (see Section 11.7). A large fraction of salt is used for winter road deicing (Fig. 11.3a and Box 12.4). However, in addition to the corrosive effects of NaCl, environmental concerns have focused on the side-effects on roadside vegetation and run-off into water sources. Increasing awareness of these problems has led to the introduction of reduced-salt road maintenance schemes (e.g. in Canada) and the use of calcium magnesium acetate in place of NaCl as a road deicing agent (see Box 12.4).

Both Na and K are involved in various electrophysiological functions in higher animals. The $[Na^+]:[K^+]$ ratio is different in intra- and extra-cellular fluids, and the

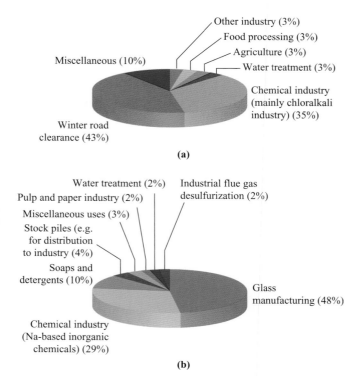

Fig. 11.3 Uses of (a) NaCl and (b) Na_2CO_3 in the US in 2009. [Data: US Geological Survey.]

concentration gradients of these ions across cell membranes are the origin of the trans-membrane potential difference that, in nerve and muscle cells, is responsible for the transmission of nerve impulses. A balanced diet therefore includes both Na^+ and K^+ salts. Potassium is also an essential plant nutrient, and K^+ salts are widely used as fertilizers (see Box 11.1). Applications of Li and Na in batteries are highlighted in Box 11.3, and the use of KO_2 in breathing masks is described in Section 11.6.

Many organic syntheses involve Li, Na or their compounds, and uses of the reagents $Na[BH_4]$ and $Li[AlH_4]$ are widespread. Alkali metals and some of their compounds also have uses in catalysts, e.g. the formation of MeOH from H_2 and CO (eq. 10.15) where doping the catalyst with Cs makes it more effective.

11.3 Physical properties

General properties

The alkali metals illustrate, more clearly than any other group of elements, the influence of increase in atomic and ionic size on physical and chemical properties. Thus, the group 1 metals are often chosen to illustrate general principles. Some physical properties of the group 1 metals are given in Table 11.1. Some important points arising from these data are listed below; see Section 7.9 for detailed discussion of the energetics of ion hydration.

- With increasing atomic number, the atoms become larger and the strength of metallic bonding (see Section 6.8) decreases.
- The effect of increasing size evidently outweighs that of increasing nuclear charge, since the ionization energies decrease from Li to Cs (see Fig. 1.16). The values of IE_2 for all the alkali metals are so high that the formation of M^{2+} ions under chemically reasonable conditions is not viable.
- Values of $E^o_{M^+/M}$ are related to energy changes accompanying the processes:

$$M(s) \longrightarrow M(g) \qquad \textit{atomization}$$
$$M(g) \longrightarrow M^+(g) \qquad \textit{ionization}$$
$$M^+(g) \longrightarrow M^+(aq) \qquad \textit{hydration}$$

and down group 1, differences in these energy changes almost cancel out, resulting in similar $E^o_{M^+/M}$ values. The lower reactivity of Li towards H_2O is *kinetic* rather than thermodynamic in origin; Li is a harder and higher melting metal, is less rapidly dispersed, and reacts more slowly than its heavier congeners.

Self-study exercise

Using data from Table 11.1, show that the enthalpy change associated with the reduction process:

$$M^+(aq) + e^- \longrightarrow M(s)$$

is $-200\,kJ\,mol^{-1}$ for $M = Na$, $-188\,kJ\,mol^{-1}$ for $M = K$, and $-189\,kJ\,mol^{-1}$ for $M = Rb$. Hence comment on the statement in the text that values of $E^o_{M^+/M}$ are similar for the group 1 metals.

In general, the chemistry of the group 1 metals is dominated by compounds containing M^+ ions. However, a small number of compounds containing the M^- ion (M = Na, K, Rb or Cs) are known (see Section 11.8), and the organometallic chemistry of the group 1 metals is a growing area that is described in Chapter 23.

Considerations of lattice energies calculated using an electrostatic model provide a satisfactory understanding for the fact that ionic compounds are central to the chemistry of Na, K, Rb and Cs. That Li shows a so-called 'anomalous' behaviour and exhibits a *diagonal relationship* to Mg can be explained in terms of similar energetic considerations (see Section 12.10).

Atomic spectra and flame tests

In the vapour state, the alkali metals exist as atoms or M_2 molecules. The strength of the M−M covalent bond decreases down the group (Table 11.1). Excitation of the outer ns^1 electron of the M atom occurs easily and emission

Table 11.1 Some physical properties of the alkali metals, M, and their ions, M^+.

Property	Li	Na	K	Rb	Cs
Atomic number, Z	3	11	19	37	55
Ground state electronic configuration	$[He]2s^1$	$[Ne]3s^1$	$[Ar]4s^1$	$[Kr]5s^1$	$[Xe]6s^1$
Enthalpy of atomization, $\Delta_a H^\circ(298\,K)\,/\,kJ\,mol^{-1}$	161	108	90	82	78
Dissociation enthalpy of M–M bond in M_2 (298 K) $/\,kJ\,mol^{-1}$	110	74	55	49	44
Melting point, mp / K	453.5	371	336	312	301.5
Boiling point, bp / K	1615	1156	1032	959	942
Standard enthalpy of fusion, $\Delta_{fus} H^\circ(mp)\,/\,kJ\,mol^{-1}$	3.0	2.6	2.3	2.2	2.1
First ionization energy, $IE_1\,/\,kJ\,mol^{-1}$	520.2	495.8	418.8	403.0	375.7
Second ionization energy, $IE_2\,/\,kJ\,mol^{-1}$	7298	4562	3052	2633	2234
Metallic radius, $r_{metal}\,/\,pm^\dagger$	152	186	227	248	265
Ionic radius, $r_{ion}\,/\,pm^\ddagger$	76	102	138	149	170
Standard enthalpy of hydration of M^+, $\Delta_{hyd} H^\circ(298\,K)\,/\,kJ\,mol^{-1}$	−519	−404	−321	−296	−271
Standard entropy of hydration of M^+, $\Delta_{hyd} S^\circ(298\,K)\,/\,J\,K^{-1}\,mol^{-1}$	−140	−110	−70	−70	−60
Standard Gibbs energy of hydration of M^+, $\Delta_{hyd} G^\circ(298\,K)\,/\,kJ\,mol^{-1}$	−477	−371	−300	−275	−253
Standard reduction potential, $E^\circ_{M^+/M}\,/\,V$	−3.04	−2.71	−2.93	−2.98	−3.03
NMR active nuclei (% abundance, nuclear spin)	^6Li (7.5, $I=1$); ^7Li (92.5, $I=\frac{3}{2}$)	^{23}Na (100, $I=\frac{3}{2}$)	^{39}K (93.3, $I=\frac{3}{2}$); ^{41}K (6.7, $I=\frac{3}{2}$)	^{85}Rb (72.2, $I=\frac{5}{2}$); ^{87}Rb (27.8, $I=\frac{3}{2}$)	^{133}Cs (100, $I=\frac{7}{2}$)

† For 8-coordinate atom in body-centred cubic metal; compare values for 12-coordinate atoms in Appendix 6.
‡ For 6-coordination.

spectra are readily observed. In Section 19.8, we describe the use of the *sodium D-line* in the emission spectrum of atomic Na for specific rotation measurements. When the salt of an alkali metal is treated with concentrated HCl (giving a volatile metal chloride) and is heated strongly in the non-luminous Bunsen flame, a characteristic flame colour is observed (Li, crimson; Na, yellow; K, lilac; Rb, red-violet; Cs, blue) and this *flame* test is used in *qualitative* analysis to identify the M^+ ion. In *quantitative* analysis, use is made of the characteristic atomic spectrum in *atomic absorption spectroscopy* (see Section 4.3).

Worked example 11.1   The Na₂ molecule

Construct an MO diagram for the formation of Na₂ from two Na atoms using only the valence orbitals and electrons of Na. Use the MO diagram to determine the bond order in Na₂.

The atomic number of Na is 11.

The ground state electronic configuration of Na is $1s^2 2s^2 2p^6 3s^1$ or $[Ne]3s^1$.

The valence orbital of Na is the $3s$.

An MO diagram for the formation of Na₂ is:

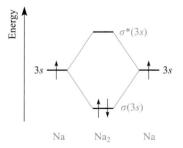

Bond order $= \frac{1}{2}[$(number of bonding electrons) − (number of antibonding electrons)$]$

Bond order in Na₂ $= \frac{1}{2} \times 2 = 1$

APPLICATIONS

Box 11.2 Keeping time with caesium

In 1993, the National Institute of Standards and Technology (NIST) brought into use a caesium-based atomic clock called NIST-7 which kept international standard time to within one second in 10^6 years. The system depends upon repeated transitions from the ground to a specific excited state of atomic Cs, and the monitoring of the frequency of the electromagnetic radiation emitted.

In 1995, the first caesium fountain atomic clock was constructed at the Paris Observatory in France. A fountain clock, NIST-F1, was introduced in 1999 in the US to function as the country's primary time and frequency standard. NIST-F1 is accurate to within one second in 80×10^6 years. The uncertainty in the measurement is continually being improved. While earlier caesium clocks observed Cs atoms at ambient temperatures, caesium fountain clocks use lasers to slow down and cool the atoms to temperatures approaching 0 K. Current atomic clock research is focusing on instruments based on optical transitions of neutral atoms or of a single ion (e.g. $^{88}Sr^+$). Progress in this area became viable after 1999 when optical counters based on femtosecond lasers (see Box 26.2) became available.

NIST-F1 caesium fountain atomic clock at the NIST laboratories in Boulder, Colorado.

Further reading

M. Chalmers (2009) *New Scientist*, vol. 201, issue 2694, p. 39 – 'Every second counts'.
P. Gill (2001) *Science*, vol. 294, p. 1666 – 'Raising the standards'.

M. Takamoto, F.-L. Hong, R. Higashi and H. Katori (2005) *Nature*, vol. 435, p. 321 – 'An optical lattice clock'.
R. Wynands and S. Weyers (2005) *Metrologia*, vol. 42, p. S64 – 'Atomic fountain clocks'.
www.nist.gov/physlab/div847/grp50/primary-frequency-standards.cfm

Self-study exercises

1. Why is it not necessary to include the $1s$, $2s$ and $2p$ orbitals and electrons in the MO description of the bonding in Na_2?

2. Use the MO diagram to determine whether Na_2 is paramagnetic or diamagnetic. [*Ans*: Diamagnetic]

See end-of-chapter problem 11.5 for an extension of these exercises.

Radioactive isotopes

In addition to the radioactivity of Fr, 0.02% of naturally occurring K consists of ^{40}K which decays according to scheme 11.2.

electron capture (proton to neutron conversion)

$^{40}_{19}K$

(11% of total decay processes) → $^{40}_{18}Ar$

β^- decay

(89% of total decay processes) → $^{40}_{20}Ca$

(11.2)

The overall half-life for both the β-decay and electron capture is 1.25×10^9 yr.

The decay of ^{40}K provides the human body with a natural source of radioactivity, albeit at very low levels. The decay from ^{40}K to ^{40}Ar is the basis of a technique for dating minerals (e.g. biotite, hornblende and volcanic rocks). When volcanic magma cools, ^{40}Ar formed from the decay of ^{40}K remains trapped in the mineral. Crushing and heating rock samples releases argon, and the amount of ^{40}Ar present can be determined by mass spectrometry. Atomic absorption spectroscopy is used to determine the ^{40}K content. The age of the mineral can be estimated from the ratio of $^{40}K : ^{40}Ar$.[†]

NMR active nuclei

Each of the alkali metals has at least one NMR active nucleus (Table 11.1), although not all nuclei are of sufficient sensitivity to permit their routine use. For examples of NMR spectroscopy utilizing *s*-block metals, see Section 4.8 and worked example 23.1.

[†] For an interesting discussion of ^{40}K–^{40}Ar dating, see: W.A. Howard (2005) *J. Chem. Educ.*, vol. 82, p. 1094.

APPLICATIONS

Box 11.3 Alkali metal ion batteries

The sodium/sulfur battery operates around 570–620 K and consists of a molten sodium anode and a liquid sulfur cathode which contains a carbon fibre matrix for conduction. The anode and cathode are separated by a solid β-alumina electrolyte (see Section 28.2). The cell reaction is:

$$2Na(1) + nS(1) \longrightarrow Na_2S_n(1) \qquad E_{cell} = 2.08 \text{ V}$$

and this is reversed when the battery is recharged by changing the polarity of the cell. In the 1990s, it appeared that sodium/sulfur batteries may have potential application in the electric vehicle (EV) market, but the high operating temperature of the sodium/sulfur battery is a drawback to the motor industry, and other battery technologies have superseded these batteries for electric and hybrid electric vehicles. Stationary sodium/sulfur batteries are used for energy storage, notably in Japan. This application follows from the fact that self-discharge from sodium/sulfur batteries occurs only at very low levels. The 2005 EXPO exhibition in Aichi, Japan featured an experimental power system incorporating solar cell and fuel cell electrical power generators and a sodium/sulfur battery system to store the energy. The use of an efficient storage system allows the balance between the generation of and demand for electrical energy to be regulated.

An important advance in battery technology has been the development of rechargeable, high energy-density lithium-ion batteries, first introduced to the commercial market in 1991. Ten billion US dollars worth of lithium-ion batteries were sold in 2008, and the market continues to grow. The lithium-ion battery has a cell potential of 3.6 V and consists of a positive $LiCoO_2$ electrode separated from a graphite electrode by a solid electrolyte across which Li^+ ions can migrate when the cell is charging. In commercial lithium-ion batteries, the electrolyte is usually $LiPF_6$ in an alkyl carbonate material. Lithium-ion batteries are manufactured in a discharged state. Solid $LiCoO_2$ adopts an α-$NaFeO_2$ structure type in which the O atoms are approximately cubic close-packed. The octahedral holes are occupied by M(I) or M'(III) (Li^+ or Co^{3+} in $LiCoO_2$) in such a way that the different metal ions are arranged in layers. During charging, Li^+ ions move out of these layers, are transported across the electrolyte, and are intercalated by the graphite (see Section 14.4). During discharge of the cell, the Li^+ ions return to the metal oxide lattice. The cell reaction can be represented as follows:

$$LiCoO_2 + 6C(graphite) \underset{\text{discharge}}{\overset{\text{charge}}{\rightleftharpoons}} LiC_6 + CoO_2$$

The cobalt centres are redox active, being oxidized from Co(III) to Co(IV) as Li^+ is removed from $LiCoO_2$. The crucial factor in lithium-ion batteries is that both electrodes are able to act as hosts for Li^+ ions. Rechargeable, lithium-ion batteries now dominate the market for small electronic devices such as laptop computers, mobile phones, iPods and MP3 players, and in electric bicycles. In 2005, Sony introduced a new generation of lithium-ion batteries (the Nexelion battery) in which the

Mercedes-Benz S400 Blue Hybrid car which utilizes a lithium-ion battery.

mixed metal oxide $Li(Ni,Mn,Co)O_2$ replaces the all-cobalt $LiCoO_2$ electrode, and a tin-based electrode replaces graphite.

A disadvantage of lithium-ion batteries containing cobalt is their relatively high cost. Current research strategies are aimed at finding replacement electrode materials both to increase battery performance and to reduce cost. Two contenders are $LiMn_2O_4$ and $LiFePO_4$. $LiMn_2O_4$ has a spinel structure (see Box 13.7) and when coupled with a graphite electrode forms a lithium-ion battery, the cell reaction of which is summarized below:

$$LiMn_2O_4 + 6C(graphite) \underset{\text{discharge}}{\overset{\text{charge}}{\rightleftharpoons}} LiC_6 + Mn_2O_4$$

Potential applications of this type of lithium-ion battery include those in hybrid electric vehicles (HEVs). Manufacturers including Toyota and Honda produce hybrid electric and plug-in electric vehicles (rechargeable from an external power supply when the car is parked) incorporating lithium-ion batteries, but the first mass-produced HEV containing a lithium-ion battery was launched by Mercedes-Benz in 2009. In the S400 Blue Hybrid, a 120 V lithium-ion battery pack powers an electric motor which works in conjunction with an internal combustion engine, the operating mode being computer controlled. A regenerative braking system (see Box 10.5) converts kinetic energy to electrical energy which is stored in the battery, and the electrical motor also recovers energy during deceleration.

Further reading on lithium-ion batteries

C.-M. Park, J.-H. Kim, H. Kim and H.-J. Sohn (2010) *Chem. Soc. Rev.*, vol. 39, p. 3115.

B. Scrosati and J. Garche (2010) *J. Power Sources*, vol. 195, p. 2419.

F.T. Wagner, B. Lakshmanan and M.F. Mathias (2010) *J. Phys. Chem. Lett.*, vol. 1, p. 2204.

11.4 The metals

Appearance

The metals Li, Na, K and Rb are silvery-white, but Cs has a golden-yellow cast. All are soft, Li the least so, and the trend is consistent with their melting points (Table 11.1). The particularly low melting point of Cs (301.5 K) means that it may be a liquid at ambient temperatures in some hot climates.

Reactivity

We have already described the behaviour of the metals in liquid NH_3 (see Section 9.6). The ultimate products are alkali metal amides (see eq. 9.28), and $LiNH_2$, $NaNH_2$ and KNH_2 are important reagents in organic synthesis. In the solid state, these amides adopt structures consisting of cubic close-packed $[NH_2]^-$ ions with M^+ ions occupying half the tetrahedral holes.

Worked example 11.2 Structure of NaNH₂

The solid state structure of $NaNH_2$ can be approximately described as consisting of an fcc arrangement of amide ions with Na^+ ions occupying half the tetrahedral holes. To which structure type (or prototype structure) does this correspond?

A face-centred cubic (i.e. cubic close-packed) arrangement of $[NH_2]^-$ ions (assuming each is spherical) corresponds to the following unit cell:

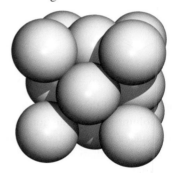

There are eight tetrahedral holes within the unit cell. The Na^+ ions occupy half of these interstitial sites:

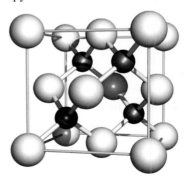

$NaNH_2$ adopts a zinc blende (ZnS) structure (compare with Fig. 6.19b).

Self-study exercises

1. Use the diagram of the unit cell for sodium amide to confirm the $1:1$ $Na^+:[NH_2]^-$ ratio.

2. Using the diagram of the unit cell of $NaNH_2$, determine the coordination number of each $[NH_2]^-$ ion. To check your answer, think how this coordination number must be related to that of an Na^+ ion.

Although Li, Na and K are stored under a hydrocarbon solvent to prevent reaction with atmospheric O_2 and water vapour, they can be handled in air, provided undue exposure is avoided; Rb and Cs should be handled in an inert atmosphere. Lithium reacts quickly with water (eq. 11.3); Na reacts vigorously, and K, Rb and Cs react violently with the ignition of H_2 produced.

$$2Li + 2H_2O \longrightarrow 2LiOH + H_2 \qquad (11.3)$$

Sodium is commonly used as a drying agent for hydrocarbon and ether solvents. Sodium should *never* be used to dry halogenated solvents (see eq. 14.47). The disposal of excess Na must be carried out with care and usually involves the reaction of Na with propan-2-ol:

This is a less vigorous, and therefore safer, reaction than that of Na with H_2O or a low molecular mass alcohol. An alternative method for disposing of small amounts of Na involves adding H_2O to a sand-filled ceramic container (e.g. plant pot) in which the metal has been buried. The conversion of Na to NaOH occurs slowly, and the NaOH reacts with the sand (i.e. SiO_2) to yield sodium silicate.[†]

All the group 1 metals react with the halogens (eq. 11.4) and H_2 when heated (eq. 11.5). The energetics of metal hydride formation are essentially like those of metal halide formation, being expressed in terms of a Born–Haber cycle (see Section 6.14).

$$2M + X_2 \longrightarrow 2MX \qquad X = halogen \qquad (11.4)$$

$$2M + H_2 \longrightarrow 2MH \qquad (11.5)$$

$$6Li + N_2 \longrightarrow 2Li_3N \qquad (11.6)$$

[†] See: H.W. Roesky (2001) *Inorg. Chem.*, vol. 40, p. 6855 – 'A facile and environmentally friendly disposal of sodium and potassium with water'.

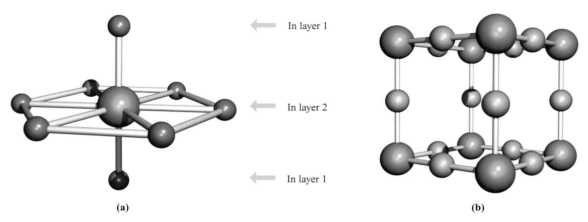

(a)   **(b)**

Fig. 11.4 (a) The solid state structure of Li_3N consists of layers of N^{3-} and Li^+ ions (ratio 1:2) alternating with layers of Li^+ ions; the latter are arranged such that they lie over the N^{3-} ions. Each N centre is in a hexagonal bipyramidal (8-coordinate) environment; there are two types of Li^+ ion, those in layer 1 are 2-coordinate, and those in layer 2 are 3-coordinate with respect to the N centres (see end-of-chapter problem 11.12). (b) The unit cell of sodium nitride; Na_3N adopts an anti-ReO_3 structure. Colour code: N, blue; Li, red; Na, orange.

Lithium reacts spontaneously with N_2, and reaction 11.6 occurs at 298 K to give red-brown, moisture-sensitive lithium nitride. Solid Li_3N has an interesting structure (Fig. 11.4a) and a high ionic conductivity (see Section 28.2). Attempts to prepare the binary nitrides of the later alkali metals were not successful until 2002. Na_3N (which is very moisture-sensitive) may be synthesized in a vacuum chamber by depositing atomic sodium and nitrogen onto a cooled sapphire substrate and then heating to room temperature. The structure of Na_3N is very different from that of Li_3N (Fig. 11.4), with Na_3N adopting an anti-ReO_3 structure (see Fig. 22.4 for ReO_3) in which the Na^+ ions are 2-coordinate and the N^{3-} ions are octahedrally sited. Reactions of the alkali metals with O_2 are discussed in Section 11.6.

Acetylides, M_2C_2, are formed when Li or Na is heated with carbon. These compounds can also be prepared by treating the metal with C_2H_2 in liquid NH_3. Reactions between K, Rb or Cs and graphite lead to a series of *intercalation compounds* MC_n (n = 8, 24, 36, 48 and 60) in which the alkali metal atoms are inserted between the layers in a graphite host lattice (see structure 14.2 and Fig. 14.4a). For a given formula, the compounds are structurally similar and exhibit similar properties, irrespective of the metal. Under high-pressure conditions, MC_{4-6} (M = K, Rb, Cs) can be formed. In contrast, the intercalation of lithium into graphite (the basis of lithium-ion battery technology, see Box 11.3) gives LiC_6, LiC_{12}, LiC_{18} and LiC_{27}. At high pressures, LiC_{2-4} can be produced. The formation of sodium–graphite intercalation compounds is more difficult. The reaction of Na vapour with graphite at high temperatures gives NaC_{64}. We return to graphite intercalation compounds in Section 14.4.

The alkali metals dissolve in Hg to give amalgams (see Box 22.3). Sodium amalgam is a liquid only when the percentage of Na is low. It is a useful reducing agent in inorganic and organic chemistry, and can be used in aqueous media because there is a large overpotential for the discharge of H_2.

An innovative method of handling alkali metals is to absorb them into silica gel, thus providing a convenient source of the metals as powerful reducing agents, e.g. in Birch reductions:

Na₂K-SG = Na₂K alloy in silica gel

Foreseeable applications of these materials are in the use of continuous-flow columns for reduction reactions in, for example, the pharmaceutical industry. The silica gel–alkali metal powders react quantitatively with water, liberating H_2. Since the powders are easily handled and stored, they have the potential to act as a 'supply-on-demand' source of H_2.[†]

11.5 Halides

The MX halides (see Chapter 6 for structures) are prepared by direct combination of the elements (eq. 11.4) and all the halides have large negative $\Delta_f H^o$ values. However, Table 11.2 shows that for X = F, values of $\Delta_f H^o$(MX)

[†] See: J.L. Dye *et al.* (2005) *J. Am. Chem. Soc.*, vol. 127, p. 9338; M. Shatnawi *et al.* (2007) *J. Am. Chem. Soc.*, vol. 129, p. 1386.

Table 11.2 Standard enthalpies of formation ($\Delta_f H^o$) and lattice energies ($\Delta_{lattice} H^o$) of alkali metal halides, MX.

M	$\Delta_f H^o(MX) / kJ\,mol^{-1}$ Halide ion size increases →				$\Delta_{lattice} H^o(MX) / kJ\,mol^{-1}$ Halide ion size increases →			
	F	Cl	Br	I	F	Cl	Br	I
Li	−616	−409	−351	−270	−1030	−834	−788	−730
Na	−577	−411	−361	−288	−910	−769	−732	−682
K	−567	−436	−394	−328	−808	−701	−671	−632
Rb	−558	−435	−395	−334	−774	−680	−651	−617
Cs	−553	−443	−406	−347	−744	−657	−632	−600

(Metal ion size increases ↓)

become *less negative* down the group, while the reverse trend is true for X = Cl, Br and I. For a given metal, $\Delta_f H^o(MX)$ always becomes less negative on going from MF to MI. These generalizations can be explained in terms of a Born–Haber cycle. Consider the formation of MX (eq. 11.7) and refer to Fig. 6.25.

$\Delta_f H^o(MX, s)$

$$= \underbrace{\{\Delta_a H^o(M, s) + IE_1(M, g)\}}_{\text{metal-dependent term}} + \underbrace{\{\Delta_a H^o(X, g) + \Delta_{EA} H(X, g)\}}_{\text{halide-dependent term}}$$

$$+ \Delta_{lattice} H^o(MX, s) \qquad (11.7)$$

For MF, the variable quantities are $\Delta_a H^o(M)$, $IE_1(M)$ and $\Delta_{lattice} H^o(MF)$, and similarly for each of MCl, MBr and MI. The sum of $\Delta_a H^o(M)$ and $IE_1(M)$ gives for the formation of Li$^+$ 681, of Na$^+$ 604, of K$^+$ 509, of Rb$^+$ 485 and of Cs$^+$ 454 kJ mol^{-1}. For the fluorides, the trend in the values of $\Delta_f H^o(MF)$ depends on the relative values of $\{\Delta_a H^o(M) + IE_1(M)\}$ and $\Delta_{lattice} H^o(MF)$ (Table 11.2), and similarly for chlorides, bromides and iodides. Inspection of the data shows that the variation in $\{\Delta_a H^o(M) + IE_1(M)\}$ is *less* than the variation in $\Delta_{lattice} H^o(MF)$, but *greater* than the variation in $\Delta_{lattice} H^o(MX)$ for X = Cl, Br and I. This is because lattice energy is proportional to $1/(r_+ + r_-)$ (see Section 6.13) and so variation in $\Delta_{lattice} H^o(MX)$ for a given halide is greatest when r_- is smallest (for F$^-$) and least when r_- is largest (for I$^-$). Considering the halides of a given metal (eq. 11.7), the small change in the term $\{\Delta_a H^o(X) + \Delta_{EA} H(X)\}$ (−249, −228, −213, −188 kJ mol^{-1} for F, Cl, Br, I respectively) is outweighed by the decrease in $\Delta_{lattice} H^o(MX)$. In Table 11.2, note that the *difference* between the values of $\Delta_f H^o(MF)$ and $\Delta_f H^o(MI)$ *decreases* significantly as the size of the M$^+$ ion increases.

The solubilities of the alkali metal halides in water are determined by a delicate balance between lattice energies and Gibbs energies of hydration (see Section 7.9 for $\Delta_{sol} G^o$ and $\Delta_{hyd} G^o$). LiF has the highest lattice energy of the group 1 metal halides and is only sparingly soluble, but solubility relationships among the other halides call for detailed discussion beyond the scope of this book.[†] The salts LiCl, LiBr, LiI and NaI are soluble in some oxygen-containing organic solvents, e.g. LiCl dissolves in THF and MeOH. Complexation of the Li$^+$ or Na$^+$ ion by the *O*-donor solvents is likely in all cases (see Section 11.8). Both LiI and NaI are very soluble in liquid NH$_3$, forming complexes; the unstable complex [Na(NH$_3$)$_4$]I has been isolated and contains a tetrahedrally coordinated Na$^+$ ion.

In the vapour state, alkali metal halides are present mainly as ion-pairs, but measurements of M−X bond distances and electric dipole moments suggest that covalent contributions to the bonding, particularly in the lithium halides, are important.

11.6 Oxides and hydroxides

Oxides, peroxides, superoxides, suboxides and ozonides

When the group 1 metals are heated in an excess of air or in O$_2$, the principal products obtained depend on the metal: lithium *oxide*, Li$_2$O (eq. 11.8), sodium *peroxide*, Na$_2$O$_2$ (eq. 11.9), and the *superoxides* KO$_2$, RbO$_2$ and CsO$_2$ (eq. 11.10).

[†] For further discussion, see: W.E. Dasent (1984) *Inorganic Energetics*, 2nd edn, Cambridge University Press, Cambridge, Chapter 5.

$$4Li + O_2 \longrightarrow 2Li_2O \qquad \textit{oxide formation} \qquad (11.8)$$

$$2Na + O_2 \longrightarrow Na_2O_2 \qquad \textit{peroxide formation} \qquad (11.9)$$

$$K + O_2 \longrightarrow KO_2 \qquad \textit{superoxide formation} \qquad (11.10)$$

The oxides Na_2O, K_2O, Rb_2O and Cs_2O can be obtained impure by using a limited air supply, but are better prepared by thermal decomposition of the peroxides or superoxides. The colours of the oxides vary from white to orange: Li_2O and Na_2O form white crystals while K_2O is pale yellow, Rb_2O yellow and Cs_2O orange. All the oxides are strong bases, the basicity increasing from Li_2O to Cs_2O. A peroxide of lithium can be obtained by the action of H_2O_2 on an ethanolic solution of $LiOH$, but it decomposes on heating. Sodium peroxide is widely used as an oxidizing agent. When pure, Na_2O_2 is colourless and the faint yellow colour usually observed is due to the presence of small amounts of NaO_2. The superoxides and peroxides contain the paramagnetic $[O_2]^-$ and diamagnetic $[O_2]^{2-}$ ions respectively (see end-of-chapter problem 11.13). Superoxides have magnetic moments of $\approx 1.73\mu_B$, consistent with one unpaired electron.

Partial oxidation of Rb and Cs at low temperatures yields *suboxides* such as Rb_9O_2 and $Cs_{11}O_3$. Their structures consist of octahedral units of metal ions with the oxygen residing at the centre. The octahedra are fused together by sharing faces (Fig. 11.5). The suboxides Rb_6O, Cs_7O and Cs_4O also contain Rb_9O_2 or $Cs_{11}O_3$ clusters. In each case, alkali metal atoms are present in the crystalline solid in addition to Rb_9O_2 or $Cs_{11}O_3$ units. Thus, more informative formulations of Rb_6O, Cs_7O and Cs_4O are $Rb_9O_2{\cdot}Rb_3$, $Cs_{11}O_3{\cdot}Cs_{10}$ and $Cs_{11}O_3{\cdot}Cs$, respectively. The formulae of the suboxide clusters are misleading in terms of the oxidation states. Each contains M^+ and O^{2-} ions, and, for example, the formula of Rb_9O_2 is better written as $(Rb^+)_9(O^{2-})_2{\cdot}5e^-$, indicating the presence of free electrons.

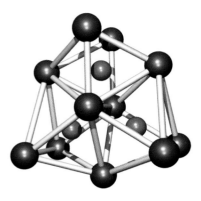

Fig. 11.5 The structure of the suboxide $Cs_{11}O_3$ consists of three oxygen-centred, face-sharing octahedral units. Colour code: Cs, blue; O, red.

The alkali metal oxides, peroxides and superoxides react with water according to eqs. 11.11–11.13. One use of KO_2 is in breathing masks where it absorbs H_2O producing O_2 for respiration and KOH, which absorbs exhaled CO_2 (reaction 11.14).

$$M_2O + H_2O \longrightarrow 2MOH \qquad (11.11)$$

$$M_2O_2 + 2H_2O \longrightarrow 2MOH + H_2O_2 \qquad (11.12)$$

$$2MO_2 + 2H_2O \longrightarrow 2MOH + H_2O_2 + O_2 \qquad (11.13)$$

$$KOH + CO_2 \longrightarrow KHCO_3 \qquad (11.14)$$

Sodium peroxide reacts with CO_2 to give Na_2CO_3, making it suitable for use in air purification in confined spaces (e.g. in submarines); KO_2 acts similarly but more effectively.

Although all the group 1 peroxides decompose on heating according to equation 11.15, their thermal stabilities depend on cation size. Li_2O_2 is the least stable peroxide, while Cs_2O_2 is the most stable. The stabilities of the superoxides with respect to decomposition to M_2O_2 and O_2 follow a similar trend.

$$M_2O_2(s) \longrightarrow M_2O(s) + \tfrac{1}{2}O_2(g) \qquad (11.15)$$

Ozonides, MO_3, containing the paramagnetic, bent $[O_3]^-$ ion (see Section 16.4), are known for all the alkali metals. The salts KO_3, RbO_3 and CsO_3 can be prepared from the peroxides or superoxides by reaction with ozone, but this method fails, or gives low yields, for LiO_3 and NaO_3. These ozonides have been prepared in liquid ammonia by the interaction of CsO_3 with an ion-exchange resin loaded with either Li^+ or Na^+ ions. The ozonides are violently explosive.

> An ***ion-exchange resin*** consists of a solid phase (e.g. a zeolite) which contains acidic or basic groups which may exchange with cations or anions, respectively, from solutions washed through the resin; an important application is in water purification (see Box 16.3).

Hydroxides

In 2008, global demand for $NaOH$ (*caustic soda*) was ≈ 50 Mt. It is manufactured in the chloralkali industry (Box 11.4), China being the largest producer. $NaOH$ is used throughout organic and inorganic chemistry wherever a cheap alkali is needed, and over half of the $NaOH$ manufactured is consumed in the chemical industry. Remaining uses are in the soap and textile industries, water treatment, aluminium manufacturing (Section 13.2) and pulp and paper manufacturing.

Solid $NaOH$ (mp 591 K) is often handled as flakes or pellets, and dissolves in water with considerable evolution of heat. Potassium hydroxide (mp 633 K) closely resembles $NaOH$ in preparation and properties. It is more soluble than $NaOH$ in EtOH, in which it produces a low concentration of

ENVIRONMENT

Box 11.4 The chloralkali industry

The *chloralkali industry* produces huge quantities of NaOH and Cl_2 by the electrolysis of *aqueous* NaCl (brine).

At the anode: $2Cl^-(aq) \longrightarrow Cl_2(g) + 2e^-$

At the cathode: $2H_2O(l) + 2e^- \longrightarrow 2[OH]^-(aq) + H_2(g)$

The anode discharges Cl_2 rather than O_2 even though, from values of E°, it appears easier to oxidize H_2O than Cl^-. This observation is a consequence of the *overpotential* required to release O_2 and is explained more fully in worked example 17.3.

Three types of electrolysis cell are available:

- the mercury cell, which employs a mercury cathode;
- the diaphragm cell, which uses an asbestos diaphragm separating the steel cathode and the graphite or platinum-coated titanium anode;
- the membrane cell, in which a cation-exchange membrane, with high permeability to Na^+ ions and low permeability to Cl^- and $[OH]^-$ ions, is placed between the anode and the cathode.

Currently, 45–50 Mt of Cl_2 is manufactured by the chloralkali process each year; this represents 95% of the global supply. For every 1 t of Cl_2 produced, 1.1 t of NaOH are also manufactured. The main producers are the US, Western Europe and Japan. Whereas the Japanese chloralkali industry operates almost entirely with the membrane cell, the US favours use of the diaphragm cell. In Europe, 46% of the industry uses the membrane cell, 34% the mercury cell and 14% the diaphragm cell. On environmental grounds, the chloralkali industry is being pressured to replace mercury and diaphragm cells by the membrane cell. In the European Union, use of the mercury-based process is being gradually phased out with a target date of 2020 for conversion of the industry to the membrane cell. However, the disposal of mercury from electrolysis cells is not trivial. The scale of the problem can be appreciated from the photograph above which shows part of the cell room in a chloralkali plant that operates using mercury cells. The export of mercury from the European Union has been banned since 2011, and the chloralkali industry must ensure the safe storage of mercury from decommissioned mercury cells, e.g. deep underground in steel cannisters.

Use of mercury and diaphragm cells is not the only environmental concern facing the industry; demand for Cl_2 has fallen in the pulp and paper industry and in the production of chlorofluorocarbons, the latter being phased out as a result of the *Montreal Protocol for the Protection of the Ozone Layer*. Nevertheless, overall demand for Cl_2 remains high, much being used in the production of chloroethene

A technician checking mercury cells in the cell room of a plant producing Cl_2 and NaOH.

(for manufacture of polyvinylchloride, PVC). Uses of Cl_2 are summarized in Fig. 17.2.

Aqueous NaOH from the electrolytic process is evaporated to give solid NaOH (caustic soda) as a white, translucent solid which is fused and cast into sticks, or made into flakes or pellets.

The chloralkali industry illustrates an interesting market problem. While the electrolysis of brine produces NaOH and Cl_2 in a *fixed molar ratio*, the markets for the two chemicals are different and unrelated. When NaOH exceeds demand, it can be stored, but storage of excess Cl_2 is more difficult. In these circumstances, the scale of production of both Cl_2 and NaOH tends to be reduced. Interestingly, prices of the two chemicals follow opposite trends. In times of recession, demand for Cl_2 falls more sharply than that of NaOH, with the result that the price of Cl_2 falls as stocks build up. Conversely, industrial demand for Cl_2 increases faster than that of NaOH when the economy is strong. Consequently, the price of the alkali falls as stocks increase. The net result is clearly important to the long-term stability of the chloralkali industry as a whole.

Further reading

N. Botha (1995) *Chemistry & Industry*, p. 832 – 'The outlook for the world chloralkali industry'.

R. Shamel and A. Udis-Kessler (2001) *Chemistry & Industry*, p. 179 – 'Critical chloralkali cycles continue'.

For up-to-date information on the European chloralkali industry, visit the website: www.eurochlor.org

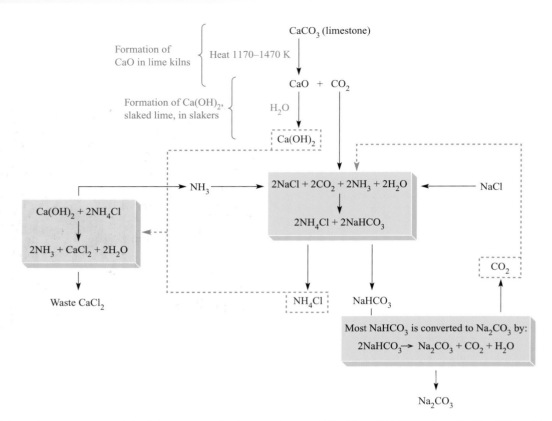

Fig. 11.6 Schematic representation of the Solvay process for the manufacture of Na_2CO_3 and $NaHCO_3$ from $CaCO_3$, NH_3 and $NaCl$. The recycling parts of the process are shown with blue, broken lines.[†]

ethoxide ions (eq. 11.16). This gives rise to the use of *etha-nolic* KOH in organic synthesis.

$$C_2H_5OH + [OH]^- \rightleftharpoons [C_2H_5O]^- + H_2O \qquad (11.16)$$

The crystal structures of the group 1 hydroxides are usually complicated, but the high-temperature form of KOH has the NaCl structure, with the $[OH]^-$ ions undergoing rotation making them pseudo-spherical.

The reactions of alkali metal hydroxides (see Section 7.4) with acids and acidic oxides call for no special mention (see end-of-chapter problem 11.23), except for reactions with CO which give metal formates (methanoates), e.g. reaction 11.17.

$$NaOH + CO \xrightarrow{450\,K} HCO_2Na \qquad (11.17)$$

Many non-metals disproportionate when treated with aqueous alkali: P_4 gives PH_3 and $[H_2PO_2]^-$, S_8 gives S^{2-} and a mixture of oxoanions, and Cl_2 reacts to give Cl^- and $[OCl]^-$ or $[ClO_3]^-$ (see also Section 17.9). Non-metals that do not form stable hydrides, and amphoteric metals (e.g. Al, eq. 11.18), react with aqueous MOH to yield H_2 and complex anions.

$$2Al + 2NaOH + 6H_2O \longrightarrow 2Na[Al(OH)_4] + 3H_2 \qquad (11.18)$$

11.7 Salts of oxoacids: carbonates and hydrogencarbonates

The properties of alkali metal salts of most oxoacids depend on the anion present and not on the cation. Thus we tend to discuss salts of oxoacids under the appropriate acid. However, we single out the carbonates and hydrogencarbonates because of their importance. Whereas Li_2CO_3 is sparingly soluble in water, the remaining carbonates of the group 1 metals are very soluble.

In many countries, sodium carbonate (soda ash) and sodium hydrogencarbonate (commonly called sodium bicarbonate) are manufactured by the Solvay process (Fig. 11.6), but this is being superseded where natural sources of the mineral *trona*, $Na_2CO_3 \cdot NaHCO_3 \cdot 2H_2O$, are available (the world's largest deposit of trona is in the Green River Basin in Wyoming, USA). The two sources are distinguished by using the terms 'natural' (refined from trona) and 'synthetic' (from the Solvay process) Na_2CO_3. Figure 11.6 shows that in the Solvay process, NH_3 can be recycled, but most waste $CaCl_2$ is dumped (e.g. into the sea) or used in winter road clearance (see

[†] See: T. Kasikowski, R. Buczkowski and M. Cichosz (2008) *Int. J. Production Economics*, vol. 112, p. 971 – 'Utilisation of synthetic soda-ash industry by-products'.

Box 12.4). In 2009, ≈ 46 Mt of sodium carbonate (natural plus synthetic) were produced worldwide and uses are summarized in Fig. 11.3b. Sodium hydrogencarbonate, although a direct product in the Solvay process, is also manufactured by passing CO_2 through aqueous Na_2CO_3 or by dissolving trona in H_2O saturated with CO_2. Its uses include those as a foaming agent, a food additive (e.g. baking powder) and an effervescent in pharmaceutical products. The Solvay company has now developed a process for using $NaHCO_3$ in pollution control, e.g. by neutralizing SO_2 or HCl in industrial and other waste emissions.

There are some notable differences between Na^+ and other alkali metal $[CO_3]^{2-}$ and $[HCO_3]^-$ salts. Whereas $NaHCO_3$ can be separated from NH_4Cl in the Solvay process by *precipitation*, the same is not true of $KHCO_3$. Hence, K_2CO_3 is produced, not via $KHCO_3$, but by the reaction of KOH with CO_2. K_2CO_3 has uses in the manufacture of certain glasses and ceramics. Among its applications, $KHCO_3$ is used as a buffering agent in water treatment and wine production. Lithium carbonate (see also Section 11.2) is only sparingly soluble in water; '$LiHCO_3$' has not been isolated. The thermal stabilities of the group 1 metal carbonates with respect to reaction 11.19 increase down the group as r_{M^+} increases, lattice energy being a crucial factor. Such a trend in stability is common to all series of oxo-salts of the alkali metals.

$$M_2CO_3 \xrightarrow{\Delta} M_2O + CO_2 \qquad (11.19)$$

The solid state structures of $NaHCO_3$ and $KHCO_3$ exhibit hydrogen bonding (see Section 10.6). In $KHCO_3$, the anions associate in pairs (Fig. 11.7a) whereas in $NaHCO_3$, infinite chains are present (Fig. 11.7b). In each case, the hydrogen bonds are asymmetrical.

Sodium silicates are of great commercial importance and are discussed in Sections 14.2 and 14.9.

11.8 Aqueous solution chemistry and macrocyclic complexes

Hydrated ions

We introduced hydrated alkali metal cations in Sections 7.7 and 7.9. Some Li^+ salts (e.g. LiF, Li_2CO_3) are sparingly soluble in water, but for large anions, the Li^+ salts are soluble while many K^+, Rb^+ and Cs^+ salts are sparingly soluble (e.g. $MClO_4$, $M_2[PtCl_6]$ for M = K, Rb or Cs).

Worked example 11.3   Salts in aqueous solutions

Starting from Rb_2CO_3, how might you prepare and isolate $RbClO_4$?

Rb_2CO_3 is soluble in water, whereas $RbClO_4$ is sparingly soluble. Therefore, a suitable method of preparation is the neutralization of Rb_2CO_3 in aqueous $HClO_4$ with the formation of $RbClO_4$ precipitate. *Caution!* Perchlorates are potentially explosive.

Self-study exercises

Answers can be determined by reading the text.

1. Would the reaction of $CsNO_3$ and perchloric acid be a convenient method of preparing $CsClO_4$?

2. Would the collection of $LiClO_4$ precipitate from the reaction in aqueous solution of Li_2CO_3 and $NaClO_4$ be a convenient way of preparing and isolating $LiClO_4$?

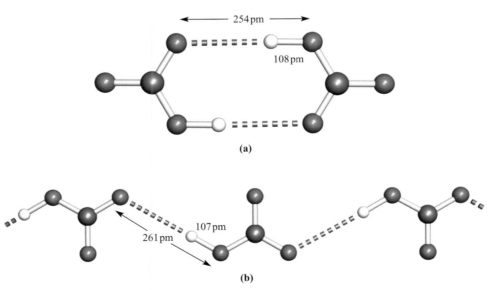

(a)

(b)

 Fig. 11.7 In the solid state, hydrogen bonding results in anion association in $NaHCO_3$ and $KHCO_3$, and the formation of (a) dimers in $NaHCO_3$ and (b) infinite chains in $KHCO_3$. Colour code: C, grey; O, red; H, white.

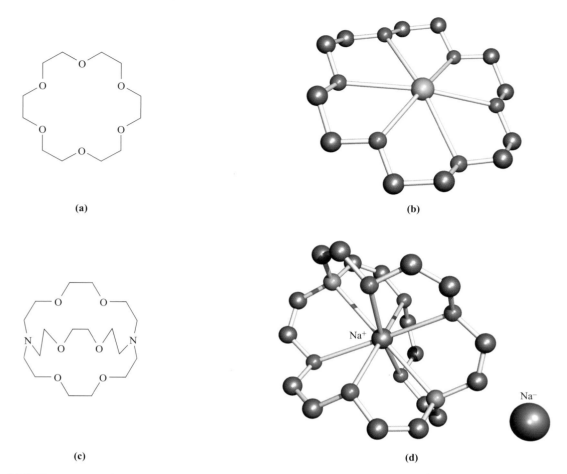

Fig. 11.8 The structures of (a) the macrocyclic polyether 18-crown-6, (b) the $[K(18\text{-crown-6})]^+$ cation for the $[Ph_3Sn]^-$ salt (X-ray diffraction) [T. Birchall *et al.* (1988) *J. Chem. Soc., Chem. Commun.*, p. 877], (c) the cryptand ligand crypt-[222], and (d) [Na(crypt-[222])]$^+$Na$^-$ (X-ray diffraction) [F.J. Tehan *et al.* (1974) *J. Am. Chem. Soc.*, vol. 96, p. 7203]. Colour code: K, orange; Na, purple; C, grey; N, blue; O, red.

3. The solubility of sodium sulfate in water, expressed in g of sodium sulfate per 100 g of water, increases from 273 to 305 K, while from 305 to 373 K, the solubility decreases slightly. What can you infer from these observations? [*Hint*: Is only one solid involved?]

In *dilute* solutions, alkali metal ions rarely form complexes, but where these are formed, e.g. with $[P_2O_7]^{4-}$ and $[EDTA]^{4-}$ (see Table 7.7), the normal order of stability constants is $Li^+ > Na^+ > K^+ > Rb^+ > Cs^+$. In contrast, when the aqueous ions are adsorbed on an *ion-exchange resin*, the order of the strength of adsorption is usually $Li^+ < Na^+ < K^+ < Rb^+ < Cs^+$. This suggests that the *hydrated ions* are adsorbed, since hydration energies decrease along this series and the total hydration interaction (i.e. primary hydration plus secondary interaction with more water molecules) is greatest for Li^+.

Complex ions

Unlike simple inorganic ligands, *polyethers* (see worked example 4.1) and, in particular, *cyclic polyethers* complex alkali metal ions quite strongly. The *crown ethers* are cyclic ethers which include 1,4,7,10,13,16-hexaoxacyclo-octadecane (Fig. 11.8a), the common name for which is 18-crown-6. This nomenclature gives the total number (C + O) and number of O atoms in the ring. Figure 11.8b shows the structure of the $[K(18\text{-crown-6})]^+$ cation; the K^+ ion is coordinated by the six *O*-donors. The radius of the cavity† inside the 18-crown-6 ring is 140 pm, and this compares with values of r_{ion} for the alkali metal ions ranging from 76 pm for Li^+ to 170 pm for Cs^+ (Table 11.1). The

---

† The concept of 'cavity size' is not as simple as it may appear; for further discussion, see the further reading list under 'Macrocyclic ligands' at the end of the chapter.

THEORY

Box 11.5 Large cations for large anions 1

Alkali metal ions encapsulated within crown ether or cryptand ligands are often used as a source of 'large cations' to aid the crystallization of salts containing large anions. An example is the compound $[K(crypt-222)]_2[C_{60}] \cdot 4C_6H_5Me$ which contains the fulleride $[C_{60}]^{2-}$. The diagram shows the unit cell of $[K(crypt-222)]_2[C_{60}] \cdot 4C_6H_5Me$; solvent molecules have been removed for clarity. The $[K(crypt-222)]^+$ cations have similar overall dimensions to the fulleride dianions, allowing the ions to pack efficiently in the crystal lattice.

Colour code: C, grey; K, purple; N, blue; O, red.

[Data from: T.F. Fassler *et al.* (1997) *Angew. Chem., Int. Ed.*, vol. 36, p. 486.]

See also: Box 24.1 – *Large cations for large anions 2*.

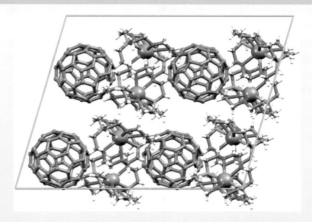

radius of the K^+ ion (138 pm) is well matched to that of the macrocycle, and stability constants for the formation of $[M(18\text{-crown-}6)]^+$ (eq. 11.20) in acetone follow the sequence $K^+ > Rb^+ > Cs^+ \approx Na^+ > Li^+$.

$$M^+ + 18\text{-crown-}6 \rightleftharpoons [M(18\text{-crown-}6)]^+ \qquad (11.20)$$

Different crown ethers have different cavity sizes, although the latter is not a fixed property because of the ability of the ligand to change conformation. Thus, the radii of the holes in 18-crown-6, 15-crown-5 and 12-crown-4 can be taken to be roughly 140, 90 and 60 pm respectively. It is, however, dangerous to assume that an $[ML]^+$ complex will fail to form simply because the size of M^+ is not matched correctly to the hole size of the macrocyclic ligand L. For example, if the radius of M^+ is slightly larger than the radius of L, a complex may form in which M^+ sits above the plane containing the donor atoms, e.g. $[Li(12\text{-crown-}4)Cl]$ (**11.1**). Alternatively a 1:2 complex $[ML_2]^+$ may result in which the metal ion is sandwiched between two ligands, e.g. $[Li(12\text{-crown-}4)_2]^+$. Note that these latter examples refer to complexes crystallized from solution.

The concept of matching ligand hole size to the size of the metal ion has played a role in discussions of the apparent selectivity of particular ligands for particular metal ions. The selectivity (such as that discussed above for $[M(18\text{-crown-}6)]^+$ complexes, eq. 11.20) is based on measured stability constants. It has, however, also been pointed out that the stability constants for $[KL]^+$ complexes are often higher than for corresponding $[ML]^+$ complexes where $M = Li$, Na, Rb or Cs, even when hole-matching is clearly not the all-important factor. An alternative explanation focuses on the fact that, when a crown ether binds M^+, the chelate rings that are formed are all 5-membered, and that the size of the K^+ ion is ideally suited to 5-membered chelate ring formation (see Section 7.12).[†] Complexes formed by such macrocyclic ligands are appreciably more stable than those formed by closely related open chain ligands (see Section 7.12).

The crown ether-complexed alkali metal ions are large and hydrophobic, and their salts tend to be soluble in *organic* solvents. For example, whereas $KMnO_4$ is water-soluble but insoluble in benzene, $[K(18\text{-crown-}6)][MnO_4]$ is soluble in benzene; mixing benzene with aqueous $KMnO_4$ leads to the purple colour being transferred from the aqueous to the benzene layer. This phenomenon is very useful in preparative organic chemistry, the *anions* being little solvated and, therefore, highly reactive.

> A *cryptand* is a polycyclic ligand containing a cavity. When the ligand coordinates to a metal ion, the complex ion is called a *cryptate*.

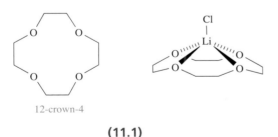

12-crown-4

(11.1)

[†] For more detailed discussion, see: R.D. Hancock (1992) *J. Chem. Educ.*, vol. 69, p. 615 – 'Chelate ring size and metal ion selection'.

Figure 11.8c shows the structure of the *cryptand* ligand 4,7,13,16,21,24-hexaoxa-1,10-diazabicyclo[8.8.8]hexacosane, commonly called cryptand-222 or crypt-222, where the 222 notation gives the number of *O*-donor atoms in each of the three chains. Cryptand-222 is an example of a *bicyclic* ligand which can *encapsulate* an alkali metal ion. Cryptands protect the complexed metal cation even more effectively than do crown ethers. They show selective coordination behaviour; cryptands-211, -221 and -222 with cavity radii of 80, 110 and 140 pm, respectively, form their most stable alkali metal complexes with Li^+, Na^+ and K^+ respectively (see Table 11.1 for r_{ion}).

$$2Na \rightleftharpoons Na^+ + Na^- \qquad (11.21)$$

The ability of crypt-222 to shift equilibrium 11.21 to the right-hand side is striking. This is observed when crypt-222 is added to Na dissolved in ethylamine, and the isolated product is the diamagnetic, golden-yellow [Na(crypt-222)]$^+$Na$^-$ (Fig. 11.8d). The solid state structure indicates that the effective radius of the *sodide* ion is ≈ 230 pm, i.e. Na$^-$ is similar in size to I$^-$. The replacement of the O atoms in crypt-222 by NMe groups generates ligand **11.2**, ideally suited to encapsulate K$^+$. Its use in place of crypt-222 has aided the study of alkalide complexes by increasing their thermal stability. Whereas [Na(crypt-222)]$^+$Na$^-$ usually has to be handled below ≈ 275 K, [K(**11.2**)]$^+$Na$^-$ and [K(**11.2**)]$^+$K$^-$ are stable at 298 K.

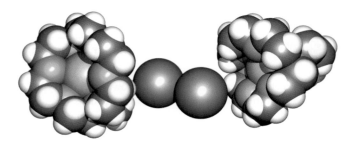

Fig. 11.9 A space-filling diagram of the [Na₂]²⁻ dimer, sandwiched between two [BaL]$^+$ cations in the complex [BaL]Na·2MeNH₂, where L is the ligand (**11.3** – H)$^-$ (see eq. 11.22). The structure was determined by X-ray diffraction; nitrogen-bonded H atoms are not shown [M.Y. Redko *et al.* (2003) *J. Am. Chem. Soc.*, vol. 125, p. 2259]. Colour code: Na, purple; Ba, orange; N, blue; C, grey; H, white.

[Na₂]²⁻, in which the Na–Na distance is 417 pm (Fig. 11.9). The dimer appears to be stabilized by N–H····Na$^-$ hydrogen-bonded interactions involving the [Ba(**11.3** – H)]$^+$ cation (see end-of-chapter problem 11.26a). The first hydrogen sodide 'H$^+$Na$^-$' was prepared using ligand **11.4** to encapsulate H$^+$, thereby protecting it and rendering it kinetically stable with respect to strong bases and alkali metals. The space-filling diagram of ligand **11.4** shows its globular nature, and illustrates how the nitrogen donor atoms are directed towards the central cavity.

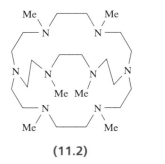

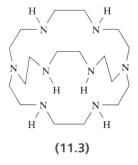

(11.2)   **(11.3)**

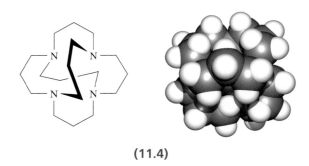

(11.4)

Replacement of O in crypt-222 by NMe rather than NH (i.e. to give ligand **11.2** rather than **11.3**) is necessary because the NH groups would react with M$^-$, liberating H₂. This is illustrated in reaction 11.22 which is carried out in liquid NH₃/MeNH₂; the Ba²⁺ ion in the product is encapsulated within the deprotonated ligand.

$$Ba + Na + \mathbf{11.3} \longrightarrow [Ba^{2+}(\mathbf{11.3} - H)^-]Na^- \qquad (11.22)$$

$$(\mathbf{11.3} - H)^- = \text{deprotonated ligand } \mathbf{11.3}$$

Despite this complication, this reaction is noteworthy for its product. In the solid state, the Na$^-$ ions pair up to give

Alkalides have also been prepared containing Rb$^-$ and Cs$^-$. In these reactions, the cryptand:metal molar ratio is 1:2. If the reaction is carried out using a greater proportion of ligand, paramagnetic black *electrides* can be isolated, e.g. [Cs(crypt-222)₂]$^+$e$^-$ in which the electron is trapped in a cavity of radius ≈ 240 pm. Electrides can also be prepared using crown ethers, and examples of crystallographically confirmed complexes are [Cs(15-crown-5)₂]$^+$e$^-$, [Cs(18-crown-6)₂]$^+$e$^-$ and [Cs(18-crown-6)(15-crown-5)]$^+$e$^-$·18-crown-6. The arrangement of the electron-containing cavities in the solid state has a profound effect on the

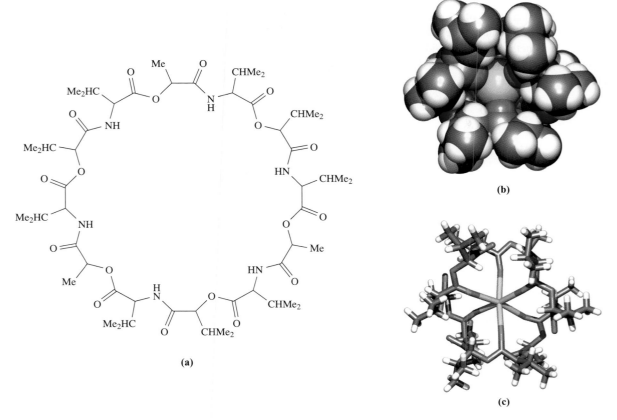

Fig. 11.10 (a) The structure of valinomycin. (b) and (c) The structure (X-ray diffraction) of [K(valinomycin)]⁺ in the salt [K(valinomycin)]$_2$[I$_3$][I$_5$] [K. Neupert-Laves *et al.* (1975) *Helv. Chim. Acta*, vol. 58, p. 432]. The stick representation illustrates the octahedral coordination of the K⁺ ion, while the space-filling representation of the [K(valinomycin)]⁺ ion which illustrates the hydrophobic exterior. Colour code: O, red; N, blue; C, grey; K⁺ ion, orange.

electrical conductivities of these materials. The conductivity of [Cs(18-crown-6)(15-crown-5)]⁺e⁻·18-crown-6 (in which the electron-cavities form rings) is $\approx 10^6$ times greater than that of either [Cs(15-crown-5)$_2$]⁺e⁻ or [Cs(18-crown-6)$_2$]⁺e⁻ (in which the free electron-cavities are organized in chains).

Cryptands have also been used to isolate crystalline LiO$_3$ and NaO$_3$ as [Li(crypt-211)][O$_3$] and [Na(crypt-222)][O$_3$] respectively, and further applications of these encapsulating ligands are in the isolation of alkali metal salts of *Zintl ions* (see Sections 9.6 and 14.7). Sodium and potassium cryptates are interesting models for biologically occurring materials involved in the transfer of Na⁺ and K⁺ across cell membranes. An example is valinomycin, a cyclic polypeptide (Fig. 11.10a). Valinomycin is present in certain microorganisms and is selective towards binding K⁺ ions. Figure 11.10 illustrates that the valinomycin ligand uses six of its carbonyl groups to octahedrally coordinate K⁺. The [K(valinomycin)]⁺ ion has a hydrophobic exterior which makes it lipid-soluble, and the complex ion can therefore be transported across the lipid bilayer of a cell membrane.[†]

11.9 Non-aqueous coordination chemistry

Many complexes of the group 1 metals prepared in non-aqueous conditions (e.g. in polar organic solvents) are known. Alkali metal ions are typically hard Lewis acids and favour coordination by hard *O*- and *N*-donor ligands. The use of macrocyclic ligands was detailed in Section 11.8, and in this section, we focus on examples of discrete molecular and polymeric species incorporating other types of *O*- and *N*-donors. The bonding in these types of compounds is considered to be predominantly ionic. The use of sterically

[†] Relevant articles are: E. Gouaux and R. MacKinnon (2005) *Science*, vol. 310, p. 1461 – 'Principles of selective ion transport in channels and pumps'; S. Varma, D. Sabo and S.B. Rempe (2007) *J. Mol. Biol.*, vol. 376, p. 13 – 'K⁺/Na⁺ selectivity in K Channels and valinomycin: over-coordination versus cavity-size constraints'.

demanding ligands favours the formation of low nuclearity complexes. A general method of synthesis is to prepare an alkali metal salt (e.g. LiCl) in the presence of a coordinating ligand such as hexamethylphosphoramide (HMPA, **11.5**) to produce aggregates, e.g. $[\{LiCl(HMPA)\}_4]$ with a cubane core (**11.6**) and $[Li_2Br_2(HMPA)_3]$ (**11.7**).

(11.5)

RO = HMPA

(11.6) **(11.7)**

Bulky alkoxides and amides also stabilize discrete complexes, e.g. $[\{KO^tBu\}_4]$ with a cubane structure, and $[\{^tBuHNLi\}_8]$ (Fig. 11.11). Amidolithium complexes of type RR'NLi in which R and R' are sterically demanding alkyl, aryl or trialkylsilyl groups exhibit diverse structures with planar Li_2N_2-rings being common building blocks in the structures. Trimethylamido complexes are known for the group 1 metals from Li to Cs, and examples include $[\{LiN(SiMe_3)_2\}_3]$, $[\{(THF)LiN(SiMe_3)_2\}_2]$, $[\{NaN(SiMe_3)_2\}_3]$, $[\{KN(SiMe_3)_2\}_2]$, $[\{RbN(SiMe_3)_2\}_2]$ and $[\{CsN(SiMe_3)_2\}_2]$. Polymeric species have also

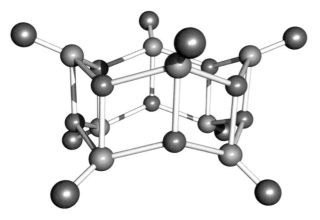

Fig. 11.11 The structure of $[\{LiNH^tBu\}_8]$ determined by X-ray diffraction; hydrogen and methyl-carbon atoms have been omitted for clarity [N.D.R. Barnett *et al.* (1996) *J. Chem. Soc., Chem. Commun.*, p. 2321]. Colour code: Li, red; N, blue; C, grey.

been structurally characterized in which the amido ligands and/or solvent molecules such as 1,4-dioxane (**11.8**) bridge alkali metal centres, e.g. $[\{NaN(SiMe_3)_2\}_\infty]$ and $[\{KN(SiMe_3)_2(\mathbf{11.8})_2\}_\infty]$.

1,4-dioxane   bridging mode of 1,4-dioxane

(11.8)

In Section 23.2, organometallic compounds of the group 1 metals are described and these include examples in which the chelating ligand $Me_2NCH_2CH_2NMe_2$ (TMEDA) plays a stabilizing role.

KEY TERMS

The following terms were introduced in this chapter. Do you know what they mean?

- ❏ amalgam
- ❏ peroxide ion
- ❏ superoxide ion

- ❏ ozonide ion
- ❏ ion-exchange (ion-exchange resin)
- ❏ crown ether

- ❏ cryptand
- ❏ alkalide
- ❏ electride

FURTHER READING

N.N. Greenwood and A. Earnshaw (1997) *Chemistry of the Elements*, 2nd edn, Butterworth-Heinemann, Oxford – Chapter 4 gives a good account of the inorganic chemistry of the group 1 metals.

M. Jansen and H. Nuss (2007) *Z. Anorg. Allg. Chem.*, vol. 633, p. 1307 – 'Ionic ozonides'.

A.G. Massey (2000) *Main Group Chemistry*, 2nd edn, Wiley, Chichester – Chapter 4 covers the chemistry of the group 1 metals.

A. Simon (1997) *Coord. Chem. Rev.*, vol. 163, p. 253 – A review which includes details of synthesis, crystallization and structures of alkali metal suboxides.

A.F. Wells (1984) *Structural Inorganic Chemistry*, 5th edn, Clarendon Press, Oxford – A well-illustrated and detailed account of the structures of alkali metal compounds.

Macrocyclic ligands

The following five references give excellent accounts of the macrocyclic effect:

J. Burgess (1999) *Ions in Solution: Basic Principles of Chemical Interactions*, 2nd edn, Horwood Publishing, Chichester, Chapter 6.

E.C. Constable (1996) *Metals and Ligand Reactivity*, revised edn, VCH, Weinheim, Chapter 6.

E.C. Constable (1999) *Coordination Chemistry of Macrocyclic Compounds*, Oxford University Press, Oxford, Chapter 5.

L.F. Lindoy (1989) *The Chemistry of Macrocyclic Ligand Complexes*, Cambridge University Press, Cambridge, Chapter 6.

A.E. Martell, R.D. Hancock and R.J. Motekaitis (1994) *Coord. Chem. Rev.*, vol. 133, p. 39.

The following reference gives an account of the coordination chemistry of alkali metal crown ether complexes:

J.W. Steed (2001) *Coord. Chem. Rev.*, vol. 215, p. 171.

Alkalides and electrides

J.L. Dye (2009) *Acc. Chem. Res.*, vol. 42, p. 1564 – 'Electrides: early examples of quantum confinement'.

J.L. Dye, M.Y. Redko, R.H. Huang and J.E. Jackson (2007) *Adv. Inorg. Chem.*, vol. 59, p. 205 – 'Role of cation complexants in the synthesis of alkalides and electrides'.

Q. Xie, R.H. Huang, A.S. Ichimura, R.C. Phillips, W.P. Pratt Jr and J.L. Dye (2000) *J. Am. Chem. Soc.*, vol. 122, p. 6971 – Report of the electride $[Rb(crypt-222)]^+ e^-$, its structure, polymorphism and electrical conductivity, with references to previous work in the area.

PROBLEMS

11.1 (a) Write down, in order, the names and symbols of the metals in group 1; check your answer by reference to the first page of this chapter. (b) Give a *general* notation that shows the ground state electronic configuration of each metal.

11.2 Explain why, for a given alkali metal, the second ionization energy is very much higher than the first.

11.3 Describe the solid state structures of (a) the alkali metals and (b) the alkali metal chlorides, and comment on trends down the group.

11.4 Discuss trends in (a) melting points, and (b) ionic radii, r_+, for the metals on descending group 1.

11.5 (a) Describe the bonding in the M_2 diatomics (M = Li, Na, K, Rb, Cs) in terms of valence bond and molecular orbital theories. (b) Account for the trend in metal–metal bond dissociation energies given in Table 11.1.

11.6 (a) Write an equation for the decay of ^{40}K by electron capture. (b) Determine the volume of gas produced when 1 g of ^{40}K decays according to this equation. (c) The decay of ^{40}K is the basis of a method for dating rock samples. Suggest how this method works.

11.7 Comment on the following observations:
(a) Li is the alkali metal that forms the nitride most stable with respect to decomposition into its elements.
(b) The mobilities of the alkali metal ions in aqueous solution follow the sequence $Li^+ < Na^+ < K^+ < Rb^+ < Cs^+$.
(c) E^o for $M^+(aq) + e^- \rightleftharpoons M(s)$ is nearly constant (see Table 11.1) for the alkali metals.

11.8 Suggest what will happen when a mixture of LiI and NaF is heated.

11.9 Very often, samples for IR spectroscopy are prepared as solid state discs by grinding the compound for analysis with an alkali metal halide. Suggest why the IR spectra of $K_2[PtCl_4]$ in KBr and KI discs might be different.

11.10 Suggest why KF is a better reagent than NaF for replacement of chlorine in organic compounds by fluorine by the autoclave reaction:

11.11 Suggest why the solubility of sodium sulfate in water increases to 305 K and then decreases.

11.12 By considering Fig. 11.4a and the packing of the units shown into an infinite lattice, show that (a) the ratio of $Li^+:N^{3-}$ ions in layer 2 is 2:1, and (b) the stoichiometry of the compound is Li_3N.

11.13 Construct approximate MO diagrams for $[O_2]^-$ and $[O_2]^{2-}$ and confirm that $[O_2]^-$ is paramagnetic, while $[O_2]^{2-}$ is diamagnetic.

11.14 What general type of reaction is equilibrium 11.21? Confirm your answer by considering the oxidation state changes involved. Give two other examples of this general type of reaction.

11.15 Write down the formulae of the following ions: (a) superoxide; (b) peroxide; (c) ozonide; (d) azide; (e) nitride; (f) sodide.

11.16 Write a brief account of the uses of the alkali metals and their compounds, with reference to relevant industrial processes.

11.17 Alkali metal cyanides, MCN, are described as *pseudohalides*. (a) Draw the structure of the cyanide ion, and give a description of its bonding. (b) Interpret the structure of NaCN if it possesses an NaCl-type structure.

11.18 Give an account of what happens when Na dissolves in liquid NH_3.

11.19 Write balanced equations for the following reactions:
(a) sodium hydride with water;
(b) potassium hydroxide with acetic acid;
(c) thermal decomposition of sodium azide;
(d) potassium peroxide with water;
(e) sodium fluoride with boron trifluoride;
(f) electrolysis of molten KBr;
(g) electrolysis of aqueous NaCl.

11.20 Suggest explanations for the following observations.
(a) Although Na_2O_2 is described as being colourless, samples of Na_2O_2 often appear to be very pale yellow.
(b) NaO_2 is paramagnetic.

11.21 (a) Explain how face-sharing between M_6O octahedra leads to compounds with stoichiometries of M_9O_2 for M = Rb, and $M_{11}O_3$ for M = Cs.
(b) The suboxide Cs_7O contains $Cs_{11}O_3$ clusters. Explain how this arises.

11.22 (a) Which of the following compounds is the least soluble in water at 298 K: Li_2CO_3, LiI, Na_2CO_3, NaOH, Cs_2CO_3, KNO_3?
(b) Which of the following compounds decompose(s) when added to water at 298 K: RbOH, $NaNO_3$, Na_2O, Li_2SO_4, K_2CO_3, LiF?
(c) Determine the solubility of Li_2CO_3 in water if $K_{sp} = 8.15 \times 10^{-4}$.

OVERVIEW PROBLEMS

11.23 Suggest products and write balanced equations for each of the following reactions; these are *not* necessarily balanced on the left-hand side.
(a) $KOH + H_2SO_4 \longrightarrow$
(b) $NaOH + SO_2 \longrightarrow$
(c) $KOH + C_2H_5OH \longrightarrow$
(d) $Na + (CH_3)_2CHOH \longrightarrow$
(e) $NaOH + CO_2 \longrightarrow$
(f) $NaOH + CO \xrightarrow{450\,K}$
(g) $H_2C_2O_4 + CsOH \longrightarrow$
(h) $NaH + BCl_3 \longrightarrow$

11.24 (a) Na_3N remained an elusive compound until 2002. Calculate a value for $\Delta_f H^{\circ}(Na_3N, s)$ using data from Appendices 8 and 10, and the following estimated values of $\Delta H(298\,K)$:

$$N(g) + 3e^- \longrightarrow N^{3-}(g)$$
$$\Delta_{EA} H = +2120\,kJ\,mol^{-1}$$

$$3Na^+(g) + N^{3-}(g) \longrightarrow Na_3N(s)$$
$$\Delta_{lattice} H^{\circ} = -4422\,kJ\,mol^{-1}$$

Comment on whether the value obtained is sufficient to indicate whether Na_3N is thermodynamically stable.

(b) The high-temperature crystalline form of $RbNH_2$ adopts a structure with a ccp array of $[NH_2]^-$ ions and Rb^+ ions occupying octahedral sites. To which structure type does this correspond? Sketch a unit cell of $RbNH_2$ and confirm the stoichiometry of the compound by considering the number of ions per unit cell.

11.25 (a) Suggest products for the reaction of Li_3N with water. Write a balanced equation for the reaction.

(b) A compound **A** was isolated from the reaction between a group 1 metal M and O_2. **A** reacts with water to give only MOH, while M reacts in a controlled manner with water giving MOH and another product, **B**. Suggest identities for M, **A** and **B**. Write equations for the reactions described. Compare the reaction of M with O_2 with those of the other group 1 metals with O_2.

11.26 (a) The crystalline product from reaction 11.22 contains $[Na_2]^{2-}$ units. Construct an MO diagram for $[Na_2]^{2-}$ and determine the bond order in this species. Comment on the result in the light of the text discussion of this species, explaining differences between the MO model and the experimental data.

(b) The enthalpies of hydration for Na^+, K^+ and Rb^+ are -404, -321 and $-296\,kJ\,mol^{-1}$ respectively. Suggest an explanation for this trend.

11.27 (a) Stability constants for the formation of $[M(18\text{-}crown\text{-}6)]^+$ complexes in acetone are given below. Comment critically on these data.

M^+	Li^+	Na^+	K^+	Rb^+	Cs^+
$\log K$	1.5	4.6	6.0	5.2	4.6

(b) Of the salts $NaNO_3$, $RbNO_3$, Cs_2CO_3, Na_2SO_4, Li_2CO_3, LiCl and LiF, which are soluble in water? Using LiCl and LiF as examples, discuss factors that contribute to the solubility of a salt.

11.28 The first list below contains the formula of a group 1 metal or metal compound. Match these to the descriptions given in the second column.

List 1	List 2
Li_3N	Reacts explosively with water, liberating H_2
NaOH	Sparingly soluble in water
Cs	Basic compound with an antifluorite structure
Cs_7O	Possesses the highest first ionization energy of the group 1 metals
Li_2CO_3	Formed by direct combination of the elements, and possesses a layer structure
$NaBH_4$	Neutralizes aqueous HNO_3 with no evolution of gas
Rb_2O	Used as a reducing agent
Li	A suboxide

INORGANIC CHEMISTRY MATTERS

11.29 Mercedes-Benz launched the first mass-produced hybrid electric vehicle containing a lithium-ion battery in 2009. (a) Explain how this battery works and show how both $LiCoO_2$ and $LiFePO_4$ function as positive electrode materials. (b) How does a plug-in electric vehicle differ from a hybrid electric vehicle?

11.30 Sodium carbonate is a vital component of the glass manufacturing industry. (a) By referring to Section 14.9, explain the role that Na^+ plays in soda-lime glass. (b) Commercial Na_2CO_3 is categorized into 'natural' and 'synthetic'. Distinguish between these and detail the process whereby synthetic Na_2CO_3 is manufactured, paying attention to any recycling processes.

Topics

12 The group 2 metals

1	2			13	14	15	16	17	18
H									He
Li	Be			B	C	N	O	F	Ne
Na	Mg			Al	Si	P	S	Cl	Ar
K	Ca	*d*-block		Ga	Ge	As	Se	Br	Kr
Rb	Sr			In	Sn	Sb	Te	I	Xe
Cs	Ba			Tl	Pb	Bi	Po	At	Rn
Fr	Ra								

12.1 Introduction

The group 2 metals – beryllium, magnesium, calcium, strontium, barium and radium – are collectively known as the
alkaline earth metals. The relationships between them are
very similar to those among the alkali metals. However, Be
stands apart from the other group 2 metals to a *greater*
extent than does Li from its homologues. For example, whereas
Li^+ and Na^+ salts (with a common counter-ion) usually crystallize with the same structure type, this is not true for Be(II)
and Mg(II) compounds. Beryllium compounds tend either to
be covalent or to contain the hydrated $[Be(OH_2)_4]^{2+}$ ion.
The high values of the enthalpy of atomization (Appendix
10) and ionization energies (Appendix 8) of the Be atom,
and the small size and consequent high charge density of a

naked Be^{2+} ion, militate against the formation of naked
Be^{2+}. It is noteworthy that Be is the only group 2 metal not
to form a stable complex with $[EDTA]^{4-}$ (see Table 7.7).

Radium is radioactive and is formed as $^{226}_{88}Ra$
(α-emitter, $t_{\frac{1}{2}} = 1622\,yr$) in the $^{238}_{92}U$ decay series. Uses of
radium-226 in cancer treatment have generally been super-
seded by other radioisotopes. The properties of radium and
its compounds can be inferred by extrapolation from those
of corresponding Ca, Sr and Ba compounds.

We have already described some aspects of the chemistry
of the group 2 elements as follows:

- ionization energies of metals (Section 1.10);
- bonding in diatomic Be_2 (Section 2.3);
- bonding schemes for $BeCl_2$ (Sections 2.8 and 5.2);
- structures of metals (Table 6.2);
- structures of halides and oxides, see CaF_2, CdI_2 and NaCl
  structures (Section 6.11);
- lattice energy treatment of disproportionation of CaF into
  Ca and CaF_2 (Section 6.16);
- solubility products, e.g. for CaF_2 (Section 7.9);
- hydration of metal ions (Section 7.9);
- saline hydrides, MH_2 (Section 10.7).

12.2 Occurrence, extraction and uses

Occurrence

Beryllium occurs mainly as the silicate mineral *beryl*,
$Be_3Al_2[Si_6O_{18}]$ (silicates, see Section 14.9). It is also
found in many natural minerals including *bertrandite*,

Fig. 12.1 Crystals of gypsum ($CaSO_4 \cdot 2H_2O$) in the Cueva de los Cristales in the Naica mine system in Mexico.

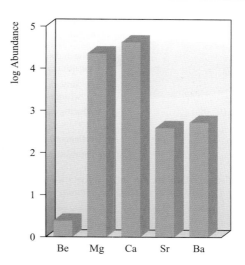

Fig. 12.2 Relative abundances in the Earth's crust of the alkaline earth metals (excluding Ra); the data are plotted on a logarithmic scale. The units of abundance are ppm.

$Be_4Si_2O_7(OH)_2$. Precious forms include *emerald* and *aquamarine*. Magnesium and calcium are the eighth and fifth most abundant elements, respectively, in the Earth's crust, and Mg is the third most abundant in the sea. The elements Mg, Ca, Sr and Ba are widely distributed in minerals and as dissolved salts in seawater. Some important minerals are *dolomite* ($CaCO_3 \cdot MgCO_3$), *magnesite* ($MgCO_3$), *olivine* (($Mg,Fe)_2SiO_4$), *carnallite* ($KCl \cdot MgCl_2 \cdot 6H_2O$), $CaCO_3$ (in the forms of *chalk*, *limestone* and *marble*), *gypsum* ($CaSO_4 \cdot 2H_2O$), *celestite* ($SrSO_4$), *strontianite* ($SrCO_3$) and *barytes* ($BaSO_4$). Figure 12.1 shows 11 m long crystals of gypsum in Mexico's Cueva de los Cristales (Cave of Crystals). Their very slow growth over hundreds of thousands of years occurred because geothermally heated water originally filling the caves provided a constant temperature of 311 K which is the transition temperature between $CaSO_4$ (anhydrite) and $CaSO_4 \cdot 2H_2O$ (gypsum).

The natural abundances of Be, Sr and Ba are far less than those of Mg and Ca (Fig. 12.2).

Extraction

Of the group 2 metals, only Mg is manufactured on a large scale. The mixed metal carbonate dolomite is thermally decomposed to a mixture of MgO and CaO, and MgO is reduced by ferrosilicon in Ni vessels (eq. 12.1). Magnesium is then removed by distillation *in vacuo*.

$$2MgO + 2CaO + FeSi \xrightarrow{1450\,K} 2Mg + Ca_2SiO_4 + Fe \quad (12.1)$$

Extraction of Mg by electrolysis of fused $MgCl_2$ is also important and is applied to the extraction of the metal from seawater. The first step is precipitation (see Table 7.4) of $Mg(OH)_2$ by addition of $Ca(OH)_2$ (*slaked lime*), produced from $CaCO_3$ (available as various calcareous deposits, see Fig. 11.6). Neutralization with

hydrochloric acid (eq. 12.2) and evaporation of water gives $MgCl_2 \cdot xH_2O$, which, after heating at 990 K, yields the anhydrous chloride. This is followed by electrolysis of molten $MgCl_2$ and solidification of Mg (eq. 12.3).

$$2HCl + Mg(OH)_2 \longrightarrow MgCl_2 + 2H_2O \quad (12.2)$$

$$\left. \begin{array}{ll} \textit{At the cathode:} & Mg^{2+}(l) + 2e^- \longrightarrow Mg(l) \\ \textit{At the anode:} & 2Cl^-(l) \longrightarrow Cl_2(g) + 2e^- \end{array} \right\} \quad (12.3)$$

Beryllium may be obtained from *beryl* by first heating with Na_2SiF_6 (eq. 12.4), extracting the water-soluble BeF_2 formed, and precipitating $Be(OH)_2$.

Beryllium is also produced from *bertrandite* or *beryl* by extraction processes which involve leaching the ores with H_2SO_4 and steam, and conversion of beryllium sulfate to $Be(OH)_2$. This is an intermediate compound in the production of Be, Be alloys and BeO. Production of the metal involves either reduction of BeF_2 with Mg, or electrolysis of $BeCl_2$ fused with NaCl.

$$Be_3Al_2[Si_6O_{18}] + 3Na_2SiF_6 \longrightarrow 3BeF_2 + 2Na_3AlF_6 + 9SiO_2 \quad (12.4)$$

The production of Ca is by electrolysis of fused $CaCl_2$ and CaF_2. The metals Sr and Ba are extracted by reduction of the corresponding oxides by Al, or by electrolysis of MCl_2 (M = Sr, Ba).

Major uses of the group 2 metals and their compounds

Caution! Beryllium and soluble barium compounds are extremely toxic.

Beryllium is one of the lightest metals known, is non-magnetic, and has a high thermal conductivity and a very

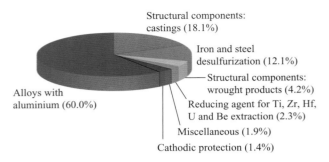

Fig. 12.3 Uses of Mg in the US in 2008 [data from US Geological Survey]; for a discussion of *cathodic protection*, see Box 8.4.

high melting point (1560 K). These properties, combined with inertness towards aerial oxidation, render it of industrial importance. It is used in the manufacture of body parts in high-speed aircraft and missiles, and in communication satellites. Because of its low electron density, Be is a poor absorber of electromagnetic radiation and, as a result, is used in X-ray tube windows. Its high melting point and low cross-section for neutron capture make Be useful in the nuclear energy industry.

Figure 12.3 summarizes the major uses of Mg. The presence of Mg in Mg/Al alloys imparts greater mechanical strength and resistance to corrosion, and improves fabrication properties. Mg/Al alloys are used in aircraft and automobile body parts and lightweight tools. Die-casting of structural components accounted for 18.1% of the consumption of primary Mg in the US in 2008, compared with 57.0% in 2004. This corresponds not to a change in use of Mg, but to a fall in total Mg consumption (Box 12.1) which is largely associated with trends in the vehicle manufacturing industry. Miscellaneous uses (Fig. 12.3) include flares, fireworks and photographic flashlights, and medical applications such as indigestion powders (*milk of magnesia*, $Mg(OH)_2$) and a purgative (*Epsom salts*, $MgSO_4 \cdot 7H_2O$). Both Mg^{2+} and Ca^{2+} ions are catalysts for diphosphate–triphosphate (see Box 15.11) transformations in biological systems; Mg^{2+} is an essential constituent of chlorophylls in green plants (see Section 12.8).

Uses of compounds of calcium far outnumber those of the metal. World production of CaO, $Ca(OH)_2$, CaO·MgO, $Ca(OH)_2 \cdot MgO$ and $Ca(OH)_2 \cdot Mg(OH)_2$ was $\approx 296\,000$ Mt in 2008 with China being by far the greatest producer. Calcium oxide (quicklime or lime) is produced by calcining limestone (see Fig. 11.6) and a major use is as a component in building mortar. Dry sand and CaO mixtures can be stored and transported. On adding water, and as CO_2 is absorbed,

ENVIRONMENT

Box 12.1 Recycling of materials: magnesium

Recycling of materials became increasingly important during the last decades of the 20th century, and continues to have a significant influence on chemical industries. A large fraction of the total Mg consumed is in the form of Al/Mg alloys (see Fig. 12.3), and recycling of Al cans necessarily means recovery of Mg. The graph below shows the variation in total consumption of primary Mg in the US from 1970 to 2008, and the increasing trend towards recovering the metal. The significant fall in total Mg consumption after 2004 is mainly associated with trends in the vehicle manufacturing industry.

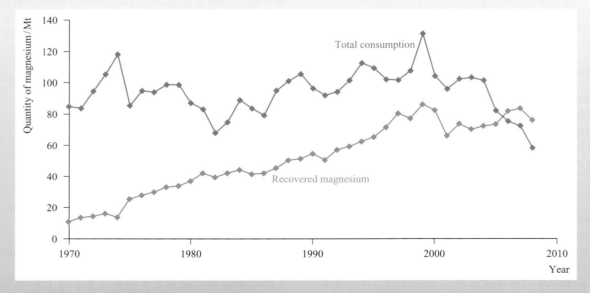

[Data from US Geological Survey]

the mortar sets as solid $CaCO_3$ (scheme 12.5). The sand in the mortar is a binding agent.

$$CaO(s) + H_2O(l) \longrightarrow Ca(OH)_2(s) \quad \Delta_r H^\circ = -65\,kJ\,mol^{-1}$$
Quicklime   Slaked lime

$$Ca(OH)_2(s) + CO_2(g) \longrightarrow CaCO_3(s) + H_2O(l)$$
$$(12.5)$$

Other important uses of lime are in the steel industry (see Box 6.1), pulp and paper manufacturing, and extraction of Mg. Calcium carbonate is in huge demand in steel, glass, cement and concrete manufacturing (see Box 14.8), and the Solvay process (Fig. 11.6). Applications of $CaCO_3$ and $Ca(OH)_2$ with environmental significance are in desulfurization processes (see Box 12.2). Large quantities of $Ca(OH)_2$ are used to manufacture bleaching powder, $Ca(OCl)_2 \cdot Ca(OH)_2 \cdot CaCl_2 \cdot 2H_2O$ (see Sections 17.2 and 17.9) and in water treatment (see eq. 12.29).

Calcium fluoride occurs naturally as the mineral fluorspar, and is commercially important as the raw material for the manufacture of HF (eq. 12.6) and F_2 (see Section 17.2). Smaller amounts of CaF_2 are used as a flux in the steel industry, for welding electrode coatings, and in glass manufacture. Prisms and cell windows made from CaF_2 are used in spectrophotometers.

$$CaF_2 + 2H_2SO_4 \longrightarrow 2HF + Ca(HSO_4)_2 \quad (12.6)$$
conc

The two mineral sources for strontium are the sulfate (*celestite*) and carbonate (*strontianite*). Trends in the commercial market for strontium have altered considerably over the last decade. Its main use was as a component ($\approx 8\%$ SrO) in colour television faceplate glass where its function was to block X-ray emissions from the cathode ray tube (CRT). However, this use of Sr has almost completely disappeared following the global uptake of flat screen televisions which incorporate only small amounts of $SrCO_3$. Current commercial demands for Sr are in pyrotechnics ($Sr(NO_3)_2$), ceramic ferrite magnets ($SrCO_3$), alloys and pigments.

Barite (or *barytes*) is the mineral form of $BaSO_4$. World production in 2008 was ≈ 8000 Mt, with China supplying over half this total. About 85% of global barite consumption is as a weighting material in oil- and gas-well drilling fluids. On a much smaller scale of application, the ability of $BaSO_4$ to stop the passage of X-rays leads to its use as a 'barium meal' in radiology for imaging the alimentary tract. Uses of Ba as a 'getter' in vacuum tubes arise from its high reactivity with gases including O_2 and N_2.

12.3 Physical properties

General properties

Selected physical properties of the group 2 elements are listed in Table 12.1. The intense radioactivity of Ra makes it impossible to obtain all the data for this element. Some general points to note from Table 12.1 are as follows:

- The general trend in decreasing values of IE_1 and IE_2 down the group (see Section 1.10) is broken by the increase in going from Ba to Ra, attributed to the *thermodynamic 6s inert pair effect* (see Box 13.4).
- High values of IE_3 preclude the formation of M^{3+} ions.
- Quoting a value of r_{ion} for beryllium assumes that the Be^{2+} ion is present in BeF_2 and BeO, a questionable assumption.
- There are no simple explanations for the irregular group variations in properties such as melting points and $\Delta_a H^\circ$.
- Values of E° for the M^{2+}/M couple are fairly constant (with the exception of Be), and can be explained in a similar way as for the group 1 metals (see Sections 8.7 and 11.3).

Flame tests

As for the alkali metals, emission spectra for the group 2 metals are readily observed and flame tests (see Section 11.3) can be used to distinguish between Ca-, Sr- and Ba-containing compounds: Ca (orange-red, but pale green when viewed through blue glass), Sr (crimson, but violet through blue glass), Ba (apple-green).

Radioactive isotopes

The isotope ^{90}Sr is a β-emitter ($t_{\frac{1}{2}} = 29.1$ yr) and a fission product of uranium. In the event of a nuclear energy plant disaster or through the dumping of nuclear waste, there is a danger that grass, and then milk, may be contaminated with ^{90}Sr and that it may be incorporated with calcium phosphate into bone.[†] For discussion of ^{226}Ra, see Section 12.1.

Refer to Section 10.3 for help if necessary.

1. $^{90}_{38}Sr$ decays by emission of a β-particle. Write an equation for the decay.

2. The product of the reaction in question 1 is also radioactive. It is a β-emitter and produces $^{90}_{40}Zr$. Use this information to confirm that your answer to question 1 is correct.

(continued)

[†] For further details, see: N. Vajda and C.-K. Kim (2010) *Appl. Radiat. Isot.*, vol. 68, p. 2306 – 'Determination of radiostrontium isotopes: A review of analytical methodology'.

Table 12.1 Some physical properties of the group 2 metals, M, and their ions, M^{2+}.

Property	Be	Mg	Ca	Sr	Ba	Ra
Atomic number, Z	4	12	20	38	56	88
Ground state electronic configuration	$[He]2s^2$	$[Ne]3s^2$	$[Ar]4s^2$	$[Kr]5s^2$	$[Xe]6s^2$	$[Rn]7s^2$
Enthalpy of atomization, $\Delta_a H^{\circ}(298\,K)/kJ\,mol^{-1}$	324	146	178	164	178	130
Melting point, mp/K	1560	923	1115	1040	1000	973
Boiling point, bp/K	≈ 3040	1380	1757	1657	1913	1413
Standard enthalpy of fusion, $\Delta_{fus}H^{\circ}(mp)/kJ\,mol^{-1}$	7.9	8.5	8.5	7.4	7.1	–
First ionization energy, $IE_1/kJ\,mol^{-1}$	899.5	737.7	589.8	549.5	502.8	509.3
Second ionization energy, $IE_2/kJ\,mol^{-1}$	1757	1451	1145	1064	965.2	979.0
Third ionization energy, $IE_3/kJ\,mol^{-1}$	14850	7733	4912	4138	3619	3300
Metallic radius, $r_{metal}/pm^{\ddagger}$	112	160	197	215	224	–
Ionic radius, r_{ion}/pm^*	27	72	100	126	142	148
Standard enthalpy of hydration of M^{2+}, $\Delta_{hyd}H^{\circ}(298\,K)/kJ\,mol^{-1}$	−2500	−1931	−1586	−1456	−1316	–
Standard entropy of hydration of M^{2+}, $\Delta_{hyd}S^{\circ}(298\,K)/J\,K^{-1}\,mol^{-1}$	−300	−320	−230	−220	−200	–
Standard Gibbs energy of hydration of M^{2+}, $\Delta_{hyd}G^{\circ}(298\,K)/kJ\,mol^{-1}$	−2410	−1836	−1517	−1390	−1256	–
Standard reduction potential, $E^{\circ}_{M^{2+}/M}/V$	−1.85	−2.37	−2.87	−2.89	−2.90	−2.92

$\ddagger$ For 12-coordinate atoms.
$*$ For 4-coordination for Be^{2+}, and 6-coordination for other M^{2+} ions.

---

3. $^{90}_{38}Sr$ is formed as a fission product of $^{235}_{92}U$. Complete the following equation and determine the second fission product:

$$^{235}_{92}U + ^{1}_{0}n \longrightarrow ^{90}_{38}Sr + ? + 3^{1}_{0}n$$

4. Why is $^{90}_{38}Sr$ considered to be especially dangerous when it is released into the environment?

12.4 The metals

Appearance

Beryllium and magnesium are greyish metals, while the remaining group 2 metals are soft and silver-coloured. The metals are malleable, ductile and quite brittle. In air, the shiny surface of each metal quickly tarnishes.

Reactivity

Beryllium and magnesium are passivated (eq. 12.7) and are kinetically inert to O_2 and H_2O at ambient temperatures.

However, Mg *amalgam* liberates H_2 from water, since no coating of oxide forms on its surface. Mg metal reacts with steam or hot water (eq. 12.8).

$$2Be + O_2 \longrightarrow \underset{\substack{\text{protective oxide} \\ \text{coating on metal}}}{2BeO} \quad (12.7)$$

$$Mg + 2H_2O \underset{\text{steam}}{\longrightarrow} Mg(OH)_2 + H_2 \quad (12.8)$$

Beryllium and magnesium dissolve readily in non-oxidizing acids. Magnesium is attacked by nitric acid, whereas beryllium reacts with dilute HNO_3 but is passivated by concentrated nitric acid. Magnesium does not react with aqueous alkali, whereas Be forms an *amphoteric* hydroxide (see Section 12.6).

The metals Ca, Sr and Ba exhibit similar chemical behaviours, generally resembling, but being slightly less reactive than, Na. They react with water and acids liberating H_2, and the similarity with Na extends to dissolution in liquid NH_3 to give blue solutions containing solvated electrons. From these solutions, it is possible to isolate hexaammines,

ENVIRONMENT

Box 12.2 Desulfurization processes to limit SO$_2$ emissions

Current awareness of the effects of environmental pollution has been instrumental in the development of *desulfurization processes*. This includes desulfurization of fossil fuels and flue gases from a variety of sources. Fossil fuels are the major source of fuel in electricity-generating power stations, and this is exemplified by the chart below which shows the source of fuel for the generation of electricty in the US in 2009:

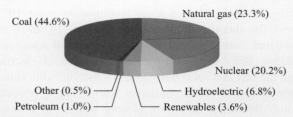

Coal (44.6%)
Natural gas (23.3%)
Nuclear (20.2%)
Other (0.5%)
Hydroelectric (6.8%)
Petroleum (1.0%)
Renewables (3.6%)

[Data: US Energy Information Administration]

The aim in a flue gas desulfurization process is to optimize the removal of SO$_2$ from emissions into the atmosphere. One important method of desulfurization in commercial operation throughout the world is based upon the neutralization reactions between Ca(OH)$_2$ or CaCO$_3$ and sulfuric acid. General equations for the neutralization reactions are:

$$Ca(OH)_2 + H_2SO_4 \longrightarrow CaSO_4 + 2H_2O$$

$$CaCO_3 + H_2SO_4 \longrightarrow CaSO_4 + H_2O + CO_2$$

Drax power station in the UK (shown opposite) operates six coal-fired boilers which produce superheated steam to drive turbines which generate electricity. About 7% of the electricity used in the UK is generated by this one power station. Different grades of coal contain varying amounts of sulfur, an upper limit being close to 2%. Each boiler at Drax power station consumes approximately 6300 tonnes of coal a day, and a 2% sulfur content by weight corresponds to 126 tonnes of sulfur being converted to SO$_2$. The photograph shows the main chimney (in the foreground) in addition to the cooling towers at Drax. When the power plant was first operational in the 1970s, gas emissions (CO$_2$, NO$_x$ and SO$_2$) left the power station through the main chimney. The chimney was fitted with a flue gas desulfurization system in the 1990s and the emitted gases now pass through absorbers containing limestone, CaCO$_3$. Slaked lime, Ca(OH)$_2$, can also be used. The reactions shown below remove >90% of the SO$_2$ produced at the Drax power station, and the desulfurization system is capable of removing up to 280,000 tonnes of SO$_2$ per year. The reactions occurring are:

$$SO_2 + H_2O \rightleftharpoons H^+ + [HSO_3]^-$$

$$H^+ + [HSO_3]^- + \tfrac{1}{2}O_2 \longrightarrow 2H^+ + [SO_4]^{2-}$$

$$2H^+ + [SO_4]^{2-} + Ca(OH)_2 \longrightarrow CaSO_4 \cdot 2H_2O$$

or

$$2H^+ + [SO_4]^{2-} + H_2O + CaCO_3 \longrightarrow CaSO_4 \cdot 2H_2O + CO_2$$

The Drax coal-fired power station in the UK uses desulfurization systems, installed between 1993 and 1996.

An advantage of the system is that CaSO$_4$·2H$_2$O, *gypsum*, is non-toxic and is not a waste product. It has a number of commercial applications, for example in the production of plaster of Paris (see Section 12.7) and cement (see Box 14.8).

An alternative way of reducing sulfur emissions is to replace coal by biomass (timber and agricultural waste in addition to specifically grown crops). In 2007, Sembcorp Biomass Power Station was commissioned as a wood-fired power station in the UK, and uses wood from sustainable sources.

Ammonia can be used as an alternative to CaCO$_3$ and Ca(OH)$_2$ in desulfurization processes. The sulfuric acid (see above) combines with NH$_3$ to give [NH$_4$]$_2$[SO$_4$]. Like gypsum, ammonium sulfate is a commercially desirable chemical, and is recycled as a fertilizer.

Further reading

V.C. Baligar, R.B. Clark, R.F. Korcak and R.J. Wright (2011) *Adv. Agron.*, vol. 111, p. 51 – 'Flue gas desulfurization product use on agricultural land'.

C. Li, Q. Zhang, N.A. Krotkov, D.G. Streets, K. He, S.-C. Tsay and J.F. Gleason, *Geophys. Res. Lett.* (2010) vol. 37, p. L08807/1 – 'Recent large reduction in sulfur dioxide emissions from Chinese power plants observed by the ozone monitoring instrument'.

D. Stirling (2000) *The Sulfur Problem: Cleaning Up Industrial Feedstocks*, Royal Society of Chemistry, Cambridge.

S. Su, B. Li, S. Cui and S. Tao (2011) *Environ. Sci. Technol.*, vol. 45, p. 8403 – 'Sulfur dioxide emissions from combustion in China: from 1990 to 2007'.

D. Wang, A. Bao, W. Kunc and W. Liss (2012) *Appl. Energy*, vol. 91, p. 341 – 'Coal power plant flue gas waste heat and water recovery'.

C.F. You and X.C. Xu (2010) *Energy*, vol. 35, p. 4467 – 'Coal combustion and its pollution control in China'.

www.draxpower.com
See also Boxes 15.7, 16.5 and 22.5.

$[M(NH_3)_6]$ (M = Ca, Sr, Ba), but these slowly decompose to amides (eq. 12.9).

$$[M(NH_3)_6] \longrightarrow M(NH_2)_2 + 4NH_3 + H_2 \quad M = Ca, Sr, Ba$$
(12.9)

When heated, all the group 2 metals combine with O_2, N_2, sulfur or halogens (eq. 12.10–12.13).

$$2M + O_2 \xrightarrow{\Delta} 2MO$$
(12.10)

$$3M + N_2 \xrightarrow{\Delta} M_3N_2$$
(12.11)

$$8M + S_8 \xrightarrow{\Delta} 8MS$$
(12.12)

$$M + X_2 \xrightarrow{\Delta} MX_2 \qquad X = F, Cl, Br, I$$
(12.13)

Differences between the first and later members of group 2 are illustrated by the formation of hydrides and carbides. When heated with H_2, Ca, Sr and Ba form saline hydrides, MH_2, but Mg reacts only under high pressure. In contrast, BeH_2 (which is polymeric, Fig. 10.15) is prepared from beryllium alkyls (see Section 23.3). Beryllium combines with carbon at high temperatures to give Be_2C which possesses an antifluorite structure (see Section 6.11). The other group 2 metals form carbides MC_2 which contain the $[C \equiv C]^{2-}$ ion, and adopt NaCl-type structures that are elongated along one axis. Be_2C reacts with water according to eq. 12.14.

$$Be_2C + 4H_2O \longrightarrow 2Be(OH)_2 + CH_4$$
(12.14)

The carbides of Ca, Mg, Sr and Ba react with water to yield C_2H_2. For CaC_2, this reaction is a means of manufacturing ethyne (eq. 12.15) in areas of the world where coal is a more important feedstock for the chemical industry than oil, e.g. South Africa and China. Calcium carbide is made from CaO, which in turn is manufactured by calcining limestone (scheme 12.16). Production and consumption of CaC_2 in China accounts for 95% of the global demand. In Europe, the US and Japan, the production of CaC_2 has declined and it is manufactured, not for conversion to ethyne but to produce the nitrogenous fertilizer calcium cyanamide (eq. 12.17).

$$CaC_2 + 2H_2O \longrightarrow C_2H_2 + Ca(OH)_2$$
(12.15)

$$\left. \begin{array}{l} CaCO_3 \xrightarrow{\Delta} CaO + CO_2 \\ CaO + 3C \xrightarrow{2300\,K} CaC_2 + CO \end{array} \right\}$$
(12.16)

$$CaC_2 + N_2 \xrightarrow{1300\,K} CaNCN + C$$
(12.17)

CaH_2 is used as a drying agent (see Box 12.3) but its reaction with water is highly exothermic.

The carbide Mg_2C_3 (which contains the linear $[C_3]^{4-}$ ion, **12.1**, isoelectronic with CO_2) is formed by heating MgC_2, or by reaction of Mg dust with pentane vapour at 950 K. Reaction of Mg_2C_3 with water produces $MeC \equiv CH$.

$$\left[C = C = C \right]^{4-}$$

(12.1)

12.5 Halides

Beryllium halides

Anhydrous beryllium halides are covalent. The fluoride, BeF_2, is obtained as a glass (sublimation point 1073 K) from the thermal decomposition of $[NH_4]_2[BeF_4]$, itself prepared from BeO and NH_3 in an excess of aqueous HF. Molten BeF_2 is virtually a non-conductor of electricity, and the fact that solid BeF_2 adopts a β-cristobalite structure (see Section 6.11) is consistent with its being a covalent solid. Beryllium difluoride is very soluble in water, the formation of $[Be(OH_2)_4]^{2+}$ (see Section 12.8) being thermodynamically favourable (Table 12.1).

Anhydrous $BeCl_2$ (mp 688 K, bp 793 K) can be prepared by reaction 12.18. This is a standard method of preparing a metal chloride that cannot be made by dehydration of hydrates obtained from aqueous media. In the case of Be, $[Be(OH_2)_4]^{2+}$ is formed and attempted dehydration of $[Be(OH_2)_4]Cl_2$ yields the hydroxide, not the chloride (eq. 12.19).

$$2BeO + CCl_4 \xrightarrow{1070\,K} 2BeCl_2 + CO_2$$
(12.18)

$$[Be(OH_2)_4]Cl_2 \xrightarrow{\Delta} Be(OH)_2 + 2H_2O + 2HCl$$
(12.19)

A *deliquescent* substance absorbs water from the surrounding air and eventually forms a liquid.

In the vapour state above 1020 K, $BeCl_2$ is monomeric and has a linear structure. At lower temperatures, the vapour also contains planar dimers. We return to the structures of gas-phase BeX_2 molecules later in the section. $BeCl_2$ forms colourless, deliquescent crystals containing infinite chains in which the coordination environment of each Be is tetrahedral. The Be–Cl distances are longer than in the monomer (Fig. 12.4). In Section 5.2, we described the bonding in monomeric $BeCl_2$ in terms of *sp* hybridization. In the polymer, each Be atom can be considered to be sp^3 hybridized and a localized σ-bonding scheme is appropriate in which each Cl donates a lone pair of electrons into an empty hybrid orbital on an adjacent Be atom (Fig. 12.4c). The formation of this chain demonstrates the Lewis acidity of beryllium dihalides. $BeCl_2$ acts as a Friedel–Crafts catalyst (i.e. like $AlCl_3$), and the formation of adducts is illustrated by $[BeF_4]^{2-}$, $[BeCl_4]^{2-}$ and $BeCl_2 \cdot 2L$ (L = ether, aldehyde, ketone).

APPLICATIONS

Box 12.3 Inorganic elements and compounds as drying agents

It is useful to distinguish between different classes of *drying agent* as being reagents that react with water either *reversibly* or *irreversibly*. The former can be regenerated, usually by heating, while the latter (sometimes classed as *dehydrating* agents) cannot. Caution is always needed when choosing a drying agent for the following reasons:

- the substance from which water is being removed may react with the drying agent;
- dehydrating agents often react vigorously with water and should not be used to dry very wet solvents, for which a pre-drying stage is appropriate;
- magnesium perchlorate, $Mg(ClO_4)_2$, although an extremely efficient drying agent, is best avoided because of the risk of explosions.

Many drying or dehydrating agents are compounds of group 1 or 2 metals. Concentrated H_2SO_4, molecular sieves and silica gel are also commonly used to absorb water. Silica gel is a porous form of silica, SiO_2, which is manufactured by coagulating acidified, aqueous solutions of sodium silicate. After washing to remove Na^+ and other ions, the precipitate is heated to drive off water. Although called a gel, silica gel is a microporous solid with a surface area of around $800\,m^2\,g^{-1}$. It is naturally colourless, but cobalt(II) salts (e.g. $[NH_4]_2[CoCl_4]$) are often added to act as an indicator. Such salts are blue in the absence of water and pink when hydrated. The photograph shows anhydrous cobalt-dyed silica gel spheres. Another highly effective dehydrating agent is phosphorus(V) oxide (see Section 15.10).

Blue cobalt-dyed silica gel spheres; these turn pink when water is absorbed.

alcohols or aldehydes), $LiAlH_4$ (for hydrocarbons and ethers) and sodium. The latter, generally extruded as wire, is extremely efficient for removing water from hydrocarbons or ethers, but reacts with, for example, alcohols, and is not suitable for drying halogenated solvents.

Agents for drying or predrying solvents

Typically, anhydrous salts that absorb water as solvate are suitable for removing water from solvents. Anhydrous $MgSO_4$, $CaCl_2$, $CaSO_4$, Na_2SO_4 and K_2CO_3 are hygroscopic and of these, $CaSO_4$ and $MgSO_4$ are particularly efficient and inert drying agents.

Drying agents that react irreversibly with H_2O

Drying agents in this category include Ca and Mg (for alcohols), CaH_2 (for a range of solvents, but not lower

Drying agents for use in desiccators and drying tubes

Suitable agents for drying samples in desiccators are anhydrous $CaCl_2$, $CaSO_4$, KOH and P_2O_5. Gases may be dried by passage through drying tubes packed with a suitable agent, but possible reaction of the gas with the drying agent must be considered. Although P_2O_5 is a common choice for use in desiccators, reaction with water results in the formation of a brown, viscous layer on the surface of the anhydrous powder, thereby curtailing its dehydrating ability.

The reaction of $BeCl_2$ with $[Ph_4P]Cl$ in a 1 : 1 molar ratio produces $[Ph_4P]_2[Be_2Cl_6]$ containing the anion shown in Fig. 12.4d. The Be–Cl bonds involved in the bridging interactions are longer (210 pm) than the terminal bonds (196 pm), consistent with the differences observed on going from polymeric $BeCl_2$ to gas-phase $BeCl_2$ (Fig. 12.4). When $BeCl_2$ reacts with two equivalents of $[Ph_4P]Cl$, $[Ph_4P]_2[BeCl_4]$ is formed which contains the tetrahedral $[BeCl_4]^{2-}$ ion.

Worked example 12.1   Lewis acidity of $BeCl_2$

Suggest a structure for a dimer of $BeCl_2$ and explain how its formation illustrates $BeCl_2$ acting as a Lewis acid.

Each Be atom can accommodate up to eight electrons in its valence shell. In a $BeCl_2$ monomer, there are only four valence electrons associated with each Be atom. Each Be atom can therefore accept one or two lone pairs of electrons,

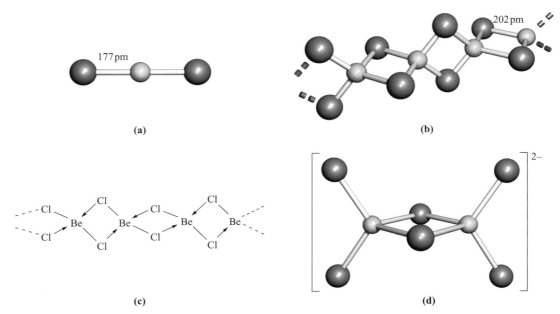

Fig. 12.4 (a) The linear structure of $BeCl_2$ in the gas phase. (b) The solid state polymeric structure of $BeCl_2$ is similar to that of BeH_2 (Fig. 10.15), although the bonding in these two compounds is *not* the same. (c) In $BeCl_2$, there are sufficient valence electrons to invoke 2c-2e Be—Cl bonds. (d) The structure of the $[Be_2Cl_6]^{2-}$ ion in $[Ph_4P]_2[Be_2Cl_6]$ determined by X-ray diffraction [B. Neumüller *et al.* (2003) *Z. Anorg. Allg. Chem.*, vol. 629, p. 2195]; the average Be–Cl terminal distance is 196 pm and the bridging Be–Cl distance is 210 pm. Colour code: Be, yellow; Cl, green.

thereby acting as a Lewis acid. Each Cl atom in monomeric $BeCl_2$ has three lone pairs of electrons. The dimer of $BeCl_2$ forms by donation of a lone pair of electrons from Cl to Be:

Each Be centre will be in a trigonal planar environment.

Self-study exercises

1. Rationalize why, on going from monomeric $BeCl_2$ to dimeric $(BeCl_2)_2$ to polymeric $(BeCl_2)_n$, the environment of the Be atom changes from linear to trigonal planar to tetrahedral.

    [*Ans.* The number of electrons in the valence shell of Be changes from four to six to eight]

2. The recrystallization of $BeCl_2$ from diethyl ether solutions leads to a Lewis acid–base adduct. Draw the likely structure of the adduct and rationalize its formation in terms of the electron-accepting properties of $BeCl_2$.

    [*Ans.* Tetrahedral $BeCl_2 \cdot 2Et_2O$; O donates a lone pair of electrons to Be]

Halides of Mg, Ca, Sr and Ba

The fluorides of Mg(II), Ca(II), Sr(II) and Ba(II) are ionic, have high melting points, and are sparingly soluble in water, the solubility increasing slightly with increasing cation size (K_{sp} for MgF_2, CaF_2, SrF_2 and $BaF_2 = 7.42 \times 10^{-11}$, 1.46×10^{-10}, 4.33×10^{-9} and 1.84×10^{-7} respectively). Whereas MgF_2 adopts a rutile structure (see Fig. 6.22), CaF_2, SrF_2 and BaF_2 crystallize with the fluorite structure (Fig. 6.19). In contrast to the behaviour of BeF_2, none of the later metal fluorides behaves as a Lewis acid.

The structures of gaseous group 2 metal fluoride and later halide molecules are the subject of ongoing theoretical interest.[†] It has been suggested that the term 'quasilinear' be used for a species for which the calculated energy difference between linear and bent structures (with a change in angle of $>20°$) is less than 4 kJ mol^{-1}. Using this definition leads to the structures given in Table 12.2. Of those compounds listed as quasilinear, $SrBr_2$ has the lowest energy barrier between a linear and bent structure. Some theoretical studies suggest only CaF_2, $CaCl_2$, $SrCl_2$ and $SrBr_2$ should be categorized as quasilinear, while in the extreme, only $SrBr_2$ should be considered quasilinear, with $CaCl_2$, $CaBr_2$, CaI_2 and SrI_2 being linear and CaF_2, $SrCl_2$ and BaI_2 being

[†] See: M. Kaupp (2001) *Angew. Chem. Int. Ed.*, vol. 40, p. 3534; M. Hargittai (2000) *Chem. Rev.*, vol. 100, p. 2233; K.J. Donald and R. Hoffmann (2006) *J. Am. Chem. Soc.*, vol. 128, p. 11236; M. Vasiliu, D. Feller, J.L. Gole and D.A. Dixon (2010) *J. Phys. Chem. A*, vol. 114, p. 9349.

Table 12.2 Structures of the monomeric group 2 metal dihalides, MX_2. The term 'quasilinear' is explained in the text.

Metal	Halide			
	F	**Cl**	**Br**	**I**
Be	Linear	Linear	Linear	Linear
Mg	Linear	Linear	Linear	Linear
Ca	Quasilinear	Quasilinear	Quasilinear	Quasilinear
Sr	Bent	Quasilinear	Quasilinear	Quasilinear
Ba	Bent	Bent	Bent	Quasilinear

bent. The most bent of the dihalides is BaF_2. It has a bond angle in the region of 110–126° (values come from a range of theoretical and experimental data) and the calculated energy to convert bent BaF_2 to a linear molecule is $\approx 21 \, kJ \, mol^{-1}$. The preference for bent structures for the heaviest metals combined with F, Cl or Br (see Table 12.2) has been explained in terms of both 'inverse (or core) polarization' and the participation of d atomic orbitals for Ca, Sr and Ba. Inverse polarization occurs when the metal ion is polarizable and is polarized by F^- or Cl^-, or to a lesser extent, by Br^-. This is represented in diagram **12.2**. The polarization is termed 'inverse' to distinguish it from the polarization of a large, polarizable *anion* by a *cation* (see Section 6.13).

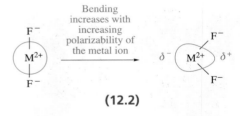

(12.2)

An alternative explanation focuses on the participation of d orbitals in the bonding in CaX_2, SrX_2 and BaX_2. Table 12.2 shows that Be and Mg form only linear gaseous dihalides. These two metals have only s and p atomic orbitals available for bonding and the best $M-X$ orbital overlap is achieved for a linear molecule. This is shown in diagram **12.3** for an np orbital on M with the out-of-phase combination of $X---X$ orbitals. For Ca, Sr and Ba, vacant $3d$, $4d$ and $5d$ orbitals, respectively, are available, but can only overlap efficiently with orbitals on the X atoms if the MX_2 molecule is bent. Two interactions must be considered as shown in diagram **12.4** (the axes are defined arbitrarily as shown). The out-of-phase combination of $X---X$ orbitals only overlaps efficiently with the d_{yz} orbital of M if the MX_2 molecule is bent; opening the molecule up to a linear shape 'switches off' this orbital interaction. Although the interaction between the metal d_{z^2} orbital and the in-phase

combination of $X---X$ orbitals is most efficient when MX_2 is linear, it is still effective when the molecule is bent (diagram **12.4**). The inverse polarization and participation of d atomic orbitals may both contribute to the problem of bent MX_2 molecules, and the explanation for the trend in shapes listed in Table 12.2 remains a matter for debate.

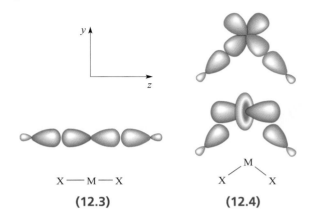

(12.3)   **(12.4)**

In addition to monomers of MX_2 being present in the vapour state, there is evidence that magnesium and calcium halides form dimers. Electron diffraction data are consistent with the presence of <5% Ca_2X_4 for calcium halides, while data at 1065 K for magnesium bromide indicate that 12% of the gaseous sample is composed of Mg_2Br_4.

Worked example 12.2   Linear *vs* bent MX_2 molecules

What shape for the gas-phase molecule SrF_2 is consistent with the VSEPR model?

Sr is in group 2 and has two valence electrons.
Each F atom provides one electron for bonding.
The valence shell of Sr in SrF_2 contains two bonding pairs of electrons and no lone pairs, therefore, by the VSEPR model SrF_2 should be a linear molecule.

Self-study exercises

1. Comment on the prediction of the VSEPR model for SrF_2 in the light of experimental observation.

[*Ans.* See text]

2. For which of the following gas-phase species is the VSEPR model in agreement with experimental observations: $BeCl_2$, BaF_2, MgF_2?   [*Ans.* See text]

3. Suggest a plausible structure for Mg_2Br_4.

[*Ans.* Like Be_2Cl_4 in worked example 12.1]

Magnesium chloride, bromide and iodide crystallize from aqueous solution as hydrates which undergo partial hydrolysis when heated. The anhydrous salts are, therefore, prepared by reaction 12.20.

$$Mg + X_2 \longrightarrow MgX_2 \qquad X = Cl, Br, I \qquad (12.20)$$

> A *hygroscopic* solid absorbs water from the surrounding air but does not become a liquid.

Anhydrous MCl_2, MBr_2 and MI_2 (M = Ca, Sr and Ba) can be prepared by dehydration of the hydrated salts. These anhydrous halides are *hygroscopic* and $CaCl_2$ (manufactured as a by-product in the Solvay process, see Fig. 11.6) is used as a laboratory drying agent (see Box 12.3), for road deicing and for dust control (see Box 12.4). In the solid state, many of the anhydrous halides possess complicated layer structures such as the CdI_2-type structure (Fig. 6.23). Most of these halides are somewhat soluble in

---

ENVIRONMENT

Box 12.4 Winter road deicing and controlling dust on roads

In Section 11.2, we described the widescale use of NaCl for winter road deicing. In 2008, the US used ≈ 23 Mt of salt for the control of ice on roads. The great advantage of NaCl is that it is cheap. The disadvantages are that it is corrosive to motor vehicles and to concrete structures such as bridges and, when the snow melts, it is carried into water courses. The environmental effects that this has on water supplies and to fish and vegetation are a cause for concern and a topic of current research. Sodium chloride acts most effectively as a deicing agent at temperatures above $-6\,°C\,(267\,K)$.

Calcium chloride is also commonly applied as a road deicing agent. Its advantage over NaCl is that, when applied as solid anhydrous $CaCl_2$, it is effective at temperatures as low as $-32\,°C\,(241\,K)$. An added benefit of using anhydrous $CaCl_2$ is that its dissolution into melted snow or ice is an exothermic process which results in further snow or ice melting. Aqueous solutions of $CaCl_2$ (sold as 'liquid $CaCl_2$') are also applied to roads. A solution that is 32% $CaCl_2$ by weight is an effective deicing agent down to $-18\,°C\,(255\,K)$. Two disadvantages of $CaCl_2$ are that it is significantly more corrosive than NaCl, and it is more expensive. One compromise is to pre-wet NaCl with $CaCl_2$ solution, and the application of 'pre-wetted salt' to roads is common practice.

While NaCl and $CaCl_2$ have been applied as winter road deicing agents for many years, their environmental disadvantages and corrosive properties make them far from ideal. The corrosive nature of chloride deicers makes them unsuitable for de-icing aircraft, and glycols are typically used for this purpose. An alternative to NaCl and $CaCl_2$ is the double salt calcium magnesium acetate (CMA), the potential of which was first recognized in the 1970s. CMA is manufactured by treating calcined dolomite (CaO.MgO) with acetic acid but, generated in this way, the product is about five times as expensive as NaCl. CMA is most efficient as a deicer above $-7\,°C$ and therefore compares favourably with NaCl. However, CMA has many advantages. It is far less corrosive than chloride deicers, exhibits a low toxicity to vegetation and aquatic wildlife, and is biodegradable. Both NaCl and $CaCl_2$ are mobile in groundwater, and about 50% of NaCl applied to roads ends up in groundwater supplies. In contrast, CMA is poorly mobile in soil and shows a low tendency to reach groundwater. Current research into cheaper routes to its manufacture include oxidation of organic

food waste and fermentation processes, e.g. from calcined dolomite and whey lactose. The latter is converted to lactic acid by the bacterium *Lactobacillus plantarum*, and then to acetic and propanoic acids by *Propionibacterium acidipropionici*.

About 21% of $CaCl_2$ produced in North America is consumed in road deicing. A further 27% is used to control dust on unpaved roads. This application arises from the hygroscopic nature of $CaCl_2$. Addition of anhydrous $CaCl_2$, flaked $CaCl_2$ (78% $CaCl_2$ and 22% moisture) or 'liquid $CaCl_2$' (which dries out *in situ*) to dusty road surfaces provides a means of trapping water, thereby helping to aggregate the dust particles. In addition to reducing dust pollution, particle aggregation helps to slow down deterioration of the road surface. Canada, for example, uses $CaCl_2$ widely on its 'dirt roads', and in 2000, ≈ 100 kt were applied across the country.

Salt or pre-wetted salt is spread in huge amounts on roads in snow-belts.

Further reading

R.E. Jackson and E.G. Jobbágy (2005) *Proc. Nat. Acad. Sci.*, vol. 102, p. 14487 – 'From icy roads to salty streams'.

P.V. Vadlani, A.P. Mathews and G.S. Karr (2008) *World J. Microbiol. Biotechnol.*, vol. 24, p. 825 – 'Low-cost propionate salt as road deicer: evaluation of cheese whey and other media constituents'.

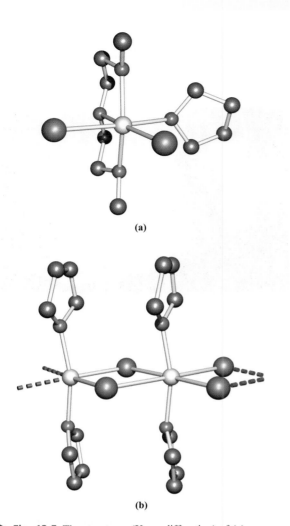

(a)

(b)

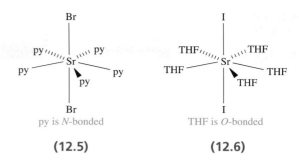

py is *N*-bonded   THF is *O*-bonded

(12.5)   **(12.6)**

12.6 Oxides and hydroxides

Oxides and peroxides

Beryllium oxide, BeO, is formed by ignition of Be or its compounds in O_2. It is an insoluble white solid which adopts a wurtzite-type structure (see Fig. 6.21). The oxides of the other group 2 metals are usually prepared by thermal decomposition of the corresponding carbonate (eq. 12.21, for which temperature T refers to $P(CO_2) = 1$ bar).

$$MCO_3 \xrightarrow{T \text{ K}} MO + CO_2 \qquad \begin{cases} M = Mg & T = 813\,K \\ Ca & 1173\,K \\ Sr & 1563\,K \\ Ba & 1633\,K \end{cases}$$

(12.21)

Figure 12.6 shows the trend in melting points of the oxides. MgO, CaO, SrO and BaO crystallize with an NaCl-type structure and the decrease in melting point reflects the decrease in lattice energy as the cation size increases (Table 12.1). The high melting point of MgO makes it suitable as a refractory material (see Box 12.5).

> ***Refractory materials*** are suitable for use in furnace linings; such a material has a high melting point, low electrical conductivity and high thermal conductivity, and is chemically inert at the high operating temperatures of the furnace.

Fig. 12.5 The structures (X-ray diffraction) of (a) [MgBr₂(diglyme)(THF)] (diglyme = MeOCH₂CH₂OCH₂CH₂OMe) [N. Metzler *et al.* (1994) *Z. Naturforsch.*, *Teil B*, vol. 49, p. 1448] and (b) [MgBr₂(THF)₂] [R. Sarma *et al.* (1977) *J. Am. Chem. Soc.*, vol. 99, p. 5289]; H atoms have been omitted. Colour code: Mg, pale grey; Br, gold; O, red; C, grey.

polar solvents such as ethers or pyridine, and a number of crystalline complexes have been isolated. Octahedral coordination has been confirmed by X-ray diffraction studies of complexes including *trans*-[MgBr₂(py)₄], *trans*-[MgBr₂(THF)₄], *cis*-[MgBr₂(diglyme)(THF)] (Fig. 12.5a) and *trans*-[CaI₂(THF)₄]. In [MgBr₂(THF)₂], octahedral coordination in the solid state is achieved by the formation of a chain structure (Fig. 12.5b); py = pyridine, THF = tetrahydrofuran (see Table 7.7). The larger sizes of the heavier metals permit higher coordination numbers, e.g. pentagonal bipyramidal *trans*-[SrBr₂(py)₅], **12.5**, and *trans*-[SrI₂(THF)₅], **12.6**. In organic chemistry, MgBr₂ is used as a catalyst for esterification reactions, and MgBr₂·2Et₂O is commercially available, being a catalyst for the conversion of aliphatic epoxides to the corresponding ketones.

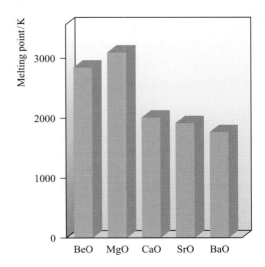

Fig. 12.6 The melting points of the group 2 metal oxides.

APPLICATIONS

Box 12.5 MgO: refractory material

When one looks for a commercially viable refractory oxide, MgO (*magnesia*) is high on the list: it has a very high melting point (3073 K), can withstand heating above 2300 K for long periods, and is relatively inexpensive. Magnesia is fabricated into bricks for lining furnaces in steelmaking. Incorporating chromium ore into the refractory bricks increases their resistance to thermal shock. Magnesia bricks are also widely used in night-storage radiators: MgO conducts heat extremely well, but also has the ability to store it. In a radiator, the bricks absorb heat which is generated by electrically heated filaments during periods of 'off-peak' consumer rates, and then radiate the thermal energy over relatively long periods.

A blast furnace used in steel manufacturing.

The action of water on MgO slowly converts it to $Mg(OH)_2$ which is sparingly soluble. Oxides of Ca, Sr and Ba react rapidly and exothermically with water, and absorb CO_2 from the atmosphere (eq. 12.5). The conversion of CaO to calcium carbide and its subsequent hydrolysis (see eqs. 12.15 and 12.16) is industrially important.

Group 2 metal peroxides, MO_2, are known for M = Mg, Ca, Sr and Ba. Attempts to prepare BeO_2 have so far failed, and there is no experimental evidence for any beryllium peroxide compound.[†] As for the group 1 metal peroxides, the stability with respect to the decomposition reaction 12.22 increases with the size of the M^{2+} ion. This trend arises from the difference between the lattice energies of MO and MO_2 (for a given M) which becomes smaller as r_+ increases. $\Delta_{lattice}H^\circ(MO, s)$ is always more negative than $\Delta_{lattice}H^\circ(MO_2, s)$ (see worked example 12.3).

$$MO_2 \longrightarrow MO + \tfrac{1}{2}O_2 \quad (M = Mg, Ca, Sr, Ba) \quad (12.22)$$

All the peroxides are strong oxidizing agents. Magnesium peroxide is manufactured by reacting $MgCO_3$ or MgO with H_2O_2, and is used in toothpaste and has environmental and agricultural applications as a slow O_2 release agent (see end-of-chapter problem 12.25). Calcium peroxide is

prepared by cautious dehydration of $CaO_2 \cdot 8H_2O$, itself made by reaction 12.23.

$$Ca(OH)_2 + H_2O_2 + 6H_2O \longrightarrow CaO_2 \cdot 8H_2O \quad (12.23)$$

The reactions of SrO and BaO with O_2 (600 K, 200 bar pressure, and 850 K, respectively) yield SrO_2 and BaO_2. Pure BaO_2 has not been isolated and the commercially available material contains BaO and $Ba(OH)_2$. Reactions of the peroxides with acids (eq. 12.24) generate H_2O_2.

$$SrO_2 + 2HCl \longrightarrow SrCl_2 + H_2O_2 \quad (12.24)$$

Worked example 12.3 Using the Kapustinskii equation

The lattice energies of SrO and SrO_2 are -3220 and $-3037\,kJ\,mol^{-1}$ respectively. (a) For what processes are these values defined? (b) Show that the relative magnitudes of these values are consistent with estimates obtained using the Kapustinskii equation.

(a) The lattice energies are negative values and therefore refer to the formation of 1 mole of crystalline lattice from gaseous ions:

$$Sr^{2+}(g) + O^{2-}(g) \longrightarrow SrO(s)$$
$$Sr^{2+}(g) + [O_2]^{2-}(g) \longrightarrow SrO_2(s)$$

† See: R.J.F. Berger, M. Hartmann, P. Pyykkö, D. Sundholm and H. Schmidbaur (2001) *Inorg. Chem.*, vol. 40, p. 2270 – 'The quest for beryllium peroxides'.

(b) This part of the problem makes use of the relationship introduced at the end of Section 6.16: the Kapustinskii equation:

$$\Delta U(0\,\text{K}) = -\frac{(1.07 \times 10^5)\upsilon |z_+||z_-|}{r_+ + r_-}$$

where: υ = number of ions in the formula of the salt

$\quad |z_+|$ = numerical charge on cation

$\quad |z_-|$ = numerical charge on anion

$\quad r_+$ = radius of cation in pm

$\quad r_-$ = radius of anion in pm

For SrO and SrO_2:

$\upsilon = 2$ in each compound

$|z_+| = 2 \qquad |z_-| = 2$

$r_+ = 126\,\text{pm}$ (see Appendix 6)

$\quad$ = constant for both compounds

The only variable is r_-.

Therefore:

$$\Delta U(0\,\text{K}) \propto -\frac{1}{126 + r_-}$$

Because the ionic radius of $[O_2]^{2-} > O^{2-}$, it follows from the equation above that $\Delta U(SrO_2)$ is less negative than $\Delta U(SrO)$. This result is in agreement with the data given in the question.

Self-study exercises

Use data from the Appendices in the book where necessary.

1. Use the Kapustinskii equation to estimate a value for the process (at 0 K):

$$SrO(s) \longrightarrow Sr^{2+}(g) + O^{2-}(g) \qquad [\textit{Ans.}\ 3218\,\text{kJ mol}^{-1}]$$

2. The values of the lattice energies of MgO, CaO and SrO are -3795, -3414 and $-3220\,\text{kJ mol}^{-1}$ respectively. Show that this trend in values is consistent with the Kapustinskii equation.

3. The *difference* between the lattice energies of CaO and CaO_2 is $270\,\text{kJ mol}^{-1}$. Will the difference between the lattice energies of MgO and MgO_2 be larger or smaller than $270\,\text{kJ mol}^{-1}$? Use the Kapustinskii equation to rationalize your answer. $\qquad$ [*Ans.* Larger]

Hydroxides

Beryllium hydroxide is amphoteric and this sets it apart from the hydroxides of the other group 2 metals which are basic. In the presence of excess $[OH]^-$, $Be(OH)_2$ behaves as a Lewis acid (eq. 12.25), forming the tetrahedral complex ion **12.7**, but $Be(OH)_2$ also reacts with acids, e.g. reaction 12.26.

(12.7)

$$Be(OH)_2 + 2[OH]^- \longrightarrow [Be(OH)_4]^{2-} \qquad (12.25)$$
$$Be(OH)_2 + H_2SO_4 \longrightarrow BeSO_4 + 2H_2O \qquad (12.26)$$

The water solubilities of $M(OH)_2$ (M = Mg, Ca, Sr, Ba) increase down the group, as do their thermal stabilities with respect to decomposition into MO and H_2O. Magnesium hydroxide acts as a weak base, whereas $Ca(OH)_2$, $Sr(OH)_2$ and $Ba(OH)_2$ are strong bases. *Soda lime* is a mixture of NaOH and $Ca(OH)_2$ and is manufactured from CaO and aqueous NaOH. Soda lime is easier to handle than NaOH and is commercially available, being used, for example, as an absorbent for CO_2, and in qualitative tests for $[NH_4]^+$ salts, amides, imides and related compounds which evolve NH_3 when heated with soda lime.

12.7 Salts of oxoacids

In this section, we give selected coverage of group 2 metal salts of oxoacids, paying attention only to compounds of special interest or importance.

Most beryllium salts of strong oxoacids crystallize as soluble hydrates. Beryllium carbonate tends to hydrolyse, giving a salt containing $[Be(OH_2)_4]^{2+}$ (see Section 12.8). $BeCO_3$ can be isolated only by precipitation under an atmosphere of CO_2. This tendency towards hydrolysis is also illustrated by the formation of *basic beryllium acetate* $[Be_4(\mu_4\text{-}O)(\mu\text{-}O_2CMe)_6]$ (rather than $Be(MeCO_2)_2$) by the action of $MeCO_2H$ on $Be(OH)_2$. Figure 12.7 shows the structure of $[Be_4(\mu_4\text{-}O)(\mu\text{-}O_2CMe)_6]$; the central oxygen atom is bonded to four tetrahedral Be centres. A similar structure is observed in the basic nitrate $[Be_4(\mu_4\text{-}O)(\mu\text{-}O_2NO)_6]$ which is formed in reaction sequence 12.27.

$$BeCl_2 \xrightarrow{\ N_2O_4\ } [NO]_2[Be(NO_3)_4] \xrightarrow{\ 323\,\text{K}\ } Be(NO_3)_2$$
$$\Big\downarrow\ 398\,\text{K} \qquad (12.27)$$
$$[Be_4(\mu_4\text{-}O)(\mu\text{-}O_2NO)_6]$$

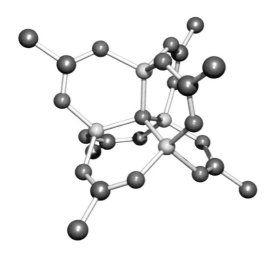

Fig. 12.7 The structure of basic beryllium acetate, $[Be_4(\mu_4\text{-}O)(\mu\text{-}O_2CMe)_6]$ (X-ray diffraction) [A. Tulinsky *et al.* (1959) *Acta Crystallogr.*, vol. 12, p. 623]; hydrogen atoms have been omitted. Colour code: Be, yellow; C, grey; O, red.

The carbonates of Mg and the later metals are sparingly soluble in water. Their thermal stabilities (eq. 12.21) increase with cation size, and this trend can be rationalized in terms of lattice energies. The metal carbonates are much more soluble in an aqueous solution of CO_2 than in water due to the formation of $[HCO_3]^-$. However, salts of the type '$M(HCO_3)_2$' have not been isolated. *Hard water* contains Mg^{2+} and Ca^{2+} ions which complex with the stearate ions in soaps, producing insoluble 'scum' in household baths and basins. *Temporary hardness* is due to the presence of hydrogencarbonate salts and can be overcome by boiling (which shifts equilibrium 12.28 to the right-hand side causing $CaCO_3$, or similarly $MgCO_3$, to precipitate) or by adding an appropriate amount of $Ca(OH)_2$ (again causing precipitation, eq. 12.29).

$$Ca(HCO_3)_2(aq) \rightleftharpoons CaCO_3(s) + CO_2(g) + H_2O(l) \tag{12.28}$$

$$Ca(HCO_3)_2(aq) + Ca(OH)_2(aq) \rightarrow 2CaCO_3(s) + 2H_2O(l) \tag{12.29}$$

Permanent hardness is caused by other Mg^{2+} and Ca^{2+} salts (e.g. sulfates). The process of *water softening* involves passing the hard water through a cation-exchange resin (see Section 11.6). Washing-machine detergents contain 'builders' that remove Mg^{2+} and Ca^{2+} ions from washing water. Polyphosphates have been used for this purpose (see Box 15.11), but zeolites (see Section 14.9) are preferred.

Calcium carbonate occurs naturally in two crystalline forms, *calcite* and the metastable *aragonite*. In calcite, the Ca^{2+} and $[CO_3]^{2-}$ ions are arranged in such as way that each Ca^{2+} ion is 6-coordinate with respect to the carbonate

O atoms, whereas in aragonite, each Ca^{2+} ion is surrounded by nine O atoms. The energy difference between them is $<5\,kJ\,mol^{-1}$ with calcite being the thermodynamically favoured form. However, aragonite is kinetically stable with respect to conversion to calcite. Aragonite can be prepared in the laboratory by precipitation of $CaCO_3$ from hot aqueous solution.

Sulfates of Mg and Ca have important applications and those of $CaSO_4$ are described in Section 16.2. $CaSO_4$ is a product of industrial desulfurization processes (see Box 12.2). Hydrated calcium sulfate ($CaSO_4\cdot2H_2O$, *gypsum*) occurs naturally (see Fig 12.1). Gypsum crystals cleave easily owing to the presence of layers which are held together by hydrogen bonding. When gypsum is heated at $\approx400\,K$, it forms the hemihydrate $CaSO_4\cdot\frac{1}{2}H_2O$ (*plaster of Paris*), and if this is mixed with water, the material expands slightly as the dihydrate is regenerated (see Box 12.6). Barium sulfate is a sparingly soluble salt ($K_{sp} = 1.07 \times 10^{-10}$) and the formation of a white precipitate of $BaSO_4$ is used as a qualitative test for the presence of sulfate ions in aqueous solution (eq. 12.30).

$$BaCl_2(aq) + [SO_4]^{2-}(aq) \rightarrow BaSO_4(s) + 2Cl^-(aq) \tag{12.30}$$

Calcium phosphate is described in Section 15.2.

A hydrate $X\cdot nH_2O$ in which $n = \frac{1}{2}$ is called a **hemihydrate**; if $n = 1\frac{1}{2}$, it is a **sesquihydrate**.

12.8 Complex ions in aqueous solution

Aqua species of beryllium

There is a high tendency to form $[Be(OH_2)_4]^{2+}$ in aqueous media. In ^{17}O-enriched water, exchange between coordinated water and solvent is slow on the NMR spectroscopic timescale, permitting the nature of the hydrated ion to be established. The tetrahedral coordination sphere (Be−O = 162.0 pm) has been established in the solid state structures of $[Be(OH_2)_4]Cl_2$ and $[Be(OH_2)_4][O_2CC\equiv CCO_2]$ (Fig. 12.8). The charge density of Be^{2+} is high and solutions of beryllium salts are acidic (see Section 7.7). Reaction 12.31 is an over-simplistic representation of the acid dissociation, since various condensation processes occur, e.g. reaction 12.32, and hydroxido-bridged species are also present.

$$[Be(OH_2)_4]^{2+} + H_2O \rightleftharpoons [Be(OH_2)_3(OH)]^+ + [H_3O]^+$$
$$pK_a = 5.40 \tag{12.31}$$

$$4[Be(OH_2)_4]^{2+} + 2H_2O$$
$$\rightleftharpoons 2[(H_2O)_3Be\text{-}O\text{-}Be(OH_2)_3]^{2+} + 4[H_3O]^+ \tag{12.32}$$

APPLICATIONS

Box 12.6 Gypsum plasters

The earliest known use of gypsum plaster was in Anatolia (part of modern-day Turkey) and Syria in about 6000 BC, and in about 3700 BC, the Egyptians used gypsum plaster in the inside of the pyramids. The building industry is the major consumer of gypsum plasters. Gypsum, $CaSO_4 \cdot 2H_2O$, is mined on a large scale worldwide, and is calcined to form the β-hemihydrate, $CaSO_4 \cdot \frac{1}{2}H_2O$. The hemihydrate is referred to as *plaster of Paris*, the name being derived from Montmartre in Paris where gypsum was quarried. Hydration of the hemihydrate with a carefully controlled amount of H_2O initially gives a slurry which hardens as $CaSO_4 \cdot 2H_2O$ crystallizes. Crystals are needle-like and it is their intergrowth that provides gypsum with its strength and suitability for the building trade. Calcined gypsum which is stored for long periods may age by absorbing water, and this affects the rehydration process. The setting process of gypsum plasters may be accelerated or slowed down by suitable additives, e.g. <0.1% of citric acid is sufficient to retard the crystallization process. Gypsum plasters suitable for applying to walls have been developed so that additives are already present with the hemihydrate. Building contractors commonly use prefabricated gypsum plasterboards and tiles. Plasterboards are fabricated by pouring a hemihydrate–water–additive slurry onto cardboard sheets ≈0.5 mm thick. After completing the lamination by applying a second sheet of cardboard, the plasterboard is dried. The incorporation of fibreglass (see Box 13.6) into plasterboards is also possible, giving fibreboard products. An advantage of gypsum plasterboards as partition walls is their degree of fire resistance.

In 2008, 159 Mt of gypsum was produced by mining worldwide. A second source referred to as 'synthetic gypsum' is produced in the flue gas desulfurization processes described in Box 12.2. The contribution made by synthetic gypsum to world supplies is growing in importance, and its production is less expensive than mining natural gypsum. In 2008, 31% of gypsum consumed in the US was synthetic in origin. Recycling gypsum is also increasing. It can be salvaged from wallboard manufacturing, building demolition and house construction; approximately 10% of wallboard supplied for building a new house is scrapped. Recycled wallboard is crushed, and separated into paper and gypsum. The latter enters the production process for new prefabricated plasterboards along with mined and synthetic gypsum. The average new home in the US contains ≥570 m² of gypsum wallboard.

The Yoshino gypsum plant in Japan: recycling gypsum wallboards is an integral part of the manufacturing process which produces new gypsum plasterboards.

Statistical data: US Geological Survey.

Aqua species of Mg^{2+}, Ca^{2+}, Sr^{2+} and Ba^{2+}

In contrast to the coordination number of four in the aquated Be^{2+} ion, each of the later group 2 metal ions can accommodate six or more water molecules in the first coordination sphere. In solution (^{17}O-labelled water), ^{17}O NMR spectroscopic data are consistent with the presence of $[Mg(OH_2)_6]^{2+}$, and crystallographic data on a range of salts confirm that this complex ion is octahedral. The $[Mg(OH_2)_6]^{2+}$ ion dissociates to some extent in aqueous solution ($pK_a = 11.44$).

In solution, the coordination number of the $[Ca(OH_2)_n]^{2+}$ ion depends on the concentration with $n \geq 6$. In the solid state, octahedral $[Ca(OH_2)_6]^{2+}$ is present in crystalline $CaCl_2 \cdot 6H_2O$ and $CaBr_2 \cdot 6H_2O$, and a number of salts containing $[Ca(OH_2)_7]^{2+}$ have been structurally characterized. The geometry of the cation varies, e.g. pentagonal bipyramidal in $[Ca(OH_2)_7][Co(EDTA)]_2$ and capped octahedral in $[Ca(OH_2)_7][Bi(EDTA)]_2 \cdot H_2O$ (see Fig. 19.8 for 7-coordinate geometries). $[Ca(OH_2)_8]^{2+}$ ions occur in $CaK[AsO_4] \cdot 8H_2O$. The large Sr^{2+} and Ba^{2+} ions can also accommodate more than six aqua ligands. The presence of $[M(OH_2)_8]^{2+}$ ions (M = Sr, Ba) in aqueous solution has been confirmed using the EXAFS technique (see Box 25.2). The single crystal structures of $[Sr(OH_2)_8][OH]_2$ and $[Ba(OH_2)_8][OH]_2$ have revealed hydrogen-bonded networks surrounding distorted square antiprismatic $[M(OH_2)_8]^{2+}$ ions. A similar coordination geometry has been observed for the $[Sr(OH_2)_8]^{2+}$ ion (Fig. 12.9) in a *host–guest complex* that involves an extensive hydrogen-bonded network with $[Sr(OH_2)_8]^{2+}$ ions in the cavities. The hydrated cations of Ca^{2+}, Sr^{2+} and Ba^{2+} do not undergo appreciable acid dissociation, and solutions of their salts derived from strong acids are neutral.

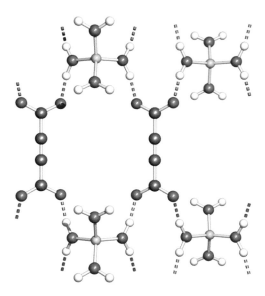

Fig. 12.8 Part of the packing diagram of $[Be(OH_2)_4][O_2CC\equiv CCO_2]$ showing hydrogen bonding between $[Be(OH_2)_4]^{2+}$ cations and $[O_2CC\equiv CCO_2]^{2-}$ anions; the structure was determined by neutron diffraction [C. Robl *et al.* (1992) *J. Solid State Chem.*, vol. 96, p. 318]. Colour code: Be, yellow; C, grey; O, red; H, white.

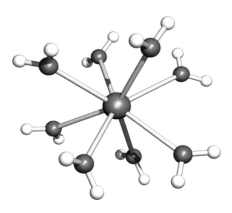

Fig. 12.9 The distorted square antiprismatic structure of the $[Sr(OH_2)_8]^{2+}$ ion present in a host–guest complex, the solid state structure of which has been determined by X-ray diffraction [M.J. Hardie *et al.* (2001) *Chem. Commun.*, p. 1850]. Colour code: Sr, gold; O, red; H, white.

In a **host–guest complex**, a molecule (the guest) occupies a cavity in the molecular structure of a larger molecular entity (the host). Intermolecular interactions are involved between the host and guest species. Examples are a metal ion within a crown ether or cryptand (ion–dipole interactions), hydrogen-bonded cages called **clathrates** that encapsulate guest molecules, and **inclusion compounds** in which guest molecules occupy channels in the structure of the host (van der Waals forces between host and guest).

Complexes with ligands other than water

The group 2 metal ions are hard acids and are preferentially coordinated by hard bases (see Table 7.9). In this section we consider complexes formed in aqueous solution in which the metal centre is coordinated by *O*- and *N*-donor ligands to give cationic species. Two important ligands are $[EDTA]^{4-}$ (see eq. 7.75) and $[P_3O_{10}]^{5-}$ (see Fig. 15.19). Both form water-soluble complexes with Mg^{2+} and the heavier metal ions, and are *sequestering agents* used in water-softening to remove Mg^{2+} and Ca^{2+} ions.

Macrocyclic ligands, including crown ethers and cryptands (see Sections 7.12 and 11.8), form stable complexes with Mg^{2+}, Ca^{2+}, Sr^{2+} and Ba^{2+}. In an analogous manner to that noted for group 1 cations, selectivity corresponding to matching of cation (Table 12.1) and ligand-cavity sizes is observed. Thus, values of the stability constants for complexation with cryptand-222 (cavity radius 140 pm) in water follow the sequence $Ba^{2+} > Sr^{2+} \gg Ca^{2+} > Mg^{2+}$. An important class of macrocyclic ligands are the porphyrins and the parent compound is shown in Fig. 12.10a. Deprotonation of the two NH groups of a porphyrin gives a dianionic porphyrinato ligand. Chlorophylls, the pigments in green plants involved in photosynthesis, are porphyrinato derivatives containing Mg^{2+} coordinated within a square planar array of the four *N*-donor atoms. The structure of chlorophyll *a* is shown in Fig. 12.10b. The extensive conjugation in the ring system means that the molecule absorbs light in the visible region (λ_{max} 660 nm) and this initiates a series of reactions involving other systems containing Mn or Fe. Note that it is the *ligand* (not Mg^{2+}) that is involved in these redox reactions.

Self-study exercise

The fullerenes C_{60} and C_{70} are discussed in Chapter 14. When these fullerenes are reduced using barium in liquid ammonia, fulleride salts containing $[Ba(NH_3)_7]^{2+}$ and $[Ba(NH_3)_9]^{2+}$ counter-ions are obtained. Suggest possible structures for these cations.

[*Ans.* Refer to Figs. 10.14c and 19.9]

12.9 Complexes with amido or alkoxy ligands

In Section 12.5, we described group 2 metal halide complexes such as *trans*-$[CaI_2(THF)_4]$ and *trans*-$[SrBr_2(py)_5]$.

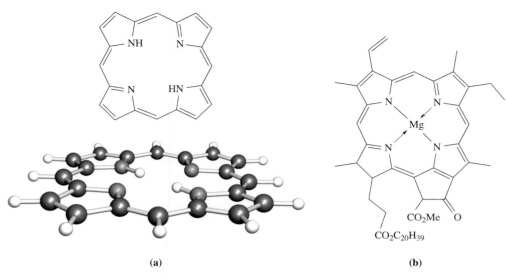

(a)

(b)

 Fig. 12.10 The structures of (a) porphyrin and (b) chlorophyll *a*.

The number of complexes of the group 2 metals with *N*- or *O*-donor ligands continues to grow, notably those incorporating sterically demanding amido or alkoxy ligands.

With the bulky bis(trimethylsilyl)amido ligand, each of the M^{2+} ions forms at least one type of complex. In the gas phase, monomeric $[Be\{N(SiMe_3)_2\}_2]$ contains a linear N−Be−N unit. In the solid state structure of $[Mg\{N(SiMePh_2)_2\}_2]$, $\angle N-Mg-N = 162.8°$, the deviation from linearity being attributed to weak dipolar interactions between the electropositive metal centre and the electron density of the aromatic rings. Coordination numbers of 3 and 4 for Mg(II), Ca(II), Sr(II) and Ba(II) are seen in dimers $[M\{N(SiMe_3)_2\}_2]_2$ or solvated monomers, e.g. tetrahedral $[Ba\{N(SiMe_3)_2\}_2(THF)_2]$. The structure of $[Ca\{N(SiMe_3)_2\}_2]_2$ is shown in Fig. 12.11a, and similar structures have been confirmed crystallographically for the analogous Mg, Sr and Ba compounds as well as for $[Mg\{N(CH_2Ph)_2\}_2]_2$.

While alkoxy derivatives of the alkaline earth metals have been known for many years, the area has undergone significant expansion since 1990. Much of this interest stems from the fact that calcium, strontium and barium alkoxides are potential precursors for high-temperature superconductors (see Chapter 28) and volatile compounds suitable for *chemical vapour deposition* (CVD) studies are being sought. Mononuclear complexes include several of the type $[M(OR)_2(THF)_3]$, e.g. $[Ca(OC_6H_2-2,6-^tBu_2-4-Me)_2(THF)_3]$. Some interesting high nuclearity species have also been isolated, including $[Ba_4(\mu_4-O)(\mu-OC_6H_2(CH_2NMe_2)_3-2,4,6)_6]$, formed by treating BaI_2 with $K[OC_6H_2(CH_2NMe_2)_3-2,4,6]$ in THF, and $[Ca_9(OCH_2CH_2OMe)_{18}(HOCH_2CH_2OMe)_2]$ (Fig. 12.11b), produced by reacting Ca metal with 2-methoxyethanol in hexane.

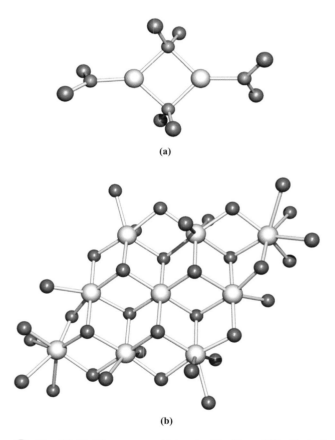

(a)

(b)

Fig. 12.11 The structures (determined by X-ray diffraction) of (a) $[Ca_2\{N(SiMe_3)_2\}_2\{\mu-N(SiMe_3)_2\}_2]$ in which the methyl groups have been omitted [M. Westerhausen *et al.* (1991) *Z. Anorg. Allg. Chem.*, vol. 604, p. 127] and (b) $[Ca_9(OCH_2CH_2OMe)_{18}-(HOCH_2CH_2OMe)_2]$ for which only the $Ca_9(\mu_3-O)_8(\mu-O)_8O_{20}$ core is shown (four of the ligands in $[Ca_9(OCH_2CH_2OMe)_{18}-(HOCH_2CH_2OMe)_2]$ are terminally attached, leaving four oxygen atoms non-coordinated to Ca^{2+} centres) [S.C. Goel *et al.* (1991) *J. Am. Chem. Soc.*, vol. 113, p. 1844]. Colour code: Ca, pale grey; O, red; N, blue; Si, gold.

12.10 Diagonal relationships between Li and Mg, and between Be and Al

In Section 11.3, we noted that the properties of Li and its compounds are often considered to be anomalous when compared with those of the later group 1 metals, and that a *diagonal relationship* exists between Li and Mg. In this section, we consider this relationship in detail and also describe a similar diagonal relationship between Be and Al. The positions of Li, Be, Mg and Al in the periodic table are shown below:

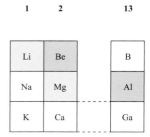

Table 12.3 lists selected physical properties of the first three elements in groups 1, 2 and 13. From a comparison of the properties of Li with those of Na and K, or of Li with Mg, it can be seen that Li resembles Mg more closely than it does the later members of group 1. A similar comparison between Be, Mg, Ca and Al leads to the conclusion that the physical properties of Be listed in Table 12.3 resemble those of Al more than they do those of the later group 2 metals. The Li^+ ion is small and highly polarizing, and this results in a degree of covalency in some of its compounds. On going from Li^+ to Be^{2+}, the ionic radius *decreases*, while on going down group 2 from Be^{2+} to Mg^{2+}, the ionic radius *increases*. The net result is that the sizes of Li^+ and Mg^{2+} are similar (Table 12.3), and we observe similar patterns in behaviour between lithium and magnesium despite the fact that they are in different groups. The diagonal relationship that exists between Li^+ and Mg^{2+} is also observed between Be^{2+} and Al^{3+}, and between the Na^+, Ca^{2+} and Y^{3+} ions.

Lithium and magnesium

Some of the chemical properties of Li that make it diagonally related to Mg rather than vertically related to the other alkali metals are:

- Lithium readily combines with N_2 to give the nitride, Li_3N; Mg reacts with N_2 to give Mg_3N_2.
- Lithium combines with O_2 to give the oxide Li_2O rather than a peroxide or superoxide (see eqs. 11.8–11.10); Mg forms MgO. The peroxides of both metals can be formed by reacting LiOH or $Mg(OH)_2$ with H_2O_2.
- Lithium and magnesium carbonates decompose readily on heating to give Li_2O and CO_2, and MgO and CO_2 respectively; down the group, the carbonates of the group 1 metals become increasingly stable with respect to thermal decomposition (see eq. 11.19 and accompanying text).
- Lithium and magnesium nitrates decompose on heating according to eqs. 12.33 and 12.34, whereas $NaNO_3$ and the later alkali metal nitrates decompose according to eq. 12.35.

$$4LiNO_3 \xrightarrow{\Delta} 2Li_2O + 2N_2O_4 + O_2 \qquad (12.33)$$

$$2Mg(NO_3)_2 \xrightarrow{\Delta} 2MgO + 2N_2O_4 + O_2 \qquad (12.34)$$

$$2MNO_3 \xrightarrow{\Delta} 2MNO_2 + O_2 \quad (M = Na, K, Rb, Cs) \qquad (12.35)$$

- The Li^+ and Mg^{2+} ions are more strongly hydrated in aqueous solution than are the ions of the later group 1 and 2 metals.
- LiF and MgF_2 are sparingly soluble in water; the later group 1 fluorides are soluble.

Table 12.3 Selected physical properties of the first three elements of groups 1, 2 and 13.

Property	Group 1			Group 2			Group 13		
	Li	**Na**	**K**	**Be**	**Mg**	**Ca**	**B**	**Al**	**Ga**
Metallic radius, r_{metal} /pm[†]	157	191	235	112	160	197	–	143	153
Ionic radius, r_{ion} /pm[‡]	76	102	138	27	72	100	–	54	62
Pauling electronegativity, χ^P	1.0	0.9	0.8	1.6	1.3	1.0	2.0	1.6	1.8
$\Delta_{atom}H^o(298\,K)$ / kJ mol^{-1}	161	108	90	324	146	178	582	330	277

[†] For 12-coordinate atoms (see also Table 11.1).
[‡] For 6-coordination except for Be, which is for 4-coordination; the ionic radius refers to M^+ for group 1, M^{2+} for group 2, and M^{3+} for group 13.

- LiOH is much less soluble in water than the other alkali metal hydroxides; $Mg(OH)_2$ is sparingly soluble.
- $LiClO_4$ is much more soluble in water than the other alkali metal perchlorates; $Mg(ClO_4)_2$ and the later group 2 metal perchlorates are very soluble.

Beryllium and aluminium

Representative chemical properties of Be that make it diagonally related to Al rather than vertically related to the later group 2 metals are:

- The Be^{2+} ion is hydrated in aqueous solution, forming $[Be(OH_2)_4]^{2+}$ in which the Be^{2+} centre significantly polarizes the already polar O−H bonds, leading to loss of H^+ (see eq. 12.31); the pK_a value of 5.4 for $[Be(OH_2)_4]^{2+}$ is close to that of $[Al(OH_2)_6]^{3+}$ ($pK_a = 5.0$) which contains a highly polarizing Al^{3+} ion (see eq. 7.34).
- Be and Al both react with aqueous alkali, liberating H_2; Mg does not react with aqueous alkali.
- $Be(OH)_2$ and $Al(OH)_3$ are amphoteric, reacting with both acids and bases (see eqs. 12.25 and 12.26 for reactions of $Be(OH)_2$, and eqs. 7.41 and 7.42 for $Al(OH)_3$); the hydroxides of the later group 2 metals are basic.
- $BeCl_2$ and $AlCl_3$ fume in moist air, reacting to give HCl.
- Both Be and Al form complex halides, hence the ability of the chlorides to act as Friedel–Crafts catalysts.

Further examples of similarities between the behaviours of Be and Al can be found by comparing their reactivities (see Sections 12.4 and 13.4).

KEY TERMS

The following terms were introduced in this chapter. Do you know what they mean?

- ❑ deliquescent
- ❑ hygroscopic
- ❑ refractory material
- ❑ permanent and temporary hardness of water
- ❑ water-softening agent (sequestering agent)
- ❑ hemihydrate
- ❑ sesquihydrate
- ❑ host–guest complex
- ❑ clathrate
- ❑ inclusion compound
- ❑ porphyrin
- ❑ amido ligand
- ❑ alkoxy ligand

FURTHER READING

K.M. Fromm (2002) *CrystEngComm*, vol. 4, p. 318 – An article that uses structural data to consider the question of ionic versus covalent bonding in group 2 metal iodide complexes.

N.N. Greenwood and A. Earnshaw (1997) *Chemistry of the Elements*, 2nd edn, Butterworth-Heinemann, Oxford – Chapter 5 gives a detailed account of the inorganic chemistry of the group 2 metals.

A.G. Massey (2000) *Main Group Chemistry*, 2nd edn, Wiley, Chichester – Chapter 5 covers the chemistry of the group 2 metals.

K.A. Walsh (2009) *Beryllium Chemistry and Processsing*, ASM International, Ohio – A book that details sources, production, chemistry and metallurgy of beryllium.

A.F. Wells (1984) *Structural Inorganic Chemistry*, 5th edn, Clarendon Press, Oxford – A full account of the structural chemistry of the group 2 metals and their compounds.

Special topics

L. Addadi, ed. (2007) *CrystEngComm*, vol. 9, issue 12 – An issue dedicated to biomineralization including that of $CaCO_3$.

K.M. Fromm and E.D. Gueneau (2004) *Polyhedron*, vol. 23, p. 1479 – 'Structures of alkali and alkaline earth metal clusters with oxygen donor ligands' (a review that includes comments on CVD).

D.L. Kepert, A.F. Waters and A.H. White (1996) *Aust. J. Chem.*, vol. 49, p. 117 – 'Synthesis and structural systematics of nitrogen base adducts of group 2 salts' (Part VIII in a series of papers covering this subject).

N.A.J.M. Sommerdijk and G. de With (2008) *Chem. Rev.*, vol. 108, p. 4499 – 'Biomimetic $CaCO_3$ mineralization using designer molecules and interfaces'.

PROBLEMS

12.1 (a) Write down, in order, the names and symbols of the metals in group 2; check your answer by reference to the first page of this chapter. (b) Give a *general* notation that shows the ground state electronic configuration of each metal.

12.2 Using data in Table 7.4, determine the relative solubilities of $Ca(OH)_2$ and $Mg(OH)_2$ and explain the relevance of your answer to the extraction of magnesium from seawater.

12.3 (a) Write an equation to show how Mg reacts with N_2 when heated. (b) Suggest how the product reacts with water.

12.4 The structure of magnesium carbide, MgC_2, is of the NaCl type, elongated along one axis. (a) Explain how this elongation arises. (b) What do you infer from the fact that there is no similar elongation in NaCN which also crystallizes with a NaCl-type structure?

12.5 Write balanced equations for the following reactions:
(a) the thermal decomposition of $[NH_4]_2[BeF_4]$;
(b) the reaction between NaCl and $BeCl_2$;
(c) the dissolution of BeF_2 in water.

12.6 (a) Suggest a likely structure for the dimer of $BeCl_2$, present in the vapour phase below 1020 K. What hybridization scheme is appropriate for the Be centres? (b) $BeCl_2$ dissolves in diethyl ether to form monomeric $BeCl_2 \cdot 2Et_2O$; suggest a structure for this compound and give a description of the bonding.

12.7 MgF_2 has a TiO_2-type structure. (a) Sketch a unit cell of MgF_2, and (b) confirm the stoichiometry of MgF_2 using the solid state structure.

12.8 Discuss the trends in data in Table 12.4.

12.9 (a) How do anhydrous $CaCl_2$ and CaH_2 function as drying agents?
(b) Compare the solid state structures and properties of $BeCl_2$ and $CaCl_2$.

Table 12.4 Data for problem 12.8.

Metal, M	$\Delta_f H^\circ$ / kJ mol^{-1}			
	MF_2	MCl_2	MBr_2	MI_2
Mg	-1113	-642	-517	-360
Ca	-1214	-795	-674	-535
Sr	-1213	-828	-715	-567
Ba	-1200	-860	-754	-602

12.10 How would you attempt to estimate the following?
(a) $\Delta_r H^\circ$ for the solid state reaction:
$$MgCl_2 + Mg \longrightarrow 2MgCl$$
(b) $\Delta_r H^\circ$ for the reaction:
$$CaCO_3(\text{calcite}) \longrightarrow CaCO_3(\text{aragonite})$$

12.11 (a) Identify the conjugate acid–base pairs in reaction 12.24. (b) Suggest how BaO_2 will react with water.

12.12 (a) Determine $\Delta_r H^\circ$ for the reactions of SrO and BaO with water, given that values of $\Delta_f H^\circ(298\,K)$ for SrO(s), BaO(s), $Sr(OH)_2(s)$, $Ba(OH)_2(s)$ and $H_2O(l)$ are -592.0, -553.5, -959.0, -944.7 and -285.5 kJ mol^{-1} respectively. (b) Compare the values of $\Delta_r H^\circ$ with that for the reaction of CaO with water (eq. 12.5), and comment on factors contributing to the trend in values.

12.13 (a) What qualitative test is used for CO_2? (b) What reaction takes place, and (c) what is observed in a positive test?

12.14 Discuss the data presented in Table 12.5.

12.15 Write a short account that justifies the so-called *diagonal relationship* between Li and Mg.

12.16 Suggest why MgO is more soluble in aqueous $MgCl_2$ solution than in pure water.

12.17 Suggest why Be^{2+} forms the tetrahedral ion $[Be(OH_2)_4]^{2+}$, while Mg^{2+} forms octahedral $[Mg(OH_2)_6]^{2+}$.

12.18 The reaction between $Ca(OH)_2$ and H$_2$**12.9** in aqueous solution leads to the formation of the complex $[Ca(OH_2)_2(\textbf{12.9})]$. This crystallizes as a centrosymmetric dimer in which each Ca^{2+} centre is 8-coordinate. This contains a central $Ca_2(\mu\text{-}O)_2$ unit in which each bridging *O*-donor involves a carboxylate group; only one oxygen atom of each carboxylate group is coordinated. Propose a structure for the centrosymmetric dimer.

(H$_2$12.9)

Table 12.5 Data for problem 12.14: log K for the formation of the complexes $[M(\text{crypt-222})]^{n+}$.

M^{n+}	Na^+	K^+	Rb^+	Mg^{2+}	Ca^{2+}	Sr^{2+}	Ba^{2+}
log K	4.2	5.9	4.9	2.0	4.1	13.0	>15

OVERVIEW PROBLEMS

12.19 Suggest explanations for the following observations.

   (a) The energy released when a mole of crystalline BaO is formed from its constituent ions is less than that released when a mole of MgO forms from its ions. (Each compound possesses an NaCl-structure.)

   (b) Despite being a covalent solid, BeF_2 is very soluble in water.

   (c) At 298 K, Be adopts an hcp lattice; above 1523 K, the coordination number of a Be atom in elemental beryllium is 8.

12.20 Comment on the following statements.

   (a) Na_2S adopts a solid state structure that is related to that of CaF_2.

   (b) $[C_3]^{4-}$, CO_2 and $[CN_2]^{2-}$ are isoelectronic species.

   (c) $Be(OH)_2$ is virtually insoluble in water, but is soluble in aqueous solutions containing excess hydroxide ions.

   (d) MgO is used as a refractory material.

12.21 Suggest products for the following reactions, and write balanced equations for the reactions. Comment on any of these reactions that are important in chemical manufacturing processes.

   (a) $CaH_2 + H_2O \longrightarrow$

   (b) $BeCl_2 + LiAlH_4 \longrightarrow$

   (c) $CaC_2 + H_2O \longrightarrow$

   (d) $BaO_2 + H_2SO_4 \longrightarrow$

   (e) $CaF_2 + H_2SO_4(conc) \longrightarrow$

   (f) $MgO + H_2O_2 \longrightarrow$

   (g) $MgCO_3 \xrightarrow{\Delta}$

   (h) Mg in air $\xrightarrow{\Delta}$

12.22 (a) A group 2 metal, **M**, dissolves in liquid NH_3, and from the solution, compound **A** can be isolated. **A** slowly decomposes to **B** with liberation of NH_3 and a gas **C**. Metal **M** gives a crimson flame test; through blue glass, the flame appears pale purple. Suggest identities for **M**, **A**, **B** and **C**.

   (b) The group 2 metal **X** occurs naturally in great abundance as the carbonate. Metal **X** reacts with cold water, forming compound **D**, which is a strong base. Aqueous solutions of **D** are used in qualitative tests for CO_2. **X** combines with H_2 to give a saline hydride that is used as a drying agent. Identify **X** and **D**. Write equations for the reaction of **X** with H_2O and of the hydride of **X** with H_2O. Explain how you would carry out a qualitative test for CO_2 using an aqueous solution of **D**.

12.23 (a) A 6-coordinate complex may be obtained by crystallizing anhydrous CaI_2 from THF solution at 253 K. In contrast, when anhydrous BaI_2 is crystallized from THF at 253 K, a 7-coordinate complex is isolated. Suggest structures for the two complexes, and comment on possible isomerism and factors that may favour one particular isomer in each case. Rationalize why CaI_2 and BaI_2 form complexes with THF that have different coordination numbers.

   (b) Which of the following compounds are sparingly soluble in water, which are soluble without reaction, and which react with water: $BaSO_4$, CaO, $MgCO_3$, $Mg(OH)_2$, SrH_2, $BeCl_2$, $Mg(ClO_4)_2$, CaF_2, $BaCl_2$, $Ca(NO_3)_2$? For the compounds that react with water, what are the products formed?

12.24 Each compound in List 1 has a matching description in List 2. Correctly match the partners. There is only one correct statement for each compound.

List 1	List 2
$CaCl_2$	Polymeric in the solid state
BeO	Soda lime
$Be(OH)_2$	Strong oxidizing agent
CaO	Used in qualitative analysis for sulfates
CaF_2	Hygroscopic solid, used for de-icing
$BaCl_2$	Amphoteric
$BeCl_2$	Quicklime
MgO_2	Crystallizes with a wurtzite-type structure
$Ca(OH)_2/NaOH$	A prototype crystal structure

INORGANIC CHEMISTRY MATTERS

12.25 Magnesium peroxide is used as a slow O_2 release agent in agriculture, ponds and lakes. It is manufactured by treating magnesium oxide or carbonate with H_2O_2. (a) Write equations for these reactions. (b) Suggest how the decomposition of magnesium peroxide depends upon pH by considering decomposition in neutral H_2O, dilute acid and dilute alkali.

12.26 Describe how sulfur-containing emissions from coal-fired power stations (a) arise, and (b) are controlled. (c) What are the products of standard desulfurization processes and how are they utilized?

12.27 Discuss how the properties of Mg lead to the following applications. (a) Professional camera bodies are made from magnesium alloys (>90% Mg). (b) Mg is used for cathodic protection of steel structures exposed to seawater. (c) Mg is used in fireworks. (d) Aluminium alloys used in the vehicle manufacturing industry contain up to 5% Mg.

12.28 World production of lime in 2008 was 296 Mt. The term 'lime' may refer to CaO (quicklime) and/or slaked lime ($Ca(OH)_2$), but is also used to encompass CaO, $Ca(OH)_2$, CaO·MgO, $Ca(OH)_2$·MgO and $Ca(OH)_2$·$Mg(OH)_2$. (a) How are CaO and CaO·MgO manufactured? (b) Describe the role of CaO in the building industry. (c) How is CaO converted to calcium carbide. Comment on recent trends in the industrial importance of this reaction. (d) In the paper industry, wood chips are converted to pulp by treatment with aqueous NaOH and Na_2S. Heating the spent liquor gives Na_2CO_3. Explain how treating this residue with $Ca(OH)_2$ followed by appropriate steps allows both NaOH and $Ca(OH)_2$ to be recovered.

Topics

13
The group 13 elements

1	2		13	14	15	16	17	18
H								He
Li	Be		**B**	C	N	O	F	Ne
Na	Mg		**Al**	Si	P	S	Cl	Ar
K	Ca	*d*-block	**Ga**	Ge	As	Se	Br	Kr
Rb	Sr		**In**	Sn	Sb	Te	I	Xe
Cs	Ba		**Tl**	Pb	Bi	Po	At	Rn
Fr	Ra							

13.1 Introduction

The elements in group 13 – boron, aluminium, gallium, indium and thallium – show a wide variation in properties: B is a non-metal, Al is a metal but exhibits many chemical similarities to B, and the later elements essentially behave as metals. The diagonal relationship between Al and Be was discussed in Section 12.10. Although the M(III) oxidation state is characteristic for elements in group 13, the M(I) state occurs for all elements except B. For Tl, the more stable oxidation state is +1. Thallium shows similarities in its chemistry to the alkali metals, Ag, Hg and Pb.

In contrast to the later elements, B forms a large number of so-called *electron-deficient* cluster compounds, the bond-ing in which poses problems within valence bond theory (see Section 13.11).

13.2 Occurrence, extraction and uses

Occurrence

The relative abundances of the group 13 elements are shown in Fig. 13.1. The main sources of boron are *borax*, $Na_2[B_4O_5(OH)_4] \cdot 8H_2O$, and *kernite*, $Na_2[B_4O_5(OH)_4] \cdot 2H_2O$, with extensive deposits being worked commercially in the Mojave Desert, California. Aluminium is the most abundant metal in the Earth's crust (Fig. 13.2), and occurs in aluminosilicates such as *clays*, *micas* and *feldspars*, in *bauxite* (hydrated oxides) and, to a lesser extent, in *cryolite*, $Na_3[AlF_6]$. Gallium occurs with aluminium in bauxite. Gallium, indium and thallium occur in trace amounts as sulfides in various minerals.

Extraction

Of the group 13 elements, Al is of the greatest commercial importance, with uses exceeding those of all metals except Fe. Figure 13.3 shows the dramatic rise in the production of Al in the US (the world's largest producer) since 1960, and emphasizes the increasing importance of aluminium recycling. Its isolation from the widely available aluminosilicate minerals is prohibitively difficult. Hence, bauxite and cryolite are the chief ores, and both are consumed in the extraction process. Crude bauxite is a mixture of oxides (impurities include Fe_2O_3, SiO_2 and TiO_2) and is purified using the Bayer process. After addition of the crude ore to hot aqueous NaOH under pressure (which causes Fe_2O_3 to

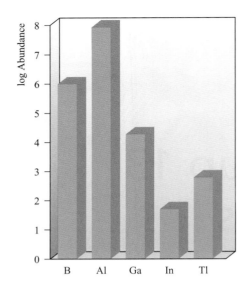

Fig. 13.1 Relative abundances of the group 13 elements in the Earth's crust. The data are plotted on a logarithmic scale. The units of abundance are parts per billion; 1 billion $= 10^9$.

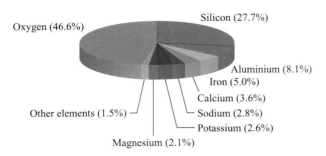

Fig. 13.2 Abundances of elements in the Earth's crusts. Aluminium is the most abundant metal. The most abundant minerals are silicates (see Section 14.9).

The first steps in the extraction of boron from borax are its conversion to boric acid (eq. 13.1) and then to the oxide (eq. 13.2).

$$Na_2[B_4O_5(OH)_4]\cdot 8H_2O + H_2SO_4$$
$$\longrightarrow 4B(OH)_3 + Na_2SO_4 + 5H_2O \quad (13.1)$$

$$2B(OH)_3 \xrightarrow{\Delta} B_2O_3 + 3H_2O \quad (13.2)$$

Boron of low purity is obtained by reduction of the oxide by Mg, followed by washing the product with alkali, hydrochloric acid and then hydrofluoric acid. The product is a very hard, black solid of low electrical conductivity which is inert towards most acids, but is slowly attacked by concentrated HNO_3 or fused alkali. Pure boron is made by the vapour-phase reduction of BBr_3 with H_2, or by pyrolysis of B_2H_6 or BI_3. At least four allotropes can be obtained under different conditions but transitions between them are extremely slow. For a discussion of the production of boron fibres, see Section 28.7.

separate), the solution is seeded with $Al_2O_3\cdot 3H_2O$ and cooled, or is treated with a stream of CO_2 to precipitate crystalline α-$Al(OH)_3$. Anhydrous Al_2O_3 (*alumina*) is produced by the action of heat. Electrolysis of molten Al_2O_3 gives Al at the cathode, but the melting point (2345 K) is high, and it is more practical and economical to use a mixture of cryolite and alumina as the electrolyte with an operating temperature for the melt of 1220 K. The extraction is expensive in terms of the electrical power required, and Al production is often associated with hydroelectric schemes.

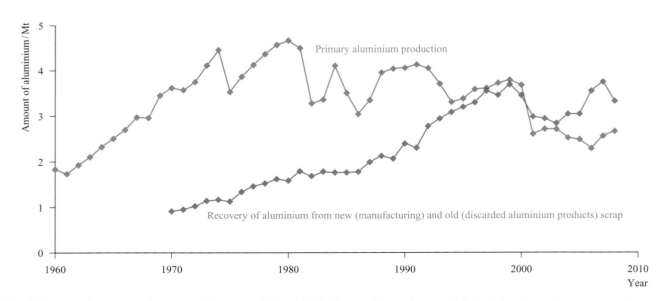

Fig. 13.3 Production of aluminium in the US between 1960 and 2008. The contribution that recycled aluminium has made to the market became increasingly important in the latter part of the twentieth century and has now overtaken primary production. [Data: US Geological Survey.]

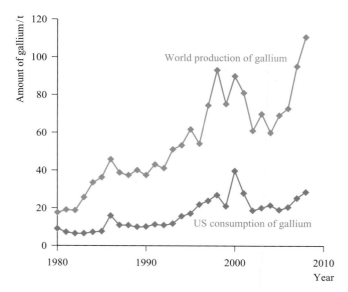

Fig. 13.4 World production (estimated) and US consumption of gallium between 1980 and 2008. [Data: US Geological Survey.]

An increase in world production of Ga over the last part of the 20th century (Fig. 13.4) coincides with increased demand for gallium arsenide (GaAs) in components for electronic equipment. Commercial end-uses for Ga are in integrated circuits and optoelectronic devices (laser diodes, light-emitting diodes, photodetectors and solar cells). The main source of Ga is crude *bauxite*, in which Ga is associated with Al. Gallium is also obtained from residues from the Zn-processing industry. The development of the electronics industry has also led to a significant increase in the demand for indium. Indium occurs in the zinc sulfide ore *sphalerite* (also called *zinc blende*, see Fig. 6.19) where, because it is a similar size to Zn, it substitutes for some of the Zn. The extraction of zinc from ZnS (see Section 21.2) therefore provides indium as a by-product. Recycling of In is becoming important. It is recovered from indium–tin oxide (ITO) in Japan, China and Korea where production of ITO (see Section 28.3) is centred. The manufacture of thin films of ITO accounts for most of the indium consumed worldwide. Thallium is obtained as a by-product of the smelting of Cu, Zn and Pb ores, although demand for the element is low.

Major uses of the group 13 elements and their compounds

The widespread applications of Al are summarized in Fig. 13.5a. Its strength can be increased by alloying with Cu or Mg. Aluminium oxide (see Section 13.7) has many important uses. *Corundum* (α-alumina) and *emery* (corundum mixed with the iron oxides *magnetite* and *haematite*) are extremely hard and are used as abrasives. Diamond is

the only naturally occurring mineral harder than corundum. Gemstones including ruby, sapphire, oriental topaz, oriental amethyst and oriental emerald result from the presence of trace metal salts in Al_2O_3, e.g. Cr(III) produces the red colour of ruby.[†] Artificial crystals can be manufactured from bauxite in furnaces, and artificial rubies are important as components in lasers. The γ-form of Al_2O_3 is used as a catalyst and as a stationary phase in chromatography. Al_2O_3 fibres are described in Section 28.7.

The two commercially most important borates are $Na_2[B_4O_5(OH)_4]\cdot8H_2O$ (*borax*) and $Na_2[B_4O_5(OH)_4]\cdot2H_2O$ (*kernite*). Figure 13.5b illustrates the applications of boron (in terms of boron oxide usage). Borosilicate glass has a high refractive index and is suitable for optical lenses. Borax has been used in pottery glazes for many centuries and remains in use in the ceramics industry. The reaction between fused borax and metal oxides is the basis for using borax as a flux in brazing. When metals are being fused together, coatings of metal oxides must be removed to ensure good metal–metal contact at the point of fusion. Boric acid, $B(OH)_3$, is used on a large scale in the glass industry, as a flame retardant (see Box 17.1), as a component in buffer solutions and is also an antibacterial agent. The use of B_2O_3 in the glass industry is described in Box 13.6. Elemental boron is used in the production of impact-resistant steels and (because [10]B has a high cross-section for neutron capture) in control rods for nuclear reactors. Amorphous boron is used in pyrotechnics, giving a characteristic green colour when it burns. The green colour probably arises from an emission from an electronically excited state of the BO_2 radical.

Gallium and indium phosphides, arsenides and antimonides have important applications in the semiconductor industry (see Sections 6.9 and 28.6; Boxes 14.2 and 23.2). They are used as transistor materials and in light-emitting diodes (LEDs) in, for example, pocket calculators; the colour of the light emitted depends on the band gap (see Table 28.5). Figure 13.4 shows that, in 2008, the US used 26% of the gallium produced worldwide. Almost all of this was consumed in the form of GaAs or GaN, and was used in LEDs, laser diodes (e.g. GaN laser diodes in DVD players), photodetectors, solar cells and integrated circuits. The application of LEDs for backlighting liquid crystal display TVs, and computer (including notebook and laptop) monitors is responsible for the recent increase in world demand for gallium. However, markets linked to the electronics industry are susceptible to fluctuation depending on world or local economies. The decrease in demand for gallium (specifically GaAs) in the US between 2000 and

[†] See: C. Degli Esposti and L. Bizzocchi (2007) *J. Chem. Educ.*, vol. 84, p. 1316 – 'Absorption and emission spectroscopy of a lasing material: ruby'.

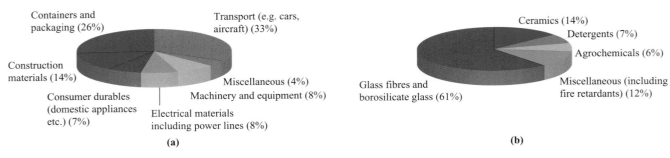

Fig. 13.5 (a) Uses of aluminium in the US in 2008; China, Russia, Canada and the US are the world's largest producers of the metal. (b) Uses of boron in the US in 2008; the data are given in terms of tons of boron oxide content. [Data: US Geological Survey.]

BIOLOGY AND MEDICINE

Box 13.1 Borax and boric acid: essentiality and toxicity

It has been recognized since 1923 that boron is an essential plant micronutrient. Other micronutrients are Mn, Zn, Cu, Mo, Fe and Cl. A deficiency in boron results in a range of problems, including die-back of terminal buds, stunted growth, hollow hearts in some vegetables, hollow stems and failure for grain to set (e.g. in wheat). Boron deficiency appears to be most prevalent in sandy conditions or in soils with a low content of organic matter, and in boron-poor soils, crop yields are diminished. Under neutral (or close to neutral) conditions, boron is available as boric acid, $B(OH)_3$, and the borate ion $[B(OH)_4]^-$. Although the exact function of boron remains undetermined, there is evidence that it plays a vital role in cell walls. The primary walls of plant cells are composed of pectic polysaccharides (with galacturonic acid being a dominant monosaccharide unit), cellulose and hemicelluloses. One of the principal pectic polysaccharides is rhamnogalacturonan II (RG-II). In 1996, it was determined that RG-II exists as a dimer which is cross-linked by a 1 : 2 borate-diol diester:

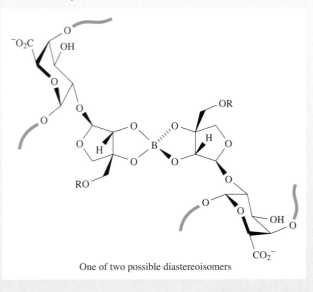

One of two possible diastereoisomers

It is thought that borate ester cross-linking of pectin is necessary for the normal growth and development of higher plants. Thus, a deficiency of boron leads to the effects described above. Application of borate fertilizers such as borax $(Na_2[B_4O_5(OH)_4] \cdot 8H_2O)$ to crops is therefore important. A balance has to be sought, however, because an excess of boron can be toxic to plants, and cereal crops are especially sensitive.

The toxicities of boric acid and borax to animal life are sufficient for them to be used as insecticides, e.g. in ant and cockroach control. Borax is also used as a fungicide; it acts by preventing the formation of fungal spores. The level of toxicity of borax is relatively low, but does cause some concern; e.g. borax and honey was, at one time, used to relieve the pain of teething in children, but this use is no longer recommended.

Further reading

L. Bolaños, K. Lukaszewski, I. Bonilla and D. Blevins (2004) *Plant Physiol. Biochem.*, vol. 42, p. 907 – 'Why boron?'.

K.H. Caffall and D. Mohnen (2009) *Carbohydr. Res.*, vol. 344, p. 1879 – 'The structure, function, and biosynthesis of plant cell wall pectic polysaccharides.'

M.A. O'Neill, S. Eberhard, P. Albersheim and A.G. Darvill (2001) *Science*, vol. 294, p. 846 – 'Requirement of borate cross-linking of cell wall rhamnogalacturonan II for *Arabidopsis* growth'.

P.P. Power and W.G. Woods (1997) *Plant and Soil*, vol. 193, p. 1 – 'The chemistry of boron and its speciation in plants'.

APPLICATIONS

Box 13.2 The unusual properties of indium–tin oxide (ITO)

Indium–tin oxide (ITO) is indium oxide doped with tin oxide. Thin films of ITO have commercially valuable properties: it is transparent, electrically conducting and reflects IR radiation. Applications of ITO are varied. It is used as a coating material for flat-panel computer displays, for coating architectural glass panels, and in electrochromic devices. Coating motor vehicle and aircraft windscreens and motor vehicle rear windows allows them to be electrically heated for de-icing purposes. A thin film of ITO (or related material) on the cockpit canopy of an aircraft such as the stealth plane renders this part of the plane radar-silent, contributing to the sophisticated design that allows the stealth plane to go undetected by radar.

By ensuring that all outer surfaces of a spacecraft are electrically conducting, the vessel is protected against build-up of electrostatic charge. The photograph shows the solar satellite *Ulysses* (a joint venture from NASA and the European Space Agency). The external surfaces of the spacecraft are covered in multi-layer insulation and a layer of electrically conducting ITO (the gold-coloured blanket). *Ulysses* was launched in 1990 and operated until June 2009, completing almost three

orbits of the Sun, providing information about the heliosphere and revealing that the strength of the solar wind is presently weakening.

The solar satellite *Ulysses*.

2001 (Fig. 3.4) can be attributed to a drop in sales of mobile phones. The largest use of indium is in thin-film coatings, e.g. laptop computers, flat panel displays and liquid crystal displays that use indium–tin oxide (ITO) coatings. In 2008, such coatings accounted for >75% of global indium consumption. Indium is also used in lead-free solders, in semiconductors, for producing seals between glass, ceramics and metals (because In has the ability to bond to non-wettable materials), and for fabricating special mirrors which reduce headlight glare. Uses of indium–tin oxide (ITO) are highlighted in Box 13.2.

Thallium sulfate was formerly used to kill ants and rats, but the extreme toxicity of Tl compounds means that they must be treated with caution. The world production of thallium (10 000 kg in 2008) is far less than that of gallium (Fig. 13.4) and indium. Important uses of Tl are in semiconducting materials in selenium rectifiers, in Tl-activated NaCl and NaI crystals in γ-radiation detectors, and in IR radiation detection and transmission equipment. The radioisotope ^{201}Tl ($t_{\frac{1}{2}} = 12.2$ d) is used for cardiovascular imaging.

13.3 Physical properties

Table 13.1 lists selected physical properties of the group 13 elements. Despite the discussion of ionization energies that follows, there is little evidence for the formation of *free* M^{3+} ions in compounds of the group 13 elements.

Electronic configurations and oxidation states

The group 13 elements have an outer electronic configuration ns^2np^1. There is a larger difference between IE_1 and IE_2 than between IE_2 and IE_3 (i.e. comparing the removal of a p with that of an s electron). The relationships between the electronic structures of the group 13 elements and those of the preceding noble gases are more complex than for the group 1 and 2 elements. For Ga and In, the electronic structures of the species formed after the removal of three valence electrons are $[Ar]3d^{10}$ and $[Kr]4d^{10}$ respectively, while for Tl, the corresponding species has the configuration $[Xe]4f^{14}5d^{10}$. Whereas for B and Al, the value of IE_4 (Table 13.1) refers to the removal of an electron from a noble gas configuration, this is not the case for the three later elements. The difference between IE_3 and IE_4 is not nearly so large for Ga, In and Tl as for B and Al. On going down group 13, the observed discontinuities in values of IE_2 and IE_3, and the differences between them (Table 13.1), originate in the failure of the d and f electrons (which have a low screening power, see Section 1.7) to compensate for the increase in nuclear charge. This failure is also reflected in the relatively small difference between values of r_{ion} for Al^{3+} and Ga^{3+}. For Tl, *relativistic effects* (see Box 13.3) are also involved.

On descending group 13, the trend in IE_2 and IE_3 shows *increases* at Ga and Tl (Table 13.1), and this leads to a marked increase in stability of the +1 oxidation state for these elements. In the case of Tl (the only salt-like trihalide of which is TlF_3), this is termed the *thermodynamic 6s inert pair effect* (see Box 13.4), so called to distinguish it from the

Table 13.1   Some physical properties of the group 13 elements, M, and their ions.

Property	B	Al	Ga	In	Tl
Atomic number, Z	5	13	31	49	81
Ground state electronic configuration	$[\text{He}]2s^2 2p^1$	$[\text{Ne}]3s^2 3p^1$	$[\text{Ar}]3d^{10}4s^2 4p^1$	$[\text{Kr}]4d^{10}5s^2 5p^1$	$[\text{Xe}]4f^{14}5d^{10}6s^2 6p^1$
Enthalpy of atomization, $\Delta_a H^\circ(298\,\text{K})\,/\,\text{kJ mol}^{-1}$	582	330	277	243	182
Melting point, mp / K	2453†	933	303	430	576.5
Boiling point, bp / K	4273	2792	2477	2355	1730
Standard enthalpy of fusion, $\Delta_{fus} H^\circ(\text{mp})\,/\,\text{kJ mol}^{-1}$	50.2	10.7	5.6	3.3	4.1
First ionization energy, $IE_1\,/\,\text{kJ mol}^{-1}$	800.6	577.5	578.8	558.3	589.4
Second ionization energy, $IE_2\,/\,\text{kJ mol}^{-1}$	2427	1817	1979	1821	1971
Third ionization energy, $IE_3\,/\,\text{kJ mol}^{-1}$	3660	2745	2963	2704	2878
Fourth ionization energy, $IE_4\,/\,\text{kJ mol}^{-1}$	25 030	11 580	6200	5200	4900
Metallic radius, r_{metal} / pm‡	–	143	153	167	171
Covalent radius, r_{cov} / pm	88	130	122	150	155
Ionic radius, r_{ion} / pm*	–	54 (Al^{3+})	62 (Ga^{3+})	80 (In^{3+})	89 (Tl^{3+}) 159 (Tl^+)
Standard reduction potential, E° (M^{3+}/M) / V	–	−1.66	−0.55	−0.34	+0.72
Standard reduction potential, E° (M^+/M) / V	–	–	−0.2	−0.14	−0.34
NMR active nuclei (% abundance, nuclear spin)	^{10}B (19.6, $I = 3$) ^{11}B (80.4, $I = \frac{3}{2}$)	^{27}Al (100, $I = \frac{5}{2}$)	^{69}Ga (60.4, $I = \frac{3}{2}$) ^{70}Ga (39.6, $I = \frac{3}{2}$)	^{113}In (4.3, $I = \frac{9}{2}$)	^{203}Tl (29.5, $I = \frac{1}{2}$) ^{205}Tl (70.5, $I = \frac{1}{2}$)

† For β-rhombohedral boron.

‡ Only the values for Al, In and Tl (the structures of which are close-packed) are strictly comparable; see text (Section 6.3) for Ga.

* There is no evidence for the existence of simple cationic boron under chemical conditions; values of r_{ion} for M^{3+} refer to 6-coordination; for Tl^+, r_{ion} refers to 8-coordination.

stereochemical inert pair effect mentioned in Section 2.8. Similar effects are seen for Pb (group 14) and Bi (group 15), for which the most stable oxidation states are +2 and +3 respectively, rather than +4 and +5. The inclusion in Table 13.1 of E° values for the M^{3+}/M and M^+/M redox couples for the later group 13 elements reflects the variable accessibility of the M^+ state within the group.

Although an oxidation state of +3 (and for Ga, In and Tl, +1) is characteristic of a group 13 element, most of the group 13 elements also form compounds in which a *formal* oxidation state of +2 is suggested, e.g. B_2Cl_4 and $GaCl_2$. However, caution is needed. In B_2Cl_4, the +2 oxidation state arises because of the presence of a B−B bond, whereas $GaCl_2$ is the mixed oxidation state species $Ga[GaCl_4]$.

THEORY

Box 13.3 Relativistic effects

Among many generalizations about heavier elements are two that depend on quantum theory for explanation:

- the ionization energies of the $6s$ electrons are anomalously high, leading to the marked stabilization of Hg(0), Tl(I), Pb(II) and Bi(III) compared with Cd(0), In(I), Sn(II) and Sb(III);
- whereas bond energies usually decrease down a group of p-block elements, they often increase down a group of d-block metals, in both the elements themselves and their compounds.

These observations can be accounted for (though often far from simply) if Einstein's theory of relativity is combined with quantum mechanics, in which case they are attributed to *relativistic effects*. We focus here on *chemical* generalizations.

According to the theory of relativity, the mass m of a particle increases from its rest mass m_0 when its velocity v approaches the speed of light, c, and m is then given by the equation:

$$m = \frac{m_0}{\sqrt{1 - \left(\frac{v}{c}\right)^2}}$$

For a one-electron system, the Bohr model of the atom (which, despite its shortcomings, gives the correct value for the ionization energy) leads to the velocity of the electron being expressed by the equation:

$$v = \frac{Ze^2}{2\varepsilon_0 nh}$$

where Z = atomic number, e = charge on the electron, ε_0 = permittivity of a vacuum, h = Planck constant.

For $n = 1$ and $Z = 1$, v is only $\approx \left(\frac{1}{137}\right)c$, but for $Z = 80$, $\frac{v}{c}$ becomes ≈ 0.58, leading to $m \approx 1.2 m_0$. Since the radius of the Bohr orbit is given by the equation:

$$r = \frac{Ze^2}{4\pi\varepsilon_0 mv^2}$$

the increase in m results in an approximately 20% contraction of the radius of the $1s$ ($n = 1$) orbital; this is called *relativistic contraction*. Other s orbitals are affected in a similar way and as a consequence, when Z is high, s orbitals have diminished overlap with orbitals of other atoms. A detailed treatment shows that p orbitals (which have a low electron density near to the nucleus) are less affected. On the other hand, d orbitals (which are more effectively screened from the nuclear charge by the contracted s and p orbitals) undergo a *relativistic expansion*; a similar argument applies to f orbitals. The relativistic contraction of the s orbitals means that for an atom of high atomic number, there is an extra energy of attraction between s electrons and the nucleus. This is manifested in higher ionization energies for the $6s$ electrons, contributing to the *thermodynamic 6s inert pair effect* which is discussed in Box 13.4. The effects of the relativistic contraction of the 6s orbital on the chemistry of Au have received particular attention (see below).

Further reading

D.J. Gorin and F.D. Toste (2007) *Nature*, vol. 446. p. 395 – 'Relativistic effects in homogeneous gold catalysis'.

P. Pyykkö (1988) *Chem. Rev.*, vol. 88, p. 563 – 'Relativistic effects in structural chemistry'.

Worked example 13.1   Thermochemistry of TlF and TlF₃

The enthalpy changes for the formation of crystalline TlF and TlF$_3$ from their component ions in the gas phase are -845 and $-5493\,\mathrm{kJ\,mol^{-1}}$, respectively. Use data from the Appendices in this book to calculate a value for the enthalpy change for the reaction:

$$\mathrm{TlF(s)} + \mathrm{F_2(g)} \longrightarrow \mathrm{TlF_3(s)}$$

Let ΔH° be the standard enthalpy change for the reaction:

$$\mathrm{TlF(s)} + \mathrm{F_2(g)} \longrightarrow \mathrm{TlF_3(s)} \qquad \text{(i)}$$

You are given enthalpy changes ($\approx$lattice energies) for TlF and TlF$_3$, i.e. for the reactions:

$$\mathrm{Tl^+(g)} + \mathrm{F^-(g)} \longrightarrow \mathrm{TlF(s)} \qquad \text{(ii)}$$

$$\mathrm{Tl^{3+}(g)} + 3\mathrm{F^-(g)} \longrightarrow \mathrm{TlF_3(s)} \qquad \text{(iii)}$$

for which lattice energies are *negative*.

Set up an appropriate thermochemical cycle that relates equations (i), (ii) and (iii):

THEORY

Box 13.4 The thermodynamic 6s inert pair effect

We confine attention here to the conversion of a metal halide MX_n into MX_{n+2}:

$$MX_n + X_2 \longrightarrow MX_{n+2}$$

In the simplest possible case, both halides are ionic solids and the energy changes involved are:

- absorption of the lattice energy of MX_n;
- absorption of $IE_{(n+1)} + IE_{(n+2)}$ to convert $M^{n+}(g)$ into $M^{(n+2)+}(g)$;
- liberation of the enthalpy of formation of $2X^-(g)$ (which is nearly constant for X = F, Cl, Br and I, see Appendices 9 and 10);
- liberation of the lattice energy of MX_{n+2}.

For a given M, the difference between the lattice energies of MX_n and MX_{n+2} is greatest for X = F, so if any saline halide MX_{n+2} is formed, it will be the fluoride. This treatment is probably a good representation of the conversion of TlF into TlF_3, and PbF_2 into PbF_4.

If, however, the halides are covalent compounds, the energy changes in the conversion are quite different. In this case, n times the M–X bond energy in MX_n and $2\Delta_f H^o(X, g)$ have to be absorbed, while $(n + 2)$ times the M–X bond energy in MX_{n+2} is liberated; $IE_{(n+1)}$ and $IE_{(n+2)}$ are not involved. The most important quantities in determining whether the conversion is possible are now the M–X bond energies in the two halides. The limited experimental data available indicate that both sets of M–X bond energies decrease along the series

F > Cl > Br > I, and that the M–X bond energy is always greater in MX_n than in MX_{n+2}. The overall result is that formation of MX_{n+2} is most likely for X = F. (The use of bond energies relative to ground-state atoms is unfortunate, but is inevitable since data are seldom available for valence state atoms. In principle, it would be better to consider the promotion energy for the change from one valence state of M to another, followed by a term representing the energy liberated when each valence state of M forms M–X bonds. However, this is beyond our present capabilities.)

The third possibility for the MX_n to MX_{n+2} conversion, and the one most likely in practice, is that MX_n is an ionic solid and MX_{n+2} is a covalent molecule. The problem now involves many more quantities and is too complicated for discussion here. Representative changes are the conversions of TlCl to $TlCl_3$, and of $PbCl_2$ to $PbCl_4$.

Finally, we must consider the effect of varying M down a group. In general, ionization energies (see Appendix 8) and lattice energies of compounds *decrease* as atomic and ionic radii (see Appendix 6) *increase*. It is where there is actually an *increase* in ionization energies, as is observed for the valence s electrons of Tl, Pb and Bi, that we get the clearest manifestations of the *thermodynamic 6s inert pair effect*. Where covalent bond formation is involved, a really satisfactory discussion of this inert pair effect is not yet possible, but the attempt at formulation of the problem can nevertheless be a rewarding exercise.

Apply Hess's law to this cycle:

$$\Delta_{lattice} H^o(TlF, s) + \Delta H^o = IE_2 + IE_3 + 2\Delta_a H^o(F, g)$$
$$+ 2\Delta_{EA} H^o(F, g)$$
$$+ \Delta_{lattice} H^o(TlF_3, s)$$

$$\Delta H^o = IE_2 + IE_3 + 2\Delta_a H^o(F, g) + 2\Delta_{EA} H^o(F, g)$$
$$+ \Delta_{lattice} H^o(TlF_3, s) - \Delta_{lattice} H^o(TlF, s)$$

Values of IE, $\Delta_a H^o$ and $\Delta_{EA} H^o$ are in Appendices 8, 10 and 9 respectively.

$$\Delta H^o = 1971 + 2878 + (2 \times 79) - (2 \times 328) - 5493 + 845$$
$$= -297 \, kJ \, mol^{-1}$$

Self-study exercises

1. For TlF(s), $\Delta_f H^o = -325 \, kJ \, mol^{-1}$. Use this value and ΔH^o for reaction (i) in the worked example to determine a value for $\Delta_f H^o(TlF_3, s)$.   [*Ans.* $-622 \, kJ \, mol^{-1}$]

2. Explain why $\Delta_{EA} H^o(F, g)$ is a negative value ($-328 \, kJ \, mol^{-1}$), while IE_1, IE_2 and IE_3 for Tl are all positive (589, 1971 and 2878 $kJ \, mol^{-1}$ respectively).
   [*Ans.* See Section 1.10]

NMR active nuclei

All the group 13 elements possess at least one isotope that is NMR active (Table 13.1). In particular, routine use is made of ^{11}B NMR spectroscopy in the characterization of B-containing compounds (e.g. Fig. 4.21). The ^{205}Tl nucleus is readily observed, and, since Tl^+ behaves similarly to Na^+ and K^+, replacement of these group 1 metal ions by Tl^+ allows ^{205}Tl NMR spectroscopy to be used to investigate Na- or K-containing biological systems.

13.4 The elements

Appearance

Impure (amorphous) boron is a brown powder, but the pure element forms shiny, silver-grey crystals. Properties including its high melting point and low electrical conductivity make B an important refractory material (see Section 12.6). Aluminium is a hard, white metal. Thermodynamically, it should react with air and water but it is resistant owing to the formation of an oxide layer, 10^{-6} to 10^{-4} mm thick. A thicker layer of Al_2O_3 can be obtained by making Al the anode in the electrolysis of H_2SO_4; the result is *anodized aluminium* which will take up dyes and pigments to produce a strong and decorative finish. Gallium is a silver-coloured metal with a particularly long liquid range (303–2477 K). Indium and thallium are soft metals, and In has the unusual property of emitting a high-pitched 'cry' when the metal is bent.

Structures of the elements

The structures of the group 13 *metals* were described in Section 6.3 and Table 6.2. The first 'allotrope' of boron to be documented was the α-tetragonal form, but this has been reformulated as a carbide or nitride, $B_{50}C_2$ or $B_{50}N_2$, the presence of C or N arising as a result of synthetic conditions. This carbidic phase is *not* the same as the boron carbide B_4C (more correctly formulated as $B_{13}C_2$) which has a structure related to that of β-rhombohedral B. The standard state of B is the β-rhombohedral form, but the structure of α-rhombohedral B makes an easier starting point in our discussion. Both the α- and β-rhombohedral allotropes contain icosahedral B_{12}-units (Figs. 13.6 and 13.7a). The bonding in elemental B is covalent, and within each B_{12}-unit, it is delocalized. We return to bonding descriptions in boron cluster compounds in Section 13.11, but for now note that the connectivity of each B atom in Figs. 13.6 and 13.7 exceeds the number of valence electrons available per B.

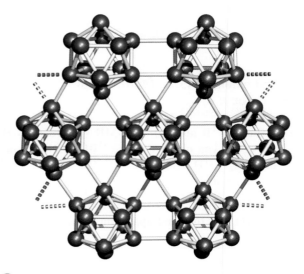

Fig. 13.6 Part of one layer of the infinite lattice of α-rhombohedral boron, showing the B_{12}-icosahedral building blocks which are covalently linked to give a rigid, infinite lattice.

α-Rhombohedral boron consists of B_{12}-icosahedra covalently linked by B–B bonds to form an infinite lattice. A readily interpretable picture of the lattice is to consider each icosahedron as an approximate sphere, and the overall structure as a ccp array of B_{12}-icosahedra, one layer of which is shown in Fig. 13.6. However, note that this is an infinite covalent lattice, as distinct from the close-packed metal lattices described in Chapter 6.

The structure of β-rhombohedral B consists of B_{84}-units, connected through B_{10}-units. Each B_{84}-unit is conveniently viewed in terms of the sub-units shown in Fig. 13.7. Their interrelationship is described in the figure caption, but an interesting point to note is the structural relationship between the B_{60}-sub-unit shown in Fig. 13.7c and the fullerene C_{60} (Fig. 14.5). The covalent lattices of both α- and β-rhombohedral B are extremely rigid, making crystalline B very hard, with a high melting point (2453 K for β-rhombohedral B).

Reactivity

Boron is inert under normal conditions except for attack by F_2. At high temperatures, it reacts with most non-metals (exceptions include H_2), most metals and with NH_3. The formations of metal borides (see Section 13.10) and boron nitride (see Section 13.8) are of particular importance.

The reactivities of the heavier group 13 elements contrast with that of the first member of the group. Aluminium readily oxidizes in air (see above). It dissolves in dilute mineral acids (e.g. reaction 13.3) but is passivated by concentrated HNO_3. Aluminium reacts with aqueous NaOH or KOH, liberating H_2 (eq. 13.4).

$$2Al + 3H_2SO_4 \xrightarrow{\text{dilute, aq}} Al_2(SO_4)_3 + 3H_2 \qquad (13.3)$$

$$2Al + 2MOH + 6H_2O \longrightarrow 2M[Al(OH)_4] + 3H_2$$
$$(M = Na, K) \qquad (13.4)$$

Reactions of Al with halogens at room temperature or with N_2 on heating give the Al(III) halides or nitride. Aluminium is often used to reduce metal oxides, e.g. in the *thermite process* (eq. 13.5) which is highly exothermic.

$$2Al + Fe_2O_3 \longrightarrow Al_2O_3 + 2Fe \qquad (13.5)$$

Gallium, indium and thallium dissolve in most acids to give salts of Ga(III), In(III) or Tl(I), but only Ga liberates H_2 from aqueous alkali. All three metals react with halogens at, or just above, 298 K. The products are of the type MX_3 with the exceptions of reactions 13.6 and 13.7.

$$2Tl + 2Br_2 \longrightarrow Tl[TlBr_4] \qquad (13.6)$$

$$3Tl + 2I_2 \longrightarrow Tl_3I_4 \qquad (13.7)$$

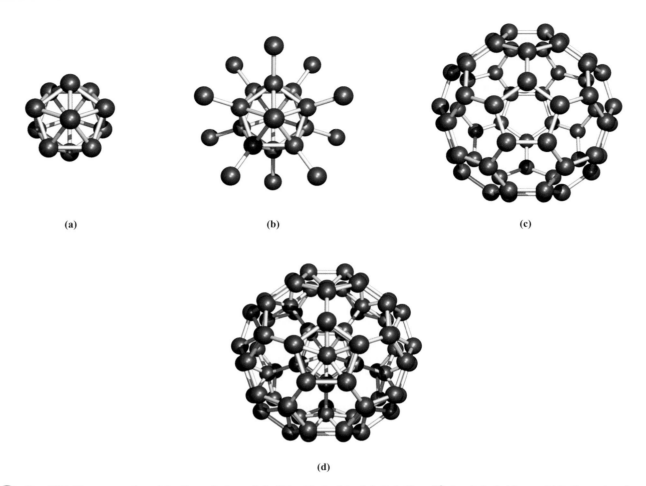

Fig. 13.7 The construction of the B_{84}-unit, the main building block of the infinite lattice of β-rhombohedral boron. (a) In the centre of the unit is a B_{12}-icosahedron, and (b) to each of these 12, another boron atom is covalently bonded. (c) A B_{60}-cage is the outer 'skin' of the B_{84}-unit. (d) The final B_{84}-unit can be described in terms of covalently bonded sub-units $(B_{12})(B_{12})(B_{60})$.

13.5 Simple hydrides

Neutral hydrides

With three valence electrons, each group 13 element might be expected to form a hydride MH_3. Although the existence of BH_3 has been established in the gas phase, its propensity to dimerize means that B_2H_6 (diborane(6), **13.1**) is, in practice, the simplest hydride of boron.

(13.1)

We have already discussed the structure of and bonding in B_2H_6 (Sections 10.7 and 5.7) and you are reminded of the presence of 3c-2e (delocalized, 3-centre 2-electron)

B−H−B interactions.[†] In worked example 13.2, the ^{11}B and ^1H NMR spectra of B_2H_6 are analysed.

Worked example 13.2   Multinuclear NMR spectroscopy: B_2H_6

Predict the (a) ^{11}B and (b) ^1H NMR spectra of B_2H_6. (c) What would you observe in the $^{11}B\{^1H\}$ NMR spectrum of B_2H_6? [^1H, 100%, $I = \frac{1}{2}$; ^{11}B, 80.4%, $I = \frac{3}{2}$.] Information needed:

• In the ^1H NMR spectrum, coupling to ^{10}B (see Table 13.1) can, to a first approximation, be ignored.[‡]

[†] For historical insight, see: P. Laszlo (2000) *Angew. Chem. Int. Ed.*, vol. 39, p. 2071 – 'A diborane story'.
[‡] For further details, see: C.E. Housecroft (1994) *Boranes and Metallaboranes: Structure, Bonding and Reactivity*, 2nd edn, Ellis Horwood, Hemel Hempstead, Chapter 3, and references cited therein.

● **A general point in the NMR spectra of boranes is that:**

$$J(^{11}B-^{1}H_{terminal}) > J(^{11}B-^{1}H_{bridge})$$

(a) First, draw the structure of B_2H_6; there is one B environment, and two H environments:

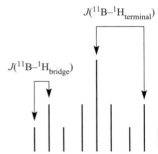

Consider the ^{11}B NMR spectrum. There is one signal, but each ^{11}B nucleus couples to two terminal ^{1}H nuclei and two bridging ^{1}H nuclei. The signal therefore appears as a triplet of triplets:

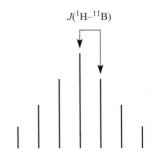

The exact nature of the observed spectrum depends upon the values of $J(^{11}B-^{1}H_{terminal})$ and $J(^{11}B-^{1}H_{bridge})$.

(b) In the ^{1}H NMR spectrum, there will be two signals, with relative integrals 2:4 (bridge H:terminal H).

Consider first the signal due to the terminal protons. For ^{11}B, $I = \frac{3}{2}$, meaning that there are four spin states with values $+\frac{3}{2}$, $+\frac{1}{2}$, $-\frac{1}{2}$ and $-\frac{3}{2}$. There is an *equal probability* that each terminal ^{1}H will 'see' the ^{11}B nucleus in each of the four spin states, and this gives rise to the ^{1}H signal being split into four equal intensity lines: a 1:1:1:1 multiplet.

Now consider the bridging protons. Each ^{1}H nucleus couples to *two* ^{11}B nuclei, and the signal will be a 1:2:3:4:3:2:1 multiplet since the combined nuclear spins of the two ^{11}B nuclei can adopt seven orientations, but not with equal probabilities:

J(^{1}H–^{11}B)

(c) The proton-decoupled ^{11}B NMR spectrum (written as the $^{11}B\{^{1}H\}$ NMR spectrum) will exhibit a singlet, since all $^{11}B-^{1}H$ coupling has been removed.

Self-study exercises

1. Refer to the spectral diagram in part (a) above. (i) Which part of the signal is the triplet due to $^{11}B-^{1}H_{terminal}$ spin–spin coupling? (ii) Indicate where else on the above diagram you could measure values of $J(^{11}B-^{1}H_{terminal})$ and $J(^{11}B-^{1}H_{bridge})$.

2. Refer to the spectral diagram in part (b) above. (i) Confirm the 1:2:3:4:3:2:1 intensities by considering the coupling to one ^{11}B nucleus and then adding in the effects of coupling to the second ^{11}B nucleus. (ii) Where else in the spectrum could you measure values of $J(^{1}H-^{11}B)$?

3. The $[BH_4]^-$ ion has a tetrahedral structure. Explain why the ^{1}H NMR spectrum exhibits a 1:1:1:1 multiplet, while the ^{11}B NMR spectrum shows a binomial quintet.
   [*Ans.* Refer to Case study 3 in Section 4.8]

Monomeric AlH_3 has been isolated at low temperature in a matrix. Evidence for the existence of Al_2H_6 (formed from laser-ablated Al atoms in a solid H_2 matrix at 3.5–6.5 K) has been obtained from vibrational spectroscopic data. The dissociation enthalpy for Al_2H_6 into $2AlH_3$ has been estimated from mass spectrometric data to be $138 \pm 20 \, kJ \, mol^{-1}$, a value similar to that for B_2H_6 going to $2BH_3$ (see after eq. 13.18). In the solid state at normal temperatures, X-ray and neutron diffraction data have shown that aluminium hydride (see eq. 3.19) consists of a 3-dimensional network in which each Al centre is octahedrally sited, being involved in six Al−H−Al 3c-2e interactions. Electron diffraction data show digallane, Ga_2H_6, is structurally similar to B_2H_6 ($Ga-H_{term} = 155 \, pm$, $Ga-H_{bridge} = 172 \, pm$, $Ga-H-Ga = 95°$). The enthalpy change for the dissociation of Ga_2H_6 to $2GaH_3$ is estimated to be $59 \pm 16 \, kJ \, mol^{-1}$, far lower than for either B_2H_6 or Al_2H_6. The existence of InH_3 was confirmed in 2004 (IR spectroscopic data for matrix-isolated InH_3), but, at present, the isolation of TlH_3 remains uncertain. Laser-ablated Tl atoms combine with H_2 in a Ne, Ar or H_2 matrix to give TlH as the dominant species. The hydrides of the group 13 elements are extremely air- and moisture-sensitive, and handling them requires the use of high vacuum techniques with all-glass apparatus.

(13.2)

Diborane(6) is an important reagent in synthetic organic chemistry, and reaction 13.8 is one convenient laboratory preparation. The structure of *diglyme*, used as the solvent, is shown in diagram **13.2**.

$$3Na[BH_4] + 4Et_2O \cdot BF_3$$

$$\xrightarrow{\text{diglyme, 298 K}} 2B_2H_6 + 3Na[BF_4] + 4Et_2O \qquad (13.8)$$

Although this reaction is standard procedure for the preparation of B_2H_6, it is not without problems. The reaction temperature must be carefully controlled because the solubility of $Na[BH_4]$ in diglyme varies significantly with temperature. Secondly, the solvent cannot easily be recycled.[†] Reaction 13.9, which uses a triglyme (**13.3**) adduct of BF_3 as precursor, produces B_2H_6 quantitatively and is an improvement on the traditional reaction 13.8. Reaction 13.9 can be applied to large-scale syntheses, and the triglyme solvent can be recycled. Tetraglyme can be used in place of triglyme.

$$3Na[BH_4] + 4(\mathbf{13.3}) \cdot BF_3$$

$$\xrightarrow{\text{triglyme, 298 K}} 2B_2H_6 + 3Na[BF_4] + 4(\mathbf{13.3}) \qquad (13.9)$$

(13.3)

Reaction 13.10 is the basis for an industrial synthesis of B_2H_6.

$$2BF_3 + 6NaH \xrightarrow{450\,K} B_2H_6 + 6NaF \qquad (13.10)$$

Diborane(6) is a colourless gas (bp 180.5 K) which is rapidly decomposed by water (eq. 13.11). Like other boron hydrides (see Section 13.11), B_2H_6 has a small positive value of $\Delta_f H^\circ$ ($+36\,kJ\,mol^{-1}$). Mixtures of B_2H_6 with air or O_2 are liable to inflame or explode (reaction 13.12).

$$B_2H_6 + 6H_2O \longrightarrow 2B(OH)_3 + 6H_2 \qquad (13.11)$$

$$B_2H_6 + 3O_2 \longrightarrow B_2O_3 + 3H_2O$$

$$\Delta_r H^\circ = -2138\,kJ \text{ per mole of } B_2H_6 \qquad (13.12)$$

Digallane, Ga_2H_6, is prepared by reaction 13.13. The product condenses at low temperature as a white solid (mp 223 K) but decomposes above 243 K.

$$(13.13)$$

Figure 13.8 summarizes some reactions of B_2H_6 and Ga_2H_6. Compared with the much studied B_2H_6, Ga_2H_6 has received far less attention. Three points should be noted:

- Ga_2H_6 is *unlike* B_2H_6 in that Ga_2H_6 rapidly decomposes to its constituent elements;
- Ga_2H_6 and B_2H_6 both react with HCl, but in the case of the borane, substitution of a terminal H by Cl is observed, whereas both terminal and bridging H atoms can be replaced in Ga_2H_6;
- Ga_2H_6 is *like* B_2H_6 in that it reacts with Lewis bases.

This last class of reaction is well documented and the examples in Fig. 13.8 illustrate two reaction types with the steric demands of the Lewis base being an important factor in determining which pathway predominates. For example, two NH_3 molecules can attack the *same* B or Ga centre, resulting in *asymmetric cleavage* of the E_2H_6 molecule. In contrast, reactions with more sterically demanding Lewis bases tend to cause *symmetric cleavage* (eq. 13.14).

$$(13.14)$$

The gallaborane $GaBH_6$ can be prepared by the reaction of $H_2Ga(\mu\text{-Cl})_2GaH_2$ (see eq. 13.13) with $Li[BH_4]$ at 250 K in the absence of air and moisture. In the gas phase, $GaBH_6$ has a molecular structure (**13.4**) analogous to those of B_2H_6 and Ga_2H_6. However, in the solid state it forms helical chains (Fig. 13.9).

(13.4)

$GaBH_6$ decomposes above 343 K (eq. 13.15), and it undergoes asymmetric cleavage (eq. 13.16). Although this reaction is carried out at low temperature, the product is stable at 298 K. Symmetric cleavage occurs when $GaBH_6$ reacts with NMe_3 or PMe_3 (eq. 13.17).

$$2GaBH_6 \xrightarrow{>343\,K} 2Ga + B_2H_6 + 3H_2 \qquad (13.15)$$

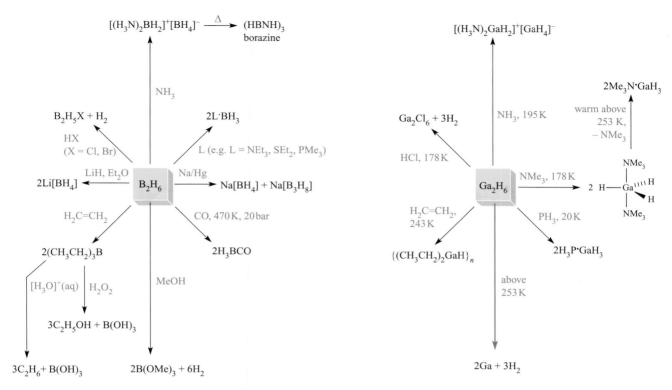

Fig. 13.8 Selected reactions of B_2H_6 and Ga_2H_6; all reactions of Ga_2H_6 must be carried out at low temperature since it decomposes above 253 K to gallium and dihydrogen. Borazine (top left-hand of the diagram) is discussed in Section 13.8.

$$GaBH_6 + 2NH_3 \xrightarrow{\ 195\,K\ } [H_2Ga(NH_3)_2]^+[BH_4]^-$$
$$(13.16)$$

$$GaBH_6 + 2EMe_3 \longrightarrow Me_3E\cdot GaH_3 + Me_3E\cdot BH_3$$
$$(E = N \text{ or } P) \quad (13.17)$$

At low temperatures, $H_2Ga(\mu\text{-Cl})_2GaH_2$ can be used as a precursor to Ga_2H_6 and $GaBH_6$, but thermal decomposition of $H_2Ga(\mu\text{-Cl})_2GaH_2$ (under vacuum at room temperature) leads to the mixed-valence compound $Ga^+[GaCl_3H]^-$. At higher temperatures, decomposition occurs according to eq. 13.18.

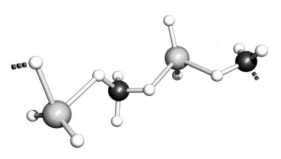

Fig. 13.9 Part of one chain of the polymeric structure of crystalline $GaBH_6$ (X-ray diffraction at 110 K) [A.J. Downs *et al.* (2001) *Inorg. Chem.*, vol. 40, p. 3484]. Colour code: B, blue; Ga, yellow; H, white.

$$2H_2Ga(\mu\text{-Cl})_2GaH_2 \longrightarrow 2Ga + Ga^+[GaCl_4]^- + 4H_2$$
$$(13.18)$$

Amine adducts of GaH_3 are of interest with respect to their use as precursors in chemical vapour deposition (CVD) (see Section 28.6). Tertiary amine adducts, $R_3N\cdot GaH_3$, dissociate, giving R_3N and GaH_3, and the latter then decomposes to Ga and H_2. Adducts of secondary and primary amines may eliminate H_2, as has been shown for $RH_2N\cdot GaH_3$ (R = Me, ^tBu) (Fig. 13.10).

Many of the reactions of B_2H_6 involve the non-isolable BH_3, and a value of 150 kJ mol^{-1} has been estimated for the dissociation enthalpy of B_2H_6 into $2BH_3$. Using this value, we can compare the Lewis acid strengths of BH_3, boron trihalides (BX_3) and boron trialkyls, and find that BH_3 lies between BX_3 and BMe_3 in behaviour towards simple Lewis bases such as NMe_3. However, only BH_3 forms adducts with CO and PF_3. Both CO and PF_3 are capable of acting as both electron donors (each using a lone pair of electrons centred on C or P respectively) and electron acceptors (using empty antibonding orbitals in CO or PF_3 respectively). Formation of $OC\cdot BH_3$ and $F_3P\cdot BH_3$ suggests that BH_3 can also act in both capacities. Its electron acceptance is readily understood in terms of an empty atomic orbital, i.e. B has four valence atomic orbitals, but only three are used for bonding in BH_3. Electron donation by BH_3 is ascribed to *hyperconjugation*

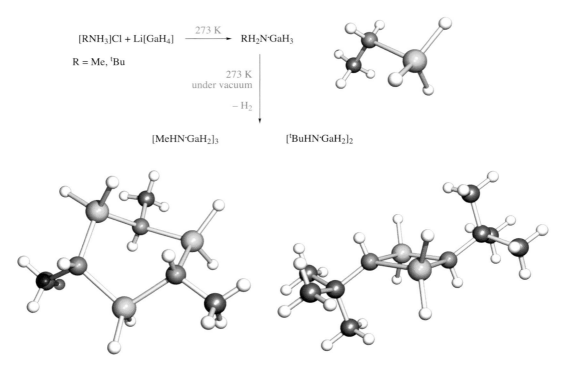

$$[RNH_3]Cl + Li[GaH_4] \xrightarrow{273 \text{ K}} RH_2N \cdot GaH_3$$

R = Me, ^{t}Bu

273 K
under vacuum

$- H_2$

$[MeHN \cdot GaH_2]_3$ $[^tBuHN \cdot GaH_2]_2$

Fig. 13.10 Formation of adducts $RH_2N \cdot GaH_3$ (R = Me, ^{t}Bu), and subsequent elimination of H_2 to give cyclic products, the size of which depends on R. The structures of the products have been determined by X-ray diffraction [S. Marchant *et al.* (2005) *Dalton Trans.*, p. 3281]. Colour code: Ga, yellow; N, blue; C, grey; H, white.

analogous to that proposed for a methyl group in organic compounds.[†]

In the solid state, the adduct $H_3N \cdot BH_3$ provides an interesting example of the so-called *dihydrogen bond* (see Fig. 10.12 and discussion). There is significant interest in the potential use of $H_3N \cdot BH_3$ as a hydrogen storage material. The adduct contains 19.6% hydrogen by weight and is non-inflammable under ambient conditions. Research efforts focus on methods of releasing H_2 from $H_3N \cdot BH_3$ (e.g. thermally, metal catalysed) but for this to be reversible, the production of the thermodynamically stable boron nitride (BN) has to be avoided. Effectively, this means that only two-thirds of the hydrogen in $H_3N \cdot BH_3$ could be utilized.[‡]

Worked example 13.3 Bonding in L·BH₃ adducts

Describe how BH_3 can behave as both an electron acceptor and an electron donor in the adduct $OC \cdot BH_3$.

First, consider the structure of $OC \cdot BH_3$:

The molecular orbitals of CO were described in Fig. 2.15. The HOMO possesses mainly carbon character; this MO is outward-pointing and is, to a first approximation, a lone pair on the C atom.

The $OC \cdot BH_3$ molecule contains a tetrahedral B atom; an sp^3 hybridization scheme is appropriate for B. Formation of the three B–H σ-bonds uses three sp^3 hybridized orbitals and the three valence electrons of B. This leaves a vacant sp^3 hybrid orbital on B that can act as an electron acceptor. The acceptance of two electrons completes an octet of electrons around the B atom:

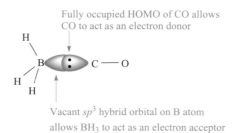

Fully occupied HOMO of CO allows
CO to act as an electron donor

Vacant sp^3 hybrid orbital on B atom
allows BH_3 to act as an electron acceptor

[†]For a discussion of hyperconjugation, see: M.B. Smith and J. March (2000) *March's Advanced Organic Chemistry: Reactions, Mechanisms and Structure*, 5th edn, Wiley, New York.
[‡]For reviews, see: A. Staubitz, A.P.M. Robertson and I. Manners (2010) *Chem. Rev.*, vol. 110, p. 4079; N.C. Smythe and J.C. Gordon (2010) *Eur. J. Inorg. Chem.*, p. 509.

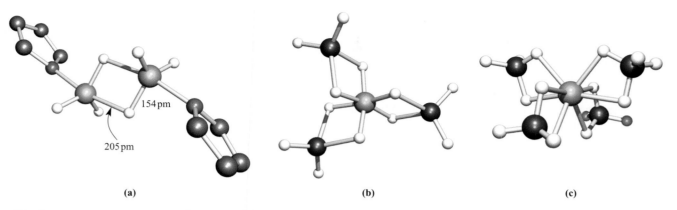

Fig. 13.11 (a) The structure of $[Al_2H_6(THF)_2]$ (X-ray diffraction at 173 K); hydrogen atoms have been omitted from the THF ligands [I.B. Gorrell *et al.* (1993) *J. Chem. Soc., Chem. Commun.*, p. 189]. (b) The structure of $[Al(BH_4)_3]$ deduced from spectroscopic studies. (c) The structure of $[Al(BH_4)_4]^-$ (X-ray diffraction) in the salt $[Ph_3MeP][Al(BH_4)_4]$ [D. Dou *et al.* (1994) *Inorg. Chem.*, vol. 33, p. 5443]. Colour code: B, blue; Al, gold; H, white; O, red; C, grey.

The LUMO of CO is a π^* orbital (Fig. 2.15). This orbital can act as an electron acceptor. Electrons can be donated from a B−H σ-bond (hyperconjugation):

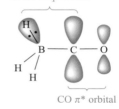

B–H σ-orbital can overlap with one lobe of the C 2p orbital

CO π^* orbital

The dominant effect is the σ-donation from CO to BH$_3$.

[***Note***: Although significantly less important than the σ-donation, the extent of the hyperconjugation is not clearly understood. See: A.S. Goldman and K. Krogh-Jespersen (1996) *J. Am. Chem. Soc.*, vol. 118, p. 12159.]

Self-study exercise

The structure of OC·BH$_3$ can be represented as illustrated below; this is one of several resonance forms that can be drawn. Rationalize the charge distribution shown in the diagram.

$$H_3B-C\equiv O^+$$

Aluminium hydride can be prepared by reaction 13.19; the solvent can be Et$_2$O, but the formation of etherate complexes $(Et_2O)_n AlH_3$ complicates the synthesis.

$$3Li[AlH_4] + AlCl_3 \longrightarrow \frac{4}{n}[AlH_3]_n + 3LiCl \qquad (13.19)$$

Above 423 K, $[AlH_3]_n$ is unstable with respect to decomposition to the elements, and this thermal instability has potential for generating thin films of Al. Aluminium hydride reacts with Lewis bases, e.g. to give Me$_3$N·AlH$_3$ (see reaction 13.26), in which the Al centre is tetrahedrally coordinated. As is general among the *p*-block elements, later elements in a group may exhibit higher coordination numbers than earlier congeners, and one example is THF·AlH$_3$, the solid state structure of which is dimeric, albeit with asymmetrical Al−H−Al bridges (Fig. 13.11a).

A number of adducts of InH$_3$ containing phosphine donors have been isolated, e.g. **13.5** and **13.6**, which are stable in the solid state at 298 K, but decompose in solution.[†]

Cy = cyclohexyl

(13.5)   **(13.6)**

The $[MH_4]^-$ ions

The syntheses and reducing properties of $[BH_4]^-$ and $[AlH_4]^-$ were described in Section 10.7, and reactions 13.8 and 13.9 showed the use of Na$[BH_4]$ (the most important salt containing the $[BH_4]^-$ ion) as a precursor to B$_2$H$_6$. Sodium

---

[†] For an overview of indium trihydride complexes, see: C. Jones (2001) *Chem. Commun.*, p. 2293. See also: S.G. Alexander and M.L. Cole (2008) *Eur. J. Inorg. Chem.*, p. 4493 – 'Lewis base adducts of heavier group 13 halohydrides – not just aspiring trihydrides!'

tetrahydridoborate(1−) is a white non-volatile crystalline solid, a typical ionic salt with an NaCl-type structure. It is stable in dry air and soluble in water, and is kinetically, rather than thermodynamically, stable in water. Although insoluble in Et_2O, it dissolves in THF and polyethers. Despite the salt-like properties of $Na[BH_4]$, derivatives with some other metals are covalent, involving M−H−B 3c-2e interactions. An example is $[Al(BH_4)_3]$ (Fig. 13.11b) in which the $[BH_4]^-$ ion behaves as a *bidentate ligand* as in structure **13.7**. In *trans*-$[V(BH_4)_2(Me_2PCH_2CH_2PMe_2)_2]$, each $[BH_4]^-$ ligand is *monodentate* (**13.8**), forming one B−H−V bridge, and in $[Zr(BH_4)_4]$, the 12-coordinate Zr(IV) centre is surrounded by four *tridentate* ligands (**13.9**). Complex formation may (eq. 13.20) or may not (eq. 13.21) be accompanied by reduction of the central metal.

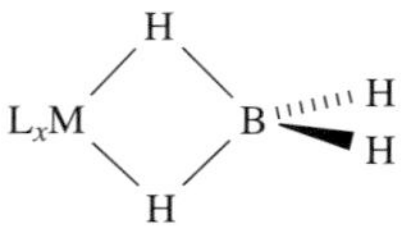

(13.7) 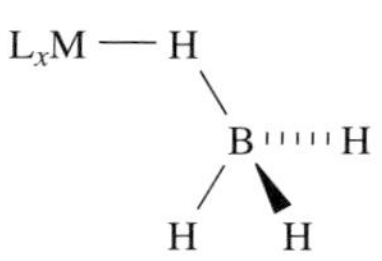　　**(13.8)** 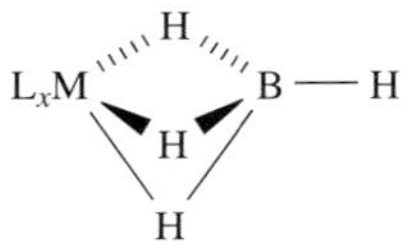　　**(13.9)**

$$2[VCl_4(THF)_2] + 10[BH_4]^-$$
$$\longrightarrow 2[V(BH_4)_4]^- + 8Cl^- + B_2H_6 + H_2 + 4THF$$
(13.20)

$$HfCl_4 + 4[BH_4]^- \longrightarrow [Hf(BH_4)_4] + 4Cl^-$$
(13.21)

Although $[Al(BH_4)_3]$ is a widely cited example of a tetrahydridoborate(1−) complex of Al(III), the first complex to be characterized by X-ray diffraction, $[Ph_3MeP][Al(BH_4)_4]$ (Fig. 13.11c), was not reported until 1994. It is prepared by reaction 13.22 and was the first example of a molecular species containing an 8-coordinate Al(III) centre; the coordination sphere is approximately dodecahedral (see Fig. 19.9).

$$[Al(BH_4)_3] + [BH_4]^- \longrightarrow [Al(BH_4)_4]^-$$
(13.22)

In solution, many covalent complexes containing the $[BH_4]^-$ ligand exhibit dynamic behaviour which may be observed on the NMR spectroscopic timescale. For example, the room temperature ^1H NMR spectrum of $[Al(BH_4)_3]$ shows only one signal.

Worked example 13.4　Dynamic behaviour of complexes containing $[BH_4]^-$

The room temperature solution ^{11}B NMR spectrum of $[Ph_3MeP][Al(BH_4)_4]$ shows a well-resolved binomial quintet ($\delta -34.2$ ppm, $J = 85$ Hz). At 298 K, the ^1H NMR spectrum of this compound exhibits signals at

δ **7.5–8.0 (multiplet), 2.8 (doublet, $J = 13$ Hz) and 0.5 (very broad) ppm. The latter signal remains broad on cooling the sample to 203 K. Interpret these data. The solid state structure of $[Al(BH_4)_4]^-$ is given in Fig. 13.11; NMR data are listed in Table 4.3.**

First, consider the solid state structure as a starting point, but remember that the NMR spectrum relates to a solution sample:

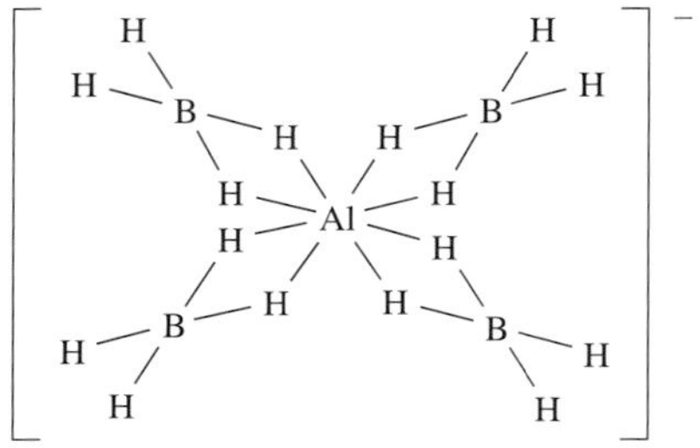

In the ^1H NMR spectrum, the multiplet at δ 7.5–8.0 ppm is assigned to the Ph protons in $[Ph_3MeP]^+$, and the doublet at δ 2.8 ppm is assigned to the Me protons which couple to the ^{31}P nucleus ($I = \frac{1}{2}$, 100%). The signal at δ 0.5 ppm must arise from the boron-attached protons.

In the solid state, each $[BH_4]^-$ ion is involved in two Al−H−B interactions. There are two H environments: terminal (8H) and bridging (8H). The observation of one broad signal for the ^1H nuclei attached to ^{11}B is consistent with a fluxional (dynamic) process which exchanges the terminal and bridging protons.

The observation of a *binomial* quintet in the ^{11}B NMR spectrum is consistent with each ^{11}B nucleus (all are in equivalent environments) coupling to four ^1H nuclei which are *equivalent* on the NMR timescale, i.e. which are undergoing a dynamic process.

Self-study exercise

The solid state structure of $H_3Zr_2(PMe_3)_2(BH_4)_5$ (compound **A**) is shown schematically below. There are four tridentate and one bidentate $[BH_4]^-$ and three bridging hydride ligands.

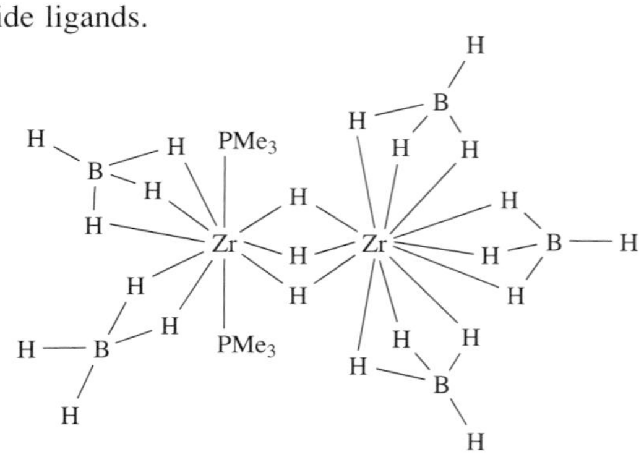

At 273 K, the solution ^{11}B NMR spectrum of **A** shows two quintets ($\delta -12.5$ ppm, $J = 88$ Hz and $\delta -9.8$ ppm,

$J = 88\,\text{Hz}$, relative integrals $3:2$). The ^{1}H NMR spectrum (273 K), exhibits a triplet ($J = 14\,\text{Hz}$, 3 H) at δ 3.96 ppm, a triplet at δ 1.0 ppm ($J = 3\,\text{Hz}$, 18 H) and two $1:1:1:1$ quartets ($J = 88\,\text{Hz}$) with integrals relative to one another of $3:2$. Interpret these spectroscopic data and explain the origin of the spin–spin couplings; see Table 4.3 for nuclear spin data.

[*Ans.* See: J.E. Gozum *et al.* (1991) *J. Am. Chem. Soc.*, vol. 113, p. 3829]

The salt Li[AlH$_4$] is a widely used reducing and hydrogenating agent. It is obtained as a white solid by reaction 13.23 or 13.24, and is stable in dry air but is decomposed by water (eq. 13.25).

$$4LiH + AlCl_3 \xrightarrow{\text{Et}_2\text{O}} 3LiCl + Li[AlH_4] \qquad (13.23)$$

$$Li + Al + 2H_2 \xrightarrow{\text{250 bar, 400 K, Et}_2\text{O}} Li[AlH_4] \qquad (13.24)$$

$$Li[AlH_4] + 4H_2O \longrightarrow LiOH + Al(OH)_3 + 4H_2 \qquad (13.25)$$

Adducts of aluminium hydride can be obtained from [AlH$_4$]$^-$ (e.g. reaction 13.26) and some of these compounds are important reducing agents and polymerization catalysts in organic chemistry.

$$3Li[AlH_4] + AlCl_3 + 4Me_3N \longrightarrow 4Me_3N\cdot AlH_3 + 3LiCl \qquad (13.26)$$

The compounds Li[EH$_4$] for E = Ga, In and Tl have been prepared at low temperatures, (e.g. reaction 13.27) but are thermally unstable.

$$4LiH + GaCl_3 \longrightarrow Li[GaH_4] + 3LiCl \qquad (13.27)$$

13.6 Halides and complex halides

Boron halides: BX$_3$ and B$_2$X$_4$

Boron trihalides are monomeric under ordinary conditions, possess trigonal planar structures (**13.10**), and are much more volatile than the corresponding compounds of Al. Boron trifluoride is a colourless gas (bp 172 K), BCl$_3$ and BBr$_3$ are colourless liquids (BCl$_3$, mp 166 K, bp 285 K; BBr$_3$, mp 227 K, bp 364 K), while BI$_3$ is a white solid (mp 316 K). Low-temperature X-ray diffraction data for BCl$_3$ and BI$_3$ show that discrete trigonal planar molecules are present in the solid state.

	B—X distance
X = F	131 pm
X = Cl	174 pm
X = Br	189 pm
X = I	210 pm

(13.10)

Equation 13.28 shows the usual synthesis of BF$_3$; excess H$_2$SO$_4$ removes the H$_2$O formed. Boron trifluoride fumes strongly in moist air and is partially hydrolysed by excess H$_2$O (eq. 13.29). With small amounts of H$_2$O at low temperatures, the adducts BF$_3\cdot$H$_2$O and BF$_3\cdot$2H$_2$O are obtained.

$$B_2O_3 + 3CaF_2 + 3H_2SO_4 \xrightarrow{\text{conc}} 2BF_3 + 3CaSO_4 + 3H_2O \qquad (13.28)$$

$$4BF_3 + 6H_2O \longrightarrow 3[H_3O]^+ + 3[BF_4]^- + B(OH)_3 \qquad (13.29)$$

Pure tetrafluoroboric acid, HBF$_4$, is *not* isolable but is commercially available in Et$_2$O solution, or as solutions formulated as [H$_3$O][BF$_4$]$\cdot$4H$_2$O. It can also be formed by reaction 13.30.

$$B(OH)_3 + 4HF \longrightarrow [H_3O]^+ + [BF_4]^- + 2H_2O \qquad (13.30)$$

Tetrafluoroboric acid is a very strong acid, and mixtures of HF and BF$_3$ are extremely strong proton donors, although not quite as strong as those of HF and SbF$_5$ (see Section 9.7). Salts containing the [BF$_4$]$^-$ ion are frequently encountered in synthetic chemistry. The [BF$_4$]$^-$ ion (like [PF$_6$]$^-$ coordinates very weakly, if at all, to metal centres and is often used as an 'innocent' anion to precipitate cations. For a discussion of the stability of KBF$_4$ with respect to KF + BF$_3$, see Section 6.16.

The [BF$_4$]$^-$ ion can be converted to [B(CN)$_4$]$^-$ in the solid state reaction 13.31. A range of salts can then be prepared from Li[B(CN)$_4$] as exemplified in reactions 13.32 and 13.33.

$$K[BF_4] + 4KCN + 5LiCl \xrightarrow[\text{no solvent}]{573\,\text{K}} Li[B(CN)_4]$$
$$+ 5KCl + 4LiF \qquad (13.31)$$

$$Li[B(CN)_4] \xrightarrow{\text{HCl, }^n\text{Pr}_3\text{N}} [^nPr_3NH][B(CN)_4] + LiCl \qquad (13.32)$$

$$[^nPr_3NH][B(CN)_4] + MOH \xrightarrow{\text{H}_2\text{O}} M[B(CN)_4] + H_2O + {}^nPr_3N$$
$$M = Na, K \qquad (13.33)$$

Self-study exercises

1. To what point group does [BF$_4$]$^-$ belong? Explain why [BF$_4$]$^-$ has two IR active T_2 vibrational modes.
   [*Ans.* See Fig. 3.16 and accompanying discussion]

2. The ^{13}C NMR spectrum of a CDCl$_3$ solution of [Bu$_4$N][B(CN)$_4$] shows (in addition to signals for solvent and [Bu$_4$N]$^+$) a $1:1:1:1$ multiplet overlying a less intense $1:1:1:1:1:1:1$ signal. Both signals are centred at δ 122.3 ppm, and coupling constants for the two multiplets are 71 and 24 Hz, respectively. Rationalize the appearance of the spectrum.
   [*Ans.* See Section 4.8, Case study 4; a figure of the spectrum can be found in E. Bernhardt *et al.* (2000) *Z. Anorg. Allg. Chem.*, vol. 626, p. 560]

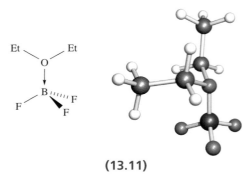

(13.11)

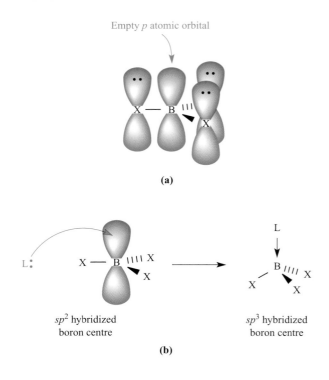

(a)

sp² hybridized boron centre   *sp³* hybridized boron centre

(b)

Fig. 13.12 (a) The formation of partial π-bonds in a trigonal planar BX_3 molecule can be considered in terms of the donation of electron density from filled p atomic orbitals on the X atoms into the empty $2p$ atomic orbital on boron. (b) Reaction of BX_3 with a Lewis base, L, results in a change from a trigonal planar (*sp²* boron centre) to tetrahedral (*sp³* boron centre) molecule.

Boron trifluoride forms a range of complexes with ethers, nitriles and amines. It is commercially available as the adduct $Et_2O \cdot BF_3$ (**13.11**). Being a liquid at 298 K, this is a convenient means of handling BF_3 which has many applications as a catalyst in organic reactions, e.g. in Friedel–Crafts alkylations and acylations.

The reactions between B and Cl_2 or Br_2 yield BCl_3 or BBr_3 respectively, while BI_3 is prepared by reaction 13.34 or 13.35. Each of BCl_3, BBr_3 and BI_3 is decomposed by water (eq. 13.36), and reacts with inorganic or organic compounds containing labile protons to eliminate HX (X = Cl, Br, I). Thus, while BF_3 forms an adduct with NH_3, BCl_3 reacts in liquid NH_3 to form $B(NH_2)_3$. The adduct $H_3N \cdot BCl_3$ can be isolated in low yield from the reaction of BCl_3 and NH_4Cl, the major product being $(ClBNH)_3$ (see eq. 13.62). The adduct is stable at room temperature in an inert atmosphere. In the solid state, $H_3N \cdot BCl_3$ adopts an ethane-like, staggered conformation and there is intermolecular hydrogen bonding involving $N-H \cdots Cl$ interactions.

$$BCl_3 + 3HI \xrightarrow{\Delta} BI_3 + 3HCl \qquad (13.34)$$

$$3Na[BH_4] + 8I_2 \longrightarrow 3NaI + 3BI_3 + 4H_2 + 4HI \qquad (13.35)$$

$$BX_3 + 3H_2O \longrightarrow B(OH)_3 + 3HX \qquad X = Cl, Br, I \qquad (13.36)$$

Unlike $[BF_4]^-$, the ions $[BCl_4]^-$, $[BBr_4]^-$ and $[BI_4]^-$ are stabilized only in the presence of large cations such as $[^nBu_4N]^+$.

In mixtures containing two or three of BF_3, BCl_3 and BBr_3, exchange of the halogen atoms occurs to yield BF_2Cl, $BFBr_2$, $BFClBr$ etc. and their formation can be monitored by using ^{11}B or ^{19}F NMR spectroscopy (see end-of-chapter problem 4.43).

The thermodynamics of adduct formation by BF_3, BCl_3 and BBr_3 have been much discussed, and reactions with NMe_3 (Lewis base L) in the gas phase show that the order of adduct stabilities is $L \cdot BF_3 < L \cdot BCl_3 < L \cdot BBr_3$. Determinations of Δ_rH° for reaction 13.37 in nitrobenzene solution reveal the same sequence.

$$py(soln) + BX_3(g) \longrightarrow py \cdot BX_3(soln) \qquad py = \qquad (13.37)$$

This sequence is the opposite of that predicted on the basis of the electronegativities of the halogens, but by considering changes in bonding during adduct formation, one can rationalize the experimental observations. In BX_3, the $B-X$ bonds contain partial π-character (Fig. 13.12a) (see Section 5.3). Reaction with a Lewis base, L, leads to a change in stereochemistry at the B centre from trigonal planar to tetrahedral and, as a result, the π-contributions to the $B-X$ bonds are lost (Fig. 13.12b). This is demonstrated by the observation that the $B-F$ bond length increases from 130 pm in BF_3 to 145 pm in $[BF_4]^-$. We can formally consider adduct formation to occur in two steps: (i) the reorganization of trigonal planar to pyramidal B, and (ii) the formation of an $L \longrightarrow B$ coordinate bond. The first step is endothermic, while the second is exothermic. The pyramidal BX_3 *intermediate* cannot be isolated and is only a *model* state. The observed ordering of adduct stabilities can now be understood in terms of the energy difference between that associated with loss of π-character (which is greatest for BF_3) and that associated with formation of the $L \longrightarrow B$ bond. Evidence for the amount of π-character in BX_3 following the sequence $BF_3 > BCl_3 > BBr_3$ comes from the fact that the increase in the $B-X$ bond distances in BX_3 (130, 176 and 187 pm for BF_3, BCl_3 and BBr_3) is greater than the increase in the values of r_{cov} for X (71, 99 and 114 pm for F, Cl and Br). It has been suggested that the presence of the π-bonding in boron trihalides is the reason why these molecules are monomeric, while the corresponding halides of the heavier group 13 elements are oligomeric (e.g. Al_2Cl_6). π-Bonding is always stronger in compounds involving first-row elements (e.g. compare the chemistries of C and Si, or N and P, in Chapters 14 and 15). An alternative explanation for the relative Lewis acid strengths of BF_3, BCl_3 and BBr_3 is that the ionic contributions to the bonding in BX_3 (see Fig. 5.10) are greatest for BF_3 and least for BBr_3. Thus, the reorganization energy associated with lengthening the B–X bonds on going from BX_3 to $L\cdot BX_3$ follows the order $BF_3 > BCl_3 > BBr_3$, making the formation of $L\cdot BF_3$ the least favourable of $L\cdot BF_3$, $L\cdot BCl_3$ and $L\cdot BBr_3$. It is significant that for *very weak* Lewis bases such as CO, little geometrical change occurs to the BX_3 unit on going from BX_3 to $OC\cdot BX_3$. In this case, the observed order of complex stability is $OC\cdot BF_3 > OC\cdot BCl_3$, consistent with the Lewis acid strength of BX_3 being controlled by the polarity of the BX_3 molecule.

X—B—B—X with four X substituents **(13.12)**   X—B—B—X staggered with X on wedge/dash **(13.13)**

(13.12)   **(13.13)**

Among the group 13 elements, B alone forms halides of the type X_2B-BX_2, although adducts of the type LX_2M-MX_2L (M = Al, Ga; L = Lewis base) are closely related compounds, e.g. see structure **13.18**. At 298 K, B_2Cl_4 is a colourless, unstable liquid, and is prepared by co-condensing BCl_3 and Cu vapours on a surface cooled with liquid N_2. B_2Cl_4 is converted to B_2F_4 (a colourless gas at 298 K) by reaction with SbF_3. The compounds B_2Br_4 and B_2I_4 are, respectively, an easily hydrolysed

liquid and a pale yellow solid. In the solid state, B_2F_4 and B_2Cl_4 are planar (D_{2h}, **13.12**), but in the vapour phase, B_2F_4 remains planar while B_2Cl_4 has a staggered structure (D_{2d}, **13.13**). B_2Br_4 adopts a staggered conformation in the vapour, liquid and solid phases. These preferences are not readily explained.

$$B_2Cl_4$$
720 K, a few minutes ↙   ↘ 373 K, in presence of CCl_4, several days

$$B_9Cl_9 \qquad\qquad B_8Cl_8 \qquad (13.38)$$

The thermal decomposition of B_2X_4 (X = Cl, Br, I) gives BX_3 and cluster molecules (Fig. 13.13) of type B_nX_n (X = Cl, $n = 8$–12; X = Br, $n = 7$–10; X = I, $n = 8$ or 9). Some degree of selectiveness can be achieved by fine tuning the reaction conditions (e.g. scheme 13.38), but this general synthetic route to these clusters is difficult. Higher yields of B_9X_9 (X = Cl, Br, I) are obtained using reactions 13.39 and 13.40 for which radical mechanisms are proposed.

$$B_{10}H_{14} + \tfrac{26}{6}C_2Cl_6$$

In a sealed tube
$$\xrightarrow{470\,K,\,2\,days} B_9Cl_9 + BCl_3 + \tfrac{26}{3}C + 14HCl$$
$$(13.39)$$

In an autoclave
$$B_{10}H_{14} + 13X_2 \xrightarrow{470\,K,\,20\,h} B_9X_9 + BX_3 + 14HX$$
$$(X = Br\ or\ I) \quad (13.40)$$

Reduction of B_9X_9 with I^- leads, first, to the radical anion $[B_9X_9]^{\cdot-}$ and then to $[B_9X_9]^{2-}$. The solid state structures of B_9Cl_9, B_9Br_9, $[Ph_4P][B_9Br_9]$ and $[Bu_4N]_2[B_9Br_9]$ have been determined and confirm that each cluster possesses a tricapped trigonal prismatic structure (Fig. 13.13c). This represents an unusual example of a main-group cluster core maintaining the same core structure along a redox series (eq. 13.41). However, each reduction step results in

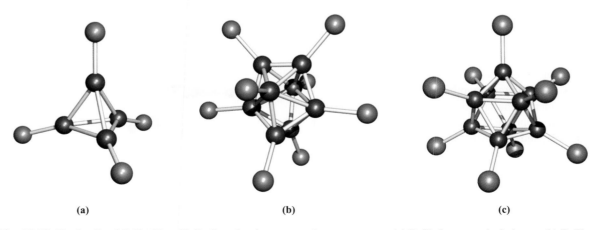

(a)   (b)   (c)

 Fig. 13.13 The family of B_nX_n (X = Cl, Br, I) molecules possess cluster structures. (a) B_4Cl_4 has a tetrahedral core, (b) B_8Cl_8 possesses a dodecahedral cluster core and (c) B_9Br_9 has a tricapped trigonal prismatic core. Colour code: B, blue; Cl, green; Br, gold.

significant changes in bond lengths within the cluster framework.

$$B_9Br_9 \xrightarrow{1e^- \text{ reduction}} [B_9Br_9]^{\bullet -} \xrightarrow{1e^- \text{ reduction}} [B_9Br_9]^{2-}$$

Retention of a trigonal tricapped prismatic cluster core

(13.41)

The cluster B_4Cl_4 can be obtained by passing an electrical discharge through BCl_3 in the presence of Hg. Figure 13.13 shows the structures of B_4Cl_4 and B_8Cl_8. Reactions of B_4Cl_4 may occur with retention of the cluster core (e.g. reaction 13.42) or its fragmentation (e.g. reaction 13.43). Reactions of B_8Cl_8 are often accompanied by cage expansion (e.g. reaction 13.44), an exception being Friedel–Crafts bromination which gives B_8Br_8.

$$B_4Cl_4 + 4Li^tBu \longrightarrow B_4{}^tBu_4 + 4LiCl \qquad (13.42)$$

$$B_4Cl_4 \xrightarrow{480\,K,\,CFCl_3} BF_3 + B_2F_4 \qquad (13.43)$$

$$B_8Cl_8 \xrightarrow{AlMe_3} B_9Cl_{9-n}Me_n \qquad n = 0-4 \qquad (13.44)$$

Analysis of the bonding in any of these clusters poses problems. If the terminal B–X bonds are considered to be localized 2c-2e interactions, then there are insufficient valence electrons remaining for a localized treatment of the B–B interactions in the B_n core. We return to this problem at the end of Section 13.11.

Al(III), Ga(III), In(III) and Tl(III) halides and their complexes

The trifluorides of Al, Ga, In and Tl are non-volatile solids, best prepared by fluorination of the metal (or one of its simple compounds) with F_2. AlF_3 is also prepared by reaction 13.45.

$$Al_2O_3 + 6HF \xrightarrow{970\,K} 2AlF_3 + 3H_2O \qquad (13.45)$$

Each trifluoride is high melting and has an infinite structure. In AlF_3, each Al centre is octahedral, surrounded by six F atoms, each of which links two Al centres. The octahedral AlF_6-unit is encountered in other Al fluorides: Tl_2AlF_5 contains polymeric chains composed of AlF_6-octahedra linked through opposite vertices (represented by either **13.14** or **13.15**), and in $TlAlF_4$ and $KAlF_4$, AlF_6 octahedra

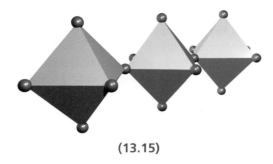

(13.15)

are linked through four vertices to form sheets. In the salt $[pyH]_4[Al_2F_{10}]\cdot4H_2O$ ([pyH]$^+$ = pyridinium ion), the anions contain two edge-sharing octahedral AlF_6-units, two representations of which are shown in structure **13.16**. Corner-sharing AlF_6-units are present in $[Al_7F_{30}]^{9-}$ which is a discrete anion (Fig. 13.14), and in $[Al_7F_{29}]^{8-}$

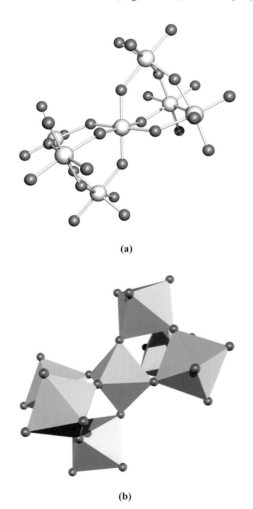

(a)

(b)

Fig. 13.14 The structure (X-ray diffraction) of the $[Al_7F_{30}]^{9-}$ anion in the salt $[NH(CH_2CH_2NH_3)_3]_2[H_3O][Al_7F_{30}]$ [E. Goreshnik *et al.* (2002) *Z. Anorg. Allg. Chem.*, vol. 628, p. 162]. (a) A 'ball-and-stick' representation of the structure (colour code: Al, pale grey; F, green) and (b) a polyhedral representation showing the corner-sharing octahedral AlF_6-units.

(13.14)

which forms polymeric chains in the compound $[NH(CH_2CH_2NH_3)_3]_2[Al_7F_{29}]\cdot 2H_2O$.

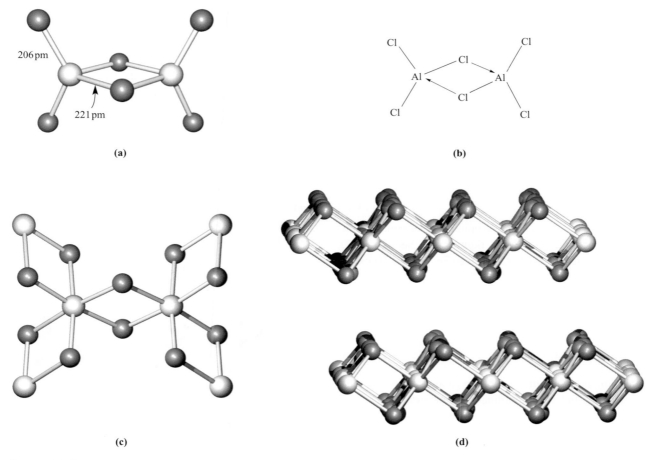

(13.16)

Cryolite, $Na_3[AlF_6]$ (see Section 13.2) occurs naturally but is also synthesized (reaction 13.46) to meet commercial needs. The solid state structure of cryolite is related to the perovskite-type structure.

$$Al(OH)_3 + 6HF + 3NaOH \longrightarrow Na_3[AlF_6] + 6H_2O \qquad (13.46)$$

Compounds MX_3 (M = Al, Ga or In; X = Cl, Br or I) are obtained by direct combination of the elements. They are relatively volatile and in the solid state possess layer structures or structures containing dimers M_2X_6. Solid $AlCl_3$ adopts a layer structure with octahedrally sited Al (Figs. 13.15c–d). The vapours consist of dimeric molecules and these are also present in solutions of the compounds in inorganic solvents. Only at high temperatures does dissociation to monomeric MX_3 occur. In the monomer, the group 13 metal is trigonal planar, but in the dimer, a tetrahedral environment results from X $\longrightarrow$ M coordinate bond formation involving a halogen lone pair of electrons (Fig. 13.15).

When water is dripped on to solid $AlCl_3$, vigorous hydrolysis occurs, but in *dilute* aqueous solution, $[Al(OH_2)_6]^{3+}$ (see eq. 7.34) and Cl^- ions are present. In coordinating solvents such as Et_2O, $AlCl_3$ forms adducts such as $Et_2O\cdot AlCl_3$, structurally analogous to **13.11**. With NH_3, AlX_3 (X = Cl, Br, I) forms $H_3N\cdot AlX_3$, and in the solid state (as for $H_3N\cdot BCl_3$) there is intermolecular hydrogen

(a)

206 pm

221 pm

(b)

(c)

(d)

Fig. 13.15 (a) The structure of Al_2Cl_6 with bond distances determined in the vapour phase; the terminal M–X bond distances are similarly shorter than the bridging distances in Al_2Br_6, Al_2I_6, Ga_2Cl_6, Ga_2Br_6, Ga_2I_6 and In_2I_6. In $AlCl_3$ monomer, the Al–Cl distances are 206 pm. (b) A representation of the bonding in Al_2Cl_6 showing the Cl lone pair donation to Al. (c) Part of one layer of crystalline $AlCl_3$ viewed from above, showing Cl atoms bridging between octahedrally sited Al atoms. (d) The layer structure of solid $AlCl_3$ viewed from the side; the upper and lower faces of each layer consist of Cl atoms, and weak van der Waals forces operate between layers. Colour code: Al, pale grey; Cl, green.

APPLICATIONS

Box 13.5 Lewis acid pigment solubilization

Applications of pigments for coatings, printing and information storage are widespread, but the fabrication of thin films of pigments is difficult because of their insoluble nature. Dyes, on the other hand, are easier to manipulate. Research at the Xerox Corporation has shown that Lewis acid complexes can be utilized to solubilize and lay down thin films of certain pigments. For example, the photosensitive perylene derivative shown below forms an adduct with $AlCl_3$:

Complex formation occurs in $MeNO_2$ solution and the solution is then applied to the surface to be coated. Washing with water removes the Lewis acid leaving a thin film of the photosensitive pigment. The Lewis acid pigment solubilization (LAPS) technique has been used to fabricate multilayer photoconductors and a range of other thin film devices.

Further reading

B.R. Hsieh and A.R. Melnyk (1998) *Chemistry of Materials*, vol. 10, p. 2313 – 'Organic pigment nanoparticle thin film devices via Lewis acid pigment solubilization'.

B.R. Hsieh and A.R. Melnyk (2001) *J. Imaging Sci. Technol.*, vol. 45, p. 37 – 'Organic pigment nanoparticle thin film devices via Lewis acid pigment solubilization and *in situ* pigment dispersions'.

bonding involving $N-H\cdots X$ interactions. (A commercial application of $AlCl_3$ adducts is highlighted in Box 13.5.) Addition of Cl^- to $AlCl_3$ yields the tetrahedral $[AlCl_4]^-$ and this reaction is important in Friedel–Crafts acylations and alkylations, the initial steps in which are summarized in eq. 13.47.

$$RC\overset{+}{\equiv}O + [AlCl_4]^- \xleftarrow{RC(O)Cl} AlCl_3 \xrightarrow{RCl} R^+ + [AlCl_4]^-$$
$$(13.47)$$

Salts of $[AlCl_4]^-$ containing asymmetric, organic cations represent one family of ionic liquids (see Section 9.12). Molten salts such as $NaCl-Al_2Cl_6$ (see Section 9.12) contain $[AlCl_4]^-$ in equilibrium with $[Al_2Cl_7]^-$ (eq. 13.48). The solid state structures of a number of salts of $[Al_2Cl_7]^-$ have been determined, and demonstrate that the ion can adopt either a staggered or eclipsed conformation (Fig. 13.16).

$$2[AlCl_4]^- \rightleftharpoons [Al_2Cl_7]^- + Cl^-$$
$$(13.48)$$

Gallium and indium trichlorides and tribromides also form adducts, but with coordination numbers of 4, 5 or 6: $[MCl_6]^{3-}$, $[MBr_6]^{3-}$, $[MCl_5]^{2-}$, $[MCl_4]^-$ and $[MBr_4]^-$ (M = Ga or In) or $L\cdot GaX_3$ or $L_3\cdot InX_3$ (L = neutral Lewis base). The square-based pyramidal structure of $[InCl_5]^{2-}$ has been confirmed by X-ray diffraction for the $[Et_4N]^+$ salt. This is not expected by VSEPR arguments, but one must bear in mind that energy differences between 5-coordinate geometries are often small and preferences can be tipped by, for example, crystal packing forces.

The Tl(III) halides are less stable than those of the earlier group 13 elements. $TlCl_3$ and $TlBr_3$ are very unstable with respect to conversion to the Tl(I) halides (eq. 13.49).

$$TlBr_3 \longrightarrow TlBr + Br_2 \qquad (13.49)$$

The compound TlI_3 is isomorphous with the alkali metal triiodides and is really thallium(I) triiodide, **13.17**. However, when treated with excess I^-, an interesting redox reaction occurs with the formation of $[TlI_4]^-$ (see Section 13.9). The decrease in stability of the higher oxidation state on going from the binary fluoride to iodide is a general feature of all metals that exhibit more than one oxidation state. For ionic compounds, this is easily explained in terms of

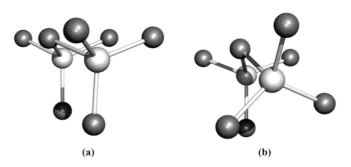

Fig. 13.16 The crystallographically determined structure of the $[Al_2Cl_7]^-$ ion. In the compound $[(C_6Me_6)_3Zr_3Cl_6][Al_2Cl_7]_2$, the anions adopt one of two different conformations: (a) an eclipsed conformation and (b) a staggered conformation [F. Stollmaier *et al.* (1981) *J. Organomet. Chem.*, vol. 208, p. 327]. Colour code: Al, grey; Cl, green.

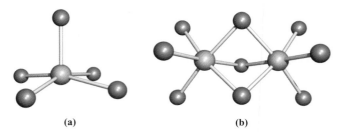

(a)   **(b)**

Fig. 13.17 (a) The structure of $[TlCl_5]^{2-}$ determined by
X-ray diffraction for the salt $[H_3N(CH_2)_5NH_3][TlCl_5]$.
[M.A. James *et al.* (1996) *Can. J. Chem.*, vol. 74, p. 1490.] (b) The
crystallographically determined structure of $[Tl_2Cl_9]^{3-}$ in
$Cs_3[Tl_2Cl_9]$. Colour code: Tl, orange; Cl, green.

lattice energies. The difference between the values of the
lattice energies for MX and MX_3 (X = halide) is greatest
for the smallest anions (see eq. 6.16).

$$Tl^+ \quad \left[I\!\!-\!\!-\!\!I\!\!-\!\!-\!\!I \right]^-$$

(13.17)

Thallium(III) exhibits coordination numbers higher than 4
in complex chlorides, prepared by addition of chloride
salts to $TlCl_3$. In $[H_3N(CH_2)_5NH_3][TlCl_5]$, a square-based
pyramidal structure for the anion has been confirmed
(Fig. 13.17a). In $K_3[TlCl_6]$, the anion has the expected octahedral structure, and in $Cs_3[Tl_2Cl_9]$, the Tl(III) centres in the
anion are also octahedral (Fig. 13.17b).

Self-study exercises

1. Using the method outlined in Section 3.4, confirm that
AlI_3 and Al_2I_6 belong to the D_{3h} and D_{2h} point groups,
respectively.

2. The IR spectrum of AlI_3 vapour has been measured in the
region 50–700 cm^{-1}. Three absorptions at 427, 147 and
66 cm^{-1} are observed, and the band at 66 cm^{-1} is also present in the Raman spectrum. Given that the absorption at
427 cm^{-1} is a stretching mode, assign the three bands and
draw diagrams to illustrate the vibrational modes.
[*Ans.* Refer to Fig. 3.14 and accompanying discussion]

Lower oxidation state Al, Ga, In and Tl halides

Aluminium(I) halides are formed in reactions of Al(III)
halides with Al at 1270 K followed by rapid cooling. Red
AlCl is also formed by treating the metal with HCl at
1170 K. However, the monohalides are unstable with
respect to disproportionation (eq. 13.50).

$$3AlX \longrightarrow 2Al + AlX_3 \qquad (13.50)$$

The reaction of AlBr with PhOMe at 77 K followed by
warming to 243 K yields $[Al_2Br_4(OMePh)_2]$, **13.18**. This
is air- and moisture-sensitive and decomposes at 298 K,
but represents a close relation of the $X_2B{-}BX_2$ compounds
described earlier. Crystals of $[Al_2I_4(THF)_2]$ (**13.19**) are
deposited from metastable AlI·THF/toluene solutions
which are formed by co-condensation of AlI with THF
and toluene. The Al–Al bond lengths in **13.18** and **13.19**
are 253 and 252 pm respectively, consistent with single
bonds ($r_{cov} = 130$ pm). Co-condensation of AlBr with THF
and toluene gives solutions from which $[Al_{22}Br_{20}(THF)_{12}]$
and $[Al_5Br_6(THF)_6]^+[Al_5Br_8(THF)_4]^-$ (Fig. 13.18) can be
isolated; aluminium metal is also deposited. The structure
of $[Al_{22}Br_{20}(THF)_{12}]$ (**13.20**) consists of an icosahedral
Al_{12}-core; an $AlBr_2(THF)$-unit is bonded to 10 of the Al
atoms, and THF donors are coordinated to the remaining
two Al atoms. The Al–Al distances within the Al_{12}-cage

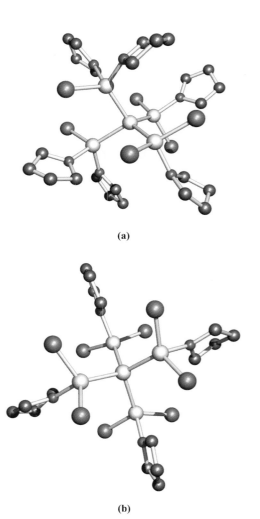

(a)

(b)

Fig. 13.18 The structures (X-ray diffraction) of
(a) $[Al_5Br_6(THF)_6]^+$ and (b) $[Al_5Br_8(THF)_4]^-$ in the
aluminium subhalide 'Al$_5$Br$_7$·5THF' [C. Klemp *et al.* (2000) *Angew.
Chem. Int. Ed.*, vol. 39, p. 3691]. Colour code: Al, pale grey; Br,
gold; O, red; C, grey.

lie in the range 265–276 pm, while the Al–Al bond lengths outside the cage are 253 pm. Formal oxidation states of 0 and +2, respectively, can be assigned to the Al atoms inside and outside the Al_{12}-cage. The compound $Ga_2Br_4py_2$ (py = pyridine) is structurally similar to **13.18** and **13.19**, and the Ga–Ga bond length of 242 pm corresponds to a single bond ($r_{cov} = 122$ pm).

(13.18)   **(13.19)**

Al represents another Al atom with terminal $AlBr_2(THF)$ group

(13.20)

Gallium(I) chloride forms when $GaCl_3$ is heated at 1370 K. It is formed in the gas phase when HCl and Ga are heated under vacuum at 1200 K, and has been characterized by IR spectroscopy as a matrix isolated species. Above 273 K, GaCl disproportionates, but solutions of GaCl in toluene and Et_2O are metastable and may be used as starting materials in a similar way to solutions containing GaBr described below. Gallium(I) bromide can also be formed at high temperatures. A pale green, insoluble powder, 'GaI', can be prepared from Ga metal and I_2 in toluene under ultrasonic conditions. This material appears to be a mixture of gallium subhalides, with $Ga_2[Ga_2I_6]$ being a major component. Co-condensation of GaBr with toluene and THF at 77 K gives metastable GaBr-containing solutions, but these disproportionate to Ga and $GaBr_3$ when warmed above 253 K. However, if $Li[Si(SiMe_3)_3]$ is added

to the solution at 195 K, low oxidation state gallium species can be isolated (eq. 13.51). The structure of $Ga_{22}\{Si(SiMe_3)_3\}_8$ consists of a central Ga atom surrounded by a Ga_{13}-cage, with eight $Ga\{Si(SiMe_3)_3\}$ groups capping the eight square faces of the Ga_{13}-cage.[†] Examples of the use of GaBr and GaI as precursors to organometallic gallium species are described in Section 23.4.[‡]

(13.51)

When $GaCl_3$ is heated with Ga, a compound of stoichiometry '$GaCl_2$' is formed, but crystallographic and magnetic data show this is $Ga^+[GaCl_4]^-$. The mixed In(I)/In(III) compound $In[InCl_4]$ is prepared in a similar way to its Ga analogue. InCl can also be isolated from the $InCl_3$/In reaction mixture and has a deformed NaCl structure, and is practically insoluble in most organic solvents.

Thallium(I) halides, TlX, are stable compounds which in some ways resemble Ag(I) halides. Thallium(I) fluoride is very soluble in water, but TlCl, TlBr and TlI are sparingly soluble. The trend in solubilities can be rationalized in terms of the increased covalent contributions in the 'ionic' lattices for the larger halides, a situation that parallels the trend for the Ag(I) halides (see Section 6.15). In the solid state, TlF has a distorted NaCl-type structure, while TlCl and TlBr adopt CsCl structures. Thallium(I) iodide is dimorphic; below 443 K, the yellow form adopts a lattice derived from an NaCl structure in which neighbouring layers are slipped with respect to each other and, above 443 K, the red form crystallizes with a CsCl-type structure. Under high pressures, TlCl, TlBr and TlI become metallic in character.

[†] For further information on metalloid Al and Ga cluster molecules, see: H. Schnöckel (2005) *Dalton Trans.*, p. 3131; A. Schnepf and H. Schnöckel (2002) *Angew. Chem. Int. Ed.*, vol. 41, p. 3532; H. Schnöckel (2008) *Chem. Rev.*, vol. 110, p. 4125.

[‡] For an overview of GaI, see: R.J. Baker and C. Jones (2005) *Dalton Trans.*, p. 1341.

13.7 Oxides, oxoacids, oxoanions and hydroxides

It is a general observation that, within the *p*-block, basic character increases down a group. Thus:

- boron oxides are exclusively acidic;
- aluminium and gallium oxides are amphoteric;
- indium and thallium oxides are exclusively basic.

Thallium(I) oxide is soluble in water and the resulting hydroxide is as strong a base as KOH.

Boron oxides, oxoacids and oxoanions

The principal oxide of boron, B_2O_3, is obtained as a vitreous solid by dehydration of boric acid at red heat (eq. 13.2), or in a crystalline form by controlled dehydration. The latter possesses a 3-dimensional, covalent structure comprising planar BO_3 units (B−O = 138 pm) which share O atoms, but which are mutually twisted with respect to each other to give a rigid lattice. Under high pressure and at 803 K, a transition to a more dense form occurs, the change in density being 2.56 to 3.11 g cm^{-3}. This second polymorph contains tetrahedral BO_4 units, which are irregular because three O atoms are shared among three BO_4 units, while one atom connects two BO_4 units. Heating B_2O_3 with B at 1273 K gives BO or $\{BO\}_x$. Its structure has not been established, but the fact that reaction with water yields $(HO)_2BB(OH)_2$ (see Fig. 13.20) suggests it contains B−B bonds. Dehydration of $(HO)_2BB(OH)_2$ regenerates BO. The solid state ^{11}B NMR spectrum of BO is also consistent with the presence of B–B bonds. Theoretical studies using DFT (see Section 4.13) suggest that 6-membered rings are the most likely structural motifs, but indicate that there is little energetic preference between a number of possible structures[†]. Trigonal planar and tetrahedral B exemplified in the polymorphs of B_2O_3 occur frequently in boron–oxygen chemistry.

The commercial importance of B_2O_3 is in its use in the borosilicate glass industry (Box 13.6). As a Lewis acid, B_2O_3 is a valuable catalyst. BPO_4 (formed by reacting B_2O_3 with P_4O_{10}) catalyses the hydration of alkenes and dehydration of amides to nitriles. The structure of BPO_4 can be considered in terms of SiO_2 (see Section 14.9) in which alternate Si atoms have been replaced by B or P atoms.

Worked example 13.5   Isoelectronic relationships

The structure of BPO_4 is derived from that of SiO_2 by replacing alternate Si atoms by B or P atoms. Explain how this description relates to the isoelectronic principle.

[†]See: F. Claeyssens, N.L. Allan, N.C. Norman and C.A. Russell (2010) *Phys. Rev. B*, vol. 82, article number 094119.

Consider the positions of B, P and Si in the periodic table:

	13	14	15
	B	C	N
	Al	**Si**	P
	Ga	Ge	As

Considering only valence electrons:

B$^-$ is isoelectronic with Si
P$^+$ is isoelectronic with Si
BP is isoelectronic with Si$_2$

Therefore, replacement of two Si atoms in the solid state structure of SiO_2 by B and P does not affect the number of valence electrons in the system.

Self-study exercises

1. Boron phosphide, BP, crystallizes with a zinc blende structure. Comment on how this relates to the structure of elemental silicon.
   [*Ans.* Look at Fig. 6.20, and consider isoelectronic relationships as above]

2. Explain why $[CO_3]^{2-}$ and $[BO_3]^{3-}$ are isoelectronic. Are they isostructural?
   [*Ans.* B$^-$ isoelectronic with C; both trigonal planar]

3. Comment on the isoelectronic and structural relationships between $[B(OMe)_4]^-$, $Si(OMe)_4$ and $[P(OMe)_4]^+$.
   [*Ans.* B$^-$, Si and P$^+$ are isoelectronic (valence electrons); all tetrahedral]

Water is taken up slowly by B_2O_3 giving $B(OH)_3$ (ortho-boric or boric acid), but above 1270 K, molten B_2O_3 reacts rapidly with steam to give $B_3O_3(OH)_3$ (metaboric acid, Fig. 13.19a). Industrially, boric acid is obtained from borax (reaction 13.1), and heating $B(OH)_3$ converts it to $B_3O_3(OH)_3$. Both boric acids have layer structures in which molecules are linked by hydrogen bonds. The slippery feel of $B(OH)_3$ and its use as a lubricant are consequences of the layers (Figs. 4.5 and 13.19b). In aqueous solution, $B(OH)_3$ behaves as a weak acid, but is a *Lewis* rather than a Brønsted acid (eq. 13.52). Ester formation with 1,2-diols leads to an increase in acid strength (eq. 13.53). The importance of borate esters in nature was highlighted in Box 13.1.

$$B(OH)_3(aq) + 2H_2O(l) \rightleftharpoons [B(OH)_4]^-(aq) + [H_3O]^+(aq)$$
$$pK_a = 9.1 \qquad (13.52)$$

APPLICATIONS

Box 13.6 B_2O_3 in the glass industry

The glass industry in Western Europe and the US accounts for over half the B_2O_3 consumed (see Fig. 13.5b). Fused B_2O_3 dissolves metal oxides to give metal borates. Fusion with Na_2O or K_2O results in a viscous molten phase, rapid cooling of which produces a glass. Fusion with appropriate metal oxides leads to coloured metal borate glasses. Borosilicate glass is of particular commercial importance. It is formed by fusing together B_2O_3 and SiO_2 (glass formers) with additives (glass modifiers), typically Na_2O, K_2O, and/or Al_2O_3. The structures of borosilicate glasses (in which Si is in a tetrahedral environment and B may be trigonal planar or tetrahedral) are complex. Details can be found in the references below.

Borosilicate glasses include *Pyrex* which is used to manufacture most laboratory glassware as well as kitchenware. It contains a high proportion of SiO_2 and exhibits a low linear coefficient of expansion. Pyrex glass can be heated and cooled rapidly without breaking, and is resistant to attack by alkalis or acids. The refractive index of Pyrex is 1.47, and if a piece of clean Pyrex glassware is immersed in a mixture of $MeOH/C_6H_6$, 16/84 by weight, it seems to 'disappear'. This gives a quick way of testing if a piece of glassware is made from Pyrex. Although the linear coefficient of expansion of silica glass is lower than that of Pyrex glass (0.8 versus 3.3), the major advantage of borosilicate over silica glass is its workability. The softening point (i.e. the temperature at which the glass can be worked and blown) of fused silica glass is 1983 K, while that of Pyrex is 1093 K.

The photograph opposite shows a borosilicate glass mirror at the University of Arizona Mirror Laboratory. The laboratory specializes in manufacturing large, lightweight mirrors for optical and infrared telescopes. Each mirror has a honeycomb design, and is constructed from borosilicate glass that is melted, moulded and spun-cast in a rotating furnace before being polished. The 8.4 metre-wide mirror shown here was the first of its type and was completed in 1997 for the Large Binocular Telescope, Mount Graham, Arizona.

Fibreglass falls into two categories: textile fibres and insulation fibreglass. Of the textile fibres, alumino-borosilicate glass has the most widespread applications. The fibres possess high tensile strength and low thermal expansion, and are used in reinforced plastics. Insulation fibreglass includes glass wool which contains $\approx$55–60% SiO_2, $\approx$3% Al_2O_3, $\approx$10–14% Na_2O, 3–6% B_2O_3 plus other components such as CaO, MgO and ZrO_2.

Researcher Roger Angel with the borosilicate glass mirror for the Large Binocular Telescope.

Further reading

J.C. Phillips and R. Kerner (2008) *J. Chem. Phys.*, vol. 128, p. 174506 – 'Structure and function of window glass and Pyrex'.

N.M. Vedishcheva, B.A. Shakhmatkim and A.C. Wright (2004) *J. Non-Cryst. Solids*, vol. 345–346, p. 39 – 'The structure of sodium borosilicate glasses: thermodynamic modelling vs. experiment'.

$$2 \quad \text{(di-tert-butyl diol)} \quad + \quad \text{(boronic acid)} \quad \rightleftharpoons \quad \left[\text{(boronate complex)} \right]^{-} + [H_3O]^{+} + 2H_2O \tag{13.53}$$

Diboronic acid, $B_2(OH)_4$, can be obtained by hydrolysis of B_2Cl_4. Like boric acid, diboronic acid crystallizes with a layer structure, each layer consisting of hydrogen-bonded molecules (Fig. 13.20).

Many borate anions exist and metal borates such as *colemanite* ($Ca[B_3O_4(OH)_3]\cdot H_2O$), *borax* ($Na_2[B_4O_5(OH)_4]\cdot 8H_2O$), *kernite* ($Na_2[B_4O_5(OH)_4]\cdot 2H_2O$) and *ulexite* ($NaCa[B_5O_6(OH)_6]\cdot 5H_2O$) occur naturally. The solid state structures of borates are well established, and Fig. 13.21 shows selected anions. In planar BO_3 groups, B–O $\approx$136 pm, but in tetrahedral BO_4 units, B–O $\approx$148 pm. This increase is similar to that observed on going from BF_3 to $[BF_4]^{-}$ (see Section 13.6) and suggests that B–O π-bonding involving O lone pairs is present in

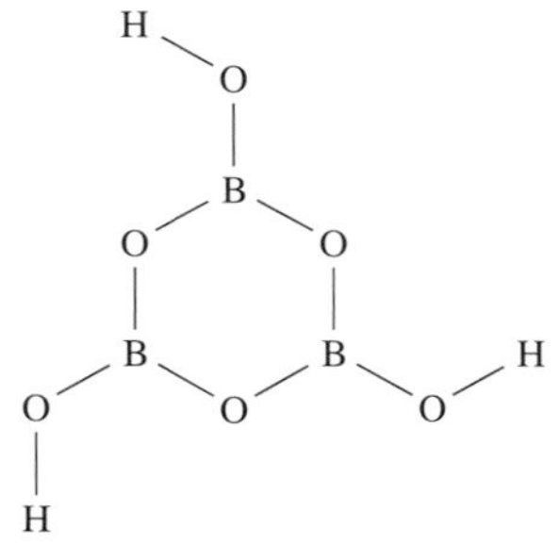

(a) 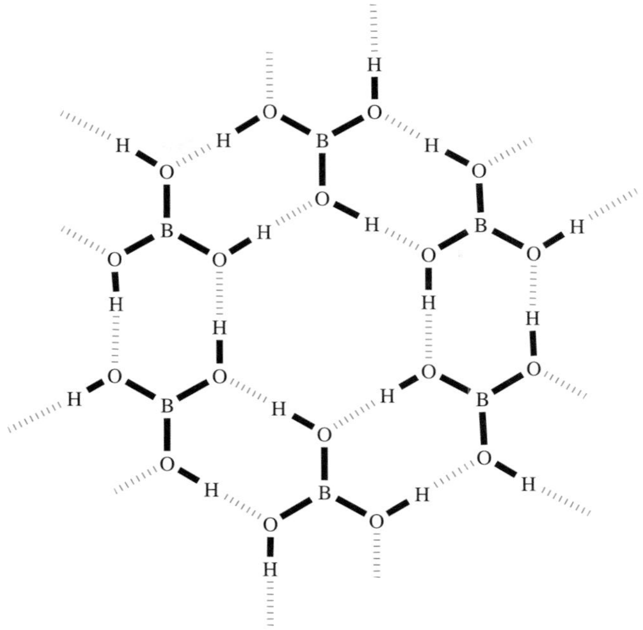   **(b)**

Fig. 13.19 (a) The structure of metaboric acid, $B_3O_3(OH)_3$. (b) Schematic representation of part of one layer of the solid state lattice of boric acid (orthoboric acid), $B(OH)_3$; covalent bonds within each molecule are highlighted in bold, and intermolecular hydrogen bonds are shown by red hashed lines. The hydrogen bonds are asymmetrical, with $O-H = 100\,pm$ and $O\cdots O = 270\,pm$.

planar BO_3 units. This is lost on going to a tetrahedral BO_4 unit. While solid state data abound, less is known about the nature of borate anions in aqueous solution. It is possible to distinguish between trigonal planar and tetrahedral B using ^{11}B NMR spectroscopy and data show that species

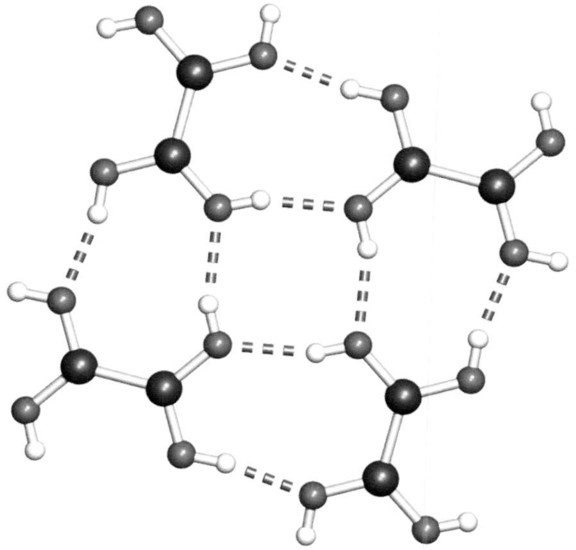

Fig. 13.20 Part of one layer of the solid state structure of $B_2(OH)_4$, determined by X-ray diffraction [R.A. Baber *et al.* (2003) *New J. Chem.*, vol. 27, p. 773]. The structure is supported by a network of hydrogen-bonded interactions. Colour code: B, blue; O, red; H, white.

containing only 3-coordinate B are unstable in solution and rapidly convert to species with 4-coordinate B. The species present in solution are also pH- and temperature-dependent.

The reactions of $B(OH)_3$ with Na_2O_2, or borates with H_2O_2, yield sodium peroxoborate (commonly known as sodium perborate). This is an important constituent of washing powders because it hydrolyses in water to give H_2O_2 and so is a bleaching agent. On an industrial scale, sodium peroxoborate is manufactured from borax by electrolytic oxidation. The solid state structure of sodium peroxoborate has been determined by X-ray diffraction and contains anion **13.21**; the compound is formulated as $Na_2[B_2(O_2)_2(OH)_4]\cdot 6H_2O$.

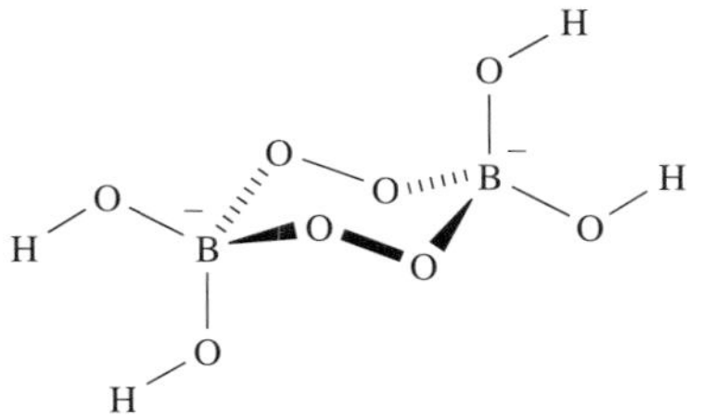

(13.21)

Aluminium oxides, oxoacids, oxoanions and hydroxides

Aluminium oxide occurs in two main forms: α-alumina (*corundum*) and γ-Al_2O_3 (*activated alumina*). The solid

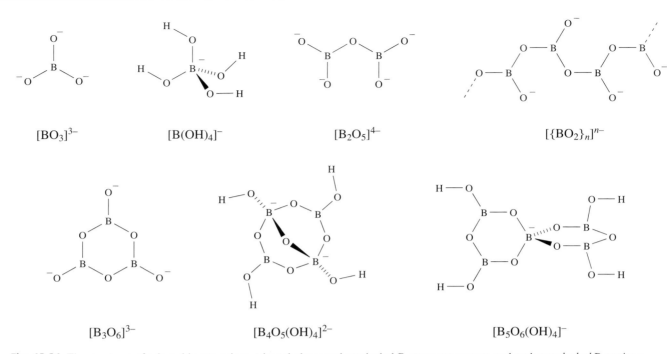

$[BO_3]^{3-}$ $[B(OH)_4]^-$ $[B_2O_5]^{4-}$ $[\{BO_2\}_n]^{n-}$

$[B_3O_6]^{3-}$ $[B_4O_5(OH)_4]^{2-}$ $[B_5O_6(OH)_4]^-$

Fig. 13.21 The structures of selected borate anions; trigonal planar and tetrahedral B atoms are present, and each *tetrahedral* B carries a negative charge. The $[B_4O_5(OH)_4]^{2-}$ anion occurs in the minerals *borax* and *kernite*. In the pyroborate ion, $[B_2O_5]^{4-}$, the B−O−B bond angle depends on the cation present, e.g. $\angle B-O-B = 153°$ in $Co_2B_2O_5$, and $131.5°$ in $Mg_2B_2O_5$.

state structure of α-Al_2O_3 consists of an hcp array of O^{2-} ions with cations occupying two-thirds of the octahedral interstitial sites. α-Alumina is extremely hard and is relatively unreactive (e.g. it is resistant to attack by acids). Its density ($4.0\,g\,cm^{-3}$) exceeds that of γ-Al_2O_3 ($3.5\,g\,cm^{-3}$) which has a defect spinel structure (see Box 13.7 and Section 20.11). The α-form is made by dehydrating $Al(OH)_3$ or $AlO(OH)$ at $\approx 1300\,K$, while dehydration of γ-$AlO(OH)$ below $720\,K$ gives γ-Al_2O_3. Both $Al(OH)_3$ and $AlO(OH)$ occur as minerals: *diaspore*, α-$AlO(OH)$, *boehmite*, γ-$AlO(OH)$, and *gibbsite*, γ-$Al(OH)_3$. α-$Al(OH)_3$ (*bayerite*) does not occur naturally but can be prepared by reaction 13.54.

$$2Na[Al(OH)_4](aq) + CO_2(g)$$
$$\longrightarrow 2Al(OH)_3(s) + Na_2CO_3(aq) + H_2O(l) \qquad (13.54)$$

The catalytic and adsorbing properties of γ-Al_2O_3, $AlO(OH)$ and $Al(OH)_3$ make this group of compounds invaluable commercially. One use of $Al(OH)_3$ is as a *mordant*, i.e. it absorbs dyes and is used to fix them to fabrics. The amphoteric nature of γ-Al_2O_3 and $Al(OH)_3$ is illustrated in reactions 13.55–13.58. Equation 13.57 shows the formation of an *aluminate* when $Al(OH)_3$ dissolves in excess alkali.

$$\gamma\text{-}Al_2O_3 + 3H_2O + 2[OH]^- \longrightarrow 2[Al(OH)_4]^- \qquad (13.55)$$

$$\gamma\text{-}Al_2O_3 + 3H_2O + 6[H_3O]^+ \longrightarrow 2[Al(OH_2)_6]^{3+} \qquad (13.56)$$

$$Al(OH)_3 + [OH]^- \longrightarrow [Al(OH)_4]^- \qquad (13.57)$$

$$Al(OH)_3 + 3[H_3O]^+ \longrightarrow [Al(OH_2)_6]^{3+} \qquad (13.58)$$

For use as the stationary phases in chromatography, acidic, neutral and basic forms of alumina are commercially available.

The electrical and/or magnetic properties of a number of mixed oxides of Al and other metals including members of the spinel family (Box 13.7) and sodium β-alumina (see Section 28.2) have extremely important industrial applications. In this section, we single out $3CaO\cdot Al_2O_3$ because of its role in cement manufacture, and because it contains a discrete aluminate ion. Calcium aluminates are prepared from CaO and Al_2O_3, the product depending on the stoichiometry of the reactants. The mixed oxide $3CaO\cdot Al_2O_3$ comprises Ca^{2+} and $[Al_6O_{18}]^{18-}$ ions and in the solid state, Ca^{2+} ions hold the cyclic anions (**13.22**) together through Ca--O interactions, the Ca^{2+} ions being in distorted octahedral environments. The oxide is a major component in Portland cement (see Box 14.8). $[Al_6O_{18}]^{18-}$ is isostructural with $[Si_6O_{18}]^{12-}$ (see Section 14.9) and the presence of these units in the solid state lattice imparts a very open

THEORY
Box 13.7 'Normal' spinel and 'inverse' spinel lattices

A large group of minerals called *spinels* have the general formula AB_2X_4 in which X is most commonly oxygen and the oxidation states of metals A and B are $+2$ and $+3$ respectively; examples include $MgAl_2O_4$ (*spinel*, after which this structural group is named), $FeCr_2O_4$ (*chromite*) and Fe_3O_4 (*magnetite*, a mixed Fe(II), Fe(III) oxide). The spinel family also includes sulfides, selenides and tellurides, and may contain metal ions in the $+4$ and $+2$ oxidation states, e.g. $TiMg_2O_4$, usually written as Mg_2TiO_4. Our discussion below focuses on spinel-type compounds containing A^{2+} and B^{3+} ions.

The spinel lattice is not geometrically simple but can be considered in terms of a cubic close-packed array of O^{2-} ions with one-eighth of the tetrahedral holes occupied by A^{2+} ions and half of the octahedral holes occupied by B^{3+} ions. The unit cell contains eight formula units, i.e. $[AB_2X_4]_8$.

Some mixed metal oxides AB_2X_4 in which at least one of the metals is a *d*-block element (e.g. $CoFe_2O_4$) possess an *inverse spinel* structure which is derived from the spinel lattice by exchanging the sites of the A^{2+} ions with half of the B^{3+} ions.

The occupation of octahedral sites may be ordered or random, and structure types cannot be simply partitioned into 'normal' or 'inverse'. A parameter λ is used to provide information about the distribution of cations in the interstitial sites of the close-packed array of X^{2-} ions; λ indicates the proportion of B^{3+} ions occupying *tetrahedral* holes. For a normal spinel, $\lambda = 0$; for an inverse spinel, $\lambda = 0.5$. Thus, for $MgAl_2O_4$, $\lambda = 0$, and for $CoFe_2O_4$, $\lambda = 0.5$. Other spinel-type compounds have values of λ between 0 and 0.5; for example, for $MgFe_2O_4$, $\lambda = 0.45$ and for $NiAl_2O_4$, $\lambda = 0.38$. We discuss factors governing the preference for a normal or inverse spinel structure in Section 20.11.

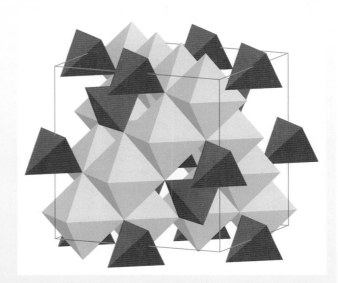

The inverse spinel structure of Fe_3O_4 showing the unit cell and the tetrahedral and octahedral environments of the Fe centres. The vertex of each tetrahedron and octahedron is occupied by an O atom.

structure which facilitates the formation of hydrates, a property crucial to the setting of cement.

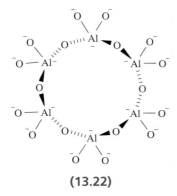

(13.22)

Self-study exercise

The conventional way of expressing the composition of mixed metal oxide minerals is in terms of the metal oxide content. Thus, Fe_3O_4 can be formulated as $FeO \cdot Fe_2O_3$ which reveals that it is a mixed iron(II)/iron(III) oxide.

The following mixed oxides are components in the cement industry. Rewrite the formulae in a way that shows the oxide compositions: $Al_2Ca_3O_6$, Ca_3SiO_5, Ca_2SiO_4, $Al_2Ca_4Fe_2O_{10}$.   [*Ans*: see Box 14.8]

Oxides of Ga, In and Tl

The oxides and related compounds of the heavier group 13 metals call for less attention than those of Al. Gallium, like Al, forms more than one polymorph of Ga_2O_3, $GaO(OH)$ and $Ga(OH)_3$, and the compounds are amphoteric. This contrasts with the basic nature of In_2O_3, $InO(OH)$ and $In(OH)_3$. Thallium is unique among the group in exhibiting an oxide for the M(I) state: Tl_2O forms when Tl_2CO_3 is heated in N_2, and it reacts with water (eq. 13.59).

$$Tl_2O + H_2O \longrightarrow 2TlOH \qquad (13.59)$$

Thallium(III) forms the oxide Tl_2O_3, but no simple hydroxide. Tl_2O_3 is insoluble in water and decomposes in acids. In concentrated NaOH solution and in the presence of $Ba(OH)_2$, the hydrated oxide $Tl_2O_3 \cdot xH_2O$ forms

$Ba_2[Tl(OH)_6]OH$. In the solid state, the $[Tl(OH)_6]^{3-}$ ions are connected to Ba^{2+} and $[OH]^-$ ions to give a structure that is related to that of K_2PtCl_6 (see Section 22.11).

13.8 Compounds containing nitrogen

The BN unit is isoelectronic with C_2 and many boron–nitrogen analogues of carbon systems exist. However useful this analogy is *structurally*, a BN group does *not* mimic a CC unit *chemically*, and reasons for this difference can be understood by considering the electronegativity values $\chi^P(B) = 2.0$, $\chi^P(C) = 2.6$ and $\chi^P(N) = 3.0$.

Nitrides

Boron nitride, BN, is a robust, chemically rather inert compound. Preparative routes include the high-temperature reactions of borax with $[NH_4]Cl$, B_2O_3 with NH_3, and $B(OH)_3$ with $[NH_4]Cl$. High-purity boron nitride can be made by reacting NH_3 with BF_3 or BCl_3. The fabrication of thin films of BN is described in Section 28.6, and the use of transmission electron microscopy (TEM) to study hollow boron nitride nanospheres is detailed in Box 13.8. The common form of boron nitride sublimes at 2603 K. It is referred to as hexagonal-BN (or α-BN) and has a layer structure consisting of hexagonal rings. The structure is similar to that of graphite. However, in α-BN, the layers are arranged so that a B atom in one layer lies directly over an N atom in the next (Fig. 13.22) and this contrasts with the staggered arrangement of alternate layers in graphite (Fig. 14.4). The B−N distances within a layer are much shorter than those between layers and, in Table 13.2, it is compared with those in other B−N species. The B−N bonds are shorter than in adducts

such as $Me_3N \cdot BBr_3$ in which a single boron–nitrogen bond can be assigned, and imply the presence of π-bonding in BN resulting from overlap between N $2p$ (occupied) and B $2p$ (vacant) orbitals orthogonal to the 6-membered rings. The *interlayer* distance of 330 pm is consistent with van der Waals interactions, and α-BN acts as a good lubricant, thus resembling graphite. Unlike graphite, α-BN is white and an insulator. This difference can be interpreted in terms of band theory (see Section 6.8), with the band gap in boron nitride being considerably greater than that in graphite because of the polarity of the B−N bond.

Heating α-BN at ≈2000 K and >50 kbar pressure in the presence of catalytic amounts of Li_3N or Mg_3N_2 converts it to a more dense polymorph, cubic-BN (or β-BN), with the zinc blende structure (see Section 6.11). Table 13.2 shows that the B−N bond distance in cubic-BN is similar to those in $R_3N \cdot BR_3$ adducts and longer than in hexagonal-BN. This further supports the existence of π-bonding within the layers of the latter. Structurally, the cubic form of BN resembles diamond (Fig. 6.20) and the two materials are almost equally hard. Crystalline cubic BN is called *borazon* and is used as an abrasive. A third polymorph of boron nitride (γ-BN) with a wurtzite-type structure is formed by compression of the layered form at ≈12 kbar.

Of the group 13 metals, only Al reacts directly with N_2 (at 1020 K) to form a nitride; AlN has a wurtzite-type structure and is hydrolysed to NH_3 by hot dilute alkali. Gallium and indium nitrides also crystallize with the wurtzite structure, and are more reactive than their B or Al counterparts. The importance of the group 13 metal nitrides, and of the related MP, MAs and MSb (M = Al, Ga, In) compounds, lies in their applications in the semiconductor industry (see also Section 23.4).

Table 13.2   Boron–nitrogen bond distances in selected neutral species; all data are from X-ray diffraction studies (≤298 K).

Species	B−N distance / pm	Comment
$Me_3N \cdot BBr_3$	160.2	Single bond
$Me_3N \cdot BCl_3$	157.5	Single bond
Cubic-BN(β-BN)	157	Single bond
Hexagonal-BN(α-BN)	144.6	Intralayer distance, see Fig. 13.22; some π-contribution
$B(NMe_2)_3$	143.9	Some π-contribution
$Mes_2\bar{B}=\overset{+}{N}H_2$	137.5	Double bond
$Mes_2\bar{B}=\overset{+}{N}=\bar{B}Mes_2{}^\dagger$	134.5	Double bond
$^tBu\bar{B}\equiv\overset{+}{N}{}^tBu$	125.8	Triple bond

† Mes = 2,4,6-$Me_3C_6H_2$.

THEORY

Box 13.8 Transmission electron microscopy (TEM)

A transmission electron microscope.

Transmission electron microscopy (TEM) is a technique for imaging the internal structures of materials, and high-resolution transmission electron microscopy (HRTEM) can achieve images on an atomic scale. Samples must be available as ultra-thin slices (<200 nm), and must be able to withstand the high vacuum under which the instrument operates. Biological samples can also be studied, but require specialized sample preparation. In a transmission electron microscope, the energy source is a beam of high-energy (usually 100–400 keV) electrons originating from a single crystal of LaB_6 or a tungsten filament that is heated to high temperatures. The electron beam is focused onto the sample by a series of condenser 'lenses' (i.e. magnetic fields, rather than the conventional lenses found in optical microscopes). Part of the beam is transmitted while the remaining electrons are scattered. The transmitted beam passes through a series of objective lenses to produce a magnified image that is projected onto a phosphor screen. Permanent images are recorded using a charge-coupled detector (CCD). A TEM image shows regions of different contrast (see the figure opposite) which can be interpreted in terms of varying structure. Darker regions correspond to parts of the sample in which the atomic mass is the highest.

TEM is highly suited to imaging *nanoscale materials*. For example, the TEM image opposite illustrates hollow spheres of boron nitride, formed from an autoclave reaction between BBr_3 and NaN_3:

$$BBr_3 + 3NaN_3 \xrightarrow{373\,K,\ 8\,h} BN + 3NaBr + 4N_2$$

The use of sodium amide instead of an azide precursor gives a room temperature synthesis:

$$BBr_3 + NaNH_2 \xrightarrow{298\,K} BN + NaBr + 2HBr$$

Other routes involve the use of magnesium boride or sodium tetrafluoridoborate as sources of boron. All are autoclave methods to boron nitride nanospheres:

$$2NH_4Cl + MgB_2 \xrightarrow{820\,K,\ 8\,h} 2BN + MgCl_2 + 4H_2$$

$$5NH_4Cl + 2MgB_2 + NaN_3 \xrightarrow{770\,K,\ 8\,h}$$
$$4BN + 2MgCl_2 + 10H_2 + NaCl + 2N_2$$

$$NaBF_4 + 3NaN_3 \xrightarrow{702\,K,\ 20\,h} BN + 4NaF + 4N_2$$

The purity and composition of the product can be determined by using X-ray photoelectron spectroscopy (see Section 4.12), and TEM is used to examine the form in which the boron nitride has been formed. The images reveal that 30–40% of the sample is composed of hollow spheres (each with a well-defined boundary) of BN. The external diameter of a sphere is in the range 100–200 nm and the wall thickness is <10 nm. Research into the formation of such hollow boron nitride spheres forms part of a wider initiative into the development of hollow nanoparticles for, for example, catalysis, drug delivery, and encapsulation of biological agents. Hollow spheres of BN, as well as boron nitride and carbon nanotubes (see Section 28.9), are also potential materials for hydrogen storage (see Box 10.2).

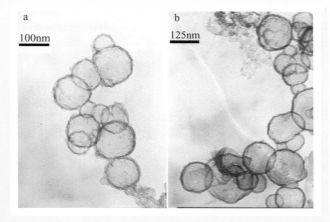

TEM images of hollow spheres of boron nitride. The 100nm and 125nm scale bars allow you to estimate the diameter of each sphere.
X. Wang *et al.* (2003) *Chem. Commun.*, p. 2688.

Further reading

E.M. Slayter and H.S. Slayter (1992) *Light and Electron Microscopy*, Cambridge University Press, Cambridge.

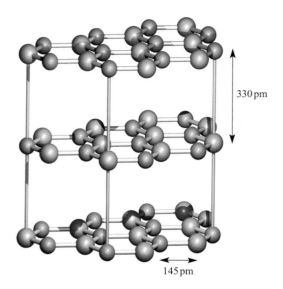

Fig. 13.22 Part of the layer structure of the common polymorph of boron nitride, BN. Hexagonal rings in adjacent layers lie over one another so that B and N atoms are eclipsed. This is emphasized by the yellow lines.

Ternary boron nitrides

Ternary boron nitrides (i.e. compounds of type $M_xB_yN_z$) are a relatively new addition to boron–nitrogen chemistry. The high-temperature reactions of hexagonal BN with Li_3N or Mg_3N_2 lead to Li_3BN_2 and Mg_3BN_3 respectively. Reaction 13.60 is used to prepare Na_3BN_2 because of the difficulty in accessing Na_3N as a starting material (see Section 11.4).

$$2Na + NaN_3 + BN \xrightarrow{\text{1300 K, 4 GPa}} Na_3BN_2 + N_2 \qquad (13.60)$$

Structural determinations for Li_3BN_2, Na_3BN_2 and Mg_3BN_3 confirm the presence of discrete $[BN_2]^{3-}$ ions, and Mg_3BN_3 is therefore better formulated as $(Mg^{2+})_3[BN_2]^{3-}(N^{3-})$. The $[BN_2]^{3-}$ ion (**13.23**) is isoelectronic and isostructural with CO_2.

$$\bar{N} = B = \bar{N}$$

(13.23)

Ternary boron nitrides containing *d*-block metal ions are not as well represented as those containing lanthanoid metals. These include $Eu_3(BN_2)_2$, $La_3[B_3N_6]$, $La_5[B_3N_6][BN_3]$ and $Ce_3[B_2N_4]$ which are formulated as involving $[BN_2]^{3-}$, $[BN_3]^{6-}$, $[B_2N_4]^{8-}$ and $[B_3N_6]^{9-}$ ions. Lanthanoid (Ln) compounds $Ln_3[B_2N_4]$ contain one conduction electron per formula unit, i.e. $(Ln^{3+})_3[B_2N_4]^{8-}(e^-)$. These nitridoborate compounds may be formed by heating

($>1670\,K$) mixtures of powdered lanthanoid metal, metal nitride and α-BN, or by metathesis reactions between Li_3BN_2 and $LaCl_3$. The ions $[BN_3]^{6-}$ and $[B_2N_4]^{8-}$ are isoelectronic analogues of $[CO_3]^{2-}$ and $[C_2O_4]^{2-}$, respectively. The B–N bonds in $[BN_3]^{6-}$ are equivalent and diagram **13.24** shows a set of resonance structures consistent with this observation. The bonding can also be described in terms of a delocalized bonding model involving π-interactions between N 2p and B 2p orbitals. Similarly, sets of resonance structures or delocalized bonding models are needed to describe the bonding in $[B_2N_4]^{8-}$ (**13.25**) and $[B_3N_6]^{9-}$ (see end-of-chapter problem 13.31c).

(13.24)

(13.25)

The solid state structures of $La_3[B_3N_6]$, $La_5[B_3N_6][BN_3]$ and $La_6[B_3N_6][BN_3]N$ show that the $[B_3N_6]^{9-}$ ion contains a 6-membered B_3N_3 ring with a chair conformation (diagram **13.26**, B atoms shown in orange). Each boron atom is in a planar environment, allowing it to participate in π-bonding to nitrogen.

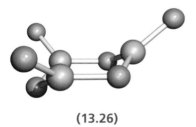

(13.26)

Self-study exercise

The compound $Mg_2[BN_2]Cl$ contains $[BN_2]^{3-}$ ions belonging to the $D_{\infty h}$ point group. The Raman spectrum of $Mg_2[BN_2]Cl$ shows one line at $1080\,cm^{-1}$. (a) What shape is the $[BN_2]^{3-}$ ion? (b) Which vibrational mode gives rise to the observed Raman line? (c) Why is this vibrational mode not IR active?

[*Ans.* See structure 13.23, and Fig. 3.11 and accompanying text]

Molecular species containing B–N or B–P bonds

We have already described the formation of B−N single bonds in adducts $R_3N \cdot BH_3$, and now we extend the discussion to include compounds with boron–nitrogen multiple bonds.

(13.27)

The hexagonal B_3N_3-motif in the layered form of boron nitride appears in a group of compounds called *borazines*. The parent compound $(HBNH)_3$, **13.27**, is isoelectronic and isostructural with benzene. It is prepared by reaction 13.61, from B_2H_6 (Fig. 13.8) or from the *B*-chloro-derivative, itself prepared from BCl_3 (eq. 13.62).

$$NH_4Cl + Na[BH_4] \xrightarrow{-NaCl, -H_2} H_3N \cdot BH_3 \xrightarrow{\Delta} (HBNH)_3$$
$$(13.61)$$

$$BCl_3 + 3NH_4Cl \xrightarrow{420\,K, C_6H_5Cl} (ClBNH)_3$$

$$\xrightarrow{Na[BH_4]} (HBNH)_3 \quad (13.62)$$

The use of an alkylammonium chloride in place of NH_4Cl in reaction 13.62 leads to the formation of an *N*-alkyl derivative $(ClBNR)_3$ which can be converted to $(HBNR)_3$ by treatment with $Na[BH_4]$.

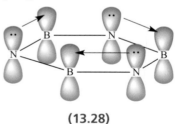

(13.28)

Borazine is a colourless liquid (mp 215 K, bp 328 K) with an aromatic odour and *physical* properties that resemble those of benzene. The B−N distances in the planar B_3N_3 ring are equal (144 pm) and close to those in the layered form of BN (Table 13.2). This is consistent with a degree of delocalization of the N lone pairs around the ring as represented in **13.28**. Structure **13.27** gives one resonance form of borazine, analogous to a Kekulé structure for benzene.[†] Despite the

---

[†] Theoretical studies suggest that the N lone pairs may be localized. See: J.J. Engelberts, R.W.A. Havenith, J.H. van Lenthe, L.W. Jenneskens and P.W. Fowler (2005) *Inorg. Chem.*, vol. 44, p. 5266.

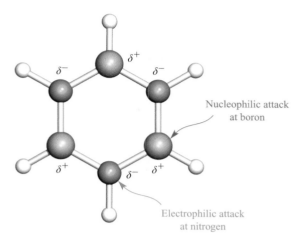

Fig. 13.23 In borazine, the difference in electronegativities of boron and nitrogen leads to a charge distribution which makes the B atoms (shown in orange) and N atoms (shown in blue), respectively, susceptible to nucleophilic and electrophilic attack.

formal charge distribution, a consideration of the relative electronegativities of B ($\chi^P = 2.0$) and N ($\chi^P = 3.0$) indicates that B is susceptible to attack by nucleophiles while N attracts electrophiles (Fig. 13.23). Thus, the reactivity of borazine contrasts sharply with that of benzene, although it must be remembered that C_6H_6 is *kinetically* inert towards the addition of, for example, HCl and H_2O. Equations 13.63 and 13.64 give representative reactions of borazine; the formula notation indicates the nature of the *B*- or *N*-substituents, e.g. $(ClHBNH_2)_3$ contains Cl attached to B.

$$(HBNH)_3 + 3HCl \longrightarrow (ClHBNH_2)_3 \qquad \textit{addition reaction}$$
$$(13.63)$$

$$(HBNH)_3 + 3H_2O \longrightarrow \{H(HO)BNH_2\}_3 \qquad \textit{addition reaction}$$
$$(13.64)$$

Each of the products of these reactions possesses a chair conformation (compare cyclohexane). Treatment of $(ClHBNH_2)_3$ with $Na[BH_4]$ leads to the formation of $(H_2BNH_2)_3$ (Fig. 13.24a).

(13.29)

Dewar borazine derivatives **13.29** can be stabilized by the introduction of sterically demanding substituents.

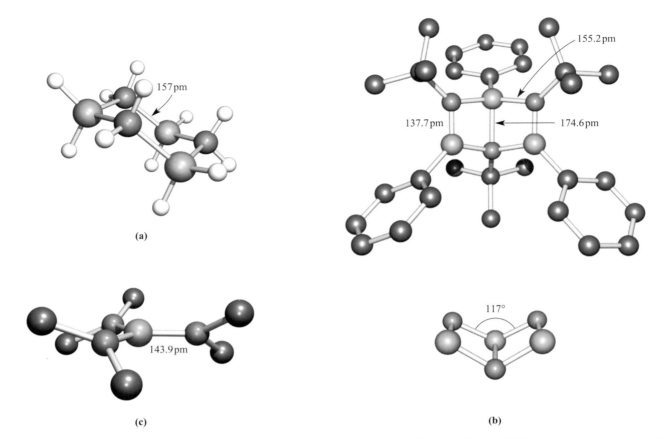

Fig. 13.24 The structures (determined by X-ray diffraction) of (a) $B_3N_3H_{12}$ [P.W.R. Corfield *et al.* (1973) *J. Am. Chem. Soc.*, vol. 95, p. 1480], (b) the Dewar borazine derivative N,N',N''-$^tBu_3$-B,B',B''-$Ph_3B_3N_3$ [P. Paetzold *et al.* (1991) *Z. Naturforsch., Teil B*, vol. 46, p. 853], (c) $B(NMe_2)_3$ [G. Schmid *et al.* (1982) *Z. Naturforsch., Teil B*, vol. 37, p. 1230, structure determined at 157 K]; H atoms in (b) and (c) have been omitted. Colour code: B, orange; N, blue; C, grey; H, white.

Figure 13.24b shows the structure of N,N',N''-$^tBu_3$-B,B',B''-$Ph_3B_3N_3$. The 'open-book' conformation of the B_3N_3 framework mimics that of the C_6-unit in Dewar benzene. By comparing the bond distances in Fig. 13.24b with those in Table 13.2, we see that the central B−N bond in **13.29** is longer than a typical single bond, the four distances of 155 pm (Fig. 13.24b) are close to those expected for single bonds, and the two remaining B−N bond lengths correspond to double bonds. Dewar borazines are prepared by cyclotrimerization of iminoboranes RBNR′ (**13.30**), although cyclooligomerization processes are not simple.[†] A family of RBNR′ compounds is known, and can be rendered kinetically stable with respect to oligomerization by the introduction of bulky substituents and/or maintaining low temperatures. For example, $^tBuBN^tBu$ (with sterically demanding *tert*-butyl groups) has a half-life of 3 days at 323 K. Iminoboranes can be made by elimination of a suit-

able species from compounds of type **13.31** (e.g. reaction 13.65) and possess very short B−N bonds (Table 13.2) consistent with triple bond character.

$$R-\overset{-}{B}\equiv\overset{+}{N}-R'$$

(13.30)

$$\begin{array}{c}R\\ \diagdown\overset{-}{}\\ B=N\\ \diagup\diagdown\\ RR'\end{array}\quad\begin{array}{c}R'\\ \overset{+}{\diagup}\\ \\ \diagdown\\ R'\end{array}$$

(13.31)

$$\begin{array}{cc}R & R'\\ \diagdown\overset{-}{}\overset{+}{\diagup}\\ B=N\\ \diagup\diagdown\\ ClSiMe_3\end{array}\xrightarrow{\Delta,\ \sim10^{-3}\ \text{bar}} Me_3SiCl + R\overset{-}{B}\equiv\overset{+}{N}R'$$

$$(13.65)$$

Compounds **13.31** can be made by reactions such as 13.66 or 13.67, and reaction 13.68 has been used to prepare Mes_2BNH_2 which has been structurally characterized. The

[†] For a detailed account, see: P. Paetzold (1987) *Adv. Inorg. Chem.*, vol. 31, p. 123.

B−N distance in Mes_2BNH_2 (Table 13.2) implies a double bond, and the planes containing the C_2B and NH_2 units are close to being coplanar as required for efficient overlap of the B and N $2p$ atomic orbitals in π-bond formation.

$$M[BH_4] + [R_2NH_2]Cl \longrightarrow H_2BNR_2 + MCl + 2H_2 \tag{13.66}$$

$$R_2BCl + R'_2NH + Et_3N \longrightarrow R_2BNR'_2 + [Et_3NH]Cl \tag{13.67}$$

$$Mes_2BF \xrightarrow[-NH_4F]{\text{Liquid } NH_3, Et_2O} Mes_2BNH_2 \tag{13.68}$$

$$Mes = mesityl$$

While considering the formation of B−N π-bonds, it is instructive to consider the structure of $B(NMe_2)_3$. As Fig. 13.24c shows, each B and N atom is in a trigonal planar environment, and the B–N bond distances indicate partial π-character (Table 13.2) as expected. On the other hand, in the solid state structure, the twisting of the NMe_2 units, which is clearly apparent in Fig. 13.24c, will militate against efficient $2p$–$2p$ atomic orbital overlap. Presumably, such twisting results from steric interactions and the observed structure of $B(NMe_2)_3$ provides an interesting example of a subtle balance of steric and electronic effects.

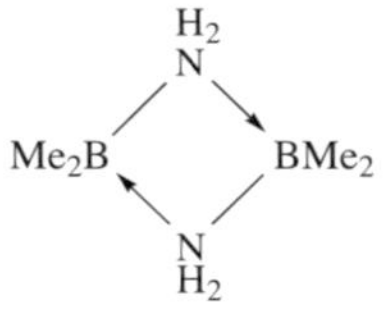

(13.32)

With less bulky substituents, compounds **13.31** readily dimerize. For example, Me_2BNH_2 forms the cyclodimer **13.32**. Whereas Me_2BNH_2 is a gas at room temperature (bp 274 K) and reacts rapidly with H_2O, dimer **13.32** has a melting point of 282 K and is kinetically stable towards hydrolysis by water.

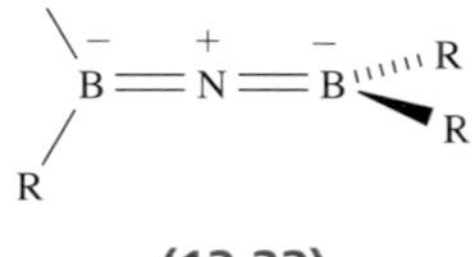

(13.33)

Compounds **13.30** and **13.31** are analogues of alkynes and alkenes respectively. Allene analogues, **13.33**, can also be prepared, e.g. scheme 13.69. Crystallographic data for $[Mes_2BNBMes_2]^-$ reveal B−N bond lengths consistent with double bond character (Table 13.2) and the presence of B−N π-bonding is further supported by the fact that the planes containing the C_2B units are mutually orthogonal as shown in structure **13.33**.

$$Mes_2BNH_2 \xrightarrow[-2^nBuH]{2^nBuLi \text{ in } Et_2O} \{Li(OEt_2)NHBMes_2\}_2$$

$$\Bigg\downarrow \begin{array}{l} 2Mes_2BF \text{ in } Et_2O \\ -2LiF \end{array}$$

$$[Li(OEt_2)_3][Mes_2BNBMes_2] \xleftarrow[-^nBuH]{^nBuLi \text{ in } Et_2O} (Mes_2B)_2NH \tag{13.69}$$

Compounds containing B−P bonds are also known, and some chemistry of these species parallels that of the B−N-containing compounds described above. However, there are some significant differences, one of the main ones being that no phosphorus-containing analogue of borazine has been isolated. Monomers of the type $R_2BPR'_2$ analogous to **13.31** are known for R and R' being bulky substituents.

At 420 K, the adduct $Me_2PH \cdot BH_3$ undergoes dehydrogenation to give $(Me_2PBH_2)_3$ as the major product and $(Me_2PBH_2)_4$ as the minor product. Structural data for the phenyl-substituted analogues of these compounds show that in the solid state, **13.34** and **13.35** adopt chair and boat–boat conformations, respectively. These cyclic compounds can also be obtained by heating $Ph_2PH \cdot BH_3$ at 400 K in the presence of a catalytic amount of the rhodium(I) compound $[Rh_2(\mu\text{-}Cl)_2(cod)_2]$ (see structure **24.22** for the ligand cod). However, if this reaction is carried out at ≤ 360 K, cyclization does not occur and the product is $Ph_2PHBH_2PPh_2BH_3$ (**13.36**).

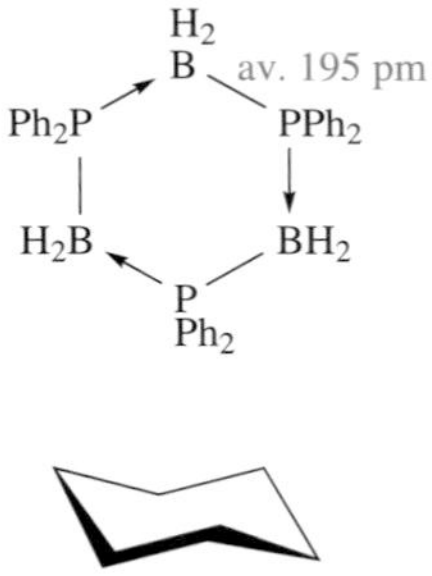

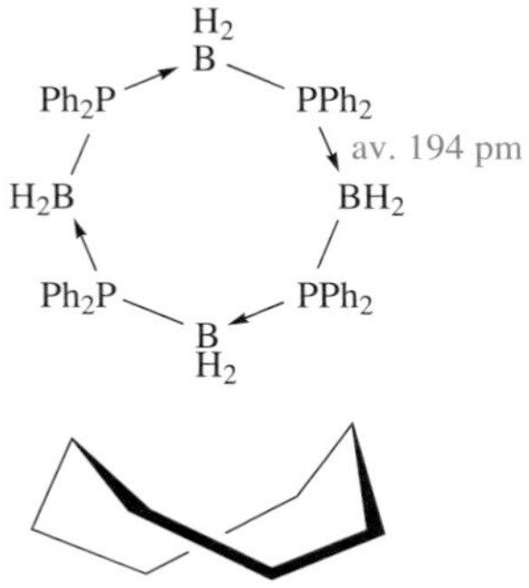

(13.34)

(13.35)

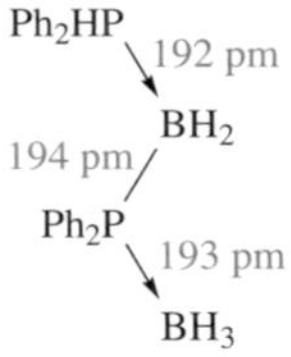

(13.36)

Self-study exercise

The reaction of LiHPhP·BH$_3$ with Me$_2$HN·BH$_2$Cl leads to a boron-containing product **A**. The highest mass peak in the mass spectrum of **A** is at $m/z = 180$. The ^{31}P NMR spectrum of a solution of **A** exhibits a broadened doublet at $\delta -54.8$ ppm (J 344 Hz), and the ^{11}B$\{^1$H$\}$ NMR spectrum shows two doublets at $\delta -12.8$ (J 70 Hz) and -41.5 ppm (J 50 Hz). The ^{1}H NMR spectrum contains multiplets in the range δ 7.77–7.34 ppm, broad signals at δ 4.7, 2.0 and 0.7 ppm, a doublet of doublets at δ 2.61 ppm (J 35, 5.8 Hz), and a doublet of sextets (J 344, 6 Hz). Suggest a structure of **A** that is consistent with the experimental data. What is the cause of the broadening of the doublet in the ^{31}P NMR spectrum?

[*Ans.* See C.A. Jaska *et al.* (2004) *Inorg. Chem.*, vol. 43, p. 1090]

Molecular species containing group 13 metal–nitrogen bonds

Coordinate M–N bond formation where M = Al, Ga, In (and to a lesser extent Tl) gives rise to a wide variety of complexes ranging from R$_3$N·GaH$_3$ (see Section 13.5) and *trans*-[GaCl$_2$(py)$_4$]$^+$ (py = pyridine) to cyclic species such as (Me$_2$AlNMe$_2$)$_2$ (**13.37**) or cages such as those in Fig. 13.25. The bond distances in these compounds are consistent with localized 2-centre 2-electron bonds.

$$
\begin{array}{c}
\text{Me}_2 \\
\text{Al} \\
\text{Me}_2\text{N} \qquad \text{NMe}_2 \\
\text{Al} \\
\text{Me}_2
\end{array}
$$

(13.37)

Compounds containing Al–N and Ga–N bonds are of interest as precursors to aluminium and gallium nitrides. Simple adducts are made by combining appropriate Lewis acids and bases. However, when primary or secondary amines react with organoaluminium, -gallium or -indium compounds, the reactions are usually accompanied by elimination of H$_2$ or an alkane to give cyclic or cage compounds (eqs. 13.70–13.72). The dependence of the reaction on temperature is illustrated by reaction 13.73 versus 13.74. In general, alkane elimination from R$_3$Al·NH$_2$R' leads to dimers and trimers of formula (R$_2$AlNHR')$_n$ (n = 2, 3) and further alkane elimination leads to higher oligomers of type (RAlNR')$_n$ ($n > 3$).

$$n\text{Na[AlH}_4] + n\text{RNH}_2 \longrightarrow (\text{HAlNR})_n + n\text{NaH} + 2n\text{H}_2 \tag{13.70}$$

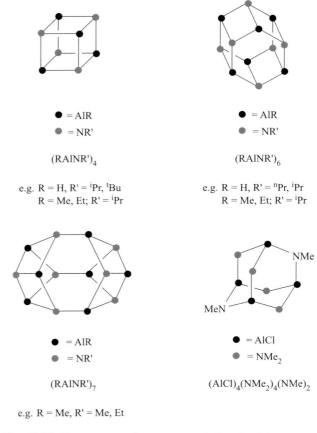

$\bullet$ = AlR $\bullet$ = NR'

(RAlNR')$_4$

e.g. R = H, R' = ^{i}Pr, ^{t}Bu
R = Me, Et; R' = ^{i}Pr

$\bullet$ = AlR $\bullet$ = NR'

(RAlNR')$_6$

e.g. R = H, R' = ^{n}Pr, ^{i}Pr
R = Me, Et; R' = ^{i}Pr

$\bullet$ = AlR $\bullet$ = NR'

(RAlNR')$_7$

e.g. R = Me, R' = Me, Et

$\bullet$ = AlCl $\bullet$ = NMe$_2$

(AlCl)$_4$(NMe$_2$)$_4$(NMe)$_2$

Fig. 13.25 The structures of some representative aluminium–nitrogen cluster compounds. Localized bonding schemes are appropriate for each cage (see end-of-chapter problem 13.24).

$$n\text{AlR}_3 + n\text{R'NH}_2 \longrightarrow (\text{RAlNR'})_n + 2n\text{RH} \tag{13.71}$$

$$n\text{GaR}_3 + n\text{R}'_2\text{NH} \longrightarrow (\text{R}_2\text{GaNR}'_2)_n + n\text{RH} \tag{13.72}$$

$$3\text{AlMe}_3 + 3\text{MeNH}_2 \xrightarrow{340\,\text{K}} (\text{Me}_2\text{AlNHMe})_3 + 3\text{CH}_4 \tag{13.73}$$

$$7\text{AlMe}_3 + 7\text{MeNH}_2 \xrightarrow{490\,\text{K}} (\text{Me}_2\text{AlNHMe})_7 + 7\text{CH}_4 \tag{13.74}$$

Compounds containing multiple M=N or M≡N bonds do not play an important role in the chemistry of the group 13 metals. The cyclic compound {MeAlN(2,6-^{i}Pr$_2$C$_6$H$_3$)}$_3$ (Fig. 13.26) has been isolated and contains a planar Al$_3$N$_3$ ring. The compound appears to be an analogue of a borazine derivative. However, there is no evidence for any significant Al–N π-contributions to the bonding (i.e. negligible 3p–2p π-overlap), even though the Al–N bonds are shorter (178 pm) than a typical single bond (185–200 pm). The sterically demanding substituents present in {MeAl-N(2,6-^{i}Pr$_2$C$_6$H$_3$)}$_3$ and other related substituents play a

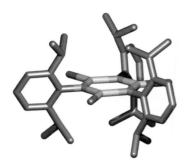

Fig. 13.26 The structure of $\{MeAlN(2,6-^iPr_2C_6H_3)\}_3$ determined by X-ray diffraction [K.M. Waggoner *et al.* (1988) *Angew. Chem. Int. Ed.*, vol. 27, p. 1699]; H atoms are omitted for clarity. Colour code: Al, yellow; N, blue; C, grey.

vital role stabilizing a range of rather unusual main group compounds (see Section 14.3 and Chapter 23).

13.9 Aluminium to thallium: salts of oxoacids, aqueous solution chemistry and complexes

Aluminium sulfate and alums

An *alum* has the general formula $M^IM^{III}(SO_4)_2\cdot12H_2O$.

The most important soluble oxosalts of Al are undoubtedly $Al_2(SO_4)_3\cdot16H_2O$ and the double sulfates $MAl(SO_4)_2\cdot12H_2O$ (*alums*). In alums, M^+ is usually K^+, Rb^+, Cs^+ or $[NH_4]^+$, but Li^+, Na^+ and Tl^+ compounds also exist. The Al^{3+} ion may be replaced by another M^{3+} ion, but its size must be comparable; possible metals are Ga, In (but not Tl), Ti, V, Cr, Mn, Fe and Co. The sulfate ion in an alum can be replaced by $[SeO_4]^{2-}$. Alums occur naturally in *alum shales*, but are well known in crystal growth experiments. Beautiful octahedral crystals are characteristic, e.g. in colourless $KAl(SO_4)_2\cdot12H_2O$ or purple $KFe(SO_4)_2\cdot12H_2O$. The purple colour of the latter arises from the presence of the $[Fe(OH_2)_6]^{3+}$ ion and, in all alums, the M^{3+} ion is octahedrally coordinated by six aqua ligands. The remaining water molecules are held in the crystal lattice by hydrogen bonds and connect the hydrated cations to the anions. Aluminium sulfate is used in water purification (see Box 16.3) for the removal of phosphate and of colloidal matter, the coagulation of which is facilitated by the high charge on the Al^{3+} cation.

Aqua ions

The M^{3+} aqua ions (M = Al, Ga, In, Tl) are acidic (see eq. 7.34) and the acidity increases down the group. Solutions of their salts are appreciably hydrolysed and salts of weak acids (e.g. carbonates and cyanides) cannot exist in aqueous solution. Solution NMR spectroscopic studies show that in acidic media, Al(III) is present as octahedral $[Al(OH_2)_6]^{3+}$, but raising the pH leads to the formation of multinuclear species such as hydrated $[Al_2(OH)_2]^{4+}$

(13.38) and $[Al_7(OH)_{16}]^{5+}$. Further increase in pH causes $Al(OH)_3$ to precipitate, and in alkaline solution, the aluminate anions $[Al(OH)_4]^-$ (tetrahedral), $[Al(OH)_6]^{3-}$ (octahedral) and $[(HO)_3Al(\mu\text{-}O)Al(OH)_3]^{2-}$ and other polymeric species are present. The aqueous solution chemistry of Ga(III) resembles that of Al(III), but the later metals are not amphoteric (see Section 13.7).

$$
\left[\begin{array}{c}
\text{H}_2\text{O} \qquad\qquad \text{OH}_2 \\
\text{H}_2\text{O}\cdots\overset{|}{\underset{|}{\text{Al}}}\cdots\overset{\text{H}}{\text{O}}\cdots\overset{|}{\underset{|}{\text{Al}}}\cdots\text{OH}_2 \\
\text{H}_2\text{O} \qquad\overset{\text{O}}{\underset{\text{H}}{}}\qquad \text{OH}_2 \\
\text{H}_2\text{O} \qquad\qquad \text{OH}_2
\end{array}\right]^{4+}
$$

(13.38)

Redox reactions in aqueous solution

The standard reduction potentials for the M^{3+}/M couples (Table 13.1) show that Al^{3+}(aq) is much less readily reduced in aqueous solution than are the later M^{3+} ions. This can be attributed, in part, to the more negative Gibbs energy of hydration of the smaller Al^{3+} ion. However, an important contributing factor (scheme 13.75) in differentiating between the values of E^o for the Al^{3+}/Al and Ga^{3+}/Ga couples is the significant increase in the sum of the first three ionization energies (Table 13.1).

$$
M^{3+}(aq) \xrightarrow{-\Delta_{hyd}H^o} M^{3+}(g) \xrightarrow{-\Sigma IE_{1-3}} M(g) \xrightarrow{-\Delta_a H^o} M(s)
$$

$$(13.75)$$

Although In(I) can be obtained in low concentration by oxidation of an In anode in dilute $HClO_4$, the solution rapidly evolves H_2 and forms In(III). A value of -0.44 V has been measured for the In^{3+}/In^+ couple (eq. 13.76).

$$
In^{3+}(aq) + 2e^- \longrightarrow In^+(aq) \qquad E^o = -0.44\,\text{V} \qquad (13.76)
$$

For the $Ga^{3+}(aq)/Ga^+(aq)$ couple, a value of $E^o = -0.75$ V has been determined and, therefore, studies of aqueous Ga^+ are rare because of the ease of oxidation of Ga^+ to Ga^{3+}. The compound $Ga^+[GaCl_4]^-$ (see the end of Section 13.6) can be used as a source of Ga^+ in aqueous solution, but it is very unstable with respect to oxidation and rapidly reduces $[I_3]^-$, aqueous Br_2, $[Fe(CN)_6]^{3-}$ and $[Fe(bpy)_3]^{3+}$.

Worked example 13.6　Potential diagrams

The potential diagram for indium in acidic solution (pH = 0) is given below with standard redox potentials given in V:

$$
In^{3+} \xrightarrow{-0.44} In^+ \xrightarrow{-0.14} In
$$
$$
\underset{E^o}{\underline{\phantom{In^{3+}\qquad\qquad\qquad\qquad In}}}
$$

Determine the value of E^o for the In^{3+}/In couple.

The most rigorous method is to determine $\Delta G^{\circ}(298\,K)$ for each step, and then to calculate E° for the In^{3+}/In couple. However, it is not necessary to evaluate ΔG° for each step; instead leave values of ΔG° in terms of the Faraday constant (see worked example 8.7).

Reduction of In^{3+} to In^{+} is a 2-electron process:

$$\Delta G^{\circ}_{1} = -[2 \times F \times (-0.44)] = +0.88F\,J\,mol^{-1}$$

Reduction of In^{+} to In is a 1-electron process:

$$\Delta G^{\circ}_{2} = -[1 \times F \times (-0.14)] = +0.14F\,J\,mol^{-1}$$

Next, find ΔG° for the reduction of In^{3+} to In:

$$\Delta G^{\circ} = \Delta G^{\circ}_{1} + \Delta G^{\circ}_{2} = +0.88F + 0.14F = +1.02F\,J\,mol^{-1}$$

Reduction of In^{3+} to In is a 3-electron process, and E° is found from the corresponding value of ΔG°:

$$E^{\circ} = -\frac{\Delta G^{\circ}}{zF} = -\frac{1.02F}{3F} = -0.34\,V$$

Self-study exercises

1. The potential diagram for gallium (at pH = 0) is as follows:

$$Ga^{3+} \xrightarrow{-0.75} Ga^{+} \xrightarrow{E^{\circ}} Ga$$
$$\underset{-0.55}{\underline{\hspace{4cm}}}$$

Calculate a value for E° for the Ga^{+}/Ga couple.
[*Ans.* $-0.15\,V$]

2. The potential diagram (at pH = 0) for thallium is as follows:

$$Tl^{3+} \xrightarrow{E^{\circ}} Tl^{+} \xrightarrow{-0.34} Tl$$
$$\underset{+0.72}{\underline{\hspace{4cm}}}$$

Determine the value of E° for the reduction of Tl^{3+} to Tl^{+}.
[*Ans.* $+1.25\,V$]

3. Construct Frost–Ebsworth diagrams for Ga, In and Tl at pH = 0. Use the diagrams to comment on (a) the relative abilities of Ga^{3+}, In^{3+} and Tl^{3+} to act as oxidizing agents under these conditions, and (b) the relative stabilities of the +1 oxidation state of each element.

E° for the reduction of Tl(III) to Tl(I) in molar $HClO_{4}$ is $+1.25\,V$, and under these conditions, Tl(III) is a powerful oxidizing agent. The value of E° is, however, dependent on the anion present and complex formed (see Section 8.3). Tl(I) (like the alkali metal ions) forms few stable complexes in aqueous solution, whereas Tl(III) is strongly complexed by a variety of anions. For example, consider the presence of Cl^{-} in solution. Whereas TlCl is fairly insoluble,

Tl(III) forms the soluble complex $[TlCl_{4}]^{-}$ and, at $[Cl^{-}] = 1\,mol\,dm^{-3}$, $E^{\circ}(Tl^{3+}/Tl^{+}) = +0.9\,V$. Thallium(III) forms a more stable complex with I^{-} than Cl^{-}, and at high $[I^{-}]$, $[TlI_{4}]^{-}$ is produced in solution even though $E^{\circ}(Tl^{3+}/Tl^{+})$ is more positive than $E^{\circ}(I_{2}/2I^{-})$ ($+0.54\,V$) and TlI is sparingly soluble. Thus, while tabulated reduction potentials for the Tl^{3+}/Tl^{+} and $I_{2}/2I^{-}$ couples might suggest that aqueous I^{-} will reduce Tl(III) to Tl(I) (see Appendix 11), in the presence of high concentrations of I^{-}, Tl(III) is stabilized. Indeed, the addition of I^{-} to solutions of TlI_{3} (see structure **13.17**), which contain $[I_{3}]^{-}$ (i.e. $I_{2} + I^{-}$), brings about reaction 13.77 oxidizing Tl(I) to Tl(III).

$$TlI_{3} + I^{-} \longrightarrow [TlI_{4}]^{-} \tag{13.77}$$

In alkaline media, Tl(I) is also easily oxidized, since TlOH is soluble in water and hydrated $Tl_{2}O_{3}$ (which is in equilibrium with Tl^{3+} and $[OH]^{-}$ ions in solution) is very sparingly soluble in water ($K_{sp} \approx 10^{-45}$).

Electrochemical data (from cyclic voltammetry and rotating disc electrode techniques, see Box 8.2) for the 2-electron reduction of Tl^{3+} to Tl^{+} in aqueous solution, are consistent with the formation of a transient intermediate Tl(II) species, $[Tl-Tl]^{4+}$, formed near the electrode.

Coordination complexes of the M^{3+} ions

A wide range of coordination complexes of the group 13 metal ions are known. Octahedral coordination is common, e.g. in $[M(acac)_{3}]$ (M = Al, Ga, In), $[M(ox)_{3}]^{3-}$ (M = Al, Ga, In) and *mer*-$[Ga(N_{3})_{3}(py)_{3}]$ (see Table 7.7 for ligand abbreviations and structures). Figure 13.27a shows the structure of $[Al(ox)_{3}]^{3-}$. The complexes $[M(acac)_{3}]$ are structurally related to $[Fe(acac)_{3}]$ (see Fig. 7.10). In Section 7.11, we discussed the influence of $[H^{+}]$ on the formation of $[Fe(acac)_{3}]$ and similar arguments apply to the group 13 metal ion complexes.

(13.39)

Deprotonation of 8-hydroxyquinoline gives the bidentate ligand **13.39** which has a number of applications. For example, Al^{3+} may be extracted into organic solvents as the octahedral complex $[Al(\mathbf{13.39})_{3}]$ providing a weighable form for the metal in gravimetric analysis.

Complexes involving macrocyclic ligands with pendant carboxylate or phosphate groups have received attention in the development of highly stable metal complexes suitable for *in vivo* applications, e.g. tumour-seeking complexes containing radioisotopes. The incorporation of ^{67}Ga (γ-emitter, $t_{\frac{1}{2}} = 3.2$ days), ^{68}Ga (β^{+}-emitter, $t_{\frac{1}{2}} = 68$ min)

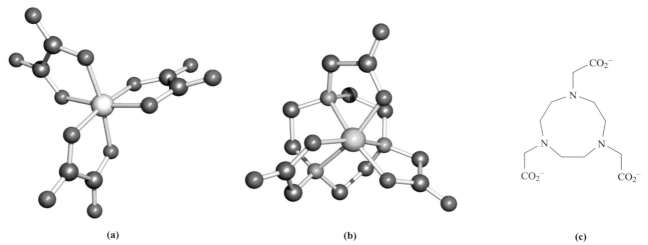

Fig. 13.27 The structures (X-ray diffraction) of (a) $[Al(ox)_3]^{3-}$ in the ammonium salt [N. Bulc *et al.* (1984) *Acta Crystallogr., Sect. C*, vol. 40, p. 1829], and (b) [GaL] [C.J. Broan *et al.* (1991) *J. Chem. Soc., Perkin Trans. 2*, p. 87] where ligand L^{3-} is shown in diagram (c). Hydrogen atoms have been omitted from (a) and (b); colour code: Al, pale grey; Ga, yellow; O, red; C, grey; N, blue.

or ^{111}In (γ-emitter, $t_{\frac{1}{2}} = 2.8$ days) into such complexes yields potential radiopharmaceuticals. Figure 13.27c shows an example of a well-studied ligand which forms very stable complexes with Ga(III) and In(III) ($\log K \geq 20$). The way in which this ligand encapsulates the M^{3+} ion with the three *N*-donor atoms forced into a *fac*-arrangement can be seen in Fig. 13.27b.

We noted in Section 13.6 that InCl is virtually insoluble in most organic solvents. In contrast, the triflate salt, $InSO_3CF_3$, dissolves in a range of solvents, making it a more convenient source of In(I). The salt can be stabilized as a crown ether complex (Fig. 13.28), the solid state structure of which reveals an In–O(triflate) distance (237 pm) that is shorter than the In–O(ether) distances (average 287 pm).

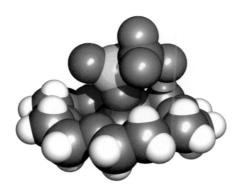

Fig. 13.28 The structure of [In(18-crown-6)][CF$_3$SO$_3$] determined by X-ray diffraction [C.G. Andrews *et al.* (2005) *Angew. Chem. Int. Ed.*, vol. 44, p. 7453]; the space-filling representation illustrates the embedding of the In(I) centre within the crown ether, and the interaction between the In$^+$ and [CF$_3$SO$_3$]$^-$ ions. Colour code: In, blue; O, red; C, grey; S, yellow; F, green.

13.10 Metal borides

Solid state metal borides are characteristically extremely hard, involatile, high melting and chemically inert materials which are industrially important with uses as refractory materials and in rocket cones and turbine blades, i.e. components that must withstand extreme stress, shock and high temperatures. The borides LaB_6 and CeB_6 are excellent thermionic electron emission sources, and single crystals are used as cathode materials in electron microscopes (see Box 13.8).

Preparative routes to metal borides are varied, as are their structures. Some may be made by direct combination of the elements at high temperatures, and others from metal oxides (e.g. reactions 13.78 and 13.79).

$$Eu_2O_3 \xrightarrow{\text{boron carbide/carbon, } \Delta} EuB_6 \tag{13.78}$$

$$TiO_2 + B_2O_3 \xrightarrow{Na, \Delta} TiB_2 \tag{13.79}$$

Metal borides may be boron- or metal-rich, and general families include MB_3, MB_4, MB_6, MB_{10}, MB_{12}, M_2B_5 and M_3B_4 (B-rich), and M_3B, M_4B, M_5B, M_3B_2 and M_7B_3 (M-rich). The formulae bear no relation to those expected on the basis of the formal oxidation states of boron and metal.

The structural diversity of these materials is so great as to preclude a full discussion here, but we can conveniently consider them in terms of the categories shown in Table 13.3, which are identified in terms of the arrangement of the B atoms within a host metal lattice. The structure of the MB_6 borides (e.g. CaB_6) is similar to a CsCl-type structure with B_6-units (Table 13.3) replacing

Table 13.3 Classification of the structures of solid state metal borides.

Description of the boron atom organization	Pictorial representation of the boron association	Examples of metal borides adopting each structure type
Isolated B atoms		Ni_3B, Mn_4B, Pd_5B_2, Ru_7B_3
Pairs of B atoms	B——B	Cr_5B_3
Chains		V_3B_4, Cr_3B_4, HfB, CrB, FeB
Linked double chains		Ta_3B_4
Sheets		MgB_2, TiB_2, CrB_2, Ti_2B_5, W_2B_5
Linked B_6 octahedra (see text)		Li_2B_6, CaB_6, LaB_6, CeB_6
Linked B_{12} icosahedra (see text; see also Fig. 13.6)		ZrB_{12}, UB_{12}

(B—B links to adjacent icosahedra are not shown)

Cl^- ions. However, the B$-$B distances *between* adjacent B_6-octahedra are similar to those *within* each unit and so a 'discrete ion' model is not appropriate. The structure type of MB_{12} (e.g. UB_{12}) can be described in terms of an NaCl structure in which the Cl^- ions are replaced by B_{12}-icosahedra (Table 13.3), but again, an ionic model is not appropriate.

Although this summary of metal borides is brief, it illustrates the complexity of structures frequently encountered in the chemistry of boron. Research interest in metal borides has been stimulated since 2001 by the discovery that MgB_2 is a superconductor with a critical temperature, T_c, of 39 K† (see Section 28.4).

13.11 Electron-deficient borane and carbaborane clusters: an introduction

This section introduces *electron-deficient clusters* containing boron, focusing on the small clusters $[B_6H_6]^{2-}$, B_5H_9 and B_4H_{10}. A comprehensive treatment of borane and carbaborane clusters is beyond the scope of this book, but more detailed accounts can be found in the references cited at the end of the chapter.

> An *electron-deficient* species possesses fewer valence electrons than are required for a localized bonding scheme. In a *cluster*, the atoms form a cage-like structure.

The pioneering work of Alfred Stock between 1912 and 1936 revealed that boron formed a range of hydrides of varying nuclearities. Since these early studies, the number of neutral and anionic boron hydrides has increased greatly, and the structures of three of the smaller boranes are shown in Fig. 13.29. The following classes of boron hydride cluster are the most commonly encountered.

- In a *closo*-cluster, the atoms form a closed, deltahedral cage and have the general formula $[B_nH_n]^{2-}$ (e.g. $[B_6H_6]^{2-}$).
- In a *nido*-cluster, the atoms form an open cage which is derived from a closed deltahedron with one vertex unoccupied; general formulae are B_nH_{n+4}, $[B_nH_{n+3}]^-$ etc. (e.g. B_5H_9, $[B_5H_8]^-$).
- In an *arachno*-cluster, the atoms form an open cage which is derived from a closed deltahedron with two vertices unoccupied; general formulae are B_nH_{n+6}, $[B_nH_{n+5}]^-$ etc. (e.g. B_4H_{10}, $[B_4H_9]^-$).
- In a *hypho*-cluster, the atoms form an open cage which is derived from a closed deltahedron with three vertices

unoccupied; this is a poorly exemplified group of compounds with general formulae B_nH_{n+8}, $[B_nH_{n+7}]^-$ etc.
- A *conjuncto*-cluster consists of two or more cages connected together through a shared atom, an external bond, a shared edge or a shared face (e.g. $\{B_5H_8\}_2$).

> A *deltahedron* is a polyhedron that possesses only *triangular* faces, e.g. an octahedron.

In the 1950s–1960s, there was considerable interest in the possibility of using boron hydrides as high-energy fuels, but in practice, it is difficult to ensure complete combustion to B_2O_3, and involatile polymers tend to block exhaust ducts. Although interest in fuel applications has faded, boranes remain a fascination to structural and theoretical chemists.

The name of a borane denotes the number of boron atoms, the number of hydrogen atoms, and the overall charge. The number of boron atoms is given by a Greek prefix (di-, tri-, tetra-, penta-, hexa- etc.), the exception being for nine and eleven, where the Latin nona- and undeca- are used. The number of hydrogen atoms is shown as an Arabic numeral in parentheses at the end of the name (see below). The charge for an ion is shown at the end of the name; the nomenclature for anions is also distinguished from that of neutral boranes (see examples below). As a prefix, the class of cluster (*closo*-, *nido*-, *arachno*-, *conjuncto*- etc.) should be stated.

- $[B_6H_6]^{2-}$ *closo*-hexahydrohexaborate(2−)
- B_4H_{10} *arachno*-tetraborane(10)
- B_5H_9 *nido*-pentaborane(9)
- B_6H_{10} *nido*-hexaborane(10)

The higher boranes can be prepared by controlled pyrolysis of B_2H_6 in the vapour phase. The pyrolysis of B_2H_6 in a hot–cold reactor (i.e. a reactor having an interface between two regions of extreme temperatures) gives, for example, B_4H_{10}, B_5H_{11} or B_5H_9 depending upon the temperature interface. Decaborane(14), $B_{10}H_{14}$, is produced by heating B_2H_6 at 453–490 K under static conditions. Such methods are complicated by the interconversion of one borane to another, and it has been desirable to seek selective syntheses. The reaction between B_2H_6 and $Na[BH_4]$ (eq. 13.80) gives $Na[B_3H_8]$ which contains the $[B_3H_8]^-$ ion (**13.40**).

(13.40)

† J. Nagamatsu, N. Nakagawa, T. Muranaka, Y. Zenitani and J. Akimitsu (2001) *Nature*, vol. 410, p. 63 – 'Superconductivity at 39 K in magnesium boride'.

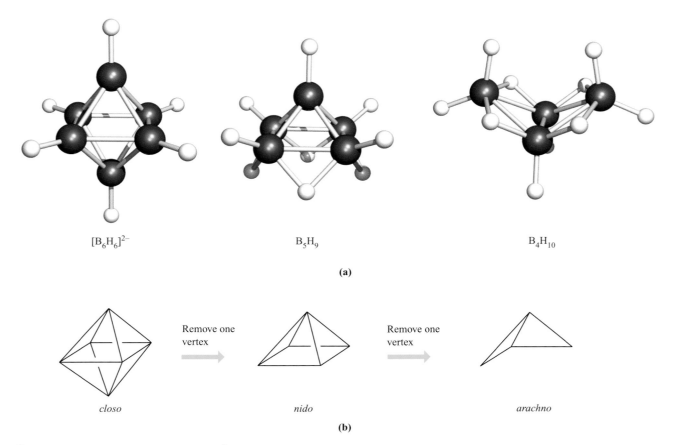

$[B_6H_6]^{2-}$ B_5H_9 B_4H_{10}

(a)

Remove one vertex   Remove one vertex

closo   *nido*   *arachno*

(b)

Fig. 13.29 (a) The structures of $[B_6H_6]^{2-}$, B_5H_9 and B_4H_{10}; colour code: B, blue; H, white. (b) Schematic representation of the derivation of *nido* (with $n = 5$) and *arachno* (with $n = 4$) cages from a parent *closo* deltahedral cage with $n = 6$.

$[B_3H_8]^-$ is a convenient precursor to B_4H_{10}, B_5H_9 and $[B_6H_6]^{2-}$ (eqs. 13.81–13.83).

$$B_2H_6 + Na[BH_4] \xrightarrow{\text{363 K in diglyme}} Na[B_3H_8] + H_2 \tag{13.80}$$

$$4Na[B_3H_8] + 4HCl \longrightarrow 3B_4H_{10} + 3H_2 + 4NaCl \tag{13.81}$$

$$5[B_3H_8]^- + 5HBr \xrightarrow{-H_2} 5[B_3H_7Br]^-$$
$$\xrightarrow{\text{373 K}} 3B_5H_9 + 4H_2 + 5Br^- \tag{13.82}$$

$$2Na[B_3H_8] \xrightarrow{\text{435 K in diglyme}} Na_2[B_6H_6] + 5H_2 \tag{13.83}$$

(Diglyme: see structure **13.2**)

The formation of $Na_2[B_6H_6]$ in reaction 13.83 competes with that of $Na_2[B_{10}H_{10}]$ and $Na_2[B_{12}H_{12}]$ (eqs. 13.84 and 13.85) and the reaction gives only low yields of $Na_2[B_6H_6]$. Starting from $Na[B_3H_8]$ prepared *in situ* by reaction 13.80, a typical molar ratio of $[B_6H_6]^{2-} : [B_{10}H_{10}]^{2-} : [B_{12}H_{12}]^{2-}$ from a combination of reactions 13.83–13.85 is $2 : 1 : 15$.

$$4Na[B_3H_8] \xrightarrow{\text{435 K in diglyme}} Na_2[B_{10}H_{10}] + 2Na[BH_4] + 7H_2 \tag{13.84}$$

$$5Na[B_3H_8] \xrightarrow{\text{435 K in diglyme}} Na_2[B_{12}H_{12}] + 3Na[BH_4] + 8H_2 \tag{13.85}$$

Higher yields of $Na_2[B_6H_6]$ are obtained by changing the *in situ* synthesis of $Na[B_3H_8]$ to reaction 13.86, followed by heating in diglyme at reflux for 36 hours.

$$5Na[BH_4] + 4Et_2O \cdot BF_3 \xrightarrow{\text{373 K in diglyme}}$$
$$2Na[B_3H_8] + 2H_2 + 3Na[BF_4] + 4Et_2O \tag{13.86}$$

The dianion $[B_6H_6]^{2-}$ has a closed octahedral B_6 cage (Fig. 13.29a) and is a *closo*-cluster. Each B atom is connected to four other B atoms within the cage, and to one terminal H. The structure of B_5H_9 (Fig. 13.29a) consists of a square-based pyramidal cage of B atoms, each of which carries one terminal H. The remaining four H atoms occupy B–H–B bridging sites around the square face of the cage. Figure 13.29a shows the structure of B_4H_{10}

which has an open framework of two edge-sharing B_3 triangles. The inner B atoms carry one terminal H each, and two terminal H atoms are bonded to each of the outer B atoms. The remaining four H atoms are involved in B−H−B bridges. X-ray diffraction data for the K^+, Cs^+, $[Li(NH_3)_4]^+$ and 1-aminoguanidinium salts have shown that the B−B bond distances in $[B_6H_6]^{2-}$ are equal (172 pm), but in B_5H_9, the unbridged B−B edges (apical–basal, 166 pm) are shorter than the H-bridged edges (basal–basal, 172 pm). The apical and basal atoms in B_5H_9 are defined in structure **13.41**. A similar situation is observed in B_4H_{10} (H-bridged edges = 186 pm, unique B−B edge = 173 pm from X-ray diffraction data). The *range* of B−B distances in these three cages is significant and, in the light of the discussion of bonding that follows, it is instructive to compare these distances with twice the covalent radius of B ($r_{cov} = 88$ pm). Longer B−B edges are observed in other clusters (e.g. 198 pm in $B_{10}H_{14}$) but are still regarded as bonding interactions.

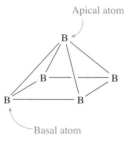

Apical atom

Basal atom

(13.41)

Formally, we can consider the structure of B_5H_9 as being related to that of $[B_6H_6]^{2-}$ by removing one vertex from the B_6 octahedral cage (Fig. 13.29b). Similarly, the B_4 cage in B_4H_{10} is related to that of B_5H_9 by the removal of another vertex. The removal of a vertex is accompanied by the addition of bridging H atoms. These observations lead us to a discussion of the bonding in boranes. The first point is that boron-containing and related clusters exhibit structures in which the bonding is *not* readily represented in terms of localized bonding models. This is in contrast to the situation in B_2H_6, $[BH_4]^-$ and $[B_3H_8]^-$ where 2c-2e and 3c-2e interactions can adequately represent the distributions of valence electrons.[†] A satisfactory solution to this problem is to consider a delocalized approach and invoke MO theory (see Box 13.9). The situation has been greatly helped by an empirical set of rules developed by Wade, Williams and Mingos. The initial *Wade's rules* can be summarized as follows, and 'parent' deltahedra are shown in Fig. 13.30:

[†] A valence bond method called *styx* rules, devised by W.N. Lipscomb, provides a means of constructing bonding networks for boranes in terms of 3c-2e B−H−B interactions, 3c-2e B−B−B interactions, 2c-2e B−B bonds, and BH_2-units, but the method is applied easily only to a limited number of clusters.

- a *closo*-deltahedral cluster cage with n vertices requires $(n + 1)$ pairs of electrons which occupy $(n + 1)$ cluster bonding MOs;
- from a 'parent' *closo*-cage with n vertices, a set of more open cages (*nido*, *arachno* and *hypho*) can be derived, each of which possesses $(n + 1)$ pairs of electrons occupying $(n + 1)$ cluster bonding MOs;
- for a parent *closo*-deltahedron with n vertices, the related *nido*-cluster has $(n - 1)$ vertices and $(n + 1)$ pairs of electrons;
- for a parent *closo*-deltahedron with n vertices, the related *arachno*-cluster has $(n - 2)$ vertices and $(n + 1)$ pairs of electrons;
- for a parent *closo*-deltahedron with n vertices, the related *hypho*-cluster has $(n - 3)$ vertices and $(n + 1)$ pairs of electrons.

In counting the number of cluster-bonding electrons available in a borane, we first formally break down the cluster into fragments and determine the number of valence electrons that each fragment can contribute for cluster bonding. A procedure is as follows.

- Determine how many {BH}-units are present (i.e. assume each B atom carries a terminal hydrogen atom); each {BH}-unit provides two electrons for cage bonding (of the three valence electrons of B, one is used to form a localized terminal B−H bond, leaving two for cluster bonding).
- Count how many additional H atoms there are; each provides one electron.
- Add up the number of electrons available from the cluster fragments and take account of any overall charge.
- The total number of electrons corresponds to $(n + 1)$ pairs of electrons, and thus, the number of vertices, n, of the parent deltahedron can be established.
- Each {BH}-unit occupies one vertex in the parent deltahedron, and from the number of vertices left vacant, the class of cluster can be determined; if vertices are non-equivalent, the first to be left vacant *tends* to be either one of highest connectivity or a 'cap' in 'capped' structures (e.g. $n = 9$ and 10 in Fig. 13.30).
- Additional H atoms are placed in bridging sites along B−B edges of an *open* face of the cluster, or in extra terminal sites, usually available if there are any B atoms of especially low connectivity.

Worked example 13.7   Using Wade's rules to rationalize a structure

Rationalize why $[B_6H_6]^{2-}$ adopts an octahedral cage.

There are six {BH}-units and no additional H atoms. Each {BH}-unit provides two valence electrons. There are two electrons from the 2− charge.

THEORY

Box 13.9 Bonding in $[B_6H_6]^{2-}$

In Section 24.5, we discuss the *isolobal principle*, and the relationship between the bonding properties of different cluster *fragments*. The bonding in boron-containing clusters and, more generally, in organometallic clusters, is conveniently dealt with in terms of molecular orbital theory. In this box, we show how the *frontier orbitals* (i.e. the highest occupied and lowest unoccupied MOs) of six BH units combine to give the seven cluster bonding MOs in $[B_6H_6]^{2-}$. This *closo*-anion has O_h symmetry:

After accounting for the localized B–H bonding orbital (σ_{BH}) and its antibonding counterpart, a BH fragment has

three orbitals remaining which are classed as its frontier orbitals:

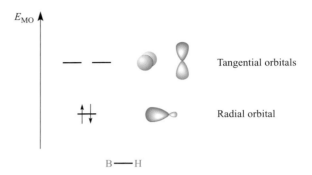

If we consider the BH fragments as being placed in the orientations shown in the structural diagram on the left, then the three frontier orbitals can be classified as one *radial* orbital (pointing into the B_6 cage) and two *tangential* orbitals (lying over the cluster surface). When the six BH-units come together, a total of (6×3) orbitals combine to give 18 MOs, seven of which possess cluster-bonding character. The interactions that give rise to these bonding MOs are shown below. The 11 non-bonding and antibonding MOs are omitted from the diagram.

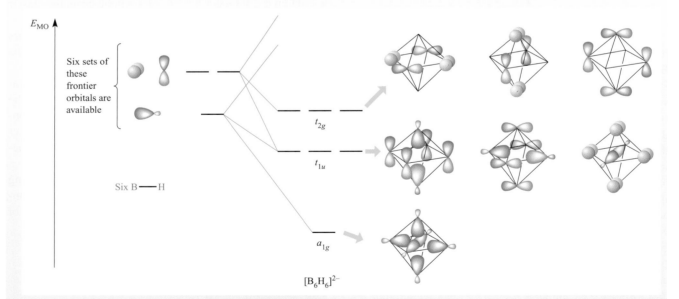

Once the molecular orbital interaction diagram has been constructed, the electrons that are available in $[B_6H_6]^{2-}$ can be accommodated in the lowest-lying MOs. Each BH unit provides two electrons, and in addition the 2– charge provides two electrons. There is, therefore, a total of seven

electron pairs available, which will completely occupy the seven bonding MOs shown in the diagram above. Relating this to Wade's rules, the MO approach shows that there are seven electron-pairs for a *closo*-cage possessing six cluster vertices.

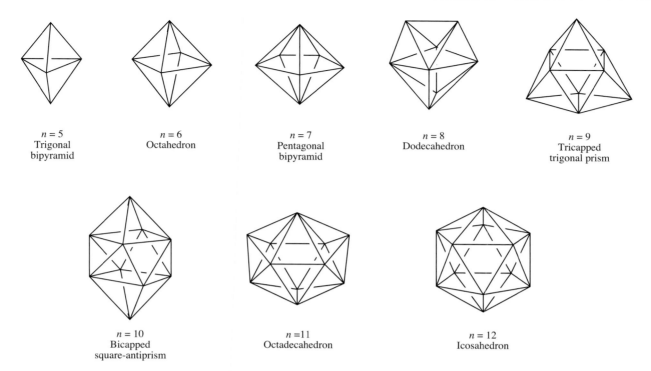

$n = 5$
Trigonal
bipyramid

$n = 6$
Octahedron

$n = 7$
Pentagonal
bipyramid

$n = 8$
Dodecahedron

$n = 9$
Tricapped
trigonal prism

$n = 10$
Bicapped
square-antiprism

$n = 11$
Octadecahedron

$n = 12$
Icosahedron

Fig. 13.30 The deltahedral cages with 5–12 vertices which are the parent cages used in conjunction with Wade's rules to rationalize borane cluster structures. As a general (but not foolproof) scheme, when removing vertices from these cages to generate *nido*-frameworks, remove a vertex of connectivity three from the trigonal bipyramid, any vertex from the octahedron or icosahedron, a 'cap' from the tricapped trigonal prism or bicapped square-antiprism, and a vertex of highest connectivity from the remaining deltahedra. See also Fig. 13.35 for 13-vertex cages.

Total number of cage-bonding electrons available
$$= (6 \times 2) + 2 = 14 \text{ electrons}$$
$$= 7 \text{ pairs}$$
Thus, $[B_6H_6]^{2-}$ has seven pairs of electrons with which to bond six $\{BH\}$-units.

This means that there are $(n + 1)$ pairs of electrons for n vertices, and so $[B_6H_6]^{2-}$ is a *closo*-cage, a six-vertex deltahedron, i.e. the octahedron is adopted (see Fig. 13.30).

Self-study exercises

Refer to Fig. 13.30.

1. Rationalize why $[B_{12}H_{12}]^{2-}$ adopts an icosahedral structure for the boron cage.

2. Show that the observed bicapped square-antiprismatic structure of the boron cage in $[B_{10}H_{10}]^{2-}$ is consistent with Wade's rules.

3. In each of the following, rationalize the observed boron cage structure in terms of Wade's rules: (a) B_5H_9 (a square-based pyramid); (b) B_4H_{10} (two edge-fused triangles, Fig. 13.29); (c) $[B_6H_9]^-$ (a pentagonal pyramid); (d) B_5H_{11} (an open network of three edge-fused triangles).

Worked example 13.8   Using Wade's rules to predict a structure

Suggest a likely structure for $[B_5H_8]^-$.

There are five $\{BH\}$-units and three additional H atoms. Each $\{BH\}$-unit provides two valence electrons. There is one electron from the $1-$ charge.
Total number of cage-bonding electrons available
$$= (5 \times 2) + 3 + 1 = 14 \text{ electrons}$$
$$= 7 \text{ pairs}$$

Seven pairs of electrons are consistent with the parent deltahedron having six vertices, i.e. $(n + 1) = 7$, and so $n = 6$.

The parent deltahedron is an octahedron and the B_5-core of $[B_5H_8]^-$ will be derived from an octahedron with one vertex left vacant:

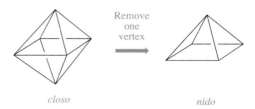

closo   *nido*

The three extra H atoms form B−H−B bridges along three of the four B−B edges of the open (square) face of the B_5-cage. The predicted structure of $[B_5H_8]^-$ is:

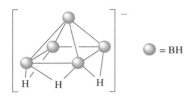

Self-study exercises

Refer to Fig. 13.30.

1. Confirm the following classifications within Wade's rules: (a) $[B_9H_9]^{2-}$, *closo*; (b) B_6H_{10}, *nido*; (c) B_4H_{10}, *arachno*; (d) $[B_8H_8]^{2-}$, *closo*; (e) $[B_{11}H_{13}]^{2-}$, *nido*.

2. Suggest likely structures for the following: (a) $[B_9H_9]^{2-}$; (b) B_6H_{10}; (c) B_4H_{10}; (d) $[B_8H_8]^{2-}$.
   [*Ans.* (a) Tricapped trigonal prism; (b) pentagonal pyramid; (c) see Fig. 13.29; (d) dodecahedron]

The types of reactions that borane clusters undergo depend upon the class and size of the cage. The clusters $[B_6H_6]^{2-}$ and $[B_{12}H_{12}]^{2-}$ provide examples of *closo*-hydroborate dianions; B_5H_9 and B_4H_{10} are examples of small *nido*- and *arachno*-boranes, respectively.

The development of the chemistry of $[B_6H_6]^{2-}$ has been relatively slow, but improved synthetic routes (see eq. 13.86 and accompanying text) have now made the dianion more accessible. The reactivity of $[B_6H_6]^{2-}$ is influenced by its ability to act as a Brønsted base ($pK_b = 7.0$). Protonation of $Cs_2[B_6H_6]$ (using HCl) yields $Cs[B_6H_7]$. This reaction is atypical of *closo*-hydroborate dianions. Furthermore, the added proton in $[B_6H_7]^-$ (**13.42**) adopts an unusual triply-bridging (μ_3) site, capping a B_3-face. Both ^1H and ^{11}B NMR spectra are consistent with the dynamic behaviour of the μ_3-H atom, which renders all six BH$_{terminal}$-units equivalent (see end-of-chapter problem 13.34a).

(13.42)   **(13.43)**

Chlorination, bromination and iodination of $[B_6H_6]^{2-}$ occur with X_2 in strongly basic solution to give mixtures of products (eq. 13.87, X = Cl, Br, I). Monofluorination of

$[B_6H_6]^{2-}$ can be achieved using XeF_2, but is complicated by protonation, the products being $[B_6H_5F]^{2-}$ and $[B_6H_5(\mu_3\text{-}H)F]^-$. By using **13.43** as the fluorinating agent, $[B_6H_5(\mu_3\text{-}H)F]^-$ is selectively formed.

$$[B_6H_6]^{2-} + nX_2 + n[OH]^- \longrightarrow$$
$$[B_6H_{(6-n)}X_n]^{2-} + nH_2O + nX^- \quad (X = Cl, Br, I)$$
$$(13.87)$$

The tendency for $[B_6H_6]^{2-}$ to gain H^+ affects the conditions under which alkylation reactions are carried out. Neutral conditions must be used, contrasting with the acidic conditions under which $[B_{10}H_{10}]^{2-}$ and $[B_{12}H_{12}]^{2-}$ are alkylated. Even so, as scheme 13.88 shows, the reaction is not straightforward.

$$[Bu_4N]_2[B_6H_6] + RX \xrightarrow[-\,[Bu_4N]X]{\text{in } CH_2Cl_2} [Bu_4N][B_6H_5(\mu_3\text{-}H)R]$$

$$\downarrow \text{CsOH in EtOH}$$

$$Cs_2[B_6H_5R] \text{ (precipitate)}$$
$$(13.88)$$

The oxidation of $[B_6H_6]^{2-}$ by dibenzoyl peroxide leads to the *conjuncto*-cluster **13.44**. Treatment of **13.44** with $Cs[O_2CMe]$ and then with CsOH removes the capping protons one by one to give $[\{B_6H_5(\mu_3\text{-}H)\}\{B_6H_5\}]^{3-}$ and then $[\{B_6H_5\}_2]^{4-}$.

$$[\{B_6H_5(\mu_3\text{-}H)\}_2]^{2-}$$

(13.44)

The chemistry of $[B_{12}H_{12}]^{2-}$ (and also of $[B_{10}H_{10}]^{2-}$) is well explored. Electrophilic substitution reactions predominate, although some reactions with nucleophiles also occur. The vertices in the icosahedral cage of $[B_{12}H_{12}]^{2-}$ (**13.45**) are all equivalent, and therefore there is no preference for the first site of substitution. The reactions of $[B_{12}H_{12}]^{2-}$ with Cl_2 and Br_2 lead to $[B_{12}H_{(12-x)}X_x]^{2-}$ ($x = 1$–12), and the rate of substitution decreases as x increases. The rate also decreases on going from X = Cl to X = Br, and the reaction is slower still for X = I. Iodination with I_2 leads to some degree of substitution, but for the formation of $[B_{12}I_{12}]^{2-}$, it is necessary to use a mixture of I_2 and ICl. Perfluorination of $[B_{12}H_{12}]^{2-}$ can be achieved by heating $K_2[B_{12}H_{12}]$ in anhydrous liquid HF at 340 K (to form $[B_{12}H_8F_4]^{2-}$), followed by treatment of the reaction mixture with 20% F_2/N_2 at 298 K.

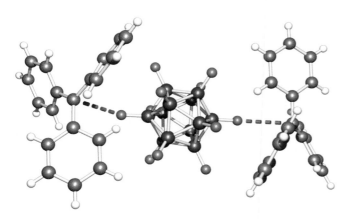

Fig. 13.31 In the solid state, the ions in [CPh₃]₂[B₁₂F₁₂] exhibit weak BF···C interactions (BF···C = 309 pm), consistent with [B₁₂F₁₂]²⁻ behaving as a weakly coordinating anion [S.V. Ivanov *et al.* (2003) *J. Am. Chem. Soc.*, vol. 125, p. 4694]. Colour code: B, blue; F, green; C, grey; H, white.

The cation can be exchanged to give a range of different salts including [CPh₃]₂[B₁₂F₁₂], the structure of which has been determined (Fig. 13.31). In Section 9.9, we introduced halogenated carbaborane anions such as $[CHB_{11}Cl_{11}]^-$ that are weak bases and extremely weakly coordinating. The cation–anion interactions in [CPh₃]₂[B₁₂F₁₂] (Fig. 13.31) are consistent with $[B_{12}F_{12}]^{2-}$ behaving as a weakly coordinating anion; each BF···C distance of 309 pm is only 11 pm less than the sum of the van der Waals radii of C and F.

Scheme 13.89 shows further examples of substitutions in $[B_{12}H_{12}]^{2-}$, and the atom numbering scheme for the cage is shown in structure **13.46**. In each reaction, the icosahedral B₁₂-cage is retained. Since CO is a 2-electron donor, its introduction in place of an H atom (which provides one electron) affects the overall charge on the cluster (scheme 13.89). The thiol $[B_{12}H_{11}(SH)]^{2-}$ (scheme 13.89) is of particular importance because of its application in treating cancer using boron neutron capture therapy (BNCT).[†]

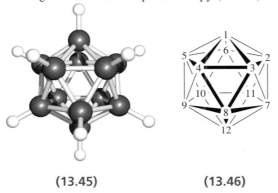

(13.45)   **(13.46)**

[†] See: I.B. Sivaev and V.V. Bregadze (2009) *Eur. J. Inorg. Chem.*, p. 1433 – 'Polyhedral boranes for medical applications: Current status and perspectives'.

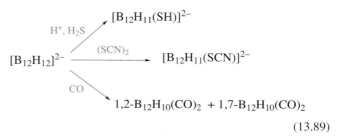

$$[B_{12}H_{12}]^{2-} \xrightarrow[\text{(SCN)}_2]{H^+, H_2S} \begin{array}{l} [B_{12}H_{11}(SH)]^{2-} \\ [B_{12}H_{11}(SCN)]^{2-} \\ 1,2\text{-}B_{12}H_{10}(CO)_2 + 1,7\text{-}B_{12}H_{10}(CO)_2 \end{array}$$

(13.89)

The reaction of [Bu₄N]₂[B₁₂H₁₂] with MeI and AlMe₃ leads first to $[B_{12}Me_{(12-x)}I_x]^{2-}$ ($x \leq 5$) and, after prolonged heating, to $[B_{12}Me_{12}]^{2-}$ and $[B_{12}Me_{11}I]^{2-}$. Scheme 13.90 shows the formation of $H_2B_{12}(OH)_{12}$ and salts of $[B_{12}(OH)_{12}]^{2-}$.

$$Cs_2[B_{12}H_{12}] \xrightarrow{30\% \, H_2O_2} Cs_2[B_{12}(OH)_{12}]$$
$$\xrightarrow[\text{HCl(aq), 423 K}]{} \xrightarrow[\text{MCl (M = Na, K, Rb)}]{}$$
$$H_2B_{12}(OH)_{12} \qquad M_2[B_{12}(OH)_{12}]$$

(13.90)

Even though $[B_{12}(OH)_{12}]^{2-}$ has 12 terminal OH groups available for hydrogen bonding, the alkali metal salts are not very soluble in water. This surprising observation can be understood by considering the solid state structures of the Na⁺, K⁺, Rb⁺ and Cs⁺ salts. These all exhibit extensive hydrogen-bonded networks as well as highly organized M⁺····OH interactions. The observed low solubilities correspond to small values of the equilibrium constant, K, for the dissolution process. Since $\ln K$ is related to $\Delta_{sol}G^{\circ}$ (see Section 7.9), it follows from the thermodynamic cycle in eq. 13.91 that the Gibbs energy of hydration is insufficient to offset the lattice energy of each salt.

$$M_2[B_{12}(OH)_{12}](s) \xrightarrow{-\Delta_{lattice}G^{\circ}} 2M^+(g) + [B_{12}(OH)_{12}]^{2-}(g)$$
$$\Delta_{sol}G^{\circ} \searrow \qquad \swarrow \Delta_{hyd}G^{\circ}$$
$$2M^+(aq) + [B_{12}(OH)_{12}]^{2-}(aq)$$

(13.91)

$H_2B_{12}(OH)_{12}$ (Fig. 13.32) is also poorly soluble in water, and this is rationalized in terms of the extensive intermolecular hydrogen bonding in the solid state. In contrast to the *Lewis* acidity of B(OH)₃ (eq. 13.52), $H_2B_{12}(OH)_{12}$ is a *Brønsted* acid. Solid $H_2B_{12}(OH)_{12}$ is a proton conductor ($1.5 \times 10^{-5}\,\Omega^{-1}\,cm^{-1}$ at 298 K). It is proposed that the protons migrate through the solid by a *Grotthuss* mechanism in which protons 'hop' between relatively stationary anions.[‡]

[‡] For an overview of the principles and properties of proton conductors, see: T. Norby (1999) *Solid State Ionics*, vol. 125, p. 1.

Fig. 13.32 The structure of $H_2B_{12}(OH)_{12}$ determined by X-ray diffraction [D.J. Stasko *et al.* (2004) *Inorg. Chem.*, vol. 43, p. 3786]. The sites of protonation of the conjugate base, $[B_{12}(OH)_{12}]^{2-}$, are on the left- and right-hand sides of the figure, respectively. Colour code: B, blue; O, red; H, white.

The reactivities of B_5H_9 and B_4H_{10} have been well explored and typical reactions are given in Figs. 13.33 and 13.34. The *nido*-B_5H_9 cluster is more reactive than *closo*-$[B_6H_6]^{2-}$, and *arachno*-B_4H_9 is more susceptible

still to reactions involving cage degradation or cleavage. For example, B_4H_{10} is hydrolysed by H_2O, while B_5H_9 is hydrolysed only slowly by water but completely by alcohols. Many reactions involving *arachno*-B_4H_{10} with Lewis bases are known and Fig. 13.34 illustrates cleavage with NH_3 (a small base) to give an ionic salt and by a more sterically demanding base to give neutral adducts. Compare these reactions with those of B_2H_6 (eq. 13.14). Carbon monoxide and PF_3, on the other hand, react with B_4H_{10} with elimination of H_2 and retention of the B_4 cage. Deprotonation of both B_4H_{10} and B_5H_9 can be achieved using NaH or KH and in each case H^+ is removed from a *bridging* site. This preference is quite general among boranes and can be rationalized in terms of redistribution of the two electrons from the B−H−B bridge into a B−B interaction upon H^+ removal. Electrophiles react with B_5H_9 (Fig. 13.33) with initial attack being at the apical B atom. Isomerizations to give the basally substituted derivatives occur but have been shown by ^{10}B labelling studies to involve B_5 cage rearrangement rather than migration of the substituent. Both B_4H_{10} and B_5H_9 react with ethyne to generate a new family of cluster compounds, the *carboranes*. Structurally, carboranes resemble boranes,

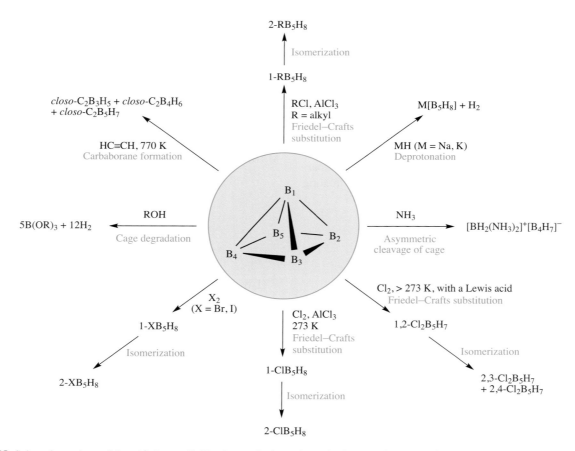

Fig. 13.33 Selected reactions of the *nido*-borane B_5H_9; the numbering scheme in the central structure is used to indicate positions of substitution in products that retain the B_5-core.

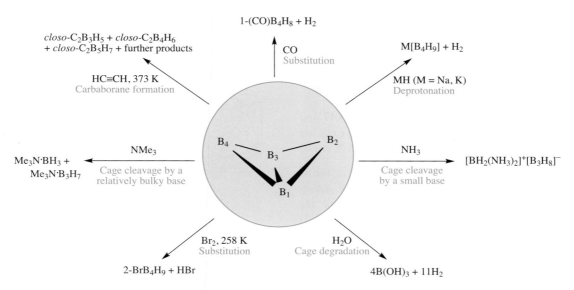

Fig. 13.34 Selected reactions of the *arachno*-borane B_4H_{10}; the numbering scheme in the central structure is used to show positions of substitution in products that retain the B_4-core.

with structures rationalized in terms of Wade's rules (a CH unit provides one more electron for bonding than a BH unit). The structures of the carbaborane products in Figs. 13.33 and 13.34 are shown in **13.47–13.49**, although in each case only one cage-isomer is illustrated; an example of the application of Wade's rules to them is given in worked example 13.9.

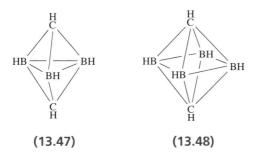

(13.47)   **(13.48)**

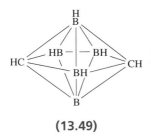

(13.49)

Worked example 13.9   Applying Wade's rules to carbaborane structures

(a) Rationalize why the cage structure of $C_2B_4H_6$ is an octahedron. (b) How many cage isomers are possible?

(a) In $C_2B_4H_6$, there are four {BH}-units, two {CH}-units and no additional H atoms.

Each {BH}-unit provides two valence electrons.
Each {CH}-unit provides three valence electrons.
Total number of cage-bonding electrons available

$$= (4 \times 2) + (2 \times 3) = 14 \text{ electrons}$$
$$= 7 \text{ pairs}$$

Thus, $C_2B_4H_6$ has seven pairs of electrons with which to bond six cluster units.

There are $(n + 1)$ pairs of electrons for n vertices, and so $C_2B_4H_6$ is a *closo*-cage, a six-vertex deltahedron, i.e. the octahedron is adopted (see Fig. 13.30).

(b) In an octahedron, all vertices are equivalent. It follows that there are two possible arrangements of the two carbon and four boron atoms, leading to two cage isomers:

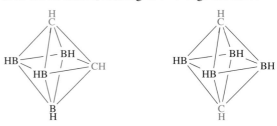

It is *not* possible to say anything about isomer preference using Wade's rules.

The deltahedra shown in Fig. 13.30 and used as 'parent deltahedra' for deriving or rationalizing structures using Wade's rules go only as far as the 12-vertex icosahedron. No single-cage hydroborate dianions $[B_nH_n]^{2-}$ are known for $n > 12$. However, in 2003, the first 13-vertex *closo*-carbaborane was reported (Fig. 13.35a). The strategy for the preparation of this compound follows two steps (scheme 13.92). First, a 12-vertex *closo*-cage is reduced and this leads to cage-opening, consistent with Wade's rules. The open face in the intermediate cluster is highlighted in scheme 13.92. In the second step, the open cage

is capped with a boron-containing fragment to generate a 13-vertex *closo*-cluster.

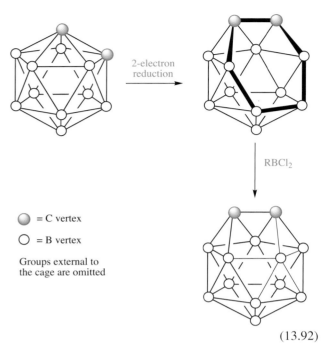

(13.92)

In practice the two C atoms must be 'tethered' together in order that the cluster does not rearrange or degrade during the reaction. In Fig. 13.35, this 'tether' corresponds to the

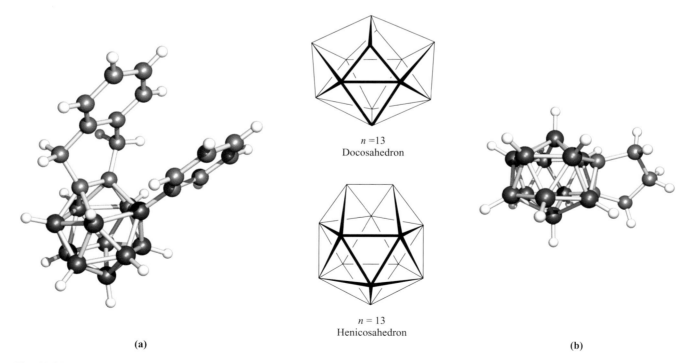

Fig. 13.35 (a) The structure (X-ray diffraction) of the 13-vertex carbaborane $1,2$-μ-$\{C_6H_4(CH_2)_2\}$-3-Ph-$1,2$-$C_2B_{11}H_{10}$ [A. Burke *et al.* (2003) *Angew. Chem. Int. Ed.*, vol. 42, p. 225]. The henicosahedron adopted by the carbaborane, and the docosahedron predicted for *closo*-$[B_{13}H_{13}]^{2-}$ (see text). (b) The structure of the 14-vertex *closo*-$1,2$-$(CH_2)_3$-$1,2$-$C_2B_{12}H_{12}$ determined by X-ray diffraction [L. Deng *et al.* (2005) *Angew. Chem. Int. Ed.*, vol. 44, p. 2128]. Colour code: B, blue; C, grey; H, white.

organic fragment that bridges the two cluster carbon atoms. The phenyl substituent attached directly to the cage labels the site at which a boron atom is introduced in the second step in scheme 13.92. Interestingly, this first example of a 13-vertex *closo*-carbaborane adopts a polyhedron which is not a deltahedron. Rather, the polyhedron is a *henicosahedron* (Fig. 13.35). This contrasts with the deltahedron (the *docosahedron*, Fig. 13.35) that has been predicted by theory to be the lowest energy structure for the hypothetical $[B_{13}H_{13}]^{2-}$.

A strategy similar to that in scheme 13.92 was used in 2005 to prepare the first 14-vertex carbaborane. In the first step, a 12-vertex, *closo*-carbaborane with adjacent carbon atoms tethered together by a $(CH_2)_3$-chain, undergoes a 4-electron reduction to form an *arachno*-cluster. The reaction of this open cage with $HBr_2B \cdot SMe_2$ follows competitive routes: (i) addition of one boron vertex with concomitant 2-electron oxidation to yield a *closo*-C_2B_{11} cluster, or (ii) addition of two boron vertices to give a *closo*-14-vertex cage. Two isomers of the latter have been observed, and structural data for one isomer confirm a bicapped hexagonal prismatic cage (Fig. 13.35b).

Before leaving this introduction to boron clusters, we return briefly to the boron halides of type B_nX_n (X = halogen). Although these have deltahedral structures, they do not 'obey' Wade's rules. Formally, by Wade's rules, each {BX}-unit in B_8X_8 provides two electrons for cage-bonding, but this approach gives an electron count (eight pairs) which is inconsistent with the observed closed dodecahedral cage (Fig. 13.13b). Similarly, B_4Cl_4 has a tetrahedral structure (Fig. 13.13a), although a simple electron count gives only four electron pairs for cluster bonding. The apparent violation of Wade's rules arises because the symmetry of the B_n-cluster-bonding MOs is appropriate to allow interaction with filled p atomic orbitals of the terminal halogens. Donation of electrons from the terminal halogen atoms to boron can occur. One must therefore be aware that, while Wade's rules are extremely useful in many instances, apparent exceptions do exist and require more in-depth bonding analyses.

KEY TERMS

The following terms were introduced in this chapter. Do you know what they mean?

- ❏ thermodynamic 6*s* inert pair effect
- ❏ relativistic effect
- ❏ mordant
- ❏ cyclodimer
- ❏ alum
- ❏ electron-deficient cluster
- ❏ deltahedron
- ❏ Wade's rules

FURTHER READING

S. Aldridge and A.J. Downs (2001) *Chem. Rev.*, vol. 101, p. 3305 – A review of hydrides of main group metals with particular reference to group 13 elements.

A.J. Downs, ed. (1993) *The Chemistry of Aluminium, Gallium, Indium and Thallium*, Kluwer, Dordrecht – Covers the chemistry and commercial aspects of these elements including applications to materials.

A.J. Downs and C.R. Pulham (1994) *Chem. Soc. Rev.*, vol. 23, p. 175 – 'The hydrides of aluminium, gallium, indium and thallium: a re-evaluation'.

R.C. Fischer and P.P. Power (2010) *Chem. Rev.*, vol. 110, p. 3877 – A review of multiple bonding involving heavier main group elements.

A.G. Massey (2000) *Main Group Chemistry*, 2nd edn, Wiley, Chichester – Chapter 7 covers the chemistry of the group 13 elements.

H.W. Roesky (2004) *Inorg. Chem.*, vol. 43, p. 7284 – 'The renaissance of aluminum chemistry' reviews recent developments in aluminium chemistry.

H.W. Roesky and S.S. Kumar (2005) *Chem. Commun.*, p. 4027 – 'Chemistry of aluminium(I)' gives an account of monomeric and tetrameric Al(I) compounds.

D.F. Shriver and M.A. Drezdon (1986) *The Manipulation of Air-sensitive Compounds*, 2nd edn, Wiley-Interscience, New York – Many compounds of the group 13 elements are extremely sensitive to air and moisture; this book gives a detailed account of methods of handling such compounds.

A.F. Wells (1984) *Structural Inorganic Chemistry*, 5th edn, Clarendon Press, Oxford – Includes a full account of the structural chemistry of the elements and compounds in group 13.

Boranes and borane clusters

M.A. Fox and K. Wade (2003) *Pure Appl. Chem.*, vol. 75, p. 1315 – 'Evolving patterns in boron cluster chemistry'.

N.N. Greenwood (1992) *Chem. Soc. Rev.*, vol. 21, p. 49 – 'Taking stock: the astonishing development of boron hydride cluster chemistry'.

N.N. Greenwood and A. Earnshaw (1997) *Chemistry of the Elements*, 2nd edn, Butterworth-Heinemann, Oxford – Chapter 6 covers boron clusters in some detail.

R.N. Grimes (2004) *J. Chem. Educ.*, vol. 81, p. 657 – A review: 'Boron clusters come of age'.

C.E. Housecroft (1994) *Boranes and Metalloboranes: Structure, Bonding and Reactivity*, 2nd edn, Ellis Horwood, Hemel Hempstead – A clear and well-illustrated introduction to borane clusters and their derivatives.

C.E. Housecroft (1994) *Clusters of the p-Block Elements*, Oxford University Press, Oxford – An introductory

survey of clusters containing *p*-block elements including boron.

W. Preetz and G. Peters (1999) *Eur. J. Inorg. Chem.*, p. 1831 – A review: 'The hexahydro-*closo*-hexaborate dianion $[B_6H_6]^{2-}$ and its derivatives'.

A. Staubitz, A.P.M. Robertson, M.E. Sloan and I. Manners (2010) *Chem. Rev.*, vol. 110, p. 4023 – 'Amine- and phosphine-borane adducts: new interest in old molecules'.

PROBLEMS

13.1 (a) Write down, in order, the names and symbols of the elements in group 13; check your answer by reference to the first page of this chapter. (b) Classify the elements in terms of metallic and non-metallic behaviour. (c) Give a *general* notation showing the ground state electronic configuration of each element.

13.2 Using the data in Table 13.1, draw a potential diagram for Tl and determine the value of $E^\circ(Tl^{3+}/Tl^+)$.

13.3 Plot a graph to show the variation in values of IE_1, IE_2 and IE_3 for the group 13 elements (Table 13.1), and plot a similar graph to show the variation in values of IE_1 and IE_2 for the group 2 metals (Table 12.1). Account for differences in trends of IE_2 for the group 2 and 13 elements.

13.4 Write equations for the following processes, involved in the extraction of the elements from their ores:

(a) the reduction of boron oxide by Mg;
(b) the result of the addition of hot aqueous NaOH to a mixture of solid Al_2O_3 and Fe_2O_3;
(c) the reaction of CO_2 with aqueous $Na[Al(OH)_4]$.

13.5 Predict the following NMR spectra: (a) the ^{11}B NMR spectrum of $[BH_4]^-$; (b) the ^{1}H NMR spectrum of $[BH_4]^-$; (c) the ^{11}B NMR spectrum of the adduct $BH_3 \cdot PMe_3$; (d) the $^{11}B\{^1H\}$ NMR spectrum of THF·BH_3. [^{1}H, 100%, $I = \frac{1}{2}$; ^{31}P, 100%, $I = \frac{1}{2}$; ^{11}B, 80.4%, $I = \frac{3}{2}$; ignore ^{10}B.]

13.6 The thermite process is shown in eq. 13.5. Determine $\Delta_r H^\circ$ for this reaction if $\Delta_f H^\circ(Al_2O_3, s, 298\,K)$ and $\Delta_f H^\circ(Fe_2O_3, s, 298\,K) = -1675.7$ and -824.2 kJ mol^{-1}, and comment on the relevance of this value to that of $\Delta_{fus} H(Fe, s) = 13.8$ kJ mol^{-1}.

13.7 Explain how, during dimerization, each BH_3 molecule acts as both a Lewis base and a Lewis acid.

13.8 Describe the bonding in Ga_2H_6 and Ga_2Cl_6, both of which have structures of the type shown in **13.50**.

X = H or Cl

(13.50)

13.9 The ordering of the relative stabilities of adducts L·BH_3 for some common adducts is, according to L: $Me_2O < THF < Me_2S < Me_3N < Me_3P < H^-$. In addition to answering each of the following, indicate how you could use NMR spectroscopy to confirm your proposals.

(a) What happens when Me_3N is added to a THF solution of THF·BH_3?
(b) Will Me_2O displace Me_3P from $Me_3P \cdot BH_3$?
(c) Is $[BH_4]^-$ stable in THF solution with respect to a displacement reaction?
(d) Suggest what may be formed when $Ph_2PCH_2CH_2PPh_2$ is added to a THF solution of THF·BH_3, the latter remaining in excess.

13.10 (a) One gallium-containing product, **A**, was obtained from the following reaction, carried out in Et_2O solvent:

The room temperature, solution ^{1}H NMR spectrum of **A** showed the following signals: δ 4.90 (s, 2H), 3.10 (t, 4H), 2.36 (t, 4H), 2.08 (s, 12H) ppm, and the ^{13}C NMR spectrum exhibited three signals at δ 61.0, 50.6 and 45.7 ppm. The highest mass peak in the mass spectrum of **A** is $m/z = 230$. Structural data for **A** reveal that the Ga atom is 5-coordinate. Suggest the likely identity of **A**, and propose its structure. (b) Compound **A** can function as a Lewis base. Rationalize why this is the case, and suggest a product for the reaction of **A** with $Me_3N \cdot GaH_3$.

13.11 The reaction of $K[B(CN)_4]$ with ClF_3 in liquid HF leads to the formation of $K[B(CF_3)_4]$. Explain why, in the ^{11}B NMR spectrum of this salt, a 13-line pattern is observed. What will be the relative intensities of the middle and outside lines of this multiplet?

13.12 The solvolysis of $K[B(CF_3)_4]$ in concentrated H_2SO_4 generates $(F_3C)_3BCO$. (a) Write a balanced equation for the solvolysis process. (b) In the gas phase, $(F_3C)_3BCO$ possesses C_3 rather than C_{3v} symmetry. Rationalize this observation, and draw a structure for the molecule which is consistent with the C_3 point group.

13.13 Suggest explanations for the following facts.

(a) $Na[BH_4]$ is very much less rapidly hydrolysed by H_2O than is $Na[AlH_4]$.

(b) The rate of hydrolysis of B_2H_6 by water vapour is given by the equation:

$$Rate \propto (P_{B_2H_6})^{\frac{1}{2}}(P_{H_2O})$$

(c) A saturated aqueous solution of boric acid is neutral to the indicator bromocresol green (pH range 3.8–5.4), and a solution of $K[HF_2]$ is acidic to this indicator. When excess boric acid is added to a solution of $K[HF_2]$, the solution becomes alkaline to bromocresol green.

13.14 Suggest likely products for the following reactions:

(a) $BCl_3 + EtOH \longrightarrow$
(b) $BF_3 + EtOH \longrightarrow$
(c) $BCl_3 + PhNH_2 \longrightarrow$
(d) $BF_3 + KF \longrightarrow$

13.15 (a) Write down the formula of cryolite. (b) Write down the formula of perovskite. (c) Cryolite is described as possessing a 3-dimensional structure closely related to that of perovskite. Suggest how this is possible when the stoichiometries of the two compounds do not appear to be compatible.

13.16 (a) Suggest structures for $[MBr_6]^{3-}$, $[MCl_5]^{2-}$ and $[MBr_4]^-$ (M = Ga or In). (b) In the salt $[Et_4N]_2[InCl_5]$, the anion has a square-based pyramidal structure, as does $[TlCl_5]^{2-}$ in the salt $[H_3N(CH_2)_5NH_3][TlCl_5]$. Comment on these observations in the light of your answer to part (a). (c) Suggest methods of preparing $[H_3N(CH_2)_5NH_3][TlCl_5]$ and $Cs_3[Tl_2Cl_9]$. (d) Explain how magnetic data enable you to distinguish between the formulations $GaCl_2$ and $Ga[GaCl_4]$ for gallium dichloride.

13.17 Comment on each of the following observations.

(a) AlF_3 is almost insoluble in anhydrous HF, but dissolves if KF is present. Passage of BF_3 through the resulting solution causes AlF_3 to reprecipitate.

(b) The Raman spectra of germanium tetrachloride, a solution of gallium trichloride in concentrated hydrochloric acid, and fused gallium dichloride contain the following lines:

	Absorption / cm^{-1}			
GeCl$_4$	134	172	396	453
GaCl$_3$/HCl	114	149	346	386
GaCl$_2$	115	153	346	380

(c) When TlI$_3$, which is isomorphous with the alkali metal triiodides, is treated with aqueous NaOH, hydrated Tl_2O_3 is quantitatively precipitated.

13.18 Figure 13.11c shows the solid state structure of the $[Al(BH_4)_4]^-$ ion, present in $[Ph_3MeP][Al(BH_4)_4]$. In the light of these structural data, account for the following observations, recorded for the compound *in solution*. (a) At 298 K, the ^1H NMR spectrum of $[Ph_3MeP][Al(BH_4)_4]$ shows one broad signal in addition to signals assigned to the cation; this pattern of signals is retained at 203 K. (b) In the ^{11}B NMR spectrum (298 K) of the same compound, a quintet is observed. (c) In the IR spectrum of $[Ph_3MeP][Al(BH_4)_4]$, absorptions due to bridging $Al-H-B$ and terminal $B-H$ interactions are both observed.

13.19 Figure 13.20 shows four hydrogen-bonded molecules of $B_2(OH)_4$. To what point group does a *single molecule* of $B_2(OH)_4$ belong?

13.20 (a) The behaviour of H_3BO_3 in aqueous solution is not typical of a mineral acid such as HCl or H_2SO_4. Illustrate, using appropriate examples, these differing behaviours. (b) The formula of borax is sometimes written as $Na_2B_4O_7 \cdot 10H_2O$. Comment critically on this representation.

13.21 Compare the physical and chemical properties of α- and γ-alumina, choosing examples that highlight why it is important not to call Al_2O_3 simply 'alumina'.

13.22 (a) Suggest products for the following reactions.

$$3EtB(OH)_2 \xrightarrow{-3H_2O}$$

$$ClB(NMe_2)_2 \xrightarrow{2Na}$$

$$K[(C_2F_5)_3BF] + SbF_5 \longrightarrow$$

(b) $PhB(OH)_2$ forms dimers in the solid state. Dimers further associate into a 3-dimensional network. Describe how this assembly is likely to arise.

13.23 Write a brief account of the bonding and reactivity of borazine which emphasizes the ways in which this compound is similar or dissimilar to benzene.

13.24 Give appropriate bonding descriptions for the aluminium–nitrogen compounds depicted in Fig. 13.25.

13.25 $GaCl_3$ reacts with $KP(H)Si^tBu_3$ (equimolar amounts) to give KCl and two isomers of a 4-membered, cyclic compound which contains 38.74% C, 7.59% H and 19.06% Cl. Suggest the identity of the product, and draw structural diagrams to illustrate the isomerism.

13.26 Use Wade's rules to suggest likely structures for B_5H_9, $[B_8H_8]^{2-}$, $C_2B_{10}H_{12}$ and $[B_6H_9]^-$. Are any cage-isomers possible?

13.27 (a) Two-electron reduction of B_5H_9 followed by protonation is a convenient route to B_5H_{11}. What structural change (and why) do you expect the B_5 cage to undergo during this reaction?

(b) Account for the fact that the solution ^{11}B NMR spectrum of $[B_3H_8]^-$ (**13.40**) exhibits one signal which is a binomial nonet.

(c) The photolysis of B_5H_9 leads to the formation of a mixture of three isomers of $B_{10}H_{16}$. The products arise from the intermolecular elimination of H_2. Suggest the nature of the product, and the reason that three isomers are formed.

13.28 Suggest likely products for the following reactions, with the stoichiometries stated:

(a) $B_5H_9 + Br_2 \xrightarrow{298\ K}$

(b) $B_4H_{10} + PF_3 \longrightarrow$

(c) $1\text{-}BrB_5H_8 \xrightarrow{KH,\ 195\ K}$

(d) $2\text{-}MeB_5H_8 \xrightarrow{ROH}$

13.29 Crystalline $Ag_2[B_{12}Cl_{12}]$ may be described as having a structure based on an anti-fluorite-type arrangement. By approximating each $[B_{12}Cl_{12}]^{2-}$ ion to a sphere, draw a diagram to represent a unit cell of $Ag_2[B_{12}Cl_{12}]$. What type of interstitial hole does each Ag^+ ion occupy in this idealized structure?

OVERVIEW PROBLEMS

13.30 (a) Write balanced equations for the reactions of aqueous Ga^+ with $[I_3]^-$, Br_2, $[Fe(CN)_6]^{3-}$ and $[Fe(bpy)_3]^{3+}$.

(b) The ^{205}Tl NMR spectrum of an acidic solution that contains Tl^{3+} and ^{13}C-enriched $[CN]^-$ ions in concentrations of 0.05 and 0.31 mol dm^{-3} respectively shows a binomial quintet ($\delta\ 3010$ ppm, $J = 5436$ Hz) and quartet ($\delta\ 2848$ ppm, $J = 7954$ Hz). Suggest what species are present in solution and rationalize your answer. (See Table 4.3 for nuclear spin data.)

13.31 (a) Comment why, in Fig. 13.1, the data are presented on a logarithmic scale. What are the relative abundances of Al (Fig. 13.1) and Mg (Fig. 12.2) in the Earth's crust?

(b) Show that the changes in oxidation states for elements undergoing redox changes in reaction 13.18 balance.

(c) The ion $[B_3N_6]^{9-}$ in $La_5(BN_3)(B_3N_6)$ possesses a chair conformation with each B atom being in an approximately trigonal planar environment (see structure **13.26**); B−N bond lengths in the ring are 148 pm, and the exocyclic B−N bond lengths average 143 pm. Draw a set of resonance structures for $[B_3N_6]^{9-}$, focusing on

those structures that you consider will contribute the most to the overall bonding. Comment on the structures you have drawn in the light of the observed structure of the ion in crystalline $La_5(BN_3)(B_3N_6)$.

13.32 (a) NMR spectroscopic data for $[HAl(BH_4)_2]_n$ are consistent with the compound existing in two forms in solution. One form is probably a dimer and the other, a higher oligomer. Each species possesses one boron environment, and in the ^{11}B NMR spectrum, each species exhibits a binomial quintet. The chemical shift of the signal for each species in the ^{27}Al NMR spectrum suggests an octahedral environment for the Al atom. Suggest a structure for the dimer $[HAl(BH_4)_2]_2$ which is consistent with these observations and comment on whether the data indicate a static or dynamic molecule.

(b) The elemental analysis for an adduct **A** is 15.2% B, 75.0% Cl, 4.2% C and 5.6% O. The ^{11}B NMR spectrum of **A** contains two singlets ($\delta\ -20.7$ and $+68.9$ ppm) with relative integrals 1:3; the signal at $\delta\ -20.7$ ppm is characteristic of a B atom in a tetrahedral environment, while that at $\delta\ +68.9$ ppm is consistent with trigonal planar boron. In the IR spectrum, there is a characteristic absorption at $2176 \, cm^{-1}$. Suggest an identity for **A** and draw its structure.

13.33 (a) What type of semiconductors are formed by doping silicon with boron or gallium? Using simple band theory, explain how the semiconducting properties of Si are altered by doping with B or Ga.

(b) An active area of research within the field of Ga^{3+} and In^{3+} coordination chemistry is the search for complexes suitable for use as radiopharmaceuticals. Suggest how ligands **13.51** and **13.52** are likely to coordinate to Ga^{3+} and In^{3+} respectively.

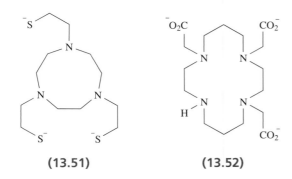

(13.51)   **(13.52)**

13.34 (a) At 297 K, the ^{11}B NMR spectrum of a CD_2Cl_2 solution of $[Ph_4As][B_6H_7]$ shows one doublet ($\delta -18.0$ ppm, $J = 147$ Hz). In the ^1H NMR spectrum, two signals are observed ($\delta -5.5$ ppm, broad; $\delta +1.1$ ppm, $1:1:1:1$ quartet). At 223 K, the ^{11}B NMR spectrum exhibits signals at $\delta -14.1$ and -21.7 ppm (relative integrals $1:1$). Lowering the temperature has little effect on the ^1H NMR spectrum. Draw the solid state structure of $[B_6H_7]^-$ and rationalize the solution NMR spectroscopic data.

(b) The reaction of Ga metal with NH_4F at 620 K liberates H_2 and NH_3 and yields an ammonium salt **X** in which gallium is in oxidation state $+3$. The solid state structure of **X** consists of discrete cations lying between sheets composed of vertex-sharing GaF_6-octahedra; sharing of vertices occurs only in one plane. Suggest an identity for **X**. Write a balanced equation for the reaction of Ga and NH_4F to give **X**. Explain with the aid of a diagram how the stoichiometry of **X** is maintained in the solid state structure.

13.35 Comment on the following statements:

(a) World gallium production increased from 11 t in 1975 to 111 t in 2008.
(b) Rubies are composed of α-Al_2O_3 (corundum) but are red in colour.
(c) The adduct $H_3N \cdot BH_3$ has possible potential as a hydrogen storage material.

13.36 Glass is manufactured by cooling a melt to produce a rigid structure without crystallization. The binary oxides B_2O_3 and SiO_2 are the primary components of borosilicate glass. Possible structural units include $B(\mu\text{-}O)_4$, $B(\mu\text{-}O)_3$, $B(O_t)_3$, $Si(\mu\text{-}O)_4$ and $Si(O_t)_4$ where μ-O and O_t are bridging O atoms and terminal O^- units, respectively. (a) Draw diagrams to show the structures of these units and how they might interconnect where appropriate, paying attention to atomic charges. Comment on the bonding in each unit. (b) Suggest other Si- or B-centred structural units that might be present in borosilicate glass. (b) Na_2O is added to glass as a modifier. Suggest what role it plays when added to a borosilicate glass containing approximately 70% SiO_2, 14% B_2O_3, 10% Na_2O and 6% Al_2O_3, by weight.

13.37 The commercial applications of boron nitride include those as an electrical insulator, a lubricant, an abrasive, a material for making crucibles for high temperature work (e.g. moulds for molten steel) and in cutting tools. Explain why BN has such diverse applications.

You should also try problem 7.36.

Topics

14

The group 14 elements

1	2		13	**14**	15	16	17	18
H								He
Li	Be		B	**C**	N	O	F	Ne
Na	Mg		Al	**Si**	P	S	Cl	Ar
K	Ca	*d*-block	Ga	**Ge**	As	Se	Br	Kr
Rb	Sr		In	**Sn**	Sb	Te	I	Xe
Cs	Ba		Tl	**Pb**	Bi	Po	At	Rn
Fr	Ra							

14.1 Introduction

The elements in group 14 – carbon, silicon, germanium, tin and lead – show a gradation from C, which is non-metallic, to Pb, which, though its oxides are amphoteric, is mainly metallic in nature. The so-called '*diagonal line*' through the *p*-block separates metallic from non-metallic elements and passes between Si and Ge, indicating that Si is non-metallic and Ge is metallic. However, this distinction is not definitive. In the solid state, Si and Ge possess a covalent diamond-type structure (see Fig. 6.20a), but their electrical resistivities (see Section 6.8) are significantly lower than that of diamond, indicating metallic behaviour. Silicon and germanium are classed as *semi-metals*[†] and we have already discussed their semi-conducting properties (see Section 6.9).

[†] Under IUPAC recommendations, the term 'semi-metal' is preferred over 'metalloid'.

All members of group 14 exhibit an oxidation state of $+4$, but the $+2$ oxidation state increases in stability as the group is descended. Carbenes exemplify the C(II) state but exist only as reaction intermediates, silicon dihalides are stable only at high temperatures, the Ge(II) and Sn(II) states are well established, and Pb(II) is more stable than the Pb(IV) state. In this respect, Pb resembles its periodic neighbours, Tl and Bi, with the inertness of the 6*s* electrons being a general feature of the last member of each of groups 13, 14 and 15 (see Box 13.4).

Carbon is essential to life on Earth (see Box 14.7), and most of its compounds lie within the remit of organic chemistry. Nonetheless, compounds of C that are formally classified as 'inorganic' abound and extend to organometallic species (see Chapters 23 and 24).

14.2 Occurrence, extraction and uses

Occurrence

Figure 14.1 illustrates the relative abundances of the group 14 elements in the Earth's crust. The two long-established crystalline allotropes of carbon, diamond and graphite, occur naturally, as does amorphous carbon (e.g. in coal). Diamonds occur in igneous rocks (e.g. in the Kimberley volcanic pipes, South Africa). Carbon dioxide constitutes only 0.04% of the Earth's atmosphere, and, although vital for photosynthesis, CO_2 is not a major source of carbon. During the 1990s, it was discovered that molecular allotropes of carbon, the *fullerenes* (see Section 14.4), occur naturally in a number of deposits in Australia, New Zealand and North America. Soot contains fullerenes and related carbon species (nanotubes, concentric fullerenes,

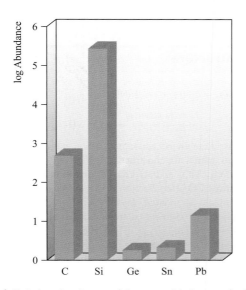

Fig. 14.1 Relative abundances of the group 14 elements in the Earth's crust. The data are plotted on a logarithmic scale. The units of abundance are parts per million (ppm).

open hemi-shells). The formation of soot under fuel-rich conditions involves growth of polycyclic aromatic hydrocarbons which aggregate to form particles. The current development of the chemistry of fullerenes and carbon nanotubes relies, however, on their laboratory synthesis.

Elemental Si does not occur naturally, but it constitutes 27.7% of the Earth's crust (Si is the second most abundant element after O, Fig. 13.2) in the form of sand, quartz, rock crystal, flint, agate and silicate minerals (see Section 14.9). In contrast, Ge makes up only 1.8 ppm of the Earth's crust, being present in trace amounts in a range of minerals (e.g. zinc ores) and in coal. The principal tin-bearing ore is *cassiterite* (SnO_2). Important ores of lead are *galena* (PbS), *anglesite* ($PbSO_4$) and *cerussite* ($PbCO_3$).

Extraction and manufacture

Sources of natural graphite are supplemented by manufactured material formed by heating powdered coke (high-temperature carbonized coal) with silica at ≈ 2800 K. Approximately 30% of diamonds for industrial use in the US are synthetic (see Box 14.4). Diamond films may be grown using a chemical vapour deposition method (see Section 28.6), and hydrothermal processes are currently being investigated.[†] The manufacture of amorphous carbon (carbon black, used in synthetic rubber) involves burning oil in a limited supply of air.

Silicon (not of high purity) is extracted from silica, SiO_2, by heating with C or CaC_2 in an electric furnace. Impure Ge

---

[†] See, for example: X.-Z. Zhao, R. Roy, K.A. Cherian and A. Badzian (1997) *Nature*, vol. 385, p. 513; R.C. DeVries (1997) *Nature*, vol. 385, p. 485; N. Yamasaki, Y. Yamasaki, K. Tohji, N. Tsuchiya, T. Hashida and K. Ioku (2006) *J. Mater. Sci.*, vol. 41, p. 1599.

can be obtained from flue dusts collected during the extraction of zinc from its ores, or by reducing GeO_2 with H_2 or C. For use in the electronic and semiconductor industries, ultra-pure Si and Ge are required, and both can be obtained by zone-melting techniques (see Box 6.3 and Section 28.6).

Tin is obtained from *cassiterite* (SnO_2) by reduction with C in a furnace (see Section 8.8), but a similar process cannot be applied to extract Pb from its sulfide ore because $\Delta_f G^\circ(CS_2, g)$ is $+67$ kJ mol^{-1}; thermodynamically viable processes involve reactions 14.1 or 14.2 at high temperatures. Both Sn and Pb are refined electrolytically. Recycling of Sn and Pb takes place on a huge scale. In the US in 2009, 1.12 Mt of secondary Pb (i.e. from recycled Pb, mainly from used lead–acid batteries) was produced and this accounted for about 80% of Pb consumed in the US.

$$\left.\begin{array}{l} 2PbS + 3O_2 \longrightarrow 2PbO + 2SO_2 \\ PbO + C \longrightarrow Pb + CO \\ \text{or} \\ PbO + CO \longrightarrow Pb + CO_2 \end{array}\right\} \qquad (14.1)$$

$$PbS + 2PbO \longrightarrow 3Pb + SO_2 \qquad (14.2)$$

Uses

Diamond is the hardest known substance, and apart from its commercial value as a gemstone, it has applications in cutting tools and abrasives (see Box 14.4). Whereas diamond has a rigid 3-dimensional structure, graphite possesses a layer structure (see Section 14.4), and this results in a remarkable difference in physical properties and applications. The properties of graphite that are exploited commercially (Fig. 14.2) are its inertness, high thermal stability, electrical and thermal conductivities (which are direction-dependent, see Section 14.4) and ability to act as a lubricant. Its thermal and electrical properties make graphite suitable as a refractory material (see Section 12.6) and for uses in batteries and fuel cells. The growing importance of fuel-cell technology has resulted in a growth in demand for high-purity graphite. Other new technologies are having an impact on the market for graphite. Graphite cloth ('flexible graphite') is a relatively

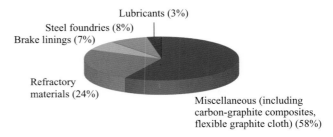

Fig. 14.2 Uses of natural graphite in the US in 2009. [Data: US Geological Survey.]

APPLICATIONS

Box 14.1 Activated charcoal: utilizing a porous structure

Activated charcoal is a finely divided form of carbon and is manufactured from organic materials (e.g. peat, wood) by heating in the presence of reagents that promote both oxidation and dehydration. Although the structure of activated charcoal is based upon that of graphite, it is highly irregular and contains non-graphitic carbon, extensive cross-linking and heteroatoms (e.g. H, N and O). The result is a pore structure with a large internal surface area: *microporous* materials exhibit pores <2 nm wide, *macroporous* refers to activated charcoals with a pore size >50 nm, and *mesoporous* materials fall in between these extremes. The largest internal surface areas are found for microporous materials ($>700 \, m^2 \, g^{-1}$). The surface of activated charcoal is hydrophobic and adsorption of small molecules occurs largely through van der Waals interactions. Non-polar molecules are adsorbed in preference to polar ones, although the adsorption properties of the charcoal are affected by the number and types of heteroatoms bound to the surface. This, in turn, reflects the method of manufacture.

Activated charcoal has widespread applications both for gas- and liquid-phase adsorptions. Gas-phase applications (adsorption of gas molecules including organic vapours) usually require surface areas of 1000 to $2000 \, m^2 \, g^{-1}$. In addition to gas purification, activated charcoal is used in the recovery of volatile organic solvents (e.g. acetone, ethanol, toluene, chlorinated hydrocarbons) and in the paints, inks, synthetic fibres, rubbers and adhesives industries.

Early large-scale applications of activated charcoal were in gas masks in World War I. Various gas-filters, including those in cooker extractors and mobile or bench-top laboratory fume-hoods, contain activated charcoal filters. About 20% of the activated charcoal that is produced is consumed in the sugar industry, where it is used as a decolouring agent. Liquid-phase applications are dominated by water purification. After removal of particulate matter, water from rivers or reservoirs is treated with ozone to kill bacteria. It then passes into rapid gravity filter tanks which contain gravel covered with a bed of granular activated charcoal. Once adsorption of contaminants including O_3 is complete, the water may be chlorinated before distribution as drinking water.

The porous structure means that activated charcoal is an excellent heterogeneous catalyst, especially when impregnated with a *d*-block metal such as palladium. On an industrial scale, it is used, for example, in the manufacture of phosgene and sulfuryl dichloride:

$$CO + Cl_2 \xrightarrow{\text{activated charcoal}} \underset{\text{phosgene}}{COCl_2}$$

$$SO_2 + Cl_2 \xrightarrow{\text{activated charcoal}} \underset{\text{sulfuryl dichloride}}{SO_2Cl_2}$$

In the laboratory, activated charcoal has catalytic uses, for example in the aerial oxidation of cobalt(II) in the presence of NH_3 and NH_4Cl to give the octahedral cobalt(III) complex $[Co(NH_3)_6]^{3+}$:

$$4CoCl_2 + O_2 + 4[NH_4]Cl + 20NH_3$$
$$\xrightarrow{\text{activated charcoal}} 4[Co(NH_3)_6]Cl_3 + 2H_2O$$

The porous skeleton of activated carbon can be used as a template on which to construct other porous materials, for example, SiO_2, TiO_2 and Al_2O_3. The oxide is first dissolved in supercritical CO_2 (see Section 9.13) and then the activated carbon template is coated in the supercritical fluid. The carbon template is removed by treatment with oxygen plasma or by calcination in air at 870 K, leaving a nanoporous ('nano' refers to the scale of the pore size) metal oxide with a macroporous structure that mimics that of the activated carbon template.

Activated charcoal is routinely used for treatment after the ingestion of poisons, especially in children. It is typically administered in single or multiple doses, each of 50 g and acts in the gut, adsorbing toxins.

Rapid gravity filter tanks at the River Itchen water treatment works in Southampton, UK.

Further reading

R.H. Bradley (2011) *Adsorp. Sci. Technol.*, vol. 29, p. 1 – 'Recent developments in the physical adsorption of toxic organic vapours by activated carbons'.

A.J. Evans (1999) *Chemistry & Industry*, p. 702 – 'Cleaning air with carbon'.

H. Wakayama, H. Itahara, N. Tatsuda, S. Inagaki and Y. Fukushima (2001) *Chemistry of Materials*, vol. 13, p. 2392 – 'Nanoporous metal oxides synthesized by the nanoscale casting process using supercritical fluids'.

D.A. Warrell (2009) *Trans. R. Soc. Trop. Med. Hyg.*, vol. 103, p. 860 – 'Researching nature's venoms and poisons'.

new product and applications are increasing. In 2004, Novoselov and Geim (awarded the Nobel Prize for Physics in 2010) succeeded in isolating a single layer of graphite. This one-atom thick sheet is called *graphene*. The high speed at which charge carriers move through graphene, its mechanical strength and optical transparency make this an exciting material, properties of which are discussed in Section 28.8.

Carbon black is of huge commercial importance and is manufactured by the partial combustion of hydrocarbons (e.g. natural gas, petroleum) under controlled conditions. Carbon black has a well-defined morphology and is composed of aggregates of particles with a graphitic microstructure. The major application of carbon black is in the reinforcement of vulcanized rubber, and about 66% ends up in vehicle tyres. Other important uses are in printing inks, paints and plastics. Charcoal (made by heating wood) and animal charcoal (produced by charring treated bones) are microcrystalline forms of graphite, supported, in the case of animal charcoal, on calcium phosphate. The adsorption properties of *activated charcoal* render it commercially important (Box 14.1). Carbon fibres of great tensile strength (formed by heating oriented organic polymer fibres at ≥ 1750 K) contain graphite crystallites oriented parallel to the fibre axis, and are used to strengthen materials such as plastics. Carbon-composites are fibre-reinforced, chemically inert materials which possess high strength, rigidity, thermal stability, high resistance to thermal shock and retain their mechanical properties at high temperature. Carbon fibre materials are discussed in detail in Section 28.7.

Silicon has major applications in the steel industry (see Box 6.1) and in the electronic and semiconductor industries (Box 14.2). Silica, SiO_2, is an extremely important commercial material. It is the main component of glass, and large quantities of sand are consumed worldwide by the building industry. Quartz glass (formed on cooling fused SiO_2) can withstand sudden temperature changes and has specialist uses. Different types of glasses are described in Section 14.9. Silica gel (an amorphous form of silica, produced by treating aqueous sodium silicate with acid) is used as a drying agent (see Box 12.3), a stationary phase in chromatography, and a heterogeneous catalyst. Alkali metals may be absorbed into silica gel, giving a convenient means of handling the metals prior to use as reducing agents (see Section 11.4). *Caution*! Inhalation of silica dusts may lead to the lung disease *silicosis*. Hydrated silica forms the exoskeletons of marine diatoms, but the role of Si in other biological systems is less well defined.[†] The applications of silicates and aluminosilicates are discussed in Section 14.9.

[†] See: J.D. Birchall (1995) *Chem. Soc. Rev.*, vol. 24, p. 351 – 'The essentiality of silicon in biology'.

The commercial demand for Ge is small but significant. Its major uses are in germanium-based polymerization catalysts for the production of polyethylene terephthalate (PET), fibre optics, infrared (night vision) optical devices, and the electronics and solar electrical industries. Applications in optical devices arise from the optical properties of GeO_2. More than 60% of the Ge used in optical devices is recycled. About 54 000 kg of Ge was used in the US in 2008. Compared with this, the demand for Sn and Pb is much greater. In 2008, the US consumed 59 000 t of Sn (primary and recycled) and 1.5 Mt of Pb (primary and recycled). Tin-plating of steel cans improves corrosion resistance and is a major use of Sn. The metal is, however, soft and tin alloys such as pewter, soldering metal, bronze and die-casting alloy have greater commercial value than pure Sn. High-quality window glass is usually manufactured by the Pilkington process which involves floating molten glass on molten tin to produce a flat surface. Tin dioxide is an opacifier used in enamels and paints (also see Section 28.4), and its applications in gas sensors are described in Box 14.11. The use of tin-based chemicals as flame retardants (see Box 17.1) is increasing in importance.

Lead is a soft metal and has been widely used in the plumbing industry, but this use has diminished as awareness of the toxicity of the metal has grown (Box 14.3). Similarly, uses of Pb in paints have been reduced, and 'environmentally friendly' lead-free fuels have replaced leaded counterparts (Fig. 14.3). Lead oxides are of great commercial importance, e.g. in the manufacture of 'lead crystal' glass. *Red lead*, Pb_3O_4, is used as a pigment and a corrosion-resistant coating for steel and iron. By far the greatest demand for lead is in lead–acid batteries. The cell reaction is a combination of half-reactions 14.3 and 14.4; a normal

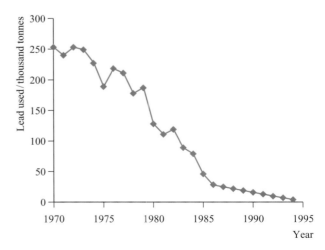

Fig. 14.3 The declining use of leaded fuels in motor vehicles is illustrated by these statistics from the US. In 1970, lead additives to fuel were at a peak. [Data: US Geological Survey.]

ENVIRONMENT

Box 14.2 Solar power: thermal and electrical

Harnessing renewable energy from the Sun is an environmentally acceptable method of producing power. Conversion via solar heat exchange panels provides thermal energy to raise the temperature of swimming pools or to provide domestic hot water. Conversion via photovoltaic systems (solar cells) produces electricity and involves the use of semiconductors. Initially, NASA's space programme was the driving force behind the development of solar cells, and applications in satellites and other space vessels remain at the cutting edge of design technology. However, we all now feel the benefits of solar cells which are used in items such as solar-powered calculators. Most solar cells are fabricated using silicon. Other semiconductors include GaAs and CdTe, but Si-based cells dominate the commercial market. They comprise solid state junction devices which are typically 200–350 µm thick. The device is constructed from an n-doped layer (which faces the sun), a p-doped layer and a metal-contact grid on the top and bottom surfaces as shown in the diagram below. The contact grids are connected by a conducting wire. When light falls on the cell, electrons at the n–p junction move from the p-type to the n-type silicon, and 'holes' (see Section 6.9) move in the opposite direction; this leads to a flow of electricity around the circuit. A typical solar cell comprises a silicon wafer of about $100\,cm^2$ surface area. The power output per cell is $\leq 0.6\,V$, and this voltage is too small to be practically useful. Cells are therefore combined in series to create a module which has an output of $\approx 12\,V$. The output is dependent upon weather conditions and number of daylight hours, and a typical module contains between 28 and 36 solar cells. For some applications, a single solar module is sufficient. For higher voltage needs, modules are connected in series to produce photovoltaic panels such as those shown in the photograph. These are commonly called solar panels, but should not be confused with the solar heat exchange panels mentioned above. Photovoltaic panels may provide stand-alone power sources or can be connected into a supply grid for storage and distribution of electricity. The efficiency of silicon-based photovoltaic cells is $\geq 20\%$. To improve efficiency, hybrid systems have been developed that combine photovoltaic technology with, for example, wind turbines.

The Grätzel cell is a new generation of solar cell that relies upon a wide band gap semiconductor (usually nanocrystalline TiO_2) in contact with an electrolyte. However, TiO_2 is optically transparent. Thus, a dye which absorbs light in the visible region is anchored onto the surface of the semiconductor,

Solar photovoltaic energy panels in Provence, France.

and when illuminated, the dye injects an electron into the conduction band of the solid. The chemistry behind these *dye-sensitized solar cells* is described in Section 28.3. Photovoltaic devices that utilize conducting organic polymers are another challenge to silicon-based solar cells. The importance of electrically conductive polymers was recognized by the award of the 2000 Nobel Prize in Chemistry to Heeger, MacDiarmid and Shirakawa. Solar cells fabricated using organic polymers are cheaper, lighter and more flexible than solid-state junction devices, but (as of 2011) they are less efficient at converting solar into electrical energy. Research efforts are continuing to improve the efficiencies of inorganic dye-sensitized solar cells and organic photovoltaics.

Further reading

J.-L. Bredas and J.R. Durrant, eds. (2009) *Acc. Chem. Res.*, vol. 42, issue 11 – A special issue containing a series of reviews on the topic of organic photovoltaics.

R. Eisenberg and D.G. Nocera, eds. (2005) *Inorg. Chem.*, vol. 44, p. 6799 – A series of cutting-edge papers published as a 'Forum on Solar and Renewable Energy'.

M. Grätzel (2003) *J. Photochem. Photobiol C*, vol. 4, p. 145 – 'Dye-sensitized solar cells'.

M.A. Green (2001) *Adv. Mater.*, vol. 13, p. 1019 – 'Crystalline silicon photovoltaic cells'.

G.J. Meyer (2010) *ACS Nano*, vol. 4, p. 4337 – 'The 2010 Millennium Technology Grand Prize: Dye-sensitized solar cells'.

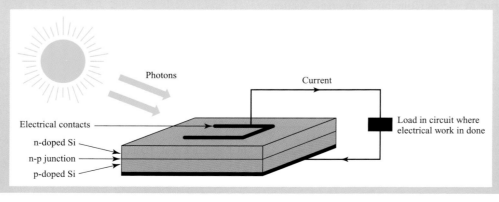

ENVIRONMENT

Box 14.3 Toxicity of lead

Lead salts are extremely toxic. The ingestion of a soluble lead salt can cause acute poisoning, and long-term exposure to a source of the metal (e.g. old water pipes, Pb-based paints) may result in chronic poisoning. Organolead(IV) compounds such as Et_4Pb, used as an anti-knock additive to leaded motor fuels, attack the nervous system. In a relevant piece of research, analysis of wines produced between 1962 and 1991 from grapes grown in roadside vineyards has shown some correlation between a decrease in Pb content and the introduction of unleaded fuels. Sequestering agents such as $[EDTA]^{4-}$ (see eq. 7.75 and accompanying text) are used to complex Pb^{2+} ions in the body, and their removal follows by natural excretion.

Joints between metals, including those in electronic components, have traditionally used SnPb solders. However, in the European Union, new environmental legislation banning the inclusion of lead, cadmium, mercury, hexavalent chromium and polybrominated flame retardants (see Box 17.1) in new electrical and electronic equipments came into force in 2006. The ban on lead-containing components has several exceptions which include lead in high-melting solders (Sn–Pb alloys containing >85% Pb) and lead in solders for network infrastructure equipment for signalling and transmission. These exemptions will be reviewed with the expectation that replacement materials will eventually be used. The introduction of the legislation has had a huge impact on the soldering of electronic components. Eutectic SnPb solder exhibits many desirable properties (e.g. low melting, easily worked and inexpensive) and it is a challenge for research and development initiatives to find alloys for lead-free solders that replicate these properties. Solders based on Sn with Ag, Bi, Cu and Zn as alloying metals are the most promising candidates. Of these SnAgCu (3–4% by weight of Ag and 0.5–0.9% by weight of Cu) solders are the most common replacement in the electronics industry. However, an SnAgCu solder is ≈ 2.5 times more expensive than an SnPb eutectic. A cheaper alternative is an SnCu solder, but its disadvantage is that it melts at a higher temperature than the SnAgCu-based materials.

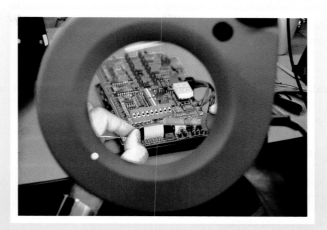

Soldering the pins of a chip onto an electronic circuit board.

Further reading

H. Black (2005) *Chem. Ind.*, issue 21, p. 22 – 'Lead-free solder'.

R.A. Goyer (1988) in *Handbook on Toxicity of Inorganic Compounds*, eds. H.G. Seiler, H. Sigel and A. Sigel, Marcel Dekker, New York, p. 359 – 'Lead'.

Y.-S. Lai, H.-M. Tong and K.-N. Tu, eds. (2009) *Microelectron. Reliab.*, vol. 49, issue 3 – A series of articles: 'Recent research advances in Pb-free solders'.

R. Lobinski *et al.* (1994) *Nature*, vol. 370, p. 24 – 'Organolead in wine'.

H. Ma and J.C. Suhling (2009) *J. Mater. Sci.*, vol. 44, p. 1141 – 'A review of mechanical properties of lead-free solders for electronic packaging'.

---

automobile 12 V battery contains six cells connected in series.

$$PbSO_4(s) + 2e^- \rightleftharpoons Pb(s) + [SO_4]^{2-}(aq) \qquad E^o = -0.36\,V \tag{14.3}$$

$$PbO_2(s) + 4H^+(aq) + [SO_4]^{2-}(aq) + 2e^-$$
$$\rightleftharpoons PbSO_4(s) + 2H_2O(l) \quad E^o = +1.69\,V \tag{14.4}$$

Lead–acid storage batteries are used not only in the automobile industry but also as power sources for industrial forklifts, mining vehicles and airport ground services, and for independent electrical power sources in, for example, hospitals.

14.3 Physical properties

Table 14.1 lists selected physical properties of the group 14 elements. A comparison with Table 13.1 shows there to be some similarities in *trends* down groups 13 and 14.

Ionization energies and cation formation

On descending group 14, the trends in ionization energies reveal two particular points:

- the relatively large increases between values of IE_2 and IE_3 for each element;
- the discontinuities (i.e. *increases*) in the trends of values of IE_3 and IE_4 at Ge and Pb.

The sums of the first four ionization energies for any element suggest that it is unlikely that M^{4+} ions are formed. For example, although both SnF_4 and PbF_4 are non-volatile solids, neither has a symmetrical 3-dimensional structure in the solid state. Both SnO_2 and PbO_2 adopt a rutile-type structure. Agreement between values of lattice energies determined using a Born–Haber cycle and calculated from an electrostatic model is good for SnO_2, but is poor for PbO_2. This suggests a degree of covalent bonding in PbO_2, rather than a formulation of $Pb^{4+}(O^{2-})_2$. Thus, values of the M^{4+} ionic radii (Table 14.1) should be treated with some caution.

Table 14.1 Some physical properties of the group 14 elements, M, and their ions.

Property	C	Si	Ge	Sn	Pb
Atomic number, Z	6	14	32	50	82
Ground state electronic configuration	$[He]2s^2 2p^2$	$[Ne]3s^2 3p^2$	$[Ar]3d^{10}4s^2 4p^2$	$[Kr]4d^{10}5s^2 5p^2$	$[Xe]4f^{14}5d^{10}6s^2 6p^2$
Enthalpy of atomization, $\Delta_a H^o(298\,K)\,/\,kJ\,mol^{-1}$	717	456	375	302	195
Melting point, mp / K	>3823[†]	1687	1211	505	600
Boiling point, bp / K	5100	2628	3106	2533	2022
Standard enthalpy of fusion, $\Delta_{fus}H^o(mp)\,/\,kJ\,mol^{-1}$	104.6	50.2	36.9	7.0	4.8
First ionization energy, $IE_1\,/\,kJ\,mol^{-1}$	1086	786.5	762.2	708.6	715.6
Second ionization energy, $IE_2\,/\,kJ\,mol^{-1}$	2353	1577	1537	1412	1450
Third ionization energy, $IE_3\,/\,kJ\,mol^{-1}$	4620	3232	3302	2943	3081
Fourth ionization energy, $IE_4\,/\,kJ\,mol^{-1}$	6223	4356	4411	3930	4083
Metallic radius, r_{metal} / pm	–	–	–	158	175
Covalent radius, r_{cov} / pm[‡]	77	118	122	140	154
Ionic radius, r_{ion} / pm[*]	–	–	53 (Ge^{4+})	74 (Sn^{4+})	78 (Pb^{4+})
				93 (Sn^{2+})	119 (Pb^{2+})
Standard reduction potential, $E^o(M^{2+}/M)$ / V	–	–	–	−0.14	−0.13
Standard reduction potential, $E^o(M^{4+}/M^{2+})$ / V	–	–	–	+0.15	+1.69[**]
NMR active nuclei (% abundance, nuclear spin)	^{13}C (1.1, $I=\frac{1}{2}$)	^{29}Si (4.7, $I=\frac{1}{2}$)	^{73}Ge (7.8, $I=\frac{9}{2}$)	^{117}Sn (7.6, $I=\frac{1}{2}$); ^{119}Sn (8.6, $I=\frac{1}{2}$)	^{207}Pb (22.6, $I=\frac{1}{2}$)

[†] For diamond.
[‡] Values for C, Si, Ge and Sn refer to diamond-type structures and thus refer to 4-coordination; the value for Pb also applies to a 4-coordinate centre.
[*] Values are for 6-coordination.
[**] This value is for the half-reaction: $PbO_2(s) + 4H^+(aq) + [SO_4]^{2-}(aq) + 2e^- \rightleftharpoons PbSO_4(s) + 2H_2O(l)$.

Aqueous solution chemistry involving cations of the group 14 elements is restricted mainly to Sn and Pb (see Section 14.13), and so Table 14.1 gives E^o values only for these metals.

Self-study exercises

1. Comment on the fact that covalent radii are listed in Table 14.1 for all the group 14 elements, but ionic radii are listed only for Ge, Sn and Pb. Why are radii for M^{4+} *and* M^{2+} listed for Sn and Pb, but not for Ge?

2. How accurate do you expect the value of r_{ion} for Sn^{4+} to be? Rationalize your answer.

3. No electrochemical data are listed in Table 14.1 for C, Si and Ge. Suggest reasons for this.

4. Explain why, for each group 14 element, the value of IE_3 is significantly larger than that of IE_2.

Some energetic and bonding considerations

Table 14.2 lists some experimentally determined values for covalent bond enthalpy terms. When we try to interpret the chemistry of the group 14 elements on the basis of such bond energies, caution is necessary for two reasons:

- many thermodynamically favourable reactions are kinetically controlled;
- in order to use bond enthalpy terms successfully, *complete* reactions must be considered.

The first point is illustrated by considering that although the combustions of CH_4 and SiH_4 are both thermodynamically favourable, SiH_4 is spontaneously inflammable in air, whereas CH_4 explodes in air only when a spark provides

Table 14.2 Some experimental covalent bond enthalpy terms ($kJ\,mol^{-1}$); the values for single bonds refer to the group 14 elements in tetrahedral environments.

C–C	C=C	C≡C	C–H	C–F	C–Cl	C–O	C=O
346	598	813	416	485	327	359	806

Si–Si			Si–H	Si–F	Si–Cl	Si–O	Si=O
226			326	582	391	466	642

Ge–Ge			Ge–H	Ge–F	Ge–Cl	Ge–O	
186			289	465	342	350	

Sn–Sn			Sn–H		Sn–Cl		
151			251		320		

					Pb–Cl		
					244		

the energy to overcome the activation barrier. In respect of the second point, consider reaction 14.5.

(14.5)

Inspection of Table 14.2 shows that $E(C-H) > E(C-Cl)$, but the fact that the H–Cl bond ($431\,kJ\,mol^{-1}$) is significantly stronger than the Cl–Cl bond ($242\,kJ\,mol^{-1}$) results in reaction 14.5 being energetically favourable.

> *Catenation* is the tendency for covalent bond formation between atoms of a given element, e.g. C–C bonds in hydrocarbons or S–S bonds in polysulfides.

The particular strength of the C–C bond contributes towards the fact that catenation in carbon compounds is common. However, it must be stressed that *kinetic* as well as thermodynamic factors may be involved, and any detailed discussion of kinetic factors is subject to complications:

- Even when C–C bond breaking is the rate-determining step, it is the bond dissociation *energy* (zero point energy) rather than the enthalpy term that is important.
- Reactions are often bimolecular processes in which bond-making and bond-breaking occur simultaneously, and in such cases, the rate of reaction may bear no relationship to the difference between bond enthalpy terms of the reactants and products.

In contrast to the later elements in group 14, C tends not to exhibit coordination numbers greater than four. While complexes such as $[SiF_6]^{2-}$ and $[Sn(OH)_6]^{2-}$ are known, carbon analogues are not. The fact that CCl_4 is kinetically inert towards hydrolysis but $SiCl_4$ is readily hydrolysed by water has traditionally been ascribed to the availability of $3d$ orbitals on Si, which can stabilize an associative transition state. This view has been challenged with the suggestion that the phenomenon is steric in origin associated purely with the lower accessibility of the C centre arising from the shorter C–Cl bonds with respect to the Si–Cl bonds.

The possible role of $(p-d)\pi$-bonding for Si and the later elements in group 14 has been a controversial issue (see Section 5.7) and we return to this in Section 14.6. On the other hand, $(p-p)\pi$-bonding leading to double and triple homonuclear bonds, which is so common in carbon chemistry, is relatively unimportant later in the group.[†] A similar situation is observed in groups 15 and 16. The

---

[†]See: R.C. Fischer and P.P. Power (2010) *Chem. Rev.*, vol. 110, p. 3877.

mesityl derivative **14.1** was the first compound containing an Si=Si bond to be characterized. In the Raman spectrum of **14.1**, an absorption at $529\,cm^{-1}$ is assigned to the $\nu(Si=Si)$ mode, and in the solid state structure, the Si–Si bond distance of 216 pm is less than twice the value of r_{cov} (2×118 pm). Such species are stabilized with respect to polymerization by the presence of bulky substituents such as mesityl (in **14.1**), CMe_3 or $CH(SiMe_3)_2$.

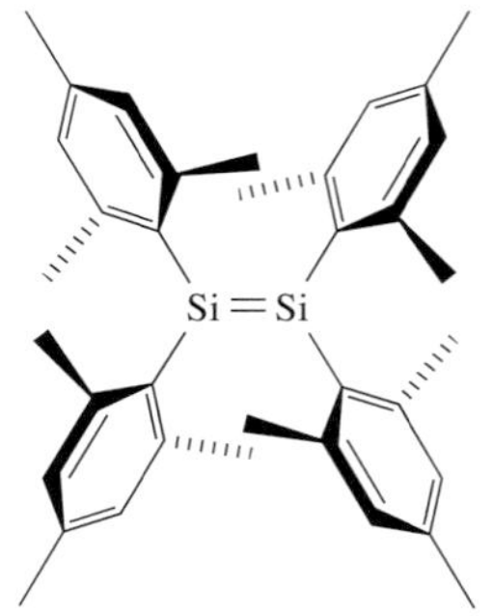

Mesityl = Mes = 1,3,5-trimethylphenyl

(14.1)

The central Si_2C_4-unit in **14.1** is planar, allowing overlap of $3p$ orbitals (orthogonal to the plane) for π-bond formation; the bulky mesityl substituents adopt a 'paddle-wheel' conformation minimizing steric interactions.[†] In contrast, theoretical studies on Si_2H_4 (mass spectrometric evidence for which has been obtained), indicate that a non-planar structure is energetically favoured. The same *trans*-bent conformation has been observed experimentally for Sn_2R_4 compounds (see Fig. 23.19 and accompanying text). Theoretical studies on the hypothetical $HSi\equiv SiH$ suggest that a non-linear structure is energetically preferred over an ethyne-like structure.

The first compound containing an Si≡Si bond was isolated and structurally characterized in 2004. The product of reaction 14.6 (the reducing agent is the intercalation compound KC_8, see structure **14.2**) is kinetically and thermodynamically stabilized by very bulky silyl substituents. In line with theoretical predictions, the disilyne has a non-linear structure. The Si≡Si bond (206 pm) is about 4% shorter than a typical Si=Si bond (214–216 pm), and about 13% shorter than a typical Si–Si bond ($2 \times r_{cov} = 236$ pm). This degree of shortening is significantly less than that observed on going from C–C to C=C to C≡C, consistent with less efficient π-overlap for Si compared with C.

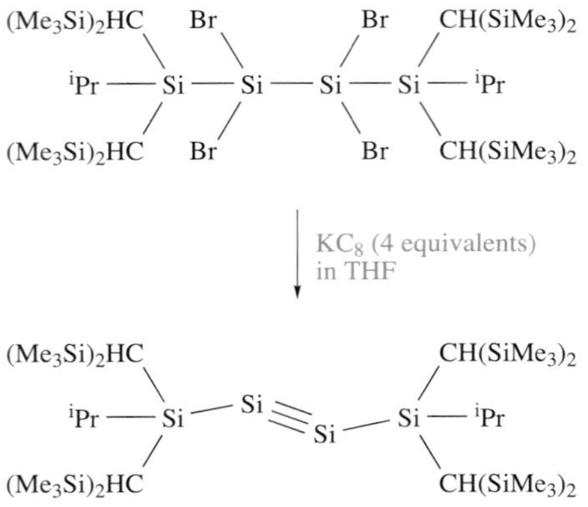

$$ (14.6) $$

$$ Si \equiv Si = 206\ \text{pm};\ \angle Si\text{–}Si\text{–}Si = 137° $$

^{29}Si NMR spectroscopy provides diagnostic data for the formation of the RSi≡SiR unit and supports the assignment of a disilyne formed by the dehalogenation of a disilene with lithium naphthalide (eq. 14.7).[‡]

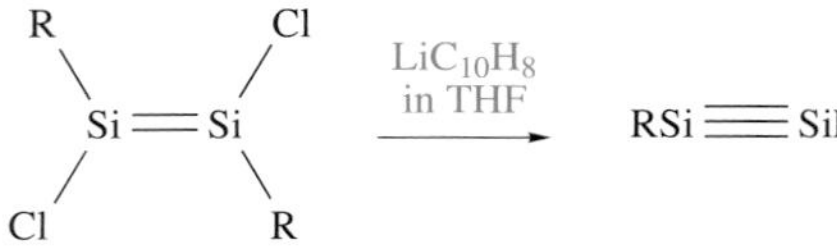

$$ (14.7) $$

$$ R = SiMe(Si^tBu_3)_2 $$

The compound $^tBu_3SiSiBr_2SiBr_2Si^tBu_3$ is related to the precursor in eq. 14.6. It reacts with $(^tBu_2MeSi)_2SiLi_2$ to generate a cyclotrisilene that may be converted into an aromatic cyclotrisilenylium cation:

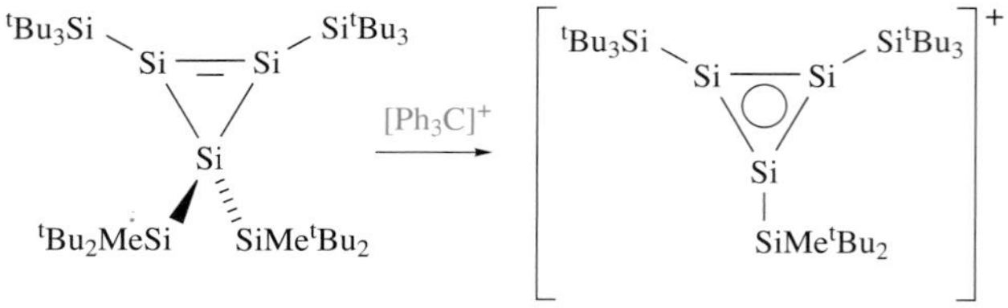

The formation of $(p–p)\pi$-bonds between C and Si is also rare. An example is shown in eq. 14.8. In 1999, the first examples of a C≡Si bond were confirmed in the gas-phase molecules HC≡SiF and HC≡SiCl. These species were detected using neutralization–reionization mass spectrometry, but have not been isolated.

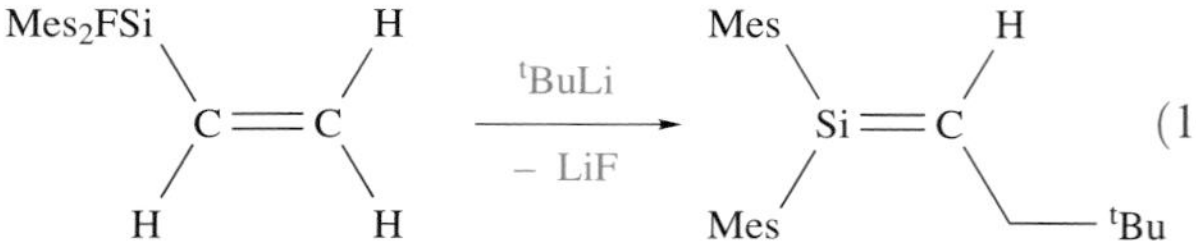

$$ (14.8) $$

The first Ge=C double bond was reported in 1987, since when a number of examples have been reported, including $Mes_2Ge=CHCH_2^tBu$ which is stable at 298 K. The formation of Ge=Ge bonds is described in Section 23.5.

[†] In a second structurally characterized polymorph, the orientations of the mesityl groups differ; see: R. Okazaki and R. West (1996) *Adv. Organomet. Chem.*, vol. 39, p. 231.

[‡] Si≡Si formation, see: N. Wiberg *et al.* (2004) *Z. Anorg. Allg. Chem.*, vol. 630, p. 1823; A. Sekiguchi *et al.* (2004) *Science*, vol. 305, p. 1755; M. Karni *et al.* (2005) *Organometallics*, vol. 24, p. 6319.

NMR active nuclei

Table 14.1 lists NMR active nuclei for the group 14 elements. Although the isotopic abundance of ^{13}C is only 1.1%, use of ^{13}C NMR spectroscopy is very important. The low abundance means that, unless a sample is isotopically enriched, satellite peaks in, for example, a ^{1}H NMR spectrum, will not be observed and application of ^{13}C as an NMR active nucleus lies in its *direct observation*. The appearance of satellite peaks due to coupling of an observed nucleus such as ^{1}H to ^{29}Si or ^{119}Sn is diagnostic (see case study 5 in Section 4.8). Direct observation of ^{29}Si nuclei is a routine means of characterizing Si-containing compounds. Tin-119 NMR spectroscopy (^{119}Sn being generally favoured over ^{117}Sn for direct observation) is also valuable. The chemical shift range is large and, as with many heteronuclei, δ values may provide an indication of coordination environments.

Mössbauer spectroscopy

The ^{119}Sn nucleus is suitable for Mössbauer spectroscopy (see Section 4.10) and isomer shift values can be used to distinguish between Sn(II) and Sn(IV) environments. The spectroscopic data may also provide information about the coordination number of the Sn centre.

Worked example 14.1   NMR spectroscopy

The ^{1}H NMR spectrum of SnMe$_4$ consists of a singlet with two superimposed doublets. The coupling constants for the doublets are 52 and 54 Hz, and the overall five-line signal exhibits an approximately 4:4:84:4:4 pattern. Use data from Table 14.1 to interpret the spectrum.

In Me$_4$Sn, all 12 protons are equivalent and one signal is expected. Sn has two NMR active nuclei: ^{117}Sn (7.6%, $I = \frac{1}{2}$) and ^{119}Sn (8.6%, $I = \frac{1}{2}$). The ^{1}H nuclei couple to the ^{117}Sn nucleus to give a doublet, and to the ^{119}Sn nucleus to give another doublet. The relative intensities of the lines in the signal reflect the abundances of the spin-active nuclei:

- 83.8% of the ^{1}H nuclei are in molecules containing isotopes of Sn that are not spin-active, and these protons give rise to a singlet.
- 7.6% of the ^{1}H nuclei are in molecules containing ^{117}Sn and these protons give rise to a doublet.
- 8.6% of the ^{1}H nuclei are in molecules containing ^{119}Sn and these protons give rise to a doublet.

The coupling constants for the doublets are 52 and 54 Hz. From the data given, it is not possible to assign these to coupling to a particular isotope. (In fact, $J(^{117}Sn-^{1}H) = 52$ Hz, and $J(^{119}Sn-^{1}H) = 54$ Hz.)

Self-study exercises

Data: see Table 14.1; ^{1}H and ^{19}F, 100%, $I = \frac{1}{2}$.

1. The ^{13}C NMR spectrum of Me$_3$SnCl contains five lines in a non-binomial pattern; the separation between the outer lines is 372 Hz. Interpret these data.
   [*Ans.* As in the worked example;
   $J(^{119}Sn-^{13}C) = 372$ Hz]

2. Apart from the chemical shift value, how do you expect well-resolved ^{1}H NMR spectra of Me$_4$Sn and Me$_4$Si to differ?
   [*Ans.* Take into account the % abundances of spin-active nuclei]

3. Explain why the ^{29}Si NMR spectrum of SiH$_3$CH$_2$F consists of a quartet ($J = 203$ Hz) of doublets ($J = 25$ Hz) of triplets ($J = 2.5$ Hz).
   [*Ans.* ^{29}Si couples to directly bonded ^{1}H, 2-bond coupling to ^{19}F, and 2-bond coupling to ^{1}H]

4. The $^{119}Sn\{^{1}H\}$ NMR spectra of the compounds Sn(CH$_2$CH$_2$CH$_2$CH$_2$SnPh$_2$R)$_4$ with R = Ph, Cl or H are as follows: R = Ph: δ −11.45, −99.36 ppm; R = Cl, δ −10.46, 17.50 ppm; R = H, δ −11.58, −136.65 ppm. In each spectrum, the signal close to δ −11 ppm is of lower intensity. (a) Assign the spectra and comment on the effect of changing the R group. (b) In the $^{13}C\{^{1}H\}$ NMR spectrum of Sn(CH$_2$CH$_2$CH$_2$CH$_2$SnPh$_3$)$_4$, signals for the directly Sn-bonded CH$_2$ groups appear at δ 8.3 and 10.7 ppm. Each signal has two pairs of satellite peaks. What is the origin of these peaks?
   [Ans. See H. Schumann *et al.* (2006) *J. Organomet. Chem.*, vol. 691, p. 1703]

14.4 Allotropes of carbon

Graphite and diamond: structure and properties

We have already described the rigid structure of diamond (Fig. 6.20a). Diamond is not the thermodynamically most stable form of the element but is *metastable*. At room temperature, the conversion of diamond into graphite is thermodynamically favoured (eq. 14.9), making graphite the standard state of C at 298 K. However, reaction 14.9 is infinitely slow.

$$C(\text{diamond}) \longrightarrow C(\text{graphite})$$

$$\Delta_r G^\circ(298 \text{ K}) = -2.9 \text{ kJ mol}^{-1} \qquad (14.9)$$

A state is **metastable** if it exists without observable change even though it is thermodynamically unstable with respect to another state.

APPLICATIONS

Box 14.4 Diamonds: natural and synthetic

The commercial value of diamonds as gemstones is well recognized, and the world production of natural gem-quality diamonds in 2008 is shown in the chart at the bottom of the page. The chart also shows the production of natural diamonds (non-gemstone quality) used for industrial purposes. Because diamond is the hardest known substance,[†] it has widespread applications as an abrasive and in cutting-tools and drill-bits. These applications extend from drill-bits for mining to diamond saws for cutting crystals into wafer-thin slices for the electronics industry. Diamond exhibits electrical, optical and thermal properties (it has the highest thermal conductivity of any material at 298 K) that make it suitable for use in corrosion and wear-resistant coatings, in heat sinks in electrical circuits, and in certain types of lenses. An application in the laboratory is in diamond anvil cells in which diamonds on the tips of pistons are compressed together. A stainless steel gasket placed between the tips of the diamonds provides a sample chamber, and pressures of up to 200 GPa (comparable with those at the centre of the Earth) can be achieved in the cell. Diamonds are transparent to IR, visible, near-UV and X-ray radiation, and therefore diamond anvil cells can be used in conjunction with spectroscopic and X-ray diffraction equipment to study high-pressure phases of minerals.

Industrial demand for diamond is met in part by synthetic diamonds. The scale of production of synthetic diamonds is significantly greater than that of mining natural material.

Part of the phase diagram for carbon is shown below:

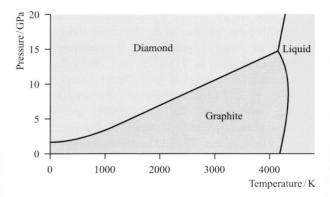

The diagram illustrates that graphite converts to diamond under high-pressure high-temperature (HPHT) conditions.

Synthetic diamonds are produced by dissolving graphite in a melted metal (e.g. Fe) and crystallizing the mixture under appropriate high P and T conditions. After being cooled, the metal is dissolved into acid, leaving synthetic diamonds of sizes ranging between ≈ 0.05 and 0.5 mm. Major uses of these industrial diamonds include grinding, honing (e.g. smoothing cylinder bores), saw-blades and polishing powders. The relative importance of synthetic diamond production (which has risen dramatically since 1950) compared with mining of the natural material is clearly seen by comparing the two charts below. The US leads the world in the manufacture of synthetic diamonds, while the main reserves of gemstone diamonds are in Africa, Australia, Canada and Russia.

[†] But see: T. Ferroir, L. Dubrovinsky, A. El Goresy, A. Simionovici, T. Nakamura and P. Gillet (2010) *Earth Plant. Sc. Lett.*, vol. 290, p. 150 – 'Carbon polymorphism in shocked meteorites: Evidence for new natural ultrahard phases'.

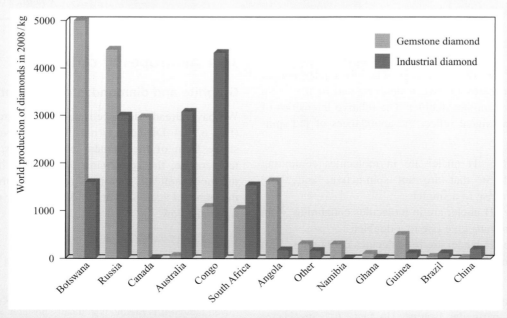

[Data: US Geological Survey using a conversion factor of 5 carats = 1 g]

Diamond has a higher density than graphite ($\rho_{\text{graphite}} = 2.25$; $\rho_{\text{diamond}} = 3.51 \, \text{g cm}^{-3}$), and this allows artificial diamonds to be made from graphite at high pressures (see Box 14.4). There are two structural modifications of graphite. The 'normal' form is α-graphite and can be converted to the β-form by grinding; a β $\longrightarrow$ α-transition occurs above 1298 K. Both forms possess layered structures and Fig. 14.4a shows 'normal' graphite. The *intra*layer C–C bond distances are equal (142 pm) while the *inter*-layer distances are 335 pm. A comparison of these distances with the values for C of $r_{\text{cov}} = 77 \, \text{pm}$ and $r_{\text{v}} = 185 \, \text{pm}$ indicates that while covalent bonding is present within each layer, only weak van der Waals interactions operate between adjacent layers. Graphite cleaves readily and is used as a lubricant. This property depends not only on the weak van der Waals forces between the layers but also on the presence of intercalated (see below) water molecules. The absorption of water vapour causes the coefficient of friction of graphite to decrease from 0.5–0.8 (the range for graphite in a vacuum) to ≈0.1. This contrasts with, for example, MoS_2 which also has a layered structure (see Box 22.6) and is an intrinsic solid state lubricant.[†] The electrical conductivity (see Section 6.8) of α-graphite is direction-dependent. In a direction parallel to the layers, the electrical resistivity is $1.3 \times 10^{-5} \, \Omega \, \text{m}$ (at 293 K) but is ≈1 Ω m in a direction perpendicular to the layers. Each C atom has four valence electrons and forms three σ-bonds, leaving one electron to participate in delocalized π-bonding. The molecular π-orbitals extend over each layer, and while the bonding MOs are fully occupied, the energy gap between them and the vacant antibonding MOs is very small, allowing the electrical conductivity in the direction *parallel* to the layers to approach that of a metal. In contrast, the electrical resistivity of diamond is $1 \times 10^{11} \, \Omega \, \text{m}$, making diamond an excellent insulator. The chemistry of a single layer of graphite (*graphene*) is detailed further in Section 28.8.

Graphite is more reactive than diamond. Graphite is oxidized by atmospheric O_2 above 970 K, whereas diamond burns at >1170 K. Graphite reacts with hot, concentrated HNO_3 to give the aromatic compound $C_6(CO_2H)_6$. Polymeric carbon monofluoride, CF_n ($n \leq 1$), is formed when F_2 reacts with graphite at 720 K (or at lower temperatures in the presence of HF), although at 970 K, the product is monomeric CF_4. The fluorine content in materials formulated as CF_n is variable and their colour varies, being white when $n \approx 1.0$. Carbon monofluoride possesses a layer structure, and is used as a lubricant, being more resistant to atmospheric oxidation at higher temperatures than graphite. Part of one layer is shown in Fig. 14.4b. In the idealized compound CF, each

[†]For a review of solid lubricants, see: H.E. Sliney (1982) *Tribiology Int.*, vol. 15, p. 303.

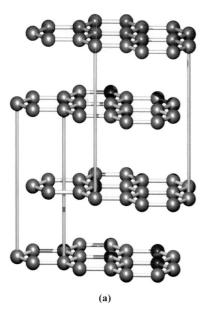

(a)

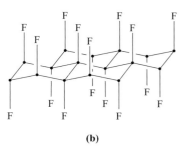

(b)

Fig. 14.4 (a) Part of the infinite layered-lattice of α-graphite ('normal' graphite); the layers are co-parallel, and atoms in *alternate* layers lie over each other. This is emphasized by the yellow lines in the diagram. (b) Part of one layer of the proposed structure of CF_n for $n = 1$.

C atom is tetrahedral; each C–C bond distance within a layer is 154 pm, and between layers it is 820 pm, i.e. more than double that in α-graphite.

Graphite: intercalation compounds

An ***intercalation compound*** results from the reversible inclusion (with no associated covalent bond formation) of a species (atom, ion or molecule) in a solid host which has a laminar structure.

Graphite possesses the remarkable property of forming many *intercalation* (*lamellar* or *graphitic*) compounds, the formation of which involves movement apart of the carbon layers and the penetration of atoms or ions between them. There are two general types of graphite intercalation compound in which the graphite layers remain planar:

- intercalation with associated reduction of the graphite by a metal (group 1, Ca, Sr, Ba, or some lanthanoid metals);
- intercalation with concomitant oxidation of the graphite by an oxidizing agent such as Br_2, HNO_3 or H_2SO_4.

At the end of Section 11.4, we described the different conditions under which Li, Na, K, Rb or Cs may be intercalated into graphite, and the application of lithium ion intercalation in lithium-ion batteries. We now look in more detail at the intercalation of potassium.

When graphite is treated with an excess of K (and unreacted metal is washed out with Hg), a paramagnetic copper-coloured material formulated as $K^+[C_8]^-$ results. The penetration of K^+ ions between the layers causes structural changes in the graphite framework: the initially staggered layers (Fig. 14.4a) become eclipsed, and the interlayer spacing increases from 335 to 540 pm. The K^+ ions lie above (or below) the centres of alternate C_6-rings, as indicated in structure **14.2** which shows a projection through the eclipsed layers.

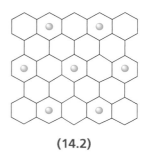

(14.2)

The electrical conductivity of KC_8 is greater than that of α-graphite, consistent with the addition of electrons to the delocalized π-system. Heating KC_8 leads to the formation of a series of decomposition products as the metal is eliminated (eq. 14.10). The structures of these materials are related, there being one, two, three, four or five carbon layers respectively between layers of K^+ ions.

$$KC_8 \xrightarrow{\Delta} KC_{24} \xrightarrow{\Delta} KC_{36} \xrightarrow{\Delta} KC_{48} \xrightarrow{\Delta} KC_{60}$$

copper-coloured   blue

$$(14.10)$$

Such alkali metal intercalates are extremely reactive, igniting in air and exploding on contact with water.

In intercalation compounds formed with strong acids in the presence of oxidizing agents, the carbon layers lose electrons and become positively charged, e.g. graphite hydrogensulfate, $[C_{24}]^+[HSO_4]^- \cdot 24H_2O$, which is produced when graphite is treated with concentrated H_2SO_4 and a little HNO_3 or CrO_3. A related product forms when the acid is $HClO_4$; in this intercalate, the planar layers of carbon atoms are 794 pm apart and are separated by $[ClO_4]^-$ ions and acid molecules. Cathodic reduction of this material, or treatment with graphite, gives a series of compounds corresponding to the sequential elimination of $HClO_4$. These materials are better electrical conductors than graphite, and this can be explained in terms of a positive-hole mechanism (see Section 6.9).

Other intercalation compounds in which graphite is oxidized include those formed with Cl_2, Br_2, ICl and halides such as KrF_2, UF_6 and $FeCl_3$. Reaction of graphite with $[O_2]^+[AsF_6]^-$ results in the formation of the salt $[C_8]^+[AsF_6]^-$.

The catalytic properties of some graphite intercalation compounds render them of practical importance; e.g. KC_8 is a hydrogenation catalyst.

Fullerenes: synthesis and structure

In 1985, Kroto, Smalley and coworkers discovered that, by subjecting graphite to laser radiation at >10 000 K, new allotropes of carbon were formed. Because of their molecular architectures, the *fullerenes* are named after architect Buckminster Fuller, known for designing geodesic domes. The family of molecular fullerenes includes C_{60}, C_{70}, C_{76}, C_{78}, C_{80} and C_{84}. Several synthetic routes to fullerenes have been developed. The standard route for preparing C_{60} and C_{70} is by the Krätschmer–Huffmann method, first reported in 1990. These two fullerenes are the major components of the graphitic soot mixture produced as graphite rods are evaporated (by applying an electrical arc between them) in a helium atmosphere at ≈130 mbar and the vapour condensed. Extraction of the soot into benzene yields a red solution from which C_{60} and C_{70} can be separated by chromatography, typically HPLC (see Section 4.2). Hexane or benzene solutions of C_{60} are magenta, while those of C_{70} are red. Both C_{60} and C_{70} are available commercially.

Figure 14.5a shows the structure of C_{60}. Although a number of X-ray diffraction studies of C_{60} have been carried out, the near-spherical shape of the molecule has led to frustrating orientational disorder (see Box 15.5) problems. The C_{60} molecule belongs to the I_h point group and consists of an approximately spherical network of atoms which are connected in 5- and 6-membered rings. All the C atoms are equivalent, and the ^{13}C NMR spectrum of C_{60} exhibits one signal (δ +143 ppm). The rings are arranged such that no 5-membered rings are adjacent to one another. Thus, C_{60} (the smallest fullerene that can be isolated as a stable species) satisfies the *isolated pentagon rule* (IPR).[†] The separation of the 5-membered rings by 6-membered rings is easily seen in the schematic representation of C_{60} shown in Fig. 14.5b which also gives a bonding scheme. Each C atom is covalently bonded to three others in an approximately trigonal planar arrangement. The relatively large surface of the 'sphere' means that there is only slight deviation from planarity at each C centre. There are two types of C–C bond: those at the

---

[†] For the origins of IPR, see: H.W. Kroto (1985) *Nature*, vol. 318, p. 354.

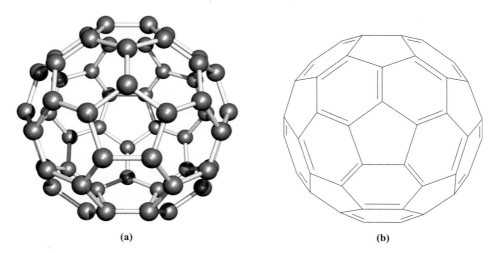

Fig. 14.5 (a) The structure of the fullerene C_{60}; the approximately spherical molecule is composed of fused 5- and 6-membered rings of carbon atoms. [X-ray diffraction at 173 K of the benzene solvate $C_{60} \cdot 4C_6H_6$, M.F. Meidine *et al.* (1992) *J. Chem. Soc., Chem. Commun.*, p. 1534.] (b) A representation of C_{60}, in the same orientation as in (a), but showing only the upper surface and illustrating the localized single and double carbon–carbon bonds.

junctions of two hexagonal rings (6,6-edges) are of length 139 pm, while those between a hexagonal and a pentagonal ring (5,6-edges) are longer, 145.5 pm. These differences indicate the presence of localized double and single bonds, and similar bonding descriptions are appropriate for other fullerene cages. We consider chemical evidence for the presence of C=C double bonds below. After C_{60}, the next smallest fullerene to satisfy IPR is C_{70}. The C_{70} molecule has D_{5h} symmetry and is approximately ellipsoidal (Fig. 14.6). It comprises 6- and 5-membered rings organized so that, as in C_{60}, 5-membered rings are never adjacent. The ^{13}C NMR spectrum of C_{70} confirms that there are five C environments in solution, consistent with the solid state structure (Fig. 14.6a).

Fullerenes: reactivity

Since efficient syntheses have been available, fullerenes (in particular C_{60}) have been the focus of an explosion of research. We provide a brief introduction to the chemical

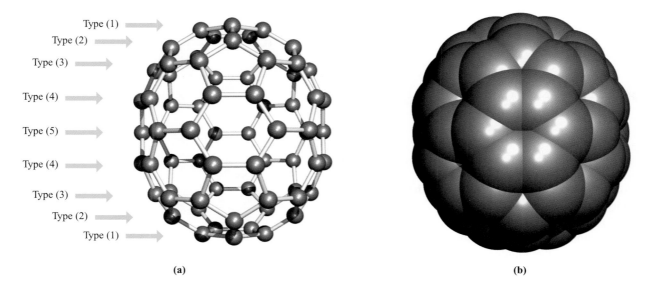

Type (1)
Type (2)
Type (3)
Type (4)
Type (5)
Type (4)
Type (3)
Type (2)
Type (1)

(a)

(b)

Fig. 14.6 The structure of C_{70} determined from an X-ray diffraction study of $C_{70} \cdot 6S_8$ [H.B. Bürgi *et al.* (1993) *Helv. Chim. Acta*, vol. 76, p. 2155]: (a) a ball-and-stick representation showing the five carbon atom types, and (b) a space-filling diagram illustrating the ellipsoidal shape of the molecule.

properties of C_{60}; organometallic derivatives are covered in Section 24.10, and the end-of-chapter reading gives more in-depth coverage.

Protonation of C_{60} has only been observed by superacids of type $HCHB_{11}R_5X_6$ (e.g. X = Cl) (see Section 9.9). The solution ^{13}C NMR spectrum of $[HC_{60}]^+$ shows a single, sharp peak, indicating that the proton migrates over the entire fullerene cage on the NMR timescale. Solid state NMR spectroscopic data (i.e. for a static structure) show that the protonated sp^3 C atom (δ 56 ppm) is directly bonded to the sp^2 cationic site (δ 182 ppm).

The structural representation in Fig. 14.5b suggests connected benzene rings. However, although C_{60} exhibits a small degree of aromatic character, its reactions tend to reflect the presence of *localized* double and single C–C bonds, e.g. C_{60} undergoes *addition* reactions. Birch reduction (i.e. Na in liquid NH_3) gives a mixture of polyhydrofullerenes (eq. 14.11) with $C_{60}H_{32}$ being the dominant product. Reoxidation to C_{60} occurs with the quinone shown. Reaction 14.12 shows a selective route to $C_{60}H_{36}$; the hydrogen-transfer agent is 9,10-dihydroanthracene (DHA). In addition to being a selective method of hydrogenation, use of 9,9',10,10'-[D_4]dihydroanthracene provides a method of selective deuteration.

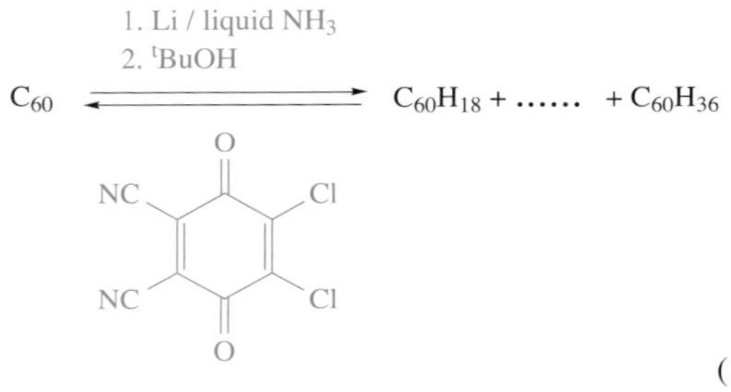

$$(14.11)$$

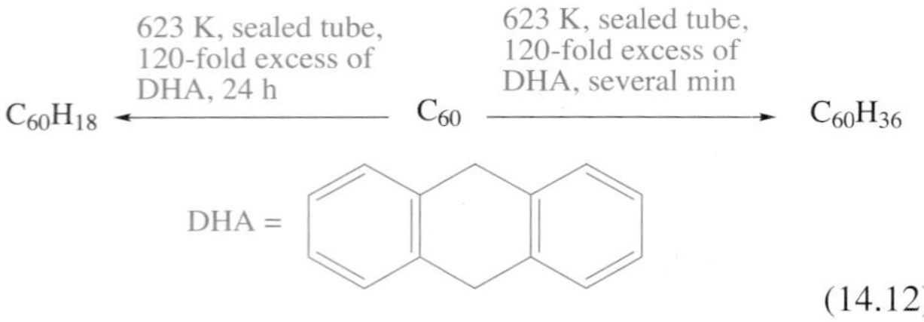

$$(14.12)$$

Additions of F_2, Cl_2 and Br_2 also occur, the degree and selectivity of halogenation depending on conditions (Fig. 14.7). Because F atoms are small, addition of F_2 to adjacent C atoms in C_{60} is possible, e.g. to form 1,2-$C_{60}F_2$. However, in the addition of Cl_2 or Br_2, the halogen atoms prefer to add to remote C atoms. Thus, in $C_{60}Br_8$ and in $C_{60}Br_{24}$ (Fig. 14.8a), the Br atoms are in 1,3- or 1,4-positions with respect to each other. Just as going from benzene to cyclohexane causes a change from a

planar to a boat- or chair-shaped ring, addition of substituents to C_{60} causes deformation of the near-spherical surface. This is illustrated in Fig. 14.8 with the structures of $C_{60}Br_{24}$, $C_{60}F_{18}$ and $C_{60}Cl_{30}$. The C_{60}-cage in $C_{60}Br_{24}$ includes both boat and chair C_6-rings. Addition of a Br to a C atom causes a change from sp^2 to sp^3 hybridization. The arrangement of the Br atoms over the surface of the C_{60} cage is such that they are relatively far apart from each other. In contrast, in $C_{60}F_{18}$ (Fig. 14.8b), the F atoms are in 1,2-positions with respect to each other and the C_{60}-cage suffers severe 'flattening' on the side associated with fluorine addition. At the centre of the flattened part of the cage lies a planar, C_6-ring (shown at the centre of the lower part of Fig. 14.8b). This ring has equal C–C bond lengths (137 pm) and has aromatic character. It is surrounded by sp^3 hybridized C atoms, each of which bears an F atom. The cage distortion is even more severe in $C_{60}Cl_{30}$. This high degree of chlorination results in the formation of two 15-membered rings of sp^3-hybridized C atoms (top and bottom in Fig. 14.8c) and a flattening of the C_{60} framework into a drum-shaped structure.

The ene-like nature of C_{60} is reflected in a range of reactions such as the additions of an O atom to give an epoxide ($C_{60}O$), and of O_3 at 257 K to yield an intermediate ozonide ($C_{60}O_3$). In hydrocarbon solvents, addition occurs at the junction of two 6-membered rings (a 6,6-bond), i.e. at a C=C bond, as shown in scheme 14.13. Loss of O_2 from $C_{60}O_3$ gives $C_{60}O$ but the structure of this product depends on the reaction conditions. At 296 K, the product is an epoxide with the O bonded across a 6,6-bond. In contrast, photolysis opens the cage and the O atom bridges a 5,6-edge (scheme 14.13).

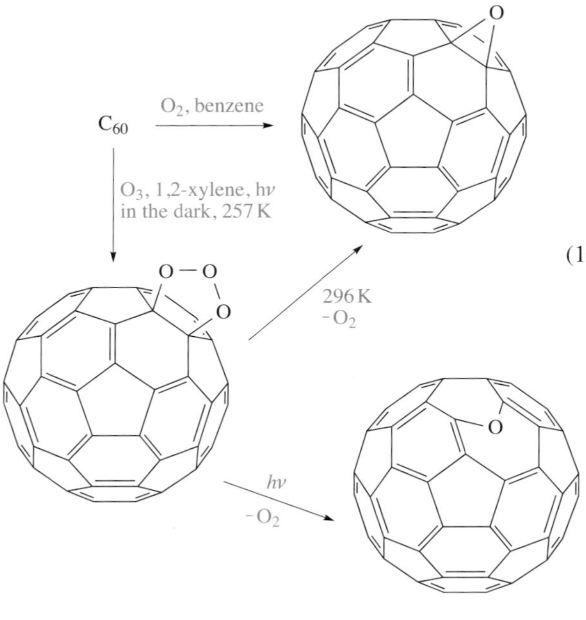

$$(14.13)$$

Other reactions typical of double-bond character include the formation of cycloaddition products (exemplified

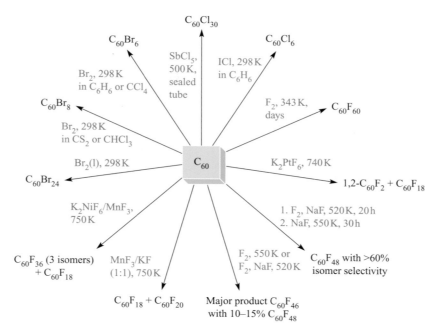

Fig. 14.7 Examples of halogenation reactions of C_{60}. Although the number of possible isomers for products $C_{60}X_n$ where $2 \leq n \leq 58$ is, at the very least, large, some of the reactions (such as fluorination using NaF and F_2) are surprisingly selective.

schematically in eq. 14.14). Such additions can be used to synthesize a wide range of derivatives of C_{60}.

(14.14)

The Diels–Alder reaction of tetrazine **14.3** with C_{60} followed by intramolecular cycloaddition and loss of N_2[†] results in the insertion of a C_2-unit into the C_{60} cage. In structure **14.3**, the two carbon atoms marked with the pink dots are those that are eventually incorporated into the C_{62} cage. Figure 14.9 shows the structure of the product, $C_{62}(C_6H_4\text{-}4\text{-Me})_2$, and confirms the presence of a 4-membered ring surrounded by four 6-membered rings.

(14.3)

Reactions of C_{60} with free radicals readily occur, e.g. photolysis of RSSR produces RS˙ which reacts with C_{60} to give $C_{60}SR˙$, although this is unstable with respect to regeneration of C_{60}. The stabilities of radical species $C_{60}Y˙$ are highly dependent on the steric demands of Y. When the reaction of ${}^tBu˙$ (produced by photolysis of a *tert*-butyl halide) with C_{60}

is monitored by EPR spectroscopy (which detects the presence of unpaired electrons, see Section 4.9), the intensity of the signal due to the radical $C_{60}{}^tBu˙$ increases over the temperature range 300–400 K. These data are consistent with equilibrium 14.15, with reversible formation and cleavage of an inter-cage C−C bond.

(14.15)

The formation of methanofullerenes, $C_{60}CR_2$, occurs by reaction at either 5,6- or 6,6-edges in C_{60}. For the

---

[†]For details of the reaction pathway, see: W. Qian *et al.* (2003) *J. Am. Chem. Soc.*, vol. 125, p. 2066.

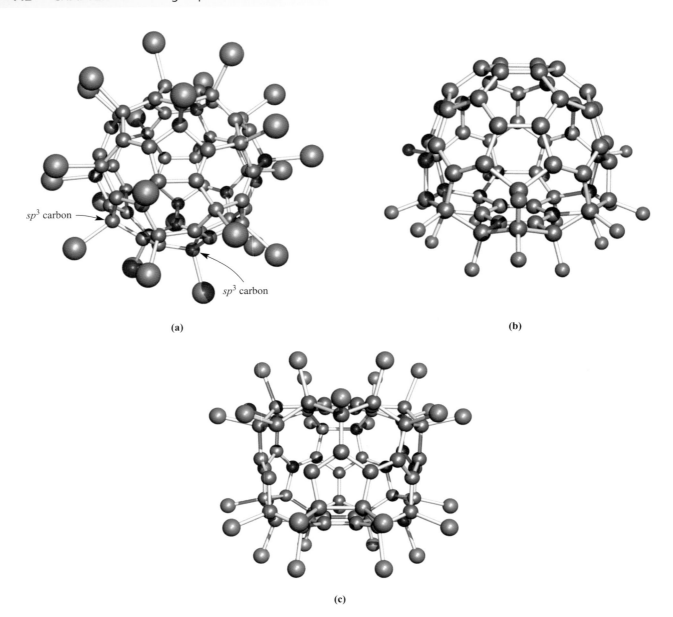

🌐 **Fig. 14.8** The structure of $C_{60}Br_{24}$ determined by X-ray diffraction at 143 K [F.N. Tebbe *et al.* (1992) *Science*, vol. 256, p. 822]. The introduction of substituents results in deformation of the C_{60} surface; compare the structure of $C_{60}Br_{24}$ with that of C_{60} in Fig. 14.5a which shows the C_{60} cage in a similar orientation. (b) The structure (X-ray diffraction at 100 K) of $C_{60}F_{18}$ [I.S. Neretin *et al.* (2000) *Angew. Chem. Int. Ed.*, vol. 39, p. 3273]. Note that the F atoms are all associated with the 'flattened' part of the fullerene cage. (c) The structure of $C_{60}Cl_{30}$ determined by X-ray diffraction [P.A. Troshin *et al.* (2005) *Angew. Chem. Int. Ed.*, vol. 44, p. 234]. Colour code: C, grey; Br, gold; F, green; Cl, green.

6,6-addition products, the product of the reaction of C_{60} with diphenylazomethane is $C_{61}Ph_2$ (eq. 14.16). Initially, structural data suggested that the reaction was an example of 'cage expansion' with the addition of the CPh_2 unit being concomitant with the cleavage of the C–C bond marked *a* in eq. 14.16. This conclusion was at odds with NMR spectroscopic data and theoretical calculations. A low-temperature X-ray diffraction study of compound **14.4** finally confirmed that 6,6-edge-bridged methanofullerenes should be described in terms of the C_{60} cage sharing a common C–C bond with a cyclopropane ring.

$$C_{60} \xrightarrow[-N_2]{Ph_2C-\overset{-}{N}\equiv\overset{+}{N}} \qquad (14.16)$$

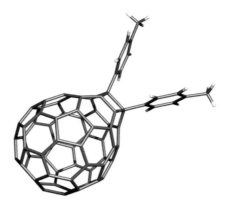

Fig. 14.9 A stick representation of the structure of $C_{62}(C_6H_4\text{-}4\text{-}Me)_2$ determined by X-ray diffraction [W. Qian *et al.* (2003) *J. Am. Chem. Soc.*, vol. 125, p. 2066].

SiMe$_3$

C
|||
C
\
C
|||
C

C$\cdots$C$\equiv$C–C$\equiv$C—SiMe$_3$

157.4 pm

(14.4)

Theoretical studies on C_{60} show that the LUMO is triply degenerate and the HOMO–LUMO separation is relatively small. It follows that reduction of C_{60} should be readily achieved. A number of charge transfer complexes have been prepared in which a suitable donor molecule transfers an electron to C_{60} as in eq. 14.17. The product in this particular reaction is of importance because, on cooling to 16 K, it becomes *ferromagnetic* (see Fig. 20.32).

Me$_2$N	NMe$_2$

Me$_2$N	NMe$_2$	$+$ C_{60} $\xrightleftharpoons[\text{Dissolve in toluene}]{\text{No solvent}}$ $[C_2(NMe_2)_4]^+\ [C_{60}]^-$

Liquid at 298 K

(14.17)

The electrochemical reduction of C_{60} results in the formation of a series of *fulleride* ions, $[C_{60}]^{n-}$ where $n = 1$–6. The mid-point potentials (obtained using cyclic voltammetry and measured with respect to the ferrocenium/

ferrocene couple, $Fc^+/Fc = 0$ V (see Box 8.2), for the reversible 1-electron steps at 213 K are given in scheme 14.18.

$C_{60} \xrightleftharpoons{-0.81\text{ V}} [C_{60}]^- \xrightleftharpoons{-1.24\text{ V}} [C_{60}]^{2-} \xrightleftharpoons{-1.77\text{ V}} [C_{60}]^{3-}$

$\xrightleftharpoons{-2.22\text{ V}} [C_{60}]^{4-} \xrightleftharpoons{-2.71\text{ V}} [C_{60}]^{5-} \xrightleftharpoons{-3.12\text{ V}} [C_{60}]^{6-}$

(14.18)

By titrating C_{60} in liquid NH_3 against an Rb/NH_3 solution (see Section 9.6) at 213 K, five successive reduction steps are observed and the $[C_{60}]^{n-}$ anions have been studied by vibrational and electronic spectroscopies. At low temperatures, some alkali metal fulleride salts of type $[M^+]_3[C_{60}]^{3-}$ become *superconducting* (see Section 28.4). The structures of the M_3C_{60} fullerides can be described in terms of M^+ ions occupying the interstitial holes in a lattice composed of close-packed, near-spherical C_{60} cages. In K_3C_{60} and Rb_3C_{60}, the $[C_{60}]^{3-}$ cages are arranged in a ccp lattice, and the cations fully occupy the octahedral and tetrahedral holes (Fig. 14.10). The temperature at which a material becomes superconducting is its *critical temperature*, T_c. Values of T_c for K_3C_{60} and Rb_3C_{60} are 18 K and 28 K respectively, and for Cs_3C_{60} (in which the C_{60} cages adopt a bcc lattice), $T_c = 40$ K. Although Na_3C_{60} is structurally related to K_3C_{60} and Rb_3C_{60}, it is not superconducting. The paramagnetic $[C_{60}]^{2-}$ anion has been isolated as the $[K(crypt\text{-}222)]^+$ salt (reaction 14.19 and Section 11.8). In the solid state, the $[C_{60}]^{2-}$ cages are arranged in layers with hexagonal packing, although the cages are well separated; $[K(crypt\text{-}222)]^+$ cations reside between the layers of fulleride anions.

$C_{60} \xrightarrow[\text{toluene/crypt-222}]{\text{DMF/K}} [K(crypt\text{-}222)]_2[C_{60}]$ (14.19)

Whereas C_{60} is readily reduced, it is difficult to oxidize. By using cyclic voltammetry (see Box 8.2) with ultra-dry solvent (CH_2Cl_2) and a supporting electrolyte with a very high oxidation resistance and low nucleophilicity ($[^nBu_4N][AsF_6]$), three reversible oxidation processes have been observed (eq. 14.20). The $[C_{60}]^{2+}$ ion is very unstable,

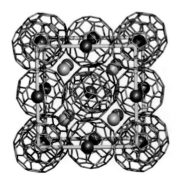

Fig. 14.10 A representation of the structures of K_3C_{60} and Rb_3C_{60} in which the $[C_{60}]^{3-}$ cages are arranged in an fcc lattice with the M^+ ions occupying all the octahedral (blue) and tetrahedral (orange) holes. The unit cell is shown in yellow.

and the third oxidation process can be studied only at low temperatures.

$$C_{60} \underset{+1.27\,V}{\rightleftharpoons} [C_{60}]^+ \underset{+1.71\,V}{\rightleftharpoons} [C_{60}]^{2+} \underset{+2.14V}{\rightleftharpoons} [C_{60}]^{3+}$$

(14.20)

The coupling of C_{60} molecules through $[2+2]$ cycloaddition to give C_{120} (**14.5**) can be achieved by a solid state reaction that involves high-speed vibration milling of C_{60} in the presence of catalytic amounts of KCN. When heated at 450 K for a short period, the C_{120} molecule dissociates into C_{60}. Under conditions of high temperature and pressure, repeated $[2+2]$ cycloadditions between C_{60} cages can lead to the formation of polymerized fullerene chains and networks. Once formed, these materials remain stable at ambient pressure and temperature, and exhibit interesting electronic and magnetic (*ferromagnetic* above room temperature, see Fig. 20.32) properties.

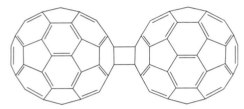

(14.5)

Endohedral metallofullerenes are a series of compounds in which metal atoms are encapsulated within a fullerene cage. The general family is denoted as $M_x@C_n$. Examples of these compounds include $Sc_2@C_{84}$, $Y@C_{82}$, $La_2@C_{80}$ and $Er@C_{60}$. In general, the larger fullerenes produce more stable compounds than C_{60}. The compounds are prepared by vaporizing graphite rods impregnated with an appropriate metal oxide or metal carbide. By use of ^{13}C and ^{139}La NMR spectroscopies, it has been shown that the two lanthanum atoms in $La_2@C_{80}$ undergo circular motion within the fullerene cage. Fullerene derivatives that possess a 'hole' have been designed so that gaseous species such as H_2 and He can enter and escape from the cage. In the example in structure **14.6**, the He atom (drawn in **14.6** with an arbitrary radius) moves in and out of the cage via the opening on the right-hand side.

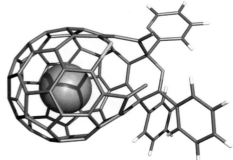

Colour code: C, grey; H, white; O, red; N, blue; S, yellow; He, orange.

(14.6)

Carbon nanotubes

Carbon *nanotubes* were discovered in 1991 and consist of elongated cages, best thought of as rolled graphene sheets. In contrast to the fullerenes, nanotubes consist of networks of fused 6-membered rings. Nanotubes are very flexible and are of importance in materials science. The end-of-chapter reading provides an entry into the area, and nanotubes are described in detail in Section 28.9.

Self-study exercise

The rate of escape of helium from compound 14.6 has been monitored by using ^{3}He NMR spectroscopy (^{3}He, $I = \frac{1}{2}$), measuring the signal integral relative to that of a known amount of ^{3}He@C_{60} added as an internal standard. The data are as follows:

Temperature / K	303	313	323	333
k / s^{-1}	4.78×10^{-6}	1.62×10^{-5}	5.61×10^{-5}	1.40×10^{-4}

[Data: C.M. Stanisky *et al.* (2005) *J. Am. Chem. Soc.*, vol. 127, p. 299]

Use the Arrhenius equation (see Box 25.1) to determine the activation energy for the escape of ^{3}He.

[*Ans.* 95.5 kJ mol^{-1}]

14.5 Structural and chemical properties of silicon, germanium, tin and lead

Structures

The solid state structures of Si, Ge, Sn and Pb and the trends from semiconductor to metal on descending the group have already been discussed:

- diamond-type structure of Si, Ge and α-Sn (Section 6.11 and Fig. 6.20);
- polymorphism of Sn (Section 6.4);
- structure of Pb (Section 6.3);
- semiconducting properties (Section 6.9).

Chemical properties

Silicon is much more reactive than carbon. At high temperatures, Si combines with O_2, F_2, Cl_2, Br_2, I_2, S_8, N_2, P_4, C and B to give binary compounds. Silicon liberates H_2 from aqueous alkali (eq. 14.21), but is insoluble in acids other than a mixture of concentrated HNO_3 and HF.

$$Si + 4[OH]^- \longrightarrow [SiO_4]^{4-} + 2H_2$$

(14.21)

On descending group 14, the electropositivity and reactivity of the elements increase. In general, Ge behaves in a similar manner to Si, but, being more electropositive, reacts with concentrated HNO_3 (forming GeO_2), and does not react with aqueous alkali. Reactions between Ge and HCl or H_2S yield $GeCl_4$ or GeS_2 respectively. Although high temperatures are needed for reactions between Sn and O_2 (to give SnO_2) or sulfur (giving SnS_2), the metal reacts readily with halogens to yield SnX_4. Tin is little affected by dilute HCl or H_2SO_4, but reacts with dilute HNO_3 (to give $Sn(NO_3)_2$ and NH_4NO_3) and with concentrated acids yielding $SnCl_2$ (from HCl) and $SnSO_4$ and SO_2 (from H_2SO_4). Hot aqueous alkali oxidizes the metal to Sn(IV) (eq. 14.22).

$$Sn + 2[OH]^- + 4H_2O \xrightarrow{-2H_2} \begin{bmatrix} OH \\ HO_{\prime\prime\prime\prime}\!\!\underset{|}{\overset{|}{Sn}}\!\!^{\backslash\backslash\backslash\backslash}OH \\ HO \qquad OH \\ OH \end{bmatrix}^{2-} \quad (14.22)$$

When finely divided, Pb is pyrophoric, but bulk pieces are passivated by coatings of, for example, PbO, and reaction with O_2 in air occurs only above $\approx 900\,K$. Lead reacts very slowly with dilute mineral acids, slowly evolves H_2 from hot concentrated HCl, and reacts with concentrated HNO_3 to give $Pb(NO_3)_2$ and oxides of nitrogen. For reactions of Pb with halogens, see Section 14.8.

Worked example 14.2   Reactivity of the group 14 elements with halogens

Write an equation for the reaction that takes place when Si is heated in F_2. The product of the reaction is a gas for which $\Delta_f H^o(298\,K) = -1615\,kJ\,mol^{-1}$. Use this value and data in Appendix 10 to calculate a value for the Si–F bond enthalpy. Compare the value obtained with that in Table 14.2.

F_2 oxidizes Si to Si(IV) and the reaction is:

$$Si(s) + 2F_2(g) \longrightarrow SiF_4(g)$$

To find the bond enthalpy term, start by writing an equation for the dissociation of gaseous SiF_4 into gaseous atoms, and then set up an appropriate thermochemical cycle that incorporates $\Delta_f H^o(SiF_4, g)$:

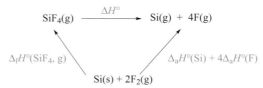

ΔH^o corresponds to the enthalpy change (gas-phase reaction) when the four Si–F bonds are broken. By Hess's law:

$$\Delta H^o + \Delta_f H^o(SiF_4, g) = \Delta_a H^o(Si) + 4\Delta_a H^o(F)$$

The atomization enthalpies are listed in Appendix 10.

$$\begin{aligned} \Delta H^o &= \Delta_a H^o(Si) + 4\Delta_a H^o(F) - \Delta_f H^o(SiF_4, g) \\ &= 456 + (4 \times 79) - (-1615) \\ &= 2387\,kJ\,mol^{-1} \end{aligned}$$

$$Si{-}F \text{ bond enthalpy} = \frac{2387}{4} = 597\,kJ\,mol^{-1}$$

This compares with a value of $582\,kJ\,mol^{-1}$ in Table 14.2.

Self-study exercises

1. Germanium reacts with F_2 to give gaseous GeF_4. Use data from Table 14.2 and Appendix 10 to estimate a value of $\Delta_f H^o(GeF_4, g)$.   [*Ans.* $-1169\,kJ\,mol^{-1}$]

2. Suggest reasons why $PbCl_2$ rather than $PbCl_4$ is formed when Pb reacts with Cl_2.   [*Ans.* See Box 13.4]

14.6 Hydrides

Although the extensive chemistry of hydrocarbons (i.e. carbon hydrides) lies outside this book, we note several points for comparisons with later group 14 hydrides:

- Table 14.2 illustrated the relative strength of a C–H bond compared with C–Cl and C–O bonds, and this trend is *not* mirrored by later elements;
- CH_4 is chlorinated with some difficulty, whereas SiH_4 reacts violently with Cl_2;
- CH_4 is stable with respect to hydrolysis, but SiH_4 is readily attacked by water;
- SiH_4 is spontaneously inflammable in air and, although it is the *kinetic* stability of CH_4 with respect to reaction with O_2 at 298 K that is crucial, values of $\Delta_c H^o$ show that combustion of SiH_4 is more exothermic than that of CH_4;
- catenation is more common for C than the later group 14 elements, and hydrocarbon families are much more diverse than their Si, Ge, Sn and Pb analogues.

Worked example 14.3   Bond enthalpies and group 14 hydrides

Suggest why catenation is more common for C than for Si, Ge and Sn. Why is this relevant to the formation of families of saturated hydrocarbon molecules?

The much higher C–C bond enthalpies (see Table 14.2) compared with those of Si–Si, Ge–Ge and Sn–Sn bonds means that the formation of compounds containing bonds between carbon atoms is thermodynamically more favourable than analogous compounds containing Si–Si, Ge–Ge and Sn–Sn bonds. On descending group 14, orbital overlap becomes less efficient as the valence orbitals become more diffuse, i.e. as the principal quantum number increases.

The backbones of saturated hydrocarbons are composed of C−C bonds, i.e. their formation depends on catenation being favourable. An additional factor that favours the formation of hydrocarbons is the strength of the C−H bonds (stronger than Si−H, Ge−H or Sn−H: see Table 14.2). On descending group 14, the hydrides become thermodynamically less stable, and the kinetic barriers to reactions such as hydrolysis of E−H bonds become lower.

Self-study exercises

1. Using bond enthalpies from Table 14.2 and Appendix 12, calculate values of ΔH° for the reactions:

$$SiH_4(g) + 4Cl_2(g) \longrightarrow SiCl_4(g) + 4HCl(g)$$

$$CH_4(g) + 4Cl_2(g) \longrightarrow CCl_4(g) + 4HCl(g)$$

   Additional data: see Appendix 10. Comment on the results. [*Ans.* −1020; −404 kJ mol^{-1}]

2. Use the fact that CH_4 is kinetically stable, but thermodynamically unstable, with respect to oxidation by O_2 at 298 K to sketch an approximate energy profile for the reaction:

$$CH_4(g) + 3O_2(g) \longrightarrow 2CO_2(g) + 2H_2O(l)$$

   Comment on the relative energy changes that you show in the diagram.
   [*Ans.* Plot *E versus* reaction coordinate, showing the relative energy levels of reactants and products; $\Delta_r H$ is negative; E_a is relatively large]

Binary hydrides

Silane, SiH_4, is formed when $SiCl_4$ or SiF_4 reacts with Li[AlH$_4$]. Industrially, SiH_4 is produced from powdered silicon. This is first treated with HCl at 620 K to give $SiHCl_3$. Passage over a catalyst (e.g. $AlCl_3$) then converts $SiHCl_3$ into SiH_4 and $SiCl_4$. Large-scale production of $SiHCl_3$ and SiH_4 is necessary because they are sources of pure Si (eqs. 14.23 and 14.29) for semiconductors (see Section 28.6). Silanes Si_nH_{2n+2} with straight or branched chains are known for $1 \leq n \leq 10$, and Fig. 14.11 compares the boiling points of the first five straight-chain silanes with their hydrocarbon analogues. Silanes are explosively inflammable in air (eq. 14.24).

$$SiH_4 \xrightarrow{\Delta} Si + 2H_2 \tag{14.23}$$

$$SiH_4 + 2O_2 \longrightarrow SiO_2 + 2H_2O \tag{14.24}$$

A mixture of SiH_4, Si_2H_6, Si_3H_8 and Si_4H_{10} along with traces of higher silanes is obtained when Mg_2Si reacts with aqueous

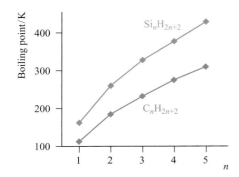

Fig. 14.11 Boiling points of the straight-chain silanes, Si_nH_{2n+2}, and hydrocarbons C_nH_{2n+2}.

acid, but the non-specificity of this synthesis renders it of little practical value. By irradiating SiH_4 with a CO_2 laser, SiH_4 can be converted selectively into Si_2H_6. Silane is a colourless gas which is insoluble in water, reacts rapidly with alkalis (eq. 14.25), and forms compounds of the type M[SiH$_3$] with Na, K (eq. 14.26), Rb and Cs. The crystalline salt K[SiH$_3$] possesses an NaCl structure and is a valuable synthetic reagent, e.g. eq. 14.27.

$$SiH_4 + 2KOH + H_2O \longrightarrow K_2SiO_3 + 4H_2 \tag{14.25}$$

$$2SiH_4 + 2K \xrightarrow{\text{in MeOCH}_2\text{CH}_2\text{OMe}} 2K[SiH_3] + H_2 \tag{14.26}$$

$$Me_3ESiH_3 + KCl \xleftarrow{\text{Me}_3\text{ECl}}{} K[SiH_3] \xrightarrow{\text{MeI}} MeSiH_3 + KI$$
$$\qquad\qquad\quad E = Si, Ge, Sn \tag{14.27}$$

Germanes Ge_nH_{2n+2} (straight and branched chain isomers) are known for $1 \leq n \leq 9$. GeH_4 is less reactive than SiH_4; it is a colourless gas (bp 184 K, dec 488 K), insoluble in water, and prepared by treating GeO_2 with Na[BH$_4$] although higher germanes are also formed. Discharges of various frequencies are finding increased use for this type of synthesis and have been used to convert GeH_4 into higher germanes, or mixtures of SiH_4 and GeH_4 into Ge_2H_6, $GeSiH_6$ and Si_2H_6. Mixed hydrides of Si and Ge, e.g. $GeSiH_6$ and $GeSi_2H_8$, are also formed when an intimate mixture of Mg_2Ge and Mg_2Si is treated with acid. Reactions between GeH_4 and alkali metals, M, in liquid NH_3 produce M[GeH$_3$], and, like [SiH$_3$]$^-$, the [GeH$_3$]$^-$ ion is synthetically useful. For example, reaction 14.28 shows an application of K[GeH$_3$] for the formation of pure $GeSiH_6$ which can be used to grow mixed Ge/Si thin films.

$$K[GeH_3] + H_3SiOSO_2CF_3 \longrightarrow GeSiH_6 + K[CF_3SO_3] \tag{14.28}$$

The reaction of $SnCl_4$ with Li[AlH$_4$] gives SnH_4 (bp 221 K) but this decomposes at 298 K into Sn and H_2. The trend in reactivities is $SiH_4 > GeH_4 < SnH_4$. Plumbane, PbH_4, has been prepared from $Pb(NO_3)_2$ and $NaBH_4$, but decomposes in less than 10 s at room temperature. The IR spectrum (the only means by which PbH_4 has been characterized) is

ENVIRONMENT

Box 14.5 Methane hydrates

A gas hydrate (an example of a *clathrate*) is a crystalline solid comprising a *host* (a 3-dimensional assembly of hydrogen-bonded H_2O molecules which form cage-like arrays) and *guest* molecules (small molecules such as CH_4 which occupy the cavities in the host lattice). The hydrates crystallize in one of three structure types: structure I (the most common), structure II or structure H. In each structure-type, water molecules form a hydrogen-bonded network made up of inter-connected cages, defined by the positions of the O atoms. A structure I hydrate has a cubic unit cell composed of dodecahe-dral (20 H_2O molecules, right-hand side below) and tetrakaide-cahedral (24 H_2O molecules, left-hand side below) cages which share pentagonal faces:

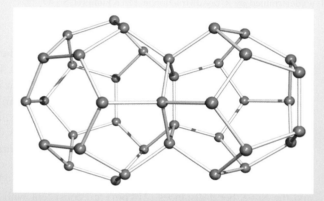

Part of the overall 3-dimensional network of the structure I hydrate is shown below in a stick representation (hydrogen atoms omitted):

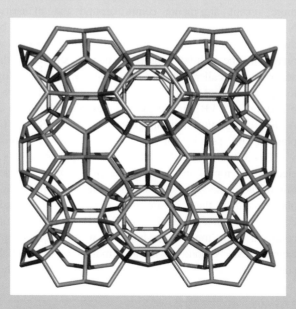

Gas hydrates occur naturally in the Arctic and in deep-sea continental margins. Their importance lies in their ability to trap gases within crystalline masses, thereby acting rather like natural gas 'storage tanks'. Under normal conditions of temperature and pressure, gas clathrates collapse and release the guest molecules, for example CH_4 is released from a gas clathrate of composition $CH_4 \cdot 6H_2O$:

$$CH_4 \cdot 6H_2O \longrightarrow CH_4 + 6H_2O$$

Although methane hydrates represent potential fuel sources, no method of commercially producing methane from them has yet been developed. Uncontrolled release of the huge amounts of CH_4 that are presently trapped inside these clathrates could add to the 'greenhouse' effect (see Box 14.7). The total amount of naturally occurring organic compound-based carbon on Earth is estimated to be about $19\,000 \times 10^{15}$ t. In addition to this, carbon occurs widely in inorganic minerals such as carbonates. The pie-chart below shows the relative importance of methane hydrates as a potential source of carbon from organic-based carbon materials.

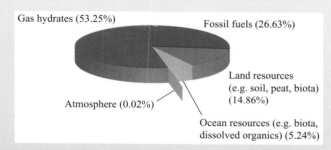

Gas hydrates (53.25%) Fossil fuels (26.63%)

Atmosphere (0.02%)

Land resources (e.g. soil, peat, biota) (14.86%)

Ocean resources (e.g. biota, dissolved organics) (5.24%)

[Data: US Geological Survey]

Further reading

A. Demirbas (2010) *Energy Convers. Manage.*, vol. 51, p. 1547 – 'Methane hydrates as potential energy resource: Part 1 – Importance, resource and recovery facilities'.

A. Demirbas (2010) *Energy Convers. Manage.*, vol. 51, p. 1562 – 'Methane hydrates as potential energy resource: Part 2 – Methane production processes from gas hydrates'.

V. Krey *et al.* (2009) *Environ. Res. Lett.*, vol. 4, article 034007 – 'Gas hydrates: entrance to a methane age or climate threat?'

W.L. Mao, C.A. Koh and E.D. Sloan (2007) *Phys. Today*, vol. 60, p. 42 – 'Clathrate hydrates under pressure'.

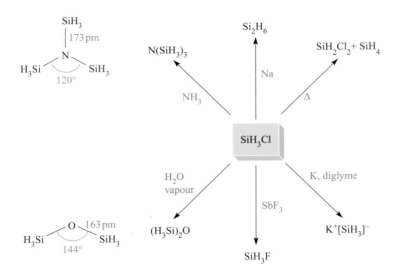

Fig. 14.12 Representative reactions of SiH_3Cl. The structures of $N(SiH_3)_3$ (determined by X-ray diffraction at 115 K) and $(H_3Si)_2O$ (determined by electron diffraction).

consistent with a tetrahedral, molecular structure. Replacement of the H atoms in PbH_4 by alkyl or aryl substituents is accompanied by increased stability (see Section 23.5).

Halohydrides of silicon and germanium

Among compounds of the type SiH_nX_{4-n} (X = halogen, $n = 1–3$), $SiHCl_3$ is of particular importance in the purification of Si in the semiconductor industry (eq. 14.29). The success of the second step in scheme 14.29 depends on the precursor being volatile. $SiHCl_3$ (mp 145 K, bp 306 K) is ideally suited to the process, as is SiH_4 (mp 88 K, bp 161 K).

$$Si(\text{impure}) + 3HCl \xrightarrow[-H_2]{620\,K} SiHCl_3$$

$$\xrightarrow[\substack{2.\,CVD\,(chemical \\ vapour\,deposition)}]{\substack{1.\,Purification \\ by\,distillation}} Si(\text{pure, polycrystalline}) \qquad (14.29)$$

Another application of $SiHCl_3$ is *hydrosilation* (eq. 14.30), a method of introducing an $SiCl_3$ group and an entry to organosilicon chemistry.

$$RCH{=}CH_2 + SiHCl_3 \longrightarrow RCH_2CH_2SiCl_3 \qquad (14.30)$$

$$SiH_4 + nHX \xrightarrow{\Delta,\,AlCl_3} SiH_{4-n}X_n + nH_2$$
$$n = 1 \text{ or } 2 \qquad (14.31)$$

The halo-derivatives SiH_2X_2 and SiH_3X (X = Cl, Br, I) can be prepared from SiH_4 (eq. 14.31) and some reactions of SiH_3Cl (bp 243 K) are shown in Fig. 14.12. The ease with which SiH_nX_{4-n} compounds hydrolyse releasing HX means that they must be handled in moisture-free conditions. The preparation and reactivity of GeH_3Cl resemble those of SiH_3Cl.

The structures of trisilylamine, $N(SiH_3)_3$, and disilyl ether, $(H_3Si)_2O$, are shown in Fig. 14.12. The NSi_3 skeleton in $N(SiH_3)_3$ is planar and the N–Si bond distance of 173 pm is shorter than the sum of the covalent radii ($\Sigma r_{cov} = 193$ pm). Similarly, in $(H_3Si)_2O$, the Si–O–Si bond angle of 144° is large (compare with 111° in Me_2O) and the Si–O bonds of 163 pm are shorter than Σr_{cov}. Trigermylamine is isostructural with $N(SiH_3)_3$, but $P(SiH_3)_3$ is pyramidal with P–Si bonds of length 225 pm. In $(H_3Si)_2S$, the Si–S–Si bond angle is 97° and the Si–S bond distances (214 pm) are consistent with a bond order of 1. For many years, these data were taken as an indication that N and O took part in $(p–d)\pi$-bonding with Si (diagram **14.7**), there being no corresponding interactions in Si–P or Si–S bonds. This explanation is outdated and more recent arguments centre around the planarity of $N(SiH_3)_3$ (and related strengthening of Si–N bonds) being due to $n(N) \longrightarrow \sigma^*(Si–H)$ electron donation, where $n(N)$ represents the non-bonding (lone pair) electrons of the N atom. This so-called *negative hyperconjugation*[†] is analogous to the donation of electrons from a *d*-block metal centre to a σ^*-orbital of a PR_3 ligand that we describe in Section 20.4. A stereoelectronic effect also contributes to $N(SiH_3)_3$ being planar. The polarity of the N–Si bonds ($\chi^P(Si) = 1.9$, $\chi^P(N) = 3.0$) is such that there are significant long-range electrostatic repulsions between the SiH_3 groups. These are minimized if the NSi_3-skeleton in $N(SiH_3)_3$ adopts a trigonal planar, rather than pyramidal, geometry. The outdated concept of $(p–d)\pi$-bonding in $N(SiH_3)_3$ should not be confused with the $(p–p)\pi$-bonding which occurs in, for example, Si=N bonds (with a formal bond order of 2) in compounds such as $^tBu_2Si{=}NSi^tBu_3$, **14.8**. Notice that in

---

[†] Negative hyperconjugation: see Y. Mo, Y. Zhang and J. Gao (1999) *J. Am. Chem. Soc.*, vol. 121, p. 5737 and references cited in this paper.

14.8 the nitrogen atom is in a *linear* environment and can be considered to have a stereochemically inactive lone pair, possibly involved in π-interactions.

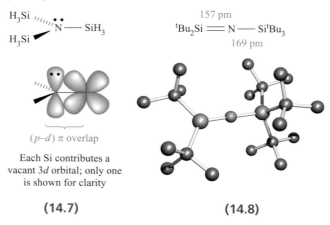

(p–d) π overlap

Each Si contributes a vacant 3d orbital; only one is shown for clarity

(14.7)   **(14.8)**

Self-study exercise

Propose a bonding scheme for $^tBu_2SiNSi^tBu_3$ (**14.8**) that is consistent with the experimentally determined structure. State clearly what hybridization schemes are appropriate.

14.7 Carbides, silicides, germides, stannides and plumbides

Carbides

Classifying carbides is not simple, but some useful categories are:

- saline (salt-like) carbides which produce mainly CH_4 when hydrolysed;
- those containing the $[C{\equiv}C]^{2-}$ ion;
- those containing the $[C{=}C{=}C]^{4-}$ ion;
- interstitial carbides;
- solid state carbides with other 3-dimensional structures;
- fulleride salts (see Section 14.4);
- endohedral metallofullerenes (see Section 14.4).

Examples of saline carbides are Be_2C (see Section 12.4 and eq. 12.14) and Al_4C_3, both made by heating the constituent elements at high temperatures. Although their solid state structures contain isolated C centres which are converted to CH_4 on reaction with H_2O, it is unlikely that the 'C^{4-}' ion is present, since the interelectronic repulsion energy would be enormous.

Carbides containing the $[C{\equiv}C]^{2-}$ (acetylide) ion include Na_2C_2, K_2C_2, MC_2 ($M = Mg$, Ca, Sr, Ba), Ag_2C_2 and Cu_2C_2. They evolve C_2H_2 when treated with water (see eq. 12.15). Calcium carbide is manufactured as a grey solid by heating CaO with coke (see eq. 12.16 and discussion), and when pure, it is colourless. It adopts a distorted

NaCl-type structure, the axis along which the $[C{\equiv}C]^{2-}$ ions are aligned being lengthened; the C–C bond distance is 119 pm, compared with 120 pm in C_2H_2. The reaction between CaC_2 and N_2 (eq. 14.32) is used to manufacture calcium cyanamide, a nitrogenous fertilizer (eq. 14.33). The cyanamide ion, **14.9**, is isoelectronic with CO_2.

$$^-N{=}C{=}N^-$$

(14.9)

$$CaC_2 + N_2 \xrightarrow{1300\,K} CaNCN + C \tag{14.32}$$

$$CaNCN + 3H_2O \longrightarrow CaCO_3 + 2NH_3 \tag{14.33}$$

Equations 14.34 and 14.35 show syntheses of Na_2C_2, Ag_2C_2 and Cu_2C_2. The group 11 carbides are heat- and shock-sensitive, and explosive when dry.

$$2NaNH_2 + C_2H_2 \xrightarrow{\text{in liquid NH}_3} Na_2C_2 + 2NH_3 \tag{14.34}$$

$$2[M(NH_3)_2]^+ + C_2H_2 \longrightarrow M_2C_2 + 2[NH_4]^+ + 2NH_3$$
$$M = Ag, Cu \tag{14.35}$$

Carbides of formula MC_2 do not necessarily contain the acetylide ion. The room temperature form of ThC_2 (Th is an actinoid metal, see Chapter 27) adopts an NaCl-type structure but is not isostructural with CaC_2. In ThC_2, the C_2-units ($d_{CC} = 133$ pm) in alternating layers lie in different orientations. The solid state structure of LaC_2 contains C_2-units with $d_{CC} = 129$ pm. Unlike CaC_2 which is an insulator, ThC_2 and LaC_2 have metallic appearances and are electrical conductors. The C–C bond lengths can be rationalized in terms of structures approximating to $Th^{4+}[C_2]^{4-}$ and $La^{3+}[C_2]^{3-}$. Compared with $[C_2]^{2-}$, the extra electrons in $[C_2]^{4-}$ and $[C_2]^{3-}$ are in antibonding MOs, thus weakening the C–C bond. However, the conducting properties and diamagnetism of ThC_2 and LaC_2 show that this is an oversimplified description since electron delocalization into a conduction band (see Section 6.8) must occur. Hydrolysis of these carbides is also atypical of a $[C_2]^{2-}$-containing species, e.g. the reaction of ThC_2 and H_2O yields mainly C_2H_2, C_2H_6 and H_2.

Carbides containing $[C{=}C{=}C]^{4-}$ are rare. They include Mg_2C_3 which liberates propyne upon hydrolysis.

The structures of the so-called *interstitial carbides* (formed by heating C with d-block metals having $r_{\text{metal}} > 130$ pm, e.g. Ti, Zr, V, Mo, W) may be described in terms of a close-packed metal lattice with C atoms occupying octahedral holes (see Fig. 6.5). In carbides of type M_2C (e.g. V_2C, Nb_2C) the metal atoms are in an hcp lattice and half of the octahedral sites are occupied. In the MC type (e.g. TiC and WC), the metal atoms adopt a ccp structure and all the octahedral holes are occupied. These interstitial carbides are important refractory materials; they are very hard and infusible, have melting points >2800 K and, in contrast to the acetylide derivatives, do not react with water. Tungsten carbide, WC, is one of the hardest substances known and is widely used in cutting tools and dies. Although TiC, WC, V_2C, Nb_2C and related compounds are commonly described as *interstitial* compounds, this does not imply weak bonding. To convert solid carbon into isolated carbon atoms is a very endothermic process and this must be compensated by the formation of strong W$-$C bonds. Similar considerations apply to interstitial nitrides (see Section 15.6).

Transition metals with $r_{\text{metal}} < 130$ pm (e.g. Cr, Fe, Co, Ni) form carbides with a range of stoichiometries (e.g. Cr_3C_2, Fe_3C) which possess complicated structures involving C$-$C bonding. In Cr_3C_2 (formed by reaction 14.36), the Cr atoms form a 3-dimensional structure of edge-sharing trigonal prisms each occupied by a C atom such that carbon chains run through the structure with C$-$C distances comparable to single bonds.

$$3Cr_2O_3 + 13C \xrightarrow[\text{presence of } H_2]{1870\,\text{K, in}} 2Cr_3C_2 + 9CO \qquad (14.36)$$

Carbides of this type are hydrolysed by water or dilute acid to give mixtures of hydrocarbons and H_2.

Silicides

The structures of the metal silicides (prepared by direct combination of the elements at high temperatures) are diverse, and a full discussion of the structures is beyond the scope of this book.[†] Some examples of their solid state structural types are:

- isolated Si atoms (e.g. Mg_2Si, Ca_2Si);
- Si_2-units (e.g. U_3Si_2);
- Si_4-units (e.g. NaSi, KSi, CsSi)
- Si_n-chains (e.g. CaSi);
- planar or puckered hexagonal networks of Si atoms (e.g. β-USi_2, $CaSi_2$);
- 3-dimensional network of Si atoms (e.g. $SrSi_2$, α-USi_2).

The Si_4-units present in the alkali metal silicides are noteworthy. The $[Si_4]^{4-}$ anion is isoelectronic with P_4 and the solid state structures of several group 1 metal silicides contain tetrahedral Si_4-units, but these are not isolated anions. The structure of Cs_4Si_4 comes close to featuring discrete, tetrahedral $[Si_4]^{4-}$ ions, but significant cation–anion interactions exist. The silicide K_3LiSi_4 possesses tetrahedral Si_4-units linked by Li^+ ions to give infinite chains, and in K_7LiSi_8, pairs of Si_4-units are connected as shown in structure **14.10** with additional interactions involving K^+ ions. The tetrahedral clusters present in M_4Si_4 (M = group 1 metal) cannot be extracted into solution, so there is a distinction between the presence of these units in

solid state silicides and the formation of Zintl ions described in the next section.

(14.10)

Silicides are hard materials, but their melting points are generally lower than those of the metal carbides. Treatment of Mg_2Si with dilute acids gives mixtures of silanes (see Section 14.6). The properties of some silicides make them useful as refractory materials (e.g. Fe_3Si and $CrSi_2$). Fe_3Si is used in magnetic tapes and disks to increase their thermal stability.

Zintl ions containing Si, Ge, Sn and Pb

Germanium, tin and lead do not form solid state binary compounds with metals. In contrast, the formation of *Zintl phases* and *Zintl ions* (see Section 9.6), which contain clusters of group 14 metal atoms, is characteristic of these elements. Historically, Zintl phases have been produced by the reduction of Ge, Sn or Pb by Na in liquid NH_3. The synthesis of $[Sn_5]^{2-}$ (eq. 9.35) typifies the preparation of Zintl ions, and the use of the encapsulating ligand crypt-222 to bind an alkali metal counter-ion (see Fig. 11.8) has played a crucial role in the development of Zintl ion chemistry. Thus, salts such as $[K(\text{crypt-222})]_2[Sn_5]$ and $[Na(\text{crypt-222})]_4[Sn_9]$ can be isolated. Modern technology allows low-temperature X-ray diffraction studies of sensitive (e.g. thermally unstable) compounds. It is therefore possible[‡] to investigate salts such as $[Li(NH_3)_4]_4[Pb_9]\cdot NH_3$ and $[Li(NH_3)_4]_4[Sn_9]\cdot NH_3$ which are formed by the direct reaction of an excess of Pb or Sn in solutions of lithium in liquid NH_3. The isolation of silicon-containing Zintl ions *in solution* was not reported until 2004. Dissolution of $K_{12}Si_{17}$ or $Rb_{12}Si_{17}$ (intermetallic compounds, known to contain $[Si_4]^{4-}$ and $[Si_9]^{4-}$ units in the solid state) in liquid NH_3 followed by the addition of crypt-222 produces red solutions from which crystals of $[K(\text{crypt-222})]_3[Si_9]$ and $[Rb(\text{crypt-222})]_3[Si_9]$ (solvated with NH_3) may be isolated. Interestingly, whereas the $M_{12}Si_{17}$ precursor contains $[Si_9]^{4-}$, the species isolated from solution is $[Si_9]^{3-}$. The presence of a mild oxidizing agent (e.g. Ph_3P or Ph_3GeCl) is needed to obtain $[Si_9]^{2-}$ (isolated as $[K(\text{18-crown-6})]_2[Si_9]$) and $[Si_5]^{2-}$ (isolated as $[Rb(\text{crypt-222})]_2[Sn_5]$).

Diamagnetic Zintl ions include $[M_4]^{4-}$ (M = Ge, Sn, Pb), $[M_5]^{2-}$ (M = Si, Sn, Pb), $[M_9]^{4-}$ (M = Ge, Sn, Pb), $[M_9]^{2-}$

[†] For further details, see: A.F. Wells (1984) *Structural Inorganic Chemistry*, 5th edn, Clarendon Press, Oxford, p. 987.

[‡] N. Korber and A. Fleischmann (2001) *J. Chem. Soc., Dalton Trans.*, p. 383.

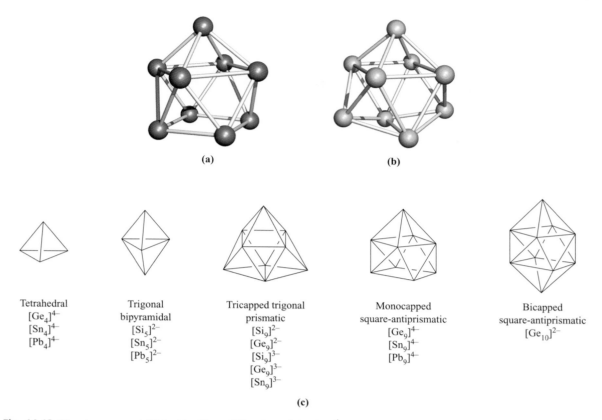

Fig. 14.13 The structures, established by X-ray diffraction, of (a) $[Sn_9]^{4-}$, determined for the salt $[Na(crypt-222)]_4[Sn_9]$ [J.D. Corbett *et al.* (1977) *J. Am. Chem. Soc.*, vol. 99, p. 3313], and (b) $[Ge_9]^{3-}$, determined for the compound $[K(crypt-222)]_3[Ge_9] \cdot PPh_3$ [C. Belin *et al.* (1991) *New J. Chem.*, vol. 15, p. 931]; for discussion of cryptand ligands including crypt-222, see Section 11.8. (c) Schematic representations of structure types for selected Zintl ions. See also Fig. 14.14.

(M = Si, Ge), $[Ge_{10}]^{2-}$ (see discussion at the end of Section 14.7), $[Sn_8Tl]^{3-}$, $[Sn_9Tl]^{3-}$ and $[Pb_2Sb_2]^{2-}$. Paramagnetic ions are exemplified by $[M_9]^{3-}$ (M = Si, Ge, Sn). The structure of $[Sn_5]^{2-}$ was shown in Fig. 9.3. Figure 14.13 shows the structures of $[Sn_9]^{4-}$ and $[Ge_9]^{3-}$, and illustrates some of the main deltahedral families of the group 14 Zintl ions. While Fig. 14.13 shows the $[M_9]^{3-}$ clusters as having tricapped trigonal prismatic structures, the cages are significantly distorted (by means of bond elongation), and lie somewhere between the tricapped trigonal prismatic and monocapped square-antiprismatic limits. Bonding in these ions is delocalized, and for the diamagnetic clusters, Wade's rules (see Section 13.11) can be used to rationalize the observed structures. Wade's rules were developed for borane clusters. A {BH}-unit contributes two electrons to cluster bonding and, similarly, a group 14 atom contributes two electrons to cluster bonding if a lone pair of electrons is localized outside the cage. Thus, in bonding terms, an Si, Ge, Sn or Pb atom can mimic a {BH}-unit. More strictly, an atom of each group 14 element is *isolobal* with a {BH}-unit (see Section 24.5).

Worked example 14.4   Structures of Zintl ions

Rationalize the structure of $[Sn_9]^{4-}$ shown in Fig. 14.13a.

There are nine Sn atoms and each provides two valence electrons, assuming that each atom carries a lone pair of electrons.

There are four electrons from the 4− charge.

Total number of cage-bonding electrons available

$$= (9 \times 2) + 4 = 22 \text{ electrons}$$
$$= 11 \text{ pairs}$$

Thus, $[Sn_9]^{4-}$ has 11 pairs of electrons with which to bond nine Sn atoms.

This means that there are $(n + 2)$ pairs of electrons for n vertices, and so $[Sn_9]^{4-}$ is a *nido*-cage, based on a 10-vertex deltahedron (see Fig. 13.30) with one vertex vacant. This corresponds to the observed structure of a monocapped square-antiprism.

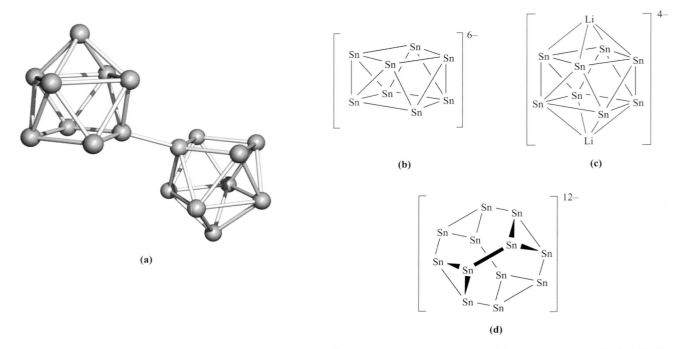

Fig. 14.14 (a) The structure (X-ray diffraction) of the $[(Ge_9)_2]^{6-}$ ion in $Cs_4[K(crypt-222)]_2[(Ge_9)_2]\cdot6en$ (en = 1,2-ethanediamine) [L. Xu *et al.* (1999) *J. Am. Chem. Soc.*, vol. 121, p. 9245]. (b) The *arachno*-$[Sn_8]^{6-}$ cluster in $Rb_4Li_2Sn_8$. (c) The solid state structure of $Rb_4Li_2Sn_8$ shows that Li^+ ions cap the open cage to give $[Li_2Sn_8]^{4-}$ (see text). (d) The open $[Sn_{12}]^{12-}$ cluster in the compound $CaNa_{10}Sn_{12}$; the cage encapsulates a Ca^{2+} ion.

Self-study exercises

1. By referring to Figs. 13.30 and 14.13c, rationalize the structures of: (a) $[Ge_4]^{4-}$; (b) $[Sn_5]^{2-}$; (c) $[Ge_9]^{2-}$; (d) $[Ge_{10}]^{2-}$.

2. Rationalize why $[Sn_5]^{2-}$ and $[Pb_5]^{2-}$ are isostructural.

3. Rationalize why $[Si_5]^{2-}$ adopts the same cluster structure as $C_2B_3H_5$. [*Hint*: Look back to worked example 13.9.]

Reaction conditions are critical to the selective formation of a Zintl ion. For example, the alloy KSn_2 reacts with crypt-222 (see Section 11.8) in 1,2-ethanediamine to give $[K(crypt-222)]_3[Sn_9]$ containing the paramagnetic $[Sn_9]^{3-}$ ion. However, reaction times must be less than two days, since longer periods favour the formation of $[K(crypt-222)]_4[Sn_9]$ containing the diamagnetic $[Sn_9]^{4-}$ ion. The paramagnetic clusters $[Sn_9]^{3-}$ and $[Ge_9]^{3-}$ both adopt distorted tricapped trigonal prismatic structures (Fig. 14.13b). When $Cs_2K[Ge_9]$ is added to a mixture of 1,2-ethanediamine and crypt-222, coupling of the $[Ge_9]^{3-}$ radicals occurs to give $Cs_4[K(crypt-222)]_2[(Ge_9)_2]$; formally, the coupling involves the oxidation of one lone pair on each $[Ge_9]^{3-}$ cage. The structure of the $[(Ge_9)_2]^{6-}$ ion (Fig. 14.14a) consists of two monocapped square-antiprismatic clusters (each with delocalized bonding) connected by a localized, 2-centre 2-electron Ge–Ge

bond. Wade's rules can be applied to each cage in $[(Ge_9)_2]^{6-}$ as follows:

- eight of the Ge atoms each carries a lone pair of electrons and provides two electrons for cluster bonding;
- the Ge atom involved in the inter-cage Ge–Ge bond contributes three electrons to cluster bonding (one electron is used for the external Ge–Ge bond);
- the 6– charge provides three electrons to each cage;
- total electron count *per cage* = 16 + 3 + 3 = 22 electrons;
- 11 pairs of electrons are available to bond nine Ge atoms, and so each cage is classed as a *nido*-cluster, consistent with the observed monocapped square-antiprism (Fig. 14.14a).

The Zintl ions shown in Fig. 14.13 are *closo*- or *nido*-clusters. The compounds $Rb_4Li_2Sn_8$ and $K_4Li_2Sn_8$, which contain *arachno*-$[Sn_8]^{6-}$ (Fig. 14.14b), have been prepared by the direct fusion of tin metal with the respective alkali metals. X-ray diffraction studies on $Rb_4Li_2Sn_8$ show that the *arachno*-$[Sn_8]^{6-}$ cluster is stabilized by interactions with Li^+ ions which effectively close up the open cage as shown in Fig. 14.14c. In addition, each Li^+ ion interacts with an Sn–Sn edge of an adjacent cluster and as a result, a network of interconnected cages is formed, with Rb^+ ions in cavities between the Zintl ions. The combination of small and large cations is an important factor in the stabilization of this system. The same strategy has been used to stabilize another open-cage Zintl ion, $[Sn_{12}]^{12-}$

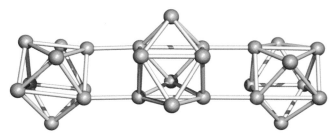

Fig. 14.15 The structure (X-ray diffraction) of the $[(Ge_9)_3]^{6-}$ ion in $[Rb(crypt-222)]_6[(Ge_9)_3]\cdot 3en$ (en = 1,2-ethanediamine) [A. Ugrinov *et al.* (2002) *J. Am. Chem. Soc.*, vol. 124, p. 10990].

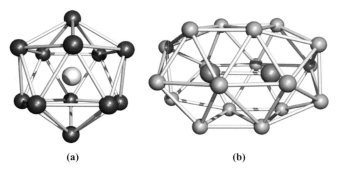

(a)   (b)

Fig. 14.16 (a) The structure of $[Pt@Pb_{12}]^{2-}$ [E.N. Esenturk *et al.* (2004) *Angew. Chem. Int. Ed.*, vol. 43, p. 2132]. (b) The structure of $[Pd_2@Ge_{18}]^{4-}$ [J.M. Goicoechea *et al.* (2005) *J. Am. Chem. Soc.*, vol. 127, p. 7676]. Both structures were determined by X-ray diffraction for $[K(crypt-222)]^+$ salts. Colour code: Pb, red; Ge, orange; Pt, pale grey; Pd, green.

(Fig. 14.14d), which is formed by fusing together stoichiometric amounts of Na, Ca and Sn. The product is $CaNa_{10}Sn_{12}$, and in the solid state, the Ca^{2+} ion provides a stabilizing effect by being sited at the centre of the $[Sn_{12}]^{12-}$ cluster. A related system in which Sr^{2+} replaces Ca^{2+} has also been prepared.

As more Zintl ions are isolated, challenges to the rationalization of the bonding within Wade's rules are encountered. For example, the oxidation of $[Ge_9]^{4-}$ using PPh_3, $AsPh_3$, As or Sb gives $[(Ge_9)_3]^{6-}$ (eqs. 14.37 and 14.38). The $[(Ge_9)_3]^{6-}$ anion (Fig. 14.15) consists of three tricapped trigonal prismatic cages, each with two elongated prism edges.

$$3Rb_4[Ge_9] + 3EPh_3 \longrightarrow Rb_6[(Ge_9)_3] + 3Rb[EPh_2] + 3RbPh$$
$$(E = P, As) \qquad (14.37)$$

$$3[Ge_9]^{4-} + 14E \longrightarrow [(Ge_9)_3]^{6-} + 2[E_7]^{3-} \quad (E = As, Sb)$$
$$(14.38)$$

Cage-coupling also occurs in a saturated 1,2-ethanediamine solution of $Rb_4[Ge_9]$; addition of 18-crown-6 leads to the formation of $[Rb(18-crown-6)]_8[(Ge_9)_4]$. The $[(Ge_9)_4]^{8-}$ ion is structurally similar to $[(Ge_9)_3]^{6-}$ (Fig. 14.15), with four Ge_9-cages connected in a linear chain of overall length 2 nm. This observation leads to the description of the system as a 'nanorod'.

In the discussion of Wade's rules in Box 13.9, we described the involvement of *radial* and *tangential* orbitals in cluster bonding in boranes. Outward-pointing radial orbitals on each B atom are involved in the formation of the external (*exo*) B−H σ-bonds. Similarly, in most Zintl ions, the lone pair of electrons that is localized on each atom is accommodated in an outward-pointing orbital. In the oxidative coupling of two $[Ge_9]^{3-}$ cages to give $[(Ge_9)_2]^{6-}$ (Fig. 14.14a), the localized single bond that joins the cages and which formally arises from the oxidation of a lone pair per cluster is radially oriented with respect to each cluster. However, in $[(Ge_9)_3]^{6-}$ (Fig. 14.15) and $[(Ge_9)_4]^{8-}$, the intercluster bonds are *not* radially related to each cluster, but lie parallel to the prism edges. In addition, the Ge−Ge bond lengths for the intercluster bonds are significantly longer in $[(Ge_9)_3]^{6-}$ and $[(Ge_9)_4]^{8-}$ than

that in $[(Ge_9)_2]^{6-}$. This suggests that the bonds that connect the cages in $[(Ge_9)_3]^{6-}$ and $[(Ge_9)_4]^{8-}$ are of bond orders less than 1 and that the bonding is not localized. It is, therefore, not possible to apply Wade's rules to each cage in this tricluster system.

Figure 14.13 showed $[Ge_{10}]^{2-}$ as a sole example of a 10-atom cluster. $[Ge_{10}]^{2-}$ was reported in 1991, but the structure suffered from a crystallographic disorder (see Box 15.5). Homonuclear Zintl ions with more than nine cage atoms can be stabilized by the inclusion of an interstitial atom. Such endohedral Zintl ions (written as $[M@E_n]^{x-}$ where M is the interstitial atom) include:

- $[Ni@Pb_{10}]^{2-}$ with a bicapped square-antiprismatic arrangement of Pb atoms;
- $[Pt@Pb_{12}]^{2-}$ with an icosahedral arrangement of Pb atoms;
- $[Pd_2@Ge_{18}]^{4-}$ with the structure shown in Fig. 14.16b.

Their syntheses (eqs. 14.39–14.41) are similar, starting from an $[E_9]^{4-}$ ion and a source of metal M in a zero oxidation state. In eq. 14.39, cod stands for cycloocta-1,5-diene (structure **24.20**).

$$\underset{\text{in 1,2-ethanediamine}}{K_4[Pb_9]} + \underset{\text{in toluene}}{Ni(cod)_2} \xrightarrow{\text{crypt-222}}$$
$$[K(crypt-222)]_2[Ni@Pb_{10}]$$
$$(14.39)$$

$$\underset{\text{in 1,2-ethanediamine}}{K_4[Pb_9]} + \underset{\text{in toluene}}{Pt(PPh_3)_4} \xrightarrow{\text{crypt-222}}$$
$$[K(crypt-222)]_2[Pt@Pb_{12}]$$
$$(14.40)$$

$$\underset{\text{in 1,2-ethanediamine}}{K_4[Ge_9]} + \underset{\text{in toluene}}{Pd(PPh_3)_4} \xrightarrow{\text{crypt-222}}$$
$$[K(crypt-222)]_4[Pd_2@Ge_{18}]$$
$$(14.41)$$

The anions $[Ni@Pb_{10}]^{2-}$ and $[Pt@Pb_{12}]^{2-}$ obey Wade's rules, provided that the central M(0) atom contributes zero valence electrons to cluster bonding. Following the procedure shown in worked example 14.4, $[Ni@Pb_{10}]^{2-}$ possesses $(10 \times 2) + 2 = 22$ cluster-bonding electrons; 11 pairs of electrons are consistent with a *closo*-cage with 10 vertices. $[Pd_2@Ge_{18}]^{4-}$ is unusual in possessing such a large, single-cage deltahedron.

14.8 Halides and complex halides

Carbon halides

Selected physical properties of the tetrahalides of C and Si are listed in Table 14.3. The carbon tetrahalides differ markedly from those of the later group 14 elements: they are inert towards water and dilute alkali and do not form complexes with metal halides. Historically, the distinction has been attributed to the absence of d orbitals in the valence shell of a C atom; look back at the electronic versus steric debate, outlined in Section 14.3. However, one must be cautious. In the case of CX_4 being inert towards attack by water, the 'lack of C d orbitals' presupposes that the reaction would proceed through a 5-coordinate intermediate (i.e. as is proposed for hydrolysis of silicon halides). Of course, it is impossible to establish the mechanism of a reaction that does not occur! Certainly, CF_4 and CCl_4 are *thermodynamically* unstable with respect to hydrolysis; compare the value of $\Delta_r G^{\circ}$ for eq. 14.42 with that of $-290\,\text{kJ}\,\text{mol}^{-1}$ for the hydrolysis of $SiCl_4$.

$$CCl_4(l) + 2H_2O(l) \longrightarrow CO_2(g) + 4HCl(aq)$$

$$\Delta_r G^{\circ} = -380\,\text{kJ}\,\text{mol}^{-1} \qquad (14.42)$$

Carbon tetrafluoride is extremely inert and may be prepared by the reaction of SiC and F_2, with the second product, SiF_4, being removed by passage through aqueous NaOH. Equation 14.43 shows a convenient laboratory-scale synthesis of CF_4 from graphite-free calcium cyanamide (see structure **14.9**); trace amounts of CsF are added to prevent the formation of NF_3.

$$CaNCN + 3F_2 \xrightarrow{\text{CsF, 298 K, 12 h}} CF_4 + CaF_2 + N_2$$

$$(14.43)$$

Uncontrolled fluorination of an organic compound usually leads to decomposition because large amounts of heat are evolved (eq. 14.44).

$$\Delta_r H^{\circ} = -480\,\text{kJ}\,\text{mol}^{-1} \qquad (14.44)$$

The preparation of a fully fluorinated organic compound tends therefore to be carried out in an inert solvent (the vaporization of which consumes the heat liberated) in a reactor packed with gold- or silver-plated copper turnings (which similarly absorb heat but may also play a catalytic role). Other methods include use of CoF_3 or AgF_2 as fluorinating agents, or electrolysis in liquid HF (see Section 9.7).

Fluorocarbons (see also Section 17.3) have boiling points close to those of the corresponding hydrocarbons but have higher viscosities. They are inert towards concentrated alkalis and acids, and dissolve only in non-polar organic solvents. Their main applications are as high-temperature lubricants. *Freons* are chlorofluorocarbons (CFCs) or chlorofluorohydrocarbons, made by partial replacement of chlorine as in, for example, the first step of scheme 14.45. Although CFCs used to be applied extensively in aerosol propellants, air-conditioners, foams for furnishings, refrigerants and solvents, concern over their role in the depletion of the ozone layer has resulted in rapid phasing out of their use (see Box 14.6).

$$CHCl_3 \xrightarrow[\text{SbCl}_5, \text{SbF}_3]{\text{HF}} CHF_2Cl \xrightarrow{970\,\text{K}} C_2F_4 + HCl \qquad (14.45)$$

Two important polymers are manufactured from chlorofluoro-compounds. The monomer for the commercially named *Teflon* or PTFE is C_2F_4 (tetrafluoroethene) which is prepared by reaction 14.45. Polymerization occurs in the presence of water with an organic peroxide catalyst. Teflon is an inert white solid, stable up to 570 K. It has widespread domestic applications, e.g. non-stick coatings for kitchenware. The monomer CF_2=CFCl is used to manufacture the commercial polymer *Kel-F*. Both Teflon

Table 14.3   Selected physical properties of the carbon and silicon tetrahalides.

Property	CF_4	CCl_4	CBr_4	CI_4	SiF_4	$SiCl_4$	$SiBr_4$	SiI_4
Melting point / K	89	250	363	444 (dec)	183	203	278.5	393.5
Boiling point / K	145	350	462.5	–	187	331	427	560.5
Appearance at 298 K	Colourless gas	Colourless liquid	Colourless solid	Dark red solid	Colourless gas, fumes in air	Colourless, fuming liquid	Colourless, fuming liquid	Colourless solid

ENVIRONMENT

Box 14.6 CFCs and the Montreal Protocol

The *ozone layer* is a stratum in the atmosphere 15–30 km above the Earth's surface. It protects life on the Earth from UV radiation originating from the Sun because O_3 absorbs strongly in the ultraviolet region of the spectrum. An effect of UV radiation on humans is skin cancer. Chlorofluorocarbons (CFCs) are atmospheric pollutants which contribute towards the depletion of the ozone layer. In 1987, the 'Montreal Protocol for the Protection of the Ozone Layer' was established and legislation was implemented to phase out the use of CFCs, e.g. in aerosol propellants, refrigerants: an almost complete phase-out of CFCs was required by 1996 for industrial nations, with developing nations following this ban by 2010. The phasing out of CFCs has affected the manufacture of asthma inhalers, large numbers of which used to use a CFC-based propellant. These inhalers have been replaced by models with hydrofluoroalkane (HFA) propellants.

CFCs are not the only ozone-depleting chemicals. Other 'Class I' ozone-depleters include CH_2ClBr, CBr_2F_2, CF_3Br, CCl_4, $CHCl_3$ and CH_3Br. In the past, methyl bromide has been used as an agricultural pest control (see Box 17.3). Alternative pesticides for soil treatment continue to be developed in order to comply with the Montreal Protocol which has banned the use of CH_3Br since 2005 (from 2015 in developing countries). Strictly controlled exceptions to the ban are currently (2010–2012) permitted. For example, the US Environmental Protection Agency allows 'critical use exemptions' where viable alternatives to CH_3Br are still unavailable.

While less harmful to the environment than CFCs, hydrochlorofluorocarbons (HCFCs) are still ozone-depleting (they are classified as 'Class II' ozone-depleters) and will be phased out by 2020. Hydrofluorocarbons appear to have little or no ozone-depleting effect and are replacing CFCs and HCFCs in refrigerants and aerosol propellants.

Loss of ozone was first detected in the stratosphere over Antarctica, and the growth of the 'ozone hole' is now monitored from satellite photographs and by using ground-based instruments. The chemical events and environmental circumstances that lead to ozone depletion over Antarctica can be summarized as follows. Initially, emissions of CFCs enter the stratosphere and are decomposed by high-energy UV radiation. Over the Antarctic, polar stratospheric clouds (containing ice with dissolved HNO_3) form in the 'polar vortex' in the exceptionally cold winter temperatures. It is on the surfaces of these clouds that HCl and $ClONO_2$ (the long-lived chlorine carriers after CFC breakdown) are converted to active forms of chlorine:

$$HCl + ClONO_2 \longrightarrow HNO_3 + Cl_2$$

$$H_2O + ClONO_2 \longrightarrow HNO_3 + HOCl$$

$$HCl + HOCl \longrightarrow H_2O + Cl_2$$

$$N_2O_5 + HCl \longrightarrow HNO_3 + ClONO$$

$$N_2O_5 + H_2O \longrightarrow 2HNO_3$$

In the Antarctic winter, sunlight is absent. Once it returns in the spring (i.e. September), photolysis of Cl_2 results in the

formation of chlorine radicals, $Cl^{\bullet}$, and their presence initiates catalytic O_3 destruction:

$$2ClO^{\bullet} \longrightarrow Cl_2O_2$$

$$Cl_2O_2 \xrightarrow{h\nu} Cl^{\bullet} + ClO_2^{\bullet}$$

$$ClO_2^{\bullet} \longrightarrow Cl^{\bullet} + O_2$$

$$2Cl^{\bullet} + 2O_3 \longrightarrow 2ClO^{\bullet} + 2O_2$$

The $ClO^{\bullet}$ goes back into the cycle of reactions, and, from the steps shown above, the overall reaction is:

$$2O_3 \longrightarrow 3O_2$$

The role of bromine can be summarized in the following reaction sequence:

$$ClO^{\bullet} + BrO^{\bullet} \longrightarrow Cl^{\bullet} + Br^{\bullet} + O_2$$

$$Cl^{\bullet} + O_3 \longrightarrow ClO^{\bullet} + O_2$$

$$Br^{\bullet} + O_3 \longrightarrow BrO^{\bullet} + O_2$$

Further information

For up-to-date information from the Environmental Protection Agency, see: http://www.epa.gov/ozone/

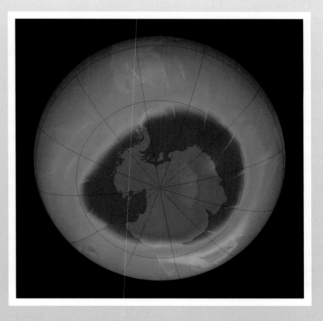

A false colour satellite photograph (taken in September 2009) of the hole in the ozone layer over Antarctica. This was the fifth largest ozone hole recorded, despite reductions in ozone depleters.

and Kel-F are used in laboratory equipment such as sealing tape and washers, parts in gas cylinder valves and regulators, coatings for stirrer bars, and sleeves for glass joints operating under vacuum.

Carbon tetrachloride (Table 14.3) is produced by chlorination of CH_4 at 520–670 K or by the reaction sequence 14.46, in which the CS_2 is recycled.

$$\left.\begin{array}{l} CS_2 + 3Cl_2 \xrightarrow{\text{Fe catalyst}} CCl_4 + S_2Cl_2 \\ CS_2 + 2S_2Cl_2 \longrightarrow CCl_4 + 6S \\ 6S + 3C \longrightarrow 3CS_2 \end{array}\right\} \quad (14.46)$$

In the past, CCl_4 was widely used as a solvent and for the chlorination of inorganic compounds. However, its high toxicity and the fact that photochemical or thermal decomposition results in the formation of $CCl_3^{\bullet}$ and $Cl^{\bullet}$ radicals has led to its manufacture and use being controlled by environmental legislation. The potentially violent reaction of CCl_4 with Na (eq. 14.47) demonstrates why sodium should never be used to dry halogenated solvents.

$$CCl_4 + 4Na \longrightarrow 4NaCl + C \qquad \Delta_r G^o = -1478 \text{ kJ mol}^{-1}$$
$$(14.47)$$

Reactions 14.48 and 14.49 give preparations of CBr_4 and CI_4 (Table 14.3). Both compounds are toxic and are easily decomposed to their elements $(\Delta_f G^o (CBr_4, s, 298 \text{ K}) = +47.7 \text{ kJ mol}^{-1})$. CI_4 decomposes slowly in the presence of H_2O, giving CHI_3 and I_2.

$$3CCl_4 + 4AlBr_3 \longrightarrow 3CBr_4 + 4AlCl_3 \qquad (14.48)$$

$$CCl_4 + 4C_2H_5I \xrightarrow{AlCl_3} CI_4 + 4C_2H_5Cl \qquad (14.49)$$

Carbonyl chloride (*phosgene*), **14.11**, is a highly toxic, colourless gas (bp 281 K) with a choking smell, and was used in World War I chemical warfare. It is manufactured by reaction 14.50, and is used industrially in the production of polycarbonates (formed from bisphenol A, **14.13** and $COCl_2$ and used, for example, in compact discs and DVDs), diisocyanates (for polyurethane polymers) and 1-naphthyl-*N*-methylcarbamate, **14.12** (for insecticides). Because of the highly toxic nature of $COCl_2$, new phosgene-free methods of industrial scale synthesis of polycarbonates are being sought.[†]

$$CO + Cl_2 \xrightarrow{\text{activated carbon catalyst}} COCl_2 \qquad (14.50)$$

[†]See: W.B. Kim, U.A. Joshi and J.S. Lee (2004) *Ind. Eng. Chem. Res.*, vol. 43, p. 1897; B. Schäffner, F. Schäffner, S.P. Verevkin and A. Börner (2010) *Chem. Rev.*, vol. 110, p. 4554.

Phosgene

(14.11)   **(14.12)**

Bisphenol A   Urea

(14.13)   **(14.14)**

Fluorination of $COCl_2$ using SbF_3 yields $COClF$ and COF_2 which, like $COCl_2$, are unstable to water, and react with NH_3 (to give urea, **14.14**) and alcohols (to give esters). Reaction of $COCl_2$ with SbF_5 yields the linear cation $[ClCO]^+$. Its presence in the condensed phase has been established by vibrational spectroscopy. Reaction between COF_2 and SbF_5, however, gives an adduct $F_2CO \cdot SbF_5$ rather than $[FCO]^+[SbF_6]^-$.

Silicon halides

Many fluorides and chlorides of Si are known, but we confine our discussion to SiF_4 and $SiCl_4$ (Table 14.3) and some of their derivatives. Silicon and Cl_2 react to give $SiCl_4$, and SiF_4 can be obtained by fluorination of $SiCl_4$ using SbF_3, or by reaction 14.51; compare with eqs. 13.28 and 15.84.

$$SiO_2 + 2H_2SO_4 + 2CaF_2 \longrightarrow SiF_4 + 2CaSO_4 + 2H_2O$$
$$(14.51)$$

Both SiF_4 and $SiCl_4$ are molecular with tetrahedral structures. They react readily with water, but the former is only partially hydrolysed (compare eqs. 14.52 and 14.53). Controlled hydrolysis of $SiCl_4$ results in the formation of $(Cl_3Si)_2O$, through the intermediate $SiCl_3OH$.

$$2SiF_4 + 4H_2O \longrightarrow SiO_2 + 2[H_3O]^+ + [SiF_6]^{2-} + 2HF$$
$$(14.52)$$

$$SiCl_4 + 2H_2O \longrightarrow SiO_2 + 4HCl \qquad (14.53)$$

The reaction between equimolar amounts of neat $SiCl_4$ and $SiBr_4$ at 298 K leads to an equilibration mixture of $SiCl_4$, $SiBrCl_3$, $SiBr_2Cl_2$, $SiBr_3Cl$ and $SiBr_4$ (see end-of-chapter problem 4.39) which can be separated by fractional

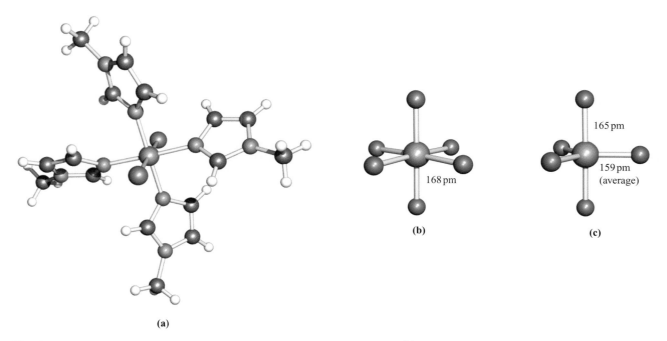

Fig. 14.17 Solid state structures (X-ray diffraction) of (a) *trans*-$[SiCl_2(MeIm)_4]^{2+}$ from the salt $[SiCl_2(MeIm)_4]Cl_2\cdot3CHCl_3$ (MeIm = *N*-methylimidazole) [K. Hensen *et al.* (2000) *J. Chem. Soc., Dalton Trans.*, p. 473], (b) octahedral $[SiF_6]^{2-}$, determined for the salt $[C(NH_2)_3]_2[SiF_6]$ [A. Waskowska (1997) *Acta Crystallogr., Sect. C*, vol. 53, p. 128] and (c) trigonal bipyramidal $[SiF_5]^-$, determined for the compound $[Et_4N][SiF_5]$ [D. Schomburg *et al.* (1984) *Inorg. Chem.*, vol. 23, p. 1378]. Colour code: Si, pink; F, green; N, blue; C, grey; Cl, green; H, white.

distillation. The Lewis base *N*-methylimidazole (MeIm) reacts with $SiCl_4$ and $SiBr_2Cl_2$ to give *trans*-$[SiCl_2(MeIm)_4]^{2+}$ (Fig. 14.17a) as the chloride and bromide salts respectively. This provides a means of stabilizing an $[SiCl_2]^{2+}$ cation.

The formation of $[SiF_6]^{2-}$, the hexafluoridosilicate ion (Fig. 14.17b), illustrates the ability of Si to act as an F^- acceptor and increase its coordination number beyond 4. Hexafluoridosilicates are best prepared by reactions of SiF_4 with metal fluorides in aqueous HF; the K^+ and Ba^{2+} salts are sparingly soluble. In aqueous solution, H_2SiF_6 behaves as a strong acid, but the pure compound has not been isolated. The $[SiF_5]^-$ ion (Fig. 14.17c) is formed in the reaction of SiO_2 with aqueous HF, and may be isolated as a tetraalkylammonium ion. Silicon tetrachloride does not react with alkali metal chlorides, although lattice energy considerations suggest that it might be possible to stabilize the $[SiCl_6]^{2-}$ ion using a very large quaternary ammonium cation.

Halides of germanium, tin and lead

There are many similarities between the tetrahalides of Ge and Si, and GeX_4 (X = F, Cl, Br or I) is prepared by direct combination of the elements. At 298 K, GeF_4 is a colourless gas, $GeCl_4$ is a colourless liquid, and GeI_4 is a red-orange solid (mp 417 K); $GeBr_4$ melts at 299 K. Each hydrolyses, liberating HX. Unlike $SiCl_4$, $GeCl_4$ accepts Cl^- (e.g. reaction 14.54).

$$GeCl_4 + 2[Et_4N]Cl \xrightarrow{\text{in } SOCl_2} [Et_4N]_2[GeCl_6] \qquad (14.54)$$

The Si(II) halides SiF_2 and $SiCl_2$ can be obtained only as unstable species (by action of SiF_4 or $SiCl_4$ on Si at ≈1500 K) which polymerize to cyclic products. In contrast, Ge forms stable dihalides; GeF_2, $GeCl_2$ and $GeBr_2$ are produced when Ge is heated with GeX_4, but the products disproportionate on heating (eq. 14.55).

$$2GeX_2 \xrightarrow{\Delta} GeX_4 + Ge \qquad (14.55)$$

Reaction between GeF_2 and F^- gives $[GeF_3]^-$. Several compounds of type $MGeCl_3$ exist where M^+ may be an alkali metal ion or a quaternary ammonium or phosphonium ion (e.g. eqs. 14.56–14.58). Crystal structure determinations for $[BzEt_3N][GeCl_3]$ (Bz = benzyl) and $[Ph_4P][GeCl_3]$ confirm the presence of well-separated trigonal pyramidal $[GeCl_3]^-$ ions. In contrast, $CsGeCl_3$ adopts a perovskite-type structure (Fig. 6.24) which is distorted at 298 K and non-distorted above 328 K. $CsGeCl_3$ belongs to a group of semiconducting compounds $CsEX_3$ (E = Ge, Sn, Pb; X = Cl, Br, I).

$$Ge(OH)_2 \xrightarrow{\text{CsCl, conc HCl}} CsGeCl_3 \qquad (14.56)$$

$$GeCl_2(1,4\text{-dioxane}) + Ph_4PCl$$
$$\longrightarrow [Ph_4P][GeCl_3] + 1,4\text{-dioxane} \qquad (14.57)$$

$$Ge + RbCl \xrightarrow{\text{in 6M HCl}} RbGeCl_3 \qquad (14.58)$$

The preference for the +2 over the +4 oxidation state increases down the group, the change being due to the thermodynamic 6s inert pair effect (Box 13.4). Whereas members of the GeX_4 family are more stable than GeX_2, PbX_2 halides are more stable than PbX_4. Tin tetrafluoride (which forms hygroscopic crystals) is prepared from $SnCl_4$ and HF. At 298 K, SnF_4 is a white solid and has a sheet structure, **14.15**, with octahedral Sn atoms. At 978 K, SnF_4 sublimes to give a vapour containing tetrahedral molecules. Lead tetrafluoride (mp 870 K) has the same solid state structure as SnF_4, and may be prepared by the action of F_2 or halogen fluorides on Pb(II) compounds, e.g. PbF_2 or $Pb(NO_3)_2$.

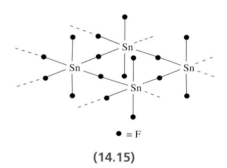

$\bullet = F$

(14.15)

F — Sn — F ... 205 pm ... 218 pm

(14.16)

Tin(II) fluoride is water-soluble and can be prepared in aqueous media. In contrast, PbF_2 is only sparingly soluble. One form of PbF_2 adopts a CaF_2-type structure (see Fig. 6.19a). The solid state structure of SnF_2 consists of puckered Sn_4F_8 rings, **14.16**, with each Sn being trigonal pyramidal consistent with the presence of a lone pair. In structures **14.15** and **14.16**, the Sn–F bridge bonds are longer than the terminal

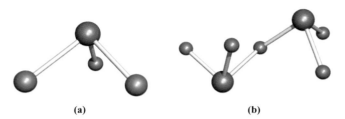

(a)   **(b)**

Fig. 14.18 The structures of (a) $[SnCl_2F]^-$ and (b) $[Sn_2F_5]^-$ from the solid state structure (X-ray diffraction) of $[Co(en)_3][SnCl_2F][Sn_2F_5]Cl$ (en, see Table 7.7); each Sn atom is in a trigonal pyramidal environment [I.E. Rakov *et al.* (1995) *Koord. Khim.*, vol. 21, p. 16]. Colour code: Sn, brown; F, small green; Cl, large green.

bonds, a feature that is common in this type of structure. Many tin fluoride compounds show a tendency to form F–Sn–F bridges in the solid state, as we illustrate later.

Tin(IV) chloride, bromide and iodide are made by combining the respective elements and resemble their Si and Ge analogues. The compounds hydrolyse, liberating HX, but hydrates such as $SnCl_4 \cdot 4H_2O$ can also be isolated. The reaction of Sn and HCl gives $SnCl_2$, a white solid which is partially hydrolysed by water. The hydrate $SnCl_2 \cdot 2H_2O$ is commercially available and is used as a reducing agent. In the solid state, $SnCl_2$ has a puckered-layer structure, but discrete, bent molecules are present in the gas phase.

The Sn(IV) halides are Lewis acids, and their ability to accept halide ions (e.g. reaction 14.59) follows the order $SnF_4 > SnCl_4 > SnBr_4 > SnI_4$.

$$2KCl + SnCl_4 \xrightarrow{\text{in presence of HCl(aq)}} K_2[SnCl_6] \qquad (14.59)$$

Similarly, $SnCl_2$ accepts Cl^- to give trigonal pyramidal $[SnCl_3]^-$, but the existence of discrete anions in the solid state is cation-dependent (see earlier discussion of $CsGeCl_3$). The $[SnF_5]^-$ ion can be formed from SnF_4, but in the solid state, it is polymeric with bridging F atoms and octahedral Sn centres. The bridging F atoms are mutually *cis* to one another. Bridge formation is similarly observed in Na^+ salts of $[Sn_2F_5]^-$ and $[Sn_3F_{10}]^{4-}$, formed by reacting NaF and SnF_2 in aqueous solution. Figure 14.18 shows the structures of the $[SnCl_2F]^-$ and $[Sn_2F_5]^-$ ions.

Lead tetrachloride is obtained as an oily liquid by the reaction of cold concentrated H_2SO_4 on $[NH_4]_2[PbCl_6]$. The latter is made by passing Cl_2 through a saturated solution of $PbCl_2$ in aqueous NH_4Cl. The ease with which $[PbCl_6]^{2-}$ is obtained is a striking example of stabilization of a higher oxidation state by complexation (see Section 8.3). In contrast, $PbCl_4$ is hydrolysed by water and decomposes to $PbCl_2$ and Cl_2 when gently heated. The Pb(II) halides are considerably more stable than their Pb(IV) analogues and are crystalline solids at 298 K. They can be precipitated by mixing aqueous solutions of soluble halide and soluble Pb(II) salts (e.g. eq. 14.60). Note that

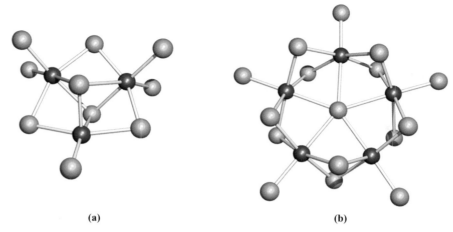

(a)   **(b)**

Fig. 14.19 The structures (X-ray diffraction) of (a) the $[Pb_3I_{10}]^{4-}$ ion in the $[^nBu_3N(CH_2)_4N^nBu_3]^{2+}$ salt [H. Krautscheid *et al.* (1999) *J. Chem. Soc., Dalton Trans.*, p. 2731] and (b) the $[Pb_5I_{16}]^{6-}$ ion in the salt $[^nBuN(CH_2CH_2)_3N^nBu]_3[Pb_5I_{16}]\cdot 4DMF$ [H. Krautscheid *et al.* (2000) *Z. Anorg. Allg. Chem.*, vol. 626, p. 3]. Colour code: Pb, blue; I, yellow.

only a few Pb(II) salts (e.g. $Pb(NO_3)_2$ and $Pb(O_2CMe)_2$) are very soluble in water.

$$Pb(NO_3)_2(aq) + 2NaCl(aq) \longrightarrow PbCl_2(s) + 2NaNO_3(aq)$$
$$(14.60)$$

Lead(II) chloride is much more soluble in hydrochloric acid than in water owing to the formation of $[PbCl_4]^{2-}$. In the solid state, $PbCl_2$ has a complicated structure with 9-coordinate Pb centres, but PbF_2 adopts the fluorite structure (Fig. 6.19a). Yellow PbI_2 adopts the CdI_2 structure (Fig. 6.23). Discrete iodidoplumbate anions such as $[Pb_3I_{10}]^{4-}$ (Fig. 14.19a), $[Pb_7I_{22}]^{8-}$, $[Pb_{10}I_{28}]^{8-}$ and $[Pb_5I_{16}]^{6-}$ (Fig. 14.19b) as well as related polymeric iodidoplumbates[†] can be formed by reacting PbI_2 and NaI in the presence of large cations such as $[R_3N(CH_2)_4NR_3]^{2+}$ (R = Me, ^nBu) or $[P(CH_2Ph)_4]^+$. The reactions can be driven towards a particular product by varying the reactant stoichiometry, reaction conditions and counter-ion. In these iodidoplumbates, the Pb(II) centres are in either octahedral or square-based pyramidal environments (Fig. 14.19).

Worked example 14.5   Group 14 halides: structure and energetics

SnF_4 sublimes at 978 K. Describe the changes that take place during sublimation and the processes that contribute to the enthalpy of sublimation.

Sublimation refers to the process:

$$SnF_4(s) \longrightarrow SnF_4(g)$$

In the solid state, SnF_4 has a sheet structure (see structure **14.15**) in which each Sn is octahedrally sited. In the gas phase, SnF_4 exists as discrete, tetrahedral molecules. During sublimation, the SnF_4 units must be released from the solid state structure, and this involves breaking Sn–F–Sn bridges and converting them into terminal Sn–F bonds. Each Sn atom goes from an octahedral to a tetrahedral environment. Enthalpy changes that take place are:

- enthalpy change associated with Sn–F bond cleavage (endothermic process);
- enthalpy change associated with the conversion of half an Sn–F–Sn bridge interaction to a terminal Sn–F bond (two of these per molecule);
- enthalpy change associated with a change in hybridization of the Sn atom as it changes from octahedral to tetrahedral, and an associated change in the Sn–F bond strength for the terminal Sn–F bonds.

Self-study exercises

1. Above 328 K, $CsGeCl_3$ adopts a perovskite structure; at 298 K, the structure is distorted, but remains based on perovskite. Does solid $CsGeCl_3$ contain discrete $[GeCl_3]^-$ ions? Explain your answer.
   [*Ans.* Refer to Fig. 6.24 and related discussion]

2. Explain why PbX_2 halides are more stable than PbX_4 halides. [*Ans.* The answer is in Box 13.4]

3. In reactions 14.54 and 14.57, which reactants are Lewis acids and which are Lewis bases? Give an explanation for your answer. What is the general name for the products?
   [*Ans.* Acid = electron acceptor; base = electron donor; adduct]

[†]See for example: H. Krautscheid, C. Lode, F. Vielsack and H. Vollmer (2001) *J. Chem. Soc., Dalton Trans.*, p. 1099.

14.9 Oxides, oxoacids and hydroxides

Oxides and oxoacids of carbon

Unlike the later elements in group 14, carbon forms stable, volatile monomeric oxides: CO and CO_2. A comment on the difference between CO_2 and SiO_2 can be made in the light of the thermochemical data in Table 14.2: the $C=O$ bond enthalpy term is *more* than twice that for the $C-O$ bond, while the $Si=O$ bond enthalpy term is *less* than twice that of the $Si-O$ bond. In rationalizing these differences, there is justification for saying that the $C=O$ bond is strengthened relative to $Si=O$ by $(p-p)\pi$ contributions. In the past, it has been argued that the $Si-O$ bond is strengthened relative to the $C-O$ bond by $(p-d)\pi$-bonding (but see comments at the end of Section 14.6). Irrespective of the interpretation of the enthalpy terms, however, the data indicate that (ignoring enthalpy and entropy changes associated with vaporization) SiO_2 is stable with respect to conversion into molecular $O=Si=O$, while (unless subjected to extreme conditions, see later) CO_2 is stable with respect to the formation of a macromolecular species containing 4-coordinate C and $C-O$ single bonds.

Carbon monoxide is a colourless gas, formed when C burns in a restricted supply of O_2. Small-scale preparations involve the dehydration of methanoic acid (eq. 14.61). CO is manufactured by reduction of CO_2 using coke heated above 1070 K or by the water–gas shift reaction (see Section 10.4). Industrially, CO is very important and we consider some relevant catalytic processes in Chapter 25. The thermodynamics of the oxidation of carbon is of immense importance in metallurgy as discussed in Section 8.8.

$$HCO_2H \xrightarrow{\text{conc } H_2SO_4} CO + H_2O \qquad (14.61)$$

Carbon monoxide is almost insoluble in water under normal conditions and does not react with aqueous NaOH, but at high pressures and temperatures, HCO_2H and $Na[HCO_2]$ are formed respectively. Carbon monoxide combines with F_2, Cl_2 and Br_2 (e.g. eq. 14.50), sulfur and selenium. The high toxicity of CO arises from the formation of a stable complex with haemoglobin (see Section 29.3) with the consequent inhibition of O_2 transport in the body. The oxidation of CO to CO_2 can be used for the quantitative analysis for CO (eq. 14.62) with the I_2 formed being titrated against thiosulfate. CO is similarly oxidized by a mixture of MnO_2, CuO and Ag_2O at ambient temperatures and this reaction is used in respirators.

$$I_2O_5 + 5CO \longrightarrow I_2 + 5CO_2 \qquad (14.62)$$

Selected physical properties of CO and CO_2 are given in Table 14.4. Bonding models are described in Sections 2.7 and 5.7. The bond in CO is the strongest known in a stable molecule and confirms the efficiency of $(p-p)\pi$-bonding between the $2p$ orbitals of C and O. However,

Table 14.4   Selected properties of CO and CO_2.

Property	CO	CO_2
Melting point / K	68	–
Boiling point / K	82	195 (sublimes)
$\Delta_f H^\circ$ (298 K) / kJ mol^{-1}	–110.5	–393.5
$\Delta_f G^\circ$ (298 K) / kJ mol^{-1}	–137	–394
Bond energy / kJ mol^{-1}	1075	806
$C-O$ bond distance / pm	112.8	116.0
Dipole moment / D	0.11	0

considerations of the bonding provide no simple explanation as to why the dipole moment of CO is so low. We described the bonding in CO using MO theory in Fig. 2.15. The HOMO of the CO molecule is predominantly an outward-pointing orbital centred on the C atom. As a result, CO acts as a donor to electron-deficient molecules such as BH_3 (see worked example 13.3). More important is the role that CO plays in organometallic chemistry, and we return to this in Chapter 24.

In an excess of O_2, C burns to give CO_2. Under normal temperatures and pressures, CO_2 exists as linear molecules with $C=O$ double bonds. Solid phases containing CO_2 molecules can be produced at low temperatures and high pressures. The most commonly encountered example is *dry ice* which is produced by first liquefying CO_2 at a pressure of 6 MPa, and then cooling the liquid CO_2 (still under pressure) to its freezing point of 195 K. Subjecting a molecular phase of CO_2 to laser-heating at 1800 K, under a pressure of 40 GPa, results in the formation of a solid phase which is structurally similar to crystalline quartz. When the pressure is reduced, the 3-dimensional structure is retained as low as 1 GPa, at which point, molecules of CO_2 reform. In 2006, a dense, amorphous form of CO_2 was prepared by compressing molecular solid CO_2 at 40–64 GPa with heating to 564 K. The amorphous, glass-like nature of this new phase has been confirmed from vibrational spectroscopic and high-intensity (i.e. using a synchrotron source) X-ray diffraction data.[†] The discovery of phases of CO_2 that exhibit SiO_2-like stuctures has stimulated much recent research interest, and the next hurdle to overcome is to find a means of maintaining these structures under ambient conditions. Dry ice readily sublimes (Table 14.4) but may be kept in insulated containers for

---

[†] Quartz-like CO_2, see: V. Iota *et al.* (1999) *Science*, vol. 283, p. 1510; amorphous silica-like CO_2, see: M. Santoro *et al.* (2006) *Nature*, vol. 441, p. 857; high-pressure phases of CO_2, see: J. Sun *et al.* (2009) *Proc. Natl. Acad. Sci. U.S.A.* vol, 106, p. 6077.

ENVIRONMENT

Box 14.7 'Greenhouse' gases

Carbon dioxide normally constitutes $\approx 0.04\%$ by volume of the Earth's atmosphere, from which it is removed and returned according to the carbon cycle:

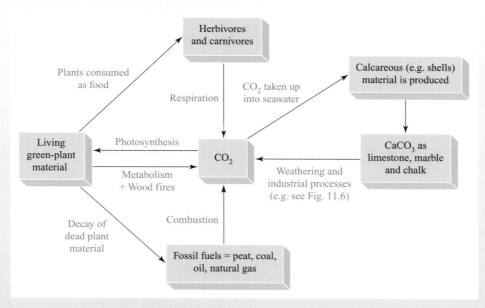

The balance is a delicate one, and the increase in combustion of fossil fuels and decomposition of limestone for cement manufacture in recent years have given rise to fears that a consequent increase in the CO_2 content of the atmosphere will lead to an 'enhanced greenhouse effect', raising the temperature of the atmosphere. This arises because the sunlight that reaches the Earth's surface has its maximum energy in the visible region of the spectrum where the atmosphere is transparent. However, the energy maximum of the Earth's thermal radiation is in the infrared, where CO_2 absorbs strongly (see Fig. 3.11). Even a small increase in the CO_2 component of the atmosphere might have serious effects because of its effects on the extent of the polar ice caps and glaciers, and because of the sensitivity of reaction rates to even small temperature changes. The danger is enhanced by the cutting down and burning of tropical rain forests which would otherwise reduce the CO_2 content of the atmosphere by photosynthesis.

The second major 'greenhouse' gas is CH_4 which is produced by the anaerobic decomposition of organic material. The old name of 'marsh gas' came about because bubbles of CH_4 escape from marshes. Flooded areas such as rice paddy fields produce large amounts of CH_4, and ruminants (e.g. cows, sheep and goats) also expel sizeable quantities of CH_4. Although the latter is a natural process, recent increases in the numbers of domestic animals around the world are naturally leading to increased release of CH_4 into the atmosphere.

The 1997 Kyoto Protocol is an international agreement that commits the industrialized countries that signed it to reducing their levels of emissions of the 'greenhouse gases' CO_2, CH_4, N_2O, SF_6, hydrofluorocarbons and perfluorocarbons (see Box 14.6). The emission targets cover all six emissions, weighted according to their global-warming potentials. For example, although emissions of SF_6 are low, it is long-lived in the atmosphere and its global warming potential is significantly higher than that of CO_2. Taking 1990 emission levels as a baseline, a target of $\approx 5\%$ reduction by 2008–2012 was set out in the Kyoto Protocol. This target is an average over all participating countries.

The term 'carbon footprint' refers to schemes for offsetting CO_2 emissions. Some of the emissions from manufacturing processes, domestic fuel consumption and transport using fossil fuels can be offset by measures such as installing solar panels, using energy-efficient lightbulbs and planting trees to partially compensate for the effects of deforestation.

Further reading

R. Conrad (2009) *Environ. Microbiol. Rep.*, vol. 1, p. 285 – 'The global methane cycle: recent advances in understanding the microbial processes involved'.

A.A. Lacis, G.A. Schmidt, D. Rind and R.A. Ruedy (2010) *Science*, vol. 330, p. 356 – 'Atmospheric CO_2: Principal control knob governing Earth's temperature'.

K.P. Shine and W.T. Sturges (2009) *Science*, vol. 315, p. 1804 – 'CO_2 is not the only gas'.

For information from the European Environment Agency, see: http://www.eea.europa.eu/

The Carbon Dioxide Information Analysis Center (CDIAC) provides up-to-date information on trends in 'greenhouse' gas emissions and global change: http://cdiac.esd.ornl.gov

Table 14.5   Selected low-temperature baths involving dry ice.[†]

Bath components	Temperature / K
Dry ice + ethane-1,2-diol	258
Dry ice + heptan-3-one	235
Dry ice + acetonitrile	231
Dry ice + ethanol	201
Dry ice + acetone	195
Dry ice + diethyl ether	173

[†] To construct a bath, add *small* pieces of solid CO_2 to the solvent. Initial sublimation of the CO_2 ceases as the bath temperature decreases to the point where solid dry ice persists. The bath temperature is maintained by occasionally adding small pieces of dry ice. See also Table 15.1.

laboratory use in, e.g., low-temperature baths (Table 14.5). *Supercritical CO_2* has become a much studied and versatile solvent (see Section 9.13). Small-scale laboratory syntheses of gaseous CO_2 usually involve reactions such as 14.63. For the industrial production of CO_2, see Fig. 11.6 and Section 10.4.

$$CaCO_3 + 2HCl \longrightarrow CaCl_2 + CO_2 + H_2O \qquad (14.63)$$

Carbon dioxide is the world's major environmental source of acid and its low solubility in water is of immense biochemical and geochemical significance. However, under aqueous conditions, H_2CO_3 is not a readily studied species.[†] In an aqueous solution of carbon dioxide, most of the solute is present as molecular CO_2 rather than as H_2CO_3, as can be seen from the value of $K \approx 1.7 \times 10^{-3}$ for the equilibrium:

$$CO_2(aq) + H_2O(l) \rightleftharpoons H_2CO_3(aq)$$

Aqueous solutions of CO_2 are only weakly acidic, but it does not follow that H_2CO_3 (carbonic acid) is a very weak acid. The value of $pK_a(1)$ for H_2CO_3 is usually quoted as 6.37. This evaluation, however, assumes that all the acid is present in solution as H_2CO_3 or $[HCO_3]^-$ when, in fact, a large proportion is present as dissolved CO_2. By taking this into account, one arrives at a $pK_a(1)$ for H_2CO_3 of ≈ 3.6. This is consistent with a value of $pK_a = 3.45 \pm 0.15$ reported in 2009 from the ultra-fast protonation of the hydrogencarbonate ion. This study provided the first observation of carbonic acid (albeit in its deuterated form) in aqueous solution. Something that is of great biological and industrial importance is the fact

[†] See: R. Ludwig and A. Kornath (2000) *Angew. Chem. Int. Ed.*, vol. 39, p. 1421 – 'In spite of the chemist's belief: Carbonic acid is surprisingly stable'; K. Adamczyk *et al.* (2009) *Science*, vol. 326, p. 1690 – 'Real-time observation of carbonic acid formation in aqueous solution'.

that combination of CO_2 with water is a relatively slow process. This can be shown by titrating a saturated solution of CO_2 against aqueous NaOH using phenolphthalein as indicator. Neutralization of CO_2 occurs by two routes. For pH < 8, the main pathway is by direct hydration (eq. 14.64), which shows pseudo-first order kinetics. At pH > 10, the main pathway is by attack of hydroxide ion (eq. 14.65). The overall rate of process 14.65 (which is first order in both CO_2 and $[OH]^-$) is greater than that of process 14.64.

$$\left.\begin{array}{l} CO_2 + H_2O \longrightarrow H_2CO_3 \\ H_2CO_3 + [OH]^- \longrightarrow [HCO_3]^- + H_2O \end{array}\right\} \begin{array}{l} \textit{slow} \\ \textit{very fast} \end{array}$$
$$(14.64)$$

$$\left.\begin{array}{l} CO_2 + [OH]^- \longrightarrow [HCO_3]^- \\ [HCO_3]^- + [OH]^- \longrightarrow [CO_3]^{2-} + H_2O \end{array}\right\} \begin{array}{l} \textit{slow} \\ \textit{very fast} \end{array}$$
$$(14.65)$$

Until 1993, there was no evidence that free carbonic acid had been isolated, though an unstable ether adduct is formed when dry HCl reacts with $NaHCO_3$ suspended in Me_2O at 243 K, and there is mass spectrometric evidence for H_2CO_3 being a product of the thermal decomposition of $[NH_4][HCO_3]$. However, IR spectroscopic data now indicate that H_2CO_3 can be isolated using a cryogenic method in which glassy MeOH solution layers of $KHCO_3$ (or Cs_2CO_3) and HCl are quenched on top of each other at 78 K and the reaction mixture warmed to 300 K. Under these conditions, and in the absence of water, H_2CO_3 can be sublimed unchanged.

(14.17)

The carbonate ion is planar and possesses D_{3h} symmetry with all C–O bonds of length 129 pm. A delocalized bonding picture involving $(p–p)\pi$-interactions is appropriate, and VB theory describes the ion in terms of three resonance structures of which one is **14.17**. The C–O bond distance in $[CO_3]^{2-}$ is longer than in CO_2 (Table 14.4) and is consistent with a formal bond order of 1.33. Most metal carbonates, other than those of the group 1 metals (see Section 11.7), are sparingly soluble in water. A general method of preparing peroxo salts can be used to convert K_2CO_3 to $K_2C_2O_6$; the electrolysis of aqueous K_2CO_3 at 253 K using a high current density produces a salt believed to contain the peroxocarbonate ion, **14.18**. An alternative route involves the reaction of CO_2 with KOH in 86% aqueous H_2O_2 at 263 K. The colour of the product is variable and probably depends upon the presence of impurities such as KO_3. The electrolytic method gives a blue material whereas the product from the second route is orange. Peroxocarbonates are also believed to be intermediates in the reactions of CO_2 with superoxides (see Section 11.6).

(14.18)   **(14.19)**

A third oxide of carbon is the suboxide C_3O_2 which is made by dehydrating malonic acid, $CH_2(CO_2H)_2$, using P_2O_5 at 430 K. At room temperature, C_3O_2 is a gas (bp 279 K), but it polymerizes above 288 K to form a red-brown paramagnetic material. The structure of C_3O_2 is usually described as 'quasilinear' because IR spectroscopic and electron diffraction data for the gaseous molecule show that the energy barrier to bending at the central C atom is only $0.37\,kJ\,mol^{-1}$, i.e. very close to the vibrational ground state. The melting point of C_3O_2 is 160 K. An X-ray diffraction study of crystals grown just below this temperature confirms that the molecules are essentially linear in the solid state (structure **14.19**). However, the data are best interpreted in terms of disordered (see Box 15.5), bent molecules with a C–C–C bond angle close to 170°, consistent with a 'quasilinear' description. The species $[OCNCO]^+$, $[NCNCN]^-$ and $[N_5]^+$ are isoelectronic with C_3O_2, but they are not isostructural with the 'quasilinear' C_3O_2. Unambiguously non-linear structures are observed for $[OCNCO]^+$ ($\angle C–N–C = 131°$ in $[OCNCO]^+[Sb_3F_{16}]^-$), the dicyanamide ion $[NCNCN]^-$ ($\angle C–N–C = 124°$ in Cs[NCNCN]), and $[N_5]^+$ (see Section 15.5).

Worked example 14.6   Lewis structures

(a) Draw a Lewis structure for linear C_3O_2. (b) Consider possible Lewis structures for linear and non-linear (bent at the central atom) $[OCNCO]^+$ and $[NCNCN]^-$. Comment on these structures in view of the following solid state data:

$[OCNCO]^+[Sb_3F_{16}]^-$ $\angle C–N–C = 131°$,
    $\angle O–C–N = 173°$, $C–O = 112\,pm$, $C–N = 125\,pm$

Cs[NCNCN] $\angle C–N–C = 124°$, $\angle N–C–N = 172°$,
    av. $C–N_{term} = 115\,pm$, av. $C–N_{centre} = 128\,pm$

(a) A Lewis structure for C_3O_2 is:

(b) Possible Lewis structures can be drawn by considering isoelectronic relationships between C and N^+, O and N^-, and N and O^+.

Therefore starting from linear C_3O_2, Lewis structures for linear $[OCNCO]^+$ and $[NCNCN]^-$ are:

However, the observed bond angles at the central atom show that the ions are non-linear in the solid state salts studied. For each ion, if a negative charge is localized on the central N atom, then a Lewis structure consistent with a non-linear structure can be drawn:

The observed bond lengths in salts of $[OCNCO]^+$ and $[NCNCN]^-$ are consistent with the above Lewis structures.

Double deprotonation of oxalic acid (see Section 7.4) gives the oxalate ion, $[C_2O_4]^{2-}$, and many oxalate salts are available commercially. The solid state structures of anhydrous alkali metal oxalates respond to an increase in the size of the metal ion. In $Li_2C_2O_4$, $Na_2C_2O_4$, $K_2C_2O_4$ and one polymorph of $Rb_2C_2O_4$, the $[C_2O_4]^{2-}$ ion is planar (**14.20**). In the second polymorph of $Rb_2C_2O_4$ and in $Cs_2C_2O_4$, the $[C_2O_4]^{2-}$ ion adopts a staggered conformation (**14.21**). Oxalate salts in general tend to exhibit planar anions in the solid state. The C–C bond length (157 pm) is consistent with a single bond and indicates that the planar structure is not a consequence of π-delocalization but is, instead, a result of intermolecular interactions in the crystal lattice.

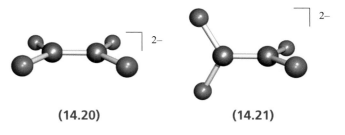

(14.20)   **(14.21)**

Silica, silicates and aluminosilicates

Silica, SiO_2, is an involatile solid and occurs in many different forms, nearly all of which possess 3-dimensional structures constructed from tetrahedral SiO_4 building blocks, often represented as in structure **14.22**. The diagram at the right-hand side of **14.22** is a polyhedral representation of the SiO_4 unit, and is commonly used when illustrating the connectivities of the building blocks in β-quartz $\underset{\text{slow}}{\overset{1143 \text{ K}}{\rightleftharpoons}}$ β-tridymite $\underset{\text{slow}}{\overset{1742 \text{ K}}{\rightleftharpoons}}$ β-cristobalite $\underset{\text{slow}}{\overset{1983 \text{ K}}{\rightleftharpoons}}$ liquid

846 K ↕ fast   393–433 K ↕ fast   473–548 K ↕ fast

α-quartz   α-tridymite   α-cristobalite

Fig. 14.20 Transition temperatures between polymorphs of SiO_2.

3-dimensional silicate structures. Each unit is connected to the next by sharing an oxygen atom to give Si−O−Si bridges. At atmospheric pressure, three polymorphs of silica exist. Each polymorph is stable within a characteristic temperature range, but possesses a low-temperature (α) and a high-temperature (β) modification (Fig. 14.20). The structure of β-cristobalite and its relationship to that of diamond was shown in Fig. 6.20. The different polymorphs of silica resemble β-cristobalite in having tetrahedral SiO_4-units, but each is made unique by exhibiting a different arrangement of these building blocks. α-Quartz has an interlinked helical chain structure and is optically active because the chain has a handedness. It is also *piezoelectric* and is therefore used in crystal oscillators and filters for frequency control and in electromechanical devices such as microphones and loudspeakers.

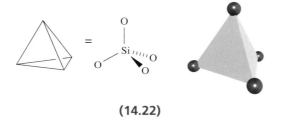

(14.22)

A *piezoelectric* crystal is one that generates an electric field (i.e. develops charges on opposite crystal faces when subjected to mechanical stress) or that undergoes some change to atomic positions when an electric field is applied to it; such crystals must lack a centre of symmetry (e.g. contain tetrahedral arrangements of atoms). Their ability to transform electrical oscillations into mechanical vibration, and vice versa, is the basis of their use in, e.g., crystal oscillators.

Transitions from one polymorph of silica to another involve initial Si−O bond cleavage and require higher temperatures than the changes between α- and β-forms of one polymorph. When liquid silica cools, it forms a non-crystalline glass consisting of a 3-dimensional structure assembled from SiO_4 tetrahedra connected in a random manner. Only a few oxides form glasses (e.g. B_2O_3, Al_2O_3, SiO_2, GeO_2, P_2O_5 and As_2O_5) since the criteria for a *random* assembly are:

- the coordination number of the non-oxygen element must be 3 or 4 (a coordination number of 2 gives a chain and greater than 4 gives too rigid a structure);

- only one O atom must be shared between any two non-oxygen atoms (greater sharing leads to too rigid an assembly);
- a flexible X−O−X bond angle around a value of ≈150°;
- free rotation about the X−O bonds.

When silica glass is heated to ≈1750 K, it becomes plastic and can be worked in an oxy-hydrogen flame. *Silica glass* apparatus is highly insensitive to thermal shock owing to the low coefficient of thermal expansion of silica. *Borosilicate glass* (see Box 13.6) contains 10–15% B_2O_3 and has a lower melting point than silica glass. Glass for windows, bottles and many other commercial uses is *soda-lime glass*. This is manufactured by fusing sand, Na_2CO_3 and limestone to give a glass that contains 70–75% SiO_2 and 12–15% Na_2O, with additional CaO and MgO. The added Na_2O modifies the silica structure by converting some Si−O−Si bridges in the silica network to terminal Si−O bonds. The Na^+ ions reside in cavities in the 3-dimensional network and are coordinated by the terminal Si−O$^-$ units. The melting point of soda-lime glass is lower than that of borosilicate glass. Recycled glass (*cullet*) now contributes significantly to the manufacture of new glass, and this trend continues to grow.

In all forms of silica mentioned so far, the Si−O bond length is ≈160 pm and the Si−O−Si bond angle ≈144°, values close to those in $(H_3Si)_2O$ (Fig. 14.12). By heating silica under very high pressure, a rutile form (see Fig. 6.22) containing 6-coordinate Si is formed in which the Si−O bond length is 179 pm (compare with the sum of $r_{cov}(Si) = 118$ pm and $r_{cov}(O) = 73$ pm). This form of silica is more dense and less reactive than ordinary forms. Silica is not attacked by acids other than HF, with which it forms $[SiF_6]^{2-}$.

Although esters of type $Si(OR)_4$ (eq. 14.66) are known, no well-defined 'silicic acid' (H_4SiO_4) has been established. In aqueous solution, there is evidence for equilibria involving silicic acids including H_4SiO_4, $H_6Si_2O_7$, H_2SiO_3 and $H_2Si_2O_5$, but condensation reactions with formation of Si−O−Si linkages occur too readily for these simple molecules to be isolated in the solid state. These same reactions are central to biomineralization processes in which marine organisms construct exoskeletons from silicic acids, i.e. the latter provide a bioavailable source of silica (Fig. 14.21).

$$SiCl_4 + 4ROH \longrightarrow Si(OR)_4 + 4HCl \qquad (14.66)$$

Normal silica is only very slowly attacked by alkali, but *silicates* are readily formed by fusion of SiO_2 and metal hydroxides, oxides or carbonates. The range of known silicates is large and they, and the *aluminosilicates* (see later), are extremely important, both in nature and for commercial and industrial purposes.

Sodium silicates of variable composition are made by heating sand (which is impure quartz containing, e.g.,

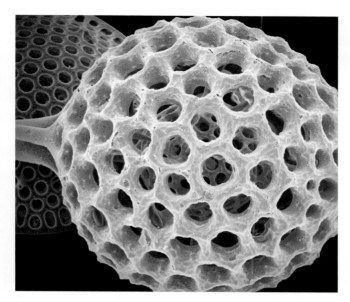

Fig. 14.21 Scanning electron micrograph (SEM) of the silica exoskeleton of the radiolarian (*Amphisphaerina radiolarian*). Radiolaria are single-celled protozoans.

iron(III) oxide) with Na_2CO_3 at $\approx 1600\,K$. If the sodium content is high ($Na:Si \approx 3.2\text{--}4:1$), the silicates are water-soluble and the resulting alkaline solution (*water glass*) contains ions such as $[SiO(OH)_3]^-$ and $[SiO_2(OH)_2]^{2-}$. Water glass is used commercially in detergents where it controls the pH and degrades fats by hydrolysis. If the Na content is low, the silicate ions consist of large polymeric species and their Na^+ salts are insoluble in water. Equilibrium between the different species is attained rapidly at pH > 10, and more slowly in less alkaline solutions.

The Earth's crust is largely composed of silica and silicate minerals, which form the principal constituents of all rocks and of the sands, clays and soils that result from degradation of rocks. Most inorganic building materials are based on silicate minerals and include natural silicates such as sandstone, granite and slate, as well as manufactured materials such as cement, concrete (see Box 14.8) and ordinary glass (see above). Clays (see Box 14.10) are used in the ceramics industry and mica is used as an electrical insulator.

It is common practice to describe silicates in terms of a purely ionic model. However, although we might *write* Si^{4+}, the 4+ charge is unlikely on ionization energy grounds and is incompatible with the commonly observed $Si-O-Si$ bond angle of $\approx 140°$. Figure 14.22 compares the ionic radii of ions commonly present in silicates; the value for the 'Si^{4+}' ion is an estimate. Since the Al^{3+} and Si^{4+} ions are similar sizes, replacement is common and leads to the formation of aluminosilicates. If Al^{3+} replaces Si^{4+}, an extra singly charged cation must be present to maintain electrical neutrality. Thus, in the feldspar *orthoclase*, $KAlSi_3O_8$, the anion $[AlSi_3O_8]^-$ is related to

SiO_2 (i.e. $[AlSi_3O_8]^-$ is isoelectronic with Si_4O_8) and $[AlSi_3O_8]^-$ possesses the structure of quartz with one-quarter of the Si replaced by aluminium. The K^+ ions occupy cavities in the relatively open lattice. Double replacements are also common, e.g. $\{Na^+ + Si^{4+}\}$ replaced by $\{Ca^{2+} + Al^{3+}\}$ (look at the radii comparisons in Fig. 14.22).

The overwhelming majority of silicates have structures based on SiO_4 tetrahedra (**14.22**) which, by sharing O atoms, assemble into small groups such as **14.23**, cyclic motifs, infinite chains, infinite layers or infinite 3-dimensional networks. Sharing an atom only involves *corners* of tetrahedra; sharing an edge would bring two O^{2-} ions too close together.

Of the metal ions most commonly occurring in silicates, the coordination numbers with respect to O^{2-} ions are 4 for Be^{2+}, 4 or 6 for Al^{3+}, 6 for Mg^{2+}, Fe^{3+} or Ti^{4+}, 6 or 8 for Na^+, and 8 for Ca^{2+}.

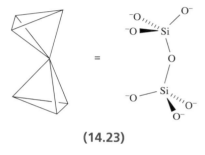

(14.23)

Figure 14.23 illustrates the structures of some silicate anions; $[Si_2O_7]^{6-}$ is shown in structure **14.23**. The simplest silicates contain the $[SiO_4]^{4-}$ ion and include Mg_2SiO_4 (*olivine*) and the β- and γ-phases of synthetic Ca_2SiO_4 ($2CaO·SiO_2$, see Box 14.8). The mineral *thortveitite*, $Sc_2Si_2O_7$ (a major source of scandium), contains discrete $[Si_2O_7]^{6-}$ ions. The cyclic ions $[Si_3O_9]^{6-}$ and $[Si_6O_{18}]^{12-}$

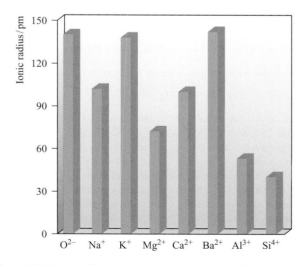

Fig. 14.22 Ionic radii of selected ions involved in silicates. These data can be used to rationalize cation replacements in silicates.

APPLICATIONS

Box 14.8 Materials chemistry: cement and concrete

Although the ancient Romans produced durable mortars using lime, volcanic ash (containing alumina) and clay, it was not until 1824 that *Portland cement* came into being. Its production was patented by Joseph Aspidin, and the name 'Portland' derives from the fact that Aspidin's cement resembles natural stone that occurs on the Isle of Portland in southwest England. Cement is manufactured and stored in a dry, powder form and, once hydrated, is used for mortar for binding bricks or stones. The combination of cement with fine and coarse aggregates (sand and stones) generates concrete.

The primary ingredients required for the manufacture of cement are limestone ($CaCO_3$) and silica (SiO_2), along with smaller quantities of alumina (Al_2O_3) and Fe_2O_3. Initially, $CaCO_3$ is calcined at 1070 K. The CaO so formed is then heated with SiO_2, Al_2O_3 and Fe_2O_3 in a rotary kiln at 1570–1720 K. At these temperatures, the material sinters (i.e. becomes partially molten) and forms a *clinker* which is composed of the mixed oxides $3CaO \cdot SiO_2$ ('C_3S'), $2CaO \cdot SiO_2$ ('C_2S'), $3CaO \cdot Al_2O_3$ ('C_3A') and $4CaO \cdot Al_2O_3 \cdot Fe_2O_3$ ('C_4AF'). (These abbreviations are in common use in the cement industry.) Phase diagrams for the system are complicated. The cooled clinker is ground to a powder and gypsum ($CaSO_4 \cdot 2H_2O$) is added; the gypsum controls the setting time of the cement (see later). A typical Portland cement has a composition within the ranges 55–60% $3CaO \cdot SiO_2$, 15–18% $2CaO \cdot SiO_2$, 2–9% $3CaO \cdot Al_2O_3$, 7–14% $4CaO \cdot Al_2O_3 \cdot Fe_2O_3$, 5–6% gypsum and <1% Na_2O or K_2O. White cements (in demand for architectural use) have a reduced iron content and contain only 1% 'C_3A' and increased amounts of 'C_3S' and 'C_2S'. Cements that will be exposed to high levels of sulfates (e.g. in ground water) must also have a reduced 'C_3A' content. This is because sulfates react with calcium aluminate hydrate with an associated 220% increase in the volume of the material, resulting in structural degradation.

The hydration of cement powder is the final step in which it is transformed into a hard material which is insoluble in water. The hydration process is exothermic, and is exemplified for $3CaO \cdot SiO_2$ as follows:

$$2(3CaO \cdot SiO_2) + 6H_2O \longrightarrow 3CaO \cdot 2SiO_2 \cdot 3H_2O + 3Ca(OH)_2$$

or, in the notation form used in the cement industry:

$$2\text{'}C_3S\text{'} + 6\text{'}H\text{'} \longrightarrow \text{'}C_3S_2H_3\text{'} + 3\text{'}CH\text{'}$$

For 'C_3S', the enthalpy of hydration is $500 \, kJ \, kg^{-1}$, compared with 250, 850 and $330 \, kJ \, kg^{-1}$ for 'C_2S', 'C_3A' and 'C_4AF', respectively. The setting period consists of several stages. The addition of water to dry cement initially causes a rapid, and highly exothermic, dissolution of ions and formation of hydrated species. This is followed by a dormant period, and then a final exothermic event which completes the cement setting process. 'C_3S', 'C_2S', 'C_3A' and 'C_4AF' set at different rates ('C_3A' > 'C_3S' > 'C_2S' ≈ 'C_4AF'). The addition of gypsum to the cement mix slows down the setting process, a point that is particularly important for 'C_3A'.

The hardening of cement (or concrete) is ultimately caused by the formation of Si–O–Si bridges. Hydrated 'C_3S' and 'C_2S' exist as extended networks, coexisting with fibrous domains, calcium hydroxide and unhydrated cement grains. This framework (usually referred to as the *cement gel*) contains pores, the total volume of which may be one-quarter of the entire material. Cement contains a number of NMR active nuclei (^1H, ^{29}Si, ^{27}Al and ^{23}Na), and solid state ^{29}Si magic angle spinning (MAS) NMR spectroscopy has been used for structural investigations.

Cement is the binding agent in concrete and constitutes 15–25% of the final material by weight. Water and cement together generate a *cement paste* which, when combined with fine and/or coarse aggregate, results in concrete. Most concrete is now produced ready-mixed and is moved to building sites in cement vehicles consisting of slowly rotating drums to prevent the concrete from setting prematurely. The quality of concrete is determined largely by the ratio of cement to water, and particles of aggregate should be completely surrounded by cement paste. The properties of a given concrete can be tuned by the presence of additives. In addition to its production from almost universally available raw materials, and its strength and adaptability, concrete is fire-resistant. It is now one of the most important building materials worldwide.

The contribution that the calcining of $CaCO_3$ makes to global CO_2 emissions is an environmental concern. As a result of the 1997 Kyoto Protocol, methods of reducing CO_2 emissions are prime objectives for cement manufacturers.

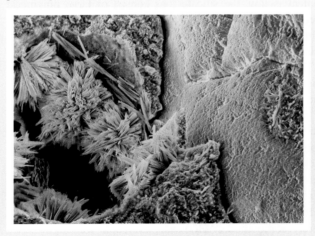

Coloured scanning electron micrograph of gypsum crystals (brown) that have formed in setting concrete (blue).

Further reading

D.C. MacLaren and M.A. White (2003) *J. Chem. Educ.*, vol. 80, p. 623 – 'Cement: its chemistry and properties'.

R. Rehan and M. Nehdi (2005) *Environ. Sci. Policy*, vol. 8, p. 105 – 'Carbon dioxide emissions and climate change: policy implications for the cement industry'.

APPLICATIONS

Box 14.9 The rise and fall of fibrous asbestos

In the commercial market, the term asbestos covers fibrous forms of the minerals actinolite, amosite, anthophyllite, chrysotile, crocidolite and tremolite. The ability of the fibres to be woven along with their heat resistance and high tensile strength led to widespread applications of asbestos in fire-proofing materials, brake linings, prefabricated boards for construction, roofing tiles and insulation. As the graph below shows, world production of asbestos was at a peak in the mid-1970s and has since declined. Most of the asbestos mined nowadays is chrysotile, and the world's leading producers are Russia, China, Brazil and Kazakhstan. Continuing applications are largely in roofing materials, gaskets and friction products including brake linings. The dramatic downturn in the use of asbestos is associated with its severe health risks. The respiratory disease asbestosis is caused by the inhalation of asbestos fibres by workers constantly exposed to them. Strict legislation controls the use of asbestos, and demolition or renovation of old buildings often reveals large amounts of asbestos, which can be cleared only under qualified specialists. In most countries, the decline in the use of asbestos is set to continue as further restrictive legislation is passed. Many countries have now banned the processing and use of asbestos. Legislation, such as that enforced in the Republic of Korea in 2009, focuses on products in which asbestos particles may break loose or come into contact with the skin. On the other hand, since 2000, consumption of asbestos has increased in parts of Asia, South America and the Commonwealth of Independent States, and this accounts for the recent upturn in world production in the graph below. Although the health risks of asbestos fibres are well known, the huge expansion of nanotechnology raises the question that similar risks may exist with carbon nanotubes, and nanofibres and nanoparticles of other materials. Risk assessment in this area is therefore critical, and is actively being pursued.

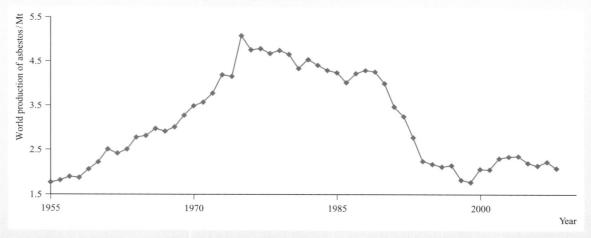

[Data: US Geological Survey]

Scanning electron micrograph (SEM) of chrysotile asbestos fibres.

Further reading

J.D. Brain, M.A. Curran, T. Donaghey and R.M. Molina (2009) *Nanotoxicology*, vol. 3, p. 174 – 'Biologic responses to nanomaterials depend on exposure, clearance and material characteristics'.

I. Fenoglio, M. Tomatis and B. Fubini (2001) *Chem. Commun.*, p. 2182 – 'Spontaneous polymerisation on amphibole asbestos: relevance to asbestos removal'.

B. Fubini and C. Otero Areán (1999) *Chem. Soc. Rev.*, vol. 28, p. 373 – 'Chemical aspects of the toxicity of inhaled mineral dusts'.

V.W. Hoyt and E. Mason (2008) *J. Chem. Health Safety*, vol. 15, p. 10 – 'Nanotechnology: Emerging health issues'.

For information from the Environmental Protection Agency on asbestos, see: http://www.epa.gov/asbestos/

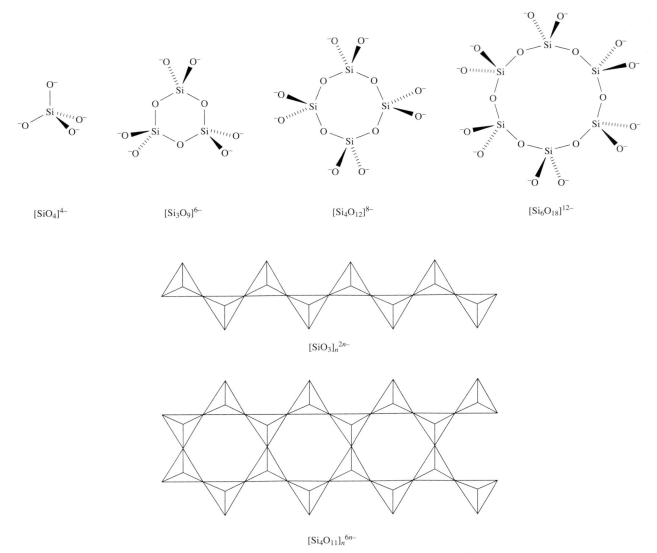

$[SiO_4]^{4-}$ $[Si_3O_9]^{6-}$ $[Si_4O_{12}]^{8-}$ $[Si_6O_{18}]^{12-}$

$[SiO_3]_n^{2n-}$

$[Si_4O_{11}]_n^{6n-}$

Fig. 14.23 Schematic representations of the structures of selected silicates. Conformational details of the rings are omitted. In the polymeric structures, each tetrahedron represents an SiO_4-unit as shown in structure **14.22**. (See also Fig. 14.25.)

occur in $Ca_3Si_3O_9$ (α-*wollastonite*) and $Be_3Al_2Si_6O_{18}$ (*beryl*) respectively, while $[Si_4O_{12}]^{8-}$ is present in the synthetic salt $K_8Si_4O_{12}$. Short-chain silicates are not common, although $[Si_3O_{10}]^{8-}$ occurs in a few rare minerals. Cage structures have been observed in some synthetic silicates and two examples are shown in Fig. 14.24.

If SiO_4 tetrahedra sharing two corners form an infinite chain, the $Si:O$ ratio is $1:3$ (Fig. 14.23). Such chains are present in $CaSiO_3$ (β-*wollastonite*) and $CaMg(SiO_3)_2$ (*diopside*, a member of the *pyroxene* group of minerals which possess $[SiO_3]_n^{2n-}$ chains). Although infinite chains are present in these minerals, the relative orientations of the chains are different. *Asbestos* (see Box 14.9) consists of a group of fibrous minerals, some of which (e.g. $Ca_2Mg_5(Si_4O_{11})_2(OH)_2$, *tremolite*) contain the double-chain silicate $[Si_4O_{11}]_n^{6n-}$ shown in Figs. 14.23 and 14.25. More extended cross-linking of chains produces

layer structures of composition $[Si_2O_5]^{2-}$; ring sizes within the layers may vary. Such sheets occur in *micas* and are responsible for the characteristic cleavage of these minerals into thin sheets. *Talc*, characterized by its softness, has the composition $Mg_3(Si_2O_5)_2(OH)_2$. The Mg^{2+} ions are sandwiched between layers comprising $[Si_2O_5]_n^{2n-}$ sheets and $[OH]^-$ ions. This produces electrically neutral sheets (Fig. 14.26) which are held together by weak interactions, allowing talc to cleave readily in a direction parallel to the sheets. A consequence of this cleavage is that talc is used as a dry lubricant, e.g. in personal care preparations.

Infinite sharing of all four oxygen atoms of the SiO_4 tetrahedra gives a composition SiO_2 (see earlier) but partial replacement of Al for Si in Si_nO_{2n} leads to the anions $[AlSi_{n-1}O_{2n}]^-$ and $[Al_2Si_{n-2}O_{2n}]^{2-}$ etc. Minerals belonging to this group include *orthoclase* ($KAlSi_3O_8$),

Box 14.10 Kaolin, smectite and hormite clays: from ceramics to natural absorbers

Crystalline clays (aluminosilicate minerals) are categorized according to structure. Clays in the *kaolin* or *china clay* group (e.g. *kaolinite*, $Al_2Si_2O_5(OH)_4$) possess sheet structures with alternating layers of linked SiO_4 tetrahedra and AlO_6 octahedra. *Smectite* clays (e.g. *sodium montmorillonite*, $Na[Al_5MgSi_{12}O_{30}(OH)_6]$) also have layer structures, with cations (e.g. Na^+, Ca^{2+}, Mg^{2+}) situated between the aluminosilicate layers. Interactions between the layers are weak, and water molecules readily penetrate the channels causing the lattice to expand; the volume of montmorillonite increases several times over as water is absorbed. *Hormite* clays (e.g. *palygorskite*) possess structures in which chains of SiO_4 tetrahedra are connected by octahedral AlO_6 or MgO_6 units; these clays exhibit outstanding adsorbent and absorbent properties.

Within industry and commerce, terms other than the mineral classifications are common. *Ball clay* is a type of kaolin particularly suited to the manufacture of ceramics. It consists of kaolinite, quartz and mica, and is usually worked from open-cast mines. In 2009, 38% of the ball clay produced in the US was used for tile manufacture, 24% for sanitary ware, and the remainder for pottery and various ceramics. When the clay consists predominantly of kaolinite, it is often referred to simply as kaolin or china clay. These clays are white and soft, and are extracted from open-cast mines using high-pressure water jets. This produces a slurry which is transferred to storage tanks where the clay is separated from feldspar and other silicate minerals. The photograph shows an aerial view of an open-cast china clay mine and groups of circular settling tanks are visible. White kaolinite is of great importance in the paper industry for coatings and as a filler, and in 2009, 62% of the kaolinite produced in the US was consumed in the paper industry. In 2008, the world production of kaolin clays (ball and china clays) was $\approx$36 Mt. In contrast, world production of smectite and hormite clays (11.7 and 3.5 Mt, respectively, in 2008) was significantly lower.

Smectite clays tend to be referred to as *bentonite*, the name deriving from the rock in which the clays occur. *Fuller's earth* is a general term used commercially to describe

hormite clays. Applications of smectite and hormite clays stem from their ability to absorb water, swelling as they do so. Drilling fluids rely on the outstanding, reversible behaviour of sodium montmorillonite as it takes in water: this property is called *thixotropy*. When static, or at low drill speeds, an aqueous suspension of the clay is highly viscous owing to the absorption of water by the lattice and the realignment of the charged aluminosilicate layers. At high drill speeds, electrostatic interactions between the layers are destroyed and the drill-fluid viscosity decreases. Fuller's earth clays are remarkably effective absorbents and two major applications are in pet litter, and in granules which can be applied to minor oil spillages (e.g. at fuel stations).

[Statistical data: US Geological Survey]

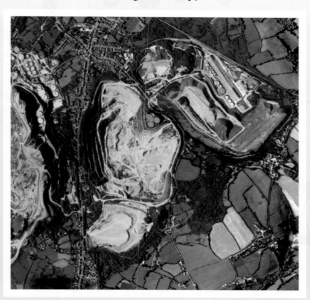

Aerial photograph of Gunheath china clay mines in Cornwall, UK.

albite ($NaAlSi_3O_8$), *anorthite* ($CaAl_2Si_2O_8$) and *celsian* ($BaAl_2Si_2O_8$). Feldspars are aluminosilicate salts of K^+, Na^+, Ca^{2+} or Ba^{2+} and constitute an important class of rock-forming minerals which include *orthoclase*, *celsian*, *albite* and *anorthite*. The *feldspathoid* minerals are related to feldspars, but have a lower silica content. An example is *sodalite*, $Na_8[Al_6Si_6O_{24}]Cl_2$ (see Box 16.4). Both the feldspars and feldspathoid minerals are anhydrous.

Zeolites constitute an important class of aluminosilicates, but their ability to absorb water makes them distinct from the feldspars and feldspathoids. In feldspars, the holes in the structure that accommodate the cations are quite small. In zeolites, the cavities are much larger and can accommo-

date not only cations but also molecules such as H_2O, CO_2, MeOH and hydrocarbons. Commercially and industrially, zeolites (both natural and synthetic) are extremely important. The Al:Si ratio varies widely among zeolites; Al-rich systems are hydrophilic and their ability to take up H_2O leads to their use as laboratory drying agents (molecular sieves). Different zeolites contain different-sized cavities and channels, permitting a choice of zeolite to effect selective molecular adsorption. Silicon-rich zeolites are hydrophobic. Catalytic uses of zeolites (see Sections 25.6 and 25.7) are widespread, e.g. the synthetic zeolite ZSM-5 with composition $Na_n[Al_nSi_{96-n}O_{192}]\cdot\approx16H_2O$ ($n < 27$) catalyses benzene alkylation, xylene isomerization

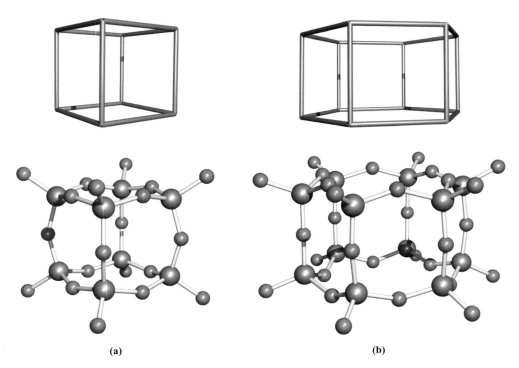

(a)   **(b)**

Fig. 14.24 The structures, elucidated by X-ray diffraction, of (a) $[Si_8O_{20}]^{8-}$, determined for the salt $[Me_4N]_8[Si_8O_{20}]\cdot65H_2O$ [M. Wiebcke *et al.* (1993) *Microporous Materials*, vol. 2, p. 55], and (b) $[Si_{12}O_{30}]^{12-}$, determined for the salt $K_{12}[\alpha\text{-cyclodextrin}]_2[Si_{12}O_{30}]\cdot36H_2O$ [K. Benner *et al.* (1997) *Angew. Chem. Int. Ed.*, vol. 36, p. 743]. The silicon atoms in (a) and (b) define a cube and hexagonal prism respectively. Colour code: Si, purple; O, red.

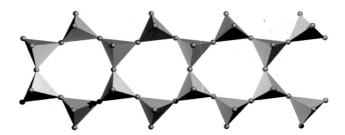

Fig. 14.25 Part of one of the double chains of general formula $[Si_4O_{11}]_n^{6n-}$ present in the mineral tremolite. Compare this representation with that in Fig. 14.23. Each red sphere represents an O atom, and each tetrahedral O$_4$-unit surrounds an Si atom.

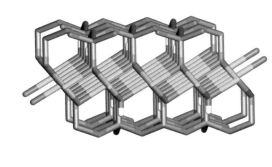

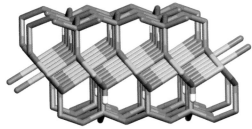

Fig. 14.26 A stick representation of part of two layers in the solid state structure of the mineral talc, $Mg_3(Si_2O_5)_2(OH)_2$. Each Mg^{2+} ion is in an octahedral environment, surrounded by six O atoms. Colour code: Si, purple; O, red; Mg, yellow; H, white.

and conversion of methanol to hydrocarbons (for motor fuels). Figure 14.27 illustrates the cavities present in zeolite H-ZSM-5.[†] Electrical neutrality upon Al-for-Si replacement can also be achieved by converting O$^-$ to a terminal OH group. These groups are strongly acidic, which means that such zeolites are excellent ion-exchange (see Section 11.6) materials and have applications in, for example, water purification and washing powders (see Section 12.7).

[†] Zeolites are generally known by acronyms that reflect the research or industrial companies of origin, e.g. ZSM stands for Zeolite Socony Mobil.

> ***Zeolites*** are crystalline, hydrated aluminosilicates that possess framework structures containing regular channels and/or cavities; the cavities contain H_2O molecules and cations (usually group 1 or 2 metal ions).

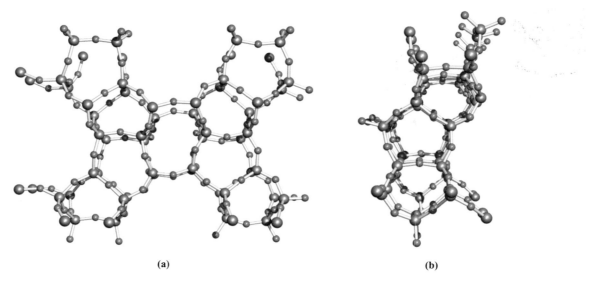

(a)   (b)

Fig. 14.27 The structure of H-ZSM-5 zeolite ($Al_{0.08}Si_{23.92}O_{48}$) is typical of a zeolite in possessing cavities which can accommodate guest molecules. (a) and (b) show two orthogonal views of the host lattice; the structure was determined by X-ray diffraction for the zeolite hosting 1,4-dichlorobenzene [H. van Koningsveld *et al.* (1996) *Acta Crystallogr.*, *Sect. B*, vol. 52, p. 140]. Colour code: (Si, Al), purple; O, red.

Oxides, hydroxides and oxoacids of germanium, tin and lead

The dioxides of Ge, Sn and Pb are involatile solids. Germanium dioxide closely resembles SiO_2, and exists in both quartz and rutile forms. It dissolves in concentrated HCl forming $[GeCl_6]^{2-}$ and in alkalis to give *germanates*. While these are not as important as silicates, it should be noted that many silicates possess germanate analogues. However, relatively few open-framework germanates (i.e. with structures related to those of zeolites) are known. Although Si and Ge are both group 14 elements, the structural building-blocks in silicates and germanates differ. Whereas silicates are composed of tetrahedral SiO_4-units (Figs. 14.23–14.27), the larger size of Ge allows it to be in GeO_4 (tetrahedral), GeO_5 (square-based pyramidal or trigonal bipyramidal) and GeO_6 (octahedral) environments.[†] For example, the germanate $[Ge_{10}O_{21}(OH)][N(CH_2CH_2NH_3)_3]$, contains 4-, 5- and 6-coordinate Ge atoms. This compound is synthesized by a hydrothermal method (such methods are used for both germanate and zeolite syntheses) using the amine $N(CH_2CH_2NH_2)_3$ to direct the assembly of the 3-dimensional network.

A *hydrothermal method* of synthesis refers to a heterogeneous reaction carried out in a closed system in an aqueous solvent with $T > 298\,K$ and $P > 1$ bar. Such reaction conditions permit the dissolution of reactants and the isolation of products that are poorly soluble under ambient conditions.

[†]Z.-E. Lin and G.-Y. Yang (2010) *Eur. J. Inorg. Chem.*, p. 2895 – 'Germanate frameworks constructed from oxo germanium cluster building units'.

Germanium monoxide is prepared by dehydration of the yellow hydrate, obtained by reaction of $GeCl_2$ with aqueous NH_3, or by heating $Ge(OH)_2$, obtained from $GeCl_2$ and water. The monoxide, which is amphoteric, is not as well characterized as GeO_2, and disproportionates at high temperature (eq. 14.67).

$$2GeO \xrightarrow{970\,K} GeO_2 + Ge \qquad (14.67)$$

Solid SnO_2 and PbO_2 adopt a rutile-type structure (Fig. 6.22). SnO_2 occurs naturally as cassiterite but can easily be prepared by oxidation of Sn. Tin(IV) oxide is unusual in being an optically transparent, electrical conductor; in Box 14.11, we look at the use of SnO_2 in resistive gas sensors. The formation of PbO_2 requires the action of powerful oxidizing agents such as alkaline hypochlorite on Pb(II) compounds. On heating, PbO_2 decomposes to PbO via a series of other oxides (eq. 14.68). In the last step in the pathway, the reaction conditions favour the decomposition of Pb_3O_4, the O_2 formed being removed. This is in contrast to the conditions used to make Pb_3O_4 from PbO (see the end of Section 14.9).

$$PbO_2 \xrightarrow{566\,K} Pb_{12}O_{19} \xrightarrow{624\,K} Pb_{12}O_{17} \xrightarrow{647\,K} Pb_3O_4 \xrightarrow{878\,K} PbO$$
$$(14.68)$$

When freshly prepared, SnO_2 is soluble in many acids (eq. 14.69) but it exhibits amphoteric behaviour and also

APPLICATIONS

Box 14.11 Sensing gases

Tin(IV) oxide is of significant commercial value because it is an optically transparent, electrical conductor. Its three main applications are in gas sensors, as a transparent conducting oxide, and as an oxidation catalyst. We focus here on the first of these uses.

Tin(IV) oxide is classed as a wide band-gap (3.6 eV) n-type semiconductor. In a stoichiometric form, SnO_2 is an insulator. However, intrinsic oxygen deficiency renders it non-stoichiometric (see Section 6.17) and leads to electrical conductance. The conductivity can be increased by doping (e.g. with Pd) which modifies the band structure (see Section 6.8). In the presence of reducing gases such as CO, the electrical conductivity of SnO_2 increases, and it is this phenomenon that is the basis for the use of SnO_2 in resistive gas sensors. Other metal oxides that show a change in electrical conductivity in response to the presence of certain gases include In_2O_3, GeO_2, TiO_2, Mn_2O_3, CuO, ZnO, WO_3, MoO_3, Nb_2O_5 and CeO_2. In commercial sensors, SnO_2 and ZnO are by far the most commonly used metal oxides. Detecting the presence of toxic gases can be carried out by, for example, IR spectroscopic means, but such techniques do not lend themselves to regular monitoring of industrial and domestic environments. Solid state gas sensors are advantageous because they can monitor levels of gases continuously, and are relatively inexpensive. Sensors that detect gases such as CO, hydrocarbons or solvent (alcohols, ketones, esters, etc.) vapours at a parts-per-million (ppm) level are now in common use in underground car parking garages, automatic ventilation systems, fire alarms and gas-leak detectors. The presence of even small amounts of the target gases results in a significant increase in the electrical conductivity of SnO_2, and this change provides a measure of the gas concentration, triggering a signal or alarm if a pre-set threshold level is detected.

The increase in electrical conductivity that arises in the presence of a reducing gas also depends on the presence of O_2. The SnO_2 in a sensor is usually in the form of a porous, thick film which has a high surface area. Adsorption of O_2 onto a SnO_2 surface draws electrons from the conduction band. Below ≈ 420 K, the adsorbed oxygen is in the form of O_2^-, while above this temperature, O^- and O^{2-} are present. The presence of paramagnetic O_2^- and O^- has been evidenced by using EPR spectroscopy (see Section 4.9). The operating temperature of an SnO_2 sensor is 450–750 K and in the presence of a reducing gas such as CO or hydrocarbon, the SnO_2 surface loses oxygen in a process that may be represented as:

$$CO(g) + O^-(surface) \longrightarrow CO_2(g) + e^-$$

The electrons that are released are conducted through the solid and return to the conduction band, thereby producing an increase in the electrical conductivity of the material. The process is illustrated schematically below where O^-(ads) represents an O^- ion adsorbed on the SnO_2 surface:

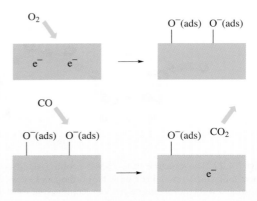

Tin(IV) oxide sensors play a major role in the commercial market and can be used to detect CO, CH_4, C_2H_5OH vapour, H_2 and NO_x. Other sensor materials include:

- ZnO, Ga_2O_3 and TiO_2/V_2O_5 for CH_4 detection;
- La_2CuO_4, Cr_2O_3/MgO and $Bi_2Fe_4O_9$ for C_2H_5OH vapour detection;
- ZnO, Ga_2O_3, ZrO_2 and WO_3 for H_2 detection;
- ZnO, TiO_2 (doped with Al and In) and WO_3 for NO_x;
- ZnO, Ga_2O_3, Co_3O_4 and TiO_2 (doped with Pt) for CO detection;
- WO_3 for O_3 detection at a parts-per-billion (ppb) level.

Further reading

E. Comini and G. Sberveglieri (2010) *Materials Today*, vol. 13, p. 36 – 'Metal oxide nanowires as chemical sensors'.

M.E. Franke, T.J. Koplin and U. Simon (2006) *Small*, vol. 2, p. 36 – 'Metal and metal oxide nanoparticles in chemiresistors: Does the nanoscale matter?'

A. Gurlo (2011) *Nanoscale*, vol. 3, p. 154 – 'Nanosensors: towards morphological control of gas sensing activity. SnO_2, In_2O_3, ZnO and WO_3 case studies'.

J. Riegel, H. Neumann and H.-W. Wiedenmann (2002) *Solid State Ionics*, vol. 152–153, p. 783 – 'Exhaust gas sensors for automotive emission control'.

M. Tiemann (2007) *Chem. Eur. J.*, vol. 13, p. 8376 – 'Porous metal oxides as gas sensors'.

A. Tricoli, M. Righettoni and A. Teleki (2010) *Angew. Chem. Int. Ed.*, vol. 49, p. 7632 – 'Semiconductor gas sensors: Dry synthesis and application'.

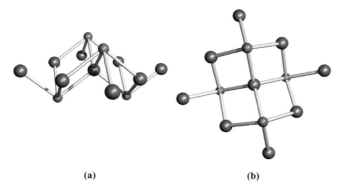

Fig. 14.28 Two views (a) from the side and (b) from above of a part of one layer of the SnO and red PbO lattices. Colour code: Sn, Pb, brown; O, red.

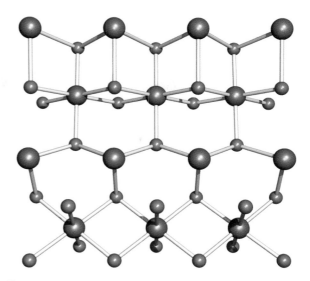

Fig. 14.29 Part of the network structure of Pb_3O_4 (i.e. $2PbO \cdot PbO_2$) showing the interconnected chains of octahedral $\{Pb^{(IV)}O_6\}$-units and trigonal pyramidal $\{Pb^{(II)}O_3\}$-units. Colour code: Pb, brown; O, red.

reacts with alkalis; reaction 14.70 occurs in strongly alkaline media to give a stannate.

$$SnO_2 + 6HCl \longrightarrow 2[H_3O]^+ + [SnCl_6]^{2-} \qquad (14.69)$$

$$SnO_2 + 2KOH + 2H_2O \longrightarrow K_2[Sn(OH)_6] \qquad (14.70)$$

In contrast, PbO_2 shows acidic (but no basic) properties, forming $[Pb(OH)_6]^{2-}$ when treated with alkali. Crystalline salts such as $K_2[Sn(OH)_6]$ and $K_2[Pb(OH)_6]$ can be isolated.

The monoxides SnO (the form which is stable under ambient conditions is blue-black in colour) and PbO (red form, *litharge*) possess layer structures in which each metal centre is at the apex of a square-based pyramidal array (Fig. 14.28). Each metal centre bears a lone pair of electrons occupying an orbital pointing towards the space between the layers, and electronic effects contribute to the preference for this asymmetric structure. Litharge is the more important form of PbO, but a yellow form also exists. While PbO can be prepared by heating the metal in air above 820 K, SnO is sensitive to oxidation and is best prepared by thermal decomposition of tin(II) oxalate. PbO can also be made by dehydrating $Pb(OH)_2$. Both SnO and PbO are amphoteric, but the oxoanions formed from them, like those from GeO, are not well characterized. Of the group 14 elements, only lead forms a mixed oxidation state oxide: Pb_3O_4 (*red lead*) is obtained by heating PbO in an excess of air at 720–770 K, and is better formulated as $2PbO \cdot PbO_2$. The solid state structure of Pb_3O_4 consists of chains of edge-sharing $\{Pb^{(IV)}O_6\}$-octahedra linked together by trigonal pyramidal $\{Pb^{(II)}O_3\}$-units (Fig. 14.29). Nitric acid reacts with Pb_3O_4 (according to eq. 14.71), while treatment with glacial acetic acid yields a mixture of $Pb(CH_3CO_2)_2$ and $Pb(CH_3CO_2)_4$, the latter compound being an important reagent in organic chemistry; the two acetate salts can be separated by crystallization.

$$Pb_3O_4 + 4HNO_3 \longrightarrow PbO_2 + 2Pb(NO_3)_2 + 2H_2O \quad (14.71)$$

14.10 Siloxanes and polysiloxanes (silicones)

Although siloxanes are often classed as organometallic compounds, they are conveniently described in this chapter because of their structural relationship to silicates. Hydrolysis of Me_nSiCl_{4-n} ($n = 1$–3) might be expected to give the derivatives $Me_nSi(OH)_{4-n}$ ($n = 1$–3). By analogy with carbon analogues, we might expect Me_3SiOH to be stable (except with respect to dehydration at higher temperatures), but would expect $Me_2Si(OH)_2$ and $MeSi(OH)_3$ to undergo dehydration to $Me_2Si=O$ and $MeSiO_2H$ respectively. However, at the beginning of Section 14.9, we indicated that an $Si=O$ bond is energetically less favourable than two $Si-O$ bonds. As a consequence, hydrolysis of Me_nSiCl_{4-n} ($n = 1$–3) yields *siloxanes* which are oligomeric products (e.g. reaction 14.72) containing the tetrahedral groups **14.24–14.26** in which each O atom represents part of an $Si-O-Si$ bridge. Diols can condense to give chains (**14.26**) or rings (e.g. **14.27**). Hydrolysis of $MeSiCl_3$ produces a cross-linked polymer.

$$2Me_3SiOH \longrightarrow Me_3SiOSiMe_3 + H_2O \qquad (14.72)$$

(structures)

14.24   **14.25**   **14.26**

(14.27)

(14.28)

Polysiloxanes (often referred to as silicones) have a range of structures and applications (see Box 14.12), and, in their manufacture, control of the polymerization is essential. The methylsilicon chlorides are co-hydrolysed, or the initial products of hydrolysis are equilibrated by heating with H_2SO_4 which catalyses the conversion of cyclic oligomers into chain polymers, bringing about redistribution of the terminal $OSiMe_3$ groups. For example, equilibration of $HOSiMe_2(OSiMe_2)_nOSiMe_2OH$ with $Me_3SiOSiMe_3$ leads to the polymer $Me_3Si(OSiMe_2)_nOSiMe_3$. Cross-linking, achieved by co-hydrolysis of Me_2SiCl_2 and $MeSiCl_3$, leads, after heating at 520 K, to silicone resins that are hard and inert. Tailoring the product so that it possesses a smaller degree of cross-linking results in the formation of silicone rubbers.

14.11 Sulfides

The disulfides of C, Si, Ge and Sn show the gradation in properties that might be expected to accompany the increasingly metallic character of the elements. Some properties of these sulfides are given in Table 14.6. Lead(IV) is too powerful an oxidizing agent to coexist with S^{2-}, and lead(IV) sulfide is not known.

Carbon disulfide is made by heating charcoal with sulfur at 1200 K, or by passing CH_4 and sulfur vapour over Al_2O_3 at 950 K. It is highly toxic (by inhalation and absorption through the skin) and extremely flammable, but is an excellent solvent which is used in the production of rayon and cellophane. Carbon disulfide is insoluble in water, but is, by a narrow margin, thermodynamically unstable with respect to hydrolysis to CO_2 and H_2S. However, this reaction has a high kinetic barrier and is very slow. Unlike CO_2, CS_2 polymerizes under high pressure to give a black solid with the chain structure **14.29**. When shaken with solutions of group 1 metal sulfides, CS_2 dissolves readily to give trithiocarbonates, M_2CS_3, which contain the $[CS_3]^{2-}$ ion **14.30**. The $[CS_3]^{2-}$ ion can also be made by the reaction of CS_2 with group 1 metal hydroxides in polar solvents. Salts of $[CS_3]^{2-}$ are readily isolated, e.g. Na_2CS_3 forms yellow needles (mp 353 K). The free acid H_2CS_3 separates as an oil when salts are treated with hydrochloric acid (eq. 14.73), and behaves as a weak acid in aqueous solution: $pK_a(1) = 2.68$, $pK_a(2) = 8.18$.

$$BaCS_3 + 2HCl \xrightarrow{273\,K} BaCl_2 + H_2CS_3 \qquad (14.73)$$

(14.29)   **(14.30)**

The action of an electric discharge on CS_2 results in the formation of C_3S_2, **14.31** (compare with **14.19**), a red liquid which decomposes at room temperature, producing a black polymer $(C_3S_2)_x$. When heated, C_3S_2 explodes. In contrast to CO, CS is a short-lived radical species which

Table 14.6   Selected properties of ES_2 (E = C, Si, Ge, Sn).

Property	CS_2	SiS_2	GeS_2	SnS_2
Melting point / K	162	1363 (sublimes)	870 (sublimes)	873 (dec.)
Boiling point / K	319	–	–	–
Appearance at 298 K	Volatile liquid, foul odour	White needle-like crystals	White powder or crystals	Golden-yellow crystals
Structure at 298 K	Linear molecule S=C=S	Solid state, chain[†]	3-Dimensional lattice with Ge_3S_3 and larger rings with shared vertices[†]	CdI_2-type structure (see Fig. 6.23)

[†] At high pressures and temperatures, SiS_2 and GeS_2 adopt a β-cristobalite lattice (see Fig. 6.20c).

APPLICATIONS

Box 14.12 Diverse applications of siloxane polymers (silicones)

Siloxane polymers (known in the commercial market as silicones) have widespread applications including personal care products, greases, sealants, varnishes, waterproofing materials, synthetic rubbers and gas-permeable membranes such as those used in soft contact lenses. Medical applications are increasing in importance, although the use of silicone breast implants is controversial because of claims of the migration of low molecular weight siloxanes from the implant into the surrounding body tissue.

Siloxane surfactants are crucial ingredients in personal care products: they are the components of shampoos and conditioners that improve the softness and silkiness of hair, and are also used in shaving foams, toothpastes, antiperspirants, cosmetics, hair-styling gels and bath oils. These applications follow from a combination of properties, e.g. low surface tension, water-solubility or dispersion in water to give emulsions, and low toxicity. Some of these properties contrast sharply with those required by polysiloxanes used in, for example, greases, sealants and rubbers. Siloxane surfactants contain a permethylated backbone which incorporates polar substituents. For example, polyether groups are hydrophilic and allow the polymers to be used in aqueous media, a prerequisite for use in shampoos and hair conditioners:

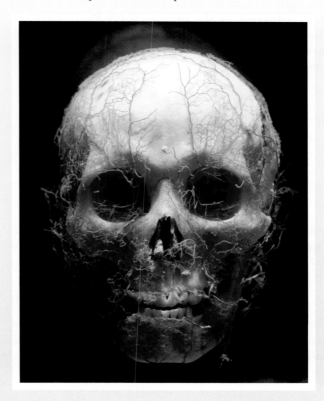

curing process, the elastomer is peeled away from the surface, giving a mould that can be used to replicate the original feature. One of the applications of silicone elastomers that has significant potential in nanoscience is *soft-lithography*. This technique was first developed in 1998 by Whitesides, and is increasingly being used for the replication of micro- and nanostructures. The structure or pattern to be copied is coated with liquid PDMS which is then cured to give a polysiloxane elastomer. When removed, the PDMS elastomer serves as a high-resolution 'stamp' for structure replication.

Veins surrounding a skeletal head at 'Bodies ... The Exhibition' which opened in New York City in 2005. The tissue in the body specimens is permanently preserved by using liquid polysiloxanes followed by a hardening process.

Silicone fluids cover a diverse range of polysiloxanes with uses that include lubricants, hydraulic fluids, water repellents, power transmission fluids and paint additives. Polydimethylsiloxane, $\{-SiMe_2O-\}_n$ (PDMS), and polymethylphenylsiloxane, $\{-SiMePhO-\}_n$, fluids are particularly important. Modifying the organic substituents is a means of tuning the properties of the polymer. For example, the introduction of phenyl substituents gives polymers that are able to withstand higher temperatures than PDMS, while the incorporation of fluoroalkyl groups leads to silicone fluids which can be used as low-temperature lubricants.

Silicone rubbers or *elastomers* are cross-linked polymers, the tensile strength of which is increased by adding a filler, usually 'fumed silica' (a form of SiO_2 with a particularly high surface area). Cross-linking takes place during curing (vulcanization) of the elastomer, which can occur at high temperatures or room temperature, depending on the polymer. For the replication of, for example, a plaster or carved wooden surface, room-temperature vulcanization is used to cross-link a polysiloxane paste which is first applied to the surface. After the

Further reading

A.B. Braunschweig, F. Huo and C.A. Mirkin (2009) *Nature Chem.*, vol. 1, p. 353 – 'Molecular printing'.

J.E. Mark, H.R. Allcock and R. West (2005) *Inorganic Polymers*, 2nd edn, Oxford University Press, Oxford – Chapter 4: 'Polysiloxanes and related polymers'.

J. Rogers and R.G. Nuzzo (2005) *Materials Today*, vol. 8, p. 50 – 'Recent progress in soft lithography'.

Y. Xia and G.M. Whitesides (1998) *Angew. Chem. Int. Ed.*, vol. 37, p. 550 – 'Soft lithography'.

decomposes at 113 K; it has been observed in the upper atmosphere.

$$S = C = C = C = S$$

(14.31)

Several salts of the $[C_2S_4]^{2-}$ anion are known and are made by, for example, reaction 14.74. The free acid $H_2C_2S_4$ (an analogue of oxalic acid) has not been isolated.

$$[CH_3CS_2]^- + 2[S_x]^{2-}$$
$$\longrightarrow [C_2S_4]^{2-} + [HS]^- + H_2S + [S_{2x-4}]^{2-} \quad (14.74)$$

In $[Et_4N]_2[C_2S_4]$, the anion has D_{2d} symmetry, i.e. the dihedral angle between the planes containing the two CS_2-units is $90°$ (structure **14.32**), whereas in $[Ph_4P]_2[C_2S_4]\cdot6H_2O$, this angle is $79.5°$. Compare these structural data with those for salts of the related oxalate ion, $[C_2O_4]^{2-}$ (structures **14.20** and **14.21**).

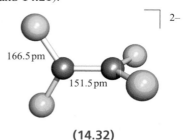

166.5 pm

151.5 pm

(14.32)

Silicon disulfide is prepared by heating Si in sulfur vapour. Both the structure of this compound (Table 14.6) and the chemistry of SiS_2 show no parallels with SiO_2; SiS_2 is instantly hydrolysed (eq. 14.75).

$$SiS_2 + 2H_2O \longrightarrow SiO_2 + 2H_2S \quad (14.75)$$

The disulfides of Ge and Sn (Table 14.6) are precipitated when H_2S is passed into acidic solutions of Ge(IV) and Sn(IV) compounds. Some sulfides have cluster structures, e.g. $[Ge_4S_{10}]^{4-}$ (**14.33**), prepared by reaction 14.76.

$$4GeS_2 + 2S^{2-} \xrightarrow{\text{Aqueous solution in presence of } Cs^+} [Ge_4S_{10}]^{4-}$$
$$(14.76)$$

Tin(IV) forms a number of thiostannates containing discrete anions, e.g. Na_4SnS_4 contains the tetrahedral $[SnS_4]^{4-}$ ion, and $Na_4Sn_2S_6$ and $Na_6Sn_2S_7$ contain anions **14.34** and **14.35** respectively.

(14.33)

(14.34)   **(14.35)**

The monosulfides of Ge, Sn and Pb are all obtained by precipitation from aqueous media. Both GeS and SnS crystallize with layer structures similar to that of black phosphorus (see Section 15.4). Lead(II) sulfide occurs naturally as galena and adopts an NaCl-type structure. Its formation as a black precipitate ($K_{sp} \approx 10^{-30}$) is observed in the qualitative test for H_2S (eq. 14.77). The colour and very low solubility of PbS suggest that it is not a purely ionic compound.

$$Pb(NO_3)_2 + H_2S \longrightarrow \underset{\text{black ppt}}{PbS} + 2HNO_3 \quad (14.77)$$

Pure PbS is a p-type semiconductor when S-rich, and an n-type when Pb-rich (the non-stoichiometric nature of solids is discussed in Section 6.17). It exhibits *photoconductivity* and has applications in photoconductive cells, transistors and photographic exposure meters.

> If a material is a *photoconductor*, it absorbs light with the result that electrons from the valence band are excited into the conducting band. Thus, the electrical conductivity increases on exposure to light.

Worked example 14.7   Tin and lead sulfides

Calculate the solubility of PbS given that $K_{sp} = 10^{-30}$. Is your answer consistent with the fact that PbS is shown as a precipitate in reaction 14.77?

K_{sp} refers to the equilibrium:

$$PbS(s) \rightleftharpoons Pb^{2+}(aq) + S^{2-}(aq)$$

$$K_{sp} = 10^{-30} = \frac{[Pb^{2+}][S^{2-}]}{[PbS]} = [Pb^{2+}][S^{2-}]$$

$$[Pb^{2+}] = [S^{2-}]$$

Therefore, making this substitution in the equation for K_{sp} gives:

$$[Pb^{2+}]^2 = 10^{-30}$$

$$[Pb^{2+}] = 10^{-15} \, \text{mol dm}^{-3}$$

Thus, the extremely low solubility means that PbS will appear as a precipitate in reaction 14.77.

1. Describe the coordination environment of each Pb^{2+} and S^{2-} ion in galena.   [*Ans.* NaCl structure; see Fig. 6.16]

2. The solubility of SnS in water is 10^{-13} mol dm^{-3}. Calculate a value for K_{sp}.   [*Ans.* 10^{-26}]

3. Lead-deficient and lead-rich PbS are p- and n-type semiconductors respectively. Explain the difference between these two types of semiconductors.
    [*Ans.* see Fig. 6.14 and accompanying discussion]

14.12 Cyanogen, silicon nitride and tin nitride

In discussing bonds formed between the group 14 elements and nitrogen, two compounds of particular importance emerge: cyanogen, C_2N_2, and silicon nitride. Tin(IV) nitride has been prepared more recently.

Cyanogen and its derivatives

The $CN^•$ radical is a *pseudo-halogen*, i.e. its chemistry resembles that of a halogen atom, X; it forms C_2N_2, HCN and $[CN]^-$, analogues of X_2, HX and X^-. Although C_2N_2 and HCN are thermodynamically unstable with respect to decomposition into their elements, hydrolysis by H_2O, and oxidation by O_2, they and $[CN]^-$ are *kinetically* stable enough for them to be well-established and much studied species.

Cyanogen, C_2N_2, is a toxic, extremely flammable gas (mp 245 K, bp 252 K) which is liable to react explosively with some powerful oxidants. Although $\Delta_f H^o(C_2N_2,$ 298 K$) = +297$ kJ mol^{-1}, pure C_2N_2 can be stored for long periods without decomposition. Reactions 14.78 and 14.79 give two syntheses of C_2N_2; reaction 14.79 illustrates the pseudo-halide like nature of $[CN]^-$ which is oxidized by Cu(II) in an analogous fashion to the oxidation of I^- to I_2. Cyanogen is manufactured by air-oxidation of HCN over a silver catalyst.

$$Hg(CN)_2 + HgCl_2 \xrightarrow{570\,K} C_2N_2 + Hg_2Cl_2 \qquad (14.78)$$

$$2CuSO_4 + 4NaCN \xrightarrow[\text{solution, } \Delta]{\text{aqueous}} C_2N_2 + 2CuCN + 2Na_2SO_4 \qquad (14.79)$$

$$N \equiv C \underset{137\ pm}{\overset{116\ pm}{-\!\!\!-\!\!\!-}} C \equiv N$$

(14.36)

Cyanogen has the linear structure **14.36** and the short C–C distance indicates considerable electron delocalization. It burns in air with a very hot, violet flame (eq. 14.80), and resembles the halogens in that it is hydrolysed by alkali (eq. 14.81) and undergoes thermal dissociation to $CN^•$ at high temperatures.

$$C_2N_2 + 2O_2 \longrightarrow 2CO_2 + N_2 \qquad (14.80)$$

$$C_2N_2 + 2[OH]^- \longrightarrow [OCN]^- + [CN]^- + H_2O \qquad (14.81)$$

$$H \underset{106.5\ pm}{\overset{115\ pm}{-\!\!\!-\!\!\!-}} C \equiv N$$

(14.37)   **(14.38)**

Hydrogen cyanide, HCN, **14.37**, is an extremely toxic and flammable, colourless volatile liquid (mp 260 K, bp 299 K) with a high dielectric constant due to strong hydrogen bonding. It has a characteristic smell of bitter almonds. The pure liquid polymerizes and in the absence of a stabilizer such as H_3PO_4, polymerization may be explosive. In the presence of traces of H_2O and NH_3, HCN forms adenine, **14.38**, and on reduction, gives $MeNH_2$. It is thought that HCN was one of the small molecules in the early atmosphere of the Earth, and played an important role in the formation of many biologically important compounds. Hydrogen cyanide is prepared on a small scale by adding acid to NaCN, and industrially by reactions 14.82 and 14.83.

$$2CH_4 + 2NH_3 + 3O_2 \xrightarrow[1250-1550\,K,\,2\,bar]{Pt/Rh,} 2HCN + 6H_2O \qquad (14.82)$$

$$CH_4 + NH_3 \xrightarrow{Pt,\ 1450-1550\,K} HCN + 3H_2 \qquad (14.83)$$

Many organic syntheses involve HCN, and it is of great industrial importance, a large fraction going into the production of 1,4-dicyanobutane (adiponitrile) for nylon manufacture, and cyanoethene (acrylonitrile) for production of acrylic fibres.

In aqueous solution, HCN behaves as a weak acid (p$K_a = 9.31$) and is slowly hydrolysed (eq. 14.84). An older name for hydrocyanic acid is prussic acid.

$$HCN + 2H_2O \longrightarrow [NH_4]^+ + [HCO_2]^- \qquad (14.84)$$

$$2HCN + Na_2CO_3 \xrightarrow{\text{aqueous solution}} 2NaCN + H_2O + CO_2 \qquad (14.85)$$

The neutralization of aqueous HCN by Na_2CO_3, $NaHCO_3$ or $Na[HCO_2]$ generates NaCN, the most important salt of the acid. It is manufactured by reaction 14.85, and has widespread uses in organic chemistry (e.g. for the formation of

BIOLOGY AND MEDICINE

Box 14.13 Hydrogen cyanide in plant material

A number of plants (e.g. cassava, sugar cane, some varieties of white clover) and fruits are natural sources of HCN. The origins of the HCN are cyanoglucosides such as *amygdalin* (e.g. in almonds, peach and apricot stones, apple seeds) and *linamarin* (in cassava). The release of HCN from certain plants (cyanogenesis) occurs in the presence of specific enzymes. For example, the enzyme *linamarase* is present in the cell walls of cassava plants. Crushing or chewing cassava root results in release of linamarase, allowing it to act on its cyanoglucoside substrate linamarin. Initially Me$_2$C(OH)CN is released, and this rapidly produces HCN. Cassava is an important root crop grown in tropical regions as a source of starch, e.g. it is used for the production of tapioca. Cassava plants may be either a sweet or a bitter variety. The HCN content ranges from 250 to 900 mg kg^{-1} depending on variety (compare this with a

lethal dose of HCN of 1 mg per kg body weight). Bitter cassava contains the greatest amounts of cyanoglucosides, and in order to produce it as a foodstuff, it must be subjected to careful treatment of shredding, pressure and heat. This cleaves the cyanoglucoside, thereby removing HCN prior to human consumption of the foodstuff. A beneficial side effect of cyanoglucosides in plants is that they act as a natural, chemical defence, for example, against insects and rodents.

Further reading

D.A. Jones (1998) *Phytochemistry*, vol. 47, p. 155 – 'Why are so many plants cyanogenic?'

Apricot stones contain amygdalin.

C–C bonds). NaCN is also used in the extraction of Ag and Au (see eq. 22.4 and Box 22.2). At 298 K, NaCN and KCN adopt an NaCl-type structure, each [CN]$^-$ ion freely rotating (or having random orientations) about a fixed point in the lattice and having an effective ionic radius of $\approx$190 pm. At lower temperatures, transitions to structures of lower symmetry occur, e.g. NaCN undergoes a cubic to hexagonal transition below 283 K. Crystals of NaCN and KCN are deliquescent, and both salts are soluble in water and are highly toxic. Fusion of KCN and sulfur gives potassium thiocyanate, KSCN.

Mild oxidizing agents convert [CN]$^-$ to cyanogen (eq. 14.79) but with more powerful oxidants such as PbO or *neutral* [MnO$_4$]$^-$, cyanate ion, **14.39**, is formed (reaction 14.86). Potassium cyanate reverts to the cyanide on heating (eq. 14.87).

$$PbO + KCN \longrightarrow Pb + K[OCN] \qquad (14.86)$$

$$2K[OCN] \xrightarrow{\Delta} 2KCN + O_2 \qquad (14.87)$$

$$O = C = N^- \longleftrightarrow {}^-O - C \equiv N$$

(14.39)

O=C=N with 121 pm, 117 pm, 128°, 99 pm, H

(14.40)

Two acids can be derived from **14.39**: HOCN (cyanic acid or hydrogen cyanate) and HNCO (isocyanic acid, **14.40**). It has been established that HOCN and HNCO are not in equilibrium with each other. Isocyanic acid (pK_a = 3.66) is obtained by heating urea (eq. 14.88) but rapidly trimerizes, although heating the trimer regenerates the monomer.

keto-Tautomer of
cyanuric acid

(14.88)

The fulminate ion, $[CNO]^-$, is an isomer of the cyanate ion. Fulminate salts can be reduced to cyanides but cannot be prepared by oxidation of them. The free acid readily polymerizes but is stable for short periods in Et_2O at low temperature. Metal fulminates are highly explosive. Mercury(II) fulminate may be prepared by reaction 14.89 and is a dangerous detonator.

$$2Na[CH_2NO_2] + HgCl_2 \longrightarrow Hg(CNO)_2 + 2H_2O + 2NaCl$$
(14.89)

(14.41)   **(14.42)**

Cyanogen chloride, **14.41** (mp 266 K, bp 286 K), is prepared by the reaction of Cl_2 with NaCN or HCN, and readily trimerizes to **14.42**, which has applications in the manufacture of dyestuffs and herbicides.

Silicon nitride

Silicon nitride, Si_3N_4, has wide applications as a ceramic and refractory material and in the form of whiskers (see Section 28.6). It is a white, chemically inert amorphous powder, which can be formed by reaction 14.90, or by combining Si and N_2 above 1650 K.

$$SiCl_4 + 4NH_3 \xrightarrow{-4HCl} Si(NH_2)_4 \xrightarrow{\Delta} Si(NH)_2 \xrightarrow{\Delta} Si_3N_4$$
(14.90)

The two most important polymorphs, α- and β-Si_3N_4, possess 3-dimensional structures containing distorted tetrahedral Si atoms, and 3-coordinate, near-planar N atoms. The detailed crystal structures and the presence of lattice defects have been the subject of debate for many years.[†] A denser, harder polymorph, γ-Si_3N_4, has been obtained by high-

---

[†]See: C.-M. Wang, X. Pan, M. Rühle, F.L. Riley and M. Mitomo (1996) *J. Mater. Sci.,* vol. 31, p. 5281; F.L. Riley (2000) *J. Am. Ceram. Soc.,* vol. 83, p. 245; A. Kuwabara, K. Matsunaga and I. Tanaka (2008) *Phys. Rev. B,* vol. 78, article 064104.

pressure, high-temperature (15 GPa, >2000 K) fabrication. This polymorph has the spinel structure (see Box 13.7): the N atoms form a cubic close-packed structure in which two-thirds of the Si atoms occupy octahedral holes and one-third occupy tetrahedral holes. The oxide spinels that we discussed in Box 13.7 contained metal ions in the +2 and +3 oxidation states, i.e. $(A^{II})(B^{III})_2O_4$. In γ-Si_3N_4, all the Si atoms are in a single (+4) oxidation state. Another refractory material is Si_2N_2O, made from Si and SiO_2 under N_2/Ar atmosphere at 1700 K. It possesses puckered hexagonal nets of alternating Si and N atoms, the sheets being linked by Si$-$O$-$Si bonds.

Tin(IV) nitride

Tin(IV) nitride, Sn_3N_4, was first isolated in 1999 from the reaction of SnI_4 with KNH_2 in liquid NH_3 at 243 K followed by annealing the solid product at 573 K. Sn_3N_4 adopts a spinel-type structure, related to that of γ-Si_3N_4. Tin(IV) nitride is the first nitride spinel found to be stable under ambient conditions.

14.13 Aqueous solution chemistry and salts of oxoacids of germanium, tin and lead

When GeO_2 is dissolved in basic aqueous solution, the solution species formed is $[Ge(OH)_6]^{2-}$. With hydrochloric acid, GeO_2 forms $[GeCl_6]^{2-}$. Although GeO_2 is reduced by H_3PO_2 in aqueous HCl solution and forms the insoluble $Ge(OH)_2$ when the solution pH is increased, it is possible to retain Ge(II) in aqueous solution under controlled conditions. Thus, 6 M aqueous HCl solutions that contain 0.2–0.4 mol dm^{-3} of Ge(II) generated *in situ* (eq. 14.91) are stable for several weeks.

$$Ge^{IV} + H_2O + H_3PO_2 \longrightarrow H_3PO_3 + Ge^{II} + 2H^+$$
(14.91)

Table 14.1 lists standard reduction potentials for the M^{4+}/M^{2+} and M^{2+}/M (M = Sn, Pb) couples. The value of $E^o(Sn^{4+}/Sn^{2+}) = +0.15$ V shows that Sn(II) salts in aqueous solution are readily oxidized by O_2. In addition, hydrolysis of Sn^{2+} to species such as $[Sn_2O(OH)_4]^{2-}$ and $[Sn_3(OH)_4]^{2+}$ is extensive. Aqueous solutions of Sn(II) salts are therefore usually acidified and complex ions are then likely to be present, e.g. if $SnCl_2$ is dissolved in dilute hydrochloric acid, $[SnCl_3]^-$ forms. In alkaline solutions, the dominant species is $[Sn(OH)_3]^-$. Extensive hydrolysis of Sn(IV) species in aqueous solution also occurs unless sufficient acid is present to complex the Sn(IV). Thus, in aqueous HCl, Sn(IV) is present as $[SnCl_6]^{2-}$. In alkaline solution at high pH, $[Sn(OH)_6]^{2-}$ is the main species and salts of this octahedral ion, e.g. $K_2[Sn(OH)_6]$, can be isolated.

In comparison with their Sn(II) analogues, Pb(II) salts are much more stable in aqueous solution with respect to hydrolysis and oxidation. The most important *soluble* salts are $Pb(NO_3)_2$ and $Pb(CH_3CO_2)_2$. The fact that many water-insoluble Pb(II) salts dissolve in a mixture of $[NH_4][CH_3CO_2]$ and CH_3CO_2H reveals that Pb(II) is strongly complexed by acetate. Most Pb(II) oxo-salts are, like the halides, sparingly soluble in water; $PbSO_4$ ($K_{sp} = 1.8 \times 10^{-8}$) dissolves in concentrated H_2SO_4.

The Pb^{4+} ion does not exist in aqueous solution, and the value of $E^o(Pb^{4+}/Pb^{2+})$ given in Table 14.1 is for the half-reaction 14.92 which forms part of the familiar lead–acid battery (see eqs. 14.3 and 14.4). For half-reaction 14.92, the fourth-power dependence of the half-cell potential upon $[H^+]$ explains why the relative stabilities of Pb(II) and Pb(IV) depend upon the pH of the solution (see Section 8.2).

$$PbO_2(s) + 4H^+(aq) + 2e^- \rightleftharpoons Pb^{2+}(aq) + 2H_2O(l)$$

$$E^o = +1.45\,V \qquad (14.92)$$

Thus, for example, PbO_2 oxidizes concentrated HCl to Cl_2, but Cl_2 oxidizes Pb(II) in alkaline solution to PbO_2. It may be noted that thermodynamically, PbO_2 should oxidize water at pH = 0, and the usefulness of the lead–acid battery depends on there being a high overpotential for O_2 evolution.

Yellow crystals of $Pb(SO_4)_2$ may be obtained by electrolysis of fairly concentrated H_2SO_4 using a Pb anode. However, in cold water, it is hydrolysed to PbO_2, as are Pb(IV) acetate and $[NH_4]_2[PbCl_6]$ (see Section 14.8). The complex ion $[Pb(OH)_6]^{2-}$ forms when PbO_2 dissolves in concentrated KOH solution, but on dilution of the solution, PbO_2 is reprecipitated.

KEY TERMS

The following terms were introduced in this chapter. Do you know what they mean?

- ❏ catenation
- ❏ metastable
- ❏ intercalation compound
- ❏ Zintl ion
- ❏ piezoelectric
- ❏ hydrothermal
- ❏ photoconductor

FURTHER READING

Group 14: general

J.S. Casas and J. Sordo, eds. (2006) *Lead: Chemistry, Analytical Aspects, Environmental Impact and Health Effects*, Elsevier, Amsterdam – Modern coverage of many aspects of the chemistry, uses and toxicity of lead.

A.G. Davies, M. Gielen, K.H. Pannell and E.R.T. Tiekink, eds. (2008) *Tin Chemistry: Fundamentals, Frontiers and Applications*, Wiley, Chichester – A detailed survey of the chemistry of tin.

N.N. Greenwood and A. Earnshaw (1997) *Chemistry of the Elements*, 2nd edn, Butterworth-Heinemann, Oxford – Chapters 8–10 describe the chemistry of the group 14 elements in detail.

P. Jutzi and U. Schubert, eds. (2003) *Silicon Chemistry: From the Atom to Extended Systems*, Wiley-VCH, Weinheim – A series of chapters covering molecular and materials chemistry of silicon.

A.G. Massey (2000) *Main Group Chemistry*, 2nd edn, Wiley, Chichester – Chapter 8 covers the chemistry of the group 14 elements.

Carbon: fullerenes and nanotubes

R.C. Haddon, ed. (2002) *Acc. Chem. Res.*, vol. 35, issue 12 – 'Carbon nanotubes' (a special issue of the journal covering different aspects of the area).

A. Hirsch and M. Brettreich (2005) *Fullerenes: Chemistry and Reactions*, Wiley-VCH, Weinheim – A detailed account of the chemistry of C_{60} and higher fullerenes.

H.W. Kroto (1992) *Angew. Chem. Int. Ed.*, vol. 31, p. 111 – 'C_{60}: Buckminsterfullerene, the celestial sphere that fell to earth'.

S. Margadonna and K. Prassides (2002) *J. Solid State Chem.*, vol. 168, p. 639 – 'Recent advances in fullerene superconductivity'.

K. Prassides, ed. (2004) *Struct. Bond.*, vol. 109 – An entire volume of this journal with the theme 'Fullerene-based materials: structures and properties'.

C.A. Reed and R.D. Bolskov (2000) *Chem. Rev.*, vol. 100, p. 1075 – 'Fulleride anions and fullerenium cations'.

D. Tasis, N. Tagmatarchis, A. Bianco and M. Prato (2006) *Chem. Rev.*, vol. 106, p. 1105 – 'Chemistry of carbon nanotubes'.

Zintl ions and other cluster compounds

J.D. Corbett (2000) *Angew. Chem. Int. Ed.*, vol. 39, p. 671 – 'Polyanionic clusters and networks of the early p-element metals in the solid state: beyond the Zintl boundary'.

T.F. Fässler (2001) *Coord. Chem. Rev.*, vol. 215, p. 347 – 'The renaissance of homoatomic nine-atom polyhedra of the heavier carbon-group elements Si–Pb'.

T.F. Fässler (2001) *Angew. Chem. Int. Ed.*, vol. 40, p. 4161 – 'Homoatomic polyhedra as structural modules in chemistry: what binds fullerenes and homonuclear Zintl ions?'

N. Korber (2009) *Angew. Chem. Int. Ed.*, vol. 48, p. 3216 – 'The shape of germanium clusters to come'.

A. Schnepf (2007) *Chem. Soc. Rev.*, vol. 36, p. 745 – 'Metalloid group 14 cluster compounds: An introduction and perspectives to this novel group of cluster compounds.'

Silicates, polysiloxanes and zeolites

J.E. Mark, H.R. Allcock and R. West (2005) *Inorganic Polymers*, 2nd edn, Oxford University Press, Oxford – Chapter 4: 'Polysiloxanes and related polymers'.

P.M. Price, J.H. Clark and D.J. Macquarrie (2000) *J. Chem. Soc., Dalton Trans.*, p. 101 – A review entitled: 'Modified silicas for clean technology'.

A.F. Wells (1984) *Structural Inorganic Chemistry*, 5th edn, Clarendon Press, Oxford – Chapter 23 contains a full account of silicate structures.

Other topics

R.C. Fischer and P.P. Power (2010) *Chem. Rev.*, vol. 110, p. 3877 – 'π-Bonding and the lone pair effect in multiple bonds involving heavier main group elements: Developments in the new millennium'.

M.S. Hill (2010) *Struct. Bond.*, vol. 136, p. 189 – 'Homocatenation of metal and metalloid main group elements'.

M.J. Hynes and B. Jonson (1997) *Chem. Soc. Rev.*, vol. 26, p. 133 – 'Lead, glass and the environment'.

S.T. Oyama (1996) *The Chemistry of Transition Metal Carbides and Nitrides*, Kluwer, Dordrecht.

W. Schnick (1999) *Angew. Chem. Int. Ed.*, vol. 38, p. 3309 – 'The first nitride spinels – New synthetic approaches to binary group 14 nitrides'.

See also Chapter 6 Further reading: Semiconductors.

PROBLEMS

14.1 (a) Write down, in order, the names and symbols of the elements in group 14. (b) Classify the elements in terms of metallic, semi-metallic or non-metallic behaviour. (c) Give a *general* notation showing the ground state electronic configuration of each element.

14.2 Comment on the trends in values of (a) melting points, (b) $\Delta_{atom}H^{\circ}(298\,K)$ and (c) $\Delta_{fus}H^{\circ}(mp)$ for the elements on descending group 14.

14.3 How does the structure of graphite account for (a) its use as a lubricant, (b) the design of graphite electrodes, and (c) the fact that diamond is the more stable allotrope at very high pressures?

14.4 Figure 14.10 shows a unit cell of K_3C_{60}. From the structural information given, confirm the stoichiometry of this fulleride.

14.5 Give four examples of reactions of C_{60} that are consistent with the presence of C=C bond character.

14.6 Comment on each of the following observations.

(a) The carbides Mg_2C_3 and CaC_2 liberate propyne and ethyne respectively when treated with water, reaction between ThC_2 and water produces mixtures composed mainly of C_2H_2, C_2H_6 and H_2, but no reaction occurs when water is added to TiC.

(b) Mg_2Si reacts with $[NH_4]Br$ in liquid NH_3 to give silane.

(c) Compound **14.43** is hydrolysed by aqueous alkali at the same rate as the corresponding Si–D compound.

(14.43)

14.7 (a) Suggest why the NSi_3 skeleton in $N(SiMe_3)_3$ is planar. (b) Suggest reasons why, at 298 K, CO_2 and SiO_2 are not isostructural. Under what conditions can phases of CO_2 with silica-like structures be made?

14.8 Predict the shapes of the following molecules or ions: (a) ClCN; (b) OCS; (c) $[SiH_3]^-$; (d) $[SnCl_5]^-$; (e) Si_2OCl_6; (f) $[Ge(C_2O_4)_3]^{2-}$; (g) $[PbCl_6]^{2-}$; (h) $[SnS_4]^{4-}$.

14.9 The observed structure of $[Sn_9Tl]^{3-}$ is a bicapped square-antiprism. (a) Confirm that this is consistent with Wade's rules. (b) How many isomers (retaining the bicapped square-antiprism core) of $[Sn_9Tl]^{3-}$ are possible?

14.10 Compare and contrast the structures and chemistries of the hydrides of the group 14 elements, and give pertinent examples to illustrate structural and chemical differences between BH_3 and CH_4, and between AlH_3 and SiH_4.

14.11 Write equations for: (a) the hydrolysis of $GeCl_4$; (b) the reaction of $SiCl_4$ with aqueous NaOH; (c) the 1:1 reaction of CsF with GeF_2; (d) the hydrolysis of SiH_3Cl; (e) the hydrolysis of SiF_4; (f) the 2:1 reaction of $[Bu_4P]Cl$ with $SnCl_4$. In each case suggest the structure of the product containing the group 14 element.

14.12 Rationalize the following signal multiplicities in the ^{119}Sn NMR spectra of some halo-anions and, where possible, use the data to distinguish between geometric isomers [^{19}F 100% $I = \frac{1}{2}$]: (a) $[SnCl_5F]^{2-}$ doublet; (b) $[SnCl_4F_2]^{2-}$ isomer A, triplet; isomer B, triplet; (c) $[SnCl_3F_3]^{2-}$ isomer A, doublet of triplets; isomer B, quartet; (d) $[SnCl_2F_4]^{2-}$ isomer A, quintet; isomer B, triplet of triplets; (e) $[SnClF_5]^{2-}$ doublet of quintets; (f) $[SnF_6]^{2-}$ septet.

14.13 What would you expect to form when:

(a) Sn is heated with concentrated aqueous NaOH;
(b) SO_2 is passed over PbO_2;
(c) CS_2 is shaken with aqueous NaOH;
(d) SiH_2Cl_2 is hydrolysed by water;
(e) four molar equivalents of $ClCH_2SiCl_3$ react with three equivalents of $Li[AlH_4]$ in Et_2O solution?

14.14 Suggest one method for the estimation of each of the following quantities:

(a) $\Delta_r H^\circ$ for the conversion:
GeO_2(quartz) $\longrightarrow$ GeO_2(rutile);
(b) the Pauling electronegativity value, χ^P, of Si;
(c) the purity of a sample of $Pb(MeCO_2)_4$ prepared in a laboratory experiment.

14.15 By referring to Fig. 8.6, deduce whether carbon could be used to extract Sn from SnO_2 at (a) 500 K; (b) 750 K; (c) 1000 K. Justify your answer.

14.16 Comment on the following observations.

(a) the pyroxenes $CaMgSi_2O_6$ and $CaFeSi_2O_6$ are isomorphous;
(b) the feldspar $NaAlSi_3O_8$ may contain up to 10% of $CaAl_2Si_2O_8$;
(c) the mineral *spodumene*, $LiAlSi_2O_6$, is isostructural with *diopside*, $CaMgSi_2O_6$, but when it is heated it is transformed into a polymorph having the quartz structure with the Li^+ ions in the interstices.

Table 14.7 Data for problem 14.17.

Compound	v_1(symmetric)/cm^{-1}	v_3(asymmetric)/cm^{-1}
I	2330	2158
II	658	1535
III	1333	2349

14.17 Table 14.7 gives values of the symmetric and asymmetric stretches of the heteronuclear bonds in CO_2, CS_2 and $(CN)_2$, although the molecules are indicated only by the labels I, II and III. (a) Assign an identity to each of I, II and III. (b) State whether the stretching modes listed in Table 14.7 are IR active or inactive.

14.18 Account for the fact that when aqueous solution of KCN is added to a solution of aluminium sulfate, a precipitate of $Al(OH)_3$ forms.

14.19 What would you expect to be the hydrolysis products of (a) cyanic acid, (b) isocyanic acid and (c) thiocyanic acid?

14.20 For solid $Ba[CSe_3]$, the vibrational wavenumbers and assignments for the $[CSe_3]^{2-}$ ion are 802 (E', stretch), 420 (A_2''), 290 (A_1') and 185 (E', deformation) cm^{-1}.
(a) Based on these assignments, deduce the shape of the $[CSe_3]^{2-}$ ion. (b) Draw diagrams to illustrate the modes of vibration of $[CSe_3]^{2-}$. (c) Which modes of vibration are IR active?

14.21 Deduce the point groups of each of the following molecular species: (a) SiF_4, (b) $[CO_3]^{2-}$, (c) CO_2, (d) SiH_2Cl_2.

14.22 By using the diagram shown below, confirm that C_{60} belongs to the I_h point group.

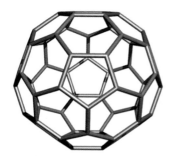

OVERVIEW PROBLEMS

14.23 (a) By using the description of the bonding in Sn_2R_4 as a guide (see Fig. 23.19), suggest a bonding scheme for a hypothetical $HSi{\equiv}SiH$ molecule with the following geometry:

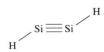

(b) Do you expect the $[FCO]^+$ ion to have a linear or bent structure? Give an explanation for your answer.

(c) The α-form of SnF_2 is a cyclotetramer. Give a description of the bonding in this tetramer and explain why the ring is non-planar.

14.24 Which description in the second list below can be correctly matched to each compound in the first list? There is only one match for each pair.

List 1	List 2
SiF_4	A semiconductor at 298 K with a diamond-type structure
Si	A Zintl ion
Cs_3C_{60}	Its Ca^{2+} salt is a component of cement
SnO	A water-soluble salt that is not decomposed on dissolution
$[Ge_9]^{4-}$	Gas at 298 K consisting of tetrahedral molecules
GeF_2	An acidic oxide
$[SiO_4]^{4-}$	An amphoteric oxide
PbO_2	Solid at 298 K with a sheet structure containing octahedral Sn centres
$Pb(NO_3)_2$	Becomes superconducting at 40 K
SnF_4	An analogue of a carbene

14.25 (a) $[SnF_5]^-$ has a polymeric structure consisting of chains with *cis*-bridging F atoms. Draw a repeat unit of the polymer. State the coordination environment of each Sn atom, and explain how the overall stoichiometry of $Sn:F = 1:5$ is retained in the polymer.

(b) Which of the salts PbI_2, $Pb(NO_3)_2$, $PbSO_4$, $PbCO_3$, $PbCl_2$ and $Pb(O_2CCH_3)_2$ are soluble in water?

(c) The IR spectrum of ClCN shows absorptions at 1917, 1060 and 230 cm^{-1}. Suggest assignments for these bands and justify your answer.

14.26 Suggest products for the following reactions; the left-hand sides of the equations are not necessarily balanced.

(a) $GeH_3Cl + NaOCH_3 \longrightarrow$

(b) $CaC_2 + N_2 \xrightarrow{\Delta}$

(c) $Mg_2Si + H_2O/H^+ \longrightarrow$

(d) $K_2SiF_6 + K \xrightarrow{\Delta}$

(e) $1,2\text{-}(OH)_2C_6H_4 + GeO_2 \xrightarrow{\text{NaOH/MeOH}}$

(f) $(H_3Si)_2O + I_2 \longrightarrow$

(g) $C_{60} \xrightarrow[]{O_3,\,257\,K\text{ in xylene}} \xrightarrow{296\,K}$

(h) $Sn \xrightarrow{\text{Hot NaOH(aq)}}$

14.27 (a) Describe the solid state structures of K_3C_{60} and of KC_8. Comment on any physical or chemical properties of the compounds that are of interest.

(b) Comment on the use of lead(II) acetate in a qualitative test for H_2S.

(c) In the $[Et_4N]^+$ salt, the $[C_2S_4]^{2-}$ ion is non-planar; the dihedral angle between the planes containing the two CS_2 groups is 90°. In contrast, in many of its salts, the $[C_2O_4]^{2-}$ ion is planar. Deduce, with reasoning, the point groups of these anions.

14.28 The reaction between a 1,2-ethanediamine solution of $K_4[Pb_9]$ and a toluene solution of $[Pt(PPh_3)_4]$ in the presence of crypt-222 leads to the formation of the platinum-centred Zintl ion $[Pt@Pb_{12}]^{2-}$, the ^{207}Pb NMR spectrum of which consists of a pseudo-triplet ($J_{^{207}Pb^{195}Pt} = 3440\,Hz$).

(a) What is the role of the crypt-222 in the reaction?

(b) Sketch the appearance of the ^{207}Pb NMR spectrum, paying attention to the relative intensities of the components of the triplet. Explain how this signal arises, and indicate on your diagram where the value of $J_{^{207}Pb^{195}Pt}$ is measured.

(c) The ^{195}Pt NMR spectrum of $[Pt@Pb_{12}]^{2-}$ is a non-binomial multiplet. Explain the origins of the coupling pattern. What is the separation (in Hz) of any pair of adjacent lines in the multiplet?

[Data: ^{207}Pb, 22.1% abundant, $I = \frac{1}{2}$; ^{195}Pt, 33.8% abundant, $I = \frac{1}{2}$]

14.29 (a) Equation 14.47 shows the reaction of Na with CCl_4. From the following data, confirm the value of $\Delta_r G^\circ = -1478\,kJ\,mol^{-1}$ at 298 K. Data: $\Delta_f G^\circ(NaCl, s) = -384\,kJ\,mol^{-1}$; $\Delta_f H^\circ(CCl_4, l) = -128.4\,kJ\,mol^{-1}$; $S^\circ(CCl_4, l) = 214\,J\,K^{-1}$ mol^{-1}; $S^\circ(C, gr) = +5.6\,J\,K^{-1}\,mol^{-1}$; $S^\circ(Cl_2, g) = 223\,J\,K^{-1}\,mol^{-1}$.

(b) Comment on similarities and differences between the structures of β-cristobalite and a non-crystalline silica glass.

INORGANIC CHEMISTRY MATTERS

14.30 Carbon monoxide detectors are in widespread use in the workplace and home. Metal oxide (e.g. SnO_2) semiconductor and electrochemical sensors are commonly used. (a) Explain how an SnO_2 sensor for CO works. (b) An electrochemical sensor employs platinum electrodes with aqueous sulfuric acid as the electrolyte. CO is oxidized at the anode. Write half equations for the anode and cathode reactions. Outline how the cell works as a quantitative sensor for CO.

14.31 Zeolite A is used as a water softener in detergents and washing powders. Industrial production combines hydrated Al_2O_3, aqueous NaOH and Na_4SiO_4. After crystallization, the product (zeolite A) is subject to a number of washing and drying processes before use. (a) Describe the general structural characteristics of a zeolite, and comment on how an aluminosilicate differs from a silicate. (b) By representing zeolite A as Na_xA, suggest how it functions as a water softener.

14.32 The glass industry manufactures millions of tonnes of glass per year. (a) Only certain element oxides form glasses. Explain why this is, giving examples of what are termed in the glass industry as 'network-forming oxides'. Which oxide is the most important starting material in commercial glasses? Explain how a glass differs from a crystalline oxide such as α-Al_2O_3 (corundum). (b) Glass can be 'modified' by adding oxides such as Na_2O or CaO. Suggest how an O^{2-} ion might interact with the original oxide network. What role will the Na^+ or Ca^{2+} ions play in the modified glass? (c) A modified glass is treated with Al_2O_3 and the reaction is represented by the following equation:

$$2[SiO_{3/2}O]^- + 2Na^+ \xrightarrow[-2SiO_2]{Al_2O_3} 2[AlO_{4/2}]^- + 2Na^+$$

Explain the meaning of the fractional notation used in the equation and show schematically how the structure of the glass is altered when Al_2O_3 is added.

14.33 Lead-acid batteries accounted for 88% of all lead consumed in the US in 2009. (a) Complete the cell reaction given below (not balanced on the left-hand side) and show that the oxidation state changes balance in the final equation:

$$PbO_2(s) + Pb(s) + H_2SO_4(aq) \longrightarrow$$

(b) What are the two half-cell reactions in a lead–acid battery? (c) The value of E°_{cell} is 2.05 V. However, a normal automobile battery operates at about 12 V. Explain why the values differ. (d) How is a lead–acid battery recharged?

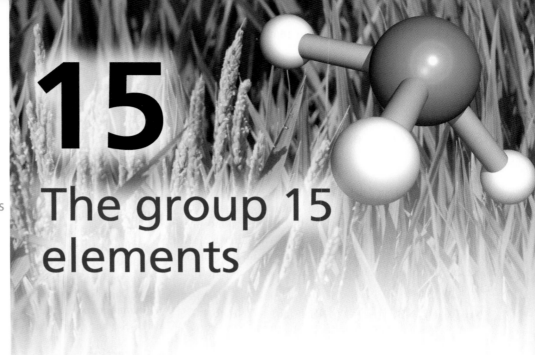

15

The group 15 elements

1	2			13	14	**15**	16	17	18
H									He
Li	Be			B	C	**N**	O	F	Ne
Na	Mg			Al	Si	**P**	S	Cl	Ar
K	Ca	*d*-block		Ga	Ge	**As**	Se	Br	Kr
Rb	Sr			In	Sn	**Sb**	Te	I	Xe
Cs	Ba			Tl	Pb	**Bi**	Po	At	Rn
Fr	Ra								

15.1 Introduction

> The group 15 elements – nitrogen, phosphorus, arsenic,
> antimony and bismuth – are called the ***pnictogens***.[†]

The rationalization of the properties of the group 15
elements (nitrogen, phosphorus, arsenic, antimony and
bismuth) and their compounds is difficult, despite there
being some general similarities in trends of the group 13,
14 and 15 elements, e.g. increase in metallic character and
stabilities of lower oxidation states on descending the
group. Although the 'diagonal' line (Fig. 7.8) can be

[†]See: G.S. Girolami (2009) *J. Chem. Educ.*, vol. 86, p. 1200 – 'Origin
of the Terms Pnictogen and Pnictide'.

drawn between As and Sb, formally separating non-metallic
and metallic elements, the distinction is not well defined and
should be treated with caution.

Very little of the chemistry of the group 15 elements is
that of simple ions. Although metal nitrides and phosphides
that react with water are usually considered to contain N^{3-}
and P^{3-} ions, electrostatic considerations make it doubtful
whether these ionic formulations are correct. The only
definite case of a simple cation in a *chemical* environment
is that of Bi^{3+}, and nearly all the chemistry of the group
15 elements involves covalently bonded compounds. The
thermochemical basis of the chemistry of such species is
much harder to establish than that of ionic compounds. In
addition, they are much more likely to be *kinetically* inert,
both to substitution reactions (e.g. NF_3 to hydrolysis,
$[H_2PO_2]^-$ to deuteration), and to oxidation or reduction
when these processes involve making or breaking covalent
bonds, as well as the transfer of electrons. Nitrogen, for
example, forms a range of oxoacids and oxoanions, and in
aqueous media can exist in all oxidation states from $+5$ to
-3, e.g. $[NO_3]^-$, N_2O_4, $[NO_2]^-$, NO, N_2O, N_2, NH_2OH,
N_2H_4, NH_3. Tables of standard reduction potentials
(usually calculated from thermodynamic data) or potential
diagrams (see Section 8.5) are of limited use in summarizing
the relationships between these species. Although they
provide information about the thermodynamics of possible reactions, they say nothing about the kinetics. Much
the same is true about the chemistry of phosphorus. The
chemistry of the first two members of group 15 is far
more extensive than that of As, Sb and Bi, and we can
mention only a small fraction of the known inorganic
compounds of N and P. In our discussions, we shall need
to emphasize *kinetic* factors more than in earlier chapters.

Arsenic is extremely toxic (Box 15.1). Like lead(II) and mercury(II), arsenic(III) is a soft metal centre and interacts with sulfur-containing residues in proteins.

15.2 Occurrence, extraction and uses

Occurrence

Figure 15.1a illustrates the relative abundances of the group 15 elements in the Earth's crust. Naturally occurring N_2 makes up 78% (by volume) of the Earth's atmosphere (Fig. 15.1b) and contains $\approx 0.36\%$ ^{15}N. The latter is useful for isotopic labelling and can be obtained in concentrated form by chemical exchange processes such as equilibria 15.1 and 15.2.

$$^{15}NH_3(g) + [^{14}NH_4]^+(aq) \rightleftharpoons [^{15}NH_4]^+(aq) + {}^{14}NH_3(g)$$
$$(15.1)$$

$$^{15}NO(g) + [^{14}NO_3]^-(aq) \rightleftharpoons [^{15}NO_3]^-(aq) + {}^{14}NO(g)$$
$$(15.2)$$

Because of the availability of N_2 in the atmosphere and its requirement by living organisms (in which N is present as proteins), the *fixing of nitrogen* in forms in which it may be assimilated by plants is of great importance. Attempts to devise synthetic nitrogen-fixation processes (see Section 29.4) that mimic the action of bacteria living in root nodules of leguminous plants have not yet been successful. However, N_2 can be fixed by other processes, e.g. its industrial conversion to NH_3 (see Section 15.5) or the conversion of metal-coordinated N_2 to NH_3 (see Section 25.4). The only natural source of nitrogen suitably 'fixed' for uptake by plants is crude $NaNO_3$ (*Chile saltpetre* or *sodanitre*) which occurs in the deserts of South America.

Phosphorus is an essential constituent of plant and animal tissue. Calcium phosphate occurs in bones and teeth, and phosphate esters of nucleotides (e.g. DNA, Fig. 10.13) are of immense biological significance (see Box 15.11). Phosphorus occurs naturally in the form of *apatites*, $Ca_5X(PO_4)_3$, the important minerals being *fluorapatite* (X = F), *chlorapatite* (X = Cl) and *hydroxyapatite* (X = OH). Major deposits of the apatite-containing ore *phosphate rock* occur in North Africa, North America, Asia and the Middle East. Although arsenic occurs in the elemental form, commercial sources of the element are *mispickel (arsenopyrite*, FeAsS), *realgar* (As_4S_4) and *orpiment* (As_2S_3). Native antimony is rare and the only commercial ore is *stibnite* (Sb_2S_3). Bismuth occurs as the element, and as the ores *bismuthinite* (Bi_2S_3) and *bismite* (Bi_2O_3).

Extraction

The industrial separation of N_2 is discussed in Section 15.4. Mining of phosphate rock takes place on a vast scale (in 2008, 161 Mt was mined worldwide), with the majority destined for the production of fertilizers (see Box 15.10) and animal feed supplements. Elemental phosphorus is extracted from phosphate rock (which approximates in composition to $Ca_3(PO_4)_2$) by heating with sand and coke in an electric furnace (eq. 15.3); phosphorus vapour distils out and is condensed under water to yield white phosphorus.

$$2Ca_3(PO_4)_2 + 6SiO_2 + 10C \xrightarrow{\approx 1700\,K} P_4 + 6CaSiO_3 + 10CO$$
$$(15.3)$$

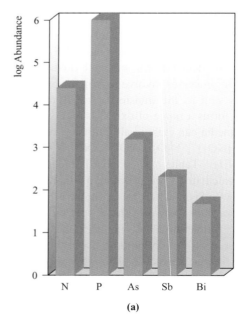

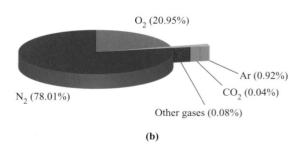

Fig. 15.1 (a) Relative abundances of the group 15 elements in the Earth's crust. Data are plotted on a logarithmic scale; units of abundance are parts per billion (1 billion $= 10^9$). (b) The main components (by percentage volume) of the Earth's atmosphere.

ENVIRONMENT

Box 15.1 Toxicity of arsenic

The toxicity of arsenic is well known, and the element features regularly in crime novels as a poison. A lethal dose is of the order of ≈ 130 mg. Arsenic occurs naturally in a number of minerals. Leaching of the element gradually occurs giving rise to arsenite, $[AsO_3]^{3-}$, and (predominately) arsenate, $[AsO_4]^{3-}$, salts in natural waters. In West Bengal (India) and Bangladesh, clean drinking water has been obtained from deep-bore wells since the 1970s. While this initially reduced the number of deaths from cholera and other diseases arising from bacteria, the large number of cases of chronic arsenic poisoning had become a major issue by 2000. In test bore wells, concentrations of dissolved arsenic were found to be at a maximum at a depth of about 30–40 m, and hence the depth of the wells is now known to be a key issue.

Despite the known toxicity of arsenic, it was used in agricultural pesticides until replaced by effective organic compounds in the second half of the 20th century. While this use of arsenic declined, its application in the form of chromated copper arsenate (CCA) in wood preservatives showed a general increase between 1980 and 2000 (see the graph below). Wood for a wide range of construction purposes has been treated under high pressure with CCA, resulting in a product with a higher resistance to decay caused by insect and larvae infestation. Typically, 1 m³ of pressure-treated wood contains approximately 0.8 kg of arsenic, and therefore the total quantities used in the construction and garden landscape businesses pose a major environmental risk. Once

pressure-treated wood is destroyed by burning, the residual ash contains high concentrations of arsenic. Wood left to rot releases arsenic into the ground. Added to this, the chromium waste from the wood preservative is also toxic.

The 2002 US Presidential Green Chemistry Challenge Awards (see Box 9.2) recognized the development of a copper-based 'environmentally advanced wood preservative' as a replacement for chromated copper arsenate. The new preservative contains a copper(II) complex and a quaternary ammonium salt. Its introduction into the market coincides with a change of policy within the wood-preserving industry: in 2003, US manufacturers initiated a change from arsenic-based products to alternative wood preservatives. This can be seen in the dramatic fall in CCA-containing wood preservatives in 2004 in the graph below. A number of other countries including Australia, Germany, Japan, Switzerland and Sweden have also banned the use of CCA.

In contrast to the toxic effects of arsenic, some prokaryotic (anaerobic) bacteria depend upon arsenic. Mono Lake, California, is a closed, saline basin (i.e. no water outlet) that is fed by freshwater streams and underwater springs including volcanic sources. In 2008, researchers discovered that cyanobacteria and photosynthetic, prokaryotic bacteria in Mono Lake use arsenic(III) compounds as their only photosynthetic electron donor. The process converts As(III) to As(V). The discovery may be relevant to an understanding of the arsenic cycle on ancient Earth in which oxygen played no role.

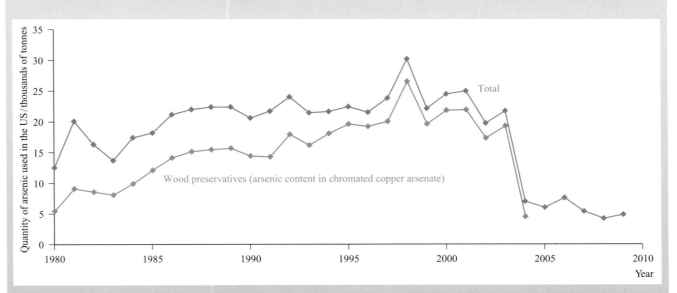

[Data: US Geological Survey; 'total' arsenic includes uses in wood preservatives, agricultural chemicals, glass, non-ferrous alloys and electronics.]

Further reading

D. Bleiwas (2000) US Geological Survey, http://minerals.usgs.gov/minerals/mflow/d00-0195/ – 'Arsenic and old waste'.

C. Cox (1991) *J. Pesticide Reform*, vol. 11, p. 2 – 'Chromated copper arsenate'.

J.S. Wang and C.M. Wai (2005) *J. Chem. Educ.*, vol. 81, p. 207 – 'Arsenic in drinking water – a global environmental problem'.

The principal source of As is FeAsS, and the element is extracted by heating (eq. 15.4) and condensing the As sublimate. An additional method is air-oxidation of arsenic sulfide ores to give As_2O_3 which is then reduced by C. As_2O_3 is also recovered on a large scale from flue dusts in Cu and Pb smelters.

$$FeAsS \xrightarrow{\Delta(\text{in absence of air})} FeS + As \qquad (15.4)$$

Antimony is obtained from stibnite by reduction using scrap iron (eq. 15.5) or by conversion to Sb_2O_3 followed by reduction with C.

$$Sb_2S_3 + 3Fe \longrightarrow 2Sb + 3FeS \qquad (15.5)$$

The extraction of Bi from its sulfide or oxide ores involves reduction with carbon (via the oxide when the ore is Bi_2S_3), but the metal is also obtained as a by-product of Pb, Cu, Sn, Ag and Au refining processes.

Self-study exercises

1. Calculate $\Delta_rG^\circ(298\text{ K})$ for the following reaction, given that values of $\Delta_fG^\circ(298\text{ K})$ for $Bi_2O_3(s)$ and $CO(g)$ are -493.7 and $-137.2\text{ kJ mol}^{-1}$, respectively:

   $$Bi_2O_3(s) + 3C(gr) \longrightarrow 2Bi(s) + 3CO(g)$$

   Comment on the answer in the light of the fact that carbon is used industrially to extract Bi from Bi_2O_3.
   [*Ans.* $+82.1\text{ kJ mol}^{-1}$]

2. Bismuth melts at 544 K, and values of $\Delta_fG^\circ(Bi_2O_3)$ can be estimated (in kJ mol^{-1}) by using the following equations:

   $T = 300{-}525\text{ K}: \quad \Delta_fG^\circ(Bi_2O_3) = -580.2 + 0.410T$
   $\qquad\qquad\qquad\qquad\qquad\quad - 0.0209T\log T$

   $T = 600{-}1125\text{ K}: \quad \Delta_fG^\circ(Bi_2O_3) = -605.5 + 0.478T$
   $\qquad\qquad\qquad\qquad\qquad\quad - 0.0244T\log T$

   For $CO(g)$, $\Delta_fG^\circ(300\text{ K}) = -137.3\text{ kJ mol}^{-1}$, and $\Delta_fG^\circ(1100\text{ K}) = -209.1\text{ kJ mol}^{-1}$. Use these data to construct a graph that shows the variation in Δ_fG° of CO and Bi_2O_3 (*in kJ per half-mole of O_2*) with temperature over the range 300–1100 K. What is the significance of the graph in terms of the extraction of Bi from Bi_2O_3 using carbon as the reducing agent?
   [*Ans.* Refer to Fig. 8.7 and related discussion]

Uses

Dinitrogen ranks as one of the most important industrial chemicals, and a large proportion of N_2 is converted to NH_3 (see Box 15.3). Gaseous N_2 is widely used to provide inert atmospheres, both industrially (e.g. in the electronics industry during the production of transistors, etc.)

Table 15.1 Selected low-temperature baths involving liquid N_2.[†]

Bath contents	Temperature / K
Liquid N_2 + cyclohexane	279
Liquid N_2 + acetonitrile	232
Liquid N_2 + octane	217
Liquid N_2 + heptane	182
Liquid N_2 + hexa-1,5-diene	132

[†] To prepare a liquid N_2 *slush bath*, liquid N_2 is poured into an appropriate solvent which is constantly stirred. See also Table 14.5.

and in laboratories. Liquid N_2 (bp 77 K) is an important coolant (Table 15.1) with applications in some freezing processes. Nitrogen-based chemicals are extremely important, and include nitrogenous fertilizers (see Box 15.3), nitric acid (see Box 15.8) and nitrate salts, explosives such as nitroglycerine (**15.1**) and trinitrotoluene (TNT, **15.2**), nitrite salts (e.g. in the curing of meat where they prevent discoloration by inhibiting oxidation of blood), cyanides and azides (e.g. in motor vehicle airbags where decomposition produces N_2 to inflate the airbag, see eq. 15.6).

(15.1)   **(15.2)**

By far the most important application of phosphorus is in phosphate fertilizers, and in Box 15.10 we highlight this use and possible associated environmental problems. Bone ash (calcium phosphate) is used in the manufacture of bone china. Most white phosphorus is converted to H_3PO_4, or to compounds such as P_4O_{10}, P_4S_{10}, PCl_3 and $POCl_3$. Phosphoric acid is industrially very important and is used on a large scale in the production of fertilizers, detergents and food additives. It is responsible for the sharp taste of many soft drinks, and is used to remove oxide and scale from the surfaces of iron and steel. Phosphorus trichloride is also manufactured on a large scale. It is a precursor to many organophosphorus compounds, including nerve agents (see Box 15.2), flame retardants (see Box 17.1) and insecticides. Phosphorus is important in steel manufacture and phosphor bronzes. Red phosphorus (see Section 15.4) is used in safety matches and in the generation of smoke (e.g. fireworks, smoke bombs).

APPLICATIONS

Box 15.2 Phosphorus-containing nerve agents

During the second half of the 20th century, the development of organophosphorus nerve agents became coupled not just with their actual use, but with the threat of potential use in acts of terrorism and during war. Nerve agents such as Sarin, Soman and VX tend to be called 'nerve gases' even though they are liquids at room temperature.

Sarin   R = Me
Soman   R = ^{t}Bu

VX

Each country signing the 1997 Chemical Weapons Convention agreed to ban the development, production, stockpiling and use of chemical weapons, and to destroy chemical weapons and associated production facilities by 2012. A problem for those involved in developing processes of destroying nerve agents is to ensure that the end-products are harmless. Sarin, for example, can be destroyed by room temperature hydrolysis using aqueous Na_2CO_3 to give NaF and the sodium salt of an organophosphate:

aqueous Na_2CO_3

$- NaF, -CO_2$

Nerve agent VX is more difficult to hydrolyse. It reacts slowly with aqueous NaOH at room temperature, so the reaction is carried out at 360 K for several hours. Hydrolysis follows the two routes shown in the scheme at the bottom of the page. The product in which the P–S bond remains intact is highly toxic. After the hydrolysis stage, the aqueous waste must be processed to render it safe.

Rapid detection of chemical warfare agents in the field is essential. However, analytical techniques such as gas or liquid chromatography are not suitable for routine use out of the laboratory. One method that has been investigated makes use of the release of HF from the hydrolysis of a fluorophosphonate compound (e.g. Sarin). The reaction is catalysed by a copper(II) complex:

$+ H_2O$ Cu(II) catalyst $+ HF$

The reaction is carried out over a thin film of porous silicon (which contains the copper(II) catalyst), the surface of which has been oxidized. As HF is produced, it reacts with the surface SiO_2 to give gaseous SiF_4:

$$SiO_2 + 4HF \longrightarrow SiF_4 + 2H_2O$$

Porous silicon is luminescent, and the above reaction results in changes in the emission spectrum of the porous silicon and provides a method of detecting the $R_2P(O)F$ agent. Other detection methods that have been investigated include application of carbon nanotubes as sensors for nerve agents.

Further reading

K.A. Joshi *et al.* (2006) *Anal. Chem.*, vol. 78, p. 331 – 'V-type nerve agent detection using a carbon nanotube-based amperometric enzyme electrode'.

H. Sohn *et al.* (2000) *J. Am. Chem. Soc.*, vol. 122, p. 5399 – 'Detection of fluorophosphonate chemical warfare agents by catalytic hydrolysis with a porous silicon interferometer'.

Y.-C. Yang, J.A. Baker and J.R. Ward (1992) *Chem. Rev.*, vol. 92, p. 1729 – 'Decontamination of chemical warfare agents'.

Y.-C. Yang (1999) *Acc. Chem. Res.*, vol. 32, p. 109 – 'Chemical detoxification of nerve agent VX'.

Major products

aqueous
NaOH

360 K

Minor products

Toxic

Table 15.2 Some physical properties of the group 15 elements and their ions.

Property	N	P	As	Sb	Bi
Atomic number, Z	7	15	33	51	83
Ground state electronic configuration	$[\text{He}]2s^2 2p^3$	$[\text{Ne}]3s^2 3p^3$	$[\text{Ar}]3d^{10}4s^2 4p^3$	$[\text{Kr}]4d^{10}5s^2 5p^3$	$[\text{Xe}]4f^{14}5d^{10}6s^2 6p^3$
Enthalpy of atomization, $\Delta_a H^o(298\,\text{K})/\text{kJ mol}^{-1}$	473‡	315	302	264	210
Melting point, mp / K	63	317	887 sublimes	904	544
Boiling point, bp / K	77	550	–	2023	1837
Standard enthalpy of fusion, $\Delta_{\text{fus}}H^o(\text{mp})/\text{kJ mol}^{-1}$	0.71	0.66	24.44	19.87	11.30
First ionization energy, $IE_1/\text{kJ mol}^{-1}$	1402	1012	947.0	830.6	703.3
Second ionization energy, $IE_2/\text{kJ mol}^{-1}$	2856	1907	1798	1595	1610
Third ionization energy, $IE_3/\text{kJ mol}^{-1}$	4578	2914	2735	2440	2466
Fourth ionization energy, $IE_4/\text{kJ mol}^{-1}$	7475	4964	4837	4260	4370
Fifth ionization energy, $IE_5/\text{kJ mol}^{-1}$	9445	6274	6043	5400	5400
Metallic radius, r_{metal} / pm	–	–	–	–	182
Covalent radius, r_{cov} / pm*	75	110	122	143	152
Ionic radius, r_{ion} / pm**	171 (N^{3-})	–	–	–	103 (Bi^{3+})
NMR active nuclei (% abundance, nuclear spin)	^{14}N (99.6, $I=1$) ^{15}N (0.4, $I=\frac{1}{2}$)	^{31}P (100, $I=\frac{1}{2}$)	^{75}As (100, $I=\frac{3}{2}$)	^{121}Sb (57.3, $I=\frac{5}{2}$) ^{123}Sb (42.7, $I=\frac{7}{2}$)	^{209}Bi (100, $I=\frac{9}{2}$)

‡ For nitrogen, $\Delta_a H^o = \frac{1}{2} \times$ dissociation energy of N_2.
* For 3-coordination.
** For 6-coordination.

Arsenic salts and arsines are extremely toxic, and uses of arsenic compounds in weedkillers, sheep- and cattle-dips, and poisons against vermin are less widespread than was once the case (see Box 15.1). Antimony compounds are less toxic, but large doses result in liver damage. Potassium antimony tartrate (*tartar emetic*) was used medicinally as an emetic and expectorant but has now been replaced by less toxic reagents. Bismuth is one of the less toxic heavy metals and compounds, such as the subcarbonate $(\text{BiO})_2\text{CO}_3$, find use in stomach remedies including treatments for peptic ulcers.

Arsenic is a doping agent in semiconductors (see Section 6.9) and GaAs has widespread uses in solid state devices and semiconductors. Other uses of As include those in alloys (e.g. it increases the strength of Pb) and in batteries. Sb_2O_3 is used in paints, adhesives and plastics, and as a flame retardant (see Box 17.1). Uses of Sb_2S_3 include those in photoelectric devices and electrophotographic recording materials, and as a flame retardant. Major uses of bismuth are in alloys (e.g. with Sn) and as Bi-containing compounds such as BiOCl in cosmetic products (e.g. creams, hair dyes and tints). Other uses are as oxidation catalysts and in high-temperature superconductors. Bi_2O_3 has many uses in the glass and ceramics industry, and for catalysts and magnets. The move towards lead-free solders (see Box 14.3) has resulted in increased use of Bi-containing solders, e.g. Sn/Bi/Ag alloys. A number of other applications are emerging in which Bi substitutes for Pb, for example in bismuth shot for game-hunting.†

15.3 Physical properties

Table 15.2 lists selected physical properties of the group 15 elements. Some observations regarding ionization energies are that:

- they increase rather sharply after removal of the *p* electrons;
- they decrease only slightly between P and As (similar behaviour to that between Al and Ga, and between Si and Ge);
- for removal of the *s* electrons, there is an increase between Sb and Bi, just as between In and Tl, and between Sn and Pb (see Box 13.4).

---

† Studies have indicated that bismuth may be not without toxic side effects: R. Pamphlett, G. Danscher, J. Rungby and M. Stoltenberg (2000) *Environ. Res. Sect. A*, vol. 82, p. 258 – 'Tissue uptake of bismuth from shotgun pellets'.

Values of $\Delta_a H^{\circ}$ decrease steadily from N to Bi, paralleling similar trends in groups 13 and 14.

Worked example 15.1 Thermochemical data for the group 15 elements

At 298 K, the values of the enthalpy changes for the processes:

$$N(g) + e^- \longrightarrow N^-(g)$$

and

$$N(g) + 3e^- \longrightarrow N^{3-}(g)$$

are ≈ 0 and $2120\,kJ\,mol^{-1}$. Comment on these data.

The ground state electronic configuration of N is $1s^2 2s^2 2p^3$ and the process:

$$N(g) + e^- \longrightarrow N^-(g)$$

involves the addition of an electron into a $2p$ atomic orbital to create a spin-paired pair of electrons. Repulsive interactions between the valence electrons of the N atom and the incoming electron would give rise to a positive enthalpy term. This is offset by a negative enthalpy term associated with the attraction between the nucleus and the incoming electron. In the case of nitrogen, these two terms essentially compensate for one another.

The process:

$$N(g) + 3e^- \longrightarrow N^{3-}(g)$$

is highly endothermic. After the addition of the first electron, electron repulsion between the N^- ion and the incoming electron is the dominant term, making the process:

$$N^-(g) + e^- \longrightarrow N^{2-}(g)$$

endothermic. Similarly, the process:

$$N^{2-}(g) + e^- \longrightarrow N^{3-}(g)$$

is highly endothermic.

Self-study exercises

1. Comment on reasons for the trend in the first five ionization energies for bismuth (703, 1610, 2466, 4370 and 5400 kJ mol^{-1}). [*Ans.* Refer to Section 1.10 and Box 13.4]

2. Give an explanation for the trend in values of IE_1 down group 15 (N, 1402; P, 1012; As, 947; Sb, 831; Bi, 703 kJ mol^{-1}). [*Ans.* Refer to Section 1.10]

3. Why is there a decrease in the values of IE_1 on going from N to O, and from P to S? [*Ans.* Refer to Section 1.10 and Box 1.6]

Table 15.3 Some covalent bond enthalpy terms (kJ mol^{-1}); the values for single bonds refer to the group 15 elements in 3-coordinate environments, and values for triple bonds are for dissociation of the appropriate diatomic molecule.

N–N 160	N=N ≈400[‡]	N≡N 946	N–H 391	N–F 272	N–Cl 193	N–O 201
P–P 209		P≡P 490	P–H 322	P–F 490	P–Cl 319	P–O 340
As–As 180			As–H 296	As–F 464	As–Cl 317	As–O 330
					Sb–Cl 312	
					Bi–Cl 280	

[‡] See text.

Bonding considerations

Analogies between groups 14 and 15 are seen if we consider certain bonding aspects. Table 15.3 lists some covalent bond enthalpy terms for group 15 elements. Data for most single bonds follow trends reminiscent of those in group 14 (Table 14.2); e.g. N forms stronger bonds with H than does P, but weaker bonds with F, Cl or O. These observations, together with the absence of stable P-containing analogues of N_2, NO, HCN, $[N_3]^-$ and $[NO_2]^+$ (**15.3–15.7**), indicate that strong $(p–p)\pi$-bonding is important only for the first member of group 15.[†]

N≡N **(15.3)**   N=O **(15.4)**   H—C≡N **(15.5)**

$^-$N=N$^+$=N$^-$ **(15.6)**   O=N$^+$=O **(15.7)**

It can be argued that differences between the chemistries of nitrogen and the heavier group 15 elements (e.g. existence of PF$_5$, AsF$_5$, SbF$_5$ and BiF$_5$, but not NF$_5$) arise from the fact that an N atom is simply too small to accommodate five atoms around it. Historically, the differences have been attributed to the availability of d-orbitals on P, As, Sb and Bi, but not on N. However, even in the presence of electronegative atoms which would lower the energy of the d-orbitals, it is now considered that these orbitals play

[†] For an account of attempts to prepare $[PO_2]^+$ by F$^-$ abstraction from $[PO_2F_2]^-$, see: S. Schneider, A. Vij, J.A. Sheehy, F.S. Tham, T. Schroer and K.O. Christe (1999) *Z. Anorg. Allg. Chem.*, vol. 627, p. 631.

no significant role in hypervalent compounds of the group 15 (and later) elements. As we saw in Chapter 5, it is possible to account for the bonding in hypervalent molecules of the p-block elements in terms of a valence set of ns and np orbitals, and you should be cautious about using sp^3d and sp^3d^2 hybridization schemes to describe trigonal bipyramidal and octahedral species of p-block elements. Although we shall show molecular structures of compounds in which P, As, Sb and Bi are in oxidation states of $+5$ (e.g. PCl_5, $[PO_4]^{3-}$, $[SbF_6]^-$), the representation of a line between two atoms does not necessarily mean the presence of a localized 2-centre 2-electron bond. Similarly, the representation of a double line between two atoms does not necessarily imply that the interaction comprises covalent σ- and π-contributions. For example, while it is often convenient to draw structures for Me_3PO and PF_5 as:

it is more realistic to show the role that charge-separated species play when one is discussing the electronic distribution in ions or molecules, i.e.

Furthermore, PF_5 should really be represented by a series of resonance structures to provide a description that accounts for the equivalence of the two axial P–F bonds and the equivalence of the three equatorial P–F bonds. When we wish to focus on the *structure* of a molecule rather than on its bonding, charge-separated representations are not always the best option because they often obscure the observed geometry. This problem is readily seen by looking at the charge-separated representation of PF_5, in which the trigonal bipyramidal structure of PF_5 is not immediately apparent.

The largest *difference* between groups 14 and 15 lies in the relative strengths of the $N\equiv N$ (in N_2) and N–N (in N_2H_4) bonds compared with those of $C\equiv C$ and C–C bonds (Tables 15.3 and 14.2). There is some uncertainty about a value for the $N=N$ bond enthalpy term because of difficulty in choosing a reference compound, but the approximate value given in Table 15.3 is seen to be *more* than twice that of the N–N bond, whereas the $C=C$ bond is significantly *less* than twice as strong as the C–C bond (Table 14.2). While N_2 is thermodynamically stable with respect to oligomerization to species containing N–N

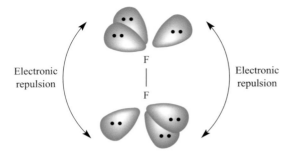

Fig. 15.2 Schematic representation of the electronic repulsion, believed to weaken the F–F bond in F_2. This represents the simplest example of a phenomenon that also occurs in N–N and O–O single bonds.

bonds, $HC\equiv CH$ is thermodynamically unstable with respect to species with C–C bonds, (see end-of-chapter problem 15.2). Similarly, the dimerization of P_2 to tetrahedral P_4 is thermodynamically favourable. The σ- and π-contributions that contribute to the very high strength of the $N\equiv N$ bond (which makes many nitrogen compounds endothermic and most of the others only *slightly* exothermic) were discussed in Section 2.3. However, the particular weakness of the N–N single bond calls for comment. The O–O ($146\,kJ\,mol^{-1}$ in H_2O_2) and F–F ($159\,kJ\,mol^{-1}$ in F_2) bonds are also very weak, much weaker than S–S or Cl–Cl bonds. In N_2H_4, H_2O_2 and F_2, the N, O or F atoms carry lone pairs, and it is believed that the N–N, O–O and F–F bonds are weakened by repulsion between lone pairs on adjacent atoms (Fig. 15.2). Lone pairs on larger atoms (e.g. in Cl_2) are further apart and experience less mutual repulsion. Each N atom in N_2 also has a non-bonding lone pair, but they are directed away from each other. Table 15.3 illustrates that N–O, N–F and N–Cl are also rather weak and, again, interactions between lone pairs of electrons can be used to rationalize these data. However, when N is singly bonded to an atom with no lone pairs (e.g. H), the bond is strong. In pursuing such arguments, we must remember that in a heteronuclear bond, extra energy contributions may be attributed to partial ionic character (see Section 2.5).

Another important difference between N and the later group 15 elements is the ability of N to take part in strong hydrogen bonding (see Sections 10.6 and 15.5). This arises from the much higher electronegativity of N ($\chi^P = 3.0$) compared with values for the later elements (χ^P values: P, 2.2; As, 2.2; Sb, 2.1; Bi, 2.0). The ability of the first row element to participate in hydrogen bonding is also seen in group 16 (e.g. O–H····O and N–H····O interactions) and group 17 (e.g. O–H····F, N–H····F interactions). For carbon, the first member of group 14, weak hydrogen bonds (e.g. C–H····O interactions) are important in the solid state structures of molecular and biological systems.

NMR active nuclei

Nuclei that are NMR active are listed in Table 15.2. Routinely, ^{31}P NMR spectroscopy is used in characterizing P-containing species; see for example Case studies 1, 2 and 4 and end-of-chapter problem 4.40 in Chapter 4. Chemical shifts are usually reported with respect to $\delta = 0$ for 85% aqueous H_3PO_4, but other reference compounds are used, e.g. trimethylphosphite, $P(OMe)_3$. The chemical shift range for ^{31}P is large.

Self-study exercise

At 307 K, the ^{31}P NMR spectrum of a CD_2Cl_2 solution containing $[PF_5(CN)]^-$ consists of a sextet (δ –157.7 ppm, J_{PF} 744 Hz). At 178 K, the ^{19}F NMR spectrum of the same anion exhibits two signals (δ –47.6 ppm, doublet of doublets; δ –75.3 ppm, doublet of quintets) from which the following coupling constants can be measured: $J_{PF(axial)}$ 762 Hz, $J_{PF(eq)}$ 741 Hz, J_{FF} 58 Hz. (a) Rationalize these observations. (b) Draw a diagram to show the ^{19}F NMR spectrum and mark on it where the values of the coupling constants may be measured.

[*Ans.* 307 K, fluxional; 178 K, static; see Section 4.8]

Radioactive isotopes

Although the only naturally occurring isotope of phosphorus is ^{31}P, 16 man-made radioactive isotopes are known. Of these, ^{32}P is the most important with its half-life of 14.3 days making it suitable as a tracer.

15.4 The elements

Nitrogen

Dinitrogen is obtained industrially by fractional distillation of liquid air, and the product contains some Ar and traces of O_2. Dioxygen can be removed by addition of a small amount of H_2 and passage over a Pt catalyst. The separation of N_2 and O_2 using gas-permeable membranes is growing in importance and is a cheaper alternative to purifying N_2 by the fractional distillation of liquid air. Compared with the latter, N_2 produced by membrane separation is less pure (typically it contains 0.5–5% O_2), and the technology is suited to the production of lower volumes of gas. Nonetheless, the use of gas-permeable membranes is well suited for applications such as the production of inert atmospheres for the storage and transport of fruit and vegetables, or for generating small

volumes or low flow-rates of N_2 for laboratory applications. Membranes are made from polymeric materials, the gas permeability of which is selective. The factors which determine this are the solubility of a given gas in the membrane and its rate of diffusion across the membrane. When the N_2/O_2 mixture passes across the surface of the membrane, O_2 permeates through the membrane, leaving the initial stream of gas enriched in the less permeable gas (N_2). Small amounts of N_2 can be prepared by thermal decomposition of sodium azide (eq. 15.6) or by reaction 15.7 or 15.8. The latter should be carried out cautiously because of the risk of explosion. Ammonium nitrite (NH_4NO_2) is potentially explosive, as is ammonium nitrate which is a powerful oxidant and a component of dynamite. In car airbags, the decomposition of NaN_3 is initiated by an electrical impulse.[†]

$$2NaN_3(s) \xrightarrow{\Delta} 2Na + 3N_2 \qquad (15.6)$$

$$NH_4NO_2(aq) \xrightarrow{\Delta} N_2 + 2H_2O \qquad (15.7)$$

$$2NH_4NO_3(s) \xrightarrow{>570\,K} 2N_2 + O_2 + 4H_2O \qquad (15.8)$$

Dinitrogen is generally unreactive. It combines slowly with Li at ambient temperatures (eq. 11.6), and, when heated, with the group 2 metals, Al, Si, Ge (Section 14.5) and many *d*-block metals. The reaction between CaC_2 and N_2 is used industrially for manufacturing the nitrogenous fertilizer calcium cyanamide (eqs. 14.32 and 14.33). Many elements (e.g. Na, Hg, S) which are inert towards N_2 do react with atomic nitrogen, produced by passing N_2 through an electric discharge. At ambient temperatures, N_2 is reduced to hydrazine (N_2H_4) by vanadium(II) and magnesium hydroxides. The reaction of N_2 with H_2 is discussed later in the chapter.

A large number of *d*-block metal complexes containing coordinated N_2 are known (see Fig. 15.9 and eqs. 22.98 and 22.99 and discussion). N_2 is isoelectronic with CO and the bonding in complexes containing the N_2 ligand can be described in a similar manner to that in metal carbonyl complexes (see Chapter 24).

Phosphorus

Phosphorus exhibits complicated allotropy; 12 forms have been reported, and these include both crystalline and amorphous forms. Crystalline white phosphorus contains tetrahedral P_4 molecules (Fig. 15.3a) in which the $P-P$

[†] A. Madlung (1996) *J. Chem. Educ.*, vol. 73, p. 347 – 'The chemistry behind the air bag'.

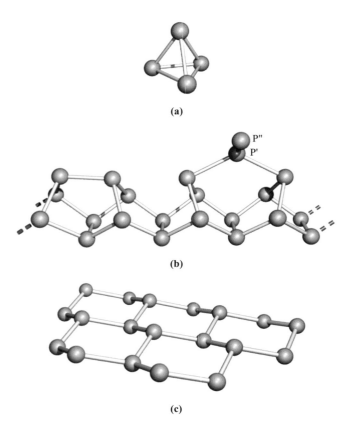

Hittorf's phosphorus (also called violet phosphorus) is a well-characterized form of the red allotrope and its complicated structure is best described in terms of interlocking chains (Fig. 15.3b). Non-bonded chains lie parallel to each other to give layers, and the chains in one layer lie at right-angles to the chains in the next layer, being connected by the P′−P″ bonds shown in Fig. 15.3b. All P−P bond distances are ≈222 pm, indicating covalent single bonds. One method of obtaining crystals of Hittorf's phosphorus is to sublime commercially available amorphous red phosphorus under vacuum in the presence of an I_2 catalyst. Under these conditions, another allotrope, fibrous red phosphorus, also crystallizes. Both Hittorf's and fibrous red phosphorus consist of the chains shown in Fig. 15.3b. Whereas in Hittorf's phosphorus pairs of these chains are linked in a mutually perpendicular orientation, in fibrous phosphorus they lie parallel to one another. Black phosphorus is the most stable allotrope and is obtained by heating white phosphorus under high pressure. Its appearance and electrical conductivity resemble those of graphite, and it possesses a double-layer lattice of puckered 6-membered rings (Fig. 15.3c); P−P distances within a layer are 220 pm and the shortest interlayer P−P distance is 390 pm. On melting, all allotropes give a liquid containing P_4 molecules, and these are also present in the vapour. Above 1070 K or at high pressures, P_4 is in equilibrium with P_2 (**15.8**).

187 pm

P ≡≡≡ P

(15.8)

Fig. 15.3 (a) The tetrahedral P_4 molecule found in white phosphorus. (b) Part of one of the chain-like arrays of atoms present in the infinite lattice of Hittorf's phosphorus; the repeat unit contains 21 atoms, and atoms P′ and P″ are equivalent atoms in adjacent chains, with chains connected through P′−P″ bonds. The same chains are also present in fibrous red phosphorus. (c) Part of one layer of puckered 6-membered rings present in black phosphorus and in the rhombohedral allotropes of arsenic, antimony and bismuth.

Most of the chemical differences between the allotropes of phosphorus are due to differences in activation energies for reactions. Black phosphorus is kinetically inert and does not ignite in air even at 670 K. Red phosphorus is intermediate in reactivity between the white and black allotropes. It is not poisonous, is insoluble in organic solvents, does not react with aqueous alkali, and ignites in air above 520 K. It reacts with halogens, sulfur and metals, but less vigorously than does white phosphorus. The latter is a soft, waxy solid which becomes yellow on exposure to light; it is very poisonous, being readily absorbed into the blood and liver. White phosphorus is soluble in benzene, PCl_3 and CS_2 but is virtually insoluble in water, and is stored under water to prevent oxidation. In moist air, it undergoes *chemiluminescent* oxidation, emitting a green glow and slowly forming P_4O_8 (see Section 15.10) and some O_3; the chain reaction involved is extremely complicated.

distances (221 pm) are used to define $r_{cov} = 110$ pm for a single bond. White phosphorus is *defined* as the standard state of the element, but is actually metastable (eq. 15.9). The lower stability of the white form probably originates in strain associated with the 60° bond angles.

$$\text{P} \xleftarrow{\;\Delta_f H° = -39.3\,\text{kJ mol}^{-1}\;} \tfrac{1}{4}\text{P}_4 \xrightarrow{\;\Delta_f H° = -17.6\,\text{kJ mol}^{-1}\;} \text{P}$$

Black   White   Red

(15.9)

White phosphorus is manufactured by reaction 15.3, and heating this allotrope in an inert atmosphere at ≈540 K produces red phosphorus. Several crystalline forms of red phosphorus exist, and all probably possess infinite lattices.[†]

[†] For details, see: H. Hartl (1995) *Angew. Chem. Int. Ed.*, vol. 34, p. 2637 – 'New evidence concerning the structure of amorphous red phosphorus'; M. Ruck *et al.* (2005) *Angew. Chem. Int. Ed.*, vol. 44, p. 7616 – 'Fibrous red phosphorus'.

A ***chemiluminescent*** reaction is one that is accompanied by the emission of light.

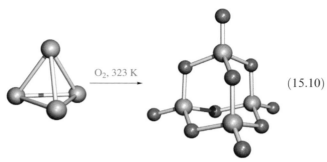

(15.10)

Above 323 K, white phosphorus inflames, yielding phosphorus(V) oxide (eq. 15.10); in a limited supply of air, P_4O_6 may form. White phosphorus combines violently with all of the halogens, giving PX_3 (X = F, Cl, Br, I) or PX_5 (X = F, Cl, Br) depending on the relative amounts of P_4 and X_2. Concentrated HNO_3 oxidizes P_4 to H_3PO_4, and with hot aqueous NaOH, reaction 15.11 occurs, some H_2 and P_2H_4 also being formed.

$$P_4 + 3NaOH + 3H_2O \longrightarrow 3NaH_2PO_2 + PH_3 \qquad (15.11)$$

$$23P_4 + 12LiPH_2 \longrightarrow 6Li_2P_{16} + 8PH_3 \qquad (15.12)$$

Reaction 15.12 yields Li_2P_{16}, while Li_3P_{21} and Li_4P_{26} can be obtained by altering the ratio of $P_4 : LiPH_2$. The structures of the phosphide ions $[P_{16}]^{2-}$, **15.9**, $[P_{21}]^{3-}$, **15.10**, and $[P_{26}]^{4-}$ are related to one chain in Hittorf's and fibrous red phosphorus (Fig. 15.3b).

(15.9)

(15.10)

Like N_2, P_4 can act as a ligand in *d*-block metal complexes. Examples of different coordination modes of P_4 are shown in structures **15.11–15.13**.

(15.11)   **(15.12)**

(15.13)

Arsenic, antimony and bismuth

Arsenic vapour contains As_4 molecules, and the unstable yellow form of solid As probably also contains these units. At relatively low temperatures, Sb vapour contains molecular Sb_4. At room temperature and pressure, As, Sb and Bi are grey solids with extended structures resembling that of black phosphorus (Fig. 15.3c). On descending the group, although intralayer bond distances increase as expected, similar increases in interlayer spacing do not occur, and the coordination number of each atom effectively changes from 3 (Fig. 15.3c) to 6 (three atoms within a layer and three in the next layer).

Arsenic, antimony and bismuth burn in air (eq. 15.13) and combine with halogens (see Section 15.7).

$$4M + 3O_2 \xrightarrow{\Delta} 2M_2O_3 \qquad M = As, Sb\ or\ Bi \qquad (15.13)$$

The elements are not attacked by non-oxidizing acids but react with concentrated HNO_3 to give H_3AsO_4 (hydrated As_2O_5), hydrated Sb_2O_5 and $Bi(NO_3)_3$ respectively, and with concentrated H_2SO_4 to produce As_4O_6, $Sb_2(SO_4)_3$ and $Bi_2(SO_4)_3$ respectively. None of the elements reacts with aqueous alkali, but As is attacked by fused NaOH (eq. 15.14).

$$2As + 6NaOH \longrightarrow 2Na_3AsO_3 + 3H_2 \qquad (15.14)$$
$$\text{sodium arsenite}$$

15.5 Hydrides

Trihydrides, EH_3 (E = N, P, As, Sb and Bi)

Each group 15 element forms a trihydride, selected properties of which are given in Table 15.4; the lack of data for BiH_3 stems from its instability. The variation in boiling points (Fig. 10.7b, Table 15.4) is one of the strongest pieces of evidence for hydrogen bond formation by nitrogen. Further evidence comes from the fact that NH_3 has a greater value of $\Delta_{vap}H^{\circ}$ and a higher surface tension than the later trihydrides. Thermal stabilities of these compounds decrease down the group (BiH_3 decomposes above 228 K), and this trend is reflected in the bond enthalpy terms (Table 15.3). Ammonia is the only trihydride to possess a negative value of $\Delta_f H^{\circ}$ (Table 15.4).

Table 15.4 Selected data for the group 15 trihydrides, EH_3.

	NH_3	PH_3	AsH_3	SbH_3	BiH_3
Name[†]	Ammonia (azane)	Phosphane (phosphine)	Arsane (arsine)	Stibane (stibine)	Bismuthane
Melting point / K	195.5	140	157	185	206
Boiling point / K	240	185.5	210.5	256	290[‡]
$\Delta_{vap}H^{\circ}(bp)/kJ\,mol^{-1}$	23.3	14.6	16.7	21.3	–
$\Delta_f H^{\circ}(298\,K)/kJ\,mol^{-1}$	−45.9	5.4	66.4	145.1	277[‡]
Dipole moment / D	1.47	0.57	0.20	0.12	–
E−H bond distance / pm	101.2	142.0	151.1	170.4	–
∠H−E−H / deg	106.7	93.3	92.1	91.6	–

[†] Ammonia is the accepted trivial name for NH_3. Azane is the IUPAC parent name for NH_3 and is used for naming derivatives. Phosphane, arsane and stibane are the IUPAC systematic names for PH_3, AsH_3 and SbH_3. Phosphine, arsine and stibine are still in use but are no longer accepted by the IUPAC.
[‡] Estimated value.

Worked example 15.2
Bond enthalpies in group 15 hydrides

Given that $\Delta_f H^{\circ}(298\,K)$ for $PH_3(g)$ is $+5.4\,kJ\,mol^{-1}$, calculate a value for the P−H bond enthalpy term in PH_3. [Other data: see Appendix 10.]

Construct an appropriate Hess cycle, bearing in mind that the P−H bond enthalpy term can be determined from the standard enthalpy of atomization of $PH_3(g)$.

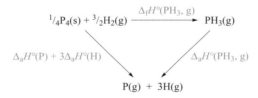

$$\Delta_f H^{\circ}(PH_3,\,g) + \Delta_a H^{\circ}(PH_3,\,g)$$
$$= \Delta_a H^{\circ}(P,\,g) + 3\Delta_a H^{\circ}(H,\,g)$$

Standard enthalpies of atomization of the elements are listed in Appendix 10.

$\Delta_a H^{\circ}(PH_3,\,g)$

$= \Delta_a H^{\circ}(P,\,g) + 3\Delta_a H^{\circ}(H,\,g) - \Delta_f H^{\circ}(PH_3,\,g)$

$= 315 + 3(218) - 5.4$

$= 963.6 = 964\,kJ\,mol^{-1}$ (to 3 sig. fig.)

P−H bond enthalpy term $= \dfrac{964}{3} = 321\,kJ\,mol^{-1}$

Self-study exercises

1. Using data from Table 15.3 and Appendix 10, calculate a value for $\Delta_f H^{\circ}(NH_3,\,g)$. [*Ans.* $-46\,kJ\,mol^{-1}$]

2. Calculate a value for the Bi−H bond enthalpy term in BiH_3 using data from Table 15.4 and Appendix 10. [*Ans.* $196\,kJ\,mol^{-1}$]

3. Use data in Table 15.4 and Appendix 10 to calculate the As−H bond enthalpy term in AsH_3. [*Ans.* $297\,kJ\,mol^{-1}$]

Ammonia is obtained by the action of H_2O on the nitrides of Li or Mg (eq. 15.15), by heating $[NH_4]^+$ salts with base (e.g. reaction 15.16), or by reducing a nitrate or nitrite in alkaline solution with Zn or Al (e.g. reaction 15.17).

$$Li_3N + 3H_2O \longrightarrow NH_3 + 3LiOH \quad (15.15)$$

$$2NH_4Cl + Ca(OH)_2 \longrightarrow 2NH_3 + CaCl_2 + 2H_2O \quad (15.16)$$

$$[NO_3]^- + 4Zn + 6H_2O + 7[OH]^- \longrightarrow NH_3 + 4[Zn(OH)_4]^{2-} \quad (15.17)$$

Trihydrides of the later elements are best made by method 15.18, or by acid hydrolysis of phosphides, arsenides, antimonides or bismuthides (e.g. reaction 15.19). Phosphane can also be made by reaction 15.20, $[PH_4]I$ being prepared from P_2I_4 (see Section 15.7).

$$ECl_3 \xrightarrow{\text{Li[AlH}_4\text{] in Et}_2\text{O}} EH_3 \qquad E = P,\ As,\ Sb,\ Bi \quad (15.18)$$

APPLICATIONS

Box 15.3 Ammonia: an industrial giant

Ammonia is manufactured on a huge scale, the major producers being China, the US, India and Russia. The graph below shows the trends for world and US production of NH_3 between 1990 and 2008.

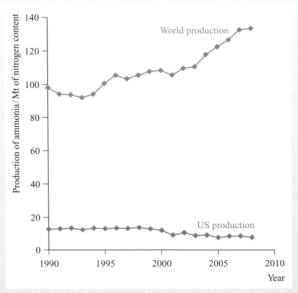

[Data: US Geological Survey]

Agriculture demands vast quantities of fertilizers to supplement soil nutrients; this is critical when the same land is used year after year for crop production. Essential nutrients are N, P, K (the three required in largest amounts), Ca, Mg and S plus

trace elements. In 2008, in the US, direct use and its conversion into other nitrogenous fertilizers accounted for ≈90% of all NH_3 produced. In addition to NH_3 itself, the nitrogen-rich compound $CO(NH_2)_2$ (urea) is of prime importance, along with $[NH_4][NO_3]$ and $[NH_4]_2[HPO_4]$ (which has the benefit of supplying both N and P nutrients); $[NH_4]_2[SO_4]$ accounts for a smaller portion of the market. The remaining 10% of NH_3 produced in the US was used in the synthetic fibre industry (e.g. nylon-6, nylon-66 and rayon), manufacture of explosives (see structures **15.1** and **15.2**), resins and miscellaneous chemicals.

Phosphorus-containing fertilizers are highlighted in Box 15.10.

Application of anhydrous NH_3 as a fertilizer. Gaseous NH_3 is released from a tank via a nozzle alongside a blade that cuts through the soil.

$$Ca_3P_2 + 6H_2O \longrightarrow 2PH_3 + 3Ca(OH)_2 \qquad (15.19)$$

$$[PH_4]I + KOH \longrightarrow PH_3 + KI + H_2O \qquad (15.20)$$

The industrial manufacture of NH_3 (Box 15.3) involves the Haber process (reaction 15.21), and the manufacture of the H_2 (see Section 10.4) required contributes significantly to the overall cost of the process.

$$N_2 + 3H_2 \rightleftharpoons 2NH_3 \quad \begin{cases} \Delta_r H^\circ(298\,K) = -92\,kJ\,mol^{-1} \\ \Delta_r G^\circ(298\,K) = -33\,kJ\,mol^{-1} \end{cases}$$

$$(15.21)$$

The Haber process is a classic application of physicochemical principles to a system in equilibrium. The decrease in

number of moles of gas means that $\Delta_r S^\circ(298\,K)$ is negative. For industrial viability, NH_3 must be formed in optimum yield *and* at a reasonable rate. Increasing the temperature increases the rate of reaction, but decreases the yield since the forward reaction is exothermic. At a given temperature, both the equilibrium yield and the reaction rate are increased by working at high pressures. The presence of a suitable catalyst (see Section 25.8) also increases the rate. The rate-determining step is the dissociation of N_2 into N atoms chemisorbed onto the catalyst. The optimum reaction conditions are $T = 723\,K$, $P = 20\,260\,kPa$, and Fe_3O_4 mixed with K_2O, SiO_2 and Al_2O_3 as the heterogeneous catalyst. The Fe_3O_4 is reduced to give the catalytically active α-Fe. The NH_3 formed is either liquefied or dissolved in H_2O to form a saturated solution of specific gravity 0.880.

Worked example 15.3   Thermodynamics of NH₃ formation

For the equilibrium:

$$\tfrac{1}{2}N_2(g) + \tfrac{3}{2}H_2(g) \rightleftharpoons NH_3(g)$$

values of $\Delta_r H^{\circ}(298\,K)$ and $\Delta_r G^{\circ}(298\,K)$ are -45.9 and $-16.4\,kJ\,mol^{-1}$, respectively. Calculate $\Delta_r S^{\circ}(298\,K)$ and comment on the value.

$$\Delta_r G^{\circ} = \Delta_r H^{\circ} - T\Delta_r S^{\circ}$$

$$\Delta_r S^{\circ} = \frac{\Delta_r H^{\circ} - \Delta_r G^{\circ}}{T}$$

$$= \frac{-45.9 - (-16.4)}{298}$$

$$= -0.0990\,kJ\,K^{-1}\,mol^{-1}$$

$$= -99.0\,J\,K^{-1}\,mol^{-1}$$

The negative value is consistent with a decrease in the number of moles of gas in going from the left- to right-hand side of the equilibrium.

Self-study exercises

These exercises all refer to the equilibrium given in the worked example.

1. Determine ln K at 298 K. [*Ans.* 6.62]

2. At 700 K, $\Delta_r H^{\circ}$ and $\Delta_r G^{\circ}$ are -52.7 and $+27.2\,kJ\,mol^{-1}$, respectively. Determine a value for $\Delta_r S^{\circ}$ under these conditions. [*Ans.* $-114\,J\,K^{-1}\,mol^{-1}$]

3. Determine ln K at 700 K. [*Ans.* -4.67]

4. Comment on your answer to question 3, given that the optimum temperature for the industrial synthesis of NH₃ is 723 K.

Ammonia is a colourless gas with a pungent odour. Table 15.4 lists selected properties and structural data for the trigonal pyramidal molecule **15.14**, the barrier to inversion for which is very low ($24\,kJ\,mol^{-1}$). Oxidation products of NH₃ depend on conditions. Reaction 15.22 occurs on combustion in O₂, but at $\approx1200\,K$ in the presence of a Pt/Rh catalyst and a contact time of $\approx1\,ms$, the less exothermic reaction 15.23 takes place. This reaction forms part of the manufacturing process for HNO₃ (see Section 15.9).

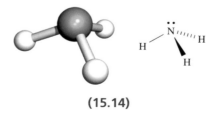

(15.14)

$$4NH_3 + 3O_2 \longrightarrow 2N_2 + 6H_2O \qquad (15.22)$$

$$4NH_3 + 5O_2 \xrightarrow{Pt/Rh} 4NO + 6H_2O \qquad (15.23)$$

The solubility of NH₃ in water is greater than that of any other gas, doubtless because of hydrogen bond formation between NH₃ and H₂O. The equilibrium constant (at 298 K) for reaction 15.24 shows that nearly all the dissolved NH₃ is *non-ionized*, consistent with the fact that even dilute solutions retain the characteristic smell of NH₃. Since $K_w = 10^{-14}$, it follows that the aqueous solutions of $[NH_4]^+$ salts of strong acids (e.g. NH₄Cl) are slightly acidic (eq. 15.25). (See worked example 7.2 for calculations relating to equilibria 15.24 and 15.25, and worked example 7.3 for the relationship between pK_a and pK_b.)

$$NH_3(aq) + H_2O(l) \rightleftharpoons [NH_4]^+(aq) + [OH]^-(aq)$$
$$K_b = 1.8 \times 10^{-5} \qquad (15.24)$$

$$[NH_4]^+(aq) + H_2O(l) \rightleftharpoons [H_3O]^+(aq) + NH_3(aq)$$
$$K_a = 5.6 \times 10^{-10} \qquad (15.25)$$

Ammonium salts are easily prepared by neutralization reactions, e.g. eq. 15.26. Industrial syntheses are carried out using the Solvay process (Fig. 11.6), or reactions 15.27 and 15.28. Both ammonium sulfate and ammonium nitrate are important fertilizers, and NH₄NO₃ is a component of some explosives (see eq. 15.8).

$$NH_3 + HBr \longrightarrow NH_4Br \qquad (15.26)$$

$$CaSO_4 + 2NH_3 + CO_2 + H_2O$$
$$\longrightarrow CaCO_3 + [NH_4]_2[SO_4] \qquad (15.27)$$

$$NH_3 + HNO_3 \longrightarrow NH_4NO_3 \qquad (15.28)$$

Detonation of NH₄NO₃ may be initiated by another explosion, and ammonium perchlorate is similarly metastable with respect to oxidation of the $[NH_4]^+$ cation by the anion; NH₄ClO₄ is used in solid rocket propellants, e.g. in the booster rockets of the space shuttle.

All perchlorate salts are potentially explosive and must be treated with extreme caution.

'Technical ammonium carbonate' (used in smelling salts) is actually a mixture of $[NH_4][HCO_3]$ and $[NH_4][NH_2CO_2]$ (ammonium carbamate). The latter is prepared by reacting NH₃ and CO₂ under pressure. It smells strongly of NH₃ because carbamic acid is an extremely weak acid (scheme 15.29). Pure carbamic acid (H_2NCO_2H) has not been isolated; the compound dissociates completely at 332 K.

$$\underbrace{[NH_4]^+(aq) + [H_2NCO_2]^-(aq)}_{\text{salt of a strong base and a weak acid}}$$

$$\rightleftharpoons NH_3(aq) + \{H_2NCO_2H(aq)\}$$
$$\Updownarrow$$
$$NH_3(aq) + CO_2(aq) \qquad (15.29)$$

Ammonium salts often crystallize with structures similar to those of the corresponding K^+, Rb^+ or Cs^+ salts. The $[NH_4]^+$ ion can be approximated to a sphere (Fig. 6.18) with $r_{ion} = 150\,pm$, a value similar to that of Rb^+. However, if, in the solid state, there is potential for hydrogen bonding involving the $[NH_4]^+$ ions, ammonium salts adopt structures unlike those of their alkali metal analogues, e.g. NH_4F possesses a wurtzite rather than an NaCl-type structure. The majority of $[NH_4]^+$ salts are soluble in water, with hydrogen bonding between $[NH_4]^+$ and H_2O being a contributing factor. An exception is $[NH_4]_2[PtCl_6]$.

Phosphane (Table 15.4) is an extremely toxic, colourless gas which is much less soluble in water than is NH_3. The P–H bond is not polar enough to form hydrogen bonds with H_2O. In contrast to NH_3, aqueous solutions of PH_3 are neutral, but in liquid NH_3, PH_3 acts as an acid (e.g. eq. 15.30).

$$K + PH_3 \xrightarrow{\text{liquid } NH_3} K^+ + [PH_2]^- + \tfrac{1}{2}H_2 \qquad (15.30)$$

The spontaneous flammability associated with PH_3 is due to the presence of P_2H_4 (see later). Phosphonium halides, PH_4X, are formed by treating PH_3 with HX but only the iodide is stable under ambient conditions. The chloride is unstable above 243 K and the bromide decomposes at 273 K. The $[PH_4]^+$ ion is decomposed by water (eq. 15.31). Phosphane acts as a Lewis base and a range of adducts (including those with low oxidation state d-block metal centres) are known. Examples include $H_3B\cdot PH_3$, $Cl_3B\cdot PH_3$, $Ni(PH_3)_4$ (decomposes above 243 K) and $Ni(CO)_2(PH_3)_2$. Combustion of PH_3 yields H_3PO_4.

$$[PH_4]^+ + H_2O \longrightarrow PH_3 + [H_3O]^+ \qquad (15.31)$$

The hydrides AsH_3 and SbH_3 resemble those of PH_3 (Table 15.4), but they are less stable with respect to decomposition into their elements. The thermal instability of AsH_3 and SbH_3 was the basis for the Marsh test. This is a classic analytical technique used in forensic science in which arsenic- or antimony-containing materials were first converted to AsH_3 or SbH_3, and the latter were then thermally decomposed (eq. 15.32). Treatment of the brown-black residue with aqueous NaOCl was used to distinguish between As (which reacted, eq. 15.33) and Sb (which did not react).

$$2EH_3(g) \xrightarrow{\Delta} 2E(s) + 3H_2(g) \qquad E = As, Sb \qquad (15.32)$$

$$5NaOCl + 2As + 3H_2O \longrightarrow 2H_3AsO_4 + 5NaCl \qquad (15.33)$$

Both AsH_3 and SbH_3 are extremely toxic gases, and SbH_3 is liable to explode. They are less basic than PH_3, but can be protonated with HF in the presence of AsF_5 or SbF_5 (eq. 15.34). The salts $[AsH_4][AsF_6]$, $[AsH_4][SbF_6]$ and $[SbH_4][SbF_6]$ form air- and moisture-sensitive crystals which decompose well below 298 K.

$$AsH_3 + HF + AsF_5 \longrightarrow [AsH_4]^+ + [AsF_6]^- \qquad (15.34)$$

Hydrides E_2H_4 (E = N, P, As)

Hydrazine, N_2H_4, is a colourless liquid (mp 275 K, bp 386 K), miscible with water and with a range of organic solvents, and is corrosive and toxic. Its vapour forms explosive mixtures with air. Although $\Delta_f H^o(N_2H_4, 298\,K) = +50.6\,kJ\,mol^{-1}$, N_2H_4 at ambient temperatures is *kinetically* stable with respect to N_2 and H_2. Alkyl derivatives of hydrazine (see eq. 15.44) have been used as rocket fuels, e.g. combined with N_2O_4 in the *Apollo* missions.[†] N_2H_4 has uses in the agricultural and plastics industries, and in the removal of O_2 from industrial water boilers to minimize corrosion (eq. 15.35).

$$N_2H_4 + O_2 \longrightarrow N_2 + 2H_2O \qquad (15.35)$$

Hydrazine is obtained by the Raschig reaction (the basis for the industrial synthesis) which involves the partial oxidation of NH_3 (eq. 15.36). Glue or gelatine is added to inhibit side-reaction 15.37 which otherwise consumes the N_2H_4 as it is formed; the additive removes traces of metal ions that catalyse reaction 15.37.

$$NH_3 + NaOCl \longrightarrow NH_2Cl + NaOH \qquad \textit{fast}$$
$$NH_3 + NH_2Cl + NaOH \longrightarrow N_2H_4 + NaCl + H_2O \qquad \textit{slow}$$
$$(15.36)$$

$$2NH_2Cl + N_2H_4 \longrightarrow N_2 + 2NH_4Cl \qquad (15.37)$$

Hydrazine is obtained commercially from the Raschig process as the monohydrate and is used in this form for many purposes. Dehydration is difficult, and direct methods to produce anhydrous N_2H_4 include reaction 15.38.

$$2NH_3 + [N_2H_5][HSO_4] \longrightarrow N_2H_4 + [NH_4]_2[SO_4] \qquad (15.38)$$

In aqueous solution, N_2H_4 usually forms $[N_2H_5]^+$ (hydrazinium) salts, but some salts of $[N_2H_6]^{2+}$ have been isolated, e.g. $[N_2H_6][SO_4]$. The pK_b values for hydrazine are given in eqs. 15.39 and 15.40, and the first step shows N_2H_4 to be a weaker base than NH_3 (eq. 15.24).

$$N_2H_4(aq) + H_2O(l) \rightleftharpoons [N_2H_5]^+(aq) + [OH]^-(aq)$$
$$K_b(1) = 8.9 \times 10^{-7} \qquad (15.39)$$

$$[N_2H_5]^+(aq) + H_2O(l) \rightleftharpoons [N_2H_6]^{2+}(aq) + [OH]^-(aq)$$
$$K_b(2) \approx 10^{-14} \qquad (15.40)$$

[†] O. de Bonn *et al.* (2001) *Z. Anorg. Allg. Chem.*, vol. 627, p. 2011 – 'Plume deposits from bipropellant rocket engines: methylhydrazinium nitrate and *N,N*-dimethylhydrazinium nitrate'.

Both N_2H_4 and $[N_2H_5]^+$ are reducing agents, and reaction 15.41 is used for the determination of hydrazine.

$$N_2H_4 + KIO_3 + 2HCl \longrightarrow N_2 + KCl + ICl + 3H_2O$$
$$(15.41)$$

The use of N_2H_4 in rocket fuels was described above. The stored energy in explosives and propellants ('high energy density materials') usually arises either from oxidation of an organic framework, or from an inherent high positive enthalpy of formation. For the hydrazinium salt $[N_2H_5]_2[\mathbf{15.15}]$ (prepared by reaction 15.42), $\Delta_f H^\circ(s, 298\,K) = +858\,kJ\,mol^{-1}$ (or $3.7\,kJ\,g^{-1}$), making $[N_2H_5]_2[\mathbf{15.15}]$ a spectacular example of a high energy density material.

$$Ba[\mathbf{15.15}] + [N_2H_5]_2[SO_4] \longrightarrow [N_2H_5]_2[\mathbf{15.15}]\cdot 2H_2O$$

$$\xrightarrow{\text{373 K, }in\ vacuo} [N_2H_5]_2[\mathbf{15.15}] \qquad (15.42)$$

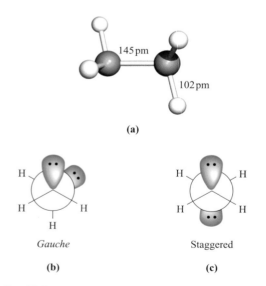

5,5'-Azotetrazolate dianion

(15.15)

Figure 15.4a shows the structure of N_2H_4. Of the conformations possible for N_2H_4, electron diffraction and IR spectroscopic data confirm that the *gauche*-form is favoured in the gas phase. The *gauche* conformation (Figs. 15.4a and 15.4b) is also adopted by P_2H_4 in the gas phase. In the solid

145 pm

102 pm

(a)

H   H
   •• ••
H   H
   H

Gauche

(b)

H   H

H   H
   ••

Staggered

(c)

Fig. 15.4 (a) The structure of N_2H_4, and Newman projections showing (b) the observed *gauche* conformation, and (c) the possible staggered conformation. An eclipsed conformation is also possible.

Fig. 15.5 The solid state structure (X-ray diffraction at 123 K) of the anion in $[Na(NH_3)_5]^+[Na(NH_3)_3(P_3H_3)]^-$ [N. Korber *et al.* (2001) *J. Chem. Soc., Dalton Trans.*, p. 1165]. Two of the three P atoms coordinate to the sodium centre (Na–P = 308 pm). Colour code: P, orange; Na, purple; N, blue; H, white.

state, P_2H_4 has a staggered conformation (Fig. 15.4c) while the related N_2F_4 exhibits both conformers. The eclipsed conformation (which would maximize lone pair–lone pair repulsions) is not observed.

Diphosphane, P_2H_4, is a colourless liquid (mp 174 K, bp 329 K), and is toxic and spontaneously inflammable. When heated, it forms higher phosphanes. Diphosphane is formed as a minor product in several reactions in which PH_3 is prepared (e.g. reaction 15.11) and may be separated from PH_3 by condensation in a freezing mixture. It exhibits no basic properties.

The $[P_3H_3]^{2-}$ ion is formed in reaction 15.43 and is stabilized by coordination to the sodium centre in $[Na(NH_3)_3(P_3H_3)]^-$. In the solid state, the H atoms in $[P_3H_3]^{2-}$ are in an all-*trans* configuration (Fig. 15.5).

$$5Na + 0.75P_4 + 11NH_3$$

$$\xrightarrow{\text{Na in liquid NH}_3\ 238\,K} [Na(NH_3)_5]^+[Na(NH_3)_3(P_3H_3)]^-$$
$$+ 3NaNH_2 \qquad (15.43)$$

Chloramine and hydroxylamine

H

N ''''''Cl

107°   103°

H

(15.16)

The reactions of NH_3 and Cl_2 (diluted with N_2) or aqueous NaOCl (the first step in reaction 15.36) yield chloramine, **15.16**, the compound responsible for the odour of water containing nitrogenous matter that has been sterilized with Cl_2. Chloramine is unstable, and violently explosive, and is usually handled in dilute solutions (e.g. in H_2O or Et_2O). Its reaction with Me_2NH (eq. 15.44) yields the rocket fuel 1,1-dimethylhydrazine.

$$NH_2Cl + 2Me_2NH \longrightarrow Me_2NNH_2 + [Me_2NH_2]Cl \quad (15.44)$$

Although dilute aqueous solutions of NH_2Cl can conveniently be handled, it is not practical to work with neat NH_2Cl because of its instability and the risk of explosion. Thus, apparently simple reactions such as the preparation and isolation of salts containing $[NH_3Cl]^+$ are not trivial. Use of pure NH_2Cl can be avoided by reaction of $(Me_3Si)_2NCl$ with HF in the presence of a strong Lewis base. As soon as NH_2Cl forms, it immediately forms $[NH_3Cl]^+$ (scheme 15.45).

$$(Me_3Si)_2NCl + 2HF \longrightarrow 2Me_3SiF + \{NH_2Cl\}$$

$$\left. \begin{array}{l} \text{Immediate reaction} \\ \text{with HF and } SbF_5 \end{array} \right| \qquad (15.45)$$

$$[NH_3Cl]^+[SbF_6]^-$$

Reaction 15.46 is one of several routes to hydroxylamine, NH_2OH, which is usually handled as a salt (e.g. the sulfate) or in aqueous solution. The free base can be obtained from its salts by treatment with NaOMe in MeOH.

$$2NO + 3H_2 + H_2SO_4$$

$$\xrightarrow{\text{platinized charcoal catalyst}} [NH_3OH]_2[SO_4] \qquad (15.46)$$

Pure NH_2OH forms white, hygroscopic crystals which melt at 306 K and explode at higher temperatures. It is a weaker base than NH_3 or N_2H_4. Many of its reactions arise from the great variety of redox reactions in which it takes part in aqueous solution, e.g. it reduces Fe(III) in acidic solution (eq. 15.47) but oxidizes Fe(II) in the presence of alkali (eq. 15.48).

$$2NH_2OH + 4Fe^{3+} \longrightarrow N_2O + 4Fe^{2+} + H_2O + 4H^+$$
$$(15.47)$$

$$NH_2OH + 2Fe(OH)_2 + H_2O \longrightarrow NH_3 + 2Fe(OH)_3$$
$$(15.48)$$

More powerful oxidizing agents (e.g. $[BrO_3]^-$) oxidize NH_2OH to HNO_3. The formation of N_2O in most oxidations of NH_2OH exemplifies dominance of kinetic over thermodynamic factors. Consideration of the potential diagram (see Section 8.5) in Fig. 15.6 shows that, on thermodynamic grounds, the expected product from the action of weak oxidizing agents on $[NH_3OH]^+$ (i.e. NH_2OH in acidic solution) would be N_2, but it seems that the reaction occurs by the steps in eq. 15.49.

$$\left. \begin{array}{l} NH_2OH \longrightarrow NOH + 2H^+ + 2e^- \\ 2NOH \longrightarrow HON{=}NOH \\ HON{=}NOH \longrightarrow N_2O + H_2O \end{array} \right\} \qquad (15.49)$$

Figure 15.6 also shows that, at pH $= 0$, $[NH_3OH]^+$ is unstable with respect to disproportionation into N_2 and $[NH_4]^+$ or $[N_2H_5]^+$. In fact, hydroxylamine does slowly decompose to N_2 and NH_3.

Worked example 15.4 Using potential and Frost–Ebsworth diagrams

(a) Use the data in Fig. 15.6 to calculate ΔG^o(298 K) for the following reduction process.

$$2[NH_3OH]^+(aq) + H^+(aq) + 2e^-$$
$$\longrightarrow [N_2H_5]^+(aq) + 2H_2O(l)$$

(b) Estimate ΔG^o(298 K) for the same process using the Frost–Ebsworth diagram in Fig. 8.4c.

(a) From the potential diagram, E^o for this half-reaction is $+1.41$ V.

$$\Delta G^o = -zFE^o$$
$$= -2 \times (96\,485 \times 10^{-3}) \times 1.41$$
$$= -272 \text{ kJ mol}^{-1}$$

(b) The gradient of the line joining the points for $[NH_3OH]^+$ and $[N_2H_5]^+ \approx \dfrac{1.9 - 0.5}{1} = 1.4 \text{ V}$

$$E^o = \frac{\text{Gradient of line}}{\text{Number of electrons transferred per mole of N}}$$
$$= \frac{1.4}{1} = 1.4 \text{ V}$$

$$\Delta G^o = -zFE^o$$
$$= -2 \times (96\,485 \times 10^{-3}) \times 1.4$$
$$= -270 \text{ kJ mol}^{-1}$$

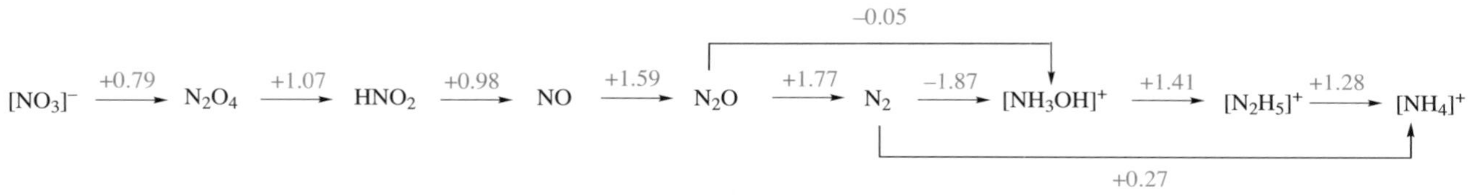

Fig. 15.6 Potential diagram for nitrogen at pH $= 0$. A Frost–Ebsworth diagram for nitrogen is given in Fig. 8.4c.

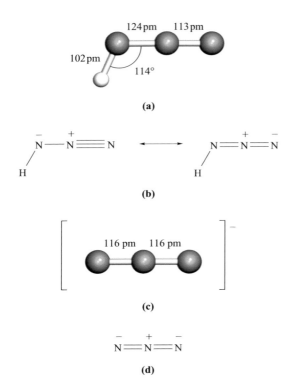

Fig. 15.7 (a) Structure of HN_3, (b) the major contributing resonance forms of HN_3, (c) the structure of the azide ion (the ion is symmetrical but bond distances vary slightly in different salts), and (d) the principal resonance structure of $[N_3]^-$. Colour code: N, blue; H, white.

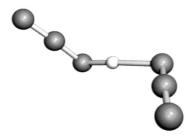

Fig. 15.8 The solid state structure (X-ray diffraction at 203 K) of the anion in $[PPh_4]^+[N_3HN_3]^-$ [B. Neumüller *et al.* (1999) *Z. Anorg. Allg. Chem.*, vol. 625, p. 1243]. Colour code: N, blue; H, white.

Self-study exercises

1. Explain how the Frost–Ebsworth diagram for nitrogen (Fig. 8.4c) illustrates that $[NH_3OH]^+$ (at pH 0) is unstable with respect to disproportionation.

    [*Ans.* See the bullet-point list in Section 8.6]

2. Use the data in Fig. 15.6 to calculate E^o for the reduction process:

    $$[NO_3]^-(aq) + 4H^+(aq) + 3e^- \longrightarrow NO(g) + 2H_2O(l)$$

    [*Ans.* +0.95 V]

3. In basic solution (pH = 14), $E^o_{[OH^-]=1}$ for the following process is +0.15 V. Calculate $\Delta G^o(298\,K)$ for the reduction process.

    $$2[NO_2]^-(aq) + 3H_2O(l) + 4e^- \rightleftharpoons N_2O(g) + 6[OH]^-(aq)$$

    [*Ans.* $-58\,kJ\,mol^{-1}$]

Further relevant problems can be found after worked example 8.8.

Hydrogen azide and azide salts

Sodium azide, NaN_3, is obtained from molten sodium amide by reaction 15.50 (or by reacting $NaNH_2$ with $NaNO_3$ at 450 K), and treatment of NaN_3 with H_2SO_4 yields hydrogen azide, HN_3.

$$2NaNH_2 + N_2O \xrightarrow{460\,K} NaN_3 + NaOH + NH_3 \quad (15.50)$$

Hydrogen azide (hydrazoic acid) is a colourless liquid (mp 193 K, bp 309 K). It is dangerously explosive ($\Delta_f H^o(l, 298\,K) = +264\,kJ\,mol^{-1}$) and highly poisonous. Aqueous solutions of HN_3 are weakly acidic (eq. 15.51).

$$HN_3 + H_2O \rightleftharpoons [H_3O]^+ + [N_3]^- \quad pK_a = 4.75 \quad (15.51)$$

The structure of HN_3 is shown in Fig. 15.7a, and a consideration of the resonance structures in Fig. 15.7b provides an explanation for the asymmetry of the NNN-unit. The azide ion is isoelectronic with CO_2, and the symmetrical structure of $[N_3]^-$ (Fig. 15.7c) is consistent with the bonding description in Fig. 15.7d. A range of azide salts is known; Ag(I), Cu(II) and Pb(II) azides, which are insoluble in water, are explosive, and $Pb(N_3)_2$ is used as an initiator for less sensitive explosives. On the other hand, group 1 metal azides decompose less violently when heated (eqs. 11.2 and 15.6). The reaction between NaN_3 and Me_3SiCl yields the covalent compound Me_3SiN_3 which is a useful reagent in organic synthesis. Reaction 15.52 occurs when Me_3SiN_3 is treated with $[PPh_4]^+[N_3]^-$ in the presence of ethanol. The $[N_3HN_3]^-$ anion in the product is stabilized by hydrogen bonding (compare with $[FHF]^-$, see Fig. 10.9). Although the position of the H atom in the anion is not known with great accuracy, structural parameters for the solid state structure of $[PPh_4][N_3HN_3]$ (Fig. 15.8) are sufficiently accurate to confirm an asymmetrical $N-H\cdots N$ interaction ($N\cdots N = 272\,pm$).

$$[PPh_4][N_3] + Me_3SiN_3 + EtOH$$
$$\longrightarrow [PPh_4][N_3HN_3] + Me_3SiOEt \quad (15.52)$$

The azide group, like $CN^{\bullet}$ (though to a lesser extent), shows similarities to a halogen and is another example of

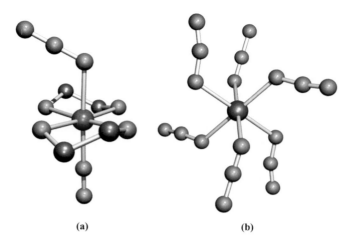

Fig. 15.9 The structures (X-ray diffraction) of (a) *trans*-[Ru(en)$_2$(N$_2$)(N$_3$)]$^+$ in the [PF$_6$]$^-$ salt (H atoms omitted) [B.R. Davis *et al.* (1970) *Inorg. Chem.*, vol. 9, p. 2768] and (b) [Sn(N$_3$)$_6$]$^{2-}$ structurally characterized as the [Ph$_4$P]$^+$ salt [D. Fenske *et al.* (1983) *Z. Naturforsch., Teil B*, vol. 38, p. 1301]. Colour code: N, blue; Ru, red; Sn, brown; C, grey.

a pseudo-halogen (see Section 14.12). However, no N$_6$ molecule (i.e. a dimer of N$_3^{\bullet}$ and so an analogue of an X$_2$ halogen) has yet been prepared.[†] Like halide ions, the azide ion acts as a ligand in a wide variety of complexes of both metals and non-metals, e.g. [Au(N$_3$)$_4$]$^-$, *trans*-[TiCl$_4$(N$_3$)$_2$]$^{2-}$, *cis*-[Co(en)$_2$(N$_3$)$_2$]$^+$, *trans*-[Ru(en)$_2$(N$_2$)(N$_3$)]$^+$ (which is also an example of a dinitrogen complex, Fig. 15.9a), [Sn(N$_3$)$_6$]$^{2-}$ (Fig. 15.9b), [Si(N$_3$)$_6$]$^{2-}$, [Sb(N$_3$)$_6$], [W(N$_3$)$_6$], [W(N$_3$)$_7$]$^-$ and [U(N$_3$)$_7$]$^{3-}$.

The reaction of HN$_3$ with [N$_2$F][AsF$_6$] (prepared by reaction 15.69) in HF at 195 K results in the formation of [N$_5$][AsF$_6$]. Designing the synthesis of [N$_5$]$^+$ was not trivial. Precursors in which the N≡N and N=N bonds are preformed are critical, but should not involve gaseous N$_2$ since this is too inert. The HF solvent provides a heat sink for the exothermic reaction, the product being potentially explosive. Although [N$_5$][AsF$_6$] was the first example of a salt of [N$_5$]$^+$ and is therefore of significant interest, it is not very stable and tends to explode. In contrast, [N$_5$][SbF$_6$] (eq. 15.53) is stable at 298 K and is relatively resistant to impact. Solid [N$_5$][SbF$_6$] oxidizes NO, NO$_2$ and Br$_2$ (scheme 15.54), but not Cl$_2$ or O$_2$.

$$[N_2F]^+[SbF_6]^- + HN_3 \xrightarrow[\text{(ii) warm to 298 K}]{\text{(i) liquid HF, 195 K}} [N_5]^+[SbF_6]^- + HF$$

$$(15.53)$$

[†]For theoretical data on N$_6$, see: T.M. Klapötke (2000) *J. Mol. Struct.*, vol. 499, p. 99; L.J. Wang, P. Warburton and P.G. Mezey (2002) *J. Phys. Chem. A*, vol. 106, p. 2748.

$$[N_5]^+[SbF_6]^- \begin{cases} \xrightarrow{NO} [NO]^+[SbF_6]^- + 2.5N_2 \\ \xrightarrow{NO_2} [NO_2]^+[SbF_6]^- + 2.5N_2 \\ \xrightarrow{Br_2} [Br_2]^+[SbF_6]^- + 2.5N_2 \end{cases}$$

$$(15.54)$$

The reaction of [N$_5$][SbF$_6$] with SbF$_5$ in liquid HF yields [N$_5$][Sb$_2$F$_{11}$], the solid state structure of which has been determined, confirming a V-shaped [N$_5$]$^+$ ion (central N–N–N angle = 111°). The N–N bond lengths are 111 pm (almost the same as in N$_2$) and 130 pm (slightly more than in MeN=NMe), respectively, for the terminal and central bonds. Resonance stabilization (structures **15.17**) provides a degree of multiple-bond character to all the N–N bonds. The three resonance structures shown in blue contain one or two terminal sextet N atoms. Their inclusion helps to account for the observed N$_{terminal}$–N–N$_{central}$ bond angles of 168°.

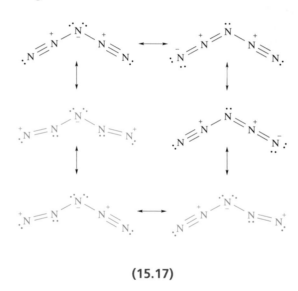

(15.17)

15.6 Nitrides, phosphides, arsenides, antimonides and bismuthides

Nitrides

Classifying nitrides is not simple, but nearly all nitrides fall into one of the following groups, although, as we have seen for the borides and carbides, some care is needed in attempting to generalize:

- saline nitrides of the group 1 and 2 metals, and aluminium;
- covalently bonded nitrides of the *p*-block elements (see Sections 13.8, 14.12 and 16.10 for BN, C$_2$N$_2$, Si$_3$N$_4$, Sn$_3$N$_4$ and S$_4$N$_4$);
- interstitial nitrides of *d*-block metals;
- pernitrides of the group 2 metals.

The classification of 'saline nitride' implies the presence of the N^{3-} ion, but this is unlikely (see Section 15.1). However, it is usual to consider Li_3N, Na_3N (see Section 11.4), Be_3N_2, Mg_3N_2, Ca_3N_2, Ba_3N_2 and AlN in terms of ionic formulations. Hydrolysis of saline nitrides liberates NH_3. Sodium nitride is very hygroscopic, and samples are often contaminated with NaOH (reaction 15.55).

$$Na_3N + 3H_2O \longrightarrow 3NaOH + NH_3 \qquad (15.55)$$

Among the nitrides of the p-block elements, Sn_3N_4 and the γ-phase of Si_3N_4 represent the first examples of spinel nitrides (see Section 14.12).

Nitrides of the d-block metals are hard, inert solids which resemble metals in appearance, and have high melting points and electrical conductivities (see Box 15.4). They can be prepared from the metal or metal hydride with N_2 or NH_3 at high temperatures. Most possess structures in which the nitrogen atoms occupy octahedral holes in a close-packed metal lattice. Full occupancy of these holes leads to the stoichiometry MN (e.g. TiN, ZrN, HfN, VN, NbN). Cubic close-packing of the metal atoms and an NaCl-type structure for the nitride MN is favoured for metals in the earliest groups of the d-block.

Pernitrides contain the $[N_2]^{2-}$ ion and are known for barium, strontium and platinum. BaN_2 is prepared from the elements under a 5600 bar pressure of N_2 at 920 K. It is structurally related to the carbide ThC_2 (see Section 14.7), and contains isolated $[N_2]^{2-}$ ions with an $N-N$ distance of 122 pm, consistent with an $N=N$ bond. The strontium nitrides SrN_2 and SrN are made from Sr_2N at 920 K under N_2 pressures of 400 and 5500 bar, respectively. The structure of SrN_2 is derived from the layered structure of Sr_2N by having half of the octahedral holes between the layers occupied by $[N_2]^{2-}$ ions. Its formation can be considered in terms of N_2 (at high pressure) oxidizing Sr from a formal oxidation state of +1.5 to +2, and concomitant reduction of N_2 to

APPLICATIONS

Box 15.4 Materials chemistry: metal and non-metal nitrides

Nitrides of the d-block metals are hard, are resistant to wear and chemical attack including oxidation, and have very high melting points. These properties render nitrides such as TiN, ZrN and HfN invaluable for protecting high-speed cutting tools. The applied coatings are extremely thin (typically $\leq 10\,\mu m$), but nonetheless significantly prolong the lifetimes of tools that operate under the toughest of work conditions. Nitride coatings can be applied using the technique of chemical vapour deposition (see Section 28.6), or by forming a surface layer of Fe_3N or Fe_4N by reacting the prefabricated steel tool with N_2.

Ceramic cutting-tool materials include alumina (α-Al_2O_3) and silicon nitride. Of these two refractory materials, Si_3N_4 has greater strength at higher temperatures, higher thermal stability, lower thermal coefficient of expansion, and higher thermal conductivity. Other properties include extremely high thermal shock resistance, and resistance to attack by oxidizing agents. Despite these advantages, Si_3N_4 is difficult to fabricate in a dense form, and additives (e.g. MgO, Y_2O_3) are required to aid the conversion of powdered Si_3N_4 to the final material (i.e. the *sintering* process). The powdered Si_3N_4 comprises mainly the α-phase which is metastable with respect to the β-form. Conversion of α- to β-Si_3N_4 occurs during sintering, and gives rise to elongated grains. These grains grow within a fine-grained matrix, giving the material a reinforced microstructure. Silicon nitride is used extensively in cutting tools (e.g. for machining cast iron), and in bearings such as in machine-tool spindles. The thermal properties of Si_3N_4 have resulted in its use in ceramic heating devices. Since the mid-1980s, silicon nitride glow plugs in diesel engines have found widespread application. 'Glow plugs' are used to heat the combustion chamber of a diesel engine, aiding ignition on a cold-start. As the photograph illustrates, Si_3N_4 heaters can

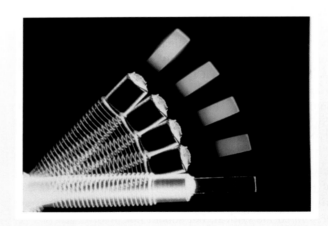

A series of photographs showing a silicon nitride (ceramic) heater going from room temperature to 600°C over a period of a few seconds.

achieve a rise in temperature from ambient to around 600°C ($\approx 900\,K$) in a few seconds. Silicon nitride heating devices are also employed for household appliances, e.g. for heating water.

Layers of TiN, ZrN, HfN or TaN are applied as diffusion barriers in semiconducting devices. The barrier layer ($\approx 100\,nm$ thick) is fabricated between the semiconducting material (e.g. GaAs or Si) and the protective metallic (e.g. Au or Ni) coating, and prevents diffusion of metal atoms into the GaAs or Si device.

For related information: see the discussions on boron nitride, silicon nitride and ceramic coatings in Section 28.6.

$[N_2]^{2-}$. At higher pressures of N_2, all the octahedral holes in the structure become occupied by $[N_2]^{2-}$ ions, and the final product, SrN, is better formulated as $(Sr^{2+})_4(N^{3-})_2(N_2^{2-})$. Platinum pernitride, PtN_2, is prepared from Pt and N_2 under extreme conditions of temperature and pressure in a diamond anvil cell (see Box 14.4). In PtN_2, the N–N bond length of 141 pm is consistent with a single bond and the compound is best formulated as a platinum(IV) compound containing $[N_2]^{4-}$ units.

Phosphides

Most elements combine with phosphorus to give solid state binary phosphides; exceptions are Hg, Pb, Sb and Te. A solid of composition BiP has been reported, but the formation of a phosphide as opposed to a mixture of elemental bismuth and phosphorus has not been confirmed.[†] Types of solid state phosphides are very varied, and simple classification is not possible. Phosphides of the d-block metals tend to be inert, metallic-looking compounds with high melting points and electrical conductivities. Their formulae are often deceptive in terms of the oxidation state of the metal and their structures may contain isolated P centres, P_2 groups, or rings, chains or layers of P atoms.

The group 1 and 2 metals form compounds M_3P and M_3P_2 respectively, which are hydrolysed by water and can be considered to be ionic. The alkali metals also form phosphides which contain groups of P atoms forming chains or cages, the cages being either $[P_7]^{3-}$ (**15.18**) or $[P_{11}]^{3-}$ (**15.19**). Lithium phosphide of stoichiometry LiP consists of helical chains and is better formulated as $Li_n[P_n]$, the $[P_n]^{n-}$ chains being isoelectronic with S_n (see Box 1.1). The P–P distances in the chains are 221 ppm, consistent with single bonds ($r_{cov} = 110$ pm). K_4P_3 contains $[P_3]^{4-}$ chains, Rb_4P_6 has planar $[P_6]^{4-}$ rings, Cs_3P_7 contains $[P_7]^{3-}$ cages, and Na_3P_{11} features $[P_{11}]^{3-}$ cages. Ba_3P_{14} and Sr_3P_{14} also contain $[P_7]^{3-}$ cages. The phosphides BaP_{10}, CuP_7, Ag_3P_{11}, MP_4 (e.g. M = Mn, Tc, Re, Fe, Ru, Os) and TlP_5 contain more extended arrays of P atoms, two examples (**15.9** and **15.10**) of which have already been mentioned.

For the preparation of metal phosphides, the most general method is to heat the metal with red phosphorus. Alkali metal phosphides may be prepared using, for example, $LiPH_2$ (eq. 15.12) or P_2H_4. The reaction of Cs with P_2H_4 (eq. 15.56) followed by recrystallization from liquid NH_3 gives $Cs_2P_4 \cdot 2NH_3$ which contains planar $[P_4]^{2-}$ rings.

$$6P_2H_4 + 10Cs \longrightarrow Cs_2P_4 + 8CsPH_2 + 4H_2 \qquad (15.56)$$

The P–P bond distances of 215 pm are shorter than a typical single bond (220 pm) but longer than a double bond (see Section 23.6). Cyclic $[P_4]^{2-}$ is a 6π-aromatic system, and the bonding is explored in the exercise below.

Self-study exercise

Three of the occupied MOs of $[P_4]^{2-}$ are shown below. The e_g orbitals are the highest lying occupied MOs. Assume that the P_4-ring is oriented in the xy-plane.

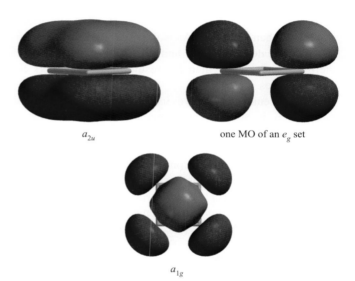

a_{2u} one MO of an e_g set

a_{1g}

(a) Assuming that $[P_4]^{2-}$ is a perfect square, use Fig. 3.10 to confirm that $[P_4]^{2-}$ belongs to the D_{4h} point group.

(b) By using the appropriate character table in Appendix 3, confirm that the symmetries of the MOs labelled a_{1g} and a_{2u} are consistent with the labels given.

(c) How many occupied MOs does $[P_4]^{2-}$ possess? (Consider only the valence electrons.)

(d) Explain why the a_{1g} orbital is classed as a σ-bonding MO.

(e) The π-orbitals of $[P_4]^{2-}$ are derived from combinations of the phosphorus $3p_z$ orbitals. How many π-MOs are there? Draw diagrams to represent these π-MOs. Draw an energy level diagram to show the relative energies of these MOs.
    [*Ans.* (c) 11; (e) $[P_4]^{2-}$ is isoelectronic with $[C_4H_4]^{2-}$, see Fig. 24.28]

(15.18)   **(15.19)**

[†]G.C. Allen *et al.* (1997) *Chem. Mater.*, vol. 9, p. 1385 –'Material of composition BiP'.

Arsenides, antimonides and bismuthides

Metal arsenides, antimonides and bismuthides can be prepared by direct combination of the metal and group 15 element. Like the phosphides, classification is not simple, and structure types vary. The coverage here is, therefore, selective.

Gallium arsenide, GaAs, is an important III–V semiconductor (see Section 28.6) and crystallizes with a zinc blende-type structure (see Fig. 6.19b). Slow hydrolysis occurs in moist air and protection of semiconductor devices from the air is essential; N_2 is often used as a 'blanket gas'. At 298 K, GaAs has a band gap of 1.42 eV, and can be used to make devices that emit light in the infrared region. Gallium arsenide exhibits a high electron mobility ($8500 \, cm^2 \, V^{-1} \, s^{-1}$, compared with a value of $1500 \, cm^2 \, V^{-1} \, s^{-1}$ for silicon). While this property, coupled with the highly desirable optical properties of GaAs, provides advantages for GaAs over Si, there are a number of disadvantages for use in devices: (i) GaAs is more expensive than Si, (ii) GaAs wafers are more brittle than those fabricated from Si, and (iii) GaAs has a lower thermal conductivity than Si, resulting in heat-sinks in GaAs devices.

Nickel arsenide, NiAs, gives its name to a prototype structure which is adopted by a number of d-block metal arsenides, antimonides, sulfides, selenides and tellurides. The structure can be described as a hexagonal close-packed (hcp) array of As atoms with Ni atoms occupying octahedral holes. Although such a description might conjure up the concept of an *ionic* lattice, the bonding in NiAs is certainly *not* purely ionic. Figure 15.10 shows a unit cell of NiAs. The placement of the Ni atoms in

octahedral holes in the hcp arrangement of As atoms means that the coordination environment of the As centres is *trigonal prismatic*. Although each Ni atom has six As neighbours at 243 pm, there are two Ni neighbours at a distance of only 252 pm (compare $r_{metal}(Ni) = 125$ pm) and there is almost certainly Ni—Ni bonding running through the structure. This is consistent with the observation that NiAs conducts electricity.

Arsenides and antimonides containing the $[As_7]^{3-}$ and $[Sb_7]^{3-}$ ions can be prepared by, for example, reactions 15.57 and 15.58. These Zintl ions are structurally related to $[P_7]^{3-}$ (**15.18**) and their bonding can be described in terms of localized 2-centre 2-electron interactions.

$$3Ba + 14As \xrightarrow{1070 \, K} Ba_3[As_7]_2 \quad (15.57)$$

$$Na/Sb \text{ alloy} \xrightarrow[\text{crypt-222}]{1,2\text{-ethanediamine,}} [Na(crypt\text{-}222)]_3[Sb_7] \quad (15.58)$$

Heteroatomic Zintl ions incorporating group 15 elements are present in the compounds $[K(crypt\text{-}222)]_2[Pb_2Sb_2]$, $[K(crypt\text{-}222)]_2[GaBi_3]$, $[K(crypt\text{-}222)]_2[InBi_3]$ and $[Na(crypt\text{-}222)]_3[In_4Bi_5]$, all of which are prepared (mostly as solvates with 1,2-ethanediamine) in a similar way to reaction 15.58. The $[Pb_2Sb_2]^{2-}$, $[GaBi_3]^{2-}$ and $[InBi_3]^{2-}$ ions are tetrahedral in shape. The $[In_4Bi_5]^{3-}$ ion adopts a monocapped square-antiprismatic structure in which the Bi atoms occupy the unique capping site and the four open-face sites. These structures are consistent with Wade's rules (see Section 13.11).[†] Examples of non-cluster $[E_n]^{x-}$ species are provided by $[Bi_2]^{2-}$, $[As_4]^{4-}$, $[Sb_4]^{4-}$ and $[Bi_4]^{4-}$. The $[Bi_2]^{2-}$ ion forms as a minor component in a 1,2-ethanediamine solution of the phase $K_5In_2Bi_4$ (made by heating a stoichiometric mixture of K, In and Bi) and may be crystallized as the salt $[K(crypt\text{-}222)]_2[Bi_2]$. The short Bi–Bi distance of 284 pm is consistent with a double bond. The $[Bi_2]^{2-}$ ion is also present in the salt $[Cs(18\text{-crown-}6)]_2[Bi_2]$. The phases M_5E_4 (M = K, Rb, Cs; E = As, Sb, Bi) are formed by heating the respective elements under vacuum and slowly cooling the mixtures. They are noteworthy because they contain $[E_4]^{4-}$ chains and 'extra' electrons, i.e. they are formulated as $[M^+]_5[E_4^{4-}][e^-]$ with the additional electron being delocalized over the structure.

The syntheses of cationic bismuth clusters were described in Section 9.12. The $[Bi_5]^{3+}$ ion may also be obtained by oxidation of Bi using $GaCl_3$ in benzene, or using AsF_5. Although $[Bi_5]^{3+}$, $[Bi_8]^{2+}$ and $[Bi_9]^{5+}$ have been known for many years, no well-characterized example of a homopolyatomic antimony cation was reported until 2004.

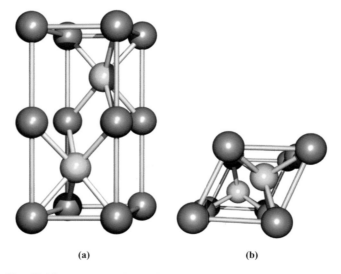

(a)   (b)

Fig. 15.10 Two views of the unit cell (defined by the yellow lines) of the nickel arsenide (NiAs) lattice; colour code: Ni, green; As, yellow. View (a) emphasizes the trigonal prismatic coordination environment of the As centres, while (b) (which views (a) from above) illustrates more clearly that the unit cell is not a cuboid.

[†]For examples of related clusters that violate Wade's rules, see: L. Xu and S.C. Sevov (2000) *Inorg. Chem.*, vol. 39, 5383.

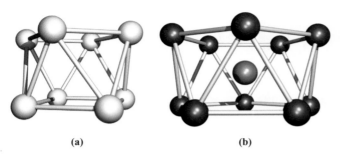

(a)   (b)

Fig. 15.11 (a) The square antiprismatic structure of $[Sb_8]^{2+}$ and $[Bi_8]^{2+}$. (b) The Pd-centred pentagonal antiprismatic structure of $[Pd@Bi_{10}]^{4+}$.

The salt $[Sb_8][GaCl_3]_2$ is formed by reducing $SbCl_3$ using $Ga^+[GaCl_4]^-$ in $GaCl_3$/benzene solution. The $[Sb_8]^{2+}$ cation is isostructural with $[Bi_8]^{2+}$ and possesses a square antiprismatic structure (Fig. 15.11a), consistent with Wade's rules (i.e. a 22 cluster-electron *arachno*-cage). Figure 15.11b shows the Pd-centred pentagonal antiprismatic cluster adopted by $[Pd@Bi_{10}]^{4+}$. This cluster is related to the endohedral Zintl ions shown in Fig. 14.16. Assuming that the palladium is a Pd(0) centre and contributes no electrons to cluster bonding, then $[Pd@Bi_{10}]^{4+}$ is a 26 cluster-electron *arachno*-cage.

Worked example 15.5 Electron counting in heteroatomic Zintl ions

Explain how Wade's rules rationalize the tetrahedral shape of $[GaBi_3]^{2-}$.

Assume that each main group element in the cluster retains a lone pair of electrons, localized outside the cluster (i.e. not involved in cluster bonding).

Electrons available for cluster bonding are as follows:

Ga (group 13) provides one electron.
Bi (group 15) provides three electrons.
The overall 2− charge provides two electrons.
Total cluster electron count $= 1 + (3 \times 3) + 2$
$= 12$ electrons.

The $[GaBi_3]^{2-}$ ion has six pairs of electrons with which to bond four atoms. $[GaBi_3]^{2-}$ is therefore classed as a *nido*-cluster, based on a 5-vertex trigonal bipyramid with one vertex missing. This is consistent with the observed tetrahedral shape:

      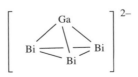

closo-trigonal bipyramid

Self-study exercises

1. Explain how Wade's rules rationalize why $[Pb_2Sb_2]^{2-}$ has a tetrahedral shape. What class of cluster is $[Pb_2Sb_2]^{2-}$? [*Ans.* 6 cluster electron pairs; *nido*]

2. Explain why the monocapped square-antiprismatic structure for $[In_4Bi_5]^{3-}$ shown below is consistent with Wade's rules. What class of cluster is $[In_4Bi_5]^{3-}$?

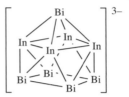

[*Ans.* 11 cluster electron pairs; *nido*]

3. In theory, would isomers be possible for tetrahedral $[Pb_2Sb_2]^{2-}$ and for tetrahedral $[InBi_3]^{2-}$?

[*Ans.* No isomers possible]

15.7 Halides, oxohalides and complex halides

Nitrogen halides

The highest molecular halides of nitrogen are of formula NX_3. Nitrogen pentahalides are not known and this has been attributed to the steric crowding of five halogen atoms around the small N atom. Important nitrogen halides are NX_3 ($X = F$, Cl), N_2F_4 and N_2F_2, selected properties for which are listed in Table 15.5. NBr_3 and NI_3 exist but are less well characterized than NF_3 and NCl_3.

Nitrogen trifluoride is made either by reaction 15.59 which must be carried out in a controlled manner, or by electrolysis of anhydrous NH_4F/HF mixtures.

$$4NH_3 + 3F_2 \xrightarrow{\text{Cu catalyst}} NF_3 + 3NH_4F \qquad (15.59)$$

NF_3 is the most stable of the trihalides of nitrogen, being the only one to have a negative value of $\Delta_f H^o$ (Table 15.5). It is a colourless gas which is resistant to attack by acids and alkalis, but is decomposed by sparking with H_2 (eq. 15.60). The resistance towards hydrolysis parallels that observed for the carbon tetrahalides (Section 14.8).

$$2NF_3 + 3H_2 \longrightarrow N_2 + 6HF \qquad (15.60)$$

(15.20)

The gas-phase structure of NF_3 is trigonal pyramidal **(15.20)**, and the molecular dipole moment is very small

Table 15.5   Selected data for nitrogen fluorides and trichloride.

	NF$_3$	NCl$_3$	N$_2$F$_4$	cis-N$_2$F$_2$	trans-N$_2$F$_2$
Melting point / K	66	<233	108.5	<78	101
Boiling point / K	144	<344; explodes at 368	199	167	162
$\Delta_f H^\circ$(298 K) / kJ mol^{-1}	−132.1	230.0	−8.4	69.5	82.0
Dipole moment / D	0.24	0.39	0.26†	0.16	0
N−N bond distance / pm	−	−	149	121	122
N−X bond distance / pm	137	176	137	141	140
Bond angles / deg	∠F–N–F 102.5	∠Cl–N–Cl 107	∠F–N–F 103 ∠N–N–F 101	∠N–N–F 114	∠N–N–F 106

† *Gauche* conformation (see Fig. 15.4).

(Table 15.5). In contrast to NH$_3$ and PF$_3$, NF$_3$ shows no donor properties.

Worked example 15.6   Dipole moments in NX$_3$ molecules

Explain why NH$_3$ is polar. In which direction does the dipole moment act?

NH$_3$ is a trigonal pyramidal molecule with a lone pair of electrons on the N atom:

The Pauling electronegativity values of N and H are 3.0 and 2.2, respectively (see Appendix 7) and, therefore, each N–H bond is polar in the sense N$^{\delta-}$–H$^{\delta+}$. The resultant molecular dipole moment is reinforced by the lone pair of electrons:

(By SI convention, the arrow representing the dipole moment points from δ^- to δ^+: see Section 2.6.)

Self-study exercises

1. Rationalize why there is a significant difference between the dipole moments of the gas-phase molecules NH$_3$ ($\mu = 1.47$ D) and NF$_3$ ($\mu = 0.24$ D).

    [*Ans.* See Example 3 in Section 2.6]

2. Account for the fact that the dipole moment of NHF$_2$ (1.92 D) is greater than that of NF$_3$ (0.24 D).

3. Suggest how the directionalities of the resulting dipole moments in NH$_3$ and NHF$_2$ differ. Give reasons for your answer.

Nitrogen trichloride is an oily, yellow liquid at 289 K, but it is highly endothermic and dangerously explosive (Table 15.5). The difference in stabilities of NF$_3$ and NCl$_3$ lies in the relative bond strengths of N−F over N−Cl, and of Cl$_2$ over F$_2$. Nitrogen trichloride can be prepared by reaction 15.61, with the equilibrium being drawn to the right-hand side by extracting NCl$_3$ into a suitable organic solvent. Diluted with air, NCl$_3$ is used for bleaching flour since hydrolysis by moisture forms HOCl (see Section 17.9). Alkalis hydrolyse NCl$_3$ according to eq. 15.62.

$$NH_4Cl + 3Cl_2 \rightleftharpoons NCl_3 + 4HCl \qquad (15.61)$$

$$2NCl_3 + 6[OH]^- \longrightarrow N_2 + 3[OCl]^- + 3Cl^- + 3H_2O \quad (15.62)$$

Nitrogen tribromide is more reactive than NCl$_3$, and explodes at temperatures as low as 175 K. It can be prepared by reaction 15.63, attempts to make it by treating NCl$_3$ with Br$_2$ being unsuccessful.

$$(Me_3Si)_2NBr + 2BrCl \xrightarrow{\text{in pentane, 186 K}} NBr_3 + 2Me_3SiCl \qquad (15.63)$$

Nitrogen triiodide has been made by reacting IF with boron nitride in CFCl$_3$. Although NI$_3$ is stable at 77 K and has been characterized by IR, Raman and ^{15}N NMR spectroscopies, it is highly explosive at higher temperatures ($\Delta_f H^\circ$(NI$_3$, g) = +287 kJ mol^{-1}). The reaction between concentrated aqueous NH$_3$ and [I$_3$]$^-$ yields NH$_3$·NI$_3$, black crystals of which are dangerously explosive ($\Delta_f H^\circ$(NH$_3$·NI$_3$, s) = +146 kJ mol^{-1}) as the compound decomposes to NH$_3$, N$_2$ and I$_2$.

The nitrogen fluorides N_2F_4 and N_2F_2 can be obtained from reactions 15.64 and 15.65. Properties of these fluorides are listed in Table 15.5, and both compounds are explosive.

$$2NF_3 \xrightarrow{Cu,\ 670\,K} N_2F_4 + CuF_2 \qquad (15.64)$$

$$2N_2F_4 + 2AlCl_3 \xrightarrow{203\,K} trans\text{-}N_2F_2 + 3Cl_2 + N_2 + 2AlF_3$$

$$\Big\downarrow 373\,K$$

$$cis\text{-}N_2F_2 \qquad (15.65)$$

The structure of N_2F_4 resembles that of hydrazine, except that both the *gauche* and *trans* (staggered) conformers (Fig. 15.4) are present in the liquid and gas phases. At temperatures above 298 K, N_2F_4 reversibly dissociates into blue $NF_2{}^{\bullet}$ radicals which undergo a wide range of reactions (e.g. eqs 15.66–15.68).

$$2NF_2 + S_2F_{10} \longrightarrow 2F_2NSF_5 \qquad (15.66)$$

$$2NF_2 + Cl_2 \longrightarrow 2NClF_2 \qquad (15.67)$$

$$NF_2 + NO \longrightarrow F_2NNO \qquad (15.68)$$

(15.21)   **(15.22)**

Dinitrogen difluoride, N_2F_2, exists in both the *trans-* and *cis*-forms (**15.21** and **15.22**), with the *cis*-isomer being thermodynamically the more stable of the two but also the more reactive. Reaction 15.65 gives a selective method of preparing *trans*-N_2F_2. Isomerization by heating gives a mixture of isomers from which *cis*-N_2F_2 can be isolated by treatment with AsF_5 (reaction 15.69).

Mixture of isomers:

$$cis\text{-}N_2F_2 \xrightarrow{AsF_5} [N_2F]^+[AsF_6]^- \xrightarrow{NaF/HF} cis\text{-}N_2F_2$$

$$trans\text{-}N_2F_2 \xrightarrow{AsF_5} \text{No reaction}$$

$$(15.69)$$

Reaction 15.69 illustrates the ability of N_2F_2 to donate F^- to *strong* acceptors such as AsF_5 and SbF_5, a reaction type shared by N_2F_4 (eqs. 15.70 and 15.71). The cation $[NF_4]^+$ is formed in reaction 15.72. We return to the properties of AsF_5 and SbF_5 later.

$$N_2F_4 + AsF_5 \longrightarrow [N_2F_3]^+[AsF_6]^- \qquad (15.70)$$

$$N_2F_4 + 2SbF_5 \longrightarrow [N_2F_3]^+[Sb_2F_{11}]^- \qquad (15.71)$$

$$NF_3 + F_2 + SbF_5 \longrightarrow [NF_4]^+[SbF_6]^- \qquad (15.72)$$

Self-study exercise

Use the data below to determine $\Delta_r H^\circ(298\,K)$ for the following reaction, and comment on why the endothermic compound $NI_3 \cdot NH_3$ forms.

$$3I_2(s) + 5NH_3(aq) \longrightarrow NI_3 \cdot NH_3(s) + 3NH_4I(aq)$$

Data: $\Delta_f H^\circ(298\,K)$: $NH_3(aq)$, -80; $NH_4I(aq)$, -188; $NI_3 \cdot NH_3(s)$, $+146$ kJ mol^{-1}.

[*Ans.* See: D. Tudela (2002) *J. Chem. Educ.*, vol. 79, p. 558]

Oxofluorides and oxochlorides of nitrogen

	X		
	F	Cl	Br
a / pm	152	198	214
b / pm	113	114	115
a / °	110	113	117

(15.23)

Several oxofluorides and oxochlorides of nitrogen are known, but all are unstable gases or volatile liquids which are rapidly hydrolysed. Nitrosyl halides FNO, ClNO and BrNO are formed in reactions of NO with F_2, Cl_2 and Br_2 respectively. Structural details for gas-phase molecules are shown in **15.23**. The short N–O bond lengths indicate triple rather than double bond character and a contribution from the left-hand resonance structure in the resonance pair **15.24** is clearly important. Crystals of FNO and ClNO have been grown from condensed samples of the compounds, and their solid state structures have been determined at 128 and 153 K, respectively. Compared with those in the gas phase, FNO molecules in the crystal have shorter (108 pm) N–O and longer (165 pm) N–F bonds. A similar trend is seen for ClNO (solid: N–O = 105 pm, N–Cl = 219 pm). These data suggest that the $[NO]^+X^-$ form in resonance pair **15.24** becomes more dominant on going from gaseous to solid XNO.

$$X = F, Cl, Br$$

(15.24)

Nitryl fluoride, FNO_2, and nitryl chloride, $ClNO_2$, are prepared, respectively, by fluorination of N_2O_4 (reaction 15.73) and oxidation of ClNO (using e.g. Cl_2O or O_3). Both are planar molecules; FNO_2 is isoelectronic with $[NO_3]^-$.

$$N_2O_4 + 2CoF_3 \xrightarrow{570\,K} 2FNO_2 + 2CoF_2 \qquad (15.73)$$

The oxohalides FNO, ClNO, FNO$_2$ and ClNO$_2$ combine with suitable fluorides or chlorides to give salts containing [NO]$^+$ or [NO$_2$]$^+$, e.g. reactions 15.74–15.76.

$$\text{FNO} + \underset{\substack{\text{halide}\\\text{acceptor}}}{\text{AsF}_5} \longrightarrow [\text{NO}]^+[\text{AsF}_6]^- \tag{15.74}$$

$$\text{ClNO} + \text{SbCl}_5 \longrightarrow [\text{NO}]^+[\text{SbCl}_6]^- \tag{15.75}$$

$$\text{FNO}_2 + \text{BF}_3 \longrightarrow [\text{NO}_2]^+[\text{BF}_4]^- \tag{15.76}$$

The main factor involved in the change from covalent to ionic halide is believed to be the enthalpy change accompanying the attachment of the halide ion to the halide acceptor.

The reaction of FNO with the powerful fluorinating agent IrF$_6$ results in the formation of the nitrogen(V) oxofluoride F$_3$NO. Above 520 K, F$_3$NO is in equilibrium with FNO and F$_2$. Resonance structures **15.25** and **15.26** can be written to depict the bonding, and the short N–O bond (116 pm) and long N–F bonds (143 pm) suggest that contributions from **15.26** (and similar structures) are important.

(15.25)   **(15.26)**

Reactions of F$_3$NO with strong F$^-$ acceptors such as BF$_3$ and AsF$_5$ yield the salts [F$_2$NO]$^+$[BF$_4$]$^-$ and [F$_2$NO]$^+$[AsF$_6$]$^-$. In the [AsF$_6$]$^-$ salt (see Box 15.5), the [F$_2$NO]$^+$ ion is planar (structure **15.27**), consistent with the formation of an N(2p)–O(2p) π-bond (diagram **15.28**).

(15.27)   **(15.28)**

Phosphorus halides

Phosphorus forms the halides PX$_3$ (X = F, Cl, Br and I) and PX$_5$ (X = F, Cl and Br); PI$_5$ is unknown. Most are made by direct combination of the elements with the product determined by which element is in excess. PF$_3$, however, must be made by reaction 15.77 and a convenient synthesis of PF$_5$ is from KPF$_6$ (see below). The halides are all

hydrolysed by water (e.g. eq. 15.78), although PF$_3$ reacts only slowly.

$$\text{PCl}_3 + \text{AsF}_3 \longrightarrow \text{PF}_3 + \text{AsCl}_3 \tag{15.77}$$

$$\text{PCl}_3 + 3\text{H}_2\text{O} \longrightarrow \text{H}_3\text{PO}_3 + 3\text{HCl} \tag{15.78}$$

X	a / pm	α / $^\circ$
F	156	96.5
Cl	204	100
Br	222	101
I	243	102

(15.29)

Each of the trihalides has a trigonal pyramidal structure, **15.29**. Phosphorus trifluoride is a very poisonous, colourless and odourless gas. It has the ability (like CO, see Section 24.2) to form complexes with metals and Lewis acids such as BH$_3$, and its toxicity arises from complex formation with haemoglobin. Protonation of PF$_3$ can be achieved when HF/SbF$_5$ is used as the acid (eq. 15.79), although an analogous reaction does not occur with AsF$_3$. [HPF$_3$][SbF$_6$]·HF is thermally unstable, but low-temperature structural data show that the tetrahedral [HPF$_3$]$^+$ ion has bond lengths of P–H = 122 and P–F = 149 pm.

$$\text{PF}_3 + \text{HF} + \text{SbF}_5(\text{excess})$$
$$\xrightarrow[\substack{\text{anhydrous HF, 77 K;}\\\text{crystallize at 213 K}}]{} [\text{HPF}_3][\text{SbF}_6]\cdot\text{HF} \tag{15.79}$$

The reaction of PF$_3$ with Me$_4$NF in MeCN gives [Me$_4$N][PF$_4$]. The [PF$_4$]$^-$ ion has a see-saw shape, consistent with the VSEPR model, i.e. the structure is derived from a trigonal bipyramid with a lone pair of electrons occupying one equatorial position. In solution, [PF$_4$]$^-$ is stereochemically non-rigid and the mechanism of F atom exchange is probably by Berry pseudo-rotation (see Fig. 4.24). When treated with an equimolar amount of water, [PF$_4$]$^-$ hydrolyses (eq. 15.80). With an excess of water, [HPF$_5$]$^-$ (**15.30**) hydrolyses to [HPO$_2$F]$^-$ (**15.31**). Further hydrolysis of [HPO$_2$F]$^-$ is not observed.

$$2[\text{PF}_4]^- + 2\text{H}_2\text{O} \xrightarrow{\text{MeCN, 293 K}} [\text{HPF}_5]^- + [\text{HPO}_2\text{F}]^- + 2\text{HF} \tag{15.80}$$

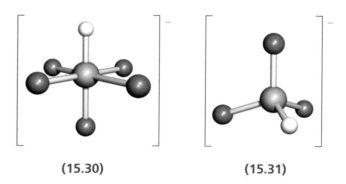

(15.30)   **(15.31)**

THEORY

Box 15.5 Crystal structure disorders: disorders involving F and O atoms

The technique of X-ray diffraction was introduced in Section 4.11, and throughout this book we have made use of the results of single-crystal structure determinations. Not all structure solutions are straightforward. Some involve disordering of atomic positions, a problem that, for example, made the elucidation of the structure of C_{60} difficult (see Section 14.4). Examples of disordered structures occur commonly in oxofluorides because the O and F atoms are similar in size and possess similar electronic properties. Thus, in a crystal containing molecules of an oxofluoride XF_xO_y, a given atomic position might be occupied by O in one molecule and by F in another molecule. The overall result is modelled by *fractional occupation* of each site by O and F. Fractional occupancies can lead to difficulties in determining true X–F and X–O bond lengths and true bond angles. The compound $[F_2NO]^+[AsF_6]^-$ represents a classic example of the problem. Although first prepared and characterized in 1969, its structure was not reported until 2001. The $[F_2NO]^+$ ions in crystalline $[F_2NO][AsF_6]$ are disordered such that the fluorine occupancy of each 'F' position is 78% and 77% respectively (rather than being 100%), and the fluorine occupancy of the 'O' position is 45% (rather than being 0%). The paper cited in the further reading below illustrates how the structural data can be treated so that meaningful N–O and N–F bond lengths and F–N–F and F–N–O bond angles are obtained. Crystalline

$[F_2NO][AsF_6]$ is composed of infinite chains of alternating cations and anions. There are close contacts between the N atom of each cation and the F atoms of adjacent $[AsF_6]^-$ ions as shown in the figure.

Colour code: N, blue; O, red; F, green; As, orange.

Further reading

A. Vij, X. Zhang and K.O. Christe (2001) *Inorg. Chem.*, vol. 40, p. 416 – 'Crystal structure of $F_2NO^+AsF_6^-$ and method for extracting meaningful geometries from oxygen/fluorine disordered crystal structures'.

For other examples of crystallographic disorders, see: Section 14.4, C_{60}; Section 14.9, C_3O_2; Section 15.13, $(NPF_2)_4$; Section 16.10, $Se_2S_2N_4$; Box 16.2, $[O_2]^-$; Fig. 23.4, Cp_2Be; Section 24.13, $(\eta^5\text{-Cp})_2Fe$.

Phosphorus trichloride is a colourless liquid (mp 179.5 K, bp 349 K) which fumes in moist air (eq. 15.78) and is toxic. Its reactions include those in scheme 15.81.

$$PCl_3 \begin{cases} \xrightarrow{O_2} POCl_3 \\ \xrightarrow{X_2 \ (X=halogen)} PCl_3X_2 \\ \xrightarrow{NH_3} P(NH_2)_3 \end{cases}$$

(15.81)

(15.32)

Single-crystal X-ray diffraction (at 109 K) data show that PF_5 has a trigonal bipyramidal structure, **15.32**. In solution, the molecule is fluxional on the NMR spectroscopic timescale and one doublet is observed in the ^{19}F NMR spectrum, i.e. all ^{19}F environments are equivalent and couple with the ^{31}P nucleus. This stereochemical non-rigidity is another example of Berry pseudo-rotation (see Fig. 4.24).

Electron diffraction data show that in the gas phase, PCl_5 has a molecular, trigonal bipyramidal structure ($P–Cl_{ax} = 214$, $P–Cl_{eq} = 202$ pm), provided that thermal dissociation into PCl_3 and Cl_2 is prevented by the presence of an excess of Cl_2. In the solid state, however, tetrahedral $[PCl_4]^+$ ($P–Cl = 197$ pm) and octahedral $[PCl_6]^-$ ($P–Cl = 208$ pm) ions are present, and the compound crystallizes with a CsCl-type structure (Fig. 6.17). In contrast, PBr_5 (which dissociates in the gas phase to PBr_3 and Br_2) crystallizes as $[PBr_4]^+Br^-$. The mixed halide PF_3Cl_2 is obtained as a gas (bp 280 K) from the reaction of PF_3 and Cl_2 and has a molecular structure with equatorial Cl atoms. However, when PCl_5 reacts with AsF_3 in $AsCl_3$ solution, the solid product $[PCl_4]^+[PF_6]^-$ (mp ≈ 403 K) is isolated. Solid PI_5 has not been isolated,[†] but the isolation of the salts $[PI_4]^+[AsF_6]^-$ (from the reaction of PI_3 and $[I_3]^+[AsF_6]^-$) and $[PI_4]^+[AlCl_4]^-$ (from the reaction between PI_3, ICl and $AlCl_3$) confirms the existence of the tetrahedral $[PI_4]^+$ ion. The reaction of PBr_3 with $[I_3][AsF_6]$ leads to a mixture of $[PBr_4][AsF_6]$, $[PBr_3I][AsF_6]$ and small amounts of $[PBr_2I_2][AsF_6]$.

[†] An estimate of $\Delta_f H^\circ([PI_4]^+I^-, s) = +180 \ kJ \ mol^{-1}$ has been made: see I. Tornieporth-Oetting *et al.* (1990) *J. Chem. Soc., Chem. Comm.*, p. 132.

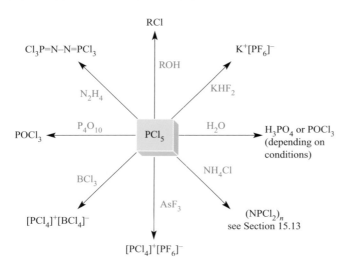

Fig. 15.12 Selected reactions of PCl_5.

Selective formation of $[PBr_4][AsF_6]$ can be achieved by treating PBr_3 with $[Br_3]^+[AsF_6]^-$.

(15.33)

Phosphorus pentafluoride is a strong Lewis acid and forms stable complexes with amines and ethers. The hexafluoridophosphate ion, $[PF_6]^-$, **15.33**, is made in aqueous solution by reacting H_3PO_4 with concentrated HF. $[PF_6]^-$ is isoelectronic and isostructural with $[SiF_6]^{2-}$ (see Fig. 14.17b). Salts such as $[NH_4][PF_6]$ are commercially available, and $[PF_6]^-$ is used to precipitate salts containing large organic or complex cations. Solid KPF_6 (prepared as in Fig. 15.12) decomposes on heating to give PF_5 and this route is a useful means of preparing PF_5. Phosphorus pentachloride is an important reagent, and is made industrially by the reaction of PCl_3 and Cl_2. Selected reactions are given in Fig. 15.12.

Of the lower halides P_2X_4, the most important is the red, crystalline P_2I_4 (mp 398 K), made by reacting white phosphorus with I_2 in CS_2. In the solid state, molecules of P_2I_4 adopt a *trans* (staggered) conformation (see Fig. 15.4). In many of its reactions, P_2I_4 undergoes P–P bond fission, e.g. hydrolysis of P_2I_4 leads to a mixture of mononuclear products including PH_3, H_3PO_4, H_3PO_3 and H_3PO_2.

(15.34)

Salts of $[P_2I_5]^+$ (**15.34**) can be obtained according to scheme 15.82. However, the $[P_2I_5]^+$ ion exists only in the solid state. The ^{31}P NMR spectra of CS_2 solutions of dissolved samples show a singlet at $\delta +178$ ppm, consistent with the presence of PI_3 rather than $[P_2I_5]^+$. In contrast, solution ^{31}P NMR spectra have been obtained for $[P_2I_5]^+$ in the presence of the $[Al\{OC(CF_3)_3\}_4]^-$ anion (see exercise 1 after worked example 15.7).

$$P_2I_4 + I_2 + EI_3 \xrightarrow{CS_2}$$
$$[P_2I_5]^+ [EI_4]^- \qquad\qquad (15.82)$$
$$2PI_3 + EI_3 \xrightarrow{CS_2} \qquad\qquad E = Al, Ga, In$$

Worked example 15.7 ^{31}P NMR spectroscopy of phosphorus halides

The $[P_3I_6]^+$ ion is formed in the reaction of P_2I_4 with PI_3 and $Ag[Al\{OC(CF_3)_3\}_4]\cdot CH_2Cl_2$. The solution ^{31}P NMR spectrum shows a triplet and a doublet with relative integrals $1:2$ ($J = 385$ Hz). Suggest a structure for $[P_3I_6]^+$ that is consistent with the NMR spectroscopic data.

First look up the spin quantum number and natural abundance of ^{31}P (Tables 4.3 or 15.2): $I = \frac{1}{2}$, 100%.

Adjacent ^{31}P nuclei will couple, and the presence of a triplet and doublet in the spectrum is consistent with a P–P–P backbone in $[P_3I_6]^+$. The terminal P atoms must be equivalent and therefore the following structure can be proposed:

Self-study exercises

1. Rationalize why the ^{31}P NMR spectrum of $[P_2I_5]^+$ contains two, equal-intensity doublets ($J = 320$ Hz).

2. Prolonged reaction between PI_3, $PSCl_3$ and powdered Zn results in the formation of P_3I_5 as one of the products. The solution ^{31}P NMR spectrum of P_3I_5 shows a doublet at $\delta +98$ ppm and a triplet at $\delta +102$ ppm. These values compare with $\delta +106$ ppm for P_2I_4. Suggest a structure for P_3I_5 and give reasoning for your answer.

   [*Ans.* See K.B. Dillon *et al.* (2001) *Inorg. Chim. Acta*, vol. 320, p. 172]

3. The solution ^{31}P NMR spectrum of $[HPF_5]^-$ consists of a 20-line multiplet from which three coupling constants can be obtained. Explain the origins of these spin–spin coupling constants in terms of the structure of $[HPF_5]^-$.

See also end-of-chapter problems 4.40, 15.32a and 15.35a.

Phosphoryl trichloride, POCl₃

(15.35)

Of the phosphorus oxohalides, the most important is $POCl_3$, prepared by reaction of PCl_3 with O_2. Phosphoryl trichloride is a colourless, fuming liquid (mp 275 K, bp 378 K), which is readily hydrolysed by water, liberating HCl. The vapour contains discrete molecules (**15.35**). Some of the many uses of $POCl_3$ are as a phosphorylating and chlorinating agent, and as a reagent in the preparation of phosphate esters. An example of its use is the basis for end-of-chapter problem 15.39b.

Arsenic and antimony halides

Arsenic forms the halides AsX_3 (X = F, Cl, Br, I) and AsX_5 (X = F, Cl). The trihalides $AsCl_3$, $AsBr_3$ and AsI_3 can be made by direct combination of the elements, and reaction 15.83 is another route to $AsCl_3$. Reaction 15.84 is used to prepare AsF_3 (mp 267 K, bp 330 K) despite the fact that AsF_3 (like the other trihalides) is hydrolysed by water; the H_2O formed in the reaction is removed with excess H_2SO_4. Glass containers are not practical for AsF_3 as it reacts with silica in the presence of moisture.

$$As_2O_3 + 6HCl \xrightarrow{\quad} 2AsCl_3 + 3H_2O \qquad (15.83)$$
$$\text{conc}$$

$$As_2O_3 + 3H_2SO_4 + 3CaF_2$$
$$\text{conc}$$
$$\xrightarrow{\quad} 2AsF_3 + 3CaSO_4 + 3H_2O \qquad (15.84)$$

In the solid, liquid and gas states, AsF_3 and $AsCl_3$ have molecular, trigonal pyramidal structures. With an appropriate reagent, AsF_3 may act as either an F^- donor or acceptor (eqs. 15.85 and 15.86). Compare this with the behaviours of BrF_3 (Section 9.10) and $AsCl_3$ (eq. 15.87) which finds some use as a non-aqueous solvent.

$$AsF_3 + KF \xrightarrow{\quad} K^+[AsF_4]^- \qquad (15.85)$$

$$AsF_3 + SbF_5 \xrightarrow{\quad} [AsF_2]^+[SbF_6]^- \qquad (15.86)$$

$$2AsCl_3 \rightleftharpoons [AsCl_2]^+ + [AsCl_4]^- \qquad (15.87)$$

The reaction of $AsCl_3$ with Me_2NH and excess HCl in aqueous solution gives $[Me_2NH_2]_3[As_2Cl_9]$ containing anion **15.36**.[†]

(15.36)

Salts containing the $[AsX_4]^+$ (X = F, Cl, Br, I) ions include $[AsF_4][PtF_6]$ and $[AsCl_4][AsF_6]$ which are stable compounds, and $[AsBr_4][AsF_6]$ and $[AsI_4][AlCl_4]$, both of which are unstable. By using the weakly coordinating anions $[AsF(OTeF_5)_5]^-$ and $[As(OTeF_5)_6]^-$ (for example, in redox reaction 15.88), it is possible to stabilize $[AsBr_4]^+$ in the solid state.

$$AsBr_3 + BrOTeF_5 + \underset{[OTeF_5]^- \text{ acceptor}}{As(OTeF_5)_5} \xrightarrow{\quad} [AsBr_4]^+[As(OTeF_5)_6]^-$$

$$(15.88)$$

The only stable pentahalide of arsenic is AsF_5 (prepared by reaction 15.89), although $AsCl_5$ can be made at 173 K by treating $AsCl_3$ with Cl_2 under UV radiation. X-ray diffraction data for $AsCl_5$ at 150 K confirm the presence of discrete, trigonal bipyramidal molecules in the solid state ($As-Cl_{ax}$ = 221 pm, $As-Cl_{eq}$ = 211 pm). If, during the preparation of $AsCl_5$, H_2O and HCl are present, the isolated, crystalline products are $[H_5O_2]_5[AsCl_6]Cl_4$ and $[H_5O_2][AsCl_6]\cdot AsOCl_3$. These are stable below 253 K and contain hydrogen-bonded $[H_5O_2]^+$ and $[AsCl_6]^-$ ions. $[H_5O_2][AsCl_6]\cdot AsOCl_3$ is the result of cocrystallization of $[H_5O_2][AsCl_6]$ and $AsOCl_3$. This provides an example of *monomeric*, tetrahedral $AsOCl_3$, whereas solid $AsOCl_3$ (made by reacting $AsCl_3$ and O_3 at 195 K) contains the dimers **15.37**; each As atom is in a trigonal bipyramidal environment.

(15.37)

At 298 K, AsF_5 is a colourless gas and has a molecular structure similar to **15.32**.

$$AsF_3 + 2SbF_5 + Br_2 \xrightarrow{\quad} AsF_5 + 2SbBrF_4 \qquad (15.89)$$

[†]For comments on the effect that cation size may have on the solid state structure of $[E_2X_9]^{3-}$ (E = As, Sb, Bi; X = Cl, Br), see: M. Wojtaś *et al.* (2002) *Z. Anorg. Allg. Chem.*, vol. 628, p. 516.

AsF$_5$ is a strong F$^-$ acceptor (e.g. reactions 15.69, 15.70 and 15.74) and many complexes containing the octahedral [AsF$_6$]$^-$ ion are known. One interesting reaction of AsF$_5$ is with metallic Bi to give [Bi$_5$][AsF$_6$]$_3$ which contains the trigonal bipyramidal cluster [Bi$_5$]$^{3+}$. Although [AsF$_6$]$^-$ is the usual species formed when AsF$_5$ accepts F$^-$, the [As$_2$F$_{11}$]$^-$ adduct has also been isolated. X-ray diffraction data for [(MeS)$_2$CSH]$^+$[As$_2$F$_{11}$]$^-$ (formed from (MeS)$_2$CS, HF and AsF$_5$) confirm that [As$_2$F$_{11}$]$^-$ is structurally like [Sb$_2$F$_{11}$]$^-$ (Fig. 15.13b).

Antimony trihalides are low melting solids, and although these contain trigonal pyramidal molecules, each Sb centre has additional, longer range, intermolecular Sb····X interactions. The trifluoride and trichloride are prepared by reacting Sb$_2$O$_3$ with concentrated HF and HCl, respectively. SbF$_3$ is a widely used fluorinating agent, e.g. converting B$_2$Cl$_4$ to B$_2$F$_4$ (Section 13.6), CHCl$_3$ to CHF$_2$Cl (eq. 14.45), COCl$_2$ to COClF and COF$_2$ (Section 14.8), SiCl$_4$ to SiF$_4$ (Section 14.8) and SOCl$_2$ to SOF$_2$ (Section 16.7). However, reactions may be complicated by SbF$_3$ acting as an oxidizing agent (eq. 15.90). Reactions between SbF$_3$ and MF (M = alkali metal) give salts which include K$_2$SbF$_5$ (containing [SbF$_5$]$^{2-}$, **15.38**), KSb$_2$F$_7$ (with discrete SbF$_3$ and [SbF$_4$]$^-$, **15.39**), KSbF$_4$ (in which the anion is [Sb$_4$F$_{16}$]$^{4-}$, **15.40**) and CsSb$_2$F$_7$ (containing [Sb$_2$F$_7$]$^-$, **15.41**).

$$3C_6H_5PCl_2 + 4SbF_3 \longrightarrow 3C_6H_5PF_4 + 2SbCl_3 + 2Sb \quad (15.90)$$

(15.38)

(15.39)

(15.40)

(15.41)

Antimony pentafluoride (mp 280 K, bp 422 K) is prepared from SbF$_3$ and F$_2$, or by reaction 15.91. In the solid state, SbF$_5$ is tetrameric (Fig. 15.13a) and the presence of Sb–F–Sb bridges accounts for the very high viscosity of the liquid. Antimony pentachloride (mp 276 K, bp 352 K)

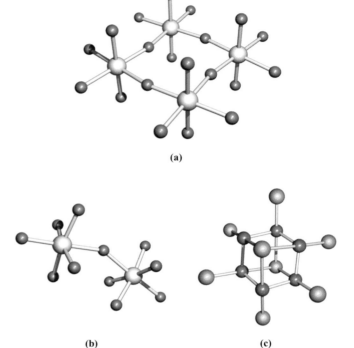

(a)

(b)　　　　　　　　**(c)**

Fig. 15.13 The solid state structures of (a) {SbF$_5$}$_4$, (b) [Sb$_2$F$_{11}$]$^-$ (X-ray diffraction) in the *tert*-butyl salt [S. Hollenstein *et al.* (1993) *J. Am. Chem. Soc.*, vol. 115, p. 7240] and (c) [As$_6$I$_8$]$^{2-}$ (X-ray diffraction) in [{MeC(CH$_2$PPh$_2$)$_3$}NiI]$_2$[As$_6$I$_8$] [P. Zanello *et al.* (1990) *J. Chem. Soc., Dalton Trans.*, p. 3761]. The bridge Sb–F bonds in {SbF$_5$}$_4$ and [Sb$_2$F$_{11}$]$^-$ are ≈15 pm longer than the terminal bonds. Colour code: Sb, silver; As, red; F, green; I, yellow.

is prepared from the elements, or by reaction of Cl$_2$ with SbCl$_3$. Liquid SbCl$_5$ contains discrete trigonal bipyramidal molecules, and these are also present in the solid between 219 K and the melting point. Like PCl$_5$ and AsCl$_5$, the axial bonds in SbCl$_5$ are longer than the equatorial bonds (233 and 227 pm for the solid at 243 K). Below 219 K, the solid undergoes a reversible change involving dimerization of the SbCl$_5$ molecules (diagram **15.42**).

(15.42)

$$SbCl_5 + 5HF \longrightarrow SbF_5 + 5HCl \quad (15.91)$$

We have already illustrated the role of SbF$_5$ as an extremely powerful fluoride acceptor (e.g. reactions 9.44, 9.55, 15.71, 15.72 and 15.86). Similarly, SbCl$_5$ is one of the strongest chloride acceptors known (e.g. reactions 15.75 and 15.92). Reactions of SbF$_5$ and SbCl$_5$ with alkali metal fluorides and chlorides yield compounds of the type $M[SbF_6]$ and $M[SbCl_6]$.

$$SbCl_5 + AlCl_3 \longrightarrow [AlCl_2]^+[SbCl_6]^- \qquad (15.92)$$

Whereas the addition of Cl^- to $SbCl_5$ invariably gives $[SbCl_6]^-$, acceptance of F^- by SbF_5 may be accompanied by further association by the formation of $Sb-F-Sb$ bridges. Thus, products may contain $[SbF_6]^-$, $[Sb_2F_{11}]^-$ (Fig. 15.13b) or $[Sb_3F_{16}]^-$ in which each Sb centre is octahedrally sited. The strength with which SbF_5 can accept F^- has led to the isolation of salts of some unusual cations, including $[O_2]^+$, $[XeF]^+$, $[Br_2]^+$, $[ClF_2]^+$ and $[NF_4]^+$. Heating $Cs[SbF_6]$ and CsF (molar ratio 1:2) at 573 K for 45 h produces $Cs_2[SbF_7]$. Vibrational spectroscopic and theoretical results are consistent with the $[SbF_7]^{2-}$ ion having a pentagonal bipyramidal structure.

When $SbCl_3$ reacts with Cl_2 in the presence of $CsCl$, dark blue Cs_2SbCl_6 precipitates. Black $[NH_4]_2[SbBr_6]$ can be similarly obtained. Since these compounds are diamagnetic, they cannot contain Sb(IV) and are, in fact, mixed oxidation state species containing $[SbX_6]^{3-}$ and $[SbX_6]^-$. The dark colours of the compounds arise from absorption of light associated with electron transfer between the two anions. The solid state structures of Cs_2SbCl_6 and $[NH_4]_2[SbBr_6]$ show similar characteristics to one another, e.g. in $[NH_4]_2[SbBr_6]$, two distinct octahedral anions are present, $[SbBr_6]^-$ (Sb−Br = 256 pm) and $[SbBr_6]^{3-}$ (Sb−Br = 279 pm); the lone pair in the Sb(III) species is stereochemically inactive.

A number of high nuclearity halo-anions of As and Sb are known which contain doubly and triply bridging X^-, e.g. $[As_6I_8]^{2-}$ (Fig. 15.13c), $[As_8I_{28}]^{4-}$, $[Sb_5I_{18}]^{3-}$ and $[Sb_6I_{22}]^{4-}$.

To what point group do the $[SbBr_6]^-$ and $[SbBr_6]^{3-}$ ions belong if they possess regular octahedral structures? Explain why one of these ions possesses a stereochemically inactive lone pair of electrons, while the other ion has no Sb-centred lone pair.

Bismuth halides

The trihalides BiF_3, $BiCl_3$, $BiBr_3$ and BiI_3 are all well characterized, but BiF_5 is the only Bi(V) halide known. All are solids at 298 K. In the vapour phase, the trihalides have molecular (trigonal pyramidal) structures. In the solid state, β-BiF_3 contains 9-coordinate Bi(III) centres. Solid $BiCl_3$ and $BiBr_3$ have molecular structures but with an additional five long $Bi \cdots X$ contacts, and in BiI_3, the Bi atoms occupy octahedral sites in an hcp array of I atoms. The trihalides can be formed by combination of the elements at high

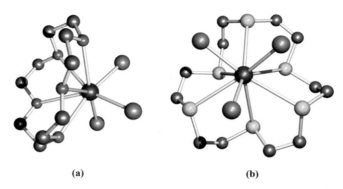

Fig. 15.14 The structures (X-ray diffraction) of (a) [BiCl₃(15-crown-5)] [N.W. Alcock *et al.* (1993) *Acta Crystallogr.*, *Sect. B*, vol. 49, p. 507] and (b) [BiCl₃L] where L = 1,4,7,10,13,16-hexathiacyclooctadecane [G.R. Willey *et al.* (1992) *J. Chem. Soc., Dalton Trans.*, p. 1339]. Note the high coordination numbers of the Bi(III) centres. Hydrogen atoms have been omitted. Colour code: Bi, blue; O, red; S, yellow; Cl, green; C, grey.

temperature. Each trihalide is hydrolysed by water to give BiOX, which are insoluble compounds with layer structures. The reaction of BiF_3 with F_2 at 880 K yields BiF_5 which is a powerful fluorinating agent. Heating BiF_5 with an excess of MF (M = Na, K, Rb or Cs) at 503–583 K for four days produces $M_2[BiF_7]$; the reactions are carried out under a low pressure of F_2 to prevent reduction of Bi(V) to Bi(III). Treatment of BiF_5 with an excess of FNO at 195 K yields $[NO]_2[BiF_7]$, but this is thermally unstable and forms $[NO][BiF_6]$ when warmed to room temperature. The $[BiF_7]^{2-}$ ion has been assigned a pentagonal bipyramidal structure on the basis of vibrational spectroscopic and theoretical data.

The trihalides are Lewis acids and form donor–acceptor complexes with a number of ethers, e.g. *fac*-[BiCl₃(THF)₃], *mer*-[BiI₃(py)₃] (py = pyridine), *cis*-[BiI₄(py)₂]⁻, [BiCl₃(py)₄] (**15.43**) and the macrocyclic ligand complexes shown in Fig. 15.14. Reactions with halide ions give species such as $[BiCl_5]^{2-}$ (square pyramidal), $[BiBr_6]^{3-}$ (octahedral), $[Bi_2Cl_8]^{2-}$ (**15.44**), $[Bi_2I_8]^{2-}$ (structurally similar to **15.44**), and $[Bi_2I_9]^{3-}$ (**15.45**). Bismuth(III) also forms some higher nuclearity halide complexes, e.g. $[Bi_4Cl_{16}]^{4-}$, as well as the polymeric species $[\{BiX_4\}_n]^{n-}$ and $[\{BiX_5\}_n]^{2n-}$. In each case, the Bi atoms are octahedrally sited.

py is *N*-bonded

(15.43)

(15.44)

(15.45)

Worked example 15.8   Redox chemistry of group 15 metal halides

In reaction 15.88, which species undergo oxidation and which reduction? Confirm that the equation balances in terms of changes in oxidation states.

The reaction to be considered is:

$$AsBr_3 + BrOTeF_5 + As(OTeF_5)_5 \longrightarrow [AsBr_4]^+[As(OTeF_5)_6]^-$$

Oxidation states:	$AsBr_3$	As, +3; Br, −1
	$BrOTeF_5$	Br, +1; Te, +6
	$As(OTeF_5)_5$	As, +5; Te, +6
	$[AsBr_4]^+$	As, +5; Br, −1
	$[As(OTeF_5)_6]^-$	As, +5; Te, +6

The redox chemistry involves As and Br. The As in $AsBr_3$ is oxidized on going to $[AsBr_4]^+$, while Br in $BrOTeF_5$ is reduced on going to $[AsBr_4]^+$.

Oxidation: As(+3) to As(+5)   Change in oxidation state = +2

Reduction: Br(+1) to Br(−1)   Change in oxidation state = −2

Therefore the equation balances in terms of oxidation state changes.

Self-study exercises

1. In reaction 15.59, which elements are oxidized and which reduced? Confirm that the reaction balances in terms of changes in oxidation states.
[*Ans.* N, oxidized; F, reduced]

2. Which elements undergo redox changes in reaction 15.62? Confirm that the equation balances in terms of the oxidation state changes.
[*Ans.* N, reduced; half of the Cl, oxidized]

3. Are reactions 15.74, 15.75 and 15.76 redox reactions? Confirm your answer by determining the oxidation states of the N atoms in the reactants and products in each equation. [*Ans.* Non-redox]

4. Confirm that reaction 15.90 is a redox process, and that the equation balances with respect to changes in oxidation states for the appropriate elements.

15.8 Oxides of nitrogen

As in group 14, the first element of group 15 stands apart in forming oxides in which (p–p)π-bonding is important. Table 15.6 lists selected properties of nitrogen oxides, excluding NO_3 which is an unstable radical. NO_2 exists in equilibrium with N_2O_4.

Dinitrogen monoxide, N_2O

Dinitrogen monoxide (Table 15.6) is usually prepared by decomposition of solid ammonium nitrate (eq. 15.93,

Table 15.6 Selected data for the oxides of nitrogen.

	N_2O	NO	N_2O_3	NO_2	N_2O_4	N_2O_5
Name	Dinitrogen monoxide[†]	Nitrogen monoxide[†]	Dinitrogen trioxide	Nitrogen dioxide	Dinitrogen tetraoxide	Dinitrogen pentaoxide
Melting point / K	182	109	173	–	262	303
Boiling point / K	185	121	277 dec.	–	294	305 sublimes
Physical appearance	Colourless gas	Colourless gas	Blue solid or liquid	Brown gas	Colourless solid or liquid, but see text	Colourless solid, stable below 273 K
$\Delta_f H^\circ(298\,K)/kJ\,mol^{-1}$	82.1 (g)	90.2 (g)	50.3 (l) 83.7 (g)	33.2 (g)	−19.5 (l) 9.2 (g)	−43.1 (s)
Dipole moment of gas-phase molecule / D	0.16	0.16	–	0.315	–	–
Magnetic properties	Diamagnetic	Paramagnetic	Diamagnetic	Paramagnetic	Diamagnetic	Diamagnetic

[†] N_2O and NO are commonly called nitrous oxide and nitric oxide, respectively.

compare reaction 15.8) but the aqueous solution reaction 15.94 is useful for obtaining a purer product. For further detail on the oxidation of NH_2OH to N_2O, see Section 15.5.

$$NH_4NO_3 \xrightarrow{450-520\,K} N_2O + 2H_2O \qquad (15.93)$$

$$NH_2OH + HNO_2 \longrightarrow N_2O + 2H_2O \qquad (15.94)$$

Dinitrogen monoxide has a faint, sweet odour. It dissolves in water to give a neutral solution, but does not react to any significant extent. The position of equilibrium 15.95 is far to the left.

$$N_2O + H_2O \rightleftharpoons H_2N_2O_2 \qquad (15.95)$$

$$\underset{113\text{ pm}\quad 119\text{ pm}}{\overset{-}{N}\!\!=\!\!\overset{+}{N}\!\!=\!\!O} \qquad\qquad \overset{+}{N}\!\!\equiv\!\!\overset{-}{N}\!\!-\!\!O$$

(15.46)   **(15.47)**

Dinitrogen monoxide is a non-toxic gas which is fairly unreactive at 298 K. The N_2O molecule is linear, and the bonding can be represented as in structure **15.46**, although the bond lengths suggest some contribution from resonance structure **15.47**. In the past, N_2O ('laughing gas') was widely used as an anaesthetic, but possible side effects coupled with the availability of a range of alternative anaesthetics have led to a significant decline in its use.[†] Nitrous oxide remains in use as a propellant in whipped cream dispensers. Its reactivity is higher at elevated temperatures; N_2O supports combustion, and reacts with $NaNH_2$ at 460 K (eq. 15.50). This reaction is used commercially to prepare NaN_3, a precursor to other azides such as $Pb(N_3)_2$ which is used as a detonator.

Nitrogen monoxide, NO

Nitrogen monoxide (Table 15.6 and Box 15.6) is made industrially from NH_3 (eq. 15.96), and on a laboratory scale by reducing HNO_3 or nitrites (e.g. KNO_2) in the presence of H_2SO_4 in aqueous solution (reaction 15.97 or 15.98).

$$4NH_3 + 5O_2 \xrightarrow{1300\,K,\ Pt\ catalyst} 4NO + 6H_2O \qquad (15.96)$$

$$[NO_3]^- + 3Fe^{2+} + 4H^+ \longrightarrow NO + 3Fe^{3+} + 2H_2O \qquad (15.97)$$

$$[NO_2]^- + I^- + 2H^+ \longrightarrow NO + H_2O + \tfrac{1}{2}I_2 \qquad (15.98)$$

Reaction 15.97 is the basis of the brown ring test for $[NO_3]^-$. After the addition of an equal volume of aqueous $FeSO_4$ to the test solution, cold concentrated H_2SO_4 is added slowly to form a separate, lower layer. If $[NO_3]^-$ is present, NO is liberated, and a brown ring forms between the two layers. The brown colour is due to the formation of $[Fe(NO)(OH_2)_5]^{2+}$, an example of one of many *nitrosyl complexes* in which NO acts as a ligand (see Sections 20.4 and 24.2). The IR spectrum of $[Fe(NO)(OH_2)_5]^{2+}$ shows an absorption at $1810\,cm^{-1}$ assigned to $\nu(NO)$ and is consistent with the formulation of an $[NO]^-$ ligand bound to Fe(III) rather than $[NO]^+$ coordinated to Fe(I). The presence of Fe(III) is also supported by Mössbauer spectroscopic data. The reaction between $[Fe(OH_2)_6]^{2+}$ and NO is discussed again in Box 26.1. The compound $[Et_4N]_5[NO][V_{12}O_{32}]$ is unusual because it contains a non-coordinated $[NO]^-$ ion. The $[V_{12}O_{32}]^{4-}$ ion (see Section 21.6) has a 'bowl-shaped' structure and acts as a 'host', trapping the $[NO]^-$ ion as a 'guest' within the cage. There are only weak van der Waals interactions between the host and guest.

$$\overset{\bullet}{N}\!\!=\!\!O$$
$$115\text{ pm}$$

(15.48)

Structure **15.48** shows that NO is a radical. Unlike NO_2, it does not dimerize unless cooled to low temperature under high pressure. In the diamagnetic solid, a dimer with a long N–N bond (218 pm) is present. It is probable that a dimer is an intermediate in reactions 15.99, for which reaction rates decrease with increasing temperature.

$$\left.\begin{array}{ll} 2NO + Cl_2 \longrightarrow 2ClNO & \text{Rate} \propto (P_{NO})^2(P_{Cl_2}) \\[4pt] 2NO + O_2 \longrightarrow 2NO_2 & \text{Rate} \propto (P_{NO})^2(P_{O_2}) \end{array}\right\} \quad (15.99)$$

The reaction of NO with O_2 is important in the manufacture of nitric acid (Section 15.9), but NO can also be oxidized directly to HNO_3 by acidified $[MnO_4]^-$. The reduction of NO depends on the reducing agent, e.g. with SO_2, the product is N_2O, but reduction with tin and acid gives NH_2OH. Although NO is thermodynamically unstable with respect to its elements (Table 15.6), it does not decompose at an appreciable rate below 1270 K, and so does not support combustion well. The positive value of Δ_fH° means that at high temperatures, the formation of NO is favoured, and this is significant during combustion of motor and aircraft fuels where NO is one of several oxides formed. These oxides are collectively described by NO_x (see Box 15.7) and contribute to the formation of smogs over large cities.

A reaction of NO that has been known since the early 1800s is that with sulfite ion to form $[O_3SNONO]^{2-}$. One resonance structure for this ion is shown in diagram **15.49**. The bond lengths for the K^+ salt are consistent with an S–N single bond, and double bond character for the N–N bond, but they also suggest some degree of multiple bond character for the N–O bonds. It is proposed that $[O_3SNONO]^{2-}$ forms by sequential addition of NO to

---

[†] See: U.R. Jahn and E. Berendes (2005) *Best Practice & Research Clinical Anaesthesiology*, vol. 19, p. 391 – 'Nitrous oxide – an outdated anaesthetic'; R. D. Sanders, J. Weimann and M. Maze (2008) *Anesthesiology*, vol. 109, p. 707 – 'Biologic effects of nitrous oxide'.

Research into the role played by NO in biological systems is an active area, and in 1992, *Science* named NO 'Molecule of the Year'. The 1998 Nobel Prize in Physiology or Medicine was awarded to Robert F. Furchgott, Louis J. Ignarro and Ferid Murad for 'their discoveries concerning nitric oxide as a signalling molecule in the cardiovascular system' (http://www.nobel.se/medicine/laureates/1998/press.html).

Nitrogen monoxide is synthesized *in vivo* from the amino acid L-arginine, and the reaction is catalysed by the haem-containing NO synthase (NOS) enzymes which are related to cytochromes P450 (see Section 29.3). There are three forms of NO synthases: neuronal (nNOS), inducible (iNOS) and endothelial (eNOS). nNOS and eNOS are classed as low-output enzymes and their activities are regulated by Ca^{2+} ions in the presence of the calcium-binding protein, calmodulin. In contrast, iNOS is a high-output enzyme and its activity does not depend on the presence of Ca^{2+} ions. L-Arginine is converted to NO and L-citrulline (see below) in a two-step reaction. This involves a 5-electron oxidation process, and requires O_2 in addition to the cofactor NADPH (nicotinamide adenine dinucleotide phosphate).

L-Arginine   L-Citrulline

The enzymic receptor for NO in the body (*guanylyl cyclase*) also contains a haem unit. In addition to these haem groups, NO may coordinate to the haem iron atom in myoglobin and haemoglobin (see Section 29.3). The structure of iron(II)

horse heart myoglobin with NO bound to the haem-iron centre has been determined by single crystal X-ray diffraction. The structure is illustrated in the left-hand figure below, with the protein shown in a ribbon representation and the haem-unit in a stick representation. The secondary structure of the protein is colour-coded, with α-helices shown in red, turns in green, and coils in silver-grey. The right-hand figure below is an enlargement of the active site. The iron(II) centre is octahedrally coordinated. The NO molecule is bound *trans* to the histidine residue (see Table 29.2) that connects the active site to the protein backbone. The structural data confirm that the Fe–N–O unit is non-linear (angle Fe–N–O = 147°).

The small molecular dimensions of NO mean that it readily diffuses through cell walls. It acts as a messenger molecule in biological systems, and appears to have an active role in mammalian functions such as the regulation of blood pressure, muscle relaxation and neuro-transmission. A remarkable property exhibited by NO is that it appears to be cytotoxic (i.e. it is able to specifically destroy particular cells) and it affects the ability of the body's immune system to kill tumour cells.

Further reading

L.E. Goodrich, F. Paulat, V.K.K. Praneeth and N. Lehnert (2010) *Inorg. Chem.*, vol. 49, p. 6293 – 'Electronic structure of heme-nitrosyls and its significance for nitric oxide reactivity, sensing, transport, and toxicity in biological systems'.

J.A. McCleverty (2004) *Chem. Rev.*, vol. 104, p. 403 – 'Chemistry of nitric oxide relevant to biology'.

R.J.P. Williams (1995) *Chem. Soc. Rev.*, vol. 24, p. 77 – 'Nitric oxide in biology: its role as a ligand'.

Reviews by the winners of the 1998 Nobel Prize for Physiology or Medicine: *Angew. Chem. Int. Ed.* (1999) vol. 38, pp. 1856, 1870, 1882.

See also Box 29.2: How the blood-sucking *Rhodnius prolixus* utilizes NO.

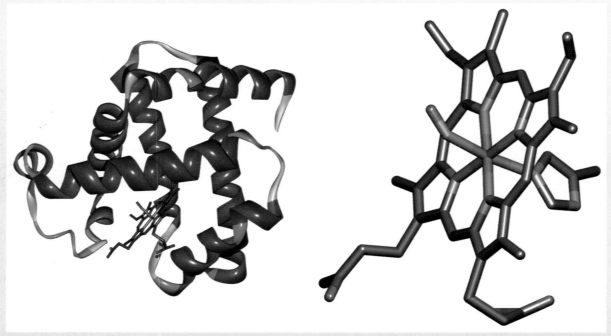

The structure of iron(II) horse heart myoglobin with bound NO, and an enlargement of the active site. Colour code for the haem-unit: Fe, green; N, blue; O, red; C, grey. [Data: D.M. Copeland *et al.* (2003) *Proteins: Structure, Function and Genetics*, vol. 53, p. 182.]

ENVIRONMENT

Box 15.7 NO$_x$: tropospheric pollutant

'NO$_x$' (pronounced 'NOX') is a combination of nitrogen oxides arising from both natural (soil emissions and lightning) and man-made sources. The major man-made culprits are vehicle and aircraft exhausts and large industrial power (e.g. electricity-generating) plants (see Fig. 25.17). The term NO$_x$ usually refers to a combination of NO and NO$_2$, and excludes N$_2$O which is naturally produced by denitrifying bacteria (see Box 15.9) and is formed in the manufacture of adipic acid. Adipic acid is one of the reagents in the industrial synthesis of Nylon-66, and is produced by the oxidation of cyclohexanol and cyclohexanone using nitric acid:

Because of the large scale of adipic acid manufacture, emissions of N$_2$O from this source are significant. Since the 1990s, measures have been put in place to prevent N$_2$O reaching the atmosphere. Both thermal destruction and catalytic (e.g. CuO/Al$_2$O$_3$) decomposition convert N$_2$O to N$_2$ and O$_2$.

In the closing years of the 20th century, a better awareness of our environment led to the regulation of exhaust emissions. Regulated emissions are CO, hydrocarbons and NO$_x$, as well as particulate matter. Catalytic converters (see Section 25.8) are now routinely used to reduce the emissions of CO, hydrocarbons and NO$_x$ from the combustion of transport fuels. A combination of Rh, Pt and Pd metals is required in a catalytic converter to catalyse the oxidation of CO and hydrocarbons, and the reduction of NO$_x$, thereby producing acceptable emissions of CO$_2$, H$_2$O and N$_2$. Rhodium metal catalyses the following reactions:

$$2NO + 2CO \longrightarrow 2CO_2 + N_2$$

$$2NO + 2H_2 \longrightarrow N_2 + 2H_2O$$

The troposphere is the region immediately above the Earth's surface and varies in depth from around 10 to 15 km. The region that follows the troposphere is the stratosphere (up to 50 km above the Earth's surface). Of the nitrogen oxides NO, NO$_2$ and N$_2$O, dinitrogen oxide is moderately unreactive and has a relatively long lifetime in the troposphere. Once N$_2$O crosses into the stratosphere, it undergoes photolytic decomposition to N$_2$ and O$_2$. The effects of NO$_x$ (i.e. NO and NO$_2$) in the troposphere are to increase HO$^{\bullet}$ and O$_3$ concentrations. While O$_3$ in the upper atmosphere acts as a barrier against UV radiation, increased levels at lower altitudes are detrimental to human lung tissue. Photochemical smog which forms over large cities consists mainly of O$_3$ and is associated with volatile

organic compounds (VOCs) and with NO$_x$ emissions from motor vehicle exhausts. Ozone formation in the troposphere usually begins with a reaction between CO or a VOC with an HO$^{\bullet}$ radical and atmospheric O$_2$, for example:

$$CO + HO^{\bullet} + O_2 \longrightarrow HO_2^{\bullet} + CO_2$$

This is followed by a sequence of radical reactions (in which we explicitly show NO and NO$_2$ as radicals) resulting in the production of O$_3^{\bullet}$.

$$HO_2^{\bullet} + ON^{\bullet} \longrightarrow HO^{\bullet} + NO_2^{\bullet}$$
$$NO_2^{\bullet} \xrightarrow{h\nu} ON^{\bullet} + O$$
$$O_2 + O \longrightarrow O_3$$

Notice that overall, neither NO nor HO$^{\bullet}$ is destroyed, and therefore each radical can re-enter the reaction chain. The photograph below shows the detrimental effects of these reactions in the production of photochemical smog over Santiago, Chile. Santiago, Mexico City and Sao Paulo are the most polluted cities in Latin America.

Smog over Santiago, Chile.

Further reading

M.G. Lawrence and P.J. Crutzen (1999) *Nature*, vol. 402, p. 167 – 'Influence of NO$_x$ emissions from ships on tropospheric photochemistry and climate'.

S. Sillman (2004) in *Treatise on Geochemistry*, eds H.D. Holland and K.K. Turekian, Elsevier, Oxford, vol. 9, p. 407 – 'Tropospheric ozone and photochemical smog'.

M.H. Thiemens and W.C. Trogler (1991) *Science*, vol. 251, p. 932 – 'Nylon production: an unknown source of atmospheric nitrous oxide'.

R.P. Wayne (2000) *Chemistry of Atmospheres*, Oxford University Press, Oxford.

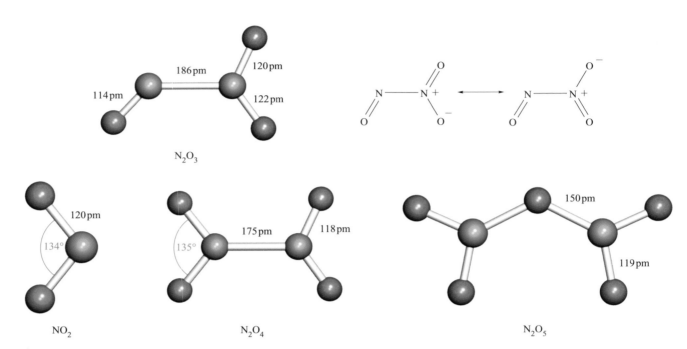

Fig. 15.15 The molecular structures of N_2O_3 (with resonance structures), NO_2, N_2O_4 and N_2O_5; molecules of N_2O_3, N_2O_4 and N_2O_5 are planar. The N–N bonds in N_2O_3 and N_2O_4 are particularly long (compare with N_2H_4, Fig. 15.4). Colour code: N, blue; O, red.

$[SO_3]^{2-}$, rather than the single-step addition of the transient dimer, ONNO.

For the K^+ salt:

S–N = 175 pm

N–N = 128 pm

N–O = 129, 132 pm

(15.49)

Reactions 15.74 and 15.75 showed the formation of salts containing the $[NO]^+$ (nitrosyl) cation. Many salts are known and X-ray diffraction data confirm an N–O distance of 106 pm, i.e. less than in NO (115 pm). A molecular orbital treatment of the bonding (see end-of-chapter problem 15.20) is consistent with this observation. In going from NO to $[NO]^+$ there is an increase in the NO vibrational frequency (from 1876 to $\approx$2300 cm^{-1}), in keeping with an increase in bond strength. All nitrosyl salts are decomposed by water (eq. 15.100).

$$[NO]^+ + H_2O \longrightarrow HNO_2 + H^+ \qquad (15.100)$$

Dinitrogen trioxide, N_2O_3

Dinitrogen trioxide (Table 15.6 and Fig. 15.15) is obtained as a dark blue liquid in reaction 15.101 at low temperatures, but even at 195 K, extensive dissociation back to NO and N_2O_4 occurs.

$$2NO + N_2O_4 \rightleftharpoons 2N_2O_3 \qquad (15.101)$$

Dinitrogen trioxide is water-soluble and is the *acid anhydride* of HNO_2, nitrous acid (eq. 15.102).

$$N_2O_3 + H_2O \longrightarrow 2HNO_2 \qquad (15.102)$$

An ***acid anhydride*** is formed when one or more molecules of acid lose one or more molecules of water.

Dinitrogen tetraoxide, N_2O_4, and nitrogen dioxide, NO_2

Dinitrogen tetraoxide and nitrogen dioxide (Table 15.6 and Fig. 15.15) exist in equilibrium 15.103, and must be discussed together.

$$N_2O_4 \rightleftharpoons 2NO_2 \qquad (15.103)$$

The solid is colourless and is diamagnetic, consistent with the presence of only N_2O_4. Dissociation of this dimer gives the brown NO_2 radical. Solid N_2O_4 melts to give a yellow liquid, the colour arising from the presence of a little NO_2. At 294 K (bp), the brown vapour contains 15% NO_2. The colour of the vapour darkens as the temperature is raised, and at 413 K dissociation of N_2O_4 is almost complete. Above 413 K, the colour lightens again as NO_2 dissociates to NO and O_2. Laboratory-scale preparations of NO_2 or N_2O_4 are usually by thermal decomposition of *dry* lead(II) nitrate (eq. 15.104). If the brown gaseous NO_2 is cooled to $\approx$273 K, N_2O_4 condenses as a yellow liquid.

$$2Pb(NO_3)_2(s) \xrightarrow{\Delta} 2PbO(s) + 4NO_2(g) + O_2(g) \qquad (15.104)$$

Dinitrogen tetraoxide is a powerful oxidizing agent (e.g. see Box 9.1) which attacks many metals, including Hg, at 298 K. The reaction of NO_2 or N_2O_4 with water gives a 1:1 mixture of nitrous and nitric acids (eq. 15.105), although nitrous acid disproportionates (see below). Because of the formation of these acids, atmospheric NO_2 is corrosive and contributes to 'acid rain' (see Box 16.5). In concentrated H_2SO_4, N_2O_4 yields the nitrosyl and nitryl cations (eq. 15.106). The reactions of N_2O_4 with halogens and uses of N_2O_4 as a non-aqueous solvent were outlined in Sections 15.7 and 9.11, respectively.

$$2NO_2 + H_2O \longrightarrow HNO_2 + HNO_3 \qquad (15.105)$$

$$N_2O_4 + 3H_2SO_4 \longrightarrow [NO]^+ + [NO_2]^+ + [H_3O]^+ + 3[HSO_4]^- \qquad (15.106)$$

The nitryl cation **15.50** is linear, compared with the bent structures of NO_2 (Fig. 15.15) and of $[NO_2]^-$ ($\angle O-N-O = 115°$).

$$O \!=\!\!=\! \overset{+}{N} \!=\!\!=\! O$$
$$115 \text{ pm}$$

(15.50)

Dinitrogen pentaoxide, N_2O_5

Dinitrogen pentaoxide (Table 15.6 and Fig. 15.15) is the acid anhydride of HNO_3 and is prepared by reaction 15.107.

$$2HNO_3 \xrightarrow[\text{(dehydrating agent)}]{P_2O_5} N_2O_5 + H_2O \qquad (15.107)$$

It forms colourless deliquescent crystals but slowly decomposes above 273 K to give N_2O_4 and O_2. In the solid state, N_2O_5 consists of $[NO_2]^+$ and $[NO_3]^-$ ions, but the vapour contains planar molecules (Fig. 15.15). A molecular form of the solid can be formed by sudden cooling of the vapour to 93 K. Dinitrogen pentaoxide reacts violently with water, yielding HNO_3, and is a powerful oxidizing agent (e.g. reaction 15.108).

$$N_2O_5 + I_2 \longrightarrow I_2O_5 + N_2 \qquad (15.108)$$

15.9 Oxoacids of nitrogen

Isomers of $H_2N_2O_2$

An aqueous solution of the sodium salt of $[N_2O_2]^{2-}$ can be made from organic nitrites by reaction 15.109 or by the reduction of $NaNO_2$ with sodium amalgam. Addition of Ag^+ leads to the precipitation of $Ag_2N_2O_2$. Treatment of

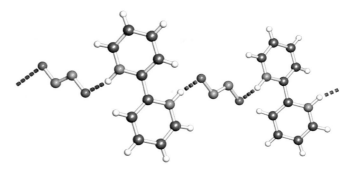

Fig. 15.16 Part of one of the hydrogen-bonded chains in the solid state structure of the 2,2'-bipyridinium salt of $[N_2O_2]^{2-}$. The structure was determined by X-ray diffraction at 173 K [N. Arulsamy *et al.* (1999) *Inorg. Chem.*, vol. 38, p. 2716].

this salt with anhydrous HCl in dry diethyl ether leads to the formation of hyponitrous acid, $H_2N_2O_2$.[†]

$$RONO + NH_2OH + 2EtONa$$
$$\longrightarrow Na_2N_2O_2 + ROH + 2EtOH \qquad (15.109)$$

Free $H_2N_2O_2$ is a weak acid. It is potentially explosive, decomposing spontaneously into N_2O and H_2O. The hyponitrite ion, $[N_2O_2]^{2-}$, exists in both the *trans-* and *cis*-forms. The *trans*-configuration is kinetically the more stable and has been confirmed in the solid state structure of $Na_2N_2O_2 \cdot 5H_2O$. The *cis*-form can be prepared as $Na_2N_2O_2$ by heating solid Na_2O with gaseous N_2O. Spectroscopic data for $H_2N_2O_2$ also indicate a *trans*-configuration (structure **15.51**). In the 2,2'-bipyridinium salt, $O\cdots H-N$ hydrogen-bonded interactions between the O atoms of the $[N_2O_2]^{2-}$ anions and the NH groups of the 2,2'-bipyridinium cations (**15.52**) lead to the formation of chains in the solid state (Fig. 15.16).

(15.51)   **(15.52)**

The reaction of acid with potassium nitrocarbamate results in the formation of nitramide (eq. 15.110) which is an isomer of hyponitrous acid. Nitramide has been structurally characterized; one N atom is trigonal planar (O_2NN) and the other is trigonal pyramidal (H_2NN). The compound is

---

[†] Although the name hyponitrous acid remains in common use, it is no longer recommended by the IUPAC. The recommendation is to use *diazenediol* (which derives from the substitution of each H atom in diazene ($HN=NH$) by an OH group), or *bis(hydroxidonitrogen)(N–N)*.

potentially explosive, and undergoes base-catalysed decomposition to N_2O and H_2O.

$$\text{nitrocarbamate ion} \xrightarrow[- CO_2]{H^+} \text{nitramide} \tag{15.110}$$

Nitrous acid, HNO_2

Nitrous acid is known only in solution and in the vapour phase. In the latter, it has structure **15.53**. It is a weak acid ($pK_a = 3.37$), but is unstable with respect to disproportionation in solution (eq. 15.111). It may be prepared *in situ* by reaction 15.112, the water-soluble reagents being chosen so as to give an insoluble metal salt ($BaSO_4$) as a product. $AgNO_2$ is insoluble but other metal nitrites are soluble in water.

$$\angle O–N–O = 111° $$
$$\angle H–O–N = 102°$$

(15.53)

$$3HNO_2 \longrightarrow 2NO + HNO_3 + H_2O \tag{15.111}$$

$$Ba(NO_2)_2 + H_2SO_4 \xrightarrow{aqu} BaSO_4 + 2HNO_2 \tag{15.112}$$

Sodium nitrite is an important reagent in the preparation of diazonium compounds, e.g. reaction 15.113 in which HNO_2 is prepared *in situ*. Alkali metal nitrates yield the nitrites when heated alone or, better, with Pb (reaction 15.114).

$$PhNH_2 \xrightarrow{NaNO_2, HCl, <273 K} [PhN_2]^+Cl^- \tag{15.113}$$

$$NaNO_3 + Pb \xrightarrow{\Delta} NaNO_2 + PbO \tag{15.114}$$

Nitrous acid can be oxidized to $[NO_3]^-$ by powerful oxidants such as acidified $[MnO_4]^-$. The products of the reduction of HNO_2 depend on the reducing agent:

- NO is formed with I^- or Fe^{2+};
- N_2O is produced with Sn^{2+};
- NH_2OH results from reduction by SO_2;
- NH_3 is formed with Zn in alkaline solution.

Kinetic rather than thermodynamic control over a reaction is illustrated by the fact that, in dilute solution, HNO_2, but not HNO_3, oxidizes I^- to I_2. Equations 15.115 show that the values of E^o_{cell} for these redox reactions are similar; nitrous acid is a faster, rather than a more powerful, oxidizing agent than dilute nitric acid.

$$I_2 + 2e^- \rightleftharpoons 2I^- \qquad E^o = +0.54\,V$$
$$[NO_3]^- + 3H^+ + 2e^- \rightleftharpoons HNO_2 + H_2O \qquad E^o = +0.93\,V$$
$$HNO_2 + H^+ + e^- \rightleftharpoons NO + H_2O \qquad E^o = +0.98\,V$$

$$\tag{15.115}$$

Nitric acid, HNO_3, and its derivatives

Nitric acid is an important industrial chemical (Box 15.8) and is manufactured on a large scale in the Ostwald process, which is closely tied to NH_3 production in the Haber–Bosch process. The first step is the oxidation of NH_3 to NO (eq. 15.23). After cooling, NO is mixed with air and absorbed in a countercurrent of water. The reactions involved are summarized in scheme 15.116. This produces HNO_3 in a concentration of $\approx 60\%$ by weight and it can be concentrated to 68% by distillation.

$$2NO + O_2 \rightleftharpoons 2NO_2$$
$$2NO_2 \rightleftharpoons N_2O_4$$
$$N_2O_4 + H_2O \longrightarrow HNO_3 + HNO_2$$
$$2HNO_2 \longrightarrow NO + NO_2 + H_2O$$
$$3NO_2 + H_2O \longrightarrow 2HNO_3 + NO$$

$$\tag{15.116}$$

Pure nitric acid can be made in the laboratory by adding H_2SO_4 to KNO_3 and distilling the product *in vacuo*. It is a colourless liquid, but must be stored below 273 K to prevent slight decomposition (eq. 15.117) which gives the acid a yellow colour.

$$4HNO_3 \longrightarrow 4NO_2 + 2H_2O + O_2 \tag{15.117}$$

Ordinary concentrated HNO_3 is the *azeotrope* containing 68% by weight of HNO_3 and boiling at 393 K. Photochemical decomposition occurs by reaction 15.117. Fuming HNO_3 is orange owing to the presence of an excess of NO_2.

An *azeotrope* is a mixture of two liquids that distils unchanged, the composition of liquid and vapour being the same. Unlike a pure substance, the composition of the azeotropic mixture depends on pressure.

In aqueous solution, HNO_3 acts as a strong acid which attacks most metals, often more rapidly if a trace of HNO_2 is present. Exceptions are Au and the *platinum-group metals* (see Section 22.9); Fe and Cr are passivated by concentrated HNO_3. Equations 9.8–9.10 illustrate HNO_3 acting as a base.

APPLICATIONS

Box 15.8 Commercial demand for HNO₃ and [NH₄][NO₃]

The industrial production of nitric acid (scheme 15.116) is carried out on a large scale and its manufacture is closely linked to that of ammonia. About 80% of all HNO_3 produced is converted into fertilizers, with $[NH_4][NO_3]$ being a key product:

$$NH_3 + HNO_3 \longrightarrow [NH_4][NO_3]$$

The commercial grade of $[NH_4][NO_3]$ contains $\approx 34\%$ nitrogen. For fertilizers, it is manufactured in the form of pellets which are easily handled. Its high solubility in water ensures efficient uptake by the soil.

Ammonium nitrate has other important applications: about 25% of the manufactured output is used directly in explosives, but its ready accessibility makes it a target for misuse, e.g. in the Oklahoma City bombing in 1995. The potentially explosive nature of $[NH_4][NO_3]$ also makes it a high-risk chemical for transportation.

Nitric acid is usually produced as an aqueous solution containing 50–68% HNO_3 by weight, and this is highly suitable for use in the fertilizer industry. However, for applications of HNO_3 as a nitrating agent in the production of, for example, explosives, acid containing >98% HNO_3 by weight is needed. Ordinary distillation is not appropriate because HNO_3 and H_2O form an azeotrope (see text). Alternative methods are dehydration using concentrated H_2SO_4, or by oxidation of NH_3 and of the NO so-formed, followed by a final oxidation step:

$$4NH_3 + 5O_2 \xrightarrow{\text{Pt/Rh}} 4NO + 6H_2O$$

$$2NO + O_2 \rightleftharpoons 2NO_2$$

$$2NO_2 \rightleftharpoons N_2O_4$$

$$2N_2O_4 + O_2 + 2H_2O \rightleftharpoons 4HNO_3$$

See also Box 15.3: Ammonia: an industrial giant.

The explosive decomposition of ammonium nitrate.

Tin, arsenic and a few *d*-block metals are converted to their oxides when treated with HNO_3, but others form nitrates. Only Mg, Mn and Zn liberate H_2 from *very dilute* nitric acid. If the metal is a more powerful reducing agent than H_2, reaction with HNO_3 reduces the acid to N_2, NH_3, NH_2OH or N_2O. Other metals liberate NO or NO_2 (e.g. reactions 15.118 and 15.119).

$$3Cu(s) + 8HNO_3(aq) \xrightarrow{\text{dilute}} 3Cu(NO_3)_2(aq) + 4H_2O(l) + 2NO(g) \quad (15.118)$$

$$Cu(s) + 4HNO_3(aq) \xrightarrow{\text{conc}} Cu(NO_3)_2(aq) + 2H_2O(l) + 2NO_2(g) \quad (15.119)$$

Large numbers of metal nitrate salts are known. Anhydrous nitrates of the group 1 metals, Sr^{2+}, Ba^{2+}, Ag^+ and Pb^{2+} are readily accessible, but for other metals, anhydrous nitrate salts are typically prepared using N_2O_4 (see Section 9.11). The preparations of anhydrous $Mn(NO_3)_2$ and $Co(NO_3)_2$ by slow dehydration of the corresponding hydrated salts using concentrated HNO_3 and phosphorus(V) oxide illustrate an alternative strategy. Nitrate salts of all metals and cations such as $[NH_4]^+$ are soluble in water. Alkali metal nitrates decompose on heating to the nitrite (reaction 15.120; see also eq. 15.114). The decomposition of NH_4NO_3 depends on the temperature (eqs. 15.8 and 15.93). Most metal nitrates decompose to the oxide when

ENVIRONMENT

Box 15.9 The nitrogen cycle, and nitrates and nitrites in waste water

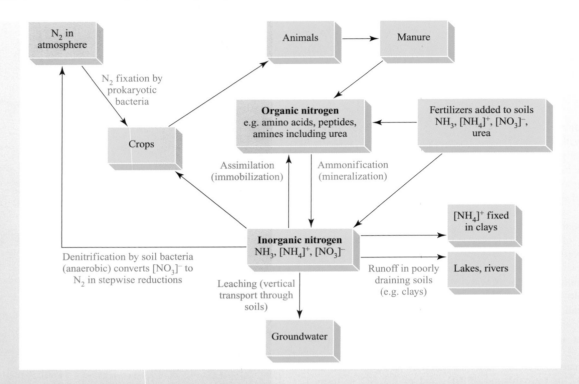

[Adapted from: D.S. Powlson and T.M. Addiscott (2004) in *Encyclopedia of Soils in the Environment*, ed. D. Hillel, Elsevier, Oxford, vol. 3, p. 21.]

The global nitrogen cycle includes chemical changes and nitrogen transport between the oceans, land and atmosphere, and involves natural and man-made sources of nitrogen. The diagram above gives a simplified nitrogen cycle, highlighting the processes that are coupled to agriculture. Sources of nitrates in groundwater (i.e. levels above those occurring naturally) include nitrate-based fertilizers and decaying organic material, as well as septic tanks, industrial effluent and waste from food processing factories.

Levels of $[NO_3]^-$ in waste water are controlled by legislation, limits being recommended by the World Health Organization, Environmental Protection Agency and European Union. Nitrites, because of their toxicity, must also be removed. One of the principal concerns arising from nitrates and nitrites in drinking water is their association with the disease *methaemoglobinaemia* ('blue baby syndrome'). This is primarily suffered by infants and results in the blood having a lower than normal O_2-carrying capacity. In the body, bacteria in the digestive system convert $[NO_3]^-$ to $[NO_2]^-$ which is able to irreversibly oxidize Fe^{2+} in haemoglobin to Fe^{3+}. The product is methaemoglobin and in this state, the iron can no longer bind O_2.

Nitrate salts are highly soluble and their removal from aqueous solution by techniques based on precipitation is not viable. Methods of nitrate removal include anion exchange, reverse osmosis (see Box 16.3), and enzymic denitrification. Ion-exchange involves passing the nitrate-containing water through a tank filled with resin beads on which chloride ions are adsorbed. In most conventional water purification systems, resins bind anions preferentially in the order $[SO_4]^{2-} > [NO_3]^- > Cl^- > [HCO_3]^- > [OH]^-$. Thus, as water containing nitrate ions passes through the resin, $[NO_3]^-$ exchanges for Cl^-, leaving $[NO_3]^-$ ions adsorbed on the surface. However, if the water contains significant amounts of sulfate, $[SO_4]^{2-}$ binds preferentially. Specialized resins must therefore be used for sulfate-rich wastes. Once the ion-exchange process has exhausted the resin of Cl^- ion, the system is regenerated by passing aqueous NaCl through the resin.

Removal of nitrates using enzymic denitrification takes advantage of the fact that certain anaerobic bacteria reduce $[NO_3]^-$ and $[NO_2]^-$ to N_2 in a sequence of steps, each involving a specific enzyme:

$$[NO_3]^- \xrightarrow{\text{nitrate reductase}} [NO_2]^- \xrightarrow{\text{nitrite reductase}} NO \xrightarrow{\text{nitric oxide reductase}} N_2O \xrightarrow{\text{nitrous oxide reductase}} N_2$$

Other methods of removing $[NO_2]^-$ involve oxidation to $[NO_3]^-$ (using $[OCl]^-$ or H_2O_2 as oxidant), and removing the $[NO_3]^-$ as detailed above, or using urea or sulfamic acid to reduce $[NO_2]^-$ to N_2 (see end-of-chapter problem 15.41).

heated (eq. 15.121), but silver and mercury(II) nitrates give the respective metal (eq. 15.122)

$$2KNO_3 \xrightarrow{\Delta} 2KNO_2 + O_2 \qquad (15.120)$$

$$2Cu(NO_3)_2 \xrightarrow{\Delta} 2CuO + 4NO_2 + O_2 \qquad (15.121)$$

$$2AgNO_3 \xrightarrow{\Delta} 2Ag + 2NO_2 + O_2 \qquad (15.122)$$

Many organic and inorganic compounds are oxidized by concentrated HNO_3, although nitrate ion in aqueous solution is usually a very *slow* oxidizing agent (see above). *Aqua regia* contains free Cl_2 and $ONCl$ and attacks Au (eq. 15.123) and Pt with the formation of chlorido complexes.

$$Au + HNO_3 + 4HCl \longrightarrow HAuCl_4 + NO + 2H_2O$$
$$\underbrace{\text{conc} \qquad\qquad \text{conc}}_{\text{aqua regia}} \qquad (15.123)$$

Aqua regia is a mixture of concentrated nitric and hydrochloric acids.

Concentrated HNO_3 oxidizes I_2, P_4 and S_8 to HIO_3, H_3PO_4 and H_2SO_4 respectively.

The molecular structure of HNO_3 is shown in Fig. 15.17a. Differences in N–O bond distances are readily understood in terms of the resonance structures shown. The nitrate ion has a trigonal planar (D_{3h}) structure and the equivalence of the

bonds may be rationalized using valence bond or molecular orbital theory (Figs. 5.25 and 15.17b). We considered an MO treatment for the bonding in $[NO_3]^-$ in Fig. 5.25 and described how interaction between the N 2p orbital and a ligand-group orbital involving in-phase O 2p orbitals gives rise to one occupied MO in $[NO_3]^-$ that has π-bonding character delocalized over all four atoms.

The hydrogen atom in HNO_3 can be replaced by fluorine by treating dilute HNO_3 or KNO_3 with F_2. The product, fluorine nitrate, **15.54**, is an explosive gas which reacts slowly with H_2O but rapidly with aqueous alkali (eq. 15.124).

(15.54)

$$2FONO_2 + 4[OH]^-$$
$$\longrightarrow 2[NO_3]^- + 2F^- + 2H_2O + O_2 \qquad (15.124)$$

The reaction of $NaNO_3$ with Na_2O at 570 K leads to the formation of Na_3NO_4 (sodium orthonitrate), and K_3NO_4 may be prepared similarly. X-ray diffraction data confirm that the $[NO_4]^{3-}$ ion is tetrahedral with N–O bond lengths of 139 pm, consistent with single bond character.

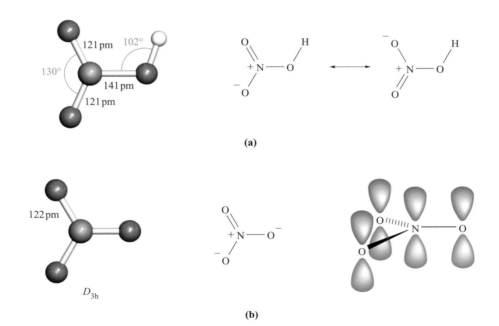

Fig. 15.17 (a) The gas-phase planar structure of HNO_3, and appropriate resonance structures. (b) The molecular structure of the planar $[NO_3]^-$ anion; the equivalence of the three N–O bonds can be rationalized by valence bond theory (one of three resonance structures is shown) or by MO theory (partial π-bonds are formed by overlap of N and O 2p atomic orbitals and the π-bonding is delocalized over the NO_3-framework as was shown in Fig. 5.25). Colour code: N, blue; O, red; H, white.

Structure **15.55** includes a valence bond picture of the bonding. The free acid H_3NO_4 is not known.

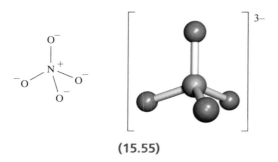

(15.55)

15.10 Oxides of phosphorus, arsenic, antimony and bismuth

Each of the group 15 elements from P to Bi forms two oxides, E_2O_3 (or E_4O_6) and E_2O_5 (or E_4O_{10}), the latter becoming less stable as the group is descended:

- E_2O_5 (E = P, As, Sb, Bi) are acidic;
- P_4O_6 is acidic;
- As_4O_6 and Sb_4O_6 are amphoteric;
- Bi_2O_3 is basic.

In addition to describing the common oxides of the group 15 elements, the section below introduces several other oxides of phosphorus.

Oxides of phosphorus

Phosphorus(III) oxide, P_4O_6, is obtained by burning white phosphorus in a restricted supply of O_2. It is a colourless, volatile solid (mp 297 K, bp 447 K) with molecular structure **15.56**. The P–O bond distances (165 pm) are consistent with single bonds, and the angles P–O–P and O–P–O are 128° and 99° respectively. The oxide is soluble in diethyl ether or benzene, but reacts with cold water (eq. 15.125).

$$P_4O_6 + 6H_2O \longrightarrow 4H_3PO_3 \qquad (15.125)$$

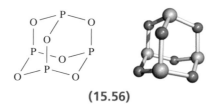

(15.56)

Each P atom in P_4O_6 carries a lone pair of electrons and P_4O_6 can therefore act as a Lewis base. Adducts with one and two equivalents of BH_3 have been reported, but the

reaction of P_4O_6 with one equivalent of $Me_2S \cdot BH_3$ followed by slow crystallization from toluene solution at 244 K gives $P_8O_{12}(BH_3)_2$ (**15.57**) rather than an adduct of P_4O_6. The solid state structure confirms that dimerization of P_4O_6 has occurred through P–O bond cleavage in structure **15.56** and reformation of P–O bonds between monomeric units. Free P_8O_{12} has not, to date, been isolated.

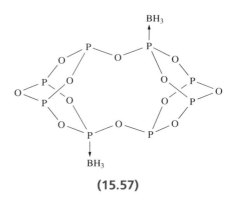

(15.57)

Oxidation of P_4O_6 with O_2 gives P_4O_{10} (see below), while ozone oxidation leads to the formation of P_4O_{18} (eq. 15.126) which has been structurally characterized. The square-based pyramidal environment of each P atom is related to that found in phosphite ozonides $(RO)_3PO_3$ (see Fig. 16.5). In solution, P_4O_{18} decomposes above 238 K with gradual release of O_2, but the decomposition of dry P_4O_{18} powder is explosive.

$$P_4O_6 \xrightarrow[\text{in } CH_2Cl_2]{O_3,\ 195\ K} P_4O_{18} \qquad (15.126)$$

The most important oxide of phosphorus is P_4O_{10} (phosphorus(V) oxide), commonly called *phosphorus pentoxide*. It can be made directly from P_4 (eq. 15.10) or by oxidizing P_4O_6. In the vapour phase, phosphorus(V) oxide contains P_4O_{10} molecules with structure **15.58**; the P–O_{bridge} and P–$O_{terminal}$ bond distances are 160 and 140 pm. When the vapour is condensed rapidly, a volatile and extremely hygroscopic solid is obtained which also contains P_4O_{10} molecules. If this solid is heated in a closed vessel for several hours and the melt maintained at a high temperature before being allowed to cool, the solid obtained is macromolecular. Three polymorphic forms exist at ordinary pressure and temperature, with the basic building block being unit **15.59**. In each polymorph, only three of the four

O atoms are involved in interconnecting the PO_4 units via P−O−P bridges. Phosphorus(V) oxide has a great affinity for water (eq. 15.127), and is the anhydride of the wide range of oxoacids described in Section 15.11. It is used as a drying agent (see Box 12.3).

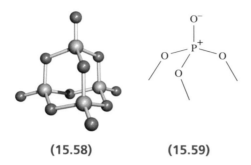

(15.58)   **(15.59)**

$$P_4O_{10} + 6H_2O \longrightarrow 4H_3PO_4 \qquad (15.127)$$

Three other oxides of phosphorus, P_4O_7 (**15.60**), P_4O_8 (**15.61**) and P_4O_9 (**15.62**), have structures that are related to those of P_4O_6 and P_4O_{10}.

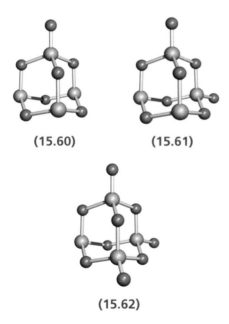

(15.60)   **(15.61)**

(15.62)

These oxides are mixed P(III)P(V) species, each centre bearing a terminal oxo group being P(V). For example, P_4O_8 is made by heating P_4O_6 in a sealed tube at 710 K, the other product being red phosphorus (eq. 15.128).

$$4P_4O_6 \xrightarrow{710\,K} 3P_4O_8 + 4P(red) \qquad (15.128)$$

Oxides of arsenic, antimony and bismuth

The normal combustion products of As and Sb are As(III) and Sb(III) oxides (eq. 15.13). The vapour and high-temperature solid polymorph of each oxide contains E_4O_6 (E = As or Sb) molecules structurally related to **15.56**. Lower temperature polymorphs have layer structures containing trigonal pyramidal As or Sb atoms. Condensation of As_4O_6 vapour above 520 K leads to the formation of As_2O_3 glass. Arsenic(III) oxide is an important precursor in arsenic chemistry and is made industrially from the sulfide (Section 15.2). Dissolution of As_2O_3 in water gives a very weakly acidic solution, and it is probable that the species present is $As(OH)_3$ (*arsenous acid*) although this has never been isolated. Crystallization of aqueous solutions of $As(OH)_3$ yields As_2O_3. Arsenic(III) oxide dissolves in aqueous alkali to give salts containing the $[AsO_2]^-$ ion, and in aqueous HCl with the formation of $AsCl_3$. The properties of Sb_2O_3 in water and aqueous alkali or HCl resemble those of As_2O_3.

Bismuth(III) oxide occurs naturally as *bismite*, and is formed when Bi combines with O_2 on heating. In contrast to earlier members of group 15, molecular species are not observed for Bi_2O_3, and the structure is more like that of a typical *metal* oxide.

Arsenic(V) oxide is most readily made by reaction 15.129 than by direct oxidation of the elements. The route makes use of the fact that As_2O_5 is the acid anhydride of arsenic acid, H_3AsO_4. In the solid state, As_2O_5 has a 3-dimensional structure consisting of As−O−As linked octahedral AsO_6 and tetrahedral AsO_4-units.

$$As_2O_3 \xrightarrow{\text{conc HNO}_3} 2H_3AsO_4 \xrightarrow{\text{dehydration}} As_2O_5 + 3H_2O$$
$$(15.129)$$

Antimony(V) oxide may be made by reacting Sb_2O_3 with O_2 at high temperatures and pressures. It crystallizes with a 3-dimensional structure in which the Sb atoms are octahedrally sited with respect to six O atoms. Bismuth(V) oxide is poorly characterized, and its formation requires the action of strong oxidants (e.g. alkaline hypochlorite) on Bi_2O_3.

15.11 Oxoacids of phosphorus

Table 15.7 lists selected oxoacids of phosphorus. This is an important group of compounds, but the acids are difficult to classify in a straightforward manner. It should be remembered that the basicity of each acid corresponds to the number of OH-groups, *and not simply to the total number of hydrogen atoms*, e.g. H_3PO_3 and H_3PO_2 are dibasic and monobasic respectively (Table 15.7). The P-attached hydrogens do not ionize in aqueous solution, and diagnostic absorptions in the IR spectra of these compounds confirm the presence of P–H bonds. The IR spectrum of aqueous H_3PO_2 exhibits absorptions at 2408, 1067 and 811 cm^{-1} assigned to the stretching, deformation and rocking modes of the PH_2 group. The band at 2408 cm^{-1}

Table 15.7   Selected oxoacids of phosphorus; older names that are still in common use are given in parentheses.

Formula	Name	Structure	pK_a values
H_3PO_2	Phosphinic acid		$pK_a = 1.24$
H_3PO_3	Phosphonic acid (phosphorous acid)		$pK_a(1) = 2.00$; $pK_a(2) = 6.59$
H_3PO_4	Phosphoric acid (orthophosphoric acid)		$pK_a(1) = 2.21$; $pK_a(2) = 7.21$; $pK_a(3) = 12.67$
$H_4P_2O_6$	Hypodiphosphoric acid		$pK_a(1) = 2.2$; $pK_a(2) = 2.8$; $pK_a(3) = 7.3$; $pK_a(4) = 10.0$
$H_4P_2O_7$	Diphosphoric acid		$pK_a(1) = 0.85$; $pK_a(2) = 1.49$; $pK_a(3) = 5.77$; $pK_a(4) = 8.22$
$H_5P_3O_{10}$	Triphosphoric acid		$pK_a(1) \leq 0$ $pK_a(2) = 0.89$; $pK_a(3) = 4.09$; $pK_a(4) = 6.98$; $pK_a(5) = 9.93$

is the most easily observed. In the IR spectrum of aqueous H_3PO_3, an absorption at $2440\,cm^{-1}$ corresponds to the P–H stretching mode.

The absorption at $2408\,cm^{-1}$ in the IR spectrum of aqueous H_3PO_2 shifts when the sample is fully deuterated. Explain why this shift occurs, and calculate the wavenumber at which the new band should be observed. [*Ans.* $1735\,cm^{-1}$]

Phosphinic acid, H_3PO_2

The reaction of white phosphorus with aqueous alkali (eq. 15.11) produces the phosphinate ion, $[H_2PO_2]^-$. By using $Ba(OH)_2$ as alkali, precipitating the Ba^{2+} ions as $BaSO_4$, and evaporating the aqueous solution, white deliquescent crystals of H_3PO_2 can be obtained. In aqueous solution, H_3PO_2 is a fairly strong monobasic acid (eq. 15.130 and Table 15.7).

$$H_3PO_2 + H_2O \rightleftharpoons [H_3O]^+ + [H_2PO_2]^- \qquad (15.130)$$

Phosphinic acid and its salts are reducing agents. $NaH_2PO_2 \cdot H_2O$ is used industrially in a non-electrochemical reductive process which reduces Ni^{2+} to Ni, and plates nickel onto, for example, steel. The so-called *electroless nickel* coatings also contain phosphorus, and the amount of P present influences the corrosion and wear-resistance properties of the coating. For example, coatings with a high P content (11–13%) exhibit enhanced resistance to attack by acids, whereas lowering the P content to <4% makes the coating more resistant to corrosion by alkalis.

When heated, H_3PO_2 disproportionates according to eq. 15.131, the products being determined by reaction temperature.

$$\left. \begin{array}{l} 3H_3PO_2 \xrightarrow{\Delta} PH_3 + 2H_3PO_3 \\ \text{or} \\ 2H_3PO_2 \xrightarrow{\Delta} PH_3 + H_3PO_4 \end{array} \right\} \qquad (15.131)$$

Intramolecular transfer of a proton to the terminal O atom in phosphinic acid produces a tautomer in which the P atom is 3-coordinate:

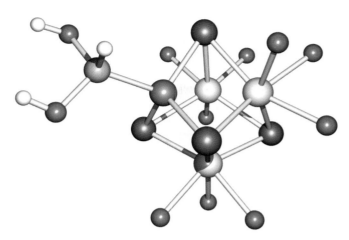

In practice, this equilibrium lies far over to the left-hand side, with a ratio $H_2PO(OH) : HP(OH)_2 > 10^{12} : 1$. $HP(OH)_2$ carries a lone pair of electrons and, in 2003, the tautomer was stabilized by coordination to $[W_3(OH_2)_9NiX_4]^{4+}$

Fig. 15.18 The structure of $[W_3(OH_2)_9NiSe_4\{PH(OH)_2\}]^{4+}$ determined by X-ray diffraction [M.N. Solokov *et al.* (2003) *Chem. Commun.*, p. 140]. The H atoms in the structure were not fully located. Colour code: P, orange; W, silver; Ni, green; Se, brown; O red; H, white.

(X = S, Se) (Fig. 15.18). The presence of one P–H bond was confirmed by the appearance of a doublet (J_{PH} 393 Hz) in the ^{31}P NMR spectrum.

Phosphonic acid, H_3PO_3

Phosphonic acid (often called *phosphorous acid*) may be crystallized from the solution obtained by adding ice-cold water to P_4O_6 (eq. 15.125) or PCl_3 (eq. 15.78). Pure H_3PO_3 forms colourless, deliquescent crystals (mp 343 K) and in the solid state, molecules of the acid (Table 15.7) are linked by hydrogen bonds to form a 3-dimensional network. In aqueous solution, it is dibasic (eqs. 15.132 and 15.133).

$$H_3PO_3(aq) + H_2O(l) \rightleftharpoons [H_3O]^+(aq) + [H_2PO_3]^-(aq) \qquad (15.132)$$

$$[H_2PO_3]^-(aq) + H_2O(l) \rightleftharpoons [H_3O]^+(aq) + [HPO_3]^{2-}(aq) \qquad (15.133)$$

Salts containing the $[HPO_3]^{2-}$ ion are called *phosphonates*. Although the name 'phosphite' remains in common use, it is a possible source of confusion since esters of type $P(OR)_3$ are also called phosphites, e.g. $P(OEt)_3$ is triethylphosphite.

Phosphonic acid is a reducing agent, but disproportionates when heated (eq. 15.134).

$$4H_3PO_3 \xrightarrow{470\,K} PH_3 + 3H_3PO_4 \qquad (15.134)$$

Hypodiphosphoric acid, $H_4P_2O_6$

The reaction between red phosphorus and NaOCl or $NaClO_2$ yields $Na_2H_2P_2O_6$. This can be converted in aqueous solution into the dihydrate of the free acid which is best formulated as $[H_3O]_2[H_2P_2O_6]$. Dehydration using P_4O_{10} gives $H_4P_2O_6$. The first indication of a P–P bonded dimer (i.e. rather than H_2PO_3) came from the observation that the acid was diamagnetic, and X-ray diffraction data for the salt $[NH_4]_2[H_2P_2O_6]$ have confirmed this structural feature. All four terminal P–O bonds are of equal length (157 pm), and the bonding description in diagram **15.63** is consistent with this observation. In keeping with our comments on hypervalent species in Section 15.3, this description is more appropriate than a pair of resonance structures, each involving one P=O and one P–O$^-$ bond. The acid is thermodynamically unstable with respect to disproportionation and reaction 15.135 occurs slowly in aqueous solution. For this reason, $H_4P_2O_6$ cannot be made by reduction of H_3PO_4 or by oxidation of H_3PO_3 in aqueous media. Hence the need to use a precursor (i.e. elemental phosphorus) in which the P–P bond is already present.

(15.63)

$$H_4P_2O_6 + H_2O \longrightarrow H_3PO_3 + H_3PO_4 \qquad (15.135)$$

Phosphoric acid, H_3PO_4, and its derivatives

Phosphoric acid is made from phosphate rock (eq. 15.136) or by hydration of P_4O_{10} (eq. 15.127).

$$Ca_3(PO_4)_2 + 3H_2SO_4 \underset{\text{conc}}{\longrightarrow} 2H_3PO_4 + 3CaSO_4 \qquad (15.136)$$

The pure acid forms deliquescent, colourless crystals (mp 315 K). It has a molecular structure (Table 15.7) with P–OH and P–O bond distances of 157 and 152 pm. In the crystalline state, extensive hydrogen bonding links H_3PO_4 molecules into a layered network. On standing, crystalline H_3PO_4 rapidly forms a viscous liquid. In this and in the commercially available 85% (by weight with water) acid, extensive hydrogen bonding is responsible for the syrupy nature of the acid. In dilute aqueous solutions, acid molecules are hydrogen-bonded to water molecules rather than to each other.

Phosphoric acid is very stable and has no oxidizing properties except at very high temperatures. Aqueous H_3PO_4 is a tribasic acid (Table 15.7) and salts containing $[H_2PO_4]^-$, $[HPO_4]^{2-}$ and $[PO_4]^{3-}$ can be isolated. Thus, three Na$^+$ salts can be prepared under suitable neutralization conditions. Two of the most commonly encountered sodium and potassium salts are $Na_2HPO_4 \cdot 12H_2O$ and KH_2PO_4. Sodium phosphates are extensively used for buffering aqueous solutions, and tri-*n*-butyl phosphate is a valuable

solvent for the extraction of metal ions from aqueous solution (see Box 7.3).

When H_3PO_4 is heated at 510 K, it is dehydrated to diphosphoric acid (eq. 15.137). Comparison of the structures of these acids (Table 15.7) shows that water is eliminated with concomitant P–O–P bridge formation. Further heating yields triphosphoric acid (eq. 15.138).

$$2H_3PO_4 \xrightarrow{\Delta} H_4P_2O_7 + H_2O \qquad (15.137)$$

$$H_3PO_4 + H_4P_2O_7 \xrightarrow{\Delta} H_5P_3O_{10} + H_2O \qquad (15.138)$$

Such species containing P–O–P bridges are commonly called *condensed phosphates* and eq. 15.139 shows the general condensation process.

$$(15.139)$$

The *controlled* hydrolysis of P_4O_{10} is sometimes useful as a means of preparing condensed phosphoric acids. In principle, the condensation of phosphate ions (e.g. reaction 15.140) should be favoured at low pH, but in practice such reactions are usually slow.

$$2[PO_4]^{3-} + 2H^+ \rightleftharpoons [P_2O_7]^{4-} + H_2O \qquad (15.140)$$

Clearly, the number of OH groups in a particular unit determines the extent of the condensation processes. In condensed phosphate anion formation, chain-terminating end groups (**15.64**) are formed from $[HPO_4]^{2-}$, chain members (**15.65**) from $[H_2PO_4]^-$, and cross-linking groups (**15.66**) from H_3PO_4.

(15.64)   **(15.65)**   **(15.66)**

In free condensed acids such as $H_5P_3O_{10}$ (Table 15.7), different phosphorus environments can be distinguished by ^{31}P NMR spectroscopy or chemical methods:

- the pK_a values for successive proton dissociations depend on the position of the OH group; terminal P atoms carry one strongly and one weakly acidic proton, while each

ENVIRONMENT

Box 15.10 Phosphate fertilizers: essential to crops but are they damaging our lakes?

Worldwide demand for fertilizers is enormous and world consumption is increasing at a rate of between 2% and 3% per year. Phosphorus is an essential plant nutrient and up to 90% (depending on the country) of phosphate rock (see Section 15.2) that is mined is consumed in the manufacture of phosphorus-containing fertilizers. Insoluble phosphate rock is treated with concentrated H_2SO_4 to generate soluble *superphosphate* fertilizers containing $Ca(H_2PO_4)_2$ mixed with $CaSO_4$ and other sulfates. Reaction between phosphate rock and H_3PO_4 gives *triple superphosphate*, mainly $Ca(H_2PO_4)_2$. Ammonium phosphate fertilizers are valuable sources of both N and P. Environmentalists are concerned about the effects that phosphates and polyphosphates from fertilizers and detergents have on the natural balance of lake populations. Phosphates in run-off water which flows into lakes contribute to the excessive growth of algae (the formation of *algal bloom* as shown in the photograph opposite) and the *eutrophication* of the lake. Algae produce O_2 during photosynthesis. However, the presence of large amounts of dead algae provides a ready food supply for aerobic organisms. The net result of excessive algal blooms, therefore, is a depletion in lakes of O_2 which in turn affects fish and other aquatic life. When water is classed as hypereutrophic, it is at an extreme end of its trophic state. The latter is qualitatively measured by monitoring the phosphorus, nitrogen and chlorophyll concentrations, and the transparency of the water. The chlorophyll concentration is determined by the amount of plant and algal growth. The transparency decreases with increased algal blooms and sedimentary particles, and is measured by lowering a Secchi disk (a disk with white and black segments) into the water until it is no longer visible. This depth is referred to as the Secchi depth. The table below shows approximate divisions between trophic states based on these measurements:

Eutrophication of a farm pond.

condensed phosphates, and the levels that must be removed before the waste can be discharged are controlled by legislation. In most cases, phosphates are removed by methods based on precipitation (this is the reverse of the situation for nitrate removal: see Box 15.9). Fe^{3+}, Al^{3+} and Ca^{2+} are most commonly used to give precipitates that can be separated by filtration. Values of K_{sp} for $FePO_4 \cdot 2H_2O$, $AlPO_4$ and $Ca_2(PO_4)_3$ are 9.91×10^{-16}, 9.84×10^{-21} and 2.07×10^{-33} respectively.

The issue of phosphates in lakes is not clear-cut: field studies indicate that adding phosphates to acid lakes (the result of acid rain pollution) stimulates plant growth, which in turn leads to a production of $[OH]^-$, which neutralizes excess acid.

Further reading

L.E. de-Bashan and Y. Bashan (2004) *Water Res.*, vol. 38, p. 4222 – 'Recent advances in removing phosphorus from wastewater and its future use as a fertilizer (1997–2003)'.

W. Davison, D.G. George and N.J.A. Edwards (1995) *Nature*, vol. 377, p. 504 – 'Controlled reversal of lake acidification by treatment with phosphate fertilizer'.

R. Gächter and B. Müller (2003) *Limnol. Oceanogr.*, vol. 48, p. 929 – 'Why the phosphorus retention of lakes does not necessarily depend on the oxygen supply to their sediment surface'.

J.H. Kinniburgh and M. Barnett (2009) *Water Environ. J.*, vol. 24, p. 107 – 'Orthophosphate concentrations in the River Thames: Reductions in the past decade'.

V.H. Smith and D.W. Schindler (2009) *Trends Ecol. Evol.*, vol. 24, p. 201 – 'Eutrophication science: Where do we go from here?'

Trophic state	[Phosphorus]/ $\mu g\ dm^{-3}$	[Nitrogen]/ $\mu g\ dm^{-3}$	[Chlorophyll]/ $\mu g\ dm^{-3}$	Secchi depth/m
Oligotrophic	<10	<400	<3	>6
Mesotrophic	10–35	400–600	3–8	6–3
Eutrophic	35–100	600–1500	8–40	3–1
Hypereutrophic	>100	>1500	>40	<1

Eutrophication can occur as a natural process, although the term is most often applied to situations that have been exacerbated by external influences.

Fertilizers are a major source of phosphates entering rivers and lakes. However, domestic and industrial waste water (e.g. from detergent manufacturing) also contains $[PO_4]^{3-}$ and

P atom in the body of the chain bears one strongly acidic group;

- cross-linking P—O—P bridges are hydrolysed by water much faster than other such units.

The simplest condensed phosphoric acid, $H_4P_2O_7$, is a solid at 298 K and can be obtained from reaction 15.137 or, in a purer form, by reaction 15.141. It is a stronger acid than H_3PO_4 (Table 15.7).

$$5H_3PO_4 + POCl_3 \longrightarrow 3H_4P_2O_7 + 3HCl \qquad (15.141)$$

The sodium salt $Na_4P_2O_7$ is obtained by heating Na_2HPO_4 at 510 K. Note the electronic and structural relationship between $[P_2O_7]^{4-}$ (in which the terminal P—O bond distances are equal) and $[Si_2O_7]^{6-}$, **14.23**. In aqueous solution, $[P_2O_7]^{4-}$ is very slowly hydrolysed to $[PO_4]^{3-}$, and the two ions can be distinguished by chemical tests, e.g. addition of Ag^+ ions precipitates white $Ag_4P_2O_7$ or pale yellow Ag_3PO_4.

The acid referred to as 'metaphosphoric acid' with an empirical formula of HPO_3 is actually a sticky mixture of polymeric acids, obtained by heating H_3PO_4 and $H_4P_2O_7$ at ≈ 600 K. More is known about the salts of these acids than about the acids themselves. For example, $Na_3P_3O_9$ can be isolated by heating NaH_2PO_4 at 870–910 K and maintaining the melt at 770 K to allow water vapour to escape. It contains the cyclic $[P_3O_9]^{3-}$ ion (*cyclo*-triphosphate ion, Fig. 15.19a) which has a chair conformation. In alkaline solution,

$[P_3O_9]^{3-}$ hydrolyses to $[P_3O_{10}]^{5-}$ (triphosphate ion, Fig. 15.19b). The salts $Na_5P_3O_{10}$ and $K_5P_3O_{10}$ (along with several hydrates) are well characterized and $Na_5P_3O_{10}$ (manufactured by reaction 15.142) is used in detergents where it acts as a water softener. However, concerns that phosphates in waste water are associated with the eutrophication of lakes and rivers (Box 15.10) have been responsible for the replacement of phosphates in detergents by zeolites (see Section 14.9). Uses of polyphosphates as sequestering agents were mentioned in Sections 12.7 and 12.8. The parent acid $H_5P_3O_{10}$ has not been prepared in a pure form, but solution titrations allow pK_a values to be determined (Table 15.7).

$$2Na_2HPO_4 + NaH_2PO_4 \xrightarrow{550-650\ K} Na_5P_3O_{10} + 2H_2O$$
$$(15.142)$$

The salt $Na_4P_4O_{12}$ may be prepared by heating $NaHPO_4$ with H_3PO_4 at 670 K and slowly cooling the melt. Alternatively, the volatile form of P_4O_{10} may be treated with ice-cold aqueous NaOH and $NaHCO_3$. Figure 15.19c shows the structure of $[P_4O_{12}]^{4-}$, in which the P_4O_4-ring adopts a chair conformation. Several salts of the $[P_6O_{18}]^{6-}$ ion (Fig. 15.19d) are also well characterized; the Na^+ salt is made by heating NaH_2PO_4 at ≈ 1000 K.

The discussion above illustrates how changes in the conditions of heating Na_2HPO_4 or NaH_2PO_4 cause

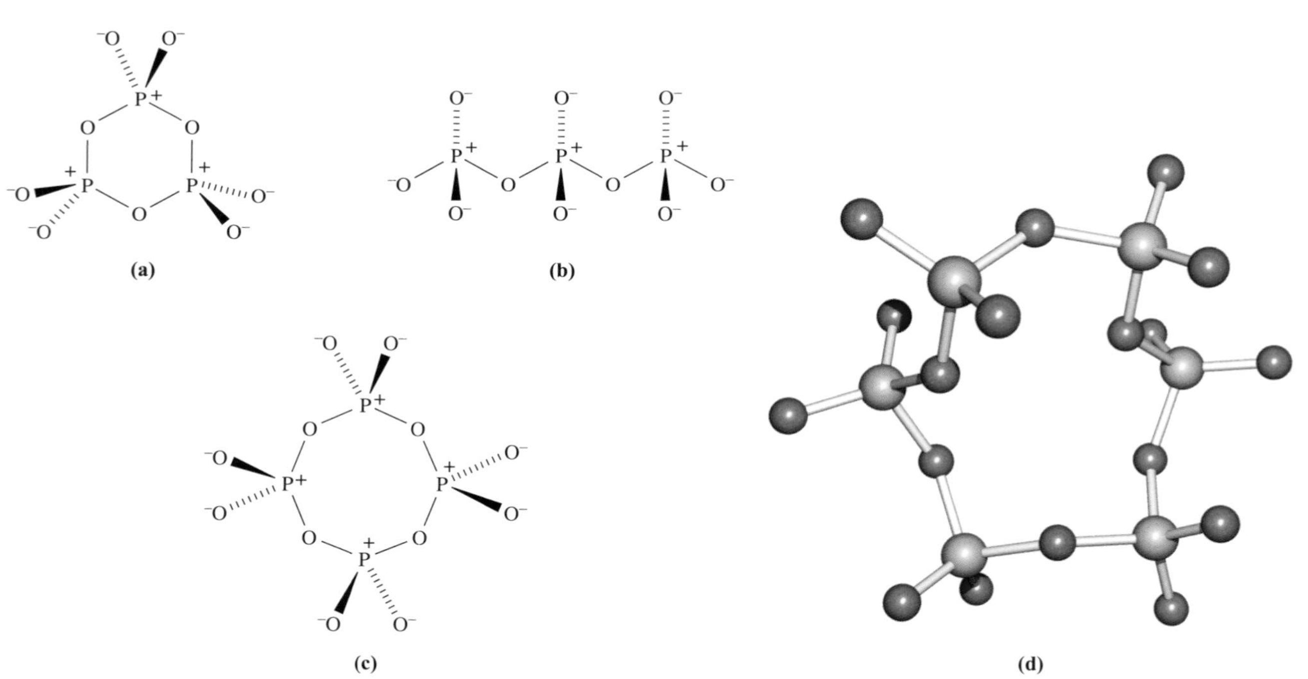

Fig. 15.19 Schematic representations of the structures of (a) $[P_3O_9]^{3-}$, (b) $[P_3O_{10}]^{5-}$ and (c) $[P_4O_{12}]^{4-}$. (d) The structure of $[P_6O_{18}]^{6-}$ (X-ray diffraction) in the compound $[Et_4N]_6[P_6O_{18}]\cdot 4H_2O$ [M.T. Averbuch-Pouchot *et al.* (1991) *Acta Crystallogr., Sect. C*, vol. 47, p. 1579]. Compare these structures with those of the isoelectronic silicates, see Fig. 14.24 and associated text. Colour code: P, orange; O, red.

BIOLOGY AND MEDICINE

Box 15.11 Biological significance of phosphates and arsenates

Phosphates play an enormously important role in biological systems. The genetic substances deoxyribonucleic acid (DNA) and ribonucleic acid (RNA) are phosphate esters (see Fig. 10.13). The form of DNA originally characterized is referred to as B-DNA and possesses a right-handed helix supported by hydrogen-bonded, complementary nucleobase pairs. Other common forms include A-DNA (right-handed) and Z-DNA (left-handed). The phosphate groups lie on the outside of the helix as shown in the left-hand diagram below.

Bones and teeth are constructed from *collagen* (fibrous protein) and single crystals of *hydroxyapatite*, $Ca_5(OH)(PO_4)_3$. Tooth decay involves acid attack on the phosphate, but the addition of fluoride ion to water supplies facilitates the formation of fluoroapatite, which is more resistant to decay.

$$Ca_5(OH)(PO_4)_3 + F^- \longrightarrow Ca_5F(PO_4)_3 + [OH]^-$$

All living cells contain *adenosine triphosphate*, ATP, which consists of adenine, ribose and triphosphate units. The structure of ATP is shown on the right below. Hydrolysis results in the loss of a phosphate group and converts ATP to ADP (adenosine diphosphate), releasing energy which is used for functions such as cell growth and muscle movement. In a simplified form:

$$[ATP]^{4-} + 2H_2O \longrightarrow [ADP]^{3-} + [HPO_4]^{2-} + [H_3O]^+$$

and, at the standard state usually employed in discussions of biochemical processes (pH 7.4 and $[CO_2] = 10^{-5}$ mol dm^{-3}),

$\Delta G \approx -40$ kJ per mole of reaction. Conversely, energy released by, for example, the oxidation of carbohydrates can be used to convert ADP to ATP (see Section 29.4); thus ATP is continually being reformed, ensuring a continued supply of stored energy in the body.

In 2011, researchers published evidence for the incorporation of arsenic in place of phosphorus (i.e. arsenate replacing phosphate) in nucleic acids and proteins in a strain of bacterium isolated from Mono Lake in California (see Box 15.1). However, the mechanisms by which these biomolecules operate are not yet known.

Further reading

J.J.R. Fraústo da Silva and R.J.P. Williams (1991) *The Biological Chemistry of the Elements*, Clarendon Press, Oxford.

C.K. Mathews, K.E. van Holde and K.G. Ahern (2000) *Biochemistry*, 3rd edn, Benjamin/Cummings, New York.

R.S. Oremland and J.F. Stolz (2003) *Science*, vol. 300, p. 939 – 'The ecology of arsenic'.

F. Wolfe-Simon *et al.* (2011) *Science*, vol. 332, p. 1163 – 'A bacterium that can grow by using arsenic instead of phosphorus'.

F. Wolfe-Simon, P.C.W. Davies and A.D. Anbar (2009) *Int. J. Astrobiol.*, vol. 8, p. 69 – 'Did nature also choose arsenic?'.

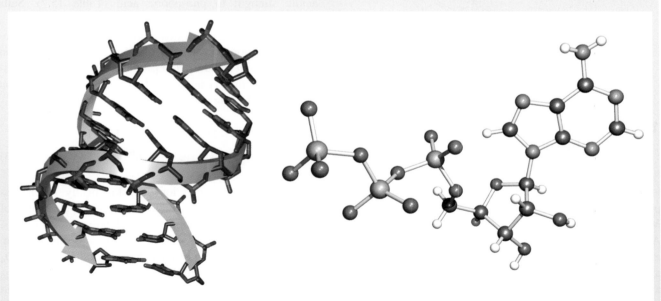

Colour code: P, orange; O, red; C, grey; N, blue; H, white.

product variation. Carefully controlled conditions are needed to obtain long-chain polyphosphates. Depending on the relative orientations of the PO_4-units, several modifications can be made. Cross-linked polyphosphates (some of which are glasses) can be made by heating NaH_2PO_4 with P_4O_{10}.

Chiral phosphate anions

Although the octahedral ion $[Sb(OH)_6]^-$ exists (see Section 15.12), the analogous phosphorus-containing anion has not been isolated. However, related anions containing chelating O,O'-donor ligands are known and we introduce them here because of their stereoselective applications. An example is anion **15.67** which has D_3 symmetry (see worked example 3.9) and is chiral (Fig. 15.20). The importance of anions of this family lies in their ability to discriminate between chiral cations.[†] We return to this in Section 19.8.

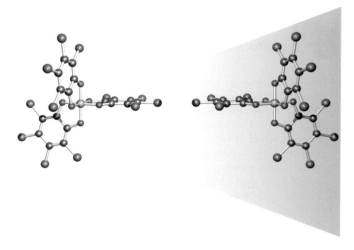

Fig. 15.20 The two enantiomers (non-superposable mirror images) of anion **15.67**. Colour code: P, orange; O, red; C, grey; Cl, green.

(15.68)

Arsenic acid, H_3AsO_4, is obtained by dissolving As_2O_5 in water or by oxidation of As_2O_3 using nitric acid (reaction 15.129). Values of $pK_a(1) = 2.25$, $pK_a(2) = 6.77$ and $pK_a(3) = 11.60$ for H_3AsO_4 show that it is of similar acidic strength to phosphoric acid (Table 15.7). Salts derived from H_3AsO_4 and containing the $[AsO_4]^{3-}$, $[HAsO_4]^{2-}$ or $[H_2AsO_4]^-$ ions can be prepared under appropriate conditions. In acidic solution, H_3AsO_4 acts as an oxidizing agent and the pH-dependence of the ease of oxidation or reduction is understood in terms of half-equation 15.143 and the relevant discussion in Section 8.2.

$$H_3AsO_4 + 2H^+ + 2e^- \rightleftharpoons H_3AsO_3 + H_2O$$
$$E^{\circ} = +0.56\,V \qquad (15.143)$$

Condensed polyarsenate ions are kinetically much less stable with respect to hydrolysis (i.e. cleavage of As−O−As bridges) than condensed polyphosphate ions, and only monomeric $[AsO_4]^{3-}$ exists in aqueous solution. Thus, $Na_2H_2As_2O_7$ can be made by dehydrating NaH_2AsO_4 at 360 K. Further dehydration (410 K) yields $Na_3H_2As_3O_{10}$ and, at 500 K, polymeric $(NaAsO_3)_n$ is formed. In the solid state, the latter contains infinite chains of tetrahedral AsO_4 units linked by As−O−As bridges. All these condensed arsenates revert to $[AsO_4]^{3-}$ on adding water.

Oxoacids of Sb(III) are not stable, and few antimonite salts are well characterized. Meta-antimonites include $NaSbO_2$

TRISPHAT

(15.67)

15.12 Oxoacids of arsenic, antimony and bismuth

'Arsenous acid' ($As(OH)_3$ or H_3AsO_3) has not been isolated. Aqueous solutions of As_2O_3 (see Section 15.10) probably contain H_3AsO_3, but metal arsenites (e.g. M_3AsO_3 where M = Ag or alkali metal) are known. Although there is little evidence for the existence of an acid of formula $As(O)OH$, salts of $[AsO_2]^-$ (meta-arsenites) are known. Sodium meta-arsenite, $NaAsO_2$ (commercially available), contains Na^+ ions and infinite chains, **15.68**, with trigonal pyramidal As(III) centres.

[†] For an overview, see: J. Lacour and V. Hebbe-Viton (2003) *Chem. Soc. Rev.*, vol. 32, p. 373 – 'Recent developments in chiral anion mediated asymmetric chemistry'.

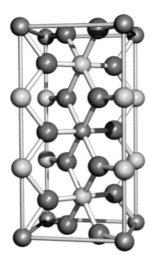

Fig. 15.21 The unit cell of $FeSb_2O_6$ which has a *trirutile* lattice; compare with the rutile unit cell in Fig. 6.22. Colour code: Sb, yellow; Fe, green; O, red; the edges of the unit cell are defined in yellow.

which can be prepared as the trihydrate from Sb_2O_3 and aqueous NaOH; the anhydrous salt has a polymeric structure. No oxoacids of Sb(V) are known, and neither is the tetrahedral anion '$[SbO_4]^{3-}$'. However, well-defined antimonates can be obtained, for example, by dissolving antimony(V) oxide in aqueous alkali and crystallizing the product. Some antimonates contain the octahedral $[Sb(OH)_6]^-$ ion, e.g. $Na[Sb(OH)_6]$ (originally formulated as $Na_2H_2Sb_2O_7 \cdot 5H_2O$) and $[Mg(OH_2)_6][Sb(OH)_6]_2$ (with the old formula of $Mg(SbO_3)_2 \cdot 12H_2O$). The remaining antimonates should be considered as mixed metal oxides. Their solid state structures

consist of 3-dimensional arrays in which Sb(V) centres are octahedrally coordinated by six O atoms and connected by Sb—O—Sb bridges, e.g. $NaSbO_3$, $FeSbO_4$, $ZnSb_2O_6$ and $FeSb_2O_6$ (Fig. 15.21).

No oxoacids of Bi are known, although some bismuthate salts are well characterized. Sodium bismuthate is an insoluble, orange solid, obtained by fusing Bi_2O_3 with NaOH in air or with Na_2O_2. It is a very powerful oxidizing agent, e.g. in the presence of acid, it oxidizes Mn(II) to $[MnO_4]^-$, and liberates Cl_2 from hydrochloric acid. Like antimonates, some of the bismuthates are better considered as mixed metal oxides. An example is the Bi(III)–Bi(V) compound $K_{0.4}Ba_{0.6}BiO_{3-x}$ ($x \approx 0.02$) which has a perovskite-type structure (Fig. 6.24) and is of interest as a Cu-free superconductor at 30 K (see Section 28.4).

15.13 Phosphazenes

Phosphazenes are a group of P(V)/N(III) compounds featuring chain or cyclic structures, and are oligomers of the hypothetical $N{\equiv}PR_2$. The reaction of PCl_5 with NH_4Cl in a chlorinated solvent (e.g. C_6H_5Cl) gives a mixture of colourless solids of formula $(NPCl_2)_n$ in which the predominant species have $n = 3$ or 4. The compounds $(NPCl_2)_3$ and $(NPCl_2)_4$ are readily separated by distillation under reduced pressure. Although eq. 15.144 summarizes the overall reaction, the mechanism is complicated. There is some evidence to support the scheme in Fig. 15.22 which illustrates the formation of the trimer.

$$n PCl_5 + n NH_4Cl \longrightarrow (NPCl_2)_n + 4n HCl \qquad (15.144)$$

$$3 PCl_5 + NH_4Cl \xrightarrow{-4HCl} [Cl_3P{=}N{=}PCl_3]^+ [PCl_6]^-$$

$$[NH_4]^+ + [PCl_6]^- \longrightarrow Cl_3P{=}NH + 3HCl$$

$$[Cl_3P{=}N{=}PCl_3]^+ + Cl_3P{=}NH \xrightarrow{-HCl} [Cl_3P{=}N{-}PCl_2{=}N{=}PCl_3]^+$$

$$\downarrow \begin{array}{c} Cl_3P{=}NH \\ -HCl \end{array}$$

$$[Cl_3P{=}N{-}PCl_2{=}N{-}PCl_2{=}N{=}PCl_3]^+$$

$$\downarrow -[PCl_4]^+$$

(a)

(b)

🌐 **Fig. 15.22** Proposed reaction scheme for the formation of the cyclic phosphazene $(NPCl_2)_3$, and the structures of (a) $[Cl_3P{=}N{-}PCl_2{=}N{=}PCl_3]^+$ and (b) $[Cl_3P{=}N{-}(PCl_2{=}N)_2{=}PCl_3]^+$. Both were determined by X-ray diffraction for the chloride salts [E. Rivard *et al.* (2004) *Inorg. Chem.*, vol. 43, p. 2765]. Colour code: P, orange; N, blue; Cl, green.

Reaction 15.144 is the traditional method of preparing $(NPCl_2)_3$, but yields are typically $\approx 50\%$. Improved yields can be obtained by using reaction 15.145. Again, although this looks straightforward, the reaction pathway is complicated and the formation of $(NPCl_2)_3$ competes with that of $Cl_3P{=}NSiMe_3$ (eq. 15.146). Yields of $(NPCl_2)_3$ can be optimized by ensuring a slow rate of addition of PCl_5 to $N(SiMe_3)_3$ in CH_2Cl_2. Yields of $Cl_3P{=}NSiMe_3$ (a precursor for phosphazene polymers, see below) are optimized if $N(SiMe_3)_3$ is added rapidly to PCl_5 in CH_2Cl_2, and hexane is then added.

$$3N(SiMe_3)_3 + 3PCl_5 \longrightarrow (NPCl_2)_3 + 9Me_3SiCl \tag{15.145}$$

$$N(SiMe_3)_3 + PCl_5 \longrightarrow Cl_3P{=}NSiMe_3 + 2Me_3SiCl \tag{15.146}$$

Reaction 15.144 can be adapted to produce $(NPBr_2)_n$ or $(NPMe_2)_n$ by using PBr_5 or Me_2PCl_3 (in place of PCl_5) respectively. The fluoro derivatives $(NPF_2)_n$ ($n = 3$ or 4) are not made directly, but are prepared by treating $(NPCl_2)_n$ with NaF suspended in MeCN or $C_6H_5NO_2$.

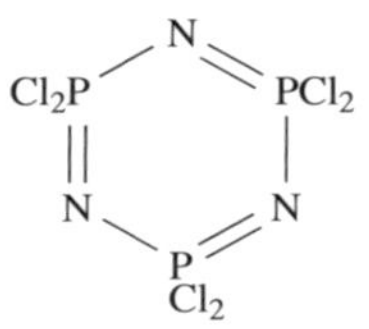

(15.69) 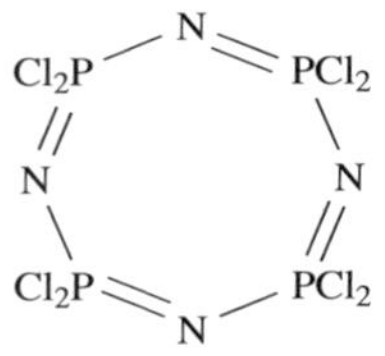   **(15.70)**

The Cl atoms in $(NPCl_2)_3$, **15.69**, and $(NPCl_2)_4$, **15.70**, readily undergo nucleophilic substitutions, e.g. the following groups can be introduced:

- F using NaF (see above);
- NH_2 using liquid NH_3;
- NMe_2 using Me_2NH;
- N_3 using LiN_3;
- OH using H_2O;
- Ph using LiPh.

Two substitution pathways are observed. If the group that first enters *decreases* the electron density on the P centre (e.g. F replaces Cl), the second substitution occurs at the *same* P atom. If the electron density *increases* (e.g. NMe_2 substitutes for Cl), then the second substitution site is at a different P centre.

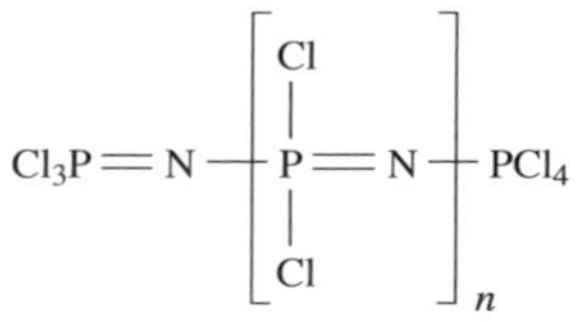

(15.71) 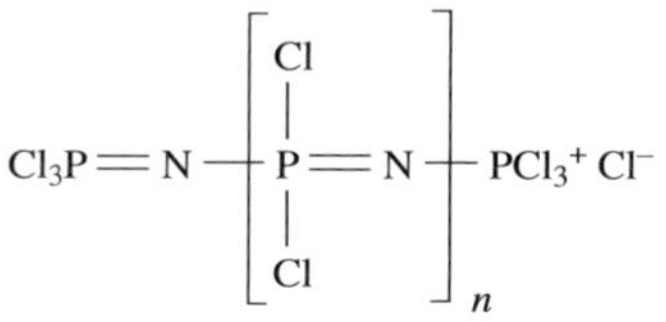   **(15.72)**

Small amounts of linear polymers are also produced in reaction 15.145, and their yield can be increased by using excess PCl_5. Such polymers may exist in either covalent (**15.71**) or ionic (**15.72**) forms. Polymers of $(NPCl_2)_3$ with molecular masses in the range 10^6, but with a wide mass distribution, result from heating molten $(NPCl_2)_3$ at 480–520 K. Room temperature cationic-polymerization can be achieved using $Cl_3P{=}NSiMe_3$ as a precursor (eq. 15.147). This leads to polymers with molecular masses around 10^5 and with a relatively small mass distribution.

$$Cl_3P{=}NSiMe_3 \xrightarrow[\text{(e.g. } PCl_5\text{) 297 K}]{\text{Cationic initiator}}$$
$$[Cl_3P{=}N(PCl_2{=}N)_nPCl_3]^+[PCl_6]^- \tag{15.147}$$

The first step in reaction 15.147 is the formation of $[Cl_3P{=}N{=}PCl_3]^+[PCl_6]^-$, which can be converted to the chloride salt by reaction 15.148. This is a convenient route to $[Cl_3P{=}N{=}PCl_3]^+Cl^-$ which is a precursor to higher polymers (e.g. eq. 15.149).

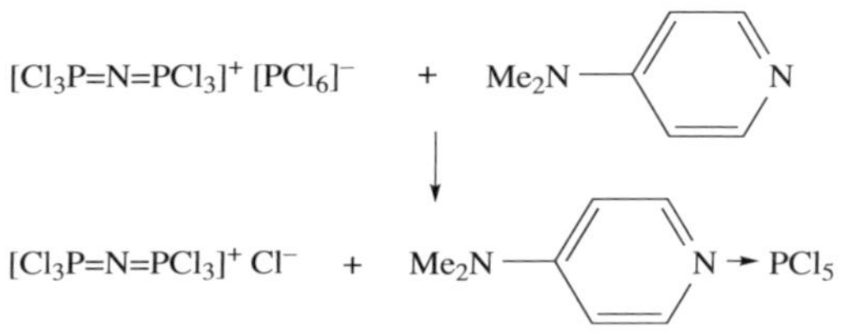

$$(15.148)$$

$$[Cl_3P{=}N{=}PCl_3]^+ Cl^- \ + \ n\,Cl_3P{=}NSiMe_3$$
$$\xrightarrow{-\,n\,Me_3SiCl} [Cl_3P{=}N{-}(PCl_2{=}N)_x{=}PCl_3]^+ Cl^-$$
$$x = 1, 2, 3$$
$$(15.149)$$

The structures of the $[Cl_3P{=}N{-}(PCl_2{=}N)_x{=}PCl_3]^+$ cations for $x = 1$ and 2 are shown in Fig. 15.22. The P–N–P bond angles in these polyphosphazenes lie in the range 134–157°, and the P–N bond distances are all similar (153–158 pm). This indicates that the bonding is delocalized, rather than the combination of double and single bonds that is traditionally drawn. The bonding is best described in terms of contributions from charge-separated resonance structures (i.e. ionic bonding), and negative hyperconjugation involving $n(N) \longrightarrow \sigma^*(P{-}Cl)$ electron donation where $n(N)$ represents the N lone pair. This is analogous to the negative hyperconjugation in $N(SiH_3)_3$ (Section 14.6).

The $[Ph_3P{=}N{=}PPh_3]^+$ ion (commonly abbreviated to $[PPN]^+$) is related to $[Cl_3P{=}N{=}PCl_3]^+$, and is often used to stabilize salts containing large anions (see Box 24.1).

The Cl atoms in the polymers are readily replaced, and this is a route to some commercially important materials. Treatment with sodium alkoxides, NaOR, yields linear polymers $[NP(OR)_2]_n$ which have water-resistant properties, and

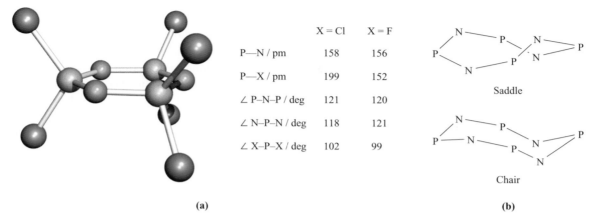

	X = Cl	X = F
P—N / pm	158	156
P—X / pm	199	152
∠ P–N–P / deg	121	120
∠ N–P–N / deg	118	121
∠ X–P–X / deg	102	99

Saddle

Chair

(a)   (b)

Fig. 15.23 (a) Structural parameters for the phosphazenes $(NPX_2)_3$ (X = Cl or F); colour code: P, orange; N, blue; X, green. (b) Schematic representations of the P_4N_4 ring conformations in $(NPF_2)_4$ (saddle conformation only) and $(NPCl_2)_4$ (saddle and chair conformations).

when R = CH_2CF_3, the polymers are inert enough for use in the construction of artificial blood vessels and organs. Many phosphazene polymers are used in fire-resistant materials (see Box 17.1).

The structures of $(NPCl_2)_3$, $(NPCl_2)_4$, $(NPF_2)_3$ and $(NPF_2)_4$ are shown in Fig. 15.23. Each of the 6-membered rings is planar, while the 8-membered rings are puckered. In $(NPF_2)_4$, the ring adopts a saddle conformation (Fig. 15.23b),[†] but two ring conformations exist for $(NPCl_2)_4$. The metastable form has a saddle conformation, while the stable form of $(NPCl_2)_4$ adopts a chair conformation (Fig. 15.23b). Although structures **15.69** and **15.70** indicate double and single bonds in the rings, crystallographic data show that the P−N bond lengths in a given ring are equal. Data for $(NPCl_2)_3$ and $(NPF_2)_3$ are given in Fig. 15.23a; in $(NPF_2)_4$, d(P−N) = 154 pm, and in the saddle and chair conformers of $(NPCl_2)_4$, d(P−N) = 157 and 156 pm respectively. The P−N bond distances are significantly shorter than expected for a P−N single bond (e.g. 177 pm in the anion in $Na[H_3NPO_3]$), indicating a degree of multiple bond character. Resonance structures **15.73** could be used to describe the bonding in the planar 6-membered rings, but both involve hypervalent P atoms.

Traditional bonding descriptions for the 6-membered rings have involved $N(2p)$–$P(3d)$ overlap, both in and perpendicular to the plane of the P_3N_3-ring. However, this model is not consistent with current opinion that phosphorus makes little or no use of its $3d$ orbitals. Structure **15.74** provides another resonance form for a 6-membered cyclophosphazene, and is consistent with the observed P−N bond equivalence, as well as the observation that the N and P atoms are subject to attack by electrophiles and nucleophiles, respectively. Theoretical results are consistent with highly polarized $P^{\delta+}$−$N^{\delta-}$ bonds and the absence of aromatic character in the P_3N_3-ring.[‡] As for the linear polyphosphazenes, both ionic bonding and negative hyperconjugation appear to contribute to the bonding in cyclic phosphazenes.

(15.74)

Self-study exercise

The azido derivative, $N_3P_3(N_3)_6$, is fully combusted according to the equation:

$$N_3P_3(N_3)_6(s) + \tfrac{15}{4}O_2(g) \longrightarrow \tfrac{3}{4}P_4O_{10}(s) + \tfrac{21}{2}N_2(g)$$

(15.73)

[†] Prior to 2001, the ring was thought to be planar; the correct conformation was previously masked by a crystallographic disorder (see Box 15.5). See: A.J. Elias *et al.* (2001) *J. Am. Chem. Soc.*, vol. 123, p. 10299.

[‡] For recent analysis of the bonding in phosphazenes, see: V. Luaña, A.M. Pendás, A. Costales, G.A. Carriedo and F.J. García-Alonso (2001) *Inorg. Chem.*, vol. 105, p. 5280; A.B. Chaplin, J.A. Harrison and P.J. Dyson (2005) *Inorg. Chem.*, vol. 44, p. 8407; L. Kapička, P. Kubáček and P. Holub (2007) *J. Mol. Struct.*, vol. 820, p. 148.

and the standard enthalpy of combustion has been determined as $-4142\,\text{kJ}\,\text{mol}^{-1}$. Calculate the value of $\Delta_fH^{\circ}(N_3P_3(N_3)_6,\ s)$ given that $\Delta_fH^{\circ}(P_4O_{10},\ s) = -2984\,\text{kJ}\,\text{mol}^{-1}$. Comment on the fact that $N_3P_3(N_3)_6$ is classed as a 'high energy density material'. What is the origin of the large difference between the value of $\Delta_fH^{\circ}(N_3P_3(N_3)_6,\ s)$ and that of $\Delta_fH^{\circ}(N_3P_3Cl_6,\ s) = -811\,\text{kJ}\,\text{mol}^{-1}$? $\qquad$ [*Ans.* $+1904\,\text{kJ}\,\text{mol}^{-1}$]

15.14 Sulfides and selenides

Sulfides and selenides of phosphorus

Sulfur–nitrogen compounds are described in Section 16.10, and in this section we look at the molecular sulfides and selenides formed by phosphorus. Although the structures of the sulfides (Fig. 15.24) appear to be closely related to those of the oxides (Section 15.10), there are some notable differences, e.g. P_4O_6 and P_4S_6 are not isostructural. The bond distances *within* the cages of all the sulfides indicate single P–P and P–S bonds. The data for P_4S_3 shown in Fig. 15.24 are typical. The terminal P–S bonds are shorter than those in the cage (e.g. 191 versus 208 pm in P_4S_{10}), and this can be rationalized in terms of a greater ionic contribution to the terminal bonds. Only some of the sulfides are prepared by direct combination of the elements. Above 570 K, white phosphorus combines with sulfur to give P_4S_{10} which is the most useful of the phosphorus sulfides. It is a thiating agent (i.e. one that introduces sulfur into a system) in organic reactions, and is a precursor to organothiophosphorus compounds. The reaction of red phosphorus with sulfur above 450 K yields P_4S_3, and P_4S_7 can also be made by direct combination under appropriate conditions. The remaining sulfides in Fig. 15.24 are made by one of the general routes:

- abstraction of sulfur using PPh_3 (e.g. reaction 15.150);
- treatment of a phosphorus sulfide with sulfur (e.g. reaction 15.151);
- treatment of a phosphorus sulfide with phosphorus (e.g. reaction 15.152);
- reaction of α- (**15.75**) or β-$P_4S_3I_2$ (**15.76**) with $(Me_3Sn)_2S$ (reaction 15.153).

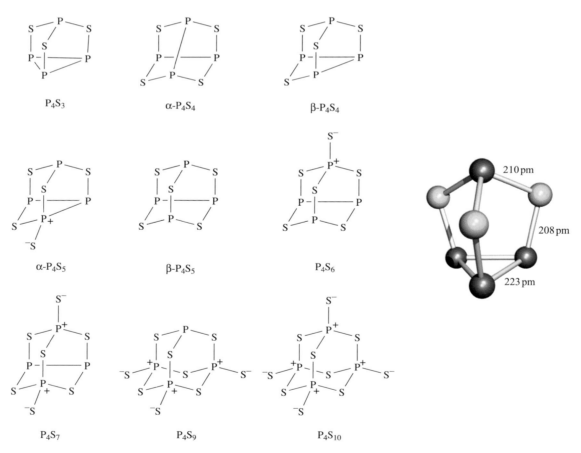

Fig. 15.24 Schematic representations of the molecular structures of phosphorus sulfides, and the structure (X-ray diffraction) of P_4S_3 [L.Y. Goh *et al.* (1995) *Organometallics*, vol. 14, p. 3886]. Colour code: S, yellow; P, brown.

There is ^{31}P NMR spectroscopic evidence that P_4S_8 has been prepared by treating P_4S_9 with PPh_3.

(15.75)   **(15.76)**

$$P_4S_7 + Ph_3P \longrightarrow P_4S_6 + Ph_3P{=}S \qquad (15.150)$$

$$P_4S_3 \xrightarrow{\text{excess sulfur}} P_4S_9 \qquad (15.151)$$

$$P_4S_{10} \xrightarrow{\text{red phosphorus}} \alpha\text{-}P_4S_5 \qquad (15.152)$$

$$\beta\text{-}P_4S_3I_2 + (Me_3Sn)_2S \longrightarrow \beta\text{-}P_4S_4 + 2Me_3SnI$$
$$\qquad (15.153)$$

Phosphorus sulfides ignite easily, and P_4S_3 is used in 'strike anywhere' matches; it is combined with $KClO_3$, and the compounds inflame when subjected to friction. (In safety matches, the head of the match contains $KClO_3$ and this reacts with red phosphorus which is combined with glass powder on the side of the match box; see end-of-chapter problem 15.42). Whereas P_4S_3 does not react with water, other phosphorus sulfides are slowly hydrolysed (e.g. reaction 15.154).

$$P_4S_{10} + 16H_2O \longrightarrow 4H_3PO_4 + 10H_2S \qquad (15.154)$$

We have already noted (Section 15.10) that, although sometimes referred to as 'phosphorus pentoxide', phosphorus(V) oxide does not exist as P_2O_5 molecules. In contrast, the vapour of phosphorus(V) sulfide contains some P_2S_5 molecules (although decomposition of the vapour to S, P_4S_7 and P_4S_3 also occurs). The phosphorus selenides P_2Se_5 and P_4Se_{10} are distinct species. Both can be made by direct combination of P and Se under appropriate conditions; P_2Se_5 is also formed by the decomposition of P_3Se_4I, and P_4Se_{10} from the reaction of P_4Se_3 and selenium at 620 K. Structure **15.77** has been confirmed by X-ray diffraction for P_2Se_5; P_4Se_{10} is isostructural with P_4S_{10} and P_4O_{10}.

(15.77)   **(15.78)**

When P_2S_5 is heated under vacuum with Cs_2S and sulfur in a $1:2:7$ molar ratio, $Cs_4P_2S_{10}$ is formed. This contains discrete $[P_2S_{10}]^{4-}$ ions (**15.78**), the terminal P–S bonds in which are shorter (201 pm) than the two in the central chain (219 pm).

Self-study exercise

P_2S_7 can be stabilized by using pyridine (py) to form an adduct: $(py)_2P_2S_7$ forms when P_4S_{10} is heated with S_8 in pyridine. The ^{31}P NMR spectrum of $(py)_2P_2S_7$ shows one signal (δ 82.2 ppm). Suggest a structure for the product and draw a resonance structure in which all atoms obey the octet rule.

[*Ans*. see C. Rotter *et al.* (2010) *Chem. Commun.*, vol. 46, p. 5024.]

Arsenic, antimony and bismuth sulfides

Arsenic and antimony sulfide ores are major sources of the group 15 elements (see Section 15.2). In the laboratory, As_2S_3 and As_2S_5 are usually precipitated from aqueous solutions of arsenite or arsenate. Reaction 15.155 proceeds when the H_2S is passed slowly through the solution at 298 K. If the temperature is lowered to 273 K and the rate of flow of H_2S is increased, the product is As_2S_5.

$$2[AsO_4]^{3-} + 6H^+ + 5H_2S \longrightarrow As_2S_3 + 2S + 8H_2O$$
$$\text{conc} \qquad (15.155)$$

Solid As_2S_3 has the same layer structure as the low-temperature polymorph of As_2O_3, but it vaporizes to give As_4S_6 molecules (see below). As_2S_5 exists in crystalline and vitreous forms, but structural details are not known. Both As_2S_3 and As_2S_5 are readily soluble in alkali metal sulfide solutions with the formation of thioarsenites and thioarsenates (e.g. eq. 15.156); acids decompose these salts, reprecipitating the sulfides.

$$As_2S_3 + 3S^{2-} \longrightarrow 2[AsS_3]^{3-} \qquad (15.156)$$

The sulfides As_4S_3 (*dimorphite*), As_4S_4 (*realgar*) and As_2S_3 (*orpiment*) occur naturally; the last two are red and golden-yellow respectively and were used as pigments in early times.[†] The arsenic sulfides As_4S_3, α-As_4S_4, β-As_4S_4 and β-As_4S_5 are structural analogues of the phosphorus sulfides in Fig. 15.24, but As_4S_6 is structurally related to P_4O_6 and As_4O_6 rather than to P_4S_6. The bond distances in α-As_4S_4 (**15.79**) are consistent with As–As and As–S single bonds, and this view of the cage allows a comparison with S_4N_4 (see Section 16.10).

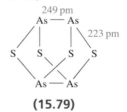

(15.79)

[†] For wider discussions of inorganic pigments, see: R.J.H. Clark and P.J. Gibbs (1997) *Chem. Commun.*, p. 1003 – 'Identification of lead(II) sulfide and pararealgar on a 13th century manuscript by Raman microscopy', and Further reading in Box 4.1.

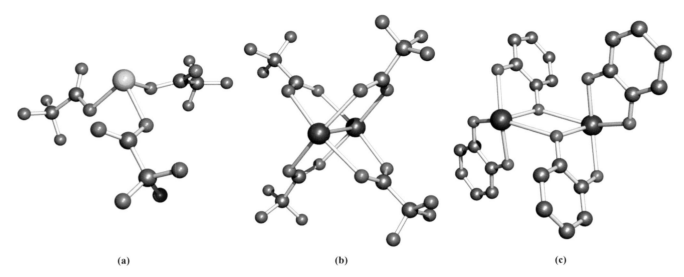

(a)   (b)   (c)

Fig. 15.25 The structures (X-ray diffraction) of (a) (*R*)-[Sb(O$_2$CCF$_3$)$_3$] [D.P. Bullivant *et al.* (1980) *J. Chem. Soc., Dalton Trans.*, p. 105], (b) Bi$_2$(O$_2$CCF$_3$)$_4$ [E.V. Dikarev *et al.* (2004) *Inorg. Chem.*, vol. 43, p. 3461] and (c) [Bi$_2$(C$_6$H$_4$O$_2$)$_4$]$^{2-}$, crystallized as a hydrated ammonium salt [G. Smith *et al.* (1994) *Aust. J. Chem.*, vol. 47, p. 1413]. Colour code: Sb, yellow; Bi, blue; O, red; F, green; C, grey.

The only well-characterized binary sulfide of Sb is the naturally occurring Sb$_2$S$_3$ (*stibnite*), which has a double-chain structure in which each Sb(III) is pyramidally sited with respect to three S atoms. The sulfide can be made by direct combination of the elements. A metastable red form can be precipitated from aqueous solution, but reverts to the stable black form on heating. Like As$_2$S$_3$, Sb$_2$S$_3$ dissolves in alkali metal sulfide solutions (see eq. 15.156). Bismuth(III) sulfide, Bi$_2$S$_3$, is isostructural with Sb$_2$S$_3$, but in contrast to its As and Sb analogues, Bi$_2$S$_3$ does not dissolve in alkali metal sulfide solutions.

15.15 Aqueous solution chemistry and complexes

Many aspects of the aqueous solution chemistry of the group 15 elements have already been covered:

- acid–base properties of NH$_3$, PH$_3$, N$_2$H$_4$ and HN$_3$ (Section 15.5);
- redox behaviour of nitrogen compounds (Section 15.5 and Fig. 15.6);
- the *brown ring test* for nitrate ion (Section 15.8);
- oxoacids (Sections 15.9, 15.11 and 15.12);
- condensed phosphates (Section 15.11);
- lability of condensed arsenates (Section 15.12);
- sequestering properties of polyphosphates (Section 15.11).

In this section we focus on the formation of aqueous solution species by Sb(III) and Bi(III). Solutions of Sb(III)

contain either hydrolysis products or complex ions. The former are commonly written as [SbO]$^+$, but by analogy with Bi(III) (see below), this is oversimplified. Complexes are formed with ligands such as oxalate, tartrate or trifluoro-acetate ions, and it is usual to observe an arrangement of donor atoms about the Sb atom that reflects the presence of a stereochemically active lone pair of electrons, e.g. in [Sb(O$_2$CCF$_3$)$_3$], the Sb(III) centre is in a trigonal pyramidal environment (Fig. 15.25a). The analogous bismuth(III) complex, Bi(O$_2$CCF$_3$)$_3$, undergoes an interesting reaction when heated with finely divided Bi in an evacuated, sealed vessel. The product, Bi$_2$(O$_2$CCF$_3$)$_4$ (Fig. 15.25b), is a rare example of a simple, bismuth(II) compound. It is diamagnetic and contains a Bi–Bi bond of length 295 pm (compare $2r_{cov} = 304$ pm). Bi$_2$(O$_2$CCF$_3$)$_4$ is decomposed by water and other polar solvents and, at $\approx$500 K, its vapour disproportionates to Bi(III) and Bi(0).

When a mixture of Bi$_2$O$_3$ and aqueous trifluoro-methanesulfonic acid is heated at reflux, crystals of [Bi(OH$_2$)$_9$][CF$_3$SO$_3$]$_3$ are obtained after cooling the solution. The [Bi(OH$_2$)$_9$]$^{3+}$ ion has a tricapped trigonal prismatic arrangement (**15.80**) of aqua ligands (compare this with the structure of [ReH$_9$]$^{2-}$ in Fig. 10.14c). However, in highly acidic aqueous media, the cation [Bi$_6$(OH)$_{12}$]$^{6+}$ is the dominant species. The six Bi(III) centres are arranged in an octahedron, but at non-bonded separations (Bi$\cdots$Bi = 370 pm), and each of the 12 Bi–Bi edges is supported by a bridging hydroxo ligand. In more alkaline solutions, [Bi$_6$O$_6$(OH)$_3$]$^{3+}$ is formed, and ultimately, Bi(OH)$_3$ is precipitated. Although there is evidence for the

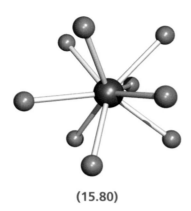

(15.80)

formation of bismuth polyoxo cations with nuclearities ranging from two to nine in alkaline solutions, few species have been isolated and structural data are sparse. An exception is the $[Bi_9(\mu_3\text{-}O)_8(\mu_3\text{-}OH)_6]^{5+}$ cation which is formed by hydrolysing $BiO(ClO_4)$ with aqueous NaOH and has been structurally characterized as the perchlorate salt. The Bi atoms in $[Bi_9(\mu_3\text{-}O)_8(\mu_3\text{-}OH)_6]^{5+}$ adopt a tricapped trigonal prismatic arrangement with O or OH groups capping the faces. The coordination geometry of Bi(III) is often influenced by the presence of a stereochemically active lone pair; e.g. in the catecholate complex $[Bi_2(C_6H_4O_2)_4]^{2-}$ (Fig. 15.25c), each Bi atom is in a square-based pyramidal environment. Figure 15.14 showed the structures of two complexes of $BiCl_3$ with macrocyclic ligands.

KEY TERMS

The following terms were introduced in this chapter. Do you know what they mean?

❑ chemiluminescent reaction ❑ acid anhydride ❑ azeotrope

FURTHER READING

D.E.C. Corbridge (1995) *Phosphorus*, 5th edn, Elsevier, Amsterdam – A book covering of all aspects of phosphorus chemistry.

J. Emsley (2000) *The Shocking Story of Phosphorus*, Macmillan, London – A readable book described as 'a biography of the devil's element'.

N.N. Greenwood and A. Earnshaw (1997) *Chemistry of the Elements*, 2nd edn, Butterworth-Heinemann, Oxford – Chapters 11–13 give a detailed account of the chemistries of the group 15 elements.

A.G. Massey (2000) *Main Group Chemistry*, 2nd edn, Wiley, Chichester – Chapter 9 covers the chemistry of the group 15 elements.

N.C. Norman, ed. (1998) *Chemistry of Arsenic, Antimony and Bismuth*, Blackie, London – A series of articles covering both inorganic and organometallic aspects of the later group 15 elements.

A.F. Wells (1984) *Structural Inorganic Chemistry*, 5th edn, Clarendon Press, Oxford – Chapters 18–20 give detailed accounts of the structures of compounds of the group 15 elements.

Specialized topics

J. Akhavan (2004) *The Chemistry of Explosives*, 2nd edn, RSC, Cambridge.

J.C. Bottaro (1996) *Chem. Ind.*, p. 249 – 'Recent advances in explosives and solid propellants'.

K. Dehnicke and J. Strähle (1992) *Angew. Chem. Int. Ed.*, vol. 31, p. 955 – 'Nitrido complexes of the transition metals'.

D.P. Gates and I. Manners (1997) *J. Chem. Soc., Dalton Trans.*, p. 2525 – 'Main-group-based rings and polymers'.

A.C. Jones (1997) *Chem. Soc. Rev.*, vol. 26, p. 101 – 'Developments in metal-organic precursors for semiconductor growth from the vapour phase'.

E. Maciá (2005) *Chem. Soc. Rev.*, vol. 34, p. 691 – 'The role of phosphorus in chemical evolution'.

J.E. Mark, H.R. Allcock and R. West (2005) *Inorganic Polymers*, 2nd edn, Oxford University Press, Oxford – Chapter 3 deals with polyphosphazenes, including applications.

S.T. Oyama (1996) *The Chemistry of Transition Metal Carbides and Nitrides*, Kluwer, Dordrecht.

G.B. Richter-Addo, P. Legzdins and J. Burstyn, eds (2002) *Chem. Rev.*, vol. 102, number 4 – A journal issue devoted to the chemistry of NO, and a source of key references for the area.

W. Schnick (1999) *Angew. Chem. Int. Ed.*, vol. 38, p. 3309 – 'The first nitride spinels – New synthetic approaches to binary group 14 nitrides'.

G. Steinhauser and T.M. Klapötke (2008) *Angew. Chem. Int. Ed.*, vol. 47, p. 3330 – '"Green" pyrotechnics: a chemists' challenge'.

PROBLEMS

15.1 What are the formal oxidation states of N or P in the following species? (a) N_2; (b) $[NO_3]^-$; (c) $[NO_2]^-$; (d) NO_2; (e) NO; (f) NH_3; (g) NH_2OH; (h) P_4; (i) $[PO_4]^{3-}$; (j) P_4O_6; (k) P_4O_{10}.

15.2 Using bond enthalpy terms from Tables 14.2 and 15.3, estimate values of $\Delta_r H^o$ for the following reactions:

   (a) $2N_2 \longrightarrow N_4$ (tetrahedral structure);
   (b) $2P_2 \longrightarrow P_4$ (tetrahedral structure);
   (c) $2C_2H_2 \longrightarrow C_4H_4$ (tetrahedrane, with a tetrahedral C_4 core).

15.3 Give a brief account of allotropy among the group 15 elements.

15.4 Write equations for the reactions of (a) water with Ca_3P_2; (b) aqueous NaOH with NH_4Cl; (c) aqueous NH_3 with $Mg(NO_3)_2$; (d) AsH_3 with an excess of I_2 in neutral aqueous solution; (e) PH_3 with KNH_2 in liquid NH_3.

15.5 Explain why (a) a dilute aqueous solution of NH_3 smells of the gas whereas dilute HCl does not retain the acrid odour of gaseous HCl, and (b) ammonium carbamate is used in smelling salts.

15.6 If (at 298 K) pK_b for NH_3 is 4.75, show that pK_a for $[NH_4]^+$ is 9.25.

15.7 Give the relevant half-equations for the oxidation of NH_2OH to HNO_3 by $[BrO_3]^-$, and write a balanced equation for the overall process.

15.8 (a) Write a balanced equation for the preparation of NaN_3 from $NaNH_2$ with $NaNO_3$. (b) Suggest a route for preparing the precursor $NaNH_2$. (c) How might NaN_3 react with $Pb(NO_3)_2$ in aqueous solution?

15.9 (a) We noted that $[N_3]^-$ is isoelectronic with CO_2. Give three other species that are also isoelectronic with $[N_3]^-$. (b) Describe the bonding in $[N_3]^-$ in terms of an MO picture.

15.10 Refer to Fig. 15.10. (a) By considering a number of unit cells of NiAs connected together, confirm that the coordination number of each Ni atom is 6. (b) How does the information contained in the unit cell of NiAs confirm the stoichiometry of the compound?

15.11 Suggest how you might confirm the conformation of N_2H_4 in (a) the gas phase and (b) the liquid phase.

15.12 In each of reactions 15.66, 15.67 and 15.68, $NF_2^{\bullet}$ reacts with another radical. What is the second radical in each reaction, and how is it formed? Draw a Lewis structure of F_2NNO (the product of reaction 15.68).

15.13 (a) Discuss structural variation among the phosphorus(III) and phosphorus(V) halides, indicating where stereochemical non-rigidity is possible. (b) On what basis is it appropriate to compare the lattice of $[PCl_4][PCl_6]$ with that of CsCl?

15.14 What might you expect to observe (at 298 K) in the ^{19}F NMR spectra of solutions containing (a) $[PF_6]^-$ and (b) $[SbF_6]^-$. Data needed are in Table 15.2.

15.15 Which of the following equations show redox reactions: eqs. 15.64, 15.70, 15.73, 15.111 and 15.123? For each redox reaction, indicate which species is being oxidized and which reduced. Confirm that the changes in oxidation states for the oxidation and reduction processes balance.

15.16 Explain whether it is possible to distinguish between the following pairs of isomers based *only* on the coupling patterns in the ^{31}P NMR spectra: (a) *cis*- and *trans*-$[PF_4(CN)_2]^-$, and (b) *mer*- and *fac*-$[PF_3(CN)_3]^-$.

15.17 Draw the structures of the possible isomers of $[PCl_2F_3(CN)]^-$, and state how many fluorine environments there are based on the structures you have drawn. At room temperature, the ^{19}F NMR spectra of CH_2Cl_2 solutions of two of the isomers exhibit two signals, while the spectrum of the third isomer shows only one signal. Account for these observations.

15.18 Suggest products for the reactions between (a) $SbCl_5$ and PCl_5; (b) KF and AsF_5; (c) NOF and SbF_5; (d) HF and SbF_5.

15.19 (a) Draw the structures of $[Sb_2F_{11}]^-$ and $[Sb_2F_7]^-$, and rationalize them in terms of the VSEPR model. (b) Suggest likely structures for the $[\{BiX_4\}_n]^{n-}$ and $[\{BiX_5\}_n]^{2n-}$ oligomers mentioned in Section 15.7.

15.20 By using an MO approach, rationalize why, in going from NO to $[NO]^+$, the bond order increases, bond distance decreases and NO vibrational wavenumber increases.

15.21 $25.0\,cm^3$ of a 0.0500 M solution of sodium oxalate $(Na_2C_2O_4)$ reacted with $24.8\,cm^3$ of a solution of $KMnO_4$, **A**, in the presence of excess H_2SO_4. $25.0\,cm^3$ of a 0.0494 M solution of NH_2OH in H_2SO_4 was boiled with an excess of iron(III) sulfate solution, and when the reaction was complete, the iron(II) produced was found to be equivalent to $24.65\,cm^3$ of solution **A**. The product **B** formed from the NH_2OH in this reaction can be assumed not to interfere with the determination of iron(II). What can you deduce about the identity of **B**?

15.22 Write a brief account that supports the statement that 'all the oxygen chemistry of phosphorus(V) is based on the tetrahedral PO_4 unit'.

15.23 Figure 15.21 shows a unit cell of $FeSb_2O_6$. (a) How is this unit cell related to the rutile-type structure? (b) Why can the solid state structure of $FeSb_2O_6$ not be described in terms of a single unit cell of the rutile-type structure? (c) What is the coordination environment of each atom type? (d) Confirm the stoichiometry of this compound using only the information provided in the unit cell diagram.

15.24 How may NMR spectroscopy be used:

(a) to distinguish between solutions of $Na_5P_3O_{10}$ and $Na_6P_4O_{13}$;

(b) to determine whether F atoms exchange rapidly between non-equivalent sites in AsF_5;

(c) to determine the positions of the NMe_2 groups in $P_3N_3Cl_3(NMe_2)_3$?

15.25 Deduce what you can about the nature of the following reactions.

(a) One mole of NH_2OH reacts with two moles of Ti(III) in the presence of excess alkali, and the Ti(III) is converted to Ti(IV).

(b) When Ag_2HPO_3 is warmed in water, all the silver is precipitated as metal.

(c) When one mole of H_3PO_2 is treated with excess I_2 in acidic solution, one mole of I_2 is reduced; on making the solution alkaline, a second mole of I_2 is consumed.

15.26 Predict the structures of (a) $[NF_4]^+$; (b) $[N_2F_3]^+$; (c) NH_2OH; (d) $SPCl_3$; (e) PCl_3F_2.

15.27 Suggest syntheses for each of the following from $K^{15}NO_3$: (a) $Na^{15}NH_2$, (b) $^{15}N_2$ and (c) $[^{15}NO][AlCl_4]$.

15.28 Suggest syntheses for each of the following from $Ca_3(^{32}PO_4)_2$: (a) $^{32}PH_3$, (b) $H_3{}^{32}PO_3$ and (c) $Na_3{}^{32}PS_4$.

15.29 $25.0\,cm^3$ of a $0.0500\,M$ solution of sodium oxalate reacted with $24.7\,cm^3$ of a solution of $KMnO_4$, **C**, in the presence of excess H_2SO_4. $25.0\,cm^3$ of a $0.0250\,M$ solution of N_2H_4 when treated with an excess of alkaline $[Fe(CN)_6]^{3-}$ solution gave $[Fe(CN)_6]^{4-}$ and a product **D**. The $[Fe(CN)_6]^{4-}$ formed was reoxidized to $[Fe(CN)_6]^{3-}$ by $24.80\,cm^3$ of solution **C**, and the presence of **D** did not influence this determination. What can you deduce about the identity of **D**?

15.30 Comment on the fact that $AlPO_4$ exists in several forms, each of which has a structure which is also that of a form of silica.

15.31 (a) Explain what is meant by *hyperconjugation* in a phosphazene such as $[Cl_3P=N-PCl_2=N=PCl_3]^+$. (b) Draw resonance structures for $[Cl_3P=N-PCl_2=N=PCl_3]^+$ which illustrate contributions to the bonding from charge-separated species.

OVERVIEW PROBLEMS

15.32 (a) The ^{31}P and ^{11}B NMR spectra of $Pr_3P\cdot BBr_3$ (Pr = *n*-propyl) exhibit a $1:1:1:1$ quartet ($J = 150\,Hz$) and a doublet ($J = 150\,Hz$), respectively. Explain the origin of these signals.

(b) Discuss the factors that contribute towards $[NH_4][PF_6]$ being soluble in water.

(c) The ionic compound $[AsBr_4][AsF_6]$ decomposes to Br_2, AsF_3 and $AsBr_3$. The proposed pathway is as follows:

$$[AsBr_4][AsF_6] \longrightarrow [AsBr_4]F + AsF_5$$

$$[AsBr_4]F \longrightarrow AsBr_2F + Br_2$$

$$AsBr_2F + AsF_5 \longrightarrow 2AsF_3 + Br_2$$

$$3AsBr_2F \longrightarrow 2AsBr_3 + AsF_3$$

Discuss these reactions in terms of redox processes and halide redistributions.

15.33 Suggest products for the following reactions; the equations are not necessarily balanced on the left-hand sides.

(a) $PI_3 + IBr + GaBr_3 \longrightarrow$

(b) $POBr_3 + HF + AsF_5 \longrightarrow$

(c) $Pb(NO_3)_2 \xrightarrow{\Delta}$

(d) $PH_3 + K \xrightarrow{\text{liquid } NH_3}$

(e) $Li_3N + H_2O \longrightarrow$

(f) $H_3AsO_4 + SO_2 + H_2O \longrightarrow$

(g) $BiCl_3 + H_2O \longrightarrow$

(h) $PCl_3 + H_2O \longrightarrow$

15.34 (a) Draw the structure of P_4S_3 and describe an appropriate bonding scheme for this molecule. Compare the structures of P_4S_{10}, P_4S_3 and P_4, and comment on the formal oxidation states of the P atoms in these species.

(b) The electrical resistivity of Bi at 273 K is $1.07 \times 10^{-6} \, \Omega \, m$. How do you expect this property to change as the temperature increases? On what grounds have you drawn your conclusion?

(c) Hydrated iron(III) nitrate was dissolved in hot HNO_3 (100%), and the solution was placed in a desiccator with P_2O_5 until the sample had become a solid residue. The pure Fe(III) product (an ionic salt $[NO_2][X]$) was collected by sublimation; crystals were extremely deliquescent. Suggest an identity for the product, clearly stating the charges on the ions. The Fe(III) centre has a coordination number of 8. Suggest how this is achieved.

15.35 (a) Predict the ^{31}P NMR spectrum of $[HPF_5]^-$ (assuming a static structure) given that $J_{PH} = 939$ Hz, $J_{PF(axial)} = 731$ Hz and $J_{PF(equatorial)} = 817$ Hz.

(b) The $[BiF_7]^{2-}$ and $[SbF_6]^{3-}$ ions have pentagonal bipyramidal and octahedral structures, respectively. Are these observations consistent with the VSEPR model?

(c) Consider the following reaction scheme (K.O. Christe (1995) *J. Am. Chem. Soc.*, vol. 117, p. 6136):

$$NF_3 + NO + 2SbF_5 \xrightarrow{420 \text{ K}} [F_2NO]^+[Sb_2F_{11}]^- + N_2$$

$$\Big\updownarrow >450 \text{ K}$$

$$F_3NO + 2SbF_5$$

$$\Big\updownarrow >520 \text{ K}$$

$$[NO]^+[SbF_6]^- \xleftarrow{SbF_5} FNO + F_2$$

$$\Big\downarrow NF_3, SbF_5$$

$$[NF_4]^+[SbF_6]^-$$

Discuss the reaction scheme in terms of redox and Lewis acid–base chemistry. Comment on the structures of, and bonding in, the nitrogen-containing species in the scheme.

15.36 (a) Sn_3N_4, γ-Si_3N_4 and γ-Ge_3N_4 are the first examples of *nitride spinels*. What is a spinel, and how do the structures of these nitrides relate to that of the oxide Fe_3O_4? Comment on any features that distinguish the nitride spinels from typical oxide analogues.

(b) The reaction between O_3 and $AsCl_3$ at 195 K leads to an As(V) compound **A**. Raman spectra of **A** in CH_2Cl_2 solution are consistent with a molecular structure with C_{3v} symmetry. However, a single-crystal X-ray diffraction study of **A** at 153 K reveals a molecular structure with C_{2h} symmetry.

Suggest an identity for **A** and rationalize the experimental data.

15.37 (a) Why does fuming nitric acid appear orange in colour?

(b) By using nitric acid as an example, explain what is meant by the term *azeotrope*.

15.38 (a) Use Wade's rules to rationalize the fact that in $[Pd@Bi_{10}]^{4+}$, the Bi atoms are arranged in a pentagonal antiprism. How is this structure related to that of $[Pd@Pb_{12}]^{2-}$?

(b) At 298 K, ammonium perchlorate decomposes according to the equation:
$$4NH_4ClO_4(s) \longrightarrow 2Cl_2(g) + 2N_2O(g) + 3O_2(g) + 8H_2O(l)$$

Determine $\Delta_r G^{\circ}(298 \text{ K})$ for this decomposition if $\Delta_f G^{\circ}(298 \text{ K})$ of $N_2O(g)$, $H_2O(l)$ and $NH_4ClO_4(s)$ are $+104$, -237 and $-89 \, kJ \, mol^{-1}$. What part does entropy play in the reaction?

15.39 (a) What would you predict would happen when equimolar amounts of NaN_3 and $NaNO_2$ react in acidic solution? How would you attempt to confirm your prediction?

(b) $POCl_3$ reacts with an excess of Me_2NH to yield compound **A** as the only phosphorus-containing product; compound **A** is miscible with water. **A** contains 40.21%C, 23.45%N and 10.12%H, and each of the solution ^1H and ^{13}C NMR spectra exhibits one signal. Equimolar amounts of **A** and RNH_2 (R = alkyl) react, eliminating dimethylamine to give **B** (shown below).

B

(i) Suggest the identity of **A**, and draw its structure, giving a resonance form in which all non-H atoms obey the octet rule. (ii) What is the origin of the miscibility of **A** with water? (iii) Write a balanced equation for the formation of **A** from $POCl_3$ and Me_2NH.

15.40 Electron diffraction and spectroscopic studies of mixed fluoro/chloro phosphorus pentahalides are consistent with trigonal bipyramidal structures in which the most electronegative halogens occupy the axial positions. Confirm that this statement is in agreement with PCl_3F_2, PCl_2F_3 and $PClF_4$ having D_{3h}, C_{2v} and C_{2v} symmetries, respectively. Draw the structure of each compound and state whether the compound is polar.

INORGANIC CHEMISTRY MATTERS

15.41 Box 15.9 deals with the nitrogen cycle and the removal of nitrates and nitrites from waste water. (a) Urea is used to reduce $[NO_2]^-$ to N_2. Write a balanced equation for the reaction of HNO_2 with urea. (b) Sulfamic acid is also used to reduce $[NO_2]^-$ to N_2 during water treatment. Give an equation for this reaction. (c) Nitrites can be removed using H_2O_2, $[OCl]^-$ or $HOCl$ as oxidants. How does the reduction potential for the following process depend on pH?

$$[NO_3]^-(aq) + 3H^+(aq) + 2e^- \rightleftharpoons HNO_2(aq) + H_2O(l)$$
$$E^\circ = +0.93V$$

Suggest products for the reduction of H_2O_2 and $HOCl$ in acidic, aqueous solution. Give equations for the reactions of H_2O_2 with $[NO_2]^-$ in acidic solution, and for $[OCl]^-$ with $[NO_2]^-$ in alkaline solution.

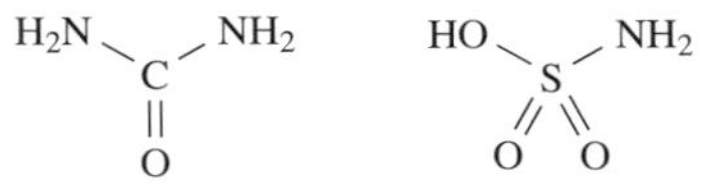

Urea   Sulfamic acid

15.42 (a) The head of a safety match contains $KClO_3$ and this reacts with red phosphorus which is combined with glass powder on the side of the match box. Write an equation for the reaction that occurs when the match is struck. Show that changes in oxidation states balance. Comment on the role of the glass powder.

(b) In a 'strike-it-anywhere' match, the match head contains P_4S_3 and $KClO_3$. Use oxidation state changes to suggest what happens when you strike the match, assuming that additional O_2 is required for combustion.

15.43 Polyphosphazenes are an important class of inorganic macromolecule and have many commercial applications, e.g. fire retardants, elastomers, fuel cell membranes, biomedical applications. The scheme below shows synthetic routes to three polymers. (a) Outline how $(NPCl_2)_n$ is produced on a large scale starting from PCl_5 and NH_4Cl. (b) What are the identities of the polymers **A–C**?

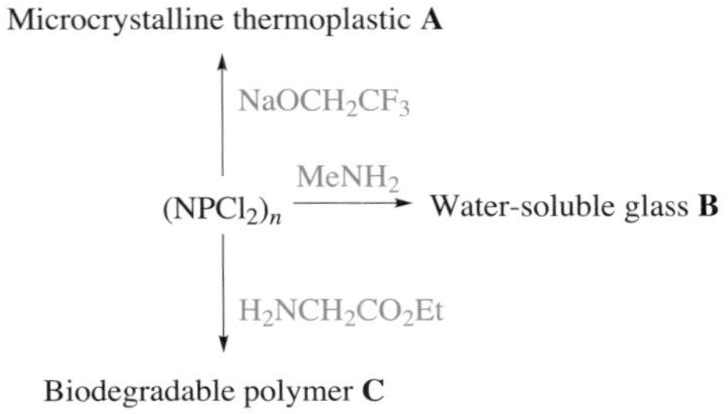

Microcrystalline thermoplastic **A**

NaOCH$_2$CF$_3$

$(NPCl_2)_n$ $\xrightarrow{\text{MeNH}_2}$ Water-soluble glass **B**

H$_2$NCH$_2$CO$_2$Et

Biodegradable polymer **C**

15.44 Over 95% of the phosphate rock mined in the US is used to manufacture phosphoric acid and phosphate-based fertilizers. The remaining phosphate rock is used for the production of white phosphorus. (a) Give equations to show how phosphate rock is converted to phosphoric acid and to white phosphorus. (b) Phosphate rock is converted to superphosphate fertilizers by treatment with concentrated H_2SO_4 and to triple superphosphate by reaction with H_3PO_4. What is the chemical composition of these fertilizers? Why are they suitable for use as fertilizers when phosphate rock itself is not? (c) What is meant by condensation of phosphates? (d) What is the role of $Na_5[P_3O_{10}]$ in detergents? Why is its use in decline, and what substitute is now favoured? Explain how the newer material carries out the role originally played by $Na_5[P_3O_{10}]$.

Topics

16

The group 16 elements

1	2		13	14	15	**16**	17	18
H								He
Li	Be		B	C	N	**O**	F	Ne
Na	Mg		Al	Si	P	**S**	Cl	Ar
K	Ca	*d*-block	Ga	Ge	As	**Se**	Br	Kr
Rb	Sr		In	Sn	Sb	**Te**	I	Xe
Cs	Ba		Tl	Pb	Bi	**Po**	At	Rn
Fr	Ra							

16.1 Introduction

> The group 16 elements – oxygen, sulfur, selenium, tellurium and polonium – are called the *chalcogens*.

Oxygen occupies so central a position in any treatment of inorganic chemistry that discussions of many of its compounds are dealt with under other elements. The decrease in non-metallic character down the group is easily recognized in the elements:

- oxygen exists only as two gaseous allotropes (O_2 and O_3);
- sulfur has many allotropes, all of which are insulators;
- the stable forms of selenium and tellurium are semi-conductors;
- polonium is a metallic conductor.

Knowledge of the chemistry of Po and its compounds is limited because of the absence of a stable isotope[†] and the difficulty of working with ^{210}Po, the most readily available isotope. Polonium-210 is produced by bombarding ^{209}Bi with thermal (or 'slow') neutrons. The neutrons are produced by fission of ^{235}U nuclei, and their kinetic energy is reduced by elastic collisions with ^{12}C or ^{2}H nuclei during passage through graphite or D_2O. Combination of ^{209}Bi with a neutron yields ^{210}Bi, which undergoes β-decay to form ^{210}Po. Polonium-210 is an intense α-emitter ($t_{1/2} = 138\,d$) which liberates $520\,kJ\,g^{-1}\,h^{-1}$. This large energy loss causes many compounds of Po to decompose; Po decomposes water, making studies of chemical reactions in aqueous solution difficult. Polonium is a metallic conductor and crystallizes in a simple cubic lattice. It forms volatile, readily hydrolysed halides $PoCl_2$, $PoCl_4$, $PoBr_2$, $PoBr_4$ and PoI_4 and complex ions $[PoX_6]^{2-}$ (X = Cl, Br, I). Polonium(IV) oxide is formed by reaction between Po and O_2 at 520 K; it adopts a fluorite-type structure (see Fig. 6.19) and is sparingly soluble in aqueous alkali.

16.2 Occurrence, extraction and uses

Occurrence

Figure 16.1 illustrates the relative abundances of the group 16 elements in the Earth's crust. Dioxygen makes up 21% of the Earth's atmosphere (see Fig. 15.1b), and 47% of the Earth's crust is composed of O-containing compounds (Fig. 13.2), e.g. water, limestone, silica, silicates, bauxite

---

[†]For a recent update, see: R. Collé, L. Laureano-Perez and I. Outola (2007) *Appl. Radiat. Isotopes*, vol. 65, p. 728 – 'A note on the half-life of ^{209}Po'.

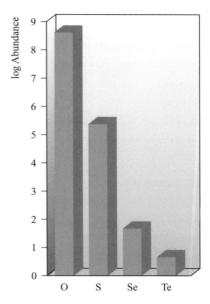

Fig. 16.1 Relative abundances of the group 16 elements (excluding Po) in the Earth's crust. The data are plotted on a logarithmic scale. The units of abundance are parts per billion (1 billion $= 10^9$). Polonium is omitted because its abundance is only 3×10^{-7} ppb, giving a negative number on the log scale.

and haematite. It is a component of innumerable compounds and is essential to life, being converted to CO_2 during respiration. Native sulfur occurs in deposits around volcanoes and hot springs, and sulfur-containing minerals include *iron pyrites* (*fool's gold*, FeS_2), *galena* (PbS), *sphalerite* or *zinc blende* (ZnS), *cinnabar* (HgS), *realgar* (As_4S_4), *orpiment* (As_2S_3), *stibnite* (Sb_2S_3), *molybdenite*

(MoS_2) and *chalcocite* (Cu_2S). Selenium and tellurium are relatively rare (Fig. 16.1). Selenium occurs in only a few minerals, while Te is usually combined with other metals, e.g. in *sylvanite* ($AgAuTe_4$).

Extraction

The Frasch process is the traditional means of extracting sulfur from natural deposits. Superheated water (440 K under pressure) is used to melt the sulfur, and compressed air then forces it to the surface. This method is now in decline and many operations have been closed. Canada and the US are the largest producers of sulfur in the world, and Fig. 16.2 shows the dramatic changes in methods of sulfur production in the US between 1980 and 2008. The trend is being followed worldwide as sulfur recovery from crude petroleum refining and natural gas production becomes the dominant production process for environmental reasons. In natural gas, the source of sulfur is H_2S which occurs in concentrations of up to 30%. Sulfur is recovered by reaction 16.1. The third method of production is labelled 'by-product from sulfuric acid' manufacture in Fig. 16.2. This refers to the direct coupling of the manufacture of sulfuric acid (only a fraction of the total produced) to the extraction of metals (e.g. copper) from sulfide ores by roasting in air. The SO_2 evolved is used for the manufacture of H_2SO_4 (see Section 16.9), and in Fig. 16.2, it is included as a source of sulfur (although elemental sulfur is never isolated).

$$2H_2S + O_2 \xrightarrow{\text{activated carbon or alumina catalyst}} 2S + 2H_2O$$

$$(16.1)$$

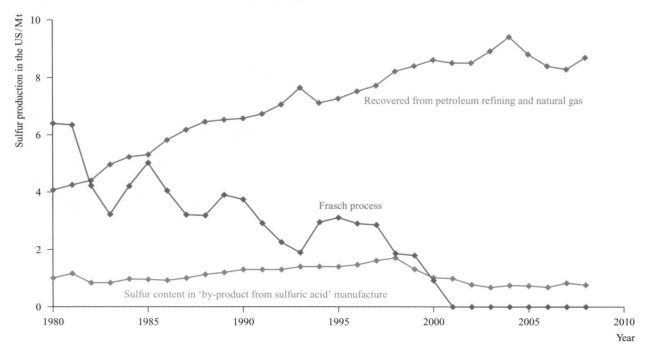

Fig. 16.2 Production of sulfur in the US from 1980 to 2008; note the increasing importance of recovery methods which have now replaced the Frasch process as a source of sulfur in the US. [Data: US Geological Survey.] See text for an explanation of 'by-product sulfuric acid'.

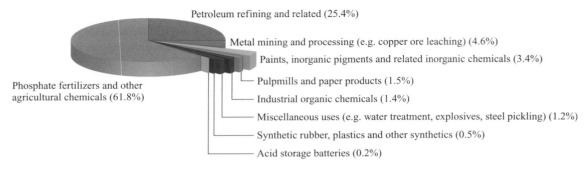

Petroleum refining and related (25.4%)

Metal mining and processing (e.g. copper ore leaching) (4.6%)

Paints, inorganic pigments and related inorganic chemicals (3.4%)

Pulpmills and paper products (1.5%)

Industrial organic chemicals (1.4%)

Miscellaneous uses (e.g. water treatment, explosives, steel pickling) (1.2%)

Synthetic rubber, plastics and other synthetics (0.5%)

Acid storage batteries (0.2%)

Phosphate fertilizers and other agricultural chemicals (61.8%)

Fig. 16.3 Uses of sulfur and sulfuric acid (by sulfur content) in the US in 2008. [Data: US Geological Survey.]

Commercial sources of Se and Te are flue dusts deposited during the refining of, for example, copper sulfide ores and from anode residues from the electrolytic refining of copper.

Uses

The chief use of O_2 is as a fuel (e.g. for oxyacetylene and hydrogen flames), as a supporter of respiration under special conditions (e.g. in air- and spacecraft), and in steel manufacturing.

Sulfur, mainly in the form of sulfuric acid, is an enormously important industrial chemical. The amount of sulfuric acid consumed by a given nation is an indicator of that country's industrial development. Figure 16.3 illustrates applications of sulfur and sulfuric acid. Sulfur is usually present in the form of an industrial *reagent* (e.g. in

APPLICATIONS

Box 16.1 Selenium as a photoconductor

The photoconductive properties of selenium are responsible for its role in photocopiers: the technique of *xerography* developed rapidly in the latter half of the 20th century. Amorphous selenium is deposited by a vaporization technique to provide a thin film ($\approx$50 μm thick) on an Al drum which is then installed in a photocopier. At the start of a photocopying run, the amorphous Se film is charged by a high-voltage corona discharge. Exposure of the amorphous Se film to light, with the image to be copied present in the light beam, creates a latent image which is produced in the form of regions of differing electrostatic potential. The image is developed using powdered toner which distributes itself over the 'electrostatic image'. The latter is then transferred to paper (again electrostatically) and fixed by heat treatment. An amorphous Se-coated photoreceptor drum has a lifetime of $\approx$100 000 photocopies. Spent drums are recycled, with some of the main recycling units being in Canada, Japan, the Philippines and several European countries. Once the mainstay of the photocopying industry, Se is gradually being replaced by organic photoreceptors, which are preferable to selenium on both performance and environmental grounds.

In the last decade or so, amorphous selenium has been applied as a photoconductor in X-ray image detectors, particularly for biomedical imaging. For this application, flat-panel detectors with large sensing areas (> 30 cm × 30 cm) have been developed. The coating of amorphous Se is typically 500 μm thick, and is deposited over a silicon thin film transistor layer. After applying an electrical potential to the surface, the detector is exposed to an X-ray beam and the electrons released are used to transmit information that ultimately provides an image.

A medical technician laminating an amorphous selenium-containing X-ray imaging plate.

Further reading

G. Belev and S.O. Kasap (2004) *J. Non-Cryst. Solids*, vol. 345/346, p. 484 – 'Amorphous selenium as an X-ray photoconductor'.

Table 16.1 Some physical properties of the group 16 elements and their ions.

Property	O	S	Se	Te	Po
Atomic number, Z	8	16	34	52	84
Ground state electronic configuration	$[He]2s^2 2p^4$	$[Ne]3s^2 3p^4$	$[Ar]3d^{10} 4s^2 4p^4$	$[Kr]4d^{10} 5s^2 5p^4$	$[Xe]4f^{14} 5d^{10} 6s^2 6p^4$
Enthalpy of atomization, $\Delta_a H^\circ(298\,\mathrm{K}) / \mathrm{kJ\,mol}^{-1}$	249‡	277	227	197	$\approx$146
Melting point, mp / K	54	388	494	725	527
Boiling point, bp / K	90	718	958	1263*	1235
Standard enthalpy of fusion, $\Delta_{fus} H^\circ(\mathrm{mp}) / \mathrm{kJ\,mol}^{-1}$	0.44	1.72	6.69	17.49	–
First ionization energy, $IE_1 / \mathrm{kJ\,mol}^{-1}$	1314	999.6	941.0	869.3	812.1
$\Delta_{EA} H^\circ_1(298\,\mathrm{K}) / \mathrm{kJ\,mol}^{-1}$ **	−141	−201	−195	−190	−183
$\Delta_{EA} H^\circ_2(298\,\mathrm{K}) / \mathrm{kJ\,mol}^{-1}$ **	+798	+640			
Covalent radius, r_{cov} / pm	73	103	117	135	–
Ionic radius, r_{ion} for X^{2-} / pm	140	184	198	211	–
Pauling electronegativity, χ^P	3.4	2.6	2.6	2.1	2.0
NMR active nuclei (% abundance, nuclear spin)	^{17}O (0.04, $I = \frac{5}{2}$)	^{33}S (0.76, $I = \frac{3}{2}$)	^{77}Se (7.6, $I = \frac{1}{2}$)	^{123}Te (0.9, $I = \frac{1}{2}$) ^{125}Te (7.0, $I = \frac{1}{2}$)	

‡ For oxygen, $\Delta_a H^\circ = \frac{1}{2} \times$ Dissociation energy of O_2.
* For amorphous Te.
** $\Delta_{EA} H^\circ_1(298\,\mathrm{K})$ is the enthalpy change associated with the process $X(g) + e^- \longrightarrow X^-(g) \approx -\Delta U(0\,\mathrm{K})$; see Section 1.10. $\Delta_{EA} H^\circ_2(298\,\mathrm{K})$ refers to the process $X^-(g) + e^- \longrightarrow X^{2-}(g)$.

H_2SO_4 in the production of superphosphate fertilizers), and it is not necessarily present in the end-product.

An important property of amorphous-Se is its ability to convert light into electricity, and selenium is used in photo-electric cells, photographic exposure meters, photocopiers and X-ray imaging detectors (Box 16.1). A major use of selenium is in the glass industry. It is used to counteract the green tint caused by iron impurities in soda-lime silica glasses, and is also added to architectural plate glass to reduce solar heat transmission. In the form of CdS_xSe_{1-x}, selenium is used as a red pigment in glass and ceramics. Crystalline Se is a semiconductor. Tellurium is used as an additive ($\leq$0.1%) to low-carbon steels in order to improve the machine qualities of the metal. This accounts for about half of the world's consumption of tellurium. Catalytic applications are also important, and other applications stem from its semiconducting properties, e.g. cadmium telluride has recently been incorporated into solar cells (see Box 14.2). However, uses of Te are limited, partly because Te compounds are readily absorbed by the body and excreted in the breath and perspiration as foul-smelling organic derivatives.

16.3 Physical properties and bonding considerations

Table 16.1 lists selected physical properties of the group 16 elements. The trend in electronegativity values has important consequences as regards the ability of O−H bonds to form hydrogen bonds. This pattern follows that in group 15. While O−H····X and X−H····O (X = O, N, F) interactions are relatively strong hydrogen bonds, those involving sulfur are weak, and typically involve a strong hydrogen-bond donor with sulfur acting as a weak acceptor (e.g. O−H····S).[†] In the case of S−H····S hydrogen bonds, the calculated hydrogen bond enthalpy is $\approx$5 kJ mol^{-1} in H_2S····H_2S, compared with $\approx$20 kJ mol^{-1} for the O−H····O hydrogen bond in H_2O····H_2O (see Table 10.4).

In comparing Table 16.1 with analogous tables in Chapters 11–15, you should note the importance of *anion*, rather than cation, formation. With the possible exception of PoO_2, there is no evidence that group 16 compounds contain

---

[†] See: T. Steiner (2002) *Angew. Chem. Int. Ed.*, vol. 41, p. 48 – 'The hydrogen bond in the solid state'.

monatomic cations. Thus Table 16.1 lists values only of the *first* ionization energies to illustrate the expected decrease on descending the group. Electron affinity data for oxygen show that reaction 16.2 for E = O is highly endothermic, and O^{2-} ions exist in ionic lattices only because of the high lattice energies of metal oxides (see Section 6.16).

$$\left.\begin{array}{l} E(g) + 2e^- \longrightarrow E^{2-}(g) \qquad (E = O, S) \\ \Delta_r H^\circ(298\,K) = \Delta_{EA}H^\circ_1(298\,K) + \Delta_{EA}H^\circ_2(298\,K) \end{array}\right\}(16.2)$$

Reaction 16.2 for E = S is also endothermic (Table 16.1), but less so than for O since the repulsion between electrons is less in the larger anion. However, the energy needed to compensate for this endothermic step tends not to be available since lattice energies for sulfides are much lower than those of the corresponding oxides because of the much greater radius of the S^{2-} ion. Consequences of this are that:

- high oxidation state oxides (e.g. MnO_2) often have no sulfide analogues;
- agreement between calculated and experimental values of lattice energies (see Section 6.15) for many *d*-block metal sulfides is much poorer than for oxides, indicating significant covalent contributions to the bonding.

Similar considerations apply to selenides and tellurides.

Worked example 16.1   Thermochemical cycles for metal oxides and sulfides

(a) **Using data from the Appendices and the value $\Delta_f H^\circ$ (ZnO, s) = −350 kJ mol^{-1}, determine the enthalpy change (at 298 K) for the process:**

$$Zn^{2+}(g) + O^{2-}(g) \longrightarrow ZnO(s)$$

(b) **What percentage contribution does $\Delta_{EA}H^\circ_2(O)$ make to the overall enthalpy change for the following process?**

$$Zn(s) + \tfrac{1}{2}O_2(g) \longrightarrow Zn^{2+}(g) + O^{2-}(g)$$

(a) Set up an appropriate Born–Haber cycle:

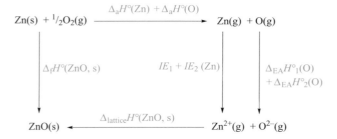

From Appendix 8, for Zn: $\quad IE_1 = 906\,\text{kJ mol}^{-1}$
$\qquad\qquad\qquad\qquad\quad IE_2 = 1733\,\text{kJ mol}^{-1}$

From Appendix 9, for O: $\quad \Delta_{EA}H^\circ_1 = -141\,\text{kJ mol}^{-1}$
$\qquad\qquad\qquad\qquad\quad \Delta_{EA}H^\circ_2 = 798\,\text{kJ mol}^{-1}$

From Appendix 10: $\qquad \Delta_a H^\circ(Zn) = 130\,\text{kJ mol}^{-1}$
$\qquad\qquad\qquad\qquad \Delta_a H^\circ(O) = 249\,\text{kJ mol}^{-1}$

From the thermochemical cycle, applying Hess's law:

$\Delta_{\text{lattice}}H^\circ(\text{ZnO, s})$

$\quad = \Delta_f H^\circ(\text{ZnO, s}) - \Delta_a H^\circ(Zn) - \Delta_a H^\circ(O) - IE_1 - IE_2$

$\qquad - \Delta_{EA}H^\circ_1 - \Delta_{EA}H^\circ_2$

$\quad = -350 - 130 - 249 - 906 - 1733 + 141 - 798$

$\quad = -4025\,\text{kJ mol}^{-1}$

(b) The process:

$$Zn(s) + \tfrac{1}{2}O_2(g) \longrightarrow Zn^{2+}(g) + O^{2-}(g)$$

is part of the Hess cycle shown in part (a). The enthalpy change for this process is given by:

$\Delta H^\circ = \Delta_a H^\circ(Zn) + \Delta_a H^\circ(O) + IE_1 + IE_2 + \Delta_{EA}H^\circ_1$

$\qquad + \Delta_{EA}H^\circ_2$

$\quad = 130 + 249 + 906 + 1733 - 141 + 798$

$\quad = 3675\,\text{kJ mol}^{-1}$

As a percentage of this,

$$\Delta_{EA}H^\circ_2 = \frac{798}{3675} \times 100 \approx 22\%$$

Self-study exercises

1. Given that $\Delta_f H^\circ(\text{Na}_2\text{O, s}) = -414\,\text{kJ mol}^{-1}$, determine the enthalpy change for the process:

$$2Na^+(g) + O^{2-}(g) \longrightarrow Na_2O(s)$$

   [*Ans.* −2528 kJ mol^{-1}]

2. What percentage contribution does $\Delta_{EA}H^\circ_2(O)$ make to the overall enthalpy change for the following process? How significant is this contribution in relation to each of the other contributions?

$$2Na(s) + \tfrac{1}{2}O_2(g) \longrightarrow 2Na^+(g) + O^{2-}(g)$$

   [*Ans.* ≈38%]

3. NaF and CaO both adopt NaCl-type structures. Consider the enthalpy changes that contribute to the overall value of $\Delta H^\circ(298\,K)$ for each of the following processes:

$$Na(s) + \tfrac{1}{2}F_2(g) \longrightarrow Na^+(g) + F^-(g)$$

$$Ca(s) + \tfrac{1}{2}O_2(g) \longrightarrow Ca^{2+}(g) + O^{2-}(g)$$

   Assess the relative role that each enthalpy contribution plays to determining the sign and magnitude of ΔH° for each process.

Table 16.2 Some covalent bond enthalpy terms (kJ mol^{-1}) for bonds involving oxygen, sulfur, selenium and tellurium.

O–O	O=O	O–H	O–C	O–F	O–Cl
146	498	464	359	190[†]	205[†]
S–S	S=S	S–H	S–C	S–F	S–Cl
266	427	366	272	326[†]	255[†]
Se–Se		Se–H		Se–F	Se–Cl
192		276		285[†]	243[†]
		Te–H		Te–F	
		238		335[†]	

[†] Values for O–F, S–F, Se–F, Te–F, O–Cl, S–Cl and Se–Cl derived from OF$_2$, SF$_6$, SeF$_6$, TeF$_6$, OCl$_2$, S$_2$Cl$_2$ and SeCl$_2$ respectively.

Some bond enthalpy terms for compounds of the group 16 elements are given in Table 16.2. In discussing groups 14 and 15, we emphasized the importance of $(p–p)\pi$-bonding for the first element in each group. We also pointed out that the failure of nitrogen to form 5-coordinate species such as NF$_5$ can be explained in terms of the N atom being too small to accommodate five F atoms around it. These factors are also responsible for some of the differences between O and its heavier congeners. For example:

- there are no stable sulfur analogues of CO and NO (although CS$_2$ and OCS are well known);
- the highest fluoride of oxygen is OF$_2$, but the later elements form SF$_6$, SeF$_6$ and TeF$_6$.

Coordination numbers above 4 for S, Se and Te can be achieved using a valence set of ns and np orbitals, with d-orbitals playing little or no role as valence orbitals (see Chapter 5). Thus, valence structures such as **16.1** can be used to represent the bonding in SF$_6$, although a set of resonance structures is required in order to rationalize the equivalence of the six S–F bonds. When describing the *structure* of SF$_6$, diagram **16.2** is more enlightening than **16.1**. Provided that you keep in mind that a line between two atoms does not represent a localized single bond, then **16.2** is an acceptable (and useful) representation of the molecule.

(16.1)   (16.2)

Similarly, while diagram **16.3** is a resonance form for H$_2$SO$_4$ which describes the S atom obeying the octet rule, structures **16.4** and **16.5** are useful for a rapid appreciation of the oxidation state of the S atom and coordination environment of the S atom. For these reasons, throughout the chapter we shall use diagrams analogous to **16.2**, **16.4** and

16.5 for hypervalent compounds of S, Se and Te. We shall also use 3-dimensional representations of the type shown in **16.6** (SF$_6$) and **16.7** (H$_2$SO$_4$) to provide structural information, but these should not be used to draw conclusions about the distribution of bonding electrons within the molecule.

(16.3)   (16.4)   (16.5)

(16.6)   (16.7)

Values in Table 16.2 illustrate the particular weakness of the O–O and O–F bonds and this can be rationalized in terms of lone pair repulsions (see Fig. 15.2). Note that O–H and O–C bonds are much stronger than S–H and S–C bonds.

NMR active nuclei and isotopes as tracers

Despite its low abundance (Table 16.1), ^{17}O has been used in studies of, for example, hydrated ions in aqueous solution and polyoxometallates (see Section 22.7).

The isotope ^{18}O is present to an extent of 0.2% in naturally occurring oxygen and is commonly used as a (non-radioactive) tracer for the element. The usual tracer for sulfur is ^{35}S, which is made by an (n,p) reaction on ^{35}Cl; ^{35}S is a β-emitter with $t_{\frac{1}{2}} = 87$ days.

Worked example 16.2   NMR spectroscopy using ^{77}Se and ^{125}Te nuclei

The solution ^{125}Te NMR spectrum of Te($cyclo$-C$_6$H$_{11}$)$_2$ at 298 K shows one broad signal. On increasing the temperature to 353 K, the signal sharpens. On cooling to 183 K, the signal splits into three signals at δ 601, 503 and 381 ppm with relative integrals of 25:14:1. Rationalize these data.

Te($cyclo$-C$_6$H$_{11}$)$_2$ contains only one Te environment, but the Te atom can be in either an equatorial or axial position of the cyclohexyl ring. This leads to three possible conformers:

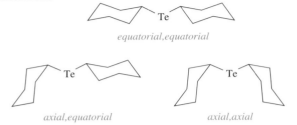

equatorial,equatorial

axial,equatorial   *axial,axial*

On steric grounds, the most favoured is the *equatorial,equatorial* conformer, and the least favoured is the *axial,axial* conformer. Signals at δ 601, 503 and 381 ppm in the low-temperature spectrum can be assigned to the *equatorial, equatorial*, *axial,equatorial* and *axial,axial* conformers respectively. At higher temperatures, the cyclohexyl rings undergo ring inversion (ring-flipping), causing the Te atom to switch between axial and equatorial positions. This interconverts the three conformers of Te($cyclo$-C$_6$H$_{11}$)$_2$. At 353 K, the interconversion is faster than the NMR timescale and one signal is observed (its chemical shift is the weighted average of the three signals observed at 183 K). On cooling from 353 to 298 K, the signal broadens, before splitting at lower temperatures.

[For a figure of the variable temperature spectra of Te($cyclo$-C$_6$H$_{11}$)$_2$, see: K. Karaghiosoff *et al.* (1999) *J. Organomet. Chem.*, vol. 577, p. 69.]

Self-study exercises

Data: see Table 16.1.

1. The reaction of SeCl$_2$ with ^{t}BuNH$_2$ in differing molar ratios leads to the formation of a series of compounds, among which are the following:

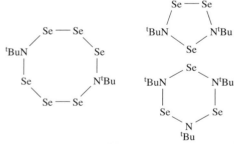

How many signals would you expect to see for each compound in the ^{77}Se NMR spectrum?
[*Ans.* See: T. Maaninen *et al.* (2000) *Inorg. Chem.*, vol. 39, p. 5341]

2. The ^{125}Te NMR spectrum (263 K) of an MeCN solution of the [Me$_4$N]$^+$ salt of [MeOTeF$_6$]$^-$ shows a septet of quartets with values of $J_{TeF} = 2630$ Hz and $J_{TeH} = 148$ Hz. The ^{19}F NMR spectrum exhibits a singlet with two satellite peaks. In the solid state, [MeOTeF$_6$]$^-$ has a pentagonal bipyramidal structure

with the MeO group in an axial position. (a) Interpret the ^{125}Te and ^{19}F NMR spectroscopic data. (b) Sketch the ^{19}F NMR spectrum and indicate where you would measure J_{TeF}.
[*Ans.* See: A. R. Mahjoub *et al.* (1992) *Angew. Chem. Int. Ed.*, vol. 31, p. 1036]

See also end-of-chapter problem 4.42.

16.4 The elements

Dioxygen

Dioxygen is obtained industrially by the liquefaction and fractional distillation of air, and is stored and transported as a liquid. Convenient laboratory preparations of O$_2$ are the electrolysis of aqueous alkali using Ni electrodes, and decomposition of H$_2$O$_2$ (eq. 16.3). A mixture of KClO$_3$ and MnO$_2$ used to be sold as 'oxygen mixture' (eq. 16.4) and the thermal decompositions of many other oxo salts (e.g. KNO$_3$, KMnO$_4$ and K$_2$S$_2$O$_8$) produce O$_2$.

$$2H_2O_2 \xrightarrow{MnO_2 \text{ or Pt catalyst}} O_2 + 2H_2O \qquad (16.3)$$

$$2KClO_3 \xrightarrow{\Delta, \; MnO_2 \text{ catalyst}} 3O_2 + 2KCl \qquad (16.4)$$

Caution! Chlorates are potentially explosive.

Dioxygen is a colourless gas, but condenses to a pale blue liquid or solid. Its bonding was described in Sections 2.2 and 2.3. In all phases, it is paramagnetic with a *triplet* ground state, i.e. the two unpaired electrons have the same spin, with the valence electron configuration being:

$$\sigma_g(2s)^2\sigma_u^*(2s)^2\sigma_g(2p_z)^2\pi_u(2p_x)^2\pi_u(2p_y)^2\pi_g^*(2p_x)^1\pi_g^*(2p_y)^1$$

This triplet ground state is designated by the term symbol $^3\Sigma_g^-$. In this state, O$_2$ is a powerful oxidizing agent (see eq. 8.28 and associated discussion) but, fortunately, the kinetic barrier is often high. If it were not, almost all organic chemistry would have to be carried out in closed systems. The O$_2$ molecule possesses two excited states that lie 94.7 and 157.8 kJ mol^{-1} above the ground state. The first excited state is a singlet state (designated by the term symbol $^1\Delta_g$) with two spin-paired electrons in the π_g^* level occupying one MO:

$$\sigma_g(2s)^2\sigma_u^*(2s)^2\sigma_g(2p_z)^2\pi_u(2p_x)^2\pi_u(2p_y)^2\pi_g^*(2p_x)^2\pi_g^*(2p_y)^0$$

In the higher excited state (singlet state, $^1\Sigma_g^+$), the two electrons occupy different MOs as in the ground state, but have *opposite* spins. The blue colour of liquid and solid O$_2$ arises from the simultaneous excitation by a single photon of two O$_2$ molecules from their ground to excited states. The associated absorption of energy corresponds to absorption of light in the red to green region of the visible part of the

THEORY

Box 16.2 Accurate determination of the O–O bond distance in $[O_2]^-$

Textbook discussions of MO theory of homonuclear diatomic molecules often consider the trends in bond distances in $[O_2]^+$, O_2, $[O_2]^-$ and $[O_2]^{2-}$ (see end-of-chapter problem 2.10) in terms of the occupancy of molecular orbitals. However, the determination of the bond distance in the superoxide ion $[O_2]^-$ has not been straightforward owing to disorder problems in the solid state and, as a result, the range of reported values for $d(O-O)$ is large. Cation exchange in liquid NH_3 has been used to isolate the salt $[1,3\text{-}(NMe_3)_2C_6H_4][O_2]_2 \cdot 3NH_3$ from $[NMe_4][O_2]$. In the solid state, each $[O_2]^-$ ion is fixed in a particular orientation by virtue of a hydrogen-bonded network.

The figure below shows the N–H····O interactions between solvate NH_3 and $[O_2]^-$, and the weak C–H····O interactions between cation methyl groups and $[O_2]^-$. Structural parameters for the hydrogen bonds indicate that the interactions are very weak. Consequently, the length of the bond in the $[O_2]^-$ anion ought not to be significantly perturbed by their presence. In $[1,3\text{-}(NMe_3)_2C_6H_4][O_2]_2 \cdot 3NH_3$, there are two crystallographically independent anions with O–O distances of 133.5 and 134.5 pm.

Further reading

H. Seyeda and M. Jansen (1998) *J. Chem. Soc., Dalton Trans.*, p. 875.

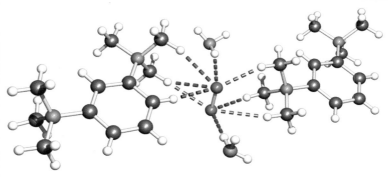

Colour code: O, red; N, blue; C, grey; H, white.

spectrum (see end-of-chapter problem 16.5). Singlet dioxygen (the $^1\Delta_g$ state) can be generated photochemically by irradiation of O_2 in the presence of an organic dye as sensitizer, or non-photochemically by reactions such as 16.5 and 16.16.[†]

$$H_2O_2 + NaOCl \longrightarrow O_2(^1\Delta_g) + NaCl + H_2O \qquad (16.5)$$

Singlet O_2 is short-lived, but extremely reactive, combining with many organic compounds, e.g. in reaction 16.6, $O_2(^1\Delta_g)$ acts as a dienophile in a Diels–Alder reaction.

$$\xrightarrow{O_2(^1\Delta_g)} \qquad (16.6)$$

At high temperatures, O_2 combines with most elements, exceptions being the halogens and noble gases, and N_2 unless under special conditions. Reactions with the group 1 metals are of particular interest, oxides, peroxides, superoxides and suboxides being possible products. Bond lengths in O_2, $[O_2]^-$ and $[O_2]^{2-}$ are 121, 134 and 149 pm (see

Box 16.2), consistent with a weakening of the bond caused by increased occupation of the π^* MOs (see Fig. 2.10).

The first ionization energy of O_2 is $1168 \, kJ \, mol^{-1}$ and it may be oxidized by very powerful oxidizing agents such as PtF_6 (eq. 16.7). The bond distance of 112 pm in $[O_2]^+$ is in keeping with the trend for O_2, $[O_2]^-$ and $[O_2]^{2-}$. Other salts include $[O_2]^+[SbF_6]^-$ (made from irradiation of O_2 and F_2 in the presence of SbF_5, or from O_2F_2 and SbF_5) and $[O_2]^+[BF_4]^-$ (eq. 16.8).

$$O_2 + PtF_6 \longrightarrow [O_2]^+[PtF_6]^- \qquad (16.7)$$

$$2O_2F_2 + 2BF_3 \longrightarrow 2[O_2]^+[BF_4]^- + F_2 \qquad (16.8)$$

The chemistry of O_2 is an enormous topic, and examples of its reactions can be found throughout this book. Its biological role is discussed in Chapter 29.

Ozone

Ozone, O_3, is usually prepared in up to 10% concentration by the action of a silent electrical discharge between two concentric metallized tubes in an apparatus called an *ozonizer*. Electrical discharges in thunderstorms convert O_2 into ozone. The action of UV radiation on O_2, or heating O_2 above 2750 K followed by rapid quenching,

---

[†]For an introduction to singlet state O_2, see: C.E. Wayne and R.P. Wayne (1996) *Photochemistry*, OUP, Oxford.

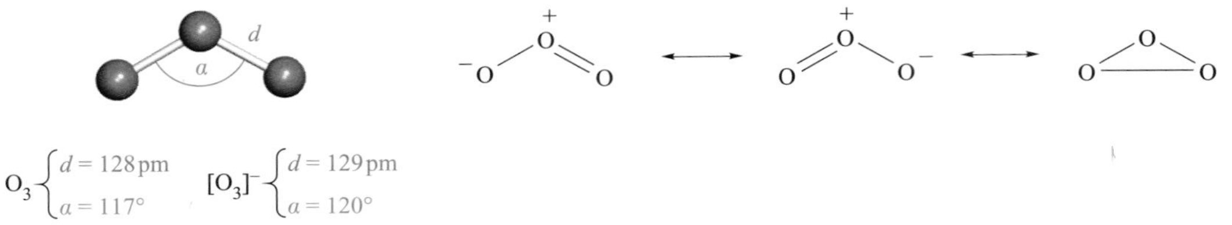

Fig. 16.4 The structures of O_3 and $[O_3]^-$, and contributing resonance structures in O_3. The O–O bond order in O_3 is taken to be 1.5.

$$O_3 \begin{cases} d = 128\,pm \\ a = 117° \end{cases} \quad [O_3]^- \begin{cases} d = 129\,pm \\ a = 120° \end{cases}$$

also produces O_3. In each of these processes, O atoms are produced and combine with O_2 molecules. Pure ozone can be separated from reaction mixtures by fractional liquefaction. The liquid is blue and boils at 163 K to give a perceptibly blue gas with a characteristic 'electric' smell. Molecules of O_3 are bent (Fig. 16.4). Ozone absorbs strongly in the UV region, and its presence in the upper atmosphere of the Earth is essential in protecting the planet's surface from over-exposure to UV radiation from the Sun (see Box 14.6).

Ozone is highly endothermic (eq. 16.9). The pure liquid is dangerously explosive, and the gas is a very powerful oxidizing agent (eq. 16.10).

$$\tfrac{3}{2}O_2(g) \longrightarrow O_3(g) \quad \Delta_f H°(O_3, g, 298\,K) = +142.7\,kJ\,mol^{-1}$$
$$(16.9)$$

$$O_3(g) + 2H^+(aq) + 2e^- \rightleftharpoons O_2(g) + H_2O(l)$$
$$E° = +2.07\,V \quad (16.10)$$

The value of $E°$ in eq. 16.10 refers to pH = 0 (see Box 8.1), and at higher pH, E becomes less positive: +1.65 V at pH = 7, and +1.24 V at pH = 14. The presence of high concentrations of alkali stabilizes O_3 both thermodynamically and kinetically. Ozone is much more reactive than O_2 (hence the use of O_3 in water purification, see Box 14.1). Reactions 16.11–16.13 typify this high reactivity, as does the reaction of ozone with alkenes to give ozonides.

$$O_3 + S + H_2O \longrightarrow H_2SO_4 \quad (16.11)$$

$$O_3 + 2I^- + H_2O \longrightarrow O_2 + I_2 + 2[OH]^- \quad (16.12)$$

$$4O_3 + PbS \longrightarrow 4O_2 + PbSO_4 \quad (16.13)$$

Potassium ozonide, KO_3 (formed in reaction 16.14), is an unstable red salt which contains the paramagnetic $[O_3]^-$ ion (Fig. 16.4). Ozonide salts are known for all the alkali metals. The compounds $[Me_4N][O_3]$ and $[Et_4N][O_3]$ have been prepared using reactions of the type shown in eq. 16.15. Ozonides are explosive, but $[Me_4N][O_3]$ is relatively stable, decomposing above 348 K (see also Sections 11.6 and 11.8).

$$2KOH + 5O_3 \longrightarrow 2KO_3 + 5O_2 + H_2O \quad (16.14)$$

$$CsO_3 + [Me_4N][O_2] \xrightarrow{\text{liquid } NH_3} CsO_2 + [Me_4N][O_3]$$
$$(16.15)$$

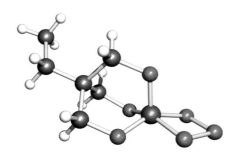

Fig. 16.5 The structure (X-ray diffraction at 188 K) of the phosphite ozonide $EtC(CH_2O)_3PO_3$ [A. Dimitrov *et al.* (2001) *Eur. J. Inorg. Chem.*, p. 1929]. Colour code: P, brown; O, red; C, grey; H, white.

Phosphite ozonides, $(RO)_3PO_3$, have been known since the early 1960s, and are made *in situ* as precursors to singlet oxygen (eq. 16.16). The ozonides are stable only at low temperatures, and it is only with the use of modern low-temperature crystallographic methods that structural data are now available. Figure 16.5 shows the structure of the phosphite ozonide prepared by the steps in scheme 16.17. In the PO_3 ring, the P–O and O–O bond lengths are 167 and 146 pm, respectively; the ring is close to planar, with a dihedral angle of 7°.

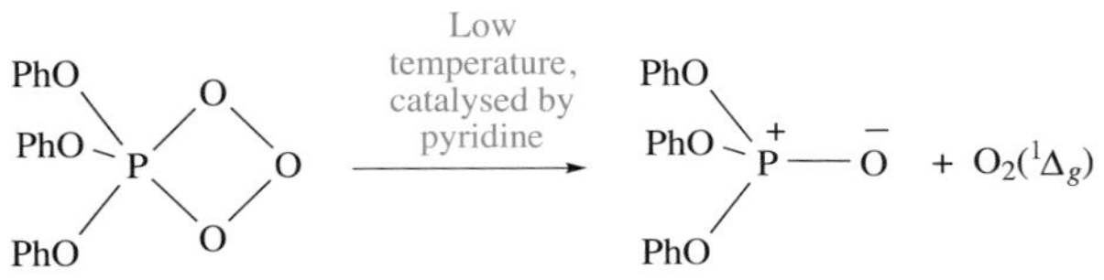

$$(16.16)$$

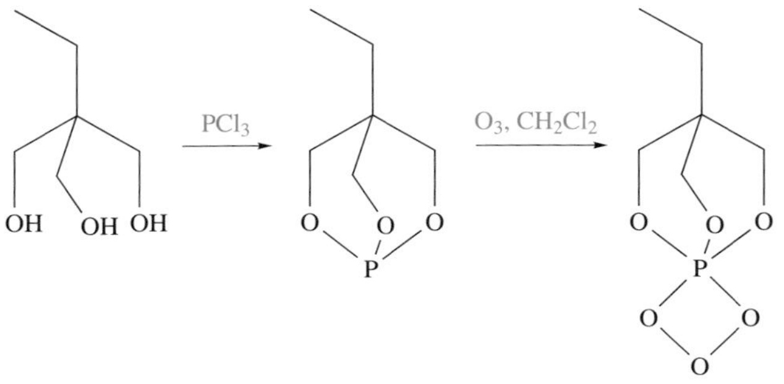

$$(16.17)$$

Sulfur: allotropes

The allotropy of sulfur is complicated, and we describe only the best-established species. The tendency for catenation (see Section 14.3) by sulfur is high and leads to the formation of both rings of varying sizes and chains. Allotropes of known structure include cyclic S_6, S_7, S_8, S_9, S_{10}, S_{11}, S_{12}, S_{18} and S_{20} (all with puckered rings, e.g. Figs. 16.6a–c) and fibrous sulfur (*catena*-S_∞, Figs. 16.6d and 3.20a). In most of these, the S−S bond distances are 206 ± 1 pm, indicative of single bond character; the S−S−S bond angles lie in the range 102–108°. The ring conformations of S_6 (chair) and S_8 (crown) are readily envisaged but other rings have more complicated conformations. The structure of S_7 (Fig. 16.6b) is noteworthy because of the wide range of S−S bond lengths (199–218 pm) and angles (101.5–107.5°). The energies of interconversion between the cyclic forms are very small.

The most stable allotrope is orthorhombic sulfur (the α-form and standard state of the element) and it occurs naturally as large yellow crystals in volcanic areas. At 367.2 K, the α-form transforms reversibly into monoclinic sulfur (β-form). Both the α- and β-forms contain S_8 rings; the density of the α-form is 2.07 g cm^{-3}, compared with 1.94 g cm^{-3} for the β-form in which the packing of the rings is less efficient. However, if single crystals of the α-form are rapidly heated to 385 K, they melt before the α ⟶ β transformation occurs. If crystallization takes place at 373 K, the S_8 rings adopt the structure of the β-form, but the crystals must be cooled rapidly to 298 K. On standing at 298 K, a β ⟶ α transition occurs within a few weeks. β-Sulfur melts at 401 K, but this is not a true melting point, since some breakdown of S_8 rings takes place, causing the melting point to be depressed.

Rhombohedral sulfur (the ρ-form) comprises S_6 rings and is obtained by the ring closure reaction 16.18. It decomposes in light to S_8 and S_{12}.

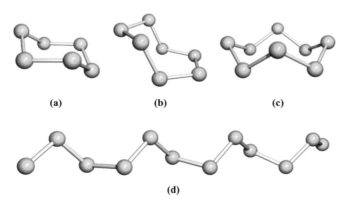

(a)   **(b)**   **(c)**

(d)

Fig. 16.6 Representations of the structures of some allotropes of sulfur: (a) S_6, (b) S_7, (c) S_8 and (d) *catena*-S_∞ (the chain continues at each end).

$$S_2Cl_2 + H_2S_4 \xrightarrow{\text{dry diethyl ether}} S_6 + 2HCl \qquad (16.18)$$

(16.8)   **(16.9)**

Similar ring closures starting from H_2S_x (**16.8**) and S_yCl_2 (**16.9**) lead to larger rings, but a more recent strategy makes use of $[(C_5H_5)_2TiS_5]$ (**16.10**) which is prepared by reaction 16.19 and contains a coordinated $[S_5]^{2-}$ ligand. The Ti(IV) complex reacts with S_yCl_2 to give *cyclo*-S_{y+5}, allowing synthesis of a series of sulfur allotropes. All the *cyclo*-allotropes are soluble in CS_2.

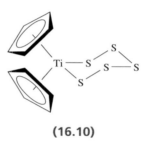

(16.10)

$$2NH_3 + H_2S + \tfrac{1}{2}S_8 \longrightarrow [NH_4]_2[S_5]$$

$$\xrightarrow{[(C_5H_5)_2TiCl_2]} [(C_5H_5)_2TiS_5] \qquad (16.19)$$

By rapidly quenching molten sulfur at 570 K in ice-water, fibrous sulfur (which is insoluble in water) is produced. Fibrous sulfur, *catena*-S_∞, contains infinite, helical chains (Fig. 16.6d) and slowly reverts to α-sulfur on standing. α-Sulfur melts to a mobile yellow liquid which darkens in colour as the temperature is raised. At 433 K, the viscosity increases enormously as S_8 rings break by homolytic S−S bond fission, giving diradicals which react together to form polymeric chains containing $\leq 10^6$ atoms. The viscosity reaches a maximum at ≈ 473 K, and then decreases up to the boiling point (718 K). At this point the liquid contains a mixture of rings and shorter chains. The vapour above liquid sulfur at 473 K consists mainly of S_8 rings, but at higher temperatures, smaller molecules predominate, and above 873 K, paramagnetic S_2 (a diradical like O_2) becomes the main species. Dissociation into atoms occurs above 2470 K.

Sulfur: reactivity

Sulfur is a reactive element. It burns in air with a blue flame to give SO_2, and reacts with F_2, Cl_2 and Br_2 (scheme 16.20).

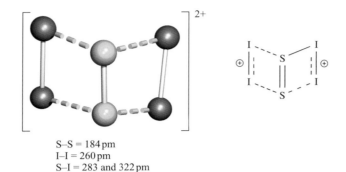

S–S = 184 pm
I–I = 260 pm
S–I = 283 and 322 pm

Fig. 16.7 The structure of $[S_2I_4]^{2+}$ determined by X-ray diffraction at low temperature for the $[AsF_6]^-$ salt [S. Brownridge *et al.* (2005) *Inorg. Chem.*, vol. 44, p. 1660], and a representation of the bonding in terms of S_2 interacting with two $[I_2]^+$. Colour code: S, yellow; I, brown.

For the syntheses of other halides and oxides, see Sections 16.7 and 16.8.

$$S_8 \begin{cases} \xrightarrow{F_2} SF_6 \\ \xrightarrow{Cl_2} S_2Cl_2 \\ \xrightarrow{Br_2} S_2Br_2 \end{cases} \qquad (16.20)$$

Sulfur does not react directly with I_2, but in the presence of AsF_5 or SbF_5 in liquid SO_2, the salts $[S_7I][EF_6]$ (E = As or Sb) are produced which contain the $[S_7I]^+$ cation (**16.11**). If an excess of I_2 is used, the products are $[S_2I_4][EF_6]_2$ (E = As or Sb). The $[S_2I_4]^{2+}$ cation has the 'open-book' structure shown in Fig. 16.7. The bonding can be considered in terms of S_2 interacting with two $[I_2]^+$ ions by means of donation of the unpaired electron in the π^* MO of each $[I_2]^+$ into a vacant MO of S_2. On the basis of the short S–S bond, stretching mode at $734\,cm^{-1}$, and theoretical investigations, it is proposed that the S–S bond order lies between 2.2 and 2.4. When treated with hot aqueous alkali, sulfur forms a mixture of polysulfides, $[S_x]^{2-}$, and polythionates (**16.12**), while oxidizing agents convert it to H_2SO_4.

(**16.11**)   (**16.12**)

Saturated hydrocarbons are dehydrogenated when heated with sulfur, and further reaction with alkenes occurs. An application of this reaction is in the vulcanization of rubber, in which soft rubber is toughened by cross-linking of the polyisoprene chains, making it suitable for use in, for example, tyres. The reactions of sulfur with CO or $[CN]^-$ yield OCS (**16.13**) or the thiocyanate ion (**16.14**), while treatment with sulfites gives thiosulfates (eq. 16.21).

$$O=C=S \qquad {}^-N=C=S \longleftrightarrow N\equiv C-S^-$$

(**16.13**)   (**16.14**)

$$Na_2SO_3 + \tfrac{1}{8}S_8 \xrightarrow{H_2O, \, 373\,K} Na_2S_2O_3 \qquad (16.21)$$

The oxidation of S_8 by AsF_5 or SbF_5 in liquid SO_2 (see Section 9.5) yields salts containing the cations $[S_4]^{2+}$, $[S_8]^{2+}$ and $[S_{19}]^{2+}$ (Fig. 16.8). In reaction 16.22, AsF_5 acts as an oxidizing agent *and* a fluoride acceptor (eq. 16.23).

$$S_8 + 3AsF_5 \xrightarrow{\text{liquid } SO_2} [S_8][AsF_6]_2 + AsF_3 \qquad (16.22)$$

$$\left. \begin{array}{l} AsF_5 + 2e^- \longrightarrow AsF_3 + 2F^- \\ AsF_5 + F^- \longrightarrow [AsF_6]^- \end{array} \right\} \qquad (16.23)$$

Two-electron oxidation of S_8 results in a change in ring conformation (Fig. 16.8a). The red $[S_8]^{2+}$ cation was originally reported as being blue, but the blue colour is now known to arise from the presence of radical impurities such as $[S_5]^+$.[†] In S_8, all the S–S bond lengths are equal (206 pm) and the distance between two S atoms across the ring from one another is greater than the sum of the van der Waals radii ($r_v = 185\,pm$). A redetermination of the structure of the $[AsF_6]^-$ salt of $[S_8]^{2+}$ (Fig. 16.8c) illustrates (i) a variation in S–S bond distances around the ring and (ii) cross-ring S–S separations that are smaller than the sum of the van der Waals radii, i.e. $[S_8]^{2+}$ exhibits *trans-annular* interactions. The most important transannular interaction corresponds to the shortest S····S contact and Fig. 16.8d shows a resonance structure that describes an appropriate bonding contribution.

The $[S_4]^{2+}$ cation is square (S–S = 198 pm) with delocalized bonding. It is isoelectronic with $[P_4]^{2-}$ (see the self-study exercise on p. 505). In $[S_{19}]^{2+}$ (Fig. 16.8e), two 7-membered, puckered rings are connected by a 5-atom chain. The positive charge can be considered to be localized on the two 3-coordinate S centres.

A cyclic species has an ***annular*** form, and a ***transannular*** interaction is one between atoms across a ring.

Selenium and tellurium

Selenium possesses several allotropes. Crystalline, red monoclinic selenium exists in three forms, each containing Se_8 rings with the crown conformation of S_8 (Fig. 16.6c). Black selenium consists of larger polymeric rings. The thermodynamically stable allotrope is grey selenium which contains infinite, helical chains (Se–Se = 237 pm), the axes of which lie parallel to one another.

Elemental selenium can be prepared by reaction 16.24. By substituting Ph_3PSe in this reaction by Ph_3PS, rings

[†] See: T.S. Cameron *et al.* (2000) *Inorg. Chem.*, vol. 39, p. 5614.

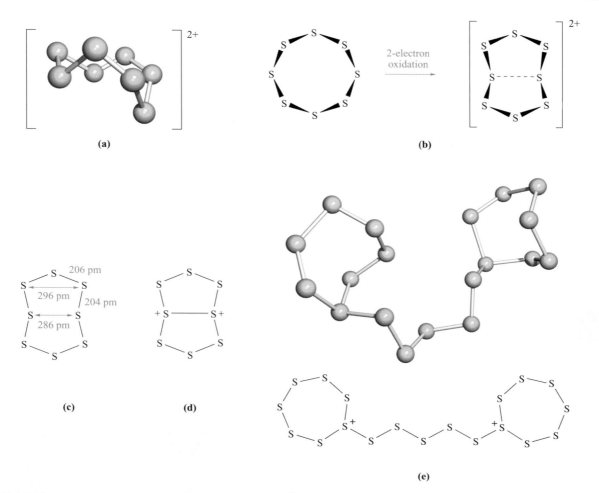

Fig. 16.8 (a) Schematic representation of the structure of $[S_8]^{2+}$. (b) The change in conformation of the ring during oxidation of S_8 to $[S_8]^{2+}$. (c) Structural parameters for $[S_8]^{2+}$ from the $[AsF_6]^-$ salt. (d) One resonance structure that accounts for the transannular interaction in $[S_8]^{2+}$. (e) The structure of the $[S_{19}]^{2+}$ cation, determined by X-ray diffraction for the $[AsF_6]^-$ salt [R.C. Burns *et al.* (1980) *Inorg. Chem.*, vol. 19, p. 1423], and a schematic representation showing the localization of positive charge.

of composition Se_nS_{8-n} ($n = 1–5$) can be produced (see end-of-chapter problem 4.42).

$$4SeCl_2 + 4Ph_3PSe \longrightarrow Se_8 + 4Ph_3PCl_2 \qquad (16.24)$$

Tellurium has only one crystalline form which is a silvery-white metallic-looking solid and is isostructural with grey selenium. The red allotropes of Se can be obtained by rapid cooling of molten Se and extraction into CS_2. The photoconductivity of Se (see Box 16.1) and Te arises because, in the solid, the band gap of 1.66 eV is small enough for the influence of visible light to cause the promotion of electrons from the filled bonding MOs to the unoccupied antibonding MOs (see Section 6.8).

Although *cyclo*-Te_8 is not known as an allotrope of the element, it has been characterized in the salt $Cs_3[Te_{22}]$ which has the composition $[Cs^+]_3[Te_6{}^{3-}][Te_8]_2$.

Although less reactive, Se and Te are chemically similar to sulfur. This resemblance extends to the formation of cations such as $[Se_4]^{2+}$, $[Te_4]^{2+}$, $[Se_8]^{2+}$ and $[Te_8]^{2+}$. The salt

$[Se_8][AsF_6]_2$ can be made in an analogous manner to $[S_8][AsF_6]_2$ in liquid SO_2 (eq. 16.22), whereas reaction 16.25 is carried out in fluorosulfonic acid (see Section 9.8). Newer methods use metal halides (e.g. $ReCl_4$ and WCl_6) as oxidizing agents, e.g. the formation of $[Te_8]^{2+}$ (eq. 16.26). Reaction 16.27 (in AsF_3 solvent) produces $[Te_6]^{4+}$, **16.15**, which has no S or Se analogue.

$$4Se + S_2O_6F_2 \xrightarrow{HSO_3F} [Se_4][SO_3F]_2 \xrightarrow{Se, HSO_3F} [Se_8][SO_3F]_2 \qquad (16.25)$$

$$2ReCl_4 + 15Te + TeCl_4 \xrightarrow{\Delta, \text{ sealed tube}} 2[Te_8][ReCl_6] \qquad (16.26)$$

$$6Te + 6AsF_5 \xrightarrow{AsF_3} [Te_6][AsF_6]_4 + 2AsF_3 \qquad (16.27)$$

The structures of $[Se_4]^{2+}$, $[Te_4]^{2+}$ and $[Se_8]^{2+}$ mimic those of their S analogues, but $[Te_8]^{2+}$ exists in two forms. In $[Te_8][ReCl_6]$, $[Te_8]^{2+}$ is structurally similar to $[S_8]^{2+}$ and

$[Se_8]^{2+}$, but in $[Te_8][WCl_6]_2$, the cation has the bicyclic structure, i.e. resonance structure **16.16** is dominant.

a lies in the range 266–269 pm
b lies in the range 306–315 pm

(16.15)

(16.16)

16.5 Hydrides

Water, H_2O

Aspects of the chemistry of water have already been covered as follows:

- the properties of H_2O (Section 7.2);
- acids, bases and ions in aqueous solution (Chapter 7);
- 'heavy water', D_2O (Section 10.3);
- comparison of the properties of H_2O and D_2O (Table 10.2);
- hydrogen bonding (Section 10.6).

Water purification is discussed in Box 16.3.

APPLICATIONS

Box 16.3 Purification of water

The simplest method for the removal of all solid solutes from water is distillation, but because of the high boiling point and enthalpy of vaporization (Table 7.1), this method is expensive. If the impurities are ionic, ion exchange is an effective (and relatively cheap) means of purification. The treatment involves the passage of water down a column of an organic resin containing acidic groups and then down a similar column containing basic groups:

$$Resin-SO_3H + M^+ + X^- \longrightarrow [Resin-SO_3]^-M^+ + H^+ + X^-$$

Acidic resin

$$[Resin-NR_3]^+[OH]^- + H^+ + X^- \longrightarrow [Resin-NR_3]^+X^- + H_2O$$

Basic resin

After treatment, *deionized water* is produced. The resins are reactivated by treatment with dilute H_2SO_4 and Na_2CO_3 solutions respectively. Reverse osmosis at high pressures is also an important process in water purification, with cellulose acetate as the usual membrane. The latter prevents the passage of dissolved solutes or insoluble impurities. The removal of nitrates is highlighted in Box 15.9.

The purification of drinking water is a complicated industrial process. Water may be abundant on the Earth, but impurities such as microorganisms, particulate materials and chemicals usually make it unfit for human consumption. Coagulation and separation methods are used to remove many particles. In the coagulation step, coagulants are dispersed throughout the water by rapid mixing. This is followed by a slower flocculation process in which coagulants and suspended impurities come together to form 'floc' which can then be separated by sedimentation. Aluminium and iron(III) salts are widely used in the coagulation stages, and the treatment relies upon the formation of polymeric species in solution. Pre-polymerized coagulants are commercially available and include polyaluminium silicate

Flocculator-clarifier units in a water treatment plant in Florida, US.

sulfate (PASS) and polyferric sulfate (PFS). About two-thirds of all $Al_2(SO_4)_3$ manufactured goes into water treatment processes, with the paper manufacturing industry consuming about a half of this amount.

Further reading

Encyclopedia of Separation Science (2000) eds C.F. Poole, M. Cooke and I.D. Wilson, Academic Press, New York: J. Irving, p. 4469 – 'Water treatment: Overview: ion exchange'; W.H. Höll, p. 4477 – 'Water treatment: Anion exchangers: ion exchange'.

J.-Q. Jiang and N.J.D. Graham (1997) *Chem. Ind.*, p. 388 – 'Pre-polymerized inorganic coagulants for treating water and waste water'.

See also Box 14.1.

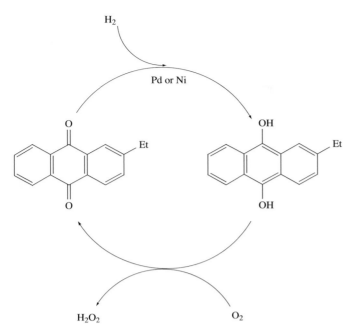

Fig. 16.9 The catalytic cycle used in the industrial manufacture of hydrogen peroxide; O_2 is converted to H_2O_2 during the oxidation of the organic alkylanthraquinol. The organic product is reduced by H_2 in a Pd- or Ni-catalysed reaction. Such cycles are discussed in detail in Chapter 25.

Hydrogen peroxide, H_2O_2

The oldest method for the preparation of H_2O_2 is reaction 16.28. The hydrolysis of peroxydisulfate (produced by electrolytic oxidation of $[HSO_4]^-$ at high current densities using Pt electrodes) has also been an important route to H_2O_2 (eq. 16.29).

$$BaO_2 + H_2SO_4 \longrightarrow BaSO_4 + H_2O_2 \qquad (16.28)$$

$$2[NH_4][HSO_4] \xrightarrow[-H_2]{\text{electrolytic oxidation}} [NH_4]_2[S_2O_8]$$

$$\xrightarrow{H_2O} 2[NH_4][HSO_4] + H_2O_2 \qquad (16.29)$$

Nowadays, H_2O_2 is manufactured by the oxidation of 2-ethylanthraquinol (or a related alkyl derivative). The H_2O_2 formed is extracted into water and the organic product is reduced back to starting material. The process is summarized in the catalytic cycle in Fig. 16.9.[†]

Some physical properties of H_2O_2 are given in Table 16.3. Like water, it is strongly hydrogen-bonded. Pure or strongly concentrated aqueous solutions of H_2O_2 readily decompose (eq. 16.30) in the presence of alkali, heavy metal ions or heterogeneous catalysts (e.g. Pt or MnO_2), and traces of complexing agents (e.g. 8-hydroxyquinoline, **16.17**) or

---

[†]For an overview of H_2O_2 production processes, see: W.R. Thiel (1999) *Angew. Chem. Int. Ed.*, vol. 38, p. 3157 – 'New routes to hydrogen peroxide: alternatives for established processes?'

Table 16.3 Selected properties of H_2O_2.

Property	
Physical appearance at 298 K	Colourless (very pale blue) liquid
Melting point / K	272.6
Boiling point / K	425 (decomposes)
$\Delta_f H^\circ(298\,K) / kJ\,mol^{-1}$	-187.8
$\Delta_f G^\circ(298\,K) / kJ\,mol^{-1}$	-120.4
Dipole moment / debye	1.57
O–O bond distance (gas phase) / pm	147.5
$\angle O-O-H$ (gas phase) / deg	95

adsorbing materials (e.g. sodium stannate, $Na_2[Sn(OH)_6]$) are often added as stabilizers.

$$H_2O_2(l) \longrightarrow H_2O(l) + \tfrac{1}{2}O_2(g)$$

$$\Delta_r H^\circ(298\,K) = -98\,kJ \text{ per mole of } H_2O_2 \quad (16.30)$$

(16.17)

Mixtures of H_2O_2 and organic or other readily oxidized materials are dangerously explosive, and H_2O_2 mixed with hydrazine has been used as a rocket propellant. A major application of H_2O_2 is in the paper and pulp industry where it is replacing chlorine as a bleaching agent (see Fig. 17.2). Other uses are as an antiseptic, in water pollution control and for the manufacture of sodium peroxoborate (see Section 13.7) and peroxocarbonates (see Section 14.9).

Figure 16.10 shows the gas-phase structure of H_2O_2 and bond parameters are listed in Table 16.3. The internal dihedral angle is sensitive to the surroundings (i.e. the extent of hydrogen bonding) being $111°$ in the gas phase, $90°$ in the solid state and $180°$ in the adduct $Na_2C_2O_4 \cdot H_2O_2$. In this last example, H_2O_2 has a *trans*-planar conformation and the O lone pairs appear to interact with the Na^+ ions. Values of the dihedral angle in organic peroxides, ROOR, show wide variations ($\approx 80°-145°$).

In aqueous solution, H_2O_2 is partially ionized (eq. 16.31), and in alkaline solution, is present as the $[HO_2]^-$ ion.

$$H_2O_2 + H_2O \rightleftharpoons [H_3O]^+ + [HO_2]^-$$

$$K_a = 2.4 \times 10^{-12} \quad (298\,K) \qquad (16.31)$$

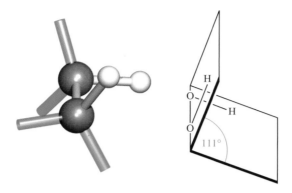

Fig. 16.10 The gas-phase structure of H_2O_2. The directions in which the lone pairs point are indicated by the green cylinders. The angle shown as $111°$ is the *internal dihedral angle*, the angle between the planes containing each OOH-unit; see Table 16.3 for other bond parameters.

Hydrogen peroxide is a powerful oxidizing agent as is seen from the standard reduction potential (at $pH = 0$) in eq. 16.32. For example, H_2O_2 oxidizes I^- to I_2, SO_2 to H_2SO_4 and (in alkaline solution) Cr(III) to Cr(VI). Powerful oxidants such as $[MnO_4]^-$ and Cl_2 will oxidize H_2O_2 (eqs. 16.33–16.35), and in alkaline solution, H_2O_2 is a good reducing agent (eq. 16.36).

$$H_2O_2 + 2H^+ + 2e^- \rightleftharpoons 2H_2O \qquad E^o = +1.78\,\text{V} \quad (16.32)$$

$$O_2 + 2H^+ + 2e^- \rightleftharpoons H_2O_2 \qquad E^o = +0.70\,\text{V} \quad (16.33)$$

$$2[MnO_4]^- + 5H_2O_2 + 6H^+ \longrightarrow 2Mn^{2+} + 8H_2O + 5O_2 \qquad (16.34)$$

$$Cl_2 + H_2O_2 \longrightarrow 2HCl + O_2 \qquad (16.35)$$

$$O_2 + 2H_2O + 2e^- \rightleftharpoons H_2O_2 + 2[OH]^- \\ E^o_{[OH^-]=1} = -0.15\,\text{V} \quad (16.36)$$

Tracer studies using ^{18}O show that in these redox reactions $H_2(^{18}O)_2$ is converted to $(^{18}O)_2$, confirming that no oxygen from the solvent (which is not labelled) is incorporated and the O–O bond is not broken.

Worked example 16.3   Redox reactions of H_2O_2 in aqueous solution

Use data from Appendix 11 to determine $\Delta G^o(298\,\text{K})$ for the oxidation of $[Fe(CN)_6]^{4-}$ by H_2O_2 in aqueous solution at $pH = 0$. Comment on the significance of the value obtained.

First, look up the appropriate half-equations and corresponding E^o values:

$$[Fe(CN)_6]^{3-}(aq) + e^- \rightleftharpoons [Fe(CN)_6]^{4-}(aq) \qquad E^o = +0.36\,\text{V}$$

$$H_2O_2(aq) + 2H^+(aq) + 2e^- \rightleftharpoons 2H_2O(l) \qquad E^o = +1.78\,\text{V}$$

The overall redox process is:

$$2[Fe(CN)_6]^{4-}(aq) + H_2O_2(aq) + 2H^+(aq) \\ \longrightarrow 2[Fe(CN)_6]^{3-}(aq) + 2H_2O(l)$$

$$E^o_{cell} = 1.78 - 0.36 = 1.42\,\text{V}$$

$$\Delta G^o(298\,\text{K}) = -zFE^o_{cell}$$

$$= -2 \times 96\,485 \times 1.42 \times 10^{-3}$$

$$= -274\,\text{kJ\,mol}^{-1}$$

The value of ΔG^o is large and negative showing that the reaction is spontaneous and will go to completion.

Self-study exercises

1. In aqueous solution at pH 14, $[Fe(CN)_6]^{3-}$ is reduced by H_2O_2. Find the relevant half-equations in Appendix 11 and calculate $\Delta G^o(298\,\text{K})$ for the overall reaction.
   [*Ans.* $-98\,\text{kJ}$ per mole of H_2O_2]

2. At pH 0, H_2O_2 oxidizes aqueous sulfurous acid. Find the appropriate half-equations in Appendix 11 and determine $\Delta G^o(298\,\text{K})$ for the overall reaction.
   [*Ans.* $-311\,\text{kJ}$ per mole of H_2O_2]

3. Is the oxidation of Fe^{2+} to Fe^{3+} by aqueous H_2O_2 (at pH 0) thermodynamically more or less favoured when the Fe^{2+} ions are in the form of $[Fe(bpy)_3]^{2+}$ or $[Fe(OH_2)_6]^{2+}$? Quantify your answer by determining $\Delta G^o(298\,\text{K})$ for each reduction.
   [*Ans.* Less favoured for $[Fe(bpy)_3]^{2+}$; $\Delta G^o = -145$; $-195\,\text{kJ}$ per mole of H_2O_2]

See also end-of-chapter problem 8.8.

Deprotonation of H_2O_2 gives $[OOH]^-$ (eq. 16.31) and loss of a second proton yields the peroxide ion, $[O_2]^{2-}$. In addition to peroxide salts such as those of the alkali metals (see Section 11.6), many peroxido complexes are known. Figure 16.11 shows two such complexes, one of which also contains the $[OOH]^-$ ion in a bridging mode. Typical O–O bond distances for coordinated peroxido groups are $\approx 140–148$ pm. Further peroxido complexes are described elsewhere in this book (e.g. Fig. 21.12 and accompanying discussion) and include models for the active centre in cytochrome c oxidase (see the end of Section 29.4).

Hydrides H_2E (E = S, Se, Te)

Selected physical data for hydrogen sulfide, selenide and telluride are listed in Table 16.4 and illustrated in Figs. 10.7 and 10.8. Hydrogen sulfide is more toxic than HCN, but because H_2S has a very characteristic odour of rotten eggs, its presence is easily detected. It is a natural

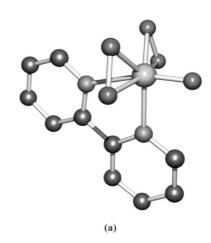

(a)

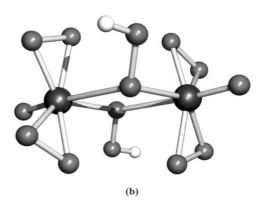

(b)

Fig. 16.11 The structures (X-ray diffraction) of (a) $[V(O_2)_2(O)(bpy)]^-$ in the hydrated ammonium salt [H. Szentivanyi *et al.* (1983) *Acta Chem. Scand., Ser. A*, vol. 37, p. 553] and (b) $[Mo_2(O_2)_4(O)_2(\mu\text{-OOH})_2]^{2-}$ in the pyridinium salt [J.-M. Le Carpentier *et al.* (1972) *Acta Crystallogr., Sect. B*, vol. 28, p. 1288]. The H atoms in the second structure were not located but have been added here for clarity. Colour code: V, yellow; Mo, dark blue; O, red; N, light blue; C, grey; H, white.

product of decaying S-containing matter, and is present in coal pits, gas wells and sulfur springs. Where it occurs in natural gas deposits, H_2S is removed by reversible absorption in a solution of an organic base and is converted to S by controlled oxidation. Figure 16.2 showed the increasing importance of sulfur recovery from natural gas as a source of commercial sulfur. In the laboratory, H_2S was historically prepared by reaction 16.37 in a Kipp's apparatus. The hydrolysis of calcium or barium sulfides (e.g. eq. 16.38) produces purer H_2S, but the gas is also commercially available in small cylinders.

$$FeS(s) + 2HCl(aq) \longrightarrow H_2S(g) + FeCl_2(aq) \qquad (16.37)$$

$$CaS + 2H_2O \longrightarrow H_2S + Ca(OH)_2 \qquad (16.38)$$

Hydrogen selenide may be prepared by reaction 16.39, and a similar reaction can be used to make H_2Te.

$$Al_2Se_3 + 6H_2O \longrightarrow 3H_2Se + 2Al(OH)_3 \qquad (16.39)$$

The enthalpies of formation of H_2S, H_2Se and H_2Te (Table 16.4) indicate that the sulfide can be prepared by direct combination of H_2 and sulfur (boiling), and is more stable with respect to decomposition into its elements than H_2Se or H_2Te.

Like H_2O, the hydrides of the later elements in group 16 have bent structures but the angles of $\approx 90°$ (Table 16.4) are significantly less than that in H_2O (105°). This suggests that the E−H bonds (E = S, Se or Te) involve p character from the central atom (i.e. little or no contribution from the valence s orbital).

In aqueous solution, the hydrides behave as weak acids (Table 16.4 and Section 7.5). The second acid dissociation constant of H_2S is $\approx 10^{-19}$ and, thus, metal sulfides are hydrolysed in aqueous solution. The only reason that many metal sulfides can be isolated by the action of H_2S on solutions of their salts is that the sulfides are extremely

Table 16.4 Selected data for H_2S, H_2Se and H_2Te.

	H_2S	H_2Se	H_2Te
Name[†]	Hydrogen sulfide	Hydrogen selenide	Hydrogen telluride
Physical appearance and general characteristics	Colourless gas; offensive smell of rotten eggs; toxic	Colourless gas; offensive smell; toxic	Colourless gas; offensive smell; toxic
Melting point / K	187.5	207	224
Boiling point / K	214	232	271
$\Delta_{vap}H°$(bp) / kJ mol^{-1}	18.7	19.7	19.2
$\Delta_f H°$(298 K) / kJ mol^{-1}	−20.6	+29.7	+99.6
$pK_a(1)$	7.04	4.0	3.0
$pK_a(2)$	19	−	−
E−H bond distance / pm	134	146	169
$\angle$H−E−H / deg	92	91	90

[†] The IUPAC names of sulfane, selane and tellane are rarely used.

insoluble. For example, a qualitative test for H_2S is its reaction with aqueous lead acetate (eq. 16.40).

$$H_2S + Pb(O_2CCH_3)_2 \longrightarrow \underset{\text{Black ppt.}}{PbS} + 2CH_3CO_2H \quad (16.40)$$

Sulfides such as CuS, PbS, HgS, CdS, Bi_2S_3, As_2S_3, Sb_2S_3 and SnS have solubility products (see Sections 7.9 and 7.10) less than $\approx 10^{-30}$ and can be precipitated by H_2S in the presence of dilute HCl. The acid suppresses ionization of H_2S, lowering the concentration of S^{2-} in solution. Sulfides such as ZnS, MnS, NiS and CoS with solubility products in the range $\approx 10^{-15}$ to 10^{-30} are precipitated only from neutral or alkaline solutions.

Protonation of H_2S to $[H_3S]^+$ can be achieved using the superacid HF/SbF_5 (see Section 9.9). The salt $[H_3S][SbF_6]$ is a white crystalline solid which reacts with quartz glass. Vibrational spectroscopic data for $[H_3S]^+$ are consistent with a trigonal pyramidal structure like that of $[H_3O]^+$. The addition of MeSCl to $[H_3S][SbF_6]$ at 77 K followed by warming of the mixture to 213 K yields $[Me_3S][SbF_6]$, which is stable below 263 K. Spectroscopic data (NMR, IR and Raman) are consistent with the presence of the trigonal pyramidal $[Me_3S]^+$ cation.

Polysulfanes

Polysulfanes are compounds of the general type H_2S_x where $x \geq 2$ (see structure **16.8**). Sulfur dissolves in aqueous solutions of group 1 or 2 metal sulfides (e.g. Na_2S) to yield polysulfide salts, (e.g. Na_2S_x). Acidification of such solutions gives a mixture of polysulfanes as a yellow oil, which can be fractionally distilled to yield H_2S_x ($x = 2$–6). An alternative method of synthesis, particularly useful for polysulfanes with $x > 6$, is by condensation reaction 16.41.

$$2H_2S + S_nCl_2 \longrightarrow H_2S_{n+2} + 2HCl \quad (16.41)$$

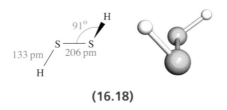

(16.18)

The structure of H_2S_2 (**16.18**) resembles that of H_2O_2 (Fig. 16.10) with an internal dihedral angle of $90.5°$ in the gas phase. All polysulfanes are thermodynamically unstable with respect to decomposition to H_2S and S. Their use in the preparation of *cyclo*-S_n species was described in Section 16.4.

16.6 Metal sulfides, polysulfides, polyselenides and polytellurides

Sulfides

Descriptions of metal sulfides already covered include:

- the zinc blende and wurtzite structures (Section 6.11, Figs. 6.19 and 6.21);
- precipitation of metal sulfides using H_2S (Section 16.5);
- sulfides of the group 14 metals (Section 14.11);
- sulfides of the group 15 elements (Section 15.14).

The group 1 and 2 metal sulfides possess the antifluorite and NaCl structure types respectively (see Section 6.11), and are typical ionic salts. However, the adoption of the NaCl-type structure (e.g. by PbS and MnS) cannot be regarded as a criterion for ionic character, as we discussed in Section 14.11. Most *d*-block metal monosulfides crystallize with the NiAs-type structure (e.g. FeS, CoS, NiS) (see Fig. 15.10) or the zinc blende or wurtzite structure (e.g. ZnS, CdS, HgS) (see Figs. 6.19 and 6.21). Metal disulfides may adopt the CdI_2 structure (e.g. TiS_2 and SnS_2 with metal(IV) centres), but others such as FeS_2 (iron pyrites) contain $[S_2]^{2-}$ ions. The latter are formally analogous to peroxides and may be considered to be salts of H_2S_2.

The blue paramagnetic $[S_2]^-$ ion is an analogue of the superoxide ion and has been detected in solutions of alkali metal sulfides in acetone or dimethyl sulfoxide. Simple salts containing $[S_2]^-$ are not known, but the blue colour of the aluminosilicate mineral *ultramarine* is due to the presence of the radical anions $[S_2]^-$ and $[S_3]^-$ (Box 16.4).

Polysulfides

Polysulfide ions $[S_x]^{2-}$ are not prepared by deprotonation of the corresponding polysulfanes. Instead, methods of synthesis include reactions 16.19 and 16.42, and the reaction of H_2S with S suspended in NH_4OH solution which yields a mixture of $[NH_4]_2[S_4]$ and $[NH_4]_2[S_5]$.

$$2Cs_2S + S_8 \xrightarrow{\text{aq medium}} 2Cs_2[S_5] \quad (16.42)$$

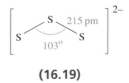

(16.19)

Polysulfides of the *s*-block metals are well established. The $[S_3]^{2-}$ ion is bent (**16.19**), but as the chain length increases, it develops a helical twist, making it chiral

ENVIRONMENT

Box 16.4 Ultramarine blues

The death mask of Tutankhamen, Cairo Museum, Egypt.

The soft, metamorphic mineral *lapis lazuli* (or *lazurite*) is a natural resource that was prized by ancient Egyptians for its blue colour and was cut, carved and polished for ornamental uses. The photograph above illustrates its use in the elaborate decoration of Tutankhamen's death mask. Natural deposits of lapis lazuli occur in, for example, Afghanistan, Iran and Siberia. Powdered lapis lazuli is a natural source of the blue pigment *ultramarine*. Lapis lazuli is related to the aluminosilicate mineral *sodalite*, $Na_8[Al_6Si_6O_{24}]Cl_2$, whose structure is shown opposite. The cavities in the aluminosilicate framework contain Na^+ cations and Cl^- anions. Partial or full replacement of Cl^- by the radical anions $[S_2]^-$ and $[S_3]^-$ results in the formation of $Na_8[Al_6Si_6O_{24}][Cl,S_n]_2$. Additional replacement of ions (e.g. $2Na^+$ for Ca^{2+}) may also occur. The presence of the chalcogenide ions gives rise to the blue pigmentation. The relative amounts of $[S_2]^-$ and $[S_3]^-$ present determine the colour of the pigment: in the UV–VIS spectrum, $[S_2]^-$ absorbs at 370 nm and $[S_3]^-$ at 595 nm. The $[S_2]^-$ and $[S_3]^-$ chromophores therefore give rise to yellow and blue colours, respectively. In some cases, red $[S_4]^-$ is present. In artificial ultramarines, the ratio of different $[S_n]^-$ ions can be controlled, so producing a range of colours from green to blue to violet to pink. Synthetic ultramarine is manufactured by heating together kaolinite (see Box 14.10), Na_2CO_3 and sulfur. This method means that SO_2 is produced, and desulfurization of the waste gases must be

carried out to meet legislative requirements. More environmentally friendly methods of production are being sought (see the reading list below).

Further reading

E. Climent-Pascual, J. Romero de Paz, J. Rodríguez-Carvajal, E. Suard and R. Sáez-Puche (2009) *Inorg. Chem.*, vol. 48, p. 6526 – 'Synthesis and characterization of the ultramarine-type analog $Na_{8-x}[Si_6Al_6O_{24}](S_2,S_3,CO_3)_{1-2}$'.

N. Gobeltz-Hautecoeur, A. Demortier, B. Lede, J.P. Lelieur and C. Duhayon (2002) *Inorg. Chem.*, vol. 41, p. 2848 – 'Occupancy of the sodalite cages in the blue ultramarine pigments'.

S. Kowalak, A. Janowska and S. Łączkowska (2004) *Catal. Today*, vol. 90, p. 167 – 'Preparation of various color ultramarine from zeolite A under environment-friendly conditions'.

D. Reinen and G.-G. Linder (1999) *Chem. Soc. Rev.*, vol. 28, p. 75 – 'The nature of the chalcogen colour centres in ultramarine-type solids'.

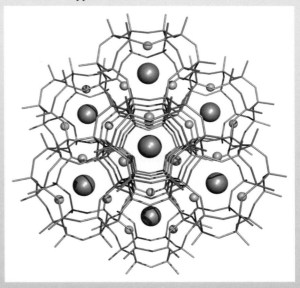

Part of the 3-dimensional structure of sodalite. The aluminosilicate framework is shown in a stick representation; Na^+ (orange) and Cl^- (green) ions occupy the cavities.

(Fig. 16.12a). The coordination chemistry of these anions leads to some complexes such as those in Figs. 16.12 and 22.26b. For chains containing four or more S atoms, the $[S_x]^{2-}$ ligand often chelates to one metal centre or bridges between two centres. The structure of $[AuS_9]^-$ (Fig. 16.12d) illustrates a case where a long chain is required to satisfy the fact that the Au(I) centre favours a linear arrangement of donor atoms.

The cyclic $[S_6]^-$ radical has been prepared by reaction 16.43. In $[Ph_4P][S_6]$, the anion adopts a chair conformation, with two S–S bonds significantly longer than the other four (structure **16.20**).

$$2[Ph_4P][N_3] + 22H_2S + 20Me_3SiN_3$$

$$\longrightarrow 2[Ph_4P][S_6] + 10(Me_3Si)_2S + 11[NH_4][N_3] + 11N_2$$

$$(16.43)$$

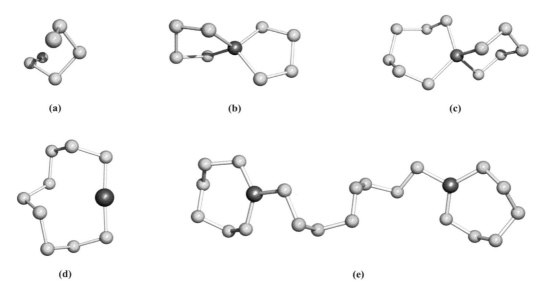

Fig. 16.12 The structures (X-ray diffraction) of (a) $[S_6]^{2-}$ in the salt $[H_3NCH_2CH_2NH_3][S_6]$ [P. Bottcher *et al.* (1984) *Z. Naturforsch., Teil B*, vol. 39, p. 416], (b) $[Zn(S_4)_2]^{2-}$ in the tetraethylammonium salt [D. Coucouvanis *et al.* (1985) *Inorg. Chem.*, vol. 24, p. 24], (c) $[Mn(S_5)(S_6)]^{2-}$ in the $[Ph_4P]^+$ salt [D. Coucouvanis *et al.* (1985) *Inorg. Chem.*, vol. 24, p. 24], (d) $[AuS_9]^-$ in the $[AsPh_4]^+$ salt [G. Marbach *et al.* (1984) *Angew. Chem. Int. Ed.*, vol. 23, p. 246], and (e) $[(S_6)Cu(\mu-S_8)Cu(S_6)]^{4-}$ in the $[Ph_4P]^+$ salt [A. Müller *et al.* (1984) *Angew. Chem. Int. Ed.*, vol. 23, p. 632]. Colour code: S, yellow.

$$\left[\; S \overset{263\ pm}{\underset{206\ pm}{\cdots}} S \cdots S \;\right]^-$$

(16.20)

Polyselenides and polytellurides

Although Se and Te analogues of polysulfanes do not extend beyond the poorly characterized H_2Se_2 and H_2Te_2, the chemistries of polyselenides, polytellurides and their metal complexes are well established. Equations 16.44–16.47 illustrate preparations of salts of $[Se_x]^{2-}$ and $[Te_x]^{2-}$; see Section 11.8 for details of crown ethers and cryptands. Solvothermal conditions may also be used for the synthesis of $[Se_x]^{2-}$ and $[Te_x]^{2-}$ anions. For example, when Cs_2CO_3 and Se react in superheated MeOH solution, the Cs_2CO_3 facilitates the disproportionation of Se to $[Se_x]^{2-}$ and oxoanions. An example of this type of approach is given in self-study exercise 3 below.

$$3Se + K_2Se_2 \xrightarrow{DMF} K_2[Se_5] \qquad (16.44)$$

$$4Se + K_2Se_2 + 2[Ph_4P]Br \longrightarrow [Ph_4P]_2[Se_6] + 2KBr \qquad (16.45)$$

$$3Se + K_2Se_2 \xrightarrow{DMF,\ 15\text{-crown-5}} [K(15\text{-crown-5})]_2[Se_5] \qquad (16.46)$$

$$2K + 3Te \xrightarrow{1,2\text{-diaminoethane, crypt-222}} [K(crypt\text{-}222)]_2[Te_3] \qquad (16.47)$$

(16.21)

(16.22)

Structurally, the smaller polyselenide and polytelluride ions resemble their polysulfide analogues, e.g. $[Te_5]^{2-}$ has structure **16.21** with a helically twisted chain. The structures of higher anions are less simple, e.g. $[Te_8]^{2-}$ (**16.22**) can be considered in terms of $[Te_4]^{2-}$ and $[Te_3]^{2-}$ ligands bound to a Te^{2+} centre. Similarly, $[Se_{11}]^{2-}$ can be described in terms of two $[Se_5]^{2-}$ ligands chelating to an Se^{2+} centre. The coordination chemistry of the $[Se_x]^{2-}$ and $[Te_x]^{2-}$

chain anions has developed significantly since 1990. Examples include $[(Te_4)Cu(\mu\text{-}Te_4)Cu(Te_4)]^{4-}$ and $[(Se_4)_2In(\mu\text{-}Se_5)In(Se_4)_2]^{4-}$ (both of which have bridging and chelating ligands), octahedral $[Pt(Se_4)_3]^{2-}$ with chelating $[Se_4]^{2-}$ ligands, $[Zn(Te_3)(Te_4)]^{2-}$, $[Cr(Te_4)_3]^{3-}$ and $[Au_2(TeSe_2)_2]^{2-}$ (**16.23**) which contains a rare example of a metal-coordinated, mixed Se/Te polychalcogenide anion.

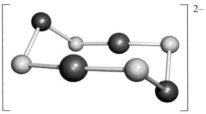

Colour code: Au, red; Te, blue; Se, yellow.

(16.23)

1. $[TeSe_2]^{2-}$ and $[TeSe_3]^{2-}$ possess bent and trigonal pyramidal structures, respectively. Rationalize these structures in terms of the available valence electrons, and draw a set of resonance structures for $[TeSe_3]^{2-}$. Ensure that each atom obeys the octet rule.

2. The $[Te(Se_5)_2]^{2-}$ ion contains two $[Se_2]^{2-}$ bidentate ligands coordinated to a square planar tellurium centre. (i) What is the oxidation state of Te in $[Te(Se_5)_2]^{2-}$? (ii) Each $TeSe_6$-ring has a chair-conformation. Draw the structure of $[Te(Se_5)_2]^{2-}$, given that the Te atom lies on an inversion centre.

3. $[MnCl_2(cyclam)]Cl$ was reacted with elemental selenium in the presence of Cs_2CO_3 in MeOH at 150 °C in a sealed glass tube for 20 h. After this period, the mixture was cooled slowly to room temperature, and crystals of product **X** formed. **X** is a manganese(II) complex and contains 21.03% C, 4.24% H and 9.81% N. (a) Suggest how the cyclam ligand coordinates to a metal ion. (b) Suggest a molecular formula and possible structure for **X**.

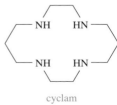

cyclam

[*Ans.* See A. Kromm *et al.* (2008) *Z. Anorg. Allg. Chem.*, vol. 634, p. 2191]

16.7 Halides, oxohalides and complex halides

In contrast to the trend found in earlier groups, the stability of the lowest oxidation state (+2) of the central atom in the halides of the group 16 elements *decreases* down the group. This is well exemplified in the halides discussed in this section. Our discussion is confined to the fluorides of O, and the fluorides and chlorides of S, Se and Te. The bromides and iodides of the later elements are similar to their chloride analogues. Compounds of O with Cl, Br and I are described in Section 17.8.

Oxygen fluorides

Oxygen difluoride, OF_2 (**16.24**), is highly toxic and may be prepared by reaction 16.48. Selected properties are given in Table 16.5. Although OF_2 is formally the anhydride of hypofluorous acid, HOF, only reaction 16.49 occurs with water and this is very slow at 298 K. With concentrated alkali, decomposition is much faster, and with steam, it is explosive.

$$F \overset{O}{\diagdown} \underset{103°}{\diagup} F \quad 141 \text{ pm}$$

(16.24)

$$2NaOH + 2F_2 \longrightarrow OF_2 + 2NaF + H_2O \quad (16.48)$$

$$H_2O + OF_2 \longrightarrow O_2 + 2HF \quad (16.49)$$

Pure OF_2 can be heated to 470 K without decomposition, but it reacts with many elements (to form fluorides and oxides) at, or slightly above, room temperature. When subjected to UV radiation in an argon matrix at 4 K, the $OF^•$ radical is formed (eq. 16.50) and on warming, the radicals combine to give dioxygen difluoride, O_2F_2.

$$OF_2 \xrightarrow{\text{UV radiation}} OF^• + F^• \quad (16.50)$$

Dioxygen difluoride may also be made by the action of a high-voltage discharge on a mixture of O_2 and F_2 at 77–90 K and 1–3 kPa pressure. Selected properties of O_2F_2 are listed in Table 16.5. The low-temperature decomposition of O_2F_2 initially yields $O_2F^•$ radicals. Even at low temperatures, O_2F_2 is an extremely powerful fluorinating agent, e.g. it inflames with S at 93 K, and reacts with BF_3 (eq. 16.8) and SbF_5 (eq. 16.51). O_2F_2 is one of the most powerful oxidative fluorinating agents known, and this is well exemplified by reactions with oxides and fluorides of uranium, plutonium and neptunium (e.g. reactions 16.52 and 16.53). These reactions occur at or below ambient temperatures, in contrast to

Table 16.5 Selected physical properties of oxygen and sulfur fluorides.

Property	OF_2	O_2F_2	S_2F_2	$F_2S=S$	SF_4	SF_6	S_2F_{10}
Physical appearance and general characteristics	Colourless (very pale yellow) gas; explosive and toxic	Yellow solid below 119 K; decomposes above 223 K	Colourless gas; extremely toxic	Colourless gas	Colourless gas; toxic; reacts violently with water	Colourless gas; highly stable	Colourless liquid; extremely toxic
Melting point / K	49	119	140	108	148	222 (under pressure)	220
Boiling point / K	128	210	288	262	233	subl. 209	303
$\Delta_f H^o$(298 K)/kJ mol^{-1}	+24.7	+18.0			−763.2	−1220.5	
Dipole moment / D	0.30	1.44			0.64	0	0
E−F bond distance / pm†	141	157.5	163.5	160	164.5 (ax) 154.5 (eq)	156	156

† For other structural data, see text.

the high temperatures required to form UF_6, PuF_6 and NpF_6 using F_2 or halogen fluorides as fluorinating agents. However, because of the high reactivity of O_2F_2, choosing appropriate reaction conditions is crucial to being able to control the reaction.

$$2O_2F_2 + 2SbF_5 \longrightarrow 2[O_2]^+[SbF_6]^- + F_2 \qquad (16.51)$$

$$NpF_4 + O_2F_2 \longrightarrow NpF_6 + O_2 \qquad (16.52)$$

$$NpO_2 + 3O_2F_2 \longrightarrow NpF_6 + 4O_2 \qquad (16.53)$$

The molecular shape of O_2F_2 (**16.25**) resembles that of H_2O_2 (Fig. 16.9) although the internal dihedral angle is smaller (87°). The very long O−F bond probably accounts for the ease of dissociation into $O_2F^•$ and $F^•$. Structures **16.26** show valence bond representations which reflect the long O−F and short O−O bonds; compare the O−O bond distance with those for O_2 and derived ions (Section 16.4) and H_2O_2 (Table 16.3).

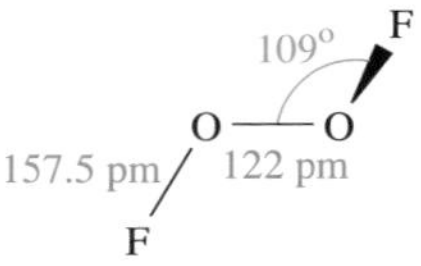

(16.25) 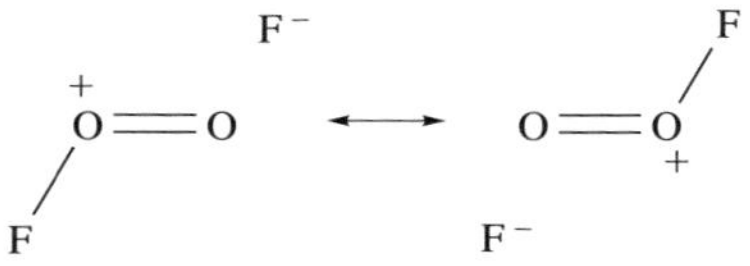   **(16.26)**

Sulfur fluorides and oxofluorides

Table 16.5 lists some properties of the most stable fluorides of sulfur. The fluorides SF_4 and S_2F_2 can be prepared from the

reaction of SCl_2 and HgF_2 at elevated temperatures. Both fluorides are highly unstable. Disulfur difluoride exists as two isomers: S_2F_2 (**16.27**) and $F_2S=S$ (**16.28**), with S_2F_2 (made from AgF and S at 398 K) readily isomerizing to $F_2S=S$. The structure of S_2F_2 is like that of O_2F_2, with an internal dihedral angle of 88°. The S−S bond distances in both isomers are very short and imply multiple bond character (compare ≈206 pm for a single S−S bond, and 184 pm in $[S_2I_4]^{2+}$, Fig. 16.7). For S_2F_2, contributions from resonance structures analogous to those shown for O_2F_2 are important, while for $S=SF_2$, we may write:

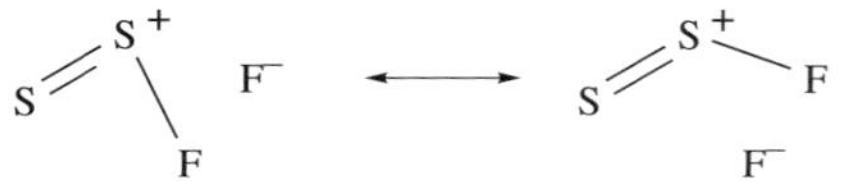

Both isomers are unstable with respect to disproportionation into SF_4 and S, and are extremely reactive, attacking glass and being rapidly hydrolysed by water and alkali (e.g. eq. 16.54).

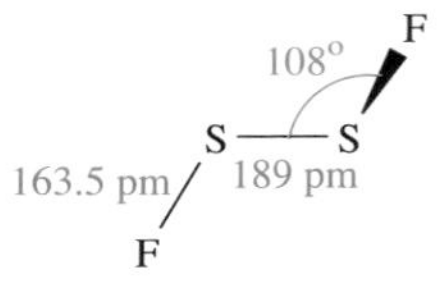

(16.27) 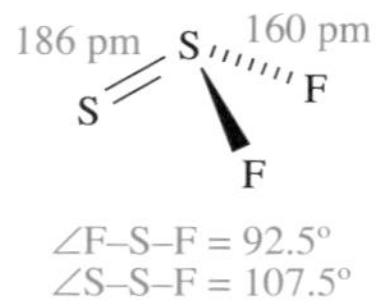   **(16.28)**

$$2S=SF_2 + 2[OH]^- + H_2O \longrightarrow \tfrac{1}{4}S_8 + [S_2O_3]^{2-} + 4HF \qquad (16.54)$$

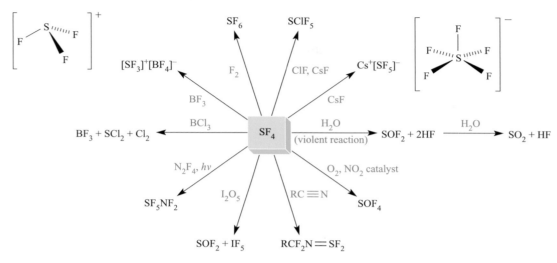

Fig. 16.13 Selected reactions of sulfur tetrafluoride.

Sulfur tetrafluoride, SF_4, is best prepared by reaction 16.55. It is commercially available and is used as a selective fluorinating agent, e.g. it converts carbonyl groups into CF_2 groups without destroying any unsaturation in the molecule. Representative reactions are shown in Fig. 16.13. SF_4 hydrolyses rapidly and must be handled in moisture-free conditions.

$$3SCl_2 + 4NaF \xrightarrow{\text{MeCN, 350 K}} SF_4 + S_2Cl_2 + 4NaCl$$

(16.55)

The structure of SF_4, **16.29**, is derived from a trigonal bipyramid and can be rationalized in terms of the VSEPR model. The $S-F_{ax}$ and $S-F_{eq}$ bond distances are quite different (Table 16.5). Oxidation by O_2 in the absence of a catalyst to form SOF_4 is slow. The structure of SOF_4, **16.30**, is related to that of SF_4, but with $S-F_{ax}$ and $S-F_{eq}$ bond distances that are close in value.

(16.29)   **(16.30)**   **(16.31)**

Among the sulfur fluorides, SF_6, **16.31**, stands out for its high stability and chemical inertness. Its bonding was discussed in Section 5.7. SF_6 is commercially available and is manufactured by burning sulfur in F_2. It has a high dielectric constant and its main use is as an electrical insulator. However, SF_6 that enters the atmosphere is long-lived, and emissions are controlled under the Kyoto Protocol (see Box 14.7). Its lack of reactivity (e.g. it is unaffected

by steam at 770 K or molten alkalis) is kinetic rather than thermodynamic in origin. The value of $\Delta_r G^{\circ}$ for reaction 16.56 certainly indicates thermodynamic spontaneity. Rather surprisingly, SF_6 has been shown to be a reactive fluorinating agent towards low-valent organometallic Ti and Zr compounds (e.g. reaction 16.57 which involves cyclopentadienyl derivatives).

$$SF_6 + 3H_2O \longrightarrow SO_3 + 6HF$$
$$\Delta_r G^{\circ}(298 \text{ K}) = -221 \text{ kJ mol}^{-1}$$

(16.56)

(16.57)

The preparation of SF_6 from S and F_2 produces small amounts of S_2F_{10} and the yield can be optimized by controlling the reaction conditions. An alternative route is reaction 16.58. Selected properties of S_2F_{10} are given in Table 16.5.

$$2SF_5Cl + H_2 \xrightarrow{h\nu} S_2F_{10} + 2HCl$$

(16.58)

(16.32)

Molecules of S_2F_{10} have the staggered structure **16.32**. The S–S bond length of 221 pm is significantly longer than the single bonds in elemental S (206 pm). It disproportionates when heated (eq. 16.59) and is a powerful oxidizing agent. An interesting reaction is that with NH_3 to yield $N\equiv SF_3$ (see structure **16.66**).

$$S_2F_{10} \xrightarrow{420\,K} SF_4 + SF_6 \qquad (16.59)$$

Many compounds containing SF_5 groups are now known, including $SClF_5$ and SF_5NF_2 (Fig. 16.13). In accord with the relative strengths of the S–Cl and S–F bonds (Table 16.2), reactions of $SClF_5$ usually involve cleavage of the S–Cl bond (e.g. reaction 16.60).

$$2SClF_5 + O_2 \xrightarrow{h\nu} F_5SOOSF_5 + Cl_2 \qquad (16.60)$$

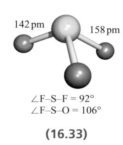

∠F–S–F = 92°
∠F–S–O = 106°

(16.33)

Sulfur forms several oxofluorides, and we have already mentioned SOF_4. Thionyl difluoride, SOF_2 (**16.33**), is a colourless gas (bp 229 K), prepared by fluorinating $SOCl_2$ using SbF_3. It reacts with F_2 to give SOF_4, and is slowly hydrolysed by water (see Fig. 16.13). The reaction of SOF_2 and $[Me_4N]F$ at 77 K followed by warming to 298 K produces $[Me_4N][SOF_3]$, the first example of a salt containing $[SOF_3]^-$. The anion rapidly hydrolyses (reaction 16.61 followed by reaction 16.62 depending on conditions) and reacts with SO_2 to give SOF_2 and $[SO_2F]^-$.

$$3[SOF_3]^- + H_2O \longrightarrow 2[HF_2]^- + [SO_2F]^- + 2SOF_2 \qquad (16.61)$$

$$4[SO_2F]^- + H_2O \longrightarrow 2[HF_2]^- + [S_2O_5]^{2-} + 2SO_2 \qquad (16.62)$$

Sulfuryl difluoride,[†] SO_2F_2 (**16.34**), is a colourless gas (bp 218 K) which is made by reaction 16.63 or 16.64.

$$SO_2Cl_2 + 2NaF \longrightarrow SO_2F_2 + 2NaCl \qquad (16.63)$$

$$Ba(SO_3F)_2 \xrightarrow{\Delta} SO_2F_2 + BaSO_4 \qquad (16.64)$$

[†] Sulfuryl difluoride is also called sulfonyl difluoride or sulfonyl fluoride.

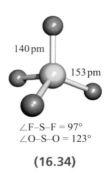

140 pm
153 pm

∠F–S–F = 97°
∠O–S–O = 123°

(16.34)

Although unaffected by water, SO_2F_2 is hydrolysed by concentrated aqueous alkali. A series of sulfuryl fluorides is known, including FSO_2OSO_2F and FSO_2OOSO_2F. The latter compound is prepared by reaction 16.65; fluorosulfonic acid (see Section 9.8) is related to the intermediate in this reaction.

$$SO_3 + F_2 \xrightarrow{AgF_2,\ 450\,K} FSO_2OF \xrightarrow{SO_3} FSO_2OOSO_2F \qquad (16.65)$$

The dissociation of FSO_2OOSO_2F at 393 K produces the brown paramagnetic radical $FSO_2O^{\bullet}$, selected reactions of which are shown in scheme 16.66.

$$(16.66)$$

The reaction of F_2 with sulfate ion yields $[FSO_4]^-$ which can be isolated as the caesium salt and is an extremely powerful oxidizing agent (eq. 16.67).

$$[FSO_4]^- + 2H^+ + 2e^- \rightleftharpoons [HSO_4]^- + HF \qquad E^o \approx +2.5\,V \qquad (16.67)$$

Self-study exercises

1. Consider structure **16.33**. Draw a set of resonance structures for SOF_2 that maintains an octet of electrons around the S atom. Comment on the structures that you have drawn, given that values of r_{cov} of S, O and F are 103, 73 and 71 pm, respectively.

2. Show that SO_2F_2 belongs to the C_{2v} point group.

3. SF_6 is a greenhouse gas. In the upper stratosphere, photolysis to SF_5 is possible. Combination of SF_5 with O_2 gives the radical F_5SO_2. Draw Lewis structures for SF_5 and F_5SO_2 showing which atom formally carries the unpaired electron in each species.

Sulfur chlorides and oxochlorides

The range of sulfur chlorides and oxochlorides (which are all hydrolysed by water) is more restricted than that of the corresponding fluorides. There are no stable chloro analogues of SF_4, SF_6 and S_2F_{10}. One example of a high oxidation state chloride is $SClF_5$, prepared as shown in Fig. 16.13.

Disulfur dichloride, S_2Cl_2, is a fuming orange liquid (mp 193 K, bp 409 K) which is toxic and has a repulsive smell. It is manufactured by passing Cl_2 through molten S, and further chlorination yields SCl_2 (a dark-red liquid, mp 195 K, dec. 332 K). Both are used industrially for the manufacture of $SOCl_2$ (scheme 16.68) and S_2Cl_2 for the vulcanization of rubber. Pure SCl_2 is unstable with respect to equilibrium 16.69.

$$2SO_2 + S_2Cl_2 + 3Cl_2 \longrightarrow 4SOCl_2$$
$$SO_3 + SCl_2 \longrightarrow SOCl_2 + SO_2$$
$$\text{(16.68)}$$

$$2SCl_2 \rightleftharpoons S_2Cl_2 + Cl_2 \qquad \text{(16.69)}$$

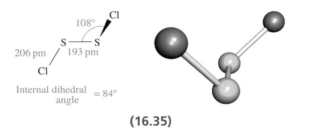

(16.35)

The structure of S_2Cl_2, **16.35**, resembles that of S_2F_2. SCl_2 is a bent molecule (S–Cl = 201 pm, $\angle Cl–S–Cl = 103°$). Decomposition of both chlorides by water yields a complicated mixture containing S, SO_2, $H_2S_5O_6$ and HCl. Equation 16.18 showed the use of S_2Cl_2 in the formation of an S_n ring. Condensation of S_2Cl_2 with polysulfanes (eq. 16.70) gives rise to chlorosulfanes that can be used, for example, in the formation of various sulfur rings (see structures **16.8** and **16.9** and discussion).

$$ClS-SCl + H\!-\!(S)_x\!-\!H + ClS-SCl$$
$$\longrightarrow ClS_{x+4}Cl + 2HCl \quad \text{(16.70)}$$

Thionyl dichloride, $SOCl_2$ (prepared, for example, by reaction 16.68 or 16.71), and sulfuryl dichloride,[‡] SO_2Cl_2 (prepared by reaction 16.72), are colourless, fuming liquids: $SOCl_2$, bp 351 K; SO_2Cl_2, bp 342 K. Their ease of hydrolysis accounts for their fuming nature, e.g. eq. 16.73.

$$SO_2 + PCl_5 \longrightarrow SOCl_2 + POCl_3 \qquad \text{(16.71)}$$

$$SO_2 + Cl_2 \xrightarrow{\text{activated charcoal}} SO_2Cl_2 \qquad \text{(16.72)}$$

$$SOCl_2 + H_2O \longrightarrow SO_2 + 2HCl \qquad \text{(16.73)}$$

[‡] Sulfuryl dichloride is also called sulfonyl dichloride or sulfonyl chloride.

The structural parameters shown for $SOCl_2$, **16.36**, and SO_2Cl_2, **16.37**, are for the gas-phase molecules.

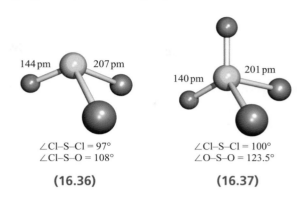

| ∠Cl–S–Cl = 97° | ∠Cl–S–Cl = 100° |
| ∠Cl–S–O = 108° | ∠O–S–O = 123.5° |

(16.36) **(16.37)**

Both $SOCl_2$ and SO_2Cl_2 are available commercially. Thionyl dichloride is used to prepare acyl chlorides (eq. 16.74) and anhydrous metal chlorides (i.e. removing water of crystallization by reaction 16.73), while SO_2Cl_2 is a chlorinating agent.

$$RCO_2H + SOCl_2 \xrightarrow{\Delta} RC(O)Cl + SO_2 + HCl \qquad \text{(16.74)}$$

Self-study exercises

1. Show that SCl_2 belongs to the C_{2v} point group.

2. Does SCl_2 possess $(3n - 5)$ or $(3n - 6)$ degrees of vibrational freedom? Rationalize your answer.
   [*Ans.* See eqs. 3.5 and 3.6]

3. By using the C_{2v} character table (Appendix 3), show that an SCl_2 molecule has A_1 and B_2 normal modes of vibration. Draw diagrams to illustrate these modes of vibration. Confirm that each mode is both IR and Raman active.
   [*Ans.* Refer to Fig. 3.12 (SCl_2 is like SO_2) and related discussion]

4. Show that S_2Cl_2 has C_2 symmetry.

Halides of selenium and tellurium

In contrast to sulfur chemistry where dihalides are well established, the isolation of dihalides of selenium and tellurium has only been achieved for $SeCl_2$ and $SeBr_2$ (reactions 16.75 and 16.76). Selenium dichloride is a thermally unstable red oil; $SeBr_2$ is a red-brown solid.

$$Se_{\text{powder}} + SO_2Cl_2 \xrightarrow{296 \text{ K}} SeCl_2 + SO_2 \qquad \text{(16.75)}$$

$$SeCl_2 + 2Me_3SiBr \xrightarrow{296 \text{ K, THF}} SeBr_2 + 2Me_3SiCl \qquad \text{(16.76)}$$

Table 16.6 lists selected properties of SeF_4, SeF_6, TeF_4 and TeF_6. Selenium tetrafluoride is a good fluorinating agent. It is a liquid at 298 K and (compared with SF_4) is relatively

Table 16.6 Selected properties of the fluorides of selenium and tellurium.

Property	SeF$_4$	SeF$_6$	TeF$_4$	TeF$_6$
Physical appearance and general characteristics	Colourless fuming liquid; toxic; violent hydrolysis	White solid at low temp.; colourless gas; toxic	Colourless solid; highly toxic	White solid at low temp.; colourless gas; foul smelling; highly toxic
Melting point / K	263.5	subl. 226	403	subl. 234
Boiling point / K	375	–	dec. 467	–
$\Delta_f H^{\circ}$(298 K) / kJ mol^{-1}		−1117.0		−1318.0
E−F bond distance for gas phase molecules / pm†	Se−F$_{ax}$ = 176.5 Se−F$_{eq}$ = 168	169	Te−F$_{ax}$ = 190 Te−F$_{eq}$ = 179	181.5

† For other structural data, see text.

convenient to handle. It is prepared by reacting SeO$_2$ with SF$_4$. Combination of F$_2$ and Se yields SeF$_6$ which is thermally stable and relatively inert. The tellurium fluorides are similarly prepared, TeF$_4$ from TeO$_2$ and SF$_4$ (or SeF$_4$), and TeF$_6$ from the elements. In the liquid and gas phases, SeF$_4$ contains discrete molecules (Fig. 16.14a) but in the solid state, significant intermolecular interactions are present. These are considerably weaker than in TeF$_4$, in which the formation of asymmetrical Te−F−Te bridges leads to a polymeric structure in the crystal (Fig. 16.14b). ^{19}F NMR spectroscopic studies of liquid SeF$_4$ have shown that the molecules are stereochemically non-rigid (see Section 4.8). The structures of SeF$_6$ and TeF$_6$ are regular octahedra. SeF$_4$ fumes in moist air, whereas SeF$_6$ resists hydrolysis. SeF$_4$ reacts with CsF to give Cs$^+$[SeF$_5$]$^-$. Further reaction with fluoride ion to give [SeF$_6$]$^{2-}$ can be achieved only by using a highly active fluoride source (the so-called 'naked' fluoride ion, accessed by using anhydrous Me$_4$NF or organic fluorides containing large counter-ions). Thus, the reaction of **16.38** with [SeF$_5$]$^-$ gives the hexamethylpiperidinium salt of [SeF$_6$]$^{2-}$. In the solid state, the [SeF$_6$]$^{2-}$ ion in this salt has a distorted octahedral structure (its symmetry lies between C_{3v} and C_{2v}). The distortion away from O_h symmetry has been attributed to the presence of a stereochemically active lone pair of electrons, but the fact that there are C−H$\cdots$F

hydrogen bonds between cations and anions means that the [SeF$_6$]$^{2-}$ ion cannot be considered to be in an isolated environment.

1,1,3,3,5,5-hexamethylpiperidinium fluoride (a source of 'naked' fluoride ion)

(16.38)

Tellurium hexafluoride is hydrolysed by water to telluric acid, H$_6$TeO$_6$ (see **16.62**), and undergoes a number of exchange reactions such as reaction 16.77. It is also a fluoride acceptor, reacting with alkali metal fluorides and [Me$_4$N]F under anhydrous conditions (eq. 16.78).

$$\text{TeF}_6 + \text{Me}_3\text{SiNMe}_2 \longrightarrow \text{Me}_2\text{NTeF}_5 + \text{Me}_3\text{SiF} \qquad (16.77)$$

$$\left.\begin{array}{l} \text{TeF}_6 + [\text{Me}_4\text{N}]\text{F} \xrightarrow{\text{MeCN, 233 K}} [\text{Me}_4\text{N}][\text{TeF}_7] \\ [\text{Me}_4\text{N}][\text{TeF}_7] + [\text{Me}_4\text{N}]\text{F} \xrightarrow{\text{MeCN, 273 K}} [\text{Me}_4\text{N}]_2[\text{TeF}_8] \end{array}\right\}$$

$$(16.78)$$

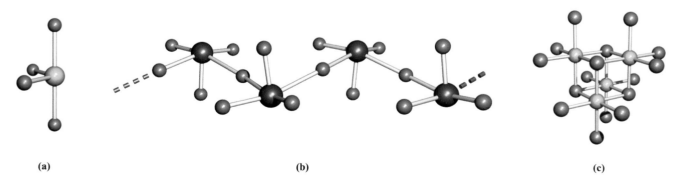

Fig. 16.14 (a) The structure of SeF$_4$ in the gas and liquid phases. (b) In the solid state, TeF$_4$ consists of polymeric chains; the Te−F−Te bridges are asymmetrical (Te−F = 208 and 228 pm). (c) The structure of the molecular Se$_4$Cl$_{16}$-unit present in crystalline SeCl$_4$. Colour code: Se, yellow; Te, blue; F and Cl, green.

The $[TeF_7]^-$ ion has a pentagonal bipyramidal structure (**16.39**) although in the solid state, the equatorial F atoms deviate slightly from the mean equatorial plane. In $[TeF_8]^{2-}$, **16.40**, vibrational spectroscopic data are consistent with the Te centre being in a square-antiprismatic environment.

$$Te-F_{ax} = 179 \text{ pm}$$
$$Te-F_{eq} = 183-190 \text{ pm}$$

(16.39)   **(16.40)**

In contrast to sulfur, Se and Te form stable tetrachlorides, made by direct combination of the elements. At 298 K, both tetrachlorides are solids (SeCl$_4$, colourless, subl. 469 K; TeCl$_4$, yellow, mp 497 K, bp 653 K) which contain tetrameric units, depicted in Fig. 16.14c for SeCl$_4$. The E–Cl (E = Se or Te) bonds within the cubane core are significantly longer than the terminal E–Cl bonds: e.g. Te–Cl = 293 (core) and 231 (terminal) pm. Thus, the structure may also be described in terms of $[ECl_3]^+$ and Cl^- ions. Stepwise removal of $[ECl_3]^+$ from the tetramers E_4Cl_{16} (E = Se, Te) occurs in the presence of chloride ion in non-polar solvents, e.g. for the first step:

$$Te_4Cl_{16} + R^+Cl^- \longrightarrow R^+[Te_3Cl_{13}]^- + TeCl_4$$

in which R^+ is a large, organic cation, or overall:

$$Te_4Cl_{16} \longrightarrow [Te_3Cl_{13}]^- \longrightarrow [Te_2Cl_{10}]^{2-} \longrightarrow [TeCl_6]^{2-}$$

> A ***cubane*** contains a central cubic (or near-cubic) arrangement of atoms.

The $[SeCl_3]^+$ and $[TeCl_3]^+$ cations are also formed in reactions with Cl^- acceptors, e.g. reaction 16.79.

$$SeCl_4 + AlCl_3 \longrightarrow [SeCl_3]^+ + [AlCl_4]^- \qquad (16.79)$$

Both SeCl$_4$ and TeCl$_4$ are readily hydrolysed by water, but with group 1 metal chlorides in the presence of concentrated HCl, yellow complexes such as $K_2[SeCl_6]$ and $K_2[TeCl_6]$ are formed. Reaction 16.80 is an alternative route to $[TeCl_6]^{2-}$, while $[SeCl_6]^{2-}$ is formed when SeCl$_4$ is dissolved in molten SbCl$_3$ (eq. 16.81).

$$TeCl_4 + 2^tBuNH_2 + 2HCl \longrightarrow 2[^tBuNH_3]^+ + [TeCl_6]^{2-}$$
$$\qquad (16.80)$$

$$2SbCl_3 + SeCl_4 \rightleftharpoons 2[SbCl_2]^+ + [SeCl_6]^{2-} \qquad (16.81)$$

The $[SeCl_6]^{2-}$ and $[TeCl_6]^{2-}$ ions usually (see below) possess *regular octahedral* structures (O_h symmetry), rather than the distorted structure (with a stereochemically active lone pair) that would be expected on the basis of the VSEPR model. It may be argued that a change from a distorted to a regular octahedral structure arises from a decrease in stereochemical activity of the lone pair as the steric crowding of the ligands increases,[†] e.g. on going from $[SeF_6]^{2-}$ to $[SeCl_6]^{2-}$. However, as we have already noted, the origins of the distortion in $[SeF_6]^{2-}$ in the solid state are not unambiguous because of the presence of cation–anion hydrogen-bonded interactions. This word of caution extends to other examples where the nature of the cation influences the structure of the anion in the *solid state*. For example, in $[H_3N(CH_2)_3NH_3][TeCl_6]$, the $[TeCl_6]^{2-}$ has approximately C_{2v} symmetry, and in $[^tBuNH_3]_2[TeBr_6]$, the $[TeBr_6]^{2-}$ ion has approximately C_{3v} symmetry. For the octahedral anions, a molecular orbital scheme can be developed (Fig. 16.15) that uses only the valence shell $4s$ and $4p$ (Se) or $5s$ and $5p$ (Te) orbitals. Combined with six Cl $3p$ orbitals, this leads to seven occupied MOs in $[ECl_6]^{2-}$ (E = Se, Te), of which four have bonding character, two have non-bonding character, and one has antibonding character. The net number of bonding MOs is therefore three, and the net E–Cl bond order is 0.5.

Tellurium forms a series of subhalides, e.g. Te$_3$Cl$_2$ and Te$_2$Cl, the structures of which can be related to the helical chains in elemental Te. When Te is oxidized to Te$_3$Cl$_2$, oxidation of one in three Te atoms occurs to give polymer **16.41**.

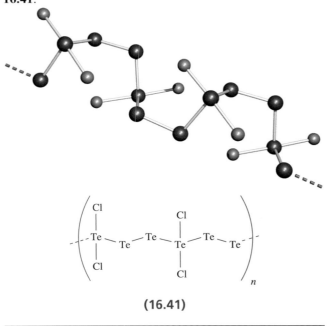

(16.41)

[†] For a fuller discussion of these ideas, see: R.J. Gillespie and P.L.A. Popelier (2001) *Chemical Bonding and Molecular Geometry*, OUP, Oxford, Chapter 9. For a theoretical investigation of $[EX_6]^{n-}$ species, see: M. Atanasov and D. Reinen (2005) *Inorg. Chem.*, vol. 44, p. 5092.

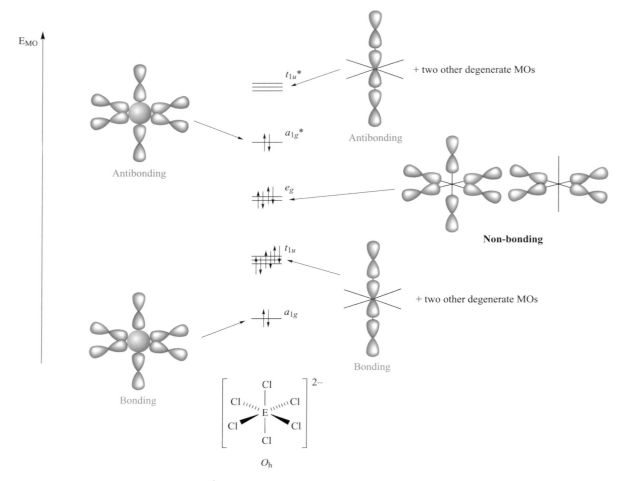

Fig. 16.15 An MO diagram for octahedral $[ECl_6]^{2-}$ (E = Se or Te) using a valence set of $4s$ and $4p$ orbitals for Se or $5s$ and $5p$ orbitals for Te. These orbitals overlap with Cl $3p$ orbitals. The diagram can be derived from that for SF_6 described in Figs. 5.27 and 5.28.

16.8 Oxides

Oxides of sulfur

The most important oxides of sulfur are SO_2 and SO_3, but there are also a number of unstable oxides. Among these are S_2O (**16.42**) and S_8O (**16.43**), made by reactions 16.82 and 16.83. The oxides S_nO ($n = 6$–10) can be prepared by reaction 16.84, exemplified for S_8O.

(16.42)   **(16.43)**

$$SOCl_2 + Ag_2S \xrightarrow{430\,K} S_2O + 2AgCl \tag{16.82}$$

$$HS_7H + SOCl_2 \longrightarrow S_8O + 2HCl \tag{16.83}$$

$$S_8 \xrightarrow{CF_3C(O)OOH} S_8O \tag{16.84}$$

Sulfur dioxide is manufactured on a large scale by burning sulfur (the most important process) or H_2S, by roasting sulfide ores (e.g. eq. 16.85), or reducing $CaSO_4$ (eq. 16.86). Desulfurization processes to limit SO_2 emissions (see Box 12.2) and reduce acid rain (see Box 16.5) are now in use. In the laboratory, SO_2 may be prepared by reaction 16.87, and it is commercially available in cylinders. Selected physical properties of SO_2 are listed in Table 16.7.

$$4FeS_2 + 11O_2 \xrightarrow{\Delta} 2Fe_2O_3 + 8SO_2 \tag{16.85}$$

$$CaSO_4 + C \xrightarrow{>1620\,K} CaO + SO_2 + CO \tag{16.86}$$

$$Na_2SO_3 + 2HCl \underset{conc}{\longrightarrow} SO_2 + 2NaCl + H_2O \tag{16.87}$$

The boiling point of SO_2 is 263 K, but it can be safely handled in a sealed tube at room temperature under its own vapour pressure. It is a good solvent with a wide

Table 16.7 Selected physical properties of SO_2 and SO_3.

Property	SO_2	SO_3
Physical appearance and general characteristics	Colourless, dense gas; pungent smell	Volatile white solid, or a liquid
Melting point / K	197.5	290
Boiling point / K	263.0	318
$\Delta_{vap}H^{\circ}(bp) / kJ\,mol^{-1}$	24.9	40.7
$\Delta_f H^{\circ}(298\,K) / kJ\,mol^{-1}$	−296.8 (SO_2, g)	−441.0 (SO_3, l)
Dipole moment / D	1.63	0
S−O bond distance / pm [†]	143	142
∠O−S−O / deg [†]	119.5	120

[†] Gas phase parameters; for SO_3, data refer to the monomer.

range of uses (see Section 9.8). Sulfur dioxide has a molecular structure (**16.44**).

S–O = 143 pm
∠O–S–O = 119.5°

(16.44)

Sulfur dioxide reacts with O_2 (see below), F_2 and Cl_2 (eq. 16.88). It also reacts with the heavier alkali metal fluorides to give metal fluorosulfites (eq. 16.89), and with CsN_3 to give the Cs^+ salts of $[SO_2N_3]^-$ (Fig. 16.16a) and $[(SO_2)_2N_3]^-$ (Fig. 16.16b). The latter is formed when CsN_3 dissolves in liquid SO_2 at 209 K. On raising the temperature to 243 K, $[(SO_2)_2N_3]^-$ loses one equivalent of SO_2 to yield $[SO_2N_3]^-$.

$$SO_2 + X_2 \longrightarrow SO_2X_2 \qquad (X = F, Cl) \qquad (16.88)$$

$$SO_2 + MF \xrightarrow{258\,K} M^+[SO_2F]^- \qquad (M = K, Rb, Cs) \qquad (16.89)$$

In aqueous solution, SO_2 is converted to only a small extent to sulfurous acid. Aqueous solutions of H_2SO_3 contain significant amounts of dissolved SO_2 (see eqs. 7.18–7.20). Sulfur dioxide is a weak reducing agent in acidic solution, and a slightly stronger one in basic media (eqs. 16.90 and 16.91).

$$[SO_4]^{2-}(aq) + 4H^+(aq) + 2e^- \rightleftharpoons H_2SO_3(aq) + H_2O(l)$$
$$E^{\circ} = +0.17\,V \qquad (16.90)$$

$$[SO_4]^{2-}(aq) + H_2O(l) + 2e^- \rightleftharpoons [SO_3]^{2-}(aq) + 2[OH]^-(aq)$$
$$E^{\circ}_{[OH^-]=1} = -0.93\,V \qquad (16.91)$$

Thus, aqueous solutions of SO_2 are oxidized to sulfate by many oxidizing agents (e.g. I_2, $[MnO_4]^-$, $[Cr_2O_7]^{2-}$ and Fe^{3+} in acidic solutions). However, if the concentration of H^+ is very high, $[SO_4]^{2-}$ can be reduced to SO_2 as in, for

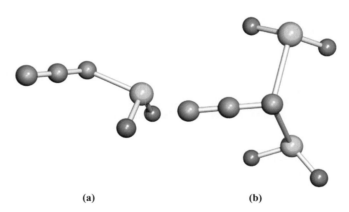

(a)   **(b)**

Fig. 16.16 (a) The structure of the azidosulfite anion, $[SO_2N_3]^-$, determined by X-ray diffraction at 173 K for the Cs^+ salt [K.O. Christe *et al.* (2002) *Inorg. Chem.*, vol. 41, p. 4275]. (b) The structure of $[(SO_2)_2N_3]^-$ determined for the Cs^+ salt by X-ray diffraction at 130 K [K.O. Christe *et al.* (2003) *Inorg. Chem.*, vol. 42, p. 416]. Colour code: N, blue; S, yellow; O, red.

ENVIRONMENT

Box 16.5 The contribution of SO$_2$ to acid rain

Despite being recognized as far back as the 1870s, the environmental problems associated with 'acid rain' came to the fore in the 1960s with the decline of fish stocks in European and North American lakes. Two of the major contributors towards acid rain are SO$_2$ and NO$_x$. (In Section 25.8, we discuss the use of catalytic converters to combat pollution due to nitrogen oxides, NO$_x$.) Although SO$_2$ emissions arise from natural sources such as volcanic eruptions, artificial sources contribute ≈90% of the sulfur in the atmosphere. Fossil fuels such as coal contain ≈2–3% sulfur and combustion produces SO$_2$. This is being countered by the recovery of sulfur from petroleum (Fig. 16.2). Sulfur dioxide is released when metal sulfide ores are roasted in the production of metals such as Co, Ni, Cu and Zn, for example:

$$Cu_2S + O_2 \xrightarrow{\Delta} 2Cu + SO_2$$

However, this source of SO$_2$ is now utilized for the production of sulfuric acid (see Fig. 16.2 and accompanying text). Once released into the atmosphere, SO$_2$ dissolves in water vapour, forming H$_2$SO$_3$ and H$_2$SO$_4$. Acid formation may take several days and involves multi-stage reactions, the outcome of which is:

$$2SO_2 + O_2 + 2H_2O \longrightarrow 2H_2SO_4$$

By the time acid rain falls to the Earth's surface, the pollutants may have travelled long distances from their industrial sources. For example, prevailing winds in Europe may carry SO$_2$ from the UK, France and Germany to Scandinavia.

The effects of acid rain can be devastating. The pH of lakes and streams is lowered, although the composition of the bedrock is significant, and in some cases provides a natural buffering effect. A second effect is that acid rain penetrating the bedrock can react with aluminosilicate minerals, or can leach heavy metal ions from the bedrock. As the acid rain makes its way through the bedrock and into waterways, it carries with it the metal pollutants. Acidified and polluted waters not only kill fish, but also affect the food chain. Acid rain falling on soils may be neutralized if the soil is alkaline, but otherwise the lowering of the pH and the leaching of plant nutrients has devastating effects on vegetation. The effects of acid rain on some building materials are all around us: crumbling gargoyles on ancient churches are a sad reminder of pollution by acid rain. The photograph illustrates the damage caused by acid rain to one of the limestone gargoyles on Notre Dame Cathedral in Paris. The cathedral was completed in 1345, and photographic records in the 20th century show that fine detail in the stone carving was still present in 1920. During the next 70 years, coincident with the growth of industrialized nations, significant corrosion occurred, caused predominantly by acid rain.

International legislation to reduce acidic gas emissions has been in operation since the 1980s. Emissions of SO$_2$ and NO$_2$ in Europe climbed steadily after 1920. Maximum emis-

A gargoyle on Notre Dame Cathedral, Paris, photographed in 1996. The damage to the limestone has been caused mainly by acid rain.

sions of 62 Mt of SO$_2$ were recorded in 1980, while emissions of NO$_2$ peaked at 29 Mt in 1990. The effects of legislation have been significant reductions in both SO$_2$ and NO$_2$ emissions with projected values of 18 and 15 Mt per year, respectively, by 2020. Recent environmental studies indicate some improvement in the state of Western European and North American streams and lakes.

For related information: see Box 12.2: Desulfurization processes to limit SO$_2$ emissions; Box 16.6: Volcanic emissions.

Further reading

T. Loerting, R.T. Kroemer and K.R. Liedl (2000) *Chem. Commun.*, p. 999 – 'On the competing hydrations of sulfur dioxide and sulfur trioxide in our atmosphere'.

D. Malakoff (2010) *Science*, vol. 330, p. 910 – 'Taking the sting out of acid rain'.

J.L. Stoddard *et al.* (1999) *Nature*, vol. 401, p. 575 – 'Regional trends in aquatic recovery from acidification in North America and Europe'.

J. Vuorenmaa (2004) *Environ. Pollut.*, vol. 128, p. 351 – 'Long-term changes of acidifying deposition in Finland (1973–2000)'.

R.F. Wright *et al.* (2005) *Environ. Sci. Technol.*, vol. 39, p. 64A – 'Recovery of acidified European surface waters'.

example, reaction 16.92; the dependence of E on $[H^+]$ is detailed in Section 8.2.

$$Cu + 2H_2SO_4 \xrightarrow{\text{conc}} SO_2 + CuSO_4 + 2H_2O \quad (16.92)$$

In the presence of concentrated HCl, SO_2 will itself act as an oxidizing agent; in reaction 16.93, the Fe(III) produced is then complexed by Cl^-.

$$\left. \begin{array}{l} SO_2 + 4H^+ + 4Fe^{2+} \longrightarrow S + 4Fe^{3+} + 2H_2O \\ Fe^{3+} + 4Cl^- \longrightarrow [FeCl_4]^- \end{array} \right\} \quad (16.93)$$

The oxidation of SO_2 by atmospheric O_2 (eq. 16.94) is very slow, but is catalysed by V_2O_5 (see Section 25.8). This is the first step in the *Contact process* for the manufacture of sulfuric acid. Operating conditions are crucial since equilibrium 16.94 shifts further towards the left-hand side as the temperature is raised, although the yield can be increased somewhat by use of high pressures of air. In practice, the industrial catalytic process operates at ≈750 K and achieves conversion factors of >98%.

$$2SO_2 + O_2 \rightleftharpoons 2SO_3 \quad \Delta_rH^\circ = -96 \text{ kJ per mole of } SO_2$$
$$(16.94)$$

Self-study exercise

For the equilibrium:

$$SO_2(g) + \tfrac{1}{2}O_2(g) \rightleftharpoons SO_3(g)$$

values of $\ln K$ are 8.04 and -1.20 at 1073 and 1373 K respectively. Determine ΔG° at each of these temperatures and comment on the significance of the data with respect to the application of this equilibrium in the first step in the manufacture of H_2SO_4.

[*Ans.* $\Delta G^\circ(1073 \text{ K}) = -71.7 \text{ kJ mol}^{-1}$;
$\Delta G^\circ(1373 \text{ K}) = +13.7 \text{ kJ mol}^{-1}$]

In the manufacture of sulfuric acid, gaseous SO_3 is removed from the reaction mixture by passage through concentrated H_2SO_4, in which it dissolves to form *oleum* (see Section 16.9). Absorption into water to yield H_2SO_4 directly is not a viable option because SO_3 reacts vigorously and very exothermically with H_2O, forming a thick mist. On a small scale, SO_3 can be prepared by heating oleum.

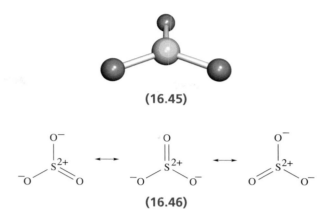

(16.45)

(16.46)

Table 16.7 lists selected physical properties of SO_3. In the gas phase, it is an equilibrium mixture of monomer (planar molecules, **16.45**, S–O = 142 pm) and trimer. Resonance structures **16.46** are consistent with three equivalent S–O bonds, and with the S atom possessing an octet of electrons. Solid SO_3 is polymorphic, with all forms containing SO_4-tetrahedra sharing two oxygen atoms. Condensation of the vapour at low temperatures yields γ-SO_3 which contains trimers (Fig. 16.17a); crystals of γ-SO_3 have an ice-like appearance. In the presence of traces of water, white crystals of β-SO_3 form; β-SO_3 consists of polymeric chains (Fig. 16.17b), as does α-SO_3 in which the chains are arranged into layers in the solid state. Differences in the thermodynamic properties of the different polymorphs are very small, although they do react with water at different

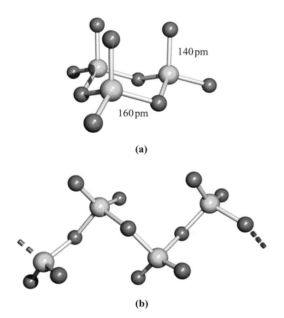

Fig. 16.17 The structures of solid state polymorphs of sulfur trioxide contain tetrahedral SO_4 units: (a) γ-SO_3 consists of trimeric units and (b) α- and β-SO_3 contain polymeric chains. Colour code: S, yellow; O, red.

rates. Sulfur trioxide is very reactive and representative reactions are given in scheme 16.95.

$$SO_3 \begin{cases} \xrightarrow{HX} HSO_3X \quad X = F, Cl \\ \xrightarrow{L} L \cdot SO_3 \quad L = \text{Lewis base, e.g. pyridine, } PPh_3 \\ \xrightarrow{H_2O} H_2SO_4 \end{cases}$$

(16.95)

Oxides of selenium and tellurium

Selenium and tellurium dioxides are white solids obtained by direct combination of the elements. The polymorph of TeO_2 so formed is α-TeO_2, whereas β-TeO_2 occurs naturally as the mineral *tellurite*. Both forms of TeO_2 contain structural units **16.47** which are connected by shared O atoms into a 3-dimensional lattice in α-TeO_2, and into a sheet structure in the β-form. The structure of SeO_2 consists of chains (**16.48**) in which the Se centres are in trigonal pyramidal environments. Whereas SeO_2 sublimes at 588 K, TeO_2 is an involatile solid (mp 1006 K). In the gas phase, SeO_2 is monomeric with structure **16.49**. Resonance structures for SeO_2 can be drawn as for SO_2 (structure **16.44**). The trends in structures of the dioxides of S, Se and Te and their associated properties (e.g. mp, volatility) reflect the increase in metallic character on descending group 16.

(16.47)   **(16.48)**   **(16.49)**

ENVIRONMENT
Box 16.6 Volcanic emissions

The eruption of a volcano is accompanied by emissions of water vapour (>70% of the volcanic gases), CO_2 and SO_2 plus lower levels of CO, sulfur vapour and Cl_2. Carbon dioxide contributes to the 'greenhouse' effect, and it has been estimated that volcanic eruptions produce ≈112 million tonnes of CO_2 per year. Levels of CO_2 in the plume of a volcano can be monitored by IR spectroscopy (the asymmetric stretching mode of the linear CO_2 molecule is observed at 2349 cm^{-1}). Ultraviolet spectroscopy is used to monitor SO_2 (it absorbs at ≈300 nm).

Mount Etna in southern Italy is classed as a 'continuously degassing' volcano and its emissions of SO_2 are among the largest of any volcano. In 1991, its SO_2 emission rate of ≈4000–5000 Mg day^{-1} was estimated to be similar to the total industrial sulfur emissions from France. Sulfur dioxide emissions are particularly damaging to the environment, since they result in the formation of acid rain. Sulfuric acid aerosols persist as suspensions in the atmosphere for long periods after an eruption. The Mount St Helens eruption occurred in May 1980. Towards the end of the eruption, the level of SO_2 in the volcanic plume was ≈2800 tonnes per day, and an emission rate of ≈1600 tonnes per day was measured in July 1980. Emissions of SO_2 (diminishing with time after the major eruption) continued for over two years, being boosted periodically by further volcanic activity.

Related discussions: see Box 12.2; Box 14.7; Box 16.5.

Explosive eruption of Mount St Helens, Washington, US on 22 July 1980.

Further reading

T. Casadevall, W. Rose, T. Gerlach, L.P. Greenland, J. Ewert, R. Wunderman and R. Symonds (1983) *Science*, vol. 221, p. 1383 – 'Gas emissions and eruptions of Mount St. Helens through 1982'.

R. von Glasgow (2010) *Proc. Nat. Acad. Sci.*, vol. 107, p. 6594 – 'Atmospheric chemistry in volcanic plumes'.

L.L. Malinconico, Jr (1979) *Nature*, vol. 278, p. 43 – 'Fluctuations in SO_2 emission during recent eruptions of Etna'.

C. Oppenheimer (2004) in *Treatise on Geochemistry*, eds H.D. Holland and K.K. Turekian, Elsevier, Oxford, vol. 3, p. 123 – 'Volcanic degassing'.

R.B. Symonds, T.M. Gerlach and M.H. Reed (2001) *J. Volcanol. Geoth. Res.*, vol. 108, p. 303 – 'Magmatic gas scrubbing: implications for volcano monitoring'.

Selenium dioxide is very toxic and is readily soluble in water to give selenous acid, H_2SeO_3. It is readily reduced, e.g. by hydrazine, and is used as an oxidizing agent in organic reactions. The α-form of TeO_2 is sparingly soluble in water, giving H_2TeO_3, but is soluble in aqueous HCl and alkali. Like SeO_2, TeO_2 is a good oxidizing agent. Like SO_2, SeO_2 and TeO_2 react with KF (see eq. 16.89). In solid $K[SeO_2F]$, weak fluoride bridges link the $[SeO_2F]^-$ ions into chains. In contrast, the tellurium analogue contains trimeric anions (structure **16.50**, see worked example 16.4). Selenium trioxide is a white, hygroscopic solid. It is difficult to prepare, being thermodynamically unstable with respect to SeO_2 and O_2 ($\Delta_fH^o(298\,K)$: $SeO_2 = -225$; $SeO_3 = -184\,kJ\,mol^{-1}$). It may be made by reaction of SO_3 with K_2SeO_4 (a salt of selenic acid). Selenium trioxide decomposes at 438 K, is soluble in water, and is a stronger oxidizing agent than SO_3. In the solid state, tetramers (**16.51**) are present.

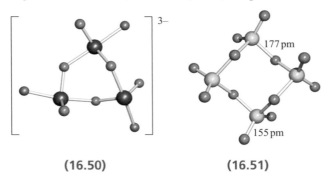

(16.50)　　　　　**(16.51)**

Tellurium trioxide (the α-form) is formed by dehydrating telluric acid (eq. 16.96). It is an orange solid which is insoluble in water but dissolves in aqueous alkali, and is a very powerful oxidizing agent. On heating above 670 K, TeO_3 decomposes to TeO_2 and O_2. Solid TeO_3 has a 3-dimensional structure in which each Te(VI) centre is octahedrally sited and connected by bridging O atoms.

$$H_6TeO_6 \longrightarrow TeO_3 + 3H_2O \qquad (16.96)$$

Worked example 16.4　Selenium and tellurium oxides and their derivatives

Structure 16.50 shows a representation of $[Te_3O_6F_3]^{3-}$. Rationalize why the coordination environment of the Te atom is *not* tetrahedral.

Apply the VSEPR model to structure **16.50**:
Te is in group 16 and has six valence electrons.

The formation of Te–F and three Te–O bonds (terminal and two bridging O atoms) adds four more electrons to the valence shell of Te.

In $[Te_3O_6F_3]^{3-}$, each Te centre is surrounded by five electron pairs, of which one is a lone pair.

Within the VSEPR model, a trigonal bipyramidal coordination environment is expected.

Self-study exercises

1. Draw a resonance structure for Se_4O_{12} (**16.51**) that is consistent with selenium retaining an octet of electrons.
   [*Hint*: See structure 16.46]

2. Explain what is meant by the phrase 'TeO_2 is dimorphic'.

3. SeO_2 is soluble in aqueous NaOH. Suggest what species are formed in solution, and write equations for their formation.　　　[*Ans.* $[SeO_3]^{2-}$ and $[HSeO_3]^-$]

4. 'TeO_2 is amphoteric'. Explain what this statement means.　　　　　　　　　[*Ans.* See Section 7.8]

16.9 Oxoacids and their salts

By way of an introduction to oxoacids, we note some generalities:

- oxoacid chemistry of sulfur resembles the complicated system of phosphorus;
- there are structural analogies between sulfates and phosphates, although fewer condensed sulfates are known;
- redox processes involving sulfur oxoanions are often slow, and thermodynamic data alone do not give a very good picture of their chemistry (compare similar situations for nitrogen- and phosphorus-containing oxoanions);
- selenium and tellurium have a relatively simple oxoacid chemistry.

Structures and pK_a values for important sulfur oxoacids are given in Table 16.8.

Dithionous acid, $H_2S_2O_4$

Although we show the structure of dithionous acid in Table 16.8, only its salts are known and these are powerful reducing agents. Dithionite is prepared by reduction of sulfite in aqueous solution (eq. 16.97) by Zn or Na amalgam and possesses eclipsed structure **16.52**.

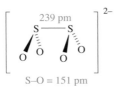

(16.52)

$$2[SO_3]^{2-} + 2H_2O + 2e^- \rightleftharpoons 4[OH]^- + [S_2O_4]^{2-}$$
$$E^o = -1.12\,V \qquad (16.97)$$

Table 16.8 Selected oxoacids of sulfur.[†]

Formula	Name	Structure[‡]	pK_a values (298 K)
$H_2S_2O_4$	Dithionous acid		pK_a(1) = 0.35; pK_a(2) = 2.45
H_2SO_3	Sulfurous acid[*]		pK_a(1) = 1.82; pK_a(2) = 6.92
H_2SO_4	Sulfuric acid		pK_a(2) = 1.92
$H_2S_2O_7$	Disulfuric acid		pK_a(1) = 3.1
$H_2S_2O_8$	Peroxydisulfuric acid		pK_a values not known with certainty.
$H_2S_2O_3$	Thiosulfuric acid		pK_a(1) = 0.6; pK_a(2) = 1.74

[†] Commonly used names have been included in this table; for systematic additive names and comments on uses of traditional names, see: *IUPAC: Nomenclature of Inorganic Chemistry* (*Recommendations 2005*), senior eds N.G. Connelly and T. Damhus, RSC Publishing, Cambridge.
[‡] See text; not all the acids can be isolated.
[*] See text for comment on structure of conjugate base.

APPLICATIONS

Box 16.7 SO$_2$ and sulfites in wine

During the fermentation process in the manufacture of wine, SO$_2$ or K$_2$S$_2$O$_5$ is added to the initial wine pressings to kill microorganisms, the presence of which results in spoilage of the wine. Molecular SO$_2$ is only used for large-scale wine production, while K$_2$S$_2$O$_5$ is the common additive in small-scale production. In acidic solution, [S$_2$O$_5$]$^{2-}$ undergoes the following reactions:

$$[S_2O_5]^{2-} + H_2O \rightleftharpoons 2[HSO_3]^-$$

$$[HSO_3]^- + H^+ \rightleftharpoons SO_2 + H_2O$$

The overall equilibrium system for aqueous SO$_2$ is:

$$SO_2 + H_2O \rightleftharpoons H^+ + [HSO_3]^- \rightleftharpoons 2H^+ + [SO_3]^{2-}$$

(These equilibria are discussed more fully with eqs. 7.18–7.20.) The position of equilibrium is pH-dependent; for the fermentation process, the pH is in the range 2.9–3.6. Only *molecular* SO$_2$ is active against microorganisms.

The first (i.e. yeast) fermentation step is followed by a bacterial fermentation step (malolactic fermentation) in which malic acid is converted to lactic acid. After this stage, SO$_2$ is added to stabilize the wine against oxidation. Adding SO$_2$ too early destroys the bacteria that facilitate malolactic fermentation. Malolactic fermentation is usually only important in red wine production.

The addition of SO$_2$ to white and red wines is handled differently. Red wines contain anthocyanin pigments, and these react with [HSO$_3$]$^-$ or [SO$_3$]$^{2-}$, resulting in a partial loss of the red coloration. Clearly, this must be avoided and means that addition of SO$_2$ to red wine must be carefully controlled. On the other hand, significantly more SO$_2$ can be added to white wine. Red wine, therefore, is less well protected by SO$_2$ against oxidation and spoilage by microorganisms than white wine, and it is essential to ensure that sugar and malic acid (food for the microbes) are removed

Red wine contains anthocyanin pigments which react with [SO$_3$]$^{2-}$ and [HSO$_3$]$^-$.

from red wine before bottling. Red wine does possess a higher phenolic content than white wine, and this acts as a built-in anti-oxidant.

Wines manufactured in the US carry a '*contains sulfites*' statement on the label. Some people are allergic to sulfites, and one possible substitute for SO$_2$ is the enzyme lysozyme. Lysozyme attacks lactic bacteria, and is used in cheese manufacture. However, it is not able to act as an anti-oxidant. A possible solution (not yet adopted by the wine industry) would be to mount a combined offensive: adding lysozyme and a reduced level of SO$_2$.

[Dr Paul Bowyer is acknowledged for assistance with the content of this box.]

The very long S–S bond in [S$_2$O$_4$]$^{2-}$ (compare $2 \times r_{cov}(S) = 206$ pm) shows it to be particularly weak and this is consistent with the observation that ^{35}S undergoes rapid exchange between [S$_2$O$_4$]$^{2-}$ and SO$_2$ in neutral or acidic solution. The presence of the [SO$_2$]$^-$ radical anion in solutions of Na$_2$S$_2$O$_4$ has been demonstrated by EPR spectroscopy. In aqueous solutions, [S$_2$O$_4$]$^{2-}$ is oxidized by air but in the absence of air, it undergoes reaction 16.98.

$$2[S_2O_4]^{2-} + H_2O \longrightarrow [S_2O_3]^{2-} + 2[HSO_3]^- \qquad (16.98)$$

Sulfurous and disulfurous acids, H$_2$SO$_3$ and H$_2$S$_2$O$_5$

Neither 'sulfurous acid' (see also Section 16.8) nor 'disulfurous acid' has been isolated as a free acid. Salts containing

the sulfite ion, [SO$_3$]$^{2-}$, are well established (e.g. Na$_2$SO$_3$ and K$_2$SO$_3$ are commercially available) and are quite good reducing agents (eq. 16.91). Applications of sulfites include those as food preservatives, e.g. an additive in wines (see Box 16.7). The [SO$_3$]$^{2-}$ ion has a trigonal pyramidal structure with delocalized bonding (S–O = 151 pm, ∠O–S–O =106°). There is evidence from ^{17}O NMR spectroscopic data that protonation of [SO$_3$]$^{2-}$ occurs to give a mixture of isomers as shown in equilibrium 16.99.

$$\left[H-OSO_2 \right]^- \rightleftharpoons \left[H-SO_3 \right]^- \qquad (16.99)$$

Although the [HSO$_3$]$^-$ ion exists in solution, and salts such as NaHSO$_3$ (used as a bleaching agent) may be isolated, evaporation of a solution of NaHSO$_3$ which has been saturated with SO$_2$ yields Na$_2$S$_2$O$_5$ (eq. 16.100).

$$2[HSO_3]^- \rightleftharpoons H_2O + [S_2O_5]^{2-} \qquad (16.100)$$

The $[S_2O_5]^{2-}$ ion is the only known derived anion of disulfurous acid and possesses structure **16.53** with a long, weak S−S bond. The bond distances given in structure **16.53** are for the K$^+$ salt.

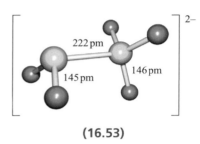

(16.53)

Dithionic acid, H$_2$S$_2$O$_6$

Dithionic acid is another sulfur oxoacid that is only known in aqueous solution (in which it behaves as a strong acid) or in the form of salts containing the dithionate, $[S_2O_6]^{2-}$, ion. Such salts can be isolated as crystalline solids and Fig. 16.18a shows the presence of a long S−S bond; the anion possesses a staggered conformation in the solid state. The dithionate ion can be prepared by controlled oxidation of $[SO_3]^{2-}$ (eqs. 16.101 and 16.102), but *not* by the reduction of $[SO_4]^{2-}$ (eq. 16.103). $[S_2O_6]^{2-}$ can be isolated as the soluble salt BaS$_2$O$_6$, which is easily converted into salts of other cations.

$$[S_2O_6]^{2-} + 4H^+ + 2e^- \rightleftharpoons 2H_2SO_3 \quad E^\circ = +0.56\,V$$
$$(16.101)$$

$$MnO_2 + 2[SO_3]^{2-} + 4H^+ \rightarrow Mn^{2+} + [S_2O_6]^{2-} + 2H_2O$$
$$(16.102)$$

$$2[SO_4]^{2-} + 4H^+ + 2e^- \rightleftharpoons [S_2O_6]^{2-} + 2H_2O$$
$$E^\circ = -0.22\,V \quad (16.103)$$

The $[S_2O_6]^{2-}$ ion is not easily oxidized or reduced, but in acidic solution it slowly decomposes according to eq. 16.104, consistent with there being a weak S−S bond.

$$[S_2O_6]^{2-} \rightarrow SO_2 + [SO_4]^{2-} \qquad (16.104)$$

Sulfuric acid, H$_2$SO$_4$

Sulfuric acid is the most important of the oxoacids of sulfur and is manufactured by the *Contact process*. The first stages of this process (conversion of SO$_2$ to SO$_3$ and formation of oleum) were described in Section 16.8. The oleum is finally diluted with water to give H$_2$SO$_4$. Pure H$_2$SO$_4$ is a colourless liquid with a high viscosity caused by extensive intermolecular hydrogen bonding. Its self-ionization and use as a non-aqueous solvent were described in Section 9.8, and selected properties given in Table 9.6. Gas-phase H$_2$SO$_4$ molecules have C_2 symmetry (Fig. 16.18b) with S−O bond distances (157 and 142 pm) that reflect two different types of S−O bond. In the solid state, hydrogen bonding between adjacent H$_2$SO$_4$ molecules results in the formation of a 3-dimensional network (Fig. 16.19). Diagram **16.54** shows a hypervalent structure for H$_2$SO$_4$, and **16.55** gives a bonding scheme in which the S atom obeys the octet rule (refer back to the discussion of bonding in Section 16.3). In the sulfate ion, all four S−O bond distances are equal (149 pm) because of charge delocalization, and in $[HSO_4]^-$, the S−OH bond distance is 156 pm and the remaining S−O bonds are of equal length (147 pm).

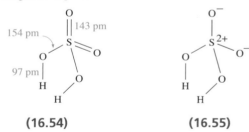

(16.54)   **(16.55)**

In aqueous solution, H$_2$SO$_4$ acts as a strong acid (eq. 16.105) but the $[HSO_4]^-$ ion is a fairly weak acid

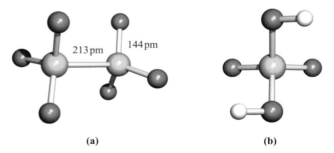

(a)   **(b)**

Fig. 16.18 (a) The structure of $[S_2O_6]^{2-}$ showing the staggered conformation; from the salt $[Zn\{H_2NNHC(O)Me\}_3][S_2O_6]\cdot 2.5H_2O$ [I.A. Krol *et al.* (1981) *Koord. Khim.*, vol. 7, p. 800]; (b) the C_2 structure of gas-phase H$_2$SO$_4$. Colour code: S, yellow; O, red; H, white.

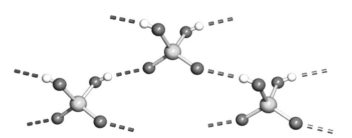

Fig. 16.19 Part of the 3-dimensional, hydrogen-bonded network of H$_2$SO$_4$ molecules in crystalline sulfuric acid. The structure was determined by X-ray diffraction at 113 K [E. Kemnitz *et al.* (1996) *Acta Crystallogr.*, Sect. C, vol. 52, p. 2665]. Colour code: S, yellow; O, red; H, white.

(eq. 16.106 and Table 16.8). Two series of salts are formed and can be isolated, e.g. $KHSO_4$ and K_2SO_4.

$$H_2SO_4 + H_2O \longrightarrow [H_3O]^+ + [HSO_4]^- \qquad (16.105)$$

$$[HSO_4]^- + H_2O \rightleftharpoons [H_3O]^+ + [SO_4]^{2-} \qquad (16.106)$$

Dilute aqueous H_2SO_4 (typically 2 M) neutralizes bases (e.g. eq. 16.107), and reacts with electropositive metals, liberating H_2, and metal carbonates (eq. 16.108).

$$H_2SO_4(aq) + 2KOH(aq) \longrightarrow K_2SO_4(aq) + 2H_2O(l) \qquad (16.107)$$

$$H_2SO_4(aq) + CuCO_3(s) \longrightarrow CuSO_4(aq) + H_2O(l) + CO_2(g) \qquad (16.108)$$

Commercial applications of sulfate salts are numerous, e.g. $(NH_4)_2SO_4$ as a fertilizer, $CuSO_4$ in fungicides, $MgSO_4 \cdot 7H_2O$ (Epsom salt) as a laxative, and hydrated $CaSO_4$ (see Boxes 12.2 and 12.7). Uses of H_2SO_4 were included in Fig. 16.3.

Concentrated H_2SO_4 is a good oxidizing agent (e.g. reaction 16.92) and a powerful dehydrating agent (see Box 12.3). Its reaction with HNO_3 is important for organic nitrations (eq. 16.109).

$$HNO_3 + 2H_2SO_4 \longrightarrow [NO_2]^+ + [H_3O]^+ + 2[HSO_4]^- \qquad (16.109)$$

Although HF/SbF_5 is a superacid, attempts to use it to protonate pure H_2SO_4 are affected by the fact that pure sulfuric acid undergoes reaction 16.110 to a small extent. The presence of the $[H_3O]^+$ ions in the HF/SbF_5 system prevents complete conversion of H_2SO_4 to $[H_3SO_4]^+$ (see worked example 16.5).

$$2H_2SO_4 \rightleftharpoons [H_3O]^+ + [HS_2O_7]^- \qquad (16.110)$$

An ingenious method of preparing a salt of $[H_3SO_4]^+$ is to use reaction 16.111 which is thermodynamically driven by the high Si−F bond enthalpy term in Me_3SiF (see Table 14.2). In the solid state structure of $[D_3SO_4]^+[SbF_6]^-$ (made by using DF in place of HF), the cation has structure **16.56** and there are extensive O−D···F interactions between cations and anions.

$$\underset{\text{a silyl ester of } H_2SO_4}{(Me_3SiO)_2SO_2} + 3HF + SbF_5$$

$$\xrightarrow{\text{liquid HF}} [H_3SO_4]^+[SbF_6]^- + 2Me_3SiF \qquad (16.111)$$

(16.56)

Worked example 16.5 Protonation of sulfuric acid

Reaction of HF/SbF_5 with H_2SO_4 does not result in complete protonation of sulfuric acid because of the presence of the $[H_3O]^+$ ions. (a) Explain the origin of the $[H_3O]^+$ ions and (b) explain how $[H_3O]^+$ interferes with attempts to use HF/SbF_5 to protonate H_2SO_4.

Pure sulfuric acid undergoes self-ionization processes. The most important is:

$$2H_2SO_4 \rightleftharpoons [H_3SO_4]^+ + [HSO_4]^-$$

and the following dehydration process also occurs:

$$2H_2SO_4 \rightleftharpoons [H_3O]^+ + [HS_2O_7]^-$$

The equilibrium constants for these processes are 2.7×10^{-4} and 5.1×10^{-5} respectively (see eqs. 9.46 and 9.47).

(b) The equilibrium for the superacid system in the absence of pure H_2SO_4 is:

$$2HF + SbF_5 \rightleftharpoons [H_2F]^+ + [SbF_6]^-$$

$[H_2F]^+$ is a stronger acid than H_2SO_4 and, in theory, the following equilibrium should lie to the right:

$$H_2SO_4 + [H_2F]^+ \rightleftharpoons [H_3SO_4]^+ + HF$$

However, a competing equilibrium is established which arises from the self-ionization process of H_2SO_4 described in part (a):

$$HF + SbF_5 + 2H_2SO_4 \rightleftharpoons [H_3O]^+ + [SbF_6]^- + H_2S_2O_7$$

Since H_2O is a stronger base than H_2SO_4, protonation of H_2O is favoured over protonation of H_2SO_4.

Self-study exercises

1. The $[H_3SO_4]^+$ ion is formed when boric acid is dissolved in oleum. Draw the structure of the anion that is produced, and write a balanced equation for the reaction.

   [*Ans.* See eq. 9.51]

2. The preparation of $[D_3SO_4]^+$ requires the use of DF. Suggest a method of preparing DF.

   [*Ans.* See eq. 17.1]

3. The methodology of reaction 16.111 has been used to protonate H_2O_2 and H_2CO_3. Write equations for these reactions and suggest structures for the protonated acids. [*Ans.* See R. Minkwitz *et al.* (1998, 1999) *Angew. Chem. Int. Ed.*, vol. 37, p. 1681; vol. 38, p. 714]

Fluoro- and chlorosulfonic acids, HSO₃F and HSO₃Cl

Fluoro- and chlorosulfonic acids, HSO_3F and HSO_3Cl, are obtained as shown in scheme 16.95, and their structures are related to that of H_2SO_4 with one OH group replaced by F or Cl. Both are colourless liquids at 298 K, and fume in moist air. HSO_3Cl reacts explosively with water. Both acids are commercially available. HSO_3F has wide applications in *superacid* systems (see Section 9.9) and as a fluorinating agent, while HSO_3Cl is used as a chlorosulfonating agent.

Polyoxoacids with S–O–S units

Although K^+ salts of the polysulfuric acids $HO_3S(OSO_2)_nOSO_3H$ ($n = 2, 3, 5, 6$) have been obtained by the reaction of SO_3 with K_2SO_4, the free acids cannot be isolated. Disulfuric and trisulfuric acids are present in oleum, i.e. when SO_3 is dissolved in concentrated H_2SO_4. The salt $[NO_2]_2[S_3O_{10}]$ has also been prepared and structurally characterized. Structure **16.57** shows $[S_3O_{10}]^{2-}$ as a representative member of this group of polyoxoanions.

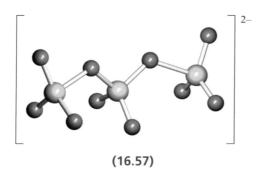

(16.57)

Peroxysulfuric acids, H₂S₂O₈ and H₂SO₅

The reaction between cold, anhydrous H_2O_2 and chlorosulfonic acid yields peroxysulfuric acid, H_2SO_5, and peroxydisulfuric acid, $H_2S_2O_8$ (eq. 16.112). However, $H_2S_2O_8$ (Table 16.8) is readily hydrolysed to H_2SO_5 (**16.58**) (eq. 16.113).

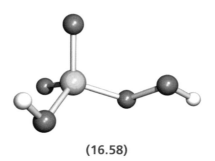

(16.58)

$$H_2O_2 \xrightarrow[-HCl]{ClSO_3H} H_2SO_5 \xrightarrow[-HCl]{ClSO_3H} H_2S_2O_8 \qquad (16.112)$$

$$H_2S_2O_8 + H_2O \xrightarrow{273\,K} H_2SO_5 + H_2SO_4 \qquad (16.113)$$

Both H_2SO_5 and $H_2S_2O_8$ are crystalline solids at 298 K. Few salts of H_2SO_5 are known, but those of $H_2S_2O_8$ are easily made by anodic oxidation of the corresponding sulfates in acidic solution at low temperatures and high current densities. Peroxydisulfates are strong oxidizing agents (eq. 16.114), and oxidations are often catalysed by Ag^+, with Ag(II) species being formed as intermediates. In acidic solutions, $[S_2O_8]^{2-}$ oxidizes Mn^{2+} to $[MnO_4]^-$, and Cr^{3+} to $[Cr_2O_7]^{2-}$.

$$[S_2O_8]^{2-} + 2e^- \rightleftharpoons 2[SO_4]^{2-} \qquad E^o = +2.01\,V \quad (16.114)$$

Peroxydisulfuric acid smells of ozone, and when $K_2S_2O_8$ is heated, a mixture of O_2 and O_3 is produced.

Thiosulfuric acid, H₂S₂O₃, and polythionates

Thiosulfuric acid may be prepared under *anhydrous* conditions by reaction 16.115, or by treatment of lead thiosulfate (PbS_2O_3) with H_2S, or sodium thiosulfate with HCl. The free acid is very unstable, decomposing at 243 K or upon contact with water.

$$H_2S + HSO_3Cl \xrightarrow{low\ temp} H_2S_2O_3 + HCl \qquad (16.115)$$

A representation of the structure of thiosulfuric acid is given in Table 16.8, but the conditions of reaction 16.115 may suggest protonation at sulfur, i.e. $(HO)(HS)SO_2$. Thiosulfate salts are far more important than the acid. Crystallization of the aqueous solution from reaction 16.116 yields $Na_2S_2O_3 \cdot 5H_2O$.

$$Na_2SO_3 + S \xrightarrow{in\ aqueous\ solution} Na_2S_2O_3 \qquad (16.116)$$

$$\left[\begin{array}{c} S \\ | \ 201\,pm \\ S \\ O \diagup \ \diagdown O \\ O \end{array} \right]^{2-}$$

147 pm

(16.59)

The thiosulfate ion, **16.59**, is a very good complexing agent for Ag^+, and $Na_2S_2O_3$ is used in photography for removing unchanged AgBr from exposed photographic film (eq. 16.117) although this use is in decline as a result of the huge growth in digital photography. In $[Ag(S_2O_3)_3]^{5-}$, each thiosulfate ion coordinates to Ag^+ through a sulfur donor atom.

$$AgBr + 3Na_2S_2O_3 \longrightarrow Na_5[Ag(S_2O_3)_3] + NaBr \qquad (16.117)$$

Most oxidizing agents (including Cl_2 and Br_2) slowly oxidize $[S_2O_3]^{2-}$ to $[SO_4]^{2-}$, and $Na_2S_2O_3$ is used to remove excess Cl_2 in bleaching processes. In contrast, I_2 rapidly oxidizes $[S_2O_3]^{2-}$ to tetrathionate; reaction 16.118 is of importance in titrimetric analysis.

$$2[S_2O_3]^{2-} + I_2 \longrightarrow [S_4O_6]^{2-} + 2I^- \qquad (16.118)$$

$$[S_4O_6]^{2-} + 2e^- \rightleftharpoons 2[S_2O_3]^{2-} \qquad E^o = +0.08\,\text{V}$$

$$I_2 + 2e^- \rightleftharpoons 2I^- \qquad E^o = +0.54\,\text{V}$$

Polythionates contain ions of type $[S_nO_6]^{2-}$ and may be prepared by condensation reactions such as those in scheme 16.119, but some ions must be made by specific routes. Polythionate ions are structurally similar and have two $\{SO_3\}^-$ groups connected by a sulfur chain (**16.60** shows $[S_5O_6]^{2-}$). Solid state structures for a number of salts show chain conformations are variable. In aqueous solution, polythionates slowly decompose to H_2SO_4, SO_2 and sulfur.

$$\left.\begin{array}{l} SCl_2 + 2[HSO_3]^- \longrightarrow [S_3O_6]^{2-} + 2HCl \\ S_2Cl_2 + 2[HSO_3]^- \longrightarrow [S_4O_6]^{2-} + 2HCl \end{array}\right\} \qquad (16.119)$$

Some compounds are known in which S atoms in a polythionate are replaced by Se or Te, e.g. $Ba[Se(SSO_3)_2]$ and $Ba[Te(SSO_3)_2]$. Significantly, Se and Te cannot replace the terminal S atoms, presumably because in their highest oxidation states they are too powerfully oxidizing and attack the remainder of the chain.

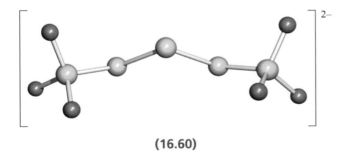

(16.60)

Oxoacids of selenium and tellurium

Selenous acid, H_2SeO_3, may be crystallized from aqueous solutions of SeO_2 and gives rise to two series of salts containing the $[HSeO_3]^-$ and $[SeO_3]^{2-}$ ions. In aqueous solution, it behaves as a weak acid: $pK_a(1) \approx 2.46$, $pK_a(2) \approx 7.31$. Heating salts of $[HSeO_3]^-$ generates diselenites containing ion **16.61**. Tellurous acid, H_2TeO_3, is not as stable as H_2SeO_3 and is usually prepared in aqueous solution where it acts as a weak acid: $pK_a(1) \approx 2.48$, $pK_a(2) \approx 7.70$. Most tellurite salts contain the $[TeO_3]^{2-}$ ion.

(16.61)

Oxidation of H_2SeO_3 with 30% aqueous H_2O_2 yields selenic acid, H_2SeO_4, which may be crystallized from the solution. In some ways it resembles H_2SO_4, being fully dissociated in aqueous solution with respect to loss of the first proton. For the second step, $pK_a = 1.92$. It is a more powerful oxidant than H_2SO_4, e.g. it liberates Cl_2 from concentrated HCl. Selenic acid dissolves gold metal, oxidizing it to Au(III). When the reaction is carried out at 520 K, the product is $Au_2(SeO_3)_2(SeO_4)$ which contains both tetrahedral $[SeO_4]^{2-}$ and trigonal pyramidal $[SeO_3]^{2-}$ ions. Reaction in the solid state between Na_2SeO_4 and Na_2O (2:1 molar equivalents) leads to $Na_6Se_2O_9$. This formula is more usefully written as $Na_{12}(SeO_6)(SeO_4)_3$, showing the presence of the octahedral $[SeO_6]^{6-}$ ion which is stabilized in the crystalline lattice by interaction with eight Na^+ ions. The $[SeO_5]^{4-}$ ion has been established in Li_4SeO_5 and Na_4SeO_5. The formula, H_6TeO_6 or $Te(OH)_6$, and properties of telluric acid contrast with those of selenic acid. In the solid, octahedral molecules (**16.62**) are present and in solution, it behaves as a weak acid: $pK_a(1) = 7.68$, $pK_a(2) = 11.29$. Typical salts include those containing $[Te(O)(OH)_5]^-$ and $[Te(O)_2(OH)_4]^{2-}$ and the presence of the $[TeO_4]^{2-}$ ion has been confirmed in the solid state structure of $Rb_6[TeO_5][TeO_4]$.

(16.62)

16.10 Compounds of sulfur and selenium with nitrogen

Sulfur–nitrogen compounds

Sulfur–nitrogen chemistry is an area that has seen major developments over the last few decades, in part because of the conductivity of the polymer $(SN)_x$. The following discussion is necessarily selective, and more detailed accounts are listed at the end of the chapter. Probably the best known of the sulfur–nitrogen compounds is tetrasulfur tetranitride, S_4N_4. It has traditionally been obtained using reaction 16.120, but a more convenient method is reaction 16.121. Tetrasulfur tetranitride is a diamagnetic orange solid (mp 451 K) which explodes when heated or struck. Pure samples are very sensitive (see exercise 1 at the end of the section). It is hydrolysed slowly by water (in which it is insoluble) and rapidly by warm alkali (eq. 16.122).

$$6S_2Cl_2 + 16NH_3 \xrightarrow{\text{CCl}_4,\ 320\,\text{K}} S_4N_4 + 12NH_4Cl + S_8$$
(16.120)

$$2\{(Me_3Si)_2N\}_2S + 2SCl_2 + 2SO_2Cl_2$$
$$\longrightarrow S_4N_4 + 8Me_3SiCl + 2SO_2 \quad (16.121)$$

$$S_4N_4 + 6[OH]^- + 3H_2O \longrightarrow [S_2O_3]^{2-} + 2[SO_3]^{2-} + 4NH_3$$
(16.122)

The structure of S_4N_4, **16.63**, is a cradle-like ring in which pairs of S atoms are brought within weak bonding distance of one another (compare with $[S_8]^{2+}$, Fig. 16.8). The S—N bond distances in S_4N_4 indicate delocalized bonding with π-contributions (compare the S—N distances of 163 pm with the sum of the S and N covalent radii of 178 pm). Transfer of charge from S to N occurs giving $S^{\delta+}$—$N^{\delta-}$ polar bonds. A resonance structure for S_4N_4 that illustrates the cross-cage S—S bonding interactions is shown in **16.64**.

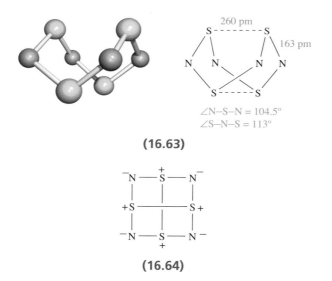

$\angle$N–S–N = 104.5°
$\angle$S–N–S = 113°

(16.63)

(16.64)

Figure 16.20 gives selected reactions of S_4N_4. Some lead to products containing S—N rings in which the cross-cage interactions of S_4N_4 are lost. Reduction (at N) gives tetrasulfur tetraimide, $S_4N_4H_4$, which has a crown-shaped ring with equal S—N bond lengths. Tetrasulfur tetraimide is one of a number of compounds in which S atoms in S_8 are formally replaced by NH groups with retention of the crown conformation; S_7NH, $S_6N_2H_2$, $S_5N_3H_3$ (along with S_4N_4 and S_8) are all obtained by treating S_2Cl_2 with NH_3.

No members of this family with adjacent NH groups in the ring are known.

(16.65)

$\angle$F–S–F = 94°
$\angle$N–S–F = 122°

(16.66)

Halogenation of S_4N_4 (at S) may degrade the ring depending on X_2 or the conditions (Fig. 16.20). The ring in $S_4N_4F_4$ has a puckered conformation quite different from that in $S_4N_4H_4$. Fluorination of S_4N_4 under appropriate conditions (Fig. 16.20) yields thiazyl fluoride, NSF, **16.65**, or thiazyl trifluoride NSF$_3$, **16.66**, which contain S≡N triple bonds (see end-of-chapter problem 16.29a). Both are pungent gases at room temperature, and NSF slowly trimerizes to $S_3N_3F_3$; note that $S_4N_4F_4$ is not made from the monomer. The structures of $S_3N_3Cl_3$ (**16.67**) and $S_3N_3F_3$ are similar. The rings exhibit only slight puckering and the S—N bond distances are equal in $S_3N_3Cl_3$ and approximately equal in the fluoro analogue. The salt $[S_3N_2Cl]^+Cl^-$ (made by heating a mixture of S_2Cl_2, sulfur and NH_4Cl) contains cation **16.68**. Oxidation of S_4N_4 with AsF_5 or SbF_5 gives $[S_4N_4][EF_6]_2$ (E = As or Sb) containing $[S_4N_4]^{2+}$. This has the planar structure **16.69** in many of its salts, but $[S_4N_4]^{2+}$ can also adopt a planar structure with alternating bond distances, or a puckered conformation. The $[S_4N_3]^+$ cation (prepared as shown in Fig. 16.20) has the planar structure **16.70** with delocalized bonding.

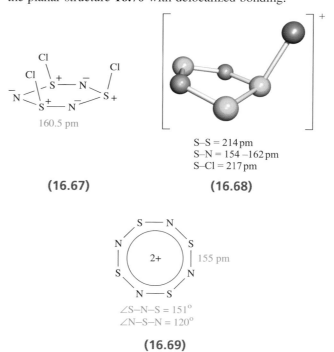

160.5 pm

S–S = 214 pm
S–N = 154 –162 pm
S–Cl = 217 pm

(16.67)   **(16.68)**

$\angle$S–N–S = 151°
$\angle$N–S–N = 120°

(16.69)

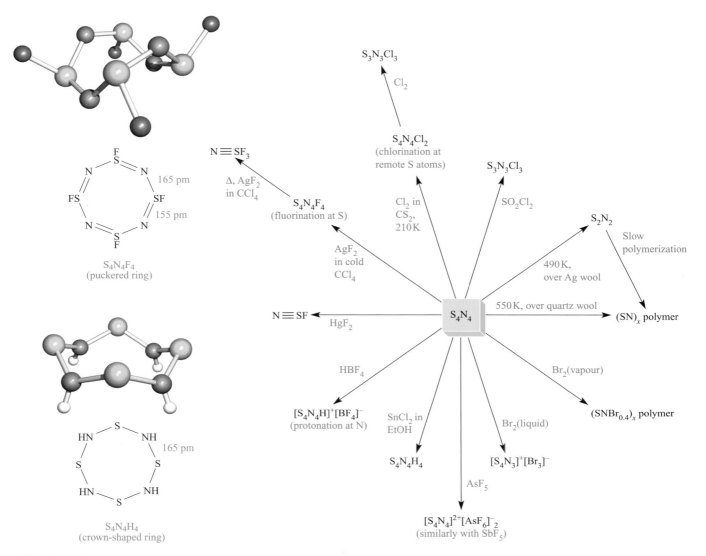

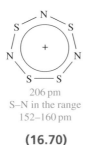

Fig. 16.20 Selected reactions of S_4N_4; the rings in $S_4N_4H_4$ and $S_4N_4F_4$ are non-planar.

206 pm
S–N in the range
152–160 pm

(16.70)

The S_4N_4 cage can be degraded to S_2N_2 (Fig. 16.20) which is isoelectronic with $[S_4]^{2+}$ (see Section 16.4). S_2N_2 is planar with delocalized bonding (S–N = 165 pm), and resonance structures are shown in **16.71**. At room temperature, this converts to the lustrous golden-yellow, fibrous polymer $(SN)_x$, which can also be prepared from S_4N_4. The polymer decomposes explosively at 520 K, but can be sublimed *in vacuo* at $\approx$410 K. It is a remarkable material, being covalently bonded but showing metallic properties:

a 1-dimensional pseudo-metal. It has an electrical conductance about one-quarter of that of mercury in the direction of the polymer chains, and at 0.3 K it becomes a super-conductor. However, the explosive nature of S_4N_4 and S_2N_2 limits commercial production of $(SN)_x$. In the solid state, X-ray diffraction data indicate that the S–N bond lengths in $(SN)_x$ alternate (159 and 163 pm) but highly precise data are still not available; the closest interchain distances are non-bonding S–S contacts of 350 pm. Structure **16.72** gives a representation of the polymer chain and the conductivity can be considered to arise from the unpaired electrons on sulfur occupying a half-filled conduction band (see Section 6.8).

(16.71)

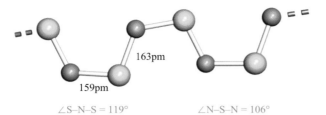

∠S–N–S = 119°   ∠N–S–N = 106°

Colour code: S, yellow; N blue

(16.72)

Equations 16.123 and 16.124 show convenient routes to $[NS_2][SbF_6]$. This product is soluble in liquid SO_2 and is readily separated from AgCl which precipitates out of solution. The $[NS_2]^+$ ion (**16.73**) is isoelectronic (in terms of valence electrons) with $[NO_2]^+$ (see structure **15.50**). The $[NS_2]^+$ ion is a useful synthon, undergoing cycloaddition reactions with, for example, alkynes, nitriles and alkenes.

$$S_3N_3Cl_3 + \tfrac{3}{8}S_8 + 3AgSbF_6 \xrightarrow{\text{liquid } SO_2} 3[NS_2][SbF_6] + 3AgCl$$
$$(16.123)$$

$$[S_3N_2Cl]Cl + \tfrac{1}{8}S_8 + 2AgSbF_6 \xrightarrow{\text{liquid } SO_2} 2[NS_2][SbF_6] + 2AgCl$$
$$(16.124)$$

$$\overset{+}{S=N=S}$$
146 pm

(16.73)

Tetraselenium tetranitride

Among the compounds formed by Se and N, we mention only Se analogues of S_4N_4. Selenium tetranitride, Se_4N_4, can be prepared by reacting $SeCl_4$ with $\{(Me_3Si)_2N\}_2Se$. It forms orange, hygroscopic crystals and is highly explosive. The structure of Se_4N_4 is like that of S_4N_4 (**16.63**) with Se–N bond lengths of 180 pm and cross-cage Se····Se separations of 276 pm (compare with $r_{cov}(Se) = 117$ pm). The reactivity of Se_4N_4 has not been as fully explored as that of S_4N_4. Reaction 16.125 is an adaptation of the synthesis of Se_4N_4 and leads to the 1,5-isomer of $Se_2S_2N_4$ (**16.74**). In the solid state structure, the S and Se atoms are disordered (see Box 15.5), making it difficult to tell whether the crystalline sample is $Se_2S_2N_4$ or a solid solution of S_4N_4 and Se_4N_4. Mass spectrometric data are consistent with the presence of $Se_2S_2N_4$, and the appearance of only one signal in the ^{14}N NMR spectrum confirms the 1,5- rather than 1,3-isomer.

$$2\{(Me_3Si)_2N\}_2S + 2SeCl_4 \longrightarrow Se_2S_2N_4 + 8Me_3SiCl$$
$$(16.125)$$

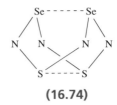

(16.74)

Self-study exercises

1. Although $\Delta_fH^\circ(S_4N_4, s, 298\,K) = +460\,kJ\,mol^{-1}$, S_4N_4 is kinetically stable under ambient conditions with respect to decomposition to the elements. (a) What is meant by 'kinetically stable'? (b) An electrically heated Pt wire can be used to initiate the explosive decomposition of S_4N_4 under a pressure of N_2. Write an equation for what happens during the reaction, and determine the enthalpy change, $\Delta_rH^\circ(298\,K)$.

2. Theoretical calculations suggest that the $S_3N_3^\bullet$ radical has D_{3h} symmetry. Deduce, with reasoning, whether the ring is planar or puckered.

3. Suggest products for the reactions of S_4N_4 with the following reagents: (a) SO_2Cl_2; (b) AsF_5; (c) $SnCl_2$ in EtOH; (d) HgF_2; (e) liquid Br_2.

[*Ans.* See Fig. 16.20]

16.11 Aqueous solution chemistry of sulfur, selenium and tellurium

As we saw earlier in the chapter, the redox reactions between compounds of S in different oxidation states are often slow, and values of E° for half-reactions are invariably obtained from thermochemical information or estimated on the basis of observed chemistry. The data in Fig. 16.21 illustrate the relative redox properties of some S-, Se- and Te-containing species. Points to note are:

- the greater oxidizing powers of selenate and tellurate than of sulfate;
- the similarities between the oxidizing powers of sulfate, selenite and tellurite;
- the instabilities in aqueous solution of H_2Se and H_2Te.

$$[SO_4]^{2-} \xrightarrow{+0.17} H_2SO_3 \xrightarrow{+0.45} S \xrightarrow{+0.14} H_2S$$

$$[SeO_4]^{2-} \xrightarrow{+1.15} H_2SeO_3 \xrightarrow{+0.74} Se \xrightarrow{-0.40} H_2Se$$

$$H_6TeO_6 \xrightarrow{+1.02} TeO_2 \xrightarrow{+0.59} Te \xrightarrow{-0.79} H_2Te$$

Fig. 16.21 Potential diagrams (values in V) for sulfur, selenium and tellurium at pH = 0.

There is little difference in energy between the various oxidation state species of sulfur, a fact that is doubtless involved in the complicated oxoacid and oxoanion chemistry of sulfur. We have already discussed some aspects of the aqueous solution chemistry of the group 16 elements:

- the ionization of the hydrides (Sections 7.5 and 16.5);
- formation of metal sulfides (Section 16.6);

- formation of polysulfide ions, e.g. $[S_5]^{2-}$ (eq. 16.42);
- oxoacids and their salts (Section 16.9);
- the oxidizing power of $[S_2O_8]^{2-}$ (eq. 16.114).

There is no cation chemistry in aqueous solution for the group 16 elements. The coordination to metal ions of oxoanions such as $[SO_4]^{2-}$ and $[S_2O_3]^{2-}$ is well established (e.g. see eq. 16.117).

KEY TERMS

The following terms were introduced in this chapter. Do you know what they mean?

❑ annular
❑ transannular interaction

❑ cubane

FURTHER READING

N.N. Greenwood and A. Earnshaw (1997) *Chemistry of the Elements*, 2nd edn, Butterworth-Heinemann, Oxford – Chapters 14–16 cover the chalcogens in detail.

A.G. Massey (2000) *Main Group Chemistry*, 2nd edn, Wiley, Chichester – Chapter 10 covers the chemistry of the group 16 elements.

A.F. Wells (1984) *Structural Inorganic Chemistry*, 5th edn, Clarendon Press, Oxford – Chapters 11–17 cover the structures of a large number of compounds of the group 16 elements.

Specialized topics

A.J. Banister and I.B. Gorrell (1998) *Adv. Mater.*, vol. 10, p. 1415 – 'Poly(sulfur nitride): the first polymeric metal'.
J. Beck (1994) *Angew. Chem. Int. Ed.*, vol. 33, p. 163 –

'New forms and functions of tellurium: from polycations to metal halide tellurides'.

M. Jansen and H. Nuss (2007) *Z. Anorg. Allg. Chem.*, vol. 633, p. 1307 – 'Ionic ozonides'.

P. Kelly (1997) *Chem. Brit.*, vol. 33, no. 4, p. 25 – 'Hell's angel: a brief history of sulfur'.

D. Stirling (2000) *The Sulfur Problem: Cleaning Up Industrial Feedstocks*, Royal Society of Chemistry, Cambridge.

W.-T. Tsai (2007) *J. Fluorine Chem.*, vol. 128, p. 1345 – 'The decomposition products of sulfur hexafluoride (SF_6): Reviews of environmental and health risk analysis'.

R.P. Wayne (2000) *Chemistry of Atmospheres*, Oxford University Press, Oxford.

PROBLEMS

16.1 (a) Write down, in order, the names and symbols of the elements in group 16; check your answer by reference to the first page of this chapter. (b) Give a *general* notation showing the ground state electronic configuration of each element.

16.2 Write an equation to represent the formation of ^{210}Po from ^{209}Bi.

16.3 Write half-equations to show the reactions involved during the electrolysis of aqueous alkali.

16.4 By considering the reactions $8E(g) \longrightarrow 4E_2(g)$ and $8E(g) \longrightarrow E_8(g)$ for $E = O$ and $E = S$, show that the formation of diatomic molecules is favoured for oxygen, whereas ring formation is favoured for sulfur. [Data: see Table 16.2.]

16.5 (a) Draw diagrams to show the occupancies of the $\pi_g{}^*$ level in the ground state and first two excited states of O_2. Does the formal bond order change upon excitation from the ground to the first excited state ($^1\Delta_g$) of O_2? (b) The $^1\Delta_g$ state of O_2 lies 94.7 kJ mol^{-1} above the ground state. Show that the simultaneous excitation of two O_2 molecules from their ground to $^1\Delta_g$ states corresponds to an absorption of light of wavelength 631 nm.

16.6 (a) Use the values of E^o for reactions 16.32 and 16.33 to show that H_2O_2 is thermodynamically unstable with respect to decomposition into H_2O and O_2. (b) '20 Volume' H_2O_2 is so called because 1 volume of the solution liberates 20 volumes of O_2 when it decomposes. If the volumes are measured at 273 K and 1 bar pressure, what is the concentration of the solution expressed in grams of H_2O_2 per dm^3?

16.7 Suggest products for the following reactions: (a) H_2O_2 and Ce^{4+} in acidic solution; (b) H_2O_2 and I^- in acidic solution. [Data needed: see Appendix 11.]

16.8 Hydrogen peroxide oxidizes $Mn(OH)_2$ to MnO_2. (a) Write an equation for this reaction. (b) What secondary reaction will occur?

16.9 Explain why *catena*-Se_∞ is chiral.

16.10 The diagrams below show two views of S_6. Confirm that this molecule has D_{3d} symmetry.

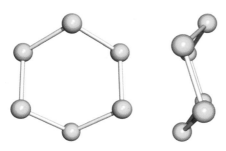

16.11 Predict the structures of (a) H_2Se; (b) $[H_3S]^+$; (c) SO_2; (d) SF_4; (e) SF_6; (f) S_2F_2.

16.12 (a) Explain why the reaction of SF_4 with BF_3 yields $[SF_3]^+$, whereas the reaction with CsF gives $Cs[SF_5]$. (b) Suggest how SF_4 might react with a carboxylic acid, RCO_2H.

16.13 The Raman spectrum of solid $[SeI_3][AsF_6]$ contains absorptions at 227, 216, 99 and $80\,cm^{-1}$ assigned to the vibrational modes of the $[SeI_3]^+$ ion. Explain the origins of the four absorptions. Draw diagrams to represent the modes of vibration, and assign a symmetry label to each mode.

16.14 Discuss the trends in (a) the O−O bond lengths in O_2 (121 pm), $[O_2]^+$ (112 pm), H_2O_2 (147.5 pm), $[O_2]^{2-}$ (149 pm) and O_2F_2 (122 pm), and (b) the S−S bond distances in S_6 (206 pm), S_2 (189 pm), $[S_4]^{2+}$ (198 pm), H_2S_2 (206 pm), S_2F_2 (189 pm), S_2F_{10} (221 pm) and S_2Cl_2 (193 pm). [Data: $r_{cov}(S) = 103$ pm.]

16.15 Comment on the following values of gas-phase dipole moments: SeF_6, 0 D; SeF_4, 1.78 D; SF_4, 0.64 D; SCl_2, 0.36 D; $SOCl_2$, 1.45 D; SO_2Cl_2, 1.81 D.

16.16 The ^{125}Te NMR spectrum of $[Me_4N][TeF_7]$ (298 K in MeCN) consists of a binomial octet ($J = 2876$ Hz), while the ^{19}F NMR spectrum exhibits a singlet with two (superimposed over the singlet), very low-intensity doublets ($J = 2876$ and 2385 Hz respectively). Rationalize these observations. [Data: see Table 16.1; ^{19}F, 100%, $I = \frac{1}{2}$.]

16.17 In the following series of compounds or ions, identify those that are isoelectronic (with respect to the valence electrons) and those that are also isostructural: (a) $[SiO_4]^{4-}$, $[PO_4]^{3-}$, $[SO_4]^{2-}$; (b) CO_2, SiO_2, SO_2, TeO_2, $[NO_2]^+$; (c) SO_3, $[PO_3]^-$, SeO_3; (d) $[P_4O_{12}]^{4-}$, Se_4O_{12}, $[Si_4O_{12}]^{8-}$.

16.18 (a) Give the structures of SO_3 and $[SO_3]^{2-}$ and rationalize the difference between them. (b) Outline the properties of aqueous solutions of SO_2 and discuss the species that can be derived from them.

16.19 (a) Draw the structures of S_7NH, $S_6N_2H_2$, $S_5N_3H_3$ and $S_4N_4H_4$, illustrating isomerism where appropriate. (The structures of hypothetical isomers with two or more adjacent NH groups should be ignored.) (b) Write a brief account of the preparation and reactivity of S_4N_4, giving the structures of the products formed in the reactions described.

16.20 Discuss the interpretation of each of the following observations.

(a) When metallic Cu is heated with concentrated H_2SO_4, in addition to $CuSO_4$ and SO_2, some CuS is formed.

(b) The $[TeF_5]^-$ ion is square pyramidal.

(c) Silver nitrate gives a white precipitate with aqueous sodium thiosulfate; the precipitate dissolves in an excess of $[S_2O_3]^{2-}$. If the precipitate is heated with water, it turns black, and the supernatant liquid then gives a white precipitate with acidified aqueous $Ba(NO_3)_2$.

16.21 Interpret the following experimental results.

(a) Sodium dithionite, $Na_2S_2O_4$ (0.0261 g) was added to an excess of ammoniacal $AgNO_3$ solution; the precipitated silver was removed by filtration, and dissolved in nitric acid. The resulting solution was found to be equivalent to $30.0\,cm^3$ 0.10 M thiocyanate solution.

(b) A solution containing 0.0725 g of $Na_2S_2O_4$ was treated with $50.0\,cm^3$ 0.0500 M iodine solution and acetic acid. After completion of the reaction, the residual I_2 was equivalent to $23.75\,cm^3$ 0.1050 M thiosulfate.

16.22 The action of concentrated H_2SO_4 on urea, $(H_2N)_2CO$, results in the production of a white crystalline solid **X** of formula H_3NO_3S. This is a monobasic acid. On treatment with sodium nitrite and dilute hydrochloric acid at 273 K, one mole of **X** liberates one mole of N_2, and on addition of aqueous $BaCl_2$, the resulting solution yields one mole of $BaSO_4$ per mole of **X** taken initially. Deduce the structure of **X**.

16.23 Write a brief account of the oxoacids of sulfur, paying attention to which species are isolable.

16.24 Give the structures of S_2O, $[S_2O_3]^{2-}$, NSF, NSF_3, $[NS_2]^+$ and S_2N_2 and rationalize their shapes.

16.25 $[NS_2][SbF_6]$ reacts with nitriles, $RC\equiv N$, to give $[X][SbF_6]$ where $[X]^+$ is a cycloaddition product.

Propose a structure for $[X]^+$ and show that it is a 6π-electron system. Do you expect the ring to be planar or puckered? Give reasons for your answer.

OVERVIEW PROBLEMS

16.26 Which description in the second list below can be correctly matched to each element or compound in the first list? There is only one match for each pair.

List 1	List 2
S_∞	A toxic gas
$[S_2O_8]^{2-}$	Readily disproportionates in the presence of Mn^{2+}
$[S_2]^-$	Reacts explosively with H_2O
S_2F_2	Exists as a tetramer in the solid state
Na_2O	A strong reducing agent, oxidized to $[S_4O_6]^{2-}$
$[S_2O_6]^{2-}$	A blue, paramagnetic species
PbS	Exists as two monomeric isomers
H_2O_2	A chiral polymer
HSO_3Cl	Crystallizes with an antifluorite structure
$[S_2O_3]^{2-}$	A black, insoluble solid
H_2S	A strong oxidizing agent, reduced to $[SO_4]^{2-}$
SeO_3	Contains a weak S–S bond, readily cleaved in acidic solution

16.27 (a) A black precipitate forms when H_2S is added to an aqueous solution of a Cu(II) salt. The precipitate redissolves when Na_2S is added to the solution. Suggest a reason for this observation.
   (b) In the presence of small amounts of water, the reaction of SO_2 with CsN_3 leads to $Cs_2S_2O_5$ as a by-product in the formation of $Cs[SO_2N_3]$. Suggest how the formation of $Cs_2S_2O_5$ arises.
   (c) The complex ion $[Cr(Te_4)_3]^{3-}$ possesses a $\Delta\lambda\lambda\lambda$-conformation. Using the information in Box 19.3, explain (i) to what the symbols Δ and λ refer, and (ii) how the $\Delta\lambda\lambda\lambda$-conformation arises.

16.28 Suggest products for the following reactions; the equations are not necessarily balanced on the left-hand sides. Draw the structures of the sulfur-containing products.
   (a) $SF_4 + SbF_5 \xrightarrow{\text{liq HF}}$
   (b) $SO_3 + HF \longrightarrow$

(c) $Na_2S_4 + HCl \longrightarrow$
(d) $[HSO_3]^- + I_2 + H_2O \longrightarrow$
(e) $[SN][AsF_6] + CsF \xrightarrow{\Delta}$
(f) $HSO_3Cl + $ anhydrous $H_2O_2 \longrightarrow$
(g) $[S_2O_6]^{2-} \xrightarrow{\text{in acidic solution}}$

16.29 (a) Structures **16.65** and **16.66** show hypervalent sulfur in NSF and NSF_3. Draw resonance structures for each molecule that retains an octet of electrons around the S atoms, and account for the three equivalent S–F bonds in NSF_3.
   (b) The enthalpies of vaporization (at the boiling point) of H_2O, H_2S, H_2Se and H_2Te are 40.6, 18.7, 19.7 and $19.2\,kJ\,mol^{-1}$. Give an explanation for the trend in these values.
   (c) Which of the following compounds undergoes significant reaction when they dissolve in water under ambient conditions: Al_2Se_3, HgS, SF_6, SF_4, SeO_2, FeS_2 and As_2S_3? Give equations to show the reactions that occur. Which of these compounds is kinetically, but not thermodynamically, stable with respect to hydrolysis?

16.30 The $[Se_4]^{2+}$ ion has D_{4h} symmetry and the Se–Se bond lengths are equal (228 pm).
   (a) Is the ring in $[Se_4]^{2+}$ planar or puckered?
   (b) Look up a value of r_{cov} for Se. What can you deduce about the Se–Se bonding?
   (c) Draw a set of resonance structures for $[Se_4]^{2+}$.
   (d) Construct an MO diagram that describes the π-bonding in $[Se_4]^{2+}$. What is the π-bond order?

16.31 (a) S_8O can be prepared by treating S_8 with CF_3CO_3H. The structure of S_8O is shown below. Explain why the addition of one O atom to the S_8 ring reduces the molecular symmetry from D_{4d} to C_1.

(b) The reaction between TeF_4 and Me_3SiCN leads to the formation of Me_3SiF and substitution products $TeF_{4-n}(CN)_n$. When the ^{125}Te NMR spectrum of a mixture of TeF_4 and less than four equivalents of Me_3SiCN is recorded at 173 K, three signals are observed: δ 1236 ppm (quintet, J 2012 Hz), δ 816 ppm (triplet, J 187 Hz), δ 332 ppm (doublet, J 200 Hz). The ^{19}F NMR spectrum of the same mixture at 173 K exhibits three signals, each with ^{125}Te satellites. Rationalize the observed spectra (including comments on fluxional species where appropriate), and suggest structures for the products.

16.32 The reaction of $TeCl_4$ with PPh_3 in THF solution in air leads to the formation of the salt $[(Ph_3PO)_2H]_2[Te_2Cl_{10}]$. Structural data reveal that each Te centre in the anion is in an approximately octahedral environment. (a) Suggest a structure for $[Te_2Cl_{10}]^{2-}$. (b) The cation $[(Ph_3PO)_2H]^+$ is derived from the phosphine oxide Ph_3PO. Suggest a structure for the cation and comment on its bonding.

INORGANIC CHEMISTRY MATTERS

16.33 HSO is an intermediate in the atmospheric oxidation of H_2S, and it has been implicated in ozone depletion. Calculated wavenumbers for the fundamental modes of vibration of HSO are 2335, 1077 and 1002 cm^{-1}. These are assigned to stretching, bending and stretching modes, respectively. Deuteration causes the fundamental modes of vibration to shift to 1704, 772 and 1041 cm^{-1}. (a) Is the HSO molecule linear or bent? (b) Draw one or more resonance structures to describe the bonding in HSO. (c) How many of the fundamental modes of vibration are IR active? (d) Which of the modes of vibration (2335, 1077 or 1002 cm^{-1}) arises from the S–H stretch? Use the deuteration data to confirm your assignment. (e) Suggest why HSO is associated with ozone depletion.

16.34 (a) North African desert dusts contain particles of calcite and dolomite. Why are these dusts important in countering acid rain in the Eastern Mediterranean? Give equations to illustrate your answer.

(b) Crystalline $CaC_2O_4 \cdot H_2O$ (log $K_{sp} = -8.6$) and $CaC_2O_4 \cdot 2H_2O$ (log $K_{sp} \approx -8.2$) accumulate on the surfaces of some lichens. This is thought to be in response to high levels of calcium ions. Zones of $CaC_2O_4 \cdot H_2O$ and $CaC_2O_4 \cdot 2H_2O$ crystals on lichens exposed to SO_2 emissions or acid rain are interrupted by growths of gypsum crystals. (i) Why are $CaC_2O_4 \cdot H_2O$ and $CaC_2O_4 \cdot 2H_2O$ suited to immobilizing excess Ca^{2+} ions? (ii) What is the origin of the gypsum crystals on lichens. Include appropriate equations for their formation.

16.35 Comment on the relevance of homonuclear bond formation between group 16 elements to the following:

(a) ozone and dioxygen;
(b) iron pyrites (fool's gold);
(c) the blue gemstone lapis lazuli;
(d) elemental Te, used as an additive in low carbon steels.

Topics

17 The group 17 elements

1	2		13	14	15	16	**17**	18
H								He
Li	Be		B	C	N	O	**F**	Ne
Na	Mg		Al	Si	P	S	**Cl**	Ar
K	Ca	*d*-block	Ga	Ge	As	Se	**Br**	Kr
Rb	Sr		In	Sn	Sb	Te	**I**	Xe
Cs	Ba		Tl	Pb	Bi	Po	**At**	Rn
Fr	Ra							

17.1 Introduction

The group 17 elements are called the *halogens*.

Fluorine, chlorine, bromine and iodine

The chemistry of fluorine, chlorine, bromine and iodine is probably better understood than that of any other group of elements except the alkali metals. This is partly because much of the chemistry of the halogens is that of singly bonded atoms or singly charged anions, and partly because of the wealth of structural and physicochemical data available for most of their compounds. The fundamental principles of inorganic chemistry are often illustrated by discussing properties of the halogens and halide compounds, and topics already discussed include:

- electron affinities of the halogens (Section 1.10);
- valence bond theory for F_2 (Section 2.2);
- molecular orbital theory for F_2 (Section 2.3);
- electronegativities of the halogens (Section 2.5);
- dipole moments of hydrogen halides (Section 2.6);
- bonding in HF by molecular orbital theory (Section 2.7);
- VSEPR model (which works well for many halide compounds, Section 2.8);
- application of the packing-of-spheres model, solid state structure of F_2 (Section 6.3);
- ionic radii (Section 6.10);
- ionic structure types: NaCl, CsCl, CaF_2, antifluorite, CdI_2 (Section 6.11);
- lattice energies: comparisons of experimental and calculated values for metal halides (Section 6.15);
- estimation of fluoride ion affinities (Section 6.16);
- estimation of standard enthalpies of formation and disproportionation, illustrated using halide compounds (Section 6.16);
- hydrogen halides as Brønsted acids (Section 7.4);
- energetics of hydrogen halide dissociation in aqueous solution (Section 7.5);
- solubilities of metal halides (Section 7.9);
- common-ion effect, exemplified by AgCl (Section 7.10);
- stability of complexes containing hard and soft metal ions and ligands, illustrated with halides of Fe(III) and Hg(II) (Section 7.13);
- redox half-cells involving silver halides (Section 8.3);
- non-aqueous solvents: liquid HF (Section 9.7);
- non-aqueous solvents: BrF_3 (Section 9.10);
- reactions of halogens with H_2 (Section 10.4);
- hydrogen bonding involving halogens (Section 10.6).

In Sections 11.5, 12.5, 13.6, 14.8, 15.7 and 16.7 we have discussed the halides of the group 1, 2, 13, 14, 15 and 16 elements respectively. Fluorides of the noble gases are discussed in Sections 18.4 and 18.5, and of the *d*- and *f*-block metals in Chapters 21, 22 and 27. In this chapter, we discuss the halogens themselves, their oxides and oxoacids, interhalogen compounds and polyhalide ions.

Astatine

Astatine is the heaviest member of group 17 and is known only in the form of radioactive isotopes, all of which have short half-lives. The longest lived isotope is ^{210}At ($t_{\frac{1}{2}} = 8.1$ h). Several isotopes are present naturally as transient products of the decay of uranium and thorium minerals; ^{218}At is formed from the β-decay of ^{218}Po, but the path competes with decay to ^{214}Pb (the dominant decay, see Fig. 27.3). Other isotopes are artificially prepared, e.g. ^{211}At (an α-emitter) from the nuclear reaction:

$$^{209}_{83}\text{Bi} + {}^{4}_{2}\text{He} \longrightarrow {}^{211}_{85}\text{At} + 2{}^{1}_{0}\text{n}.$$

The ^{211}At isotope may be separated by vacuum distillation. In general, At is chemically similar to iodine. Tracer studies (which are the only sources of information about the element) show that At_2 is less volatile than I_2, is soluble in organic solvents, and is reduced by SO_2 to At^- which can be coprecipitated with AgI or TlI. Hypochlorite, $[\text{ClO}]^-$, or peroxydisulfate, $[\text{S}_2\text{O}_8]^{2-}$, oxidizes astatine to an anion that is carried by $[\text{IO}_3]^-$ (e.g. coprecipitation with AgIO_3) and is therefore probably $[\text{AtO}_3]^-$. Less powerful oxidizing agents such as Br_2 also oxidize astatine, probably to $[\text{AtO}]^-$ or $[\text{AtO}_2]^-$.

Self-study exercises

1. The preparation of ^{211}At can be described by the abbreviated nuclear reaction $^{209}_{83}\text{Bi}(\alpha, 2n)^{211}_{85}\text{At}$. Explain what this means.

2. Explain what happens when $^{218}_{84}\text{Po}$ loses a β-particle.

[*Ans.* See Section 10.3]

17.2 Occurrence, extraction and uses

Occurrence

Figure 17.1 shows the relative abundances of the group 17 elements in the Earth's crust and in seawater. The major natural sources of fluorine are the minerals *fluorspar* (*fluorite*, CaF_2), *cryolite* ($\text{Na}_3[\text{AlF}_6]$) and *fluorapatite*, ($\text{Ca}_5\text{F}(\text{PO}_4)_3$) (see Section 15.2 and Box 15.11). The importance of cryolite lies in its being an *aluminium* ore (see Section 13.2). Sources of chlorine are closely linked to those of Na and K (see Section 11.2): *rock salt* (NaCl),

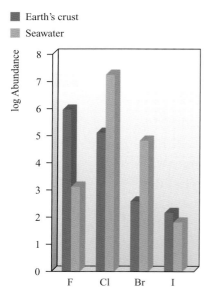

Fig. 17.1 Relative abundances of the halogens (excluding astatine) in the Earth's crust and seawater. The data are plotted on a logarithmic scale. The units of abundance are parts per billion (1 billion = 10^9).

sylvite (KCl) and *carnallite* ($\text{KCl·MgCl}_2\text{·6H}_2\text{O}$). Seawater is one source of Br_2 (Fig. 17.1), but significantly higher concentrations of Br^- are present in salt lakes and natural brine wells (see Box 17.3). The natural abundance of iodine is less than that of the lighter halogens. It occurs as iodide ion in seawater and is taken up by seaweed, from which it may be extracted. Impure Chile saltpetre (*caliche*) contains up to 1% sodium iodate and this has become an important source of I_2. Brines associated with oil and salt wells are of increasing importance.

Extraction

Most fluorine-containing compounds are made using HF, the latter being prepared from fluorite by reaction 17.1. In 2009, ≈85% of CaF_2 consumed in the US was converted into HF. Hydrogen fluoride is also recycled from Al manufacturing processes and from petroleum alkylation processes, and re-enters the supply chain. Difluorine is strongly oxidizing and must be prepared industrially by electrolytic oxidation of F^- ion. The electrolyte is a mixture of anhydrous molten KF and HF, and the electrolysis cell contains a steel or copper cathode, ungraphitized carbon anode, and a Monel metal (Cu/Ni) diaphragm which is perforated below the surface of the electrolyte, but not above it, thus preventing the H_2 and F_2 products from recombining. As electrolysis proceeds, the HF content of the melt is renewed by adding dry gas from cylinders.

$$\text{CaF}_2 + \text{H}_2\text{SO}_4 \xrightarrow{\text{conc}} \text{CaSO}_4 + 2\text{HF} \qquad (17.1)$$

Dichlorine is one of the most important industrial chemicals in the world, and is manufactured by the chloralkali process

($\approx 95\%$ of the world's supply, see Box 11.4) and by the Downs process (see Fig. 11.2). In 2009, the US and Europe manufactured 8.4 and 10.1 Mt of Cl_2, respectively. Higher demand for Cl_2 in Europe in 2009 was associated with increased production of polyvinyl chloride (PVC). The manufacture of Br_2 involves oxidation of Br^- by Cl_2, with air being swept through the system to remove Br_2. Similarly, I^- in brines is oxidized to I_2. The extraction of I_2 from $NaIO_3$ involves controlled reduction by SO_2; complete reduction yields NaI.

Uses

The nuclear fuel industry uses large quantities of F_2 in the production of UF_6 for fuel enrichment processes and this is the major use of F_2. Reprocessing of spent nuclear fuels involves both recovery of uranium and separation of ^{235}U from fission products. Short-lived radionuclides decay during a period of fuel storage (*pond storage*). After this, uranium is converted in the soluble salt $[UO_2][NO_3]_2$, and then into UF_6 (see Box 7.3).

Industrially, the most important F-containing compounds are HF, BF_3, CaF_2 (as a flux in metallurgy), synthetic cryo-lite (see reaction 13.46) and chlorofluorocarbons (CFCs, see Box 14.6). Water fluoridation was introduced in many developed countries during the mid-20th century, in order to reduce occurrences of dental caries, especially in children. The fluoridation agents are H_2SiF_6 and Na_2SiF_6, the former being a by-product of the manufacture of phosphoric acid from phosphate rock (see eq. 15.136). Phosphate rock contains fluoride impurities and, in the presence of SiO_2, phosphoric acid production results in the formation of gaseous HF and SiF_4. Scrubbing these emissions with water yields H_2SiF_6. Since the 1990s, there have been concerns that long-term uptake of fluoridated water may be linked to higher cancer risks.[†] Although the fluoridation of many water supplies continues, the introduction of toothpastes containing NaF or sodium monofluorophosphate, and the use of fluoridized salt (in Germany, France and Switzerland), provide other means of fighting tooth decay.

Figure 17.2a summarizes the major uses of chlorine. Chlorinated organic compounds, including 1,2-dichloroethene and vinyl chloride for the polymer industry, are hugely important. Dichlorine was widely used as a bleach in the paper and pulp industry, but environmental legislations have resulted in changes (Fig. 17.2b). Chlorine dioxide, ClO_2 (an 'elemental chlorine-free' bleaching agent) is favoured over Cl_2 because it does not produce toxic effluents.[‡] In addition to pulp bleaching, ClO_2 is increasingly used for the treatment of drinking water. However, because ClO_2 is unstable as a

[†] See: P.H.C. Harrison (2005) *J. Fluorine Chem.*, vol. 126, p. 1448 – 'Fluoride in water: a UK perspective'; D. Fagin (2008) *Sci. Am.*, vol. 298, issue 1, p. 74 – 'Second thoughts about fluoride'.
[‡] For a discussion of methods of cleaning up contaminated groundwater, including the effects of contamination by chlorinated solvent waste, see: B. Ellis and K. Gorder (1997) *Chem. Ind.*, p. 95.

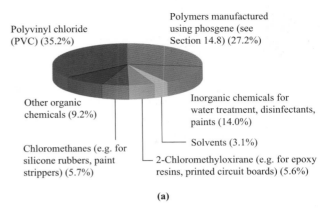

Polyvinyl chloride (PVC) (35.2%)

Polymers manufactured using phosgene (see Section 14.8) (27.2%)

Other organic chemicals (9.2%)

Inorganic chemicals for water treatment, disinfectants, paints (14.0%)

Solvents (3.1%)

Chloromethanes (e.g. for silicone rubbers, paint strippers) (5.7%)

2-Chloromethyloxirane (e.g. for epoxy resins, printed circuit boards) (5.6%)

(a)

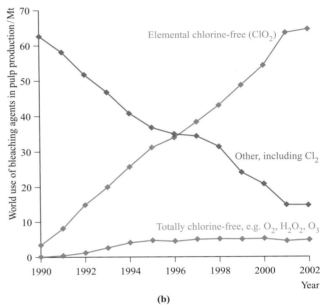

(b)

Fig. 17.2 (a) Industrial uses of Cl_2 in Europe in 2009 [data: www.eurochlor.org]. (b) The trends in uses of bleaching agents in the pulp industry between 1990 and 2002; ClO_2 has replaced Cl_2. Both elemental chlorine-free and totally chlorine-free agents comply with environmental legislations [data: Alliance for Environmental Technology].

compressed gas, it must be produced on site from either $NaClO_3$ or $NaClO_2$ (e.g. reactions 17.2 and 17.3).[*]

$$2NaClO_3 + 2NaCl + 2H_2SO_4 \longrightarrow 2ClO_2 + Cl_2 + 2Na_2SO_4 + 2H_2O \quad (17.2)$$

$$5NaClO_2 + 4HCl \longrightarrow 4ClO_2 + 5NaCl + 2H_2O \quad (17.3)$$

The manufacture of bromine- and iodine-containing organic compounds is a primary application of these halogens. Other uses include those of iodide salts (e.g. KI) and silver bromide in the photographic industry (although this is diminishing with the use of digital cameras), and bromine-based organic compounds as flame retardants (Box 17.1). Applications of iodine and its compounds are described in Box 17.2.

[*] See: G. Gordon and A.A. Rosenblatt (2005) *Ozone: Science and Engineering*, vol. 27, p. 203 – 'Chlorine dioxide: the current state of the art'.

APPLICATIONS

Box 17.1 Flame retardants

The incorporation of flame retardants into plastics, textiles, electronic equipment (e.g. printed circuit boards) and other materials is big business, and demand in the US reached approximately US\$ 1.3 billion in 2008. The chart on the right shows the split between the three main categories of flame retardants in the US in 2003. A range of brominated organics is used commercially, the most important being:

- the perbrominated diphenyl ether $(C_6Br_5)_2O$ (abbreviated as deca-BDE in the commercial market);
- 1,2,5,6,9,10-hexabromocyclododecane (HBCD);
- tetrabromobisphenol A (TBBPA, see below);
- octabromodiphenyl ether (octa-BDE);
- pentabromodiphenyl ether (penta-BDE);
- polybrominated biphenyl derivatives (PBBs);
- brominated polymers (e.g. epoxy resins, polycarbonates).

TBBPA

Deca-BDE is used as a flame retardant in textiles and in plastics for electrical and electronic equipment including housings for computers and television sets; HBCD is applied in thermal insulation foam (e.g. in the building trade) and in textile coatings. TBBPA has its main application as a flame retardant in printed circuit boards and related electronic equipment. In 2004, brominated flame retardants accounted for $\approx 50\%$ of the bromine consumed in the US. There is, however, growing awareness of the bioaccumulation of bromine-containing fire retardants, and, in the last decade, many studies have been carried out to assess the levels of these chemicals in the environment. The European Union has enforced legislation arising from concerns about the side effects (including hormone-related effects and the production of bromodioxins). In the EU, the production of polybrominated biphenyls (PBBs) has been banned since 2000, and since July 2006, the use of penta-BDE and octa-BDE in new electrical and electronic equipment has been prohibited. Following EU risk assessment of TBBPA in 2008, no restrictions in its use as a flame retardant have been put in place.

Phosphorus-based flame retardants include tris(1,3-dichloroisopropyl) phosphate, used in polyurethane foams and polyester resins. Once again, there is debate concerning toxic side-effects of such products: although these flame retardants may save lives, they produce noxious fumes during a fire.

Many inorganic compounds are used as flame retardants; for example

- Sb_2O_3 is used in PVC, and in aircraft and motor vehicles; scares that Sb_2O_3 in cot mattresses may be the cause of 'cot deaths' appear to have subsided;
- $Ph_3Sb(OC_6Cl_5)_2$ is added to polypropene;
- borates, exemplified by:

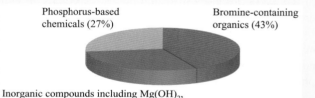

are used in polyurethane foams, polyesters and polyester resins;
- $ZnSnO_3$ has applications in PVC, thermoplastics, polyester resins and certain resin-based gloss paints.

Tin-based flame retardants appear to have a great potential future: they are non-toxic, apparently producing none of the hazardous side-effects of the widely used phosphorus-based materials.

Phosphorus-based chemicals (27%) Bromine-containing organics (43%)

Inorganic compounds including $Mg(OH)_2$, $ZnSnO_3$, Sb_2O_3 and borates (30%)

[Data: *Additives for Polymers* (2005), issue 3, p. 12]

Further reading

K. Harley *et al.* (2010) *Environ. Health Persp.*, vol. 118, p. 699 – 'PBDE concentrations in women's serum and fecundability'.

P.C. Hartmann, D. Bürgi and W. Giger (2004) *Chemosphere*, vol. 57, p. 781 – 'Organophosphate flame retardants and plasticizers in indoor air'.

R.J. Law *et al.* (2006) *Chemosphere*, vol. 64, p. 187 – 'Levels and trends of brominated flame retardants in the European environment'.

R.J. Letcher, ed. (2003) *Environ. Int.*, vol. 29, issue 6, pp. 663–885 – A themed issue of the journal entitled: 'The state-of-the-science and trends of brominated flame retardants in the environment'.

A. Marklund, B. Andersson and P. Haglund (2005) *Environ. Sci. Technol.*, vol. 39, p. 7423 – 'Organophosphorus flame retardants and plasticizers in Swedish sewage treatment plants'.

APPLICATIONS

Box 17.2 Iodine: from X-ray contrast agents to disinfectants and catalytic uses

The annual output of iodine is significantly lower than that of chlorine or bromine, but, nonetheless, it has a wide range of important applications. Determining accurate data for end-uses of iodine is difficult because many iodine-containing intermediate compounds are marketed before the final application is reached. One major application of certain iodine-containing compounds is as X-ray contrast agents. Such agents are radio-opaque (i.e. they prevent X-rays from penetrating) and are used to assist the diagnosis of disorders of the heart, central nervous system, gall bladder, urinary tract and other organs. For example, following an injection of an iodine-based radio-opaque contrast agent, an X-ray examination of the kidneys, ureters and bladder results in an image called an intravenous pyelogram (IVP). The photograph below shows an IVP of a patient in whom a kidney stone has passed into the ureter but could not be passed into the bladder. The resulting obstruction causes dilation of the ureter and renal pelvis.

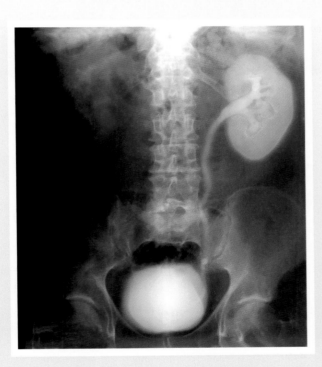

An intravenous pyelogram (IVP) imaged by using X-rays and an iodine-based contrast agent.

An important application of iodine itself is as a biocide and disinfectant. Since I_2 is insoluble in water, solubilizing agents such as poly-N-vinyl-2-pyrrolidine (PVP) are required to prepare commercially useful aqueous iodine-containing solutions. PVP/I_2 antiseptic solutions are marketed under a number of trade names including Betadine. Although PVP/I_2

has been used as an antiseptic since the 1950s, the nature of the interaction between PVP and I_2 is not fully elucidated. Electronic, Raman and IR spectroscopic data are consistent with the initial formation of a charge transfer complex (see Section 17.4) followed by release of I^- ion and formation of $[I_3]^-$. When I^- is released, the formation of $[PVP-I]^+$ has been proposed but not proven. An alternative proposal is the formation of $[PVPH]^+[I_3]^-$ with a proton hydrogen bonded between two ketone oxygen atoms in the polymer backbone. Uses of I_2 as a disinfectant range from wound antiseptics and disinfecting skin before surgery to maintaining germ-free swimming pools and water supplies.

PVP

At an industrial level, the square planar iodido-complexes cis-$[Rh(CO)_2I_2]^-$ and cis-$[Ir(CO)_2I_2]^-$ are the catalysts for the Monsanto and Cativa acetic acid and Tennessee–Eastman acetic anhydride processes, discussed in detail in Section 25.4. Application of iodine as a stabilizer includes its incorporation into nylon used in carpet and tyre manufacture. Iodized animal feed supplements are responsible for reduced instances of goitre (enlarged thyroid gland) which are otherwise prevalent in regions where the iodine content of soil and drinking water is low; iodized hen feeds increase egg production. Iodine is usually added to feeds in the form of $[H_3NCH_2CH_2NH_3]I_2$, KI, $Ca(IO_3)_2$ or $Ca(IO_4)_2$. Among dyes that have a high iodine content is erythrosine B (food red-colour additive E127) which is added to carbonated soft drinks, gelatins and cake icings. The use of ^{131}I as a medical radioisotope is described at the end of Section 17.3.

Erythrosine B contains 58% iodine aqueous solutions: λ_{max} = 525 nm

Iodine is essential for life and a deficiency results in a swollen thyroid gland; 'iodized salt' (NaCl with added I^-) provides us with iodine supplement.

17.3 Physical properties and bonding considerations

Table 17.1 lists selected physical properties of the group 17 elements (excluding astatine). Most of the differences between fluorine and the later halogens can be attributed to the:

- inability of F to exhibit any oxidation state other than -1 in its compounds, formally, an exception is $[F_3]^-$;
- relatively small size of the F atom and F^- ion;
- low dissociation energy of F_2 (Figs. 15.2 and 17.3);
- higher oxidizing power of F_2;
- high electronegativity of fluorine.

The last factor is *not* a rigidly defined quantity. However, it is useful in rationalizing such observations as the anomalous physical properties of, for example, HF (see

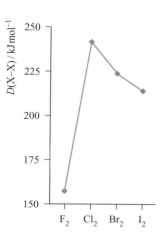

Fig. 17.3 The trend in X−X bond energies for the first four halogens.

Section 10.6), the strength of F-substituted carboxylic acids, the deactivating effect of the CF_3 group in electrophilic aromatic substitutions, and the non-basic character of NF_3 and $(CF_3)_3N$ (see end-of-chapter problem 17.4).

Table 17.1 Some physical properties of fluorine, chlorine, bromine and iodine.

Property	F	Cl	Br	I
Atomic number, Z	9	17	35	53
Ground state electronic configuration	$[He]2s^2 2p^5$	$[Ne]3s^2 3p^5$	$[Ar]3d^{10}4s^2 4p^5$	$[Kr]4d^{10}5s^2 5p^5$
Enthalpy of atomization, $\Delta_a H^\circ$(298 K)/kJ mol^{-1} [†]	79	121	112	107
Melting point, mp/K	53.5	172	266	387
Boiling point, bp/K	85	239	332	457.5
Standard enthalpy of fusion of X_2, $\Delta_{fus}H^\circ$(mp)/kJ mol^{-1}	0.51	6.40	10.57	15.52
Standard enthalpy of vaporization of X_2, $\Delta_{vap}H^\circ$(bp)/kJ mol^{-1}	6.62	20.41	29.96	41.57
First ionization energy, IE_1/kJ mol^{-1}	1681	1251	1140	1008
$\Delta_{EA}H_1^\circ$(298 K)/kJ mol^{-1} [‡]	-328	-349	-325	-295
$\Delta_{hyd}H^\circ(X^-, g)$/kJ mol^{-1}	-504	-361	-330	-285
$\Delta_{hyd}S^\circ(X^-, g)$/J K^{-1} mol^{-1}	-150	-90	-70	-50
$\Delta_{hyd}G^\circ(X^-, g)$/kJ mol^{-1}	-459	-334	-309	-270
Standard reduction potential, $E^\circ(X_2/2X^-)$/V	$+2.87$	$+1.36$	$+1.09$	$+0.54$
Covalent radius, r_{cov}/pm	71	99	114	133
Ionic radius, r_{ion} for X^-/pm [*]	133	181	196	220
van der Waals radius, r_v/pm	135	180	195	215
Pauling electronegativity, χ^P	4.0	3.2	3.0	2.7

[†] For each element X, $\Delta_a H^\circ = \frac{1}{2} \times$ Dissociation energy of X_2.
[‡] $\Delta_{EA}H_1^\circ$(298 K) is the enthalpy change associated with the process $X(g) + e^- \longrightarrow X^-(g) \approx -$(electron affinity); see Section 1.10.
[*] Values of r_{ion} refer to a coordination number of 6 in the solid state.

Fluorine forms no high oxidation state compounds (e.g. there are no analogues of $HClO_3$ and Cl_2O_7). When F is attached to another atom, Y, the Y−F bond is usually *stronger* than the corresponding Y−Cl bond (e.g. Tables 14.2, 15.3 and 16.2). If atom Y possesses no lone pairs, or has lone pairs but a large r_{cov}, then the Y−F bond is much stronger than the corresponding Y−Cl bond (e.g. C−F versus C−Cl, Table 14.2). Consequences of the small size of the F atom are that high coordination numbers can be achieved in molecular fluorides YF_n, and good overlap of atomic orbitals between Y and F leads to short, strong bonds, reinforced by ionic contributions when the difference in electronegativities of Y and F is large. The volatility of covalent F-containing compounds (e.g. fluorocarbons, see Section 14.8) originates in the weakness of the inter-molecular van der Waals or London dispersion forces. This, in turn, can be correlated with the low polarizability and small size of the F atom. The small ionic radius of F^- leads to high coordination numbers in saline fluorides, high lattice energies and highly negative values of $\Delta_f H^o$ for these compounds, as well as a large negative standard enthalpy and entropy of hydration of the ion (Table 17.1).

Worked example 17.1   Saline halides

For the process:

$$Na^+(g) + X^-(g) \longrightarrow NaX(s)$$

values of $\Delta H^o(298\,K)$ are −910, −783, −732 and −682 kJ mol^{-1} **for** $X^- = F^-$, Cl^-, Br^- **and** I^-, **respectively. Account for this trend.**

The process above corresponds to the formation of a crystalline lattice from gaseous ions, and $\Delta H^o(298\,K) \approx \Delta U(0\,K)$.

The Born–Landé equation gives an expression for $\Delta U(0\,K)$ assuming an electrostatic model and this is appropriate for the group 1 metal halides:

$$\Delta U(0\,K) = -\frac{LA|z_+||z_-|e^2}{4\pi\varepsilon_0 r_0}\left(1 - \frac{1}{n}\right)$$

NaF, NaCl, NaBr and NaI all adopt an NaCl structure, therefore A (the Madelung constant) is constant for this series of compounds.

The only variables in the equation are r_0 (internuclear distance) and n (Born exponent, see Table 6.3).

The term $\left(1 - \frac{1}{n}\right)$ varies little since n varies only from 7 for NaF to 9.5 for NaI.

The internuclear distance $r_0 = r_{cation} + r_{anion}$ and, since the cation is constant, varies only as a function of r_{anion}.

Therefore, the trend in values of $\Delta U(0\,K)$ can be explained in terms of the trend in values of r_{anion}.

$$\Delta U(0\,K) \propto -\frac{1}{\text{constant} + r_{anion}}$$

r_{anion} follows the trend $F^- < Cl^- < Br^- < I^-$, and therefore, $\Delta U(0\,K)$ has the most negative value for NaF.

Self-study exercises

1. What is meant by 'saline', e.g. saline fluoride?
    [*Ans.* See Section 10.7]

2. The alkali metal fluorides, MgF_2 and the heavier group 2 metal fluorides adopt NaCl, rutile and fluorite structures, respectively. What are the coordination numbers of the metal ion in each case?
    [*Ans.* See Figs. 6.16, 6.19a and 6.22]

3. Given the values (at 298 K) of $\Delta_f H^o(SrF_2, s) = -1216$ kJ mol^{-1} and $\Delta_f H^o(SrBr_2, s) = -718$ kJ mol^{-1}, calculate values for $\Delta_{lattice} H^o(298\,K)$ for these compounds using data from the Appendices. Comment on the relative magnitudes of the values.
    [*Ans.* SrF_2, −2496 kJ mol^{-1}; $SrBr_2$, −2070 kJ mol^{-1}]

In Section 16.3, we pointed out the importance of anion, rather than cation, formation in group 15. As expected, this is even more true in group 16. Table 17.1 lists values of the first ionization energies simply to show the expected decrease down the group. Although none of the halogens has yet been shown to form a discrete and stable monocation X^+, complexed or solvated I^+ is established, e.g. in $[I(py)_2]^+$ (Fig. 17.4), $[Ph_3PI]^+$ (see Section 17.4) and, apparently, in solutions obtained from reaction 17.4.

$$I_2 + AgClO_4 \xrightarrow{\text{Et}_2\text{O}} AgI + IClO_4 \qquad (17.4)$$

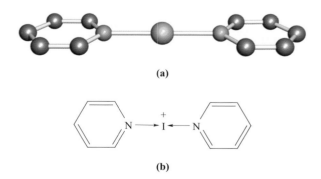

(a)

(b)

Fig. 17.4 (a) The structure of $[I(py)_2]^+$ (determined by X-ray crystallography) from the salt $[I(py)_2][I_3]\cdot 2I_2$ [O. Hassel *et al.* (1961) *Acta Chem. Scand.*, vol. 15, p. 407]; (b) A representation of the bonding in the cation. Colour code: I, gold; N, blue; C, grey.

The corresponding Br- and Cl-containing species are less stable, though they are probably involved in aromatic bromination and chlorination reactions in aqueous media.

The electron affinity of F is out of line with the trend observed for the later halogens (Table 17.1). Addition of an electron to the small F atom is accompanied by greater electron–electron repulsion than is the case for Cl, Br and I, and this probably explains why the process is less exothermic than might be expected on chemical grounds. Methods of accessing 'naked' fluoride ion are of considerable current interest. Structure **16.38** showed one example and its use in the preparation of the $[SeF_6]^{2-}$ ion. Two other sources of 'naked' F^- are Me_4NF and Me_4PF, and the increased reactivity of such fluoride ions, free of interactions with other species, promises to be useful in a wide range of reactions. This has a direct parallel with use of $[K(18\text{-crown-}6)][MnO_4]$ in organic chemistry as a highly reactive form of $[MnO_4]^-$ ion (see Section 11.8).

As we consider the chemistry of the halogens, it will be clear that there is an increasing trend towards higher oxidation states down the group. This is well exemplified among the interhalogen compounds (Section 17.7).

NMR active nuclei and isotopes as tracers

Although F, Cl, Br and I all possess spin active nuclei, in practice only ^{19}F (100%, $I = \frac{1}{2}$) is used routinely. Fluorine-19 NMR spectroscopy is a valuable tool in the elucidation of structures and reaction mechanisms of F-containing compounds; see case studies 1 and 5 and the discussion of stereochemically non-rigid species in Section 4.8.

ENVIRONMENT

Box 17.3 Bromine: resources and commercial demand

Bromine occurs as bromide salts in seawater, salt lakes and natural brines. World reserves are plentiful. The major producers of Br_2 draw on brines from Arkansas and Michigan in the US, and from the Dead Sea in Israel, and the chart below indicates the extent to which these countries dominate the world market. The Dead Sea is the deepest saline lake in the world, containing large reserves of NaCl, $MgCl_2$ and $CaCl_2$ with smaller quantities of bromide salts.

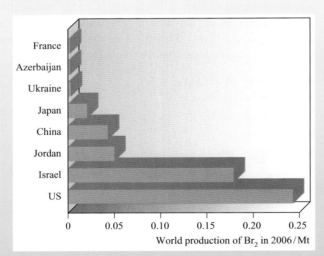

[Data: US Geological Survey]

Environmental issues, however, are likely to have a dramatic effect on the commercial demand for Br_2. We have already mentioned the call to phase out some bromine-based flame retardants (Box 17.1) and this significantly affects the demand for Br_2.

Crystalline salt deposits in the Dead Sea, Israel.

The commercial market for Br_2 has already been hit by the switch from leaded to unleaded motor vehicle fuels. Leaded fuels contained 1,2-$C_2H_4Br_2$ as an additive to facilitate the release of lead (formed by decomposition of the anti-knock agent Et_4Pb) as a volatile bromide. 1,2-Dibromoethane is also used as a nematocide and fumigant, and CH_3Br is a widely applied fumigant for soil. Bromomethane, however, falls in the category of a potential ozone depleter and its use has effectively been banned by the Montreal Protocol since 2005 (2015 in developing countries). Strictly controlled 'critical exceptions' to the ban are permitted (see Box 14.6).

Further reading

B. Reuben (1999) *Chemistry & Industry*, p. 547 – 'An industry under threat?'

Self-study exercises

In each example, use the VSEPR model to help you.

1. In the solution ^{19}F NMR spectrum (at 298 K) of $[BrF_6]^+[AsF_6]^-$, the octahedral cation gives rise to two overlapping, equal intensity 1:1:1:1 quartets $(J(^{19}F^{79}Br) = 1578$ Hz; $J(^{19}F^{80}Br) = 1700$ Hz). What can you deduce about the nuclear spins of ^{79}Br and ^{80}Br? Sketch the spectrum and indicate where you would measure the coupling constants.

[*Ans.* See R.J. Gillespie *et al.* (1974) *Inorg. Chem.*, vol. 13, p. 1230]

2. The room temperature ^{19}F NMR spectrum of MePF$_4$ shows a doublet $(J = 965$ Hz), whereas that of $[MePF_5]^-$ exhibits a doublet $(J = 829$ Hz) of doublets $(J = 33$ Hz) of quartets $(J = 9$ Hz), and a doublet $(J = 675$ Hz) of quintets $(J = 33$ Hz). Rationalize these data, and assign the coupling constants to $^{31}P-^{19}F$, $^{19}F-^{19}F$ or $^{19}F-^1H$ spin–spin coupling.

[*Ans.* MePF$_4$, trigonal bipyramidal, fluxional; $[MePF_5]^-$, octahedral, static]

See also end-of-chapter problems 4.43, 4.45, 14.12, 15.14, 15.24b, 16.16 and 17.10, and self-study exercises after worked examples 14.1 and 16.2.

Artificial isotopes of F include ^{18}F (β^+ emitter, $t_{\frac{1}{2}} = 1.83$ h) and ^{20}F (β^- emitter, $t_{\frac{1}{2}} = 11.0$ s). The former is the longest lived radioisotope of F and is used as a radioactive tracer, for example, in *positron emission tomography* (PET). This is a medical imaging technique that uses radionuclides (e.g. ^{18}F, ^{11}C, ^{13}N, ^{15}O) which decay by loss of a positron (β^+ particle). When a positron collides with an electron, the particles are annihilated and two γ-rays of equal energy, but travelling in opposite directions, are emitted. The γ-rays leaving a patient's body are detected by a PET scanner. ^{18}F-Fluorodeoxyglucose is widely used in PET to monitor glucose metabolism which is perturbed by the presence of cancerous tumours.[†] The ^{20}F isotope has application in F dating of bones and teeth; these usually contain apatite (see Section 15.2 and Box 15.11) which is slowly converted to fluorapatite when the mineral is buried in the soil. By using the technique of *neutron activation analysis*, naturally occurring ^{19}F is converted to ^{20}F by neutron bombardment; the radioactive decay of the latter is then monitored, allowing the amount of ^{19}F originally present in the sample to be determined. An alternative method of fluorine dating makes use of a fluoride ion-selective electrode.[‡]

[†] K. Wechalekar, B. Sharma and G. Cook (2005) *Clin. Radiol.*, vol. 60, p. 1143 – 'PET/CT in oncology – a major advance'.
[‡] For details of this technique, see: M.R. Schurr (1989) *J. Archaeol. Sci.*, vol. 16, p. 265 – 'Fluoride dating of prehistoric bones by ion selective electrode'.

An electrode that is sensitive to the concentration of a specific ion is called an ***ion-selective electrode***. A common example is a pH meter, the electrode in which is sensitive to H^+ ions.

The uses of radioisotopes (e.g. ^{131}I) in medicine are extremely important. Iodine-131 is a β^--emitter and also emits γ-radiation. If a patient ingests ^{131}I (usually as a solution of ^{131}I-labelled NaI), the isotope quickly accumulates in the thyroid gland. By using a γ-camera to record the emitted γ-radiation, a γ-camera scan (a *scintigram*) is recorded which reveals the size and state of the gland. The half-life of ^{131}I is 8 days, and so the dose administered decays relatively quickly. The half-life of a radionuclide used in nuclear medicine must be long enough to allow for the preparation of the radiopharmaceutical and its administration to the patient, but must be short enough so as to minimize the patient's exposure to radiation. It is also important that the radioisotope decays to a daughter nuclide that is not itself hazardous to the patient. In the case of ^{131}I, decay is to a stable (naturally occurring) isotope of xenon:

$$^{131}_{53}I \longrightarrow {}^{131}_{54}Xe + \beta^-$$

17.4 The elements

Difluorine

Difluorine is a pale yellow gas with a characteristic smell similar to that of O_3 or Cl_2. It is extremely corrosive, being easily the most reactive element known. Difluorine is handled in Teflon or special steel vessels,[*] although glass (see below) apparatus can be used if the gas is freed of HF by passage through sodium fluoride (eq. 17.5).

$$NaF + HF \longrightarrow Na[HF_2] \qquad (17.5)$$

The synthesis of F_2 cannot be carried out in aqueous media because F_2 decomposes water, liberating ozonized oxygen (i.e. O_2 containing O_3). The oxidizing power of F_2 is apparent from the E^o value listed in Table 17.1. The decomposition of a few high oxidation state metal fluorides generates F_2, but the only efficient alternative to the electrolytic method used industrially (see Section 17.2) is reaction 17.6. Difluorine is commercially available in cylinders, making laboratory synthesis generally unnecessary.

$$K_2[MnF_6] + 2SbF_5 \xrightarrow{420 K} 2K[SbF_6] + MnF_2 + F_2$$
$$(17.6)$$

Difluorine combines directly with all elements except O_2, N_2 and the lighter noble gases; reactions tend to be very

[*] See, for example, R.D. Chambers and R.C.H. Spink (1999) *Chem. Commun.*, p. 883 – 'Microreactors for elemental fluorine'.

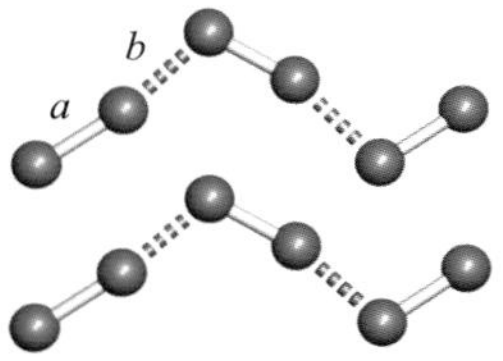

	Intramolecular distance for molecule in the gaseous state / pm	Intramolecular distance, a / pm	Intermolecular distance *within* a layer, b / pm	Intermolecular distance *between* layers / pm
Cl	199	198	332	374
Br	228	227	331	399
I	267	272	350	427

Fig. 17.5 Part of the solid state structures of Cl_2, Br_2 and I_2 in which molecules are arranged in stacked layers, and relevant intramolecular and intermolecular distance data.

violent. Combustion in compressed F_2 (*fluorine bomb calorimetry*) is a suitable method for determining values of $\Delta_f H^\circ$ for many binary metal fluorides. However, many metals are passivated by the formation of a layer of non-volatile metal fluoride. Silica is thermodynamically unstable with respect to reaction 17.7, but, unless the SiO_2 is powdered, the reaction is slow provided that HF is absent; the latter sets up chain reaction 17.8.

$$SiO_2 + 2F_2 \longrightarrow SiF_4 + O_2 \qquad (17.7)$$

$$\left.\begin{array}{l} SiO_2 + 4HF \longrightarrow SiF_4 + 2H_2O \\ 2H_2O + 2F_2 \longrightarrow 4HF + O_2 \end{array}\right\} \qquad (17.8)$$

The high reactivity of F_2 arises partly from the low bond dissociation energy (Fig. 17.3) and partly from the strength of the bonds formed with other elements (see Section 17.3).

Dichlorine, dibromine and diiodine

Dichlorine is a pale green-yellow gas with a characteristic odour. Inhalation causes irritation of the respiratory system and liquid Cl_2 burns the skin. Reaction 17.9 can be used for small-scale synthesis, but, like F_2, Cl_2 may be purchased in cylinders for laboratory use.

$$MnO_2 + 4HCl \longrightarrow MnCl_2 + Cl_2 + 2H_2O \qquad (17.9)$$
conc

Dibromine is a dark orange, volatile liquid (the only liquid non-metal at 298 K) but is often used as the aqueous solution 'bromine water'. Skin contact with liquid Br_2 results in burns, and Br_2 vapour has an unpleasant smell and causes eye and respiratory irritation. At 298 K, I_2 forms dark purple crystals which sublime readily at 1 bar pressure into a purple vapour.

In the crystalline state, Cl_2, Br_2 or I_2 molecules are arranged in layers (Fig. 17.5). The molecules Cl_2 and Br_2 have *intra*molecular distances which are the same as in the vapour (compare these distances with $2r_{cov}$, Table 17.1). *Inter*molecular distances for Cl_2 and Br_2 are also listed in Fig. 17.5. The distances within a layer are shorter than $2r_v$ (Table 17.1), suggesting some degree of interaction

between the X_2 molecules. The shortest *inter*molecular $X\cdots X$ distance *between* layers is significantly longer. In solid I_2, the *intra*molecular I–I bond distance is longer than in a gaseous molecule, and the lowering of the bond order (i.e. decrease in intramolecular bonding) is offset by a degree of *inter*molecular bonding within each layer (Fig. 17.5). It is significant that solid I_2 possesses a metallic lustre and exhibits appreciable electrical conductivity at higher temperatures. Under very high pressure I_2 becomes a metallic conductor.

Chemical reactivity decreases steadily from Cl_2 to I_2, notably in reactions of the halogens with H_2, P_4, S_8 and most metals. The values of E° in Table 17.1 indicate the decrease in oxidizing power along the series $Cl_2 > Br_2 > I_2$, and this trend is the basis of the methods of extraction of Br_2 and I_2 described in Section 17.2. Notable features of the chemistry of iodine which single it out among the halogens are that it is more easily:

- oxidized to high oxidation states;
- converted to stable salts containing I in the +1 oxidation state (e.g. Fig. 17.4).

Charge transfer complexes

A ***charge transfer complex*** is one in which a donor and acceptor interact *weakly* together with some transfer of electronic charge, usually facilitated by the acceptor.

The observed colours of the halogens arise from an electronic transition from the highest occupied π^* MO to the lowest unoccupied σ^* MO (see Fig. 2.10). The HOMO–LUMO energy gap decreases in the order $F_2 > Cl_2 > Br_2 > I_2$, leading to a progressive shift in the absorption maximum from the near-UV to the red region of the visible spectrum. Dichlorine, dibromine and diiodine dissolve unchanged in many organic solvents (e.g. saturated hydrocarbons, CCl_4). However in, for example, ethers, ketones and pyridine, which contain donor atoms, Br_2 and I_2 (and Cl_2 to a smaller extent) form *charge transfer complexes* with the halogen σ^*

MO acting as the acceptor orbital. In the extreme, complete transfer of charge could lead to heterolytic bond fission as in the formation of $[I(py)_2]^+$ (Fig. 17.4 and eq. 17.10).

$$2py + 2I_2 \longrightarrow [I(py)_2]^+ + [I_3]^- \qquad (17.10)$$

Solutions of I_2 in donor solvents, such as pyridine, ethers or ketones, are brown or yellow. Even benzene acts as a donor, forming charge transfer complexes with I_2 and Br_2. The colours of these solutions are noticeably different from those of I_2 or Br_2 in cyclohexane (a non-donor). Whereas amines, ketones and similar compounds donate electron density through a σ lone pair, benzene uses its π-electrons. This is apparent in the relative orientations of the donor (benzene) and acceptor (Br_2) molecules in Fig. 17.6b. The fact that solutions of the charge transfer complexes are

coloured means that they absorb in the visible region of the spectrum ($\approx$400–750 nm), but the electronic spectrum also contains an intense absorption in the UV region ($\approx$230–330 nm) arising from an electronic transition from the solvent–X_2 occupied bonding MO to a vacant anti-bonding MO. This is the so-called *charge transfer band*. Many charge transfer complexes can be isolated in the solid state and examples are given in Fig. 17.6. In complexes in which the donor is weak, e.g. C_6H_6, the $X-X$ bond distance is unchanged (or nearly so) by complex formation. Elongation as in 1,2,4,5-$(EtS)_4C_6H_2\cdot(Br_2)_2$ (compare the $Br-Br$ distance in Fig. 17.6c with that for free Br_2, in Fig. 17.5) is consistent with the involvement of a good donor. It has been estimated from theoretical calculations that -0.25 negative charges are transferred from

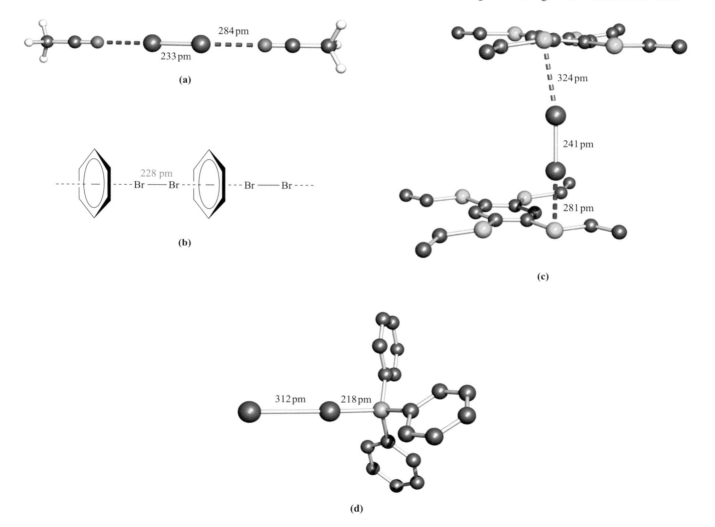

(a)

228 pm

(b)

(c)

(d)

🌐 **Fig. 17.6** Some examples of charge transfer complexes involving Br_2; the crystal structure of each has been determined by X-ray diffraction: (a) $2MeCN\cdot Br_2$ [K.-M. Marstokk *et al.* (1968) *Acta Crystallogr., Sect. B*, vol. 24, p. 713]; (b) schematic representation of the chain structure of $C_6H_6\cdot Br_2$; (c) 1,2,4,5-$(EtS)_4C_6H_2\cdot(Br_2)_2$ in which Br_2 molecules are sandwiched between layers of 1,2,4,5-$(EtS)_4C_6H_2$ molecules; interactions involving only one Br_2 molecule are shown and H atoms are omitted [H. Bock *et al.* (1996) *J. Chem. Soc., Chem. Commun.*, p. 1529]; (d) $Ph_3P\cdot Br_2$ [N. Bricklebank *et al.* (1992) *J. Chem. Soc., Chem. Commun.*, p. 355]. Colour code: Br, brown; C, grey; N, blue; S, yellow; P, orange; H, white.

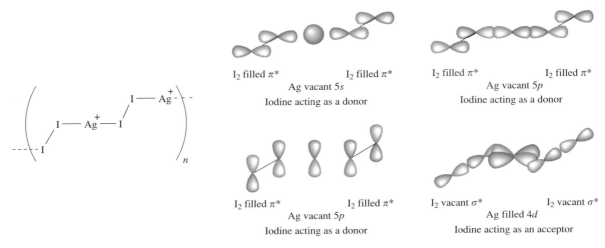

Fig. 17.7 Part of the chain structure of polymeric $[(AgI_2)_n]^{n+}$. Proposed bonding scheme for $[(AgI_2)_n]^{n+}$ illustrating the ability of I_2 to act as both a charge donor and a charge acceptor.

$1,2,4,5\text{-}(EtS)_4C_6H_2$ to Br_2. Different degrees of charge transfer are also reflected in the relative magnitudes of Δ_rH given for reactions 17.11. Further evidence for the weakening of the $X-X$ bond comes from vibrational spectroscopic data, e.g. a shift for $\bar{\nu}(X-X)$ from $215\,cm^{-1}$ in I_2 to $204\,cm^{-1}$ in $C_6H_6 \cdot I_2$.

$$\left.\begin{array}{ll} C_6H_6 + I_2 \longrightarrow C_6H_6 \cdot I_2 & \Delta_rH = -5\,kJ\,mol^{-1} \\ C_2H_5NH_2 + I_2 \longrightarrow C_2H_5NH_2 \cdot I_2 & \Delta_rH = -31\,kJ\,mol^{-1} \end{array}\right\}$$
$$(17.11)$$

Figure 17.6d shows the solid state structure of $Ph_3P \cdot Br_2$. $Ph_3P \cdot I_2$ has a similar structure ($I-I = 316\,pm$). In CH_2Cl_2 solution, $Ph_3P \cdot Br_2$ ionizes to give $[Ph_3PBr]^+Br^-$ and, similarly, Ph_3PI_2 forms $[Ph_3PI]^+I^-$ or, in the presence of excess I_2, $[Ph_3PI]^+[I_3]^-$. The formation of complexes of this type is not easy to predict:

- the reaction of Ph_3Sb with Br_2 or I_2 is an oxidative addition yielding Ph_3SbX_2, **17.1**;
- Ph_3AsBr_2 is an As(V) compound, whereas $Ph_3As \cdot I_2$, $Me_3As \cdot I_2$ and $Me_3As \cdot Br_2$ are charge transfer complexes of the type shown in Fig. 17.6d.[†]

(17.1)

[†] For insight into the complexity of this problem, see, for example, N. Bricklebank, S.M. Godfrey, H.P. Lane, C.A. McAuliffe, R.G. Pritchard and J.-M. Moreno (1995) *J. Chem. Soc., Dalton Trans.*, p. 3873.

The nature of the products from reaction 17.12 are dependent on the solvent and the R group in R_3P. Solid state structure determinations exemplify products of type $[R_3PI]^+[I_3]^-$ (e.g. $R = {}^nPr_2N$, solvent $= Et_2O$) and $[(R_3PI)_2I_3]^+[I_3]^-$ (e.g. $R = Ph$, solvent $= CH_2Cl_2$; $R = {}^iPr$, solvent $= Et_2O$). Structure **17.2** shows the $[({}^iPr_3PI)_2I_3]^+$ cation in $[({}^iPr_3PI)_2I_3][I_3]$.

$$R_3P + 2I_2 \longrightarrow R_3PI_4 \qquad (17.12)$$

(17.2)

Although there are many examples of X_2 molecules acting as electron acceptors, the role of X_2 as a donor is less well exemplified. Two examples of I_2 acting as an electron donor towards a metal centre are $Rh_2(O_2CCF_3)_4I_2 \cdot I_2$ and the Ag^+-containing polymer (made by reaction 17.13) shown in Fig. 17.7. The $I-I$ bond distance of close to $267\,pm$ indicates that the bond order is 1 (see Fig. 17.5). In the silver complex, the bonding scheme shown in Fig. 17.7 has been proposed. Charge donation from the filled π^* orbitals (the degenerate HOMO of I_2, see Fig. 2.10) to low-lying, empty $5s$ and $5p$ orbitals on the Ag^+ centre strengthens the $I-I$ bond, while back-donation of charge from a filled Ag $4d$ orbital to the empty σ^* MO of I_2 weakens the $I-I$ bond. In the Raman spectrum of

$[(AgI_2)_n][SbF_6]_n$, the value of $208\,cm^{-1}$ for $\bar{\nu}(I-I)$ is slightly lower than in I_2 ($215\,cm^{-1}$). A valence bond approach to the bonding in $[(AgI_2)_n]^{n+}$ is the subject of end-of-chapter problem 17.27.

$$AgMF_6 + I_2 \xrightarrow{\text{in liquid } SO_2} \frac{1}{n}[(AgI_2)_n][MF_6]_n \quad (M = As, Sb)$$
$$(17.13)$$

Clathrates

Dichlorine, dibromine and diiodine are sparingly soluble in water. By freezing aqueous solutions of Cl_2 and Br_2, solid hydrates of approximate composition $X_2 \cdot 8H_2O$ may be obtained. These crystalline solids, called *clathrates* (see Section 12.8), consist of hydrogen-bonded structures with X_2 molecules occupying cavities in the 3-dimensional network. An example is $1,3,5\text{-}(HO_2C)_3C_6H_3 \cdot 0.16Br_2$ which consists of a hydrogen-bonded 3-dimensional network of benzene-1,3,5-tricarboxylic acid molecules with Br_2 molecules hosted within some of the cavities in the framework.

17.5 Hydrogen halides

All the hydrogen halides, HX, are gases at 298 K with sharp, acid smells. Selected properties are given in Table 17.2. Direct combination of H_2 and X_2 to form HX (see eqs. 10.25–10.27 and accompanying discussion) can be used synthetically only for the chloride and bromide. Hydrogen fluoride is prepared by treating suitable metal fluorides with concentrated H_2SO_4 (e.g. reaction 17.14) and analogous reactions are also a convenient means of making

HCl. Analogous reactions with bromides and iodides result in partial oxidation of HBr or HI to Br_2 or I_2 (reaction 17.15), and synthesis is thus by reaction 17.16 with PX_3 prepared *in situ*.

$$CaF_2 + 2H_2SO_4 \xrightarrow{\text{conc}} 2HF + Ca(HSO_4)_2 \quad (17.14)$$

$$2HBr + H_2SO_4 \xrightarrow{\text{conc}} Br_2 + 2H_2O + SO_2 \quad (17.15)$$

$$PX_3 + 3H_2O \longrightarrow 3HX + H_3PO_3 \quad X = Br \text{ or } I \quad (17.16)$$

Some aspects of the chemistry of the hydrogen halides have already been covered:

- liquid HF (Section 9.7);
- solid state structure of HF (Fig. 10.9);
- hydrogen bonding and trends in boiling points, melting points and $\Delta_{vap}H^\circ$ (Section 10.6);
- formation of the $[HF_2]^-$ ion (Section 9.7; eq. 10.31 and accompanying discussion);
- Brønsted acid behaviour in aqueous solution and energetics of acid dissociation (Sections 7.4 and 7.5).

Hydrogen fluoride is an important reagent for the introduction of F into organic and other compounds (e.g. reaction 14.45 in the production of CFCs). It differs from the other hydrogen halides in being a weak acid in aqueous solution ($pK_a = 3.45$). This is in part due to the high $H-F$ bond dissociation enthalpy (Table 7.2 and Section 7.5). At high concentrations, the acid strength increases owing to

Table 17.2   Selected properties of the hydrogen halides.

Property	HF	HCl	HBr	HI
Physical appearance at 298 K	Colourless gas	Colourless gas	Colourless gas	Colourless gas
Melting point / K	189	159	186	222
Boiling point / K	293	188	207	237.5
$\Delta_{fus}H^\circ$(mp) / kJ mol^{-1}	4.6	2.0	2.4	2.9
$\Delta_{vap}H^\circ$(bp) / kJ mol^{-1}	34.0	16.2	18.0	19.8
$\Delta_f H^\circ$(298 K) / kJ mol^{-1}	−273.3	−92.3	−36.3	+26.5
$\Delta_f G^\circ$(298 K) / kJ mol^{-1}	−275.4	−95.3	−53.4	+1.7
Bond dissociation energy / kJ mol^{-1}	570	432	366	298
Bond length / pm	92	127.5	141.5	161
Dipole moment / D	1.83	1.11	0.83	0.45

the stabilization of F^- by formation of $[HF_2]^-$, **17.3** (scheme 17.17 and Table 10.4).

$$\left[F \text{—} H \text{—} F \right]^-$$

(17.3)

$$HF(aq) + H_2O(l) \rightleftharpoons [H_3O]^+(aq) + F^-(aq)$$
$$F^-(aq) + HF(aq) \rightleftharpoons [HF_2]^-(aq) \quad K = \frac{[HF_2^-]}{[HF][F^-]} = 0.2$$

(17.17)

The formation of $[HF_2]^-$ (see Section 9.7) is also observed when HF reacts with group 1 metal fluorides. $M[HF_2]$ salts are stable at room temperature, and structural data allow a realistic assessment of the strength of the F–H–F hydrogen-bonded interaction in the $[HF_2]^-$ ion. Analogous compounds are formed with HCl, HBr and HI only at low temperatures.

17.6 Metal halides: structures and energetics

All the halides of the alkali metals have NaCl or CsCl structures (Figs. 6.16 and 6.17) and their formation may be considered in terms of the Born–Haber cycle (see Section 6.14). In Section 11.5, we discussed trends in lattice energies of these halides, and showed that lattice energy is proportional to $1/(r_+ + r_-)$. We can apply this relationship to see why, for example, CsF is the best choice of alkali metal fluoride to effect the halogen exchange reaction 17.18.

$$\ce{>C-Cl + MF -> >C-F + MCl}$$

(17.18)

In the absence of solvent, the energy change associated with reaction 17.18 involves:

- the difference between the C–Cl and C–F bond energy terms (*not* dependent on M);
- the difference between the electron affinities of F and Cl (*not* dependent on M);
- the difference in lattice energies between MF and MCl (dependent on M).

The last difference is approximately proportional to:

$$\frac{1}{(r_{M^+} + r_{Cl^-})} - \frac{1}{(r_{M^+} + r_{F^-})}$$

which is always negative because $r_{F^-} < r_{Cl^-}$. The term above approaches zero as r_{M^+} increases. Thus, reaction 17.18 is favoured most for $M^+ = Cs^+$.

A few other monohalides possess the NaCl or CsCl structure, e.g. AgF, AgCl, and we have already discussed (Section 6.15) that these silver(I) halides exhibit significant covalent character. The same is true for CuCl, CuBr, CuI and AgI which possess the wurtzite structure (Fig. 6.21).

Most metal difluorides crystallize with CaF_2 (Fig. 6.19) or rutile (Fig. 6.22) structures, and for most of these, a simple ionic model is appropriate (e.g. CaF_2, SrF_2, BaF_2, MgF_2, MnF_2 and ZnF_2). With slight modification, this model also holds for other d-block difluorides. Chromium(II) chloride adopts a *distorted* rutile structure type, but other first row d-block metal dichlorides, dibromides and diiodides possess $CdCl_2$ or CdI_2 structures (see Fig. 6.23 and accompanying discussion). For these dihalides, neither purely electrostatic nor purely covalent models are satisfactory. Dihalides of the heavier d-block metals are considered in Chapter 22.

Metal trifluorides are crystallographically more complex than the difluorides, but symmetrical 3-dimensional structures are commonly found. Many contain octahedral (sometimes distorted) metal centres, e.g. AlF_3 (Section 13.6), VF_3 and MnF_3. For trichlorides, tribromides and triiodides, layer structures predominate. Among the tetrafluorides, a few have 3-dimensional structures, e.g. the two polymorphs of ZrF_4 possess, respectively, corner-sharing square-antiprismatic and dodecahedral ZrF_8 units. Most metal tetrahalides are either volatile molecular species (e.g. $SnCl_4$, $TiCl_4$) or contain rings or chains with M–F–M bridges (e.g. SnF_4, **14.14**). Metal–halogen bridges are longer than terminal bonds. Metal pentahalides may possess chain or ring structures (e.g. NbF_5, RuF_5, SbF_5, Fig. 15.13a) or molecular structures (e.g. $SbCl_5$), while metal hexahalides are molecular and octahedral (e.g. UF_6, MoF_6, WF_6, WCl_6). In general, an increase in oxidation state results in a structural change along the series 3-dimensional ionic → layer or polymer → molecular.

For metals exhibiting variable oxidation states, the relative thermodynamic stabilities of two ionic halides that contain a common halide ion but differ in the oxidation state of the metal (e.g. AgF and AgF_2) can be assessed using Born–Haber cycles. In such a reaction as 17.19, if the increase in ionization energies (e.g. $M \to M^+$ versus $M \to M^{2+}$) is approximately offset by the difference in lattice energies of the compounds, the two metal halides will be of about equal stability. This commonly happens with d-block metal halides.

$$MX + \tfrac{1}{2}X_2 \longrightarrow MX_2 \tag{17.19}$$

Worked example 17.2 Thermochemistry of metal fluorides

The lattice energies of CrF_2 and CrF_3 are -2921 and $-6040 \, kJ \, mol^{-1}$ respectively. **(a)** Calculate values of $\Delta_f H^{\circ}(298 \, K)$ for $CrF_2(s)$ and $CrF_3(s)$, and comment on the stability of these compounds with respect to $Cr(s)$ and $F_2(g)$. **(b)** The third ionization energy of Cr is large and positive. What factor offsets this and results in the standard enthalpies of formation of CrF_2 and CrF_3 being of the same order of magnitude?

(a) Set up a Born–Haber cycle for each compound; data needed are in the Appendices. For CrF_2 this is:

$$Cr(s) + F_2(g) \xrightarrow{\Delta_a H^{\circ}(Cr) + 2\Delta_a H^{\circ}(F)} Cr(g) + 2F(g)$$

$$\Delta_f H^{\circ}(CrF_2, s) \qquad IE_1 + IE_2 \, (Cr) \qquad 2\Delta_{EA} H^{\circ}(F)$$

$$CrF_2(s) \xleftarrow{\Delta_{lattice} H^{\circ}(CrF_2, s)} Cr^{2+}(g) + 2F^-(g)$$

$$\begin{aligned}
\Delta_f H^{\circ}(CrF_2, s) &= \Delta_a H^{\circ}(Cr) + 2\Delta_a H^{\circ}(F) + \Sigma IE(Cr) \\
&\quad + 2\Delta_{EA} H^{\circ}(F) + \Delta_{lattice} H^{\circ}(CrF_2, s) \\
&= 397 + 2(79) + 653 + 1591 + 2(-328) - 2921 \\
&= -778 \, kJ \, mol^{-1}
\end{aligned}$$

A similar cycle for CrF_3 gives:

$$\begin{aligned}
\Delta_f H^{\circ}(CrF_3, s) &= \Delta_a H^{\circ}(Cr) + 3\Delta_a H^{\circ}(F) + \Sigma IE(Cr) \\
&\quad + 3\Delta_{EA} H^{\circ}(F) + \Delta_{lattice} H^{\circ}(CrF_3, s) \\
&= 397 + 3(79) + 653 + 1591 + 2987 \\
&\quad + 3(-328) - 6040 \\
&= -1159 \, kJ \, mol^{-1}
\end{aligned}$$

The large negative values of $\Delta_f H^{\circ}(298 \, K)$ for both compounds show that the compounds are stable with respect to their constituent elements.

(b) $IE_3(Cr) = 2987 \, kJ \, mol^{-1}$

There are two negative terms that help to offset this: $\Delta_{EA} H^{\circ}(F)$ and $\Delta_{lattice} H^{\circ}(CrF_3, s)$. Note also that:

$$\Delta_{lattice} H^{\circ}(CrF_3, s) - \Delta_{lattice} H^{\circ}(CrF_2, s) = -3119 \, kJ \, mol^{-1}$$

and this term alone effectively cancels the extra energy of ionization required on going from Cr^{2+} to Cr^{3+}.

Self-study exercises

1. Values of $\Delta_{lattice} H^{\circ}$ for MnF_2 and MnF_3 (both of which are stable with respect to their elements at 298 K) are -2780 and $-6006 \, kJ \, mol^{-1}$. The third ionization energy of Mn is $3248 \, kJ \, mol^{-1}$. Comment on these data.

2. $\Delta_f H^{\circ}(AgF_2, \, s)$ and $\Delta_f H^{\circ}(AgF, \, s) = -360$ and $-205 \, kJ \, mol^{-1}$. Calculate values of $\Delta_{lattice} H^{\circ}$ for each compound. Comment on the results in the light of the fact that the values of $\Delta_f H^{\circ}$ for AgF_2 and AgF are of the same order of magnitude.

   [*Ans.* AgF, $-972 \, kJ \, mol^{-1}$; AgF_2, $-2951 \, kJ \, mol^{-1}$]

17.7 Interhalogen compounds and polyhalogen ions

Interhalogen compounds

Properties of interhalogen compounds are listed in Table 17.3. All are prepared by direct combination of elements, and where more than one product is possible, the outcome of the reaction is controlled by temperature and relative proportions of the halogens. Reactions of F_2 with the later halogens at ambient temperature and pressure give ClF, BrF_3 or IF_5, but increased temperatures give ClF_3, ClF_5, BrF_5 and IF_7. For the formation of IF_3, the reaction between I_2 and F_2 is carried out at 228 K. Table 17.3 shows clear trends among the four families of compounds XY, XY_3, XY_5 and XY_7:

- F is always in oxidation state -1;
- highest oxidation states for X reached are $Cl < Br < I$;
- combination of the later halogens with *fluorine* leads to the highest oxidation state compounds.

The structural families are **17.4–17.7** and are consistent with the VSEPR model (see Section 2.8). Angle α in **17.5** is $87.5°$ in ClF_3 and $86°$ in BrF_3. In each of ClF_5, BrF_5 and IF_5, the X atom lies just below the plane of the four F atoms. In this series of compounds, angle α in structure **17.6** lies between $90°$ (in ClF_5) and $81°$ (in IF_5). Among the interhalogens, ICl_3 is unusual in dimerizing. It possesses structure **17.8** and the planar I environments are rationalized in terms of the VSEPR model.

$$X \!\!-\!\! Y$$

(T-shaped structure with axial and equatorial Y atoms, angle α)

T-shaped

(17.4)   **(17.5)**

Table 17.3 Properties of interhalogen compounds.

Compound	Appearance at 298 K	Melting point / K	Boiling point / K	$\Delta_f H^o(298\,K)^{**}$ / kJ mol^{-1}	Dipole moment for gas-phase molecule / D	Bond distances in gas-phase molecules except for IF$_3$ and I$_2$Cl$_6$ / pm§
ClF	Colourless gas	117	173	−50.3	0.89	163
BrF	Pale brown gas	≈240‡	≈293‡	−58.5	1.42	176
BrCl	†	–	–	+14.6	0.52	214
ICl	Red solid	300 (α) 287 (β)	≈373*	−23.8	1.24	232
IBr	Black solid	313	389*	−10.5	0.73	248.5
ClF$_3$	Colourless gas	197	285	−163.2	0.6	160 (eq), 170 (ax)
BrF$_3$	Yellow liquid	282	399	−300.8	1.19	172 (eq), 181 (ax)
IF$_3$	Yellow solid	245 (dec)	–	≈ −500	–	187 (eq), 198 (ax)§§
I$_2$Cl$_6$	Orange solid	337 (sub)	–	−89.3	0	238 (terminal)§§ 268 (bridge)
ClF$_5$	Colourless gas	170	260	−255	–	172 (basal), 162 (axial)
BrF$_5$	Colourless liquid	212.5	314	−458.6	1.51	178 (basal), 168 (axial)
IF$_5$	Colourless liquid	282.5	373	−864.8	2.18	187 (basal), 185 (axial)
IF$_7$	Colourless gas	278 (sub)	–	−962	0	186 (eq), 179 (ax)

† Exists only in equilibrium with dissociation products: $2BrCl \rightleftharpoons Br_2 + Cl_2$.
‡ Significant disproportionation means values are approximate.
* Some dissociation: $2IX \rightleftharpoons I_2 + X_2$ (X = Cl, Br).
** Values quoted for the state observed at 298 K.
§ See structures **17.3–17.7**.
§§ Solid state (X-ray diffraction) data.

Square-based pyramid **(17.6)**

Pentagonal bipyramid **(17.7)**

(17.8)

In a series XY$_n$ in which the oxidation state of X increases, the X−Y bond enthalpy term decreases, e.g. for the Cl−F bonds in ClF, ClF$_3$ and ClF$_5$, they are 257, 172 and 153 kJ mol^{-1} respectively.

The most stable of the diatomic molecules are ClF and ICl. At 298 K, IBr dissociates somewhat into its elements, while BrCl is substantially dissociated (Table 17.3). Bromine monofluoride readily disproportionates (eq. 17.20), and IF is unstable at room temperature with respect to reaction 17.21.

$$3BrF \longrightarrow Br_2 + BrF_3 \qquad (17.20)$$
$$5IF \longrightarrow 2I_2 + IF_5 \qquad (17.21)$$

In general, the diatomic interhalogens exhibit properties intermediate between their parent halogens. However, where the electronegativities of X and Y differ significantly, the X−Y bond is stronger than the mean of the X−X and Y−Y bond strengths (see eqs. 2.10 and 2.11). Consistent with this is the observation that, if $\chi^P(X) \ll \chi^P(Y)$, the X−Y bond lengths (Table 17.3) are shorter than the mean of $d(X-X)$ and $d(Y-Y)$. Figure 17.5 illustrated that in the solid state, molecules of Cl_2, Br_2 and I_2 form zigzag chains which stack in layers. The structure of crystalline IBr is similar (Fig. 17.8a). Within each chain, the intermolecular I⋯Br distances (316 pm) are significantly shorter than the sum of the van der Waals radii of I and Br (410 pm). In contrast, solid ClF is composed of ribbons of molecules which are supported by Cl⋯Cl interactions (Fig. 17.8b). The Cl⋯Cl distances (307 pm) are appreciably shorter than twice $r_v(Cl)$. Two polymorphs (α- and β-forms) of crystalline ICl have been structurally characterized. Each form comprises chains of molecules, with each chain consisting of alternating non-equivalent ICl units (I−Cl = 235 and 244 pm). Figure 17.8c shows part of one chain of β-ICl and illustrates that short Cl⋯I and I⋯I interactions are present. The I⋯I distances of

306 pm are particularly significant, being closer to the sum of r_{cov} (266 pm) than to the sum of r_v (430 pm). This suggests that ICl tends towards forming I_2Cl_2 dimers in the solid state.

Chlorine monofluoride is commercially available and acts as a powerful fluorinating and oxidizing agent (e.g. reaction 17.22). Oxidative addition of ClF to SF_4 was shown in Fig. 16.13. It may behave as a fluoride donor (eq. 17.23) or as a fluoride acceptor (eq. 17.24). The structures of $[Cl_2F]^+$ (**17.9**) and $[ClF_2]^-$ (**17.10**) can be rationalized using the VSEPR model. Iodine monochloride and monobromide are less reactive than ClF, but of importance is the fact that, *in polar solvents*, ICl is a source of I^+ and iodinates aromatic compounds.

$$W + 6ClF \longrightarrow WF_6 + 3Cl_2 \tag{17.22}$$

$$2ClF + AsF_5 \longrightarrow [Cl_2F]^+[AsF_6]^- \tag{17.23}$$

$$ClF + CsF \longrightarrow Cs^+[ClF_2]^- \tag{17.24}$$

$$\left[\begin{array}{c} Cl \\ Cl \diagup \diagdown F \end{array} \right]^+ \qquad \left[F - Cl - F \right]^-$$

(17.9)   **(17.10)**

With the exception of I_2Cl_6, the higher interhalogens contain F and are extremely reactive, exploding or reacting violently with water or organic compounds; ClF_3 even ignites asbestos. Despite these hazards, they are valuable fluorinating agents, e.g. the highly reactive ClF_3 converts metals, metal chlorides and metal oxides to metal fluorides. One of its main uses is in nuclear fuel reprocessing for the formation of UF_6 (reaction 17.25).

$$U + 3ClF_3 \xrightarrow{\Delta} UF_6 + 3ClF \tag{17.25}$$

Reactivity decreases in the general order $ClF_n > BrF_n > IF_n$, and within a series having common halogens, the compound with the highest value of n is the most reactive, e.g. $BrF_5 > BrF_3 > BrF$. In line with these trends is the use of IF_5 as a relatively mild fluorinating agent in organic synthesis.

We have already discussed the self-ionization of BrF_3 and its use as a non-aqueous solvent (see Section 9.10). There is some evidence for the self-ionization of IF_5 (eq. 17.26), but little to support similar processes for other interhalogens.

$$2IF_5 \rightleftharpoons [IF_4]^+ + [IF_6]^- \tag{17.26}$$

Reactions 17.23 and 17.24 showed the fluoride donor and acceptor abilities of ClF. All the higher interhalogens undergo similar reactions, although ClF_5 does not form stable complexes at 298 K with alkali metal fluorides. However, it does react with CsF or $[Me_4N]F$ at low

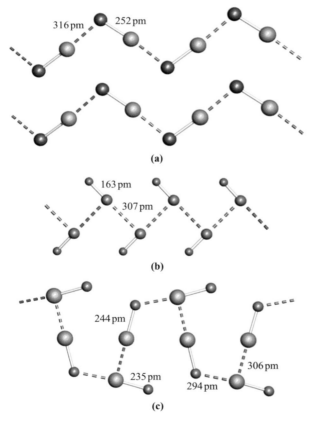

Fig. 17.8 The solid state structures (determined by X-ray diffraction) of (a) IBr [L.N. Swink *et al.* (1968) *Acta Crystallogr., Sect. B*, vol. 24, p. 429], (b) ClF (determined at 85 K) in which the chains are supported by short Cl⋯Cl contacts [R. Boese *et al.* (1997) *Angew. Chem. Int. Ed.*, vol. 36, p. 1489] and (c) β-ICl [G.B. Carpenter *et al.* (1962) *Acta Crystallogr.*, vol. 15, p. 360]. Colour code: F and Cl, green; Br, brown; I, gold.

temperatures to give salts containing $[ClF_6]^-$. Examples of F^- donation and acceptance by interhalogens are given in eqs. 9.42 and 17.27–17.31.

$$NOF + ClF_3 \longrightarrow [NO]^+[ClF_4]^- \qquad (17.27)$$

$$CsF + IF_7 \longrightarrow Cs^+[IF_8]^- \qquad (17.28)$$

$$IF_3 \xrightarrow{[Me_4N]F} [Me_4N]^+[IF_4]^- \xrightarrow{[Me_4N]F} [Me_4N]^+{}_2[IF_5]^{2-} \qquad (17.29)$$

$$ClF_3 + AsF_5 \longrightarrow [ClF_2]^+[AsF_6]^- \qquad (17.30)$$

$$IF_5 + 2SbF_5 \longrightarrow [IF_4]^+[Sb_2F_{11}]^- \qquad (17.31)$$

The choice of a large cation (e.g. Cs^+, $[NMe_4]^+$) for stabilizing $[XY_n]^-$ anions follows from lattice energy considerations; see also Boxes 11.5 and 24.1. Thermal decomposition of salts of $[XY_n]^-$ leads to the halide salt of highest lattice energy, e.g. reaction 17.32.

$$Cs[ICl_2] \xrightarrow{\Delta} CsCl + ICl \qquad (17.32)$$

Whereas $[IF_6]^+$ can be made by treating IF_7 with a fluoride acceptor (e.g. AsF_5), $[ClF_6]^+$ or $[BrF_6]^+$ must be made from ClF_5 or BrF_5 using an extremely powerful oxidizing agent because ClF_7 and BrF_7 are not known. Reaction 17.33 illustrates the use of $[KrF]^+$ to oxidize Br(V) to Br(VII). $[ClF_6]^+$ can be prepared in a similar reaction, or by using PtF_6 as oxidant. However, PtF_6 is not a strong enough oxidizing agent to oxidize BrF_5. In reaction 17.34, the active oxidizing species is $[NiF_3]^+$.[†] This cation is formed *in situ* in the $Cs_2[NiF_6]/AsF_5/HF$ system, and is a more powerful oxidative fluorinating agent than PtF_6.

$$[KrF]^+[AsF_6]^- + BrF_5 \longrightarrow [BrF_6]^+[AsF_6]^- + Kr \qquad (17.33)$$

$$Cs_2[NiF_6] + 5AsF_5 + XF_5 \qquad (17.34)$$
$$\xrightarrow[\substack{213\,K\ warmed \\ to\ 263\,K}]{anhydrous\ HF} [XF_6][AsF_6] + Ni(AsF_6)_2 + 2CsAsF_6$$
$$(X = Cl, Br)$$

Reaction 17.35 further illustrates the use of a noble gas fluoride in interhalogen synthesis. Unlike reaction 17.29, this route to $[Me_4N][IF_4]$ avoids the use of the thermally unstable IF_3.

$$2XeF_2 + [Me_4N]I \xrightarrow{242\,K;\ warm\ to\ 298\,K} [Me_4N][IF_4] + 2Xe \qquad (17.35)$$

[†] For details of the formation of $[NiF_3]^+$, see: T. Schroer and K.O. Christe (2001) *Inorg. Chem.*, vol. 40, p. 2415.

X = Cl, Br, I

(17.11)

On the whole, the observed structures of interhalogen anions and cations (Table 17.4) are in accord with the VSEPR model, but $[BrF_6]^-$ is regular octahedral, indicating the presence of a stereochemically inactive lone pair. Raman spectroscopic data suggest that $[ClF_6]^-$ is isostructural with $[BrF_6]^-$. On the other hand, the vibrational spectrum of $[IF_6]^-$ shows it is not regular octahedral; however, on the ^{19}F NMR timescale, $[IF_6]^-$ is stereochemically non-rigid. The difference between the structures of $[BrF_6]^-$ and $[IF_6]^-$ may be rationalized in terms of the difference in size of the central atom (see Section 16.7, $[SeF_6]^{2-}$ versus $[SeCl_6]^{2-}$). However, a word of caution: the solid state structure of $[Me_4N][IF_6]$ reveals the presence of loosely bound $[I_2F_{12}]^{2-}$ dimers (Fig. 17.9a).

Of particular interest in Table 17.4 is $[IF_5]^{2-}$. Only two examples of pentagonal planar XY_n species are known, the other being $[XeF_5]^-$ (see Section 18.4). In salts such as $[BrF_2][SbF_6]$, $[ClF_2][SbF_6]$ and $[BrF_4][Sb_2F_{11}]$, there is significant cation–anion interaction. Diagram **17.12** focuses on the Br environment on the solid state structure

Table 17.4 Structures of selected interhalogens and derived anions and cations. Each is consistent with VSEPR theory.

Shape	Examples
Linear	$[ClF_2]^-$, $[IF_2]^-$, $[ICl_2]^-$, $[IBr_2]^-$
Bent	$[ClF_2]^+$, $[BrF_2]^+$, $[ICl_2]^+$
T-shaped[†]	ClF_3, BrF_3, IF_3, ICl_3
Square planar	$[ClF_4]^-$, $[BrF_4]^-$, $[IF_4]^-$, $[ICl_4]^-$
See-saw, **17.11**	$[ClF_4]^+$, $[BrF_4]^+$, $[IF_4]^+$
Square-based pyramidal	ClF_5, BrF_5, IF_5
Pentagonal planar	$[IF_5]^{2-}$
Octahedral	$[ClF_6]^+$, $[BrF_6]^+$, $[IF_6]^+$
Pentagonal bipyramidal	IF_7
Square antiprismatic	$[IF_8]^-$

[†] Low-temperature X-ray diffraction data show that solid ClF_3 contains discrete T-shaped molecules, but in solid BrF_3 and IF_3 there are intermolecular X–F····X bridges resulting in coordination spheres not unlike those in $[BrF_4]^-$ and $[IF_5]^{2-}$.

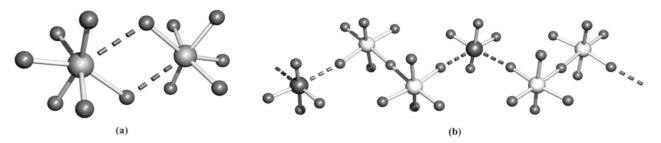

Fig. 17.9 (a) The structure of the $[I_2F_{12}]^{2-}$ dimer present in $[Me_4N][IF_6]$ (determined by X-ray diffraction); the I–F bridge distances are 211 and 282 pm [A.R. Mahjoub *et al.* (1991) *Angew. Chem. Int. Ed.*, vol. 30, p. 323]. Colour code: I, gold; F, green. (b) The solid state structure of $[BrF_4][Sb_2F_{11}]$ (determined by X-ray diffraction) consists of infinite chains of $[BrF_4]^+$ and $[Sb_2F_{11}]^-$ ions supported by Br$\cdots$F interactions (Br$\cdots$F = 235 and 241 pm, Sb–F_{term} = 184 pm, Sb–F_{br} = 190 pm, Br–F_{ax} = 173 pm, Br–F_{eq} = 166 pm) [A. Vij *et al.* (2002) *Inorg. Chem.*, vol. 41, p. 6397]. Colour code: Br, brown; Sb, silver; F, green.

of $[BrF_2][SbF_6]$, and Fig. 17.9b shows part of a chain of alternating cations and anions in crystalline $[BrF_4][Sb_2F_{11}]$.

$$(17.12)$$

229 pm

169 pm

93°

Self-study exercises

1. ^{127}I (100%, $I = \frac{5}{2}$) is a quadrupolar nucleus (see Section 4.8), but under certain circumstances (e.g. highly symmetrical environment), narrow line widths can be observed for signals in a ^{127}I NMR spectrum. Rationalize why the ^{127}I NMR spectrum of a liquid HF solution of $[IF_6][Sb_3F_{16}]$ exhibits a well-resolved binomial septet.

   [*Ans.* See J.F. Lehmann *et al.* (2004) *Inorg. Chem.*, vol. 43, p. 6905]

2. Draw the structure of an $[Sb_2F_{11}]^-$ anion which possesses D_{4h} symmetry. Indicate on the diagram where the C_4 axis and σ_h plane lie. Compare this structure with the structures of the $[Sb_2F_{11}]^-$ ions shown in Figs. 15.13b and 17.9. Comment on how the anion is able to change its conformation, and what factors might affect the structure.

Bonding in $[XY_2]^-$ ions

In Section 5.7, we used molecular orbital theory to describe the bonding in XeF_2, and developed a picture which gave a bond order of $\frac{1}{2}$ for each Xe–F bond. In terms of valence electrons, XeF_2 is isoelectronic with $[ICl_2]^-$, $[IBr_2]^-$, $[ClF_2]^-$ and related anions, and all have linear structures. The bonding in these anions can be viewed as being similar to that in XeF_2, and suggests weak X–Y bonds. This is in

contrast to the localized hypervalent picture that emerges from a structure such as **17.13**. Evidence for weak bonds comes from the X–Y bond lengths (e.g. 255 pm in $[ICl_2]^-$ compared with 232 in gas phase ICl) and from X–Y bond stretching wavenumbers (e.g. 267 and 222 cm^{-1} for the symmetric and asymmetric stretches of $[ICl_2]^-$ compared with 384 cm^{-1} in ICl).

$$(17.13)$$

Polyhalogen cations

In addition to the interhalogen cations described earlier, homonuclear cations $[Br_2]^+$, $[I_2]^+$, $[Cl_3]^+$, $[Br_3]^+$, $[I_3]^+$, $[Br_5]^+$, $[I_5]^+$ and $[I_4]^{2+}$ are well established. The cations $[Br_2]^+$ and $[I_2]^+$ can be obtained by oxidation of the corresponding halogen (eqs. 17.36, 17.37 and 9.15).

$$Br_2 + SbF_5 \xrightarrow{BrF_5} [Br_2]^+[Sb_3F_{16}]^- \qquad (17.36)$$

$$2I_2 + S_2O_6F_2 \xrightarrow{HSO_3F} 2[I_2]^+[SO_3F]^- \qquad (17.37)$$

On going from X_2 to the corresponding $[X_2]^+$, the bond shortens consistent with the loss of an electron from an antibonding orbital (see Figs. 2.7 and 2.10). In $[Br_2]^+[Sb_3F_{16}]^-$, the Br–Br distance is 215 pm, and in $[I_2]^+[Sb_2F_{11}]^-$ the I–I bond length is 258 pm (compare values of X_2 in Fig. 17.5). Correspondingly, the stretching wavenumber increases, e.g. 368 cm^{-1} in $[Br_2]^+$ compared with 320 cm^{-1} in Br_2. The cations are paramagnetic, and $[I_2]^+$ dimerizes at 193 K to give $[I_4]^{2+}$ (**17.14**). In reaction 17.38, AsF_5 acts as the oxidizing agent, and as a source of the counter-ion. When the oxidant is SbF_5, it is possible to isolate

$[I_4]^{2+}[Sb_3F_{14}]^-[SbF_6]^-$. The $[Sb_3F_{14}]^-$ ion (**17.15**) contains one Sb(III) and two Sb(V) centres, and can be considered as comprising a $[SbF_2]^+$ cation linked to two $[SbF_6]^-$ ions.

$$2I_2 + 3AsF_5 \xrightarrow{\text{in liquid SO}_2} [I_4][AsF_6]_2 + AsF_3 \qquad (17.38)$$

(17.14)

(17.15)

The cations $[Cl_3]^+$, $[Br_3]^+$ and $[I_3]^+$ are bent (**17.16**) as expected from the VSEPR model, and the X—X bond lengths are similar to those in gaseous X_2, consistent with single bonds. Reactions 17.39 and 17.40 may be used to prepare salts of $[Br_3]^+$ and $[I_3]^+$, and use of a higher concentration of I_2 in the I_2/AsF_5 reaction leads to the formation of $[I_5]^+$ (see reaction 9.15). The $[I_5]^+$ and $[Br_5]^+$ ions are structurally similar (**17.17**) with $d(\text{X–X})_{\text{terminal}} < d(\text{X–X})_{\text{non-terminal}}$, e.g. in $[I_5]^+$, the distances are 264 and 289 pm.

(17.16)   X = Cl, Br, I

(17.17)   X = Br, I

$$3Br_2 + 2[O_2]^+[AsF_6]^- \longrightarrow 2[Br_3]^+[AsF_6]^- + 2O_2 \qquad (17.39)$$

$$3I_2 + 3AsF_5 \xrightarrow{\text{in liquid SO}_2} 2[I_3]^+[AsF_6]^- + AsF_3 \qquad (17.40)$$

Even using extremely powerful oxidizing agents such as $[O_2]^+$, it has not proved possible (so far) to obtain the free $[Cl_2]^+$ ion by oxidizing Cl_2. When Cl_2 reacts with $[O_2]^+[SbF_6]^-$ in HF at low temperature, the product is $[Cl_2O_2]^+$ (**17.18**) which is best described as a charge transfer complex of $[Cl_2]^+$ and O_2. With IrF_6 as oxidant, reaction 17.41 takes place. The blue $[Cl_4][IrF_6]$ decomposes at 195 K to give salts of $[Cl_3]^+$, but X-ray diffraction data at 153 K show that the $[Cl_4]^+$ ion is structurally analogous to **17.14** (Cl–Cl = 194 pm, Cl····Cl = 294 pm).

$$2Cl_2 + IrF_6 \xrightarrow[<193 \text{ K}]{\text{anhydrous HF}} [Cl_4]^+[IrF_6]^- \qquad (17.41)$$

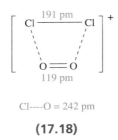

Cl----O = 242 pm

(17.18)

Polyhalide anions

Of the group 17 elements, iodine forms the largest range of homonuclear polyhalide ions, which include $[I_3]^-$, $[I_4]^{2-}$, $[I_5]^-$, $[I_7]^-$, $[I_8]^{2-}$, $[I_9]^-$, $[I_{10}]^{4-}$, $[I_{12}]^{2-}$, $[I_{16}]^{2-}$, $[I_{16}]^{4-}$, $[I_{22}]^{4-}$, $[I_{26}]^{3-}$ and $[I_{29}]^{3-}$. The $[F_3]^-$ ion has not been isolated under ambient conditions, but has been produced at 4 K by laser-ablation of metals with F_2 under conditions of electron capture.[†] Both $[Cl_3]^-$ and $[Br_3]^-$ are well established, but the $[I_3]^-$ ion is the most important member of this family. It is formed when I_2 is dissolved in aqueous solutions containing iodide ion. It has a linear structure, and in the solid state, the two I–I bond lengths may be equal (e.g. 290 pm in $[Ph_4As][I_3]$) or dissimilar (e.g. 283 and 303 pm in $Cs[I_3]$). The latter indicates something approaching an $[I–I\cdots I]^-$ entity (compare I–I = 266 pm in I_2). In the higher polyiodide ions, different I–I bond distances point to the structures being described in terms of association between I_2, I^- and $[I_3]^-$ units as examples in Fig. 17.10 show. This reflects their origins, since the higher polyiodides are formed upon crystallization of solutions containing I_2 and I^-. Details of the solid state structures of the anions are cation-dependent, e.g. although usually V-shaped, linear $[I_5]^-$ has also been observed in the solid state. The reaction between HI and pentafluorobenzyldibenzylamine produces compound **17.19**. The $[I_3]^-$ ion is essentially symmetrical (I–I = 290 and 292 pm). In the $[I_4]^{2-}$ ion, the central I–I distance is 276 pm, while the outer distances are 336 pm, indicating that, in this solid state structure, the $[I_4]^{2-}$ ion comprises an I_2 molecule interacting weakly with two I^- ions.

(17.19)

[†] See: S. Riedel, T. Köchner, X. Wang and L. Andrews (2010) *Inorg. Chem.*, vol. 49, p. 7156.

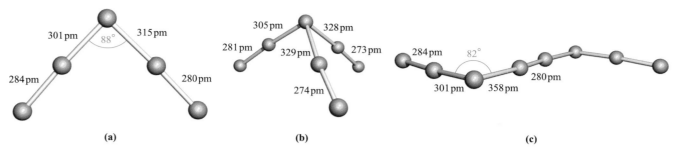

Fig. 17.10 The structures (X-ray diffraction) of (a) $[I_5]^-$ in $[Fe(S_2CNEt_2)_3][I_5]$ [C.L. Raston *et al.* (1980) *J. Chem. Soc., Dalton Trans.*, p. 1928], (b) $[I_7]^-$ in $[Ph_4P][I_7]$ [R. Poli *et al.* (1992) *Inorg. Chem.*, vol. 31, p. 3165], and (c) $[I_8]^{2-}$ in $[C_{10}H_8S_8]_2[I_3][I_8]_{0.5}$ [M.A. Beno *et al.* (1987) *Inorg. Chem.*, vol. 26, p. 1912].

Compared with polyiodides, fewer polybromide ions have been characterized. Many salts involving $[Br_3]^-$ are known, and the association in the solid state of $[Br_3]^-$ and Br^- has been observed to give rise to the linear species **17.20**. The $[Br_8]^{2-}$ ion is structurally analogous to $[I_8]^{2-}$ (Fig. 17.10c) with Br–Br bond distances that indicate association between Br_2 and $[Br_3]^-$ units in the crystal.

$$\left[Br\!\!-\!\!Br\!\!-\!\!Br \text{-----} Br\!\!-\!\!Br\!\!-\!\!Br \text{-----} Br \right]^{3-}$$

(17.20)

Polyiodobromide ions are exemplified by $[I_2Br_3]^-$ and $[I_3Br_4]^-$. In the 2,2'-bipyridinium salt, $[I_2Br_3]^-$ is V-shaped like $[I_5]^-$ (Fig. 17.10a), while in the $[Ph_4P]^+$ salt, $[I_3Br_4]^-$ resembles $[I_7]^-$ (Fig. 17.10b). Both $[I_2Br_3]^-$ and $[I_3Br_4]^-$ can be described as containing IBr units linked by a Br^- ion.

17.8 Oxides and oxofluorides of chlorine, bromine and iodine

Earlier in the book (e.g. Sections 15.3 and 16.3), we described how heavier elements of the *p*-block can expand their coordination numbers beyond 4 but still retain an octet of electrons in the valence shell. In this section, we discuss oxides and oxofluorides of Cl, Br and I, and in the majority of these compounds, halogen atoms are in an oxidation state of +3, +5 or +7. In terms of bonding, we may draw charge-separated species and sets of resonance structures to represent the bonding in many of these molecules. For example, the bonding in Br_2O_3 and FClO may be described as follows (lone pairs are shown explicitly only on the high oxidation state halogen atom):

While these structures show each atom obeying the octet rule, drawing them becomes cumbersome for many species.

Furthermore, for halogen atoms in oxidation state +7, charge separated structures place an unusually high positive formal charge on the central atom. We have therefore chosen to illustrate the structures of high oxidation state oxides and oxofluorides of Cl, Br and I showing hypervalent halogen atoms. Throughout the section, you should keep in mind that the representation of a 'line' between two atoms in a structural diagram does not necessarily imply a 2-centre 2-electron interaction.

Oxides

Oxygen fluorides were described in Section 16.7. Iodine is the only halogen to form an oxide which is *thermodynamically stable* with respect to decomposition into its elements (eq. 17.42). The chlorine and bromine oxides are hazardous materials which tend to explode.

$$I_2 + \tfrac{5}{2}O_2 \longrightarrow I_2O_5 \qquad \Delta_f H^\circ(298\,K) = -158.1\,kJ\,mol^{-1}$$

(17.42)

Chlorine oxides, although not difficult to prepare, are all liable to decompose explosively. Far less is known about the oxides of Br (which are very unstable) than those of chlorine and iodine, although Br_2O_3 (**17.21**) and Br_2O_5 (**17.22**) have been unambiguously prepared (scheme 17.43) and structurally characterized. The Br(V) centres are trigonal pyramidal and in Br_2O_5, the BrO_2 groups are eclipsed. The oxide Br_2O may be made by reaction 17.44 and has the non-linear structure **17.23**.

$$Br_2 \xrightarrow{O_3,\ 195\,K} Br_2O_3 \xrightarrow{O_3,\ 195\,K} Br_2O_5 \qquad (17.43)$$
Brown \qquad Orange \qquad Colourless

$$H_2O + 2BrOTeF_5 \xrightarrow{in\ CCl_3F,\ 195\,K} Br_2O + 2HOTeF_5$$

(17.44)

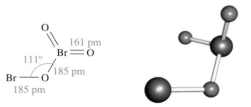

(17.21)

$O=Br$ ... 188 pm ... O ... $Br=O$ 161 pm, 121°

(17.22)

X	d / pm	a / °
Cl	170	111
Br	186	114

(17.23)

Dichlorine monoxide, Cl_2O (**17.23**), is obtained as a yellow-brown gas by action of Cl_2 on mercury(II) oxide or moist sodium carbonate (eqs. 17.45 and 17.46). Cl_2O liquefies at ≈ 277 K, and explodes on warming. It hydrolyses to hypochlorous acid (eq. 17.47), and is formally the anhydride of this acid (see Section 15.8).

$$2Cl_2 + 3HgO \longrightarrow Cl_2O + Hg_3O_2Cl_2 \qquad (17.45)$$

$$2Cl_2 + 2Na_2CO_3 + H_2O \longrightarrow 2NaHCO_3 + 2NaCl + Cl_2O \qquad (17.46)$$

$$Cl_2O + H_2O \longrightarrow 2HOCl \qquad (17.47)$$

$Cl-O = 147$ pm; $\angle O-Cl-O = 117°$

(17.24)

Chlorine dioxide, ClO_2 (**17.24**) is a yellow gas (bp 283 K), and is produced in the highly dangerous reaction between potassium chlorate, $KClO_3$, and concentrated H_2SO_4. Reaction 17.48 is a safer method of synthesis, and reactions 17.2 and 17.3 showed two of the commercial methods used to make ClO_2. ClO_2 is used to bleach flour and wood pulp and for water treatment. Its application as a bleach in the paper industry has increased (see Fig. 17.2).

$$2KClO_3 + 2H_2C_2O_4 \longrightarrow K_2C_2O_4 + 2ClO_2 + 2CO_2 + 2H_2O \qquad (17.48)$$

Despite being a radical, ClO_2 shows no tendency to dimerize. It dissolves unchanged in water, but is slowly hydrolysed to HCl and $HClO_3$, a reaction that involves the $ClO^\bullet$ radical. In alkaline solution, hydrolysis is rapid (eq. 17.49). Ozone reacts with ClO_2 at 273 K to form Cl_2O_6, a dark red liquid which is also made by reaction 17.50.

$$2ClO_2 + 2[OH]^- \longrightarrow [ClO_3]^- + [ClO_2]^- + H_2O \qquad (17.49)$$

$$ClO_2F + HClO_4 \longrightarrow Cl_2O_6 + HF \qquad (17.50)$$

(17.25)

Reaction 17.50, and the hydrolysis of Cl_2O_6 to chlorate and perchlorate, suggest that it has structure **17.25** and is the mixed anhydride of $HClO_3$ and $HClO_4$. The IR spectrum of matrix-isolated Cl_2O_6 is consistent with a molecular structure with two inequivalent Cl centres. The solid, however, contains $[ClO_2]^+$ and $[ClO_4]^-$ ions. Cl_2O_6 is unstable with respect to decomposition into ClO_2, $ClOClO_3$ and O_2, and, with H_2O, reaction 17.51 occurs. The oxide $ClOClO_3$ is the mixed acid anhydride of HOCl and $HClO_4$, and is made by reaction 17.52.

$$Cl_2O_6 + H_2O \longrightarrow HClO_4 + HClO_3 \qquad (17.51)$$

$$Cs[ClO_4] + ClSO_3F \longrightarrow Cs[SO_3F] + ClOClO_3 \qquad (17.52)$$

The anhydride of perchloric acid is Cl_2O_7 (**17.26**), an oily, explosive liquid (bp ≈ 353 K), which is made by dehydrating $HClO_4$ using phosphorus(V) oxide at low temperatures.

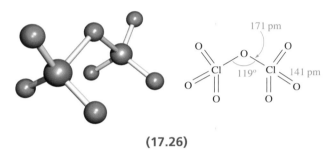

171 pm, 119°, 141 pm

(17.26)

In contrast to Br_2O_5 which is thermally unstable, I_2O_5 is stable to 573 K. It is a white, hygroscopic solid, prepared by dehydration of iodic acid (HIO_3). The reaction is reversed when I_2O_5 dissolves in water (eq. 17.53). I_2O_5 is used in analysis for CO (see eq. 14.62).

$$I_2O_5 + H_2O \longrightarrow 2HIO_3 \qquad (17.53)$$

In the solid state, I_2O_5 is structurally related to Br_2O_5 (**17.22**), with the difference that it has a staggered conformation, probably as a result of extensive intermolecular interactions ($I\cdots O \leq 223$ pm).

Oxofluorides

Several families of halogen oxides with X$-$F bonds exist: FXO_2 (X = Cl, Br, I), FXO_3 (X = Cl, Br, I), F_3XO (X = Cl, Br, I), F_3XO_2 (X = Cl, I) and F_5IO. The thermally unstable FClO is also known. Their structures are consistent with the VSEPR model (**17.27–17.32**).

(17.27)   **(17.28)**   **(17.29)**

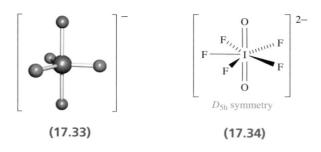

(17.30)   **(17.31)**   **(17.32)**

Chloryl fluoride, $FClO_2$, is a colourless gas (bp 267 K) and can be prepared by reacting F_2 with ClO_2. It hydrolyses to $HClO_3$ and HF, and acts as a fluoride donor towards SbF_5 (eq. 17.54) and a fluoride acceptor with CsF (eq. 17.55).

$$FClO_2 + SbF_5 \longrightarrow [ClO_2]^+[SbF_6]^- \qquad (17.54)$$

$$CsF + FClO_2 \longrightarrow Cs^+[F_2ClO_2]^- \qquad (17.55)$$

Perchloryl fluoride, $FClO_3$ (bp 226 K, $\Delta_f H^\circ(298\,K) = -23.8\,kJ\,mol^{-1}$) is surprisingly stable and decomposes only above 673 K. It can be prepared by reaction 17.56, or by treating $KClO_3$ with F_2.

$$KClO_4 + 2HF + SbF_5 \longrightarrow FClO_3 + KSbF_6 + H_2O \qquad (17.56)$$

Alkali attacks $FClO_3$ only slowly, even at 500 K. Perchloryl fluoride is a mild fluorinating agent and has been used in the preparation of fluorinated steroids. It is also a powerful oxidizing agent at elevated temperatures, e.g. it oxidizes SF_4 to SF_6. Reaction 17.57 illustrates its reaction with an organic nucleophile. In contrast to $FClO_2$, $FClO_3$ does not behave as a fluoride donor or acceptor.

$$C_6H_5Li + FClO_3 \longrightarrow LiF + C_6H_5ClO_3 \qquad (17.57)$$

The reaction between F_2 and Cl_2O at low temperatures yields F_3ClO (mp 230 K, bp 301 K, $\Delta_f H^\circ(g, 298\,K) = -148\,kJ\,mol^{-1}$) which decomposes at 570 K to ClF_3 and O_2. Reactions of F_3ClO with CsF and SbF_5 show its ability to accept or donate F^-, producing $[F_4ClO]^-$ and $[F_2ClO]^+$ respectively.

The compound $FBrO_3$ is one of only a few examples of bromine in oxidation state +7. It is made by treating $K[BrO_4]$ with AsF_5 or SbF_5 in liquid HF. $FBrO_3$ is more reactive than $FClO_3$ and is hydrolysed more rapidly than $FClO_3$ by aqueous alkali at room temperature (eq. 17.58). Unlike $FClO_3$, $FBrO_3$ acts as a fluoride acceptor (eq. 17.59) to give *trans*-$[F_2BrO_3]^-$ (**17.33**).

$$FBrO_3 + 2[OH]^- \longrightarrow [BrO_4]^- + H_2O + F^- \qquad (17.58)$$

$$MF + FBrO_3 \longrightarrow M[F_2BrO_3] \qquad e.g.\ M^+ = Cs^+,\ [NO]^+ \qquad (17.59)$$

(17.33)   **(17.34)**

D_{5h} symmetry

The only representative of the neutral F_5XO family of oxofluorides is F_5IO, produced when IF_7 reacts with water. It does not readily undergo further reaction with H_2O. One reaction of note is that of F_5IO with $[Me_4N]F$ in which the pentagonal bipyramidal ion $[F_6IO]^-$ is formed; X-ray diffraction data show that the oxygen atom is in an axial site and that the equatorial F atoms are essentially coplanar, in contrast to the puckering observed in IF_7 (see Section 2.8). The pentagonal pyramidal $[F_5IO]^{2-}$ is formed as the Cs^+ salt when CsF, I_2O_5 and IF_5 are heated at 435 K. The stoichiometry of the reaction must be controlled to prevent $[F_4IO]^-$ being formed as the main product. Transfer of F^- from $[Me_4N]F$ to F_3IO_2 (**17.30**) leads to a mixture of *cis*- and *trans*-$[F_4IO_2]^-$. The *trans*-isomer reacts further to give $[F_5IO_2]^{2-}$, and this gives a means of isolating the pure *cis*-isomer of $[F_4IO_2]^-$ from the *cis*/*trans* mixture. Vibrational spectroscopic data are consistent with $[F_5IO_2]^{2-}$ possessing D_{5h} symmetry (**17.34**).

Self-study exercises

1. Rationalize each of the following structures in terms of the VSEPR model.

2. Confirm that the $[IOF_5]^{2-}$ ion (the structure is given above) has C_{5v} symmetry.

3. To what point groups do the following fluorides belong: BrF_5, $[BrF_4]^-$, $[BrF_6]^+$? Assume that each structure is regular. [*Ans.* C_{4v}; D_{4h}; O_h]

4. In the IR spectra of isotopically labelled $F^{35}ClO_2$ and $F^{37}ClO_2$, bands at 630 and 622 cm^{-1}, respectively, are assigned to the Cl–F stretching mode. What is the origin of the shift in wavenumber? Confirm that the observed shift is consistent with that expected.

17.9 Oxoacids and their salts

Hypofluorous acid, HOF

Fluorine is unique among the halogens in forming no species in which it has a formal oxidation state other than −1. The only known oxoacid is hypofluorous acid, HOF, which is unstable and does not ionize in water but reacts according to eq. 17.60; no salts are known. It is obtained by passing F_2 over ice at 230 K (eq. 17.61) and condensing the gas produced. At 298 K, HOF decomposes rapidly (eq. 17.62). The solid state structure of HOF has been determined by X-ray diffraction at 113 K, and shows that HOF molecules associate into chains through $O-H\cdots O$ hydrogen bonds.

$$HOF + H_2O \longrightarrow H_2O_2 + HF \qquad (17.60)$$

$$F_2 + H_2O \xrightarrow{230\,K} HOF + HF \qquad (17.61)$$

$$2HOF \longrightarrow 2HF + O_2 \qquad (17.62)$$

Oxoacids of chlorine, bromine and iodine

Table 17.5 lists the families of oxoacids known for Cl, Br and I. The hypohalous acids, HOX, are obtained in aqueous solution by reaction 17.63 (compare reactions 17.45 and 17.47).

$$2X_2 + 3HgO + H_2O \longrightarrow Hg_3O_2X_2 + 2HOX \qquad (17.63)$$

In contrast to HOF, HOCl, HOBr and HOI are unknown as isolated compounds, but in aqueous solution they act as weak acids (pK_a values: HOCl, 4.53; HOBr, 8.69; HOI, 10.64). Hypochlorite salts such as NaOCl, KOCl and $Ca(OCl)_2$ (eq. 17.64) can be isolated. NaOCl can be crystallized from a solution obtained by electrolysing aqueous NaCl in such a way that the Cl_2 liberated at the anode mixes with the NaOH produced at the cathode. Hypochlorites are powerful oxidizing agents and in the presence of alkali convert $[IO_3]^-$ to $[IO_4]^-$, Cr^{3+} to $[CrO_4]^{2-}$, and

even Fe^{3+} to $[FeO_4]^{2-}$. Bleaching powder is a non-deliquescent mixture of $CaCl_2$, $Ca(OH)_2$ and $Ca(OCl)_2$ and is manufactured by the action of Cl_2 on $Ca(OH)_2$. NaOCl is a bleaching agent and disinfectant.

$$2CaO + 2Cl_2 \longrightarrow Ca(OCl)_2 + CaCl_2 \qquad (17.64)$$

All hypohalites are unstable with respect to disproportionation (eq. 17.65). At 298 K, the reaction is slow for $[OCl]^-$, fast for $[OBr]^-$ and very fast for $[OI]^-$. Sodium hypochlorite disproportionates in hot aqueous solution (eq. 17.66), and the passage of Cl_2 through *hot* aqueous alkali yields chlorate and chloride salts rather than hypochlorites. Hypochlorite solutions decompose by reaction 17.67 in the presence of cobalt(II) compounds as catalysts.

$$3[OX]^- \longrightarrow [XO_3]^- + 2X^- \qquad (17.65)$$

$$3NaOCl \longrightarrow NaClO_3 + 2NaCl \qquad (17.66)$$

$$2[OX]^- \longrightarrow 2X^- + O_2 \qquad (17.67)$$

Like HOCl, chlorous acid, $HClO_2$, is not isolable but is known in aqueous solution and is prepared by reaction 17.68. It is a relatively weak acid ($pK_a = 2.0$). Sodium chlorite (used as a bleach) is made by reaction 17.69. The chlorite ion has the bent structure **17.35**.

$$Ba(ClO_2)_2 + H_2SO_4(aq) \longrightarrow 2HClO_2(aq) + BaSO_4(s)$$
$$\text{suspension} \qquad (17.68)$$

$$Na_2O_2 + 2ClO_2 \longrightarrow 2NaClO_2 + O_2 \qquad (17.69)$$

Cl—O = 157 pm; ∠O–Cl–O = 111°

(17.35)

Alkaline solutions of chlorites persist unchanged over long periods, but in the presence of acid, $HClO_2$ disproportionates according to eq. 17.70.

$$5HClO_2 \longrightarrow 4ClO_2 + H^+ + Cl^- + 2H_2O \qquad (17.70)$$

Table 17.5 Oxoacids of chlorine, bromine and iodine.

Oxoacids of chlorine		Oxoacids of bromine		Oxoacids of iodine	
Hypochlorous acid	HOCl	Hypobromous acid	HOBr	Hypoiodous acid	HOI
Chlorous acid	HOClO ($HClO_2$)				
Chloric acid	$HOClO_2$ ($HClO_3$)	Bromic acid	$HOBrO_2$ ($HBrO_3$)	Iodic acid	$HOIO_2$ (HIO_3)
Perchloric acid	$HOClO_3$ ($HClO_4$)	Perbromic acid	$HOBrO_3$ ($HBrO_4$)	Periodic acid	$HOIO_3$ (HIO_4)
				Orthoperiodic acid	$(HO)_5IO$ (H_5IO_6)

Fig. 17.11 In the solid state, molecules of HIO_3 form hydrogen-bonded chains. The structure was determined by neutron diffraction [K. Staahl (1992) *Acta Chem. Scand.*, vol. 46, p. 1146]. Colour code: I, gold; O, red; H, white.

Chloric and bromic acids, $HClO_3$ and $HBrO_3$, are both strong acids but cannot be isolated as pure compounds. The aqueous acids can be made by reaction 17.71 (compare with reaction 17.68).

$$Ba(XO_3)_2 + H_2SO_4 \longrightarrow BaSO_4 + 2HXO_3 \quad (X = Cl, Br) \tag{17.71}$$

Iodic acid, HIO_3, is a stable, white solid at room temperature, and is produced by reacting I_2O_5 with water (eq. 17.53) or by the oxidation of I_2 with nitric acid. Crystalline iodic acid contains trigonal pyramidal HIO_3 molecules connected into chains by extensive hydrogen bonding (Fig. 17.11). In aqueous solution it is a fairly strong acid ($pK_a = 0.77$).

Chlorates are strong oxidizing agents. Commercially, $NaClO_3$ is used for the manufacture of ClO_2 (eq. 17.2), and is used as a weedkiller, although this application has been banned in the EU since 2009. $KClO_3$ has applications in fireworks and safety matches. Chlorates are produced by electrolysis of brine at 340 K, allowing the products to mix efficiently (scheme 17.72); chlorate salts are crystallized from the mixture. Anodic oxidation of $[OCl]^-$ produces further $[ClO_3]^-$.

Electrolysis: $\quad 2Cl^- \longrightarrow Cl_2 + 2e^-$

$\qquad\qquad\quad 2H_2O + 2e^- \longrightarrow H_2 + 2[OH]^-$

Mixing and disproportionation:

$\qquad\qquad Cl_2 + 2[OH]^- \longrightarrow Cl^- + [OCl]^- + H_2O$

$\qquad\qquad 3[OCl]^- \longrightarrow [ClO_3]^- + 2Cl^-$

$$\tag{17.72}$$

Bromates are made by, for example, reaction 17.73 under alkaline conditions. Reaction 17.74 is a convenient synthesis of KIO_3.

$$KBr + 3KOCl \longrightarrow KBrO_3 + 3KCl \tag{17.73}$$

$$2KClO_3 + I_2 \longrightarrow 2KIO_3 + Cl_2 \tag{17.74}$$

Potassium bromate and iodate are commonly used in volumetric analysis. Very pure KIO_3 is easily obtained, and reaction 17.75 is used as a source of I_2 for the standardization of thiosulfate solutions (reaction 16.118).

$$[IO_3]^- + 5I^- + 6H^+ \longrightarrow 3I_2 + 3H_2O \tag{17.75}$$

$$\begin{array}{c} O \diagup\!\!\!\!\overset{X}{\diagdown} O \\ | \\ O^- \end{array}$$

X = Cl, Br, I

(17.36)

Halate ions are trigonal pyramidal (**17.36**) although, in the solid state, some metal iodates contain infinite structures in which two O atoms of each iodate ion bridge two metal centres.[†] The thermal decomposition of alkali metal chlorates follows reaction 17.76, but in the presence of a suitable catalyst, $KClO_3$ decomposes to give O_2 (eq. 16.4). Some iodates (e.g. KIO_3) decompose when heated to iodide and O_2, but others (e.g. $Ca(IO_3)_2$) give oxide, I_2 and O_2. Bromates behave similarly and the interpretation of these observations is a difficult problem in energetics and kinetics.

$$4[ClO_3]^- \longrightarrow 3[ClO_4]^- + Cl^- \quad \textit{Caution: risk of explosion!} \tag{17.76}$$

Perchloric acid is the only oxoacid of Cl that can be isolated, and its vapour state structure is shown in Fig. 17.12a. Discrete molecules are also present in crystalline, anhydrous $HClO_4$, the structure having been determined at 113 K. At 298 K, $HClO_4$ is a colourless liquid

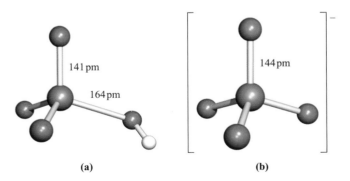

(a)   (b)

Fig. 17.12 Structures of (a) perchloric acid (vapour state), in which one Cl−O bond is unique, and (b) perchlorate ion, in which all Cl−O bonds are equivalent. Colour code: Cl, green; O, red; H, white.

[†] For further discussion, see: A.F. Wells (1984) *Structural Inorganic Chemistry*, 5th edn, Clarendon Press, Oxford, pp. 327–337.

(bp 363 K with some decomposition). It is made by heating $KClO_4$ with concentrated H_2SO_4 under reduced pressure. Pure perchloric acid is liable to explode when heated or in the presence of organic material, but in dilute solution, $[ClO_4]^-$ is very difficult to reduce despite the reduction potentials (which provide thermodynamic but not kinetic data) shown in eqs. 17.77 and 17.78. Zinc, for example, merely liberates H_2, and iodide ion has no action. Reduction to Cl^- can be achieved by Ti(III) in acidic solution or by Fe(II) in the presence of alkali.

$$[ClO_4]^- + 2H^+ + 2e^- \rightleftharpoons [ClO_3]^- + H_2O$$
$$E^o = +1.19\,V \quad (\text{at pH 0}) \qquad (17.77)$$

$$[ClO_4]^- + 8H^+ + 8e^- \rightleftharpoons Cl^- + 4H_2O$$
$$E^o = +1.39\,V \quad (\text{at pH 0}) \qquad (17.78)$$

Perchloric acid is an extremely strong acid in aqueous solution (see Table 7.3). Although $[ClO_4]^-$ (Fig. 17.12b) does form complexes with metal cations, the tendency to do so is less than for other common anions. Consequently, $NaClO_4$ solution is a standard medium for the investigation of ionic equilibria in aqueous systems, e.g. it is used as a supporting electrolyte in electrochemical experiments (see Box 8.2). Alkali metal perchlorates can be obtained by disproportionation of chlorates (eq. 17.76) under carefully controlled conditions; traces of impurities can catalyse decomposition to chloride and O_2. *Perchlorate salts are potentially explosive and must be handled with particular care.* For example, solid NH_4ClO_4 decomposes at 298 K according to eq. 17.79, and mixtures of ammonium perchlorate and aluminium are standard missile propellants.

$$4NH_4ClO_4(s) \longrightarrow 2Cl_2(g) + 2N_2O(g)$$
$$+ 3O_2(g) + 8H_2O(l) \qquad (17.79)$$

When heated, $KClO_4$ gives KCl and O_2, apparently without intermediate formation of $KClO_3$. Silver perchlorate, like silver salts of some other very strong acids (e.g. $AgBF_4$, $AgSbF_6$ and AgO_2CCF_3), is soluble in many organic solvents including C_6H_6 and Et_2O owing to complex formation between Ag^+ and the organic molecules.

The best method of preparation of perbromate ion is by reaction 17.80. Cation exchange (see Section 11.6) can be used to give $HBrO_4$ in solution, but the anhydrous acid has not been isolated.

$$[BrO_3]^- + F_2 + 2[OH]^- \longrightarrow [BrO_4]^- + 2F^- + H_2O$$
$$(17.80)$$

Potassium perbromate has been structurally characterized and contains tetrahedral $[BrO_4]^-$ ions (Br−O = 161 pm). Thermochemical data show that $[BrO_4]^-$ (half-reaction 17.81) is a slightly stronger oxidizing agent than $[ClO_4]^-$ or $[IO_4]^-$ under the same conditions. However, oxidations

by $[BrO_4]^-$ (as for $[ClO_4]^-$) are slow in dilute neutral solution, but more rapid at higher acidities.

$$[BrO_4]^- + 2H^+ + 2e^- \rightleftharpoons [BrO_3]^- + H_2O$$
$$E^o = +1.76\,V \quad (\text{at pH 0}) \qquad (17.81)$$

Several different periodic acids and periodates are known; Table 17.5 lists periodic acid, HIO_4 and orthoperiodic acid, H_5IO_6 (compare with H_6TeO_6, Section 16.9). Oxidation of KIO_3 by hot alkaline hypochlorite yields $K_2H_3IO_6$ which is converted to KIO_4 by nitric acid. Treatment of KIO_4 with concentrated alkali yields $K_4H_2I_2O_{10}$, and dehydration of this at 353 K leads to $K_4I_2O_9$. Apart from $[IO_4]^-$ (**17.37**) and $[IO_5]^{3-}$ and $[HIO_5]^{2-}$ (which are square-based pyramidal), periodic acids and periodate ions feature octahedral I centres, e.g. H_5IO_6 (**17.38**), $[H_2I_2O_{10}]^{4-}$ (**17.39**) and $[I_2O_9]^{4-}$ (**17.40**). The presence of five OH groups per molecule in H_5IO_6 leads to extensive intermolecular hydrogen bonding in the solid state, resulting in a 3-dimensional network (Fig. 17.13a). The solid state structure of anhydrous HIO_4 contrasts with that of $HClO_4$. While the latter consists of discrete molecules, solid HIO_4 contains chains of edge-sharing octahedra (Fig. 17.13b); the chains interact with one another through O–H$\cdots$O hydrogen bonds.

I−O = 178 pm

(17.37)

I−O(terminal) = 178 pm
I−OH = 189 pm

(17.38)

I−O(terminal) = 181 pm
I−O(bridge) = 200 pm
I−OH = 198 pm

(17.39)

I−O(terminal) = 177 pm
I−O(bridge) = 201 pm

(17.40)

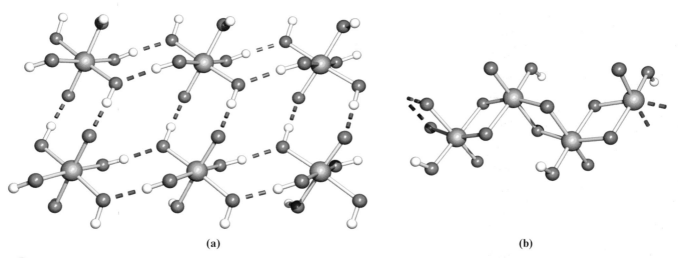

Fig. 17.13 (a) Part of the 3-dimensional hydrogen-bonded network in crystalline H_5IO_6 (determined by neutron diffraction) [Y.D. Feikema (1966) *Acta Crystallogr.*, vol. 20, p. 765]. (b) Part of one chain of *cis*-edge sharing octahedra in crystalline HIO_4 (determined by combined X-ray and neutron diffraction) [T. Kraft *et al.* (1997) *Angew. Chem. Int. Ed.*, vol. 36, p. 1753]. Colour code: I, gold; O, red; H, white.

The relationships between these ions in aqueous solution may be expressed by equilibria 17.82, and aqueous solutions of periodates are therefore not simple systems.

$$\left.\begin{array}{l}
[H_3IO_6]^{2-} + H^+ \rightleftharpoons [IO_4]^- + 2H_2O \\
2[H_3IO_6]^{2-} \rightleftharpoons 2[HIO_5]^{2-} + 2H_2O \\
2[HIO_5]^{2-} \rightleftharpoons [H_2I_2O_{10}]^{4-} \rightleftharpoons [I_2O_9]^{4-} + H_2O
\end{array}\right\} \quad (17.82)$$

Orthoperiodic acid is obtained by electrolytic oxidation of iodic acid, or by adding concentrated nitric acid to $Ba_5(IO_6)_2$, prepared by reaction 17.83.

$$5Ba(IO_3)_2 \xrightarrow{\Delta} Ba_5(IO_6)_2 + 4I_2 + 9O_2 \quad (17.83)$$

Heating H_5IO_6 dehydrates it, first to $H_4I_2O_9$, and then to HIO_4. In aqueous solution, both H_5IO_6 ($pK_a = 3.3$) and HIO_4 ($pK_a = 1.64$) behave as rather weak acids. Periodate oxidizes iodide (eq. 17.84) *rapidly* even in neutral solution (compare the actions of chlorate and bromate). It liberates ozonized O_2 from hot acidic solution, and oxidizes Mn(II) to $[MnO_4]^-$ in acidic solution (eq. 17.85).

$$[IO_4]^- + 2I^- + H_2O \longrightarrow [IO_3]^- + I_2 + 2[OH]^- \quad (17.84)$$

$$5[IO_4]^- + 2Mn^{2+} + 3H_2O \longrightarrow 5[IO_3]^- + 2[MnO_4]^- + 6H^+ \quad (17.85)$$

17.10 Aqueous solution chemistry

In this section, we are mainly concerned with redox processes in aqueous solution; see Section 17.1 for a list of relevant topics already covered in the book. Values of E^o for half-reactions 17.86 can be measured directly for X = Cl, Br and I (Table 17.1) and their magnitudes are determined by the X—X bond energies (Fig. 17.3), the electron affinities of the halogen atoms (Table 17.1) and the standard Gibbs energies of hydration of the halide ions (Table 17.1). This can be seen from scheme 17.87; for X = Br and I, an additional vaporization stage is needed for the element.

$$\tfrac{1}{2}X_2 + e^- \rightleftharpoons X^- \quad (17.86)$$

$$\tfrac{1}{2}X_2 \longrightarrow X(g) \longrightarrow X^-(g) \longrightarrow X^-(aq) \quad (17.87)$$

Dichlorine is a more powerful oxidizing agent in aqueous media than Br_2 or I_2, partly because of a more negative enthalpy of formation of the anion but, more importantly, because the Cl^- ion (which is smaller than Br^- or I^-) interacts more strongly with solvent molecules. (In solid salt formation, the lattice energy factor similarly explains why chloride salts are more exothermic than corresponding bromides or iodides.)

Since F_2 liberates ozonized O_2 from water, the value of E^o for half-reaction 17.86 has no physical reality, but a value of $+2.87\,V$ can be estimated by comparing the energy changes for each step in scheme 17.87 for X = F and Cl, and hence deriving the difference in E^o for half-equation 17.86 for X = F and Cl. Most of the difference between these E^o values arises from the much more negative value of $\Delta_{hyd}G^o$ of the smaller F^- ion (Table 17.1).

Diiodine is much more soluble in aqueous iodide solutions than in water. At low concentrations of I_2, eq. 17.88 describes the system; K can be found by partitioning I_2 between the aqueous layer and a solvent immiscible with water (e.g. CCl_4).

$$I_2 + I^- \rightleftharpoons [I_3]^- \qquad K \approx 10^2\ (298\,K) \quad (17.88)$$

Potential diagrams (partly calculated from thermochemical data) for Cl, Br and I are given in Fig. 17.14. Because

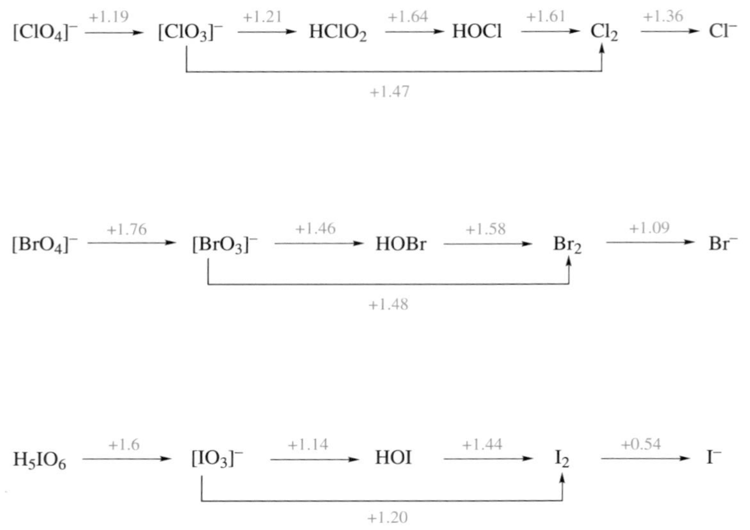

$$[ClO_4]^- \xrightarrow{+1.19} [ClO_3]^- \xrightarrow{+1.21} HClO_2 \xrightarrow{+1.64} HOCl \xrightarrow{+1.61} Cl_2 \xrightarrow{+1.36} Cl^-$$
$$[ClO_3]^- \xrightarrow{\hspace{4cm}+1.47\hspace{4cm}} Cl_2$$

$$[BrO_4]^- \xrightarrow{+1.76} [BrO_3]^- \xrightarrow{+1.46} HOBr \xrightarrow{+1.58} Br_2 \xrightarrow{+1.09} Br^-$$
$$[BrO_3]^- \xrightarrow{\hspace{4cm}+1.48\hspace{4cm}} Br_2$$

$$H_5IO_6 \xrightarrow{+1.6} [IO_3]^- \xrightarrow{+1.14} HOI \xrightarrow{+1.44} I_2 \xrightarrow{+0.54} I^-$$
$$[IO_3]^- \xrightarrow{\hspace{4cm}+1.20\hspace{4cm}} I_2$$

Fig. 17.14 Potential diagrams for chlorine, bromine and iodine at pH = 0. A Frost–Ebsworth diagram for chlorine (pH = 0) is given in Fig. 17.15 in end-of-chaper problem 17.21.

several of the oxoacids are weak, the effects of $[H^+]$ on values of some of the reduction potentials are quite complicated. For example, the disproportionation of hypochlorite to chlorate and chloride could be written as equilibrium 17.89 without involving protons.

$$3[OCl]^- \rightleftharpoons [ClO_3]^- + 2Cl^- \tag{17.89}$$

However, the fact that HOCl is a weak acid, while $HClO_3$ and HCl are strong ones (see Table 7.3) means that, in the presence of hydrogen ions, $[OCl]^-$ is protonated and this affects the position of equilibrium 17.89: HOCl is more stable with respect to disproportionation than $[OCl]^-$. On the other hand, the disproportionation of chlorate into perchlorate and chloride is realistically represented by equilibrium 17.90. From the data in Fig. 17.14, this reaction is easily shown to be thermodynamically favourable (see end-of-chapter problem 17.22b). Nevertheless, the reaction does not occur in aqueous solution owing to kinetic factors.

$$4[ClO_3]^- \rightleftharpoons 3[ClO_4]^- + Cl^- \tag{17.90}$$

Another example of the limitations of the data in Fig. 17.14 is the inference that O_2 should oxidize I^- and Br^- at pH 0. Further, the fact that Cl_2 rather than O_2 is evolved when hydrochloric acid is electrolysed is a consequence of the high overpotential for O_2 evolution at most surfaces (see worked example 17.3). Despite some limitations, Fig. 17.14 does provide some useful information: for example, the more powerful oxidizing properties of periodate and perbromate than of perchlorate when these species are being reduced to halate ions, and the more weakly oxidizing powers of iodate and iodine than of the other halates or halogens respectively.

The fact that Fig. 17.14 refers only to *specific conditions* is illustrated by considering the stability of I(I). Hypoiodous acid is unstable with respect to disproportionation into $[IO_3]^-$ and I_2, and is therefore not formed when $[IO_3]^-$ acts as an oxidant in aqueous solution. However, in hydrochloric acid, HOI undergoes reaction 17.91.

$$HOI + 2HCl \rightarrow [ICl_2]^- + H^+ + H_2O \tag{17.91}$$

Under these conditions, the potential diagram becomes:

$$[IO_3]^- \xrightarrow{+1.23} [ICl_2]^- \xrightarrow{+1.06} I_2$$

and I(I) is now stable with respect to disproportionation.

Worked example 17.3   The effects of overpotentials

Explain why, when aqueous HCl is electrolysed, the anode discharges Cl_2 (or a mixture of Cl_2 and O_2) rather than O_2 even though standard electrode potentials (at pH 0, see Appendix 11) indicate that H_2O is more readily oxidized than Cl_2.

For the anode reaction (i.e. the reverse of that shown in Appendix 11), the relevant half-reactions are:

$$2Cl^-(aq) \rightarrow Cl_2(g) + 2e^- \qquad E^o = -1.36\,V$$

$$2H_2O(l) \rightarrow O_2(g) + 4H^+(aq) + 4e^- \qquad E^o = -1.23\,V$$

The second half-reaction originates from the electrolysis of water:

$$2H_2O(l) \longrightarrow 2H_2(g) + O_2(g)$$

The spontaneous process is actually the *reverse* reaction (i.e. formation of H_2O from H_2 and O_2) and for this at pH 7, $E_{cell} = 1.23\,V$ (see the self-study exercises below). In order to drive the electrolysis of H_2O, the electrical power source must be able to supply a minimum of 1.23 V. In practice, however, this potential is insufficient to cause the electrolysis of H_2O and an additional potential (the *overpotential*) is needed. The size of the overpotential depends on several factors, one being the nature of the electrode surface. For Pt electrodes, the overpotential for the electrolysis of H_2O is $\approx 0.60\,V$. Thus, in practice, Cl_2 (or a mixture of Cl_2 and O_2) is discharged from the anode during the electrolysis of aqueous HCl.

Self-study exercises

1. For the following process, $E^o = 0\,V$. Calculate E at pH 7.

$$2H^+(aq) + 2e^- \rightleftharpoons H_2(g) \qquad [\textit{Ans.} -0.41\,V]$$

2. For the process below, $E^o = +1.23\,V$. Determine E at pH 7.

$$O_2(g) + 4H^+(aq) + 4e^- \rightleftharpoons 2H_2O(l) \qquad [\textit{Ans.} +0.82\,V]$$

3. Using your answers to the first two exercises, calculate E_{cell} at pH 7 for the overall reaction:

$$2H_2(g) + O_2(g) \longrightarrow 2H_2O(l) \qquad [\textit{Ans.} 1.23\,V]$$

KEY TERMS

The following terms were introduced in this chapter. Do you know what they mean?

- ❏ ion-selective electrode
- ❏ ozonized oxygen
- ❏ charge transfer complex
- ❏ charge transfer band
- ❏ polyhalide ion

FURTHER READING

R.E. Banks, ed. (2000) *Fluorine Chemistry at the Millennium*, Elsevier Science, Amsterdam – Covers many aspects of fluorine chemistry including metal fluorides, noble gas fluorides, biological topics and nuclear fuels.

N.N. Greenwood and A. Earnshaw (1997) *Chemistry of the Elements*, 2nd edn, Butterworth-Heinemann, Oxford – Chapter 17 covers the halogens in detail.

A.G. Massey (2000) *Main Group Chemistry*, 2nd edn, Wiley, Chichester – Chapter 11 covers the chemistry of the group 17 elements.

A.G. Sharpe (1990) *J. Chem. Educ.*, vol. 67, p. 309 – A review of the solvation of halide ions and its chemical significance.

A.F. Wells (1984) *Structural Inorganic Chemistry*, 5th edn, Clarendon Press, Oxford – Chapter 9 gives a detailed account of inorganic halide structures.

A.A. Woolf (1981) *Adv. Inorg. Chem. Radiochem.*, vol. 24, p. 1 – A review of the thermochemistry of fluorine compounds.

Special topics

E.H. Appelman (1973) *Acc. Chem. Res.*, vol. 6, p. 113 – 'Nonexistent compounds: two case histories'; deals with the histories of the perbromates and hypofluorous acid.

A.J. Blake, F.A. Devillanova, R.O. Gould, W.S. Li, V. Lippolis, S. Parsons, C. Radek and M. Schröder (1998) *Chem. Soc. Rev.*, vol. 27, p. 195 – 'Template self-assembly of polyiodide networks'.

B.J. Finlayson-Pitts (2010) *Anal. Chem.*, vol. 82, p. 770 – 'Halogens in the troposphere'.

M. Hargittai (2000) *Chem. Rev.*, vol. 100, p. 2233 – 'Molecular structure of metal halides'.

I.V. Nikitin (2008) *Russ. Chem. Rev.*, vol. 77, p. 739 – 'Halogen monoxides'.

K. Seppelt (1997) *Acc. Chem. Res.*, vol. 30, p. 111 – 'Bromine oxides'.

P.H. Svensson and L. Kloo (2003) *Chem. Rev.*, vol. 103, p. 1649 –'Synthesis, structure and bonding in polyiodide and metal iodide–iodine systems'.

PROBLEMS

17.1 (a) What is the collective name for the group 17 elements? (b) Write down, in order, the names and symbols of these elements; check your answer by reference to the first two pages of this chapter. (c) Give a *general* notation showing the ground state electronic configuration of each element.

17.2 (a) Write equations to show the reactions involved in the extraction of Br_2 and I_2 from brines.
(b) What reactions occur in the Downs process, and why must the products of the process be kept apart?
(c) In the electrolysis cell used for the industrial preparation of F_2, a diaphragm is used to separate the products. Give an equation for the reaction that would occur in the absence of the diaphragm and describe the nature of the reaction.

17.3 For a given atom Y, the Y−F bond is usually stronger than the corresponding Y−Cl bond. An exception is when Y is oxygen (Table 16.2). Suggest a reason for this observation.

17.4 Give explanations for the following observations.
(a) pK_a values for CF_3CO_2H and CH_3CO_2H are 0.23 and 4.75, respectively.
(b) The dipole moment of a gas phase NH_3 molecule is 1.47 D, but that of NF_3 is 0.24 D.
(c) In electrophilic substitution reactions in monosubstituted aryl compounds C_6H_4X, X = Me is activating and *ortho-* and *para*-directing, whereas $X = CF_3$ is deactivating and *meta*-directing.

17.5 Briefly discuss the trends in boiling points and values of $\Delta_{vap}H^{\circ}$ listed in Table 17.2 for the hydrogen halides.

17.6 Use values of r_{cov} (Table 17.1) to estimate the X−Y bond lengths of ClF, BrF, BrCl, ICl and IBr. Compare the answers with values in Fig. 17.8 and Table 17.3, and comment on the validity of the method of calculation.

17.7 Suggest products for the following reactions (which are not balanced):
(a) $AgCl + ClF_3 \longrightarrow$
(b) $ClF + BF_3 \longrightarrow$
(c) $CsF + IF_5 \longrightarrow$
(d) $SbF_5 + ClF_5 \longrightarrow$
(e) $Me_4NF + IF_7 \longrightarrow$
(f) $K[BrF_4] \overset{\Delta}{\longrightarrow}$

17.8 Discuss the role of halide acceptors in the formation of interhalogen cations and anions.

17.9 Predict the structures of (a) $[ICl_4]^-$, (b) $[BrF_2]^+$, (c) $[ClF_4]^+$, (d) IF_7, (e) I_2Cl_6, (f) $[IF_6]^+$, (g) BrF_5.

17.10 (a) Assuming *static* structures and no observed coupling to the central atom, what would you expect to see in the ^{19}F NMR spectra of BrF_5 and $[IF_6]^+$? (b) Do you expect these spectra to be temperature-dependent?

17.11 Discuss the interpretation of each of the following observations:
(a) Al_2Cl_6 and I_2Cl_6 are not isostructural.
(b) Thermal decomposition of $[Bu_4N][ClHI]$ yields $[Me_4N]I$ and HCl.
(c) 0.01 M solutions of I_2 in *n*-hexane, benzene, ethanol and pyridine are violet, purple, brown and yellow respectively. When 0.001 mol of pyridine is added to $100\,cm^3$ of each of the solutions of I_2 in *n*-hexane, benzene and ethanol, all become yellow.

17.12 The UV–VIS spectrum of a solution of I_2 and hexamethylbenzene in hexane exhibits absorptions at 368 and 515 nm. Assign these absorptions to electronic transitions, and explain how each transition arises.

17.13 The electronic spectra of mixtures of CH_2Cl_2 solutions (each $0.993\,mmol\,dm^{-3}$) of I_2 and the donor D shown in the diagram on the next page were recorded for different volume ratios of the two solutions. Values of the absorbance for the absorption at $\lambda_{max} = 308\,nm$ are as follows:

Volume ratio $I_2 : D$	Absorbance
0 : 10	0.000
1 : 9	0.056
2 : 8	0.097
3 : 7	0.129
4 : 6	0.150
5 : 5	0.164
6 : 4	0.142
7 : 3	0.130
8 : 2	0.103
9 : 1	0.070
10 : 0	0.000

[Data: A.J. Blake *et al.* (1997) *J. Chem. Soc., Dalton Trans.*, p. 1337.]

Donor, D

(a) Suggest how compound D might interact with I_2.

(b) Use the data in the table to establish the stoichiometry of the complex formed between D and I_2. Why can the absorbance data be used for this purpose?

(c) In the Raman spectrum of the complex, a band at $162 \, cm^{-1}$ is assigned to the I_2 stretching mode. Explain why this value is shifted from that of $215 \, cm^{-1}$ for I_2 itself.

17.14 Suggest likely structures for (a) $[F_2ClO_2]^-$, (b) $FBrO_3$, (c) $[ClO_2]^+$, (d) $[F_4ClO]^-$.

17.15 (a) Give equations to show the effect of temperature on the reaction between Cl_2 and aqueous NaOH.

(b) In neutral solution 1 mol $[IO_4]^-$ reacts with excess I^- to produce 1 mol I_2. On acidification of the resulting solution, a further 3 mol I_2 is liberated. Derive equations for the reactions which occur under these conditions.

(c) In strongly alkaline solution containing an excess of barium ions, a solution containing $0.01587 \, g$ of I^- was treated with $0.1 \, M$ $[MnO_4]^-$ until a pink colour persisted in the solution; $10.0 \, cm^3$ was required. Under these conditions, $[MnO_4]^-$ was converted into the sparingly soluble $BaMnO_4$. What is the product of the oxidation of iodide?

17.16 (a) Give descriptions of the bonding in ClO_2 and $[ClO_2]^-$ (**17.24** and **17.35**), and rationalize the differences in Cl–O bond lengths. (b) Rationalize why $KClO_4$ and $BaSO_4$ are isomorphous.

17.17 Suggest products for the following (which are not balanced):

(a) $[ClO_3]^- + Fe^{2+} + H^+ \longrightarrow$

(b) $[IO_3]^- + [SO_3]^{2-} \longrightarrow$

(c) $[IO_3]^- + Br^- + H^+ \longrightarrow$

17.18 Describe in outline how you would attempt:

(a) to determine the equilibrium constant and standard enthalpy change for the aqueous solution reaction:

$$Cl_2 + H_2O \rightleftharpoons HCl + HOCl$$

(b) to show that the oxide I_4O_9 (reported to be formed by reaction between I_2 and O_3) reacts with water according to the reaction:

$$I_4O_9 + 9H_2O \longrightarrow 18HIO_3 + I_2$$

(c) to show that when alkali metal atoms and Cl_2 interact in a solidified noble gas matrix at very low temperatures, the ion $[Cl_2]^-$ is formed.

17.19 Discuss the interpretation of each of the following observations:

(a) Although the hydrogen bonding in HF is stronger than that in H_2O, water has much the higher boiling point.

(b) Silver chloride and silver iodide are soluble in saturated aqueous KI, but insoluble in saturated aqueous KCl.

17.20 Explain why:

(a) $[NH_4]F$ has the wurtzite structure, unlike other ammonium halides which possess the CsCl or NaCl lattice depending on temperature;

(b) $[PH_4]I$ is the most stable of the $[PH_4]^+X^-$ halides with respect to decomposition to PH_3 and HX.

17.21 Figure 17.15 shows a Frost–Ebsworth diagram for chlorine.

(a) How is this diagram related to the potential diagram for chlorine in Fig. 17.14?

(b) Which is the most thermodynamically favoured species in Fig. 17.15? Explain how you reach your conclusion.

(c) State, with reasons, which species in the figure is the best oxidizing agent.

(d) Why is it important to state the pH value in the caption to Fig. 17.15?

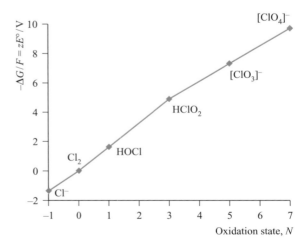

Fig. 17.15 A Frost–Ebsworth diagram for chlorine at pH $= 0$.

OVERVIEW PROBLEMS

17.22 (a) The reaction of CsF, I_2O_5 and IF_5 at 435 K leads to Cs_2IOF_5. When the amount of CsF is halved, the product is $CsIOF_4$. Write balanced equations for the reactions. Are they redox reactions?

(b) Using data in Fig. 17.14, calculate ΔG^o(298 K) for the reaction:

$$4[ClO_3]^-(aq) \rightleftharpoons 3[ClO_4]^-(aq) + Cl^-(aq)$$

Comment on the fact that the reaction does not occur at 298 K.

(c) Chlorine dioxide is the major bleaching agent in the pulp industry. While some statistics for bleaching agents list ClO_2, others give $NaClO_3$ instead. Suggest reasons for this difference.

17.23 (a) BrO has been detected in the emission gases from volcanoes (N. Bobrowski *et al.* (2003) *Nature*, vol. 423, p. 273). Construct an MO diagram for the formation of BrO from Br and O atoms. Comment on any properties and bonding features of BrO that you can deduce from the diagram.

(b) $[Cl_2O_2]^+$ is approximately planar and is described as a charge transfer complex of $[Cl_2]^+$ and O_2. By considering the HOMOs and LUMOs of $[Cl_2]^+$ and O_2, suggest what orbital interactions are involved in the charge transfer.

17.24 (a) Comment on the fact that HOI disproportionates in aqueous solution at pH 0, but in aqueous HCl at pH 0, iodine(I) is stable with respect to disproportionation.

(b) The solid state structure of $[ClF_4][SbF_6]$ reveals the presence of ions, but asymmetrical Cl–F–Sb bridges result in infinite zigzag chains running through the lattice. The Cl atoms are in pseudo-octahedral environments. Draw the structures of the separate ions present in $[ClF_4][SbF_6]$, and use the structural description to illustrate part of one of the infinite chains.

17.25 Which description in the second list below can be correctly matched to each element or compound in the first list? There is only one match for each pair.

List 1	List 2
$HClO_4$	Weak acid in aqueous solution
CaF_2	Charge transfer complex
I_2O_5	Solid contains octahedrally sited chloride ion
ClO_2	Strong acid in aqueous solution
$[BrF_6]^+$	Contains a halogen atom in a square planar coordination environment
$[IF_6]^-$	Its formation requires the use of an extremely powerful oxidative fluorinating agent
HOCl	Anhydride of HIO_3
$C_6H_6 \cdot Br_2$	Adopts a prototype structure
ClF_3	Possesses a distorted octahedral structure
RbCl	Used in the nuclear fuel industry to fluorinate uranium
I_2Cl_6	Radical

17.26 (a) How many degrees of vibrational freedom does each of ClF_3 and BF_3 possess? The IR spectrum of ClF_3 in an argon matrix exhibits six absorptions, whereas that of BF_3 has only three. Explain why the spectra differ in this way.

(b) Which of the following compounds are potentially explosive and must be treated with caution: ClO_2, $KClO_4$, KCl, Cl_2O_6, Cl_2O, Br_2O_3, HF, CaF_2, ClF_3 and BrF_3? State particular conditions under which explosions may occur. Are other serious hazards associated with any of the compounds in the list?

17.27 (a) Figure 17.7 showed the structure of $[(AgI_2)_n]^{n+}$ and an MO scheme for the bonding. The bonding may also be represented using the valence bond approach. The diagram below illustrates the positive charge localized on Ag$^+$ centres. Use this as a starting point to draw a set of resonance structures which illustrate I$_2$ acting as a charge donor. How does this compare with a VB scheme for the bonding in [I$_5$]$^+$?

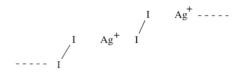

(b) When I$_2$ reacts with SbF$_5$ in liquid SO$_2$, the compound [I$_4$][Sb$_3$F$_{14}$][SbF$_6$] is formed. Explain what happens in this reaction, and draw the structures of the ions present in the product. Assign oxidation states to each atom in the product.

INORGANIC CHEMISTRY MATTERS

17.28 Both HBr and BrO (see problem 17.23) are present in volcanic plumes. A model for reactions in the plume involves the following sequence:

$$BrO + HO_2 \longrightarrow HOBr + O_2 \qquad (i)$$
$$HOBr + HBr \longrightarrow Br_2 + H_2O \qquad (ii)$$
$$Br_2 \overset{h\nu}{\longrightarrow} 2Br \qquad (iii)$$
$$Br + O_3 \longrightarrow BrO + O_2 \qquad (iv)$$

(a) Briefly discuss the types of reactions shown above.

(b) Discuss the roles of ClO and Cl in ozone depletion over Antarctica. Comment on any similarities between these reactions and those occurring in the volcanic plume.

17.29 Fluoride ions are added to drinking water, toothpaste and drugs used to treat osteoporosis. However, an excess of F$^-$ is toxic to the body. The design of water-soluble, selective fluoride ion receptors is therefore topical. One group of receptors that has been investigated consists of the boron-containing species shown below (Mes = mesityl = 2,4,6-Me$_3$C$_6$H$_2$):

(a) Explain in detail how compound **A** binds F$^-$.

(b) Suggest why the stability constant of the complex formed between **B** and F$^-$ is greater than that formed between **A** and F$^-$.

(c) Compounds **C** and **D** form zwitter-ionic complexes with F$^-$. Explain what this means.

(d) The ^{1}H NMR spectrum of the complex formed between **C** and F$^-$ shows two signals for the CH$_2$ protons: δ 3.82 (d, J_{HH} = 12.9 Hz) and 6.50 (dd, J_{HH} = 12.9 Hz, J_{HF} = 9.2 Hz). Draw the structure for a complex that is consistent with this observation, and also explains the high stability of the complex.

(e) Which of the receptors would you expect to form water-soluble complexes with F$^-$?

17.30 Comment on the following statements, giving equations to illustrate relevant reactions:

(a) Iodine supplements for the human body comprise I$_2$/KI.

(b) HF is used commercially to etch silica glass.

(c) Mixtures of solid ammonium perchlorate and powdered aluminium are used as rocket propellants.

BMes$_3$

BMes$_2$

BPh$_2$

A   **B**

BMes$_2$ BMes$_2$

$^+$NMe$_3$ $^+$NMe$_3$

C   **D**

Topics

Occurrence and extraction
Applications
Physical properties
Compounds of xenon
Compounds of argon, krypton and radon

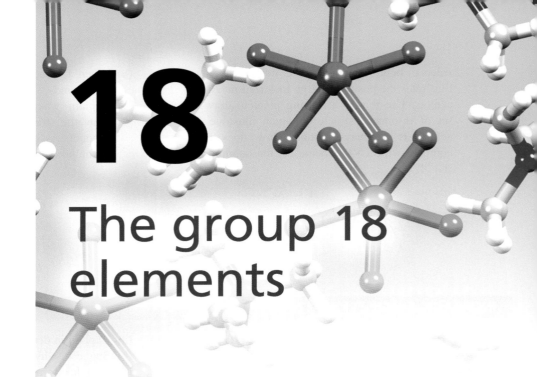

18

The group 18 elements

1	2		13	14	15	16	17	**18**
H								He
Li	Be		B	C	N	O	F	Ne
Na	Mg		Al	Si	P	S	Cl	Ar
K	Ca	*d*-block	Ga	Ge	As	Se	Br	Kr
Rb	Sr		In	Sn	Sb	Te	I	Xe
Cs	Ba		Tl	Pb	Bi	Po	At	Rn
Fr	Ra							

18.1 Introduction

> The group 18 elements (helium, neon, argon, krypton, xenon and radon) are called the **noble gases**.

This section gives a brief, partly historical, introduction to the group 18 elements, the ground state electronic configurations of which tend to suggest chemical inertness. Until 1962, the chemistry of the noble gases was restricted to a few very unstable species such as $[HHe]^+$, $[He_2]^+$, $[ArH]^+$, $[Ar_2]^+$ and $[HeLi]^+$ formed by the combination of an ion and an atom under highly energetic conditions, and detected spectroscopically. Molecular orbital theory provides a simple explanation of why diatomic species such as He_2 and Ne_2 are not known. In He_2, bonding and antibonding MOs are fully occupied (Section 2.3). However, in a monocation such as $[Ne_2]^+$, the highest

energy MO is *singly* occupied, meaning that there is a *net bonding* interaction. Thus, the bond energies in $[He_2]^+$, $[Ne_2]^+$ and $[Ar_2]^+$ are 126, 67 and 104 $kJ\,mol^{-1}$, respectively, but no stable compounds containing these cations have been isolated. Although $[Xe_2]^+$ has been known for some years and characterized by Raman spectroscopy ($\bar{\nu}(XeXe) = 123\,cm^{-1}$), it was only in 1997 that $[Xe_2][Sb_4F_{21}]$ (prepared from $[XeF][Sb_2F_{11}]$ and HF/SbF_5, see Section 9.9) was crystallographically characterized. Discrete $[Xe_2]^+$ ions (**18.1**) are present in the solid state of $[Xe_2][Sb_4F_{21}]$, although there are weak Xe····F interactions. The Xe−Xe bond is extremely long, the longest structurally confirmed homonuclear bond between main group elements. Under a 30–50 bar pressure of Xe and in the presence of excess SbF_5, the $[Sb_4F_{11}]^-$ salt of green $[Xe_2]^+$ transforms into a salt of blue $[Xe_4]^+$. Reducing the pressure of Xe gas reverses the reaction. The $[Xe_4]^+$ ion has been characterized by Raman spectroscopy and the observed absorption at $110\,cm^{-1}$ agrees with that calculated for the stretch of the central Xe−Xe bond in a linear cation with structure **18.2**.

$$\left[Xe \underset{309\text{ pm}}{\overline{\hspace{1.5cm}}} Xe \right]^+ \qquad \left[Xe \overset{353\text{ pm}}{\overline{\hspace{1cm}}} Xe \underset{319\text{ pm}}{\overline{\hspace{1cm}}} Xe \overset{353\text{ pm}}{\overline{\hspace{1cm}}} Xe \right]^+$$

(18.1)   **(18.2)**

When H_2O is frozen in the presence of Ar, Kr or Xe at high pressures, clathrates (see Box 14.5 and Section 12.8) of limiting composition $Ar·6H_2O$, $Kr·6H_2O$ and $Xe·6H_2O$ are obtained. The noble gas atoms are guests within hydrogen-bonded host lattices. Other noble gas-containing clathrates include $3.5Xe·8CCl_4·136D_2O$

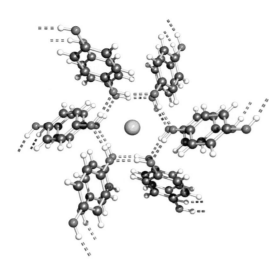

Fig. 18.1 Part of the solid state structure of tris(β-hydroquinone) xenon clathrate showing the arrangement of hydrogen-bonded organic molecules around a xenon atom [T. Birchall *et al.* (1989) *Acta Crystallogr., Sect. C*, vol. 45, p. 944]. Colour code: Xe, yellow; C, grey; O, red; H, white.

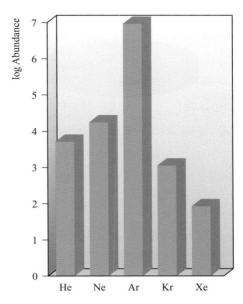

Fig. 18.2 Relative abundances of the noble gases (excluding radon, the abundance of which is 1×10^{-12} ppb) in the Earth's atmosphere. The data are plotted on a logarithmic scale. The units of abundance are parts per billion by volume (1 billion $= 10^9$).

and $0.866Xe \cdot 3[1,4\text{-}(HO)_2C_6H_4]$ (Fig. 18.1). This type of system is well established, but note that no *chemical* change occurs to the noble gas atoms upon formation of the clathrate.

The first indication that Xe was not chemically inert came in 1962 from work of Neil Bartlett when the reaction between Xe and PtF_6 gave a compound formulated as '$XePtF_6$' (see Section 6.16). A range of species containing Xe chemically bonded to other elements (most commonly F or O) is now known. Compounds of Kr are limited to KrF_2 and its derivatives. In principle, there should be many more compounds of Rn. However, the longest lived isotope, ^{222}Rn, has a half-life of 3.8 d and is an intense α-emitter (which leads to decomposition of its compounds), and, in practice, information about the chemistry of Rn is very limited.

18.2 Occurrence, extraction and uses

Occurrence

After hydrogen, He is the second most abundant element in the universe. It occurs to an extent of $\leq 7\%$ by volume in natural gas from sources in the US and Canada, and this origin is doubtless from the radioactive decay of heavier elements (α-particle $= {}^4_2He$). Helium is also found in various minerals containing α-emitting unstable isotopes. Helium was first detected spectroscopically in the Sun's atmosphere. Nuclear fusion reactions taking place in the Sun start at temperatures above 10^7 K, and the following reactions are

believed to be the main source of the Sun's energy ($β^+ =$ positron, $ν_e =$ neutrino):

$$^1_1H + {}^1_1H \longrightarrow {}^2_1H + β^+ + ν_e$$

$$^1_1H + {}^2_1H \longrightarrow {}^3_2He + γ$$

$$^3_2He + {}^3_2He \longrightarrow {}^4_2He + 2{}^1_1H$$

Figure 18.2 shows the relative abundances of the noble gases in the Earth's atmosphere. Argon is present to an extent of 0.92% by volume in the Earth's atmosphere (Fig. 15.1b). Radon is formed by decay of ^{226}Ra in the ^{238}U decay chain (see Fig. 27.3), and poses a serious health hazard in uranium mines, being linked to cases of lung cancer.[†]

Extraction

In terms of commercial production, He and Ar are the two most important noble gases. Helium is extracted from natural gas by liquefaction of other gases present (He has the lowest boiling point of all the elements), leaving gaseous He which is removed by pumping. Neon is extracted as a by-product when air is liquefied, being left behind as the only gas. Argon has almost the same boiling

---

[†] Development of lung cancer apparently associated with radon emissions is a more general cause for concern: P. Phillips, T. Denman and S. Barker (1997) *Chem. Brit.*, vol. 33, number 1, p. 35 – 'Silent, but deadly'; J. Woodhouse (2002) in *Molecules of Death*, eds R.H. Warding, G.B. Steventon and S.C. Mitchell, Imperial College Press, London, p. 197 – 'Radon'.

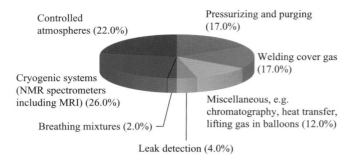

Controlled atmospheres (22.0%)

Pressurizing and purging (17.0%)

Welding cover gas (17.0%)

Cryogenic systems (NMR spectrometers including MRI) (26.0%)

Miscellaneous, e.g. chromatography, heat transfer, lifting gas in balloons (12.0%)

Breathing mixtures (2.0%)

Leak detection (4.0%)

Fig. 18.3 Uses of helium in the US in 2009. The total consumption of 'grade A' helium in the US in 2009 was $47 \times 10^6 \, m^3$. [Data: US Geological Survey.]

point as O_2 (Ar, 87 K; O_2, 90 K) and the two gases remain together during the fractionation of liquid air. The O_2/Ar mixture can be partially separated by further fractionation. The crude Ar is mixed with H_2 and sparked to remove O_2 as H_2O, excess H_2 being removed by passage over hot CuO. Krypton and xenon are usually separated from O_2 by selective absorption on charcoal.

Uses

Figure 18.3 summarizes the main uses of helium. Both helium and argon are used to provide inert atmospheres, for example for arc-welding (see Box 18.1) and during the

growth of single Si or Ge crystals for the semiconductor industry (see Box 6.3). Argon is also used in laboratory inert atmosphere ('dry' or 'glove') boxes for handling air-sensitive compounds. Being very light and non-inflammable, He is used to inflate the tyres of large aircraft, and in balloons including weather balloons and NASA's unmanned suborbital research balloons. Liquid He is an important coolant and is used in highfield NMR spectrometers including those used in medical imaging (see Box 4.3). The superconductivity of metals cooled to the temperature of liquid He suggests that the latter may become important in power transmission. An O_2/He mixture is used in place of O_2/N_2 for deep-sea divers. Helium is much less soluble in blood than N_2, and does not cause 'the bends' when the pressure is released on surfacing. Helium is also used as a heat-transfer agent in gas-cooled nuclear reactors, for which it has the advantages of being non-corrosive and of not becoming radioactive under irradiation. Neon, krypton and xenon are used in electric discharge signs (e.g. for advertising) and Ar is contained in metal filament bulbs to reduce evaporation from the filament.

18.3 Physical properties

Some physical properties of the group 18 elements are listed in Table 18.1. Of particular significance is the fact that the

APPLICATIONS

Box 18.1 Protective inert gases for metal arc-welding

The high-temperature conditions under which metal arc-welding takes place would, in the absence of protective gases, lead to reaction between molten metal and atmospheric gases including O_2 and N_2. Noble gases such as He and Ar are an obvious choice for the protective blanket, but these may be mixed with an active ingredient such as CO_2 (or H_2) to provide an oxidizing (or reducing) component to the protective layer. Of He and Ar, the latter is of greater industrial importance and is used in welding CrNi alloy steels and a range of metals. Argon is denser than He (1.78 versus $0.18 \, kg \, m^{-3}$ at 273 K) and so gives better protection. High-purity Ar (>99.99%) is commercially available and such levels of purity are essential when dealing with metals such as Ti, Ta and Nb which are extremely prone to attack by O_2 or N_2 during arc-welding.

An engineer using a metal inert gas welding torch in the manufacture of an exhaust for a racing car.

APPLICATIONS

Box 18.2 Xenon in 21st century space propulsion systems

In October 1998, at the start of its New Millennium Program, NASA launched a new space probe called *Deep Space 1* (DS1), designed to test new technologies with potential applications in future solar exploration. One of the revolutionary technologies on this flight was a xenon-based ion propulsion system, ten times more efficient than any other used prior to the DS1 mission. The system operates by using a solar power source, and ionizes Xe gas contained in a chamber, at one end of which is a pair of metal grids charged at 1280 V. A xenon-ion beam is produced as ions are ejected through the grids at $\approx 145\,000\ \mathrm{km\,h^{-1}}$, and the resultant thrust is used to propel DS1 through space. Since the fuel is Xe gas (and only 81 kg is required for an approximately two-year mission), an advantage of the system, in addition to the efficient thrust, is that DS1 is smaller and lighter than previous unmanned spacecraft.

Deep Space 1 was taken out of service in December 2001. During its three years in space, it trialled a number of new technologies and proved the future potential of the xenon propulsion system. After *DS1*, the next spacecraft to be propelled by xenon-ion thrusters was *SMART-1* (SMART = Small Missions for Advanced Research in Technology). This was launched by the European Space Agency in 2003 for a lunar mission, and maintained an orbit around the moon from 2004 to 2006. The Japanese Institute of Space and Astronautical Science (ISAS) has also developed a xenon ion propulsion system for application in its MUSES-C asteroid sample return mission. NASA is developing a new generation of ion propulsion systems as part of its Evolutionary Xenon Thruster (NEXT) programme. These thrusters are for use in outer planet missions (e.g. Neptune orbiter, Titan explorer, Saturn ring observer) and inner solar system missions. NASA's Dawn spacecraft was launched in 2007 on a mission to asteroid Vesta (arrival 2011) and Dwarf planet Ceres (scheduled arrival 2015), the two largest bodies in the asteroid belt between Mars and Jupiter. The mission will acquire data about the early history of the solar system and is scheduled to last for eight years.

Computer illustration of SMART-1 lunar spacecraft. Two large solar panels collect the energy from the Sun needed to ionize the onboard xenon fuel.

Further reading

K. Nishiyama and H. Kunuaka (2006) *Thin Solid Films*, vol. 506–507, p. 588 – '20-cm ECR plasma generator for xenon ion propulsion'.

nmp.nasa.gov/ds1
dawn.jpl.nasa.gov/
www.isas.jaxa.jp
www.esa.int/esaMI/SMART-1

---

noble gases have the highest ionization energies of the elements in their respective periods (Fig. 1.16), but there is a decrease in values on descending the group (Fig. 6.26). The extremely low values of $\Delta_{\mathrm{fus}}H^{\circ}$ and $\Delta_{\mathrm{vap}}H^{\circ}$ correspond to the weak van der Waals interactions between the atoms, and the increase in values of $\Delta_{\mathrm{vap}}H^{\circ}$ down the group is due to increased interatomic interactions as atomic size and polarizability increase.

The properties of He deserve special note. It can diffuse through rubber and most glasses. Below 2.18 K, ordinary liquid ^{4}He (but not ^{3}He) is transformed into liquid He(II) which has the remarkable properties of a thermal conductivity 600 times that of copper, and a viscosity approaching

zero. Liquid He(II) forms films only a few hundred atoms thick which flow up and over the side of the containing vessel.

NMR active nuclei

In the NMR spectroscopic characterization of Xe-containing compounds, use is made of ^{129}Xe, with a natural abundance of 26.4% and $I = \frac{1}{2}$. Although direct observation of ^{129}Xe is possible, the observation of satellite peaks in, for example, ^{19}F NMR spectra of xenon fluorides is a valuable diagnostic tool as is illustrated for $[\mathrm{XeF_5}]^{-}$ in Case study 5, Section 4.8. For a potential clinical application of ^{129}Xe, see Box 4.3.

Table 18.1 Some physical properties of the group 18 elements (noble gases).

Property	He	Ne	Ar	Kr	Xe	Rn
Atomic number, Z	2	10	18	36	54	86
Ground state electronic configuration	$1s^2$	$[He]2s^22p^6$	$[Ne]3s^23p^6$	$[Ar]3d^{10}4s^24p^6$	$[Kr]4d^{10}5s^25p^6$	$[Xe]4f^{14}5d^{10}6s^26p^6$
Melting point, mp / K	$-^{\dagger}$	24.5	84	116	161	202
Boiling point, bp / K	4.2	27	87	120	165	211
Standard enthalpy of fusion, $\Delta_{fus}H^{\circ}(mp)/kJ\,mol^{-1}$	–	0.34	1.12	1.37	1.81	–
Standard enthalpy of vaporization, $\Delta_{vap}H^{\circ}(bp)/kJ\,mol^{-1}$	0.08	1.71	6.43	9.08	12.62	18.0
First ionization energy, $IE_1/kJ\,mol^{-1}$	2372	2081	1521	1351	1170	1037
Van der Waals radius, r_v/pm	99	160	191	197	214	–

† Helium cannot be solidified under any conditions of temperature and pressure.

Worked example 18.1   NMR spectroscopy of xenon-containing compounds

Reaction of XeF_4 and $C_6F_5BF_2$ at 218 K yields $[C_6F_5XeF_2][BF_4]$. (a) Use the VSEPR model to suggest a structure for $[C_6F_5XeF_2]^+$. (b) The ^{129}Xe NMR spectrum of $[C_6F_5XeF_2][BF_4]$ consists of a triplet ($J = 3892$ Hz), and the ^{19}F NMR spectrum shows a three-line signal (relative intensities $\approx 1:5.6:1$), three multiplets and a singlet. The relative integrals of the five signals are $2:2:1:2:4$. Rationalize these data.

(a) Xe has eight valence electrons.

The positive charge can be formally localized on Xe, leaving seven valence electrons.

Each F atom provides one electron to the valence shell of Xe.

The C_6F_5 group is bonded through carbon to Xe and provides one electron to the valence shell of Xe.

Total number of electrons in the valence shell of Xe = 10.

The parent shape for $[C_6F_5XeF_2]^+$ is a trigonal bipyramid with the two lone pairs in the equatorial plane to minimize lone pair–lone pair repulsions. For steric reasons, the C_6F_5 group is expected to lie in the equatorial plane with the plane of the aryl ring orthogonal to the plane containing the XeF_2 unit. The expected structure is T-shaped:

(b) The triplet in the ^{129}Xe NMR spectrum of $[C_6F_5XeF_2][BF_4]$ shows a large coupling constant (3892 Hz) and arises from coupling between ^{129}Xe and the two equivalent, directly bonded ^{19}F nuclei.

There are four F environments in $[C_6F_5XeF_2]^+$ (*ortho*, *meta* and *para*-F atoms in the aryl group and the two equivalent F atoms bonded to Xe, with a ratio $2:2:1:2$, respectively. The signals for the aryl F atoms appear as multiplets because of ^{19}F–^{19}F coupling between non-equivalent F atoms. There are four equivalent F atoms in the $[BF_4]^-$ ion leading to a singlet; coupling to ^{11}B is not observed. Only the directly bonded ^{19}F nuclei couple to ^{129}Xe ($I = \frac{1}{2}$, 26.4%). The signal in the ^{19}F NMR spectrum assigned to these F atoms appears as a singlet with satellites for the 26.4% of the ^{19}F bonded to spin-active ^{129}Xe. The relative intensities $1:5.6:1$ correspond to 26.4% of the signal split into a doublet (see Fig. 4.23).

Self-study exercises

Nuclear spin data: see Tables 4.3 and 18.1.

1. The reaction of $CF_2=CFBF_2$ with XeF_2 gives the $[BF_4]^-$ salt of the following cation:

The solution ^{129}Xe NMR spectrum of the compound exhibits an eight-line multiplet with lines of equal intensity. Account for this observation.

[*Ans*. See H.-J. Frohn *et al.* (1999)
Chem. Commun., p. 919]

2. What would you expect to see in the ^{19}F NMR spectrum of XeF$_4$, the structure of which is consistent with the VSEPR model?

[*Ans*. Similar to Fig. 4.23 (experimental data:
δ 317 ppm, $J = 3895$ Hz)]

18.4 Compounds of xenon

Fluorides

The most stable Xe compounds are the colourless fluorides XeF$_2$, XeF$_4$ and XeF$_6$ (Table 18.2). Upon irradiation with UV light, Xe reacts with F$_2$ at ambient temperature to give XeF$_2$. The rate of formation is increased by using HF as a catalyst and pure XeF$_2$ can be prepared by this method. Xenon difluoride may also be made by action of an electrical discharge on a mixture of Xe and F$_2$, or by passing these gases through a short nickel tube at 673 K. The latter method gives a mixture of XeF$_2$ and XeF$_4$, and the yield of XeF$_4$ is optimized by using a 1:5 Xe:F$_2$ ratio. With an NiF$_2$ catalyst, the reaction proceeds at a lower temperature, and even at 393 K, XeF$_6$ can be formed under these same conditions. It is not possible to prepare XeF$_4$ free of XeF$_2$ and/or XeF$_6$. Similarly, XeF$_6$ always forms with contamination by the lower fluorides. Separation of XeF$_4$ from a mixture involves preferential complexation of XeF$_2$ and XeF$_6$ (eq. 18.1) and the XeF$_4$ is then removed *in vacuo*, while separation of XeF$_6$ involves reaction 18.2 followed by thermal decomposition of the complex.

$$\left.\begin{array}{l} \text{XeF}_2 \\ \text{XeF}_4 \\ \text{XeF}_6 \end{array}\right\} \xrightarrow{\text{excess AsF}_5 \text{ in liq. BrF}_5} \left\{\begin{array}{l} [\text{Xe}_2\text{F}_3]^+[\text{AsF}_6]^- \\ \text{XeF}_4 \\ [\text{XeF}_5]^+[\text{AsF}_6]^- \end{array}\right. \quad (18.1)$$

$$\text{XeF}_6 + 2\text{NaF} \longrightarrow \text{Na}_2[\text{XeF}_8] \quad (18.2)$$

All the fluorides sublime *in vacuo*, and all are readily decomposed by water, XeF$_2$ very slowly, and XeF$_4$ and XeF$_6$, rapidly (eqs. 18.3–18.5 and 18.14).

$$2\text{XeF}_2 + 2\text{H}_2\text{O} \longrightarrow 2\text{Xe} + 4\text{HF} + \text{O}_2 \quad (18.3)$$

$$6\text{XeF}_4 + 12\text{H}_2\text{O} \longrightarrow 2\text{XeO}_3 + 4\text{Xe} + 24\text{HF} + 3\text{O}_2 \quad (18.4)$$

$$\text{XeF}_6 + 3\text{H}_2\text{O} \longrightarrow \text{XeO}_3 + 6\text{HF} \quad (18.5)$$

All three fluorides are powerful oxidizing and fluorinating agents, the relative reactivities being XeF$_6$ > XeF$_4$ > XeF$_2$. The difluoride is available commercially and is widely used for fluorinations, e.g. eqs. 17.35, 18.6 and 18.7. At 298 K, XeF$_6$ reacts with silica (preventing the handling of XeF$_6$ in silica glass apparatus, eq. 18.8) and with H$_2$, while XeF$_2$ and XeF$_4$ do so only when heated.

$$\text{S} + 3\text{XeF}_2 \xrightarrow{\text{anhydrous HF}} \text{SF}_6 + 3\text{Xe} \quad (18.6)$$

$$2\text{Ir} + 5\text{XeF}_2 \xrightarrow{\text{anhydrous HF}} 2\text{IrF}_5 + 5\text{Xe} \quad (18.7)$$

$$2\text{XeF}_6 + \text{SiO}_2 \longrightarrow 2\text{XeOF}_4 + \text{SiF}_4 \quad (18.8)$$

Table 18.2   Selected properties of XeF$_2$, XeF$_4$ and XeF$_6$.

Property	XeF$_2$	XeF$_4$	XeF$_6$
Melting point / K	413	390	322
$\Delta_f H^\circ$(s, 298 K)/kJ mol^{-1}	-163	-267	-338
$\Delta_f H^\circ$(g, 298 K)/kJ mol^{-1}	-107	-206	-279
Calculated $\Delta_f H^\circ$(298 K)/kJ mol$^{-1\dagger}$	-100.0 ± 1.3	-182.1 ± 4.2	-244.0 ± 8.4
Mean Xe$-$F bond enthalpy term/kJ mol^{-1}	133	131	126
Xe$-$F bond distance / pm	200‡	195‡	189*
Molecular shape	Linear	Square planar	Octahedral

† The results of high-level computational studies indicate that the currently available experimental thermochemical data may be too negative. As a consequence, the Xe$-$F bond enthalpy terms may also be in error [D.A. Dixon, W.A. de Jong, K.A. Peterson, K.O. Christe and G.J. Schrobilgen (2005) *J. Am. Chem. Soc.*, vol. 127, p. 8627].
‡ Neutron diffraction.
* Gas-phase electron diffraction.

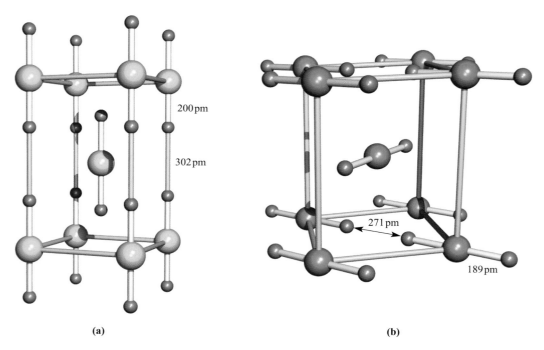

(a)   **(b)**

Fig. 18.4 Unit cells (yellow lines) of (a) XeF_2 and (b) β-KrF_2 showing the arrangements and close proximity of molecular units. Colour code: Xe, yellow; Kr, red; F, green.

The structures of the xenon halides are consistent with the VSEPR model. The XeF_2 molecule is linear, but in the solid state, there are significant intermolecular interactions (Fig. 18.4a). The XeF_4 molecule is square planar (D_{4h}). Discrete molecules are present in solid XeF_4, but there are extensive intermolecular F····Xe interactions with an average distance of 324 pm, significantly less than the sum of the van der Waals radii (349 pm). In the vapour state, the vibrational spectrum of XeF_6 indicates C_{3v} symmetry, i.e. an octahedron distorted by a stereochemically active lone pair in the centre of one face (**18.3**), but the molecule is readily converted into other configurations. High-level theoretical studies indicate that the energies of C_{3v} and O_h structures for XeF_6 are very similar, and that the lone pair is highly fluxional. Solid XeF_6 exists in a number of crystalline forms which were reinvestigated in 2006.[†] The two highest temperature modifications are formed by crystallization at 303 K over several days or by rapid sublimation, respectively. Although crystallographically distinct, the two forms are similar at the molecular level and are formulated as $\{[XeF_5]^+F^-\}_3 \cdot XeF_6$ (**18.4**). The room temperature crystalline form of XeF_6 comprises both tetramers and hexamers, each made up of square pyramidal $[XeF_5]^+$ units bridged by F^- ions: $\{[XeF_5]^+F^-\}_4$ and $\{[XeF_5]^+F^-\}_6$. The low-temperature structural modifications contain tetramers, $\{[XeF_5]^+F^-\}_4$. Crystallization of

XeF_6 from anhydrous HF produces $[XeF_5]_2[HF_2]_2 \cdot HF$ which contains dimeric units of $[XeF_5]^+$ ions bridged by $[HF_2]^-$ ions (**18.5**). Thus, in crystalline XeF_6, the most common building block is the $[XeF_5]^+$ unit.

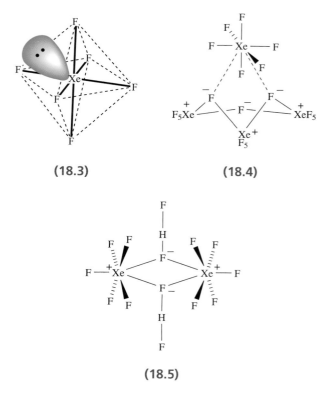

(18.3)   **(18.4)**

(18.5)

[†] S. Hoyer, T. Emmler and K. Seppelt (2006) *J. Fluorine Chem.*, vol. 127, p. 1415 – 'The structure of xenon hexafluoride in the solid state'.

The bonding in XeF_2 and XeF_4 can be described in terms of using only s and p valence orbitals. We showed in Fig. 5.30 that the net bonding in linear XeF_2 can be considered in terms of the overlap of a $5p$ orbital on the Xe atom with an out-of-phase combination of F $2p$ orbitals (a σ_u-orbital). This gives a formal bond order of $\frac{1}{2}$ per Xe−F bond. A similar bonding scheme can be developed for square planar XeF_4. The net σ-bonding orbitals are shown in diagram **18.6**. These are fully occupied, resulting in a formal bond order of $\frac{1}{2}$ per Xe−F bond.

(18.6)

If the $[XeF]^+$ ion (see below) is taken to contain a single bond, then the fact that its bond distance of 184–190 pm (depending on the salt) is noticeably shorter than those in XeF_2 and XeF_4 (Table 18.2) is consistent with a model of 3c-2e interactions in the xenon fluorides. Further support for low bond orders in XeF_2 and XeF_4 comes from the fact that the strengths of the Xe−F bonds in XeF_2, XeF_4 and XeF_6 are essentially the same (Table 18.2, but see table footnote), in contrast to the significant decrease noted (Section 17.7) along the series $ClF > ClF_3 > ClF_5$.

Xenon difluoride reacts with F^- acceptors. With pentafluorides such as SbF_5, AsF_5, BrF_5, NbF_5 and IrF_5, it forms three types of complex: $[XeF]^+[MF_6]^-$, $[Xe_2F_3]^+[MF_6]^-$ and $[XeF]^+[M_2F_{11}]^-$, although in the solid state, there is evidence for cation–anion interaction through the formation of Xe−F−M bridges. The $[Xe_2F_3]^+$ cation has structure **18.7**. A number of complexes formed between XeF_2 and metal tetrafluorides have been reported, e.g. the reactions of XeF_2 with RuF_5 or CrF_4 give $[XeF]^+[RuF_6]^-$ (**18.8**) and polymeric $[XeF]^+[CrF_5]^-$ (**18.9**), respectively. In these and related compounds, cation–anion interactions in the solid state are significant.

$\alpha = 140\text{--}160°$ depending on counter-ion

(18.7)

(18.8)   **(18.9)**

The reaction of $[XeF][AsF_6]$ with a stoichiometric amount of H_2O in HF solution produces $[H_3O][AsF_6]\cdot2XeF_2$ and $[Xe_3OF_3][AsF_6]$, but is highly sensitive to reaction conditions. The $[Xe_3OF_3]^+$ cation is the first example of a xenon(II) oxofluoride. It possesses structure **18.10** although there is an O/F atom disorder in the solid state (see Box 15.5), making it difficult to determine Xe–O/F bond lengths for the central bonds in **18.10**. The terminal Xe–F bond distances are close to those in XeF_2 (200 pm).

(18.10)

Xenon hexafluoride acts as an F^- donor to numerous pentafluorides, giving complexes of types $[XeF_5]^+[MF_6]^-$, $[XeF_5]^+[M_2F_{11}]^-$ (for M = Sb or V) and $[Xe_2F_{11}]^+[MF_6]^-$. The $[XeF_5]^+$ ion (average Xe−F = 184 pm) is isoelectronic and isostructural with IF_5 (**17.6**), but in solid state salts, there is evidence for fluoride bridge formation between cations and anions. The $[Xe_2F_{11}]^+$ cation can be considered as $[F_5Xe\cdots F\cdots XeF_5]^+$ in the same way that $[Xe_2F_3]^+$ can be written as $[FXe\cdots F\cdots XeF]^+$. The compounds $[XeF_5][AgF_4]$ and $[Xe_2F_{11}]_2[NiF_6]$ contain Ag(III) and Ni(IV) respectively, and are prepared from XeF_6, the metal(II) fluoride and KrF_2. In these cases, XeF_6 is not strong enough to oxidize Ag(II) to Ag(III) or Ni(II) to Ni(IV), and KrF_2 is employed as the oxidizing agent. The range of Xe−F bond distances in $[Xe_2F_{11}]_2[NiF_6]$ (Fig. 18.5) illustrates the $[F_5Xe\cdots F\cdots XeF_5]^+$ nature of the cation and the longer $F\cdots Xe$ contacts between anion and cations. Xenon tetrafluoride is much less reactive than XeF_2 with F^- acceptors. Among the few complexes formed is $[XeF_3]^+[Sb_2F_{11}]^-$. The $[XeF_3]^+$ cation (**18.11**) is isostructural with ClF_3 (**17.5**).

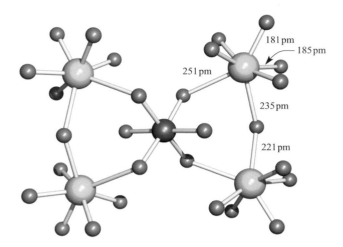

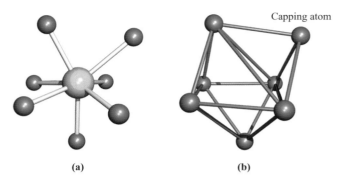

Fig. 18.6 (a) The structure of $[XeF_7]^-$, determined by X-ray diffraction for the caesium salt [A. Ellern *et al.* (1996) *Angew. Chem. Int. Ed. Engl.*, vol. 35, p. 1123]; (b) the capped octahedral arrangement of the F atoms in $[XeF_7]^-$. Colour code: Xe, yellow; F, green.

Fig. 18.5 The structure of $[Xe_2F_{11}]_2[NiF_6]$ determined by X-ray diffraction [A. Jesih *et al.* (1989) *Inorg. Chem.*, vol. 28, p. 2911]. The environment about each Xe centre is similar to that in the solid state $[XeF_6]_4$ (**18.3**). Colour code: Xe, yellow; Ni, blue; F, green.

gives $[NO_2]^+[Xe_2F_{13}]^-$, the solid state structure of which reveals that the anion can be described as an adduct of $[XeF_7]^-$ and XeF_6 (structure **18.13**).

$$\text{(18.13)}$$

(18.11)   (18.12)

Chlorides

Both XeF_4 and XeF_6 act as F^- acceptors. The ability of XeF_4 to accept F^- to give $[XeF_5]^-$ has been observed in reactions with CsF and $[Me_4N]F$. The $[XeF_5]^-$ ion (**18.12**) is one of only two pentagonal planar species known, the other being the isoelectronic $[IF_5]^{2-}$ (Section 17.7). Equation 18.9 shows the formations of $[XeF_7]^-$ and $[XeF_8]^{2-}$ (which has a square-antiprismatic structure with a stereochemically inactive lone pair). The salts $Cs_2[XeF_8]$ and $Rb_2[XeF_8]$ are the most stable compounds of Xe yet made, and decompose only when heated above 673 K.

$$MF + XeF_6 \longrightarrow M[XeF_7] \xrightarrow{\Delta,\, -XeF_6} M_2[XeF_8] \qquad (18.9)$$
$$M = Rb,\, Cs,\, NO$$

Structural information on $[XeF_7]^-$ has been difficult to obtain because of its ready conversion into $[XeF_8]^{2-}$. Recrystallization of freshly prepared $Cs[XeF_7]$ from liquid BrF_5 yields crystals suitable for X-ray diffraction studies. The $[XeF_7]^-$ ion has a capped octahedral structure (Fig. 18.6a) with Xe–F = 193 and 197 pm in the octahedron and Xe–F = 210 pm to the capping F atom. The coordination sphere defined by the seven F atoms is shown in Fig. 18.6b; the octahedral part is significantly distorted. The reaction between NO_2F and excess XeF_6

Xenon dichloride has been detected by matrix isolation. It is obtained on condensing the products of a microwave discharge in a mixture of Cl_2 and a large excess of Xe at 20 K. Fully characterized compounds containing Xe–Cl bonds are rare, and most also contain Xe–C bonds (see the end of Section 18.4). The $[XeCl]^+$ ion is formed as the $[Sb_2F_{11}]^-$ salt on treatment of $[XeF]^+[SbF_6]^-$ in anhydrous HF/SbF_5 with $SbCl_5$. In the solid state (data collected at 123 K), cation–anion interactions are observed in $[XeCl][Sb_2F_{11}]$ (**18.14**). The Xe–Cl bond length is the shortest known to date. At 298 K, $[XeCl][Sb_2F_{11}]$ decomposes according to eq. 18.10.

$$\text{(18.14)}$$

$$2[XeCl][Sb_2F_{11}] \longrightarrow Xe + Cl_2 + [XeF][Sb_2F_{11}] + 2SbF_5$$
$$(18.10)$$

Oxides

Equations 18.4 and 18.5 showed the formation of XeO_3 by hydrolysis of XeF_4 and XeF_6. Solid XeO_3 forms colourless crystals and is dangerously explosive ($\Delta_f H^o(298\,K) = +402\,kJ\,mol^{-1}$). The solid contains trigonal pyramidal molecules (**18.15**). Xenon trioxide is only weakly acidic and its aqueous solution is virtually non-conducting. Reactions of XeO_3 and MOH (M = K, Rb, Cs) produce xenates (eq. 18.11) which slowly disproportionate in solution (eq. 18.12).

$$Xe—O = 176\,pm$$
$$\angle O–Xe–O = 103^\circ$$

(18.15)

$$XeO_3 + MOH \longrightarrow M[HXeO_4] \qquad (18.11)$$

$$2[HXeO_4]^- + 2[OH]^- \longrightarrow [XeO_6]^{4-} + Xe + O_2 + 2H_2O$$
$$\text{perxenate}$$
$$(18.12)$$

Aqueous $[XeO_6]^{4-}$ is formed when O_3 is passed through a dilute solution of XeO_3 in alkali. Insoluble salts such as $Na_4XeO_6\cdot8H_2O$ and Ba_2XeO_6 may be precipitated, but perxenic acid 'H_4XeO_6' (a weak acid in aqueous solution) has not been isolated. The perxenate ion is a powerful oxidant and is rapidly reduced in aqueous acid (eq. 18.13). Oxidations such as Mn(II) to $[MnO_4]^-$ occur instantly in acidic media at 298 K.

$$[XeO_6]^{4-} + 3H^+ \longrightarrow [HXeO_4]^- + \tfrac{1}{2}O_2 + H_2O \qquad (18.13)$$

Xenon tetraoxide is prepared by the slow addition of concentrated H_2SO_4 to Na_4XeO_6 or Ba_2XeO_6. It is a pale yellow, highly explosive solid ($\Delta_f H^o(298\,K) = +642\,kJ\,mol^{-1}$) which is a very powerful oxidizing agent. Tetrahedral XeO_4 molecules (**18.16**) are present in the gas phase.

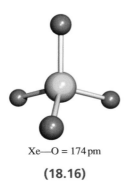

$$Xe—O = 174\,pm$$

(18.16)

Oxofluorides and oxochlorides

Oxofluorides are known for Xe(II) (structure **18.10**), Xe(IV), Xe(VI) and Xe(VIII): $XeOF_2$, $XeOF_4$, XeO_2F_2, XeO_2F_4 and XeO_3F_2. Their structures are consistent with VSEPR theory: see end-of-chapter problem 18.8. The 1:1 reaction of XeF_4 and H_2O in liquid HF yields $XeOF_2$, isolated as a pale yellow solid which decomposes explosively at 273 K. In contrast to reaction 18.5, *partial* hydrolysis of XeF_6 (eq. 18.14) gives $XeOF_4$ (a colourless liquid, mp 227 K), which can be converted to XeO_2F_2 by reaction 18.15. Reaction 18.16 is used to prepare XeO_3F_2 which can be separated *in vacuo*. Further reaction between XeO_3F_2 and XeF_6 yields XeO_2F_4.

$$XeF_6 + H_2O \longrightarrow XeOF_4 + 2HF \qquad (18.14)$$
$$XeO_3 + XeOF_4 \longrightarrow 2XeO_2F_2 \qquad (18.15)$$
$$XeO_4 + XeF_6 \longrightarrow XeOF_4 + XeO_3F_2 \qquad (18.16)$$

The stable salts $M[XeO_3F]$ (M = K or Cs) are obtained from MF and XeO_3, and contain infinite chain anions with F^- ions bridging XeO_3 groups. Similar complexes are obtained from CsCl or RbCl with XeO_3 but these contain linked $[XeO_3Cl_2]^{2-}$ anions as shown in **18.17**.

(18.17)

Other compounds of xenon

Members of a series of compounds of the type FXeA where, for example, A^- is $[OClO_3]^-$, $[OSO_2F]^-$, $[OTeF_5]^-$ or $[O_2CCF_3]^-$ have been prepared by the highly exothermic elimination of HF between XeF_2 and HA. Further loss of HF leads to XeA_2 (e.g. eq. 18.17). Elimination of HF also drives the reaction of XeF_2 with $HN(SO_3F)_2$ to yield $FXeN(SO_3F)_2$, a relatively rare example of Xe–N bond formation.

$$XeF_2 + HOSO_2F \xrightarrow{-HF} FXeOSO_2F \xrightarrow[-HF]{HOSO_2F} Xe(OSO_2F)_2$$
$$\textbf{(18.18)} \qquad\qquad (18.17)$$

(18.18)

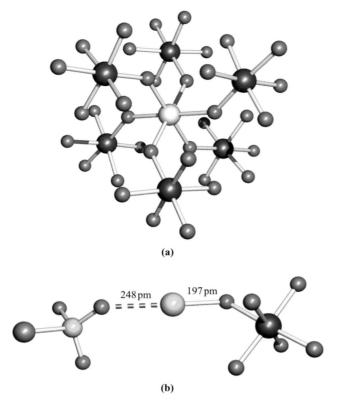

(a)

248 pm   197 pm

(b)

Fig. 18.7 The structure of (a) the $[Sb(OTeF_5)_6]^-$ anion and (b) the SO_2ClF adduct of the $[XeOTeF_5]^+$ cation in $[XeOTeF_5][Sb(OTeF_5)_6] \cdot SO_2ClF$ (determined by X-ray diffraction) [H.P.A. Mercier *et al.* (2005) *Inorg. Chem.*, vol. 44, p. 49]. Colour code: Xe and S, yellow; O, red; Cl and F, green; Te, blue; Sb, silver.

For Xe(II), a linear coordination environment is typical and in the solid state, salts containing $[XeF]^+$, $[XeOTeF_5]^+$ or related cations exhibit significant cation---anion interactions as shown in structures **18.7–18.9** and **18.14**. Similarly, in $[XeOTeF_5][AsF_6]$, the second coordination site on the Xe centre is occupied by an F atom of $[AsF_6]^-$ ($Xe \cdots F = 224$ pm). In an attempt to produce a salt containing a 'naked' cation, use has been made of the $[Sb(OTeF_5)_6]^-$ anion (Fig. 18.7a). This is a very weakly coordinating anion because the negative charge is spread out over 30 F atoms; $[Sb(OTeF_5)_6]^-$ is

also sterically demanding. $[XeOTeF_5][Sb(OTeF_5)_6]$ is prepared by reaction 18.18, in which one equivalent of $Xe(OTeF_5)_2$ acts as an oxidant. Although the structure of $[XeOTeF_5][Sb(OTeF_5)_6]$ confirms an absence of cation---anion interactions, the Xe(II) centre persists in being 2-coordinate by virtue of association with the SO_2ClF solvent (Fig. 18.7b).

$$2Xe(OTeF_5)_2 + Sb(OTeF_5)_3 \xrightarrow[\text{in liquid } SO_2ClF]{253\ K}$$
$$[XeOTeF_5][Sb(OTeF_5)_6] + Xe \quad (18.18)$$

Xenon–carbon bond formation is now well exemplified, and many products contain fluorinated aryl substituents, e.g. $(C_6F_5CO_2)Xe(C_6F_5)$, $[(2,6-F_2C_5H_3N)XeC_6F_5]^+$ (Fig. 18.8a), $[(2,6-F_2C_6H_3)Xe][BF_4]$ (Fig. 18.8b), $[(2,6-F_2C_6H_3)Xe][CF_3SO_3]$ and $[(MeCN)Xe(C_6F_5)]^+$. The degree of interaction between the Xe centre and non-carbon donor (i.e. F, O or N) in these species varies. Some species are best described as containing Xe in a linear environment (e.g. Fig. 18.8a) and others tend towards containing an $[RXe]^+$ cation (e.g. Fig. 18.8b). The compounds C_6F_5XeF and $(C_6F_5)_2Xe$ are obtained using the reactions in scheme 18.19. Stringent safety precautions must be taken when handling such compounds; $(C_6F_5)_2Xe$ decomposes explosively above 253 K.

$$Me_3SiC_6F_5 + XeF_2 \longrightarrow Me_3SiF + C_6F_5XeF$$
$$\Big\downarrow Me_3SiC_6F_5 \quad (18.19)$$
$$Me_3SiF + (C_6F_5)_2Xe$$

The $[C_6F_5XeF_2]^+$ ion (formed as the $[BF_4]^-$ salt from $C_6F_5BF_2$ and XeF_4, see worked example 18.1) is an extremely powerful oxidative-fluorinating agent, e.g. it converts I_2 to IF_5.

The use of a difluoroborane, RBF_2, precursor has proved to be a successful strategy for alkyl, alkenyl and alkynyl derivatives of xenon(II). Xenon–carbon(alkene) and

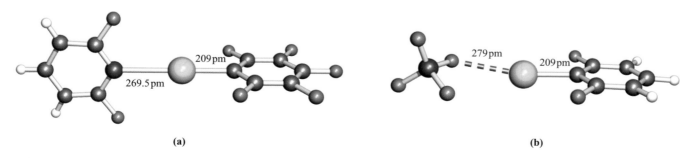

(a)   269.5 pm   209 pm

(b)   279 pm   209 pm

Fig. 18.8 The structures (X-ray diffraction) of (a) $[(2,6-F_2C_5H_3N)Xe(C_6F_5)]^+$ in the $[AsF_6]^-$ salt [H.J. Frohn *et al.* (1995) *Z. Naturforsch., Teil B*, vol. 50, p. 1799] and (b) $[(2,6-F_2C_6H_3)Xe][BF_4]$ [T. Gilles *et al.* (1994) *Acta Crystallogr., Sect. C*, vol. 50, p. 411]. Colour code: Xe, yellow; N, blue; B, blue; C, grey; F, green; H, white.

Xe–C(alkyne) bond formation is illustrated by reactions 18.20 and 18.21.

$$\text{(18.20)}$$

$$\text{(18.21)}$$

Compounds containing linear C–Xe–Cl units are now known, the first examples being C_6F_5XeCl (eq. 18.22) and $[(C_6F_5Xe)_2Cl]^+$ (eq. 18.23 and structure **18.19**).

$$[C_6F_5Xe]^+[AsF_6]^- + \quad 4\text{-}ClC_5H_4N{\cdot}HCl$$

4-chloropyridine hydrochloride

$$\xrightarrow[\text{195 K}]{\text{CH}_2\text{Cl}_2} C_6F_5XeCl + [4\text{-}ClC_5H_4NH]^+[AsF_6]^-$$

$$\text{(18.22)}$$

$$2[C_6F_5Xe]^+[AsF_6]^- + 6Me_3SiCl$$

$$\xrightarrow[\text{195 K}]{\text{CH}_2\text{Cl}_2} [(C_6F_5Xe)_2Cl]^+[AsF_6]^- + 6Me_3SiF$$

$$+ AsCl_3 + Cl_2 \quad \text{(18.23)}$$

(18.19)

The first detection of compounds containing metal–xenon bonds ($Fe(CO)_4Xe$ and $M(CO)_5Xe$ with M = Cr, Mo, W) was in the 1970s and involved matrix isolation studies. Since 2000, a number of fully isolated and characterized compounds containing Au–Xe or Hg–Xe bonds have been known, but even the most stable of these compounds decomposes at ≈ 298 K with loss of Xe. Their isolation depends upon the solvent and counter-ion being a weaker base than Xe(0). The first example was the square planar $[AuXe_4]^{2+}$ cation (av. Au–Xe = 275 pm). It is produced when AuF_3 is reduced to Au(II) in anhydrous HF/SbF_5 in the presence of Xe (eq. 18.24).

$$AuF_3 + 6Xe + 3H^+$$

$$\xrightarrow[\text{warm to 298 K}]{\text{HF/SbF}_5, \text{77 K,}} [AuXe_4]^{2+} + [Xe_2]^{2+} + 3HF$$

$$\text{(18.24)}$$

Removal of Xe from $[AuXe_4][Sb_2F_{11}]_2$ under vacuum at 195 K leads to $[cis\text{-}AuXe_2][Sb_2F_{11}]_2$. The *cis*-description arises as a result of Au–F–Sb bridge formation in the solid state (diagram **18.20**). The *trans*-isomer of $[AuXe_2]^{2+}$ is formed by reacting finely divided Au with XeF_2 in HF/SbF_5 under a pressure of Xe, but if the pressure is lowered, the product is the Au(II) complex $[XeAuFAuXe][SbF_6]_3$. The +2 oxidation state is rare for gold (see Section 22.12).

Au–Xe = 266, 267 pm   Au–F = 218, 224 pm

(18.20)

The acid strength of the HF/SbF_5 system can be lowered by reducing the amount of SbF_5 relative to HF. Under these conditions, crystals of the Au(III) complex **18.21** (containing *trans*-$[AuXe_2F]^{2+}$) are isolated from the reaction of XeF_2, Au and Xe. Bond formation between Xe(0) and Au(I) is exemplified in $[F_3AsAuXe]^+$ (reaction 18.25). In crystalline $[F_3AsAuXe][Sb_2F_{11}]$, the ions are essentially discrete species, with the shortest cation---anion contact being that shown in Fig. 18.9.

$$[F_3AsAu]^+[SbF_6]^- + Xe \xrightarrow[\text{HF/SbF}_5]{\text{in anhydrous}} [F_3AsAuXe]^+[Sb_2F_{11}]^-$$

$$\text{(18.25)}$$

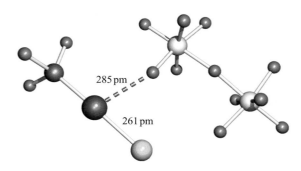

Fig. 18.9 The structure of $[F_3AsAuXe][Sb_2F_{11}]$ (X-ray diffraction, 173 K) shows the Xe(0) centre bonded to linear Au(I), and only weak cation---anion interactions; the shortest contact is shown [I.-C. Hwang *et al.* (2003) *Angew. Chem. Int. Ed.*, vol. 42, p. 4392]. Colour code: Xe, yellow; Au, red; As, brown; Sb, silver; F, green.

(18.21)

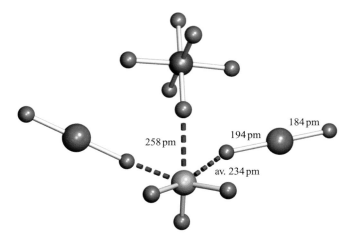

18.5 Compounds of argon, krypton and radon

The chemistry of argon is still in its infancy. Photolysis of HF in a solid argon matrix results in the formation of HArF, which has been identified by comparing the IR spectra of $^1H^{40}ArF$, $^1H^{36}ArF$ and $^2H^{40}ArF$. Theoretical studies suggest that the formation of Ar–C and Ar–Si bonds should be possible.

The only binary compound containing Kr is KrF_2. It is a colourless solid which decomposes >250 K, and is best prepared by UV irradiation of a mixture of Kr and F_2 (4 : 1 molar ratio) at 77 K. Krypton difluoride is dimorphic. The low-temperature phase, α-KrF_2, is isomorphous with XeF_2 (Fig. 18.4a). The structure of the β-form of KrF_2 is shown in Fig. 18.4b. The phase transition from β- to α-KrF_2 occurs below 193 K. Krypton difluoride is much less stable than XeF_2. It is rapidly hydrolysed by water (in an analogous manner to reaction 18.3), and dissociates into Kr and F_2 at 298 K ($\Delta_f H^\circ(298\,K) = +60.2\,kJ\,mol^{-1}$). We have already exemplified the use of KrF_2 as a powerful oxidizing agent in the syntheses of $[XeF_5][AgF_4]$ and $[Xe_2F_{11}]_2[NiF_6]$ (Section 18.4).

Krypton difluoride reacts with a number of pentafluorides, MF_5 (typically in anhydrous HF or BrF_5 at low temperature), to form $[KrF]^+[MF_6]^-$ (M = As, Sb, Bi, Ta), $[KrF]^+[M_2F_{11}]^-$ (M = Sb, Ta, Nb) and $[Kr_2F_3]^+[MF_6]^-$ (M = As, Sb, Ta). In the solid state, the $[KrF]^+$ ion in $[KrF]^+[MF_6]^-$ (M = As, Sb, Bi) is strongly associated with the anion (e.g. structure **18.22**). The $[Kr_2F_3]^+$ ion (**18.23**)[†] is structurally similar to $[Xe_2F_3]^+$ (**18.7**). The oxidizing and fluorinating powers of KrF_2 are illustrated by its reaction with metallic gold to give $[KrF]^+[AuF_6]^-$ (eq. 18.26).

$$7KrF_2 + 2Au \longrightarrow 2[KrF][AuF_6] + 5Kr \qquad (18.26)$$

In the examples above, KrF_2 reacts with Lewis acids that are strong enough F^- acceptors to abstract F^-. In reaction

---

[†]For details of variation of bond lengths and angles in $[Kr_2F_3]^+$ with the salt, see J.F. Lehmann *et al.* (2001) *Inorg. Chem.*, vol. 40, p. 3002.

Fig. 18.10 The structure of $[BrOF_2][AsF_6]\cdot2KrF_2$ (X-ray diffraction, 100 K) showing Br····F interactions between the structural units [D.S. Brock *et al.* (2010) *J. Am. Chem. Soc.*, vol. 132, p. 3533]. Colour code: Kr, red; Br, gold; As, brown; O, red; F, green.

18.26, KrF_2 also acts as a very powerful oxidizing agent. An unusual example of the interaction of KrF_2 with a *p*-block element without oxidation of the latter or transfer of F^-, is observed in the reaction of KrF_2 with $[BrOF_2][AsF_6]$ (eq. 18.27). The product is stable at low temperature for several days and the solid state structure (Fig. 18.10) confirms adduct formation between KrF_2 and the bromine(V) atom.

$$[BrOF_2][AsF_6]^- + 2KrF_2 \xrightarrow[\text{/195 K}]{\text{anhydrous HF,}} [BrOF_2][AsF_6]\cdot2KrF_2$$

$$(18.27)$$

Few compounds are known that contain Kr bonded to elements other than F. The reactions between KrF_2, RC≡N (e.g. R = H, CF_3) and AsF_5 in liquid HF or BrF_5 yield $[(RCN)KrF]^+[AsF_6]^-$ with Kr–N bond formation, and Kr–O bond formation has been observed in the reaction of KrF_2 and $B(OTeF_5)_3$ to give $Kr(OTeF_5)_2$.

(18.22)

(18.23)

Radon is oxidized by halogen fluorides (e.g. ClF, ClF_3) to the non-volatile RnF_2. The latter is reduced by H_2 at 770 K, and is hydrolysed by water in an analogous manner to XeF_2 (eq. 18.3). As we mentioned in Section 18.1, little chemistry of radon has been explored.

FURTHER READING

K.O. Christe (2001) *Angew. Chem. Int. Ed.*, vol. 40, p. 1419 – An overview of recent developments: 'A renaissance in noble gas chemistry'.

W. Grochala (2007) *Chem. Soc. Rev.*, vol. 36, p. 1632 – A 'critical review' of compounds of the noble gases.

I. Hargittai and D.K. Menyhárd (2010) *J. Mol. Struct.*, vol. 978, p. 136 – The VSEPR model applied to noble gas compounds.

J.F. Lehmann, H.P.A. Mercier and G.J. Schrobilgen (2002) *Coord. Chem. Rev.*, vol. 233–234, p. 1 – A comprehensive review: 'The chemistry of krypton'.

J.F. Liebman and C.A. Deakyne (2003) *J. Fluorine Chem.*, vol. 121, p. 1 – A review of noble gas compounds, emphasizing interrelations and interactions with fluorine-containing species.

A.G. Massey (2000) *Main Group Chemistry*, 2nd edn, Wiley, Chichester – Chapter 12 covers the chemistry of the group 18 elements.

K. Seppelt (2003) *Z. Anorg. Allg. Chem.*, vol. 629, p. 2427 – 'Metal–xenon complexes'.

PROBLEMS

18.1 (a) What is the collective name for the group 18 elements? (b) Write down, in order, the names and symbols of these elements; check your answer by reference to the first page of this chapter. (c) What common feature does the ground state electronic configuration of each element possess?

18.2 Construct MO diagrams for He_2 and $[He_2]^+$ and rationalize why the former is not known but the latter may be detected.

18.3 Confirm that the observed gas-phase structures of XeF_2, XeF_4 and XeF_6 are consistent with the VSEPR model.

18.4 Rationalize the structure of $[XeF_8]^{2-}$ (a square antiprism) in terms of the VSEPR model.

18.5 How would you attempt to determine values for (a) $\Delta_f H^\circ (XeF_2, 298\,K)$ and (b) the $Xe-F$ bond energy in XeF_2?

18.6 Why is $XeCl_2$ likely to be much less stable than XeF_2?

18.7 How may the standard enthalpy of the unknown salt Xe^+F^- be estimated?

18.8 Predict the structures of $[XeO_6]^{4-}$, $XeOF_2$, $XeOF_4$, XeO_2F_2, XeO_2F_4 and XeO_3F_2.

18.9 Suggest products for the following reactions (which are not necessarily balanced on the left-hand sides):

(a) $CsF + XeF_4 \longrightarrow$
(b) $SiO_2 + XeOF_4 \longrightarrow$
(c) $XeF_2 + SbF_5 \longrightarrow$
(d) $XeF_6 + [OH]^- \longrightarrow$
(e) $KrF_2 + H_2O \longrightarrow$

18.10 Write a brief account of the chemistry of the xenon fluorides.

18.11 (a) The reaction of XeF_2 with RuF_5 at 390 K results in the formation of a compound, the Raman spectrum of which is similar to that of $CsRuF_6$ but with an additional band at $600\,cm^{-1}$. Rationalize these data. (b) When the product of the reaction in part (a) reacts with excess F_2 at 620 K, a compound of molecular formula $RuXeF_{11}$ is formed. The compound is monomeric in the solid state. Propose a structure for this product.

18.12 The reaction of $F_2C=CClBF_2$ with XeF_2 gives a product **A** for which the NMR spectroscopic data are as follows: ^{19}F NMR δ/ppm −64.3 (s + d, J 8 Hz, 1F), −75.9 (s + d, J 138 Hz, 1F), −148.1 (non-binomial quartet, J 11 Hz); ^{129}Xe NMR δ/ ppm −3550 (dd, J 8 Hz, 138 Hz) (s = singlet, d = doublet, dd = doublet of doublets). Rationalize the data and suggest the identity of **A**.

18.13 Equation 18.25 showed the preparation of $[F_3AsAuXe][Sb_2F_{11}]$ from $[F_3AsAu][SbF_6]$. Solid $[F_3AsAu][SbF_6]$ contains a distorted $[SbF_6]^-$ ion; one Sb–F bond is 193 pm long, and five are in the range 185–189 pm. The Au centre interacts with the F atom of the long Sb–F bond (Au–F = 212 pm, compared with 203 pm calculated for the hypothetical $[AuF_2]^-$ ion). Suggest why $[F_3AsAu][SbF_6]$ was chosen as the precursor to $[F_3AsAuXe]^+$, rather than a route involving reduction of AuF_3 in anhydrous HF/SbF_5 in the presence of Xe.

OVERVIEW PROBLEMS

18.14 (a) The ^{19}F NMR spectrum of $[Kr_2F_3][SbF_6]$ in BrF_5 at 207 K contains a doublet ($J = 347$ Hz) and triplet ($J = 347$ Hz) assigned to the cation. Explain the origin of these signals.

(b) Give examples that illustrate the role of E–F–Xe and E–F–Kr bridge formation (E = any element) in the solid state. To what extent does bridge formation occur betwen cations and anions, and how does it affect the description of a solid as containing discrete ions?

18.15 Suggest products for the following reactions, which are not necessarily balanced on the left-hand side:

(a) $KrF_2 + Au \longrightarrow$

(b) $XeO_3 + RbOH \longrightarrow$

(c) $[XeCl][Sb_2F_{11}] \xrightarrow{298\,K}$

(d) $KrF_2 + B(OTeF_5)_3 \longrightarrow$

(e) $C_6F_5XeF + Me_3SiOSO_2CF_3 \longrightarrow$

(f) $[C_6F_5XeF_2]^+ + C_6F_5I \longrightarrow$

18.16 By referring to the following literature source, assess the safety precautions required when handling XeO_4: M. Gerken and G.J. Schrobilgen (2002) *Inorg. Chem.*, vol. 41, p. 198.

18.17 The vibrational modes of KrF_2 are at 590, 449 and 233 cm^{-1}. Explain why only the bands at 590 and 233 cm^{-1} are observed in the IR spectrum of gaseous KrF_2.

18.18 Use MO theory to rationalize why the Xe–F bond strength in $[XeF]^+$ is greater than in XeF_2.

INORGANIC CHEMISTRY MATTERS

18.19 High-field NMR spectrometers, including those used for magnetic resonance imaging in hospitals, contain magnets with superconducting coils, e.g. NbTi which becomes superconducting at 9.5 K. (a) Why is liquid helium used to cool the magnet? (b) For He, $\Delta_{vap}H(bp) = 0.1\,kJ\,mol^{-1}$. To what process does this refer and why is the value so low? (c) Suggest why the liquid helium tank in an NMR spectrometer is surrounded by a tank of liquid N_2.

18.20 Discharge lamps are used throughout the world for lighting and advertising. Such a lamp consists of a sealed tube with a metal electrode at each end, and contains a gas (e.g. He, Ar, Ne) or vapour (Na, Hg). The atoms are excited by an electrical discharge. Explain the origin of the pale yellow glow emitted by a helium discharge lamp.

Topics

Ground state electronic configurations
Physical properties
Reactivity of the metals
Characteristic properties
Electroneutrality principle
Kepert model
Coordination numbers
Isomerism

19

d-Block metal chemistry: general considerations

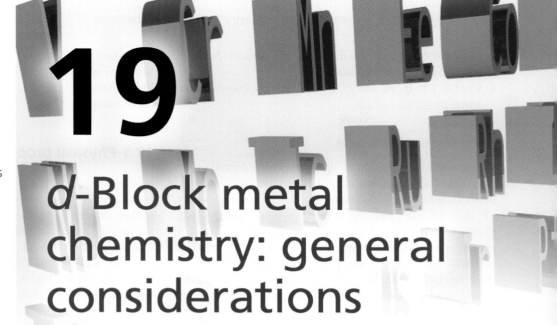

1–2	**3**	**4**	**5**	**6**	**7**	**8**	**9**	**10**	**11**	**12**	13–18
s-block											*p*-block
	Sc	Ti	V	Cr	Mn	Fe	Co	Ni	Cu	Zn	
	Y	Zr	Nb	Mo	Tc	Ru	Rh	Pd	Ag	Cd	
	La	Hf	Ta	W	Re	Os	Ir	Pt	Au	Hg	

19.1 Topic overview

In Chapters 19–22, we discuss the chemistry of the *d*-block metals, covering first some general principles including magnetic and electronic spectroscopic properties. We move then to a systematic coverage of the metals and their inorganic compounds. The organometallic chemistry of the *d*-block metals is covered in Chapter 24 after an account of *p*-block organometallic chemistry in Chapter 23. We have already discussed some aspects of the *d*-block metals:

- ground state electronic configurations (Table 1.3);
- trends in first ionization energies (Fig. 1.16 and Section 1.10);
- structures of bulk metals (Section 6.3);
- polymorphism (Section 6.4);
- metallic radii (Section 6.5);
- trends in melting points and $\Delta_a H^\circ$(298 K) (Section 6.6);
- alloys and intermetallic compounds (Section 6.7);
- metallic bonding including electrical resistivity (Section 6.8 and Fig. 6.10);
- aquated cations: formation and acidic properties (Section 7.7);

- solubilities of ionic salts and common-ion effect (Sections 7.9 and 7.10);
- stability constants for metal complexes (Section 7.12);
- selected ligand structures and abbreviations (Table 7.7);
- an introduction to coordination complexes (Section 7.11);
- redox chemistry in aqueous solution, including potential diagrams and Frost–Ebsworth diagrams (Chapter 8);
- stereoisomerism (Section 2.9);
- chiral molecules (Section 3.8);
- binary metal hydrides (Section 10.7).

19.2 Ground state electronic configurations

d-Block metals versus transition elements

The three rows of *d*-block metals are shown in the schematic periodic table at the beginning of the chapter. The term 'transition elements (metals)' is also widely used. However, the group 12 metals (Zn, Cd and Hg) are not always classified as transition metals.[†] The elements in the *f*-block (see Chapter 27) have, in the past, been called *inner transition elements*. Throughout our discussions, we shall use the terms *d*-block and *f*-block metals, so being consistent with the use of the terms *s*-block and *p*-block elements in earlier chapters. Three further points should be noted:

- each group of *d*-block metals consists of three members and is called a *triad*;

[†] *IUPAC Nomenclature of Inorganic Chemistry* (Recommendations 2005), senior eds N.G. Connelly and T. Damhus, RSC Publishing, Cambridge, p. 51.

● metals of the second and third rows are sometimes called the *heavier d-block metals*;
● Ru, Os, Rh, Ir, Pd and Pt are collectively known as the *platinum-group metals*.

Electronic configurations

To a first approximation, the observed ground state electronic configurations of the first, second and third row *d*-block metal atoms correspond to the progressive filling of the 3*d*, 4*d* and 5*d* atomic orbitals respectively (Table 1.3). However, there are minor deviations from this pattern, e.g. in the first row, the ground state of chromium is $[Ar]4s^1 3d^5$ rather than $[Ar]4s^2 3d^4$. The reasons for these deviations are beyond the scope of this book: we should need to know both the energy difference between the 3*d* and 4*s* atomic orbitals when the nuclear charge is 24 (the atomic number of Cr) and the interelectronic interaction energies for each of the $[Ar]4s^1 3d^5$ and $[Ar]4s^2 3d^4$ configurations. Fortunately, M^{2+} and M^{3+} *ions* of the *first row* *d*-block metals all have electronic configurations of the general form $[Ar]3d^n$, and so the comparative chemistry of these metals is largely concerned with the consequences of the successive filling of the 3*d* orbitals. For metals of the second and third rows, the picture is more complicated, and a systematic treatment of their chemistry cannot be given. The emphasis in this and the next chapter is therefore on the first row metals, but we shall include some material that illustrates ways in which the heavier metals differ from their lighter congeners.

An important point that must not be forgotten is that *d*-block metal atoms are, of course, *many-electron species*,

and when we discuss, for example, radial distribution functions of the *nd* atomic orbitals, we refer to hydrogen-like atoms and, therefore, the discussion is extremely approximate.

19.3 Physical properties

In this section, we consider physical properties of the *d*-block metals (see cross-references in Section 19.1 for further details). An extended discussion of properties of the heavier metals is given in Section 22.1. Nearly all the *d*-block metals are hard, ductile and malleable, with high electrical and thermal conductivities. With the exceptions of Mn, Zn, Cd and Hg, at room temperature, the metals possess one of the typical metal structures (see Table 6.2). The metallic radii (r_{metal}) for 12-coordination (Table 6.2 and Fig. 19.1) are much smaller than those of the *s*-block metals of comparable atomic number. Figure 19.1 also illustrates that values of r_{metal}:

● show little variation across a given row of the *d*-block;
● are greater for second and third row metals than for first row metals;
● are similar for the second and third row metals in a given triad.

This last observation is due to the so-called *lanthanoid contraction*: the steady decrease in size along the 14 lanthanoid metals between La and Hf (see Section 27.3).

Metals of the *d*-block are (with the exception of the group 12 metals) much harder and less volatile than those of the *s*-block. The trends in enthalpies of atomization

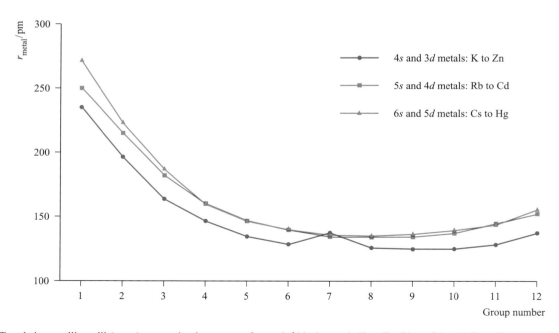

Fig. 19.1 Trends in metallic radii (r_{metal}) across the three rows of *s*- and *d*-block metals K to Zn, Rb to Cd, and Cs to Hg.

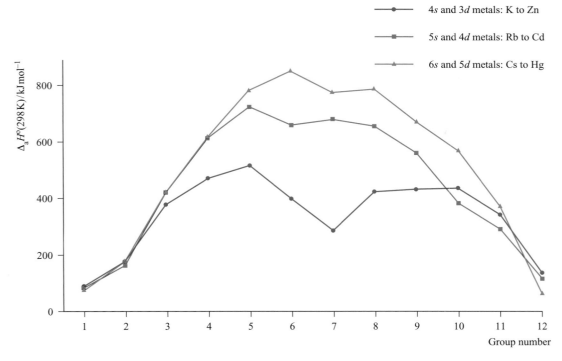

Fig. 19.2 Trends in standard enthalpies of atomization, $\Delta_a H^\circ (298\,K)$, across the three rows of s- and d-block metals K to Zn, Rb to Cd, and Cs to Hg.

(Table 6.2) are shown in Fig. 19.2. Metals in the second and third rows generally possess higher enthalpies of atomization than the corresponding elements in the first row. This is a substantial factor in accounting for the far greater occurrence of metal–metal bonding in compounds of the heavier d-block metals compared with their first row congeners. In general, Fig. 19.2 shows that metals in the centre of the d-block possess higher values of $\Delta_a H^\circ (298\,K)$ than early or late metals. However, one must be careful in comparing metals with different structure types and this is particularly true of manganese (see Section 6.3).

The first ionization energies (IE_1) of the d-block metals in a given period (Fig. 1.16 and Appendix 8) are higher than those of the preceding s-block metals. Figure 1.16 shows that across each of the periods K to Kr, Rb to Xe, and Cs to Rn, the variation in values of IE_1 is small across the d-block and far greater among the s- and p-block elements. Within each period, the overall trend for the d-block metals is for the ionization energies to increase, but many small variations occur. Chemical comparisons between metals from the s- and d-blocks are complicated by the number of factors involved. Thus, all $3d$ metals have values of IE_1 (Fig. 1.16) and IE_2 larger than those of calcium, and all except zinc have higher values of $\Delta_a H^\circ$ (Fig. 19.2) than calcium. These factors make the metals less reactive than calcium. However, since all known M^{2+} ions of the $3d$ metals are smaller than Ca^{2+}, lattice and solvation energy effects (see Chapters 6 and 7) are more favourable for the

Table 19.1 Standard reduction potentials (298 K) for some metals in the first long period; the concentration of each aqueous solution is $1\,mol\,dm^{-3}$.

Reduction half-equation	E° / V
$Ca^{2+}(aq) + 2e^- \rightleftharpoons Ca(s)$	-2.87
$Ti^{2+}(aq) + 2e^- \rightleftharpoons Ti(s)$	-1.63
$V^{2+}(aq) + 2e^- \rightleftharpoons V(s)$	-1.18
$Cr^{2+}(aq) + 2e^- \rightleftharpoons Cr(s)$	-0.91
$Mn^{2+}(aq) + 2e^- \rightleftharpoons Mn(s)$	-1.19
$Fe^{2+}(aq) + 2e^- \rightleftharpoons Fe(s)$	-0.44
$Co^{2+}(aq) + 2e^- \rightleftharpoons Co(s)$	-0.28
$Ni^{2+}(aq) + 2e^- \rightleftharpoons Ni(s)$	-0.25
$Cu^{2+}(aq) + 2e^- \rightleftharpoons Cu(s)$	$+0.34$
$Zn^{2+}(aq) + 2e^- \rightleftharpoons Zn(s)$	-0.76

$3d$ metal ions. In practice, it turns out that, in the formation of species containing M^{2+} ions, all the $3d$ metals are thermodynamically less reactive than calcium, and this is consistent with the standard reduction potentials listed in Table 19.1. However, interpretation of observed chemistry based on these E° data is not always straightforward, since the formation of a coherent surface film of metal oxide often renders a metal less reactive than expected (see Section 19.4). A few *d*-block metals are very powerful reducing agents, e.g. E^o for the Sc^{3+}/Sc couple ($-2.08\,V$) is more negative than that for Al^{3+}/Al ($-1.66\,V$).

Worked example 19.1   Reduction potentials of the first row *d*-block metals

In what way does the value of E^o for the $Fe^{2+}(aq)/Fe(s)$ couple depend on the first two ionization energies of $Fe(g)$?

E^o for the $Fe^{2+}(aq)/Fe(s)$ couple refers to the reduction process:

$$Fe^{2+}(aq) + 2e^- \rightleftharpoons Fe(s)$$

relative to the reduction:

$$2H^+(aq) + 2e^- \rightleftharpoons H_2(g)$$

The sum of the first and second ionization energies, IE_1 and IE_2, refers to the process:

$$Fe(g) \longrightarrow Fe^{2+}(g)$$

The entropy changes on ionization are negligible compared with the enthalpy changes. Therefore, IE_1 and IE_2 may be approximated to Gibbs energy changes.

In order to relate the processes, construct a thermochemical cycle:

$$
\begin{array}{ccc}
Fe^{2+}(aq) + 2e^- & \xrightarrow{\;\Delta G^o{}_1\;} & Fe(s) \\
{\scriptstyle\Delta_{hyd}G^o}\uparrow & & \downarrow{\scriptstyle\Delta_a G^o} \\
Fe^{2+}(g) & \xleftarrow{\;IE_1 + IE_2\;} & Fe(g)
\end{array}
$$

$\Delta_{hyd}G^o$ is the Gibbs energy change for the hydration of a mole of gaseous Fe^{2+} ions. This cycle illustrates the contribution that the ionization energies of Fe make to $\Delta G^o{}_1$, the Gibbs energy change associated with the reduction of $Fe^{2+}(aq)$. This in turn is related to $E^o{}_{Fe^{2+}/Fe}$ by the equation:

$$\Delta G^o{}_1 = -zFE^o$$

where $F = 96\,485\,C\,mol^{-1}$ and $z = 2$.

Self-study exercises

Use the data in Table 19.1 for these questions.

1. Which of the metals Cu and Zn will liberate H_2 from dilute hydrochloric acid?   [*Ans.* See Section 8.2]

2. Calculate a value of $\Delta G^o(298\,K)$ for the reaction:

$$Zn(s) + 2H^+(aq) \longrightarrow Zn^{2+}(aq) + H_2(g)$$

   Is the result consistent with your answer to question 1?
       [*Ans.* $-147\,kJ\,mol^{-1}$]

3. A polished Cu rod is placed in an aqueous solution of $Zn(NO_3)_2$. In a second experiment, a polished Zn rod is placed in an aqueous solution of $CuSO_4$. Does anything happen to (a) the Cu rod and (b) the Zn rod? Quantify your answers by calculating appropriate values of $\Delta G^o(298\,K)$.   [*Ans.* See Section 8.2]

19.4 The reactivity of the metals

In Chapters 21 and 22 we look at individual elements of the *d*-block in detail. However, a few points are given here as an overview. In general, the metals are moderately reactive and combine to give binary compounds when heated with dioxygen, sulfur or the halogens (e.g. reactions 19.1–19.3), product stoichiometry depending, in part, on the available oxidation states (see below). Combination with H_2, B, C or N_2 may lead to interstitial hydrides (Section 10.7), borides (Section 13.10), carbides (Section 14.7) or nitrides (Section 15.6).

$$Os + 2O_2 \xrightarrow{\;\Delta\;} OsO_4 \qquad (19.1)$$

$$Fe + S \xrightarrow{\;\Delta\;} FeS \qquad (19.2)$$

$$V + \frac{n}{2}X_2 \xrightarrow{\;\Delta\;} VX_n \quad (X = F, n = 5; X = Cl, n = 4;$$
$$X = Br, I, n = 3) \qquad (19.3)$$

Most *d*-block metals should, on thermodynamic grounds (e.g. Table 19.1), liberate H_2 from acids but, in practice, many do not since they are passivated by a thin surface coating of oxide or by having a high dihydrogen overpotential, or both. Silver, gold and mercury (i.e. late, second and third row metals) are, even in the thermodynamic sense, the least reactive metals known. For example, gold is not oxidized by atmospheric O_2 or attacked by acids, except by a 3:1 mixture of concentrated HCl and HNO_3 (*aqua regia*).

19.5 Characteristic properties: a general perspective

In this section, we introduce properties that are characteristic of *d*-block metal compounds. More detailed discussion follows in Chapter 20.

Colour

The colours of *d*-block metal compounds are a characteristic feature of species with ground state electronic configurations other than d^0 and d^{10}. For example, $[Cr(OH_2)_6]^{2+}$ is sky-blue, $[Mn(OH_2)_6]^{2+}$ very pale pink,

Table 19.2 The visible part of the electromagnetic spectrum.

Colour of light *absorbed*	Approximate wavelength ranges / nm	Corresponding wavenumbers (approximate values) / cm^{-1}	Colour of light *transmitted*, i.e. complementary colour of the absorbed light	In a 'colour wheel' representation,[†] complementary colours are in opposite sectors
Red	700–620	14 300–16 100	Green	
Orange	620–580	16 100–17 200	Blue	
Yellow	580–560	17 200–17 900	Violet	
Green	560–490	17 900–20 400	Red	
Blue	490–430	20 400–23 250	Orange	
Violet	430–380	23 250–26 300	Yellow	

[†] When an electronic spectrum exhibits more than one absorption in the visible region, the simplicity of the colour wheel does not hold.

$[Co(OH_2)_6]^{2+}$ pink, $[MnO_4]^-$ intense purple and $[CoCl_4]^{2-}$ dark blue. In contrast, complexes of Sc(III) (d^0) or Zn(II) (d^{10}) are colourless unless the ligands contain a chromophore that absorbs in the visible region.

> A **chromophore** is the group of atoms in a molecule responsible for the absorption of electromagnetic radiation.

The fact that many of the observed colours are of *low intensity* is consistent with the colour originating from electronic '*d–d*' transitions. If we are dealing with an isolated gas-phase ion, such transitions are forbidden by the Laporte selection rule (eq. 19.4 where l is the orbital quantum number). The pale colours observed in complexes indicate that the probability of a transition occurring is low. Table 19.2 shows relationships between the wavelength of light absorbed and observed colours.

$$\Delta l = \pm 1 \qquad \text{(Laporte selection rule)} \qquad (19.4)$$

The intense colours of species such as $[MnO_4]^-$ have a different origin, namely *charge transfer* absorptions or emissions. The latter are *not* subject to selection rule 19.4 and are always more intense than electronic transitions between different d orbitals. We return to electronic spectra in Section 20.7.

Paramagnetism

The occurrence of *paramagnetic* compounds of d-block metals is common and arises from the presence of unpaired electrons. This phenomenon can be investigated using electron paramagnetic resonance (EPR) spectroscopy (see Section 4.9). It also leads to signal broadening and anomalous chemical shift values in NMR spectra (see Box 4.2).

Complex formation

d-Block metal ions readily form complexes, with complex formation often being accompanied by a change in colour and sometimes a change in the intensity of colour. Equation 19.5 shows the effect of adding concentrated HCl to aqueous cobalt(II) ions.

$$[Co(OH_2)_6]^{2+} + 4Cl^- \longrightarrow [CoCl_4]^{2-} + 6H_2O \qquad (19.5)$$

pale pink   dark blue

The formation of such complexes is analogous to the formation of those of s- and p-block metals and discussed in previous chapters, e.g. $[K(18\text{-crown-}6)]^+$, $[Be(OH_2)_4]^{2+}$, *trans*-$[SrBr_2(py)_5]$, $[AlF_6]^{3-}$, $[SnCl_6]^{2-}$ and $[Bi_2(O_2C_6H_4)_4]^{2-}$.

Self-study exercises

For the answers, refer to Table 7.7.

1. Many ligands in complexes have common abbreviations. Give the full names of the following ligands: en, THF, phen, py, $[acac]^-$, $[ox]^{2-}$.

2. Draw the structures of the following ligands. Indicate the potential donor atoms in and the denticity of each ligand: en, $[EDTA]^{4-}$, DMSO, dien, bpy, phen.

Variable oxidation states

The occurrence of variable oxidation states and, often, the interconversion between them, is a characteristic of most d-block metals. Exceptions are in groups 3 and 12 as Table 19.3 illustrates. In group 12, the $+1$ oxidation state is found for species containing (or formally containing) the $[M_2]^{2+}$ unit. This is extremely common for Hg, but is

Table 19.3 Oxidation states of the *d*-block metals; the most stable states are marked in blue. Tabulation of zero oxidation states refers to their appearance in *compounds* of the metal. In organometallic compounds, oxidation states of less than zero are encountered (see Chapter 23). An oxidation state enclosed in [ ] is rare.

Sc	Ti	V	Cr	Mn	Fe	Co	Ni	Cu	Zn
	0	0	0	0	0	0	0	[0]	
		1	1	1	1	1	1	1	[1]
	2	2	2	2	2	2	2	2	2
3	3	3	3	3	3	3	3	3	
	4	4	4	4	4	4	4	[4]	
		5	5	5					
			6	6	6				
				7					

Y	Zr	Nb	Mo	Tc	Ru	Rh	Pd	Ag	Cd
			0	0	0	0	0		
						1		1	[1]
	2	2	2	[2]	2	2	2	2	2
3	3	3	3	3	3	3		3	
	4	4	4	4	4	4	4		
		5	5	5	5	5			
			6	6	6	6			
				7	7				
					8				

La	Hf	Ta	W	Re	Os	Ir	Pt	Au	Hg
			0	0	0	0	0		
			1	1		1		1	1
	2	2	2	2	2	2	2	[2]	2
3	3	3	3	3	3	3		3	
	4	4	4	4	4	4	4		
		5	5	5	5	5	5	5	
			6	6	6	6	6		
				7	7				
					8				

rare for Zn and Cd (see Sections 21.13 and 22.13). A comparison between the available oxidation states for a given metal (Table 19.3) and the electronic configurations listed in Table 1.3 is instructive. As expected, metals that display the greatest number of different oxidation states occur in or near the middle of a *d*-block row. Two cautionary notes (illustrated by *d*- and *f*-block metal compounds) should be made:

- The apparent oxidation state deduced from a molecular or empirical formula may be misleading. For example, (i) LaI_2 is a metallic conductor and is best formulated as $La^{3+}(I^-)_2(e^-)$, and (ii) $MoCl_2$ contains metal cluster units with metal–metal bonds and is formally $[Mo_6Cl_8]^{4+}(Cl^-)_4$. Metal–metal bond formation becomes more important for the heavier metals.
- There are many metal compounds in which it is impossible to assign oxidation states unambiguously, e.g. in the complexes $[Ti(bpy)_3]^{n-}$ ($n = 0, 1, 2$), there is evidence that the negative charge is localized on the bpy ligands (see Table 7.7) not the metal centres, and in nitrosyl complexes, the NO ligand may donate one or three electrons (see Sections 20.4 and 24.2).

19.6 Electroneutrality principle

Pauling's ***electroneutrality principle*** is an *approximate method* of estimating the charge distribution in molecules and complex ions. It states that the distribution of charge in a molecule or ion is such that the charge on any single atom is within the range +1 to −1 (ideally close to zero).

Consider the complex ion $[Co(NH_3)_6]^{3+}$. Figure 19.3a gives a representation of the complex which indicates that the coordinate bonds are formed by lone pair donation from the ligands to the Co(III) centre. It implies transfer of charge from ligand to metal, and Fig. 19.3b shows the resulting charge distribution. This is clearly unrealistic, since the cobalt(III) centre becomes more negatively charged than would be favourable given its electropositive nature. At the other extreme, we could consider the bonding in terms of a wholly ionic model (Fig. 19.3c): the 3+ charge remains localized on the cobalt ion and the six NH_3 ligands remain neutral. However, this model is also flawed. Experimental evidence shows that the $[Co(NH_3)_6]^{3+}$ complex ion remains as an entity in aqueous solution, and the electrostatic interactions implied by the ionic model are unlikely to be

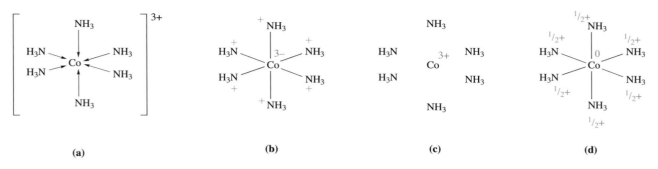

Fig. 19.3 The complex cation $[Co(NH_3)_6]^{3+}$: (a) a conventional diagram showing the donation of lone pairs of electrons from ligands to metal ion; (b) the charge distribution that results from a 100% covalent model of the bonding; (c) the charge distribution that results from a 100% ionic model of the bonding; and (d) the approximate charge distribution that results from applying the electroneutrality principle.

strong enough to allow this to happen. Thus, neither of the extreme bonding models is appropriate.

If we now apply the electroneutrality principle to $[Co(NH_3)_6]^{3+}$, then, ideally, the net charge on the metal centre should be zero. That is, the Co^{3+} ion may accept a total of *only three electrons* from the six ligands, thus giving the charge distribution shown in Fig. 19.3d. The electroneutrality principle results in a bonding description for the $[Co(NH_3)_6]^{3+}$ ion which is 50% ionic (or 50% covalent).

Self-study exercises

1. In $[Fe(CN)_6]^{3-}$, a realistic charge distribution results in each ligand carrying a charge of $-\frac{2}{3}$. In this model, what charge does the Fe centre carry and why is this charge consistent with the electroneutrality principle?

2. If the bonding in $[CrO_4]^{2-}$ were described in terms of a 100% ionic model, what would be the charge carried by the Cr centre? Explain how this charge distribution can be modified by the introduction of covalent character into the bonds.

19.7 Coordination numbers and geometries

In this section, we give an overview of the coordination numbers and geometries found within d-block metal compounds. It is impossible to give a comprehensive account, and several points should be borne in mind:

- most examples in this section involve mononuclear complexes, and in complexes with more than one metal centre, structural features are often conveniently described in terms of individual metal centres (e.g. in polymer **19.4**, each Pd(II) centre is in a square planar environment);
- although coordination environments are often described in terms of *regular* geometries such as those in Table 19.4, in

practice they are often distorted, for example as a consequence of steric effects;

- detailed discussion of a particular geometry usually involves bond lengths and angles determined in the solid state and these may be affected by crystal packing forces;
- where the energy difference between different possible structures is small (e.g. for 5- and 8-coordinate complexes), fluxional behaviour in solution may be observed; the small energy difference may also lead to the observation of different structures in the solid state, e.g. in salts of $[Ni(CN)_5]^{3-}$ the shape of the anion depends upon the cation present and in $[Cr(en)_3][Ni(CN)_5] \cdot 1.5H_2O$, *both* trigonal bipyramidal and square-based pyramidal structures are present.

The description of coordination numbers in this section does not include ionic lattices, but instead focuses on mononuclear species in which the metal centre is covalently bonded to the atoms in the coordination sphere. The metal–ligand bonding in complexes can generally be considered in terms of σ-donor ligands interacting with a metal centre which acts as a σ-acceptor. This may, in some complexes, be augmented with interactions involving π-donor ligands (with the metal as a π-acceptor) or π-acceptor ligands (with the metal as a π-donor). For a preliminary discussion of stereochemistry, it is not necessary to detail the metal–ligand bonding but we shall find it useful to draw attention to the electronic configuration of the metal centre. The reasons for this will become clear in Chapter 20, but for now you should remember that both steric and electronic factors are involved in dictating the coordination geometry around a metal ion.

It is difficult to provide generalizations about the trends in coordination *number* within the d-block. However, it is useful to bear the following points in mind:

- sterically demanding ligands favour low coordination numbers at metal centres;
- high coordination numbers are most likely to be attained with small ligands and large metal ions;

Table 19.4 Coordination geometries; each describes the arrangement of the donor atoms that surround the metal centre. Note that for some coordination numbers, more than one possible arrangement of donor atoms exists.

Coordination number	Arrangement of donor atoms around metal centre	Less common arrangements
2	Linear	
3	Trigonal planar	Trigonal pyramidal
4	Tetrahedral; square planar	
5	Trigonal bipyramidal; square-based pyramidal	
6	Octahedral	Trigonal prismatic
7	Pentagonal bipyramidal	Monocapped trigonal prismatic; monocapped octahedral
8	Dodecahedral; square antiprismatic; hexagonal bipyramidal	Cube; bicapped trigonal prismatic
9	Tricapped trigonal prismatic	

- the size of a metal ion decreases as the formal charge increases, e.g. $r(Fe^{3+}) < r(Fe^{2+})$;
- low coordination numbers will be favoured by metals in high oxidation states with π-bonding ligands.

The Kepert model

For many years after the classic work of Werner which laid the foundations for the correct formulation of *d*-block metal complexes,[†] it was assumed that a metal in a given oxidation state would have a fixed coordination number and geometry. In the light of the success (albeit not universal success) of the VSEPR model in predicting the shapes of molecular species of the *p*-block elements (see Section 2.8), we might reasonably expect the structures of the complex ions $[V(OH_2)_6]^{3+}$ (d^2), $[Mn(OH_2)_6]^{3+}$ (d^4), $[Co(OH_2)_6]^{3+}$ (d^6), $[Ni(OH_2)_6]^{2+}$ (d^8) and $[Zn(OH_2)_6]^{2+}$ (d^{10}) to vary as the electronic configuration of the metal ion changes. However, each of these species has an octahedral arrangement of ligands (**19.1**). Thus, it is clear that the VSEPR model is not applicable to *d*-block metal complexes.

$$\left[\begin{array}{c} OH_2 \\ H_2O_{\prime\prime\prime\prime\prime} | \prime\prime\prime\prime\prime OH_2 \\ M \\ H_2O \quad OH_2 \\ | \\ OH_2 \end{array} \right]^{n+}$$

(19.1)

[†]Alfred Werner was the first to recognize the existence of coordination complexes and was awarded the 1913 Nobel Prize in Chemistry; see http://nobelprize.org.

We turn instead to the *Kepert model*, in which the metal lies at the centre of a sphere and the ligands are free to move over the surface of the sphere. The ligands are considered to repel one another in a similar manner to the point charges in the VSEPR model. However, unlike the VSEPR model, that of Kepert *ignores non-bonding electrons*. Thus, the coordination geometry of a *d*-block species is considered by Kepert to be *independent* of the ground state electronic configuration of the metal centre, and so species of type $[ML_n]$, $[ML_n]^{m+}$ and $[ML_n]^{m-}$ have the *same* coordination geometry.

> The **Kepert model** rationalizes the shapes of *d*-block metal complexes $[ML_n]$, $[ML_n]^{m+}$ or $[ML_n]^{m-}$ by considering the repulsions between the groups L. Lone pairs of electrons are ignored. For coordination numbers between 2 and 6, the following arrangements of donor atoms are predicted:
>
> 2 linear
> 3 trigonal planar
> 4 tetrahedral
> 5 trigonal bipyramidal *or* square-based pyramidal
> 6 octahedral

Table 19.4 lists coordination environments associated with coordination numbers between 2 and 9. Some, but not all, of these ligand arrangements are in accord with the Kepert model. For example, the coordination sphere in $[Cu(CN)_3]^{2-}$ is predicted by the Kepert model to be trigonal planar (**19.2**). Indeed, this is what is found experimentally. The other option in Table 19.4 is trigonal pyramidal, but this does not minimize interligand repulsions. One of the most important classes of structure for which the Kepert model does not predict the correct answer is that of the square planar complex, and here electronic effects are usually the controlling factor (see Section 20.3). Another factor that may lead to a breakdown of the Kepert model is the inherent constraint of a ligand. For example:

- the four nitrogen donor atoms of a porphyrin ligand (Fig. 12.10a) are confined to a square planar array;
- *tripodal ligands* such as **19.3** have limited flexibility which means that the donor atoms are not necessarily free to adopt the positions predicted by Kepert;
- macrocyclic ligands (see Section 11.8) are less flexible than open chain ligands.

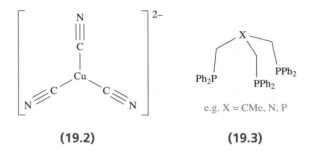

(19.2)   **(19.3)**

e.g. X = CMe, N, P

> A *tripodal* ligand (e.g. **19.3**) is one containing three arms, each with a donor atom, which radiate from a central atom or group; this central point may itself be a donor atom.

Coordination numbers in the solid state

In the remaining part of this section, we give a systematic outline of the occurrence of different coordination numbers and geometries in *solid state d*-block metal complexes. A general word of caution: molecular formulae can be misleading in terms of coordination number. For example in

CdI_2 (Fig. 6.23), each Cd centre is octahedrally sited, and molecular halides or pseudohalides (e.g. $[CN]^-$) may contain M−X−M bridges and exist as oligomers, e.g. α-$PdCl_2$ is polymeric (**19.4**).

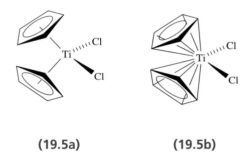

(19.4)

A further ambiguity arises when the bonding mode of a ligand can be described in more than one way. This often happens in organometallic chemistry, for example with cyclopentadienyl ligands as discussed in Chapter 24. The nomenclature introduced in Box 19.1 assists, but there is still the question of whether to consider, for example, an $[\eta^5\text{-}C_5H_5]^-$ ligand as occupying one or five sites in the coordination sphere of a metal atom: thus, the coordination number of the Ti(IV) centre in $[(\eta^5\text{-}C_5H_5)_2TiCl_2]$ may be represented as either **19.5a** or **19.5b**.

(19.5a)   **(19.5b)**

THEORY

Box 19.1 η-Nomenclature for ligands

In organometallic chemistry in particular, but also in coordination chemistry, use of the Greek prefix η (eta) is encountered; the letter is accompanied by a superscript number (e.g. η^3). This prefix describes the number of atoms in a ligand which directly interact with the metal centre, the *hapticity* of the ligand. For example, the cyclopentadienyl ligand, $[C_5H_5]^-$ or

Cp^-, is versatile in its modes of bonding, and examples include those shown below. Note the different ways of representing the η^3- and η^5-modes.

In coordination chemistry, the η terminology is used for ligands such as $[O_2]^{2-}$ (peroxido ligand), as exemplified in structure 21.3.

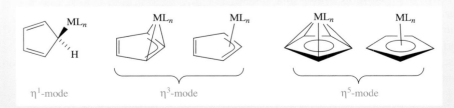

η^1-mode η^3-mode η^5-mode

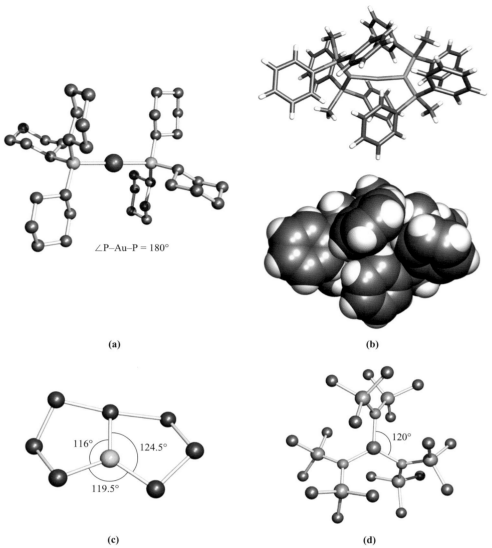

∠P–Au–P = 180°

(a)

(b)

116°   124.5°

119.5°

(c)

120°

(d)

Fig. 19.4 Examples of 2- and 3-coordinate structures (X-ray diffraction data): (a) $[Au\{P(cyclo\text{-}C_6H_{11})_3\}_2]^+$ in the chloride salt [J.A. Muir *et al.* (1985) *Acta Crystallogr., Sect. C*, vol. 41, p. 1174], (b) stick and space-filling representations of the 2-coordinate iron(II) complex $[Fe\{N(SiMePh_2)_2\}_2]$ [R.A. Bartlett *et al.* (1987) *J. Am. Chem. Soc.*, vol. 109, p. 7563], (c) $[AgTe_7]^{3-}$ in the salt $[Et_4N][Ph_4P]_2[AgTe_7]$ [J.M. McConnachie *et al.* (1993) *Inorg. Chem.*, vol. 32, p. 3201], and (d) $[Fe\{N(SiMe_3)_2\}_3]$ [M.B. Hursthouse *et al.* (1972) *J. Chem. Soc., Dalton Trans.*, p. 2100]. Hydrogen atoms are omitted for clarity; colour code: Au, red; Ag, yellow; Fe, green; C, grey; P, orange; Te, dark blue; Si, pink; N, light blue.

Coordination number 2

Examples of coordination number 2 are uncommon, being generally restricted to Cu(I), Ag(I), Au(I) and Hg(II), all d^{10} ions. Examples include $[CuCl_2]^-$, $[Ag(NH_3)_2]^+$, $[Au(CN)_2]^-$, $(R_3P)AuCl$, $[Au(PR_3)_2]^+$ (R = alkyl or aryl, Fig. 19.4a) and $Hg(CN)_2$, in each of which the metal centre is in a linear environment. However, in the solid state, the Cu(I) centre in $K[Cu(CN)_2]$ is 3-coordinate by virtue of cyanido-bridge formation (see structure 21.71). Bulky amido ligands, e.g. $[N(SiR_3)_2]^-$, are often associated with *low* coordination numbers. For example, in $[Fe\{N(SiMePh_2)_2\}_2]$ (∠N–Fe–N = 169°, Fig. 19.4b), the sterically demanding amido groups force a 2-coordinate

environment on a metal centre that usually prefers to be surrounded by a greater number of ligands.

Coordination number 3

3-Coordinate complexes are not common. Usually, trigonal planar structures are observed, and examples involving d^{10} metal centres include:

- Cu(I) in $[Cu(CN)_3]^{2-}$ (**19.2**), $[Cu(CN)_2]^-$ (see above), $[Cu(SPMe_3)_3]^+$;
- Ag(I) in $[AgTe_7]^{3-}$ (Fig. 19.4c), $[Ag(PPh_3)_3]^+$;
- Au(I) in $[Au\{PPh(C_6H_{11})_2\}_3]^+$;
- Hg(II) in $[HgI_3]^-$, $[Hg(SPh_3)_3]^-$;
- Pt(0) in $[Pt(PPh_3)_3]$, $[Pt(P^tBu_2H)_3]$.

Sterically demanding amido ligands have been used to stabilize complexes containing 3-coordinate metal ions, e.g. $[Fe\{N(SiMe_3)_2\}_3]$ (Fig. 19.4d). In the solid state, $[Y\{N(SiMe_3)_2\}_3]$ and $[Sc\{N(SiMe_3)_2\}_3]$ possess *trigonal pyramidal* metal centres ($\angle N–Y–N = 115°$ and $\angle N–Sc–N = 115.5°$), but it is likely that crystal packing effects cause the deviation from planarity. The fact that in the gas phase $[Sc\{N(SiMe_3)_2\}_3]$ contains a trigonal planar Sc(III) centre supports this proposal.

p-Block chemistry has a number of examples of T-shaped molecules (e.g. ClF_3) in which stereochemically active lone pairs play a crucial role. *d*-Block metal complexes do not mimic this behaviour, although ligand constraints (e.g. the bite angle of a chelate) may distort a 3-coordinate structure away from the expected trigonal planar structure.

Coordination number 4

4-Coordinate complexes are extremely common, with a tetrahedral arrangement of donor atoms being the most frequently observed. The tetrahedron is sometimes 'flattened', distortions being attributed to steric or crystal packing effects or, in some cases, electronic effects. Tetrahedral complexes for d^3 ions are rarely, if ever, encountered. Complex **19.6** exemplifies the stabilization of 4-coordinate Cr^{3+} (d^3) using a tripodal ligand. This coordination geometry is enforced by the ligand, and the Cr–N distances in the 'trigonal plane' are shorter (188 pm) than the axial distance (224 pm).

(19.6)

Tetrahedral complexes for d^4 ions have been stabilized only with bulky amido ligands, e.g. $[M(NPh_2)_4]$ and $[M\{N(SiMe_3)_2\}_3Cl]$ for M = Hf or Zr. Simple tetrahedral species include:

- d^0: $[VO_4]^{3-}$, $[CrO_4]^{2-}$, $[MoS_4]^{2-}$, $[WS_4]^{2-}$, $[MnO_4]^-$, $[TcO_4]^-$, RuO_4, OsO_4;
- d^1: $[MnO_4]^{2-}$, $[TcO_4]^{2-}$, $[ReO_4]^{2-}$, $[RuO_4]^-$;
- d^2: $[FeO_4]^{2-}$, $[RuO_4]^{2-}$;
- d^5: $[FeCl_4]^-$, $[MnCl_4]^{2-}$;
- d^6: $[FeCl_4]^{2-}$, $[FeI_4]^{2-}$;
- d^7: $[CoCl_4]^{2-}$;
- d^8: $[NiCl_4]^{2-}$, $[NiBr_4]^{2-}$;
- d^9: $[CuCl_4]^{2-}$ (distorted);
- d^{10}: $[ZnCl_4]^{2-}$, $[HgBr_4]^{2-}$, $[CdCl_4]^{2-}$, $[Zn(OH)_4]^{2-}$, $[Cu(CN)_4]^{3-}$, $[Ni(CO)_4]$.

The solid state structures of apparently simple anions may in fact be polymeric (e.g. the presence of fluoride bridges in $[CoF_4]^{2-}$ and $[NiF_4]^{2-}$ leads to a layered structure with octahedral metal centres) or may be cation-dependent (e.g. discrete tetrahedral $[MnCl_4]^{2-}$ ions are present in the Cs^+ and $[Me_4N]^+$ salts, but a polymeric structure with Mn−Cl−Mn bridges is adopted by the Na^+ salt).

Square planar complexes are rarer than tetrahedral, and are often associated with d^8 configurations where electronic factors strongly favour a square planar arrangement (see Section 20.3), e.g. $[PdCl_4]^{2-}$, $[PtCl_4]^{2-}$, $[AuCl_4]^-$, $[AuBr_4]^-$, $[RhCl(PPh_3)_3]$ and $trans$-$[IrCl(CO)(PPh_3)_2]$. The classification of distorted structures such as those in $[Ir(PMePh_2)_4]^+$ and $[Rh(PMe_2Ph)_4]^+$ (Fig. 19.5a) may be ambiguous, but in this case, the fact that each metal ion is d^8 suggests that steric crowding causes deviation from a square planar arrangement (not from a tetrahedral one). The $[Co(CN)_4]^{2-}$ ion is a rare example of a square planar d^7 complex.

Coordination number 5

The limiting structures for 5-coordination are the trigonal bipyramid and square-based pyramid. In practice, many structures lie between these two extremes, the energy difference between trigonal bipyramidal and square-based pyramidal structures usually being small (see Section 4.8). Among simple 5-coordinate complexes are trigonal bipyramidal $[CdCl_5]^{3-}$, $[HgCl_5]^{3-}$ and $[CuCl_5]^{3-}$ (d^{10}) and a series of square-based pyramidal oxido- or nitrido-complexes in which the oxido or nitrido ligand occupies the axial site:

- d^0: $[NbCl_4(O)]^-$;
- d^1: $[V(acac)_2(O)]$, $[WCl_4(O)]^-$ (**19.7**), $[TcCl_4(N)]^-$ (**19.8**), $[TcBr_4(N)]^-$;
- d^2: $[TcCl_4(O)]^-$, $[ReCl_4(O)]^-$.

(19.7)　　　　**(19.8)**

The formulae of some complexes may misleadingly suggest '5-coordinate' metal centres: e.g. Cs_3CoCl_5 is actually $Cs_3[CoCl_4]Cl$.

5-Coordinate structures are found for many compounds with polydentate amine, phosphane or arsane ligands. Of particular interest among these are complexes containing tripodal ligands (**19.3**) in which the central atom is a donor atom. This makes the ligand ideally suited to occupy one axial and the three equatorial sites of a trigonal bipyramidal complex as in $[CoBr\{N(CH_2CH_2NMe_2)_3\}]^+$, $[Rh(SH)\{P(CH_2CH_2PPh_2)_3\}]$ and $[Zn\{N(CH_2CH_2NH_2)_3\}Cl]^+$ (Fig. 19.5b). On the other

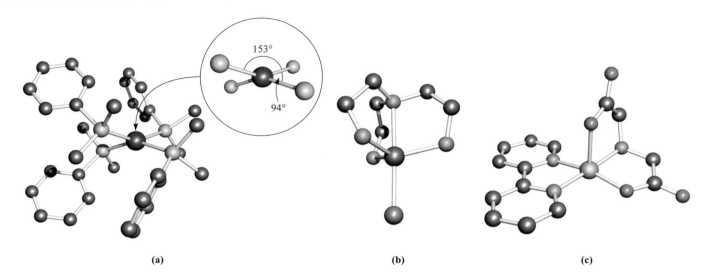

(a)   **(b)**   **(c)**

Fig. 19.5 Examples of 4- and 5-coordinate structures (X-ray diffraction data): (a) in $[Rh(PMe_2Ph)_4]^+$, the steric demands of the ligands distort the structure from the square planar structure expected for this d^8 metal centre [J.H. Reibenspies *et al.* (1993) *Acta Crystallogr., Sect. C*, vol. 49, p. 141], (b) $[Zn\{N(CH_2CH_2NH_2)_3\}Cl]^+$ in the $[Ph_4B]^-$ salt [R.J. Sime *et al.* (1971) *Inorg. Chem.*, vol. 10, p. 537], and (c) $[Cu(bpy)\{NH(CH_2CO_2)_2\}]$, crystallized as the hexahydrate [R.E. Marsh *et al.* (1995) *Acta Crystallogr., Sect. B*, vol. 51, p. 300]. Hydrogen atoms are omitted for clarity; colour code: Rh, dark blue; P, orange; Zn, brown; Cl, green; N, light blue; Cu, yellow; O, red; C, grey.

hand, the conformational constraints of the ligands may result in a preference for a square-based pyramidal complex in the solid state, e.g. $[Cu(bpy)\{NH(CH_2CO_2)_2\}]\cdot6H_2O$ (Fig. 19.5c).

Coordination number 6

For many years after Werner's proof from stereochemical studies that many 6-coordinate complexes of chromium and cobalt had octahedral structures (see Box 21.7), it was believed that no other form of 6-coordination occurred, and a vast amount of data from X-ray diffraction studies seemed to support this. Eventually, however, examples of trigonal prismatic coordination were confirmed.

The regular or nearly regular octahedral coordination sphere is found for all electronic configurations from d^0 to d^{10}, e.g. $[TiF_6]^{2-}$ (d^0), $[Ti(OH_2)_6]^{3+}$ (d^1), $[V(OH_2)_6]^{3+}$ (d^2), $[Cr(OH_2)_6]^{3+}$ (d^3), $[Mn(OH_2)_6]^{3+}$ (d^4), $[Fe(OH_2)_6]^{3+}$ (d^5), $[Fe(OH_2)_6]^{2+}$ (d^6), $[Co(OH_2)_6]^{2+}$ (d^7), $[Ni(OH_2)_6]^{2+}$ (d^8), $[Cu(NO_2)_6]^{4-}$ (d^9) and $[Zn(OH_2)_6]^{2+}$ (d^{10}). There are distinctions between what are termed *low-spin* and *high-spin* complexes (see Section 20.1): where the distinction is meaningful, the examples listed above are high-spin complexes, but many octahedral low-spin complexes are also known, e.g. $[Mn(CN)_6]^{3-}$ (d^4), $[Fe(CN)_6]^{3-}$ (d^5), $[Co(CN)_6]^{3-}$ (d^6). Octahedral complexes of d^4 and d^9 metal ions tend to be *tetragonally distorted*, i.e. they are elongated or squashed. This is an electronic effect called *Jahn–Teller distortion* (see Section 20.3).

While the vast majority of 6-coordinate complexes containing simple ligands are octahedral, there is a small group of d^0 or d^1 metal complexes in which the metal centre is in a trigonal prismatic or distorted trigonal prismatic environment. The octahedron and trigonal prism are closely related, and can be described in terms of two triangles which are staggered (**19.9**) or eclipsed (**19.10**).

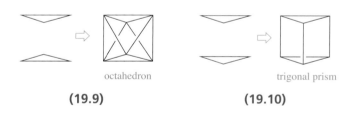

octahedron   trigonal prism

(19.9)   **(19.10)**

The complexes $[ReMe_6]$ (d^1), $[TaMe_6]^-$ (d^0) and $[ZrMe_6]^{2-}$ (d^0) contain regular trigonal prismatic (D_{3h}) metal centres, while in $[MoMe_6]$ (d^0), $[WMe_6]$ $(d^0$, Fig. 19.6a), $[NbMe_6]^-$ (d^0) and $[TaPh_6]^-$ (d^0) the coordination environment is distorted trigonal prismatic (C_{3v}). The common feature of the ligands in these complexes is that they are σ-donors, with no π-donating or π-accepting properties. In $[Li(TMEDA)]_2[Zr(SC_6H_4\text{-}4\text{-}Me)_6]$ (TMEDA = $Me_2NCH_2CH_2NMe_2$), the $[Zr(SC_6H_4\text{-}4\text{-}Me)_6]^{2-}$ ion also has a distorted trigonal prismatic structure. Although thiolate ligands are usually weak π-donor ligands, it has been suggested that the cation–anion interactions in crystalline $[Li(TMEDA)]_2[Zr(SC_6H_4\text{-}4\text{-}Me)_6]$ result in the RS^- ligands behaving only as σ-donors. Another related group of trigonal prismatic d^0, d^1 or d^2 metal complexes contains

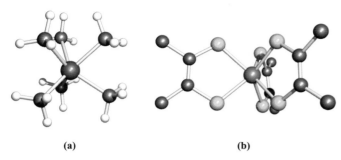

Fig. 19.6 The trigonal prismatic structures of (a) $[WMe_6]$ [V. Pfennig *et al.* (1996) *Science*, vol. 271, p. 626] and (b) $[Re(S_2C_2Ph_2)_3]$, only the *ipso*-C atoms of each Ph ring are shown [R. Eisenberg *et al.* (1966) *Inorg. Chem.*, vol. 5, p. 411]. Hydrogen atoms are omitted from (b); colour code: W, red; Re, green; C, grey; S, yellow; H, white.

the dithiolate ligands, **19.11**, and includes $[Mo(S_2C_2H_2)_3]$ and $[Re(S_2C_2Ph_2)_3]$ (Fig. 19.6b). We return to σ-donor and π-donor ligands in Section 20.4, and to the question of octahedral versus trigonal prismatic complexes in Box 20.3.

R——C——S⁻
      ‖
R——C——S⁻

e.g. R = H, Ph

(19.11)

The complexes $[WL_3]$, $[TiL_3]^{2-}$, $[ZrL_3]^{2-}$ and $[HfL_3]^{2-}$ (L is **19.12**) also possess trigonal prismatic structures. For a regular trigonal prism, angle α in **19.13** is $0°$ and this is observed for $[TiL_3]^{2-}$ and $[HfL_3]^{2-}$. In $[ZrL_3]^{2-}$, $\alpha = 3°$, and in $[WL_3]$, $\alpha = 15°$. Formally, $[WL_3]$ contains W(0) and is a d^6 complex, while $[ML_3]^{2-}$ (M = Ti, Zr, Hf) contains the metal in a -2 oxidation state. However, theoretical results for $[WL_3]$ indicate that negative charge is transferred on to the ligands. In the extreme case, the ligands can be formulated as L^{2-} and the metal as a d^0 centre.[†]

(19.12)

(19.13)

[†] For a detailed discussion, see: P. Rosa, N. Mézailles, L. Ricard, F. Mathey and P. Le Floch (2000) *Angew. Chem. Int. Ed.*, vol. 39, p. 1823 and references in this paper.

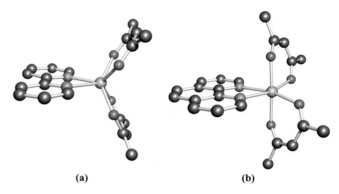

Fig. 19.7 The solid state structures of (a) $[Mn(acac)_2(bpy)]$ (trigonal prismatic) [R. van Gorkum *et al.* (2005) *Eur. J. Inorg. Chem.*, p. 2255] and (b) $[Mn(acac)_2(phen)]$ (octahedral) [F.S. Stephens (1977) *Acta Crystallogr., Sect. B*, vol. 33, p. 3492]. Hydrogen atoms have been omitted; colour code: Mn, orange; N, blue; O, red; C, grey.

The solid state structures of $[Mn(acac)_2(bpy)]$ (trigonal prismatic, Fig. 19.7a) and $[Mn(acac)_2(phen)]$ (octahedral, Fig. 19.7b) provide an example in which crystal packing forces appear to dictate the difference in ligand arrangement. The energy difference between the two structures is calculated to be very small, and the preference for a trigonal prism in $[Mn(acac)_2(bpy)]$ is observed only in the solid state.

Coordination number 7

High coordination numbers (≥ 7) are observed most frequently for ions of the early second and third row d-block metals and for the lanthanoids and actinoids, i.e. r_{cation} must be relatively large (see Chapter 27). Figure 19.8a shows the arrangement of the donor atoms for the three idealized 7-coordinate structures. In the capped trigonal prism, the 'cap' is over one of the square faces of the prism. In reality, there is much distortion from these idealized structures, and this is apparent for the example of a capped octahedral complex shown in Fig. 19.8b. The anions in $[Li(OEt_2)]^+[MoMe_7]^-$ and $[Li(OEt_2)]^+[WMe_7]^-$ are further examples of capped octahedral structures. A problem in the chemical literature is that the distortions may lead to ambiguity in the way in which a given structure is described. Among binary metal halides and pseudohalides, 7-coordinate structures are exemplified by the pentagonal bipyramidal ions $[V(CN)_7]^{4-}$ (d^2) and $[NbF_7]^{3-}$ (d^1). In the ammonium salt, $[ZrF_7]^{3-}$ (d^0) is pentagonal bipyramidal, but in the guanidinium salt, it has a monocapped trigonal prismatic structure (Fig. 19.8c). Further examples of monocapped trigonal prismatic complexes are $[NbF_7]^{2-}$ and $[TaF_7]^{2-}$ (d^0). 7-Coordinate complexes containing oxido ligands may favour pentagonal bipyramidal structures with the oxido group in an axial site, e.g. $[Nb(O)(ox)_3]^{3-}$,

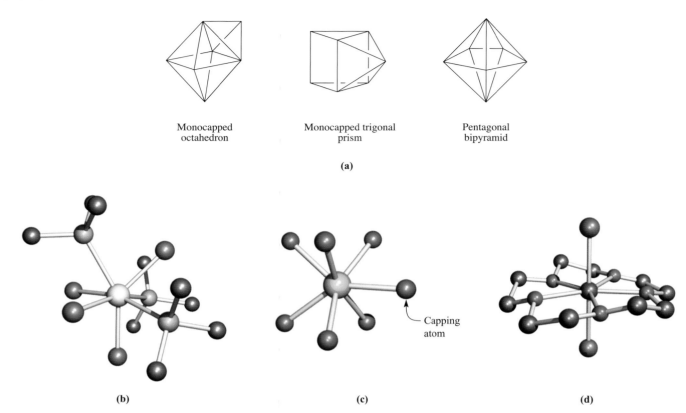

Monocapped octahedron

Monocapped trigonal prism

Pentagonal bipyramid

(a)

(b)

(c)

Capping atom

(d)

Fig. 19.8 (a) The coordination spheres defined by the donor atoms in idealized 7-coordinate structures. Examples of 7-coordinate complexes (X-ray diffraction data): (b) the capped octahedral structure of $[TaCl_4(PMe_3)_3]$ [F.A. Cotton *et al.* (1984) *Inorg. Chem.*, vol. 23, p. 4046], (c) the capped trigonal prismatic $[ZrF_7]^{3-}$ in the guanidinium salt [A.V. Gerasimenko *et al.* (1985) *Koord. Khim.*, vol. 11, p. 566], and (d) the pentagonal bipyramidal cation in $[ScCl_2(15\text{-crown-5})]_2[CuCl_4]$ with the crown ether occupying the equatorial plane [N.R. Strel'tsova *et al.* (1992) *Zh. Neorg. Khim.*, vol. 37, p. 1822]. Hydrogen atoms have been omitted for clarity; colour code: Ta, silver; Cl, green; P, orange; Zr, yellow; F, green; Sc, brown; C, grey; O, red.

$[Nb(O)(OH_2)_2(ox)_2]^-$ and $[Mo(O)(O_2)_2(ox)]^{2-}$ (all d^0). In this last example, two peroxido ligands are present, each in an η^2 mode (**19.14**). Macrocyclic ligands containing five donor atoms (e.g. 15-crown-5) may dictate that the coordination geometry is pentagonal bipyramidal as shown in Fig. 19.8d.

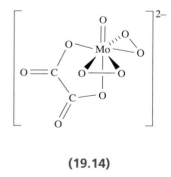

(19.14)

Coordination number 8

As the number of vertices in a polyhedron increases, so does the number of possible structures (Fig. 19.9a). Probably, the best known 8-vertex polyhedron is the cube, (**19.15**), but this

is hardly ever observed as an arrangement of donor atoms in complexes. The few examples include the anions in the actinoid complexes $Na_3[PaF_8]$, $Na_3[UF_8]$ and $[Et_4N]_4[U(NCS\text{-}N)_8]$. Steric hindrance between ligands can be reduced by converting a cubic into a square antiprismatic arrangement, i.e. on going from **19.15** to **19.16**.

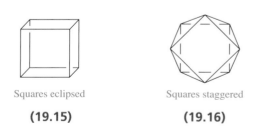

Squares eclipsed

Squares staggered

(19.15)

(19.16)

Square antiprismatic coordination environments occur in $[Zr(acac)_4]$ (d^0) and in the anions in the salts $Na_3[TaF_8]$ (d^0), $K_2[ReF_8]$ (d^1) and $K_2[H_3NCH_2CH_2NH_3][Nb(ox)_4]$ (d^1) (Fig. 19.9b). Specifying the counter-ion is important since the energy difference between 8-coordinate structures tends to be small with the result that the preference between two structures may be altered by crystal packing forces in

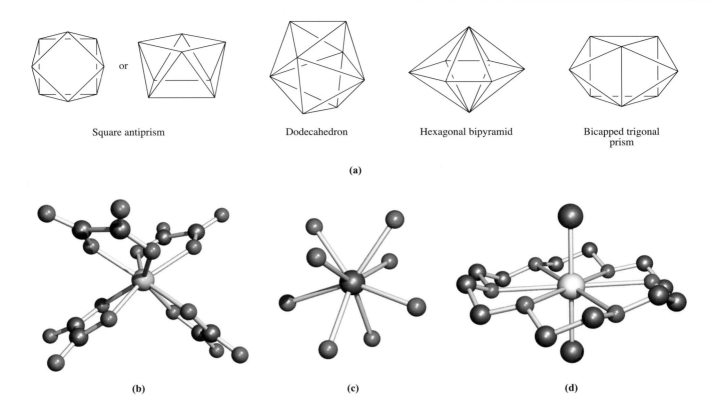

(a)

(b)   **(c)**   **(d)**

Fig. 19.9 (a) The coordination spheres defined by the donor atoms in idealized 8-coordinate structures. The left-hand drawing of the square antiprism emphasizes that the two square faces are mutually staggered. Examples of 8-coordinate complexes (X-ray diffraction): (b) the square antiprismatic structure of $[Nb(ox)_4]^{4-}$ in the salt $K_2[H_3NCH_2CH_2NH_3][Nb(ox)_4] \cdot 4H_2O$ [F.A. Cotton *et al.* (1987) *Inorg. Chem.*, vol. 26, p. 2889]; (c) the dodecahedral ion $[Y(OH_2)_8]^{3+}$ in the salt $[Y(OH_2)_8]Cl_3 \cdot (15\text{-crown-5})$ [R.D. Rogers *et al.* (1986) *Inorg. Chim. Acta*, vol. 116, p. 171]; and (d) $[CdBr_2(18\text{-crown-6})]$ with the macrocyclic ligand occupying the equatorial plane of a hexagonal bipyramid [A. Hazell (1988) *Acta Crystallogr., Sect. C*, vol. 44, p. 88]. Hydrogen atoms have been omitted for clarity; colour code: Nb, yellow; O, red; Y, brown; Cd, silver; C, grey; Br, brown.

two different salts. Examples are seen in a range of salts of $[Mo(CN)_8]^{3-}$, $[W(CN)_8]^{3-}$, $[Mo(CN)_8]^{4-}$ or $[W(CN)_8]^{4-}$ which possess square antiprismatic or dodecahedral structures depending on the cation. Further examples of dodecahedral complexes include $[Y(OH_2)_8]^{3+}$ (Fig. 19.9c) and a number of complexes with bidentate ligands: $[Mo(O_2)_4]^{2-}$ (d^0), $[Ti(NO_3)_4]$ (d^0), $[Cr(O_2)_4]^{3-}$ (d^1), $[Mn(NO_3)_4]^{2-}$ (d^5) and $[Fe(NO_3)_4]^-$ (d^5).

The hexagonal bipyramid is a rare coordination environment, but may be favoured in complexes containing a hexadentate macrocyclic ligand, for example $[CdBr_2(18\text{-crown-6})]$, Fig. 19.9d. A bicapped trigonal prism is another option for 8-coordination, but is only rarely observed, e.g. in $[ZrF_8]^{4-}$ (d^0) and $[La(acac)_3(OH_2)_2] \cdot H_2O$ (d^0).

Coordination number 9

The anions $[ReH_9]^{2-}$ and $[TcH_9]^{2-}$ (both d^0) provide examples of 9-coordinate species in which the metal centre is in a tricapped trigonal prismatic environment (see Fig. 10.14c). A coordination number of 9 is most often associated with yttrium, lanthanum and the *f*-block elements. The tricapped trigonal prism is the only *regular* arrangement of donor atoms yet observed, e.g. in $[Sc(OH_2)_9]^{3+}$, $[Y(OH_2)_9]^{3+}$ and $[La(OH_2)_9]^{3+}$.

Coordination numbers of 10 and above

It is always dangerous to draw conclusions on the basis of the non-existence of structure types, but, from data available at the present time, it seems that a coordination of ≥ 10 is generally confined to the *f*-block metal ions (see Chapter 27). Lanthanum exhibits coordination numbers of 10 and 12, e.g. in $[La(NO_3\text{-}O,O')_2(OH_2)_6]^+$ (Fig. 19.10a) and $[La(NO_3\text{-}O,O')_6]^{3-}$. In both complexes, the nitrate ions are bidentate, as indicated by the nomenclature in the formulae. However, although La is in group 3, it is usually classed with the lanthanoid, rather than *d*-block, metals. Within the *d*-block, complexes containing $[BH_4]^-$ and related ligands provide rare examples of coordination numbers >9. For example, in $[Hf(BH_4)_4]$ and $[Zr(MeBH_3)_4]$ the ligands are tridentate (see structure **13.9**) and the metal centres

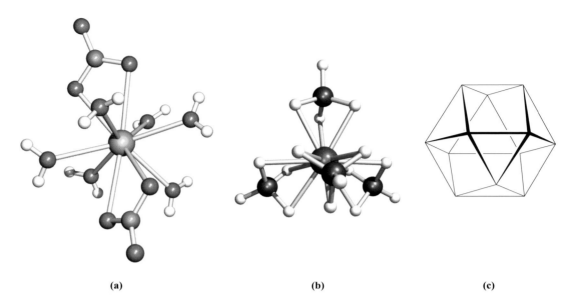

(a)   **(b)**   **(c)**

Fig. 19.10 (a) The structure (X-ray diffraction) of the $[La(NO_3\text{-}O,O')_2(OH_2)_6]^+$ cation in the nitrate salt [J.C. Barnes (2006) Private communication to the CSD]. Colour code: La, gold; O, red; N, blue; H, white. (b) The structure of $[Hf(BH_4)_4]$ determined by neutron diffraction at low temperature [R.W. Broach *et al.* (1983) *Inorg. Chem.*, vol. 22, p. 1081]. Colour code: Hf, red; B, blue; H, white. (c) The 12-vertex cubeoctahedral coordination sphere of the Hf(IV) centre in $[Hf(BH_4)_4]$.

are 12-coordinate. Figures 19.10b and c show the structure of $[Hf(BH_4)_4]$ and the cubeoctahedral arrangement of the 12 hydrogen atoms around the metal centre. The same coordination environment is found in $[Zr(MeBH_3)_4]$.

19.8 Isomerism in *d*-block metal complexes

In this book so far, we have not had cause to mention isomerism very often, and most references have been to *trans*- and *cis*-isomers, e.g. *trans*-$[CaI_2(THF)_4]$ (Section 12.5) and the *trans*- and *cis*-isomers of N_2F_2 (Section 15.7). These are *diastereoisomers* (see Section 2.9).

> **Stereoisomers** possess the same connectivity of atoms, but differ in the spatial arrangement of atoms or groups. Examples include *trans*- and *cis*-isomers, and *mer*- and *fac*-isomers. If the stereoisomers are *not* mirror images of one another, they are called **diastereoisomers**. Stereoisomers that *are* mirror images of one another are called **enantiomers**.

Self-study exercises

All the answers can be found by reading Section 2.9.

1. Draw possible structures for the square planar complexes $[PtBr_2(py)_2]$ and $[PtCl_3(PEt_3)]^-$ and give names to distinguish between any isomers that you have drawn.

2. In $[Ru(CO)_4(PPh_3)]$, the Ru centre is in a trigonal bipyramidal environment. Draw the structures of possible isomers and give names to distinguish between them.

3. Draw the structures and name the isomers of octahedral $[CrCl_2(NH_3)_4]^+$.

4. Octahedral $[RhCl_3(OH_2)_3]$ has two isomers. Draw their structures and give them distinguishing names.

Figure 19.11 classifies the types of isomers exhibited by coordination complexes. In the rest of this section, we introduce *structural* (or *constitutional*) *isomerism*, followed by a discussion of *enantiomers*.

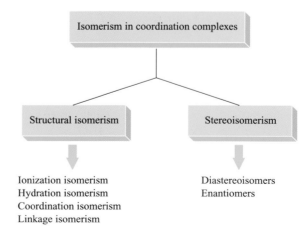

Fig. 19.11 Classification of types of isomerism in metal complexes.

Structural isomerism: ionization isomers

Ionization isomers result from the interchange of an anionic ligand within the first coordination sphere with an anion outside the coordination sphere.

Examples of ionization isomers are violet $[Co(NH_3)_5Br][SO_4]$ (prepared by reaction scheme 19.6) and red $[Co(NH_3)_5(SO_4)]Br$ (prepared by reaction sequence 19.7). These isomers can be readily distinguished by appropriate qualitative tests for *ionic* sulfate or bromide, respectively (eqs. 19.8 and 19.9). The isomers are also easily distinguished by IR spectroscopy. The free $[SO_4]^{2-}$ ion belongs to the T_d point group. The two T_2 vibrational modes are IR active (see Fig. 3.16) and strong absorptions are observed at $1104\,cm^{-1}$ (stretch) and $613\,cm^{-1}$ (deformation). In $[Co(NH_3)_5(SO_4)]Br$, the sulfate ion acts as a monodentate ligand and the symmetry of the $[SO_4]^{2-}$ group is lowered with respect to the free ion. As a result, the IR spectrum of $[Co(NH_3)_5(SO_4)]Br$ shows three absorptions (1040, 1120 and $970\,cm^{-1}$) arising from stretching modes of the coordinated $[SO_4]^{2-}$ ligand.

$$CoBr_2 \xrightarrow{[NH_4]Br,\ NH_3,\ O_2} [Co(NH_3)_5(OH_2)]Br_3$$

$$\downarrow \Delta$$

$$[Co(NH_3)_5Br]Br_2$$

$$\downarrow Ag_2SO_4$$

$$[Co(NH_3)_5Br][SO_4] \qquad (19.6)$$

$$[Co(NH_3)_5Br]Br_2 \xrightarrow{conc\ H_2SO_4} [Co(NH_3)_5(SO_4)][HSO_4]$$

$$\downarrow BaBr_2$$

$$[Co(NH_3)_5(SO_4)]Br \qquad (19.7)$$

$$BaCl_2(aq) + [SO_4]^{2-}(aq) \longrightarrow \underset{\text{white ppt}}{BaSO_4(s)} + 2Cl^-(aq) \qquad (19.8)$$

$$AgNO_3(aq) + Br^-(aq) \longrightarrow \underset{\substack{\text{pale yellow} \\ \text{ppt}}}{AgBr(s)} + [NO_3]^-(aq) \qquad (19.9)$$

Structural isomerism: hydration isomers

Hydration isomers result from the interchange of H_2O and another ligand between the first coordination sphere and the ligands outside it.

The classic example of hydrate isomerism is that of the compound of formula $CrCl_3 \cdot 6H_2O$. Green crystals of chromium(III) chloride formed from a hot solution obtained by reducing chromium(VI) oxide with concentrated hydrochloric acid are $[Cr(OH_2)_4Cl_2]Cl \cdot 2H_2O$. When this is dissolved in water, the chloride ions in the complex are slowly replaced by water to give blue-green $[Cr(OH_2)_5Cl]Cl_2 \cdot H_2O$ and finally violet $[Cr(OH_2)_6]Cl_3$. The complexes can be distinguished by precipitation of the *free* chloride ion using aqueous silver nitrate (eq. 19.10).

$$AgNO_3(aq) + Cl^-(aq) \longrightarrow \underset{\text{white ppt}}{AgCl(s)} + [NO_3]^-(aq) \qquad (19.10)$$

Structural isomerism: coordination isomerism

Coordination isomers are possible only for salts in which both cation and anion are complex ions. The isomers arise from interchange of ligands between the two metal centres.

Examples of coordination isomers are:

- $[Co(NH_3)_6][Cr(CN)_6]$ and $[Cr(NH_3)_6][Co(CN)_6]$;
- $[Co(NH_3)_6][Co(NO_2)_6]$ and $[Co(NH_3)_4(NO_2)_2][Co(NH_3)_2(NO_2)_4]$;
- $[Pt^{II}(NH_3)_4][Pt^{IV}Cl_6]$ and $[Pt^{IV}(NH_3)_4Cl_2][Pt^{II}Cl_4]$.

Structural isomerism: linkage isomerism

Linkage isomers may arise when one or more of the ligands can coordinate to the metal ion in more than one way, e.g. in $[SCN]^-$ (**19.17**), both the N and S atoms are potential donor sites. Such a ligand is *ambidentate*.

$$[S = C = N]^-$$

(19.17)

Because $[SCN]^-$ is ambidentate, the complex $[Co(NH_3)_5(NCS)]^{2+}$ has two isomers which are distinguished by using the following nomenclature:

- in $[Co(NH_3)_5(NCS\text{-}N)]^{2+}$, the thiocyanate ligand coordinates through the nitrogen donor atom;
- in $[Co(NH_3)_5(NCS\text{-}S)]^{2+}$, the thiocyanate ion is bonded to the metal centre through the sulfur atom.

Scheme 19.11 shows how linkage isomers of $[Co(NH_3)_5(NO_2)]^{2+}$ can be prepared.

$$[Co(NH_3)_5Cl]Cl_2 \xrightarrow{dil\ NH_3(aq)} [Co(NH_3)_5(OH_2)]Cl_3$$

$$\downarrow NaNO_2 \qquad\qquad\qquad \downarrow NaNO_2,\ conc\ HCl$$

$$\underset{red}{[Co(NH_3)_5(NO_2\text{-}O)]Cl_2} \underset{UV}{\overset{warm\ HCl\ or\ spontaneous}{\rightleftharpoons}} \underset{yellow}{[Co(NH_3)_5(NO_2\text{-}N)]Cl_2}$$

$$(19.11)$$

In this example, the complexes $[Co(NH_3)_5(NO_2\text{-}O)]^{2+}$ and $[Co(NH_3)_5(NO_2\text{-}N)]^{2+}$ can be distinguished by using IR spectroscopy. For the *O*-bonded ligand, characteristic

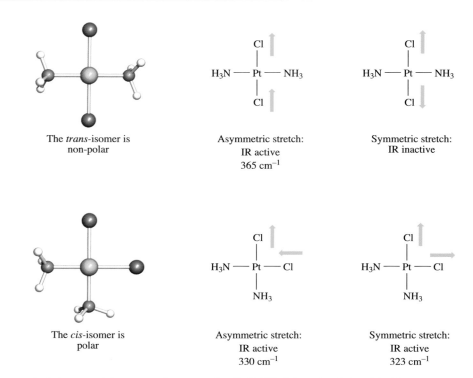

Fig. 19.12 The *trans*- and *cis*-isomers of the square planar complex $[PtCl_2(NH_3)_2]$ can be distinguished by IR spectroscopy. The selection rule for an IR active vibration is that it must lead to a *change in molecular dipole moment* (see Section 3.7).

absorption bands at 1065 and 1470 cm^{-1} are observed, while for the *N*-bonded ligand, the corresponding vibrational wavenumbers are 1310 and 1430 cm^{-1}.

The DMSO ligand (dimethylsulfoxide, **19.18**) can coordinate to metal ions through either the *S*- or *O*-donor atom. These modes can be distinguished by using IR spectroscopy: $\bar{\nu}_{SO}$ for free DMSO is 1055 cm^{-1}, for *S*-bonded DMSO, $\bar{\nu}_{SO} = 1080–1150$ cm^{-1}, and for *O*-bonded DMSO, $\bar{\nu}_{SO} = 890–950$ cm^{-1}. An example of the interconversion of linkage isomers involving the DMSO ligand is shown in scheme 19.12. The isomerization also involves a *trans–cis* rearrangement of chlorido-ligands.

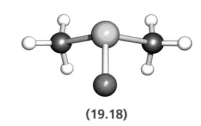

(19.18)

$$Me_2(O)S\text{,,,,}\underset{\underset{Cl}{|}}{\overset{\overset{Cl}{|}}{Ru}}\text{,,,,,}S(O)Me_2 \quad \underset{h\nu}{\overset{heat}{\rightleftharpoons}} \quad Me_2(O)S\text{,,,,}\underset{\underset{Cl}{|}}{\overset{\overset{S(O)Me_2}{|}}{Ru}}\text{,,,,,}Cl$$

$$Me_2(O)S \quad \quad \quad Me_2(O)S \quad \quad \quad OSMe_2$$

(19.12)

Stereoisomerism: diastereoisomers

Distinguishing between *cis*- and *trans*-isomers of a square planar complex or between *mer*- and *fac*-isomers of an octahedral complex is most unambiguously confirmed by structural determinations using single-crystal X-ray diffraction. Vibrational spectroscopy (applications of which were introduced in Section 3.7) may also be of assistance. For example, Fig. 19.12 illustrates that the asymmetric stretch for the PtCl$_2$ unit in $[Pt(NH_3)_2Cl_2]$ is IR active for both the *trans*- and *cis*-isomers, but the symmetric stretch is IR active only for the *cis*-isomer. In square planar complexes containing phosphane ligands, the ^{31}P NMR spectrum may be particularly diagnostic, as is illustrated in Box 19.2.

The existence of ions or molecules in different structures (e.g. trigonal bipyramidal and square-based pyramidal $[Ni(CN)_5]^{3-}$) is just a special case of diastereoisomerism. In the cases of, for example, tetrahedral and square planar $[NiBr_2(PBzPh_2)_2]$ (Bz = benzyl), the two forms can be distinguished by the fact that they exhibit different magnetic properties as we discuss in Section 20.10. To complicate matters, square planar $[NiBr_2(PBzPh_2)_2]$ may exist as either *trans*- or *cis*-isomers.

Stereoisomerism: enantiomers

A pair of ***enantiomers*** consists of two molecular species which are non-superposable mirror images of each other; see also Section 3.8.

THEORY

Box 19.2 *Trans-* and *cis-*isomers of square planar complexes: an NMR spectroscopic probe

In Section 4.8, we described how *satellite peaks* may arise in some NMR spectra. In square planar platinum(II) complexes containing two phosphane (PR_3) ligands, the ^{31}P NMR spectrum of the complex provides valuable information about the *cis-* or *trans-*arrangement of the ligands. Platinum possesses six naturally occurring isotopes (Appendix 5) but only one, ^{195}Pt, is NMR active. ^{195}Pt is 33.8% abundant and has a nuclear spin quantum number of value $I = \frac{1}{2}$. In a ^{31}P NMR spectrum of a complex such as $[PtCl_2(PPh_3)_2]$, there is spin–spin coupling between the ^{31}P and ^{195}Pt nuclei which gives rise to satellite peaks.

If the PR_3 ligands are mutually *trans*, the value of $J_{PPt} \approx 2000$–$2500\,Hz$, but if the ligands are *cis*, the coupling constant is much larger, ≈ 3000–$3500\,Hz$. While the values vary somewhat, comparison of the ^{31}P NMR spectra of *cis-* and *trans-*isomers of a given complex enables the isomers to be assigned. For example, for *cis-* and *trans-*$[PtCl_2(P^nBu_3)_2]$, values of J_{PPt} are 3508 and 2380 Hz, respectively. The figure on the right shows a 162 MHz ^{31}P NMR spectrum of *cis-*$[PtCl_2(P^nBu_3)_2]$, simulated using experimental data; (the chemical shift reference is 85% aqueous H_3PO_4).

Similar diagnostic information can be obtained from NMR spectroscopy for square planar complexes containing metal centres with spin-active isotopes. For example, rhodium is monotopic (i.e. 100% of one isotope) with ^{103}Rh having $I = \frac{1}{2}$. In square planar rhodium(I) complexes containing two phosphane ligands, values of J_{PRh} are ≈ 160–$190\,Hz$ for a *cis-*arrangement and ≈ 70–$90\,Hz$ for a *trans-*arrangement. Thus, the ^{31}P NMR spectrum of a complex of the type $[RhCl(PR_3)_2L]$ (L = neutral ligand) exhibits a *doublet* with a J_{PRh} coupling constant characteristic of a particular isomer.

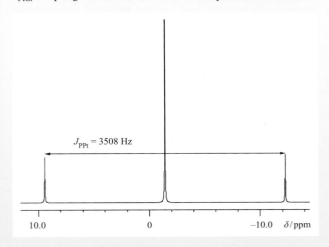

The occurrence of enantiomers (optical isomerism) is concerned with *chirality*, and some important terms relating to chiral complexes are defined in Box 19.3. Enantiomers of a coordination compound most often occur when chelating ligands are involved. Figure 19.13a shows $[Cr(acac)_3]$, an octahedral *tris-chelate* complex, and Fig. 19.13b shows *cis-*$[Co(en)_2Cl_2]^+$, an octahedral *bis-chelate* complex. In this case, only the *cis-*isomer possesses enantiomers; the *trans-*isomer is achiral. Enantiomers are distinguished by using the labels Δ and Λ (see Box 19.3).

Self-study exercises

1. Explain why *cis-*$[Co(en)_2Cl_2]^+$ is chiral while *trans-*$[Co(en)_2Cl_2]^+$ is achiral.

2. A chiral molecule lacks an inversion centre and a plane of symmetry. Use these criteria to show that species belonging to the C_2 and D_3 point groups are chiral. [*Hint*: see Appendix 3.]

3. The diagrams below represent two tetrahedral, bischelate complexes. Explain in terms of symmetry elements why **A** is achiral, but **B** is chiral. Draw the structure of the other enantiomer of **B**.

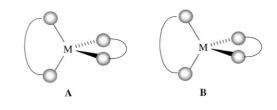

Chiral molecules rotate the plane of polarized light (Fig. 19.14). This property is known as *optical activity*. Enantiomers rotate the light to equal extents, but in opposite directions, the dextrorotatory (*d*) enantiomer to the right and the laevorotatory (*l*) enantiomer to the left. The amount of rotation *and* its sign depend upon the wavelength of the incident light. The observation of optical activity depends upon *chemical* properties of the chiral molecule. If the two enantiomers interconvert rapidly to give an equilibrium mixture containing equal amounts of the two forms, no overall rotation occurs. A mixture of equal amounts of two enantiomers is called a *racemate*.

THEORY

Box 19.3 Definitions and notation for chiral complexes

Chirality was introduced in Section 3.8. Here, we collect together some terms that are frequently encountered in discussing optically active complexes.

Enantiomers are a pair of stereoisomers that are non-superposable mirror images.

Diastereoisomers are stereoisomers that are not enantiomers.

(+) and (−) prefixes: the specific rotation of enantiomers is equal and opposite, and a useful means of distinguishing between enantiomers is to denote the *sign* of $[\alpha]_D$. Thus, if two enantiomers of a compound A have $[\alpha]_D$ values of $+12°$ and $-12°$, they are labelled (+)-A and (−)-A.

***d* and *l* prefixes**: sometimes (+) and (−) are denoted by *dextro-* and *laevo-* (derived from the Latin for right and left) and these refer to right- and left-handed rotation of the plane of polarized light respectively; *dextro* and *laevo* are generally abbreviated to *d* and *l*.

The $+/-$ or d/l notation is not a direct descriptor of the *absolute configuration* of an enantiomer (the arrangement of the substituents or ligands) for which the following prefixes are used.

***R* and *S* prefixes**: the convention for labelling chiral carbon atoms (tetrahedral with four different groups attached) uses *sequence rules* (also called the Cahn–Ingold–Prelog notation). The four groups attached to the chiral carbon atom are prioritized according to the atomic number of the attached atoms, highest priority being assigned to highest atomic number, and the molecule then viewed down the C−X vector, where X has the lowest priority. The *R*- and *S*-labels for the enantiomers refer to a clockwise (*rectus*) and anticlockwise (*sinister*) sequence of the prioritized atoms, working from high to low. Example: CHClBrI, view down the C−H bond:

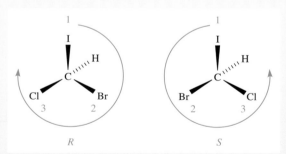

This notation is used for chiral organic ligands, and also for tetrahedral complexes.

Δ and Λ prefixes: enantiomers of octahedral complexes containing three equivalent bidentate ligands (tris-chelate complexes) are among those that are distinguished using Δ (delta) and Λ (lambda) prefixes. The octahedron is viewed down a 3-fold axis, and the chelates then define either a right- or a left-handed helix. The enantiomer with right-handedness is labelled Δ, and that with left-handedness is Λ.

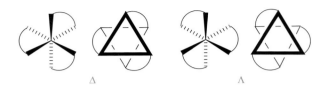

δ and λ prefixes: the situation with chelating ligands is often more complicated than the previous paragraph suggests. Consider the chelation of 1,2-diaminoethane to a metal centre. The 5-membered ring so formed is not planar but adopts an envelope conformation. This is most easily seen by taking a Newman projection along the C−C bond of the ligand. Two enantiomers are possible and are distinguished by the prefixes δ and λ.

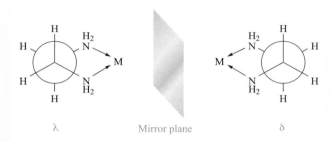

P and M descriptors: a helical, propeller or screw-shaped structure (e.g. S_n has a helical chain) can be right- or left-handed and is termed *P* ('plus') or *M* ('minus'), respectively. This is illustrated with (*P*)- and (*M*)-hexahelicene:

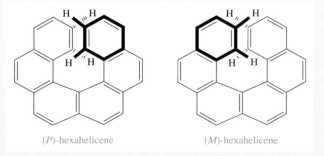

(*P*)-hexahelicene   (*M*)-hexahelicene

For detailed information, see:

IUPAC Nomenclature of Inorganic Chemistry (*Recommendations 2005*), senior eds N.G. Connelly and T. Damhus, RSC Publishing, Cambridge, p. 189.

Basic terminology of stereochemistry: *IUPAC Recommendations 1996* (1996) *Pure Appl. Chem.*, vol. 68, p. 2193.

A. von Zelewsky (1996) *Stereochemistry of Coordination Compounds*, Wiley, Chichester.

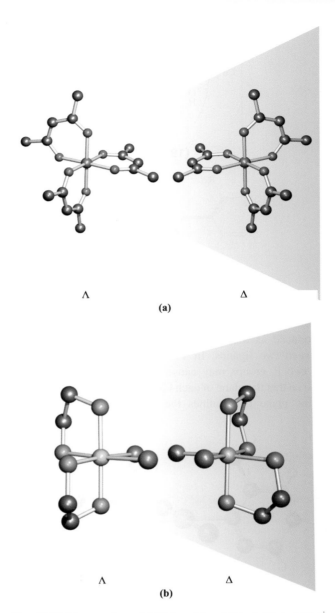

Λ   Δ

(a)

Λ   Δ

(b)

Fig. 19.13 The complexes (a) $[Cr(acac)_3]$ and (b) *cis*-$[Co(en)_2Cl_2]^+$ are chiral. The two enantiomers are non-superposable mirror images of one another. Hydrogen atoms are omitted from the diagrams for clarity. Colour code: Cr, green; Co yellow; Cl, green; N, blue; O, red; C, grey.

The rotation, α, is measured in a *polarimeter* (Fig. 19.14). In practice, the amount of rotation depends upon the wavelength of the light, temperature and the concentration of compound present in solution. The *specific rotation*, $[\alpha]$, for a chiral compound in solution is given by eq. 19.13.

$$[\alpha] = \frac{\alpha}{c \times \ell} \tag{19.13}$$

where: α = observed rotation

ℓ = path length of solution in polarimeter (in dm)

c = concentration (in g cm^{-3})

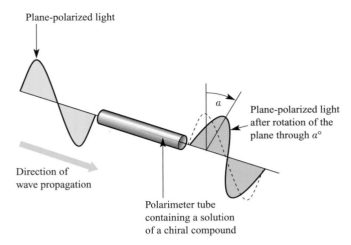

Fig. 19.14 One enantiomer of a chiral compound rotates the plane of linearly polarized light through a characteristic angle, $\alpha°$; the instrument used to measure this rotation is called a polarimeter. The direction indicated (a clockwise rotation as you view the light as it emerges from the polarimeter) is designated as $+\alpha°$. The other enantiomer of the same compound rotates the plane of polarized light through an angle $-\alpha°$.

Light of a single frequency is used for specific rotation measurements and a common choice is the *sodium D-line* in the emission spectrum of atomic sodium. The specific rotation at this wavelength is denoted as $[\alpha]_D$.

Pairs of enantiomers such as Δ- and Λ-$[Cr(acac)_3]$ differ only in their action on polarized light. However, for ionic complexes such as $[Co(en)_3]^{3+}$, there is the opportunity to form salts with a chiral counter-ion A^-. These salts now contain two different types of chirality: the Δ- or Λ-chirality at the metal centre and the $(+)$ or $(-)$ chirality of the anion. Four combinations are possible of which the pair $\{\Delta\text{-}(+)\}$ and $\{\Lambda\text{-}(-)\}$ is enantiomeric as is the pair $\{\Delta\text{-}(-)\}$ and $\{\Lambda\text{-}(+)\}$. However, with a given anion chirality, the pair of salts $\{\Delta\text{-}(-)\}$ and $\{\Lambda\text{-}(-)\}$ are diastereoisomers (see Box 19.3) and may differ in the packing of the ions in the solid state, and separation by fractional crystallization is often possible.

Octahedral, phosphorus-containing anions (see p. 534) that are tris-chelates are chiral, and are used for resolving enantiomers and as NMR shift reagents.[†] We exemplify this with TRISPHAT (**19.19**). The presence of the electron-withdrawing substituents in TRISPHAT increases the configurational stability of the anion, i.e. the rate of interconversion of enantiomers is slow. Thus, salts of Δ- and Λ-TRISPHAT can be prepared enantiomerically pure. An example of its use is the resolution of the enantiomers of *cis*-$[Ru(phen)_2(NCMe)_2]^{2+}$ (**19.20**). This

---

[†] For relevent reviews, see: J. Lacour and V. Hebbe-Viton (2003) *Chem. Soc. Rev.*, vol. 32, p. 373 – 'Recent developments in chiral anion mediated asymmetric chemistry'; J. Lacour (2010) *C. R. Chimie*, vol. 13, p. 985 – 'Chiral hexacoordinated phosphates: From pioneering studies to modern uses in stereochemistry'.

Δ-TRISPHAT

(19.19)

(19.20)

(19.21)

(19.22)

may conveniently be prepared as the $[CF_3SO_3]^-$ salt which is a racemate. Anion exchange using enantiomerically pure $[\Delta\text{-TRISPHAT}]^-$ gives a mixture of $[\Delta\text{-}$**19.20**$][\Delta\text{-TRISPHAT}]_2$ and $[\Lambda\text{-}$**19.20**$][\Delta\text{-TRISPHAT}]_2$ which can be separated by chromatography. The salts $[\Delta\text{-}$**19.20**$][\Delta\text{-TRISPHAT}]_2$ and $[\Lambda\text{-}$**19.20**$][\Delta\text{-TRISPHAT}]_2$ are diastereoisomers. When dissolved in non-coordinating or poorly solvating solvents, they form diastereoisomeric ion-pairs which can be distinguished in the ^1H NMR spectrum. For example, in CD_2Cl_2 solution, the signal for the methyl group of the MeCN ligands occurs at δ 2.21 ppm for $[\Delta\text{-}$**19.20**$][\Delta\text{-TRISPHAT}]_2$, and at δ 2.24 ppm for $[\Lambda\text{-}$**19.20**$][\Delta\text{-TRISPHAT}]_2$. This illustrates the use of TRISPHAT as a *diamagnetic* NMR chiral shift reagent. We look at the use of *paramagnetic* NMR shift reagents in Box 27.3.

The first purely inorganic complex to be resolved into its enantiomers was $[CoL_3]^{6+}$ (**19.21**) in which each L^+ ligand is the complex *cis*-$[Co(NH_3)_4(OH)_2]^+$ which chelates through the two *O*-donor atoms.[†]

Chirality is not usually associated with square planar complexes but there are some special cases where chirality is introduced as a result of, for example, steric interactions between two ligands. In **19.22**, steric repulsions between the two R groups may cause the aromatic substituents to twist so that the plane of each C_6-ring is no longer orthogonal to the plane that defines the square planar environment

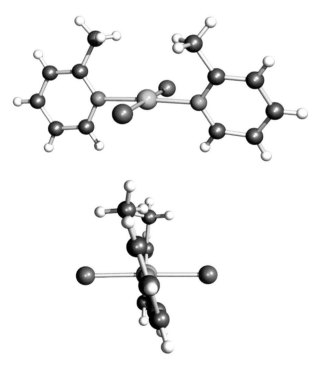

Fig. 19.15 Two views of the structure (X-ray diffraction) of *trans*-$[PdCl_2(2\text{-Mepy})_2]$ (2-Mepy = 2-methylpyridine) showing the square planar environment of the Pd(II) centre and the mutual twisting of the 2-methylpyridine ligands. The torsion angle between the rings is 18.6° [M.C. Biagini (1999) *J. Chem. Soc., Dalton Trans.*, p. 1575]. Colour code: Pd, yellow; N, blue; Cl, green; C, grey; H, white.

[†]For elucidation of a related species, see: W.G. Jackson, J.A. McKeon, M. Zehnder and M. Neuburger (2004) *Chem. Commun.*, p. 2322 – 'The rediscovery of Alfred Werner's second hexol'.

around M. Such a twist is defined by the torsion angle A–B–C–D in structure **19.22**, and renders the molecule chiral. The chirality can be recognized in terms of a handedness, as in a helix, and the terms *P* and *M* (see Box 19.3) can be used to distinguish between related chiral molecules. If, in **19.22**, the sequence rules priority of R is higher than R' (e.g. R = Me,

R' = H), then a positive torsion angle corresponds to *P*-chirality. An example is *trans*-[PdCl$_2$(2-Mepy)$_2$] (2-Mepy = 2-methylpyridine), for which the *P*-isomer is shown in Fig. 19.15. End-of-chapter problem 19.28 is concerned with the incorporation of chiral ligands into square-planar platinum(II) complexes.

KEY TERMS

The following terms were introduced in this chapter. Do you know what they mean?

- *d*-block metal
- transition element
- platinum-group metal
- chromophore
- electroneutrality principle
- Kepert model
- tripodal ligand
- structural isomerism
- stereoisomerism
- ionization isomerism
- hydration isomerism
- linkage isomerism
- enantiomer
- diastereoisomer
- specific rotation
- racemate
- resolution of enantiomers

FURTHER READING

S. Alvarez (2005) *Dalton Trans.*, p. 2209 – 'Polyhedra in (inorganic) chemistry' gives a systematic survey of polyhedra with examples from inorganic chemistry.

M.C. Biagini, M. Ferrari, M. Lanfranchi, L. Marchiò and M.A. Pellinghelli (1999) *J. Chem. Soc., Dalton Trans.*, p. 1575 – An article that illustrates chirality of square planar complexes.

M. Gerloch and E.C. Constable (1994) *Transition Metal Chemistry: The Valence Shell in d-Block Chemistry*, VCH, Weinheim – An introductory and very readable text.

J.M. Harrowfield and S.B. Wild (1987) *Comprehensive Coordination Chemistry*, eds G. Wilkinson, R.D. Gillard and J.A. McCleverty, Pergamon, Oxford, vol. 1, Chapter 5 – An excellent overview: 'Isomerism in coordination chemistry'.

C.E. Housecroft (1999) *The Heavier d-Block Metals: Aspects of Inorganic and Coordination Chemistry*, Oxford University Press, Oxford – A short textbook which highlights differences between the first row and the heavier *d*-block metals.

J.A. McCleverty (1999) *Chemistry of the First-row Transition Metals*, Oxford University Press, Oxford – A valuable introduction to metals and solid compounds, solution species, high and low oxidation state species and bio-transition metal chemistry.

D.M.P. Mingos (1998) *Essential Trends in Inorganic Chemistry*, Oxford University Press, Oxford – Chapter 5 deals with trends among the *d*- and *f*-block elements.

G. Seeber, B.E.F. Tiedemann and K.N. Raymond (2006) *Top. Curr. Chem.*, vol. 265, p. 147 – 'Supramolecular chirality in coordination chemistry' takes chiral systems beyond mononuclear complexes to supramolecular assemblies.

D. Venkataraman, Y. Du, S.R. Wilson, K.A. Hirsch, P. Zhang and J.S. Moore (1997) *J. Chem. Educ.*, vol. 74, p. 915 – An article entitled: 'A coordination geometry table of the *d*-block elements and their ions'.

M.J. Winter (1994) *d-Block Chemistry*, Oxford University Press, Oxford – An introductory text concerned with the principles of the *d*-block metals.

PROBLEMS

Ligand abbreviations: see Table 7.7.

19.1 Comment on (a) the observation of variable oxidation states among elements of the *s*- and *p*-blocks, and (b) the statement that 'variable oxidation states are a characteristic feature of any *d*-block metal'.

19.2 (a) Write down, in order, the metals that make up the first row of the *d*-block and give the ground state valence electronic configuration of each element. (b) Which triads of metals make up groups 4, 8 and 11? (c) Which metals are collectively known as the platinum-group metals?

19.3 Comment on the reduction potential data in Table 19.1.

19.4 By referring to relevant sections earlier in the book, write a brief account of the formation of hydrides, borides, carbides and nitrides of the *d*-block metals.

19.5 Give a brief overview of properties that characterize a *d*-block metal.

19.6 Suggest why (a) high coordination numbers are not usual for first row *d*-block metals, (b) in early *d*-block metal complexes the combination of a high oxidation state and high coordination number is common, and (c) in first row *d*-block metal complexes, high oxidation states are stabilized by fluorido or oxido ligands.

19.7 For each of the following complexes, give the oxidation state of the metal and its d^n configuration:

(a) $[Mn(CN)_6]^{4-}$; (b) $[FeCl_4]^{2-}$; (c) $[CoCl_3(py)_3]$; (d) $[ReO_4]^-$; (e) $[Ni(en)_3]^{2+}$; (f) $[Ti(OH_2)_6]^{3+}$; (g) $[VCl_6]^{3-}$; (h) $[Cr(acac)_3]$.

19.8 Within the Kepert model, what geometries do you associate with the following coordination numbers: (a) 2; (b) 3; (c) 4; (d) 5; (e) 6?

19.9 Show that the trigonal bipyramid, square-based pyramid, square antiprism and dodecahedron belong to the point groups D_{3h}, C_{4v}, D_{4d} and D_{2d} respectively.

19.10 (a) In the solid state, $Fe(CO)_5$ possesses a trigonal bipyramidal structure. How many carbon environments are there? (b) Explain why only one signal is observed in the ^{13}C NMR spectrum of solutions of $Fe(CO)_5$, even at low temperature.

19.11 Structures **19.23–19.25** show bond angle data (determined by X-ray diffraction) for some complexes with low coordination numbers. Comment on these data, suggesting reasons for deviations from regular geometries.

(19.23) **(19.24)**

(19.25)

19.12 Suggest a structure for the complex $[CuCl(\mathbf{19.26})]^+$ assuming that all donor atoms are coordinated to the Cu(II) centre.

(19.26)

19.13 What chemical tests would you use to distinguish between (a) $[Co(NH_3)_5Br][SO_4]$ and $[Co(NH_3)_5(SO_4)]Br$, and (b) $[CrCl_2(OH_2)_4]Cl \cdot 2H_2O$ and $[CrCl(OH_2)_5]Cl_2 \cdot H_2O$? (c) What is the relationship between these pairs of compounds? (d) What isomers are possible for $[CrCl_2(OH_2)_4]^+$?

19.14 (a) Give formulae for compounds that are coordination isomers of $[Co(bpy)_3]^{3+}[Fe(CN)_6]^{3-}$. (b) What other types of isomerism could be exhibited by any of the complex ions given in your answer to part (a)?

19.15 What isomers would you expect to exist for the platinum(II) compounds:
(a) $[Pt(H_2NCH_2CHMeNH_2)_2]Cl_2$, and
(b) $[Pt(H_2NCH_2CMe_2NH_2)(H_2NCH_2CPh_2NH_2)]Cl_2$?

19.16 How many different forms of $[Co(en)_3]^{3+}$ are possible in principle? Indicate how they are related as enantiomers or diastereoisomers.

19.17 State the types of isomerism that may be exhibited by the following complexes, and draw structures of the isomers: (a) $[Co(en)_2(ox)]^+$, (b) $[Cr(ox)_2(OH_2)_2]^-$, (c) $[PtCl_2(PPh_3)_2]$, (d) $[PtCl_2(Ph_2PCH_2CH_2PPh_2)]$ and (e) $[Co(en)(NH_3)_2Cl_2]^+$.

19.18 Using spectroscopic methods, how would you distinguish between the pairs of isomers (a) *cis*- and *trans*-$[PdCl_2(PPh_3)_2]$, (b) *cis*- and *trans*-$[PtCl_2(PPh_3)_2]$ and (c) *fac*- and *mer*-$[RhCl_3(PMe_3)_3]$?

19.19 Structure **19.27** shows the ligand tpy (2,2':6',2''-terpyridine). What conformational changes does the ligand undergo when it coordinates to a metal ion? Comment on possible isomer formation in the following complexes: (a) $[Ru(py)_3Cl_3]$, (b) $[Ru(bpy)_2Cl_2]^+$, and (c) $[Ru(tpy)Cl_3]$.

(19.27)

19.20 The conjugate base of **19.28** forms the complex [CoL₃] which has *mer*- and *fac*-isomers. (a) Draw the structures of these isomers, and explain why the labels *mer* and *fac* are used. (b) What other type of isomerism does [CoL₃] exhibit? (c) When a freshly prepared sample of [CoL₃] is chromatographed, two fractions, **A** and **B**, are collected. The ^{19}F NMR spectrum of **A** exhibits a singlet, while that of **B** shows three signals with relative integrals of $1:1:1$. Rationalize these data.

HL

(19.28)

19.21 The reaction of [RuCl₂(PPh₃)(dppb)] with phen leads to the loss of PPh₃ and the formation of an octahedral complex, **X**.

dppb

The solution ^{31}P{^{1}H} NMR spectrum of a freshly made sample of **X** shows a singlet at δ 33.2 ppm. The sample is left standing in the light for a few hours, after which time the ^{31}P{^{1}H} NMR spectrum is again recorded. The signal at δ 33.2 ppm has diminished in intensity, and two doublets at δ 44.7 and 32.4 ppm (relative integrals $1:1$, each signal with $J = 31$ Hz) have appeared. Rationalize these data.

19.22 One isomer of [PdBr₂(NH₃)₂] is unstable with respect to a second isomer, and the isomerization process can be followed by IR spectroscopy. The IR spectrum of the first isomer shows absorptions at 480 and 460 cm^{-1} assigned to ν(PdN) modes. During isomerization, the band at 460 cm^{-1} gradually disappears and that at 480 cm^{-1} shifts to 490 cm^{-1}. Rationalize these data.

19.23 Consider the following reaction in which $[P_3O_{10}]^{5-}$ (see Fig. 15.19) displaces the carbonate ion to give a mixture of linkage isomers:

(a) Suggest possible coordination modes for the $[P_3O_{10}]^{5-}$ ion in the products, given that an octahedral metal centre is retained. (b) How might the products formed in the reaction be influenced by the pH of the solution?

OVERVIEW PROBLEMS

19.24 (a) In each of the following complexes, determine the overall charge, n, which may be positive or negative: $[Fe^{II}(bpy)_3]^n$, $[Cr^{III}(ox)_3]^n$, $[Cr^{III}F_6]^n$, $[Ni^{II}(en)_3]^n$, $[Mn^{II}(ox)_2(OH_2)_2]^n$, $[Zn^{II}(py)_4]^n$, $[Co^{III}Cl_2(en)_2]^n$.

(b) If the bonding in $[MnO_4]^-$ were 100% ionic, what would be the charges on the Mn and O atoms? Is this model realistic? By applying Pauling's electroneutrality principle, redistribute the charge in $[MnO_4]^-$ so that Mn has a resultant charge of $+1$. What are the charges on each O atom? What does this charge distribution tell you about the degree of covalent character in the Mn–O bonds?

19.25 (a) Which of the following octahedral complexes are chiral: *cis*-[CoCl₂(en)₂]$^+$, [Cr(ox)₃]$^{3-}$, *trans*-[PtCl₂(en)₂]$^{2+}$, [Ni(phen)₃]$^{2+}$, [RuBr₄(phen)]$^-$, *cis*-[RuCl(py)(phen)₂]$^+$?

(b) The solution ^{31}P NMR spectrum of a mixture of isomers of the square planar complex [Pd(SCN)₂(Ph₂PCH₂PPh₂)] shows one broad signal at 298 K. At 228 K, two singlets and two doublets ($J = 82$ Hz) are observed and the relative integrals of these signals are solvent-dependent. Draw the structures of the possible isomers of [Pd(SCN)₂(Ph₂PCH₂PPh₂)] and rationalize the NMR spectroscopic data.

19.26 (a) Explain why complex **19.29** is chiral.

(19.29)

(b) In each of the following reactions, the left-hand sides are balanced. Suggest possible products and give the structures of each complex formed.

$$AgCl(s) + 2NH_3(aq) \longrightarrow$$

$$Zn(OH)_2(s) + 2KOH(aq) \longrightarrow$$

(c) What type of isomerism relates the Cr(III) complexes [Cr(en)$_3$][Cr(ox)$_3$] and [Cr(en)(ox)$_2$] [Cr(en)$_2$(ox)]?

19.27 (a) The following complexes each possess one of the structures listed in Table 19.4. Use the point group to deduce each structure: [ZnCl$_4$]$^{2-}$ (T_d); [AgCl$_3$]$^{2-}$ (D_{3h}); [ZrF$_7$]$^{3-}$ (C_{2v}); [ReH$_9$]$^{2-}$ (D_{3h}); [PtCl$_4$]$^{2-}$ (D_{4h}); [AuCl$_2$]$^-$ ($D_{\infty h}$).

(b) How does the coordination environment of Cs$^+$ in CsCl differ from that of typical, discrete 8-coordinate complexes? Give examples to illustrate the latter, commenting on factors that may influence the preference for a particular coordination geometry.

INORGANIC CHEMISTRY MATTERS

19.28 Interactions between DNA and metal complexes are the basis for the use of square-planar platinum(II)-containing anti-cancer drugs. (a) Explain how the interaction of right-handed DNA with chiral complexes leads to diastereoisomeric species. (b) How does the replacement of the two NH$_3$ ligands in cisplatin (see below) by two PhMeCHNH$_2$ ligands affect the chirality of the complex?

cisplatin

(c) The bidentate ligands drawn below may be used to prepare analogues of cisplatin. Draw the structures of the complexes formed and indicate all asymmetric centres. For a given ligand, which pairs of complexes are related by being enantiomers or diastereoisomers?

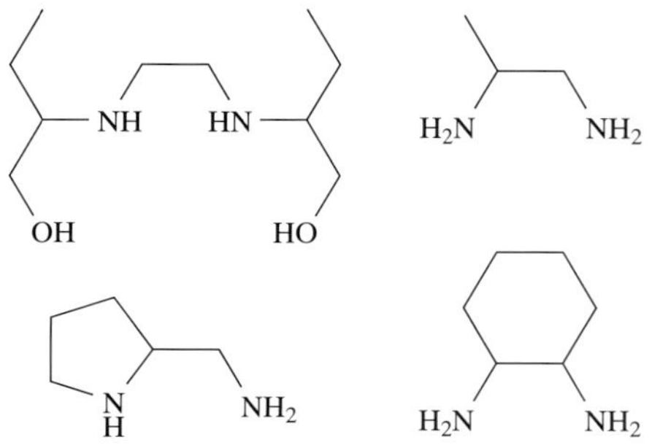

Topics

Valence bond model

Crystal field theory

Spectrochemical series

Crystal field stabilization
    energy

Molecular orbital theory

Microstates and term
    symbols

Electronic absorption
    and emission spectra

Nephelauxetic effect

Magnetic properties

Thermodynamic aspects

20

d-Block metal chemistry: coordination complexes

20.1 Introduction

In this chapter, we discuss complexes of the *d*-block metals and consider bonding theories that rationalize experimental facts such as electronic spectra and magnetic properties. Most of the discussion centres on first row *d*-block metals, for which theories of bonding are most successful. The bonding in *d*-block metal complexes is not fundamentally different from that in other compounds, and we shall show applications of valence bond theory, the electrostatic model and molecular orbital theory.

Fundamental to discussions about *d*-block chemistry are the 3*d*, 4*d* or 5*d* orbitals for the first, second or third row *d*-block metals, respectively. We introduced *d*-orbitals in Section 1.6, and showed that a *d*-orbital is characterized by having a value of the quantum number $l = 2$. The conventional representation of a set of five degenerate *d*-orbitals is shown in Fig. 20.1b.[†] The lobes of the d_{yz}, d_{xy} and d_{xz} orbitals point *between* the Cartesian axes and each orbital lies in one of the three planes defined by the axes. The $d_{x^2 - y^2}$ orbital is related to d_{xy}, but the lobes of the $d_{x^2 - y^2}$ orbital point *along* (rather than between) the x and y axes. We could envisage being able to draw two more atomic orbitals which are related to the $d_{x^2 - y^2}$ orbital, i.e. the $d_{z^2 - y^2}$ and $d_{z^2 - x^2}$ orbitals (Fig. 20.1c). However, this would give a total of six *d*-orbitals. For $l = 2$, there are only five real solutions to the Schrödinger equation ($m_l = +2, +1, 0, -1, -2$). The problem is solved by taking a linear combination of the $d_{z^2 - x^2}$ and $d_{z^2 - y^2}$ orbitals. This means that the two orbitals

---

[†] Although we refer to the *d* orbitals in these 'pictorial' terms, it is important not to lose sight of the fact that these orbitals are *not real* but merely mathematical solutions of the Schrödinger wave equation (see Section 1.5).

are combined (Fig. 20.1c), with the result that the fifth real solution to the Schrödinger equation corresponds to what is traditionally labelled the d_{z^2} orbital (although this is actually shorthand notation for $d_{2z^2 - y^2 - x^2}$).

The fact that three of the five *d*-orbitals have their lobes directed *between* the Cartesian axes, while the other two are directed along these axes (Fig. 20.1b), is a key point in the understanding of bonding models for and physical properties of *d*-block metal complexes. As a consequence of there being a distinction in their directionalities, the *d* orbitals in the presence of ligands are split into groups of different energies, the type of splitting and the magnitude of the energy differences depending on the arrangement and nature of the ligands. Magnetic properties and electronic absorption spectra, both of which are observable properties, reflect the splitting of *d* orbitals.

High- and low-spin states

In Section 19.5, we stated that paramagnetism is a characteristic of some *d*-block metal compounds. In Section 20.10 we consider magnetic properties in detail, but for now, let us simply state that magnetic data allow us to determine the number of unpaired electrons. In an isolated first row *d*-block metal ion, the 3*d* orbitals are degenerate and the electrons occupy them according to Hund's rules: e.g. diagram **20.1** shows the arrangement of six electrons.

$$\uparrow\downarrow \quad \uparrow \quad \uparrow \quad \uparrow \quad \uparrow$$

(20.1)

However, magnetic data for a range of octahedral d^6 *complexes* show that they fall into two categories: paramagnetic

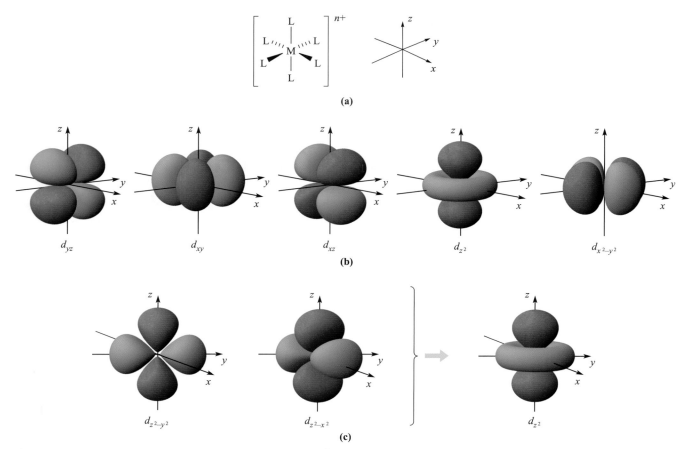

Fig. 20.1 (a) The six M—L vectors of an octahedral complex $[ML_6]^{n+}$ can be defined to lie along the x, y and z axes. (b) The five d orbitals; the d_{z^2} and $d_{x^2-y^2}$ atomic orbitals point directly along the axes, but the d_{xy}, d_{yz} and d_{xz} atomic orbitals point between them. (c) The formation of a d_{z^2} orbital from a linear combination of $d_{x^2-y^2}$ and $d_{z^2-x^2}$ orbitals. The orbitals have been generated using the program *Orbital Viewer* [David Manthey, www.orbitals.com/orb/index.html].

or diamagnetic. The former are called *high-spin complexes* and correspond to those in which, despite the d orbitals being split, there are still four unpaired electrons. The diamagnetic d^6 complexes are termed *low-spin* and correspond to those in which electrons are doubly occupying three orbitals, leaving two unoccupied. High- and low-spin complexes exist for octahedral d^4, d^5, d^6 and d^7 metal complexes. As shown above, for a d^6 configuration, low-spin corresponds to a diamagnetic complex and high-spin to a paramagnetic one. For d^4, d^5 and d^7 configurations, both high- and low-spin complexes of a given configuration are paramagnetic, but with different numbers of unpaired electrons. Magnetic properties of d-block metal complexes are described in detail in Section 20.10.

20.2 Bonding in *d*-block metal complexes: valence bond theory

Hybridization schemes

Although VB theory (see Sections 2.1, 2.2 and 5.2) in the form developed by Pauling in the 1930s is not much used now in discussing d-block metal complexes, the

terminology and many of the ideas have been retained and some knowledge of the theory remains useful. In Section 5.2, we described the use of sp^3d, sp^3d^2 and sp^2d hybridization schemes in trigonal pyramidal, square-based pyramidal, octahedral and square planar molecules. Applications of these hybridization schemes to describe the bonding in d-block metal complexes are given in Table 20.1. An *empty* hybrid orbital on the metal centre can accept a pair of electrons from a ligand to form a σ-bond. The choice of particular p or d atomic orbitals may depend on the definition of the axes with respect to the molecular framework, e.g. in linear ML_2, the M–L vectors are usually defined to lie along the z axis. We have included the cube in Table 20.1 only to point out the required use of an f orbital.

The limitations of VB theory

This short section on VB theory is included for historical reasons, and we illustrate the limitations of the VB model by considering octahedral complexes of Cr(III) (d^3) and Fe(III) (d^5) and octahedral, tetrahedral and square planar complexes of Ni(II) (d^8). The atomic orbitals required

Table 20.1 Hybridization schemes for the σ-bonding frameworks of different geometrical configurations of ligand donor atoms.

Coordination number	Arrangement of donor atoms	Orbitals hybridized	Hybrid orbital description	Example
2	Linear	s, p_z	sp	$[Ag(NH_3)_2]^+$
3	Trigonal planar	s, p_x, p_y	sp^2	$[HgI_3]^-$
4	Tetrahedral	s, p_x, p_y, p_z	sp^3	$[FeBr_4]^{2-}$
4	Square planar	$s, p_x, p_y, d_{x^2-y^2}$	sp^2d	$[Ni(CN)_4]^{2-}$
5	Trigonal bipyramidal	$s, p_x, p_y, p_z, d_{z^2}$	sp^3d	$[CuCl_5]^{3-}$
5	Square-based pyramidal	$s, p_x, p_y, p_z, d_{x^2-y^2}$	sp^3d	$[Ni(CN)_5]^{3-}$
6	Octahedral	$s, p_x, p_y, p_z, d_{z^2}, d_{x^2-y^2}$	sp^3d^2	$[Co(NH_3)_6]^{3+}$
6	Trigonal prismatic	$s, d_{xy}, d_{yz}, d_{xz}, d_{z^2}, d_{x^2-y^2}$ or $s, p_x, p_y, p_z, d_{xz}, d_{yz}$	sd^5 or sp^3d^2	$[ZrMe_6]^{2-}$
7	Pentagonal bipyramidal	$s, p_x, p_y, p_z, d_{xy}, d_{x^2-y^2}, d_{z^2}$	sp^3d^3	$[V(CN)_7]^{4-}$
7	Monocapped trigonal prismatic	$s, p_x, p_y, p_z, d_{xy}, d_{xz}, d_{z^2}$	sp^3d^3	$[NbF_7]^{2-}$
8	Cubic	$s, p_x, p_y, p_z, d_{xy}, d_{xz}, d_{yz}, f_{xyz}$	sp^3d^3f	$[PaF_8]^{3-}$
8	Dodecahedral	$s, p_x, p_y, p_z, d_{z^2}, d_{xy}, d_{xz}, d_{yz}$	sp^3d^4	$[Mo(CN)_8]^{4-}$
8	Square antiprismatic	$s, p_x, p_y, p_z, d_{xy}, d_{xz}, d_{yz}, d_{x^2-y^2}$	sp^3d^4	$[TaF_8]^{3-}$
9	Tricapped trigonal prismatic	$s, p_x, p_y, p_z, d_{xy}, d_{xz}, d_{yz}, d_{z^2}, d_{x^2-y^2}$	sp^3d^5	$[ReH_9]^{2-}$

for hybridization in an octahedral complex of a first row *d*-block metal are the $3d_{z^2}$, $3d_{x^2-y^2}$, $4s$, $4p_x$, $4p_y$ and $4p_z$ (Table 20.1). These orbitals must be *unoccupied* so as to be available to accept six pairs of electrons from the ligands. The $Cr^{3+}(d^3)$ ion has three unpaired electrons and these are accommodated in the $3d_{xy}$, $3d_{xz}$ and $3d_{yz}$ orbitals:

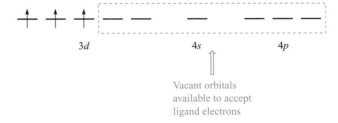

With the electrons from the six ligands included and a hybridization scheme applied for an octahedral complex, the diagram becomes:

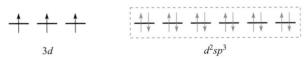

This diagram is appropriate for all octahedral Cr(III) complexes because the three $3d$ electrons always singly occupy different orbitals.

For octahedral Fe(III) complexes (d^5), we must account for the existence of both high- and low-spin complexes. The electronic configuration of the free Fe^{3+} ion is:

For a low-spin octahedral complex such as $[Fe(CN)_6]^{3-}$, we can represent the electronic configuration by means of the following diagram where the electrons shown in red are donated by the ligands:

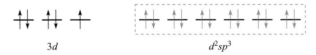

For a high-spin octahedral complex such as $[FeF_6]^{3-}$, the five $3d$ electrons occupy the five $3d$ atomic orbitals (as in the free ion shown above) and the two d orbitals required for the sp^3d^2 hybridization scheme must come from the $4d$ set. With the ligand electrons included, valence bond theory describes the bonding as follows, leaving three empty $4d$ atomic orbitals (not shown):

However, this scheme is unrealistic because the $4d$ orbitals are at a significantly higher energy than the $3d$ atomic orbitals.

Nickel(II) (d^8) forms paramagnetic tetrahedral and octahedral complexes, and diamagnetic square planar complexes. Bonding in a tetrahedral complex can be represented as follows (electrons donated by the four ligands are shown in red):

An octahedral nickel(II) complex can be described by the diagram:

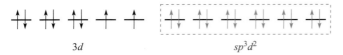

in which the three empty 4*d* atomic orbitals are not shown. For diamagnetic square planar nickel(II) complexes, valence bond theory gives the following picture:

Valence bond theory may rationalize stereochemical and magnetic properties, but only at a simplistic level. It can say *nothing* about electronic spectroscopic properties or about the kinetic inertness (see Section 26.2) that is a characteristic of the low-spin *d*⁶ configuration. Furthermore, the model implies a distinction between high- and low-spin complexes that is actually misleading. Finally, it cannot tell us *why* certain ligands are associated with the formation of high- (or low-)spin complexes. These limitations of VB theory necessitate that we must approach the bonding in *d*-block metal complexes in other ways.

20.3 Crystal field theory

A second approach to the bonding in complexes of the *d*-block metals is *crystal field theory*. This is an *electrostatic*

model and simply uses the ligand electrons to create an electric field around the metal centre. Ligands are considered as point charges and there are *no* metal–ligand covalent interactions.

The octahedral crystal field

Consider a first row metal cation, M^{n+}, surrounded by six ligands placed on the Cartesian axes at the vertices of an octahedron (Fig. 20.1a). Each ligand is treated as a negative point charge and there is an electrostatic attraction between the metal ion and ligands. However, there is also a repulsive interaction between electrons in the *d* orbitals and the ligand point charges. *If* the electrostatic field (the *crystal field*) were spherical, then the energies of the five *d* orbitals would be raised (destabilized) by the same amount. However, since the d_{z^2} and $d_{x^2-y^2}$ atomic orbitals point *directly at* the ligands while the d_{xy}, d_{yz} and d_{xz} atomic orbitals point *between* them, the d_{z^2} and $d_{x^2-y^2}$ atomic orbitals are destabilized to a greater extent than the d_{xy}, d_{yz} and d_{xz} atomic orbitals (Fig. 20.2). Thus, with respect to their energy in a spherical field (the *barycentre*, a kind of 'centre of gravity'), the d_{z^2} and $d_{x^2-y^2}$ atomic orbitals are destabilized while the d_{xy}, d_{yz} and d_{xz} atomic orbitals are stabilized.

Crystal field theory is an electrostatic model which predicts that the *d* orbitals in a metal complex are not degenerate. The pattern of splitting of the *d* orbitals depends on the crystal field, this being determined by the arrangement and type of ligands.

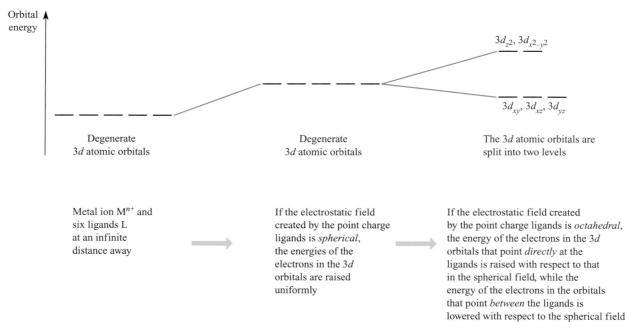

Fig. 20.2 The changes in the energies of the electrons occupying the 3*d* orbitals of a first row M^{n+} ion when the latter is in an octahedral crystal field. The energy changes are shown in terms of the orbital energies. Similar diagrams can be drawn for second (4*d*) and third (5*d*) row metal ions.

THEORY

Box 20.1 A reminder about symmetry labels

The two sets of d orbitals in an octahedral field are labelled e_g and t_{2g} (Fig. 20.3). In a tetrahedral field (Fig. 20.8), the labels become e and t_2. The symbols t and e refer to the degeneracy of the level:

- a triply degenerate level is labelled t;
- a doubly degenerate level is labelled e.

The subscript g means *gerade* and the subscript u means *ungerade*. *Gerade* and *ungerade* designate the behaviour of the wavefunction under the operation of *inversion*, and denote the *parity* (even or odd) of an orbital.

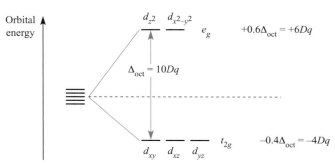

Point A is related to A' by passing through the centre of inversion. Similarly, B is related to B'.

The u and g labels are applicable *only* if the system possesses a centre of symmetry (centre of inversion) and thus are used for the octahedral field, but not for the tetrahedral one:

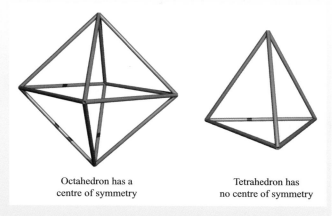

Octahedron has a centre of symmetry

Tetrahedron has no centre of symmetry

For more on the origins of symmetry labels: see Chapter 5.

From the O_h character table (Appendix 3), it can be deduced (see Chapter 5) that the d_{z^2} and $d_{x^2-y^2}$ orbitals have e_g symmetry, while the d_{xy}, d_{yz} and d_{xz} orbitals possess t_{2g} symmetry (Fig. 20.3). The energy separation between them is Δ_{oct} ('delta oct') or $10Dq$. The overall stabilization of the t_{2g} orbitals equals the overall destabilization of the e_g set. Thus, the two orbitals in the e_g set are raised by $0.6\Delta_{oct}$ with respect to the barycentre while the three in the t_{2g} set are lowered by $0.4\Delta_{oct}$. Figure 20.3 also shows these energy differences in terms of $10Dq$. Both Δ_{oct} and $10Dq$ notations are in common use, but we use Δ_{oct} in this book.[†] The stabilization and destabilization of the t_{2g} and e_g sets, respectively, are given *in terms of* Δ_{oct}. The magnitude of Δ_{oct} is determined by the *strength of the crystal field*, the two extremes being called *weak field* and *strong field* (eq. 20.1).

$$\Delta_{oct}(\text{weak field}) < \Delta_{oct}(\text{strong field}) \qquad (20.1)$$

It is a merit of crystal field theory that, in principle at least, values of Δ_{oct} can be evaluated from electronic absorption spectroscopic data (see Section 20.7). Consider the d^1 complex $[Ti(OH_2)_6]^{3+}$, for which the ground state is represented by diagram **20.2** or the notation $t_{2g}^{1}e_g^{0}$.

$$\begin{array}{c} \underline{\quad} \ \underline{\quad} \quad e_g \\[4pt] \underline{\uparrow} \ \underline{\quad} \ \underline{\quad} \quad t_{2g} \end{array}$$

(20.2)

The absorption spectrum of the ion (Fig. 20.4) exhibits one broad band for which $\lambda_{max} = 20\,300\ \text{cm}^{-1}$ corresponding to an energy change of $243\ \text{kJ mol}^{-1}$. (The conversion is $1\ \text{cm}^{-1} = 11.96 \times 10^{-3}\ \text{kJ mol}^{-1}$.) The absorption results from a change in electronic configuration from $t_{2g}^{1}e_g^{0}$ to $t_{2g}^{0}e_g^{1}$, and the value of λ_{max} (see Fig. 20.16) gives a measure of Δ_{oct}. For systems with more than one d electron, the evaluation of Δ_{oct} is more complicated. It is important to remember that Δ_{oct} is an *experimental* quantity.

Factors governing the magnitude of Δ_{oct} (Table 20.2) are the identity and oxidation state of the metal ion and the nature of the ligands. We shall see later that Δ parameters

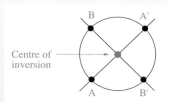

Fig. 20.3 Splitting of the d orbitals in an octahedral crystal field, with the energy changes measured with respect to the barycentre, the energy level shown by the hashed line.

Orbital energy

d_{z^2} $d_{x^2-y^2}$ e_g $+0.6\Delta_{oct} = +6Dq$

$\Delta_{oct} = 10Dq$

d_{xy} d_{xz} d_{yz} t_{2g} $-0.4\Delta_{oct} = -4Dq$

[†] The notation Dq has mathematical origins in crystal field theory. We prefer the use of Δ_{oct} because of its experimentally determined origins (see Section 20.7).

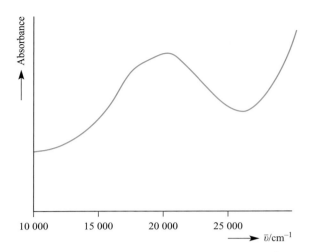

Fig. 20.4 The electronic absorption spectrum of $[Ti(OH_2)_6]^{3+}$ in aqueous solution.

Table 20.2 Values of Δ_{oct} for some *d*-block metal complexes.

Complex	Δ/cm^{-1}	Complex	Δ/cm^{-1}
$[TiF_6]^{3-}$	17 000	$[Fe(ox)_3]^{3-}$	14 100
$[Ti(OH_2)_6]^{3+}$	20 300	$[Fe(CN)_6]^{3-}$	35 000
$[V(OH_2)_6]^{3+}$	17 850	$[Fe(CN)_6]^{4-}$	33 800
$[V(OH_2)_6]^{2+}$	12 400	$[CoF_6]^{3-}$	13 100
$[CrF_6]^{3-}$	15 000	$[Co(NH_3)_6]^{3+}$	22 900
$[Cr(OH_2)_6]^{3+}$	17 400	$[Co(NH_3)_6]^{2+}$	10 200
$[Cr(OH_2)_6]^{2+}$	14 100	$[Co(en)_3]^{3+}$	24 000
$[Cr(NH_3)_6]^{3+}$	21 600	$[Co(OH_2)_6]^{3+}$	18 200
$[Cr(CN)_6]^{3-}$	26 600	$[Co(OH_2)_6]^{2+}$	9 300
$[MnF_6]^{2-}$	21 800	$[Ni(OH_2)_6]^{2+}$	8 500
$[Fe(OH_2)_6]^{3+}$	13 700	$[Ni(NH_3)_6]^{2+}$	10 800
$[Fe(OH_2)_6]^{2+}$	9 400	$[Ni(en)_3]^{2+}$	11 500

are also defined for other ligand arrangements (e.g. Δ_{tet}). For octahedral complexes, Δ_{oct} increases along the following *spectrochemical series* of ligands. The $[NCS]^-$ ion may coordinate through the *N*- or *S*-donor (distinguished in red below) and accordingly, it has two positions in the series:

$$I^- < Br^- < [NCS]^- < Cl^- < F^- < [OH]^- < [ox]^{2-}$$

$$\approx H_2O < [NCS]^- < NH_3 < en < bpy < phen < [CN]^- \approx CO$$

weak field ligands ――――――――→ strong field ligands increasing Δ_{oct}

The spectrochemical series is reasonably general. Ligands with the same donor atoms are close together in the series. If we consider octahedral complexes of *d*-block metal ions, a number of points arise which can be illustrated by the following examples:

● the complexes of Cr(III) listed in Table 20.2 illustrate the effects of different ligand field strengths for a given M^{n+} ion;

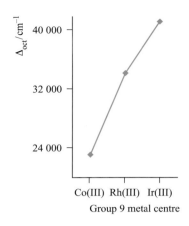

Fig. 20.5 The trend in values of Δ_{oct} for the complexes $[M(NH_3)_6]^{3+}$ where M = Co, Rh, Ir.

● the complexes of Fe(II) and Fe(III) in Table 20.2 illustrate that for a given ligand and a given metal, Δ_{oct} increases with increasing oxidation state;
● where analogous complexes exist for a series of M^{n+} metals ions (constant *n*) in a triad, Δ_{oct} increases significantly down the triad (e.g. Fig. 20.5);
● for a given ligand and a given oxidation state, Δ_{oct} varies *irregularly* across the first row of the *d*-block, e.g. over the range 8000 to 14 000 cm^{-1} for the $[M(OH_2)_6]^{2+}$ ions.

Trends in values of Δ_{oct} lead to the conclusion that metal ions can be placed in a spectrochemical series which is independent of the ligands:

$$Mn(II) < Ni(II) < Co(II) < Fe(III) < Cr(III) < Co(III)$$

$$< Ru(III) < Mo(III) < Rh(III) < Pd(II) < Ir(III) < Pt(IV)$$

――――――――――――――――→

increasing field strength

Spectrochemical series are empirical generalizations and simple crystal field theory *cannot* account for the magnitudes of Δ_{oct} values.

Crystal field stabilization energy: high- and low-spin octahedral complexes

We now consider the effects of different numbers of electrons occupying the *d* orbitals in an octahedral crystal field. For a d^1 system, the ground state corresponds to the configuration t_{2g}^1 (**20.2**). With respect to the barycentre, there is a stabilization energy of $-0.4\Delta_{oct}$ (Fig. 20.3). This is the so-called *crystal field stabilization energy*, *CFSE*.[†] For a d^2 ion, the ground state configuration is t_{2g}^2 and the CFSE = $-0.8\Delta_{oct}$ (eq. 20.2). A d^3 ion (t_{2g}^3) has a CFSE = $-1.2\Delta_{oct}$.

$$CFSE = -(2 \times 0.4)\Delta_{oct} = -0.8\Delta_{oct} \qquad (20.2)$$

――――――――――

[†] The sign convention used here for CFSE follows the thermodynamic convention.

(20.3)   **(20.4)**

For a ground state d^4 ion, two arrangements are possible: the four electrons may occupy the t_{2g} set with the configuration t_{2g}^4 (**20.3**), or may singly occupy four d orbitals, $t_{2g}^3 e_g^1$ (**20.4**). Configuration **20.3** corresponds to a low-spin arrangement, and **20.4** to a high-spin case. The preferred configuration is that with the lower energy and depends on whether it is energetically preferable to pair the fourth electron or promote it to the e_g level. Two terms contribute to the electron-pairing energy, P, which is the energy required to transform two electrons with parallel spin in different degenerate orbitals into spin-paired electrons in the same orbital:

- the loss in the *exchange energy* (see Box 1.7) which occurs upon pairing the electrons;
- the coulombic repulsion between the spin-paired electrons.

For a given d^n configuration, the CFSE is the *difference* in energy between the d electrons in an octahedral crystal field and the d electrons in a spherical crystal field (see Fig. 20.2). To exemplify this, consider a d^4 configuration. In a spherical crystal field, the d orbitals are degenerate and each of four orbitals is singly occupied. In an octahedral crystal field, eq. 20.3 shows how the CFSE is determined for a high-spin d^4 configuration (**20.4**).

$$\text{CFSE} = -(3 \times 0.4)\Delta_{\text{oct}} + 0.6\Delta_{\text{oct}} = -0.6\Delta_{\text{oct}} \quad (20.3)$$

For a low-spin d^4 configuration (**20.3**), the CFSE consists of two terms: the four electrons in the t_{2g} orbitals give rise to a

$-1.6\Delta_{\text{oct}}$ term, and a pairing energy, P, must be included to account for the spin-pairing of two electrons. Now consider a d^6 ion. In a spherical crystal field (Fig. 20.2), one d orbital contains spin-paired electrons, and each of four orbitals is singly occupied. On going to the high-spin d^6 configuration in the octahedral field ($t_{2g}^4 e_g^2$), no change occurs to the number of spin-paired electrons and the CFSE is given by eq. 20.4.

$$\text{CFSE} = -(4 \times 0.4)\Delta_{\text{oct}} + (2 \times 0.6)\Delta_{\text{oct}} = -0.4\Delta_{\text{oct}} \quad (20.4)$$

For a low-spin d^6 configuration ($t_{2g}^6 e_g^0$) the six electrons in the t_{2g} orbitals give rise to a $-2.4\Delta_{\text{oct}}$ term. Added to this is a pairing energy term of $2P$ which accounts for the spin-pairing associated with the two pairs of electrons in excess of the one in the high-spin configuration. Table 20.3 lists values of the CFSE for all d^n configurations in an octahedral crystal field. Inequalities 20.5 and 20.6 show the requirements for high- or low-spin configurations. Inequality 20.5 holds when the crystal field is weak, whereas expression 20.6 is true for a strong crystal field. Figure 20.6 summarizes the preferences for low- and high-spin d^5 octahedral complexes.

For high-spin: $\Delta_{\text{oct}} < P$ (20.5)

For low-spin: $\Delta_{\text{oct}} > P$ (20.6)

We can now relate types of ligand with a preference for high- or low-spin complexes. Strong field ligands such as $[CN]^-$ favour the formation of low-spin complexes, while weak field ligands such as halides tend to favour high-spin complexes. However, we cannot predict whether high- or low-spin complexes will be formed unless we have accurate values of Δ_{oct} and P. On the other hand, with some experimental knowledge in hand, we can make some comparative

Table 20.3 Octahedral crystal field stabilization energies (CFSE) for d^n configurations; pairing energy, P, terms are included where appropriate (see text). High- and low-spin octahedral complexes are shown only where the distinction is appropriate.

d^n	High-spin = weak field		Low-spin = strong field	
	Electronic configuration	CFSE	Electronic configuration	CFSE
d^1	$t_{2g}^1 e_g^0$	$-0.4\Delta_{\text{oct}}$		
d^2	$t_{2g}^2 e_g^0$	$-0.8\Delta_{\text{oct}}$		
d^3	$t_{2g}^3 e_g^0$	$-1.2\Delta_{\text{oct}}$		
d^4	$t_{2g}^3 e_g^1$	$-0.6\Delta_{\text{oct}}$	$t_{2g}^4 e_g^0$	$-1.6\Delta_{\text{oct}} + P$
d^5	$t_{2g}^3 e_g^2$	0	$t_{2g}^5 e_g^0$	$-2.0\Delta_{\text{oct}} + 2P$
d^6	$t_{2g}^4 e_g^2$	$-0.4\Delta_{\text{oct}}$	$t_{2g}^6 e_g^0$	$-2.4\Delta_{\text{oct}} + 2P$
d^7	$t_{2g}^5 e_g^2$	$-0.8\Delta_{\text{oct}}$	$t_{2g}^6 e_g^1$	$-1.8\Delta_{\text{oct}} + P$
d^8	$t_{2g}^6 e_g^2$	$-1.2\Delta_{\text{oct}}$		
d^9	$t_{2g}^6 e_g^3$	$-0.6\Delta_{\text{oct}}$		
d^{10}	$t_{2g}^6 e_g^4$	0		

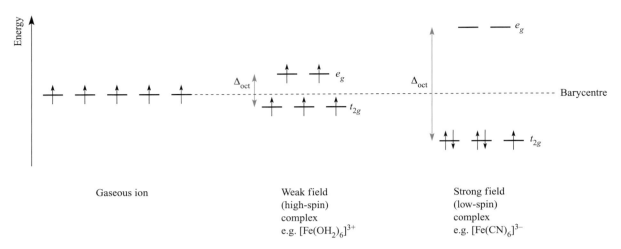

Fig. 20.6 The occupation of the $3d$ orbitals in weak and strong field Fe^{3+} (d^5) complexes.

predictions. For example, if we know from magnetic data that $[Co(OH_2)_6]^{3+}$ is low-spin, then from the spectrochemical series we can say that $[Co(ox)_3]^{3-}$ and $[Co(CN)_6]^{3-}$ will be low-spin. The only common high-spin cobalt(III) complex is $[CoF_6]^{3-}$.

Self-study exercises

All questions refer to ground state electronic configurations.

1. Draw energy level diagrams to represent a high-spin d^6 electronic configuration for an octahedral (O_h) complex. Confirm that the diagram is consistent with a value of CFSE $= -0.4\Delta_{oct}$.

2. Why does Table 20.3 not list high- and low-spin cases for all d^n configurations?

3. Explain why the CFSE for a low-spin d^5 configuration contains a $2P$ term (Table 20.3).

4. Given that $[Co(OH_2)_6]^{3+}$ is low-spin, explain why it is possible to predict that $[Co(bpy)_3]^{3+}$ is also low-spin.

Jahn–Teller distortions

Octahedral complexes of d^9 and high-spin d^4 ions are often distorted, e.g. CuF_2 (the solid state structure of which contains octahedrally sited Cu^{2+} centres, see Section 21.12) and $[Cr(OH_2)_6]^{2+}$, so that two metal–ligand bonds (axial) are different lengths from the remaining four (equatorial). This is shown in structures **20.5** (elongated octahedron) and **20.6** (compressed octahedron).[†] For a high-spin d^4

[†] Other distortions may arise and these are exemplified for Cu(II) complexes in Section 21.12.

ion, one of the e_g orbitals contains one electron while the other is vacant. If the singly occupied orbital is the d_{z^2}, most of the electron density in this orbital will be concentrated between the cation and the two ligands on the z axis. Thus, there will be greater electrostatic repulsion associated with these ligands than with the other four, and therefore the complex suffers elongation (**20.5**). Conversely, occupation of the $d_{x^2-y^2}$ orbital would lead to elongation along the x and y axes as in structure **20.6**. A similar argument can be put forward for the d^9 configuration in which the two orbitals in the e_g set are occupied by one and two electrons respectively. Electron-density measurements confirm that the electronic configuration of the Cr^{2+} ion in $[Cr(OH_2)_6]^{2+}$ is *approximately* $d_{xy}{}^1 d_{yz}{}^1 d_{xz}{}^1 d_{z^2}{}^1$. The corresponding effect when the t_{2g} set is unequally occupied is expected to be very much smaller since the orbitals are not pointing directly at the ligands. This expectation is usually, but not invariably, confirmed experimentally. Distortions of this kind are called *Jahn–Teller* or *tetragonal distortions*.

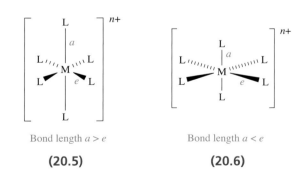

Bond length $a > e$ Bond length $a < e$

(20.5)   **(20.6)**

The *Jahn–Teller theorem* states that any non-linear molecular system in a degenerate electronic state will be unstable and will undergo distortion to form a system of lower symmetry and lower energy, thereby removing the degeneracy.

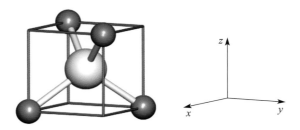

Fig. 20.7 The relationship between a tetrahedral ML₄ complex and a cube; the cube is readily related to a Cartesian axis set. The ligands lie *between* the x, y and z axes. Compare this with an octahedral complex, where the ligands lie on the axes.

The observed tetragonal distortion of an octahedral $[ML_6]^{n+}$ complex is accompanied by a change in symmetry (O_h to D_{4h}) and a splitting of the e_g and t_{2g} sets of orbitals (see Fig. 20.10). Elongation of the complex (**20.5**) is accompanied by the stabilization of each d orbital that has a z component, while the d_{xy} and $d_{x^2-y^2}$ orbitals are destabilized.

The tetrahedral crystal field

Now we consider the tetrahedral crystal field. Figure 20.7 shows a convenient way of relating a tetrahedron to a Cartesian axis set. With the complex in this orientation, none of the metal d orbitals points exactly at the ligands, but the d_{xy}, d_{yz} and d_{xz} orbitals come nearer to doing so than the d_{z^2} and $d_{x^2-y^2}$ orbitals. For a regular tetrahedron, the splitting of the d orbitals is therefore inverted compared with that for a regular octahedral structure, and the energy difference (Δ_{tet}) is smaller. If all other things are equal (and of course, they never are), the relative splittings Δ_{oct} and Δ_{tet} are related by eq. 20.7.

$$\Delta_{tet} = \tfrac{4}{9}\Delta_{oct} \approx \tfrac{1}{2}\Delta_{oct} \qquad (20.7)$$

Figure 20.8 compares crystal field splitting for octahedral and tetrahedral fields. Remember, the subscript g in the symmetry labels (see Box 20.1) is not needed in the tetrahedral case.

Since Δ_{tet} is significantly smaller than Δ_{oct}, tetrahedral complexes are high-spin. Also, since smaller amounts of energy are needed for $t_2 \longleftarrow e$ transitions (tetrahedral) than for $e_g \longleftarrow t_{2g}$ transitions (octahedral), corresponding octahedral and tetrahedral complexes often have different colours. (The notation for electronic transitions is given in Section 4.7.)

> *Tetrahedral complexes* are almost invariably *high-spin*.

While one can anticipate that tetrahedral complexes will be high-spin, the effects of a strong field ligand which also lowers the symmetry of the complex can lead to a low-spin 'distorted tetrahedral' system. This is a rare situation, and is observed in the cobalt(II) complex shown in Fig. 20.9. The lowering in symmetry from a model T_d CoL₄ complex to C_{3v} CoL₃X results in the change in orbital energy levels (Fig. 20.9). If the a_1 orbital is sufficiently stabilized and the e set is significantly destabilized, a low-spin system is energetically favoured.

Jahn–Teller effects in tetrahedral complexes are illustrated by distortions in d^9 (e.g. $[CuCl_4]^{2-}$) and high-spin d^4 complexes. A particularly strong structural distortion is observed in $[FeO_4]^{4-}$ (see structure 21.33).

The square planar crystal field

A square planar arrangement of ligands can be formally derived from an octahedral array by removal of two *trans*-ligands (Fig. 20.10). If we remove the ligands lying along the z axis, then the d_{z^2} orbital is greatly stabilized; the energies of the d_{yz} and d_{xz} orbitals are also lowered (Fig. 20.10). The fact that square planar d^8 complexes such as $[Ni(CN)_4]^{2-}$ are diamagnetic is a consequence of the relatively large energy difference between the d_{xy} and $d_{x^2-y^2}$ orbitals. Worked example 20.1 shows an experimental means (other than single-crystal X-ray diffraction) by which square planar and tetrahedral d^8 complexes can be distinguished.

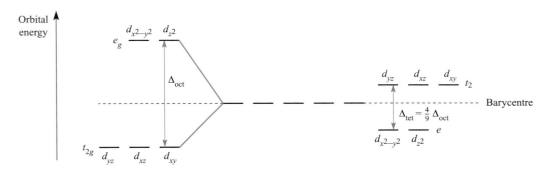

Fig. 20.8 Crystal field splitting diagrams for octahedral (left-hand side) and tetrahedral (right-hand side) fields. The splittings are referred to a common barycentre. See also Fig. 20.2.

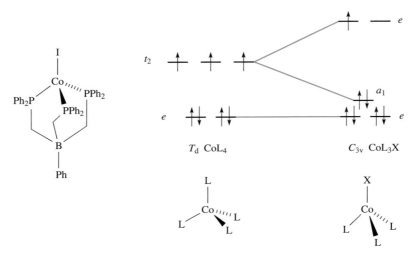

Fig. 20.9 [PhB(CH₂PPh₂)₃CoI] is a rare example of a low-spin, distorted tetrahedral complex. The tripodal tris(phosphane) is a strong-field ligand.

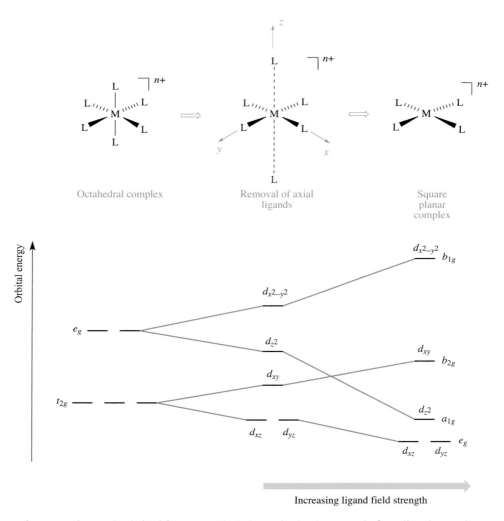

Fig. 20.10 A square planar complex can be derived from an octahedral complex by the removal of two ligands, e.g. those on the _z_ axis. The intermediate stage is analogous to a Jahn–Teller distorted (elongated) octahedral complex.

Worked example 20.1   Square planar and tetrahedral d^8 complexes

The d^8 complexes $[Ni(CN)_4]^{2-}$ and $[NiCl_4]^{2-}$ are square planar and tetrahedral respectively. Will these complexes be paramagnetic or diamagnetic?

Consider the splitting diagrams shown in Fig. 20.11. For $[Ni(CN)_4]^{2-}$ and $[NiCl_4]^{2-}$, the eight electrons occupy the d orbitals as follows:

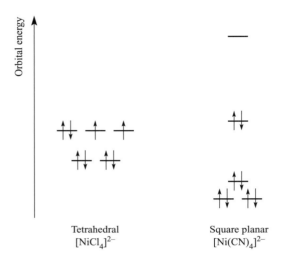

Thus, $[NiCl_4]^{2-}$ is paramagnetic while $[Ni(CN)_4]^{2-}$ is diamagnetic.

Self-study exercises

The answer to each question is closely linked to the theory in worked example 20.1.

1. The complexes $[NiCl_2(PPh_3)_2]$ and $[PdCl_2(PPh_3)_2]$ are paramagnetic and diamagnetic respectively. What does this tell you about their structures?

2. The anion $[Ni(SPh)_4]^{2-}$ is tetrahedral. Explain why it is paramagnetic.

3. Diamagnetic *trans*-$[NiBr_2(PEtPh_2)_2]$ converts to a form which is paramagnetic. Suggest a reason for this observation.

Although $[NiCl_4]^{2-}$ (d^8) is tetrahedral and paramagnetic, $[PdCl_4]^{2-}$ and $[PtCl_4]^{2-}$ (also d^8) are square planar and diamagnetic. This difference is a consequence of the larger crystal field splitting observed for second and third row metal ions compared with their first row congener. Palladium(II) and platinum(II) complexes are invariably square planar.

Second and third row metal d^8 complexes (e.g. Pt(II), Pd(II), Rh(I), Ir(I)) are invariably *square planar*.

Other crystal fields

Figure 20.11 shows crystal field splittings for some common geometries with the relative splittings of the d orbitals with respect to Δ_{oct}. By using these splitting diagrams, it is possible to rationalize the magnetic properties of a given complex (see Section 20.9). However, a word of caution: Fig. 20.11 refers to ML_x complexes containing *like* ligands, and so *only* applies to simple complexes.

Crystal field theory: uses and limitations

Crystal field theory can bring together structures, magnetic properties and electronic properties, and we shall expand upon the last two topics later in the chapter. Trends in CFSEs provide some understanding of thermodynamic and kinetic aspects of d-block metal complexes (see Sections 20.11–20.13 and 26.4). Crystal field theory is surprisingly useful when one considers its simplicity. However, it has limitations. For example, although we can interpret the contrasting magnetic properties of high- and low-spin octahedral complexes on the basis of the positions of weak- and strong-field ligands in the spectrochemical series, crystal field theory provides no explanation as to *why* particular ligands are placed where they are in the series.

20.4 Molecular orbital theory: octahedral complexes

In this section, we consider another approach to the bonding in metal complexes: the use of molecular orbital theory. In contrast to crystal field theory, the molecular orbital model considers covalent interactions between the metal centre and ligands.

Complexes with *no* metal–ligand π-bonding

We illustrate the application of MO theory to d-block metal complexes first by considering an octahedral complex such as $[Co(NH_3)_6]^{3+}$ in which metal–ligand σ-bonding is dominant. In the construction of an MO energy level diagram for such a complex, many approximations are made and the result is only *qualitatively* accurate. Even so, the results are useful for an understanding of metal–ligand bonding.

By following the procedures detailed in Chapter 5, an MO diagram can be constructed to describe the bonding in an O_h $[ML_6]^{n+}$ complex. For a first row metal, the valence shell atomic orbitals are $3d$, $4s$ and $4p$. Under O_h

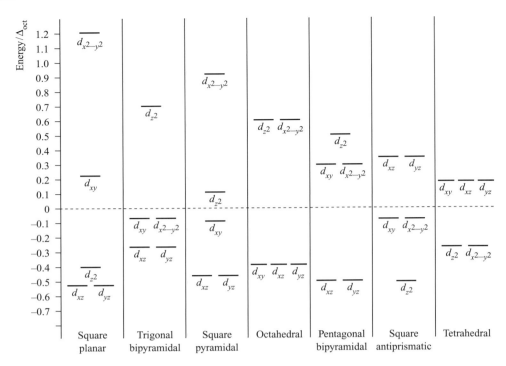

Fig. 20.11 Crystal field splitting diagrams for some common fields referred to a common barycentre. Splittings are given with respect to Δ_{oct}.

symmetry (see Appendix 3), the *s* orbital has a_{1g} symmetry, the *p* orbitals are degenerate with t_{1u} symmetry, and the *d* orbitals split into two sets with e_g (d_{z^2} and $d_{x^2-y^2}$ orbitals) and t_{2g} (d_{xy}, d_{yz} and d_{xz} orbitals) symmetries, respectively (Fig. 20.12). Each ligand, L, provides one orbital and derivation of the ligand group orbitals for the O_h L_6 fragment is analogous to those for the F_6 fragment in SF_6 (see Fig. 5.27, eqs. 5.26–5.31 and accompanying text). These LGOs have a_{1g}, t_{1u} and e_g symmetries (Fig. 20.12). Symmetry matching between metal orbitals and LGOs allows the construction of the MO diagram shown in Fig. 20.13. Combinations of the metal and ligand orbitals generate six bonding and six anti-bonding molecular orbitals. The metal d_{xy}, d_{yz} and d_{xz} atomic orbitals have t_{2g} symmetry and are non-bonding (Fig. 20.13). The overlap between the ligand and metal *s* and *p* orbitals is greater than that involving the metal *d* orbitals, and so the a_{1g} and t_{1u} MOs are stabilized to a greater extent than the e_g MOs. In an octahedral complex *with no π-bonding*, the energy difference between the t_{2g} and e_g^* levels corresponds to Δ_{oct} in crystal field theory (Fig. 20.13).

Having constructed the MO diagram in Fig. 20.13, we are able to describe the bonding in a range of octahedral σ-bonded complexes. For example:

- in low-spin $[Co(NH_3)_6]^{3+}$, 18 electrons (six from Co^{3+} and two from each ligand) occupy the a_{1g}, t_{1u}, e_g and t_{2g} MOs;
- in high-spin $[CoF_6]^{3-}$, 18 electrons are available, 12 occupying the a_{1g}, t_{1u} and e_g MOs, four the t_{2g} level, and two the e_g^* level.

Whether a complex is high- or low-spin depends upon the energy separation of the t_{2g} and e_g^* levels. *Notionally*, in a σ-bonded octahedral complex, the 12 electrons supplied by the ligands are considered to occupy the a_{1g}, t_{1u} and e_g orbitals. Occupancy of the t_{2g} and e_g^* levels corresponds to the number of valence electrons of the metal ion, just as in crystal field theory. The molecular orbital model of bonding in octahedral complexes gives much the same results as crystal field theory. It is when we move to complexes with M–L π-bonding that distinctions between the models emerge.

Complexes with metal–ligand π-bonding

The metal d_{xy}, d_{yz} and d_{xz} atomic orbitals (the t_{2g} set) are non-bonding in an $[ML_6]^{n+}$, σ-bonded complex (Fig. 20.13) and these orbitals may overlap with ligand orbitals of the correct symmetry to give π-interactions (Fig. 20.14). Although π-bonding between metal and ligand *d* orbitals is sometimes considered for interactions between metals and phosphane ligands (e.g. PR_3 or PF_3), it is more realistic to consider the roles of ligand σ*-orbitals as the acceptor orbitals.[†] Two types of ligand must be differentiated: π-donor and π-acceptor ligands.

[†] For further discussion, see: A.G. Orpen and N.G. Connelly (1985) *J. Chem. Soc., Chem. Commun.*, p. 1310; T. Leyssens, D. Peeters, A.G. Orpen and J.N. Harvey (2007) *Organometallics*, vol. 26, p. 2637. See also the discussion of *negative hyperconjugation* at the end of Section 14.6.

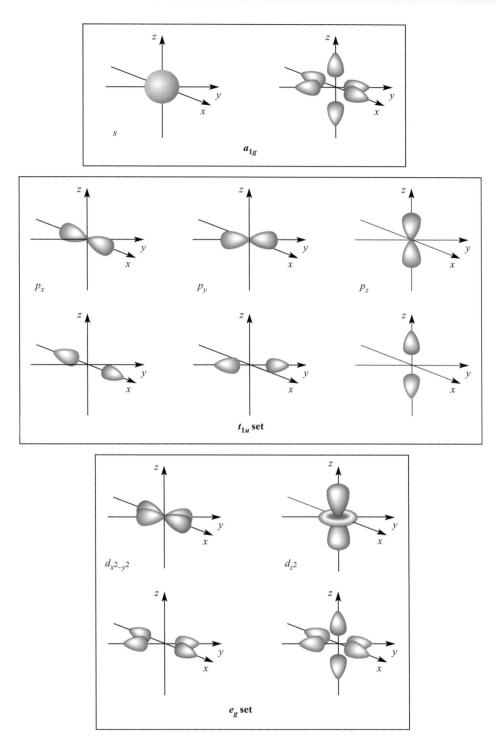

Fig. 20.12 Metal atomic orbitals s, p_x, p_y, p_z, $d_{x^2-y^2}$, d_{z^2} matched by symmetry with ligand group orbitals for an octahedral (O_h) complex with only σ-bonding.

A **π-donor ligand** donates electrons to the metal centre in an interaction that involves a filled ligand orbital and an empty metal orbital.

A **π-acceptor ligand** accepts electrons from the metal centre in an interaction that involves a filled metal orbital and an empty ligand orbital.

π-Donor ligands include Cl^-, Br^- and I^- and the metal–ligand π-interaction involves transfer of electrons from filled ligand p orbitals to the metal centre (Fig. 20.14a). Examples of π-acceptor ligands are CO, N_2, NO and alkenes, and the metal–ligand π-bonds arise from the

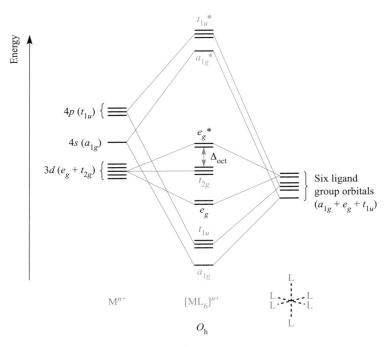

Fig. 20.13 An approximate MO diagram for the formation of $[ML_6]^{n+}$ (where M is a first row metal) using the ligand group orbital approach; the orbitals are shown pictorially in Fig. 20.12. The bonding only involves M–L σ-interactions.

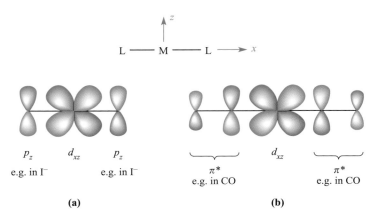

Fig. 20.14 π-Bond formation in a linear L–M–L unit in which the metal and ligand donor atoms lie on the x axis: (a) between metal d_{xz} and ligand p_z orbitals as for L = I$^-$, an example of a π-donor ligand, and (b) between metal d_{xz} and ligand π^*-orbitals as for L = CO, an example of a π-acceptor ligand.

back-donation of electrons from the metal centre to vacant antibonding orbitals on the ligand (for example, Fig. 20.14b). π-Acceptor ligands can stabilize low oxidation state metal complexes (see Chapter 24). Figure 20.15 shows partial MO diagrams which describe metal–ligand π-interactions in octahedral complexes; the metal *s* and *p* orbitals which are involved in σ-bonding (see Fig. 20.13) have been omitted. Figure 20.15a shows the interaction between a metal ion and six π-donor ligands; electrons are omitted from the diagram, and we return to them later. The ligand group π-orbitals (see Box 20.2) are filled and lie above,

but relatively close to, the ligand σ-orbitals, and interaction with the metal d_{xy}, d_{yz} and d_{xz} atomic orbitals leads to bonding (t_{2g}) and antibonding (t_{2g}^*) MOs. The energy separation between the t_{2g}^* and e_g^* levels corresponds to Δ_{oct}. Figure 20.15b shows the interaction between a metal ion and six π-acceptor ligands. The vacant ligand π^*-orbitals lie significantly higher in energy than the ligand σ-orbitals. Orbital interaction leads to bonding (t_{2g}) and antibonding (t_{2g}^*) MOs as before, but now the t_{2g}^* MOs are at high energy and Δ_{oct} is identified as the energy separation between the t_{2g} and e_g^* levels (Fig. 20.15b).

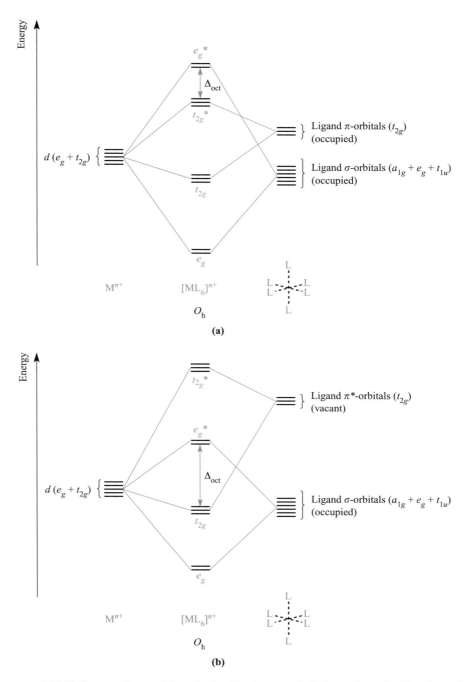

Fig. 20.15 Approximate partial MO diagrams for metal–ligand π-bonding in an octahedral complex: (a) with π-donor ligands and (b) with π-acceptor ligands. In addition to the MOs shown, σ-bonding in the complex involves the a_{1g} and t_{1u} MOs (see Fig. 20.13). Electrons are omitted from the diagram, because we are dealing with a general M^{n+} ion. Compared with Fig. 20.13, the energy scale is expanded.

Although Figs. 20.13 and 20.15 are qualitative, they reveal important differences between octahedral $[ML_6]^{n+}$ complexes containing σ-donor, π-donor and π-acceptor ligands:

- Δ_{oct} *decreases* on going from a σ-complex to one containing π-donor ligands;
- for a complex with π-donor ligands, increased π-donation stabilizes the t_{2g} level and destabilizes the $t_{2g}{}^{*}$, thus decreasing Δ_{oct};

- Δ_{oct} values are relatively large for complexes containing π-acceptor ligands, and such complexes are likely to be low-spin;
- for a complex with π-acceptor ligands, increased π-acceptance stabilizes the t_{2g} level, increasing Δ_{oct}.

The above points are consistent with the positions of the ligands in the spectrochemical series; π-donors such as I^-

THEORY

Box 20.2 The t_{2g} set of ligand π-orbitals for an octahedral complex

Figure 20.15 shows *three* ligand group π-orbitals and you may wonder how these arise from the combination of six ligands, especially since we show a simplistic view of the π-interactions in Fig. 20.14. In an octahedral $[ML_6]^{n+}$ complex with six π-donor or -acceptor ligands lying on the x, y and z axes, each ligand provides *two* π-orbitals, e.g. for ligands on the x axis, both p_y and p_z orbitals are available for π-bonding. Now consider just one plane containing four ligands of the octahedral complex, e.g. the xz plane. Diagram (a) below shows a ligand group orbital (LGO) comprising the p_z orbitals of two ligands and the p_x orbitals of the other two. Diagram (b) shows how the LGO in (a) combines with the metal d_{xz} orbital to give a bonding MO, while (c) shows the antibonding combination.

Three LGOs of the type shown in (a) can be constructed, one in each plane, and these can, respectively, overlap with the metal d_{xy}, d_{yz} and d_{xz} atomic orbitals to give the t_{2g} and t_{2g}^* MOs shown in Fig. 20.15.

Self-study exercise

Show that, under O_h symmetry, the LGO in diagram (a) belongs to a t_{2g} set.

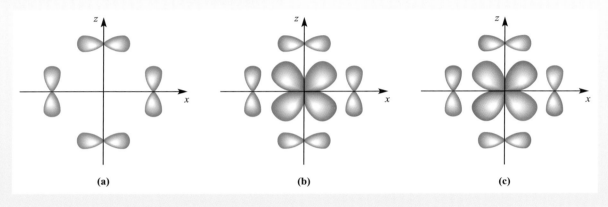

(a)   (b)   (c)

and Br^- are weak-field, while π-acceptor ligands such as CO and $[CN]^-$ are strong-field ligands.

Let us complete this section by considering the occupancies of the MOs in Figs. 20.15a and b. Six π-donor ligands provide 18 electrons (12 σ- and six π-electrons) and these can *notionally* be considered to occupy the a_{1g}, t_{1u}, e_g and t_{2g} orbitals of the complex. The occupancy of the t_{2g}^* and e_g^* levels corresponds to the number of valence electrons of the metal ion. Six π-acceptor ligands provide 12 electrons (i.e. 12 σ-electrons since the π^*-ligand orbitals are empty) and, *formally*, we can place these in the a_{1g}, t_{1u} and e_g orbitals of the complex. The number of electrons supplied by the metal centre then corresponds to the occupancy of the t_{2g} and e_g^* levels. Since occupying *antibonding* MOs lowers the metal–ligand bond order, it follows that, for example, octahedral complexes with π-accepting ligands will not be favoured for metal centres with d^7, d^8, d^9 or d^{10} configurations. This last point brings us back to some fundamental observations in experimental inorganic chemistry: *d*-block metal organometallic and related complexes tend to obey the *effective atomic number rule* or *18-electron rule*. Worked example 20.2 illustrates this rule, and we return to its applications in Chapter 24.

A low oxidation state organometallic complex contains π-acceptor ligands and the metal centre tends to acquire 18 electrons in its valence shell (the ***18-electron rule***), thus filling the valence orbitals, e.g. Cr in $Cr(CO)_6$, Fe in $Fe(CO)_5$, and Ni in $Ni(CO)_4$.

Worked example 20.2   18-Electron rule

Show that $Cr(CO)_6$ obeys the 18-electron rule.

The Cr(0) centre has six valence electrons.

CO is a π-acceptor ligand, and each CO ligand is a 2-electron donor.

The total electron count at the metal centre in $Cr(CO)_6 = 6 + (6 \times 2) = 18$.

Self-study exercises

1. Show that the metal centre in each of the following obeys the 18-electron rule: (a) $Fe(CO)_5$; (b) $Ni(CO)_4$; (c) $[Mn(CO)_5]^-$; (d) $Mo(CO)_6$.

2. (a) How many electrons does a PPh_3 ligand donate? (b) Use your answer to (a) to confirm that the Fe centre in $Fe(CO)_4(PPh_3)$ obeys the 18-electron rule.

3. What is the oxidation state of each metal centre in the complexes in question (1)?

[*Ans.* (a) 0; (b) 0; (c) -1; (d) 0]

In applying the 18-electron rule, one clearly needs to know the number of electrons donated by a ligand, e.g. CO is a 2-electron donor. An ambiguity arises over NO groups in complexes. Nitrosyl complexes fall into two classes:

- NO as a 3-electron donor: crystallographic data show linear M—N—O (observed range $\angle$M—N—O = 165–180°) and short M—N and N—O bonds indicating multiple bond character; IR spectroscopic data give $\nu(NO)$ in the range 1650–1900 cm^{-1}; the bonding mode is represented as **20.7** with the N atom taken to be sp hybridized.

- NO as a 1-electron donor: crystallographic data reveal a bent M—N—O group (observed range $\angle$M—N—O $\approx$ 120–140°), and N—O bond length typical of a double bond; IR spectroscopic data show $\nu(NO)$ in the range 1525–1690 cm^{-1}; the bonding mode is represented as **20.8** with the N atom considered as sp^2 hybridized.

$$M{=}N{=}\overset{..}{\underset{..}{O}}\colon \;\longleftrightarrow\; \colon\!M{-}N{\equiv}O\colon \qquad\qquad M{-}N$$

(20.7)   **(20.8)**

Although the 18-electron rule is quite widely obeyed for low oxidation state organometallic compounds containing π-acceptor ligands, it is useless for higher oxidation state metals. This is clear from examples of octahedral complexes cited in Section 19.7, and can be rationalized in terms of the

---

THEORY

Box 20.3 Octahedral versus trigonal prismatic d^0 and d^1 metal complexes

In Section 19.7, we stated that there is a small group of d^0 or d^1 metal complexes in which the metal centre is in a trigonal prismatic (e.g. $[TaMe_6]^-$ and $[ZrMe_6]^{2-}$) or distorted trigonal prismatic (e.g. $[MoMe_6]$ and $[WMe_6]$) environment. The methyl groups in these d^0 complexes form M–C σ-bonds, and 12 electrons are available for the bonding: one electron from each ligand and six electrons from the metal, including those from the negative charge where applicable. (In counting electrons, we assume a zero-valent metal centre: see Section 24.3.) The qualitative energy level diagram drawn opposite shows that, in a model MH_6 complex with an octahedral structure, these 12 electrons occupy the a_{1g}, e_g and t_{1u} MOs. Now consider what happens if we change the geometry of the model MH_6 complex from octahedral to trigonal prismatic. The point group changes from O_h to D_{3h}, and as a consequence, the properties of the MOs change as shown in the figure. The number of electrons stays the same, but there is a net gain in energy. This stabilization explains why d^0 (and also d^1) complexes of the MMe_6 type show a preference for a trigonal prismatic structure. However, the situation is further complicated because of the observation that $[MoMe_6]$ and $[WMe_6]$, for example, exhibit structures with C_{3v} symmetry (i.e. distorted trigonal prismatic): three of the M–C bonds are normal but three are elongated and have smaller angles between them. This distortion can also be explained in terms of MO theory, since additional orbital stabilization for the 12-electron system is achieved with respect to the D_{3h} structure.

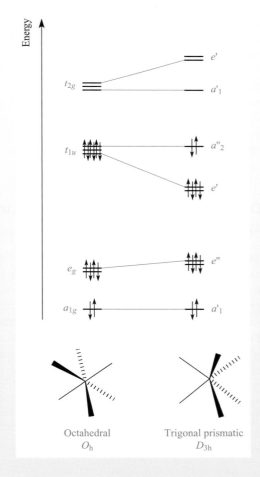

Further reading

K. Seppelt (2003) *Acc. Chem. Res.*, vol. 36, p. 147 – 'Non-octahedral structures'.

smaller energy separations between bonding and anti-bonding orbitals illustrated in Figs. 20.13 and 20.15a compared with that in Fig. 20.15b.

We could extend our arguments to complexes such as $[CrO_4]^{2-}$ and $[MnO_4]^-$ showing how π-donor ligands help to stabilize high oxidation state complexes. However, for a valid discussion of these examples, we need to construct new MO diagrams appropriate to tetrahedral species. To do so would not provide much more insight than we have gained from considering the octahedral case, and interested readers are directed to more specialized texts.[†]

20.5 Ligand field theory

Although we shall not be concerned with the mathematics of ligand field theory, it is important to comment upon it briefly since we shall be using ligand field stabilization energies (LFSEs) later in this chapter.

> ***Ligand field theory*** is an extension of crystal field theory which is freely parameterized rather than taking a localized field arising from point charge ligands.

Ligand field, like crystal field, theory is *confined* to the role of *d* orbitals, but unlike the crystal field model, the ligand field approach is *not* a purely electrostatic model. It is a freely parameterized model, and uses Δ_{oct} and *Racah parameters* (to which we return later) which are obtained from electronic spectroscopic (i.e. *experimental*) data. Most importantly, although (as we showed in the last section) it is possible to approach the bonding in *d*-block metal complexes by using molecular orbital theory, it is *incorrect* to state that ligand field theory is simply the application of MO theory.[‡]

20.6 Describing electrons in multi-electron systems

In crystal field theory, we consider repulsions between *d*-electrons and ligand electrons, but ignore interactions between *d*-electrons on the metal centre. This is actually an aspect of a more general question about how we describe the interactions between electrons in multi-electron systems. We will now show why simple electron configurations such as $2s^2 2p^1$ or $4s^2 3d^2$ do not uniquely define the arrangement of the electrons. This leads us to an introduction of term

symbols for free atoms and ions. For the most part, use of these symbols is confined to our discussions of the electronic spectra of *d*- and *f*-block complexes. In Section 1.7, we showed how to assign a set of quantum numbers to a given electron. For many purposes, this level of discussion is adequate. However, for an understanding of electronic spectra, a more detailed discussion is required. Before studying this section, you should review Box 1.4.

Quantum numbers *L* and *M*$_L$ for multi-electron species

In the answer to worked example 1.7, we ignored a complication. In assigning quantum numbers to the four $2p$ electrons, how do we indicate whether the last electron is in an orbital with $m_l = +1$, 0 or -1? This, and related questions, can be answered only by considering the interaction of electrons, primarily by means of the *coupling* of magnetic fields generated by their spin or orbital motion: hence the importance of spin and orbital angular momentum (see Section 1.6).

For any system containing more than one electron, the energy of an electron with principal quantum number *n* depends on the value of *l*, and this also determines the orbital angular momentum which is given by eq. 20.8 (see Box 1.4).

$$\text{Orbital angular momentum} = \left(\sqrt{l(l+1)}\right)\frac{h}{2\pi} \qquad (20.8)$$

The energy and the orbital angular momentum of a multi-electron species are determined by a new quantum number, *L*, which is related to the values of *l* for the individual electrons. Since the orbital angular momentum has magnitude *and* $(2l + 1)$ spatial orientations with respect to the *z* axis (i.e. the number of values of m_l), *vectorial* summation of individual *l* values is necessary. The value of m_l for any electron denotes the component of its orbital angular momentum, $m_l(h/2\pi)$, along the *z* axis (see Box 1.4). Summation of m_l values for individual electrons in a multi-electron system therefore gives the resultant orbital magnetic quantum number M_L:

$$M_L = \sum m_l$$

Just as m_l may have the $(2l + 1)$ values $l, (l-1) \dots 0 \dots -(l-1)$, $-l$, so M_L can have $(2L + 1)$ values $L, (L-1) \dots 0 \dots -(L-1)$, $-L$. If we can find all possible values of M_L for a multi-electron species, we can determine the value of *L* for the system.

As a means of cross-checking, it is useful to know what values of *L* are possible. The allowed values of *L* can be determined from *l* for the individual electrons in the multi-electron system. For two electrons with values of l_1 and l_2:

$$L = (l_1 + l_2), (l_1 + l_2 - 1), \dots |l_1 - l_2|$$

[†] For application of MO theory to geometries other than octahedral, see Chapter 9 in: J.K. Burdett (1980) *Molecular Shapes: Theoretical Models of Inorganic Stereochemistry*, Wiley, New York.

[‡] For a more detailed introduction to ligand field theory, see: M. Gerloch and E.C. Constable (1994) *Transition Metal Chemistry: The Valence Shell in d-Block Chemistry*, VCH, Weinheim, pp. 117–120; also see the further reading list at the end of the chapter.

The *modulus* sign around the last term indicates that $|l_1 - l_2|$ may only be zero or a positive value. As an example, consider a p^2 configuration. Each electron has $l = 1$, and so the allowed values of L are 2, 1 or 0. Similarly, for a d^2 configuration, each electron has $l = 2$, and so the allowed values of L are 4, 3, 2, 1 or 0. For systems with three or more electrons, the electron–electron coupling must be considered in sequential steps: couple l_1 and l_2 as above to give a resultant L, and then couple L with l_3, and so on.

Energy states for which $L = 0, 1, 2, 3, 4...$ are known as S, P, D, F, $G...$ terms, respectively. These are analogous to the $s, p, d, f, g...$ labels used to denote atomic orbitals with $l = 0$, 1, 2, 3, 4... in the 1-electron case. By analogy with eq. 20.8, eq. 20.9 gives the resultant orbital angular momentum for a multi-electron system.

$$\text{Orbital angular momentum} = \left(\sqrt{L(L+1)}\right)\frac{h}{2\pi} \qquad (20.9)$$

Quantum numbers S and M_S for multi-electron species

Now let us move from the orbital quantum number to the spin quantum number. In Section 1.6, we stated that the spin quantum number, s, determines the magnitude of the spin angular momentum of an electron and has a value of $\frac{1}{2}$. For a 1-electron species, m_s is the magnetic spin angular momentum and has a value of $+\frac{1}{2}$ or $-\frac{1}{2}$. We now need to define the quantum numbers S and M_S for multi-electron species. The spin angular momentum for a multi-electron species is given by eq. 20.10, where S is the total spin quantum number.

$$\text{Spin angular momentum} = \left(\sqrt{S(S+1)}\right)\frac{h}{2\pi} \qquad (20.10)$$

The quantum number M_S is obtained by algebraic summation of the m_s values for individual electrons:

$$M_S = \sum m_s$$

For a system with n electrons, each having $s = \frac{1}{2}$, possible values of S fall into two series depending on the total number of electrons:

- $S = \frac{1}{2}, \frac{3}{2}, \frac{5}{2} ...$ for an odd number of electrons;
- $S = 0, 1, 2 ...$ for an even number of electrons.

S cannot take negative values. The case of $S = \frac{1}{2}$ clearly corresponds to a 1-electron system, for which values of m_s are $+\frac{1}{2}$ or $-\frac{1}{2}$, and values of M_S are also $+\frac{1}{2}$ or $-\frac{1}{2}$. For each value of S, there are $(2S + 1)$ values of M_S:

Allowed values of M_S: $S, (S-1), ... -(S-1), -S$

Thus, for $S = 0$, $M_S = 0$, for $S = 1$, $M_S = 1, 0$ or -1, and for $S = \frac{3}{2}$, $M_S = \frac{3}{2}, \frac{1}{2}, -\frac{1}{2}$ or $-\frac{3}{2}$.

Microstates and term symbols

With sets of quantum numbers in hand, the electronic states (*microstates*) that are possible for a given electronic configuration can be determined. This is best achieved by constructing a table of microstates, remembering that:

- no two electrons may possess the same set of quantum numbers (the Pauli exclusion principle);
- only *unique* microstates may be included.

Let us start with the case of two electrons in s orbitals. There are two general electronic configurations which describe this: ns^2 and $ns^1n's^1$. Our aim is to determine the possible arrangements of electrons within these two configurations. This will give us a general result which relates all ns^2 states (regardless of n) and another which relates all $ns^1n's^1$ states (regardless of n and n'). An extension of these results leads to the conclusion that a single electronic configuration (e.g. $2s^2 2p^2$) does *not* define a unique arrangement of electrons.

Case 1: ns^2 configuration

An electron in an s atomic orbital must have $l = 0$ and $m_l = 0$, and for each electron, m_s can be $+\frac{1}{2}$ or $-\frac{1}{2}$. The ns^2 configuration is described in Table 20.4. Applying the Pauli exclusion principle means that the two electrons in a given microstate must have different values of m_s, i.e. ↑ and ↓ in one row in Table 20.4. A second arrangement of electrons is given in Table 20.4, but now we must check whether this is the same as or different from the first arrangement. We cannot physically distinguish the electrons, so must use sets of quantum numbers to decide if the microstates (i.e. rows in the table) are the same or different:

- first microstate: $l = 0$, $m_l = 0$, $m_s = +\frac{1}{2}$; $l = 0$, $m_l = 0$, $m_s = -\frac{1}{2}$;
- second microstate: $l = 0$, $m_l = 0$, $m_s = -\frac{1}{2}$; $l = 0$, $m_l = 0$, $m_s = +\frac{1}{2}$.

The microstates are identical (the electrons have simply been switched around) and so one microstate is discounted. Hence, for the ns^2 configuration, only one microstate is possible. The values of M_S and M_L are obtained by

Table 20.4 Table of microstates for an ns^2 configuration; an electron with $m_s = +\frac{1}{2}$ is denoted as ↑, and an electron with $m_s = -\frac{1}{2}$ is denoted as ↓. The two microstates are identical and so one row can be discounted (see text for explanation).

First electron: $m_l = 0$	Second electron: $m_l = 0$	$M_L = \Sigma m_l$	$M_S = \Sigma m_s$	
↑	↓	0	0	$L = 0, S = 0$
↓	↑			

Table 20.5 Table of microstates for an $ns^1n's^1$ configuration. An electron with $m_s = +\frac{1}{2}$ is denoted as ↑, and an electron with $m_s = -\frac{1}{2}$ as ↓. Each row in the table corresponds to a different microstate.

First electron: $m_l = 0$	Second electron: $m_l = 0$	$M_L = \Sigma\, m_l$	$M_S = \Sigma\, m_s$	
↑	↑	0	+1	$\left.\begin{array}{l} \\ \\ \\ \end{array}\right\} L = 0,\ S = 1$
↑	↓	0	0	
↓	↓	0	−1	

reading across the table. The result in Table 20.4 is represented as a *term symbol* which has the form $^{(2S+1)}L$, where $(2S+1)$ is called the *multiplicity* of the term:

$$\text{Multiplicity of the term} \longrightarrow {}^{(2S+1)}L \longleftarrow \begin{cases} L = 0 & S \text{ term} \\ L = 1 & P \text{ term} \\ L = 2 & D \text{ term} \\ L = 3 & F \text{ term} \\ L = 4 & G \text{ term} \end{cases}$$

Terms for which $(2S + 1) = 1, 2, 3, 4 \ldots$ (corresponding to $S = 0, \frac{1}{2}, 1, \frac{3}{2} \ldots$) are called *singlet, doublet, triplet, quartet* ... terms, respectively. Hence, the ns^2 configuration in Table 20.4 corresponds to a ^1S term (a 'singlet S term').[†]

Case 2: $ns^1n's^1$ configuration

Table 20.5 shows allowed microstates for an $ns^1n's^1$ configuration. It is important to check that the three microstates are indeed different from one another:

- first microstate: $l = 0$, $m_l = 0$, $m_s = +\frac{1}{2}$; $l = 0$, $m_l = 0$, $m_s = +\frac{1}{2}$;
- second microstate: $l = 0$, $m_l = 0$, $m_s = +\frac{1}{2}$; $l = 0$, $m_l = 0$, $m_s = -\frac{1}{2}$;
- third microstate: $l = 0$, $m_l = 0$, $m_s = -\frac{1}{2}$; $l = 0$, $m_l = 0$, $m_s = -\frac{1}{2}$.

Values of M_S and M_L are obtained by reading across the table. Values of L and S are obtained by fitting the values of M_S and M_L to the series:

$M_L : L, (L-1) \ldots 0, \ldots -(L-1), -L$
$M_S : S, (S-1) \ldots -(S-1), -S$

and are shown in the right-hand column of Table 20.5. A value of $S = 1$ corresponds to a multiplicity of $(2S + 1) = 3$. This gives rise to a ^3S term (a 'triplet S term').

Self-study exercises

1. Show that an s^1 configuration corresponds to a ^2S term.

2. Show that a d^1 configuration corresponds to a ^2D term.

3. In Table 20.5, why is there not a microstate in which the first electron has $m_s = -\frac{1}{2}$, and the second electron has $m_s = +\frac{1}{2}$?

The quantum numbers *J* and *M*ᴶ

Before moving to further examples, we must address the interaction between the total angular orbital momentum, L, and the total spin angular momentum, S. To do so, we define the total angular momentum quantum number, J. Equation 20.11 gives the relationship for the total angular momentum for a multi-electron species.

$$\text{Total angular momentum} = \left(\sqrt{J(J+1)}\right)\frac{h}{2\pi} \quad (20.11)$$

The quantum number J takes values $(L + S)$, $(L + S - 1) \ldots |L - S|$, and these values fall into the series $0, 1, 2 \ldots$ or $\frac{1}{2}, \frac{3}{2}, \frac{5}{2} \ldots$ (like j for a single electron, J for the multi-electron system must be positive or zero). It follows that there are:

$(2S + 1)$ possible values of J for $S < L$;

$(2L + 1)$ possible values of J for $L < S$.

The value of M_J denotes the component of the total angular momentum along the z axis. Just as there are relationships between S and M_S, and between L and M_L, there is one between J and M_J:

Allowed values of $M_J : J, (J - 1) \ldots -(J - 1), -J$

The method of obtaining J from L and S is based on *LS* (or *Russell–Saunders*) *coupling*, i.e. *spin–orbit coupling*. Although it is the only form of coupling of orbital and spin angular momentum that we shall consider in this book, it is not valid for all elements (especially those with high atomic numbers). In an alternative method of coupling, l and s for all the individual electrons are first combined to give j, and the individual j values are combined in a *j–j coupling* scheme.[‡] The difference in coupling schemes arises from whether the spin–orbit interaction is greater or smaller than the orbit–orbit and spin–spin interactions between electrons.

We are now in a position to write *full* term symbols which include information about S, L and J. The notation for a full term symbol is:

$$\text{Multiplicity of the term} \longrightarrow {}^{(2S+1)}L_J \begin{array}{l} \nearrow\ S, P, D, F, G \ldots \text{ term} \\ \searrow\ J \text{ value} \end{array}$$

A term symbol $^3P_0$ ('triplet P zero') signifies a term with $L = 1$, $(2S + 1) = 3$ (i.e. $S = 1$), and $J = 0$. Different values

---

[†] S is used for the resultant spin quantum number as well as a term with $L = 0$, but, in practice, this double usage rarely causes confusion.

[‡] For details of *j–j* coupling, see: M. Gerloch (1986) *Orbitals, Terms and States*, Wiley, Chichester, p. 74; H. Orofino and R.B. Faria (2010) *J. Chem. Educ.*, vol. 87, p. 1451.

of J denote different *levels* within the term, i.e. $^{(2S+1)}L_{J_1}$, $^{(2S+1)}L_{J_2}\ldots$, for example:

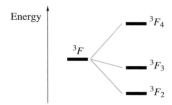

The degeneracy of any J level is $(2J+1)$. This follows from the allowed M_J values being $J, (J-1)\ldots-(J-1), -J$. The J levels have different energies and we illustrate their importance when we discuss magnetic properties (see Fig. 20.28). In inorganic chemistry, it is often sufficient to write the term symbol without the J value, and refer simply to a $^{(2S+1)}L$ term as in the ns^2 and $ns^1n's^1$ examples that we described earlier.

Ground states of elements with $Z=1-10$

In this section, we look in detail at the electronic ground states of atoms with $Z=1$ to 10. This allows you to practise writing tables of microstates, placing the microstates in groups so as to designate terms, and finally assigning ground or excited states. An understanding of this process is essential before we can proceed to a discussion of electronic spectroscopy. An important point to note is that *only electrons in open (incompletely filled) shells (e.g. ns^1, np^2, nd^4) contribute to the term symbol.*

When constructing tables of microstates, it is all too easy to write down a duplicate set, or to miss a microstate. The ns^2 and $ns^1n'p^1$ examples above are relatively simple, but for other systems, it is useful to follow a set of guidelines. Book-keeping of microstates is extremely important, if extremely tedious!

Follow these 'rules' when constructing a ***table of microstates***:

1. Write down the electron configuration (e.g. d^2).
2. Ignore closed shell configurations (e.g. ns^2, np^6, nd^{10}) as these will always give a $^1S_0$ term. This is totally symmetric and makes no contribution to the angular momentum.
3. Determine the number of microstates: for x electrons in a sub-level of $(2l+1)$ orbitals, this is given by:[†]

$$\frac{\{2(2l+1)\}!}{x!\{2(2l+1)-x\}!}$$

4. Tabulate microstates by m_l and m_s, and sum to give M_L and M_S on each row. Check that the number of microstates in the table is the same as that expected from rule (3).
5. Collect the microstates into groups based on values of M_L.

[†] The ! sign means *factorial*: $x! = x \times (x-1) \times (x-2) \ldots \times 1$.

Hydrogen ($Z=1$)

The electronic configuration for an H atom in its ground state is $1s^1$. For one electron in an s orbital ($l=0$):

$$\text{Number of microstates} = \frac{\{2(2l+1)\}!}{x!\{2(2l+1)-x\}!}$$
$$= \frac{2!}{1! \times 1!} = 2$$

The table of microstates is as follows:

$m_l = 0$	$M_L = \Sigma m_l$	$M_S = \Sigma m_s$	
↑	0	$+\frac{1}{2}$	$\left.\right\} L = 0, S = \frac{1}{2}$
↓	0	$-\frac{1}{2}$	

Since $S = \frac{1}{2}$, the multiplicity of the term, $(2S+1)$, is 2 (a doublet term). Since $L = 0$, this is a ^2S term. To determine J, look at the values: use $J = (L+S), (L+S-1) \ldots |L-S|$. The only possible value of J is $\frac{1}{2}$, so the complete term symbol for the H atom is $^2S_{1/2}$.

Helium ($Z=2$)

The electronic configuration of a ground state He atom is $1s^2$ ($l=0$) and hence the table of microstates is like that in Table 20.4:

$m_l = 0$	$m_l = 0$	$M_L = \Sigma m_l$	$M_S = \Sigma m_s$	
↑	↓	0	0	$\left.\right\} L = 0, S = 0$

Since $M_L = 0$ and $M_S = 0$, it follows that $L = 0$ and $S = 0$. The only value of J is 0, and so the term symbol is $^1S_0$.

Lithium ($Z=3$)

Atomic Li has the ground state electronic configuration $1s^22s^1$. Since only the $2s^1$ configuration contributes to the term symbol, the term symbol for Li is the same as that for H (both in their ground states): $^2S_{1/2}$.

Beryllium ($Z=4$)

The ground state electronic configuration of Be is $1s^22s^2$, and contains only closed configurations. Therefore, the term symbol for the ground state of Be is like that of He: $^1S_0$.

Boron ($Z=5$)

When we consider boron ($1s^22s^22p^1$), a new complication arises. Only the $2p^1$ configuration contributes to the term symbol, but because there are three distinct p orbitals ($m_l = +1$, 0 or -1), the p^1 configuration cannot be represented by a unique term symbol. For one electron in a p orbital ($l = 1$):

$$\text{Number of microstates} = \frac{\{2(2l+1)\}!}{x!\{2(2l+1)-x\}!}$$

$$= \frac{6!}{1! \times 5!} = 6$$

A table of microstates for the $2p^1$ configuration is as follows:

$m_l = +1$	$m_l = 0$	$m_l = -1$	M_L	M_S	
↑			$+1$	$+\frac{1}{2}$	
	↑		0	$+\frac{1}{2}$	$L=1, S=\frac{1}{2}$
		↑	-1	$+\frac{1}{2}$	
		↓	-1	$-\frac{1}{2}$	
	↓		0	$-\frac{1}{2}$	$L=1, S=\frac{1}{2}$
↓			$+1$	$-\frac{1}{2}$	

The microstates fall into two sets with $M_L = +1, 0, -1$, and therefore with $L = 1$ (a P term); $S = \frac{1}{2}$ and so $(2S+1) = 2$ (a doublet term). J can take values $(L+S), (L+S-1) \ldots |L-S|$, and so $J = \frac{3}{2}$ or $\frac{1}{2}$. The term symbol for boron may be $^2P_{3/2}$ or $^2P_{1/2}$.

Providing that Russell–Saunders coupling holds, the relative energies of the terms for a given configuration can be found by stating Hund's rules in a formal way:

For the **relative energies of terms** for a given electronic configuration:

1. The term with the highest spin multiplicity has the lowest energy.

2. If two or more terms have the same multiplicity (e.g. ^3F and ^3P), the term having the highest value of L has the lowest energy (e.g. ^3F is lower than ^3P).

3. For terms having the same multiplicity and the same values of L (e.g. $^3P_0$ and $^3P_1$), the level with the lowest value of J is the lowest in energy if the sub-level is less than half-filled (e.g. p^2), and the level with the highest value of J is the more stable if the sub-level is more than half-filled (e.g. p^4). If the level is half-filled with maximum spin multiplicity (e.g. p^3 with $S = \frac{3}{2}$), L must be zero, and $J = S$.

For boron, there are two terms to consider: $^2P_{3/2}$ or $^2P_{1/2}$. These are both doublet terms, and both have $L = 1$. For the p^1 configuration, the p level is less than half-filled, and therefore the ground state level is the one with the lower value of J, i.e. $^2P_{1/2}$.

Carbon ($Z = 6$)

The electron configuration of carbon is $1s^2 2s^2 2p^2$, but only the $2p^2$ ($l = 1$) configuration contributes to the term symbol:

$$\text{Number of microstates} = \frac{\{2(2l+1)\}!}{x!\{2(2l+1)-x\}!}$$

$$= \frac{6!}{2! \times 4!} = 15$$

Table 20.6 Table of microstates for a p^2 configuration. An electron with $m_s = +\frac{1}{2}$ is denoted as ↑, and an electron with $m_s = -\frac{1}{2}$ by ↓.

$m_l = +1$	$m_l = 0$	$m_l = -1$	M_L	M_S	
↑↓			2	0	
↑	↓		1	0	
	↑↓		0	0	$L=2, S=0$
	↓	↑	-1	0	
		↑↓	-2	0	
↑	↑		1	1	
↑		↑	0	1	
	↑	↑	-1	1	
↓	↑		1	0	
↓		↑	0	0	$L=1, S=1$
	↓	↑	-1	0	
↓	↓		1	-1	
↓		↓	0	-1	
	↓	↓	-1	-1	
↑		↓	0	0	$L=0, S=0$

The table of microstates for a p^2 configuration is given in Table 20.6. The microstates have been grouped according to values of M_L and M_S. Remember that values of L and S are derived by looking for sets of M_L and M_S values:

Allowed values of M_L: $L, (L-1), \ldots, 0, \ldots -(L-1), -L$
Allowed values of M_S: $S, (S-1), \ldots -(S-1), -S$

There is no means of telling which entry with $M_L = 0$ and $M_S = 0$ should be assigned to which term (or similarly, how entries with $M_L = 1$ and $M_S = 0$, or $M_L = -1$ and $M_S = 0$ should be assigned). *Indeed, it is not meaningful to do so.* Term symbols are now assigned as follows:

- $L = 2$, $S = 0$ gives the singlet term, ^1D; J can take values $(L+S), (L+S-1) \ldots |L-S|$, so only $J = 2$ is possible; the term symbol is $^1D_2$.
- $L = 1$, $S = 1$ corresponds to a triplet term; possible values of J are 2, 1, 0 giving the terms $^3P_2$, $^3P_1$ and $^3P_0$.
- $L = 0$, $S = 0$ corresponds to a singlet term, and only $J = 0$ is possible; the term symbol is $^1S_0$.

The predicted energy ordering (from the rules above) is $^3P_0 < {}^3P_1 < {}^3P_2 < {}^1D_2 < {}^1S_0$, and the ground state is the $^3P_0$ term.

Nitrogen to neon ($Z = 7$–10)

A similar treatment for the nitrogen atom shows that the $2p^3$ configuration gives rise to ^4S, ^2P and ^2D terms. For the $2p^4$ configuration (oxygen), we introduce a useful simplification by considering the $2p^4$ case in terms of microstates arising from two positrons. This follows from the fact that a positron has the same spin and angular momentum properties as an electron, and differs only in charge. Hence, the terms arising from the np^4 and np^2 configurations are the same. Similarly, np^5 is equivalent to np^1. This positron or *positive hole* concept is very useful and we shall later extend it to nd configurations.

1. Show that the terms for the $3s^2 3p^2$ configuration of Si are $^1D_2$, $^3P_2$, $^3P_1$, $^3P_0$ and $^1S_0$, and that the ground term is $^3P_0$.

2. Show that the ground term for the $2s^2 2p^5$ configuration of an F atom is $^2P_{3/2}$.

3. Confirm that a p^3 configuration has 20 possible microstates.

4. Show that the $2s^2 2p^3$ configuration of nitrogen leads to ^4S, ^2D and ^2P terms, and that the ground term is $^4S_{3/2}$.

The d^2 configuration

Finally in this section, we move to d electron configurations. With $l = 2$, and up to 10 electrons, tables of microstates soon become large. We consider only the d^2 configuration for which:

$$\text{Number of microstates} = \frac{\{2(2l+1)\}!}{x!\{2(2l+1) - x\}!}$$

$$= \frac{10!}{2! \times 8!} = 45$$

Table 20.7 shows the 45 microstates which have been arranged according to values of M_L and M_S. Once again, remember that for microstates such as those with $M_L = 0$ and $M_S = 0$, there is no means of telling which entry should be assigned to which term. The terms arising from the microstates in Table 20.7 are determined as follows:

- $L = 3$, $S = 1$ gives a ^3F term with J values of 4, 3 or 2 ($^3F_4$, $^3F_3$, $^3F_2$);
- $L = 4$, $S = 0$ gives a ^1G term with only $J = 4$ possible ($^1G_4$);
- $L = 2$, $S = 0$ gives a ^1D term with only $J = 2$ possible ($^1D_2$);
- $L = 1$, $S = 1$ gives a ^3P term with J values of 2, 1 or 0 ($^3P_2$, $^3P_1$, $^3P_0$);
- $L = 0$, $S = 0$ gives a ^1S term with only $J = 0$ possible ($^1S_0$).

The relative energies of these terms are determined by considering Hund's rules. The terms with the highest spin multiplicity are the ^3F and ^3P, and of these, the term with higher value of L has the lower energy. Therefore, ^3F is the ground term. The remaining terms are all singlets and so their relative energies depend on the values of L. Hund's rules therefore predict the energy ordering of the terms for a d^2 configuration to be $^3F < ^3P < ^1G < ^1D < ^1S$.

The d^2 configuration is less than a half-filled level and so, if we include the J values, a more detailed description of the predicted ordering of the terms is $^3F_2 < ^3F_3 < ^3F_4 < ^3P_0 < ^3P_1 < ^3P_2 < ^1G_4 < ^1D_2 < ^1S_0$. We return to this ordering when we discuss Racah parameters in Section 20.7, and magnetism (see Fig. 20.24).

Further explanations for the answers can be found by reading Section 20.6.

1. Set up a table of microstates for a d^1 configuration and show that the term symbol is ^2D, and that the ground term is $^2D_{3/2}$.

2. Explain why a value of $S = 1$ corresponds to a triplet state.

3. The terms for a d^2 configuration are ^1D, ^3F, ^1G, ^3P and ^1S. Which is the ground state term? Rationalize your answer.

4. Explain why a d^9 configuration has the same ground state term as a d^1 configuration.

5. Set up a table of microstates for a d^5 configuration, considering *only* those microstates with the highest possible spin multiplicity (the *weak field limit*). Show that the term symbol for the ground term is $^6S_{5/2}$.

20.7 Electronic spectra: absorption

Spectral features

A characteristic feature of many d-block metal complexes is their colours, which arise because they absorb light in the visible region (e.g. Fig. 20.4). Studies of electronic spectra of metal complexes provide information about structure and bonding, although interpretation of the spectra is not always straightforward. Absorptions arise from transitions between electronic energy levels:

- transitions between metal-centred orbitals possessing d-character ('d–d' transitions);
- transitions between metal- and ligand-centred MOs which transfer charge from metal to ligand or ligand to metal (charge transfer bands).

Electronic absorption spectra and the notation for electronic transitions were introduced in Section 4.7, along with the Beer–Lambert law which relates absorbance to the concentration of the solution. The molar extinction coefficient, ε_{max}, is determined from the Beer–Lambert law (eq. 20.12) and indicates the intensity of an absorption.

Table 20.7 Table of microstates for a d^2 configuration. An electron with $m_s = +\frac{1}{2}$ is denoted as ↑, and an electron with $m_s = -\frac{1}{2}$ by ↓.

$m_l = +2$	$m_l = +1$	$m_l = 0$	$m_l = -1$	$m_l = -2$	M_L	M_S	
↑	↑				+3	+1	
↑		↑			+2	+1	
↑			↑		+1	+1	
↑				↑	0	+1	
	↑			↑	−1	+1	
		↑		↑	−2	+1	
			↑	↑	−3	+1	
↑	↓				+3	0	
↑		↓			+2	0	
↑			↓		+1	0	$L = 3, S = 1$
↑				↓	0	0	
	↑			↓	−1	0	
		↑		↓	−2	0	
			↑	↓	−3	0	
↓	↓				+3	−1	
↓		↓			+2	−1	
↓			↓		+1	−1	
↓				↓	0	−1	
	↓			↓	−1	−1	
		↓		↓	−2	−1	
			↓	↓	−3	−1	
↑↓					+4	0	
↓	↑				+3	0	
↓		↑			+2	0	
↓			↑		+1	0	
↓				↑	0	0	$L = 4, S = 0$
	↓			↑	−1	0	
		↓		↑	−2	0	
			↓	↑	−3	0	
				↑↓	−4	0	
	↑↓				+2	0	
	↑	↓			+1	0	
	↑		↓		0	0	$L = 2, S = 0$
		↑	↓		−1	0	
			↑↓		−2	0	
	↑	↑			+1	+1	
	↑		↑		0	+1	
		↑	↑		−1	+1	
	↓	↑			+1	0	
	↓		↑		0	0	$L = 1, S = 1$
		↓	↑		−1	0	
	↓	↓			+1	−1	
	↓		↓		0	−1	
		↓	↓		−1	−1	
		↑↓			0	0	$L = 0, S = 0$

Values of ε_{max} range from close to zero to $>10\,000\,\mathrm{dm^3\,mol^{-1}\,cm^{-1}}$ (Table 20.8).

$$\varepsilon_{max} = \frac{A_{max}}{c \times \ell} \qquad (\varepsilon_{max}\text{ in dm}^3\text{ mol}^{-1}\text{ cm}^{-1}) \qquad (20.12)$$

An absorption band is characterized by both the wavelength, λ_{max}, of the absorbed electromagnetic radiation and ε_{max}. An absorption spectrum may be represented as a plot of absorbance (A) against wavelength (Fig. 20.16), ε against wavelength, A against wavenumber ($\bar{\nu}$), or ε against wavenumber (see Section 4.7). Wavelength is usually quoted in nm and wavenumber in cm^{-1}.

$$\bar{\nu} = \frac{1}{\lambda} = \frac{\nu}{c}$$

400 nm corresponds to 25 000 cm^{-1}; 200 nm corresponds to 50 000 cm^{-1}.

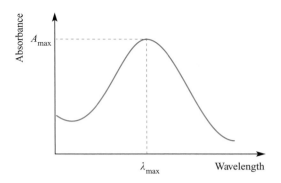

Fig. 20.16 Absorptions in the electronic spectrum of a molecule or molecular ion are often broad, and cover a range of wavelengths. The absorption is characterized by values of λ_{max} and ε_{max} (see eq. 20.12).

Some important points (for which explanations will be given later in the section) are that the electronic absorption spectra of:

- d^1, d^4, d^6 and d^9 complexes consist of one broad absorption;
- d^2, d^3, d^7 and d^8 complexes consist of three broad absorptions;
- d^5 complexes consist of a series of very weak, relatively sharp absorptions.

Self-study exercises

1. Show that $20\,000\,cm^{-1} = 500\,nm$.

2. Figure 20.4 shows the absorption spectrum of aqueous $[Ti(OH_2)_6]^{3+}$ as a plot of A against $\bar{\nu}$. How will the appearance of the plot change if it is redrawn as (a) A against λ, and (b) ε against $\bar{\nu}$? (c) What other information do you require to generate a plot of ε against $\bar{\nu}$?

Charge transfer absorptions

In Section 17.4, we introduced charge transfer bands in the context of their appearance in the UV region of the spectra of halogen-containing charge transfer complexes. In metal complexes, intense absorptions (typically in the UV or visible part of the electronic spectrum) may arise from ligand-centred n–π^* or π–π^* transitions, or from the transfer of electronic charge between ligand and metal orbitals. The latter fall into two categories:

- transfer of an electron from an orbital with primarily ligand character to one with primarily metal character (ligand-to-metal charge transfer, LMCT).
- transfer of an electron from an orbital with primarily metal character to one with primarily ligand character (metal-to-ligand charge transfer, MLCT).

Charge transfer transitions are not restricted by the selection rules that govern 'd–d' transitions (see later). The probability of these electronic transitions is therefore high, and the absorption bands are therefore intense (Table 20.8).

Since electron transfer from metal to ligand corresponds to metal oxidation and ligand reduction, an MLCT transition occurs when a ligand that is easily reduced is bound to a metal centre that is readily oxidized. Conversely, LMCT occurs when a ligand that is easily oxidized is bound to a metal centre (usually one in a high oxidation state) that is readily reduced. There is, therefore, a correlation between the energies of charge transfer absorptions and the electrochemical properties of metals and ligands.

Ligand-to-metal charge transfer may give rise to absorptions in the UV or visible region of the electronic spectrum. One of the most well-known examples is observed for $KMnO_4$. The deep purple colour of aqueous solutions of $KMnO_4$ arises from an intense LMCT absorption in the visible part of the spectrum (Fig. 20.17). This transition corresponds to the promotion of an electron from an orbital that is mainly oxygen lone pair in character to a low-lying, mainly Mn-centred orbital. The following series of complexes illustrate the effects of the metal, ligand and oxidation state of the metal on the position (λ_{max}) of the LMCT band:

- $[MnO_4]^-$ (528 nm), $[TcO_4]^-$ (286 nm), $[ReO_4]^-$ (227 nm);
- $[CrO_4]^{2-}$ (373 nm), $[MoO_4]^{2-}$ (225 nm), $[WO_4]^{2-}$ (199 nm);
- $[FeCl_4]^{2-}$ (220 nm), $[FeBr_4]^{2-}$ (244 nm);
- $[OsCl_6]^{3-}$ (282 nm), $[OsCl_6]^{2-}$ (370 nm).

Table 20.8 Typical ε_{max} values for electronic absorptions. A large ε_{max} corresponds to an intense absorption and, if the absorption is in the visible region, a highly coloured complex.

Type of transition	Typical ε_{max} / $dm^3\,mol^{-1}\,cm^{-1}$	Example
Spin-forbidden 'd–d'	<1	$[Mn(OH_2)_6]^{2+}$ (high-spin d^5)
Laporte-forbidden, spin-allowed 'd–d'	1–10 10–1000	Centrosymmetric complexes, e.g. $[Ti(OH_2)_6]^{3+}$ (d^1) Non-centrosymmetric complexes, e.g. $[NiCl_4]^{2-}$
Charge transfer (fully allowed)	1000–50 000	$[MnO_4]^-$

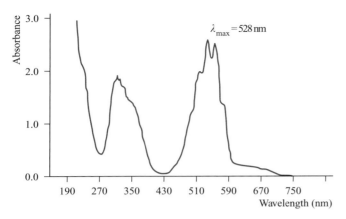

Fig. 20.17 Part of the electronic absorption spectrum of an aqueous solution of $KMnO_4$. Both absorptions arise from LMCT, but it is the band at 528 nm that gives rise to the observed purple colour. Very dilute solutions (here, 1.55×10^{-3} mol dm^{-3}) must be used so that the absorptions remain within the absorption scale.

Across the first two series above, the LMCT band moves to lower wavelength (higher energy) as the metal centre becomes harder to reduce (see Fig. 22.14). The values of the absorption maxima for $[FeX_4]^{2-}$ with different halido-ligands illustrate a shift to longer wavelength (lower energy) as the ligand becomes easier to oxidize (I$^-$ easier than Br$^-$, easier than Cl$^-$). Finally, a comparison of two osmium complexes that differ only in the oxidation state of the metal centre illustrates that the observed ordering of the λ_{max} values is consistent with Os(IV) being easier to reduce than Os(III).

Metal-to-ligand charge transfer typically occurs when the ligand has a vacant, low-lying π^* orbital, for example, CO (see Fig. 2.15), py, bpy, phen and other heterocyclic, aromatic ligands. Often, the associated absorption occurs in the UV region of the spectrum and is not responsible for producing intensely coloured species. In addition, for ligands where a ligand-centred $\pi^* \leftarrow \pi$ transition is possible (e.g. heterocyclic aromatics such as bpy), the MLCT band may be obscured by the $\pi^* \leftarrow \pi$ absorption. For $[Fe(bpy)_3]^{2+}$ and $[Ru(bpy)_3]^{2+}$, the MLCT bands appear in the visible region at 520 and 452 nm, respectively. These are both metal(II) complexes, and the metal *d*-orbitals are relatively close in energy to the ligand π^* orbitals, giving rise to an MLCT absorption energy corresponding to the visible part of the spectrum.

Self-study exercises

1. Explain why aqueous solutions of $[MnO_4]^-$ are purple whereas those of $[ReO_4]^-$ are colourless.

2. Explain the origin of an absorption at 510 nm ($\varepsilon = 11\,000$ dm^3 mol^{-1} cm^{-1}) in the electronic absorption spectrum of $[Fe(phen)_3]^{2+}$ (phen, see Table 7.7).

Selection rules

As we saw in Section 20.6, electronic energy levels are labelled with term symbols. For the most part, we shall use the simplified form of these labels, omitting the *J* states. Thus, the term symbol is written in the general form:

Multiplicity of the term $\longrightarrow$ $(2S + 1)_L \longleftarrow$
$\begin{cases} L = 0 & S \text{ term} \\ L = 1 & P \text{ term} \\ L = 2 & D \text{ term} \\ L = 3 & F \text{ term} \\ L = 4 & G \text{ term} \end{cases}$

Electronic transitions between energy levels obey the following selection rules:

Spin selection rule: $\Delta S = 0$

Transitions may occur from singlet to singlet, or from triplet to triplet states, and so on, but a change in spin multiplicity is *forbidden*.

Laporte selection rule: There must be a change in parity:

allowed transitions:	$g \leftrightarrow u$	
forbidden transitions:	$g \leftrightarrow g$	$u \leftrightarrow u$

This leads to the selection rule:

$\Delta l = \pm 1$

and, thus, *allowed* transitions are $s \rightarrow p, p \rightarrow d, d \rightarrow f$; *forbidden* transitions are $s \rightarrow s, p \rightarrow p, d \rightarrow d, f \rightarrow f, s \rightarrow d, p \rightarrow f$ etc.

Since these selection rules *must* be *strictly obeyed*, why do many *d*-block metal complexes exhibit '*d*–*d*' bands in their electronic absorption spectra?

A spin-forbidden transition becomes 'allowed' if, for example, a singlet state mixes to some extent with a triplet state. This is possible by *spin–orbit coupling* (see Section 20.6) but for first row metals, the degree of mixing is small and so bands associated with 'spin-forbidden' transitions are very weak (Table 20.8). Spin-allowed '*d*–*d*' transitions remain Laporte-forbidden and their observation is explained by a mechanism called '*vibronic coupling*'. An octahedral complex possesses a centre of symmetry, but molecular vibrations result in its temporary loss. At an instant when the molecule does *not* possess a centre of symmetry, mixing of *d* and *p* orbitals can occur. Since the lifetime of the vibration ($\approx 10^{-13}$ s) is longer than that of an electronic transition ($\approx 10^{-18}$ s), a '*d*–*d*' transition involving an orbital of mixed *pd* character can occur although the absorption is still relatively weak (Table 20.8). In a molecule which is non-centrosymmetric (e.g. tetrahedral), *p*–*d* mixing can occur to a greater extent and so the probability of '*d*–*d*' transitions is greater than in a centrosymmetric complex. This leads to tetrahedral complexes being more intensely coloured than octahedral complexes.

Worked example 20.3   Spin-allowed and spin-forbidden transitions

Explain why an electronic transition for high-spin $[Mn(OH_2)_6]^{2+}$ is spin-forbidden, but for $[Co(OH_2)_6]^{2+}$ is spin-allowed.

$[Mn(OH_2)_6]^{2+}$ is high-spin d^5 Mn(II):

A transition from a t_{2g} to e_g orbital is impossible without breaking the spin selection rule: $\Delta S = 0$.

$[Co(OH_2)_6]^{2+}$ is a high-spin d^7 Co(II) complex:

A transition from a t_{2g} to e_g orbital can occur without violating the spin selection rule.
NB: Transitions in both complexes are Laporte-forbidden.

Self-study exercises

1. Write down the spin selection rule.   [*Ans.* See p. 690]

2. What is the d^n configuration and the spin multiplicity of the ground state of (a) a Ti^{3+} and (b) a V^{3+} ion?
   [*Ans.* (a) d^1; doublet; (b) d^2; triplet]

3. Why is a transition from a t_{2g} to an e_g orbital spin allowed in $[V(OH_2)_6]^{3+}$?
   [*Ans.* Triplet to triplet; see question 1]

Electronic absorption spectra of octahedral and tetrahedral complexes

Electronic spectroscopy is a complicated topic and we shall restrict our discussion to high-spin complexes. This corresponds to the *weak field* limit. We begin with the electronic absorption spectrum of an octahedral d^1 ion, exemplified by $[Ti(OH_2)_6]^{3+}$. The spectrum of $[Ti(OH_2)_6]^{3+}$ (Fig. 20.4) exhibits one broad band. However, close inspection shows the presence of a shoulder indicating that the absorption is actually two closely spaced bands (see below). The term symbol for the ground state of Ti^{3+} (d^1, one electron with $L = 2$, $S = \frac{1}{2}$) is ^2D. In an octahedral field, this is split into

$^2T_{2g}$ and ^2E_g terms separated by an energy Δ_{oct}. More generally, it can be shown from group theory that, in an octahedral or tetrahedral field, D, F, G, H and I, but not S and P, terms split. (Lanthanoid metal ions provide examples of ground state H and I terms: see Table 27.3.)

Term	Components in an octahedral field
S	A_{1g}
P	T_{1g}
D	$T_{2g} + E_g$
F	$A_{2g} + T_{2g} + T_{1g}$
G	$A_{1g} + E_g + T_{2g} + T_{1g}$
H	$E_g + T_{1g} + T_{1g} + T_{2g}$
I	$A_{1g} + A_{2g} + E_g + T_{1g} + T_{2g} + T_{2g}$

Similar splittings occur in a tetrahedral field, but the g labels are no longer applicable (see Box 20.1).

The splittings arise because the S, P, D, F and G terms refer to a degenerate set of d orbitals. In an octahedral field, this splits into the t_{2g} and e_g sets of orbitals (Fig. 20.2). For the d^1 ion (term symbol ^2D), there are therefore two possible configurations: $t_{2g}^1 e_g^0$ or $t_{2g}^0 e_g^1$, and these give rise to the $^2T_{2g}$ (ground state) and ^2E_g (excited state) terms. The energy separation between these states increases with increasing field strength (Fig. 20.18). The electronic absorption spectrum of Ti^{3+} arises from a transition from the T_{2g} to the E_g level. The energy of the transition depends on the field strength of the ligands in the octahedral Ti(III) complex. The observation that the absorption spectrum of $[Ti(OH_2)_6]^{3+}$ (Fig. 20.4) consists of two bands, rather than one, can be

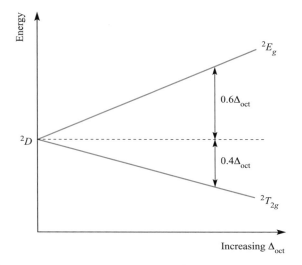

Fig. 20.18 Energy level diagram for a d^1 ion in an octahedral field.

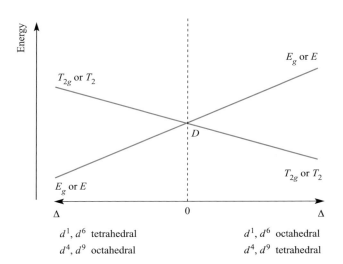

Fig. 20.19 Orgel diagram for d^1, d^4 (high-spin), d^6 (high-spin) and d^9 ions in octahedral (for which T_{2g} and E_g labels are relevant) and tetrahedral (E and T_2 labels) fields. In contrast to Fig. 20.18, multiplicities are not stated because they depend on the d^n configuration.

rationalized in terms of a Jahn–Teller effect in the excited state, $t_{2g}^0 e_g^1$. Single occupancy of the e_g level results in a lowering of the degeneracy, although the resultant energy separation between the two orbitals is small. A corresponding effect in the t_{2g} set is even smaller and can be ignored. Hence, two transitions from ground to excited state are possible. These are close in energy and the spectrum exhibits two absorptions which are of similar wavelength.

For the d^9 configuration (e.g. Cu^{2+}) in an octahedral field (actually, a rare occurrence because of Jahn–Teller effects which lower the symmetry), the ground state of the free ion (^2D) is again split into $^2T_{2g}$ and ^2E_g terms, but, in contrast to the d^1 ion (Fig. 20.18), the ^2E_g term is lower than the $^2T_{2g}$ term. The d^9 and d^1 configurations are related by a *positive hole* concept: d^9 is derived from a d^{10} configuration by replacing one electron by a positive hole. Thus, whereas the d^1 configuration contains one electron, d^9 contains one 'hole' (see Section 20.6). For a d^9 ion in an octahedral field, the splitting diagram is an inversion of that for the octahedral d^1 ion. This relationship is shown in Fig. 20.19 (an *Orgel diagram*) where the right-hand side describes the octahedral d^1 case and the left-hand side describes the octahedral d^9 ion.

Just as there is a relationship between the d^1 and d^9 configurations, there is a similar relationship between the d^4 and d^6 configurations. Further, we can relate the four configurations in an octahedral field as follows. In the weak-field limit, a d^5 ion is high-spin and spherically symmetric, and in this latter regard, d^0, d^5 and d^{10} configurations are analogous. Addition of one electron to the high-spin d^5 ion to give a d^6 configuration mimics going from a d^0 to d^1 configuration. Likewise, going from d^5 to d^4 by adding a positive hole mimics going from d^{10}

to d^9. The result is that the Orgel diagrams for octahedral d^1 and d^6 ions are the same, as are the diagrams for octahedral d^4 and d^9 (Fig. 20.19).

Figure 20.19 also shows that the diagram for a d^1 or d^9 ion is inverted by going from an octahedral to tetrahedral field. Because the Orgel diagram uses a single representation for octahedral *and* tetrahedral fields, it is not possible to indicate that $\Delta_{tet} = \frac{4}{9}\Delta_{oct}$. Tetrahedral d^4 and d^6 ions can also be represented on the same Orgel diagram.

Finally, Fig. 20.19 shows that for each of the octahedral and tetrahedral d^1, d^4, d^6 and d^9 ions, only one electronic transition from a ground to excited state is possible:

- for octahedral d^1 and d^6, the transition is $E_g \longleftarrow T_{2g}$
- for octahedral d^4 and d^9, the transition is $T_{2g} \longleftarrow E_g$
- for tetrahedral d^1 and d^6, the transition is $T_2 \longleftarrow E$
- for tetrahedral d^4 and d^9, the transition is $E \longleftarrow T_2$

Each transition is *spin-allowed* (no change in total spin, S) and the electronic absorption spectrum of each ion exhibits one band. For the sake of completeness, the notation for the transitions given above should include spin multiplicities, $2S + 1$, e.g. for octahedral d^1, the notation is $^2E_g \longleftarrow {}^2T_{2g}$, and for high-spin, octahedral d^4 it is $^5T_{2g} \longleftarrow {}^5E_g$.

In an analogous manner to grouping d^1, d^4, d^6 and d^9 ions, we can consider together d^2, d^3, d^7 and d^8 ions in octahedral and tetrahedral fields. In order to discuss the electronic spectra of these ions, the terms that arise from the d^2 configuration must be known. In an absorption spectrum, we are concerned with electronic transitions from the ground state to one or more excited states. Transitions are possible from one excited state to another, but their probability is so low that they can be ignored. Two points are particularly important:

- selection rules restrict electronic transitions to those between terms with the *same* multiplicity;
- the ground state will be a term with the highest spin multiplicity (Hund's rules, see Section 20.6).

In order to work out the terms for the d^2 configuration, a table of microstates (Table 20.7) must be constructed. However, for interpreting electronic spectra, the table can be simplified because we need concern ourselves only with the terms of maximum spin multiplicity. This corresponds to a *weak field limit*. For the d^2 ion, we therefore focus on the ^3F and ^3P (triplet) terms. These are summarized in Table 20.9, with the corresponding microstates represented only in terms of electrons with $m_s = +\frac{1}{2}$. It follows from the rules given in Section 20.6 that the ^3F term is expected to be lower in energy than the ^3P term. In an octahedral field, the ^3P term does not split, and is labelled $^3T_{1g}$. The ^3F term splits into $^3T_{1g}$, $^3T_{2g}$ and $^3A_{2g}$ terms. The $^3T_{1g}(F)$ term corresponds to a $t_{2g}^2 e_g^0$ arrangement and is triply degenerate because there are three ways of placing two electrons (with parallel spins) in any two of the d_{xy},

Table 20.9 A shorthand table of microstates for a d^2 configuration; only a high-spin case (weak field limit) is considered, and each electron has $m_s = +\frac{1}{2}$. The microstates are grouped so as to show the derivation of the ^3F and ^3P terms. Table 20.7 provides the complete table of microstates for a d^2 ion.

$m_l = +2$	$m_l = +1$	$m_l = 0$	$m_l = -1$	$m_l = -2$	M_L	
↑	↑				+3	
↑		↑			+2	
↑			↑		+1	
↑				↑	0	$^3F\ (L = 3)$
			↑	↑	−1	
		↑		↑	−2	
			↑	↑	−3	
	↑	↑			+1	
↑			↑		0	$^3P\ (L = 1)$
		↑	↑		−1	

d_{yz} and d_{xz} orbitals. The $^3A_{2g}$ term corresponds to $t_{2g}^0 e_g^2$ arrangement (singly degenerate). The $^3T_{2g}$ and $^3T_{1g}(P)$ terms equate with a $t_{2g}^1 e_g^1$ configuration; the lower energy $^3T_{2g}$ term arises from placing two electrons in orbitals lying in mutually perpendicular planes, e.g. $(d_{xy})^1 (d_{z^2})^1$, while the higher energy $^3T_{1g}(P)$ term arises from placing two electrons in orbitals lying in the same plane e.g. $(d_{xy})^1 (d_{x^2-y^2})^1$. The energies of the $^3T_{1g}(F)$, $^3T_{2g}$, $^3A_{2g}$ and $^3T_{1g}(P)$ terms are shown on the right-hand side of Fig. 20.20; note the effect of increasing field strength. Starting from this diagram and using the same arguments as for the d^1, d^4, d^6 and d^9 ions, we can derive the complete Orgel diagram shown in Fig. 20.20.

At increased field strengths, the lines describing the $T_{1g}(F)$ and $T_{1g}(P)$ terms (or T_1, depending on whether we are dealing with octahedral or tetrahedral cases) curve away from one another; there is interaction between terms of the same symmetry and they are not allowed to cross (the *non-crossing rule*). From Fig. 20.20, we can see why three absorptions are observed in the electronic spectra of d^2, d^3, d^7 and d^8 octahedral and tetrahedral complexes. The transitions are from the ground to excited states, and are all spin-allowed, e.g. for an octahedral d^3 ion, the allowed transitions are $^4T_{2g} \leftarrow {}^4A_{2g}$, $^4T_{1g}(F) \leftarrow {}^4A_{2g}$ and $^4T_{1g}(P) \leftarrow {}^4A_{2g}$. Figure 20.21 illustrates spectra for octahedral nickel(II) (d^8) complexes.

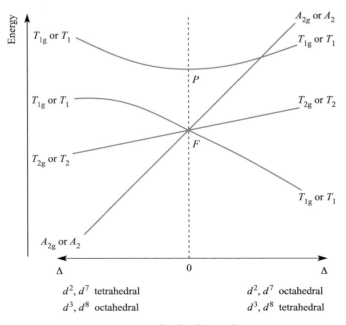

Fig. 20.20 Orgel diagram for d^2, d^3, d^7 and d^8 ions (high-spin) in octahedral (for which T_{1g}, T_{2g} and A_{2g} labels are relevant) and tetrahedral (T_1, T_2 and A_2 labels) fields. Multiplicities are not stated because they depend on the d^n configuration, e.g. for the octahedral d^2 ion, $^3T_{1g}$, $^3T_{2g}$ and $^3A_{2g}$ labels are appropriate.

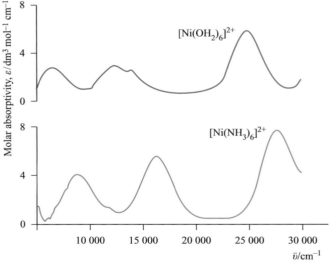

Fig. 20.21 Electronic spectra of $[\text{Ni(OH}_2)_6]^{2+}$ (0.101 mol dm^{-3}) and $[\text{Ni(NH}_3)_6]^{2+}$ (0.315 mol dm^{-3} in aqueous NH$_3$ solution) showing three absorption bands. Values of the molar absorptivity, ε, are related to absorbance by the Beer–Lambert law (eq. 20.12). [This figure is based on data provided by Christian Reber; see: M. Triest, G. Bussière, H. Bélisle and C. Reber (2000) *J. Chem. Educ.*, vol. 77, p. 670; http://jchemed.chem.wisc.edu/JCEWWW/Articles/JCENi/JCENi.html]

For the high-spin d^5 configuration, all transitions are *spin-forbidden* and '*d–d*' transitions that are observed are between the ^6S ground state and quartet states (three unpaired electrons). Associated absorptions are extremely weak.

Worked example 20.4   Electronic absorption spectra

The electronic absorption spectrum of an aqueous solution of $[Ni(en)_3]^{2+}$ exhibits broad absorptions with $\lambda_{max} \approx 325$, 550 and 900 nm. (a) Suggest assignments for the electronic transitions. (b) Which bands are in the visible region?

(a) $[Ni(en)_3]^{2+}$ is a Ni(II), d^8 complex. From the Orgel diagram in Fig. 20.20 the three transitions can be assigned; *lowest wavelength* corresponds to *highest energy* transition:

900 nm assigned to $^3T_{2g} \leftarrow {}^3A_{2g}$
550 nm assigned to $^3T_{1g}(F) \leftarrow {}^3A_{2g}$
325 nm assigned to $^3T_{1g}(P) \leftarrow {}^3A_{2g}$

(b) Visible region spans ≈ 400–750 nm, so only the 550 nm absorption falls in this range (see Table 19.2). The band at 325 nm may tail into the visible region.

Self-study exercises

1. Of the three absorptions in $[Ni(en)_3]^{2+}$, which is closest to the UV end of the spectrum? [*Ans.* Look at Appendix 4]

2. Does the notation $^3T_{2g} \leftarrow {}^3A_{2g}$ indicate an absorption or an emission band? [*Ans.* Look in Section 4.7]

3. Why are the three transitions for $[Ni(en)_3]^{2+}$ (a) spin-allowed, and (b) Laporte-forbidden?

Interpretation of electronic absorption spectra: use of Racah parameters

For a d^1 configuration, the energy of the absorption band in an electronic spectrum gives a direct measure of Δ_{oct}. Figure 20.22a shows the splitting of the ^2D term in an octahedral field into the T_{2g} and E_g levels, the difference in energy being Δ_{oct}. The dependence of this splitting on the ligand field strength is represented in the Orgel diagram in Fig. 20.18. For electron configurations other than d^1, the situation is more complicated. For example, from Fig. 20.20, we expect the electronic spectrum of an octahedral d^2, d^3, d^7 or d^8 ion to consist of three absorptions arising from d–d transitions. How do we determine a

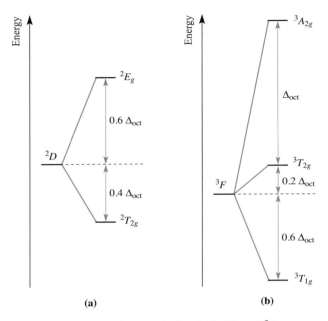

Fig. 20.22 (a) Splitting in an octahedral field of the (a) ^2D term arising from a d^1 configuration, and (b) ^3F term arising from a d^2 configuration. The ^3P term for the d^2 configuration does not split. The diagram applies only to the weak field limit.

value of Δ_{oct} from such a spectrum? For a given electron configuration, the energies of the terms are given by equations involving *Racah parameters* (A, B and C) which allow for electron–electron repulsions. These parameters are used in addition to Δ_{oct} to quantify the description of the spectrum. For example, the ^3F, ^3P, ^1G, ^1D and ^1S terms arise from a d^2 configuration, and their energies are as follows:

Energy of $^1S = A + 14B + 7C$

Energy of $^1D = A - 3B + 2C$

Energy of $^1G = A + 4B + 2C$

Energy of $^3P = A + 7B$

Energy of $^3F = A - 8B$

The actual energies of the terms can be determined spectroscopically[†] and the ordering is found to be $^3F < {}^1D < {}^3P < {}^1G < {}^1S$. (Compare this with a predicted ordering from Hund's rules of $^3F < {}^3P < {}^1G < {}^1D < {}^1S$: see Section 20.6.) Racah parameters for a given system can be evaluated using the above equations. The values of the parameters depend on ion size (B and C become larger with decreasing ionic radius), and usually, the ratio $C/B \approx 4$. From the equations above, the energy difference between

---

[†] Note that this involves the observation of transitions involving both triplet and singlet states.

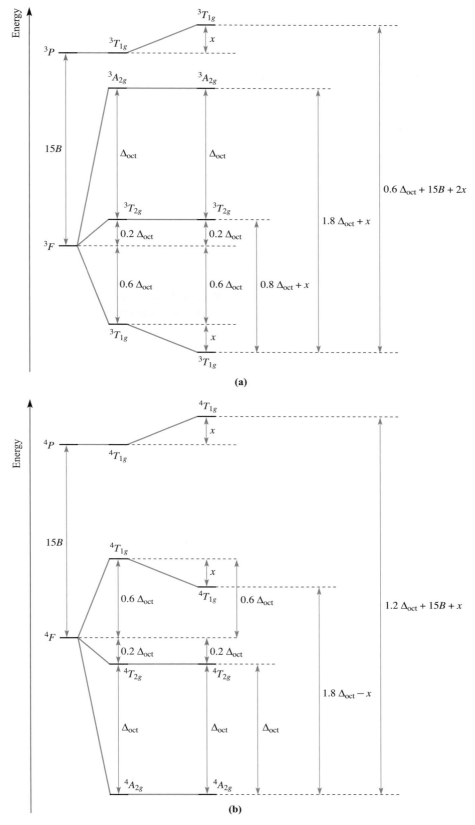

Fig. 20.23 Splitting in an octahedral field of (a) the ^3F and ^3P terms arising from a d^2 configuration, and (b) the ^4F and ^4P terms arising from a d^3 configuration. The initial splitting of the energy levels for the d^2 ion follows from Fig. 20.22b, and the splitting for the d^3 configuration is the inverse of this; x is the energy perturbation caused by mixing of the $T_{1g}(F)$ and $T_{1g}(P)$ terms. The three energy separations marked on the right-hand side of each diagram can be related to energies of transitions observed in electronic spectra of d^2 and d^3 ions. B is a Racah parameter (see eq. 20.13). The diagram applies only to the weak field limit.

the terms of maximum spin multiplicity, ^3P and ^3F, requires the use only of the Racah parameter B. This is true for P and F terms of a common multiplicity arising from other d^n configurations (eq. 20.13).

Energy difference between
$$\begin{aligned}P \text{ and } F \text{ terms} &= (A + 7B) - (A - 8B)\\ &= 15B\end{aligned} \qquad (20.13)$$

Now let us look in detail at the analysis of an electronic absorption spectrum for a d^2 ion in an octahedral field (*weak field ligands*). We have already seen that there are three possible transitions from the ground state: $^3T_{2g} \leftarrow {}^3T_{1g}(F)$, $^3T_{1g}(P) \leftarrow {}^3T_{1g}(F)$ and $^3A_{2g} \leftarrow {}^3T_{1g}(F)$ (Fig. 20.20). Figure 20.22b shows the splitting of the ^3F term, and how the energy separations of the T_{1g}, T_{2g} and A_{2g} levels are related to Δ_{oct}. There is, however, a complication. Figure 20.22b ignores the influence of the ^3P term which lies above the ^3F. As we have already noted, terms of the same symmetry (e.g. $T_{1g}(F)$ and $T_{1g}(P)$) interact, and this causes the 'bending' of the lines in the Orgel diagram in Fig. 20.20. Hence, a perturbation, x, has to be added to the energy level diagram in Fig. 20.22b, and this is represented in Fig. 20.23a. From eq. 20.13, the energy separation of the ^3P and ^3F terms is $15B$. The energies of the $^3T_{2g} \leftarrow {}^3T_{1g}(F)$, $^3T_{1g}(P) \leftarrow {}^3T_{1g}(F)$ and $^3A_{2g} \leftarrow {}^3T_{1g}(F)$ transitions can be written in terms of Δ_{oct}, B and x. Provided that all three absorptions are observed in the spectrum, Δ_{oct}, B and x can be determined. Unfortunately, one or more absorptions may be hidden under a charge transfer band, and in the next section, we describe an alternative strategy for obtaining values of Δ_{oct} and B.

Figure 20.23b shows the splitting of terms for a d^3 configuration (again, for a *limiting weak field* case). The initial splitting of the ^4F term is the inverse of that for the ^3F term in Fig. 20.23a (see Fig. 20.20 and related discussion). Mixing of the $^4T_{1g}(F)$ and $^4T_{1g}(P)$ terms results in the perturbation x shown in the figure. After taking this into account, the energies of the $^4T_{2g} \leftarrow {}^4A_{2g}$, $^4T_{1g}(F) \leftarrow {}^4A_{2g}$ and $^4T_{1g}(P) \leftarrow {}^4A_{2g}$ transitions are written in terms of Δ_{oct}, B and x. Once again, provided that all three bands are observed in the electron spectrum, these parameters can be determined.

The energy level diagram in Fig. 20.23a describes not only an octahedral d^2 configuration, but also an octahedral d^7, a tetrahedral d^3 and a tetrahedral d^8 (with appropriate changes to the multiplicities of the term symbols). Similarly, Fig. 20.23b describes octahedral d^3 and d^8, and tetrahedral d^2 and d^7 configurations.

It must be stressed that the above procedure works *only for the limiting case in which the field strength is very weak*. It cannot, therefore, be applied widely for the determination of Δ_{oct} and B for d-block metal complexes. Further development of this method is beyond the

scope of this book,[†] and a more general approach is the use of Tanabe–Sugano diagrams.

Interpretation of electronic absorption spectra: Tanabe–Sugano diagrams

A more advanced treatment of the energies of electronic states is found in *Tanabe–Sugano diagrams*. The energy of the ground state is taken to be zero for all field strengths, and the energies of all other terms and their components are plotted with respect to the ground term. If there is a change in ground term as the field strength increases, a discontinuity appears in the diagram. Figure 20.24 shows the Tanabe–Sugano diagram for the d^2 configuration in an octahedral field. Notice that the energy and field strength are both expressed in terms of the Racah parameter B. Application of Tanabe–Sugano diagrams is illustrated in worked example 20.5.

Worked example 20.5   Application of Tanabe–Sugano diagrams

Aqueous solutions of $[V(OH_2)_6]^{3+}$ show absorptions at 17 200 and 25 600 cm^{-1} assigned to the $^3T_{2g} \leftarrow {}^3T_{1g}(F)$ and $^3T_{1g}(P) \leftarrow {}^3T_{1g}(F)$ transitions. Estimate values of B and Δ_{oct} for $[V(OH_2)_6]^{3+}$.

$[V(OH_2)_6]^{3+}$ is a d^2 ion and the Tanabe–Sugano diagram in Fig. 20.24 is therefore appropriate. An important point to recognize is that with the diagram provided, only approximate values of B and Δ_{oct} can be obtained.

Let the transition energies be $E_2 = 25\,600\,cm^{-1}$ and $E_1 = 17\,200\,cm^{-1}$.

Values of transition energies cannot be read directly from the Tanabe–Sugano diagram, but ratios of energies can be obtained since:

$$\frac{\left(\dfrac{E_2}{B}\right)}{\left(\dfrac{E_1}{B}\right)} = \frac{E_2}{E_1}$$

From the observed absorption data:

$$\frac{E_2}{E_1} = \frac{25\,600}{17\,200} = 1.49$$

We now proceed by trial and error, looking for the value of $\dfrac{\Delta_{oct}}{B}$ which corresponds to a ratio:

$$\frac{\left(\dfrac{E_2}{B}\right)}{\left(\dfrac{E_1}{B}\right)} = 1.49$$

[†] For a full account of the treatment, see: A.B.P. Lever (1984) *Inorganic Electronic Spectroscopy*, Elsevier, Amsterdam.

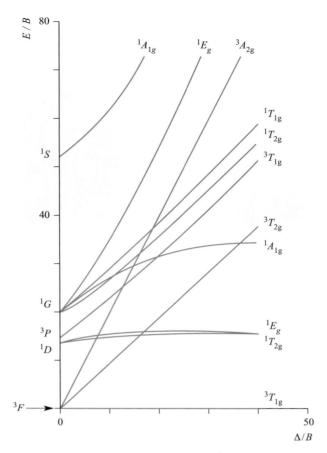

Fig. 20.24 Tanabe–Sugano diagram for the d^2 configuration in an octahedral field.

Trial points:

when $\dfrac{\Delta_{oct}}{B} = 20$, $\dfrac{\left(\dfrac{E_2}{B}\right)}{\left(\dfrac{E_1}{B}\right)} \approx \dfrac{32}{18} = 1.78$

when $\dfrac{\Delta_{oct}}{B} = 30$, $\dfrac{\left(\dfrac{E_2}{B}\right)}{\left(\dfrac{E_1}{B}\right)} \approx \dfrac{41}{28} = 1.46$

when $\dfrac{\Delta_{oct}}{B} = 29$, $\dfrac{\left(\dfrac{E_2}{B}\right)}{\left(\dfrac{E_1}{B}\right)} \approx \dfrac{40.0}{26.9} = 1.49$

This is an *approximate* answer but we are now able to estimate B and Δ_{oct} as follows:

- when $\dfrac{\Delta_{oct}}{B} = 29$, we have $\dfrac{E_2}{B} \approx 40.0$, and since $E_2 = 25\,600\ cm^{-1}$, $B \approx 640\ cm^{-1}$;
- when $\dfrac{\Delta_{oct}}{B} = 29$, $\dfrac{E_1}{B} \approx 26.9$, and since $E_1 = 17\,200$ cm^{-1}, $B \approx 640\ cm^{-1}$.

Substitution of the value of B into $\dfrac{\Delta_{oct}}{B} = 29$ gives an estimate of $\Delta_{oct} \approx 18\,600\ cm^{-1}$.

Accurate methods involving mathematical expressions can be used. These can be found in the advanced texts listed in the further reading at the end of the chapter.

Self-study exercises

1. Why are the two values of B obtained above self-consistent?

2. For $[Ti(OH_2)_6]^{3+}$, a value of Δ_{oct} can be determined directly from λ_{max} in the electronic spectrum. Why is this not possible for $[V(OH_2)_6]^{3+}$, and for most other octahedral ions?

20.8 Electronic spectra: emission

The energy of the absorbed radiation corresponds to the energy of a transition from ground to excited state. Selection rules for electronic spectroscopy only allow transitions between states of the same *multiplicity* (see Sections 20.6 and 20.7). Thus, excitation may occur from a singlet ground state (S_0 in photochemical notation) to the singlet first excited state (S_1). Decay of the excited state back to the ground state may take place by:

- radiative decay (i.e. the emission of electromagnetic radiation),
- non-radiative decay in which thermal energy is lost, or
- non-radiative intersystem crossing to a triplet state (T_1 in Fig. 20.25 represents the lowest energy triplet state).

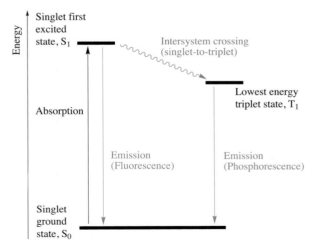

Fig. 20.25 Emission processes are represented using a Jablonski diagram. Radiative and non-radiative transitions are depicted by straight and wavy arrows, respectively.

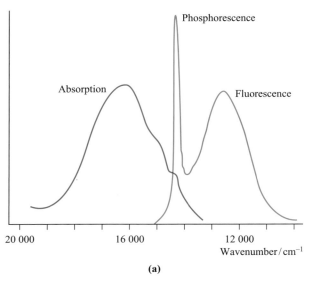

Fig. 20.26 (a) The absorption (at 298 K) and emission (at 77 K) spectra of $[Cr\{OC(NH_2)_2\}_6]^{3+}$. [Adapted from: G.B. Porter and H.L. Schläfer (1963) *Z. Phys.Chem.*, vol. 38, p. 227 with permission.] (b) A ruby laser emitting from a laser splitter.

These processes compete with each other. Emission without a change in multiplicity is called *fluorescence*, while *phosphorescence* refers to an emission in which there is a change in multiplicity. The excitation and decay processes are represented using a *Jablonski diagram* in which radiative transitions are represented by straight arrows and non-radiative transitions by wavy arrows (Fig. 20.25). It follows from the Jablonski diagram that the wavelength of light emitted in a phosphorescence will be to longer wavelength (red-shifted) than the absorbed radiation. In fluorescence, the emitted light is also red-shifted, but to a much smaller extent; while the absorption (Fig. 20.25) involves the ground vibrational state of S_0 and an excited vibrational state of S_1, the emission is from the lowest vibrational state of S_1. Since phosphorescence involves a spin-forbidden transition, the lifetime of the excited state is often relatively long (nanoseconds to microseconds or longer). Fluorescence lifetimes (typically between singlet states) are shorter and usually lie in the picosecond to nanosecond range.

> *Luminescence* refers to the spontaneous emission of radiation from an electronically excited species and covers both *fluorescence* and *phosphorescence*.
>
> Emission without a change in multiplicity is called *fluorescence*, and *phosphorescence* is emission in which the multiplicity changes.

Luminescence from *d*-block metal complexes was first observed for chromium(III) complexes. This is illustrated in Fig. 20.26a with the absorption and emission spectra of $[Cr\{OC(NH_2)_2\}_6]^{3+}$ ($OC(NH_2)_2$ = urea). In a typical fluorescence experiment, the absorption spectrum is initially recorded. The sample is then irradiated with light corresponding to λ_{max} in the absorption spectrum, and the emission spectrum is recorded. $[Cr\{OC(NH_2)_2\}_6]^{3+}$ exhibits both fluorescence and phosphorescence, and the broad and sharp band shapes in Fig. 20.26 are typical. The ground and first excited states of Cr^{3+} ion (d^3) in O_h symmetry are the triplet $^4A_{2g}$ and $^4T_{2g}$, respectively (see Fig. 20.20). Fluorescence arises from a $^4T_{2g} \longrightarrow {}^4A_{2g}$ transition. Intersystem crossing populates the singlet ^2E_g excited state, and phosphorescence arises from a $^2E_g \longrightarrow {}^4A_{2g}$ transition. The red colour of rubies is due to the presence of trace amounts of Cr^{3+} ions in Al_2O_3, each Cr^{3+} ion residing in an octahedral CrO_6 environment. Synthetic ruby (Al_2O_3 doped with $\approx 0.05\%$ by weight of Cr^{3+}) rods are used in ruby lasers, and Fig. 20.26b shows the emission at 694.3 nm corresponding to the $^2E_g \longrightarrow {}^4A_{2g}$ transition following excitation by a continuous or pulsed source of light.

20.9 Evidence for metal–ligand covalent bonding

The nephelauxetic effect

In metal complexes, there is evidence for sharing of electrons between metal and ligand. Pairing energies are lower in complexes than in gaseous M^{n+} ions, indicating that interelectronic repulsion is less in complexes and that the *effective* size of the metal orbitals has increased. This is the *nephelauxetic effect*.

> *Nephelauxetic* means (electron) 'cloud expanding'.

Table 20.10 Selected values of h and k which are used to parameterize the nephelauxetic series; worked example 20.6 shows their application.

Metal ion	k	Ligands	h
Co(III)	0.35	6 Br⁻	2.3
Rh(III)	0.28	6 Cl⁻	2.0
Co(II)	0.24	6 [CN]⁻	2.0
Fe(III)	0.24	3 en	1.5
Cr(III)	0.21	6 NH₃	1.4
Ni(II)	0.12	6 H₂O	1.0
Mn(II)	0.07	6 F⁻	0.8

For complexes with a common metal ion, it is found that the nephelauxetic effect of ligands varies according to a series independent of metal ion:

$$F^- < H_2O < NH_3 < en < [ox]^{2-} < [NCS]^- < Cl^- < [CN]^- < Br^- < I^-$$

increasing nephelauxetic effect ⟶

A nephelauxetic series for metal ions (independent of ligands) is as follows:

$$Mn(II) < Ni(II) \approx Co(II) < Mo(II) < Re(IV) < Fe(III)$$
$$< Ir(III) < Co(III) < Mn(IV)$$

increasing nephelauxetic effect ⟶

The nephelauxetic effect can be parameterized and the values shown in Table 20.10 used to estimate the reduction in electron–electron repulsion upon complex formation. In eq. 20.14, the interelectronic repulsion in the complex is the Racah parameter B; B_0 is the interelectronic repulsion in the gaseous M^{n+} ion.

$$\frac{B_0 - B}{B_0} \approx h_{\text{ligands}} \times k_{\text{metal ion}} \qquad (20.14)$$

The worked example and exercises below illustrate how to apply eq. 20.14.

Worked example 20.6   The nephelauxetic series

Using data in Table 20.10, estimate the reduction in the interelectronic repulsion in going from the gaseous Fe^{3+} ion to $[FeF_6]^{3-}$.

The reduction in interelectronic repulsion is given by:

$$\frac{B_0 - B}{B_0} \approx h_{\text{ligands}} \times k_{\text{metal ion}}$$

In Table 20.10, values of h refer to an octahedral set of ligands.

For $[FeF_6]^{3-}$:

$$\frac{B_0 - B}{B_0} \approx 0.8 \times 0.24 = 0.192$$

Therefore, the reduction in interelectronic repulsion in going from the gaseous Fe^{3+} ion to $[FeF_6]^{3-}$ is $\approx 19\%$.

Self-study exercises

Refer to Table 20.10.

1. Show that the reduction in interelectronic repulsion in going from the gaseous Ni^{2+} ion to $[NiF_6]^{4-}$ is $\approx 10\%$.

2. Estimate the reduction in interelectronic repulsion on going from gaseous Rh^{3+} to $[Rh(NH_3)_6]^{3+}$.
   [*Ans.* $\approx 39\%$]

EPR spectroscopy

Further proof of electron sharing comes from electron paramagnetic resonance (EPR) spectroscopy. As we described in Section 4.9, if a metal ion carrying an unpaired electron is linked to a ligand containing nuclei with nuclear spin quantum number $I \neq 0$, hyperfine splitting of the EPR signal is observed, showing that the orbital occupied by the electron has both metal and ligand character, i.e. there is metal–ligand covalent bonding. An example is the EPR spectrum of $Na_2[IrCl_6]$ (paramagnetic low-spin d^5) recorded for a solid solution in $Na_2[PtCl_6]$ (diamagnetic low-spin d^6); this was a classic EPR experiment, reported in 1953.[†]

20.10 Magnetic properties

Magnetic susceptibility and the spin-only formula

We begin the discussion of magnetochemistry with the so-called *spin-only formula*, an *approximation* that has limited, but useful, applications.

Paramagnetism arises from unpaired electrons. Each electron has a magnetic moment with one component associated with the spin angular momentum of the electron and (except when the quantum number $l = 0$) a second component associated with the orbital angular momentum. For many complexes of first row d-block metal ions we can ignore the second component and the magnetic moment, μ, can be regarded as being determined by the

---

[†] See: J. Owen and K.W.H. Stevens (1953) *Nature*, vol. 171, p. 836.

THEORY

Box 20.4 Magnetic susceptibility

It is important to distinguish between the magnetic susceptibilities χ, χ_g and χ_m.

• Volume susceptibility is χ and is dimensionless.

• Gram susceptibility is $\chi_g = \dfrac{\chi}{\rho}$ where ρ is the density of the sample; the units of χ_g are $m^3\,kg^{-1}$.

• Molar susceptibility is $\chi_m = \chi_g M$ (where M is the molecular mass of the compound) and has SI units of $m^3\,mol^{-1}$.

---

number of unpaired electrons, n (eqs. 20.15 and 20.16). The two equations are related because the total spin quantum number $S = \dfrac{n}{2}$.

$$\mu(\text{spin-only}) = 2\sqrt{S(S+1)} \tag{20.15}$$

$$\mu(\text{spin-only}) = \sqrt{n(n+2)} \tag{20.16}$$

The *effective magnetic moment*, μ_{eff}, can be obtained from the experimentally measured *molar magnetic susceptibility*, χ_m (see Box 20.4), and is expressed in Bohr magnetons (μ_B) where $1\mu_B = eh/4\pi m_e = 9.27 \times 10^{-24}\,J\,T^{-1}$. Equation 20.17 gives the relationship between μ_{eff} and χ_m. Using SI units for the constants, this expression reduces to eq. 20.18 in which χ_m is in $cm^3\,mol^{-1}$. In the laboratory, the continued use of Gaussian units in magnetochemistry means that *irrational susceptibility* is the measured quantity and eq. 20.19 is therefore usually applied.[†]

$$\mu_{eff} = \sqrt{\frac{3k\chi_m T}{L\mu_0\mu_B{}^2}} \tag{20.17}$$

where k = Boltzmann constant; L = Avogadro number; μ_0 = vacuum permeability; T = temperature in kelvin

$$\mu_{eff} = 0.7977\sqrt{\chi_m T} \tag{20.18}$$

$$\mu_{eff} = 2.828\sqrt{\chi_m T} \quad \text{(for use with Gaussian units)} \tag{20.19}$$

Several methods can be used to measure χ_m, e.g. the *Gouy balance* (Fig. 20.27), the *Faraday balance* (which operates in a similar manner to the Gouy balance) and a more modern technique using a *SQUID* (see Section 28.4). The Gouy method makes use of the interaction between unpaired electrons and a magnetic field. A diamagnetic material is repelled by a magnetic field whereas a paramagnetic material is attracted into it. The compound for study is placed in a glass tube, suspended from a balance on which the weight of the sample is recorded. The tube is placed so that one end of the sample lies at the point of

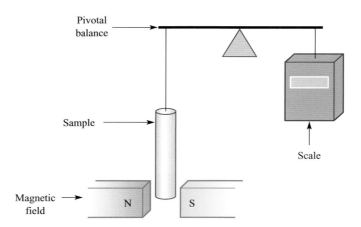

Fig. 20.27 Schematic representation of a Gouy balance.

maximum magnetic flux in an electromagnetic field while the other end is at a point of low flux. Initially the magnet is switched off, but upon applying a magnetic field, paramagnetic compounds are drawn into it by an amount that depends on the number of unpaired electrons. The change in weight caused by the movement of the sample into the field is recorded, and from the associated force it is possible to calculate the magnetic susceptibility of the compound. The effective magnetic moment is then derived using eq. 20.19.

For metal complexes in which the spin quantum number S is the same as for the isolated gaseous metal ion, the spin-only formula (eq. 20.15 or 20.16) can be applied to find the number of unpaired electrons. Table 20.11 lists examples in which measured values of μ_{eff} correlate fairly well with those derived from the spin-only formula; note that all the metal ions are from the *first row* of the *d*-block. Use of the spin-only formula allows the number of unpaired electrons to be determined and gives information about oxidation state of the metal and whether the complex is low- or high-spin.

The use of magnetic data to assist in the assignments of coordination geometries is exemplified by the difference

---

[†] Units in magnetochemistry are non-trivial; for detailed information, see: I. Mills *et al.* (1993) *IUPAC: Quantities, Units and Symbols in Physical Chemistry*, 2nd edn, Blackwell Science, Oxford.

Table 20.11   Spin-only values of μ_{eff} compared with approximate ranges of observed magnetic moments for high-spin complexes of first row d-block ions.

Metal ion	d^n configuration	S	μ_{eff}(spin-only)$/\mu_B$	Observed values of μ_{eff}/μ_B
Sc^{3+}, Ti^{4+}	d^0	0	0	0
Ti^{3+}	d^1	$\frac{1}{2}$	1.73	1.7–1.8
V^{3+}	d^2	1	2.83	2.8–3.1
V^{2+}, Cr^{3+}	d^3	$\frac{3}{2}$	3.87	3.7–3.9
Cr^{2+}, Mn^{3+}	d^4	2	4.90	4.8–4.9
Mn^{2+}, Fe^{3+}	d^5	$\frac{5}{2}$	5.92	5.7–6.0
Fe^{2+}, Co^{3+}	d^6	2	4.90	5.0–5.6
Co^{2+}	d^7	$\frac{3}{2}$	3.87	4.3–5.2
Ni^{2+}	d^8	1	2.83	2.9–3.9
Cu^{2+}	d^9	$\frac{1}{2}$	1.73	1.9–2.1
Zn^{2+}	d^{10}	0	0	0

between tetrahedral and square planar d^8 species, e.g. Ni(II), Pd(II), Pt(II), Rh(I) and Ir(I). Whereas the greater crystal field splitting for the second and third row metal ions invariably leads to square planar complexes, nickel(II) is found in both tetrahedral and square planar environments. Square planar Ni(II) complexes are diamagnetic, whereas tetrahedral Ni(II) species are paramagnetic (see worked example 20.1).

Worked example 20.7   Magnetic moments: spin-only formula

At room temperature, the observed value of μ_{eff} for $[Cr(en)_3]Br_2$ is 4.75 μ_B. Is the complex high- or low-spin? (Ligand abbreviations: see Table 7.7.)

$[Cr(en)_3]Br_2$ contains the octahedral $[Cr(en)_3]^{2+}$ complex and a Cr^{2+} (d^4) ion. Low-spin will have two unpaired electrons ($n = 2$), and high-spin will have four ($n = 4$).

Assume that the spin-only formula is valid (first row metal, octahedral complex):

$$\mu(\text{spin-only}) = \sqrt{n(n+2)}$$

For low-spin: $\mu(\text{spin-only}) = \sqrt{8} = 2.83$

For high-spin: $\mu(\text{spin-only}) = \sqrt{24} = 4.90$

The latter is close to the observed value, and is consistent with a high-spin complex.

Self-study exercises

1. Given that (at 293 K) the observed value of μ_{eff} for $[VCl_4(MeCN)_2]$ is 1.77 μ_B, deduce the number of

unpaired electrons and confirm that this is consistent with the oxidation state of the V atom.

2. At 298 K, the observed value of μ_{eff} for $[Cr(NH_3)_6]Cl_2$ is 4.85 μ_B. Confirm that the complex is high-spin.

3. At 300 K, the observed value of μ_{eff} for $[V(NH_3)_6]Cl_2$ is 3.9 μ_B. Confirm that this corresponds to what is expected for an octahedral d^3 complex.

Spin and orbital contributions to the magnetic moment

By no means do all paramagnetic complexes obey the spin-only formula and caution must be exercised in its use. It is often the case that moments arising from *both* the spin and orbital angular momenta contribute to the observed magnetic moment. Details of the Russell–Saunders coupling scheme to obtain the total angular momentum quantum number, J, from quantum numbers L and S are given in Section 20.6, along with notation for term symbols $^{(2S+1)}L_J$. The energy difference between adjacent states with J values of J' and $(J' + 1)$ is given by the expression $(J' + 1)\lambda$ where λ is called the *spin–orbit coupling constant*. For the d^2 configuration, for example, the ^3F term in an octahedral field is split into $^3F_2$, $^3F_3$ and $^3F_4$, the energy differences between successive pairs being 3λ and 4λ respectively. In a magnetic field, each state with a different J value splits again to give $(2J + 1)$ different levels separated by $g_J\mu_B B_0$ where g_J is a constant called the Landé splitting factor and B_0 is the magnetic field. It is the very small energy differences between these levels with which EPR spectroscopy is concerned and g-values are measured using this technique (see Section 4.9). The overall splitting pattern for a d^2 ion is shown in Fig. 20.28.

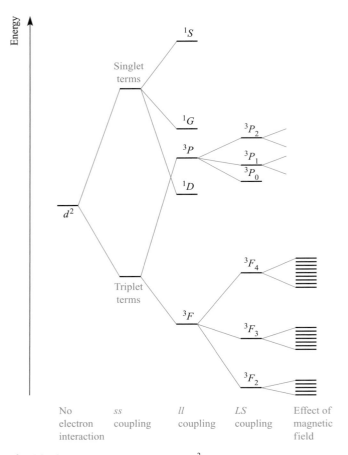

Fig. 20.28 Splitting of the terms of a d^2 ion (not to scale). See Section 20.6 for derivations of the term symbols.

The value of λ varies from a fraction of a cm^{-1} for the very lightest atoms to a few thousand cm^{-1} for the heaviest ones. The extent to which states of different J values are populated at ambient temperature depends on how large their separation is compared with the thermal energy available, kT; at 300 K, $kT \approx 200\,cm^{-1}$ or $2.6\,kJ\,mol^{-1}$. It can be shown theoretically that if the separation of energy

levels is large, the magnetic moment is given by eq. 20.20. Strictly, this applies only to free-ion energy levels, but it gives values for the magnetic moments of lanthanoid ions (for which λ is typically $1000\,cm^{-1}$) that are in good agreement with observed values (see Section 27.4).

$$\left. \begin{aligned} \mu_{\text{eff}} &= g_J \sqrt{J(J+1)} \\ \text{where} \quad g_J &= 1 + \left(\frac{S(S+1) - L(L+1) + J(J+1)}{2J(J+1)} \right) \end{aligned} \right\}$$
(20.20)

For *d*-block metal ions, eq. 20.20 gives results that correlate poorly with experimental data (Tables 20.11 and 20.12). For many (but not all) first row metal ions, λ is very small and the spin and orbital angular momenta of the electrons operate independently. For this case, the van Vleck formula (eq. 20.21) has been derived. Strictly, eq. 20.21 applies to free ions but, in a complex ion, the crystal field partly or fully *quenches* the orbital angular momentum. However, in practice, there is generally a poor fit between values of μ_{eff} calculated from eq. 20.21 and those observed (compare data in Tables 20.12 and 20.11).

$$\mu_{\text{eff}} = \sqrt{4S(S+1) + L(L+1)}$$
(20.21)

If there is *no contribution* from orbital motion, then eq. 20.21 reduces to eq. 20.22 which is the spin-only formula we met earlier. Any ion for which $L = 0$ (e.g. high-spin d^5 Mn^{2+} or Fe^{3+} in which each orbital with $m_l = +2, +1, 0, -1, -2$ is singly occupied, giving $L = 0$) should, therefore, obey eq. 20.22.

$$\mu_{\text{eff}} = \sqrt{4S(S+1)} = 2\sqrt{S(S+1)}$$
(20.22)

However, some other complex ions also obey the spin-only formula (Tables 20.11 and 20.12). In order for an electron to have orbital angular momentum, it must be possible to

Table 20.12 Calculated magnetic moments for first row *d*-block metal ions in high-spin complexes at ambient temperatures. Compare these values with those observed (Table 20.11).

Metal ion	Ground term	μ_{eff} / μ_B calculated from eq. 20.20	μ_{eff} / μ_B calculated from eq. 20.21	μ_{eff} / μ_B calculated from eq. 20.22
Ti^{3+}	$^2D_{3/2}$	1.55	3.01	1.73
V^{3+}	$^3F_2$	1.63	4.49	2.83
V^{2+}, Cr^{3+}	$^4F_{3/2}$	0.70	5.21	3.87
Cr^{2+}, Mn^{3+}	$^5D_0$	0	5.50	4.90
Mn^{2+}, Fe^{3+}	$^6S_{5/2}$	5.92	5.92	5.92
Fe^{2+}, Co^{3+}	$^5D_4$	6.71	5.50	4.90
Co^{2+}	$^4F_{9/2}$	6.63	5.21	3.87
Ni^{2+}	$^3F_4$	5.59	4.49	2.83
Cu^{2+}	$^2D_{5/2}$	3.55	3.01	1.73

Table 20.13 Spin–orbit coupling coefficients, λ, for selected first row d-block metal ions.

Metal ion	Ti^{3+}	V^{3+}	Cr^{3+}	Mn^{3+}	Fe^{2+}	Co^{2+}	Ni^{2+}	Cu^{2+}
d^n configuration	d^1	d^2	d^3	d^4	d^6	d^7	d^8	d^9
$\lambda\,/\,cm^{-1}$	155	105	90	88	-102	-177	-315	-830

transform the orbital it occupies into an entirely equivalent and degenerate orbital by rotation. The electron is then effectively rotating about the axis used for the rotation of the orbital. In an octahedral complex, for example, the three t_{2g} orbitals can be interconverted by rotations through 90°. Thus, an electron in a t_{2g} orbital has orbital angular momentum. The e_g orbitals, having different shapes, cannot be interconverted and so electrons in e_g orbitals never have angular momentum. There is, however, another factor that needs to be taken into account: if all the t_{2g} orbitals are singly occupied, an electron in, say, the d_{xz} orbital cannot be transferred into the d_{xy} or d_{yz} orbital because these already contain an electron having the same spin quantum number as the incoming electron. If all the t_{2g} orbitals are doubly occupied, electron transfer is also impossible. It follows that in high-spin octahedral complexes, orbital contributions to the magnetic moment are important only for the configurations t_{2g}^1, t_{2g}^2, $t_{2g}^4 e_g^2$ and $t_{2g}^5 e_g^2$. For tetrahedral complexes, it is similarly shown that the configurations that give rise to an orbital contribution are $e^2 t_2^1$, $e^2 t_2^2$, $e^4 t_2^4$ and $e^4 t_2^5$. These results lead us to the conclusion that an octahedral high-spin d^7 complex should have a magnetic moment greater than the spin-only value of $3.87\,\mu_B$ but a tetrahedral d^7 complex should not. However, the observed values of μ_{eff} for $[Co(OH_2)_6]^{2+}$ and $[CoCl_4]^{2-}$ are 5.0 and $4.4\,\mu_B$ respectively, i.e. *both* complexes have magnetic moments greater than μ(spin-only). The third factor involved is *spin–orbit coupling*.

Spin–orbit coupling is a complicated subject. We introduced LS (or Russell–Saunders) coupling in Section 20.6, and Fig. 20.28 shows the effects of LS coupling on the energy level diagram for a d^2 configuration. The extent of spin–orbit coupling is quantified by the constant λ, and for the d^2 configuration in Fig. 20.28, the energy differences between the $^3F_2$ and $^3F_3$ levels, and between the $^3F_3$ and $^3F_4$ levels, are 3λ and 4λ, respectively (see earlier). As a result of spin–orbit coupling, mixing of terms occurs. Thus, for example, the $^3A_{2g}$ ground term of an octahedral d^8 ion (Fig. 20.20) mixes with the higher $^3T_{2g}$ term. The extent of mixing is related to Δ_{oct} and to the spin–orbit coupling constant, λ. Equation 20.23 is a modification of the spin-only formula which takes into account spin–orbit coupling. Although the relationship depends on Δ_{oct}, it also applies to tetrahedral complexes. Equation 20.23 applies only to ions having A or E ground terms

(Figs. 20.19 and 20.20). This simple approach is not applicable to ions with a T ground term.

$$\mu_{eff} = \mu(\text{spin-only})\left(1 - \frac{\alpha\lambda}{\Delta_{oct}}\right)$$

$$= \sqrt{n(n+2)}\left(1 - \frac{\alpha\lambda}{\Delta_{oct}}\right) \tag{20.23}$$

where: λ = spin–orbit coupling constant

$\alpha = 4$ for an A ground term

$\alpha = 2$ for an E ground term

Some values of λ are given in Table 20.13. Note that λ is positive for less than half-filled shells and negative for shells that are more than half-filled. Thus, spin–orbit coupling leads to:

- $\mu_{eff} > \mu(\text{spin-only})$ for d^6, d^7, d^8 and d^9 ions;
- $\mu_{eff} < \mu(\text{spin-only})$ for d^1, d^2, d^3 and d^4 ions.

Worked example 20.8   Magnetic moments: spin–orbit coupling

Calculate a value for μ_{eff} for $[Ni(en)_3]^{2+}$ taking into account spin–orbit coupling. Compare your answer with μ(spin-only) and the value of $3.16\,\mu_B$ observed experimentally for $[Ni(en)_3][SO_4]$. [Data: see Tables 20.2 and 20.13.]

Octahedral Ni(II) (d^8) has a $^3A_{2g}$ ground state. Equation needed:

$$\mu_{eff} = \mu(\text{spin-only})\left(1 - \frac{4\lambda}{\Delta_{oct}}\right)$$

$$\mu(\text{spin-only}) = \sqrt{n(n+2)} = \sqrt{8} = 2.83$$

From Table 20.2: $\Delta_{oct} = 11\,500\,cm^{-1}$
From Table 20.13: $\lambda = -315\,cm^{-1}$

$$\mu_{eff} = 2.83\left(1 + \frac{4 \times 315}{11\,500}\right) = 3.14\,\mu_B$$

The calculated value is significantly larger than μ(spin-only) as expected for a d^n configuration with a more than half-full shell. It agrees well with the experimental value.

Self-study exercises

Use data in Tables 20.2 and 20.13.

1. Calculate a value for μ_{eff} for $[\text{Ni(NH}_3)_6]^{2+}$ taking into account spin–orbit coupling. [*Ans.* $3.16\mu_B$]

2. Calculate a value for μ_{eff} for $[\text{Ni(OH}_2)_6]^{2+}$ taking into account spin–orbit coupling. [*Ans.* $3.25\mu_B$]

An important point is that spin–orbit coupling is generally large for second and third row *d*-block metal ions and this leads to large discrepancies between μ(spin-only) and observed values of μ_{eff}. The d^1 complexes *cis*-$[\text{NbBr}_4(\text{NCMe})_2]$ and *cis*-$[\text{TaCl}_4(\text{NCMe})_2]$ illustrate this clearly. Nb and Ta are second and third row group 5 metals, and the room temperature values of μ_{eff} for *cis*-$[\text{NbBr}_4(\text{NCMe})_2]$ and *cis*-$[\text{TaCl}_4(\text{NCMe})_2]$ are 1.27 and 0.45 μ_B, respectively. These data compare with a calculated μ(spin-only) of 1.73 μ_B.

The effects of temperature on μ_{eff}

So far, we have ignored the effects of temperature on μ_{eff}. If a complex obeys the Curie law (eq. 20.24), then μ_{eff} is independent of temperature. This follows from a combination of eqs. 20.18 and 20.24.

$$\chi = \frac{C}{T} \tag{20.24}$$

where: C = Curie constant
$\qquad\quad T$ = temperature in K

However, the Curie law is rarely obeyed and so it is essential to state the temperature at which a value of μ_{eff} has been measured. For second and third row *d*-block metal ions in particular, quoting *only* a room temperature value of μ_{eff} is usually meaningless. When spin–orbit coupling is large, μ_{eff} is highly dependent on T. For a given electronic configuration, the influence of temperature on μ_{eff} can be seen from a *Kotani plot* of μ_{eff} against kT/λ where k is the Boltzmann constant, T is the temperature in K, and λ is the spin–orbit coupling constant. Remember that λ is small for first row metal ions, is large for a second row metal ion, and is even larger for a third row ion. Figure 20.29 shows a Kotani plot for a t_{2g}^4 configuration. Four points are indicated on the curve and correspond to typical values of $\mu_{\text{eff}}(298\,\text{K})$ for complexes of Cr(II) and Mn(III) from the first row, and Ru(IV) and Os(IV) from the second and third rows respectively. Points to note from these data are:

• the points corresponding to $\mu_{\text{eff}}(298\,\text{K})$ for the first row metal ions lie on the near-horizontal part of the curve, and so changing the temperature has little effect on μ_{eff};

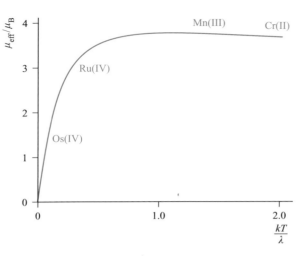

Fig. 20.29 Kotani plot for a t_{2g}^4 configuration; λ is the spin–orbit coupling constant. Typical values of $\mu_{\text{eff}}(298\,\text{K})$ for Cr(II), Mn(III), Ru(IV) and Os(IV) are indicated on the curve.

• the points relating to $\mu_{\text{eff}}(298\,\text{K})$ for the heavier metal ions lie on parts of the curve with steep gradients, and so μ_{eff} is sensitive to changes in temperature; this is especially true for Os(IV) which lies on the steepest part of the curve.

Spin crossover

The choice between a low- and high-spin configuration for d^4, d^5, d^6 and d^7 complexes is not always unique and a *spin crossover* sometimes occurs. This may be initiated by a change in pressure (e.g. a low- to high-spin crossover for $[\text{Fe(CN)}_5(\text{NH}_3)]^{3-}$ at high pressure) or temperature (e.g. octahedral $[\text{Fe(phen)}_2(\text{NCS-}N)_2]$, octahedral $[\text{Fe(20.9)}_2]$ and the square-based pyramidal complex **20.10** undergo low- to high-spin crossovers at 175, 391 and 180 K respectively). The change in the value of μ_{eff} which accompanies the spin crossover may be gradual or abrupt (Fig. 20.30).[†]

(20.9)

* = coordination site

[†] For a review of spin crossover in Fe(II) complexes, see: P. Gütlich, Y. Garcia and H.A. Goodwin (2000) *Chem. Soc. Rev.*, vol. 29, p. 419. An application of spin crossover is described in 'Molecules with short memories': O. Kahn (1999) *Chem. Brit.*, vol. 35, number 2, p. 24.

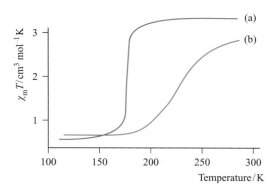

Fig. 20.30 The dependence of the observed values of μ_{eff} on temperature for (a) $[Fe(phen)_2(NCS-N)_2]$ where low- to high-spin crossover occurs abruptly at 175 K, and (b) $[Fe(btz)_2(NCS-N)_2]$ where low- to high-spin crossover occurs more gradually. Ligand abbreviations are defined in the figure [data: J.-A. Real *et al.* (1992) *Inorg. Chem.*, vol. 31, p. 4972].

(20.10)

In addition to magnetic measurements, Mössbauer spectroscopy can be used to study spin-crossover transitions. Isomer shifts of iron complexes are sensitive not only to oxidation state (see Section 4.9) but also to spin state. Figure 20.31 shows the Mössbauer spectra of $[Fe\{HC(3,5-Me_2pz)_3\}_2]I_2$ over a temperature range from 295 to 4.2 K. Each spectrum is characterized by a 'split peak' which is described in terms of the isomer shift, δ, and the quadrupole splitting, ΔE_Q. At 295 K, the iron(II) centre is high-spin ($\delta = 0.969$ mm s^{-1}, $\Delta E_Q = 3.86$ mm s^{-1}). On cooling, the complex undergoes a change to a low-spin state, and at 4.2 K, the transition is complete. The lowest spectrum in Fig. 20.31 arises from low-spin $[Fe\{HC(3,5-Me_2pz)_3\}_2]I_2$ ($\delta = 0.463$ mm s^{-1}, $\Delta E_Q = 0.21$ mm s^{-1}). At intermediate temperatures, the Mössbauer spectroscopic data are fitted to a mixture of low- and high-spin complexes (exemplified in Fig. 20.31 by the spectrum at 166 K).

Ferromagnetism, antiferromagnetism and ferrimagnetism

Whenever we have mentioned magnetic properties so far, we have assumed that metal centres have no interaction with each other (Fig. 20.32a). This is true for substances where the paramagnetic centres are well separated from each other by diamagnetic species. Such systems are said to be *magnetically dilute* (see Section 4.9). For a paramagnetic material, the magnetic susceptibility, χ, is inversely proportional to temperature. This is expressed by the Curie law

(eq. 20.24 and Fig. 20.33a). When the paramagnetic species are very close together (as in the bulk metal) or are separated by a species that can transmit magnetic interactions (as in many *d*-block metal oxides, fluorides and chlorides), the metal centres may interact (*couple*) with one another. The interaction may give rise to *ferromagnetism* or *antiferromagnetism* (Figs. 20.32b and 20.32c).

> In a ***ferromagnetic*** material, large domains of magnetic dipoles are aligned in the same direction. In an ***antiferromagnetic*** material, neighbouring magnetic dipoles are aligned in opposite directions.

Ferromagnetism leads to greatly enhanced paramagnetism as in iron metal at temperatures of up to 1043 K (the *Curie temperature*, T_C), above which thermal energy is sufficient to overcome the alignment and normal paramagnetic behaviour prevails. Above the Curie temperature, a ferromagnetic material obeys the Curie–Weiss law (eq. 20.25). This is represented graphically in Fig. 20.33b which illustrates that, on cooling a sample, ferromagnetic ordering (i.e. the change from paramagnetic to ferromagnetic domains, Fig. 20.32a to 20.32b) occurs at the Curie temperature, T_C. In many cases, the Weiss constant equals the Curie temperature, and the Curie–Weiss law can be written as in eq. 20.26.

$$\chi = \frac{C}{T - \theta} \qquad \text{Curie–Weiss law} \qquad (20.25)$$

where: θ = Weiss constant

   C = Curie constant

$$\chi = \frac{C}{T - T_C} \qquad (20.26)$$

where: T_C = Curie temperature

Antiferromagnetism occurs below the *Néel temperature*, T_N. As the temperature decreases, less thermal energy is available and the paramagnetic susceptibility falls

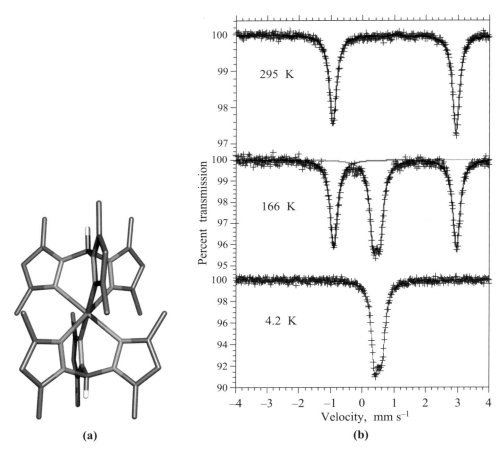

Fig. 20.31 (a) The structure (X-ray diffraction) of the [Fe{HC(3,5-Me$_2$pz)$_3$}$_2$]$^{2+}$ cation in which HC(3,5-Me$_2$pz)$_3$ is a tripodal ligand related to **20.9** (pz = pyrazoyl) [D.L. Reger *et al.* (2002) *Eur. J. Inorg. Chem.*, p. 1190]. (b) Mössbauer spectra of crystallized [Fe{HC(3,5-Me$_2$pz)$_3$}$_2$]I$_2$ at 295, 166 and 4.2 K obtained during cooling of the sample. The data points are shown by black crosses, and the data are fitted to the curves that are shown. [Gary J. Long is acknowledged for providing the spectra.]

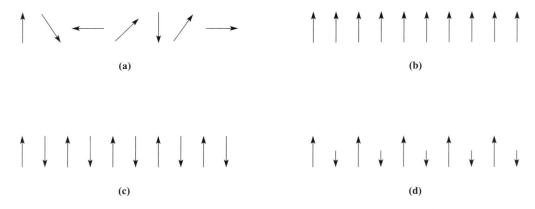

Fig. 20.32 Representations of (a) paramagnetism, (b) ferromagnetism, (c) antiferromagnetism and (d) ferrimagnetism.

rapidly. The dependence of magnetic susceptibility on temperature for an antiferromagnetic material is shown in Fig. 20.33c. The classic example of antiferromagnetism is MnO which has an NaCl-type structure and a Néel temperature of 118 K. Neutron diffraction is capable of distinguishing between sets of atoms having opposed magnetic moments and reveals that the unit cell of MnO at 80 K is double the one at 293 K. This indicates that in the conventional unit cell (Fig. 6.16), metal atoms at adjacent corners have opposed moments at 80 K and that the cells must be stacked to produce the 'true' unit cell. More complex behaviour may occur if some moments are systematically aligned so as to oppose others, but relative numbers or relative values of the moments are such as to lead to a

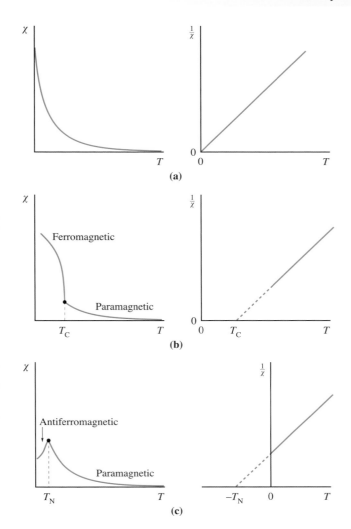

Fig. 20.33 The dependence of the magnetic susceptibility, χ, and of $^1/\chi$ on temperature for (a) a paramagnetic material, (b) a ferromagnetic material and (c) an antiferromagnetic material. The temperatures T_C and T_N are the Curie and Néel temperatures, respectively.

finite resultant magnetic moment: this is *ferrimagnetism* and is represented schematically in Fig. 20.32d.

When a bridging ligand facilitates the coupling of electron spins on adjacent metal centres, the mechanism is one of *superexchange*. This is shown schematically in diagram **20.11**, in which the unpaired metal electrons are represented in red.

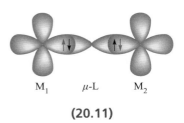

(20.11)

In a *superexchange* pathway, the unpaired electron on the first metal centre, M_1, interacts with a spin-paired pair of electrons on the bridging ligand with the result that the unpaired electron on M_2 is aligned in an antiparallel manner with respect to that on M_1.

20.11 Thermodynamic aspects: ligand field stabilization energies (LFSE)

Trends in LFSE

So far, we have considered Δ_{oct} (or Δ_{tet}) only as a quantity derived from electronic spectroscopy and representing the energy required to transfer an electron from a t_{2g} to an e_g level (or from an e to t_2 level). However, chemical significance can be attached to these values. Table 20.3 showed the variation in crystal field stabilization energies (CFSE) for high- and low-spin octahedral systems. The trend for high-spin systems is restated in Fig. 20.34, where it is compared with that for a tetrahedral field, Δ_{tet} being expressed as a fraction of Δ_{oct} (see eq. 20.7). Note the change from CFSE to LFSE in moving from Table 20.3 to Fig. 20.34. This reflects the fact that we are now dealing with *ligand field theory* and *ligand field stabilization energies*. In the discussion that follows, we consider relationships between observed trends in LFSE values and selected thermodynamic properties of high-spin compounds of the *d*-block metals.

Lattice energies and hydration energies of M^{n+} ions

Figure 20.35 shows a plot of experimental lattice energy data for metal(II) chlorides of first row *d*-block elements. In each salt, the metal ion is high-spin and lies in an octahedral environment in the solid state.[†] The 'double hump' in Fig. 20.35 is reminiscent of that in Fig. 20.34, albeit with respect to a reference line which shows a general increase in lattice energy as the period is crossed. Similar plots can be obtained for species such as MF_2, MF_3 and $[MF_6]^{3-}$, but for each series, only limited data are available and complete trends cannot be studied.

Water is a weak-field ligand and $[M(OH_2)_6]^{2+}$ ions of the first row metals are high-spin. The relationship between absolute enthalpies of hydration of M^{2+} ions (see Section 7.9) and d^n configuration is shown in Fig. 20.36, and again we see the same 'double-humped' appearance of Figs. 20.34 and 20.35.

For each plot in Figs. 20.35 and 20.36, deviations from the reference line joining the d^0, d^5 and d^{10} points may be taken

---

[†] Strictly, a purely electrostatic model does not hold for chlorides, but we include them because more data are available than for fluorides, for which the electrostatic model is more appropriate.

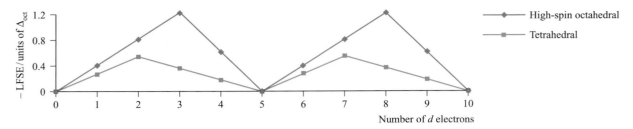

Fig. 20.34 Ligand field stabilization energies as a function of Δ_{oct} for high-spin octahedral systems and for tetrahedral systems; Jahn–Teller effects for d^4 and d^9 configurations have been ignored.

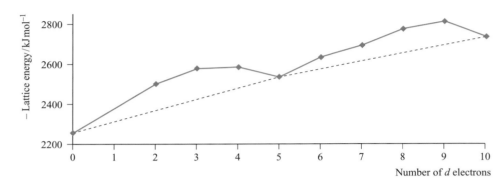

Fig. 20.35 Lattice energies (derived from Born–Haber cycle data) for MCl_2 where M is a first row *d*-block metal; the point for d^0 corresponds to $CaCl_2$. Data are not available for scandium where the stable oxidation state is $+3$.

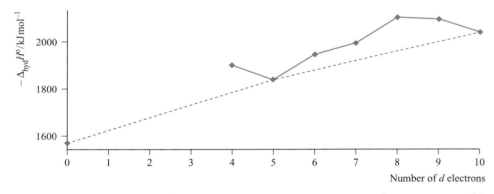

Fig. 20.36 Absolute enthalpies of hydration of the M^{2+} ions of the first row metals; the point for d^0 corresponds to Ca^{2+}. Data are not available for Sc^{2+}, Ti^{2+} and V^{2+}.

as measures of 'thermochemical LFSE' values. In general, the agreement between these values and those calculated from the values of Δ_{oct} derived from electronic spectroscopic data are fairly close. For example, for $[Ni(OH_2)_6]^{2+}$, the values of LFSE(thermochemical) and LFSE(spectroscopic) are 120 and $126\,kJ\,mol^{-1}$ respectively. The latter comes from an evaluation of $1.2\Delta_{oct}$ where Δ_{oct} is determined from the electronic spectrum of $[Ni(OH_2)_6]^{2+}$ to be $8500\,cm^{-1}$. We have to emphasize that this level of agreement is fortuitous. If we look more closely at the problem, we note that only *part* of the measured hydration enthalpy can be attributed to the first coordination sphere of six H_2O molecules, and, moreover, the definitions of LFSE(thermochemical) and

LFSE(spectroscopic) are not strictly equivalent. In conclusion, interesting and useful though discussions of 'double-humped' graphs are in dealing with trends in the thermodynamics of high-spin complexes, it is important to note that they are never more than *approximations*. It is crucial to remember that LFSE terms are only *small parts* of the total interaction energies (generally $<10\%$).

Octahedral versus tetrahedral coordination: spinels

Figure 20.34 indicates that, if all other factors are equal, d^0, high-spin d^5 and d^{10} ions should have no electronically imposed preference between tetrahedral and octahedral coordination, and that the strongest preference for octahedral coordination should be found for d^3, d^8 and low-spin d^6 ions. Is there any unambiguous evidence for these preferences?

The distribution of metal ions between tetrahedral and octahedral sites in a spinel (see Box 13.7) can be rationalized in terms of LFSEs. In a normal spinel $A^{II}B_2^{III}O_4$ the tetrahedral sites are occupied by the A^{2+} ions and the octahedral sites by B^{3+} ions: $(A^{II})^{tet}(B^{III})_2^{oct}O_4$. In an inverse spinel, the distribution is $(B^{III})^{tet}(A^{II}B^{III})^{oct}O_4$. For spinel itself, $A = Mg$, $B = Al$. If at least one of the cations is from the d-block, the inverse structure is frequently (though by no means always) observed: $Zn^{II}Fe_2^{III}O_4$, $Fe^{II}Cr_2^{III}O_4$ and $Mn^{II}Mn_2^{III}O_4$ are normal spinels while $Ni^{II}Ga_2^{III}O_4$, $Co^{II}Fe_2^{III}O_4$ and $Fe^{II}Fe_2^{III}O_4$ are inverse spinels. To account for these observations we first note the following:

- the Madelung constants for the spinel and inverse spinel lattices are usually nearly equal;
- the charges on the metal ions are independent of environment (an assumption);
- Δ_{oct} values for complexes of M^{3+} ions are significantly greater than for corresponding complexes of M^{2+} ions.

Consider compounds with normal spinel structures: in $Zn^{II}Fe_2^{III}O_4$ (d^{10} and d^5), LFSE = 0 for each ion; in $Fe^{II}Cr_2^{III}O_4$ (d^6 and d^3), Cr^{3+} has a much greater LFSE in an octahedral site than does high-spin Fe^{2+}; in $Mn^{II}Mn_2^{III}O_4$ (d^5 and d^4), only Mn^{3+} has any LFSE and this is greater in an octahedral than a tetrahedral site. Now consider some inverse spinels: in $Ni^{II}Ga_2^{III}O_4$, only Ni^{2+} (d^8) has any LFSE and this is greater in an octahedral site; in each of $Co^{II}Fe_2^{III}O_4$ (d^7 and d^5) and $Fe^{II}Fe_2^{III}O_4$ (d^6 and d^5), LFSE = 0 for Fe^{3+} and so the preference is for Co^{2+} and Fe^{2+} respectively to occupy octahedral sites. While this argument is impressive, we must note that observed structures do not always agree with LFSE expectations, e.g. $Fe^{II}Al_2^{III}O_4$ is a normal spinel.

20.12 Thermodynamic aspects: the Irving–Williams series

In aqueous solution, water is replaced by other ligands (eq. 20.27, and see Table 7.7) and the position of equilibrium will be related to the difference between two LFSEs, since Δ_{oct} is ligand-dependent.

$$[Ni(OH_2)_6]^{2+} + [EDTA]^{4-} \rightleftharpoons [Ni(EDTA)]^{2-} + 6H_2O$$
$$(20.27)$$

Table 20.14 lists overall stability constants (see Section 7.12) for $[M(en)_3]^{2+}$ and $[M(EDTA)]^{2-}$ high-spin complexes for d^5 to d^{10} first row M^{2+} ions. For a given ligand and cation charge, ΔS° should be nearly constant along the series and the variation in $\log\beta_n$ should approximately parallel the trend in values of $-\Delta H^\circ$. Table 20.14 shows that the trend from d^5 to d^{10} follows a 'single hump', with the ordering of $\log\beta_n$ for the high-spin ions being:

$$Mn^{2+} < Fe^{2+} < Co^{2+} < Ni^{2+} < Cu^{2+} > Zn^{2+}$$

This is called the *Irving–Williams series* and is observed for a wide range of ligands. The trend is a 'hump' that peaks at Cu^{2+} (d^9) and not at Ni^{2+} (d^8) as might be expected from a consideration of LFSEs (Fig. 20.34). While the variation in LFSE values is a contributing factor, it is not the sole arbiter. Trends in stability constants should bear a relationship to trends in ionic radii (see Appendix 6). The pattern in values of r_{ion} for 6-coordinate high-spin ions is:

$$Mn^{2+} > Fe^{2+} > Co^{2+} > Ni^{2+} < Cu^{2+} < Zn^{2+}$$

We might expect r_{ion} to decrease from Mn^{2+} to Zn^{2+} as Z_{eff} increases, but once again we see a dependence on the d^n configuration with Ni^{2+} being smallest. In turn, this predicts the highest value of $\log\beta_n$ for Ni^{2+}. Why, then, are copper(II) complexes so much more stable than might be expected? The answer lies in the Jahn–Teller distortion that a d^9 complex suffers. The six metal–ligand bonds are not of equal length and thus the concept of a 'fixed' ionic radius for Cu^{2+} is not valid. In an elongated complex (structure **20.5**) such as $[Cu(OH_2)_6]^{2+}$, there are four short and two long Cu–O bonds. Plots of stepwise stability constants for the displacement of H_2O by NH_3 ligands in $[Cu(OH_2)_6]^{2+}$ and $[Ni(OH_2)_6]^{2+}$ are shown in Fig. 20.37. For the first four substitution steps, complex stability is greater for Cu^{2+} than Ni^{2+}, reflecting the formation of four short (strong) Cu–N bonds. The value of $\log K_5$ for Cu^{2+} is consistent with the formation of a weak (axial) Cu–N bond; $\log K_6$ cannot be measured in aqueous solution. The magnitude of the overall stability constant for complexation of Cu^{2+} is dominated by values of K_n for the first four steps and the thermodynamic favourability of

Table 20.14 Overall stability constants for selected high-spin d-block metal complexes.

Metal ion	Mn^{2+}	Fe^{2+}	Co^{2+}	Ni^{2+}	Cu^{2+}	Zn^{2+}
$\log\beta_3$ for $[M(en)_3]^{2+}$	5.7	9.5	13.8	18.6	18.7	12.1
$\log\beta$ for $[M(EDTA)]^{2-}$	13.8	14.3	16.3	18.6	18.7	16.1

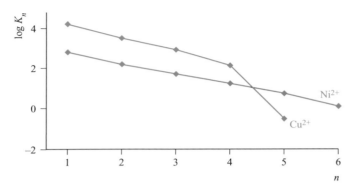

Fig. 20.37 Stepwise stability constants ($\log K_n$) for the displacement of H_2O by NH_3 from $[Ni(OH_2)_6]^{2+}$ (d^8) and $[Cu(OH_2)_6]^{2+}$ (d^9).

these displacement steps is responsible for the position of Cu^{2+} in the Irving–Williams series.

20.13 Thermodynamic aspects: oxidation states in aqueous solution

In the preceding sections, we have, with some degree of success, attempted to rationalize irregular trends in some thermodynamic properties of the first row *d*-block metals. Now we consider the variation in E° values for equilibrium 20.28 (Table 19.1 and Fig. 20.38). The more negative the value of E°, the less easily M^{2+} is reduced.

$$M^{2+}(aq) + 2e^- \rightleftharpoons M(s) \qquad (20.28)$$

This turns out to be a difficult problem. Water is relatively easily oxidized or reduced, and the range of oxidation states on which measurements can be made under aqueous conditions is therefore restricted, e.g. Sc(II) and Ti(II) would liberate H_2. Values of $E^\circ(M^{2+}/M)$ are related (see

Fig. 8.5) to energy changes accompanying the processes:

$M(s) \longrightarrow M(g)$	*atomization* $(\Delta_a H^\circ)$
$M(g) \longrightarrow M^{2+}(g)$	*ionization* $(IE_1 + IE_2)$
$M^{2+}(g) \longrightarrow M^{2+}(aq)$	*hydration* $(\Delta_{hyd} H^\circ)$

In crossing the first row of the *d*-block, the general trend is for $\Delta_{hyd} H^\circ$ to become more negative (Fig. 20.36). There is also a successive increase in the sum of the first two ionization energies albeit with discontinuities at Cr and Cu (Fig. 20.39). Values of $\Delta_a H^\circ$ vary erratically and over a wide range with a particularly low value for zinc (Table 6.2). The net effect of all these factors is an irregular variation in values of $E^\circ(M^{2+}/M)$ across the row, and it is clearly not worth discussing the relatively small variations in LFSEs.

Consider now the variations in $E^\circ(M^{3+}/M^{2+})$ across the row. The enthalpy of atomization is no longer relevant and we are concerned only with trends in the third ionization energy (Table 20.15) and the hydration energies of M^{2+} and M^{3+}. Experimental values for $E^\circ(M^{3+}/M^{2+})$ (Table 20.15) are restricted to the middle of the series; Sc(II) and Ti(II) would reduce water while Ni(III), Cu(III) and Zn(III) would oxidize it. In general, larger values of IE_3 correspond to more positive E° values. This suggests that a steady increase in the difference between the hydration energies of M^{3+} and M^{2+} (which would become larger as the ions become smaller) is outweighed by the variation in IE_3. The only pair of metals for which the change in E° appears out of step is vanadium and chromium. The value of IE_3 for Cr is $165\,kJ\,mol^{-1}$ greater than for V and so it is harder to oxidize *gaseous* Cr^{2+} than V^{2+}. In aqueous solution however, Cr^{2+} is a more powerful reducing agent than V^{2+}. These oxidations correspond to changes in electronic configuration of $d^3 \longrightarrow d^2$ for V and $d^4 \longrightarrow d^3$ for Cr. The V^{2+}, V^{3+}, Cr^{2+} and Cr^{3+} hexaaqua ions are high-spin.

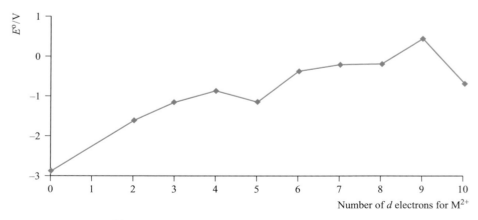

Fig. 20.38 The variation in values of $E^\circ(M^{2+}/M)$ as a function of d^n configuration for the first row metals; the point for d^0 corresponds to M = Ca.

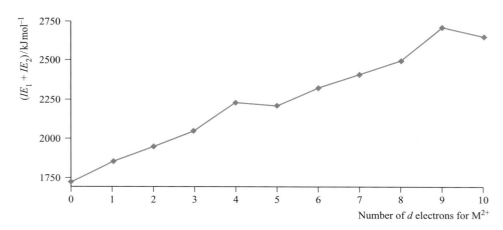

Fig. 20.39 The variation in the sum of the first and second ionization energies as a function of d^n configuration for the first row metals; the point for d^0 corresponds to M = Ca.

Table 20.15   Standard reduction potentials for the equilibrium $M^{3+}(aq) + e^- \rightleftharpoons M^{2+}(aq)$ and values of the third ionization energies.

M	V	Cr	Mn	Fe	Co
E° / V	−0.26	−0.41	+1.54	+0.77	+1.92
$IE_3 / kJ\,mol^{-1}$	2827	2992	3252	2962	3232

Oxidation of V^{2+} is accompanied by a *loss* of LFSE (Table 20.3), while there is a *gain* in LFSE (i.e. more negative) upon oxidation of Cr^{2+} (minor consequences of the Jahn–Teller effect are ignored). Using values of Δ_{oct} from Table 20.2, these changes in LFSE are expressed as follows:

Change in LFSE on oxidation of V^{2+} is

$$-(1.2 \times 12\,400) \text{ to } -(0.8 \times 17\,850)$$
$$= -14\,880 \text{ to } -14\,280\,cm^{-1}$$
$$= +600\,cm^{-1}$$

Change in LFSE on oxidation of Cr^{2+} is

$$-(0.6 \times 14\,100) \text{ to } -(1.2 \times 17\,400)$$
$$= -8460 \text{ to } -20\,880\,cm^{-1}$$
$$= -12\,420\,cm^{-1}$$

The gain in LFSE upon formation of Cr^{3+} corresponds to $\approx 150\,kJ\,mol^{-1}$ and largely cancels out the effect of the third ionization energy. Thus, the apparent anomaly of $E^{\circ}(Cr^{3+}/Cr^{2+})$ can be mostly accounted for in terms of LFSE effects – a considerable achievement in view of the simplicity of the theory.

KEY TERMS

The following terms were introduced in this chapter. Do you know what they mean?

- high-spin
- low-spin
- crystal field theory
- Δ_{oct}, Δ_{tet} ...
- weak-field ligand
- strong-field ligand
- spectrochemical series
- crystal field stabilization energy (CFSE)
- pairing energy
- Jahn–Teller distortion
- π-donor ligand
- π-acceptor ligand
- 18-electron rule
- 'd–d' transition
- quantum numbers for multi-electron species
- term symbol
- table of microstates
- Russell–Saunders coupling
- charge transfer absorption
- MLCT
- LMCT
- λ_{max} and ε_{max} for an absorption band
- selection rule: $\Delta S = 0$
- selection rule: $\Delta l = \pm 1$
- vibronic coupling
- Orgel diagram
- Racah parameter
- Tanabe–Sugano diagram
- nephelauxetic effect
- magnetic susceptibility
- effective magnetic moment
- spin-only formula
- Gouy balance
- spin–orbit coupling constant
- Curie law
- Kotani plot
- spin crossover
- ferromagnetism
- antiferromagnetism
- ferrimagnetism
- superexchange
- ligand field stabilization energy (LFSE)

FURTHER READING

Texts that complement the present treatment

I.B. Bersuker (1996) *Electronic Structure and Properties of Transition Metal Compounds*, Wiley, New York.

M. Gerloch and E.C. Constable (1994) *Transition Metal Chemistry: the Valence Shell in d-Block Chemistry*, VCH, Weinheim.

J.E. Huheey, E.A. Keiter and R.L. Keiter (1993) *Inorganic Chemistry*, 4th edn, Harper Collins, New York, Chapter 11.

W.L. Jolly (1991) *Modern Inorganic Chemistry*, 2nd edn, McGraw-Hill, New York, Chapters 15, 17 and 18.

S.F.A. Kettle (1996) *Physical Inorganic Chemistry*, Spektrum, Oxford.

Term symbols and symmetry labels

P. Atkins and J. de Paula (2010) *Atkins' Physical Chemistry*, 9th edn, Oxford University Press, Oxford – Chapter 11 gives a good introduction to term symbols.

M.L. Campbell (1996) *J. Chem. Educ.*, vol. 73, p. 749 – 'A systematic method for determining molecular term symbols for diatomic molecules' is an extremely good summary of a related topic not covered in this book.

M. Gerloch (1986) *Orbitals, Terms and States*, Wiley, Chichester – A detailed, but readable, account of state symbols which includes *j–j* coupling.

D.W. Smith (1996) *J. Chem. Educ.*, vol. 73, p. 504 – 'Simple treatment of the symmetry labels for the *d–d* states of octahedral complexes'.

Crystal and ligand field theories, electronic spectra and magnetism: advanced texts

B.N. Figgis (1966) *Introduction to Ligand Fields*, Interscience, New York.

B.N. Figgis and M.A. Hitchman (2000) *Ligand Field Theory and its Applications*, Wiley-VCH, New York.

M. Gerloch (1983) *Magnetism and Ligand Field Analysis*, Cambridge University Press, Cambridge.

M. Gerloch and R.C. Slade (1973) *Ligand Field Parameters*, Cambridge University Press, Cambridge.

D.A. Johnson and P.G. Nelson (1999) *Inorg. Chem.*, vol. 38, p. 4949 – 'Ligand field stabilization energies of the hexaaqua 3+ complexes of the first transition series'.

A.F. Orchard (2003) *Magnetochemistry*, Oxford University Press, Oxford – A general account of the subject.

E.I. Solomon and A.B.P. Lever, eds (1999) *Inorganic Electronic Structure and Spectroscopy, Vol. 1 Methodology; Vol. 2 Applications and Case Studies*, Wiley, New York.

PROBLEMS

20.1 Outline how you would apply crystal field theory to explain why the five *d*-orbitals in an octahedral complex are not degenerate. Include in your answer an explanation of the 'barycentre'.

20.2 The absorption spectrum of $[Ti(OH_2)_6]^{3+}$ exhibits a band with $\lambda_{max} = 510$ nm. What colour of light is absorbed and what colour will aqueous solutions of $[Ti(OH_2)_6]^{3+}$ appear?

20.3 Draw the structures of the following ligands, highlight the donor atoms and give the likely modes of bonding (e.g. monodentate): (a) en; (b) bpy; (c) $[CN]^-$; (d) $[N_3]^-$; (e) CO; (f) phen; (g) $[ox]^{2-}$; (h) $[NCS]^-$; (i) PMe_3.

20.4 Arrange the following ligands in order of increasing field strength: Br^-, F^-, $[CN]^-$, NH_3, $[OH]^-$, H_2O.

20.5 For which member of the following pairs of complexes would Δ_{oct} be the larger and why: (a) $[Cr(OH_2)_6]^{2+}$ and $[Cr(OH_2)_6]^{3+}$; (b) $[CrF_6]^{3-}$ and $[Cr(NH_3)_6]^{3+}$; (c) $[Fe(CN)_6]^{4-}$ and $[Fe(CN)_6]^{3-}$; (d) $[Ni(OH_2)_6]^{2+}$ and $[Ni(en)_3]^{2+}$; (e) $[MnF_6]^{2-}$ and $[ReF_6]^{2-}$; (f) $[Co(en)_3]^{3+}$ and $[Rh(en)_3]^{3+}$?

20.6 (a) Explain why there is no distinction between low- and high-spin arrangements for an octahedral d^8 metal ion. (b) Discuss the factors that contribute to the preference for forming either a high- or a low-spin d^4 complex. (c) How would you distinguish experimentally between the two configurations in (b)?

20.7 Verify the CFSE values in Table 20.3.

20.8 In each of the following complexes, rationalize the number of observed unpaired electrons (stated after the formula): (a) $[Mn(CN)_6]^{4-}$ (1); (b) $[Mn(CN)_6]^{2-}$ (3); (c) $[Cr(en)_3]^{2+}$ (4); (d) $[Fe(ox)_3]^{3-}$ (5); (e) $[Pd(CN)_4]^{2-}$ (0); (f) $[CoCl_4]^{2-}$ (3); (g) $[NiBr_4]^{2-}$ (2).

20.9 (a) Explain the forms of the *d* orbital splitting diagrams for trigonal bipyramidal and square pyramidal complexes of formula ML_5 shown in Fig. 20.11. (b) What would you expect concerning the magnetic properties of such complexes of Ni(II)?

20.10 (a) What do you understand by the *nephelauxetic effect*? (b) Place the following ligands in order of increasing nephelauxetic effect: H_2O, I^-, F^-, en, $[CN]^-$, NH_3.

20.11 Discuss each of the following observations:
(a) The $[CoCl_4]^{2-}$ ion is a regular tetrahedron but $[CuCl_4]^{2-}$ has a flattened tetrahedral structure.
(b) The electronic absorption spectrum of $[CoF_6]^{3-}$ contains two bands with maxima at 11 500 and 14 500 cm^{-1}.

20.12 The $3p^2$ configuration of an Si atom gives rise to the following terms: $^1S_0$, $^3P_2$, $^3P_1$, $^3P_0$ and $^1D_2$. Use Hund's rules to predict the relative energies of these terms, giving an explanation for your answer.

20.13 With reference to the ^3F, ^1D, ^3P, ^1G and ^1S terms of a d^2 configuration, explain how you can use term symbols to gain information about allowed electronic transitions.

20.14 What term or terms arise from a d^{10} configuration, and what is the ground state term? Give an example of a first row d-block metal ion with this configuration.

20.15 What are the limitations of the Russell–Saunders coupling scheme?

20.16 Deduce possible J values for a ^3F term. What is the degeneracy of each of these J levels, and what happens when a magnetic field is applied? Sketch an energy level diagram to illustrate your answer, and comment on its significance to EPR spectroscopy.

20.17 In an octahedral field, how will the following terms split, if at all: (a) ^2D, (b) ^3P and (c) ^3F?

20.18 (a) Set up a table of microstates to show that the ground term for the d^1 ion is the singlet ^2D. What are the components of this term in a tetrahedral field? (b) Repeat the process for a d^2 ion and show that the ground and excited terms are the ^3F and ^3P. What are the components of these terms in tetrahedral and octahedral fields?

20.19 (a) On Fig. 20.21, convert the wavenumber scale to nm. (b) Which part of the scale corresponds to the visible range? (c) What would you predict are the colours of $[Ni(OH_2)_6]^{2+}$ and $[Ni(NH_3)_6]^{2+}$. (d) Are the spectra in Fig. 20.21 consistent with the relative positions of H_2O and NH_3 in the spectrochemical series?

20.20 (a) How many 'd–d' bands would you expect to find in the electronic absorption spectrum of an octahedral Cr(III) complex? (b) Account for the observation that the colour of $trans$-$[Co(en)_2F_2]^+$ is less intense than those of cis-$[Co(en)_2F_2]^+$ and $trans$-$[Co(en)_2Cl_2]^+$.

20.21 Comment on the following statements concerning electronic absorption spectra.
(a) $[OsCl_6]^{3-}$ and $[RuCl_6]^{3-}$ exhibit LMCT bands at 282 and 348 nm, respectively.
(b) $[Fe(bpy)_3]^{2+}$ is expected to exhibit an MLCT rather than an LMCT absorption.

20.22 Rationalize why the absorption spectrum of an aqueous solution of $[Ti(OH_2)_6]^{2+}$ (stable under acidic conditions) exhibits two well-separated bands (430 and 650 nm) assigned to 'd–d' transitions, whereas that of an aqueous solution of $[Ti(OH_2)_6]^{3+}$ consists of one absorption ($\lambda_{max} = 490$ nm) with a shoulder (580 nm).

20.23 Describe how you could use Fig. 20.23 to determine Δ_{oct} and the Racah parameter B from the energies of absorptions observed in the spectrum of an octahedral d^3 ion. What are the significant limitations of this method?

20.24 The electronic absorption spectrum of $[Co(OH_2)_6]^{2+}$ exhibits bands at 8100, 16 000 and 19 400 cm^{-1}.
(a) Assign these bands to electronic transitions.
(b) The value of Δ_{oct} for $[Co(OH_2)_6]^{2+}$ listed in Table 20.2 is 9300 cm^{-1}. What value of Δ_{oct} would you obtain using the diagram in Fig. 20.23b? Why does the calculated value not match that in Table 20.2?

20.25 Values of the Racah parameter B for free gaseous Cr^{3+}, Mn^{2+} and Ni^{2+} ions are 918, 960 and 1041 cm^{-1}, respectively. For the corresponding hexaaqua ions, values of B are 725, 835 and 940 cm^{-1}. Suggest a reason for the reduction in B on forming each complex ion.

20.26 Find x in the formulae of the following complexes by determining the oxidation state of the metal from the experimental values of μ_{eff}: (a) $[VCl_x(bpy)]$, 1.77 μ_B; (b) $K_x[V(ox)_3]$, 2.80 μ_B; (c) $[Mn(CN)_6]^{x-}$, 3.94 μ_B. What assumption have you made and how valid is it?

20.27 Explain why in high-spin octahedral complexes, orbital contributions to the magnetic moment are only important for d^1, d^2, d^6 and d^7 configurations.

20.28 The observed magnetic moment for $K_3[TiF_6]$ is 1.70 μ_B. (a) Calculate μ(spin-only) for this complex. (b) Why is there a difference between calculated and observed values?

20.29 Comment on the observations that octahedral Ni(II) complexes have magnetic moments in the range 2.9–3.4 μ_B, tetrahedral Ni(II) complexes have moments up to ≈ 4.1 μ_B, and square planar Ni(II) complexes are diamagnetic.

20.30 For which of the following ions would you expect the spin-only formula to give reasonable estimates of the magnetic moment: (a) $[Cr(NH_3)_6]^{3+}$, (b) $[V(OH_2)_6]^{3+}$, (c) $[CoF_6]^{3-}$? Rationalize your answer.

20.31 Which of the following ions are diamagnetic: (a) $[Co(OH_2)_6]^{3+}$, (b) $[CoF_6]^{3-}$, (c) $[NiF_6]^{2-}$, (d) $[Fe(CN)_6]^{3-}$, (e) $[Fe(CN)_6]^{4-}$, (f) $[Mn(OH_2)_6]^{2+}$? Rationalize your answer.

20.32 (a) Using data from Appendix 6, plot a graph to show how the ionic radii of high-spin, 6-coordinate M^{2+} ions of the first row of the *d*-block vary with the d^n configuration. Comment on factors that contribute to the observed trend. (b) Briefly discuss other properties of these metal ions that show related trends.

20.33 Values of Δ_{oct} for $[Ni(OH_2)_6]^{2+}$ and high-spin $[Mn(OH_2)_6]^{3+}$ have been evaluated spectroscopically as 8500 and 21 000 cm^{-1} respectively. Assuming that these values also hold for the corresponding oxide lattices, predict whether $Ni^{II}Mn_2^{III}O_4$ should have the normal or inverse spinel structure. What factors might make your prediction unreliable?

20.34 Discuss each of the following observations:
(a) Although $Co^{2+}(aq)$ forms the tetrahedral complex $[CoCl_4]^{2-}$ on treatment with concentrated HCl, $Ni^{2+}(aq)$ does not form a similar complex.
(b) E^o for the half-reaction:
$$[Fe(CN)_6]^{3-} + e^- \rightleftharpoons [Fe(CN)_6]^{4-}$$
depends on the pH of the solution, being most positive in strongly acidic medium.
(c) E^o for the Mn^{3+}/Mn^{2+} couple is much more positive than that for Cr^{3+}/Cr^{2+} or Fe^{3+}/Fe^{2+}.

OVERVIEW PROBLEMS

20.35 (a) Explain clearly why, under the influence of an octahedral crystal field, the energy of the d_{z^2} orbital is raised whereas that of the d_{xz} orbital is lowered. State how the energies of the other three *d* orbitals are affected. With respect to what are the orbital energies raised or lowered?
(b) What is the expected ordering of values of Δ_{oct} for $[Fe(OH_2)_6]^{2+}$, $[Fe(CN)_6]^{3-}$ and $[Fe(CN)_6]^{4-}$? Rationalize your answer.
(c) Would you expect there to be an orbital contribution to the magnetic moment of a tetrahedral d^8 complex? Give an explanation for your answer.

20.36 (a) Which of the following complexes would you expect to suffer from a Jahn–Teller distortion: $[CrI_6]^{4-}$, $[Cr(CN)_6]^{4-}$, $[CoF_6]^{3-}$ and $[Mn(ox)_3]^{3-}$? Give reasons for your answers.
(b) $[Et_4N]_2[NiBr_4]$ is paramagnetic, but $K_2[PdBr_4]$ is diamagnetic. Rationalize these observations.
(c) Using a simple MO approach, explain what happens to the energies of the metal *d* orbitals on the formation of a σ-bonded complex such as $[Ni(NH_3)_6]^{2+}$.

20.37 Ligand **20.12** forms an octahedral complex, $[Fe(\mathbf{20.12})_3]^{2+}$. (a) Draw diagrams to show what isomers are possible. (b) $[Fe(\mathbf{20.12})_3]Cl_2$ exhibits spin crossover at 120 K. Explain clearly what this statement means.

(20.12)

20.38 (a) The values of ε_{max} for the most intense absorptions in the electronic spectra of $[CoCl_4]^{2-}$ and $[Co(OH_2)_6]^{2+}$ differ by a factor of about 100. Comment on this observation and state which complex you expect to exhibit the larger value of ε_{max}.
(b) In the electronic absorption spectrum of a solution containing $[V(OH_2)_6]^{3+}$, two bands are observed at 17 200 and 25 600 cm^{-1}. No absorption for the $^3A_{2g} \leftarrow {}^3T_{1g}(F)$ transition is observed. Suggest a reason for this, and assign the two observed absorptions.
(c) Red crystalline $[NiCl_2(PPh_2CH_2Ph)_2]$ is diamagnetic. On heating to 387 K for 2 hours, a blue-green form of the complex is obtained, which has a magnetic moment of $3.18\mu_B$ at 295 K. Suggest an explanation for these observations and draw structures for the complexes, commenting on possible isomerism.

20.39 (a) A Kotani plot for the t_{2g}^1 configuration consists of a curve similar to that in Fig. 20.29, but levelling off at $\mu_{eff} \approx 1.8\mu_B$ when $kT/\lambda \approx 1.0$. Suggest two metal ions that you might expect to possess room temperature values of μ_{eff} (i) on the near horizontal part of the curve and (ii) on the steepest part of the curve with $\mu_{eff} < 0.5$. For the four metal ions you have chosen, how do you expect μ_{eff} to be affected by an increase in temperature?

(b) Classify the following ligands as being σ-donor only, π-donor and π-acceptor: F^-, CO and NH_3. For each ligand, state what orbitals are involved in σ- or π-bond formation with the metal ion in an octahedral complex. Give diagrams to illustrate the overlap between appropriate metal orbitals and ligand group orbitals.

20.40 (a) Explain the origins of MLCT and LMCT absorptions in the electronic spectra of d-block metal complexes. Give examples to illustrate your answer.

(b) Explain what information can be obtained from a Tanabe–Sugano diagram.

INORGANIC CHEMISTRY MATTERS

20.41 The structure of phthalocyanine is shown below:

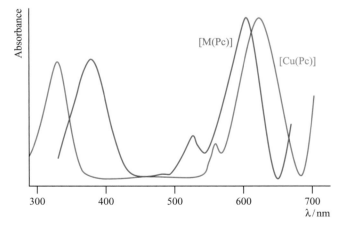

H₂Pc

The complex [Cu(Pc)] is an important commercial pigment and its electronic absorption spectrum is shown in Fig. 20.40. The absorption spectrum represented by the green line in Fig. 20.40 arises from another metal(II) phthalocyanine complex, [M(Pc)]. (a) Suggest how the ligand H₂Pc binds Cu^{2+}, and draw the structure of [Cu(Pc)]. Comment on its formation in terms of the chelate effect. (b) What colour is the [Cu(Pc)] pigment? Rationalize your answer. (c) The absorptions around 600–650 nm in Fig. 20.40 arise from $\pi^* \leftarrow \pi$ transitions. Explain what this notation means and the origins of the orbitals involved. (d) Explain how the absorption spectrum of [M(Pc)] gives rise to a green pigment. (e) When Cu^{2+} forms a complex with the perchlorinated analogue of H₂Pc, there is a red shift in the lowest energy absorption band compared to that in the spectrum of [Cu(Pc)]. How does this affect the colour of the pigment? (f) Inkjet printing dyes based on [Cu(Pc)] contain sulfonate substituents. Suggest a reason for this.

20.42 A crucial component of a dye-sensitized solar cell (DSC) is the sensitizer. The latter captures photons which are converted to electric current in the cell. A typical dye is cis-[Ru(L)₂(NCS–N)₂] where L is the bpy derivative shown below:

CO₂H

N

N

HO₂C

(a) What conformational change does the bpy ligand undergo when it binds to Ru^{2+}? (b) An ideal dye should absorb light over the whole visible range, and the extinction coefficient should be high over the whole absorption range. Explain why this is. (c) The electronic transition associated with the absorption of cis-[Ru(L)₂(NCS-N)₂] in the visible region involves an electron in the metal t_{2g} level and a low-lying π^* MO. What is the origin of the acceptor orbital, and what is the general name for this type of transition? Relate the type of transition to the redox properties of ruthenium(II).

Fig. 20.40 The electronic absorption spectrum of copper(II) phthalocyanine, [Cu(Pc)], (red trace) and an absorption spectrum of a different metal(II) phthalocyanine, [M(Pc)] (green trace). [Based on Figure 4 in: P. Gregory in *Comprehensive Coordination Chemistry II*, 2004, Elsevier, Chapter 9.12, p. 549.].

21

d-Block metal chemistry: the first row metals

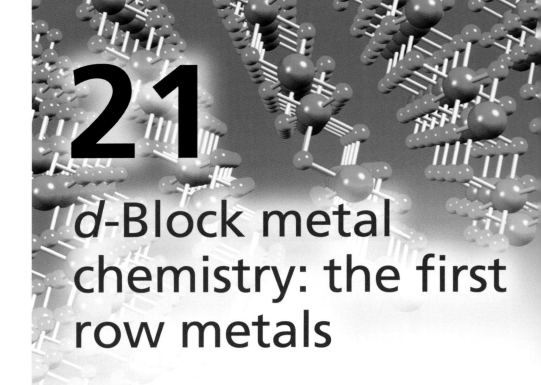

1–2	3	4	5	6	7	8	9	10	11	12	13–18
s-block											*p*-block
	Sc	Ti	V	Cr	Mn	Fe	Co	Ni	Cu	Zn	
	Y	Zr	Nb	Mo	Tc	Ru	Rh	Pd	Ag	Cd	
	La	Hf	Ta	W	Re	Os	Ir	Pt	Au	Hg	

21.1 Introduction

The chemistry of the first row *d*-block metals is best considered separately from that of the second and third row metals for several reasons, including the following:

- the chemistry of the first member of a triad is distinct from that of the two heavier metals, e.g. Zr and Hf have similar chemistries but that of Ti differs;
- electronic absorption spectra and magnetic properties of many complexes of the first row metals can often be rationalized using crystal or ligand field theory, but effects of spin–orbit coupling are more important for the heavier metals (see Sections 20.9 and 20.10);
- complexes of the heavier metal ions show a wider range of coordination numbers than those of their first row congeners;
- trends in oxidation states (Table 19.3) are not consistent for all members of a triad, e.g. although the *maximum* oxidation state of Cr, Mo and W is +6, its stability is greater for Mo and W than for Cr;
- metal–metal bonding is more important for the heavier metals than for those in the first row.

The emphasis of this chapter is on inorganic and coordination chemistry; organometallic complexes are discussed in Chapter 24.

21.2 Occurrence, extraction and uses

Figure 21.1 shows the relative abundances of the first row *d*-block metals in the Earth's crust. *Scandium* occurs as a rare component in a range of minerals. Its main source is *thortveitite* $(Sc,Y)_2Si_2O_7$ (a rare mineral found in Scandinavia and Japan), and it can also be extracted from residues in uranium processing. Uses of scandium are limited; it is a component in high-intensity lights.

The main ore of *titanium* is ilmenite $(FeTiO_3)$, and it also occurs as three forms of TiO_2 (*anatase*, *rutile* and *brookite*) and *perovskite* $(CaTiO_3$, Fig. 6.24). The structures of anatase, rutile and brookite differ as follows: whereas the structure of rutile (Fig. 6.22) is based on an hcp array of O^{2-} ions with half the octahedral holes occupied by Ti(IV) centres, those of anatase and brookite contain ccp arrays of O^{2-} ions. Titanium is present in meteorites, and rock samples from the *Apollo 17* lunar mission contain $\approx 12\%$ of Ti. Production of Ti involves conversion of rutile or ilmenite to $TiCl_4$ (by heating in a stream of Cl_2 at 1200 K in the presence of coke) followed by reduction using Mg. Titanium(IV) oxide is also purified via $TiCl_4$ in the 'chloride process' (see Box 21.3). Titanium metal is resistant to corrosion at ambient temperatures, and is lightweight and strong, making it valuable as a component in alloys, e.g. in aircraft construction. Superconducting magnets (used, for example, in MRI equipment, see Box 4.3) contain NbTi multicore conductors.

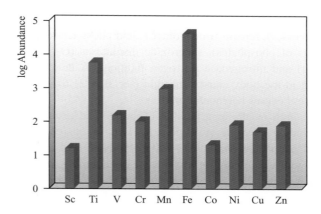

Fig. 21.1 Relative abundances of the first row *d*-block metals in the Earth's crust. The data are plotted on a logarithmic scale, and the units of abundance are parts per million (ppm).

Vanadium occurs in several minerals including *vanadinite* ($Pb_5(VO_4)_3Cl$), *carnotite* ($K_2(UO_2)_2(VO_4)_2 \cdot 3H_2O$), *roscoelite* (a vanadium-containing mica) and the polysulfide *patronite* (VS_4). It also occurs in phosphate rock (see Section 15.2) and in some crude oils. It is not mined directly and extraction of vanadium is associated with that of other metals. Roasting vanadium ores with Na_2CO_3 gives water-soluble $NaVO_3$ and from solutions of this salt, the sparingly soluble $[NH_4][VO_3]$ can be precipitated. This is heated to give V_2O_5, reduction of which with Ca yields V. The steel industry consumes about 85% of world supplies of V and *ferrovanadium* (used for toughening steels) is made by reducing a mixture of V_2O_5 and Fe_2O_3 with Al; steel–vanadium alloys are used for spring and high-speed cutting-tool steels. Vanadium(V) oxide is used as a catalyst in the oxidations of

SO_2 to SO_3 (see Section 25.7) and of naphthalene to phthalic acid.

The major ore of *chromium* is *chromite* ($FeCr_2O_4$) which has a normal spinel structure (see Box 13.7 and Section 20.11). Chromite is reduced with carbon to produce *ferrochromium* for the steel industry; stainless steels contain Cr to increase their corrosion resistance (see Box 6.2). For the production of Cr metal, chromite is fused with Na_2CO_3 in the presence of air (eq. 21.1) to give water-soluble Na_2CrO_4 and insoluble Fe_2O_3. Extraction with water followed by acidification with H_2SO_4 gives a solution from which $Na_2Cr_2O_7$ can be crystallized. Equations 21.2 and 21.3 show the final two stages of production.

$$4FeCr_2O_4 + 8Na_2CO_3 + 7O_2$$
$$\longrightarrow 8Na_2CrO_4 + 2Fe_2O_3 + 8CO_2 \quad (21.1)$$
$$Na_2Cr_2O_7 + 2C \xrightarrow{\Delta} Cr_2O_3 + Na_2CO_3 + CO \quad (21.2)$$
$$Cr_2O_3 + 2Al \xrightarrow{\Delta} Al_2O_3 + 2Cr \quad (21.3)$$

The corrosion resistance of Cr leads to its widespread use as a protective coating (*chromium plating*). The metal is deposited by electrolysing aqueous $Cr_2(SO_4)_3$, produced by dissolving Cr_2O_3 in H_2SO_4. After the steel industry, the next major consumer of Cr ($\approx$25%) is the chemical industry where applications include pigments (e.g. chrome yellow), tanning agents, mordants, catalysts and oxidizing agents. Chromite is used as a refractory material (see Section 12.6), e.g. in refractory bricks and furnace linings. Chromium compounds are toxic; chromates are corrosive to skin.

ENVIRONMENT

Box 21.1 Chromium: resources and recycling

About 95% of the world's reserve base of chromium ore lies in South Africa and Kazakhstan. The bar chart illustrates the dominance of South Africa in world chromite output.

Industrial nations in Europe and North America must rely on a supply of chromium ore from abroad, the US consuming $\approx$10% of world output. Because chromium is such a vital metal to the economy, government stockpiles in the US are considered an important strategy to ensure supplies during periods of military activity. Chromium ore is converted to chromium ferroalloys (for stainless steel and other alloys), chromite-containing refractory materials and chromium-based chemicals. The most important commercial applications of the latter are for pigments, leather tanning and wood preservation.

Recycling of stainless steel scrap as a source of Cr is an important secondary source. In 2008, the US supply of chromium consisted of 15% from government and industry

stockpiles, 67% from imports, and 18% from recycled material.

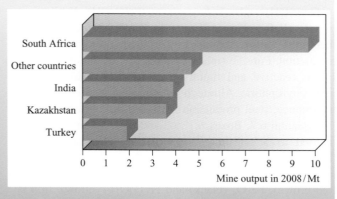

[Data: US Geological Survey.]

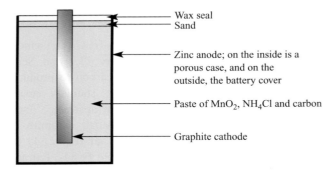

Fig. 21.2 Schematic representation of the dry battery cell ('acid' version).

Several oxides of *manganese* occur naturally, the most important being *pyrolusite* (β-MnO_2). South Africa and Ukraine hold 80% and 10%, respectively, of the world's ore reserves. Little recycling of Mn currently takes place. Manganese nodules containing up to 24% of the metal have been discovered on the ocean bed. The main use of the element is in the steel industry. Pyrolusite is mixed with Fe_2O_3 and reduced with coke to give *ferromanganese* (≈80% Mn). Almost all steels contain some Mn; those with a high Mn content (up to 12%) possess very high resistance to shock and wear and are suitable for crushing, grinding and excavating machinery. Manganese metal is produced by the electrolysis of $MnSO_4$ solutions. Manganese(IV) oxide is used in dry cell batteries. Figure 21.2 shows the Leclanché cell (the 'acid' cell); in the long-life 'alkaline' version, NaOH or KOH replaces NH_4Cl. The strong oxidizing power of $KMnO_4$ makes this an important chemical (see Box 21.4). Manganese is an essential trace element for plants, and small amounts of $MnSO_4$ are added to fertilizers.

Iron is the most important of all metals and is the fourth most abundant element in the Earth's crust. The Earth's core is believed to consist mainly of iron and it is the main constituent of metallic meteorites. The chief ores are *haematite* (α-Fe_2O_3), *magnetite* (Fe_3O_4), *siderite* ($FeCO_3$), *goethite* (α-Fe(O)OH) and *lepidocrocite* (γ-Fe(O)OH). While *iron pyrites* (FeS_2) and *chalcopyrite* ($CuFeS_2$) are common, their high sulfur contents render them unsuitable for Fe production. Pure Fe (made by reduction of the oxides with H_2) is reactive and rapidly corrodes, and finely divided iron is pyrophoric. Although *pure* iron is not of commercial importance, steel production is carried out on a huge scale (see Section 6.7, Boxes 6.1, 6.2 and 8.4). α-Iron(III) oxide is used as a polishing and grinding agent and in the formation of ferrites (see Section 21.9). Iron oxides are important commercial pigments: α-Fe_2O_3 (red), γ-Fe_2O_3 (red-brown), Fe_3O_4 (black) and Fe(O)OH (yellow). Iron is of immense biological importance (see Chapter 29), and is present in, for example, haemoglobin and myoglobin

(O_2 carriers), ferredoxins and cytochromes (redox processes), ferritin (iron storage), acid phosphatase (hydrolysis of phosphates), superoxide dismutases (O_2 dismutation) and nitrogenase (nitrogen fixation). A deficiency of iron in the body causes anaemia (see Box 21.6), while an excess causes haemochromatosis.

Cobalt occurs as a number of sulfide and arsenide ores including *cobaltite* (CoAsS) and *skutterudite* ((Co,Ni)As_3) which contains planar As_4-units. Production of the metal generally relies on the fact that it often occurs in ores of other metals (e.g. Ni, Cu and Ag) and the final processes involve reduction of Co_3O_4 with Al or C followed by electrolytic refining. Pure Co is brittle but it is commercially important in special steels, alloyed with Al, Fe and Ni (*Alnico* is a group of carbon-free alloys) in permanent magnets, and in the form of hard, strong, corrosion-resistant non-ferrous alloys (e.g. with Cr and W) which are important in the manufacture of jet engines and aerospace components. Cobalt compounds are widely used as pigments (blue hues in porcelain, enamels and glass, see Box 21.8), catalysts and as additives to animal feeds. Vitamin B_{12} is a cobalt complex, and a range of enzymes require B_{12} coenzymes.

Like cobalt, *nickel* occurs as sulfide and arsenide minerals, e.g. *pentlandite*, (Ni,Fe)$_9S_8$. Roasting such ores in air gives nickel oxide which is then reduced to the metal using carbon. The metal is refined electrolytically or by conversion to Ni(CO)$_4$ followed by thermal decomposition (eq. 21.4). This is the *Mond process*, which is based on the fact that Ni forms a carbonyl derivative more readily than any other metal.

$$Ni + 4CO \underset{423-573\,K}{\overset{323\,K}{\rightleftharpoons}} Ni(CO)_4 \qquad (21.4)$$

Nickel is used extensively in alloys, notably in stainless steel, other corrosion-resistant alloys such as *Monel metal* (68% Ni and 32% Cu), and coinage metals. Electroplated Ni provides a protective coat for other metals. Nickel has widespread use in batteries; recently, this has included the production of 'environmentally friendly' nickel–metal hydride batteries (see Box 10.5) which out-perform NiCd cells (eq. 21.5) as rechargeable sources of power in portable appliances.

$$\begin{array}{ll} \textit{Anode:} & Cd + 2[OH]^- \longrightarrow Cd(OH)_2 + 2e^- \\ \textit{Cathode:} & NiO(OH) + H_2O + e^- \longrightarrow Ni(OH)_2 + [OH]^- \end{array} \Big\}$$
$$(21.5)$$

Nickel is an important catalyst, e.g. for the hydrogenation of unsaturated organic compounds and in the water–gas shift reaction (see Section 10.4). *Raney nickel* is prepared by treating a NiAl alloy with NaOH and is a spongy material (pyrophoric when dry) which is a highly active catalyst.

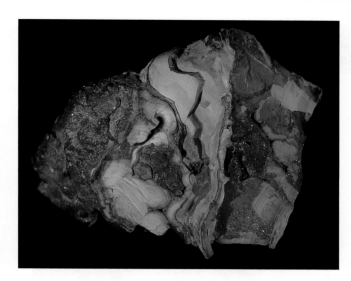

Fig. 21.3 A piece of unpolished malachite, $Cu_2(OH)_2CO_3$.

Recycling of nickel is becoming increasingly important with the major source being austenitic stainless steel (see Box 6.2). In the US, approximately 40% of nickel is recycled.

Copper is, by a considerable margin, the least reactive of the first row metals and occurs native in small deposits in several countries. The chief ore is *chalcopyrite* ($CuFeS_2$). Other ores include *chalcanthite* ($CuSO_4·5H_2O$), *atacamite* ($Cu_2(OH)_3Cl$), *malachite* ($Cu_2(OH)_2CO_3$) (Fig. 21.3), *azurite* ($Cu_3(OH)_2(CO_3)_2$) and *cuprite* (Cu_2O). Polished malachite is widely used for decorative purposes. Traditional Cu production involves roasting chalcopyrite in a limited air supply to give Cu_2S and FeO. The latter is removed by combination with silica to form a slag, and Cu_2S is converted to Cu by reaction 21.6. However, over the last two decades, methods which avoid SO_2 emissions have been introduced (Box 21.2).

$$Cu_2S + O_2 \longrightarrow 2Cu + SO_2 \qquad (21.6)$$

Electrolytic purification of Cu is carried out by constructing a cell with impure Cu as the anode, clean Cu as the cathode and $CuSO_4$ as electrolyte. During electrolysis, Cu is transferred from anode to cathode, yielding high-purity metal (e.g. suitable for electrical wiring, a major use) and a deposit under the anode from which metallic Ag and Au can be extracted. Recycling of copper is important (Box 21.2). Being corrosion-resistant, Cu is in demand for water and steam piping and is used on the exterior of buildings, e.g. roofing and flashing, where long-term exposure results in a green patina of basic copper sulfate or carbonate. Alloys of Cu such as brass (Cu/Zn) (see Section 6.7), bronze (Cu/Sn), nickel silver (Cu/Zn/Ni) and

coinage metal (Cu/Ni) are commercially important. Copper(II) sulfate is used extensively as a fungicide. Copper has a vital biochemical role (see Section 29.4), e.g. in cytochrome oxidase (involved in reduction of O_2 to H_2O) and haemocyanin (an O_2-carrying copper protein in arthropods). Copper compounds have numerous catalytic uses, and analytical applications include the biuret test and use of Fehling's solution (see Section 21.12).

The principal ores of *zinc* are *sphalerite* (*zinc blende*, ZnS, see Fig. 6.19), *calamine* (*hemimorphite*, $Zn_4Si_2O_7(OH)_2·H_2O$) and *smithsonite* ($ZnCO_3$). Extraction from ZnS involves roasting in air to give ZnO followed by reduction with carbon. Zinc is more volatile (bp 1180 K) than most metals and can be separated by rapid chilling (to prevent reversing the reaction) and purified by distillation or electrolysis. Recycling of Zn has grown in importance, providing a secondary source of the metal. Figure 21.4 summarizes major uses of Zn. It is used to galvanize steel (see Section 6.7 and Box 8.4), and Zn alloys are commercially important, e.g. brass (Cu/Zn) and nickel silver (Cu/Zn/Ni). Dry cell batteries use zinc as the anode (Fig. 21.2). A recent development is that of the zinc–air battery. The cell reactions are shown in scheme 21.7, and spent batteries can be regenerated at specialized recycling centres, or *in situ* in specially designed regenerative zinc–air fuel cells. During regeneration, the cell reactions in scheme 21.7 are reversed.[†] Zinc–air batteries or rechargeable zinc–air fuel cells are used as back-up emergency power supplies, and show potential for use in electrically powered vehicles.

At the anode: $Zn + 4[OH]^- \longrightarrow [Zn(OH)_4]^{2-} + 2e^-$

$[Zn(OH)_4]^{2-} \longrightarrow ZnO + 2[OH]^- + H_2O$

At the cathode: $O_2 + 2H_2O + 4e^- \longrightarrow 4[OH]^-$

Overall: $2Zn + O_2 \longrightarrow 2ZnO$

$$(21.7)$$

Zinc oxide is widely used in skin creams and talcum powders, and is an ingredient in sunscreen lotions for protection against UV radiation. One of its major applications is in the rubber industry, where it lowers the vulcanization temperature and facilitates faster vulcanization (see Section 16.4). Both ZnO and ZnS are used as white pigments, although for most purposes TiO_2 is superior (see Box 21.3 and Section 28.5).

[†] For further details, see: J. Goldstein, I. Brown and B. Koretz (1999) *J. Power Sources*, vol. 80, p. 171 – 'New developments in the Electric Fuel Ltd. zinc/air system'; S.I. Smedley and X.G. Zhang (2007) *J. Power Sources*, vol. 165, p. 897 – 'A regenerative zinc–air fuel cell'.

ENVIRONMENT

Box 21.2 Copper: resources and recycling

The resources of copper on the Earth's surface have recently been re-estimated. About 550 million tonnes of copper are thought to be present in bedrock minerals and deep-sea nodules. The main copper ore for traditional mining is chalcopyrite ($CuFeS_2$). The conventional extraction process involves smelting and produces large quantities of SO_2 (see Box 16.5). In the 1980s, a new copper extraction method was introduced that uses H_2SO_4 from the smelting process to extract Cu from copper ores other than those used in traditional mining, e.g. azurite ($Cu_3(OH)_2(CO_3)_2$) and malachite ($Cu_2(OH)_2CO_3$). Copper is extracted in the form of aqueous $CuSO_4$. This is mixed with an organic solvent, chosen so that it can extract Cu^{2+} ions by exchanging H^+ for Cu^{2+}, thus producing H_2SO_4 which is recycled back into the leaching stage of the operation. The aqueous-to-organic phase change separates Cu^{2+} ions from impurities. Acid is again added, releasing Cu^{2+} into an aqueous phase which is electrolysed to produce copper metal. The overall process is known as leach–solvent extraction–electrowinnning (SX/EW) and operates at ambient temperatures. It is an environmentally friendly, hydrometallurgical process, but because it relies on H_2SO_4, it is currently coupled to conventional smelting of sulfide ores. In South America, >40% of Cu is currently (in 2011) extracted by the SX/EW process.

In regions where sulfide ores predominate, copper is leached using bacteria. Naturally occurring bacteria called *Acidithiobacillus thiooxidans* oxidize sulfide to sulfate ion, and this bioleaching process now works in parallel with SX/EW as a substitute for a significant fraction of conventional smelting operations.

Among metals, consumption of Cu is exceeded only by steel and Al. The recovery of Cu from scrap metal is an essential part of copper-based industries, e.g. in 2009 in the US, recycled metal constituted ≈35% of the Cu supply. Worldwide mine production in 2009 was 15.8 Mt, with 34% originating from Chile, 8% from the US and 7.5% from Peru (the world's leading producers). Recycling of the metal is important for environmental reasons: dumping of waste leads to pollution, e.g. of water supplies. In the electronics industry, solutions of NH_3–NH_4Cl in the presence of O_2 are used to etch Cu in printed circuit boards. The resulting Cu(II) waste is subjected to a process analogous to SX/EW described above. The waste is first treated with an organic solvent XH which is a compound

Bioleaching of copper from copper sulfide ores at the Skouriotissa copper mine in Cyprus.

of the type RR'C(OH)C(NOH)R'', the conjugate base of which can function as a ligand:

$$[Cu(NH_3)_4]^{2+}(aq) + 2XH(org)$$
$$\longrightarrow CuX_2(org) + 2NH_3(aq) + 2NH_4^+(aq)$$

where aq and org represent the aqueous and organic phases respectively. Treatment with H_2SO_4 follows:

$$CuX_2 + H_2SO_4 \longrightarrow CuSO_4 + 2XH$$

and then Cu is reclaimed by electrolytic methods:

At the cathode: $Cu^{2+}(aq) + 2e^- \longrightarrow Cu(s)$

Further reading

C.L. Brierley (2008) *Trans. Nonferrous Met. Soc. China*, vol. 18, p. 1302 – 'How will biomining be applied in the future?'

Md.E. Hoque and O.J. Philip (2011) *Mater. Sci. Eng. C*, vol. 31, p. 57 – 'Biotechnological recovery of heavy metals from secondary sources – An overview'.

J. Lee, S. Acar, D.L. Doerr and J.A. Brierley (2011) *Hydrometallurgy*, vol. 105, p. 213 – 'Comparative bioleaching and mineralogy of composited sulfide ores containing enargite, covellite and chalcocite by mesophilic and thermophilic microorganisms'.

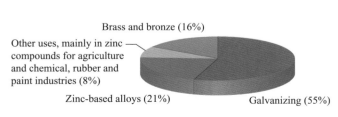

Fig. 21.4 Uses of zinc in the US in 2009. [Data: US Geological Survey.]

21.3 Physical properties: an overview

Physical data for the first row metals have already been discussed earlier in the book, but Table 21.1 summarizes selected physical properties. Additional data are tabulated as follows:

● metal structure types (Table 6.2);

Table 21.1 Selected physical properties of the metals of the first row of the *d*-block.

Property	Sc	Ti	V	Cr	Mn	Fe	Co	Ni	Cu	Zn
Atomic number, Z	21	22	23	24	25	26	27	28	29	30
Physical appearance of pure metal	Soft; silver-white; tarnishes in air	Hard; lustrous silver coloured	Soft; ductile; bright white	Hard; blue-white	Hard; lustrous silver-blue	Quite soft; malleable; lustrous, white	Hard; brittle; lustrous blue-white	Hard; malleable and ductile; grey-white	Malleable and ductile; reddish	Brittle at 298 K; malleable 373–423 K; lustrous blue-white
Melting point / K	1814	1941	2183	2180	1519	1811	1768	1728	1358	693
Boiling point / K	3104	3560	3650	2945	2235	3023	3143	3005	2840	1180
Ground state valence electronic configuration (core = [Ar]):										
Atom	$4s^2 3d^1$	$4s^2 3d^2$	$4s^2 3d^3$	$4s^1 3d^5$	$4s^2 3d^5$	$4s^2 3d^6$	$4s^2 3d^7$	$4s^2 3d^8$	$4s^1 3d^{10}$	$4s^2 3d^{10}$
M^+	$4s^1 3d^1$	$4s^2 3d^1$	$3d^4$	$3d^5$	$4s^1 3d^5$	$4s^1 3d^6$	$3d^8$	$3d^9$	$3d^{10}$	$4s^1 3d^{10}$
M^{2+}	$3d^1$	$3d^2$	$3d^3$	$3d^4$	$3d^5$	$3d^6$	$3d^7$	$3d^8$	$3d^9$	$3d^{10}$
M^{3+}	[Ar]	$3d^1$	$3d^2$	$3d^3$	$3d^4$	$3d^5$	$3d^6$	$3d^7$	$3d^8$	$3d^9$
Enthalpy of atomization, $\Delta_a H^\circ (298\,K) / kJ\,mol^{-1}$	378	470	514	397	283	418	428	430	338	130
First ionization energy, $IE_1 / kJ\,mol^{-1}$	633.1	658.8	650.9	652.9	717.3	762.5	760.4	737.1	745.5	906.4
Second ionization energy, $IE_2 / kJ\,mol^{-1}$	1235	1310	1414	1591	1509	1562	1648	1753	1958	1733
Third ionization energy, $IE_3 / kJ\,mol^{-1}$	2389	2653	2828	2987	3248	2957	3232	3395	3555	3833
Metallic radius, $r_{metal} / pm^\dagger$	164	147	135	129	137	126	125	125	128	137
Electrical resistivity $(\rho) \times 10^8 / \Omega\,m$ (at 273 K)‡	56*	39	18.1	11.8	143	8.6	5.6	6.2	1.5	5.5
	Sc	Ti	V	Cr	Mn	Fe	Co	Ni	Cu	Zn

† Metallic radius for 12-coordinate atom.
‡ See eq. 6.3 for relationship between electrical resistivity and resistance.
* At 290–300 K.

- values of ionic radii, r_{ion}, which depend on charge, geometry and whether the ion is high- or low-spin (Appendix 6);
- standard reduction potentials, $E^\circ (M^{2+}/M)$ and $E^\circ (M^{3+}/M^{2+})$ (see Tables 19.1 and 20.15 and Appendix 11).

For electronic absorption spectroscopic data (e.g. Δ_{oct} and spin–orbit coupling constants) and magnetic moments, relevant sections in Chapter 20 should be consulted.

21.4 Group 3: scandium

The metal

In its chemistry, Sc shows a greater similarity to Al than to the heavier group 3 metals; E° values are given for comparison in eq. 21.8.

$$M^{3+}(aq) + 3e^- \rightleftharpoons M(s) \quad \begin{cases} M = Al, \ E^\circ = -1.66\,V \\ M = Sc, \ E^\circ = -2.08\,V \end{cases}$$

(21.8)

Scandium metal dissolves in both acids and alkalis, and combines with halogens. It reacts with N_2 at high temperatures to give ScN which is hydrolysed by water. Scandium normally shows one stable oxidation state in its compounds, Sc(III). However, reactions of $ScCl_3$ and Sc at high temperatures lead to a number of subhalides (e.g. Sc_7Cl_{10} and Sc_7Cl_{12}).

Scandium(III)

Direct combination of Sc and a halogen gives anhydrous ScF_3 (water-insoluble white solid), $ScCl_3$ and $ScBr_3$ (soluble white solids) and ScI_3 (moisture-sensitive yellow solid). The fluoride crystallizes with the ReO_3 structure (Fig. 21.5) in which each Sc centre is octahedrally sited. In each of $ScCl_3$, $ScBr_3$ and ScI_3, the Sc atoms occupy two thirds of the octahedral sites in an hcp array of halogen atoms (i.e. a BiI_3-type structure). On reaction with MF (M = Na, K, Rb, NH_4), ScF_3 forms water-soluble complexes $M_3[ScF_6]$ containing octahedral $[ScF_6]^{3-}$.

Addition of aqueous alkali to solutions of Sc(III) salts precipitates ScO(OH) which is isostructural with AlO(OH). In the presence of excess $[OH]^-$, ScO(OH) redissolves as $[Sc(OH)_6]^{3-}$. Dehydration of ScO(OH) yields Sc_2O_3.

The coordination chemistry of Sc(III) is far more limited than that of the other first row *d*-block metal ions and is generally restricted to hard donors such as N and O. Coordination numbers of 6 are favoured, e.g. $[ScF_6]^{3-}$, $[Sc(bpy)_3]^{3+}$, *mer*-$[ScCl_3(OH_2)_3]$, *mer*-$[ScCl_3(THF)_3]$ and $[Sc(acac)_3]$. Among complexes with higher coordination numbers are $[ScF_7]^{4-}$ (pentagonal bipyramid), $[ScCl_2(15\text{-crown-}5)]^+$ (Fig. 19.8d), $[Sc(NO_3)_5]^{2-}$ (see end of Section 9.11) and $[Sc(OH_2)_9]^{3+}$ (tricapped trigonal prism). Bulky amido ligands stabilize low coordination numbers, e.g. $[Sc\{N(SiMe_3)_2\}_3]$.

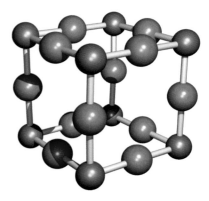

Fig. 21.5 Unit cell of ReO_3, a prototype structure; Re atoms are shown in brown and O atoms in red. This structure type is adopted by ScF_3 and FeF_3.

21.5 Group 4: titanium

The metal

Titanium does not react with alkalis (cold or hot) and does not dissolve in mineral acids at room temperature. It is attacked by hot HCl, forming Ti(III) and H_2, and hot HNO_3 oxidizes the metal to hydrous TiO_2. Titanium wire dissolves in aqueous HF with vigorous liberation of H_2 and the formation of green-yellow solutions containing Ti(IV) and Ti(II) (eq. 21.9).

$$2Ti + 6HF \longrightarrow [TiF_6]^{2-} + Ti^{2+} + 3H_2 \qquad (21.9)$$

Titanium reacts with most non-metals at high temperatures. With C, O_2, N_2 and halogens X_2, it forms TiC, TiO_2 (see Fig. 6.22), TiN (see Section 15.6) and TiX_4 respectively. With H_2, it forms 'TiH_2' but this has a wide non-stoichiometric range, e.g. $TiH_{1.7}$. The binary hydrides, carbide (see Section 14.7), nitride and borides (see Section 13.10) are all inert, high-melting, refractory materials.

In its compounds, Ti exhibits oxidation states of +4 (by far the most stable), +3, +2 and, rarely, 0.

Titanium(IV)

Titanium(IV) halides can be formed from the elements. Industrially, $TiCl_4$ is prepared by reacting TiO_2 with Cl_2 in the presence of carbon and this reaction is also used in the purification of TiO_2 in the 'chloride process' (see Box 21.3). Titanium(IV) fluoride is a hygroscopic white solid which forms HF on hydrolysis. The vapour contains tetrahedral TiF_4 molecules. Solid TiF_4 consists of Ti_3F_{15}-units in which the Ti atoms are octahedrally sited; the corner-sharing octahedra (Fig. 21.6) are linked through the F_a atoms (shown in Fig. 21.6a) to generate isolated columns in an infinite array. $TiCl_4$ and $TiBr_4$ hydrolyse more readily than TiF_4. At 298 K, $TiCl_4$ is a colourless liquid (mp 249 K, bp 409 K) and $TiBr_4$ a yellow solid. The tetraiodide is a red-brown hygroscopic solid which sublimes *in vacuo* at 473 K to a red vapour. Tetrahedral molecules are present in the solid and vapour phases of $TiCl_4$, $TiBr_4$ and TiI_4. Each tetrahalide acts as a Lewis acid. $TiCl_4$ is the most important, being used with $AlCl_3$ in Ziegler–Natta catalysts for alkene polymerization (see Section 25.8) and as a catalyst in a variety of other organic reactions. The Lewis acidity of $TiCl_4$ is seen in complex formation. It combines with tertiary amines and phosphanes to give octahedral complexes such as $[TiCl_4(NMe_3)_2]$ and $[TiCl_4(PEt_3)_2]$. Salts containing $[TiCl_6]^{2-}$ are made in thionyl chloride solution since they are hydrolysed by water. In contrast, salts of $[TiF_6]^{2-}$ can be prepared in aqueous media. With the diarsane **21.1** (see scheme 23.89), the dodecahedral complex $[TiCl_4(\mathbf{21.1})_2]$ is formed. Reaction of N_2O_5 with $TiCl_4$ yields anhydrous

APPLICATIONS

Box 21.3 Commercial demand for TiO$_2$

Titanium dioxide has wide industrial applications as a brilliant white pigment and its applications as a pigment in the US in 2009 are shown in the chart opposite. This commercial application arises from the fact that fine particles scatter incident light extremely strongly; even crystals of TiO$_2$ possess a very high refractive index ($\mu = 2.70$ for rutile, 2.55 for anatase). Historically, Pb(II) compounds were used as pigments in paints but the associated health hazards make lead undesirable; TiO$_2$ has negligible health risks. Two manufacturing methods are used:

- the *sulfate process* produces TiO$_2$ in the form of rutile and anatase;
- the *chloride process* produces rutile.

The raw material for the sulfate process is ilmenite, FeTiO$_3$; treatment with H$_2$SO$_4$ at 420–470 K yields Fe$_2$(SO$_4$)$_3$, TiOSO$_4$ and some FeSO$_4$. The Fe$_2$(SO$_4$)$_3$ is reduced and separated as FeSO$_4$·7H$_2$O by a crystallization process. Hydrolysis of TiOSO$_4$ yields hydrated TiO$_2$ which is subsequently dehydrated to give TiO$_2$:

$$TiOSO_4 + (n+1)H_2O \xrightarrow{\text{aqu. alkali}} TiO_2 \cdot nH_2O + H_2SO_4$$

$$TiO_2 \cdot nH_2O \xrightarrow{\Delta} TiO_2 + nH_2O$$

Sulfuric acid is removed by neutralization with CaCO$_3$ to produce gypsum as a by-product:

$$CaCO_3 + H_2SO_4 + H_2O \longrightarrow CaSO_4 \cdot 2H_2O + CO_2$$

Gypsum is recycled into the building trade (see Boxes 12.6 and 14.8). TiO$_2$ produced by the sulfate process is in the form of *anatase* unless seed crystals of *rutile* are introduced in the final stages of production. Rutile ore occurs naturally in, for example, apatite veins in Norway, and is the raw material for the chloride process. Initially, TiO$_2$ ore is converted to TiCl$_4$ by treatment with Cl$_2$ and C at 1200 K. Oxidation by O$_2$ at $\approx$1500 K yields pure rutile:

$$TiO_2 + 2Cl_2 + C \longrightarrow TiCl_4 + CO_2$$
crude

$$TiCl_4 + O_2 \longrightarrow TiO_2 + 2Cl_2$$
pure

Scanning electron microscopy (SEM) image of TiO$_2$ flakes. Magnification ×900.

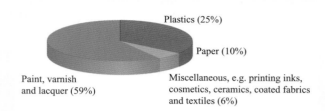

Plastics (25%)

Paper (10%)

Paint, varnish and lacquer (59%)

Miscellaneous, e.g. printing inks, cosmetics, ceramics, coated fabrics and textiles (6%)

[Data: US Geological Survey.]

The Cl$_2$ from the second step is recycled for use in the initial chlorination step.

Originally, the sulfate process was the more industrially important process, but since the early 1990s, the chloride process has been favoured on both financial and environmental grounds. Both processes are in current use.

Titanium dioxide is a wide band-gap semiconductor and is an excellent photocatalyst for the photomineralization of water, i.e. the degradation of pollutants in water is catalysed by TiO$_2$ in the presence of UV radiation. Pollutants which can be successfully destroyed include a wide range of hydrocarbons and halogenated organic compounds as well as some herbicides, pesticides and dyes. The semiconducting properties of TiO$_2$ have also led to its being used as a gas sensor for detection of Me$_3$N emitted from decaying fish. Other uses of TiO$_2$ include applications in cosmetics and ceramics, and in anodes for various electrochemical processes. TiO$_2$ is used as a UV filter in suncreams and for this application, control over particle size is important since the optimum light scattering occurs when the TiO$_2$ particle diameter is 180–220 nm.

While only a fraction of a percent of the world's demand for TiO$_2$, the application of nanocrystalline TiO$_2$ (anatase) in Grätzel's dye-sensitized solar cells is becoming increasingly important. A film of the TiO$_2$ wide band-gap semiconductor is coated onto a transparent fluorine-doped tin oxide conducting glass. A redox active dye (which absorbs light over as wide a range of the visible spectrum as possible) is then adsorbed onto the TiO$_2$ surface. Excitation of the dye as a photon is absorbed is accompanied by injection of an electron into the conduction band of the semiconductor (see Section 28.3).

Further reading

X. Chen and S.S. Mao (2006) *J. Nanosci. Nanotechno.*, vol. 6, p. 906 – 'Synthesis of titanium dioxide (TiO$_2$) nanomaterials'.

U. Diebold (2003) *Surf. Sci. Rep.*, vol. 48, p. 53 – 'The surface science of titanium dioxide'.

G.J. Meyer (2010) *ACS NANO*, vol. 4, p. 4337 – 'The 2010 Millennium Technology Grand Prize: Dye-sensitized solar cells'.

A. Mills, R.H. Davies and D. Worsley (1993) *Chem. Soc. Rev.*, vol. 22, p. 417 – 'Water purification by semiconductor photocatalysis'.

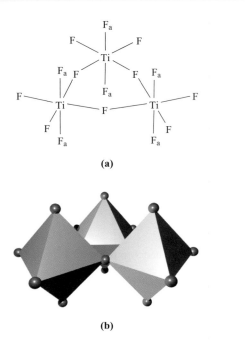

(a)

(b)

Fig. 21.6 The solid state structure of TiF_4 consists of columnar stacks of corner-sharing octahedra. The building blocks are Ti_3F_{15}-units shown here in (a) schematic representation and (b) polyhedral representation; F atoms are shown in green. [Data: H. Bialowons *et al.* (1995) *Z. Anorg. Allg. Chem.*, vol. 621, p. 1227.]

[Ti(NO$_3$)$_4$] in which the Ti(IV) centre is in a dodecahedral environment (Fig. 21.7a).

(21.1)

The commercial importance of TiO_2 is described in Box 21.3, and the structure of its rutile form was shown in Fig. 6.22. Although it may be formulated as $Ti^{4+}(O^{2-})_2$, the very high value of the sum of the first four ionization energies of the metal ($8797 \, kJ \, mol^{-1}$) makes the validity of the ionic model doubtful. Dry TiO_2 is difficult to dissolve in acids, but the hydrous form (precipitated by adding base to solutions of Ti(IV) salts) dissolves in HF, HCl and H_2SO_4 giving fluorido, chlorido and sulfato complexes respectively. There is no simple aqua ion of Ti^{4+}. The reaction of TiO_2 with CaO at 1620 K gives the *titanate* $CaTiO_3$. Other members of this group include $BaTiO_3$ and $FeTiO_3$ (*ilmenite*). The $M^{II}TiO_3$ titanates are *mixed oxides* and do *not* contain $[TiO_3]^{2-}$ ions. The structure type depends on the size of M^{2+}: if it is large (e.g. M = Ca), a perovskite lattice is favoured (Fig. 6.24) but

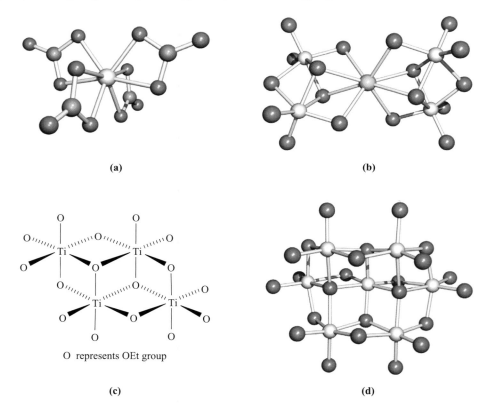

(a)

(b)

O represents OEt group

(c)

(d)

Fig. 21.7 (a) The structure of Ti(NO$_3$)$_4$ (X-ray diffraction) showing the dodecahedral environment of the Ti atom; compare with Fig. 19.9 [C.D. Garner *et al.* (1966) *J. Chem. Soc., A*, p. 1496]; (b) the structure of [Ca{Ti$_2$(OEt)$_9$}$_2$] (X-ray diffraction); Et groups are omitted [E.P. Turevskaya *et al.* (1994) *J. Chem. Soc., Chem. Commun.*, p. 2303]; (c) the tetrameric structure of [Ti(OEt)$_4$] i.e. [Ti$_4$(OEt)$_{16}$] with ethyl groups omitted for clarity; (d) the structure of [Ti$_7$(μ$_4$-O)$_2$(μ$_3$-O)$_2$(OEt)$_{20}$] (X-ray diffraction); Et groups are omitted [R. Schmid *et al.* (1991) *J. Chem. Soc., Dalton Trans.*, p. 1999]. Colour code: Ti, pale grey; O, red; N, blue; Ca, yellow.

if M^{2+} is similar in size to Ti(IV), a corundum structure (see Section 13.7), in which M(II) and Ti(IV) replace two Al(III) centres, is preferred, e.g. ilmenite. Above 393 K, $BaTiO_3$ has the perovskite structure, but at lower temperatures it transforms successively into three phases, each of which is a *ferroelectric*, i.e. the phase has an electric dipole moment even in the absence of an external magnetic field. This arises because the small Ti(IV) centre tends to lie off-centre in the octahedral O_6-hole (Fig. 6.24). Application of an electric field causes all such ions to be drawn to the same side of the holes and leads to a great increase in specific permittivity; thus, barium titanates are used in capacitors. Application of pressure to one side of a $BaTiO_3$ crystal causes the Ti^{4+} ions to migrate, generating an electric current (the piezoelectric effect, see Section 14.9), and this property makes $BaTiO_3$ suitable for use in electronic devices such as microphones. Interest in perovskite-phases such as $BaTiO_3$ and $CaTiO_3$ has led to investigations of solid state materials such as $[M\{Ti_2(OEt)_9\}_2]$ (M = Ba or Ca) (Fig. 21.7b) derived from reactions of alkoxides of Ti(IV) and Ba or Ca. Titanium alkoxides are widely used in waterproofing fabrics and in heat-resistant paints. Thin films of TiO_2 are used in capacitors and can be deposited using Ti(IV) alkoxides such as $[Ti(OEt)_4]$. The ethoxide is prepared from $TiCl_4$ and Na[OEt] (or from $TiCl_4$, dry NH_3 and EtOH) and has a tetrameric structure (Fig. 21.7c) in which each Ti is octahedrally sited. Larger structures which retain TiO_6 'building-blocks' can be assembled. For example, reaction of $[Ti(OEt)_4]$ with anhydrous EtOH at 373 K gives $[Ti_{16}O_{16}(OEt)_{32}]$, while $[Ti_7O_4(OEt)_{20}]$ (Fig. 21.7d) is the product if basic $CuCO_3$ is present. Similar structures are observed for vanadates (Section 21.6), molybdates and tungstates (Section 22.7).

(21.2)   **(21.3)**

The reaction of TiO_2 and $TiCl_4$ at 1320 K in a fluidized bed produces $[Cl_3Ti(\mu\text{-}O)TiCl_3]$ which reacts with $[Et_4N]Cl$ to give $[Et_4N]_2[TiOCl_4]$. The $[TiOCl_4]^{2-}$ ion **(21.2)** has a square-based pyramidal structure with the oxido ligand in the apical position. A number of peroxido complexes of Ti(IV) are known and include products of reactions between TiO_2 in 40% HF and 30% H_2O_2; at pH 9 the product is $[TiF_2(\eta^2\text{-}O_2)_2]^{2-}$ while at pH 6, $[TiF_5(\eta^2\text{-}O_2)]^{3-}$ is formed. The dinuclear species $[Ti_2F_6(\mu\text{-}F)_2(\eta^2\text{-}O_2)_2]^{4-}$ **(21.3)** is made by treating $[TiF_6]^{2-}$ with 6% H_2O_2 at pH 5.

Titanium(III)

Titanium(III) fluoride is prepared by passing H_2 and HF over Ti or its hydride at 970 K. TiF_3 is a blue solid (mp 1473 K) with a structure related to ReO_3 (Fig. 21.5). The trichloride exists in four forms (α, β, γ and δ). The α-form (a violet solid) is prepared by reducing $TiCl_4$ with H_2 above 770 K and has a layer structure with Ti atoms in octahedral sites. The brown β-form is prepared by heating $TiCl_4$ with trialkyl aluminium compounds; it is fibrous and contains face-sharing $TiCl_6$ octahedra. The trichloride is commercially available. It is used as a catalyst in alkene polymerization (see Section 25.7) and is a powerful reducing agent. In air, $TiCl_3$ is readily oxidized, and disproportionates above 750 K (eq. 21.10).

$$2TiCl_3 \longrightarrow TiCl_4 + TiCl_2 \qquad (21.10)$$

Titanium tribromide is made by heating $TiBr_4$ with Al, or by reaction of BBr_3 with $TiCl_3$; it is a grey solid with a layer structure analogous to α-$TiCl_3$. Reduction of TiI_4 with Al gives violet TiI_3. Both $TiBr_3$ and TiI_3 disproportionate when heated >600 K. The magnetic moment of TiF_3 (1.75 μ_B at 300 K) is consistent with one unpaired electron per metal centre. However, magnetic data for $TiCl_3$, $TiBr_3$ and TiI_3 indicate significant Ti–Ti interactions in the solid state. For $TiCl_3$, the magnetic moment at 300 K is 1.31 μ_B and $TiBr_3$ is only weakly paramagnetic.

When aqueous solutions of Ti(IV) are reduced by Zn, the purple aqua ion $[Ti(OH_2)_6]^{3+}$ is obtained (see eq. 7.35 and Fig. 20.4). This is a powerful reductant (eq. 21.11) and aqueous solutions of Ti(III) must be protected from aerial oxidation.

$$[TiO]^{2+}(aq) + 2H^+(aq) + e^- \rightleftharpoons Ti^{3+}(aq) + H_2O(l)$$
$$E^o = +0.1\,V \qquad (21.11)$$

In alkaline solution (partly because of the involvement of H^+ in redox equilibrium 21.11, and partly because of the low solubility of the product), Ti(III) compounds liberate H_2 from H_2O and are oxidized to TiO_2. In the absence of air, alkali precipitates hydrous Ti_2O_3 from solutions of $TiCl_3$. Dissolution of this oxide in acids gives salts containing $[Ti(OH_2)_6]^{3+}$, e.g. $[Ti(OH_2)_6]Cl_3$ and $CsTi(SO_4)_2 \cdot 12H_2O$, the latter being isomorphous with other alums (see Section 13.9).

Titanium(III) oxide is made by reducing TiO_2 with Ti at high temperatures. It is a purple-black, insoluble solid with the corundum structure (see Section 13.7) and exhibits a transition from semiconductor to metallic character on heating above 470 K or doping with, for example, V(III). Uses of Ti_2O_3 include those in thin film capacitors.

Complexes of Ti(III) usually have octahedral structures, e.g. $[TiF_6]^{3-}$, $[TiCl_6]^{3-}$, $[Ti(CN)_6]^{3-}$, *trans*-$[TiCl_4(THF)_2]^-$, *trans*-$[TiCl_4(py)_2]^-$, *mer*-$[TiCl_3(THF)_3]$, *mer*-$[TiCl_3(py)_3]$ and $[Ti\{(H_2N)_2CO\text{-}O\}_6]^{3+}$, and magnetic moments close to the spin-only values. Examples of 7-coordinate complexes include $[Ti(EDTA)(OH_2)]^-$ and $[Ti(OH_2)_3(ox)_2]^-$.

Low oxidation states

Titanium(II) chloride, bromide and iodide can be prepared by thermal disproportionation of TiX_3 (eq. 21.10) or by reaction 21.12. They are red or black solids which adopt the CdI_2 structure (Fig. 6.23).

$$TiX_4 + Ti \xrightarrow{\Delta} 2TiX_2 \qquad (21.12)$$

With water, $TiCl_2$, $TiBr_2$ and TiI_2 react violently, liberating H_2 as Ti(II) is oxidized. Equation 21.9, however, shows that the Ti^{2+} ion can be formed in aqueous solution under appropriate conditions. Either Ti or $TiCl_3$ dissolved in aqueous HF gives a mixture of $[TiF_6]^{2-}$ and $[Ti(OH_2)_6]^{2+}$. The former can be precipitated as $Ba[TiF_6]$ or $Ca[TiF_6]$, and the remaining $[Ti(OH_2)_6]^{2+}$ ion (d^2) exhibits an electronic absorption spectrum with two bands at 430 and 650 nm, which is similar to that of the isoelectronic $[V(OH_2)_6]^{3+}$ ion.

Titanium(II) oxide is manufactured by heating TiO_2 and Ti *in vacuo*. It is a black solid and a metallic conductor which adopts a defect NaCl-type structure: at room temperature one-sixth of both anion and cation sites are unoccupied, i.e. a Schottky defect. The oxide also exists as a non-stoichiometric compound with compositions in the range $TiO_{0.82}$–$TiO_{1.23}$. Conducting properties of the first row metal(II) oxides are compared in Section 28.2.

Reduction of $TiCl_3$ with Na/Hg, or of $TiCl_4$ with Li in THF and 2,2'-bipyridine leads to violet $[Ti(bpy)_3]$. Formally this contains Ti(0), but results of MO calculations and spectroscopic studies indicate that electron delocalization occurs such that the complex should be considered as $[Ti^{3+}(bpy^-)_3]$; see also the end of Section 19.5 and discussion of complexes containing ligand **19.12** in Section 19.7.

Self-study exercises

1. The structure of TiO_2 (rutile) is a 'prototype structure'. What does this mean? What are the coordination environments of the Ti and O centres? Give two other examples of compounds that adopt the same structure as TiO_2.
   [*Ans.* See Fig. 6.22 and discussion]

2. The pK_a value for $[Ti(OH_2)_6]^{3+}$ is 3.9. To what equilibrium does this value relate? How does the strength of aqueous $[Ti(OH_2)_6]^{3+}$ as an acid compare with those of $MeCO_2H$, $[Al(OH_2)_6]^{3+}$, HNO_2 and HNO_3?
   [*Ans.* See eqs. 7.35, and 7.9, 7.14, 7.13 and 7.34]

3. What is the electronic configuration of the Ti^{3+} ion? Explain why the electronic absorption spectrum of $[Ti(OH_2)_6]^{3+}$ consists of an absorption with a shoulder rather than a single absorption.
   [*Ans.* See Section 20.7, after worked example 20.3]

4. The electronic absorption spectrum of $[Ti(OH_2)_6]^{2+}$ consists of two bands assigned to '*d–d*' transitions. Is this consistent with what is predicted from the Orgel diagram shown in Fig. 20.20? Comment on your answer.

21.6 Group 5: vanadium

The metal

In many ways, V metal is similar to Ti. Vanadium is a powerful reductant (eq. 21.13) but is passivated by an oxide film.

$$V^{2+} + 2e^- \rightleftharpoons V \qquad E^o = -1.18\,V \qquad (21.13)$$

The metal is insoluble in non-oxidizing acids (except HF) and alkalis, but is attacked by HNO_3, *aqua regia* and peroxydisulfate solutions. On heating, V reacts with halogens (eq. 21.14) and combines with O_2 to give V_2O_5, and with B, C and N_2 to yield solid state materials (see Sections 13.10, 14.7 and 15.6).

$$V \begin{cases} \xrightarrow{F_2} VF_5 \\ \xrightarrow{Cl_2} VCl_4 \\ \xrightarrow{X_2} VX_3 \quad (X = Br\ or\ I) \end{cases} \qquad (21.14)$$

The normal oxidation states of vanadium are $+5$, $+4$, $+3$ and $+2$. Oxidation state 0 occurs in a few compounds with π-acceptor ligands, e.g. $V(CO)_6$ (see Chapter 24).

Vanadium(V)

The only binary halide of vanadium(V) is VF_5 (eq. 21.14). It is a volatile white solid which is readily hydrolysed and is a powerful fluorinating agent. In the gas phase, VF_5 exists as trigonal bipyramidal molecules but the solid has a polymeric structure (**21.4**). The salts $K[VF_6]$ and $[Xe_2F_{11}][VF_6]$ are made by reacting VF_5 with KF or XeF_6 (at 250 K) respectively.

(21.4)

The oxohalides VOX_3 (X = F or Cl) are made by halogenation of V_2O_5. Reaction of VOF_3 with $(Me_3Si)_2O$ yields VO_2F, and treatment of $VOCl_3$ with Cl_2O gives VO_2Cl.

The oxohalides are hygroscopic and hydrolyse readily. Both VO_2F and VO_2Cl decompose on heating (eq. 21.15).

$$3VO_2X \xrightarrow{\Delta} VOX_3 + V_2O_5 \qquad (X = F \text{ or } Cl) \qquad (21.15)$$

Pure V_2O_5 is an orange or red powder depending on its state of division, and is manufactured by heating $[NH_4][VO_3]$ (eq. 21.16).

$$2[NH_4][VO_3] \xrightarrow{\Delta} V_2O_5 + H_2O + 2NH_3 \qquad (21.16)$$

Vanadium(V) oxide is amphoteric, being sparingly soluble in water but dissolving in alkalis to give a wide range of vanadates, and in strong acids to form complexes of $[VO_2]^+$. The species present in vanadium(V)-containing solutions depend on the pH:

pH 14 $[VO_4]^{3-}$

$[VO_3(OH)]^{2-}$ in equilibrium with $[V_2O_7]^{4-}$

$[V_4O_{12}]^{4-}$

pH 6 $[H_nV_{10}O_{28}]^{(6-n)-}$

V_2O_5

pH 0 $[VO_2]^+$

This dependence can be expressed in terms of a series of equilibria such as eqs. 21.17–21.23.

$$[VO_4]^{3-} + H^+ \rightleftharpoons [VO_3(OH)]^{2-} \qquad (21.17)$$

$$2[VO_3(OH)]^{2-} \rightleftharpoons [V_2O_7]^{4-} + H_2O \qquad (21.18)$$

$$[VO_3(OH)]^{2-} + H^+ \rightleftharpoons [VO_2(OH)_2]^- \qquad (21.19)$$

$$4[VO_2(OH)_2]^- \rightleftharpoons [V_4O_{12}]^{4-} + 4H_2O \qquad (21.20)$$

$$10[V_3O_9]^{3-} + 15H^+ \rightleftharpoons 3[HV_{10}O_{28}]^{5-} + 6H_2O \qquad (21.21)$$

$$[HV_{10}O_{28}]^{5-} + H^+ \rightleftharpoons [H_2V_{10}O_{28}]^{4-} \qquad (21.22)$$

$$[H_2V_{10}O_{28}]^{4-} + 14H^+ \rightleftharpoons 10[VO_2]^+ + 8H_2O \qquad (21.23)$$

Isopolyanions (homopolyanions) are complex metal oxoanions (polyoxometallates) of type $[M_xO_y]^{n-}$, e.g. $[V_{10}O_{28}]^{6-}$ and $[Mo_6O_{19}]^{2-}$. A ***heteropolyanion*** contains a hetero atom, e.g. $[PW_{12}O_{40}]^{3-}$.

The formation of polyoxometallates is a characteristic of V, Mo, W (see Section 22.7) and, to a lesser extent, Nb, Ta and Cr. Characterization of solution species is aided by ^{17}O and ^{51}V NMR spectroscopies, and solid state structures for a range of salts are known. The structural chemistry of V_2O_5 and vanadates is complicated and only a brief survey is given here. The structure of V_2O_5 consists of layers of edge-sharing, approximately square-based pyramids (**21.5**). Each V centre is bonded to one O at 159 pm (apical site and not shared), one O at 178 pm (shared with one other V) and two O at 188 pm and one at 202 pm (shared with two other V atoms). Salts of $[VO_4]^{3-}$ (*ortho-vanadates*) contain discrete tetrahedral ions, and those of $[V_2O_7]^{4-}$ (*pyrovanadates*) also contain discrete anions (Fig. 21.8a); $[V_2O_7]^{4-}$ is isoelectronic and isostructural with $[Cr_2O_7]^{2-}$. The ion $[V_4O_{12}]^{4-}$ has a cyclic structure (Fig. 21.8b). Anhydrous salts of $[VO_3]^-$ (*metavanadates*) contain infinite chains of vertex-sharing VO_4 units (Figs. 21.8c and d). However, this structure type is not common to all metavanadates, e.g. in $KVO_3 \cdot H_2O$ and $Sr(VO_3)_2 \cdot 4H_2O$ each V is bonded to five O atoms in a double-chain structure. The $[V_{10}O_{28}]^{6-}$ anion exists in solution (at appropriate pH) and has been characterized in the solid state in, for example, $[H_3NCH_2CH_2NH_3]_3[V_{10}O_{28}] \cdot 6H_2O$ and $[^iPrNH_3]_6[V_{10}O_{28}] \cdot 4H_2O$ (Fig. 21.8e). It consists of 10 VO_6 octahedral units with two μ_6-O, four μ_3-O, 14 μ-O and eight terminal O atoms. Crystalline salts of $[HV_{10}O_{28}]^{5-}$, $[H_2V_{10}O_{28}]^{4-}$ and $[H_3V_{10}O_{28}]^{3-}$ have also been isolated and the anions retain the framework shown in Fig. 21.8e. Examples of isopolyanions of vanadium with open ('bowl-shaped') structures are known, e.g. $[V_{12}O_{32}]^{4-}$, and these may act as 'hosts' to small molecules. In $[Ph_4P]_4[V_{12}O_{32}] \cdot 4MeCN \cdot 4H_2O$, one MeCN molecule resides partially within the cavity of the anion, while an $[NO]^-$ ion is encapsulated in $[Et_4N]_5[NO][V_{12}O_{32}]$.

Reduction of yellow $[VO_2]^+$ in acidic solution yields successively blue $[VO]^{2+}$, green V^{3+} and violet V^{2+}. The potential and Frost–Ebsworth diagrams in Fig. 21.9 show that all oxidation states of vanadium in aqueous solution are stable with respect to disproportionation.

(21.5)

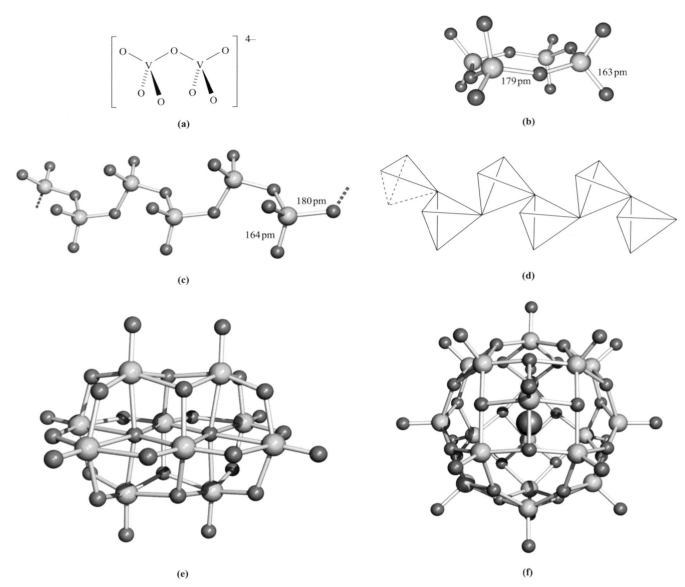

Fig. 21.8 (a) The structure of the $[V_2O_7]^{4-}$ anion consists of two tetrahedral units sharing a common oxygen atom; (b) the structure of $[V_4O_{12}]^{4-}$ in the salt $[Ni(bpy)_3]_2[V_4O_{12}]\cdot11H_2O$ (X-ray diffraction) [G.-Y. Yang *et al.* (1998) *Acta Crystallogr.*, *Sect. C*, vol. 54, p. 616]; (c) infinite chains of corner-sharing tetrahedral VO_4 units are present in anhydrous metavanadates; this shows part of one chain in [*n*-$C_6H_{13}NH_3][VO_3]$ (an X-ray diffraction determination) [P. Roman *et al.* (1991) *Mater. Res. Bull.*, vol. 26, p. 19]; (d) the structure of the metavanadate shown in (c) can be represented as a chain of corner-sharing tetrahedra, each tetrahedron representing a VO_4 unit; (e) the structure of $[V_{10}O_{28}]^{6-}$ in the salt $[^iPrNH_3]_6[V_{10}O_{28}]\cdot4H_2O$ (X-ray diffraction) [M.-T. Averbuch-Pouchot *et al.* (1994) *Eur. J. Solid State Inorg. Chem.*, vol. 31, p. 351]; (f) in $[Et_4N]_5[V_{18}O_{42}I]$ (X-ray diffraction), the $[V_{18}O_{42}]^{4-}$ ion contains square-based pyramidal VO_5 units and the cage encapsulates I^- [A. Müller *et al.* (1997) *Inorg. Chem.*, vol. 36, p. 5239]. Colour code: V, yellow; O, red; I, purple.

Vanadium(IV)

The highest chloride of vanadium is VCl_4 (eq. 21.14). It is a toxic, red-brown liquid (mp 247 K, bp 421 K) and the liquid and vapour phases contain tetrahedral molecules (**21.6**). It readily hydrolyses to $VOCl_2$ (see below), and at 298 K, slowly decomposes (eq. 21.24). The reaction of VCl_4 with anhydrous HF gives lime-green VF_4 (solid at 298 K) which is also formed with VF_5 when V reacts with F_2. On heating, VF_4 disproportionates (eq. 21.25) in contrast to the behaviour of VCl_4 (eq. 21.24).

(21.6)

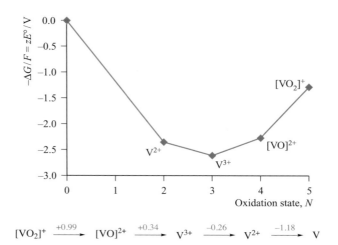

$$[VO_2]^+ \xrightarrow{+0.99} [VO]^{2+} \xrightarrow{+0.34} V^{3+} \xrightarrow{-0.26} V^{2+} \xrightarrow{-1.18} V$$

Fig. 21.9 Potential (lower) and Frost–Ebsworth diagrams for vanadium at pH 0.

$$2VCl_4 \longrightarrow 2VCl_3 + Cl_2 \qquad (21.24)$$

$$2VF_4 \xrightarrow{\geq 298\,K} VF_5 + VF_3 \qquad (21.25)$$

The structure of solid VF_4 consists of fluorine bridged VF_6-units. Four VF_6-units are linked by V−F−V bridges to give tetrameric rings (as in CrF_4, structure **21.14**) and these motifs are connected through additional fluorine bridges to form layers. Reaction between VF_4 and KF in anhydrous HF gives $K_2[VF_6]$ containing octahedral $[VF_6]^{2-}$. Vanadium(IV) bromide is known but decomposes at 250 K to VBr_3 and Br_2.

The green oxochloride $VOCl_2$ (prepared from V_2O_5 and VCl_3) is polymeric and has a temperature-dependent magnetic moment ($1.40\,\mu_B$ at 296 K, $0.95\,\mu_B$ at 113 K); it decomposes on heating (eq. 21.26).

$$2VOCl_2 \xrightarrow{650\,K} VOCl_3 + VOCl \qquad (21.26)$$

(21.7)

Figure 21.9 shows that vanadium(V) is quite a powerful oxidant, and only mild reducing agents (e.g. SO_2) are needed to convert V(V) to V(IV). In aqueous solution, V(IV) is present as the hydrated vanadyl ion $[VO]^{2+}$ **(21.7)** of which many salts are known. Anhydrous $V(O)SO_4$ is manufactured by reducing a solution of V_2O_5 in H_2SO_4 with $H_2C_2O_4$. Blue solid $V(O)SO_4$ has a polymeric structure with vertex-sharing VO_6 octahedra linked by sulfate groups. The hydrate $V(O)SO_4 \cdot 5H_2O$ contains octahedrally sited V(IV) involving one oxido ligand (V−O = 159 pm) and five other O atoms (from sulfate and four H_2O) at 198–222 pm. The reaction of V_2O_5 and Hacac (see Table 7.7) gives blue $[VO(acac)_2]$ which has a square-based pyramidal structure **(21.8)**. This readily forms complexes with *N*-donor ligands which occupy the site *trans* to the oxido ligand. The salt $[NH_4]_2[VOCl_4]$ can be obtained by crystallization of a solution of $VOCl_3$ and $[NH_4]Cl$ in hydrochloric acid. The $[VOCl_4]^{2-}$ ion has a square-based pyramidal structure with the oxido ligand in the apical site. This preference is seen throughout related derivatives containing the $[VO]^{2+}$ unit. Its presence is detected by a characteristic IR spectroscopic absorption around $980\,cm^{-1}$ (the corresponding value for a V−O single bond is $\approx 480\,cm^{-1}$).

(21.8)

Vanadium(IV) oxide, VO_2, is prepared by heating V_2O_5 with $H_2C_2O_4$. It crystallizes with a rutile-type structure (Fig. 6.22) which is distorted at 298 K so that pairs of V(IV) centres are alternately 262 and 317 pm apart. The shorter distance is consistent with metal–metal bonding. This polymorph is an insulator, but above 343 K, the electrical conductivity and magnetic susceptibility of VO_2 increase as the regular rutile structure is adopted. Vanadium(IV) oxide is blue but shows thermochromic behaviour.

> The colour of a ***thermochromic*** compound is temperature dependent. The phenomenon is called ***thermochromism***.

Vanadium(IV) oxide is amphoteric, dissolving in non-oxidizing acids to give $[VO]^{2+}$ and in alkalis to form homopolyanions such as $[V_{18}O_{42}]^{12-}$, the Na^+ and K^+ salts of which can be isolated by heating $V(O)SO_4$ and MOH (M = Na or K) in water at pH 14 in an inert atmosphere. The structure of $[V_{18}O_{42}]^{12-}$ consists of square-based pyramidal VO_5-units, the apical O atoms of which are terminal (i.e. V=O units) while basal O atoms are involved in V−O−V bridges to build an almost spherical cage. Related anions such as $[V_{18}O_{42}]^{4-}$, $[V_{18}O_{42}]^{5-}$ and $[V_{18}O_{42}]^{6-}$ formally contain V(IV) and V(V) centres. The cavity in $[V_{18}O_{42}]^{n-}$ is able to accommodate an *anionic* guest as in $[V_{18}O_{42}I]^{5-}$ (Fig. 21.8f) or $[H_4V_{18}O_{42}X]^{9-}$ (X = Cl, Br, I).

Vanadium(III)

The trihalides VF_3, VCl_3, VBr_3 and VI_3 are all known. The yellow-green, insoluble trifluoride is made from V and HF at 500 K. Vanadium(III) chloride is a violet, hygroscopic solid which dissolves in water without decomposition to give $[V(OH_2)_6]Cl_3$. Anhydrous VCl_3 is made by decomposition of VCl_4 at 420 K (eq. 21.24), but above 670 K, it disproportionates to VCl_4 and VCl_2. Reaction of VCl_3 with BBr_3, or V with Br_2, yields VBr_3, a green-black, water-soluble solid which disproportionates to VBr_2 and VBr_4. The brown, hygroscopic VI_3 is made from V with I_2, and decomposes above 570 K to VI_2 and I_2. Each of the solid trihalides adopts a structure in which the V(III) centres occupy two thirds of the octahedral sites in an hcp array of halogen atoms (i.e. a BiI_3 prototype structure).

(21.9)

Vanadium(III) forms a variety of octahedral complexes, e.g. *mer*-$[VCl_3(THF)_3)]$ and *mer*-$[VCl_3(^tBuNC-C)_3]$, which have magnetic moments close to the spin-only value for a d^2 ion. The $[VF_6]^{3-}$ ion is present in simple salts such as K_3VF_6, but various extended structures are observed in other salts. The reaction of CsCl with VCl_3 at 1000 K produces $Cs_3[V_2Cl_9]$. $[V_2Cl_9]^{3-}$ (**21.9**) is isomorphous with $[Cr_2Cl_9]^{3-}$ and consists of two face-sharing octahedra with *no* metal–metal interaction. Examples of complexes with higher coordination numbers are known, e.g. $[V(CN)_7]^{4-}$ (pentagonal bipyramidal) made from VCl_3 and KCN in aqueous solution and isolated as the K^+ salt.

The oxide V_2O_3 (which, like Ti_2O_3, adopts the corundum structure, see Section 13.7) is made by partial reduction of V_2O_5 using H_2, or by heating (1300 K) V_2O_5 with vanadium. It is a black solid which, on cooling, exhibits a metal–insulator transition at 155 K. The oxide is basic, dissolving in acids to give $[V(OH_2)_6]^{3+}$. The hydrated oxide may be precipitated by adding alkali to green solutions of vanadium(III) salts. The $[V(OH_2)_6]^{3+}$ ion is present in alums such as $[NH_4]V(SO_4)_2 \cdot 12H_2O$ formed by electrolytic reduction of $[NH_4][VO_3]$ in sulfuric acid.

Vanadium(II)

Green VCl_2 is made from VCl_3 and H_2 at 770 K and is converted to blue VF_2 by reaction with HF and H_2. VCl_2 can also be obtained from VCl_3 as described above, and

similarly brown-red VBr_2 and violet VI_2 can be produced from VBr_3 and VI_3, respectively. Vanadium(II) fluoride crystallizes with a rutile-type structure (Fig. 6.22) and becomes antiferromagnetic below 40 K. VCl_2, VBr_2 and VI_2 (all paramagnetic) possess CdI_2 layer structures (Fig. 6.23). The dihalides are water-soluble.

Vanadium(II) is present in aqueous solution as the violet, octahedral $[V(OH_2)_6]^{2+}$ ion. It can be prepared by reduction of vanadium in higher oxidation states electrolytically or using zinc amalgam. It is strongly reducing, being rapidly oxidized on exposure to air. Compounds such as Tutton salts contain $[V(OH_2)_6]^{2+}$; e.g. $K_2V(SO_4)_2 \cdot 6H_2O$ is made by adding K_2SO_4 to an aqueous solution of VSO_4 and forms violet crystals.

A ***Tutton salt*** has the general formula $[M^I]_2M^{II}(SO_4)_2 \cdot 6H_2O$ (compare with an ***alum***, Section 13.9).

Vanadium(II) oxide is a grey, metallic solid and is obtained by reduction of higher oxides at high temperatures. It is non-stoichiometric, varying in composition from $VO_{0.8}$ to $VO_{1.3}$, and possesses an NaCl (Fig. 6.16) or defect NaCl structure (see Section 6.17). Conducting properties of the first row metal(II) oxides are compared in Section 28.2.

Simple vanadium(II) complexes include $[V(CN)_6]^{4-}$, the K^+ salt of which is made by reducing $K_4[V(CN)_7]$ with K metal in liquid NH_3. The magnetic moment of 3.5 μ_B for $[V(CN)_6]^{4-}$ is close to the spin-only value of 3.87 μ_B. Octahedral $[V(NCMe)_6]^{2+}$ has been isolated in the $[ZnCl_4]^{2-}$ salt from the reaction of VCl_3 with Et_2Zn in MeCN. Treatment of $VCl_2 \cdot 4H_2O$ with phen gives $[V(phen)_3]Cl_2$, for which $\mu_{eff} = 3.82 \mu_B$ (300 K), consistent with octahedral d^3.

Self-study exercises

1. The magnetic moment of a green salt $K_n[VF_6]$ is 2.79 μ_B at 300 K. With what value of *n* is this consistent?

   [*Ans. n* = 3]

2. The octahedral complex $[VL_3]$ where HL = $CF_3COCH_2COCH_3$ (related to Hacac) exists as *fac* and *mer* isomers in solution. Draw the structures of these isomers, and comment on further isomerism exhibited by this complex.
   [*Ans.* See structures 2.36 and 2.37, Fig. 3.20b, Section 19.8]

3. The electronic absorption spectrum of $[VCl_4(bpy)]$ shows an asymmetric band: $\lambda_{max} = 21\,300\,cm^{-1}$ with a shoulder at $17\,400\,cm^{-1}$. Suggest an explanation for this observation.

   [*Ans.* d^1, see Fig. 20.4 and discussion]

4. The vanadate $[V_{14}O_{36}Cl]^{5-}$ is an open cluster with a guest Cl^- ion. (a) What are the formal oxidation states of the V centres? (b) The electronic absorption spectrum exhibits intense charge transfer bands. Explain why these are likely to arise from LMCT rather than MLCT transitions.

[*Ans.* (a) 2 V(IV) and 12 V(V)]

21.7 Group 6: chromium

The metal

At ordinary temperatures, Cr metal is resistant to chemical attack (although it dissolves in dilute HCl and H_2SO_4). This inertness is kinetic rather than thermodynamic in origin as the Cr^{2+}/Cr and Cr^{3+}/Cr couples in Fig. 21.10 show. Nitric acid renders Cr passive, and Cr is resistant to alkalis. At higher temperatures the metal is reactive: it decomposes steam and combines with O_2, halogens, and most other non-metals. Borides, carbides and nitrides (see Sections 13.10, 14.7 and 15.6) exist in various phases (e.g. CrN, Cr_2N, Cr_3N, Cr_3N_2) and are inert materials (e.g. CrN is used in wear-resistant coatings). The black sulfide Cr_2S_3 is formed by direct combination of the elements on heating. Other sulfides are formed by reactions other than direct combination of Cr and S_8, e.g. CrS is formed by thermal decomposition of Cr_2S_3.

The main oxidation states of chromium are +6, +3 and +2. A few compounds of Cr(V) and Cr(IV) are known, but are unstable with respect to disproportionation. Chromium(0) is stabilized by π-acceptor ligands (see Chapter 24).

Chromium(VI)

No halides of chromium(VI) have been isolated. Early reports of CrF_6 have since been shown to be incorrect, the vibrational spectrum being due not to CrF_6, but to CrF_5. However, the oxohalides CrO_2F_2 and CrO_2Cl_2 are known. Fluorination of CrO_3 with SeF_4, SF_4 or HF yields CrO_2F_2 (violet crystals, mp 305 K), while CrO_2Cl_2 (red liquid, mp 176 K, bp 390 K) is prepared by heating a mixture of $K_2Cr_2O_7$, KCl and concentrated H_2SO_4. Chromyl chloride is an oxidant and chlorinating agent. It has a molecular structure (**21.10**) and is light-sensitive and readily hydrolysed (eq. 21.27). If CrO_2Cl_2 is added to a concentrated

KCl solution, $K[CrO_3Cl]$ precipitates. Structure **21.11** shows the $[CrO_3Cl]^-$ ion.

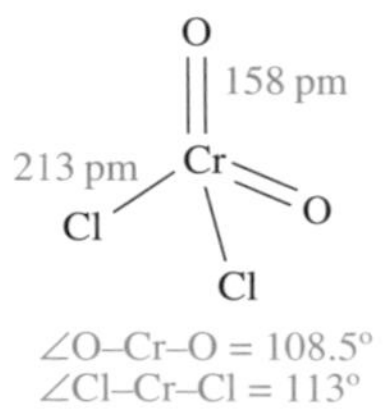

O $\quad \parallel$ 158 pm

213 pm $\diagdown$ Cr $=$ O

Cl $\diagup$

Cl

∠O–Cr–O = 108.5°
∠Cl–Cr–Cl = 113°

(21.10)

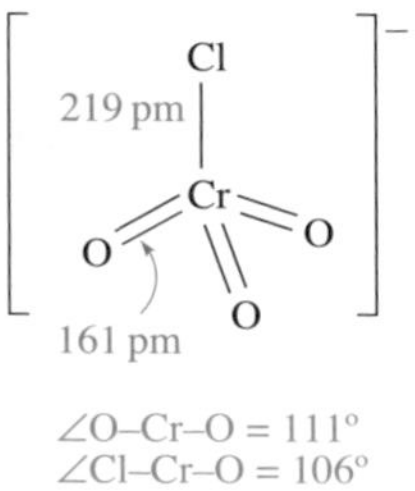

Cl $\mid$ 219 pm

Cr

O $=$ $\diagup$ O

$\diagup$ O 161 pm

∠O–Cr–O = 111°
∠Cl–Cr–O = 106°

(21.11)

$$2CrO_2Cl_2 + 3H_2O \longrightarrow [Cr_2O_7]^{2-} + 4Cl^- + 6H^+ \quad (21.27)$$

Chromium(VI) oxide ('chromic acid'), CrO_3, separates as a purple-red solid when concentrated H_2SO_4 is added to a solution of a dichromate(VI) salt. It is a powerful oxidant with uses in organic synthesis. It melts at 471 K and at slightly higher temperatures decomposes to Cr_2O_3 and O_2 with CrO_2 formed as an intermediate. The solid state structure of CrO_3 consists of chains of corner-sharing tetrahedral CrO_4 units (as in Fig. 21.8d).

Chromium(VI) oxide dissolves in base to give yellow solutions of $[CrO_4]^{2-}$. This is a weak base and forms $[HCrO_4]^-$ and then H_2CrO_4 as the pH is lowered (H_2CrO_4: $pK_a(1) = 0.74$; $pK_a(2) = 6.49$). In solution, these equilibria are complicated by the formation of orange dichromate(VI), $[Cr_2O_7]^{2-}$ (eq. 21.28).

$$2[HCrO_4]^- \rightleftharpoons [Cr_2O_7]^{2-} + H_2O \quad (21.28)$$

Further condensation occurs at high $[H^+]$ to give $[Cr_3O_{10}]^{2-}$ and $[Cr_4O_{13}]^{2-}$. The structures (determined for solid state salts) of $[Cr_2O_7]^{2-}$ and $[Cr_3O_{10}]^{2-}$ are shown in Fig. 21.11. Like $[CrO_4]^{2-}$, they contain tetrahedral CrO_4 units and the chains in the di- and trinuclear species

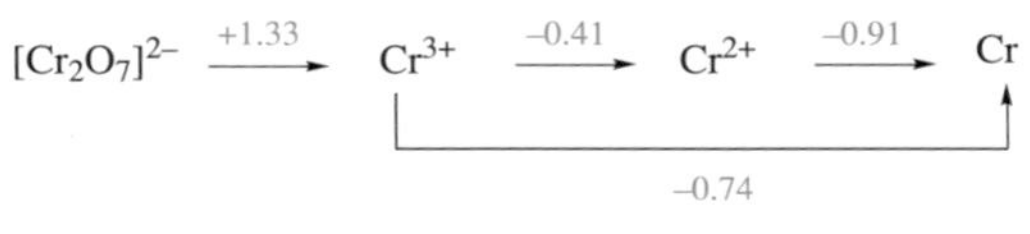

$[Cr_2O_7]^{2-} \xrightarrow{+1.33} Cr^{3+} \xrightarrow{-0.41} Cr^{2+} \xrightarrow{-0.91} Cr$

$\underbrace{\qquad\qquad}_{-0.74}$

Fig. 21.10 Potential diagram for chromium at pH 0. A Frost–Ebsworth diagram for Cr is shown in Fig. 8.4a.

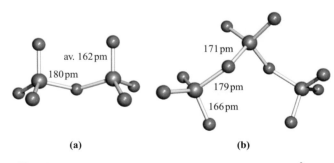

(a)   (b)

Fig. 21.11 Structures (X-ray diffraction) of (a) $[Cr_2O_7]^{2-}$ in the 2-amino-5-nitropyridinium salt [J. Pecaut *et al.* (1993) *Acta Crystallogr., Sect. B*, vol. 49, p. 277], and (b) $[Cr_3O_{10}]^{2-}$ in the guanidinium salt [A. Stepien *et al.* (1977) *Acta Crystallogr., Sect. B*, vol. 33, p. 2924]. Colour code: Cr, green; O, red.

contain corner-sharing tetrahedra (i.e. as in CrO_3). The $[Cr_4O_{13}]^{2-}$ ion has a related structure. Higher species are not observed and thus chromates do not mimic vanadates in their structural complexity.

Complex formation by Cr(VI) requires strong π-donor ligands such as O^{2-} or $[O_2]^{2-}$. When H_2O_2 is added to an acidified solution of a chromate(VI) salt, the product (formed as a solution species) is a deep violet-blue complex which contains both oxido and peroxido ligands (eq. 21.29).

$$[CrO_4]^{2-} + 2H^+ + 2H_2O_2 \longrightarrow [Cr(O)(O_2)_2] + 3H_2O$$
$$(21.29)$$

In aqueous solution, $[Cr(O)(O_2)_2]$ rapidly decomposes to Cr(III) and O_2. An ethereal solution is more stable and, from it, the pyridine adduct $[Cr(O)(O_2)_2(py)]$ may be isolated. In the solid state, $[Cr(O)(O_2)_2(py)]$ contains an approximate pentagonal pyramidal arrangement of donor atoms with the oxido ligand in the apical site (Figs. 21.12a and b). If each peroxido ligand is considered to occupy one rather than two coordination sites, then the coordination environment is tetrahedral (Fig. 21.12c). This and related compounds (which are explosive when dry) have uses as oxidants in organic syntheses. Like other Cr(VI) compounds, $[Cr(O)(O_2)_2(py)]$ has a very small paramagnetic susceptibility (arising from coupling of the diamagnetic ground state with excited states). The action of H_2O_2 on neutral or slightly acidic solutions of $[Cr_2O_7]^{2-}$ (or reaction between $[Cr(O)(O_2)_2]$ and alkalis) yields diamagnetic, dangerously explosive, red-violet salts of $[Cr(O)(O_2)_2(OH)]^-$. Imido ligands $[RN]^{2-}$ may formally replace oxido groups in Cr(VI) species, e.g. $[Cr(N^tBu)_2Cl_2]$ is structurally related to CrO_2Cl_2.

Chromium(VI) in acidic solution is a powerful oxidizing agent (eq. 21.30), but reactions are often slow. Both $Na_2Cr_2O_7$ and $K_2Cr_2O_7$ are manufactured on a large scale; $K_2Cr_2O_7$ is less soluble in water than $Na_2Cr_2O_7$. Both are widely used as oxidants in organic syntheses. Commercial applications include those in tanning, corrosion inhibitors and insecticides. The use of 'chromated copper arsenate' in wood preservatives is being discontinued on environmental grounds (see Box 15.1). Potassium dichromate(VI) is used in titrimetric analysis (e.g. reaction 21.31) and the colour change accompanying reduction of $[Cr_2O_7]^{2-}$ to Cr^{3+} is the basis for some types of breathalyser units in which ethanol is oxidized to acetaldehyde. Sodium chromate(VI), also an important oxidant, is manufactured by reaction 21.32.

$$[Cr_2O_7]^{2-} + 14H^+ + 6e^- \rightleftharpoons 2Cr^{3+} + 7H_2O$$

orange   green[†]

$$E^o = +1.33\,V \qquad (21.30)$$

$$[Cr_2O_7]^{2-} + 14H^+ + 6Fe^{2+} \longrightarrow 2Cr^{3+} + 7H_2O + 6Fe^{3+}$$
$$(21.31)$$

$$Na_2Cr_2O_7 + 2NaOH \longrightarrow 2Na_2CrO_4 + H_2O \qquad (21.32)$$

Chromium(VI) compounds are highly toxic (suspected carcinogens) and must be stored away from combustible materials; violent reactions occur with some organic compounds.

Chromium(V) and chromium(IV)

Unlike CrF_6, CrF_5 is well established. It is a red, volatile solid (mp 303 K), formed by direct combination of the elements at $\approx$570 K. The vapour is yellow and contains distorted trigonal bipyramidal CrF_5 molecules. It is a strong oxidizing and fluorinating agent. For Cr(V), the fluoride is the only halide known. Pure CrF_4 can be made by fluorination of Cr using HF/F_2 under solvothermal conditions. The pure material is violet, but the colour of samples prepared by different routes varies (green, green-black, brown) with descriptions being affected by the presence of impurities. In the vapour, CrF_4 exists as a tetrahedral molecule. Solid CrF_4 is dimorphic. In α-CrF_4, pairs of edge-sharing CrF_6-octahedra (**21.12** and **21.13**) assemble into columns through Cr$-$F$-$Cr bridges involving the atoms marked F_a in structure **21.12**. In β-CrF_4, Cr_4F_{20}-rings (**21.14**) are connected through the apical

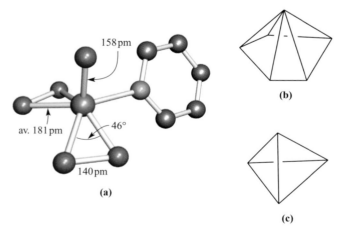

Fig. 21.12 (a) The structure of $[Cr(O)(O_2)_2(py)]$ determined by X-ray diffraction [R. Stomberg (1964) *Ark. Kemi*, vol. 22, p. 29]; colour code: Cr, green; O, red; N, blue; C, grey. The coordination environment can be described as (b) pentagonal pyramidal or (c) tetrahedral (see text).

[†] The green colour is due to a sulfato complex, H^+ being supplied as sulfuric acid; $[Cr(OH_2)_6]^{3+}$ is violet, see later.

F_a atoms to generate columns. Compare the structures of α- and β-CrF_4 with that of solid TiF_4 (Fig. 21.6).

(21.12)   **(21.13)**

(21.14)

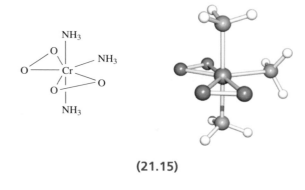

(21.15)

Self-study exercises

1. The solid state structure of $[XeF_5]^+[CrF_5]^-$ contains infinite chains of distorted CrF_6 octahedra connected through *cis*-vertices. Draw part of the chain, ensuring that the 1:5 Cr:F stoichiometry is maintained.

2. Assuming that the cations in $[XeF_5]^+[CrF_5]^-$ are discrete, what geometry for each cation would be consistent with the VSEPR model?

[For the answers to both exercises, see: K. Lutar *et al.* (1998) *Inorg. Chem.*, vol. 37, p. 3002]

Chromium(IV) chloride and bromide have been prepared but are unstable.

Chromium(IV) oxide, CrO_2, is usually made by controlled decomposition of CrO_3. It is a brown-black solid which has the rutile structure and is a metallic conductor (compare with VO_2). It is ferromagnetic and is widely used in magnetic recording tapes.

When an acidic solution in which $[Cr_2O_7]^{2-}$ is oxidizing propan-2-ol is added to aqueous $MnSO_4$, MnO_2 is precipitated, although acidified $[Cr_2O_7]^{2-}$ alone does not effect this oxidation. This observation is evidence for the participation of Cr(V) or Cr(IV) in dichromate(VI) oxidations. Under suitable conditions, it is possible to isolate salts of $[CrO_4]^{3-}$ and $[CrO_4]^{4-}$. For example, dark blue Sr_2CrO_4 is produced by heating $SrCrO_4$, Cr_2O_3 and $Sr(OH)_2$ at 1270 K, and dark green Na_3CrO_4 results from reaction of Na_2O, Cr_2O_3 and Na_2CrO_4 at 770 K. Complexes of chromium(V) may be stabilized by π-donor ligands, e.g. $[CrF_6]^-$, $[CrOF_4]^-$, $[CrOF_5]^{2-}$, $[CrNCl_4]^{2-}$ and $[Cr(N^tBu)Cl_3]$. Peroxido complexes containing $[Cr(O_2)_4]^{3-}$ are obtained by reaction of chromate(V) with H_2O_2 in alkaline solution; $[Cr(O_2)_4]^{3-}$ has a dodecahedral structure. These salts are explosive but are less dangerous than the Cr(VI) peroxido complexes. The explosive Cr(IV) peroxido complex $[Cr(O_2)_2(NH_3)_3]$ **(21.15)** is formed when $[Cr_2O_7]^{2-}$ reacts with aqueous NH_3 and H_2O_2. A related complex is $[Cr(O_2)_2(CN)_3]^{3-}$.

Chromium(III)

The +3 oxidation state is the most stable for chromium in its compounds and octahedral coordination dominates for Cr(III) centres. Table 20.3 shows the large LFSE associated with the octahedral d^3 configuration, and Cr(III) complexes are generally kinetically inert (see Section 26.2).

Anhydrous $CrCl_3$ (red-violet solid, mp 1425 K) is made from the metal and Cl_2, and is converted to green CrF_3 by heating with HF at 750 K. Solid CrF_3 is isostructural with VF_3, and $CrCl_3$ adopts a BiI_3 structure. The dark green tribromide and triiodide can be prepared from Cr and the respective halogen and are isostructural with $CrCl_3$. Chromium trifluoride is sparingly soluble and may be precipitated as the hexahydrate. The formation of $CrCl_3·6H_2O$ and its hydrate isomerism were described in Section 19.8. Although pure $CrCl_3$ is insoluble in water, addition of a trace of Cr(II) (e.g. $CrCl_2$) results in dissolution. The fast redox reaction between Cr(III) in the $CrCl_3$ lattice and Cr(II) in solution is followed by rapid substitution of Cl^- by H_2O at the solid surface since Cr(II) is labile (see Chapter 26).

Chromium(III) oxide is made by combination of the elements at high temperature, by reduction of CrO_3, or by reaction 21.33. It has the corundum structure (Section 13.7) and is semiconducting and antiferromagnetic ($T_N = 310$ K). Commercially Cr_2O_3 is used in abrasives and is an important green pigment. The dihydrate (*Guignet's green*) is used in paints. Traces of Cr(III) in Al_2O_3 give rise to the red colour of rubies (see Section 20.8).

$$[NH_4]_2[Cr_2O_7] \xrightarrow{\Delta} Cr_2O_3 + N_2 + 4H_2O \qquad (21.33)$$

Large numbers of mononuclear, octahedral Cr(III) complexes are known with magnetic moments close to the spin-only value of $3.87\,\mu_B$ (Table 20.11). The electronic absorption spectra of octahedral d^3 complexes contain three absorptions due to '*d–d*' transitions (see Fig. 20.20). Selected examples of octahedral chromium(III) complexes are $[Cr(acac)_3]$, $[Cr(ox)_3]^{3-}$, $[Cr(en)_3]^{3+}$, $[Cr(bpy)_3]^{3+}$, *cis*- and *trans*-$[Cr(en)_2F_2]^+$, *trans*-$[CrCl_2(MeOH)_4]^+$, $[Cr(CN)_6]^{3-}$ and $[Cr(NH_3)_2(S_5)_2]^-$ ($[S_5]^{2-}$ is bidentate; see Fig. 16.12 for related structures). Complex halides include $[CrF_6]^{3-}$, $[CrCl_6]^{3-}$ and $[Cr_2Cl_9]^{3-}$. Violet $Cs_3[Cr_2Cl_9]$ is made by reaction 21.34. $[Cr_2Cl_9]^{3-}$ is isostructural with $[V_2Cl_9]^{3-}$ (**21.9**) and magnetic data are consistent with the presence of three unpaired electrons per Cr(III) centre, i.e. *no* Cr–Cr interaction.

$$3CsCl + 2CrCl_3 \xrightarrow{\text{in a melt}} Cs_3[Cr_2Cl_9] \qquad (21.34)$$

Pale violet $[Cr(OH_2)_6]^{3+}$ is obtained in aqueous solution when $[Cr_2O_7]^{2-}$ is reduced by SO_2 or by EtOH and H_2SO_4 below 200 K. The commonest salt containing $[Cr(OH_2)_6]^{3+}$ is chrome alum, $KCr(SO_4)_2 \cdot 12H_2O$. $[Cr(OH_2)_6]^{3+}$ has also been structurally characterized in the solid state in a number of salts, e.g. $[Me_2NH_2][Cr(OH_2)_6][SO_4]_2$ (av. Cr–O = 196 pm). From aqueous solutions of Cr(III) salts, alkali precipitates Cr_2O_3 which dissolves to give $[Cr(OH_2)_6]^{3-}$. The hexaaqua ion is quite acidic (p$K_a \approx 4$) and hydroxido-bridged species are present in solution (see eq. 7.38 and accompanying discussion). Figure 21.13 shows the structure of $[Cr_2(OH_2)_8(\mu\text{-}OH)_2]^{4+}$. Addition of NH_3 to aqueous solutions of $[Cr(OH_2)_6]^{3+}$ results in the slow formation of ammine

complexes; it is preferable to use Cr(II) precursors since substitution is faster in Cr(II) than Cr(III) (see Chapter 26). The dinuclear complex **21.16** is reversibly converted to the oxido-bridged **21.17** in the presence of alkali (eq. 21.35).

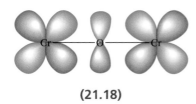

The two Cr(III) (d^3) centres in complex **21.17** are antiferromagnetically coupled and this is rationalized in terms of $(d–p)\pi$-bonding involving Cr d and O p orbitals (diagram **21.18**). Weak antiferromagnetic coupling also occurs between the Cr(III) centres in trinuclear complexes of type $[Cr_3L_3(\mu\text{-}O_2CR)_6(\mu_3\text{-}O)]^+$ (Fig. 21.14).

(21.18)

Chromium(II)

Anhydrous CrF_2, $CrCl_2$ and $CrBr_2$ are made by reacting Cr with HX (X = F, Cl, Br) at >850 K; CrI_2 is formed by heating Cr and I_2. The fluoride and chloride adopt distorted rutile structures (Fig. 6.22), while $CrBr_2$ and CrI_2 crystallize with distorted CdI_2 structures (Fig. 6.23). The distortions arise from the Jahn–Teller effect (high-spin d^4). Crystals of $CrCl_2$ are colourless but dissolve in water to give blue solutions of the strongly reducing hexaaqua ion. Solutions of $[Cr(OH_2)_6]^{2+}$ are usually obtained by dissolving Cr in acids or by reduction (Zn amalgam or electrolytically) of Cr(III)-containing solutions. Hydrated salts such as $Cr(ClO_4)_2 \cdot 6H_2O$, $CrCl_2 \cdot 4H_2O$ and $CrSO_4 \cdot 7H_2O$ may be isolated from solution, but cannot be dehydrated without decomposition.

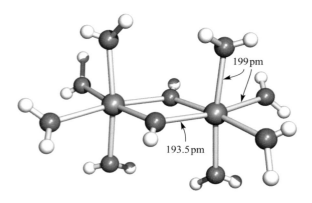

Fig. 21.13 The structure of $[Cr_2(OH_2)_8(\mu\text{-}OH)_2]^{4+}$ determined by X-ray diffraction for the mesitylene-2-sulfonate salt; the *non-bonded* Cr····Cr separation is 301 pm [L. Spiccia *et al.* (1987) *Inorg. Chem.*, vol. 26, p. 474]. Colour code: Cr, green; O, red; H, white.

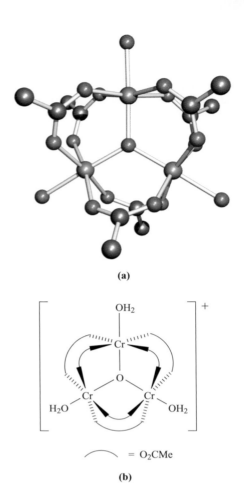

(a)

OH$_2$

Cr

O

H$_2$O   Cr   Cr   OH$_2$

$^+$

⌒ = O$_2$CMe

(b)

Fig. 21.14 A representative member of the [Cr$_3$L$_3$(μ-O$_2$CR)$_6$(μ_3-O)]$^+$ family of complexes: (a) the structure of [Cr$_3$(OH$_2$)$_3$(μ-O$_2$CMe)$_6$(μ_3-O)]$^+$ (X-ray diffraction) in the hydrated chloride salt [C.E. Anson *et al.* (1997) *Inorg. Chem.*, vol. 36, p. 1265], and (b) a schematic representation of the same complex. In (a), the H atoms are omitted for clarity; colour code: Cr, green; O, red; C, grey.

Self-study exercises

1. CrI$_2$ adopts a distorted CdI$_2$ structure. What is the environment about each Cr(II) centre?

[*Ans.* See Fig. 6.23; Cr replaces Cd]

2. In CrBr$_2$, four Cr−Br distances are 254 pm and two are 300 pm. What is the *d* electron configuration of the Cr centre? Explain the origin of the difference in bond lengths.   [*Ans. d*^{4}; see Section 20.3]

For the Cr^{3+}/Cr^{2+} couple, $E^{\circ} = -0.41$ V, and Cr(II) compounds slowly liberate H$_2$ from water, as well as undergo oxidation by O$_2$ (see worked example 8.4). The potential diagram in Fig. 21.10 shows that Cr(II) compounds are just

stable with respect to disproportionation. The study of the oxidation of Cr^{2+} species has played an important role in establishing the mechanisms of redox reactions (see Chapter 26).

Complexes of Cr(II) include halide anions such as [CrX$_3$]$^-$, [CrX$_4$]$^{2-}$, [CrX$_5$]$^{3-}$ and [CrX$_6$]$^{4-}$. Despite the range of formulae, the Cr(II) centres in the solids are usually octahedrally sited, e.g. [CrCl$_3$]$^-$ consists of chains of distorted face-sharing octahedra, the distortion being a Jahn–Teller effect. Some of these salts show interesting magnetic properties. For example, salts of [CrCl$_4$]$^{2-}$ show *ferromagnetic coupling* (as opposed to antiferromagnetic coupling which is a more common phenomenon, see Section 20.10) at low temperatures, with T_C values in the range 40–60 K; communication between the metal centres is through Cr−Cl−Cr bridging interactions.

Cyanido complexes of Cr(II) include [Cr(CN)$_6$]$^{4-}$ and [Cr(CN)$_5$]$^{3-}$. K$_4$[Cr(CN)$_6$] may be prepared in aqueous solution, but only in the presence of excess cyanide ion; octahedral [Cr(CN)$_6$]$^{4-}$ is low-spin. The reaction of [Cr$_2$(μ-O$_2$CMe)$_4$] (see below) with [Et$_4$N][CN] leads to the formation of [Et$_4$N]$_3$[Cr(CN)$_5$]. In the solid state, both trigonal bipyramidal and square-based pyramidal [Cr(CN)$_5$]$^{3-}$ ions are present. The small energy difference between the 5-coordinate structures has also been observed for [Ni(CN)$_5$]$^{3-}$. At 300 K, [Cr(CN)$_5$]$^{3-}$ exhibits an effective magnetic moment of 4.90 μ_B, consistent with high-spin Cr(II). The [CN]$^-$ ion is a strong-field ligand, and so [Cr(CN)$_5$]$^{3-}$ represents a rare example of a high-spin cyanido complex, one other being [Mn(CN)$_4$]$^{2-}$. Theoretical data (a combination of ligand field theory and DFT) indicate that for the 5-coordinate [Cr(CN)$_5$]$^{3-}$, the promotion energy associated with a change in spin state is smaller than the spin-pairing energy and this leads to a high-spin complex. In contrast, for octahedral [Cr(CN)$_6$]$^{4-}$, the reverse is true and the complex is low-spin.†

Chromium–chromium multiple bonds

Chromium(II) carboxylates are dimers of general formula [Cr$_2$(μ-O$_2$CR)$_4$] or [Cr$_2$L$_2$(μ-O$_2$CR)$_4$] and are examples of *d*-block metal complexes that involve metal–metal multiple bonding. For example, red [Cr$_2$(OH$_2$)$_2$(μ-O$_2$CMe)$_4$] is precipitated when aqueous CrCl$_2$ is added to saturated aqueous Na[MeCO$_2$]. Figure 21.15 shows the structures of [Cr$_2$(μ-O$_2$CC$_6$H$_2$-2,4,6-^{i}Pr$_3$)$_4$] and [Cr$_2$(py)$_2$(μ-O$_2$CMe)$_4$]. The significant difference between these two compounds is the presence of axial ligands, i.e. the pyridine ligands in the latter complex. Even when no axial ligands are present, association can occur in the solid state as is observed in [Cr$_2$(μ-O$_2$CMe)$_4$] (**21.19**). In [Cr$_2$(μ-O$_2$CC$_6$H$_2$-2,4,6-^{i}Pr$_3$)$_4$], the steric demands of the aryl substituents prevent

---

† For details, see: R.J. Deeth (2006) *Eur. J. Inorg. Chem.*, p. 2551 – 'A theoretical rationale for the formation, structure and spin state of pentacyanochromate(II)'.

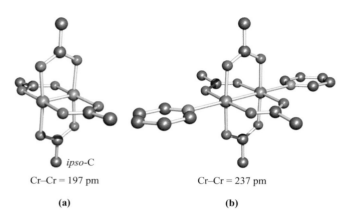

ipso-C

Cr–Cr = 197 pm   Cr–Cr = 237 pm

(a)   **(b)**

Fig. 21.15 The structures (X-ray diffraction) of (a) [Cr$_2$(μ-O$_2$CC$_6$H$_2$-2,4,6-^{i}Pr$_3$)$_4$] with only the *ipso*-C atoms of the aryl substituents shown [F.A. Cotton *et al.* (2000) *J. Am. Chem. Soc.*, vol. 122, p. 416] and (b) [Cr$_2$(py)$_2$(μ-O$_2$CMe)$_4$] with H atoms omitted for clarity [F.A. Cotton *et al.* (1980) *Inorg. Chem.*, vol. 19, p. 328]. Colour code: Cr, green; O, red; C, grey; N, blue.

association and the solid contains discrete molecules (Fig. 21.15a).

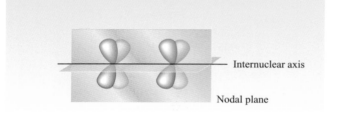

(21.19)

Compounds of the type [Cr$_2$(μ-O$_2$CR)$_4$] and [Cr$_2$L$_2$(μ-O$_2$CR)$_4$] (Fig. 21.15) are *diamagnetic*, possess *short* Cr–Cr bonds (cf. 258 pm in Cr metal), and have eclipsed ligand conformations. These properties are consistent with the Cr(II) *d* electrons being involved in *quadruple* bond formation. For the bridging ligands in [Cr$_2$(μ-O$_2$CR)$_4$] to be eclipsed is less surprising than in complexes with monodentate ligands, e.g. [Re$_2$Cl$_8$]$^{2-}$ (see Section 22.8), but the observation is a key feature in the description of the metal–metal quadruple bond. The bonding in [Cr$_2$(μ-O$_2$CR)$_4$] can be described as shown in Fig. 21.16. The Cr atoms are defined to lie on the *z* axis, and each Cr atom uses four (*s*, *p$_x$*, *p$_y$* and *d$_{x^2-y^2}$*)† of its nine atomic orbitals to form Cr–O bonds. Now allow mixing of the *p$_z$* and *d$_{z^2}$* orbitals to give two hybrid orbitals directed along the *z* axis. Each Cr atom has four orbitals available for metal–

metal bonding: *d$_{xz}$*, *d$_{yz}$*, *d$_{xy}$* and one *p$_z$d$_{z^2}$* hybrid, with the second *p$_z$d$_{z^2}$* hybrid being non-bonding and pointing outwards from the Cr–Cr-unit (see below). Figure 21.16a shows that overlap of the metal *p$_z$d$_{z^2}$* hybrid orbitals leads to a σ-bond, while *d$_{xz}$–d$_{xz}$* and *d$_{yz}$–d$_{yz}$* overlap gives a degenerate pair of π-orbitals. Finally, overlap of the *d$_{xy}$* orbitals gives rise to a δ-bond. The degree of overlap follows the order σ > π > δ and Fig. 21.16b shows an approximate energy level diagram for the σ, π, δ, σ*, π* and δ* MOs. Each Cr(II) centre provides four electrons for Cr–Cr bond formation and these occupy the MOs in Fig. 21.16b to give a σ^{2}π^{4}δ^2 configuration, i.e. a quadruple bond. A consequence of this bonding picture is that the δ component forces the two CrO$_4$-units to be eclipsed. The red colour of [Cr$_2$(μ-O$_2$CMe)$_4$] (λ$_{max}$ = 520 nm, see Table 19.2) and related complexes can be understood in terms of the δ–δ* energy gap and a σ^{2}π^{4}δ^{1}δ^{*1} ◂── σ^{2}π^{4}δ^2 transition.

A **δ-bond** is formed by the face-on overlap of two *d$_{xz}$* (or two *d$_{yz}$*, or two *d$_{xy}$*) orbitals. The resultant MO possesses *two* nodal planes that contain the internuclear axis:

Internuclear axis

Nodal plane

This bonding description for [Cr$_2$(μ-O$_2$CR)$_4$] leaves a non-bonding, outward-pointing *p$_z$d$_{z^2}$* hybrid orbital per Cr atom (**21.20**). Complex formation with donors such as H$_2$O and pyridine (Fig. 21.15b) occurs by donation of a lone pair of electrons into each vacant orbital. The Cr–Cr bond length increases significantly when axial ligands are introduced, e.g. 197 to 239 pm on going from [Cr$_2$(μ-O$_2$CC$_6$H$_2$-2,4,6-^{i}Pr$_3$)$_4$] to [Cr$_2$(MeCN)$_2$(μ-O$_2$CC$_6$H$_2$-2,4,6-^{i}Pr$_3$)$_4$].

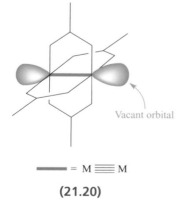

Vacant orbital

━━ = M ≡≡≡ M

(21.20)

The chromium–chromium quadruple bond is typically considered to be strong, and this is supported by the fact that reactions of [Cr$_2$(μ-O$_2$CR)$_4$] with Lewis bases generate

---

† The choice of the *d$_{x^2-y^2}$* orbital for Cr–O bond formation is arbitrary. The *d$_{xy}$* could also have been used, leaving the *d$_{x^2-y^2}$* orbital available for metal–metal bonding.

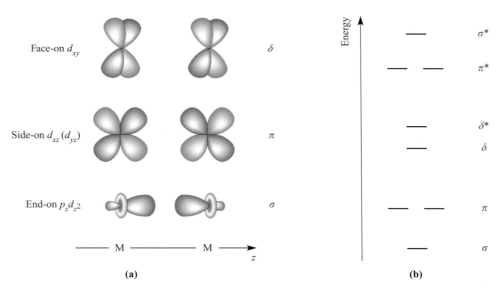

Fig. 21.16 (a) The formation of σ, π and δ components of a metal–metal quadruple bond by overlap of appropriate metal orbitals. *Both the* d_{xz} and d_{yz} atomic orbitals are used to form π-bonds, and the d_{xy} atomic orbital is used for δ-bond formation. (b) Approximate energy levels of the metal–metal bonding and antibonding MOs. This figure is relevant for M_2L_8 or $M_2(\mu\text{-}L)_4$ type complexes.

adducts without loss of the metal–metal bonding interaction. However, there are examples in which the bond is readily cleaved. [Li(THF)]$_4$[Cr$_2$Me$_8$] contains the [Cr$_2$Me$_8$]$^{4-}$ ion, in which the Cr–Cr distance is 198 pm in the solid state. Treatment with the chelating ligand Me$_2$NCH$_2$CH$_2$NMe$_2$ (TMEDA) results in [Li(TMEDA)]$_2$[CrMe$_4$]. A second example of a weak quadruple bond is found in the amidato complex **21.21**. In the solid state, **21.21** is diamagnetic and the Cr–Cr bond length is 196 pm. However, in benzene solution, **21.21** dissociates to give paramagnetic monomers [CrL$_2$].

metallic, we include it here because it exemplifies a metal–metal bond with a bond order of five. Compound **21.22** formally contains two Cr(I) (d^5) centres, and extremely bulky organo-ligands are used to protect the Cr$_2$-core (Fig. 21.17). Structural data show that the Cr–Cr bond is very short (183.5 pm) and magnetic data indicate the presence of strongly coupled d^5–d^5 bonding electrons. These observations are consistent with the presence of a chromium–chromium quintuple bond. This can be described in terms of the orbital interactions shown in Fig. 21.16 plus an additional δ-bond arising from the face-on overlap of two $d_{x^2-y^2}$ orbitals. The quintuple bond therefore has a $\sigma^2\pi^4\delta^4$ configuration. The organo substituent used to stabilize this system is from the same family of R groups used to stabilize E$_2$R$_2$ type compounds, where E is a heavy group 13 or 14 element (see Chapter 23).

an amidato ligand

Cr ━━━ Cr represents Cr ≡≡≡ Cr

(21.21)

The diagrams in Fig. 21.16 are relevant for complexes of the type M_2L_8 or $M_2(\mu\text{-}L)_4$ in which one of the metal d orbitals is reserved for metal–ligand bonding. However, if the number of ligands is reduced, more metal orbitals and more metal valence electrons become available for metal–metal bonding. Although compound **21.22** is organo-

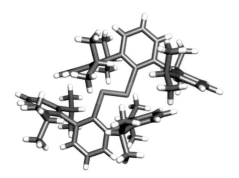

Fig. 21.17 The structure (X-ray diffraction) of [Cr$_2$L$_2$] in [Cr$_2$L$_2$]·MeC$_6$H$_5$, where HL = 2,6-bis(2,6-di-isopropylphenyl)benzene [T. Nguyen *et al.* (2005) *Science*, vol. 310, p. 844]. Colour code: Cr, green; C, grey; H, white.

Cr $\equiv$ Cr = quintuple bond

(21.22)

Self-study exercise

A quick method of working out the number of electrons available for metal–metal bonding in a complex such as $[Cr_2Me_8]^{4-}$ is to 'remove' each ligand along with an appropriate charge, write down the formula of the 'metal core', and hence determine the number of valence electrons remaining. For example, removal of $8Me^-$ ligands from $[Cr_2Me_8]^{4-}$ leaves $[Cr_2]^{4+}$ which has $(2 \times 6) - 4 = 8$ valence electrons. These electrons occupy the MOs shown in Fig. 21.16b: $\sigma^2\pi^4\delta^2$. In this method, the removal of *anionic* methyl groups takes care of the electrons required for Cr–C bond formation. Carry out this same exercise for $[Cr_2(\mu\text{-}O_2CMe)_4]$, complex **21.21** and $[Re_2Cl_8]^{2-}$ to confirm the presence of a quadruple bond in each complex.

21.8 Group 7: manganese

The metal

Metallic Mn is slowly attacked by water and dissolves readily in acids (e.g. eq. 21.36). The finely divided metal is pyrophoric in air, but the bulk metal is not attacked unless heated (eq. 21.37). At elevated temperatures, it combines with most non-metals, e.g. N_2 (eq. 21.38), halogens (eq. 21.39), C, Si and B (see Sections 13.10, 14.7 and 15.6).

$$Mn + 2HCl \longrightarrow MnCl_2 + H_2 \tag{21.36}$$

$$3Mn + 2O_2 \xrightarrow{\Delta} Mn_3O_4 \tag{21.37}$$

$$3Mn + N_2 \xrightarrow{\Delta} Mn_3N_2 \tag{21.38}$$

$$Mn + Cl_2 \xrightarrow{\Delta} MnCl_2 \tag{21.39}$$

Manganese exhibits the widest range of oxidation states of any of the first row *d*-block metals. The lowest states are stabilized by π-acceptor ligands, usually in organometallic complexes (see Chapter 24). However, dissolution of Mn powder in air-free aqueous NaCN gives the Mn(I) complex $Na_5[Mn(CN)_6]$.

This section describes Mn(II)–Mn(VII) species. A potential diagram for manganese was given in Fig. 8.2 and a Frost–Ebsworth diagram was shown in Fig. 8.3. On going from Cr to Mn, there is an abrupt change in the stability with respect to oxidation of M^{2+} (eq. 21.40). The difference in E° values arises from the much higher third ionization energy of Mn (see Table 21.1). All oxidation states above Mn(II) are powerful oxidizing agents.

$$M^{3+}(aq) + e^- \rightleftharpoons M^{2+}(aq) \quad \begin{cases} M = Mn, \; E^\circ = +1.54\,V \\ M = Cr, \; E^\circ = -0.41\,V \end{cases}$$
$$\tag{21.40}$$

Manganese(VII)

Binary halides of Mn(VII) have not been isolated. The oxohalides MnO_3F and MnO_3Cl may be made by reacting $KMnO_4$ with HSO_3X ($X = F$ or Cl) at low temperature. Both are powerful oxidants and decompose explosively at room temperature. Both MnO_3F and MnO_3Cl have molecular (C_{3v}) structures. The oxido and imido, $[RN]^{2-}$, groups are isoelectronic, and compounds of the type $Mn(NR)_3Cl$ have been prepared by reacting a complex of $MnCl_3$ with $RNH(SiMe_3)$. The chlorido ligand in $Mn(NR)_3Cl$ can be substituted by a range of anions (Fig. 21.18).

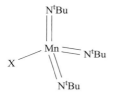

e.g. $X = Cl$, O_2CMe, OC_6F_5

(a)

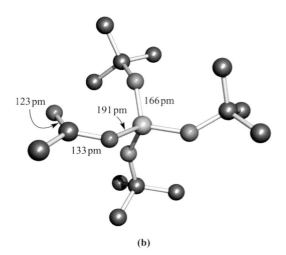

(b)

Fig. 21.18 (a) Examples of manganese(VII) imido complexes. (b) The structure of $[Mn(N^tBu)_3(O_2CMe)]$ determined by X-ray diffraction [A.A. Danopoulos *et al.* (1994) *J. Chem. Soc., Dalton Trans.*, p. 1037]. Hydrogen atoms are omitted for clarity; colour code: Mn, orange; N, blue; O, red; C, grey.

APPLICATIONS

Box 21.4 KMnO₄: a powerful oxidant at work

About 0.05 Mt per year of KMnO$_4$ are manufactured worldwide. Although this amount does not compete with those of inorganic chemicals such as CaO, NH$_3$, TiO$_2$ and the major mineral acids, the role of KMnO$_4$ as an oxidizing agent is nonetheless extremely important. The photograph shows the vigorous reaction that occurs when propan-1,2,3-triol (glycerol) is dripped onto crystals of KMnO$_4$:

$$4 \; \text{glycerol} + 14KMnO_4 \longrightarrow 7K_2CO_3 + 7Mn_2O_3 + 5CO_2 + 16H_2O$$

In addition to oxidations of organic compounds in industrial manufacturing processes, KMnO$_4$ is used in water purification where it is preferable to Cl$_2$ for two reasons: it does not affect the taste of the water, and MnO$_2$ (produced on reduction) is a coagulant for particulate impurities. The oxidizing power of KMnO$_4$ is also applied to the removal of impurities, for example in the purification of MeOH, EtOH, MeCO$_2$H and NC(CH$_2$)$_4$CN (a precursor in nylon manufacturing). Some commercial bleaching processes use KMnO$_4$, e.g. bleaching some cotton fabrics, jute fibres and beeswax.

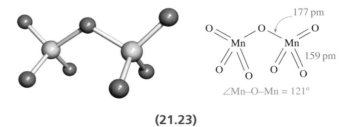

The reaction of KMnO₄ with propan-1,2,3-triol (glycerol).

Manganese(VII) chemistry is dominated by the manganate(VII) ion (permanganate). The potassium salt, KMnO$_4$, is a strong oxidizing agent and is corrosive to human tissue. It is manufactured on a large scale (see Box 21.4) by conversion of MnO$_2$ to K$_2$MnO$_4$ followed by electrolytic oxidation. In analytical chemistry, Mn determination involves oxidation of Mn(II) to [MnO$_4$]$^-$ by bismuthate, periodate or peroxydisulfate. Solid KMnO$_4$ forms dark purple-black crystals and is isostructural with KClO$_4$. Tetrahedral [MnO$_4$]$^-$ ions have equivalent bonds (Mn−O = 163 pm). Aqueous solutions of KMnO$_4$ deposit MnO$_2$ on standing. Although KMnO$_4$ is insoluble in benzene, the addition of the cyclic ether 18-crown-6 results in the formation of the soluble [K(18-crown-6)][MnO$_4$] (see Section 11.8). Potassium permanganate is intensely coloured owing to ligand-to-metal charge transfer (see Fig. 20.17). It also shows weak temperature-independent paramagnetism arising from the coupling of the diamagnetic ground state of [MnO$_4$]$^-$ with paramagnetic excited states under the influence of a magnetic field.

The free acid HMnO$_4$ can be obtained by low-temperature evaporation of its aqueous solution (made by ion exchange). It is a violent oxidizing agent and explodes above 273 K. The anhydride of HMnO$_4$ is Mn$_2$O$_7$, made by the action of concentrated H$_2$SO$_4$ on pure KMnO$_4$. It is a green, hygroscopic, highly explosive liquid, unstable above 263 K (eq. 21.41) and has molecular structure **21.23**.

$$2Mn_2O_7 \xrightarrow{>263\,K} 4MnO_2 + 3O_2 \qquad (21.41)$$

∠Mn–O–Mn = 121°

(21.23)

Equations 21.42–21.44 show reductions of [MnO$_4$]$^-$ to Mn(VI), Mn(IV) and Mn(II) respectively.

$$[MnO_4]^-(aq) + e^- \rightleftharpoons [MnO_4]^{2-}(aq)$$
$$E^o = +0.56\,V \qquad (21.42)$$

$$[MnO_4]^-(aq) + 4H^+ + 3e^- \rightleftharpoons MnO_2(s) + 2H_2O$$
$$E^o = +1.69\,V \qquad (21.43)$$

$$[MnO_4]^-(aq) + 8H^+ + 5e^- \rightleftharpoons Mn^{2+}(aq) + 4H_2O$$
$$E^o = +1.51\,V \qquad (21.44)$$

The H^+ concentration plays an important part in influencing which reduction takes place (see Section 8.2). Although many reactions of $KMnO_4$ can be understood by considering redox potentials, kinetic factors are also important. Permanganate at pH 0 should oxidize water, but in practice the reaction is extremely slow. It should also oxidize $[C_2O_4]^{2-}$ at room temperature, but reaction 21.45 is very slow unless Mn^{2+} is added or the temperature is raised.

$$2[MnO_4]^- + 16H^+ + 5[C_2O_4]^{2-}$$
$$\longrightarrow 2Mn^{2+} + 8H_2O + 10CO_2 \quad (21.45)$$

Many studies have been made on the mechanism of such reactions and, as in oxidations by $[Cr_2O_7]^{2-}$, it has been shown that intermediate oxidation states are involved.

Manganese(VI)

No binary halides of Mn(VI) have been isolated, and the only oxohalide is MnO_2Cl_2 (**21.24**). It is prepared by reducing $KMnO_4$ with SO_2 at low temperature in HSO_3Cl, and is a brown liquid which readily hydrolyses and decomposes at 240 K.

(21.24)

Salts of dark green $[MnO_4]^{2-}$ are made by fusing MnO_2 with group 1 metal hydroxides in the presence of air, or by reaction 21.46. This oxidation may be reversed by reaction 21.47.

$$4[MnO_4]^- + 4[OH]^- \longrightarrow 4[MnO_4]^{2-} + 2H_2O + O_2$$
$$(21.46)$$

$$2[MnO_4]^{2-} + Cl_2 \longrightarrow 2[MnO_4]^- + 2Cl^- \qquad (21.47)$$

Manganate(VI) is unstable with respect to disproportionation (eq. 21.48) in the presence of even weak acids such as H_2CO_3 and is therefore not formed in the reduction of acidified $[MnO_4]^-$.

$$3[MnO_4]^{2-} + 4H^+ \longrightarrow 2[MnO_4]^- + MnO_2 + 2H_2O$$
$$(21.48)$$

The $[MnO_4]^{2-}$ ion is tetrahedral ($Mn-O = 166$ pm), and K_2MnO_4 is isomorphous with K_2CrO_4 and K_2SO_4. At

298 K, the magnetic moment of K_2MnO_4 is $1.75\,\mu_B$ (d^1). The tetrahedral anion $[Mn(N^tBu)_4]^{2-}$ (an imido analogue of $[MnO_4]^{2-}$) is made by treating $Mn(N^tBu)_3Cl$ with $Li[NH^tBu]$.

Manganese(V)

Although studies of the MnF_3/F_2 system indicate the existence of MnF_5 in the gas phase, binary halides of Mn(V) have not been isolated. The only oxohalide is $MnOCl_3$ (**21.25**) which is made by reacting $KMnO_4$ with $CHCl_3$ in HSO_3Cl. Above 273 K, $MnOCl_3$ decomposes, and in moist air, it hydrolyses to $[MnO_4]^{3-}$. Salts of $[MnO_4]^{3-}$ are blue and moisture-sensitive; the most accessible are $K_3[MnO_4]$ and $Na_3[MnO_4]$, made by reduction of $[MnO_4]^-$ in concentrated aqueous KOH or NaOH at 273 K. Solutions of $[MnO_4]^{3-}$ must be strongly alkaline to prevent disproportionation which occurs readily in weakly alkaline (eq. 21.49) or acidic (eq. 21.50) media.

(21.25)

$$2[MnO_4]^{3-} + 2H_2O \longrightarrow [MnO_4]^{2-} + MnO_2 + 4[OH]^-$$
$$(21.49)$$

$$3[MnO_4]^{3-} + 8H^+ \longrightarrow [MnO_4]^- + 2MnO_2 + 4H_2O \quad (21.50)$$

The tetrahedral structure of $[MnO_4]^{3-}$ has been confirmed in the solid state in $Na_{10}Li_2(MnO_4)_4$. The Mn−O bonds are longer (170 pm) than in manganate(VI) or manganate(VII). Magnetic moments of $[MnO_4]^{3-}$ salts are typically $\approx 2.8\,\mu_B$.

1. Values of Δ_{tet} for $[MnO_4]^{3-}$, $[MnO_4]^{2-}$ and $[MnO_4]^-$ have been estimated from electronic absorption spectroscopic data to be 11 000, 19 000 and 26 000 cm^{-1} respectively. Comment on this trend.

   [*Ans.* See discussion of trends in Table 20.2]

2. Values of μ_{eff} for K_2MnO_4 and K_3MnO_4 are 1.75 and $2.80\,\mu_B$ (298 K) respectively, while $KMnO_4$ is diamagnetic. Rationalize these observations.

   [*Ans.* Relate to d^n configuration; see Table 20.11]

3. Explain why $KMnO_4$ is intensely coloured, whereas $KTcO_4$ and $KReO_4$ are colourless.

   [*Ans.* See Section 20.7]

Manganese(IV)

The only binary halide of Mn(IV) is MnF_4, prepared from the elements. It is an unstable blue solid which decomposes at ambient temperatures (eq. 21.51). Crystalline MnF_4 is dimorphic. The building blocks in α-MnF_4 are tetramers like those in VF_4 and β-CrF_4 (**21.14**). However, in these three metal fluorides, the assembly of the tetramers differs and in α-MnF_4, they are linked to give a 3-dimensional network.

$$2MnF_4 \longrightarrow 2MnF_3 + F_2 \qquad (21.51)$$

Manganese(IV) oxide is polymorphic and often non-stoichiometric. Only the high-temperature β-form has the stoichiometry MnO_2 and adopts a rutile structure (Fig. 6.22). It acts as an oxidizing agent when heated with concentrated acids (e.g. reaction 21.52).

$$MnO_2 + 4HCl \xrightarrow[\text{conc}]{\Delta} MnCl_2 + Cl_2 + 2H_2O \qquad (21.52)$$

Hydrated forms of MnO_2 are extremely insoluble and are often obtained as dark black-brown precipitates in redox reactions involving $[MnO_4]^-$ (eq. 21.43) when the $[H^+]$ is insufficient to allow reduction to Mn^{2+}.

The reaction of Mn_2O_3 with $CaCO_3$ at 1400 K yields Ca_2MnO_4, which formally contains $[MnO_4]^{4-}$. However, Ca_2MnO_4 crystallizes with a layer structure in which each Mn(IV) centre is in an octahedral MnO_6 environment; isolated $[MnO_4]^{4-}$ ions are not present.

The coordination chemistry of Mn(IV) is limited. Mononuclear complexes include $[Mn(CN)_6]^{2-}$ and $[MnF_6]^{2-}$. The cyanido complex is made by oxidizing $[Mn(CN)_6]^{3-}$ and has a magnetic moment of 3.94 μ_B. Salts of $[MnF_6]^{2-}$ also have values of μ_{eff} close to the spin-only value of 3.87 μ_B. $[MnF_6]^{2-}$ is prepared by fluorinating mixtures of chlorides or by reducing $[MnO_4]^-$ with H_2O_2 in aqueous HF. Reaction 21.53 shows the first viable non-electrolytic method of producing F_2.

$$K_2[MnF_6] + 2SbF_5 \xrightarrow{\Delta} MnF_2 + 2K[SbF_6] + F_2 \quad (21.53)$$

The structure of K_2MnF_6 is a prototype for some AB_2X_6 systems (e.g. Cs_2FeF_6 and K_2PdF_6). It is best considered as a close-packed array of K^+ and F^- ions in an alternating cubic–hexagonal sequence. The Mn^{4+} centres occupy some of the octahedral holes such that they are surrounded by six F^- ions giving $[MnF_6]^{2-}$ ions present in the lattice. Closely related structure types are K_2GeF_6 and K_2PtCl_6 in which the K^+ and X^- ions in each compound form hcp or ccp arrays respectively.[†]

[†] For detailed descriptions of these structure types, see A.F. Wells (1984) *Structural Inorganic Chemistry* 5th edn, OUP, Oxford, p. 458.

Self-study exercises

1. Calculate μ(spin-only) for $[Mn(CN)_6]^{2-}$.

   [*Ans.* 3.87 μ_B]

2. Explain why orbital contributions to the magnetic moments of $[MnF_6]^{2-}$ and $[Mn(CN)_6]^{2-}$ are not important.

   [*Ans.* Electronic configuration $t_{2g}{}^3$; see Section 20.10]

3. In the electronic absorption spectrum of $[Mn(CN)_6]^{2-}$, one might expect to see three absorptions arising from spin-allowed transitions. What would be the assignments of these transitions?

   [*Ans.* See Fig. 20.20 and discussion]

The enzyme Photosystem II (PSII) is responsible for the conversion of H_2O to O_2 during photosynthesis. Reaction 21.54 takes place in the oxygen evolving centre (OEC) in PSII, the active site consisting of a cubane-like Mn_3CaO_4-unit linked to an Mn spike (Fig. 21.19a). The structure was elucidated in 2004 through an X-ray diffraction study of PSII isolated from the cyanobacterium *Thermosynechococcus elongatus*.[‡]

$$2H_2O \longrightarrow O_2 + 4H^+ + 4e^- \qquad (21.54)$$

Electron transfer involves the four Mn centres undergoing a sequence of redox steps, the fully oxidized and reduced states being $\{Mn^{IV}{}_3Mn^{III}\}$ and $\{Mn^{III}{}_3Mn^{II}\}$ respectively. The proposed catalytic cycle (called a Kok cycle) by which reaction 21.54 is achieved is shown below, where each of the intermediates S_0 to S_4 represents the Mn_4-unit in different oxidation states:

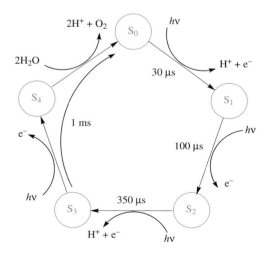

[‡] See: K.N. Ferreira, T.M. Iverson, K. Maghlaoui, J. Barber and S. Iwata (2004) *Science*, vol. 303, p. 1831; J. Barber (2008) *Inorg. Chem.*, vol. 47, p. 1700.

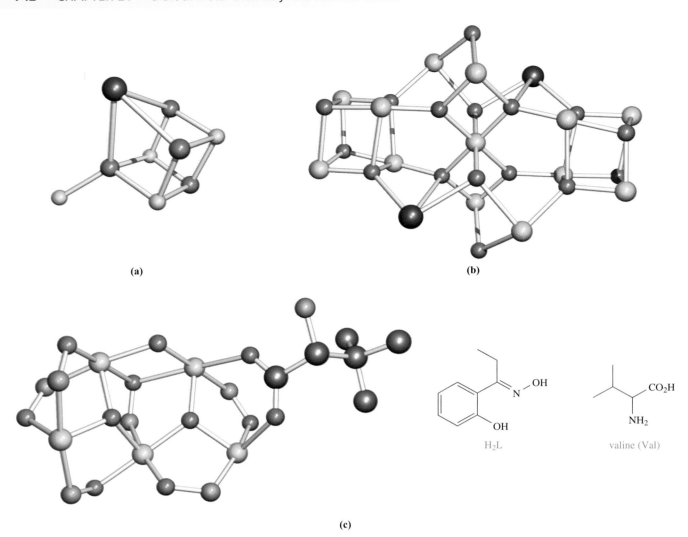

(a)

(b)

(c)

H_2L

valine (Val)

Fig. 21.19 (a) The structure (X-ray diffraction) of the cubane-like Mn_4CaO_4 active site in the oxygen evolving centre (OEC) in Photosystem II [K.N. Ferreira *et al.* (2004) *Science*, vol. 303, p. 1831]. High oxidation state manganese-containing cluster model compounds: the structures (X-ray diffraction) of (b) the $\{Mn_{13}Ca_2O_{16}\}$-core in $[Mn_{13}Ca_2O_{10}(OH)_2(OMe)_2(O_2CPh)_{18}(OH_2)_4]\cdot10MeCN$ [A. Mishra *et al.* (2005) *Chem. Commun.*, p. 54], and (c) the $\{Mn_5(\mu_3-O)_2(\mu-O)_2(\mu-NO)_6(Val)\}$-core in $[Mn_5O_2(OMe)L_6(Val)]\cdot Val\cdot1.5H_2O$ [C. Kozoni *et al.* (2009) *Dalton Trans.*, p. 9117] and the structures of the H_2L and amino acid ligands in the complex. Colour code: Mn, yellow; O, red; Ca, dark blue; N, blue; C, grey.

A variety of discrete molecular complexes has been developed to study the chemistry of the water-splitting centre in PSII. Two examples are shown in Fig. 21.19. The first is $[Mn_{13}Ca_2O_{10}(OH)_2(OMe)_2(O_2CPh)_{18}(OH_2)_4]$ which contains a $\{Mn_{13}Ca_2O_{16}\}$-core comprising two linked cubane units (Fig. 21.19b). In PSII, the $\{Mn_4CaO_4\}$ active site is linked to the protein backbone of the enzyme through amino acid residues (aspartic acid, glutamic acid and histidine). Figure 21.19c shows the cluster core of $[Mn_5O_2(OMe)L_6(Val)]$ which models the interaction between an amino acid (valine) and an $[Mn^{IV}_2 Mn^{III}_3]$-unit.

Manganese(III)

The only binary halide of Mn(III) is the red-purple MnF_3 which is made by the action of F_2 on Mn(II) halides at 520 K. It is thermally stable but is immediately hydrolysed by water. The solid state structure of MnF_3 is related to those of TiF_3, VF_3, CrF_3, FeF_3 and CoF_3 but is Jahn–Teller distorted (high-spin d^4). In MnF_3 there are *three* pairs of Mn–F distances (179, 191 and 209 pm) rather than the distortions shown in structures **20.5** and **20.6**. At room temperature, the magnetic moment of MnF_3 is 4.94 μ_B, but on cooling, MnF_3 becomes antiferromagnetic ($T_N = 43$ K) (see Section 20.10).

The black oxide Mn_2O_3 (the α-form) is obtained when MnO_2 is heated at 1070 K or (in the hydrous form) by oxidation of Mn(II) in alkaline media. At higher temperatures, it forms Mn_3O_4, a normal spinel ($Mn^{II}Mn^{III}_2O_4$, see Box 13.7) but with the Mn(III) centres being Jahn–Teller distorted. The Mn atoms in α-Mn_2O_3 are in distorted octahedral MnO_6 sites (elongated, diagram **20.5**). The structure differs from the corundum structure adopted by Ti_2O_3, V_2O_3 and Cr_2O_3. Whereas Mn_2O_3 is *antiferromagnetic* below 80 K, Mn_3O_4 is *ferrimagnetic* below 43 K.

Most complexes of Mn(III) are octahedral, high-spin d^4 and are Jahn–Teller distorted. The red aqua ion $[Mn(OH_2)_6]^{3+}$ can be obtained by electrolytic oxidation of aqueous Mn^{2+} and is present in the alum $CsMn(SO_4)_2 \cdot 12H_2O$. Surprisingly, the $[Mn(OH_2)_6]^{3+}$ ion shows no Jahn–Teller distortion, at least down to 78 K. In aqueous solution, $[Mn(OH_2)_6]^{3+}$ is appreciably hydrolysed (see Section 7.7) and polymeric cations are present. It is also unstable with respect to disproportionation (eq. 21.55) as expected from the potentials in Figs. 8.2 and 8.3; it is less unstable in the presence of high concentrations of Mn^{2+} or H^+ ions.

$$2Mn^{3+} + 2H_2O \longrightarrow Mn^{2+} + MnO_2 + 4H^+ \qquad (21.55)$$

The Mn^{3+} ion is stabilized by hard ligands including F^-, $[PO_4]^{3-}$, $[SO_4]^{2-}$ or $[C_2O_4]^{2-}$. The pink colour sometimes seen before the end of the permanganate–oxalate titration (eq. 21.45) is due to an oxalato complex of Mn(III). The salt $Na_3[MnF_6]$ is made by heating NaF with MnF_3, and reaction of MnO_2 with KHF_2 in aqueous HF gives $K_3[MnF_6]$. Both salts are violet and have magnetic moments of 4.9 μ_B (298 K), consistent with the spin-only value for high-spin d^4. Reaction of NaF with MnF_3 in aqueous HF yields pink $Na_2[MnF_5]$ which contains chains of distorted octahedral Mn(III) centres (**21.26**) in the solid state. Salts of $[MnF_4]^-$ also crystallize with the Mn centres in Jahn–Teller distorted octahedral sites, e.g. $CsMnF_4$ has a layer structure (**21.27**). However, in salts of $[MnCl_5]^{2-}$ for which solid state data are available, discrete square-based pyramidal anions are present. Contrasting structures are also observed in the related complexes $[Mn(N_3)(acac)_2]$ and $[Mn(NCS-N)(acac)_2]$; whereas the azido ligand presents two nitrogen donors to adjacent Mn(III) centres to produce a chain polymer, the thiocyanate ligand binds only through the hard *N*-donor leaving the soft *S*-donor uncoordinated (Fig. 21.20). The complex $[Mn(acac)_3]$ (obtained from $MnCl_2$ and $[acac]^-$ followed by oxidation with $KMnO_4$) is also of structural interest. It is dimorphic, crystallizing in one form with an elongated octahedral coordination sphere (**20.5**) while in the other, it is compressed (**20.6**).

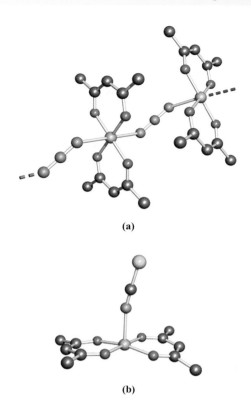

(a)

(b)

Fig. 21.20 The structures (X-ray diffraction) of the Mn(III) complexes (a) $[Mn(N_3)(acac)_2]$ which forms polymeric chains [B.R. Stults *et al.* (1975) *Inorg. Chem.*, vol. 14, p. 722] and (b) $[Mn(NCS-N)(acac)_2]$ [B.R. Stults *et al.* (1979) *Inorg. Chem.*, vol. 18, p. 1847]. Hydrogen atoms are omitted for clarity; colour code: Mn, orange; C, grey; O, red; N, blue; S, yellow.

The only well-known low-spin complex of Mn(III) is the dark red $K_3[Mn(CN)_6]$, made from KCN and $K_3[MnF_6]$ or by oxidation of $K_4[Mn(CN)_6]$ using 3% H_2O_2. As expected for low-spin d^4, $[Mn(CN)_6]^{3-}$ has a regular octahedral structure (Mn$-$C = 198 pm).

Self-study exercises

1. Explain why $[MnF_6]^{3-}$ is Jahn–Teller distorted, but $[Mn(CN)_6]^{3-}$ is not.

   [*Ans.* See structures **20.5** and **20.6** and discussion]

2. Write down expressions for the CFSE of high- and low-spin octahedral Mn^{3+} in terms of Δ_{oct} and the pairing energy, *P*.   [*Ans.* See Table 20.3]

3. Green solutions of $[Mn(OH_2)_6]^{3+}$ contain $[Mn(OH_2)_5(OH)]^{2+}$ and $[Mn_2(OH_2)_8(\mu\text{-}OH)_2]^{4+}$. Explain how these species arise, and include equations for appropriate equilibria. How might $[Mn(OH_2)_6]^{3+}$ be stabilized in aqueous solution?

   [*Ans.* See Section 7.7]

(21.26)   **(21.27)**

Manganese(II)

Manganese(II) salts are obtained from MnO_2 by a variety of methods. The soluble $MnCl_2$ and $MnSO_4$ result from heating MnO_2 with the appropriate concentrated acid (eqs. 21.52 and 21.56). The sulfate is commercially made by this route (MnO_2 being supplied as the mineral pyrolusite) and is commonly encountered as the hydrate $MnSO_4 \cdot 5H_2O$.

$$2MnO_2 + 2H_2SO_4 \xrightarrow{\ \Delta\ } 2MnSO_4 + O_2 + 2H_2O \qquad (21.56)$$
$$\text{conc}$$

Insoluble $MnCO_3$ is obtained by precipitation from solutions containing Mn^{2+}; however, the carbonate so obtained contains hydroxide. Pure $MnCO_3$ can be made by reaction of manganese(II) acetate or hydroxide with supercritical CO_2 (see Section 9.13).

Manganese(II) salts are characteristically very pale pink or colourless. For the d^5 Mn^{2+} ion in an octahedral high-spin complex, '*d–d*' transitions are both spin- and Laporte-forbidden (see Section 20.7). Although the electronic absorption spectrum of $[Mn(OH_2)_6]^{2+}$ does contain several absorptions, they are all weaker by a factor of $\approx 10^2$ than those arising from spin-allowed transitions of other first row metal ions. The weak absorptions observed for Mn^{2+} arise from promotion of an electron to give various excited states containing only three unpaired electrons.

All four halides of Mn(II) are known. Hydrates of MnF_2 and $MnBr_2$ are prepared from $MnCO_3$ and aqueous HF or HBr and the anhydrous salts are then obtained by dehydration. The chloride is prepared by reaction 21.52, and MnI_2 results from direct combination of the elements. The fluoride adopts a rutile structure (Fig. 6.22) in the solid state, while $MnCl_2$, $MnBr_2$ and MnI_2 possess the CdI_2 layer structure (Fig. 6.23).

The reduction of a higher oxide of manganese (e.g. MnO_2 or Mn_2O_3) with H_2 at elevated temperature gives MnO, which is also obtained by thermal decomposition of manganese(II) oxalate. Green MnO adopts an NaCl structure and its anti-ferromagnetic behaviour was discussed in Section 20.10. The conductivity of metal(II) oxides is described in Section 28.2. Manganese(II) oxide is a basic oxide, insoluble in water but dissolving in acids to give pale pink solutions containing $[Mn(OH_2)_6]^{2+}$. The oxidation of Mn(II) compounds in acidic solution requires a powerful oxidant such as periodate, but in alkaline media, oxidation is easier because hydrous Mn_2O_3 is far less soluble than $Mn(OH)_2$. Thus, when alkali is added to a solution of a Mn(II) salt in the presence of air, the white precipitate of $Mn(OH)_2$ that initially forms rapidly darkens owing to atmospheric oxidation.

Large numbers of Mn(II) complexes exist. This oxidation state is stable with respect to both oxidation and reduction

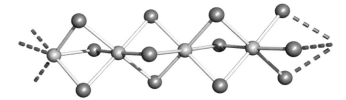

Fig. 21.21 Part of one of the infinite chains of face-sharing octahedra present in the lattice of $[Me_2NH_2][MnCl_3]$; the structure was determined by X-ray diffraction [R.E. Caputo *et al.* (1976) *Phys. Rev. B*, vol. 13, p. 3956]. Colour code: Mn, orange; Cl, green.

(Fig. 8.3), and in high-spin complexes. The lack of any LFSE means that Mn^{2+} does not favour a particular arrangement of ligand donor atoms. Manganese(II) halides form a range of complexes. Reaction of MnF_2 with MF (e.g. M = Na, K, Rb) gives $M[MnF_3]$ salts which adopt the perovskite structure (Fig. 6.24); discrete $[MnF_3]^-$ ions are not present. Heating a 1:2 ratio of $MnF_2:KF$ at 950 K gives $K_2[MnF_4]$ which has an extended structure containing MnF_6 octahedra connected by Mn–F–Mn bridges. Discrete anions are, again, *not* present in salts of $[MnCl_3]^-$, e.g. $[Me_2NH_2][MnCl_3]$ crystallizes with infinite chains of face-sharing $MnCl_6$ octahedra (Fig. 21.21). Structural determinations for several compounds which appear to be salts of $[MnCl_5]^{3-}$ reveal significant cation-dependence. The green-yellow Cs_3MnCl_5 contains discrete tetrahedral $[MnCl_4]^{2-}$ and Cl^- ions, whereas pink $[(H_3NCH_2CH_2)_2NH_2][MnCl_5]$ has an extended structure containing corner-sharing $MnCl_6$ octahedra. The salt $K_4[MnCl_6]$ contains discrete octahedral anions, and in green-yellow $[Et_4N]_2[MnCl_4]$ and $[PhMe_2(PhCH_2)N]_2[MnCl_4]$, isolated tetrahedral anions are present. The presence of the tetrahedral $[MnCl_4]^{2-}$ ion leads to complexes that are rather more intensely coloured than those containing related octahedral species (see Section 20.7). Tetrahedral $[Mn(CN)_4]^{2-}$ (a rare example of a high-spin cyanido complex) results from the photoinduced, reductive decomposition of $[Mn(CN)_6]^{2-}$. As a solid, the yellow salt $[N(PPh_3)_2]_2[Mn(CN)_4]$ is fairly stable in air. It is also stable in dry, aprotic solvents (e.g. MeCN), but hydrolyses in protic solvents.

The reactions of $MnCl_2$, $MnBr_2$ and MnI_2 with, for example, *N*-, *O*-, *P*- or *S*-donor ligands have led to the isolation of a wide variety of complexes. A range of coordination geometries is observed as the following examples show ($H_2pc = \mathbf{21.28}$; tpy $= \mathbf{21.29}$; Hpz $= \mathbf{21.30}$:

- tetrahedral: $[MnCl_2(OPPh_3)_2]$, $[Mn(N_3)_4]^{2-}$, $[Mn(Se_4)_2]^{2-}$;
- square planar: $[Mn(pc)]$;
- trigonal bipyramidal: $[MnBr_2\{OC(NHMe)_2\}_3]$, $[MnBr\{N(CH_2CH_2NMe_2)_3\}]^+$, $[MnI_2(THF)_3]$;

- octahedral: *trans*-[MnBr$_2$(Hpz)$_4$], *cis*-[Mn(bpy)$_2$(NCS-N)$_2$], *cis*-[MnCl$_2$(HOCH$_2$CH$_2$OH)$_2$], [MnI(THF)$_5$]$^+$, *mer*-[MnCl$_3$(OH$_2$)$_3$]$^-$, [Mn(tpy)$_2$]$^{2+}$, [Mn(EDTA)]$^{2-}$;
- 7-coordinate: [Mn(EDTA)(OH$_2$)]$^{2-}$, *trans*-[Mn(**21.31**)(OH$_2$)$_2$]$^{2+}$;
- square-antiprism: [Mn(**21.32**)$_2$]$^{2+}$;
- dodecahedral: [Mn(NO$_3$-*O,O'*)$_4$]$^{2-}$.

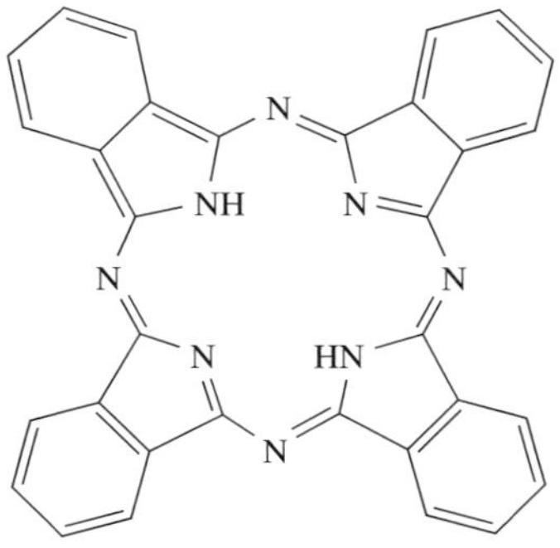

H$_2$pc = phthalocyanine

(21.28)

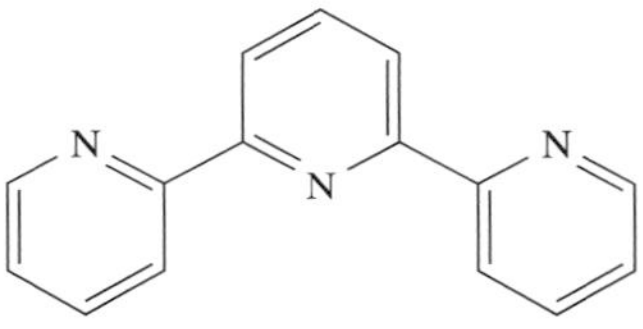

tpy = 2,2':6',2"-terpyridine
(free ligand has the *trans,trans*-configuration shown; coordinated ligand has a *cis,cis*-configuration)

(21.29)

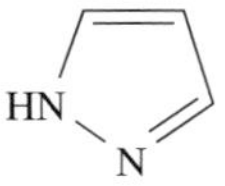

Hpz = 1*H*-pyrazole   15-crown-5   12-crown-4

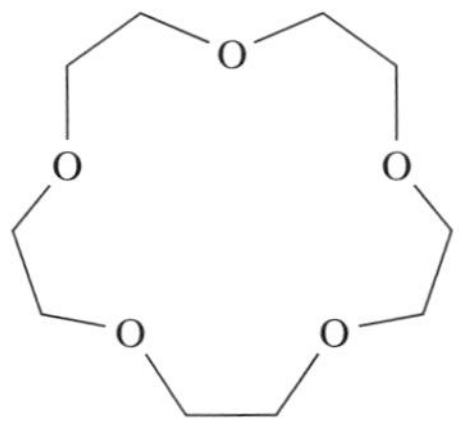

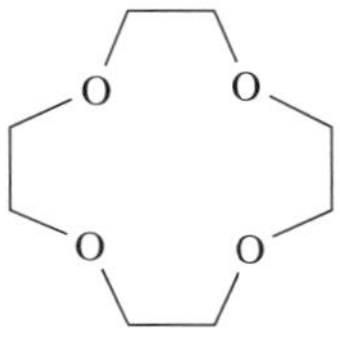

(21.30)   **(21.31)**   **(21.32)**

The only common low-spin complex of Mn(II) is the blue, efflorescent K$_4$[Mn(CN)$_6$]·3H$_2$O ($\mu_{eff} = 2.18 \mu_B$) which is prepared in aqueous solution from MnCO$_3$ and KCN. Conversion of K$_4$[Mn(CN)$_6$] to K$_3$[Mn(CN)$_6$] occurs readily, the presence of the cyanido ligands significantly destabilizing Mn(II) with respect to Mn(III) (see Section 8.3).

Efflorescence is the loss of water from a hydrated salt.

Manganese(I)

Manganese(I) is typically stabilized by π-acceptor ligands in organometallic derivatives, but several compounds deserve a mention here. When Mn powder is dissolved in air-free aqueous NaCN, the Mn(I) complex Na$_5$[Mn(CN)$_6$] is formed, the oxidizing agent being water. The low-spin d^6 [Mn(CN)$_6$]$^{5-}$ ion can also be made by reducing [Mn(CN)$_6$]$^{4-}$ with Na or K amalgam, again in the absence of O$_2$. The [Mn(OH$_2$)$_3$(CO)$_3$]$^+$ ion is the first example of a mixed aqua/carbonyl complex containing a first row *d*-block metal. It is an analogue of [Tc(OH$_2$)$_3$(CO)$_3$]$^+$, *in vivo* applications of which are described in Box 22.7. Evidence for [Mn(OH$_2$)$_3$(CO)$_3$]$^+$ existing as the *fac*-isomer comes from the ν(CO) region of the IR spectrum; the observation of two absorptions (2051 and 1944 cm^{-1}, assigned to the A_1 and E vibrational modes, respectively) is consistent with C_{3v} symmetry.

In Chapter 23, we describe many examples of the use of sterically demanding ligands. Structure **21.22** and Fig. 21.17 illustrate the use of a highly bulky ligand to stabilize a Cr–Cr quintuple bond. Reaction 21.57 shows how an Mn(I) complex is stabilized by the use of a bulky β-diketiminato ligand. The product of this reaction is the first example of a 3-coordinate, Mn(I) compound. The molecule formally contains an {Mn$_2$}$^{2+}$ core (Mn–Mn = 272 pm) and magnetic data are consistent with a rare example of a high-spin Mn(I) (d^6) complex in which there is antiferromagnetic coupling between the metal centres.

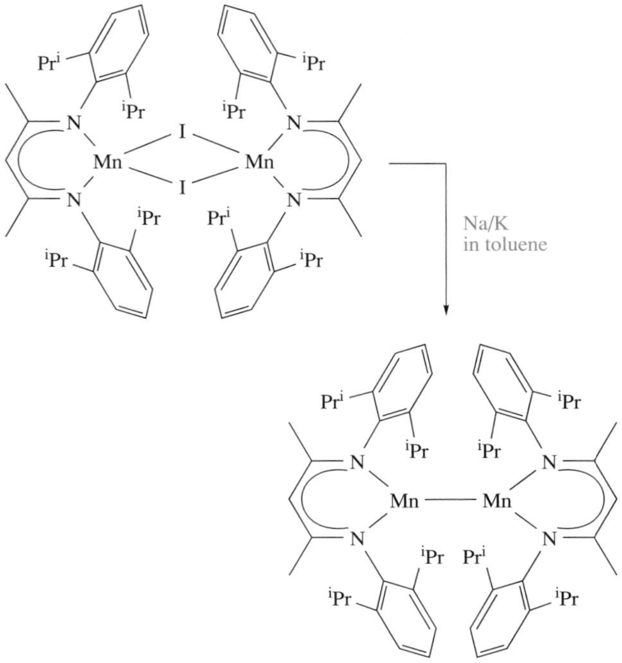

(21.57)

21.9 Group 8: iron

The metal

Finely divided Fe is pyrophoric in air, but the bulk metal oxidizes in dry air only when heated. In moist air, Fe rusts, forming a hydrated oxide $Fe_2O_3 \cdot xH_2O$. Rusting is an electrochemical process (Box 8.4 and eq. 21.58) and occurs only in the presence of O_2, H_2O and an electrolyte. The latter may be water, but is more effective if it contains dissolved SO_2 (e.g. from industrial pollution) or NaCl (e.g. from sea-spray or salt-treated roads). Diffusion of the ions formed in reaction 21.58 deposits $Fe(OH)_2$ at places between the points of attack and this is further oxidized to hydrated iron(III) oxide.

$$\left. \begin{array}{l} 2Fe \longrightarrow 2Fe^{2+} + 4e^- \\ O_2 + 2H_2O + 4e^- \longrightarrow 4[OH]^- \end{array} \right\} \quad (21.58)$$

Iron reacts with halogens at 470–570 K to give FeF_3, $FeCl_3$, $FeBr_3$ and FeI_2, respectively. The metal dissolves in dilute mineral acids to yield Fe(II) salts, but concentrated HNO_3 and other powerful oxidizing agents make it passive. Iron metal is unaffected by alkalis. When powdered iron and sulfur are heated together, FeS is produced. The formation of iron carbides and alloys is crucial to the steel industry (see Boxes 6.1 and 6.2 and Section 6.7).

Most of the chemistry of Fe involves Fe(II) or Fe(III), with Fe(IV) and Fe(VI) known in a small number of compounds; Fe(V) is rare. Lower formal oxidation states occur with π-acceptor ligands (see Chapter 24).

Iron(VI), iron(V) and iron(IV)

In iron chemistry, Mössbauer spectroscopy is widely used to gain information about the oxidation state and/or spin state of the Fe centres (see Section 4.10 and Fig. 20.31). The highest oxidation states of iron are found in compounds of $[FeO_4]^{2-}$, $[FeO_4]^{3-}$, $[FeO_4]^{4-}$ and $[FeO_3]^{2-}$ although these free ions are not necessarily present. Salts of $[FeO_4]^{2-}$ can be made by hypochlorite oxidation of Fe(III) salts in the presence of alkali (eq. 21.59). They contain discrete tetrahedral ions and are paramagnetic with magnetic moments corresponding to two unpaired electrons. The Na^+ and K^+ salts are deep red-purple and are soluble in water; aqueous solutions decompose (eq. 21.60) but alkaline solutions are stable. Ferrate(VI) is a powerful oxidant (eq. 21.61).

$$Fe_2O_3 + 3[OCl]^- + 4[OH]^- \longrightarrow 2[FeO_4]^{2-} + 3Cl^- + 2H_2O \quad (21.59)$$

$$4[FeO_4]^{2-} + 6H_2O \longrightarrow 4FeO(OH) + 8[OH]^- + 3O_2 \quad (21.60)$$

$$[FeO_4]^{2-} + 8H^+ + 3e^- \rightleftharpoons Fe^{3+} + 4H_2O \quad E^o = +2.20\,V \quad (21.61)$$

The reaction of K_2FeO_4 with KOH in O_2 at 1000 K gives K_3FeO_4, a rare example of an Fe(V) salt.

Iron(IV) ferrates include Na_4FeO_4 (made from Na_2O_2 and $FeSO_4$), Sr_2FeO_4 (prepared by heating Fe_2O_3 and SrO in the presence of O_2) and Ba_2FeO_4 (made from BaO_2 and $FeSO_4$). Na_4FeO_4 and Ba_2FeO_4 contain discrete $[FeO_4]^{4-}$ ions. The high-spin d^4 configuration of Fe(IV) in $[FeO_4]^{4-}$ leads to a Jahn–Teller distortion, reducing the symmetry from T_d to approximately D_{2d} (structure **21.33** with structural data for the Na^+ salt).

average Fe–O = 181 pm

(21.33)

In aqueous solution Na_4FeO_4 disproportionates (eq. 21.62).

$$3Na_4FeO_4 + 5H_2O \longrightarrow Na_2FeO_4 + Fe_2O_3 + 10NaOH \quad (21.62)$$

Compounds formally containing $[FeO_3]^{2-}$ are actually mixed metal oxides; $CaFeO_3$, $SrFeO_3$ and $BaFeO_3$ crystallize with the perovskite structure (Fig. 6.24).

Attempts to stabilize Fe in high oxidation states using fluorido ligands have met with limited success. The reaction of Cs_2FeO_4 with F_2 (40 bar, 420 K) gives Cs_2FeF_6 along with $CsFeF_4$ and Cs_3FeF_6. In the solid state, Cs_2FeF_6 adopts a K_2MnF_6 structure (see Section 21.8, Mn(IV)). There is current interest in the coordination chemistry of Fe(IV) since Fe(IV) intermediates may be present in bioinorganic processes involving cytochromes P-450 (see Section 29.3). However, the number of Fe(IV) complexes so far isolated and structurally characterized is small. The coordination environment is octahedral or square-based pyramidal, and ligands that stabilize Fe(IV) include dithiocarbamates (Fig. 21.22), dithiolates as in $[Fe(PMe_3)_2(1,2-S_2C_6H_4)_2]$, porphyrins and phthalocyanines.

Self-study exercises

1. Explain why $[FeO_4]^{4-}$ (structure **21.33**) suffers from a Jahn–Teller distortion. The distortion is particularly strong. Is this expected?

   [*Ans.* See discussion of the tetrahedral crystal field in Section 20.3]

2. Typically, values of μ_{eff} for salts of $[FeO_4]^{2-}$ lie in the range 2.8–3.0 μ_B. Show that this is consistent with a μ(spin-only) value for tetrahedral Fe(VI) and comment on why orbital contributions to the magnetic moment are not expected.

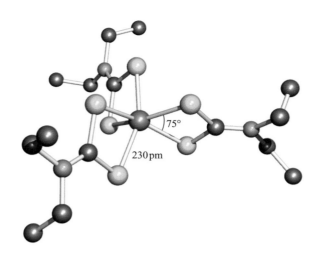

 Fig. 21.22 The structure (X-ray diffraction) of the iron(IV) complex $[Fe(S_2CNEt_2)_3]^+$ in the $[I_5]^-$ salt [C.L. Raston *et al.* (1980) *J. Chem. Soc., Dalton Trans.*, p. 1928]. Hydrogen atoms are omitted; colour code: Fe, green; S, yellow; C, grey; N, blue.

3. SrFeO$_3$ crystallizes with a perovskite structure. What are the coordination environments of Sr, Fe and O?

[*Ans.* Relate to CaTiO$_3$ in Fig. 6.24]

4. (a) The Fe(IV) compound Ba$_3$FeO$_5$ contains discrete ions in the solid state. Suggest what ions are present.
(b) Ba$_3$FeO$_5$ is paramagnetic down to 5 K. Illustrate how the molar magnetic susceptibility varies over the temperature range 5–300 K.

[*Ans.* See J.L. Delattre *et al.* (2002) *Inorg. Chem.*, vol. 41, p. 2834]

Iron(III)

The old name for iron(III) is *ferric*. Iron(III) fluoride, chloride and bromide are made by heating Fe with the halogen. The fluoride is a white, involatile solid isostructural with ScF$_3$ (Fig. 21.5). In the solid state, FeCl$_3$ adopts the BiI$_3$ structure but the gas phase (bp 588 K) contains discrete molecules, dimers below 970 K and monomers above 1020 K. Anhydrous FeCl$_3$ forms hygroscopic dark green or black crystals. It dissolves in water to give strongly acidic solutions (see below) from which the orange-brown hydrate FeCl$_3 \cdot 6H_2O$ (properly formulated as *trans*-[FeCl$_2$(OH$_2$)$_4$]Cl·2H$_2$O) can be crystallized. The trichloride is a useful precursor in Fe(III) chemistry, and both anhydrous FeCl$_3$ and FeBr$_3$ are used as Lewis acid catalysts in organic synthesis. Anhydrous FeBr$_3$ forms deliquescent, red-brown, water-soluble crystals; the solid adopts a BiI$_3$ structure, but in the gas phase, molecular dimers are present. Iron(III) iodide readily decomposes (eq. 21.63) but, under inert conditions, it can be isolated from reaction 21.64.

$$2FeI_3 \longrightarrow 2FeI_2 + I_2 \qquad (21.63)$$

$$2Fe(CO)_4I_2 + I_2 \xrightarrow{h\nu} 2FeI_3 + 8CO \qquad (21.64)$$

Iron(III) oxide exists in a number of forms. The paramagnetic α-form (a red-brown solid or grey-black crystals) occurs as the mineral *haematite* and adopts a corundum structure (see Section 13.7) with octahedrally sited Fe(III) centres. The β-form is produced by hydrolysing FeCl$_3 \cdot 6H_2O$, or by chemical vapour deposition (CVD, see Section 28.6) at 570 K from iron(III) trifluoroacetylacetonate. On annealing at 770 K, a β ⟶ α phase change occurs. The γ-form is obtained by careful oxidation of Fe$_3$O$_4$ and crystallizes with an extended

---

APPLICATIONS

Box 21.5 The super-iron battery

The MnO$_2$–Zn dry battery is a major contributor to the commercial supply of batteries. In the long-life 'alkaline' version, the lifetime of the battery is mainly dependent on the lifetime of the MnO$_2$ cathode. Prolonging the lifetimes of batteries which are used, for example, in implanted pacemakers has obvious advantages, and the use of the Fe(VI) compounds K$_2$FeO$_4$, BaFeO$_4$ and SrFeO$_4$ as cathodic materials has been investigated with promising results. The so-called 'super-iron battery' contains, for example, K$_2$FeO$_4$ as a replacement for MnO$_2$ in the alkaline dry battery. The reduction of Fe(VI) to Fe(III):

$$[FeO_4]^{2-} + \tfrac{5}{2}H_2O + 3e^- \longrightarrow \tfrac{1}{2}Fe_2O_3 + 5[OH]^-$$

provides a high-capacity source of cathodic charge and the $[FeO_4]^{2-}$-for-MnO$_2$ cathode replacement leads to an increase

in the energy capacity of the battery of more than 50%. The cell reaction of the super-iron battery is:

$$2K_2FeO_4 + 3Zn \longrightarrow Fe_2O_3 + ZnO + 2K_2ZnO_2$$

and a further advantage of the system is that it is rechargeable.

Further reading

S. Licht, B. Wang and S. Ghosh (1999) *Science*, vol. 285, p. 1039 – 'Energetic iron(VI) chemistry: The super-iron battery'.

S. Licht and R. Tel-Vered (2004) *Chem. Commun.*, p. 628 – 'Rechargeable Fe(III/VI) super-iron cathodes'.

S. Licht and X. Yu (2008) *ACS Symposium Series*, vol. 985, p. 197 – 'Recent advances in Fe(VI) charge storage and super-iron batteries'.

structure in which the O^{2-} ions adopt a ccp array and the Fe^{3+} ions randomly occupy octahedral and tetrahedral holes. γ-Fe_2O_3 is ferromagnetic and is used in magnetic data storage. Iron(III) oxide is insoluble in water but can be dissolved with difficulty in acids. Several hydrates of Fe_2O_3 exist, and when Fe(III) salts are dissolved in alkali, the red-brown gelatinous precipitate that forms is *not* $Fe(OH)_3$ but $Fe_2O_3 \cdot H_2O$ (also written as Fe(O)OH). The precipitate is soluble in acids giving $[Fe(OH_2)_6]^{3+}$, and in concentrated aqueous alkalis, $[Fe(OH)_6]^{3-}$ is present. Several forms of Fe(O)OH exist and consist of chain structures with edge-sharing FeO_6 octahedra. The minerals *goethite* and *lepidocrocite* are α- and γ-Fe(O)OH respectively.

Mixed metal oxides derived from Fe_2O_3 and of general formula $M^{II}Fe^{III}_2O_4$ or $M^IFe^{III}O_2$ are commonly known as *ferrites* despite the absence of discrete oxoanions. They include compounds of commercial importance by virtue of their magnetic properties, e.g. electromagnetic devices for information storage; for discussion of the magnetic properties of mixed metal oxides, see Chapter 28. Spinel and inverse spinel structures adopted by $M^{II}Fe^{III}_2O_4$ oxides were described in Box 13.7 and Section 20.11, e.g. $MgFe_2O_4$ and $NiFe_2O_4$ are inverse spinels while $MnFe_2O_4$ and $ZnFe_2O_4$ are normal spinels. Some oxides of the $M^IFe^{III}O_2$ type adopt structures that are related to NaCl (e.g. $LiFeO_2$, in which the Li^+ and Fe^{3+} ions occupy Na^+ sites and O^{2-} ions occupy Cl^- sites, Fig. 6.16). Among the $M^IFe^{III}O_2$ group of compounds, $CuFeO_2$ and $AgFeO_2$ are noteworthy in being semiconductors. Other ferrites exist with more complex structures: permanent magnets are made using $BaFe_{12}O_{19}$, and the *iron garnet* family includes $Y_3Fe_5O_{12}$ (yttrium iron garnet, YIG) which is used as a microwave filter in radar equipment.

When Fe_2O_3 is heated at 1670 K, it converts to black Fe_3O_4 ($Fe^{II}Fe^{III}_2O_4$) which also occurs as the mineral *magnetite*, and possesses an inverse spinel structure (see Box 13.7). Its ferrimagnetic behaviour (see Fig. 20.32) makes Fe_3O_4 commercially important, e.g. it is used in magnetic toner in photocopiers. Mixtures of Fe_3O_4 and γ-Fe_2O_3 are used in magnetic data storage, and this market competes with that of CrO_2 (see Section 21.7).

Self-study exercises

1. Spinel and inverse spinel structures are based on cubic close-packed (ccp) arrangements of O^{2-} ions. Draw a representation of a unit cell of a ccp arrangement of O^{2-} ions. How many octahedral and tetrahedral holes are there in this unit cell?   [*Ans.* See Section 6.2]

2. Refer to the diagram drawn in question 1. If half of the octahedral and one-eighth of the tetrahedral holes are filled with Fe^{3+} and Zn^{2+} ions respectively, show that the resultant oxide has the formula $ZnFe_2O_4$.

3. The inverse spinel structure of magnetite can be described as follows. Starting with a ccp arrangement of O^{2-} ions, one-quarter of the octahedral holes are filled with Fe^{3+} ions and one-quarter with Fe^{2+} ions; one-eighth of the tetrahedral holes are occupied with Fe^{3+} ions. Show that this corresponds to a formula of Fe_3O_4, and that the compound is charge-neutral.

The chemistry of Fe(III) is well researched and among many commercially available starting materials are the chloride (see above), perchlorate, sulfate and nitrate. *Hazard: Perchlorates are potentially explosive.* Anhydrous $Fe(ClO_4)_3$ is a yellow solid, but it is commercially available as a hydrate $Fe(ClO_4)_3 \cdot xH_2O$ with variable water content. The hydrate is prepared from aqueous $HClO_4$ and $Fe_2O_3 \cdot H_2O$ and, depending on contamination with chloride, may be pale violet (<0.005% chloride content) or yellow. Iron(III) sulfate (made by oxidation of $FeSO_4$ with concentrated H_2SO_4) is purchased as the hydrate $Fe_2(SO_4)_3 \cdot 5H_2O$. The nitrate is available as $Fe(NO_3)_3 \cdot 9H_2O$ (correctly formulated as $[Fe(OH_2)_6][NO_3]_3 \cdot 3H_2O$) which forms colourless or pale violet deliquescent crystals. It is made by reaction of iron oxides with concentrated HNO_3. The violet hexahydrate, $Fe(NO_3)_3 \cdot 6H_2O$ (correctly written as $[Fe(OH_2)_6][NO_3]_3$), can be obtained by reaction of $Fe_2O_3 \cdot H_2O$ with HNO_3. The octahedral $[Fe(OH_2)_6]^{3+}$ ion is also present in crystals of the violet alum $[NH_4]Fe(SO_4)_2 \cdot 12H_2O$ (see Section 13.9). These Fe(III) salts are all water-soluble, dissolving to give brown-yellow solutions due to hydrolysis of $[Fe(OH_2)_6]^{3+}$ (eqs. 7.36 and 7.37); solution species include $[(H_2O)_5FeOFe(OH_2)_5]^{4+}$ (**21.34**) which has a *linear* Fe–O–Fe bridge indicative of $(d–p)\pi$-bonding involving Fe d and O p orbitals. The structural characterization of **21.34** has been achieved by hydrogen-bonded association of this cation with the crown ether 18-crown-6 (Fig. 21.23)

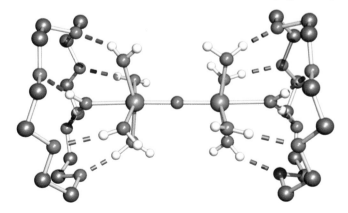

Fig. 21.23 The structure of $[(H_2O)_5Fe(\mu\text{-}O)Fe(OH_2)_5]^{4+} \cdot (18\text{-}crown\text{-}6)_2$ present in crystalline $[(H_2O)_5Fe(\mu\text{-}O)Fe(OH_2)_5][ClO_4]_4 \cdot (18\text{-}crown\text{-}6)_2 \cdot 2H_2O$. The structure was determined by X-ray diffraction at 173 K [P.C. Junk *et al.* (2002) *J. Chem. Soc., Dalton Trans.*, p. 1024]; hydrogen atoms are omitted from the crown ethers. Colour code: Fe, green; O, red; C, grey; H, white.

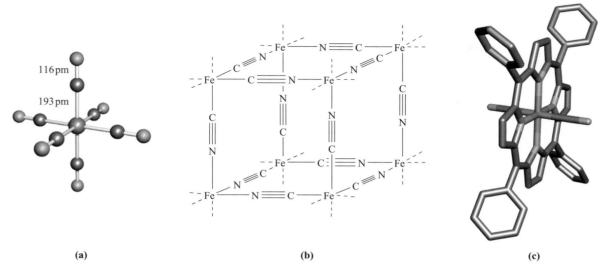

Fig. 21.24 Examples of iron(III) cyanido complexes: (a) the structure of $[Fe(CN)_6]^{3-}$ in the salt $Cs[NH_4]_2[Fe(CN)_6]$ (X-ray diffraction) [D. Babel (1982) *Z. Naturforsch.*, *Teil B*, vol. 37, p. 1534], (b) one-eighth of the unit cell of $KFe[Fe(CN)_6]$ (the K^+ ions occupy the cavities and are omitted from the figure), and (c) the structure (determined by X-ray diffraction) of $[Fe(CN)_2(TPP)]$ where $H_2TPP = 5,10,15,20$-tetraphenyl-21*H*,23*H*-porphyrin (see Fig. 12.10 for parent porphyrin) [W.R. Scheidt *et al.* (1980) *J. Am. Chem. Soc.*, vol. 102, p. 3017]. Hydrogen atoms are omitted from (c); colour code in (a) and (c): Fe, green; N, blue, C, grey.

or 15-crown-5 (**21.31**). The average $Fe-O_{bridge}$ and $Fe-O_{aqua}$ bond distances in **21.34** are 179 and 209 pm. The magnetic moment of $5.82\,\mu_B$ for $[Fe(OH_2)_6]^{3+}$ is close to the spin-only value for high-spin d^5.

(21.34)

The $[Fe(CN)_6]^{3-}$ ion (Fig. 21.24a) contains low-spin Fe(III) ($\mu_{eff} = 2.25\,\mu_B$) and is made by oxidation of $[Fe(CN)_6]^{4-}$, e.g. by reaction 21.65 or electrolytically. The cyanido ligands in $[Fe(CN)_6]^{3-}$ are more labile than in $[Fe(CN)_6]^{4-}$ and cause the former to be more toxic than the latter.

$$2K_4[Fe(CN)_6] + Cl_2 \longrightarrow 2K_3[Fe(CN)_6] + 2KCl \qquad (21.65)$$

The ruby-red salt $K_3[Fe(CN)_6]$ (potassium hexacyanido-ferrate(III) or ferricyanide) is commercially available. It is an oxidizing agent although $[Fe(CN)_6]^{3-}$ is less powerful an oxidant than $[Fe(OH_2)_6]^{3+}$ (see Section 8.3). Addition of $[Fe(CN)_6]^{3-}$ to aqueous Fe^{2+} gives the deep blue complex *Turnbull's blue* and this reaction is used as a qualitative test for Fe^{2+}. Conversely, if $[Fe(CN)_6]^{4-}$ is added to

aqueous Fe^{3+}, the deep blue complex *Prussian blue* is produced.[†] Both Prussian blue and Turnbull's blue are hydrated salts of formula $Fe^{III}_4[Fe^{II}(CN)_6]_3 \cdot xH_2O$ ($x \approx 14$), and related to them is $KFe[Fe(CN)_6]$, *soluble Prussian blue*. In the solid state, these complexes possess extended structures containing cubic arrangements of Fe^{n+} centres linked by $[CN]^-$ bridges (Fig. 21.24b). The Fe^{3+} cations are high-spin, and $[Fe(CN)_6]^{4-}$ contains low-spin Fe(II). The deep blue colour is the result of electron transfer between Fe(II) and Fe(III). $K_2Fe[Fe(CN)_6]$, which contains only Fe(II), is white. Electron transfer can be prevented by shielding the cation as in the compound $[Fe^{II}L_2]_3[Fe^{III}(CN)_6]_2 \cdot 2H_2O$ (*Ukrainian red*) shown in Fig. 21.25a. Figure 21.24b shows part of the unit cell of $KFe[Fe(CN)_6]$; each Fe^{n+} is in an octahedral environment, either FeC_6 or FeN_6. Turnbull's blue, Prussian blue and Berlin green ($Fe^{III}[Fe^{III}(CN)_6]$) have been widely used in inks and dyes.

Figure 21.24b shows the ability of $[CN]^-$ to act as a bridging ligand and a number of polymeric materials containing either Fe(III) or Fe(II) as well as other metal centres have been made utilizing this property. An example is $[Ni(en)_2]_3[Fe(CN)_6]_2 \cdot 2H_2O$, the solid state structure of which (Fig. 21.25b) consists of interconnected helical chains in which octahedral Ni^{2+} and Fe^{3+} centres are

---

[†] Prussian blue celebrated its 300th birthday in 2005: S.K. Ritter (2005) *Chem. Eng. News*, vol. 83, issue 18, p. 32 – 'Prussian blue: Still a hot topic'.

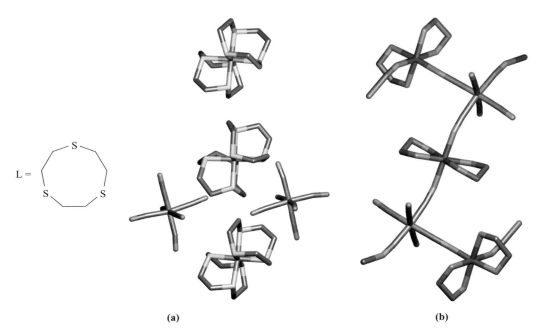

(a)   **(b)**

Fig. 21.25 (a) The structure (X-ray diffraction) of $[FeL_2]_3[Fe(CN)_6]_2\cdot2H_2O$ (L is defined in the scheme in the figure) in which the Fe(II) and Fe(III) centres are remote from each other ('valence trapped') [V.V. Pavlishchuk *et al.* (2001) *Eur. J. Chem.*, p. 297]. (b) Part of the polymeric structure (X-ray diffraction) of $[Ni(en)_2]_3[Fe(CN)_6]_2\cdot2H_2O$ in which Fe^{3+} ions are in $Fe(CN-C)_6$ environments and Ni^{2+} ions are in $Ni(CN-N)_2(en)_2$ sites [M. Ohba *et al.* (1994) *J. Am. Chem. Soc.*, vol. 116, p. 11566]. Hydrogen atoms are omitted; colour code: Fe, green; Ni, red; N, blue; S, yellow; C, grey.

connected by bridging $[CN]^-$ ligands. The latter facilitate electronic communication between the metal centres resulting in a ferromagnetic material (i.e. one in which the magnetic spins are aligned in the same direction, see Fig. 20.32). This is an example of a so-called *molecule-based magnet*. The design and assembly of such materials from paramagnetic building-blocks (both inorganic and organic) have undergone significant development over the past decade.[†]

Large numbers of Fe(III) complexes are known, and octahedral coordination is common. Examples of simple complexes (see Table 7.7 for ligand abbreviations) include:

- high-spin octahedral: $[Fe(OH_2)_6]^{3+}$, $[FeF_6]^{3-}$, $[Fe(ox)_3]^{3-}$, $[Fe(acac)_3]$;

[†] For examples of molecule-based magnets involving cyanido-bridging ligands, see: M. Ohba and H. Ōkawa (2000) *Coord. Chem. Rev.*, vol. 198, p. 313 – 'Synthesis and magnetism of multi-dimensional cyanide-bridged bimetallic assemblies'; S.R. Batten and K.S. Murray (2003) *Coord. Chem. Rev.*, vol. 246, p. 103 – 'Structure and magnetism of coordination polymers containing dicyanamide and tricyanomethanide'; M. Pilkington and S. Decurtins (2004) *Comprehensive Coordination Chemistry II*, eds J.A. McCleverty and T.J. Meyer, Elsevier, Oxford, vol. 7, p. 177 – 'High nuclearity clusters: clusters and aggregates with paramagnetic centers: Cyano and oxalato bridged systems'; S. Wang, X.-H. Ding, J.-L. Zuo, X.-Z. You and W. Huang (2011) *Coord. Chem. Rev.*, vol. 255, p. 1713 – 'Tricyanometalate molecular chemistry: A type of versatile building blocks for the construction of cyano-bridged molecular architectures'.

- low-spin octahedral: $[Fe(CN)_6]^{3-}$, $[Fe(bpy)_3]^{3+}$, $[Fe(phen)_3]^{3+}$, $[Fe(en)_3]^{3+}$;
- 7-coordinate: $[Fe(EDTA)(OH_2)]^-$.

The octahedral complex $[Fe(NH_3)_6]^{3+}$ can be prepared in liquid NH_3, but it has low stability in aqueous solutions, decomposing with loss of NH_3. Both bpy and phen stabilize Fe(II) more than they do Fe(III). This is ascribed to the existence of relatively low-lying π^* MOs on the ligands, allowing them to function as π-acceptors. In aqueous solution, both $[Fe(bpy)_3]^{3+}$ and $[Fe(phen)_3]^{3+}$ are more readily reduced than the hexaaqua ion (eqs. 21.66 and 21.67).

$$[Fe(bpy)_3]^{3+} + e^- \rightleftharpoons [Fe(bpy)_3]^{2+} \qquad E^o = +1.03\,V$$
blue   red
$$(21.66)$$

$$[Fe(phen)_3]^{3+} + e^- \rightleftharpoons [Fe(phen)_3]^{2+} \quad E^o = +1.12\,V$$
blue   red
$$(21.67)$$

The addition of thiocyanate to aqueous solutions of Fe^{3+} produces a blood-red coloration due to the formation of $[Fe(OH_2)_5(SCN-N)]^{2+}$. Complete exchange of ligands to give $[Fe(SCN-N)_6]^{3-}$ is best carried out in non-aqueous media.

Iron(III) favours *O*-donor ligands and stable complexes such as the green $[Fe(ox)_3]^{3-}$ and red $[Fe(acac)_3]$ are commonly encountered. Iron(III) porphyrinato complexes are of relevance for modelling haem-proteins (see Section 29.3) and there is interest in reactions of these complexes with, for example, CO, O_2, NO and $[CN]^-$. The N_4-donor set of a porphyrinato ligand is confined to a plane and this restriction forces the Fe(III) centre to be in a square planar environment with respect to the macrocycle. Other ligands may then enter in axial sites above and below the FeN_4-plane to give either square-based pyramidal or octahedral complexes (Fig. 21.24c).

Low coordination numbers can be stabilized by interaction with amido ligands, e.g. $[Fe\{N(SiMe_3)_2\}_3]$ (Fig. 19.4d).

Self-study exercises

1. In Fig. 21.23, the oxido-bridge atom lies on an inversion centre. Explain what this means. [*Ans*. See Section 3.2]

2. For $[Fe(tpy)Cl_3]$ (tpy is structure **21.29**), $\mu_{eff} = 5.85 \, \mu_B$ at 298 K. Comment on why there is no orbital contribution to the magnetic moment, and determine the number of unpaired electrons. Why does this complex exist only in the *mer*-form?

[*Ans*. See Section 20.10; see Fig. 2.18 and consider flexibility of ligand]

3. In $[Fe(CN)_6]^{3-}$, does the CN^- ligand act as a π-donor or a π-acceptor ligand? Explain how the ligand properties lead to $[Fe(CN)_6]^{3-}$ being low-spin.

[*Ans*. See Fig. 20.15b and discussion]

4. In the caption to Fig. 21.24b, why is the structure described as being 'one-eighth of the unit cell of $KFe[Fe(CN)_6]$' rather than being a complete unit cell?

Iron(II)

The old name for iron(II) is *ferrous*. Anhydrous FeF_2, $FeCl_2$ and $FeBr_2$ can be prepared by reaction 21.68, while FeI_2 is made by direct combination of the elements.

$$Fe + 2HX \xrightarrow{\Delta} FeX_2 + H_2 \qquad (X = F, Cl, Br) \qquad (21.68)$$

Iron(II) fluoride is a sparingly soluble, white solid with a distorted rutile structure (Fig. 6.22); the environment around the high-spin Fe(II) centre (d^6) is surprisingly irregular with 4F at 212 pm and 2F at 198 pm. In the gas phase, FeF_2 is monomeric. Iron(II) chloride forms white, hygroscopic, water-soluble crystals and adopts a $CdCl_2$ structure (see Section 6.11). In the gas phase of $FeCl_2$,

BIOLOGY AND MEDICINE

Box 21.6 Iron complexes fight anaemia

In Chapter 29, the crucial role that iron plays in biological systems is discussed in detail. Anaemia, in which the body suffers from a deficiency of iron, leads to a general state of lethargy and weakness. Iron is usually administered orally to a patient as iron supplement tablets containing an Fe(II) or Fe(III) salt. Iron(II) salts are more typical because they exhibit better solubilities than Fe(III) salts at physiological pH, but Fe(III) has the advantage that, unlike Fe(II), it is not susceptible to oxidation in aqueous solution. Among compounds which are in common use are iron(III) chloride, iron(II) sulfate, iron(II) fumarate, iron(II) succinate and iron(II) gluconate; the structures of fumaric acid, succinic acid and gluconic acid are shown below.

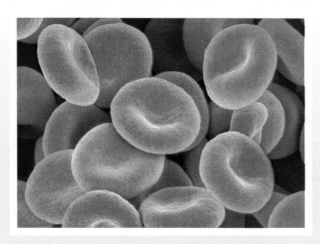

Colorized scanning electron micrograph of red blood cells (magnified $\times 5000$).

Fumaric acid   Succinic acid   Gluconic acid

monomers and dimers (**21.35**) are present. The pale green hydrate $FeCl_2 \cdot 4H_2O$, properly formulated as octahedral $[FeCl_2(OH_2)_4]$, is a convenient precursor in Fe(II) chemistry. The hexahydrate (which loses water readily) can be obtained by recrystallizing $FeCl_2$ from water below 285 K. The reaction of $FeCl_2$ with Et_4NCl in acetone yields the air-sensitive $[Et_4N]_2[Fe_2Cl_6]$ containing anion **21.36**.

$$\text{(structures)}$$

(21.35)   (21.36)

Iron(II) bromide is a deliquescent, yellow or brown solid and adopts a CdI_2 structure. It is very soluble in water and forms hydrates $FeBr_2 \cdot xH_2O$ where $x = 4$, 6 or 9 depending on crystallization conditions. Dark violet FeI_2 has a CdI_2 layer structure, and is hygroscopic and light-sensitive; it forms a green tetrahydrate. All the halides or their hydrates are commercially available, as are salts such as the perchlorate, sulfate and $[NH_4]_2Fe[SO_4]_2 \cdot 6H_2O$. Iron(II) sulfate is a common source of Fe(II) and is available as the blue-green $FeSO_4 \cdot 7H_2O$, an old name for which is *green vitriol*. Like most hydrated Fe(II) salts, it dissolves in water to give $[Fe(OH_2)_6]^{2+}$, the electronic absorption spectrum and magnetic moment of which are consistent with high-spin d^6. The salt $[NH_4]_2Fe[SO_4]_2 \cdot 6H_2O$ is an important source of Fe^{2+} because, in the solid state, it is kinetically more stable towards oxidation than most Fe(II) salts.

Iron(II) oxide is a black, insoluble solid with an NaCl structure above its Curie temperature (200 K). The FeO lattice suffers defects because it is always deficient in Fe (see Section 6.17). Below 200 K, FeO undergoes a phase change and becomes antiferromagnetic. It can be made *in vacuo* by thermal decomposition of iron(II) oxalate but the product must be cooled rapidly to prevent disproportionation (eq. 21.69).

$$4FeO \longrightarrow Fe_3O_4 + Fe \qquad (21.69)$$

White $Fe(OH)_2$ is precipitated by adding alkali to solutions of Fe(II) salts but it rapidly absorbs O_2, turning dark green, then brown. The products are a mixed Fe(II)Fe(III) hydroxide and $Fe_2O_3 \cdot H_2O$. Iron(II) hydroxide dissolves in acids, and from concentrated NaOH solutions, the blue-green $Na_4[Fe(OH)_6]$ can be crystallized.

An interesting distinction between iron(II) oxides and sulfides is that, whereas FeO has an analogue in FeS, there is no peroxide analogue of FeS_2 (*iron pyrites*). The sulfide FeS is made by heating together the elements; it is found in lunar rock samples and adopts an NiAs structure (Fig. 15.10). Reaction of FeS with hydrochloric acid used to be a common laboratory synthesis of H_2S (eq. 16.37). Iron

pyrites is $Fe^{2+}(S_2)^{2-}$ and contains low-spin Fe(II) in a distorted NaCl structure.

The coordination chemistry of Fe(II) is well developed and only a brief introduction to simple species is given here. Iron(II) halides combine with gaseous NH_3 to give salts of $[Fe(NH_3)_6]^{2+}$ but this decomposes in aqueous media, precipitating $Fe(OH)_2$. In aqueous solutions, $[Fe(OH_2)_6]^{2+}$ is unstable with respect to oxidation, although, as we saw above, double salts such as $[NH_4]_2Fe[SO_4]_2 \cdot 6H_2O$ are more stable. Displacement of the ligands in $[Fe(OH_2)_6]^{2+}$ leads to a range of complexes. We have already discussed the stabilization of Fe(II) by bpy and phen (eqs. 21.66 and 21.67). Oxidation of red $[Fe(phen)_3]^{2+}$ to blue $[Fe(phen)_3]^{3+}$ is more difficult than that of $[Fe(OH_2)_6]^{2+}$ to $[Fe(OH_2)_6]^{3+}$, and hence arises the use of $[Fe(phen)_3][SO_4]$ as a redox indicator. Both $[Fe(phen)_3]^{2+}$ and $[Fe(bpy)_3]^{2+}$ are low-spin d^6 and diamagnetic. $[Fe(CN)_6]^{4-}$ is also low-spin. The latter, like $[Fe(CN)_6]^{3-}$ (Fig. 21.24a), is octahedral but the Fe–C bonds in the Fe(II) species are shorter (192 pm) than those in the Fe(III) complex. This provides support for stronger Fe–C π-bonding in the lower oxidation state complex. However, the C–N bond lengths and stretching frequencies differ little between $[Fe(CN)_6]^{3-}$ and $[Fe(CN)_6]^{4-}$. There are many known *mono*substitution products of $[Fe(CN)_6]^{4-}$. Sodium nitropentacyanidoferrate(II) (*sodium nitroprusside*), $Na_2[Fe(CN)_5(NO)] \cdot 2H_2O$, is made by reaction 21.70 or 21.71; among its uses are those as an anti-hypertensive drug (it acts as a vasodilator through release of NO) and as a standard reference for ^{57}Fe Mössbauer spectroscopy.

$$[Fe(CN)_6]^{4-} + 4H^+ + [NO_3]^-$$
$$\longrightarrow [Fe(CN)_5(NO)]^{2-} + CO_2 + [NH_4]^+ \qquad (21.70)$$

$$[Fe(CN)_6]^{4-} + H_2O + [NO_2]^-$$
$$\longrightarrow [Fe(CN)_5(NO)]^{2-} + [CN]^- + 2[OH]^- \qquad (21.71)$$

Nitrogen monoxide is a radical (structure **15.48**), but $Na_2[Fe(CN)_5(NO)]$ is diamagnetic. In the complex, the N–O distance of 113 pm is shorter, and the stretching wavenumber of 1947 cm^{-1} higher, than in free NO. Thus, the complex is formulated as containing an $[NO]^+$ ligand. The addition of S^{2-} to $[Fe(CN)_5(NO)]^{2-}$ produces the red $[Fe(CN)_5(NOS)]^{4-}$ and this is the basis of a sensitive test for S^{2-}. Similarly, reaction with $[OH]^-$ gives $[Fe(CN)_5(NO_2)]^{4-}$ (eq. 21.72). For details of $[Fe(NO)(OH_2)_5]^{2+}$, see Section 15.8.

$$[Fe(CN)_5(NO)]^{2-} + 2[OH]^- \longrightarrow [Fe(CN)_5(NO_2)]^{4-} + H_2O \qquad (21.72)$$

The active sites of NiFe and Fe-only hydrogenase enzymes (see Figs. 29.19 and 29.21) contain $Fe(CO)_x(CN)_y$

coordination units, and there is active interest in studying model Fe(II) compounds containing both CO and $[CN]^-$ ligands. Besides being a good π-acceptor ligand, $[CN]^-$ is a strong σ-donor and can stabilize carbonyl complexes of Fe(II). More commonly, we associate CO with low oxidation state (≤ 0) compounds (see Chapter 24). The reaction of CO with $FeCl_2$ suspended in MeCN, followed by addition of $[Et_4N][CN]$ leads to salts of $[Fe(CN)_5(CO)]^{3-}$ and *trans-* and *cis-*$[Fe(CN)_4(CO)_2]^{2-}$. Alternatively, *trans-*$[Fe(CN)_4(CO)_2]^{2-}$ can be made by adding $[CN]^-$ to an aqueous solution of $FeCl_2\cdot 4H_2O$ under an atmosphere of CO, while the same reaction yields $Na_3[Fe(CN)_5(CO)]$ if five equivalents of NaCN are used. Reaction of NaCN with $Fe(CO)_4I_2$ yields the Na^+ salt of *fac-*$[Fe(CO)_3(CN)_3]^-$, and further addition of $[CN]^-$ leads to the formation of *cis-*$[Fe(CN)_4(CO)_2]^{2-}$. Exchanging a CO ligand in $Fe(CO)_5$ by $[CN]^-$ is described in Section 24.7.

In addition to the hexaaqua ion, high-spin Fe(II) complexes include $[Fe(en)_3]^{2+}$. Its magnetic moment of $5.45\,\mu_B$ is larger than the spin-only value of $4.90\,\mu_B$ and reflects orbital contributions for the configuration $t_{2g}^4 e_g^2$. Although iron(II) favours an octahedral arrangement of donor atoms, there are some tetrahedral complexes, for example $[FeCl_4]^{2-}$ (eq. 21.73), $[FeBr_4]^{2-}$, $[FeI_4]^{2-}$ and $[Fe(SCN)_4]^{2-}$.

$$2MCl + FeCl_2 \xrightarrow{\text{in EtOH}} M_2[FeCl_4] \quad (M = \text{group 1 metal})$$
$$(21.73)$$

The amido complex $[Fe\{N(SiMePh_2)_2\}_2]$ is an unusual example of 2-coordinate Fe(II) (see Section 19.7).

Self-study exercises

1. Rationalize why $[Fe(OH_2)_6]^{2+}$ and $[Fe(CN)_6]^{4-}$, both octahedral Fe(II) complexes, are paramagnetic and diamagnetic, respectively.

    [*Ans.* See Section 20.3 and Table 20.3]

2. Explain why there is an orbital contribution to the magnetic moment of $[Fe(en)_3]^{2+}$.

    [*Ans.* See Section 20.10]

3. The value of $\log \beta_6$ for $[Fe(CN)_6]^{4-}$ is 32.[†] Calculate a value for $\Delta G^o(298\,K)$ for the process:

    $$Fe^{2+}(aq) + 6[CN]^-(aq) \longrightarrow [Fe(CN)_6]^{4-}(aq)$$

    [*Ans.* $-183\,kJ\,mol^{-1}$]

4. To which point group does *fac-*$[Fe(CO)_3(CN)_3]^-$ belong? How many $v(CO)$ and how many $v(CN)$ absorptions are expected in the IR spectrum of *fac-*$[Fe(CO)_3(CN)_3]^-$?

    [*Ans.* See Table 3.5 and the associated self-study exercises]

[†] This system was reassessed in 2003: W.N. Perera and G. Hefter (2003) *Inorg. Chem.*, vol. 42, p. 5917.

5. Confirm that *trans-*$[Fe(CN)_4(CO)_2]$ belongs to the D_{4h} point group. Explain why the IR spectrum of *trans-*$[Fe(CN)_4(CO)_2]$ contains one $v(CO)$ absorption and one $v(CN)$ band.

Iron in low oxidation states

Low oxidation states of iron are typically associated with organometallic compounds and will be discussed mainly in Chapter 24. However, several iron-containing nitrosyl complexes deserve a mention here. The NO molecule is a radical and, as was described in Section 20.5, the M–N–O unit in a complex is either linear or bent depending on whether the ligand behaves as a 3- or a 1-electron donor, respectively. Formally, the nitrosyl ligand is classified as behaving as $[NO]^-$ (bent) or $[NO]^+$ (linear), but in many cases, the oxidation state of the metal in a nitrosyl complex remains ambiguous. Equation 21.74 shows the formation of the tetrahedral $Fe(NO)_3I$ in which each Fe–N–O bond angle is $166°$. Loss of NO from $Fe(NO)_3I$ occurs under vacuum to give $(ON)_2Fe(\mu-I)_2Fe(NO)_2$.

$$Fe(CO)_4I_2 + 3NO \longrightarrow Fe(NO)_3I + \tfrac{1}{2}I_2 + 4CO \quad (21.74)$$

Attempts to prepare binary iron nitrosyl complexes by treatment of $Fe(CO)_3Cl$ with $AgPF_6$ or $AgBF_4$ lead to $Fe(NO)_3(\eta^1\text{-}PF_6)$ and $Fe(NO)_3(\eta^1\text{-}BF_4)$ rather than salts containing 'naked' $[Fe(NO)_3]^+$ ions; the 'η^1' notation indicates that the anions coordinate through the metal centre through one F atom.

21.10 Group 9: cobalt

The metal

Cobalt is less reactive than Fe (e.g. see eq. 21.75); Co does not react with O_2 unless heated, although when very finely divided, it is pyrophoric. It dissolves slowly in dilute mineral acids (e.g. reaction 21.76), but concentrated HNO_3 makes it passive; alkalis have no effect on the metal.

$$M^{2+}(aq) + 2e^- \longrightarrow M(s) \quad \begin{cases} M = Fe,\ E^o = -0.44\,V \\ M = Co,\ E^o = -0.28\,V \end{cases}$$
$$(21.75)$$

$$Co + H_2SO_4 \longrightarrow CoSO_4 + H_2 \quad (21.76)$$

Cobalt reacts at 520 K with F_2 to give CoF_3, but with Cl_2, Br_2 and I_2, CoX_2 is formed. Even when heated, cobalt does not react with H_2 or N_2, but it does combine with B, C (see Section 14.7), P, As and S.

The trend in decreasing stability of high oxidation states on going from Mn to Fe continues along the row (Table 19.3). Cobalt(IV) is the highest oxidation state but it is of far less importance than Co(III) and Co(II). Cobalt(I) and lower oxidation states are stabilized in organometallic species by π-acceptor ligands (see Chapter 24). Among Co(I) complexes containing only phosphane ligands is tetrahedral $[Co(PMe_3)_4]^+$.

Cobalt(IV)

Few Co(IV) species have been established. Yellow $Cs_2[CoF_6]$ is obtained by fluorination of a mixture of CsCl and $CoCl_2$ at 570 K. The fact that $[CoF_6]^{2-}$ (d^5) is *low-spin* contrasts with the *high-spin* nature of $[CoF_6]^{3-}$ (d^6) and the difference reflects the increase in Δ_{oct} with increasing oxidation state. Cobalt(IV) oxide (made by oxidizing Co(II) using alkaline hypochlorite) is poorly defined. Several mixed oxides are known: Ba_2CoO_4 and M_2CoO_3 (M = K, Rb, Cs).

Cobalt(III)

There are few *binary* compounds of Co(III) and only a limited number of Co(III) compounds are commercially available. The only binary halide is brown CoF_3 which is isostructural with FeF_3. It is used as a fluorinating agent, e.g. for preparing perfluorinated organics, and is corrosive and an oxidant. The reaction of N_2O_5 with CoF_3 at 200 K gives the dark green, anhydrous $Co(NO_3)_3$ which has a molecular structure with three bidentate $[NO_3]^-$ groups bound to octahedral Co(III).

Although reports of Co_2O_3 are found in the literature, the anhydrous compound probably does not exist. In contrast, the mixed oxidation state Co_3O_4 ($Co^{II}Co^{III}_2O_4$) is well established and is formed when Co is heated in O_2. The insoluble, grey-black Co_3O_4 crystallizes with a normal spinel structure (Box 13.7) containing high-spin Co^{2+} in tetrahedral holes and low-spin Co^{3+} in octahedral holes. It is, therefore, a worse electrical conductor than Fe_3O_4, in which both high-spin Fe^{2+} and high-spin Fe^{3+} are present in the same octahedral environment. A hydrated oxide is precipitated when excess alkali reacts with most Co(III) compounds, or on aerial oxidation of aqueous suspensions of $Co(OH)_2$. Mixed metal oxides $MCoO_2$, where M is an alkali metal, can be made by heating mixtures of the oxides and consist of layer structures built of edge-sharing CoO_6 octahedra with M^+ ions in interlayer sites. Of particular significance is $LiCoO_2$ which is used in lithium-ion batteries (see Box 11.3).

The blue, low-spin $[Co(OH_2)_6]^{3+}$ ion can be prepared *in situ* by electrolytic oxidation of aqueous $CoSO_4$ in acidic solution at 273 K. A more convenient method for routine use is to dissolve solid $[Co(NH_3)_6][Co(CO_3)_3]$ (a stable, sparingly soluble salt, made according to scheme 21.77) in aqueous nitric or perchloric acid. The acid is chosen so as

to ensure precipitation of $[Co(NH_3)_6]X_3$ (eq. 21.78), leaving $[Co(OH_2)_6]^{3+}$ in solution.

$$Co(NO_3)_2 \cdot 6H_2O(aq) \xrightarrow{\text{excess } [HCO_3]^-,\ H_2O_2} [Co(CO_3)_3]^{3-}(aq)$$

$$\downarrow [Co(NH_3)_6]Cl_3(aq)$$

$$[Co(NH_3)_6][Co(CO_3)_3](s)$$

green crystals

$$(21.77)$$

$$[Co(NH_3)_6][Co(CO_3)_3] + 6HX(aq) \qquad X^- = ClO_4^-, NO_3^-$$

$$\downarrow$$

$$[Co(OH_2)_6]^{3+}(aq) + 3X^-(aq) + 3CO_2(g) + [Co(NH_3)_6]X_3(s)$$

$$(21.78)$$

The $[Co(OH_2)_6]^{3+}$ ion is a powerful oxidant (eq. 21.79) and is unstable in aqueous media, decomposing to Co(II) with the liberation of ozonized O_2. The $[Co(OH_2)_6]^{3+}$ ion is best isolated as the sparingly soluble blue alum $CsCo(SO_4)_2 \cdot 12H_2O$, although this decomposes within hours on standing. Complex formation with, for example, bpy, NH_3, RNH_2 or $[CN]^-$ greatly stabilizes Co(III) as eqs. 21.80–21.83 illustrate.

$$Co^{3+}(aq) + e^- \rightleftharpoons Co^{2+}(aq) \qquad E^\circ = +1.92\,V \qquad (21.79)$$

$$[Co(bpy)_3]^{3+} + e^- \rightleftharpoons [Co(bpy)_3]^{2+}$$
$$E^\circ = +0.31\,V \qquad (21.80)$$

$$[Co(NH_3)_6]^{3+} + e^- \rightleftharpoons [Co(NH_3)_6]^{2+}$$
$$E^\circ = +0.11\,V \qquad (21.81)$$

$$[Co(en)_3]^{3+} + e^- \rightleftharpoons [Co(en)_3]^{2+} \qquad E^\circ = -0.26\,V \quad (21.82)$$

$$[Co(CN)_6]^{3-} + H_2O + e^- \rightleftharpoons [Co(CN)_5(OH_2)]^{3-} + [CN]^-$$
$$E^\circ = -0.83\,V \qquad (21.83)$$

Replacing aqua by ammine ligands, for example, results in a dramatic change in E° (eqs. 21.79 and 21.81) and shows that the overall stability constant of $[Co(NH_3)_6]^{3+}$ is $\approx 10^{30}$ greater than that of $[Co(NH_3)_6]^{2+}$. Much of this difference arises from LFSEs:

• Δ_{oct} for the ammine complex is greater than for the aqua complex in both oxidation states (Table 20.2);
• both Co(II) complexes are high-spin whereas both Co(III) complexes are low-spin (Table 20.3).

Cobalt(III) complexes (d^6) are usually low-spin octahedral and *kinetically inert* (see Section 26.2). The latter means that ligands are not labile and so preparative methods of Co(III) complexes usually involve oxidation of

THEORY

Box 21.7 Alfred Werner

Alfred Werner (working at the University of Zürich) was awarded the Nobel Prize for Chemistry in 1913 for his pioneering work that began to unravel the previous mysteries of the compounds formed between *d*-block metal ions and species such as H_2O, NH_3 and halide ions. A famous problem that led to Werner's theory of coordination concerns the fact that $CoCl_3$ forms a series of complexes with NH_3:

- violet $CoCl_3 \cdot 4NH_3$
- green $CoCl_3 \cdot 4NH_3$
- purple $CoCl_3 \cdot 5NH_3$
- yellow $CoCl_3 \cdot 6NH_3$

and that addition of $AgNO_3$ precipitates different amounts of AgCl per equivalent of Co(III). Thus, one equivalent of $CoCl_3 \cdot 6NH_3$ reacts with an excess of $AgNO_3$ to precipitate *three* equivalents of AgCl, one equivalent of $CoCl_3 \cdot 5NH_3$ precipitates *two* equivalents of AgCl, while one equivalent of either green or violet $CoCl_3 \cdot 4NH_3$ precipitates only *one* equivalent of AgCl. Werner realized that any Cl^- precipitated was free chloride ion and that any other chloride was held in the compound in some other way. The crucial conclusion that Werner drew was that in all these cobalt(III) compounds, the metal was intimately associated with six ligands (NH_3 molecules or Cl^- ions), and that only the remaining Cl^- ions behaved as 'normal' ions, free to react with Ag^+:

$$Ag^+(aq) + Cl^-(aq) \longrightarrow AgCl(s)$$

Werner referred to the oxidation state of the metal ion as its 'primary valence' and to what we now call the coordination number as its 'secondary valence'. The compounds $CoCl_3 \cdot 6NH_3$, $CoCl_3 \cdot 5NH_3$ and $CoCl_3 \cdot 4NH_3$ were thus reformulated as $[Co(NH_3)_6]Cl_3$, $[Co(NH_3)_5Cl]Cl_2$ and $[Co(NH_3)_4Cl_2]Cl$. This picture contrasted greatly with earlier ideas such as the 'chain theory' of Danish chemist Sophus Mads Jørgensen.

Werner's studies went on to show that the numbers of ions in solution (determined from conductivity measurements) were consistent with the formulations $[Co(NH_3)_6]^{3+}[Cl^-]_3$, $[Co(NH_3)_5Cl]^{2+}[Cl^-]_2$ and $[Co(NH_3)_4Cl_2]^+Cl^-$. The fact that $[Co(NH_3)_4Cl_2]Cl$ existed as two *isomers* (the green and violet forms) was a key to the puzzle of the shape of the $[Co(NH_3)_4Cl_2]^+$ complex. The possible *regular* arrangements for six ligands are planar hexagonal, octahedral and trigonal prismatic. There are three ways of arranging the ligands in $[Co(NH_3)_4Cl_2]^+$ in a hexagon:

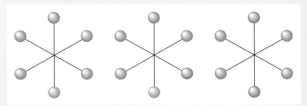

There are two ways for an octahedral arrangement (what we now call *cis*- and *trans*-isomers):

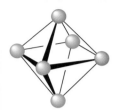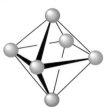

and three for a trigonal prismatic arrangement:

From the fact that only two isomers of $[Co(NH_3)_4Cl_2]Cl$ had been isolated, Werner concluded that $[Co(NH_3)_4Cl_2]^+$ had an octahedral structure, and, by analogy, so did other complexes containing six ligands. Werner's work extended well beyond this one system and his contributions to the groundwork of the understanding of coordination chemistry were immense.

Alfred Werner (1866–1919).

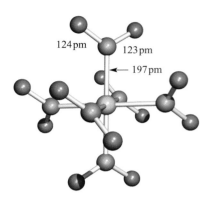

Fig. 21.26 The structure (X-ray diffraction) of $[Co(NO_2-N)_6]^{3-}$ in the salt $Li[Me_4N]_2[Co(NO_2-N)_6]$ [R. Bianchi *et al.* (1996) *Acta Crystallogr.*, *Sect. B*, vol. 52, p. 471]. Colour code: Co, yellow; N, blue; O, red.

the corresponding or related Co(II) species, often *in situ*. For example:

- oxidation by PbO_2 of aqueous Co^{2+} in the presence of excess oxalate gives $[Co(ox)_3]^{3-}$;
- action of excess $[NO_2]^-$ and acid on aqueous Co^{2+} gives $[Co(NO_2-N)_6]^{3-}$ (Fig. 21.26); some $[NO_2]^-$ acts as oxidant and NO is liberated;
- reaction between $Co(CN)_2$ and excess KCN in aqueous solution with *in situ* oxidation gives yellow $K_3[Co(CN)_6]$ (the intermediate Co(II) species is $[Co(CN)_5]^{3-}$ or $[Co(CN)_5(OH_2)]^{3-}$, see later);
- reaction of aqueous $CoCl_2$ with bpy and Br_2 gives $[Co(bpy)_3]^{3+}$;
- aerial oxidation of aqueous $CoCl_2$ in the presence of NH_3 and $[NH_4]Cl$ gives purple $[Co(NH_3)_5Cl]Cl_2$ containing cation **21.37**.

$$\begin{bmatrix} & Cl & \\ H_3N_{\prime\prime\prime\prime\prime\prime} & | & \prime\prime\prime\prime\prime NH_3 \\ & Co & \\ H_3N & | & NH_3 \\ & NH_3 & \end{bmatrix}^{2+}$$

Co–N = 197 pm; Co–Cl 229 pm

(21.37)

The identity of the product may depend on reaction conditions and in the last example, if charcoal is added as a catalyst, the isolated complex is $[Co(NH_3)_6]Cl_3$ containing the $[Co(NH_3)_6]^{3+}$ ion. Similarly, the preparation of orange-red $[Co(en)_3]Cl_3$ requires careful control of reaction conditions (eq. 21.84).

$$\begin{array}{ccc} & CoCl_2 & \\ \text{en, en.HCl, } O_2 \nearrow & & \searrow \text{en.HCl, } O_2 \\ \text{aqueous solution} & & \text{acidic solution} \\ \downarrow & & \downarrow \\ [Co(en)_3]Cl_3 & & trans\text{-}[Co(en)_2Cl_2]Cl \end{array}$$

$$(21.84)$$

The $[Co(en)_3]^{3+}$ ion is frequently used to precipitate large anions, and the kinetic inertness of the d^6 ion allows its enantiomers to be separated. The green *trans*-$[Co(en)_2Cl_2]Cl$ is isolated from reaction 21.84 as the salt *trans*-$[Co(en)_2Cl_2]Cl·2H_2O·HCl$ but this loses HCl on heating. It can be converted to the racemic red *cis*-$[Co(en)_2Cl_2]Cl$ by heating an aqueous solution and removing the solvent. Enantiomers of *cis*-$[Co(en)_2Cl_2]^+$ can be separated using a chiral anion such as (1*S*)- or (1*R*)-3-bromocamphor-8-sulfonate. In aqueous solution, one Cl^- ligand in $[Co(en)_2Cl_2]^+$ is replaced by H_2O to give $[Co(en)_2Cl(OH_2)]^{2+}$. Because ligand substitutions in Co(III) complexes are so slow, these species have been the subject of many kinetic studies (see Chapter 26).

The $[Co(CN)_6]^{3-}$ ion is so stable that if a solution of $K_3[Co(CN)_5]$ containing excess KCN is heated, H_2 is evolved and $K_3[Co(CN)_6]$ is formed. In this reaction, the hydrido complex $[Co(CN)_5H]^{3-}$ is an intermediate. It can be obtained almost quantitatively (reversible reaction 21.85) and can be precipitated as $Cs_2Na[Co(CN)_5H]$.

$$2[Co^{II}(CN)_5]^{3-} + H_2 \rightleftharpoons 2[Co^{III}(CN)_5H]^{3-} \qquad (21.85)$$

The $[Co(CN)_5H]^{3-}$ ion is an effective homogeneous hydrogenation catalyst for alkenes. The process is summarized in eq. 21.86, with reaction 21.85 regenerating the catalyst.

$$\left. \begin{array}{l} [Co(CN)_5H]^{3-} + CH_2=CHX \\ \qquad \longrightarrow [Co(CN)_5CH_2CH_2X]^{3-} \\ [Co(CN)_5CH_2CH_2X]^{3-} + [Co(CN)_5H]^{3-} \\ \qquad \longrightarrow CH_3CH_2X + 2[Co(CN)_5]^{3-} \end{array} \right\}$$

$$(21.86)$$

$$\begin{bmatrix} & CN & & \\ NC & | & CN & \\ & Co & & \\ NC & | & & CN \\ & CN & O\text{---}O & CN \\ & & & Co \\ & NC & | & CN \\ & & CN & \end{bmatrix}^{6-}$$

145 pm

(21.38)

By aerial oxidation of $[Co^{II}(CN)_5]^{3-}$ in aqueous cyanide solution, it is possible to isolate the diamagnetic peroxido complex $[(CN)_5Co^{III}OOCo^{III}(CN)_5]^{6-}$ (**21.38**) which can be precipitated as the brown potassium salt. Oxidation of $K_6[(CN)_5CoOOCo(CN)_5]$ using Br_2 leads to the paramagnetic, red $K_5[(CN)_5CoOOCo(CN)_5]$. The structure of $[(CN)_5CoOOCo(CN)_5]^{5-}$ resembles that of **21.38**, except that the O–O distance is 126 pm, indicating that oxidation takes place at the peroxido bridge and not at a metal centre. Thus, $[(CN)_5CoOOCo(CN)_5]^{5-}$ is a superoxido complex retaining two Co(III) centres. The ammine complexes $[(H_3N)_5CoOOCo(NH_3)_5]^{4+}$ and $[(H_3N)_5CoOOCo(NH_3)_5]^{5+}$ (which have been isolated as the brown nitrate and green chloride salts respectively) are similar, containing peroxido and superoxido ligands respectively; the peroxido complex is stable in solution only in the presence of >2 M NH_3.

One of the few examples of a high-spin Co(III) complex is $[CoF_6]^{3-}$. The blue K^+ salt (obtained by heating $CoCl_2$, KF and F_2) has a magnetic moment of 5.63 μ_B.

Cobalt(II)

In contrast to Co(III), Co(II) forms many simple compounds and all four Co(II) halides are known. Reaction of anhydrous $CoCl_2$ with HF at 570 K gives sparingly soluble, pink CoF_2 which crystallizes with the rutile structure (see Fig. 6.22). Blue $CoCl_2$ is made by combination of the elements and has a $CdCl_2$ structure (see Section 6.11). It turns pink on exposure to moisture and readily forms hydrates. The dark pink hexahydrate is commercially available and is a common starting material in Co(II) chemistry. The di- and tetrahydrates can also be crystallized from aqueous solutions of $CoCl_2$. Crystalline $CoCl_2 \cdot 6H_2O$ contains *trans*-$[CoCl_2(OH_2)_4]$, connected to the extra water molecules through a hydrogen-bonded network. In contrast, the structure of $CoCl_2 \cdot 4H_2O$ consists of hydrogen-bonded *cis*-$[CoCl_2(OH_2)_4]$ molecules, while $CoCl_2 \cdot 2H_2O$ contains chains of edge-sharing octahedra (structure **21.39**).

(21.39)

In aqueous solutions of all forms of $CoCl_2$, the major species are $[Co(OH_2)_6]^{2+}$, $[CoCl(OH_2)_5]^+$ and $[CoCl_4]^{2-}$, with minor amounts of $[CoCl_2(OH_2)_4]$ and $[CoCl_3(OH_2)]^-$. Green $CoBr_2$ (made by heating Co and Br_2) is dimorphic, adopting either the $CdCl_2$ or CdI_2 structure. It is water-soluble and can be crystallized as the purple-blue dihydrate or red hexahydrate. Heating Co metal with HI produces blue-black CoI_2 which adopts a CdI_2 layer structure. The red hexahydrate $CoI_2 \cdot 6H_2O$ can be crystallized from aqueous solutions. Both $CoBr_2 \cdot 6H_2O$ and $CoI_2 \cdot 6H_2O$ contain the octahedral $[Co(OH_2)_6]^{2+}$ ion in the solid state, as do a number of hydrates, e.g. $CoSO_4 \cdot 6H_2O$, $Co(NO_3)_2 \cdot 6H_2O$ and $Co(ClO_4)_2 \cdot 6H_2O$. Aqueous solutions of most simple Co(II) salts contain $[Co(OH_2)_6]^{2+}$ (see below).

Cobalt(II) oxide is an olive-green, insoluble solid but its colour may vary depending on its dispersion. It is best obtained by thermal decomposition of the carbonate or nitrate in the absence of air, and has the NaCl structure; CoO is used as a pigment in glasses and ceramics (see Box 21.8). When heated in air at 770 K, CoO converts to Co_3O_4.

The sparingly soluble $Co(OH)_2$ may be pink or blue, with the pink form being the more stable. Freshly precipitated blue $Co(OH)_2$ turns pink on standing. The change in colour is presumably associated with a change in coordination about the Co(II) centre. Cobalt(II) hydroxide is amphoteric and dissolves in hot, concentrated alkalis to give salts of $[Co(OH)_4]^{2-}$ (**21.40**).

(21.40)

Whereas the coordination chemistry of Co^{3+} is essentially that of octahedral complexes, that of Co^{2+} is structurally variable since LFSEs for the d^7 configuration do not tend to favour a particular ligand arrangement. The variation in coordination geometries is shown in the following examples:

- linear: $[Co\{N(SiMe_3)_2\}_2]$;
- trigonal planar: $[Co\{N(SiMe_3)_2\}_2(PPh_3)]$, $[Co\{N(SiMe_3)_2\}_3]^-$;
- tetrahedral: $[Co(OH)_4]^{2-}$, $[CoCl_4]^{2-}$, $[CoBr_4]^{2-}$, $[CoI_4]^{2-}$, $[Co(NCS-N)_4]^{2-}$, $[Co(N_3)_4]^{2-}$, $[CoCl_3(NCMe)]^-$;
- square planar: $[Co(CN)_4]^{2-}$, $[Co(pc)]$ ($H_2pc = $ **21.28**);
- trigonal bipyramidal: $[Co\{N(CH_2CH_2PPh_2)_3\}(SMe)]^+$;
- square-based pyramidal: $[Co(CN)_5]^{3-}$;
- octahedral: $[Co(OH_2)_6]^{2+}$, $[Co(NH_3)_6]^{2+}$, $[Co(en)_3]^{2+}$;
- pentagonal bipyramidal: $[Co(15\text{-crown-}5)L_2]^{2+}$ ($L = H_2O$ or MeCN; see **21.45**);
- dodecahedral: $[Co(NO_3-O,O')_4]^{2-}$ (Fig. 21.28c).

Aqueous solutions of simple salts usually contain $[Co(OH_2)_6]^{2+}$ but there is evidence for the existence of equilibrium 21.87, although $[Co(OH_2)_6]^{2+}$ is by far the dominant species; speciation in aqueous $CoCl_2$ was discussed earlier.

$$[Co(OH_2)_6]^{2+} \rightleftharpoons [Co(OH_2)_4]^{2+} + 2H_2O \qquad (21.87)$$
$$\text{octahedral} \qquad\qquad \text{tetrahedral}$$

Whereas $[Co(OH_2)_6]^{2+}$ is a stable complex, $[Co(NH_3)_6]^{2+}$ is easily oxidized (eqs. 21.79 and 21.81). The same is true of amine complexes. $[Co(en)_3]^{2+}$ can be prepared from $[Co(OH_2)_6]^{2+}$ and en in an inert atmosphere and is usually made *in situ* when needed. The $[Co(bpy)_3]^{2+}$ ion is stable enough to be isolated in a number of salts, e.g. orange $[Co(bpy)_3]Cl_2 \cdot 2H_2O \cdot EtOH$ which has been

APPLICATIONS

Box 21.8 Cobalt blues

Blue glass and ceramic glazes and enamels are in high demand for decorative wear, and the source of colour is very often a cobalt-based pigment. Cobalt(II) oxide is the form that is incorporated into the molten glass, but initial sources vary. Black Co_3O_4 is transformed in $\approx 93\%$ yield to CoO at $\approx 1070\,K$. Purple $CoCO_3$ can also be used as the raw material but has lower conversion yields. Only very small amounts of the oxide are required to obtain a discernible blue pigment. Variations in colour are achieved by combining with other oxides, e.g. purple shades result if manganese oxide is added. Cobalt oxide is also used to counter the yellow colouring in glazes that arises from iron impurities. Blue pigmentation can also be obtained using $(Zr,V)SiO_4$ (see Section 28.5).

While the importance of cobalt-based pigments in ceramics is well established, it has also been shown that thin films of Co_3O_4 provide an effective coating for solar collectors that operate at high temperatures. The properties of black Co_3O_4 that make it suitable for this application are its high solar absorbance and low IR emittance.

Related material: see Box 14.2 – Solar power: thermal and electrical.

Late 19th century English earthenware vases covered in a cobalt glaze.

crystallographically characterized (Co−N = 213 pm). Among the stable complexes of Co(II) is tetrahedral $[CoX_4]^{2-}$ (X = Cl, Br, I). Addition of concentrated HCl to solutions of pink $[Co(OH_2)_6]^{2+}$ produces the intensely blue $[CoCl_4]^{2-}$. Many salts of $[CoCl_4]^{2-}$ are known; of note is Cs_3CoCl_5 which is actually $Cs_3[CoCl_4]Cl$ and does *not* contain $[CoCl_5]^{3-}$. Both $[Co(OH_2)_6]^{2+}$ and $[CoCl_4]^{2-}$, like most Co(II) complexes, are high-spin with magnetic moments higher than the spin-only value. Typically, for high-spin Co^{2+}, μ_{eff} lies in the range 4.3–5.2 μ_B for octahedral complexes and 4.2–4.8 μ_B for tetrahedral species. Among other tetrahedral complexes is $[Co(NCS-N)_4]^{2-}$, isolated in blue $[Me_4N]_2[Co(NCS-N)_4]$ ($\mu_{eff} = 4.40\,\mu_B$) and $K_2[Co(NCS-N)_4]\cdot4H_2O$ ($\mu_{eff} = 4.38\,\mu_B$). The insoluble mercury(II) salt of $[Co(NCS-N)_4]^{2-}$ is the standard calibrant for magnetic susceptibility measurements. By using cation **21.41**, it has been possible to isolate a red salt of the octahedral $[Co(NCS-N)_6]^{4-}$.

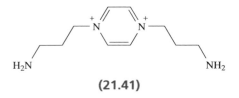

(21.41)

The ability of chlorido ligands to bridge between two metal centres allows the formation of dinuclear species

such as $[Co_2Cl_6]^{2-}$ (Fig. 21.27a), as well as higher nuclearity complexes such as polymer **21.39**. The complex $[CoCl_2(py)_2]$ exists in two modifications: one is monomer **21.42** containing a tetrahedral Co(II) centre, while the other contains edge-sharing octahedra in polymer **21.43**. Equation 21.88 summarizes the formation of $[CoCl_2(py)_2]$ and $[CoCl_2(py)_2]_n$. Similar tetrahedral–octahedral interconversions are seen for some Ni(II) complexes of type L_2NiX_2 where X^- has the propensity for bridge formation (see Section 21.11).

$$CoCl_2 + 2py \longrightarrow \underset{\text{blue}}{[CoCl_2(py)_2]} \underset{\Delta,\ 390\,K}{\overset{\text{polymerizes on standing in air}}{\rightleftharpoons}} \underset{\text{violet}}{[CoCl_2(py)_2]_n}$$

(21.88)

py = pyridine

(21.42)   **(21.43)**

Heating a solution of $CoCl_2$ in THF at reflux produces the dark blue $[Co_4Cl_2(\mu\text{-}Cl)_6(THF)_6]$ in which bridging chlorido ligands support the tetranuclear framework (Fig. 21.27b).

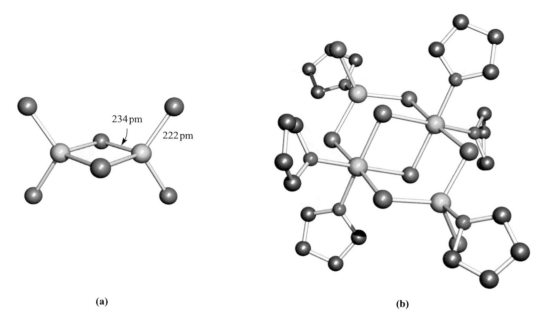

(a)   (b)

Fig. 21.27 The structures (X-ray diffraction) of (a) $[Co_2Cl_6]^{2-}$ in the salt $[Co(15\text{-crown-}5)(NCMe)_2][Co_2Cl_6]$; the cation is shown in structure **21.45** [O.K. Kireeva *et al.* (1992) *Polyhedron*, vol. 11, p. 1801] and (b) $[Co_4Cl_2(\mu\text{-Cl})_6(THF)_6]$ [P. Sobota *et al.* (1993) *Polyhedron*, vol. 12, p. 613]. Hydrogen atoms are omitted; colour code: Co, yellow; Cl, green; C, grey; O, red.

Two Co(II) centres are octahedrally coordinated and two are in 4-coordinate environments. At 300 K, the magnetic moment is 4.91 μ_B, typical of isolated high-spin Co(II) centres. On lowering the temperature to 4.2 K, the value of μ_{eff} increases to 7.1 μ_B. Such behaviour indicates ferromagnetic coupling between the metal centres which are able to communicate through the bridging ligands (see Section 20.10).

Chloride is just one example of a ligand which may coordinate to a metal centre in a terminal or bridging mode; other ligands may be equally versatile. For example, $[Co(acac)_2]$ is prepared from $CoCl_2$, Hacac and $Na[O_2CMe]$ in aqueous methanol. In the solid state, the blue anhydrous salt is tetrameric with a structure related to that of the trimer $[\{Ni(acac)_2\}_3]$ (see Fig. 21.29b).

Low-spin cyanido complexes of Co(II) provide examples of square-based pyramidal and square planar species. The addition of an excess of $[CN]^-$ to aqueous Co^{2+} yields $[Co(CN)_5]^{3-}$. That this is formed in preference to $[Co(CN)_6]^{4-}$ (which has not been isolated) can be understood by considering Fig. 20.15b. For the strong-field cyanido ligands, Δ_{oct} is large and for a hypothetical octahedral d^7 complex, partial occupancy of the $e_g{}^*$ MOs would be unfavourable since it would impart significant *antibonding* character to the complex. The brown $K_3[Co(CN)_5]$ is paramagnetic, but a violet, diamagnetic salt $K_6[Co_2(CN)_{10}]$ has also been isolated. The $[Co_2(CN)_{10}]^{6-}$ ion, **21.44**, possesses a Co–Co single bond and a staggered conformation; it is isoelectronic and isostructural with $[Mn_2(CO)_{10}]$ (see

Fig. 24.10). By using the large cation $[(Ph_3P)_2N]^+$, it has been possible to isolate a salt of the square planar complex $[Co(CN)_4]^{2-}$ (Fig. 21.28a). This is an unusual example of a square planar Co(II) species where the geometry is *not* imposed by the ligand. In complexes such as $[Co(pc)]$, the phthalocyanine ligand (**21.28**) has a rigid framework and forces the coordination environment to be square planar.

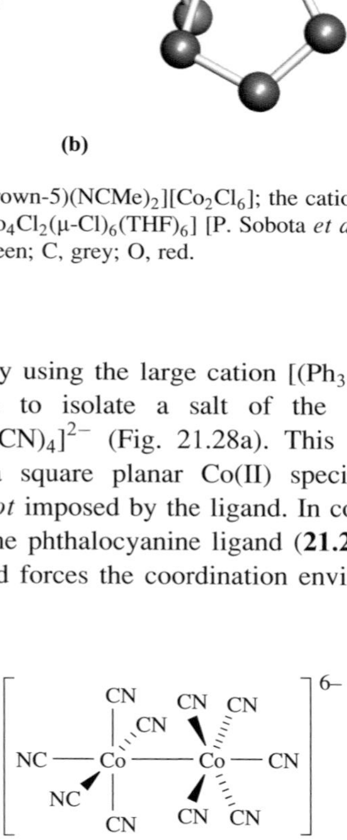

$$\left[\begin{array}{c} \text{21.44} \end{array} \right]^{6-}$$

(21.44)

The highest coordination numbers for Co(II) are 7 and 8. The effects of a coordinatively restricted macrocyclic ligand give rise to pentagonal bipyramidal structures for $[Co(15\text{-crown-}5)(NCMe)_2]^{2+}$ (**21.45**) and $[Co(15\text{-crown-}5)(OH_2)_2]^{2+}$. Larger macrocycles are more flexible, and in the complex $[Co(\textbf{21.46})]^{2+}$, the S_6-donor set is octahedrally arranged. Figure 21.28b shows the solid state structure of $[Co(12\text{-crown-}4)(NO_3)_2]$ in which the Co(II) centre is 7-coordinate. In $[Co(NO_3)_4]^{2-}$, a dodecahedral arrangement of donor atoms is observed, although as Fig. 21.28c shows, each $[NO_3]^-$ ligand is bound asymmetrically with one oxygen donor interacting more strongly than the

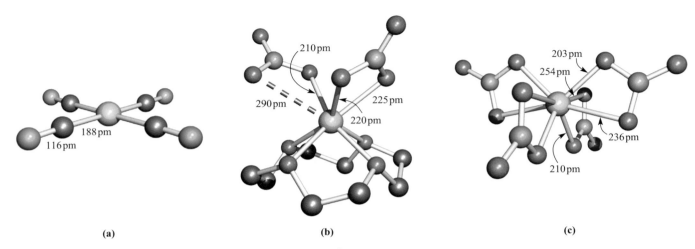

Fig. 21.28 The structures (X-ray diffraction) of (a) $[Co(CN)_4]^{2-}$ in the salt $[(Ph_3P)_2N]_2[Co(CN)_4]\cdot4DMF$; there is also a *weak* interaction with a solvate molecule in an axial site [S.J. Carter *et al.* (1984) *J. Am. Chem. Soc.*, vol. 106, p. 4265]; (b) $[Co(12\text{-}crown\text{-}4)(NO_3)_2]$ [E.M. Holt *et al.* (1981) *Acta Crystallogr.*, *Sect. B*, vol. 37, p. 1080]; and (c) $[Co(NO_3)_4]^{2-}$ in the $[Ph_4As]^+$ salt [J.G. Bergman *et al.* (1966) *Inorg. Chem.*, vol. 5, p. 1208]. Hydrogen atoms are omitted; colour code: Co, yellow; N, blue; C, grey; O, red.

other. These nitrato complexes illustrate that caution is sometimes needed in interpreting coordination geometries and a further example concerns $[LCoX]^+$ complexes where L is the tripodal ligand $N(CH_2CH_2PPh_2)_3$. For $X^- = [MeS]^-$ or $[EtO(O)_2S]^-$, the Co(II) centre in $[LCoX]^+$ is 5-coordinate (**21.47**) with a Co–N distance of 213 or 217 pm, respectively. However, for $X^- = Cl^-$, Br^- or I^-, there is only a weak interaction between the nitrogen and metal centre (**21.48**) with Co$\cdots$N in the range 268–273 pm. These data refer to the *solid state* and say nothing about solution species.

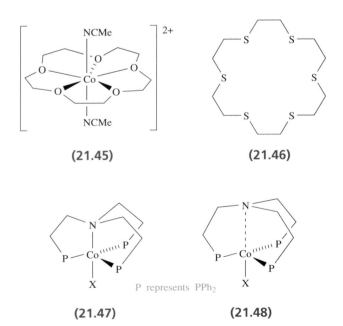

(21.45)

(21.46)

(21.47)

(21.48)

P represents PPh₂

Self-study exercises

1. For octahedral Co^{2+}, what is the ground state term that arises from the $t_{2g}^5 e_g^2$ electronic configuration?
   [*Ans.* $^4T_{1g}$; see Fig. 20.20]

2. The electronic absorption spectrum of $[Co(OH_2)_6]^{2+}$ shows absorptions at 8100, 16 000 and 19 400 cm^{-1}. The middle band is assigned to the transition $^4T_{1g}(P) \longleftarrow {}^4T_{1g}(F)$. Assign the remaining two transitions.
   [*Ans.* See Fig. 20.20]

3. For tetrahedral Co^{2+}, what is the ground state electronic configuration, and to what ground state term does this correspond?
   [*Ans.* $e^4 t_2^3$; $^4A_2$]

4. Explain why, rather than using the spin-only formula, the magnetic moments of tetrahedral Co^{2+} complexes may be estimated using the following equation:

$$\mu_{\text{eff}} = 3.87\left(1 - \frac{4\lambda}{\Delta_{\text{oct}}}\right)$$

21.11 Group 10: nickel

The metal

The reactivity of Ni metal resembles that of Co (e.g. eq. 21.89). It is attacked by dilute mineral acids, made passive by concentrated HNO_3, and is resistant to aqueous alkalis.

$$M^{2+}(aq) + 2e^- \longrightarrow M(s) \qquad \begin{cases} M = Ni, \ E^o = -0.25\,V \\ M = Co, \ E^o = -0.28\,V \end{cases}$$

$$(21.89)$$

The bulk metal is oxidized by air or steam only at high temperatures, but *Raney nickel* (see Section 21.2) is pyrophoric. Nickel reacts with F_2 to give a coherent coating of NiF_2 which prevents further attack; hence the use of nickel and its alloy *Monel metal* (68% Ni and 32% Cu) in apparatus for handling F_2 or xenon fluorides. With Cl_2, Br_2 and I_2, Ni(II) halides are formed. At elevated temperatures, Ni reacts with P, S and B and a range of different phosphide (see Section 15.6), sulfide and boride (see Section 13.10) phases are known.

Nickel(II) is far the most important oxidation state for the metal (Table 19.3). Low oxidation states are most common in organometallic species (Chapter 24), but other Ni(0) species include $[Ni(PF_3)_4]$ and $[Ni(CN)_4]^{4-}$. Yellow $K_4[Ni(CN)_4]$ is made by reduction of $K_2[Ni(CN)_4]$ in liquid NH_3 using excess K, but oxidizes immediately on exposure to air.

Nickel(IV) and nickel(III)

The formation of nickel(IV) requires the use of extremely strong oxidants. $K_2[NiF_6]$ is prepared from $NiCl_2$, F_2 and KCl. The salt $[Xe_2F_{11}]_2[NiF_6]$ (Fig. 18.5) is made from XeF_2, KrF_2 and NiF_2. Octahedral $[NiF_6]^{2-}$ is diamagnetic (low-spin d^6) and the red K^+ salt crystallizes with the $K_2[PtF_6]$ structure (see Mn(IV), Section 21.8). Above 620 K, $K_2[NiF_6]$ decomposes to $K_3[NiF_6]$. Salts of $[NiF_6]^{2-}$ are powerful oxidants, and $[NF_4]_2[NiF_6]$ has been used as an oxidizing agent in some solid propellants. It decomposes on heating according to eq. 21.90. Nickel(IV) fluoride can be prepared from $K_2[NiF_6]$ and BF_3 or AsF_5, but is unstable above 208 K (eq. 21.91).

$$[NF_4]_2[NiF_6] \xrightarrow{\text{anhydrous HF}} 2NF_3 + NiF_2 + 3F_2 \qquad (21.90)$$

$$2NiF_4 \longrightarrow 2NiF_3 + F_2 \qquad (21.91)$$

Nickel(IV) is present in $KNiIO_6$, formally a salt of $[IO_6]^{5-}$ (see Section 17.9). This compound is formed by oxidation of $[Ni(OH_2)_6]^{2+}$ by $[S_2O_8]^{2-}$ in the presence of $[IO_4]^-$. The structure of $KNiIO_6$ can be considered as an hcp array of O atoms with K, Ni and I occupying octahedral sites.

Impure NiF_3 is made by reaction 21.91. It is a black solid, and is a strong fluorinating agent, but decomposes when heated (eq. 21.92).

$$2NiF_3 \xrightarrow{\Delta} 2NiF_2 + F_2 \qquad (21.92)$$

Reaction of $NiCl_2$, KCl and F_2 produces violet $K_3[NiF_6]$. Octahedral $[NiF_6]^{3-}$ is low-spin d^7 $(t_{2g}^6 e_g^1)$ and shows the expected Jahn–Teller distortion.

The black hydrous oxide Ni(O)OH is obtained by alkaline hypochlorite oxidation of aqueous Ni(II) salts is used in NiCd rechargeable batteries (eq. 21.5). It is a strong oxidizing agent, liberating Cl_2 from hydrochloric acid. Mixed metal oxides of Ni(IV) include $BaNiO_3$ and $SrNiO_3$, which are isostructural and contain chains of face-sharing NiO_6 octahedra.

Nickel(III) is a very good oxidizing agent, but is stabilized by σ-donor ligands. Complexes include $[Ni(1,2\text{-}S_2C_6H_4)_2]^-$ (**21.49**) and $[NiBr_3(PEt_3)_2]$ (**21.50**). The latter has a magnetic moment of $1.72\,\mu_B$, indicative of low-spin Ni(III); the solid compound is stable for only a few hours. Other ligands used to stabilize Ni(III) include porphyrins and aza-macrocycles. In $[Ni(\textbf{21.51})]$, each set of three N-donors and three O-donors is in a *fac*-arrangement about an octahedral Ni(III) centre.

(21.49)

(21.50)

(21.51)

Nickel(II)

Nickel(II) fluoride is made by fluorination of $NiCl_2$, and is a yellow solid with a rutile structure (Fig. 6.22). Both NiF_2 and its green tetrahydrate are commercially available. Anhydrous $NiCl_2$, $NiBr_2$ and NiI_2 are made by direct combination of the elements. $NiCl_2$ and NiI_2 adopt a $CdCl_2$ structure, while $NiBr_2$ has a CdI_2 structure (see Section 6.11). The chloride is a useful precursor in Ni(II) chemistry and can be purchased as the yellow anhydrous salt or green hydrate. The hexahydrate contains the $[Ni(OH_2)_6]^{2+}$ ion in the solid state, but the dihydrate (obtained by partial dehydration of $NiCl_2 \cdot 6H_2O$) has a polymeric structure analogous to **21.39**. Anhydrous $NiBr_2$ is yellow and can be crystallized as a number of hydrates. Black NiI_2 forms a green hexahydrate.

The water-insoluble, green NiO is obtained by thermal decomposition of $NiCO_3$ or $Ni(NO_3)_2$ and crystallizes with the NaCl structure. Thin amorphous films of NiO exhibiting electrochromic behaviour (see Box 22.4) may be deposited by CVD (*chemical vapour deposition*, see Section 28.6) starting from [Ni(acac)$_2$]. Nickel(II) oxide is antiferromagnetic (T_N = 520 K), and its conducting properties are discussed in Section 28.3. Nickel(II) oxide is basic, reacting with acids, e.g. reaction 21.93.

$$NiO + H_2SO_4 \longrightarrow NiSO_4 + H_2O \qquad (21.93)$$

Oxidation of NiO by hypochlorite yields Ni(O)OH (see earlier). Aerial oxidation converts NiS to Ni(S)OH, a fact that explains why, although NiS is not precipitated in acidic solution, after exposure to air it is insoluble in dilute acid. Addition of [OH]$^-$ to aqueous solutions of Ni^{2+} precipitates green Ni(OH)$_2$ which has a CdI$_2$ structure; it is used in NiCd batteries (eq. 21.5). Nickel(II) hydroxide is insoluble in aqueous NaOH except at very high hydroxide concentrations when it forms soluble Na$_2$[Ni(OH)$_4$]. Ni(OH)$_2$ is soluble in aqueous NH$_3$ with formation of [Ni(NH$_3$)$_6$]$^{2+}$. The pale green basic carbonate, 2NiCO$_3\cdot$3Ni(OH)$_2\cdot$4H$_2$O, forms when Na$_2$CO$_3$ is added to aqueous Ni^{2+} and it is this carbonate that is usually bought commercially.

A range of coordination geometries is observed for nickel(II) complexes with coordination numbers from 4 to 6 being common. Octahedral and square planar geometries are most usual. Examples include:

- tetrahedral: [NiCl$_4$]$^{2-}$, [NiBr$_4$]$^{2-}$, [Ni(NCS-*N*)$_4$]$^{2-}$;
- square planar: [Ni(CN)$_4$]$^{2-}$, [Ni(Hdmg)$_2$] (H$_2$dmg = dimethylglyoxime);
- trigonal bipyramidal: [Ni(CN)$_5$]$^{3-}$ (cation-dependent), [NiCl{N(CH$_2$CH$_2$NMe$_2$)$_3$}]$^+$;
- square-based pyramidal: [Ni(CN)$_5$]$^{3-}$ (cation-dependent);
- octahedral: [Ni(OH$_2$)$_6$]$^{2+}$, [Ni(NH$_3$)$_6$]$^{2+}$, [Ni(bpy)$_3$]$^{2+}$, [Ni(en)$_3$]$^{2+}$, [Ni(NCS-*N*)$_6$]$^{4-}$, [NiF$_6$]$^{4-}$.

Some structures are complicated by interconversions between square planar and tetrahedral, or square planar and octahedral coordination as we discuss later. In addition, the potential of some ligands to bridge between metal centres may cause ambiguity. For example, alkali metal salts of [NiF$_3$]$^-$, [NiF$_4$]$^{2-}$ and [NiCl$_3$]$^-$ crystallize with extended structures, whereas salts of [NiCl$_4$]$^{2-}$ and [NiBr$_4$]$^{2-}$ contain discrete tetrahedral anions. The compounds KNiF$_3$ and CsNiF$_3$ are obtained by cooling melts containing NiF$_2$ and MHF$_2$. KNiF$_3$ has a perovskite structure (Fig. 6.24) and is antiferromagnetic, while CsNiF$_3$ possesses chains of face-sharing NiF$_6$ octahedra and is ferrimagnetic. A similar chain structure is adopted by CsNiCl$_3$. The antiferromagnetic K$_2$NiF$_4$ contains layers of corner-sharing octahedral NiF$_6$ units (Fig. 21.29a) separated by K$^+$ ions.

In Section 21.10, we noted that [Co(acac)$_2$] is tetrameric. Similarly, [Ni(acac)$_2$] oligomerizes, forming trimers (Fig. 21.29b) in which [acac]$^-$ ligands are in chelating and bridging modes. Reaction of [{Ni(acac)$_2$}$_3$] with

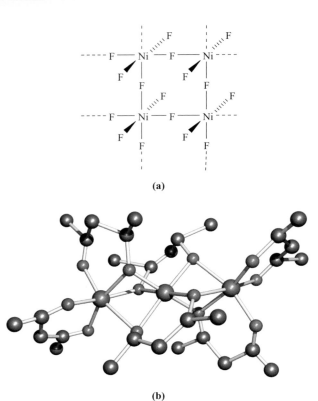

Fig. 21.29 (a) Representation of part of a layer of corner-sharing NiF$_6$ octahedra in K$_2$NiF$_4$. (b) The structure of [{Ni(acac)$_2$}$_3$] (X-ray diffraction) with H atoms omitted [G.J. Bullen *et al.* (1965) *Inorg. Chem.*, vol. 4, p. 456]. Colour code: Ni, green; C, grey; O, red.

aqueous AgNO$_3$ yields Ag[Ni(acac)$_3$] containing the octahedral [Ni(acac)$_3$]$^-$ ion.

Solid, hydrated nickel(II) salts and their aqueous solutions usually contain green [Ni(OH$_2$)$_6$]$^{2+}$, the electronic absorption spectrum of which was shown in Fig. 20.21 with that of [Ni(NH$_3$)$_6$]$^{2+}$. Salts of the latter are typically blue, giving violet solutions. In aqueous solution, [Ni(NH$_3$)$_6$]$^{2+}$ is stable only in the presence of excess NH$_3$ without which species such as [Ni(NH$_3$)$_4$(OH$_2$)$_2$]$^{2+}$ form. The violet chloride, bromide or perchlorate salts of [Ni(en)$_3$]$^{2+}$ are obtained as racemates, the cation being kinetically labile (see Section 26.2). The octahedral complexes *trans*-[Ni(ClO$_4$-*O*,*O'*)$_2$(NCMe)$_2$] (**21.52**) and *trans*-[Ni(ClO$_4$-*O*)$_2$(py)$_4$] illustrate the ability of perchlorate ions to act as bidentate or monodentate ligands respectively. The latter complex is discussed again later.

(21.52)

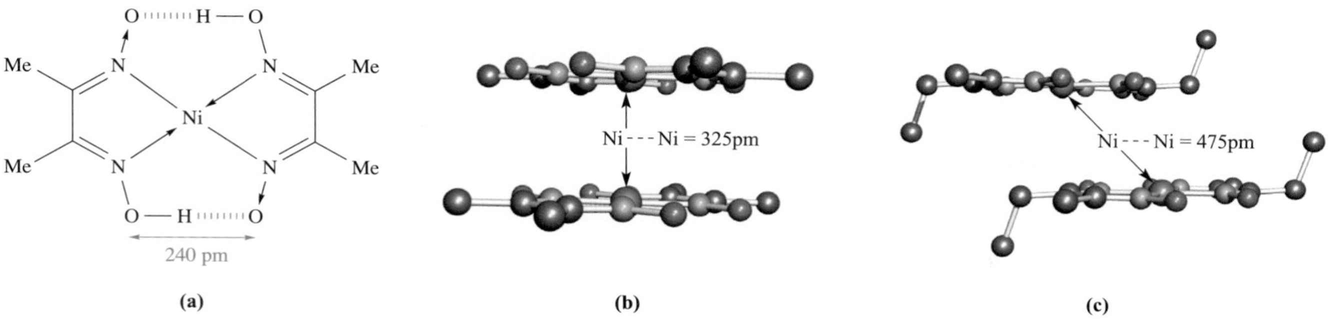

(a)

(b)

(c)

Fig. 21.30 (a) Representation of the square planar structure of bis(dimethylglyoximato)nickel(II), [Ni(Hdmg)$_2$]; (b) in the solid state, molecules of [Ni(Hdmg)$_2$] pack in vertical columns with relatively short Ni····Ni distances [X-ray diffraction data: D.E. Williams *et al.* (1959) *J. Am. Chem. Soc.*, vol. 81, p. 755]; but, (c) in bis(ethylmethylglyoximato)nickel(II), the packing is not so efficient [X-ray diffraction data: E. Frasson *et al.* (1960) *Acta Crystallogr.*, vol. 13, p. 893]. Hydrogen atoms are omitted; colour code: Ni, green; N, blue; O, red; C, grey.

Magnetic moments of *octahedral* Ni(II) complexes are usually close to the spin-only value of 2.83 μ_B. In contrast, *tetrahedral* complexes possess magnetic moments $\approx 4\,\mu_B$ due to orbital contributions (see Section 20.10), and *square planar* complexes such as [Ni(CN)$_4$]$^{2-}$ (eq. 21.94) are *diamagnetic*. These differences in magnetic moments are invaluable in providing information about the coordination geometry in a Ni(II) complex.

$$[Ni(OH_2)_6]^{2+} + 4[CN]^- \longrightarrow [Ni(CN)_4]^{2-} + 6H_2O \quad (21.94)$$
$$\text{yellow}$$

The red square planar complex bis(dimethylglyoximato)-nickel(II), [Ni(Hdmg)$_2$]† (Fig. 21.30a), is used for gravimetric determination of nickel; Ni(II) is precipitated along with Pd(II) when the ligand H$_2$dmg in weakly ammoniacal solution is used as a reagent. The specificity for Ni^{2+} arises from the low solubility of [Ni(Hdmg)$_2$], *not* its high stability constant. All complexes of type [M(Hdmg)$_2$], where M^{2+} is a first row *d*-block metal ion, have stability constants of the same order. The low solubility of [Ni(Hdmg)$_2$] can be rationalized in terms of its solid state structure. Strong hydrogen bonding links the two ligands (Fig. 21.30a) and plays a role in determining a square planar structure. As a consequence of the molecular framework being planar, molecules in the crystalline solid are able to assemble into 1-dimensional stacks such that intermolecular Ni····Ni separations are 325 pm (Fig. 21.30b; but contrast [Cu(Hdmg)$_2$] in Section 21.12). Bis(ethylmethylglyoximato)-nickel(II) has a related structure, but the bulkier ligand forces the molecules to pack less efficiently

(Fig. 21.30c). The fact that the latter complex is more soluble than [Ni(Hdmg)$_2$] supports a structure–solubility relationship.

For some Ni(II) complexes, there is only a small energy difference between structure types. In Section 19.7, we stated that *both* trigonal bipyramidal and square-based pyramidal [Ni(CN)$_5$]$^{3-}$ ions (eq. 21.95) are present in crystals of [Cr(en)$_3$][Ni(CN)$_5$]·1.5H$_2$O. In the anhydrous salt, however, the anions are square-based pyramidal. It is impossible to give a simple interpretation of these observations which may be attributed to a 'subtle balance of steric and electronic effects'.

$$[Ni(CN)_4]^{2-} + \text{excess } [CN]^- \longrightarrow [Ni(CN)_5]^{3-} \quad (21.95)$$

The preference between different 4- and 6-coordination geometries for a number of Ni(II) systems is often marginal and examples are as follows.

Octahedral-planar

- [Ni(ClO$_4$)$_2$(py)$_4$] exists in a blue, paramagnetic *trans*-octahedral form and as a yellow diamagnetic salt containing square planar [Ni(py)$_4$]$^{2+}$ ions.
- Salicylaldoxime (2-HOC$_6$H$_4$CH=NOH) reacts with Ni(II) to give colourless crystals of the square planar complex **21.53**, but on dissolving in pyridine, a green solution of the paramagnetic octahedral [Ni(2-OC$_6$H$_4$CH=NOH)$_2$(py)$_2$] forms.

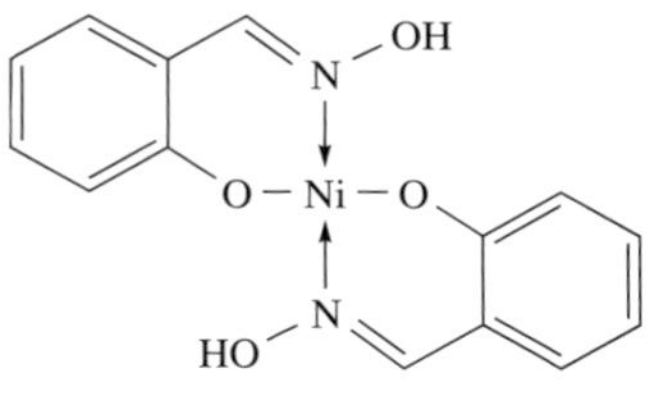

(21.53)

---

† For an introduction to the use of [Ni(Hdmg)$_2$] and related complexes in template syntheses of macrocycles, see: E.C. Constable (1999) *Coordination Chemistry of Macrocyclic Compounds*, OUP, Oxford (Chapter 4).

Tetrahedral-planar

● Halides of type NiL_2X_2 are generally planar when $L = $ trialkylphosphane, but tetrahedral when L is triarylphosphane; when $X = Br$ and $L = PEtPh_2$ or $P(CH_2Ph)Ph_2$, both forms are known (scheme 21.96).

$$NiBr_2 + PEtPh_2$$

↓ ethanol

$$[NiBr_2(PEtPh_2)_2]$$
green, tetrahedral
$\mu_{eff} = 3.18\,\mu_B$

(21.96)

slow isomerization at 298 K ↕ dissolve in CS_2

$$[NiBr_2(PEtPh_2)_2]$$
brown, square planar
diamagnetic

Nickel(I)

Nickel(I) complexes are rather rare, but this oxidation state is thought to be involved in the catalytic function of nickel-containing enzymes such as [NiFe]-hydrogenase (see Fig. 29.19). Dark red $K_4[Ni_2(CN)_6]$ can be prepared by Na amalgam reduction of $K_2[Ni(CN)_4]$. It is diamagnetic and the anion has structure **21.54** in which the $Ni(CN)_3$ units are mutually perpendicular. The reaction of $K_4[Ni_2(CN)_6]$ with water liberates H_2 and forms $K_2[Ni(CN)_4]$.

$$\left[\begin{array}{c} CN \\ | \\ NC-Ni-Ni-CN \\ | \\ CN \end{array} \begin{array}{c} CN \\ NC \end{array} \right]^{4-}$$

(21.54)

Self-study exercises

1. Sketch and label an Orgel diagram for an octahedral d^8 ion. Include the multiplicities in the term symbols.
   [*Ans.* See Fig. 20.20; multiplicity $= 3$]

2. Why are tetrahedral Ni(II) complexes paramagnetic whereas square planar complexes are diamagnetic? Give an example of each type of complex.
   [*Ans.* See worked example 20.1]

3. Draw the structure of H_2dmg. Explain how the presence of intramolecular hydrogen bonding in $[Ni(Hdmg)_2]$ results in a preference for a square planar over tetrahedral structure.
   [*Ans.* See Fig. 21.30a; $O-H\cdots O$ not possible in tetrahedral structure]

THEORY

Box 21.9 Copper: from antiquity to present day

5000–4000 BC	Copper metal used in tools and utensils; heat is used to make the metal malleable.
4000–2000 BC	In Egypt, copper is cast into specific shapes; bronze (an alloy with tin) is made; first copper mining in Asia Minor, China and North America.
2000 BC–0	Bronze weapons are introduced; bronze is increasingly used in decorative pieces.
0–AD 200	Brass (an alloy of copper and zinc) is developed.
AD 200–1800	A period of little progress.
1800–1900	Deposits of copper ores in Michigan, US, are mined, increasing dramatically the US output and availability of the metal. The presence of copper in plants and animals is first discovered.
1900–1960	The electrical conducting properties of copper are discovered and, as a result, many new applications.
1960 onwards	North American production continues to increase but the world market also reaps the benefits of mining in many other countries, in particular Chile; copper recycling becomes important. From mid-1980s, materials such as $YBa_2Cu_3O_{7-x}$ are discovered to be high-temperature superconductors (see Section 28.4).

Table adapted from: R.R. Conry and K.D. Karlin (1994) 'Copper: inorganic & coordination chemistry' in *Encyclopedia of Inorganic Chemistry*, ed. R.B. King, Wiley, Chichester, vol. 2, p. 829.

21.12 Group 11: copper

The metal

Copper is the least reactive of the first row metals. It is not attacked by non-oxidizing acids in the absence of air (eq. 21.97), but it reacts with hot concentrated sulfuric acid (eq. 21.98) and with HNO_3 of all concentrations (eqs. 15.118 and 15.119).

$$Cu^{2+} + 2e^- \rightleftharpoons Cu \qquad\qquad E^o = +0.34\,V \qquad (21.97)$$

$$Cu + 2H_2SO_4 \xrightarrow{\text{conc}} SO_2 + CuSO_4 + 2H_2O \qquad (21.98)$$

In the presence of air, Cu reacts with many dilute acids (the green patina on roofs in cities is basic copper sulfate) and also dissolves in aqueous NH_3 to give $[Cu(NH_3)_4]^{2+}$. When heated strongly, Cu combines with O_2 (eq. 21.99).

$$2Cu + O_2 \xrightarrow{\Delta} 2CuO \xrightarrow{>1300\,K} Cu_2O + \tfrac{1}{2}O_2 \qquad (21.99)$$

Heating Cu with F_2, Cl_2 or Br_2 produces the corresponding dihalide.

Copper is the only first row d-block metal to exhibit a *stable* +1 oxidation state. In aqueous solution, Cu(I) is unstable by a relatively small margin with respect to Cu(II) and the metal (eqs. 21.97, 21.100 and 21.101).

$$Cu^+ + e^- \rightleftharpoons Cu \qquad E^o = +0.52\,V \qquad (21.100)$$

$$Cu^{2+} + e^- \rightleftharpoons Cu^+ \qquad E^o = +0.15\,V \qquad (21.101)$$

This disproportionation is usually fast, but when aqueous Cu(I) is prepared by reduction of Cu(II) with V(II) or Cr(II), decomposition in the absence of air takes several hours. Copper(I) can be stabilized by the formation of an insoluble compound (e.g. CuCl) or a complex (e.g. $[Cu(CN)_4]^{3-}$) (see Section 8.4). The stable oxidation state may depend on reaction conditions: e.g. when Cu powder reacts with aqueous $AgNO_3$, reaction 21.102 takes place, but in MeCN reaction 21.103 occurs.

$$Cu + 2Ag^+ \xrightarrow{\text{aq. solution}} Cu^{2+} + 2Ag \qquad (21.102)$$

$$Cu + [Ag(NCMe)_4]^+ \rightarrow [Cu(NCMe)_4]^+ + Ag \qquad (21.103)$$

Copper(0) is rarely stabilized. The unstable $Cu_2(CO)_6$ has been isolated in a matrix at low temperature. The highest oxidation state attained for copper is +4.

Copper(IV) and copper(III)

Copper(IV) is rare. It exists in the red Cs_2CuF_6 which is made by fluorinating $CsCuCl_3$ at 520 K. The $[CuF_6]^{2-}$ ion is low-spin d^7 and has a Jahn–Teller distorted octahedral structure. Copper(IV) oxide has been prepared in a matrix

by vaporizing the metal and co-depositing it with O_2. Spectroscopic data are consistent with a linear structure, $O{=}Cu{=}O$.

High-pressure fluorination of a mixture of CsCl and $CuCl_2$ gives $Cs_3[CuF_6]$. Green $K_3[CuF_6]$ is similarly prepared and has a magnetic moment of 3.01 μ_B indicative of octahedral Cu(III). The diamagnetic compounds $K[CuO_2]$ and $K_7[Cu(IO_6)_2]$ contain square planar Cu(III) (structures **21.55** and **21.56**).

(21.55)

(21.56)

(21.57)

Ligands that stabilize Cu(III) include 1,2-dithiooxalate. Reaction of $[C_2O_2S_2]^{2-}$ with $CuCl_2$ produces $[Cu^{II}(C_2O_2S_2)_2]^{2-}$, oxidation of which by $FeCl_3$ gives $[Cu^{III}(C_2O_2S_2)_2]^-$ (**21.57**). This readily undergoes a photo-induced two-electron intramolecular transfer, cleaving one of the C–C bonds and releasing two equivalents of SCO.

Probably the most important use of Cu(III) species is in high-temperature superconductors such as $YBa_2Cu_3O_{7-x}$ ($x \approx 0.1$) which are discussed in Chapter 28.

Copper(II)

Cupric is the old name for copper(II). Throughout copper(II) chemistry, Jahn–Teller distortions are observed as predicted for an octahedral d^9 ion, although the degree of distortion varies considerably.

White CuF_2 (made, like $CuCl_2$ and $CuBr_2$, from the elements) has a distorted rutile structure (Fig. 6.22) with elongated CuF_6-units (four Cu–F = 193 pm, two Cu–F = 227 pm). In moist air, CuF_2 turns blue as it forms the dihydrate. Copper(II) chloride forms yellow or brown deliquescent crystals and forms the green-blue $CuCl_2{\cdot}2H_2O$ on standing in moist air. The structure of anhydrous $CuCl_2$

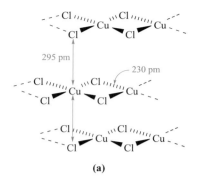

(a)

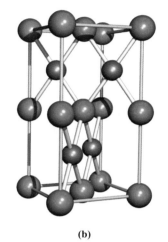

(b)

Fig. 21.31 (a) Representation of the solid state structure of CuCl$_2$ in which chains stack to place each Cu(II) centre in a distorted octahedral environment; (b) the *cooperite* (PtS) structure adopted by CuO with Cu^{2+} (square planar) and O^{2-} (distorted tetrahedral) centres shown in brown and red respectively. The edges of the unit cell are defined by the yellow lines.

(Fig. 21.31a) consists of chains so stacked that each Cu(II) centre is in a distorted octahedral site. In solid CuCl$_2$·2H$_2$O (**21.58**), *trans*-square planar molecules are arranged so that there are weak intermolecular Cu····Cl interactions. Above 570 K, CuCl$_2$ decomposes to CuCl and Cl$_2$. Black CuBr$_2$ has a distorted CdI$_2$ structure (Fig. 6.23). Copper(II) iodide is not known.

$a = 193; b = 228; c = 291$ pm

(21.58)

Black CuO is made by heating the elements (eq. 21.99) or by thermal decomposition of solid Cu(NO$_3$)$_2$ or CuCO$_3$

(eq. 21.104). Its structure consists of square planar CuO$_4$ units linked by bridging O atoms into chains; these lie in a criss-cross arrangement so that each O atom is in a distorted tetrahedral site. Figure 21.31b shows a unit cell of this lattice which is an example of the *cooperite* (PtS) structure type. Below 225 K, CuO is antiferromagnetic. One use of CuO is as a black pigment in ceramics.

$$CuCO_3 \xrightarrow{\Delta} CuO + CO_2 \qquad (21.104)$$

Blue Cu(OH)$_2$ precipitates when [OH]$^-$ is added to aqueous solutions of Cu^{2+}. Cu(OH)$_2$ dissolves in acids and also in concentrated aqueous alkalis in which an ill-defined hydroxido species is formed. Copper(II) hydroxide is readily dehydrated to CuO.

Aqueous solutions of Cu^{2+} contain the [Cu(OH$_2$)$_6$]$^{2+}$ ion and this has been isolated in several salts including Cu(ClO$_4$)$_2$·6H$_2$O and the Tutton salt [NH$_4$]$_2$Cu[SO$_4$]$_2$·6H$_2$O (see Section 21.6). The solid state structures of both salts reveal distortions of [Cu(OH$_2$)$_6$]$^{2+}$ such that there are *three* pairs of Cu—O distances, e.g. in Cu(ClO$_4$)$_2$·6H$_2$O the Cu—O bond lengths are 209, 216 and 228 pm. Crystals of the blue hydrated sulfate CuSO$_4$·5H$_2$O (*blue vitriol*) contain square planar [Cu(OH$_2$)$_4$]$^{2+}$ units with two sulfate O atoms completing the remaining sites in an elongated octahedral coordination sphere. The solid state structure consists of a hydrogen-bonded assembly which incorporates the non-coordinated H$_2$O molecules. The pentahydrate loses water in stages on heating (eq. 21.105 and self-study exercise 2 after worked example 4.2) and finally forms the white, hygroscopic anhydrous CuSO$_4$.

$$CuSO_4 \cdot 5H_2O \xrightarrow[-2H_2O]{300\,K} CuSO_4 \cdot 3H_2O$$
$$\xrightarrow[-2H_2O]{380\,K} CuSO_4 \cdot H_2O \xrightarrow[-H_2O]{520\,K} CuSO_4$$
$$(21.105)$$

Copper(II) sulfate and nitrate are commercially available and, in addition to uses as precursors in Cu(II) chemistry, they are used as fungicides, e.g. *Bordeaux mixture* contains CuSO$_4$ and Ca(OH)$_2$ and when added to water forms a basic copper(II) sulfate which acts as the antifungal agent. Copper(II) nitrate is widely used in the dyeing and printing industries. It forms hydrates Cu(NO$_3$)$_2$·xH$_2$O where $x = 2.5$, 3 or 6. The blue hexahydrate readily loses water at 300 K to give green Cu(NO$_3$)$_2$·3H$_2$O. Anhydrous Cu(NO$_3$)$_2$ is made from Cu and N$_2$O$_4$: reaction 9.78 followed by decomposition of [NO][Cu(NO$_3$)$_3$] so formed. The solid state structure of α-Cu(NO$_3$)$_2$ consists of Cu(II) centres linked into an infinite lattice by bridging [NO$_3$]$^-$

ligands (**21.59**). At 423 K, the solid volatilizes *in vacuo* giving molecular $Cu(NO_3)_2$ (**21.60**).

(21.59)

(21.60)

The salt $Cu(O_2CMe)_2 \cdot H_2O$ is dimeric and is structurally similar to $[Cr_2(OH_2)_2(\mu-O_2CMe)_4]$ (see Fig. 21.15 for structure type) but lacks the strong metal–metal bonding. The distance between the two Cu centres of 264 pm is greater than in the bulk metal (256 pm). The magnetic moment of 1.4 μ_B per Cu(II) centre (i.e. less than μ(spin-only) of 1.73 μ_B) suggests that in $[Cu_2(OH_2)_2(\mu-O_2CMe)_4]$ there is only weak antiferromagnetic coupling between the unpaired electrons. On cooling, the magnetic moment decreases. These observations can be explained in terms of the two unpaired electrons giving a singlet ground state ($S = 0$) and a low-lying triplet excited state ($S = 1$) which is thermally populated at 298 K but which becomes less populated as the temperature is lowered (see Section 20.6 for singlet and triplet states).

Vast numbers of copper(II) complexes are known and this discussion covers only simple species. Jahn–Teller distortions are generally observed (d^9 configuration). Halido complexes include $[CuCl_3]^-$, $[CuCl_4]^{2-}$ and $[CuCl_5]^{3-}$ but the solid state structures of species possessing these stoichiometries are highly dependent on the counter-ions. For example, $[Ph_4P][CuCl_3]$ contains dimers (**21.61**), whereas $K[CuCl_3]$ and $[Me_3NH]_3[CuCl_3][CuCl_4]$ contain chains of distorted, face-sharing octahedra (Fig. 21.32a). The latter salt also contains discrete tetrahedral $[CuCl_4]^{2-}$ ions. $[PhCH_2CH_2NH_2Me]_2[CuCl_4]$ crystallizes in two forms, one with distorted tetrahedral and the other with square planar $[CuCl_4]^{2-}$ ions. The salt $[NH_4]_2[CuCl_4]$ has a polymeric structure containing distorted octahedral Cu(II) centres. Dimeric $[Cu_2Cl_8]^{4-}$ (with edge-sharing trigonal bipyramidal Cu(II) centres) may be stabilized by very bulky cations, e.g. $[M(en)_3]_2[Cu_2Cl_8]Cl_2$ (M = Co, Rh or Ir, Fig. 21.32b). The $[CuCl_5]^{3-}$ ion is trigonal bipyramidal in the Cs^+ and $[Me_3NH]^+$ salts, but in $[\mathbf{21.62}][CuCl_5]$, it is square-based pyramidal.

(21.61)   **(21.62)**

Complexes containing *N*- and *O*-donor ligands are very common, and coordination numbers of 4, 5 and 6 predominate. We have already mentioned the aqua species $[Cu(OH_2)_6]^{2+}$ and $[Cu(OH_2)_4]^{2+}$. When NH_3 is added to aqueous Cu^{2+}, only four aqua ligands in $[Cu(OH_2)_6]^{2+}$ are replaced (see Section 20.12), but salts of $[Cu(NH_3)_6]^{2+}$ can be made in liquid NH_3. $[Cu(en)_3]^{2+}$ is formed in very concentrated aqueous solutions of 1,2-ethanediamine. Deep blue aqueous $[Cu(NH_3)_4](OH)_2$ (formed when $Cu(OH)_2$ is dissolved in aqueous NH_3) has the remarkable property of dissolving cellulose, and if the resulting solution

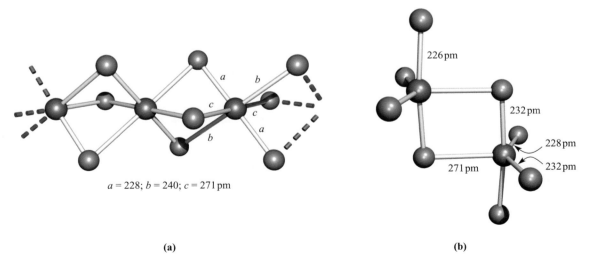

$a = 228; b = 240; c = 271$ pm

226 pm
232 pm
228 pm
271 pm   232 pm

(a)   **(b)**

Fig. 21.32 The structures (X-ray diffraction) of (a) polymeric $[CuCl_3]_n^{n-}$ in the salt $[Me_3NH]_3[CuCl_3][CuCl_4]$; the $[CuCl_4]^{2-}$ ion in this salt is tetrahedral [R.M. Clay *et al.* (1973) *J. Chem. Soc., Dalton Trans.*, p. 595] and (b) the $[Cu_2Cl_8]^{4-}$ ion in the salt $[Rh(en)_3]_2[Cu_2Cl_8]Cl_2 \cdot 2H_2O$ [S.K. Hoffmann *et al.* (1985) *Inorg. Chem.*, vol. 24, p. 1194]. Colour code: Cu, brown; Cl, green.

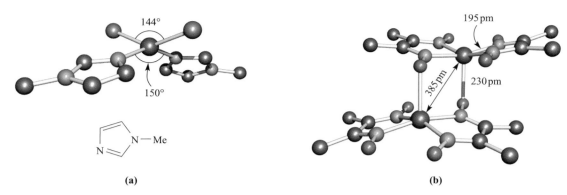

Fig. 21.33 (a) The flattened tetrahedral structure of [CuCl$_2$(Meim)$_2$] (determined by X-ray diffraction) and a schematic representation of the *N*-methylimidazole (Meim) ligand [J.A.C. van Ooijen *et al.* (1979) *J. Chem. Soc., Dalton Trans.*, p. 1183]; (b) [Cu(Hdmg)$_2$] forms dimers in the solid state, in contrast to [Ni(Hdmg)$_2$] (Fig. 21.30); structure determined by X-ray diffraction [A. Vaciago *et al.* (1970) *J. Chem. Soc. A*, p. 218]. Colour code: Cu, brown; N, blue; Cl, green; O, red; C, grey.

is squirted into acid, the synthetic fibre *rayon* is produced as cellulose is precipitated. The reaction is historically important as a means of producing rayon. Further examples of complexes with *N*- and *O*-donor ligands are:

- tetrahedral (flattened): [Cu(NCS-*N*)$_4$]$^{2-}$; [CuCl$_2$(Meim)$_2$] (Fig. 21.33a);
- square planar: [Cu(ox)$_2$]$^{2-}$; *cis*- and *trans*-[Cu(H$_2$NCH$_2$CO$_2$)$_2$]; [Cu(en)(NO$_3$-*O*)$_2$];
- trigonal bipyramidal: [Cu(NO$_3$-*O*)$_2$(py)$_3$] (equatorial nitrates); [Cu(CN){N(CH$_2$CH$_2$NH$_2$)$_3$}]$^+$ (axial cyanide);
- square-based pyramidal: [Cu(NCS-*N*)(**21.63**)]$^+$ (ligand **21.63** is tetradentate in the basal sites); [Cu(OH$_2$)(phen)(**21.64**)] (apical H$_2$O), [CuCl$_2$(OH$_2$)$_2$(MeOH)] (apical MeOH, *trans* Cl in the basal sites);
- octahedral: [Cu(HOCH$_2$CH$_2$OH)$_3$]$^{2+}$; [Cu(bpy)$_3$]$^{2+}$; [Cu(phen)$_3$]$^{2+}$; *trans*-[CuCl(OH$_2$)(en)$_2$]$^+$; *trans*-[Cu(BF$_4$)$_2$(en)$_2$] (see below).

(21.63)   **(21.64)**

Jahn–Teller distortions are apparent in many complexes. In [Cu(bpy)$_3$]$^{2+}$, the distortion is particularly severe with equatorial Cu–N bonds of 203 pm, and axial distances of 223 and 245 pm. The complex *trans*-[Cu(BF$_4$)$_2$(en)$_2$] illustrates the ability of [BF$_4$]$^-$ to act as a monodentate ligand; the long Cu–F bonds (256 pm) indicate rather weak Cu–F interactions. In Section 21.11, we described the structure of [Ni(Hdmg)$_2$]; [Cu(Hdmg)$_2$] also exhibits hydrogen bonding between the ligands but, in the solid

state, molecules are associated in *pairs* with the coordination sphere being square-based pyramidal (Fig. 21.33b).

A practical application of the coordination of *N*,*O*-donors to Cu(II) is the *biuret test* for peptides and proteins. Compounds containing peptide linkages form a violet complex ($\lambda_{max} = 540$ nm) when treated in NaOH solution with a few drops of aqueous CuSO$_4$. The general form of the complex can be represented by that of **21.65**, in which the ligand is the doubly deprotonated form of biuret, H$_2$NC(O)NHC(O)NH$_2$.

(21.65)

When a Cu(II) salt is treated with excess KCN at room temperature, cyanogen is evolved and the copper reduced (eq. 21.106). However, in aqueous methanol at low temperatures, violet square planar [Cu(CN)$_4$]$^{2-}$ forms.

$$2Cu^{2+} + 4[CN]^- \longrightarrow 2CuCN(s) + C_2N_2 \qquad (21.106)$$

Some copper-containing complexes are studied as models for bioinorganic systems (see Chapter 29).

Copper(I)

Cuprous is the old name for copper(I). The Cu$^+$ ion has a d^{10} configuration and salts are diamagnetic and colourless except when the counter-ion is coloured or when charge transfer absorptions occur in the visible region, e.g. in red Cu$_2$O.

Copper(I) fluoride is not known although the CuF unit is stabilized in the tetrahedral complex [CuF(PPh$_3$)$_3$]. CuCl, CuBr and CuI are white solids and are made by reduction of a Cu(II) salt in the presence of halide ions, e.g. CuBr forms when SO$_2$ is bubbled through an aqueous solution of CuSO$_4$ and KBr. Copper(I) chloride has a zinc blende structure (see Fig. 6.19). The γ-forms of CuBr and CuI adopt the zinc blende structure but convert to the β-forms (wurtzite structure, Fig. 6.21) at 660 and 690 K respectively. Values of K_{sp}(298 K) for CuCl, CuBr and CuI are 1.72×10^{-7}, 6.27×10^{-9} and 1.27×10^{-12}. Copper(I) iodide precipitates when any Cu(II) salt is added to KI solution (eq. 21.107).

$$2Cu^{2+} + 4I^- \longrightarrow 2CuI + I_2 \qquad (21.107)$$

Anions and ligands available in solution strongly influence the relative stabilities of Cu(I) and Cu(II) species. The very low solubility of CuI is crucial to reaction 21.107 which occurs despite the fact that the E^o values of the Cu^{2+}/Cu$^+$ and I$_2$/I$^-$ couples are $+0.15$ and $+0.54$ V respectively. However, in the presence of 1,2-ethanediamine or tartrate, which form stable complexes with Cu^{2+}, I$_2$ oxidizes CuI.

Copper(I) cyanide (eq. 21.106) is commercially available. This polymorph converts to a high-temperature form at 563 K. Both polymorphs contain chains (**21.66**). In the high-temperature form, the chains are linear (as in AgCN and AuCN), but in the low-temperature form, each chain adopts an unusual 'wave-like' configuration. The two polymorphs can be interconverted at room temperature by use of aqueous KBr under the conditions shown in scheme 21.108.

$$\cdots\cdots Cu - C \equiv N \longrightarrow Cu - C \equiv N \cdots\cdots$$

(21.66)

CuCN
Low-temperature polymorph
$\xrightarrow{\text{KBr(aq), 298 K}}$
$\xleftarrow{\text{Limited H}_2\text{O}}$
KCu$_2$(CN)$_2$Br.H$_2$O

KBr(aq), 298 K ↑ | Large amounts of H$_2$O ↓

CuCN
High-temperature polymorph   (21.108)

Copper(I) hydride is obtained by reduction of Cu(II) salts with H$_3$PO$_2$ and crystallizes with the wurtzite structure. It decomposes when treated with acids, liberating H$_2$.

Red copper(I) oxide may be made by oxidation of Cu (reaction 21.99), but is more readily obtained by reduction of Cu(II) compounds in alkaline media. When Fehling's solution (Cu^{2+} in aqueous alkaline sodium tartrate) is added to a reducing sugar such as glucose, Cu$_2$O

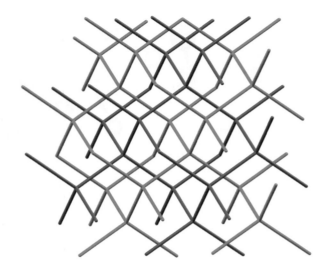

Fig. 21.34 The solid state structure of Cu$_2$O (cuprite) consists of two interpenetrating 3-dimensional networks (shown in red and blue). Each network has an anti-cristobalite structure.

precipitates. This is a qualitative test for reducing sugars. The solid state structure of Cu$_2$O is based on an anti-β-cristobalite (SiO$_2$, Fig. 6.20c) i.e. with Cu(I) in linear sites and O^{2-} in tetrahedral sites. Because the Cu$_2$O framework is particularly open, the crystal consists of two interpenetrating frameworks (Fig. 21.34), and the Cu$_2$O, *cuprite*, structure is a structural prototype. Copper(I) oxide is used as a red pigment in ceramics, porcelain glazes and glasses. Cu$_2$O has fungicidal properties and is added to certain paints as an antifouling agent. It is insoluble in water, but dissolves in aqueous NH$_3$ to give colourless [Cu(NH$_3$)$_2$]$^+$ (**21.67**); the solution readily absorbs O$_2$ and turns blue as [Cu(NH$_3$)$_4$]$^{2+}$ forms.

$$\left[H_3N \longrightarrow Cu \longleftarrow NH_3 \right]^+$$

(21.67)

In acidic solutions, Cu$_2$O disproportionates (eq. 21.109).

$$Cu_2O + H_2SO_4 \longrightarrow CuSO_4 + Cu + H_2O \qquad (21.109)$$

Complex **21.67** illustrates a linear environment for Cu(I). The most common geometry is tetrahedral, and 3-coordinate species also occur. Halide complexes exhibit great structural diversity and the identity of the cation is often crucial in determining the structure of the anion. For example, [CuCl$_2$]$^-$ (formed when CuCl dissolves in concentrated HCl) may occur as discrete, linear anions (**21.68**) or as a polymer with tetrahedral Cu(I) centres (**21.69**). Trigonal planar [CuCl$_3$]$^{2-}$ has been isolated, e.g. in [Me$_4$P]$_2$[CuCl$_3$], but association into discrete, halido-bridged anions is also possible, e.g. [Cu$_2$I$_4$]$^{2-}$ (**21.70**), [Cu$_2$Br$_5$]$^{3-}$ (Fig. 21.35a) and [Cu$_4$Br$_6$]$^{2-}$ (Fig. 21.35b). An unusual *linear* Cu−Br−Cu bridge links

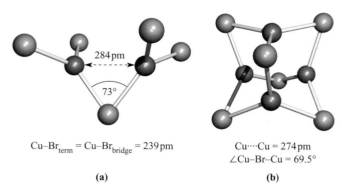

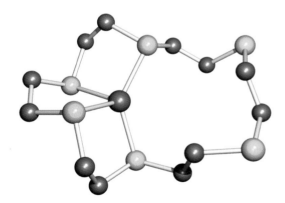

(21.70)

(21.71)

Fig. 21.35 The structures (X-ray diffraction) of (a) $[Cu_2Br_5]^{3-}$ in the $[Me_4N]^+$ salt [M. Asplund *et al.* (1985) *Acta Chem. Scand., Ser. A*, vol. 39, p. 47] and (b) $[Cu_4Br_6]^{2-}$ in the $[^nPr_4N]^+$ salt [M. Asplund *et al.* (1984) *Acta Chem. Scand. Ser. A*, vol. 38, p. 725]. In both, the Cu(I) centres are in trigonal planar environments and in $[Cu_4Br_6]^{2-}$, the copper atoms are in a tetrahedral arrangement; the Cu····Cu distances are longer than in the bulk metal. Colour code: Cu, brown; Br, pink.

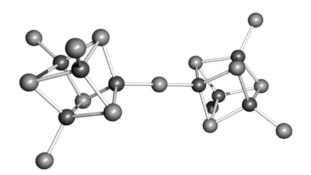

Fig. 21.36 The structure (X-ray diffraction at 203 K) of the mixed-valence $[Cu_8Br_{15}]^{6-}$ ion in the compound $[MePh_3P]_6[Cu_8Br_{15}]$ [G.A. Bowmaker *et al.* (1999) *Inorg. Chem.*, vol. 38, p. 5476]. Colour code: Cu, brown; Br, pink.

two cubane-like sub-units in the mixed-valence anion $[Cu_8Br_{15}]^{6-}$ (Fig. 21.36). This ion formally contains one Cu(II) and seven Cu(I) centres, but structural and EPR spectroscopic properties and theoretical calculations are consistent with delocalized bonding. Complexation between Cu(I) and $[CN]^-$ can lead to $[Cu(CN)_2]^-$ (polymeric **21.71** as in the K^+ salt), $[Cu(CN)_3]^{2-}$ (trigonal planar, **19.2**) or $[Cu(CN)_4]^{3-}$ (tetrahedral).

Copper(I) is a soft metal centre (Table 7.9) and tends to interact with soft donor atoms such as S and P, although complex formation with *O*- and *N*-donor ligands is well documented. Many complexes with *S*-donor ligands are known, and the propensity of sulfur to form bridges leads to many multinuclear complexes, e.g. $[(S_6)Cu(\mu\text{-}S_8)Cu(S_6)]^{4-}$ (Fig. 16.12), $[Cu_4(SPh)_6]^{2-}$ (which is structurally related to $[Cu_4Br_6]^{2-}$ with $[SPh]^-$ replacing Br^- bridges), and $[\{Cu(S_2O_3)_2\}_n]$ (structurally related to **21.69** with *S*-bonded thiosulfates replacing Cl^- bridges). We have seen several times in this chapter how macrocyclic ligands may impose unusual coordination numbers on metal ions, or, if the ring is large enough, may wrap around a metal ion, e.g. in $[Co(\mathbf{21.46})]^{2+}$. In $[Cu(\mathbf{21.46})]^+$ (Fig. 21.37), the preference for the Cu^+ ion to be tetrahedrally coordinated means that it interacts with only four of the six donor atoms of the macrocycle.

Cu–Br$_{term}$ = Cu–Br$_{bridge}$ = 239 pm

(a)

Cu····Cu = 274 pm
∠Cu–Br–Cu = 69.5°

(b)

$\begin{bmatrix} Cl - Cu - Cl \end{bmatrix}^-$

Cu–Cl = 209 pm

(21.68)

Cu–Cl = 235 pm

(21.69)

Fig. 21.37 The structure of $[Cu(\mathbf{21.46})]^+$ (ligand **21.46** is an S_6-macrocycle) determined by X-ray diffraction for the $[BF_4]^-$ salt; the Cu^+ is in a distorted tetrahedral environment [J.R. Hartman *et al.* (1986) *J. Am. Chem. Soc.*, vol. 108, p. 1202]. Hydrogen atoms are omitted; colour code: Cu, brown; S, yellow; C, grey.

Self-study exercises

1. 'Octahedral' Cu(II) complexes are often described as having a (4 + 2)-coordination pattern. Suggest the origin of this description.

   [*Ans.* See structure 20.5 and discussion]

2. Values of log K_n for the displacement of H_2O ligands in $[Cu(OH_2)_6]^{2+}$ by NH_3 ligands are 4.2, 3.5, 2.9, 2.1 and -0.52 for $n = 1, 2, 3, 4$ and 5 respectively. A value for $n = 6$ cannot be measured in aqueous solution. Comment on these data.

   [*Ans.* See Fig. 20.37 and discussion]

3. CuO adopts a cooperite structure. Confirm the stoichiometry of the compound from the unit cell shown in Fig. 21.31b.

4. The $[Cu_3Cl_{12}]^{6-}$ ion contains one tetragonally distorted, octahedral Cu(II) centre and two tetrahedral Cu(II) centres. The ion is centrosymmetric. Draw the structure of the anion, and comment on what is meant by a 'tetragonally distorted octahedral' environment.

5. In liquid NH_3, the standard reduction potentials of the couples Cu^{2+}/Cu^+ and $Cu^+/Cu(s)$ (relative to $H^+/H_2(g)$) are $+0.44$ and $+0.36$ V, respectively. These values are $+0.15$ and $+0.52$ V under aqueous conditions. Calculate K for the equilibrium:

   $$2Cu^+ \rightleftharpoons Cu^{2+} + Cu(s)$$

   in liquid NH_3 and in aqueous solution at 298 K, and comment on the significance of the results.

   [*Ans.* K(liquid NH_3)=0.045; K(aq) $= 1.8 \times 10^6$]

21.13 Group 12: zinc

The metal

Zinc is not attacked by air or water at room temperature, but the hot metal burns in air and decomposes steam, forming ZnO. Zinc is much more reactive than Cu (compare eqs. 21.110 and 21.97), liberating H_2 from dilute mineral acids and from alkalis (eq. 21.111). With hot concentrated sulfuric acid, reaction 21.112 occurs. The products of reactions with HNO_3 depend on temperature and acid concentration. On heating, Zn reacts with all the halogens to give ZnX_2, and combines with elemental S and P.

$$Zn^{2+} + 2e^- \rightleftharpoons Zn \qquad E^o = -0.76\,V \qquad (21.110)$$

$$Zn + 2NaOH + 2H_2O \longrightarrow Na_2[Zn(OH)_4] + H_2 \qquad (21.111)$$

$$Zn + 2H_2SO_4 \xrightarrow{\text{hot, conc}} ZnSO_4 + SO_2 + 2H_2O \qquad (21.112)$$

The first (Sc) and last (Zn) members of the first row of the *d*-block exhibit a more restricted range of oxidation states than the other metals, and the chemistry of Zn is essentially confined to that of Zn(II). The $[Zn_2]^{2+}$ ion (analogues of which are well established for the heavier group 10 metals) has only been established in a yellow diamagnetic glass obtained by cooling a solution of metallic Zn in molten $ZnCl_2$. It rapidly disproportionates (eq. 21.113).

$$[Zn_2]^{2+} \longrightarrow Zn^{2+} + Zn \qquad (21.113)$$

Since the electronic configuration of Zn^{2+} is d^{10}, compounds are colourless and diamagnetic. There is no LFSE associated with the d^{10} ion and, as the discussion below shows, no particular geometry is preferred for Zn^{2+}. There are some similarities with Mg, and many compounds of Zn are isomorphous with their Mg analogues.

Zinc(II)

Binary halides are best made by action of HF, HCl, Br_2 or I_2 on hot Zn. ZnF_2 is also prepared by thermal decomposition of $Zn(BF_4)_2$. The vapours of the halides contain linear molecules. Solid ZnF_2 adopts a rutile structure (Fig. 6.22) and has a high lattice energy and melting point. Evidence for significant covalent character is apparent in the structures and properties of $ZnCl_2$, $ZnBr_2$ and ZnI_2 which possess layer structures, have lower melting points than ZnF_2 (Fig. 21.38) and are soluble in a range of organic solvents. The water solubility of ZnF_2 is low, but $ZnCl_2$, $ZnBr_2$ and ZnI_2 are highly soluble. Uses of $ZnCl_2$ are varied, e.g. in some fireproofings, wood preservation, as an astringent, in deodorants and, combined with NH_4Cl, as a soldering flux.

Zinc hydride is made by reaction 21.114 (or from LiH and $ZnBr_2$) and is a fairly stable solid at 298 K.

$$ZnI_2 + 2NaH \xrightarrow{THF} ZnH_2 + 2NaI \qquad (21.114)$$

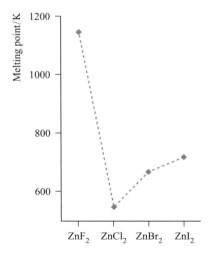

Fig. 21.38 Trend in melting points of the zinc halides.

Zinc is of great commercial significance and ZnO (made from Zn and O_2) is the most important compound of zinc (see Section 21.2). It is a white solid with the wurtzite structure (Fig. 6.21) at 298 K. It turns yellow on heating and in this form is a semiconductor owing to loss of oxygen and production of some interstitial Zn atoms. Zinc oxide is amphoteric, dissolving in acids to give solutions containing $[Zn(OH_2)_6]^{2+}$ or derivatives thereof (some anions coordinate to Zn^{2+}). Hydrolysis of $[Zn(OH_2)_6]^{2+}$ occurs to give various solution species resulting from H^+ loss. In alkalis, ZnO forms zincates such as $[Zn(OH)_4]^{2-}$ (**21.72**). This ion also forms when $Zn(OH)_2$ dissolves in aqueous alkalis. Zinc hydroxide is water-insoluble; there are five polymorphs of which ε-$Zn(OH)_2$ (distorted β-cristobalite structure, Fig. 6.20c) is thermodynamically the most stable.

(21.72)

Zinc sulfide occurs naturally as the minerals *zinc blende* and, more rarely, *wurtzite*. These are structural prototypes (see Section 6.11). It is a light-sensitive white solid and, on exposure to cathode- or X-rays, it fluoresces and is used in fluorescent paints and radar screens. Adding Cu to ZnO results in a green phosphorescence after exposure to light, and other colour variations are achieved by using different additives. The conversion of ZnS to ZnO by roasting in air is the commercial method of producing the oxide.

Other Zn(II) compounds that are commercially available include the carbonate, sulfate and nitrate. The sulfate is very soluble in water. Crystals of $ZnSO_4 \cdot 7H_2O$ form on evaporating solutions from reactions of Zn, ZnO, $Zn(OH)_2$ or $ZnCO_3$ with aqueous H_2SO_4. Dehydration initially occurs on heating, followed by decomposition (eq. 21.115).

$$ZnSO_4 \cdot 7H_2O \xrightarrow[-7H_2O]{520\,K} ZnSO_4 \xrightarrow{1020\,K} ZnO + SO_3$$

$$(21.115)$$

Insoluble $ZnCO_3$ occurs naturally as *smithsonite*, but the mineral tends to be coloured owing to the presence of, for example, Fe(II). The carbonate is usually purchased as the basic salt $ZnCO_3 \cdot 2Zn(OH)_2 \cdot xH_2O$ and is used in calamine lotion.

Zinc nitrate can be obtained as one of several hydrates, of which $Zn(NO_3)_2 \cdot 6H_2O$ is the most common. Anhydrous $Zn(NO_3)_2$ is made from Zn and N_2O_4 since heating the hydrates yields hydroxy salts. The hexahydrates of $Zn(NO_3)_2$ and $Zn(ClO_4)_2$ contain octahedral $[Zn(OH_2)_6]^{2+}$ in the solid state. Similarly, it is possible to isolate salts

containing $[Zn(NH_3)_6]^{2+}$ from reactions done in liquid NH_3, e.g. $ZnCl_2 \cdot 6NH_3$. However, in aqueous solution, $[Zn(NH_3)_6]^{2+}$ exists in equilibrium with tetrahedral $[Zn(NH_3)_4]^{2+}$. Equation 9.25 showed the formation of $[Zn(NH_2)_4]^{2-}$. Basic zinc acetate $[Zn_4(\mu_4\text{-}O)(\mu\text{-}O_2CMe)_6]$ is isostructural with its Be(II) analogue (Fig. 12.7), but is more readily hydrolysed in water. Another salt of interest is $Zn(acac)_2 \cdot H_2O$ (**21.73**) in which the coordination of Zn^{2+} is square-based pyramidal.

(21.73)

Our discussion of Zn(II) compounds has introduced complexes including $[Zn(OH_2)_6]^{2+}$, $[Zn(NH_3)_6]^{2+}$, $[Zn(NH_3)_4]^{2+}$, $[Zn(OH)_4]^{2-}$ and $[Zn(acac)_2(OH_2)]$, exemplifying octahedral, tetrahedral and square-based pyramidal coordination. Large numbers of Zn(II) complexes are known (some interest arises from developing models for Zn-containing bioinorganic systems, see Chapter 29) and coordination numbers of 4 to 6 are the most common. Zinc(II) is a borderline hard/soft ion and readily complexes with ligands containing a range of donor atoms, e.g. hard *N*- and *O*- and soft *S*-donors.

(21.74)   **(21.75)**

Tetrahedral $[ZnCl_4]^{2-}$ and $[ZnBr_4]^{2-}$ can be formed from $ZnCl_2$ and $ZnBr_2$ and many salts are known. Salts of $[ZnI_4]^{2-}$ are stabilized using large cations. Crystallographic data for '$[ZnCl_3]^-$' salts usually reveal the presence of $[Zn_2Cl_6]^{2-}$ (**21.74**), and in coordinating solvents, tetrahedral $[ZnCl_3(solv)]^-$ is present. Salts such as $K[ZnCl_3] \cdot H_2O$ contain $[ZnCl_3(OH_2)]^-$ (**21.75**) in the solid state. A similar picture is true for '$[ZnBr_3]^-$' and '$[ZnI_3]^-$' salts; both $[Zn_2Br_6]^{2-}$ and $[Zn_2I_6]^{2-}$ have been confirmed in the solid state.

The structure of $Zn(CN)_2$ is an *anticuprite* lattice with $[CN]^-$ groups bridging between tetrahedral Zn(II) centres, and two interpenetrating networks. (See Fig. 21.34 for the cuprite structure.) In contrast, $[Zn(CN)_4]^{2-}$ exists as discrete tetrahedral ions, as do $[Zn(N_3)_4]^{2-}$ and $[Zn(NCS\text{-}N)_4]^{2-}$. Just as it is possible to isolate both $[Zn(NH_3)_4]^{2+}$ and

$[Zn(NH_3)_6]^{2+}$, pairs of tetrahedral $[ZnL_2]^{2+}$ and octahedral $[ZnL_3]^{2+}$ complexes (L = en, bpy, phen) are also known.

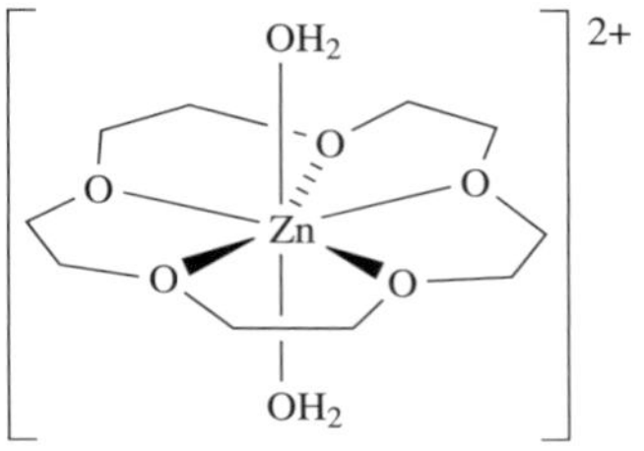

(21.76)   **(21.77)**

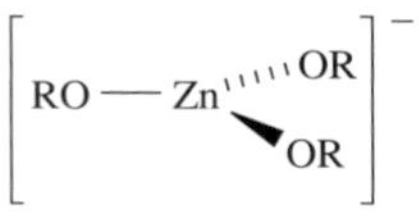

Examples of high coordination numbers for Zn^{2+} are rare, but include pentagonal bipyramidal $[Zn(15\text{-crown-}5)(OH_2)_2]^{2+}$ (**21.76**), and dodecahedral $[Zn(NO_3)_4]^{2-}$ (structurally similar to $[Co(NO_3)_4]^{2-}$, Fig. 21.28c).

By using a sterically demanding aryloxide ligand, it is possible to isolate a 3-coordinate (trigonal planar) Zn(II) complex, structure **21.77**.

Zinc(I)

In eq. 21.57, we illustrated the use of a β-diketiminato ligand, L, to stabilize Mn(I) in a dinuclear complex. The same ligand is capable of stabilizing the Zn_2L_2 complex shown in scheme 21.116. The complex formally contains a $\{Zn_2\}^{2+}$ core, in which the Zn–Zn bond distance is 236 pm. This is the second example of a compound containing a Zn–Zn bond, the first being the organometallic species $(\eta^5\text{-}C_5Me_5)_2Zn_2$ (Zn–Zn = 230.5 pm, see Fig. 24.23).

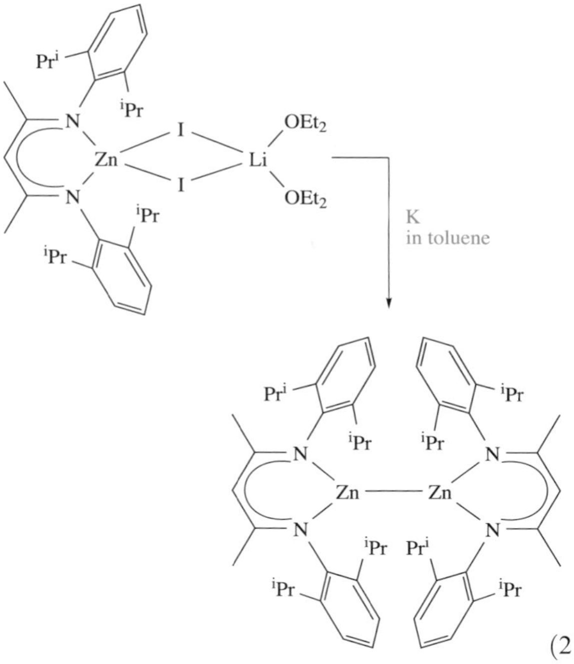

(21.116)

Self-study exercises

1. Explain why Zn(II) compounds are diamagnetic, irrespective of the coordination environment of the Zn^{2+} ion.   [*Ans. d^{10} and see Figs. 20.8 and 20.11*]

2. Do you expect Zn^{2+} to form stable, octahedral complexes with π-acceptor ligands? Give reasons for your answer.
[*Ans. See end of Section 20.4*]

KEY TERMS

The following terms were introduced in this chapter. Do you know what they mean?

- ❏ isopolyanion
- ❏ polyoxometallate
- ❏ homopolyanion
- ❏ heteropolyanion
- ❏ thermochromic

FURTHER READING

See also further reading suggested for Chapters 19 and 20.

F.A. Cotton (2000) *J. Chem. Soc., Dalton Trans.*, p. 1961 – 'A millennial overview of transition metal chemistry'.
F.A. Cotton, G. Wilkinson, M. Bochmann and C. Murillo (1999) *Advanced Inorganic Chemistry*, 6th edn, Wiley

Interscience, New York – One of the best detailed accounts of the chemistry of the *d*-block metals.
J. Emsley (1998) *The Elements*, 3rd edn, Oxford University Press, Oxford – An invaluable source of data for the elements.

N.N. Greenwood and A. Earnshaw (1997) *Chemistry of the Elements*, 2nd edn, Butterworth-Heinemann, Oxford – A very good account including historical, technological and structural aspects; the metals in each triad are treated together.

J. McCleverty (1999) *Chemistry of the First-row Transition Metals*, Oxford University Press, Oxford – An introductory text dealing with the metals Ti to Cu.

S. Riedel and M. Kaupp (2009) *Coord. Chem. Rev.*, vol. 253, p. 606 – 'The highest oxidation states of the transition metal elements'.

J. Silver (ed.) (1993) *Chemistry of Iron*, Blackie, London – A series of articles covering different facets of the chemistry of iron.

A.F. Wells (1984) *Structural Inorganic Chemistry*, 5th edn, Clarendon Press, Oxford – An excellent source for detailed structural information of, in particular, binary compounds.

PROBLEMS

21.1 Write out, in sequence, the first row *d*-block elements and give the valence electronic configuration of each metal and of its M^{2+} ion.

21.2 Comment on the variation in oxidation states of the first row metals.

21.3 In the complex $[Ti(BH_4)_3(MeOCH_2CH_2OMe)]$, the Ti(III) centre is 8-coordinate. Suggest modes of coordination for the ligands.

21.4 Comment on each of the following observations. (a) Li_2TiO_3 forms a continuous range of solid solutions with MgO. (b) When $TiCl_3$ is heated with concentrated aqueous NaOH, H_2 is evolved.

21.5 An acidified solution of $0.1000 \, mol \, dm^{-3}$ ammonium vanadate ($25.00 \, cm^3$) was reduced by SO_2 and, after boiling off excess reductant, the blue solution remaining was found to require addition of $25.00 \, cm^3$ $0.0200 \, mol \, dm^{-3}$ $KMnO_4$ to give a pink colour to the solution. Another $25.00 \, cm^3$ portion of the vanadate solution was shaken with Zn amalgam and then immediately poured into excess of the ammonium vanadate solution; on titration of the resulting solution with the $KMnO_4$ solution, $74.5 \, cm^3$ of the latter was required. Deduce what happened in these experiments.

21.6 Give equations to describe what happens to VBr_3 on heating.

21.7 The magnetic moment of $[NH_4]V(SO_4)_2 \cdot 12H_2O$ is $2.8 \, \mu_B$ and the electronic absorption spectrum of an aqueous solution contains absorptions at $17\,800$, $25\,700$ and $34\,500 \, cm^{-1}$. Explain these observations.

21.8 Suggest the formula and structure of the mononuclear complex formed between Cr^{3+} and ligand **21.78**. Comment on possible isomerism.

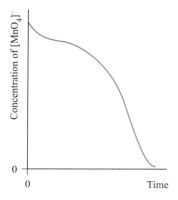

(21.78)

21.9 Use data from Appendix 11 to predict qualitatively the outcome of the following experiment at 298 K: Cr is dissolved in excess of molar $HClO_4$ and the solution is shaken in air.

21.10 Figure 21.39 shows the change in concentration of $[MnO_4]^-$ with time during a reaction with acidified oxalate ions. (a) Suggest a method of monitoring the reaction. (b) Explain the shape of the curve.

Fig. 21.39 Figure for problem 21.10.

21.11 Comment on the modes of bonding of the ligands in the Mn(II) complexes listed at the end of Section 21.8, drawing attention to any conformational restrictions.

21.12 How would you (a) distinguish between the formulations $Cu^{II}Fe^{II}S_2$ and $Cu^{I}Fe^{III}S_2$ for the mineral *chalcopyrite*, (b) show that Fe^{3+} is a hard cation, and (c) show that the blue compound precipitated when a solution of $[MnO_4]^-$ in concentrated aqueous KOH is reduced by $[CN]^-$ contains Mn(V)?

21.13 Give equations for the following reactions: (a) heating Fe with Cl_2; (b) heating Fe with I_2; (c) solid $FeSO_4$ with concentrated H_2SO_4; (d) aqueous Fe^{3+} with $[SCN]^-$; (e) aqueous Fe^{3+} with $K_2C_2O_4$; (f) FeO with dilute H_2SO_4; (g) aqueous $FeSO_4$ and NaOH.

21.14 How would you attempt to (a) estimate the crystal field stabilization energy of FeF_2, and (b) determine the overall stability constant of $[Co(NH_3)_6]^{3+}$ in aqueous solution given that the overall formation constant for $[Co(NH_3)_6]^{2+}$ is 10^5, and:

$$Co^{3+}(aq) + e^- \rightleftharpoons Co^{2+}(aq) \qquad E^o = +1.92\,V$$

$$[Co(NH_3)_6]^{3+}(aq) + e^- \rightleftharpoons [Co(NH_3)_6]^{2+}(aq)$$
$$E^o = +0.11\,V$$

21.15 Suggest why Co_3O_4 adopts a normal rather than inverse spinel structure.

21.16 Give explanations for the following observations. (a) The complex $[Co(en)_2Cl_2]_2[CoCl_4]$ has a room temperature magnetic moment of 3.71 μ_{eff}. (b) The room temperature magnetic moment of $[CoI_4]^{2-}$ (e.g. 5.01 μ_B for the $[Et_4N]^+$ salt) is larger than that of salts of $[CoCl_4]^{2-}$.

21.17 (a) When $[CN]^-$ is added to aqueous Ni^{2+} ions, a green precipitate forms; if excess KCN is added, the precipitate dissolves to give a yellow solution and at high concentrations of $[CN]^-$, the solution becomes red. Suggest an explanation for these observations. (b) If the yellow compound from part (a) is isolated and reacted with Na in liquid NH_3, a red, air-sensitive, diamagnetic product can be isolated. Suggest its identity.

21.18 Treatment of an aqueous solution of $NiCl_2$ with $H_2NCHPhCHPhNH_2$ gives a blue complex ($\mu_{eff} = 3.30\,\mu_B$) which loses H_2O on heating to form a yellow, diamagnetic compound. Suggest explanations for these observations and comment on possible isomerism in the yellow species.

21.19 Give equations for the following reactions: (a) aqueous NaOH with $CuSO_4$; (b) CuO with Cu in concentrated HCl at reflux; (c) Cu with concentrated HNO_3; (d) addition of aqueous NH_3 to a precipitate of $Cu(OH)_2$; (e) $ZnSO_4$ with aqueous NaOH followed by addition of excess NaOH; (f) ZnS with dilute HCl.

21.20 (a) Compare the solid state structures of $[M(Hdmg)_2]$ for M = Ni and Cu and comment on the fact that $[Cu(Hdmg)_2]$ is more soluble in water than is $[Ni(Hdmg)_2]$. (b) Suggest the likely structural features of $[Pd(Hdmg)_2]$.

21.21 Copper(II) chloride is not completely reduced by SO_2 in concentrated HCl solution. Suggest an explanation for this observation and state how you would try to establish if the explanation is correct.

21.22 When the ligands do not sterically control the coordination geometry, do 4-coordinate complexes of (a) Pd(II), (b) Cu(I) and (c) Zn(II) prefer to be square planar or tetrahedral? Explain your answer. In the absence of crystallographic data, how could you distinguish between a square planar or tetrahedral structure for a Ni(II) complex?

21.23 Write down formulae for the following ions: (a) manganate(VII); (b) manganate(VI); (c) dichromate(VI); (d) vanadyl; (e) vanadate (*ortho* and *meta*); (f) hexacyanidoferrate(III). Give an alternative name for manganate(VII).

21.24 Give a brief account of the variation in properties of binary oxides of the first row *d*-block metals on going from Sc to Zn.

21.25 Give an overview of the formation of halido complexes of type $[MX_n]^{m-}$ by the first row *d*-block metal ions, noting in particular whether discrete ions are present in the solid state.

21.26 When iron(II) oxalate (oxalate = ox^{2-}) is treated with H_2O_2, H_2ox and K_2ox, a green compound **X** is obtained. **X** reacts with aqueous NaOH to give hydrated Fe_2O_3, and is decomposed by light with production of iron(II) oxalate, K_2ox and CO_2. Analysis of **X** shows it contains 11.4% Fe and 53.7% ox^{2-}. Deduce the formula of **X** and write equations for its reaction with alkali and its photochemical decomposition. State, with reasons, whether you would expect **X** to be chiral.

21.27 Dimethyl sulfoxide (DMSO) reacts with cobalt(II) perchlorate in EtOH to give a pink compound **A** which is a 1:2 electrolyte and has a magnetic moment of 4.9 μ_B. Cobalt(II) chloride also reacts with DMSO, but in this case the dark blue product, **B**, is a 1:1 electrolyte, and the magnetic moment of **B** is 4.6 μ_B per Co centre. Suggest a formula and structure for **A** and **B**.

21.28 When H_2S is passed into a solution of copper(II) sulfate acidified with H_2SO_4, copper(II) sulfide precipitates. When concentrated H_2SO_4 is heated with metallic Cu, the principal sulfur-containing product is SO_2 but a residue of copper(II) sulfide is also formed. Account for these reactions.

OVERVIEW PROBLEMS

21.29 (a) Write an equation to represent the discharge of an alkaline electrolyte cell containing a Zn anode and $BaFeO_4$ cathode.

(b) The first charge transfer band for $[MnO_4]^-$ occurs at $18\,320\,cm^{-1}$, and that for $[MnO_4]^{2-}$ at $22\,940\,cm^{-1}$. Explain the origin of these absorptions, and comment on the trend in relative energies on going from $[MnO_4]^{2-}$ to $[MnO_4]^-$.

(c) Explain why FeS_2 adopts a NaCl structure rather than a structure in which the cation:anion ratio is 1:2.

21.30 (a) The value of μ_{eff} for $[CoF_6]^{3-}$ is $5.63\,\mu_B$. Explain why this value does not agree with the value for μ calculated from the spin-only formula.

(b) By using a simple MO approach, rationalize why one-electron oxidation of the bridging ligand in $[(CN)_5CoOOCo(CN)_5]^{6-}$ leads to a shortening of the O–O bond.

(c) Salts of which of the following complex ions might be expected to be formed as racemates: $[Ni(acac)_3]^-$, $[CoCl_3(NCMe)]^-$, *cis*-$[Co(en)_2Cl_2]^+$, *trans*-$[Cr(en)_2Cl_2]^+$?

21.31 (a) The electronic absorption spectrum of $[Ni(DMSO)_6]^{2+}$ $(DMSO = Me_2SO)$ exhibits three absorptions at 7728, 12970 and $24\,038\,cm^{-1}$. Assign these absorptions.

(b) CuF_2 has a distorted rutile structure (four $Cu–F = 193\,pm$ and two $Cu–F = 227\,pm$ per Cu centre); $[CuF_6]^{2-}$ and $[NiF_6]^{3-}$ are distorted octahedral ions. Explain the origins of these distortions.

(c) Dissolution of vanadium metal in aqueous HBr leads to a complex '$VBr_3 \cdot 6H_2O$'. X-ray diffraction data reveal that the compound contains a complex cation containing a centre of symmetry. Suggest a formulation for the compound, and a structure for the cation.

21.32 The complex $[V_2L_4]$, where HL is diphenylformamidine, is diamagnetic. Each L^- ligand acts as a bridging, *N*,*N'*-donor such that the complex is structurally similar to complexes of the type $[Cr_2(O_2CR)_4]$. (a) Describe a bonding scheme

for the $[V_2]^{4+}$ core and derive the formal metal–metal bond order in $[V_2L_4]$. (b) The reaction of $[V_2L_4]$ with KC_8 in THF results in the formation of $K(THF)_3[V_2L_4]$. What is the role of KC_8 in this reaction? (c) Do you expect the V–V bond length to increase or decrease on going from $[V_2L_4]$ to $K(THF)_3[V_2L_4]$? Rationalize your answer.

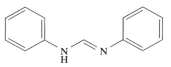

Diphenylformamidine (HL)

21.33 (a) The ligand 1,4,7-triazacyclononane, L, forms the nickel complexes $[NiL_2]_2[S_2O_6]_3 \cdot 7H_2O$ and $[NiL_2][NO_3]Cl \cdot H_2O$. X-ray diffraction data for these complexes reveal that in the cation in $[NiL_2][NO_3]Cl \cdot H_2O$, the Ni–N bond lengths lie in the range 209–212 pm, while in $[NiL_2]_2[S_2O_6]_3 \cdot 7H_2O$, two Ni–N bonds (mutually *trans*) are of length 211 pm and the remaining Ni–N bonds are in the range 196–199 pm. Rationalize these data.

1,4,7-triazacyclononane

(b) Suggest why some reports of the properties of low-spin $[Fe(bpy)_3]^{2+}$ state that its salts possess very low magnetic moments.

(c) The ligand HL can be represented as follows:

What is the term given to these forms of HL? The conjugate base of HL forms the complexes *mer*-$[VL_3]^-$ and $[V(Me_2NCH_2CH_2NMe_2)L_2]$. Draw the structure of *mer*-$[VL_3]^-$, and the structures of the possible isomers of $[V(Me_2NCH_2CH_2NMe_2)L_2]$.

INORGANIC CHEMISTRY MATTERS

21.34 Vanadium(IV) complexes act as mimics for insulin, a hormone secreted by the pancreas. Among the complexes being studied is [VOL$_2$] in which HL is maltol:

O

OH

O

maltol

Figure 21.40 shows the pH dependence of the species in aqueous solution containing [VO]$^{2+}$ and HL in a 1 : 2 ratio. (a) Suggest a structure for [VOL$_2$]. (b) For which ion is [VO]$^{2+}$ an abbreviation? (c) Rationalize the shapes of the curves in Fig. 21.40 and suggest structures for the species present. (d) Why are studies such as that summarized in Fig. 21.40 important in the development of anti-diabetes drugs?

21.35 The tanning process in the manufacture of leather relies upon the interaction of Cr^{3+} with the fibrous protein collagen. Although glycine and L-proline are the most important amino acids (see Table 29.2) in collagen, glutamic acid (pK_a = 3.8) and aspartic acid (pK_a = 4.2) are also present. During tanning, the pH of an aqueous solution of Cr(OH)SO$_4$ is lowered from $\approx$2.8 to 3.8. (a) What chromium(III) ion is present in aqueous solution at very low pH and why is H$^+$ needed to stabilize this species? (b) In the absence of collagen, aqueous solutions of Cr^{3+} at pH 3.8 contain linear, tri- and tetranuclear

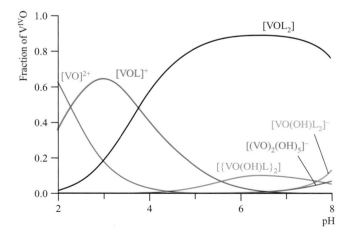

Fig. 21.40 Speciation curves for complexes formed in an aqueous solution of vanadium(IV) and maltol (HL) with a metal : ligand ratio of 1 : 2. [Redrawn with permission from T. Kiss *et al.* (2000) *J. Inorg. Biochem.*, vol. 80, p. 65, Elsevier.]

species. Suggest structures for these species. (c) A study [A.D. Covington *et al.* (2001) *Polyhedron*, vol. 20, p. 461] of the interaction of chromium(III) with collagen states that, at pH 3.8, carboxylate groups should compete with hydroxido ligands for chromium. The study concludes that the predominant chromium-containing species bound to leather is a linear oligomer with nuclearity 2 or 3. Using this information, suggest how Cr^{3+} interacts with collagen and indicate how Cr^{3+} may be involved in cross-linking of collagen fibres. (d) A major concern in the tanning industry is to prevent toxic waste arising from the oxidation of Cr(III) to Cr(VI). Chromium(VI) may be present as [Cr$_2$O$_7$]$^{2-}$ and two other anions, depending upon the pH. Write equilibria to show how the three Cr(VI) species are interrelated, and suggest a method to remove them from waste water.

21.36 The compound shown below is a formazan dye usually referred to as 'zincon'. It is used to detect Zn^{2+} and Cu^{2+} ions:

OH   HO$_2$C

HO$_3$S

N   HN

N   N

(a) Suggest how the ligand binds to Zn^{2+} or Cu^{2+}, and comment on the role of pH in determining the overall charge of the complex. (b) Why does the ligand include a SO$_3$H substituent? (c) Zincon itself absorbs at 463 nm. Suggest how the absorption arises. (d) When zincon binds Cu^{2+}, the absorption at 463 nm is replaced by one at 600 nm. Why does this make zincon an easy method of detection for Cu^{2+} ions? (e) The copper(II) complex of zincon can be used as a sensor for [CN]$^-$ ions in aqueous solution. Addition of [CN]$^-$ results in the disappearance of the absorption at 600 nm and reappearance of the absorption at 463 nm. Outline the chemical changes occurring in solution.

Try also end-of-chapter problems: 8.32, 8.33, 8.35 and 8.36.

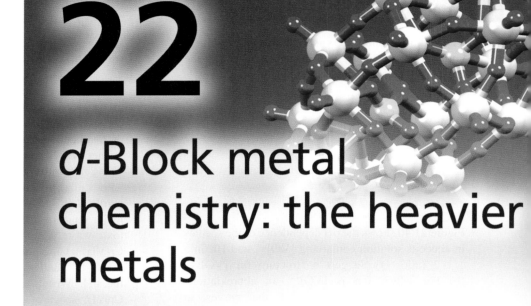

Topics

Occurrence and extraction
Applications
Physical properties
Inorganic and coordination chemistry of the second and the third row metals

22
d-Block metal chemistry: the heavier metals

1–2	3	4	5	6	7	8	9	10	11	12	13–18
s-block											*p*-block
	Sc	Ti	V	Cr	Mn	Fe	Co	Ni	Cu	Zn	
	Y	Zr	Nb	Mo	Tc	Ru	Rh	Pd	Ag	Cd	
	La	Hf	Ta	W	Re	Os	Ir	Pt	Au	Hg	

22.1 Introduction

Chapter 21 dealt with descriptive chemistry of the first row *d*-block metals and, in this chapter, we focus on the second and third row metals (the *heavier metals*). Reasons for discussing the lighter and heavier metals separately were given in Section 21.1.

Lanthanum, La, is commonly classified with the lanthanoids (see Fig. 1.14) even though 'lanthanoid' means 'like lanthanum' and La is strictly a group 3 metal. Because of the chemical similarity of La to the elements Ce−Lu, they are considered together in Chapter 27. The only mention of La in this chapter is its occurrence.

22.2 Occurrence, extraction and uses

Figure 22.1 shows the relative abundances of the second and third row *d*-block metals. Compared with the first row metals (Fig. 21.1), the abundances of some of the heavier metals are very low, e.g. Os, 1×10^{-4} ppm and Ir, 6×10^{-6} ppm; Tc does not occur naturally. *Yttrium* and *lanthanum* are similar to the lanthanoids and occur with them in nature. The major yttrium and lanthanum ores are *monazite* (a mixed metal phosphate, $(Ce,La,Nd,Pr,Th,Y...)PO_4$) and *bastnäsite* $((Ce,La,Y...)CO_3F)$; their composition varies, e.g. an 'yttrium-rich' mineral might contain $\leq 1\%$ Y, a 'lanthanum-rich' one up to 35% La. The extraction of yttrium involves conversion to YF_3 or YCl_3 followed by reduction with Ca or K respectively; the separation of lanthanoid metals is described in Section 27.5. The most important use of yttrium is in phosphors (high purity Y_2O_3 and YVO_4) for television and computer displays. Although traditional uses in TVs containing cathode-ray tubes have declined, phosphors remain in use in some flat screen displays, e.g. plasma TVs. Yttrium is also used in corrosion-resistant alloys, and in the formation of yttrium garnets for microwave filters and synthetic gemstones (yttrium aluminium garnets, YAG, $Al_5Y_3O_{12}$).

Zirconium is the next most abundant *d*-block metal in the Earth's crust after Fe, Ti and Mn, and is present to quite a large extent in lunar rock samples collected in the Apollo missions. Zirconium and *hafnium* occur naturally together and are hard to separate. Hf is rarer than Zr, 5.3 and 190 ppm, respectively, of the Earth's crust. The main ores are *baddeleyite* (ZrO_2), *zircon* $((Zr,Hf)SiO_4, <2\%$ Hf) and *alvite* $((Zr,Hf)SiO_4 \cdot xH_2O, <2\%$ Hf). Extraction of Zr involves reduction of ZrO_2 by Ca, or conversion of ZrO_2 to K_2ZrF_6 (by treatment with K_2SiF_6) followed by reduction. Both Zr and Hf can be produced from zircon by reaction sequence 22.1. The mixture of metals so obtained is used for strengthening steel.

$$MO_2 \xrightarrow{CCl_4, 770\,K} MCl_4 \xrightarrow{Mg\ under\ Ar,\ 1420\,K} M$$

$$(M = Zr\ or\ Hf) \quad (22.1)$$

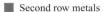

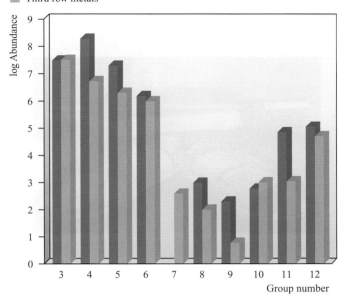

Fig. 22.1 Relative abundances of the second and third row *d*-block metals in the Earth's crust. The data are plotted on a logarithmic scale, and the units of abundance are parts per 10^9. Technetium (group 7) does not occur naturally.

Zirconium has a high corrosion resistance and low cross-section for neutron capture and is used for cladding fuel rods in water-cooled nuclear reactors. For this application, Zr must be free of Hf, which is a very good neutron absorber. The main use of pure Hf is in nuclear reactor control rods. Zirconium and hafnium compounds possess similar lattice energies and solubilities, and their complexes have similar stabilities. This means that separation techniques (e.g. ion exchange, solvent extraction) encounter the same problems as those of the lanthanoids. Very pure metals can be obtained by zone refining (see Box 6.3) or by thermal decomposition of the iodides on a hot metal filament. Zirconium compounds have a range of catalytic applications. Uses of ZrO_2 are described in Section 22.5. In Box 15.4, we highlighted applications of Hf and Zr nitrides.

Niobium (formerly called *columbium*) and *tantalum* occur together in the mineral *columbite* $(Fe,Mn)(Nb,Ta)_2O_6$. When the mineral is Nb-rich, it is called *niobite* and when Ta-rich, *tantalite*. Fusion of the ore with alkali gives poly-niobates and -tantalates, and further treatment with dilute acid yields Nb_2O_5 and Ta_2O_5. One method of separation utilizes the more basic character of Ta: at a controlled concentration of HF and KF in aqueous solution, the oxides are converted to $K_2[NbOF_5]$ and $K_2[TaF_7]$. The former is more water-soluble than the latter. The modern separation technique is fractional extraction from aqueous HF solution into

methyl isobutyl ketone. Niobium is used in the manufacture of tough steels and superalloys which are used in the aerospace industry, e.g. in frameworks designed for the Gemini space program (the forerunner to the Apollo moon missions). Superconducting magnets (e.g. in MRI equipment, see Box 4.3) contain NbTi metallic multicore conductors. Tantalum is very high melting (mp 3290 K) and extremely resistant to corrosion by air and water. It is used in corrosion-resistant alloys, e.g. for construction materials in the chemical industry. The inertness of the metal makes it suitable for use in surgical appliances including prostheses. Tantalum has wide application in the manufacture of electronic components, in particular, capacitors that are used in mobile phones and personal computers.

The German for tungsten is *wolfram*, hence the symbol W. Although *molybdenum* and *tungsten* compounds are usually isomorphous, the elements occur separately. The major Mo-containing ore is *molybdenite* (MoS_2) and the metal is extracted by reactions 22.2. Tungsten occurs in *wolframite* ($(Fe,Mn)WO_4$) and *scheelite* ($CaWO_4$) and scheme 22.3 shows typical extraction processes.

$$MoS_2 \xrightarrow{\Delta\,(870\,K)\,in\,air} MoO_3 \xrightarrow{H_2,\,870\,K} Mo \qquad (22.2)$$

$$\left.\begin{array}{l} (Fe,Mn)WO_4 \xrightarrow{Na_2CO_3,\,fusion} \underset{insoluble}{(Fe,Mn)_2O_3} + \underset{soluble}{Na_2WO_4} \\[2mm] Na_2WO_4 \xrightarrow{HCl} WO_3 \xrightarrow{H_2,\,870\,K} W \end{array}\right\}$$
$$(22.3)$$

Molybdenum is very hard and high melting (mp 2896 K), and tungsten has the highest melting point (3695 K) of all metals (Table 6.2). Both metals are used in the manufacture of toughened steels (for which wolframite can be reduced directly by Al). Tungsten carbides have extensive use in cutting tools and abrasives. A major use of W metal is in electric light bulb filaments including low-energy halogen lamps. Molybdenum has an essential role in biological systems (see Section 29.1).

Technetium is an artificial element, available as ^{99}Tc (a β-particle emitter, $t_{\frac{1}{2}} = 2.13 \times 10^5$ yr) which is isolated from fission product wastes by oxidation to $[TcO_4]^-$. Separation employs solvent extraction and ion-exchange methods. The $[TcO_4]^-$ ion is the common precursor in technetium chemistry. Technetium metal can be obtained by H_2 reduction of $[NH_4][TcO_4]$ at high temperature. The principal use of Tc compounds is in nuclear medicine where they are important imaging agents (see Box 22.7). *Rhenium* is rare and occurs in small amounts in Mo ores. During roasting (first step in eq. 22.2), volatile Re_2O_7 forms and is deposited in flue dusts. It is dissolved in water and precipitated as $KReO_4$. The two major uses of Re are in petroleum-reforming catalysts and as a component of high-temperature superalloys. Such alloys are used in, for example, heating

ENVIRONMENT

Box 22.1 Environmental catalysts

The platinum-group metals Rh, Pd and Pt play a vital role in keeping the environment devoid of pollutants originating from vehicle exhausts. They are present in catalytic converters (which we discuss in detail in Section 25.8) where they catalyse the conversion of hydrocarbon wastes, CO and NO_x (see Box 15.7) to CO_2, H_2O and N_2. In 2008, the manufacture of catalytic converters used 81% of the rhodium, 47% of palladium and 44% of platinum consumed worldwide. The growth rate of environmental catalyst manufacture by companies such as Johnson Matthey in the UK is driven by legislative measures for the control of exhaust emissions. Regulations in force in the US and Europe have had a major impact on the levels of emissions and have improved the quality of urban air. Tighter control of vehicle emissions has now been introduced in most parts of Asia.

For more details of catalytic converters, see Section 25.8.

Catalytic converters showing interior design. Small particles (≈ 1600 pm diameter) of Pd, Pt or Rh are dispersed on a support such as γ-alumina.

---

elements, thermocouples and filaments for photographic flash equipment and mass spectrometers.

The *platinum-group metals* (Ru, Os, Rh, Ir, Pd and Pt) are rare (Fig. 22.1) and expensive, and occur together either native or in sulfide ores of Cu and Ni. World production of platinum-group metals is dominated by South Africa (59% of world output in 2008) and Russia (26%), with mines in the US, Canada and Zimbabwe producing most of the remainder. The main source of *ruthenium* is from wastes from Ni refining, e.g. from *pentlandite*, (Fe,Ni)S. *Osmium* and *iridium* occur in *osmiridium*, a native alloy with variable composition: 15–40% osmium and 80–50% iridium. *Rhodium* occurs in *native platinum* and in *pyrrhotite* ores ($Fe_{1-n}S$, $n = 0$–0.2, often with $\leq 5\%$ Ni). Native platinum is of variable composition but may contain as much as 86% Pt, other constituents being Fe, Ir, Os, Au, Rh, Pd and Cu. The ore is an important source of *palladium* which is also a side-product of Cu and Zn refining. Besides being obtained native, *platinum* is extracted from *sperrylite* ($PtAs_2$). Extraction and separation methods for the six metals are interlinked, solvent extraction and ion-exchange methods being used.[†] The metals are important heterogeneous catalysts, e.g. Pd for hydrogenation and dehydrogenation, Pt for NH_3 oxidation and hydrocarbon reforming, and Rh and Pt for catalytic converters (see Box 22.1). Uses of Ru and Rh include alloying with Pt and Pd to increase their hardness for use in, for example, the manufacture of electrical components

(e.g. electrodes and thermocouples) and laboratory crucibles. Osmium and iridium have few commercial uses. They are employed to a limited extent as alloying agents; an IrOs alloy is used in pen-nibs. Palladium is widely used in the electronics industry (in printed circuits and multilayer ceramic capacitors). The ability of Pd to absorb large amounts of H_2 (see Section 10.7) leads to it being used in the industrial purification of H_2. Platinum is particularly inert: Pt electrodes[‡] have laboratory applications (e.g. in the standard hydrogen and pH electrodes), and the metal is widely used in electrical wires, thermocouples and jewellery. Platinum-containing compounds such as *cisplatin* (**22.1**) and *carboplatin* (**22.2**) are antitumour drugs, and we discuss these further in Box 22.9.

(22.1)

(22.2)

Silver and *gold* occur native, and in sulfide, arsenide and telluride ores, e.g. argentite (Ag_2S) and *sylvanite* (($Ag,Au)Te_2$). Silver is usually worked from the residues of Cu, Ni or Pb refining and, like Au, can be extracted from all its ores by reaction 22.4, the cyanido complex being reduced to the metal by Zn.[*]

[†] For further discussion, see: P.A. Tasker, P.G. Plieger and L.C. West (2004) in *Comprehensive Coordination Chemistry II*, eds J.A. McCleverty and T.J. Meyer, Elsevier, Oxford, vol. 9, p. 759 – 'Metal complexes for hydrometallurgy and extraction'.

[‡] Microelectrodes are a relatively new innovation; see: G. Denuault (1996) *Chem. & Ind.*, p. 678.
[*] Extraction of gold, see: J. Barrett and M. Hughes (1997) *Chem. Brit.*, vol. 33, issue 6, p. 23 – 'A golden opportunity'.

ENVIRONMENT

Box 22.2 Treatment of cyanide waste

The toxicity of $[CN]^-$ was brought to public attention early in 2000 when a huge spillage of cyanide (originating from gold extraction processes at the Aurul gold mine in Baia Mare, Romania) entered the River Danube and surrounding rivers in Eastern Europe, devastating fish stocks and other river life.

The high toxicity of $[CN]^-$ makes it essential for cyanide-containing waste produced by industry to be treated. Several methods are used. For dilute solutions of cyanide, destruction using hypochlorite solution is common:

$$[CN]^- + [OCl]^- + H_2O \longrightarrow ClCN + 2[OH]^-$$

$$ClCN + 2[OH]^- \longrightarrow Cl^- + [OCN]^- + H_2O \quad (\text{at pH} > 11)$$

$$[OCN]^- + 2H_2O \longrightarrow [NH_4]^+ + [CO_3]^{2-} \quad (\text{at pH} < 7)$$

The operation must be further modified to take into account the large amounts of Cl^- produced. An alternative method is oxidation by H_2O_2:

$$[CN]^- + H_2O_2 \longrightarrow [OCN]^- + H_2O$$

Older methods such as formation of $[SCN]^-$ or complexation to give $[Fe(CN)_6]^{4-}$ are no longer favoured.

The removal of dead fish from the Tisza River in Hungary following the cyanide spill from the Aurul gold mine, Romania, on 9 February 2000.

$$4M + 8[CN]^- + 2H_2O + O_2 \longrightarrow 4[M(CN)_2]^- + 4[OH]^-$$

$$(M = Ag, Au) \quad (22.4)$$

Although the use of cyanide is currently the most important means of extracting gold from its ores, its toxicity (Box 22.2) is a clear disadvantage. Other methods of extraction are therefore being considered, for example the use of ligands such as thiourea, thiocyanate and thiosulfate, which form water-stable gold complexes. Native gold typically contains 85–95% Au with Ag as the second constituent. Silver is used in soldering alloys, high-capacity batteries, electrical equipment and printed circuits. Silver salts were used extensively in the photographic industry, but their importance has declined as the digital camera market has expanded. Silver iodide (in the form of flares or ground-sited acetone–AgI generators) is used in cloud seeding to control rain patterns in certain regions. Gold has been worked since ancient civilization, not only in the usual yellow form, but as red, purple or blue *colloidal gold*. Modern uses of colloidal gold are in electron micro-scope imaging, staining of microscope slides and as colouring agents, e.g. reduction of Au(III) with $SnCl_2$ yields *purple of Cassius*, used in the manufacture of ruby glass. Uses of gold include coinage, the electronics industry and jewellery; *carat* indicates the gold content (24 carat = pure gold). Some gold compounds are used as anti-arthritic drugs. Recycling of Ag and Au (as with other precious metals) is an important way of conserving resources.

Cadmium occurs as the rare mineral *greenockite* (CdS), but the metal is isolated almost entirely from zinc ores, CdS occurring (<0.5%) in ZnS. Being more volatile than Zn, Cd can be collected in the first stage of the distillation of the metal. Cadmium has a relatively low melting point (594 K) and is used as an alloying agent in low-melting alloys. The main use of cadmium is in NiCd batteries (see eq. 21.5). Cadmium selenide and telluride are semiconductors and are employed in the electronics industry. CdTe has potential application in solar cells, although the market currently makes greatest use of

BIOLOGY AND MEDICINE

Box 22.3 Mercury: a highly toxic, liquid metal

The low melting point (234 K) of Hg results in its being a unique metal. Its high thermal expansion coefficient makes it a suitable liquid for use in thermometers, and it has widespread application in barometers, diffusion pumps and in Hg switches in electrical apparatus. The use of mercury cells in the chloralkali process is gradually being phased out (see Box 11.4). Some other metals dissolve in mercury to give *amalgams*; their uses are varied, for example:

- Cd/Hg amalgam is a component in the Weston cell (see **cell 22.5**);
- Na/Hg amalgam is a convenient source of Na as a reducing agent;
- silver amalgam ($\approx$50% Hg, 35% Ag, 13% Sn, 2% Cu by weight) is used for silver fillings in dentistry.

Despite these uses, Hg poses a serious health risk, as do its compounds (e.g. Me_2Hg). Mercury has a low enthalpy of vaporization (59 kJ mol^{-1}), and even below its boiling point (630 K) its volatility is high. At 293 K, a drop of liquid Hg vaporizes at a rate of 5.8 μg h^{-1} cm^{-2}, and at its saturation point the surrounding air contains 13 mg m^{-3}, a level far in excess of safe limits. Similarly, amalgams are a source of Hg vapour, and those in tooth fillings release toxic vapour directly into the human body. Research has shown that brushing teeth and chewing increases the vaporization process. The toxicity is now well established, and steps have been taken in some countries to phase out the use of Hg in dental fillings.

Mercury enters the environment from both industrial and natural sources. A continuously degassing volcano such as Mount Etna emits significant quantities of Hg (for Mount Etna, the rate of Hg emission is about 27 Mg per year). Once inorganic mercury enters water-courses and sediments of lakes, bacteria which normally reduce sulfates convert the metal into the form of methyl mercury, $[MeHg]^+$. In this form, mercury(II) progresses along the food chain, eventually accumulating in fish. The accumulation occurs because the rate of uptake exceeds the rate of mercury excretion by animals, and species high up the food chain (large fish and fish-eating birds and mammals, including man) may accumulate potentially toxic levels of mercury. Methyl mercury is lipophilic and is able to cross the blood–brain barrier. Mercury vapour, Hg(0), that enters the body accumulates in the kidneys, brain and testicles. It is oxidized to Hg(II) and, along with methyl mercury, is readily coordinated by soft sulfur donors present in proteins. The end result of mercury poisoning is severe damage to the central nervous system. One of the reasons why the toxicity is so high is that its retention time in body tissue is especially long, $\approx$65 days in the kidneys. The effects of Hg poisoning were referred to by Lewis Carroll in *Alice in Wonderland* – the hat-making occupation of the Mad Hatter brought him into regular contact with $Hg(NO_3)_2$ which was used in the production of felt hats.

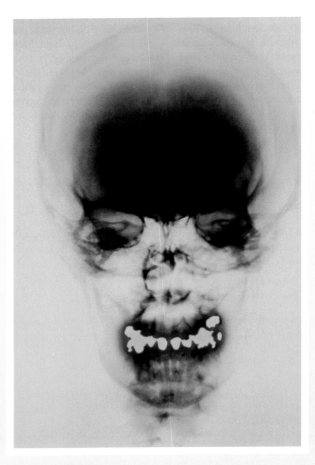

A colour-enhanced dental X-ray showing silver/mercury amalgam fillings (yellow).

Further reading

M.B. Blayney, J.S. Winn and D.W. Nierenberg (1997) *Chem. Eng. News*, vol. 75, May 12 issue, p. 7 – 'Handling dimethyl mercury'.

S.A. Counter and L.H. Buchanan (2004) *Toxicol. Appl. Pharm.*, vol. 198, p. 209 – 'Mercury exposure in children: a review'.

N.J. Langford and R.E. Ferner (1999) *J. Human Hypertension*, vol. 13, p. 651 – 'Toxicity of mercury'.

L. Magos and T.W. Clarkson (2006) *Ann. Clin. Biochem.*, vol. 43, p. 257 – 'Overview of the clinical toxicity of mercury'.

A. Sigel, H. Sigel and R.K.O. Sigel, eds. (2010) *Metal Ions in Life Sciences*, vol. 7, RSC Publishing, Cambridge – A series of reviews on the theme of 'Organometallics in the environment and toxicology'.

M.J. Vimy (1995) *Chem. Ind.*, p. 14 – 'Toxic teeth: The chronic mercury poisoning of modern man'.

Si-based cells. The Weston standard cell (cell 22.5) uses a Cd/Hg amalgam cathode, but use of this cell is declining. Cadmium is toxic and environmental legislation in the European Union and US in particular has led to a reduction in its use. Cadmium used in NiCd batteries can be recycled, but its use in other areas is expected to decrease.

$$Cd(Hg) \mid CdSO_4, H_2O \vdots Hg_2SO_4 \mid Hg \qquad (22.5)$$

The symbol Hg is derived from *hydrargyrum* (Latin) meaning 'liquid silver'. The major source of *mercury* is cinnabar (HgS), from which the metal is extracted by roasting in air (eq. 22.6).

$$HgS + O_2 \longrightarrow Hg + SO_2 \qquad (22.6)$$

Mercury has many uses but is a cumulative poison (see Box 22.3).

22.3 Physical properties

Some physical properties of the heavier *d*-block metals have already been discussed or tabulated:

- trends in first ionization energies (Fig. 1.16);
- ionization energies (Appendix 8);
- metallic radii (Table 6.2 and Fig. 19.1);
- values of $\Delta_a H^\circ$ (Table 6.2);
- lattice types (Table 6.2);
- an introduction to electronic absorption and emission spectra, and magnetism (Chapter 20).

For convenience, selected physical properties are listed in Table 22.1.

The electronic configurations of the ground state M(g) atoms change rather irregularly with increasing atomic number, more so than for the first row metals (compare Tables 22.1 and 21.1). The *nd* and $(n+1)s$ atomic orbitals are closer in energy for $n = 4$ or 5 than for $n = 3$. For ions of the first row metals, the electronic configuration is generally d^n and this brings a certain amount of order to discussions of the properties of M^{2+} and M^{3+} ions. Simple M^{n+} cations of the heavier metals are rare, and it is not possible to discuss their chemistry in terms of simple redox couples (e.g. M^{3+}/M^{2+}) as we did for most of the first row metals.

The atomic numbers of pairs of second and third row metals (except Y and La) differ by 32 units and there is an appreciable difference in electronic energy levels and, therefore, electronic spectra and ionization energies. Within a triad, the first ionization energy is generally higher for the third than for the first and second metals, but the reverse is often true for removal of subsequent electrons. Even where a pair of second and third row metal compounds are isostructural, there are often quite significant differences in stability with respect to oxidation and reduction.

Figure 22.2 shows that, with the exception of Hg (group 12), the heavier metals have higher values of $\Delta_a H^\circ$ than their first row congeners. This is a consequence of the greater spatial extent of *d* orbitals with an increase in principal quantum number, and greater orbital overlap: $5d$–$5d > 4d$–$4d > 3d$–$3d$. The trend corresponds to the fact that, compared with the first row metals, the heavier metals exhibit many more compounds containing M–M bonds. Figure 22.2 also shows that the highest values of $\Delta_a H^\circ$ occur for metals in the middle of a row. Among the heavier metals, there are numerous multimetal species with metal–metal bonding and these are discussed later in the chapter. Many low oxidation state *metal carbonyl clusters* also exist (see Chapter 24).

It is difficult to discuss satisfactorily the relative stabilities of oxidation states (Table 19.3). The situation is complicated by the fact that low oxidation states for the heavier metals are stabilized in organometallic complexes, while in non-organometallic species, the stability of *higher* oxidation states tends to *increase* down a group. Consider group 6. Tungsten forms the stable WF_6 and WCl_6, while CrF_6 is not known. Although CrO_3 and chromate(VI) ions are powerful oxidizing agents, WO_3, tungstate(VI) species and molybdenum analogues are not readily reduced. In general, the stability of high oxidation states increases for a given triad in the sequence first row $\ll$ second row $<$ third row metals. Two factors appear important in the stabilization of third row high oxidation state compounds (e.g. AuF_5 and ReF_7) which have no second or first row counterparts:

- easier promotion of electrons for the $5d$ metals compared with $4d$ or $3d$ metals;
- better orbital overlap for $5d$ orbitals (or those with $5d$ character) than for $4d$ or $3d$ orbitals.

In comparing pairs of compounds such as MoF_6 and WF_6, or RuF_6 and OsF_6, the M–F bonds are stronger for the third than second row metal, and the symmetric stretching wavenumber and force constant are higher. Relativistic effects (see Box 13.3) are also important for the third row metals.

Effects of the lanthanoid contraction

Table 22.1 shows that pairs of metals in a triad (Zr and Hf, Nb and Ta etc.) are of similar radii. This is due to the *lanthanoid contraction*: the steady decrease in size along the series of lanthanoid metals Ce–Lu which lie between La and Hf in the third row of the *d*-block. The similarity extends to values of r_{ion} (where meaningful) and r_{cov} (e.g. the M–O distances in the high-temperature forms of ZrO_2 and HfO_2 differ by less than 1 pm) and to many pairs of second and third row compounds being isomorphous. Properties that depend mainly on atom or ion size (e.g. lattice energies, solvation energies, complex stability constants) are nearly the same for corresponding pairs of $4d$ and $5d$ metal compounds. Pairs of metals often occur naturally together (e.g. Zr and Hf, Nb and Ta) and are difficult to separate (Section 22.2).

Table 22.1 Selected physical properties of the second and third row *d*-block metals.

Second row	Y	Zr	Nb	Mo	Tc**	Ru	Rh	Pd	Ag	Cd
Atomic number, Z	39	40	41	42	43	44	45	46	47	48
Physical appearance of pure metal	Soft; silver-white	Hard; lustrous; silver-coloured	Soft; shiny; silver-white	Hard; lustrous; silver-coloured; often encountered as grey powder	Silver; often encountered as grey powder	Hard; lustrous; silver-white	Hard; lustrous; silver-white	Grey-white; malleable and ductile; strength increased by cold-working	Lustrous; silver-white	Soft; blue-white; ductile
Ground state valence electronic configuration (core = [Kr])	$5s^2 4d^1$	$5s^2 4d^2$	$5s^1 4d^4$	$5s^1 4d^5$	$5s^2 4d^5$	$5s^1 4d^7$	$5s^1 4d^8$	$5s^0 4d^{10}$	$5s^1 4d^{10}$	$5s^2 4d^{10}$
Melting point / K	1799	2128	2750	2896	2430	2607	2237	1828	1235	594
Boiling point / K	3611	4650	5015	4885	5150	4173	4000	3413	2485	1038
Enthalpy of atomization, $\Delta_a H^\circ$ (298 K) / kJ mol^{-1}	423	609	721	658	677	651	556	377	285	112
Metallic radius, r_{metal} / pm[†]	182	160	147	140	135	134	134	137	144	152
Electrical resistivity (ρ) × 10^8 / Ω m (at 273 K)[‡]	59.6*	38.8	15.2	4.9	–	7.1	4.3	9.8	1.5	6.8

Third row	La	Hf	Ta	W	Re	Os	Ir	Pt	Au	Hg
Atomic number, Z	57	72	73	74	75	76	77	78	79	80
Physical appearance of pure metal	Soft; silver-white; tarnishes in air	Lustrous; silver-coloured; ductile	Hard; shiny; silver-coloured; ductile	Lustrous; silver-white; often encountered as grey powder	Silver-grey; often encountered as grey powder	Very hard; lustrous; blue-white; dense§	Very hard; brittle; lustrous; silver-coloured; dense§	Lustrous; silver-coloured; malleable; ductile	Soft; yellow; malleable; ductile	Liquid at 298 K; silver-coloured
Ground state valence electronic configuration (core = [Xe]$4f^{14}$)	$6s^2 5d^1$	$6s^2 5d^2$	$6s^2 5d^3$	$6s^2 5d^4$	$6s^2 5d^5$	$6s^2 5d^6$	$6s^2 5d^7$	$6s^1 5d^9$	$6s^1 5d^{10}$	$6s^2 5d^{10}$
Melting point / K	1193	2506	3290	3695	3459	3306	2719	2041	1337	234
Boiling point / K	3730	5470	5698	5930	5900	5300	4403	4100	3080	630
Enthalpy of atomization, $\Delta_a H^\circ$ (298 K) / kJ mol^{-1}	423	619	782	850	774	787	669	566	368	61
Metallic radius, r_{metal} / pm[†]	188	159	147	141	137	135	136	139	144	155
Electrical resistivity (ρ) × 10^8 / Ω m (at 273 K)[‡]	61.5*	30.4	12.2	4.8	17.2	8.1	4.7	9.6	2.1	94.1

[†] Metallic radius for 12-coordinate atom.
[‡] See eq. 6.3 for relationship between electrical resistivity and resistance.
* At 290–300 K.
** Technetium is radioactive (see text).
§ Osmium and iridium are the densest elements known (22.59 and 22.56 g cm^{-3} respectively).

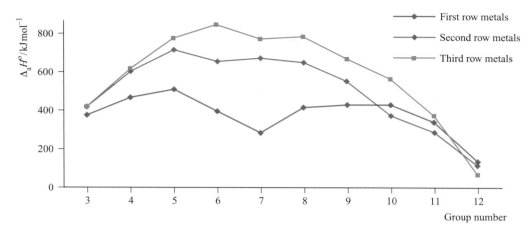

Fig. 22.2 Trends in the values of standard enthalpies of atomization (298 K) of the *d*-block metals; values are given in Tables 21.1 and 22.1.

Coordination numbers

Consistent with the increase in size in going from a first row to later metals in a triad, the heavier metals tend to show higher coordination numbers. The common range is 4 to 9 with the highest numbers being especially prevalent for metals in groups 3–5.

NMR active nuclei

Several of the metals have spin-active nuclei and this sometimes allows *direct* observation using NMR spectroscopy, e.g. ^{89}Y has a shift range >1000 ppm and ^{89}Y NMR spectroscopy is valuable for characterizing yttrium-containing compounds. In general, it is more convenient to make use of the coupling of metal nuclei to more easily observed nuclei such as ^{1}H, ^{13}C or ^{31}P. Some examples of nuclei with $I = \frac{1}{2}$ are ^{89}Y (100% abundant), ^{103}Rh (100%), ^{183}W (14.3%), ^{107}Ag (51.8%), ^{109}Ag (48.2%), ^{195}Pt (33.8%) and ^{187}Os (1.6%). Coupling to isotopes present in <100% abundance gives rise to satellite peaks (see Section 4.8, Fig. 4.23 and Box 19.2).

22.4 Group 3: yttrium

The metal

Bulk yttrium metal is passivated by an oxide layer and is quite stable in air. Metal turnings ignite if heated >670 K (eq. 22.7). Yttrium reacts with halogens (eq. 22.8) and combines with most other non-metals. The reaction between Y and H_2 under pressure was described in Section 10.7. Yttrium reacts slowly with cold water and dissolves in dilute acids (half-equation 22.9), liberating H_2.

$$4Y + 3O_2 \xrightarrow{\Delta} 2Y_2O_3 \qquad (22.7)$$

$$2Y + 3X_2 \xrightarrow{\Delta} 2YX_3 \qquad (X = F, Cl, Br, I) \qquad (22.8)$$

$$Y^{3+} + 3e^- \rightleftharpoons Y \qquad E^\circ = -2.37\,V \qquad (22.9)$$

The chemistry of yttrium is that of the +3 oxidation state, the formation of lower hydrides being an exception.

Yttrium(III)

The halides YF_3, YCl_3, YBr_3 and YI_3 are white solids. The fluoride is water-insoluble, but YCl_3, YBr_3 and YI_3 are soluble. In solid YF_3, each Y atom is 9-coordinate (distorted tricapped trigonal prismatic), while both YCl_3 and YI_3 have layer structures (e.g. YI_3 adopts a BiI_3-structure) with 6-coordinate Y centres. Yttrium(III) chloride forms a hexahydrate, $YCl_3 \cdot 6H_2O$, which is correctly formulated as $[YCl_2(OH_2)_6]^+Cl^-$. Reaction of YCl_3 with KCl gives $K_3[YCl_6]$ containing the octahedral $[YCl_6]^{3-}$ ion. In contrast to ScF_3 which forms $[ScF_6]^{3-}$, YF_3 forms no complex ions of this type.

The white oxide Y_2O_3 is insoluble in water but dissolves in acids. It is used in ceramics, optical glasses and refractory materials (see also Section 22.2). The mixed metal oxide $YBa_2Cu_3O_7$ is a member of a family of materials that become superconducting upon cooling. These so-called *high-temperature superconductors* are discussed further in Section 28.4. Yttrium(III) hydroxide is a colourless solid, in which each Y^{3+} ion is in a tricapped trigonal prismatic YO_9 environment. The hydroxide is water-insoluble and exclusively basic.

In the coordination chemistry of Y^{3+}, coordination numbers of 6 to 9 are usual. Crystalline salts containing the aqua ions $[Y(OH_2)_8]^{3+}$ (dodecahedral, Fig. 19.9c) and $[Y(OH_2)_9]^{3+}$ (tricapped trigonal prismatic) have been structurally characterized. The Y^{3+} ion is 'hard' and in its complexes favours hard *N*- and *O*- donors, e.g. *trans*-$[YCl_4(THF)_2]^-$ (octahedral), *trans*-$[YCl_2(THF)_5]^+$ (pentagonal bipyramidal), $[Y(OH_2)_7(pic)]^{2+}$ (8-coordinate, Hpic = **22.3**), $[Y(NO_3\text{-}O,O')_3(OH_2)_3]$ (irregular 9-coordinate) and $[Y(NO_3)_5]^{2-}$ (see end of Section 9.11). Reaction 22.10 yields a rare example of 3-coordinate Y(III). In the solid state, $[Y\{N(SiMe_3)_2\}_3]$ has a trigonal pyramidal rather

OH
O$_2$N   NO$_2$

NO$_2$

Hpic = picric acid

(22.3)

than planar structure but this is probably caused by crystal packing effects (see Section 19.7).

$$YCl_3 + 3Na[N(SiMe_3)_2] \longrightarrow [Y\{N(SiMe_3)_2\}_3] + 3NaCl$$
(22.10)

22.5 Group 4: zirconium and hafnium

The metals

In a finely divided form, Hf and Zr metals are pyrophoric, but the bulk metals are passivated. The high corrosion resistance of Zr is due to the formation of a dense layer of inert ZrO_2. The metals are not attacked by dilute acids (except HF) unless heated, and aqueous alkalis have no effect even when hot. At elevated temperatures, Hf and Zr combine with most non-metals (e.g. eq. 22.11).

$$MCl_4 \xleftarrow{Cl_2, \Delta} M \xrightarrow{O_2, \Delta} MO_2 \quad (M = Hf \text{ or } Zr) \quad (22.11)$$

More is known about the chemistry of Zr than Hf, the former being more readily available (see Section 22.2).

Much of the chemistry concerns Zr(IV) and Hf(IV), the lower oxidation states being less stable with respect to oxidation than the first group member, Ti(III). In aqueous solutions, only M(IV) is stable although not as M^{4+}, even though tables of data may quote half-equation 22.12; solution species (see below) depend upon conditions.

$$M^{4+} + 4e^- \rightleftharpoons M \quad \begin{cases} M = Zr, & E^\circ = -1.70\,V \\ M = Hf, & E^\circ = -1.53\,V \end{cases} \quad (22.12)$$

Stabilization of low oxidation states of Zr and Hf by π-acceptor ligands is discussed in Chapter 24.

Zirconium(IV) and hafnium(IV)

The halides MX_4 (M = Zr, Hf; X = F, Cl, Br, I; see Fig. 22.16), formed by direct combination of the elements, are white solids with the exception of orange-yellow ZrI_4 and HfI_4. The solids possess infinite structures ($ZrCl_4$, $ZrBr_4$, ZrI_4 and HfI_4 contain chains of edge-sharing octahedra) but the vapours contain tetrahedral molecules. Zirconium(IV) fluoride is dimorphic. The α-form consists of a network of F-bridged ZrF_8 square antiprisms and converts ($>720\,K$) to β-ZrF_4 in which each Zr centre is dodecahedrally sited. Ultra-pure ZrF_4 for use in optical fibres and IR spectrometer parts is made by treatment of $[Zr(BH_4)_4]$ (see Section 13.5) with HF and F_2. The chlorides, bromides

and iodides are water-soluble, but hydrolyse to MOX_2. Water reacts with ZrF_4 to give $[F_3(H_2O)_3Zr(\mu\text{-}F)_2Zr(OH_2)_3F_3]$. Both ZrF_4 and $ZrCl_4$ form highly electrically conducting materials with graphite, e.g. the reaction of ZrF_4, F_2 and graphite gives $C_nF(ZrF_4)_m$ ($n = 1\text{–}100$, $m = 0.0001\text{–}0.15$). The Lewis acidity of the halides is seen in the formation of complexes such as $HfCl_4 \cdot 2L$ (L = NMe_3, THF) and in the use of $ZrCl_4$ as a Lewis acid catalyst.

Oxides of Zr(IV) and Hf(IV) are produced by direct combination of the elements or by heating MCl_4 with H_2O followed by dehydration. The white oxides are isostructural and adopt extended structures in which Zr and Hf centres are 7-coordinate. Zirconium(IV) oxide is inert, and is used as an opacifier in ceramics and enamels and as an additive to synthetic apatites (see Section 15.2 and Box 15.11) used in dentistry. Pure ZrO_2 undergoes a phase change at 1370 K which results in cracking of the material, and for use in, for example, refractory materials, the higher temperature cubic phase is stabilized by adding MgO or CaO. Crystals of *cubic zirconia* (see Section 6.17) are commercially important as artificial diamonds. The addition of $[OH]^-$ to any water-soluble Zr(IV) compound produces the white amorphous $ZrO_2 \cdot xH_2O$. There is no true Zr(IV) hydroxide.

In aqueous acidic solution, Zr(IV) compounds are present as partly hydrolysed species, e.g. $[Zr_3(OH)_4]^{8+}$ and $[Zr_4(OH)_8]^{8+}$. From solutions of $ZrCl_4$ in dilute HCl, '$ZrOCl_2 \cdot 8H_2O$' can be crystallized; this is tetrameric, $[Zr_4(OH)_8(OH_2)_{16}]Cl_8 \cdot 12H_2O$, and contains $[Zr_4(OH)_8(OH_2)_{16}]^{8+}$ (Fig. 22.3a) in which each Zr is dodecahedrally sited.

The high coordination numbers exhibited in some apparently simple compounds of Zr(IV) and Hf(IV) extend to their complexes (e.g. see Fig. 19.10), hard fluorido and oxygen-donor ligands being favoured, e.g.:

- pentagonal bipyramidal: $[ZrF_7]^{3-}$ (Fig. 22.3b, e.g. Na^+, K^+ salts, structure is cation-dependent), $[HfF_7]^{3-}$, (e.g. K^+ salt, eq. 22.13), $[F_4(H_2O)Zr(\mu\text{-}F)_2Zr(OH_2)F_4]^{2-}$;
- capped trigonal prismatic: $[ZrF_7]^{3-}$ (Fig. 22.3c, e.g. $[NH_4]^+$ salt, structure is cation-dependent);
- dodecahedral: $[Zr(NO_3\text{-}O,O')_4]$ (eq. 22.14), $[Zr(ox)_4]^{4-}$;
- square antiprismatic: $[Zr(acac)_4]$ (Fig. 22.3d).

$$HfF_4 + 3KF \xrightarrow{\Delta \text{ in sealed Pt tube}} K_3[HfF_7] \quad (22.13)$$

$$ZrCl_4 + 4N_2O_5$$
$$\xrightarrow{\text{anhydrous conditions}} [Zr(NO_3\text{-}O,O')_4] + 4NO_2Cl \quad (22.14)$$

$$ZrCl_4 + 2CsCl \xrightarrow{1070\,K,\text{ in sealed SiO}_2\text{ tube}} Cs_2[ZrCl_6] \quad (22.15)$$

The $[ZrCl_6]^{2-}$ ion (eq. 22.15) is octahedral; colourless $Cs_2[ZrCl_6]$ adopts a $K_2[PtCl_6]$ structure (see Pt(IV) in Section 22.11) and is used as an image intensifier in X-ray

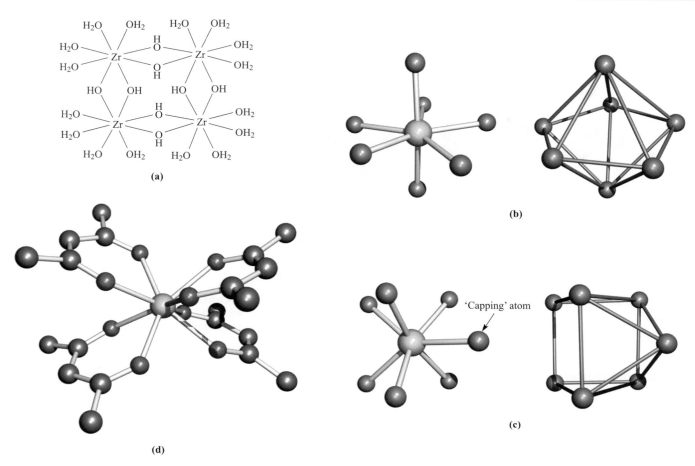

Fig. 22.3 (a) Representation of the structure of $[Zr_4(OH)_8(OH_2)_{16}]^{8+}$ in $[Zr_4(OH)_8(OH_2)_{16}]Cl_8 \cdot 12H_2O$, (b) the pentagonal bipyramidal structure (X-ray diffraction) of $[ZrF_7]^{3-}$ in $[H_3N(CH_2)_2NH_2(CH_2)_2NH_3][ZrF_7]$ [V.V. Tkachev *et al.* (1993) *Koord. Khim.*, vol. 19, p. 288], (c) the monocapped trigonal prismatic structure (X-ray diffraction) of $[ZrF_7]^{3-}$ in the guanidinium salt [A.V. Gerasimenko *et al.* (1985) *Koord. Khim.*, vol. 11, p. 566], and (d) the square antiprismatic structure (X-ray diffraction) of $[Zr(acac)_4]$ [W. Clegg (1987) *Acta Crystallogr.*, *Sect. C*, vol. 43, p. 789]. Hydrogen atoms in (d) have been omitted; colour code: Zr, yellow; C, grey; O, red; F, green.

imaging. A number of oxido complexes with square-based pyramidal structures are known, e.g. $[M(O)(ox)_2]^{2-}$ (M = Hf, Zr, **22.4**) and $[Zr(O)(bpy)_2]^{2+}$. Lower coordination numbers are stabilized by amido ligands, e.g. tetrahedral $[M(NPh_2)_4]$ and $[M\{N(SiMe_3)_2\}_3Cl]$ (M = Hf, Zr).

$$\left[\begin{array}{c} O \\ O = \overset{O}{\underset{\underset{O}{\parallel}}{\overset{O_{\prime\prime\prime\prime}}{M}}}{\overset{\prime\prime\prime\prime O}{\diagdown}} \\ O \diagup \quad \diagdown O \end{array} \right]^{2-}$$

M = Hf, Zr

(22.4)

Lower oxidation states of zirconium and hafnium

The blue or black halides ZrX_3, ZrX_2 and ZrX (X = Cl, Br, I) are obtained by reduction of ZrX_4, e.g. heating Zr and $ZrCl_4$ in a sealed Ta tube gives ZrCl or $ZrCl_3$ depending on temperature. The corresponding hafnium chlorides are prepared similarly, e.g. eqs. 22.16 and 22.17.

$$Hf + HfCl_4 \xrightarrow{\text{1070 K, in sealed Ta tube}} HfCl \qquad (22.16)$$

$$Hf + HfCl_4 \xrightarrow{\text{720 K, in sealed Ta tube}} HfCl_3 \qquad (22.17)$$

The monohalides have layer structures consisting of sheets of metal and halogen atoms sequenced XMMX...XMMX... and are metallic conductors in a direction *parallel* to the layers. Compare this with the conductivity of graphite (see Section 14.4). The di- and trihalides disproportionate (eqs. 22.18 and 22.19).

$$2MCl_2 \longrightarrow M + MCl_4 \qquad (M = Hf, Zr) \qquad (22.18)$$

$$2MCl_3 \longrightarrow MCl_2 + MCl_4 \qquad (M = Hf, Zr) \qquad (22.19)$$

In general, there is no aqueous chemistry of M(I), M(II) and M(III), but exceptions are some hexazirconium clusters which are water-stable.[†]

[†] X. Xie and T. Hughbanks (2000) *Inorg. Chem.*, vol. 39, p. 555 – 'Reduced zirconium halide clusters in aqueous solution'.

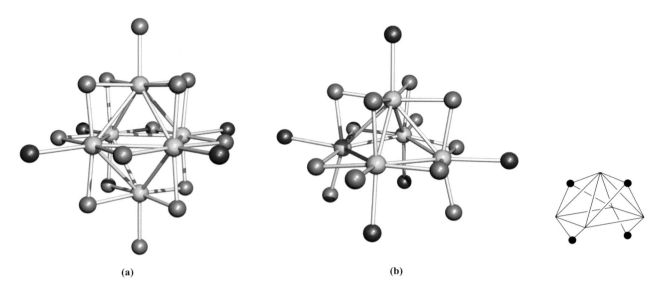

(a) **(b)**

Fig. 22.4 The structures (determined by X-ray diffraction) of (a) $[Zr_6Cl_{14}(P^nPr_3)_4]$ [F.A. Cotton *et al.* (1992) *Angew. Chem. Int. Ed.*, vol. 31, p. 1050] and (b) $[Zr_5Cl_{12}(\mu\text{-}H)_2(\mu_3\text{-}H)_2(PMe_3)_5]$ [F.A. Cotton *et al.* (1994) *J. Am. Chem. Soc.*, vol. 116, p. 4364]. Colour code: Zr, yellow; Cl, green; P, red; Me and ^nPr groups are omitted. The inset in (b) shows the μ-H and μ_3-H positions (black dots) with respect to the Zr_5 framework.

Zirconium clusters

In this section, we introduce the first group of cluster compounds of the heavier *d*-block metals in which the external ligands are halides. Octahedral M_6 frameworks are present in most of these clusters, but, in contrast to similar group 5 and 6 species (Sections 22.6 and 22.7), most zirconium clusters are stabilized by an *interstitial atom*, e.g. Be, B, C or N.

Heating Zr powder, $ZrCl_4$ and carbon in a sealed Ta tube above 1000 K produces $Zr_6Cl_{14}C$. Under similar reaction conditions and with added alkali metal halides, clusters such as $Cs_3[Zr_6Br_{15}C]$, $K[Zr_6Br_{13}Be]$ and $K_2[Zr_6Br_{15}B]$ are formed. In the solid state, these octahedral Zr_6 clusters are connected by bridging halido ligands to generate extended structures. The formulae may be written to show the connectivity, e.g. writing $[Zr_6Br_{15}B]^{2-}$ as $[\{Zr_6(\mu\text{-}Br)_{12}B\}Br_{6/2}]^{2-}$ indicates that Zr_6 clusters are connected into a 3-dimensional network by six doubly-bridging Br atoms, three 'belonging' to each cluster.[†] Some of these clusters can be 'cut-out' from the 3-dimensional network, for example, by working in ionic liquid media (see Fig. 9.9c and accompanying text) or, in some cases, by dissolving the solid state precursors in aqueous solution. Salts such as $[H_3O]_4[Zr_6Cl_{18}B]$ and $[H_3O]_5[Zr_6Cl_{18}Be]$ have been isolated under aqueous conditions, and are stabilized in the presence of acid.

In contrast to the high-temperature syntheses of Zr_6X clusters (X = B, C, N) described above, reduction of $ZrCl_4$ by Bu_3SnH followed by addition of PR_3 gives discrete clusters such as $[Zr_6Cl_{14}(P^nPr_3)_4]$ (Fig. 22.4a, Zr–Zr = 331–337 pm) and $[Zr_5Cl_{12}(\mu\text{-}H)_2(\mu_3\text{-}H)_2(PMe_3)_5]$ (Fig. 22.4b, Zr–Zr = 320–354 pm). Varying the reaction conditions leads to clusters such as $[Zr_6Cl_{14}(PMe_3)H_4]$, $[Zr_6Cl_{18}H_4]^{3-}$ and $[Zr_6Cl_{18}H_5]^{4-}$.[‡]

22.6 Group 5: niobium and tantalum

The metals

The properties of Nb and Ta (and of corresponding pairs of their compounds) are similar. At high temperatures, both are attacked by O_2 (eq. 22.20) and the halogens (eq. 22.21) and combine with most non-metals.

$$4M + 5O_2 \xrightarrow{\Delta} 2M_2O_5 \qquad (M = Nb, Ta) \qquad (22.20)$$

$$2M + 5X_2 \xrightarrow{\Delta} 2MX_5$$
$$(M = Nb, Ta; X = F, Cl, Br, I) \qquad (22.21)$$

The metals are passivated by the formation of oxide coatings, giving them high corrosion resistance. They are inert towards non-oxidizing acids, HF and HF/HNO_3 being two of the few reagents to attack them under ambient conditions. Fused alkalis react with Nb and Ta at high temperatures.

The chemistry of Nb and Ta is predominantly that of the +5 oxidation state. The heavier group 5 metals differ from V (see Section 21.6) in the relative instability of their lower

---

[†] The nomenclature is actually more complicated, but more informative, e.g. R.P. Ziebarth and J.D. Corbett (1985) *J. Am. Chem. Soc.*, vol. 107, p. 4571.

[‡] For a discussion of the location of H atoms in these and related Zr_6 cages, see: L. Chen, F.A. Cotton and W.A. Wojtczak (1997) *Inorg. Chem.*, vol. 36, p. 4047.

oxidation states, their failure to form simple ionic compounds, and the inertness of the M(V) oxides. In contrast to V, it is not meaningful to assign ionic radii to Nb and Ta in their lower oxidation states since they tend to form hexanuclear clusters with metal–metal bonding (see later). For M(V), radii of 64 pm are usually tabulated for 'Nb^{5+}' and 'Ta^{5+}', but these are unreal quantities since Nb(V) and Ta(V) compounds are essentially covalent.

Niobium(V) and tantalum(V)

Niobium(V) and tantalum(V) halides (white MF_5, yellow MCl_5, yellow-red MBr_5 and yellow-brown MI_5) are volatile, air- and moisture-sensitive solids made by reaction 22.21. The chlorides and bromides are also made by halogenation of M_2O_5. NbI_5 is produced commercially by reaction of $NbCl_5$, I_2 and HI, and TaI_5 by treating $TaCl_5$ with BI_3. Each halide is monomeric (trigonal bipyramidal) in the gas phase, but the solid fluorides are tetrameric (**22.5**), while solid MCl_5, MBr_5 and MI_5 consist of dimers (**22.6**). The M–F–M bridges in tetramer **22.5** are linear and the M–F_{bridge} bonds are longer (and weaker) than M–$F_{terminal}$ (206 vs 177 pm for M = Nb). Similarly, in dimer **22.6**, M–X_{bridge} > M–$X_{terminal}$.

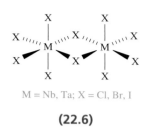

M = Nb, Ta; X = Cl, Br, I

(22.6)

The halides NbF_5, TaF_5, $NbCl_5$ and $TaCl_5$ are useful starting materials in the chemistry of these metals. They are Friedel–Crafts catalysts and the Lewis acidity of NbF_5 and TaF_5 is apparent in reaction 22.22 (which takes place in non-aqueous media, see Section 9.10), in the formation of related salts and other complexes (see later), and in the ability of a TaF_5/HF mixture to act as a superacid (see Section 9.9).

$$MF_5 + BrF_3 \longrightarrow [BrF_2]^+[MF_6]^- \quad (M = Nb, Ta) \quad (22.22)$$

The oxohalides MOX_3 and MO_2X (M = Nb, Ta; X = F, Cl, Br, I) are prepared by halogenation of M_2O_5, or reaction of MX_5 with O_2 under controlled conditions. The oxohalides are monomeric in the vapour and polymeric in the solid: $NbOCl_3$ is representative with gas-phase monomer (**22.7**) and solid-phase polymer (**22.8**) which contains oxygen-bridged Nb_2Cl_6-units. Oxoanions include octahedral $[MOX_5]^{2-}$ (M = Nb, Ta; X = F, Cl), $[MOCl_4]^-$ (eq. 22.23), and $[Ta_2OX_{10}]^{2-}$ (X = F, Cl; Fig. 22.5a). The linearity of the bridge in $[Ta_2OX_{10}]^{2-}$ indicates multiple bond character (refer to Fig. 22.19).

$$MOCl_3 + ONCl \longrightarrow [NO]^+[MOCl_4]^- \quad (M = Nb, Ta)$$
$$(22.23)$$

The structure of $[Nb(OH_2)(O)F_4]^-$ (Fig. 22.5b) shows how oxido and aqua ligand O atoms can be distinguished from

M = Nb, Ta

(22.5)

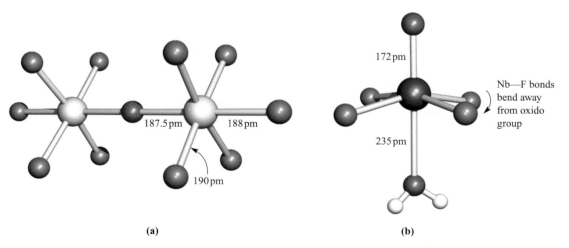

(a)   (b)

Fig. 22.5 The structures (determined by X-ray diffraction for the $[Et_4N]^+$ salts) of (a) $[Ta_2OF_{10}]^{2-}$ [J.C. Dewan *et al.* (1977) *J. Chem. Soc., Dalton Trans.*, p. 978] and (b) $[Nb(OH_2)(O)F_4]^-$ [N.G. Furmanova *et al.* (1992) *Kristallografiya*, vol. 37, p. 136]. Colour code: Ta, pale grey; Nb, blue; F, green; O, red; H, white.

the Nb—O bond lengths; it is not always possible to locate H atoms in X-ray diffraction studies (see Section 4.11).

(22.7)   **(22.8)**

Hydrolysis of $TaCl_5$ with H_2O produces the hydrated oxide $Ta_2O_5 \cdot xH_2O$. $Nb_2O_5 \cdot xH_2O$ is best formed by boiling $NbCl_5$ in aqueous HCl. Heating the hydrates yields anhydrous Nb_2O_5 and Ta_2O_5 which are dense, inert white solids. Various polymorphs of Nb_2O_5 exist, with NbO_6 octahedra being the usual structural unit; the structures of both metal(V) oxides are complicated networks. Uses of Nb_2O_5 include those as a catalyst, in ceramics and in humidity sensors. Both Nb_2O_5 and Ta_2O_5 are insoluble in acids except concentrated HF, but dissolve in molten alkalis. If the resultant melts are dissolved in water, salts of niobates (precipitated below $\approx$pH 7) and tantalates (precipitated below $\approx$pH 10) can be isolated, e.g. $K_8[Nb_6O_{19}] \cdot 16H_2O$ and $[Et_4N]_6[Nb_{10}O_{28}] \cdot 6H_2O$. The $[Nb_6O_{19}]^{8-}$ ion consists of six MO_6 octahedral units with shared O atoms; it is isoelectronic and isostructural with $[Mo_6O_{19}]^{2-}$ and $[W_6O_{19}]^{2-}$ (see Fig. 22.9c). The $[Nb_{10}O_{28}]^{6-}$ ion is isostructural with $[V_{10}O_{28}]^{6-}$ (Fig. 21.8e) and contains octahedral building blocks as in $[Nb_6O_{19}]^{8-}$.

Heating Nb_2O_5 or Ta_2O_5 with group 1 or group 2 metal carbonates at high temperatures (e.g. Nb_2O_5 with Na_2CO_3 at 1650 K in a Pt crucible) yields mixed metal oxides such as $LiNbO_3$, $NaNbO_3$, $LiTaO_3$, $NaTaO_3$ and $CaNb_2O_6$. The $M'MO_3$ compounds crystallize with perovskite structures (Fig. 6.24), and exhibit ferroelectric and piezoelectric properties (see Section 14.9) which lead to uses in electro-optical and acoustic devices.

The coordination chemistry of Nb(V) and Ta(V) is well developed and there is a close similarity in the complexes formed by the two metals. Complexes with hard donors are favoured. Although 6-, 7- and 8-coordinate complexes are the most common, lower coordination numbers are observed, e.g. in $[Ta(NEt_2)_5]$ (trigonal bipyramidal), $[Nb(NMe_2)_5]$ and $[NbOCl_4]^-$ (square-based pyramidal). The Lewis acidity of the pentahalides, especially NbF_5 and TaF_5, leads to the formation of salts such as $Cs[NbF_6]$ and $K[TaF_6]$ (octahedral anions), $K_2[NbF_7]$ and $K_2[TaF_7]$ (capped trigonal prismatic anions), $Na_3[TaF_8]$ and $Na_3[NbF_8]$ (square antiprismatic anions) and $[^nBu_4N][M_2F_{11}]$ (eq. 22.24 and structure **22.9**).

$$MF_5 \xrightarrow{[^nBu_4N][BF_4]} [^nBu_4N][MF_6] \xrightarrow{MF_5} [^nBu_4N][M_2F_{11}]$$

$$(M = Nb, Ta) \qquad (22.24)$$

M = Nb, Ta

(22.9)

Other complexes include:

- octahedral: $[Nb(OH_2)(O)F_4]^-$ (Fig. 22.5b), $[Nb(NCS-N)_6]^-$, $[NbF_5(OEt_2)]$, mer-$[NbCl_3(O)(NCMe)_2]$;
- intermediate between octahedral and trigonal prismatic: $[Nb(SCH_2CH_2S)_3]^-$;
- pentagonal bipyramidal: $[Nb(OH_2)_2(O)(ox)_2]^-$ (**22.10**); $[Nb(O)(ox)_3]^{3-}$ (oxido ligand in an axial site);
- dodecahedral: $[M(\eta^2\text{-}O_2)_4]^{3-}$ (M = Nb, Ta), $[Nb(\eta^2\text{-}O_2)_2(ox)_2]^{3-}$;
- square antiprismatic: $[Ta(\eta^2\text{-}O_2)_2F_4]^{3-}$.

(For explanation of the η-nomenclature, see Box 19.1.)

(22.10)

Self-study exercises

1. The solution ^{19}F NMR spectrum of $[^nBu_4N][Ta_2F_{11}]$ at 173 K shows three signals: a doublet of quintets ($J = 165$ and 23 Hz, respectively), a doublet of doublets ($J = 23$ and 42 Hz) and a signal consisting of 17 lines with relative intensities close to 1:8:28:56:72:72:84:120:142:120:84:72:72:56:28:8:1. Rationalize these data.

    [*Ans.* See S. Brownstein (1973) *Inorg. Chem.*, vol. 12, p. 584]

2. The anion $[NbOF_6]^{3-}$ has C_{3v} symmetry. Suggest a structure for this ion.

    [*Ans.* See Fig. 19.8a; O atom in unique site]

Niobium(IV) and tantalum(IV)

With the exception of TaF_4, all halides of Nb(IV) and Ta(IV) are known. They are dark solids, prepared by reducing the respective MX_5 by heating with metal M or Al. Niobium(IV) fluoride is paramagnetic (d^1) and isostructural with SnF_4 (**14.15**). In contrast, MCl_4, MBr_4 and MI_4 are diamagnetic (or weakly paramagnetic) consistent with the pairing of metal atoms in the solid state. The structures of $NbCl_4$ and NbI_4 consist of edge-sharing distorted NbX_6 octahedra (**22.11**) with alternating Nb−Nb distances (303 and 379 pm in $NbCl_4$; 331 and 436 pm in NbI_4). The solid state structure of $TaCl_4$ is similar, with alternating Ta–Ta distances of 299 and 379 pm.

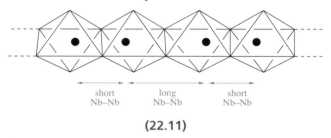

short
Nb−Nb

long
Nb−Nb

short
Nb−Nb

(22.11)

The tetrahalides are readily oxidized in air (e.g. NbF_4 to NbO_2F) and disproportionate on heating (reaction 22.25).

$$2TaCl_4 \xrightarrow{\Delta} TaCl_5 + TaCl_3 \qquad (22.25)$$

Blue-black NbO_2 is formed by reduction of Nb_2O_5 at 1070 K using H_2 or NH_3. It has a rutile structure, distorted by pairing of Nb atoms (Nb−Nb = 280 pm). Heating Nb or Ta with elemental sulfur produces the metal(IV) sulfides (NbS_2 and TaS_2) which possess layer structures. Both compounds are polymorphic. The normal phase of NbS_2 comprises layers in which each Nb atom is in a trigonal prismatic environment. The layer structure of TaS_2 resembles that of CdI_2 (Fig. 6.23), but other phases of TaS_2 are known. TaS_2 is commercially available and exhibits lubricating properties similar to those of MoS_2 (see Box 22.6). One important property of the layered metal sulfides is their ability to form intercalation compounds by accommodating guest molecules or ions between the layers. For example, TaS_2 intercalates Li^+ ions and this is the basis for the use of TaS_2 and similar layered MS_2 solids as electrode materials in lithium ion-based batteries (see Section 28.2).

A range of Nb(IV) and Ta(IV) complexes are formed by reactions of MX_4 (X = Cl, Br, I) with Lewis bases containing *N*-, *P*-, *As*-, *O*- or *S*-donors, or by reduction of MX_5 in the presence of a ligand. Coordination numbers are typically 6, 7 or 8. For example, some structures confirmed for the *solid state* are:

- octahedral: *trans*-$[TaCl_4(PEt_3)_2]$, *cis*-$[TaCl_4(PMe_2Ph)_2]$;
- capped octahedral: $[TaCl_4(PMe_3)_3]$ (Fig. 19.8b);
- capped trigonal prismatic: $[NbF_7]^{3-}$ (eq. 22.26);
- dodecahedral: $[Nb(CN)_8]^{4-}$;
- square antiprismatic: $[Nb(ox)_4]^{4-}$.

$$4NbF_5 + Nb + 15KF \longrightarrow 5K_3[NbF_7] \qquad (22.26)$$

Lower oxidation state halides

Of the lower oxidation state compounds of Nb and Ta, we focus on halides. The compounds MX_3 (M = Nb, Ta and X = Cl, Br) are prepared by reduction of MX_5 and are quite inert solids. NbF_3 and TaF_3 crystallize with the ReO_3 structure (see Fig. 21.5).

• = Cl

(22.12)

A range of halides with M_3 or M_6 frameworks exist, but all have extended structures with the metal cluster units connected by bridging halides. The structure of Nb_3Cl_8 is represented in **22.12**, but of the nine outer Cl atoms shown, six are shared between two adjacent units, and three between three (see worked example 22.1). Alternatively, the structure can be considered in terms of an hcp array of Cl atoms with three-quarters of the octahedral holes occupied by Nb atoms such that they form Nb_3 triangles. Reduction of Nb_3I_8 (structurally analogous to Nb_3Cl_8) with Nb in a sealed tube at 1200 K yields Nb_6I_{11}. The formula can be written as $[Nb_6I_8]I_{6/2}$ indicating that $[Nb_6I_8]^{3+}$ units are connected by iodides shared between two clusters. (The ionic formulation is purely a formalism.) The $[Nb_6I_8]^{3+}$ cluster consists of an octahedral Nb_6-core, each face of which is iodido-capped (Fig. 22.6a). The clusters are connected into a network by bridges (Fig. 22.6c). Two other families of halides are M_6X_{14} (e.g. Nb_6Cl_{14}, Ta_6Cl_{14}, Ta_6I_{14}) and M_6X_{15} (e.g. Nb_6F_{15}, Ta_6Cl_{15}, Ta_6Br_{15}). Their formulae can be written as $[M_6X_{12}]X_{4/2}$ or $[M_6X_{12}]X_{6/2}$ showing that they contain cluster units $[M_6X_{12}]^{2+}$ and $[M_6X_{12}]^{3+}$ respectively (Fig. 22.6b). The clusters are connected into either a 3-dimensional network (M_6X_{15}, Fig. 22.6c) or 2-dimensional sheet (M_6X_{14}, Fig. 22.6d).

Magnetic data show that the subhalides exhibit metal–metal bonding. The magnetic moment of Nb_3Cl_8 is 1.86 μ_B per Nb_3-unit (298 K) indicating one unpaired electron. This can be rationalized as follows:

- 3 Nb atoms provide 15 electrons (Nb s^2d^3);
- 8 Cl atoms provide 8 electrons (this is irrespective of the Cl bonding mode because bridge formation invokes coordinate bonds using Cl lone pairs);
- the total number of valence electrons is 23;
- 22 electrons are used in 8 Nb−Cl and 3 Nb−Nb single bonds;
- 1 electron is left over.

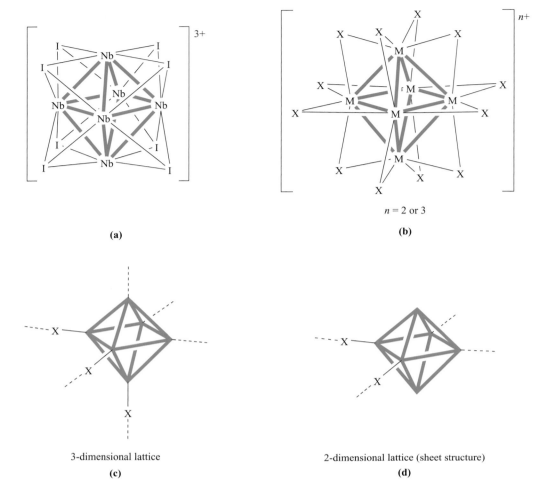

(a)   **(b)**

n = 2 or 3

3-dimensional lattice

(c)

2-dimensional lattice (sheet structure)

(d)

Fig. 22.6 Representations of the structures of (a) the $[Nb_6I_8]^{3+}$ unit found in Nb_6I_{11} and (b) the $[M_6X_{12}]^{n+}$ ($n = 2$ or 3) unit found in compounds of type M_6X_{14} and M_6X_{15} for M = Nb or Ta, X = halide. The cluster units are connected into (c) a 3-dimensional network or (d) a 2-dimensional sheet by bridging halides (see text).

Compounds of the type M_6X_{14} are diamagnetic, while M_6X_{15} compounds have magnetic moments corresponding to one unpaired electron per M_6-cluster. If we consider M_6X_{14} to contain an $[M_6X_{12}]^{2+}$ unit, there are eight pairs of valence electrons remaining after allocation of 12 M—X bonds, giving a bond order of two-thirds per M—M edge (12 edges). In M_6X_{15}, after allocating electrons to 12 M—X single bonds, the $[M_6X_{12}]^{3+}$ unit has 15 valence electrons for M—M bonding. The observed paramagnetism indicates that one unpaired electron remains unused. The magnetic moment (per hexametal unit) of Ta_6Br_{15}, for example, is temperature-dependent: $\mu_{eff} = 2.17\,\mu_B$ at 623 K, 1.73 μ_B at 222 K and 1.34 μ_B at 77 K.

There is also a family of discrete clusters $[M_6X_{18}]^{n-}$ (M = Nb, Ta; X = Cl, Br, I). For example, the reaction of Nb_6Cl_{14} with KCl at 920 K produces $K_4[Nb_6Cl_{18}]$. The $[Nb_6Cl_{18}]^{4-}$ ion is oxidized by I_2 to $[Nb_6Cl_{18}]^{3-}$ or by Cl_2 to $[Nb_6Cl_{18}]^{2-}$. The $[M_6X_{18}]^{n-}$ ions are structurally similar (Fig. 22.7) and relationships between the structure of this

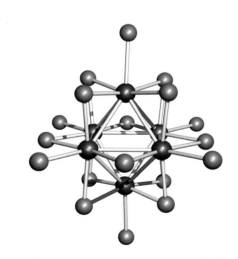

Fig. 22.7 The structure (X-ray diffraction) of $[Nb_6Cl_{18}]^{3-}$ in the $[Me_4N]^+$ salt [F.W. Koknat *et al.* (1974) *Inorg. Chem.*, vol. 13, p. 295]. Colour code: Nb, blue; Cl, green.

discrete ion, that of the $[M_6Cl_{12}]^{n+}$ ion (Fig. 22.6b) and of the Zr_6 clusters (e.g. Fig. 22.4a) are clear.

Worked example 22.1   Structures of halides of Nb

Part of the solid state structure of Nb_3Cl_8 is shown below. Explain how this structure is consistent with the stoichiometry of the compound.

The diagram above represents part of an extended structure. The 'terminal' Cl atoms are shared between units: 6 are shared between 2 units, and 3 are shared between 3 units.

Per Nb_3 unit, the number of Cl atoms

$$= 4 + (6 \times \tfrac{1}{2}) + (3 \times \tfrac{1}{3}) = 8$$

Thus, the stoichiometry of the compound $= Nb_3Cl_8$.

Self-study exercises

The answers to these questions can be found by reading the last subsection.

1. The solid state structure of NbI_4 consists of edge-shared octahedra. Explain how this description is consistent with the stoichiometry of the compound.

2. The formula of Nb_3I_{11} can be written as $[Nb_3I_8]I_{6/2}$. Explain how this can be translated into a description of the solid state structure of the compound.

22.7 Group 6: molybdenum and tungsten

The metals

The properties of Mo and W are similar. Both have very high melting points and enthalpies of atomization (Table 6.2 and Fig. 22.2). The metals are not attacked in air at 298 K, but react with O_2 at high temperatures to give MO_3, and are readily oxidized by the halogens (see later). Even at 298 K, oxidation to M(VI) occurs with F_2 (eq. 22.27, see Fig. 22.16). Sulfur reacts with Mo or W

(e.g. eq. 22.28). Other sulfide phases of Mo are produced under different conditions.

$$M + 3F_2 \longrightarrow MF_6 \qquad (M = Mo, W) \qquad (22.27)$$

$$M + 2S \xrightarrow{\Delta} MS_2 \qquad (M = Mo, W) \qquad (22.28)$$

The metals are inert towards most acids but are rapidly attacked by fused alkalis in the presence of oxidizing agents.

Molybdenum and tungsten exhibit a range of oxidation states (Table 19.3) although simple mononuclear species are not known for all states. The extensive chemistry of Cr(II) and Cr(III) (see Section 21.7) has no counterpart in the chemistries of the heavier group 6 metals, and, in contrast to Cr(VI), Mo(VI) and W(VI) are poor oxidizing agents. Since $W^{3+}(aq)$ is not known, no reduction potential for the W(VI)/W(III) couple can be given. Equations 22.29 and 22.30 compare the Cr and Mo systems at pH 0.

$$[Cr_2O_7]^{2-} + 14H^+ + 6e^- \rightleftharpoons 2Cr^{3+} + 7H_2O \quad E^{\circ} = +1.33\,V$$
$$(22.29)$$

$$H_2MoO_4 + 6H^+ + 3e^- \rightleftharpoons Mo^{3+} + 4H_2O \qquad E^{\circ} = +0.1\,V$$
$$(22.30)$$

Molybdenum and tungsten compounds are usually isomorphous and essentially isodimensional.

Molybdenum(VI) and tungsten(VI)

The hexafluorides are formed by reaction 22.27, or by reactions of MoO_3 with SF_4 (sealed vessel, 620 K) and WCl_6 with HF or SbF_3. Both MoF_6 (colourless liquid, bp 307 K) and WF_6 (pale yellow, volatile liquid, bp 290 K) have molecular structures (**22.13**) and are readily hydrolysed. The only other hexahalides that are well established are the dark blue WCl_6 and WBr_6. The former is made by heating W or WO_3 with Cl_2 and has an octahedral molecular structure. WBr_6 (also molecular) is best made by reaction 22.31. Both WCl_6 and WBr_6 readily hydrolyse. Reactions of WF_6 with Me_3SiCl, or WCl_6 with F_2, yield mixed halides, e.g. *cis-* and *trans-*WCl_2F_4 and *mer-* and *fac-*WCl_3F_3.

M = Mo, W   M = Mo, W

(22.13)   **(22.14)**

$$W(CO)_6 + 3Br_2 \longrightarrow WBr_6 + 6CO \qquad (22.31)$$

While MoF_6 and WF_6 are octahedral, the isoelectronic molecules $MoMe_6$ and WMe_6 adopt distorted trigonal

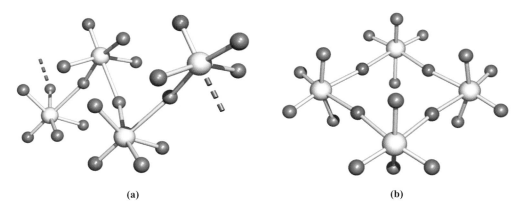

(a)   **(b)**

Fig. 22.8 (a) Part of one of the infinite chains that constitute the solid state structure of $MoOF_4$, and (b) one of the tetrameric units present in crystalline WOF_4. Colour code: Mo or W, pale grey; O, red; F, green.

prismatic structures (Box 20.3). Theoretical studies at the DFT level (see Section 4.13) show that there is only a low energy barrier to interconversion of octahedral and trigonal prismatic structures for MoF_6 and WF_6. Since the F atoms in an MF_6 molecule are equivalent, it is difficult to prove whether interconversion occurs in practice. However, the solution ^{19}F NMR spectra of $MF_5(OC_6F_5)$ (**22.14**) are temperature dependent, consistent with stereo-chemically non-rigid molecules. Moreover, in $WF_5(OC_6F_5)$, the retention of coupling between ^{19}F and ^{183}W nuclei between the low and high-temperature limiting spectra confirms that the fluxional process occurs without W–F bond cleavage.[†]

Self-study exercise

WF_6 reacts with PMe_3 or PMe_2Ph to form the 7-coordinate complex $WF_6(PMe_3)$ or $WF_6(PMe_2Ph)$. In the solid state, these possess capped trigonal prismatic and capped octahedral structures, respectively. Each of the solution ^{31}P and ^{19}F NMR spectra of $WF_6(PMe_3)$ exhibits one signal ($J_{PF} = 74\,Hz$), as does each of the ^{31}P and ^{19}F NMR spectra of $WF_6(PMe_2Ph)$. What conclusions can you draw from these data?
[*Ans.* S. El-Kurdi *et al.* (2010) *Chem. Eur. J.*, vol. 16, p. 595.]

Oxohalides MOX_4 (M = Mo, X = F, Cl; M = W, X = F, Cl, Br) and MO_2X_2 (M = Mo, W; X = F, Cl, Br) can be made by a variety of routes, e.g. eq. 22.32. Reactions of MO_3 with CCl_4 yield MO_2Cl_2; WO_2Cl_2 decomposes on heating (eq. 22.33). The oxohalides readily hydrolyse.

$$\left.\begin{array}{l} M + O_2 + F_2 \\ MOCl_4 + HF \\ MO_3 + F_2 \end{array}\right\} \longrightarrow MOF_4 \quad (M = Mo, W) \qquad (22.32)$$

$$2WO_2Cl_2 \xrightarrow{450-550\,K} WO_3 + WOCl_4 \qquad (22.33)$$

The solids do not contain monomeric units, e.g. $MoOF_4$ contains chains of $MoOF_5$ octahedra linked by Mo–F–Mo bridges (Fig. 22.8a). In WOF_4, W–O–W bridges are present within tetrameric units (Fig. 22.8b). The layer structure of WO_2Cl_2 is related to that of SnF_4 (**14.15**); each layer comprises bridged WO_4Cl_2 units (**22.15**) and the lattice is able to act as an intercalation host.

(22.15)

The most important compounds of Mo(VI) and W(VI) are the oxides and the molybdate and tungstate anions. White MoO_3 (mp 1073 K) is usually made by reaction 22.34, and yellow WO_3 (mp 1473 K) by dehydration of tungstic acid (see below). Both oxides are commercially available.

$$MoS_2 \xrightarrow{roast\ in\ air} MoO_3 \qquad (22.34)$$

The structure of MoO_3 consists of layers of linked MoO_6 octahedra. The arrangement of the MoO_6 units is complex and results in a unique 3-dimensional network. Several polymorphs of WO_3 exist, all based on the ReO_3 structure (Fig. 21.5). Thin films of WO_3 are used in electrochromic windows (Box 22.4). Neither MoO_3 nor WO_3 reacts with

[†] For further details, see: G.S. Quiñones, G. Hägele and K. Seppelt (2004) *Chem. Eur. J.*, vol. 10, p. 4755 – 'MoF_6 and WF_6: non-rigid molecules?'

APPLICATIONS

Box 22.4 Electrochromic 'smart' windows

Walls of windows are commonplace in modern office blocks: use of electrochromic glass improves energy efficiency and working environment.

An electrochromic material is one which changes colour when an electrical potential difference is applied across the material. Turning off the external voltage supply reverses the colour change. Applications of electrochromic films include the manufacture of 'smart' windows, i.e. windows that can be reversibly made darker by applying a voltage stimulus. Such windows are able to modify the amount of daylight entering a building, and can be used to regulate solar heat transmission. Electrochromic devices are also used in vehicle mirrors and sunroof panels. The use of flexible polymers in place of glass substrates for the electrochromic films increases the range of applications.

Metal oxides containing redox active metals (e.g. WO_3, MoO_3, Ta_2O_5, Nb_2O_5, IrO_2) are the key inorganic materials applied to the production of electrochromic films, and of these oxides, WO_3 (either crystalline or amorphous) is the most important. A film of pure WO_3 is transparent and a colour change to blue arises from the reversible formation of a lithium tungsten bronze (see eq. 22.42 and associated text). A typical electrochemical film design is shown below. It consists of two outer layers of electrically conducting glass (usually SnO_2:F or In_2O_3:Sn, see Section 28.3). Between these are the active electrode (WO_3) and the counter electrode (e.g. $Li_xV_2O_5$), and sandwiched between them is a solid (polymer) electrolyte which conducts Li^+ ions. The counter electrode acts as a store for Li^+ ions. When a small potential ($\approx 1.5\,V$) is applied across the cell, Li^+ ions migrate from the counter electrode, through the electrolyte to the active electrode. The uptake of Li^+ ions by WO_3 under a potential bias is represented by the equation:

$$WO_3 + xLi^+ + xe^- \rightleftharpoons Li_xWO_3$$
colourless   blue

The reaction is fully reversible. For each Li^+ ion incorporated into Li_xWO_3, one W^{VI} centre is reduced to W^V. The above equation therefore becomes:

$$W^{VI}O_3 + xLi^+ + xe^- \rightleftharpoons Li_xW^{VI}_{1-x}W^V_xO_3$$
colourless   blue

As the equilibria above show, WO_3 is a *cathodic* electrochromic material, i.e. the darkening of the electrochromic film occurs at the cathode in the cell. In contrast, IrO_2 is an *anodic* electrochromic material, and the reversible colour change of the electrochromic device depends on proton rather than lithium ion migration:

$$H_xIrO_2 \rightleftharpoons IrO_2 + xH^+ + xe^-$$
colourless   dark blue

The two oxidation states of iridium are Ir(IV) and Ir(III):

$$H_xIr^{IV}_{1-x}Ir^{III}_xO_2 \rightleftharpoons Ir^{IV}O_2 + xH^+ + xe^-$$
colourless   dark blue

Further reading

D.T. Gillaspie, R.C. Tenent and A.C. Dillon (2010) *J. Mater. Chem.*, vol. 20, p. 9585 – 'Metal-oxide films for electrochromic applications: present technology and future directions'.

C.G. Granqvist (2008) *Pure Appl. Chem.*, vol. 80, p. 2489 – 'Electrochromics for energy efficiency and indoor comfort'.

G.A. Niklasson and C.G. Granqvist (2007) *J. Mater. Chem.*, vol. 17, p. 127 – 'Electrochromics for smart windows: Thin films of tungsten oxide and nickel oxide, and devices based on these'.

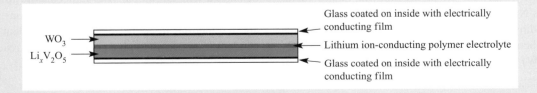

WO_3 — Glass coated on inside with electrically conducting film
$Li_xV_2O_5$ — Lithium ion-conducting polymer electrolyte
Glass coated on inside with electrically conducting film

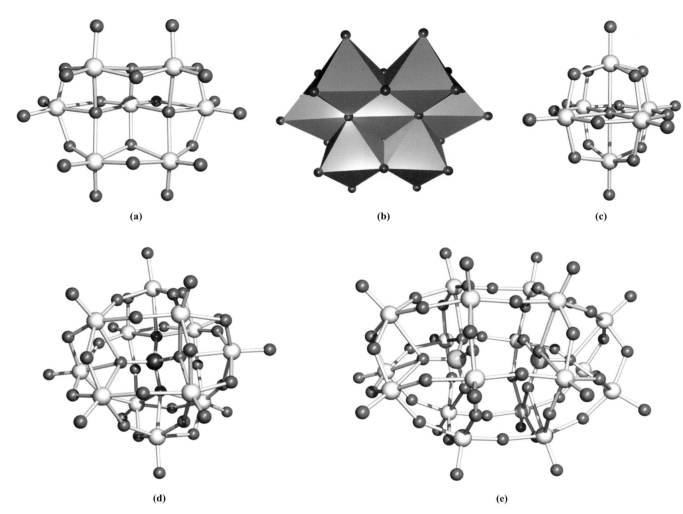

Fig. 22.9 (a) The structure of $[Mo_7O_{24}]^{6-}$ in the $[H_3N(CH_2)_2NH_2(CH_2)_2NH_3]^{3+}$ salt [P. Roman *et al.* (1992) *Polyhedron*, vol. 11, p. 2027]; (b) the $[Mo_7O_{24}]^{6-}$ ion represented in terms of seven octahedral building blocks (these can be generated in diagram (a) by connecting the O atoms); (c) the structure of $[W_6O_{19}]^{2-}$ determined for the $[W(CN^tBu)_7]^{2+}$ salt [W.A. LaRue *et al.* (1980) *Inorg. Chem.*, vol. 19, p. 315]; (d) the structure of the α-Keggin ion $[SiMo_{12}O_{40}]^{4-}$ in the guanidinium salt (the Si atom is shown in dark blue) [H. Ichida *et al.* (1980) *Acta Crystallogr., Sect. B*, vol. 36, p. 1382]; (e) the structure of $[H_3S_2Mo_{18}O_{62}]^{5-}$ (in the $[^nBu_4N]^+$ salt) formed by reducing the α-Dawson anion $[S_2Mo_{18}O_{62}]^{4-}$ (H atoms are omitted) [R. Neier *et al.* (1995) *J. Chem. Soc., Dalton Trans.*, p. 2521]. Colour code: Mo and W, pale grey; O, red; Si, blue; S, yellow.

acids, but in aqueous alkali, $[MO_4]^{2-}$ or polyoxometallate ions are produced. The chemistry of molybdates and tungstates is complicated and the uses of the homo- and heteropolyanions are extremely varied.[†] The simplest molybdate(VI) and tungstate(VI) ions are $[MoO_4]^{2-}$ and $[WO_4]^{2-}$, many salts of which are known. Alkali metal salts such as Na_2MoO_4 and Na_2WO_4 (commercially available as the dihydrates and useful starting materials in this area of chemistry) are made by dissolving MO_3 (M = Mo, W) in aqueous alkali metal hydroxide. From

strongly acidic solutions of these molybdates and tungstates, it is possible to isolate yellow 'molybdic acid' and 'tungstic acid'. Crystalline molybdic and tungstic acids are formulated as $MoO_3·2H_2O$ and $WO_3·2H_2O$, and possess layer structures consisting of corner-sharing $MO_5(OH_2)$ octahedra with additional H_2O molecules residing between the layers. In crystalline salts, the $[MO_4]^{2-}$ ions are discrete and tetrahedral. In acidic media and dependent on the pH, condensation occurs to give polyanions, e.g. reaction 22.35.

$$7[MoO_4]^{2-} + 8H^+ \longrightarrow [Mo_7O_{24}]^{6-} + 4H_2O \qquad pH \approx 5$$
$$(22.35)$$

Structural features of $[Mo_7O_{24}]^{6-}$ (Fig. 22.9a), which are in common with other polynuclear molybdates and tungstates, are that:

[†] For overviews of applications, see: D.-L. Long, R. Tsunashima and L. Cronin (2010) *Angew. Chem. Int. Ed.*, vol. 49, p. 1736; A. Dolbecq, E. Dumas, C.R. Mayer and P. Mialane (2010) *Chem. Rev.*, vol. 110, p. 6009; A. Proust, R. Thouvenot and P. Gouzerh (2008) *Chem. Commun.*, p. 1837.

APPLICATIONS

Box 22.5 Catalytic applications of MoO$_3$ and molybdates

Molybdenum-based catalysts are used to facilitate a range of organic transformations including benzene to cyclohexane, ethylbenzene to styrene, and propene to acetone.

Acrylonitrile (used in the manufacture of acrylic fibres, resins and rubbers) is produced commercially on a large scale by the reaction:

$$+ \ {}^3\!/_2 O_2 \ + \ NH_3$$

$$\downarrow \ Bi_2O_3/MoO_3 \text{ catalyst}$$

$$+ \ 3H_2O$$

Propene is also the precursor to acrolein (acrylaldehyde):

$$+ \ O_2$$

$$\downarrow \ Bi_2O_3/MoO_3 \text{ catalyst}$$

$$+ \ H_2O$$

The two manufacturing processes are together known as the SOHIO (Standard Oil of Ohio Company) process. The bismuth–molybdate catalyst functions by providing intimately associated Bi−O and Mo=O sites. The Bi−O sites are involved in abstracting α-hydrogen (see structure **24.47**) while the Mo=O sites interact with the incoming alkene, and are involved in activation of NH$_3$ and in C−N bond formation.

In Box 12.2, we described methods of desulfurizing emission gases. A combination of MoO$_3$ and CoO supported on activated alumina acts as an effective catalyst for the desulfurization of petroleum and coal-based products. This catalyst system has wide application in a process that contributes significantly to reducing SO$_2$ emissions.

Further reading

R.K. Grasselli (1986) *J. Chem. Educ.*, vol. 63, p. 216 – 'Selective oxidation and ammoxidation of olefins by heterogeneous catalysis'.

J. Haber and E. Lalik (1997) *Catal. Today*, vol. 33, p. 119 – 'Catalytic properties of MoO$_3$ revisited'.

T.A. Hanna (2004) *Coord. Chem. Rev.*, vol. 248, p. 429 – 'The role of bismuth in the SOHIO process'.

C. Limberg (2007) *Top. Organomet. Chem.* (2007) vol. 22, p. 79 – 'The SOHIO process as an inspiration for molecular organometallic chemistry'.

- the cage is supported by oxygen bridges and there is no metal–metal bonding;
- the cage is constructed from octahedral MO$_6$-units connected by shared oxygen atoms.

As a consequence of this last point, the structures may be represented in terms of linked octahedra, in much the same way that silicate structures are depicted by linked tetrahedra (see structure **14.22** and Fig. 14.23). Figure 22.9b shows such a representation for $[Mo_7O_{24}]^{6-}$; each vertex corresponds to an O atom in Fig. 22.9a. By controlling the pH or working in non-aqueous media, salts of other molybdates and tungstates can be isolated. One of the simplest is $[M_6O_{19}]^{2-}$ (M = Mo, W) which is isostructural with $[M_6O_{19}]^{8-}$ (M = Nb, Ta) and possesses the *Lindqvist structure* (Fig. 22.9c). For tungsten, the solution system is more complicated than for molybdenum, and involves equilibria with W$_7$, W$_{10}$, W$_{11}$ and W$_{12}$ species. The lowest nuclearity anion, $[W_7O_{24}]^{6-}$, is isostructural with $[Mo_7O_{24}]^{6-}$. Salts can be isolated by careful control of pH, and under non-aqueous conditions salts of polytungstates unknown in aqueous solution can be crystallized.

Heteropolyanions have been well studied and have many applications, e.g. as catalysts. Two families are especially important:

- the α-Keggin anions,[†] $[XM_{12}O_{40}]^{n-}$ (M = Mo, W; e.g. X = P or As, $n = 3$; X = Si, $n = 4$; X = B, $n = 5$);
- the α-Dawson anions, $[X_2M_{18}O_{62}]^{n-}$ (M = Mo, W; e.g. X = P or As, $n = 6$).

Equations 22.36 and 22.37 show typical syntheses of α-Keggin ions. All ions are structurally similar (Fig. 22.9d) with the hetero-atom tetrahedrally sited in the centre of the polyoxometallate cage. The construction of the cage from oxygen-linked, octahedral MO$_6$-units is apparent by studying Fig. 22.9d.

$$[HPO_4]^{2-} + 12[WO_4]^{2-} + 23H^+ \longrightarrow [PW_{12}O_{40}]^{3-} + 12H_2O \tag{22.36}$$

$$[SiO_3]^{2-} + 12[WO_4]^{2-} + 22H^+ \longrightarrow [SiW_{12}O_{40}]^{4-} + 11H_2O \tag{22.37}$$

[†] The prefix α distinguishes the structural type discussed here from other isomers; the first example, $[PMo_{12}O_{40}]^{3-}$, was reported in 1826 by Berzelius, and was structurally elucidated using X-ray diffraction in 1933 by J.F. Keggin.

α-Dawson anions of Mo are formed spontaneously in solutions containing $[MoO_4]^{2-}$ and phosphates or arsenates at appropriate pH, but formation of corresponding tungstate species is slower and requires an excess of phosphate or arsenate. The α-Dawson cage structure can be viewed as the condensation of two α-Keggin ions with loss of six MO_3-units (compare Figs. 22.9d and 22.9e). The structure shown in Fig. 22.9e is that of $[H_3S_2Mo_{18}O_{62}]^{5-}$, a protonated product of the 4-electron reduction of the α-Dawson ion $[S_2Mo_{18}O_{62}]^{4-}$. Apart from bond length changes, the cage remains unaltered by the addition of electrons. Similarly, reduction of α-Keggin ions occurs without gross structural changes. Reduction converts some of the M(VI) to M(V) centres and is accompanied by a change in colour to intense blue. Hence, reduced Keggin and Dawson anions are called *heteropoly blues*.

Heteropolyanions with incomplete cages, *lacunary* anions, may be made under controlled pH conditions, e.g. at pH ≈ 1, $[PW_{12}O_{40}]^{3-}$ can be prepared (eq. 22.36) while at pH ≈ 2, $[PW_{11}O_{39}]^{7-}$ is formed. Lacunary ions act as ligands by coordination through terminal *O*-atoms. Complexes include $[PMo_{11}VO_{40}]^{4-}$, $[(PW_{11}O_{39})Ti(\eta^5\text{-}C_5H_5)]^{4-}$ and $[(PW_{11}O_{39})Rh_2(O_2CMe)_2(DMSO)_2]^{5-}$.

The formation of mononuclear complexes by Mo(VI) and W(VI) is limited. Simple complexes include octahedral $[WOF_5]^-$ and *cis*-$[MoF_4O_2]^{2-}$. Salts of $[MoF_7]^-$ (eq. 22.38) and $[MoF_8]^{2-}$ (eq. 22.39) have been isolated.

$$MoF_6 + [Me_4N]F \xrightarrow{\text{MeCN}} [Me_4N][MoF_7] \qquad (22.38)$$

$$MoF_6 + 2KF \xrightarrow{\text{in IF}_5} K_2[MoF_8] \qquad (22.39)$$

The peroxido ligand, $[O_2]^{2-}$, forms a range of complexes with Mo(VI) and W(VI), e.g. $[M(O_2)_2(O)(ox)]^{2-}$ (M = Mo, W) is pentagonal bipyramidal (**22.16**) and $[Mo(O_2)_4]^{2-}$ is dodecahedral. Some peroxido complexes of Mo(VI) catalyse the epoxidation of alkenes.

(22.16)

Molybdenum(V) and tungsten(V)

The known pentahalides are yellow MoF_5, yellow WF_5, black $MoCl_5$, dark green WCl_5 and black WBr_5, all solids at 298 K. The pentafluorides are made by heating MoF_6 with Mo (or WF_6 with W, eq. 22.40), but both disproportionate on heating (eq. 22.41).

$$5WF_6 + W \xrightarrow{\Delta} 6WF_5 \qquad (22.40)$$

$$2MF_5 \xrightarrow{\Delta,\, T\,\text{K}} MF_6 + MF_4 \qquad (22.41)$$

M = Mo, $T > 440$ K; M = W, $T > 320$ K

Direct combination of the elements under controlled conditions gives $MoCl_5$ and WCl_5. The pentafluorides MoF_5 and WF_5 are tetrameric in the solid, isostructural with NbF_5 and TaF_5 (**22.5**). $MoCl_5$ and WCl_5 are dimeric and structurally similar to $NbCl_5$ and $TaCl_5$ (**22.6**). Each pentahalide is paramagnetic, indicating little or no metal–metal interaction.

Tungsten bronzes contain M(V) and M(VI) (see Box 22.4) and are formed by vapour-phase reduction of WO_3 by alkali metals, reduction of Na_2WO_4 by H_2 at 800–1000 K, or by reaction 22.42.

$$\frac{x}{2}Na_2WO_4 + \frac{3-2x}{3}WO_3 + \frac{x}{6}W \xrightarrow{1120\,\text{K}} Na_xWO_3$$

$$(22.42)$$

Tungsten bronzes are inert materials M_xWO_3 ($0 < x < 1$) with defect perovskite structures (Fig. 6.24). Their colour depends on x: golden for $x \approx 0.9$, red for $x \approx 0.6$, violet for $x \approx 0.3$. Bronzes with $x > 0.25$ exhibit metallic conductivity owing to a band-like structure associated with W(V) and W(VI) centres in the lattice. Those with $x < 0.25$ are semiconductors (see Section 6.8). Similar compounds are formed by Mo, Ti and V.[†]

Our discussion of complexes of Mo(V) and W(V) is restricted to selected mononuclear species. Octahedral coordination is common, for example $[MoF_6]^-$ (eq. 22.43), $[WF_6]^-$ and $[MoCl_6]^-$. 8-Coordinate W(V) is found in $[WF_8]^{3-}$ (eq. 22.44).

$$MoF_6 \text{ (in excess)} \xrightarrow{\text{KI in liquid SO}_2} K[MoF_6] \qquad (22.43)$$

$$W(CO)_6 \xrightarrow{\text{KI in IF}_5} K_3[WF_8] \qquad (22.44)$$

Treatment of WCl_5 with concentrated HCl leads to $[WOCl_5]^{2-}$, and $[WOBr_5]^{2-}$ forms when $[W(O)_2(ox)_2]^{3-}$ reacts with aqueous HBr. Dissolution of $[MoOCl_5]^{2-}$ in aqueous acid produces yellow $[Mo_2O_4(OH_2)_6]^{2+}$ (**22.17**) which is diamagnetic, consistent with a Mo−Mo single bond. A number of complexes $[MoCl_3L_2]$ are known, e.g. $WOCl_3(THF)_2$ (a useful starting material since the THF ligands are labile), $[WOCl_3(PEt_3)_2]$ (**22.18**) and $[MoOCl_3(bpy)]$. High coordination numbers are observed in $[Mo(CN)_8]^{3-}$ and $[W(CN)_8]^{3-}$, formed by oxidation of $[M(CN)_8]^{4-}$ using Ce^{4+} or $[MnO_4]^-$. The coordination geometries are cation-dependent, illustrating the small

---

[†] See for example: C.X. Zhou, Y.X. Wang, L.Q. Yang and J.H. Lin (2001) *Inorg. Chem.*, vol. 40, p. 1521 – 'Syntheses of hydrated molybdenum bronzes by reduction of MoO_3 with $NaBH_4$'; X.K. Hu, Y.T. Qian, Z.T. Song, J.R. Huang, R. Cao and J.Q. Xiao (2008) *Chem. Mater.*, vol. 20, p. 1527 – 'Comparative study on MoO_3 and H_xMoO_3 nanobelts: structure and electric transport'.

energy difference between dodecahedral and square anti-prismatic structures.

$$\begin{bmatrix} & O & & O & \\ H_2O_{\prime\prime\prime\prime,} & \| & ,\prime\prime\prime O_{\prime\prime\prime\prime,} & \| & ,\prime\prime\prime OH_2 \\ & Mo & & Mo & \\ H_2O & | & O & | & OH_2 \\ & H_2O & & OH_2 & \end{bmatrix}^{2+}$$

(22.17)

$$\begin{array}{c} O \\ Et_3P_{\prime\prime\prime,} \| ,\prime\prime\prime Cl \\ W \\ Cl \quad | \quad PEt_3 \\ Cl \end{array}$$

(22.18)

Molybdenum(IV) and tungsten(IV)

Binary halides MX_4 are established for M = Mo, W and X = F, Cl and Br; WI_4 exists but is not well characterized. Equations 22.45 and 22.46 show representative syntheses.

$$MoO_3 \xrightarrow{H_2, 720\,K} MoO_2 \xrightarrow{CCl_4, 520\,K} MoCl_4 \qquad (22.45)$$

$$WCl_6 \xrightarrow{W(CO)_6, \text{reflux in chlorobenzene}} WCl_4 \qquad (22.46)$$

Tungsten(IV) fluoride is polymeric, and a polymeric structure for MoF_4 is consistent with Raman spectroscopic data. Three polymorphs of $MoCl_4$ exist: α-$MoCl_4$ has the $NbCl_4$ structure (**22.11**) and, at 520 K, transforms to the β-form containing cyclic Mo_6Cl_{24} units (Fig. 22.10a). The structure of the third polymorph is unknown. Tungsten(IV) chloride (structurally like α-$MoCl_4$) is a useful starting material in W(IV) and lower oxidation state

chemistry. All the tetrahalides are air- and moisture-sensitive.

Reduction of MO_3 (M = Mo, W) by H_2 yields MoO_2 and WO_2 which adopt rutile structures (Fig. 6.22), distorted (as in NbO_2) by pairing of metal centres; in MoO_2, Mo−Mo distances are 251 and 311 pm. The oxides do not dissolve in non-oxidizing acids. Molybdenum(IV) sulfide (eq. 22.28) has a layer structure and is used as a lubricant (Box 22.6).

Molybdenum(IV) is stabilized in acidic solution as red $[Mo_3(\mu_3\text{-}O)(\mu\text{-}O)_3(OH_2)_9]^{4+}$ (Fig. 22.10b) which is formed by reduction of $Na_2[MoO_4]$ or oxidation of $[Mo_2(OH_2)_8]^{4+}$.

The halido complexes $[MX_6]^{2-}$ (M = Mo, W; X = F, Cl, Br) are known although $[WF_6]^{2-}$ has been little studied. By adjusting the conditions of reaction 22.43 (i.e. taking a 1:2 molar ratio $MoF_6:I^-$, and removing I_2 as it is formed), $K_2[MoF_6]$ can be isolated. Salts of $[MoCl_6]^{2-}$ can be made starting from $MoCl_5$, e.g. $[NH_4]_2[MoCl_6]$ by heating $MoCl_5$ with NH_4Cl. Many salts of $[WCl_6]^{2-}$ are known (e.g. reaction 22.47) but the ion decomposes on contact with water.

$$2M[WCl_6] \xrightarrow{550\,K} M_2[WCl_6] + WCl_6$$
$$(M = \text{group 1 metal}) \qquad (22.47)$$

Reduction of H_2WO_4 using Sn in HCl in the presence of K_2CO_3 leads to $K_4[W_2(\mu\text{-}O)Cl_{10}]$; the anion is structurally like $[Ta_2(\mu\text{-}O)F_{10}]^{2-}$ (Fig. 22.5a).

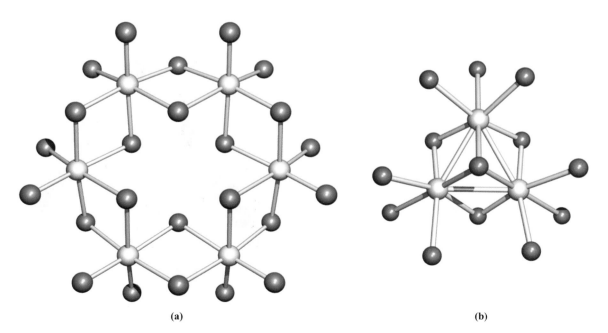

(a)   **(b)**

Fig. 22.10 (a) The β-form of $MoCl_4$ consists of cyclic Mo_6Cl_{24} units. The structure was determined by X-ray diffraction [U. Müller (1981) *Angew. Chem.*, vol. 93, p. 697]. Colour code: Mo, pale grey; Cl, green. (b) The structure of $[Mo_3(\mu_3\text{-}O)(\mu\text{-}O)_3(OH_2)_9]^{4+}$ determined by X-ray diffraction for the hydrated $[4\text{-}MeC_6H_4SO_3]^-$ salt; H atoms are omitted from the terminally bound H_2O ligands. Mo−Mo distances are in the range 247–249 pm. [D.T. Richens *et al.* (1989) *Inorg. Chem.*, vol. 28, p. 1394]. Colour code: Mo, pale grey; O, red.

APPLICATIONS

Box 22.6 MoS₂: a solid lubricant

After purification and conversion into appropriate grade powders, the mineral *molybdenite*, MoS_2, has widespread commercial applications as a solid lubricant. It is applied to reduce wear and friction, and is able to withstand high-temperature working conditions. The lubricating properties are a consequence of the solid state layer structure (compare with graphite). Within each layer, each Mo centre is in a trigonal prismatic environment and each S atom bridges three Mo centres (in the figure, the colour code is Mo, pale grey, and S, yellow). The upper and lower surfaces of each layer consist entirely of S atoms, and there are only weak van der Waals forces operating between S−Mo−S slabs. Applications of MoS_2 lubricants range from engine oils and greases (used in engineering equipment) to coatings on sliding fitments.

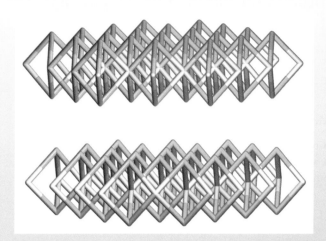

Octahedral geometries are common for complexes of Mo(IV) and W(IV), syntheses of which often involve ligand-mediated reduction of the metal centre, e.g. reactions 22.48 and 22.49.

$$WOCl_4 + 3Ph_3P \longrightarrow \textit{trans-}[WCl_4(PPh_3)_2] + Ph_3PO \tag{22.48}$$

$$MoCl_5 \xrightarrow{\text{excess py or bpy}} \begin{cases} MoCl_4(py)_2 \\ MoCl_4(bpy) \end{cases} \tag{22.49}$$

The salt $K_4[Mo(CN)_8]\cdot 2H_2O$ was the first example (in 1939) of an 8-coordinate (dodecahedral) complex. However, studies on a range of salts of $[Mo(CN)_8]^{4-}$ and $[W(CN)_8]^{4-}$ reveal cation dependence, both dodecahedral and square antiprismatic anions being found. The $K_4[M(CN)_8]$ salts are formed by reactions of K_2MO_4, KCN and KBH_4 in the presence of acetic acid. The $[M(CN)_8]^{4-}$ ions are kinetically inert with respect to ligand substitution (see Section 26.2), but can be oxidized to $[M(CN)_8]^{3-}$ as described earlier.

Molybdenum(III) and tungsten(III)

All the binary halides of Mo(III) and W(III) are known except for WF_3. The Mo(III) halides are made by reducing a halide of a higher oxidation state. Reduction of $MoCl_5$ with H_2 at 670 K gives $MoCl_3$ which has a layer structure similar to $CrCl_3$ but distorted and rendered diamagnetic by pairing of metal atoms (Mo−Mo = 276 pm). The 'W(III) halides' contain M_6 clusters and are prepared by controlled halogenation of a lower halide (see eqs. 22.55 and 22.56). W_6Cl_{18} (Fig. 22.11) has also been made by reducing WCl_4 using graphite in a silica tube at 870 K.

In contrast to Cr(III) (see Section 21.7), mononuclear complexes of Mo(III) and W(III) (especially the latter) are rare, there being an increased tendency for M−M bonding for the M(III) state. Electrolytic reduction of MoO_3 in concentrated HCl yields $[MoCl_5(OH_2)]^{2-}$ and $[MoCl_6]^{3-}$, the red K^+ salts of which are stable in dry air but are readily hydrolysed to $[Mo(OH_2)_6]^{3+}$, one of the few simple aqua ions of the heavier metals. By changing the reaction conditions, $[Mo_2Cl_9]^{3-}$ is formed in place of $[MoCl_6]^{3-}$, but reduction of WO_3 in concentrated HCl always gives $[W_2Cl_9]^{3-}$; $[WX_6]^{3-}$ has not been isolated. Both $[MoF_6]^{3-}$ and $[MoCl_6]^{3-}$ are

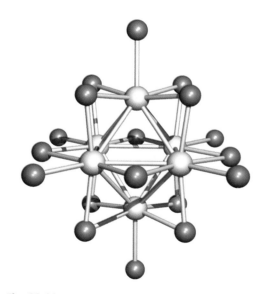

Fig. 22.11 The structure of W_6Cl_{18}, determined by X-ray diffraction [S. Dill *et al.* (2004) *Z. Anorg. Allg. Chem.*, vol. 630, p. 987]. The W−W bond distances are all close to 290 pm. Colour code: W, pale grey; Cl, green.

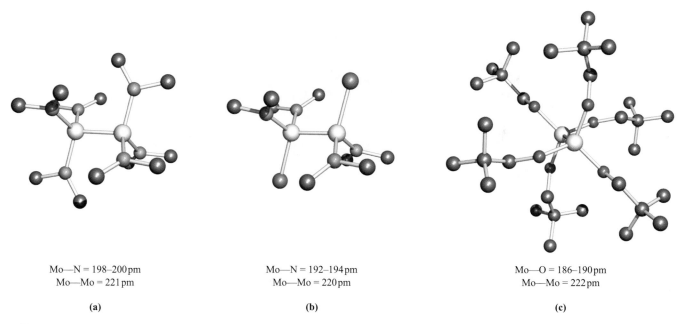

Mo—N = 198–200 pm
Mo—Mo = 221 pm

(a)

Mo—N = 192–194 pm
Mo—Mo = 220 pm

(b)

Mo—O = 186–190 pm
Mo—Mo = 222 pm

(c)

Fig. 22.12 The staggered structures (X-ray diffraction) of (a) $Mo_2(NMe_2)_6$ [M.H. Chisholm *et al.* (1976) *J. Am. Chem. Soc.*, vol. 98, p. 4469], (b) $Mo_2Cl_2(NMe_2)_4$ [M. Akiyama *et al.* (1977) *Inorg. Chem.*, vol. 16, p. 2407] and (c) $Mo_2(OCH_2{}^tBu)_6$ [M.H. Chisholm *et al.* (1977) *Inorg. Chem.*, vol. 16, p. 1801]. Hydrogen atoms are omitted for clarity; colour code: Mo, pale grey; N, blue; O, red; C, grey; Cl, green.

paramagnetic with magnetic moments close to $3.8\,\mu_B$ ($\approx \mu$(spin-only) for d^3). The $[M_2X_9]^{3-}$ ions adopt structure **22.19**; magnetic data and M—M distances (from crystalline salts) are consistent with metal–metal bonding. The $[W_2Cl_9]^{3-}$ ion is diamagnetic, indicating a W≡W triple bond consistent with the short bond length of 242 pm. Oxidation (eq. 22.50) to $[W_2Cl_9]^{2-}$ causes the W–W bond to lengthen to 254 pm consistent with a lower bond order of 2.5.

$$2[W_2Cl_9]^{3-} + Cl_2 \longrightarrow 2[W_2Cl_9]^{2-} + 2Cl^- \qquad (22.50)$$

$$\left[\begin{array}{c} X \quad X \quad X \\ X\cdots M \!\!=\!\! M\cdots X \\ X \quad X \quad X \end{array} \right]^{3-}$$

M = Mo, W; X = Cl, Br (see text)

(22.19)

In $Cs_3[Mo_2X_9]$, the Mo—Mo bond lengths are 266 pm (X = Cl) and 282 pm (X = Br). These data and magnetic moments at 298 K of $0.6\,\mu_B$ (X = Cl) and $0.8\,\mu_B$ (X = Br) per Mo, indicate significant Mo—Mo interaction but with a bond order <3. Contrast this with $[Cr_2X_9]^{3-}$, in which there is no Cr—Cr bonding (Section 21.7).

In Mo(III) and W(III) chemistry, Mo≡Mo and W≡W triple bonds ($\sigma^2\pi^4$, see Fig. 21.16) are common, and derivatives with amido and alkoxy ligands have received much attention, e.g. as precursors for solid state materials.

Reaction of $MoCl_3$ (or $MoCl_5$) or WCl_4 with $LiNMe_2$ gives $Mo_2(NMe_2)_6$ or $W_2(NMe_2)_6$ respectively. Both possess staggered structures (Fig. 22.12a) with M—M bond lengths of 221 (Mo) and 229 pm (W) typical of triple bonds. The orientations of the NMe_2 groups in the solid state suggest that the M—N bonds contain metal d–nitrogen p π-contributions. A staggered conformation, short Mo—Mo bond and shortened Mo—N bonds are also observed in $Mo_2Cl_2(NMe_2)_4$ (Fig. 22.12b); this and the W analogue are made by reacting $M_2(NMe_2)_6$ with Me_3SiCl. The air- and moisture-sensitive $M_2(NMe_2)_6$ and $M_2Cl_2(NMe_2)_4$ (M = Mo, W) are precursors for many derivatives including alkoxy compounds (eq. 22.51 and Fig. 22.12c); $[W_2(OR)_6]$ compounds are less stable than their Mo analogues. An extensive chemistry of alkoxy derivatives has been developed.[†]

$$Mo_2(NMe_2)_6 \xrightarrow{ROH} Mo_2(OR)_6 \qquad (22.51)$$

(R must be bulky, e.g. ^tBu, ^iPr, $CH_2{}^tBu$)

Molybdenum(II) and tungsten(II)

With the exception of organometallic and cyanido complexes, few mononuclear species are known for Mo(II)

[†] For example, see M.H. Chisholm (1995) *Chem. Soc. Rev.*, vol. 24, 79; M.H. Chisholm (1996) *J. Chem. Soc., Dalton Trans.*, p. 1781; M.H. Chisholm and A.M. Macintosh (2005) *Chem. Rev.*, vol. 105, p. 2949; M.H. Chisholm and Z. Zhou (2004) *J. Mater. Chem.*, vol. 14, p. 3081.

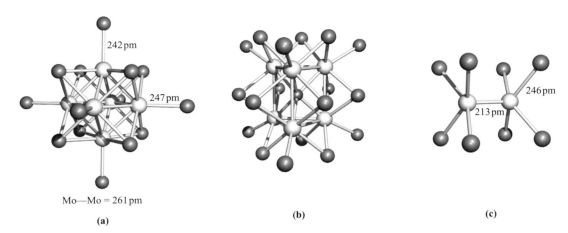

242 pm

247 pm

Mo—Mo = 261 pm

(a)

(b)

246 pm

213 pm

(c)

Fig. 22.13 The structures (X-ray diffraction) of (a) $[Mo_6Cl_{14}]^{2-}$ in the $[Ph_4P]^+$ salt [M.A. White *et al.* (1994) *Acta Crystallogr., Sect. C*, vol. 50, p. 1087], (b) $W_6Cl_{18}C$ [Y.-Q. Zheng *et al.* (2003) *Z. Anorg. Allg. Chem.*, vol. 629, p. 1256] and (c) $[Mo_2Cl_8]^{4-}$ in the compound $[H_3NCH_2CH_2NH_3]_2[Mo_2Cl_8]\cdot2H_2O$ [J.V. Brenic *et al.* (1969) *Inorg. Chem.*, vol. 8, p. 2698]. Colour code: Mo or W, pale grey; Cl, green; C, dark grey.

and W(II). The pentagonal bipyramidal $[Mo(CN)_7]^{5-}$ ion is formed by reducing $[MoO_4]^{2-}$ using H_2S in the presence of $[CN]^-$. In the capped trigonal prismatic $[MoBr(CN^tBu)_6]^+$ (eq. 22.52), the Br^- ligand occupies the capping site.

$$[MoBr_2(CO)_4]+6\,^tBuNC \longrightarrow [MoBr(CN^tBu)_6]Br + 4CO \tag{22.52}$$

The binary M(II) (M = Mo, W) halides are made from the higher halides, e.g. WCl_2 by disproportionation of WCl_4 at $\approx700\,K$ or from reduction of WCl_6 by H_2 at $\approx700\,K$, and $MoCl_2$ by fusion of Mo with $MoCl_3$, $MoCl_4$ or $MoCl_5$. The structures of the dihalides consist of $[M_6X_8]^{4+}$ clusters (structurally like $[Nb_6I_8]^{3+}$, Fig. 22.6a) with each M atom bonded to an additional X atom. The clusters are connected into a 2-dimensional layer structure by M−X−M bridges, i.e. $[M_6X_8]X_2X_{4/2}$. Reactions such as 22.53 and 22.54 produce salts containing discrete $[M_6X_{14}]^{2-}$ ions (Fig. 22.13a). The diamagnetism of $[M_6X_{14}]^{2-}$ is consistent with M−M bonding, and M−M single bonds can be allocated by following a similar electron-counting procedure as for $[Nb_6X_{12}]^{2+}$ and $[Ta_6X_{12}]^{2+}$ (see Section 22.6). Octahedral $Mo_6(\mu_3\text{-}X)_8$ (usually X = S, Se, Te) cluster units are the building blocks in so-called *Chevrel phases*, $M_xMo_6X_8$ (M = group 1 or 2 metal, or *p*-, *d*- or *f*-block metal). These materials exhibit interesting electronic (in particular, superconducting) properties and are discussed further in Section 28.4.

$$MoCl_2 \xrightarrow{\text{Et}_4\text{NCl in dil. HCl, EtOH}} [Et_4N]_2[Mo_6Cl_{14}] \tag{22.53}$$

$$MoBr_2 \xrightarrow{\text{CsBr, IBr, 325 K}} Cs_2[Mo_6Br_{14}] \tag{22.54}$$

While the Mo(II) halides are not readily oxidized, WCl_2 and WBr_2 (eqs. 22.55 and 22.56) are oxidized to give products

containing $[W_6(\mu\text{-}Cl)_{12}]^{6+}$ or $[W_6(\mu_3\text{-}Br)_8]^{6+}$ clusters, terminal halides and bridging $[Br_4]^{2-}$ units. The formulae of the products in the equations below indicate whether the clusters are discrete or linked.

$$[W_6Cl_8]Cl_2Cl_{4/2} \xrightarrow{Cl_2,\,373\,K} [W_6Cl_{12}]Cl_6 \tag{22.55}$$

$$[W_6Br_8]Br_2Br_{4/2} \xrightarrow{Br_2,\,T<420\,K} \begin{cases} [W_6Br_8]Br_6 \\ [W_6Br_8]Br_4(Br_4)_{2/2} \\ [W_6Br_8]Br_2(Br_4)_{4/2} \end{cases} \tag{22.56}$$

When WBr_2 (i.e. $[W_6Br_8]Br_2Br_{4/2}$) is heated with AgBr *in vacuo* with a temperature gradient of $925/915\,K$, the products are yellow-green $Ag_2[W_6Br_{14}]$ and black-brown $Ag[W_6Br_{14}]$. Both silver salts contain discrete anions, structurally similar to $[Mo_6Cl_{14}]^{2-}$ (Fig. 22.13a). The difference in colour of the compounds is characteristic of W in different oxidation states: $[W_6Br_{14}]^{2-}$ contains W in oxidation state $+2$, whereas $[W_6Br_{14}]^-$ is formulated as $[W^{II}_5W^{III}Br_{14}]^-$.

In contrast to the more usual octahedral tungsten halide clusters, $W_6Cl_{16}C$ (formed in the reaction of W, WCl_5 and CCl_4 *in vacuo* with a temperature gradient of $1030/870\,K$) contains an example of a carbon-centred trigonal prismatic cluster unit. The cluster units are connected into a 2-dimensional sheet (**22.20**) and a formulation of $[W_6Cl_{12}C]Cl_2Cl_{4/2}$ is appropriate. The related compound $W_6Cl_{18}C$ can also be isolated and consists of discrete trigonal prismatic, carbon-centred clusters (Fig. 22.13b). When WCl_6 is reduced by Bi at $670\,K$ in the presence of CCl_4, a black solid is produced (proposed to be $W_6Cl_{16}C$) from which discrete $[W_6Cl_{18}C]^{2-}$ clusters can be isolated by the addition of $[Bu_4N]Cl$. Reduction of WCl_6 by Bi at $770\,K$ in the presence of NaN_3 leads to the formation of

Na[$W_6Cl_{18}N$]. The [$W_6Cl_{18}C$]$^{2-}$ and [$W_6Cl_{18}N$]$^-$ anions contain interstitial C and N, respectively, and are structurally analogous to $W_6Cl_{18}C$ (Fig. 22.13b).

(22.20)

Compounds containing an {Mo≡Mo}$^{4+}$ unit are well exemplified in Mo(II) chemistry. In contrast, the chemistry of the {W≡W}$^{4+}$ unit is less extensive. Two reasons for this observation are that complexes containing a {W_2}$^{4+}$ core are more prone to oxidation than those based on {Mo_2}$^{4+}$, and that suitable precursors to {Mo_2}$^{4+}$-containing complexes are more abundant than those to analogous tungsten-containing compounds. A description of Mo≡Mo and W≡W quadruple bonds in terms of σ, π and δ components is analogous to that of a Cr≡Cr bond (see Section 21.7 and Fig. 21.16), and the effect of the δ component in forcing the ligands to be eclipsed is illustrated by the structure of [Mo_2Cl_8]$^{4-}$ (Fig. 22.13c). This is made in the reaction sequence 22.57, the intermediate acetate $Mo_2(\mu$-$O_2CMe)_4$ (**22.21**) being a useful synthon in this area of chemistry, e.g. reaction 22.58. Replacement of the Cl$^-$ ligands in [Mo_2Cl_8]$^{4-}$ yields a range of derivatives. Equation 22.59 gives examples, one of which involves concomitant oxidation of the {Mo≡Mo}$^{4+}$ core. Derivatives containing [$MeSO_3$]$^-$ or [CF_3SO_3]$^-$ bridges are useful precursors and can be used to prepare the highly reactive [$Mo_2(NCMe)_8$]$^{4+}$ (eq. 22.60).

= quadruple bond

(22.21)

$$Mo(CO)_6 \xrightarrow{MeCO_2H} Mo_2(\mu\text{-}O_2CMe)_4 \xrightarrow{KCl \text{ in } HCl} [Mo_2Cl_8]^{4-}$$

(22.57)

$$Mo_2(\mu\text{-}O_2CMe)_4 \xrightarrow{LiMe \text{ in } Et_2O} Li_4[Mo_2Me_8]\cdot 4Et_2O$$

(22.58)

$$[Mo_2Cl_8]^{4-} \xrightarrow[\text{in presence of } O_2]{[HPO_4]^{2-}} [Mo_2(\mu\text{-}HPO_4)_4]^{2-}$$

(22.59)

$$\downarrow [SO_4]^{2-}$$

$$[Mo_2(\mu\text{-}SO_4)_4]^{4-}$$

$$[Mo_2(\mu\text{-}O_3SCF_3)_2(OH_2)_4]^{2+} \xrightarrow{MeCN} [Mo_2(NCMe)_8]^{4+}$$

(22.60)

Each Mo centre in these Mo_2 derivatives possesses a vacant orbital (as in structure **21.20**), but forming Lewis base adducts is not facile; '[$Mo_2(\mu$-$O_2CMe)_4(OH_2)_2$]' has not been isolated, although the oxidized species [$Mo_2(\mu$-$SO_4)_4(OH_2)_2$]$^{3-}$ and [$Mo_2(\mu$-$HPO_4)_4(OH_2)_2$]$^{2-}$ are known. An unstable adduct [$Mo_2(\mu$-$O_2CMe)_4(py)_2$] results from addition of pyridine to [$Mo_2(\mu$-$O_2CMe)_4$], and a more stable one can be made by using [$Mo_2(\mu$-$O_2CCF_3)_4$].

Not all the derivatives mentioned above contain Mo≡Mo bonds, e.g. oxidation occurs in reaction 22.59 in the formation of [$Mo_2(\mu$-$HPO_4)_4$]$^{2-}$. Table 22.2 lists the Mo−Mo bond lengths in selected compounds, and the bond orders follow from the energy level diagram in Fig. 21.16b; e.g. [Mo_2Cl_8]$^{4-}$ has a $\sigma^2\pi^4\delta^2$ configuration (Mo≡Mo), but [$Mo_2(HPO_4)_4$]$^{2-}$ and [$Mo_2(HPO_4)_4(OH_2)_2$]$^{2-}$ are $\sigma^2\pi^4$ (Mo≡Mo). Oxidation of the [M_2]$^{4+}$ (M = Mo or W) core is most facile when the bridging ligand is **22.22**. Dissolution of [Mo_2(**22.22**)$_4$] in CH_2Cl_2 results in a 1-electron oxidation of the [Mo_2]$^{4+}$ core and formation of [Mo_2(**22.22**)$_4$Cl]. In contrast, when [W_2(**22.22**)$_4$] dissolves in a chloroalkane solvent, it is oxidized directly to [W_2(**22.22**)$_4$Cl$_2$], i.e. [W_2]$^{6+}$. Gas-phase photoelectron spectroscopic data (see Section 4.12) show that initial ionization of [W_2(**22.22**)$_4$] requires only 339 kJ mol^{-1}. Just how low this value is can be appreciated by comparing it with a value of $IE_1 = 375.7$ kJ mol^{-1} for Cs, the element with the lowest first ionization energy (see Fig. 1.16). [W_2(**22.22**)$_4$] can be prepared in a convenient 2-step procedure, starting from the readily available and air-stable W(CO)$_6$:

$$W(CO)_6 \xrightarrow[\text{in 1,2-dichlorobenzene}]{} [W_2(\textbf{22.22})_4Cl_2] \xrightarrow{K, THF} [W_2(\textbf{22.22})_4]$$

Conjugate base of hexahydropyrimidopyrimidine [hpp]$^-$

(22.22)

Table 22.2 Mo−Mo bond lengths and orders in selected dimolybdenum species.

Compound or ion[†]	Mo−Mo bond distance / pm	Mo−Mo bond order	Notes
$Mo_2(\mu\text{-}O_2CMe)_4$	209	4.0	Structure **22.21**
$Mo_2(\mu\text{-}O_2CCF_3)_4$	209	4.0	Analogous to **22.21**
$[Mo_2Cl_8]^{4-}$	214	4.0	Fig. 22.13c
$[Mo_2(\mu\text{-}SO_4)_4]^{4-}$	211	4.0	Analogous to **22.21**
$[Mo_2(\mu\text{-}SO_4)_4(OH_2)_2]^{3-}$	217	3.5	Contains axial H_2O ligands
$[Mo_2(\mu\text{-}HPO_4)_4(OH_2)_2]^{2-}$	223	3.0	Contains axial H_2O ligands

[†] Data for anions refer to K^+ salts; contrast Fig. 22.13c where the $[Mo_2Cl_8]^{4-}$ parameters refer to the $[H_3NCH_2CH_2NH_3]^{2+}$ salt. For an overview of Mo≡Mo bond lengths, see: F.A. Cotton, L.M. Daniels, E.A. Hillard and C.A. Murillo (2002) *Inorg. Chem.*, vol. 41, p. 2466.

Self-study exercises

1. Rationalize why the Mo−Mo bond length increases (≈ 7 pm) when $[Mo_2(\mu\text{-}O_2CR)_4]$ ($R = 2,4,6\text{-}^iPr_3C_6H_2$) undergoes a 1-electron oxidation.
   [*Ans.* See: F.A. Cotton *et al.* (2002) *Inorg. Chem.*, vol. 41, p. 1639]

2. Rationalize why the two $MoCl_4$-units in $[Mo_2Cl_8]^{4-}$ are eclipsed. [*Ans.* See Fig. 21.16 and discussion]

3. Confirm that $W_2Cl_4(NH_2Cy)_4$ (Cy = cyclohexyl) possesses a ditungsten quadruple bond. The $W_2Cl_4N_4$ core in $W_2Cl_4(NH_2Cy)_4$ is non-centrosymmetric. Which of the following is the structure of this core?

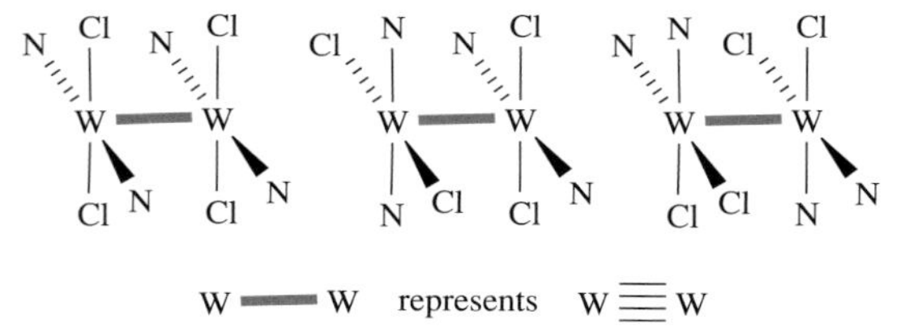

22.8 Group 7: technetium and rhenium

The metals

The heavier group 7 metals, Tc and Re, are less reactive than Mn. Technetium does not occur naturally (see Section 22.2). The bulk metals tarnish slowly in air, but more finely divided Tc and Re burn in O_2 (eq. 22.61) and react with the halogens (see below). Reactions with sulfur give TcS_2 and ReS_2.

$$4M + 7O_2 \xrightarrow{650\,K} 2M_2O_7 \qquad (M = Tc, Re) \qquad (22.61)$$

The metals dissolve in oxidizing acids (e.g. conc HNO_3) to give $HTcO_4$ (pertechnetic acid) and $HReO_4$ (perrhenic acid), but are insoluble in HF or HCl.

Technetium and rhenium exhibit oxidation states from 0 to +7 (Table 19.3), although M(II) and lower states are stabilized by π-acceptor ligands such as CO (see Box 22.7) and will not be considered further in this section. The chemistry of Re is better developed than that of Tc, but interest in the latter has expanded with the current use of its compounds in nuclear medicine. There are significant differences between the chemistries of Mn and the heavier group 7 metals:

- a comparison of the potential diagrams in Fig. 22.14 with that for Mn (Fig. 8.2) shows that $[TcO_4]^-$ and $[ReO_4]^-$ are significantly more stable with respect to reduction than $[MnO_4]^-$;
- the heavier metals have less cationic chemistry than manganese;
- a tendency for M−M bond formation leads to higher nuclearity species being important for the heavier metals.

High oxidation states of technetium and rhenium: M(VII), M(VI) and M(V)

Rhenium reacts with F_2 to give yellow ReF_6 and ReF_7 (see Fig. 22.16) depending on conditions, and ReF_5 is made by

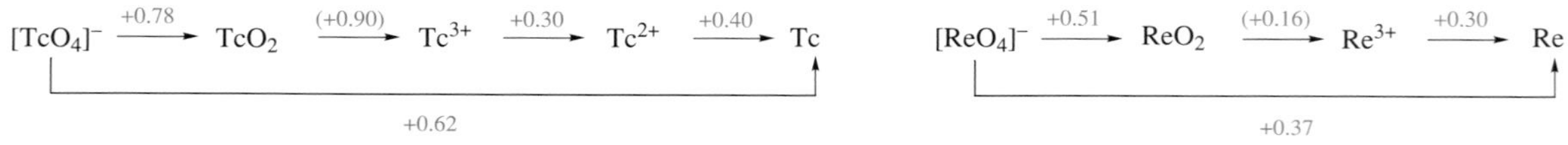

Fig. 22.14 Potential diagrams for technetium and rhenium in aqueous solution at pH 0; compare with the diagram for manganese in Fig. 8.2.

reaction 22.62. Direct combination of Tc and F_2 leads to TcF_6 and TcF_5; TcF_7 is not known.[†]

$$ReF_6 \xrightarrow{\text{over W wire, 870 K}} ReF_5 \qquad (22.62)$$

For the later halogens, combination of the elements at appropriate temperatures affords $TcCl_6$, $ReCl_6$, $ReCl_5$ and $ReBr_5$. The high oxidation state halides are volatile solids which are hydrolysed by water to $[MO_4]^-$ and MO_2 (e.g. eq. 22.63).

$$3ReF_6 + 10H_2O \longrightarrow 2HReO_4 + ReO_2 + 18HF \qquad (22.63)$$

(22.23)   **(22.24)**

The fluorides ReF_7, ReF_6 and TcF_6 are molecular with pentagonal bipyramidal, **22.23**, and octahedral structures. $ReCl_6$ is probably a molecular monomer, but $ReCl_5$ (a useful precursor in Re chemistry) is a dimer (**22.24**).

Oxohalides are well represented:

- M(VII): $TcOF_5$, $ReOF_5$, TcO_2F_3, ReO_2F_3, ReO_2Cl_3, ReO_3F, TcO_3F, TcO_3Cl, TcO_3Br, TcO_3I, ReO_3Cl, ReO_3Br;
- M(VI): $TcOF_4$, $ReOF_4$, ReO_2F_2, $TcOCl_4$, $ReOCl_4$, $ReOBr_4$;
- M(V): $ReOF_3$, $TcOCl_3$.

They are prepared by reacting oxides with halogens, or halides with O_2, or by reactions such as 22.64 and 22.65. ReO_2Cl_3 is prepared by treating Re_2O_7 with an excess of BCl_3. Whereas $ReOF_5$ can be prepared by the high-temperature reaction between ReO_2 and F_2, the Tc analogue must be made by reaction 22.66 because the reaction of F_2 and TcO_2 gives TcO_3F.

$$ReCl_5 + 3Cl_2O \longrightarrow ReO_3Cl + 5Cl_2 \qquad (22.64)$$

$$\left.\begin{array}{l} Tc_2O_7 + 4HF \longrightarrow 2TcO_3F + [H_3O]^+ + [HF_2]^- \\ TcO_3F + XeF_6 \longrightarrow TcO_2F_3 + XeOF_4 \end{array}\right\} \qquad (22.65)$$

$$2TcO_2F_3 + 2KrF_2 \xrightarrow{\text{in anhydrous HF}} 2TcOF_5 + 2Kr + O_2 \qquad (22.66)$$

Few oxohalides have been structurally characterized in the solid state. $ReOCl_4$ (**22.25**) and $TcOF_5$ (**22.26**) are molecular. ReO_2Cl_3 is dimeric with each Re atom in a distorted octahedral environment (**22.27**), while TcO_2F_3 is polymeric with oxido groups *trans* to bridging F atoms (**22.28**). X-ray

diffraction data for $K[Re_2O_4F_7]\cdot2ReO_2F_3$ show that ReO_2F_3 adopts a polymeric structure analogous to TcO_2F_3. The oxofluorides $TcOF_4$ and $ReOF_4$ also have polymeric structures with O atoms *trans* to M−F−M bridges. In SO_2ClF solution, ReO_2F_3 exists as an equilibrium mixture of a cyclic trimer (**22.29**) and tetramer; the Tc analogue is present only as the trimer.

(22.25)   **(22.26)**

Re−Cl$_{\text{term}}$ = 227 pm   Re−Cl$_{\text{bridge}}$ = 259 pm

(22.27)

(22.28)   **(22.29)**

Self-study exercises

1. Rationalize why the ^{19}F NMR spectrum of a solution of $TcOF_5$ in SO_2ClF at 163 K exhibits a doublet and a quintet ($J = 75$ Hz). What will be the relative integrals of these signals?   [*Hint*: See structure **22.26**]

2. The reaction of $TcOF_5$ with SbF_5 gives $[Tc_2O_2F_9]^+[Sb_2F_{11}]^-$. Suggest a structure for the cation.   [*Ans.* See N. LeBlond *et al.* (2000) *Inorg. Chem.*, vol. 39, p. 4494]

3. Assuming a static structure, predict what you would expect to see in the solution ^{19}F NMR spectrum of the following anion:

[*Ans.* See W.J. Casteel, Jr *et al.* (1999) *Inorg. Chem.*, vol. 38, p. 2340]

[†] Calculations suggest TcF_7 might be prepared: S. Riedel, M. Renz and M. Kaupp (2007) *Inorg. Chem.*, vol. 46, p. 5734 – 'High-valent technetium fluorides. Does TcF_7 exist?'.

4. To obtain pure TcO_3F, the equilibrium:

$$[TcO_4]^- + 3HF \rightleftharpoons TcO_3F + [H_3O]^+ + 2F^-$$

is driven to the right-hand side by adding BiF_5. Suggest how this works.

[*Ans*. See J. Supel *et al.* (2007) *Inorg. Chem.*, vol. 46, p. 5591]

A number of imido analogues of the oxohalides ($[RN]^{2-}$ is isoelectronic with O^{2-}) have been structurally characterized and include tetrahedral $Re(N^tBu)_3Cl$ and trigonal bipyramidal $Re(N^tBu)_2Cl_3$ (**22.30**). Reduction of $Tc(NAr)_3I$ using Na leads to the dimer $[(ArN)_2Tc(\mu\text{-}NAr)_2Tc(NAr)_2]$ when Ar = 2,6-$Me_2C_6H_3$, but when Ar is the bulkier 2,6-$^iPr_2C_6H_3$, the product is $Tc_2(NAr)_6$ with an ethane-like configuration. Each dimer contains a Tc—Tc single bond.

(**22.30**) (**22.31**)

The yellow, volatile oxides M_2O_7 (M = Tc, Re) form when the metals burn in O_2. The volatility of Re_2O_7 is used in manufacturing Re (see Section 22.2). In the solid and vapour states, Tc_2O_7 is molecular with a linear Tc—O—Tc bridge (**22.31**). In the vapour, Re_2O_7 has a similar structure but the solid adopts a complex layer structure. The oxides are the anhydrides of $HTcO_4$ and $HReO_4$ and dissolve in water (eq. 22.67) to give solutions containing $[TcO_4]^-$ (pertechnetate) and $[ReO_4]^-$ (perrhenate). Pertechnetate and perrhenate salts are the commonest starting materials in Tc and Re chemistries.

$$M_2O_7 + H_2O \longrightarrow 2HMO_4 \qquad (M = Tc, Re) \qquad (22.67)$$

Pertechnetic and perrhenic acids are strong acids. Crystalline $HReO_4$ (yellow), $HReO_4 \cdot H_2O$ and $HTcO_4$ (dark red) have been isolated; crystalline $HReO_4 \cdot H_2O$ consists of a hydrogen-bonded network of $[H_3O]^+$ and $[ReO_4]^-$ ions. The acids react with H_2S to precipitate M_2S_7 (eq. 22.68) in striking contrast to the reduction of $[MnO_4]^-$ to Mn^{2+} by H_2S.

$$2HMO_4 + 7H_2S \longrightarrow M_2S_7 + 8H_2O \qquad (M = Tc, Re)$$
$$(22.68)$$

The $[TcO_4]^-$ and $[ReO_4]^-$ ions are tetrahedral and isostructural with $[MnO_4]^-$. Whereas $[MnO_4]^-$ is intense purple due to a ligand-to-metal charge transfer absorption in the visible region, $[ReO_4]^-$ and $[TcO_4]^-$ are colourless because the corresponding LMCT band is in the UV region (see Section 20.7).

Solvated or complexed $[TcO_3]^+$ and $[ReO_3]^+$ are both known, e.g. cation **22.32**. The closest example of a free $[TcO_3]^+$ ion is found in the fluorosulfonate salt. The solid state structure of $[TcO_3][SO_3F]$ contains trigonal pyramidal $[TcO_3]^+$ (Tc–O = 168 pm), but there are three short $O_3Tc^+ \cdots OSO_2F^-$ contacts (Tc $\cdots$ O = 224–228 pm) placing the Tc atom in a distorted octahedral environment.

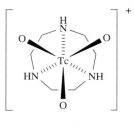

(**22.32**)

Rhenium(VI) oxide, ReO_3, is made by reducing Re_2O_7 with CO. TcO_3 has not been isolated. Red ReO_3 crystallizes with a cubic structure (Fig. 21.5) which is a prototype structure. ReO_3 is a metallic-like electrical conductor owing to delocalization of the d^1 electrons. No reaction between ReO_3 and H_2O, dilute acids or alkalis occurs, but reaction 22.69 occurs with concentrated alkalis.

$$3ReO_3 + 2[OH]^- \xrightarrow[\text{conc}]{\Delta} 2[ReO_4]^- + ReO_2 + H_2O \qquad (22.69)$$

For the +5 oxidation state, the only oxide or oxohalide is the blue Re_2O_5 but it is unstable with respect to disproportionation.

Technetium(VII) and rhenium(VII) form a series of hydride complexes (neutron diffraction data are essential for accurate H location) including the tricapped trigonal prismatic $[TcH_9]^{2-}$, $[ReH_9]^{2-}$ (see Section 10.7) and $[ReH_7(Ph_2PCH_2CH_2PPh_2\text{-}P,P')]$. Some hydrido complexes contain coordinated η^2-H_2 (**22.33**) with a 'stretched' H–H bond, e.g. in $[ReH_5(\eta^2\text{-}H_2)\{P(4\text{-}C_6H_4Me)_3\}_2]$ two H atoms are separated by 136 pm, the next shortest H $\cdots$ H separation being 175 pm.

Longer bond than in free H_2

(**22.33**)

A few halido complexes are known for M(VI) and M(V): square antiprismatic $[ReF_8]^{2-}$ (formed from KF and ReF_6), $[ReF_6]^-$ (from reduction of ReF_6 with KI in liquid SO_2), $[TcF_6]^-$ (from TcF_6 and CsCl in IF_5), $[ReCl_6]^-$ (in the salt $[PCl_4]_3[Re^VCl_6][Re^{IV}Cl_6]$ formed from the reaction of $ReCl_5$ and PCl_5) and $[Re_2F_{11}]^-$ (**22.34**, formed as the $[Re(CO)_6]^+$ salt when an excess of ReF_6 reacts with $Re_2(CO)_{10}$ in anhydrous HF).

Eclipsed, with a linear bridge

(22.34)

Complexes of M(VII), M(VI) and M(V) (M = Tc, Re) are dominated by oxido and nitrido species, with octahedral and square-based pyramidal (oxido or nitrido ligand in the apical site) structures being common. Complexes of M(V) outnumber those of the higher oxidation states, with square-based pyramidal structures usually favoured. Examples include:

- octahedral M(VII): *fac*-[ReO$_3$L]$^+$ (L = **22.35**, see structure **22.32**), *fac*-[ReO$_3$Cl(phen)], [TcNCl(η^2-O$_2$)$_2$]$^-$;
- octahedral M(VI): [ReOCl$_5$]$^-$, *trans*-[TcN(OH$_2$)Br$_4$]$^-$, *mer*-[TcNCl$_3$(bpy)];
- square-based pyramidal M(VI): [TcNCl$_4$]$^-$, [TcNBr$_4$]$^-$;
- octahedral M(V): [ReOCl$_5$]$^{2-}$, [ReOCl$_4$(py)]$^-$, *trans*-[TcO$_2$(en)$_2$]$^+$, *trans*-[TcO$_2$(py)$_4$]$^+$, *cis*-[TcNBr(bpy)$_2$]$^+$;
- square-based pyramidal M(V): [ReOCl$_4$]$^-$, [TcOCl$_4$]$^-$, [TcO(ox)$_2$]$^-$;
- pentagonal bipyramidal, rare for Re(V): complex **22.36**;
- square antiprismatic Re(V): [Re(CN)$_8$]$^{3-}$.

X = NH or S

(22.35)   **(22.36)**

The development of technetium agents for imaging the brain, heart and kidneys has prompted the study of a range of Tc(V) oxido complexes, many of which are square-based pyramidal and contain a tetradentate ligand, often a mixed *S*- and *N*-donor. The oxido ligand occupies the apical site. Complexes **22.37** and **22.38** (in their ^{99m}Tc forms, see Box 22.7) are examples of radiopharmaceuticals used as kidney and brain imaging agents respectively. Complex **22.39** (containing the metastable radioisotope ^{99m}Tc) is marketed as the heart imaging agent Myoview. The ethoxyethyl substituents render the complex lipophilic, a requirement for biodistribution (uptake into the heart and clearance from the blood and liver).[†]

[†] For review articles, see: S.R. Banerjee, K.P. Maresca, L. Francesconi, J. Valliant, J.W. Babich and J. Zubieta (2005) *Nucl. Med. Biol.*, vol. 32, p. 1; R. Alberto (2005) *Top. Curr. Chem.*, vol. 252, p. 1; T. Mindt, H. Struthers, E. Garcia-Garayoa, D. Desbois and R. Schibli (2007) *Chimia*, vol. 61, p. 725; M.D. Bartholoma, A.S. Louie, J.F. Valliant and J. Zubieta (2010) *Chem. Rev.*, vol. 110, p. 2903. See also Box 22.7 further reading.

(22.37)   **(22.38)**

(22.39)

Technetium(IV) and rhenium(IV)

The reaction of Tc$_2$O$_7$ with CCl$_4$ at 670 K (or heating Tc and Cl$_2$) gives TcCl$_4$ as a moisture-sensitive, red solid. The halides ReX$_4$ (X = F, Cl, Br, I) are all known. Blue ReF$_4$ forms when ReF$_5$ is reduced by H$_2$ over a Pt gauze, and black ReCl$_4$ is made by heating ReCl$_5$ and Re$_3$Cl$_9$. Solid TcCl$_4$ and ReCl$_4$ are polymeric but not isostructural. TcCl$_4$ adopts chain structure **22.40** and has a magnetic moment of 3.14 μ_B (298 K) per Tc(IV) centre. In ReCl$_4$, dimers are linked into zigzag chains by chlorido-bridges (**22.41**) and the short Re–Re distance is consistent with metal–metal bonding (compare **22.41** with **22.19**). The salt [PCl$_4$]$^+$[Re$_2$Cl$_9$]$^-$ is formed by reducing ReCl$_5$ using PCl$_3$ at 373–473 K under a stream of N$_2$. The salt contains discrete ions. [Re$_2$Cl$_9$]$^-$ adopts a structure analogous to **22.19**, and the Re–Re distance of 272 pm is consistent with a single bond. When PCl$_5$ is heated with ReCl$_4$ at 570 K under vacuum, the product is [PCl$_4$]$_2$[Re$_2$Cl$_{10}$]. The structure of the [Re$_2$Cl$_{10}$]$^{2-}$ ion is similar to the ReCl$_5$ dimer (**22.24**); neither has a direct Re–Re bond.

Re–Re = 273 pm

(22.40)   **(22.41)**

Most radioisotopes used in nuclear medicine are artificial. They are produced in nuclear reactors (e.g. ^{89}Sr, ^{57}Co), cyclotrons (e.g. ^{11}C, ^{18}F) or specialized generators such as a molybdenum-technetium generator. The metastable radioisotope ^{99m}Tc has a half-life of 6 hours and is important for medical imaging. It is a decay product of ^{99}Mo ($t_{1/2}$ = 2.8 days), which is itself man-made, being produced in a nuclear reactor. The radioactive decay of ^{99}Mo to ^{99m}Tc and the much longer lived ^{99}Tc is summarized below:

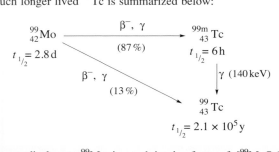

For medical use, ^{99}Mo is used in the form of $[^{99}MoO_4]^{2-}$ adsorbed on alumina in a 'cold kit' generator. The commercial assembly of the latter is shown in the photograph opposite. Decay of $[^{99}MoO_4]^{2-}$ gives $[^{99m}TcO_4]^-$ which is selectively eluted from the generator and combined with an appropriate ligand to give a complex suitable for injection into a patient. The ^{99m}Tc-containing complexes are designed to target tumour cells. As ^{99m}Tc decays to ^{99}Tc, γ-radiation of energy 140 keV is emitted. This energy falls within the range ($\approx$100–200 keV) compatible with modern γ-detectors. The emitted γ-radiation can be recorded in the form of a γ-camera scan or *scintigram*. These 2-dimensional images are used to evaluate both primary tumours and metastases. (Metastasis is the growth of a secondary malignant tumour from a primary cancer.) Three-dimensional images are obtained by using *single photon emission computed tomography* (SPECT). This technique enables a radiographer to obtain an image by using only micromolar or nanomolar concentrations of the ^{99m}Tc tracer. ^{99m}Tc is invaluable for diagnostic imaging, and with the development of a suite of technetium-containing complexes, it is now possible to image the heart, liver, kidneys, brain and bone.

In developing new techniques of tumour imaging with radioisotopes, one goal is to label single-chain antibody fragments which may efficiently target tumours. The complex $[^{99m}Tc(OH_2)_3(CO)_3]^+$:

Clean-room assembly of technetium-99m generators. $[^{99}MoO_4]^{2-}$ is adsorbed on alumina in a 'cold kit' generator, and radioactive decay produces $[^{99m}TcO_4]^-$.

can be used to label single-chain antibody fragments which carry *C*-terminal histidine tags. High activities are achieved (90 mCi mg^{-1}) and, *in vivo*, the technetium-labelled fragments are very stable. This technique appears to have a high potential for application in clinical medicine. The original method of preparing $[^{99m}Tc(OH_2)_3(CO)_3]^+$ involved the reaction between $[^{99m}TcO_4]^-$ and CO at 1 bar pressure in aqueous NaCl at pH 11. For commercial radiopharmaceutical kits, use of gaseous CO is inconvenient and solid, air-stable sources of CO are desirable. Potassium boranocarbonate, $K_2[H_3BCO_2]$ (made from $H_3B \cdot THF$/CO and ethanolic KOH), is ideal: it acts as both a source of CO and a reducing agent, and reacts with $[^{99m}TcO_4]^-$ under buffered, aqueous conditions to give $[^{99m}Tc(OH_2)_3(CO)_3]^+$.

Further reading

R. Alberto (2009) *Eur. J. Inorg. Chem.*, p. 21 – 'The chemistry of technetium-water complexes within the manganese triad: Challenges and perspectives'.

R. Alberto (2010) *Top. Organomet. Chem.*, vol. 32, p. 219 – 'Organometallic radiopharmaceuticals'.

R. Alberto, K. Ortner, N. Wheatley, R. Schibli and A.P. Schubiger (2001) *J. Am. Chem. Soc.*, vol. 123, p. 3135 – 'Synthesis and properties of boranocarbonate: a convenient *in situ* CO source for the aqueous preparation of $[^{99m}Tc(OH_2)_3(CO)_3]^+$'.

R. Waibel *et al.* (1999) *Nature Biotechnol.*, vol. 17, p. 897 – 'Stable one-step technetium-99m labelling of His-tagged recombinant proteins with a novel Tc(I)-carbonyl complex'.

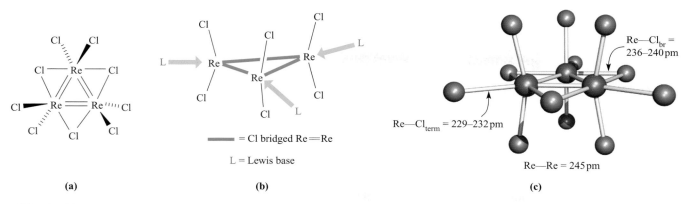

Fig. 22.15 Schematic representations of (a) the structure of Re_3Cl_9 (interactions between units occur in the solid, see text) and (b) the sites of addition of Lewis bases to Re_3Cl_9. (c) The structure (X-ray diffraction) of $[Re_3Cl_{12}]^{3-}$ in the $[Me_3NH]^+$ salt [M. Irmler *et al.* (1991) *Z. Anorg. Allg. Chem.*, vol. 604, p. 17]; colour code: Re, brown; Cl, green.

The oxides TcO_2 and ReO_2 are made by thermal decomposition of $[NH_4][MO_4]$ or reduction of M_2O_7 by M or H_2. Both adopt rutile structures (Fig. 6.22), distorted by pairing of metal centres as in MoO_2. With O_2, TcO_2 is oxidized to Tc_2O_7, and with H_2 at 770 K, reduction of TcO_2 to the metal occurs.

Reduction of $KReO_4$ using I^- in concentrated HCl produces $K_4[Re_2(\mu\text{-}O)Cl_{10}]$. The $[Re_2(\mu\text{-}O)Cl_{10}]^{4-}$ anion has a linear Re–O–Re bridge with Re–O π-character (Re–O = 186 pm) and is structurally related to $[W_2(\mu\text{-}O)Cl_{10}]^{4-}$ and $[Ru_2(\mu\text{-}O)Cl_{10}]^{4-}$ (see Fig. 22.17). The octahedral complexes $[MX_6]^{2-}$ (M = Tc, Re; X = F, Cl, Br, I) are all known and are probably the most important M(IV) complexes. The ions $[MX_6]^{2-}$ (X = Cl, Br, I) are formed by reducing $[MO_4]^-$ (e.g. by I^-) in concentrated HX. Reactions of $[MBr_6]^{2-}$ with HF yield $[MF_6]^{2-}$. The chlorido complexes (e.g. as K^+ or $[Bu_4N]^+$ salts) are useful starting materials in Tc and Re chemistries, but both are readily hydrolysed in water. In aqueous solution, $[TcCl_6]^{2-}$ is in equilibrium with $[TcCl_5(OH_2)]^-$, and complete hydrolysis gives TcO_2. Halide exchange between $[ReI_6]^{2-}$ and HCl leads to *fac*-$[ReCl_3I_3]^{2-}$, *cis*- and *trans*-$[ReCl_4I_2]^{2-}$ and $[ReCl_5I]^{2-}$. In most complexes, octahedral coordination for Re(IV) and Tc(IV) is usual, e.g. *cis*-$[TcCl_2(acac)_2]$, *trans*-$[TcCl_4(PMe_3)_2]$, $[Tc(NCS\text{-}N)_6]^{2-}$, $[Tc(ox)_3]^{2-}$, *trans*-$[ReCl_4(PPh_3)_2]$, $[ReCl_5(OH_2)]^-$, $[ReCl_5(PEt_3)]^-$, $[ReCl_4(bpy)]$ and *cis*-$[ReCl_4(THF)_2]$. A notable exception is the pentagonal bipyramidal $[Re(CN)_7]^{3-}$ ion which is made as the $[Bu_4N]^+$ salt by heating $[Bu_4N]_2[ReCl_6]$ with $[Bu_4N][CN]$ in DMF.

Self-study exercise

The hydrolysis of $TcCl_4$ immediately gives TcO_2. However, *cis*-$[TcCl_4(OH_2)_2]$ can be isolated as the solvate $[TcCl_4(OH_2)_2]\cdot 2C_4H_8O_2$ from a solution of $TcCl_4$ in dioxane ($C_4H_8O_2$) containing trace amounts of water. Suggest what intermolecular interactions are responsible for stabilizing $[TcCl_4(OH_2)_2]$, and indicate how *cis*-$[TcCl_4(OH_2)_2]\cdot 2C_4H_8O_2$ can adopt a chain-like structure in the solid state.

dioxane

[*Ans.* See E. Yegen *et al.* (2005) *Chem. Commun.*, p. 5575]

Technetium(III) and rhenium(III)

For the +3 oxidation state, metal–metal bonding becomes important. Rhenium(III) halides (X = Cl, Br, I) are trimeric, M_3X_9. No Tc(III) halide or ReF_3 is known. Rhenium(III) chloride is an important precursor in Re(III) chemistry and is made by heating $ReCl_5$. Its structure (Fig. 22.15a) consists of an Re_3 triangle (Re–Re = 248 pm), each edge being chlorido-bridged; the terminal Cl atoms lie above and below the metal framework. In the solid, two-thirds of the terminal Cl atoms are involved in weak bridging interactions to Re atoms of adjacent molecules. Rhenium(III) chloride is diamagnetic, and Re=Re double bonds are allocated to the metal framework, i.e. the (formally) $\{Re_3\}^{9+}$ core contains 12 valence electrons (Re, s^2d^5) which are used for metal–metal bonding. Lewis bases react with Re_3Cl_9 (or $Re_3Cl_9(OH_2)_3$) to give complexes of type $Re_3Cl_9L_3$ (Fig. 22.15b). $Re_3Cl_9(OH_2)_3$ can be isolated from aqueous solutions of the chloride at 273 K. Scheme 22.70 shows further examples of Lewis base additions.

$$Re_3Cl_9(py)_3 \xleftarrow{\text{py}} Re_3Cl_9 \xrightarrow{\text{PR}_3} Re_3Cl_9(PR_3)_3 \qquad (22.70)$$

The reaction of MCl with Re_3Cl_9 gives $M[Re_3Cl_{10}]$, $M_2[Re_3Cl_{11}]$ or $M_3[Re_3Cl_{12}]$ depending upon the conditions, for example reactions 22.71 and 22.72. Figure 22.15c shows the structure of the $[Re_3Cl_{12}]^{3-}$ ion.

$$Re_3Cl_9 \xrightarrow{\text{excess CsCl, conc HCl}} Cs_3[Re_3Cl_{12}] \qquad (22.71)$$

$$Re_3Cl_9 \xrightarrow{\text{[Ph}_4\text{As]Cl, dil HCl}} [Ph_4As]_2[Re_3Cl_{11}] \qquad (22.72)$$

The diamagnetic $[Re_2Cl_8]^{2-}$ was the first example of a species containing a metal–metal quadruple bond. It is made by reducing $[ReO_4]^-$ using H_2 or $[HPO_2]^{2-}$ and is isostructural with $[Mo_2Cl_8]^{4-}$ (Fig. 22.13c) with a Re−Re distance of 224 pm. Salts of $[Re_2Cl_8]^{2-}$ are blue ($\lambda_{max} = 700$ nm) arising from a $\sigma^2\pi^4\delta^1\delta^{*1} \longleftarrow \sigma^2\pi^4\delta^2$ transition (Fig. 21.16). Reactions of $[Re_2Cl_8]^{2-}$ include ligand displacements and redox processes. With Cl_2, $[Re_2Cl_9]^-$ is formed (i.e. oxidation and Cl^- addition). Reaction 22.73 shows the reaction of carboxylates with $[Re_2Cl_8]^{2-}$; the reaction can be reversed by treatment with HCl.

$$[Re_2Cl_8]^{2-} + 4[RCO_2]^- \longrightarrow [Re_2(\mu\text{-}O_2CR)_4Cl_2] + 6Cl^- \qquad (22.73)$$

$$\textbf{(22.42)}$$

= quadruple bond

(22.42)

When $[Re_2Cl_8]^{2-}$ reacts with bidentate phosphanes (e.g. reaction 22.74), the $\{Re_2\}^{6+}$ core with a $\sigma^2\pi^4\delta^2$ configuration (Re≣Re) is reduced to a $\{Re_2\}^{4+}$ unit ($\sigma^2\pi^4\delta^2\delta^{*2}$, Re≡Re). The change might be expected to lead to an increase in the Re−Re bond length, but in fact it stays the same (224 pm). The introduction of the bridging ligands counters the decrease in bond order by 'clamping' the Re atoms together.

$$[Re_2Cl_8]^{2-} + 2Ph_2PCH_2CH_2PPh_2$$

$$\xrightarrow[\text{reduction}]{-Cl^-} [Re_2Cl_4(\mu\text{-}Ph_2PCH_2CH_2PPh_2)_2] \qquad (22.74)$$

The $[Tc_2Cl_8]^{2-}$ ion is also known but is less stable than $[Re_2Cl_8]^{2-}$. The paramagnetic $[Tc_2Cl_8]^{3-}$ ($\sigma^2\pi^4\delta^2\delta^{*1}$, Tc−Tc = 211 pm, eclipsed ligands) is easier to isolate

than $[Tc_2Cl_8]^{2-}$ ($\sigma^2\pi^4\delta^2$, Tc−Tc = 215 pm, eclipsed ligands). The *increase* in Tc−Tc distance of 4 pm in going from $[Tc_2Cl_8]^{3-}$ to $[Tc_2Cl_8]^{2-}$ is not readily rationalized. Reduction of the $\{Tc_2\}^{6+}$ core occurs when $[Tc_2Cl_8]^{2-}$ undergoes reaction 22.75. The product (also made from $Tc^{II}_2Cl_4(PR_3)_4$ and $HBF_4\cdot OEt_2$) is expected to have a staggered arrangement of ligands consistent with the change from $\sigma^2\pi^4\delta^2$ to $\sigma^2\pi^4\delta^2\delta^{*2}$, and this has been confirmed for the related $[Tc_2(NCMe)_8(OSO_2CF_3)_2]^{2+}$.

$$[Tc_2Cl_8]^{2-} \xrightarrow{HBF_4\cdot OEt_2, \text{ in MeCN}} [Tc_2(NCMe)_{10}]^{4+} \qquad (22.75)$$

Mononuclear complexes of Re(III) and Tc(III) are quite well exemplified (often with π-acceptor ligands stabilizing the +3 oxidation state) and octahedral coordination is usual, e.g. $[Tc(acac)_2(NCMe)_2]^+$, $[Tc(acac)_3]$, $[Tc(NCS\text{-}N)_6]^{3-}$, *mer*-$[Tc(Ph_2PCH_2CH_2CO_2)_3]$, *mer,trans*-$[ReCl_3(NCMe)(PPh_3)_2]$. 7-Coordination has been observed in $[ReBr_3(CO)_2(bpy)]$ and $[ReBr_3(CO)_2(PMe_2Ph)_2]$. Simple aqua ions such as $[Tc(OH_2)_6]^{3+}$ are not known, although, stabilized by CO, it is possible to prepare the Tc(I) species $[Tc(OH_2)_3(CO)_3]^+$ (see Box 22.7).

Technetium(I) and rhenium(I)

The chemistry of Tc(I) and Re(I) complexes has increased in importance with their application as, or models for, diagnostic imaging agents (see Box 22.7). The M(I) centre is stabilized by using π-acceptor ligands, e.g. CO and RNC. Reduction of $[TcO_4]^-$ by $[S_2O_4]^{2-}$ in alkaline aqueous EtOH in the presence of an isocyanide, RNC, gives octahedral $[Tc(CNR)_6]^+$. With R = CH_2CMe_2OMe, this species is lipophilic and the ^{99m}Tc complex is marketed under the trade name of Cardiolite as a heart imaging agent. Since $[S_2O_4]^{2-}$ is not a strong enough reducing agent to convert $[ReO_4]^-$ to $[Re(CNR)_6]^+$ in the presence of RNC, these rhenium(I) complexes are made by treating $[ReOCl_3(PPh_3)_2]$ with an excess of RNC.

Another group of important Tc(I) and Re(I) complexes are those containing the *fac*-$M(CO)_3^+$ unit. A convenient synthesis of *fac*-$[M(CO)_3X_3]$ involves treatment of $[MO_4]^-$ or $[MOCl_4]^-$ (M = Tc or Re) with $BH_3\cdot THF/CO$ (see Box 22.7). The CO ligands exhibit a strong *trans*-effect (see Section 26.3) and labilize the Cl^- ligands in *fac*-$[M(CO)_3Cl_3]^{2-}$. Consequently, solvent molecules, including H_2O, readily exchange with Cl^- to give *fac*-$[M(CO)_3(solv)_3]^+$. While the *fac*-$M(CO)_3^+$ unit is inert with respect to ligand substitution, solvent molecules exchange with a variety of ligands. *fac*-$[Tc(CO)_3(OH_2)_3]^+$ is of increasing significance in radiopharmaceutical applications.

Self-study exercises

1. Show that fac-$[Tc(CO)_3(CN)_3]^{2-}$ belongs to the C_{3v} point group.

2. A sample of fac-$[Tc(CO)_3(CN)_3]^{2-}$ was made using $K^{13}CN$, and was labelled to an extent of $\approx 70\%$. Rationalize why the ^{99}Tc NMR spectrum of this complex shows a quartet (J 186 Hz), superimposed on a less intense triplet. Are any other signals expected in the spectrum?
   [*Ans.* See P. Kurz *et al.* (2004) *Inorg. Chem.*, vol. 43, p. 3789]

22.9 Group 8: ruthenium and osmium

The metals

Like all platinum-group metals, Ru and Os are relatively noble. Osmium powder reacts slowly with O_2 at 298 K to give the volatile OsO_4 (the bulk metal requires heating to 670 K). Ruthenium metal is passivated by a coating of non-volatile RuO_2 and reacts further with O_2 only at temperatures above 870 K. Both metals react with F_2 and Cl_2 when heated (see below), and are attacked by mixtures of HCl and oxidizing agents, and by molten alkalis.

Table 19.3 shows the range of oxidation states exhibited by the group 8 metals. In this section we consider oxidation states from +2 to +8. The lower states are stabilized by π-acceptor ligands and are covered in Chapter 24. Consistent with trends seen for earlier second and third row metals, Ru and Os form some compounds with metal–metal multiple bonds.

High oxidation states of ruthenium and osmium: M(VIII), M(VII) and M(VI)

Despite the fact that the maximum oxidation state of Ru and Os is +8 (e.g. in RuO_4 and OsO_4), the only binary halides formed for the high oxidation states are RuF_6 (eq. 22.76) and OsF_6 (eq. 22.77). The formation of OsF_7 has been claimed but not proven (Fig. 22.16).

$$2RuF_5 + F_2 \xrightarrow{\text{500 K, 50 bar}} 2RuF_6 \qquad (22.76)$$

$$Os + 3F_2 \xrightarrow{\text{500 K, 1 bar}} OsF_6 \qquad (22.77)$$

Ruthenium(VI) fluoride is an unstable brown solid. OsF_6 is a volatile yellow solid with a molecular (octahedral) structure. Neutron powder diffraction data for OsF_6 reveal that the four equatorial Os−F bonds are slightly shorter than the apical bonds, providing evidence for a small Jahn–Teller effect, consistent with the t_{2g}^{2} ground state electronic

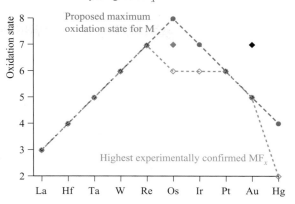

Fig. 22.16 Highest oxidation state $5d$ metal fluorides showing confirmed experimental data, erroneous or unconfirmed data, and the possible maximum oxidation state that could be attained by the metal. [Adapted from: S. Riedel and M. Kaupp (2006) *Inorg. Chem.*, vol. 45, p. 10497.]

configuration for Os(VI). Metal carbonyl cations (see Section 24.4) are rare but in superacid media, OsF_6 reacts with CO to give the osmium(II) complex $[Os(CO)_6]^{2+}$ (eq. 22.78).

$$OsF_6 + 4SbF_5 + 8CO \xrightarrow[\text{in HF/SbF}_5]{\text{300 K, 1.5 bar CO}}$$

$$[Os(CO)_6][Sb_2F_{11}]_2 + 2COF_2 \quad (22.78)$$

Several oxofluorides of Os(VIII), Os(VII) and Os(VI) are known, but $RuOF_4$ is the only example for Ru. All are very moisture-sensitive. Yellow-green $RuOF_4$ can be prepared from RuO_2 and argon-diluted F_2 at 720 K. This form of $RuOF_4$ is polymeric, with octahedral units **22.43** linked into helical chains (see Fig. 22.17a). If the synthesis is carried out with neat F_2 at 570 K, a yellow form of $RuOF_4$ is produced. Crystallographic data for this polymorph confirm the presence of loosely held dimers **22.44** ($Ru \cdots F = 234$ pm compared with $Ru–F = 184$ to 192 pm).

yellow-green form of $RuOF_4$

(22.43)

yellow form of $RuOF_4$

(22.44)

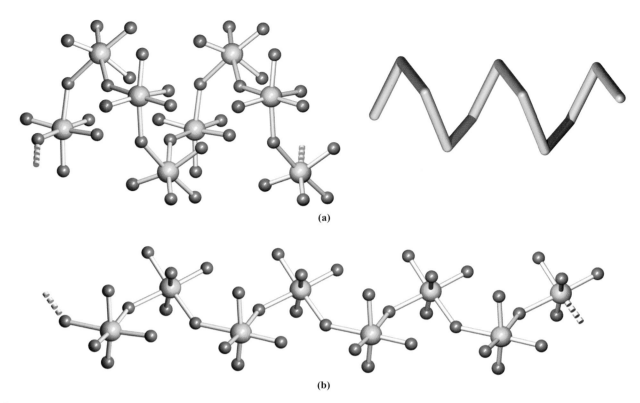

(a)

(b)

Fig. 22.17 Structures (X-ray diffraction) of the two polymorphs of OsOF$_4$. (a) Form I consists of helical polymer chains, and the structure of the polymeric form of RuOF$_4$ is analogous. (b) In form II of OsOF$_4$, each chain has a flattened, zigzag backbone. Colour code: Os, yellow; O, red; F, green. [Data: structural data were provided by Konrad Seppelt; from H. Shorafa *et al.* (2007) *Z. Anorg. Allg. Chem.*, vol. 633, p. 543.]

Deep blue OsOF$_4$ is best prepared[†] by lengthy heating of OsO$_4$ with OsF$_6$. Like RuOF$_4$, OsOF$_4$ crystallizes in two modifications: a helical polymer (Fig. 22.17a) and a closely related polymeric structure in which the chain is flattened (Fig. 22.17b). OsOCl$_4$ can be prepared from OsO$_4$ and BCl$_3$ but is very unstable. Crystalline OsOCl$_4$ consists of loosely bound dimers like those of yellow RuOF$_4$ (**22.44**).

The only oxofluoride that contains Os(VII) is green OsOF$_5$. This may be made by heating OsF$_6$ and OsO$_4$ at ≈ 500 K, and crystalline OsOF$_5$ consists of monomeric molecules. The compound 'OsO$_2$F$_3$' does not contain Os(VII), but is the mixed oxidation state species OsO$_3$F$_2$·OsOF$_4$.[‡]

Red *cis*-OsO$_2$F$_4$ forms when OsO$_4$ reacts with HF and KrF$_2$ at 77 K. Yellow OsO$_3$F$_2$ (made from F$_2$ and OsO$_4$) is also molecular in the gas phase (**22.45**) but is polymeric in the solid with Os$-$F$-$Os bridges connecting *fac*-octahedral units. Scheme 22.79 illustrates the ability of OsO$_3$F$_2$ to act as a fluoride acceptor.

$$OsO_3F_2 \xrightarrow[\text{warm to 213 K}]{\text{NOF, 195 K}} [NO]^+[fac\text{-}OsO_3F_3]^-$$

$$\downarrow \begin{array}{l} \text{[Me}_4\text{N]F in} \\ \text{anhydrous HF} \end{array}$$

$$[Me_4N]^+[fac\text{-}OsO_3F_3]^-$$

(22.79)

(22.45)

(22.46)

M = Ru, Os; *d* = 171 pm (gas phase)

Both Ru and Os form toxic, volatile, yellow oxides MO$_4$ (RuO$_4$ mp 298 K, bp 403 K; OsO$_4$ mp 313 K, bp 403 K)[†] but RuO$_4$ is more readily reduced than OsO$_4$. Osmium(VIII) oxide ('osmic acid') is made from Os and O$_2$ (see above), but the formation of RuO$_4$ requires acidified [IO$_4$]$^-$ or [MnO$_4$]$^-$ oxidation of RuO$_2$ or RuCl$_3$. Both tetraoxides have penetrating ozone-like odours; they are sparingly

---

[†] For discussion of problems associated with literature reports of OsOF$_4$, see: H. Shorafa and K. Seppelt (2007) *Z. Anorg. Allg. Chem.*, vol. 633, p. 543.
[‡] For detailed discussion, see: H. Shorafa and K. Seppelt (2006) *Inorg. Chem.*, vol. 45, p. 7929.

[†] The literature contains differing values for OsO$_4$: see Y. Koda (1986) *J. Chem. Soc., Chem. Commun.*, p. 1347.

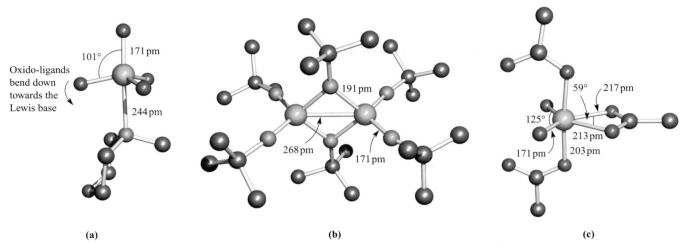

Fig. 22.18 The structures (X-ray diffraction) of (a) the adduct formed between *N*-methylmorpholine and OsO$_4$ [A.J. Bailey *et al.* (1997) *J. Chem. Soc., Dalton Trans.*, p. 3245], (b) [Os$_2$(N^{t}Bu)$_4$(μ-N^{t}Bu)$_2$]$^{2+}$ in the [BF$_4$]$^-$ salt [A.A. Danopoulos *et al.* (1991) *J. Chem. Soc., Dalton Trans.*, p. 269] and (c) [OsO$_2$(O$_2$CMe)$_3$]$^-$ in the solvated K$^+$ salt [T. Behling *et al.* (1982) *Polyhedron*, vol. 1, p. 840]. Colour code: Os, yellow; O, red; C, grey; N, blue.

soluble in water but soluble in CCl$_4$. The oxides are isostructural (molecular structure **22.46**). Ruthenium(VIII) oxide is thermodynamically unstable with respect to RuO$_2$ and O$_2$ (eq. 22.80) and is liable to explode. It is a very powerful oxidant, reacting violently with organic compounds. Osmium(VIII) oxide is used as an oxidizing agent in organic synthesis (e.g. converting alkenes to 1,2-diols) and as a biological stain, but its ease of reduction and its volatility make it dangerous to the eyes. Reaction 22.80 occurs on heating for M = Os.

$$MO_4 \longrightarrow MO_2 + O_2 \qquad (M = Ru, Os) \qquad (22.80)$$

Osmium(VIII) oxide forms adducts with Lewis bases such as Cl$^-$, 4-phenylpyridine and *N*-morpholine. The adducts are distorted trigonal bipyramidal with the oxido ligands in the equatorial and one axial site (Fig. 22.18a). OsO$_4$ acts as a fluoride acceptor, reacting with [Me$_4$N]F at 298 K to give [Me$_4$N][OsO$_4$F], and with two equivalents of [Me$_4$N]F at 253 K to yield [Me$_4$N]$_2$[*cis*-OsO$_4$F$_2$].

When RuO$_4$ dissolves in aqueous alkali, O$_2$ is evolved and [RuO$_4$]$^-$ forms; in concentrated alkali, reduction proceeds to [RuO$_4$]$^{2-}$ (eq. 22.81). K$_2$RuO$_4$ can also be made by fusing Ru with KNO$_3$ and KOH.

$$4[RuO_4]^- + 4[OH]^- \longrightarrow 4[RuO_4]^{2-} + 2H_2O + O_2 \qquad (22.81)$$

$$\left[\begin{array}{c} OH \\ | \\ O = Ru \underset{\displaystyle O}{\overset{\displaystyle O}{\diagup}} \\ | \\ OH \end{array} \right]^{2-}$$

(22.47)

$$\left[\begin{array}{c} O \\ \| \\ O \diagdown \hspace{-0.3em} \underset{\displaystyle O}{Os} \diagup \hspace{-0.3em} OH \\ OH \end{array} \right]^{2-}$$

(22.48)

Both [RuO$_4$]$^-$ and [RuO$_4$]$^{2-}$ are powerful oxidants but can be stabilized in solution by pH control under non-reducing conditions. In solid state salts, [RuO$_4$]$^-$ (d^1) has a flattened tetrahedral structure (Ru–O = 179 pm), but crystals of 'K$_2$[RuO$_4$]·H$_2$O' are actually K$_2$[RuO$_3$(OH)$_2$] containing anion **22.47**. In contrast to its action on RuO$_4$, alkali reacts with OsO$_4$ to give *cis*-[OsO$_4$(OH)$_2$]$^{2-}$ (**22.48**) which is reduced to *trans*-[OsO$_2$(OH)$_4$]$^{2-}$ by EtOH. Anion **22.48** has been isolated in crystalline Na$_2$[OsO$_4$(OH)$_2$]·2H$_2$O. Reaction 22.82 gives K[Os(N)O$_3$] which contains tetrahedral [Os(N)O$_3$]$^-$ (**22.49**), isoelectronic and isostructural with OsO$_4$. The IR spectrum of [Os(N)O$_3$]$^-$ (C_{3v}) contains bands at 871 and 897 cm^{-1} (asymmetric and symmetric $\nu_{Os=O}$, respectively) and 1021 cm^{-1} ($\nu_{Os\equiv N}$); this compares with absorptions at 954 and 965 cm^{-1} for OsO$_4$ (T_d, see Fig. 3.16).

$$OsO_4 + NH_3 + KOH \longrightarrow K[Os(N)O_3] + 2H_2O \qquad (22.82)$$

$$OsO_4 \xrightarrow{(Me_3Si)NH^tBu} Os(N^tBu)_4 \qquad (22.83)$$

Reaction 22.83 gives an imido analogue of OsO$_4$. The tetrahedral shape is retained and the Os–N bond lengths of 175 pm are consistent with double bonds. Sodium amalgam reduces Os(N^{t}Bu)$_4$ to the Os(VI) dimer Os$_2$(N^{t}Bu)$_4$(μ-N^{t}Bu)$_2$ (Os–Os = 310 pm) and subsequent oxidation gives [Os$_2$(N^{t}Bu)$_4$(μ-N^{t}Bu)$_2$]$^{2+}$ (Fig. 22.18b), a rare example of an Os(VII) complex. Trigonal planar Os(NAr)$_3$ is stabilized against dimerization if the aryl group, Ar, is very bulky, e.g. 2,6-^{i}Pr$_2$C$_6$H$_3$.

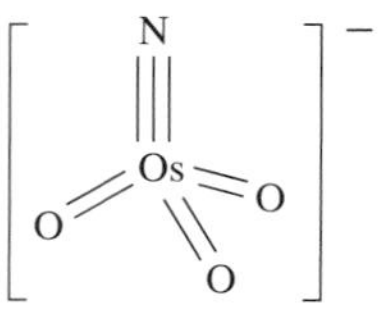

(22.49) 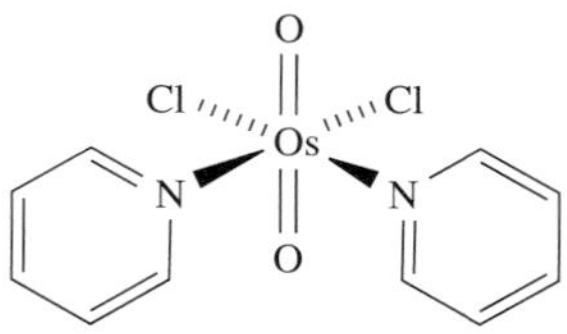   **(22.50)**

Complexes of M(VIII) and M(VII) are few, e.g. $[Os(N)O_3]^-$ (see above), but are well exemplified for M(VI), particularly for M = Os, with oxido, nitrido or imido ligands commonly present, e.g.

- tetrahedral: $[OsO_2(S_2O_3\text{-}S)_2]^{2-}$;
- square-based pyramidal: $[RuNBr_4]^-$, $[OsNBr_4]^-$;
- octahedral: $[OsO_2(O_2CMe)_3]^-$ (distorted, Fig. 22.18c), *trans*-$[OsO_2Cl_4]^{2-}$, *trans*-$[RuO_2Cl_4]^{2-}$, $[OsO_2Cl_2(py)_2]$ (**22.50**), *trans*-$[OsO_2(en)_2]^{2+}$.

Worked example 22.2 Osmium(VI) compounds

Rationalize why salts of *trans*-$[OsO_2(OH)_4]^{2-}$ are diamagnetic.

$[OsO_2(OH)_4]^{2-}$ contains Os(VI) and therefore has a d^2 configuration.

The structure of *trans*-$[OsO_2(OH)_4]^{2-}$ is:

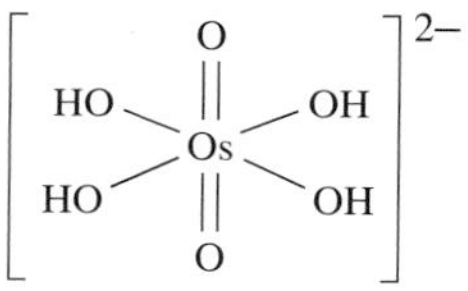

An octahedral (O_h) d^2 complex would be paramagnetic, but in $[OsO_2(OH)_4]^{2-}$, the axial Os−O bonds are shorter than the equatorial Os−O bonds. The complex therefore suffers from a tetragonal distortion and, consequently, the d orbitals split as follows, assuming that the z axis is defined to lie along the O=Os=O axis:

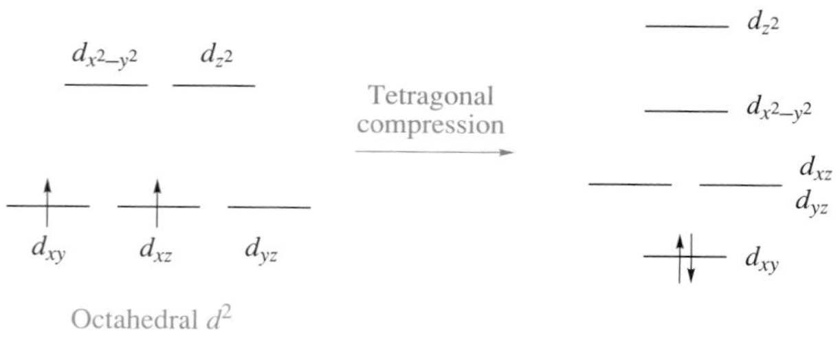

The complex is therefore diamagnetic.

Self-study exercises

1. Rationalize why OsF_6 suffers only a *small* Jahn–Teller effect.

   [*Ans.* See 'Jahn–Teller distortions' in Section 20.3]

2. Suggest why the high oxidation state compounds of Os are dominated by those containing oxido, nitrido and fluorido ligands.

   [*Ans.* All π-donor ligands; see Section 20.4]

3. Comment on the fact that, at 300 K, μ_{eff} for OsF_6 is 1.49 μ_B.

   [*Ans.* See discussion of Kotani plots in Section 20.10]

Ruthenium(V), (IV) and osmium(V), (IV)

Green RuF_5 and OsF_5 (readily hydrolysed solids) are made by reactions 22.84 and 22.85 and are tetrameric like NbF_5 (**22.5**) but with non-linear bridges. Black $OsCl_5$ is the only other halide of the M(V) state and is made by reducing and chlorinating OsF_6 with BCl_3. It is dimeric, analogous to $NbCl_5$ (**22.6**).

$$2Ru + 5F_2 \xrightarrow{570\,K} 2RuF_5 \qquad (22.84)$$

$$OsF_6 \xrightarrow{I_2,\,IF_5,\,328\,K} OsF_5 \qquad (22.85)$$

For the M(IV) state, RuF_4, OsF_4, $OsCl_4$ (two polymorphs) and $OsBr_4$ are known and are polymeric. The fluorides are made by reducing higher fluorides, and $OsCl_4$ and $OsBr_4$ by combining the elements at high temperature and, for $OsBr_4$, high pressure.

In contrast to iron, the lowest oxides formed by the heavier group 8 metals are for the M(IV) state. Both RuO_2 and OsO_2 adopt a rutile structure (Fig. 6.22); these oxides are far less important than RuO_4 and OsO_4.

The electrochemical oxidation of $[Ru(OH_2)_6]^{2+}$ in aqueous solution produces a Ru(IV) species. Its formulation as $[Ru_4O_6(OH_2)_{12}]^{4+}$ (or a protonated form depending on pH) is consistent with ^{17}O NMR spectroscopic data and, of the two proposed structures **22.51** and **22.52**, the latter is supported by EXAFS studies (see Box 25.2).

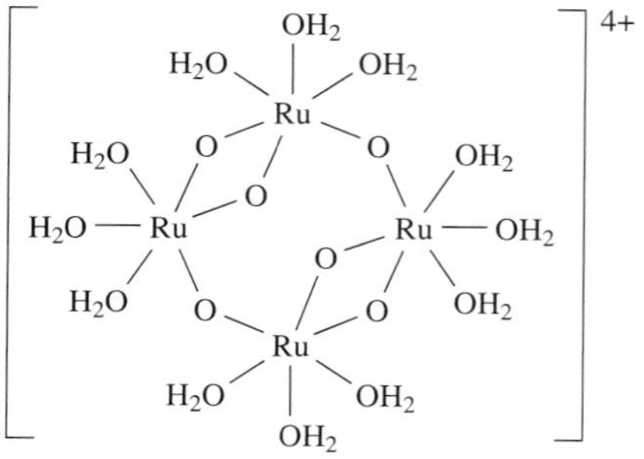

(22.51)

(22.52)

Octahedral halido complexes of Ru(V) and Os(V) are represented by $[MF_6]^-$ (M = Ru, Os) and $[OsCl_6]^-$. $K[OsF_6]$, for example, can be made by reduction of OsF_6 with KBr in anhydrous HF. The Os(V) anions [fac-$OsCl_3F_3]^-$, cis-$[OsCl_4F_2]^-$ and $trans$-$[OsCl_4F_2]^-$ are made by oxidation of the analogous Os(IV) dianions using $KBrF_4$ or BrF_3. For the +4 oxidation state, all the $[MX_6]^{2-}$ ions are known except $[RuI_6]^{2-}$. Various synthetic routes are used, e.g. $[RuCl_6]^{2-}$ can be made by heating Ru, Cl_2 and an alkali metal chloride, or by oxidizing $[RuCl_6]^{3-}$ with Cl_2. The salt $K_2[RuCl_6]$ has a magnetic moment (298 K) of 2.8 μ_B, close to μ(spin-only) for a low-spin d^4 ion but the value is temperature dependent. For $K_2[OsCl_6]$, the value of 1.5 μ_B arises from the greater spin–orbit coupling constant for the $5d$ metal ion (see Fig. 20.29 and discussion). Mixed M(IV) halido complexes are produced by halogen exchange. In reaction 22.86, the products are

formed by stepwise substitution, the position of F^- entry being determined by the stronger $trans$-effect (see Section 26.3) of the chlorido ligand.

$$[OsCl_6]^{2-} \xrightarrow{BrF_3} [OsCl_5F]^{2-} + cis\text{-}[OsCl_4F_2]^{2-}$$
$$+ fac\text{-}[OsCl_3F_3]^{2-} + cis\text{-}[OsCl_2F_4]^{2-}$$
$$+ [OsClF_5]^{2-} + [OsF_6]^{2-} \qquad (22.86)$$

The reduction of OsO_4 by Na_2SO_3 in aqueous H_2SO_4 containing Cl^- produces $[OsCl_5(OH_2)]^-$ in addition to $[OsCl_6]^{2-}$ and $[\{Cl_3(HO)(H_2O)Os\}_2(\mu\text{-}OH)]^-$.

Reaction of RuO_4 in aqueous HCl in the presence of KCl gives K^+ salts of $[Ru^{IV}_2OCl_{10}]^{4-}$, $[Ru^{III}Cl_5(OH_2)]^{2-}$ and $[Ru^{III}Cl_6]^{3-}$. Each Ru(IV) centre in $[Ru_2OCl_{10}]^{4-}$ is octahedrally sited and the Ru−O−Ru bridge is linear (Fig. 22.19a). Salts of $[Ru_2OCl_{10}]^{4-}$ are diamagnetic. This is rationalized by considering the formation of two 3-centre π-interactions (Fig. 22.19b) involving the d_{xz} and d_{yz} atomic orbitals of the two low-spin Ru(IV) centres (each of configuration $d_{xy}{}^2 d_{xz}{}^1 d_{yz}{}^1$) and the filled p_x and p_y atomic orbitals of the O atom. In addition to the π and π^* MOs, four non-bonding MOs result from combinations of the d_{xy}, d_{xz} and d_{yz} orbitals (Fig. 22.19b). These are fully occupied in $[Ru_2OCl_{10}]^{4-}$. The same MO diagram can be used to describe the bonding in the related anions $[Os_2OCl_{10}]^{4-}$ (two d^4 metal centres), $[W_2OX_{10}]^{4-}$ (X = Cl, Br; d^2), $[Re_2OCl_{10}]^{4-}$ (d^3) and $[Ta_2OX_{10}]^{2-}$ (X = F, Cl; d^0). Changes in d^n configuration only affect

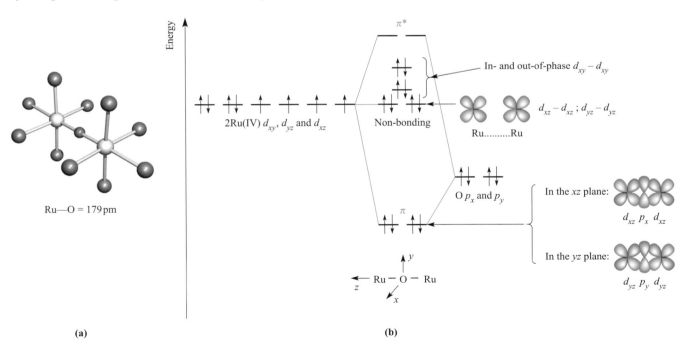

Fig. 22.19 (a) The structure of $[Ru_2(\mu\text{-}O)Cl_{10}]^{4-}$ determined by X-ray diffraction for the histaminium salt [I.A. Efimenko *et al.* (1994) *Koord. Khim.*, vol. 20, p. 294]; colour code: Ru, pale grey; Cl, green; O, red. (b) A partial MO diagram for the interaction between the d_{xy}, d_{xz} and d_{yz} atomic orbitals of the Ru(IV) centres and the p_x and p_y atomic orbitals of the O atom to give two bonding, two antibonding and four non-bonding MOs; the non-bonding MOs are derived from combinations of d orbitals with no oxygen contribution. Relative orbital energies are approximate, and the non-bonding MOs lie close together.

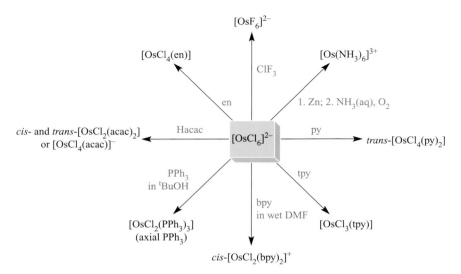

Fig. 22.20 Representative complex-forming reactions starting from $[OsCl_6]^{2-}$. Note that reduction to Os(III) occurs in three reactions, and to Os(II) in one. See Table 7.7 for ligand abbreviations; tpy = 2,2':6',2''-terpyridine.

the occupancy of the non-bonding MOs, leaving the metal–oxygen π-bonding MOs occupied. The diamagnetic $[Ru_2(\mu\text{-}N)Cl_8(OH_2)_2]^{3-}$ (**22.53**) is a nitrido-bridged analogue of $[Ru_2OCl_{10}]^{4-}$, and Ru–N distances of 172 pm indicate strong π-bonding; it is made by reducing $[Ru(NO)Cl_5]^{2-}$ with $SnCl_2$ in HCl.

(22.53)

While a number of dinuclear Ru(IV) complexes containing bridging nitrido ligands are known, mononuclear complexes are rare. The conversion of azide to nitride (eq. 22.87) has a high activation barrier and usually involves thermolysis or photolysis. However, $[N_3]^-$ is a useful precursor to an $[N]^{3-}$ ligand when coupled with the reducing power of Ru(II) in the absence of π-acceptor ligands and in the presence of π-donors. This is exemplified by reaction 22.88, in which the amido group acts as a π-donor.

$$[N_3]^- + 2e^- \longrightarrow N^{3-} + N_2 \tag{22.87}$$

$$\text{(22.88)}$$

Although the coordination chemistry of Ru(IV) and Os(IV) is varied, halido complexes are dominant. Complexes of Os(IV) outnumber those of Ru(IV). Hexahalido complexes are common precursors (Fig. 22.20). Apart from those already described, examples with mixed ligands include octahedral $trans$-$[OsBr_4(AsPh_3)_2]$, $[OsX_4(acac)]^-$ (X = Cl, Br, I), $[OsX_4(ox)]^{2-}$ (X = Cl, Br, I), cis-$[OsCl_4(NCS\text{-}N)_2]^{2-}$ and $trans$-$[OsCl_4(NCS\text{-}S)_2]^{2-}$. Two complexes of special note for their stereochemistries are the pentagonal bipyramidal $[RuO(S_2CNEt_2)_3]^-$ (**22.54**) and the square planar $trans$-$[Ru(PMe_3)_2(NR)_2]$ in which R is the bulky 2,6-^{i}Pr$_2$C$_6$H$_3$.

(22.54)

Osmium(IV) complexes containing terminal Os=O units are rather rare, and it is more usual for a bridging mode to be adopted as in $[Os_2OCl_{10}]^{4-}$. One example of a mononuclear species is shown in scheme 22.89, in which the Os(IV) dioxido derivative is obtained by activation of molecular O_2 via an Os(VI) intermediate. The square planar geometry of the Os(IV) product is unusual for a d^4 configuration. The precursor in scheme 22.89 is prepared by treating $OsCl_3 \cdot xH_2O$ with P^iPr_3, and has a distorted 6-coordinate coordination sphere.

$$(22.89)$$

Ruthenium(III) and osmium(III)

All the binary halides RuX_3 are known, but for Os, only $OsCl_3$ and OsI_3 have been established. OsF_4 is the lowest fluoride of Os. Reduction of RuF_5 with I_2 gives RuF_3, a brown solid isostructural with FeF_3. Reactions 22.90–22.92 show preparations of $RuCl_3$, $RuBr_3$ and RuI_3. The chloride is commercially available as a hydrate of variable composition '$RuCl_3 \cdot xH_2O$' ($x \approx 3$) and is an important starting material in Ru(III) and Ru(II) chemistry.

$$Ru_3(CO)_{12} \xrightarrow{Cl_2,\ 630\,K\ under\ N_2} \beta\text{-}RuCl_3 \xrightarrow{Cl_2,\ >720\,K} \alpha\text{-}RuCl_3$$
$$(22.90)$$

$$Ru \xrightarrow{Br_2,\ 720\,K,\ 20\,bar} RuBr_3 \qquad (22.91)$$

$$RuO_4 \xrightarrow{aq\ HI} RuI_3 \qquad (22.92)$$

The α-forms of $RuCl_3$ and $OsCl_3$ are isostructural with α-$TiCl_3$ (see Section 21.5), while β-$RuCl_3$ has the same structure as $CrCl_3$ (see Section 21.7). Extended structures with octahedral Ru(III) are adopted by $RuBr_3$ and RuI_3.

There are no binary oxides or oxoanions for Ru(III), Os(III) or lower oxidation states. No simple aqua ion of Os(III) has been established, but octahedral $[Ru(OH_2)_6]^{3+}$ can be obtained by aerial oxidation of $[Ru(OH_2)_6]^{2+}$ and has been isolated in the alum (see Section 13.9) $CsRu(SO_4)_2 \cdot 12H_2O$ and the salt $[Ru(OH_2)_6][4\text{-}MeC_6H_4SO_3]_3 \cdot 3H_2O$. The Ru–O bond length of 203 pm is shorter than in $[Ru(OH_2)_6]^{2+}$ (212 pm). In aqueous solution, $[Ru(OH_2)_6]^{3+}$ is acidic (compare eq. 22.93 with eq. 7.36 for Fe^{3+}) and it is less readily reduced than $[Fe(OH_2)_6]^{3+}$ (eq. 22.94).

$$[Ru(OH_2)_6]^{3+} + H_2O \rightleftharpoons [Ru(OH_2)_5(OH)]^{2+} + [H_3O]^+$$
$$pK_a \approx 2.4 \qquad (22.93)$$

$$[M(OH_2)_6]^{3+} + e^- \rightleftharpoons [M(OH_2)_6]^{2+}$$
$$\begin{cases} M = Ru, & E^\circ = +0.25\,V \\ M = Fe, & E^\circ = +0.77\,V \end{cases} \qquad (22.94)$$

Substitution in Ru(III) complexes (low-spin d^5) is slow (see Chapter 26) and all members of the series $[RuCl_n(OH_2)_{6-n}]^{(n-3)-}$, including isomers, have been characterized. Aerial oxidation of $[Ru(NH_3)_6]^{2+}$ (see below) gives $[Ru(NH_3)_6]^{3+}$ (eq. 22.95).

$$[Ru(NH_3)_6]^{3+} + e^- \rightleftharpoons [Ru(NH_3)_6]^{2+} \qquad E^\circ = +0.10\,V$$
$$(22.95)$$

Halido complexes $[MX_6]^{3-}$ are known for M = Ru, X = F, Cl, Br, I and M = Os, X = Cl, Br, I. The anions $[RuCl_5(OH_2)]^{2-}$ and $[RuCl_6]^{3-}$ are made in the same reaction as $[Ru_2OCl_{10}]^{4-}$ (see above). In aqueous solution, $[RuCl_6]^{3-}$ is rapidly aquated to $[RuCl_5(OH_2)]^{2-}$. The anion $[Ru_2Br_9]^{3-}$ can be made by treating $[RuCl_6]^{3-}$ with HBr. The ions $[Ru_2Br_9]^{3-}$, $[Ru_2Cl_9]^{3-}$ (reaction 22.96) and $[Os_2Br_9]^{3-}$ adopt structure **22.55**. The Ru–Ru distances of 273 (Cl) and 287 pm (Br) along with magnetic moments of 0.86 (Cl) and 1.18 μ_B (Br) suggest a degree of Ru–Ru bonding, a conclusion supported by theoretical studies.

M = Ru, X = Cl, Br
M = Os, X = Br

(22.55)

$$K_2[RuCl_5(OH_2)] \xrightarrow{520\,K} K_3[Ru_2Cl_9] \qquad (22.96)$$

$$[OsCl_6]^{2-} \xrightarrow{MeCO_2H,\ (MeCO)_2O} [Os_2(\mu\text{-}O_2CMe)_4Cl_2]$$
$$\xrightarrow[X = Cl,\ Br,\ I]{HX(g),\ EtOH} [Os_2X_8]^{2-} \qquad (22.97)$$

The anions $[Os_2X_8]^{2-}$ (X = Cl, Br, I) are made in reaction sequence 22.97. In diamagnetic $[Os_2X_8]^{2-}$ and $[Os_2(\mu\text{-}O_2CMe)_4Cl_2]$, the electronic configuration of the Os_2-unit is (from Fig. 21.16) $\sigma^2 \pi^4 \delta^2 \delta^{*2}$ corresponding to an Os≡Os triple bond. Since the δ^* MO is occupied, the influence of the δ-bond is lost and so no electronic factor restricts the orientation of the ligands (compare the eclipsed orientation of the ligands in $[Re_2Cl_8]^{2-}$ and $[Mo_2Cl_8]^{4-}$ which contain an M≡M). Crystal structures for several salts of $[Os_2Cl_8]^{2-}$ show different ligand arrangements (Fig. 22.21) and this is true also for $[Os_2Br_8]^{2-}$. For $[Os_2I_8]^{2-}$, steric factors appear to favour a staggered arrangement. Ruthenium(III) forms a number of acetate complexes. The reaction of $RuCl_3 \cdot xH_2O$ with $MeCO_2H$ and $MeCO_2Na$ yields paramagnetic $[Ru_3(OH_2)_3(\mu\text{-}O_2CMe)_6(\mu_3\text{-}O)]^+$ (structurally analogous to the Cr(III) species in Fig. 21.14) which is reduced

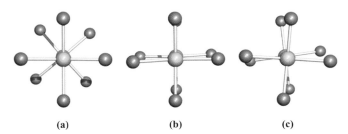

Fig. 22.21 Differences in energy between the arrangement of the chlorido ligands (staggered, eclipsed or somewhere in between) in $[Os_2Cl_8]^{2-}$ are small. The solid state structure of $[Os_2Cl_8]^{2-}$ (viewed along the Os−Os bond) in (a) the $[^nBu_4N]^+$ salt [P.A. Agaskar *et al.* (1986) *J. Am. Chem. Soc.*, vol. 108, p. 4850] ($[Os_2Cl_8]^{2-}$ is also staggered in the $[Ph_3PCH_2CH_2PPh_3]^{2+}$ salt), (b) the $[MePh_3P]^+$ salt [F.A. Cotton *et al.* (1990) *Inorg. Chem.*, vol. 29, p. 3197] and (c) the $[(Ph_3P)_2N]^+$ salt (structure determined at 83 K) [P.E. Fanwick *et al.* (1986) *Inorg. Chem.*, vol. 25, p. 4546]. In the $[(Ph_3P)_2N]^+$ salt, an eclipsed conformer is also present.

by PPh$_3$ to give the mixed-valence complex $[Ru_3(PPh_3)_3(\mu-O_2CMe)_6(\mu_3-O)]$.

Both Ru(III) and Os(III) form a range of octahedral complexes with ligands other than those already mentioned. Ru(III) complexes outnumber those of Os(III), the reverse of the situation for the M(IV) state, reflecting the relative stabilities Os(IV) > Ru(IV) but Ru(III) > Os(III). $[Os(CN)_6]^{3-}$ may be prepared by electrochemical oxidation of $[Os(CN)_6]^{4-}$, but reduction back to the Os(II) complex readily occurs. Oxidation of $[Ru(CN)_6]^{4-}$ using Ce(IV) gives $[Ru(CN)_6]^{3-}$, but isolation of its salts from aqueous solutions requires rapid precipitation. This is best achieved using $[Ph_4As]^+$ as the counter-ion. Examples of other mononuclear complexes include $[Ru(acac)_3]$, $[Ru(ox)_3]^{3-}$, $[Ru(en)_3]^{3+}$, *cis*-$[RuCl(OH_2)(en)_2]^{2+}$, *cis*-$[RuCl_2(bpy)_2]^+$, $[RuCl_4(bpy)]^-$, *trans*-$[RuCl(OH)(py)_4]^+$, *mer*-$[RuCl_3(DMSO-S)_2(DMSO-O)]$, $[Ru(NH_3)_5(py)]^{3+}$, *mer*-$[OsCl_3(py)_3]$, $[Os(acac)_3]$, $[Os(en)_3]^{3+}$, *trans*-$[OsCl_2(PMe_3)_4]^+$ and *trans*-$[OsCl_4(PEt_3)_2]^-$.

The Ru(III) compounds [HIm][*trans*-RuCl$_4$(Im)(DMSO-*S*)] and [HInd][*trans*-RuCl$_4$(Ind)$_2$] (Im = imidazole, Ind = indazole, Fig. 22.22) have completed phase I clinical trials as anti-cancer drugs. The former complex selectively targets metastases of solid tumours. The range of ruthenium-based compounds that exhibits anti-cancer activity includes organometallic ruthenium(II) complexes.[†]

Ruthenium(II) and osmium(II)

Binary halides of Ru(II) and Os(II) are not well characterized and there are no oxides. Heating the metal with S gives MS$_2$ (M = Ru, Os) which contain $[S_2]^{2-}$ and adopt

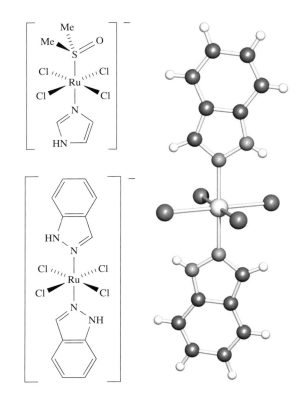

Fig. 22.22 The structures of the anions in the ruthenium(III) anti-cancer agents [HIm][*trans*-RuCl$_4$(Im)(DMSO-*S*)] and [HInd][*trans*-RuCl$_4$(Ind)$_2$] (Im = imidazole, Ind = indazole). The right-hand structure shows the [*trans*-RuCl$_4$(Ind)$_2$]$^-$ anion in the salt [H(Ind)$_2$][*trans*-RuCl$_4$(Ind)$_2$]. [X-ray diffraction data: E. Reisner *et al.* (2004) *Inorg. Chem.*, vol. 43, p. 7083.] Colour code: Ru, pale grey; Cl, green; N, blue; C, dark grey; H, white.

a pyrite structure (see Section 21.9). Most of the chemistry of Ru(II) and Os(II) concerns complexes, all of which are diamagnetic, low-spin d^6 and, with a few exceptions, are octahedral. We saw in Section 20.3 that values of Δ_{oct} (for a set of related complexes) are greater for second and third row metals than for the first member of the triad, and low-spin complexes are favoured. A vast number of Ru(II) complexes are known and we can give only a brief introduction.

The hydrido anions $[RuH_6]^{4-}$ and $[OsH_6]^{4-}$ (analogous to $[FeH_6]^{4-}$, Fig. 10.14b) are formed by heating the metal with MgH$_2$ or BaH$_2$ under a pressure of H$_2$. There are no simple halido complexes. H$_2$ or electrochemical reduction of RuCl$_3$·xH$_2$O in MeOH produces blue solutions (*ruthenium blues*) which, despite their synthetic utility for preparing Ru(II) complexes, have not been fully characterized. The blue species present have been variously formulated, but cluster anions seem likely.

Substitution reactions involving Ru(II) or Os(II) are affected by the kinetic inertness of the low-spin d^6 ion (see Section 26.2), and methods of preparation of M(II) complexes often start from higher oxidation states, e.g. RuCl$_3$·xH$_2$O or $[OsCl_6]^{2-}$. Reducing aqueous solutions of

---

[†]See: W.H. Ang and P.J. Dyson (2006) *Eur. J. Inorg. Chem.*, p. 4003; I. Bratsos, S. Jedner, T. Gianferrara and E. Alessio (2007) *Chimia*, vol. 61, p. 692 and other articles in this issue of *Chimia*; A. Levina, A. Mitra and P. A. Lay (2009) *Metallomics*, vol. 1, p. 458; G. Süss-Fink (2010) *Dalton Trans.*, vol. 39, p. 1673.

$RuCl_3 \cdot xH_2O$ from which Cl^- has been precipitated by Ag^+ produces $[Ru(OH_2)_6]^{2+}$; there is no Os(II) analogue. In air, $[Ru(OH_2)_6]^{2+}$ readily oxidizes (eq. 22.94) but is present in Tutton salts (see Section 21.6) $M_2Ru(SO_4)_2 \cdot 6H_2O$ (M = Rb, NH_4). Its structure has been determined in the salt $[Ru(OH_2)_6][4\text{-}MeC_6H_4SO_3]_2$ (see discussion of $[Ru(OH_2)_6]^{3+}$). Under 200 bar pressure of N_2, $[Ru(OH_2)_6]^{2+}$ reacts to give $[Ru(OH_2)_5(N_2)]^{2+}$. The related $[Ru(NH_3)_5(N_2)]^{2+}$ (which can be isolated as the chloride salt and is structurally similar to **22.57**) is formed either by reaction scheme 22.98 or by N_2H_4 reduction of aqueous solutions of $RuCl_3 \cdot xH_2O$.[†]

$$[Ru(NH_3)_5(OH_2)]^{3+} \xrightarrow{\text{Zn/Hg}} [Ru(NH_3)_5(OH_2)]^{2+}$$

$$\xrightarrow[-H_2O]{N_2,\ 100\ \text{bar}} [Ru(NH_3)_5(N_2)]^{2+} \qquad (22.98)$$

The cation $[(H_3N)_5Ru(\mu\text{-}N_2)Ru(NH_3)_5]^{4+}$ (**22.56**) forms when $[Ru(NH_3)_5(OH_2)]^{2+}$ reacts with $[Ru(NH_3)_5(N_2)]^{2+}$, or when aqueous $[Ru(NH_3)_5Cl]^{2+}$ is reduced by Zn amalgam under N_2. Reduction of $[OsCl_6]^{2-}$ with N_2H_4 gives $[Os(NH_3)_5(N_2)]^{2+}$ (**22.57**) which can be oxidized or converted to the bis(N_2) complex (eq. 22.99); note the presence of the π-acceptor ligand to stabilize Os(II).

$$[Os(NH_3)_5(N_2)]^{2+} \begin{cases} \xrightarrow[-2H_2O]{HNO_2} cis\text{-}[Os(NH_3)_4(N_2)_2]^{2+} \\ \\ \xrightarrow{Ce^{4+}} [Os(NH_3)_5(N_2)]^{3+} \end{cases} \qquad (22.99)$$

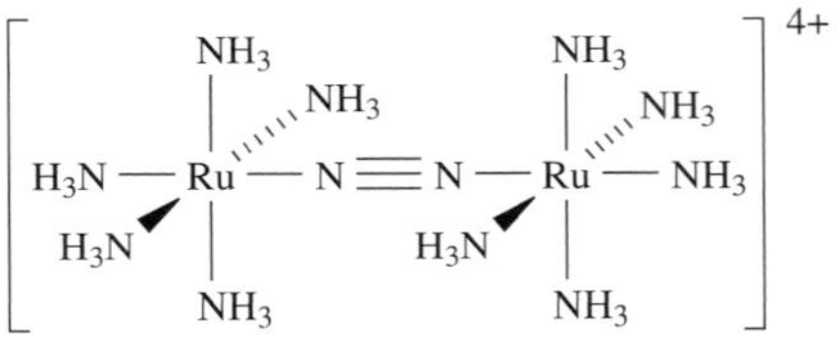

N–N = 112 pm   Ru–N$_{bridge}$ = 193 pm
Ru–N(NH$_3$) = 212–214 pm

(22.56)

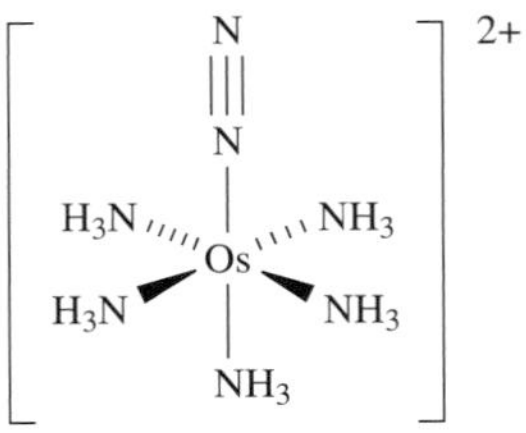

N–N = 112 pm   Os–N(N$_2$) = 184 pm
Os–N(NH$_3$) = 214–215 pm

(22.57)

[†] Much of the interest in metal complexes containing N_2 ligands arises from the possibility of reducing the ligand to NH_3: see Y. Nishibayashi, S. Iwai and M. Hidai (1998) *Science*, vol. 279, p. 540; R.R. Schrock (2008) *Angew. Chem. Int. Ed.*, vol. 47, p. 5512; L.D. Field (2010) *Nature Chem.*, vol. 2, p. 520; N. Hazari (2010) *Chem. Soc. Rev.*, vol. 39, p. 4044; J.L. Crossland and D.R. Tyler (2010) *Coord. Chem. Rev.*, vol. 254, p. 1883.

Most dinitrogen complexes decompose when gently heated, but those of Ru, Os and Ir can be heated to 370–470 K. Although the bonding in a terminal, linear $M-N\equiv N$ unit can be described in a similar manner to a terminal $M-C\equiv O$ unit, the bridging modes of N_2 and CO are different as shown in **22.58**. Coordination of CO to metals is described in Section 24.2.

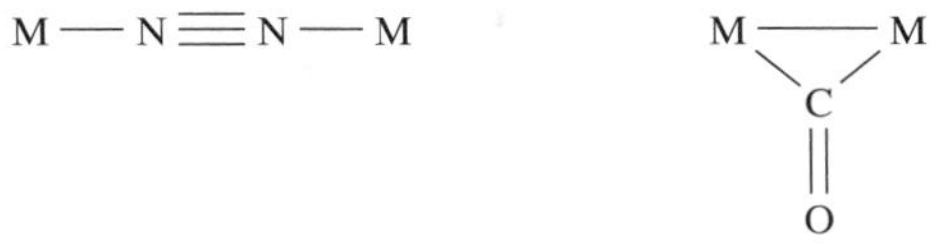

Typical μ-N_2 mode of bonding   Typical μ-CO mode of bonding

(22.58)

The complex $[Ru(NH_3)_6]^{2+}$ (which oxidizes in air, eq. 22.95) is made by reacting $RuCl_3 \cdot xH_2O$ with Zn dust in concentrated NH_3 solution. The analogous Os(II) complex may be formed in liquid NH_3, but is unstable. The reaction of HNO_2 with $[Ru(NH_3)_6]^{2+}$ gives the nitrosyl complex $[Ru(NH_3)_5(NO)]^{3+}$ in which the Ru–N–O angle is close to 180°. Numerous mononuclear nitrosyl complexes of ruthenium are known. In each of $[Ru(NH_3)_5(NO)]^{3+}$, $[RuCl_5(NO)]^{2-}$, $[RuCl(bpy)_2(NO)]^{2+}$, *mer,trans*-$[RuCl_3(PPh_3)_2(NO)]$ and $[RuBr_3(Et_2S)(Et_2SO)(NO)]$ (Fig. 22.23a), the Ru–N–O unit is linear and an Ru(II) state is formally assigned. Without prior knowledge of structural and spectroscopic properties of nitrosyl complexes (see Section 20.4), the oxidation state of the metal centre remains ambiguous, for example in $[RuCl(NO)_2(PPh_3)_2]$ (Fig. 22.23b). Stable ruthenium nitrosyl complexes are formed during the extraction processes for the recovery of uranium and plutonium from nuclear wastes, and are difficult to remove; [106]Ru is a fission product from uranium and plutonium and the use of HNO_3 and TBP (see Box 7.3) in the extraction process facilitates the formation of Ru(NO)-containing complexes. While complexes containing NO ligands are well known, $[Ru(NH_3)_5(N_2O)]^{2+}$ is presently the only example of an isolated complex containing an N_2O ligand. The ligand coordinates to the Ru(II) centre through a N atom. There is significant interest in the coordination chemistry of N_2O, owing to its relevance to biological denitrification (see Box 15.9) in which the enzyme nitrous oxide reductase (the active site in which is a Cu_4 cluster unit) catalyses the final reduction step, i.e. conversion of N_2O to N_2.

The tris-chelates $[Ru(en)_3]^{2+}$, $[Ru(bpy)_3]^{2+}$ (Fig. 10.3) and $[Ru(phen)_3]^{2+}$ are made in a similar manner to $[Ru(NH_3)_6]^{2+}$. The $[Ru(bpy)_3]^{2+}$ complex is widely studied as a photosensitizer. It absorbs light at 452 nm to give an excited singlet state $^1\{[Ru(bpy)_3]^{2+}\}^*$ (Fig. 22.24), which results from transfer of an electron from the Ru(II) centre to a bpy π^*-orbital, i.e. the excited

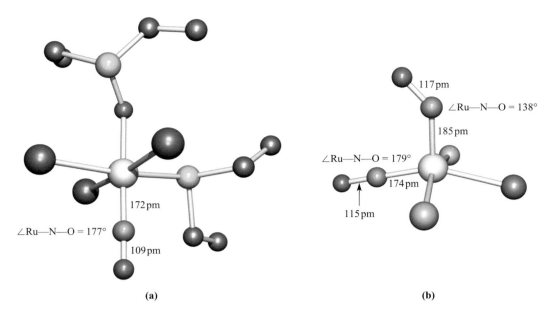

(a)   (b)

Fig. 22.23 The structures (X-ray diffraction) of (a) $[RuBr_3(Et_2S)(Et_2SO)(NO)]$ [R.K. Coll *et al.* (1987) *Inorg. Chem.*, vol. 26, p. 106] and (b) $[RuCl(NO)_2(PPh_3)_2]$ (only the P atoms of the PPh$_3$ groups are shown) [C.G. Pierpont *et al.* (1972) *Inorg. Chem.*, vol. 11, p. 1088]. Hydrogen atoms are omitted in (a); colour code: Ru, pale grey; Br, brown; Cl, green; O, red; N, blue; S, yellow; P, orange; C, grey.

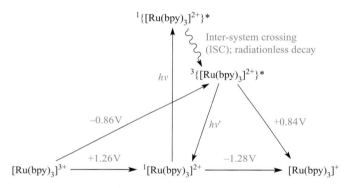

Fig. 22.24 $[Ru(bpy)_3]^{2+}$ (low-spin d^6 is a singlet state) absorbs light to give an excited state which rapidly decays to a longer-lived excited state, $^3\{[Ru(bpy)_3]^{2+}\}^*$. This state can decay by emission or can undergo electron transfer. Standard reduction potentials are given for 1-electron processes involving $[Ru(bpy)_3]^{2+}$ and $^3\{[Ru(bpy)_3]^{2+}\}^*$.

state may be considered to contain Ru(III), two bpy and one [bpy]$^-$. The singlet excited state rapidly decays to a triplet excited state,[†] the lifetime of which in aqueous solution at 298 K is 600 ns, long enough to allow redox activity to occur. The standard reduction potentials in Fig. 22.24 show that the excited $^3\{[Ru(bpy)_3]^{2+}\}^*$ state is both a

[†] For a detailed review, see: A. Juris, V. Balzani, F. Barigelletti, S. Campagna, P. Belser and A. von Zelewsky (1988) *Coord. Chem. Rev.*, vol. 84, p. 85 – 'Ru(II) polypyridine complexes: photophysics, photochemistry, electrochemistry and chemiluminescence'. For an introduction to photochemical principles, see C.E. Wayne and R.P. Wayne (1996) *Photochemistry*, Oxford University Press Primer Series, Oxford.

better oxidant *and* reductant than the ground $[Ru(bpy)_3]^{2+}$ state. In neutral solution, for example, H_2O can be oxidized or reduced by the excited complex. In practice, the system works only in the presence of a *quenching agent* such as methyl viologen (paraquat), $[MV]^{2+}$ (**22.59**) and a sacrificial donor, D, which reduces $[Ru(bpy)_3]^{3+}$ to $[Ru(bpy)_3]^{2+}$ (scheme 22.100) as described in Section 10.4.

$$\overset{+}{MeN}\text{—}\langle\text{aryl}\rangle\text{—}\langle\text{aryl}\rangle\text{—}\overset{+}{NMe}$$

(22.59)

$$\text{(scheme)}$$

(22.100)

Many low oxidation state complexes of Ru and Os including those of Ru(II) and Os(II) are stabilized by PR$_3$ (π-acceptor) ligands. Treatment of RuCl$_3\cdot x$H$_2$O with PPh$_3$ in EtOH/HCl at reflux gives *mer*-$[RuCl_3(PPh_3)_3]$ or, with excess PPh$_3$ in MeOH at reflux, $[RuCl_2(PPh_3)_3]$. Reaction with H$_2$ converts

$[RuCl_2(PPh_3)_3]$ to $[HRuCl(PPh_3)_3]$ which is a hydrogenation catalyst for alk-1-enes (see Section 25.5). Both $[RuCl_2(PPh_3)_3]$ and $[HRuCl(PPh_3)_3]$ have square-based pyramidal structures (**22.60** and **22.61**).

(22.60)   **(22.61)**

Mixed-valence ruthenium complexes

Equation 22.97 showed the formation of the Os(III) complex $[Os_2(\mu\text{-}O_2CMe)_4Cl_2]$. For ruthenium, the scenario is different, and in reaction 22.101, the product is an Ru(II)/Ru(III) polymer (**22.62**).

$$RuCl_3 \cdot xH_2O \xrightarrow{\quad MeCO_2H, (MeCO)_2O \quad} [Ru_2(\mu\text{-}O_2CMe)_4Cl]_n$$

$$(22.101)$$

—— = multiple bond (see text)

(22.62)

Complex **22.62** formally possesses a $\{Ru_2\}^{5+}$ core and from Fig. 21.16 we would predict a configuration of $\sigma^2\pi^4\delta^2\delta^{*2}\pi^{*1}$. However, the observed paramagnetism corresponding to three unpaired electrons is consistent with the π^* level lying at lower energy than the δ^*, i.e. $\sigma^2\pi^4\delta^2\pi^{*2}\delta^{*1}$. This reordering is reminiscent of the σ–π crossover amongst first row diatomics (Fig. 2.10) and illustrates the importance of utilizing *experimental facts* when constructing and interpreting qualitative MO diagrams. With three unpaired electrons, a $[Ru_2(\mu\text{-}O_2CR)_4]^+$ ion has an $S = \frac{3}{2}$ ground state and is therefore an excellent candidate as a building block for the assembly of *molecule-based magnetic materials* (see also Fig. 21.25b and associated text). Coordination polymers in which $[Ru_2(\mu\text{-}O_2CMe)_4]^+$ ions are connected by organic bridging ligands (e.g. $[N(CN)_2]^-$, $[C(CN)_3]^-$, **22.63** and **22.64**) typically show weak antiferromagnetic coupling rather than long-range ferromagnetic ordering. By using $[M(CN)_6]^{3-}$ (M = Cr,

Fe), $[Ru_2(\mu\text{-}O_2CR)_4]^+$ units can be connected into 3-dimensional (for R = Me) or 2-dimensional (for R = ^tBu) networks which become magnetically ordered at low temperatures.

phenazine   TCNQ

(22.63)   **(22.64)**

The *Creutz–Taube* cation $[(H_3N)_5Ru(\mu\text{-}pz')Ru(NH_3)_5]^{5+}$ (pz' = pyrazine) is a member of the series of cations **22.65** (eq. 22.102).

$$[Ru(NH_3)_5(OH_2)]^{2+}$$

$n = 4, 5, 6$

(22.65)

When the charge is 4+ or 6+, the complexes are Ru(II)/Ru(II) or Ru(III)/Ru(III) species respectively. For $n = 5$, a mixed-valence Ru(II)/Ru(III) species might be formulated but spectroscopic and structural data show the Ru centres are equivalent with charge delocalization across the pyrazine bridge. Such electron transfer (see Section 26.5) is not observed in all related species. For example, $[(bpy)_2ClRu(\mu\text{-}pz')RuCl(bpy)_2]^{3+}$ exhibits an intervalence charge transfer absorption in its electronic spectrum indicating an Ru(II)/Ru(III) formulation. The complex $[(H_3N)_5Ru^{III}(\mu\text{-}pz')Ru^{II}Cl(bpy)_2]^{4+}$ is similar.

Self-study exercises

1. *rac-cis*-$[Ru(bpy)_2(DMSO\text{-}S)Cl]^+$ can be separated into its enantiomers by HPLC using a chiral stationary phase. (i) Draw the structure of Δ-*cis*-$[Ru(bpy)_2(DMSO\text{-}S)Cl]^+$ and suggest why it does not racemize under normal conditions. (ii) Outline the principles of HPLC. [*Ans.* See Section 4.2]

2. Triazine acts as a bridging ligand between $[Ru_2(O_2CPh)_4]$ molecules to give a 2-dimensional coordination polymer. Suggest how triazine coordinates to $[Ru_2(O_2CPh)_4]$, and predict the structure of the repeat motif that appears in one layer of the structure of the crystalline product.

triazine

[*Ans.* See S. Furukawa *et al.* (2005) *Chem. Commun.*, p. 865]

22.10 Group 9: rhodium and iridium

The metals

Rhodium and iridium are unreactive metals. They react with O_2 or the halogens only at high temperatures (see below) and neither is attacked by *aqua regia*. The metals dissolve in fused alkalis. For Rh and Ir, the range of oxidation states (Table 19.3) and the stabilities of the highest ones are less than for Ru and Os. The most important states are Rh(III) and Ir(III), i.e. d^6 which is invariably low-spin, giving diamagnetic and kinetically inert complexes (see Section 26.2).

High oxidation states of rhodium and iridium: M(VI) and M(V)

Rhodium(VI) and iridium(VI) occur only in black RhF_6 and yellow IrF_6 (see Fig. 22.16), formed by heating the metals with F_2 under pressure and quenching the volatile products. Both RhF_6 and IrF_6 are octahedral monomers. The pentafluorides are made by direct combination of the elements (eq. 22.103) or by reduction of MF_6, and are moisture-sensitive (reaction 22.104) and very reactive. They are tetramers, structurally analogous to NbF_5 (**22.5**).

$$2RhF_5 \xleftarrow{M = Rh, 520\,K, 6\,bar} 2M + 5F_2 \xrightarrow{M = Ir, 650\,K} 2IrF_5$$
(22.103)

$$IrF_5 \xrightarrow{H_2O} IrO_2 \cdot xH_2O + HF + O_2$$
(22.104)

For M(V) and M(VI), no binary compounds with the heavier halogens and no oxides are known. Iridium(VI) fluoride is the precursor to $[Ir(CO)_6]^{3+}$, the only example to date of a tripositive, binary metal carbonyl cation. Compare reaction 22.105 (reduction of IrF_6 to $[Ir(CO)_6]^{3+}$) with reaction 22.78 (reduction of OsF_6 to $[Os(CO)_6]^{2+}$).

$$2IrF_6 + 12SbF_5 + 15CO$$
$$\xrightarrow[\text{in } SbF_5]{320\,K, \,1\,bar\,CO} 2[Ir(CO)_6][Sb_2F_{11}]_3 + 3COF_2 \quad (22.105)$$

Salts of octahedral $[MF_6]^-$ (M = Rh, Ir) can be made in HF or interhalogen solvents (reaction 22.106). On treatment with water, they liberate O_2 forming Rh(IV) and Ir(IV) compounds.

$$RhF_5 + KF \xrightarrow{HF \text{ or } IF_5} K[RhF_6]$$
(22.106)

A number of Ir(V) hydrido complexes are known, e.g. $[IrH_5(PMe_3)_2]$.

Rhodium(IV) and iridium(IV)

The unstable fluorides are the only established neutral halides of Rh(IV) and Ir(IV), and no oxohalides are known. The reaction of $RhBr_3$ or $RhCl_3$ with BrF_3 yields RhF_4. IrF_4 is made by reduction of IrF_6 or IrF_5 with Ir, but above 670 K, IrF_4 disproportionates (eq. 22.107). Before 1965, reports of 'IrF_4' were erroneous and actually described IrF_5.

$$8IrF_5 + 2Ir \xrightarrow{670\,K} 10IrF_4 \xrightarrow{>670\,K} 5IrF_3 + 5IrF_5$$
(22.107)

Iridium(IV) oxide forms when Ir is heated with O_2 and is the only well-established oxide of Ir. It is also made by controlled hydrolysis of $[IrCl_6]^{2-}$ in alkaline solution. Heating Rh and O_2 gives Rh_2O_3 (see below) unless the reaction is carried out under high pressure, in which case RhO_2 is obtained. Rutile structures (Fig. 6.22) are adopted by RhO_2 and IrO_2.

The series of paramagnetic (low-spin d^5) anions $[MX_6]^{2-}$ with M = Rh, X = F, Cl and M = Ir, X = F, Cl, Br, can be made, but the Ir(IV) species are the more stable. $[RhF_6]^{2-}$ and $[RhCl_6]^{2-}$ (eqs. 22.108 and 22.109) are hydrolysed to RhO_2 by an excess of H_2O. White alkali metal salts of $[IrF_6]^{2-}$ are made by reaction 22.110. $[IrF_6]^{2-}$ is stable in neutral or acidic solution but decomposes in alkali.

$$2KCl + RhCl_3 \xrightarrow{BrF_3} K_2[RhF_6]$$
(22.108)

$$CsCl + [RhCl_6]^{3-} \xrightarrow{Cl_2, \text{aqu.}} Cs_2[RhCl_6]$$
(22.109)

$$M[IrF_6] \xrightarrow{H_2O} M_2[IrF_6] + IrO_2 + O_2$$
$$(M = Na, K, Rb, Cs) \quad (22.110)$$

Salts of $[IrCl_6]^{2-}$ are common starting materials in Ir chemistry. Alkali metal salts are made by chlorinating a mixture of MCl and Ir. $Na_2[IrCl_6] \cdot 3H_2O$, $K_2[IrCl_6]$ and the acid $H_2[IrCl_6] \cdot xH_2O$ (*hexachloridoiridic acid*) are commercially available. The $[IrCl_6]^{2-}$ ion is quantitatively reduced (eq. 22.111) by KI or $[C_2O_4]^{2-}$ and is used as an oxidizing agent in some organic reactions. In alkaline solution, $[IrCl_6]^{2-}$ decomposes, liberating O_2, but the reaction is reversed in strongly acidic solution (see Section 8.2). In its reactions, $[IrCl_6]^{2-}$ is often reduced to Ir(III) (scheme 22.112), but reaction with Br^- yields $[IrBr_6]^{2-}$.

$$[IrCl_6]^{2-} + e^- \rightleftharpoons [IrCl_6]^{3-} \qquad E^o = +0.87\,V \qquad (22.111)$$

$$[IrCl_6]^{2-} \begin{cases} \xrightarrow{[CN]^-} [Ir(CN)_6]^{3-} \\ \xrightarrow{NH_3} [Ir(NH_3)_5Cl]^{2+} \xrightarrow{NH_3} [Ir(NH_3)_6]^{3+} \\ \xrightarrow{Et_2S} [IrCl_3(SEt_2)_3] \end{cases} \quad (22.112)$$

Octahedral coordination is usual for Ir(IV). Complexes with O-donors are relatively few, and include $[Ir(OH)_6]^{2-}$ (the red K^+ salt is made by heating $Na_2[IrCl_6]$ with KOH), $[Ir(NO_3)_6]^{2-}$ (formed by treating $[IrBr_6]^{2-}$ with N_2O_5) and $[Ir(ox)_3]^{2-}$ (made by oxidizing $[Ir(ox)_3]^{3-}$). Complexes with group 15 donors include $[IrCl_4(phen)]$, $[IrCl_2H_2(P^iPr_3)_2]$ (**22.66**) and *trans*-$[IrBr_4(PEt_3)_2]$.

$$\begin{array}{c} P^iPr_3 \\ | \\ H_{\,\prime\prime\prime\prime}\!\!-\!\!Ir\!\!-\!\!^{\backslash\backslash\backslash\backslash}Cl \\ Cl\diagup \;|\; \diagdown H \\ | \\ P^iPr_3 \end{array}$$

(22.66)

Rhodium(III) and iridium(III)

Binary halides MX_3 for M = Rh, Ir and X = Cl, Br and I can be made by heating mixtures of the appropriate elements. Reactions 22.113 and 22.114 show routes to MF_3. Direct reaction of M and F_2 leads to higher fluorides (e.g. eq. 22.103).

$$RhCl_3 \xrightarrow{F_2,\,750\,K} RhF_3 \qquad (22.113)$$

$$Ir + IrF_6 \xrightarrow{750\,K} 2IrF_3 \qquad (22.114)$$

Anhydrous $RhCl_3$ and α-$IrCl_3$ adopt layer structures and are isomorphous with $AlCl_3$. Brown α-$IrCl_3$ converts to the red β-form at 870–1020 K. Water-soluble $RhCl_3 \cdot 3H_2O$ (dark red) and $IrCl_3 \cdot 3H_2O$ (dark green) are commercially available, being common starting materials in Rh and Ir chemistry. Figure 22.25 shows selected complex formations starting from $IrCl_3 \cdot 3H_2O$. In particular, note the formation of $[Ir(bpy)_2(bpy\text{-}C,N)]^{2+}$: this contains a 2,2'-bipyridine ligand which has undergone *orthometallation*. As the structure in Fig. 22.25 illustrates, deprotonation of 2,2'-bipyridine in the 6-position occurs to give the $[bpy\text{-}C,N]^-$ ligand. This leaves an uncoordinated N atom which can be protonated as is observed in $[Ir(bpy)_2(Hbpy\text{-}C,N)]^{3+}$.

The oxide Ir_2O_3 is known only as an impure solid. Rhodium(III) oxide is well characterized, and is made by heating the elements at ordinary pressure or by thermal decomposition of $Rh(NO_3)_3$ (eq. 22.115). Several polymorphs of Rh_2O_3 are known; α-Rh_2O_3 has a corundum structure (see Section 13.7).

$$4Rh(NO_3)_3 \cdot 6H_2O$$
$$\xrightarrow{1000\,K} 2Rh_2O_3 + 24H_2O + 12NO_2 + 3O_2 \quad (22.115)$$

In the presence of aqueous $HClO_4$, octahedral $[Rh(OH_2)_6]^{3+}$ can be formed but it hydrolyses (eq. 22.116). Crystalline $Rh(ClO_4)_3 \cdot 6H_2O$ contains $[Rh(OH_2)_6]^{3+}$, i.e. it should be formulated as $[Rh(OH_2)_6][ClO_4]_3$. The $[Ir(OH_2)_6]^{3+}$ ion exists in aqueous solutions in the presence of concentrated $HClO_4$. The hexaaqua ions are present in the crystalline alums $CsM(SO_4)_2 \cdot 12H_2O$ (M = Rh, Ir).

$$[Rh(OH_2)_6]^{3+} + H_2O \rightleftharpoons [Rh(OH_2)_5(OH)]^{2+} + [H_3O]^+$$
$$pK_a = 3.33 \qquad (22.116)$$

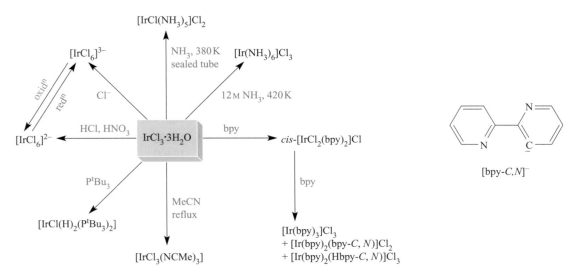

Fig. 22.25 Selected reactions of $IrCl_3 \cdot xH_2O$. In the complexes $[Ir(bpy)_2(bpy\text{-}C,N)]^{2+}$ and $[Ir(bpy)_2(Hbpy\text{-}C,N)]^{3+}$, the ligands coordinating in a C,N-mode have undergone *orthometallation* in which a C—H bond has been broken and a C^- coordination site formally created.

When $Rh_2O_3 \cdot H_2O$ is dissolved in a limited amount of aqueous HCl, $RhCl_3 \cdot 3H_2O$ (better written as $[RhCl_3(OH_2)_3]$) forms. All members of the series $[RhCl_n(OH_2)_{6-n}]^{(3-n)+}$ ($n = 0$–6) are known and can be made in solution by reaction of $[Rh(OH_2)_6]^{3+}$ with Cl^- or by substitution starting from $[RhCl_6]^{3-}$ (see end-of-chapter problem 26.10). Interconversions involving $[Rh(OH_2)_6]^{3+}$ and $[RhCl_6]^{3-}$ are given in scheme 22.117.

$$(22.117)$$

Reduction of $[IrCl_6]^{2-}$ by SO_2 yields $[IrCl_6]^{3-}$ (Fig. 22.25) which hydrolyses in H_2O to $[IrCl_5(OH_2)]^{2-}$ (isolated as the green $[NH_4]^+$ salt), $[IrCl_4(OH_2)_2]^-$ and $[IrCl_3(OH_2)_3]$. Reaction of $[Ir(OH_2)_6]^{3+}$ with $[22.67]Cl_3$ in aqueous Cs_2SO_4 produces $[22.67][IrCl_2(OH_2)_4][SO_4]_2$ containing cation 22.68 for which $pK_a(1) = 6.31$.

(22.67)   **(22.68)**

Routes to ammine complexes of Ir(III) are shown in Fig. 22.25. For Rh(III), it is more difficult to form $[Rh(NH_3)_6]^{3+}$ than $[Rh(NH_3)_5Cl]^{2+}$ (eq. 22.118). Reaction of $RhCl_3 \cdot 3H_2O$ with Zn dust and aqueous NH_3 gives $[Rh(NH_3)_5H]^{2+}$.

$$(22.118)$$

Large numbers of octahedral Rh(III) and Ir(III) complexes exist, and common precursors include $[IrCl_6]^{3-}$ (e.g. Na^+, K^+ or $[NH_4]^+$ salts), $[RhCl(NH_3)_5]Cl_2$, $[Rh(OH_2)(NH_3)_5][ClO_4]_3$ (made by treating $[RhCl(NH_3)_5]Cl_2$ with $AgClO_4$) and *trans*-$[RhCl_2(py)_4]^+$ (made from $RhCl_3 \cdot 3H_2O$ and pyridine). Schemes 22.119 and 22.120 give selected examples.

$$(22.119)$$

$$(22.120)$$

Rhodium(III) and iridium(III) form complexes with both hard and soft donors and examples (in addition to those above and in Fig. 22.25) include:

- *N*-donors: $[Ir(NO_2)_6]^{3-}$, *cis*-$[RhCl_2(bpy)_2]^+$, $[Rh(bpy)_2(phen)]^{3+}$, $[Rh(bpy)_3]^{3+}$, $[Rh(en)_3]^{3+}$;
- *O*-donors: $[Rh(acac)_3]$, $[Ir(acac)_3]$, $[Rh(ox)_3]^{3-}$;
- *P*-donors: *fac*- and *mer*-$[IrH_3(PPh_3)_3]$, $[RhCl_4(PPh_3)_2]^-$, $[RhCl_2(H)(PPh_3)_2]$;
- *S*-donors: $[Ir(NCS-S)_6]^{3-}$ (Fig. 22.26a), *mer*-$[IrCl_3(SEt_2)_3]$, $[Ir(S_6)_3]^{3-}$ (Fig. 22.26b).

Both metal ions form $[M(CN)_6]^{3-}$. Linkage isomerization is exhibited by $[Ir(NH_3)_5(NCS)]^{2+}$, i.e. both $[Ir(NH_3)_5(NCS-N)]^{2+}$ and $[Ir(NH_3)_5(NCS-S)]^{2+}$ can be isolated. The nitrite ligand in $[Ir(NH_3)_5(NO_2)]^{2+}$ undergoes a change from *O*- to *N*-coordination in alkaline solution.

Self-study exercises

1. $[Rh_2Cl_9]^{3-}$ and $[Rh_2Br_9]^{3-}$ possess face-sharing octahedral structures. Heating a propylene carbonate solution of the $[Bu_4N]^+$ salts of $[Rh_2Cl_9]^{3-}$ and $[Rh_2Br_9]^{3-}$ results in a mixture of $[Rh_2Cl_nBr_{9-n}]^{3-}$ ($n = 0$–9) in which all possible species are present. Suggest an experimental technique that can be used to detect these species. Assuming retention of the face-sharing octahedral structure, draw the structures of all possible isomers for $n = 5$. [*Ans.* See: J.-U. Vogt *et al.* (1995) *Z. Anorg. Allg. Chem.*, vol. 621, p. 186]

2. Comment on factors that affect the trend in the values of Δ_{oct} tabulated below.

Complex	Δ_{oct}/cm^{-1}	Complex	Δ_{oct}/cm^{-1}
$[Rh(OH_2)_6]^{3+}$	25 500	$[Rh(CN)_6]^{3-}$	44 400
$[RhCl_6]^{3-}$	19 300	$[RhBr_6]^{3-}$	18 100
$[Rh(NH_3)_6]^{3+}$	32 700	$[Rh(NCS-S)_6]^{3-}$	19 600

[*Ans.* See Table 20.2 and discussion]

Rhodium(II) and iridium(II)

Mononuclear Rh(II) and Ir(II) complexes are relatively rare. The chemistry of Rh(II) is quite distinct from that of Ir(II) since dimers of type $[Rh_2(\mu\text{-}L)_4]$ (e.g. $L^- = RCO_2^-$) and

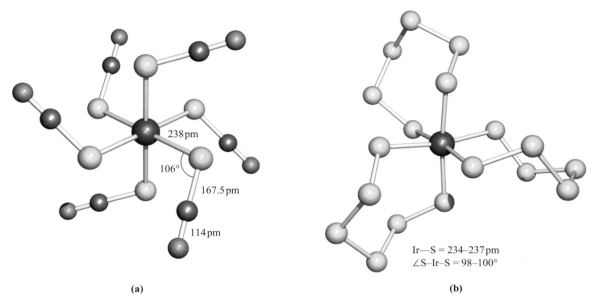

Fig. 22.26 The structures (X-ray diffraction) of (a) $[Ir(NCS-S)_6]^{3-}$ in the $[Me_4N]^+$ salt [J.-U. Rohde *et al.* (1998) *Z. Anorg. Allg. Chem.*, vol. 624, p. 1319] and (b) $[Ir(S_6)_3]^{3-}$ in the $[NH_4]^+$ salt [T.E. Albrecht-Schmitt *et al.* (1996) *Inorg. Chem.*, vol. 35, p. 7273]. Colour code: Ir, red; S, yellow; C, grey; N, blue.

$[Rh_2(\mu\text{-}L)_4L'_2]$ are well known but Ir analogues are rare. The best-known Rh(II) dimers contain carboxylate bridges (Figs. 22.27a and 22.27b); other bridging ligands include $[RC(O)NH]^-$ and $[RC(O)S]^-$. The dimers $[Rh_2(\mu\text{-}O_2CMe)_4L_2]$ (L = MeOH or H_2O) are made by reactions 22.121 and 22.122. The axial ligands can be removed by heating *in vacuo*, or replaced (e.g. reaction 22.123).

$$RhCl_3\cdot3H_2O$$

$$\xrightarrow[\text{MeOH reflux}]{MeCO_2H / Na[MeCO_2]} [Rh_2(\mu\text{-}O_2CMe)_4(MeOH)_2]$$

$$(22.121)$$

$$[NH_4]_3[RhCl_6] \xrightarrow[\text{EtOH, }H_2O]{MeCO_2H} [Rh_2(\mu\text{-}O_2CMe)_4(OH_2)_2]$$

$$(22.122)$$

$$[Rh_2(\mu\text{-}O_2CMe)_4(OH_2)_2] \begin{cases} \xrightarrow{py} [Rh_2(\mu\text{-}O_2CMe)_4(py)_2] \\ \xrightarrow{SEt_2} [Rh_2(\mu\text{-}O_2CMe)_4(SEt_2)_2] \end{cases}$$

$$(22.123)$$

Figure 22.27c shows the structure of $[Rh_2(\mu\text{-}O_2CMe)_4(OH_2)_2]$, and related complexes are similar. If the axial ligand has a second donor atom appropriately oriented, polymeric chains in which L' bridges $[Rh_2(\mu\text{-}L)_4]$ units can result, e.g. when L' = phenazine (**22.63**) or **22.69**. Each dimer formally contains an $\{Rh_2\}^{4+}$ core which (from Fig. 21.16) has a $\sigma^2\pi^4\delta^2\delta^{*2}\pi^{*4}$ configuration, and a Rh–Rh *single* bond. Compare this with the multiply-bonded Mo(II), Re(III) and Os(III) dimers discussed earlier.

(22.69)

The only example of an $[Ir_2(\mu\text{-}L)_4]$ dimer containing an $\{Ir_2\}^{4+}$ core and an Ir–Ir single bond (252 pm) occurs for $L^- = $ **22.70**.

Me — N,N'-di-4-tolylformamidine — Me

Conjugate base of *N,N'*-di-4-tolylformamidine

(22.70)

Rhodium(I) and iridium(I)

The +1 oxidation state of Rh and Ir (d^8) is stabilized by π-acceptor ligands such as phosphanes, with square planar, and to a lesser extent, trigonal bipyramidal coordination being favoured. Being low oxidation state species, it may be appropriate to consider the bonding in terms of the 18-electron rule (Section 20.4). In fact, most Rh(I) complexes are square planar, 16-electron species and some, such as $[RhCl(PPh_3)_3]$ (*Wilkinson's catalyst*, **22.71**), have important applications in homogeneous catalysis (see Chapter 25). Preparation of $[RhCl(PPh_3)_3]$ involves reduction of Rh(III) by PPh_3 (eq. 22.124). Other $[RhCl(PR_3)_3]$

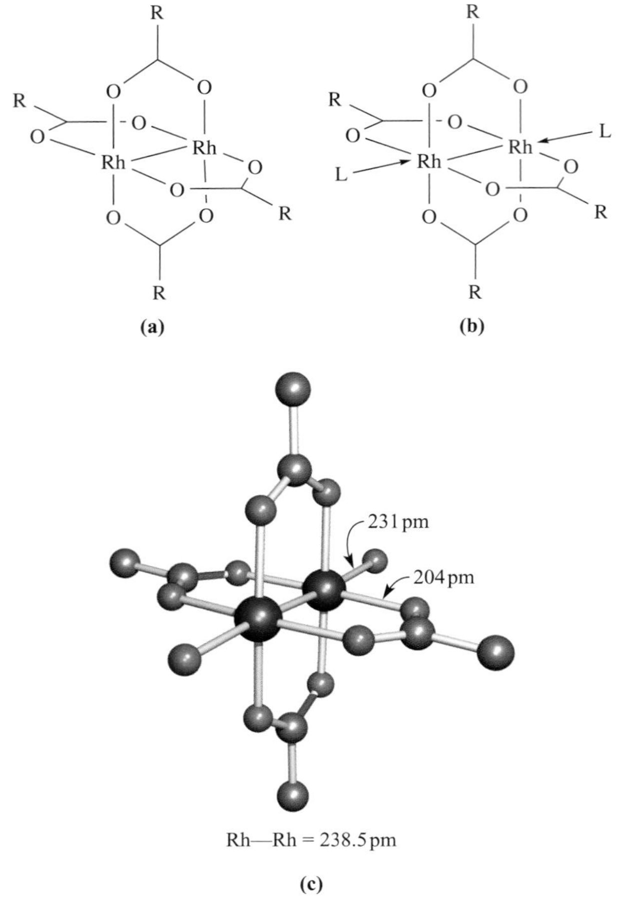
(a)
(b)
231 pm
204 pm
Rh—Rh = 238.5 pm
(c)

Fig. 22.27 Schematic representations of two families of Rh(II) carboxylate dimers: (a) [$Rh_2(\mu\text{-}O_2CR)_4$] and (b) [$Rh_2(\mu\text{-}O_2CR)_4L_2$]. (c) The structure of [$Rh_2(\mu\text{-}O_2CMe)_4(OH_2)_2$] (H atoms omitted) determined by X-ray diffraction [F.A. Cotton *et al.* (1971) *Acta Crystallogr., Sect. B*, vol. 27, p. 1664]; colour code: Rh, blue; C, grey; O, red.

complexes are made by routes such as 22.125. Alkene complexes like that in this reaction are described in Chapter 24. Starting from [$RhCl(PPh_3)_3$], it is possible to make a variety of square planar complexes in which phosphane ligands remain to stabilize the Rh(I) centre, e.g. scheme 22.126. Treatment of [$RhCl(PPh_3)_3$] with $TlClO_4$ yields the perchlorate salt of the trigonal planar cation [$Rh(PPh_3)_3$]$^+$.

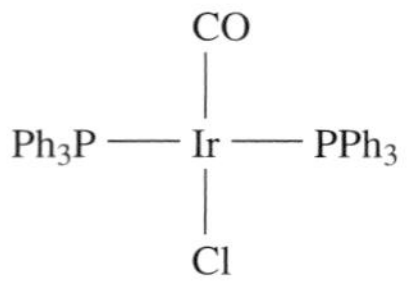

(22.71)

$$RhCl_3{\cdot}3H_2O + 6PPh_3 \xrightarrow{\text{EtOH, reflux}} [RhCl(PPh_3)_3] + \ldots \qquad (22.124)$$

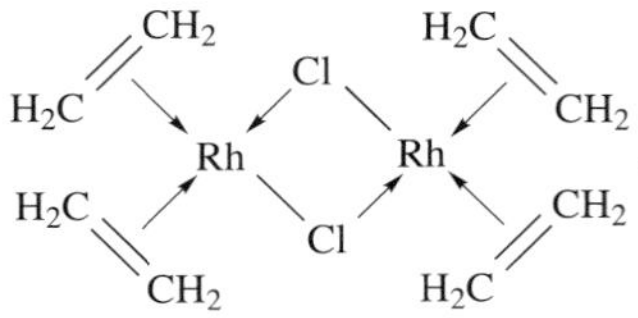

$$+ 6PR_3 \longrightarrow 2[RhCl(PR_3)_3] + 4C_2H_4 \qquad R \neq Ph \qquad (22.125)$$

$$[RhCl(PPh_3)_3] \begin{cases} \xrightarrow{NaBH_4} [RhH(PPh_3)_3] \\ \xrightarrow[MeCN]{[Me_3O][BF_4]} [Rh(MeCN)(PPh_3)_3][BF_4] \\ \xrightarrow{RCO_2H} [Rh(O_2CR\text{-}O)(PPh_3)_3] \end{cases} \qquad (22.126)$$

The square planar Ir(I) complex *trans*-[$IrCl(CO)(PPh_3)_2$] (*Vaska's compound*, **22.72**) is strictly organometallic since it contains an Ir–C bond, but it is an important precursor in Ir(I) chemistry. Both *trans*-[$IrCl(CO)(PPh_3)_2$] and [$RhCl(PPh_3)_3$] undergo many oxidative addition reactions (see Section 24.9) in which the M(I) centre is oxidized to M(III).

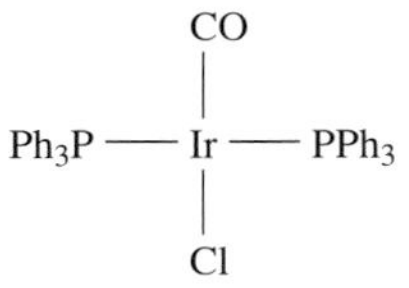

(22.72)

22.11 Group 10: palladium and platinum

The metals

At 298 K, bulk Pd and Pt are resistant to corrosion. Palladium is more reactive than Pt, and at high temperatures is attacked by O_2, F_2 and Cl_2 (eq. 22.127).

$$PdO \xleftarrow{O_2,\,\Delta} Pd \xrightarrow{Cl_2,\,\Delta} PdCl_2 \qquad (22.127)$$

Palladium dissolves in hot oxidizing acids (e.g. HNO_3), but both metals dissolve in *aqua regia* and are attacked by molten alkali metal oxides.

The dominant oxidation states are M(II) and M(IV), but the M(IV) state is more stable for Pt than Pd. Within a given oxidation state, Pd and Pt resemble each other with the exception of their behaviour towards oxidizing and reducing agents. Palladium(II) and platinum(II) form almost exclusively low-spin, square planar complexes. This contrasts with the wide range of high- and low-spin 4- and 6-coordinate nickel(II) complexes, with 4-coordinate including both square planar and tetrahedral geometries (see Section 21.11).

The highest oxidation states: M(VI) and M(V)

The M(VI) and M(V) states are confined to platinum fluorides (reactions 22.128 and 22.129, see Fig. 22.16); PtF_5 readily disproportionates to PtF_4 and PtF_6.

$$PtCl_2 \xrightarrow{F_2,\,620\,K} PtF_5 \qquad (22.128)$$

$$Pt \xrightarrow[\text{2. rapid quenching}]{\text{1. }F_2,\,870\,K} PtF_6 \qquad (22.129)$$

Platinum(V) fluoride is a tetramer (Fig. 22.28). PtF_6 is a red solid and has a molecular structure consisting of octahedral molecules. Neutron powder diffraction data confirm little deviation from ideal O_h symmetry. The hexafluoride is a very powerful oxidizing agent (eq. 22.130, and see Section 6.16) and attacks glass. The oxidizing power of the third row d-block hexafluorides (for those that exist) follows the sequence $PtF_6 > IrF_6 > OsF_6 > ReF_6 > WF_6$.

$$O_2 + PtF_6 \longrightarrow [O_2]^+[PtF_6]^- \qquad (22.130)$$

In anhydrous HF, PtF_6 reacts with CO to give $[Pt^{II}(CO)_4]^{2+}[Pt^{IV}F_6]^{2-}$ (eq. 22.131), while in liquid SbF_5, reaction 22.132 occurs.

$$2PtF_6 + 7CO \xrightarrow[\text{223 K, warm to 298 K}]{\text{1 bar CO, anhydrous HF}} [Pt(CO)_4][PtF_6] + 3COF_2 \quad (22.131)$$

$$PtF_6 + 6CO + 4SbF_5 \xrightarrow[\text{300 K}]{\text{1 bar CO, liquid }SbF_5} [Pt(CO)_4][Sb_2F_{11}]_2 + 2COF_2 \quad (22.132)$$

The fluorides PdF_5 and PdF_6 have not been confirmed, but $[PdF_6]^-$ can be made by reaction 22.133.

$$PdF_4 + KrF_2 + O_2 \longrightarrow [O_2]^+[PdF_6]^- + Kr \qquad (22.133)$$

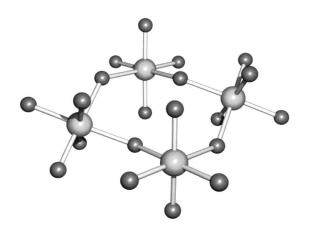

Fig. 22.28 The tetrameric structure of PtF_5 (X-ray diffraction, B.G. Müller *et al.* (1992) *Eur. J. Solid State Inorg. Chem.*, vol. 29, p. 625). Colour code: Pt, yellow; F, green.

Palladium(IV) and platinum(IV)

The only tetrahalide of Pd(IV) is PdF_4, a diamagnetic, red solid made from the elements at 570 K. The compound 'PdF$_3$' (also formed from Pd and F_2) is actually $Pd^{II}[Pd^{IV}F_6]$. Both Pd centres in this solid state compound are octahedrally sited, i.e. a rare example of octahedral Pd(II). All the Pt(IV) halides are known, and $PtCl_4$ and $PtBr_4$ are formed by reactions of the halogens with Pt. Treatment of $PtCl_2$ with F_2 ($T < 475\,K$) gives PtF_4 (compare reaction 22.128). In $PtCl_4$, $PtBr_4$ and PtI_4, the metal is octahedrally sited as shown in **22.73**. In PdF_4 and PtF_4, the connectivity is similar but results in a 3-dimensional structure.

$$X = Cl,\, Br,\, I$$

(22.73)

Hydrated PtO_2 is made by hydrolysing $[PtCl_6]^{2-}$ in boiling aqueous Na_2CO_3; heating converts it to the black anhydrous oxide. Above 920 K, PtO_2 decomposes to the elements. The hydrated oxide dissolves in NaOH as $Na_2[Pt(OH)_6]$ and in aqueous HCl as $H_2[PtCl_6]$ (hexachloridoplatinic acid). The latter is an important starting material in synthesis and has catalytic applications. Water hydrolyses $H_2[PtCl_6]$ to $H[PtCl_5(OH_2)]$ and $[PtCl_4(OH_2)_2]$; the reaction is reversed by adding HCl.

In their complexes, Pd(IV) and Pt(IV) are low-spin, octahedral and diamagnetic (d^6). The full range of halido complexes $[MX_6]^{2-}$ is known (e.g. eqs. 22.134–22.136), in contrast to PdF_4 being the only neutral Pd(IV) halide. The $[MX_6]^{2-}$ ions are stabilized by large cations.

$$M \xrightarrow{\text{aqua regia; KCl}} K_2[MCl_6] \qquad M = Pd,\, Pt \qquad (22.134)$$

$$K_2[PtCl_6] \xrightarrow{BrF_3} K_2[PtF_6] \qquad (22.135)$$

$$PtCl_4 \xrightarrow{HCl} H_2[PtCl_6] \begin{cases} \xrightarrow{KCl} K_2[PtCl_6] \\ \xrightarrow{KI} K_2[PtI_6] \end{cases} \qquad (22.136)$$

The greater *kinetic* inertness (see Section 26.2) of the Pt(IV) complexes is illustrated by the fact that $K_2[PdF_6]$ is decomposed in the air by moisture, but $K_2[PtF_6]$ can be crystallized from boiling water even though $[PtF_6]^{2-}$ is *thermodynamically* unstable with respect to hydrolysis. The solid state structure of $K_2[PtCl_6]$ is a structure prototype. It can be derived from the CaF_2 structure (Fig. 6.19a) by replacing

Ca^{2+} by octahedral $[PtCl_6]^{2-}$ ions, and F^- by K^+ ions. For details of $K_2[PtH_6]$, see Section 10.7.

The variety of Pd(IV) complexes is far less than that of Pt(IV), and their syntheses usually involve oxidation of a related Pd(II) species, e.g. reaction 22.137. When a chelating ligand such as bpy or $Me_2PCH_2CH_2PMe_2$ is present, the complex is constrained to being *cis*, e.g. **22.74**. Palladium(IV) complexes are of limited stability.

(22.74)

$$trans\text{-}[PdCl_2(NH_3)_2] + Cl_2 \longrightarrow trans\text{-}[PdCl_4(NH_3)_2]$$

$$(22.137)$$

Platinum(IV) forms a wide range of thermodynamically and kinetically inert octahedral complexes, and ammine complexes, for example, have been known since the days of Werner (see Box 21.7). In liquid NH_3 at 230 K, $[NH_4]_2[PtCl_6]$ is converted to $[Pt(NH_3)_6]Cl_4$. *Trans*-$[PtCl_2(NH_3)_4]^{2+}$ is made by oxidative addition of Cl_2 to $[Pt(NH_3)_4]^{2+}$, and, as for Pd, oxidative addition is a general strategy for Pt(II) ⟶ Pt(IV) conversions. Amine complexes include the optically active $[Pt(en)_3]^{4+}$ and *cis*-$[PtCl_2(en)_2]^{2+}$, both of which can be resolved. Although the range of ligands coordinating to Pt(IV) covers soft and hard donors (see Table 7.9), some ligands such as phosphanes tend to reduce Pt(IV) to Pt(II).

Of note is $[Pt(NH_3)_4]_2[\mathbf{22.75}]\cdot 9H_2O$, formed when aqueous $[Pt(NH_3)_4][NO_3]_2$ reacts with $K_3[Fe(CN)_6]$. Localized Fe(II) and Pt(IV) centres in $[\mathbf{22.75}]^{4-}$ have been assigned on the basis of magnetic, electrochemical and EPR spectroscopic data, and the complex exhibits an intense Fe(II) ⟶ Pt(IV) charge transfer absorption at 470 nm.

(22.75)

and Pt(III) species. Both $PtCl_3$ and $PtBr_3$ are mixed-valence compounds. The compounds of empirical formulae $Pt(NH_3)_2Br_3$ (**22.76**) and $Pt(NEtH_2)_4Cl_3\cdot H_2O$ (*Wolffram's red salt*, **22.77**) contain halide-bridged chains; extra Cl^- in the lattice of the latter balance the 4+ charge. Such mixed-valence compounds possess intense colours due to inter-valence charge transfer absorptions. Partially oxidized $[Pt(CN)_4]^{2-}$ salts are described under platinum(II).

(22.76)

octahedral Pt(IV)   square planar Pt(II)

$L = NH_3$

(22.77)

octahedral Pt(IV)   square planar Pt(II)

$L = EtNH_2$

Palladium(III) and platinum(III) dimers that are structurally related to the Rh(II) dimers discussed earlier (Fig. 22.27) include $[Pd_2(\mu\text{-}SO_4\text{-}O,O')_4(OH_2)_2]^{2-}$, $[Pd_2(\mu\text{-}O_2CMe)_4(OH_2)_2]^{2+}$, $[Pt_2(\mu\text{-}HSO_4\text{-}O,O')_2(SO_4\text{-}O,O')_2]$ and $[Pt_2(SO_4\text{-}O,O')_4(OH_2)_2]$. Each formally contains a $\{Pd_2\}^{6+}$ or $\{Pt_2\}^{6+}$ core (isoelectronic with $\{Rh_2\}^{4+}$) and an M–M single bond.

The *platinum blues*[†] are mixed-valence complexes containing discrete Pt_n chains (Fig. 22.29). They are formed by hydrolysis of *cis*-$[PtCl_2(NH_3)_2]$ or *cis*-$[PtCl_2(en)]$ in aqueous $AgNO_3$ (i.e. replacing Cl^- by H_2O and precipitating AgCl) followed by treatment with *N,O*-donors such as pyrimidines, uracils or the compounds shown in Fig. 22.29b. Figures 22.29a and 22.29c show two examples; each formally contains a $\{Pt_4\}^{9+}$ core which can be considered as $(Pt^{III})(Pt^{II})_3$. EPR spectroscopic data show that the unpaired electron is delocalized over the Pt_4-chain. Interest in platinum blues lies in the fact that some exhibit anti-tumour activity.

Palladium(III), platinum(III) and mixed-valence complexes

We saw above that 'PdF$_3$' is the mixed valence $Pd[PdF_6]$, and this cautionary note extends to some other apparently Pd(III)

[†] For related complexes, see: C. Tejel, M.A. Ciriano and L.A. Oro (1999) *Chem. Eur. J.*, vol. 5, p. 1131 – 'From platinum blues to rhodium and iridium blues'; B. Lippert (2007) *Chimia*, vol. 61, p. 732 – 'Platinum pyrimidine blues: Still a challenge to bioinorganic chemists and a treasure for coordination chemists'.

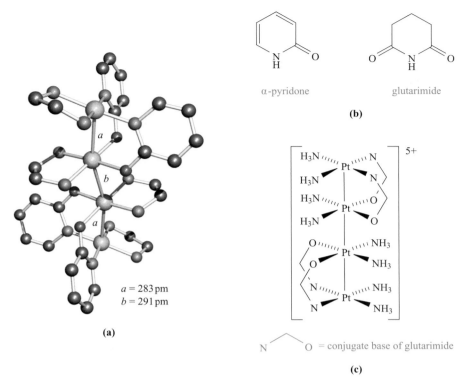

(b)

α-pyridone   glutarimide

(c)

N ⌄ O = conjugate base of glutarimide

Fig. 22.29 (a) The structure (X-ray diffraction) of the cation in the platinum blue $[Pt_4(en)_4(\mu\text{-}L)_4][NO_3]_5 \cdot H_2O$ where HL = α-pyridone. Hydrogen atoms have been omitted; colour code: Pt, yellow; N, blue; O, red; C, grey [T.V. O'Halloran *et al.* (1984) *J. Am. Chem. Soc.*, vol. 106, p. 6427]. (b) Examples of *N,O*-donor ligands present in platinum blues. (c) Schematic representation of the platinum blue $[Pt_4(NH_3)_8(\mu\text{-}L)_4]^{5+}$ where HL = glutarimide.

Self-study exercises

1. The anion in $K_4[Pt_4(SO_4)_5]$ can be written in the form $[Pt_4(SO_4)_4(SO_4)_{2/2}]^{4-}$. What does this notation tell you about the solid state structure of the complex?
   [*Ans.* See M. Pley *et al.* (2005) *Eur. J. Inorg. Chem.*, p. 529]

2. The structure of $PtCl_3$ consists of Pt_6Cl_{12} clusters with square planar Pt centres, and chains of edge-sharing *cis*-$[PtCl_2Cl_{4/2}]$-octahedra. (a) Explain how to interpret the formula *cis*-$[PtCl_2Cl_{4/2}]$ to give a structural diagram of part of one chain. (b) Assign oxidation states to the Pt centres in the two structural units. (c) Confirm that the overall stoichiometry is $PtCl_3$.
   [*Ans.* See H.G. von Schnering *et al.* (2004) *Z. Anorg. Allg. Chem.*, vol. 630, p. 109]

Palladium(II) and platinum(II)

In Section 20.3, we discussed the increase in crystal field splitting on descending group 10, and explained why Pd(II) and Pt(II) complexes favour a square planar arrangement of donor atoms.

All the halides of Pd(II) and Pt(II) except PtF_2 are known. Reaction of Pd and F_2 gives 'PdF_3' (see above) which is reduced to violet PdF_2 by SeF_4. Unusually for Pd(II), PdF_2 is paramagnetic $(\mu = 1.90\ \mu_B)$ and each Pd(II) centre is octahedrally sited in a rutile structure (Fig. 6.22). The other dihalides are diamagnetic (low-spin d^8) and contain square planar M(II) centres in polymeric structures. Heating Pd and Cl_2 gives $PdCl_2$. The α-form is a polymer (**22.78**) and above 820 K, α-$PdCl_2$ converts to the β-form which contains hexameric units (**22.79**). Palladium(II) bromide is made from the elements, and PdI_2 by heating $PdCl_2$ with HI. Direct combination of Pt and a halogen affords $PtCl_2$, $PtBr_2$ and PtI_2. $PtCl_2$ is dimorphic like $PdCl_2$.

(22.78)   **(22.79)**

Black, water-insoluble PdI_2 dissolves in a solution containing CsI and I_2, and the compounds $Cs_2[PdI_4]\cdot I_2$ and $Cs_2[PdI_6]$ can be crystallized from the solution. Pressure converts $Cs_2[PdI_4]\cdot I_2$ to $Cs_2[PdI_6]$, a process that is facilitated by the presence of chains (**22.80**) in the solid state structure of $Cs_2[PdI_4]\cdot I_2$. The structure of $Cs_2[PdI_6]$ is like that of $K_2[PtCl_6]$.

(22.80)

Black PdO, formed by heating Pd and O_2, is the only well-established oxide of Pd. In contrast, PtO_2 is the only well-characterized oxide of Pt. Dissolution of PdO in perchloric acid gives $[Pd(OH_2)_4][ClO_4]_2$ containing a diamagnetic, square planar tetraaqua ion. The $[Pt(OH_2)_4]^{2+}$ ion is made in solution by treating $[PtCl_4]^{2-}$ with aqueous $AgClO_4$. Both aqua ions are considerably better oxidizing agents than aqueous Ni^{2+} (eq. 22.138), but neither $[Pd(OH_2)_4]^{2+}$ nor $[Pt(OH_2)_4]^{2+}$ is very stable. Solid $[Pt(OH_2)_4][SbF_6]_2$, containing the square planar $[Pt(OH_2)_4]^{2+}$ cation, has been isolated by dissolving $PtO\cdot H_2O$ in $[H_3O][SbF_6]$ followed by recrystallization from anhydrous HF.

$$M^{2+} + 2e^- \rightleftharpoons M \quad \begin{cases} M = Ni,\ E^\circ = -0.25\,V \\ M = Pd,\ E^\circ = +0.95\,V \quad (22.138) \\ M = Pt,\ E^\circ = +1.18\,V \end{cases}$$

Palladium(II) and platinum(II) form a wealth of square planar complexes. A tendency for Pt−Pt interactions (i.e. for the heaviest group 10 metal) is quite often observed. The mechanisms of substitution reactions in Pt(II) complexes and the *trans*-effect have been much studied (see Section 26.3). For the discussion that follows, it is important to note that mutually *trans* ligands exert an effect on one another, and this dictates the order in which ligands are displaced and, therefore, the products of substitution reactions. *A word of caution:* do not confuse *trans*-effect with *trans*-influence (see Box 22.8).

Important families of complexes with monodentate ligands include $[MX_4]^{2-}$ (e.g. X = Cl, Br, I, CN, SCN-*S*), $[MX_2L_2]$ (e.g. X = Cl, Br; L = NH_3, NR_3, RCN, py, PR_3, SR_2; or X = CN; L = PR_3) and $[ML_4]^{2+}$ (e.g. L = PR_3, NH_3, NR_3, MeCN). For $[MX_2L_2]$, *trans*- and *cis*-isomers may exist and, in the absence of X-ray diffraction data, IR spectroscopy can be used to distinguish between *cis*- and *trans*-$[MX_2Y_2]$ species (Fig. 19.12). Isomers of, for example, $[PtCl_2(PR_3)_2]$ can also be distinguished using ^{31}P NMR spectroscopy (Box 19.2). Equations 22.139–22.141 show the formation of some ammine complexes,

the choice of route for *cis*- and *trans*-$[PtCl_2(NH_3)_2]$ arising from the *trans*-effect (see above).

Isomerization of *cis*- to *trans*-$[PtCl_2(NH_3)_2]$ occurs in solution.

$$[PtCl_6]^{2-} \xrightarrow{SO_2;\ HCl} H_2PtCl_4 \xrightarrow{conc\ NH_3,\ \Delta} [Pt(NH_3)_4]Cl_2$$
$$(22.139)$$

$$[PtCl_4]^{2-} + 2NH_3 \xrightarrow{aq\ soln} cis\text{-}[PtCl_2(NH_3)_2] + 2Cl^-$$
$$(22.140)$$

$$[Pt(NH_3)_4]^{2+} + 2HCl$$
$$\xrightarrow{aq\ soln} trans\text{-}[PtCl_2(NH_3)_2] + 2[NH_4]^+ \quad (22.141)$$

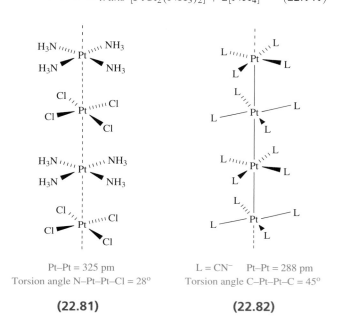

Pt–Pt = 325 pm
Torsion angle N–Pt–Pt–Cl = 28°

L = CN^- Pt–Pt = 288 pm
Torsion angle C–Pt–Pt–C = 45°

(22.81)   **(22.82)**

Magnus's green salt, $[Pt(NH_3)_4][PtCl_4]$ is prepared by precipitation from colourless $[Pt(NH_3)_4]Cl_2$ and pink $[PtCl_4]^{2-}$. It contains chains of alternating cations and anions (**22.81**) with significant Pt−Pt interactions, and this structural feature leads to the change in colour on going from the constituent ions to the solid salt. However, the Pt−Pt distance is not as short as in the partially oxidized $[Pt(CN)_4]^{2-}$ (see below), and $[Pt(NH_3)_4][PtCl_4]$ is, therefore, not a metallic conductor.

The cyanido complexes $[Pd(CN)_4]^{2-}$ and $[Pt(CN)_4]^{2-}$ are very stable. $K_2[Pt(CN)_4]\cdot 3H_2O$ can be isolated from the reaction of $K_2[PtCl_4]$ with KCN in aqueous solution. Aqueous solutions of $K_2[Pt(CN)_4]$ are colourless, but the hydrate forms yellow crystals. Similarly, other salts are colourless in solution but form coloured crystals. The colour change arises from stacking (non-eclipsed) of square planar anions in the solid, although the Pt⋯⋯Pt separations are significantly larger (e.g. 332 pm in yellow-green

THEORY

Box 22.8 The *trans*-influence

Consider a square planar complex which contains a *trans* L−M−L' arrangement:

$$
\begin{array}{ccc}
 & L & \\
 & | & \\
X & \!\!-\!\!-\!\!- M \!\!-\!\!-\!\!- & X \\
 & | & \\
 & L' &
\end{array}
$$

Ligands L and L' compete with each other for electron density because the formation of M−L and M−L' bonds uses the same metal orbitals, e.g. d_{z^2} and p_z if L and L' lie on the z axis. The existence of a *ground state trans*-influence (i.e. the influence that L has on the M−L' bond in the ground state of the complex) is established by comparing the solid state structural, and vibrational and NMR spectroscopic data for series of related complexes. Structural data are exemplified by the following series of square planar Pt(II) complexes; H^- exerts a strong *trans*-influence and as a consequence, the Pt−Cl bond in $[PtClH(PEtPh_2)_2]$ is relatively long and weak:

$\begin{bmatrix} & Cl & \\ & \| & \\ Cl\!-\!Pt\!-\!Cl \\ & \| & \\ & Cl & \end{bmatrix}^{2-}$	$\begin{bmatrix} H_2C\!=\!CH_2 \\ \downarrow \\ Cl\!-\!Pt\!-\!Cl \\ {}_{b}\;\; \|_{a} \\ Cl \end{bmatrix}^{-}$	$\begin{array}{c} PMe_3 \\ \| \\ Cl\!-\!Pt\!-\!PMe_3 \\ \| \\ Cl \end{array}$	$\begin{array}{c} PEtPh_2 \\ \| \\ Cl\!-\!Pt\!-\!H \\ \| \\ PEtPh_2 \end{array}$
	(in Zeise's salt)		
Pt−Cl / pm 231.6	$a = 232.7$ $b = 230.5$	237	242

IR and ^1H NMR spectroscopic data for a series of square planar complexes *trans*-$[PtXH(PEt_3)_2]$ are as follows:

X^-	$[CN]^-$	I^-	Br^-	Cl^-
$\bar{\nu}(Pt-H) \, / \, cm^{-1}$	2041	2156	2178	2183
$\delta(^1H \text{ for } Pt-H) \, ppm$	−7.8	−12.7	−15.6	−16.8

Values of $\bar{\nu}(Pt-H)$ show that the Pt−H bond is weakest for $X^- = [CN]^-$ and the *trans*-influence of the X^- ligands follows the order $[CN]^- > I^- > Br^- > Cl^-$. The signal for the hydride in the ^1H NMR spectrum moves to lower frequency (higher field) with a decrease in the *trans*-influence of X^-. The *trans*-influence is not unique to square planar complexes, and may be observed wherever ligands are mutually *trans*, e.g. in octahedral species.

The *trans*-influence is *not* the same as the *trans*-effect. The *trans*-influence is a ground state phenomenon, while the *trans*-effect is a kinetic effect (see Section 26.3). The two effects are sometimes distinguished by the names *structural trans-effect* and *kinetic trans-effect*.

Further reading

K.M. Andersen and A.G. Orpen (2001) *Chem. Commun.*, p. 2682 – 'On the relative magnitudes of *cis* and *trans* influences in metal complexes'.

B.J. Coe and S.J. Glenwright (2000) *Coord. Chem. Rev.*, vol. 203, p. 5 – '*Trans*-effects in octahedral transition metal complexes'.

M. Melnik and C.E. Holloway (2006) *Coord. Chem. Rev.*, vol. 250, p. 2261 – 'Stereochemistry of platinum coordination compounds'.

$Ba[Pt(CN)_4] \cdot 4H_2O$ and 309 pm in violet $Sr[Pt(CN)_4] \cdot 3H_2O$) than in Pt metal (278 pm). When $K_2[Pt(CN)_4]$ is partially oxidized with Cl_2 or Br_2, bronze complexes of formula $K_2[Pt(CN)_4]X_{0.3} \cdot 2.5H_2O$ (X = Cl, Br) are obtained. These contain isolated X^- ions and stacks of staggered $[Pt(CN)_4]^{2-}$ ions (**22.82**) with short Pt−Pt separations, and are good 1-dimensional metallic conductors. The conductivity arises from electron delocalization along the Pt_n chain, the centres no longer being localized Pt(II) after partial oxidation.

Some salts do contain non-stacked $[Pt(CN)_4]^{2-}$ ions, e.g. $[PhNH_3]_2[Pt(CN)_4]$.

The paucity of complexes with O-donors arises because Pd(II) and Pt(II) are soft metal centres (see Table 7.9) and prefer soft donors such as S or P. Although $[Pt(OH_2)_4][SbF_6]_2$ has been structurally characterized by X-ray diffraction (see earlier in the section), the tetraaqua ions of Pd(II) and Pt(II) are relatively unstable. Reaction of $[PtCl_4]^{2-}$ with KOH and excess Hacac gives monomeric

BIOLOGY AND MEDICINE

Box 22.9 Platinum-containing drugs for cancer treatment

Cisplatin is the square planar complex *cis*-[PtCl$_2$(NH$_3$)$_2$] and its capacity to act as an anti-tumour drug has been known since the 1960s. It is used to treat bladder and cervical tumours, as well as testicular and ovarian cancers, but patients may suffer from side-effects of nausea and kidney damage. Carboplatin shows similar anti-tumour properties and has the advantage of producing fewer side-effects than cisplatin. The drugs operate by interacting with guanine (G) bases in strands of DNA (Fig. 10.13), with the *N*-donors of the nucleotide base coordinating to the Pt(II) centre; GG intra-strand cross-links are formed by cisplatin. Cisplatin, carboplatin and oxaliplatin are approved for medical use by the FDA (the US Food and Drug Administration).

One Pt(II) complex to undergo phase II clinical trials is the triplatinum species shown below. The complex is significantly more active than cisplatin, and is capable of forming inter-strand cross-links involving three base pairs in DNA. The complex is targeted at the treatment of lung, ovarian and gastric cancers.

Satraplatin is one of the few Pt(IV) complexes to reach clinical trials. It has completed phase III trials but is not (as of early 2011), FDA approved.

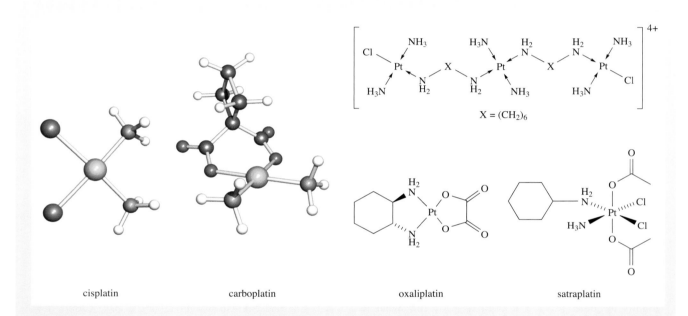

cisplatin   carboplatin   oxaliplatin   satraplatin

Further reading

M.-H. Baik, R.A. Friesner and S.J. Lippard (2003) *J. Am. Chem. Soc.*, vol. 125, p. 14082 – 'Theoretical study of cisplatin binding to purine bases: Why does cisplatin prefer guanine over adenine?'

T.W. Hambley (2001) *J. Chem. Soc., Dalton Trans.*, p. 2711 – 'Platinum binding to DNA: Structural controls and consequences'.

L. Kelland (2007) *Nature Rev. Cancer*, vol. 7, p. 573 – 'The resurgence of platinum-based cancer chemotherapy'.

A.V. Klein and T.W. Hambley (2009) *Chem. Rev.*, vol. 109, p. 4911 – 'Platinum drug distribution in cancer cells and tumors'.

B. Lippert, ed. (2000) *Cisplatin*, Wiley-VCH, Weinheim.

J.B. Mangrum and N.P. Farrell (2010) *Chem. Commun.*, vol. 46, p. 6640 – 'Excursions in polynuclear platinum DNA binding'.

S.H. van Rijt and P.J. Sadler (2009) *Drug Discov. Today*, vol. 14, p. 1089 – 'Current applications and future potential for bioinorganic chemistry in the development of anticancer drugs'.

S. van Zutphen and J. Reedijk (2005) *Coord. Chem. Rev.*, vol. 249, p. 2845 – 'Targeting platinum anti-tumour drugs: Overview of strategies employed to reduce systemic toxicity'.

[Pt(acac)$_2$]. Palladium(II) and platinum(II) acetates are trimeric and tetrameric respectively. The Pd atoms in [Pd(O$_2$CMe)$_2$]$_3$ are arranged in a triangle with each Pd$\cdots$Pd (non-bonded, 310–317 pm) bridged by two [MeCO$_2$]$^-$ ligands giving square planar coordination. In [Pt(O$_2$CMe)$_2$]$_4$, the Pt atoms form a square (Pt–Pt bond lengths = 249 pm) with two [MeCO$_2$]$^-$ bridging each edge. Palladium(II) acetate is an important industrial catalyst for the conversion of ethene to vinyl acetate.

Zeise's salt, K[PtCl$_3$(η^2-C$_2$H$_4$)], is a well-known organometallic Pt(II) complex and is discussed in Section 24.10.

Self-study exercises

1. Explain why 4-coordinate Pt(II) complexes are usually diamagnetic.

    [*Ans.* See Section 20.1 and following text]

2. Each of the mononuclear complexes **A** and **B** has the molecular formula C$_{12}$H$_{30}$Cl$_2$P$_2$Pt. The ^{31}P spectrum of each complex shows a singlet (δ –11.8 ppm for **A**, δ –3.1 ppm for **B**) superimposed on a doublet. For **A**, J_{PtP} = 2400 Hz, and for **B**, J_{PtP} = 3640 Hz. Suggest structures for **A** and **B**, and explain how the spectra and the different coupling constants arise.

    [*Ans.* See Box 19.2]

Platinum(−II)

The relativistic contraction of the 6s orbital (see Box 13.3) is greatest for Pt and Au. Among the properties that this influences† are enthalpies of electron attachment. In contrast to other d-block metals, Pt and Au possess *negative* values for the enthalpy of attachment of the first electron at 298 K. For reaction 22.142, $\Delta_{EA}H^{\circ}$ = –205 kJ mol^{-1}, a value comparable with that of sulfur ($\Delta_{EA}H^{\circ}$(S, g) = –201 kJ mol^{-1}) and greater than that for oxygen (–141 kJ mol^{-1}). Since the chalcogens readily form X^{2-} ions, it has been reasoned that Pt^{2-} may also be formed.‡

$$Pt(g) + e^- \longrightarrow Pt^-(g) \qquad (22.142)$$

The first ionization energy of Cs is the lowest of any element (see Fig. 1.16). Cs reacts with Pt sponge (a porous form of the metal with a large surface area) at 973 K followed by slow cooling to give crystals of Cs$_2$Pt. The red colour and transparency of the crystals provide evidence for a distinct band gap, supporting complete charge separation and the

formation of Cs$^+$ and Pt^{2-} ions. In the solid state, each Pt^{2-} ion lies within a tricapped trigonal prismatic array of Cs$^+$ ions. Barium also reacts with Pt at high temperatures, and forms Ba$_2$Pt which is formulated as (Ba^{2+})$_2$Pt^{2-}·2e$^-$. The free electrons in this system result in metallic behaviour.

22.12 Group 11: silver and gold

The metals

Silver and gold are generally inert, and are not attacked by O$_2$ or non-oxidizing acids. Silver dissolves in HNO$_3$ and liberates H$_2$ from concentrated HI owing to the formation of stable iodido complexes. Where sulfide (e.g. as H$_2$S) is present, Ag tarnishes as a surface coating of Ag$_2$S forms. Gold dissolves in concentrated HCl in the presence of oxidizing agents due to the formation of chlorido complexes (eq. 22.143).

$$[AuCl_4]^- + 3e^- \rightleftharpoons Au + 4Cl^- \qquad E^{\circ}_{[Cl^-] = 1} = +1.00\,V$$
$$(22.143)$$

Both metals react with halogens (see below), and gold dissolves in liquid BrF$_3$, forming [BrF$_2$]$^+$[AuF$_4$]$^-$. The dissolution of Ag and Au in cyanide solutions in the presence of air is used in their extraction from crude ores (eq. 22.4).

Stable oxidation states for the group 11 metals differ: in contrast to the importance of Cu(II) and Cu(I), silver has only one stable oxidation state, Ag(I), and for gold, Au(III) and Au(I) are dominant, with Au(III) being the more stable. Relativistic effects (see Box 13.3) are considered to be important in stabilizing Au(III). As we have already noted, discussing oxidation states of the heavy d-block metals in terms of independently obtained physicochemical data is usually impossible owing to the absence of *IE* values and the scarcity of simple ionic compounds or aqua ions. Data are more plentiful for Ag than for many of the heavier metals, and some comparisons with Cu are possible. Although the enthalpy of atomization of Ag < Cu (Table 22.3), the greater ionic radius for the silver ion along with relevant ionization energies (Table 22.3) make Ag more noble than Cu (eqs. 22.144–22.146). Gold is more noble still (eq. 22.147).

Table 22.3 Selected physical data for Cu and Ag.

Quantity	Cu	Ag
IE_1 / kJ mol^{-1}	745.5	731.0
IE_2 / kJ mol^{-1}	1958	2073
IE_3 / kJ mol^{-1}	3555	3361
$\Delta_a H^{\circ}$(298 K) / kJ mol^{-1}	338	285

---

† For a discussion of the consequences of relativistic effects, see: P. Pyykkö (1988) *Chem. Rev.*, vol. 88, p. 563; P. Pyykkö (2004) *Angew. Chem. Int. Ed.*, vol. 43, p. 4412; P. Pyykkö (2008) *Chem. Soc. Rev.*, vol. 37, p. 1967.
‡ See: A. Karpov, J. Nuss and U. Wedig (2003) *Angew. Chem. Int. Ed.*, vol. 42, p. 4818.

$$Ag^+ + e^- \rightleftharpoons Ag \qquad E^o = +0.80\,V \qquad (22.144)$$

$$Cu^+ + e^- \rightleftharpoons Cu \qquad E^o = +0.52\,V \qquad (22.145)$$

$$Cu^{2+} + e^- \rightleftharpoons Cu^+ \qquad E^o = +0.34\,V \qquad (22.146)$$

$$Au^+ + e^- \rightleftharpoons Au \qquad E^o = +1.69\,V \qquad (22.147)$$

Gold(V) and silver(V)

Gold(V) is found only in AuF_5 and $[AuF_6]^-$ (eqs. 22.148 and 22.149). AuF_5 is highly reactive and possesses dimeric structure **22.83** in the solid state.

$$Au + O_2 + 3F_2 \xrightarrow{670\,K} [O_2]^+[AuF_6]^- \xrightarrow{430\,K} AuF_5 + O_2 + \tfrac{1}{2}F_2 \qquad (22.148)$$

$$2Au + 7KrF_2 \xrightarrow[-5\,Kr]{293\,K} 2[KrF]^+[AuF_6]^-$$

$$\xrightarrow{335\,K} 2AuF_5 + 2Kr + 2F_2 \qquad (22.149)$$

(22.83)

It has been reported that AuF_5 reacts with atomic F to produce AuF_7. However, the results of DFT calculations (see Section 4.13) show that elimination of F_2 from AuF_7 is a highly exothermic reaction with a low activation energy. Hence, the report of AuF_7 is probably in error (see Fig. 22.16).[†]

Gold(III) and silver(III)

For gold, AuF_3, $AuCl_3$ and $AuBr_3$ are known, but AgF_3 is the only high oxidation state halide of silver. It is made in anhydrous HF by treating $K[AgF_4]$ with BF_3, $K[AgF_4]$ being prepared from fluorination of a KCl and AgCl mixture. Red AgF_3 is diamagnetic (d^8) and isostructural with AuF_3. Diamagnetic $K[AuF_4]$ contains square planar anions. Gold(III) fluoride is made from Au with F_2 (1300 K, 15 bar) or by reaction 22.150. It is a polymer consisting of helical chains. Part of one chain is shown in **22.84**. Each Au(III) centre is in a square planar environment (Au–F = 191–203 pm), with additional weak Au---F interactions (269 pm) between chains.

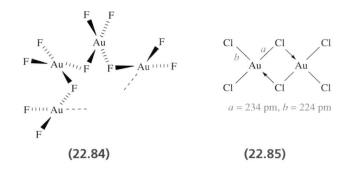

(22.84)   **(22.85)**

$$Au \xrightarrow{BrF_3} [BrF_2][AuF_4] \xrightarrow{330\,K} AuF_3 \qquad (22.150)$$

Red $AuCl_3$ and brown $AuBr_3$ (made by direct combination of the elements) are diamagnetic, planar dimers (**22.85**). In hydrochloric acid, $AuCl_3$ forms $[AuCl_4]^-$. This reacts with Br^- to give $[AuBr_4]^-$, but with I^- to yield AuI and I_2. The acid $HAuCl_4 \cdot xH_2O$ (tetrachloridoauric acid), its bromido analogue, $K[AuCl_4]$ and $AuCl_3$ are commercially available and are valuable starting materials in Au(III) and Au(I) chemistry.

The hydrated oxide $Au_2O_3 \cdot H_2O$ is precipitated by alkali from solutions of $Na[AuCl_4]$, and reacts with an excess of $[OH]^-$ to give $[Au(OH)_4]^-$. $Au_2O_3 \cdot H_2O$ is the only established oxide of gold. For Ag, the thermodynamic stability of the oxides is lowest for Ag(III): Ag_2O_3 is thermodynamically unstable with respect to decomposition to Ag_2O and O_2. The oxide, AgO, is the mixed oxidation state $Ag^{III}Ag^IO_2$ (see later).[†]

Limited numbers of Ag(III) complexes are known. Examples are paramagnetic CsK_2AgF_6 and diamagnetic $K[AgF_4]$. Numerous gold(III) complexes have been made, and square planar coordination (d^8 metal centre) predominates. The anions $[AuX_4]^-$ (X = F, Cl, Br, see above) can be made by oxidation of Au metal (e.g. eq. 22.143). The unstable $[AuI_4]^-$ is made by treating $[AuCl_4]^-$ with anhydrous, liquid HI. Other simple complexes include $[Au(CN)_4]^-$ (from $[AuCl_4]^-$ with $[CN]^-$), $[Au(NCS\text{-}S)_4]^-$, $[Au(N_3)_4]^-$ and $[Au(NO_3\text{-}O)_4]^-$ (eq. 22.151). Complexes of type R_3PAuCl_3 can be made by oxidative addition of Cl_2 to Ph_3PAuCl.

$$Au \xrightarrow{N_2O_5} [NO_2][Au(NO_3)_4] \xrightarrow{KNO_3} K[Au(NO_3)_4] \qquad (22.151)$$

Most compounds which may appear to contain gold(II) are mixed-valence compounds, e.g. '$AuCl_2$' is actually the tetramer $(Au^I)_2(Au^{III})_2Cl_8$ (**22.86**), and $CsAuCl_3$ is $Cs_2[AuCl_2][AuCl_4]$. Both compounds contain square

---

[†] See: S. Riedel and M. Kaupp (2006) *Inorg. Chem.*, vol. 45, p. 1228 – 'Has AuF_7 been made?'

[†]For a thermodynamic treatment, see: D. Tudela (2008) *J. Chem. Educ.*, vol. 85, p. 863.

BIOLOGY AND MEDICINE

Box 22.10 Bactericidal effects of silver sols

Solutions of silver sols (i.e. colloidal dispersions of Ag in aqueous solution) have some application as bactericidal agents. The active agent is Ag^+, with the metabolism of bacteria being disrupted by its presence. A silver sol exhibits a large metal surface area, and oxidation by atmospheric O_2 occurs to some extent to give Ag_2O. While this is only sparingly soluble in water, the concentration of Ag^+ in solution is sufficient to provide the necessary bactericidal effects. For example, Johnson & Johnson market the dressing 'Actisorb Silver' which consists of activated carbon impregnated with metallic silver, manufactured as a carbonized fabric enclosed in nylon fibres. The photograph shows a coloured scanning electron micrograph (SEM) of an 'Actisorb Silver' wound dressing. The charcoal and silver can be seen as the black layer while the nylon fibres appear in white. The dressing is designed not only to kill bacteria, but also to absorb toxins and to minimize the smell of the wound. Such dressings are used on infected wounds (e.g. fungal lesions and leg ulcers). Over-exposure to Ag, however, results in argyria: this is a darkening of the skin, caused by absorption of metallic Ag, which cannot be medically reversed.

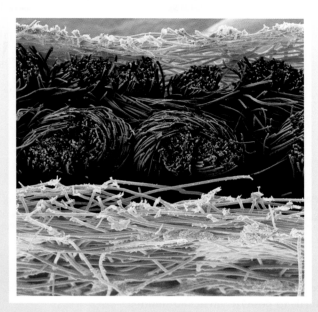

An SEM image of a silver-impregnated 'Actisorb Silver' wound dressing.

---

planar Au(III) and linear Au(I), and their dark colours arise from charge transfer between Au(I) and Au(III).

$$(22.86)$$

Gold(II) and silver(II)

True gold(II) compounds are rare (see above) and are represented by $[AuXe_4]^{2+}$ (eq. 18.24), and *trans*- and *cis*-$[AuXe_2]^{2+}$ (structure 18.20). In anhydrous HF/SbF_5, AuF_3 is reduced or partially reduced to give $Au_3F_8 \cdot 2SbF_5$, $Au_3F_7 \cdot 3SbF_5$ or $[Au(HF)_2][SbF_6]_2 \cdot 2HF$ depending on reaction conditions. For many years, $AuSO_4$ was formulated as the mixed-valence compound $Au^IAu^{III}(SO_4)_2$, but in 2001, a crystal structure determination confirmed it to be an Au(II) compound containing an $[Au_2]^{4+}$ unit (Fig. 22.30a). This dinuclear core is present in a range of complexes that formally contain Au(II). However, Fig. 22.30b shows a rare example of a *mononuclear* Au(II) complex; in the solid state, a Jahn–Teller distortion is observed as expected for a d^9 electronic configuration (Au$-$S$_{axial}$ = 284 pm, Au$-$S$_{equatorial}$ = 246 pm).

Silver(II) is stabilized in the compounds $Ag^{II}M^{IV}F_6$ (M = Pt, Pd, Ti, Rh, Sn, Pb) in which each Ag(II) and M(IV) centre is surrounded by six octahedrally arranged F atoms. Brown AgF_2 is obtained by reacting F_2 and Ag at 520 K, but is instantly decomposed by H_2O. AgF_2 is paramagnetic (Ag^{2+}, d^9) but the magnetic moment of 1.07 μ_B reflects antiferromagnetic coupling. In solid AgF_2, the environments of Ag^{2+} centres are Jahn–Teller distorted (elongated) octahedral, Ag$-$F = 207 and 259 pm. The $[AgF]^+$ ion has been characterized in $[AgF]^+[AsF_6]^-$ which, in anhydrous HF, undergoes partial disproportionation to give $[AgF]^+_2[AgF_4]^-[AsF_6]^-$ (eq. 22.152). Crystalline $[AgF]_2[AgF_4][AsF_6]$ consists of polymeric $[AgF]_n^{n+}$ chains with linear Ag(II), square planar $[Ag^{III}F_4]^-$ ions and octahedral $[AsF_6]^-$ ions.

$$4[AgF][AsF_6] \xrightarrow[\text{HF}]{\text{anhydrous}} Ag[AsF_6] + [AgF]_2[AgF_4][AsF_6]$$

$$+ 2AsF_5 \qquad (22.152)$$

The black solid of composition AgO which is precipitated when $AgNO_3$ is warmed with persulfate solution is diamagnetic and contains Ag(I) (with two O nearest neighbours) and Ag(III) (4-coordinate). However, when AgO dissolves in aqueous $HClO_4$, the paramagnetic $[Ag(OH_2)_4]^{2+}$ ion is formed. This ion (eq. 22.153), AgO and Ag(II) complexes

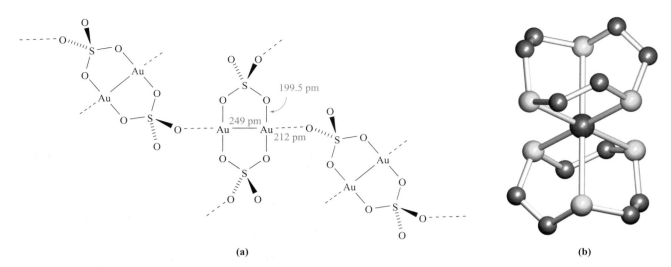

(a)   **(b)**

Fig. 22.30 (a) Schematic representation of part of one chain in the solid state structure of $AuSO_4$, which contains $[Au_2]^{4+}$ units; the $[SO_4]^{2-}$ act as both bridging and monodentate ligands. (b) The structure (X-ray diffraction) of $[AuL_2]^{2+}$ (L = 1,4,7-trithiacyclononane, see Fig. 22.31) determined for the $[BF_4]^-$ salt [A.J. Blake *et al.* (1990) *Angew. Chem. Int. Ed.*, vol. 29, p. 197]; H atoms are omitted; colour code: Au, red; S, yellow; C, grey.

are very powerful oxidizing agents, e.g. AgO converts Mn(II) to $[MnO_4]^-$ in acidic solution.

$$Ag^{2+} + e^- \rightleftharpoons Ag^+ \qquad E^o = +1.98\,V \qquad (22.153)$$

Silver(II) complexes can be precipitated from aqueous solution of Ag(I) salts by using a very strong oxidizing agent in the presence of an appropriate ligand. They are paramagnetic and usually square planar. Examples include $[Ag(py)_4]^{2+}$, $[Ag(bpy)_2]^{2+}$ and $[Ag(bpy)(NO_3\text{-}O)_2]$.

Self-study exercises

1. The compound $AgRhF_6$ is prepared from $RhCl_3$, Ag_2O (2:1 ratio) and F_2 at 770 K for 15 days. For Ag, Rh and F, what oxidation state changes occur in this reaction?

2. Justify why the following compound is classed as containing Au(II).

Me   Me
  \ /
   P
   |
Cl — Au — Au — Cl
   |
   P
  / \
Me   Me

3. The reaction of Au(CO)Cl with $AuCl_3$ gives a diamagnetic product formulated as '$AuCl_2$'. Comment on these results.
   [*Ans.* See: D.B. Dell'Amico *et al.* (1977) *J. Chem. Soc., Chem. Commun.*, p. 31]

Gold(I) and silver(I)

Many Ag(I) salts are familiar laboratory reagents. They are nearly always anhydrous and (except for AgF, $AgNO_3$ and $AgClO_4$) are usually sparingly soluble in water. Topics already covered are:

* solubilities of Ag(I) halides (Section 7.9);
* common-ion effect, exemplified with AgCl (Section 7.10);
* half-cells involving Ag(I) halides (Section 8.3);
* Frenkel defects illustrated by the structure of AgBr (Section 6.17).

Yellow AgF can be made from the elements or by dissolving AgO in HF. It adopts an NaCl structure (Fig. 6.16) as do AgCl and AgBr. Precipitation reactions 22.154 are used to prepare AgCl (white), AgBr (pale yellow) and AgI (yellow); for K_{sp} values, see Table 7.4.

$$AgNO_3(aq) + X^-(aq) \longrightarrow AgX(s) + [NO_3]^-(aq)$$
$$(X = Cl, Br, I) \qquad (22.154)$$

Silver(I) iodide is polymorphic. The stable form at 298 K and 1 bar pressure, γ-AgI, has a zinc blende structure (Fig. 6.19). At high pressures, this converts to δ-AgI with an NaCl structure, the Ag−I distance increasing from 281 to 304 pm. Between 409 and 419 K, the β-form exists with a wurtzite structure (Fig. 6.21). Above 419 K, α-AgI becomes a fast ion electrical conductor (see Section 28.2), the conductivity at the transition temperature increasing by a factor of ≈4000. In this form, the I^- ions occupy positions in a CsCl structure (Fig. 6.17) but the much smaller Ag^+ ions move freely between sites of 2-, 3- or 4-coordination

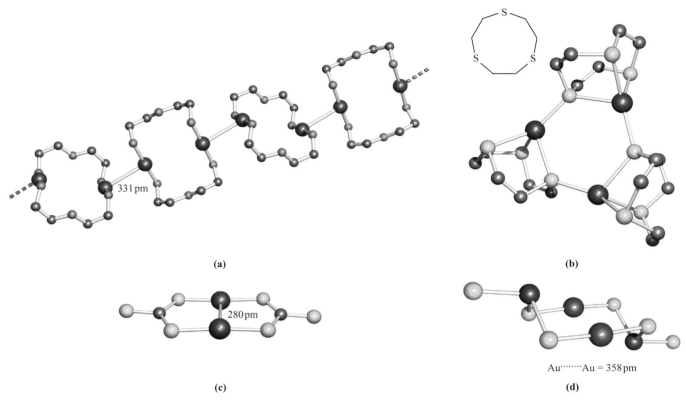

Fig. 22.31 The structures (X-ray diffraction) of (a) the cation in $[Au_2(H_2NCH_2CH_2NHCH_2CH_2NH_2)_2][BF_4]_2$ (part of one chain in which dimers are connected by weak Au–Au interactions is shown) [J. Yau *et al.* (1995) *J. Chem. Soc., Dalton Trans.*, p. 2575], (b) the trimer $[Ag_3L_3]^{3+}$ where L is the sulfur-containing macrocycle 1,4,7-trithiacyclononane shown in the inset [H.-J. Kuppers *et al.* (1987) *Angew. Chem. Int. Ed.*, vol. 26, p. 575], (c) the planar $[Au_2(CS_3)_2]^{2-}$ (from the $[(Ph_3P)_2N]^+$ salt) [J. Vicente *et al.* (1995) *J. Chem. Soc., Chem. Commun.*, p. 745], and (d) $[Au_2(TeS_3)_2]^{2-}$ (from the $[Me_4N]^+$ salt) [D.-Y. Chung *et al.* (1995) *Inorg. Chem.*, vol. 34, p. 4292]. Hydrogen atoms are omitted for clarity; colour code: Au, red; Ag, dark blue; S, yellow; N, light blue; Te, dark blue; C, grey.

among the easily deformed I^- ions. The high-temperature form of Ag_2HgI_4 shows similar behaviour.

Although gold(I) fluoride has not been isolated, it has been prepared by laser ablation of Au metal in the presence of SF_6. From its microwave spectrum, an equilibrium Au–F bond length of 192 pm has been determined from rotational constants. Yellow AuCl, AuBr and AuI can be made by reactions 22.155 and 22.156. Overheating AuCl and AuBr results in decomposition to the elements. Crystalline AuCl, AuBr and AuI possess zigzag chain structures (**22.87**). The halides disproportionate when treated with H_2O. Disproportionation of Au(I) (eq. 22.157) does not parallel that of Cu(I) to Cu and Cu(II).

Au–I = 262 pm; $\angle$Au-I-Au = 72°

(22.87)

$$AuX_3 \xrightarrow{\Delta} AuX + X_2 \quad (X = Cl, Br) \quad (22.155)$$

$$2Au + I_2 \xrightarrow{\Delta} 2AuI \quad (22.156)$$

$$3Au^+ \longrightarrow Au^{3+} + 2Au \quad (22.157)$$

Silver(I) oxide is precipitated by adding alkali to solutions of Ag(I) salts. It is a brown solid which decomposes above 423 K. Aqueous suspensions of Ag_2O are alkaline and absorb atmospheric CO_2. In alkalis, Ag_2O dissolves, forming $[Ag(OH)_2]^-$. No gold(I) oxide has been confirmed.

In complexes of gold(I), linear coordination is usual, although Au····Au interactions in the solid state are a common feature (Fig. 22.31a). Trigonal planar and tetrahedral complexes are also found. For Ag(I), linear and tetrahedral complexes are common, but the metal ion can tolerate a range of environments and coordination numbers from 2 to 6 (the latter is rare) are well established. Both Ag(I) and Au(I) favour soft donor atoms, and there are many complexes with M–P and M–S bonds, including some thiolate complexes with intriguing structures (Fig. 16.12d and Fig. 22.31b).

Dissolving Ag_2O in aqueous NH_3 gives the linear $[Ag(NH_3)_2]^+$, but in liquid NH_3 tetrahedral $[Ag(NH_3)_4]^+$ forms. Trigonal planar $[Ag(NH_3)_3]^+$ can be isolated as the nitrate. Equation 22.4 showed the formation of $[M(CN)_2]^-$ (M = Ag, Au) during metal extraction. The cyanido complexes are also made by dissolving MCN in aqueous KCN. Both AgCN and AuCN are linear polymers (**22.88**).

BIOLOGY AND MEDICINE

Box 22.11 Gold complexes in medicine

Gold(I) complexes containing thiolate, [RS]⁻, ligands are used as *disease-modifying antirheumatic drugs* in the treatment of rheumatoid arthritis. Crucial to their application is the lability of the ligands: gold(I)-containing drugs are *prodrugs*, i.e. they are not themselves pharmacologically active, but instead undergo ligand exchange to produce an active drug *in vivo*. The mode of action is not fully determined, but it is proposed that the soft *S*-donors of cysteine residues in proteins displace soft *S*-donor ligands in the prodrug. This can be represented by the equilibrium:

$$RSAuL + RS'H \rightleftharpoons R'SAuL + RSH$$

However, a structural (X-ray diffraction) study of the interaction of Et_3PAuCl with the enzyme cyclophilin reveals that a histidine residue (see Table 29.2) displaces Cl^-. Unexpectedly, the gold(I) unit is bound by a histidine *N*-donor in preference to the available cysteine *S*-donors.

Three important rheumatoid arthritic drugs are Auranofin, Myochrysine and Solganol. Auranofin is orally administered, and has been approved by the US Food and Drug Administration (FDA) since 1976. It is lipophilic and is readily absorbed by the body from the intestine after ingestion. Auranofin is typically used to treat adult patients who show no response to non-steroidal anti-inflammatory drugs (NSAIDs).

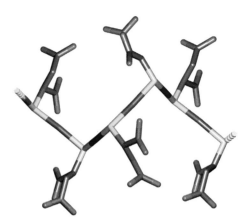

Auranofin

Whereas Auranofin is monomeric, the gold(I) rheumatoid arthritic prodrugs Solganol and Myochrysine are polymeric (see end-of-chapter problem 22.33) and are administered by injection to both adults and children.

Solganol

Myochrysine

The polymeric structure of Myochrysine:

Colour code: Au, red; S, yellow; O, red; C, grey (H atoms omitted)
[Data: R. Bau (1998) *J. Am. Chem. Soc.*, vol. 120, p. 9380]

Thioneins are small protein molecules in which about one-third of the amino acids are cysteine residues. Nature has evolved thioneins to transport and remove toxic metal ions such as Hg(II) and Cd(II), but, like the latter, Au(I) is soft and readily interacts with thioneins. As a result, thioneins are implicated in cell resistance to gold(I) drugs.

The use of Au(I) complexes to control rheumatoid arthritis is well established, but more recently, it has been recognized that some Au(III) complexes have the potential to function as anti-cancer drugs. There are electronic and structural parallels between Au(III) and Pt(II): both are d^8 metal ions with a preference for square planar coordination. Building upon the medicinal successes of cisplatin, carboplatin and oxaliplatin (see Box 22.9), development of Au(III)-containing anti-cancer agents is now an active area of research. Possible candidates include those shown below:

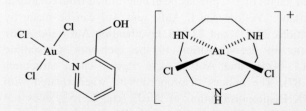

For many of the Au(III) compounds investigated, a problem is their stability under physiological conditions.

Further reading

H.E. Abdou *et al.* (2009) *Coord. Chem. Rev.*, vol. 253, p. 1661 – 'Structures and properties of gold(I) complexes of interest in biochemical applications'.

C. Gabbiani, A. Casini and L. Messori (2007) *Gold Bull.*, vol. 40, p. 73 – 'Gold(III) compounds as anticancer drugs'.

I. Ott (2009) *Coord. Chem. Rev.*, vol. 253, p. 1670 – 'On the medicinal chemistry of gold complexes as anticancer drugs'.

Their solid state structures suffer from disorder problems, but total neutron diffraction[†] has been used to give the accurate structural data shown in diagram **22.88**. The fact that the Au−C/N distance is smaller than the Ag−C/N bond length is attributed to relativistic effects. The same phenomenon is observed in the discrete, linear $[Au(CN)_2]^-$ and $[Ag(CN)_2]^-$ ions.

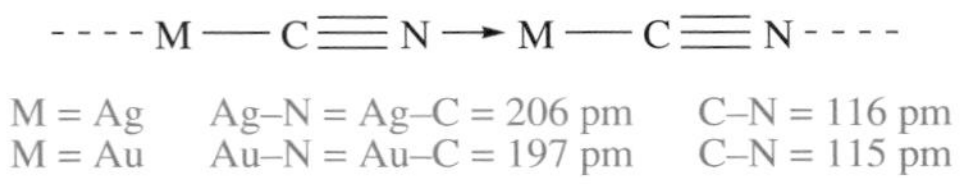

$$----M----C\equiv N\rightarrow M----C\equiv N----$$

M = Ag   Ag–N = Ag–C = 206 pm   C–N = 116 pm
M = Au   Au–N = Au–C = 197 pm   C–N = 115 pm

(22.88)

In contrast to the bridging mode of $[CN]^-$ in $[Ag(CN)_2]^-$, the complex $[Ag(CN)(NH_3)]$ crystallizes with discrete linear molecules. At room temperature, $[Ag(CN)(NH_3)]$ rapidly loses NH_3, decomposing to AgCN.

Dissolution of AgX in aqueous halide solutions produces $[AgX_2]^-$ and $[AgX_3]^{2-}$. In aqueous solutions, the ions $[AuX_2]^-$ (X = Cl, Br, I) are unstable with respect to disproportionation but can be stabilized by adding excess X^- (eq. 22.158).

$$3[AuX_2]^- \rightleftharpoons [AuX_4]^- + 2Au + 2X^- \qquad (22.158)$$

Routes to Au(I) complexes often involve reduction of Au(III) as illustrated by the formation of R_3PAuCl and R_2SAuCl species (eqs. 22.159 and 22.160).

$$[AuCl_4]^- + 2R_3P \rightarrow [R_3PAuCl] + R_3PCl_2 + Cl^- \quad (22.159)$$

$$[AuCl_4]^- + 2R_2S + H_2O$$
$$\rightarrow [R_2SAuCl] + R_2SO + 2H^+ + 3Cl^- \qquad (22.160)$$

Molecules of R_3PAuCl and R_2SAuCl (for which many examples with different R groups are known) contain linear Au(I), but aggregation in the solid state by virtue of Au⋯⋯Au interactions (similar to those in Fig. 22.31a) is often observed. Other than the expectation of a linear Au(I) environment, structures may be hard to predict. For example in $[Au_2(CS_3)_2]^{2-}$ (made from $[Au(SH)_2]^-$ and CS_2), there is a short Au−Au contact, but in $[Au_2(TeS_3)_2]^{2-}$ (made from AuCN and $[TeS_3]^{2-}$), the Au(I) centres are out of bonding range (Figs. 22.31c, d).

Gold(−I) and silver(−I)

Relativistic effects (see Box 13.3) have a profound influence on the ability of gold to exist in the −1 oxidation state.[‡] The

[†] See: S.J. Hibble, A.C. Hannon and S.M. Cheyne (2003) *Inorg. Chem.*, vol. 42, p. 4724; S.J. Hibble, S.M. Cheyne, A.C. Hannon and S.G. Eversfield (2002) *Inorg. Chem.*, vol. 41, p. 1042.
[‡] For relevant discussions of relativistic effects, see footnote references on p. 833

enthalpy of attachment of the first electron to Au (eq. 22.161) is −223 kJ mol⁻¹, a value that lies between those for iodine (−295 kJ mol⁻¹) and sulfur (−201 kJ mol⁻¹).

$$Au(g) + e^- \rightarrow Au(g) \qquad (22.161)$$

The choice of metal for the reduction of Au to Au^- is Cs, the latter having the lowest first ionization energy of any element. Caesium also reduces Pt (see the end of Section 22.11). Caesium auride, CsAu, can be formed from the elements at 490 K. It adopts a CsCl structure (see Fig. 6.17) and is a semiconductor with a band gap of 2.6 eV. Golden-brown CsAu dissolves in liquid NH_3 to give yellow solutions from which a blue ammoniate $CsAu\cdot NH_3$ can be crystallized. The Cs^+ ion in CsAu can be exchanged for $[Me_4N]^+$ using an ion-exchange resin. Crystalline $[Me_4N]^+Au^-$ is isostructural with $[Me_4N]^+Br^-$.

Although the argentide anion, Ag^-, has not yet been isolated in a crystalline compound, spectroscopic and electrochemical data have provided evidence for its formation in liquid NH_3.

22.13 Group 12: cadmium and mercury

The metals

Cadmium is chemically very like Zn, and any differences are attributable to the larger sizes of the Cd atom and Cd^{2+} ion versus Zn. Among the group 12 metals, Hg is distinct. It bears some resemblance to Cd, but in many respects is very like Au and Tl. It has been suggested that the relative inertness of Hg towards oxidation is a manifestation of the thermodynamic $6s$ inert pair effect (see Box 13.4).

Cadmium is a reactive metal and dissolves in non-oxidizing and oxidizing acids, but unlike Zn, it does not dissolve in aqueous alkali. In moist air, Cd slowly oxidizes, and when heated in air, it forms CdO. When heated, Cd reacts with the halogens and sulfur.

Mercury is less reactive than Zn and Cd. It is attacked by oxidizing (but not non-oxidizing) acids, with products dependent on conditions, e.g. with dilute HNO_3, Hg forms $Hg_2(NO_3)_2$ (containing $[Hg_2]^{2+}$, see below) but with concentrated HNO_3, the product is $Hg(NO_3)_2$. Reaction of the metal with hot concentrated H_2SO_4 gives $HgSO_4$ and SO_2. Mercury reacts with the halogens (eqs. 22.162 and 22.163). It combines with O_2 at 570 K to give HgO, but at higher temperatures HgO decomposes back to the elements, and, if sulfur is present, HgS is produced rather than the oxide.

$$Hg + X_2 \xrightarrow{\Delta} HgX_2 \qquad (X = F, Cl, Br) \qquad (22.162)$$

$$3Hg + 2I_2 \xrightarrow{\Delta} HgI_2 + Hg_2I_2 \qquad (22.163)$$

Mercury dissolves many metals to give *amalgams* (see Box 22.3). In the Na−Hg system, for example, Na_3Hg_2,

NaHg and NaHg$_2$ have been characterized. Solid Na$_3$Hg$_2$ contains square $[Hg_4]^{6-}$ units (Hg$-$Hg $=$ 298 pm), the structure and stability of which have been rationalized in terms of aromatic character.

For cadmium, the +2 oxidation state is of most importance, but compounds of Hg(I) and Hg(II) are both well known. Mercury is unique among the group 12 metals in forming a stable $[M_2]^{2+}$ ion. There is evidence for $[Zn_2]^{2+}$ and $[Cd_2]^{2+}$ in metal–metal halide melts, and Cd$_2$[AlCl$_4$] has been isolated from a molten mixture of Cd, CdCl$_2$ and AlCl$_3$. However, it is not possible to obtain $[Zn_2]^{2+}$ and $[Cd_2]^{2+}$ in aqueous solution. Species which formally contain Zn$_2^{2+}$ cores are known (see 'Zinc(I)' in Section 21.13). Force constants (60, 110 and 250 N m^{-1} for M $=$ Zn, Cd and Hg calculated from Raman spectra of $[M_2]^{2+}$) show that the bond in $[Hg_2]^{2+}$ is stronger than those in $[Zn_2]^{2+}$ and $[Cd_2]^{2+}$. However, given that Hg has the lowest value of $\Delta_a H^\circ$ of all the *d*-block metals (Table 6.2), the stability of $[Hg_2]^{2+}$ (**22.89**) is difficult to rationalize. Other polycations of mercury are known; $[Hg_3]^{2+}$ (**22.90**) is formed as the $[AlCl_4]^-$ salt in molten Hg, HgCl$_2$ and AlCl$_3$, and $[Hg_4]^{2+}$ (**22.91**) is produced as the $[AsF_6]^-$ salt from reaction of Hg with AsF$_5$ in liquid SO$_2$.

$$\left[\text{Hg} \text{—} \text{Hg} \right]^{2+} \qquad \left[\text{Hg} \text{—} \text{Hg} \text{—} \text{Hg} \right]^{2+}$$
$$\text{253 pm} \qquad\qquad\qquad \text{255 pm}$$

$$\textbf{(22.89)} \qquad\qquad \textbf{(22.90)}$$

$$\overset{\text{262 pm}}{\left[\text{Hg} \text{—} \underset{\text{259 pm}}{\text{Hg}} \text{—} \text{Hg} \text{—} \text{Hg} \right]^{2+}}$$

$$\textbf{(22.91)}$$

Ionization energies decrease from Zn to Cd but increase from Cd to Hg (Table 22.4). Whatever the origin of the high ionization energies for Hg, it is clear that they far outweigh the small change in $\Delta_a H^\circ$ and make Hg a noble metal. The reduction potentials in Table 22.4 reveal the relative electropositivities of the group 12 metals.

Table 22.4 Selected physical data for the group 12 metals.

Quantity	Zn	Cd	Hg
IE_1/kJ mol^{-1}	906.4	867.8	1007
IE_2/kJ mol^{-1}	1733	1631	1810
$\Delta_a H^\circ$(298 K)/kJ mol^{-1}	130	112	61
E° (M^{2+}/M)/V	−0.76	−0.40	+0.85
r_{ion} for M^{2+}/pm†	74	95	101

† For Hg, the value is based on the structure of HgF$_2$, one of the few mercury compounds with a typical ionic lattice.

Since much of the chemistries of Cd and Hg are distinct, we deal with the two metals separately. In making this decision, we are effectively saying that the consequences of the lanthanoid contraction are of minor significance for the heavier metals of the last group of the *d*-block.

Cadmium(II)

All four Cd(II) halides are known. The action of HF on CdCO$_3$ gives CdF$_2$, and of gaseous HCl on Cd (720 K) yields CdCl$_2$. CdBr$_2$ and CdI$_2$ are formed by direct combination of the elements. White CdF$_2$ adopts a CaF$_2$ structure (Fig. 6.19), while CdCl$_2$ (white), CdBr$_2$ (pale yellow) and CdI$_2$ (white) have layer structures (see Section 6.11). The fluoride is sparingly soluble in water, while the other halides are readily soluble, giving solutions containing aquated Cd^{2+} and a range of halido complexes, e.g. CdI$_2$ dissolves to give an equilibrium mixture of $[Cd(OH_2)_6]^{2+}$, $[Cd(OH_2)_5I]^+$, $[CdI_3]^-$ and $[CdI_4]^{2-}$, while 0.5 M aqueous CdBr$_2$ contains $[Cd(OH_2)_6]^{2+}$, $[Cd(OH_2)_5Br]^+$, $[Cd(OH_2)_5Br_2]$, $[CdBr_3]^-$ and $[CdBr_4]^{2-}$. In contrast to Zn^{2+}, the stability of halido complexes of Cd^{2+} increases from F$^-$ to I$^-$, i.e. Cd^{2+} is a softer metal centre than Zn^{2+} (Table 7.9).

Cadmium(II) oxide (formed by heating Cd in O$_2$, and varying in colour from green to black) adopts an NaCl structure. It is insoluble in H$_2$O and alkalis, but dissolves in acids, i.e. CdO is more basic than ZnO. Addition of dilute alkali to aqueous solutions of Cd^{2+} precipitates white Cd(OH)$_2$, and this dissolves only in *concentrated* alkali to give $[Cd(OH)_4]^{2-}$ (contrast $[Zn(OH)_4]^{2-}$, **21.72**). Equation 22.5 showed the role of Cd(OH)$_2$ in NiCd cells. Yellow CdS (the stable α-form has a wurtzite structure, Fig. 6.21) is commercially important as a pigment and phosphor. CdSe and CdTe are semiconductors (see Section 22.2).

In aqueous solutions, $[Cd(OH_2)_6]^{2+}$ is present and is weakly acidic (eq. 22.164); in concentrated solutions, aquated $[Cd_2(OH)]^{3+}$ is present.

$$[Cd(OH_2)_6]^{2+} + H_2O \rightleftharpoons [Cd(OH_2)_5(OH)]^+ + H^+$$
$$pK_a \approx 9 \qquad (22.164)$$

In aqueous NH$_3$, tetrahedral $[Cd(NH_3)_4]^{2+}$ is present, but at high concentrations, $[Cd(NH_3)_6]^{2+}$ forms. The lack of LFSE for Cd^{2+} (d^{10}) means that a range of coordination geometries is observed. Coordination numbers of 4, 5 and 6 are most common, but higher coordination numbers can be forced upon the metal centre by using macrocyclic ligands. Examples of complexes include:

- tetrahedral: $[CdCl_4]^{2-}$, $[Cd(NH_3)_4]^{2+}$, $[Cd(en)_2]^{2+}$;
- trigonal bipyramidal: $[CdCl_5]^{3-}$;
- octahedral: $[Cd(DMSO-O)_6]^{2+}$, $[Cd(en)_3]^{2+}$, $[Cd(acac)_3]^-$, $[CdCl_6]^{4-}$;
- hexagonal bipyramidal: $[CdBr_2(18\text{-crown-}6)$ (see Section 11.8).

As we have seen previously, formulae can be deceptive in terms of structure, e.g. $[Cd(NH_3)_2Cl_2]$ is polymeric with octahedral Cd^{2+} and bridging chlorido ligands (**22.92**).

(22.92)

Mercury(II)

All four Hg(II) halides can be prepared from the elements. A fluorite structure (Fig. 6.19) is adopted by HgF_2 ($Hg-F = 225$ pm). It is completely hydrolysed by H_2O (eq. 22.165).

$$HgF_2 + H_2O \longrightarrow HgO + 2HF \qquad (22.165)$$

The chloride and bromide are volatile solids, soluble in H_2O (in which they are un-ionized), EtOH and Et_2O. The solids contain HgX_2 units packed to give distorted octahedral Hg(II) centres (two long $Hg-X$ contacts to adjacent molecules). Below 400 K, HgI_2 is red with a layer structure, and above 400 K is yellow with HgI_2 molecules assembled into a lattice with distorted octahedral metal centres. The vapours contain linear HgX_2 molecules with bond distances of 225, 244 and 261 pm for X = Cl, Br and I respectively. Figure 22.32 shows the trend in solubilities of the halides; for HgI_2, $K_{sp} = 2.82 \times 10^{-29}$.

Mercury(II) oxide exists in both a yellow form (formed by heating Hg in O_2 or by thermal decomposition of $Hg(NO_3)_2$) and a red form (prepared by precipitation from alkaline solutions of Hg^{2+}). Both have infinite chain structures (**22.93**) with linear Hg(II). The thermal decomposition of HgO (eq. 22.166) led to the discovery of O_2 by Priestley in 1774.

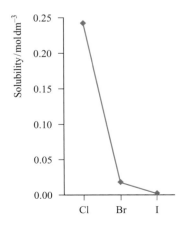

Fig. 22.32 The trend in solubilities of Hg(II) halides in water; HgF_2 decomposes.

$$2HgO \xrightarrow{>670\,K} 2Hg + O_2 \qquad (22.166)$$

Although the oxide dissolves in acids, it is only weakly basic. In aqueous solution, Hg(II) salts that are ionized (e.g. $Hg(NO_3)_2$ and $HgSO_4$) are hydrolysed to a considerable extent and many basic salts are formed, e.g. $HgO \cdot HgCl_2$ and $[O(HgCl)_3]Cl$ (a substituted oxonium salt). Solid $Hg(OH)_2$ is unknown. However, $[Hg(OH)][NO_3] \cdot H_2O$ (hydrated 'basic mercury(II) nitrate') can be isolated. In the solid state, this contains zigzag chains (**22.94**) to which H_2O molecules are loosely connected.

(22.93)   **(22.94)**

In its complexes, Hg(II) (d^{10}) exhibits coordination numbers of 2 to 6. Like Cd^{2+}, Hg^{2+} is a soft metal centre (see Table 7.8 and discussion) and coordination to *S*-donors is especially favoured. Complex chlorides, bromides and iodides are formed in aqueous solution, and the tetrahedral $[HgI_4]^{2-}$ is particularly stable. A solution of $K_2[HgI_4]$ (*Nessler's reagent*) gives a characteristic brown compound, $[Hg_2N]^+I^-$, on treatment with NH_3 and is used in determination of NH_3. In solid $[Hg_2N]I$, the $[Hg_2N]^+$ cations assemble into an infinite network related to that of β-cristobalite (Fig. 6.20c) and containing linear Hg(II). Reaction 22.167 shows the formation of its hydroxide.

$$2HgO + NH_3 \xrightarrow{aq} Hg_2N(OH) + H_2O \qquad (22.167)$$

The salt $[Hg(NH_3)_2]Cl_2$ (eq. 22.168) contains linear $[Hg(NH_3)_2]^{2+}$ ions and dissolves in aqueous NH_3 to give $[Hg(NH_2)]Cl$ which contains polymeric chains (**22.95**).

$$HgCl_2 + 2NH_3(g) \longrightarrow [Hg(NH_3)_2]Cl_2 \qquad (22.168)$$

(22.95)

Examples of Hg(II) complexes illustrating different coordination environments (see Table 7.7 for ligand abbreviations) include:

- linear: $[Hg(NH_3)_2]^{2+}$, $[Hg(CN)_2]$, $[Hg(py)_2]^{2+}$, $[Hg(SEt)_2]$;
- trigonal planar: $[HgI_3]^-$;
- tetrahedral: $[Hg(en)_2]^{2+}$, $[Hg(NCS-S)_4]^{2-}$, $[HgI_4]^{2-}$, $[Hg(S_4-S,S')_2]^{2-}$, $[Hg(Se_4-Se,Se')_2]^{2-}$, $[Hg(phen)_2]^{2+}$;

- trigonal bipyramidal: $[HgCl_2(tpy)]$, $[HgCl_2(dien)]$, $[HgCl_5]^{3-}$;
- square-based pyramidal: $[Hg(OH_2)L]^{2+}$ (L = **22.96**);
- octahedral: $[Hg(en)_3]^{2+}$, *fac*-$[HgL_2]^{2+}$ (L = **22.97**);
- square antiprism: $[Hg(NO_2\text{-}O,O')_4]^{2-}$.

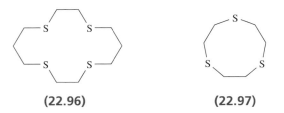

(22.96)   (22.97)

Mercury(I)

The chemistry of Hg(I) is that of the $[Hg_2]^{2+}$ unit which contains an Hg–Hg single bond (**22.89**). The general method of preparation of Hg(I) compounds is by the action of Hg metal on Hg(II) compounds, e.g. reaction 22.169 in which Hg_2Cl_2 (*calomel*) is freed from $HgCl_2$ by washing with hot water. The *standard calomel electrode* (see Box 8.3) is a reference electrode (eq. 22.170) consisting of a Pt wire dipping into Hg in contact with Hg_2Cl_2 and immersed in 1 M KCl solution. This electrode is more convenient to use than the standard hydrogen electrode which requires a source of purified gas.

$$HgCl_2 + Hg \xrightarrow{\Delta} Hg_2Cl_2 \quad (22.169)$$

$$Hg_2Cl_2 + 2e^- \rightleftharpoons 2Hg + 2Cl^-$$

$$E^o = +0.268\,V \text{ (in 1 M aq KCl)} \quad (22.170)$$

Potential diagrams for Hg are shown in scheme 22.171, and the data in acidic solution illustrate that the disproportionation of Hg(I) (eq. 22.172) has a small and positive ΔG^o value at 298 K.

$$(22.171)$$

```
Hg²⁺ ──+0.91──▶ [Hg₂]²⁺ ──+0.77──▶ Hg      pH = 0
      └──────────+0.84──────────▶
HgO ──+0.10──▶ Hg                          pH = 14
```

$$[Hg_2]^{2+} \rightleftharpoons Hg^{2+} + Hg \quad K = 4.3 \times 10^{-3}\ (298\,K) \quad (22.172)$$

Reagents that form insoluble Hg(II) salts or stable Hg(II) complexes upset equilibrium 22.172 and decompose Hg(I) salts, e.g. addition of $[OH]^-$, S^{2-} or $[CN]^-$ results in formation of Hg and HgO, HgS or $[Hg(CN)_4]^{2-}$. The Hg(I)

compounds Hg_2O, Hg_2S and $Hg_2(CN)_2$ are not known. Mercury(II) forms more stable complexes than the larger $[Hg_2]^{2+}$ and relatively few Hg(I) compounds are known. The most important are the halides (**22.98**).[†] Whereas Hg_2F_2 decomposes to Hg, HgO and HF on contact with water, the later halides are sparingly soluble.

X —— Hg —— Hg —— X	X	Hg–Hg / pm
	F	251
	Cl	252
	Br	258
	I	269

(22.98)

Basic mercury(I) has been stabilized in the salt $[Hg_2(OH)][BF_4]$, made by reaction 22.173. In the solid state, zigzag chains (**22.99**) are connected together through weak Hg---O interactions to form layers, with $[BF_4]^-$ ions occupying the spaces between adjacent layers.

$$HgO + Hg + HBF_4 \xrightarrow{H_2O} [Hg_2(OH)][BF_4] \quad (22.173)$$

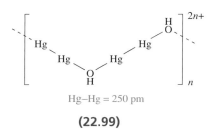

Hg–Hg = 250 pm

(22.99)

Other Hg(I) salts include $Hg_2(NO_3)_2$, Hg_2SO_4 and $Hg_2(ClO_4)_2$. The nitrate is commercially available as the dihydrate, the solid state structure of which contains $[(H_2O)HgHg(OH_2)]^{2+}$ cations. Scheme 22.174 summarizes some reactions of hydrated $Hg_2(NO_3)_2$.

$$Hg_2(NO_3)_2 \begin{cases} \xrightarrow{KSCN} Hg_2(SCN)_2 \xrightarrow{decomp.} Hg(SCN)_2 + Hg \\ \xrightarrow{NaN_3} Hg_2(N_3)_2 \text{ (explosive)} \\ \xrightarrow{H_2SO_4} Hg_2SO_4 \end{cases}$$

$$(22.174)$$

Crystalline, anhydrous $Hg_2(NO_3)_2$ can be prepared by drying $Hg_2(NO_3)_2 \cdot 2H_2O$ over concentrated H_2SO_4.

[†] Theoretical data cast doubt on the reliability of the Hg–Hg bond lengths for X = Br and I; see: M.S. Liao and W.H.E. Schwarz (1997) *J. Alloy. Compd.*, vol. 246, p. 124.

FURTHER READING

See also further reading suggested for Chapters 19 and 20.

M.H. Chisholm and A.M. Macintosh (2005) *Chem. Rev.*, vol. 105, p. 2949 – 'Linking multiple bonds between metal atoms: Clusters, dimers of 'dimers', and higher ordered assemblies'.

F.A. Cotton, G. Wilkinson, M. Bochmann and C. Murillo (1999) *Advanced Inorganic Chemistry*, 6th edn, Wiley Interscience, New York – One of the best detailed accounts of the chemistry of the *d*-block metals.

F.A. Cotton, C.A. Murillo and R.A. Walton (2005) *Multiple Bonds between Metal Atoms*, 3rd edn, Springer, New York.

S.A. Cotton (1997) *Chemistry of Precious Metals*, Blackie, London – Covers descriptive inorganic chemistry (including σ-bonded organometallic complexes) of the heavier group 8, 9, 10 and 11 metals.

A. Dolbecq, E. Dumas, C.R. Mayer and P. Mialane (2010) *Chem. Rev.*, vol. 110, p. 6009 – 'Hybrid organic-inorganic polyoxometallate compounds: From structural diversity to applications'.

J. Emsley (1998) *The Elements*, 3rd edn, Oxford University Press, Oxford – An invaluable source of data for the elements.

N.N. Greenwood and A. Earnshaw (1997) *Chemistry of the Elements*, 2nd edn, Butterworth-Heinemann, Oxford – A very good account including historical, technological and structural aspects; the metals in each triad are treated together.

G.J. Hutchings, M. Brust and H. Schmidbaur, eds. (2008) *Chem. Soc. Rev.*, Issue 9 – A series of reviews entitled: 'Gold: chemistry, materials and catalysis issue'.

C.E. Housecroft (1999) *The Heavier d-Block Metals: Aspects of Inorganic and Coordination Chemistry*, Oxford University Press, Oxford – An introductory text including chapters on aqueous solution species, structure, M–M bonded dimers and clusters, and polyoxometallates.

J.A. McCleverty and T.J. Meyer, eds (2004) *Comprehensive Coordination Chemistry II*, Elsevier, Oxford – Up-to-date reviews of the coordination chemistry of the *d*-block metals are included in volumes 4–6.

F. Mohr, ed. (2009) *Gold Chemistry: Applications and Future Directions in the Life Sciences*, Wiley-VCH, Weinheim – An overview of the chemistry of gold and applications of its compounds.

M.J. Molski and K. Seppelt (2009) *Dalton Trans.*, p. 3379 – 'The transition metal hexafluorides'.

E.A. Seddon and K.R. Seddon (1984) *The Chemistry of Ruthenium*, Elsevier, Amsterdam – An excellent, well-referenced account of the chemistry of Ru.

A.F. Wells (1984) *Structural Inorganic Chemistry*, 5th edn, Clarendon Press, Oxford – An excellent source for detailed structural information of, in particular, binary compounds.

PROBLEMS

22.1 (a) Write out the first row *d*-block metals in sequence and then complete each triad of metals. (b) Between which two metals is the series of lanthanoid metals?

22.2 Briefly discuss trends in (a) metallic radii and (b) values of $\Delta_a H^\circ(298\,K)$ for the *d*-block metals.

22.3 (a) Estimate the value of $\Delta_f H^\circ(WCl_2)$ assuming it to be an ionic compound. Comment on any assumptions made. [Data needed in addition to those in Tables 21.1, 22.1 and the Appendices: $\Delta_f H^\circ(CrCl_2) = -397\,kJ\,mol^{-1}$.] (b) What does your answer to (a) tell you about the likelihood of WCl_2 being ionic?

22.4 Comment on the following observations:
(a) The density of HfO_2 (9.68 g cm^{-3}) is much greater than that of ZrO_2 (5.73 g cm^{-3}).
(b) NbF_4 is paramagnetic but $NbCl_4$ and $NbBr_4$ are essentially diamagnetic.

22.5 Suggest products in the following reactions: (a) CsBr heated with $NbBr_5$ at 383 K; (b) KF and TaF_5 melted together; (c) NbF_5 with bpy at 298 K. (d) Comment on the structures of the group 5 metal halides in the starting materials and give possible structures for the products.

22.6 TaS_2 crystallizes with a layer structure related to that of CdI_2, whereas FeS_2 adopts a distorted NaCl structure. Why would you not expect TaS_2 and FeS_2 to crystallize with similar structure types?

22.7 Comment on the observation that $K_3[Cr_2Cl_9]$ is strongly paramagnetic but $K_3[W_2Cl_9]$ is diamagnetic.

22.8 (a) Interpret the formula $[Mo_6Cl_8]Cl_2Cl_{4/2}$ in structural terms, and show that the formula is consistent with the stoichiometry $MoCl_2$. (b) Show that the $[W_6Br_8]^{4+}$ cluster can be considered to contain W–W single bonds.

22.9 Give a short account of Tc(V) and Re(V) oxido species.

22.10 Briefly summarize similarities and differences between Mn and Tc chemistries.

22.11 Draw the structure of $[Re_2Cl_8]^{2-}$. Discuss the metal–metal bonding in the anion and its consequences on ligand orientation.

22.12 Suggest reasons for the variation in Re—Re bond lengths in the following species: $ReCl_4$ (273 pm), Re_3Cl_9 (249 pm), $[Re_2Cl_8]^{2-}$ (224 pm), $[Re_2Cl_9]^-$ (270 pm) and $[Re_2Cl_4(\mu\text{-}Ph_2PCH_2CH_2PPh_2)_2]$ (224 pm).

22.13 When $K_2[OsCl_4]$ is heated with NH_3 under pressure, compound **A** of composition $Os_2Cl_5H_{24}N_9$ is isolated. Treatment of a solution of **A** with HI precipitates a compound in which three of the five chlorines have been replaced by iodine. Treating 1 mmol of **A** with KOH releases 9 mmol NH_3. Compound **A** is diamagnetic and none of the stronger absorption bands in the IR spectrum is Raman active. Suggest a structure for **A** and account for the diamagnetism.

22.14 Give an account of the halides of Ru and Os.

22.15 (a) Give an account of the methods of synthesis of Rh(IV) and Ir(IV) halides and halido anions. (b) Reaction of $[IrCl_6]^{2-}$ with PPh_3 and $Na[BH_4]$ in EtOH gives $[IrH_3(PPh_3)_3]$. Give the structures of the isomers of this complex and suggest how you would distinguish between them using NMR spectroscopy.

22.16 $[Ir(CN)_6]^{3-}$ has a regular octahedral structure. For $K_3[Ir(CN)_6]$, the wavenumbers corresponding to the C≡N stretching modes are 2167 (A_{1g}), 2143 (E_g) and 2130 (T_{1u}) cm^{-1}. (a) To which point group does $[Ir(CN)_6]^{3-}$ belong? (b) What would you observe in the IR spectrum of $K_3[Ir(CN)_6]$ in the region between 2200 and 2000 cm^{-1}?

22.17 When $RhBr_3$ in the presence of $MePh_2As$ is treated with H_3PO_2, a monomeric compound **X** is formed. **X** contains 2 Br and 3 $MePh_2As$ per Rh, and is a non-electrolyte. Its IR spectrum has a band at 2073 cm^{-1}, and the corresponding band if the complex is made using D_3PO_2 in a deuterated solvent is 1483 cm^{-1}. Spectrophotometric titration of **X** with Br_2 shows that one molecule of **X** reacts with one molecule of Br_2; treating the product with excess mineral acid regenerates $RhBr_3$. What can you conclude about the products?

22.18 (a) Compare the structures of β-$PdCl_2$ and $[Nb_6Cl_{12}]^{2+}$. (b) Discuss, with examples, the existence (or not) of Pt(III) species. (c) Discuss the variation in stereochemistries of Ni(II), Pd(II) and Pt(II) complexes.

22.19 (a) Describe the methods by which *cis*- and *trans*-$[PtCl_2(NH_3)_2]$ can be distinguished from each other and from $[Pt(NH_3)_4][PtCl_4]$. (b) Another possible isomer would be $[(H_3N)_2Pt(\mu\text{-}Cl)_2Pt(NH_3)_2]Cl_2$. What diagnostic data would enable you to rule out its formation?

22.20 Suggest products in the reactions of $K_2[PtCl_4]$ with (a) excess KI; (b) aqueous NH_3; (c) phen; (d) tpy; (e) excess KCN. What are the expected structures of these products?

22.21 Complexes of the type $[PtCl_2(R_2P(CH_2)_nPR_2)]$ may be monomeric or dimeric. Suggest factors that might influence this preference and suggest structures for the complexes.

22.22 Comment on each of the following observations:
(a) Unlike $[Pt(NH_3)_4][PtCl_4]$, $[Pt(EtNH_2)_4][PtCl_4]$ has an electronic absorption spectrum that is the sum of those of the constituent ions.
(b) AgI is readily soluble in saturated aqueous $AgNO_3$, but AgCl is not.
(c) When $Hg(ClO_4)_2$ is shaken with liquid Hg, the ratio [Hg(I)]/[Hg(II)] in the resulting solution is independent of the value of [Hg(II)].

22.23 Discuss the variation in stable oxidation states for the group 11 metals, using examples of metal halides, oxides and complexes to illustrate your answer.

22.24 'The group 12 metals differ significantly from the *d*-block metals in groups 4–11'. Discuss this statement.

22.25 The ligand shown on the next page, 16-S-4, forms the complex $[Hg(16\text{-}S\text{-}4)]^{2+}$. The solution ^1H NMR spectrum of $[Hg(16\text{-}S\text{-}4)][ClO_4]_2$ consists of two signals at δ 3.40 and 2.46 ppm with relative integrals of 2 : 1. From the spectrum, the following coupling constants can be measured: $J_{^1H^1H} = 6.0$ Hz, $J_{^1H(\alpha)^{199}Hg} = 93.6$ Hz. [Data: ^{199}Hg: $I = \frac{1}{2}$, 16.6%]

(a) Explain why complex formation between Hg(II) and *S*-donor ligands is particularly favoured.
(b) What coordination number do you expect for the Hg(II) centre in $[Hg(16\text{-}S\text{-}4)]^{2+}$? On what basis have you made your choice?

(c) Sketch the ^1H NMR spectrum of [Hg(16-S-4)-[ClO$_4$]$_2$.

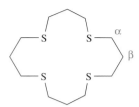

22.26 Studies of the heavier *d*-block metals are often used to introduce students to (a) metal–metal bonding, (b) high coordination numbers, (c) metal halido clusters and (d) polyoxometallates. Write an account of each topic, and include examples that illustrate why the first row metals are not generally as relevant as their heavier congeners for discussing these topics.

OVERVIEW PROBLEMS

22.27 (a) The reaction of ReCl$_4$ and PCl$_5$ at 570 K under vacuum gives [PCl$_4$]$_2$[Re$_2$Cl$_{10}$]. However, when ReCl$_5$ reacts with an excess of PCl$_5$ at 520 K, the products are [PCl$_4$]$_3$[ReCl$_6$]$_2$ and Cl$_2$. Comment on the nature of [PCl$_4$]$_3$[ReCl$_6$]$_2$ and write equations for both reactions, paying attention to the oxidation states of P and Re.

(b) The ^{19}F NMR spectrum of [Me$_4$N][*fac*-OsO$_3$F$_3$] exhibits one signal with satellites ($J = 32$ Hz). What is the origin of the satellite peaks? Sketch the spectrum and indicate clearly the nature of the coupling pattern. Show where J is measured.

22.28 (a) 'The salt [NH$_4$]$_3$[ZrF$_7$] contains discrete ions with 7-coordinate Zr(IV). On the other hand, in a compound formulated as [NH$_4$]$_3$[HfF$_7$], Hf(IV) is octahedral'. Comment on this statement and suggest possible structures for [ZrF$_7$]$^{3-}$.

(b) ^{93}Nb NMR spectroscopy has provided evidence for halide exchange when NbCl$_5$ and NbBr$_5$ are dissolved in MeCN. What would be the basis for such evidence?

22.29 (a) Figure 22.33 shows eight corner-sharing ReO$_6$ octahedra in the solid-state structure of ReO$_3$. From this, derive a diagram to show the unit cell of ReO$_3$. Explain the relationship between

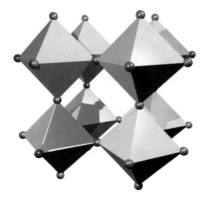

Fig. 22.33 Figure for problem 22.29a.

your diagram and that in Fig. 21.5, and confirm the stoichiometry of the oxide from the unit cell diagram.

(b) A qualitative test for [PO$_4$]$^{3-}$ is to add an excess of an acidified aqueous solution of ammonium molybdate to an aqueous solution of the phosphate. A yellow precipitate forms. Suggest a possible identity for the precipitate and write an equation for its formation.

22.30 (a) Rationalize why each of the following is diamagnetic: [Os(CN)$_6$]$^{4-}$, [PtCl$_4$]$^{2-}$, OsO$_4$ and *trans*-[OsO$_2$F$_4$]$^{2-}$.

(b) Solution ^{77}Se and ^{13}C NMR spectra for the octahedral anions in the compounds [Bu$_4$N]$_3$[Rh(SeCN)$_6$] and [Bu$_4$N]$_3$[*trans*-Rh(CN)$_2$(SeCN)$_4$] are tabulated below. Assign the spectra and explain the origin of the observed coupling patterns. [Additional data: see Table 4.3]

Anion	$\delta\,^{77}Se$ ppm	$\delta\,^{13}C$ ppm
[Rh(SeCN)$_6$]$^{3-}$	−32.7 (doublet, $J = 44$ Hz)	111.2 (singlet)
[*trans*-Rh(CN)$_2$(SeCN)$_4$]$^{3-}$	−110.7 (doublet, $J = 36$ Hz)	111.4 (singlet) 136.3 (doublet, $J = 36$ Hz)

22.31 (a) The complex shown over the page is the first example of a Pd(IV) complex containing a nitrosyl ligand (see also structure **20.9** for another view of the tridentate ligand). On the basis of the assignment of an oxidation state of +4 for Pd, what formal charge does the nitrosyl ligand carry? In view of your answer, comment on the fact that structural and spectroscopic data for the complex include the following parameters: $\angle Pd-N-O = 118°$, $N-O = 115\,pm$, $\bar{\nu}(NO) = 1650\,cm^{-1}$ (a strong absorption).

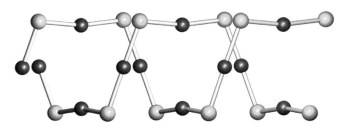

(b) The reaction of equimolar equivalents of $[Bu_4N]_2[C_2O_4]$ with $[cis\text{-}Mo_2(\mu\text{-}L)_2(MeCN)_4][BF_4]_2$ where L^- is a formamidine ligand closely related to **22.70** leads to a neutral compound **A** which is a so-called 'molecular square'. Bearing in mind the structure of $[C_2O_4]^{2-}$, suggest a structure for **A**. This compound might also be considered as a $[4+4]$ assembly. What experimental techniques would be useful in distinguishing compound **A** from a possible $[3+3]$ product?

INORGANIC CHEMISTRY MATTERS

22.32 Comment on the following statements in terms of the properties of the elements mentioned.

(a) For many decades, tungsten has been used to make filaments in incandescent light bulbs. Tungsten is used in preference to copper even though the electrical resistivity of tungsten is greater than that of copper.

(b) Incandescent light bulbs are filled with a gas such as Ar or Xe.

(c) A halogen lamp contains a tungsten filament in a quartz bulb filled with halogen gas. The halogen is Br_2 or I_2, but not F_2 or Cl_2. The lifetime of the filament is prolonged with respect to that in an incandescent bulb.

22.33 Myochrysine contains the thiomalate ligand and is an anti-arthritic drug and has a polymeric structure in the solid state:

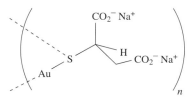

The backbone of the polymer is a helical chain of alternating Au and S atoms, and pairs of chains form double helices (Fig. 22.34). (a) Is the double helix in Fig. 22.34 chiral? Rationalize your answer. (b) Why is the thiolate ligand chiral? (c) Starting with a racemic mixture of (*R*)- and (*S*)-thiomalate, two structurally distinct double helices are observed in the unit cell of Myochrysine. Explain why this

Fig. 22.34 Part of one double helix defined by the gold and sulfur atoms in the solid state structure of Myochrysine. [Data: R. Bau (1998) *J. Am. Chem. Soc.*, vol. 120, p. 9380.]

is. (d) The shortest $Au \cdots Au$ contacts between the two chains in the double helix in Fig. 22.34 are 323 pm. Comment on the importance of such interactions in Au(I) chemistry. (e) Myochrysine is a prodrug. Explain what this means, and outline a possible means for uptake of Au(I) by the body.

22.34 (a) What is the fundamental difference between electrochromic, thermochromic and photochromic materials? (b) WO_3 is widely used in electrochromic materials. Explain why WO_3 is suited to this application. Give a brief description of how an electrochromic window based on WO_3 works. (c) Both WO_3- and IrO_2-based electrochromic glasses switch between colourless and dark blue. How do the ways in which the colour changes occur differ in the two glasses?

22.35 Two complexes that have entered clinical trials as anti-cancer drugs are $[HIm][RuCl_4(im)(DMSO)]$ and $[HInd][RuCl_4(Ind)_2]$. In the search for related active complexes, $[RuCl_2(DMSO)_2(Biim)]$ and $[RuCl_3(DMSO)(Biim)]$ have been tested; the latter is more cytotoxic against selected human cancer cell lines than the former.

indazole (Ind) imidazole (Im)

2,2′-biimidazole (Biim)

(a) What is the oxidation state of the ruthenium in each of the four complexes? (b) Suggest how each of the ligands Im, Ind and Biim coordinates to ruthenium. (c) Draw the structures of possible isomers of each complex.

Try also end-of-chapter problem 20.42: ruthenium(II) photosensitizers.

Topics

23

Organometallic compounds of s- and p-block elements

1	2		13	14	15	16	17	18
H								He
Li	Be		B	C	N	O	F	Ne
Na	Mg		Al	Si	P	S	Cl	Ar
K	Ca	d-block	Ga	Ge	As	Se	Br	Kr
Rb	Sr		In	Sn	Sb	Te	I	Xe
Cs	Ba		Tl	Pb	Bi	Po	At	Rn
Fr	Ra							

23.1 Introduction

This chapter provides an introduction to the large area of the organometallic chemistry of s- and p-block elements.

> An *organometallic* compound contains one or more metal–carbon bonds.

Compounds containing M–C bonds where M is an s-block element are readily classified as being organometallic. However, when we come to the p-block, the trend from metallic to non-metallic character means that a discussion of strictly *organometallic* compounds would ignore compounds of the semi-metals and synthetically important organoboron compounds. For the purposes of this chapter, we have broadened the definition of an organometallic compound to include species with B–C, Si–C, Ge–C, As–C, Sb–C, Se–C or Te–C bonds. Compounds

containing Xe–C bonds are covered in Chapter 18. Also relevant to this chapter is the earlier discussion of fullerenes (see Section 14.4). Quite often compounds containing, for example, Li–N or Si–N bonds are included in discussions of organometallics, but we have chosen to incorporate these in Chapters 11–15. We do not detail applications of main group organometallic compounds in organic synthesis. Abbreviations for the organic substituents mentioned in this chapter are defined in Appendix 2.

23.2 Group 1: alkali metal organometallics

Organic compounds such as terminal alkynes (RC≡CH) which contain relatively acidic hydrogen atoms form salts with the alkali metals, e.g. reactions 23.1, 23.2 and 14.34.

$$2EtC\equiv CH + 2Na \longrightarrow 2Na^+[EtC\equiv C]^- + H_2 \qquad (23.1)$$

$$MeC\equiv CH + K[NH_2] \longrightarrow K^+[MeC\equiv C]^- + NH_3 \qquad (23.2)$$

Similarly, in reaction 23.3, the acidic CH_2 group in cyclopentadiene can be deprotonated to prepare the cyclopentadienyl ligand which is synthetically important in organometallic chemistry (see also Chapter 24). Na[Cp] can also be made by direct reaction of Na with C_5H_6. Na[Cp] is pyrophoric in air, but its air-sensitivity can be lessened by complexing the Na^+ ion with 1,2-dimethoxyethane (dme). In the solid state, [Na(dme)][Cp] is polymeric (Fig. 23.1).

$$[C_5H_5]^- \text{ or } Cp^-$$

(23.3)

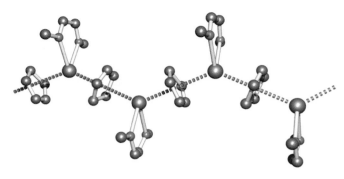

Fig. 23.1 Part of a chain that makes up the polymeric structure of [Na(dme)][Cp] (dme = 1,2-dimethoxyethane); the zigzag chain is emphasized by the hashed, red line. The structure was determined by X-ray diffraction [M.L. Coles *et al.* (2002) *J. Chem. Soc., Dalton Trans.*, p. 896]. Hydrogen atoms have been omitted for clarity; colour code: Na, purple; O, red; C, grey.

> A *pyrophoric* material is one that burns spontaneously when exposed to air.

Colourless alkyl derivatives of Na and K may be obtained by *transmetallation* reactions starting from mercury dialkyls (eq. 23.4).

$$HgMe_2 + 2Na \longrightarrow 2NaMe + Hg \qquad (23.4)$$

Organolithium compounds are of particular importance among the group 1 organometallics. They may be synthesized by treating an organic halide, RX, with Li (eq. 23.5) or by metallation reactions (eq. 23.6) using *n*-butyllithium which is commercially available as solutions in hydrocarbon (e.g. hexane) solvents.

$$^{n}BuCl + 2Li \xrightarrow{\text{hydrocarbon solvent}} {^{n}BuLi} + LiCl \qquad (23.5)$$

$$^{n}BuLi + C_6H_6 \longrightarrow {^{n}BuH} + C_6H_5Li \qquad (23.6)$$

Solvent choices for reactions involving organometallics of the alkali metals are critical. For example, $^{n}BuLi$ is decomposed by Et_2O to give ^{n}BuH, C_2H_4 and LiOEt. Organolithium, -sodium or -potassium reagents are considerably more reactive in solution in the presence of certain diamines (e.g. $Me_2NCH_2CH_2NMe_2$), and we return to this point later.

Alkali metal organometallics are extremely reactive and must be handled in air- and moisture-free environments; NaMe, for example, burns explosively in air.[†]

Lithium alkyls and aryls are more stable thermally than the corresponding compounds of the heavier group 1 metals (though they ignite spontaneously in air) and mostly differ from them in being soluble in hydrocarbons and other non-polar organic solvents and in being liquids or solids of low melting points. Sodium and potassium alkyls are insoluble in most organic solvents and, when stable enough with respect to thermal decomposition, have

[†] A useful source of reference is: D.F. Shriver and M.A. Drezdon (1986) *The Manipulation of Air-sensitive Compounds*, Wiley, New York.

fairly high melting points. In the corresponding benzyl and triphenylmethyl compounds, $Na^+[PhCH_2]^-$ and $Na^+[Ph_3C]^-$ (eq. 23.7), the negative charge in the organic anions can be delocalized over the aromatic systems, thus enhancing stability. The salts are red in colour.

$$NaH + Ph_3CH \longrightarrow Na^+[Ph_3C]^- + H_2 \qquad (23.7)$$

Sodium and potassium also form intensely coloured salts with many aromatic compounds (e.g. reaction 23.8). In reactions such as this, the oxidation of the alkali metal involves the transfer of one electron to the aromatic system producing a paramagnetic *radical anion*.

$$Na + \text{Naphthalene} \xrightarrow[\text{or THF}]{\text{Liquid } NH_3} Na^+[C_8H_{10}]^- \qquad (23.8)$$

Sodium naphthalide (deep blue)

> A *radical anion* is an anion that possesses an unpaired electron.

Lithium alkyls are polymeric both in solution and in the solid state. Table 23.1 illustrates the extent to which MeLi, $^{n}BuLi$ and $^{t}BuLi$ aggregate in solution. In an $(RLi)_4$ tetramer, the Li atoms form a tetrahedral unit, while in an $(RLi)_6$ hexamer, the Li atoms define an octahedron. Figures 23.2a and 23.2b show the structure of $(MeLi)_4$; the average Li$-$Li bond length is 261 pm compared with 267 pm in Li_2 (see Table 2.1). The bonding in lithium alkyls is the subject of end-of-chapter problem 23.2. Figures 23.2c and d show the structure of the Li_6C_6-core of $(LiC_6H_{11})_6$ ($C_6H_{11} = $ *cyclohexyl*); six Li$-$Li bond distances lie in the range 295–298 pm, while the other six are significantly shorter (238–241 pm). The presence of such aggregates in solution can be determined by using multinuclear NMR spectroscopy. Lithium possesses two spin-active isotopes (see Section 4.8 and Table 11.1) and the solution structures of lithium alkyls

Table 23.1 Degree of aggregation of selected lithium alkyls at room temperature (unless otherwise stated).

Compound	Solvent	Species present
MeLi	Hydrocarbons	$(MeLi)_6$
MeLi	Ethers	$(MeLi)_4$
$^{n}BuLi$	Hydrocarbons	$(^{n}BuLi)_6$
$^{n}BuLi$	Ethers	$(^{n}BuLi)_4$
$^{n}BuLi$	THF at low temperature	$(^{n}BuLi)_4 \rightleftharpoons 2(^{n}BuLi)_2$
$^{t}BuLi$	Hydrocarbons	$(^{t}BuLi)_4$
$^{t}BuLi$	Et_2O	Mainly solvated $(^{t}BuLi)_2$
$^{t}BuLi$	THF	Mainly solvated $^{t}BuLi$

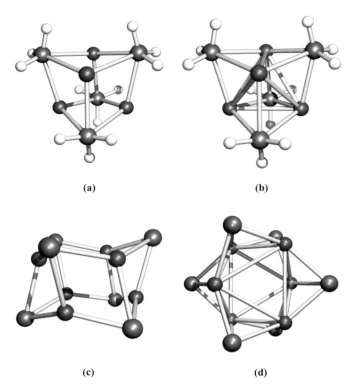

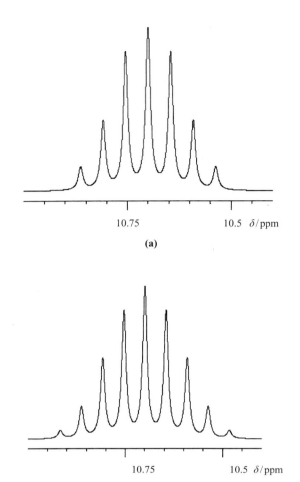

(a)

(b)

Fig. 23.2 (a) The structure of $(MeLi)_4$ (X-ray diffraction) for the perdeuterated compound [E. Weiss *et al.* (1990) *Chem. Ber.*, vol. 123, p. 79]; the Li atoms define a tetrahedral array while the Li_4C_4-unit forms a distorted cube. For clarity, the Li–Li interactions are not shown in (a) but diagram (b) shows these additional interactions. (c) The Li_6C_6-core of $(LiC_6H_{11})_6$ (X-ray diffraction) [R. Zerger *et al.* (1974) *J. Am. Chem. Soc.*, vol. 96, p. 6048]; the Li_6C_6-core can be considered as a distorted hexagonal prism with Li and C atoms at alternate corners. (d) An alternative view of the structure of the Li_6C_6-core of $(LiC_6H_{11})_6$ which also shows the Li–Li interactions (these were omitted from (c) for clarity); the Li atoms define an octahedral array. Colour code: Li, red; C, grey; H, white.

can be studied using ^6Li, ^7Li and ^{13}C NMR spectroscopies as worked example 23.1 illustrates. The alkyls of Na, K, Rb and Cs crystallize with extended structures (e.g. KMe adopts the NiAs structure, Fig. 15.10) or are amorphous solids.

Worked example 23.1   NMR spectroscopy of $(^tBuLi)_4$

The structure of $(^tBuLi)_4$ is similar to that of $(MeLi)_4$ shown in Fig. 23.2a, but with each H atom replaced by a methyl group. The 75 MHz ^{13}C NMR spectrum of a sample of $(^tBuLi)_4$, prepared from ^6Li metal, consists of two signals, one for the methyl carbons and one for the quaternary carbon atoms. The signal for the quaternary carbons is shown alongside and opposite: (a) at 185 K and (b) at 299 K. Explain how these signals arise.

[Data: for ^6Li, $I = 1$.]

First, note that the lithium present in the sample is ^6Li, and this is spin-active ($I = 1$). The multiplet nature of the signals arises from ^{13}C–^6Li spin–spin coupling.

Multiplicity of signal (number of lines) $= 2nI + 1$

Consider Fig. 23.2a with each H atom replaced by an Me group to give $(^tBuLi)_4$. The quaternary C atoms are those bonded to the Li centres, and, *in the static structure*, each ^{13}C nucleus can couple with *three* adjacent and equivalent ^6Li nuclei.

Multiplicity of signal $= (2 \times 3 \times 1) + 1 = 7$

This corresponds to the seven lines (a septet) observed in figure (a) for the low-temperature spectrum. Note that the pattern is *non-binomial*. At 299 K, a nonet is observed (non-binomial).

Multiplicity of signal $= (2 \times n \times 1) + 1 = 9$

$$n = 4$$

This means that the molecule is fluxional, and each quaternary ^{13}C nucleus 'sees' four equivalent ^6Li nuclei

Fig. 23.3 Part of one polymeric chain of $[(^nBuLi)_4 \cdot TMEDA]_\infty$ found in the solid state; the structure was determined by X-ray diffraction. Only the first carbon atom of each ^nBu chain is shown, and all H atoms are omitted for clarity. TMEDA molecules link $(^nBuLi)_4$ units together through the formation of Li–N bonds [N.D.R. Barnett *et al.* (1993) *J. Am. Chem. Soc.*, vol. 115, p. 1573]. Colour code: Li, red; C, grey; N, blue.

on the NMR spectroscopic timescale. We can conclude that at 185 K, the molecule possesses a static structure but as the temperature is raised to 299 K, sufficient energy becomes available to allow a fluxional process to occur which exchanges the ^tBu groups.

For a full discussion, see: R.D. Thomas *et al.* (1986) *Organometallics*, vol. 5, p. 1851.

[For details of NMR spectroscopy: see Section 4.8. Case study 4 in this section is concerned with a non-binomial multiplet.]

Self-study exercises

1. From the data above, what would you expect to see in the ^{13}C NMR spectrum at 340 K?

[*Ans.* Non-binomial nonet]

2. The ^{13}C NMR spectrum of $(^tBuLi)_4$ at 185 K is called the 'limiting low-temperature spectrum'. Explain what this means.

Amorphous alkali metal alkyls such as ^nBuNa are typically insoluble in common solvents, but are solubilized by the chelating ligand TMEDA (**23.1**).† Addition of this ligand may break down the aggregates of lithium alkyls to give lower nuclearity complexes, e.g. $[^nBuLi \cdot TMEDA]_2$, **23.2**. However, detailed studies have revealed that this system is far from simple, and under different conditions, it is possible to isolate crystals of either $[^nBuLi \cdot TMEDA]_2$ or $[(^nBuLi)_4 \cdot TMEDA]_\infty$ (Fig. 23.3). In the case of $(MeLi)_4$, the addition of TMEDA does not lead to cluster breakdown, and an X-ray diffraction study of $(MeLi)_4 \cdot 2TMEDA$ confirms the presence of tetramers and amine molecules in the solid state.

(23.1)   **(23.2)**

Solutions of TMEDA-complexed organoalkali metal reagents provide convenient homogeneous systems for metallations. For example, the metallation of benzene (reaction 23.6) proceeds more efficiently if $^nBuLi \cdot TMEDA$ is used in place of ^nBuLi. Alkylbenzenes are metallated by $^nBuLi \cdot TMEDA$ at the alkyl group in preference to a ring position. Thus the reaction between $C_6H_5CH_3$ and $^nBuLi \cdot TMEDA$ in hexane (303 K, 2 hours) gives $C_6H_5CH_2Li$ as the regioselective (92%) product. Metallation at the *ortho-* and *meta-*ring sites occurs under these conditions to an extent of only 2 and 6%, respectively. It is possible, however, to reverse the regioselectivity[‡] in favour of the *meta-*position by using the heterometallic reagent **23.3** which is prepared from ^nBuNa, $^nBu_2Mg$, 2,2,3,3-tetramethylpiperidine and TMEDA.

(23.3)

---

† The abbreviation TMEDA comes from the non-IUPAC name N,N,N',N'-tetramethylethylenediamine.

‡ For details of this unexpected observation, see: P.C. Andrikopoulis *et al.* (2005) *Angew. Chem. Int. Ed.*, vol. 44, p. 3459.

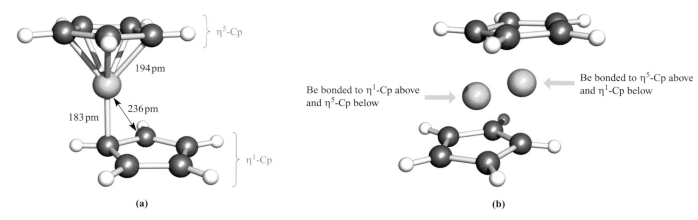

Fig. 23.4 (a) The solid state structure of Cp_2Be determined by X-ray diffraction at 128 K [K.W. Nugent *et al.* (1984) *Aust. J. Chem.*, vol. 37, p. 1601]. (b) The same structure showing the two equivalent sites over which the Be atom is disordered. Colour code: Be, yellow; C, grey; H, white.

A *regioselective* reaction is one that could proceed in more than one way but is observed to proceed only, or predominantly, in one way.

Organolithium compounds (in particular MeLi and ^nBuLi) are of great importance as synthetic reagents. Among the many uses of organolithium alkyls and aryls are the conversions of boron trihalides to organoboron compounds (eq. 23.9) and similar reactions with other *p*-block halides (e.g. $SnCl_4$).

$$3^nBuLi + BCl_3 \longrightarrow {^nBu_3B} + 3LiCl \qquad (23.9)$$

Lithium alkyls are important catalysts in the synthetic rubber industry for the stereospecific polymerization of alkenes.

23.3 Group 2 organometallics

Beryllium

Beryllium alkyls and aryls are best made by reaction types 23.10 and 23.11 respectively. They are hydrolysed by water and inflame in air.

$$HgMe_2 + Be \xrightarrow{383\,K} Me_2Be + Hg \qquad (23.10)$$

$$2PhLi + BeCl_2 \xrightarrow{Et_2O} Ph_2Be + 2LiCl \qquad (23.11)$$

In the vapour phase, Me_2Be is monomeric, with a linear C−Be−C unit (Be−C = 170 pm). The bonding was described in Section 5.2. The solid state structure is polymeric (**23.4**), and resembles that of $BeCl_2$ (Fig. 12.4b). However, whereas the bonding in $BeCl_2$ can be described in terms of a localized bonding scheme (Fig. 12.4c), there are insufficient valence electrons available in $(Me_2Be)_n$ for an analogous bonding picture. Instead, 3c-2e bonds are invoked as described for BeH_2 (see Fig. 10.15 and associated text). Higher alkyls are progressively polymerized to a lesser

extent, and the *tert*-butyl derivative is monomeric under all conditions.

(23.4)

$$2Na[Cp] + BeCl_2 \longrightarrow Cp_2Be + 2NaCl \qquad (23.12)$$

Reaction 23.12 leads to the formation of Cp_2Be, and in the solid state, the structure (Fig. 23.4a) is in accord with the description $(\eta^1\text{-}Cp)(\eta^5\text{-}Cp)Be$. Electron diffraction and spectroscopic studies of Cp_2Be in the gas phase have provided conflicting views of the structure, but current data indicate that it resembles that found in the solid state rather than the $(\eta^5\text{-}Cp)_2Be$ originally proposed. Furthermore, the solid state structure is not as simple as Fig. 23.4a shows. The Be atom is *disordered* (see Box 15.5) over two equivalent sites shown in Fig. 23.4b. Variable temperature NMR spectroscopic studies show that Cp_2Be is fluxional both in the solid state and in solution. In the solid state, an activation energy of 36.9 kJ mol^{-1} has been experimentally determined for the 'molecular inversion' in which the two Cp rings effectively exchange between η^1 and η^5-coordination modes. In solution, each of the ^1H and ^{13}C NMR spectra shows only one signal even as low as 138 K, indicating that a fluxional process makes all the proton and all the carbon environments equivalent. The compound $(C_5HMe_4)_2Be$ can be prepared at room temperature from $BeCl_2$ and $K[C_5HMe_4]$. In the solid state at 113 K, it is structurally similar to Cp_2Be although, in $(C_5HMe_4)_2Be$, the Be atom is not disordered. Solution ^1H NMR spectroscopic data for $(C_5HMe_4)_2Be$

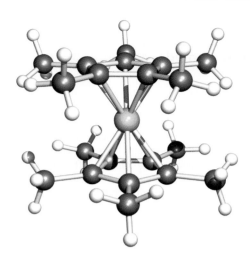

Fig. 23.5 The solid state structure (X-ray diffraction at 113 K) of $(\eta^5\text{-}C_5Me_5)_2Be$ [M. del Mar Conejo *et al.* (2000) *Angew. Chem. Int. Ed.*, vol. 39, p. 1949]. Colour code: Be, yellow; C, grey; H, white.

are consistent with the molecule being fluxional down to 183 K. The fully methylated derivative $(C_5Me_5)_2Be$ is made by reaction 23.13. In contrast to Cp_2Be and $(C_5HMe_4)_2Be$, $(C_5Me_5)_2Be$ possesses a *sandwich* structure in which the two C_5-rings are coparallel and staggered (Fig. 23.5), i.e. the compound is formulated as $(\eta^5\text{-}C_5Me_5)_2Be$.

$$2K[C_5Me_5] + BeCl_2 \xrightarrow[388\,K]{Et_2O/toluene} (C_5Me_5)_2Be + 2KCl$$
$$(23.13)$$

In a *sandwich complex*, the metal centre lies between two π-bonded hydrocarbon (or derivative) ligands. Complexes of the type $(\eta^5\text{-}Cp)_2M$ are called **metallocenes**.

We consider bonding schemes for complexes containing Cp^- ligands in Box 23.1.

Self-study exercise

Give a bonding description for the interaction between a metal atom and an η^1-Cp ring, e.g. in a general complex $L_nMH(\eta^1\text{-}Cp)$. Is the M–C interaction localized or delocalized?

Magnesium

Alkyl and aryl magnesium halides (Grignard reagents, represented by the formula RMgX) are extremely well known on account of their uses in synthetic organic chemistry. The general preparation of a Grignard reagent (eq. 23.14) requires initial activation of the metal, e.g. by addition of I_2.

$$Mg + RX \xrightarrow{Et_2O} RMgX \qquad (23.14)$$

Transmetallation of a suitable organomercury compound is a useful means of preparing pure Grignard reagents (eq. 23.15), and transmetallation 23.16 can be used to synthesize compounds of type R_2Mg.

$$Mg + RHgBr \longrightarrow Hg + RMgBr \qquad (23.15)$$

$$Mg + R_2Hg \longrightarrow Hg + R_2Mg \qquad (23.16)$$

Although eqs. 23.14–23.16 show the magnesium organometallics as simple species, this is an oversimplification. Two-coordination at Mg in R_2Mg is only observed in the solid state when the R groups are especially bulky, e.g. $Mg\{C(SiMe_3)_3\}_2$ (Fig. 23.6a). Grignard reagents are generally solvated, and crystal structure data show that

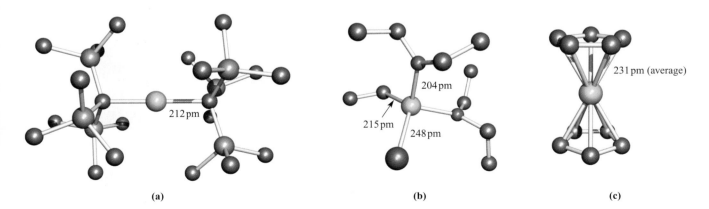

(a) (b) (c)

Fig. 23.6 The solid state structures, determined by X-ray diffraction, of (a) $Mg\{C(SiMe_3)_3\}_2$ [S.S. Al-Juaid *et al.* (1994) *J. Organomet. Chem.*, vol. 480, p. 199], (b) $EtMgBr \cdot 2Et_2O$ [L.J. Guggenberger *et al.* (1968) *J. Am. Chem. Soc.*, vol. 90, p. 5375], and (c) Cp_2Mg in which each ring is in an η^5-mode and the two rings are mutually staggered [W. Bunder *et al.* (1975) *J. Organomet. Chem.*, vol. 92, p. 1]. Hydrogen atoms have been omitted for clarity; colour code: Mg, yellow; C, grey; Si, pink; Br, brown; O, red.

THEORY

Box 23.1 Bonding in cyclopentadienyl complexes: η^5-mode

If all five C atoms of the cyclopentadienyl ring interact with the metal atom, the bonding is most readily described in terms of an MO scheme. Once the σ-bonding framework of the $[Cp]^-$ ligand has been formed, there is one $2p_z$ atomic orbital per C atom remaining, and five combinations are possible. The MO diagram below shows the formation of $(\eta^5\text{-}Cp)BeH$ (C_{5v}), a model compound that allows us to see how the $[\eta^5\text{-}Cp]^-$ ligand interacts with an s- or p-block metal fragment. For the formation of the $[BeH]^+$ fragment, we can use an sp hybridization scheme. One sp hybrid points at the H atom and the other points at the Cp ring. Using the methods from Chapter 5, the orbitals of the $[BeH]^+$ unit are classified as having a_1 or e_1 symmetry within the C_{5v} point group. To work out the π-orbitals of the $[Cp]^-$ ligand, we first determine how many C $2p_z$ orbitals are unchanged by each symmetry operation in the C_{5v} point group (Appendix 3). The resultant row of characters is:

E	$2C_5$	$2C_5^2$	$5\sigma_v$
5	0	0	1

This row can be obtained by adding the rows of characters for the A_1, E_1 and E_2 representations in the C_{5v} character table. Thus, the five π-orbitals of $[Cp]^-$ possess a_1, e_1 and e_2 symmetries. By applying the methods described in Chapter 5, the wavefunctions for these orbitals can be determined. The orbitals are shown schematically on the left-hand side of the diagram. The MO diagram is constructed by matching the symmetries of the fragment orbitals. Mixing can occur between the two a_1 orbitals of the $[BeH]^+$ fragment. Four bonding MOs (a_1 and e_1) result. The e_2 $[Cp]^-$ orbitals are non-bonding with respect to Cp–BeH interactions. (Antibonding MOs have been omitted from the diagram.) Eight electrons are available to occupy the a_1 and e_1 MOs. Representations of the a_1, e_1 and e_2 MOs are shown at the right-hand side of the figure: the e_1 set possesses Be–C bonding character, while both a_1 MOs have Be–C and Be–H bonding character.

Bonding in cyclopentadienyl complexes of d-block metals (see Chapter 24) can be described in a similar manner but must allow for the participation of metal d-orbitals.

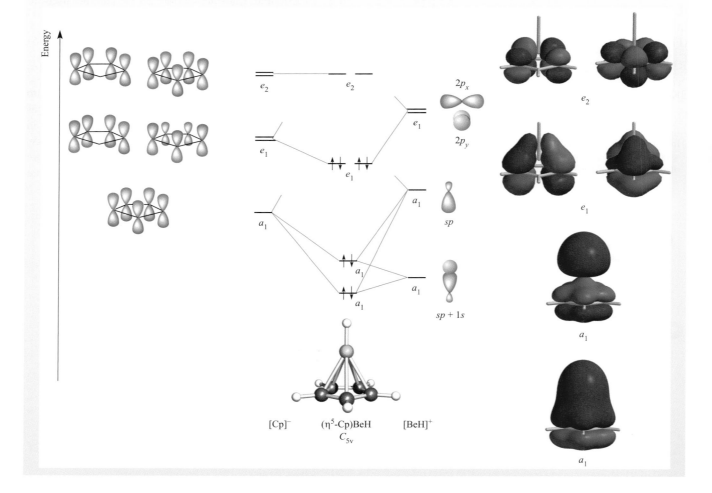

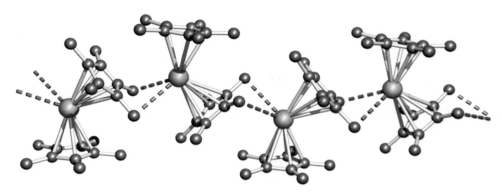

Fig. 23.7 Part of a chain in the polymeric structure (X-ray diffraction 118 K) of (η^5-$C_5Me_5)_2$Ba illustrating the bent metallocene units [R.A. Williams *et al.* (1988) *J. Chem. Soc., Chem Commun.*, p. 1045]. Hydrogen atoms have been omitted; colour code: Ba, orange; C, grey.

the Mg centre is typically tetrahedrally sited, e.g. in EtMgBr·2Et$_2$O (Fig. 23.6b) and PhMgBr·2Et$_2$O. A few examples of 5- and 6-coordination have been observed, e.g. in **23.5** where the macrocyclic ligand imposes the higher coordination number on the metal centre. The preference for an octahedral structure can be controlled by careful choice of the organic substituent, e.g. complex **23.6**. The introduction of two or more bidentate ligands into the octahedral coordination sphere leads to the possibility of *stereoisomerism*, e.g. **23.6** is chiral (see Sections 3.8 and 19.8). Enantiomerically pure Grignard reagents have potential for use in stereoselective organic synthesis. Solutions of Grignard reagents may contain several species, e.g. RMgX, R$_2$Mg, MgX$_2$, RMg(μ-X)$_2$MgR, which are further complicated by solvation. The positions of equilibria between these species are markedly dependent on concentration, temperature and solvent. Strongly donating solvents favour monomeric species in which they coordinate to the metal centre.

(23.5)   **(23.6)**

In contrast to its beryllium analogue, Cp$_2$Mg has the structure shown in Fig. 23.6c, i.e. two η^5-cyclopentadienyl ligands, and is structurally similar to ferrocene (see Section 24.13). The reaction between Mg and C$_5$H$_6$ yields Cp$_2$Mg, which is decomposed by water; the compound is therefore often inferred to be ionic and, indeed, significant ionic character is suggested by the long Mg−C bonds in the solid state and also by IR and Raman spectroscopic data.

Calcium, strontium and barium

The heavier group 2 metals are highly electropositive, and metal–ligand bonding is generally considered to be predominantly ionic. Nonetheless, this remains a topic for debate and theoretical investigation. While Cp$_2$Be and Cp$_2$Mg are monomeric and are soluble in hydrocarbon solvents, Cp$_2$Ca, Cp$_2$Sr and Cp$_2$Ba are polymeric and are insoluble in ethers and hydrocarbons. Increasing the steric demands of the substituents on the C$_5$-rings leads to structural changes in the solid state and to changes in solution properties, e.g. (C$_5$Me$_5$)$_2$Ba is polymeric, {1,2,4-(SiMe$_3$)$_3$C$_5$H$_2$}$_2$Ba is dimeric and (^{i}Pr$_5$C$_5$)$_2$Ba is monomeric. Oligomeric metallocene derivatives of Ca^{2+}, Sr^{2+} and Ba^{2+} typically exhibit bent C$_5$−M−C$_5$ units (Fig. 23.7 and see the end of Section 23.5), but in (^{i}Pr$_5$C$_5$)$_2$Ba, the C$_5$-rings are coparallel. The ^{i}Pr$_5$C$_5$-rings are very bulky, and sandwich the Ba^{2+} ion protectively, making (^{i}Pr$_5$C$_5$)$_2$Ba air-stable. The 1990s saw significant development of the organometallic chemistry of the heavier group 2 metals, with one driving force being the search for precursors for use in chemical vapour deposition (see Chapter 28). Some representative synthetic methodologies are given in eqs. 23.17–23.20, where M = Ca, Sr or Ba.[†]

$$Na[C_5R_5] + MI_2 \xrightarrow[\text{(e.g. THF, Et}_2\text{O)}]{\text{ether}} NaI + (C_5R_5)MI(\text{ether})_x$$
                                                                (23.17)

$$2C_5R_5H + M\{N(SiMe_3)_2\}_2$$
$$\xrightarrow{\text{toluene}} (C_5R_5)_2M + 2NH(SiMe_3)_2 \quad (23.18)$$

$$3K[C_5R_5] + M(O_2SC_6H_4\text{-}4\text{-Me})_2$$
$$\xrightarrow{\text{THF}} K[(C_5R_5)_3M](THF)_3 + 2K[O_2SC_6H_4\text{-}4\text{-Me}]$$
                                                                (23.19)

[†]For greater detail, see: T.P. Hanusa (2000) *Coord. Chem. Rev.*, vol. 210, p. 329; W.D. Buchanan, D.G. Allis and K. Ruhlandt-Senge (2010) *Chem. Commun.*, vol. 46, p. 4449.

$(C_5R_5)CaN(SiMe_3)_2(THF) + HC≡CR'$

$$\xrightarrow{\text{toluene}} (C_5R_5)(THF)Ca(\mu\text{-}C≡CR')_2Ca(THF)(C_5R_5)$$

$$(23.20)$$

Worked example 23.2   Cyclopentadienyl complexes of Ca²⁺, Sr²⁺ and Ba²⁺

In the solid state, $(\eta^5\text{-}1,2,4\text{-}(SiMe_3)_3C_5H_2)SrI(THF)_2$ exists as dimers, each with an inversion centre. Suggest how the dimeric structure is supported and draw a diagram to show the structure of the dimer.

The iodide ligands have the potential to bridge between two Sr centres.

When drawing the structure, ensure that the two halves of the dimer are related by an inversion centre, *i* (see Section 3.2):

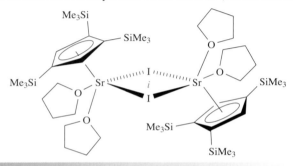

Self-study exercises

1. '$(\eta^5\text{-}C_5{}^iPr_4H)CaI$' can be stabilized in the presence of THF as a THF complex. However, removal of coordinated THF by heating results in the reaction:

$$2(\eta^5\text{-}C_5{}^iPr_4H)CaI \longrightarrow (\eta^5\text{-}C_5{}^iPr_4H)_2Ca + CaI_2$$

Comment on these observations.

2. The reaction of BaI_2 with $K[1,2,4\text{-}(SiMe_3)_3C_5H_2]$ yields a compound **A** and an ionic salt. The solution ^1H NMR spectrum of **A** shows singlets at δ 6.69 (2H), 0.28 (18H) and 0.21 (9H) ppm. Suggest an identity for **A** and assign the ^1H NMR spectrum.

[For more information and answers, see: M.J. Harvey *et al.* (2000) *Organometallics*, vol. 19, p. 1556.]

23.4 Group 13

Boron

The following aspects of organoboron compounds have already been discussed:

- reactions of alkenes with B_2H_6 to give R_3B compounds (see Fig. 13.8);

- the preparation of $B_4{}^tBu_4$ (eq. 13.42);
- organoboranes which contain B−N bonds (Section 13.8).

Organoboranes of type R_3B can be prepared by reaction 23.21, or by the hydroboration reaction mentioned above.

$$Et_2O{\cdot}BF_3 + 3RMgX \longrightarrow R_3B + 3MgXF + Et_2O$$

$$(R = \text{alkyl or aryl}) \quad (23.21)$$

Trialkylboranes are monomeric and inert towards water, but are pyrophoric. Triaryl compounds are less reactive. Both sets of compounds contain planar 3-coordinate B and act as Lewis acids towards amines and carbanions (see also Sections 13.5 and 13.6). Reaction 23.22 shows an important example: sodium tetraphenylborate is water-soluble but salts of larger monopositive cations (e.g. K^+) are insoluble. This makes $Na[BPh_4]$ useful in the precipitation of large metal ions.

$$BPh_3 + NaPh \longrightarrow Na[BPh_4] \quad (23.22)$$

Compounds of the types R_2BCl and $RBCl_2$ can be prepared by transmetallation reactions (e.g. eq. 23.23) and are synthetically useful (e.g. reactions 13.67 and 23.24).

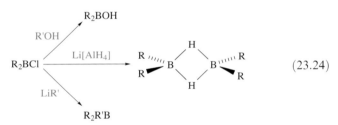

$$(23.23)$$

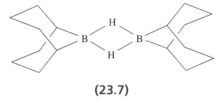

$$(23.24)$$

The bonding in $R_2B(\mu\text{-}H)_2BR_2$ can be described in a similar manner to that in B_2H_6 (see Section 5.7). An important member of this family is **23.7**, commonly known as 9-BBN,[†] which is used for the regioselective reduction of ketones, aldehydes, alkynes and nitriles.

(23.7)

By using bulky organic substituents (e.g. mesityl = 2,4,6-$Me_3C_6H_2$), it is possible to stabilize compounds of type $R_2B−BR_2$. These should be contrasted with $X_2B−BX_2$

[†] The systematic name for 9-BBN is 9-borabicyclo[3.3.1]nonane.

where X = halogen or NR_2 in which there is $X \longrightarrow B$ π-overlap (see Sections 13.6 and 13.8). Two-electron reduction of R_2B-BR_2 gives $[R_2B=BR_2]^{2-}$, an isoelectronic analogue of an alkene. The planar B_2C_4 framework has been confirmed by X-ray diffraction for $Li_2[B_2(2,4,6-Me_3C_6H_2)_3Ph]$, although there is significant interaction between the B=B unit and two Li^+ centres. The shortening of the B−B bond on going from $B_2(2,4,6-Me_3C_6H_2)_3Ph$ (171 pm) to $[B_2(2,4,6-Me_3C_6H_2)_3Ph]^{2-}$ (163 pm) is less than might be expected and this observation is attributed to the large Coulombic repulsion between the two B^- centres.

Aluminium

Aluminium alkyls can be prepared by the transmetallation reaction 23.25, or from Grignard reagents (eq. 23.26). On an industrial scale, the direct reaction of Al with a terminal alkene and H_2 (eq. 23.27) is employed.

$$2Al + 3R_2Hg \longrightarrow 2R_3Al + 3Hg \qquad (23.25)$$

$$AlCl_3 + 3RMgCl \longrightarrow R_3Al + 3MgCl_2 \qquad (23.26)$$

$$Al + \tfrac{3}{2}H_2 + 3R_2C=CH_2 \longrightarrow (R_2CHCH_2)_3Al \qquad (23.27)$$

Reactions between Al and alkyl halides yield alkyl aluminium halides (eq. 23.28). Note that 23.8 is in equilibrium with $[R_2Al(\mu-X)_2AlR_2]$ and $[RXAl(\mu-X)_2AlRX]$ via a redistribution reaction, but 23.8 predominates in the mixture.

$$2Al + 3RX \longrightarrow$$

(23.8)

$$Al + \tfrac{3}{2}H_2 + 2R_3Al \longrightarrow 3R_2AlH \qquad (23.29)$$

Alkyl aluminium hydrides are obtained by reaction 23.29. These compounds, although unstable to both air and water, are important catalysts for the polymerization of alkenes and other unsaturated organic compounds. We describe the commercially important role of alkyl aluminium derivatives as co-catalysts in Ziegler–Natta alkene polymerization in Section 25.8.[†]

(23.9)

[†] For overviews, see: G. Wilke (2003) *Angew. Chem. Int. Ed.*, vol. 42, p. 5000 – 'Fifty years of Ziegler catalysts: Consequences and development of an invention'; W.M. Alley, I.K. Hamdemir, K.A. Johnson and R.G. Finke (2010) *J. Mol. Catal. A*, vol. 315, p. 1 – 'Ziegler-type hydrogenation catalysts made from group 8–10 transition metal precatalysts and AlR_3 cocatalysts: A critical review of the literature'.

Earlier we noted that R_3B compounds are monomeric. In contrast, aluminium trialkyls form dimers. Although this resembles the behaviour of the halides discussed in Section 13.6, there are differences in bonding. Trimethylaluminium (mp 313 K) possesses structure 23.9 and a bonding description similar to that in B_2H_6 is appropriate. The fact that $Al-C_{bridge} > Al-C_{terminal}$ is consistent with 3c-2e bonding in the Al−C−Al bridges, but with 2c-2e terminal bonds. Equilibria between dimer and monomer exist in solution, with the monomer becoming more favoured as the steric demands of the alkyl group increase. Mixed alkyl halides also dimerize as exemplified in structure 23.8, but with particularly bulky R groups, the monomer (with trigonal planar Al) is favoured, e.g. (2,4,6-$^tBu_3C_6H_2$)$AlCl_2$ (Fig. 23.8a). Triphenylaluminium also exists as a dimer, but in the mesityl derivative (mesityl = 2,4,6-$Me_3C_6H_2$), the steric demands of the substituents stabilize the monomer. Figure 23.8b shows the structure of $Me_2Al(\mu-Ph)_2AlMe_2$, and the orientations of the bridging phenyl groups are the same as in $Ph_2Al(\mu-Ph)_2AlPh_2$. This orientation is sterically favoured and places each *ipso*-carbon atom in an approximately tetrahedral environment.

> The *ipso*-carbon atom of a phenyl ring is the one to which the substituent is attached; e.g. in PPh_3, the *ipso*-C of each Ph ring is bonded to P.

In dimers containing $RC\equiv C$-bridges, a different type of bonding operates. The structure of $Ph_2Al(PhC\equiv C)_2AlPh_2$ (23.10) shows that the alkynyl bridges lean over towards one of the Al centres. This is interpreted in terms of their behaving as σ,π-ligands: each forms one Al−C σ-bond and interacts with the second Al centre by using the $C\equiv C$ π-bond. Thus, each alkynyl group is able to provide three electrons (one σ- and two π-electrons) for bridge bonding in contrast to one electron being supplied by an alkyl or aryl group; the bonding is shown schematically in 23.11.

(23.10)

(23.11)

Trialkylaluminium derivatives behave as Lewis acids, forming a range of adducts, e.g. $R_3N\cdot AlR_3$, $K[AlR_3F]$, $Ph_3P\cdot AlMe_3$ and more exotic complexes such as that shown in Fig. 23.8c. Each adduct contains a tetrahedrally

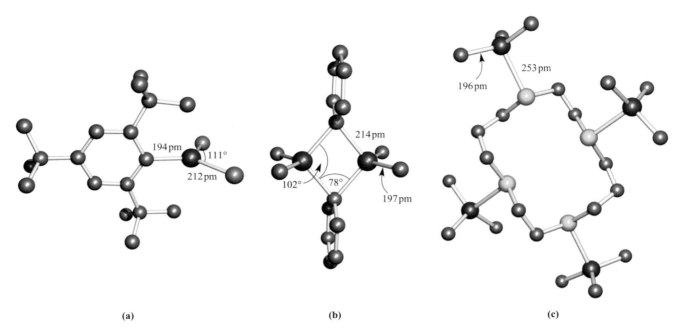

(a)   **(b)**   **(c)**

Fig. 23.8 The solid state structures (X-ray diffraction) of (a) $(2,4,6\text{-}^{t}Bu_3C_6H_2)AlCl_2$ [R.J. Wehmschulte *et al.* (1996) *Inorg. Chem.*, vol. 35, p. 3262], (b) $Me_2Al(\mu\text{-Ph})_2AlMe_2$ [J.F. Malone *et al.* (1972) *J. Chem. Soc., Dalton Trans.*, p. 2649], and (c) the adduct $L \cdot (AlMe_3)_4$ where L is the sulfur-containing macrocyclic ligand 1,4,8,11-tetrathiacyclotetradecane [G.H. Robinson *et al.* (1987) *Organometallics*, vol. 6, p. 887]. Hydrogen atoms are omitted for clarity; colour code: Al, blue; C, grey; Cl, green; S, yellow.

sited Al atom. Trialkylaluminium compounds are stronger Lewis acids than either R_3B or R_3Ga, and the sequence for group 13 follows the trend $R_3B < R_3Al > R_3Ga > R_3In > R_3Tl$. Most adducts of $AlMe_3$ with nitrogen donors (e.g. $Me_3N \cdot AlMe_3$, **23.12**, **23.13** and **23.14**) are air- and moisture-sensitive, and must be handled under inert atmospheres. One way of stabilizing the system is by the use of an internal amine as in **23.15**. Alternatively, complex **23.16** (containing a bicyclic diamine) can be handled in air, having a hydrolytic stability comparable to that of $LiBH_4$.

The first $R_2Al\text{-}AlR_2$ derivative (reported in 1988) was prepared by potassium reduction of the sterically hindered $\{(Me_3Si)_2CH\}_2AlCl$. The Al–Al bond distance in $\{(Me_3Si)_2CH\}_4Al_2$ is 266 pm (compare $r_{cov} = 130$ pm) and the Al_2C_4 framework is *planar*, despite this being a singly bonded compound. A related compound is $(2,4,6\text{-}^{i}Pr_3C_6H_2)_4Al_2$ (Al–Al = 265 pm) but here the Al_2C_4 framework is non-planar (angle between the two AlC_2 planes = 45°). One-electron reduction of Al_2R_4 ($R = 2,4,6\text{-}^{i}Pr_3C_6H_2$) gives the radical anion $[Al_2R_4]^{-}$ with a formal Al–Al bond order of 1.5. Consistent with the presence of a π-contribution, the Al–Al bond is shortened upon reduction to 253 pm for $R = (Me_3Si)_2CH$, and 247 pm for $R = 2,4,6\text{-}^{i}Pr_3C_6H_2$. In both anions, the Al_2R_4 frameworks are essentially planar. In theory, a dialane $R_2Al\text{-}AlR_2$, **23.17**, possesses an isomer, **23.18**, and such a species is exemplified by $(\eta^5\text{-}C_5Me_5)Al\text{-}Al(C_6F_5)_3$. The Al–Al bond (259 pm) in this compound is shorter than in compounds of type $R_2Al\text{-}AlR_2$ and this is consistent with the ionic contribution made to the Al–Al interaction in isomer **23.18**.

Me_2N NMe_2

Me_3Al $AlMe_3$

TMEDA·$(AlMe_3)_2$

(23.12)

$$\text{pyridine} \cdot AlMe_3$$

(23.13)

$$\text{pyrazine} \cdot 2AlMe_3$$

(23.14)

(23.15)

$Me_3Al \leftarrow N \qquad N \rightarrow AlMe_3$

(23.16)

$$R_2Al\text{-}AlR_2$$

(23.17)

$$R\text{-}Al^{+}\text{-}Al^{-}R_3$$

(23.18)

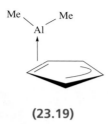

(23.19)

The reaction between cyclopentadiene and Al_2Me_6 gives $CpAlMe_2$ which is a volatile solid. In the gas phase, it is monomeric with an η^2-Cp bonding mode (**23.19**). This effectively partitions the cyclopentadienyl ring into alkene and allyl parts, since only two of the five π-electrons are donated to the metal centre. In the solid state, the molecules interact to form polymeric chains (Fig. 23.9a). The related compound Cp_2AlMe is monomeric with an η^2-mode in the solid state (Fig. 23.9b). In solution, Cp_2AlMe and $CpAlMe_2$ are highly fluxional. A small energy difference between the different modes of bonding of the cyclopentadienyl ligand is also observed in the compounds $(C_5H_5)_3Al$ (i.e. Cp_3Al), $(1,2,4\text{-}Me_3C_5H_2)_3Al$ and $(Me_4C_5H)_3Al$. In solution, even at low temperature, these are stereochemically non-rigid, with negligible energy differences between η^1-, η^2-, η^3- and η^5-modes of bonding. In the solid state, the structural parameters are

Fig. 23.10 The structure of $[(\eta^5\text{-}C_5Me_5)Al]_4$ (determined by X-ray diffraction at 200 K); $Al–Al = 277$ pm, and average $Al–C = 234$ pm [Q. Yu *et al.* (1999) *J. Organomet. Chem.*, vol. 584, p. 94]. Hydrogen atoms are omitted; colour code: Al, blue; C, grey.

consistent with the descriptions:

- $(\eta^2\text{-}C_5H_5)(\eta^{1.5}\text{-}C_5H_5)_2Al$ and $(\eta^2\text{-}C_5H_5)(\eta^{1.5}\text{-}C_5H_5)(\eta^1\text{-}C_5H_5)Al$ for the two independent molecules present in the crystal lattice;
- $(\eta^5\text{-}1,2,4\text{-}Me_3C_5H_2)(\eta^1\text{-}1,2,4\text{-}Me_3C_5H_2)_2Al$;
- $(\eta^1\text{-}Me_4C_5H)_3Al$.

These examples illustrate the non-predictable nature of these systems.

Compounds of the type R_3Al contain aluminium in oxidation state $+3$, while Al_2R_4 formally contains Al(II). The reduction of $[(\eta^5\text{-}C_5Me_5)XAl(\mu\text{-}X)]_2$ (X = Cl, Br, I) by Na/K alloy gives $[(\eta^5\text{-}C_5Me_5)Al]_4$, with the yield being the highest for X = I, corresponding to the lowest Al–X bond enthalpy. $[(\eta^5\text{-}C_5Me_5)Al]_4$ contains a tetrahedral cluster of Al atoms (Fig. 23.10) and is formally an aluminium(I) compound. Stabilization of this and related compounds requires the presence of bulky cyclopentadienyl ligands. It has not been possible to isolate monomeric $(\eta^5\text{-}C_5R_5)Al$.[†]

Gallium, indium and thallium

Since 1980, interest in organometallic compounds of Ga, In and Tl has grown, mainly because of their potential use as precursors to semiconducting materials such as GaAs and InP. Volatile compounds can be used in the growth of thin films by MOCVD (*metal organic chemical vapour deposition*) or MOVPE (*metal organic vapour phase epitaxy*)

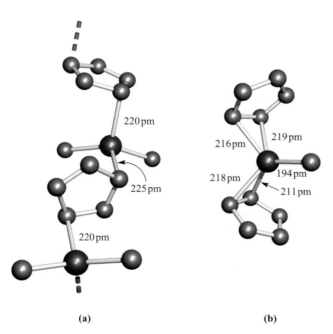

(a)   **(b)**

Fig. 23.9 The solid state structures (X-ray diffraction) of (a) polymeric $CpAlMe_2$ [B. Tecle *et al.* (1982) *Inorg. Chem.*, vol. 21, p. 458], and (b) monomeric $(\eta^2\text{-}Cp)_2AlMe$ [J.D. Fisher *et al.* (1994) *Organometallics*, vol. 13, p. 3324]. Hydrogen atoms are omitted; colour code: Al, blue; C, grey.

[†] For insight into the development of organoaluminium(I) compounds, see: H.W. Roesky (2004) *Inorg. Chem.*, vol. 43, p. 7284 – 'The renaissance of aluminum chemistry'.

APPLICATIONS
Box 23.2 III–V semiconductors

The so-called III–V semiconductors derive their name from the old group numbers for groups 13 and 15, and include AlAs, AlSb, GaP, GaAs, GaSb, InP, InAs and InSb. Of these, GaAs is of the greatest commercial interest. Although Si is probably the most important commercial semiconductor, a major advantage of GaAs over Si is that the charge carrier mobility is much greater. This makes GaAs suitable for high-speed electronic devices. Another important difference is that GaAs exhibits a fully allowed electronic transition between valence and conduction bands (i.e. it is a *direct* band gap semiconductor) whereas Si is an *indirect* band gap semiconductor. The consequence of this difference is that GaAs (and, similarly, the other III–V semiconductors) are more suited than Si for use in optoelectronic devices, since light is emitted more efficiently. The III–Vs have important applications in light-emitting diodes (LEDs). III–V semiconductors are discussed in detail in Section 28.6.

A technician handling a gallium arsenide wafer in a clean-room facility in the semiconductor industry.

Related information

Box 14.2 – Solar power: thermal and electrical

---

techniques (see Section 28.6). Precursors include Lewis base adducts of metal alkyls, e.g. $Me_3Ga \cdot NMe_3$ and $Me_3In \cdot PEt_3$. Reaction 23.30 is an example of the thermal decomposition of gaseous precursors to form a semiconductor which can be deposited in thin films (see Box 23.2).

$$Me_3Ga(g) + AsH_3(g) \xrightarrow{1000-1150\,K} GaAs(s) + 3CH_4(g)$$
$$(23.30)$$

Gallium, indium and thallium trialkyls, R_3M, can be made by use of Grignard reagents (reaction 23.31), RLi (eq. 23.32) or R_2Hg (eq. 23.33), although a variation in strategy is usually needed to prepare triorganothallium derivatives (e.g. reaction 23.34) since R_2TlX is favoured in reactions 23.31 or 23.32. The Grignard route is valuable for the synthesis of triaryl derivatives. A disadvantage of the Grignard route is that $R_3M \cdot OEt_2$ may be the isolated product.

$$MBr_3 + 3RMgBr \xrightarrow{Et_2O} R_3M + 3MgBr_2 \qquad (23.31)$$

$$MCl_3 + 3RLi \xrightarrow{\text{hydrocarbon solvent}} R_3M + 3LiCl \qquad (23.32)$$

$$2M + 3R_2Hg \longrightarrow 2R_3M + 3Hg \quad (\text{not for } M = Tl) \qquad (23.33)$$

$$2MeLi + MeI + TlI \longrightarrow Me_3Tl + 2LiI \qquad (23.34)$$

Trialkyls and triaryls of Ga, In and Tl are monomeric (trigonal planar metal centres) in solution and the gas phase. In the solid state, monomers are essentially present, but close intermolecular contacts are important in most structures. In trimethylindium, the formation of long In····C interactions (Fig. 23.11a) means that the structure can be described in terms of cyclic tetramers. Each In centre also forms an additional weak In····C interaction (356 pm) with the C atom of an adjacent tetramer to give an infinite network. The solid state structures of Me_3Ga and Me_3Tl resemble that of Me_3In. Within the planar Me_3Ga and Me_3Tl molecules, the average Ga–C and Tl–C bond distances are 196 and 230 pm, respectively. Within the tetrameric units, the Ga····C and Tl····C separations are 315 and 316 pm, respectively. Intermolecular interactions are also observed in, for example, crystalline Ph_3Ga, Ph_3In and $(PhCH_2)_3In$. Figure 23.11b shows one molecule of $(PhCH_2)_3In$, but each In atom interacts weakly with carbon atoms of phenyl rings of adjacent molecules. Dimer formation is observed in $Me_2Ga(\mu\text{-}C\equiv CPh)_2GaMe_2$ (Fig. 23.11c), and the same bonding description that we outlined for $R_2Al(\mu\text{-}C\equiv CPh)_2AlR_2$ (**23.10** and **23.11**) is appropriate.

Triorganogallium, indium and thallium compounds are air- and moisture-sensitive. Hydrolysis initially yields the linear $[R_2M]^+$ ion (which can be further hydrolysed), in contrast to the inertness of R_3B towards water and the formation of

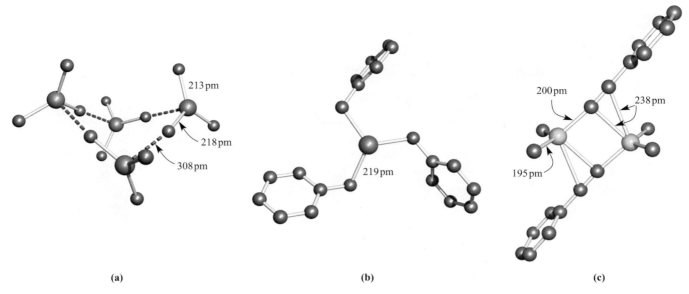

Fig. 23.11 The solid state structures (X-ray diffraction) of (a) Me$_3$In for which one of the tetrameric units (see text) is shown [A.J. Blake *et al.* (1990) *J. Chem. Soc., Dalton Trans.*, p. 2393], (b) (PhCH$_2$)$_3$In [B. Neumuller (1991) *Z. Anorg. Allg. Chem.*, vol. 592, p. 42], and (c) Me$_2$Ga(μ-C≡CPh)$_2$GaMe$_2$ [B. Tecle *et al.* (1981) *Inorg. Chem.*, vol. 20, p. 2335]. Hydrogen atoms are omitted for clarity; colour code: In, green; Ga, yellow; C, grey.

Al(OH)$_3$) from R$_3$Al. The [R$_2$Tl]$^+$ cation is also present in R$_2$TlX (X = halide), and the ionic nature of this compound differs from the covalent character of R$_2$MX for the earlier group 13 elements. Numerous adducts R$_3$M·L (L = Lewis base) are known in which the metal centre is tetrahedrally sited, e.g. Me$_3$Ga·NMe$_3$, Me$_3$Ga·NCPh, Me$_3$In·OEt$_2$, Me$_3$In·SMe$_2$, Me$_3$Tl·PMe$_3$, [Me$_4$Tl]$^-$. In compound **23.20**, donation of the lone pair comes from within the organic moiety; the GaC$_3$-unit is planar since the ligand is not flexible enough for the usual tetrahedral geometry to be adopted.

(23.20)

Species of type [E$_2$R$_4$] (single E–E bond) and [E$_2$R$_4$]$^-$ (E–E bond order 1.5) can be prepared for Ga and In provided that R is especially bulky (e.g. R = (Me$_3$Si)$_2$CH, 2,4,6-^{i}Pr$_3$C$_6$H$_2$), and reduction of [(2,4,6-^{i}Pr$_3$C$_6$H$_2$)$_4$Ga$_2$] to [(2,4,6-^{i}Pr$_3$C$_6$H$_2$)$_4$Ga$_2$]$^-$ is accompanied by a shortening of the Ga–Ga bond from 252 to 234 pm, consistent with an increase in bond order (1 to 1.5). By using even bulkier substituents, it is possible to prepare gallium(I) compounds RGa (**23.21**) starting from gallium(I) iodide. Monomeric structures have been confirmed for compound **23.21** with R' = R'' = ^{i}Pr, and with R' = H, R'' = ^{i}Pr. The latter

structure is shown in Fig. 23.12a. The space-filling representation emphasizes how the sterically demanding substituents protect the Ga atom. Compound **23.21** with R' = R'' = H crystallizes as the weakly bound dimer **23.22**, reverting to a monomer when dissolved in cyclohexane. The Ga–Ga bond in **23.22** possesses a bond order of less than 1. Reduction of **23.22** by Na leads to Na$_2$[RGaGaR], in which the dianion retains the *trans*-bent geometry of **23.22**. The Ga–Ga bond length is 235 pm, significantly shorter than in **23.22**. The salt Na$_2$[RGaGaR] was first prepared from the reaction of

R' = ^{i}Pr, ^{t}Bu or H
R'' = H or ^{i}Pr

(23.21)

Ga–Ga = 263 pm

(23.22)

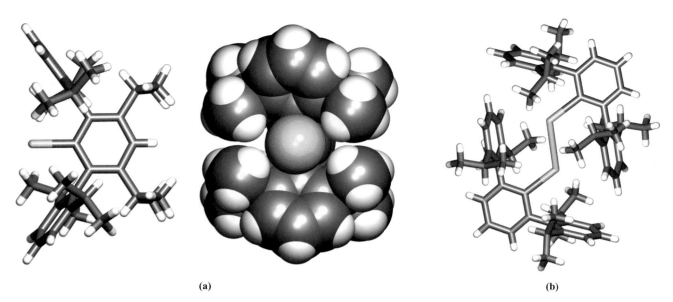

Fig. 23.12 (a) Two views of the structure (X-ray diffraction) of $\{C_6H\text{-}2,6\text{-}(C_6H_3\text{-}2,6\text{-}^iPr_2)_2\text{-}3,5\text{-}^iPr_2\}$Ga in stick and space-filling representations [Z. Zhu *et al.* (2009) *Chem. Eur. J.*, vol. 15, p. 5263]. Colour code: Ga, yellow; C, grey; H, white. (b) The structure of $\{C_6H_3\text{-}2,6\text{-}(C_6H_3\text{-}2,6\text{-}^iPr_2)_2\}_2Tl_2$ determined by X-ray diffraction [R.J. Wright *et al.* (2005) *J. Am. Chem. Soc.*, vol. 127, p. 4794]. The figure illustrates the steric crowding of the organic substituents around the central Tl_2 core; Tl–Tl = 309 pm, and angle C–Tl–Tl = 119.7°. Colour code: Tl, yellow; C, grey; H, white.

RGaCl$_2$ and Na in Et$_2$O, and it has been proposed that $[RGaGaR]^{2-}$ contains a gallium–gallium triple bond. The nature of this bonding has been the subject of intense theoretical interest. On the one hand, there is support for a Ga≡Ga formulation, while on the other, it is concluded that factors such as Ga–Na$^+$–Ga interactions contribute to the short Ga–Ga distance. More recent experimental observations indicate that the Ga–Ga interaction in Na$_2$[RGaGaR] is best described as consisting of a single bond, augmented both by Ga–Na$^+$–Ga inter-actions and by the weak interaction that is present in the precursor **23.22**.[†]

The Ga(I) chemistry described above must be compared with the following observations for In and Tl, in which the nature of the organic group R plays a critical role. The reaction of InCl with LiR when R = C$_6$H$_3$-2,6-(C$_6$H$_2$-2,4,6-^{i}Pr$_3$)$_2$ yields RIn, an analogue of compound **23.21** with R' = ^{i}Pr. The monomeric nature of RIn in the solid state has been confirmed by X-ray diffraction data. However, when R = C$_6$H$_3$-2,6-(C$_6$H$_3$-2,6-^{i}Pr$_2$)$_2$, monomeric RIn exists in cyclo-hexane solutions, but dimeric RInInR is present in the solid state. The thallium analogue of compound **23.22** is prepared by reaction 23.35. Dimeric RTlTlR (Fig. 23.12b) has the same *trans*-bent structure as its gallium and indium analogues. In hydrocarbon solvents, the dimer dissociates into monomers,

consistent with a weak Tl–Tl bond. The monomer can be stabilized by formation of the adduct RTl·B(C$_6$F$_5$)$_3$. If R is C$_6$H$_3$-2,6-(C$_6$H$_3$-2,6-Me$_2$)$_2$, reaction of LiR with TlCl produces RTl in solution, but this crystallizes as a trimer containing a triangular Tl$_3$ unit. Overall, these data illustrate how subtle changes in the organic group can lead to (not readily predictable) structural variations of the organometallic species in solution and the solid state.

$$2LiR + 2TlCl \xrightarrow[-LiCl]{Et_2O,\ 195\ K} 2RTl \longrightarrow RTlTlR$$

R = C$_6$H$_3$-2,6-(C$_6$H$_3$-2,6-^{i}Pr$_2$)$_2$ in solid state

$$(23.35)$$

The 2,6-dimesitylphenyl substituent is also extremely steri-cally demanding, and reduction of (2,6-Mes$_2$C$_6$H$_3$)GaCl$_2$ with Na yields Na$_2$[(2,6-Mes$_2$C$_6$H$_3$)$_3$Ga$_3$]. The [(2,6-Mes$_2$C$_6$H$_3$)$_3$Ga$_3$]$^{2-}$ anion possesses the cyclic structure (**23.23**) and is a 2π-electron aromatic system.

[†] For further details, see: J. Su, X.-W. Li, R.C. Crittendon and G.H. Robinson (1997) *J. Am. Chem. Soc.*, vol. 119, p. 5471; G.H. Robinson (1999) *Acc. Chem. Res.*, vol. 32, p. 773; N.J. Hardman, R.J. Wright, A.D. Phillips and P.P. Power (2003) *J. Am. Chem. Soc.*, vol. 125, p. 2667.

Mes = 2,4,6-Me$_3$C$_6$H$_2$

(23.23)

In eq. 13.51, we illustrated the use of the metastable GaBr as a precursor to multinuclear Ga-containing species. Gallium(I) bromide has also been used as a precursor to a number of organogallium clusters. For example, one of the products of the reaction of GaBr with $(Me_3Si)_3CLi$ in toluene at 195 K is **23.24**.

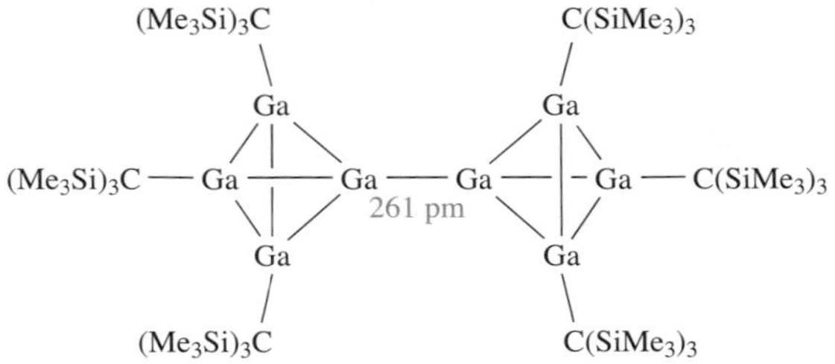

average Ga–Ga in tetrahedral cluster = 263 pm

(23.24)

Worked example 23.3   Reactions of $\{(Me_3Si)_3C\}_4E_4$ (E = Ga or In)

The reaction of the tetrahedral cluster $\{(Me_3Si)_3C\}_4Ga_4$ with I_2 in boiling hexane results in the formation of $\{(Me_3Si)_3CGaI\}_2$ and $\{(Me_3Si)_3CGaI_2\}_2$. In each compound there is only one Ga environment. Suggest structures for these compounds and state the oxidation state of Ga in the starting material and products.

The starting cluster is a gallium(I) compound:

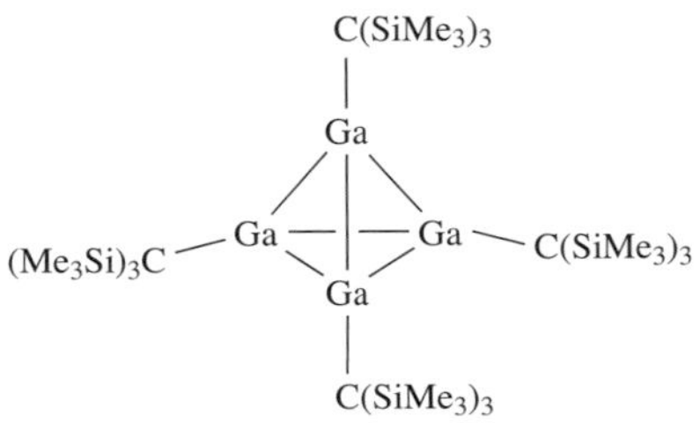

I_2 oxidizes this compound and possible oxidation states are Ga(II) (e.g. in a compound of type $R_2Ga–GaR_2$) and Ga(III). $\{(Me_3Si)_3CGaI\}_2$ is related to compounds of type $R_2Ga–GaR_2$; steric factors may contribute towards a non-planar conformation:

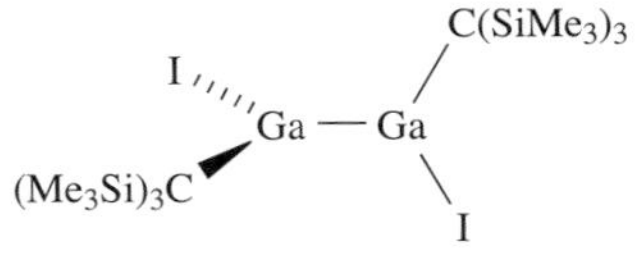

Further oxidation by I_2 results in the formation of the Ga(III) compound $\{(Me_3Si)_3CGaI_2\}_2$ and a structure consistent with equivalent Ga centres is:

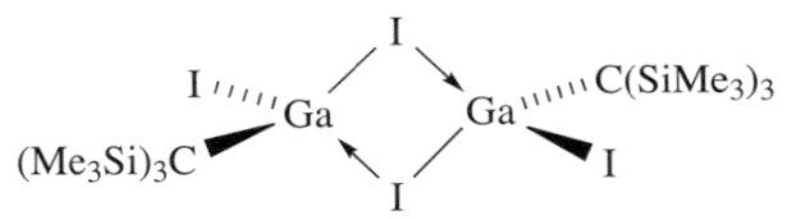

Self-study exercises

1. The Br_2 oxidation of $\{(Me_3Si)_3C\}_4In_4$ leads to the formation of the In(II) compound $\{(Me_3Si)_3C\}_4In_4Br_4$ in which each In atom retains a tetrahedral environment. Suggest a structure for the product.

2. $\{(Me_3Si)_3CGaI\}_2$ represents a Ga(II) compound of type $R_2Ga_2I_2$. However, 'Ga_2I_4', which may appear to be a related compound, is ionic. Comment on this difference.

3. A staggered conformation is observed in the solid state for $\{(Me_3Si)_3CGaI\}_2$. It has been suggested that a contributing factor may be hyperconjugation involving Ga–I bonding electrons. What acceptor orbital is available for hyperconjugation, and how does this interaction operate?

[*Ans.* W. Uhl *et al.* (2003) *Dalton Trans.*, p. 1360.]

Cyclopentadienyl complexes illustrate the increase in stability of the M(I) oxidation state as group 13 is descended, a consequence of the thermodynamic $6s$ inert pair effect (see Box 13.4). Cyclopentadienyl derivatives of Ga(III) which have been prepared (eqs. 23.36 and 23.37) and structurally characterized include Cp_3Ga and $CpGaMe_2$.

$$GaCl_3 + 3Li[Cp] \longrightarrow Cp_3Ga + 3LiCl \qquad (23.36)$$

$$Me_2GaCl + Na[Cp] \longrightarrow CpGaMe_2 + NaCl \qquad (23.37)$$

The structure of $CpGaMe_2$ resembles that of $CpAlMe_2$ (Fig. 23.9a), and Cp_3Ga is monomeric with three η^1-Cp groups bonded to trigonal planar Ga (Fig. 23.13a). The In(III) compound Cp_3In is prepared from NaCp and $InCl_3$, but is structurally different from Cp_3Ga. Solid Cp_3In contains polymeric chains in which each In atom is distorted tetrahedral (Fig. 23.13b).

The reaction of $(\eta^5\text{-}C_5Me_5)_3Ga$ with HBF_4 results in the formation of $[(C_5Me_5)_2Ga]^+[BF_4]^-$. In solution, the C_5Me_5 groups are fluxional down to 203 K, but in the solid state the complex is a dimer (**23.25**) containing $[(\eta^1\text{-}C_5Me_5)(\eta^3\text{-}C_5Me_5)Ga]^+$ ions. The structure of $[(C_5Me_5)_2Ga]^+$ contrasts with that of $[(C_5Me_5)_2Al]^+$, in which the C_5-rings are coparallel.

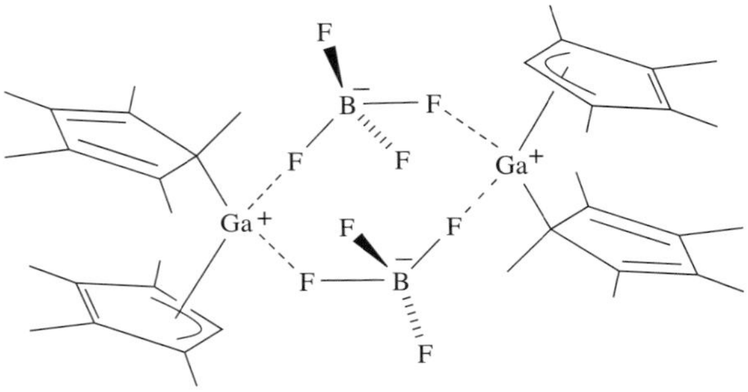

(23.25)

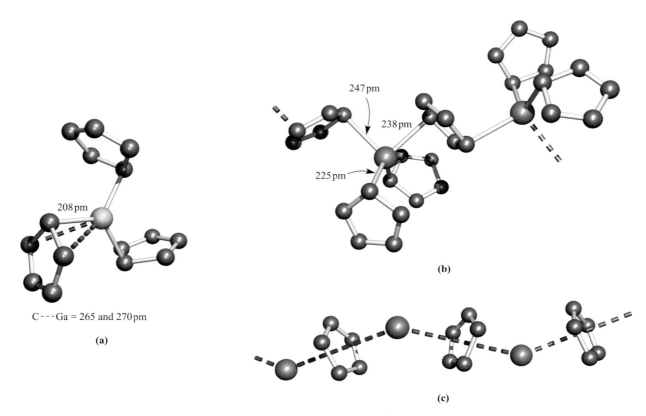

C⋯Ga = 265 and 270 pm

(a)

(b)

(c)

Fig. 23.13 The solid state structures (X-ray diffraction) of (a) monomeric (η^1-Cp)$_3$Ga [O.T. Beachley *et al.* (1985) *Organometallics*, vol. 4, p. 751], (b) polymeric Cp$_3$In [F.W.B. Einstein *et al.* (1972) *Inorg. Chem.*, vol. 11, p. 2832] and (c) polymeric CpIn [O.T. Beachley *et al.* (1988) *Organometallics*, vol. 7, p. 1051]; the zigzag chain is emphasized by the red hashed line. Hydrogen atoms are omitted for clarity; colour code: Ga, yellow; In, green; C, grey.

We saw earlier that gallium(I) halides can be used to synthesize RGa compounds (**23.21**). Similarly, metastable solutions of GaCl have been used to prepare (C$_5$Me$_5$)Ga by reactions with (C$_5$Me$_5$)Li or (C$_5$Me$_5$)$_2$Mg. An alternative route is the reductive dehalogenation of (C$_5$Me$_5$)GaI$_2$ using potassium with ultrasonic activation. In the gas phase and in solution, (C$_5$Me$_5$)Ga is monomeric, but in the solid state, hexamers are present.

On moving down group 13, the number of M(I) cyclopentadienyl derivatives increases, with a wide range being known for Tl(I). The condensation of In vapour (at 77 K) onto C$_5$H$_6$ gives CpIn, and CpTl is readily prepared by reaction 23.38.

$$C_5H_6 + TlX \xrightarrow{\text{KOH/H}_2\text{O}} CpTl + KX \qquad \text{e.g. X = halide}$$
$$(23.38)$$

Both CpIn and CpTl are monomeric in the gas phase, but in the solid, they possess the polymeric chain structure shown in Fig. 23.13c. The cyclopentadienyl derivatives (C$_5$R$_5$)M (M = In, Tl) are structurally diverse in the solid state, e.g. for R = PhCH$_2$ and M = In or Tl, 'quasi-dimers' **23.26** are present (there may or may not be a meaningful metal–metal interaction), and (η^5-C$_5$Me$_5$)In forms hexameric clusters. An important reaction of (η^5-

C$_5$Me$_5$)In is with CF$_3$SO$_3$H which gives the triflate salt of In$^+$. The salt In[O$_2$SCF$_3$] is air-sensitive and hygroscopic but is, nonetheless, a convenient source of indium(I) as an alternative to In(I) halides.

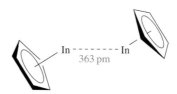

PhCH$_2$ substituents omitted

(23.26)

One use of CpTl is as a cyclopentadienyl transfer reagent to *d*-block metal ions, but it can also act as an acceptor of Cp$^-$, reacting with Cp$_2$Mg to give [Cp$_2$Tl]$^-$. This can be isolated as the salt [CpMgL][Cp$_2$Tl] upon the addition of the chelating ligand L = Me$_2$NCH$_2$CH$_2$NMeCH$_2$CH$_2$NMe$_2$. The anion [Cp$_2$Tl]$^-$ is isoelectronic with Cp$_2$Sn and possesses a structure in which the η^5-Cp rings are mutually tilted. The tilt angle (defined in Fig. 23.14c for the structurally related Cp$_2$Si) is 157°. Although this ring orientation implies the presence of a stereochemically active lone pair, it has been shown theoretically that there is only a small energy difference

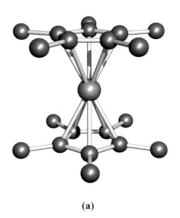
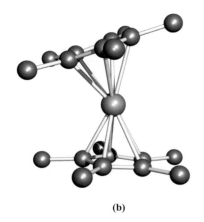
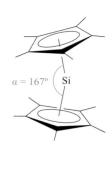

(a)   **(b)**   **(c)**

 Fig. 23.14 The solid state structure of $(\eta^5\text{-}C_5Me_5)_2Si$ contains two independent molecules. (a) In the first molecule, the cyclopentadienyl rings are co-parallel, while (b) in the other molecule they are mutually tilted; (c) the tilt angle is measured as angle α [P. Jutzi *et al.* (1986) *Angew. Chem. Int. Ed.*, vol. 25, p. 164]. Hydrogen atoms are omitted for clarity; colour code: Si, pink; C, grey.

$(3.5\,kJ\,mol^{-1})$ between this structure and one in which the η^5-Cp rings are parallel (i.e. as in Fig. 23.14a). We return to this point at the end of the next section.

23.5 Group 14

Organo-compounds of the group 14 elements include some important commercial products, including poly-siloxanes (*silicones*) discussed in Section 14.10 and

Box 14.12. Organotin compounds are employed as poly-vinylchloride (PVC) stabilizers (against degradation by light and heat), antifouling paints on ships, wood preservatives and agricultural pesticides (see Box 23.3). Leaded motor fuels contain the anti-knock agent Et_4Pb, although this use has declined on environmental grounds (see Fig. 14.3). Several general properties of the organo-derivatives of the group 14 elements, E, are as follows:

- in most compounds, the group 14 element is tetravalent;
- the E−C bonds are generally of low polarity;

APPLICATIONS

Box 23.3 Commercial uses and environmental problems of organotin compounds

Organotin(IV) compounds have a wide range of applications, with catalytic and biocidal properties being of particular importance. The compounds below are selected examples:

- $^nBu_3Sn(OAc)$ (produced by reacting $^nBu_3SnCl$ and NaOAc) is an effective fungicide and bactericide; it also has applications as a polymerization catalyst.
- $^nBu_2Sn(OAc)_2$ (from $^nBu_2SnCl_2$ and NaOAc) is used as a polymerization catalyst and a stabilizer for PVC.
- $(cyclo\text{-}C_6H_{11})_3SnOH$ (formed by alkaline hydrolysis of the corresponding chloride) and $(cyclo\text{-}C_6H_{11})_3Sn(OAc)$ (produced by treating $(cyclo\text{-}C_6H_{11})_3SnOH$ with AcOH) are used widely as insecticides in fruit orchards and vineyards.
- $^nBu_3SnOSn^nBu_3$ (formed by aqueous NaOH hydrolysis of $^nBu_3SnCl$) has uses as an algicide, fungicide and wood-preserving agent.
- $^nBu_3SnCl$ (a product of the reaction of $^nBu_4Sn$ and $SnCl_4$) is a bactericide and fungicide.
- Ph_3SnOH (formed by base hydrolysis of Ph_3SnCl) is used as an agricultural fungicide for crops such as potatoes, sugar beet and peanuts.
- The cyclic compound $(^nBu_2SnS)_3$ (formed by reacting $^nBu_2SnCl_2$ with Na_2S) is used as a stabilizer for PVC.

Tributyltin derivatives have been used as antifouling agents, applied to the underside of ships' hulls to prevent the build-up of, for example, barnacles. Global legislation now bans or greatly restricts the use of organotin-based antifouling agents on environmental grounds. Environmental risks associated with the uses of organotin compounds as pesticides, fungicides and PVC stabilizers are also a cause for concern and are the subject of regular assessments. The toxicity of organotin compounds to aquatic life follows the order triorganotin > diorganotin > monoorganotin species.

Further reading

M.A. Champ (2003) *Marine Pollut. Bull.*, vol. 46, p. 935 – 'Economic and environmental impacts on ports and harbors from the convention to ban harmful marine antifouling systems'.

K.A. Dafforn, J.A. Lewis and E.L. Johnston (2011) *Marine Pollut. Bull.*, vol. 62, p. 453 – 'Antifouling strategies: History and regulation, ecological impacts and mitigation'.

A.G. Davies (2010) *J. Chem. Res.*, vol. 34, p. 181 – 'Organotin compounds in technology and industry'.

- their stability towards all reagents decreases from Si to Pb;
- in contrast to the group 13 organometallics, derivatives of the group 14 elements are less susceptible to nucleophilic attack.

Silicon

Silicon tetraalkyl and tetraaryl derivatives (R_4Si), as well as alkyl or aryl silicon halides (R_nSiCl_{4-n}, $n = 1–3$) can be prepared by reaction types 23.39–23.43. Note that variation in stoichiometry provides flexibility in synthesis, although the product specificity may be influenced by steric requirements of the organic substituents. Reaction 23.39 is used industrially (the *Rochow* or *Direct process*).

$$n MeCl + Si/Cu \xrightarrow[\text{alloy}]{573\,K} Me_n SiCl_{4-n} \qquad (23.39)$$

$$SiCl_4 + 4RLi \longrightarrow R_4Si + 4LiCl \qquad (23.40)$$

$$SiCl_4 + RLi \longrightarrow RSiCl_3 + LiCl \qquad (23.41)$$

$$SiCl_4 + 2RMgCl \xrightarrow{Et_2O} R_2SiCl_2 + 2MgCl_2 \qquad (23.42)$$

$$Me_2SiCl_2 + {}^tBuLi \longrightarrow {}^tBuMe_2SiCl + LiCl \qquad (23.43)$$

The structures of the products of reactions 23.39–23.43 are all similar: monomeric, with tetrahedrally sited Si and resembling their C analogues.

Silicon–carbon single bonds are relatively strong ($318\,kJ\,mol^{-1}$) and R_4Si derivatives possess high thermal stabilities. The stability of the Si–C bond is further illustrated by the fact that chlorination of Et_4Si gives $(ClCH_2CH_2)_4Si$, in contrast to the chlorination of R_4Ge or R_4Sn which yields R_nGeCl_{4-n} or R_nSnCl_{4-n} (see eq. 23.53). An important reaction of Me_nSiCl_{4-n} ($n = 1–3$) is hydrolysis to produce polysiloxanes (e.g. eq. 23.44 and see Section 14.10 and Box 14.12).

$$\left.\begin{array}{l} Me_3SiCl + H_2O \longrightarrow Me_3SiOH + HCl \\ 2Me_3SiOH \longrightarrow Me_3SiOSiMe_3 + H_2O \end{array}\right\} \qquad (23.44)$$

(23.27)

The reaction of Me_3SiCl with NaCp leads to **23.27**, in which the cyclopentadienyl group is η^1. Related η^1-complexes include $(\eta^1$-$C_5Me_5)_2SiBr_2$ which reacts with anthracene/potassium to give the diamagnetic *silylene* $(\eta^5$-$C_5Me_5)_2Si$. In the solid state, two independent molecules are present (Fig. 23.14) which differ in the relative orientations of the cyclopentadienyl rings. In one molecule, the two C_5-rings are parallel and staggered (compare Cp_2Mg) whereas in the other, they are tilted. We return to this at the end of Section 23.5. Reaction 23.45 shows the formation of the proton-transfer agent

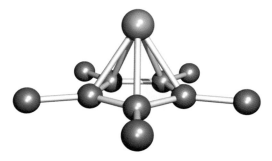

Fig. 23.15 The structure of the $[(\eta^5$-$C_5Me_5)Si]^+$ cation present in the compound $[(\eta^5$-$C_5Me_5)Si][B(C_6F_5)_4]$, determined by X-ray diffraction [P. Jutzi *et al.* (2004) *Science*, vol. 305, p. 849]. Hydrogen atoms are omitted; colour code: Si, pink; C, grey.

$[C_5Me_5H_2]^+[B(C_6F_5)_4]^-$. This reagent removes one of the $[\eta^5$-$C_5Me_5]^-$ ligands from $(\eta^5$-$C_5Me_5)_2Si$ (reaction 23.46) to give the $[(\eta^5$-$C_5Me_5)Si]^+$ cation (Fig. 23.15). This cation is important in being the only stable derivative of the $[HSi]^+$ ion which has been observed in the solar spectrum, and is thought to be present in interstellar space.

$$[H(OEt_2)_2]^+[B(C_6F_5)_4]^- \qquad (23.45)$$

$[B(C_6F_5)_4]^-$

$$[C_5Me_5H_2]^+[B(C_6F_5)_4]^- + (\eta^5\text{-}C_5Me_5)_2Si$$
$$\xrightarrow{CH_2Cl_2} [(\eta^5\text{-}C_5Me_5)Si]^+[B(C_6F_5)_4]^- + 2Me_5C_5H \qquad (23.46)$$

The reactions between R_2SiCl_2 and alkali metals or alkali metal naphthalides give *cyclo*-$(R_2Si)_n$ by loss of Cl^- and Si–Si bond formation. Bulky R groups favour small rings (e.g. $(2,6\text{-}Me_2C_6H_3)_6Si_3$ and ${}^tBu_6Si_3$) while smaller R substituents encourage the formation of large rings (e.g. $Me_{12}Si_6$, $Me_{14}Si_7$ and $Me_{32}Si_{16}$). Reaction 23.47 is designed to provide a specific route to a particular ring size.

$$Ph_2SiCl_2 + Li(SiPh_2)_5Li \longrightarrow cyclo\text{-}Ph_{12}Si_6 + 2LiCl \qquad (23.47)$$

In Section 9.9, we introduced the weakly coordinating carborane anions $[CHB_{11}R_5X_6]^-$ (R = H, Me, Cl and X = Cl, Br, I), and Fig. 9.6c showed the structure of ${}^iPr_3Si(CHB_{11}H_5Cl_6)$ which approaches that of the ion-pair $[{}^iPr_3Si]^+[CHB_{11}H_5Cl_6]^-$. The reaction of the strong electrophile $Et_3Si(CHB_{11}Me_5Br_6)$ with $Mes_3Si(CH_2CH=CH_2)$ (Mes = mesityl) gives $[Mes_3Si][CHB_{11}Me_5Br_6]$. In the solid state, this contains well-separated $[Mes_3Si]^+$ and $[CHB_{11}Me_5Br_6]^-$ ions, giving the first example of a free silylium ion. The trigonal planar Si centre is consistent with sp^2 hybridization, but steric hindrance between the mesityl groups leads to a deviation from the overall planarity that would be needed to maximize C $2p$–Si $3p$ π-overlap (diagram **23.28**). The actual arrangement of the mesityl groups (**23.29**) balances steric and electronic requirements.

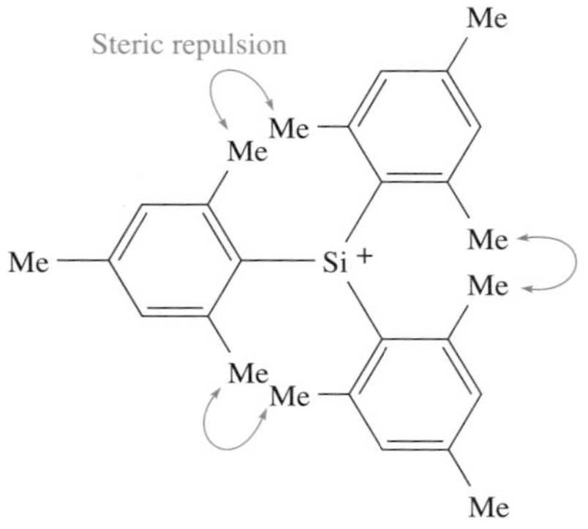

Steric repulsion

Planar arrangement of aryl groups would optimize π-bonding

(23.28)

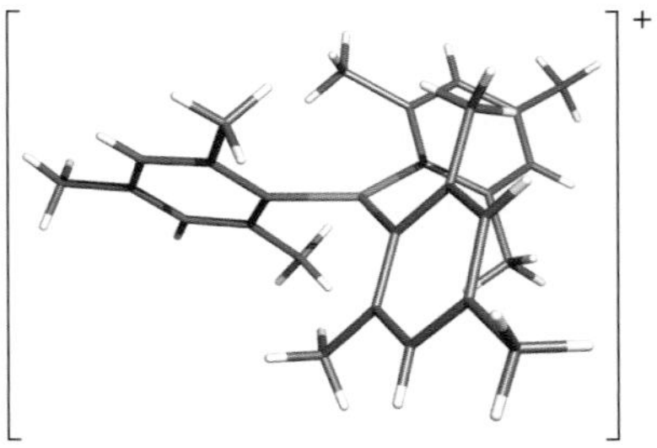

(23.29)

Silylenes, R_2Si (analogues of carbenes), can be formed by a variety of methods, for example, the photolysis of cyclic or linear organopolysilanes. As expected, R_2Si species are highly reactive, undergoing many reactions analogous to those typical of carbenes. Stabilization of R_2Si can be achieved by using sufficiently bulky substituents, and electron diffraction data confirm the bent structure of $\{(Me_3Si)_2HC\}_2Si$ ($\angle C$–Si–C = 97°).

In Section 14.3, we discussed the use of bulky substituents to stabilize $R_2Si=SiR_2$ and $RSi\equiv SiR$ compounds. The sterically demanding 2,4,6-$^iPr_3C_6H_2$ group has been used to stabilize **23.30**, the first example of a compound containing conjugated Si=Si bonds. An unusual feature of **23.30** is the preference for the s-*cis* conformation in both solution and the solid state.

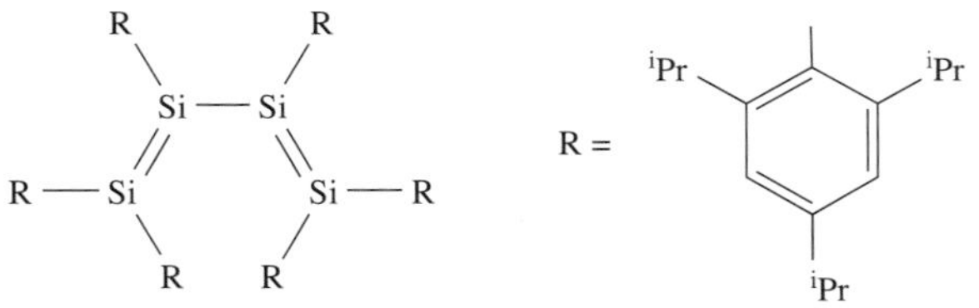

(23.30)

The spatial arrangement of two conjugated double bonds about the central single bond is described as being **s-cis** and **s-trans**, defined as follows:

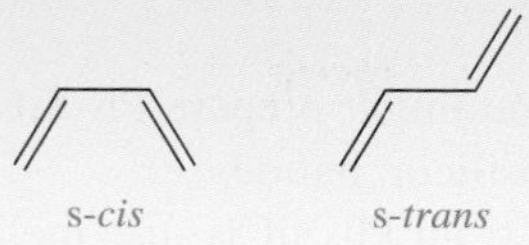

s-*cis*   s-*trans*

Worked example 23.4   Organosilicon hydrides

The reaction of Ph_2SiH_2 with potassium metal in 1,2-dimethoxyethane (DME) in the presence of 18-crown-6 yields a salt of $[Ph_3SiH_2]^-$ in which the hydride ligands are *trans* to each other. The salt has the formula $[X][Ph_3SiH_2]$. The solution ^{29}Si NMR spectrum shows a triplet ($J = 130\,Hz$) at δ −74 ppm. Explain the origin of the triplet. What signals arising from the anion would you expect to observe in the solution ^1H NMR spectrum of $[X][Ph_3SiH_2]$?

First, draw the expected structure of $[Ph_3SiH_2]^-$. The question states that the hydride ligands are *trans*, and a trigonal bipyramidal structure is consistent with the VSEPR model:

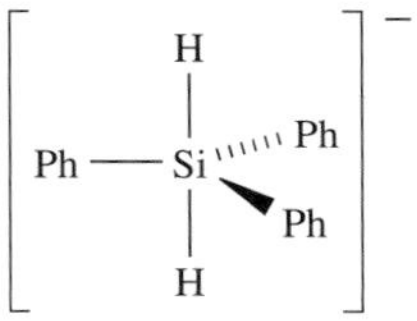

In the ^{29}Si NMR spectrum, the triplet arises from coupling of the ^{29}Si nucleus to two equivalent ^1H ($I = \frac{1}{2}$) nuclei.

Signals in the ^1H NMR spectrum that can be assigned to $[Ph_3SiH_2]^-$ arise from the phenyl and hydride groups. The three Ph groups are equivalent (all equatorial) and, in theory, give rise to three multiplets (δ 7–8 ppm) for *ortho*-, *meta*- and *para*-H atoms. In practice, these signals may overlap. The equivalent hydride ligands give rise to one signal. Silicon has one isotope that is NMR active: ^{29}Si, 4.7%, $I = \frac{1}{2}$ (see Table 14.1). We know from the ^{29}Si NMR spectrum that there is spin–spin coupling between the directly bonded ^{29}Si and ^1H nuclei. Considering these protons in the ^1H NMR spectrum, 95.3% of the protons are attached to non-spin active Si and give rise to a singlet; 4.7% are attached to ^{29}Si and give rise to a doublet ($J = 130\,Hz$). The signal will appear as a small doublet superimposed on a singlet (see Fig. 4.23).

These questions refer to the experiment described in the worked example.

1. Suggest how you might prepare Ph_2SiH_2 starting from a suitable organosilicon halide.

   [*Ans.* Start from Ph_2SiCl_2; use method of eq. 10.39]

2. Draw the structure of 18-crown-6. What is its role in this reaction? Suggest an identity for cation $[X]^+$.

   [*Ans.* See Fig. 11.8 and discussion]

[For the original literature, see: M.J. Bearpark *et al.* (2001) *J. Am. Chem. Soc.*, vol. 123, p. 7736.]

Germanium

There are similarities between the methods of preparation of compounds with Ge−C and Si−C bonds: compare reaction 23.48 with 23.39, 23.49 with 23.41, 23.50 with 23.42, and 23.51 with the synthesis of $Me_3Si(\eta^1\text{-}Cp)$.

$$nRCl + Ge/Cu \xrightarrow{\Delta} R_nGeCl_{4-n} \quad R = \text{alkyl or aryl}$$
$$(23.48)$$

$$GeCl_4 + RLi \longrightarrow RGeCl_3 + LiCl \quad\quad (23.49)$$

$$GeCl_4 + 4RMgCl \xrightarrow{Et_2O} R_4Ge + 4MgCl_2 \quad (23.50)$$

$$R_3GeCl + Li[Cp] \longrightarrow R_3Ge(\eta^1\text{-}Cp) + LiCl \quad (23.51)$$

Tetraalkyl and tetraaryl germanium compounds possess monomeric structures with tetrahedrally sited germanium. They are thermally stable and tend to be chemically inert. Halogenation requires a catalyst (eqs. 23.52 and 23.53). Chlorides can be obtained from the corresponding bromides or iodides by halogen exchange (eq. 23.54). The presence of halo-substituents increases reactivity (e.g. eq. 23.55) and makes the halo-derivatives synthetically more useful than R_4Ge compounds.

$$2Me_4Ge + SnCl_4 \xrightarrow{AlCl_3} 2Me_3GeCl + Me_2SnCl_2 \quad (23.52)$$

$$R_4Ge + X_2 \xrightarrow{AlX_3} R_3GeX + RX \quad (X = Br, I) \quad (23.53)$$

$$R_3GeBr + AgCl \longrightarrow R_3GeCl + AgBr \quad\quad (23.54)$$

$$R_3GeX \xrightarrow{KOH/EtOH, H_2O} R_3GeOH \quad\quad (23.55)$$

A simple method of preparing $RGeCl_3$ (R = alkyl or alkenyl) is by the passage of $GeCl_4$ and RCl vapours over grains of Ge heated at 650–800 K. The reaction proceeds by intermediate carbene-like $GeCl_2$ which inserts into a C–Cl bond (eq. 23.56).

$$Ge + GeCl_4 \longrightarrow 2\{Cl_2Ge:\} \xrightarrow{2RCl} 2RGeCl_3 \quad (23.56)$$

The availability of Ge(II) halides (see Section 14.8) means that the synthesis of $(\eta^5\text{-}C_5R_5)_2Ge$ derivatives does not require a reduction step as was the case for the silicon analogues described above. Reaction 23.57 is a general route to $(\eta^5\text{-}C_5R_5)_2Ge$, which exist as monomers in the solid, solution and vapour states.

$$GeX_2 + 2Na[C_5R_5] \longrightarrow (\eta^5\text{-}C_5R_5)_2Ge + 2NaX$$
$$(X = Cl, Br) \quad (23.57)$$

X-ray diffraction studies for Cp_2Ge and $\{\eta^5\text{-}C_5(CH_2Ph)_5\}_2Ge$ confirm the bent structure type illustrated in Figs. 23.14b and c for $(\eta^5\text{-}C_5Me_5)_2Si$. However, in $\{\eta^5\text{-}C_5Me_4(SiMe_2{}^tBu)\}_2Ge$, the two C_5-rings are coparallel and mutually staggered. The preferences for tilted versus coparallel rings are discussed further at the end of Section 23.5. Reaction 23.58 generates $[(\eta^5\text{-}C_5Me_5)Ge]^+$ which is structurally analogous to $[(\eta^5\text{-}C_5Me_5)Si]^+$ (Fig. 23.15). However, $[(\eta^5\text{-}C_5Me_5)Ge]^+$ (like $[(\eta^5\text{-}C_5Me_5)Sn]^+$ and $[(\eta^5\text{-}C_5Me_5)Pb]^+$) exists in the presence of more nucleophilic counter-ions than does $[(\eta^5\text{-}C_5Me_5)Si]^+$, consistent with the increasing stability of the +2 oxidation state on descending group 14.

$$(\eta^5\text{-}C_5Me_5)_2Ge + HBF_4\cdot Et_2O$$
$$\longrightarrow [(\eta^5\text{-}C_5Me_5)Ge][BF_4] + C_5Me_5H + Et_2O$$
$$(23.58)$$

Organogermanium(II) compounds are a growing family. Germylenes (R_2Ge) include the highly reactive Me_2Ge which can be prepared by reaction 23.59. Photolysis reaction 23.60 shows a general strategy to form R_2Ge.

$$Me_2GeCl_2 + 2Li \xrightarrow{THF} Me_2Ge + 2LiCl \quad (23.59)$$

$$Me(GeR_2)_{n+1}Me \xrightarrow{h\nu} R_2Ge + Me(GeR_2)_nMe \quad (23.60)$$

Using very sterically demanding R groups can stabilize the R_2Ge species. Thus, compound **23.31** is stable at room temperature. The bent structure of $\{(Me_3Si)_2HC\}_2Ge$ has been confirmed by electron diffraction ($\angle C-Ge-C = 107°$).

(23.31)

Double bond formation between C and Ge was mentioned in Section 14.3, and the formation of Ge=Ge bonds to give digermenes can be achieved (eqs. 23.61 and 23.62) if particularly bulky substituents (e.g. 2,4,6-Me$_3$C$_6$H$_2$, 2,6-Et$_2$C$_6$H$_3$, 2,6-^{i}Pr$_2$C$_6$H$_3$) are used to stabilize the system.

$$2RR'GeCl_2 \xrightarrow{\text{LiC}_{10}\text{H}_8, \text{ DME}} RR'Ge=GeRR' + 4LiCl$$

LiC$_{10}$H$_8$ = lithium naphthalide

(23.61)

$$2R_2Ge\{C(SiMe_3)_3\}_2 \xrightarrow{h\nu} 2\{R_2Ge:\} + (Me_3Si)_3CC(SiMe_3)_3$$

$$\downarrow$$

$$R_2Ge=GeR_2 \qquad (23.62)$$

Data for several structurally characterized digermenes confirm a non-planar Ge$_2$C$_4$-framework analogous to that observed for distannenes discussed in the next section (see Fig. 23.19). Digermenes are stable in the solid state in the absence of air and moisture, but in solution they show a tendency to dissociate into R$_2$Ge, the extent of dissociation depending on R. With 2,4,6-^{i}Pr$_3$C$_6$H$_2$ as substituent, R$_2$Ge=GeR$_2$ remains as a dimer in solution and can be used to generate a tetragermabuta-1,3-diene (scheme 23.63). The precursors are made *in situ* from R$_2$Ge=GeR$_2$ by treatment with Li or with Li followed by 2,4,6-Me$_3$C$_6$H$_2$Br.

(23.63)

Conditions are critical in the above reaction since prolonged reaction of R$_2$Ge=GeR$_2$ (R = 2,4,6-^{i}Pr$_3$C$_6$H$_2$) with Li in 1,2-dimethoxyethane (DME) results in the formation of **23.32**.

(**23.32**)

The reduction of R$_2$GeGeR$_2$ to [R$_2$GeGeR$_2$]$^{2-}$ is more difficult than conversion of RGeGeR to [RGeGeR]$^{2-}$ (see

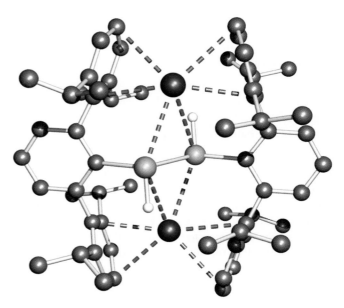

Fig. 23.16 The solid state structure (determined by X-ray diffraction) of K$_2$[Ge$_2$H$_2$\{C$_6$H$_3$-2,6-(C$_6$H$_3$-2,6-^{i}Pr$_2$)$_2$\}$_2$] with H atoms omitted except for those attached to the Ge atoms [A.F. Richards *et al.* (2004) *J. Am. Chem. Soc.*, vol. 126, p. 10530]. Hashed lines highlight the stabilizing influence of the K$^+$ ions. Colour code: Ge, orange; K, blue; C, grey; H, white.

below), but it can be achieved with careful choice of substituents and reaction conditions (eq. 23.64). The solid state structures of these salts illustrate the influential role of the metal ions. Not only do the ions interact with the Ge$_2$ unit (shown for the K$^+$ salt in Fig. 23.16), but the geometry of the central Ge$_2$H$_2$C$_2$-unit varies with M$^+$. The Li$^+$ salt contains a planar Ge$_2$H$_2$C$_2$-unit, whereas it is *trans*-trigonal pyramidal in the K$^+$ salt (Fig. 23.16). In the Na$^+$ salt, the H atoms bridge the Ge–Ge bond.

$$RHGeGeHR \xrightarrow[\text{M = Na or K in toluene}]{\text{M = Li in Et}_2\text{O/THF}} M_2[RHGeGeHR]$$

(23.64)

R=C$_6$H$_3$-2,6-(C$_6$H$_3$-2,6-^{i}Pr$_2$)$_2$

The formation of RGeGeR has been achieved by using the extremely bulky substituent R = 2,6-(2,6-^{i}Pr$_2$C$_6$H$_3$)$_2$C$_6$H$_3$. The solid state structure of RGeGeR shows a *trans*-bent conformation with a C–Ge–Ge bond angle of 129° and Ge–Ge bond length of 228.5 pm. Theoretical studies suggest a Ge–Ge bond order of ≈2.5. RGeGeR is formed by reduction of RGeCl using Li, Na or K. However, the conditions must be carefully controlled, otherwise the predominant products are the singly and doubly reduced derivatives [RGeGeR]$^-$ (a radical anion) and [RGeGeR]$^{2-}$. Analogous reactions occur when RSnCl is reduced. In both K[RGeGeR] and Li$_2$[RGeGeR], a *trans*-bent geometry is observed, as in RGeGeR. Each cation is involved in significant interactions with the anion (i.e. as in Fig. 23.16).

Tin

Some features that set organotin chemistry apart from organosilicon or organogermanium chemistries are the:

- greater accessibility of the +2 oxidation state;
- greater range of possible coordination numbers;
- presence of halide bridges (see Section 14.8).

Reactions 23.65–23.67 illustrate synthetic approaches to R_4Sn compounds, and organotin halides can be prepared by routes equivalent to reactions 23.39 and 23.48, redistribution reactions from anhydrous $SnCl_4$ (eq. 23.68), or from Sn(II) halides (eq. 23.69). Using R_4Sn in excess in reaction 23.68 gives a route to R_3SnCl. Reaction 23.66 is used industrially for the preparation of tetrabutyltin and tetraoctyltin. Commercial applications of organotin compounds are highlighted in Box 23.3.

$$4RMgBr + SnCl_4 \longrightarrow R_4Sn + 4MgBrCl \qquad (23.65)$$

$$3SnCl_4 + 4R_3Al \xrightarrow{R'_2O} 3R_4Sn + 4AlCl_3 \qquad (23.66)$$

$$^nBu_2SnCl_2 + 2\,^nBuCl + 4Na \longrightarrow \,^nBu_4Sn + 4NaCl \qquad (23.67)$$

$$R_4Sn + SnCl_4 \xrightarrow{298\,K} R_3SnCl + RSnCl_3 \xrightarrow{500\,K} 2R_2SnCl_2 \qquad (23.68)$$

$$SnCl_2 + Ph_2Hg \longrightarrow Ph_2SnCl_2 + Hg \qquad (23.69)$$

Tetraorganotin compounds tend to be colourless liquids or solids which are quite stable to attack by water and air. The ease of cleavage of the Sn−C bonds depends upon the R group, with Bu_4Sn being relatively stable. In moving to the organotin halides, reactivity increases and the chlorides are useful as precursors to a range of organotin derivatives. Figure 23.17 gives selected reactions of R_3SnCl. The structures of R_4Sn compounds are all similar with the Sn centre being tetrahedral. However, the presence of halide groups leads to significant variation in solid state structure owing to the possibility of Sn−X−Sn bridge formation. In the solid state, Me_3SnF molecules are connected into zigzag chains by asymmetric, bent Sn−F−Sn bridges (**23.33**), each Sn being in a trigonal bipyramidal arrangement. The presence of bulky substituents may result in either a straightening of the $\cdots$ Sn−F−Sn−F $\cdots$ backbone (e.g. in Ph_3SnF) or in a monomeric structure (e.g. in $\{(Me_3Si)_3C\}Ph_2SnF$). In $(Me_3SiCH_2)_3SnF$ (Fig. 23.18a), the Me_3SiCH_2 substituents are very bulky, and the Sn−F distances are much longer than the sum of the covalent radii. Solid state ^{119}Sn NMR spectroscopy and measurements of the

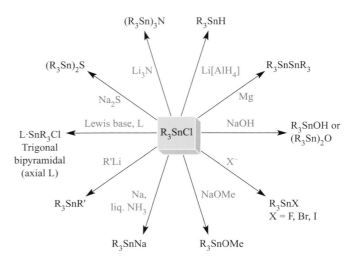

Fig. 23.17 Selected reactions of R_3SnCl; products such as R_3SnH, R_3SnNa and R_3SnSnR_3 are useful starting materials in organotin chemistry.

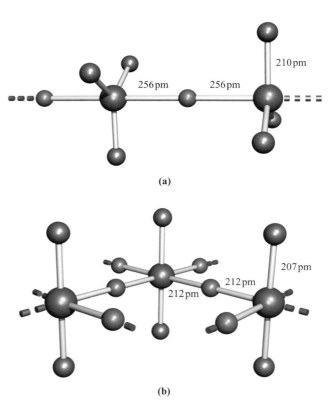

Fig. 23.18 The structures (X-ray diffraction) of (a) $(Me_3SiCH_2)_3SnF$ (only the methylene C atom of each Me_3SiCH_2 group is shown) in which the Sn−F distances are long and indicate the presence of $[(Me_3SiCH_2)_3Sn]^+$ cations interacting with F^- anions to give chains [L.N. Zakharov *et al.* (1983) *Kristallografiya*, vol. 28, p. 271], and (b) Me_2SnF_2 in which Sn−F−Sn bridge formation leads to the generation of sheets [E.O. Schlemper *et al.* (1966) *Inorg. Chem.*, vol. 5, p. 995]. Hydrogen atoms are omitted for clarity; colour code: Sn, brown; C, grey; F, green.

^{119}Sn–^{19}F spin–spin coupling constants provide a useful means of deducing the extent of molecular association in the absence of crystallographic data. Difluoro derivatives R_2SnF_2 tend to contain octahedral Sn in the solid state. In Me_2SnF_2, sheets of interconnected molecules are present (Fig. 23.18b). The tendency for association is less for the later halogens (F > Cl > Br > I). Thus, $MeSnBr_3$ and $MeSnI_3$ are monomeric, and, in contrast to Me_2SnF_2, Me_2SnCl_2 forms chains of the type shown in **23.34**. It has also been noted that the structure may be temperature dependent. Thus, at 108 K, crystalline (*cyclo*-$C_6H_{11})_3$SnCl consists of chains with asymmetrical bridges; between 108 and 298 K, changes in the Sn–Cl bond length and the intermolecular separation (structures **23.35**) suggest a transition to a structure containing discrete molecules. Figure 23.17 illustrates the ability of R_3SnCl to act as a Lewis acid. Similarly, salts of, for example, $[Me_2SnF_4]^{2-}$ may be prepared and contain discrete octahedral anions.

At 108 K: $a = 246.6$ pm
$b = 300.8$ pm

At 298 K: $a = 241.5$ pm
$b = 329.8$ pm

(23.35)

The reaction of $(2,4,6\text{-}^iPr_3C_6H_2)_3SnCH_2CH=CH_2$ with a strong electrophile in the presence of the weakly coordinating anion $[B(C_6F_5)_4]^-$ is a successful method of preparing a stannylium cation, $[R_3Sn]^+$. The approach parallels that used to prepare $[Mes_3Si]^+$, but stabilization of the stannylium ion requires more sterically demanding aryl groups. The structure of $[(2,4,6\text{-}^iPr_3C_6H_2)_3Sn]^+$ is similar to that of $[Mes_3Si]^+$ (**23.29**).

Tin(II) organometallics of the type R_2Sn, which contain Sn—C σ-bonds, are stabilized only if R is sterically demanding. For example, reaction of $SnCl_2$ with $Li[(Me_3Si)_2CH]$ gives $\{(Me_3Si)_2CH\}_2Sn$ which is monomeric in solution and dimeric in the solid state. The dimer (Fig. 23.19a) does *not* possess a planar Sn_2C_4 framework (i.e. it is *not* analogous to an alkene) and the Sn—Sn bond distance (276 pm) is too great to be consistent with a

(23.33)

210 pm
240 pm

(23.34)

240 pm
354 pm

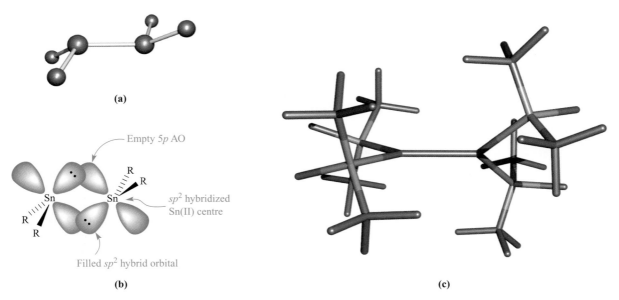

(a)

Empty 5*p* AO

sp^2 hybridized Sn(II) centre

Filled sp^2 hybrid orbital

(b)

(c)

Fig. 23.19 (a) Representation of the structure of an R_2SnSnR_2 compound which possesses a non-planar Sn_2C_4 framework, and (b) proposed bonding scheme involving sp^2 hybridized tin, and overlap of occupied sp^2 hybrid orbitals with empty 5*p* atomic orbitals to give a weak Sn=Sn double bond. (c) The structure of the distannene $\{^tBu_2MeSi\}_4Sn_2$ (determined by X-ray diffraction) [T. Fukawa *et al.* (2004) *J. Am. Chem. Soc.*, vol. 126, p. 11758]. For clarity, the structure is shown in a tube representation with H atoms omitted; colour code: Sn, brown; Si, pink; C, grey.

normal double bond. A bonding model involving overlap of filled sp^2 hybrids and vacant $5p$ atomic orbitals (Fig. 23.19b) has been suggested. Distannenes typically exist as dimers in the solid state, but dissociate into monomeric stannylenes in solution (eq. 23.70). However, $\{^tBu_2MeSi\}_4Sn_2$ (Fig. 23.19c) is an exception. The Sn–Sn bond is particularly short (267 pm), and each Sn centre is trigonal planar (sp^2 hybridized). The Sn_2Si_4-unit is twisted, and the deviation from planarity probably arises from steric factors. The presence of Sn=Sn double bond character in the solution species is exemplified by the cycloaddition reaction 23.71. The bonding scheme used for most distannenes (Fig. 23.19b) is not appropriate for $\{^tBu_2MeSi\}_4Sn_2$ and, instead, the bonding is viewed in terms of a σ-interaction supplemented by an out-of-plane p–p π-interaction.[†]

(23.70)

distannene in the solid state

stannylene in solution

(23.71)

$R = {}^tBu_2MeSi$

The formation of the *trans*-bent RSnSnR (**23.36**) is achieved by using extremely bulky R groups. The Sn–Sn bond length is 267 pm and angle C–Sn–Sn is 125° and, as for the Ge analogue, theoretical results indicate that the bond order is ≈2.5.

(23.36)

Cyclopentadienyl Sn(II) derivatives $(\eta^5\text{-}C_5R_5)_2Sn$ can be prepared by reaction 23.72.

$$2Na[Cp] + SnCl_2 \longrightarrow (\eta^5\text{-}Cp)_2Sn + 2NaCl \qquad (23.72)$$

[†] For comments on bonding schemes in distannenes, see: V.Ya. Lee *et al.* (2006) *J. Am. Chem. Soc.*, vol. 126, p. 11643.

The structures of $(\eta^5\text{-}C_5R_5)_2Sn$ with various R groups form a series in which the tilt angle α (defined in Fig. 23.14 for $(\eta^5\text{-}C_5R_5)_2Si$) increases as the steric demands of R increase: α = 125° for R = H, 144° for R = Me, 180° for R = Ph. We consider the structures of group 14 metallocenes again at the end of Section 23.5. Under appropriate conditions, Cp_2Sn reacts with Cp^- to yield $[(\eta^5\text{-}Cp)_3Sn]^-$. This last reaction shows that Cp_2Sn can function as a Lewis *acid*.

Organotin(IV) hydrides such as $^nBu_3SnH$ (prepared by $LiAlH_4$ reduction of the corresponding $^nBu_3SnCl$) are widely used as reducing agents in organic synthesis. In contrast, the first organotin(II) hydride, RSnH, was reported only in 2000. It is made by reacting $^iBu_2AlH$ with RSnCl where R is the sterically demanding substituent shown in **23.37**. In the solid state, dimers (**23.37**) supported by hydride bridges (Sn···Sn = 312 pm) are present. The orange solid dissolves in Et_2O, hexane or toluene to give blue solutions, indicating that RSnH monomers exist in solution. This conclusion is based on the electronic spectroscopic properties ($\lambda_{max} = 608$ nm) which are similar to those of monomeric R_2Sn compounds.

(23.37)

Worked example 23.5 Organotin compounds

The reaction of $\{(Me_3Si)_3C\}Me_2SnCl$ with one equivalent of ICl gives compound A. Use the mass spectrometric and ^1H NMR spectroscopic data below to suggest an identity for A. Suggest what product might be obtained if an excess of ICl is used in the reaction.

A: δ 0.37 ppm (27H, s, $J(^{29}Si–^1H) = 6.4$ Hz); δ 1.23 ppm (3H, s, $J(^{117}Sn–^1H)$, $J(^{119}Sn–^1H) = 60, 62$ Hz). No parent peak observed in the mass spectrum; highest mass peak $m/z = 421$.

The ^1H NMR spectroscopic data show the presence of two proton environments in a ratio of 27:3. These integrals, along with the coupling constants, suggest the retention of an $(Me_3Si)_3C$ group and *one* Me substituent bonded directly to Sn. Iodine monochloride acts as a chlorinating agent, and one Me group is replaced by Cl. The mass spectrometric data are consistent with a molecular formula of $\{(Me_3Si)_3C\}MeSnCl_2$, with the peak at $m/z = 421$ arising from the ion $[\{(Me_3Si)_3C\}SnCl_2]^+$, i.e. the parent ion with loss of Me.

With an excess of ICl, the expected product is $\{(Me_3Si)_3C\}SnCl_3$.

Self-study exercises

These questions refer to the experiment described above. Additional data: see Table 14.1.

1. Use Appendix 5 to deduce how the peak at $m/z = 421$ in the mass spectrum confirms the presence of *two* Cl atoms in **A**.   [*Hint*: Refer to Section 1.3]

2. Sketch the appearance of the ^1H NMR signal at δ 1.23 ppm in the spectrum of **A** and indicate where you would measure $J(^{117}Sn–^1H)$ and $J(^{119}Sn–^1H)$.
                    [*Hint*: Refer to Fig. 4.23]

3. In what coordination geometry do you expect the Sn atom to be sited in compound **A**?   [*Ans.* Tetrahedral]

[For further information, see S.S. Al-Juaid *et al.* (1998) *J. Organomet. Chem.*, vol. 564, p. 215.]

Lead

Tetraethyllead (made by reaction 23.73 or by electrolysis of $NaAlEt_4$ or $EtMgCl$ using a Pb anode) was formerly widely used as an anti-knock agent in motor fuels. However, for environmental reasons, the use of leaded fuels has declined (see Fig. 14.3).

$$4NaPb + 4EtCl \xrightarrow[\text{alloy}]{\approx 373\,K \text{ in an autoclave}} Et_4Pb + 3Pb + 4NaCl \tag{23.73}$$

Laboratory syntheses of R_4Pb compounds include the use of Grignard reagents (eqs. 23.74 and 23.75) or organolithium compounds (eqs. 23.76 and 23.77). High-yield routes to $R_3Pb–PbR_3$ involve the reactions of R_3PbLi (see below) with R_3PbCl.

$$2PbCl_2 + 4RMgBr \xrightarrow{Et_2O} 2\{R_2Pb\} + 4MgBrCl$$
$$\downarrow$$
$$R_4Pb + Pb \tag{23.74}$$

$$3PbCl_2 + 6RMgBr$$
$$\xrightarrow{Et_2O,\ 253\,K} R_3Pb–PbR_3 + Pb + 6MgBrCl \tag{23.75}$$

$$2PbCl_2 + 4RLi \xrightarrow{Et_2O} R_4Pb + 4LiCl + Pb \tag{23.76}$$

$$R_3PbCl + R'Li \longrightarrow R_3R'Pb + LiCl \tag{23.77}$$

Alkyllead chlorides can be prepared by reactions 23.78 and 23.79, and these routes are favoured over treatment of R_4Pb with X_2, the outcome of which is hard to control.

$$R_4Pb + HCl \longrightarrow R_3PbCl + RH \tag{23.78}$$

$$R_3PbCl + HCl \longrightarrow R_2PbCl_2 + RH \tag{23.79}$$

Compounds of the R_4Pb and R_6Pb_2 families possess monomeric structures with tetrahedral Pb centres as exemplified by the cyclohexyl derivative in Fig. 23.20a. The number of Pb derivatives that have been structurally studied is less than for the corresponding Sn-containing compounds. For the organolead halides, the presence of bridging halides is again a common feature giving rise to increased coordination numbers at the metal centre, e.g. in Me_3PbCl (Fig. 23.20b). Monomers are favoured if the organic substituents are sterically demanding as in $(2,4,6-Me_3C_6H_2)_3PbCl$. We mentioned above the use of R_3PbLi reagents. The first structurally characterized member of this group was 'Ph_3PbLi', isolated as the monomeric complex **23.38**.

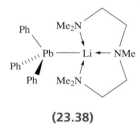

(23.38)

Tetraalkyl and tetraaryl lead compounds are inert with respect to attack by air and water at room temperature. Thermolysis leads to radical reactions such as those shown in scheme 23.80, which will be followed by further radical reaction steps.

$$\left.\begin{aligned} Et_4Pb &\longrightarrow Et_3Pb^\bullet + Et^\bullet \\ 2Et^\bullet &\longrightarrow n\text{-}C_4H_{10} \\ Et_3Pb^\bullet + Et^\bullet &\longrightarrow C_2H_4 + Et_3PbH \\ Et_3PbH + Et_4Pb &\longrightarrow H_2 + Et_3Pb^\bullet \\ &\qquad\qquad + Et_3PbCH_2CH_2^\bullet \end{aligned}\right\} \tag{23.80}$$

The chloride group in R_3PbCl can be replaced to give a range of R_3PbX species (e.g. $X^- = [N_3]^-$, $[NCS]^-$, $[CN]^-$, $[OR']^-$). Where X^- has the ability to bridge, polymeric structures are observed in the solid state. Both R_3PbN_3 and R_3PbNCS are strong Lewis acids and form adducts such as $[R_3Pb(N_3)_2]^-$. The reaction of Ph_3PbCl with $Na[Cp]$ gives $Ph_3Pb(\eta^1\text{-}Cp)$; structure **23.39** has been confirmed by X-ray diffraction and it is significant that the distance $Pb–C_{Cp} > Pb–C_{Ph}$. This is consistent with a

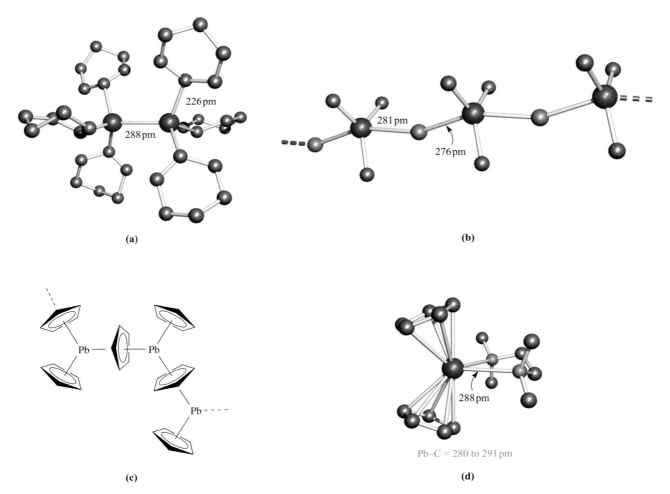

(a)

(b)

(c)

(d)

Pb–C = 280 to 291 pm

Fig. 23.20 The solid state structures of (a) $Pb_2(C_6H_{11})_6$ [X-ray diffraction: N. Kleiner *et al.* (1985) *Z. Naturforsch., Teil B*, vol. 40, p. 477], (b) Me_3PbCl [X-ray diffraction: D. Zhang *et al.* (1991) *Z. Naturforsch., Teil A*, vol. 46, p. 337], (c) Cp_2Pb (schematic diagram), and (d) $(\eta^5\text{-}Cp)_2Pb(Me_2NCH_2CH_2NMe_2)$ [X-ray diffraction: M.A. Beswick *et al.* (1996) *J. Chem. Soc., Chem. Commun.*, p. 1977]. Hydrogen atoms are omitted for clarity; colour code: Pb, red; C, grey; Cl, green; N, blue.

weaker $Pb-C_{Cp}$ bond, and preferential bond cleavage is observed, e.g. in scheme 23.81.

$$Ph_3PbO_2CMe \xleftarrow{\text{MeCO}_2\text{H}} Ph_3Pb(\eta^1\text{-}Cp) \xrightarrow{\text{PhSH}} Ph_3PbSH$$

(23.81)

(23.39)

Cyclopentadienyl derivatives of Pb(II), $(\eta^5\text{-}C_5R_5)_2Pb$, can be prepared by reactions of a Pb(II) salt (e.g. acetate or chloride) with $Na[C_5R_5]$ or $Li[C_5R_5]$. The $(\eta^5\text{-}C_5R_5)_2Pb$ compounds are generally sensitive to air, but the presence of bulky R groups increases their stability. The solid state structure of Cp_2Pb consists of polymeric chains

(Fig. 23.20c), but in the gas phase, discrete $(\eta^5\text{-}Cp)_2Pb$ molecules are present which possess the bent structure shown for $(\eta^5\text{-}C_5Me_5)_2Si$ in Fig. 23.14b. Other $(\eta^5\text{-}C_5R_5)_2Pb$ compounds which have been studied in the solid state are monomers. Bent structures (as in Fig. 23.14b) are observed for R = Me or $PhCH_2$ for example, but in $\{\eta^5\text{-}C_5Me_4(Si^tBuMe_2)\}_2Pb$ where the organic groups are especially bulky, the C_5-rings are coparallel (see the end of Section 23.5). Cp_2Pb (like Cp_2Sn) can act as a Lewis *acid*. It reacts with the Lewis bases $Me_2NCH_2CH_2NMe_2$ and $4,4'\text{-}Me_2bpy$ (**23.40**) to form the adducts $(\eta^5\text{-}Cp)_2Pb\cdot L$ where L is the Lewis base. Figure 23.20d shows the solid state structure of $(\eta^5\text{-}Cp)_2Pb\cdot Me_2NCH_2CH_2NMe_2$, and the structure of $(\eta^5\text{-}Cp)_2Pb\cdot(4,4'\text{-}Me_2bpy)$ is similar. Further evidence for Lewis acid behaviour comes from the reaction of $(\eta^5\text{-}Cp)_2Pb$ with Li[Cp] in the presence of the crown ether 12-crown-4, which gives $[Li(12\text{-crown-}4)]_2[Cp_9Pb_4][Cp_5Pb_2]$. The structures of $[Cp_9Pb_4]^-$ and $[Cp_5Pb_2]^-$ consist of fragments of the polymeric chain of

Cp$_2$Pb (see Fig. 23.20c), e.g. in [Cp$_5$Pb$_2$]$^-$, one Cp$^-$ ligand bridges between the two Pb(II) centres and the remaining four Cp$^-$ ligands are bonded in an η^5-mode, two to each Pb atom.

(23.40)

Diarylplumbylenes, R$_2$Pb, in which the Pb atom carries a lone pair of electrons, can be prepared by the reaction of PbCl$_2$ with RLi provided that R is suitably sterically demanding. The presence of monomers in the solid state has been confirmed for R = 2,4,6-(CF$_3$)$_3$C$_6$H$_2$ and 2,6-(2,4,6-Me$_3$C$_6$H$_2$)$_2$C$_6$H$_3$. Dialkyl derivatives are represented by {(Me$_3$Si)$_2$CH}$_2$Pb. The association of R$_2$Pb units to form R$_2$Pb=PbR$_2$ depends critically on R as the following examples illustrate. Crystalline {(Me$_3$Si)$_3$Si}RPb with R = 2,3,4-Me$_3$-6-^{t}BuC$_6$H and 2,4,6-(CF$_3$)$_3$C$_6$H$_2$, contain dimers in which the Pb···Pb distances are 337 and 354 pm, respectively. These separations are too long to be consistent with the presence of Pb=Pb bonds. The product in scheme 23.82 is monomeric in the gas phase and solution. In the solid, it is dimeric with a Pb–Pb bond length of 305 pm, indicative of a Pb=Pb bond. The ligand-exchange reaction 23.83 leads to a product with an even shorter Pb–Pb bond (299 pm). The bonding in R$_2$Pb=PbR$_2$ can be described in an analogous manner to that shown for R$_2$Sn=SnR$_2$ in Fig. 23.19.

2PbCl$_2$ + 4RMgBr $\xrightarrow[\text{- MgCl}_2/\text{MgBr}_2]{\substack{163\text{ K,}\\ \text{warm to }293\text{ K}}}$ 2R$_2$Pb (solution)

$$\updownarrow$$

(23.82)

R = 2,4,6-^{i}Pr$_3$C$_6$H$_2$

(solid)

R$_2$Pb + R'$_2$Pb $\longrightarrow$ 2RR'Pb (solution)

$$\updownarrow$$

(23.83)

R = 2,4,6-^{i}Pr$_3$C$_6$H$_2$
R' = (Me$_3$Si)$_3$Si

RR'Pb=PbRR' (solid)

When the Grignard reagent in scheme 23.82 is changed to 2,4,6-Et$_3$C$_6$H$_2$MgBr, the crystalline product is **23.41**, whereas with 2,4,6-Me$_3$C$_6$H$_2$MgBr, **23.42** is isolated. The formation of **23.41** can be suppressed by carrying out the reaction in the presence of dioxane. In this case, (2,4,6-Et$_3$C$_6$H$_2$)$_2$Pb trimerizes to **23.43**, in which the Pb–Pb bond distances are 318 ppm. These rather long bonds, along with the orientations of the R groups (Fig. 23.21a), lead to a bonding description involving donation of the lone pair

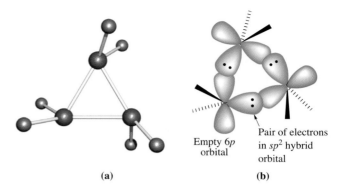

Fig. 23.21 (a) The structure of the Pb$_3$C$_6$ core in {(2,4,6-Et$_3$C$_6$H$_2$)$_2$Pb}$_3$, determined by X-ray diffraction [F. Stabenow *et al.* (2003) *J. Am. Chem. Soc.*, vol. 125, p. 10172]. Colour code: Pb, red; C, grey. (b) A bonding description for the Pb–Pb interactions in {(2,4,6-Et$_3$C$_6$H$_2$)$_2$Pb}$_3$.

from one R$_2$Pb: unit into the empty 6p orbital of the adjacent unit (Fig. 23.21b). These reaction systems are complicated, and changes in the R group and in the ratio of starting materials may result in disproportionation of PbX$_2$ (eq. 23.84), or give salts of [Pb(PbR$_3$)$_3$]$^-$ (**23.44**).

3PbCl$_2$ + 6RMgBr $\longrightarrow$ R$_3$Pb–PbR$_3$ + Pb + 6MgBrCl (23.84)

R = 2,4,6-Et$_3$C$_6$H$_2$

(23.41)

R = 2,4,6-Me$_3$C$_6$H$_2$

(23.42)

R = 2,4,6-Et$_3$C$_6$H$_2$

(23.43)

R = Ph or PhC$_6$H$_4$

(23.44)

The reaction of RPbBr (R = 2,6-(2,6-^{i}Pr$_2$C$_6$H$_3$)$_2$C$_6$H$_3$) with LiAlH$_4$ leads to RPbPbR (**23.45**) (eq. 23.85). This is not analogous, either in structure or bonding, to RGeGeR and

RSnSnR (see structure **23.36**). In RPbPbR, the Pb–Pb distance is consistent with a single bond, and each Pb atom is considered to have a sextet of electrons (one lone and two bonding pairs). The Pb–Pb bond length is close to that in trimer **23.43**.

$$2RPbBr \xrightarrow{\text{LiAlH}_4} 2RPbH \xrightarrow{-H_2} RPbPbR \qquad (23.85)$$

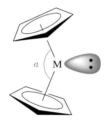

$$R = 2,6\text{-}(2,6\text{-}^iPr_2C_6H_3)_2C_6H_3$$

(23.45)

Coparallel and tilted C$_5$-rings in group 14 metallocenes

The first group 14 metallocenes to be characterized were $(\eta^5\text{-}C_5H_5)_2Sn$ and $(\eta^5\text{-}C_5H_5)_2Pb$, and in both compounds, the C$_5$-rings are mutually tilted. This observation was originally interpreted in terms of the presence of a stereochemically active lone pair of electrons as shown in structure **23.46**.

(23.46)

However, as the examples in Section 23.5 have shown, not all group 14 metallocenes exhibit structures with tilted C$_5$-rings. For example, in each of $(\eta^5\text{-}C_5Ph_5)_2Sn$, $\{\eta^5\text{-}C_5Me_4(SiMe_2{}^tBu)\}_2Ge$ and $(\eta^5\text{-}C_5{}^iPr_3H_2)_2Pb$, the two C$_5$-rings are coparallel. Trends such as that along the series $(\eta^5\text{-}C_5H_5)_2Sn$ (tilt angle $\alpha = 125°$), $(\eta^5\text{-}C_5Me_5)_2Sn$ ($\alpha = 144°$) and $(\eta^5\text{-}C_5Ph_5)_2Sn$ (coparallel rings) have been explained in terms of steric factors: as the inter-ring steric repulsions increase, angle α in **23.46** increases, and the final result is a rehybridization of the metal orbitals, rendering the lone pair stereochemically inactive. It is, however, difficult to rationalize the occurrence of *both* tilted and coparallel forms of $(\eta^5\text{-}C_5Me_5)_2Si$ (Fig. 23.14) using steric arguments. Furthermore, the preference for coparallel rings in the solid state for $\{\eta^5\text{-}C_5Me_4(SiMe_2{}^tBu)\}_2Pb$ and $(\eta^5\text{-}C_5{}^iPr_3H_2)_2Pb$, in contrast to a tilted structure for $(\eta^5\text{-}C_5{}^iPr_5)_2Pb$ ($\alpha = 170°$), cannot be rationalized in terms of inter-ring steric interactions. The situation is further complicated by the fact that as one

descends group 14, there is an increased tendency for the lone pair of electrons to be accommodated in an ns orbital and to become stereochemically inactive. A final point for consideration is that, although polymeric, the group 2 metallocenes $(\eta^5\text{-}Cp)_2M$ (M = Ca, Sr, Ba) exhibit bent C$_5$–M–C$_5$ units: here, there is no lone pair of electrons to affect the structure. Taking all current data into consideration, it is necessary to reassess (i) the stereochemical role of the lone pair of electrons in $(\eta^5\text{-}C_5R_5)_2M$ compounds (M = group 14 metal) and (ii) the role of inter-ring steric interactions as factors that contribute to the preference for coparallel or tilted C$_5$-rings. Theoretical studies indicate that the difference in energy between the two structures for a given molecule is small: $\approx 1-12\,\text{kJ mol}^{-1}$ depending on ring substituents. Crystal-packing forces have been suggested as a contributing factor, but further studies are required to provide a definitive explanation.[†]

23.6 Group 15

Bonding aspects and E=E bond formation

Our discussion of organometallic compounds of group 15 covers As, Sb and Bi. There is an extensive chemistry of compounds with N–C or P–C bonds, but much of this belongs within the remit of organic chemistry, although amines and phosphanes (e.g. R_3E, $R_2E(CH_2)_nER_2$ where E = N or P) are important ligands in inorganic complexes. In both cases, the group 15 element acts as a σ-donor, and in the case of phosphorus, also as a π-acceptor (see Section 20.4).

On descending group 15, the E–E and E–C bond enthalpy terms both decrease (e.g. see Table 15.3). In Section 15.3, we emphasized differences in bonding between nitrogen and the later elements, and illustrated that $(p-p)\pi$-bonding is important for nitrogen but not for the heavier elements. Thus, nitrogen chemistry provides many compounds of type $R_2N=NR_2$, but for most R groups the analogous $R_2E=ER_2$ compounds (E = P, As, Sb or Bi) are unstable with respect to oligomerization to give cyclic compounds such as Ph_6P_6. Only by the use of especially bulky substituents is double bond formation for the later elements made possible, with the steric hindrance preventing oligomerization. Thus, several compounds with P=P, P=As, As=As, P=Sb, Sb=Sb, Bi=Bi and P=Bi are known and possess *trans*-configurations as shown in structure **23.47**. The bulky substituents that have played a major role in enabling RE=ER compounds to be stabilized are $2,4,6\text{-}^tBu_3C_6H_2$, $2,6\text{-}(2,4,6\text{-}Me_3C_6H_2)_2C_6H_3$ and $2,6\text{-}(2,4,6\text{-}^iPr_3C_6H_2)_2C_6H_3$. Along the series RE=ER for E = P, As, Sb and Bi and R = $2,6\text{-}(2,4,6\text{-}Me_3C_6H_2)_2C_6H_3$, the E=E bond length increases (198.5 pm,

---

[†] For further discussion, see: S.P. Constantine, H. Cox, P.B. Hitchcock and G.A. Lawless (2000) *Organometallics*, vol. 19, p. 317; J.D. Smith and T.P. Hanusa (2001) *Organometallics*, vol. 20, p. 3056; V.M. Rayón and G. Frenking (2002) *Chem. Eur. J.*, vol. 8, p. 4693.

E = P; 228 pm, E = As; 266 pm, E = Sb; 283 pm, E = Bi) and the E−E−C bond angle decreases (110°, E = P; 98.5°, E = As; 94°, E = Sb; 92.5°, E = Bi). Methylation of RP=PR (R = 2,4,6-^{t}Bu$_3$C$_6$H$_2$) to give **23.48** can be achieved, but only if a 35-fold excess of methyl trifluoromethanesulfonate is used. We return to single bond formation between As, Sb and Bi atoms later.

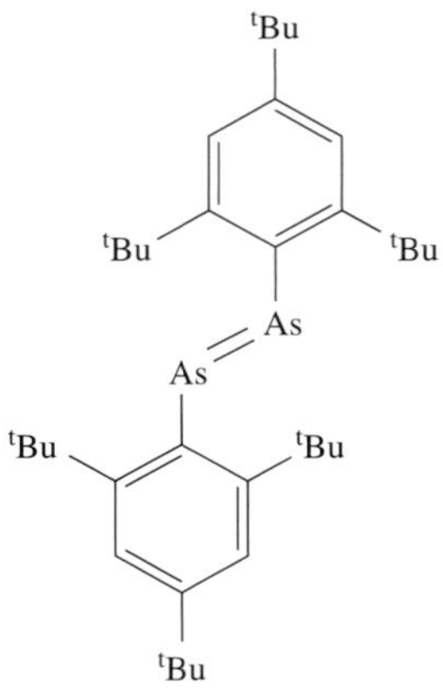

(23.47) 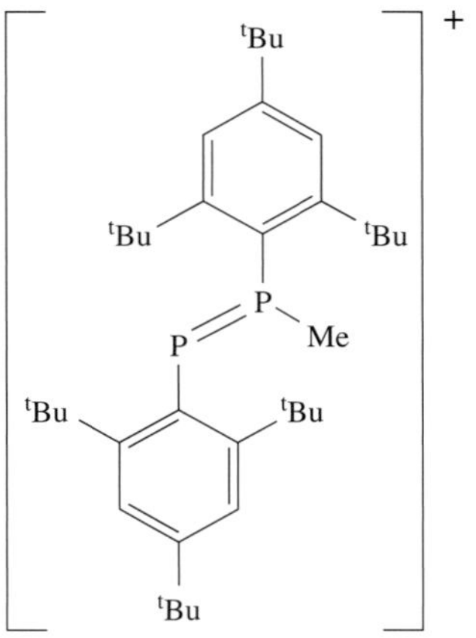   **(23.48)**

Arsenic, antimony and bismuth

Organometallic compounds of As(III), Sb(III) and Bi(III) can be prepared from the respective element and organo halides (reaction 23.86) or by use of Grignard reagents (eq. 23.87) or organolithium compounds. Treatment of organo halides (e.g. those from reaction 23.86) with R'Li gives RER'$_2$ or R$_2$ER' (e.g. eq. 23.88).

$$2As + 3RBr \xrightarrow{\text{in presence of Cu, } \Delta} RAsBr_2 + R_2AsBr$$
(23.86)

$$EX_3 + 3RMgX \xrightarrow{\text{ether solvent}} R_3E + 3MgX_2 \qquad (23.87)$$

$$R_2AsBr + R'Li \longrightarrow R_2AsR' + LiBr \qquad (23.88)$$

Scheme 23.89 shows the formation of an organoarsane that is commonly used as a chelating ligand for heavy metals, the soft As donors being compatible with soft metal centres (see Table 7.9).

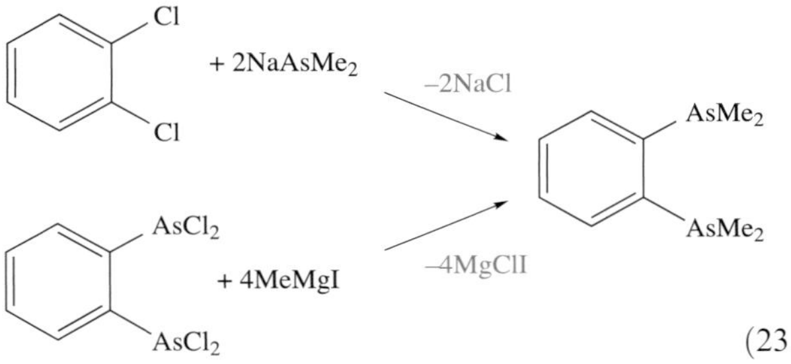

(23.89)

Metal(V) derivatives, R$_5$E, cannot be prepared from the corresponding pentahalides, but may be obtained by oxidation of R$_3$E followed by treatment with RLi (e.g. eq. 23.90). The same strategy can be used to form, for example, Me$_2$Ph$_3$Sb (reaction 23.91).

$$R_3As + Cl_2 \longrightarrow R_3AsCl_2 \xrightarrow[-2LiCl]{2RLi} R_5As \qquad (23.90)$$

$$Ph_3SbCl_2 + 2MeLi \xrightarrow{\text{Et}_2\text{O, 195 K}} Me_2Ph_3Sb + 2LiCl$$
(23.91)

The oxidative addition of R'X (R = alkyl) to R$_3$E produces R$_3$R'EX, with the tendency of R$_3$E to undergo this reaction decreasing in the order As > Sb ≫ Bi, and I > Br > Cl. Further, conversion of R$_3$X to R$_3$R'EX by this route works for R = alkyl or aryl when E = As, but not for R = aryl when E = Sb. Compounds of the type R$_3$EX$_2$ are readily prepared as shown in eq. 23.90, and R$_2$EX$_3$ derivatives can be made by addition of X$_2$ to R$_2$EX (E = As, Sb; X = Cl, Br).

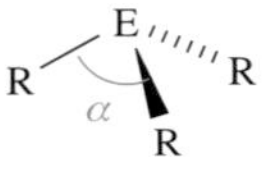

(23.49) 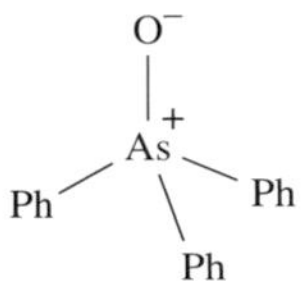   **(23.50)**

Compounds of the type R$_3$E are sensitive to oxidation by air but resist attack by water. They are more stable when R = aryl (compared to alkyl), and stability for a given series of triaryl derivatives decreases in the order R$_3$As > R$_3$Sb > R$_3$Bi. All R$_3$E compounds structurally characterized to date are trigonal pyramidal, and the C−E−C angle α in **23.49** decreases for a given R group in the order As > Sb > Bi. Hydrogen peroxide oxidizes Ph$_3$As to Ph$_3$AsO, for which **23.50** is a bonding representation. Ph$_3$SbO is similarly prepared or can be obtained by heating Ph$_3$Sb(OH)$_2$. Triphenylbismuth oxide is made by oxidation of Ph$_3$Bi or hydrolysis of Ph$_3$BiCl$_2$. The ready formation of these oxides should be compared with the relative stability of Ph$_3$P with respect to oxidation, the ready oxidation of Me$_3$P, and the use of Me$_3$NO as an oxidizing agent. (See Section 15.3 for a discussion of the bonding in hypervalent compounds of the group 15 elements.) Triphenylarsenic oxide forms a monohydrate which exists as a hydrogen-bonded dimer (**23.51**) in the solid state. Ph$_3$SbO crystallizes in several modifications which contain either monomers or polymers, and has a range of catalytic uses in organic chemistry, e.g. oxirane polymerization, and reactions between amines and acids to give amides. The reaction of Ph$_3$AsO with PhMgX leads to the salts [Ph$_4$As]X (X=Cl, Br, I). These salts are commercially available and are widely used to provide a large cation for the stabilization of salts containing large anions (see Box 24.1).

(23.51)

The ability of R_3E to act as a Lewis base decreases down group 15. *d*-Block metal complexes involving R_3P ligands are far more numerous than those containing R_3As and R_3Sb (see Section 24.2), and only a few complexes containing R_3Bi ligands have been structurally characterized, e.g. $Cr(CO)_5(BiPh_3)$ (**23.52**) and $[(\eta^5\text{-Cp})Fe(CO)_2(BiPh_3)]^+$. Adducts are also formed between R_3E or R_3EO (E = As,

Sb) and Lewis acids such as boron trifluoride (Fig. 23.22b), and in Section 17.4, we described complexes formed between Ph_3E (E = P, As, Sb) and halogens.

(23.52)

Compounds of type R_5E (E = As, Sb, Bi) adopt either a trigonal bipyramidal or square-based pyramidal structure. In the solid state, Me_5Sb, Me_5Bi, $(4\text{-MeC}_6H_4)_5Sb$ and the solvated compound $Ph_5Sb \cdot \frac{1}{2}C_6H_{12}$ are trigonal bipyramidal, while unsolvated Ph_5Sb and Ph_5Bi are square-based

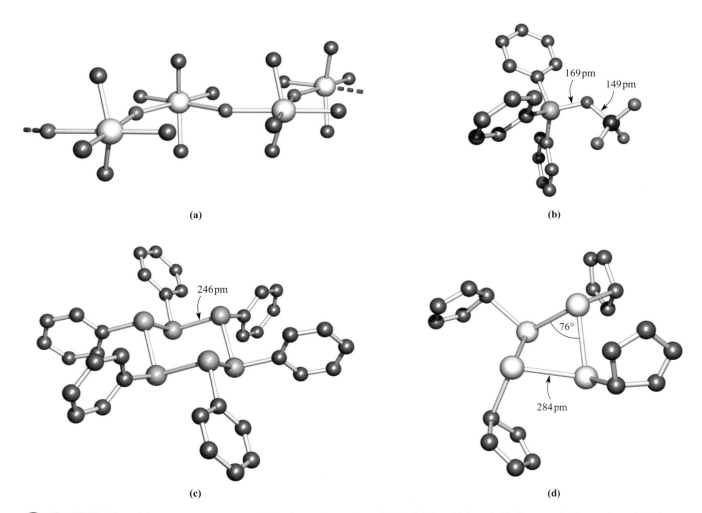

(a)

(b)

(c)

(d)

Fig. 23.22 The solid state structures (X-ray diffraction) of (a) polymeric Me_4SbF in which each Sb(V) centre is distorted octahedral [W. Schwarz *et al.* (1978) *Z. Anorg. Allg. Chem.*, vol. 444, p. 105], (b) $Ph_3AsO \cdot BF_3$ [N. Burford *et al.* (1990) *Acta Crystallogr., Sect. C*, vol. 46, p. 92], (c) Ph_6As_6 in which the As_6 adopts a chair conformation [A.L. Rheingold *et al.* (1983) *Organometallics*, vol. 2, p. 327], and (d) $(\eta^1\text{-C}_5Me_5)_4Sb_4$ with methyl groups omitted for clarity [O.M. Kekia *et al.* (1996) *Organometallics*, vol. 15, p. 4104]. Hydrogen atoms are omitted for clarity; colour code: Sb, silver; As, orange; C, grey; F, green; B, blue; O, red.

pyramidal. Electron diffraction studies on gaseous Me_5As and Me_5Sb confirm trigonal bipyramidal structures. In solution, the compounds are highly fluxional on the NMR timescale, even at low temperatures. The fluxional process involves ligand exchange via the interconversion of trigonal bipyramidal and square-based pyramidal structures (see Fig. 4.24). For $(4\text{-}MeC_6H_4)_5Sb$ in $CHFCl_2$ solvent, a barrier of $\approx 6.5\,kJ\,mol^{-1}$ to ligand exchange has been determined from ^1H NMR spectroscopic data.

On heating, R_5E compounds decompose, with the thermal stability decreasing down the group, e.g. Ph_5As is more thermally stable than Ph_5Sb than Ph_5Bi. The decomposition products vary and, for example, Ph_5Sb decomposes to Ph_3Sb and $PhPh$, while Me_5As gives Me_3As, CH_4 and C_2H_4. Cleavage of an $E-C$ bond in R_5E compounds occurs upon treatment with halogens, Brønsted acids or Ph_3B (eqs. 23.92–23.94). Both Me_5Sb and Me_5Bi react with MeLi in THF (eq. 23.95) to give salts containing the octahedral ions $[Me_6E]^-$.

$$Ph_5E + Cl_2 \longrightarrow Ph_4ECl + PhCl \qquad (23.92)$$

$$Ph_5E + HCl \longrightarrow Ph_4ECl + PhH \qquad (23.93)$$

$$Ph_5E + Ph_3B \longrightarrow [Ph_4E]^+[BPh_4]^- \qquad (23.94)$$

$$Me_5Sb + MeLi \xrightarrow{Et_2O,\ THF} [Li(THF)_4]^+[Me_6Sb]^- \qquad (23.95)$$

The monohalides R_4EX tend to be ionic for $X = Cl$, Br or I, i.e. $[R_4E]^+X^-$, but among the exceptions is Ph_4SbCl which crystallizes as discrete trigonal bipyramidal molecules. The fluorides possess covalent structures. In the solid state Me_4SbF forms polymeric chains (Fig. 23.22a) while $MePh_3SbF$ exists as trigonal bipyramidal molecules **23.53**. For the di- and trihalides there is also structural variation, and ionic, discrete molecular and oligomeric structures in the solid state are all exemplified, e.g. Me_3AsBr_2 is ionic and contains the tetrahedral $[Me_3AsBr]^+$ ion, Ph_3BiCl_2 and Ph_3SbX_2 ($X = F$, Cl, Br or I) are trigonal bipyramidal molecules with axial X atoms, Ph_2SbCl_3 is dimeric (**23.54**), while Me_2SbCl_3 exists in two structural forms, one ionic $[Me_4Sb]^+[SbCl_6]^-$ and the other a covalent dimer.

Ph
|
Me — Sb''''' Ph
|
F
(23.53)

Ph   Ph
| |
Cl''''' | '''''Cl''''' | '''''Cl
       Sb   Sb
Cl | Cl | Cl
       Ph   Ph
(23.54)

The family of R_2E-ER_2 compounds has grown significantly since 1980 and those structurally characterized by X-ray diffraction include Ph_4As_2, Ph_4Sb_2 and Ph_4Bi_2. All possess the staggered conformation shown in **23.55** for the C_4E_2 core with values of α and β of $103°$ and $96°$ for Ph_4As_2, $94°$ and $94°$ for Ph_4Sb_2, and $98°$ and $91°$ for Ph_4Bi_2. As expected, the $E-E$ bond length increases: $246\,pm$ in Ph_4As_2, $286\,pm$ in Ph_4Sb_2, and $298\,pm$ in Ph_4Bi_2. Equation 23.96 gives a typical preparative route. Some R_4Sb_2 and R_4Bi_2 (but not R_4As_2) derivatives are *thermochromic* (see Section 21.6).

$$2R_2BiCl + 2Na \xrightarrow[\text{2. 1,2-dichloroethane}]{\text{1. liquid } NH_3} R_4Bi_2 + 2NaCl$$

$$(23.96)$$

(23.55)

Ligand redistribution in liquid Me_2SbBr (no solvent) leads to the formation of the salt $[Me_3SbSbMe_2]_2[MeBr_2Sb(\mu\text{-}Br)_2SbBr_2Me]$ which contains ions **23.56** and **23.57**. The proposed pathway is given in scheme 23.97. The eclipsed conformation of cation **23.56** is probably determined by close cation–anion interactions in the solid state.

$$2Me_2SbBr \longrightarrow Me_3Sb + MeSbBr_2$$

$$2Me_2SbBr + 2Me_3Sb + 2MeSbBr_2 \rightleftharpoons [\mathbf{23.56}]_2[\mathbf{23.57}]$$

$$(23.97)$$

Me
\ 282 pm
  Sb —— Sb
Me / Me   ''' Me
     Me
(23.56) $^+$

Me
|
Br'''' Sb '''''Br'''''Sb'''''Br
Br   Br   Br
          Me
(23.57) $^{2-}$

Worked example 23.6   Application of the VSEPR model

Confirm that the octahedral structure of $[Ph_6Bi]^-$ (formed in a reaction analogous to 23.95) is consistent with the VSEPR model.

Bi has five electrons in its valence shell and the negative charge in $[Ph_6Bi]^-$ supplies one more.

Each Ph group supplies one electron to the valence shell of Bi in $[Ph_6Bi]^-$.

Total valence electron count $= 5 + 1 + 6 = 12$

The six pairs of electrons correspond to an octahedral structure within the VSEPR model, and this is consistent with the observed structure.

Self-study exercises

1. Show that the tetrahedral and trigonal pyramidal Sb centres in cation **23.56** are consistent with the VSEPR model. What does this assume about the localization of the positive charge?

2. Confirm that the structure of anion **23.57** is consistent with the VSEPR model. Comment on the preference for this structure over one in which the Me groups are on the same side of the planar Sb_2Br_6-unit.

3. Show that the octahedral centres in $Ph_4Sb_2Cl_6$ (**23.54**) are consistent with the VSEPR model.

Reduction of organometal(III) dihalides (e.g. $RAsCl_2$) with sodium or magnesium in THF, or reduction of $RAs(O)(OH)_2$ acids (reaction 23.98), gives *cyclo*-$(RE)_n$, where $n = 3$–6. Figure 23.22c shows the structure of Ph_6As_6 which illustrates the typical trigonal pyramidal environment for the group 15 element. Two crystalline polymorphs of $(\eta^1-C_5Me_5)_4Sb_4$ are known, differing in details of the molecular geometry and crystal packing; one structure is

noteworthy for its acute Sb–Sb–Sb bond angles (Fig. 23.22d). Reaction 23.99 is an interesting example of the formation of a *cyclo*-As_3 species, the organic group being tailor-made to encourage the formation of the 3-membered ring. A similar reaction occurs with $MeC(CH_2SbCl_2)_3$.

$$6PhAs(O)(OH)_2 \xrightarrow{H_3PO_2} Ph_6As_6 \qquad (23.98)$$

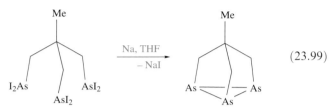

$$\qquad (23.99)$$

The lithiation (using BuLi) of Ph_2SbH in THF leads to $Ph_2SbLi(THF)_3$ which can be isolated as a crystalline solid. In contrast, lithiation of $PhSbH_2$ in $Me_2NCH_2CH_2NMe_2$ results in Sb–Sb bond formation and the $[Sb_7]^{3-}$ ion. This provides a convenient method of preparing this Zintl ion (see Section 15.6).

Organometallic chemistry involving cyclopentadienyl ligands is less important in group 15 than for groups 1, 2, 13 and 14. We have already mentioned $(\eta^1-C_5Me_5)_4Sb_4$ (Fig. 23.22d). Other compounds for which solid state structures contain $\eta^1-C_5R_5$ substituents include $(\eta^1-Cp)_3Sb$ (eq. 23.100) and $(\eta^1-C_5Me_5)AsCl_2$ (Fig. 23.23a, prepared

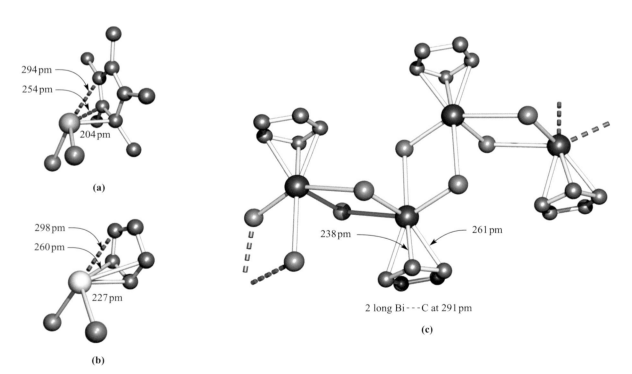

Fig. 23.23 The structures (X-ray diffraction) of (a) monomeric $(\eta^1-C_5Me_5)AsCl_2$ [E.V. Avtomonov *et al.* (1996) *J. Organomet. Chem.*, vol. 524, p. 253], (b) monomeric $(\eta^3-C_5H_5)SbCl_2$ [W. Frank (1991) *J. Organomet. Chem.*, vol. 406, p. 331], and (c) polymeric $(\eta^3-C_5H_5)BiCl_2$ [W. Frank (1990) *J. Organomet. Chem.*, vol. 386, p. 177]. Hydrogen atoms are omitted for clarity; colour code: As, orange; Sb, silver; Bi, blue; C, grey; Cl, green.

by ligand redistribution between $(\eta^1\text{-}C_5Me_5)_3As$ and $AsCl_3$). The derivatives Cp_nSbX_{3-n} (X = Cl, Br, I; n = 1, 2) are prepared by treating $(\eta^1\text{-}Cp)_3Sb$ with SbX_3. $CpBiCl_2$ is made by reaction 23.101.

$$Sb(NMe_2)_3 + 3C_5H_6 \xrightarrow{\text{Et}_2\text{O, 193 K}} (\eta^1\text{-}Cp)_3Sb + 3Me_2NH$$
$$(23.100)$$

$$BiCl_3 + Na[Cp] \xrightarrow{\text{Et}_2\text{O, 203 K}} CpBiCl_2 + NaCl \qquad (23.101)$$

In solution, the cyclopentadienyl rings in this type of compound are fluxional. In the solid state, crystallographic data (where available) reveal significant variation in bonding modes as examples in Fig. 23.23 illustrate. Consideration of the E–C bond distances leads to the designations of η^1 or η^3. Reaction 23.102 gives one of the few η^5-cyclopentadienyl derivatives of a heavier group 15 element so far prepared. The $[(\eta^5\text{-}C_5Me_5)_2As]^+$ ion is isoelectronic with $(\eta^5\text{-}C_5Me_5)_2Ge$ and possesses the same bent structure illustrated for $(\eta^5\text{-}C_5Me_5)_2Si$ in Fig. 23.14b.

$$(\eta^1\text{-}C_5Me_5)_2AsF + BF_3 \longrightarrow [(\eta^5\text{-}C_5Me_5)_2As]^+[BF_4]^-$$
$$(23.102)$$

23.7 Group 16

Our discussion of organo-compounds of group 16 elements is confined to selenium and tellurium. There are also vast numbers of organic compounds containing C–O or C–S bonds, and some relevant inorganic topics already covered are:

- oxides and oxoacids of carbon (Section 14.9);
- sulfides of carbon (Section 14.11).

Selenium and tellurium

The organic chemistry of selenium and tellurium is an expanding area of research, and one area of active interest is that of 'organic metals'. For example, the tetraselenafulvalene **23.58** acts as an electron donor to the tetracyano derivative **23.59** and 1:1 complexes formed between these, and between related molecules, crystallize with stacked structures and exhibit high electrical conductivities.

TMTSeF
(tetramethyltetraselenafulvalene)

(23.58)

TCNQ
(tetracyanoquinodimethane)

(23.59)

Organic derivatives of Se(II) include R_2Se (prepared by reaction 23.103) and RSeX (X = Cl or Br, prepared by reaction 23.104). Routes to R_2Te and R_2Te_2 compounds are shown in schemes 23.105 and 23.106. It is harder to isolate RTeX compounds than their Se analogues, but they can be stabilized by coordination to a Lewis base.

$$Na_2Se + 2RCl \longrightarrow R_2Se + 2NaCl \qquad (23.103)$$

$$R_2Se_2 + X_2 \longrightarrow 2RSeX \qquad (X = Cl, Br) \qquad (23.104)$$

$$Te + RLi \longrightarrow RTeLi \xrightarrow{\text{RBr}} R_2Te \qquad (23.105)$$

$$Na_2Te_2 + 2RX \longrightarrow R_2Te_2 + 2NaX \qquad (23.106)$$

Diselenides R_2Se_2 are readily made by treating Na_2Se_2 with RX, and have non-planar structures, e.g. for Ph_2Se_2 in the solid state, the dihedral angle (see Fig. 16.10) is $82°$ and the Se–Se bond length is 229 pm. The reaction of Ph_2Se_2 with I_2 leads, not to RSeI, but to the charge transfer complex **23.60** (see Section 17.4). In contrast, the reaction with Ph_2Te_2 leads to the tetramer $(PhTeI)_4$ (**23.61**).

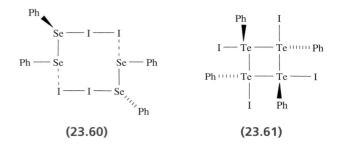

(23.60)   **(23.61)**

Dimethylselenide and telluride react with Cl_2, Br_2 and I_2 to give Me_2SeX_2 and Me_2TeX_2. The solid state structure of Me_2TeCl_2 is based on a trigonal bipyramid in accord with the VSEPR model and this is typical of R_2TeX_2 (**23.62**) compounds. What was at one time labelled as the β-form of Me_2TeI_2 is now known to be $[Me_3Te]^+[MeTeI_4]^-$, with a trigonal pyramidal cation and square-based pyramidal anion; I–Te···I bridges result in each Te centre being in a distorted octahedral environment in the solid state.

(23.62)

The oxidative addition of X_2 to RSeX (X = Cl or Br) leads to $RSeX_3$. Tellurium analogues such as $MeTeCl_3$ can be prepared by treating Me_2Te_2 with Cl_2 or by reacting $TeCl_4$ with Me_4Sn. Reaction 23.107 yields the pyrophoric compound Me_4Te which can be oxidized to Me_4TeF_2 using XeF_2; Ph_4Te can similarly be converted to $cis\text{-}Ph_4TeF_2$. Reaction of Me_4TeF_2 with Me_2Zn yields Me_6Te. The phenyl analogue, Ph_6Te, can be prepared by reaction 23.108, and treatment with Cl_2 converts Ph_6Te to Ph_5TeCl. Abstraction of chloride from the latter compound gives $[Ph_5Te]^+$ (eq. 23.109) which (in the $[B(C_6F_5)_4]^-$ salt) has a square-based pyramidal structure. Ph_6Te is thermally stable, but $[Ph_5Te]^+$ decomposes to $[Ph_3Te]^+$ (eq. 23.110).

$$TeCl_4 + 4MeLi \xrightarrow{Et_2O, \, 195 \, K} Me_4Te + 4LiCl \qquad (23.107)$$

$$Ph_4TeF_2 + 2PhLi \xrightarrow{298 \, K} Ph_6Te + 2LiF \qquad (23.108)$$

$$Ph_5TeCl \xrightarrow[\text{2. LiB(C}_6\text{F}_5)_4]{\text{1. AgSO}_3\text{CF}_3} [Ph_5Te]^+[B(C_6F_5)_4]^- \qquad (23.109)$$

$$[Ph_5Te]^+[B(C_6F_5)_4]^- \xrightarrow{420 \, K} [Ph_3Te]^+[B(C_6F_5)_4]^- + (C_6H_5)_2 \qquad (23.110)$$

KEY TERMS

The following terms were introduced in this chapter. Do you know what they mean?

- ❏ organometallic compound
- ❏ pyrophoric
- ❏ radical anion

- ❏ regioselective
- ❏ metallocene
- ❏ sandwich complex

- ❏ s-*cis* and s-*trans* conformations

FURTHER READING

General sources

Ch. Elschenbroich and A. Salzer (1992) *Organometallics*, 2nd edn, Wiley-VCH, Weinheim – An excellent text which covers both main group and transition metal organometallic chemistry.

G. Wilkinson, F.G.A. Stone and E.W. Abel, eds (1982) *Comprehensive Organometallic Chemistry*, Pergamon, Oxford – Volumes 1 and 2 provide detailed coverage of the organometallic compounds of groups 1, 2, 13, 14 and 15; the reviews include hundreds of literature references up to 1981.

G. Wilkinson, F.G.A. Stone and E.W. Abel, eds (1995) *Comprehensive Organometallic Chemistry II*, Pergamon, Oxford – Volumes 1 and 2 update the information from the above edition, covering groups 1, 2, 13, 14 and 15 for the period 1982–1994.

D.M.P Mingos and R.H. Crabtree, eds (2007) *Comprehensive Organometallic Chemistry III*, Elsevier, Oxford – Volumes 2 and 3 update the information from the above edition for groups 1, 2, 13, 14 and 15.

Specialized topics

K. Abersfelder and D. Scheschkewitz (2010) *Pure Appl. Chem.*, vol. 82, p. 595 – 'Synthesis of homo- and heterocyclic silanes via intermediates with Si=Si bonds'.

H.J. Breunig (2005) *Z. Anorg. Allg. Chem.*, vol. 631, p. 621 – 'Organometallic compounds with homonuclear bonds between bismuth atoms'.

W.D. Buchanan, D.G. Allis and K. Ruhlandt-Senge (2010) *Chem. Commun.*, vol. 46, p. 4449 – 'Synthesis and stabilization – advances in organoalkaline earth metal chemistry'.

P.H.M. Budzelaar, J.J. Engelberts and J.H. van Lenthe (2003) *Organometallics*, vol. 22, p. 1562 – 'Trends in cyclopentadienyl–main group-metal bonding'.

A.G. Davies (2004) *Organotin Chemistry*, 2nd edn, Wiley-VCH, Weinheim – This book includes an up-to-date coverage of the preparation and reactions of organotin compounds.

R. Fernández and E. Carmona (2005) *Eur. J. Inorg. Chem.*, p. 3197 – 'Recent developments in the chemistry of beryllocenes'.

R. Gleiter and D.B. Werz (2010) *Chem. Rev.*, vol. 110, p. 4447 – 'Alkynes between main group elements: From dumbbells via rods to squares and tubes'.

T.P. Hanusa (2000) *Coord. Chem. Rev.*, vol. 210, p. 329 – 'Non-cyclopentadienyl organometallic compounds of calcium, strontium and barium'.

T.P. Hanusa (2002) *Organometallics*, vol. 21, p. 2559 – 'New developments in the cyclopentadienyl chemistry of the alkaline-earth metals'.

P. Jutzi and N. Burford (1999) *Chem. Rev.*, vol. 99, p. 969 – 'Structurally diverse π-cyclopentadienyl complexes of the main group elements'.

P. Jutzi and G. Reumann (2000) *J. Chem. Soc., Dalton Trans.*, p. 2237 – 'Cp* Chemistry of main-group elements' (Cp* = C_5Me_5).

P.R. Markies, O.S. Akkerman, F. Bickelhaupt, W.J.J. Smeets and A.L. Spek (1991) *Adv. Organomet. Chem.*, vol. 32, p. 147 – 'X-ray structural analysis of organomagnesium compounds'.

N.C. Norman, ed. (1998) *Chemistry of Arsenic, Antimony and Bismuth*, Blackie, London – This book includes chapters dealing with organo-derivatives.

P.P. Power (2007) *Organometallics*, vol. 26, p. 4362 – 'Bonding and reactivity of heavier group 14 element alkyne analogues'.

P.P. Power (2010) *Nature*, vol. 463, p. 171 – 'Main group elements as transition metals'.

H.W. Roesky (2004) *Inorg. Chem.*, vol. 43, p. 7284 – 'The renaissance of aluminum chemistry'.

A. Schnepf (2004) *Angew. Chem. Int. Ed.*, vol. 43, p. 664 – 'Novel compounds of elements of group 14: ligand stabilized clusters with ''naked'' atoms'.

D.F. Shriver and M.A. Drezdon (1986) *The Manipulation of Air-sensitive Compounds*, Wiley, New York – An excellent text dealing with inert atmosphere techniques.

Y. Wang and G.H. Robinson (2007) *Organometallics*, vol. 26, p. 2 – 'Organometallics of the group 13 M–M bond (M = Al, Ga, In) and the concept of metalloaromaticity'.

Y. Wang and G.H. Robinson (2009) *Chem. Commun.*, p. 5201 – 'Unique homonuclear multiple bonding in main group compounds'.

PROBLEMS

23.1   Suggest products of the following reactions:

(a) $MeBr + 2Li \xrightarrow{Et_2O}$

(b) $Na + (C_6H_5)_2 \xrightarrow{THF}$

(c) $^nBuLi + H_2O \longrightarrow$

(d) $Na + C_5H_6 \longrightarrow$

23.2   Whether the bonding in lithium alkyls is predominantly ionic or covalent is still a matter for debate. Assuming a covalent model, use a hybrid orbital approach to suggest a bonding scheme for $(MeLi)_4$. Comment on the bonding picture you have described.

23.3   Describe the gas-phase and solid state structures of Me_2Be and discuss the bonding in each case. Compare the bonding with that in BeH_2 and $BeCl_2$.

23.4   Suggest products of the following reactions, which are *not* necessarily balanced on the left-hand side:

(a) $Mg + C_5H_6 \longrightarrow$

(b) $MgCl_2 + LiR \longrightarrow$

(c) $RBeCl \xrightarrow{LiAlH_4}$

23.5   The compound $(Me_3Si)_2C(MgBr)_2 \cdot nTHF$ is monomeric. Suggest a value for n and propose a structure for this Grignard reagent.

23.6   (a) For the equilibrium $Al_2R_6 \rightleftharpoons 2AlR_3$, comment on the fact that values of K are 1.52×10^{-8} for R = Me, and 2.3×10^{-4} for $R = Me_2CHCH_2$. (b) Describe the bonding in Al_2Me_6, Al_2Cl_6 and $Al_2Me_4(\mu\text{-}Cl)_2$.

23.7   Suggest products of the following reactions, which are *not* necessarily balanced on the left-hand side:

(a) $Al_2Me_6 + H_2O \longrightarrow$

(b) $AlR_3 + R'NH_2 \longrightarrow$

(c) $Me_3SiCl + Na[C_5H_5] \longrightarrow$

(d) $Me_2SiCl_2 + Li[AlH_4] \longrightarrow$

23.8   (a) Discuss the variation in structure for the group 13 trialkyls and triaryls. (b) Comment on features of interest in the solid state structures of $[Me_2(PhC_2)Ga]_2$ and $[Ph_3Al]_2$.

23.9   The conversion of $(\eta^1\text{-}C_5Me_5)_2SiBr_2$ to $(\eta^5\text{-}C_5Me_5)_2Si$ is achieved using anthracene/potassium. Outline the role of this reagent.

23.10   Suggest the nature of the solid state structures of (a) Ph_2PbCl_2, (b) Ph_3PbCl, (c) $(2,4,6\text{-}Me_3C_6H_2)_3PbCl$, and (d) $[PhPbCl_5]^{2-}$. In each case, state the expected coordination environment of the Pb centre.

23.11   Suggest products when Et_3SnCl reacts with the following reagents: (a) H_2O; (b) $Na[Cp]$; (c) Na_2S; (d) $PhLi$; (e) Na.

23.12   (a) In what ways do the solid state structures of $(\eta^5\text{-}C_5R_5)_2Sn$ for R = H, Me and Ph differ? (b) In the solid state structure of $(\eta^5\text{-}C_5Me_5)_2Mg$, the two cyclopentadienyl rings are parallel; however, for M = Ca, Sr and Ba, the rings are tilted with respect to one another. Say what you can about this observation.

23.13   The reaction of InBr with an excess of $HCBr_3$ in 1,4-dioxane $(C_4H_8O_2)$ leads to compound **A** which is an adduct of 1,4-dioxane and contains 21.4% In. During the reaction, the indium is oxidized. The ^1H NMR spectrum of **A** shows signals at δ 5.36 ppm (singlet) and δ 3.6 ppm (multiplet) in a ratio 1:8. Treatment of **A** with two molar equivalents of InBr followed by addition of $[Ph_4P]Br$ yields the salt **B** which contains 16.4% In and 34.2% Br. The ^1H NMR spectrum of **B** exhibits signals in the range δ 8.01–7.71 ppm and a singlet at δ 0.20 ppm with relative integrals of 60:1. Suggest identities for **A** and **B**.

23.14   Discuss the bonding between the central *p*-block elements in the following compounds and give the expected arrangements of the organic substituents with respect to the central E_2-unit:

(a) $[(2,4,6\text{-}Me_3C_6H_2)_2BB(2,4,6\text{-}Me_3C_6H_2)Ph]^{2-}$

(b) $[(2,4,6\text{-}^iPr_3C_6H_2)_2GaGa(2,4,6\text{-}^iPr_3C_6H_2)_2]^-$

(c) $\{(Me_3Si)_2CH\}_2SnSn\{CH(SiMe_3)_2\}_2$

(d) $^tBu_3GeGe^tBu_3$

(e) $(Me_3Si)_3CAsAsC(SiMe_3)_3$

23.15 Suggest products when Me_3Sb reacts with the following reagents: (a) B_2H_6; (b) H_2O_2; (c) Br_2; (d) Cl_2 followed by treatment with MeLi; (e) MeI; (f) Br_2 followed by treatment with Na[OEt].

23.16 Write a brief account of how the changes in available oxidation states for elements, E, in groups 13 to 15 affect the families of organoelement compounds of type R_nE that can be formed.

23.17 Give methods of synthesis for the following families of compound, commenting where appropriate on limitations in the choice of R: (a) R_4Ge; (b) R_3B; (c) $(C_5R_5)_3Ga$; (d) *cyclo*-$(R_2Si)_n$; (e) R_5As; (f) R_4Al_2; (g) R_3Sb.

23.18 Give a short account of the structural variation observed for cyclopentadienyl derivatives Cp_nE of the heavier *p*-block elements.

23.19 Write a brief account of the use of sterically demanding substituents in the stabilization of compounds containing E−E and E=E bonds where E is a *p*-block metal or semi-metal.

23.20 Write a short account describing methods of formation of metal–carbon bonds for metals in the *s*- and *p*-block.

23.21 By using specific examples, illustrate how heteronuclear NMR spectroscopy can be used for the routine characterization of main group organometallic compounds. [Tables 4.3, 11.1, 13.1, 14.1 and 15.2 provide relevant nuclear spin data.]

23.22 The structures of $R_2E=ER_2$ molecules where E is C, Si, Ge or Sn are usually of type **A** or **B** shown below:

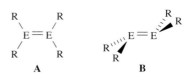

The bonding in the E_2-units is described in terms of the interaction of two triplet R_2E centres in **A**, and the interaction of two singlet R_2E centres in **B**. Explain the origins of these descriptions.

23.23 Give examples of the synthetic utility of the $[B(C_6F_5)_4]^-$ and $[CHB_{11}Me_5Br_6]^-$ anions, and rationalize the choice of these anions in the examples that you describe.

OVERVIEW PROBLEMS

23.24 (a) In 1956, it was concluded on the basis of dipole moment measurements that Cp_2Pb did not contain coparallel C_5-rings. Explain how this conclusion follows from such measurements.

(b) X-ray diffraction studies at 113 K show that two cyclopentadienyl complexes of beryllium can be formulated as $(\eta^5\text{-}C_5HMe_4)(\eta^1\text{-}C_5HMe_4)Be$ and $(\eta^5\text{-}C_5Me_5)_2Be$ respectively. The solution ^1H NMR spectrum at 298 K of $(C_5HMe_4)_2Be$ exhibits singlets at δ 1.80, 1.83 and 4.39 ppm (relative integrals 6:6:1), whereas that of $(C_5Me_5)_2Be$ shows one singlet at δ 1.83 ppm. Draw diagrams to represent the solid state structures of the compounds and rationalize the solution NMR spectroscopic data.

23.25 Treatment of $(2,4,6\text{-}^tBu_3C_6H_2)P=P(2,4,6\text{-}^tBu_3C_6H_2)$ with CF_3SO_3Me gives a salt **A** as the only product. The ^{31}P NMR spectrum of the precursor contains a singlet (δ +495 ppm), while that of the product exhibits two doublets (δ +237 and +332 ppm, $J = 633$ Hz). Compound **A** reacts with MeLi to give two isomers of **B** which are in equilibrium in solution. The solution ^{31}P NMR spectrum of **B** at 298 K shows one broad signal. On cooling to

213 K, two signals at δ −32.4 and −35.8 ppm are observed. From the solid state structures of **A** and one isomer of **B**, the P−P bond lengths are 202 and 222 pm. Suggest identities for **A** and **B**, and draw their structures which show the geometry at each P atom. Comment on the nature of the isomerism in **B**.

23.26 (a) Suggest how Na will react with $MeC(CH_2SbCl_2)_3$.

(b) Comment on aspects of the bonding in the following compound:

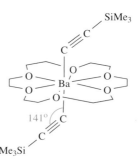

(c) Cp_2Ba and $(C_5Me_5)_2Ba$ both have polymeric structures in the solid state. However, whereas Cp_2Ba is insoluble in common organic solvents,

$(C_5Me_5)_2Ba$ is soluble in aromatic solvents. In contrast to $(C_5Me_5)_2Ba$, $(C_5Me_5)_2Be$ is monomeric. Suggest a reason for these observations.

23.27 The reactions of $(\eta^5\text{-}C_5Me_5)GeCl$ with $GeCl_2$ or $SnCl_2$ lead to the compound $[A]^+[B]^-$ or $[A]^+[C]^-$ respectively. The solution ^1H NMR spectrum of $[A][C]$ contains a singlet at δ 2.14 ppm, and the ^{13}C NMR spectrum shows two signals at δ 9.6 and 121.2 ppm. The mass spectra of the compounds exhibit a common peak at $m/z = 209$.
(a) Suggest identities for $[A][B]$ and $[A][C]$.
(b) Assign the ^{13}C NMR spectrum. (c) The peak at $m/z = 209$ is not a single line. Why is this?
(d) What structures do you expect $[B]^-$ and $[C]^-$ to adopt? (e) Describe the bonding in $[A]^+$.

23.28 (a) The reaction between $BiCl_3$ and 3 equivalents of $EtMgCl$ yields compound $\mathbf{X}$ as the organo-product. Two equivalents of BiI_3 react with 1 equivalent of $\mathbf{X}$ to produce 3 equivalents of compound $\mathbf{Y}$. In the solid state, $\mathbf{Y}$ has a polymeric structure consisting of chains in which each Bi centre is in a square-based pyramidal environment. Suggest identities for $\mathbf{X}$ and $\mathbf{Y}$, and draw possible structures for part of a chain in crystalline $\mathbf{Y}$.

(b) The reaction between $TeCl_4$ and 4 equivalents of $LiC_6H_4\text{-}4\text{-}CF_3$ (LiAr) in Et_2O leads to Ar_6Te, Ar_3TeCl and Ar_2Te as the isolated products. Suggest a pathway by which the reaction may take place that accounts for the products.

(c) The reaction of $R'SbCl_2$ with RLi (R = 2-$Me_2NCH_2C_6H_4$, $R' = CH(SiMe_3)_2$) leads to $RR'SbCl$. In the solid state, $RR'SbCl$ has a molecular structure in which the Sb centre is 4-coordinate; $RR'SbCl$ is chiral. Suggest a structure for $RR'SbCl$ and draw structures of the two enantiomers.

23.29 The following equilibrium has been studied by ^{119}Sn NMR and Mössbauer spectroscopies:

$R = C_6H_3\text{-}2,6\text{-}(C_6H_2\text{-}2,4,6\text{-}^iPr_3)_2$

The ^{119}Sn Mössbauer spectrum of a solid sample of $RSnSnRPh_2$ at 78 K provided evidence for the presence of three different tin environments. When $RSnSnRPh_2$ dissolves in toluene, a red solution is obtained, and at room temperature, the ^{119}Sn NMR spectrum of this solution shows one broad signal at δ 1517 ppm. This chemical shift is similar to that observed for $Sn(C_6H\text{-}2\text{-}^tBu\text{-}4,5,6\text{-}Me_3)_2$. On cooling the solution to 233 K, the signal at δ 1517 ppm gradually sharpens and at the same time, two new signals appear at δ 246 and 2857 ppm. Each of these new signals shows coupling $J(^{119}Sn\text{–}^{117}Sn/^{119}Sn) = 7237$ Hz. Cooling further to 213 K results in the disappearance of the signal at δ 1517 ppm, and at this point, the solution is green in colour. A colour change from green to red is observed when the solution is warmed up to room temperature. Rationalize these observations.

23.30 Experimentally determined analytical data for $PhSeCl_3$ are C, 27.5; H, 1.8; Cl, 39.9%. An X-ray diffraction study of $PhSeCl_3$ shows that it forms polymeric chains in the solid state, with each Se centre in a square-based pyramidal environment with the Ph group in the axial position. (a) To what extent are elemental analytical data useful in characterizing a new compound? (b) Draw part of the polymeric chain from the solid state structure, paying attention to the overall stoichiometry of the compound. (c) Around one Se centre in the solid state structure of $PhSeCl_3$, the observed Se–Cl bond distances are 220, 223, 263 and 273 pm. Comment on these values in the light of your answer to part (b).

INORGANIC CHEMISTRY MATTERS

23.31 The International Marine Organization is implementing a ban on the use of tributyltin compounds in anti-fouling paints on ships. The $[Bu_3Sn]^+$ cation is leached from paints into the water where it undergoes biodegradation by marine bacteria to $[Bu_2Sn]^{2+}$ and $[BuSn]^{3+}$. Tributyltin compounds (TBT) are more toxic than di- or monobutyltin derivatives (DBT and MBT). The first order rate constants for TBT, DBT and MBT are 0.33, 0.36 and 0.63 y^{-1}, respectively. Determine the half-life of each species in the water.

23.32 Comment on:

(a) the use of trimethylorganometallics in the manufacture of III–V semiconductors;

(b) the application of R_3Al compounds as catalysts.

23.33 Yeasts, fungi and bacteria are able to convert inorganic arsenic (arsenous and arsenic acids) to organoarsenic species. The pathway can be represented as follows, although not all species have been isolated:

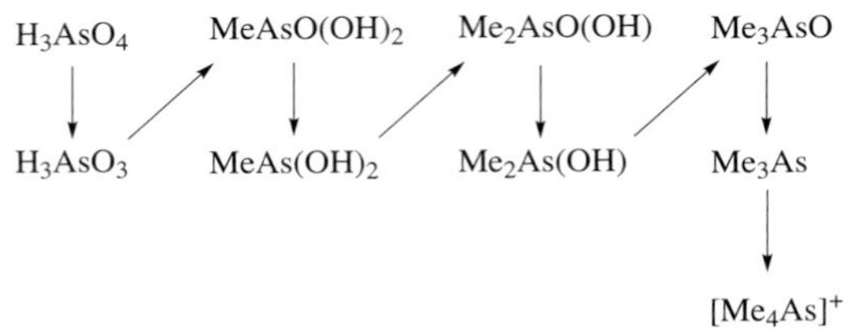

$$H_3AsO_4 \quad MeAsO(OH)_2 \quad Me_2AsO(OH) \quad Me_3AsO$$

$$H_3AsO_3 \quad MeAs(OH)_2 \quad Me_2As(OH) \quad Me_3As$$

$$[Me_4As]^+$$

The source of the methyl group is *S*-adenosylmethionine. Glutathione functions as a reducing agent and undergoes reversible oxidative coupling to a disulfide.

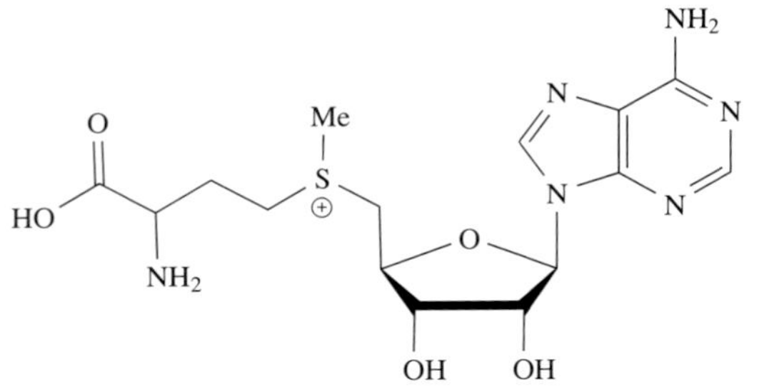

S-adenosylmethionine

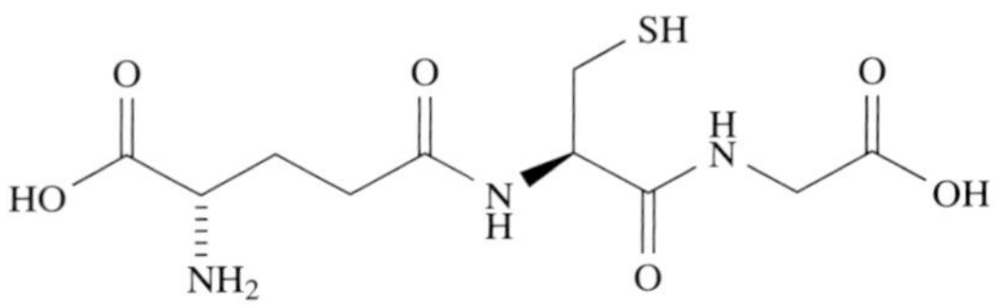

Glutathione

(a) By representing glutathione as RSH, write a half-equation to describe its oxidative coupling, showing how glutathione acts as a reducing agent.

(b) Write a half-equation to describe the reduction of Me$_2$AsO(OH).

(c) Discuss the pathway drawn above in terms of redox processes and state in what form the methyl group is transferred from *S*-adenosylmethionine to arsenic.

24

Organometallic compounds of *d*-block elements

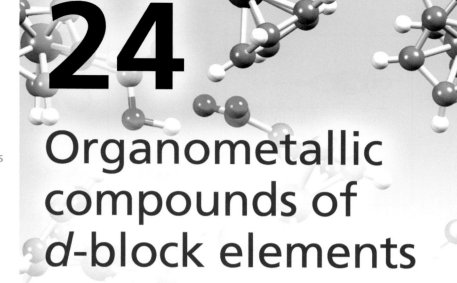

1–2	3	4	5	6	7	8	9	10	11	12	13–18
s-block											*p*-block
	Sc	Ti	V	Cr	Mn	Fe	Co	Ni	Cu	Zn	
	Y	Zr	Nb	Mo	Tc	Ru	Rh	Pd	Ag	Cd	
	La	Hf	Ta	W	Re	Os	Ir	Pt	Au	Hg	

24.1 Introduction

Organometallic chemistry of the *s*- and *p*-block elements was described in Chapter 23, and we now extend the discussion to organometallic compounds containing *d*-block metals. This topic covers a huge area of chemistry, and we can only provide an introduction to it, emphasizing the fundamental families of complexes and reactions.

In the previous chapters, we introduced compounds containing σ-bonds or π-interactions between a metal centre and a cyclopentadienyl ligand. We also introduced examples of 3-electron donor bridging ligands, e.g. halides (**23.8**) and alkynyls (**23.11**), and 2-electron alkene donors, e.g. **23.19**.

The *hapticity of a ligand* is the number of atoms that are directly bonded to the metal centre (see Boxes 19.1 and 23.1). Structures **24.1a** and **24.1b** show two representations of an $[\eta^5\text{-C}_5\text{H}_5]^-$ (cyclopentadienyl, Cp$^-$) ligand. For

clarity in the diagrams in this chapter, we adopt **24.1b** and similar representations for π-ligands such as $\eta^3\text{-C}_3\text{H}_5$ and $\eta^6\text{-C}_6\text{H}_6$.

24.2 Common types of ligand: bonding and spectroscopy

In this section, we introduce some of the most common ligands found in organometallic complexes. Many other ligands are related to those discussed below, and bonding descriptions can be developed by comparison with the ligands chosen for detailed coverage.

σ-Bonded alkyl, aryl and related ligands

In complexes such as WMe$_6$, [MoMe$_7$]$^-$, TiMe$_4$ and MeMn(CO)$_5$, the M$-$C$_{\text{Me}}$ bond can be described as a localized 2c-2e interaction, i.e. it parallels that for the $[\eta^1\text{-Cp}]^-$ ligand (see Box 23.1). The same bonding description is applicable to the Fe$-$C$_{\text{Ph}}$ bond in **24.2** and the Fe$-$C$_{\text{CHO}}$ bond in **24.3**.

(24.1a)   (24.1b)

(24.2)   (24.3)

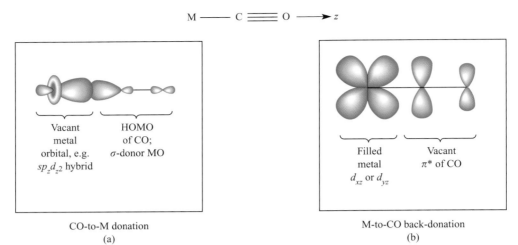

Fig. 24.1 Components of metal–carbonyl bonding: (a) the M−CO σ-bond, and (b) the M−CO π-interaction which leads to back-donation of charge from metal to carbonyl. The orbital labels are examples, and assume that the M, C and O atoms lie on the *z* axis.

Carbonyl ligands

The bonding in octahedral $M(CO)_6$ complexes was described in Section 20.4 using a molecular orbital approach, but it is also convenient to give a simple picture to describe the bonding in one M−CO interaction. Figure 24.1a shows the σ-interaction between the highest occupied molecular orbital of CO (which has predominantly C character, Fig. 2.15) and a vacant orbital on the metal centre (e.g. an $sp_zd_{z^2}$ hybrid). As a result of this interaction, electronic charge is donated from the CO ligand to the metal. Figure 24.1b shows the π-interaction that leads to *back-donation* of charge from metal to ligand; compare Fig. 24.1b with Fig. 20.14b. This 'donation/back-donation' bonding picture is the *Dewar–Chatt–Duncanson model.* Carbon monoxide is a *weak* σ-*donor* and a *strong* π-*acceptor* (or π-*acid*) and population of the CO π*-MO weakens and lengthens the C−O bond while also enhancing M−C bonding. Resonance structures **24.4** for the MCO unit also indicate a lowering of the C−O bond order as compared with free CO.

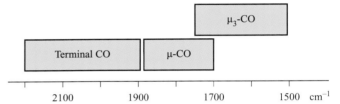

Fig. 24.2 Approximate regions in the IR spectrum in which absorptions assigned to C−O stretches observed for different carbonyl bonding modes; there is often overlap between the regions, e.g. see Table 24.1.

In multinuclear metal species, CO ligands may adopt terminal (**24.5**) or bridging (**24.6** and **24.7**) modes. Other modes are known, e.g. semi-bridging (part way between **24.5** and **24.6**) and mode **24.8**.

Evidence for a lowering of the C−O bond order on coordination comes from structural and spectroscopic data. In the IR spectrum of free CO, an absorption at $2143\,cm^{-1}$ is assigned to the C−O stretching mode and typical changes in vibrational wavenumber, $\bar{\nu}$, on going to metal carbonyl complexes are illustrated in Fig. 24.2. Absorptions due to C−O stretching modes are strong and easily observed. The lower the value of $\bar{\nu}_{CO}$, the weaker the C−O bond and this indicates greater back-donation of charge from metal to CO. Table 24.1 lists data for two sets of isoelectronic metal carbonyl complexes. On going from $Ni(CO)_4$ to $[Co(CO)_4]^-$ to $[Fe(CO)_4]^{2-}$, the additional negative charge is delocalized onto the ligands, causing a decrease in $\bar{\nu}_{CO}$. A similar effect is seen along the series $[Fe(CO)_6]^{2+}$, $[Mn(CO)_6]^+$, $Cr(CO)_6$ and $[V(CO)_6]^-$. The increased back-donation is also reflected in values of $\bar{\nu}_{MC}$, e.g. $416\,cm^{-1}$ for $[Mn(CO)_6]^+$, and $441\,cm^{-1}$ for $Cr(CO)_6$.[†]

$$: \overset{-}{M} - C \equiv \overset{+}{O} : \longleftrightarrow M = C = \overset{..}{\underset{..}{O}}:$$

(24.4)

> The interplay of donation and back-donation of electronic charge between a metal and π-acceptor ligand is an example of a *synergic effect.*

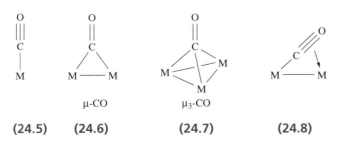

O ‖‖ C │ M	O ‖ C ⟋ ⟍ M — M	O ‖ C ⟋ │ ⟍ M ⟍ │ ⟋ M M	O ⫽ C ⟋ ↓ M — M
	μ-CO	μ₃-CO	
(24.5)	**(24.6)**	**(24.7)**	**(24.8)**

[†] For detailed discussions of IR spectroscopy in metal carbonyls, see: K. Nakamoto (1997) *Infrared and Raman Spectra of Inorganic and Coordination Compounds*, Part B, 5th edn, Wiley, New York, p. 126; S.F.A. Kettle, E. Diana, R. Rossetti and P.L. Stanghellini (1998) *J. Chem. Educ.*, vol. 75, p. 1333 – 'Bis(dicarbonyl-π-cyclopentadienyliron): a solid-state vibrational spectroscopic lesson'.

Table 24.1 Infrared spectroscopic data: values of $\bar{\nu}_{CO}$ for isoelectronic sets of tetrahedral $M(CO)_4$ and octahedral $M(CO)_6$ complexes.

Complex	$Ni(CO)_4$	$[Co(CO)_4]^-$	$[Fe(CO)_4]^{2-}$	$[Fe(CO)_6]^{2+}$	$[Mn(CO)_6]^+$	$Cr(CO)_6$	$[V(CO)_6]^-$
$\bar{\nu}_{CO}/cm^{-1}$	2060	1890	1790	2204	2101	1981	1859

Carbonyl ligand environments can also be investigated using ^{13}C NMR spectroscopy, although systems are often fluxional (e.g. $Fe(CO)_5$, see Fig. 4.24 and discussion) and information about specific CO environments may therefore be masked. Some useful points are that:

- typical ^{13}C NMR shifts for metal carbonyl ^{13}C nuclei are $\delta +170$ to $+240$ ppm;
- within a series of analogous compounds containing metals from a given triad, the ^{13}C NMR signals for the CO ligands shift to lower frequency, e.g. in the ^{13}C NMR spectra of $Cr(CO)_6$, $Mo(CO)_6$ and $W(CO)_6$, signals are at $\delta +211$, $+201$ and $+191$ ppm respectively;
- for a given metal, signals for μ-CO ligands occur at higher frequency (more positive δ value) than those for terminal carbonyls.

In keeping with the typical weakening of the C−O bond on going from free CO to coordinated CO, X-ray diffraction data show a lengthening of the C−O bond. In CO, the C−O bond length is 112.8 pm, whereas typical values in metal carbonyls for terminal and μ-CO are 117 and 120 pm respectively.

The traditional bonding model for an M−CO interaction emphasizes OC ⟶ M σ-donation and significant M ⟶ CO π-back-donation leading to C−O bond weakening and a concomitant lowering of $\bar{\nu}_{CO}$. However, there are a growing number of isolable metal carbonyl complexes in which $\bar{\nu}_{CO}$ is *higher* than in free CO (i.e. $>2143\ cm^{-1}$), the C−O bond distance is shorter than in free CO (i.e. <112.8 pm), and the M−C bonds are relatively long.[†] Members of this group include the following cations (many are salts of $[SbF_6]^-$ or $[Sb_2F_{11}]^-$, see eqs. 22.78, 22.105 and 24.24) and, in each case, the metal–carbonyl bonding is dominated by the OC ⟶ M σ-component:

- tetrahedral $[Cu(CO)_4]^+$, $\bar{\nu}_{CO} = 2184\ cm^{-1}$, C−O = 111 pm;
- square planar $[Pd(CO)_4]^{2+}$, $\bar{\nu}_{CO} = 2259\ cm^{-1}$, C−O = 111 pm;
- square planar $[Pt(CO)_4]^{2+}$, $\bar{\nu}_{CO} = 2261\ cm^{-1}$, C−O = 111 pm;
- octahedral $[Fe(CO)_6]^{2+}$, $\bar{\nu}_{CO} = 2204\ cm^{-1}$, C−O = 110 pm;
- octahedral $[Ir(CO)_6]^{3+}$, $\bar{\nu}_{CO} = 2268\ cm^{-1}$, C−O = 109 pm.

An analysis of more than 20 000 crystal structures of d-block metal carbonyl complexes[‡] confirms a clear correlation between C−O and M−C bond distances, i.e. as the M−C bond distance decreases, the C−O bond distance (d_{CO}) increases. Ninety per cent of the structural data fall into a region in which $117.0\ pm > d_{CO} > 112.8\ pm$, and for these M−CO interactions, the M−C σ- and π-bonding contributions are approximately in balance. For 4%, π-bonding dominates and $d_{CO} > 117.0\ pm$, while for 6%, σ-bonding and ionic contributions dominate and $d_{CO} < 112.8\ pm$.

Self-study exercises

Answers to the following problems can be found by reading Section 3.7. Character tables are given in Appendix 3.

1. The vibrational wavenumbers for the ν_{CO} modes in $[V(CO)_6]^-$ are 2020 (A_{1g}), 1894 (E_g) and 1859 (T_{1u}) cm^{-1}. Explain why only one of these modes is IR active.

2. Confirm that $Mn(CO)_5Cl$ belongs to the C_{4v} point group. The vibrational wavenumbers for the ν_{CO} modes in $Mn(CO)_5Cl$ are 2138 (A_1), 2056 (E) and 2000 (A_1) cm^{-1}. Use the C_{4v} character table to confirm that all three modes are IR active.

3. The IR spectrum of a salt of $[Fe(CO)_4]^{2-}$ (T_d) shows an absorption at 1788 cm^{-1}, assigned to the T_2 mode. Sketch a diagram to show the mode of vibration that corresponds to this absorption.

Hydride ligands

The term *hydride ligand* suggests $H^{\delta-}$ and is consistent with the charge distribution expected for an H atom attached to an electropositive metal centre. However, the properties of H ligands depend on environment and in many organometallic complexes, hydrido ligands behave as protons, being removed by base (eq. 24.1) or introduced by treatment with acid (reaction 24.2).

$$HCo(CO)_4 + H_2O \longrightarrow [Co(CO)_4]^- + [H_3O]^+ \quad (24.1)$$

$$[HFe(CO)_4]^- + H^+ \longrightarrow H_2Fe(CO)_4 \quad (24.2)$$

[†] For detailed discussion, see: S.H. Strauss (2000) *J. Chem. Soc., Dalton Trans.*, p. 1; H. Willner and F. Aubke (1997) *Angew. Chem. Int. Ed.*, vol. 36, p. 2403.

[‡] R.K. Hocking and T.W. Hambley (2003) *Chem. Commun.*, p. 1516 – 'Structural insights into transition-metal carbonyl bonding'.

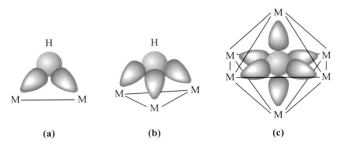

(a)   **(b)**   **(c)**

Fig. 24.3 Overlap of the H $1s$ atomic orbital with (a) two or (b) three appropriate metal hybrid orbitals to form μ-H and μ_3-H bridges. (c) For an interstitial H atom within an octahedral M_6-cage, a delocalized description involves the overlap of the H $1s$ atomic orbital with six appropriate metal orbitals to give a 7c-2e interaction.

Hydride ligands can adopt terminal, bridging or (in metal clusters) interstitial modes of bonding (**24.9–24.12**). A localized 2c-2e M–H bond is an appropriate description for a terminal hydride, delocalized 3c-2e or 4c-2e interactions describe μ-H and μ_3-H interactions respectively (Figs. 24.3a and 24.3b), and a 7c-2e interaction is appropriate for an interstitial hydride in an octahedral cage (Fig. 24.3c).

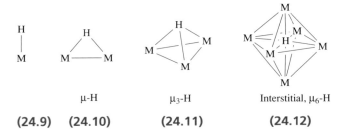

μ-H μ_3-H   Interstitial, μ_6-H

(24.9)   **(24.10)**   **(24.11)**   **(24.12)**

Locating hydride ligands by X-ray diffraction is difficult (see Section 4.11). X-rays are diffracted by *electrons* and the electron density in the region of the M–H bond is dominated by the heavy atoms. Neutron diffraction can be used, but this is an expensive and less readily available technique. In IR spectra, absorptions due to ν_{MH} modes are generally weak. Proton NMR spectroscopy is the routine way of observing metal hydrides in diamagnetic compounds. In ^{1}H NMR spectra, signals due to metal hydrides usually occur in the approximate range δ −8 to −30 ppm, although it is not easy to distinguish between terminal and bridging modes. The chemical shifts of interstitial hydrides are less diagnostic, and may occur at high frequency, e.g. δ +16.4 ppm in $[(\mu_6\text{-H})Ru_6(CO)_{18}]^-$. Spin–spin coupling to spin-active metal nuclei such as ^{103}Rh (100% abundant, $I = \frac{1}{2}$), ^{183}W (14.3%, $I = \frac{1}{2}$) or ^{195}Pt (33.8%, $I = \frac{1}{2}$) gives valuable structural information, as does that to nuclei such as ^{31}P. Typical values of J_{PH} for a *cis*-arrangement (**24.13**) are 10–15 Hz, compared with $\geq$30 Hz for *trans*-coupling (**24.14**).

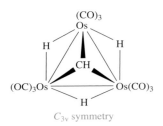

cis   *trans*

(24.13)   **(24.14)**

Examples of stereochemically non-rigid hydride complexes are common (e.g. in the tetrahedral cluster $[H_3Ru_4(CO)_{12}]^-$) and variable temperature NMR spectroscopic studies are routinely carried out.

Self-study exercise

^{187}Os has $I = \frac{1}{2}$ and is 1.64% abundant. In the ^{1}H NMR spectrum of $H_3Os_3(CO)_9CH$ (see below) in CDCl$_3$, the metal hydride signal appears as a singlet at δ −19.58 ppm, flanked by two, low-intensity doublets. Observed coupling constants are $J(^{187}\text{Os}-^1\text{H}) = 27.5$ Hz and $J(^1\text{H}-^1\text{H}) = 1.5$ Hz. Sketch the region of the spectrum that exhibits the hydride signal and rationalize the observed coupling pattern.

C_{3v} symmetry

[*Ans.* See J.S. Holmgren *et al.* (1985) *J. Organomet. Chem.*, vol. 284, p. C5]

Phosphane and related ligands

Monodentate organophosphanes[†] may be tertiary (PR$_3$), secondary (PR$_2$H) or primary (PRH$_2$) and are usually *terminally* bound; PF$_3$ behaves similarly. Bridging modes can be adopted by $[PR_2]^-$ (**24.15**) or $[PR]^{2-}$ (**24.16**) ligands. Since 1990, examples of *bridging* PR$_3$ ligands have been known. To date, these typically involve the late *d*-block metals Rh and Pd. An early example is **24.17** which exhibits μ_3-PF$_3$ in addition to bidentate R$_2$PCH$_2$PR$_2$ ligands (see below).

[†]Phosphane is the IUPAC name for PH$_3$; organophosphanes are compounds of type RPH$_2$, R$_2$PH and R$_3$P. The older names of phosphine and organophosphines remain in common use but are deemed obsolete by the IUPAC.

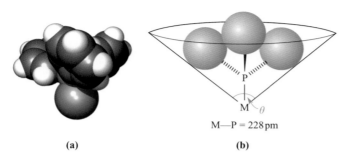

(a)   **(b)**

M—P = 228 pm

Fig. 24.4 (a) A space-filling diagram of an FePPh$_3$ unit; the phenyl groups adopt a 'paddle-wheel' arrangement; colour code: Fe, green; P, orange; C, grey; H, white. (b) Schematic representation of the measurement of the Tolman cone angle, θ, for a PR$_3$ ligand; each circle represents the spatial extent of an R group.

The series of dirhodium complexes involving μ-PR$_3$, μ-Sb^{i}Pr$_3$ (e.g. **24.18**) or μ-AsMe$_3$ ligands is steadily growing.

(24.15)   **(24.16)**

(24.17)   **(24.18)**

Phosphanes are σ-donor and π-acceptor ligands (see Section 20.4) and related to them are arsanes (AsR$_3$), stibanes (SbR$_3$) and phosphites (P(OR)$_3$). The extent of σ-donation and π-acceptance depends on the substituents, e.g. PR$_3$ (R = alkyl) is a poor π-acceptor, whereas PF$_3$ is a poor σ-donor and as strong a π-acceptor as CO. The π-accepting properties of some PR$_3$ ligands follow the ordering:

$$PF_3 > P(OPh)_3 > P(OMe)_3 > PPh_3 > PMe_3 > P^tBu_3$$

Infrared spectroscopic data can be used to determine this sequence: a ligand *trans* to a CO affects the M $\longrightarrow$ CO back-donation and, therefore, $\bar{\nu}_{CO}$, e.g. in octahedral Mo(CO)$_3$(PF$_3$)$_3$, $\bar{\nu}_{CO} = 2090$ and $2055\,\text{cm}^{-1}$, compared with 1937 and 1841 cm^{-1} in Mo(CO)$_3$(PPh$_3$)$_3$.

The steric requirements of a PR$_3$ ligand depend on the R groups. Ligands such as PPh$_3$ (Fig. 24.4a) or P^{t}Bu$_3$ are

Table 24.2 Tolman cone angles for selected phosphane and phosphite ligands.

Ligand	Tolman cone angle / deg	Ligand	Tolman cone angle / deg
P(OMe)$_3$	107	PPh$_3$	145
PMe$_3$	118	P(4-MeC$_6$H$_4$)$_3$	145
PMe$_2$Ph	122	P^{i}Pr$_3$	160
PHPh$_2$	126	P(3-MeC$_6$H$_4$)$_3$	165
P(OPh)$_3$	128	P(cyclo-C$_6$H$_{11}$)$_3$	170
PEt$_3$	132	P^{t}Bu$_3$	182
P^{n}Bu$_3$	132	P(2-MeC$_6$H$_4$)$_3$	194
PMePh$_2$	136	P(2,4,6-Me$_3$C$_6$H$_2$)$_3$	212

sterically demanding while others such as PMe$_3$ are less so. The steric requirements are assessed using the *Tolman cone angle*,[†] found by estimating the angle of a cone that has the metal atom at its apex and encompasses the PR$_3$ ligand taking the van der Waals surfaces of the H atoms as its boundary (Fig. 24.4b). Table 24.2 lists Tolman cone angles for selected ligands.

The variation in electronic and steric effects in PR$_3$ and related ligands can significantly alter the reactivities of complexes in a series in which the only variant is the phosphane ligand. Many polydentate phosphanes are known, two of the more common being **24.19** (dppm) and **24.20** (dppe).[‡] The modes of bonding of polydentate phosphanes depend on the flexibility of the backbone of the ligand. For example, dppm is ideally suited to bridge between two adjacent M centres, whereas dppe is found in chelating and bridging modes, or may act as a monodentate ligand with one P atom uncoordinated. Assigning the bonding mode is aided by ^{31}P NMR spectroscopy. Coordination of a P atom shifts its ^{31}P NMR resonance to higher frequency, e.g. the signal in the ^{31}P NMR spectrum of free PPh$_3$ is at $\delta -6$ ppm, compared with $\delta +20.6$ ppm for W(CO)$_5$(PPh$_3$).

Ph$_2$P   PPh$_2$ Ph$_2$P   PPh$_2$
dppm   dppe

(24.19)   **(24.20)**

[†] For a full discussion, see C.A. Tolman (1977) *Chem. Rev.*, vol. 77, p. 313.
[‡] The abbreviations dppm and dppe originate from the old names bis(diphenylphosphino)methane and bis(diphenylphosphino)ethane, but the IUPAC names are methylenebis(diphenylphosphane) and ethane-1,2-diylbis(diphenylphosphane).

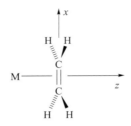

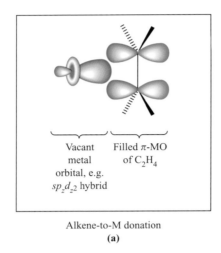

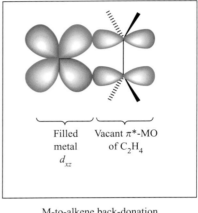

Fig. 24.5 Components of metal–alkene bonding: (a) donation of electrons from the alkene π-MO to a suitable metal d orbital or hybrid and (b) back-donation of electrons from metal to alkene π^* MO.

Self-study exercise

The ligands L and L' shown below react with $PdCl_2$ to give L_2PdCl_2 and $(L')ClPd(\mu\text{-}Cl)_2Pd(\mu\text{-}Cl)_2PdCl(L')$, respectively. Suggest why a mononuclear complex is formed in only one case.

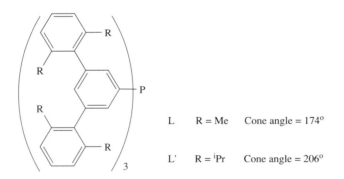

| L | R = Me | Cone angle = 174° |
| L' | R = ^{i}Pr | Cone angle = 206° |

[*Ans.* See Y. Ohzu *et al.* (2003) *Angew. Chem. Int. Ed.*, vol. 42, p. 5714]

π-Bonded organic ligands

Alkenes, $R_2C{=}CR_2$, tend to bond to metal centres in a 'side-on' (i.e. η^2) manner and behave as 2-electron donors. The

metal–ligand bonding can be described in terms of the Dewar–Chatt–Duncanson model (Fig. 24.5). The C=C π-bonding MO acts as an electron donor, while the π^*-MO is an electron acceptor. Populating the π^*-MO leads to:

- C–C bond lengthening, e.g. 133.9 pm in C_2H_4 *vs* 144.5 pm in $(\eta^5\text{-}Cp)Rh(\eta^2\text{-}C_2H_4)(PMe_3)$;
- a lowering of the absorption in the vibrational spectrum due to the stretching of the C=C bond, e.g. 1623 cm^{-1} in free C_2H_4 *vs* 1551 cm^{-1} in $Fe(CO)_4(\eta^2\text{-}C_2H_4)$.

The extent of back-donation to $R_2C{=}CR_2$ is influenced by the nature of R, and is enhanced by electron-withdrawing groups such as CN. In the extreme, the π-contribution to the C–C bond is completely removed and the complex becomes a *metallacyclopropane* ring. Structures **24.21a** and **24.21b** show limiting bonding schemes. In **24.21a**, alkene $\longrightarrow$ M donation of charge is dominant, while in **24.21b**, π-back-donation has fully populated the alkene π^*-MO, reducing the C–C bond order to one. On going from **24.21a** to **24.21b**, the alkene C atoms rehybridize from sp^2 to sp^3, M–C σ-bonds are formed, and the alkene substituents bend away from the metal (Fig. 24.6a). Comparisons of X-ray diffraction data for series of complexes provide evidence for these structural changes.

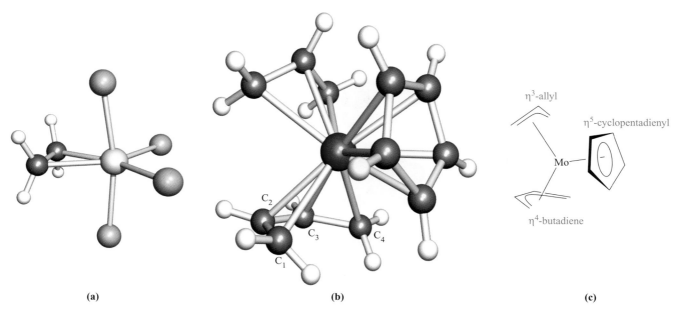

(a)   (b)   (c)

Fig. 24.6 (a) The structure (X-ray diffraction) of Ru(η^2-C_2H_4)(PMe$_3$)$_4$ illustrating the non-planarity of the coordinated ethene ligand (C−C = 144 pm); only the P atoms of the PMe$_3$ ligands are shown [W.-K. Wong *et al.* (1984) *Polyhedron*, vol. 3, p. 1255], (b) the structure (X-ray diffraction) of Mo(η^3-C_3H_5)(η^4-$CH_2CHCHCH_2$)(η^5-C_5H_5) [L.-S. Wang *et al.* (1997) *J. Am. Chem. Soc.*, vol. 119, p. 4453], and (c) a schematic representation of Mo(η^3-C_3H_5)(η^4-$CH_2CHCHCH_2$)(η^5-C_5H_5). Colour code: Ru, pale grey; Mo, red; C, dark grey; P, orange; H, white.

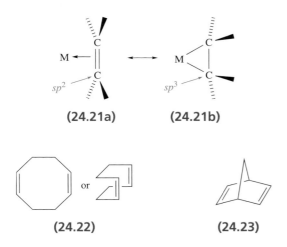

(24.21a)   (24.21b)

(24.22)   (24.23)

The bonding description for a coordinated alkene can be extended to other unsaturated organic ligands. Polyalkenes may be non-conjugated or conjugated. In complexes of non-conjugated systems (e.g. cycloocta-1,5-diene (cod), **24.22**, or 2,5-norbornadiene (nbd), **24.23**), the metal–ligand bonding is analogous to that for isolated alkene groups. For complexes of conjugated polyenes such as buta-1,3-diene, a delocalized bonding picture is appropriate. Figure 24.7a shows the four π-molecular orbitals of buta-1,3-diene. These MOs can be derived using the procedures described in Section 5.5. *cis*-Buta-1,3-diene (i.e. the *free* ligand) has C_{2v} symmetry. We define the z axis to coincide with the C_2 axis, and the molecule to lie in the yz plane. (This axis set is not that used in Fig. 24.7b. Instead, a

convenient axis set is chosen to describe the metal orbitals in the complex.) After C−H and C−C σ-bond formation, each C atom has one $2p$ orbital for π-bonding. The number of these $2p$ orbitals unchanged by each symmetry operation in the C_{2v} point group is given by the following row of characters:

E	C_2	$\sigma_v(xz)$	$\sigma_v'(yz)$
4	0	0	−4

Since there are four $2p$ orbitals, there will be four π-MOs, and from the C_{2v} character table, the row of characters above is reproduced by taking the sum of two A_2 and two B_1 representations. The π-orbitals therefore have a_2 or b_1 symmetry, and schematic representations are shown in Fig. 24.7a. In Fig. 24.7b, their symmetries (see caption to Fig. 24.7) are matched to available metal orbitals. Two combinations lead to ligand $\longrightarrow$ M donations, and two to M $\longrightarrow$ ligand back-donation. The interactions involving ψ_2 and ψ_3 weaken bonds C_1−C_2 and C_3−C_4, while strengthening C_2−C_3. The extent of ligand donation or metal back-donation depends on the metal, substituents on the diene, and other ligands present. Structure **24.24** shows the C−C bond lengths in free buta-1,3-diene, and examples of complexes include Fe(CO)$_3$(η^4-C_4H_6), in which all three C−C bonds in the coordinated diene are 145 pm, and Mo(η^3-C_3H_5)(η^4-C_4H_6)(η^5-C_5H_5) (Fig. 24.6b and c), in which the buta-diene ligand has C−C bond lengths of 142 (C_1−C_2), 138

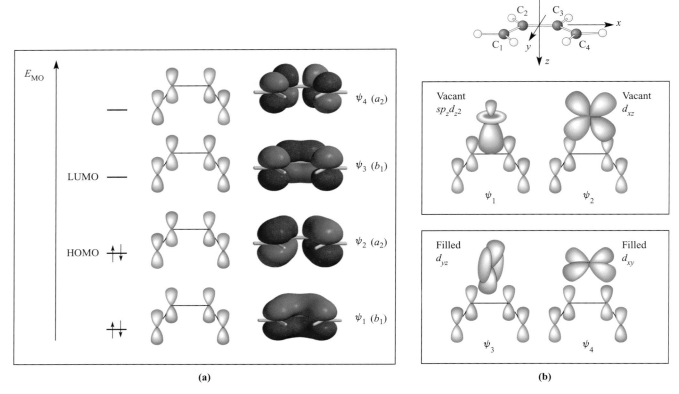

Fig. 24.7 (a) The four π-MOs of buta-1,3-diene (the energy scale is arbitrary); the symmetry labels apply to C_{2v} buta-1,3-diene with the C and H atoms lying in the *yz* plane. These symmetry labels are not applicable to the ligand in a complex of other symmetry. (b) Axis definition for a metal–buta-1,3-diene complex and the combinations of metal and ligand orbitals that lead to transfer of charge from a 1,3-diene to metal (top diagram) and from metal to 1,3-diene (lower diagram).

(C_2-C_3) and 141 pm (C_3-C_4). Just as for alkene coordination, we can draw two limiting resonance structures (**24.25**) for a buta-1,3-diene (or other 1,3-diene) complex.[†]

(24.24)

(24.25)

(24.26)

The allyl ligand, $[C_3H_5]^-$ (**24.26**), coordinates in an η^3-mode, using the two occupied π-MOs (bonding and non-bonding) as donors and the π^*-MO as an acceptor

(Fig. 24.8). Allyl can also be considered as $[C_3H_5]^\bullet$ (see later). Similar schemes can be developed for cyclobutadiene (η^4-C_4H_4), cyclopentadienyl (η^5-C_5H_5, see Box 23.1), benzene (η^6-C_6H_6) and related ligands as we discuss later in the chapter.

In solution, complexes with π-bonded organic ligands are often fluxional, with rotation of the ligand being a common dynamic process (see structure **24.55**, Fig. 24.19 and scheme 24.98). Variable-temperature NMR spectroscopy is used to study such phenomena.

Nitrogen monoxide

Nitrogen monoxide is a radical (see Section 15.8). Its bonding closely resembles that of CO, and can be represented as in Fig. 2.15 with the addition of one electron into a $\pi^*(2p)$ orbital. The NO molecule can bind to a low oxidation state metal atom in a similar way to CO, and once coordinated NO is known as a *nitrosyl* ligand. However, unlike CO, terminally bound NO can adopt two different bonding modes: linear or bent (see the end of Section 20.4). In the linear mode (**24.27**), NO donates

[†] For a critical discussion of the C–C bond lengths in Mn(η^4-$C_4H_6)_2$(CO) and related complexes, see: G.J. Reiß and S. Konietzny (2002) *J. Chem. Soc., Dalton Trans.*, p. 862.

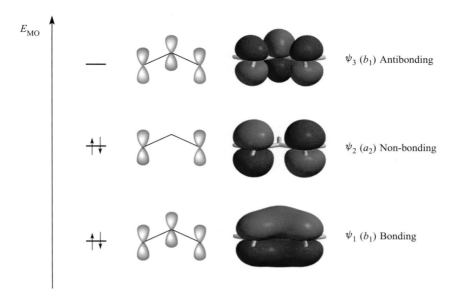

Fig. 24.8 The three π-MOs of the allyl anion, $[C_3H_5]^-$ (the energy scale is arbitrary); the symmetry labels apply to C_{2v} allyl with the C and H atoms lying in the yz plane. These symmetry labels are not applicable to the ligand in a complex of other symmetry.

three electrons to the metal centre, and the ligand behaves as a π-acceptor. Experimentally, a 'linear' MNO unit may have M–N–O bond angles in the range 165–180°, and in the IR spectrum, the vibrational wavenumber for the ν_{NO} mode lies in the approximate range 1650–1900 cm^{-1}. In the bent mode (**24.28**), NO donates one electron to the metal centre. Bent nitrosyl ligands are characterized by having M–N–O bond angles in the range 120–140°, and in the IR spectrum, the ν_{NO} absorption typically lies between 1525 and 1690 cm^{-1}.

$$M{=}N{=}\overset{..}{\underset{..}{O}} \longleftrightarrow :M{-}N{\equiv}\overset{..}{O}: \qquad M{-}\overset{\displaystyle \overset{..}{O}:}{\underset{..}{N}}$$

$$\textbf{(24.27)} \qquad\qquad\qquad \textbf{(24.28)}$$

Series of related carbonyl and nitrosyl complexes exist, e.g. $Fe(CO)_5$ and $Fe(CO)_2(NO)_2$, and $Cr(CO)_6$ and $Cr(NO)_4$. The differences in numbers of ligands can be rationalized by applying the 18-electron rule (see the exercises below).

Nitrosyl ligands may also adopt bridging modes, acting as a 3-electron donor (**24.29**).

$$\underset{\textbf{(24.29)}}{\overset{\displaystyle \overset{\displaystyle O}{\|}}{\underset{\displaystyle \cdot\overset{..}{N}\cdot}{}} \qquad \overset{\displaystyle \overset{\displaystyle O}{\|}}{\underset{\displaystyle \underset{\displaystyle M{-}M}{N}}{}}}$$

(24.29)

Self-study exercises

1. Use the 18-electron rule (see Sections 20.4 and 24.3) to explain why Cr(0) binds six CO ligands, i.e. $Cr(CO)_6$, but only four NO ligands, i.e. $Cr(NO)_4$.

2. Show that both $Fe(CO)_5$ and $Fe(CO)_2(NO)_2$ obey the 18-electron rule.

3. Explain why you would expect the Fe–N–O units to be linear in $Fe(NO)_3Cl$.

4. Explain why 'linear NO' may be considered as an NO^+ ligand.

5. The IR spectrum of $Os(NO)_2(PPh_3)_2$ exhibits an absorption at 1600 cm^{-1} assigned to ν_{NO}. This lies on the border between the ranges for bent and linear OsNO units. Suggest why you might conclude that the bonding mode is linear.

Dinitrogen

The molecules N_2 and CO are isoelectronic, and the bonding description in Fig. 24.1 can be qualitatively applied to N_2 complexes (see Section 22.9), although it must be remembered that the MOs of N_2 have equal atomic orbital contributions from each atom. Complexes of N_2 are not as stable as those of CO, and far fewer examples are known. Terminal M–N≡N units are linear (like a terminal M–C≡O), but bridging N_2 ligands do not mimic bridging CO groups (see structure **22.58** and discussion).

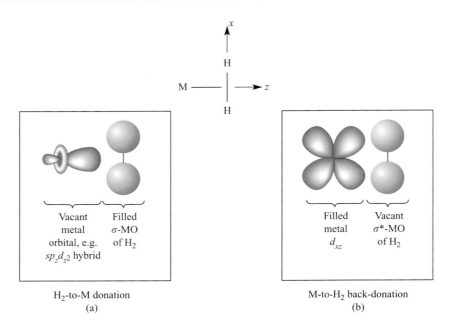

Fig. 24.9 Components of metal–dihydrogen bonding: (a) H_2-to-M donation using the H_2 σ-bonding MO and (b) M-to-H_2 back-donation into the H_2 σ^*-MO. The axis set is defined in the top diagram.

Self-study exercise

The isotopically labelled compound $Ni(^{14}N_2)_4$ has been formed in an N_2 matrix. The IR spectrum shows an absorption close to $2180\,cm^{-1}$. This band also appears in the Raman spectrum of $Ni(^{14}N_2)_4$, along with a band at $2251\,cm^{-1}$. The latter is absent in the IR spectrum. Both absorptions arise from coordinated N_2 stretching modes. Use these data to deduce the point group of $Ni(^{14}N_2)_4$. Sketch diagrams to show the modes of vibration and assign symmetry labels to them.

[*Ans.* Refer to Fig. 3.16 and accompanying text]

Dihydrogen

We have already mentioned dihydrogen complexes of Re (e.g. **22.33**) and noted the presence of a 'stretched' H—H bond. Other examples include $W(CO)_3(\eta^2\text{-}H_2)(P^iPr_3)_2$, $[OsH(\eta^2\text{-}H_2)\{P(OEt)_3\}_4]^+$, $Cr(CO)_5(\eta^2\text{-}H_2)$, $W(CO)_5(\eta^2\text{-}H_2)$ and $[Re(CO)_5(\eta^2\text{-}H_2)]^+$. The H_2 molecule only has available a σ-MO (electron donor orbital) and a σ^*-MO (acceptor). Both metal–ligand interactions shown in Fig. 24.9 weaken the H—H bond, and coordination readily leads to H—H cleavage (see Section 24.7). In dihydrogen complexes (**24.30**), the H—H bond distance is usually 80–100 pm. This compares to an H---H separation of ≥150 pm in a metal dihydride complex (**24.31**). Complexes containing $M(\eta^2\text{-}H_2)$ units but with H–H

distances of 110–150 pm (so-called 'stretched' H–H bonds) are also known.[†] Accurate determinations of H–H bond distances require neutron diffraction data. However, by preparing complexes containing HD (e.g. **24.32**), ^1H NMR spectroscopic data can be used to confirm the presence of an H–D (or, by analogy, an H–H) bond. The ^2H (D) nucleus possesses $I = 1$ and, therefore, $^1H-^2H$ coupling is observed. For example, the ^1H NMR spectrum of complex **24.32** shows a 1:1:1 multiplet, with $J_{HD} = 35.8$ Hz. This is smaller than that observed for free HD ($J_{HD} = 43$ Hz), but larger than the value for a metal dihydride complex ($J_{HD} \approx 2$–3 Hz). An estimate of the H–D bond length can be obtained by using the empirical relationship 24.3.

$$d(\text{H–H}) = 144 - (1.68 \times J_{HD}) \qquad (24.3)$$

where: d is in pm, J_{HD} is in Hz

(24.30)   **(24.31)**   **(24.32)**

[†] For an overview, see: D.M. Heinekey, A. Lledós and J.M. Lluch (2004) *Chem. Soc. Rev.*, vol. 33, p. 175 – 'Elongated dihydrogen complexes: What remains of the H–H bond?'

24.3 The 18-electron rule

In Section 20.4, we applied molecular orbital theory to octahedral complexes containing π-acceptor ligands and gave a rationale for the fact that *low oxidation state organometallic complexes tend to obey the 18-electron rule.* This rule often breaks down for early and late *d*-block metals as examples later in the chapter show: 16-electron complexes are common for e.g. Rh(I), Ir(I), Pd(0) and Pt(0). The majority of organometallic compounds with metals from the middle of the *d*-block obey the 18-electron rule and its application is useful, for example, in checking proposed structures. For electron-counting purposes, it is convenient to treat all ligands as *neutral* entities as this avoids the need to assign an oxidation state to the metal centre. However, one must not lose sight of the fact that this is a *formalism*. For example, in the synthesis of cyclopentadienyl derivatives, a common precursor is the salt $Na^+[Cp]^-$. Ferrocene, Cp_2Fe, may be formulated as an Fe(II) compound containing $[Cp]^-$ ligands, but for electron counting, it is convenient to consider the combination of an Fe(0) centre (group 8, 8 valence electrons) and two neutral $Cp^{\bullet}$ ligands (5-electron donor) giving an 18-electron complex (**24.33**). Of course, the same result is obtained if a formal oxidation state of $+2$ is assigned to the metal: Fe(II) (6 valence electrons) and two Cp^- ligands (6-electron donor). In this book we count valence electrons in terms of a zero oxidation state metal centre.

Electron count:

Fe(0) = 8 valence electrons
$2Cp^{\bullet} = 2 \times 5$ valence electrons
Total = 18 electrons

(24.33)

The number of valence electrons for a zero oxidation state metal centre is equal to the group number (e.g. Cr, 6; Fe, 8; Rh, 9). Some commonly encountered ligands[†] donate the following numbers of valence electrons:

- 1-electron donor: $H^{\bullet}$ (in any bonding mode), and terminal $Cl^{\bullet}$, $Br^{\bullet}$, $I^{\bullet}$, $R^{\bullet}$ (e.g. R = alkyl or Ph) or $RO^{\bullet}$;
- 2-electron donor: CO, PR_3, $P(OR)_3$, $R_2C=CR_2$ (η^2-alkene), R_2C: (carbene);
- 3-electron donor: η^3-$C_3H_5^{\bullet}$ (allyl radical), RC (carbyne), μ-$Cl^{\bullet}$, μ-$Br^{\bullet}$, μ-$I^{\bullet}$, μ-$R_2P^{\bullet}$ (**24.34**);
- 4-electron donor: η^4-diene (e.g. **24.24**), η^4-C_4R_4 (cyclobutadienes);
- 5-electron donor: η^5-$C_5H_5^{\bullet}$ (as in **24.33**), μ_3-$Cl^{\bullet}$, μ_3-$Br^{\bullet}$, μ_3-$I^{\bullet}$, μ_3-$RP^{\bullet}$;

[†] Notation for bridging ligands: see Section 7.7.

- 6-electron donor: η^6-C_6H_6 (and other η^6-arenes, e.g. η^6-C_6H_5Me);
- 1- or 3-electron donor: NO (**24.27** and **24.28**).

$$M \diagdown \overset{X}{} \diagup M' \qquad \text{e.g. X = Cl, Br, I, } PR_2$$

(24.34)

Counting electrons provided by bridging ligands, metal–metal bonds and net charges requires care. When bridging between two metal centres, an $X^{\bullet}$ (X = Cl, Br, I) or $R_2P^{\bullet}$ ligand uses the unpaired electron and one lone pair to give an interaction formally represented by structure **24.34**, i.e. one electron is donated to M, and two to M'. In a doubly bridged species such as $(CO)_2Rh(\mu\text{-}Cl)_2Rh(CO)_2$, the μ-Cl atoms are equivalent as are the Rh atoms, and the two Cl bridges together contribute three electrons to each Rh. A bridging $H^{\bullet}$ provides only one electron *in total*, shared between the metal atoms it bridges, e.g. in $[HFe_2(CO)_8]^-$, (**24.35**). Example **24.35** also illustrates that the formation of an M−M single bond provides each M atom with one extra electron.

Electron count:

Fe(0) = 8 electrons
3 terminal CO = 3×2 electrons
2 μ-CO = 2×1 electron per Fe
Fe–Fe bond = 1 electron per Fe
H = $^1/_2$ electron per Fe
1− charge = $^1/_2$ electron per Fe
Total = 18 electrons per Fe

(24.35)

Worked example 24.1   18-Electron rule

Confirm that the Cr centre in $[(\eta^6\text{-}C_6H_6)Cr(CO)_3]$ obeys the 18-electron rule, but Rh in $[(CO)_2Rh(\mu\text{-}Cl)_2Rh(CO)_2]$ does not.

Cr(0) (group 6) contributes 6 electrons
η^6-C_6H_6 contributes 6 electrons
3 CO contribute $3 \times 2 = 6$ electrons

Total = 18 electrons

Rh(0) (group 9) contributes 9 electrons
μ-Cl contributes 3 electrons (1 to one Rh and 2 to the other Rh)
2 CO contribute $2 \times 2 = 4$ electrons

Total per Rh = 16 electrons

Self-study exercises

1. Confirm that the Fe centres in $H_2Fe(CO)_4$ and $[(\eta^5\text{-}C_5H_5)Fe(CO)_2]^-$ obey the 18-electron rule.

2. Show that $Fe(CO)_4(\eta^2\text{-}C_2H_4)$, $HMn(CO)_3(PPh_3)_2$ and $[(\eta^6\text{-}C_6H_5Br)Mn(CO)_3]^+$ contain 18-electron metal centres.

3. Show that $[Rh(PMe_3)_4]^+$ contains a 16-electron metal centre. Comment on whether this violation of the 18-electron rule is expected.

Worked example 24.2   18-Electron rule: metal–metal bonding

Metal–metal bonding in multinuclear species is not always clear-cut. *Solely on the basis of the 18-electron rule,* **suggest whether $(\eta^5\text{-}Cp)Ni(\mu\text{-}PPh_2)_2Ni(\eta^5\text{-}Cp)$ might be expected to contain a metal–metal bond.**

The formula is instructive in terms of drawing a structure, except in respect of an M–M bond. Thus, we can draw an initial structure:

Now count the valence electrons around each metal centre: Ni(0) (group 10) contributes 10 electrons; $\eta^5\text{-}Cp^{\bullet}$ gives 5 electrons. Two $\mu\text{-}PPh_2$ contribute 6 electrons, 3 per Ni.
Total per Ni = 18 electrons.
Conclusion: each Ni atom obeys the 18-electron rule and no Ni–Ni bond is required.

Note that in all such examples, a prediction about the presence or not of the M–M bond *assumes* that the 18-electron rule is obeyed. Bridging ligands often play a very important role in supporting a dimetal framework.

Self-study exercises

1. Use the 18-electron rule to rationalize why manganese carbonyl forms a dimer $Mn_2(CO)_{10}$ containing an Mn–Mn bond, rather than a monomer $Mn(CO)_5$.

2. The presence of an Fe–Fe bond in the compound $(\eta^5\text{-}Cp)(CO)Fe(\mu\text{-}CO)_2Fe(CO)(\eta^5\text{-}Cp)$ has been a controversial topic. *Solely on the basis of the 18-electron rule,* show that an Fe–Fe bond is expected. What does your conclusion depend on? Are your assumptions infallible?

24.4 Metal carbonyls: synthesis, physical properties and structure

Table 24.3 lists many of the stable, neutral, *d*-block metal carbonyl compounds containing six or fewer metal atoms. A range of unstable carbonyls have been obtained by *matrix isolation*: the action of CO on metal atoms in a noble gas matrix at very low temperatures or the photolysis of stable metal carbonyls under similar conditions. Among species made this way are $Ti(CO)_6$, $Pd(CO)_4$, $Pt(CO)_4$, $Cu_2(CO)_6$, $Ag_2(CO)_6$, $Cr(CO)_4$, $Mn(CO)_5$, $Zn(CO)_3$, $Fe(CO)_4$, $Fe(CO)_3$ and $Ni(CO)_3$ (those of Cr, Mn, Fe and Ni being fragments formed by decomposition of stable carbonyls). In the rest of this section, we discuss compounds isolable at ordinary temperatures.

Synthesis and physical properties

The carbonyls $Ni(CO)_4$ and $Fe(CO)_5$ (both highly toxic) are the only ones normally obtained by action of CO on the finely divided metal. Formation of $Ni(CO)_4$ (eq. 21.4) occurs at 298 K and 1 bar pressure, but $Fe(CO)_5$ is made under 200 bar CO at 420–520 K. Most other simple metal carbonyls are prepared by *reductive carbonylation,* i.e. action of CO and a reducing agent (which may be excess CO) on a metal oxide, halide or other compound (e.g. reactions 24.4–24.12). Yields are often poor and we have not attempted to write stoichiometric reactions; for the preparation of $[Tc(H_2O)_3(CO)_3]^+$, see Box 22.7.

$$VCl_3 + Na + CO \xrightarrow[\text{in diglyme}]{420\,K,\,150\,bar} [Na(diglyme)_2][V(CO)_6]$$

$$\downarrow \text{HCl, Et}_2O \qquad (24.4)$$

$$V(CO)_6$$

$$CrCl_3 + CO + Li[AlH_4] \xrightarrow[\text{in Et}_2O]{390\,K,\,70\,bar} Cr(CO)_6 + LiCl + AlCl_3 \tag{24.5}$$

$$MoCl_5 + CO + AlEt_3 \xrightarrow{373\,K,\,200\,bar} Mo(CO)_6 + AlCl_3 \tag{24.6}$$

$$WCl_6 + Fe(CO)_5 \xrightarrow{373\,K} W(CO)_6 + FeCl_2 \tag{24.7}$$

$$OsO_4 + CO \xrightarrow{520\,K,\,350\,bar} Os(CO)_5 + CO_2 \tag{24.8}$$

$$Co(O_2CMe)_2 \cdot 4H_2O \xrightarrow[\text{in acetic anhydride}]{CO/H_2\,(4:1),\,200\,bar,\,430\,K} Co_2(CO)_8 \tag{24.9}$$

$$RuCl_3 \cdot xH_2O + CO \xrightarrow[\text{in MeOH}]{400\,K,\,50\,bar} Ru_3(CO)_{12} \tag{24.10}$$

Table 24.3 Neutral, low-nuclearity ($\leq M_6$) metal carbonyls of the *d*-block metals (dec. = decomposes).

Group number	5	6	7	8	9	10
First row metals	$V(CO)_6$ Dark blue solid; paramagnetic; dec. 343 K	$Cr(CO)_6$ White solid; sublimes *in vacuo*; dec. 403 K	$Mn_2(CO)_{10}$ Yellow solid; mp 427 K	$Fe(CO)_5$ Yellow liquid; mp 253 K; bp 376 K	$Co_2(CO)_8$ Air-sensitive, orange-red solid; mp 324 K	$Ni(CO)_4$ Colourless, volatile liquid; highly toxic vapour; bp 316 K
				$Fe_2(CO)_9$ Golden crystals; mp 373 K (dec.)	$Co_4(CO)_{12}$ Air-sensitive, black solid	
				$Fe_3(CO)_{12}$ Dark green solid; dec. 413 K	$Co_6(CO)_{16}$ Black solid; slowly dec. in air	
Second row metals		$Mo(CO)_6$ White solid; sublimes *in vacuo*	$Tc_2(CO)_{10}$ White solid; slowly dec. in air; mp 433 K	$Ru(CO)_5$ Colourless liquid; mp 251 K; dec. in air at 298 K to $Ru_3(CO)_{12}$ + CO	$Rh_4(CO)_{12}$ Red solid; >403 K dec. to $Rh_6(CO)_{16}$	
				$Ru_3(CO)_{12}$ Orange solid; mp 427 K; sublimes *in vacuo*	$Rh_6(CO)_{16}$ Black solid; dec. >573 K	
Third row metals		$W(CO)_6$ White solid; sublimes *in vacuo*	$Re_2(CO)_{10}$ White solid; mp 450 K	$Os(CO)_5$ Yellow liquid; mp 275 K	$Ir_4(CO)_{12}$ Slightly air-sensitive yellow solid; mp 443 K	
				$Os_3(CO)_{12}$ Yellow solid; mp 497 K	$Ir_6(CO)_{16}$ Red solid	

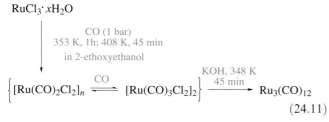

$$\left\{ [Ru(CO)_2Cl_2]_n \underset{}{\overset{CO}{\rightleftharpoons}} [Ru(CO)_3Cl_2]_2 \right\} \xrightarrow[45\ min]{KOH,\ 348\ K} Ru_3(CO)_{12}$$

$$(24.11)$$

$$OsO_4 + CO \xrightarrow[\text{in MeOH}]{400\ K,\ \leq 200\ bar} Os_3(CO)_{12} \qquad (24.12)$$

Diiron nonacarbonyl, $Fe_2(CO)_9$, is usually made by photolysis of $Fe(CO)_5$ (eq. 24.13), while $Fe_3(CO)_{12}$ is obtained by several methods, e.g. oxidation of $[HFe(CO)_4]^-$ using MnO_2.

$$2Fe(CO)_5 \xrightarrow{h\nu} Fe_2(CO)_9 + CO \qquad (24.13)$$

Some metal carbonyls including $M(CO)_6$ (M = Cr, Mo, W), $Fe(CO)_5$, $Fe_2(CO)_9$, $Fe_3(CO)_{12}$, $Ru_3(CO)_{12}$, $Os_3(CO)_{12}$

and $Co_2(CO)_8$ are commercially available. All carbonyls are thermodynamically unstable with respect to oxidation in air, but the rates of oxidation vary: $Co_2(CO)_8$ reacts under ambient conditions, $Fe(CO)_5$ and $Ni(CO)_4$ are also easily oxidized (their vapours forming explosive mixtures with air), but $M(CO)_6$ (M = Cr, Mo, W) does not oxidize unless heated. Table 24.3 lists some physical properties of some of the more common metal carbonyls. Note the increased importance of M−M bonding as one descends groups 8 and 9: e.g. whereas $Co_2(CO)_8$ is stable, $Rh_2(CO)_8$ is unstable with respect to $Rh_4(CO)_{12}$. The latter can also be formed by reaction 24.14, and above 400 K, it decomposes to $Rh_6(CO)_{16}$. Reactions 24.15 and 24.16 are routes to $Ir_4(CO)_{12}$ and $Ir_6(CO)_{16}$.

$$(CO)_2Rh(\mu\text{-Cl})_2Rh(CO)_2 \xrightarrow[\text{in hexane, NaHCO}_3]{CO,\ 1\ bar,\ 298\ K} Rh_4(CO)_{12}$$

$$(24.14)$$

Box 24.1 Large cations for large anions: 2

Metal cluster anions are usually stabilized in salts that contain large cations; common choices are $[Ph_4P]^+$, $[Ph_4As]^+$, $[^nBu_4N]^+$ and $[(Ph_3P)_2N]^+$. Compatibility between cation and anion sizes is important. The figure shows part of the packing diagram for the salt $[Ph_4P]_2[Ir_8(CO)_{22}]$; the H atoms of the Ph rings are omitted for clarity. The diagram illustrates how well the large cations pack with the cluster anions, and this is essential for the stabilization and crystallization of such salts.

See also: Box 11.5 – Large cations for large anions: 1.

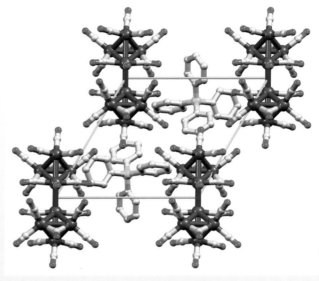

Colour coding: Ir, blue; C, grey; O, red; P, orange. The unit cell is shown in yellow. [Data from: F. Demartin *et al.* (1981) *J. Chem. Soc., Chem. Commun.*, p. 528.]

$$Na_3[IrCl_6] \xrightarrow[\text{2. base}]{\text{1. CO, 1 bar, in MeOH under reflux}} Ir_4(CO)_{12}$$

(24.15)

$$[Et_4N]_2[Ir_6(CO)_{15}] \xrightarrow{CF_3SO_3H \text{ under CO}} Ir_6(CO)_{16}$$

(24.16)

Metal carbonyl *clusters* containing four or more metal atoms are made by a variety of methods. Osmium forms a range of binary compounds and pyrolysis of $Os_3(CO)_{12}$ yields a mix of products (eq. 24.17) which can be separated by chromatography.

$$Os_3(CO)_{12} \xrightarrow{483\,K} Os_5(CO)_{16}$$
$$+ Os_6(CO)_{18} + Os_7(CO)_{21} + Os_8(CO)_{23}$$
$$\text{major product}$$

(24.17)

Metal carbonyl anions can be derived by reduction, e.g. reactions 24.18–24.23. Dimers such as $Mn_2(CO)_{10}$ and $Co_2(CO)_8$ undergo simple cleavage of the M−M bond, but in other cases, reduction is accompanied by an increase in metal nuclearity. In reactions 24.18 and 24.23, $Na[C_{10}H_8]$ (sodium naphthalide) is made from Na and

naphthalene; both $Na[C_{10}H_8]$ and $K[C_{10}H_8]$ are powerful reducing agents.

$$Fe(CO)_5 \xrightarrow{Na[C_{10}H_8]} Na_2[Fe(CO)_4] \quad (24.18)$$

$$Mn_2(CO)_{10} + 2Na \longrightarrow 2Na[Mn(CO)_5] \quad (24.19)$$

$$Co_2(CO)_8 + 2Na \longrightarrow 2Na[Co(CO)_4] \quad (24.20)$$

$$Ru_3(CO)_{12} \xrightarrow{Na, THF, \Delta} [Ru_6(CO)_{18}]^{2-} \quad (24.21)$$

$$Os_3(CO)_{12} \xrightarrow{Na, diglyme, \Delta} [Os_6(CO)_{18}]^{2-} \quad (24.22)$$

$$Ni(CO)_4 \xrightarrow{Na[C_{10}H_8]} [Ni_5(CO)_{12}]^{2-} + [Ni_6(CO)_{12}]^{2-}$$

(24.23)

The salt $Na_2[Fe(CO)_4]$ (eq. 24.18) is *Collman's reagent*, and has numerous synthetic applications. It is very air-sensitive and is best prepared *in situ*. In reactions 24.21–24.23, Na^+ salts are the initial products, but the large cluster anions are isolated as salts of large cations such as $[(Ph_3P)_2N]^+$, $[Ph_4P]^+$ or $[Ph_4As]^+$ (see Box 24.1).

The use of superacid media has been central to developing synthetic routes to isolable salts of metal carbonyl cations. Two examples are $[Os(CO)_6]^{2+}$ and $[Ir(CO)_6]^{3+}$ (eqs. 22.78 and 22.105), both isolated as the $[Sb_2F_{11}]^-$ salts. Both syntheses involve reduction of high oxidation state

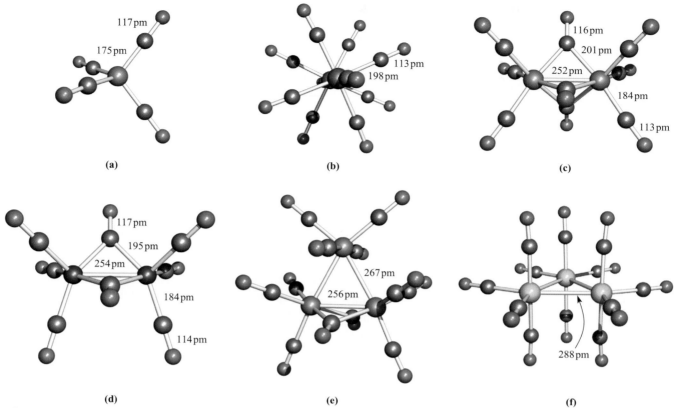

Fig. 24.10 The structures (X-ray diffraction) of (a) $[Fe(CO)_4]^{2-}$ in the K^+ salt [R.G. Teller *et al.* (1977) *J. Am. Chem. Soc.*, vol. 99, p. 1104], (b) $Re_2(CO)_{10}$ showing the staggered configuration also adopted by $Mn_2(CO)_{10}$ and $Tc_2(CO)_{10}$ [M.R. Churchill *et al.* (1981) *Inorg. Chem.*, vol. 20, p. 1609], (c) $Fe_2(CO)_9$ [F.A. Cotton *et al.* (1974) *J. Chem. Soc., Dalton Trans.*, p. 800], (d) $Co_2(CO)_8$ [P.C. Leung *et al.* (1983) *Acta Crystallogr., Sect. B*, vol. 39, p. 535], (e) $Fe_3(CO)_{12}$ [D. Braga *et al.* (1994) *J. Chem. Soc., Dalton Trans.*, p. 2911], and (f) $Os_3(CO)_{12}$ which is isostructural with $Ru_3(CO)_{12}$ [M.R. Churchill *et al.* (1977) *Inorg. Chem.*, vol. 16, p. 878]. Colour code: Fe, green; Re, brown; Co, blue; Os, yellow; C, grey; O, red.

metal fluorides (OsF_6 and IrF_6, respectively) and a similar method is used to prepare $[Pt(CO)_4]^{2+}$ (eq. 24.24). In contrast, $[Co(CO)_5]^+$ is made by oxidation of $Co_2(CO)_8$ (eq. 24.25); the oxidizing agent is probably $[H_2F]^+$. The superacid HF/BF_3 is used to produce $[BF_4]^-$ salts of $[M(CO)_6]^{2+}$ (M = Fe, Ru, Os) (eqs. 24.26 and 24.27).

$$PtF_6 + 6CO + 4SbF_5 \xrightarrow[\text{in liquid } SbF_5]{298-323 \text{ K, 1 bar CO}}$$
$$[Pt(CO)_4][Sb_2F_{11}]_2 + 2COF_2 \quad (24.24)$$

$$Co_2(CO)_8 + 2(CF_3)_3BCO + 2HF \xrightarrow[\text{in liquid HF}]{298 \text{ K, 2 bar CO}}$$
$$2[Co(CO)_5][(CF_3)_3BF] + H_2 \quad (24.25)$$

$$Fe(CO)_5 + XeF_2 + CO + 2BF_3 \xrightarrow[\text{77 to 298 K}]{HF/BF_3, \text{ CO}}$$
$$[Fe(CO)_6][BF_4]_2 + Xe \quad (24.26)$$

$$M_3(CO)_{12} \xrightarrow[\text{195 to 298 K}]{\text{liquid HF, } F_2} cis\text{-}[M(CO)_4F_2] \xrightarrow[\text{298 K}]{HF/BF_3, \text{ CO}}$$
M = Ru, Os
$$[M(CO)_6][BF_4]_2 \quad (24.27)$$

Structures

Mononuclear metal carbonyls possess the following structures (bond distances are for the solid state)[†]:

- linear: $[Au(CO)_2]^+$ (Au−C = 197 pm);
- square planar: $[Rh(CO)_4]^+$ (Rh−C = 195 pm), $[Pd(CO)_4]^{2+}$ (Pd−C = 199 pm), $[Pt(CO)_4]^{2+}$ (Pd−C = 198 pm);
- tetrahedral: $Ni(CO)_4$ (Ni−C = 182 pm), $[Cu(CO)_4]^+$ (Cu−C = 196 pm), $[Co(CO)_4]^-$ (Co−C = 175 pm), $[Fe(CO)_4]^{2-}$ (Fig. 24.10a);

[†] Electron diffraction data for gaseous $Fe(CO)_5$ are $Fe−C_{axial} = 181$ pm and $Fe−C_{equ} = 184$ pm, see: B.W. McClelland, A.G. Robiette, L. Hedberg and K. Hedberg (2001) *Inorg. Chem.*, vol. 40, p. 1358.

- trigonal bipyramidal: $Fe(CO)_5$ ($Fe-C_{axial} = 181$ pm, $Fe-C_{equ} = 180$ pm), $[Co(CO)_5]^+$ ($Co-C_{axial} = 183$ pm, $Co-C_{equ} = 185$ pm), $[Mn(CO)_5]^-$ in most salts ($Mn-C_{axial} = 182$ pm, $Mn-C_{equ} = 180$ pm);
- square-based pyramidal: $[Mn(CO)_5]^-$ in the $[Ph_4P]^+$ salt ($Mn-C_{apical} = 179$ pm, $Mn-C_{basal} = 181$ pm);
- octahedral: $V(CO)_6$ ($V-C = 200$ pm), $Cr(CO)_6$ ($Cr-C = 192$ pm), $Mo(CO)_6$ ($Mo-C = 206$ pm), $W(CO)_6$ ($W-C = 207$ pm), $[Fe(CO)_6]^{2+}$ ($Fe-C = 191$ pm), $[Ru(CO)_6]^{2+}$ ($Ru-C = 202$ pm), $[Os(CO)_6]^{2+}$ ($Os-C = 203$ pm), $[Ir(CO)_6]^{3+}$ ($Ir-C = 203$ pm).

With the exception of $V(CO)_6$, each obeys the 18-electron rule. The 17-electron count in $V(CO)_6$ suggests the possibility of dimerization to '$V_2(CO)_{12}$' containing a V–V bond, but this is sterically unfavourable. A mononuclear carbonyl of Mn would, like $V(CO)_6$, be a radical, but now, dimerization occurs and $Mn_2(CO)_{10}$ is the lowest nuclearity neutral binary carbonyl of Mn. A similar situation arises for cobalt: '$Co(CO)_4$' is a 17-electron species and the lowest nuclearity binary carbonyl is $Co_2(CO)_8$. The group 7 dimers $Mn_2(CO)_{10}$, $Tc_2(CO)_{10}$ and $Re_2(CO)_{10}$ are isostructural and have staggered arrangements of carbonyls (Fig. 24.10b); the M–M bond is unbridged and longer (Mn–Mn = 290 pm, Tc–Tc = 303 pm, Re–Re = 304 pm) than twice the metallic radius (see Tables 21.1 and 22.1). In $Fe_2(CO)_9$ (Fig. 24.10c), three CO ligands bridge between the Fe centres. Each Fe atom obeys the 18-electron rule if an Fe–Fe bond is present and this is consistent with the observed diamagnetism of the complex. Even so, many theoretical studies have been carried out to investigate the presence (or not) of Fe–Fe bonding in $Fe_2(CO)_9$. Figure 24.10d shows the *solid state* structure of $Co_2(CO)_8$. When solid $Co_2(CO)_8$ is dissolved in hexane, the IR spectrum changes. The spectrum of the solid contains bands assigned to terminal *and* bridging CO ligands, but in hexane, only absorptions due to terminal carbonyls are seen. This is explained by the equilibrium in scheme **24.36** and solid state ^{13}C NMR spectroscopic data show that terminal–bridge CO exchange occurs even in *solid* $Co_2(CO)_8$.

2. Confirm that in *each* isomer of $Co_2(CO)_8$ shown in diagram **24.36**, each Co centre obeys the 18-electron rule.

3. Does the 18-electron rule allow you to assign the structure shown in Fig. 24.10c to $Fe_2(CO)_9$ in preference to a structure of the type $(CO)_4Fe(\mu-CO)Fe(CO)_4$?

Each group 8 metal forms a trinuclear binary carbonyl $M_3(CO)_{12}$ containing a triangular framework of metal atoms. However, the arrangement of CO ligands in $Fe_3(CO)_{12}$ (Fig. 24.10e) differs from that in $Ru_3(CO)_{12}$ and $Os_3(CO)_{12}$ (Fig. 24.10f). The latter contain equilateral M_3 triangles and four terminal CO per metal, whereas in the solid state, $Fe_3(CO)_{12}$ contains an isosceles Fe_3 triangle with one Fe–Fe edge (the shortest) bridged by two CO ligands. Each M atom in $Fe_3(CO)_{12}$, $Ru_3(CO)_{12}$ and $Os_3(CO)_{12}$ obeys the 18-electron rule. The solution ^{13}C NMR spectrum of $Fe_3(CO)_{12}$ exhibits one resonance even as low as 123 K showing that the molecule is fluxional. The process can be described in terms of exchange of terminal and bridging CO ligands, or by considering the tilting of the Fe_3-unit within a shell of CO ligands. X-ray data collected at several temperatures show that $Fe_3(CO)_{12}$ also undergoes a dynamic process in the solid state. This illustrates that the CO_{term}–CO_{bridge} (CO_{term} = terminal CO ligand) exchange is a low-energy process. This is one of many such examples.

The group 9 carbonyls $Co_4(CO)_{12}$ and $Rh_4(CO)_{12}$ (Fig. 24.11a) are isostructural, with three μ-CO ligands arranged around the edges of one face of the M_4 tetrahedron. In $Ir_4(CO)_{12}$, all ligands are terminal (Fig. 24.11b). Each group 9 metal forms a hexanuclear carbonyl, $M_6(CO)_{16}$, in which the metal atoms form an octahedral cluster. In $Co_6(CO)_{16}$, $Rh_6(CO)_{16}$ and the red isomer of $Ir_6(CO)_{16}$, each M atom has two CO_{term} and there are four μ_3-CO (Fig. 24.11c). A black isomer of $Ir_6(CO)_{16}$ has been isolated and in the solid state has 12 CO_{term} and four μ-CO (Fig. 24.11d). Other octahedral carbonyl clusters include $[Ru_6(CO)_{18}]^{2-}$ and $[Os_6(CO)_{18}]^{2-}$, but in contrast $Os_6(CO)_{18}$ has a bicapped tetrahedral structure (**24.37**). This is an example of a *condensed polyhedral* cluster.

OC $\quad$ $\underset{C}{\overset{O}{C}}$ $\quad$ CO
OC$^{\text{''''''}}$Co — Co$^{\text{''''''}}$CO $\rightleftharpoons$ OC — Co — Co — CO
OC $\quad$ $\underset{O}{\overset{C}{C}}$ $\quad$ CO $\qquad$ OC $\quad$ OC $\quad$ CO

$\qquad$ OC $\quad$ CO $\quad$ CO

C_{2v} symmetry $\qquad\qquad\qquad$ D_{3d} symmetry

(24.36)

(24.37)

Self-study exercises

1. Confirm that each Tc centre in $Tc_2(CO)_{10}$ obeys the 18-electron rule.

In a ***condensed polyhedral cluster***, two or more polyhedral cages are fused together through atom, edge or face sharing.

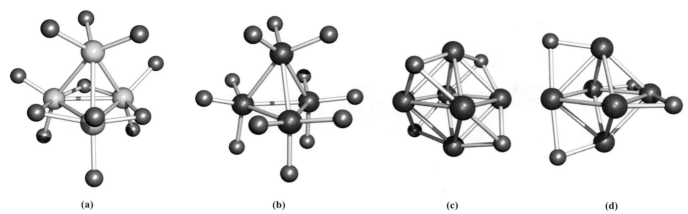

(a)   **(b)**   **(c)**   **(d)**

Fig. 24.11 The structures (X-ray diffraction) of (a) $Rh_4(CO)_{12}$ which is isostructural with $Co_4(CO)_{12}$ [C.H. Wei (1969) *Inorg. Chem.*, vol. 8, p. 2384], (b) $Ir_4(CO)_{12}$ [M.R. Churchill *et al.* (1978) *Inorg. Chem.*, vol. 17, p. 3528], (c) the red isomer of $Ir_6(CO)_{16}$ and (d) the black isomer of $Ir_6(CO)_{16}$ [L. Garlaschelli *et al.* (1984) *J. Am. Chem. Soc.*, vol. 106, p. 6664]. In (a) and (b), O atoms have been omitted for clarity. In (c) and (d), the terminal CO and the O atoms of the bridging CO ligands have been omitted; each Ir has two CO_{term}. Colour code: Rh, yellow; Ir, red; C, grey.

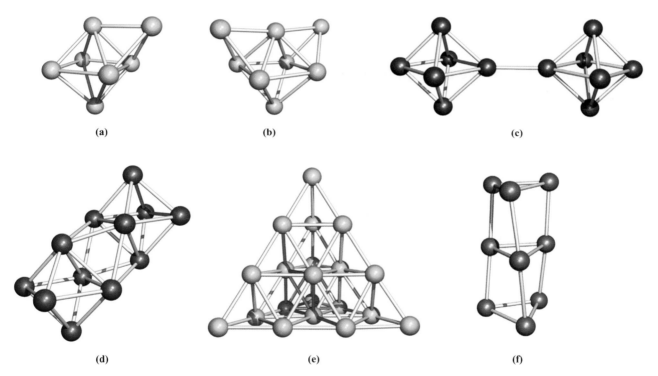

(a)   **(b)**   **(c)**

(d)   **(e)**   **(f)**

Fig. 24.12 The structures (X-ray diffraction) of the metal cores in (a) $Os_7(CO)_{21}$ [C.R. Eady *et al.* (1977) *J. Chem. Soc., Chem. Commun.*, p. 385], (b) $[Os_8(CO)_{22}]^{2-}$ in the $[(Ph_3P)_2N]^+$ salt [P.F. Jackson *et al.* (1980) *J. Chem. Soc., Chem. Commun.*, p. 60], (c) $[Rh_{12}(CO)_{30}]^{2-}$ in the $[Me_4N]^+$ salt [V.G. Albano *et al.* (1969) *J. Organomet. Chem.*, vol. 19, p. 405], (d) $[Ir_{12}(CO)_{26}]^{2-}$ in the $[Ph_4P]^+$ salt [R.D. Pergola *et al.* (1987) *Inorg. Chem.*, vol. 26, p. 3487], (e) $[Os_{20}(CO)_{40}]^{2-}$ in the $[^nBu_4P]^+$ salt [L.H. Gade *et al.* (1994) *J. Chem. Soc., Dalton Trans.*, p. 521], and (f) $[Pt_9(CO)_{18}]^{2-}$ in the $[Ph_4P]^+$ salt [J.C. Calabrese *et al.* (1974) *J. Am. Chem. Soc.*, vol. 96, p. 2614]. Colour code: Os, yellow; Rh, blue; Ir, red; Pt, brown.

The syntheses of high-nuclearity metal carbonyl clusters are not readily generalized,[†] and we focus only on the structures of selected species. For seven or more metal

atoms, metal carbonyl clusters tend to be composed of condensed (or less often, linked) tetrahedral or octahedral units. The group 10 metals form a series of clusters containing stacked triangles, e.g. $[Pt_9(CO)_{18}]^{2-}$ and $[Pt_{15}(CO)_{30}]^{2-}$. Figure 24.12 shows the metal cores of representative clusters. In $[Os_{20}(CO)_{40}]^{2-}$ (Fig. 24.12e), the Os atoms form a ccp

[†] For further details, see for example: C.E. Housecroft (1996) *Metal–Metal Bonded Carbonyl Dimers and Clusters*, Oxford University Press, Oxford.

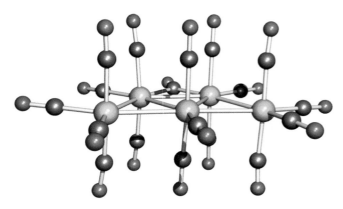

Fig. 24.13 The structure (X-ray diffraction) of $Os_5(CO)_{18}$, in which the Os atoms form a planar 'raft' [W. Wang *et al.* (1992) *J. Chem. Soc., Chem. Commun.*, p. 1737]. Colour code: Os, yellow; C, grey; O, red.

arrangement. Some metal carbonyls possess 'raft' structures, i.e. the metal atoms form planar arrangements of edge-sharing triangles, e.g. $Os_5(CO)_{18}$ (Fig. 24.13).

24.5 The isolobal principle and application of Wade's rules

In Section 13.11, we introduced *Wade's rules* to rationalize the structures of borane and related clusters. This method of counting electrons can be extended to simple organometallic clusters by making use of the *isolobal relationship* between cluster fragments.

> Two cluster fragments are ***isolobal*** if they possess the same frontier orbital characteristics: same symmetry, same number of electrons available for cluster bonding, and *approximately* the same energy.

Figure 24.14 shows the frontier MOs (i.e. those close to and including the HOMO and LUMO) of BH and C_{3v} $M(CO)_3$ (M = Fe, Ru, Os) fragments. In Box 13.9, we considered how the frontier orbitals of six BH combined to give the cluster bonding MOs in $[B_6H_6]^{2-}$ (a process that can be extended to other clusters). Now we look at why it is that BH and some organometallic fragments can be regarded as being similar in terms of cluster bonding. Features of significance in Fig. 24.14 are that the BH and C_{3v} $M(CO)_3$ fragments have three frontier MOs with matching symmetries and containing the same number of electrons. The ordering of the MOs is not important. The BH and C_{3v} $M(CO)_3$ (M = Fe, Ru, Os) fragments are *isolobal* and their relationship allows BH units in borane clusters to be replaced (in theory and sometimes in practice, although syntheses are not as simple as this formal replacement suggests) by $Fe(CO)_3$, $Ru(CO)_3$ or $Os(CO)_3$ fragments. Thus, for example, we can go from $[B_6H_6]^{2-}$ to $[Ru_6(CO)_{18}]^{2-}$. Wade's rules categorize $[B_6H_6]^{2-}$ as a 7 electron pair *closo*-cluster, and similarly, $[Ru_6(CO)_{18}]^{2-}$ is a *closo*-species. Both are predicted to have (and have in practice) octahedral cages.

Moving to the left or right of group 8 removes or adds electrons to the frontier MOs shown in Fig. 24.14. Removing or adding a CO ligand removes or adds two electrons. (The frontier MOs also change, but this is unimportant if we are simply counting electrons.) Changing the ligands similarly alters the number of electrons available. Equation 24.28 shows how the number of electrons provided by a given fragment can be determined and Table 24.4 applies this to selected fragments. These numbers are used *within the Wade approach*, also known as *polyhedral skeletal electron pair theory* (PSEPT).

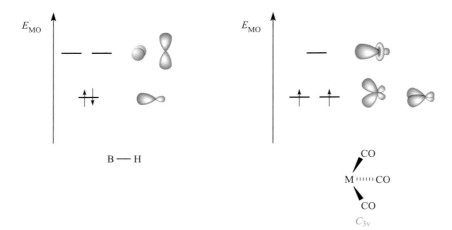

Fig. 24.14 The frontier MOs of a BH unit and a C_{3v} (i.e. 'conical') $M(CO)_3$ (M = Fe, Ru, Os) group. For the BH unit, the occupied MO is an *sp* hybrid; for $M(CO)_3$, the orbitals are represented by *pd* or *spd* hybrids. These orbitals combine with those of other cluster fragments to give cluster-bonding, non-bonding and antibonding MOs (see Box 13.9).

Table 24.4 The number of electrons (x in eq. 24.28) provided for cluster bonding by selected fragments; η^5-C_5H_5 = η^5-Cp.

Cluster fragment	Group 6: Cr, Mo, W	Group 7: Mn, Tc, Re	Group 8: Fe, Ru, Os	Group 9: Co, Rh, Ir
$M(CO)_2$	-2	-1	0	1
$M(CO)_3$	0	1	2	3
$M(CO)_4$	2	3	4	5
$M(\eta^5\text{-}C_5H_5)$	-1	0	1	2
$M(\eta^6\text{-}C_6H_6)$	0	1	2	3
$M(CO)_2(PR_3)$	0	1	2	3

$$x = v + n - 12 \tag{24.28}$$

where: x = number of cluster-bonding electrons provided by a fragment;

v = number of valence electrons from the metal atom;

n = number of valence electrons provided by the ligands

Worked example 24.3   Application of Wade's rules (PSEPT)

(a) Rationalize why $Rh_4(CO)_{12}$ has a tetrahedral core.
(b) What class of cluster is $Ir_4(CO)_{12}$?

[If you are unfamiliar with Wade's rules, first review Section 13.11.]
(a) Break the formula of $Rh_4(CO)_{12}$ down into convenient units and determine the number of cluster-bonding electrons.

- Each {$Rh(CO)_3$}-unit provides 3 cluster-bonding electrons.
- Total number of electrons available in $Rh_4(CO)_{12} = (4 \times 3) = 12$ electrons = 6 pairs.
- Thus, $Rh_4(CO)_{12}$ has 6 pairs of electrons with which to bond 4 cluster units.
- There are $(n + 2)$ pairs of electrons for n vertices, and so $Rh_4(CO)_{12}$ is a *nido*-cage; the parent deltahedron is a trigonal bipyramid, and thus $Rh_4(CO)_{12}$ is expected to be tetrahedral.

(b) Rh and Ir are both in group 9 and so $Ir_4(CO)_{12}$ is also a *nido*-cluster.

This example illustrates an important point about the use of such electron-counting schemes: *no information about the positions of the ligands can be obtained.* Although Wade's rules rationalize why $Rh_4(CO)_{12}$ and $Ir_4(CO)_{12}$ both have tetrahedral cores, they say nothing about the fact that the ligand arrangements are different (Figs. 24.11a and b).

Self-study exercises

1. Using Wade's approach, rationalize why $Co_4(CO)_{12}$ has a tetrahedral core.

2. Using PSEPT, rationalize why $[Fe_4(CO)_{13}]^{2-}$ has a tetrahedral Fe_4 core.

3. The cluster $Co_2(CO)_6C_2H_2$ has a tetrahedral Co_2C_2 core. How many electrons does each CH unit contribute to cluster bonding? Show that there are six electron pairs available for cluster bonding.

The diversity of cage structures among metal clusters is greater than that of boranes. Wade's rules were developed for boranes and extension of the rules to rationalize the structures of high-nuclearity metal clusters is limited. Boranes tend to adopt rather open structures, and there are few examples of BH units in capping positions. However, application of the *capping principle* does allow satisfactory rationalization of some condensed cages such as $Os_6(CO)_{18}$ (**24.37**).

Within the remit of **Wade's rules (PSEPT)**, the addition of one or more **capping units** to a deltahedral cage requires no additional bonding electrons. A capping unit is a cluster fragment placed over the *triangular* face of a central cage.

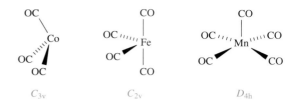

C_{3v} C_{2v} D_{4h}

Thus, for example, $Co_4(CO)_{12}$, $Co_3(CO)_9CH$, $Co_2(CO)_6C_2H_2$ and C_4H_4 form an isolobal series. Isolobal relationships have a *theoretical* premise and tell us nothing about methods of cluster synthesis.

24.6 Total valence electron counts in *d*-block organometallic clusters

The structures of many polynuclear organometallic species are not conveniently described in terms of Wade's rules, and an alternative approach is to consider the *total valence electron count*, also called the *Mingos cluster valence electron count*.

Single cage structures

Each low oxidation state *metal cluster* cage possesses a characteristic number of valence electrons (ve) as Table 24.5 shows. We shall not describe the MO basis for these numbers, but merely apply them to rationalize observed structures. Look back to Section 24.2 for the numbers of electrons donated by ligands. Any organometallic complex with a triangular M_3 framework requires 48 valence electrons, for example:

- $Ru_3(CO)_{12}$ has $(3 \times 8) + (12 \times 2) = 48$ ve;
- $H_2Ru_3(CO)_8(\mu\text{-}PPh_2)_2$ has
  $(2 \times 1) + (3 \times 8) + (8 \times 2) + (2 \times 3) = 48$ ve;
- $H_3Fe_3(CO)_9(\mu_3\text{-}CMe)$ has
  $(3 \times 1) + (3 \times 8) + (9 \times 2) + (1 \times 3) = 48$ ve.

Similarly, clusters with tetrahedral or octahedral cages require 60 or 86 valence electrons respectively, for example:

- $Ir_4(CO)_{12}$ has $(4 \times 9) + (12 \times 2) = 60$ ve;
- $(\eta^5\text{-}Cp)_4Fe_4(\mu_3\text{-}CO)_4$ has
  $(4 \times 5) + (4 \times 8) + (4 \times 2) = 60$ ve;
- $Rh_6(CO)_{16}$ has $(6 \times 9) + (16 \times 2) = 86$ ve;
- $Ru_6(CO)_{17}C$ has $(6 \times 8) + (17 \times 2) + 4 = 86$ ve.

The last example is of a cage containing an *interstitial atom* (see structure **24.12**) which *contributes all of its valence electrons* to cluster bonding. An interstitial C atom contributes four electrons, a B atom three, an N or P atom five, and so on.

A difference of two between the valence electron counts in Table 24.5 corresponds to a 2-electron reduction (adding two electrons) or oxidation (removing two electrons). This formally corresponds to breaking or making

---

Rationalize why $Os_6(CO)_{18}$ adopts structure **24.37** rather than an octahedral cage.

- $Os_6(CO)_{18}$ can be broken down into 6 $Os(CO)_3$ fragments.
- Each $\{Os(CO)_3\}$-unit provides 2 cluster-bonding electrons.
- Total number of electrons available in $Os_6(CO)_{18} = (6 \times 2) = 12$ electrons = 6 pairs.
- Thus, $Os_6(CO)_{18}$ has 6 pairs of electrons with which to bond 6 cluster units.
- This corresponds to a monocapped structure, the parent deltahedron being one with 5 vertices, i.e. a trigonal bipyramid:

- The monocapped trigonal bipyramid is the same as a bicapped tetrahedron (**24.37**).
- A *closo*-octahedral cage requires 7 pairs of electrons, and $Os_6(CO)_{18}$ has insufficient electrons for this structure.

Self-study exercises

1. The Os_7 core of $Os_7(CO)_{21}$ is a capped octahedron. Show that this is consistent with the PSEPT capping principle.

2. Use the capping principle to account for the fact that $[Os_8(CO)_{22}]^{2-}$ has a bicapped octahedral structure.

Using the isolobal principle, one can relate clusters that contain fragments having analogous orbital properties. Some isolobal pairs of metal carbonyl and hydrocarbon fragments are:

- $Co(CO)_3$ (C_{3v}) and CH (provides three orbitals and three electrons);
- $Fe(CO)_4$ (C_{2v}) and CH_2 (provides two orbitals and two electrons);
- $Mn(CO)_5$ (D_{4h}) and CH_3 (provides one orbital and one electron).

Table 24.5 Characteristic total valence electron counts for selected low oxidation state metal clusters.

Cluster framework	Diagrammatic representation of the cage	Valence electron count
Triangle		48
Tetrahedron		60
Butterfly, or planar raft of four atoms		62
Square		64
Trigonal bipyramid		72
Square-based pyramid		74
Octahedron		86
Trigonal prism		90

an M–M edge in the cage. For example, going from the 60-electron tetrahedron to a 62-electron 'butterfly' involves breaking one edge of the M_4 cage, and vice versa:

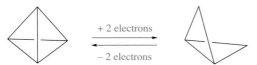

This can be applied more widely than for the structures listed in Table 24.5. Thus, the linear arrangement of the

Os atoms in $Os_3(CO)_{12}I_2$ can be rationalized in terms of its 50 valence electron count, i.e. formally, the addition of two electrons to a 48-electron triangle:

$$\begin{array}{ccc} & & Os(CO)_4I \\ & & | \\ (CO)_4 & & Os(CO)_4 \\ Os & \xrightarrow{\ +2\ electrons\ } & | \\ (OC)_4Os \text{——} Os(CO)_4 & & Os(CO)_4I \end{array}$$

48 ve $Os_3(CO)_{12}$ 50 ve $Os_3(CO)_{12}I_2$

| **Worked example 24.5   An application of total valence electron counts** |

Suggest what change in cluster structure might accompany the reaction:

$$[Co_6(CO)_{15}N]^- \longrightarrow [Co_6(CO)_{13}N]^- + 2CO$$

Both clusters contain an interstitial N atom which contributes 5 electrons to cluster bonding. The negative charge contributes 1 electron.
Total valence electron count for

$$[Co_6(CO)_{15}N]^- = (6 \times 9) + (15 \times 2) + 5 + 1 = 90$$

Total valence electron count for

$$[Co_6(CO)_{13}N]^- = (6 \times 9) + (13 \times 2) + 5 + 1 = 86$$

i.e. the loss of two CO ligands corresponds to a loss of 4 electrons, and a change from a trigonal prism to octahedral Co_6-cage.

| **Self-study exercises** |

1. $[Ru_6(CO)_{17}B]^-$ and $[Os_6(CO)_{18}P]^-$ contain interstitial B and P atoms respectively. Account for the fact that while $[Ru_6(CO)_{17}B]^-$ has an octahedral M_6 core, $[Os_6(CO)_{18}P]^-$ adopts a trigonal prismatic core.
   [*Ans.* 86 ve; 90 ve]

2. Rationalize why $Os_3(CO)_{12}$ has a triangular Os_3 core but in $Os_3(CO)_{12}Br_2$, the Os atoms are in a linear arrangement.
   [*Ans.* 48 ve; 50 ve]

3. In $Os_4(CO)_{16}$, the Os atoms are arranged in a square, but in $Os_4(CO)_{14}$ they form a tetrahedral cluster. Rationalize this observation.

Condensed cages

Structure **24.37** showed one type of *condensed cluster*. The sub-cluster units are connected either through shared M atoms, M−M edges or M_3 faces. The total valence electron count for a condensed structure is equal to the total number of electrons required by the sub-cluster units *minus* the electrons associated with the shared unit. The numbers to *subtract* are:

- 18 electrons for a shared M atom;
- 34 electrons for a shared M−M edge;
- 48 electrons for a shared M_3 face.

Examples of these families of condensed polyhedral clusters are $Os_5(CO)_{19}$ (atom-sharing, **24.38**), $H_2Os_5(CO)_{16}$ (edge-sharing, **24.39**) and $H_2Os_6(CO)_{18}$ (face-sharing, **24.40**).

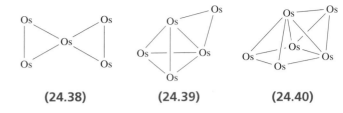

(24.38)   (24.39)   (24.40)

| **Worked example 24.6   Electron counts in condensed cluster structures** |

$Os_6(CO)_{18}$ has structure **24.37**, i.e. three face-sharing tetrahedra. Show that this structure is consistent with the number of valence electrons available.

can be represented as three face-sharing tetrahedra:

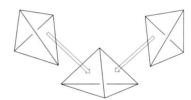

Valence electron count for three tetrahedra $= 3 \times 60 = 180$
For each shared face, subtract 48 electrons.
Valence electron count for the bicapped tetrahedron $= 180 - (2 \times 48) = 84$

The number of valence electrons available in $Os_6(CO)_{18} = (6 \times 8) + (18 \times 2) = 84$

Thus, the observed structure is consistent with the number of valence electrons available.

| **Self-study exercises** |

1. While $[Os_6(CO)_{18}]^{2-}$ has an octahedral Os_6 core, that in $Os_6(CO)_{18}$ is a capped trigonal bipyramid. Use total valence electron counts to rationalize this difference.

2. The core in $Os_5(CO)_{19}$ is shown in structure **24.38**. Show that this shape is consistent with the total valence electron count of the cluster.

3. $[Os_6(CO)_{18}]^{2-}$ has an octahedral Os_6-core, but in $H_2Os_6(CO)_{18}$, the Os_6-unit is a capped square-based pyramid. Comment on this difference in terms of the total number of valence electrons available for cluster bonding.

Limitations of total valence counting schemes

For some clusters such as Rh_x species, the number of electrons available may not match the number apparently required by the structure adopted. Two examples in rhodium carbonyl chemistry are $[Rh_5(CO)_{15}]^-$ and $[Rh_9(CO)_{19}]^{3-}$. The former possesses 76 valence electrons and yet has a trigonal bipyramidal Rh_5-core, for which 72 electrons are usual. However, a look at the Rh−Rh bond lengths reveals that six edges are in the range 292–303 pm, while three are 273–274 pm, indicating that the extra electrons have caused bond lengthening. In $[Rh_9(CO)_{19}]^{3-}$, 122 electrons are available but the Rh_9-core consists of two face-sharing octahedra for which 124 electrons are required by the scheme outlined above.[†] An example of an unexpected cluster structure is found for $[H_5Re_6(CO)_{24}]^-$. Rather than adopt a closed-cluster structure, the Re_6-unit in $[H_5Re_6(CO)_{24}]^-$ possesses a cyclohexane-like ring with a chair conformation (**24.41**). Each Re centre obeys the 18-electron rule (each $Re(CO)_4$ unit has $7 + (4 \times 2)$ valence electrons, two Re−Re bonds per Re provide 2 electrons, and the five H atoms with the $1-$ charge provide 1 electron per Re), but the preference for an open- rather than closed-cluster structure cannot be predicted.

$$\left[\begin{array}{c} (OC)_4Re \underset{(OC)_4Re}{\overset{(CO)_4}{\underset{Re}{\overset{Re}{\diagup}}}} \underset{(CO)_4}{\overset{Re(CO)_4}{\underset{Re(CO)_4}{}}} \end{array} \right]^-$$

= Re–H–Re

(24.41)

These are but three examples of the limitations of electron-counting schemes. As more clusters are structurally characterized, further exceptions arise providing yet more challenges for the theorist.

Self-study exercise

X-ray crystallography confirms that $[Rh_6(P^iPr_3)_6H_{12}]^{2+}$ possesses an octahedral Rh_6-cage. Show that this cluster is 10 electrons short of the expected electron count.

24.7 Types of organometallic reactions

In this section, we introduce the main types of ligand transformations that take place at metal centres in organometallic compounds:

- ligand substitution;
- oxidative addition (including orthometallation);

[†] For detailed discussion, see: D.M.P. Mingos and D.J. Wales (1990) *Introduction to Cluster Chemistry*, Prentice Hall, Englewood Cliffs, NJ.

- reductive elimination;
- alkyl and hydrogen migration;
- β-hydrogen elimination;
- α-hydrogen abstraction.

Substitution of CO ligands

The substitution of a CO ligand by another 2-electron donor (e.g. PR_3) may occur by photochemical or thermal activation, either by direct reaction of the metal carbonyl and incoming ligand, or by first replacing a CO by a more labile ligand such as THF or MeCN. An example of the latter is the formation of $Mo(CO)_5(PPh_3)$ (eq. 24.29) which is most effectively carried out by first making the THF (**24.42**) adduct *in situ*.

(24.42)

$$Mo(CO)_6 \xrightarrow[-CO]{\text{in THF}, \, h\nu} Mo(CO)_5(THF)$$

$$\xrightarrow[-THF]{PPh_3} Mo(CO)_5(PPh_3) \qquad (24.29)$$

The substitution steps are *dissociative* (see Chapter 26). The outgoing ligand leaves, creating a 16-electron metal centre which is *coordinatively unsaturated*. The entry of a new 2-electron ligand restores the 18-electron count. Competition between ligands for coordination to the 16-electron centre may be countered by having the incoming ligand (L in eq. 24.30) present in excess.

$$M(CO)_n \xrightarrow{-CO} \{M(CO)_{n-1}\} \underset{CO}{\overset{L}{\rightleftarrows}} \begin{array}{l} M(CO)_{n-1}L \\ M(CO)_n \end{array} \qquad (24.30)$$

In reaction 24.31, the incoming ligand provides four electrons and displaces two CO ligands. Multiple substitution by 2-electron donors is exemplified by reaction 24.32.

$$Fe(CO)_5 + \underset{\text{or } \Delta, \, 20 \, bar}{\overset{h\nu}{\diagup\!\!\!\diagup}}$$

$$Fe(CO)_3(\eta^4\text{-}CH_2CHCHCH_2) + 2CO \qquad (24.31)$$

$$Fe(CO)_5 + \text{excess } PMe_3 \xrightarrow{h\nu} Fe(CO)_3(PMe_3)_2 + 2CO$$

$$\text{(24.43)} \qquad (24.32)$$

$$\underset{\text{PMe}_3}{\overset{\text{PMe}_3}{\left| \begin{array}{c} \\ OC - Fe^{\cdots CO} \\ \diagdown CO \\ \end{array} \right.}}$$

$$\left[\underset{\text{CN}}{\overset{\text{CN}}{\left| \begin{array}{c} \\ OC - Fe^{\cdots CO} \\ \diagdown CO \\ \end{array} \right.}} \right]^{2-}$$

(24.43)   **(24.44)**

Mixed carbonyl/cyanido complexes of iron were described in Section 21.9. Their importance lies in their use as biomimetic models for [FeFe] and [NiFe]-hydrogenases (see Figs. 29.19 and 29.21). Photolysis of $Fe(CO)_5$ with $[Et_4N]CN$ leads to substitution reaction 24.33, and introduces cyanido ligands in axial positions (structure **24.44**).

$$Fe(CO)_5 + 2[Et_4N]CN \xrightarrow{\text{in MeCN, } h\nu}$$
$$[Et_4N]_2[\textit{trans}\text{-}Fe(CO)_3(CN)_2] + 2CO \quad (24.33)$$

Oxidative addition

Oxidative addition reactions are very important in organometallic synthesis. Oxidative addition involves:

- the addition of a molecule XY with cleavage of the X−Y single bond (eq. 24.34), addition of a multiply-bonded species with reduction in the bond order and formation of a metallacycle (eq. 24.35), addition of a C−H bond in an *orthometallation* step (eq. 24.36) or a similar addition;
- oxidation of the metal centre by two units;
- increase in metal coordination number by 2.

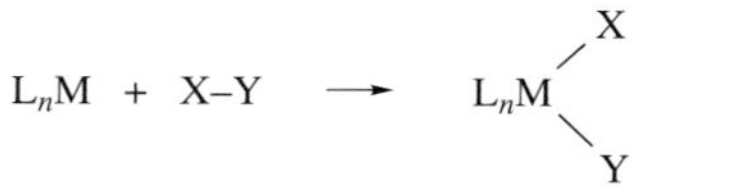

$$L_nM + X\text{–}Y \longrightarrow L_nM\diagup_{\diagdown Y}^{\diagup X} \quad (24.34)$$

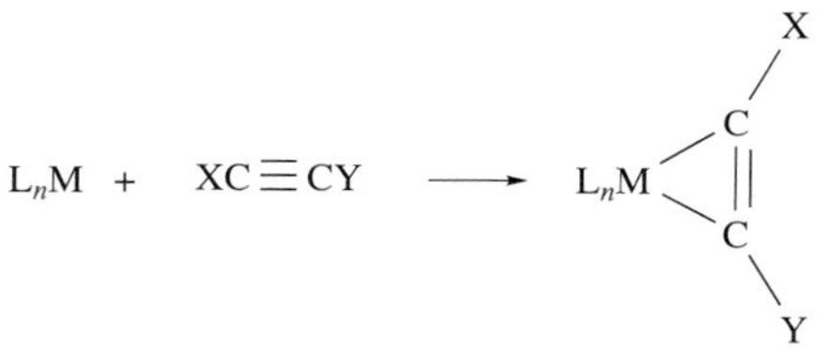

$$L_nM + XC\equiv CY \longrightarrow L_nM \quad (24.35)$$

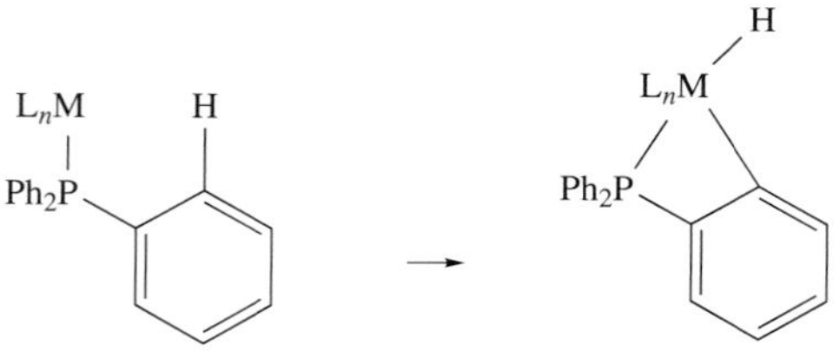

$$(24.36)$$

Addition of O_2 to give an η^2-peroxido complex is related to reaction type 24.35. Each addition in eqs. 24.34–24.36 occurs *at a 16-electron metal centre*, taking it to an 18-electron centre in the product. Most commonly, the precursor has a d^8 or d^{10} configuration, e.g. Rh(I), Ir(I), Pd(0), Pd(II), Pt(0), Pt(II), and the metal must have an accessible higher oxidation state, e.g. Rh(III). If the starting compound contains an 18-electron metal centre, oxidative addition cannot occur without loss of a 2-electron ligand as in reaction 24.37.

$$Os(CO)_5 + H_2 \xrightarrow{\Delta, \text{ 80 bar}} H_2Os(CO)_4 + CO \quad (24.37)$$

Many examples of the addition of small molecules (e.g. H_2, HX, RX) are known. The reverse of oxidative addition is *reductive elimination*, e.g. reaction 24.38, in which an acyl substituent is converted to an aldehyde.

$$H_2Co\{C(O)R\}(CO)_3 \longrightarrow HCo(CO)_3 + RCHO \quad (24.38)$$

Oxidative addition initially gives a *cis*-addition product, but ligand rearrangements can occur and the isolated product may contain the added groups mutually *cis* or *trans*.

Alkyl and hydrogen migrations

Reaction 24.39 is an example of *alkyl migration*.

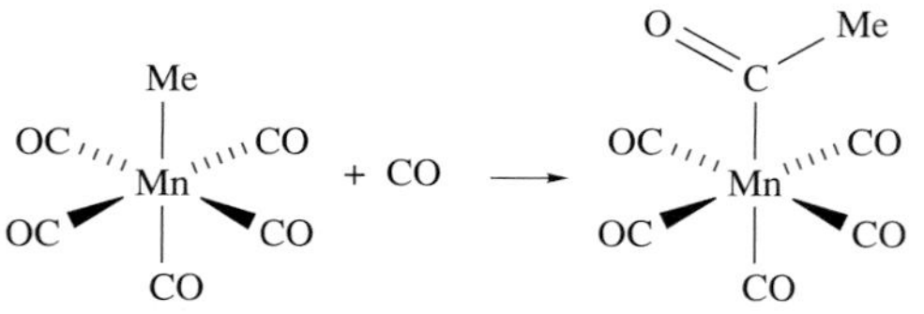

$$(24.39)$$

The reaction is also called *CO insertion* since the incoming CO molecule seems to have been inserted into the $Mn\text{–}C_{Me}$ bond: this name is misleading. If reaction 24.39 is carried out using ^{13}CO, *none* of the incoming ^{13}CO ends up in the acyl group or in the position *trans* to the acyl group; the isolated product is **24.45**.

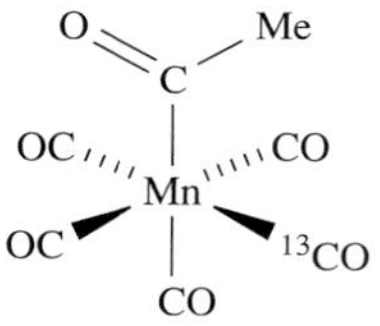

(24.45)

Reaction 24.39 involves the *intramolecular* transfer of an alkyl group to the C atom of a CO group which is *cis* to the original alkyl site. The incoming CO occupies the coordination site vacated by the alkyl group. Scheme 24.40 summarizes the process.

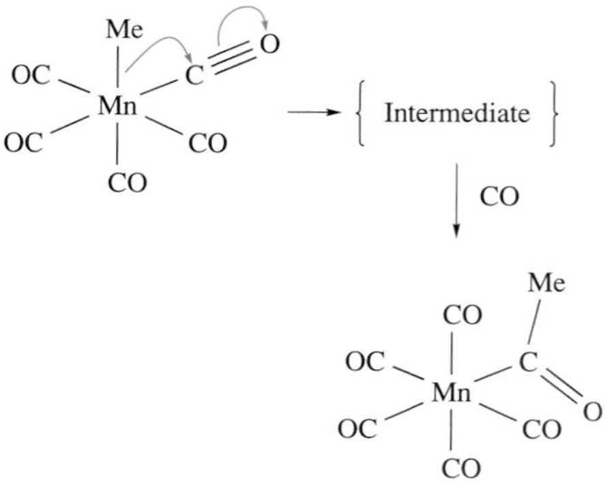

$$(24.40)$$

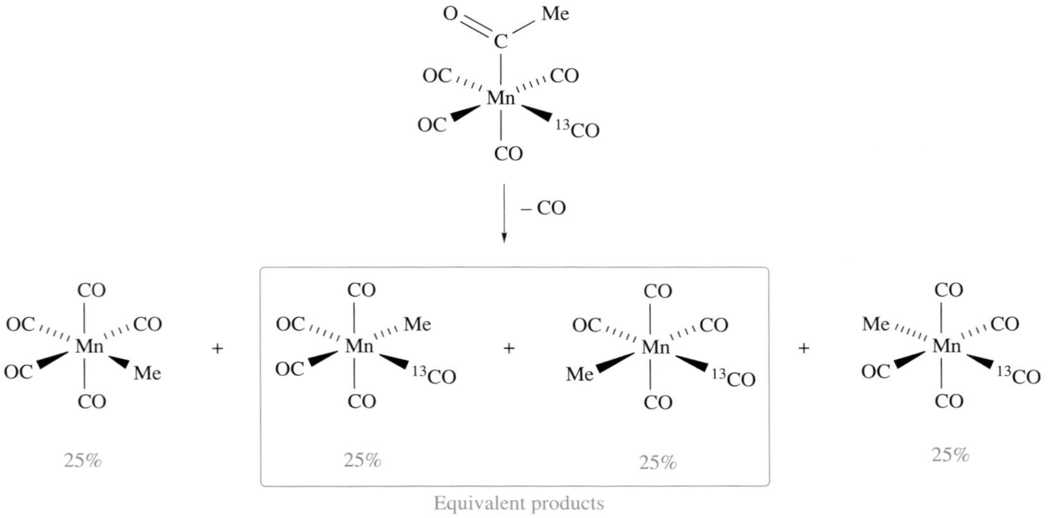

Fig. 24.15 The distribution of products from the decarbonylation of $Mn(CO)_4(^{13}CO)\{C(O)Me\}$ provides evidence for the migration of the Me group rather than movement of a CO molecule.

Scheme 24.40 implies that the intermediate is a coordinatively unsaturated species. In the presence of a solvent, S, such a species would probably be stabilized as $Mn(CO)_4(COMe)(S)$. In the absence of solvent, a 5-coordinate intermediate is likely to be stereochemically non-rigid (see Fig. 4.24 and discussion) and this is inconsistent with the observation of a selective *cis*-relationship between the incoming CO and acyl group. It has been concluded from the results of theoretical studies that the intermediate is stabilized by an *agostic* Mn$-$H$-$C interaction (structure **24.46**), the presence of which locks the stereochemistry of the system.[†]

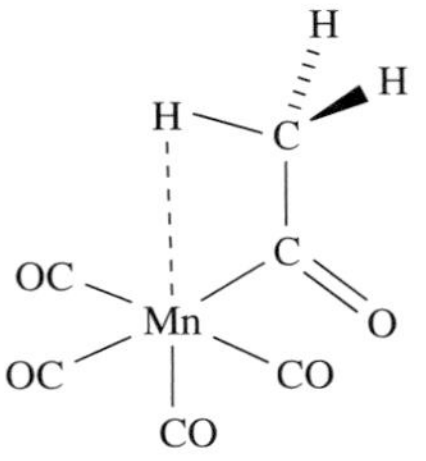

(24.46)

An ***agostic*** M$-$H$-$C interaction is a 3-centre 2-electron interaction between a metal centre, M, and a C$-$H bond in a ligand attached to M (e.g. structure **24.46**).

The migration of the methyl group is reversible and the *decarbonylation* reaction has been studied with the ^{13}C-labelled compound; the results are shown in Fig. 24.15.

[†] For a more detailed discussion, see: A. Derecskei-Kovacs and D.S. Marynick (2000) *J. Am. Chem. Soc.*, vol. 122, p. 2078.

The distribution of the products is consistent with the migration of the Me group, and not with a mechanism that involves movement of the 'inserted' CO. The reaction products can be monitored using ^{13}C NMR spectroscopy.

The 'insertion of CO' into M$-$C$_{alkyl}$ bonds is well exemplified in organometallic chemistry, and one industrial example (eq. 24.41) is a step in the Monsanto process for the production of acetic acid (see Section 25.5).

$$[Rh(Me)(CO)_2I_3]^- + CO \longrightarrow [Rh(CMeO)(CO)_2I_3]^-$$

(24.41)

Alkyl migrations are not confined to the formation of acyl groups, and, for example, 'alkene insertion' involves the conversion of a coordinated alkene to a σ-bonded alkyl group. Equation 24.42 shows the migration of an *H atom*. Related alkyl migrations occur and result in *carbon chain growth* as exemplified in Fig. 25.16.

(24.42)

β-Hydrogen elimination

The reverse of reaction 24.42 is a *β-elimination* step. It involves the transfer of a β-H atom (structure **24.47**) from the alkyl group to the metal and the conversion of the σ-alkyl group to a π-bonded alkene, i.e. a C$-$H bond is activated. For β-elimination to occur, the metal centre must be unsaturated, with a vacant coordination site *cis* to the alkyl group (eq. 24.43). The first step is thought to involve a cyclic intermediate **24.48** with an agostic M−H−C interaction.

(24.47)   **(24.48)**

$$L_nMCH_2CH_2R \longrightarrow L_nMH(\eta^2\text{-}RCH=CH_2)$$
$$\longrightarrow L_nMH + RCH=CH_2 \qquad (24.44)$$

β-Elimination is responsible for the decomposition of some metal alkyl complexes (eq. 24.44), but the reaction may be hindered or prevented by:

- steric hindrance;
- having a coordinatively *saturated* metal centre as in $(\eta^5\text{-}C_5H_5)Fe(CO)_2Et$;
- preparing an alkyl derivative which does not possess a β-hydrogen atom.

Examples of σ-bonded alkyl groups that cannot undergo β-elimination because they lack a β-H atom are Me, CH_2CMe_3, CH_2SiMe_3 and CH_2Ph. Thus, methyl derivatives cannot decompose by a β-elimination route and are usually more stable than their ethyl analogues. This does not mean that methyl derivatives are necessarily stable. The coordinatively unsaturated $TiMe_4$ decomposes at 233 K, but the stability can be increased by the formation of 6-coordinate adducts such as $Ti(bpy)Me_4$ and $Ti(Me_2PCH_2CH_2PMe_2)Me_4$.

α-Hydrogen abstraction

Early *d*-block metal complexes containing one or two α-hydrogen atoms (see **24.47**) may undergo *α-hydrogen abstraction* to yield *carbene* (*alkylidene*, **24.49**) or *carbyne* (*alkylidyne*, **24.50**) complexes. The solid state structure of the product of reaction 24.45 confirms differences in the Ta−C bond lengths: 225 pm for Ta−C_{alkyl} and 205 pm for Ta−$C_{carbene}$.

$$L_nM=CR_2 \qquad\qquad L_nM\equiv CR$$

(24.49)   **(24.50)**

(24.45)

Abstraction of a second α-H atom gives a carbyne complex (e.g. reaction 24.46). Other routes to carbenes and carbynes are described in Section 24.12.

$$WCl_6 \xrightarrow[-CMe_4,\ -LiCl]{LiCH_2CMe_3} W(\equiv CCMe_3)(CH_2CMe_3)_3 \quad (24.46)$$

Summary

A basic knowledge of the reaction types described in this section allows us to proceed to a discussion of the chemistry of selected organometallic complexes and (in Chapter 25) catalysis. Oxidative additions and alkyl migrations in particular are very important in the catalytic processes used in the manufacture of many organic chemicals. Selected important organometallic compounds used as catalysts are summarized in Box 24.2.

24.8 Metal carbonyls: selected reactions

The degradation of $Ni(CO)_4$ or $Fe(CO)_5$ to the respective metal and CO is a means of manufacturing high-purity nickel and iron. The thermal decomposition of $Ni(CO)_4$ is used in the Mond process to refine nickel (see eq. 21.4). Iron powder for use in magnetic cores in electronic components is produced by thermally decomposing $Fe(CO)_5$. The Fe particles act as nucleation centres for the production of particles up to 8 μm in diameter.

Reactions 24.18–24.23 illustrated conversions of neutral carbonyl compounds to carbonylate anions. Reduction by Na is typically carried out using Na/Hg amalgam. With Na in liquid NH_3, highly reactive anions can be formed (eqs. 24.47–24.50).

$$Cr(CO)_4(Me_2NCH_2CH_2NMe_2\text{-}N,N')$$
$$\xrightarrow{Na,\ liquid\ NH_3} Na_4[Cr(CO)_4] \qquad (24.47)$$

$$M_3(CO)_{12} \xrightarrow[low\ temp.]{Na,\ liquid\ NH_3} Na_2[M(CO)_4]$$
$$M = Ru, Os \qquad (24.48)$$

$$Ir_4(CO)_{12} \xrightarrow{Na,\ THF,\ CO\ (1\ bar)} Na[Ir(CO)_4] \qquad (24.49)$$

APPLICATIONS

Box 24.2 Homogeneous catalysts

Many of the reaction types discussed in Section 24.7 are represented in the catalytic processes described in Chapter 25. Unsaturated (16-electron) metal centres play an important role in catalytic cycles. Selected catalysts or catalyst precursors are summarized below.

Homogeneous catalyst	Catalytic application
$RhCl(PPh_3)_3$	Alkene hydrogenation
cis-$[Rh(CO)_2I_2]^-$	Monsanto acetic acid synthesis; Tennessee–Eastman acetic anhydride process
$HCo(CO)_4$	Hydroformylation; alkene isomerization
$HRh(CO)_4$	Hydroformylation (only for certain branched alkenes)[†]
$HRh(CO)(PPh_3)_3$	Hydroformylation
$[Ru(CO)_2I_3]^-$	Homologation of carboxylic acids
$[HFe(CO)_4]^-$	Water–gas shift reaction
$(\eta^5\text{-}C_5H_5)_2TiMe_2$	Alkene polymerization
$(\eta^5\text{-}C_5H_5)_2ZrH_2$	Hydrogenation of alkenes and alkynes
$Pd(PPh_3)_4$	Many laboratory applications including the Heck reaction

[†] $HRh(CO)_4$ is more active than $HCo(CO)_4$ in hydroformylation, but shows a lower regioselectivity (see eq. 25.5 and discussion).

$$Na[Ir(CO)_4] \xrightarrow[\substack{2.\ \text{liquid NH}_3,\ 195\,K, \\ \text{warm to } 240\,K}]{1.\ \text{Na, HMPA, } 293\,K} Na_3[Ir(CO)_3]$$

$$HMPA = (Me_2N)_3PO \tag{24.50}$$

The IR spectra (see Section 24.2) of highly charged anions exhibit absorptions for the terminal CO ligands in regions usually characteristic of bridging carbonyls, e.g. 1680 and $1471\,cm^{-1}$ for $[Mo(CO)_4]^{4-}$, and $1665\,cm^{-1}$ for $[Ir(CO)_3]^{3-}$.

The action of alkali on $Fe(CO)_5$ (eq. 24.51) gives $[HFe(CO)_4]^-$ (**24.51**). Nucleophilic attack by $[OH]^-$ on a CO ligand is followed by Fe–H bond formation and elimination of CO_2. The $[HFe(CO)_4]^-$ ion has a variety of synthetic uses.

$$Fe(CO)_5 + 3NaOH \xrightarrow{H_2O} Na[HFe(CO)_4] + Na_2CO_3 + H_2O \tag{24.51}$$

(24.51)   **(24.52)**

Hydrido ligands can be introduced by various routes including protonation (eqs. 24.2 and 24.52), reaction with H_2 (eqs. 24.53 and 24.54) and action of $[BH_4]^-$ (eqs. 24.55 and 24.56).

$$Na[Mn(CO)_5] \xrightarrow{\text{H}_3\text{PO}_4, \text{ in THF}} HMn(CO)_5 \tag{24.52}$$

$$Mn_2(CO)_{10} + H_2 \xrightarrow{\text{200 bar, 470 K}} 2HMn(CO)_5 \tag{24.53}$$

$$Ru_3(CO)_{12} + H_2 \xrightarrow{\text{in boiling octane}} (\mu\text{-H})_4Ru_4(CO)_{12} \tag{24.54}$$

$$Cr(CO)_6 \xrightarrow{\text{Na[BH}_4]} [(OC)_5Cr(\mu\text{-H})Cr(CO)_5]^- \tag{24.55}$$

$$Ru_3(CO)_{12} \xrightarrow{\text{Na[BH}_4] \text{ in THF}} [HRu_3(CO)_{11}]^- \tag{24.56}$$

(24.52)

Reactions 24.57–24.59 illustrate preparations of selected metal carbonyl halides (see Section 24.9) from binary carbonyls.

$$Fe(CO)_5 + I_2 \longrightarrow Fe(CO)_4I_2 + CO \tag{24.57}$$

$$Mn_2(CO)_{10} + X_2 \longrightarrow 2Mn(CO)_5X \quad X = Cl, Br, I \tag{24.58}$$

$$M(CO)_6 + [Et_4N]X \longrightarrow [Et_4N][M(CO)_5X] + CO$$
$$M = Cr, Mo, W; \ X = Cl, Br, I \tag{24.59}$$

Large numbers of derivatives are formed by displacement of CO by other ligands (see eqs. 24.29–24.33 and discussion). Whereas substitution by *tertiary* phosphane ligands usually gives terminal ligands, the introduction of a *secondary* or *primary* phosphane into a carbonyl complex creates the possibility of oxidative addition of a P–H bond. Where a second metal centre is present, the formation of a bridging R_2P ligand is possible (reaction 24.60).

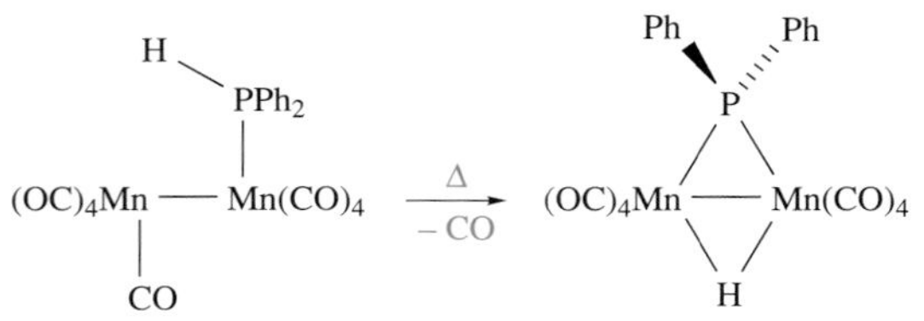

$$(24.60)$$

We saw earlier that CO displacement can be carried out photolytically or thermally, and that activation of the starting compound (as in reaction 24.29) may be necessary. In multinuclear compounds, activation of one site can control the degree of substitution, e.g. $Os_3(CO)_{11}(NCMe)$ is used as an *in situ* intermediate during the formation of monosubstituted derivatives (eq. 24.61).

$$Os_3(CO)_{12} \xrightarrow{\text{MeCN, Me}_3\text{NO}} Os_3(CO)_{11}(NCMe)$$

$$\xrightarrow{L \text{ (e.g. PR}_3)} Os_3(CO)_{11}L \qquad (24.61)$$

In the first step of the reaction, Me_3NO oxidizes CO to CO_2, liberation of which leaves a vacant coordination site that is occupied temporarily by the labile MeCN ligand. This method can be applied to higher nuclearity clusters to achieve control over otherwise complex reactions.

Displacement of CO by a nitrosyl ligand alters the electron count and, for an 18-electron centre to be retained, one-for-one ligand substitution cannot occur. Reaction 24.62 shows the conversion of octahedral $Cr(CO)_6$ to tetrahedral $Cr(NO)_4$, in which NO is a 3-electron donor.

$$Cr(CO)_6 + \text{excess NO} \longrightarrow Cr(NO)_4 + 6CO \qquad (24.62)$$

Reactions of metal carbonyls with unsaturated organic ligands are discussed in later sections.

Self-study exercises

1. The reaction of $Mn_2(CO)_{10}$ with Na/Hg can be monitored by IR spectroscopy in the region $2100-1800\,cm^{-1}$. The starting material and product show absorptions at 2046, 2015 and $1984\,cm^{-1}$, and at 1896 and $1865\,cm^{-1}$, respectively. Suggest a likely product of the reaction. Check that the Mn centre in the product obeys the 18-electron rule. Account for the changes in the IR spectra. [*Ans.* $Na[Mn(CO)_5]$, see Section 24.2]

2. Photolysis of a benzene solution of *trans*-$W(N_2)_2$(dppe)$_2$ (dppe $= Ph_2PCH_2CH_2PPh_2$) with Ph_2PH gives *trans*-$WH(PPh_2)$(dppe)$_2$ in which the W–P_{PPh_2} bond distance of 228 pm is consistent with a double bond. (a) How is a coordinated N_2 ligand related to metal-bound CO? (b) Explain what types of reaction have occurred during the conversion of *trans*-$W(N_2)_2$(dppe)$_2$ to *trans*-$WH(PPh_2)$(dppe)$_2$. [*Ans.* (a) See Section 24.2; (b) see: L.D. Field *et al.* (1998) *J. Organomet. Chem.*, vol. 571, p. 195]

24.9 Metal carbonyl hydrides and halides

Methods of preparing selected hydrido complexes were given in eqs. 24.2 and 24.52–24.56. Selected properties of the mononuclear complexes $HMn(CO)_5$, $H_2Fe(CO)_4$ and $HCo(CO)_4$ are given in Table 24.6. $HCo(CO)_4$ is an industrial catalyst (see Section 25.5). Metal hydrides play an important role in organometallic chemistry, and scheme 24.63 illustrates some ligand transformations involving M–H bonds. Each reaction of L_nMH with an alkene or alkyne is an H atom migration. This type of reaction can be extended to CH_2N_2 (**24.53**), H atom migration to which results in the loss of N_2 and formation of a metal-bound methyl group.

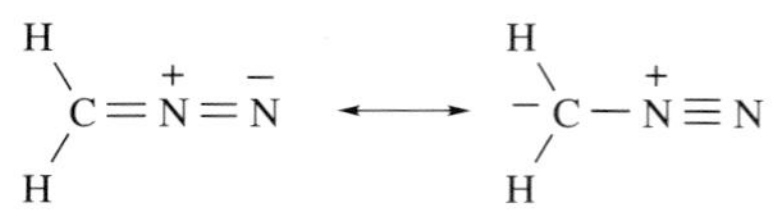

(24.53)

Table 24.6 Selected properties of $HMn(CO)_5$, $H_2Fe(CO)_4$ and $HCo(CO)_4$.

Property	HMn(CO)$_5$	H$_2$Fe(CO)$_4$	HCo(CO)$_4$
Physical appearance at 298 K	Colourless liquid	Yellow liquid	Yellow liquid
Stability	Stable up to 320 K	Dec. $\geq$253 K	Dec. >247 K (mp)
pK_a values	15.1	p$K_a(1) = 4.4$ p$K_a(2) = 14.0$	<0.4
^{1}H NMR δ/ ppm	−10.7	−11.2	−7.9

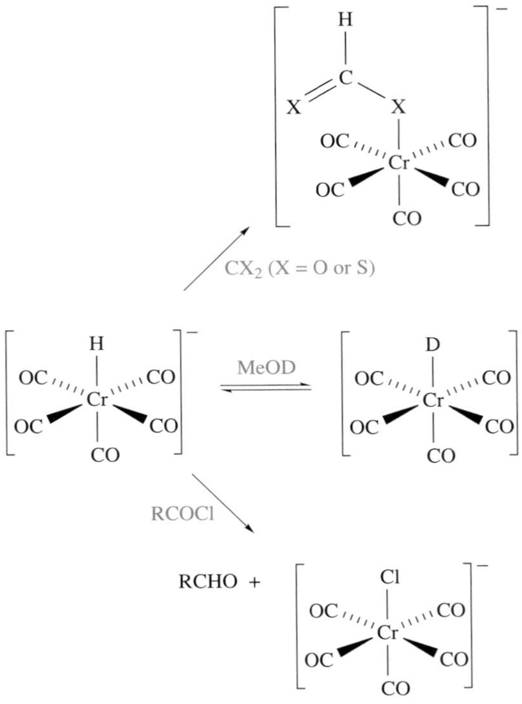

CX$_2$ (X = O or S)

MeOD

RCOCl

RCHO +

Fig. 24.16 Selected reactions of [HCr(CO)$_5$]$^-$.

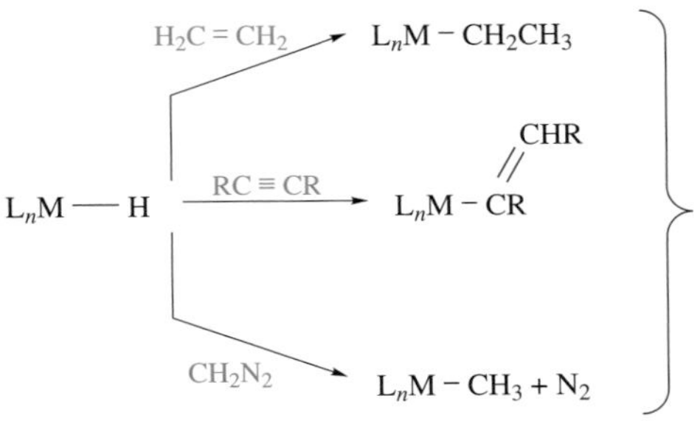

$$H_2C=CH_2 \rightarrow L_nM-CH_2CH_3$$

$$RC\equiv CR \rightarrow L_nM-CR \overset{CHR}{\|}$$

$$L_nM-H$$

$$CH_2N_2 \rightarrow L_nM-CH_3 + N_2 \qquad (24.63)$$

Mononuclear hydrido carbonyl anions include [HFe(CO)$_4$]$^-$ and [HCr(CO)$_5$]$^-$, both of which can be made by the action of hydroxide on the parent metal carbonyl (eqs. 24.51 and 24.64). Selected reactions of [HCr(CO)$_5$]$^-$ are shown in Fig. 24.16.[†]

$$Cr(CO)_6 + 2KOH \xrightarrow[298\,K,\,10\,min]{CH_2Cl_2/EtOH} K[HCr(CO)_5] + KHCO_3 \qquad (24.64)$$

Methods of forming carbonyl halides include starting from binary metal carbonyls (eqs. 24.57–24.59) or metal halides (eqs. 24.65–24.67).

$$RhCl_3 \cdot xH_2O \xrightarrow{CO,\,373\,K} (CO)_2Rh(\mu\text{-}Cl)_2Rh(CO)_2 \qquad (24.65)$$

[†] For an overview of reactions and role of [HCr(CO)$_5$]$^-$ in homogeneous catalysis, see: J.-J. Brunet (2000) *Eur. J. Inorg. Chem.*, p. 1377.

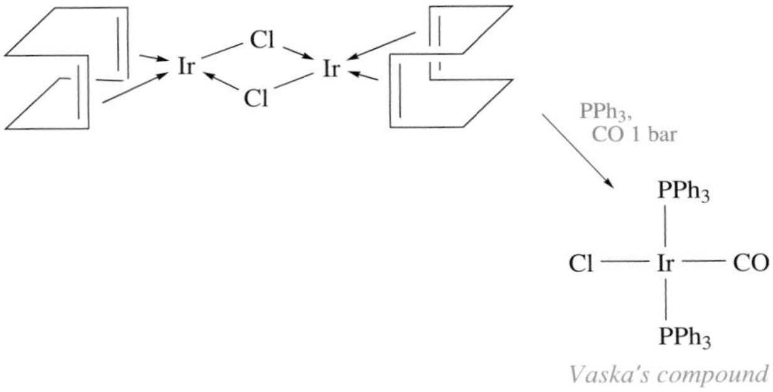

PPh$_3$, CO 1 bar

Vaska's compound (24.66)

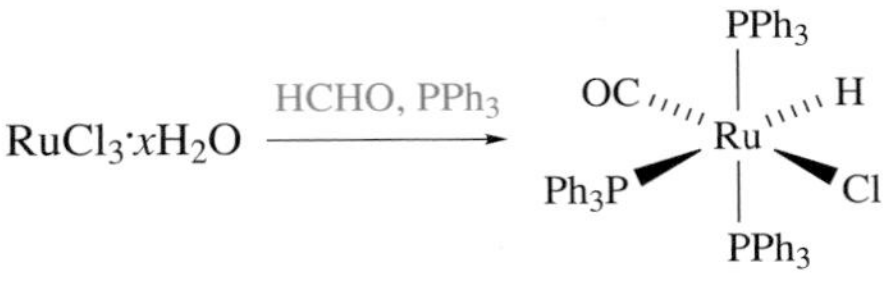

$$RuCl_3 \cdot xH_2O \xrightarrow{HCHO,\,PPh_3} \qquad (24.67)$$

$$RuCl_3 \cdot xH_2O \xrightarrow[\substack{1.\,3\,h,\,398\,K\\2.\,7\,h,\,298\,K\\3.\,extracted\,with\,hot\,THF}]{in\,MeOCH_2CH_2OH,\,CO\,(1\,bar)} \qquad (24.68)$$

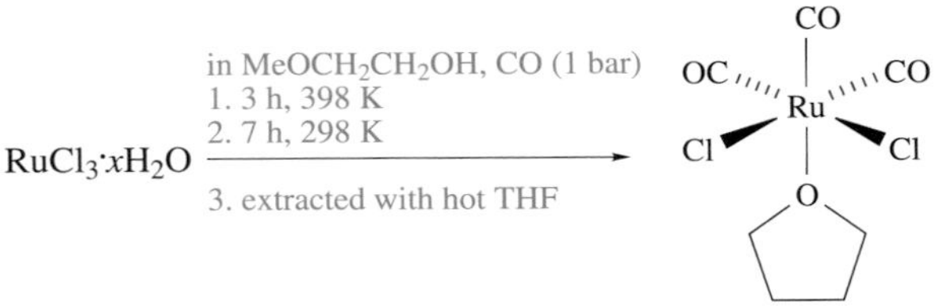

The product of reaction 24.68 is a useful precursor to Ru(PPh$_3$)$_3$(CO)$_2$ (eq. 24.69). This 18-electron complex readily loses one PPh$_3$ ligand to give an unsaturated Ru centre to which H$_2$, R$_2$C=CR$_2$, RC≡CR and related molecules easily add.

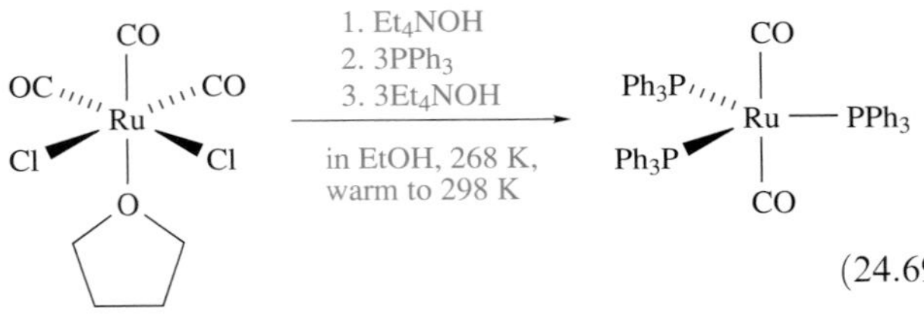

$$\xrightarrow[\substack{in\,EtOH,\,268\,K,\\warm\,to\,298\,K}]{\substack{1.\,Et_4NOH\\2.\,3PPh_3\\3.\,3Et_4NOH}} \qquad (24.69)$$

The proposed mechanism of reaction 24.69 involves initial attack by HO$^-$ at one CO ligand, followed by ligand substitution by PPh$_3$, deprotonation of the CO$_2$H ligand, and then dissociation of CO$_2$ and 2Cl$^-$. The addition of two PPh$_3$ ligands regenerates an 18-electron centre (Fig. 24.17).

The 16-electron halide complexes *cis*-[Rh(CO)$_2$I$_2$]$^-$ and *trans*-[Ir(CO)Cl(PPh$_3$)$_2$] (Vaska's compound) undergo many oxidative addition reactions and have important catalytic applications (see Chapter 25). Vaska's compound readily takes up O$_2$ to give the peroxido complex **24.54**. The product of reaction 24.67 is a catalyst precursor for alkene hydrogenation.

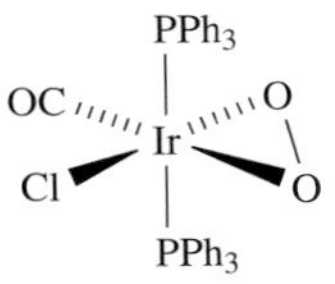

(24.54)

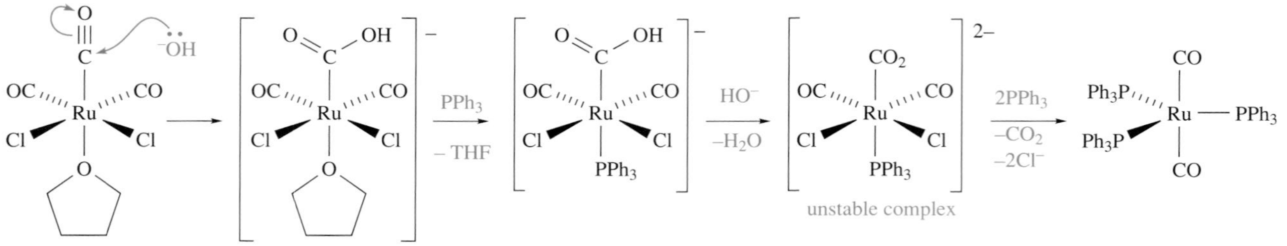

Fig. 24.17 Proposed mechanism for the conversion of *fac*-Ru(CO)$_3$Cl$_2$(THF) to Ru(CO)$_2$(PPh$_3$)$_3$.

24.10 Alkyl, aryl, alkene and alkyne complexes

σ-Bonded alkyl and aryl ligands

Simple σ-bonded organic derivatives of low oxidation state *d*-block metals are generally more reactive than analogous main group metal species. The origin is kinetic rather than thermodynamic: the availability of vacant 3*d* atomic orbitals in titanium alkyl complexes means that they (except TiMe$_4$) readily undergo β-elimination to give alkene complexes (see Section 24.7).

Alkyl and aryl derivatives can be made by reactions such as 24.70–24.78, the last being an example of an oxidative addition to a 16-electron complex. Choice of alkylating agent can affect the course of the reaction. For example, whereas LiMe is suitable in reaction 24.70, its use instead of ZnMe$_2$ in reaction 24.72 would reduce MoF$_6$.

$$TiCl_4 + 4LiMe \xrightarrow{\text{in Et}_2\text{O, 193 K}} TiMe_4 + 4LiCl \qquad (24.70)$$

$$WCl_6 + 3Al_2Me_6 \longrightarrow WMe_6 + 3Al_2Me_4Cl_2 \qquad (24.71)$$

$$MoF_6 + 3ZnMe_2 \xrightarrow{\text{Et}_2\text{O, 143 K}} MoMe_6 + 3ZnF_2 \qquad (24.72)$$

$$MoMe_6 \xrightarrow[\text{in Et}_2\text{O, 258 K}]{\text{excess LiMe}} [Li(Et_2O)]^+[MoMe_7]^- \qquad (24.73)$$

$$ScCl_3(THF)_3 + 3PhLi \xrightarrow[\text{273 K}]{\text{Et}_2\text{O/THF,}}$$
$$ScPh_3(THF)_2 + THF + 3LiCl \qquad (24.74)$$

$$Na_2[Fe(CO)_4] + EtBr \longrightarrow Na[Fe(CO)_4Et] + NaBr \qquad (24.75)$$

$$Na[Mn(CO)_5] + PhCH_2Cl \longrightarrow Mn(CO)_5(CH_2Ph) + NaCl \qquad (24.76)$$

$$Li[Mn(CO)_5] + PhC(O)Cl \xrightarrow{-LiCl} Mn(CO)_5\{C(O)Ph\}$$
$$\xrightarrow{\Delta} Mn(CO)_5Ph + CO \qquad (24.77)$$

$$cis\text{-}[Rh(CO)_2I_2]^- + MeI \longrightarrow mer\text{-}[Rh(CO)_2I_3Me]^- \qquad (24.78)$$

Hexamethyltungsten (eq. 24.71) was the first example of a discrete trigonal prismatic complex (Box 20.3). It is highly reactive in air and is potentially explosive.

Alkene ligands

Alkene complexes are often made by displacement of CO or halide ion by an alkene. The formation of *Zeise's salt*,[†] K[PtCl$_3$(η^2-C$_2$H$_4$)] (reaction 24.79), is catalysed by SnCl$_2$ with [PtCl$_3$(SnCl$_3$)]$^{2-}$ being the intermediate. The [PtCl$_3$(η^2-C$_2$H$_4$)]$^-$ ion (Fig. 24.18a) contains a square planar (or pseudo-square planar) Pt(II) centre and in the solid state, the ethene ligand lies perpendicular to the coordination 'square plane', thereby minimizing steric interactions.

$$K_2[PtCl_4] + H_2C=CH_2$$
$$\xrightarrow{SnCl_2} K[PtCl_3(\eta^2\text{-}C_2H_4)] + KCl \qquad (24.79)$$

Addition of an alkene to 16-electron metal complexes is exemplified by reaction 24.80. Ethene readily dissociates from Ir(CO)Cl(η^2-C$_2$H$_4$)(PPh$_3$)$_2$, but the related complex Ir(CO)Cl(η^2-C$_2$(CN)$_4$)(PPh$_3$)$_2$ is very stable.

$$trans\text{-}Ir(CO)Cl(PPh_3)_2 + R_2C=CR_2$$
$$\longrightarrow Ir(CO)Cl(\eta^2\text{-}C_2R_4)(PPh_3)_2 \qquad (24.80)$$

More recent additions to the family of alkene complexes are fullerene derivatives such as Mo(η^2-C$_{60}$)(CO)$_5$, W(η^2-C$_{60}$)(CO)$_5$ (eq. 24.81), Rh(CO)(η^2-C$_{60}$)(H)(PPh$_3$)$_2$, Pd(η^2-C$_{60}$)(PPh$_3$)$_2$ (Fig. 24.18b) and (η^5-Cp)$_2$Ti(η^2-C$_{60}$). The C$_{60}$ cage (see Section 14.4) functions as a polyene

---

[†] For an account of the achievements of William C. Zeise including the discovery of Zeise's salt, see: D. Seyferth (2001) *Organometallics*, vol. 20, p. 2.

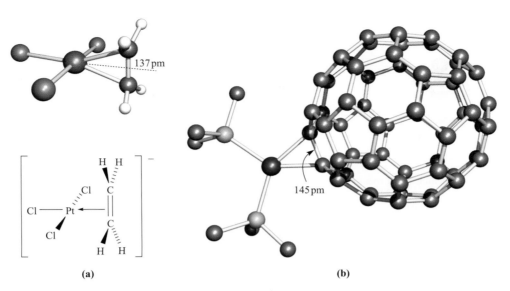

Fig. 24.18 (a) The structure of the anion in Zeise's salt, $K[PtCl_3(\eta^2\text{-}C_2H_4)]$. The Pt(II) centre can be regarded as being square planar as indicated in the schematic representation [neutron diffraction: R.A. Love *et al.* (1975) *Inorg. Chem.*, vol. 14, p. 2653]. (b) The structure of $Pd(\eta^2\text{-}C_{60})(PPh_3)_2$; for clarity, only the *ipso*-C atoms of the Ph rings are shown [X-ray diffraction: V.V. Bashilov *et al.* (1993) *Organometallics*, vol. 12, p. 991]. Colour code: Pt, brown; Pd, blue; C, grey; Cl, green; P, orange; H, white.

with localized C=C bonds, and in $C_{60}\{Pt(PEt_3)_2\}_6$, six C=C bonds (remote from one another) in the C_{60} cage have undergone addition. Reaction 24.82 illustrates C_{60}-for-ethene substitution (the 16-electron centre is retained), and reaction 24.83 shows addition to Vaska's compound (a 16- to 18-electron conversion). Equation 24.84 shows the formation of a fullerene complex of titanium, by fullerene displacement of a coordinated alkyne.

$$M(CO)_6 + C_{60} \xrightarrow{h\nu \text{ or sunlight}} M(\eta^2\text{-}C_{60})(CO)_5 + CO$$
$$M = Mo, W \tag{24.81}$$

$$Pt(\eta^2\text{-}C_2H_4)(PPh_3)_2 + C_{60} \xrightarrow{-C_2H_4} Pt(\eta^2\text{-}C_{60})(PPh_3)_2 \tag{24.82}$$

$$\textit{trans-}Ir(CO)Cl(PPh_3)_2 + C_{60}$$
$$\longrightarrow Ir(CO)Cl(\eta^2\text{-}C_{60})(PPh_3)_2 \quad (24.83)$$

$$(\eta^5\text{-}Cp)_2Ti(\eta^2\text{-}Me_3SiC\equiv CSiMe_3) + C_{60}$$
$$\longrightarrow (\eta^5\text{-}Cp)_2Ti(\eta^2\text{-}C_{60}) + Me_3SiC\equiv CSiMe_3 \tag{24.84}$$

Reactions of alkenes with metal carbonyl clusters may give simple substitution products such as $Os_3(CO)_{11}(\eta^2\text{-}C_2H_4)$ or may involve oxidative addition of one or more C–H bonds. Reaction 24.85 illustrates the reaction of $Ru_3(CO)_{12}$ with $RHC=CH_2$ (R = alkyl) to give isomers

of $H_2Ru_3(CO)_9(RCCH)$ in which the organic ligand acts as a 4-electron donor (one π- and two σ-interactions).

$$\tag{24.85}$$

In solution, alkene complexes are often fluxional, with rotation occurring as shown in Fig. 24.19. The model compound in the figure contains a mirror plane passing through M, L and L'. The limiting *low*-temperature ^1H spectrum shows one resonance for H_1 and H_4, and another due to H_2 and H_3, i.e. a static picture of the molecule. On raising the temperature, the molecule gains sufficient energy for the alkene to rotate and the limiting *high*-temperature spectrum contains one resonance since H_1,

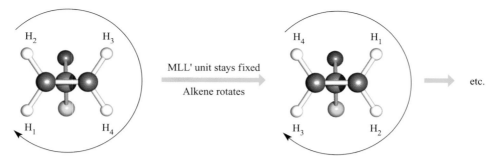

Fig. 24.19 Schematic representation of the rotation of an η^2-C_2H_4 ligand in a complex MLL'$(\eta^2$-$C_2H_4)$. The complex is viewed along a line connecting the centre of the C—C bond and the M atom; the M atom is shown in green, and ligands L and L' in red and yellow. Because L and L' are *different*, rotation of the alkene interchanges the environments of H_1 and H_4 with H_3 and H_2.

H_2, H_3 and H_4 become equivalent on the NMR timescale. In $(\eta^5$-Cp)Rh$(\eta^2$-$C_2H_4)_2$, two alkene proton signals are observed at 233 K (the different H environments are blue and black respectively in **24.55**). At 373 K, the proton environments become equivalent on the NMR spectroscopic timescale as each alkene ligand rotates about the metal–ligand coordinate bond.

(24.55)

Coordinated alkenes may be displaced by other ligands (eq. 24.82). Unlike free alkenes, which undergo electrophilic additions, *coordinated* alkenes are susceptible to nucleophilic attack and many reactions of catalytic importance involve this pathway (see Chapter 25). Reaction 24.86 shows that addition of a nucleophile, R^-, leads to a σ-bonded complex. The mechanism may involve direct attack at one alkene C atom, or attack at the $M^{\delta+}$ centre followed by alkyl migration (see Section 24.7).

(24.86)

Applications in organic synthesis make particular use of $[(\eta^5$-Cp)Fe(CO)$_2(\eta^2$-alkene)]$^+$ complexes and those involving alkenes coordinated to Pd(II) centres. The latter is the basis for the Wacker process (see Fig. 25.2) and for the Heck reaction (eq. 24.87). The Heck reaction is a palladium-catalysed C–C bond forming reaction between an aryl or vinyl halide and an activated alkene in the presence of base. In 2010, Richard Heck shared the Nobel Prize in Chemistry with Ei-ichi Negishi and Akira Suzuki 'for palladium-catalysed cross couplings in organic synthesis'.

(24.87)

When Pd(OAc)$_2$ is used as catalyst in Heck reactions with an ionic liquid (see Section 9.12) as solvent, the ionic liquid stabilizes clusters of Pd atoms producing particles of diameter 1.0–1.6 nm. (The dimensions are measured using transmission electron microscopy, see Box 13.8.) The induction periods of these Heck reactions correlate with the time taken for the Pd nanoparticles to form, consistent with the Pd clusters being the active catalysts. Compared with Pd(OAc)$_2$-catalysed reactions carried out in organic media, product separation and catalyst recycling in ionic liquids are more efficient. Nowadays, catalyst recycling is of prime importance (see Section 25.6), and the use of ionic liquids rather than conventional organic solvents is often advantageous. Ionic liquids used in Heck reactions include the following:

Often, nucleophilic addition to coordinated alkenes is regioselective, with the nucleophile attacking at the C atom carrying the *least* electron-withdrawing group. In reactions 24.88 and 24.89, a strongly electron-releasing group is exemplified by EtO, and an electron-withdrawing group by CO_2Me (Nu = nucleophile).

(24.88)

(24.89)

> **Coordinated alkenes are susceptible to nucleophilic attack.** This reaction has important applications in synthetic organic chemistry and catalysis.

Self-study exercises

1. Explain how the O_2CMe group directs nucleophilic addition as shown in reaction 24.89.

2. Why does the product of eq. 24.86 not undergo β-hydride elimination?   [*Ans.* See text under eq. 24.44]

Alkyne ligands

Many mono- and polynuclear organometallic complexes involving alkyne ligands are known. An alkyne $RC\equiv CR$ has *two* fully occupied π-MOs and may act as a 2- or 4-electron donor. The bonding in a monometallic alkyne complex can be described in a similar manner to that in an alkene complex (see Section 24.2), but allowing for the participation of the two orthogonal π-MOs. A typical $C\equiv C$ bond length in a free alkyne is 120 pm and, in complexes, this lengthens to ≈124–137 pm depending on the mode of bonding. In **24.56**, the C–C bond length (124 pm) is

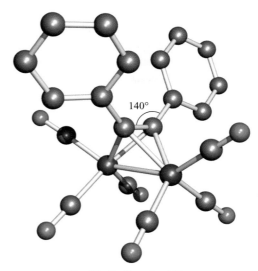

C—C in Co_2C_2-unit = 136 pm

Fig. 24.20 The structure (X-ray diffraction) of $Co_2(CO)_6(C_2Ph_2)$ [D. Gregson *et al.* (1983) *Acta Crystallogr.*, *Sect. C*, vol. 39, p. 1024]. Hydrogen atoms are omitted for clarity; colour code: Co, blue; C, grey; O, red.

consistent with a weakened triple bond; the alkyne lies perpendicular to the $PtCl_2L$-plane and occupies one site in the square planar coordination sphere of the Pt(II) centre. A similar example is $[PtCl_3(\eta^2\text{-}C_2Ph_2)]^-$ (eq. 24.90). In **24.57**, the alkyne acts as a 4-electron donor, forming a metallacycle. The C–C bond length (132 pm) is consistent with a double bond. A decrease in the alkyne $C\text{-}C\text{-}C_R$ bond angle accompanies the change in bonding mode on going from **24.56** to **24.57**. The addition of an alkyne to $Co_2(CO)_8$ (eq. 24.91) results in the formation of a Co_2C_2 cluster (Fig. 24.20) in which the alkyne C–C bond is lengthened to 136 pm.

$$K_2[PtCl_4] \xrightarrow[\text{(see Figure 11.8)}]{\text{18-crown-6}} {}^1/_2[K(18\text{-crown-6})]_2[Pt_2Cl_6]$$

$$\Big\downarrow PhC\equiv CPh$$

$$[K(18\text{-crown-6})]^+[PtCl_3(\eta^2\text{-}C_2Ph_2)]^-$$

(24.90)

$$Co_2(CO)_8 + PhC\equiv CPh \longrightarrow Co_2(CO)_6(C_2Ph_2) + 2CO$$

(24.91)

L = 4-MeC₆H₄NH₂ — L = $4\text{-MeC}_6H_4NH_2$

(24.56)   **(24.57)**

The reactions between alkynes and multinuclear metal carbonyls give various product types, with alkyne coupling and alkyne–CO coupling often being observed, e.g. reaction 24.92, in which the organic ligand in the product is a 6-electron donor (two σ- and two π-interactions). Predicting the outcome of such reactions is difficult.

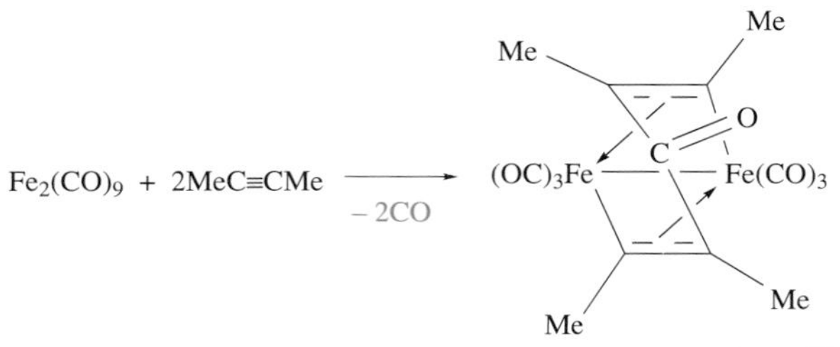

$$Fe_2(CO)_9 + 2MeC{\equiv}CMe \xrightarrow{-2CO} \quad (24.92)$$

In solution, π-bonded alkynes of the type in structure **24.56** undergo rotations analogous to those of alkenes.

Self-study exercises

1. The solution ^1H NMR spectrum of [K(18-crown-6)] [PtCl$_3$(η^2-MeC$\equiv$CMe)] exhibits a singlet at δ 3.60 ppm and a pseudo-triplet at δ 2.11 ppm (J 32.8 Hz). Assign the signals and explain the origin of the 'pseudo-triplet'. Sketch the pseudo-triplet and show where the coupling constant of 32.8 Hz is measured.

2. [K(18-crown-6)][PtCl$_3$(η^2-MeC$\equiv$CMe)] reacts with ethene to give [K(18-crown-6)][**X**]. The ^1H and ^{13}C NMR spectra of the product are as follows: ^1H NMR: δ/ppm 3.63 (singlet), 4.46 (pseudo-triplet, J 64.7 Hz); ^{13}C NMR δ/ppm 68.0 (pseudo-triplet, J 191.8 Hz), 70.0 (singlet). Suggest the identity of [**X**]$^-$, and assign the spectra.

[*Ans. to both questions*: See D. Steinborn *et al.* (1995) *Inorg. Chim. Acta*, vol. 234, p. 47; see Box 19.2]

24.11 Allyl and buta-1,3-diene complexes

Allyl and related ligands

π-Allyl and related complexes can be prepared by reactions such as 24.93–24.97. The last two reactions illustrate formation of allyl ligands by deprotonation of coordinated propene, and protonation of coordinated buta-1,3-diene respectively. Reactions 24.94 and 24.95 are examples of pathways that go via σ-bonded intermediates (e.g. **24.58**) which eliminate CO.

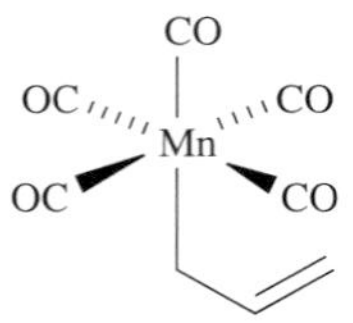

(24.58)

$$NiCl_2 + 2C_3H_5MgBr$$
$$\xrightarrow{\text{in } Et_2O, \ 263\,K} Ni(\eta^3\text{-}C_3H_5)_2 + 2MgBrCl \quad (24.93)$$

$$Na[Mn(CO)_5] + H_2C{=}CHCH_2Cl$$
$$\longrightarrow Mn(\eta^3\text{-}C_3H_5)(CO)_4 + CO + NaCl \quad (24.94)$$

$$Na[(\eta^5\text{-}Cp)Mo(CO)_3] + H_2C{=}CHCH_2Cl$$
$$\xrightarrow{h\nu, \text{ in THF}} (\eta^5\text{-}Cp)Mo(\eta^3\text{-}C_3H_5)(CO)_2 + NaCl + CO \quad (24.95)$$

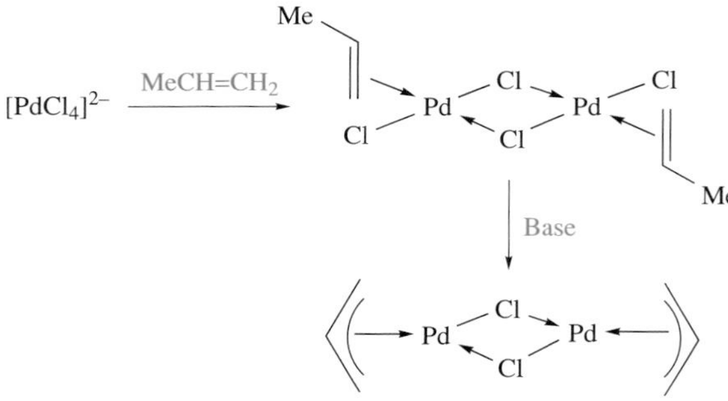

$$(24.96)$$

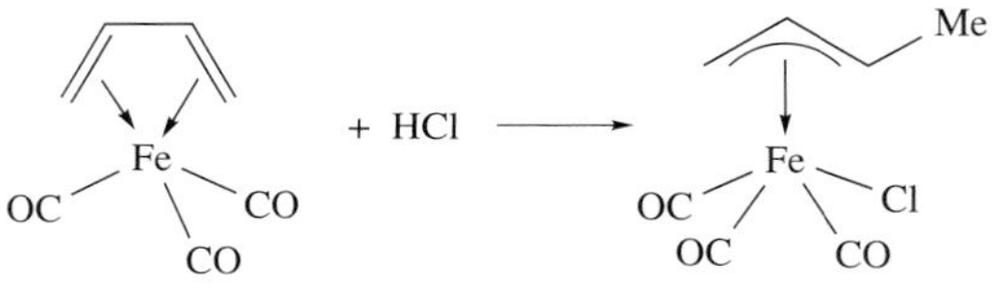

$$(24.97)$$

In Fig. 24.8, we showed the three π-MOs that the π-allyl ligand uses in bonding to a metal centre. In Mo(η^3-C$_3$H$_5$)(η^4-C$_4$H$_6$)(η^5-C$_5$H$_5$) (Fig. 24.6b), the central and two outer Mo–C bond lengths in the Mo(η^3-C$_3$) unit are different (221 and 230 pm respectively). This is a typical observation for π-allyl ligands, e.g. 198 and 203 pm respectively for the central and outer Ni–C bonds in Ni(η^3-C$_3$H$_5$)$_2$ (Fig. 24.21). In the latter, the two allyl ligands are staggered. In Fig. 24.6b and 24.21, note the orientations of the H atoms with respect to the metal centres. The two H atoms in each terminal CH$_2$ group of a coordinated η^3-C$_3$H$_5$ ligand are non-equivalent. In solution, however, they are often equivalent on the NMR spectroscopic timescale, and this can be rationalized in terms of the η^3–η^1–η^3 (i.e. π–σ–π) rearrangement shown in scheme 24.98. An η^3–η^1 rearrangement also features in some reactions of allyl ligands.

APPLICATIONS

Box 24.3 Molecular wires

The ability of chemists to design molecules for electronic applications is becoming a reality and 'molecular wires' are a topical area of research. A molecular wire is a molecule capable of transporting charge carriers from one end of the wire to the other. Molecules with extended conjugated systems are prime candidates for molecular wires since the conjugation provides the necessary electronic communication between atomic centres. The molecule must also possess a small band gap. (For details on charge carriers and band gaps, see Section 6.9.)

Although commercial applications are still future goals, much progress has been made in the design of potential molecular wires. Molecules so far studied have included organic molecules with conjugated alkyne functionalities, porphyrins connected by alkyne units, chains of connected thiophenes, and organometallic complexes. Examples of the latter are shown below:

Further reading

D.K. James and J.M. Tour (2005) *Top. Curr. Chem.*, vol. 257, p. 33 – 'Molecular wires'.

H. Lang, R. Packheiser and B. Walfort (2006) *Organometallics*, vol. 25, p. 1836 – 'Organometallic π-tweezers, NCN pincers, and ferrocenes as molecular "Tinkertoys" in the synthesis of multiheterometallic transition-metal complexes'.

N.J. Long and C.K. Williams (2003) *Angew. Chem. Int. Ed.*, vol. 42, p. 2586 – 'Metal alkynyl σ complexes: synthesis and materials'.

N. Robertson and C.A. McGowan (2003) *Chem. Soc. Rev.*, vol. 32, p. 96 – 'A comparison of potential molecular wires as components for molecular electronics'.

W.-Y. Wong (2007) *Dalton Trans.*, p. 4495 – 'Luminescent organometallic poly(aryleneethynylene)s: functional properties towards implications in molecular optoelectronics'.

W.-Y. Wong and P.D. Harvey (2010) *Macromol. Rapid Comm.*, vol. 31, p. 671 – 'Recent progress on the photonic properties of conjugated organometallic polymers built upon the *trans*-bis(*para*-ethynylbenzene)bis(phosphine)platinum(II) chromophore and related derivatives'.

B. Xi and T. Ren (2009) *Compt. Rend. Chim.*, vol. 12, p. 321 – 'Wire-like diruthenium σ-alkynyl compounds and charge mobility therein'.

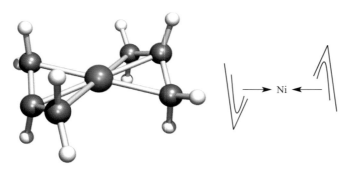

Fig. 24.21 The structure (neutron diffraction at 100 K) of $Ni(\eta^3-C_3H_5)_2$ and a schematic representation of the complex [R. Goddard *et al.* (1985) *Organometallics*, vol. 4, p. 285]. Colour code: Ni, green; C, grey; H, white.

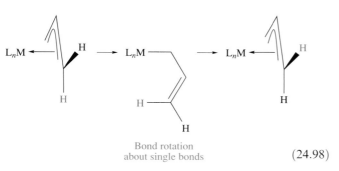

$$\text{Bond rotation about single bonds} \qquad (24.98)$$

Bis(allyl)nickel (Fig. 24.21) is one of the best known allyl complexes, but it is pyrophoric, and decomposes above 293 K. Bulky substituents can be used to stabilize analogues of $Ni(\eta^3-C_3H_5)_2$. Thus, $Ni\{\eta^3-1,3-(Me_3Si)_2C_3H_3\}_2$ is kinetically stable in air for up to several days, and decomposes only when heated above 373 K.

Buta-1,3-diene and related ligands

Photolysis of $Fe(CO)_5$ with buta-1,3-diene gives complex **24.59**, an orange liquid which loses CO at room temperature to give **24.60**, an air-stable yellow solid. The coordinated diene is difficult to hydrogenate, and does not undergo Diels–Alder reactions which are characteristic of conjugated dienes. Structural data for **24.60** confirm that the Fe atom is equidistant from each C atom of the ligand. Bonding schemes for the metal–ligand interaction were discussed in Section 24.2.

$$\underset{\text{Fe(CO)}_4}{\diagup} \qquad \underset{\substack{\text{Fe}\\(\text{CO})_3}}{\diagup}$$

(24.59)   **(24.60)**

Iron tricarbonyl complexes of 1,3-dienes (e.g. cyclohexa-1,3-diene) play an important role in organic synthesis. The complexes are stable under a variety of reaction conditions,

and iron carbonyls are inexpensive. The $Fe(CO)_3$ group acts as a protecting group for the diene functionality (e.g. against additions to the C=C bonds), allowing reactions to be carried out on other parts of the organic molecule as illustrated by reaction 24.99.

$$(24.99)$$

The presence of the $Fe(CO)_3$ group also permits reactions with *nucleophiles* to be carried out at the diene functionality with control of the stereochemistry, the nucleophile being able to attack only on the side of the coordinated diene *remote* from the metal centre. The organic ligand can be removed in the final step of the reaction.[†]

24.12 Carbene and carbyne complexes

In Section 24.7, we introduced carbene and carbyne complexes when we discussed α-hydrogen abstraction. Equations 24.45 and 24.46 exemplified methods of preparation. Carbenes can also be made by nucleophilic attack on a carbonyl C atom followed by alkylation (eq. 24.100).

$$W(CO)_6 \xrightarrow{MeLi} (OC)_5W=C\overset{O^-}{\underset{Me}{\diagdown}} \xrightarrow[\substack{e.g.\ R\ =\\ Me,\ Et}]{[R_3O]^+} (OC)_5W=C\overset{OR}{\underset{Me}{\diagdown}}$$

$$(24.100)$$

Compounds of the type formed in reactions such as 24.100 are called *Fischer-type carbenes*. They possess a low oxidation state metal, a heteroatom (O in this example) and an *electrophilic* carbene centre (i.e. subject to attack by nucleophiles, e.g. reaction 24.101). Resonance pair **24.61** gives a bonding description for a Fischer-type carbene complex.

$$(OC)_5W=C\overset{OR}{\underset{Me}{\diagdown}} \xrightarrow{PhLi} (OC)_5W=C\overset{Ph}{\underset{Me}{\diagdown}} \qquad (24.101)$$

[†] For detailed discussion of the use of metal carbonyl 1,3-diene complexes in organic synthesis, see: L.R. Cox and S.V. Ley (1998) *Chem. Soc. Rev.*, vol. 27, p. 301 – 'Tricarbonyl complexes: an approach to acyclic stereocontrol'; W.A. Donaldson and S. Chaudhury (2009) *Eur. J. Org. Chem.*, p. 3831 – 'Recent applications of acyclic (diene)iron complexes and (dienyl)iron cations in organic synthesis'.

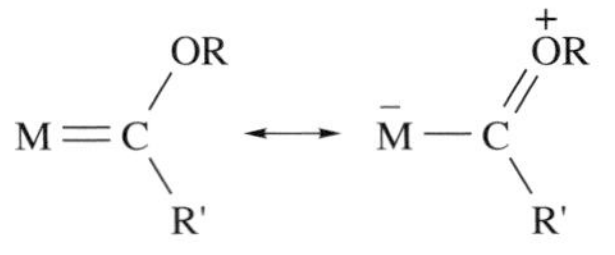

(24.61) 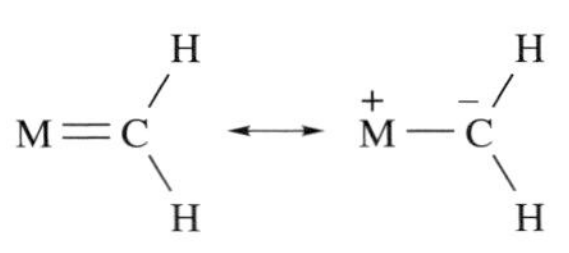   **(24.62)**

In contrast, *Schrock-type carbenes* are made by reactions such as 24.45, contain an early *d*-block metal in a high oxidation state, and show nucleophilic character (i.e. susceptible to attack by electrophiles, e.g. reaction 24.102). Resonance pair **24.62** describes a Schrock-type carbene complex.

$$(\eta^5\text{-Cp})_2\text{MeTa}=\text{CH}_2 + \text{AlMe}_3$$
$$\longrightarrow (\eta^5\text{-Cp})_2\text{Me}\overset{+}{\text{Ta}}-\text{CH}_2\overset{-}{\text{Al}}\text{Me}_3 \qquad (24.102)$$

The $M-C_{carbene}$ bonds in both Fischer- and Schrock-type complexes are *longer* than typical $M-C_{CO(term)}$ bonds, but *shorter* than typical $M-C$ single bonds, e.g. in$(OC)_5Cr=C(OMe)Ph$, $Cr-C_{carbene} = 204$ pm and $Cr-C_{CO} = 188$ pm. This implies some degree of $(d-p)$ π-character as indicated by resonance structures **24.61** and **24.62**. The π-system can be extended to the heteroatom in the Fischer-type system as shown in diagram **24.63**.

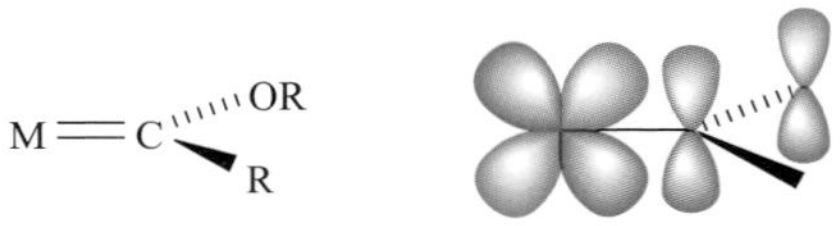

(24.63)

N-Heterocyclic carbenes (**24.64**) have become important ligands in organometallic chemistry, notable examples being Grubbs' second and third generation catalysts. Synthetic routes to *N*-heterocyclic carbene complexes (which are stable to air) typically involve salt metathesis or elimination reactions. Reaction 24.103 shows the use of a 1,3-dialkylimidazolium salt as a precursor, and scheme 24.104 illustrates conversions relevant to Grubbs' catalysts (see next page).

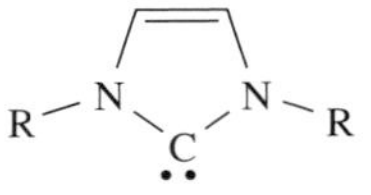

(24.64)

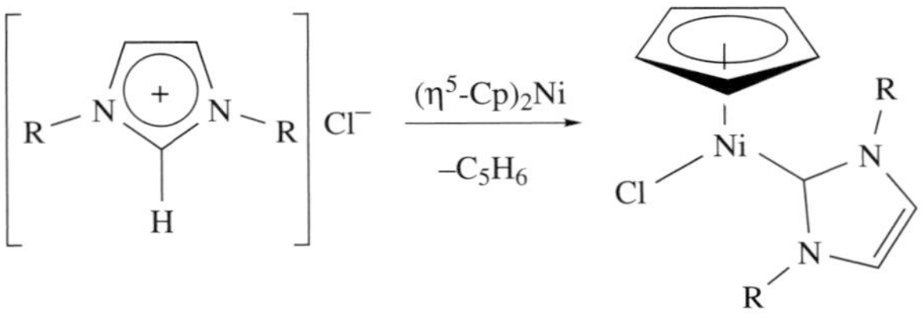

(24.103)

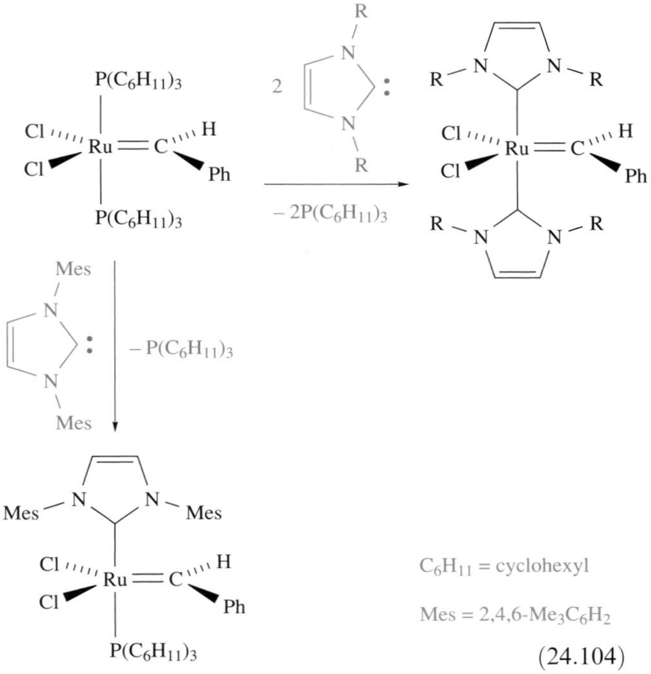

C_6H_{11} = cyclohexyl

Mes = 2,4,6-Me$_3$C$_6$H$_2$

(24.104)

N-Heterocyclic carbenes are good σ-donors. The $M-C_{carbene}$ bond distances (typically >210 pm) are longer than those in Fischer- or Schrock-type carbene complexes. This implies that the metal-to-carbene back-bonding that is characteristic of Fischer- and Schrock-type carbenes is not as important in *N*-heterocyclic carbene complexes.

One route to a carbyne (alkylidyne) complex is reaction 24.46. Equation 24.105 illustrates the initial method of Fischer. The abstraction of an α-H atom from a Schrock-type carbene yields the corresponding carbyne complex (eq. 24.106).

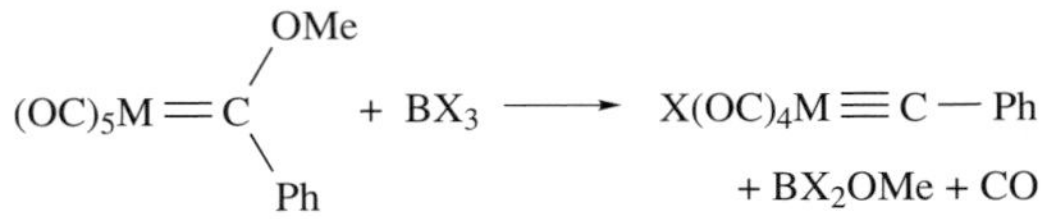

(24.105)

M = Cr, Mo, W   X = Cl, Br, I

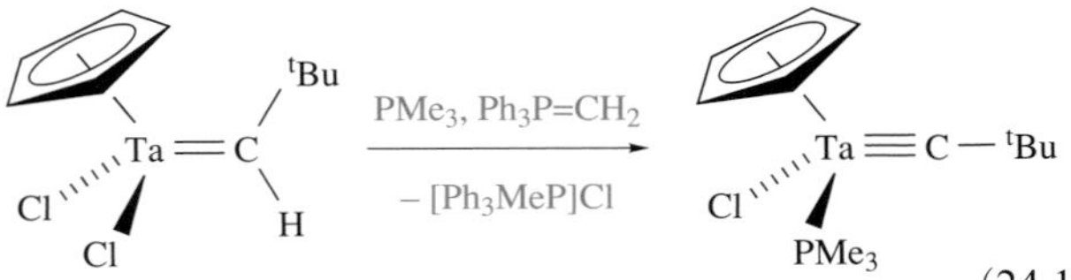

(24.106)

An $M-C_{carbyne}$ bond is typically shorter than an $M-C_{CO(term)}$ bond, e.g. structure **24.65**. The multiple bonding can be considered in terms of an *sp* hybridized $C_{carbyne}$ atom, with one $M-C$ σ-interaction (using an sp_z hybrid) and two π-interactions (using the $C_{carbyne}$ $2p_x$ and $2p_y$ atomic orbitals overlapping with the metal d_{xz} and d_{yz} atomic orbitals).

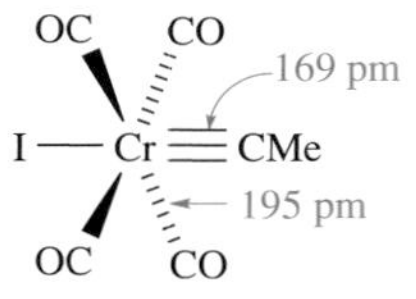

(24.65)

Alkylidyne (carbyne) complexes containing μ_3-CR groups interacting with a triangle of metal atoms include $Co_3(CO)_9(\mu_3\text{-CMe})$ and $H_3Ru_3(CO)_9(\mu_3\text{-CMe})$, and we considered the bonding in such compounds in terms of the isolobal principle in Section 24.5. Reactions 24.107 and 24.108 illustrate methods of introducing μ_3-CR groups into clusters. The precursor in reaction 24.108 is unsaturated and contains an Os=Os bond which undergoes additions. The intermediate in this reaction contains a bridging carbene group which undergoes oxidative addition of a C−H bond on heating.

$$Co_2(CO)_8 \xrightarrow{\text{MeCCl}_3} Co_3(CO)_9CMe \qquad (24.107)$$

$$H_2Os_3(CO)_{10} \xrightarrow[-N_2]{\text{CH}_2\text{N}_2} H_2Os_3(CO)_{10}(\mu\text{-CH}_2)$$

$$\xrightarrow[-CO]{\Delta} H_3Os_3(CO)_9(\mu_3\text{-CH}) \qquad (24.108)$$

In mononuclear carbyne complexes, the M≡C bond undergoes addition reactions, e.g. addition of HCl and alkynes.

Structures **24.66** (Grubbs' catalyst), **24.67** (Grubbs' second generation catalyst), **24.68** (Grubbs' third-generation catalyst) and **24.69** (Schrock-type complex) show three important carbene compounds that are used as catalysts in *alkene* (*olefin*) *metathesis*, i.e. metal-catalysed reactions in which C=C bonds are redistributed.[†] Examples include ring-opening metathesis polymerization (ROMP) and ring-closing metathesis (RCM). The importance of alkene metathesis (see Section 25.3) was acknowledged by the award of the 2005 Nobel Prize in Chemistry to Robert H. Grubbs, Richard R. Schrock and Yves Chauvin 'for the development of the metathesis method in organic synthesis'.

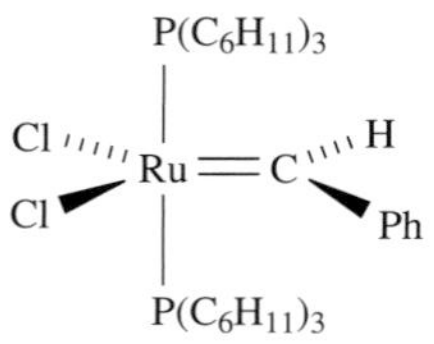

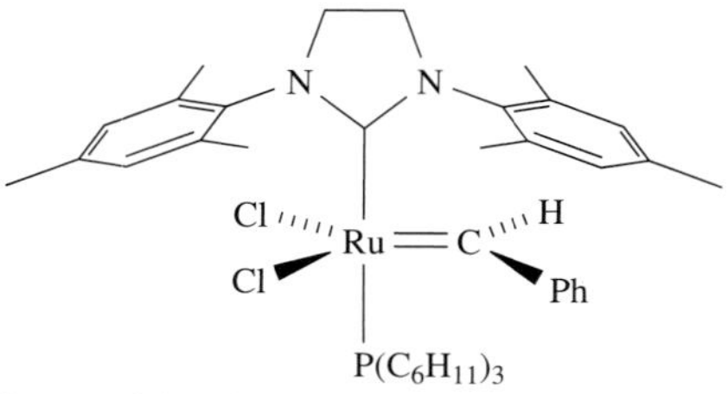

C_6H_{11} = cyclohexyl

Grubbs' catalyst

(24.66)

Grubbs' second generation catalyst

(24.67)

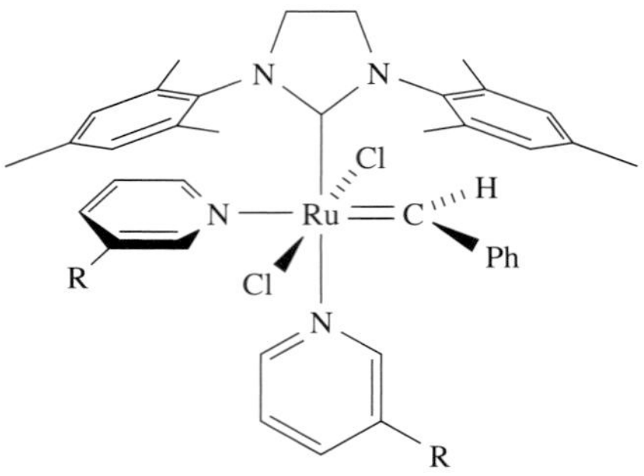

R = H or Br
Grubbs' third generation catalyst

(24.68)

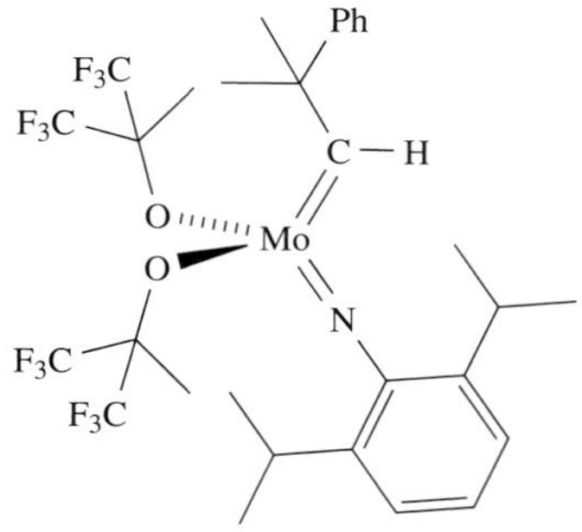

Schrock-type complex

(24.69)

24.13 Complexes containing η^5-cyclopentadienyl ligands

The cyclopentadienyl ligand was discussed in Chapter 23 and in Sections 24.1 and 24.2. Now we look at examples of some of its more important *d*-block metal complexes.

In a *sandwich complex*, the metal centre lies between two π-bonded hydrocarbon (or derivative) ligands. Complexes of the type $(\eta^5\text{-Cp})_2M$ are called *metallocenes*.

Ferrocene and other metallocenes

The best-known cyclopentadienyl complex is the *sandwich compound* ferrocene, $(\eta^5\text{-Cp})_2Fe$. It is a diamagnetic, orange solid (mp 446 K) which obeys the 18-electron rule (structure **24.33**). In the gas phase, the two cyclopentadienyl rings are *eclipsed* (**24.70**) but the solid exists in several phases in which the rings are co-parallel but in different orientations.

[†] This definition is taken from: T.M. Trnka and R.H. Grubbs (2001) *Acc. Chem. Res.*, vol. 34, p. 18 – 'The development of L_2X_2Ru=CHR olefin metathesis catalysts: an organometallic success story'.

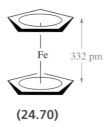

(24.70)

Solving the structure has been hampered by disorder problems. The barrier to rotation of the two rings is low and at 298 K, there is motion even in the solid state. In derivatives of ferrocene with substituents on the Cp rings, the barrier to rotation is higher, and in $(\eta^5\text{-}C_5Me_5)_2Fe$, the two C_5 rings are staggered in both the gas and solid states. The bonding in $(\eta^5\text{-Cp})_2Fe$ can be described in terms of the interactions between the π-MOs of the ligands (see Box 23.1) and the metal $3d$ atomic orbitals (see end-of-chapter problem 24.26). Ferrocene is oxidized (e.g. by I_2 or

$FeCl_3$) to the paramagnetic, blue ferrocenium ion, $[(\eta^5\text{-Cp})_2Fe]^+$. Equation 24.109 gives E° relative to the standard hydrogen electrode, but the Fc^+/Fc couple is commonly used as a convenient internal, secondary reference electrode (i.e. E° is defined as 0 V for reference purposes, see Box 8.2).

$$[(\eta^5\text{-Cp})_2Fe]^+ + e^- \rightleftharpoons (\eta^5\text{-Cp})_2Fe$$

$$\underset{Fc^+}{} \qquad \underset{Fc}{}$$

$$E^\circ = +0.40\,V \qquad (24.109)$$

Metallocenes of the first row metals are known for V(II), Cr(II), Mn(II), Fe(II), Co(II), Ni(II) and Zn(II). Reaction 24.110 is a general synthesis for all except $(\eta^5\text{-Cp})_2V$ (where the starting chloride is VCl_3) and $(\eta^5\text{-Cp})_2Zn$ (prepared by reaction 24.111). Reaction 24.112 gives an alternative synthesis for ferrocene and nickelocene. Titanocene, $(\eta^5\text{-Cp})_2Ti$, is a 14-electron, paramagnetic species. It is highly reactive and has not been isolated, although it can be made *in situ* by treating $(\eta^5\text{-Cp})_2TiCl_2$ with Mg.

BIOLOGY AND MEDICINE

Box 24.4 Biosensors for glucose oxidase

The iron-centred redox properties of a ferrocene derivative similar to that shown below are the basis of the ExacTech™ and Precision QID™ blood glucose meters manufactured by Medisense Inc. The pen-like construction of the ExacTech™ meter is shown in the photograph and its function is to measure glucose levels in people with diabetes. The iron centre facilitates electron transfer between glucose and glucose oxidase, and a glucose level reading is obtained in about 30 seconds. The lifetime of one pen-meter is about 4000 readings, and one advantage of the design is its ease of use, making it particularly suitable for use by children with diabetes.

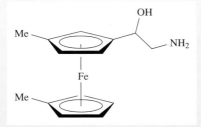

Blood glucose being tested using an ExacTech™. A drop of blood is placed in a chemically coated strip which activates the digital display of the sensor.

Further reading

N.J. Forrow and S.J. Walters (2004) *Biosens. Bioelectron.*, vol. 19, p. 763 – 'Transition metal half-sandwich complexes as redox mediators to glucose oxidase'.

M.J. Green and H.A.O. Hill (1986) *J. Chem. Soc., Faraday Trans. 1*, vol. 82, p. 1237.

H.A.O. Hill (1993) 'Bioelectrochemistry: making use of the electrochemical behaviour of proteins', *NATO ASI Ser., Ser. C*, vol. 385, p. 133.

Sterically demanding substituents are needed to stabilize the system, e.g. $(\eta^5\text{-}C_5H_4SiMe_3)_2Ti$ (scheme 24.113).

$$MCl_2 + 2Na[Cp] \longrightarrow (\eta^5\text{-}Cp)_2M + 2NaCl \qquad (24.110)$$

$$Zn\{N(SiMe_3)_2\}_2 + 2C_6H_5 \xrightarrow{Et_2O,\ 298\ K} (\eta^5\text{-}Cp)_2Zn \\ + 2(Me_3Si)_2NH \qquad (24.111)$$

$$MCl_2 + 2C_5H_6 + 2Et_2NH \longrightarrow (\eta^5\text{-}Cp)_2M + 2[Et_2NH_2]Cl \\ (M = Fe, Ni) \qquad (24.112)$$

$$2Li[C_5H_4SiMe_3] + TiCl_3 \xrightarrow[-2LiCl]{} (\eta^5\text{-}C_5H_4SiMe_3)_2TiCl \\ \xrightarrow[-NaCl]{Na/Hg} (\eta^5\text{-}C_5H_4SiMe_3)_2Ti$$
$$(24.113)$$

The complexes $(\eta^5\text{-}Cp)_2V$ (air-sensitive, violet solid), $(\eta^5\text{-}Cp)_2Cr$ (air-sensitive, red solid), $(\eta^5\text{-}Cp)_2Mn$ (brown solid, pyrophoric when finely divided), $(\eta^5\text{-}Cp)_2Co$ (very air-sensitive, black solid) and $(\eta^5\text{-}Cp)_2Ni$ (green solid) are paramagnetic; $(\eta^5\text{-}Cp)_2Cr$ and $(\eta^5\text{-}Cp)_2Ni$ have two unpaired electrons. The 19-electron complex $(\eta^5\text{-}Cp)_2Co$ is readily oxidized to the 18-electron $[(\eta^5\text{-}Cp)_2Co]^+$, yellow salts of which are air-stable. Nickelocene is a 20-electron complex and in its reactions (Fig. 24.22) often relieves this situation, forming 18-electron complexes. The 19-electron cation $[(\eta^5\text{-}Cp)_2Ni]^+$ forms when $[(\eta^5\text{-}Cp)_2Ni]$ reacts with $[H(OEt_2)_2][B(3,5\text{-}(CF_3)_2C_6H_3)_4]$ (*Brookhart's acid*). In both crystalline $[(\eta^5\text{-}Cp)_2Ni]$ and $[(\eta^5\text{-}Cp)_2Ni]^+[B(3,5\text{-}(CF_3)_2C_6H_3)_4]^-$, the cyclopentadienyl rings are mutually eclipsed. Manganocene, unlike the other metallocenes, is dimorphic. The room-temperature form is polymeric and structurally similar to $(\eta^5\text{-}Cp)_2Pb$ (Fig. 23.20c), while the high-temperature form is structurally related to ferrocene. Zincocene (air-sensitive, colourless solid) is diamagnetic and, in the solid state, is structurally similar to $(\eta^5\text{-}C_5H_5)_2Pb$ (Fig. 23.20c). The compound $(\eta^5\text{-}C_5Me_5)_2Zn_2$ was (in 2004) the first example of a *dimetallocene*. It formally contains a $\{Zn_2\}^{2+}$ core, and is made by reaction 24.114. The ratio of products depends on reaction conditions, but when carried out in Et_2O at 263 K, the reaction gives $(\eta^5\text{-}C_5Me_5)_2Zn_2$ (Fig. 24.23) as the dominant product. The Zn–Zn bond length of 230.5 pm is consistent with a metal–metal bonding interaction (see also reaction 21.116).

$$(\eta^5\text{-}C_5Me_5)_2Zn + Et_2Zn \longrightarrow (\eta^5\text{-}C_5Me_5)_2Zn_2 \\ + (\eta^5\text{-}C_5Me_5)ZnEt$$
$$(24.114)$$

The chemistry of ferrocene dominates that of the other metallocenes. It is commercially available and large numbers of derivatives are known. The rings in

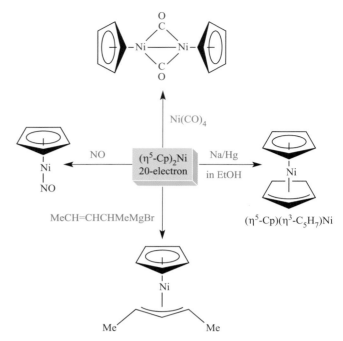

Fig. 24.22 Selected reactions of nickelocene, $(\eta^5\text{-}Cp)_2Ni$.

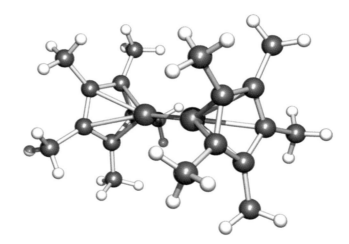

Fig. 24.23 The solid state structure (X-ray diffraction at 173 K) of $(\eta^5\text{-}C_5Me_5)_2Zn_2$ [I. Resa *et al.* (2004) *Science*, vol. 305, p. 1136]. The Zn–Zn distance is 230.5 pm. Colour code: Zn, brown; C, grey; H, white.

$(\eta^5\text{-}Cp)_2Fe$ possess aromatic character, and selected reactions are shown in Fig. 24.24. Protonation occurs at the Fe(II) centre and this is indicated by the appearance of a signal at $\delta -2.1$ ppm in the ^1H NMR spectrum of $[(\eta^5\text{-}Cp)_2FeH]^+$. The regioselective double deprotonation of each cyclopentadienyl ring in ferrocene occurs when $(\eta^5\text{-}Cp)_2Fe$ is treated with an appropriate 'superbase'. Such a base is produced by combining certain lithium or heavier group 1 metal amides with certain magnesium amides in the presence of a Lewis base cosolvent. This combination of reactants provides both a deprotonating agent and a so-called *inverse crown ether* to stabilize the

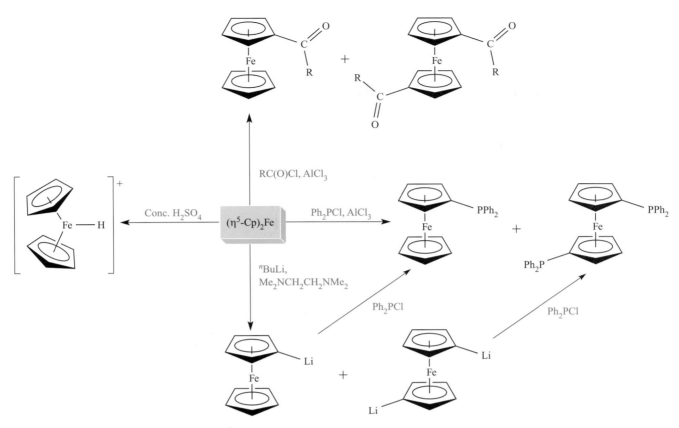

Fig. 24.24 Selected reactions of ferrocene, $(\eta^5\text{-Cp})_2\text{Fe}$.

deprotonated species. The term *inverse crown ether* arises from the fact that each Lewis base donor of a conventional crown ether is replaced by a Lewis acid (in this case, Na^+ and Mg^{2+}).[†] In reaction 24.115, the $^i\text{Pr}_2\text{N}^-$ anions formally remove H^+ from $(\eta^5\text{-Cp})_2\text{Fe}$, while the *s*-block metal ions and N atoms form the inverse crown host which stabilizes the $[(\eta^5\text{C}_5\text{H}_3)_2\text{Fe}]^{4-}$ ion (Fig. 24.25).

$$(\eta^5\text{-C}_5\text{H}_5)_2\text{Fe} + 8^i\text{Pr}_2\text{NH} + 4\text{BuNa} + 8\text{Bu}_2\text{Mg} \longrightarrow$$

$$\{(\eta^5\text{-C}_5\text{H}_3)_2\text{Fe}\}\text{Na}_4\text{Mg}_4(\text{N}^i\text{Pr}_2)_8 + 12\text{BuH} \qquad (24.115)$$

The exchange of an η^5-Cp ring for an η^6-arene ligand (eq. 24.116) is accompanied by a change in overall charge so that the Fe atom remains an 18-electron centre.

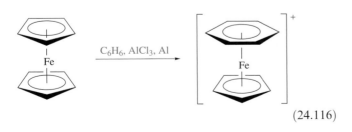

$$(24.116)$$

[†] For a review of inverse crown chemistry, see: R.E. Mulvey (2001) *Chem. Commun.*, p. 1049.

Fig. 24.25 The structure of $\{(\eta^5\text{-C}_5\text{H}_3)_2\text{Fe}\}\text{Na}_4\text{Mg}_4(\text{N}^i\text{Pr}_2)_8$ (X-ray diffraction at 160 K) in which the $[(\eta^5\text{-C}_5\text{H}_3)_2\text{Fe}]^{4-}$ anion is stabilized within the $[\text{Na}_4\text{Mg}_4(\text{N}^i\text{Pr}_2)_8]^{4+}$ host [P.C. Andrikopoulos *et al.* (2004) *J. Am. Chem. Soc.*, vol. 126, p. 11612]. Colour code: Fe, green; C, grey; H, white; N, blue; Na, purple; Mg, yellow.

The reaction of $(\eta^5\text{-Cp})(\eta^5\text{-C}_5\text{H}_4\text{I})\text{Fe}$ with $^n\text{BuLi}$ followed by treatment with half an equivalent of ZnBr_2 gave compound **A**. This was immediately reacted with hexaiodobenzene to yield **B**, the mass spectrum of which showed a parent ion at $m/z = 1182$. The solution ^1H NMR spectrum of **B** exhibited three signals: δ/ppm 4.35 (m, 12H), 4.32 (m, 12H), 3.99 (s, 30 H). In the $^{13}\text{C}\{^1\text{H}\}$ NMR spectrum of **B**, five signals were observed. Deduce the structure of **B**, and rationalize the observed NMR spectroscopic data. Suggest a structure for compound **A**.

[*Ans.* See Y.Yu *et al.* (2006) *Chem. Commun.*, p. 2572]

$(\eta^5\text{-Cp})_2\text{Fe}_2(\text{CO})_4$ and derivatives

Reactions between metal carbonyls and cyclopentadiene usually yield mixed ligand complexes, e.g. Fe(CO)_5 reacts with C_5H_6 to give $(\eta^5\text{-Cp})_2\text{Fe}_2(\text{CO})_4$. Two isomers of $(\eta^5\text{-Cp})_2\text{Fe}_2(\text{CO})_4$ exist, *cis* (**24.71**) and *trans* (**24.72**), and both have been confirmed in the solid state. The Fe−Fe bond length (253 pm) is consistent with a single bond giving each Fe centre 18 electrons.

cis-isomer	*trans*-isomer
(24.71)	**(24.72)**

In solution at 298 K, both the *cis*- and *trans*-forms are present and the terminal and bridging ligands exchange by an intramolecular process. Above 308 K, *cis* ⟶ *trans* isomerism occurs, probably through an unbridged intermediate. The *cis*-isomer can be obtained by crystallization at low temperatures.

The dimer $(\eta^5\text{-Cp})_2\text{Fe}_2(\text{CO})_4$ is commercially available and is a valuable starting material in organometallic chemistry. Reactions with Na or halogens cleave the Fe−Fe bond (eqs. 24.117 and 24.118) giving useful organometallic reagents, reactions of which are exemplified in eqs. 24.119−24.122.

$$(\eta^5\text{-Cp})_2\text{Fe}_2(\text{CO})_4 + 2\text{Na} \longrightarrow 2\text{Na}[(\eta^5\text{-Cp})\text{Fe(CO)}_2]$$

$$(24.117)$$

$$(\eta^5\text{-Cp})_2\text{Fe}_2(\text{CO})_4 + \text{X}_2 \longrightarrow 2(\eta^5\text{-Cp})\text{Fe(CO)}_2\text{X}$$

$$\text{X} = \text{Cl}, \text{Br}, \text{I} \quad (24.118)$$

$$\text{Na}[(\eta^5\text{-Cp})\text{Fe(CO)}_2] + \text{RCl} \longrightarrow (\eta^5\text{-Cp})\text{Fe(CO)}_2\text{R} + \text{NaCl}$$

$$\text{e.g. R} = \text{alkyl} \quad (24.119)$$

$$(24.120)$$

$$(24.121)$$

$$(24.122)$$

1. Give two synthetic routes to ferrocene.

[*Ans.* See eqs. 24.110 and 24.112]

2. Explain what is meant in the main text by the *regioselective* deprotonation of ferrocene to give $[(\eta^5\text{-C}_5\text{H}_3)_2\text{Fe}]^{4-}$.

3. At 295 K, the IR spectrum of a solution of $(\eta^5\text{-C}_5\text{H}_4{}^i\text{Pr})_2\text{Ti}$ shows absorptions arising from C–H vibrational modes. When the sample is cooled to 195 K under an atmosphere of N_2, new absorptions at 1986 and 2090 cm^{-1} appear. These bands disappear when the sample is warmed to 295 K. Rationalize these observations.

[*Ans.* See T.E. Hanna *et al.* (2004) *J. Am. Chem. Soc.*, vol. 126, p. 14688]

APPLICATIONS

Box 24.5 Enantioselectivity in the preparation of the herbicide *S*-Metolachlor

Each year, Novartis manufactures ≈10 000 tonnes of the herbicide *S*-Metolachlor, the synthesis of which is shown in the scheme below. It is used to control the growth of grass and broadleaved weeds among agricultural crops. The (*R*)-enantiomer of Metolachlor is inactive as a herbicide, and so if the chemical is applied as a racemate, half of the applied chemical is ineffective. The key to enantioselectivity is the first step of *asymmetric hydrogenation*.

80% enantiomeric excess

S-Metolachlor

The hydrogenation step is catalysed by an iridium(I) complex containing the chiral ferrocenyl bisphosphane ligand shown below. The ligand coordinates to the Ir(I) centre through the two *P*-donor atoms, and the complete catalyst system comprises Ir(I), the ferrocenyl ligand, I⁻ and H_2SO_4. The 80% enantiomeric excess (ee) is not as high as would be required for, say, chiral drug synthesis, but is adequate for the production of a herbicide. Chiral catalysts play a vital role in directing asymmetric syntheses, and the % ee is highly sensitive to the choice of chiral ligand; 'ee' is explained in Section 25.5. The manufacture of *S*-Metolachlor provides an example of an industrial application of one particular chiral ferrocenyl bisphosphane ligand.

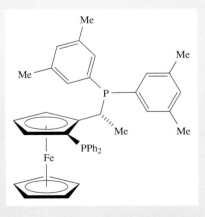

Herbicide application to a field of mid-growth cotton plants in Arkansas, US.

Further reading

There is more about asymmetric syntheses in Section 25.5.

T.J. Colacot and N.S. Hosmane (2005) *Z. Anorg. Allg. Chem.*, vol. 631, p. 2659 – 'Organometallic sandwich compounds in homogeneous catalysis: an overview'.

D.L. Lewis *et al.* (1999) *Nature*, vol. 401, p. 898 – 'Influence of environmental changes on degradation of chiral pollutants in soils'.

A. Togni (1996) *Angew. Chem. Int. Ed.*, vol. 35, p. 1475 – 'Planar–chiral ferrocenes: synthetic methods and applications'.

Box 24.6 Zirconocene derivatives as catalysts

The development of Ziegler–Natta-type catalysts (see Section 25.8) has, since the 1980s, included the use of zirconocene derivatives. In the presence of methylaluminoxane $[MeAl(\mu\text{-}O)]_n$ as a co-catalyst, compounds **A**, **B** and **C** (shown below) are active catalysts for propene polymerization. Compounds **A** and **B** are chiral because of the relative orientations of the two halves of the organic ligand. A racemic mixture of **A** facilitates the formation of *isotactic* polypropene, while use of catalyst **C** results in *syndiotactic* polypropene (see Section 25.8 for definitions of syndiotactic, isotactic and atactic). If

$(\eta^5\text{-}Cp)_2ZrCl_2$ is used, *atactic* polypropene is produced. Such catalysts are commercially available. The active species in the catalyst system is a cation of the general type $[Cp_2ZrR]^+$, and such cations are used directly as polymerization catalysts. Formed by protonolysis, oxidative Zr–R cleavage, or abstraction of R^- from Cp_2ZrR_2 (e.g. R = Me), $[Cp_2ZrR]^+$ reagents are active catalysts *without* the need for addition of the methylaluminoxane co-catalyst.

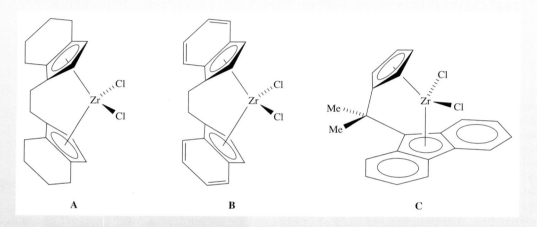

A   **B**   **C**

Zirconocene derivatives are used to catalyse a range of organic hydrogenation and C–C bond-forming reactions. In the presence of methylaluminoxane, chiral complex **A** catalyses asymmetric hydrogenations (see Section 25.5), with the active species being a cationic zirconium hydrido complex.

Further reading

Comprehensive Organometallic Chemistry III (2007), eds R.H. Crabtree and D.M.P. Mingos, Elsevier, Oxford, vol. 4, chapter 4.09, p. 1005.
Encyclopedia of Reagents in Organic Synthesis (1995), ed. L.A. Paquette, Wiley, Chichester, vol. 4, p. 2445.

24.14 Complexes containing η^6- and η^7-ligands

η^6-Arene ligands

Arenes such as benzene and toluene can act as 6π-electron donors as illustrated in eqs. 24.116 and 24.122. A wide range of arene complexes exist, and sandwich complexes can be made by co-condensation of metal and arene vapours (eq. 24.123) or by reaction 24.124.

$$Cr(g) + 2C_6H_6(g) \xrightarrow[\text{warm to 298 K}]{\text{co-condense on to surface at 77 K;}}$$

$$(\eta^6\text{-}C_6H_6)_2Cr \qquad (24.123)$$

$$\left.\begin{array}{l} 3CrCl_3 + 2Al + AlCl_3 + 6C_6H_6 \\ \quad \longrightarrow 3[(\eta^6\text{-}C_6H_6)_2Cr]^+[AlCl_4]^- \\ 2[(\eta^6\text{-}C_6H_6)_2Cr]^+ + 4[OH]^- + [S_2O_4]^{2-} \\ \quad \longrightarrow 2(\eta^6\text{-}C_6H_6)_2Cr + 2H_2O + 2[SO_3]^{2-} \end{array}\right\} \quad (24.124)$$

The group 6 metals form air-sensitive 18-electron complexes $(\eta^6\text{-}C_6H_6)_2M$ (M = Cr, Mo, W). In the solid state, the two benzene rings in $(\eta^6\text{-}C_6H_6)_2Cr$ are eclipsed (**24.73**). The C–C bonds are equal in length (142 pm) and slightly longer than in free benzene (140 pm). The bonding can be described in terms of the interaction

E_{MO}

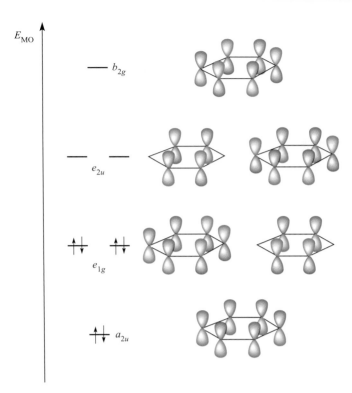

Fig. 24.26 The π-molecular orbitals of C_6H_6; the energy scale is arbitrary. The symmetry labels apply to D_{6h} C_6H_6; these labels are not applicable to the ligand in a complex of other symmetry.

between the π-MOs of the ligands (Fig. 24.26) and the metal $3d$ atomic orbitals, with the occupied ligand π-MOs acting as donors and the vacant MOs functioning as acceptors.

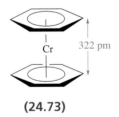

322 pm

(24.73)

Surprisingly, the brown Cr complex is easily oxidized by I_2 to the 17-electron, air-stable yellow $[(\eta^6\text{-}C_6H_6)_2Cr]^+$. The ease of oxidation precludes $(\eta^6\text{-}C_6H_6)_2Cr$ from undergoing electrophilic substitution reactions. Electrophiles oxidize $(\eta^6\text{-}C_6H_6)_2Cr$ to $[(\eta^6\text{-}C_6H_6)_2Cr]^+$ which does not react further. The lithiated derivative $(\eta^6\text{-}C_6H_5Li)_2Cr$ can be made by reaction of $(\eta^6\text{-}C_6H_6)_2Cr$ with ^nBuLi (compare with the lithiation of ferrocene, Fig. 24.24) and is a precursor to other derivatives.

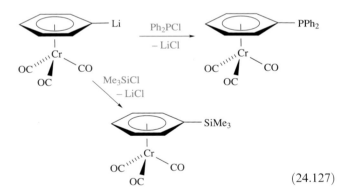

(24.74)

The reaction of $Cr(CO)_6$ or $Cr(CO)_3(NCMe)_3$ with benzene gives the *half-sandwich complex* $(\eta^6\text{-}C_6H_6)Cr(CO)_3$ **(24.74)**, and related complexes can be made similarly. The $Cr(CO)_3$ unit in $(\eta^6\text{-arene})Cr(CO)_3$ complexes withdraws electrons from the arene ligand making it *less* susceptible to electrophilic attack than the free arene, but *more* susceptible to attack by nucleophiles (reaction 24.125).

$(\eta^6\text{-}C_6H_5Cl)Cr(CO)_3 + NaOMe$

$\longrightarrow (\eta^6\text{-}C_6H_5OMe)Cr(CO)_3 + NaCl \qquad (24.125)$

As in $(\eta^6\text{-}C_6H_6)_2Cr$, the benzene ligand in $(\eta^6\text{-}C_6H_6)Cr(CO)_3$ can be lithiated (eq. 24.126) and then derivatized (scheme 24.127). The reactivity of half-sandwich complexes is not confined to sites within the π-bonded ligand: eq. 24.128 illustrates substitution of a CO ligand for PPh_3.

$(\eta^6\text{-}C_6H_6)Cr(CO)_3$

$\xrightarrow[\text{$-^nBuH$}]{^nBuLi, TMEDA} (\eta^6\text{-}C_6H_5Li)Cr(CO)_3 \qquad (24.126)$

(24.127)

$(\eta^6\text{-}C_6H_6)Cr(CO)_3 + PPh_3$

$\xrightarrow{h\nu} (\eta^6\text{-}C_6H_6)Cr(CO)_2(PPh_3) + CO \qquad (24.128)$

Cycloheptatriene and derived ligands

Cycloheptatriene (**24.75**) can act as a 6π-electron donor, and in its reaction with $Mo(CO)_6$, it forms $(\eta^6\text{-}C_7H_8)Mo(CO)_3$. The solid state structure of this complex (Fig. 24.27a) confirms that the ligand coordinates as a triene, the ring

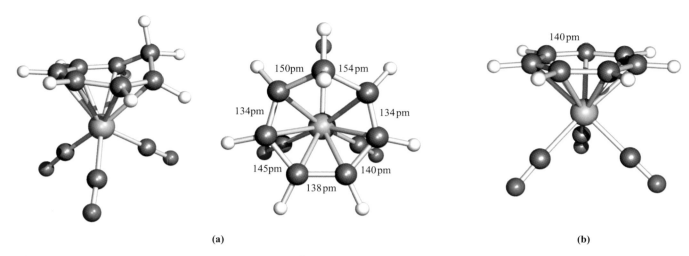

Fig. 24.27 The structures (X-ray diffraction) of (a) $[(\eta^6\text{-}C_7H_8)Mo(CO)_3]$ [J.D. Dunitz *et al.* (1960) *Helv. Chim. Acta*, vol. 43, p. 2188] and (b) $[(\eta^7\text{-}C_7H_7)Mo(CO)_3]^+$ in the $[BF_4]^-$ salt [G.R. Clark *et al.* (1973) *J. Organomet. Chem.*, vol. 50, p. 185]. Colour code: Mo, orange; C, grey; O, red; H, white.

being folded with the CH_2 group bent away from the metal centre. Reaction 24.129 shows the abstraction of H^- from coordinated $\eta^6\text{-}C_7H_8$ to give the planar $[\eta^7\text{-}C_7H_7]^+$ ion, **24.76** (the cycloheptatrienylium cation),[†] which has an aromatic π-system and retains the ability of cycloheptatriene to act as a 6π-electron donor.

(24.75)   **(24.76)**

$$\text{(24.129)}$$

The planarity of the $[C_7H_7]^+$ ligand has been confirmed in the structure of $[(\eta^7\text{-}C_7H_7)Mo(CO)_3]^+$ (Fig. 24.27b). All the C–C bond lengths are close to 140 ppm in contrast to the variation observed in $(\eta^6\text{-}C_7H_8)Mo(CO)_3$ (Fig. 24.27a).

In the complex $(\eta^4\text{-}C_7H_8)Fe(CO)_3$, cycloheptatriene acts as a diene, giving the Fe(0) centre its required 18 electrons. Equation 24.130 shows that deprotonation generates a coordinated $[C_7H_7]^-$ ligand which bonds in an η^3 manner, allowing the metal to retain 18 electrons. At room temperature, the $[C_7H_7]^-$ ligand is fluxional, and on the NMR timescale, the $Fe(CO)_3$ unit effectively 'visits' every carbon atom.

[†] The non-systematic name for the cycloheptatrienylium cation is the *tropylium cation*.

$$\text{(24.130)}$$

In $[C_7Me_7][BF_4]$, the cation is *non-planar* as a result of steric hindrance between the methyl groups. The introduction of methyl substituents affects the way in which $[C_7R_7]^+$ (R = H or Me) coordinates to a metal centre. Schemes 24.131 and 24.132 show two related reactions that lead to different types of products. The C_7-ring adopts an η^7-mode in the absence of steric crowding, and an η^5-mode when the methyl groups are sterically congested. The differing numbers of EtCN or CO ligands in the products in the two schemes are consistent with the W centre satisfying the 18-electron rule.

$$\text{(24.131)}$$

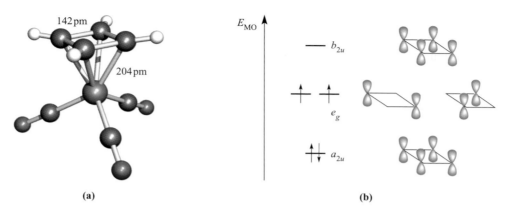

Fig. 24.28 (a) The structure (X-ray diffraction) of $(\eta^4\text{-}C_4H_4)Fe(CO)_3$ [P.D. Harvey *et al.* (1988) *Inorg. Chem.*, vol. 27, p. 57]. (b) The π-molecular orbitals of C_4H_4 in which the ligand geometry is as in its complexes, i.e. a *square* C_4 framework; the symmetry labels apply to D_{4h} C_4H_4; these labels are not applicable to the ligand in a complex of other symmetry. Colour code: Fe, green; C, grey; O, red; H, white.

A C_4H_4 ligand with the geometry found in its complexes, i.e. a *square* C_4 framework, has the π-MOs shown in Fig. 24.28b and is *paramagnetic*. However, $(\eta^4\text{-}C_4H_4)$ $Fe(CO)_3$ is *diamagnetic* and this provides evidence for pairing of electrons between ligand and metal: a C_{3v} $Fe(CO)_3$ fragment also has two unpaired electrons (Fig. 24.14).

Cyclobutadiene complexes can also be formed by the cycloaddition of alkynes as in reaction 24.134.

$$2PdCl_2(NCPh)_2 + 4PhC{\equiv}CPh$$

$$\longrightarrow (\eta^4\text{-}C_4Ph_4)ClPd(\mu\text{-}Cl)_2PdCl(\eta^4\text{-}C_4Ph_4) \qquad (24.134)$$

In its reactions, *coordinated* cyclobutadiene exhibits aromatic character, undergoing electrophilic substitution, e.g. Friedel–Crafts acylation. A synthetic application of $(\eta^4\text{-}C_4H_4)Fe(CO)_3$ in organic chemistry is as a stable source of cyclobutadiene. Oxidation of the complex releases the ligand, making it available for reaction with, for example, alkynes as in scheme 24.135.

(24.132)

24.15 Complexes containing the η^4-cyclobutadiene ligand

Cyclobutadiene, C_4H_4, is anti-aromatic (i.e. it does not have $4n + 2$ π-electrons) and readily polymerizes. However, it can be stabilized by coordination to a low oxidation state metal centre. Yellow crystalline $(\eta^4\text{-}C_4H_4)Fe(CO)_3$ is made by reaction 24.133 and its solid state structure (Fig. 24.28a) shows that (in contrast to the free ligand in which the double bonds are localized) the C—C bonds in coordinated C_4H_4 are of equal length.

(24.133)

Dewar-benzene derivative   (24.135)

KEY TERMS

The following terms were introduced in this chapter. Do you know what they mean?

- ❏ organometallic compound
- ❏ hapticity of a ligand
- ❏ Dewar–Chatt–Duncanson model
- ❏ synergic effect
- ❏ Tolman cone angle
- ❏ 18-electron rule
- ❏ condensed polyhedral cluster
- ❏ isolobal principle
- ❏ polyhedral skeletal electron pair theory (PSEPT)

- ❏ capping principle (within Wade's rules)
- ❏ total valence electron counts (for metal frameworks)
- ❏ reductive carbonylation
- ❏ ligand substitution
- ❏ oxidative addition
- ❏ orthometallation
- ❏ reductive elimination
- ❏ alkyl and hydrogen migration

- ❏ CO insertion
- ❏ β-hydrogen elimination
- ❏ agostic M−H−C interaction
- ❏ α-hydrogen abstraction
- ❏ carbene (alkylidene)
- ❏ carbyne (alkylidyne)
- ❏ sandwich complex
- ❏ metallocene
- ❏ half-sandwich complex

FURTHER READING

M. Bochmann (1994) *Organometallics 1: Complexes with Transition Metal–Carbon σ-Bonds*, Oxford University Press, Oxford – This and the companion book (see below) give a concise introduction to organometallic chemistry.

M. Bochmann (1994) *Organometallics 2: Complexes with Transition Metal–Carbon π-Bonds*, Oxford University Press, Oxford – see above.

P.J. Chirik (2010) *Organometallics*, vol. 29, p. 1500 – 'Group 4 transition metal sandwich complexes: Still fresh after almost 60 years'.

R.H. Crabtree and D.M.P. Mingos, eds (2007) *Comprehensive Organometallic Chemistry III*, Elsevier, Oxford – An update of the previous editions (see under G. Wilkinson *et al.*) covering the literature from 1993.

Ch. Elschenbroich (2005) *Organometallics*, 3rd edn, Wiley-VCH, Weinheim – An excellent text which covers both main group and transition metal organometallic chemistry.

G. Frenking (2001) *J. Organomet. Chem.*, vol. 635, p. 9 – An assessment of the bonding in *d*-block metal complexes including carbonyls which considers the relative importance of σ and π, as well as electrostatic, contributions to the metal–ligand bonds.

G. Gasser, I. Ott and N. Metzler-Nolte (2011) *J. Med. Chem.*, vol. 54, p. 3 – 'Organometallic anticancer compounds'.

A.F. Hill (2002) *Organotransition Metal Chemistry*, Royal Society of Chemistry, Cambridge – A detailed and well-organized, basic text that complements our coverage in this chapter.

S. Komiya, ed. (1997) *Synthesis of Organometallic Compounds: A Practical Guide*, Wiley-VCH, Weinheim – A book emphasizing methods of synthesis and handling of air-sensitive compounds.

G. Parkin (2010) *Struct. Bond.*, vol. 136, p. 113 – 'Metal–metal bonding in bridging hydride and alkyl compounds'.

P.L. Pauson (1993) 'Organo-iron compounds' in *Chemistry of Iron*, ed. J. Silver, Blackie Academic, Glasgow, p. 73 – A good summary of ferrocene chemistry and of other organoiron complexes.

W. Scherer and G.S. McGrady (2004) *Angew. Chem. Int. Ed.*, vol. 43, p. 1782 – 'Agostic interactions in d^0 metal alkyl complexes'.

R.R. Schrock (2001) *J. Chem. Soc., Dalton Trans.*, p. 2541 – An overview of 'Transition metal–carbon multiple bonds'.

R.R. Schrock (2005) *Chem. Commun.*, p. 2773 – 'High oxidation state alkylidene and alkylidyne complexes'.

P. Štěpnička (2008) *Ferrocenes: Ligands, Materials and Biomolecules*, Wiley, Chichester – Excellent survey that covers ferrocene compounds and their applications.

A. Togni and R.L. Halterman, eds (1998) *Metallocenes*, Wiley-VCH, Weinheim – A two-volume book covering synthesis, reactivity and applications of metallocenes.

H. Werner (2004) *Angew. Chem. Int. Ed.*, vol. 43, p. 938 – 'The way into the bridge: A new bonding mode of tertiary phosphanes, arsanes and stibanes'.

G. Wilkinson, F.G.A. Stone and E.W. Abel, eds (1982) *Comprehensive Organometallic Chemistry*, Pergamon, Oxford – A series of volumes reviewing the literature up to ≈1981.

G. Wilkinson, F.G.A. Stone and E.W. Abel, eds (1995) *Comprehensive Organometallic Chemistry II*, Pergamon, Oxford – An update of the previous set of volumes which provides an excellent entry into the literature.

H. Willner and F. Aubke (1997) *Angew. Chem. Int. Ed.*, vol. 36, p. 2403 – A review of binary carbonyl cations of metals in groups 8 to 12.

Q. Xu (2002) *Coord. Chem. Rev.*, vol. 231, p. 83 –'Metal carbonyl cations: Generation, characterization and catalytic application'.

Organometallic clusters of the *d*-block metals

C. Femoni, M.C. Iapalucci, F. Kaswalder, G. Longoni and S. Zacchini (2006) *Coord. Chem. Rev.*, vol. 250, p. 1580 – 'The possible role of metal carbonyl clusters in nanoscience and nanotechnologies'.

C.E. Housecroft (1996) *Metal–Metal Bonded Carbonyl Dimers and Clusters*, Oxford University Press, Oxford.

D.M.P. Mingos and D.J. Wales (1990) *Introduction to Cluster Chemistry*, Prentice Hall, Englewood Cliffs, NJ.

D.F. Shriver, H.D. Kaesz and R.D. Adams, eds (1990) *The Chemistry of Metal Cluster Complexes*, VCH, New York.

Fluxionality in organometallic complexes and uses of NMR spectroscopy

I.D. Gridnev (2008) *Coord. Chem. Rev.*, vol. 252, p. 1798 – 'Sigmatropic and haptotropic rearrangements in organometallic chemistry'.

W. von Phillipsborn (1999) *Chem. Soc. Rev.*, vol. 28, p. 95 – 'Probing organometallic structure and reactivity by transition metal NMR spectroscopy'.

PROBLEMS

24.1 (a) Explain the meaning of the following notations: μ-CO; μ_4-PR; η^5-C_5Me_5; η^4-C_6H_6; μ_3-H. (b) Why can the cyclopentadienyl and CO ligands be regarded as being versatile in their bonding modes? (c) Is PPh_3 a 'versatile ligand'?

24.2 What is a synergic effect, and how does it relate to metal–carbonyl bonding?

24.3 Comment on the following:

(a) Infrared spectra of $[V(CO)_6]^-$ and $Cr(CO)_6$ show absorptions at 1859 and 1981 cm^{-1} respectively assigned to ν_{CO}, and 460 and 441 cm^{-1} assigned to ν_{MC}.

(b) The Tolman cone angles of PPh_3 and $P(4\text{-}MeC_6H_4)_3$ are both 145°, but that of $P(2\text{-}MeC_6H_4)_3$ is 194°.

(c) Before reaction with PPh_3, $Ru_3(CO)_{12}$ may be treated with Me_3NO in MeCN.

(d) In the complex $[Os(en)_2(\eta^2\text{-}C_2H_4)(\eta^2\text{-}C_2H_2)]^{2+}$ the $Os-C_{ethyne}-H_{ethyne}$ bond angle is 127°.

24.4 (a) Draw a structure corresponding to the formula $[(CO)_2Ru(\mu\text{-}H)(\mu\text{-}CO)(\mu\text{-}Me_2PCH_2PMe_2)_2Ru(CO)_2]^+$. (b) The ^1H NMR spectrum of the complex in part (a) contains a quintet centred at $\delta -10.2$ ppm. Assign the signal and explain the origin of the observed multiplicity.

24.5 The solution ^1H NMR spectrum of the tetrahedral cluster $[(\eta^5\text{-}C_5Me_4SiMe_3)_4Y_4(\mu\text{-}H)_4(\mu_3\text{-}H)_4(THF)_2]$ exhibits the following signals at room temperature: δ/ppm 0.53 (s, 36H), 1.41 (m, 8H), 2.25 (s, 24H), 2.36 (s, 24H), 3.59 (m, 8H), 4.29 (quintet, $J_{^{89}Y^1H}$ 15.3 Hz, 8H). Assign the signals in the spectrum, and rationalize the appearance of the signal at δ 4.29 ppm. [Data: ^{89}Y, 100% abundant, $I = \frac{1}{2}$.]

24.6 The structure of $(\mu_3\text{-}H)_4Co_4(\eta^5\text{-}C_5M_4Et)_4$ was determined by single crystal X-ray diffraction in 1975, and by neutron diffraction in 2004. In both structure determinations, the bridging H atoms were located. To what extent can *precise* locations for these H atoms be given using single crystal X-ray and neutron diffraction techniques? Give reasons for your answer.

24.7 Consider the following compound:

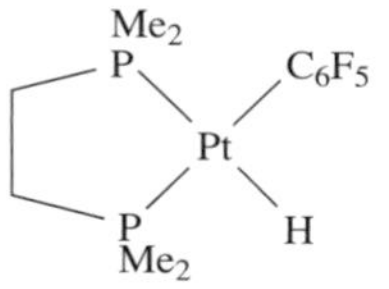

Predict the appearance of the signal assigned to the metal hydride in the ^1H NMR spectrum of this compound given the following coupling constants: $J_{^1H^{31}P(cis)}$ 17 Hz, $J_{^1H^{31}P(trans)}$ 200 Hz, $J_{^1H^{195}Pt}$ 1080 Hz. Ignore long-range $^1H-^{19}F$ coupling. [Data: ^{31}P, 100% abundant, $I = \frac{1}{2}$; ^{195}Pt, 33.8% abundant, $I = \frac{1}{2}$.]

24.8 Rationalize the following observations.

(a) On forming $[IrBr(CO)(\eta^2\text{-}C_2(CN)_4)(PPh_3)_2]$, the C−C bond in $C_2(CN)_4$ lengthens from 135 to 151 pm.

(b) During the photolysis of $Mo(CO)_5(THF)$ with PPh_3, a signal in the ^{31}P NMR spectrum at $\delta -6$ ppm disappears and is replaced by one at $\delta +37$ ppm.

(c) On going from $Fe(CO)_5$ to $Fe(CO)_3(PPh_3)_2$, absorptions in the IR spectrum at 2025 and 2000 cm^{-1} are replaced by a band at 1885 cm^{-1}.

24.9 Draw a bonding scheme (similar to that in Fig. 24.7b) for the interaction of an η^3-allyl ligand with a low oxidation state metal centre.

24.10 Show that the metal centres in the following complexes obey the 18-electron rule:
(a) $(\eta^5\text{-Cp})Rh(\eta^2\text{-}C_2H_4)(PMe_3)$
(b) $(\eta^3\text{-}C_3H_5)_2Rh(\mu\text{-Cl})_2Rh(\eta^3\text{-}C_3H_5)_2$
(c) $Cr(CO)_4(PPh_3)_2$
(d) $Fe(CO)_3(\eta^4\text{-}CH_2CHCHCH_2)$
(e) $Fe_2(CO)_9$
(f) $[HFe(CO)_4]^-$
(g) $[(\eta^5\text{-Cp})CoMe(PMe_3)_2]^+$
(h) $RhCl(H)_2(\eta^2\text{-}C_2H_4)(PPh_3)_2$

24.11 Reaction of $Fe(CO)_5$ with $Na_2[Fe(CO)_4]$ in THF gives a salt $Na_2[\mathbf{A}]$ and CO. The Raman spectrum of $[Et_4N]_2[\mathbf{A}]$ shows an absorption at $160\,cm^{-1}$ assigned to an unbridged Fe−Fe bond. Suggest an identity and structure for $[\mathbf{A}]^{2-}$.

24.12 Suggest possible structures for the cation in $[Fe_2(NO)_6][PF_6]_2$ and state how you would attempt to distinguish between them experimentally.

24.13 Comment on the following observations:
(a) In the IR spectrum of free $MeCH{=}CH_2$, $\bar{\nu}_{C=C}$ comes at $1652\,cm^{-1}$, but in the complex $K[PtCl_3(\eta^2\text{-}MeCH{=}CH_2)]$, the corresponding absorption is at $1504\,cm^{-1}$.
(b) At 303 K, the ^1H NMR spectrum of $(\eta^5\text{-Cp})(\eta^1\text{-Cp})Fe(CO)_2$ shows two singlets.

24.14 Use Wade's rules (PSEPT) to suggest structures for $Os_7(CO)_{21}$ and $[Os_8(CO)_{22}]^{2-}$.

24.15 For each of the following clusters, confirm that the total valence electron count is consistent with the metal cage framework adopted: (a) $[Ru_6(CO)_{18}]^{2-}$, octahedron; (b) $H_4Ru_4(CO)_{12}$, tetrahedron; (c) $Os_5(CO)_{16}$, trigonal bipyramid; (d) $Os_4(CO)_{16}$, square; (e) $Co_3(CO)_9(\mu_3\text{-CCl})$, triangle; (f) $H_2Os_3(CO)_9(\mu_3\text{-PPh})$, triangle; (g) $HRu_6(CO)_{17}B$, octahedron; (h) $Co_3(\eta^5\text{-Cp})_3(CO)_3$, triangle; (i) $Co_3(CO)_9Ni(\eta^5\text{-Cp})$, tetrahedron.

24.16 (a) $Os_5(CO)_{18}$ has a metal framework consisting of three edge-sharing triangles (a *raft* structure). Show that the valence electron count for this raft is consistent with the number available.
(b) Figure 24.29 shows the metal core of $[Ir_8(CO)_{22}]^{2-}$. What would be an appropriate electron-counting scheme for this cluster?

24.17 Suggest products in the following reactions, and give likely structures for the products: (a) $Fe(CO)_5$ irradiated with C_2H_4; (b) $Re_2(CO)_{10}$ with Na/Hg; (c) $Na[Mn(CO)_5]$ with ONCl; (d) $Na[Mn(CO)_5]$ with H_3PO_4; (e) $Ni(CO)_4$ with PPh_3.

24.18 In Section 24.7, we stated that the distribution of the products in Fig. 24.15 is consistent with the migration of the Me group, and not with a mechanism that

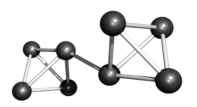

Fig. 24.29 Figure for problem 24.16b.

involves movement of the 'inserted' CO. Confirm that this is true by determining the distribution of products for the CO insertion mechanism and comparing it with that for the Me migration mechanism.

24.19 Illustrate, with examples, what is meant by (a) oxidative addition, (b) reductive elimination, (c) α-hydrogen abstraction, (d) β-hydrogen elimination, (e) alkyl migration and (f) orthometallation.

24.20 The reaction of $Cr(CO)_6$ with $Ph_2P(CH_2)_4PPh_2$ leads to the formation of two products, **A** and **B**. The ^{31}P NMR spectrum of **A** shows two signals ($\delta +46.0$ and -16.9 ppm, relative integrals 1 : 1), while that of **B** exhibits one signal ($\delta +46.2$ ppm). The IR spectra of **A** and **B** are almost identical in the region 2200–1900 cm^{-1}, with bands at 2063, 1983 and 1937 cm^{-1}. Suggest identities for **A** and **B** and explain why three absorptions are observed in the IR spectrum of each compound.

24.21 In the Heck reaction (eq. 24.87), the active catalyst is $Pd(PPh_3)_2$. Write equations to show (a) oxidative addition of PhBr to $Pd(PPh_3)_2$ to give **A**, (b) addition of $CH_2{=}CHCO_2Me$ to **A** followed by migration of the Ph group to give the σ-bonded alkyl derivative **B**, and (c) β-hydride elimination to generate the Pd(II) complex **C** and free alkene **D**.

24.22 Discuss the following statements:
(a) Complexes $Fe(CO)_3L$ where L is a 1,3-diene have applications in organic synthesis.
(b) The fullerenes C_{60} and C_{70} form a range of organometallic complexes.
(c) $Mn_2(CO)_{10}$ and C_2H_6 are related by the isolobal principle.

24.23 Explain why scheme 24.98 is invoked to explain the equivalence of the H atoms in each terminal CH_2 group of an η^3-allyl ligand, rather than a process involving rotation about the metal–ligand coordination axis.

24.24 Explain the difference between a Fischer-type carbene and a Schrock-type carbene.

24.25 The reaction of 1,3-dimethylimidazolium iodide (shown on the next page) with one equivalent of

KO^{t}Bu in THF, followed by addition of one equivalent of Ru$_3$(CO)$_{12}$ leads to product **A**. The IR spectrum of **A** has several strong absorptions between 2093 and 1975 cm^{-1}, and the solution ^{1}H NMR spectrum of **A** exhibits singlets at δ 7.02 and 3.80 ppm (relative integrals 1 : 3). (a) What role does KO^{t}Bu play in the reaction? (b) What is the likely identity of **A**? (c) Draw a possible structure of **A** and comment on possible isomers.

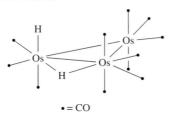

24.26 With reference to Box 23.1, develop a qualitative bonding scheme for (η^5-Cp)$_2$Fe.

24.27 Suggest products in the following reactions: (a) excess FeCl$_3$ with (η^5-Cp)$_2$Fe; (b) (η^5-Cp)$_2$Fe with PhC(O)Cl in the presence of AlCl$_3$; (c) (η^5-Cp)$_2$Fe with toluene in the presence of Al and AlCl$_3$; (d) (η^5-Cp)Fe(CO)$_2$Cl with Na[Co(CO)$_4$].

24.28 In the reaction of ferrocene with MeC(O)Cl and AlCl$_3$, how could one distinguish between the products Fe(η^5-C$_5$H$_4$C(O)Me)$_2$ and (η^5-Cp)Fe(η^5-C$_5$H$_4$C(O)Me) by methods other than elemental analysis and X-ray crystallography?

24.29 The reaction of [(C$_6$Me$_6$)RuCl$_2$]$_2$ (**A**) with C$_6$Me$_6$ in the presence of AgBF$_4$ gives [(C$_6$Me$_6$)$_2$Ru][BF$_4$]$_2$ containing cation **B**. Treatment of this compound with Na in liquid NH$_3$ yields a neutral Ru(0) complex, **C**. Suggest structures for **A**, **B** and **C**.

24.30 (a) Suggest structures for the complexes LFe(CO)$_3$ where L = 2,5-norbornadiene (**24.23**) or cyclo-heptatriene. (b) How is the bonding mode of the cycloheptatriene ligand affected on going from LFe(CO)$_3$ to LMo(CO)$_3$? (c) For L = cyclo-heptatriene, what product would you expect from the reaction of LMo(CO)$_3$ and [Ph$_3$C][BF$_4$]?

24.31 Describe the bonding in (η^4-C$_4$H$_4$)Fe(CO)$_3$, accounting for the diamagnetism of the complex.

OVERVIEW PROBLEMS

24.32 Comment on each of the following statements.

   (a) Re$_2$(CO)$_{10}$ adopts a staggered conformation in the solid state, whereas [Re$_2$Cl$_8$]$^{2-}$ adopts an eclipsed conformation.
   (b) In anions of type [M(CO)$_4$]$^{n-}$, $n = 1$ for M = Co, but $n = 2$ for M = Fe.
   (c) The reaction of benzoyl chloride with [(Ph$_3$P)$_2$N][HCr(CO)$_5$] which has first been treated with MeOD, produces PhCDO.

24.33 (a) Confirm that H$_2$Os$_3$(CO)$_{11}$ has sufficient valence electrons to adopt a triangular metal framework. Do the modes of bonding of the CO and H ligands affect the total valence electron count? Comment on the fact that H$_2$Os$_3$(CO)$_{10}$ also has a triangular Os$_3$-core.
   (b) The ^{1}H NMR spectrum of H$_2$Os$_3$(CO)$_{11}$ in deuterated toluene at 183 K shows two major signals (relative integrals 1:1) at δ −10.46 and −20.25 ppm; both are doublets with $J = 2.3$ Hz. The signals are assigned to the terminal and bridging H atoms, respectively, in the structure shown below:

The ^{1}H NMR spectrum also exhibits two pairs of low-intensity signals: δ −12.53 and −18.40 ppm (both doublets, $J = 17.1$ Hz) and δ −8.64 and −19.42 ppm (no coupling resolved). These signals are assigned to two other isomers of H$_2$Os$_3$(CO)$_{11}$. From other NMR spectroscopic experiments, it is possible to show that the two H atoms in each isomer are attached to the same Os centre. Suggest structures for the minor isomers that are consistent with the NMR spectroscopic data.

24.34 (a) The cluster H$_3$Os$_6$(CO)$_{16}$B contains an interstitial B atom and has an Os$_6$ cage derived from a pentagonal bipyramid with one equatorial vertex missing. Comment on this structure in terms of both Wade's rules and a total valence electron count for the cluster.
   (b) Give a description of the bonding in [Ir(CO)$_6$]$^{3+}$ and compare it with that in the isoelectronic compound W(CO)$_6$. How would you expect the IR spectra of these species to differ in the carbonyl stretching region?

24.35 Reduction of Ir$_4$(CO)$_{12}$ with Na in THF yields the salt Na[Ir(CO)$_x$] (**A**) which has a strong absorption in its IR spectrum (THF solution) at 1892 cm^{-1}. Reduction of **A** with Na in liquid NH$_3$, followed by addition of Ph$_3$SnCl and Et$_4$NBr, gives [Et$_4$N][**B**] as the iridium-containing product; CO is lost during the reaction. Elemental analysis of [Et$_4$N][**B**] shows that it contains 51.1% C, 4.55% H and 1.27% N. The IR spectrum of [Et$_4$N][**B**] shows one strong absorption in the carbonyl region at 1924 cm^{-1}, and the solution ^{1}H NMR spectrum exhibits multiplets between δ 7.1 and 7.3 ppm (30H), a quartet at δ 3.1 ppm (8H) and a triplet at δ 1.2 ppm (12H). Suggest structures for **A** and [**B**]$^-$. Comment on possible isomerism in [**B**]$^-$ and the preference for a particular isomer.

24.36 Suggest possible products for the following reactions:

(a)

[Ph$_3$C][BF$_4$]

(b)

PhC≡CPh

(c)

AgBF$_4$ in MeCN

(d)

+

hν

(e)

HCl

(f) (OC)$_5$W=C(Ph)(OMe)

BBr$_3$

INORGANIC CHEMISTRY MATTERS

24.37 Ferroquine has passed clinical phase II trials as an antimalarial drug. Both enantiomers of ferroquine are equally active *in vitro*. Explain why the molecule is chiral, and draw the structure of (*S*)-ferroquinone.

(*R*)-Ferroquine

24.38 Around 10 Mt per year of acetic acid are manufactured worldwide and ≈25% of this is produced using the Cativa^{TM} process. The reaction:

$$MeOH + CO \longrightarrow MeCO_2H$$

is catalysed by *cis*-[IrI$_2$(CO)$_2$]$^-$. Methanol is first converted to MeI by reaction with HI, and the catalyst undergoes oxidative addition of MeI to give a *fac*-**A**. In the presence of an I$^-$ abstractor, substitution of I$^-$ for CO leads to *fac*-**B**, which undergoes methyl migration. Reaction with I$^-$

results in elimination of MeCOI and regeneration of the catalyst. (a) What is the stereochemistry of *cis*-[IrI$_2$(CO)$_2$]$^-$? (b) Show what happens during the oxidative addition of MeI to *cis*-[IrI$_2$(CO)$_2$]$^-$, and give the structure of *fac*-**A**. (c) Draw the structure of *fac*-**B** and describe the mechanism of the methyl migration step. (d) How does the valence electron count and the oxidation state of the Ir atom change during the catalytic cycle, starting and ending with the catalyst?

24.39 Ruthenium(II) complexes of the general type shown below are potential anticancer drugs:

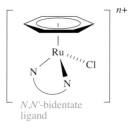

N,N'-bidentate ligand

The cytotoxicity of such complexes relies upon the replacement of the chlorido ligand by H$_2$O, and is pH dependent. (a) Write an equilibrium that defines the K_a value of the hydrolysis product of the above complex. (b) How does the introduction of electron-withdrawing substituents into the arene ring affect the pK_a value of the complex? (c) Suggest why the cytotoxicity is pH dependent.

24.40 The compound drawn below is an example of a ferrocenophane:

It undergoes ring-opening polymerization (ROP) to yield a high molecular weight polymer in which the Cp rings in each ferrocene are parallel to one another. Potential applications of these polymers include those in materials and nano-sciences. (a) Starting from Cp_2Fe, suggest a synthesis of the compound shown above. (b) Give a reaction scheme for the ROP reaction, showing the repeat unit of the polymer. Comment on the driving force for the polymerization.

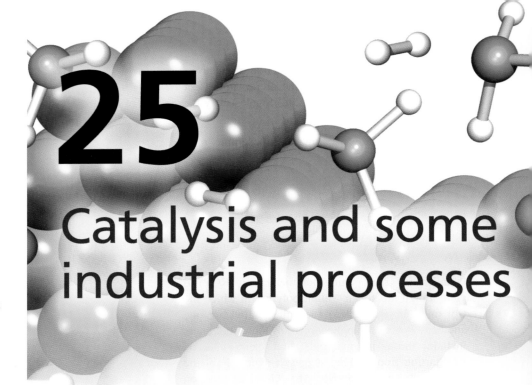

25

Catalysis and some industrial processes

25.1 Introduction and definitions

Numerous applications of catalysts in small-scale synthesis and the industrial production of chemicals have been described in this book. Now we discuss catalysis in detail, focusing on commercial applications. Catalysts containing *d*-block metals are of immense importance to the chemical industry: they provide cost-effective syntheses, and control the specificity of reactions that might otherwise give mixed products. The chemical industry (including fuels) is worth hundreds of billions of US dollars per year.[†] The search for new catalysts is one of the major driving forces behind organometallic research, and the chemistry in many parts of this chapter can be understood in terms of the reaction types introduced in Chapter 24. Current research also includes the development of environmentally friendly 'green chemistry', e.g. the use of supercritical CO_2 (scCO_2, see Section 9.13) as a medium for catalysis.[‡]

> A *catalyst* is a substance that alters the rate of a reaction without appearing in any of the products of that reaction; it may speed up or slow down a reaction. For a reversible reaction, a catalyst alters the rate at which equilibrium is attained; it does *not* alter the position of equilibrium.

The term *catalyst* is often used to encompass both the *catalyst precursor* and the *catalytically active species*. A catalyst precursor is the substance added to the reaction, but it may undergo loss of a ligand such as CO or PPh_3 before it is available as the catalytically active species.

Although one tends to associate a catalyst with *increasing* the rate of a reaction, a *negative catalyst* slows down a reaction. Some reactions are internally catalysed (*autocatalysis*) once the reaction is under way, e.g. in the reaction of $[C_2O_4]^{2-}$ with $[MnO_4]^-$, the Mn^{2+} ions formed catalyse the forward reaction.

> In an *autocatalytic reaction*, one of the products is able to catalyse the reaction.

Catalysts fall into two categories, homogeneous and heterogeneous, depending on their relationship to the phase of the reaction in which they are involved.

> A *homogeneous catalyst* is in the same phase as the components of the reaction that it is catalysing.
> A *heterogeneous catalyst* is in a different phase from the components of the reaction for which it is acting.

25.2 Catalysis: introductory concepts

Energy profiles for a reaction: catalysed versus non-catalysed

A catalyst operates by allowing a reaction to follow a different pathway from that of the non-catalysed reaction. If the activation barrier is lowered, then the reaction proceeds more rapidly. Figure 25.1 illustrates this for a

---

[†] For an overview of the growth of catalysis in industry during the 20th century, see: G.W. Parshall and R.E. Putscher (1986) *J. Chem. Educ.*, vol. 63, p. 189. For insight into the size of the chemical markets in the US and worldwide, see: W.J. Storck (2006) *Chem. Eng. News*, January 9 issue, p. 12; (2010) *Chem. Eng. News*, July 5 issue, p. 54.
[‡] For example, see: W. Leitner (2002) *Acc. Chem. Res.*, vol. 35, p. 746 – 'Supercritical carbon dioxide as a green reaction medium for catalysis'; I.P. Beletskaya and L.M. Kustov (2010) *Russ. Chem. Rev.*, vol. 79, p. 441 – 'Catalysis as an important tool of green chemistry'.

THEORY

Box 25.1 Energy and Gibbs energy of activation: E_a and $\Delta G^\ddagger$

The Arrhenius equation:

$$\ln k = \ln A - \frac{E_a}{RT} \quad \text{or} \quad k = A\,e^{\left(\frac{-E_a}{RT}\right)}$$

is often used to relate the rate constant, k, of a reaction to the activation energy, E_a, and to the temperature, T (in K). In this equation, A is the pre-exponential factor, and $R =$ molar gas constant. The activation energy is often approximated to $\Delta H^\ddagger$, but the exact relationship is:

$$E_a = \Delta H^\ddagger + RT$$

The energy of activation, $\Delta G^\ddagger$, is related to the rate constant by the equation:

$$k = \frac{k'T}{h}\,e^{\left(\frac{-\Delta G^\ddagger}{RT}\right)}$$

where k' = Boltzmann's constant, h = Planck's constant.

In Section 26.2 we discuss activation parameters, including $\Delta H^\ddagger$ and $\Delta S^\ddagger$, and show how these can be determined from an Eyring plot (Fig. 26.2) which derives from the equation above relating k to $\Delta G^\ddagger$.

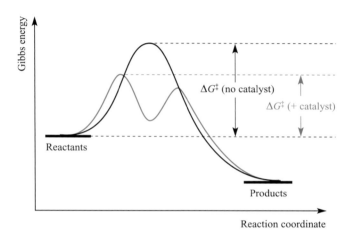

Fig. 25.1 A schematic representation of the reaction profile of a reaction without and with a catalyst. The pathway for the catalysed reaction has two steps, and the first step is rate determining.

A *catalytic cycle* consists of a series of stoichiometric reactions (often reversible) that form a closed loop. The catalyst must be regenerated so that it can participate in the cycle of reactions more than once.

For a catalytic cycle to be efficient, the intermediates must be short-lived. The downside of this for understanding the mechanism is that short lifetimes make studying a cycle difficult. Experimental probes are used to investigate the kinetics of a catalytic process, isolate or trap the intermediates, attempt to monitor intermediates in solution, or devise systems that model individual steps so that the product of the model-step represents an intermediate in the cycle. In the latter, the 'product' can be characterized by conventional techniques (e.g. NMR and IR spectroscopies, X-ray diffraction, mass spectrometry). For many cycles, however, the mechanisms are not firmly established.

reaction that follows a single step when it is non-catalysed, but a 2-step path when a catalyst is added. Each step in the catalysed route has a characteristic Gibbs energy of activation, $\Delta G^\ddagger$, but the step that matters with respect to the rate of reaction is that with the higher barrier. For the catalysed pathway in Fig. 25.1, the first step is the rate-determining step. (See Box 25.1 for the relevant equations for and relationship between E_a and $\Delta G^\ddagger$.) Values of $\Delta G^\ddagger$ for the controlling steps in the catalysed and non-catalysed routes are marked in Fig. 25.1. A crucial aspect of the catalysed pathway is that it must not pass through an energy minimum *lower* than the energy of the products. Such a minimum would be an 'energy sink', and would lead to the pathway yielding different products from those desired.

Catalytic cycles

A catalysed reaction pathway is usually represented by a *catalytic cycle*.

Self-study exercises

These exercises review types of organometallic reactions and the 18-electron rule.

1. What type of reaction is the following, and by what mechanism does it occur?

$$Mn(CO)_5Me + CO \longrightarrow Mn(CO)_5(COMe)$$

[*Ans.* See eq. 24.40]

2. Which of the following compounds contain a 16-electron metal centre: (a) $Rh(PPh_3)_3Cl$; (b) $HCo(CO)_4$; (c) $Ni(\eta^3\text{-}C_3H_5)_2$; (d) $Fe(CO)_4(PPh_3)$; (e) $[Rh(CO)_2I_2]^-$?

[*Ans.* (a), (c), (e)]

3. Write an equation to show β-hydrogen elimination from $L_nMCH_2CH_2R$.

[*Ans.* See eq. 24.44]

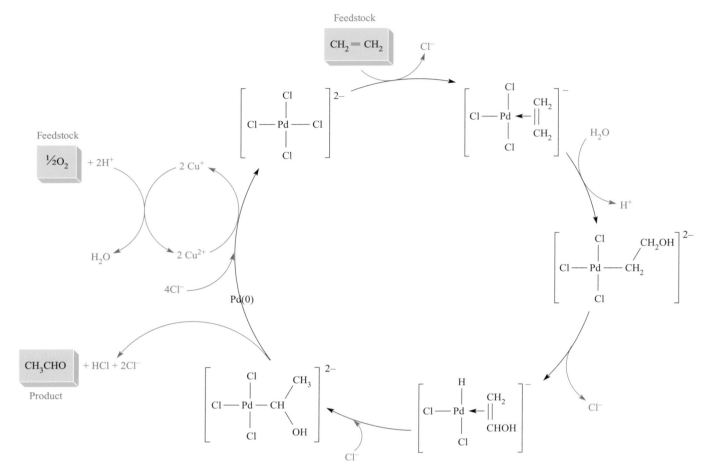

Fig. 25.2 Catalytic cycle for the Wacker process. For simplicity, we have ignored the role of coordinated H_2O, which replaces Cl^- *trans* to the alkene.

4. What is meant by 'oxidative addition'? Write an equation for the oxidative addition of H_2 to $RhCl(PPh_3)_3$.

   [*Ans.* See eq. 24.34 and associated text]

5. What type of reaction is the following, and what, typically, is the mechanism for such reactions?

   $$Mo(CO)_5(THF) + PPh_3 \longrightarrow Mo(CO)_5(PPh_3) + THF$$

   [*Ans.* See eq. 24.29 and associated text]

We now study one cycle in detail to illustrate the notations. Figure 25.2 shows a simplified catalytic cycle for the Wacker process which converts ethene to acetaldehyde (eq. 25.1). The process was developed in the 1950s and although it is not of great industrial significance nowadays, it provides a well-studied example for close examination.

$$CH_2{=}CH_2 + \tfrac{1}{2}O_2 \xrightarrow{\text{[PdCl}_4]^{2-}\text{ catalyst}} CH_3CHO \quad (25.1)$$

The *feedstocks* for the industrial process are highlighted along with the final product in Fig. 25.2. The catalyst in the Wacker process contains palladium: through most of

the cycle, the metal is present as Pd(II) but is reduced to Pd(0) as CH_3CHO is produced. We now work through the cycle, considering each step in terms of the organometallic reaction types discussed in Section 24.7.

The first step involves substitution by $CH_2{=}CH_2$ in $[PdCl_4]^{2-}$ (eq. 25.2). At the top of Fig. 25.2, the arrow notation shows $CH_2{=}CH_2$ entering the cycle and Cl^- leaving. One Cl^- is then replaced by H_2O, but we ignore this in Fig. 25.2.

$$[PdCl_4]^{2-} + CH_2{=}CH_2 \longrightarrow [PdCl_3(\eta^2\text{-}C_2H_4)]^- + Cl^-$$
$$(25.2)$$

The next step involves nucleophilic attack by H_2O with loss of H^+. Recall that coordinated alkenes are susceptible to nucleophilic attack (see eq. 24.86). In the third step, β-elimination occurs and formation of the Pd–H bond results in loss of Cl^-. This is followed by attack by Cl^- with H atom migration to give a σ-bonded $CH(OH)CH_3$ group. Elimination of CH_3CHO, H^+ and Cl^- with reduction of Pd(II) to Pd(0) occurs in the last step. To keep the cycle going, Pd(0) is now oxidized by Cu^{2+} (eq. 25.3). The secondary cycle in Fig. 25.2 shows the reduction of Cu^{2+} to Cu^+ and reoxidation of the latter by O_2 in the presence of H^+ (eq. 25.4).

$$Pd + 2Cu^{2+} + 8Cl^- \longrightarrow [PdCl_4]^{2-} + 2[CuCl_2]^- \quad (25.3)$$

$$2[CuCl_2]^- + \tfrac{1}{2}O_2 + 2HCl \longrightarrow 2CuCl_2 + 2Cl^- + H_2O \quad (25.4)$$

If the whole cycle in Fig. 25.2 is considered with species 'in' balanced against species 'out', the *net reaction* is reaction 25.1.

Choosing a catalyst

A reaction is not usually catalysed by a unique species and a number of criteria must be considered when choosing the most effective catalyst, especially for a commercial process. Moreover, altering a catalyst in an industrial plant already in operation may be costly (e.g. a new plant design may be required) and the change must be guaranteed to be financially viable. Apart from the changes in reaction conditions that the use of a catalyst may bring about (e.g. pressure and temperature), other factors that must be considered are:

- the concentration of catalyst required;
- the catalytic turnover;
- the selectivity of the catalyst to the desired product;
- how often the catalyst needs renewing.

The **catalytic turnover number** (TON) is the number of moles of product per mole of catalyst. This number indicates the number of catalytic cycles for a given process, e.g. after 2 h, the TON was 2400.

The **catalytic turnover frequency** (TOF) is the catalytic turnover per unit time: the number of moles of product per mole of catalyst per unit time, e.g. the TOF was $20\,min^{-1}$.

Defining the catalytic turnover number and frequency is not without problems. For example, if there is more than one product, one should distinguish between values of the total TON and TOF for all the catalytic products, and specific values for individual products. The term catalytic turnover number is usually used for batch processes, whereas catalytic turnover frequency is usually applied to continuous processes (flow reactors).

Now we turn to the question of selectivity, and the conversion of propene to an aldehyde provides a good example. Equation 25.5 shows the four possible products that may result from the reaction of propene with CO and H_2 (*hydroformylation*; see also Section 25.5).

The following ratios are important:

- the $n : i$ ratio of the aldehydes (*regioselectivity* of the reaction);
- the aldehyde : alcohol ratio for a given chain (*chemoselectivity* of the reaction).

The choice of catalyst can have a significant effect on these ratios. For reaction 25.5, a cobalt carbonyl catalyst (e.g. $HCo(CO)_4$) gives $\approx 80\%$ C_4-aldehyde, 10% C_4-alcohol and $\approx 10\%$ other products, and an $n : i$ ratio $\approx 3 : 1$. For the same reaction, various rhodium catalysts with phosphane co-catalysts can give an $n : i$ ratio of between 8 : 1 and 16 : 1, whereas ruthenium cluster catalysts show a high chemoselectivity to aldehydes with the regioselectivity depending on the choice of cluster, e.g. for $Ru_3(CO)_{12}$, $n : i \approx 2 : 1$, and for $[HRu_3(CO)_{11}]^-$, $n : i \approx 74 : 1$. Where the hydroformylation catalyst involves a bisphosphane ligand (e.g. $Ph_2PCH_2CH_2PPh_2$, dppe), the ligand bite angle (see structure **7.16**) can significantly influence the product distribution. For example, the $n : i$ ratios in the hydroformylation of hex-1-ene catalysed by a Rh(I)-bisphosphane complex are ≈ 2.1, 12.1 and 66.5 as the bite angle of the bisphosphane ligand increases along the series:[†]

Bite angle: 84.4°   107.6°   112.6°

Although a diagram such as Fig. 25.2 shows a catalyst being regenerated and passing once more around the cycle, in practice, catalysts eventually become exhausted or are *poisoned*, e.g. by impurities in the feedstock.

25.3 Homogeneous catalysis: alkene (olefin) and alkyne metathesis

In Section 24.12, we introduced *alkene (olefin) metathesis*, i.e. metal-catalysed reactions in which C=C bonds are redistributed. The importance of alkene and alkyne metathesis was recognized by the award of the 2005 Nobel Prize in Chemistry to Yves Chauvin, Robert H. Grubbs and Richard R. Schrock 'for the development of the metathesis method in organic synthesis'. Examples of alkene metathesis are shown in Fig. 25.3. The Chauvin mechanism for metal-catalysed alkene metathesis involves a metal alkylidene species and a series of [2 + 2]-cycloadditions and

---

[†] For further discussion of the effects of ligand bite angles on catalyst efficiency and selectivity, see: P. Dierkes and P.W.N.M. van Leeuwen (1999) *J. Chem. Soc., Dalton Trans.*, p. 1519.

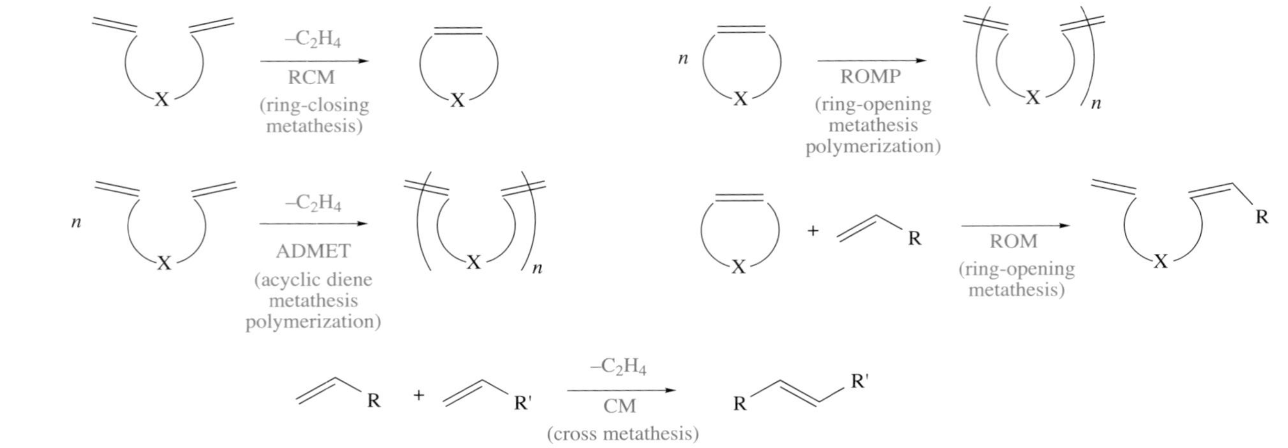

Fig. 25.3 Examples of alkene (olefin) metathesis reactions with their usual abbreviations.

cycloreversions (Fig. 25.4). Scheme 25.6 shows the mechanism for alkyne metathesis which involves a high oxidation state metal alkylidyne complex, $L_nM\equiv CR$.

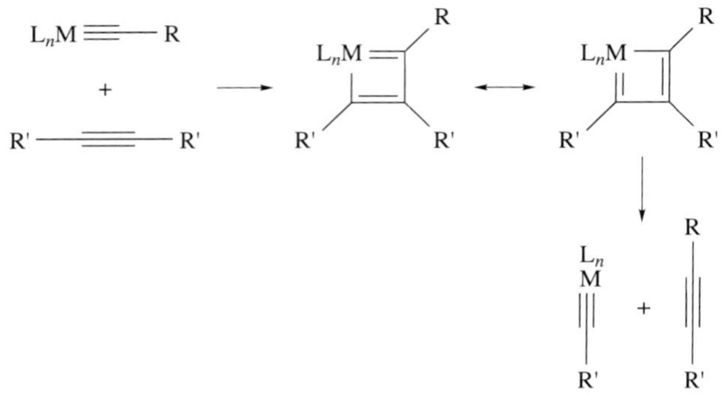

(25.6)

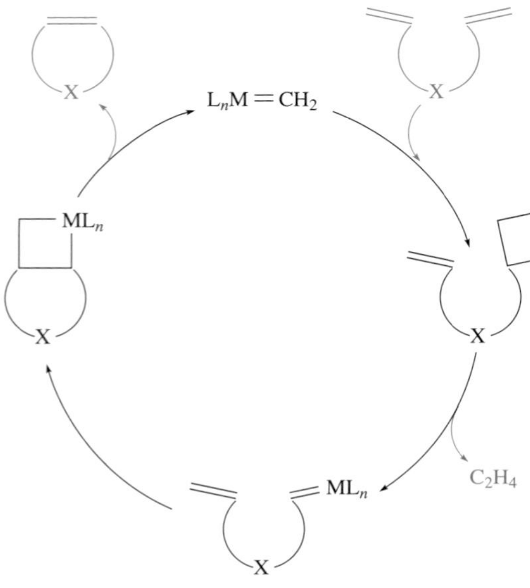

Fig. 25.4 A catalytic cycle for ring-closure metathesis (RCM) showing the Chauvin mechanism which involves [2 + 2]-cycloadditions and cycloreversions.

The catalysts that have played a dominant role in the development of this area of chemistry are those designed by Schrock (e.g. catalysts **25.1** and **25.2**) and Grubbs (catalysts **25.3** and **25.4**). Catalyst **25.3** is the traditional 'Grubbs' catalyst', and related complexes are also used. The 'second-generation' catalyst **25.4** exhibits higher catalytic activities in alkene metathesis reactions. Catalysts **25.1–25.4** are commercially available. There are around 15 modifications of Grubbs' catalysts which are optimized for different catalytic roles. This includes the recent 'third-generation' catalyst (see structure **24.68**).

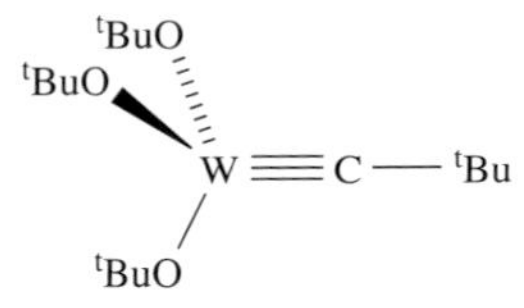

Schrock catalyst for alkyne metathesis

(25.1)

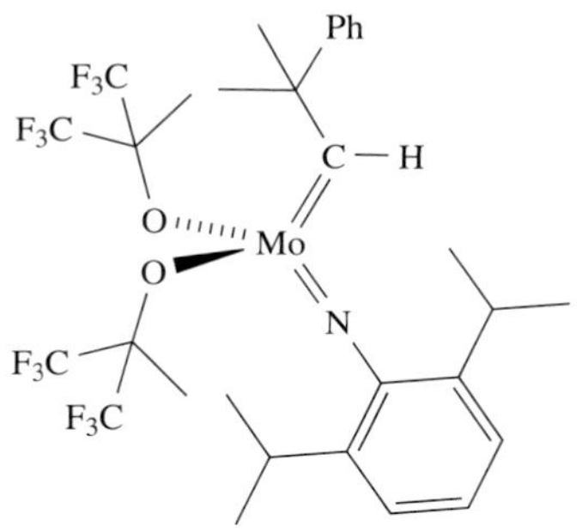

Example of a Schrock-type catalyst for alkene metathesis

(25.2)

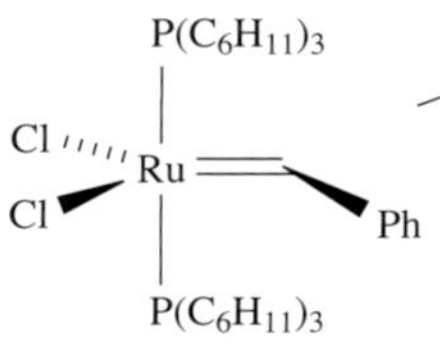

C_6H_{11} = cyclohexyl

(25.3)

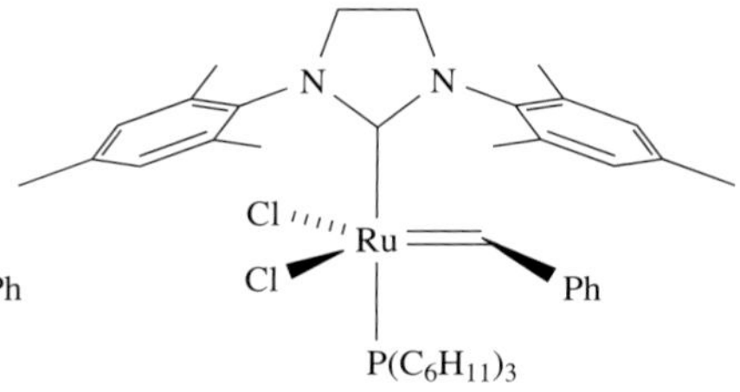

(25.4)

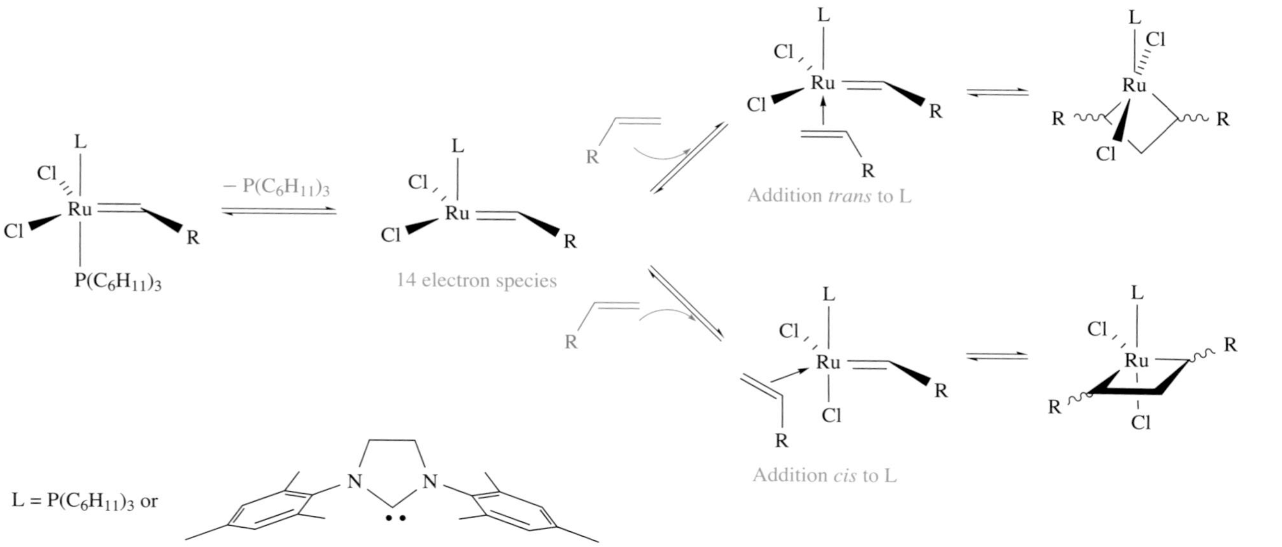

Fig. 25.5 Initial steps in the mechanism of alkene metathesis involving first and second generation Grubbs' catalysts. Two possibilities for the formation of the metallocyclobutane intermediates are shown.

In Grubbs' catalysts, tricyclohexylphosphane is chosen in preference to other PR_3 ligands because its steric hindrance and strongly electron-donating properties lead to enhanced catalytic activity. The first step in the mechanism of alkene metathesis involving Grubbs' catalysts is the dissociation of a $P(C_6H_{11})_3$ ligand to give a coordinatively unsaturated, 14-electron species (Fig. 25.5). The choice of the phosphane ligand is crucial for this initiation step: PR_3 ligands that are less sterically demanding than $P(C_6H_{11})_3$ bind too strongly to Ru, whereas those that are more bulky than $P(C_6H_{11})_3$ are too labile and a stable starting complex is not formed. The activated complex now enters the catalytic cycle by binding an alkene. This may coordinate to the Ru centre either *cis* or *trans* to $P(C_6H_{11})_3$ (first generation catalyst) or the *N*-heterocyclic carbene ligand (second generation catalyst). In keeping with the general Chauvin mechanism, the next step involves formation of metallocyclic intermediates (Fig. 25.5).[†]

A great advantage of Grubbs' catalysts is that they are tolerant of a large range of functional groups, thus permitting their widespread application. We highlight a laboratory example that combines coordination chemistry with the use of catalyst **25.3**: the synthesis of a *catenate*.

A *catenand* is a molecule containing two interlinked chains. A *catenate* is a related molecule that contains a coordinated metal ion.

[†] For elucidation of the mechanisms see, for example: R.H. Grubbs (2004) *Tetrahedron*, vol. 60, p. 7117; D.R. Anderson, D.D. Hickstein, D.J. O'Leary and R.H. Grubbs (2006) *J. Am. Chem. Soc.*, vol. 128, p. 8386; A.G. Wenzel and R.H. Grubbs (2006) *J. Am. Chem. Soc.*, vol. 128, p. 16048.

Topologically, the chemical assembly of a catenand is non-trivial because it requires one molecular chain to be threaded through another. Molecule **25.5** contains two terminal alkene functionalities and can also act as a bidentate ligand by using the *N,N'*-donor set.

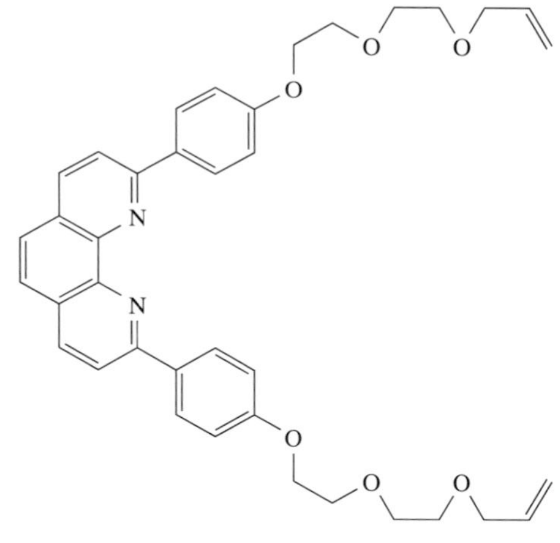

(25.5)

The complex $[Cu(\mathbf{25.5})_2]^+$ is shown schematically at the top of eq. 25.7. The tetrahedral Cu^+ centre acts as a template, fixing the positions of the two ligands with the central phenanthroline units orthogonal to one another. Ring closure of *each* separate ligand can be achieved by treating $[Cu(\mathbf{25.5})_2]^+$ with Grubbs' catalyst, and the result is the formation of a catenate, shown schematically as the product in eq. 25.7. The relative orientations of the two coordinated ligands in $[Cu(\mathbf{25.5})_2]^+$ is important if competitive reactions between *different* ligands are to be minimized.

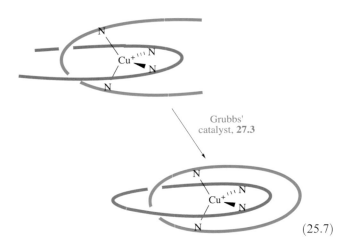

Grubbs'
catalyst, **27.3**

(25.7)

Self-study exercise

Ligand L_1 reacts with $Ru(DMSO)_4Cl_2$ in MeCN to give $[RuL_1(NCMe)_2]^{2+}$. Reaction of this complex with ligand L_2, followed by treatment with first generation Grubbs' catalyst, results in the formation of a catenate. (a) Draw a scheme for the reaction, paying attention to the coordination environment and stereochemistry of the Ru centre. (b) What type of alkene metathesis reaction is involved in the last step? (c) What complications can arise in this type of reaction?

Ligand L_1

Ligand L_2

[*Ans*: See P. Mobian *et al.* (2003) *J. Am. Chem. Soc.*, vol. 125, p. 2016; P. Mobian *et al.* (2003) *Helv. Chim. Acta*, vol. 86, p. 4195]

25.4 Homogeneous catalytic reduction of N_2 to NH_3

In nature, the fixation of nitrogen by bacteria involves the reduction of N_2 to NH_3 (eq. 25.8) catalysed by an iron- and molybdenum-containing nitrogenase (see Section 29.4). In contrast to this natural process, the industrial production of NH_3 (eq. 25.9) requires harsh conditions and a heterogeneous catalyst (see Section 25.8). Given the massive scale on which NH_3 is manufactured, the conversion of N_2 to NH_3 using a homogeneous catalyst under ambient conditions is a goal that many chemists have tried to achieve.

$$N_2 + 8H^+ + 8e^- \rightleftharpoons 2NH_3 + H_2 \qquad (25.8)$$

$$N_2 + 3H_2 \rightleftharpoons 2NH_3 \qquad (25.9)$$

Since nature depends on FeMo-nitrogenase, complexes containing these metals are of particular interest in terms of investigating N_2 to NH_3 conversion. Complexes of type **25.6** have been a starting point for a number of studies involving intermediates such as **25.7** and **25.8**. However, such interconversions produce only moderate yields of NH_3 when **25.8** is protonated.

(25.6)   **(25.7)**   **(25.8)**

Despite the large number of dinitrogen metal complexes known, their use for the catalytic production of NH_3 has not been an easy target to achieve. In 2003, Schrock reported the catalytic reduction of N_2 to NH_3 at a single Mo centre, carried out at room temperature and pressure. The reduction is selective (it does not give any N_2H_4). The catalyst is represented in Fig. 25.6a in the state in which N_2 is bound. The tripodal ligand $[N(CH_2CH_2NR)_3]^{3-}$ shown bound to the Mo(III) centre is designed to maximize steric crowding around the active metal site, creating a pocket in which small-molecule transformations occur. The substituents R increase the solubility of the complexes shown in Fig. 25.6b. Each step in the proposed catalytic cycle involves either proton or electron transfer. Of the intermedi-

N
‖
N
|
N

RN — Mo^{III}⋯N
    ⟍ NR
N

R =

Prⁱ   ⁱPr   Prⁱ   ⁱPr

ⁱPr   ⁱPr

(a)

$$Mo^{III}-N\equiv N \xrightarrow{e^-} Mo^{IV}-N=\bar{N} \xrightarrow{H^+} Mo^{IV}-N=NH \xrightarrow{H^+} \{Mo^{VI}=N-NH_2\}^+ \xrightarrow{e^-} Mo^V=N-NH_2 \xrightarrow{H^+} \{Mo^V=N-NH_3\}^+$$

$$N_2 \quad Mo^{III}$$

$$\Big\downarrow e^-, -NH_3$$

$$Mo^{VI}\equiv N$$

$$NH_3 \quad \Big\downarrow H^+$$

$$Mo^{III}-NH_3 \xleftarrow{e^-} \{Mo^{IV}-NH_3\}^+ \xleftarrow{H^+} Mo^{IV}-NH_2 \xleftarrow{e^-} \{Mo^V-NH_2\}^+ \xleftarrow{H^+} Mo^V=NH \xleftarrow{e^-} \{Mo^{VI}=NH\}^+$$

(b)

Fig. 25.6 (a) Dinitrogen bound to the single Mo(III) centre in the complex that is the starting point for the catalytic conversion of N_2 in NH_3 at room temperature and pressure. (b) The proposed scheme in which six protons and six electrons generate two equivalents of NH_3 from one equivalent of N_2. The complex shown in part (a) is abbreviated to $Mo^{III}N_2$, and so on.

ates shown, eight have been fully characterized.[†] In practice, a heptane solution of the complex $Mo^{III}N_2$ (defined in Fig. 25.6) is treated with an excess of 2,6-dimethylpyridinium ion (**25.9**) as the proton source and $(\eta^5\text{-}C_5Me_5)_2Cr$ (**25.10**) as the electron source. Decamethylchromocene is a very strong reducing agent, undergoing 1-electron oxidation to $[(\eta^5\text{-}C_5Me_5)_2Cr]^+$. The reagents must be added in a slow and controlled manner. Under these conditions, the efficiency of NH_3 formation from N_2 is ≈60%.

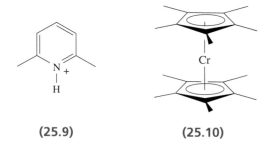

(25.9)   **(25.10)**

Although this example of the catalytic conversion of N_2 to NH_3 under ambient conditions in a well-defined molecular system remains at the research stage, it establishes that such conversions are possible.

[†] For further details, see: R.R. Schrock (2005) *Acc. Chem. Res.*, vol. 38, 955; W.W. Weare *et al.* (2006) *Proc. Nat. Acad. Sci.*, vol. 103, p. 17099; T. Kupfer and R.R. Schrock (2009) *J. Am. Chem. Soc.*, vol. 131, p. 12829; M.R. Reithofer, R.R. Schrock and P. Müller (2010) *J. Am. Chem. Soc.*, vol. 132, p. 8349; T. Munisamy and R.R. Schrock (2012) *Dalton Trans.*, vol. 41, p. 130.

25.5 Homogeneous catalysis: industrial applications

In this section, we describe selected homogeneous catalytic processes that are of industrial importance. Many more processes are applied in industry and detailed accounts can be found in the suggested reading at the end of the chapter. Two advantages of homogeneous over heterogeneous catalysis are the relatively mild conditions under which many processes operate, and the selectivity that can be achieved. A disadvantage is the need to separate the catalyst at the end of a reaction in order to recycle it, e.g. in the hydroformylation process, volatile $HCo(CO)_4$ can be removed by flash evaporation. The use of polymer supports or biphasic systems (Section 25.6) makes catalyst separation easier, and the development of such species is an active area of current research.

Throughout this section, the role of *coordinatively unsaturated 16-electron species* (see Section 24.7) and the ability of the metal centre to change coordination number (essential requirements of an active catalyst) should be noted.

Alkene hydrogenation

The most widely used procedures for the hydrogenation of alkenes nearly all employ heterogeneous catalysts, but for certain specialized purposes, homogeneous catalysts are used. Although addition of H_2 to a double bond is thermodynamically favoured (eq. 25.10), the kinetic barrier is high and a catalyst is required for the reaction to be carried out at a viable rate without the need for high temperatures and pressures.

$$CH_2=CH_2 + H_2 \longrightarrow C_2H_6 \quad \Delta G^o = -101 \, kJ \, mol^{-1}$$
$$(25.10)$$

(25.11)   **(25.12)**

Wilkinson's catalyst (**25.11**) has been widely studied, and in its presence alkene hydrogenation can be carried out at 298 K and 1 bar H_2 pressure. The red, 16-electron Rh(I) complex **25.11** can be prepared from $RhCl_3$ and PPh_3, and is commonly used in benzene/ethanol solution, in which it dissociates to some extent (equilibrium 25.11). A solvent molecule (solv) fills the fourth site in $RhCl(PPh_3)_2$ to give $RhCl(PPh_3)_2(solv)$. The 14-electron $RhCl(PPh_3)_2$ (or its solvated analogue) is the active catalyst for the hydrogenation of alkenes. Dimerization of $RhCl(PPh_3)_2$ to

25.12 leads to a catalytically inactive species, and may occur when the concentrations of H_2 and alkene are low (e.g. at the end of a batch process).

$$RhCl(PPh_3)_3 \rightleftharpoons RhCl(PPh_3)_2 + PPh_3 \quad K = 1.4 \times 10^{-4}$$
$$(25.11)$$

The *cis*-oxidative addition of H_2 to $RhCl(PPh_3)_3$ (left-hand side of Fig. 25.7) yields a coordinatively unsaturated 16-electron species (eq. 25.12).

$$\underset{\text{14-electron}}{RhCl(PPh_3)_2} + H_2 \rightleftharpoons \underset{\text{16-electron}}{RhCl(H)_2(PPh_3)_2} \qquad (25.12)$$

The addition of an alkene to $RhCl(H)_2(PPh_3)_2$ is probably the rate-determining step of the catalytic cycle shown in Fig. 25.7. The stereochemistry of octahedral $RhCl(H)_2(PPh_3)_2(\eta^2\text{-alkene})$ is such that the alkene is *cis* with respect to the two *cis*-hydrido ligands. Hydrogen migration then occurs to give a σ-bonded alkyl ligand, followed by reductive elimination of an alkane and regeneration of the active catalyst. The process is summarized in Fig. 25.7, the role of the solvent being ignored. The scheme shown should not be taken as being unique. For example, for some alkenes, experimental data suggest that $RhCl(PPh_3)_2(\eta^2\text{-alkene})$ is an intermediate. Other catalysts

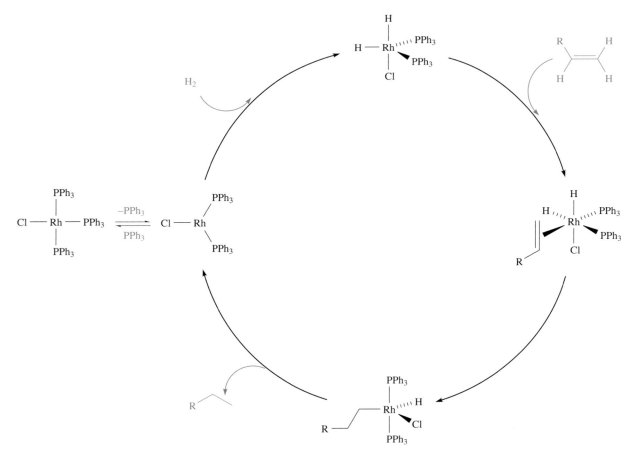

Fig. 25.7 Catalytic cycle for the hydrogenation of $RCH=CH_2$ using Wilkinson's catalyst, $RhCl(PPh_3)_3$.

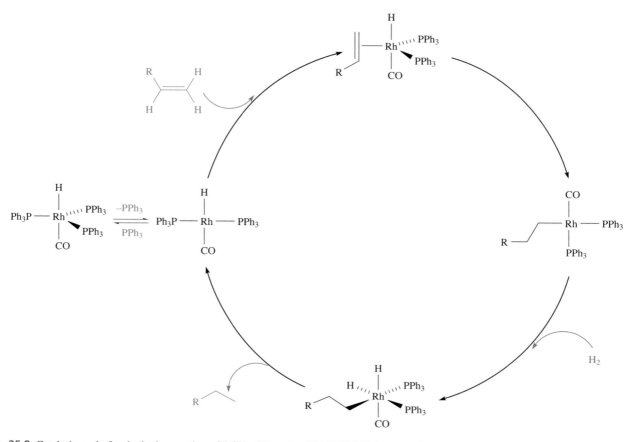

Fig. 25.8 Catalytic cycle for the hydrogenation of RCH=CH$_2$ using HRh(CO)(PPh$_3$)$_3$ as catalyst.

such as HRuCl(PPh$_3$)$_3$ and HRh(CO)(PPh$_3$)$_3$ (which loses PPh$_3$ to give an active 16-electron complex) react with alkene, rather than H$_2$, in the first step in the catalytic cycle. Figure 25.8 summarizes the route by which HRh(CO)(PPh$_3$)$_3$ catalyses the hydrogenation of an alkene. The rate-determining step is the oxidative addition of H$_2$ to the σ-bonded alkyl complex.

Substrates for hydrogenation catalysed by Wilkinson's catalyst include alkenes, dienes, allenes, terpenes, butadiene rubbers, antibiotics, steroids and prostaglandins. Significantly, ethene actually poisons its own conversion to ethane, and catalytic hydrogenation using RhCl(PPh$_3$)$_3$ cannot be applied in this case. For effective catalysis, the size of the alkene is important. The rate of hydrogenation is hindered by sterically demanding alkenes (Table 25.1). Many useful *selective* hydrogenations can be achieved, e.g. reaction 25.13.

Biologically active compounds usually have at least one *asymmetric centre* and dramatic differences in the activities of different enantiomers of chiral drugs are commonly observed. Whereas one enantiomer may be an effective therapeutic drug, the other may be inactive or highly toxic as was the case with

Table 25.1 Rate constants for the hydrogenation of alkenes (at 298 K in C$_6$H$_6$) in the presence of Wilkinson's catalyst.[†]

Alkene	$k / \times 10^{-2} \, \text{dm}^3 \, \text{mol}^{-1} \, \text{s}^{-1}$
Phenylethene (styrene)	93.0
Dodec-1-ene	34.3
Cyclohexene	31.6
Hex-1-ene	29.1
2-Methylpent-1-ene	26.6
1-Methylcyclohexene	0.6

[†] For further data, see: F.H. Jardine, J.A. Osborn and G. Wilkinson (1967) *J. Chem. Soc. A*, p. 1574.

(25.13)

Table 25.2 Observed % ee of the product of the hydrogenation of $CH_2=C(CO_2H)(NHCOMe)$ using Rh(I) catalysts containing different chiral bisphosphanes.

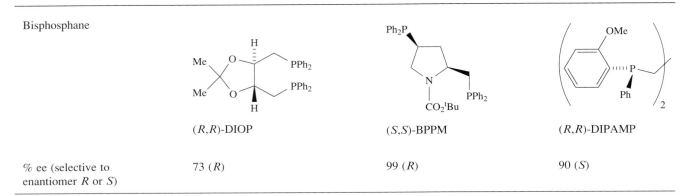

Bisphosphane	(R,R)-DIOP	(S,S)-BPPM	(R,R)-DIPAMP
% ee (selective to enantiomer R or S)	73 (R)	99 (R)	90 (S)

thalidomide.[†] *Asymmetric synthesis* is therefore an active field of research.

> **Asymmetric synthesis** is an enantioselective synthesis and its efficiency can be judged from the **enantiomeric excess** (ee):
>
> $$\% \text{ ee} = \left(\frac{|R - S|}{|R + S|}\right) \times 100$$
>
> where R and S = relative quantities of R and S enantiomers. An enantiomerically pure compound has 100% enantiomeric excess (100% ee). In **asymmetric catalysis**, the catalyst is chiral.

If hydrogenation of an alkene can, in principle, lead to enantiomeric products, then the alkene is *prochiral* (see end-of-chapter problem 25.6a). If the catalyst is *achiral* (as $RhCl(PPh_3)_3$ is), then the product of hydrogenation of the prochiral alkene is a racemate: i.e. starting from a prochiral alkene, there is an equal chance that the σ-alkyl complex formed during the catalytic cycle (Fig. 25.7) will be an *R*- or an *S*-enantiomer. If the catalyst is *chiral*, it should favour the formation of one or other of the *R*- or *S*-enantiomers, thereby making the hydrogenation enantioselective. *Asymmetric hydrogenations* can be carried out by modifying Wilkinson's catalyst, introducing a chiral phosphane or chiral bidentate bisphosphane, e.g. (R,R)-DIOP (defined in Table 25.2). By varying the chiral catalyst, hydrogenation of a given prochiral alkene proceeds with differing enantiomeric selectivities as exemplified in Table 25.2. An early triumph of the application of asymmetric alkene hydrogenation to drug manufacture was the production of

the alanine derivative L-DOPA (**25.13**), which is used in the treatment of Parkinson's disease.[‡] The anti-inflammatory drug Naproxen (active in the (*S*)-form) is prepared by chiral resolution or by asymmetric hydrogenation of a prochiral alkene (reaction 25.14); enantiopurity is essential, since the (*R*)-enantiomer is a liver toxin.

L-DOPA

(25.13)

(*S*)-BINAP

(25.14)

H_2

$Ru\{(S)\text{-BINAP}\}Cl_2$ catalyst

(*S*)-BINAP = (**25.14**)

Naproxen

(25.14)

[†] See, for example: E. Thall (1996) *J. Chem. Educ.*, vol. 73, p. 481 – 'When drug molecules look in the mirror'; S.C. Stinson (1998) *Chem. Eng. News*, 21 Sept. issue, p. 83 – 'Counting on chiral drugs'; H. Caner, E. Groner, L. Levy and I. Agranat (2004) *Drug Discovery Today*, vol. 9, p. 105 – 'Trends in the development of chiral drugs'.

[‡] For further details, see: W.A. Knowles (1986) *J. Chem. Educ.*, vol. 63, p. 222 – 'Application of organometallic catalysis to the commercial production of L-DOPA'.

Self-study exercise

Which of the following ligands are chiral? For each chiral ligand, explain how the chirality arises.

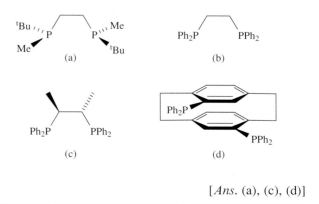

(a)

(b)

(c)

(d)

[*Ans.* (a), (c), (d)]

Monsanto and Cativa acetic acid syntheses

The conversion of MeOH to $MeCO_2H$ (eq. 25.15) is carried out on a huge industrial scale, and 60% of the world's acetyls are manufactured using the Monsanto and Cativa processes. Currently, ≈ 7 Mt per year of acetic acid are consumed worldwide, with the formation of vinyl acetate (**25.15**) being the most important commercial end use. Vinyl acetate is the precursor to polyvinylacetate (PVA, **25.16**).

PVA

(25.15)   **(25.16)**

$$MeOH + CO \longrightarrow MeCO_2H \qquad (25.15)$$

Before 1970, acetic acid was manufactured by the BASF process utilizing cobalt-based catalysts, and high temperatures and pressures. Replacement of this procedure by the Monsanto process brought advantages of milder conditions and greater selectivity (Table 25.3). The Monsanto process uses a rhodium-based catalyst, and involves two interrelated catalytic cycles (Fig. 25.9 with M = Rh). In the left-hand cycle in Fig. 25.9, MeOH is converted to MeI, which then enters the right-hand cycle by oxidative addition to the catalyst, *cis*-$[Rh(CO)_2I_2]^-$, which is a 16-electron complex. This addition is the rate-determining step in the process. It is followed by methyl migration and Fig. 25.9 shows the product of this step to be a 5-coordinate, 16-electron species. However, it is more likely to be an 18-electron complex, either dimer **25.17**, or $[Rh(CO)(COMe)I_3(solv)]$ where solv represents a solvent molecule. EXAFS studies (see Box 25.2) in THF solution indicate a dimer is present at 253 K, but a solvated monomer at 273 K. The next step in the cycle in Fig. 25.9 is addition of CO (or replacement of the solvent molecule in $[Rh(CO)(COMe)I_3(solv)]$ by CO) to give an 18-electron, octahedral complex which eliminates MeCOI. This enters the left-hand cycle in Fig. 25.9 and is converted to $MeCO_2H$.

(25.17)

The yields of products in any industrial manufacturing process must be optimized. One difficulty in the Monsanto process is the oxidation of *cis*-$[Rh(CO)_2I_2]^-$ by HI (eq. 25.16). The product easily loses CO, precipitating RhI_3 thereby removing the catalyst from the system (eq. 25.17). Operating under a pressure of CO prevents this last detrimental step and, as eq. 25.18 shows, reverses the effects of reaction 25.16. Adding small amounts of H_2 prevents oxidation of Rh(I) to Rh(III).

$$[Rh(CO)_2I_2]^- + 2HI \longrightarrow [Rh(CO)_2I_4]^- + H_2 \qquad (25.16)$$

$$[Rh(CO)_2I_4]^- \longrightarrow RhI_3(s) + 2CO + I^- \qquad (25.17)$$

$$[Rh(CO)_2I_4]^- + CO + H_2O \longrightarrow [Rh(CO)_2I_2]^- + 2HI + CO_2 \qquad (25.18)$$

Between 1995 and 2000, BP Chemicals commercialized and began to operate the Cativa process for the production of acetic acid. The catalyst is *cis*-$[Ir(CO)_2I_2]^-$ in the presence of a ruthenium-based promoter (e.g. $Ru(CO)_4I_2$) or an iodide promoter (a molecular iodide, e.g. InI_3). Catalyst

Table 25.3 Comparison of conditions and selectivities of the BASF, Monsanto and Cativa processes for the manufacture of acetic acid (eq. 25.15).

Conditions	BASF (Co-based catalyst)	Monsanto (Rh-based catalyst)	Cativa (Ir-based catalyst)
Temperature / K	500	453	453
Pressure / bar	500–700	35	20–40
Selectivity / %	90	>99	>99

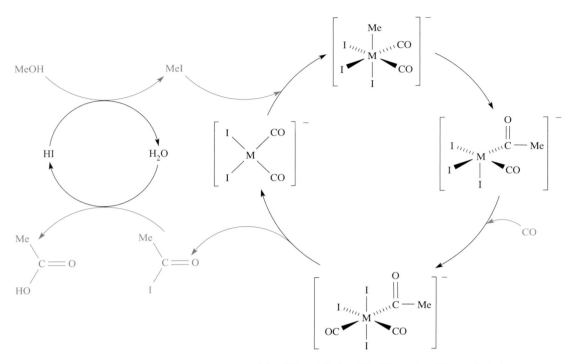

Fig. 25.9 The two interrelated catalytic cycles in the Monsanto (M = Rh) and Cativa (M = Ir) acetic acid manufacturing processes.

precursors include $IrCl_3$ and H_2IrCl_6. The catalytic cycle for the Cativa process (Fig. 25.9 with M = Ir) is essentially the same as for the Monsanto process. However, oxidative addition of MeI to cis-$[Ir(CO)_2I_2]^-$ is faster than to cis-$[Rh(CO)_2I_2]^-$, and this step is not rate-determining in the Cativa process (compare discussion above for the Monsanto process). The increased strength of the metal–ligand bonds on going from Rh to Ir (see exercise below) results in the rate-determining step being methyl migration. The rate of this step can be increased by the addition of an I^- abstractor, and this results in methyl migration occurring in the 5-coordinate $[Ir(CO)_2I_2Me]$ rather than in the 6-coordinate $[Ir(CO)_2I_3Me]^-$.

Self-study exercise

On going from Rh to Ir, metal–ligand bonding becomes stronger. Explain how the following data provide evidence for this.

	ν_{CO} / cm^{-1}	
cis-$[Rh(CO)_2I_2]^-$	2059	1988
cis-$[Ir(CO)_2I_2]^-$	2046	1968
cis,fac-$[Rh(CO)_2I_3Me]^-$	2104	2060
cis,fac-$[Ir(CO)_2I_3Me]^-$	2098	2045

An important advantage of the Cativa over Monsanto process is the fact that precipitation of $IrCl_3$ does not

occur as readily as precipitation of $RhCl_3$ (see eq. 25.17). A second advantage is that CO_2 emissions are $\approx 30\%$ lower in the Cativa than in the Monsanto process. The similarities between the two routes (Fig. 25.9) means that acetic acid manufacturing plants built to operate the Monsanto process can be retrofitted so as to switch production to the more advantageous Cativa process.

Tennessee–Eastman acetic anhydride process

The Tennessee–Eastman acetic anhydride process converts methyl acetate to acetic anhydride (eq. 25.19) and has been in commercial use since 1983.

$$MeCO_2Me + CO \longrightarrow (MeCO)_2O \qquad (25.19)$$

It closely resembles the Monsanto process but uses $MeCO_2Me$ in place of MeOH. cis-$[Rh(CO)_2I_2]^-$ remains the catalyst and the oxidative addition of MeI to cis-$[Rh(CO)_2I_2]^-$ is still the rate-determining step. One pathway can be described by adapting Fig. 25.9 with M = Rh, replacing:

- MeOH by $MeCO_2Me$;
- H_2O by $MeCO_2H$;
- $MeCO_2H$ by $(MeCO)_2O$.

However, a second pathway (Fig. 25.10) in which LiI replaces HI is extremely important for efficiency of the process. The final product is formed by the reaction of acetyl iodide and lithium acetate. Other alkali metal iodides do not function as well as LiI, e.g. replacing LiI by NaI slows the reaction by a factor of ≈ 2.5.

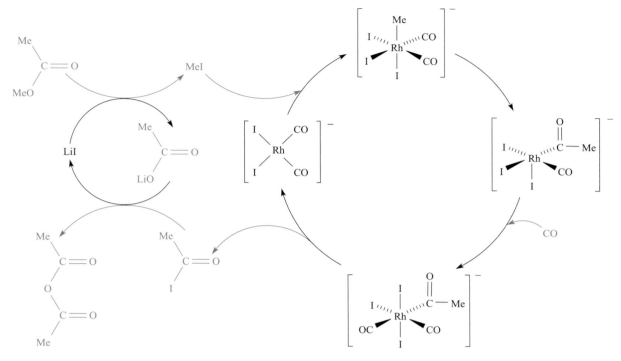

Fig. 25.10 Catalytic cycle for the Tennessee–Eastman acetic anhydride process.

Self-study exercises

1. With reference to Fig. 25.10, explain what is meant by the term 'coordinatively unsaturated'.

2. What features of $[Rh(CO)_2I_2]^-$ allow it to act as an active catalyst?

3. In Fig. 25.10, which step is an oxidative addition?
   [*Answers*: Refer to the discussion of the Monsanto process, and Section 24.7]

Hydroformylation (Oxo-process)

Hydroformylation (or the Oxo-process) is the conversion of alkenes to aldehydes (reaction 25.20). It is catalysed by cobalt and rhodium carbonyl complexes and has been exploited as a manufacturing process since World War II.

$$\text{R}\overset{\text{CO, H}_2}{\longrightarrow} \text{linear (}n\text{-isomer)} + \text{branched (}i\text{-isomer)}$$

(25.20)

Cobalt-based catalysts were the first to be employed. Under the conditions of the reaction (370–470 K, 100–400 bar),

$Co_2(CO)_8$ reacts with H_2 to give $HCo(CO)_4$. The latter is usually represented in catalytic cycles as the precursor to the coordinatively unsaturated (i.e. active) species $HCo(CO)_3$. As eq. 25.20 shows, hydroformylation can generate a mixture of linear and branched aldehydes, and the catalytic cycle in Fig. 25.11 accounts for both products. All steps (except for the final release of the aldehyde) are reversible. To interpret the catalytic cycle, start with $HCo(CO)_3$ at the top of Fig. 25.11. Addition of the alkene is the first step and this is followed by CO addition and accompanying H migration and formation of a σ-bonded alkyl group. At this point, the cycle splits into two routes depending on which C atom is involved in Co–C bond formation. The two pathways are shown as the inner and outer cycles in Fig. 25.11. In each, the next step is alkyl migration, followed by oxidative addition of H_2 and the transfer of one H atom to the alkyl group to give elimination of the aldehyde. The inner cycle eliminates a linear aldehyde, while the outer cycle produces a branched isomer. Two major complications in the process are the hydrogenation of aldehydes to alcohols, and alkene isomerization (which is also catalysed by $HCo(CO)_3$). The first of these problems (see eq. 25.5) can be controlled by using H_2 : CO ratios greater than 1 : 1 (e.g. 1.5 : 1). The isomerization problem (regioselectivity) can be addressed by using other catalysts (see below) or can be turned to advantage by purposely preparing mixtures of isomers for separation at a later stage. Scheme 25.21 illustrates the distribution of products formed when oct-1-ene undergoes

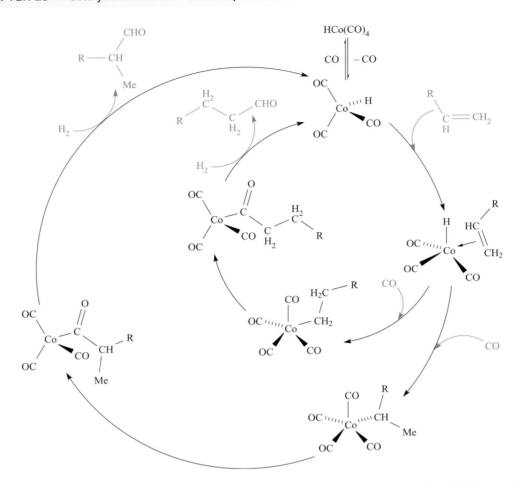

Fig. 25.11 Competitive catalytic cycles in the hydroformylation of alkenes to give linear (inner cycle) and branched (outer cycle) aldehydes.

hydroformylation at 423 K, 200 bar, and with a 1:1 H_2:CO ratio.

(25.21)

Just as we saw that the rate of hydrogenation was hindered by sterically demanding alkenes (Table 25.1), so too is the

Table 25.4 Rate constants for the hydroformylation of selected alkenes at 383 K in the presence of the active catalytic species $HCo(CO)_3$.

Alkene	$k / \times 10^{-5}\,s^{-1}$
Hex-1-ene	110
Hex-2-ene	30
Cyclohexene	10
Oct-1-ene	109
Oct-2-ene	31
2-Methylpent-2-ene	8

rate of hydroformylation affected by steric constraints, as is illustrated by the data in Table 25.4.

Other hydroformylation catalyst precursors that are used industrially are $HCo(CO)_3(PBu_3)$ (which, like $HCo(CO)_4$, must lose CO to become coordinatively unsaturated)

Table 25.5 A comparison of the operating conditions for and selectivities of three commercial hydroformylation catalysts. The formulae given are for the catalyst precursors.

	$HCo(CO)_4$	$HCo(CO)_3(PBu_3)$	$HRh(CO)(PPh_3)_3$
Temperature / K	410–450	450	360–390
Pressure / bar	250–300	50–100	30
Regioselectivity $n:i$ ratio (see eq. 25.5)	$\approx 3:1$	$\approx 9:1$	$>10:1$
Chemoselectivity (aldehyde predominating over alcohol)	High	Low	High

and $HRh(CO)(PPh_3)_3$ (which loses PPh_3 to give the catalytically active $HRh(CO)(PPh_3)_2$). Data in Table 25.5 compare the operating conditions for, and selectivities of, these catalysts with those of $HCo(CO)_4$. The Rh(I) catalyst is particularly selective towards aldehyde formation, and under certain conditions the $n:i$ ratio is as high as 20:1. An excess of PPh_3 prevents reactions 25.22 which occur in the presence of CO. The products of reactions 25.22 are also hydroformylation catalysts, but lack the selectivity of $HRh(CO)(PPh_3)_2$. The parent phosphane complex, $HRh(PPh_3)_3$, is inactive towards hydroformylation, and while $RhCl(PPh_3)_3$ is active, Cl^- acts as an inhibitor.

$$HRh(CO)(PPh_3)_2 + CO \rightleftharpoons HRh(CO)_2(PPh_3) + PPh_3$$
$$HRh(CO)_2(PPh_3) + CO \rightleftharpoons HRh(CO)_3 + PPh_3$$

$$(25.22)$$

Self-study exercises

1. Interpret the data in eq. 25.21 into a form that gives an $n:i$ ratio for the reaction. [*Ans.* $\approx 1.9:1$]

2. Draw out a catalytic cycle for the conversion of pent-1-ene to hexanal using $HRh(CO)_4$ as the catalyst precursor. [*Ans.* See inner cycle in Fig. 25.11, replacing Co by Rh]

Alkene oligomerization

The Shell Higher Olefins Process (SHOP) uses a nickel-based catalyst to oligomerize ethene. The process is designed to be flexible, so that product distributions meet consumer demand. The process is complex, but Fig. 25.12 gives a simplified catalytic cycle and indicates the form in which the nickel catalyst probably operates. Alkene addition is followed by hydrogen (first step) or alkyl (later steps) migration and formation of a σ-bonded alkyl group. This leaves a coordinatively unsaturated metal centre that can again undergo alkene addition. If β-hydride elimination

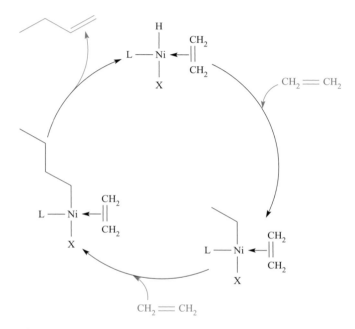

Fig. 25.12 Simplified catalytic cycle illustrating the oligomerization of ethene using a nickel-based catalyst; L = phosphane, X = electronegative group.

occurs, an alkene that contains a longer carbon chain than the starting alkene is produced.

25.6 Homogeneous catalyst development

The development of new catalysts is an important research topic, and in this section we briefly introduce some areas of current interest.

Polymer-supported catalysts

Attaching homogeneous metal catalysts to polymer supports retains the advantages of mild operating conditions and selectivity usually found for conventional homogeneous catalysts, while aiming to overcome the difficulties of catalyst separation. Types of support include polymers with a high degree of cross-linking and with large surface areas, and microporous polymers (low degree of cross-linking) which swell when they are placed in solvents. A common method of attaching the catalyst to the polymer is to functionalize the polymer with a ligand that can then be used to coordinate to, and hence bind, the catalytic metal centre. Equation 25.23 gives a schematic representation of the use of a chlorinated polymer to produce phosphane groups supported on the polymer surface. Scheme 25.24 illustrates application of the phosphane-functionalized surface to attach a Rh(I) catalyst. This system catalyses the carbonylation of MeOH in the presence of a MeI promoter, and therefore has relevance to the Monsanto process (Fig. 25.9).

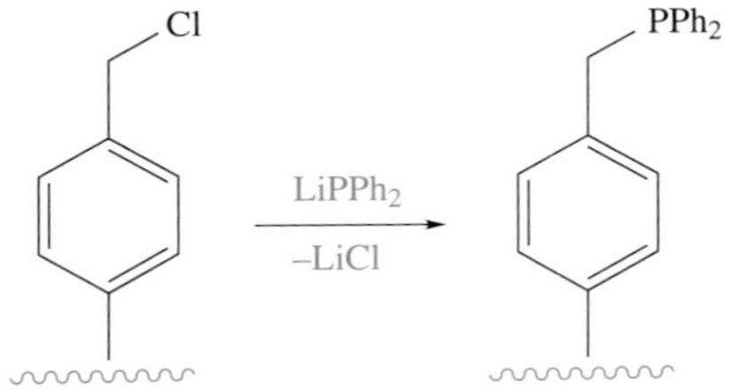

(25.23)

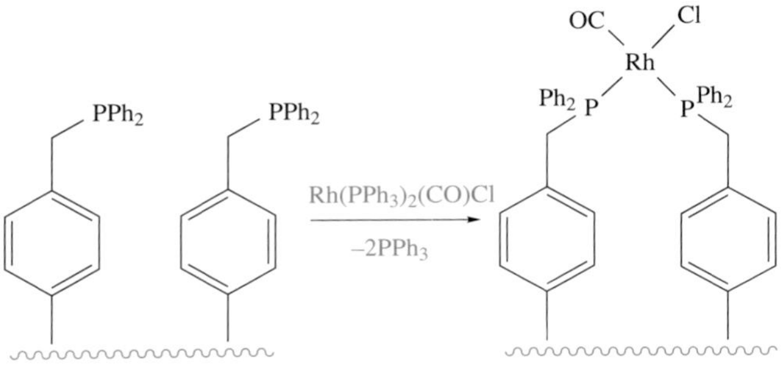

(25.24)

Alternatively, some polymers can bind the catalyst directly, e.g. poly-2-vinylpyridine (made from monomer **25.18**) is suitable for application in the preparation of hydroformylation catalysts (eq. 25.25).

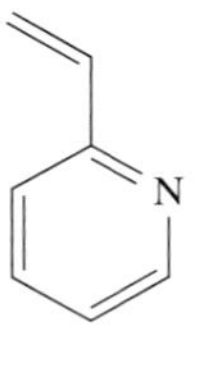

(25.18)

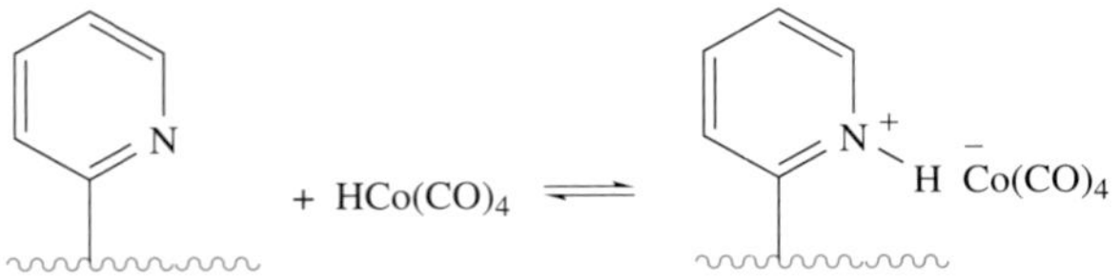

(25.25)

Hydroformylation catalysts can also be made by attaching the cobalt or rhodium carbonyl residues to a phosphane-functionalized surface through phosphane-for-carbonyl substitution. The chemo- and regioselectivities observed for the supported homogeneous catalysts are typically quite different from those of their conventional analogues.

While much progress has been made in this area, leaching of the metal into solution (which partly defeats the advantages gained with regard to catalyst separation) is a common problem.

Biphasic catalysis

Biphasic catalysis addresses the problem of catalyst separation. One strategy uses a water-soluble catalyst. This is retained in an aqueous layer that is immiscible with the organic medium in which the reaction takes place. Intimate contact between the two solutions is achieved during the catalytic reaction, after which the two liquids are allowed to settle and the catalyst-containing layer separated by decantation. Many homogeneous catalysts are hydrophobic and so it is necessary to introduce ligands that will bind to the metal but that carry hydrophilic substituents. Among ligands that have met with success is **25.19**: e.g. the reaction of an excess of **25.19** with $[Rh_2(nbd)_2(\mu\text{-Cl})_2]$ (**25.20**) gives a species, probably $[RhCl(\mathbf{25.19})_3]^{3+}$, which catalyses the hydroformylation of hex-1-ene to aldehydes (at 40 bar, 360 K) in 90% yield with an $n:i$ ratio of 4:1. An excess of the ligand in the aqueous phase stabilizes the catalyst and increases the $n:i$ ratio to $\approx 10:1$.

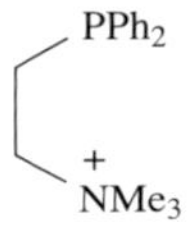

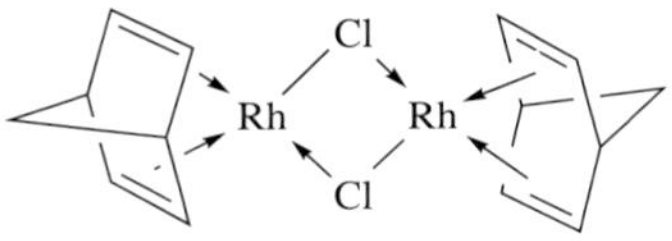

(25.19)   **(25.20)**

Much work has been carried out with the P-donor ligand **25.21** which can be introduced into a variety of organometallic complexes by carbonyl or alkene displacement. For example, the water-soluble complex $HRh(CO)(\mathbf{25.21})_3$ is a hydroformylation catalyst precursor. Conversion of hex-1-ene to heptanal proceeds with 93% selectivity for the n-isomer, a higher selectivity than is shown by $HRh(CO)(PPh_3)_3$ under conventional homogeneous catalytic conditions. A range of alkene hydrogenations are catalysed by $RhCl(\mathbf{25.21})_3$ and it is particularly efficient and selective for the hydrogenation of hex-1-ene.

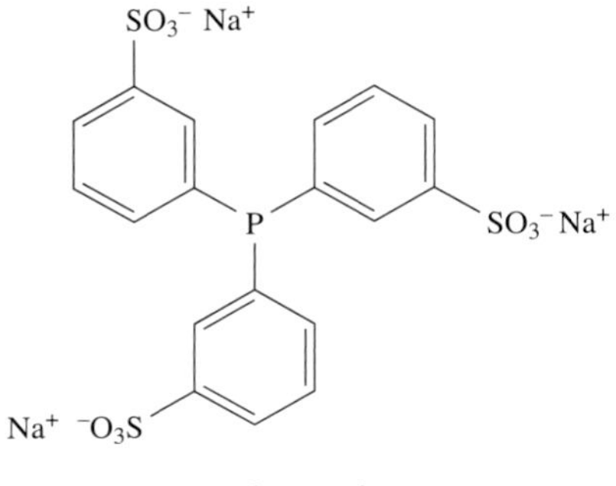

(25.21)

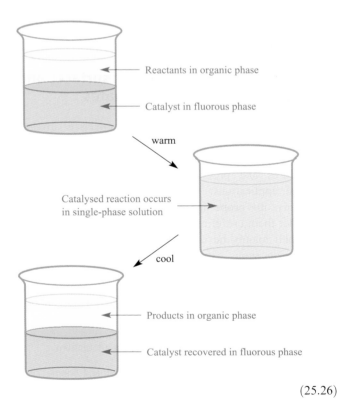

(25.22)

Biphasic asymmetric hydrogenation has also been developed using water-soluble chiral bisphosphanes such as **25.22** coordinated to Rh(I). With PhCH=C(CO$_2$H)(NH-C(O)Me) as substrate, hydrogenation takes place with 87% ee, and similar success has been achieved for related systems.

A second approach to biphasic catalysis uses a fluorous (i.e. perfluoroalkane) phase instead of an aqueous phase. There is an important difference between the higher C$_n$ perfluoroalkanes used in fluorous biphasic catalysis and the low-boiling CFCs that have been phased out under the Montreal Protocol (see Box 14.6). The principle of fluorous biphasic catalysis is summarized in scheme 25.26.

Reactants in organic phase

Catalyst in fluorous phase

warm

Catalysed reaction occurs in single-phase solution

cool

Products in organic phase

Catalyst recovered in fluorous phase

(25.26)

At room temperature, most fluorous solvents are immiscible with other organic solvents, but an increase in temperature typically renders the solvents miscible. The reactants are initially dissolved in a non-fluorinated, organic solvent and the catalyst is present in the fluorous phase. Raising the temperature of the system creates a single phase in which the catalysed reaction occurs. On cooling, the solvents, along with the products and catalyst, separate. Catalysts with suitable solubility properties can be designed by incorporating fluorophilic substituents such as C$_6$F$_{13}$ or C$_8$F$_{17}$. For example, the hydroformylation catalyst HRh(CO)(PPh$_3$)$_3$ has been adapted for use in fluorous media by using the phosphane ligand **25.23** in place of PPh$_3$. Introducing fluorinated substituents alters the electronic properties of the ligand. If the metal centre in the catalyst 'feels' this change, its catalytic properties are likely to be affected. Placing a spacer between the metal and the fluorinated substituent can minimize these effects. Thus, in phosphane ligand **25.24** (which is a derivative of PPh$_3$), the aromatic ring helps to shield the P atom from the effects of the electronegative F atoms. Although the use of the biphasic system allows the catalyst to be recovered and recycled, leaching of the Rh into the non-fluorous phase occurs over a number of catalytic cycles.

(25.23)

(25.24)

Although the biphasic catalysts described above appear analogous to those discussed in Section 25.5, it does not follow that the mechanisms by which the catalysts operate for a given reaction are similar.

Self-study exercises

1. Give an example of how PPh$_3$ can be converted into a hydrophilic catalyst.

2. The ligand (L):

forms the complex [Rh(CO)$_2$L]$^+$, which catalyses the hydrogenation of styrene in a water/heptane system.

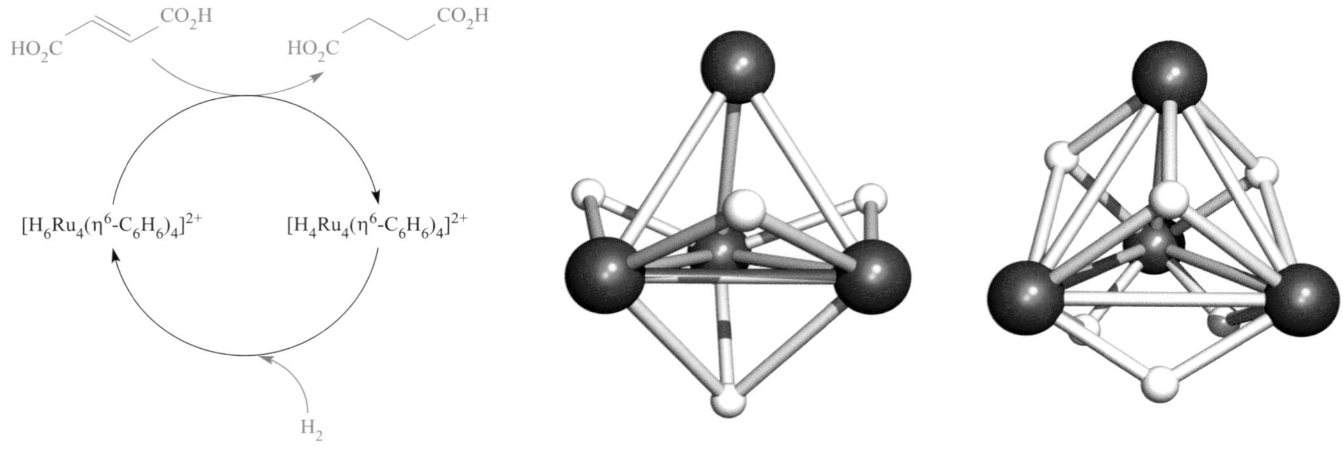

Fig. 25.13 (a) Catalytic cycle for the hydrogenation of fumaric acid by $[H_4(\eta^6-C_6H_6)_4Ru_4]^{2+}$; (b) H_4Ru_4 core of $[H_4(\eta^6-C_6H_6)_4Ru_4]^{2+}$; and (c) H_6Ru_4 core of $[H_6(\eta^6-C_6H_6)_4Ru_4]^{2+}$, both determined by X-ray diffraction [G. Meister *et al.* (1994) *J. Chem. Soc., Dalton Trans.*, p. 3215]. ^{1}H NMR spectroscopic data suggest that $[H_6(\eta^6-C_6H_6)_4Ru_4]^{2+}$ may contain an H_2 ligand and four hydrido ligands. Colour code in (b) and (c): Ru, red; H, white.

Suggest how L coordinates to the Rh centre. Explain how the catalysed reaction would be carried out, and comment on the advantages of the biphasic system over using a single solvent.

[*Ans.* See C. Bianchini *et al.* (1995) *Organometallics*, vol. 14, p. 5458]

d-Block organometallic clusters as homogeneous catalysts

Over the past 40 years, much effort has been put into investigating the use of *d*-block organometallic clusters as homogeneous catalysts, and eqs. 25.27–25.29 give examples of small-scale catalytic reactions. Note that in reaction 25.27, insertion of CO is into the O−H bond. In contrast, in the Monsanto process using $[Rh(CO)_2I_2]^-$ catalyst, CO insertion is into the C−OH bond (eq. 25.15).

$$MeOH + CO \xrightarrow[\text{400 bar, 470 K}]{Ru_3(CO)_{12}} \underset{\text{90\% selectivity}}{MeOCHO} \quad (25.27)$$

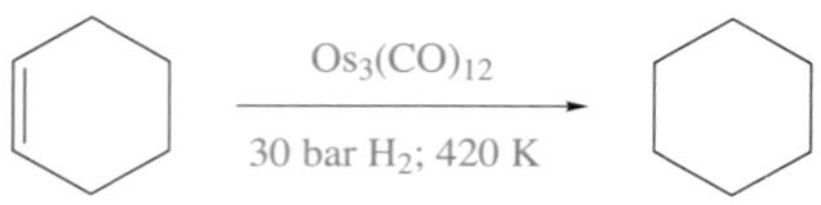

$$\xrightarrow[\text{30 bar } H_2;\ 420\ K]{Os_3(CO)_{12}} \quad (25.28)$$

$$\xrightarrow[\text{7 bar } H_2;\ 390\ K]{(\eta^5-Cp)_4Fe_4(CO)_4}$$

+ other isomers

84% selectivity

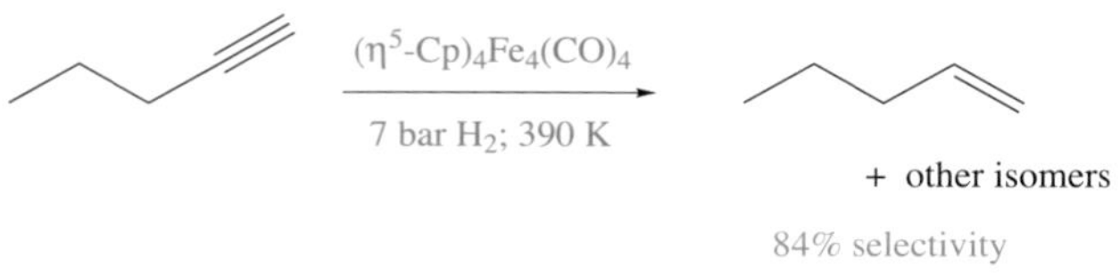

$$(25.29)$$

A promising development in the area is the use of *cationic* clusters. $[H_4(\eta^6-C_6H_6)_4Ru_4]^{2+}$ catalyses the reduction of fumaric acid, the reaction being selective to the C=C bond and leaving the carboxylic acid units intact (Fig. 25.13).

Despite the large of amount of work that has been carried out in the area and the wide range of examples now known, it would appear that there are no industrial applications of molecular cluster catalysts.

25.7 Heterogeneous catalysis: surfaces and interactions with adsorbates

The majority of industrial catalytic processes involve *heterogeneous catalysis* and Table 25.6 gives selected examples. Conditions are generally harsh, with high temperatures and pressures. Before describing specific industrial applications, we introduce some terminology and discuss the properties of metal surfaces and zeolites that render them useful as heterogeneous catalysts.

We shall mainly be concerned with reactions of gases over heterogeneous catalysts. Molecules of reactants are *adsorbed* on to the catalyst surface, undergo reaction and the products are *desorbed*. Interaction between the adsorbed species and surface atoms may be of two types: physisorption or chemisorption.

> *Physisorption* involves weak van der Waals interactions between the surface and the adsorbate.
> *Chemisorption* involves the formation of chemical bonds between surface atoms and the adsorbed species.

The process of adsorption activates molecules, either by cleaving bonds or by weakening them. The dissociation of a diatomic molecule such as H_2 on a metal surface is

Table 25.6 Examples of industrial processes that use heterogeneous catalysts.

Industrial manufacturing process	Catalyst system
NH$_3$ synthesis (Haber process)[‡]	Fe on SiO$_2$ and Al$_2$O$_3$ support
Water–gas shift reaction[*]	Ni, iron oxides
Catalytic cracking of heavy petroleum distillates	Zeolites (see Section 25.8)
Catalytic reforming of hydrocarbons to improve octane number[**]	Pt, Pt–Ir and other Pt-group metals on acidic alumina support
Methanation (CO $\longrightarrow$ CO$_2$ $\longrightarrow$ CH$_4$)	Ni on support
Ethene epoxidation	Ag on support
HNO$_3$ manufacture (Haber–Bosch process)[***]	Pt–Rh gauzes

[‡] See Section 15.5.
[*] See eqs. 10.13 and 10.14.
[**] The octane number is increased by increasing the ratio of branched or aromatic hydrocarbons to straight-chain hydrocarbons. The 0–100 octane number scale assigns 0 to *n*-heptane and 100 to 2,2,4-trimethylpentane.
[***] See Section 15.9.

represented schematically in eq. 25.30. Bond formation does not have to be with a single metal atom as we illustrate later. Bonds in molecules, e.g. C−H, N−H, are similarly activated.

$$H \text{——} H$$

$$\xrightarrow{\hspace{1cm}} \quad \begin{matrix} H & & H \\ | & & | \\ \text{—M—M—M—} & & \text{—M—M—M—} \end{matrix} \qquad (25.30)$$

The balance between the contributing bond energies is a factor in determining whether or not a particular metal will facilitate bond fission in the adsorbate. However, if metal–adsorbate bonds are especially strong, it becomes energetically less favourable for the adsorbed species to leave the surface, and this blocks adsorption sites, reducing catalytic activity.

The adsorption of CO on metal surfaces has been thoroughly investigated. Analogies can be drawn between the interactions of CO with metal atoms on a surface and those in organometallic complexes (see Section 24.2), i.e. both terminal and bridging modes of attachment are possible, and IR spectroscopy can be used to study adsorbed CO. Upon interaction with a surface metal atom, the C−O bond is weakened in much the same way as shown in Fig. 24.1. The extent of weakening depends not only on the mode of interaction with the surface but also on the surface coverage. In studies of the adsorption of CO on a Pd(111)[†] surface, it is found that the enthalpy of adsorption of CO becomes less negative as more of the surface is covered with adsorbed molecules. An abrupt decrease in the amount of heat evolved per mole of adsorbate is

observed when the surface is half-occupied by a *monolayer*. At this point, significant reorganization of the adsorbed molecules is needed to accommodate still more. Changes in the mode of attachment of CO molecules to the surface alter the strength of the C−O bond and the extent to which the molecule is activated.

Diagrams of hcp, fcc or bcc metal lattices such as we showed in Fig. 6.2 imply 'flat' metal surfaces. In practice, a surface contains imperfections such as those illustrated in Fig. 25.14. The *kinks* on a metal surface are extremely important for catalytic activity, and their presence increases the rate of catalysis. In a close-packed lattice, sections of 'flat' surface contain M$_3$ triangles (**25.25**), while a step possesses a line of M$_4$ 'butterflies' (see Table 24.5), one of which is shown in blue in structure **25.26**. Both can accommodate adsorbed species in sites which can be mimicked by discrete metal clusters. This has led to the *cluster-surface analogy* (see Section 25.9).

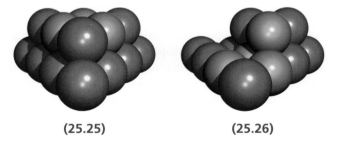

(25.25)   (25.26)

The design of metal catalysts has to take into account not only the available surface but also the fact that the catalytically active platinum-group metals (see Section 22.2) are rare and expensive. There can also be the problem that extended exposure to the metal surface may result in side reactions. In many commercial catalysts, including motor

[†] The notations (111), (110), (101)... are Miller indices and define the crystal planes in the metal lattice.

THEORY

Box 25.2 Some experimental techniques used in surface science

In much of this book, we have been concerned with studying species that are soluble and subjected to solution techniques (see Chapter 4) such as NMR and electronic spectroscopy, or with structural data obtained from X-ray or neutron diffraction studies of *single crystals* or electron diffraction studies of gases. The investigation of solid surfaces requires specialist techniques, many of which have been developed relatively recently. Selected examples are listed in the table below.

For further details of solid state techniques, see:
J. Evans (1997) *Chem. Soc. Rev.*, vol. 26, p. 11 – 'Shining light on metal catalysts'.
J. Evans (2006) *Phys. Chem. Chem. Phys.*, vol. 8, p. 3045 – 'Brilliant opportunities across the spectrum'.
G.A. Somorjai and Y. Li (2010) *Introduction to Surface Chemistry and Catalysis*, 2nd edn, Wiley, New Jersey.
A.R. West (1999) *Basic Solid State Chemistry*, 2nd edn, Wiley, Chichester.

A false colour image obtained using scanning tunnelling microscopy (STM) of iron atoms arranged in an oval on a corrugated copper surface.

Acronym	Technique	Application and description of technique
AES	Auger electron spectroscopy	Study of surface composition
EXAFS	Extended X-ray absorption fine structure	Estimation of internuclear distances around a central atom
FTIR	Fourier transform infrared spectroscopy	Study of adsorbed species
HREELS	High-resolution electron energy loss spectroscopy	Study of adsorbed species
LEED	Low-energy electron diffraction	Study of structural features of the surface and of adsorbed species
SIMS	Secondary ion mass spectrometry	Study of surface composition
STM	Scanning tunnelling microscopy	Obtaining images of a surface and adsorbed species at an atomic level
XANES	X-ray absorption near edge spectroscopy	Study of oxidation states of surface atoms
XRD	X-ray diffraction	Investigation of phases and particle sizes
XPS (ESCA)	X-ray photoelectron spectroscopy (electron spectroscopy for chemical analysis)	Study of surface composition and oxidation states of surface atoms

vehicle catalytic converters, small metal particles (e.g. 1600 pm in diameter) are dispersed on a support such as γ-alumina (*activated alumina*, see Section 13.7) which has a large surface area. Using a support of this type means that a high percentage of the metal atoms are available for catalysis. In some cases, the support itself may beneficially modify the properties of the catalyst. For example, in hydrocarbon reforming (Table 25.6), the metal and support operate together:

• the platinum-group metal catalyses the conversion of an alkane to alkene;
• isomerization of the alkene is facilitated by the acidic alumina surface;

• the platinum-group metal catalyses the conversion of the isomerized alkene to an alkane which is more highly branched than the starting hydrocarbon.

As well as having roles as supports for metals, silica and alumina are used directly as heterogeneous catalysts. A major application is in the catalytic cracking of heavy petroleum distillates. Very fine powders of silica and γ-alumina possess a huge surface area of $\approx 900 \, m^2 \, g^{-1}$. Large surface areas are a key property of zeolite catalysts (see Section 14.9), the selectivity of which can be tuned by varying the sizes, shapes and Brønsted acidity of their cavities and channels. We discuss these properties more fully in Section 25.8.

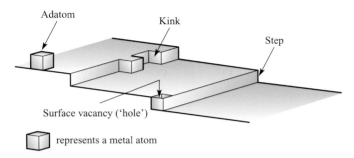

Fig. 25.14 A schematic representation of typical features of a metal surface. [Based on a figure from *Encyclopedia of Inorganic Chemistry* (1994), ed. R.B. King, vol. 3, p. 1359, Wiley, Chichester.]

25.8 Heterogeneous catalysis: commercial applications

In this section, we describe selected commercial applications of heterogeneous catalysts. The examples have been chosen to illustrate a range of catalyst types, as well as the development of motor vehicle catalytic converters.

Alkene polymerization: Ziegler–Natta catalysis and metallocene catalysts

The 1963 the Nobel Prize in Chemistry was awarded to Karl Ziegler and Giulio Natta 'for their discoveries in the field of the chemistry and technology of high polymers'. The polymerization of alkenes by heterogeneous Ziegler–Natta catalysis is of vast importance to the polymer industry. In 1953, Ziegler discovered that, in the presence of certain heterogeneous catalysts, ethene was polymerized to high-molecular-mass polyethene at relatively low pressures. In 1954, Natta showed that polymers formed using these catalytic conditions were *stereoregular*. When a terminal alkene, $RCH=CH_2$, polymerizes, the R groups in a linear polymer can be arranged as shown in Fig. 25.15. Consider polypropene in which R = Me. In the *isotactic* polymer, the methyl groups are all on the same side of the carbon chain. This gives a stereoregular polymer in which the chains pack efficiently, giving a crystalline material. *Syndiotactic* polypropene (Fig. 25.15, R = Me) is also of commercial value: the Me groups are regularly arranged on alternating sides of the carbon backbone. In contrast, *atactic* polymer contains a random arrangement of R groups and is soft and elastic.

First generation Ziegler–Natta catalysts were made by reacting $TiCl_4$ with Et_3Al to precipitate β-$TiCl_3 \cdot x AlCl_3$ which was converted to γ-$TiCl_3$. While the latter catalysed the production of isotactic polypropene, its selectivity and efficiency required significant improvement. A change in the method of catalyst preparation generated the δ-form of $TiCl_3$ which is stereoselective below 373 K. The co-catalyst,

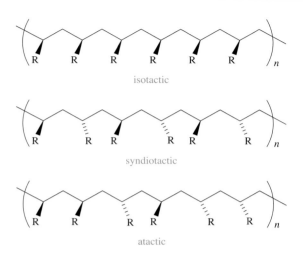

Fig. 25.15 The arrangement of R substituents in isotactic, syndiotactic and atactic linear polymers.

Et_2AlCl, in these systems is essential, its role being to alkylate Ti atoms on the catalyst surface. In third generation catalysts (used since the 1980s), $TiCl_4$ is supported on anhydrous $MgCl_2$, and Et_3Al is used for alkylation. Surface Ti(IV) is reduced to Ti(III) before coordination of the alkene (see below). The choice of $MgCl_2$ as the substrate arises from the close similarity between the crystal structures of $MgCl_2$ and β-$TiCl_3$. This allows *epitaxial* growth of $TiCl_4$ (or $TiCl_3$ after reduction) on $MgCl_2$.

> *Epitaxial* growth of a crystal on a substrate crystal is such that the growth follows the crystal axis of the substrate.

Alkene polymerization is catalysed at a surface Ti(III) centre in which there is a terminal Cl atom and a vacant coordination site. The *Cossee–Arlman mechanism* is the accepted pathway of the catalytic process and a simplified representation of the mechanism is shown in Fig. 25.16. Coordinatively unsaturated $TiCl_5$ units are the catalytically active sites. In the first step, the surface Cl atom is replaced by an ethyl group. It is crucial that the alkyl group is *cis* to the vacant coordination site to facilitate alkyl migration in the third step. In the second step, the alkene binds to Ti(III) and this is followed by alkyl migration. The repetition of these last two steps results in polymer growth. In propene polymerization, the stereoselective formation of isotactic polypropene is thought to be controlled by the catalyst's surface structure which imposes restrictions on the possible orientations of the coordinated alkene relative to the metal-attached alkyl group. Growth of the polymer is terminated by β-hydride elimination (the metal-bound H atom produced is transferred to an incoming alkene molecule to give a surface-bound alkyl group), or by reaction with H_2. The latter can be used to control the length

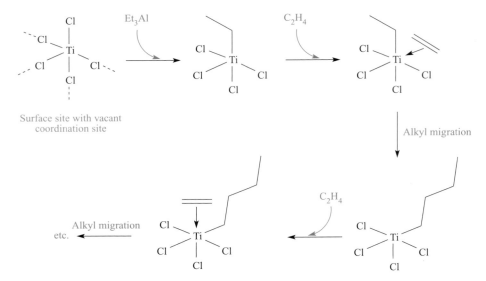

Fig. 25.16 A schematic representation of alkene polymerization on the surface of a Ziegler–Natta catalyst; the vacant coordination site must be *cis* to the coordinated alkyl group.

of the polymer chain. Heterogeneous $TiCl_3/Et_3Al$ or $MgCl_2/TiCl_4/Et_3Al$ catalysts are used industrially for the manufacture of isotactic polymers, e.g. polypropene. Only small quantities of syndiotactic polymers are produced by this route.

In addition to Ziegler–Natta catalysts, the modern polymer industry uses group 4 metallocene catalysts (see Box 24.6). Their development began in the 1970s with the observation that $(\eta^5\text{-}C_5H_5)_2MX_2$ (M = Ti, Zr, Hf) in the presence of methylaluminoxane $[MeAl(\mu\text{-}O)]_n$ catalysed the polymerization of propene. The stereospecificity of the catalysts was gradually improved (e.g. by changing the substituents on the cyclopentadienyl ring), and metallocene-based catalysts entered the commercial market in the 1990s. Although metallocenes can be used as homogeneous catalysts, for industrial purposes they are immobilized on SiO_2, Al_2O_3 or $MgCl_2$. Advantages of metallocenes over traditional Ziegler–Natta catalysts include the facts that, by changing the structure of the metallocene, the properties of the polymer may be tailored, narrow molar mass distributions can be obtained, and copolymers can be produced. Highly isotactic polypropene (e.g. using catalyst **25.27**) or syndiotactic polymers (e.g. using catalyst **25.28**) are manufactured, as well as block polymers with highly isotactic blocks or with purposely introduced irregularities (e.g. to lower the melting point). For example, isotactic polypropene with a melting point of 419 K and a molar mass of $\approx 33 \times 10^4 \, g \, mol^{-1}$ can be produced using catalyst **25.27**, whereas the product using **25.29** as catalyst melts at 435 K and has a molar mass of $\approx 99 \times 10^4 \, g \, mol^{-1}$. Note that each of metallocenes **25.27**–**25.29** contains a bridging group (CMe_2 or $SiMe_2$) that ties the cyclopentadienyl rings together and holds them in an open conformation.

Changing the tilt-angle between the rings is (in addition to the ring substitution pattern) a way of tuning catalytic behaviour.

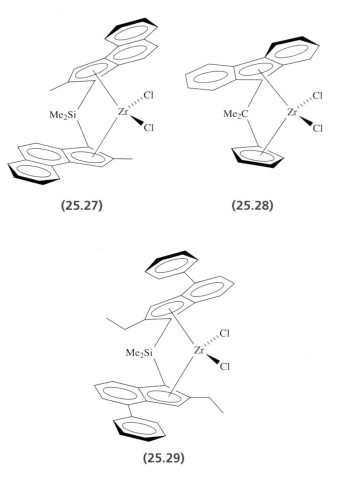

(25.27)   **(25.28)**

(25.29)

Most metallocene catalysts are active only in the presence of an $[MeAl(\mu\text{-}O)]_n$ cocatalyst. This alkylates the group 4 metal and also removes a chlorido-ligand, thereby creating a coordinatively unsaturated, cationic metal centre. The pathway for chain growth follows the Cossee–Arlman mechanism (Fig. 25.16 and eq. 25.31).

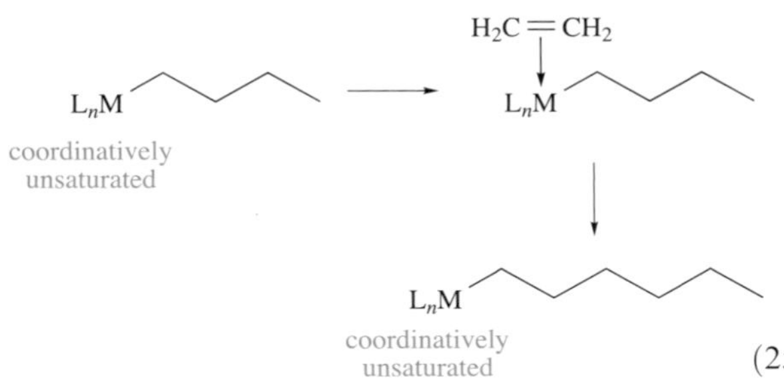

$$(25.31)$$

Self-study exercise

Propene polymerization by the Ziegler–Natta process can be summarized as follows.

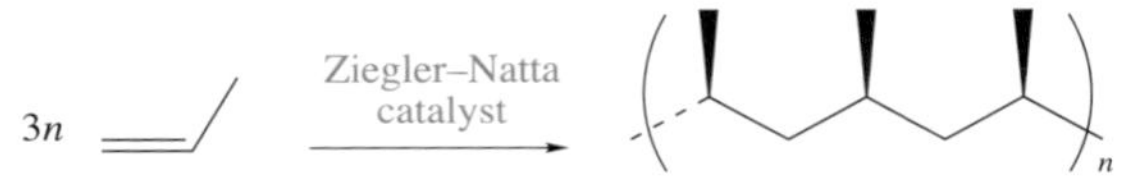

Comment on the type of polymer produced and the need for selectivity for this form of polypropene.

Fischer–Tropsch carbon chain growth

Scheme 25.32 summarizes the Fischer–Tropsch (FT) reaction, i.e. the conversion of synthesis gas (see Section 10.4) into hydrocarbons. A range of catalysts can be used (e.g. Ru, Ni, Fe, Co) but Fe and Co are currently favoured.

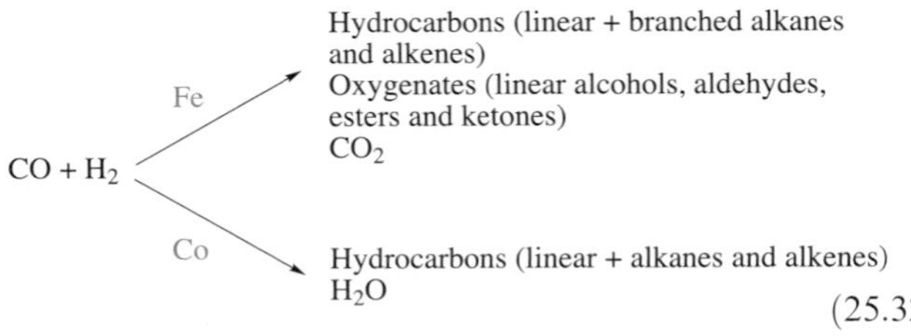

$$(25.32)$$

If petroleum is cheap and readily available, the FT process is not commercially viable and in the 1960s, many industrial plants were closed. In South Africa, the *Sasol process* continues to use H_2 and CO as feedstocks. Changes in the availability of oil reserves affect the views of industry as regards its feedstocks, and research interest in the FT reaction continues to be high. New initiatives in South Africa, Malaysia, New Zealand and the Netherlands are developing FT-based 'gas-to-liquid' fuels which use natural gas or biomass as the raw feedstock and convert it to liquid fuel.

The product distribution, including carbon chain length, of an FT reaction can be controlled by choice of catalyst, reactor design and reaction conditions. The addition of promoters such as group 1 or 2 metal salts (e.g. K_2CO_3) affects the selectivity of a catalyst. The exact mechanism by which the FT reaction occurs is not known, and many model studies have been carried out using discrete metal clusters (see Section 25.9). The original mechanism proposed by Fischer and Tropsch involved the adsorption of CO, C–O bond cleavage to give a surface carbide, and hydrogenation to produce CH_2 groups which then polymerized. Various mechanisms have been put forward, and the involvement of a surface-bound CH_3 group has been debated. Any mechanism (or series of pathways) must account for the formation of surface carbide, graphite and CH_4, and the distribution of organic products shown in scheme 25.32. Current opinion favours CO dissociation on the catalyst surface to give surface C and O and, in the presence of adsorbed H atoms (eq. 25.30), the formation of surface CH and CH_2 units and release of H_2O. If CO dissociation and subsequent formation of CH_x groups is efficient (as it is on Fe), the build-up of CH_x units leads to reaction between them and to the growth of carbon chains. The types of processes that might be envisaged on the metal surface are represented in scheme 25.33. Reaction of the surface-attached alkyl chain would release an alkane. If it undergoes β-elimination, an alkene is released.

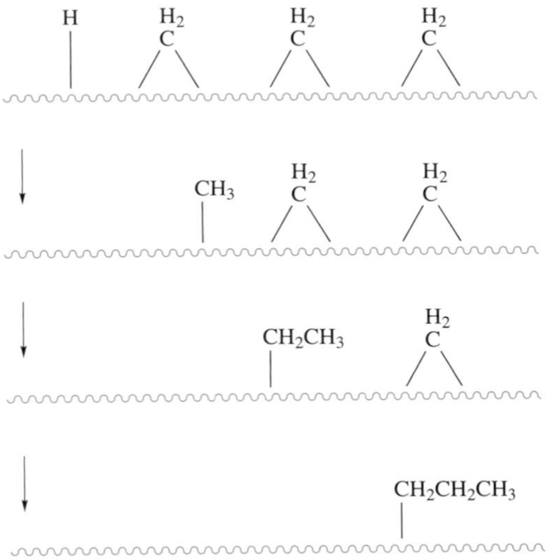

$$(25.33)$$

It has also been suggested that vinylic species are involved in FT chain growth, and that combination of surface-bound CH and CH_2 units to give $CH=CH_2$ may be followed by successive incorporation of CH_2 units alternating with alkene isomerization as shown in scheme 25.34. Release of a terminal alkene results if reaction of the adsorbate is with H instead of CH_2.

(25.34)

Haber process

The vast scale on which the industrial production of NH_3 is carried out and its growth over the latter part of the 20th century was illustrated in Box 15.3. In eq. 15.21 and the accompanying discussion, we described the manufacture of NH_3 using a heterogeneous catalyst. Now we focus on the mechanism of the reaction and on catalyst performance.

Without a catalyst, the reaction between N_2 and H_2 occurs only slowly, because the activation barrier for the dissociation of N_2 and H_2 in the gas phase is very high. In the presence of a suitable catalyst such as Fe, dissociation of N_2 and H_2 to give adsorbed atoms is facile, with the energy released by the formation of M–N and M–H bonds more than offsetting the energy required for N≡N and H–H fission. The adsorbates then readily combine to form NH_3 which desorbs from the surface. The rate-determining step is the dissociative adsorption of N_2 (eq. 25.35). The notation '(ad)' refers to an adsorbed atom.

$$N_2(g)$$

$$\longrightarrow \quad N(ad) \quad N(ad) \qquad (25.35)$$

Catalyst surface

Dihydrogen is similarly adsorbed (eq. 25.30), and the surface reaction continues as shown in scheme 25.36 with

gaseous NH_3 finally being released. Activation barriers for each step are relatively low.

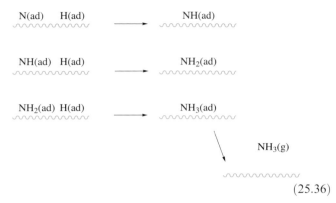

(25.36)

Metals other than Fe catalyse the reaction between N_2 and H_2, but the rate of formation of NH_3 is metal-dependent. High rates are observed for Fe, Ru and Os. Since the rate-determining step is the chemisorption of N_2, a high activation energy for this step, as is observed for late d-block metals (e.g. Co, Rh, Ir, Ni and Pt), slows down the overall formation of NH_3. Early d-block metals such as Mo and Re chemisorb N_2 efficiently, but the M–N interaction is strong enough to favour retention of the adsorbed atoms. This blocks surface sites and inhibits further reaction. The catalyst used industrially is active α-Fe which is produced by reducing Fe_3O_4 mixed with K_2O (an *electronic promoter* which improves catalytic activity), SiO_2 and Al_2O_3 (*structural promoters* which stabilize the catalyst's structure). High-purity (often synthetic) magnetite and the catalyst promoters are melted electrically and then cooled. This stage distributes the promoters homogeneously within the catalyst. The catalyst is then ground to an optimum grain size. High-purity materials are essential since some impurities poison the catalyst. Dihydrogen for the Haber process is produced as synthesis gas (Section 10.4), and contaminants such as H_2O, CO, CO_2 and O_2 are *temporary catalyst poisons*. Reduction of the Haber process catalyst restores its activity, but over-exposure of the catalyst to oxygen-containing compounds decreases the efficiency of the catalyst irreversibly. A 5 ppm CO content in the H_2 supply (see eqs. 10.13 and 10.14) decreases catalyst activity by $\approx 5\%$ per year. The performance of the catalyst depends critically on the operating temperature of the NH_3 converter, and a 770–790 K range is optimal.

Self-study exercises

1. Write equations to show how H_2 is manufactured for use in the Haber process. [*Ans.* See eqs. 10.13 and 10.14]

2. The catalytic activity of various metals with respect to the reaction of N_2 and H_2 to give NH_3 varies in the order Pt < Ni < Rh $\approx$ Re < Mo < Fe < Ru $\approx$ Os.

What factors contribute towards this trend?

[*Ans.* See text in this section]

3. In 2009, 130 Mt of NH_3 (the mass is in terms of nitrogen content) were manufactured worldwide. Production has increased dramatically over the last 40 years. Account for the scale of production in terms of the uses of NH_3.

[*Ans.* See Box 15.3]

Production of SO_3 in the Contact process

Production of sulfuric acid, ammonia and phosphate rock (see Section 15.2) heads the inorganic chemical and mineral industries in the US. The oxidation of SO_2 to SO_3 (equ. 25.37) is the first step in the Contact process, and in Section 16.8 we discussed how the yield of SO_3 depends on temperature and pressure. At ordinary temperatures, the reaction is too slow to be commercially viable, while at very high temperatures, equilibrium 25.37 shifts to the left, decreasing the yield of SO_3.

$$2SO_2 + O_2 \rightleftharpoons 2SO_3 \quad \Delta_r H^\circ = -96 \text{ kJ per mole of } SO_2 \tag{25.37}$$

Use of a catalyst increases the rate of the forward reaction 25.37, and active catalysts are Pt, V(V) compounds and iron oxides. Modern manufacturing plants for SO_3 use a V_2O_5 catalyst on an SiO_2 carrier (which provides a large surface area) with a K_2SO_4 promoter. The catalyst system contains 4–9% by weight of V_2O_5. Passage of the reactants through a series of catalyst beds is required to obtain an efficient conversion of SO_2 to SO_3, and an operating temperature of 690–720 K is optimal. Since oxidation of SO_2 is exothermic and since temperatures >890 K degrade the catalyst, the $SO_2/SO_3/O_2$ mixture must be cooled between leaving one catalyst bed and entering the next. Although the $V_2O_5/SiO_2/K_2SO_4$ system is introduced as a solid catalyst, the operating temperatures are such that the catalytic oxidation of SO_2 occurs in a liquid melt on the surface of the silica carrier.

(25.30)

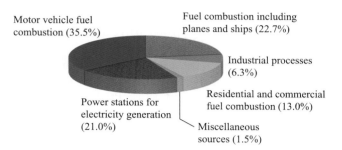

Fig. 25.17 Sources of NO_x emissions in the US. [Data: Environmental Protection Agency (2005).]

The mechanism of catalysis is complicated and has not been fully established. Initially, the liquid catalyst takes up large amounts of SO_2, and the accepted working model for the catalytic system is represented as $M_2S_2O_7$–M_2SO_4–V_2O_5/O_2–SO_2–SO_3–N_2 (M = Na, K, Rb, Cs). At normal operating temperatures, $[V(O)_2(SO_4)]^-$, the complex **25.30**, and related vanadium(V) oligomers are formed. Complex **25.30** in particular is considered to be catalytically active, while any V(III) or V(IV) species are thought to be catalytically inactive. One proposal suggests that complex **25.30** activates O_2, facilitating the oxidation of SO_2 to SO_3. The direct reaction of **25.30** with SO_2 to yield SO_3 results in reduction of V(V) to V(IV) and the formation of a catalytically inactive species. Much work remains to elucidate the details of the Contact process.

Catalytic converters

Environmental concerns have grown during the past few decades (see, for example, Box 10.2), and to the general public, the use of motor vehicle catalytic converters is well known. Regulated exhaust emissions[†] comprise CO, hydrocarbons and NO_x (see Section 15.8). The radical NO is one of several species that act as catalysts for the conversion of O_3 to O_2 and is considered to contribute to depletion of the ozone layer. Although industrial processes and the generation of electricity (see Box 12.2) contribute to NO_x emissions,[‡] the combustion of transport fuels is the major source (Fig. 25.17). A typical catalytic converter is ≥90% efficient in reducing emissions. In 2005, European regulations called for emission levels of CO, hydrocarbons and NO_x to be ≤1.0, 0.10 and 0.08 g km^{-1}, respectively, for passenger cars with petrol engines. The toughest regulations to meet are those laid down in California (the Super Ultra Low Emissions Vehicle, SULEV, standards). SULEV

[†] For reports on the current status of motor vehicle emission control, see: M.V. Twigg (2003) *Platinum Metals Rev.*, vol. 47, p. 157; M.V. Twigg and P.R. Phillips (2009) *Platinum Metals Rev.*, vol. 53, p. 27.

[‡] Shell and Bayer are among companies that have introduced processes to eliminate industrial NO_x emissions: *Chemistry & Industry* (1994) p. 415 – 'Environmental technology in the chemical industry'.

regulates emission levels of CO, hydrocarbons, NO_x and particulate matter to ≤ 0.62, 0.006, 0.012 and 0.006 $g\,km^{-1}$, respectively.

A catalytic converter consists of a honeycomb ceramic structure coated in finely divided Al_2O_3 (the *washcoat*). Fine particles of catalytically active Pt, Pd and Rh are dispersed within the cavities of the washcoat and the whole unit is contained in a stainless steel vessel placed in sequence in the vehicle's exhaust pipe. As the exhaust gases pass through the converter at high temperatures, redox reactions 25.38–25.42 occur (C_3H_8 is a representative hydrocarbon). Under legislation, the only acceptable emission products are CO_2, N_2 and H_2O.

$$2CO + O_2 \longrightarrow 2CO_2 \tag{25.38}$$

$$C_3H_8 + 5O_2 \longrightarrow 3CO_2 + 4H_2O \tag{25.39}$$

$$2NO + 2CO \longrightarrow 2CO_2 + N_2 \tag{25.40}$$

$$2NO + 2H_2 \longrightarrow N_2 + 2H_2O \tag{25.41}$$

$$C_3H_8 + 10NO \longrightarrow 3CO_2 + 4H_2O + 5N_2 \tag{25.42}$$

Whereas CO and hydrocarbons are oxidized, the destruction of NO_x involves its reduction. Modern catalytic converters have a 'three-way' system which promotes both oxidation and reduction. Pd and Pt catalyse reactions 25.38 and 25.39, while Rh catalyses reactions 25.40 and 25.41, and Pt catalyses reaction 25.42.

The efficiency of the catalyst depends, in part, on metal particle size, typically 1000–2000 pm diameter. Over a period of time, the high temperatures needed for the operation of a catalytic converter cause ageing of the metal particles with a loss of their optimal size and a decrease in the efficiency of the catalyst. Constant high-temperature running also transforms the γ-Al_2O_3 support into a phase with a lower surface area, again reducing catalytic activity. To counter degradation of the support, group 2 metal oxide stabilizers are added to the alumina. Catalytic converters operate only with unleaded fuels; lead additives bind to the alumina washcoat, deactivating the catalyst.

In order to achieve the regulatory emission standards, it is crucial to control the air:fuel ratio as it enters the catalytic converter: the optimum ratio is 14.7:1. If the air:fuel ratio exceeds 14.7:1, extra O_2 competes with NO for H_2 and the efficiency of reaction 25.41 is lowered. If the ratio is less than 14.7:1, oxidizing agents are in short supply and CO, H_2 and hydrocarbons compete with each other for NO and O_2. The air:fuel ratio is monitored by a sensor fitted in the exhaust pipe; the sensor measures O_2 levels and sends an electronic signal to the fuel injection system or carburettor to adjust the air:fuel ratio as necessary. Catalytic converter design also includes a CeO_2/Ce_2O_3 system to store oxygen. During 'lean' periods of vehicle running, O_2 can be 'stored' by reaction 25.43. During 'rich' periods

when extra oxygen is needed for hydrocarbon and CO oxidation, CeO_2 is reduced (eq. 25.44 shows oxidation of CO).

$$2Ce_2O_3 + O_2 \longrightarrow 4CeO_2 \tag{25.43}$$

$$2CeO_2 + CO \longrightarrow Ce_2O_3 + CO_2 \tag{25.44}$$

A catalytic converter cannot function immediately after the 'cold start' of an engine. At its 'light-off' temperature (typically 620 K), the catalyst operates at 50% efficiency but during the 90–120 s lead time, exhaust emissions are not controlled. Several methods have been developed to counter this problem, e.g. electrical heating of the catalyst using power from the vehicle's battery.

The development of catalytic converters has recently encompassed the use of zeolites, e.g. Cu-ZSM-5 (a copper-modified ZSM-5 system), but at the present time, and despite some advantages such as low light-off temperatures, zeolite-based catalysts have not shown themselves to be sufficiently durable for their use in catalytic converters to be commercially viable.

Zeolites as catalysts for organic transformations: uses of ZSM-5

For an introduction to zeolites, see Fig. 14.27 and the accompanying discussion. Many natural and synthetic zeolites are known, and it is the presence of well-defined cavities and/or channels, the dimensions of which are comparable with those of small molecules, that makes them invaluable as catalysts and molecular sieves. Zeolites are environmentally 'friendly' and the development of industrial processes in which they can replace less acceptable acid catalysts is advantageous. In this section, we focus on catalytic applications of synthetic zeolites such as ZSM-5 (structure-type code MFI, Fig. 25.18); the latter is silicon-rich with composition $Na_n[Al_nSi_{96-n}O_{192}] \cdot \approx 16H_2O$ ($n < 27$).[†] When H^+ replaces Na^+, the zeolite is referred to as HZSM-5 and this is highly catalytically active (see below). Within the aluminosilicate framework of ZSM-5 lies a system of interlinked channels. One set can be seen in Fig. 25.18, but the channels can be represented in the form of a structure such as **25.31**. In ZSM-5, for example, there are two sets of channels running through the structure, one of cross-section $\approx 540 \times 560$ pm and the other of cross-section $\approx 510 \times 540$ pm. The *effective pore size* is comparable to the *kinetic molecular diameter* of a molecule such as 2-methylpropane or benzene, leading to the *shape-selective* properties of zeolite catalysts. The effective pore size differs from that determined crystallographically because it takes into account the flexibility of the zeolite framework as a function of temperature. Similarly,

---

[†] Structures of zeolites can be viewed and manipulated using the website: http://www.iza-structure.org/databases/

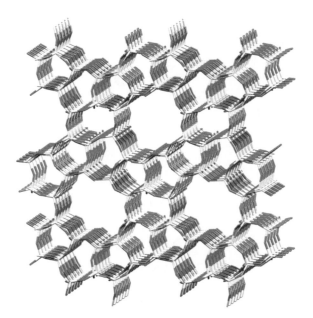

Fig. 25.18 Part of the aluminosilicate framework of synthetic zeolite ZSM-5 (structure-type MFI). Colour code: Si/Al, pale grey; O, red.

the kinetic molecular diameter allows for the molecular motions of species entering the zeolite channels or cavities.

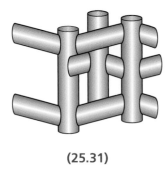

(25.31)

The high catalytic activity of zeolites arises from the Brønsted acidity of Al sites, represented in resonance pair **25.32**. The Si:Al ratio affects the number of such sites and acid strength of the zeolite.

Zeolite catalysts are important in the catalytic cracking of heavy petroleum distillates. Their high selectivities and high

rates of reactions, coupled with reduced coking effects, are major advantages over the activities of the alumina/silica catalysts that zeolites have replaced. Ultrastable Y (USY) zeolites are usually chosen for catalytic cracking because their use leads to an increase in the gasoline (motor fuels) octane number. It is essential that the catalyst is robust enough to withstand the conditions of the cracking process. Both USY and ZSM-5 (used as a co-catalyst because of its shape-selective properties) meet this requirement. The shape-selectivity of ZSM-5 is also crucial to its activity as a catalyst in the conversion of methanol to hydrocarbon fuels. The growth of carbon chains is restricted by the size of the zeolite channels and this gives a selective distribution of hydrocarbon products. In the 1970s, Mobil developed the MTG (methanol-to-gasoline) process in which ZSM-5 catalysed the conversion of MeOH to a mixture of higher ($> C_5$) alkanes, cycloalkanes and aromatics. Equations 25.45–25.47 show the initial dehydration of methanol (in the gas phase) to give dimethyl ether, followed by representative dehydrations leading to hydrocarbons. Such processes are commercially viable only when petroleum prices are high. This was the case in the 1970s and 1980s, and the MTG process was run by Mobil during the 1980s in New Zealand.

$$2CH_3OH \xrightarrow{\text{ZSM-5 catalyst}} CH_3OCH_3 + H_2O \quad (25.45)$$

$$2CH_3OCH_3 + 2CH_3OH \xrightarrow{\text{ZSM-5 catalyst}} C_6H_{12} + 4H_2O \quad (25.46)$$

$$3CH_3OCH_3 \xrightarrow{\text{ZSM-5 catalyst}} C_6H_{12} + 3H_2O \quad (25.47)$$

Recent advances have shown zeolites are effective in catalysing the direct conversion of synthesis gas to motor fuels. The MTO (methanol-to-olefins) process converts MeOH to C_2–C_4 alkenes and is also catalysed by ZSM-5. The development of a gallium-modified ZSM-5 catalyst (Ga-ZSM-5) has provided an efficient catalyst for the production of aromatic compounds from mixtures of C_3 and C_4 alkanes (commonly labelled LPG).

Zeolites are replacing acid catalysts in a number of manufacturing processes. One of the most important is the alkylation of aromatics. The Mobil–Badger process for producing C_6H_5Et from C_6H_6 and C_2H_4 provides the

(25.32)

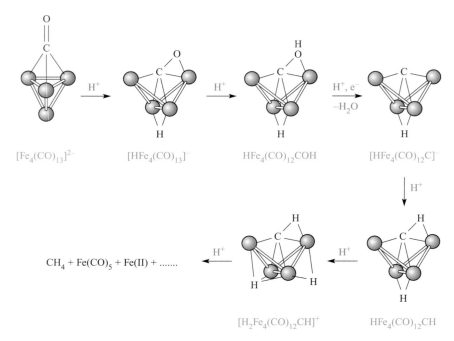

Fig. 25.19 The proton-induced conversion of a cluster-bound CO ligand to CH_4: a cluster model for catalysed hydrogenation of CO on an Fe surface. Each green sphere represents an $Fe(CO)_3$ unit.

precursor for styrene (and hence polystyrene) manufacture. The isomerization of 1,3- to 1,4-dimethylbenzene (xylenes) is also catalysed on the acidic surface of ZSM-5, presumably with channel shape and size playing an important role in the observed selectivity.

Self-study exercises

1. What are the similarities and differences between the structures of a feldspar mineral and a zeolite?

2. How does a zeolite function as a Lewis acid catalyst?

3. Give two examples of the commercial application of ZSM-5 as a catalyst.

[*Answers*: See Sections 14.9 and 25.8]

25.9 Heterogeneous catalysis: organometallic cluster models

One of the driving forces behind organometallic cluster research is to model metal-surface catalysed processes such as the Fischer–Tropsch reaction. The *cluster-surface analogy* assumes that discrete organometallic clusters containing *d*-block metal atoms are realistic models for the bulk metal. In many small clusters, the arrangements

of the metal atoms mimic units from close-packed arrays, e.g. the M_3-triangle and M_4-butterfly in structures **25.25** and **25.26**. The success of modelling studies has been limited, but a well-established and much-cited result is that shown in Fig. 25.19.[†]

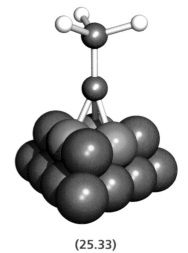

(25.33)

[†] For further details, see M.A. Drezdon, K.H. Whitmire, A.A. Bhattacharyya, W.-L. Hsu, C.C. Nagel, S.G. Shore and D.F. Shriver (1982) *J. Am. Chem. Soc.*, vol. 104, p. 5630 – 'Proton induced reduction of CO to CH_4 in homonuclear and heteronuclear metal carbonyls'.

Model studies involve transformations of organic fragments which are proposed as surface intermediates, but do not necessarily address a complete sequence of reactions as is the case in Fig. 25.19. For example, metal-supported ethylidyne units (**25.33**) are proposed as intermediates in the Rh- or Pt-catalysed hydrogenation of ethene, and there has been much interest in the chemistry of M_3-clusters such as $H_3Fe_3(CO)_9CR$, $H_3Ru_3(CO)_9CR$ and $Co_3(CO)_9CR$ which contain ethylidyne or other alkylidyne units. In the presence of base, $H_3Fe_3(CO)_9CMe$ undergoes reversible deprotonation and loss of H_2 (eq. 25.48), perhaps providing a model for an organic fragment transformation on a metal surface.

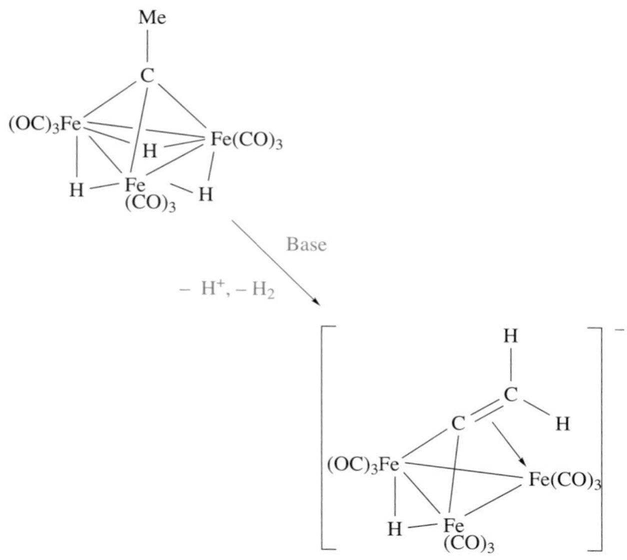

(25.48)

KEY TERMS

The following terms have been used in this chapter. Do you know what they mean?

- catalyst
- catalyst precursor
- autocatalytic
- homogeneous catalyst
- heterogeneous catalyst
- catalytic cycle
- catalytic turnover number
- catalytic turnover frequency
- alkene metathesis
- alkyne metathesis
- Grubbs' catalyst
- Schrock-type catalysts
- Chauvin mechanism
- catenand

- catenate
- coordinatively unsaturated
- Wilkinson's catalyst
- asymmetric hydrogenation
- prochiral
- enantiomeric excess
- Monsanto acetic acid process
- Cativa acetic acid process
- Tennessee–Eastman acetic anhydride process
- hydroformylation (Oxo-process)
- chemoselectivity and regioselectivity (with respect to hydroformylation)
- biphasic catalysis

- physisorption
- chemisorption
- adsorbate
- alkene polymerization
- Ziegler–Natta catalysis
- Cossee–Arlman mechanism
- Fischer–Tropsch reaction
- Haber process
- Contact process
- catalytic converter
- zeolite

FURTHER READING

General texts

G.P. Chiusoli and P.M. Maitlis (eds) (2008) *Metal-catalysis* in *Industrial Organic Processes*, Royal Society of Chemistry, Cambridge – A detailed book that covers C–O and C–C bond formation, hydrogenation, syntheses involving CO, alkene metathesis, and polymerization as well as general aspects of catalysis.

B. Cornils and W.A. Hermann (eds) (2002) *Applied Homogeneous Catalysis with Organometallic Compounds*, 2nd edn, Wiley-VCH, Weinheim (3 volumes) – This detailed 3-volume edition covers applications of catalysts and their development.

F.A. Cotton, G. Wilkinson, M. Bochmann and C. Murillo (1999) *Advanced Inorganic Chemistry*, 6th edn, Wiley Interscience, New York – Chapter 22 gives a full account of the homogeneous catalysis of organic reactions by *d*-block metal compounds.

Homogeneous catalysis

B. Alcaide, P. Almendros and A. Luna (2009) *Chem. Rev.*, vol. 109, p. 3817 – 'Grubbs' ruthenium-carbene beyond the metathesis reaction: Less conventional non-metathetic utility'. A review which looks at new applications of Grubbs' catalysts.

A. Fürstner (2000) *Angew. Chem. Int. Ed.*, vol. 39, p. 3012 – 'Olefin metathesis and beyond': a review that considers catalyst design and applications in alkene metathesis.

A. Fürstner and P.W. Davies (2005) *Chem. Commun.*, p. 2307 – A review: 'Alkyne metathesis'.

R.H. Grubbs (2004) *Tetrahedron*, vol. 60, p. 7117 – 'Olefin metathesis' gives an overview of the development and mechanistic details of Grubbs' catalysts.

A. Haynes (2007) in *Comprehensive Organometallic Chemistry III*, eds R.H. Crabtree and D.M.P. Mingos, Elsevier, Oxford, vol. 7, p. 427 – 'Commercial applications of iridium complexes in homogeneous catalysis': A review dealing with modern industrial processes utilizing Ir-based homogeneous catalysts.

A. Haynes (2010) *Adv. Catal.*, vol. 53, p. 1 – 'Catalytic methanol carbonylation': An up-to-date review of the Monsanto and Cativa processes including background information.

M.J. Krische and Y. Sun (eds) (2007) *Acc. Chem. Res.*, vol. 40, issue 12 – A special issue containing reviews on the theme of hydrogenation and transfer hydrogenation.

P.W.N.M. van Leeuwen and Z. Freixa (2007) in *Comprehensive Organometallic Chemistry III*, eds R.H. Crabtree and D.M.P. Mingos, Elsevier, Oxford, vol. 7, p. 237 - 'Application of rhodium complexes in homogeneous catalysis with carbon monoxide': A review of Rh-based homogeneous catalysts in the hydroformylation of alkenes and carbonylation of methanol.

W.E. Piers and S. Collins (2007) in *Comprehensive Organometallic Chemistry III*, eds R.H. Crabtree and D.M.P. Mingos, Elsevier, Oxford, vol. 1, p. 141 – 'Mechanistic aspects of olefin-polymerization catalysis': Detailed mechanistic review that includes homogeneous group 4 metallocene catalysts.

R.R. Schrock and A.H. Hoveyda (2003) *Angew. Chem. Int. Ed.*, vol. 42, p. 4592 – 'Molybdenum and tungsten imido alkylidene complexes as efficient olefin-metathesis catalysts'.

C.M. Thomas and G. Süss-Fink (2003) *Coord. Chem. Rev.*, vol. 243, p. 125 – 'Ligand effects in the rhodium-catalyzed carbonylation of methanol'.

T.M. Trnka and R.H. Grubbs (2001) *Acc. Chem. Res.*, vol. 34, p. 18 – 'The development of $L_2X_2Ru=CHR$ olefin metathesis catalysts: an organometallic success story': An insight into Grubbs' catalysts by their discoverer.

Heterogeneous catalysis including specific industrial processes

L.L. Böhm (2003) *Angew. Chem. Int. Ed.*, vol. 42, p. 5010 – 'The ethylene polymerization with Ziegler catalysts: fifty years after the discovery'.

M.E. Dry (2002) *Catal. Today*, vol. 71, p. 227 – 'The Fischer–Tropsch process: 1950–2000'.

G. Ertl, H. Knözinger, F. Schüth and J. Weitkamp (eds) (2008) *Handbook of Heterogeneous Catalysis*, 2nd edn. (8 volumes), Wiley-VCH, Weinheim – An encyclopedic account of heterogeneous catalysis.

P. Galli and G. Vecellio (2004) *J. Polym. Sci.*, vol. 42, p. 396 – 'Polyolefins: The most promising large-volume materials for the 21st century'.

J. Grunes, J. Zhu and G.A. Somorjai (2003) *Chem. Commun.*, p. 2257 – 'Catalysis and nanoscience'.

J.F. Haw, W. Song, D.M. Marcus and J.B. Nicholas (2003) *Acc. Chem. Res.*, vol. 36, p. 317 – 'The mechanism of methanol to hydrocarbon catalysis'.

A. de Klerk (2007) *Green Chem.*, vol. 9, p. 560 – 'Environmentally friendly refining: Fischer–Tropsch *versus* crude oil': A comparison of the refining of crude oil and the products of the FT process.

O.B. Lapina, B.S. Bal'zhinimaev, S. Boghosian, K.M. Eriksen and R. Fehrmann (1999) *Catal. Today*, vol. 51, p. 469 – 'Progress on the mechanistic understanding of SO_2 oxidation catalysts'.

S.C. Larsen (2007) *J. Phys. Chem. C*, vol. 111, p. 18464 – 'Nanocrystalline zeolites and zeolite structures: Synthesis, characterization, and applications'.

R. Schlögl (2003) *Angew. Chem. Int. Ed.*, vol. 42, p. 2004 – 'Catalytic synthesis of ammonia – A "never-ending story"?'

G.A. Somorjai, A.M. Contreras, M. Montano and R.M. Rioux (2006) *Proc. Natl. Acad. Sci.*, vol. 103, p. 10577 – 'Clusters, surfaces, and catalysis'.

G. Wilke (2003) *Angew. Chem. Int. Ed.*, vol. 42, p. 5000 – 'Fifty years of Ziegler catalysts: Consequences and development of an invention'.

Industrial processes: general

J. Hagen (2006) *Industrial Catalysis*, 2nd edn., Wiley-VCH, Weinheim – Covers both homogeneous and heterogeneous catalysis, including catalyst production, testing and development.

Ullmann's Encyclopedia of Industrial Inorganic Chemicals and Products (1998) Wiley-VCH, Weinheim – Six volumes with detailed accounts of industrial processes involving inorganic chemicals.

R.I. Wijngaarden and K.R. Westerterp (1998) *Industrial Catalysts*, Wiley-VCH, Weinheim – A book that focuses on practical aspects of applying catalysts in industry.

Biphasic catalysis

L.P. Barthel-Rosa and J.A. Gladysz (1999) *Coord. Chem. Rev.*, vol. 190–192, p. 587 – 'Chemistry in fluorous media: a user's guide to practical considerations in the application of fluorous catalysts and reagents'.

B. Cornils and W.A. Hermann (eds) (1998) *Aqueous-phase Organometallic Catalysis: Concepts and Applications*, Wiley-VCH, Weinheim – A detailed account.

A.P. Dobbs and M.R. Kimberley (2002) *J. Fluorine Chem.*, vol. 118, p. 3 – 'Fluorous phase chemistry: A new industrial technology'.

N. Pinault and D.W. Bruce (2003) *Coord. Chem. Rev.*, vol. 241, p. 1 – 'Homogeneous catalysts based on water-soluble phosphines'.

D.M. Roundhill (1995) *Adv. Organomet. Chem.*, vol. 38, p. 155 – 'Organotransition-metal chemistry and homogeneous catalysis in aqueous solution'.

E. de Wolf, G. van Koten and B.-J. Deelman (1999) *Chem. Soc. Rev.*, vol. 28, p. 37 – 'Fluorous phase separation techniques in catalysis'.

Polymer-supported catalysts

B. Clapham, T.S. Reger and K.D. Janda (2001) *Tetrahedron*, vol. 57, p. 4637 – 'Polymer-supported catalysis in synthetic organic chemistry'.

PROBLEMS

25.1 (a) Analyse the catalytic cycle shown in Fig. 25.20, identifying the types of reactions occurring. (b) Why does this process work best for R' = vinyl, benzyl or aryl groups?

25.2 Give equations that illustrate each of the following processes. Define any abbreviations used.

(a) Cross-metathesis between two alkenes.
(b) Alkyne metathesis catalysed by a high oxidation state metal alkylidyne complex $L_nM\equiv CR$.
(c) ROMP.

25.3 Suggest a suitable catalyst for the following reaction, and outline the initial steps in the mechanism of the reaction:

25.4 The isomerization of alkenes is catalysed by $HCo(CO)_3$ and Fig. 25.21 shows the relevant catalytic cycle. (a) $HCo(CO)_4$ is a catalyst precursor; explain what this means. (b) Give a fuller description of what is happening in each of the steps shown in Fig. 25.21.

25.5 Outline the catalytic processes involved in the manufacture of acetic acid (Monsanto process) and acetic anhydride (Tennessee–Eastman process), and compare the catalytic pathways.

25.6 (a) Of the following alkenes, which are prochiral: PhHC=CHPh, PhMeC=CHPh, H_2C=CHPh, H_2C=C(CO$_2$H)(NHC(O)Me)?
(b) If an asymmetric hydrogenation proceeds with 85% ee favouring the *R*-enantiomer, what is the percentage of each enantiomer formed?

25.7 (a) Assuming some similarity between the mechanism of hydroformylation using $HCo(CO)_4$ and $HRh(CO)(PPh_3)_3$ as catalysts, propose a mechanism for the conversion of RCH=CH$_2$ to RCH$_2$CH$_2$CHO and explain what is happening in each step.
(b) 'The regioselectivity of the hydroformylation of RCH=CH$_2$ catalysed by $HRh(CO)(PPh_3)_3$ drops when the temperature is increased'. Explain what is meant by this statement.

25.8 The hydroformylation of pent-2-ene using $Co_2(CO)_8$ as the catalyst was found to give rise to three aldehydes in a ratio 35:12:5. Show how the three products arose, and suggest which was formed in the most and which in the least amount.

25.9 (a) The hydrogenation of propene is catalysed by RhCl(PPh$_3$)$_3$ or HRh(CO)(PPh$_3$)$_3$. Outline the mechanisms by which these reactions occur,

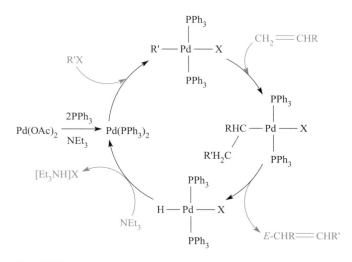

Fig. 25.20 Catalytic cycle for use in problem 25.1.

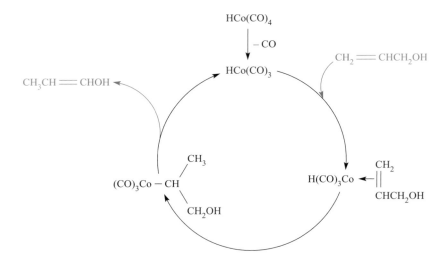

Fig. 25.21 Catalytic cycle for use in problem 25.4.

indicating clearly what the active catalyst is in each case.

(b) HRuCl(PPh$_3$)$_3$ is a very active catalyst for the hydrogenation of alkenes. However, at high catalyst concentrations and in the absence of sufficient H$_2$, orthometallation of the catalyst may accompany alkene hydrogenation. Write a reaction scheme to illustrate this process, and comment on its effect on the activity of the catalyst.

25.10 (a) Ligand **25.19** is used in biphasic catalysis. The IR spectrum of Fe(CO)$_4$(PPh$_3$) shows strong absorptions at 2049, 1975 and 1935 cm^{-1}, while that of [Fe(CO)$_4$(**25.19**)]$^+$ exhibits bands at 2054, 1983 and 1945 cm^{-1}. What can you deduce from these data?

(b) Which of the complexes [X][Ru(**25.34**)$_3$] in which X$^+$ = Na$^+$, [^{n}Bu$_4$N]$^+$ or [Ph$_4$P]$^+$ might be suitable candidates for testing in biphasic catalysis using aqueous medium for the catalyst?

(25.34)

25.11 Give a brief discussion of the use of homogeneous catalysis in selected industrial manufacturing processes.

25.12 For the catalysed hydrocyanation of buta-1,3-diene:

$$CH_2=CHCH=CH_2 \xrightarrow{HCN} NC(CH_2)_4CN$$

(a step in the manufacture of nylon-6,6), the catalyst precursor is NiL$_4$ where L = P(OR)$_3$. Consider the addition of only the first equivalent of HCN.

(a) Some values of K for:

$$NiL_4 \rightleftharpoons NiL_3 + L$$

are 6×10^{-10} for R = 4-MeC$_6$H$_4$, 3×10^{-5} for

R = ^{i}Pr and 4×10^{-2} for R = 2-MeC$_6$H$_4$. Comment on the trend in values and on the relevance of these data to the catalytic process.

(b) The first three steps in the proposed catalytic cycle are the addition of HCN to the active catalyst, loss of L, and the addition of buta-1,3-diene with concomitant H migration. Draw out this part of the catalytic cycle.

(c) Suggest the next step in the cycle, and discuss any complications.

25.13 H$_2$Os$_3$(CO)$_{10}$ (**25.35**) catalyses the isomerization of alkenes:

$$RCH_2CH=CH_2 \longrightarrow E\text{-}RCH=CHMe + Z\text{-}RCH=CHMe$$

(a) By determining the cluster valence electron count for H$_2$Os$_3$(CO)$_{10}$ deduce what makes this cluster an effective catalyst.

(b) Propose a catalytic cycle that accounts for the formation of the products shown.

(25.35)

25.14 Describe briefly why a clean nickel surface (fcc structure) should not be regarded as comprising a perfect close-packed array of atoms. Indicate the arrangements of atoms that an adsorbate might encounter on the surface, and suggest possible modes of attachment for CO.

25.15 (a) What advantages are there to using Rh supported on γ-Al$_2$O$_3$ as a catalyst rather than the bulk metal?

(b) In a catalytic converter, why is a combination of platinum-group metals used?

25.16 The forward reaction in eq. 25.37 is exothermic. What are the effects of (a) increased pressure and (b) increased temperature on the yield of SO_3? (c) In trying to optimize both the yield and rate of formation of SO_3, what problem does the Contact process encounter and how is it overcome?

25.17 (a) Outline how the gaseous reaction between N_2 and H_2 proceeds in the presence of a heterogeneous catalyst, and state why a catalyst is needed for the commercial production of NH_3.

(b) Suggest why V and Pt are poor catalysts for the reaction between N_2 and H_2, and give a possible reason why Os (although it is a good catalyst) is not used commercially.

25.18 (a) Summarize the structural features of importance in a Ziegler–Natta catalyst comprising $TiCl_4$ supported on $MgCl_2$.

(b) What is the role of the ethyl aluminium compounds which are added to the catalyst?

(c) Explain how a Ziegler–Natta catalyst facilitates the conversion of ethene to a representative oligomer.

25.19 (a) Why is it easier to investigate the Cossee–Arlman mechanism using metallocene alkene polymerization catalysts rather than Ziegler–Natta catalysts?

(b) The zirconium complex shown below is an active catalyst for the polymerization of $RCH=CH_2$. Draw a scheme to illustrate the mechanism of this reaction, assuming that it follows the Cossee–Arlman pathway.

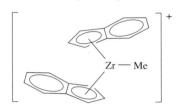

25.20 Give a brief discussion of the use of heterogeneous catalysis in selected industrial manufacturing processes.

25.21 Comment on each of the following:

(a) Zeolite 5A (effective pore size 430 pm) is used to separate a range of n- and iso-alkanes.

(b) Zeolite ZSM-5 catalyses the isomerization of 1,3- to 1,4-$Me_2C_6H_4$ (i.e. m- to p-xylene), and the conversion of C_6H_6 to EtC_6H_5.

25.22 Summarize the operation of a three-way catalytic converter, including comments on (a) the addition of cerium oxides, (b) the light-off temperature, (c) optimum air–fuel ratios and (d) catalyst ageing.

OVERVIEW PROBLEMS

25.23 Ligand **25.36** has been designed for use in Ru-based catalysts for hydrogenation reactions in an EtOH/hexane solvent system. These solvents separate into two phases upon the addition of a small amount of water.

(a) For what types of hydrogenations would this catalyst be especially useful? Rationalize your answer.

(b) Ligand **25.36** is related to BINAP (**25.14**) but has been functionalized. Suggest a reason for this functionalization.

RO OR   RO OR
      OR   RO
O   O
HN   NH

Ph₂P   PPh₂

R = CH₂C₆H₂-3,4,5-(OC₁₀H₂₁)₃

(25.36)

25.24 (a) One proposed method for removing NO from motor vehicle emissions is by catalytic reduction using NH_3 as the reducing agent. Bearing in mind the regulated, allowed emissions, write a balanced equation for the redox reaction and show that the oxidation state changes balance.

(b) In the presence of Grubbs' catalyst, compound **25.37** undergoes a selective ring-closure metathesis to give a bicyclic product **A**. Draw the structure of a 'first generation' Grubbs' catalyst. Suggest the identity of **A**, giving reasons for your choice. Write a balanced equation for the conversion of **25.37** to **A**.

O

O

(25.37)

25.25 The catalyst $[Rh(Ph_2PCH_2CH_2PPh_2)]^+$ can be prepared by the reaction of $[Rh(nbd)(Ph_2PCH_2CH_2PPh_2)]^+$ (nbd = **25.38**) with two equivalents of H_2. In coordinating solvents, $[Rh(Ph_2PCH_2CH_2PPh_2)]^+$, in the form of a solvated complex $[Rh(Ph_2PCH_2CH_2PPh_2)(solv)_2]^+$, catalyses the hydrogenation of $RCH=CH_2$.

(a) Draw the structure of $[Rh(nbd)(Ph_2PCH_2CH_2PPh_2)]^+$ and suggest what happens when this complex reacts with H_2.

(b) Draw the structure of $[Rh(Ph_2PCH_2CH_2PPh_2)(solv)_2]^+$, paying attention to the expected coordination environment of the Rh atom.

(c) Given that the first step in the mechanism is the substitution of one solvent molecule for the alkene, draw a catalytic cycle that accounts for the conversion of $RCH=CH_2$ to RCH_2CH_3. Include a structure for each intermediate complex and give the electron count at the Rh centre in each complex.

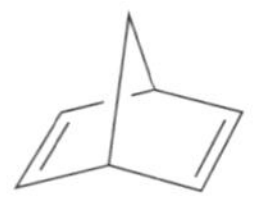

(25.38)

25.26 There is much current interest in 'dendritic' molecules, i.e. those with 'branched arms' that diverge from a central core. The supported dendritic catalyst **25.39** can be used in hydroformylation reactions, and shows high selectivity for branched over linear aldehyde products.

(a) Is **25.39** likely to be the active catalytic species? Rationalize your answer.

(b) What advantages does **25.39** have over a mononuclear hydroformylation catalyst such as $HRh(CO)_2(PPh_3)_2$?

(c) Give a general scheme for the hydroformylation of pent-1-ene (ignoring intermediates in the catalytic cycle) and explain what is meant by 'selectivity for branched over linear aldehyde products'.

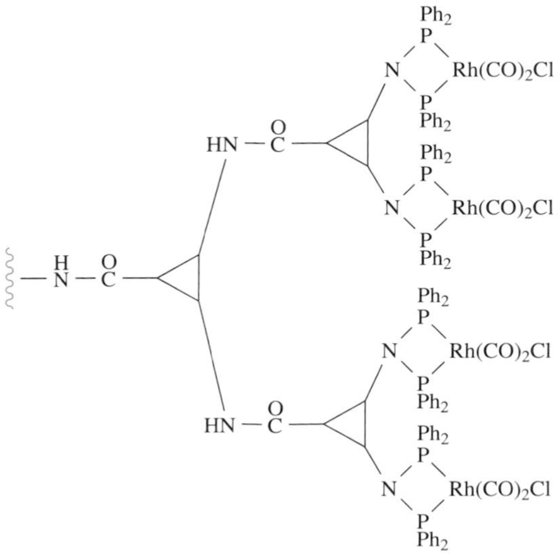

(25.39)

INORGANIC CHEMISTRY MATTERS

25.27 The first step in the Cativa process is the reaction between MeI and cis-$[Ir(CO)_2I_2]^-$. However, the catalyst may also react with HI and this step initiates a water gas shift reaction that competes with the main catalytic cycle. (a) What chemical is manufactured in the Cativa process? Why is this product of industrial importance? (b) Why is HI present in the system? (c) Give an equation for the water gas shift reaction, and state conditions typically used in industry. (d) Figure 25.22 shows the competitive catalytic cycle described above. Suggest identities for species **A**, **B**, **C** and **D**. What type of reaction is the conversion of cis-$[Ir(CO)_2I_2]^-$ to **A**? What changes in iridium oxidation state occur on going around the catalytic cycle, and what is the electron count in each iridium complex?

25.28 What roles do inorganic catalysts play in the following manufacturing processes: (a) production of aldehydes from alkenes, (b) polymerization of propene, (c) production of acetic anhydride, (d) hydrogenation of compound **25.40** to produce the drug (S)-Naproxen? State whether homogeneous or heterogeneous catalysts are used.

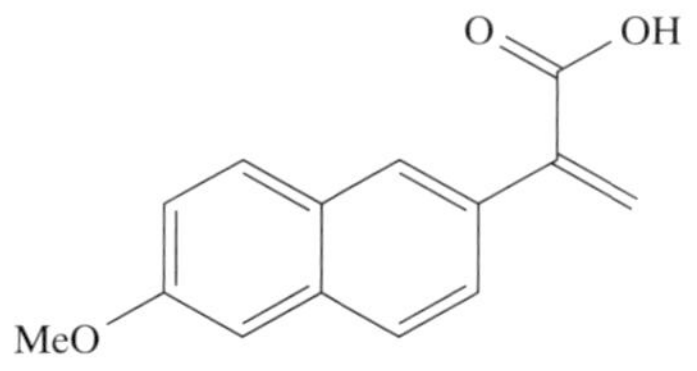

(25.40)

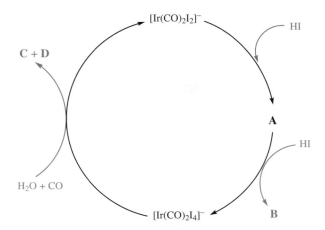

Fig. 25.22 Proposed catalytic cycle for the water gas shift reaction that is a competitive route in the Cativa process.

25.29 Measures taken to control atmospheric pollution include (a) scrubbing industrial waste gases to remove SO_2, and (b) reduction of NO in motor vehicle emissions. Explain how these are achieved and write balanced equations for relevant reactions. Are they catalytic processes?

25.30 In 2008, the US manufactured 32.4 Mt of sulfuric acid. The importance of H_2SO_4 is reflected in the fact that its consumption in a given country is a direct indicator of that country's industrial growth. (a) Give examples of commercial uses of H_2SO_4. (b) Starting from relevant feedstocks, outline how sulfuric acid is produced on an industrial scale, paying attention to reaction conditions. (c) How does the manufacture of 'by-product sulfuric acid' differ from the process you have described in part (b)?

26

d-block metal complexes: reaction mechanisms

26.1 Introduction

We have already mentioned some aspects of inorganic reaction mechanisms: *kinetically inert* metal centres such as Co(III) (Section 21.10) and organometallic reaction types (Section 24.7). Now, we discuss in more detail the mechanisms of ligand substitution and electron-transfer reactions in coordination complexes. For the substitution reactions, we confine our attention to square planar and octahedral complexes, for which kinetic data are plentiful.

A proposed mechanism *must* be consistent with all experimental facts. A mechanism cannot be proven, since another mechanism may also be consistent with the experimental data.

26.2 Ligand substitutions: some general points

In a *ligand substitution* reaction:

$$ML_xX + Y \longrightarrow ML_xY + X$$

X is the *leaving group* and Y is the *entering group*.

Kinetically inert and labile complexes

Metal complexes that undergo reactions with $t_{\frac{1}{2}} \leq 1$ min are described as being *kinetically labile*. If the reaction takes significantly longer than this, the complex is *kinetically inert*.

There is no connection between the *thermodynamic* stability of a complex and its lability towards substitution. For example, values of $\Delta_{\text{hyd}}G^{\circ}$ for Cr^{3+} and Fe^{3+} are almost equal, yet $[Cr(OH_2)_6]^{3+}$ (d^3) undergoes substitution slowly and $[Fe(OH_2)_6]^{3+}$ (high-spin d^5) rapidly. Similarly, although the overall formation constant of $[Hg(CN)_4]^{2-}$ is greater than that of $[Fe(CN)_6]^{4-}$, the Hg(II) complex rapidly exchanges $[CN]^-$ with isotopically labelled cyanide, while exchange is extremely slow for $[Fe(CN)_6]^{4-}$. The kinetic inertness of d^3 and low-spin d^6 octahedral complexes is in part associated with crystal field effects (see Section 26.4).

Figure 26.1 illustrates the range of rate constants, k, for the exchange of a water molecule in the first coordination sphere of $[M(OH_2)_x]^{n+}$ with one outside this coordination shell (eq. 26.1). The rate constant is defined according to eq. 26.2.[†]

$$[M(OH_2)_x]^{n+} + H_2O \overset{k}{\rightleftharpoons} [M(OH_2)_{x-1}(OH_2)]^{n+} + H_2O$$
(26.1)

$$\text{Rate of water exchange} = xk[M(OH_2)_x{}^{n+}]$$
(26.2)

Figure 26.1 also gives the average residence time ($\tau = {}^1/_k$) of an H_2O ligand in the first coordination sphere of a metal ion. The $[Ir(OH_2)_6]^{3+}$ ion lies at the slow-exchange extreme limit, with $\tau = 9.1 \times 10^9$ s $= 290$ years (at 298 K). At the other extreme, water exchange for the alkali metal ions is rapid, with $[Cs(OH_2)_8]^+$ being the most labile ($\tau = 2 \times 10^{10+}$s). Trends in the labilities of the main group metal ions (shown in pink in Fig. 26.1) can be understood in terms of the surface charge density and the coordination

---

[†] In rate equations, [ ] stands for 'concentration of' and should not be confused with use of square brackets around formulae of complexes in other contexts. For this reason, we omit [ ] in formulae in most reaction equations in this chapter.

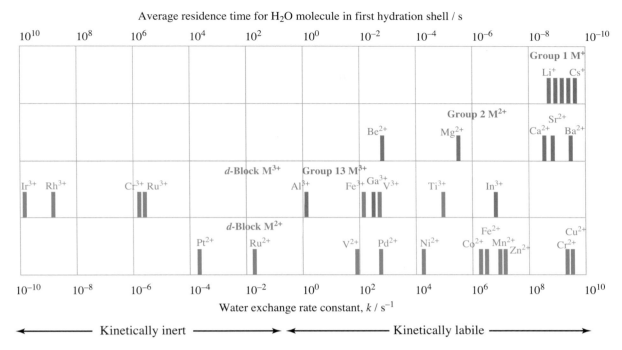

Fig. 26.1 Water exchange rate constants and average residence times for water molecules in the first coordination sphere of aquated metal ions at 298 K. Group 1, 2 and 13 metal ions are shown in pink, and *d*-block metal ions in blue. [Based on S.F. Lincoln (2005) *Helv. Chim. Acta*, vol. 88, p. 523 (Figure 1).]

number of the metal ion. On descending a given group, the rate of water exchange increases as:

- the metal ion increases in size;
- the coordination number increases;
- the surface charge density decreases.

Rates of water exchange for the group 1 metal ions vary over a small range from $[Li(OH_2)_6]^+$ (least labile) to $[Cs(OH_2)_8]^+$ (most labile). For the group 2 metal ions, k varies from $\approx 10^3 \, s^{-1}$ for $[Be(OH_2)_4]^{2+}$ to $\approx 10^9 \, s^{-1}$ for $[Ba(OH_2)_8]^{2+}$. Each group 13 M^{3+} forms a hexaaqua ion, and values of k range from $\approx 1 \, s^{-1}$ for $[Al(OH_2)_6]^{3+}$ to $\approx 10^7 \, s^{-1}$ for $[In(OH_2)_6]^{3+}$, consistent with the increase in ionic radius from 54 pm (Al^{3+}) to 80 pm (In^{3+}). The lanthanoid M^{3+} ions (not included in Fig. 26.1) are all larger than the group 13 M^{3+} ions and exhibit high coordination numbers. They are all relatively labile with $k > 10^7$. For the $[Eu(OH_2)_7]^{2+}$ ion, the average residence time for a water molecule in the first coordination sphere is only $2.0 \times 10^{-10} \, s$, and its lability is comparable to that of $[Cs(OH_2)_8]^+$.

Figure 26.1 illustrates that the rates of water exchange for the *d*-block M^{2+} and M^{3+} ions span a much greater range than do those of the group 1, 2 and 13 metal ions. The kinetic inertness of d^3 (e.g. $[Cr(OH_2)_6]^{3+}$ in Fig. 26.1) and low-spin d^6 (e.g. $[Rh(OH_2)_6]^{3+}$ and $[Ir(OH_2)_6]^{3+}$) can be understood in terms of crystal field theory. More generally, the 20 orders of magnitude covered by values of k for the *d*-block metal ions follow from the different nd-electron

configurations and crystal field effects. In Section 26.4, we consider water exchange reactions in more detail.

Stoichiometric equations say nothing about mechanism

The processes that occur in a reaction are not necessarily obvious from the stoichiometric equation. For example, reaction 26.3 might suggest a mechanism involving the direct substitution of coordinated $[CO_3]^{2-}$ by H_2O.

$$[(H_3N)_5Co(CO_3)]^+ + 2[H_3O]^+$$
$$\longrightarrow [(H_3N)_5Co(OH_2)]^{3+} + CO_2 + 2H_2O \qquad (26.3)$$

However, use of $H_2^{18}O$ as solvent shows that all the oxygen in the aqua complex is derived from carbonate, and scheme 26.4 shows the proposed pathway of the reaction.

$$[(H_3N)_5Co(OCO_2)]^+ + [H_3O]^+ \longrightarrow \left[(H_3N)_5Co \underset{CO_2}{\overset{H}{\longrightarrow}} O \right]^{2+} + H_2O$$

$$\downarrow$$

$$[(H_3N)_5Co(OH)]^{2+} + CO_2$$

$$\downarrow [H_3O]^+$$

$$[(H_3N)_5Co(OH_2)]^{3+} + H_2O$$

$$(26.4)$$

Types of substitution mechanism

In inorganic substitutions, the limiting mechanisms are *dissociative* (*D*), in which the intermediate has a lower coordination number than the starting complex (eq. 26.5), and *associative* (*A*), in which the intermediate has a higher coordination number (eq. 26.6).[†]

> *Dissociative* and *associative* reaction mechanisms involve two-step pathways and an *intermediate*.

$$
\left.
\begin{array}{l}
ML_xX \longrightarrow \underset{\text{intermediate}}{ML_x} + \underset{\substack{\text{leaving}\\\text{group}}}{X} \\[2ex]
ML_x + \underset{\substack{\text{entering}\\\text{group}}}{Y} \longrightarrow ML_xY
\end{array}
\right\} \quad \textit{dissociative } (D) \quad (26.5)
$$

$$
\left.
\begin{array}{l}
ML_xX + \underset{\substack{\text{entering}\\\text{group}}}{Y} \longrightarrow \underset{\text{intermediate}}{ML_xXY} \\[2ex]
ML_xXY \longrightarrow ML_xY + \underset{\substack{\text{leaving}\\\text{group}}}{X}
\end{array}
\right\} \quad \textit{associative } (A) \quad (26.6)
$$

> An *intermediate* occurs at a local energy minimum; it can be detected and, sometimes, isolated. A *transition state* occurs at an energy maximum, and cannot be isolated.

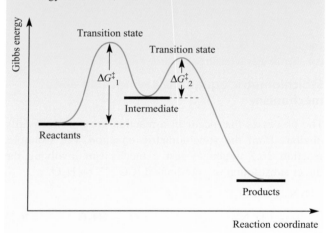

In most metal complex substitution pathways, bond formation between the metal and entering group is thought to be *concurrent* with bond cleavage between the metal and leaving group (eq. 26.7). This is the *interchange* (*I*) mechanism.

$$
ML_xX + \underset{\substack{\text{entering}\\\text{group}}}{Y} \longrightarrow \underset{\substack{\text{transition}\\\text{state}}}{Y\cdots ML_x\cdots X} \longrightarrow ML_xY + \underset{\substack{\text{leaving}\\\text{group}}}{X}
$$

$$(26.7)$$

In an *I* mechanism, there is no intermediate but various transition states are possible. Two types of interchange mechanisms can be identified:

- *dissociative interchange* (*I*$_d$), in which bond breaking dominates over bond formation;
- *associative interchange* (*I*$_a$), in which bond formation dominates over bond breaking.

In an *I*$_a$ mechanism, the reaction rate shows a dependence on the entering group. In an *I*$_d$ mechanism, the rate shows only a very small dependence on the entering group. It is usually difficult to distinguish between *A* and *I*$_a$, *D* and *I*$_d$, and *I*$_a$ and *I*$_d$ processes.

> An *interchange* (*I*) mechanism is a concerted process in which there is *no intermediate* species with a coordination number different from that of the starting complex.

Activation parameters

The diagram opposite which distinguishes between a transition state and an intermediate also shows the Gibbs energy of activation, $\Delta G^\ddagger$, for each step in the 2-step reaction path. Enthalpies and entropies of activation, $\Delta H^\ddagger$ and $\Delta S^\ddagger$, obtained from temperature dependence of rate constants, can shed light on mechanisms. Equation 26.8 (the Eyring equation) gives the relationship between the rate constant, temperature and activation parameters. A linearized form of this relationship is given in eq. 26.9.[*]

$$
k = \frac{k'T}{h} e^{\left(-\frac{\Delta G^\ddagger}{RT}\right)} = \frac{k'T}{h} e^{\left(-\frac{\Delta H^\ddagger}{RT} + \frac{\Delta S^\ddagger}{R}\right)} \quad (26.8)
$$

$$
\ln\left(\frac{k}{T}\right) = \frac{-\Delta H^\ddagger}{RT} + \ln\left(\frac{k'}{h}\right) + \frac{\Delta S^\ddagger}{R} \quad (26.9)
$$

where k = rate constant, T = temperature (K), $\Delta H^\ddagger$ = enthalpy of activation (J mol^{-1}), $\Delta S^\ddagger$ = entropy of activation (J K^{-1} mol^{-1}), R = molar gas constant, k' = Boltzmann constant, h = Planck constant[**]

From eq. 26.9, a plot of $\ln(k/T)$ against $1/T$ (an *Eyring plot*) is linear; the activation parameters $\Delta H^\ddagger$ and $\Delta S^\ddagger$ can be determined as shown in Fig. 26.2.

Values of $\Delta S^\ddagger$ are particularly useful in distinguishing between associative and dissociative mechanisms. A *large negative value* of $\Delta S^\ddagger$ is indicative of an *associative* mechanism, i.e. there is a decrease in entropy as the entering group associates with the starting complex. However, caution is needed. Solvent reorganization can result in negative values of $\Delta S^\ddagger$ even for a dissociative mechanism,

---

[†] The terminology for inorganic substitution mechanisms is not the same as for organic nucleophilic substitutions. Since readers will already be familiar with the S$_N$1 (unimolecular) and S$_N$2 (bimolecular) notation, it may be helpful to note that the *D* mechanism corresponds to S$_N$1, and *I*$_a$ to S$_N$2.

[*] For critical comments on the use of eqs. 26.8 and 26.9, see: G. Lente, I. Fábián and A. Poë (2005) *New Journal of Chemistry*, vol. 29, p. 759 – 'A common misconception about the Eyring equation'.

[**] Physical constants: see inside back cover of this book.

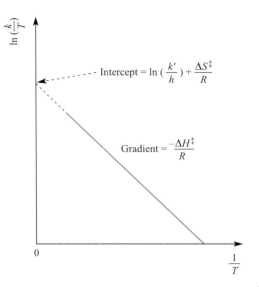

Fig. 26.2 An Eyring plot allows the activation parameters $\Delta H^{\ddagger}$ and $\Delta S^{\ddagger}$ to be determined from the temperature dependence of the rate constant; the dotted part of the line represents an extrapolation. See eq. 26.9 for definitions of quantities.

and hence the qualifier that $\Delta S^{\ddagger}$ should be *large* and negative to indicate an associative pathway.

The pressure dependence of rate constants leads to a measure of the *volume of activation*, $\Delta V^{\ddagger}$ (eq. 26.10).

$$\frac{d(\ln k)}{dP} = \frac{-\Delta V^{\ddagger}}{RT}$$

or, in integrated form:

$$\ln\left(\frac{k_{(P_1)}}{k_{(P_2)}}\right) = \frac{-\Delta V^{\ddagger}}{RT}(P_1 - P_2)$$

(26.10)

where k = rate constant; P = pressure; $\Delta V^{\ddagger}$ = volume of activation ($cm^3 mol^{-1}$); R = molar gas constant; T = temperature (K).

A reaction in which the transition state has a greater volume than the initial state shows a positive $\Delta V^{\ddagger}$, whereas a negative $\Delta V^{\ddagger}$ corresponds to the transition state being compressed relative to the reactants. After allowing for any change in volume of the solvent (which is important if solvated ions

are involved), the sign of $\Delta V^{\ddagger}$ should, in principle, distinguish between an associative and a dissociative mechanism.

> A large negative value of $\Delta V^{\ddagger}$ indicates an associative mechanism; a positive value suggests that the mechanism is dissociative.

Self-study exercise

As an alternative to eq. 26.9, the following linearized form of the Eyring equation can be derived from eq. 26.8:

$$T \times \ln\frac{k}{T} = T\left(\ln\frac{k'}{h} + \frac{\Delta S^{\ddagger}}{R}\right) - \frac{\Delta H^{\ddagger}}{R}$$

What graph would you construct to obtain a linear plot? How would you use this plot to obtain values of $\Delta H^{\ddagger}$ and $\Delta S^{\ddagger}$? [*Ans*. See G. Lente *et al.* (2005) *New. J. Chem.*, vol. 29, p. 759]

26.3 Substitution in square planar complexes

Complexes with a d^8 configuration often form square planar complexes (see Section 20.3), especially when there is a large crystal field: Rh(I), Ir(I), Pt(II), Pd(II), Au(III). However, 4-coordinate complexes of Ni(II) may be tetrahedral or square planar. The majority of kinetic work on square planar systems has been carried out on Pt(II) complexes because the rate of ligand substitution is conveniently slow. Although data for Pd(II) and Au(III) complexes indicate similarity between their substitution mechanisms and those of Pt(II) complexes, one *cannot justifiably assume* a similarity in kinetics among a series of structurally related complexes undergoing similar substitutions.

Rate equations, mechanism and the *trans*-effect

The consensus of opinion, based on a large body of experimental work, is that nucleophilic substitution reactions in square planar Pt(II) complexes normally proceed by *associative* mechanisms (A or I_a). Negative values of $\Delta S^{\ddagger}$

Table 26.1 Activation parameters for substitution in selected square planar complexes (see Table 7.7 for ligand abbreviations).

Reactants	$\Delta H^{\ddagger} / kJ\,mol^{-1}$	$\Delta S^{\ddagger} / J\,K^{-1}\,mol^{-1}$	$\Delta V^{\ddagger} / cm^3\,mol^{-1}$
$[Pt(dien)Cl]^+ + H_2O$	+84	−63	−10
$[Pt(dien)Cl]^+ + [N_3]^-$	+65	−71	−8.5
trans-$[PtCl_2(PEt_3)_2] + py$	+14	−25	−14
trans-$[PtCl(NO_2)(py)_2] + py$	+12	−24	−9

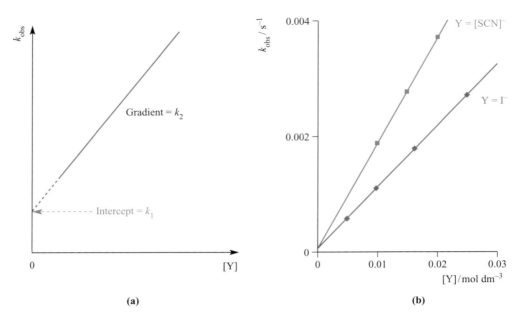

Fig. 26.3 (a) Determination of the k_1 and k_2 rate constants (eq. 26.14) from the observed rate data for ligand substitution in a square planar complex; Y is the entering ligand. The dotted part of the line represents an extrapolation. (b) Plots of k_{obs} against concentration of the entering group for the reactions of *trans*-[PtCl$_2$(py)$_2$] with [SCN]$^-$ or with I$^-$; both reactions were carried out in MeOH and so there is a common intercept. [Data from: U. Belluco *et al.* (1965) *J. Am. Chem. Soc.*, vol. 87, p. 241.]

and $\Delta V^{\ddagger}$ support this proposal (Table 26.1). The observation that the rate constants for the displacement of Cl$^-$ by H$_2$O in [PtCl$_4$]$^{2-}$, [PtCl$_3$(NH$_3$)]$^-$, [PtCl$_2$(NH$_3$)$_2$] and [PtCl(NH$_3$)$_3$]$^+$ are similar suggests an associative mechanism, since a dissociative pathway would be expected to show a significant dependence on the charge on the complex.

Reaction 26.11 shows the substitution of X by Y in a square planar Pt(II) complex.

$$PtL_3X + Y \longrightarrow PtL_3Y + X \qquad (26.11)$$

The usual form of the experimental rate law is given by eq. 26.12 indicating that the reaction proceeds simultaneously by two routes.

$$Rate = -\frac{d[PtL_3X]}{dt} = k_1[PtL_3X] + k_2[PtL_3X][Y] \qquad (26.12)$$

Reaction 26.11 would usually be studied under pseudo-first order conditions, with Y (as well as the solvent, S) in vast excess. This means that, since $[Y]_t \approx [Y]_0$, and $[S]_t \approx [S]_0$ (where the subscripts represent time t and time zero), we can rewrite eq. 26.12 in the form of eq. 26.13 where k_{obs} is the observed rate constant and is related to k_1 and k_2 by eq. 26.14.

$$Rate = -\frac{d[PtL_3X]}{dt} = k_{obs}[PtL_3X] \qquad (26.13)$$

$$k_{obs} = k_1 + k_2[Y] \qquad (26.14)$$

Carrying out a series of reactions with various concentrations of Y (always under pseudo-first order conditions) allows k_1 and k_2 to be evaluated (Fig. 26.3a). Figure 26.3b shows the effect of changing the entering group Y while maintaining a common solvent. Rate constant k_2 depends on Y, and values of k_2 are determined from the gradients of the lines in Fig. 26.3b. These lines pass through a common intercept, equal to k_1. If the kinetic runs are repeated using a different solvent, a different common intercept is observed.

The contributions of the two terms in eq. 26.12 to the overall rate reflect the relative dominance of one pathway over the other. The k_2 term arises from an associative mechanism involving attack by Y on PtL$_3$X in the rate-determining step, and when Y is a good nucleophile, the k_2 term is dominant. The k_1 term might appear to indicate a concurrent dissociative pathway. However, experiment shows that the k_1 term becomes dominant if the reaction is carried out in polar solvents, and its contribution diminishes in apolar solvents. This indicates solvent participation, and eq. 26.12 is more fully written in the form of eq. 26.15, in which S is the solvent. Since S is in vast excess, its concentration is effectively constant during the reaction (i.e. pseudo-first order conditions) and so, comparing eqs. 26.12 and 26.15, $k_1 = k_3[S]$.

$$Rate = -\frac{d[PtL_3X]}{dt} = k_3[PtL_3X][S] + k_2[PtL_3X][Y] \qquad (26.15)$$

When the solvent is a potential ligand (e.g. H_2O), it competes with the entering group Y in the rate-determining step of the reaction, and X can be displaced by Y or S. Substitution of S by Y then occurs in a *fast* step, i.e. *non-rate determining*. The two competing pathways by which reaction 26.11 occurs are shown in scheme 26.16.

$$PtL_3X + Y \xrightarrow{k_2} PtL_3Y + X$$

competes with:

$$PtL_3X + S \xrightarrow{k_1} PtL_3S + X$$
$$PtL_3S + Y \xrightarrow{fast} PtL_3Y + S$$

$$(26.16)$$

A further point in favour of both the k_1 and k_2 terms being associative is that *both* rate constants decrease when the steric demands of Y or L increase.

In the majority of reactions, substitution at square planar Pt(II) is *stereoretentive*: the entering group takes the coordination site previously occupied by the leaving group. An A or I_a mechanism involves a 5-coordinate intermediate or transition state and, since the energy difference between different 5-coordinate geometries is small, one would expect rearrangement of the 5-coordinate species to be facile unless, for example, it is sterically hindered (A or I_a) or its lifetime is too short (I_a). The stereochemical retention can be envisaged as shown in Fig. 26.4 (in which we ignore any part played by the solvent). Why does Fig. 26.4 specifically show a *trigonal bipyramidal* species as the intermediate or transition state? To answer this, we must consider additional experimental data:

The choice of leaving group in a square planar complex is determined by the nature of the ligand *trans* to it; this is the **trans-effect** and is **kinetic** in origin.

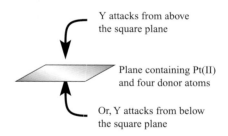

Y attacks from above the square plane

Plane containing Pt(II) and four donor atoms

Or, Y attacks from below the square plane

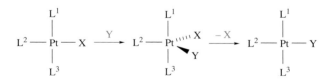

Fig. 26.4 Initial attack by the entering group at a square planar Pt(II) centre is from above or below the plane. Nucleophile Y then coordinates to give a trigonal bipyramidal species which loses X with retention of stereochemistry.

Reactions 26.17 and 26.18 illustrate the *trans*-effect in operation: *cis*- and *trans*-$[PtCl_2(NH_3)_2]$ are prepared *specifically* by different substitution routes.[†]

$$(26.17)$$

$$(26.18)$$

One contributing factor to the *trans*-effect is the *trans-influence* (see Box 22.8). The second factor, which addresses the *kinetic* origin of the *trans*-effect, is that of shared π-electron density in the 5-coordinate transition state or intermediate as shown in Fig. 26.5: ligand L^2 is

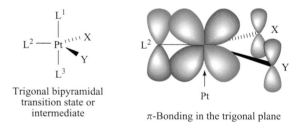

Trigonal bipyramidal transition state or intermediate

π-Bonding in the trigonal plane

Fig. 26.5 In the trigonal plane of the 5-coordinate transition state or intermediate (see Fig. 26.4), a π-bonding interaction can occur between a metal d orbital (e.g. d_{xy}) and suitable orbitals (e.g. p atomic orbitals, or molecular orbitals of π-symmetry) of ligand L^2 (the ligand *trans* to the leaving group), X (the leaving group) and Y (the entering group). Note that ligands may not necessarily contribute to the π-bonding scheme, e.g. NH_3.

[†] The use of the terms *trans-effect* and *trans-influence* in different textbooks is not consistent, and may cause confusion; attention should be paid to specific definitions.

trans to the leaving group, X, in the initial square planar complex and is also *trans* to the entering group, Y, in the final square planar complex (Fig. 26.4). These three ligands and the metal centre can communicate electronically through π-bonding *only* if they all lie in the *same plane* in the transition state or intermediate. This implies that the 5-coordinate species must be trigonal bipyramidal rather than square-based pyramidal. If L^2 is a strong π-acceptor (e.g. CO), it will stabilize the transition state by accepting electron density that the incoming nucleophile donates to the metal centre, and will thereby facilitate substitution at the site *trans* to it. The general order of the *trans*-effect (i.e. the ability of ligands to direct *trans*-substitution) spans a factor of about 10^6 in rates and is:

$$H_2O \approx [OH]^- \approx NH_3 \approx py < Cl^- < Br^- < I^- \approx [NO_2]^-$$
$$< Ph^- < Me^- < PR_3 \approx H^- \ll CO \approx [CN]^- \approx C_2H_4$$

Experimental rates of substitution are affected by both the ground state *trans*-influence and the kinetic *trans*-effect, and rationalizing the sequence above in terms of individual factors is difficult. There is no close connection between the relative magnitudes of the *trans*-influence and *trans*-effect. However, the π-bonding scheme in Fig. 26.5 does help to explain the very strong *trans*-directing abilities of CO, $[CN]^-$ and ethene.

The *trans*-effect is useful in devising syntheses of Pt(II) complexes, e.g. selective preparations of *cis*- and *trans*-isomers of $[PtCl_2(NH_3)_2]$ (schemes 26.17 and 26.18) and of $[PtCl_2(NH_3)(NO_2)]^-$ (schemes 26.19 and 26.20).

(26.19)

(26.20)

Finally, we should note that a small *cis*-effect does exist, but is usually of far less importance than the *trans*-effect.

Ligand nucleophilicity

If one studies how the rate of substitution by Y in a given complex depends on the entering group, then for most reactions at Pt(II), the rate constant k_2 (eq. 26.12) increases in the order:

$$H_2O < NH_3 \approx Cl^- < py < Br^- < I^- < [CN]^- < PR_3$$

This is called the *nucleophilicity sequence* for substitution at square planar Pt(II) and the ordering is consistent with Pt(II) being a soft metal centre (see Table 7.9). A *nucleophilicity parameter*, n_{Pt}, is defined by eq. 26.22 where k_2' is the rate constant for reaction 26.21 with Y = MeOH (i.e. for Y = MeOH, $n_{Pt} = 0$).

$$trans\text{-}[PtCl_2(py)_2] + Y \longrightarrow trans\text{-}[PtCl(py)_2Y]^+ + Cl^-$$

(26.21)

(The equation is written assuming Y is a neutral ligand.)

$$n_{Pt} = \log \frac{k_2}{k_2'} \quad \text{or} \quad n_{Pt} = \log k_2 - \log k_2' \quad (26.22)$$

Values of n_{Pt} vary considerably (Table 26.2) and illustrate the dependence of the rate of substitution on the nucleophilicity of the entering group. There is no correlation between n_{Pt} and the strength of the nucleophile as a Brønsted base.

> The *nucleophilicity parameter*, n_{Pt}, describes the dependence of the rate of substitution in a square planar Pt(II) complex on the nucleophilicity of the entering group.

Table 26.2 Values of n_{Pt} for entering ligands, Y, in reaction 26.21; values are relative to n_{Pt} for MeOH = 0 and are measured at 298 K.[†]

Ligand	Cl^-	NH_3	py	Br^-	I^-	$[CN]^-$	PPh_3
n_{Pt}	3.04	3.07	3.19	4.18	5.46	7.14	8.93

[†] For further data, see: R.G. Pearson, H. Sobel and J. Songstad (1968) *J. Am. Chem. Soc.*, vol. 90, p. 319.

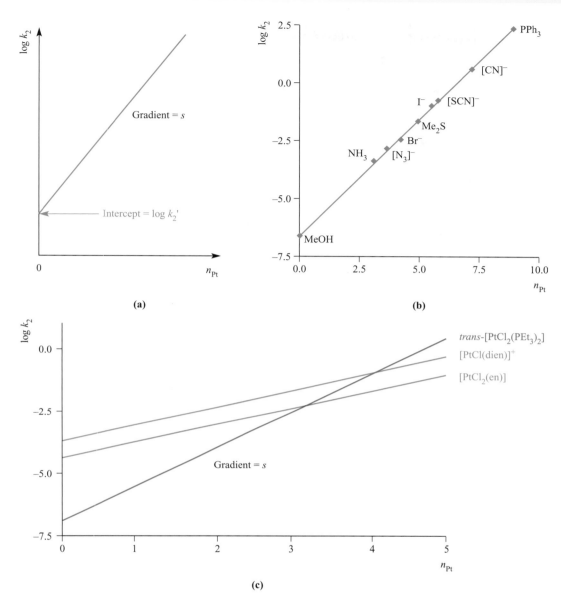

Fig. 26.6 (a) The nucleophilicity discrimination factor, s, for a particular square planar Pt(II) complex can be found from a plot of $\log k_2$ (the second order rate constant, see eq. 26.12) against n_{Pt} (the nucleophilicity parameter, see eq. 26.22). Experimental results are plotted in this way in graph (b) which shows data for the reaction of *trans*-$[PtCl_2(py)_2]$ with different nucleophiles in MeOH at 298 or 303 K. [Data from: R.G. Pearson *et al.* (1968) *J. Am. Chem. Soc.*, vol. 90, p. 319.] (c) Plots of $\log k_2$ against n_{Pt} for three square planar Pt(II) complexes; each plot is of the same type as in graph (b). The gradient of each line gives s, the nucleophilicity discrimination factor, for that particular complex. [Data from: U. Belluco *et al.* (1965) *J. Am. Chem. Soc.*, vol. 87, p. 241.]

If we now consider substitution reactions of nucleophiles with other Pt(II) complexes, linear relationships are found between values of $\log k_2$ and n_{Pt} as illustrated in Fig. 26.6. For the general reaction 26.11 (in which the ligands L do *not* have to be identical), eq. 26.23 is defined where s is the *nucleophilicity discrimination factor* and k_2' is the rate constant when the nucleophile is MeOH.

$$\log k_2 = s(n_{Pt}) + \log k_2' \qquad (26.23)$$

For a given substrate, s can be found from the gradient of a line in Fig. 26.6. Each complex has a characteristic value of s, and selected values are listed in Table 26.3. The relatively small value of s for $[Pt(dien)(OH_2)]^{2+}$ indicates that this complex does not discriminate as much between entering ligands as, for example, does *trans*-$[PtCl_2(PEt_3)_2]$; i.e. $[Pt(dien)(OH_2)]^{2+}$ is generally more reactive towards substitution than other complexes in the table, consistent with the fact that H_2O is a good leaving group.

Table 26.3 Nucleophilic discrimination factors, *s*, for selected square planar Pt(II) complexes. (See Table 7.7 for ligand abbreviations.)

Complex	*s*
trans-[PtCl$_2$(PEt$_3$)$_2$]	1.43
trans-[PtCl$_2$(AsEt$_3$)$_2$]	1.25
trans-[PtCl$_2$(py)$_2$]	1.0
[PtCl$_2$(en)]	0.64
[PtBr(dien)]$^+$	0.75
[PtCl(dien)]$^+$	0.65
[Pt(dien)(OH$_2$)]$^{2+}$	0.44

The ***nucleophilicity discrimination factor***, *s*, is a characteristic of a given square planar Pt(II) complex and describes how sensitive the complex is to variation in the nucleophilicity of the entering ligand.

Self-study exercises

1. Explain why the reaction of [Pt(NH$_3$)$_4$]$^{2+}$ with Cl$^-$ leads to *trans*-[Pt(NH$_3$)$_2$Cl$_2$] with no *cis*-isomer.

[*Ans.* See eq. 26.18 and accompanying text]

2. Suggest why the complex shown below undergoes water exchange at a rate 10^7 times faster than [Pt(OH$_2$)$_4$]$^{2+}$.

[*Ans.* See U. Frey *et al.* (1998) *Inorg. Chim. Acta*, vol. 269, p. 322]

3. For the reaction:

trans-[Pt(PEt$_3$)$_2$X(R)] + CN$^-$ ⟶
　　trans-[Pt(PEt$_3$)$_2$CN(R)] + X$^-$　(X = Cl or Br)

the *relative* rates of substitution (at 303 K) are 1 : 21 : 809 for R = 2,4,6-Me$_3$C$_6$H$_2$, 2-MeC$_6$H$_4$ and C$_6$H$_5$, respectively. If the starting complex is *cis*-[Pt(PEt$_3$)$_2$X(R)], the *relative* rates of CN$^-$ for X$^-$ substitution (at 303 K) are 1 : 7900 : 68 600 for R = 2,4,6-Me$_3$C$_6$H$_2$, 2-MeC$_6$H$_4$ and C$_6$H$_5$, respectively. Rationalize these data.

[*Ans.* See Table 3 and discussion in: G. Faraone *et al.* (1974) *J. Chem. Soc., Dalton Trans.*, p. 1377]

26.4 Substitution and racemization in octahedral complexes

Most studies of the mechanism of substitution in octahedral metal complexes have been concerned with Werner-type complexes. Organometallic complexes have entered the research field more recently. Among the former, the popular candidates for study have been Cr(III) (d^3) and low-spin Co(III) (d^6) species. These complexes are kinetically inert and their rates of reaction are relatively slow and readily followed by conventional techniques. Both Rh(III) and Ir(III) (both low-spin d^6) also undergo very slow substitution reactions. There is no universal mechanism by which octahedral complexes undergo substitution, and so care is needed when tackling the interpretation of kinetic data.

Water exchange

The exchange of coordinated H$_2$O by isotopically labelled water has been investigated for a wide range of octahedral [M(OH$_2$)$_6$]$^{n+}$ species (Co^{3+} is not among these because it is unstable in aqueous solution, see Section 21.10). Reaction 26.1, where M is an *s*-, *p*- or *d*-block metal, can be studied by using ^{17}O NMR spectroscopy (eq. 26.24), and rate constants can thus be determined (Fig. 26.1).

$$[M(OH_2)_6]^{n+} + H_2(^{17}O) \longrightarrow [M(OH_2)_5\{(^{17}O)H_2\}]^{n+} + H_2O$$
(26.24)

As was pointed out in Section 26.2, for M^{2+} and M^{3+} ions of the *d*-block metals, data for reaction 26.24 indicate a correlation between rate constants and electronic configuration. Table 26.4 lists activation volumes for reaction 26.24 with selected first row *d*-block metal ions. The change from negative to positive values of $\Delta V^{\ddagger}$ indicates a change

Table 26.4 Volumes of activation for water exchange reactions (eq. 26.24).

Metal ion	High-spin d^n configuration	$\Delta V^{\ddagger} / cm^3\,mol^{-1}$
V^{2+}	d^3	−4.1
Mn^{2+}	d^5	−5.4
Fe^{2+}	d^6	+3.7
Co^{2+}	d^7	+6.1
Ni^{2+}	d^8	+7.2
Ti^{3+}	d^1	−12.1
V^{3+}	d^2	−8.9
Cr^{3+}	d^3	−9.6
Fe^{3+}	d^5	−5.4

THEORY

Box 26.1 Reversible binding of NO to $[Fe(OH_2)_6]^{2+}$: an example of the use of flash photolysis

In Section 15.8, we described the complex $[Fe(NO)(OH_2)_5]^{2+}$ in association with the brown ring test for the nitrate ion. The binding of NO is reversible:

$$[Fe(OH_2)_6]^{2+} + NO \underset{}{\overset{K_{NO}}{\rightleftharpoons}} [Fe(OH_2)_5(NO)]^{2+} + H_2O$$

and the formation of $[Fe(OH_2)_5(NO)]^{2+}$ can be monitored by the appearance in the electronic spectrum of absorptions at 336, 451 and 585 nm with $\varepsilon_{max} = 440$, 265 and 85 $dm^3\,mol^{-1}\,cm^{-1}$, respectively. At 296 K, in a buffered solution at pH = 5.0, the value of the equilibrium constant $K_{NO} = 1.15 \times 10^3$. The IR spectrum of $[Fe(OH_2)_5(NO)]^{2+}$ has an absorption at 1810 cm^{-1} assigned to $\nu(NO)$, and this is consistent with the formulation of $[Fe^{III}(OH_2)_5(NO^-)]^{2+}$.

The kinetics of the reversible binding of NO to $[Fe(OH_2)_6]^{2+}$ can be followed by using *flash photolysis* and monitoring changes in the absorption spectrum. Irradiation of $[Fe(OH_2)_5(NO)]^{2+}$ at a wavelength of 532 nm results in rapid dissociation of NO and loss of the absorptions at 336, 451 and 585 nm, i.e. the equilibrium above moves to the left-hand side. Following the 'flash', the equilibrium re-establishes itself within 0.2 ms (at 298 K) and the rate at which $[Fe(OH_2)_5(NO)]^{2+}$ reforms can be determined from the reap-

pearance of three characteristic absorptions. The observed rate constant, k_{obs}, is $3.0 \times 10^4\,s^{-1}$. Under pseudo-first order conditions (i.e. with $[Fe(OH_2)_6]^{2+}$ in large excess), the rate constants for the forward and back reactions can be determined:

$$[Fe(OH_2)_6]^{2+} + NO \underset{k_{off}}{\overset{k_{on}}{\rightleftharpoons}} [Fe(OH_2)_5(NO)]^{2+} + H_2O$$

$$k_{obs} = k_{on}[Fe(OH_2)_6^{2+}] + k_{off}$$

in which the square brackets now stand for concentration.

At a given temperature, values of k_{on} and k_{off} can be found from the gradient and intercept of a linear plot of the variation of k_{obs} with the concentration of $[Fe(OH_2)_6]^{2+}$: at 298 K, $k_{on} = (1.42 \pm 0.04) \times 10^6\,dm^3\,mol^{-1}\,s^{-1}$ and $k_{off} = 3240 \pm 750\,s^{-1}$.

Further reading

A. Wanat, T. Schneppensieper, G. Stochel, R. van Eldik, E. Bill and K. Wieghardt (2002) *Inorg. Chem.*, vol. 41, p. 4 – 'Kinetics, mechanism and spectroscopy of the reversible binding of nitric oxide to aquated iron(II). An undergraduate text book reaction revisited'.

from associative to dissociative mechanism, and suggests that bond making becomes less (and bond breaking more) important on going from a d^3 to d^8 configuration. For the M^{3+} ions in Table 26.4, values of $\Delta V^\ddagger$ suggest an associative mechanism. Where data are available, an associative process appears to operate for second and third row metal ions, consistent with the idea that larger metal centres may facilitate association with the entering ligand.

First order rate constants, k, for reaction 26.24 vary greatly among the first row d-block metals (all high-spin M^{n+} in the hexaaqua ions):

- Cr^{2+} (d^4) and Cu^{2+} (d^9) are kinetically very labile ($k \geq 10^9\,s^{-1}$);
- Cr^{3+} (d^3) is kinetically inert ($k \approx 10^{-6}\,s^{-1}$);
- Mn^{2+} (d^5), Fe^{2+} (d^6), Co^{2+} (d^7) and Ni^{2+} (d^8) are kinetically labile ($k \approx 10^4$ to $10^7\,s^{-1}$);
- V^{2+} (d^2) has $k \approx 10^2\,s^{-1}$, i.e. considerably less labile than the later M^{2+} ions.

Although one can relate some of these trends to CFSE effects as we discuss below, charge effects are also important, e.g. compare $[Mn(OH_2)_6]^{2+}$ ($k = 2.1 \times 10^7\,s^{-1}$) and $[Fe(OH_2)_6]^{3+}$ ($k = 1.6 \times 10^2\,s^{-1}$), both of which are high-spin d^5.

The rates of water exchange (Fig. 26.1) in high-spin hexaaqua ions follow the sequences:

$$V^{2+} < Ni^{2+} < Co^{2+} < Fe^{2+} < Mn^{2+} < Zn^{2+} < Cr^{2+} < Cu^{2+}$$

and

$$Cr^{3+} \ll Fe^{3+} < V^{3+} < Ti^{3+}$$

For a series of ions of the same charge and about the same size undergoing the same reaction by the same mechanism, we may reasonably suppose that collision frequencies and values of $\Delta S^\ddagger$ are approximately constant, and that variations in rate will arise from variation in $\Delta H^\ddagger$. Let us assume that the latter arise from loss or gain of CFSE (see Table 20.3) on going from the starting complex to the transition state: *a loss of CFSE means an increase in the activation energy for the reaction and hence a decrease in its rate*. The splitting of the d orbitals depends on the coordination geometry (Figs. 20.8 and 20.11), and we can calculate the change in CFSE on the formation of a transition state. Such calculations make assumptions that are actually unlikely to be valid (e.g. constant M−L bond lengths), but for *comparative* purposes, the results should have some meaning. Table 26.5 lists results of such calculations for high-spin octahedral M^{2+} complexes going to either 5- or 7-coordinate transition states; this provides a model for both dissociative and associative processes. For either model, and despite the simplicity of crystal field theory, there is moderately good qualitative agreement between the calculated order of lability and that observed. The particular lability of Cr^{2+}(aq) (high-spin d^4) and Cu^{2+}(aq) (d^9) can be attributed to Jahn–Teller distortion which results in

Table 26.5 Changes in CFSE (ΔCFSE) on converting a high-spin octahedral complex into a square-based pyramidal (for a dissociative process) or pentagonal bipyramidal (for an associative process) transition state, other factors remaining constant (see text).

| Metal ion (high-spin) | d^n | ΔCFSE / Δ_{oct} | | Metal ion (high-spin) | d^n | ΔCFSE / Δ_{oct} | |
		Square-based pyramidal	Pentagonal bipyramidal			Square-based pyramidal	Pentagonal bipyramidal
Sc^{2+}	d^1	+0.06	+0.13	Fe^{2+}	d^6	+0.06	+0.13
Ti^{2+}	d^2	+0.11	+0.26	Co^{2+}	d^7	+0.11	+0.26
V^{2+}	d^3	−0.20	−0.43	Ni^{2+}	d^8	−0.20	−0.43
Cr^{2+}	d^4	+0.31	−0.11	Cu^{2+}	d^9	+0.31	−0.11
Mn^{2+}	d^5	0	0	Zn^{2+}	d^{10}	0	0

weakly bound axial ligands (see structure **20.5** and discussion).

The Eigen–Wilkins mechanism

Water exchange is always more rapid than substitutions with other entering ligands. Let us now consider reaction 26.25.

$$ML_6 + Y \longrightarrow products \qquad (26.25)$$

The mechanism may be associative (A or I_a) or dissociative (D or I_d), and it is not at all easy to distinguish between these, even though the rate laws are different. An associative mechanism involves a 7-coordinate intermediate or transition state and, sterically, an associative pathway seems less likely than a dissociative one. Nevertheless, activation volumes do sometimes indicate an associative mechanism (see Table 26.4). However, *for most ligand substitutions in octahedral complexes, experimental evidence supports dissociative pathways.* Two limiting cases are often observed for general reaction 26.25:

- at high concentrations of Y, the rate of substitution is independent of Y, pointing to a dissociative mechanism;
- at low concentrations of Y, the rate of reaction depends on Y and ML_6, suggesting an associative mechanism.

These apparent contradictions are explained by the *Eigen–Wilkins mechanism.*

> The ***Eigen–Wilkins mechanism*** applies to ligand substitution in an octahedral complex. An ***encounter complex*** is first formed between substrate and entering ligand in a pre-equilibrium step, and this is followed by loss of the leaving ligand in the rate-determining step.

Consider reaction 26.25. The first step in the Eigen–Wilkins mechanism is the diffusing together of ML_6 and Y to form a *weakly bound encounter complex* (equilibrium 26.26).

$$ML_6 + Y \xrightleftharpoons{K_E} \underset{\text{encounter complex}}{\{ML_6, Y\}} \qquad (26.26)$$

Usually, the rate of formation of $\{ML_6,Y\}$ and the back-reaction to ML_6 and Y are much faster than the subsequent conversion of $\{ML_6,Y\}$ to products. Thus, the formation of $\{ML_6,Y\}$ is a *pre-equilibrium*. The equilibrium constant, K_E, can rarely be determined experimentally, but it can be estimated using theoretical models. The rate-determining step in the Eigen–Wilkins mechanism is step 26.27 with a rate constant k. The overall rate law is eq. 26.28.

$$\{ML_6, Y\} \xrightarrow{k} products \qquad (26.27)$$

$$Rate = k[\{ML_6,Y\}] \qquad (26.28)$$

The concentration of $\{ML_6,Y\}$ cannot be measured directly, and we must make use of an estimated value of K_E[†] which is related to $[\{ML_6,Y\}]$ by eq. 26.29.

$$K_E = \frac{[\{ML_6,Y\}]}{[ML_6][Y]} \qquad (26.29)$$

The *total* concentration of ML_6 and $\{ML_6,Y\}$ in eq. 26.26 is measurable because it is the initial concentration of the complex; let this be $[M]_{total}$ (eq. 26.30). Thus, we have expression 26.31 for $[ML_6]$.

$$\left.\begin{array}{l} [M]_{total} = [ML_6] + [\{ML_6,Y\}] \\ [M]_{total} = [ML_6] + K_E[ML_6][Y] \\ \qquad\quad = [ML_6](1 + K_E[Y]) \end{array}\right\} \qquad (26.30)$$

$$[ML_6] = \frac{[M]_{total}}{1 + K_E[Y]} \qquad (26.31)$$

[†] K_E can be estimated using an electrostatic approach: for details of the theory, see R.G. Wilkins (1991) *Kinetics and Mechanism of Reactions of Transition Metal Complexes*, 2nd edn, Wiley-VCH, Weinheim, p. 206.

Table 26.6 Rate constants, k, for reaction 26.34; see eq. 26.28 for the rate law.

Entering ligand, Y	NH_3	py	$[MeCO_2]^-$	F^-	$[SCN]^-$
$k \times 10^{-4} / s^{-1}$	3	3	3	0.8	0.6

We can now rewrite rate equation 26.28 in the form of eq. 26.32 by substituting for $[\{ML_6,Y\}]$ (from eq. 26.29) and then for $[ML_6]$ (from eq. 26.31).

$$\text{Rate} = \frac{kK_E[M]_{\text{total}}[Y]}{1 + K_E[Y]} \qquad (26.32)$$

This equation looks complicated, but at *low concentrations of Y* where $K_E[Y] \ll 1$, eq. 26.32 approximates to eq. 26.33, a second order rate equation in which k_{obs} is the observed rate constant.

$$\text{Rate} = kK_E[M]_{\text{total}}[Y] = k_{\text{obs}}[M]_{\text{total}}[Y] \qquad (26.33)$$

Since k_{obs} can be measured experimentally, and K_E can be estimated theoretically, k can be estimated from the expression $k = k_{\text{obs}}/K_E$ which follows from eq. 26.33. Table 26.6 lists values of k for reaction 26.34 for various entering ligands. The fact that k varies so little is consistent with an I_d mechanism. If the pathway were associative, the rate would depend more significantly on the nature of Y.

$$[Ni(OH_2)_6]^{2+} + Y \longrightarrow [Ni(OH_2)_5Y]^{2+} + H_2O \qquad (26.34)$$

> The substitution of an uncharged ligand (e.g. H_2O) by an anionic ligand (e.g. Cl^-) is called ***anation***.

At a *high concentration of Y* (e.g. when Y is the solvent), $K_E[Y] \gg 1$, and eq. 26.32 approximates to eq. 26.35, a first order rate equation with *no dependence* on the entering ligand. The value of k can be measured directly ($k_{\text{obs}} = k$).

$$\text{Rate} = k[M]_{\text{total}} \qquad (26.35)$$

The water exchange reaction 26.24 exemplifies a case where the entering ligand is the solvent.

Let us now look further at *experimental* trends that are consistent with dissociative (D or I_d) mechanisms for substitution in octahedral complexes. An I_d mechanism is supported in very many instances.

The rate of ligand substitution usually depends on the *nature of the leaving ligand*.

$$[Co(NH_3)_5X]^{2+} + H_2O \rightleftharpoons [Co(NH_3)_5(OH_2)]^{3+} + X^- \qquad (26.36)$$

For reaction 26.36, the rate of substitution increases with X^- in the following order:

$$[OH]^- < [N_3]^- \approx [NCS]^- < [MeCO_2]^- < Cl^-$$
$$< Br^- < I^- < [NO_3]^-$$

This trend correlates with the M−X bond strength (the stronger the bond, the slower the rate) and is consistent

with the rate-determining step involving bond breaking in a dissociative step. We can go one step further: a plot of $\log k$ (where k is the rate constant for the forward reaction 26.36) against $\log K$ (where K is the equilibrium constant for reaction 23.36) is linear *with a gradient of 1.0* (Fig. 26.7). Equations 26.37 and 26.38 relate $\log k$ and $\log K$ to $\Delta G^{\ddagger}$ (Gibbs energy of activation) and ΔG (Gibbs energy of reaction), respectively. It follows that the linear relationship between $\log k$ and $\log K$ represents a linear relationship between $\Delta G^{\ddagger}$ and ΔG, a so-called *linear free energy relationship* (LFER).[†]

$$\Delta G^{\ddagger} \propto - \log k \qquad (26.37)$$

$$\Delta G \propto - \log K \qquad (26.38)$$

The interpretation of the LFER in Fig. 26.7 in mechanistic terms is that the transition state is closely related to the product $[Co(NH_3)_5(OH_2)]^{3+}$, and, therefore, the transition state involves, at most, only a weak Co⋯⋯X interaction. This is consistent with a dissociative (D or I_d) process.

Stereochemistry of substitution

Although most substitutions in octahedral complexes involve D or I_d pathways, we consider the stereochemical implications only of the D mechanism since this involves a 5-coordinate species which we can readily visualize (eq. 26.39).

$$(26.39)$$

The aquation (hydrolysis) reactions of *cis*- and *trans*-$[CoX(en)_2Y]^+$:

$$[CoX(en)_2Y]^+ + H_2O \longrightarrow [Co(OH_2)(en)_2Y]^{2+} + X^-$$

have been extensively studied. If the mechanism is *limiting dissociative* (D), a 5-coordinate intermediate must be involved (scheme 26.39). It follows that the stereochemistry of $[Co(OH_2)(en)_2Y]^{2+}$ must be independent of the leaving group X^-, and will depend on the structure of the intermediate. Starting with *cis*-$[CoX(en)_2Y]^+$, Fig. 26.8 shows that a square-based pyramidal intermediate leads to retention of stereochemistry. For a trigonal bipyramidal

---

[†] LFERs can also use $\ln k$ and $\ln K$, but it is common practice to use log–log relationships. Note that free energy is the same as Gibbs energy.

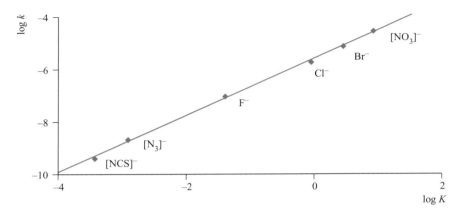

Fig. 26.7 Plot of $\log k$ against $\log K$ for selected leaving groups in reaction 26.36. [Data from: A. Haim (1970) *Inorg. Chem.*, vol. 9, p. 426.]

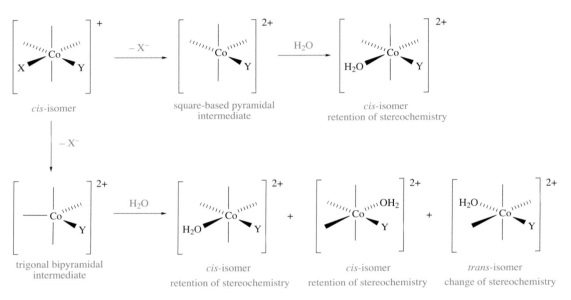

Fig. 26.8 The possible pathways for substitution of a ligand in an octahedral Co(III) complex involving a 5-coordinate intermediate. The leaving group is X^-, and the entering group is exemplified by H_2O.

intermediate, the entering group can attack at one of three positions between pairs of ligands in the equatorial plane. Figure 26.8 shows that this will give a mixture of *cis*- and *trans*-products in an approximately 2:1 ratio. Table 26.7 gives isomer distributions for the products of the spontaneous reactions of H_2O with *cis*- and *trans*-$[CoX(en)_2Y]^+$ where the leaving group is X^-. By comparing pairs of data for complexes with the same Y^- but different leaving groups ($X^- = Cl^-$ or Br^-), one concludes that the stereochemistry of aquation of *cis*- or *trans*-$[CoX(en)_2Y]^+$ is essentially independent of the leaving group. The lack of a large leaving-group effect is consistent with the dissociation of X^- being well on its way by the first transition state, i.e. the latter closely resembles the 5-coordinate intermediate.

We have already seen in Section 4.8 that there is little energy difference between trigonal bipyramidal and square-based pyramidal structures, and that 5-coordinate complexes therefore tend to undergo ligand rearrange-

ments. The 5-coordinate intermediates in the aquation reactions must therefore be very short lived, since the addition of water in the closing step of the reaction is faster than any internal square-based pyramidal–trigonal bipyramidal rearrangement. This is evidenced by the fact that, for example, any specific *cis*- and *trans*-$[CoX(en)_2Y]^+$ pair does not give a common *cis*-/*trans*-$[Co(OH_2)(en)_2Y]^{2+}$ product distribution.

Self-study exercise

The aquation reaction of Λ-*cis*-$[Co(en)_2Cl_2]^+$ leads to *cis*- and *trans*-$[Co(OH_2)(en)_2Cl]^{2+}$, the *cis*-isomer retaining its Λ-configuration. Explain why this observation indicates that a trigonal bipyramidal–square-based pyramidal rearrangement is not competitive in terms of rate with the aquation reaction.

Table 26.7 The isomer distributions in the reactions of *cis*- and *trans*-[CoX(en)$_2$Y]$^+$ with H$_2$O at 298 K.

cis-[CoX(en)$_2$Y]$^+$ + H$_2$O $\longrightarrow$ [Co(OH$_2$)(en)$_2$Y]$^{2+}$ + X$^-$			*trans*-[CoX(en)$_2$Y]$^+$ + H$_2$O $\longrightarrow$ [Co(OH$_2$)(en)$_2$Y]$^{2+}$ + X$^-$		
Y$^-$	**X$^-$**	**% of *cis*-product†**	**Y$^-$**	**X$^-$**	**% of *trans*-product‡**
[OH]$^-$	Cl$^-$	84	[OH]$^-$	Cl$^-$	30
[OH]$^-$	Br$^-$	85	[OH]$^-$	Br$^-$	29
Cl$^-$	Cl$^-$	75	Cl$^-$	Cl$^-$	74
Br$^-$	Br$^-$	73.5	Br$^-$	Br$^-$	84.5
[N$_3$]$^-$	Cl$^-$	86	[N$_3$]$^-$	Cl$^-$	91
[N$_3$]$^-$	Br$^-$	85	[N$_3$]$^-$	Br$^-$	91
[NO$_2$]$^-$	Cl$^-$	100	[NO$_2$]$^-$	Cl$^-$	100
[NO$_2$]$^-$	Br$^-$	100	[NO$_2$]$^-$	Br$^-$	100
[NCS]$^-$	Cl$^-$	100	[NCS]$^-$	Cl$^-$	58.5
[NCS]$^-$	Br$^-$	100	[NCS]$^-$	Br$^-$	57

†Remaining % is *trans*-product.
‡Remaining % is *cis*-product.
[Data: W.G. Jackson and A.M. Sargeson (1978) *Inorg. Chem.*, vol. 17, p. 1348; W.G. Jackson (1986) in *The Stereochemistry of Organometallic and Inorganic Compounds*, ed. I. Bernal, Elsevier, Amsterdam, vol. 1, Chapter 4, p. 255.]

Base-catalysed hydrolysis

Substitution reactions of Co(III) ammine complexes are catalysed by [OH]$^-$, and for reaction 26.40, the rate law is eq. 26.41.

$$[Co(NH_3)_5X]^{2+} + [OH]^- \longrightarrow [Co(NH_3)_5(OH)]^{2+} + X^-$$
$$(26.40)$$

$$\text{Rate} = k_{obs}[Co(NH_3)_5X^{2+}][OH^-] \quad (26.41)$$

That [OH]$^-$ appears in the rate equation shows it has a rate-determining role. However, this is *not* because [OH]$^-$ attacks the metal centre but rather because it deprotonates a coordinated NH$_3$ ligand. Steps 26.42–26.44 show the *conjugate–base mechanism* (*Dcb* or S$_N$1*cb* mechanism). A pre-equilibrium is first established, followed by loss of X$^-$ to give the reactive amido species **26.1**, and, finally, formation of the product in a fast step.

$$\left[\begin{array}{c} NH_3 \\ | \\ H_2N = Co^{\cdots\cdots} NH_3 \\ | \\ NH_3 \\ \end{array} NH_3 \right]^{2+}$$

(26.1)

$$[Co(NH_3)_5X]^{2+} + [OH]^- \underset{}{\overset{K}{\rightleftharpoons}} [Co(NH_3)_4(NH_2)X]^+ + H_2O$$
$$(26.42)$$

$$[Co(NH_3)_4(NH_2)X]^+ \xrightarrow{k_2} [Co(NH_3)_4(NH_2)]^{2+} + X^-$$
$$(26.43)$$

$$[Co(NH_3)_4(NH_2)]^{2+} + H_2O \xrightarrow{\text{fast}} [Co(NH_3)_5(OH)]^{2+}$$
$$(26.44)$$

If the equilibrium constant for equilibrium 26.42 is K, then the rate law consistent with this mechanism is given by eq. 26.45 (see end-of-chapter problem 26.12). If $K[OH^-] \ll 1$, then eq. 26.45 simplifies to eq. 26.41 where $k_{obs} = Kk_2$.

$$\text{Rate} = \frac{Kk_2[Co(NH_3)_5X^{2+}][OH^-]}{1 + K[OH^-]} \quad (26.45)$$

Two observations that are consistent with (but cannot rigidly establish) the conjugate–base mechanism are:

- the entry of competing nucleophiles (for example azide) is base-catalysed in exactly the same way as the hydrolysis reaction, showing that [OH]$^-$ acts as a base and not as a nucleophile;
- the exchange of H (in the NH$_3$) for D in alkaline D$_2$O is much faster than the rate of base hydrolysis.

The first point above is demonstrated by performing the base hydrolysis of [Co(NH$_3$)$_5$Cl]$^{2+}$ in the presence of [N$_3$]$^-$ (a competing nucleophile). This experiment produces [Co(NH$_3$)$_5$(OH)]$^{2+}$ and [Co(NH$_3$)$_5$(N$_3$)]$^{2+}$ in relative proportions that are independent of the concentration of [OH]$^-$, at a fixed concentration of [N$_3$]$^-$. This result is consistent with the facts that in the hydrolysis reaction, the nucleophile is H$_2$O and that [OH]$^-$ acts as a base. The second point above is demonstrated by the Green–Taube experiment which shows that a conjugate–base mechanism operates: when base

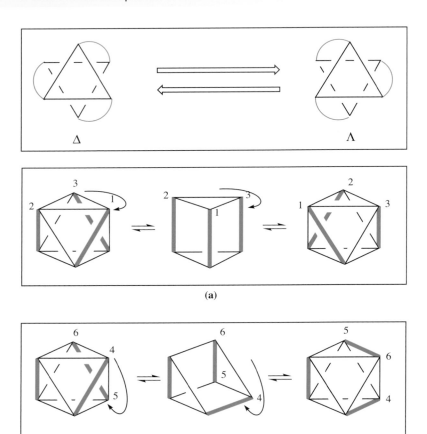

Fig. 26.9 Twist mechanisms for the interconversion of Δ and Λ enantiomers of $M(L-L)_3$: (a) the Bailar twist and (b) the Ray–Dutt twist. The chelating L–L ligands are represented by the red lines (see also Box 19.3).

hydrolysis (with a fixed concentration of $[OH]^-$) of $[Co(NH_3)_5X]^{2+}$ (X = Cl, Br, NO_3) is carried out in a mixture of $H_2(^{16}O)$ and $H_2(^{18}O)$, it is found that the ratio of $[Co(NH_3)_5(^{16}OH)]^{2+}$ to $[Co(NH_3)_5(^{18}OH)]^{2+}$ is constant and independent of X^-. This provides strong evidence that the entering group is H_2O, and not $[OH]^-$, at least in the cases of the leaving groups being Cl^-, Br^- and $[NO_3]^-$.

Isomerization and racemization of octahedral complexes

Although the octahedron is stereochemically rigid, loss of a ligand gives a 5-coordinate species which can undergo Berry pseudo-rotation (see Fig. 4.24). Although, earlier in this chapter, we discussed cases where the assumption is that such rearrangement does *not* occur, if the lifetime of the intermediate is long enough, it provides a mechanism for isomerization (e.g. eq. 26.46). Such isomerization is related to mechanisms already described.

$$trans\text{-}[MX_4Y_2] \xrightarrow{-Y} \{MX_4Y\}$$

$$\xrightarrow{Y} trans\text{-}[MX_4Y_2] + cis\text{-}[MX_4Y_2] \quad (26.46)$$

Our main concern in this section is the racemization of chiral complexes $M(L-L)_3$ and $cis\text{-}M(L-L)_2XY$ containing symmetrical or asymmetrical chelating ligands, L–L, and monodentate ligands, X and Y.

For $[Ni(bpy)_3]^{2+}$ and $[Ni(phen)_3]^{2+}$, the rates of exchange with ^{14}C-labelled ligands are the same as the rates of racemization. This is consistent with a dissociative process (eq. 26.47) in which the intermediate is racemic, or racemizes faster than recombination with L–L.

$$M(L-L)_3 \underset{-S}{\overset{\text{Solvent, S}}{\rightleftharpoons}} M(L-L)_2S_2 + L-L \quad (26.47)$$

Such a dissociative mechanism is rare, and kinetic data are usually consistent with an intramolecular process, e.g. for $[Cr(ox)_3]^{3-}$, $[Co(ox)_3]^{3-}$ (low-spin) and $[Fe(bpy)_3]^{2+}$ (low-spin), the rate of racemization exceeds that of ligand exchange.[†] Two intramolecular mechanisms are possible: a twist mechanism, or the cleavage and reformation of the M–L bond of *one end* of the bidentate ligand. Alternative twist mechanisms (the *Bailar* and *Ray–Dutt twists*) for the interconversion of enantiomers of $M(L-L)_3$ are shown in Fig. 26.9. Each transition state is a trigonal prism and the

---

[†] Ligand abbreviations: see Table 7.7.

mechanisms differ only in which pair of opposing triangular faces twist with respect to each other. The ligands remain coordinated throughout. It is proposed that the racemization of $[Ni(en)_3]^{2+}$ occurs by a twist mechanism.

The second intramolecular mechanism for racemization involves the dissociation of *one* donor atom of a bidentate ligand to give a 5-coordinate species which may undergo rearrangement within the time that the donor atom remains uncoordinated. Scheme 26.48 summarizes the available pathways for the interconversion of Δ and Λ enantiomers of $M(L{-}L)_3$.

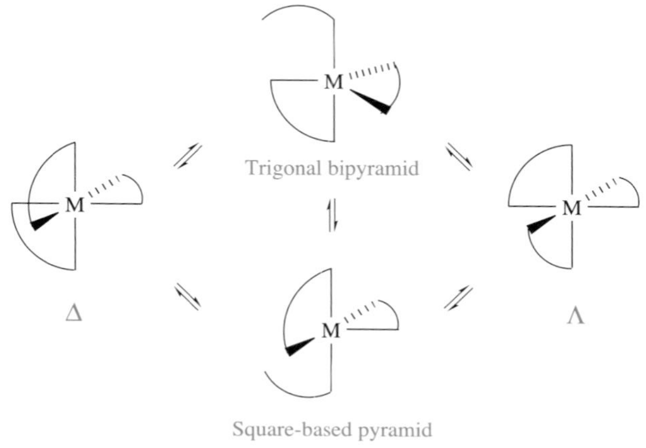

(26.48)

In aqueous solution, racemization of tris-oxalato complexes is faster than exchange of ox^{2-} by two H_2O ligands, suggesting that the two processes are mechanistically different. For $[Rh(ox)_3]^{3-}$ (**26.2**), the non-coordinated O atoms exchange with ^{18}O (from labelled H_2O) faster than do the coordinated O atoms, the rate for the latter being comparable to the rate of racemization. This is consistent with a mechanism involving dissociation of one end of the ox^{2-} ligand, both for isotope exchange of the coordinated O, and for racemization.

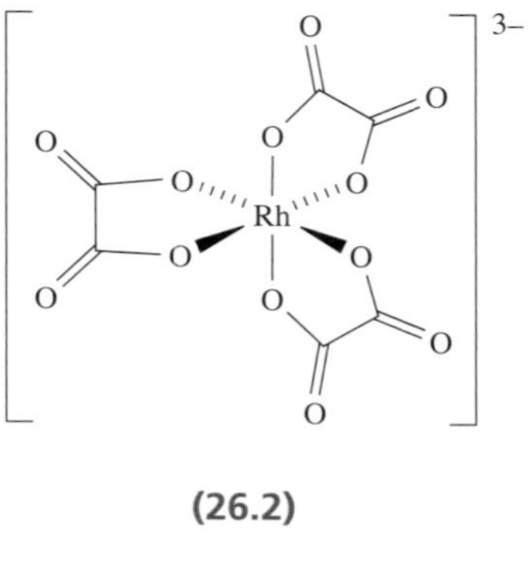

(**26.2**)

If the chelating ligand is asymmetrical (i.e. has two different donor groups), geometrical isomerization is possible as well as racemization, making the kinetics of the system more difficult to interpret. Similarly, racemization of complexes of the type *cis*-$M(L{-}L)_2XY$ is compli-

cated by competing isomerization. The kinetics of these systems are dealt with in more advanced texts.

26.5 Electron-transfer processes

The simplest redox reactions involve *only* the transfer of electrons, and can be monitored by using isotopic tracers, e.g. reaction 26.49.

$$[^{56}Fe(CN)_6]^{3-} + [^{59}Fe(CN)_6]^{4-}$$
$$\longrightarrow [^{56}Fe(CN)_6]^{4-} + [^{59}Fe(CN)_6]^{3-} \qquad (26.49)$$

If $[^{54}MnO_4]^-$ is mixed with unlabelled $[MnO_4]^{2-}$, it is found that however rapidly $[MnO_4]^{2-}$ is precipitated as $BaMnO_4$, incorporation of the label has occurred. In the case of electron transfer between $[Os(bpy)_3]^{2+}$ and $[Os(bpy)_3]^{3+}$, the rate of electron transfer can be measured by studying the loss of optical activity (reaction 26.50).

$$(+)[Os(bpy)_3]^{2+} + (-)[Os(bpy)_3]^{3+}$$
$$\rightleftharpoons (-)[Os(bpy)_3]^{2+} + (+)[Os(bpy)_3]^{3+} \qquad (26.50)$$

Electron-transfer processes fall into two classes, defined by Taube: *outer-sphere* and *inner-sphere mechanisms*.

In an *outer-sphere mechanism*, electron transfer occurs *without a covalent linkage* being formed between the reactants. In an *inner-sphere mechanism*, electron transfer occurs via a *covalently bound bridging ligand*.

In some cases, kinetic data readily distinguish between outer- and inner-sphere mechanisms, but in many reactions, rationalizing the data in terms of a mechanism is not straightforward.

Inner-sphere mechanism

In 1953, Taube (who received the Nobel Prize for Chemistry in 1983) made the classic demonstration of an inner-sphere reaction on a skilfully chosen system (reaction 26.51) in which the reduced forms were substitutionally labile and the oxidized forms were substitutionally inert.

$$[Co(NH_3)_5Cl]^{2+} + [Cr(OH_2)_6]^{2+} + 5[H_3O]^+ \longrightarrow$$
low-spin Co(III)   high-spin Cr(II)
non-labile   labile

$$[Co(OH_2)_6]^{2+} + [Cr(OH_2)_5Cl]^{2+} + 5[NH_4]^+$$
high-spin Co(II)   Cr(III)
labile   non-labile   (26.51)

All the Cr(III) produced was in the form of $[Cr(OH_2)_5Cl]^{2+}$, and tracer experiments in the presence of excess, unlabelled Cl^- showed that all the chlorido ligand in $[Cr(OH_2)_5Cl]^{2+}$ originated from $[Co(NH_3)_5Cl]^{2+}$. Since the Co centre could not have lost Cl^- before reduction, and Cr could not have gained Cl^- after oxidation, the transferred Cl^- must have been bonded to both metal centres during the reaction. Intermediate **26.3** is consistent with these observations.

(26.3)

In the above example, Cl^- is transferred between metal centres. Such transfer is often (but not necessarily) observed. In the reaction between $[Fe(CN)_6]^{3-}$ and $[Co(CN)_5]^{3-}$, the intermediate **26.4** (which is stable enough to be precipitated as the Ba^{2+} salt) is slowly hydrolysed to $[Fe(CN)_6]^{4-}$ and $[Co(CN)_5(OH_2)]^{2-}$ without transfer of the bridging ligand. Common bridging ligands in inner-sphere mechanisms include halides, $[OH]^-$, $[CN]^-$, $[NCS]^-$, pyrazine (**26.5**) and 4,4'-bipyridine (**26.6**). Pyrazine acts as an electron-transfer bridge in the Creutz–Taube cation and related species (see structure 22.65 and discussion).

(26.4)

(26.5)   **(26.6)**

The steps of an *inner-sphere mechanism* are bridge formation, electron transfer and bridge cleavage.

Equations 26.52–26.54 illustrate the inner-sphere mechanism for reaction 26.51. The product $[Co(NH_3)_5]^{2+}$ adds H_2O and then hydrolyses in a fast step to give $[Co(OH_2)_6]^{2+}$.

THEORY

Box 26.2 Timescales of experimental techniques for studying electron-transfer reactions

In Section 4.8, we discussed fluxional processes in relation to the timescales of NMR and IR spectroscopies. A range of techniques is now available to probe electron-transfer reactions, and the recent development of femtosecond (fs) and pico-

second (ps) flash photolysis methods now allows investigations of extremely rapid reactions. For his studies of transition states of chemical reactions using femtosecond spectroscopy, Ahmed H. Zewail was awarded the 1999 Nobel Prize for Chemistry.

Stopped-flow

ESR spectroscopy

NMR spectroscopy

Pulse radiolysis   Isotope separation

Nanosecond flash photolysis

Femtosecond
and picosecond flash photolysis   Spectrophotometry

10^{-12} 10^{-10} 10^{-8} 10^{-6} 10^{-4} 10^{-2} 1 10^2 Time/s

For details of experimental methods, see the 'further reading' section at the end of the chapter.

For information on femtochemistry, see:

F. Carbone, B. Barwick, O.-H. Kwon, H.S. Park, J.S. Baskin and A.H. Zewail (2009) *Chem. Phys. Lett.*, vol. 468, p. 107 – 'EELS femtosecond resolved in 4D ultrafast electron microscopy'.

M. Dantus and A. Zewail, eds (2004) *Chem. Rev.*, issue 4 – A special issue containing reviews dealing with different aspects of femtochemistry.

S.T. Park, D.J. Flannigan and A.H. Zewail (2011) *J. Am. Chem. Soc.*, vol. 133, p. 1730 – 'Irreversible chemical reactions visualized in space and time with 4D electron microscopy'.

J.C. Williamson, J. Cao, H. Ihee, H. Frey and A.H. Zewail (1997) *Nature*, vol. 386, p. 159 – 'Clocking transient chemical changes by ultrafast electron diffraction'.

A.H. Zewail (2000) *Angew. Chem. Int. Ed.*, vol. 39, p. 2586 – 'Femtochemistry: Atomic-scale dynamics of the chemical bond using ultrafast lasers'.

$$[Co^{III}(NH_3)_5Cl]^{2+} + [Cr^{II}(OH_2)_6]^{2+} \underset{k_{-1}}{\overset{k_1}{\rightleftharpoons}}$$

$$[(H_3N)_5Co^{III}(\mu\text{-}Cl)Cr^{II}(OH_2)_5]^{4+} + H_2O \qquad (26.52)$$

$$[(H_3N)_5Co^{III}(\mu\text{-}Cl)Cr^{II}(OH_2)_5]^{4+} \underset{k_{-2}}{\overset{k_2}{\rightleftharpoons}}$$

$$[(NH_3)_5Co^{II}(\mu\text{-}Cl)Cr^{III}(OH_2)_5]^{4+} \qquad (26.53)$$

$$[(H_3N)_5Co^{II}(\mu\text{-}Cl)Cr^{III}(OH_2)_5]^{4+} \underset{k_{-3}}{\overset{k_3}{\rightleftharpoons}}$$

$$[Co^{II}(NH_3)_5]^{2+} + [Cr^{III}(OH_2)_5Cl]^{2+} \qquad (26.54)$$

Most inner-sphere processes exhibit second order kinetics overall, and interpreting the data is seldom simple. Any one of bridge formation, electron transfer or bridge cleavage can be rate determining. In the reaction between $[Fe(CN)_6]^{3-}$ and $[Co(CN)_5]^{3-}$, the rate-determining step is the breaking of the bridge, but it is common for the electron transfer to be the rate-determining step. For bridge formation to be rate determining, the substitution required to form the bridge must be slower than electron transfer. This is not so in reaction 26.52: substitution in $[Cr(OH_2)_6]^{2+}$ (high-spin d^4) is very rapid, and the rate-determining step is electron transfer. However, if $[Cr(OH_2)_6]^{2+}$ is replaced by $[V(OH_2)_6]^{2+}$ (d^3), then the rate constant for reduction is similar to that for water exchange. This is also true for the reactions between $[V(OH_2)_6]^{2+}$ and $[Co(NH_3)_5Br]^{2+}$ or $[Co(CN)_5(N_3)]^{3-}$, indicating that the bridging group has little effect on the rate and that the rate-determining step is the ligand substitution required for bridge formation (the rate depending on the *leaving group*, H_2O) (see Section 26.4).

For reaction 26.55 with a range of ligands X, the rate-determining step is electron transfer, and the rates of reaction depend on X (Table 26.8). The increase in k along the series F^-, Cl^-, Br^-, I^- correlates with increased ability of the halide to act as a bridge. The value of k for $[OH]^-$ is

Table 26.8 Second order rate constants for reaction 26.55 with different bridging X ligands.

Bridging ligand, X	$k\,/\,dm^3\,mol^{-1}\,s^{-1}$
F^-	2.5×10^5
Cl^-	6.0×10^5
Br^-	1.4×10^6
I^-	3.0×10^6
$[N_3]^-$	3.0×10^5
$[OH]^-$	1.5×10^6
H_2O	0.1

similar to that for Br^-, but for H_2O, k is very small and is also pH-dependent. This observation is consistent with H_2O not being the bridging species at all, but rather $[OH]^-$, its availability in solution varying with pH.

$$[Co(NH_3)_5X]^{2+} + [Cr(OH_2)_6]^{2+} + 5[H_3O]^+ \longrightarrow$$

$$[Co(OH_2)_6]^{2+} + [Cr(OH_2)_5X]^{2+} + 5[NH_4]^+ \qquad (26.55)$$

Thiocyanate can coordinate through either the N- or S-donor, and the reaction of $[Co(NH_3)_5(NCS\text{-}S)]^{2+}$ (**26.7**) with $[Cr(OH_2)_6]^{2+}$ leads to the linkage isomers $[Cr(OH_2)_5(NCS\text{-}N)]^{2+}$ (70%) and $[Cr(OH_2)_5(NCS\text{-}S)]^{2+}$ (30%). The results are explained in terms of different bridge structures. If the free N-donor in **26.7** bonds to the Cr(II) centre to give bridge **26.8**, then the reaction proceeds to form $[Cr(OH_2)_5(NCS\text{-}N)]^{2+}$. Alternatively, bridge structure **26.9** gives the green $[Cr(OH_2)_5(NCS\text{-}S)]^{2+}$. This is unstable and isomerizes to the purple $[Cr(OH_2)_5(NCS\text{-}N)]^{2+}$.

(26.7)

(26.8)

(26.9)

Conjugated organic anions (e.g. ox^{2-}) lead to faster inner-sphere reactions than non-conjugated anions (e.g. succinate, $^-O_2CCH_2CH_2CO_2^-$). In the reaction of $[Fe(CN)_5(OH_2)]^{3-}$ with $[Co(NH_3)_5(\textbf{26.10})]^{3+}$ in which the spacer X in **26.10** is varied, the reaction is fast when X provides a conjugated bridge allowing efficient electron transfer, and is slower for short, saturated bridges such as CH_2. However, rapid electron transfer is also observed when the spacer is very flexible, even when it is a saturated (insulating) chain. This observation is consistent with the metal centres being brought in closer contact and a change to an outer-sphere mechanism.

X = CH_2, CH_2CH_2, $CH_2CH_2CH_2$,
CH=CH, C(O)

(26.10)

Table 26.9 Second order rate constants, k, for some outer-sphere redox reactions at 298 K in aqueous solution.

	Reaction	k / dm^3 mol^{-1} s^{-1}
No net chemical reaction (self-exchange)	$[Fe(bpy)_3]^{2+} + [Fe(bpy)_3]^{3+} \longrightarrow [Fe(bpy)_3]^{3+} + [Fe(bpy)_3]^{2+}$	$>10^6$
	$[Os(bpy)_3]^{2+} + [Os(bpy)_3]^{3+} \longrightarrow [Os(bpy)_3]^{3+} + [Os(bpy)_3]^{2+}$	$>10^6$
	$[Co(phen)_3]^{2+} + [Co(phen)_3]^{3+} \longrightarrow [Co(phen)_3]^{3+} + [Co(phen)_3]^{2+}$	40
	$[Fe(OH_2)_6]^{2+} + [Fe(OH_2)_6]^{3+} \longrightarrow [Fe(OH_2)_6]^{3+} + [Fe(OH_2)_6]^{2+}$	3
	$[Co(en)_3]^{2+} + [Co(en)_3]^{3+} \longrightarrow [Co(en)_3]^{3+} + [Co(en)_3]^{2+}$	10^{-4}
	$[Co(NH_3)_6]^{2+} + [Co(NH_3)_6]^{3+} \longrightarrow [Co(NH_3)_6]^{3+} + [Co(NH_3)_6]^{2+}$	10^{-6}
Net chemical reaction	$[Os(bpy)_3]^{2+} + [Mo(CN)_8]^{3-} \longrightarrow [Os(bpy)_3]^{3+} + [Mo(CN)_8]^{4-}$	2×10^9
	$[Fe(CN)_6]^{4-} + [Fe(phen)_3]^{3+} \longrightarrow [Fe(CN)_6]^{3-} + [Fe(phen)_3]^{2+}$	10^8
	$[Fe(CN)_6]^{4-} + [IrCl_6]^{2-} \longrightarrow [Fe(CN)_6]^{3-} + [IrCl_6]^{3-}$	4×10^5

Outer-sphere mechanism

When *both* reactants in a redox reaction are *kinetically inert*, electron transfer must take place by a *tunnelling* or *outer-sphere* mechanism. For a reaction such as 26.49, $\Delta G^{\circ} \approx 0$, but activation energy is needed to overcome electrostatic repulsion between ions of like charge, to stretch or shorten bonds so that they are equivalent in the transition state (see below), and to alter the solvent sphere around each complex.

In a *self-exchange reaction*, the left- and right-hand sides of the equation are identical. Only electron transfer, and no net chemical reaction, takes place.

The *Franck–Condon approximation* states that a molecular electronic transition is much faster than a molecular vibration.

The rates of outer-sphere self-exchange reactions vary considerably as illustrated in Table 26.9. Clearly, the reactants must approach closely for the electron to migrate from reductant to oxidant. This reductant–oxidant pair is called the *encounter* or *precursor complex*. When electron transfer occurs, there is an important restriction imposed upon it by the *Franck–Condon approximation* (see Section 20.7). Consider a self-exchange reaction of the type:

$$[ML_6]^{2+} + [ML_6]^{3+} \rightleftharpoons [ML_6]^{3+} + [ML_6]^{2+}$$

There is no overall reaction and therefore $\Delta G^{\circ} = 0$, and $K = 1$. Why do reactions of this type have widely differing reaction rates? It is usually the case that the M–L bond lengths in the M(III) complex are shorter than those in the corresponding M(II) complex. Consider now a hypothetical situation: what happens if an electron is transferred from the vibrational ground state of $[ML_6]^{2+}$ to the vibrational ground state of $[ML_6]^{3+}$, each with its characteristic M–L distance? The Franck–Condon approximation states that electronic transitions are far faster than nuclear motion. It follows that the loss of an electron from $[ML_6]^{2+}$ generates $[ML_6]^{3+}$ in a vibrationally excited state with an elongated M–L bond. Similarly, gain of an electron by $[ML_6]^{3+}$ produces $[ML_6]^{2+}$ in a vibrationally excited state with a compressed M–L bond. Both of these then relax to the equilibrium geometries with energy loss. If this description were correct, we would have a situation that disobeys the first law of thermodynamics. How can a reaction with $\Delta G^{\circ} = 0$ continually lose energy as the electron is transferred between $[ML_6]^{2+}$ and $[ML_6]^{3+}$? The answer, of course, is that it cannot.

The electron transfer can only take place when the M–L bond distances in the M(II) and M(III) states are the same, i.e. the bonds in $[ML_6]^{2+}$ must be compressed and those in $[ML_6]^{3+}$ must be elongated (Fig. 26.10). This is described as a Franck–Condon restriction. The activation energy required to reach these vibrational excited states varies according to the system, and hence the self-exchange rate constants vary. In the case of $[Fe(bpy)_3]^{2+}$ and $[Fe(bpy)_3]^{3+}$, both complexes are low-spin, and the Fe–N bond distances are 197 and 196 pm, respectively. Electron transfer involves only a change from t_{2g}^5 (Fe^{3+}) to t_{2g}^6 (Fe^{2+}) and vice versa. Thus, the rate of electron transfer is fast ($k > 10^6$ dm^3 mol^{-1} s^{-1}). The greater the changes in bond length

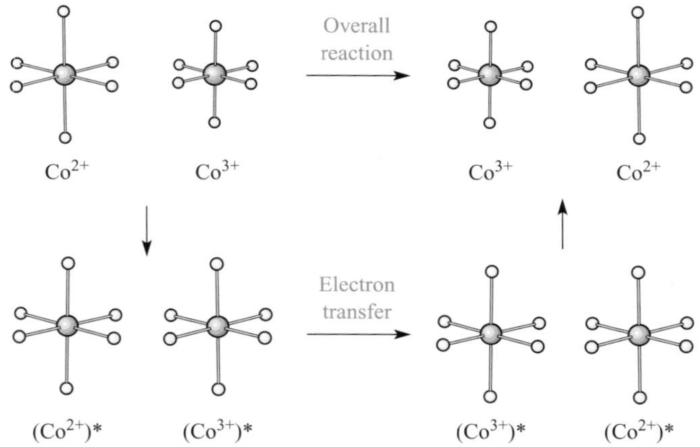

Fig. 26.10 The outer-sphere mechanism: when the reactants have differing bond lengths, vibrationally excited states with equal bond lengths must be formed in order to allow electron transfer to occur.

required to reach the encounter complex, the slower the rate of electron transfer. For example, the rate of electron transfer between $[Ru(NH_3)_6]^{2+}$ (Ru–N = 214 pm, low-spin d^6) and $[Ru(NH_3)_6]^{3+}$ (Ru–N = 210 pm, low-spin d^5) is $10^4 \, dm^3 \, mol^{-1} \, s^{-1}$.

Electron transfer between $[Co(NH_3)_6]^{2+}$ (Co–N = 211 pm) and $[Co(NH_3)_6]^{3+}$ (Co–N = 196 pm) requires not only changes in bond lengths, but also a change in spin state: $[Co(NH_3)_6]^{2+}$ is high-spin d^7 ($t_{2g}^5 e_g^2$) and $[Co(NH_3)_6]^{3+}$ is low-spin d^6 ($t_{2g}^6 e_g^0$). Transfer of an electron between the excited states shown in Fig. 26.10 leads to a configuration of $t_{2g}^5 e_g^1$ for $\{[Co(NH_3)_6]^{3+}\}*$ and $t_{2g}^6 e_g^1$ for $\{[Co(NH_3)_6]^{2+}\}*$. These are electronically excited states, each of which must undergo a spin state change to attain the ground state configuration. The activation energy for the self-exchange reaction therefore has contributions from both changes in bond lengths and changes in spin states. In such cases, the activation energy is high and the rate of electron transfer is slow ($k \approx 10^{-6} \, dm^3 \, mol^{-1} \, s^{-1}$).

Table 26.9 illustrates another point: self-exchange between $[Co(phen)_3]^{2+}$ and $[Co(phen)_3]^{3+}$ is much faster than between $[Co(NH_3)_6]^{2+}$ and $[Co(NH_3)_6]^{3+}$ or $[Co(en)_3]^{2+}$ and $[Co(en)_3]^{3+}$ (all three exchange processes are between high-spin Co(II) and low-spin Co(III)). This is consistent with the ability of phen ligands to use their π-orbitals to facilitate the intermolecular migration of an electron from one ligand to another, and phen complexes tend to exhibit fast rates of self-exchange.

The self-exchange reactions listed in Table 26.9 all involve cationic species in aqueous solution. The rates of these reactions are typically *not* affected by the nature and concentration of the anion present in solution. On the other hand, the rate of electron transfer between anions in aqueous solution generally depends on the cation and its concentration. For example, the self-exchange reaction

between $[Fe(CN)_6]^{3-}$ and $[Fe(CN)_6]^{4-}$ with K^+ as the counter-ion proceeds along a pathway that is catalysed by the K^+ ions. It has been shown[†] that, by adding the macrocyclic ligand 18-crown-6 or crypt-[222] to complex the K^+ ions (see Fig. 11.8), the K^+-catalysed pathway is replaced by a cation-independent mechanism. The rate constant that is often quoted for the $[Fe(CN)_6]^{3-}/[Fe(CN)_6]^{4-}$ self-exchange reaction is of the order of $10^4 \, dm^3 \, mol^{-1} \, s^{-1}$, whereas the value of k determined for the cation-independent pathway is $2.4 \times 10^2 \, dm^3 \, mol^{-1} \, s^{-1}$, i.e. ≈ 100 times smaller. This significant result indicates that caution is needed in the interpretation of rate constant data for electron-transfer reactions between complex anions.

The accepted method of testing for an outer-sphere mechanism is to apply *Marcus–Hush* theory[‡] which relates kinetic and thermodynamic data for two self-exchange reactions with data for the *cross-reaction* between the self-exchange partners, e.g. reactions 26.56–26.58.

$$[ML_6]^{2+} + [ML_6]^{3+} \longrightarrow [ML_6]^{3+} + [ML_6]^{2+}$$

Self-exchange 1   (26.56)

$$[M'L_6]^{2+} + [M'L_6]^{3+} \longrightarrow [M'L_6]^{3+} + [M'L_6]^{2+}$$

Self-exchange 2   (26.57)

$$[ML_6]^{2+} + [M'L_6]^{3+} \longrightarrow [ML_6]^{3+} + [M'L_6]^{2+}$$

Cross-reaction   (26.58)

[†] See: A. Zahl, R. van Eldik and T.W. Swaddle (2002) *Inorg. Chem.*, vol. 41, p. 757.

[‡] For fuller treatments of Marcus–Hush theory, see the further reading at the end of the chapter; Rudolph A. Marcus received the Nobel Prize for Chemistry in 1992.

For each self-exchange reaction, $\Delta G^{\circ} = 0$. The Gibbs energy of activation, $\Delta G^{\ddagger}$, for a self-exchange reaction can be written in terms of four contributing factors (eq. 26.59).

$$\Delta G^{\ddagger} = \Delta_{w} G^{\ddagger} + \Delta_{o} G^{\ddagger} + \Delta_{s} G^{\ddagger} + RT \ln \frac{k'T}{hZ} \quad (26.59)$$

where T = temperature in K
$\quad\quad R$ = molar gas constant
$\quad\quad k'$ = Boltzmann constant
$\quad\quad h$ = Planck constant
$\quad\quad Z$ = effective collision frequency in solution
$\quad\quad\quad \approx 10^{11} \, \text{dm}^3 \, \text{mol}^{-1} \, \text{s}^{-1}$

The contributions in this equation arise as follows:

- $\Delta_{w} G^{\ddagger}$ is the energy associated with bringing the reductant and oxidant together and includes the work done to counter electrostatic repulsions;
- $\Delta_{o} G^{\ddagger}$ is the energy associated with changes in bond distances;
- $\Delta_{s} G^{\ddagger}$ arises from rearrangements within the solvent spheres;
- the final term accounts for the loss of translational and rotational energy on formation of the encounter complex.

Although we shall not delve into the theory, it is possible to calculate the terms on the right-hand side of eq. 26.59, and thus to estimate values of $\Delta G^{\ddagger}$ for self-exchange reactions. The rate constant, k, for the self-exchange can then be calculated using eq. 26.60. The results of such calculations have been checked against much experimental data, and the validity of the theory is upheld.

$$k = \kappa Z \, e^{\left(-\Delta G^{\ddagger}/RT\right)} \quad (26.60)$$

where κ (the transmission coefficient) ≈ 1
$Z \approx 10^{11} \, \text{dm}^3 \, \text{mol}^{-1} \, \text{s}^{-1}$ (see eq. 26.59)

Now consider reactions 26.56–26.58, and let the rate and thermodynamic parameters be designated as follows:

- k_{11} and $\Delta G^{\ddagger}_{11}$ for self-exchange 1;
- k_{22} and $\Delta G^{\ddagger}_{22}$ for self-exchange 2;
- k_{12} and $\Delta G^{\ddagger}_{12}$ for the cross-reaction; the equilibrium constant is K_{12}, and the standard Gibbs energy of reaction is ΔG°_{12}.

The Marcus–Hush equation (which we shall not derive) is given by expression 26.61 and applies to outer-sphere mechanisms.

$$k_{12} = (k_{11} k_{22} K_{12} f_{12})^{1/2} \quad (26.61)$$

where f_{12} is defined by the relationship

$$\log f_{12} = \frac{(\log K_{12})^2}{4 \log \left(\dfrac{k_{11} k_{22}}{Z^2} \right)}$$

and Z is the collision frequency (see eq. 26.59)

Equation 26.62 gives a logarithmic form of eq. 26.61. Often, $f \approx 1$ and so $\log f \approx 0$, allowing this term to be neglected in some cases. Thus, eq. 26.63 is an approximate form of the Marcus–Hush equation.

$$\log k_{12} = 0.5 \log k_{11} + 0.5 \log k_{22} + 0.5 \log K_{12} + 0.5 \log f_{12} \quad (26.62)$$

$$\log k_{12} \approx 0.5 \log k_{11} + 0.5 \log k_{22} + 0.5 \log K_{12} \quad (26.63)$$

Values of k_{11}, k_{22}, K_{12} and k_{12} can be obtained experimentally, or k_{11} and k_{22} theoretically (see above); K_{12} is determined from E_{cell} (see Section 8.2). If the value of k_{12} calculated from eq. 26.63 agrees with the experimental value, this provides strong evidence that the cross-reaction proceeds by an outer-sphere mechanism. Deviation from eq. 26.63 indicates that another mechanism is operative.

Worked example 26.1 Marcus–Hush theory: a test for an outer-sphere mechanism

For the reaction:

$$[\mathbf{Ru(NH_3)_6}]^{2+} + [\mathbf{Co(phen)_3}]^{3+}$$
$$\longrightarrow [\mathbf{Ru(NH_3)_6}]^{3+} + [\mathbf{Co(phen)_3}]^{2+}$$

the observed rate constant is $1.5 \times 10^4 \, \text{dm}^3 \, \text{mol}^{-1} \, \text{s}^{-1}$ and the equilibrium constant is 2.6×10^5. The rate constants for the self-exchange reactions $[\mathbf{Ru(NH_3)_6}]^{2+}/[\mathbf{Ru(NH_3)_6}]^{3+}$ and $[\mathbf{Co(phen)_3}]^{2+}/[\mathbf{Co(phen)_3}]^{3+}$ are 8.2×10^2 and $40 \, \text{dm}^3 \, \text{mol}^{-1} \, \text{s}^{-1}$ respectively. Are these data consistent with an outer-sphere mechanism for the cross-reaction?

The approximate form of the Marcus–Hush equation is:

$$k_{12} \approx (k_{11} k_{22} K_{12})^{1/2} \quad \text{(or its log form)}$$

Calculate k_{12} using this equation:

$$k_{12} \approx [(8.2 \times 10^2)(40)(2.6 \times 10^5)]^{1/2}$$
$$\approx 9.2 \times 10^4 \, \text{dm}^3 \, \text{mol}^{-1} \, \text{s}^{-1}$$

This is in quite good agreement with the observed value of $1.5 \times 10^4 \, \text{dm}^3 \, \text{mol}^{-1} \, \text{s}^{-1}$, and suggests that the mechanism is outer-sphere electron transfer.

Self-study exercise

For the reaction given above, use the values of $k_{12} = 1.5 \times 10^4 \, \text{dm}^3 \, \text{mol}^{-1} \, \text{s}^{-1}$, $K_{12} = 2.6 \times 10^5$, and k for the self-exchange reaction $[\text{Ru(NH}_3)_6]^{2+}/[\text{Ru(NH}_3)_6]^{3+}$ to estimate a value of k for the self-exchange $[\text{Co(phen)}_3]^{2+}/[\text{Co(phen)}_3]^{3+}$. Comment on the agreement between your value and the observed value of $40 \, \text{dm}^3 \, \text{mol}^{-1} \, \text{s}^{-1}$.

[*Ans.* $\approx 1.1 \, \text{dm}^3 \, \text{mol}^{-1} \, \text{s}^{-1}$]

By using the relationships in eqs. 26.37 and 26.38, we can write eq. 26.63 in terms of Gibbs energies (eq. 26.64).

$$\Delta G^{\ddagger}_{12} \approx 0.5\Delta G^{\ddagger}_{11} + 0.5\Delta G^{\ddagger}_{22} + 0.5\Delta G^{o}_{12} \qquad (26.64)$$

In a series of related redox reactions in which one reactant is the same, a plot of $\Delta G^{\ddagger}_{12}$ against ΔG^{o}_{12} is linear with a gradient of 0.5 if an outer-sphere mechanism is operative.

An important application of Marcus–Hush theory is in bioinorganic electron-transfer systems.[†] For example, cytochrome c is an electron-transfer metalloprotein (see Section 29.4) and contains haem-iron as either Fe(II) or Fe(III). Electron transfer from one Fe centre to another is *long range*, the electron *tunnelling* through the protein.

Model systems have been devised to investigate electron transfer between cytochrome c and molecular complexes such as $[Ru(NH_3)_6]^{2+}$, and kinetic data are consistent with Marcus theory, indicating outer-sphere processes. For electron transfer in both metalloproteins and the model systems, the distance between the metal centres is significantly greater than for transfer between two simple metal complexes, e.g. up to 2500 pm. The rate of electron transfer decreases exponentially with increasing distance, r, between the two metal centres (eq. 26.65, where β is a parameter which depends on the molecular environment).

$$\text{Rate of electron transfer} \propto e^{-\beta r} \qquad (26.65)$$

KEY TERMS

The following terms have been introduced in this chapter. Do you know what they mean?

- leaving group
- entering group
- kinetically inert
- kinetically labile
- associative mechanism, A
- dissociative mechanism, D
- interchange mechanism, I_a or I_d
- intermediate
- transition state
- rate-determining step
- fast step
- activation parameters
- volume of activation, $\Delta V^{\ddagger}$
- stereoretentive
- *trans*-effect
- nucleophilicity sequence
- nucleophilicity parameter
- nucleophilicity discrimination factor
- Eigen–Wilkins mechanism
- encounter complex
- pre-equilibrium
- anation
- linear free energy relationship, LFER
- conjugate–base mechanism, Dcb
- Bailar twist mechanism
- Ray–Dutt twist mechanism
- outer-sphere mechanism
- inner-sphere mechanism
- Franck–Condon approximation
- self-exchange mechanism
- cross-reaction
- Marcus–Hush theory (fundamental principles)

FURTHER READING

For an introduction to rate laws

P. Atkins and J. de Paula (2009) *Atkins' Physical Chemistry*, 9th edn, Oxford University Press, Oxford – Chapter 21 gives a detailed account.

C.E. Housecroft and E.C. Constable (2010) *Chemistry*, 4th edn, Prentice Hall, Harlow – Chapter 15 provides a basic introduction.

Kinetics and mechanisms of inorganic and organometallic reactions

J.D. Atwood (1997) *Inorganic and Organometallic Reaction Mechanisms*, 2nd edn, Wiley-VCH, Weinheim – One of the most readable texts dealing with coordination and organometallic reaction mechanisms.

F. Basolo and R.G. Pearson (1967) *Mechanisms of Inorganic Reactions*, Wiley, New York – A classic book in the field of inorganic mechanisms.

J. Burgess (1999) *Ions in Solution*, Horwood Publishing Ltd, Chichester – Chapters 8–12 introduce inorganic kinetics in a clear and informative manner.

R.W. Hay (2000) *Reaction Mechanisms of Metal Complexes*, Horwood Publishing Ltd, Chichester – Includes excellent coverage of substitution reactions, and isomerization, racemization and redox processes.

R.B. Jordan (1998) *Reaction Mechanisms of Inorganic and Organometallic Systems*, 2nd edn, Oxford University Press, New York – A detailed text which includes experimental methods, photochemistry and bioinorganic systems.

S.F.A. Kettle (1996) *Physical Inorganic Chemistry*, Spektrum, Oxford – Chapter 14 gives an excellent introduction and includes photokinetics.

[†] For further discussion, see: R.G. Wilkins (1991) *Kinetics and Mechanism of Reactions of Transition Metal Complexes*, 2nd edn, Wiley-VCH, Weinheim, p. 285; J.J.R. Fraústo da Silva and R.J.P. Williams (1991) *The Biological Chemistry of the Elements*, Clarendon Press, Oxford, p. 105.

A.G. Lappin (1994) *Redox Mechanisms in Inorganic Chemistry*, Ellis Horwood, Chichester – A comprehensive review of redox reactions in inorganic chemistry, including multiple electron transfer and some aspects of bioinorganic chemistry.

T.W. Swaddle (2010) in *Physical Inorganic Chemistry: Reactions, Processes and Applications*, ed A. Bakac, Wiley, Hoboken, Ch. 8 – 'Ligand substitution dynamics in metal complexes'.

M.L. Tobe and J. Burgess (1999) *Inorganic Reaction Mechanisms*, Addison Wesley Longman, Harlow – A comprehensive account of inorganic mechanisms.

R.G. Wilkins (1991) *Kinetics and Mechanism of Reactions of Transition Metal Complexes*, 2nd edn, Wiley-VCH, Weinheim – An excellent and detailed text which includes experimental methods.

More specialized reviews

J. Burgess and C.D. Hubbard (2003) *Adv. Inorg. Chem.*, vol. 54, p. 71 – 'Ligand substitution reactions'.

B.J. Coe and S.J. Glenwright (2000) *Coord. Chem. Rev.*, vol. 203, p. 5 – '*Trans*-effects in octahedral transition metal complexes' (includes both structural and kinetic *trans*-effects).

R.J. Cross (1985) *Chem. Soc. Rev.*, vol. 14, p. 197 – 'Ligand substitution reactions in square planar molecules'.

R. van Eldik (1999) *Coord. Chem. Rev.*, vol. 182, p. 373 – 'Mechanistic studies in coordination chemistry'.

L. Helm, G.M. Nicolle and A.E. Merbach (2005) *Adv. Inorg. Chem.*, vol. 57, p. 327 – 'Water and proton exchange processes on metal ions'.

M.H.V. Huynh and T.J. Meyer (2007) *Chem. Rev.*, vol. 107, p. 5004 – 'Proton-coupled electron transfer'.

W.G. Jackson (2002) *Inorganic Reaction Mechanisms*, vol. 4, p. 1 – 'Base catalysed hydrolysis of aminecobalt(III) complexes: From the beginnings to the present'.

S.F. Lincoln (2005) *Helv. Chim. Acta*, vol. 88, p. 523 – 'Mechanistic studies of metal aqua ions: A semi-historical perspective'.

S.F. Lincoln, D.T. Richens and A.G. Sykes (2004) in *Comprehensive Coordination Chemistry II*, eds J.A. McCleverty and T.J. Meyer, Elsevier, Oxford, vol. 1, p. 515 – 'Metal aqua ions' covers substitution reactions.

R.A. Marcus (1986) *J. Phys. Chem.*, vol. 90, p. 3460 – 'Theory, experiment and reaction rates: a personal view'.

J. Reedijk (2008) *Platinum Metals Rev.*, vol. 52, p. 2 – 'Metal–ligand exchange kinetics in platinum and ruthenium complexes. Significance for effectiveness as anticancer drugs'.

D.T. Richens (2005) *Chem. Rev.*, vol. 105, p. 1961 – 'Ligand substitution reactions at inorganic centers'.

S.V. Rosokha and J.K. Kochi (2008) *Acc. Chem. Res.*, vol. 41, p. 641 – 'Fresh look at electron-transfer mechanisms via the donor/acceptor bindings in the critical encounter complex'.

G. Stochel and R. van Eldik (1999) *Coord. Chem. Rev.*, vol. 187, p. 329 – 'Elucidation of inorganic reaction mechanisms through volume profile analysis'.

H. Taube (1984) *Science*, vol. 226, p. 1028 – 'Electron transfer between metal complexes: Retrospective' (Nobel Prize for Chemistry lecture).

PROBLEMS

26.1 Review what is meant by the following terms:
(a) elementary step,
(b) rate-determining step,
(c) activation energy,
(d) intermediate,
(e) transition state,
(f) rate equation,
(g) zero, first and second order rate laws,
(h) nucleophile.

26.2 Sketch reaction profiles for the reaction pathways described in eqs. 26.5 and 26.6. Comment on any significant features including activation energies.

26.3 Discuss evidence to support the proposal that substitution in square planar complexes is an associative process.

26.4 Under pseudo-first order conditions, the variation of k_{obs} with [py] for reaction of square planar $[Rh(cod)(PPh_3)_2]^+$ (2×10^{-4} mol dm^{-3}, cod = **24.22**) with pyridine is as follows:

[py] / mol dm^{-3}	0.006 25	0.0125	0.025	0.05
k_{obs} / s^{-1}	27.85	30.06	34.10	42.04

Show that the data are consistent with the reaction proceeding by two competitive routes, indicate what these pathways are, and determine values of the rate constants for each pathway. [Data: H. Krüger *et al.* (1987) *J. Chem. Educ.*, vol. 64, p. 262.]

26.5 (a) The *cis*- and *trans*-isomers of $[PtCl_2(NH_3)(NO_2)]^-$ are prepared by reaction sequences 26.19 and 26.20 respectively. Rationalize the observed differences in products in these routes. (b) Suggest the products of the reaction of $[PtCl_4]^{2-}$ with PEt_3.

26.6   (a) Suggest a mechanism for the reaction:

$trans\text{-}[PtL_2Cl_2] + Y \longrightarrow trans\text{-}[PtL_2ClY]^+ + Cl^-$

(b) If the intermediate in your mechanism is sufficiently long-lived, what complication might arise?

26.7   The reaction of $trans\text{-}[Pt(PEt_3)_2PhCl]$ with the strong nucleophile thiourea (tu) in MeOH follows a 2-term rate law with $k_{obs} = k_1 + k_2[\text{tu}]$. A plot of k_{obs} against [tu] is linear with the line passing close to the origin. Rationalize these observations.

thiourea

26.8   Second order rate constants, k_2, for the reaction of $trans\text{-}[Pt(PEt_3)_2Ph(MeOH)]^+$ with pyridine (py) in MeOH to give $trans\text{-}[Pt(PEt_3)_2Ph(py)]^+$ vary with temperature as shown below. Use the data to determine the activation enthalpy and activation entropy for the reaction.

T / K	288	293	298	303	308
$k_2/\text{dm}^3\,\text{mol}^{-1}\,\text{s}^{-1}$	3.57	4.95	6.75	9.00	12.1

[Data: R. Romeo *et al.* (1974) *Inorg. Chim. Acta*, vol. 11, p. 231.]

26.9   For the reaction:

$[Co(NH_3)_5(OH_2)]^{3+} + X^- \longrightarrow [Co(NH_3)_5X]^{2+} + H_2O$

it is found that:

$$\frac{d[Co(NH_3)_5X^{2+}]}{dt} = k_{obs}[Co(NH_3)_5(OH_2)^{3+}][X^-]$$

and for $X^- = Cl^-$, $\Delta V^{\ddagger}$ is positive. Rationalize these data.

26.10   (a) Rationalize the formation of the products in the following sequence of reactions:

$[Rh(OH_2)_6]^{3+} \xrightarrow[-H_2O]{Cl^-} [RhCl(OH_2)_5]^{2+}$

$\xrightarrow[-H_2O]{Cl^-} trans\text{-}[RhCl_2(OH_2)_4]^+$

$\xrightarrow[-H_2O]{Cl^-} mer\text{-}[RhCl_3(OH_2)_3]$

$\xrightarrow[-H_2O]{Cl^-} trans\text{-}[RhCl_4(OH_2)_2]^-$

(b) Suggest methods of preparing $[RhCl_5(OH_2)]^{2-}$, $cis\text{-}[RhCl_4(OH_2)_2]^-$ and $fac\text{-}[RhCl_3(OH_2)_3]$.

26.11   What reason can you suggest for the sequence $Co > Rh > Ir$ in the rates of anation of $[M(OH_2)_6]^{3+}$ ions?

26.12   Derive rate law 26.45 for the mechanism shown in steps 26.42–26.44.

26.13   Suggest a mechanism for the possible racemization of tertiary amines $NR_1R_2R_3$. Is it likely that such molecules can be resolved?

26.14   The rate of racemization of $[CoL_3]$ where $HL = \textbf{26.11a}$ is approximately the same as its rate of isomerization into $[CoL'_3]$ where $HL' = \textbf{26.11b}$. What can you deduce about the mechanisms of these reactions?

(26.11a)   (26.11b)

26.15   Substitution of H_2O in $[Fe(OH_2)_6]^{3+}$ by thiocyanate is complicated by proton loss. By considering the reaction scheme below, derive an expression for $-\dfrac{d[SCN^-]}{dt}$ in terms of the equilibrium and rate constants, $[Fe(OH_2)_6^{3+}]$, $[SCN^-]$, $[Fe(OH_2)_5(SCN)^{2+}]$ and $[H^+]$.

26.16   Rationalize the observation that when the reaction:

$[Co(NH_3)_4(CO_3)]^+$

$\xrightarrow{[H_3O]^+, H_2O} [Co(NH_3)_4(OH_2)_2]^{3+} + CO_2$

is carried out in $H_2(^{18}O)$, the water in the complex contains equal proportions of $H_2(^{18}O)$ and $H_2(^{16}O)$.

26.17   Two twist mechanisms for the rearrangement of $\Delta\text{-}M(L-L)_3$ to $\Lambda\text{-}M(L-L)_3$ are shown in Fig. 26.9. The initial diagrams in (a) and (b) are identical; confirm that the enantiomers formed in (a) and (b) are also identical.

26.18 The rate constants for racemization (k_r) and dissociation (k_d) of $[FeL_3]^{4-}$ ($H_2L = \mathbf{26.12}$) at several temperatures, T, are given in the table. (a) Determine $\Delta H^{\ddagger}$ and $\Delta S^{\ddagger}$ for each reaction. (b) What can you deduce about the mechanism of racemization?

T/K	288	294	298	303	308
$k_r \times 10^5 / s^{-1}$	0.5	1.0	2.7	7.6	13.4
$k_d \times 10^5 / s^{-1}$	0.5	1.0	2.8	7.7	14.0

[Data from: A. Yamagishi (1986) *Inorg. Chem.*, vol. 25, p. 55.]

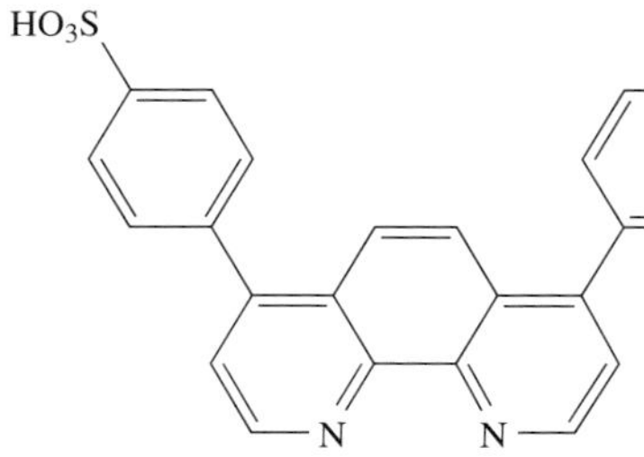

(26.12)

26.19 The reaction:

$$[Cr(NH_3)_5Cl]^{2+} + NH_3 \longrightarrow [Cr(NH_3)_6]^{3+} + Cl^-$$

in liquid NH_3 is catalysed by KNH_2. Suggest an explanation for this observation.

26.20 Give an example of a reaction that proceeds by an inner-sphere mechanism. Sketch reaction profiles for inner-sphere electron-transfer reactions in which the rate-determining step is (a) bridge formation, (b) electron transfer and (c) bridge cleavage. Which profile is most commonly observed?

26.21 Discuss, with examples, the differences between inner- and outer-sphere mechanisms, and state what is meant by a self-exchange reaction.

26.22 Account for the relative values of the rate constants for the following electron-transfer reactions in aqueous solution:

Reaction number	Reactants	$k / dm^3 mol^{-1} s^{-1}$
I	$[Ru(NH_3)_6]^{3+} + [Ru(NH_3)_6]^{2+}$	10^4
II	$[Co(NH_3)_6]^{3+} + [Ru(NH_3)_6]^{2+}$	10^{-2}
III	$[Co(NH_3)_6]^{3+} + [Co(NH_3)_6]^{2+}$	10^{-6}

For which reactions is $\Delta G^{\circ} = 0$?

26.23 (a) If, in an electron-transfer process, there is both electron and ligand transfer between reagents, what can you conclude about the mechanism? (b) Explain why very fast electron transfer between low-spin octahedral Os(II) and Os(III) in a self-exchange reaction is possible.

OVERVIEW PROBLEMS

26.24 Suggest products in the following ligand substitution reactions. Where the reaction has two steps, specify a product for each step. Where more than one product could, in theory, be possible, rationalize your choice of preferred product.

(a) $[PtCl_4]^{2-} \xrightarrow{NH_3} \xrightarrow{NH_3}$

(b) *cis*-$[Co(en)_2Cl_2]^+ + H_2O \longrightarrow$

(c) $[Fe(OH_2)_6]^{2+} + NO \longrightarrow$

26.25 (a) The reaction:

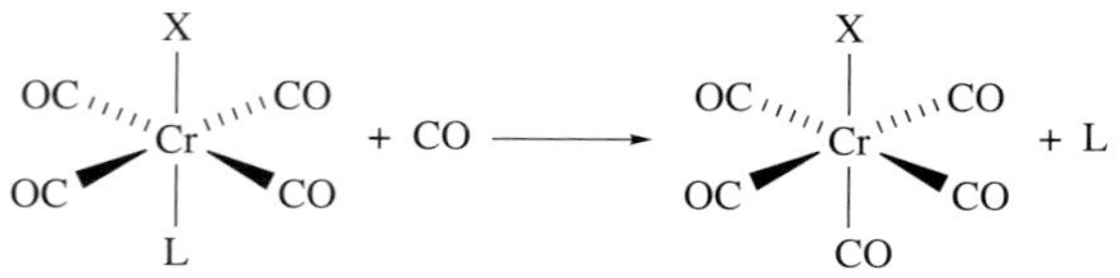

occurs by a dissociative mechanism and the first order rate constants, k_1, vary with the nature of substituent X as follows: $CO < P(OMe)_3 \approx P(OPh)_3 < P^nBu_3$.
Comment on these data.

(b) The ligand, L, shown below forms the complex $[PtLCl]^+$ which reacts with pyridine to give $[PtL(py)]^{2+}$.

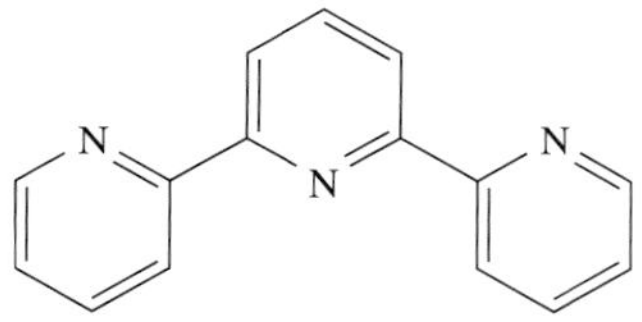

The observed rate constant, k_{obs}, can be written as:

$$k_{obs} = k_1 + k_2[pyridine]$$

What conformational change must ligand L make before complex formation? Explain the origins of the two terms in the expression for k_{obs}.

26.26 Suggest *two* experimental methods by which the kinetics of the following reaction might be monitored:

Comment on factors that contribute towards the suitability of the methods suggested.

26.27 (a) The reaction of *cis*-[PtMe$_2$(Me$_2$SO)(PPh$_3$)] with pyridine leads to *cis*-[PtMe$_2$(py)(PPh$_3$)] and the rate of reaction shows no dependence on the concentration of pyridine. At 298 K, the value of $\Delta S^{\ddagger}$ is 24 J K^{-1} mol^{-1}. Comment on these data.

(b) For the reaction:

$$[Co(NH_3)_5X]^{2+} + [Cr(OH_2)_6]^{2+} + 5[H_3O]^+ \longrightarrow$$
$$[Co(OH_2)_6]^{2+} + [Cr(OH_2)_5X]^{2+} + 5[NH_4]^+$$

rate constants for X = Cl$^-$ and I$^-$ are 6.0×10^5 and 3.0×10^6 dm^3 mol^{-1} s^{-1}, respectively. Suggest how the reactions proceed and state which step in the reaction is the rate-determining one. Comment on why the rate constants for X$^-$ = Cl$^-$ and I$^-$ differ.

26.28 Consider the following reaction that takes place in aqueous solution; L, X and Y are general ligands.

$$Co^{III}L_5X + Y \longrightarrow Co^{III}L_5Y + X$$

Discuss the possible competing pathways that exist and the factors that favour one pathway over another. Write a rate equation that takes into account the pathways that you discuss.

INORGANIC CHEMISTRY MATTERS

26.29 The structure cisplatin is shown below:

$$Cl - Pt - NH_3$$

with Cl above and NH$_3$ below the Pt.

Despite its success as an anticancer drug, the mechanism by which the drug targets DNA in the body is not fully understood, although it is known that the nucleobase guanine (see Fig. 10.13) binds more readily to Pt(II) than the other nucleobases in DNA. Among model studies reported is that of the reactions of cisplatin and three related complexes with L-histidine and 1,2,4-triazole (Nu). The ligand substitutions occur in two, reversible steps:

$$[PtLCl_2] + Nu \underset{k_1}{\overset{k_2}{\rightleftharpoons}} [PtL(Nu)Cl]^+ + Cl^-$$

$$[PtL(Nu)Cl]^+ + Nu \underset{k_3}{\overset{k_4}{\rightleftharpoons}} [PtL(Nu)_2]^{2+} + Cl^-$$

where L = (NH$_3$)$_2$, en, 1,2-diaminocyclohexane (dach), the deprotonated form, [MeCys]$^-$, of *S*-methyl-L-cysteine (see Table 29.2 for L-cysteine). The reactions were investigated at 310 K and pH 7.2 under pseudo-first order conditions.

Rate constants for the reactions are as follows:

	Nu = L-histidine		Nu = 1,2,4-triazole	
	$10^3 k_2$ /dm^3 mol^{-1}s^{-1}	$10^4 k_1$/s^{-1}	$10^2 k_2$ /dm^3 mol^{-1}s^{-1}	$10^3 k_1$/s^{-1}
cis-[Pt(NH$_3$)$_2$Cl$_2$]	8.0 ± 0.3	4.5 ± 0.4	12.0 ± 0.4	7.4 ± 0.5
[Pt(en)Cl$_2$]	7.9 ± 0.7	1.8 ± 0.1	9.9 ± 0.1	7.3 ± 0.1
[Pt(dach)Cl$_2$]	6.4 ± 0.2	2.6 ± 0.1	5.9 ± 0.1	1.2 ± 0.1
[Pt(MeCys)Cl$_2$]$^-$	352 ± 6	99 ± 1	454 ± 2	3.0 ± 0.2

	Nu = L-histidine		Nu = 1,2,4-triazole	
	$10^4 k_4$ /dm^3 mol^{-1}s^{-1}	$10^6 k_3$/s^{-1}	$10^3 k_4$ /dm^3 mol^{-1}s^{-1}	$10^4 k_3$/s^{-1}
cis-[Pt(NH$_3$)$_2$Cl$_2$]	11 ± 1	20 ± 1	12.8 ± 0.2	8.1 ± 0.2
[Pt(en)Cl$_2$]	11 ± 1	76 ± 1	11 ± 1	1.1 ± 0.1
[Pt(dach)Cl$_2$]	4.8 ± 0.4	2.2 ± 0.6	6.4 ± 0.5	1.0 ± 0.6
[Pt(MeCys)Cl$_2$]$^-$	33 ± 1	58 ± 2	24 ± 1	0.7 ± 0.01

[Data: J. Bogojeski *et al.* (2010) *Eur. J. Inorg. Chem.*, p. 5439.]

(a) Draw the structures of [Pt(en)Cl$_2$], [Pt(dach)Cl$_2$] and [Pt(MeCys)Cl$_2$]$^-$. (b) Suggest why L-histidine and 1,2,4-triazole were chosen as nucleophiles in this study. (c) Why were conditions of 310 K and pH 7.2 used? (d) Discuss the kinetic data, paying attention to and suggesting explanations for the trends in the relative rates of substitution.

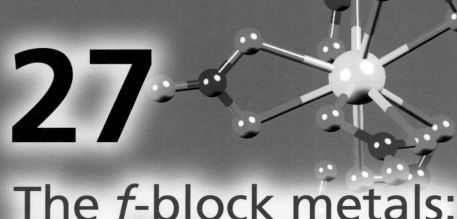

Topics

27

The *f*-block metals: lanthanoids and actinoids

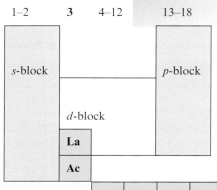

Ce	Pr	Nd	Pm	Sm	Eu	Gd	Tb	Dy	Ho	Er	Tm	Yb	Lu
Th	Pa	U	Np	Pu	Am	Cm	Bk	Cf	Es	Fm	Md	No	Lr

27.1 Introduction

In this chapter we look at *f*-block metals (Table 27.1) and their compounds. There are two series of metals: the *lanthanoids* (the 14 elements that follow lanthanum in the periodic table) and the *actinoids* (the 14 elements following actinium).[†] Scandium, yttrium, lanthanum and the lanthanoids are together called the *rare earth metals*. Although La and Ac are strictly group 3 metals, the chemical similarity of La to the elements Ce−Lu, and of Ac to Th−Lr, means that La is commonly classified with the lanthanoids, and Ac with the actinoids.

[†] The IUPAC recommends the names lanthanoid and actinoid in preference to lanthanide and actinide; the ending '-ide' usually implies a negatively charged ion.

The symbol Ln is often used to refer generically to the elements La−Lu.

The lanthanoids resemble each other much more closely than do the members of a row of *d*-block metals. The chemistry of the actinoids is more complicated, and in addition, only Th and U have naturally occurring isotopes. Studies of the *transuranium elements* (those with $Z > 92$) require specialized techniques. The occurrence of artificial isotopes among the *f*-block elements can be seen from Appendix 5: all the actinoids are unstable with respect to radioactive decay (see Tables 27.4 and 27.7), although the half-lives of the most abundant isotopes of thorium and uranium (^{232}Th and ^{238}U, $t_{\frac{1}{2}} = 1.4 \times 10^{10}$ and 4.6×10^9 yr respectively) are so long that for many purposes their radioactivity can be neglected.

Table 27.1 Lanthanum, actinium and the *f*-block elements. Ln is used as a general symbol for the metals La–Lu.

Element name	Symbol	Z	Ground state electronic configuration				Radius / pm	
			Ln	Ln^{2+}	Ln^{3+}	Ln^{4+}	Ln	Ln$^{3+ \dagger}$
Lanthanum	La	57	[Xe]$6s^2 5d^1$	[Xe]$5d^1$	[Xe]$4f^0$		188	116
Cerium	Ce	58	[Xe]$4f^1 6s^2 5d^1$	[Xe]$4f^2$	[Xe]$4f^1$	[Xe]$4f^0$	183	114
Praseodymium	Pr	59	[Xe]$4f^3 6s^2$	[Xe]$4f^3$	[Xe]$4f^2$	[Xe]$4f^1$	182	113
Neodymium	Nd	60	[Xe]$4f^4 6s^2$	[Xe]$4f^4$	[Xe]$4f^3$		181	111
Promethium	Pm	61	[Xe]$4f^5 6s^2$	[Xe]$4f^5$	[Xe]$4f^4$		181	109
Samarium	Sm	62	[Xe]$4f^6 6s^2$	[Xe]$4f^6$	[Xe]$4f^5$		180	108
Europium	Eu	63	[Xe]$4f^7 6s^2$	[Xe]$4f^7$	[Xe]$4f^6$		199	107
Gadolinium	Gd	64	[Xe]$4f^7 6s^2 5d^1$	[Xe]$4f^7 5d^1$	[Xe]$4f^7$		180	105
Terbium	Tb	65	[Xe]$4f^9 6s^2$	[Xe]$4f^9$	[Xe]$4f^8$	[Xe]$4f^7$	178	104
Dysprosium	Dy	66	[Xe]$4f^{10} 6s^2$	[Xe]$4f^{10}$	[Xe]$4f^9$	[Xe]$4f^8$	177	103
Holmium	Ho	67	[Xe]$4f^{11} 6s^2$	[Xe]$4f^{11}$	[Xe]$4f^{10}$		176	102
Erbium	Er	68	[Xe]$4f^{12} 6s^2$	[Xe]$4f^{12}$	[Xe]$4f^{11}$		175	100
Thulium	Tm	69	[Xe]$4f^{13} 6s^2$	[Xe]$4f^{13}$	[Xe]$4f^{12}$		174	99
Ytterbium	Yb	70	[Xe]$4f^{14} 6s^2$	[Xe]$4f^{14}$	[Xe]$4f^{13}$		194	99
Lutetium	Lu	71	[Xe]$4f^{14} 6s^2 5d^1$	[Xe]$4f^{14} 5d^1$	[Xe]$4f^{14}$		173	98

Element name	Symbol	Z	Ground state electronic configuration			Radius / pm	
			M	M^{3+}	M^{4+}	M$^{3+ \ddagger}$	M$^{4+ \ddagger}$
Actinium	Ac	89	[Rn]$6d^1 7s^2$	[Rn]$5f^0$		111	99
Thorium	Th	90	[Rn]$6d^2 7s^2$	[Rn]$5f^1$	[Rn]$5f^0$		94
Protactinium	Pa	91	[Rn]$5f^2 7s^2 6d^1$	[Rn]$5f^2$	[Rn]$5f^1$	104	90
Uranium	U	92	[Rn]$5f^3 7s^2 6d^1$	[Rn]$5f^3$	[Rn]$5f^2$	103	89
Neptunium	Np	93	[Rn]$5f^4 7s^2 6d^1$	[Rn]$5f^4$	[Rn]$5f^3$	101	87
Plutonium	Pu	94	[Rn]$5f^6 7s^2$	[Rn]$5f^5$	[Rn]$5f^4$	100	86
Americium	Am	95	[Rn]$5f^7 7s^2$	[Rn]$5f^6$	[Rn]$5f^5$	98	85
Curium	Cm	96	[Rn]$5f^7 7s^2 6d^1$	[Rn]$5f^7$	[Rn]$5f^6$	97	85
Berkelium	Bk	97	[Rn]$5f^9 7s^2$	[Rn]$5f^8$	[Rn]$5f^7$	96	83
Californium	Cf	98	[Rn]$5f^{10} 7s^2$	[Rn]$5f^9$	[Rn]$5f^8$	95	82
Einsteinium	Es	99	[Rn]$5f^{11} 7s^2$	[Rn]$5f^{10}$	[Rn]$5f^9$		
Fermium	Fm	100	[Rn]$5f^{12} 7s^2$	[Rn]$5f^{11}$	[Rn]$5f^{10}$		
Mendelevium	Md	101	[Rn]$5f^{13} 7s^2$	[Rn]$5f^{12}$	[Rn]$5f^{11}$		
Nobelium	No	102	[Rn]$5f^{14} 7s^2$	[Rn]$5f^{13}$	[Rn]$5f^{12}$		
Lawrencium	Lr	103	[Rn]$5f^{14} 7s^2 6d^1$	[Rn]$5f^{14}$	[Rn]$5f^{13}$		

†Ionic radius is for an 8-coordinate ion.
‡Ionic radius is for a 6-coordinate ion.

Table 27.2 Oxidation states of actinium and the actinoids. The most stable states are shown in bold.

Ac	Th	Pa	U	Np	Pu	Am	Cm	Bk	Cf	Es	Fm	Md	No	Lr
						2			2	2	2	2	**2**	
3			3	3	3	**3**	**3**	**3**	**3**	**3**	**3**	**3**	3	**3**
	4	4	4	4	**4**	4	4	4	4					
		5	5	**5**	5	5								
			6	6	6	6								
				7	7									

The ***transuranium elements*** are those with atomic number higher than that of uranium ($Z > 92$).

27.2 *f*-Orbitals and oxidation states

For an *f*-orbital, the quantum numbers are $n = 4$ or 5, $l = 3$ and $m_l = +3, +2, +1, 0, -1, -2, -3$; a set of *f*-orbitals is 7-fold degenerate. *f*-Orbitals are *ungerade*.

A set of *f*-orbitals is 7-fold degenerate and there is more than one way to represent them. You will encounter both the *general* and *cubic sets* of *f*-orbitals. The *cubic set* is commonly used and is readily related to tetrahedral, octahedral and cubic ligand fields. The cubic set comprises the f_{x^3}, f_{y^3}, f_{z^3}, f_{xyz}, $f_{z(x^2-y^2)}$, $f_{y(z^2-x^2)}$ and $f_{x(z^2-y^2)}$ atomic orbitals. Figure 27.1 shows representations of the f_{z^3} and f_{xyz} orbitals and indicates how the remaining five atomic orbitals are related to them.[†] In Fig. 27.1b, each of the eight lobes of the f_{xyz} orbital points towards one corner of a cube. Each *f* orbital contains three nodal planes.

The valence shell of a lanthanoid element contains $4f$ orbitals and that of an actinoid, $5f$ atomic orbitals. The ground state electronic configurations of the *f*-block elements are listed in Table 27.1. A $4f$ atomic orbital has no radial node, whereas a $5f$ atomic orbital has one radial node (see Section 1.6). A crucial difference between the $4f$ and $5f$ orbitals is the fact that the $4f$ atomic orbitals are deeply buried and $4f$ electrons are not available for covalent bonding. Usually for a lanthanoid metal, M, ionization beyond the M^{3+} ion is not energetically possible. This leads to a characteristic $+3$ oxidation state across the whole row from La to Lu.

The elements La to Lu are characterized by the $+3$ oxidation state, and the chemistry is mostly that of the Ln^{3+} ion.

[†] Three-dimensional representations of the cubic *f* orbitals can be viewed using the following website: http://winter.group.shef.ac.uk/orbitron/

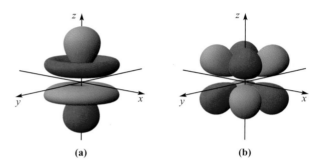

(a)   **(b)**

Fig. 27.1 The 'cubic set' of *f*-orbitals: (a) f_{z^3} and (b) f_{xyz}. The f_{x^3} and f_{y^3} orbitals are like f_{z^3} but point along the *x* and *y* axes respectively. The $f_{z(x^2-y^2)}$, $f_{y(z^2-x^2)}$ and $f_{x(z^2-y^2)}$ look like the f_{xzy} atomic orbital but, with respect to the latter, are rotated by 45° about the *z*, *y* and *x* axes respectively. The orbitals have been generated using the program *Orbital Viewer* [David Manthey, www.orbitals.com/orb/index.html]

The known oxidation states of the actinoids are shown in Table 27.2. The existence of at least two oxidation states for nearly all these metals implies that the successive ionization energies (see Appendix 8) probably differ by less than they do for the lanthanoids. For the higher oxidation states, covalent bonding must certainly be involved. This may occur either because the $5f$ atomic orbitals extend further from the nucleus than do the $4f$ atomic orbitals and are available for bonding, or because the energy separations between the $5f$, $6d$, $7s$ and $7p$ atomic orbitals are sufficiently small that appropriate valence states for covalent bonding are readily attained. Evidence that $5f$ atomic orbitals have a greater spatial extent than $4f$ atomic orbitals comes from the fine structure of the EPR spectrum of UF_3 (in a CaF_2 lattice) which arises from interaction of the electron spin of the U^{3+} ion and the F^- ions (see Section 4.9). NdF_3 (the corresponding lanthanoid species) shows no such effect.

Table 27.2 shows that a wide range of oxidation states is exhibited by the earlier actinoids, but from Cm to Lr, the elements resemble the lanthanoids. This follows from the lowering in energy of the $5f$ atomic orbitals on crossing the period and the stabilization of the $5f$ electrons.

27.3 Atom and ion sizes

The lanthanoid contraction

> The *lanthanoid contraction* is the steady decrease in size along the series of elements La−Lu.

The overall decrease in atomic and ionic radii (Table 27.1) from La to Lu has major consequences for the chemistry of the third row of *d*-block metals (see Section 22.3). The contraction is similar to that observed in a period of *d*-block metals and is attributed to the same effect: the imperfect shielding of one electron by another in the same sub-shell. However, the shielding of one $4f$ electron by another is less than for one *d* electron by another, and as the nuclear charge increases from La to

Lu, there is a fairly regular decrease in the size of the $4f^n$ sub-shell.

The ionic radii for the lanthanoids in Table 27.1 refer to 8-coordinate ions, and those for the actinoids to 6-coordination. The values should only be used as a guide. They increase with increasing coordination number and are by no means absolute values.

Coordination numbers

We introduced coordination numbers in Section 19.7. The large size of the lanthanoid and actinoid metals means that in their complexes, high coordination numbers (>6) are common, and Fig. 19.10a illustrated the 10-coordinate $[La(NO_3\text{-}O,O')_2(OH_2)_6]^+$ ion. The splitting of the degenerate

APPLICATIONS

Box 27.1 Neodymium lasers

The word *laser* stands for 'light amplification by stimulated emission of radiation'. A laser produces beams of monochromated, very intense radiation in which the radiation waves are coherent. The principle of a laser is that of *stimulated emission*: an excited state can decay spontaneously to the ground state by emitting a photon, but in a laser, the emission is stimulated by an incoming photon of the same energy as the emission. The advantages of this over spontaneous emission are that (i) the energy of emission is exactly defined, (ii) the radiation emitted is in phase with the radiation used to stimulate it, and (iii) the emitted radiation is coherent with the stimulating radiation. Further, because their properties are identical, the *emitted as well as the stimulating* radiation can stimulate further decay, and so on, i.e. the stimulating radiation has been *amplified*.

A neodymium laser consists of a YAG rod (see Section 22.2) containing a low concentration of Nd^{3+}. At each end of the rod is a mirror, one of which can also transmit radiation. An initial irradiation from an external source *pumps* the system, exciting the Nd^{3+} ions which then spontaneously relax to the longer-lived $^4F_{3/2}$ excited state (see diagram below). That the lifetime of the $^4F_{3/2}$ is relatively long is essential, allowing there to be a *population inversion* of ground and excited states. Decay to the $^4I_{11/2}$ state is the *laser transition*, and is stimulated by a photon of the correct energy. As the diagram shows, the neodymium laser is a *four level laser*. The mirror system in the laser

allows the radiation to be reflected between the ends of the rod until a high-intensity beam is eventually emitted. The wavelength of the emission from the neodymium laser is usually 1064 nm (i.e. in the infrared), but *frequency doubling* can give lasers emitting at 532 nm.

Among the many uses of YAG–Nd lasers are those for etching, cutting and welding metals. High-power lasers are used for cutting sheet metal, for example in the car and ship manufacturing industries. Cutting relies on the metal being heated to a sufficiently high temperature by energy supplied from the laser. The photograph shows a YAG–Nd laser being used to cut a steel plate during evaluation tests for the car industry.

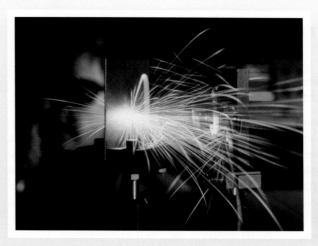

Cutting a steel plate with a YAG–Nd laser.

Further reading

P. Atkins and J. de Paula (2009) *Atkins' Physical Chemistry*, 9th edn, Oxford University Press, Oxford, Chapter 13.

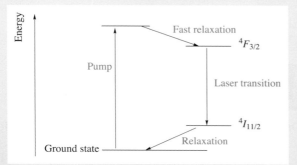

set of *f* orbitals in crystal fields in small ($\Delta_{oct} \approx 1 \, kJ \, mol^{-1}$) and crystal field stabilization considerations are of minor importance in lanthanoid and actinoid chemistry. Preferences between different coordination numbers and geometries are usually controlled by steric effects.

1. Explain why the metallic radii of Ru and Os are similar, whereas the value of r_{metal} for Fe is smaller than r_{metal} for Ru. [*Ans.* See Section 22.3]

2. Comment, with reasoning, on how you expect the trend in radii for the lanthanoid M^{3+} ions between La^{3+} and Lu^{3+} to vary. [*Ans.* See Table 27.1 and discussion of lanthanoid contraction]

3. Why is a discussion of the trend of ionic radii for the first row *d*-block metal ions less simple than a discussion of that of the Ln^{3+} ions? [*Ans.* See entries for Sc–Zn in Appendix 6, and for Ln^{3+} ions in Table 27.1]

4. The coordination environment of Nd^{3+} in $[Nd(CO_3)_4(OH_2)]^{5-}$ is a monocapped square antiprism. What is the coordination number of the Nd^{3+} ion? Suggest how this coordination number is attained, and sketch a possible structure for $[Nd(CO_3)_4(OH_2)]^{5-}$. [*Ans.* See W. Runde *et al.* (2000) *Inorg. Chem.*, vol. 39, p. 1050.]

27.4 Spectroscopic and magnetic properties

Electronic spectra and magnetic moments: lanthanoids

You should refer to Section 20.6 for term symbols for free atoms and ions. The interpretation of the electronic spectra of $4f^n$ ions is based on the principles outlined for *d*-block metal ions (Section 20.7) but there are important differences. For the lanthanoids, spin–orbit coupling is more important than crystal field splitting, and terms differing only in *J* values are sufficiently different in energy to be separated in the electronic spectrum. Further, since $l = 3$ for an *f* electron, m_l may be 3, 2, 1, 0, −1, −2 or −3, giving rise to high values of *L* for some f^n ions: e.g. for the configuration f^2, application of Hund's rules gives the ground state (with $L = 5$, $S = 1$) as $^3H_4$. Since *S*, *P*, *D*, *F* and *G* terms are also possible, many of them with different positive values of *J*, the number of possible transitions is large, even after taking into account the limita-

tions imposed by selection rules. As a result, spectra of Ln^{3+} ions often contain large numbers of absorptions. Since the $4f$ electrons are well shielded and not affected by the environment of the ion, bands arising from *f–f* transitions are sharp (rather than broad like *d–d* absorptions) and their positions in the spectrum are little affected by complex formation. Intensities of the absorptions are low, indicating that the probabilities of the *f–f* transitions are low, i.e. little *d–f* mixing. Absorptions due to $4f$–$5d$ transitions are broad and *are* affected by ligand environment. Small amounts of some lanthanoid salts are used in phosphors for television tubes (see *luminescence* on p. 1009) because of the sharpness of their electronic transitions.

> In the electronic spectra of lanthanoid metal ions, absorptions due to *f–f* transitions are *sharp*, but bands due to $4f$–$5d$ transitions are *broad*.

Typical colours of Ln^{3+} ions in aqueous solution are listed in Table 27.3. Usually (but not invariably) f^n and f^{14-n} species have similar colours.

The bulk magnetic moments (see Section 20.10) of Ln^{3+} ions are given in Table 27.3. In general, experimental values agree well with those calculated from eq. 27.1. This is based on the assumption of Russell–Saunders coupling (see Section 20.6) and large spin–orbit coupling constants, as a consequence of which only the states of lowest *J* value are populated. This is *not* true for Eu^{3+}, and not quite true for Sm^{3+}. For Eu^{3+} (f^6), the spin–orbit coupling constant λ is $\approx 300 \, cm^{-1}$, only slightly greater than kT ($\approx 200 \, cm^{-1}$). The ground state of the f^6 ion is $^7F_0$ (which is diamagnetic, since $J = 0$), but the states $^7F_1$ and $^7F_2$ are also populated to some extent and give rise to the observed magnetic moment. As expected, at low temperatures, the moment of Eu^{3+} approaches zero. The variation of μ with *n* (number of unpaired electrons) in Table 27.3 arises from the operation of Hund's third rule (see Section 20.6): $J = L - S$ for a shell less than half full but $J = L + S$ for a shell more than half full. Accordingly, *J* and g_J for ground states are both larger in the second half than the first half of the lanthanoid series.

$$\mu_{eff} = g_J \sqrt{J(J+1)} \tag{27.1}$$

where: $g_J = 1 + \left(\dfrac{S(S+1) - L(L+1) + J(J+1)}{2J(J+1)} \right)$

Worked example 27.1 Determining the term symbol for the ground state of an Ln^{3+} ion

Determine the term symbol for the ground state of Ho^{3+}.

Refer to Section 20.6 for a review of term symbols. Two general points should be noted:

Table 27.3   Colours of aqua complexes of La^{3+} and Ln^{3+}, and observed and calculated magnetic moments for the M^{3+} ions.

Metal ion	Colour	Ground state electronic configuration	Ground state term symbol	Magnetic moment, μ (298 K) / μ_B	
				Calculated from eq. 27.1	Observed
La^{3+}	Colourless	$[Xe]4f^0$	$^1S_0$	0	0
Ce^{3+}	Colourless	$[Xe]4f^1$	$^2F_{5/2}$	2.54	2.3–2.5
Pr^{3+}	Green	$[Xe]4f^2$	$^3H_4$	3.58	3.4–3.6
Nd^{3+}	Lilac	$[Xe]4f^3$	$^4I_{9/2}$	3.62	3.5–3.6
Pm^{3+}	Pink	$[Xe]4f^4$	$^5I_4$	2.68	2.7
Sm^{3+}	Yellow	$[Xe]4f^5$	$^6H_{5/2}$	0.84	1.5–1.6
Eu^{3+}	Pale pink	$[Xe]4f^6$	$^7F_0$	0	3.4–3.6
Gd^{3+}	Colourless	$[Xe]4f^7$	$^8S_{7/2}$	7.94	7.8–8.0
Tb^{3+}	Pale pink	$[Xe]4f^8$	$^7F_6$	9.72	9.4–9.6
Dy^{3+}	Yellow	$[Xe]4f^9$	$^6H_{15/2}$	10.63	10.4–10.5
Ho^{3+}	Yellow	$[Xe]4f^{10}$	$^5I_8$	10.60	10.3–10.5
Er^{3+}	Rose pink	$[Xe]4f^{11}$	$^4I_{15/2}$	9.58	9.4–9.6
Tm^{3+}	Pale green	$[Xe]4f^{12}$	$^3H_6$	7.56	7.1–7.4
Yb^{3+}	Colourless	$[Xe]4f^{13}$	$^2F_{7/2}$	4.54	4.4–4.9
Lu^{3+}	Colourless	$[Xe]4f^{14}$	$^1S_0$	0	0

- The term symbol for the ground state of an atom or ion is given by $^{(2S+1)}L_J$, and the value of L (the total angular momentum) relates to the term symbols as follows:

L	0	1	2	3	4	5	6
Term symbol	S	P	D	F	G	H	I

- From Hund's third rule (Section 20.6), the value of J for the ground state is given by $(L - S)$ for a sub-shell that is *less* than half-filled, and by $(L + S)$ for a sub-shell that is *more* than half-filled.

Now consider Ho^{3+}. Ho^{3+} has an f^{10} electronic configuration. The f orbitals have values of m_l of $-3, -2, -1, 0, +1, +2, +3$ and the lowest energy arrangement (by Hund's rules, Section 20.6) is:

m_l	+3	+2	+1	0	-1	-2	-3
	↑↓	↑↓	↑↓	↑	↑	↑	↑

There are 4 unpaired electrons.

Total spin quantum number, $S = 4 \times \frac{1}{2} = 2$

Spin multiplicity, $2S + 1 = 5$

Resultant orbital quantum number,

$L = $ sum of m_l values

$= (2 \times 3) + (2 \times 2) + (2 \times 1) - 1 - 2 - 3$

$= 6$

This corresponds to an I state.

The highest value of the resultant inner quantum number, $J = (L + S) = 8$

Therefore, the term symbol for the ground state of Ho^{3+} is $^5I_8$.

Self-study exercises

1. Confirm that the term symbol for the ground state of Ce^{3+} is $^2F_{5/2}$.

2. Confirm that Er^{3+} has a term symbol for the ground state of $^4I_{15/2}$.

3. Why do La^{3+} and Lu^{3+} both have the term symbol $^1S_0$ for the ground state?

Worked example 27.2   Calculating the effective magnetic moment of a lanthanoid ion

Calculate a value for the effective magnetic moment, μ_{eff}, of Ce^{3+}.

The value of μ_{eff} can be calculated using eq. 27.1:

$$\mu_{\text{eff}} = g\sqrt{J(J+1)}$$

where

$$g = 1 + \left(\frac{S(S+1) - L(L+1) + J(J+1)}{2J(J+1)} \right)$$

Ce^{3+} has an f^1 electronic configuration.

$$S = 1 \times \tfrac{1}{2} = \tfrac{1}{2}$$

$$L = 3 \quad \text{(see worked example 27.1)}$$

The sub-shell is less than half-filled, therefore

$$J = (L - S) = 3 - \tfrac{1}{2} = \tfrac{5}{2}$$

$$g = 1 + \left(\frac{S(S+1) - L(L+1) + J(J+1)}{2J(J+1)} \right)$$

$$= 1 + \left(\frac{(\tfrac{1}{2} \times \tfrac{3}{2}) - (3 \times 4) + (\tfrac{5}{2} \times \tfrac{7}{2})}{2(\tfrac{5}{2} \times \tfrac{7}{2})} \right)$$

$$= \tfrac{6}{7}$$

$$\mu_{\text{eff}} = g\sqrt{J(J+1)}$$

$$= \tfrac{6}{7}\sqrt{(\tfrac{5}{2} \times \tfrac{7}{2})}$$

$$= 2.54 \, \mu_B$$

ENVIRONMENT

Box 27.2 Rare earth metals: resources and demand

The chart below shows the estimated world's reserves of rare earth metals. Although large resources are available, world mine production has been almost entirely in China. However, for a number of reasons, mining of reserves in the US, Australia and other countries is now (2011) becoming economically viable, and competitive markets for the rare earth metals are expected to replace the China-dominated market.

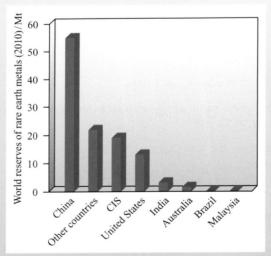

[CIS stands for Commonwealth of Independent States; data: US Geological Survey]

The principal rare earth metal-bearing ores are *bastnäsite* (in China and the US) and *monazite* (in Australia, Brazil, China, the US, India, Malaysia, South Africa, Sri Lanka and Thailand).

Bastnäsite is a mixed metal carbonate fluoride, $(M,M'...)CO_3F$. The composition varies with the source of the mineral, but the dominant component is cerium ($\approx$50%), followed by lanthanum (20–30%), neodymium (12–20%) and praseodymium ($\approx$5%). Each of the other rare earth metals (except for promethium which does not occur naturally) typically occurs to an extent of <1%. Monazite, $(M,M'...)PO_4$, is the second most important rare earth metal-containing ore and is also rich in cerium. Cerium is actually more abundant than copper in the Earth's crust.

The demand for the rare earth metals increased over the last two decades of the 20th century, and the demand for cerium oxides in motor vehicle catalytic converters (see eqs. 25.43 and 25.44) is a major contributing factor. Applications of rare earth metals in the US in 2009 were as chemical and petroleum refining catalysts (36%), metallurgical uses including alloys (21%), catalytic converters in automobiles (13%), glass polishing and ceramics (9%), rare-earth phosphors for lighting, radar, computer monitors and televisions (8%), permanent magnets (7%), electronic and other uses (6%). Alloys include those used in nickel–metal hydride batteries which contain $LaNi_5$ or a related rare earth metal-containing alloy for hydride storage (see Box 10.5). Only small amounts of the rare earth metals are recycled, most originating from scrapped permanent magnets.

The demand for the lanthanoid metals is expected to increase with the growing demands for pollution control catalysts in motor vehicles and rechargeable batteries. Along with lithium-ion batteries, nickel–metal hydride (NiMH) batteries find increasing applications in mobile phones, laptop computers and other portable electronic devices such as MP3 players.

Self-study exercises

1. Comment on why the spin-only formula is not appropriate for estimating values of μ_{eff} for lanthanoid metal ions.

2. Confirm that the estimated value of μ_{eff} for Yb^{3+} is $4.54\,\mu_B$.

3. Eu^{3+} has an f^6 electronic configuration and yet the calculated value of μ_{eff} is 0. Explain how this result arises.

Luminescence of lanthanoid complexes

Fluorescence, phosphorescence and luminescence were introduced in Section 20.8. Irradiation with UV light of many Ln^{3+} complexes causes them to fluoresce. In some species, low temperatures are required to observe this. Their fluorescence leads to the use of lanthanoids in phosphors in televisions and fluorescent lighting. The origin of the fluorescence is $4f$–$4f$ transitions, no transitions being possible for f^0, f^7 (spin-forbidden) and f^{14}. Irradiation produces Ln^{3+} in an excited state which decays to the ground state either with emission of energy (observed as fluorescence) or by a non-radiative pathway. The ions that are commercially important for their emitting properties are Eu^{3+} (red emission) and Tb^{3+} (green emission).

Electronic spectra and magnetic moments: actinoids

The spectroscopic and magnetic properties of actinoids are complicated and we mention them only briefly. Absorptions due to $5f$–$5f$ transitions are weak, but they are somewhat broader and more intense (and considerably more dependent on the ligands present) than those due to $4f$–$4f$ transitions. The interpretation of electronic spectra is made difficult by the large spin–orbit coupling constants (about twice those of the lanthanoids) as a result of which the Russell–Saunders coupling scheme partially breaks down.

Magnetic properties show an overall similarity to those of the lanthanoids in the variation of magnetic moment with the number of unpaired electrons, but values for isoelectronic lanthanoid and actinoid species, e.g. Np(VI) and Ce(III), Np(V) and Pr(III), Np(IV) and Nd(III), are lower for the actinoids, indicating partial quenching of the orbital contribution by crystal field effects.

27.5 Sources of the lanthanoids and actinoids

Occurrence and separation of the lanthanoids

All the lanthanoids except Pm occur naturally. The most stable isotope of promethium, ^{147}Pm (β-emitter, $t_{\frac{1}{2}} = 2.6\,yr$) is formed as a product of the fission of heavy nuclei and is obtained in mg amounts from products of nuclear reactors.

Bastnäsite and *monazite* are the main ores for La and the lanthanoids. All the metals (excluding Pm) can be obtained from *monazite*, a mixed phosphate $(Ce,La,Nd,Pr,Th,Y\ldots)PO_4$. *Bastnäsite*, $(Ce,La\ldots)CO_3F$, is a source of the lighter lanthanoids. The first step in extraction of the metals from monazite is removal of phosphate and thorium. The ore is heated with caustic soda, and, after cooling, Na_3PO_4 is dissolved in water. The residual hydrated Th(IV) and Ln(III) oxides are treated with hot, aqueous HCl; ThO_2 is not dissolved, but the Ln(III) oxides give a solution of MCl_3 (M = La, Ce...) which is then purified. Starting from bastnäsite, the ore is treated with dilute HCl to remove $CaCO_3$, and then converted to an aqueous solution of MCl_3 (M = La, Ce...). The similarity in ion size and properties of the lanthanoids make separation difficult. Modern methods of separating the lanthanoids involve solvent extraction using $(^nBuO)_3PO$ (see Box 7.3) or ion exchange (see Section 11.6).

A typical cation-exchange resin is sulfonated polystyrene or its Na^+ salt. When a solution containing Ln^{3+} ions is poured on to a resin column, the cations exchange with the H^+ or Na^+ ions (eq. 27.2).

$$Ln^{3+}(aq) + 3H^+(resin) \rightleftharpoons Ln^{3+}(resin) + 3H^+(aq) \quad (27.2)$$

The equilibrium distribution coefficient between the resin and the aqueous solution ($[Ln^{3+}(resin)]/[Ln^{3+}(aq)]$) is large for all the ions, but is nearly constant. The resin-bound Ln^{3+} ions are now removed using a complexing agent such as $EDTA^{4-}$ (see eq. 7.75). The formation constants of the $EDTA^{4-}$ complexes of the Ln^{3+} ions increase regularly from $10^{15.3}$ for La^{3+} to $10^{19.2}$ for Lu^{3+}. If a column on which all the Ln^{3+} ions have been absorbed is eluted with dilute aqueous H_4EDTA, and the pH adjusted to 8 using NH_3, Lu^{3+} is preferentially complexed, then Yb^{3+}, and so on. By using a long ion-exchange column, 99.9% pure components can be separated (Fig. 27.2).

The actinoids

Of the actinoids, only uranium and thorium occur naturally in significant quantities. The naturally occurring isotopes of U and Th (^{238}U, 99.275%, $t_{\frac{1}{2}} = 4.46 \times 10^9\,yr$; ^{235}U, 0.720%, $t_{\frac{1}{2}} = 7.04 \times 10^8\,yr$; ^{234}U, 0.005%, $t_{\frac{1}{2}} = 2.45 \times 10^5\,yr$; ^{232}Th, 100%, $t_{\frac{1}{2}} = 1.4 \times 10^{10}\,yr$) are all radioactive and their decay chains give rise to isotopes of actinium and protactinium:

$$^{238}_{92}U \xrightarrow{-\alpha} {}^{234}_{90}Th \xrightarrow{-\beta^-} {}^{234}_{91}Pa \xrightarrow{-\beta^-} {}^{234}_{92}U \xrightarrow{-\alpha} {}^{230}_{90}Th \longrightarrow \longrightarrow$$

$$^{235}_{92}U \xrightarrow{-\alpha} {}^{231}_{90}Th \xrightarrow{-\beta^-} {}^{231}_{91}Pa \xrightarrow{-\alpha} {}^{227}_{89}Ac \xrightarrow{-\beta^-} {}^{227}_{90}Th \longrightarrow \longrightarrow$$

$$^{232}_{90}Th \xrightarrow{-\alpha} {}^{228}_{88}Ra \xrightarrow{-\beta^-} {}^{228}_{89}Ac \xrightarrow{-\beta^-} {}^{228}_{90}Th \xrightarrow{-\alpha} {}^{224}_{88}Ra \longrightarrow \longrightarrow$$

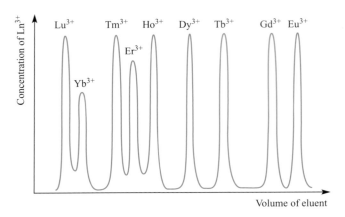

Fig. 27.2 A representation of the order in which $EDTA^{4-}$ complexes of the heavier lanthanoids are eluted from a cation-exchange column.

Radioactive decay by loss of an α-particle (i.e. loss of $[^4_2He]^{2+}$) leads to a decrease in the mass number by four and a decrease in the atomic number by two, and is accompanied by emission of γ-radiation. Decay by loss of a β-particle (i.e. loss of an electron from the nucleus) results in an increase in the atomic number by one but leaves the mass number unchanged. Figure 27.3 shows the decay chain for $^{238}_{92}U$ which ultimately produces the stable isotope $^{206}_{82}Pb$, and the properties of the nuclides in the series are given in Table 27.4.

Radioactive decay of any nuclide follows first order kinetics, but subsequent decay of the daughter nuclide results in more complicated overall kinetics. For a first order process:

$$\text{Rate of decay} = -\frac{dN}{dt} = kN$$

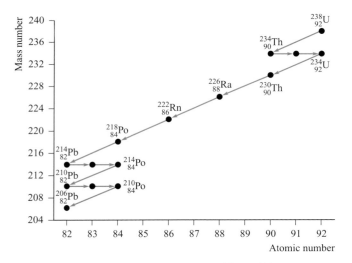

Fig. 27.3 The radioactive decay series from $^{238}_{92}U$ to $^{206}_{82}Pb$. Only the last nuclide in the series, $^{206}_{82}Pb$, is stable with respect to further decay. [Exercise: Three nuclides are not labelled. What are their identities?]

Table 27.4 The natural radioactive decay series from $^{238}_{92}U$ to $^{206}_{82}Pb$ (see Fig. 27.3) (yr = year; d = day; min = minute; s = second).

Nuclide	Symbol	Particle emitted	Half-life
Uranium-238	$^{238}_{92}U$	α	4.46×10^9 yr
Thorium-234	$^{234}_{90}Th$	$β^-$	24.1 d
Protactinium-234	$^{234}_{91}Pa$	$β^-$	1.18 min
Uranium-234	$^{234}_{92}U$	α	2.48×10^5 yr
Thorium-230	$^{230}_{90}Th$	α	8.0×10^4 yr
Radium-226	$^{226}_{88}Ra$	α	1.62×10^3 yr
Radon-222	$^{222}_{86}Rn$	α	3.82 d
Polonium-218	$^{218}_{84}Po$	α	3.05 min
Lead-214	$^{214}_{82}Pb$	$β^-$	26.8 min
Bismuth-214	$^{214}_{83}Bi$	$β^-$	19.7 min
Polonium-214	$^{214}_{84}Po$	α	1.6×10^{-4} s
Lead-210	$^{210}_{82}Pb$	$β^-$	19.4 yr
Bismuth-210	$^{210}_{83}Bi$	$β^-$	5.0 d
Polonium-210	$^{210}_{84}Po$	α	138 d
Lead-206	$^{206}_{82}Pb$	none	Non-radioactive

the integrated form of which is:

$$\ln N - \ln N_0 = -kt$$

where: N = number of nuclei at time t, N_0 = number of nuclei at time $t = 0$, k = first order rate constant

Since Pa and Ac are only formed as radioactive intermediates in decay chains, the amounts occurring naturally are so small that they, along with the remaining actinoids, are artificially produced by nuclear reactions (see below). Radiation hazards of all but Th and U lead to technical difficulties in studying actinoid compounds, and conventional experimental techniques are not generally applicable.

Uranium and thorium are isolated from natural sources. Thorium is extracted from monazite as ThO_2 (see above), and the most important source of uranium is *pitchblende* (U_3O_8). The uranium ore is heated with H_2SO_4 in the presence of an oxidizing agent to give the sulfate salt of the uranyl cation, $[UO_2]^{2+}$, which is separated on an anion-exchange resin, eluting with HNO_3 to give $[UO_2][NO_3]_2$. After further work-up, the uranium is precipitated as the oxido-peroxido complex $UO_2(O_2) \cdot 2H_2O$ or as 'yellow cake' (approximate composition $[NH_4]_2[U_2O_7]$). Thermal decomposition gives yellow UO_3 which is converted to UF_4 (see Box 7.3). Reduction of UF_4 with Mg yields U metal.

The isotopes ^{227}Ac and ^{231}Pa can be isolated from the decay products of ^{235}U in pitchblende, but are better synthesized by nuclear reactions 27.3 and 27.4.

$$^{226}Ra \xrightarrow{(n,\gamma)} {}^{227}Ra \xrightarrow{-\beta^-} {}^{227}Ac \qquad (27.3)$$

$$^{230}Th \xrightarrow{(n,\gamma)} {}^{231}Th \xrightarrow{-\beta^-} {}^{231}Pa \qquad (27.4)$$

The nuclides ^{239}Np and ^{239}Pu are decay products of ^{239}U, itself synthesized by neutron capture by ^{238}U (see Box 7.3). Lengthy irradiation of ^{239}Pu in a nuclear pile leads to the successive formation of small quantities of ^{240}Pu, ^{241}Pu, ^{242}Pu and ^{243}Pu. The last is a β^--emitter ($t_{\frac{1}{2}} = 5$ h) and decays to ^{243}Am ($t_{\frac{1}{2}} = 7400$ yr) which gives ^{244}Cm by sequence 27.5.

$$^{243}Am \xrightarrow{(n,\gamma)} {}^{244}Am \xrightarrow{-\beta^-} {}^{244}Cm \qquad (27.5)$$

Both ^{243}Am and ^{244}Cm are available on a 100 g scale, and multiple neutron capture followed by β^--decay yields milligram amounts of ^{249}Bk, ^{252}Cf, ^{253}Es and ^{254}Es, plus microgram amounts of ^{257}Fm. The transuranium elements (i.e. the elements that follow U in the periodic table, Table 27.5) extend beyond the actinoid row ($Z = 89$–103) and have all been discovered since 1940. By 1955, the periodic table extended to mendelevium and, by 1997, to meitnerium. In mid-2011, the number of elements in the periodic table stood at 117 (see the periodic table at the front of the book), although the discoveries of those elements with $Z = 113$, 115 and 118 have not (as of mid-2011) been approved by the IUPAC.[†] In 2003 and 2004, the IUPAC approved the names darmstadtium and roentgenium for elements 110 and 111, respectively, and in 2010, element 112 was named copernicium. The artificially made elements with $Z = 113$–116 and 118 are currently called ununtrium ('oneonethree'), ununquadium ('oneonefour') ... as listed in Table 27.5. This method of naming newly discovered elements is used until actual names have been approved by the IUPAC. All of these 'new' elements have been produced synthetically by the bombardment of very heavy nuclides with particles such as neutrons (see Box 7.3), $^{12}_{6}C^{n+}$ ions or $^{18}_{8}O^{n+}$ ions, for example:

$$^{249}_{97}Bk + {}^{18}_{8}O \longrightarrow {}^{260}_{103}Lr + {}^{4}_{2}He + 3{}^{1}_{0}n$$

$$^{248}_{96}Cm + {}^{18}_{8}O \longrightarrow {}^{261}_{104}Rf + 5{}^{1}_{0}n$$

The target actinoid nuclides have relatively long half-lives ($^{249}_{97}$Bk, $t_{\frac{1}{2}} = 300$ d; $^{248}_{96}$Cm, $t_{\frac{1}{2}} = 3.5 \times 10^{5}$ yr). The scale on which these transmutations is carried out is extremely small, and has been described as 'atom-at-a-time'

Table 27.5 The transuranium elements. The names are those agreed by the IUPAC.

Z	Name of element	Symbol
93	Neptunium	Np
94	Plutonium	Pu
95	Americium	Am
96	Curium	Cm
97	Berkelium	Bk
98	Californium	Cf
99	Einsteinium	Es
100	Fermium	Fm
101	Mendelevium	Md
102	Nobelium	No
103	Lawrencium	Lr
104	Rutherfordium	Rf
105	Dubnium	Db
106	Seaborgium	Sg
107	Bohrium	Bh
108	Hassium	Hs
109	Meitnerium	Mt
110	Darmstadtium	Ds
111	Roentgenium	Rg
112	Copernicium	Cn
113	Ununtrium	Uut
114	Ununquadium	Uuq
115	Ununpentium	Uup
116	Ununhexium	Uuh
118	Ununoctium	Uuo

chemistry.[‡] Studying the product nuclides is extremely difficult, because of the tiny quantities of materials produced and their short half-lives ($^{260}_{103}$Lr, $t_{\frac{1}{2}} = 3$ min; $^{261}_{104}$Rf, $t_{\frac{1}{2}} = 65$ s).

27.6 Lanthanoid metals

Lanthanum and the lanthanoids, except Eu, crystallize in one or both of the cubic or hexagonal close-packed structures. Eu has a bcc structure and the value of r_{metal} given in Table 27.1 can be adjusted to 205 pm for 12-coordination (see Section 6.5). It is important to notice in Table 27.1 that Eu and Yb have much larger metallic radii than the other lanthanoids, implying that Eu and Yb (which have well-defined lower oxidation states) contribute fewer electrons to M−M bonding. This is consistent with the lower values of $\Delta_a H^{\circ}$: Eu and Yb, 177 and 152 kJ mol^{-1} respectively, compared with the other lanthanoids (Table 27.6). The metal with the next lowest enthalpy of atomization after Eu and Yb is Sm. Like Eu and Yb, Sm has a well-defined

---

[†] See: R.C. Barber, P.J. Karol, H. Nakahara, E. Vardaci and E.W. Vogt (2011) *Pure Appl. Chem.*, vol. 83, p. 1485 and p. 1801 – 'Discovery of the elements with atomic numbers greater than or equal to 113' (IUPAC Technical Report).

[‡] See: D.C. Hoffmann and D.M. Lee (1999) *J. Chem. Educ.*, vol. 76, p. 331; W.D. Loveland, D. Morrissey and G.T. Seaborg (2005) *Modern Nuclear Chemistry*, Wiley, Weinheim.

Table 27.6 Standard enthalpies of atomization and hydration[†] (at 298 K) of the Ln^{3+} ions, sums of the first three ionization energies of $Ln(g)$, and standard reduction potentials for the lanthanoid metal ions.

Metal	$\Delta_a H^\circ(Ln)/kJ\,mol^{-1}$	$IE_1 + IE_2 +$ $IE_3/kJ\,mol^{-1}$	$\Delta_{hyd}H^\circ(Ln^{3+},$ $g)/kJ\,mol^{-1}$	$E^\circ_{Ln^{3+}/Ln}/V$	$E^\circ_{Ln^{2+}/Ln}/V$
La	431	3455	−3278	−2.38	
Ce	423	3530	−3326	−2.34	
Pr	356	3631	−3373	−2.35	−2.0
Nd	328	3698	−3403	−2.32	−2.1
Pm	348	3741	−3427	−2.30	−2.2
Sm	207	3873	−3449	−2.30	−2.68
Eu	177	4036	−3501	−1.99	−2.81
Gd	398	3750	−3517	−2.28	
Tb	389	3792	−3559	−2.28	
Dy	290	3899	−3567	−2.30	−2.2
Ho	301	3924	−3613	−2.33	−2.1
Er	317	3934	−3637	−2.33	−2.0
Tm	232	4045	−3664	−2.32	−2.4
Yb	152	4195	−3724	−2.19	−2.76
Lu	428	3886	−3722	−2.28	

[†]Values of $\Delta_{hyd}H^\circ(M^{3+}, g)$ are taken from: L.R. Morss (1976) *Chem. Rev.*, vol. 76, p. 827.

lower oxidation state, but unlike Eu and Yb, Sm shows no anomaly in its metallic radius. Further, Eu and Yb, but not Sm, form blue solutions in liquid NH_3 due to reaction 27.6 (see Section 9.6).

$$Ln \xrightarrow{\text{liquid } NH_3} Ln^{2+} + 2e^- (solv) \qquad Ln = Eu, Yb$$
$$(27.6)$$

All the lanthanoids are soft white metals. The later metals are passivated by an oxide coating and are kinetically more inert than the earlier metals. Values of E° for half-reaction 27.7 lie in the range −1.99 to −2.38 V (Table 27.6).

$$Ln^{3+}(aq) + 3e^- \rightleftharpoons Ln(s) \qquad (27.7)$$

The small variation in values of $E^\circ_{Ln^{3+}/Ln}$ can be understood by using the methods described in Section 8.7. The following thermochemical cycle illustrates the factors that contribute to the enthalpy change for the reduction of $Ln^{3+}(aq)$ to $Ln(s)$:

$$Ln^{3+}(aq) + 3e^- \xrightarrow{\Delta H^\circ} Ln(s)$$
$$-\Delta_{hyd}H^\circ(Ln^{3+}, g) \downarrow \qquad \qquad \downarrow -\Delta_a H^\circ(Ln)$$
$$Ln^{3+}(g) \xrightarrow{-IE_1 - IE_2 - IE_3} Ln(g)$$

Values of ΔH° (defined above) for the lanthanoid metals can be determined using the data in Table 27.6. The trend in

$\Delta_{hyd}H^\circ(Ln^{3+}, g)$ is a consequence of the lanthanoid contraction, and offsets the general increase in the sum of the ionization energies from La to Lu. The variations in $\Delta_{hyd}H^\circ(Ln^{3+}, g)$, ΣIE and $\Delta_a H^\circ(Ln)$ effectively cancel out, and values of ΔH° are similar for all the metals with the exception of europium. The *trend* in values of E° follows from the trend in ΔH°. However, actual E° values must (i) be determined from ΔG° rather than ΔH° and (ii) be related to E° (defined as 0 V) for the reduction of $H^+(aq)$ to $\frac{1}{2}H_2(g)$ (see Tables 8.2 and 8.3 and related discussion).

In addition to values of E° for the Ln^{3+}/Ln couple, Table 27.6 lists values of E° for half-reaction 27.8, this being of greatest importance for Sm, Eu and Yb.

$$Ln^{2+}(aq) + 2e^- \rightleftharpoons Ln(s) \qquad (27.8)$$

As a consequence of the negative reduction potentials (Table 27.6), all the metals liberate H_2 from dilute acids or steam. They burn in air to give Ln_2O_3 with the exception of Ce which forms CeO_2. When heated, lanthanoids react with H_2 to give a range of compounds between metallic (i.e. conducting) hydrides LnH_2 (best formulated as $Ln^{3+}(H^-)_2(e^-)$) and saline hydrides LnH_3. Non-stoichiometric hydrides are typified by 'GdH_3' which actually has compositions in the range $GdH_{2.85-3}$. Europium forms only EuH_2. The alloy $LaNi_5$ is a potential 'hydrogen storage vessel' (see Section 10.7 and Box 10.2) since it reversibly absorbs H_2 (eq. 27.9).

$$LaNi_5 \xrightleftharpoons[413\,K,\,-H_2]{H_2} LaNi_5H_x \qquad x \approx 6 \qquad (27.9)$$

The carbides Ln_2C_3 and LnC_2 are formed when the metals are heated with carbon. The LnC_2 carbides adopt the same structure as CaC_2 (see Section 14.7), but the $C-C$ bonds (128 pm) are significantly lengthened (119 pm in CaC_2). They are metallic conductors and are best formulated as $Ln^{3+}[C_2]^{2-}(e^-)$. Lanthanoid borides were discussed in Section 13.10. Halides are described below.

Self-study exercises

1. Using data from Table 27.6, determine $\Delta H^{\circ}(298\,K)$ for the half-reaction:

   $Gd^{3+}(aq) + 3e^- \rightleftharpoons Gd(s)$ [*Ans.* $-631\,kJ\,mol^{-1}$]

2. Write down the relationship between ΔG° and E°.

   [*Ans.* See eq. 8.9]

3. Assuming that the sign and magnitude of ΔG° for the reduction of $Gd^{3+}(aq)$ can be approximated to those of ΔH°, explain why $E^{\circ}_{Ln^{3+}/Ln}$ is negative (Table 27.6) even though $\Delta H^{\circ} = -631\,kJ\,mol^{-1}$.

   [*Ans.* See Table 8.2 and accompanying discussion]

27.7 Inorganic compounds and coordination complexes of the lanthanoids

The discussion in this section is necessarily selective. Most of the chemistry concerns the +3 oxidation state, with Ce(IV) being the only stable +4 state (eq. 27.10).

$$Ce^{4+} + e^- \rightleftharpoons Ce^{3+} \qquad E^{\circ} = +1.72\,V \qquad (27.10)$$

The +2 oxidation state is well defined for Eu, Sm and Yb and values of E° for half-reaction 27.8 are listed in Table 27.6. The estimated E° values for the Sm^{3+}/Sm^{2+} and Yb^{3+}/Yb^{2+} couples are -1.5 and $-1.1\,V$ respectively, indicating that Sm(II) and Yb(II) are highly unstable with respect to oxidation even by water. For the Eu^{3+}/Eu^{2+} couple, the E° value ($-0.35\,V$) is similar to that for Cr^{3+}/Cr^{2+} ($-0.41\,V$), and colourless Eu(II) solutions can be used for chemical studies if air is excluded.

Halides

Reactions of F_2 with Ln give LnF_3 for all the metals and, for Ce, Pr and Tb, also LnF_4. CeF_4 can also be made by reaction 27.11, or at room temperature in anhydrous HF (eq. 27.12). Improved routes to PrF_4 and TbF_4 (eqs. 27.13 and 27.14) occur slowly, but quantitatively.

$$CeO_2 \xrightarrow{F_2,\,XeF_2,\,ClF_3} CeF_4 \qquad (27.11)$$

$$CeF_3 \xrightarrow[298\,K,\,6\,days]{F_2\text{ in liquid HF}} CeF_4 \qquad (27.12)$$

$$Pr_6O_{11} \xrightarrow[298\,K,\,UV\text{ radiation, 11 days}]{F_2\text{ in liquid HF}} PrF_4 \qquad (27.13)$$

$$Tb_4O_7 \xrightarrow[298\,K,\,UV\text{ radiation, 25 days}]{F_2\text{ in liquid HF}} TbF_4 \qquad (27.14)$$

With Cl_2, Br_2 and I_2, LnX_3 are formed. However, the general route to LnX_3 is by reaction of Ln_2O_3 with aqueous HX. This gives the hydrated halides, $LnX_3(OH_2)_x$ ($x = 6$ or 7). The anhydrous trichloride is usually made by dehydrating $LnCl_3(OH_2)_x$ using $SOCl_2$ or NH_4Cl. Thermal dehydration of $LnCl_3(OH_2)_x$ results in the formation of oxochlorides. The thermal dehydration of $LaI_3 \cdot 9H_2O$ leads to polymeric $[LaIO]_n$. Dehydration of hydrates is, therefore, not always a straightforward method of preparing anhydrous LnX_3. Anhydrous LnX_3 compounds are important precursors for organometallic lanthanoid metal derivatives (e.g. reactions 27.20, 27.27 and 27.31), and one approach to water-free derivatives is to prepare solvated complexes such as $LnI_3(THF)_n$. Powdered metal reacts with I_2 in THF to give $[LnI_2(THF)_5]^+[LnI_4(THF)_2]^-$ (Ln = Nd, Sm, Gd, Dy, Er, Tm) or $[LnI_3(THF)_4]$ (Ln = La, Pr). Each of La^{3+} and Pr^{3+} can accommodate three I^- ions in a 7-coordinate molecular complex (Fig. 27.4a), whereas the smaller, later Ln^{3+} ions form solvated ion-pairs (Figs. 27.4b and c). The solid state structures of LnX_3 contain Ln(III) centres with high coordination numbers, and as $r_{M^{3+}}$ decreases across the row, the coordination number decreases. In crystalline LaF_3, each La^{3+} centre is 11-coordinate in a pentacapped trigonal prismatic environment. The chlorides $LnCl_3$ for Ln = La to Gd possess the UCl_3 structure. This is a structural prototype containing tricapped trigonal prismatic metal centres. For Ln = Tb to Lu, $LnCl_3$ adopts an $AlCl_3$ layer structure with octahedral Ln(III).

Reaction 27.15 gives metallic diiodides with high electrical conductivities. As with the lanthanoid metal dihydrides, these diiodides are actually $Ln^{3+}(I^-)_2(e^-)$. Saline LnX_2 (Ln = Sm, Eu, Yb; X = F, Cl, Br, I) can be formed by reducing LnX_3 (e.g. with H_2) and are true Ln(II) compounds. SmI_2 is available commercially and is an important 1-electron reducing agent (half-eq. 27.16) in organic synthesis.

$$Ln + 2LnI_3 \longrightarrow 3LnI_2 \qquad Ln = La, Ce, Pr, Gd \qquad (27.15)$$

$$Sm^{3+}(aq) + e^- \rightleftharpoons Sm^{2+}(aq) \qquad E^{\circ} = -1.55\,V \qquad (27.16)$$

Compounds such as $KCeF_4$, $NaNdF_4$ and Na_2EuCl_5 are made by fusion of group 1 metal fluorides and LnF_3. These are *double salts* and do not contain complex anions.

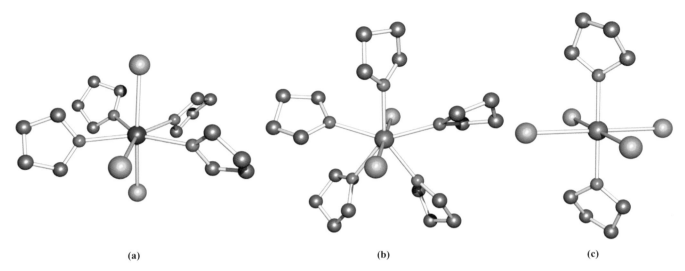

Fig. 27.4 The structures (X-ray diffraction) of (a) pentagonal bipyramidal [PrI$_3$(THF)$_4$], (b) the pentagonal bipyramidal cation in [GdI$_2$(THF)$_5$][GdI$_4$(THF)$_2$], and (c) the octahedral anion in [GdI$_2$(THF)$_5$][GdI$_4$(THF)$_2$] [K. Izod *et al.* (2004) *Inorg. Chem.*, vol. 43, p. 214]. Hydrogen atoms have been omitted from the figures. Colour code: Pr, blue; Gd, green; I, gold; O, red; C, grey.

Several discrete hexahalido anions of Ln(II) are known, e.g. [YbI$_6$]$^{4-}$ and [PrI$_6$]$^{3-}$ (see Section 9.12).

Self-study exercises

1. CeF$_4$ crystallizes with an α-ZrF$_4$ structure. What is the coordination number of each Ce^{4+} centre in the solid state?   [*Ans.* See Section 22.5]

2. In GdCl$_3$·6H$_2$O, each Gd^{3+} centre is 8-coordinate. Suggest how this is achieved.
    [*Hint*: Compare with CrCl$_3$·6H$_2$O, Section 19.8]

Hydroxides and oxides

Lanthanum hydroxide, though sparingly soluble, is a strong base and absorbs CO$_2$, giving the carbonate. Base strength and solubility decrease on crossing the lanthanoid series, and Yb(OH)$_3$ and Lu(OH)$_3$ dissolve in hot concentrated NaOH (eq. 27.17).

$$Ln(OH)_3 + 3[OH]^- \longrightarrow [Ln(OH)_6]^{3-} \qquad Ln = Yb, Lu$$
(27.17)

Cerium(III) hydroxide is a white solid, and in air, it slowly forms yellow Ce(OH)$_4$. Most oxides Ln$_2$O$_3$ are formed by thermal decomposition of oxoacid salts, e.g. reaction 27.18, but Ce, Pr and Tb give higher oxides by this method and H$_2$ is used to reduce the latter to Ln$_2$O$_3$.

$$4Ln(NO_3)_3 \xrightarrow{\Delta} 2Ln_2O_3 + 12NO_2 + 3O_2$$
(27.18)

The reaction of Nd$_2$O$_3$ and oleum (see Section 16.8) at 470 K results in the formation of Nd(S$_2$O$_7$)(HSO$_4$), a rare example of a disulfate of a rare earth metal.

Complexes of Ln(III)

The coordination and organometallic chemistries of the lanthanoid metals are rapidly growing areas of research, and in Section 9.12, we described the recent use of ionic liquids as a medium for the synthesis of new *f*-block metal complexes.[†] Since most complexes are paramagnetic, routine characterization by NMR spectroscopic methods is not usually possible. Thus, compound characterization tends to rely on X-ray diffraction studies. The paramagnetic nature of lanthanoid complexes has, however, been turned to advantage in their application as NMR shift reagents (Box 27.3) and MRI contrast agents (Box 4.3).

The Ln^{3+} ions are hard and show a preference for F$^-$ and *O*-donor ligands, e.g. in complexes with [EDTA]$^{4-}$ (Section 27.5), [Yb(OH)$_6$]$^{3-}$ (eq. 27.17) and in β-diketonate complexes (Box 27.3). In their aqua complexes, the Ln^{3+} ions are typically 9-coordinate, and a tricapped trigonal prismatic structure has been confirmed in crystalline salts such as [Pr(OH$_2$)$_9$][OSO$_3$Et]$_3$ and [Ho(OH$_2$)$_9$][OSO$_3$Et]$_3$. High coordination numbers are the norm in complexes of Ln^{3+}, with the highest exhibited by the early lanthanoids. Examples include:[‡]

- 12-coordinate: [La(NO$_3$-*O*,*O*')$_6$]$^{3-}$, [La(OH$_2$)$_2$(NO$_3$-*O*,*O*')$_5$]$^{2-}$;
- 11-coordinate: [La(OH$_2$)$_5$(NO$_3$-*O*,*O*')$_3$], [Ce(OH$_2$)$_5$(NO$_3$-*O*,*O*')$_3$], [Ce(15-crown-5)(NO$_3$-*O*,*O*')$_3$], [La(15-crown-5)(NO$_3$-*O*,*O*')$_3$] (Fig. 27.5a);
- 10-coordinate: [Ce(CO$_3$-*O*,*O*')$_5$]$^{6-}$ (Ce^{4+}, Fig. 27.5b), [Nd(NO$_3$-*O*,*O*')$_5$]$^{2-}$, [Eu(18-crown-6)(NO$_3$-*O*,*O*')$_2$]$^+$;

[†]See: K. Binnemans (2007) *Chem. Rev.*, vol. 107, p. 2592; A.-V. Mudring and S. Tang (2010) *Eur. J. Inorg. Chem.*, p. 2569.
[‡] Ligand abbreviations: see Table 7.7.

THEORY

Box 27.3 Lanthanoid shift reagents in NMR spectroscopy

The magnetic field experienced by a proton is very different from that of the applied field when a paramagnetic metal centre is present. This results in the δ range over which the ^{1}H NMR spectroscopic signals appear being larger than in a spectrum of a related diamagnetic complex (see Box 4.2). Signals for protons *close* to the paramagnetic metal centre are significantly *shifted* and this has the effect of 'spreading out' the spectrum. Values of coupling constants are generally not much changed.

^{1}H NMR spectra of large organic compounds including proteins or of mixtures of diastereoisomers, for example, are often difficult to interpret and assign due to overlapping of signals. This is particularly true when the spectrum is recorded on a lowfield (e.g. 100 or 250 MHz) instrument. Highfield NMR spectrometers (up to 1000 MHz = 1 GHz, as of 2011) provide increased sensitivity and spectroscopic signal dispersion. However, low- and midfield NMR instruments remain in daily use for research, and in these cases, paramagnetic NMR shift reagents may be used to disperse overlapping signals. Lanthanoid metal complexes are routinely employed as NMR shift reagents. The addition of a small amount of a shift reagent to a solution of an organic compound can lead to an equilibrium being established between the free and coordinated organic species. The result is that signals due to the organic species which originally overlapped, spread out, and the spectrum becomes easier to interpret. The europium(III) complex shown below is a commercially available shift reagent (Resolve-Al^{TM}), used, for example, to resolve mixtures of diastereoisomers.

The 9-coordinate complexes $[Ln(DPA)_3]^{3-}$ (Ln^{3+} is a general lanthanoid ion and H_3DPA is pyridine-2,6-dicarboxylic acid) bind effectively to proteins and can be used as paramagnetic shift reagents for protein NMR spectroscopic studies. The structure of the $[La(DPA)_3]^{3-}$ ion is shown below.

[Colour code: La, green; N, blue; O, red; C, grey; H, white; X-ray diffraction data: J.M. Harrowfield *et al.* (1995) *Aust. J. Chem.*, vol. 48, p. 807.]

Further reading

S.P Babailov (2008) *Prog. Nucl. Mag. Res. Spec.*, vol. 52, p. 1 – 'Lanthanide paramagnetic probes for NMR spectroscopic studies of molecular conformational dynamics in solution: Applications to macrocyclic molecules'.

R. Rothchild (2000) *Enantiomer*, vol. 5, p. 457 – 'NMR methods for determination of enantiomeric excess'.

T.J. Wenzel (2000) *Trends Org. Chem.*, vol. 8, p. 51 – 'Lanthanide-chiral solvating agent couples as chiral NMR shift reagents'.

- 9-coordinate: $[Sm(NH_3)_9]^{3+}$ (tricapped trigonal prismatic), $[Ln(EDTA)(OH_2)_3]^{-}$ (Ln = La, Ce, Nd, Sm, Eu, Gd, Tb, Dy, Ho), $[CeCl_2(18\text{-crown-}6)(OH_2)]^{+}$, $[PrCl(18\text{-crown-}6)(OH_2)_2]^{2+}$, $[LaCl_3(18\text{-crown-}6)]$ (Fig. 27.5c), $[Eu(tpy)_3]^{3+}$ (tpy = **27.1**), $[Nd(OH_2)(CO_3\text{-}O,O')_4]^{5-}$;
- 8-coordinate: $[Pr(NCS\text{-}N)_8]^{5-}$ (between cubic and square antiprismatic);
- 7-coordinate: **27.2**;
- 6-coordinate: $[Sm(pyr)_6]^{3-}$ (Hpyr = **27.3**), *cis*-$[GdCl_4(THF)_2]^{-}$, $[Ln(\beta\text{-ketonate})_3]$ (see Box 27.3).

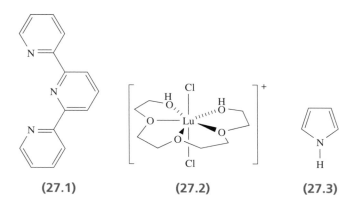

(27.1)   **(27.2)**   **(27.3)**

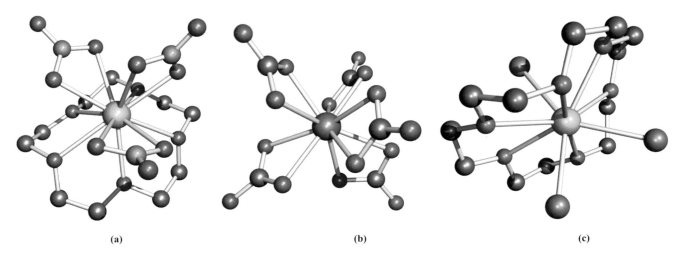

(a)   **(b)**   **(c)**

Fig. 27.5 The structures (X-ray diffraction) of (a) [La(15-crown-5)(NO$_3$-O,O')$_3$] [R.D. Rogers _et al._ (1990) _J. Crystallogr. Spectrosc. Res._, vol. 20, p. 389], (b) [Ce(CO$_3$-O,O')$_5$]$^{6-}$ in the guanidinium salt [R.E. Marsh _et al._ (1988) _Acta Crystallogr._, Sect. B, vol. 44, p. 77], and (c) [LaCl$_3$(18-crown-6)] [R.D. Rogers _et al._ (1993) _Inorg. Chem._, vol. 32, p. 3451]. Hydrogen atoms have been omitted for clarity; colour code: La, yellow; Ce, green; C, grey; N, blue; Cl, green; O, red.

The variation found in coordination geometries for a given high coordination number is consistent with the argument that spatial requirements of a ligand, and coordination restrictions of multidentate ligands, are controlling factors. The 4f atomic orbitals are deeply buried and play little role in metal–ligand bonding. Thus, the 4f^n configuration is not a controlling influence on the coordination number. Recent development of MRI contrast agents (see Box 4.3) has made studies of Gd^{3+} complexes containing polydentate ligands with _O_- and _N_-donors important.

Lower coordination numbers can be stabilized by using aryloxy or amido ligands, for example:

- 5-coordinate: **27.4, 27.5**;
- 3-coordinate: [Ln{N(SiMe$_3$)$_2$}$_3$] (Ln = Ce, Pr, Nd, Sm, Eu, Dy, Er, Yb).

R$_2$N — Nd$^{\prime\prime\prime\prime\prime}NR_2$
$\,\,\,\,\,\,\,\,\,\,\,\,\,\,\,\,\,$ NR$_2$

R = SiHMe$_2$

(27.4)

RO — Sm$^{\prime\prime\prime\prime\prime}$OR
$\,\,\,\,\,\,\,\,\,\,\,\,\,\,\,\,\,$ OR

R = 3,5-^{i}Pr$_2$C$_6$H$_3$

(27.5)

Three-coordinate bis(trimethylsilyl)amido complexes are made by reaction 27.19. In the solid state, the LnN$_3$-unit in each Ln{N(SiMe$_3$)$_2$}$_3$ complex is trigonal pyramidal (Fig. 27.6a). The preference for a trigonal pyramidal over

a trigonal planar structure is independent of the size of the Ln^{3+} ion, and DFT studies (see Section 4.13) indicate that the observed ligand arrangement is stabilized by Ln---C–Si interactions (Fig. 27.6b) and by the participation of metal _d_-orbitals in bonding.[†] In Sm{N(SiMe$_3$)$_2$}$_3$, the close Sm---C contacts shown in Fig. 27.6b are significantly shorter (300 pm) than the sum of the van der Waals radii of Sm and C ($\approx$400 pm).

$$LnCl_3 + 3MN(SiMe_3)_2 \longrightarrow Ln\{N(SiMe_3)_2\}_3 + 3MCl$$
$$(M = Na, K) \qquad (27.19)$$

27.8 Organometallic complexes of the lanthanoids

Despite the extreme air and moisture sensitivity of organolanthanoid compounds, this is a rapidly expanding research area. An exciting aspect of organolanthanoid chemistry is the discovery of many efficient catalysts for organic transformations (see Box 27.4). In contrast to the extensive carbonyl chemistry of the _d_-block metals (see Sections 24.4 and 24.9), lanthanoid metals do not form complexes with CO under normal conditions. Unstable carbonyls such as Nd(CO)$_6$ have been prepared by matrix isolation. Since organolanthanoids are usually air- and moisture-sensitive and may be pyrophoric, handling the compounds under inert atmospheres is essential.[‡]

[†] For a detailed discussion of structure and bonding in Sm{N(SiMe$_3$)$_2$}$_3$, see: E.D. Brady _et al._ (2003) _Inorg. Chem._, vol. 42, p. 6682.
[‡] Inert atmosphere techniques, see: D.F. Shriver and M.A. Drezdon (1986) _The Manipulation of Air-sensitive Compounds_, Wiley, New York.

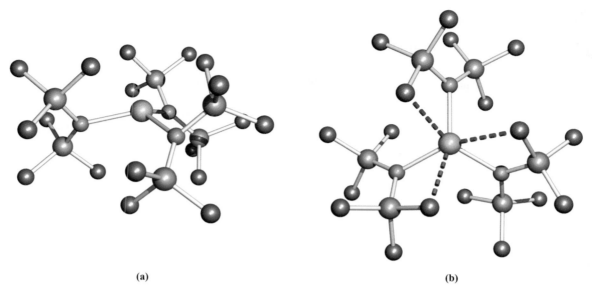

(a) **(b)**

Fig. 27.6 The solid state structure of $Sm\{N(SiMe_3)_2\}_3$ determined by X-ray diffraction [E.D. Brady *et al.* (2003) *Inorg. Chem.*, vol. 42, p. 6682]. The figure highlights (a) the trigonal pyramidal Sm centre, and (b) the close $Sm\cdots C_{methyl}$ contacts that contribute to the observed ligand arrangement. Hydrogen atoms are omitted. Colour code: Sm, orange; N, blue; Si, pink; C, grey.

σ-Bonded complexes

Reaction 27.20 shows a general method of forming Ln−C σ-bonds. In order to stabilize LnR_3, bulky alkyl substituents must be used. $LuMe_3$ may be prepared by an alternative strategy (see below).

$$LnCl_3 + 3LiR \longrightarrow LnR_3 + 3LiCl \qquad (27.20)$$

In the presence of excess LiR and with R groups that are not too sterically demanding, reaction 27.20 may proceed further to give $[LnR_4]^-$ or $[LnR_6]^{3-}$ (eqs. 27.21 and 27.22).

$$YbCl_3 + 4^tBuLi \xrightarrow{THF,\ 218\,K} [Li(THF)_3]^+[Yb^tBu_4]^- + 3LiCl \qquad (27.21)$$

$$LuCl_3 + 6MeLi \xrightarrow{DME,\ 195\,K} [Li(DMEA)]_3^+[LuMe_6]^{3-} + 3LiCl$$
$$DME = 1,2\text{-dimethoxyethane} \qquad (27.22)$$

In the solid state, $[LuMe_6]^{3-}$ is octahedral (Lu−C = 253 pm) and analogues for all the lanthanoids except Eu are known. In these reactions, a coordinating solvent such as DME or $Me_2NCH_2CH_2NMe_2$ (TMEDA) is needed to stabilize the product with a *solvated* Li^+ ion. In some cases, the solvent coordinates to the lanthanoid

metal, e.g. $TmPh_3(THF)_3$ (reaction 27.23 and structure **27.6**).

$$Tm + HgPh_2 \xrightarrow[\text{in presence of } TmCl_3]{\text{in THF,}} TmPh_3(THF)_3 \qquad (27.23)$$

Tm–C = 242 pm

(27.6)

The amido complexes $Ln(NMe_2)_3$ (Ln = La, Nd, Lu) are prepared by reaction 27.24 and are stabilized by association with LiCl. Reaction of $Ln(NMe_2)_3$ with Me_3Al gives **27.7**. For Ln = Lu, treatment of **27.7** with THF results in isolation of polymeric $[LnMe_3]_n$ (eq. 27.25).

$$LnCl_3(THF)_x + 3LiNMe_2 \longrightarrow Ln(NMe_2)_3 \cdot 3LiCl + xTHF \qquad (27.24)$$

$$Ln\{(\mu\text{-Me})_2AlMe_2\}_3 + 3THF \longrightarrow \tfrac{1}{n}[LnMe_3]_n + 3Me_3Al\cdot THF \qquad (27.25)$$

APPLICATIONS

Box 27.4 Organolanthanoid complexes as catalysts

One of the driving forces behind the study of organolanthanoid complexes is the ability of some of them to act as highly effective catalysts in organic transformations including hydrogenation, hydrosilylation, hydroboration and hydroamination reactions and the cyclization and polymerization of alkenes. The availability of a range of different lanthanoid metals coupled with a variety of ligands provides a means of systematically altering the properties of a series of organometallic complexes. In turn, this leads to controlled variation in their catalytic behaviour, including selectivity.

The presence of an $(\eta^5\text{-}C_5R_5)Ln$ or $(\eta^5\text{-}C_5R_5)_2Ln$ unit in an organolanthanoid complex is a typical feature, and often $R = Me$. When $R = H$, complexes tend to be poorly soluble in hydrocarbon solvents and catalytic activity is typically low. Hydrocarbon solvents are generally used for catalytic reactions because coordinating solvents (e.g. ethers) bind to the Ln^{3+} centre, hindering association of the metal with the desired organic substrate. In designing a potential catalyst, attention must be paid to the accessibility of the metal centre to the substrate. Dimerization of organolanthanoid complexes via bridge formation is a characteristic feature. This is a disadvantage in a catalyst because the metal centre is less accessible to a substrate than in a monomer. An inherent problem of the $(\eta^5\text{-}C_5R_5)_2Ln$-containing systems is that the steric demands of substituted cyclopentadienyl ligands may hinder the catalytic activity of the metal centre. One strategy to retain an accessible Ln centre is to increase the tilt angle between two $\eta^5\text{-}C_5R_5$ units by attaching them together as illustrated below:

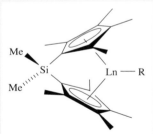

Examples of organic transformations that are catalysed by organolanthanoid complexes are given below. A significant point is that only *mild* reaction conditions are required in many reactions.

- *Hydrogenation*:

H₂, 1 bar
in cyclopentane, 298 K
$(\eta^5\text{-}C_5Me_5)_2SmCH(TMS)_2$

TMS = Me₃Si

Addition of H₂ is stereochemically specific

- *Hydrosilylation*:

PhSiH₃, in benzene, 298 K
$(\eta^5\text{-}C_5Me_5)_2YCH(TMS)_2$

TMS = Me₃Si

SiH₂Ph

H

- *Hydroamination*:

in toluene, 298 K
$(\eta^5\text{-}C_5Me_5)_2LaCH(TMS)_2$

TMS = Me₃Si

NH₂   NH

- *Cyclization with hydrosilylation*:

PhSiH₃, in pentane
$(\eta^5\text{-}C_5Me_5)_2LuMe$

SiH₂Ph

- *Hydrogenation with cyclization*:

H₂, 1 bar
in pentane, 298 K
$(\eta^5\text{-}C_5Me_5)_2YMe(THF)$

Selectivity in product formation is important, and this issue is addressed in detail in the articles listed below.

Further reading

S. Hong and T.J. Marks (2004) *Acc. Chem. Res.*, vol. 37, p. 673 – 'Organolanthanide-catalyzed hydroamination'.

Z. Hou and Y. Wakatsuki (2002) *Coord. Chem. Rev.*, vol. 231, p. 1 – 'Recent developments in organolanthanide polymerization catalysts'.

P.A. Hunt (2007) *Dalton Trans.*, p. 1743 – 'Organolanthanide mediated catalytic cycles: A computational perspective'.

K. Mikami, M. Terada and H. Matsuzawa (2002) *Angew. Chem. Int. Ed.*, vol. 41, p. 3555 – 'Asymmetric catalysis by lanthanide complexes'.

G.A. Molander and J.A.C. Romero (2002) *Chem. Rev.*, vol. 102, p. 2161 – 'Lanthanocene catalysts in selective organic synthesis'.

(27.7)

Complexes containing σ-bonded $-C\equiv CR$ groups have been prepared by a number of routes, e.g. reaction 27.26.

$$[Lu^tBu_4(THF)_4]^- \xrightarrow[-^tBuH]{HC\equiv C^tBu \text{ in THF}} [Lu(C\equiv C^tBu)_4]^-$$

$$(27.26)$$

Cyclopentadienyl complexes

Many organolanthanoids contain cyclopentadienyl ligands and reaction 27.27 is a general route to Cp_3Ln.

$$LnCl_3 + 3NaCp \longrightarrow Cp_3Ln + 3NaCl \qquad (27.27)$$

The solid state structures of Cp_3Ln compounds vary with Ln, e.g. Cp_3Tm and Cp_3Yb are monomeric, while Cp_3La, Cp_3Pr and Cp_3Lu are polymeric. Adducts with donors such as

THF, pyridine and MeCN are readily formed, e.g. tetrahedral $(\eta^5\text{-Cp})_3Tb(NCMe)$ and $(\eta^5\text{-Cp})_3Dy(THF)$, and trigonal bipyramidal $(\eta^5\text{-Cp})_3Pr(NCMe)_2$ (axial MeCN groups). The complexes $(\eta^5\text{-C}_5Me_5)_3Sm$ and $(\eta^5\text{-C}_5Me_5)_3Nd$ are 1-electron reductants: $(\eta^5\text{-C}_5Me_5)_3Ln$ reduces $Ph_3P{=}Se$ to PPh_3 and forms $(\eta^5\text{-C}_5Me_5)_2Ln(\mu\text{-Se})_nLn(\eta^5\text{-C}_5Me_5)_2$ $(Ln = Sm, \ n = 1; \ Ln = Nd, \ n = 2)$ and $(C_5Me_5)_2$. The reducing ability is attributed to the severe steric congestion in $(\eta^5\text{-C}_5Me_5)_3Ln$, the reducing agent being the $[C_5Me_5]^-$ ligand.

By altering the $LnCl_3 : NaCp$ ratio in reaction 27.27, $(\eta^5\text{-C}_5H_5)_2LnCl$ and $(\eta^5\text{-C}_5H_5)LnCl_2$ can be isolated. However, crystallographic data reveal more complex structures than these formulae suggest: e.g. $(\eta^5\text{-C}_5H_5)ErCl_2$ and $(\eta^5\text{-C}_5H_5)YbCl_2$ crystallize from THF as adducts **27.8**, $(\eta^5\text{-C}_5H_5)_2YbCl$ and $(\eta^5\text{-C}_5H_5)_2ErCl$ are dimeric (Fig. 27.7a), and $(\eta^5\text{-C}_5H_5)_2DyCl$ consists of polymeric chains (Fig. 27.7b).

Ln = Er, Yb

(27.8)

Schemes 27.28 and 27.29 show some reactions of $[(\eta^5\text{-C}_5H_5)_2LuCl]_2$ and $[(\eta^5\text{-C}_5H_5)_2YbCl]_2$. Coordinating solvents are often incorporated into the products and can cause bridge cleavage as in reaction 27.30.

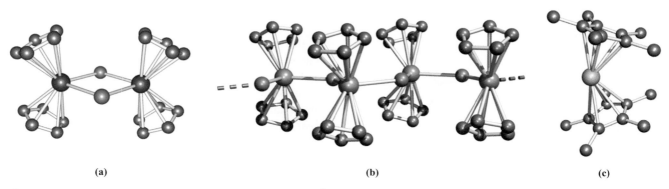

(a) (b) (c)

Fig. 27.7 The structures (X-ray diffraction) of (a) dimeric $[(\eta^5\text{-C}_5H_5)_2ErCl]_2$ [W. Lamberts *et al.* (1987) *Inorg. Chim. Acta*, vol. 134, p. 155], (b) polymeric $(\eta^5\text{-C}_5H_5)_2DyCl$ [W. Lamberts *et al.* (1987) *Inorg. Chim. Acta*, vol. 132, p. 119] and (c) the bent metallocene $(\eta^5\text{-C}_5Me_5)Sm$ [W.J. Evans *et al.* (1986) *Organometallics*, vol. 5, p. 1285]. Hydrogen atoms are omitted for clarity; colour code: Er, red; Dy, pink; Sm, orange; Cl, green; C, grey.

$$LuCl_3 + 2NaCp + LiR \xrightarrow{\text{THF, low temperature}} (\eta^5\text{-Cp})_2LuR$$

$$R = CH_2Ph, CH_2{}^tBu, 4\text{-}MeC_6H_4 \quad (27.31)$$

Use of the pentamethylcyclopentadienyl ligand (more sterically demanding than the $[C_5H_5]^-$ ligand) in organolanthanoid chemistry has played a major role in the development of this field (see Box 27.4). An increase in the steric demands of the $[C_5R_5]^-$ ligand stabilizes derivatives of the earlier lanthanoid metals. For example, the reaction of $Na[C_5H^iPr_4]$ with $YbCl_3$ in 1,2-dimethoxyethane (DME) leads to the formation of the *monomeric* complex **27.9**.

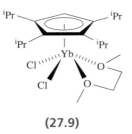

(27.9)

In contrast, the reactions of $LaCl_3$ or $NdCl_3$ with two molar equivalents of $Na[C_5H^iPr_4]$ in THF followed by recrystallization from Et_2O lead to complexes **27.10**, characterized in the solid state. In these dimeric species, there is association between $[(\eta^5\text{-}C_5H^iPr_4)_2MCl_2]^-$ ions and solvated Na^+.

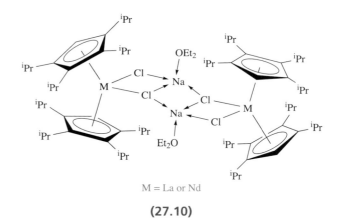

M = La or Nd

(27.10)

Lanthanoid(II) metallocenes have been known for Sm, Eu and Yb since the 1980s and are stabilized by using the bulky $[C_5Me_5]^-$ ligand (eqs. 27.32–27.34). The products are obtained as solvates. The desolvated metallocenes have *bent* structures in the solid state (Fig. 27.7c) rather than a ferrocene-like structure. For each of Sm, Eu and Yb, the most convenient route to $(\eta^5\text{-}C_5Me_5)_2Ln$ starts from $LnI_2 \cdot n$THF.

$$2Na[C_5Me_5] + YbCl_2 \xrightarrow{\text{THF}} (\eta^5\text{-}C_5Me_5)_2Yb + 2NaCl$$
$$(27.32)$$

(27.28)

(27.29)

(27.30)

Compounds of the type $(\eta^5\text{-Cp})_2LnR$ (isolated as THF adducts) can be made directly from $LnCl_3$, e.g. for Lu in reaction 27.31.

$$2K[C_5Me_5] + SmI_2 \xrightarrow{\text{THF}} (\eta^5\text{-}C_5Me_5)_2Sm + 2KI \tag{27.33}$$

$$2C_5Me_5H + Eu \xrightarrow{\text{liquid NH}_3} (\eta^5\text{-}C_5Me_5)_2Eu + H_2 \tag{27.34}$$

The first Tm(II) organometallic complex was reported in 2002. Its stabilization requires a more sterically demanding C_5R_5 substituent than is needed for the Sm(II), Eu(II) and Yb(II) metallocenes. Reaction 27.35 shows the synthesis of $\{\eta^5\text{-}C_5H_3\text{-}1,3\text{-}(SiMe_3)_2\}_2Tm(THF)$ and the use of an argon atmosphere is essential. Reaction 27.36 illustrates the effects on the reaction of using the $[C_5Me_5]^-$ ligand in place of $[C_5H_3\text{-}1,3\text{-}(SiMe_3)_2]^-$.

$$\begin{array}{l} 2K[C_5H_3\text{-}1,3\text{-}(SiMe_3)_2] \\ + \ TmI_2(THF)_3 \end{array} \xrightarrow[\text{Ar}]{\text{THF}} \tag{27.35}$$

$$\begin{array}{l} 2K[C_5Me_5] \\ + \ TmI_2(THF)_2 \end{array} \xrightarrow[\text{Ar}]{\text{THF}} \tag{27.36}$$

Thulium(II) ⟶ Thulium(III)

Bis(arene) derivatives

The co-condensation at 77 K of $1,3,5\text{-}^tBu_3C_6H_3$ with Ln metal vapour yields the bis(arene) derivatives $(\eta^6\text{-}1,3,5\text{-}^tBu_3C_6H_3)_2Ln$. The complexes are thermally stable for Ln = Nd, Tb, Dy, Ho, Er and Lu, but unstable for Ce, Eu, Tm and Yb.

Complexes containing the η^8-cyclooctatetraenyl ligand

In Chapter 24, we described organometallic sandwich and half-sandwich complexes containing π-bonded ligands with hapticities ≤7, e.g. $[(\eta^7\text{-}C_7H_7)Mo(CO)_3]^+$. The larger sizes of the lanthanoids permit the formation of sandwich complexes with the planar, octagonal $[C_8H_8]^{2-}$

ligand (see eq. 27.59). Lanthanoid(III) chlorides react with $K_2C_8H_8$ to give $[(\eta^8\text{-}C_8H_8)_2Ln]^-$ (Ln = La, Ce, Pr, Sm, Tb, Yb). Cerium also forms the Ce(IV) complex $(\eta^8\text{-}C_8H_8)_2Ce$ (**27.11**), an analogue of uranocene (see Section 27.11). For the lanthanoids with a stable +2 oxidation state, K^+ salts of $[(\eta^8\text{-}C_8H_8)_2Ln]^{2-}$ (Ln = Sm, Eu, Yb) are isolable.

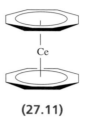

(27.11)

27.9 The actinoid metals

The artificial nature (see Section 27.5) of all but Th and U among the actinoid metals affects the extent of knowledge of their properties, and this is reflected in the varying amounts of information that we give for each metal. The instability of the actinoids with respect to radioactive decay was described in Section 27.5, and Table 27.7 lists data for the longest-lived isotope of each element. All the actinoids are highly toxic, the ingestion of long-lived α-emitters such as ^{231}Pa being extremely hazardous. Extremely small doses are lethal.

Actinium is a soft metal which glows in the dark. It is readily oxidized to Ac_2O_3 in moist air, and liberates H_2 from H_2O. *Thorium* is relatively stable in air, but is attacked slowly by H_2O and rapidly by steam or dilute HCl. On heating, Th reacts with H_2 to give ThH_2, halogens to give ThX_4, and N_2 and C to give nitrides and carbides. Thorium forms alloys with a range of metals (e.g. Th_2Zn, $CuTh_2$). *Protactinium* is ductile and malleable, is not corroded by air, but reacts with O_2, H_2 and halogens when heated (scheme 27.37), and with concentrated HF, HCl and H_2SO_4.

$$Pa \begin{cases} \xrightarrow{O_2, \Delta} Pa_2O_5 \\ \xrightarrow{H_2, \Delta} PaH_3 \\ \xrightarrow{I_2, \Delta} PaI_5 \end{cases} \tag{27.37}$$

Uranium corrodes in air; it is attacked by water and dilute acids but not alkali. Scheme 27.38 gives selected reactions. With O_2, UO_2 is produced, but on heating, U_3O_8 forms.

$$U \begin{cases} \xrightarrow{H_2, \Delta} UH_3 \\ \xrightarrow{F_2, \Delta} UF_6 \\ \xrightarrow{Cl_2, \Delta} UCl_4 + UCl_5 + UCl_6 \\ \xrightarrow{H_2O, \ 373 \ K} UO_2 \end{cases} \tag{27.38}$$

Table 27.7 Half-lives and decay modes of the longest-lived isotopes of actinium and the actinoids.

Longest-lived isotope	Half-life	Decay mode	Longest-lived isotope	Half-life	Decay mode
^{227}Ac	21.8 yr	β^-	^{247}Bk	1.4×10^3 yr	α, γ
^{232}Th	1.4×10^{10} yr	α, γ	^{251}Cf	9.0×10^2 yr	α, γ
^{231}Pa	3.3×10^4 yr	α, γ	^{252}Es	1.3 yr	α
^{238}U	4.5×10^9 yr	α, γ	^{257}Fm	100 d	α, γ
^{237}Np	2.1×10^6 yr	α, γ	^{258}Md	52 d	α
^{244}Pu	8.2×10^7 yr	α, γ	^{259}No	58 min	α
^{243}Am	7.4×10^3 yr	α, γ	^{262}Lr	3 min	α
^{247}Cm	1.6×10^7 yr	α, γ			

Neptunium is a reactive metal which quickly tarnishes in air. It reacts with dilute acids liberating H_2, but is not attacked by alkali.

A **nuclear fission** reaction such as:

$$^{235}_{2}U + {}^1_0n \longrightarrow \text{fission products} + x\,{}^1_0n + \text{energy}$$

may result in a branching chain reaction because each neutron formed can initiate another nuclear reaction. If this involves a mass of $^{235}_{2}U$ larger than the **critical mass**, a violent explosion occurs, liberating enormous amounts of energy.

Despite the fact that the critical mass of *plutonium* is <0.5 kg and it is extremely toxic, its uses as a nuclear fuel and explosive make it a much-studied element. It reacts with O_2, steam and acids, but is inert towards alkali. On heating, Pu combines with many non-metals to give, for example, PuH_2, PuH_3, $PuCl_3$, PuO_2 and Pu_3C_2. *Americium* is a very intense α- and γ-emitter. It tarnishes slowly in dry air, reacts with steam and acids, and on heating forms binary compounds with a range of non-metals. *Curium* corrodes rapidly in air; only minute quantities can be handled (<20 mg in controlled conditions). *Berkelium* and *californium* behave similarly to Cm, being attacked by air and acids, but not by alkali. Curium and the later elements are handled only in specialized research laboratories.

In the remaining sections, we focus on the chemistries of thorium and uranium (the actinoids for which the most extensive chemistries have been developed) and plutonium.

Self-study exercises

1. What happens to the mass number and atomic number of a nuclide as it undergoes decay by (a) α-particle or (b) β-particle emission? [*Ans*. See Section 27.5]

2. Identify the products in the following radioactive decay sequence:

$$^{238}_{92}U \xrightarrow{-\alpha\text{-particle}} ? \xrightarrow{-\beta\text{-particle}} ?$$

[*Ans*. $^{234}_{90}$Th; $^{234}_{91}$Pa]

3. Identify the second nuclide formed in the reaction:

$$^{235}_{92}U + {}^1_0n \longrightarrow {}^{92}_{36}U + ? + 2\,{}^1_0n$$

[*Ans*. $^{142}_{56}$Ba]

4. Identify the second nuclide formed in the reaction:

$$^{235}_{92}U + {}^1_0n \longrightarrow {}^{141}_{55}Cs + ? + 2\,{}^1_0n$$

[*Ans*. $^{93}_{37}$Rb]

27.10 Inorganic compounds and coordination complexes of thorium, uranium and plutonium

Thorium

The chemistry of thorium largely concerns Th(IV) and, in aqueous solution, there is no evidence for any other oxidation state. The E° value for the Th^{4+}/Th couple is -1.9 V.

Thorium(IV) halides are made by direct combination of the elements. White ThF_4, $ThCl_4$ and $ThBr_4$, and yellow ThI_4 crystallize with lattices in which Th(IV) is 8-coordinate. Reaction of ThI_4 with Th yields ThI_2 and ThI_3 (both polymorphic) which are metallic conductors and are formulated as $Th^{4+}(I^-)_2(e^-)_2$ and $Th^{4+}(I^-)_3(e^-)$ respectively. Thorium(IV) fluoride is insoluble in water and aqueous alkali metal fluoride solutions, but a large number of double or complex fluorides can be made by direct combination of their constituents. Their structures are complicated, e.g. $[NH_4]_3[ThF_7]$ and $[NH_4]_4[ThF_8]$ contain infinite $[ThF_7]_n^{3n-}$ chains consisting of edge-sharing tricapped trigonal prismatic Th(IV). Thorium(IV) chloride is soluble in water, and a range of salts containing the discrete, octahedral $[ThCl_6]^{2-}$ are known (reaction 27.39).

$$ThCl_4 + 2MCl \longrightarrow M_2ThCl_6 \quad \text{e.g. } M = K, Rb, Cs$$

$$(27.39)$$

White ThO_2 is made by thermal decomposition of $Th(ox)_2$ or $Th(NO_3)_4$ and adopts a CaF_2 structure (Fig. 6.19). It is precipitated in neutral or even weakly acidic solution. Nowadays, ThO_2 has application as a Fischer–Tropsch catalyst, but historically, its property of emitting a blue glow when heated led to its use in incandescent gas mantles. As expected from the high formal charge on the metal centre, aqueous solutions of Th(IV) salts contain hydrolysis products such as $[ThOH]^{3+}$, $[Th(OH)_2]^{2+}$. The addition of alkali to these solutions gives a gelatinous, white precipitate of $Th(OH)_4$ which is converted to ThO_2 at >700 K.

Coordination complexes of Th(IV) characteristically exhibit high coordination numbers, and hard donors such as oxygen are preferred, for example:

- 12-coordinate: $[Th(NO_3\text{-}O,O')_6]^{2-}$ (Fig. 27.8), $[Th(NO_3\text{-}O,O')_5(OPMe_3)_2]^-$;
- 10-coordinate: $[Th(CO_3\text{-}O,O')_5]^{6-}$;
- 9-coordinate (tricapped trigonal prismatic): $[ThCl_2(OH_2)_7]^{2+}$;
- 8-coordinate (dodecahedral): $[ThCl_4(OSPh_2)_4]$, $\alpha\text{-}[Th(acac)_4]$, $[ThCl_4(THF)_4]$;
- 8-coordinate (square antiprismatic): $\beta\text{-}[Th(acac)_4]$;
- 8-coordinate (cubic): $[Th(NCS\text{-}N)_8]^{4-}$;
- 7-coordinate: $[ThCl_4(NMe_3)_3]$.

Lower coordination numbers can be stabilized by using amido or aryloxy ligands. In reaction 27.40, the bis(silyl)-amido ligands are too bulky to allow the last chlorido

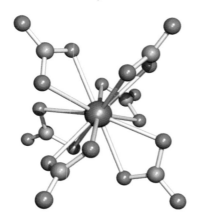

Fig. 27.8 The structure (X-ray diffraction) of the 12-coordinate $[Th(NO_3\text{-}O,O')_6]^{2-}$ in the 2,2'-bipyridinium salt [M.A. Khan *et al.* (1984) *Can. J. Chem.*, vol. 62, p. 850]. Colour code: Th, brown; N, blue; O, red.

group to be replaced. Reactions 27.41 and 27.42 illustrate that steric control dictates whether $Th(OR)_4$ is stabilized with or without other ligands in the coordination sphere.

$$ThCl_4 + 3LiN(SiMe_3)_2 \longrightarrow ThCl\{N(SiMe_3)_2\}_3 + 3LiCl$$
$$\text{tetrahedral}$$

$$(27.40)$$

$$ThI_4 + 4KO^tBu \xrightarrow{\text{pyridine/THF}} cis\text{-}Th(py)_2(O^tBu)_4 + 4KCl$$
$$\text{octahedral}$$

$$(27.41)$$

$$ThI_4 + 4KOC_6H_3\text{-}2,6\text{-}^tBu_2 \longrightarrow Th(OC_6H_3\text{-}2,6\text{-}^tBu_2)_4 + 4KI$$
$$\text{tetrahedral}$$

$$(27.42)$$

Uranium

Uranium exhibits oxidation states from $+3$ to $+6$, although U(IV) and U(VI) are the more common. The key starting point for the preparation of many uranium compounds is UO_2 and scheme 27.43 shows the syntheses of fluorides and chlorides. The fluoride UF_5 is made by controlled reduction of UF_6 but readily disproportionates to UF_4 and UF_6.

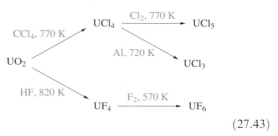

$$(27.43)$$

UCl_4 and $UI_3(THF)_4$ are useful starting materials in uranium(IV) and uranium(III) chemistries, respectively. $UI_3(THF)_4$ can be made by reaction 27.44. The use of toxic mercury can be avoided by preparing UI_3 or $UI_3(THF)_4$ from U and I_2 using specialized vacuum line techniques.[†]

$$2U + 3HgI_2 \xrightarrow{\text{THF}} 2UI_3(THF)_4 + 3Hg \quad (27.44)$$

Uranium hexafluoride is a colourless, volatile solid with a vapour pressure of 1 bar at 329 K. It is of great importance in the separation of uranium isotopes (see Box 7.3). The solid and vapour consist of octahedral UF_6 molecules (U−F = 199 pm). The hexafluoride is immediately hydrolysed by H_2O (eq. 27.45) and is a vigorous fluorinating agent. Treatment of UF_6 with BCl_3 gives the unstable, molecular UCl_6.

$$UF_6 + H_2O \longrightarrow UOF_4 + 2HF \quad (27.45)$$

The sparingly soluble, green UF_4 is an inert solid (mp 1309 K) with an extended structure containing 8-coordinate

---

[†] For details of the apparatus and method, see: W.J. Evans, S.A. Kozimor, J.W. Ziller, A.A. Fagin and M.N. Bochkarev (2005) *Inorg. Chem.*, vol. 44, p. 3993.

U(IV). Solid UCl$_4$ also contains 8-coordinate U, but UCl$_5$ is a dimer (**27.12**); the latter disproportionates on heating. The halides accept X$^-$ to give complexes such as NaUF$_7$, Cs$_2$UCl$_6$ and [NH$_4$]$_4$UF$_8$; the alkali metal salts adopt extended structures with U$-$F$-$U interactions giving U in high coordination environments.

(27.12)

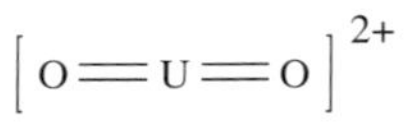

$$\left[O = U = O \right]^{2+}$$

(27.13)

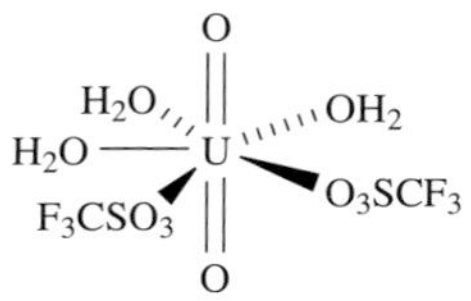

(27.14)

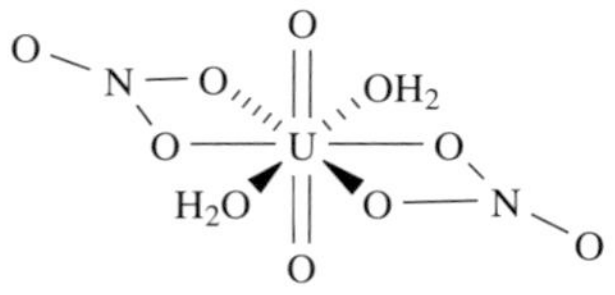

(27.15)

The oxide UO$_3$ is polymorphic and all forms decompose to the mixed oxidation state U$_3$O$_8$ on heating. Most acids dissolve UO$_3$ to give yellow solutions containing the uranyl ion (**27.13**), present as a complex, e.g. in aqueous

solution, **27.13** exists as an aqua ion, and the perchlorate salt of the pentagonal bipyramidal [UO$_2$(OH$_2$)$_5$]$^{2+}$ has been isolated. The [UO$_2$]$^{2+}$ ion is also present in many solid compounds including the alkaline earth uranates (e.g. BaUO$_4$) which are best described as mixed metal oxides. Uranyl salts of oxoacids include [UO$_2$][NO$_3$]$_2$·6H$_2$O (see Box 7.3), [UO$_2$][MeCO$_2$]$_2$·2H$_2$O and [UO$_2$][CF$_3$SO$_3$]$_2$·3H$_2$O, and coordination of the oxido-ligands and water commonly places the U(IV) centre in a 7- or 8-coordinate environment as in **27.14** and **27.15**. In aqueous solution, the [UO$_2$]$^{2+}$ ion is partially hydrolysed to species such as [U$_2$O$_5$]$^{2+}$ and [U$_3$O$_8$]$^{2+}$. In aqueous alkaline solution, the species present depend on the concentrations of both [UO$_2$]$^{2+}$ and [OH]$^-$. Investigations of complexes formed between [UO$_2$]$^{2+}$ and [OH]$^-$ are difficult because of the precipitation of U(VI) as salts such as Na$_2$UO$_4$ and Na$_2$U$_2$O$_7$. However, if Me$_4$NOH is used in place of an alkali metal hydroxide, it is possible to isolate salts of octa-hedral *trans*-[UO$_2$(OH)$_4$]$^{2-}$. The [UO$_2$]$^{2+}$ ion is hard and forms a more stable complex with F$^-$ than with the later halides. Figure 27.9 gives a potential diagram for uranium at pH = 0. Reduction of [UO$_2$]$^{2+}$ first gives [UO$_2$]$^+$, but this is somewhat unstable with respect to the disproportio-nation reaction 27.46. Since protons are involved in this reaction, the position of the equilibrium is pH-dependent. Uranium(V) can be stabilized with respect to disproportio-nation by complexing with F$^-$ as [UF$_6$]$^-$.

$$2[UO_2]^+ + 4H^+ \rightleftharpoons [UO_2]^{2+} + U^{4+} + 2H_2O \qquad (27.46)$$

Uranium metal liberates H$_2$ from acids to give the claret-coloured U^{3+} which is a powerful reducing agent

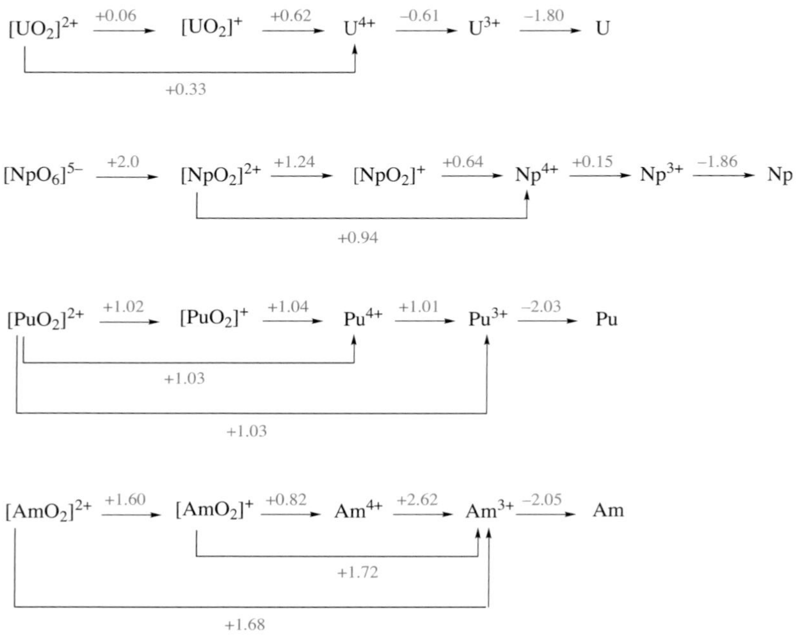

Fig. 27.9 Potential diagram for uranium at pH = 0, and comparative diagrams for Np, Pu and Am.

(Fig. 27.9). The U^{4+} ion is rapidly oxidized to $[UO_2]^{2+}$ by Cr(VI), Ce(IV) or Mn(VII) but oxidation by air is slow. The U^{4+}/U^{3+} and $[UO_2]^{2+}/[UO_2]^+$ redox couples are reversible, but the $[UO_2]^+/U^{4+}$ couple is not: the first two involve only electron transfer, but the last couple involves a structural reorganization around the metal centre.

Whereas the coordination chemistry of thorium is concerned with only the +4 oxidation state, that of uranium covers oxidation states +3 to +6. For U(VI), the linear $[UO_2]^{2+}$ unit is generally present and *trans*-octahedral, pentagonal bipyramidal and hexagonal bipyramidal complexes are usual. For other oxidation states, the coordination polyhedron is essentially determined by the spatial requirements of the ligands rather than electronic factors, and the large size of the U centre allows high coordination numbers to be attained. Complexes involving different oxidation states and coordination numbers include:

- 14-coordinate: $[U(\eta^3\text{-}BH_4)_4(THF)_2]$;
- 12-coordinate: $[U(NO_3\text{-}O,O')_6]^{2-}$, $[U(\eta^3\text{-}BH_3Me)_4]$;[†]
- 11-coordinate: $[U(\eta^3\text{-}BH_4)_2(THF)_5]^+$;
- 9-coordinate: $[U(NCMe)_9]^{3+}$ (tricapped trigonal prismatic), $[UCl_3(18\text{-crown-6})]^+$, $[UBr_2(OH_2)_5(MeCN)_2]^+$, $[U(OH_2)(ox)_4]^{4-}$;
- 8-coordinate: $[UCl_3(DMF)_5]^+$, $[UCl(DMF)_7]^{3+}$, $[UCl_2(acac)_2(THF)_2]$, $[UO_2(18\text{-crown-6})]^{2+}$, $[UO_2(NO_3\text{-}O,O')_2(NO_3\text{-}O)_2]^{2-}$, $[UO_2(\eta^2\text{-}O_2)_3]^{4-}$;
- 7-coordinate: $[U(N_3)_7]^{3-}$ (both monocapped octahedral and pentagonal bipyramidal complexes), $UO_2Cl_2(THF)_3$, $[UO_2Cl(THF)_4]^+$, $[UO_2(OSMe_2)_5]^{2+}$;
- 6-coordinate: *trans*-$[UO_2X_4]^{2-}$ (X = Cl, Br, I).

Lower coordination numbers are observed in alkoxy derivatives having sterically demanding substituents, e.g. $U(OC_6H_3\text{-}2,6\text{-}^tBu_2)_4$ (preparation analogous to reaction 27.42). The U(III) complex $U(OC_6H_3\text{-}2,6\text{-}^tBu_2)_3$ is probably monomeric. It is oxidized to $UX(OC_6H_3\text{-}2,6\text{-}^tBu_2)_3$ (X = Cl, Br, I; oxidant = PCl_5, CBr_4, I_2 respectively) in which tetrahedral U(IV) has been confirmed for X = I. Coordinating solvents tend to lead to increased coordination numbers as in reaction 27.47.

$$UCl_4 + 2LiOC^tBu_3 \xrightarrow{THF} \qquad (27.47)$$

[†] $\eta^3\text{-}[BH_3Me]^-$ is like $\eta^3\text{-}[BH_4]^-$: see structure **13.9**.

Self-study exercises

1. In $[UO_2I_2(OH_2)_2]$, the U atom lies on an inversion centre. Draw the structure of this complex.
   [*Ans.* See M.-J. Crawford *et al.* (2003) *J. Am. Chem. Soc.*, vol. 125, p. 11778]

2. What is the oxidation state and likely coordination number of the U centre in $[UO_2(phen)_3]^{2+}$? Suggest possible coordination geometries compatible with this coordination number.
   [*Ans.* See J.-C. Berthet *et al.* (2003) *Chem. Commun.*, p. 1660]

Plutonium

Oxidation states from +3 to +7 are available to plutonium, although the +7 state is known in only a few salts, e.g. Li_5PuO_6 has been prepared by heating Li_2O and PuO_2 in O_2. Hence, the potential diagram in Fig. 27.9 shows only oxidation states of +3 to +6. The chemistry of the +6 oxidation state is predominantly that of $[PuO_2]^{2+}$ although this is less stable with respect to reduction than $[UO_2]^{2+}$. The most stable oxide is PuO_2, formed when the nitrates or hydroxides of Pu in any oxidation state are heated in air. Although Pu forms PuF_6, it decomposes to PuF_4 and F_2, in contrast to the relative stability of UF_6. The highest binary chloride of Pu is $PuCl_3$, although $Cs_2[Pu^{IV}Cl_6]$ can be formed from CsCl, $PuCl_3$ and Cl_2 at 320 K.

In aqueous solution, $[PuO_2]^+$ is thermodynamically unstable (but only just) with respect to disproportionation reaction 27.48.

$$2[PuO_2]^+(aq) + 4H^+(aq)$$
$$\longrightarrow [PuO_2]^{2+}(aq) + Pu^{4+}(aq) + 2H_2O \qquad (27.48)$$

The closeness of the first three reduction potentials in the reduction of $[PuO_2]^{2+}$ (Fig. 27.9) is significant. If PuO_2 is dissolved in an excess of $HClO_4$ (an acid containing a very weakly coordinating anion) at 298 K, the solution at equilibrium contains Pu(III), Pu(IV), Pu(V) and Pu(VI). In redox systems involving Pu, however, equilibrium is not always attained rapidly. As for uranium, couples involving only electron transfer (e.g. $[PuO_2]^{2+}/[PuO_2]^+$) are rapidly reversible, but those also involving oxygen transfer (e.g. $[PuO_2]^+/Pu^{4+}$) are slower. Since hydrolysis and complex formation (the extents of which increase with increasing ionic charge, i.e. $[PuO_2]^+ < [PuO_2]^{2+} < Pu^{3+} < Pu^{4+}$) may also complicate the situation, the study of equilibria and kinetics in solutions of plutonium compounds is difficult.

The conventional means of entering plutonium chemistry is to dissolve the metal in aqueous HCl, HClO₄ or HNO₃. This generates a solution containing Pu(III). However, chloride and nitrate ions have the potential for coordination to the metal centre, and whereas $[ClO_4]^-$ is only weakly coordinating (see above), perchlorate salts have the disadvantage of being potentially explosive. A recent approach is to dissolve Pu metal in triflic acid (trifluoromethane sulfonic acid, CF_3SO_3H) to give Pu(III) as the isolable, crystalline salt $[Pu(OH_2)_9][CF_3SO_3]_3$. In the solid state, $[Pu(OH_2)_9]^{3+}$ has a tricapped trigonal prismatic structure with Pu−O bond distances of 247.6 (prism) and 257.4 pm (cap).

Self-study exercises

1. Complete the following scheme, inserting the missing nuclides and mode of decay:

$$^{238}_{92}U \xrightarrow{^{1}_{0}n} ? \xrightarrow{-\beta} ? \xrightarrow{?} {}^{239}_{94}Pu$$

2. PuO₂ crystallizes with a CaF₂-type structure. What are the coordination numbers of Pu and O in this structure?
[*Ans.* See Fig. 6.19]

3. When ^{239}Pu metal turnings are dissolved in MeCN in the presence of three equivalents of AgPF₆, a salt of $[Pu(NCMe)_9]^{3+}$ is isolated. What role does the AgPF₆ play in the reaction? Suggest a structure for $[Pu(NCMe)_9]^{3+}$.
[*Ans.* See A.E. Enriquez *et al.* (2003) *Chem. Commun.*, p. 1892]

27.11 Organometallic complexes of thorium and uranium

Although organometallic complexes are known for all the early actinoids, compounds of Th and U far exceed those of the other metals. In addition to radioactive properties, organoactinoids are air-sensitive and inert atmosphere techniques are required for their handling.

σ-Bonded complexes

Some difficulty was originally encountered in preparing homoleptic σ-bonded alkyl or aryl complexes of the actinoids, but (as for the lanthanoids, Section 27.8) use of the chelating ligand TMEDA ($Me_2NCH_2CH_2NMe_2$) was the key to stabilizing the Li⁺ salt of $[ThMe_7]^{3-}$ (eq. 27.49 and Fig. 27.10a). Similarly, hexaalkyls of type $Li_2UR_6 \cdot 7TMEDA$ have been isolated.

$$ThCl_4 + \text{excess MeLi} \xrightarrow{Et_2O, \text{ TMEDA}} [Li(TMEDA)]_3[ThMe_7]$$
(27.49)

$$(Me_3Si)_2HC \underset{CH(SiMe_3)_2}{\overset{U\cdots\cdots CH(SiMe_3)_2}{\diagdown}}$$

(27.16)

Bulky alkyl groups are also a stabilizing influence as illustrated by the isolation of $U\{CH(SiMe_3)_2\}_3$ (reaction 27.50). The solid contains *trigonal pyramidal* molecules (**27.16**). There are three short U---C_methyl contacts (309 pm),

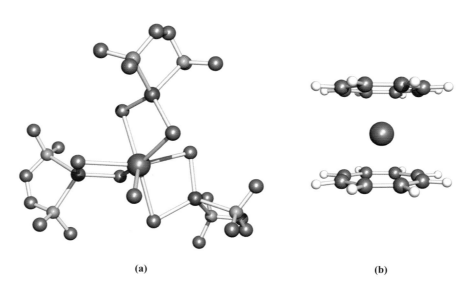

(a) **(b)**

Fig. 27.10 The structures (X-ray diffraction) of (a) $[Li(TMEDA)]_3[ThMe_7]$ showing the role of the TMEDA in stabilizing the structure (H atoms are omitted) [H. Lauke *et al.* (1984) *J. Am. Chem. Soc.*, vol. 106, p. 6841] and (b) $(\eta^8\text{-}C_8H_8)_2Th$ [A. Avdeef *et al.* (1972) *Inorg. Chem.*, vol. 11, p. 1083]. Colour code: Th, brown; Li, red; C, grey; N, blue; H, white.

and these may contribute to the deviation from planarity as in $Ln\{N(SiMe_3)_2\}_3$ (Fig. 27.6).

$$U(OC_6H_3\text{-}2,6\text{-}^tBu_2)_3 + 3LiCH(SiMe_3)_2$$

$$\longrightarrow U\{CH(SiMe_3)_2\}_3 + 3LiOC_6H_3\text{-}2,6\text{-}^tBu_2 \quad (27.50)$$

Alkyl derivatives are more stable if the actinoid metal is also bound to cyclopentadienyl ligands and reactions 27.51–27.53 show general methods of synthesis where M = Th or U.

$$(\eta^5\text{-}Cp)_3MCl + RLi \xrightarrow{Et_2O} (\eta^5\text{-}Cp)_3MR + LiCl \quad (27.51)$$

$$(\eta^5\text{-}Cp)_3MCl + RMgX \xrightarrow{THF} (\eta^5\text{-}Cp)_3MR + MgClX \quad (27.52)$$

$$(\eta^5\text{-}C_5Me_5)_2MCl_2 + 2RLi \xrightarrow{Et_2O} (\eta^5\text{-}C_5Me_5)_2MR_2 + 2LiCl \quad (27.53)$$

Cyclopentadienyl derivatives

Cyclopentadienyl derivatives are plentiful among organo-metallic complexes of Th(IV), Th(III), U(IV) and U(III), and reactions 27.54–27.57 give methods of synthesis for the main families of compounds (M = Th, U).

$$MCl_4 + 4KCp \xrightarrow{C_6H_6} (\eta^5\text{-}Cp)_4M + 4KCl \quad (27.54)$$

$$MX_4 + 3NaCp \xrightarrow{THF} (\eta^5\text{-}Cp)_3MX + 3NaX$$

$$X = Cl, Br, I \quad (27.55)$$

$$MX_4 + TlCp \xrightarrow{THF} (\eta^5\text{-}Cp)MX_3(THF)_2 + TlX$$

$$X = Cl, Br \quad (27.56)$$

$$(\eta^5\text{-}Cp)_3MCl + NaC_{10}H_8$$

$$\xrightarrow{THF} (\eta^5\text{-}Cp)_3M(THF) + NaCl + C_{10}H_8 \quad (27.57)$$

$(NaC_{10}H_8 = sodium\ naphthalide)$

Compounds of type $(\eta^5\text{-}Cp)_2MX_2$ are usually subject to a redistribution reaction such as 27.58 unless sterically hindered as in $(\eta^5\text{-}C_5Me_5)_2ThCl_2$ and $(\eta^5\text{-}C_5Me_5)_2UCl_2$.

$$2(\eta^5\text{-}Cp)_2UCl_2 \xrightarrow{THF} (\eta^5\text{-}Cp)_3UCl + (\eta^5\text{-}Cp)UCl_3(THF)_2 \quad (27.58)$$

Colourless $(\eta^5\text{-}Cp)_4Th$ and red $(\eta^5\text{-}Cp)_4U$ are monomeric in the solid state with pseudo-tetrahedral structures, **27.17** (Th−C = 287 pm, U−C = 281 pm). Tetrahedral structures are also observed for $(\eta^5\text{-}Cp)_3MX$ and $(\eta^5\text{-}Cp)_3M(THF)$ derivatives, while $(\eta^5\text{-}Cp)MX_3(THF)_2$ is octahedral. How to describe the metal–ligand bonding in these and other Cp derivatives of the actinoids is the subject of much theoretical debate. The current picture suggests involvement of the metal $6d$ atomic orbitals with the $5f$ orbitals being fairly unperturbed. Relativistic effects (see Box 13.3) also work in favour of a bonding role for the $6d$ rather than $5f$ atomic orbitals. Covalent contributions to the bonding appear to be present in Th(IV) and U(IV) cyclopentadienyl complexes, but for Th(III) and U(III), it is suggested that the bonding is mainly ionic.

A range of organometallic species can be made starting from $(\eta^5\text{-}Cp)_3ThCl$ and $(\eta^5\text{-}Cp)_3UCl$, and Fig. 27.11 shows selected reactions of $(\eta^5\text{-}Cp)_3UCl$. The hetero-metallic complex $(\eta^5\text{-}Cp)_3UFe(CO)_2(\eta^5\text{-}Cp)$ contains an unbridged U−Fe bond.

M = Th, U

(27.17)

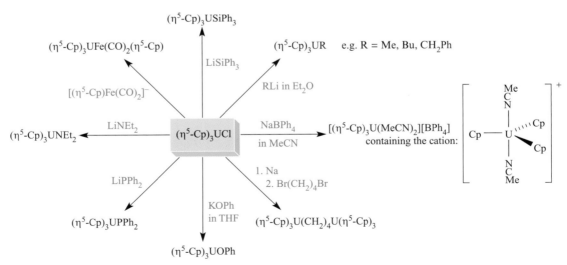

Fig. 27.11 Selected reactions of $(\eta^5\text{-}Cp)_3UCl$.

Self-study exercise

The sterically congested compound $(\eta^5\text{-}C_5Me_5)_3U$ reacts with two equivalents of PhCl to give $(\eta^5\text{-}C_5Me_5)_2UCl_2$, Ph_2 and $(C_5Me_5)_2$. This reaction is referred to as a 'sterically induced reduction'. Write a balanced equation for the overall reaction. What is being reduced in the reaction, and which two species undergo oxidation?

[*Ans.* see W.J. Evans *et al.* (2000) *J. Am. Chem. Soc.*, vol. 122, p. 12019]

Complexes containing the η^8-cyclooctatetraenyl ligand

As we have seen, the large U(IV) and Th(IV) centres accommodate up to four η^5-Cp$^-$ ligands, and ferrocene-like complexes are not observed. However, with the large $[C_8H_8]^{2-}$ ligand (reaction 27.59), sandwich complexes are formed by reaction 27.60.

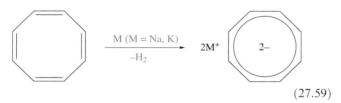

$$(27.59)$$

$$MCl_4 + 2K_2C_8H_8 \longrightarrow (\eta^8\text{-}C_8H_8)_2M + 4KCl$$

$$M = Th, U \qquad (27.60)$$

The green $(\eta^8\text{-}C_8H_8)_2U$ (*uranocene*) and yellow $(\eta^8\text{-}C_8H_8)_2Th$ (*thorocene*) are isostructural (Fig. 27.10b, mean Th$-$C $= 270$ pm and U$-$C $= 265$ pm). Bonding in these metallocenes is much studied by theorists, with arguments mirroring those discussed above for cyclopentadienyl

derivatives. Uranocene is flammable in air, but does not react with H_2O at 298 K. $(\eta^8\text{-}C_8H_8)_2Th$ is air-sensitive, is attacked by protic reagents and explodes when red hot.

(27.18)

Reaction of ThCl$_4$ with $(\eta^8\text{-}C_8H_8)_2Th$ in THF yields the half-sandwich $(\eta^8\text{-}C_8H_8)ThCl_2(THF)_2$, **27.18**, but the analogous U(IV) species is made by reaction 27.61, and the iodido derivative by reaction 27.62.

$$UCl_4 + C_8H_8 + 2NaH$$
$$\xrightarrow{\text{THF}} (\eta^8\text{-}C_8H_8)UCl_2(THF)_2 + 2NaCl + H_2 \quad (27.61)$$

$$(\eta^8\text{-}C_8H_8)_2U + I_2 \xrightarrow{\text{THF}} (\eta^8\text{-}C_8H_8)UI_2(THF)_2 + C_8H_8$$
$$(27.62)$$

The halides are useful synthons in this area of chemistry, e.g. reactions 27.63 and 27.64.

$$(\eta^8\text{-}C_8H_8)UCl_2(THF)_2 + 2NaN(SiMe_3)_2$$
$$\xrightarrow{\text{THF}} (\eta^8\text{-}C_8H_8)U\{N(SiMe_3)_2\}_2 + 2NaCl \quad (27.63)$$

$$(\eta^8\text{-}C_8H_8)UI_2(THF)_2 + 3LiCH_2SiMe_3$$
$$\xrightarrow{\text{THF}} [Li(THF)_3]^+[(\eta^8\text{-}C_8H_8)U\{CH_2SiMe_3\}_3]^- + 2LiI$$
$$(27.64)$$

KEY TERMS

The following terms have been introduced in this chapter. Do you know what they mean?

- ❑ lanthanoid
- ❑ actinoid
- ❑ transuranium element
- ❑ *f*-orbital
- ❑ lanthanoid contraction
- ❑ critical mass

FURTHER READING

H.C. Aspinall (2001) *Chemistry of f-Block Elements*, Gordon and Breach Scientific Publications, Amsterdam – An introductory general account of lanthanoids and actinoids.

S.A. Cotton (2006) *Lanthanide and Actinide Chemistry*, Wiley, New York – A good general introduction to the *f*-block elements.

A. Døssing (2005) *Eur. J. Inorg. Chem.*, p. 1425 – A review: 'Luminescence from lanthanide(3+) ions in solution'.

D.C. Hoffmann and D.M. Lee (1999) *J. Chem. Educ.*, vol. 76, p. 331 – 'Chemistry of the heaviest elements – one atom at a time' is an excellent article covering the development and future prospects of 'atom-at-a-time' chemistry of the transuranium elements.

D.C. Hoffmann, A. Ghiorso and G.T. Seaborg (2001) *The Transuranium People: The Inside Story*, Imperial College Press, London – A personalized account of the discovery and chemistry of the heaviest elements.

N. Kaltsoyannis and P. Scott (1999) *The f Elements*, Oxford University Press, Oxford – An OUP 'Primer' that complements the coverage in this chapter.

S.F.A. Kettle (1996) *Physical Inorganic Chemistry*, Spektrum, Oxford – Chapter 11 gives an excellent introduction to orbital, spectroscopic and magnetic properties of the *f*-block elements.

S.T. Liddle and D.P. Mills (2009) *Dalton Trans.*, p. 5592 – A short review entitled 'Metal–metal bonds in *f*-element chemistry'.

D. Parker (2004) *Chem. Soc. Rev.*, vol. 33, p. 156 – A review: 'Excitement in *f* block: structure, dynamics and function of nine-coordinate chiral lanthanide complexes in aqueous solution'.

P.W. Roesky (2003) *Z. Anorg. Allg. Chem.*, vol. 629, p. 1881 – A review: 'Bulky amido ligands in rare earth chemistry – synthesis, structures, and catalysis'.

G.T. Seaborg (1995) *Acc. Chem. Res.*, vol. 28, p. 257 – A review by one of the pioneers and Nobel Prize winner in the field: 'Transuranium elements: Past, present and future'.

G.T. Seaborg and W.D. Loveland (1990) *The Elements Beyond Uranium*, Wiley, New York – A text that covers syntheses of the elements, properties, experimental techniques and applications.

K. Thompson and C. Orvig, eds. (2006) *Chem. Soc. Rev.*, vol. 35, p. 499 – A themed issue of this journal focusing on the applications of lanthanoids in medicine.

Organometallic complexes

F.G.N. Cloke (1995) 'Zero oxidation state complexes of scandium, yttrium and the lanthanide elements' in *Comprehensive Organometallic Chemistry II*, eds G. Wilkinson, F.G.A. Stone and E.W. Abel, Pergamon, Oxford, vol. 4, p. 1.

S.A. Cotton (1997) *Coord. Chem. Rev.*, vol. 160, p. 93 – 'Aspects of the lanthanide–carbon σ-bond'.

F.T. Edelmann (2007) 'Complexes of group 3 and lanthanide elements' in *Comprehensive Organometallic Chemistry III*, eds R.H. Crabtree and D.M.P. Mingos, Elsevier, Oxford, vol. 4, p. 1.

F.T. Edelmann (2007) 'Complexes of actinide elements' in *Comprehensive Organometallic Chemistry III*, eds R.H. Crabtree and D.M.P. Mingos, Elsevier, Oxford, vol. 4, p. 191.

W.J. Evans (1985) *Adv. Organomet. Chem.*, vol. 24, p. 131 – 'Organometallic lanthanide chemistry'.

C.J. Schaverien (1994) *Adv. Organomet. Chem.*, vol. 36, p. 283 – 'Organometallic chemistry of the lanthanides'.

PROBLEMS

27.1 (a) What is the *lanthanoid contraction*? (b) Explain how the lanthanoids can be separated from their ores.

27.2 Use Hund's rules to derive the ground state of the Ce^{3+} ion, and calculate its magnetic moment. (The spin–orbit coupling constant for Ce^{3+} is $1000\,cm^{-1}$ and so the population of states other than the ground state can be neglected at $298\,K$.)

27.3 Show that the stability of a lanthanoid dihalide LnX_2 with respect to disproportionation into LnX_3 and Ln is greatest for $X = I$.

27.4 How would you attempt to show that a given lanthanoid diiodide, LnI_2, has saline rather than metallic character?

27.5 Comment on each of the following observations:

(a) ΔH° for the formation of $[Ln(EDTA)(OH_2)_x]^-$ ($x = 2$ or 3) in aqueous solution is nearly constant for all Ln and is almost zero.

(b) The value of E° for the Ce(IV)/Ce(III) couple (measured at pH 0) decreases along the series of acids $HClO_4$, HNO_3, H_2SO_4, HCl.

(c) $BaCeO_3$ has a perovskite structure.

27.6 Comment on the observations that the electronic spectra of lanthanoid complexes contain many absorptions some of which are weak and sharp and similar to those of the gas-phase metal ions, and some of which are broad and are affected by the ligands present.

27.7 Discuss the variation in coordination numbers among complexes of the $4f$ metals.

27.8 The reactions of $Ln(NCS)_3$ with $[NCS]^-$ under varying conditions lead to discrete anions such as $[Ln(NCS)_6]^{3-}$, $[Ln(NCS)_7(OH_2)]^{4-}$ and $[Ln(NCS)_7]^{4-}$. What can you say about possible structures for these species?

27.9 (a) Give a brief account of the formation of $Ln-C$ σ-bonds and of complexes containing cyclopentadienyl ligands, and comment on the roles of coordinating solvents.

(b) Suggest products for the reactions of $SmCl_3$ and SmI_2 with $K_2C_8H_8$.

27.10 (a) By considering Fig. 27.9, suggest a method for the separation of Am from U, Np and Pu.

(b) What would you expect to happen when a solution of $NpO_2(ClO_4)_2$ in $1 M$ $HClO_4$ is shaken with Zn amalgam and the resulting liquid decanted from the amalgam and aerated?

27.11 $25.00 cm^3$ of a solution **X** containing $21.4 g$ U(VI) dm^{-3} was reduced with Zn amalgam, decanted from the amalgam, and after being aerated for 5 min, was titrated with $0.1200 mol dm^{-3}$ Ce(IV) solution; $37.5 cm^3$ of the latter was required for reoxidation of the uranium to U(VI). Solution **X** $(100 cm^3)$ was then reduced and aerated as before, and treated with an excess of dilute aqueous KF. The resulting precipitate (after drying at $570 K$) weighed $2.826 g$. Dry O_2 was passed over the precipitate at $1070 K$, after which the solid product weighed $1.386 g$. This product was dissolved in water, the fluoride in the solution precipitated as PbClF, $2.355 g$ being obtained. Deduce what you can concerning the chemical changes in these experiments.

27.12 Suggest likely products in the following reactions: (a) UF_4 with F_2 at $570 K$; (b) Pa_2O_5 with $SOCl_2$ followed by heating with H_2; (c) UO_3 with H_2 at $650 K$; (d) heating UCl_5; (e) UCl_3 with $NaOC_6H_2$-2,4,6-Me_3.

27.13 What structural features would you expect in the solid state of (a) $Cs_2[NpO_2(acac)_3]$, (b) $Np(BH_4)_4$, (c) the guanidinium salt of $[ThF_3(CO_3)_3]^{5-}$, (d) $Li_3[LuMe_6] \cdot 3DME$, (e) $Sm\{CH(SiMe_3)_2\}_3$ and (f) a complex that analyses as having the composition $[UO_2][CF_3SO_3]_2 \cdot 2(18\text{-crown-}6) \cdot 5H_2O$?

27.14 Identify isotopes **A–F** in the following sequence of nuclear reactions:

(a) $^{238}U \xrightarrow{(n, \gamma)} A \xrightarrow{-\beta^-} B \xrightarrow{-\beta^-} C$

(b) $D \xrightarrow{-\beta^-} E \xrightarrow{(n, \gamma)} {}^{242}Am \xrightarrow{-\beta^-} F$

27.15 Identify the starting isotopes **A–E** in each of the following syntheses of transactinoid elements:

(a) $A + {}_2^4He \longrightarrow {}_{101}^{256}Md + n$

(b) $B + {}_8^{16}O \longrightarrow {}_{102}^{255}No + 5n$

(c) $C + {}_5^{11}B \longrightarrow {}_{103}^{256}Lr + 4n$

(d) $D + {}_8^{18}O \longrightarrow {}_{104}^{261}Rf + 5n$

(e) $E + {}_8^{18}O \longrightarrow {}_{106}^{263}Sg + 4n$

27.16 Discuss the following statements:

(a) Thorium forms iodides of formulae ThI_2, ThI_3 and ThI_4.

(b) In the solid state, salts of $[UO_2]^{2+}$ contain a linear cation.

(c) Reactions of NaOR with UCl_4 lead to monomeric $U(OR)_4$ complexes.

27.17 (a) What Th-containing products would you expect from the reactions of $(\eta^5\text{-Cp})_3ThCl$ with (i) $Na[(\eta^5\text{-Cp})Ru(CO)_2]$, (ii) LiCHMeEt, (iii) $LiCH_2Ph$? (b) What advantage does $(\eta^5\text{-}C_5Me_5)_2ThCl_2$ have over $(\eta^5\text{-Cp})_2ThCl_2$ as a starting material? (c) How might $(\eta^5\text{-}C_5Me_5)UI_2(THF)_3$ react with $K_2C_8H_8$?

27.18 (a) Suggest a method of preparing $U(\eta^3\text{-}C_3H_5)_4$. (b) How might $U(\eta^3\text{-}C_3H_5)_4$ react with HCl? (c) $(\eta^5\text{-}C_5Me_5)(\eta^8\text{-}C_8H_8)ThCl$ is dimeric, but its THF adduct is a monomer. Draw the structures of these compounds, and comment on the role of coordinating solvents in stabilizing other monomeric organothorium and organouranium complexes.

27.19 Discuss the following:

(a) Many actinoid oxides are non-stoichiometric, but few lanthanoid oxides are.

(b) The ion $[NpO_6]^{5-}$ can be made in aqueous solution only if the solution is strongly alkaline.

(c) A solution containing Pu(IV) undergoes negligible disproportionation in the presence of an excess of molar H_2SO_4.

27.20 Give a short account of aspects of the organometallic compounds formed by the lanthanoids and actinoids and highlight major differences between families of organometallic complexes of the *d*- and *f*-block metals.

OVERVIEW PROBLEMS

27.21 Comment on each of the following statements.

(a) Ln^{2+} complexes are strong reducing agents.

(b) In the solid state, $Cp_2YbF(THF)$ exists as a bridged dimer, while $Cp_2YbCl(THF)$ and $Cp_2YbBr(THF)$ are monomeric.

(c) In $[Th(NO_3\text{-}O,O')_3\{(C_6H_{11})_2SO\}_4]^+[Th(NO_3\text{-}O,O')_5\{(C_6H_{11})_2SO\}_2]^-$, the sulfoxide ligands are O-rather than S-bonded.

27.22 (a) The reaction of $ScCl_3 \cdot nTHF$ with one equivalent of ligand **27.19** yields a neutral compound **A** in which the metal is octahedrally sited. **A** reacts with three equivalents of MeLi to give **B**. Suggest structures for **A** and **B**. What is the oxidation state of the metal in each compound?

(27.19)

(b) The complex $[(\eta^5\text{-}C_5H_5)_2La\{C_6H_3\text{-}2,6\text{-}(CH_2NMe_2)_2\}]$ is 5-coordinate. Suggest, with reasoning, a structure for the complex.

27.23 (a) Table 27.3 lists the 'calculated' value of μ_{eff} for Eu^{3+} as 0. On what basis is this value calculated? Explain why *observed* values of μ_{eff} for Eu^{3+} are greater than zero.

(b) The complex $UO_2Cl_2(THF)_3$ contains *one* labile THF ligand and readily forms a diuranium complex, **A**, that contains 7-coordinate U(VI) with *trans*-UO_2 units. **A** is a precursor to a number of mononuclear complexes. For example, one mole of **A** reacts with four moles of $K[O\text{-}2,6\text{-}{}^tBu_2C_6H_3]$ to give two moles of **B**, and with four moles of Ph_3PO eliminating all

coordinated THF to yield two moles of **C**. Suggest identities for **A**, **B** and **C** and state the expected coordination environment of the U(VI) centre in each product.

27.24 (a) Compound **27.20** reacts with MeLi with loss of CH_4 to give **A**. When **A** reacts with $TbBr_3$, a terbium-containing complex **B** is formed, the mass spectrum of which shows an envelope of peaks at m/z 614 as the highest mass peaks. Suggest identities for **A** and **B**, and give a possible structure for **B**. Explain how the appearance of envelope of peaks at m/z 614 in the mass spectrum confirms the number of Br atoms in the product (*hint*: see Appendix 5).

(27.20)

(b) Ligand **27.21** in a mixed EtOH/MeOH solvent system extracts Pu(IV) from aqueous HNO_3. The 10-coordinate complex $[Pu(\mathbf{27.21})_2(NO_3)_2]^{2+}$ has been isolated from the EtOH/MeOH extractant as a nitrate salt. Suggest how ligand **27.21** might coordinate to Pu(IV), and state how you expect the coordination number of 10 to be achieved.

(27.21)

INORGANIC CHEMISTRY MATTERS

27.25 Vasovist® is the tradename of a Gd(III) complex which was the first intravascular contrast agent (see Box 4.3) approved in the EU for use in magnetic resonance angiography. Interactions between the lipophilic domain of the contrast agent and human serum albumin allow Vasovist® to bind reversibly to the protein. Vasovist® has the formula $Na_3[GdL(OH_2)]$ where the ligand L^{n-} is derived from H_nL shown in Fig. 27.12. (a) What is the value of n in L^{n-} and H_nL? (b) Explain why one part of the ligand is described as being lipophilic. (c) In $Na_3[GdL(OH_2)]$, the Gd^{3+} ion is 9-coordinate. Suggest a possible coordination environment for the metal ion, and indicate how

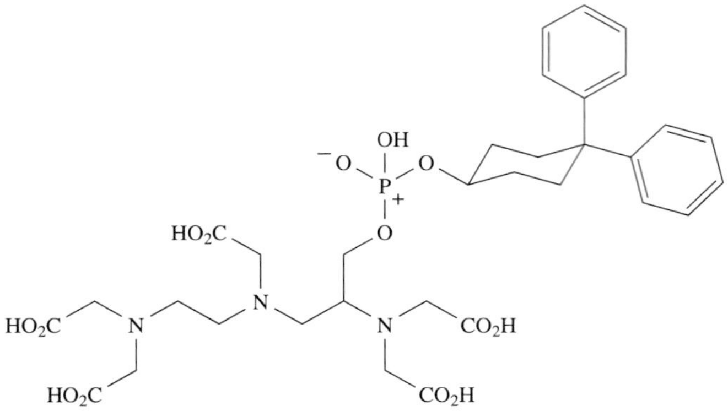

Fig. 27.12 For problem 27.25: the structure of ligand H$_n$L.

L^{n-} is likely to coordinate to Gd^{3+}. (d) Why is H$_n$L chiral? (e) Starting from (*R*)-H$_n$L, four diastereoisomers of [GdL(OH$_2$)]$^{3-}$ are possible: $\Delta R,R$, $\Delta R,S$, $\Lambda R,R$, $\Lambda R,S$. How do these arise? (f) The FAB mass spectrum of Na$_3$[GdL(OH$_2$)] shows peak envelopes at *m/z* 981, 959, 937 and 915. Suggest assignments for these peaks.

27.26 Lanthanide metal ions, Ln^{3+}, can exchange with, and mimic the function of, Ca^{2+} ions in the human body. Lanthanum carbonate is administered as chewable tablets under the tradename of Fosrenol$^{®}$ to patients with particularly high levels of phosphate ions in the blood. However, there are significant gastrointestinal side-effects, and more soluble forms of lanthanide-containing drugs are being sought. Recent research in this field has investigated the reactions of Ln(NO$_3$)$_3$·6H$_2$O with Hma (defined below) in the presence of NaOH or Et$_3$N.

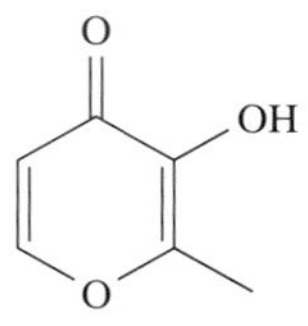

Hma

(a) Why are ligands with *O*-donor atoms chosen to bind Ln^{3+}? (b) What is the reason for adding base to the reaction mixture? (c) The highest mass peaks in the electrospray ionization mass spectra of the complexes formed with La^{3+} and Eu^{3+} are at *m/z* 537 and 551, respectively. How do these peaks arise? (d) Elemental analytical data for the Eu^{3+} complex are C 39.65, H 3.14%. What can you deduce from these data? Suggest a structure for the europium(III) complex. (e) Complexes were formed with La^{3+}, Eu^{3+}, Gd^{3+}, Tb^{3+} and Yb^{3+}, but ^{1}H NMR spectroscopic data were only reported for the La^{3+} complex. Suggest a reason for this. (f) The Ln^{3+} complexes were subjected to hydroxyapatite-binding studies. Comment on the reasons for carrying out such investigations.

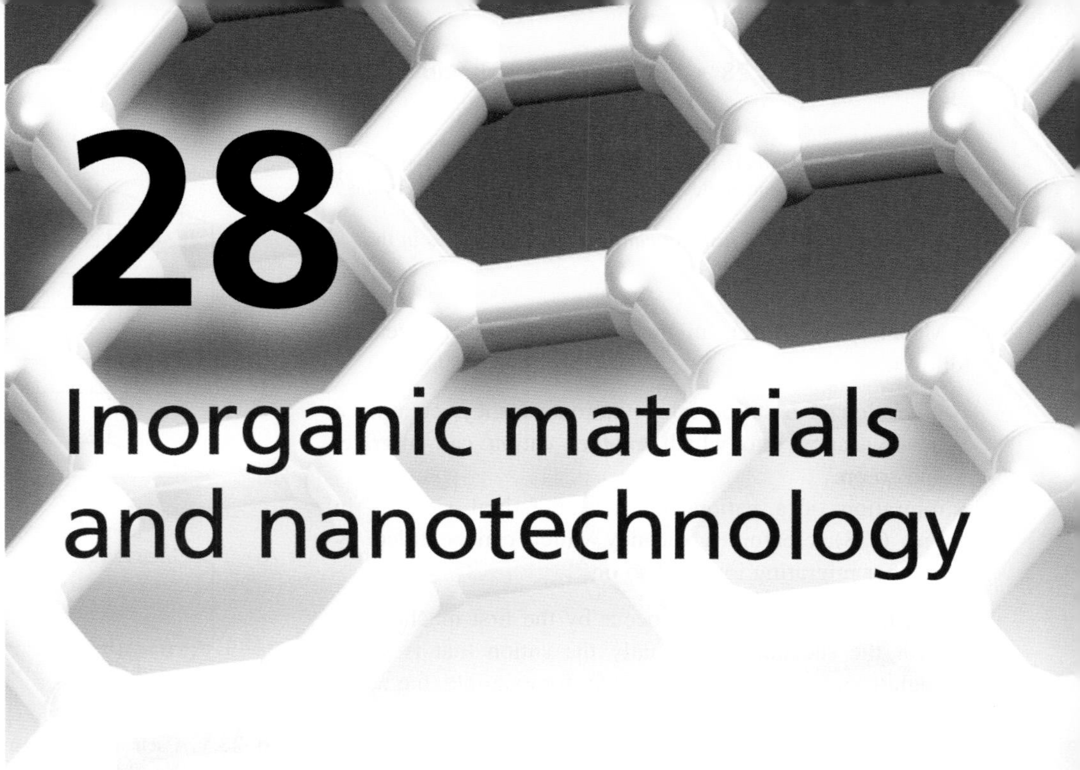

Topics

Ion conductors
Transparent conducting
  oxides
Superconductors
Chemical vapour
  deposition
Inorganic fibres
Graphene
Carbon nanotubes

28

Inorganic materials and nanotechnology

28.1 Introduction

There is intense current interest in developing new inorganic materials, and solid state, polymer and nano-chemistries are 'hot' areas of research. We have already encountered many structural aspects of the solid state and have exemplified applications of solid state materials, e.g. magnetic properties of metal oxides (Chapters 21 and 22), semiconductors (Chapter 6) and heterogeneous catalysts (Chapter 25). The following topics appeared in Chapter 6:

- structures of metals;
- polymorphism;
- alloys;
- band theory;
- semiconductors;
- prototype ionic lattices;
- lattice energies and their applications in inorganic chemistry;
- lattice defects;
- colour centres.

With the exception of semiconducting materials, these will not be discussed further here. In Section 20.10, we introduced some concepts of magnetism including *ferromagnetism*, *antiferromagnetism* and *ferrimagnetism*. Although these properties are important in materials chemistry, it is beyond the scope of this book to take this topic further.

The topics chosen for inclusion in this chapter reflect areas of active interest and elaborate upon some topics that have been given only brief mention in earlier chapters. In describing the chemistries of the *d*- and *f*-block metals in Chapters 21, 22 and 27, we included many examples of solid state compounds. We now look further at electrically conducting and superconducting materials. At various points in the book, we have mentioned colour pigments in ceramic materials when describing applications of inorganic compounds (mainly oxides). Section 28.5 looks at the colouring of ceramics in more detail. We describe *chemical vapour deposition* (CVD) for the formation of thin films of materials and its application in the semiconductor industry, and in the final two sections, we discuss inorganic fibres, graphene sheets and carbon nanotubes. Throughout the chapter, we emphasize commercial applications in order to exemplify the role that inorganic chemistry plays in technological developments. In contrast to other chapters, applications are not highlighted specifically in boxed material.

28.2 Electrical conductivity in ionic solids

Ionic *solids* usually have a high electrical resistance (low conductivity, see Section 6.8) and the conductivity is significant only when the compound is molten. The presence of defects in an ionic solid decreases the resistance, e.g. cubic zirconia stabilized with CaO (see above) is a *fast-ion conductor*, the conductivity arising from the migration of O^{2-} ions. An increased concentration of defects can be introduced by heating a solid to a high temperature and then cooling it rapidly. Since more defects are present at high temperatures, the effect of quenching the solid is to 'freeze' the defect concentration present at elevated temperature.

The presence of defects in a crystal lattice facilitates ion migration and enhances electrical conductivity (i.e. lowers the resistance).

The conductivity of a **fast-ion conductor** typically lies in the range 10^{-1} to $10^3 \ \Omega^{-1} \ m^{-1}$.

Mechanisms of ion migration can be categorized as follows:

- migration of a cation into a cation vacancy, creating a new vacancy into which another cation can migrate, and so on;
- migration of a cation into an interstitial site (as in Fig. 6.28), creating a vacancy which can be filled by another migrating ion, and so on.

Anion migration could also occur by the first mechanism, but for the second, it is usually the cation that is small enough to occupy an interstitial site, for example, the tetrahedral holes in an NaCl-type structure.

For an ionic solid to be a fast-ion conductor, it must meet some or all of the following criteria:

- it must contain mobile ions;
- the charges on the ions should be low (multiply-charged ions are less mobile than singly-charged ions);
- it must contain vacant holes between which the ions can move;
- the holes must be interconnected;
- the activation energy for the movement of an ion from one hole to the next must be low;
- the anions in the solid should be polarizable.

Sodium and lithium ion conductors

Current developments in battery technology, electrochromic devices (see Boxes 11.3 and 22.4) and research into electrically powered vehicles make use of solid electrolytes. The sodium/sulfur battery contains a solid β-alumina electrolyte. The name *β-alumina* is misleading since it is prepared by the reaction of Na_2CO_3, $NaNO_3$, NaOH and Al_2O_3 at 1770 K and is a non-stoichiometric compound of approximate composition $Na_2Al_{22}O_{34}$ (or $Na_2O \cdot 11Al_2O_3$), always containing an excess of Na^+; we therefore refer to this material as *Na β-alumina*. Equation 28.1 shows the half-reactions that occur in the sodium/sulfur battery. Na^+ ions produced at the anode migrate through the Na β-alumina electrolyte and combine with the polysulfide anions formed at the cathode (eq. 28.2). The reactions are reversed when the cell is recharged.

At the anode: $\quad Na \longrightarrow Na^+ + e^-$
At the cathode: $\quad nS + 2e^- \longrightarrow [S_n]^{2-}$ $\qquad$ (28.1)

$2Na^+ + [S_n]^{2-} \longrightarrow Na_2S_n$ $\qquad$ (28.2)

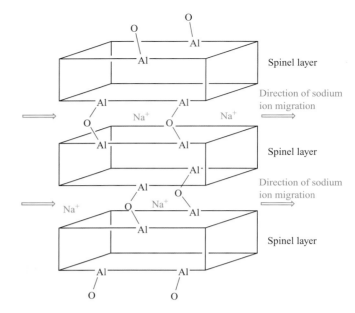

Fig. 28.1 A schematic representation of part of the structure of Na β-alumina ($Na_2O \cdot 11Al_2O_3$) in which Na^+ ions are mobile between bridged layers of Al_2O_3 spinel structure. Spinels were introduced in Box 13.7.

The Na β-alumina acts as a *sodium ion conductor*. The key to this property lies in its structure which consists of spinel-type layers 1123 pm thick, with Na^+ ions occupying the interlayer spaces (Fig. 28.1). The conductivity of Na β-alumina ($3 \ \Omega^{-1} \ m^{-1}$) arises from the ability of the Na^+ ions to migrate through the gaps between the spinel layers. It therefore conducts in one plane through the crystal in the same way that graphite conducts only in the plane parallel to the carbon-containing planes (Fig. 14.4a). Although the conductivity of Na β-alumina is small compared with a metal (Fig. 6.10), it is large compared with typical ionic solids (e.g. $10^{-13} \ \Omega^{-1} \ m^{-1}$ for solid NaCl). The Na^+ ions in Na β-alumina can be replaced by cations such as Li^+, K^+, Cs^+, Rb^+, Ag^+ and Tl^+. However, the conductivities of these materials are lower than that of Na β-alumina: the match between the size of the Na^+ ions and the interlayer channels in the host lattice leads to the most efficient cation mobility. The conductivities of Na β-alumina and selected cation and anion conductors exhibiting relatively high (i.e. in the context of ionic solids) conductivities are compared in Fig. 28.2.

The fact that silver iodide is an ion conductor was first observed in 1914 by Tubandt and Lorentz. They noted that when a current was passed between Ag electrodes separated by solid AgI, the mass of the electrodes changed. At room temperature, AgI can exist in either the β- or γ-form. On heating to 419 K, both phases transform to α-AgI, and this is accompanied by a dramatic increase in ion conductivity. In α-AgI, the I^- ions are in a body-

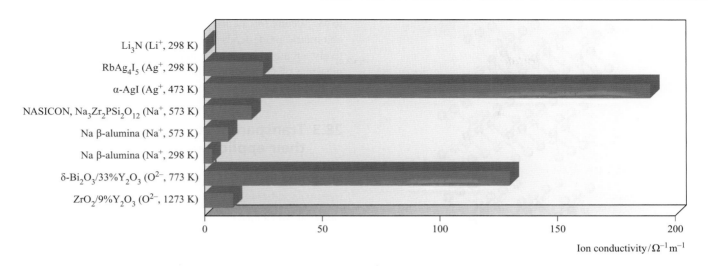

Fig. 28.2 Ion conductivities of selected ionic solids at the temperature stated. The ion given in parentheses after the solid electrolyte is the ion conductor.

centred cubic arrangement. The Ag^+ ions randomly occupy tetrahedral holes (Fig. 28.3). For each unit cell, there are 12 tetrahedral holes, only some of which need to be occupied to maintain a $1:1\ Ag^+:I^-$ stoichiometry. Two factors contribute to the high ionic conductivity of α-AgI:

- the migration of Ag^+ ions between tetrahedral holes has a low activation energy ($\approx 4.8\ kJ\ mol^{-1}$);
- Ag^+ ions are polarizing and I^- ions are highly polarizable.

The second most highly ion-conducting solid in Fig. 28.2 is Bi_2O_3 doped with Y_2O_3. Solid Bi_2O_3 has intrinsic O^{2-} vacancies. Above 1003 K, the stable phase is δ-Bi_2O_3 and the presence of disordered O^{2-} vacancies leads to a high ion (O^{2-}) conductivity. By doping Bi_2O_3 with Y_2O_3, the δ-form becomes stable at lower temperatures.

Materials of composition $Na_{1+x}Zr_2P_{3-x}Si_xO_{12}$ ($0 \leq x \leq 3$) are known as *Na super-ionic conductors* (NASICON) and also have potential application in sodium/sulfur batteries. They comprise solid solutions of $NaZr_2(PO_4)_3$ (the host lattice) and $Na_4Zr_2(SiO_4)_3$. The former adopts a structure composed of corner-sharing ZrO_6 octahedra and PO_4 tetrahedra which generate a network through which channels run. Once $Na_4Zr_2(SiO_4)_3$ is incorporated to give a solid solution, the Na^+ ion conductivity increases and is optimized to $20\ \Omega^{-1}\ m^{-1}$ (at 573 K) when $x = 2$.

Solid electrolytes with applications in lithium batteries include Li_7NbO_6, $Li_{12}Ti_{17}O_{40}$, Li_8ZrO_6 and Li_3N which are *Li^+ ion conductors* (see also Box 11.3). In Li_7NbO_6, one-eighth of the Li^+ sites are vacant, rendering the solid a good ionic conductor. Lithium nitride has the layer structure shown in Fig. 11.4a. A 1–2% deficiency of Li^+ ions in the hexagonal layers leads to Li^+ conduction *within* these layers, the interlayer Li^+ ion sites remaining fully occupied. Lithium nitride is used as the solid electrolyte in cells containing a Li electrode coupled with a TiS_2, TaS_2 or other metal sulfide electrode. The metal sulfide must possess a layered structure as shown for TaS_2 in Fig. 28.4. During battery discharge, Li^+ ions flow through the solid Li_3N barrier and into the MS_2 solid. This acts as a host lattice (eq. 28.5), intercalating the Li^+ ions between the layers.

$$xLi^+ + TiS_2 + xe^- \xrightleftharpoons[\text{charge}]{\text{discharge}} Li_xTiS_2 \qquad (28.3)$$

Self-study exercise

A solid oxide fuel cell consists of a cathode (at which O_2 from the air is reduced to O^{2-}) and an anode (at which H_2 is oxidized), and an electrolyte. The cell operates at $\approx 1300\ K$. Explain why stabilized zirconia ZrO_2 doped with CaO or Y_2O_3 is a suitable material for the electrolyte. What is the product at the anode?

[*Ans.* See Section 28.2, Fig. 28.2 and Box 10.2]

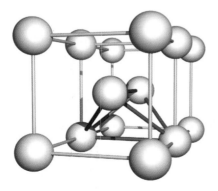

Fig. 28.3 Two unit cells of body-centred packed ions, showing one tetrahedral hole (red lines).

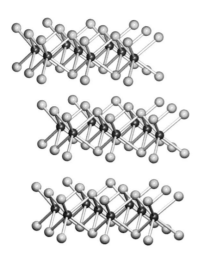

Fig. 28.4 Part of the layered structure of TaS_2. Colour code: Ta, red; S, yellow.

d-Block metal(II) oxides

In Chapter 21, we described the chemistry of the first row *d*-block metal(II) oxides, but said little about their electrical conductivity properties. The oxides TiO, VO, MnO, FeO, CoO and NiO adopt NaCl-type structures, but are non-stoichiometric, being metal-deficient as exemplified for TiO and FeO in Section 6.17. In TiO and VO, there is overlap of the metal t_{2g} orbitals giving rise to a partially occupied band (Fig. 28.5) and, as a result, TiO and VO are electrically conducting. In contrast, MnO is an insulator at 298 K but is semiconducting (see Section 6.9) at higher temperatures. FeO, CoO and NiO behave similarly, their conductivities (which are low at 298 K) increasing with temperature. The conductivity is explained in terms of an *electron-hopping* mechanism, in which an electron moves from an M^{2+}

to M^{3+} centre (recall that the metal-deficiency in the non-stoichiometric oxide leads to the presence of M^{3+} sites), effectively creating a positive hole. Heating the oxide in the presence of O_2 results in further M^{2+} ions being oxidized. In turn, this facilitates electron migration through the solid.

28.3 Transparent conducting oxides and their applications in devices

Sn-doped In_2O_3 (ITO) and F-doped SnO_2 (FTO)

> A ***transparent conducting oxide*** (TCO) is a semiconducting material with a high electrical conductivity and a high optical transparency.

Transparent conducting oxides are remarkable materials that combine high electrical conductivity ($\approx 10^4 \ \Omega^{-1} \ cm^{-1}$) with high optical transparency (>80% transmission of visible light). They are used commercially as transparent electrodes in flat-screen displays, electrochromic windows (see Box 22.4) and other opto-electronic devices. Two of the most important TCOs are tin-doped In_2O_3 (ITO) and fluorine-doped SnO_2 (FTO) in which charge carriers are created by replacing In(III) by Sn(IV), or O^{2-} by F^-, respectively. The charge carrier concentration is high (e.g. in ITO, the concentration of Sn(IV) centres is $\approx 10^{20} \ cm^{-3}$). This level of doping produces a *degenerate semiconductor* in which the Fermi level has moved into the conduction band of the host, resulting in metallic-like electrical conductivity. In order to achieve optical transmission up to the near-UV, the band gap must be $\geq 3 \ eV$. Transparent conducting oxides are basic components of photovoltaic and solid state lighting devices, and are exemplifed below by dye-sensitized solar cells (DSCs or Grätzel cells), organic light-emitting diodes (OLEDs) and light-emitting electrochemical cells (LECs).

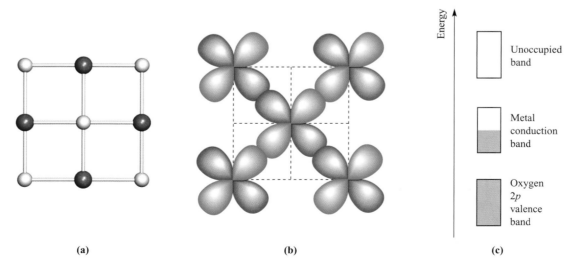

Fig. 28.5 (a) One face of the unit cell of TiO; colour code: Ti, pale grey; O, red. (b) Overlap of the Ti t_{2g} orbitals occurs and leads to (c) the formation of a partly filled metal conduction band.

Self-study exercise

Oxygen atom vacancies in In_2O_3 result in the oxide being an n-type semiconductor but the electrical conductivity is greatly increased by doping with SnO_2. How do the charge carriers differ in In_2O_3 and ITO?

Dye-sensitized solar cells (DSCs)

Traditional photovoltaic devices are silicon-based (see Box 14.2). In the Grätzel or dye-sensitized solar cell, first developed in the early 1990s, solar energy is converted to electrical energy using a wide-band gap semiconductor in contact with an electrolyte. The semiconductor is usually nanocrystalline anatase (a phase of TiO_2) which is optically transparent, and an inorganic dye (the *sensitizer*) is bound onto the semiconductor surface to absorb photons. Figure 28.6 shows the structures of two commonly used dyes in which the carboxylic acid or carboxylate acts as the surface-binding group. A typical DSC (Fig. 28.7) is constructed from two glass plates, each coated with a thin film of a TCO. One plate is the working electrode and is coated with a layer of mesoporous TiO_2 nanoparticles which support the dye. The second glass plate is the counter electrode and is coated with a thin film of platinum

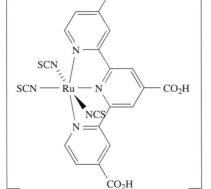

Fig. 28.6 Two ruthenium(II) complexes which are used in dye-sensitized solar cells (DSCs). The upper complex is known as 'N3', and $[Bu_4N]^+$ salts of the lower complex are called the 'black dye'.

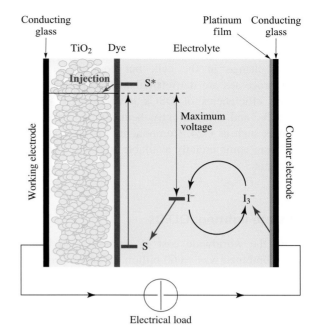

Fig. 28.7 Schematic representation of a dye-sensitized solar cell. S and S* are the ground and photoexcited states of the sensitizer. The red arrows represent electron transfer. After electron injection into the conduction band of the semiconductor, the reduced form of the dye is regenerated by oxidation of I^- in the electrolyte, and the reduced form of the latter is regenerated at the counter electrode by electrons passed around the outer circuit.

which acts as an electrical contact. The two plates are sandwiched together and the gap between them filled with an electrolyte; a typical choice contains the I^-/I_3^- redox couple. The sensitizer is designed to absorb light across as wide a spectrum as possible. The electronic absorption spectrum of the so-called 'black dye' (Fig. 28.6) has an extremely broad MLCT band which covers most of the visible region.[†] Upon irradiation, the dye is promoted from the ground state, S, to an excited state, S*, and injects an electron into the conduction band of the semiconductor (left-hand side of Fig. 28.7). After electron injection, the dye has been oxidized, i.e. the dyes shown in Fig. 28.6 contain Ru(III). The Ru(II) state is regenerated by reduction with I^-:

$$2Ru^{3+} + 3I^- \longrightarrow 2Ru^{2+} + I_3^-$$

Diffusion of I_3^- ions to the counter electrode (see below) must be rapid. Electron transport through the semiconductor to the working electrode must be faster than charge recombination at the semiconductor–sensitizer interface. Electrons flow around the outer circuit, performing electrical work. On reaching the counter electrode, electrons are transferred into the electrolyte, regenerating I^- (right-hand side of Fig. 28.7):

$$I_3^- + 2e^- \longrightarrow 3I^-$$

[†] The spectrum is pH dependent, see: Md.K. Nazeeruddin *et al.* (2001) *J. Am. Chem. Soc.*, vol. 123, p. 1613.

The I$^-$ ions must diffuse rapidly to the TiO$_2$ surface to be available for reduction of the oxidized dye. The device converts solar energy to electrical energy without a net chemical change in the cell, but competitive electron transfer processes are detrimental to the efficiency of the cell. The efficiencies of DSCs have (in 2011) reached about 11%, and are currently being commercialized by companies such as G24. Components for the construction of DSCs are commercially available from companies such as Solaronix.

Solid state lighting: OLEDs

In 2010, the worldwide cost of domestic and industrial lighting combined was ≈€60 billion. In developed nations, around 20% of electricity is used for lighting, but much of the energy in conventional (incandescent) lighting is lost as heat. Legislation aims to increase the efficiency of lighting, and the market is currently moving from incandescent to halogen and solid state lighting. It is estimated that in 2013, the global market for solid state lighting will reach €14 billion.

Solid state lighting is currently dominated by light-emitting diodes (LEDs, see Section 28.6). LEDs are semiconductor junction devices constructed from inorganic materials such as GaAs$_{1-x}$P$_x$ and In$_x$Ga$_{1-x}$As. In contrast, organic light-emitting diodes (OLEDs) which are now entering the market are fabricated from organic materials and/or metal complexes containing organic ligands. The performances of both LEDs and OLEDs are steadily increasing. LEDs are now manufactured in all the primary colours, giving a means of producing both coloured and white light. OLED displays can be produced as flat panels (Fig. 28.8) or may be printed onto substrates. They have the advantage over traditional LEDs of working at low driving voltage, and being flexible and compatible with large areas.

Fig. 28.8 OLED lighting is a very promising new and efficient flat light source based on organic compounds. The white lighting panel in the photograph is 15 × 15 cm, and was developed by chemists in Europe.

An *electroluminescent* material emits light when a direct current is passed through the material and electrons (injected from the cathode) and holes (injected from the anode) recombine.

An OLED device (Fig. 28.9a) contains one or more layers of neutral, electroluminescent materials which are sandwiched between a pair of electrodes, one of which must

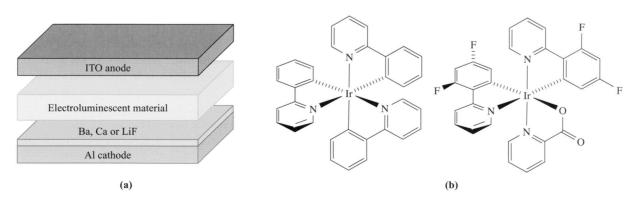

(a)   **(b)**

Fig. 28.9 (a) Schematic diagram of the components of an OLED. In the device, the layers are in contact with each other. The central layer is 60–100 nm thick. (b) Examples of coordination complexes used in OLEDs.

be optically transparent (i.e. a TCO). The material for the anode (usually ITO) is chosen so that its work function (i.e. the minimum energy required to remove an electron from the Fermi level) is close in energy to that of the HOMO of the electroluminescent material. The cathode must have a low work function that matches the energy of the LUMO of the electroluminescent material. The need for a low work function means that reactive metals such as Ca or Ba are used for the cathode and, hence, the OLED device must be encapsulated to prevent degradation in the air. When a direct current is passed through the device, electrons are injected from the cathode and populate the LUMO of the material. At the anode, electrons are removed from the HOMO. This can also be considered in terms of the injection of 'holes' into the HOMO level. Electrons and holes migrate through the material by a hopping mechanism:

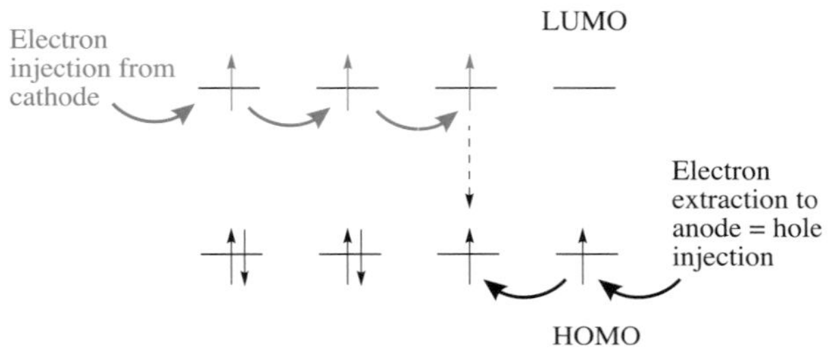

When an injected electron meets an injected hole (hashed arrow in the diagram above), recombination occurs leading to an excited state (*exciton*) which decays to its ground state in a partially radiative process. The light is emitted through the transparent TCO layer. Efficient electroluminescence can be achieved by using multiple organic layers; this overcomes the problem that individual compounds tend to transport one type of charge carrier (electrons or holes) more effectively than another. When an electron and a hole recombine, they may possess the same or opposite spin. The exciton formed is therefore in either a singlet ($S = 0$) or triplet ($S = 1$) state and, statistically, the singlet : triplet ratio is 1 : 3. A disadvantage of OLEDs containing only π-conjugated organic molecules or polymers is that only the singlet state excitons decay radiatively (fluorescence). Phosphorescence in these materials (which involves a change in spin multiplicity, see Section 20.8) is very inefficient. In order to make use of the triplet excitons, neutral inorganic complexes containing a heavy metal such as those shown in Fig. 28.9b are incorporated into the host organic matrix. The triplet state of the complex must lie at lower energy than that of the organic material, allowing triplet excitons from the matrix to migrate to the complex molecules. Due to the presence of the heavy metals in these complexes, spin–orbit coupling (see Section 20.6) leads to mixing of singlet and triplet states, and the triplet excitons are able to undergo radiative decay. OLEDs may be designed to be a single colour

emitter, but white-light sources (Fig. 28.8) can be produced by combining different wavelength emitters.

Currently, OLEDs are mainly used to prepare high-resolution, power-efficient displays for small hand-held applications, e.g. mobile phones. The preparation of larger screens for computer and TV applications is in progress. Another use of OLEDs is to generate white light for general lighting. For such applications, the device requirements are more demanding and, therefore, emissive molecules with higher efficiencies and stabilities are required.

Solid state lighting: LECs

Light-emitting electrochemical cells (LECs) differ from OLEDs in that they use ionic (rather then neutral) electroluminescent materials. The simplest LEC consists of two electrodes (the anode is a TCO) with the electroluminescent material between them. The first LECs were described in the mid-1990s by Heeger, who, along with MacDiarmid and Shirakawa, was awarded the 2000 Nobel Prize in Chemistry 'for the discovery and development of conductive polymers'. A conventional LEC consists of a semiconducting, luminescent polymer to which an ionic salt such as $Li[O_3SCF_3]$ is added. So that the ions are mobile within the cell, the electroluminescent polymer must either be conductive or is mixed with an ion-transport polymer (e.g. polyethylene oxide). Unlike an OLED, the operation of a LEC is not dependent upon the work functions of the electrodes. Thus, air-stable cathode materials (Au, Ag or, typically, Al) can be used; (contrast the requirements of an OLED). When an electric current is passed through the device, mobile ionic species migrate towards the charged electrodes. The host material becomes doped producing an n-type (electron carriers) semiconductor near the cathode and a p-type (hole carriers) material near the anode. The recombination process that leads to emission of light is similar to that described above for an OLED, but the charge carriers are different.

Recently developed LECs rely upon redox-active, emissive *d*-block metal cationic complexes as a single, electroluminescent component (Fig. 28.10a). Iridium(III) complexes are the most popular choices and are able to transport electrons and holes by successive reduction and oxidation processes. In a complex such as $[Ir(bpy)(ppy)_2]^+$ (Fig. 28.10b, Hppy = 2-phenylpyridine), the HOMO is metal-centred and the LUMO comprises ligand π^*-orbitals. Applying an electrical bias across the device leads to injection of electrons from the cathode into the LUMO of the complex, and injection of holes from the anode into the HOMO. Electrons and holes migrate through the material by a hopping mechanism analogous to that described above for OLEDs. Recombination results in an emission, the energy (and, therefore, colour) of which depends on the HOMO–LUMO gap of the complex.

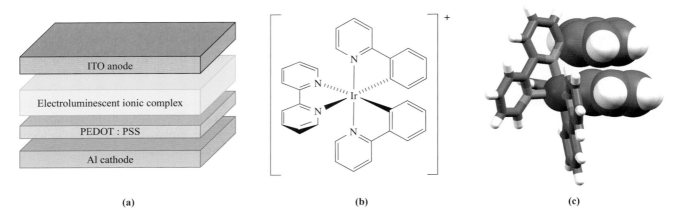

(a)   (b)   (c)

Fig. 28.10 (a) Schematic diagram of the components of a LEC; in the device, the layers are in contact with each other. PEDOT : PSS = poly(3,4-ethylenedioxythiophene) : poly(styrene sulfonate) and is a conducting polymer. The emissive layer also contains small amounts of an ionic liquid (e.g. 1-butyl-3-methylimidazolium hexafluoridophosphate, see Section 9.12) which is added to reduce the time to reach the maximum luminance of the device. (b) An example of a coordination complex used in LECs. (c) The use of face-to-face π-stacking to stabilize the ground and excited states of the complex in a LEC.

A problem of LECs based on *d*-block metal complexes is their limited stability. Expansion of the metal coordination sphere can occur upon formation of the exciton. This allows a water molecule to bind to the Ir(III) ion resulting in the formation of a new complex with consequent quenching of the emission. Introduction of a phenyl substituent into the 6-position of the bpy ligand in $[Ir(bpy)(ppy)_2]^+$ gives the complex shown in Fig. 28.10c. The pendant phenyl substituent engages in a face-to-face π-stacking interaction with the cyclometallated phenyl ring of a $[ppy]^-$ ligand. The strength of this interaction is sufficient to stabilize both the ground and excited states of the complex, and leads to dramatically increased device lifetimes.[†]

28.4 Superconductivity

Superconductors: early examples and basic theory

> A *superconductor* is a material, the electrical resistance of which drops to zero when it is cooled below its *critical temperature,* T_c. A perfect superconductor exhibits perfect diamagnetic behaviour.

Superconductivity was discovered in 1911 by Onnes who was awarded the Nobel Prize for Physics in 1913. Onnes observed that, upon cooling mercury to 4.2 K, its electrical resistance dropped abruptly to zero. Two properties characterize a superconductor: on cooling to its *critical temperature*, T_c, a superconductor loses all electrical resistance, and it becomes a perfect diamagnetic material. If an external

magnetic field is applied to a superconductor, the applied magnetic field is completely excluded (Fig. 28.11a). The latter (the *Meissner effect*) is detected experimentally by an unusual phenomenon: if a permanent magnet is placed on top of a superconductor as it is cooled, at T_c the former rises to become suspended in mid-air above the superconducting material (Fig. 28.12). If the external magnetic field being applied to the superconductor is steadily increased, a point is reached when the magnetic flux penetrates the material (Fig. 28.11b) and the superconducting properties are lost. If the transition is sharp, the superconductor is classed as a Type I (or soft) superconductor, and the field strength at which the transition occurs is the *critical magnetic field* (H_c). A Type II (or hard) superconductor undergoes a gradual transition from

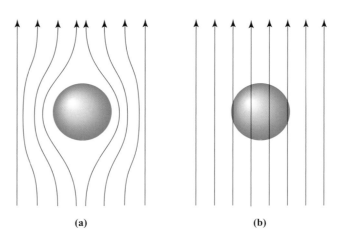

(a)   (b)

Fig. 28.11 The behaviour of magnetic flux lines when (a) a superconducting material and (b) a normal, paramagnetic material is placed in a magnetic field.

[†] See: H. J. Bolink *et al.* (2008) *Adv. Mater.*, vol. 20, p. 3910; S. Graber *et al.* (2008) *J. Am. Chem. Soc.*, vol. 130, p. 14944.

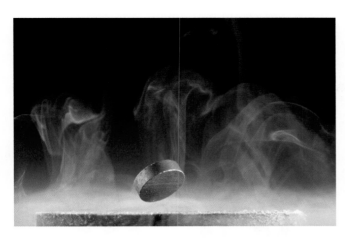

Fig. 28.12 An illustration of the Meissner effect.

superconductor to normal conductor. The lowest field strength at which magnetic flux penetration begins is H_{1c}. At some higher value, H_{2c}, the material is no longer superconducting. Between H_{1c} and H_{2c}, a 'mixed-state' region is observed.

In addition to the critical magnetic field strength, the *critical current* carried by a superconducting wire is an important characteristic. When a current flows through a wire, a magnetic field is produced. An increase in the current will, in the case of a superconducting wire, eventually produce a magnetic field of strength equal to H_c. This current is called the *critical current*. For practical purposes, a superconductor should exhibit a high critical current density.

A range of metals, alloys and metallic compounds are Type I superconductors (Table 28.1). However, to put the practical limitations of working with the materials listed in

Table 28.1 Selected superconducting metals, alloys and metallic compounds.

Element or alloy	T_c / K	Compound or alloy	T_c / K
Al	1.17	AuPb$_2$	3.15
α-Hg	4.15	InPb	6.65
In	3.41	Ir$_2$Th	6.50
Nb	9.25	Nb$_2$SnV	9.8
Ru	0.49	CuS	1.62
Sn	3.72	Nb$_3$Sn	18
Ti	0.40	TiO	0.58
Zn	0.85	SnO	3.81
Al$_2$Y	0.35	(SN)$_x$	0.26

Table 28.1 (and others including the superconducting fullerides described in Section 14.4) into perspective, you should compare the values of T_c with the boiling points of available coolants, e.g. liquid He (4.2 K), H$_2$ (20.1 K) and N$_2$ (77 K). The low values of T_c limit the possible applications of these materials, and illustrate why the so-called *high-temperature superconductors* described below have significant potential in terms of applications.

Superconductivity is usually described in terms of Bardeen–Cooper–Schrieffer (BSC) theory and we describe this model only in simple terms.[†] A *Cooper pair* consists of two electrons of opposite spin and momentum which are brought together by a cooperative effect involving the positively charged nuclei in the vibrating crystal lattice. Cooper pairs of electrons (which are present in the Fermi level, see Section 6.8) remain as bound pairs below the critical temperature, T_c, and their presence gives rise to resistance-free conductivity. The model holds for the earliest known superconductors. Although Cooper pairs may still be significant for the cuprates (*high-temperature superconductors*) described below, new theories are required. To date, no complete explanation for the conducting properties of high-temperature superconductors has been forthcoming[‡].

High-temperature superconductors

Since 1987, cuprate superconductors with $T_c > 77$ K have been the centre of intense interest. One of the first to be discovered was YBa$_2$Cu$_3$O$_7$ made by reaction 28.4. The oxygen content of the final material depends on reaction conditions (e.g. temperature and pressure).

$$4BaCO_3 + Y_2(CO_3)_3 + 6CuCO_3$$
$$\xrightarrow{1220\,K} 2YBa_2Cu_3O_{7-x} + 13CO_2 \qquad (28.4)$$

Selected *high-temperature superconductors* are listed in Table 28.2. The oxidation state of the Cu centres in YBa$_2$Cu$_3$O$_7$ can be inferred by assuming fixed oxidation states of +3, +2 and −2 for Y, Ba and O respectively; the result indicates a mixed Cu(II)/Cu(III) compound. A similar result is obtained for some other materials listed in Table 28.2. High-temperature superconductors have two structural features in common:

- Their structures are related to that of perovskite. Fig. 6.24 showed a 'ball-and-stick' representation of this structure.

[†] For a greater depth discussion of BSC (Bardeen–Cooper–Schrieffer) theory, see: J. Bardeen, L.N. Cooper and J.R. Schrieffer (1957) *Phys. Rev.*, vol. 108, p. 1175; A.R. West (1999) *Basic Solid State Chemistry*, 2nd edn, Wiley-VCH, Weinheim.
[‡] For recent theoretical models, see: A.S. Alexandrov (2011) *J. Supercond. Nov. Magn.*, vol. 24, p. 13 – 'Key pairing interaction in cuprate superconductors'; M.R. Norman (2011) *Science*, vol. 332, p. 196 – 'The challenge of unconventional superconductivity'.

Table 28.2 Selected high-temperature superconductors with $T_c >$ 77 K.

Compound	T_c / K	Compound	T_c / K
$YBa_2Cu_3O_7$	93	$Tl_2CaBa_2Cu_2O_8$	119
$YBa_2Cu_4O_8$	80	$Tl_2Ca_2Ba_2Cu_3O_{10}$	128
$Y_2Ba_4Cu_7O_{15}$	93	$TlCaBa_2Cu_2O_7$	103
$Bi_2CaSr_2Cu_2O_8$	92	$TlCa_2Ba_2Cu_3O_8$	110
$Bi_2Ca_2Sr_2Cu_3O_{10}$	110	$Tl_{0.5}Pb_{0.5}Ca_2Sr_2Cu_3O_9$	120
$HgBa_2Ca_2Cu_3O_8$	135	$Hg_{0.8}Tl_{0.2}Ba_2Ca_2Cu_3O_{8.33}$	138

The same structure is depicted in Fig. 28.13a, but with the octahedral coordination spheres of the Ti centres shown in polyhedral representation.

- They always contain layers of stoichiometry CuO_2; these may be planar (Fig. 28.13b) or puckered.

The incorporation of the two structural building blocks is illustrated in Fig. 28.14 which shows a unit cell of

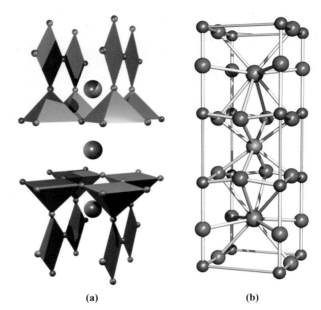

(a)   **(b)**

Fig. 28.14 A unit cell of $YBa_2Cu_3O_7$. (a) A representation showing coordination polyhedra for the Cu centres (square planar and square-based pyramidal); the Y^{3+} and Ba^{2+} ions are shown in blue and green respectively. (b) The unit cell drawn using a 'ball-and-stick' representation; colour code: Cu, brown; Y, blue; Ba, green; O, red.

$YBa_2Cu_3O_7$. The unit cell of $YBa_2Cu_3O_7$ can be considered in terms of three stacked perovskite unit cells. Taking the prototype perovskite to be $CaTiO_3$, then on going from $CaTiO_3$ to $YBa_2Cu_3O_7$, Ba^{2+} and Y^{3+} ions substitute for Ca^{2+}, while Cu centres substitute for Ti(IV). Compared with the structure derived by stacking three perovskite unit cells, the structure of $YBa_2Cu_3O_7$ is oxygen-deficient. This leads to the Cu coordination environments being square planar or square-based pyramidal (Fig. 28.14a), the Ba^{2+} ions being 10-coordinate (Fig. 28.14b), and each Y^{3+} ion being in a cubic environment. The structure is readily described in terms of sheets, and the unit cell in Fig. 28.14 can be represented schematically as layer structure **28.3**. Other high-temperature superconductors can be described in similar fashion, e.g. $Tl_2Ca_2Ba_2Cu_3O_{10}$ (containing Tl^{3+}, Ca^{2+} and Ba^{2+} centres) is composed of layer sequence **28.4**.

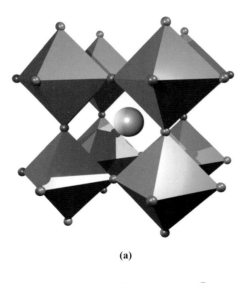

(a)

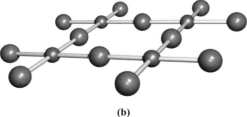

(b)

Fig. 28.13 (a) A unit cell of perovskite, $CaTiO_3$, using a polyhedral representation for the coordination environments of the Ti centres; an O atom (red) lies at each vertex of the octahedra, and the Ca^{2+} ion is shown in grey. (b) Part of a layer of stoichiometry CuO_2 which forms a building block in all cuprate high-temperature superconductors; colour code: Cu, brown; O, red.

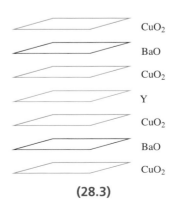

CuO₂
BaO
CuO₂
Y
CuO₂
BaO
CuO₂

(28.3)

The non-CuO_2 oxide layers in the cuprate superconductors are isostructural with layers from an NaCl structure, and so the structures are sometimes described in terms of perovskite and rock salt layers.

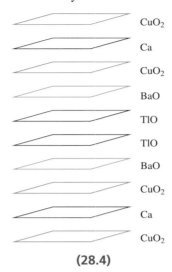

(28.4)

A full discussion of the bonding and origins of superconductivity in these cuprate materials is beyond the scope of this book, but several important points are as follows. It is the CuO_2 layers that are responsible for the superconducting properties, while the other layers in the lattice act as sources of electrons. The arrangement of the layers is an important factor in controlling the superconductivity. Taking the square planar Cu centres to be Cu(II) gives a d^9 configuration with the unpaired electron in a $d_{x^2-y^2}$ orbital. The energies of the $3d$ and $2p$ atomic orbitals are sufficiently close to allow significant orbital mixing, and a band structure is appropriate. The half-filled band is then tuned electronically by the effects of the 'electron sinks' which make up the neighbouring layers in the lattice.

Iron-based superconductors

Superconducting materials consisting of Fe-containing layered structures were first discovered in 2008.[†] The parent compound LaOFeAs (Fig. 28.15) adopts a layer structure, each layer being composed of linked O-centred La_4 or Fe-centred As_4 tetrahedra. The Fe–Fe distances of 285 pm are consistent with significant Fe–Fe bonding (r_{metal} for Fe = 126 pm). LaOFeAs is not a superconductor, but when 5–11% of the O^{2-} sites are doped with F^- ions to give $LaO_{1-x}F_xFeAs$, the material becomes superconducting. The highest T_c (26 K) is observed with 11% doped sites. By substituting La for other lanthanoid metals, values of $26 K < T_c \leq 55 K$ are achieved. Fluorine-doping in the La_2O_2 layers introduces charge

[†] See: Y. Kamihara, T. Watanabe, M. Hirano and H. Hosono (2008) *J. Am. Chem. Soc.*, vol. 130, p. 3296.

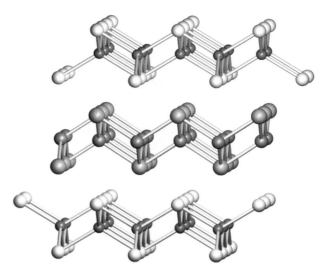

Fig. 28.15 Part of the 3-dimensional structure of LaOFeAs. The unit cell contains two LaOFeAs formula units and the structure can be considered to consist of alternating La_2O_2 and Fe_2As_2 layers. Colour code: La, pale grey; O, red; Fe, green; As, orange.

carriers into the Fe_2As_2 layers and the conduction carriers are restricted to these layers.

The five families of Fe-containing superconductors are:

- $FeSe_{1-x}Te_x$ with $T_c \leq 16 K$;
- MFeAs (M = Li, Na) with $T_c \leq 18 K$;
- $Ba_xK_{1-x}Fe_2As_2$ (Ba^{2+} sites doped with K^+) with $T_c = 38 K$ for $x = 0.4$;
- $MO_{1-x}F_xFeAs$ (M = La, Sm, Nd and O^{2-} sites doped with F^-) with $26 K \leq T_c \leq 55 K$;
- Fe_2As_2 or Fe_2P_2/perovskite materials, e.g. $Sr_4Sc_2O_6Fe_2P_2$ ($T_c = 17 K$), $Ca_4(Mg,Ti)_3O_yFe_2As_2$ ($T_c > 40 K$).

The structures of the undoped parent compounds are related. FeSe adopts the same layered structure as PbO (Fig. 14.28). In LiFeAs, the Li^+ ions are located between layers which are formally negatively charged, i.e. $[FeAs]^-$. Figure 28.16

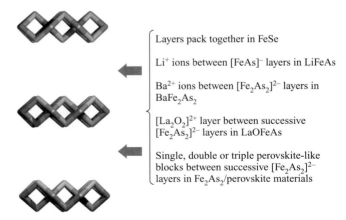

Layers pack together in FeSe

Li^+ ions between $[FeAs]^-$ layers in LiFeAs

Ba^{2+} ions between $[Fe_2As_2]^{2-}$ layers in $BaFe_2As_2$

$[La_2O_2]^{2+}$ layer between successive $[Fe_2As_2]^{2-}$ layers in LaOFeAs

Single, double or triple perovskite-like blocks between successive $[Fe_2As_2]^{2-}$ layers in Fe_2As_2/perovskite materials

Fig. 28.16 The structural relationship between the five families of Fe-containing superconductors containing FeAs or Fe_2As_2 building blocks.

summarizes the relationships between the five families of superconducting materials. In the perovskite-based materials, $[Fe_2P_2]^{2-}$ layers may replace the $[Fe_2As_2]^{2-}$ layers shown in Fig. 28.16. General formulae are $M_{n+1}M'_nO_{3n-1-y}Fe_2P_2$, $M_{n+1}M'_nO_{3n-1-y}Fe_2As_2$, $M_{n+2}M'_nO_{3n-y}Fe_2P_2$ and $M_{n+2}M'_nO_{3n-y}Fe_2As_2$ where M = Ca, Sr, Ba, M' = Mg, Al, Sc, Ti, V, Cr, Co etc. and y is a variable. As in the cuprate superconductors, there is considerable scope for tuning the critical temperature by varying the metals and O content in the perovskite blocks.

Chevrel phases

Chevrel phases are ternary metal chalcogenides (most commonly sulfides) of general formula $M_xMo_6X_8$ (M = group 1 or 2 metal, or p-, d- or f-block metal; X = S, Se, Te). They can be prepared by heating the constituent elements at ≈ 1300 K in evacuated, sealed tubes. $Mo_6(\mu_3\text{-}X)_8$ clusters are common to all Chevrel phases. Figure 28.17 shows the structure of the $Mo_6(\mu_3\text{-}S)_8$ building block in $PbMo_6S_8$, and the way in which the units are connected together. In the solid state, the Mo_6X_8 units are tilted with respect to one another, generating an extended structure that contains cavities of different sizes. Metal ions such as Pb^{2+} occupy the larger cavities (Fig. 28.17b). The smaller holes may be occupied in Chevrel phases containing small cations (e.g. Li^+). Metal ions within the range 96 pm $\leq r_{ion} \leq$ 126 pm can be accommodated. The framework has a degree of flexibility, and the unit cell dimensions vary with both X and M. The physical properties of the material also depend on X and M.

The binary compound Mo_6S_8 is metastable and cannot be synthesized by direct combination of the elements. However, it can be made by removal of metal M from $M_xMo_6S_8$ by electrochemical or chemical oxidation (e.g. by treatment with I_2 or aqueous HCl, eq. 28.5). Other metals can then be intercalated into the Mo_6S_8 framework,

Table 28.3 Examples of superconducting Chevrel phases.

Chevrel phase	T_c / K	Chevrel phase	T_c / K
$PbMo_6S_8$	15.2	$TlMo_6Se_8$	12.2
$SnMo_6S_8$	14.0	$LaMo_6Se_8$	11.4
$Cu_{1.8}Mo_6S_8$	10.8	$PbMo_6Se_8$	6.7
$LaMo_6S_8$	5.8	$Cu_2Mo_6Se_8$	5.6

M acting as an electron donor and the host lattice accepting up to four electrons per Mo_6S_8 cluster. For example, intercalation of lithium (in a reaction of Mo_6S_8 with BuLi) transfers four electrons to give $Li_4Mo_6S_8$, intercalation of an f-block metal leads to $M^{III}Mo_6S_8$, and in $Pb^{II}Mo_6S_8$, the Mo_6S_8 cluster has accepted two electrons.

$$M^{II}Mo_6S_8 + 2HCl \longrightarrow Mo_6S_8 + H_2 + MCl_2 \qquad (28.5)$$

The reversible intercalation of metal ions and their high mobility in the solid suggest possible applications of Chevrel phases as electrode materials (compare with the properties exhibited by layered sulfides, eq. 28.3). There is particular interest in the fact that most Chevrel phases are Type II superconductors, and Table 28.3 lists values of T_c for selected materials. Not only does $PbMo_6S_8$ have a relatively high T_c (i.e. compared with superconducting metals and alloys, Table 28.1), it also exhibits a very high critical flux density ($B_{c2} \approx 50$ T). These properties make $PbMo_6S_8$ suitable for high-field applications, although its critical current density (like those of other Chevrel phases) is too low for $PbMo_6S_8$ to find industrial uses. The network of Mo_6-clusters is responsible for the superconducting properties of Chevrel phases. The Mo $4d$ electrons are localized within the Mo_6-octahedral clusters, and this results in a band structure in which bands comprising Mo $4d$ character lie close to the Fermi level. Within BSC theory, this band structure is consistent with relatively high critical temperatures. In MMo_6X_8 phases, transfer of electrons from M to the Mo_6-clusters modifies the electronic properties of the material, resulting in the observed variation in values of T_c.

Superconducting properties of MgB₂

Although magnesium boride, MgB_2, has been known since the 1950s, it was only in 2001 that its superconducting properties ($T_c = 39$ K) were discovered.[†] Solid MgB_2 has hexagonal symmetry and consists of layers of Mg and B atoms (Fig. 28.18). The arrangement of the B atoms in each B_n sheet in MgB_2 mimics that of the C atoms in

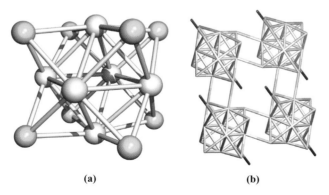

Fig. 28.17 (a) The structure of the Mo_6S_8 building block in Chevrel phases of general formula $M_xMo_6S_8$. (b) Part of the extended structure of the Chevrel phase $PbMo_6S_8$. Mo_6S_8 cages are interconnected through Mo–S and S–Pb–S interactions. Colour code: Mo, pale grey; S, yellow; Pb, red.

(a)   (b)

[†] See: J. Nagamatsu, N. Nakagawa, T. Muranaka, Y. Zenitani and J. Akimitsu (2001) *Nature*, vol. 410, p. 63.

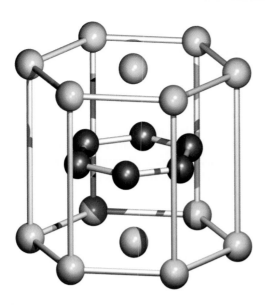

 Fig. 28.18 A repeat unit in the solid state structure of MgB_2. Colour code: Mg, yellow; B, blue.

graphite, but (unlike adjacent C_n layers in graphite, Fig. 14.4) the atoms in successive B_n layers lie directly over one another. The Mg atoms form close-packed layers sandwiched between the sheets of B atoms. The Mg atoms in MgB_2 are considered to be ionized and the boron framework (formally negatively charged) is therefore isoelectronic with graphite. In terms of band theory, MgB_2 exhibits two π-bands, one occupied (electron type) and one unoccupied (hole type). The σ-bands (which are well below the Fermi level in graphite) cross the Fermi level in MgB_2, leading to unfilled σ-bands (holes) which contribute to the unusual properties of this material. No other metal boride has yet been shown to have a T_c as high as that of MgB_2. MgB_2 is a Type II superconductor. Although the onset of superconductivity for MgB_2 occurs at a much lower temperature than that for the cuprate superconductors, the simple, layered structure of MgB_2 makes this superconductor of particular interest.

In the last 10 years, rapid progress has been made in the production of MgB_2 wires, tapes and thin films, and an MgB_2-based MRI magnet was developed by ASG Superconductors in 2006.[†] Fabrication of thin films of MgB_2 by chemical vapour deposition (CVD, see Section 28.6) uses B_2H_6 with H_2 carrier gas as a source of B which then reacts with Mg vapour under Ar atmosphere at ≈ 950 K. It is essential to exclude O_2 as this oxidizes Mg. Exclusion of C is also important: replacement of B by C atoms in MgB_2 decreases the value of T_c. On the other hand, carbon doping may be desirable in some cases as it enhances the critical current density of MgB_2.

[†] R. Penco and G. Grasso (2007) *IEEE Trans. Appl. Supercond.*, vol. 17, p. 2291 – 'Recent development of MgB_2-based large scale applications'.

Applications of superconductors

Commercial applications of high-temperature superconductors are now well established. The majority of magnetic resonance imaging scanners (see Box 4.3) rely on superconducting magnets with flux densities of 0.5–2.0 T. Currently, NbTi ($T_c = 9.5$ K) multicore conductors are used, but replacement by high-temperature superconductors would be financially beneficial. The Large Hadron Collider particle accelerator at CERN, which began operation in 2008, depends upon 10 000 NbTi superconducting magnets (fabricated as highly uniform cables) which are housed in cryostats (Fig. 28.19) containing ≈ 130 t of He to maintain the operating temperature of 1.9 K. The use of superconducting magnets produces flux densities >8 T, and the two counter-rotating proton beams in the accelerator reach energies of 7×10^{12} eV (7 TeV).

The combination of two superconductors separated by a thin oxide barrier which is a weak insulator makes up a *Josephson junction*, a device that is very sensitive to magnetic fields. Among applications of Josephson junctions is their role in SQUID (*superconducting quantum interference device*) systems for measuring magnetic susceptibilities. The extreme sensitivity of a SQUID allows it to be used to measure very weak biomagnetic signals such as those originating from the brain, and naval vessels equipped with SQUIDs have increased sensitivity to detect undersea mines.

Superconductors have been applied to develop train systems that operate with *magnetic-levitation* (MAGLEV) in which the train effectively travels ≈ 10 mm above its tracks, i.e. virtually frictionless motion. The first commercial train came into service in Shanghai in 2003 and can reach speeds of 440 km h^{-1}.

For the development of applications for superconductors, two obstacles in particular have to be surmounted. The first

Fig. 28.19 Technician inspecting the Large Hadron Collider which runs in a 27 km tunnel at CERN, near Geneva, Switzerland. The superconducting magnets are housed in the blue pipe-like cryostat.

is that the material must be cooled to low temperatures to attain T_c. As higher temperature superconductors are developed, this has become less of a major drawback, but still militates against the use of superconductors in conventional settings. The second problem is one of fabrication. When prepared as a bulk material, the cuprate superconductors have unacceptably low critical current densities, i.e. the superconductivity is lost after the material has carried only a limited amount of current. The origin of the problem is the presence of grain boundaries in the solid and can be overcome by preparing thin films using, for example, CVD (see Section 28.6) or *texturing* the material (i.e. alignment of crystallites) through specialized crystallization techniques or mechanical working. Even with the advances that have been made so far, the application of superconductors for bulk power transmission remains a long way in the future.

Self-study exercises

1. Superconducting magnets in high-field NMR spectrometers are routinely made from an NbTi alloy, and the magnet has to be cooled. What cooling agent would you use, and why? Suggest reasons why high-temperature superconductors are not currently used in NMR spectrometers.

2. In 1911, Onnes reported the first superconducting metal, mercury. T_c for Hg is 4.15 K. Sketch a graph of what Onnes observed upon cooling Hg below 4.5 K, given that the resistance of the sample in the experiment was 1.3 Ω at 4.5 K.

28.5 Ceramic materials: colour pigments

A ***ceramic*** material is a hard, high melting solid which is usually chemically inert.

Ceramic materials are commonplace in everyday life, e.g. floor and wall tiles, crockery, wash-basins, baths and decorative pottery and tiles. The cuprate high-temperature superconductors discussed above are ceramic materials. Many ceramic materials consist of metal oxides or silicates, and the addition of white and coloured pigments is a huge industrial concern. In earlier chapters, we mentioned the use of several metal oxides (e.g. CoO and TiO_2, Boxes 21.3 and 21.8) as colour pigments. One of the factors that has to be taken into account when choosing a pigment is the need for it to withstand the high firing temperatures involved in the manufacture of ceramics. This is in contrast to the introduction of pigments into, for example, fabrics.

White pigments (opacifiers)

An ***opacifier*** is a glaze additive that makes an otherwise transparent glaze opaque.

The most important commercial opacifiers in ceramic materials are TiO_2 (in the form of anatase) and $ZrSiO_4$ (zircon). While SnO_2 is also highly suitable, its use is not as cost effective as that of TiO_2 and $ZrSiO_4$, and it is retained only for specialist purposes. Zirconium(IV) oxide is also an excellent opacifier but is more expensive than $ZrSiO_4$. Fine particles of these pigments scatter incident light extremely strongly: the refractive indices of anatase, $ZrSiO_4$, ZrO_2 and SnO_2 are 2.5, 2.0, 2.2 and 2.1 respectively. The firing temperature of the ceramic material determines whether or not TiO_2 is a suitable pigment for a particular application. Above 1120 K, anatase converts to rutile, and although rutile also has a high refractive index ($\mu = 2.6$), the presence of relatively large particles of rutile prevents it from functioning as an effective opacifier. Anatase is therefore useful only if working temperatures do not exceed the phase transition temperature. Zircon is amenable to use at higher firing temperatures. It can be added to the molten glaze and precipitates as fine particles dispersed in the glaze as it is cooled.

Adding colour

Cation substitution in a host lattice such as ZrO_2, TiO_2, SnO_2 or $ZrSiO_4$ is a means of altering the colour of a pigment. The substituting cation must have one or more unpaired electrons so as to give rise to an absorption in the visible region (see Section 20.7). Yellow pigments used to colour ceramics include $(Zr,V)O_2$ (which retains the structure of *baddeleyite*, the monoclinic form of ZrO_2 in which the metal is 7-coordinate), $(Sn,V)O_2$ (with a V-doped *cassiterite* structure) and $(Zr,Pr)SiO_4$ (with a *zircon* lattice doped with $\approx 5\%$ Pr). Blue pigmentation can be obtained using $(Zr,V)SiO_4$ and this is routinely used when high-temperature firing is required. Cobalt oxide-based pigments produce a more intense blue coloration than vanadium-doped zirconia, but are unsuitable for use at high temperatures. The content of cobalt oxide needed in a blue ceramic is ≈ 0.4–0.5% Co.

Spinels (AB_2O_4) (see Box 13.7) are an important class of oxide for the manufacture of brown and black pigments for ceramics. The three spinels $FeCr_2O_4$, $ZnCr_2O_4$ and $ZnFe_2O_4$ are structurally related, forming a family in which Fe^{2+} or Zn^{2+} ions occupy tetrahedral sites, while Cr^{3+} or Fe^{3+} ions are octahedrally sited. In nature, cation substitution occurs to produce, for example, black crystals of the mineral *franklinite* $(Zn,Mn,Fe)(Fe,Mn)_2O_4$ which has a variable composition. In the ceramics industry, spinels for use as pigments are prepared by heating together suitable metal oxides in appropriate stoichiometric ratios so as to control the cation substitution in a parent spinel. In $(Zn,Fe)(Fe,Cr)_2O_4$, a range of brown shades can be obtained by varying the cation site compositions. For the commercial market, reproducibility of shade of colour is, of course, essential.

28.6 Chemical vapour deposition (CVD)

The development of *chemical vapour deposition* has been closely tied to the need to deposit thin films of a range of metals and inorganic materials for use in, for example, semi-conducting devices, ceramic coatings and electrochromic materials. Table 28.4 lists some applications of selected thin film materials. Part of the challenge of the successful production of thin films is to find suitable molecular pre-cursors, and there is much research interest in this area. We illustrate CVD by focusing on the deposition of specific materials including semiconductors. In any industrial CVD process, reactor design is crucial to the efficiency of the deposition, and it should be recognized that *the diagrams given of CVD reactors are highly schematic.*

Chemical vapour deposition (CVD) is the delivery (by uniform mass transport) of a volatile precursor or precursors to a heated surface on which reaction takes place to deposit a thin film of the solid product; the surface must be hot enough to permit reaction but cool enough to allow solid deposition. Multilayer deposition is also possible.

Metal–organic chemical vapour deposition (MOCVD) refers specifically to use of metal–organic precursors.

In *plasma-enhanced CVD*, a plasma (an ionized gas) is used to facilitate the formation of a film, either by treatment of the substrate before deposition, or by assisting molecular dissociation.

High-purity silicon for semiconductors

Although Ge was the first semiconductor to be used commercially, it is Si that now leads the world market. Germanium has been replaced, not only by Si, but by a range of recently developed semiconducting materials. All silicon semiconductors are manufactured by CVD. In Box 6.3, we described the Czochralski process for obtaining single crystals of pure silicon. The silicon used for the crystal growth must itself be of high purity and a purification stage is needed after the manufacture of Si from SiO_2 (reaction 28.6). Crude silicon is first converted to the volatile $SiHCl_3$ which is then converted back to a higher purity grade of Si (eq. 28.7) by using CVD.

$$SiO_2 + 2C \xrightarrow{\Delta} Si + 2CO \qquad (28.6)$$

$$3HCl + Si \underset{1400\,K}{\overset{620\,K}{\rightleftharpoons}} SiHCl_3 + H_2 \qquad (28.7)$$

Figure 28.20 illustrates the industrial CVD procedure: $SiHCl_3$ and H_2 pass into the reaction vessel where they

Table 28.4 Some applications of selected thin film materials; see also Table 28.6.

Thin film	Applications
Al_2O_3	Oxidation resistance
AlN	High-powered integrated circuits; acoustic devices
C (diamond)	Cutting tools and wear-resistant coatings; heat sink in laser diodes; optical components
CdTe	Solar cells
CeO_2	Optical coatings; insulating films
GaAs	Semiconducting devices; electrooptics; (includes solar cells)
GaN	Light-emitting diodes (LED)
$GaAs_{1-x}P_x$	Light-emitting diodes (LED)
$LiNbO_3$	Electrooptic ceramic
NiO	Electrochromic devices
Si	Semiconductors, many applications of which include solar cells
Si_3N_4	Diffusion barriers and inert coatings in semiconducting devices
SiO_2	Optical wave guides
SnO_2	Sensors for reducing gases, e.g. H_2, CO, CH_4, NO_x
TiC	Wear resistance
TiN	Friction reduction
W	Metal coatings on semiconducting integrated circuits
WO_3	Electrochromic windows
ZnS	Infrared windows

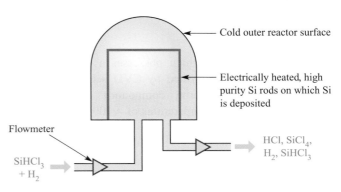

Fig. 28.20 Schematic representation of the CVD set-up used to deposit high-purity silicon by thermal decomposition of $SiHCl_3$.

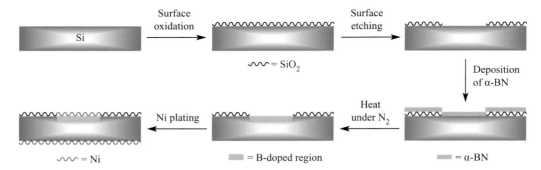

Fig. 28.21 Boron doping of silicon using α-BN.

come into contact with a high-purity silicon surface, electrically heated to 1400 K. Back-reaction 28.7 is highly endothermic and occurs on the Si surface to deposit additional Si (mp = 1687 K). No deposition occurs on the vessel walls because these are kept cold, devoid of the heat energy needed to facilitate the reaction between $SiHCl_3$ and H_2. A secondary product of the deposition reaction is $SiCl_4$ (eq. 28.8), some of which reacts with H_2 to give more $SiHCl_3$. The remainder leaves with the exhaust gases[†] and finds use in the manufacture of silica.

$$4SiHCl_3 + 2H_2 \longrightarrow 3Si + 8HCl + SiCl_4 \qquad (28.8)$$

A more recently developed CVD process starts with SiH_4 (eq. 28.9), which is first prepared from $SiHCl_3$ by scheme 28.10.

$$SiH_4 \xrightarrow{\Delta} Si + 2H_2 \qquad (28.9)$$

$$\left.\begin{array}{l} 2SiHCl_3 \longrightarrow SiH_2Cl_2 + SiCl_4 \\ 2SiH_2Cl_2 \longrightarrow SiH_3Cl + SiHCl_3 \\ 2SiH_3Cl \longrightarrow SiH_4 + SiH_2Cl_2 \end{array}\right\} \qquad (28.10)$$

The high-grade silicon produced by CVD is virtually free of B or P impurities, and this is essential despite the fact that doping with B or P is routine (Fig. 28.21). Careful tuning of the properties of n- or p-type semiconductors (see Section 6.9) depends on the *controlled* addition of B, Al, P or As during their manufacture.

α-Boron nitride

Thin films of α-BN (which possesses a layer structure, Fig. 13.22) can be deposited by CVD using reactions of NH_3 with volatile boron compounds such as BCl_3 (eq. 28.11) or BF_3 at temperatures of ≈1000 K.

$$BCl_3 + NH_3 \xrightarrow{\Delta} BN + 3HCl \qquad (28.11)$$

An important application of such films is in doping silicon to generate a p-type semiconductor (Fig. 28.21). The semiconductor-grade silicon is first oxidized to provide a layer of SiO_2 which is then etched. Deposition of a thin film of α-BN provides contact between Si and α-BN within the etched zones. By heating under N_2, B atoms from the film diffuse into the silicon to give the desired p-type semiconductor which is finally plated with a thin film of nickel (see 'Metal deposition' below).

α-Boron nitride films have a range of other applications which make use of the material's hardness, resistance to oxidation and insulating properties.

Silicon nitride and carbide

The preparation and structure of Si_3N_4 were discussed at the end of Section 14.12. Its uses as a refractory material are widespread, as are its applications in the microelectronics industry and solar cell construction. Thin films of Si_3N_4 can be prepared by reacting SiH_4 or $SiCl_4$ with NH_3 (eq. 14.90), or $SiCl_4$ with N_2H_4. Films deposited using $(\eta^5\text{-}C_5Me_5)SiH_3$ (**28.5**) as a precursor with a *plasma-enhanced CVD* technique have the advantage of low carbon contamination. The precursor is made by reduction of $(\eta^5\text{-}C_5Me_5)SiCl_3$ using $Li[AlH_4]$ and is an air- and heat-stable volatile compound, ideal for CVD.

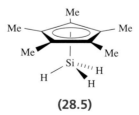

(28.5)

Silicon carbide (*carborundum*) has several polymorphs. The β-form adopts the wurtzite structure (Fig. 6.21). It is extremely hard, resists wear, withstands very high temperatures, has a high thermal conductivity and a low coefficient of thermal expansion, and has long been used as a refractory material and abrasive powder. Recent development of suitable CVD methods has made possible the deposition

---

[†] For an assessment of the treatment of waste volatiles from the semiconductor industry, see: P.L. Timms (1999) *J. Chem. Soc., Dalton Trans.*, p. 815.

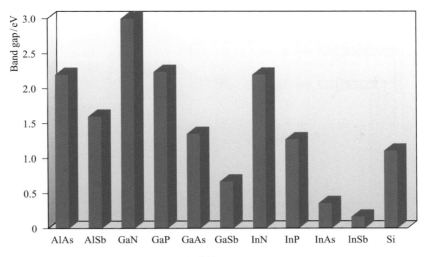

Fig. 28.22 Band gaps (at 298 K) of the III–V semiconductors and of Si.

of β-SiC of >99.9% purity. Suitable precursors are alkyl-silanes, alkylchlorosilanes, or alkanes with chlorosilanes. Silicon carbide is a IV–IV semiconductor (band gap = 2.98 eV) which has particular application for high-frequency devices and for systems operating at high temperatures. Thin films exhibit excellent reflective properties and are used for manufacturing mirrors for laser radar systems, high-energy lasers, synchrotron X-ray equipment and astronomical telescopes. Silicon carbide is also used for blue light-emitting diodes (LEDs). Silicon carbide fibres are described in Section 28.7.

III–V Semiconductors

The name III–V semiconductor derives from the old group numbering of groups 13 (III) and 15 (V). Aluminium nitride (AlN) is an insulator, and GaN and InN are wide band gap semiconductors (Fig. 28.22). The important III–V semiconductors comprise AlAs, AlSb, GaP, GaAs, GaSb, InP, InAs and InSb, and of these GaAs is the most important commercially. The band gaps of these materials are compared with that of Si (1.10 eV) in Fig. 28.22. Silicon leads the commercial market as a semiconducting material, with GaAs lying in second place. GaAs plays an important role in optoelectronics, information and mobile phone technologies. Although GaAs and InP possess similar band gaps to Si, they exhibit higher electron mobilities, making them of great commercial value for high-speed computer circuitry. Ternary materials are also important, e.g. $GaAs_{1-x}P_x$ is used in LEDs in pocket calculator, digital watch and similar displays. The colour of the emitted light depends on the band gap (Table 28.5). In such devices, the semiconductor converts electrical energy into optical energy.

Thin films of GaAs are deposited commercially using CVD techniques by reactions such as 28.12. Slow hydro-lysis of GaAs in moist air means that films must be protectively coated.

$$Me_3Ga + AsH_3 \xrightarrow{900\ K} GaAs + 3CH_4 \quad (28.12)$$

The commercial production of $GaAs_{1-x}P_x$ requires the epitaxial growth of the crystalline material on a substrate.

Epitaxial growth of a crystal on a substrate crystal is such that the growth follows the crystal axis of the substrate.

Figure 28.23 gives a representation of an apparatus used to deposit $GaAs_{1-x}P_x$. The operating temperature is typically ≈1050 K and H_2 is used as a carrier gas. Gallium (mp 303 K, bp 2477 K) is held in a vessel within the reactor. It reacts with the incoming dry HCl to give GaCl which then disproportionates (scheme 28.13), providing Ga at the substrate.

Table 28.5 The dependence of the wavelength, λ, of the emitted radiation from $GaAs_{1-x}P_x$ on the composition of the material.

x in $GaAs_{1-x}P_x$	Substrate	λ / nm	Observed colour or region of spectrum
0.10	GaAs	780	Infrared
0.39	GaAs	660	Red
0.55	GaP	650	Red
0.65	GaP	630	Orange
0.75	GaP	610	Orange
0.85	GaP	590	Yellow

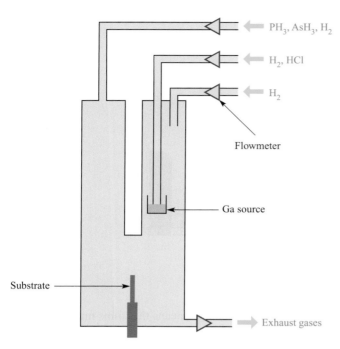

Fig. 28.23 Schematic representation of the CVD assembly used for the epitaxial growth of $GaAs_{1-x}P_x$; H_2 is the carrier gas.

$$\left.\begin{array}{l} 2Ga + 2HCl \longrightarrow 2GaCl + H_2 \\ 3GaCl \longrightarrow 2Ga + GaCl_3 \end{array}\right\} \quad (28.13)$$

The proportions of the group 15 hydrides entering the reactor can be varied as required. They thermally decompose by reaction 28.14 giving elemental components for the ternary semiconductor at the substrate surface. High-purity reagents are essential for the deposition of films that are of acceptable commercial grade.

$$2EH_3 \longrightarrow 2E + 3H_2 \qquad E = As \text{ or } P \quad (28.14)$$

Table 28.5 illustrates how the variation in semiconductor composition affects the colour of light emitted from a $GaAs_{1-x}P_x$-containing LED. Dopants can be added to the semiconductor by injecting a volatile dopant-precursor into the PH_3 and AsH_3 gas inflow. For an n-type semiconductor, H_2S or Et_2Te may be used, providing S or Te atom dopants.

Mobile phones incorporate multilayer III–V epitaxial heterojunction bipolar transistor wafers such as that illustrated in Fig. 28.24. The p–n junctions on either side of the base layer are a crucial feature of semiconductor devices. In the wafer shown in Fig. 28.24 (and in other similar wafers), the p-type base layer must be highly doped to provide high-frequency performance. Choice of dopant is critical, e.g. use of a Zn dopant (see below) results in its diffusion into the emitting n-type layers. This problem has been overcome by doping with C which exhibits a low diffusion coefficient; C-doped wafers have been used commercially since the early 1990s.

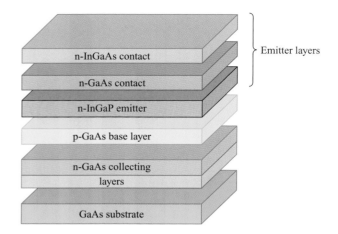

Fig. 28.24 Typical components in a multilayer heterojunction bipolar transistor wafer, each deposited by CVD. In the device, the layers are in contact with each other.

The narrow band gap of InSb (Fig. 28.22) means that InSb can be used as a photodetector within a wavelength region of 2–5 μm (i.e. in the infrared). Such IR detectors have military applications.

Self-study exercises

1. Why are amine adducts of GaH_3 of interest as possible precursors in CVD?
   [*Ans.* See Fig. 13.10 and discussion]

2. What can you say about the band structures of AlN, GaAs and Si? [*Ans.* See Section 6.8]

Metal deposition

The use of volatile molecular, often organometallic, precursors for the deposition of thin films of metals for contacts and wiring in electrical devices (i.e. semiconductor–metal connections) and as sources of dopants in semiconductors is an important part of modern manufacturing processes. The general strategy is to choose a volatile organometallic complex which can be thermally decomposed on the substrate, depositing the metal film and liberating organic products which can be removed in the exhaust gases. The use of methyl derivatives as precursors often leads to higher than acceptable carbon contamination of the deposited metal film, and for this reason other substituents tend to be preferred.

Aluminium is deposited by MOCVD using R_3Al (e.g. R = Et) despite the fact that these compounds are pyrophoric. Vanadium films can be deposited by reaction 28.15.

$$VCl_4 + 2H_2 \xrightarrow{1450\,K} V + 4HCl \quad (28.15)$$

Nickel films can be deposited from $Ni(CO)_4$, but temperature control is important since above 470 K, there is a high tendency for the deposition of carbon impurities. Other suitable precursors include $(\eta^5\text{-}Cp)_2Ni$ and $Ni(acac)_2$.

Gallium arsenide can be doped with Sn by using tin(IV) alkyl derivatives such as Me_4Sn and Bu_4Sn, although the former tends to result in carbon contamination. Zinc is added as a dopant to, for example, AlGaAs (to give a p-type semiconductor) and can be introduced by adding appropriate amounts of Et_2Zn to the volatile precursors for the ternary semiconductor (Me_3Al, Me_3Ga and AsH_3). Silicon, GaAs and InP may be doped with Er, and Cp_3Er is a suitable precursor; similarly, Cp_3Yb is used to dope InP with Yb.

Ceramic coatings

The development of CVD techniques has enabled rapid progress in the commercialization of applying ceramic coatings to carbide tools used for cutting steel. Wear-resistant coatings of thickness ≈ 5–$10\,\mu m$ are now usually added to heavy-duty cutting tools to prolong their lifetime and allow the tools to operate at significantly higher cutting speeds. Multilayers can readily be applied using CVD, and the method is amenable to coating non-uniform surfaces.

A coating of Al_2O_3 provides resistance against abrasion and oxidation, and can be deposited by the reaction at a substrate (≈ 1200–$1500\,K$) of $AlCl_3$, CO_2 and H_2. Abrasion resistance is also provided by TiC, while TiN gives a barrier against friction. The volatile precursors used for TiC are $TiCl_4$, CH_4 and H_2, and TiN is deposited using $TiCl_4$, N_2 and H_2, both at temperatures $> 1000\,K$. In general, nitride layers can be deposited using volatile metal chlorides, with H_2 and N_2 as the molecular precursors. Of particular importance for wear-resistant coatings are nitrides of Ti, Zr and Hf.

Perovskites and cuprate superconductors

Table 28.6 lists applications of some of the most commercially important mixed metal, perovskite-type oxides, and illustrates that it is the dielectric, ferroelectric, piezoelectric (see Section 14.9) and pyroelectric properties of these materials that are exploited in the electronics industry.

Ferroelectric means the spontaneous alignment of electric dipoles caused by interactions between them; domains form in an analogous manner to the domains of magnetic dipoles in a *ferromagnetic* material (see Fig. 20.32 and related discussion).

The industrial fabrication of electronic devices containing perovskite-type metal oxides traditionally involves the preparation of powdered materials which are then cast as required. However, there is great interest at the research level in developing techniques for thin film deposition and in this section we consider the use of CVD methods.

Reaction 28.16 is one conventional method of preparing $BaTiO_3$. A second route (used industrially) involves the preparation of $BaTiO(ox)_2 \cdot 4H_2O$ (ox = oxalate) from $BaCl_2$, $TiCl_4$, H_2O and H_2ox, followed by thermal decomposition (scheme 28.17).

$$TiO_2 + BaCO_3 \xrightarrow{\Delta} BaTiO_3 + CO_2 \tag{28.16}$$

$$BaTiO(ox)_2 \cdot 4H_2O \xrightarrow[\text{dehydrate}]{400\,K} BaTiO(ox)_2$$

$$\downarrow 600\,K, -CO, -CO_2$$

$$\tfrac{1}{2} BaTi_2O_5 + \tfrac{1}{2} BaCO_3$$

$$\downarrow 900\,K, -CO_2$$

$$BaTiO_3 \qquad (28.17)$$

(28.6)   **(28.7)**

It is also possible to deposit $BaTiO_3$ by using CVD (Fig. 28.25), the source of Ti being the alkoxide $Ti(O^iPr)_4$ and of Ba, a β-ketonate complex such as BaL_2 where $L^- = \mathbf{28.6}$. A typical reactor temperature for $BaTiO_3$ deposition is $\approx 500\,K$, and substrates that have been used include MgO, Si and Al_2O_3. Although often formulated as

Table 28.6 Electronic applications of selected perovskite-type mixed metal oxides.

Mixed metal oxide	Properties of the material	Electronic applications
$BaTiO_3$	Dielectric	Sensors; dielectric amplifiers; memory devices
$Pb(Zr,Ti)O_3$	Dielectric; pyroelectric; piezoelectric	Memory devices; acoustic devices
La-doped $Pb(Zr,Ti)O_3$	Electrooptic	Optical memory displays
$LiNbO_3$	Piezoelectric; electrooptic	Optical memory displays; acoustic devices; wave guides; lasers; holography
$K(Ta,Nb)O_3$	Pyroelectric; electrooptic	Pyrodetector; wave guides; frequency doubling

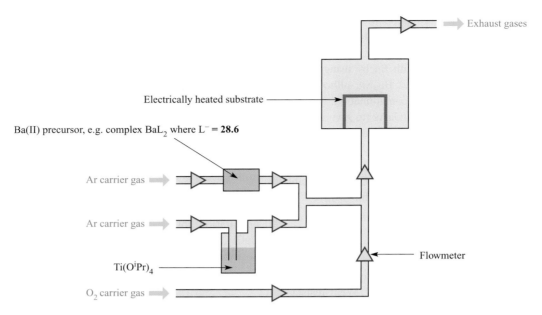

Fig. 28.25 Schematic representation of a CVD set-up used for the deposition of the perovskite $BaTiO_3$.

'BaL_2', the precursor is not so simple and its exact formulation depends on its method of preparation, e.g. adducts such as $BaL_2 \cdot (MeOH)_3$ and $[BaL_2(OEt_2)]_2$, the tetramer Ba_4L_8, and the species $Ba_5L_9(OH_2)_3(OH)$. Any increased degree of oligomerization is accompanied by a decrease in volatility, a fact that militates against the use of the precursor in CVD. Complexes containing fluorinated β-ketonate ligands such as **28.7** possess higher volatilities than related species containing non-fluorinated ligands, but unfortunately their use in CVD experiments leads to thin films of $BaTiO_3$ contaminated with fluoride.

So far, we have illustrated the formation of binary (e.g. GaAs, TiC) and ternary (e.g. $GaAs_{1-x}P_x$, $BaTiO_3$) systems through the combination of two or three volatile precursors in the CVD reactor. A problem that may be encountered is how to control the stoichiometry of the deposited material. In some cases, controlling the ratios of the precursors works satisfactorily, but in other cases, better control is achieved by trying to find a *single* precursor. There is active research in this area and it is illustrated by the formation of $LiNbO_3$ from the alkoxide precursor $LiNb(OEt)_6$. The ceramic $LiNbO_3$ is used commercially for a range of electronic purposes (Table 28.6) and is conventionally prepared by reaction 28.18 or 28.19.

$$Li_2CO_3 + Nb_2O_5 \xrightarrow{\Delta} 2LiNbO_3 + CO_2 \qquad (28.18)$$

$$Li_2O + Nb_2O_5 \xrightarrow{fuse} 2LiNbO_3 \qquad (28.19)$$

In order to develop an appropriate CVD method for depositing $LiNbO_3$ from $LiNb(OEt)_6$, one major problem has to be overcome: the volatility of bulk $LiNb(OEt)_6$ is low, and hence an aerosol-type system is used to introduce the molecular precursor into the CVD reactor. Solid $LiNb(OEt)_6$ is dissolved in toluene and the solution converted into a fine mist using ultrasonic radiation. In the first part of the reactor (550 K), the mist volatilizes and is transported in a flow of the carrier gas into a higher temperature region containing the substrate on which thermal decomposition of $LiNb(OEt)_6$ occurs to give $LiNbO_3$. Such results for the formation of ternary (or more complex) ceramic materials and the development of *aerosol-assisted CVD* may have a potential for commercial application in the future.

The explosion of interest in cuprate superconductors (see Section 28.4) during the last two decades has led to active research interest into ways of depositing these materials as thin films. For example, CVD precursors and conditions for the deposition of $YBa_2Cu_3O_7$ have included BaL_2, CuL_2 and YL_3 ($L^- = $ **28.6**) with He/O_2 carrier gas, and an $LaAlO_3$ substrate at 970 K. Superconducting MgB_2 thin films are produced by annealing CVD-deposited B in Mg vapour (see Section 28.4).

28.7 Inorganic fibres

A *fibre* (inorganic or organic) usually has a diameter <0.25 mm, a length-to-diameter ratio ≥10:1, and a cross-sectional area $<5 \times 10^{-3}$ mm^2; *whiskers* are included in this category.

Fibrous asbestos (a layer silicate which occurs naturally, see Section 14.9) was used for much of the 20th century as an insulating material. However, exposure to asbestos fibres causes lung damage (see Box 14.9) and alternative insulating materials have entered the commercial market. Certain forms of asbestos that do not utilize fibres of length 5–20 μm remain in use, e.g. in brake linings. Glass fibres have a wide range of applications, two of the major ones being insulation and reinforcement of other materials such as plastics. Aluminoborosilicate glass fibres are the most commonly employed. Alumino-lime silicate glass fibres are suited to acid-resistant needs, and when a high tensile strength material is required, aluminosilicate glass is generally appropriate. While the use of glass fibres for insulation is widespread, high-temperature working requires materials such as Al_2O_3 or ZrO_2.

We limit our main discussion in this section to B, C, SiC and Al_2O_3 fibres which can be employed for high-temperature (>1300 K) operations. Much of today's fibre technology stems from the development of new, low-density, high tensile strength materials for air and space travel. Boron fibres were among the first to be developed, with carbon and silicon carbide fibres entering and dominating the market more recently. Silicon carbide has the advantage over both B and C fibres in that it is resistant to oxidation at high temperatures, oxidizing in air only above ≈1250 K.

Boron fibres

Boron fibres can be manufactured by CVD, boron being deposited on a heated tungsten substrate (1550 K) by reaction 28.20. The reactor is schematically represented in Fig. 28.26. The tungsten substrate is drawn through the reactor, making the boron fibre production a continuous process. The proportion of H_2 and BCl_3 that interacts in the reactor is low and unchanged gases are recycled after first separating them from HCl.

$$2BCl_3 + 3H_2 \xrightarrow{\Delta} 2B + 6HCl \qquad (28.20)$$

A final step in manufacture is to coat the fibre with SiC or B_4C. This provides protection against reactions with other elements at high operating temperatures and ensures that the fibre retains its tensile strength at elevated temperature. Typically, the W wire substrate has an 8 μm diameter, the diameter of the boron fibre ≈150 μm, and the SiC or B_4C coating is ≈4 μm thick.

Carbon fibres

Since 1970, the commercial production of carbon fibres has risen dramatically. Where the low weight of a construction material is crucial, carbon-fibre reinforced polymers are now dominating the market. Body-parts for modern military aircraft contain ≤50% by weight of carbon-fibre reinforced composites in place of aluminium. This trend is also being followed in modern commercial aircraft design. The performance of Formula 1 racing cars has been greatly enhanced by turning to body parts constructed from carbon-fibre reinforced materials. The body of the 16.4 Super Sport model of the Bugatti Veyron (Fig. 28.27) is manufactured from carbon fibre, and with an 8.0 litre, 16 cylinder engine, the car achieved a record speed of 431 km h^{-1} in 2010. Carbon fibres are characterized by being stiff but brittle, and have a low density and high tensile strength. The high resistance to thermal shock arises from a high thermal conductivity but low coefficient of thermal expansion.

A number of different grades of carbon fibre are manufactured, but all are made by the thermal degradation of a polymeric organic precursor. Commercial production of carbon fibres uses three carbon-containing precursors:

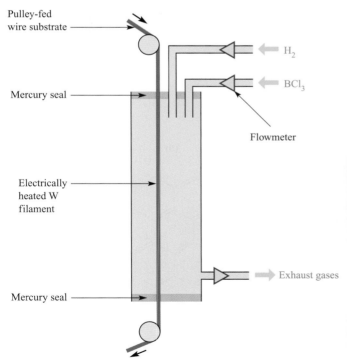

Fig. 28.26 Schematic representation of the assembly used for the manufacture of boron fibres by CVD using a tungsten substrate.

Fig. 28.27 The Bugatti Veyron Super Sport (exhibited at the Paris Motor Show in 2010) comprises a carbon fibre body.

pitch, rayon and polyacrylonitrile (PAN). The earliest type of carbon fibres were manufactured from highly crystalline rayon (cellulose). Rayon fibres are heat-treated ($\approx$500–700 K) in air, followed by heat treatment at 1300 K in an inert atmosphere. These two processes remove H and O as H_2O, CO, CO_2 and CH_4 and produce a graphite-like structure. Carbon fibres derived from rayon possess a relatively low density ($\approx$1.7 g cm^{-3}, compared with 2.26 g cm^{-3} for graphite) and a low tensile strength. Such fibres have limited uses and are not suitable for structural applications.

Pitch is the residue left after distillation of crude petroleum or coal tar. It has a high carbon content and is a cheap starting material. Pitch consists of a mixture of high molecular mass aromatic and cyclic aliphatic hydrocarbons.

The peripheries of the aromatics (e.g. with $M_r > 1000$) often carry long aliphatic chains (Fig. 28.28a), but the structure of pitches from naturally occurring organics is very variable. Heat treatment of aromatic-rich pitches at $\approx$750 K produces a *mesophase* (a liquid crystalline material). This is melt-spun into fibres which, after thermosetting, are *carbonized* by heating at $\geq$1300 K. This latter stage expels H and O as CO_2, H_2O, CO and CH_4. At this stage, the fibres are composed of *graphene sheets* (Fig. 28.28b, see Section 28.8). Heat treatment of this material produces an ordered graphite-like structure with a density of 2.20 g cm^{-3}, slightly lower than that of pure graphite (2.26 g cm^{-3}). During this process, S and N impurities are also removed. PAN-based carbon fibres are manufactured from polyacrylonitrile and may, depending on grade, retain a low nitrogen content.

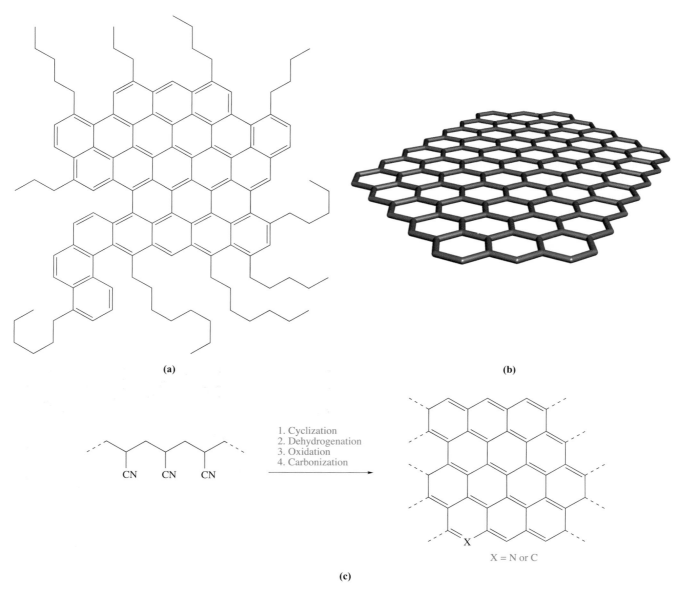

(a)

(b)

1. Cyclization
2. Dehydrogenation
3. Oxidation
4. Carbonization

X = N or C

(c)

 Fig. 28.28 (a) A representation of the type of aromatic molecule present in pitch. (b) Part of a graphene sheet (see Section 28.8). (c) Schematic representation of the formation of PAN-based carbon fibres.

Fig. 28.29 Final production stages of carbon fibres showing fibres being wound on to spools.

Their method of production is summarized in Fig. 28.28c in which atom X represents an arbitrary N content. Processing conditions (heat treatment in particular) determine the mechanical properties of the carbon fibres. Both pitch- and PAN-based carbon fibres (Fig. 28.29) are stronger and have a higher modulus of elasticity (Young's modulus) than those derived from rayon. They therefore have wider applications. Carbon fibres usually require a protective coating to provide resistance to reaction with other elements at elevated temperature.

Carbon-fibre composite materials played an important role in NASA's development of the space shuttle programme (1981–2011). Reinforced *carbon–carbon composites* make up the nose cone and wing leading edges of the space shuttle to provide the resistance to thermal shock and stress required for re-entry into the Earth's atmosphere. Carbon–carbon composites are a particular group of carbon-fibre reinforced materials in which both the bulk material and fibres are carbon. The manufacturing process for carbon–carbon composites starts by impregnating a graphitized rayon fabric with a phenolic resin and then subjecting the material to heat treatment to convert the phenolic resin to carbon. The next stage is impregnation with furfuryl alcohol (**28.8**) followed by heat treatment to convert this component to carbon. Three cycles of this process result in the desired composite material. The composite must be coated with SiC to render it resistant to oxidation. This coating is generated by heating the composite in contact with a mixture of Al_2O_3, SiO_2 and SiC in a furnace. Final impregnation with tetraethyl orthosilicate seals any surface imperfections.

(28.8)

Silicon carbide fibres

The resistance of SiC to high-temperature working and oxidation makes it a valuable advanced material. Fibres of β-SiC are produced by CVD using $R_{4-x}SiCl_x$ precursors or an alkane and chlorosilane in a reactor similar to that in Fig. 28.26. Fibres marketed under the tradename of *Nicalon* are produced by a melt-spinning process. This begins with reactions 28.21 and 28.22, the products of which are pyrolysed to give a carbosilane polymer (scheme 28.23).

$$n Me_2SiCl_2 \xrightarrow{\text{Na}} (Me_2Si)_n \qquad (28.21)$$

$$n Me_2SiCl_2 \xrightarrow{\text{Li}} cyclo\text{-}(Me_2Si)_6 \qquad (28.22)$$

$(Me_2Si)_n$

720 K

720 K

$cyclo\text{-}(Me_2Si)_6$

melt spinning → polycarbosilane fibre ~20 mm diameter

Heat in air

cross-linked fibres with SiO_2 coating

>1800 K

β-SiC fibre ~15 mm diameter

(28.23)

'Melt-spinning' involves heating the polymer until molten and forcing the melt through an appropriately sized aperture. The extruded fibres solidify, but at this stage they are fragile. A curing process is therefore required. After radiation to initiate a radical process, heating in air produces both a coating of SiO_2 and Si–O–Si cross-linkages (Fig. 28.30). This step strengthens the polymer, makes it water-insoluble and alters its properties so that it will not melt when pyrolysed in the final manufacturing stage (scheme 28.23). As a consequence of the precursors and the curing process, Nicalon fibres contain an excess of carbon and have a significant oxygen content (typically $SiC_{1.34}O_{0.36}$). Fibres with improved mechanical properties (Young's modulus $\geq$ that of steel) can be manufactured by modifying the processes shown in scheme 28.23. For example, HI-Nicalon and HI-Nicalon-S have typical compositions of $SiC_{1.39}O_{0.01}$ and $SiC_{1.05}$, respectively. The latter (close to stoichiometric silicon carbide) exhibits excellent high-temperature properties in addition to high tensile strength. Silicon carbide fibres can be made into rope, woven into cloth or used to reinforce a SiC matrix. They find wide applications, including in military aircraft components.

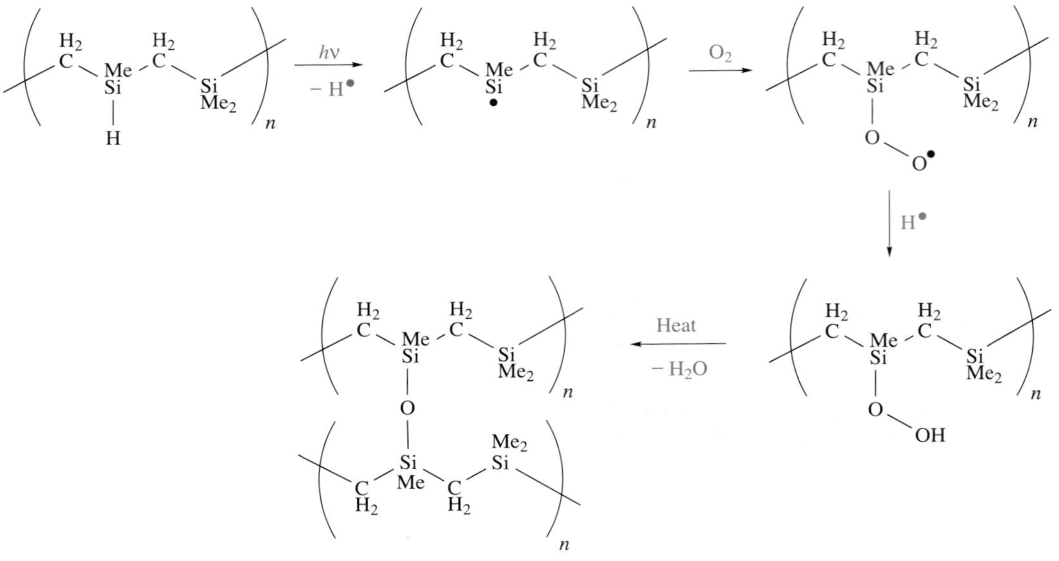

Fig. 28.30 Reactions involved in the radiation and thermal curing processes in the production of cross-linked polycarbosilane polymers.

Alumina fibres

Alumina fibres (often with silica content) are produced commercially on a large scale. Their high tensile strength, flexibility and inertness make them valuable in, for example, rope, thread (suitable for cloth manufacture), insulating material and electrical-cable coverings. A number of different manufacturing methods are in operation for the production of alumina–silica fibres, depending on the type of fibre and also the manufacturer. Polycrystalline Al_2O_3 fibres can be formed by extruding hydrated alumina slurries through suitable nozzles and then heating the extruded material. As an example of a fibre with silica content, continuous fibres containing 15% SiO_2 by weight are manufactured starting from Et_3Al. This is subject to partial hydrolysis to give a polymeric material which is dissolved along with an alkyl silicate in a suitable solvent. The viscous solution is amenable to fibre production by gel-spinning; the fibres so formed are heated (*calcined*) to convert the material into alumina–silica. Further heating results in the formation of a polycrystalline material.

28.8 Graphene

Nanoparticles and nanotechnology are well-established terms in the scientific community, in industry and within popular science. To be classed as 'nanoscale', an object should have at least one dimension of the order of 10^{-9} m (1 nm). To put this into the context of an atom, the covalent and van der Waals radii of carbon are 0.077 nm (77 pm) and 0.185 nm (185 pm), respectively.

In 2010, Geim and Novoselov were awarded the Nobel Prize in Physics for 'groundbreaking experiments regarding

the two-dimensional material graphene'. Graphene is a single sheet (i.e. one atom thick, Fig. 28.28b) cut out from a graphite lattice. In practice, this can be achieved (initially in 2004) by exfoliating flakes from a block of highly oriented pyrolytic graphite using adhesive tape and transferring them to an SiO_2 substrate. An optical microscope is used to distinguish single graphene sheets from multiple-layer flakes.[†] The strengths of the C–C σ- and π-bonds stabilize the single sheet structure.

> ***Graphene*** is a single layer, one atom thick, of the graphite structure.

> ***Highly oriented pyrolytic graphite*** (HOPG) is a graphite that exhibits a high degree of crystallographic orientation such that the c-axis (the axis perpendicular to the carbon sheets) of the crystallites has an angular range of <1°.

> The process by which layers of a multi-layered material are separated from one another is called ***exfoliation.***

Graphene exhibits remarkable physical properties, quite distinct from those of graphite, and has an exciting future as a material for advanced electronics (Fig. 28.31). Graphene is virtually optically transparent ($\approx 98\%$). It is a gapless semiconductor with an electrical resistivity so small that the charge carriers (electrons) travel through the sheet, behaving as massless elementary particles with relativistic speeds. In a normal semiconductor, the quantum mechanical behaviour of an electron is described in terms

---

[†] See: A.K. Geim and A.H. MacDonald (2007) *Physics Today*, August issue, p. 35.

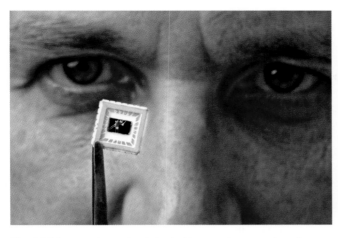

Fig. 28.31 Physicist Andre Geim holds a graphene transistor in a pair of tweezers.

of the Schrödinger equation, and the energy of the electron is proportional to the square of its momentum. In graphene, the symmetry of the structure results in the charge carriers behaving as Dirac fermions, and there is a linear relationship between the energy of an electron and its momentum. The current density in graphene is around $10^8 \, \text{A cm}^{-2}$ which is two to three orders of magnitude greater than that of copper.

Graphene in the form of thin ribbons (so-called graphene nanoribbons) exhibits electronic and mechanical properties that depend upon the width and edge structure of the ribbon. Starting from a graphene sheet, two different edge structures called the zigzag or armchair edge result depending upon the direction of a cut through the sheet (Fig. 28.32). Transmission electron microscopy (TEM) has been used to observe the graphene edge structure at atomic resolution. Movement of C atoms caused by

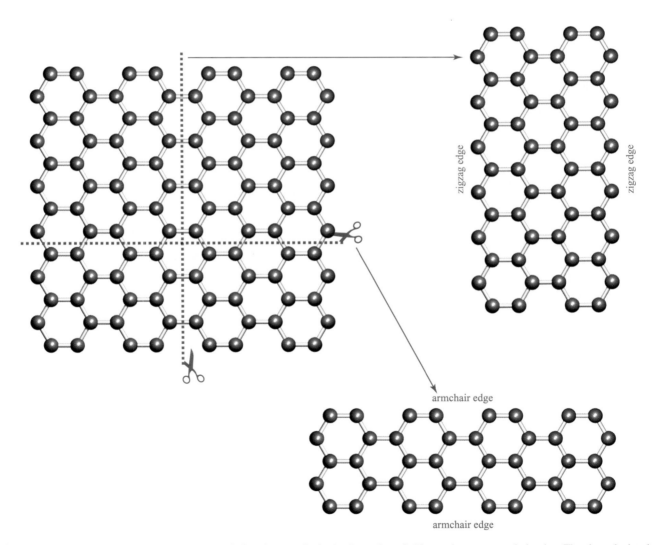

Fig. 28.32 Cutting a graphene sheet in orthogonal directions results in the formation of either a zigzag or armchair edge. The sheet depicted could also be cut diagonally to give zigzag or armchair edges.

collisions with high energy electrons during the TEM analysis results in reconstruction of the edge and the formation of a new edge structure involving pentagons and heptagons[†]:

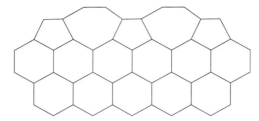

The chemistry of graphene is still in its infancy. Chemical functionalization of graphene is a means of:

- opening the band gap and tuning the electronic properties;
- solubilizing graphene to allow reaction chemistry to be more widely explored;
- solubilizing graphene to permit solution-processing of devices based on graphene.

Hydrogenation of graphene involves the formation of C–H bonds and the conversion of sp^2 to sp^3 C, thus disrupting the planar structure of the sheet. In principle, complete hydrogenation produces graphane, $(CH)_x$. Graphene can be hydrogenated using atomic H in low-pressure plasma conditions. Annealing at 720 K in Ar atmosphere regenerates graphene. Catalytic hydrogenation has been achieved by passing H_2 over a Ni/Al_2O_3 catalyst at 1100 K to give $H^{\bullet}$ radicals which react with graphene. Raman and IR spectroscopic data provide evidence for the formation of C–H bonds.

An alternative approach to opening the band gap of graphene is to start with graphene oxide which is formed when graphite undergoes acid oxidation, e.g. using $KClO_3$ and HNO_3, or $KMnO_4$ and H_2SO_4. The oxidized graphene sheets that result contain carboxylate, epoxy and hydroxyl substituents, making the sheets hydrophilic and soluble in a range of solvents. Although graphene oxide is an insulator and of no use for electronic applications, its controlled reduction decreases the band gap, allowing the formation of hybrid materials between the extremes of graphene and graphene oxide. Reduction can be carried out using $NaBH_4$ or N_2H_4. Graphene oxide is also a starting point for the preparation of other functionalized graphenes, e.g. conversion of CO_2H to CO_2R or CONHR groups.

A method of solubilizing graphene is to capitalize on the π-stacking interactions between graphene and pyrenes (**28.9**) or perylenes (**28.10**) bearing hydrophilic substituents, e.g. carboxylic acid groups. Thus, graphene can be stripped from graphite and stabilized in solution by stirring graphite in appropriately functionalized pyrene or perylene solvents.

[†] See: P. Koskinen, S. Malola and H. Häkkinen (2009) *Phys. Rev. B*, vol. 80, article 073401; Ç.Ö. Girit *et al.* (2009) *Science*, vol. 323, p. 1705.

Pyrene

Perylene

(28.9)   **(28.10)**

Between the beginning of 2008 and mid-2011, >14 000 publications on the chemistry and physics of graphene and its derivatives appeared in the scientific literature, indicating the excitement that this material has generated. The scale-up of graphene production (e.g. by epitaxal growth of graphene sheets on a silicon carbide wafer) is necessary before commercial applications such as field effect transistors, sensors, transparent electrically conducting films and composite materials can be achieved.

28.9 Carbon nanotubes

Carbon nanotubes are some of the best known nanoparticles. Their semiconducting or metallic properties, combined with high tensile strength, have resulted in an explosion of interest at both the research and commercial levels. The discovery of carbon nanotubes (in 1991) and research into their synthesis and properties are closely related to those of fullerenes and of carbon fibres. A carbon nanotube which has its ends capped is essentially an elongated fullerene. Carbon nanotubes are synthesized by electrical arc discharge between graphite rods (as for fullerenes, see Section 14.4), laser vaporization of graphite, or CVD methods. Use of the arc discharge method favours the formation of multi-walled carbon nanotubes (MWNT), but if a catalyst (e.g. Fe, Ni) is present in the graphite, single-walled carbon nanotubes (SWNT) can be produced. SWNTs are formed by the laser vaporization technique, and gram-scale syntheses can be carried out. These nanotubes usually form in bundles (Fig. 28.33) as a result of the van der Waals forces that operate between the surfaces of adjacent tubes. Suitable volatile precursors for CVD are ethyne and ethene, with Fe, Ni or Co catalysts to promote the formation of SWNTs. Purification of carbon nanotubes involves oxidation (with HNO_3 or heating in air) to remove amorphous carbon and metal catalysts, followed by sonication (the use of high-frequency sound waves to disperse the insoluble nanotubes in solution) and the use of chromatographic techniques. Treatment with oxidizing media (e.g. HNO_3/H_2SO_4) is a means of 'cutting' long carbon nanotubes into shorter lengths, and also introduces carbonyl and carboxylate functionalities (see later). Carbon nanotubes are commercially available, typical diameters being 1–100 nm (SWNT ≪ MWNT) and lengths of the order of micrometres. Further progress

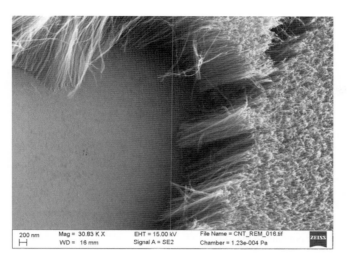

Fig. 28.33 Field-emission scanning electron micrograph (SEM) of bundles of carbon nanotubes grown on an Si wafer. Magnification is × 35 000. The sample was provided by Dr Teresa de los Arcos, Department of Physics, University of Basel. SEM image: Dr A. Wirth-Heller, FHNW, Basel.

remains to be made in synthetic methodology for the selective formation of SWNTs and MWNTs.

A SWNT is a rolled graphene sheet (Fig. 28.28b) with the C atoms connected to form a single, hollow tube. Three classes of SWNT are defined according to the vectors shown in Fig. 28.34a. Starting at a C atom (0, 0), a vector drawn in the zigzag direction is defined as having an angle

$\theta = 0°$. A vector perpendicular to this direction defines the axis of the zigzag carbon nanotube. The circumference of the zigzag nanotube is determined by the number of C_6 rings, n, through which the first vector cuts. For example, in Fig. 28.34a, $n = 5$. Starting at a C atom (0, 0), a vector with $\theta = 30°$ defines the open end of an armchair nanotube, and the axis of this tube is defined by a vector perpendicular to the first, as shown in Fig. 28.34a. Any carbon nanotube is characterized by a vector C_h (eq. 28.24). For an armchair tube, the direction of the vector C_h is fixed with $\theta = 30°$, but the magnitude depends on the number of C_6 rings through which the vector cuts. This is the sum of vectors $na_1 + ma_2$, their magnitude and direction being defined in the diagram in eq. 28.24. An armchair carbon nanotube is prefixed by a label (n, m), e.g. a (6,6)-armchair nanotube corresponds to a vector $C_h = 6a_1 + 6a_2$.

$$C_h = na_1 + ma_2 \qquad\qquad\qquad (28.24)$$

Figures 28.34b and 28.34c show part of a (5,5)-armchair and (10,0)-zigzag SWNT, respectively. The zigzag and armchair SWNTs are the limiting structures and each is achiral. The vector C_h is called the *chiral vector* and any SWNT with a value of θ between 0 and 30° is chiral. The three classes

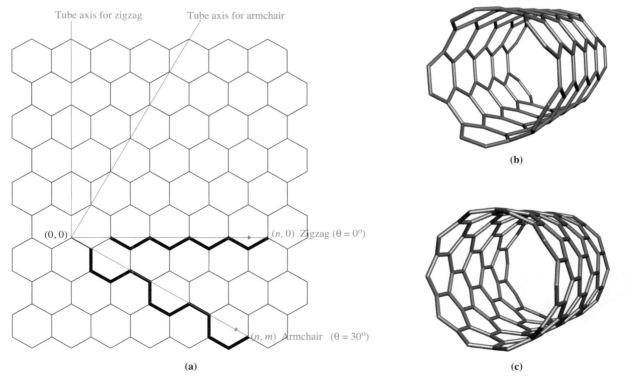

Fig. 28.34 (a) Vectors on a graphene sheet that define achiral zigzag and armchair carbon nanotubes. The angle θ is defined as being 0° for the zigzag structure. The bold lines define the shape of the open ends of the tube. (b) An example of an armchair carbon nanotube. (c) An example of a zigzag carbon nanotube.

of SWNT are therefore:

- $(n, 0)$-zigzag ($\theta = 0°$);
- (n, m)-armchair ($\theta = 30°$);
- (n, m)-chiral ($0° < \theta < 30°$).

Self-study exercise

Draw a graphene sheet of size 8×8 C_6 rings. Construct a vector triangle that defines $C_h = 5n + 5m$, and show that the end of the SWNT that this defines corresponds to that in Fig. 28.34b.

In a batch of newly synthesized carbon nanotubes, a significant proportion of carbon nanotubes are capped, i.e. the open ends of the tubes shown in Fig. 28.34 are capped by hemispherical units, formally derived from fullerenes. By removing atoms from C_{60} (Fig. 14.5), end-capping units compatible with an armchair SWNT (Fig. 28.35a) or zigzag SWNT (Fig. 28.35b) can be generated. As noted earlier, arc discharge methods of synthesis tend to favour the formation of MWNTs. These consist of concentric tubes, packed one inside another.

The physical properties that make carbon nanotubes exciting materials of the future are their electrical conductivity and mechanical properties. The graphene sheets from which SWNTs and MWNTs are made are robust, and carbon nanotubes are among the stiffest and strongest materials known. The conducting properties of carbon nanotubes depend on their structure, specifically the relationship between the tube axis and graphene sheet (Fig. 28.34a). The band structure of the material changes with (n, m). As a result, armchair nanotubes are metallic conductors, while zigzag and chiral carbon nanotubes may be semiconductors or metallic conductors. Future applications of carbon nanotubes are predicted to include nanoelectronic and optoelectronic devices, microelectrodes, sensors and polymer composites. Short carbon nanotubes may be suitable for use as components in microelectronic devices.

The tendency for SWNTs to form bundles (Fig. 28.33) results in their being insoluble in water and organic solvents, although they can be dispersed in solvents by sonication. Modification of nanotube surfaces is one way of minimizing

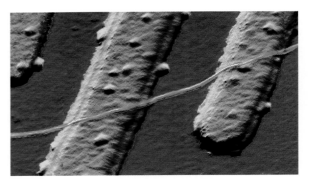

Fig. 28.36 A coloured atomic force micrograph of a carbon nanotube 'wire' (shown in blue) lying over platinum electrodes (in yellow). The diameter of the carbon nanotube is 1.5 nm (1500 pm), corresponding to the 'wire' being 10 atoms wide (r_{cov} C = 77 pm).

their aggregation, and there is much current research activity aimed at functionalizing carbon nanotubes. In addition to this leading to manipulation of single tubes, functionalization results in enhanced solubility, more diverse chemical properties and the possibility of attaching tubes to substrates. Many modified carbon nanotubes have been investigated, and three general strategies can be followed:

- functionalization by C–X covalent bond formation;
- modification based on van der Waals interactions between a carbon nanotube and another molecular species;
- using the tube as a host to generate endohedral species (compare this with endohedral fullerenes, e.g. structure 14.6).

Characterization of the functionalized materials is not trivial, and spectroscopic methods (e.g. ^{13}C NMR, IR, Raman) and microscopy (e.g. atomic force microscopy (AFM, Fig. 28.36) and scanning electron microscopy (SEM)) are among the techniques used. Selected examples of functionalization methods are described below.

Figure 28.37 illustrates fluorination of the walls of a carbon nanotube, followed by further functionalization by reactions with Grignard or organolithium reagents. The figure shows 1,2-addition of F_2, but theoretical studies suggest there is little energetic difference between 1,2- and 1,4-additions. The change from sp^2 to sp^3 hybridization that accompanies an addition reaction (see also Fig. 14.8) means that π-conjugation is lost in the region of these C atoms and this affects the electrical conductivity of the carbon nanotube. In the limiting case, the material becomes an insulator.

Acid oxidation using concentrated HNO_3 and H_2SO_4 under sonication is used to purify and shorten carbon nanotubes. This treatment introduces CO and CO_2H functionalities along the walls and at the ends of the tubes, and this is a valuable starting point for further functionalization (e.g. eqs. 28.25 and 28.26). Reaction at an open end of the nanotube converts a C–H into a C–CO_2H group without loss of aromatic π-character.

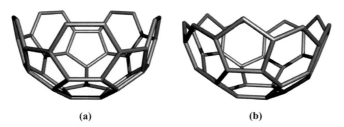

Fig. 28.35 Examples of capping units that covalently bond to open carbon nanotubes to yield closed tubes. The examples shown are portions of a C_{60} molecule and are compatible with (a) an armchair and (b) a zigzag carbon nanotube.

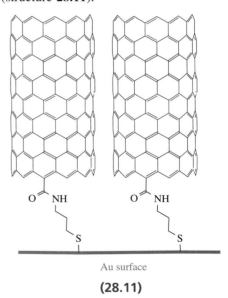

Fig. 28.37 Schematic representation of fluorination of a carbon nanotube, and reaction of a fluoro-substituted derivative with organolithium and Grignard reagents.

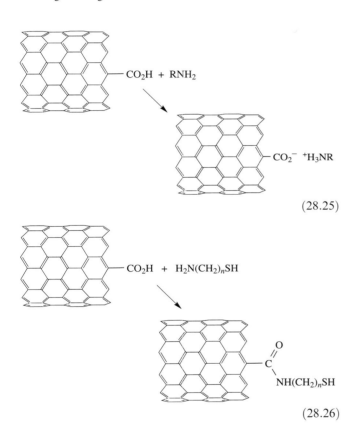

(28.25)

(28.26)

Tethering ordered assemblies of carbon nanotubes to surfaces is an important step towards producing materials suitable for applications in microelectronic devices. Thiol derivatives can be anchored to gold surfaces, and derivatives of the type formed in reaction 28.26 are suitable for this purpose (structure **28.11**).

(28.11)

KEY TERMS

The following terms and abbreviations have been introduced in this chapter. Do you know what they mean?

- ❏ cation or anion conductor
- ❏ fast ion conductor
- ❏ TCO
- ❏ DSC
- ❏ electroluminescent
- ❏ OLED
- ❏ LEC

- ❏ superconductor
- ❏ ceramic material
- ❏ opacifier
- ❏ chemical vapour deposition (CVD)
- ❏ metal–organic chemical vapour deposition (MOCVD)
- ❏ epitaxial growth

- ❏ ferroelectric
- ❏ inorganic fibre
- ❏ graphene
- ❏ highly oriented pyrolytic graphite
- ❏ exfoliation
- ❏ carbon nanotube

FURTHER READING

General and introductory texts

M. Ladd (1994) *Chemical Bonding in Solids and Fluids*, Ellis Horwood, Chichester.

U. Schubert and N. Hüsing (2000) *Synthesis of Inorganic Materials*, Wiley-VCH, Weinheim.

L. Smart and E. Moore (2005) *Solid State Chemistry: An Introduction*, 3rd edn, CRC Press, Taylor & Francis, Boca Raton.

A.R. West (1999) *Basic Solid State Chemistry*, 2nd edn, Wiley-VCH, Weinheim – An introductory text which includes structures and bonding in solids, and electrical, magnetic and optical properties.

More specialized articles

A.K. Cheetham and P. Day, eds (1992) *Solid State Chemistry*, Clarendon Press, Oxford – Two volumes covering techniques (vol. 1) and compounds (vol. 2) in detail.

D.R. Dreyer, S. Park, C.W. Bielawski and R.S. Ruoff (2010) *Chem. Soc. Rev.*, vol. 39, p. 228 – 'The chemistry of graphene oxide'.

R.A. Eppler (1998) 'Ceramic colorants' in *Ullmann's Encyclopedia of Industrial Inorganic Chemicals and Products*, Wiley-VCH, Weinheim, vol. 2, p. 1069 – Describes types and applications of ceramic pigments.

E. Fortunato, D. Ginley, H. Hosono and D.C. Paine (2007) *MRS Bulletin*, vol. 32, p. 242 – 'Transparent conducting oxides for photovoltaics'.

L. Gherghel, C. Kübel, G. Lieser, H.-J. Räder and K. Müllen (2002) *J. Am. Chem. Soc.*, vol. 124, p. 13130 – 'Pyrolysis in the mesophase: A chemist's approach toward preparing carbon nano- and microparticles'.

M. Grätzel (2003) *J. Photochem. Photobiol. C*, vol. 4, p. 145 – 'Dye-sensitized solar cells'.

A.C. Grimsdale, J. Wu and K. Müllen (2005) *Chem. Commun.*, p. 2197 – 'New carbon-rich materials for electronics, lithium battery, and hydrogen storage applications'.

A.C. Grimsdale and K. Müllen (2005) *Angew. Chem. Int. Ed.*, vol. 44, p. 5592 – 'The chemistry of organic nanomaterials'.

S. Guo and S. Dong (2011) *Chem. Soc. Rev.*, vol. 40, p. 2644 – 'Graphene nanosheet: synthesis, molecular engineering, thin film, hybrids, and energy and analytical applications'.

A. Gurevich (2011) *Nature Mater.*, vol. 10, p. 255 – 'To use or not to use cool superconductors?'

A.N. Khlobystov, D.A. Britz and G.A.D. Briggs (2005) *Acc. Chem. Res.*, vol. 38, p. 901 – 'Molecules in carbon nanotubes'.

T.T. Kodas and M. Hampden-Smith, eds (1994) *The Chemistry of Metal CVD*, VCH, Weinheim – Covers the deposition of a range of metals from organometallic precursors.

K.P. Loh, Q. Bao, P.K. Ang and J. Yang (2010) *J. Mater. Chem.*, vol. 20, p. 2277 – 'The chemistry of graphene'.

G.J. Meyer (2010) *ACS NANO*, vol. 4, p. 4337 – 'The 2010 Millennium Technology Grand Prize: Dye-sensitized solar cells'.

J. Shinar and R. Shinar (2010) in *Comprehensive Nanoscience and Technology*, eds D.L. Andrews, G.D. Scholes and G.P. Wiederrecht, Elsevier, Oxford, vol. 1, p. 73 – 'An overview of organic light-emitting diodes and their applications'.

D. Tasis, N. Tagmatarchis, A. Bianco and M. Prato (2006) *Chem. Rev.*, vol. 106, p. 1105 – 'Chemistry of carbon nanotubes'.

M.E. Thompson, P.E. Djurovich, S. Barlow and S. Marder (2007) in *Comprehensive Organometallic Chemistry III*, eds R.H. Crabtree and D.M.P. Mingos, Elsevier, Oxford, vol. 12, p. 101 – 'Organometallic complexes for optoelectronic applications'.

M.S. Whittingham (2004) *Chem. Rev.*, vol. 104, p. 4271 – 'Lithium batteries and cathode materials'.

C.H. Winter and D.M. Hoffman, eds (1999) *Inorganic Materials Synthesis*, Oxford University Press, Oxford – A detailed coverage which includes inorganic thin films.

Y. Zhu, S. Murali, W. Cai, X. Li, J.W. Suk, J.R. Potts and R.S. Ruoff (2010) *Adv. Mater.*, vol. 22, p. 3906 – 'Graphene and graphene oxide: Synthesis, properties and applications'.

PROBLEMS

28.1 If Ag electrodes are placed in contact with and on either side of a piece of bulk AgI (mp 831 K) heated at 450 K, and current is passed through the cell for a given period, it is found that one electrode gains mass and the other loses mass. Rationalize these observations.

28.2 Comment on the following values of electrical conductivities: Na β-alumina, $3 \times 10^{-2}\,\Omega^{-1}\,cm^{-1}$ (at 298 K); Li_3N, $5 \times 10^{-3}\,\Omega^{-1}\,cm^{-1}$ (at 298 K); NaCl, $10^{-15}\,\Omega^{-1}\,cm^{-1}$ (at 300 K). Would you expect these values to be direction-independent with respect to the crystal under study?

28.3 A recently developed solid state battery consists of lithium and V_6O_{13} electrodes separated by a solid polymer electrolyte. Suggest how this battery might operate.

28.4 Discuss the variation in electrical conductivities along the series TiO, VO, MnO, FeO, CoO and NiO.

28.5 (a) What is meant by a *degenerate semiconductor* and why is this relevant to TCOs?
(b) Give examples of TCOs, detailing the doping that is used in the materials.

28.6 (a) Outline the principle of operation of a dye-sensitized solar cell.
(b) Why should an ideal dye in a DSC absorb light over the whole visible range with high values of ε_{max}?

28.7 If only the fluorescent excited states are harvested for light emission in an OLED, the maximum efficiency of the device is 25%. Explain why this is. What can be done to overcome the problem?

28.8 (a) What is electroluminescence?
(b) Draw a schematic representation of an OLED device that involves the use of the complex shown below. Describe the function of each of the layers in the diagram that you have drawn.

28.9 (a) The structure of $YBa_2Cu_3O_7$ can be described as consisting of rock salt and perovskite layers. Describe the origin of this description.
(b) Why is the potential replacement of NbTi by high-temperature superconducting components in MRI equipment of commercial interest?

28.10 (a) Describe the structure of the Chevrel phase $PbMo_6S_8$.
(b) What gives rise to the superconductivity exhibited by $PbMo_6S_8$?

28.11 Explain what is meant by 'doping' using as your examples (a) MgO doping of ZrO_2, (b) LaF_3 doping of CaF_2, (c) B doping of Si, and (d) As doping of Si.

28.12 (a) Describe the layered structure of FeSe.
(b) How are the structures of NaFeAs and LaOFeAs related to that of FeSe?
(c) Whereas LaOFeAs is not superconducting, $LaO_{1-x}F_xFeAs$ is a superconductor and the value of T_c depends on x. Comment on this statement.

28.13 (a) Describe the relationship between the structures of MgB_2 and graphite.
(b) How does the electronic structure of MgB_2 differ from that of graphite, and how does this affect the properties of the materials?

28.14 Suggest likely products in the following reactions; (the reactions as shown are not necessarily balanced):
(a) $xLiI + V_2O_5 \xrightarrow{\Delta}$
(b) $CaO + WO_3 \xrightarrow{\Delta}$
(c) $SrO + Fe_2O_3 \xrightarrow{\Delta, \text{ in presence of } O_2}$

28.15 Suggest possible solid state precursors for the formation of the following compounds by pyrolysis reactions: (a) $BiCaVO_5$; (b) the Mo(VI) oxide $CuMo_2YO_8$; (c) Li_3InO_3; (d) $Ru_2Y_2O_7$.

28.16 Give a brief outline of a typical CVD process and give examples of its use in the semiconductor industry.

28.17 Briefly discuss each of the following.
(a) Precursors for, and composition and uses of, CVD wear-resistant coatings.
(b) The production of GaAs thin films.
(c) The advantages of using LEDs over traditional glass-reflector cat's eyes for road-lane markings.
(d) Problems in developing CVD methods for the deposition of perovskite and cuprate superconductors.

28.18 Rationalize the following statement: a graphene sheet consists of two interpenetrating triangular sublattices, and the unit cell of graphene contains two C atoms.

28.19 Graphene is insoluble in common solvents. Describe ways in which the material can be solubilized.

28.20 The properties of graphene nanoribbons depend upon their edge structure. What are the different edge structures and how do they arise?

28.21 (a) Explain how single-walled carbon nanotubes are classified in zigzag and armchair tubes.
(b) What inherent properties of single-walled carbon nanotubes make those formed by arc discharge or laser vaporization methods unsuitable for immediate applications?
(c) Give examples of how the problems noted in (b) can be overcome.

OVERVIEW PROBLEMS

28.22 (a) Describe the structure of lithium nitride and explain how it is able to function as a lithium ion conductor. The structures of Li_3P and Li_3As are analogous to that of the nitride. How do you expect the degree of ionic character in these compounds to vary?

(b) Epitaxial MgB_2 films can be grown from B_2H_6 and Mg vapour at temperatures up to 1030 K. Explain the meaning of 'epitaxial' and state what particular properties the films possess.

28.23 (a) MOCVD with $Al(O^iPr)_3$ as the precursor can be used to deposit α-Al_2O_3. Outline the principle of MOCVD, commenting on the required properties of the precursors.

(b) Fibres of InN can be grown at 476 K by the following reaction; nano-sized metal droplets act as catalytic sites for the formation of the crystalline fibres.

$$2H_2NNMe_2 + In^tBu_2(N_3)$$
$$\longrightarrow InN + 2Me_2NH + 2^tBuH + 2N_2$$

When $^tBu_3In$ replaces $In^tBu_2(N_3)$, only amorphous products and metallic In are obtained. What is the likely role of the 1,1-dimethylhydrazine in the reaction, and what appears to be the primary source of nitrogen for the InN? Group 13 nitrides have applications in blue/violet LED displays. What controls the wavelength of emitted light in compounds of this type?

28.24 (a) At 670 K, CaF_2 (mp = 1691 K) doped with 1% NaF has an electrical conductivity of $0.1\,\Omega^{-1}\,m^{-1}$. Suggest how this conductivity arises.

(b) The value of T_c for $YBa_2Cu_3O_7$ is 93 K. Sketch the change in electrical resistivity as a function of temperature as $YBa_2Cu_3O_7$ is cooled from 300 to 80 K. How does the shape of this graph differ from those that describe the change in resistivity with temperature for a typical metal and a typical semiconductor?

INORGANIC CHEMISTRY MATTERS

28.25 Legislation requires the phasing out of incandescent lighting. How does inorganic chemistry contribute to the development of solid state lighting?

28.26 Give a summary of how conventional superconductors have impacted on our lives. Include both scientific and more general applications.

28.27 Describe the construction and operation of a lithium-ion battery. Give examples of applications including those in the motor vehicle industry.

28.28 Describe how carbon fibres are manufactured. Summarize the properties and applications of carbon fibres and carbon–carbon composites.

Topics

29

The trace metals of life

29.1 Introduction

When one considers the chemistry of biological processes, the boundary between inorganic and organic chemistry is blurred. The *bulk biological* elements that are essential to all life include C, H, N, O (the four most abundant elements in biological systems) along with Na, K, Mg, Ca, P, S and Cl. The fundamental elements that make up the building blocks of biomolecules (e.g. amino acids, peptides, carbohydrates, proteins, lipids and nucleic acids) are C, H, N and O, with P playing its part in, for example, ATP and DNA (see Box 15.11) and S being the key to the coordinating abilities of cysteine residues in proteins. The roles of the less abundant, but nonetheless essential, elements include osmotic control and nerve action (Na, K and Cl), Mg^{2+} in chlorophyll (see Section 12.8), Mg^{2+}-containing enzymes involved in phosphate hydrolysis, structural functions of Ca^{2+} (e.g. bones, teeth, shells) and triggering actions of Ca^{2+} (e.g. in muscles). The *trace metals* are V, Cr, Mn, Fe, Co, Ni, Cu, Zn and Mo, while *trace non-metals* comprise B, Si, Se, F and I. Their essentiality to life can be summarized as follows:

- V: accumulated by a few organisms (see Box 29.1), and has been shown to be essential for growth in rats and chicks;
- Cr: essential (see Table 29.1);
- Mn, Fe, Cu, Ni, Zn: essential to all organisms (see Table 29.1);
- Co: essential to mammals and many other organisms (see Table 29.1);
- Mo: essential to all organisms (see Table 29.1) although green algae may be an exception;

- B: essential to green algae and higher plants, but its exact role is unknown (see Box 13.1);
- Si: exoskeletons of marine diatoms composed of hydrated silica (see Fig. 14.21), but its role in other biological systems is less well defined;[†]
- Se: essential to mammals and some higher plants;
- F: its role is not fully established but its deficiency causes dental caries;
- I: essential to many organisms.

Despite their crucial role in life, the trace metals make up only a tiny fraction of the human body-weight (Table 29.1). In this chapter we look at the ways in which living systems store metals, and the manner in which trace metal ions take part in the transport of molecules such as O_2, electron transfer processes and catalysis. It is assumed that the reader has already studied Chapters 19 and 20, and is familiar with the general principles of *d*-block coordination chemistry: a study of the trace metals in biological systems is *applied coordination chemistry*.

Research progress in bioinorganic chemistry has been greatly assisted in recent years by the development of methods to solve protein structures using X-ray diffraction and NMR spectroscopy. Readers are encouraged to make use of the Protein Data Bank (PDB) to update the information given in this chapter; information is available using the worldwide web (http://www.rcsb.org/pdb). In figure captions in this chapter, PDB codes are given for protein structures drawn using data from the Protein Data Bank.

[†] See: J.D. Birchall (1995) *Chem. Soc. Rev.*, vol. 24, p. 351 – 'The essentiality of silicon in biology'.

BIOLOGY AND MEDICINE

Box 29.1 The specialists: organisms that store vanadium

Storage and transport of vanadium are specialized affairs. Just why certain organisms accumulate high levels of vanadium is unknown and the biological functions of this trace metal have yet to be established.

The fungus *Amanita muscaria* (the deadly poisonous *fly agaric* toadstool) contains ≥400 times more vanadium than is typical of plants, and the amount present does not reflect the vanadium content of the soil in which the fungus grows. *Amanita muscaria* takes up the metal by using the conjugate base of (S,S)-2,2′-(hydroxyimino)dipropionic acid (H$_3$L) to transport and store the trace metal as the V(IV) complex $[VL_2]^{2-}$, *amavadin*.

(S,S)-2,2′-(hydroxyimino)dipropionic acid

The formation of a complex of 'naked' V(IV) is in contrast to the more common occurrence of complexes containing $[VO]^{2+}$ (see Section 21.6). The structure of the amavadin derivative Λ-$[V(HL)_2]\cdot H_3PO_4\cdot H_2O$ has been solved. The complex contains five chiral centres, one of which is the V(IV) centre. The latter is 8-coordinate and each HL^{2-} ligand acts as an N,O,O',O''-donor. The N–O-unit coordinates in a side-on (η^2) manner. *Amanita muscaria* contains a 1:1 mixture of the Λ- and Δ-forms of amavadin. Amavadin undergoes a reversible 1-electron oxidation without a change in structure, and this observation may be significant in view of a possible role in electron transfer.

The levels of vanadium present in some ocean-dwelling ascidians, such as the sea squirt *Ascidia nigra*, are extraordinarily high, up to 10^7 times greater than in the surrounding water. The metal is taken up from seawater (where it is typically present ≈ 1.1–1.8×10^{-3} ppm) in the form of $[VO_4]^{3-}$ and is stored in vacuoles in specialized blood cells called *vanadocytes*. Here it is reduced to V^{3+} or $[VO]^{2+}$ by the polyphenolic blood pigment *tunichrome*. (Note the structural relationship between tunichrome and L-DOPA, **25.13**.) Storage of vanadium must involve the formation of V^{3+} or $[VO]^{2+}$ complexes, but the nature of these species is not known.

The fly agaric fungus (*Amanita muscaria*).

Further information

R.E. Berry, E.M. Armstrong, R.L. Beddoes, D. Collison, S.N. Ertok, M. Helliwell and C.D. Garner (1999) *Angew. Chem. Int. Ed.*, vol. 38, p. 795 – 'The structural characterization of amavadin'.

C.D. Garner, E.M. Armstrong, R.E. Berry, R.L. Beddoes, D. Collison, J.J.A. Cooney, S.N. Ertok and M. Helliwell (2000) *J. Inorg. Biochem.*, vol. 80, p. 17 – 'Investigations of amavadin'.

T. Hubregtse, E. Neeleman, T. Maschmeyer, R.A. Sheldon, U. Hanefeld and I.W.C.E. Arends (2005) *J. Inorg. Biochem.*, vol. 99, p. 1264 – 'The first enantioselective synthesis of the amavadin ligand and its complexation to vanadium'.

D. Rehder (1991) *Angew. Chem. Int. Ed.*, vol. 30, p. 148 – 'The bioinorganic chemistry of vanadium'.

Tunichrome

Table 29.1 Mass of each trace metal present in an average 70 kg human, and a summary of where the trace metals are found and their biological roles.

Metal	Mass / mg	Biological roles
V	0.11	Enzymes (nitrogenases, haloperoxidases)
Cr	14	Claimed (not yet proven) to be essential in glucose metabolism in higher mammals
Mn	12	Enzymes (phosphatase, mitochondrial superoxide dismutase, glycosyl transferase); photoredox activity in Photosystem II (see eq. 21.54 and discussion)
Fe	4200	Electron-transfer systems (Fe–S proteins, cytochromes); O_2 storage and transport (haemoglobin, myoglobin, haemerythrin); Fe storage (ferritin, transferritin); Fe transport proteins (siderophores); in enzymes (e.g. nitrogenases, hydrogenases, oxidases, reductases)
Co	3	Vitamin B_{12} coenzyme
Ni	15	Enzymes (urease, some hydrogenases)
Cu	72	Electron transfer systems (blue copper proteins); O_2 storage and transport (haemocyanin); Cu transport proteins (ceruloplasmin)
Zn	2300	Acts as a Lewis acid (e.g. in hydrolysis processes involving carboxypeptidase, carbonic anhydrase, alcohol dehydrogenase); structural roles
Mo	5	Enzymes (nitrogenases, reductases, hydroxylases)

Amino acids, peptides and proteins: some terminology

In this chapter, we refer to polypeptides and proteins, and we now give a brief résumé of some of the terminology needed.[†]

A **peptide** is formed by condensation of two amino acids.
A **polypeptide** contains ≥ 10 amino acid residues.
A **protein** is a high molecular mass polypeptide.

A *polypeptide* in Nature is formed by the condensation, in varying sequences, of the 20 naturally occurring α-amino acids. Structure **29.1** gives the general formula of an amino acid and **29.2** shows a peptide link formed after the condensation of two amino acid residues. A peptide chain has an *N-terminus* (corresponding to an NH_2 group) and a *C-terminus* (corresponding to a CO_2H group). The names, abbreviations and structures of the 20 most common, naturally occurring amino acids are listed in Table 29.2.

All but glycine are chiral, but Nature is specific in the enantiomers that it uses.

(29.1)   **(29.2)**

Proteins are high molecular mass polypeptides with complex structures. The sequence of amino acids gives the primary structure of the protein, while the secondary and tertiary structures reveal the spatial properties of the peptide chain. The secondary structure takes into account the folding of polypeptide chains into domains called α-*helices*, β-*sheets*, *turns* and *coils*. In the ribbon representations of the protein structures illustrated in this chapter, the same colour coding is used to differentiate between

[†] For a more detailed account, see, for example: J. McMurry (2004) *Organic Chemistry*, 6th edn, Brooks/Cole, Pacific Grove, Chapter 26.

Table 29.2 The 20 most common, naturally occurring amino acids.

Name of amino acid	Abbreviation for amino acid residue (abbreviation used in sequence specification)	Structure	Acidic, neutral or basic
L-Alanine	Ala (A)		Neutral
L-Arginine	Arg (R)		Basic
L-Asparagine	Asn (N)		Neutral
L-Aspartic acid	Asp (D)		Acidic
L-Cysteine	Cys (C)		Neutral
L-Glutamic acid	Glu (E)		Acidic
L-Glutamine	Gln (Q)		Neutral
Glycine	Gly (G)		Neutral
L-Histidine	His (H)		Basic
L-Isoleucine	Ile (I)		Neutral

Table 29.2 (continued)

Name of amino acid	Abbreviation for amino acid residue (abbreviation used in sequence specification)	Structure	Acidic, neutral or basic
L-Leucine	Leu (L)		Neutral
L-Lysine	Lys (K)		Basic
L-Methionine	Met (M)		Neutral
L-Phenylalanine	Phe (F)		Neutral
L-Proline	Pro (P)		Neutral
L-Serine	Ser (S)		Neutral
L-Threonine	Thr (T)		Neutral
L-Tryptophan	Trp (W)		Neutral
L-Tyrosine	Tyr (Y)		Neutral
L-Valine	Val (V)		Neutral

these features: α-helices are shown in red, β-sheets in pale blue, turns in green and coils in silver-grey. Haemoglobin, myoglobin and most metalloenzymes are *globular proteins* in which the polypeptide chains are coiled into near-spherical structures. The *prosthetic group* in a protein is an additional, non-amino acid component of a protein which is essential for the biological activity of the protein. We shall be concerned with prosthetic groups containing metal centres, e.g. haem is the prosthetic group in haemoglobin and myoglobin. The proteins that we discuss contain metals (*metalloproteins*) and the form of the protein with the metal removed is called the *apoprotein*; the prefix *apo-* before a particular protein (e.g. ferritin and apoferritin) signifies the metal-free species. The difference between a protein and the corresponding apoprotein is analogous to that between a metal complex and the corresponding free ligand.

Self-study exercises

1. The basic groups in the side-chains of His and Lys have pK_b values of 7.9 and 3.5, respectively. Show that the corresponding K_b values are 1.3×10^{-8} and 3.2×10^{-4} and give equations to show the equilibria to which they refer. What effect does this have on the protonation states of these amino acids at pH 7.0?

2. Human serum albumin contains disulfide bridges originating from the oxidative coupling of Cys residues. Write a half-equation to represent this reaction.

29.2 Metal storage and transport: Fe, Cu, Zn and V

Living organisms require ways of storing and transporting trace metals, and storing the metal in a non-toxic form is clearly critical. Consider Fe, the most important trace metal in humans. Table 29.1 gives the average mass of Fe present in a 70 kg human, and this level needs to be maintained through a dietary intake (typically 6–40 mg per day) offsetting loss through, for example, bleeding. There is no excretory loss of Fe, a phenomenon not shared by other metals present in the body. The amount of Fe stored in the body far exceeds that taken in per day, but only a very small fraction of the iron in the body is actually in use at any one time; the mammalian system is very effective at recycling Fe. Whereas we can discuss the storage and transport of Fe in some detail, less information is currently available about the storage and transport of other trace metals.

Iron storage and transport

In mammals, the task of transferring iron from dietary sources to haemoglobin (see Section 29.3) initially involves the absorption of Fe(II) after passage through the stomach, followed by uptake into the blood in the form of the Fe(III)-containing metalloproteins *transferrins*. Iron is transported as transferrin to protein 'storage vessels' until it is required for incorporation into haemoglobin. In mammals, iron is stored mainly in the liver (typically 250–1400 ppm of Fe is present), bone marrow and spleen in the form of *ferritin*, a water-soluble metalloprotein. *Apoferritin* has been isolated from, for example, horse spleen and has a molecular weight of ≈445 000. X-ray diffraction studies confirm that it consists of 24 equivalent units. Each unit consists of a four-helix bundle which is >5 nm in length (Fig. 29.1a). These units are arranged so as to form a hollow shell (Fig. 29.1b), the cavity of which has a diameter of ≈8000 pm. In *ferritin*, this cavity contains up to 4500 high-spin Fe^{3+} centres in the form of a *microcrystalline* oxidohydroxidophosphate of composition $(FeO \cdot OH)_8(FeO \cdot H_2PO_4)$. Results of an EXAFS (see Box 25.2) study indicate that this core comprises double layers of approximately close-packed O^{2-} and $[OH]^-$ ions, with interstitial sites between the layers occupied by Fe(III) centres. Adjacent [OFeO]-triple layer blocks are only weakly associated with each other. The phosphate groups in the iron-containing core appear to function as terminators and linking groups to the protein shell.

While the structures of apoferritin and ferritin are fairly well established, the manner in which iron is transported in and out of the protein cavity is still under investigation. It is proposed that iron enters as Fe^{2+} and is oxidized once inside the protein. The formation of the crystalline core is an example of *biomineralization* and it is a remarkable achievement of evolution that iron can be stored in mammals effectively as hydrated iron(III) oxide, i.e. in a form closely related to rust!

As we illustrate throughout this chapter, studying appropriate model compounds gives insight into related, but more complicated, bioinorganic systems. The synthesis of large iron-oxido clusters from mono- and dinuclear precursors is of research interest in relation to modelling the formation of the core of ferritin, and reactions 29.1 and 29.2 give two examples. The product of reaction 29.1 is a mixed oxidation state iron species ($Fe^{III}_4Fe^{II}_8$). The Fe_6O_{14}-core of the product of reaction 29.2 is shown in Fig. 29.2. For the model complex to mimic the characteristics of iron(III)-containing ferritin, it should contain an $Fe^{III}_xO_y$-core surrounded by an organic shell. The latter should contain C, H, N and O to reproduce the protein chains, and appropriate ligands include **29.3** (H_3L) in the model complexes $[Fe_{19}(\mu_3\text{-}O)_6(\mu_3\text{-}OH)_6(\mu\text{-}OH)_8L_{10}(OH_2)_{12}]^+$ and $[Fe_{17}(\mu_3\text{-}O)_4(\mu_3\text{-}OH)_6(\mu\text{-}OH)_{10}L_8(OH_2)_{12}]^{3+}$.

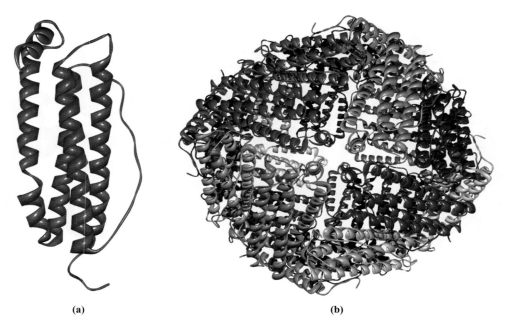

(a) **(b)**

Fig. 29.1 (a) One of the 24 equivalent units (a four-helix bundle) that are present in the protein shell of ferritin. (b) The structure of the protein shell in ferritin (isolated from the bull frog, PDB code: 1MFR) which shows the polypeptide chains in 'ribbon' representation.

(29.3)

$$Fe(OAc)_2 + LiOMe$$

in presence of O_2 | in MeOH

$$Fe_{12}(OAc)_3(\mu\text{-}OAc)_3(MeOH)_4(\mu\text{-}OMe)_8(\mu_3\text{-}OMe)_{10}(\mu_6\text{-}O)_2 \quad (29.1)$$

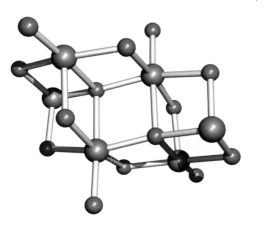

Fig. 29.2 A model for the biomineralization of ferritin. The Fe_6O_{14}-core of $[Fe_6(OMe)_4(\mu\text{-}OMe)_8(\mu_4\text{-}O)_2L_2]^{2+}$ $(L = N(CH_2CH_2NH_2)_3)$ determined by X-ray diffraction [V.S. Nair *et al.* (1992) *Inorg. Chem.*, vol. 31, p. 4048]. Colour code: Fe, green; O, red.

$$Fe(O_3SCF_3)_2 + L \xrightarrow{\text{in MeOH}}$$

$$[Fe_6(OMe)_4(\mu\text{-}OMe)_8(\mu_4\text{-}O)_2L_2][O_3SCF_3]_2 \quad (29.2)$$

$$\text{where } L = N(CH_2CH_2NH_2)_3$$

The *transferrins* are *glycoproteins* (i.e. compounds of proteins and carbohydrates) and include *serum transferrin*, *lactoferrin* (present in milk) and *ovotransferrin* (present in egg white). In humans, serum transferrin transports ≈40 mg of iron per day to the bone marrow. It contains a single polypeptide chain (molecular weight of ≈80 000) coiled in such a way as to contain two pockets suitable for binding Fe^{3+}. Each pocket presents hard *N*- and *O*-donor atoms to the metal centre, but the presence of a $[CO_3]^{2-}$ ligand is essential. The structure of human serum transferrin and details of the coordination environment of the Fe^{3+} ion are shown in Fig. 29.3. The stability constant for the Fe^{3+} complex is very high ($\log \beta = 28$ at pH 7.4), making transferrin extremely efficient as an iron transporting and scavenging agent in the body. The exact mechanism by which the Fe^{3+} enters and leaves the cavity has not been elucidated, but protonation of the carbonate ion and a change in conformation of the protein chain are probably involved.

Aerobic microorganisms also require iron, but cannot simply absorb it from their aqueous environment since Fe^{3+} is precipitated as $Fe(OH)_3$ ($K_{sp} = 2.64 \times 10^{-39}$). Evolution has provided these organisms with *O*-donor polydentate ligands called *siderophores* which scavenge for iron. Examples of siderophores are the anions derived

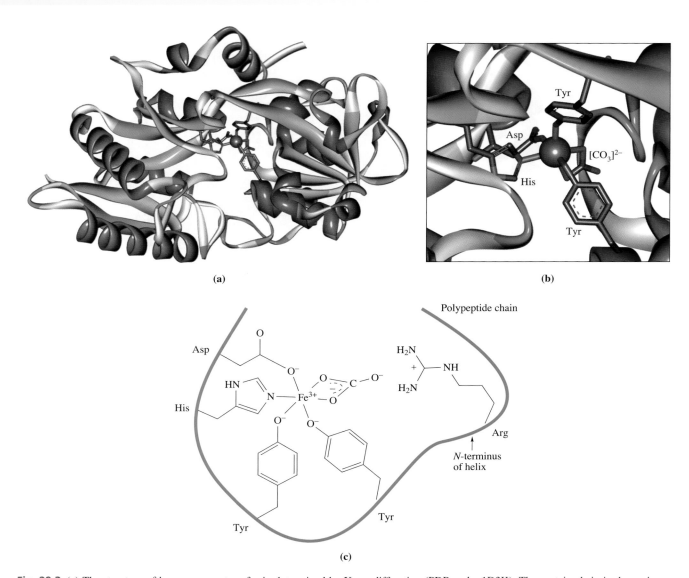

(a)

(b)

(c)

Fig. 29.3 (a) The structure of human serum transferrin determined by X-ray diffraction (PDB code: 1D3K). The protein chain is shown in ribbon representation and the coordination sphere of the Fe^{3+} ion is depicted in ball (for Fe^{3+}) and stick (for the amino acid residues and $[CO_3]^{2-}$ ion) representations. (b) An enlargement of the coordination environment of Fe^{3+}. Colour code: Fe, green; N, blue; O, red; C, grey. (c) Schematic representation of the Fe^{3+} binding site in transferrin; the coordinated $[CO_3]^{2-}$ points towards the positively charged Arg residue and the *N*-terminus of a helix.

from *enterobactin* (Fig. 29.4a), *desferrichrome* (Fig. 29.5a) and *desferrioxamine* (Fig. 29.5b). Enterobactin, H_6Ent, is derived from three L-serine residues, each carrying a 2,3-dihydroxybenzoyl group. The deprotonated form, Ent^{6-}, binds Fe^{3+} to give the complex $[Fe(Ent)]^{3-}$ in which Fe^{3+} is in an approximately octahedral environment. Spectroscopic data (electronic and circular dichroism spectra) show that the Λ-complex is formed diastereoselectively (see Box 19.3). Bacterial $[Fe(Ent)]^{3-}$ is scavenged by the protein siderocalin which occurs in the mammalian

immune system and functions as an antibacterial agent.[†] Figure 29.4b shows the interaction between siderocalin and $[Fe(Ent)]^{3-}$. Much information about $[Fe(Ent)]^{3-}$ comes from studies of model compounds. The model ligand **29.4** is closely related to enterobactin and gives a complex with

[†] See: R.J. Abergel, M.C. Clifton, J.C. Pizarro, J.A. Warner, D.K. Shuh, R.K. Strong and K.N. Raymond (2008) *J. Am. Chem. Soc.*, vol. 130, p. 11524.

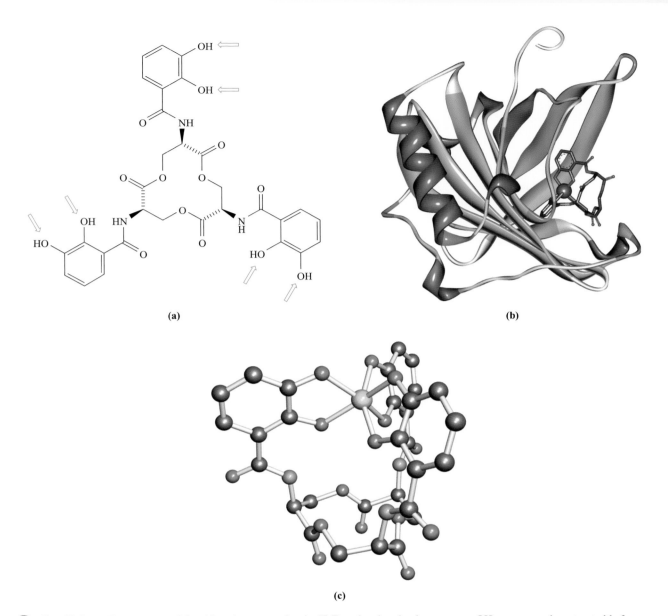

Fig. 29.4 (a) The structure of the siderophore enterobactin, H_6Ent, showing the donor atoms; OH groups are deprotonated before coordination to Fe^{3+}. (b) The structure (determined by X-ray diffraction) of the protein siderocalin complexed with $[Fe(Ent)]^{3-}$ (PDB code: 3BY0). Colour code for the complex: Fe, green; N, blue; O, red; C, grey. (c) The structure of the vanadium(IV) complex $[V(Ent)]^{2-}$ determined by X-ray diffraction of the K^+ salt [T.B. Karpishin *et al.* (1993) *J. Am. Chem. Soc.*, vol. 115, p. 1842]. Hydrogen atoms are omitted for clarity; colour code: V, yellow; C, grey; O, red; N, blue.

Fe^{3+} for which log β is close to the value for iron(III) enterobactin. The V(IV) complex of enterobactin (reaction 29.3) has been structurally characterized by X-ray diffraction, and although the radius of a V(IV) centre (58 pm) is smaller than that of Fe(III) (65 pm), the gross structural features of the Fe(III) and V(IV) complexes should be similar. The three 'arms' of the ligand lie above the central macrocycle allowing each arm to act as an *O,O'*-donor

(Fig. 29.4c). The 6-coordinate V(IV) centre is in an environment described as trigonal prismatic with a twist angle of 28° (see structures **19.9**, **19.10** and **19.13**).

$$[V(O)(acac)_2] + \underset{\text{enterobactin}}{H_6Ent} + 4KOH$$
$$\xrightarrow{\text{MeOH}} K_2[V(Ent)] + 2K[acac] + 5H_2O \qquad (29.3)$$

see **21.8**

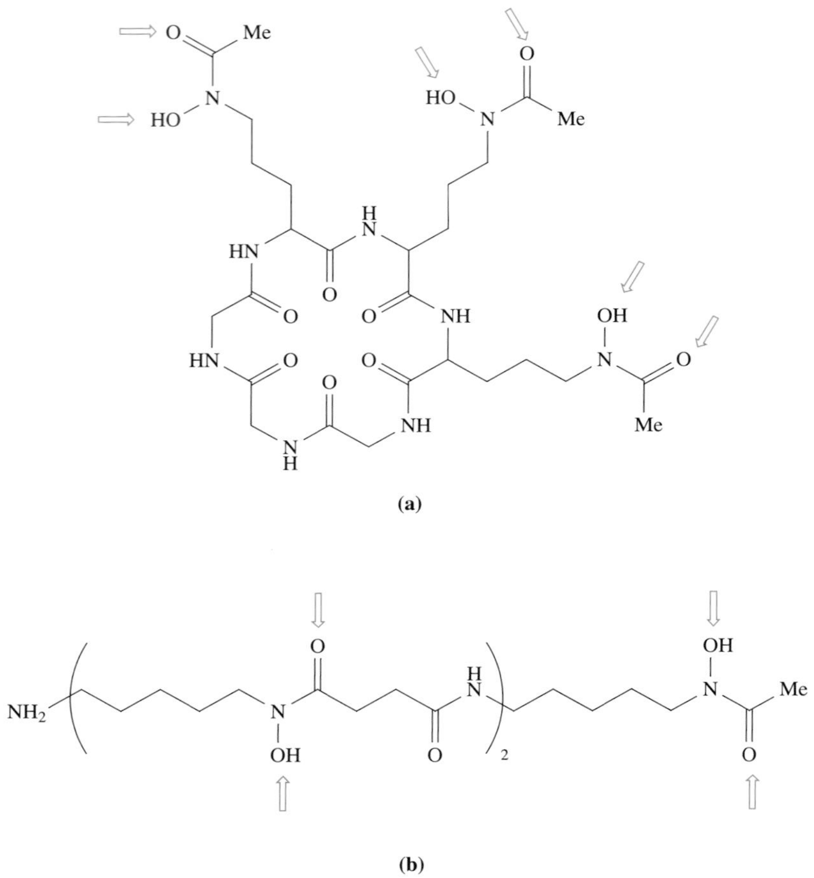

(a)

(b)

Fig. 29.5 The structures of the siderophores (a) desferrichrome and (b) desferrioxamine, showing the donor atoms; OH groups are deprotonated before coordination to Fe^{3+}.

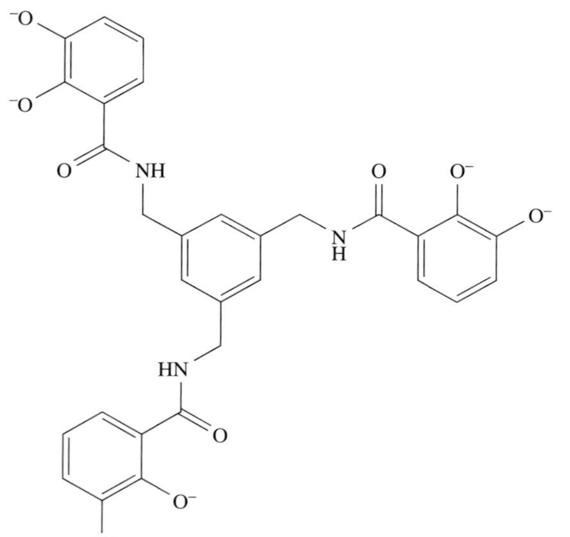

(29.4)

High-spin Fe^{3+} complexes of the siderophores are kinetically labile. If Fe^{3+} is exchanged for Cr^{3+}, kinetically inert complexes are obtained which can be studied in solution as models for the Fe^{3+} complexes.

The complexes that transport iron in mammals and microorganisms have very high overall stability constants (see above) and, although exact mechanisms have not been elucidated, it is reasonable to propose that reduction to Fe^{2+} is required for its release since the stability constant for the Fe^{2+} complex is orders of magnitude lower than that for the Fe^{3+} complex.

Self-study exercises

1. Explain why high-spin Fe^{3+} complexes of siderophores are kinetically labile whereas analogous model complexes containing Cr^{3+} are kinetically inert.

   [*Ans.* See Section 26.2]

2. The coordination of Fe^{3+} to the deprotonated form, Ent^{6-}, of enterobactin gives only the Λ-complex. Why is this? What would you expect to observe if you were to use the unnatural diastereoisomer of Ent^{6-} with an (R,R,R)-stereochemistry?

Metallothioneins: transporting some toxic metals

Transporting soft metal centres is important in protection against toxic metals such as Cd^{2+} and Hg^{2+}. Complexation requires soft ligands, which are provided by Nature in the form of cysteine residues (Table 29.2) in *thioneins*. The metal complexes that thioneins form are called *metallothioneins*. Thioneins also bind Cu^+ and Zn^{2+}, but their active role in transporting these metals in mammals has not been confirmed. Thioneins are small proteins containing ≈ 62 amino acids, about one-third of which is cysteine. The Cys residues are either adjacent to each other or separated by one other amino acid residue, thus providing pockets of *S*-donor sites ideally suited to scavenging soft metal ions. Both Cd and Hg have NMR active nuclei (the most important are ^{113}Cd, 12% abundance, $I = \frac{1}{2}$; ^{199}Hg, 17% abundance, $I = \frac{1}{2}$) and the application of NMR spectroscopy to probe the coordination sites in Cd- and Hg-containing metallothioneins has greatly aided structural determination.

The presence of Hg^{2+}, Cd^{2+}, Cu^+ and Zn^{2+} induces the production of thioneins in the liver and kidneys of mammals. Between 4 and 12 metal centres can be bound by one thionein; Zn^{2+}, Hg^{2+}, Cd^{2+} centres are likely to be in tetrahedral environments, while Cu^+ may be 3-coordinate. The structure of the Cd/Zn-containing metallothionein isoform II from rat liver has been determined by X-ray diffraction, and Fig. 29.6a illustrates the folded protein chain consisting of 61 amino acid residues of which 20 are Cys groups. One Cd^{2+} and two Zn^{2+} centres are bound in one pocket of the folded chain, and four Cd^{2+} in the other (Figs. 29.6b and 29.6c).

Thiolate and related complexes are studied as models for metallothioneins. For example, the Cu(I)-containing metallothionein in yeast has been modelled by

$[Cu_4(SPh)_6]^{2-}$ (**29.5**), while model studies on canine liver cuprothionein have utilized complex **29.6** in which the Cys residues are 'replaced' by thiourea ligands. Among Cd_xS_y-containing clusters studied as models for Cd^{2+}-containing metallothioneins is $[Cd_3(SC_6H_2{}^iPr_3)_7]^-$ (**29.7**).

(29.5)

(29.6)

$$SR = S{=}C\begin{smallmatrix} NH_2 \\ \\ NH_2 \end{smallmatrix}$$

(29.7)

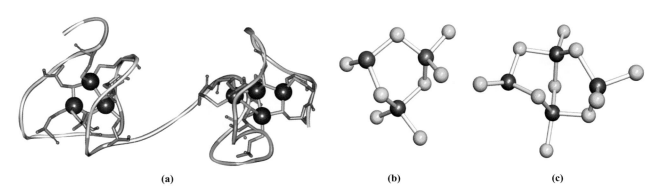

(a)   (b)   (c)

Fig. 29.6 (a) The backbone (folded to give two pockets) of the polypeptide chain in metallothionein isoform II from rat liver (PDB code: 4MT2). Each pocket contains a multinuclear metal unit coordinated by cysteine residues. The left-hand cluster contains one Cd and two Zn atoms, and the right-hand cluster consists of four Cd atoms. Details of the clusters: (b) $CdZn_2S_8$ and (c) Cd_4S_{10}. Colour code: Zn, red; Cd, blue; S, yellow.

29.3 Dealing with O_2

Haemoglobin and myoglobin

In mammals, O_2 (taken in by respiration) is carried in the bloodstream by *haemoglobin* and is stored in the tissues in *myoglobin*. Both haemoglobin and myoglobin are *haem-iron proteins*. Myoglobin has a molecular weight of $\approx 17\,000$ and is a monomer with a protein chain consisting of 153 amino acid residues. Haemoglobin has a molecular weight of $\approx 64\,500$ and is a tetramer (Fig. 29.7a). The

protein chain in myoglobin and in each chain of haemoglobin contains a protoporphyrin IX group (see Fig. 12.10a for porphyrin) which, together with a histidine residue tethered to the protein backbone, contains an Fe centre. A porphyrin ring containing an Fe centre is called a *haem group* and the one present in haemoglobin is shown in Fig. 29.7b. The Fe(II) centre is in a square-based pyramidal environment when in its 'rest state', also referred to as the deoxy-form. When O_2 binds to the haem group, it enters *trans* to the His residue to give an octahedral species (**29.8**) (see later discussion).

$$
\begin{array}{c}
N(His) \\
| \\
(Porph)N \,/\!/\!/\!/_{\cdots}\!\!\underset{|}{\overset{}{Fe(III)}}\!\!{\cdots}\!\backslash\! N(Porph) \\
(Porph)N \blacktriangleleft \quad \blacktriangleright N(Porph) \\
O {\cdots\cdots} O
\end{array}
$$

(29.8)

Although each of the four units in haemoglobin contains a haem group, the four groups do not operate independently of each other: the binding (and release) of O_2 is a *cooperative* process. Figure 29.8 compares the affinity of haemoglobin (tetramer with four haem units) and myoglobin (monomer with one haem unit) for O_2. The blue curve in Fig. 29.8 illustrates that haemoglobin has a low affinity for O_2 at

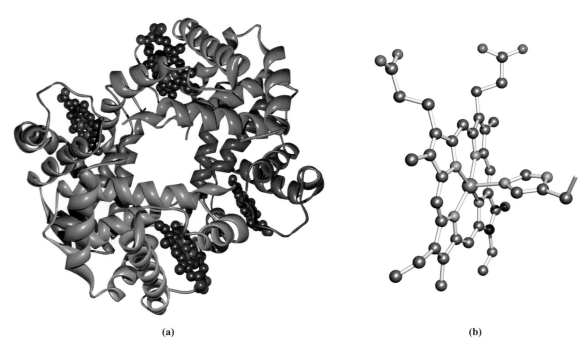

(a)   (b)

Fig. 29.7 (a) The structure of haemoglobin (PDB code: 1B86) shown in a ribbon representation. The four sub-units, each containing a haem unit, are shown in different colours. (b) The structure of the haem unit in its rest state. The Fe(II) centre is coordinated by a protoporphyrin IX ligand and a histidine residue; the non-terminated stick represents the connection to the protein backbone. Hydrogen atoms are omitted for clarity. Colour code: Fe, green; C, grey; N, blue; O, red.

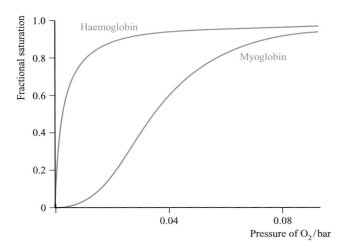

Fig. 29.8 Dioxygen binding curves for haemoglobin and myoglobin illustrating the cooperativity effects in haemoglobin. [Redrawn from: D.B. Kim-Shapiro (2004), *Free Radical Bio. Med.*, vol. 36, p. 402, Elsevier.]

low pressures of O$_2$ (e.g. in mammalian tissues), but binds O$_2$ avidly at higher O$_2$ pressure. The highest pressure of O$_2$ in a mammal is in the lungs where O$_2$ binds to haemoglobin (see end-of-chapter problem 29.27). As haemoglobin binds successive O$_2$ molecules, the affinity of the 'vacant' haem groups for O$_2$ *increases* such that the affinity for the fourth site is $\approx$300 times that of the first haem unit. The 'cooperativity' can be rationalized in terms of communication between the haem groups arising from conformational changes in the protein chains. Consider the haem group in its rest state in Fig. 29.7b: it contains high-spin Fe(II) lying $\approx$40 pm out of the plane of the N,N',N'',N'''-donor set of the porphyrin group and is drawn towards the His residue. The high-spin Fe(II) centre is apparently too large to fit within the plane of the four N-donor atoms. When O$_2$ enters the sixth coordination site, the iron centre (now low-spin Fe^{3+}, see below) moves into the plane of the porphyrin ring and pulls the His residue with it. This in turn perturbs not only the protein chain to which the His group is attached, but also the other three protein sub-units, and a cooperative process triggers the other haem units to successively bind O$_2$ more avidly. When O$_2$ is released from haemoglobin to myoglobin, the loss of the first O$_2$ molecule triggers the release of the remaining three. Myoglobin does not exhibit this cooperative effect (Fig. 29.8) since it comprises only one protein chain. When bound in either haemoglobin or myoglobin, the O$_2$ molecule resides in a sterically protected cavity. The importance of this becomes clear when we look at model compounds.

Many years of research activity have gone into reaching our current level of understanding of O$_2$ uptake by myoglobin and haemoglobin. Various proposals have been put forward to describe the nature of the iron centre and

of the O$_2$ species in the oxy-forms of these proteins. Some model studies have involved the reactions of O$_2$ with certain Co(II) Schiff base† complexes. Reactions such as that represented in eq. 29.4 yield Co(III) compounds in which the O$_2$ molecule is bound 'end-on' to the metal centre. In this complex, the Co$-$O$-$O bond angle is $\approx$125° and the O$-$O bond length $\approx$126 pm (compare values of 121 pm in O$_2$ and 134 pm in [O$_2$]$^-$, see Box 16.2).

$$\text{Co(II)} \qquad \xrightarrow{\text{O}_2} \qquad \text{Co(III)}$$

L = monodentate ligand, e.g. py

(29.4)

The Co(III) complex formed in reaction 29.4 can be considered to contain coordinated [O$_2$]$^-$, but the presence of the axial base, L, is crucial to the formation of the monomeric product. In its absence, a dicobalt species with a Co$-$O$-$O$-$Co peroxido-bridge (i.e. analogous to those discussed in Section 21.10) is formed.

A logical ligand to model the active sites in myoglobin and haemoglobin is one derived from porphyrin. Tetraphenylporphyrin (H$_2$tpp, **29.9**) is readily available, but the reaction of the Fe(II) complex [Fe(tpp)$_2$] with O$_2$ leads to a peroxido-bridged Fe(III) complex (eq. 29.5).

(29.9)

---

† A Schiff base is an imine formed by condensation of a primary amine and a carbonyl compound.

$$(29.5)$$

(29.10)

Interaction with the second iron centre can be prevented by using a porphyrin ligand with bulky substituents. An example is ligand **29.10**, a so-called 'picket-fence' porphyrin. An example of a model complex containing [Fe(**29.10**)] with an azido ligand bound to the iron(II) centre is shown in Fig. 29.9. Studies of such models provide information about the properties of high-spin iron(II) porphyrinato complexes. The four substituents in ligand **29.10** form a cavity, and reaction 29.6 shows the

binding of O_2 within this cavity. The axial ligand is 1-methylimidazole which is structurally similar to a His residue. The system clearly resembles the iron environment in haemoglobin (compare Fig. 29.7).

$$(29.6)$$

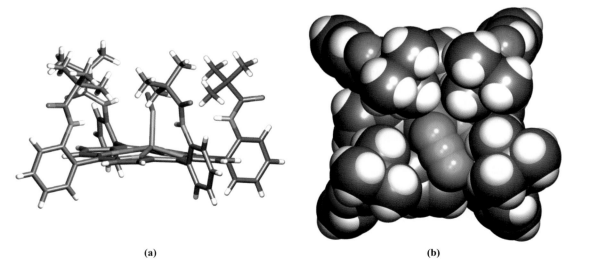

(a)   **(b)**

Fig. 29.9 The structure (determined by X-ray diffraction) of a picket-fence porphyrin model compound with an azido ligand bound to the Fe(II) centre: (a) a stick representation illustrating the protection of the [N₃]⁻ ligand by the four 'fence-posts' of the porphyrin ligand, and (b) a space-filling model looking down into the cavity. [Data: I. Hachem *et al.* (2009) *Polyhedron*, vol. 28, p. 954.] Colour code: Fe, green; N, blue; O, red; C, grey; H, white.

The solid state structure of the product of reaction 29.6 has been determined by X-ray diffraction and confirms an end-on, bent coordination mode of the O$_2$ group. The O−O bond length is 125 pm and the Fe−O−O bond angle is 136°. The vibrational spectrum of the complex exhibits an absorption at 1159 cm^{-1} assigned to ν(O−O), and, when compared with values of ν(O−O) of 1560 cm^{-1} for O$_2$, ≈1140 cm^{-1} for [O$_2$]$^-$ and ≈800 cm^{-1} for [O$_2$]$^{2-}$, it suggests the presence of an [O$_2$]$^-$ ligand. Oxyhaemoglobin and oxymyoglobin are characterized by values of ν(O−O) = 1107 and 1103 cm^{-1}, respectively. The current model for O$_2$ binding to the low-spin Fe(II) centre in haemoglobin and myoglobin is that coordination is accompanied by electron transfer, oxidizing high-spin Fe(II) to low-spin Fe(III) and reducing O$_2$ to [O$_2$]$^-$. Both low-spin Fe(III) (d^5) and [O$_2$]$^-$ contain an unpaired electron, and the fact that the oxy-forms of the proteins are diamagnetic can be understood in terms of antiferromagnetic coupling between the Fe(III) centre and [O$_2$]$^-$ ligand (see Section 20.10).

In binding to a haem group, O$_2$ acts as a π-acceptor ligand (see Section 20.4). It is not surprising, therefore, that other π-acceptor ligands can take the place of O$_2$ in haemoglobin or myoglobin, and this is the basis of the toxicity of CO. Cyanide, however, although a π-acceptor ligand, favours higher oxidation state metal centres and binds to Fe(III) in *cytochromes* (see Section 29.4); [CN]$^-$ poisoning is *not* caused by [CN]$^-$ blocking the O$_2$-binding sites in haemoglobin.

Self-study exercises

1. Construct an MO diagram for the formation of O$_2$ from two O atoms. Use the diagram to support the statement in the text that ν(O−O) in O$_2$ > ν(O−O) in [O$_2$]$^-$ > ν(O−O) in [O$_2$]$^{2-}$.   [*Ans.* See Fig. 2.10]

2. When CO binds to haemoglobin, which MOs of CO are involved in Fe−C bond formation?
    [*Ans.* The HOMO and LUMO in Fig. 2.15]

Haemocyanin

Haemocyanins are O$_2$-carrying copper-containing proteins in molluscs (e.g. whelks, snails, squid) and arthropods (e.g. lobsters, crabs, shrimps, horseshoe crabs, scorpions), and although the name suggests the presence of a haem group, haemocyanins are *not* haem proteins. Haemocyanins isolated from arthropods and molluscs are hexameric (M_r per unit ≈75 000), while those from molluscs possess 10 or 20 sub-units, each with M_r ≈350 000 to 450 000. The deoxy-form of a haemocyanin is colourless and contains Cu(I), while O$_2$ binding results in the blue Cu(II) form. The structures of a deoxyhaemocyanin (isolated from the

spiny lobster) and an oxyhaemocyanin (isolated from the Atlantic horseshoe crab) have been confirmed. The folded protein chain of one sub-unit of the deoxy-form is shown in Fig. 29.10a. Buried within the metalloprotein are two adjacent Cu(I) centres (Cu⋯⋯Cu = 354 pm, i.e. non-bonded), each of which is bound by three histidine residues (Fig. 29.10b and structure **29.11**).

(29.11)

The active site of the structurally characterized oxyhaemocyanin is shown in Fig. 29.10c. The Cu$_2$(His)$_6$-unit (Cu⋯⋯Cu = 360 pm) resembles that in the deoxy-form. The O$_2$ unit is bound in a bridging mode with an O−O bond length of 140 pm, typical of that found in peroxide complexes. The O$_2$-binding site is formulated as Cu(II)−[O$_2$]$^{2-}$−Cu(II), i.e. electron transfer accompanies O$_2$ binding. Resonance Raman spectroscopic data are consistent with this formulation: ν(O−O) ≈ 750 cm^{-1} compared with ≈800 cm^{-1} for [O$_2$]$^{2-}$. The Cu(II) centres are strongly antiferromagnetically coupled, with the μ-[O$_2$]$^{2-}$ ligand being involved in a *superexchange* mechanism (see Section 20.10).

Many model compounds have been studied in attempts to understand the binding of O$_2$ in haemocyanin, and often involve imidazole or pyrazole derivatives to represent His residues. In the light of the crystallographic data (Fig. 29.10), one model that closely resembles oxyhaemocyanin is the peroxido dicopper(II) complex (**29.12**) in which each Cu(II) centre is coordinated by an isopropyl-derivatized hydridotris(pyrazolyl)borato ligand. Like oxyhaemocyanin, complex **29.12** is diamagnetic as a result of antiferromagnetically coupled Cu(II) centres. The Raman spectrum of **29.12** shows an absorption at 741 cm^{-1} assigned to ν(O−O) which agrees well with the value for oxyhaemocyanin. However, O$_2$ binding in complex **29.12** is irreversible. In contrast, model complex **29.13** releases O$_2$ in MeCN/CH$_2$Cl$_2$ at 353 K under vacuum. When O$_2$ is added at room temperature, complex **29.13** is regenerated.

(29.12)

BIOLOGY AND MEDICINE

Box 29.2 The specialists: how the blood-sucking *Rhodnius prolixus* utilizes NO

Nitrophorins are haem proteins which are present in the salivary glands of the blood-sucking insect *Rhodnius prolixus* which lives in South and Central America. Binding of NO to the Fe(III) centre in nitrophorin (NP1) is reversible, and depends on pH. Crucial to the process of blood-sucking by *Rhodnius prolixus* is the fact that NO binds 10 times more tightly at pH 5 (i.e. the pH of the saliva within the insect) than at pH 7 (i.e. the physiological pH of the victim). Once insect saliva is released into the victim, NO is released, causing expansion of the blood vessels (vasodilation) and inhibiting blood clotting. In response to being bitten, the victim releases histamine to aid healing of the wound.

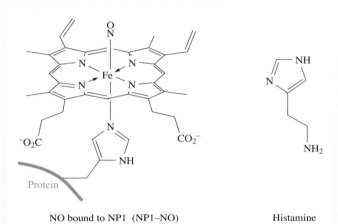

NO bound to NP1 (NP1–NO)   Histamine

within a pocket in the protein comprising β-sheets and is tethered to the protein chain by a histidine residue. X-Ray crystallographic data are available on NP1–NO and its $[CN]^-$ analogue and confirm that NO and $[CN]^-$ bind to the haem Fe. However, in the X-ray beam, photoreduction of NP1–NO readily occurs, and it is difficult to assess whether or not the observed bent coordination mode of the NO ligand is a consequence of photoreduction. The Fe–N–O bond angles differ between different structural determinations. For the cyanido complex, the Fe–C–N bond angle is 173°. The NO or $[CN]^-$ ligand is 'lodged' in a pocket of the protein chain between two leucine residues (see Table 29.2). Structural data for the histamine complex show that this same protein pocket hosts the histamine ligand, indicating that NO and histamine compete for the same binding site. At physiological pH, the haem unit in NP1 binds histamine ≈100 times more strongly than NO. This should both aid the dissociation of NO and inhibit the role of histamine, both of which work in favour of the attacking *Rhodnius prolixus*.

The assassin bug *Rhodnius prolixus* sucking blood from a human.

Further information

J.F. Andersen (2010) *Toxicon*, vol. 56, p. 1120 – 'Structure and mechanism in salivary proteins from blood-feeding arthropods'.

M.A. Hough, S.V. Antonyuk, S. Barbieri, N. Rustage, A.L. McKay, A.E. Servid, R.R. Eady, C.R. Andrew and S.S. Hasnain (2011) *J. Mol. Biol.*, vol. 405, p. 395 – 'Distal-to-proximal NO conversion in hemoproteins: The role of the proximal pocket'.

F.A. Walker (2005) *J. Inorg. Biochem.*, vol. 99, p. 216 – 'Nitric oxide interaction with insect nitrophorins and thoughts on the electronic configuration of the {FeNO}[6] complex'.

A. Weichsel, J.F. Andersen, D.E. Champagne, F.A. Walker and W.R. Montfort (1998) *Nature Struct. Biol.*, vol. 5, p. 304 – 'Crystal structures of a nitric oxide transport protein from a blood-sucking insect'.

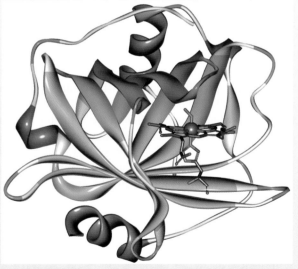

β-Sheets are shown in pale blue; colour code for ball and stick-structure: Fe, green; N, blue; O, red; C, grey.

The structure above (determined by X-ray diffraction, PDB code: 2OFR) is that of a nitrophorin from *Rhodnius prolixus* complexed with NO at pH 5.6. The haem unit is bound

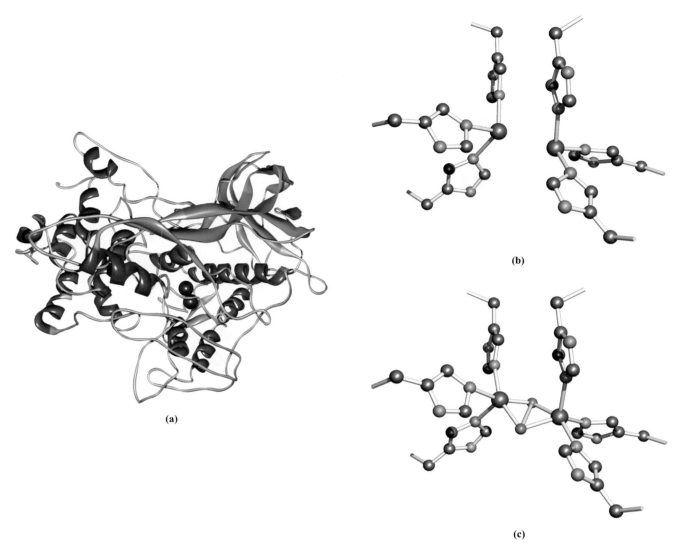

Fig. 29.10 The structure of deoxyhaemocyanin from the spiny lobster (*Panulirus interruptus*) (PDB code: 1HCY): (a) the backbone of the protein chain and the positions of the two Cu(I) centres, and (b) the active site in which the two Cu(I) centres are bound by histidine residues. (c) The O$_2$-binding site in oxyhaemocyanin from the Atlantic horseshoe crab (*Limulus polyphemus*). Hydrogen atoms are omitted; colour code: Cu, brown; C, grey; O, red; N, blue.

(29.13)

Haemerythrin

In marine invertebrates such as annelids (segmented earthworms), molluscs and arthropods (see above), O$_2$ is trans-ported by *haemerythrin*, a non-haem Fe-containing protein. In the blood, the metalloprotein ($M_r \approx 108\,000$) consists of eight sub-units, each with 113 amino acid residues and a diiron-active site. In tissues, fewer sub-units make up the metalloprotein. Unlike haemoglobin, haemerythrin exhibits no cooperativity between the sub-units during O$_2$ binding.

The structures of the deoxy- and oxy-forms of haemery-thrin have been determined crystallographically (Fig. 29.11). In the deoxy-form, a hydroxido-bridged [Fe(II)]$_2$ unit is present as shown in structure **29.14** (see Table 29.2); the dotted lines represent connections into the protein backbone. The two Fe(II) centres in deoxyhaemerythrin are strongly anti-ferromagnetically coupled through the Fe−O−Fe bridge.

The left-hand Fe(II) centre in **29.14** is coordinatively unsaturated and adds O$_2$ to give oxyhaemerythrin

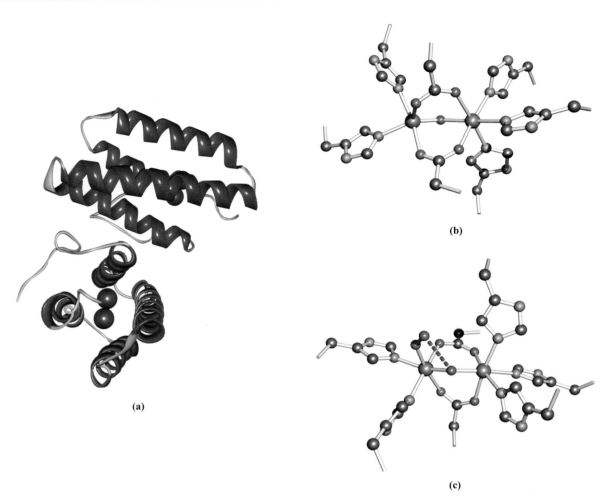

Fig. 29.11 (a) Two sub-units in the metalloprotein deoxyhaemerythrin from the sipunculid worm (*Themiste dyscrita*) (PDB code: 1HMD); the backbone of the protein chains are shown in ribbon representation and the position of the Fe_2 unit is shown. (b) The active site in which the two Fe(II) centres are bound by histidine, glutamate and aspartate residues. (c) The O_2-binding site in oxyhaemerythrin from *Themiste dyscrita*. The red hashed line represents a hydrogen-bonded interaction (see eq. 29.7). Hydrogen atoms are omitted; colour code: Fe, green; C, grey; O, red; N, blue.

(Fig. 29.11c). The hydroxyl H atom in **29.14** participates in O_2 binding, becoming part of an $[HO_2]^-$ ligand, but remaining associated with the μ-oxido group by hydrogen bond formation (eq. 29.7).

(29.14)

(29.7)

Many model studies have focused on *methaemerythrin*, i.e. the oxidized Fe(III)–Fe(III) form of haemerythrin which contains an oxido (rather than hydroxido) bridge. Methaemerythrin does not bind O$_2$, but does interact with ligands such as [N$_3$]$^-$ and [SCN]$^-$. Reaction 29.8 makes use of the hydridotris(pyrazolyl)borato ligand, [HBpz$_3$]$^-$, to model three His residues. The product (**29.15**) contains antiferromagnetically coupled Fe(III) centres.

(29.15)

$$Fe(ClO_4)_3 + Na[O_2CMe] + K[HBpz_3]$$

$$\longrightarrow [Fe_2(HBpz_3)_2(\mu\text{-}O_2CMe)_2(\mu\text{-}O)] \qquad (29.8)$$

Cytochromes P-450

> ***Oxygenases*** are enzymes that insert oxygen into other molecules; a ***monooxygenase*** inserts one oxygen atom, and a ***dioxygenase*** inserts two.

The cytochromes P-450 are metalloenzymes which function as monooxygenases, catalysing the insertion of oxygen into a C$-$H bond of an aromatic or aliphatic hydrocarbon, i.e. the conversion of RH to ROH:

$$RH + O_2 + 2H^+ + 2e^- \longrightarrow ROH + H_2O$$

Two examples of the biological utilization of this reaction are in drug metabolism and steroid synthesis. The oxygen atom originates from O$_2$: one O atom is inserted into the organic substrate and one atom is reduced to H$_2$O.

(29.16)

The active site in a cytochrome P-450 is a haem unit. An iron protoporphyrin(IX) complex (Fig. 29.7b) is covalently bound to the protein through an Fe–S$_{cysteine}$ bond. This has been confirmed crystallographically for cytochrome P-450 complexed with (1S)-camphor (Fig. 29.12). The active site contains a 5-coordinate Fe(III) centre, schematically

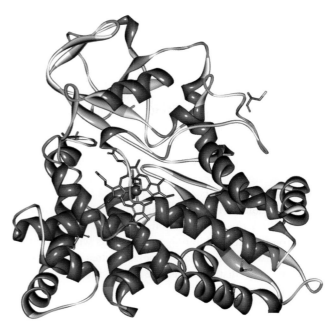

Fig. 29.12 The structure of cytochrome P-450 from the bacterium *Pseudomonas putida* (PDB code: 1AKD). The structure was determined for cytochrome P-450 complexed with (1S)-camphor, but this is omitted from the figure. The protein chain is shown in ribbon representation, with cysteine (Cys) residues highlighted in stick representation. One Cys residue is bound to the Fe(III) centre of the iron protoporphyrin(IX) unit (shown in a ball-and-stick representation with colour code: Fe, green; O, red; C, grey; N, blue; S, yellow).

represented by structure **29.16**. In its rest state, cytochrome P-450 contains a low-spin Fe(III) centre. Carbon monoxide adducts of cytochromes P-450 absorb at 450 nm and this is the origin of the name of the enzyme. It is proposed that the catalytic cycle for the conversion of RH to ROH follows the sequence of steps:

- binding of the organic substrate RH to the active site of the metalloenzyme and loss of a bound H$_2$O ligand;
- 1-electron reduction of low-spin Fe(III) to low-spin Fe(II);
- binding of O$_2$ to give an adduct, followed by 1-electron transfer from iron to produce an Fe(III)-peroxido complex;
- acceptance of another electron to give an {Fe(III)$-$O$-$O$^-$} species which is protonated to {Fe(III)$-$O$-$OH};
- further protonation and loss of H$_2$O leaving an {Fe(IV)$=$O} species with the porphyrin ring formally a radical cation;
- transfer of the oxido O atom to the bound RH substrate and release of ROH with concomitant binding of an H$_2$O ligand to the active site of the metalloenzyme which once again contains low-spin Fe(III).

The insertion of O into the C$-$H bond of RH is thought to involve a radical pathway.

29.4 Biological redox processes

In this section we look at ways in which Nature carries out redox chemistry with reference to blue copper proteins, iron–sulfur proteins and cytochromes. The redox steps in Photosystem II were outlined in the discussion accompanying eq. 21.54. We have already discussed two topics of prime importance to electron transfer in Nature. The first is the way in which the reduction potential of a metal redox couple such as Fe^{3+}/Fe^{2+} can be tuned by altering the ligands coordinated to the metal centre. Look back at the values of E^o for Fe^{3+}/Fe^{2+} redox couples listed in Table 8.1. The second is the discussion of *Marcus–Hush theory* in Section 26.5. This theory applies to electron transfer in bioinorganic systems where communication between redox active metal centres may be over relatively long distances as we shall illustrate in the following examples.

Blue copper proteins

There are three classes of copper centres in blue copper proteins:

- A Type 1 centre is characterized by an intense absorption in the electronic spectrum with $\lambda_{max} \approx 600\,nm$, and $\varepsilon_{max} \approx 100$ times greater than that of aqueous Cu^{2+}. The absorption is assigned to charge transfer from a cysteine ligand to Cu^{2+}. In the EPR spectrum (Cu^{2+} has one unpaired electron), narrow hyperfine splitting is observed (see Section 4.9).
- A Type 2 centre exhibits electronic spectroscopic characteristics typical of Cu^{2+}, and the EPR spectrum is typical of a Cu^{2+} centre in a simple coordination complex.
- A Type 3 centre exhibits an absorption with $\lambda_{max} \approx 330\,nm$ and exists as a pair of Cu(II) centres which are antiferromagnetically coupled to give a diamagnetic system. Hence, there is no EPR spectroscopic signature. The Cu_2-unit can function as a 2-electron transfer centre and is involved in the reduction of O_2.

Blue copper proteins contain a minimum of one Type 1 Cu centre, and those in this class include *plastocyanins* and *azurins*. Plastocyanins are present in higher plants and blue-green algae, where they transport electrons between Photosystems I and II (see above). The protein chain in a plastocyanin comprises between 97 and 104 amino acid residues (most typically 99) and has $M_r \approx 10\,500$. Azurins occur in some bacteria and are involved in electron transport in the conversion of $[NO_3]^-$ to N_2. Typically, the protein chain contains 128 or 129 amino acid residues ($M_r \approx 14\,600$).

Single-crystal structural data have provided valuable information about blue copper proteins containing Type 1 Cu centres. Figure 29.13a shows a representation of the folded protein chain of spinach plastocyanin. The Cu(II) centre lies within a pocket in the chain, bound by a Cys, a Met and two His residues (Fig. 29.13b). The S(Met) atom is significantly further away from the Cu(II) centre than is S(Cys). Figure 29.13c shows the backbone of the protein chain in azurin isolated from the bacterium *Pseudomonas putida*. The coordination environment of the Cu(II) centre resembles that in plastocyanin with Cu−S (Met) > Cu−S(Cys), but in addition, an O atom from an adjacent Gly residue is involved in a weak coordinate interaction (Fig. 29.13d). Structural studies have also been carried out on the reduced forms of plastocyanin and azurin. In each case, the coordination sphere remains the same except for changes in the Cu−L bond lengths. Typically, the bonds lengthen by 5–10 pm on going from Cu(II) to Cu(I). The observed coordination spheres can be considered as suiting *both* Cu(I) and Cu(II) (see Section 21.12) and thus facilitate rapid electron transfer. It should be noted, however, that in each structure discussed above, *three* donor atoms are more closely bound than the remaining donors and this indicates that binding of Cu(I) is more favourable than that of Cu(II). This is supported by the high reduction potentials (measured at pH 7) of plastocyanin (+370 mV) and azurin (+308 mV).

Oxidases are enzymes that use O_2 as an electron acceptor.

Multicopper blue copper proteins include *ascorbate oxidase* and *laccase*. These are metalloenzymes that catalyse the reduction of O_2 to H_2O (eq. 29.9) and, at the same time, an organic substrate (e.g. a phenol) undergoes a 1-electron oxidation. The overall scheme can be written in the form of eq. 29.10; $R^{\bullet}$ undergoes polymerization.

$$O_2 + 4H^+ + 4e^- \rightleftharpoons 2H_2O \tag{29.9}$$

$$4RH + O_2 \longrightarrow 4R^{\bullet} + 2H_2O \tag{29.10}$$

Spectroscopic data are consistent with the presence of all three types of copper site in ascorbate oxidase and laccase, and this was confirmed crystallographically in 1992 for ascorbate oxidase, isolated from courgettes (zucchini, *Cucurbita pepo medullosa*). Figure 29.14 shows one unit of ascorbate oxidase in which four Cu(II) centres are accommodated within the folds of the protein chain. Three Cu centres form a triangular array (non-bonded Cu⋯Cu separations of 340 pm for the bridged interaction, and 390 pm for the remaining two Cu⋯Cu distances). The fourth Cu atom (a Type 1 centre) is a significant distance away (>1200 pm), but indirectly connected to the Cu_3 unit by the protein chain. The coordination sphere of the Type 1 centre is similar to that in the oxidized form of plastocyanin (compare Fig. 29.14c with Fig. 29.13b) with the metal bound by one Met residue (Cu−S = 290 pm),

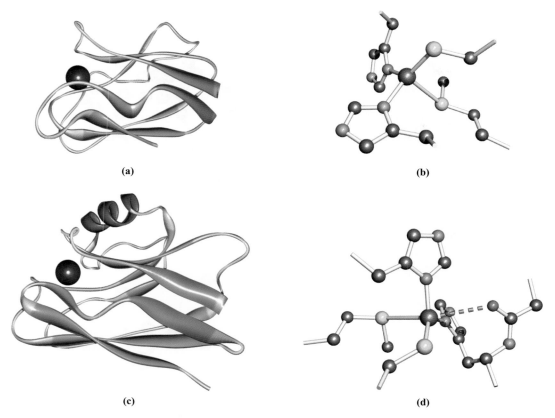

Fig. 29.13 (a) and (b) The structure of spinach plastocyanin (PDB code: 1AG6): (a) the backbone of the protein chain showing the position of the Cu(II) centre and (b) the coordination sphere of the Cu(II) centre, consisting of one methionine, one cysteine and two histidine residues. (c) and (d) The structure of azurin from *Pseudomonas putida* (PDB code: 1NWO): (c) the backbone of the protein chain showing the position of the Cu(II) centre and (d) the Cu(II) centre, coordinated by a methionine, a cysteine and two histidine residues; one O atom from the glycine residue adjacent to one of the histidines interacts weakly with the metal centre (the red hashed line). Hydrogen atoms are omitted; colour code: Cu, brown; S, yellow; C, grey; N, blue; O, red.

one Cys residue (Cu–S $= 213$ pm) and two His groups. The Cu_3-unit lies within eight His residues (Fig. 29.14b), and can be subdivided into Type 2 and Type 3 Cu centres. The Type 2 centre is coordinated by two His groups and either an H_2O or an $[OH]^-$ ligand (the experimental data cannot distinguish between them). The Type 3 centre consists of two Cu atoms bridged by either an O^{2-} or an $[OH]^-$ ligand. Magnetic data show that these Cu centres are antiferromagnetically coupled. Reduction of O_2 occurs at a Type 2/Type 3 Cu_3 site, with the remote Type 1 Cu centre acting as the main electron acceptor, removing electrons from the organic substrate. Details of the mechanism are not understood.

Laccase has been isolated from lacquer trees (e.g. *Rhus vernifera*) and from various fungi. The crystal structure of laccase obtained from the fungus *Trametes versicolor* was reported in 2002 and confirms the presence of a trinuclear copper site containing Type 2 and Type 3 copper atoms, and a monocopper (Type 1) site. The structure of the trinuclear copper site is similar to that in ascorbate oxidase (Fig. 29.14). However, the Type 1 copper atom in laccase is 3-coordinate (trigonal planar and bound by one Cys and two His residues) and lacks the axial ligand

present in the Type 1 copper centre in ascorbate oxidase. The absence of the axial ligand is thought to be responsible for tuning the reduction potential of the metalloenzyme. Laccases function over a wide range of potentials: $+500$ mV (versus a normal hydrogen electrode) is characteristic of a 'low-potential laccase' and $+800$ mV is typical for a 'high-potential laccase'. Laccase from *Trametes versicolor* belongs to the latter class.

The mitochondrial electron-transfer chain

Mitochondria are the sites in cells where raw, biological fuels are converted into energy.

Before continuing the discussion of specific electron-transfer systems, we take a look at the *mitochondrial electron-transfer chain*, i.e. the chain of redox reactions that occurs in living cells. This allows us to appreciate how the different systems discussed later fit together. Each system transfers one or more electrons and operates within a small range of reduction potentials as illustrated in Fig. 29.15. Diagrams **29.17** and **29.18** show the structures of the coenzymes $[NAD]^+$ and FAD, respectively.

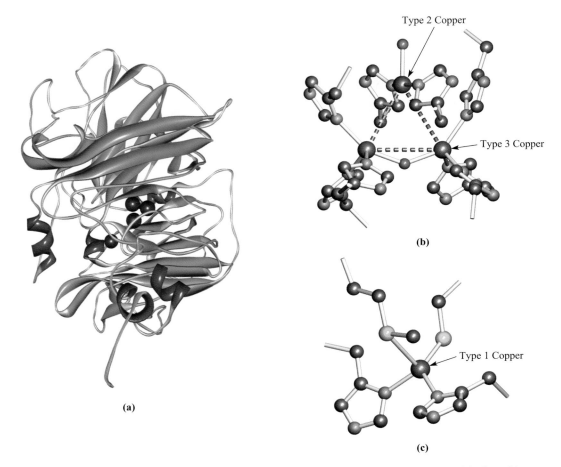

(a)

(b)

(c)

Fig. 29.14 (a) A ribbon representation of one unit of ascorbate oxidase isolated from courgettes (zucchini, *Cucurbita pepo medullosa*) (PDB code: 1AOZ). The positions of the Type 1 (on the left), Type 2 and Type 3 Cu atoms are shown. (b) Details of the tricopper unit. Each Type 3 Cu centre is bound to the protein backbone by three His residues, and the Type 2 Cu is coordinated by two His residues. (c) The Type 1 Cu centre is coordinated by a Cys, a Met and two His residues. Hydrogen atoms are omitted; colour code: Cu, brown; S, yellow; C, grey; N, blue; O, red.

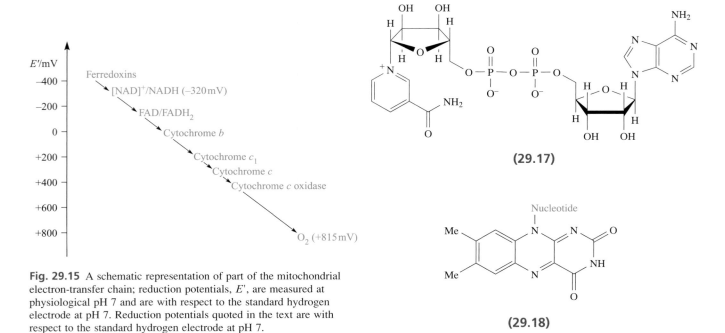

Fig. 29.15 A schematic representation of part of the mitochondrial electron-transfer chain; reduction potentials, E', are measured at physiological pH 7 and are with respect to the standard hydrogen electrode at pH 7. Reduction potentials quoted in the text are with respect to the standard hydrogen electrode at pH 7.

(29.17)

(29.18)

At one end of the chain in Fig. 29.15, *cytochrome c oxidase* catalyses the reduction of O_2 to H_2O (eq. 29.9 for which $E' = +815\,mV$). The E' scale (applicable to measurements at pH 7) in Fig. 29.15 extends to $-414\,mV$, which corresponds to reaction 29.11 at pH 7, and this range of potentials corresponds to those accessible under physiological conditions.

$$2H^+ + 2e^- \rightleftharpoons H_2 \qquad (29.11)$$

Most redox reactions involving organic molecules occur in the range $0\,mV > E' > -400\,mV$. The oxidation of a biological 'fuel' (e.g. carbohydrate) involves reactions in which electrons are passed through members of the electron transport chain until eventually H_2 and the electrons enter the $[NAD]^+$/NADH couple. Electron transfer in *steps* utilizing redox couples provided by the metal centres in metalloproteins is an essential feature of biological systems. There is a mismatch, however: oxidations and reductions of organic molecules typically involve 2-electron processes, whereas redox changes at metal centres involve 1-electron steps. The mediators in the electron transport chain are *quinones*, organic molecules which can undergo *both* 1- and 2-electron processes (eq. 29.12).

$$(29.12)$$

At several points in the mitochondrial electron-transfer chain, the release of energy is coupled to the synthesis of ATP from ADP (see Box 15.11), and this provides a means of storing energy in living cells.

Iron–sulfur proteins

The existence of iron–sulfur proteins in our present *oxidizing* environment has to be attributed to the fact that, during a stage in evolution, the environment was a *reducing* one.[†] Iron–sulfur proteins are of relatively low molecular weight and

---

[†] For a fuller discussion, see J.J.R. Fraústo da Silva and R.J.P. Williams (2001) *The Biological Chemistry of the Elements*, 2nd edn., OUP, Oxford.

contain high-spin Fe(II) or Fe(III) coordinated tetrahedrally by four *S*-donors. The latter are either S^{2-} (i.e. discrete sulfide ions) or Cys residues attached to the protein backbone. The sulfide (but not the cysteine) sulfur can be liberated as H_2S by the action of dilute acid. The FeS_4 centres occur singly in *rubredoxins*, but are combined into di-, tri- or tetrairon units in *ferredoxins*. The biological functions of iron–sulfur proteins include electron-transfer processes, nitrogen fixation, catalytic sites in hydrogenases, and oxidation of NADH to $[NAD]^+$ in mitochondria (Fig. 29.15).

> *Hydrogenases* are enzymes that catalyse the reaction:
>
> $$2H^+ + 2e^- \longrightarrow H_2$$

The simplest iron–sulfur proteins are *rubredoxins* ($M_r \approx 6000$) which are present in bacteria. Rubredoxins contain single FeS_4 centres in which all the *S*-donors are from Cys residues. Figure 29.16 shows the structure of the rubredoxin isolated from the bacterium *Clostridium pasteurianum*. The metal site lies in a pocket of the folded protein chain. The four Fe–S(Cys) bonds are of similar length (227–235 pm) and the S–Fe–S bond angles lie in the range 103–113°. The reduction potential for the Fe^{3+}/Fe^{2+} couple is sensitive to the conformation of the protein chain forming the pocket in which the FeS_4-unit lies. Consequently, a range of reduction potentials has been observed depending on the exact origin of the rubredoxin, but all are close to 0 V, e.g. $-58\,mV$ for rubredoxin from *Clostridium pasteurianum*. Rubredoxins function as 1-electron transfer sites, with the iron centre shuttling between Fe(II) and Fe(III). Upon oxidation, the Fe–S bond lengths shorten by $\approx 5\,pm$.

Ferredoxins occur in bacteria, plants and animals and are of several types:

- [2Fe–2S] ferredoxins contain two Fe centres, bridged by two S^{2-} ligands with the tetrahedral coordination sphere of each metal completed by two Cys residues (Figs. 29.17a and b);
- [3Fe–4S] ferredoxins contain three Fe and four S^{2-} centres arranged in an approximately cubic framework with one corner vacant; this unit is connected to the protein backbone by Cys residues (Fig. 29.17c);
- [4Fe–4S] resemble [3Fe–4S] ferredoxins, but contain an additional FeS(Cys) group which completes the approximately cubic cluster core (Fig. 29.17d).

The advantage of ferredoxins over rubredoxins in terms of redox chemistry is that by combining several Fe centres in close proximity, it is possible to access a greater range of reduction potentials. Different conformations of the protein pockets which surround the Fe_xS_y clusters affect the detailed structural features of the cluster cores and, thus, their reduction potentials, e.g. $-420\,mV$ for spinach [2Fe–2S] ferredoxin, and $-270\,mV$ for adrenal [2Fe–2S]

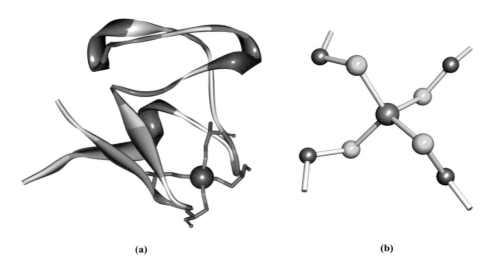

(a)   **(b)**

Fig. 29.16 (a) A ribbon representation of the metalloprotein rubredoxin from the bacterium *Clostridium pasteurianum* (PDB code: 1B13). The position of the Fe atom in the active site is shown. (b) Detail of the active site showing the tetrahedral arrangement of the Cys residues that bind the Fe centre. Hydrogen atoms are omitted; colour code: Fe, green; S, yellow; C, grey.

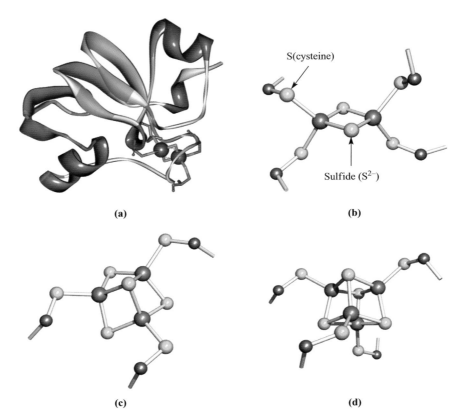

Fig. 29.17 (a) The structure of the ferredoxin metalloenzyme from spinach (*Spinacia oleracea*) and determined by X-ray diffraction (PDB code: 1A70). The protein backbone is shown in ribbon representation, and the active site in ball (Fe atoms) and stick (Cys residues) representations. (b)–(d) The iron–sulfur units from ferredoxins, structurally characterized by X-ray diffraction: (b) the [2Fe–2S] ferredoxin from spinach (*Spinacia oleracea*), (c) the [3Fe–4S] ferredoxin from the bacterium *Azotobacter vinelandii*, and (d) the [4Fe–4S] ferredoxin from the bacterium *Chromatium vinosum*. Hydrogen atoms are omitted; colour code: Fe, green; S, yellow; C, grey.

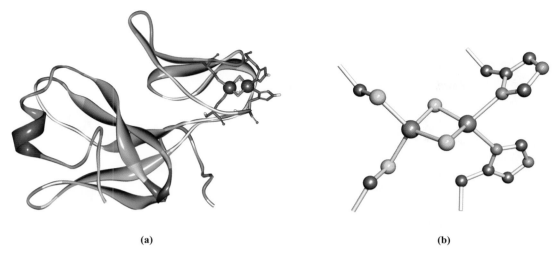

(a)   (b)

Fig. 29.18 (a) The structure (shown in ribbon representation) of Rieske protein from spinach (*Spinacia oleracea*) chloroplast (PDB code: 1RFS). The position of the Fe-containing active site is shown. (b) Detail of the [2Fe−2S] active site in which one Fe atom is coordinated by two Cys residues and the second is bound by two His residues. Hydrogen atoms are omitted; colour code: Fe, green; S, yellow; C, grey; N, blue.

ferredoxin. A [2Fe–2S] ferredoxin acts as a 1-electron transfer centre, going from an Fe(II)/Fe(II) state in the reduced form to an Fe(II)/Fe(III) state when oxidized and vice versa. Evidence for the localized, mixed valence species comes from EPR spectroscopic data.

A [4Fe–4S] ferredoxin also transfers one electron, and typical reduction potentials lie around −300 to −450 mV corresponding to the half-reaction 29.13. A [4Fe–4S] ferredoxin containing four Fe(II) centres is never accessed in biology.

$$2Fe(III)\cdot2Fe(II) + e^- \rightleftharpoons Fe(III)\cdot3Fe(II) \qquad (29.13)$$

The two species represented in eq. 29.13 do not actually possess localized Fe(II) and Fe(III) centres, rather the electrons are delocalized over the cluster core. One could envisage further oxidation to species that are formally 3Fe(III)·Fe(II) and 4Fe(III). Whereas the latter is never accessed under physiological conditions, 3Fe(III)·Fe(II) is the oxidized form of HIPIP (*high-potential protein*). Thus, 2Fe(III)·2Fe(II) is the reduced form of HIPIP or the oxidized form of ferredoxin. In contrast to the reduction potentials of ferredoxins, those of HIPIPs are *positive*, e.g. +360 mV for HIPIP isolated from the bacterium *Chromatium vinosum*. Within a given metalloprotein, redox reactions involving two electrons which effectively convert a ferredoxin into HIPIP do *not* occur.

Although we have focused on individual structural units in rubredoxins, ferredoxins and HIPIPs, some metalloproteins contain more than one Fe$_x$S$_y$ unit. For example, the ferredoxin isolated from *Azotobacter vinelandii* contains both [4Fe–4S] and [3Fe–4S] units, with the closest Fe⋯⋯Fe separation between units being ≈930 pm.

Oxygenic photosynthesis involves the cytochrome b_6f complex which is made up of sub-units including cytochrome f containing one c haem, cytochrome b_6 with two b haems, and Rieske protein which is a high-potential protein containing a [2Fe–2S] cluster. The latter is distinguished from a [2Fe–2S] ferredoxin by having one Fe centre bound by two His (rather than Cys) residues (Fig. 29.18). Rieske protein is the electron-transfer site in the oxidation of plastoquinol (a hydroquinone) to plastosemiquinone, during which protons are released. Rieske protein isolated from spinach chloroplasts has a *positive* reduction potential (+290 mV), contrasting with *negative* values for [2Fe–2S] ferredoxins. The difference is attributed to the His versus Cys coordination of one Fe centre.

Metabolism in microorganisms relies on using H$_2$ as a reducing agent and on converting H$^+$ to H$_2$ at the end of the electron-transfer chain. Three types of hydrogenases that catalyse these reactions have been identified in anaerobic bacteria. Both [NiFe]-hydrogenases and [FeFe]-hydrogenases (also called [Fe-only]-hydrogenases) contain iron–sulfur clusters. The active site of [Fe]-hydrogenase (found in methanogenic, single-celled microorganisms called archaea) contains a single Fe centre and the enzyme possesses no iron–sulfur clusters. We focus below on [NiFe]- and [FeFe]-hydrogenases which are more widely distributed among microorganisms than [Fe]-hydrogenase.

The structure of [NiFe]-hydrogenase from the bacterium *Desulfovibrio gigas* (*D. gigas*) has been crystallographically determined. It consists of two protein sub-units. The smaller unit contains one [3Fe–4S] and two [4Fe–4S] clusters. Pairs of adjacent clusters are ≈1200 pm apart and the three clusters form an electron-transfer pathway from the active site (which is located in the larger sub-unit, Fig. 29.19a) to the

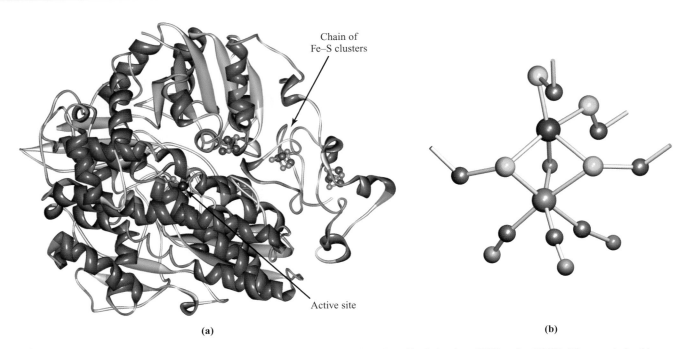

Chain of
Fe–S clusters

Active site

(a)

(b)

Fig. 29.19 (a) The structure of [NiFe]-hydrogenase from the bacterium *Desulfovibrio gigas* (PDB code: 1FRV). The protein backbone is shown in ribbon representation. The [3Fe–4S] and two [4Fe–4S] clusters are located in the smaller of the two sub-units of the protein and the active site is buried within the larger sub-unit. The Fe, S and Ni atoms in the [Fe–S] clusters and active site are shown as spheres: Fe, green; S, yellow; Ni, blue. (b) The structure of the active site in [NiFe]-hydrogenase from *D. gigas*. Colour code: Fe, green; Ni, dark blue; S, yellow; C, grey; O, red; N, blue. Each non-terminated stick represents the connection of a coordinated cysteine residue to the protein backbone.

surface of the enzyme. The active site is ≈1300 pm away from the nearest [4Fe–4S] cluster, a distance that is compatible with electron transfer (see eq. 26.65 and discussion). The dimetallic unit in the active site (Fig. 29.19b) is tethered to the protein backbone by four cysteine residues. Two are terminally bound to the Ni atom, and two bridge the NiFe unit. The identity of the ligands in the $Fe(CO)(CN)_2$ group shown in Fig. 29.19b is supported by IR spectroscopic data. Because CO and $[CN]^-$ are strong-field ligands, the Fe(II) centre is low-spin in both the oxidized and reduced forms of the enzyme. The Fe atom is not redox active during the enzymic process. The bridging ligand shown as O in Fig. 29.19b is probably a μ-peroxido ligand in the oxidized (non-active) form of the hydrogenase. The enzyme may be activated by loss of this bridging ligand. Figure 29.19b shows Ni to be coordinatively unsaturated. Both the vacant site on Ni and the bridging site between Ni and Fe are thought to be involved in binding hydride, H_2 and/or protons. Crystallographic data reveal the presence of an Mg^{2+} ion close to the active site. The ion is octahedrally sited and is bound by H_2O molecules and amino acid residues, but its role is not fully understood. [NiFe]-Hydrogenases from the bacteria *D. fructosovorans* and *D. desulfuricans* possess similar structures to that from *D. gigas*. A report of an Fe-bound SO ligand in the [NiFe]-hydrogenase from *D. vulgaris* now appears erroneous; this too contains an $Fe(CO)(CN)_2$ unit in the active site.

The crystal structures of the [FeFe]-hydrogenases from the bacteria *D. desulfuricans* and *Clostridium pasteurianum* (*C. pasteurianum*) have been determined. The enzyme from *C. pasteurianum* (Fig. 29.20) is monomeric and contains one [2Fe–2S] and three [4Fe–4S] clusters in

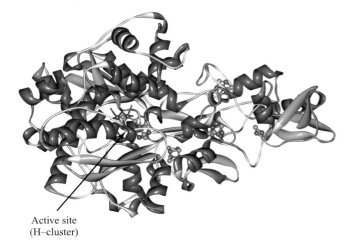

Active site
(H–cluster)

Fig. 29.20 The structure of the [FeFe]-hydrogenase from the bacterium *C. pasteurianum* (PDB code: 3C8Y). The protein backbone is shown in ribbon representation, and the Fe and S atoms in the [Fe–S] clusters and active site are shown as spheres. The H-cluster (active site) is the left-hand cluster highlighted in the diagram. Colour code: Fe, green; S, yellow; C, grey; O, red; N, blue. See Fig. 29.21 for an enlargement of the active site.

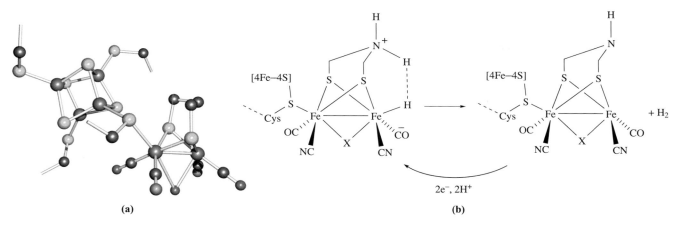

Fig. 29.21 (a) The structure of the H-cluster in the [FeFe]-hydrogenase. The Fe_4S_4-cluster has four associated Cys residues, one of which bridges to the Fe_2S_2-unit. The right-hand Fe atom is coordinatively unsaturated (but see text). Colour code: Fe, green; S, yellow; C, grey; N, blue; O, red. Each non-terminated stick represents the connection of a coordinated amino acid to the protein backbone. (b) A proposed scheme illustrating the role of a bridging amino group in enzymic H_2 evolution.

addition to the so-called H-cluster which is the active site. The latter consists of a [4Fe–4S] cluster connected directly to a [2Fe–2S] unit by a bridging Cys residue (Fig. 29.21a). Three of the Fe atoms in the [4Fe–4S] cluster are bound by Cys residues to the protein backbone. The bridging Cys residue is the only means by which the [2Fe–2S] unit is tethered to the protein chain (Fig. 29.21a). The [FeFe]-hydrogenase from *D. desulfuricans* is dimeric and each of the two protein sub-units contains two [4Fe–4S] clusters in addition to the H-cluster. Although the major structural features of the H-cluster have been elucidated (Fig. 29.21a), several ambiguities remain. The two S atoms in the [2Fe–2S] unit are the termini of a bridging group shown in Fig. 29.21a as a $SCH_2CH_2CH_2S$ unit. However, the bridge could also be SCH_2NHCH_2S or SCH_2OCH_2S since it is difficult to distinguish between CH_2, NH or O using only crystallographic data. It has been proposed that the presence of an amine group would give a site suited to a catalytic role in proton delivery and H_2 evolution (Fig. 29.21b). On the other hand, a combined crystallographic and theoretical investigation of the H-cluster in [FeFe]-hydrogenase from *C. pasteurianum* supports the presence of an SCH_2OCH_2S bridge.[†] Each Fe atom in the [2Fe–2S] unit in the H-cluster carries two terminal ligands assigned as CO and $[CN]^-$. These assignments are supported by IR spectroscopic data. An additional ligand (shown as an O atom in Fig. 29.21a) forms an asymmetrical bridge between the two Fe atoms. The identity of this ligand is uncertain, but H_2O has been proposed for the hydrogenase isolated from *D. desulfuricans*. In the structure of the enzyme from *C. pasteurianum*, the bridging ligand has been assigned as CO. The Fe centre at the right-hand side of Fig. 29.21a is

proposed to be the primary catalytic centre at which H^+ is reduced to H_2. In the structure of [FeFe]-hydrogenase from *D. desulfuricans*, this Fe site is coordinatively unsaturated (Fig. 29.21a). In contrast, this 'vacant' Fe site in the hydrogenase from *C. pasteurianum* is occupied by a terminal H_2O ligand. The differences in structural details of the active sites in the [FeFe]-hydrogenases from *C. pasteurianum* and *D. desulfuricans* are rationalized in terms of the former being an oxidized or resting state, while the latter represents a reduced state. A possible proton pathway within the enzyme involves Lys and Ser residues (see Table 29.2) in the protein backbone. Although a Lys residue is not directly coordinated to the active Fe centre, one is hydrogen-bonded to the Fe-bound $[CN]^-$ ligand. It has been established that the addition of CO inhibits enzyme activity. Crystallographic data confirm that the CO binds at the Fe site which is coordinatively unsaturated in the native enzyme.

Since the late 1990s, there has been a surge of research interest in designing and studying suitable model compounds for [NiFe]- and [FeFe]-hydrogenases. This has included Fe(II) compounds containing both CO and $[CN]^-$ ligands (see 'Iron(II)' in Section 21.9) and compounds such as **29.19** and **29.20**. Structurally, complex **29.19** closely resembles the active site of [FeFe]-hydrogenase (Fig. 29.21), but attempts to study reactions of **29.19** with H^+ lead to the formation of insoluble and catalytically inactive polymeric material. On the other hand, complex **29.20** is an active catalyst for proton reduction. Figure 29.22 shows the synthesis of a model for the complete H-cluster from [FeFe]-hydrogenase. This model complex[‡] catalyses the reduction of H^+ to H_2.

[†] A.S. Pandey, T.V. Harris, L.J. Giles, J.W. Peters and R.K. Szilagyi (2008) *J. Am. Chem. Soc.*, vol. 130, p. 4533 - 'Dithiomethylether as a ligand in the hydrogenase H-cluster'.

[‡] For details, see: C. Tard, X. Liu, S.K. Ibrahim, M. Bruschi, L. De Gioia, S.C. Davies, X. Yang, L.-S. Wang, G. Sawers and C.J. Pickett (2005) *Nature*, vol. 433, p. 610.

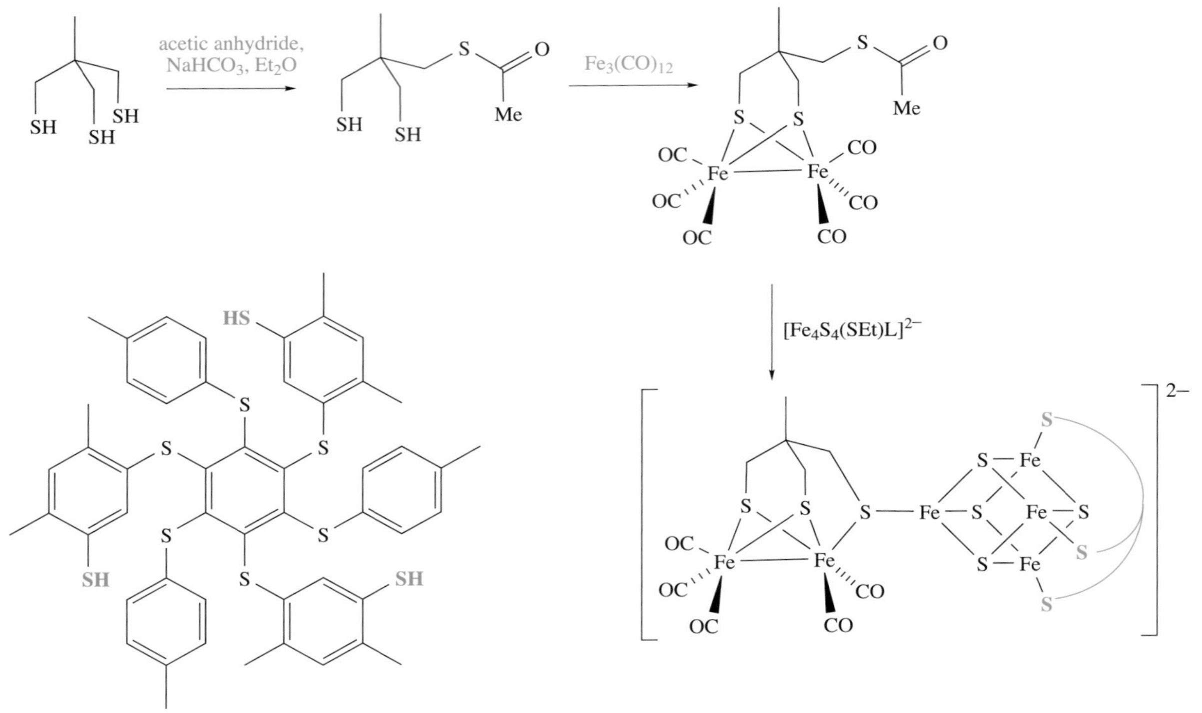

Fig. 29.22 Synthesis of a complex that models the complete H-cluster in [FeFe]-hydrogenase. The ligand L^{3-} forms a protective 'umbrella' over the [4Fe–4S] cubane.

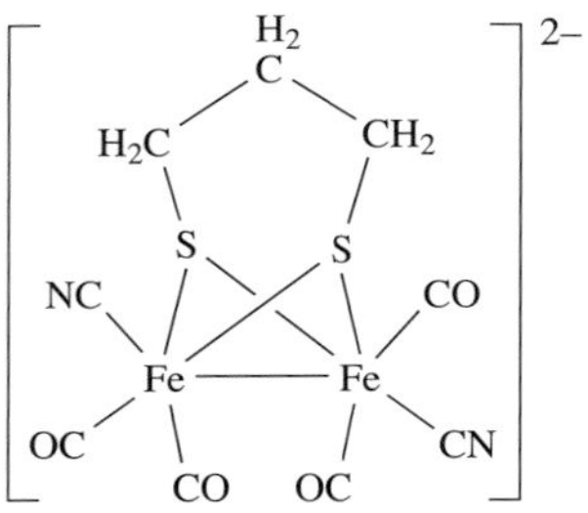

(29.19) 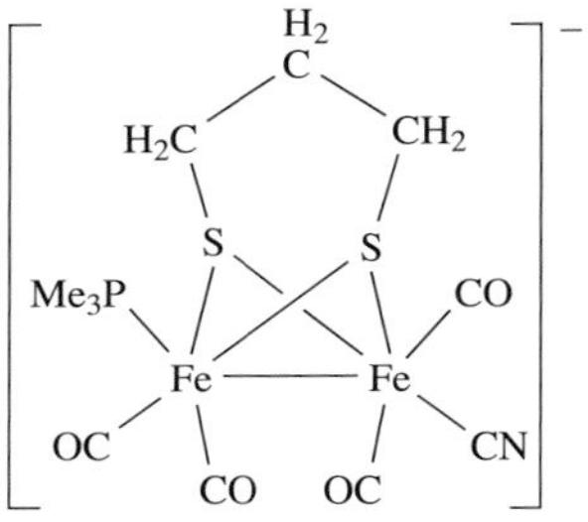   (29.20)

Nitrogen fixation by bacteria involves the reduction of N_2 to NH_3 (eq. 29.14) catalysed by *nitrogenases*. Concomitant with this process is the hydrolysis of ATP which is an energy-releasing process.

$$N_2 + 8H^+ + 8e^- \longrightarrow 2NH_3 + H_2 \qquad (29.14)$$

Studies of nitrogenase proteins from the bacteria *Azotobacter vinelandii* and *C. pasteurianum* have provided structural details of the proteins involved. Two metalloproteins make up the nitrogenase system: an Fe protein which couples the hydrolysis of ATP to electron transfer, and an FeMo protein which is responsible for binding N_2.

The dual role of these proteins can be summarized in three steps:

● reduction of Fe protein;
● 1-electron transfer from the Fe protein to FeMo protein in a process which also involves ATP hydrolysis;
● electron and H^+ transfer to N_2.

The Fe protein is a dimer and contains one [4Fe–4S] ferredoxin cluster held by Cys residues between the two halves of the protein. The ferredoxin site is relatively exposed on the surface of the protein. The FeMo protein contains two different Fe-containing clusters called the P-cluster and the FeMo cofactor. Both are buried within the protein. Details of their structures have been revealed through X-ray crystallography. In its reduced state, the P-cluster (Fig. 29.23a) consists of two [4Fe–4S] units with one S atom in common. The [4Fe–4S] cubanes are also bridged by two Cys residues, and each cubane is further connected to the protein backbone by two terminal Cys residues. The P-cluster acts as an intermediate in electron transfer from the Fe protein to the FeMo cofactor. This redox chemistry brings about structural changes in the P-cluster. On going from a reduced to oxidized state, the P-cluster opens up, replacing two Fe–S(shared atom) interactions with Fe–O(serine) and Fe–N(amide-backbone)

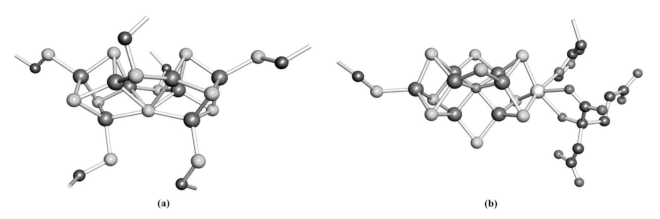

Fig. 29.23 The structures of the two types of cluster unit present in the nitrogenase molybdenum–iron protein isolated from *Azotobacter vinelandii*: (a) the P-cluster in its reduced state and (b) the FeMo cofactor. Colour code: Fe, green; Mo, pale grey; S, yellow; C, grey; N, blue; O, red. Each non-terminated stick represents the connection of a coordinated amino acid residue to the protein backbone.

bonds. The structure of the FeMo cofactor (Fig. 29.23b) has been determined through increasingly higher resolution crystal structures. It consists of a [4Fe−3S] unit connected by three bridging S atoms to a [3Fe−1Mo−3S] unit. A 6-coordinate, central atom (detected for the first time in 2002)[†] completes the cubane motif of each unit. Unambiguous assignment of this atom based on crystallographic electron density data is difficult. Possible atoms are C, N and O and, of these, the favoured candidate is N. This assignment is supported by theoretical studies. How (or, even, whether) the presence of this central atom is connected to the conversion of N_2 to NH_3 in nitrogenase is, as yet, unknown. Research results[‡] are consistent with N_2 (as well as hydrazine and small alkynes) interacting with a specific FeS site in the central part of the FeMo cofactor. Atoms Fe2 and Fe6 (structure **29.21**) are the favoured sites for N_2 binding.[*] The Mo centre in the FeMo cofactor is

approximately octahedral. It is bound to the protein backbone by a His residue and is also coordinated by a bidentate homocitrate ligand. The closest distance between metal centres in the two metal clusters in the FeMo protein is ≈ 1400 pm, a separation which is amenable to electron transfer (see eq. 26.65 and discussion). The way in which the Fe and FeMo proteins act together to catalyse the conversion of N_2 to NH_3 has yet to be established.

Before leaving iron–sulfur proteins, we must mention the important contributions that model studies have made, in particular before protein X-ray structural data were available. For discrete clusters of the type formed by reaction 29.15 and shown in diagram **29.22**, it is possible to investigate magnetic, electronic spectroscopic and electrochemical properties, record [57]Fe Mössbauer spectra (see Section 4.10) and determine accurate structural data by X-ray diffraction. Working with metalloproteins is, of course, far more difficult.

(29.21)

Fe2   Fe6

$$FeCl_3 + NaOMe + NaHS + PhCH_2SH$$
$$\longrightarrow Na_2[Fe_4S_4(SCH_2Ph)_4] \qquad (29.15)$$

(29.22)

[†] See: O. Einsle, F.A. Tezcan, S.L.A. Andrade, B. Schmid, M. Yoshida, J.B. Howard and D.C. Rees (2002) *Science*, vol. 297, p. 1696 – 'Nitrogenase MoFe-protein at 1.16 Å resolution: A central ligand in the FeMo-cofactor'.
[‡] For details, see: P.C. Dos Santos, R.Y. Igarashi, H.-I. Lee, B.M. Hoffman, L.C. Seefeldt and D.R. Dean (2005) *Acc. Chem. Res.*, vol. 38, p. 208.
[*] See: I. Dance (2006) *Biochemistry*, vol. 45, p. 6328 – 'Mechanistic significance of the preparatory migration of hydrogen atoms around the FeMo-coactive site of nitrogenase'.

Model compound **29.22** and related complexes contain high-spin Fe centres. Formally there are two Fe(II) and two Fe(III), but spectroscopic data are consistent with four equivalent metal centres and, therefore, delocalization of electrons within the cage.

Cytochromes

Figure 29.15 showed *cytochromes* to be vital members of the mitochondrial electron-transfer chain. They are also essential components in plant chloroplasts for photosynthesis. Cytochromes are haem proteins, and the ability of the iron centre to undergo reversible Fe(III) ⇌ Fe(II) changes allows them to act as 1-electron transfer centres. Many different cytochromes are known, with the reduction potential for the Fe^{3+}/Fe^{2+} couple being tuned by the surrounding protein environment. Cytochromes belong to various families, e.g. cytochromes *a*, cytochromes *b* and cytochromes *c*, which are denoted according to the substituents on the haem group. In O_2-carrying haem proteins, the 'rest state' contains a 5-coordinate Fe(II) centre which becomes 6-coordinate after O_2 uptake. In contrast, the electron-transfer cytochromes *b* and *c* contain 6-coordinate Fe which is present as either Fe(II) or Fe(III). There is little change in ligand conformation as the redox change occurs. Figure 29.24 shows the structure of cytochrome *c* isolated from horse heart. Compare the haem structure with that in haemoglobin (Fig. 29.7). In cytochrome *c*, the haem unit is bound to the protein backbone through axial His and Met residues, and through two Cys residues which are covalently linked to the porphyrin ring.

In the mitochondrial electron-transfer chain, cytochrome *c* accepts an electron from cytochrome c_1 and then transfers

it to cytochrome *c* oxidase (eq. 29.16). Ultimately, the electron is used in the 4-electron reduction of O_2 (see below). The oxidized forms of the cytochromes in eq. 29.16 contain Fe(III), and the reduced forms contain Fe(II).

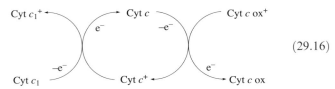

$$(29.16)$$

It is proposed that an electron is transferred by tunnelling through one of the exposed edges of the haem unit (recall that the porphyrin ring is conjugated). In relation to this, it is instructive to look at the arrangement of the haem units in cytochrome *c*554, a tetrahaem protein isolated from the bacterium *Nitrosomonas europaea* and essential to the nitrification pathway: NH_3 is converted to NH_2OH (catalysed by *ammonia monooxygenase*) which is then oxidized to $[NO_3]^-$ (catalysed by *hydroxylamine oxidoreductase*). The role of cytochrome *c*554 is to accept pairs of electrons from hydroxylamine oxidoreductase and transfer them, via cytochrome *c*552, to terminal oxidases. The crystal structure of cytochrome *c*554 shows that the four haem units are arranged in pairs such that the porphyrin rings are approximately parallel, and have overlapping edges. Adjacent pairs are then approximately perpendicular to each other (Fig. 29.25). Such arrangements have been observed in other multi-haem cytochromes and are presumably set up to provide efficient electron-transfer pathways between the edges of the haem groups.

The exact nature of the metal sites in cytochrome *c* oxidase was resolved in 1995. This terminal member of

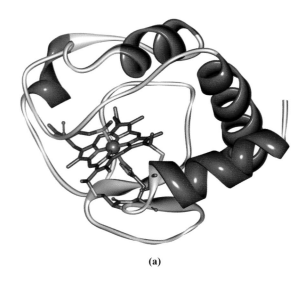

(a)

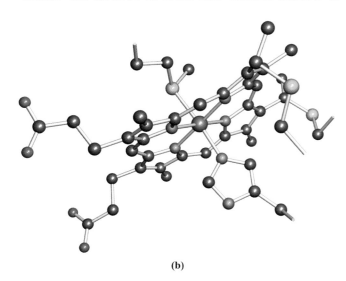

(b)

Fig. 29.24 (a) The protein chain (shown in a ribbon representation) of horse heart cytochrome *c*, showing the position of the haem unit. (b) The coordination sphere of the iron site showing the residues (Met, His and two Cys) which are covalently linked to the protein chain. Hydrogen atoms have been omitted; colour code: Fe, green; S, yellow; N, blue; C, grey; O, red. The 'broken sticks' represent connections to the protein backbone. (PDB code: 1HRC.)

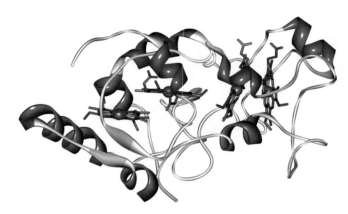

Fig. 29.25 Cytochrome *c*554 isolated from *Nitrosomonas europaea* (PDB code: 1BVB): the protein chain shown in a ribbon representation and the four haem units are shown in stick representation with Fe atoms as green balls. The Fe····Fe distances between haem units are ≈950 pm, 1220 pm and 920 pm.

the mitochondrial electron-transfer chain catalyses the reduction of O_2 to H_2O (eq. 29.9), and contains four active metal centres (Cu_A, Cu_B, haem *a* and haem a_3) which couple electron transfer to proton pumping. Electron transfer involves the Cu_A and haem *a* sites, electrons being transferred from cytochrome *c* (eq. 29.16) to Cu_A and then to haem *a*. Haem a_3 and Cu_B provide the site for O_2 binding and O_2 to H_2O conversion, and are involved in pumping H^+ (four per O_2 molecule) across the mitochondrial inner membrane. Until 1995, proposals for the identity of the metal sites were based largely on spectroscopic data and the fact that the Cu_B····Fe(haem a_3) centres were strongly antiferromagnetically coupled. The latter suggested the possible presence of a bridging ligand. Crystallographic data have now cleared the uncertainty, revealing the following structural features:

- Fe(haem *a*) is 6-coordinate with His residues in the axial sites;
- Cu_A is a dicopper site bridged by Cys residues, with a Cu_2S_2 core that is not unlike that in a [2Fe–2S] ferredoxin;
- the 3-coordinate Cu_B and 5-coordinate Fe(haem a_3) lie ≈450 pm apart and are *not* connected by a bridging ligand.

Figure 29.26 shows the active metal sites in the oxidized form of cytochrome *c* oxidase and the spatial relationship between them. They lie within a protein which has $M_r \approx 20\,000$ and is made up of 13 different polypeptide sub-units. Detailed structural studies of the protein chains have shown that a hydrogen-bonded system which incorporates residues in the protein backbone, haem propanoate side chains, and a His residue bound to Cu_A may provide an electron-transfer 'highway' between Cu_A and haem *a*.

Many model systems have been developed to aid our understanding of electron transfer and O_2 binding by

cytochromes. The initial step in the catalytic cycle involving cytochrome *c* oxidase is O_2 binding to the reduced state of the Fe(haem a_3)/Cu_B active site; this contains high-spin Fe(II) and Cu(I). Spectroscopic and mechanistic data suggest that, initially, the O_2 molecule interacts with Cu_B, and that this is followed by the formation of a haem–superoxide complex of type Fe(haem a_3)O_2/Cu_B containing Fe(III) and Cu(I). The Fe^{III}–O_2^- complex then evolves into an Fe^{IV}=O (oxido) species. The involvement of a peroxido intermediate of the type Fe^{III}–O_2^-–Cu^{II} has not been excluded, and most model systems have focused on Fe–O_2–Cu or related peroxido complexes. Structure **29.23** shows a model for this system.[†] The reaction of **29.23** with O_2 has been monitored using electronic spectroscopy, and the formation of a 1:1 complex has been confirmed. The resonance Raman spectrum of the complex exhibits an absorption at $570\,cm^{-1}$ assigned to ν(Fe−O) that shifts to $544\,cm^{-1}$ when isotopically labelled $^{18}O_2$ is used as the source of dioxygen. This absorption is characteristic of a porphyrin Fe-bound superoxide ligand.

(29.23)

[†] See: J.P. Collman, C.J. Sunderland, K.E. Berg, M.A. Vance and E.I. Solomon (2003) *J. Am. Chem. Soc.*, vol. 125, p. 6648 – 'Spectroscopic evidence for a heme–superoxide/Cu(I) intermediate in a functional model for cytochrome *c* oxidase'.

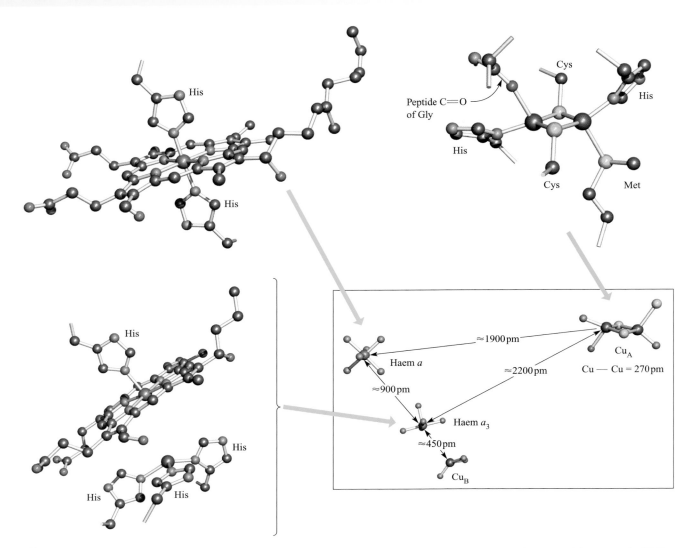

Fig. 29.26 The Cu_A, Cu_B, haem a and haem a_3 sites in cytochrome c oxidase extracted from bovine (*Bos taurus*) heart muscle (PDB code: 1OCC). The lower right-hand diagram shows the relative positions and orientations of the metal sites within the protein; an enlargement of each site shows details of the ligand spheres. Hydrogen atoms have been omitted; colour code: Cu, brown; Fe, green; S, yellow; N, blue; C, grey; O, red.

A second example of a cytochrome c oxidase model system[†] involves the reaction of a 1 : 1 mixture of complexes [Fe(**29.24**)] and [Cu(**29.25**)]$^+$ with O_2. Initially O_2 binds to [Fe^{II}(**29.24**)] with concomitant transfer of an electron from Fe to O_2 (i.e. oxidation of Fe(II) to Fe(III), and reduction of O_2 to O_2^-). [(**29.24**)Fe^{III}–O_2^-] then reacts with [Cu(**29.25**)]$^+$ to give the bridging peroxido complex [(**29.24**)Fe^{III}–O_2^{2-}–Cu^{II}(**29.25**)]$^+$. Slow transformation of [(**29.24**)Fe^{III}–O_2^{2-}–Cu^{II}(**29.25**)]$^+$ to [(**29.24**)Fe^{III}–O^{2-}–Cu^{II}(**29.25**)]$^+$ then follows. This bridging oxido species has

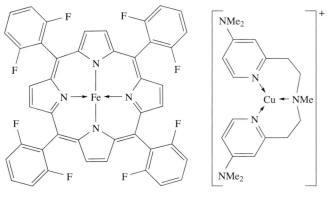

[Fe^{II}(29.24)]   **[Cu^{I}(29.25)]$^+$**

[†]See: E. Kim *et al.* (2003) *Proc. Natl. Acad. Sci.*, vol. 100, p. 3623 – 'Superoxo, μ-peroxo, and μ-oxo complexes from heme/O_2 and heme-Cu/O_2 reactivity: copper ligand influences on cytochrome c oxidase models'.

been structurally characterized and contains a bent Fe–O–Cu unit ($\angle$Fe–O–Cu = 143.4°).

Finally in this section, we note that it is the strong binding of [CN]$^-$ to Fe(III) in cytochromes that renders cyanide toxic.

In the complex formed between complex **29.23** and O$_2$, isotopic labelling of the O$_2$ causes a shift in the absorption assigned to ν(Fe−O). Explain why this shift occurs.

[*Ans.* See Section 4.6]

29.5 The Zn^{2+} ion: Nature's Lewis acid

In this section we focus on the Zn(II)-containing enzymes *carbonic anhydrase II* and *carboxypeptidases A and G2*. These are somewhat different from other systems so far described in this chapter. Zinc(II) is not a redox active centre, and so cannot take part in electron-transfer processes. It is, however, a hard metal centre (see Table 7.9) and is ideally suited to coordination by *N*- and *O*-donors. It is also highly polarizing, and the activity of Zn(II)-containing metalloenzymes depends on the Lewis acidity of the metal centre.

Carbonic anhydrase II

Human carbonic anhydrase II (CAII) is present in red blood cells and catalyses the reversible hydration of CO$_2$

(reaction 29.17). This process is slow ($k = 0.037\,\text{s}^{-1}$) but is fundamental to the removal of CO$_2$ from actively metabolizing sites. CAII increases the rate of hydrolysis by a factor of $\approx 10^7$ at physiological pH.

$$H_2O + CO_2 \rightleftharpoons [HCO_3]^- + H^+ \qquad (29.17)$$

The metalloprotein (Fig. 29.27) consists of 260 amino acids and contains a Zn^{2+} ion bound by three His residues in a pocket $\approx 1500\,\text{pm}$ deep. The tetrahedral coordination sphere is completed by a hydroxide ion or water molecule (Fig. 29.28). The peptide chain environment around the active site is crucial to the catalytic activity of the site: the Zn^{2+}-bound [OH]$^-$ ligand is hydrogen bonded to an adjacent glutamic acid residue, and to the OH group of an adjacent threonine residue (see Table 29.2). Next to the Zn^{2+} centre lies a hydrophobic pocket which 'captures' CO$_2$. The catalytic cycle by which CO$_2$ is hydrolysed is shown in Fig. 29.28b. After release of [HCO$_3$]$^-$, the coordinated H$_2$O ligand must be deprotonated in order to regenerate the active site, and the proton is transferred via a hydrogen-bonded network to a His residue (*non*-coordinated to Zn^{2+}) within the catalytic pocket.

The active site in CAII has been modelled using a hydridotris(pyrazolyl)borato ligand (**29.26**) to mimic the three histidine residues that bind Zn^{2+} in the metalloenzyme. Because Zn^{2+} is a d^{10} metal ion, it tolerates a range of coordination geometries. However, hydridotris(pyrazolyl)borato ligands are tripodal (see Section 19.7) and can force tetrahedral coordination in a complex of type [Zn(**29.26**)X]. The hydroxido complex **29.27** is one of a

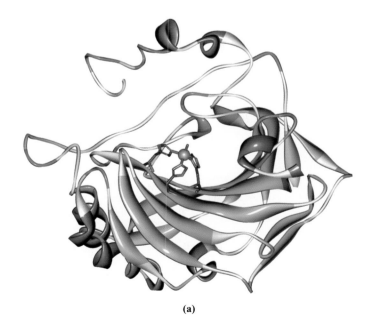

(a)

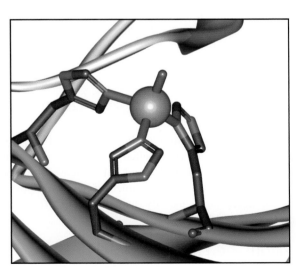

(b)

Fig. 29.27 (a) The structure of human carbonic anhydrase II determined by X-ray diffraction (PDB code: 4CAC). The protein chain is shown in ribbon representation. The active site contains three His residues and one H$_2$O molecule coordinated to a Zn^{2+} ion. (b) Enlargement of the active site. Colour code: Zn, yellow; N, blue; C, grey; O, red.

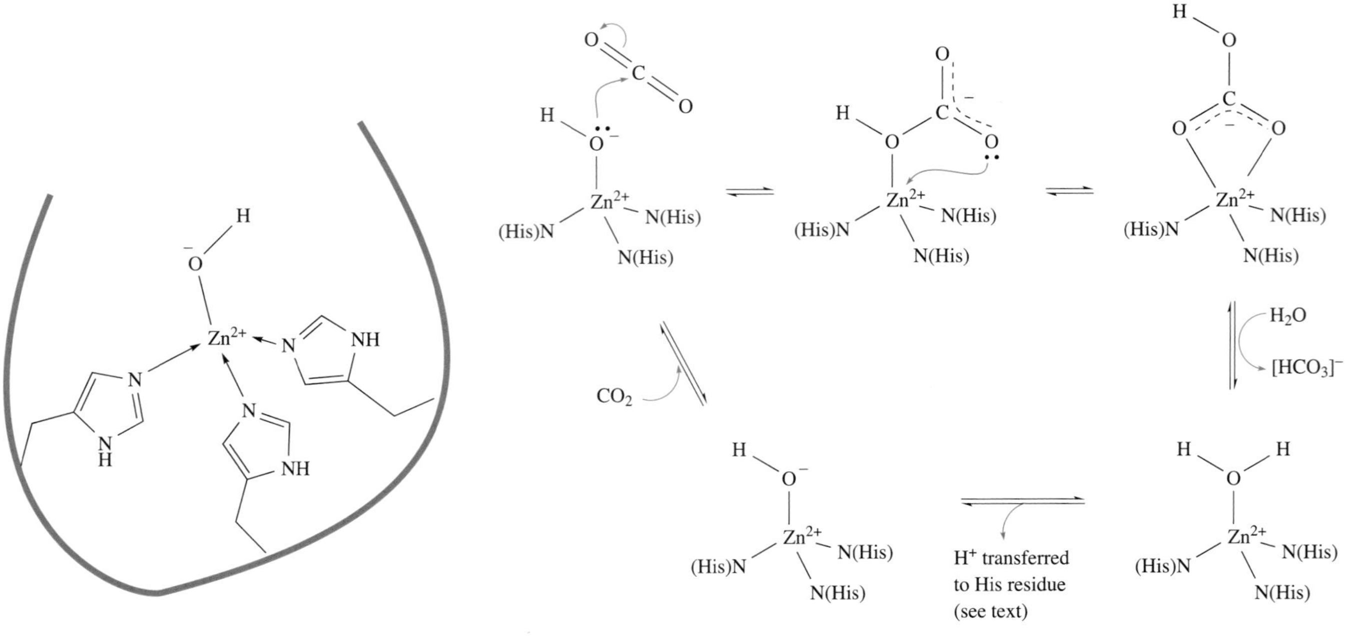

Fig. 29.28 (a) Schematic representation of the active site in human carbonic anhydrase II (CAII). (b) The catalytic cycle for the hydration of CO_2 catalysed by CAII.

series of hydridotris(pyrazolyl)borato complexes that have been studied as models for the active site in CAII.

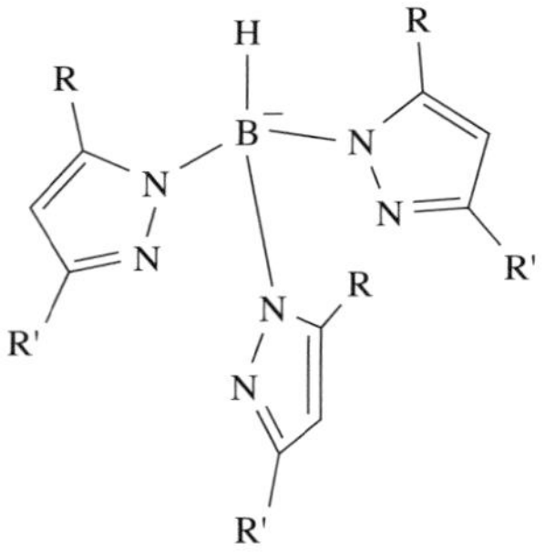

(29.26) 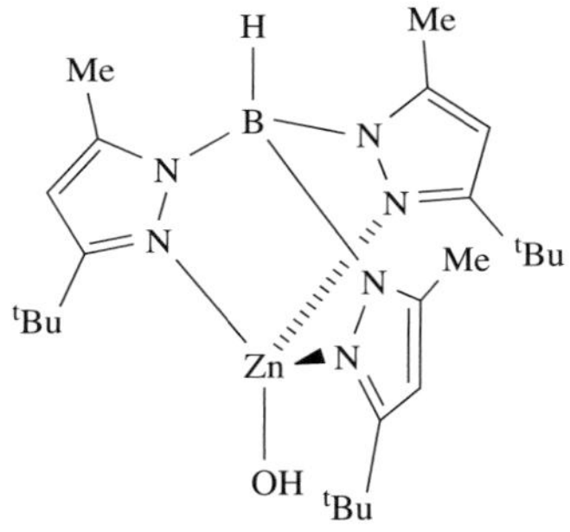   **(29.27)**

The reversible protonation of the coordinated $[OH]^-$ ligand in CAII (Fig. 29.28a) is modelled by the reaction of complex **29.27** with $(C_6F_5)_3B(OH_2)$ and subsequent deprotonation with Et_3N (eq. 29.18). The choice of acid is important as the conjugate base generally displaces the $[OH]^-$ ligand as in reaction 29.19.

$$LZn(OH) + (C_6F_5)_3B(OH_2)$$
(29.27)
$$\underset{Et_3N}{\rightleftharpoons} [LZn(OH_2)]^+[(C_6F_5)_3B(OH)]^- \qquad (29.18)$$

$$LZn(OH) + HX \longrightarrow LZnX + H_2O \qquad (29.19)$$
(29.27)

Complex **29.27** reacts with CO_2 (eq. 29.20) and catalyses oxygen exchange between CO_2 and H_2O (eq. 29.21). The latter reaction is also catalysed by carbonic anhydrase.

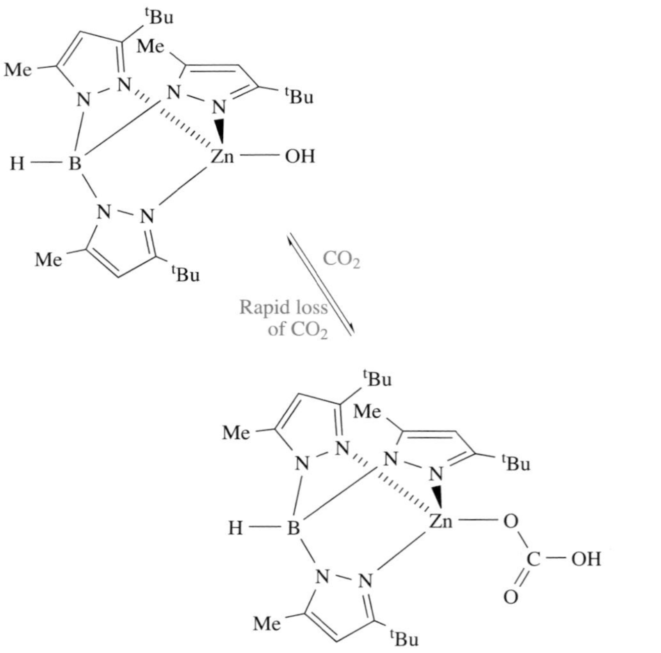

$$(29.20)$$

$$CO_2 + H_2^{17}O \rightleftharpoons CO(^{17}O) + H_2O \qquad (29.21)$$

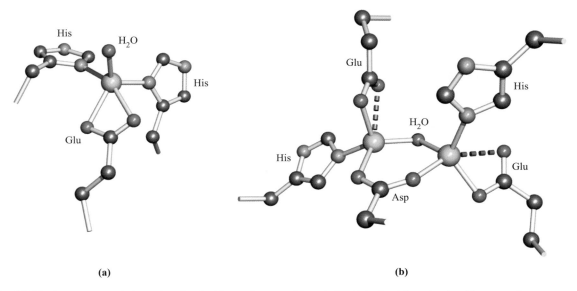

Fig. 29.29 The structures of the active sites in (a) α-carboxypeptidase A (CPA) isolated from bovine (*Bos taurus*) pancreas, and (b) carboxypeptidase G2 (CPG2) isolated from *Pseudomonas* sp.; see Table 29.2 for amino acid abbreviations. The 'broken' sticks represent connections to the protein backbone. Colour code: Zn, yellow; C, grey; O, red; N, blue.

Carboxypeptidase A

Carboxypeptidase A (CPA) is a pancreatic metalloenzyme which catalyses the cleavage of a peptide link in a polypeptide chain. The site of cleavage is specific in two ways: it occurs at the *C*-terminal amino acid (eq. 29.22), and it exhibits a high selectivity for substrates in which the *C*-terminal amino acid contains a large aliphatic or Ph substituent. The latter arises from the presence, near to the active site, of a hydrophobic pocket in the protein which is compatible with the accommodation of, for example, a Ph group (see below).

$$(29.22)$$

Carboxypeptidase A is monomeric ($M_r \approx 34\,500$) and exists in three forms (α, β and γ) which contain 307, 305 and 300 amino acids respectively. Near the surface of the protein lies a pocket in which a Zn^{2+} ion is bound to the protein backbone by one bidentate Glu and two His residues. A 5-coordinate coordination sphere is completed by a water molecule (Fig. 29.29a).

The mechanism by which the CPA-catalysed peptide-link cleavage occurs has drawn much research attention, and the pathway that is currently favoured is illustrated in a schematic form in Fig. 29.30. In the first step, the peptide to be cleaved is 'manoeuvred' into position close to the Zn^{2+} site; the dominant substrate–protein interactions involved at this stage (Fig. 29.30a) are:

- salt-bridge formation between the *C*-terminal carboxylate group of the substrate and residue Arg-145[†] which is positively charged;
- intermolecular interactions between the non-polar group R' and residues in a hydrophobic pocket of the protein chain.

These interactions may be supplemented by hydrogen bond formation (shown in Fig. 29.30a) between the OH group of Tyr-248 and the N−H group indicated in the figure, and between Arg-127 and the C=O group adjacent to the peptide cleavage site. This latter interaction polarizes the carbonyl group, activating it towards nucleophilic attack. The nucleophile is the H_2O ligand coordinated to Zn^{2+}. The Lewis acidity of the metal ion polarizes the O−H bonds, and (although this is not a unique proposal) it is likely that the carboxylate group of Glu-270 assists in the process by removing H^+ from the H_2O ligand (Fig. 29.30b). Figure 29.30c shows the next step in the proposed mechanism: the cleavage of the peptide C−N bond for which H^+ is probably provided by Glu-270. It appears likely that the second H^+ required for the formation of the NH_3^+ group on the departing terminal amino acid comes from the terminal CO_2H group of the remaining portion of the substrate (Fig. 29.30d). Figure 29.30c shows Glu-72 bound in a monodentate manner to the Zn^{2+}

[†] We have not previously included residue numbers, but do so in this discussion for the sake of clarity. Residues are numbered sequentially along the protein chain.

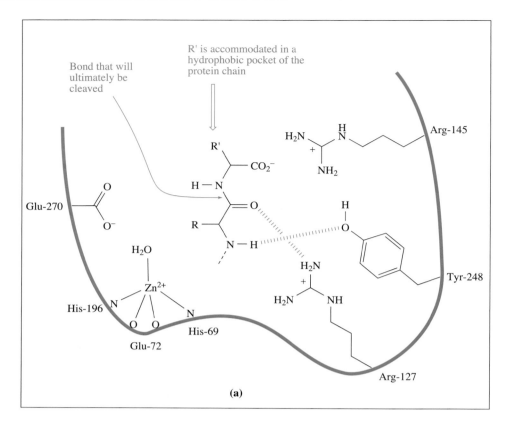

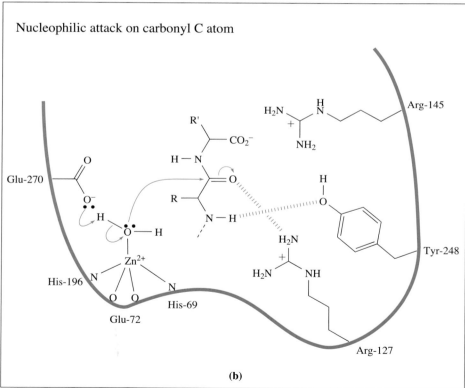

Fig. 29.30 Schematic representation of the generally accepted mechanism for the CPA-catalysed cleavage of a *C*-terminal peptide link; see Fig. 29.29a for a more detailed diagram of the coordination sphere of the Zn^{2+} ion. The red line represents the protein chain; only residues mentioned in the discussion are shown. The diagrams do not imply whether a mechanism is concerted or not.

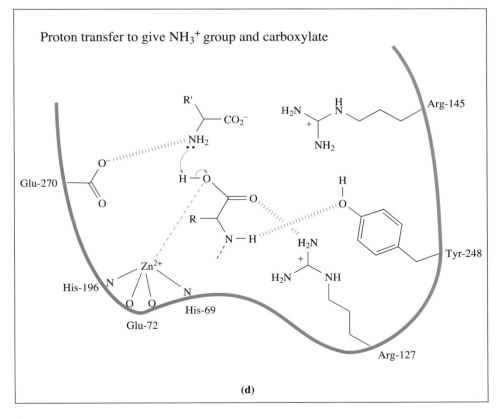

Fig. 29.30 continued

centre, whereas in the rest state, a bidentate mode has been confirmed (Fig. 29.30a). A change from a bi- to monodentate coordination appears to be associated with the formation of the $Zn^{2+}\cdots O\cdots H(Arg\text{-}127)$ interaction illustrated in Fig. 29.30c, the Zn^{2+} ion being able to move towards Arg-127 as the interaction develops. To complete the catalytic cycle, an H_2O ligand refills the vacant site on the Zn^{2+} centre. Details of this mechanism are based upon a range of data including kinetic and molecular mechanics studies and investigations of Co^{2+} substituted species (see below).

Carboxypeptidase G2

The carboxypeptidase family of enzymes also includes carboxypeptidase G2 (CPG2) which catalyses the cleavage of *C*-terminal glutamate from folate (**29.28**) and related compounds such as methotrexate (in which NH_2 replaces the OH group in the pterin group, and NMe replaces NH in the 4-amino benzoic acid unit).

Folic acid is required for growth, and the growth of tumours can be inhibited by using cancer treatment drugs which reduce the levels of folates. Structural data for the enzyme CPG2 have provided valuable information which should assist design of such drugs. Carboxypeptidase G2 (isolated from bacteria of *Pseudomonas* spp.) is a dimeric protein with $M_r \approx 41\,800$ per unit. Each monomer contains two domains, one containing the active site and one intimately involved in dimerization. Unlike carboxypeptidase A, the active site of CPG2 contains two Zn(II) centres, separated by 330 pm and bridged by an Asp residue and a water molecule (Fig. 29.29b). Each Zn^{2+} ion is further coordinated by His and Glu residues of the protein chain to give a tetrahedral environment. The pocket containing the Zn_2-unit also contains arginine and lysine residues (Table 29.2) which may be involved in binding the substrate molecule, positioning it correctly for interaction with the catalytic site.

(29.28)

Cobalt-for-zinc ion substitution

A practical disadvantage of working with metalloproteins containing Zn^{2+} is the d^{10} configuration of the ion. The metal site cannot be probed by using UV-VIS or EPR spectroscopies or by magnetic measurements. Such methods were especially important before protein crystallography became a widely applied technique. Studies involving Co^{2+}-for-Zn^{2+} substitution provide a metal centre that is amenable to investigation by spectroscopic and magnetic techniques (Co^{2+} is a d^7 ion), the choice of Co^{2+} being because:

- the ionic radii of Co^{2+} and Zn^{2+} are about the same;
- Co^{2+} can tolerate similar coordination environments to Zn^{2+};
- it is often possible to replace Zn^{2+} in a protein by Co^{2+} without greatly perturbing the protein conformation.

A typical method of metal ion substitution is shown in scheme 29.23 in which the ligand L removes Zn^{2+} by complexation.

$$[PZn^{2+}] \xrightarrow{L} [AP] \xrightarrow{Co^{2+}} [PCo^{2+}] \qquad (29.23)$$

P = protein in the metalloprotein; AP = inactive apoprotein

(29.29)   **(29.30)**

For example, treatment of carbonic anhydrase with **29.29** (or its conjugate base) results in the removal of Zn^{2+} and the formation of the catalytically inactive apoprotein. Reaction of the apoprotein with Co^{2+} gives a cobalt-substituted enzyme, $[PCo^{2+}]$, which catalyses the hydration of CO_2. Similarly, the Zn^{2+} ion can be removed from carboxypeptidase A by treatment with bpy (**29.30**), and after insertion of Co^{2+}, the model metalloenzyme $[PCo^{2+}]$ is found to be active (actually more so than native carboxypeptidase A) with respect to peptide cleavage. Investigations can be carried out with $[PCo^{2+}]$ which are impossible with native zinc enzymes, e.g. electronic spectroscopic data provide insight into coordination geometries, and monitoring the electronic spectrum as a function of pH indicates whether ligands such as H_2O are deprotonated or not.

KEY TERMS

The following terms have been introduced in this chapter. Do you know what they mean?

- trace metals
- polypeptide
- protein
- metalloprotein
- apoprotein
- ferritin
- transferrin
- siderophore
- metallothionein
- haem-protein
- haemoglobin
- myoglobin

- haemocyanin
- haemerythrin
- blue copper proteins
- oxidase
- hydrogenase
- plastocyanin
- azurin
- ascorbate oxidase
- laccase
- mitochondrial electron-transfer chain
- rubredoxin

- ferredoxins
- nitrogenase
- cytochrome
- cytochrome *c*
- cytochrome *c* oxidase
- carbonic anhydrase II
- carboxypeptidase A
- carboxypeptidase G2

FURTHER READING

Bioinorganic chemistry is a fast-moving area and readers interested in the area are advised to update the following reading list by consulting major chemical journals, in particular *Angew. Chem. Int. Ed., Chem. Commun., J. Am. Chem. Soc., Nature, Science, Nature Struct. Biol.* and *Structure*. The following series of books provides up-to-date information on all aspects of bioinorganic chemistry: *Metal Ions in Life Sciences*, eds. A. Sigel, H. Sigel and R.K.O. Sigel, vols. 1–4 (2006–2008) Wiley, Chichester; vols. 5–9 (2009–2011) RSC, Cambridge, UK; vol. 10 onwards (2012–) Springer, Dordrecht.

General sources

I. Bertini, H.B. Gray, S.J. Lippard and J.S. Valentine (1994) *Bioinorganic Chemistry*, University Science Books, Mill Valley – An excellent and detailed text, one of the best currently available.

I. Bertini, H.B. Gray, E.I. Stiefel and J.S. Valentine, eds. (2007) *Biological Inorganic Chemistry*, University Science Books, Sausalito – An up-to-date source of detailed information on essential aspects of bioinorganic chemistry.

J.A. Cowan (1997) *Inorganic Biochemistry: An Introduction*, 2nd edn, Wiley-VCH, New York – An up-to-date text covering a wider range of topics than in this chapter and including case studies.

D.E. Fenton (1995) *Biocoordination Chemistry*, Oxford University Press, Oxford – A clearly written, introductory text.

J.J.R. Fraústo da Silva and R.J.P. Williams (1991) *The Biological Chemistry of the Elements*, Oxford University Press, Oxford – An excellent, detailed text.

W. Kaim and B. Schwederski (1994) *Bioinorganic Chemistry: Inorganic Elements in the Chemistry of Life*, Wiley-VCH, Weinheim – A detailed text covering the roles of inorganic elements in living organisms, as well as applications in chemotherapy.

S.J. Lippard and J.M. Berg (1994) *Principles of Bioinorganic Chemistry*, University Science Books, Mill Valley – One of the primary texts dealing with bioinorganic chemistry.

More specialized articles including model compounds

C.A. Blindauer and P.J. Sadler (2005) *Acc. Chem. Res.*, vol. 38, p. 62 –'How to hide zinc in a small protein'.

D.W. Christianson and C.A. Fierke (1996) *Acc. Chem. Res.*, vol. 29, p. 331 – 'Carbonic anhydrase: Evolution of the zinc binding site by Nature and by design'.

A.L. de Lacey, V.M. Fernández and M. Rousset (2005) *Coord. Chem. Rev.*, vol. 249, p. 1596 – 'Native and mutant nickel–iron hydrogenases: Unravelling structure and function'.

C.L. Drennan and J.W. Peters (2003) *Curr. Opin. Struct. Biol.*, vol. 13, p. 220 – 'Surprising cofactors in metalloenzymes'.

M.C. Feiters, A.E. Rowan and R.J.M. Nolte (2000) *Chem. Soc. Rev.*, vol. 29, p. 375 – 'From simple to supramolecular cytochrome P450 mimics'.

J.C. Fontecilla-Camps, A. Volbeda, C. Cavazza and Y. Nicolet (2007) *Chem. Rev.*, vol. 107, p. 4273 – 'Structure/function relationships of [NiFe]- and [FeFe]-hydrogenases'.

D. Garner, J. McMaster, E. Raven and P. Walton, eds (2005) *Dalton Trans.*, issue 21 – A collection of articles from a

Dalton Discussion: 'Metals: Centres of biological activity'.

S.V. Kryatov, E.V. Rybak-Akimova and S. Schindler (2005) *Chem. Rev.*, vol. 105, p. 2175 – 'Kinetics and mechanisms of formation and reactivity of non-heme iron oxygen intermediates'.

X. Liu and E.C. Theil (2005) *Acc. Chem. Res.*, vol. 38, p. 167 – 'Ferritins: Dynamic management of biological iron and oxygen chemistry'.

S.S. Mansy and J.A. Cowan (2004) *Acc. Chem. Res.*, vol. 37, p. 719 – 'Iron–sulfur cluster biosynthesis: Toward an understanding of cellular machinery and molecular mechanism'.

J.A. McCleverty and T.J. Meyer, eds (2004) *Comprehensive Coordination Chemistry II*, Elsevier, Oxford – Volume 8 is entitled *Bio-coordination Chemistry* and contains chapters on all the topics covered in this text.

G. Parkin (2004) *Chem. Rev.*, vol. 104, p. 699 – 'Synthetic analogues relevant to the structure and function of zinc enzymes'.

K.N. Raymond, E.A. Dertz and S.S. Kim (2003) *Proc. Natl. Acad. Sci.*, vol. 100, p. 3584 – 'Enterobactin: An archetype for microbial iron transport'.

N. Romero-Isart and M. Vašák (2002) *J. Inorg. Biochem.*, vol. 88, p. 388 – 'Advances in the structure and chemistry of metallothioneins'.

P. Roos and N. Jakubowski, eds. (2011) *Metallomics*, vol. 3, issue 4, p. 305 – A themed issue of the journal with a series of articles dealing with cytochromes.

K. Shikama (2006) *Prog. Biophys. Mol. Biol.*, vol. 91, p. 83 – 'Nature of the FeO_2 bonding in myoglobin and hemoglobin: A new molecular paradigm'.

E.I. Solomon and R.G. Hadt (2011) *Coord. Chem. Rev.*, vol. 255, p. 774 – 'Recent advances in understanding blue copper proteins'.

M. Sommerhalter, R.L. Lieberman and A.C. Rosenzweig (2005) *Inorg. Chem.*, vol. 44, p. 770 – 'X-ray crystallography and biological metal centres: is seeing believing?'

C. Tard and C.J. Pickett (2009) *Chem. Rev.*, vol. 109, p. 2245 – 'Structural and functional analogues of the active sites of the [Fe], [NiFe]-, and [FeFe]-hydrogenases'.

R. van Eldik, ed. (2005) *Chem. Rev.*, vol. 105, issue 6 – A special issue with the theme of inorganic and bioinorganic mechanisms.

W.-D. Woggon (2005) *Acc. Chem. Res.*, vol. 38, p. 127 – 'Metalloporphyrins as active site analogues – lessons from enzymes and enzyme models'.

PROBLEMS

29.1 Give brief descriptions of the following: (a) peptide; (b) naturally occurring amino acids; (c) metalloprotein; (d) apoprotein; (e) haem unit.

29.2 Give an account of the storage and transport of metalloproteins in mammals. How does the uptake of iron by aerobic microorganisms differ from that in mammals?

29.3 $[CrL_3]^{3-}$ where $H_2L = 1,2\text{-}(HO)_2C_6H_4$ is a model complex for enterobactin. How is the model related to enterobactin, and what is the reason for chromium-for-iron substitution?

29.4 Comment on the following observations:

(a) Thioneins bind Cd^{2+} in cysteine-rich pockets.
(b) $[Cu_4(SPh)_6]^{2-}$ is a model for the Cu-containing metallothionein in yeast.
(c) Imidazole and tris(pyrazolyl)borate derivatives are often used to model histidine-binding sites.

29.5 (a) Briefly describe the mode of binding of O_2 to the iron centre in one haem unit of haemoglobin. (b) What are 'picket fence' porphyrins and why are they used in model studies of O_2 binding to myoglobin or haemoglobin? (c) The binding of O_2 to haemoglobin exhibits a 'cooperativity' effect. What is meant by this statement? (d) Why is the change from deoxyhaemoglobin to the oxy-form accompanied by a decrease in the observed magnetic moment?

29.6 Compare the modes of binding of O_2 to the metal centres in (a) myoglobin, (b) haemerythrin and (c) haemocyanin. Indicate what supporting experimental evidence is available for the structures you describe.

29.7 Differentiate between Type 1, Type 2 and Type 3 copper centres in blue copper proteins, giving both experimental and structural distinctions.

29.8 Describe the structure of the copper site in plastocyanin and discuss the features of both the metal centre and metal-binding site that allow it to function as an electron-transfer site.

29.9 Ascorbate oxidase contains four copper centres. Discuss their coordination environments, and classify the centres as Type 1, 2 or 3. What is the function of ascorbate oxidase and how do the copper centres facilitate this function?

29.10 Comment on the following observations:

(a) 'Blue copper proteins' are not always blue.

(b) Two different metalloproteins, both containing [4Fe–4S] ferredoxins bound to the protein chain by Cys ligands, exhibit reduction potentials of +350 and +490 mV.

(c) The toxicity of CO is associated with binding to haemoglobin, but that of $[CN]^-$ is not.

29.11 What is the mitochondrial electron-transfer chain, and what role do quinones play in the chain?

29.12 Model compounds are often used to model iron–sulfur proteins. Comment on the applicability of the following models, and on the data given.

(a) $[Fe(SPh)_4]^{2-}$ as a model for rubredoxin; observed values of μ_{eff} are 5.85 μ_B for the oxidized form of the model compound and 5.05 μ_B for the reduced form.

(b) $[Fe_2(\mu\text{-}S)_2(SPh)_4]^{2-}$ as a model for the active site in spinach ferredoxin.

(c) Compound **29.31** as a model for part of the active sites in nitrogenase. The Mössbauer spectrum of **29.31** is consistent with equivalent Fe centres, each with an oxidation state of 2.67.

(29.31)

29.13 For a [4Fe–4S] protein, the following series of redox reactions are possible; each step is a 1-electron reduction or oxidation:

$$4Fe(III) \rightleftharpoons 3Fe(III)\cdot Fe(II) \rightleftharpoons 2Fe(III)\cdot 2Fe(II)$$
$$\rightleftharpoons Fe(III)\cdot 3Fe(II) \rightleftharpoons 4Fe(II)$$

(a) Which of these couples are accessible under physiological conditions? (b) Which couple represents the HIPIP system? (c) How do the redox potentials of the HIPIP and [4Fe–4S] ferredoxin system differ and how does this affect their roles in the mitochondrial electron-transfer chain?

29.14 Comment on the similarities and differences between a [2Fe–2S] ferredoxin and Rieske protein, in terms of both structure and function.

29.15 (a) Outline the similarities and differences between the haem units in deoxymyoglobin and cytochrome *c*. (b) What function does cytochrome *c* perform in mammals?

29.16 (a) What is the function of cytochrome *c* oxidase? (b) Describe the four active metal-containing sites in cytochrome *c* oxidase and the proposed way in which they work together to fulfil the role of the metalloprotein.

29.17 Give an explanation for the following observations (part d assumes Box 29.2 has been studied):

(a) both haemoglobin and cytochromes contain haem-iron;

(b) cytochrome *c* oxidase contains more than one metal centre;

(c) each sub-unit in deoxyhaemoglobin contains 5-coordinate Fe(II), but in cytochrome *c*, the Fe centre is always 6-coordinate;

(d) nitrophorin (NP1) reversibly binds NO.

29.18 Discuss the role of Zn^{2+} as an example of a Lewis acid at work in a biological system.

29.19 The hydrolysis of the acid anhydride **29.32** by $[OH]^-$ is catalysed by Zn^{2+} ions. The rate equation is of the form:

$$\text{Rate} = k[\mathbf{29.32}][Zn^{2+}][OH^-]$$

(29.32)

It is also known that the addition of Zn^{2+} does not accelerate hydrolysis by H_2O or attack by other nucleophiles. Suggest a mechanism for this reaction.

29.20 Why is metal substitution used to investigate the metal binding site in carbonic anhydrase? Discuss the type of information that might be forthcoming from such a study.

OVERVIEW PROBLEMS

29.21 Compound **29.33**, H_4L, is a model for the siderophore desferrioxamine. It binds Fe^{3+} to give the complex [Fe(HL)]. What features does **29.33**

have in common with desferrioxamine? Suggest a reason for the choice of the macrocyclic unit in ligand **29.33**. Suggest a structure for [Fe(HL)].

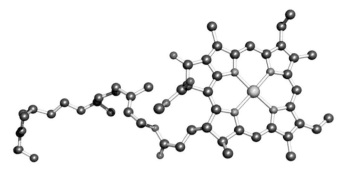

(29.33)

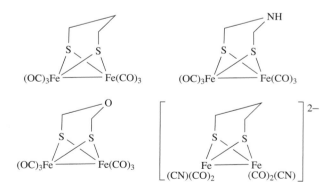

Fig. 29.31 Structure for problem 29.25b. Colour code: Mg, yellow; C, grey; O, red; N, blue.

29.22 (a) The structure of a bacterial protein reported in 2001 showed that the active site contains a $Zn_4(Cys)_9(His)_2$ cluster. To what family does this metalloprotein belong, and why is the binding site atypical?

(b) Cytochrome P-450 is a monooxygenase. Outline its function, paying attention to the structure of the active site. Construct a catalytic cycle that describes the monooxygenation of an organic substrate RH.

29.23 Compound **29.34** reacts with $Zn(ClO_4)_2 \cdot 6H_2O$ to give a complex $[Zn(\mathbf{29.34})(OH)]^+$ that is a model for the active site of carbonic anhydrase. Suggest a structure for this complex. What properties does **29.34** possess that (a) mimic the coordination site in carbonic anhydrase and (b) control the coordination geometry around the Zn^{2+} ion in the model complex?

(29.34)

29.24 (a) Comment on the relevance of studying complexes such as $[Fe(CN)_4(CO)_2]^{2-}$ and $[Fe(CO)_3(CN)_3]^-$ as models for the active sites of [NiFe]- and [FeFe]-hydrogenases.

(b) Describe the structure of the FeMo cofactor in nitrogenase. Until 2002, when a central ligand was located in the FeMo cofactor, it was suggested that N_2 binding might take place at 3-coordinate iron sites. Explain why this proposal is no longer plausible.

29.25 (a) Whereas the stability constant, K, for the equilibrium:

$$\text{Haemoglobin} + O_2 \rightleftharpoons (\text{Haemoglobin})(O_2)$$

is of the order of 10, that for the equilibrium:

$$(\text{Haemoglobin})(O_2)_3 + O_2 \rightleftharpoons (\text{Haemoglobin})(O_2)_4$$

is of the order of 3000. Rationalize this observation.

(b) Photosystem II operates in conjunction with cytochrome $b_6 f$. The crystal structure of cytochrome $b_6 f$ from the alga *Chlamydomonas reinhardtii* has been determined, and one of the cofactors present in this cytochrome is shown in Fig. 29.31. What is the function of Photosystem II? Identify the cofactor shown in Fig. 29.31.

29.26 (a) The compounds shown below are models for the active site of [FeFe]-hydrogenase. How are the models related to the active site and what problems does a crystallographer face when trying to identify the active site?

(b) The molecular mass of myoglobin is ≈ 16950. The positive mode electrospray mass spectrum of deoxymyoglobin shows peaks at m/z 1413, 1304, 1212, 1131, 1060, 998, 942, 893, 848 and 808. Suggest assignments for the peaks and explain why the mass difference between successive peaks is not a constant value.

INORGANIC CHEMISTRY MATTERS

29.27 (a) What is meant by the cooperative binding of O_2 by haemoglobin? (b) Use Fig. 29.8 to explain what would happen in your body if your blood contained myoglobin in place of haemoglobin.

29.28 Which organisms utilize: haemoglobin; [NiFe]-hydrogenases; rubredoxins; plastocyanins? Describe the active centre in each metalloprotein and the role that it plays.

29.29 Cereal grains contain high levels of phytic acid. Why is phytic acid an inhibitor of iron uptake by the human body?

OPO$_3$H$_2$

H$_2$O$_3$PO,,,,   ,,,OPO$_3$H$_2$

H$_2$O$_3$PO   OPO$_3$H$_2$

OPO$_3$H$_2$

Phytic acid

29.30 Sources of Cd include welding emissions, electronic components and NiCd batteries. EU and US legislations are now in place to reduce our exposure to Cd. Once absorbed into the body, Cd is targeted by metallothioneins, cysteine, the tripeptide glutathione and the protein albumin (Fig. 29.32) and is transported in the blood to the liver. (a) What features are highlighted in ball-and-stick representation in Fig. 29.32? Suggest how they originate in terms of the amino acid residues present in the protein. (b) Why are metallothioneins, cysteine, glutathione and albumin suited to binding Cd^{2+}? (c) Give examples of other metal ions transported by metallothioneins.

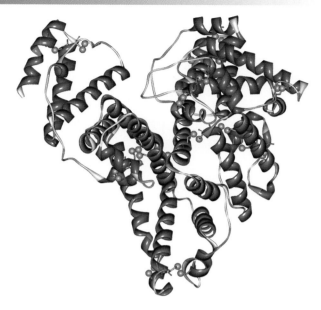

Fig. 29.32 The structure of human serum albumin determined by X-ray diffraction (PDB code: 3JRY). The sites highlighted in ball-and-stick representation are the focus of problem 29.30.

Appendices

Appendix 1
Greek letters with pronunciations

Upper case letter	Lower case letter	Pronounced
A	α	alpha
B	β	beta
Γ	γ	gamma
Δ	δ	delta
E	ε	epsilon
Z	ζ	zeta
H	η	eta
Θ	θ	theta
I	ι	iota
K	κ	kappa
Λ	λ	lambda
M	μ	mu
N	ν	nu
Ξ	ξ	xi
O	ο	omicron
Π	π	pi
P	ρ	rho
Σ	σ	sigma
T	τ	tau
Υ	υ	upsilon
Φ	φ	phi
X	χ	chi
Ψ	ψ	psi
Ω	ω	omega

Appendix 2
Abbreviations and symbols for quantities and units

For ligand structures, see Table 7.7. Where a symbol has more than one meaning, the context of its use should make the meaning clear. For further information on SI symbols and names of units, see: *Quantities, Units and Symbols in Physical Chemistry* (1993) IUPAC, 2nd edn, Blackwell Science, Oxford.

a	cross-sectional area	bpy	2,2'-bipyridine
a_i	relative activity of a component i	Bq	becquerel (unit of radioactivity)
a_0	Bohr radius of the H atom	^{n}Bu	*n*-butyl
A	ampere (unit of current)	^{t}Bu	*tert*-butyl
A	absorbance	c	coefficient (in wavefunctions)
A	frequency factor (in Arrhenius equation)	c	concentration (of solution)
A	Madelung constant	c	speed of light
A	mass number (of an atom)	c-C_6H_{11}	cyclohexyl
A	hyperfine coupling constant (EPR)	C	Curie constant
A_r	relative atomic mass	C	coulomb (unit of charge)
$A(\theta,\phi)$	angular wavefunction	Ci	curie (non-SI unit of radioactivity)
AAS	atomic absorption spectroscopy	C_n	*n*-fold rotation axis
Å	ångstrom (non-SI unit of length, used for bond distances)	ccp	cubic close-packed
acacH	acetylacetone	CFC	chlorofluorocarbon
ADP	adenosine diphosphate	CFSE	crystal field stabilization energy
Ala	alanine	cm	centimetre (unit of length)
aq	aqueous	cm^3	cubic centimetre (unit of volume)
Arg	arginine	cm^{-1}	reciprocal centimetre (wavenumber)
Asn	asparagine	conc	concentrated
Asp	aspartic acid	Cp	cyclopentadienyl
atm	atmosphere (non-SI unit of pressure)	cr	crystal
ATP	adenosine triphosphate	CT	charge transfer
ax	axial	CVD	chemical vapour deposition
B	magnetic field strength	Cys	cysteine
B	Racah parameter	d	bond distance or internuclear separation
bar	bar (unit of pressure)	d-	dextro- (see Box 19.3)
bcc	body-centred cubic	d	day (non-SI unit of time)
bp	boiling point	D	bond dissociation enthalpy

$\bar{D}$	average bond dissociation enthalpy	FID	free induction decay
D mechanism	dissociative mechanism	FT	Fourier transform
D	debye (non-SI unit of electric dipole moment)	G	Gibbs energy
Dcb mechanism	conjugate–base mechanism	g	gas
dec	decomposition	g	gram (unit of mass)
DHA	9,10-dihydroanthracene	g	Landé g-factor
dien	1,4,7-triazaheptane (see Table 7.7)	Gln	glutamine
dil	dilute	Glu	glutamic acid
dm^3	cubic decimetre (unit of volume)	Gly	glycine
DME	dimethoxyethane	H	enthalpy
DMF	N,N-dimethylformamide	H	magnetic field
$dmgH_2$	dimethylglyoxime	H_c	critical magnetic field of a superconductor
DMSO	dimethylsulfoxide	h	Planck constant
DNA	deoxyribonucleic acid	h	hour (non-SI unit of time)
E	energy	hcp	hexagonal close-packed
E	identity operator	HIPIP	high-potential protein
E	bond enthalpy term	His	histidine
e	charge on the electron	HMPA	hexamethylphosphoramide (see structure **11.5**)
e^-	electron	HOMO	highest occupied molecular orbital
EA	electron affinity	Hz	hertz (unit of frequency)
E_a	activation energy	$h\nu$	high-frequency radiation (for a photolysis reaction)
E_{cell}	electrochemical cell potential	I	nuclear spin quantum number
E°	standard reduction potential	i	centre of inversion
$EDTAH_4$	N,N,N',N'-ethylenediaminetetraacetic acid (see Table 7.7)	I_a mechanism	associative interchange mechanism
en	1,2-ethanediamine (see Table 7.7)	I_d mechanism	dissociative interchange mechanism
EPR	electron paramagnetic resonance	IE	ionization energy
eq	equatorial	Ile	isoleucine
ESR	electron spin resonance	IR	infrared
Et	ethyl	IUPAC	International Union of Pure and Applied Chemistry
eV	electron volt	j	inner quantum number
EXAFS	extended X-ray absorption fine structure	J	joule (unit of energy)
F	Faraday constant	J	spin–spin coupling constant
FAD	flavin adenine dinucleotide	J	total (resultant) inner quantum number
fcc	face-centred cubic	k	force constant

k	rate constant	m^3	cubic metre (unit of volume)
k	Boltzmann constant	m_e	electron rest mass
K	kelvin (unit of temperature)	m_i	molality
K	equilibrium constant	m_i^o	standard state molality
K_a	acid dissociation constant	m_l	magnetic quantum number
K_b	base dissociation constant	M_L	total (resultant) orbital magnetic quantum number
K_c	equilibrium constant expressed in terms of concentrations	m_s	magnetic spin quantum number
K_p	equilibrium constant expressed in terms of partial pressures	M_S	magnetic spin quantum number for the multi-electron system
K_{self}	self-ionization constant	M_r	relative molecular mass
K_{sp}	solubility product constant	Me	methyl
K_w	self-ionization constant of water	Mes	mesityl ($2,4,6\text{-}Me_3C_6H_2$)
kg	kilogram (unit of mass)	Met	methionine
kJ	kilojoule (unit of energy)	min	minute (non-SI unit of time)
kPa	kilopascal (unit of pressure)	MLCT	metal-to-ligand charge transfer
L	Avogadro's number	MO	molecular orbital
L	total (resultant) orbital quantum number	MOCVD	metal–organic chemical vapour deposition
L	ligand	mol	mole (unit of quantity)
l	liquid	mp	melting point
l	length	Mt	megatonne
l	orbital quantum number	MWNT	multi-walled (carbon) nanotube
l-	laevo- (see Box 19.3)	N	normalization factor
ℓ	path length	N	number of nuclides
LCAO	linear combination of atomic orbitals	n	neutron
LED	light-emitting diode	n	Born exponent
Leu	leucine	n	number of (e.g. moles)
LFER	linear free energy relationship	n	principal quantum number
LFSE	ligand field stabilization energy	n	nucleophilicity parameter
LGO	ligand group orbital	$[NAD]^+$	nicotinamide adenine dinucleotide
LMCT	ligand-to-metal charge transfer	NASICON	Na super ionic conductor
Ln	lanthanoid	nm	nanometre (unit of length)
LUMO	lowest unoccupied molecular orbital	NMR	nuclear magnetic resonance
Lys	lysine	OLED	organic light-emitting diode
M	molarity	oxH_2	oxalic acid
m	mass	P	pressure
m	metre (unit of length)	Pa	pascal (unit of pressure)

PES	photoelectron spectroscopy	s	second (unit of time)
Ph	phenyl	s	solid
Phe	phenylalanine	s	spin quantum number
phen	1,10-phenanthroline	s	nucleophilicity discrimination factor
pK_a	$-\log K_a$	S_n	n-fold improper rotation axis
pm	picometre (unit of length)	$S_N 1cb$ mechanism	conjugate–base mechanism
ppb	parts per billion	Ser	serine
ppm	parts per million	soln	solution
ppt	precipitate	solv	solvated; solvent
Pr	propyl	SQUID	superconducting quantum interference device
^iPr	*iso*-propyl	SWNT	single-walled (carbon) nanotube
Pro	proline	T	tesla (unit of magnetic flux density)
PVC	polyvinylchloride	T	temperature
py	pyridine	T_c	critical temperature of a superconductor
pzH	pyrazole	T_C	Curie temperature
q	point charge	T_N	Néel temperature
Q	reaction quotient	t	tonne (metric)
R	general alkyl or aryl group	t	time
R	molar gas constant	$t_{\frac{1}{2}}$	half-life
R	Rydberg constant	THF	tetrahydrofuran
R	resistance	Thr	threonine
R-	sequence rules for an enantiomer (see Box 19.3)	TMEDA	N,N,N',N'-tetramethylethylenediamine
r	radial distance	TMS	tetramethylsilane
r	radius	TOF	catalytic turnover frequency
$R(r)$	radial wavefunction	TON	catalytic turnover number
r_{cov}	covalent radius	$tppH_2$	tetraphenylporphyrin
r_{ion}	ionic radius	tpy	2,2':6,2''-terpyridine
r_{metal}	metallic radius	trien	1,4,7,10-tetraazadecane (see Table 7.7)
r_v	van der Waals radius	Trp	tryptophan
RDS	rate-determining step	Tyr	tyrosine
RF	radiofrequency	U	internal energy
S	entropy	u	atomic mass unit
S	overlap integral	UV	ultraviolet
S	total spin quantum number	UV–VIS	ultraviolet-visible
S	screening (or shielding) constant	V	potential difference
S-	sequence rules for an enantiomer (see Box 19.3)	V	volume

V	volt (unit of potential difference)	ΔH°	standard enthalpy change
v	vapour	$\Delta H^{\ddagger}$	enthalpy change of activation
v	velocity	$\Delta_{a} H$	enthalpy change of atomization
Val	valine	$\Delta_{c} H$	enthalpy change of combustion
VB	valence bond	$\Delta_{EA} H$	enthalpy change associated with the gain of an electron
ve	valence electrons (in electron counting)	$\Delta_{f} H$	enthalpy change of formation
VIS	visible	$\Delta_{fus} H$	enthalpy change of fusion
VSEPR	valence-shell electron-pair repulsion	$\Delta_{hyd} H$	enthalpy change of hydration
		$\Delta_{lattice} H$	enthalpy change for the formation of an ionic lattice
[X]	concentration of X		
XRD	X-ray diffraction	$\Delta_{r} H$	enthalpy change of reaction
yr	year (non-SI unit of time)	$\Delta_{sol} H$	enthalpy change of solution
z	number of moles of electrons transferred in an electrochemical cell	$\Delta_{solv} H$	enthalpy change of solvation
Z	atomic number	$\Delta_{vap} H$	enthalpy change of vaporization
Z	effective collision frequency in solution	ΔG°	standard Gibbs energy change
Z_{eff}	effective nuclear charge	$\Delta G^{\ddagger}$	Gibbs energy of activation
$\lvert z_{-}\rvert$	modulus of the negative charge	$\Delta_{f} G$	Gibbs energy change of formation
$\lvert z_{+}\rvert$	modulus of the positive charge	$\Delta_{r} G$	Gibbs energy change of reaction
ZSM-5	a type of zeolite (see Section 25.8)	ΔS	entropy change
α	polarizability of an atom or ion	ΔS°	standard entropy change
$[\alpha]$	specific rotation	$\Delta S^{\ddagger}$	entropy change of activation
β	stability constant	$\Delta U(0\,\text{K})$	internal energy change at $0\,\text{K}$
β^{-}	beta-particle	$\Delta V^{\ddagger}$	volume of activation
β^{+}	positron	ε	molar extinction (or absorption) coefficient
δ	chemical shift	ε_{max}	molar extinction coefficient corresponding to an absorption maximum (in an electronic spectrum)
δ-	label for an enantiomer (see Box 19.3)	ε_{0}	permittivity of a vacuum
δ^{-}	partial negative charge	ε_{r}	relative permittivity (dielectric constant)
δ^{+}	partial positive charge	η	hapticity of a ligand (see Box 19.1)
Δ	change in	λ-	label for an enantiomer (see Box 19.3)
Δ-	label for enantiomer with right-handedness (see Box 19.3)	λ	spin–orbit coupling constant
$\triangle$	heat (in a pyrolysis reaction)	λ	wavelength
Δ_{oct}	octahedral crystal field splitting energy	λ_{max}	wavelength corresponding to an absorption maximum (in an electronic spectrum)
Δ_{tet}	tetrahedral crystal field splitting energy	Λ-	label for enantiomer with left-handedness (see Box 19.3)

μ	electric dipole moment	° or $\ominus$	standard state
μ	reduced mass	‡	(called a '*double dagger*') activated complex; transition state
μ	refractive index	°	degree
μ(spin only)	spin-only magnetic moment	>	is greater than
μ_B	Bohr magneton	≫	is much greater than
μ_{eff}	effective magnetic moment	<	is less than
μ_i	chemical potential of component i	≪	is much less than
$\mu_i{}^\circ$	standard chemical potential of i	≥	is greater than or equal to
μ-	bridging ligand	≤	is less than or equal to
ν	total number of particles produced per molecule of solute	≈	is approximately equal to
ν	frequency	=	is equal to
$\bar{\nu}$	wavenumber	≠	is not equal to
ν_e	neutrino	⇌	equilibrium
ρ	density	∝	is proportional to
σ	mirror plane	×	multiplied by
τ_1	spin relaxation time (in NMR spectroscopy)	∞	infinity
χ	magnetic susceptibility	±	plus or minus
χ_m	molar magnetic susceptibility	$\sqrt{}$	square root of
χ	electronegativity	$\sqrt[3]{}$	cube root of
χ^{AR}	Allred–Rochow electronegativity	$|x|$	modulus of x
χ^M	Mulliken electronegativity	$\sum$	summation of
χ^P	Pauling electronegativity	Δ	change in (for example, ΔH is 'change in enthalpy')
ψ	wavefunction	$\angle$	angle
Ω	ohm (unit of resistance)	log	logarithm to base 10 ($\log_{10}$)
2c-2e	2-centre 2-electron	ln	natural logarithm, i.e. logarithm to base e ($\log_e$)
3c-2e	3-centre 2-electron	$\int$	integral of
(+)-	label for specific rotation of an enantiomer (see Box 19.3)	$\dfrac{d}{dx}$	differential with respect to x
(−)-	label for specific rotation of an enantiomer (see Box 19.3)	$\dfrac{\partial}{\partial x}$	partial differential with respect to x

Appendix 3
Selected character tables

The character tables given in this appendix are for some commonly encountered point groups. Complete tables are available in many physical and theoretical chemistry texts, e.g. see Chapter 3 reading list.

C_1	E
A	1

C_s	E	σ_h		
A'	1	1	x, y, R_z	x^2, y^2, z^2, xy
A''	1	-1	z, R_x, R_y	yz, xz

C_2	E	C_2		
A	1	1	z, R_z	x^2, y^2, z^2, xy
B	1	-1	x, y, R_x, R_y	yz, xz

C_{2v}	E	C_2	$\sigma_v(xz)$	$\sigma_v'(yz)$		
A_1	1	1	1	1	z	x^2, y^2, z^2
A_2	1	1	-1	-1	R_z	xy
B_1	1	-1	1	-1	x, R_y	xz
B_2	1	-1	-1	1	y, R_x	yz

C_{3v}	E	$2C_3$	$3\sigma_v$		
A_1	1	1	1	z	$x^2 + y^2, z^2$
A_2	1	1	-1	R_z	
E	2	-1	0	$(x, y)\,(R_x, R_y)$	$(x^2 - y^2, xy)\,(xz, yz)$

C_{4v}	E	$2C_4$	C_2	$2\sigma_v$	$2\sigma_d$		
A_1	1	1	1	1	1	z	$x^2 + y^2, z^2$
A_2	1	1	1	-1	-1	R_z	
B_1	1	-1	1	1	-1		$x^2 - y^2$
B_2	1	-1	1	-1	1		xy
E	2	0	-2	0	0	$(x, y)(R_x, R_y)$	(xz, yz)

C_{5v}	E	$2C_5$	$2C_5^2$	$5\sigma_v$		
A_1	1	1	1	1	z	$x^2 + y^2, z^2$
A_2	1	1	1	-1	R_z	
E_1	2	$2\cos 72°$	$2\cos 144°$	0	$(x, y)(R_x, R_y)$	(xz, yz)
E_2	2	$2\cos 144°$	$2\cos 72°$	0		$(x^2 - y^2, xy)$

D_2	E	$C_2(z)$	$C_2(y)$	$C_2(x)$		
A	1	1	1	1		x^2, y^2, z^2
B_1	1	1	-1	-1	z, R_z	xy
B_2	1	-1	1	-1	y, R_y	xz
B_3	1	-1	-1	1	x, R_x	yz

D_3	E	$2C_3$	$3C_2$		
A_1	1	1	1		$x^2 + y^2, z^2$
A_2	1	1	-1	z, R_z	
E	2	-1	0	$(x, y)(R_x, R_y)$	$(x^2 - y^2, xy)(xz, yz)$

D_{2h}	E	$C_2(z)$	$C_2(y)$	$C_2(x)$	i	$\sigma(xy)$	$\sigma(xz)$	$\sigma(yz)$		
A_g	1	1	1	1	1	1	1	1		x^2, y^2, z^2
B_{1g}	1	1	-1	-1	1	1	-1	-1	R_z	xy
B_{2g}	1	-1	1	-1	1	-1	1	-1	R_y	xz
B_{3g}	1	-1	-1	1	1	-1	-1	1	R_x	yz
A_u	1	1	1	1	-1	-1	-1	-1		
B_{1u}	1	1	-1	-1	-1	-1	1	1	z	
B_{2u}	1	-1	1	-1	-1	1	-1	1	y	
B_{3u}	1	-1	-1	1	-1	1	1	-1	x	

D_{3h}	E	$2C_3$	$3C_2$	σ_h	$2S_3$	$3\sigma_v$		
A_1'	1	1	1	1	1	1		$x^2 + y^2, z^2$
A_2'	1	1	-1	1	1	-1	R_z	
E'	2	-1	0	2	-1	0	(x, y)	$(x^2 - y^2, xy)$
A_1''	1	1	1	-1	-1	-1		
A_2''	1	1	-1	-1	-1	1	z	
E''	2	-1	0	-2	1	0	(R_x, R_y)	(xz, yz)

D_{4h}	E	$2C_4$	C_2	$2C_2'$	$2C_2''$	i	$2S_4$	σ_h	$2\sigma_v$	$2\sigma_d$		
A_{1g}	1	1	1	1	1	1	1	1	1	1		x^2+y^2, z^2
A_{2g}	1	1	1	−1	−1	1	1	1	−1	−1	R_z	
B_{1g}	1	−1	1	1	−1	1	−1	1	1	−1		x^2-y^2
B_{2g}	1	−1	1	−1	1	1	−1	1	−1	1		xy
E_g	2	0	−2	0	0	2	0	−2	0	0	(R_x, R_y)	(xz, yz)
A_{1u}	1	1	1	1	1	−1	−1	−1	−1	−1		
A_{2u}	1	1	1	−1	−1	−1	−1	−1	1	1	z	
B_{1u}	1	−1	1	1	−1	−1	1	−1	−1	1		
B_{2u}	1	−1	1	−1	1	−1	1	−1	1	−1		
E_u	2	0	−2	0	0	−2	0	2	0	0	(x, y)	

D_{2d}	E	$2S_4$	C_2	$2C_2'$	$2\sigma_d$		
A_1	1	1	1	1	1		x^2+y^2, z^2
A_2	1	1	1	−1	−1	R_z	
B_1	1	−1	1	1	−1		x^2-y^2
B_2	1	−1	1	−1	1	z	xy
E	2	0	−2	0	0	$(x, y)(R_x, R_y)$	(xz, yz)

D_{3d}	E	$2C_3$	$3C_2$	i	$2S_6$	$3\sigma_d$		
A_{1g}	1	1	1	1	1	1		x^2+y^2, z^2
A_{2g}	1	1	−1	1	1	−1	R_z	
E_g	2	−1	0	2	−1	0	(R_x, R_y)	$(x^2-y^2, xy), (xz, yz)$
A_{1u}	1	1	1	−1	−1	−1		
A_{2u}	1	1	−1	−1	−1	1	z	
E_u	2	−1	0	−2	1	0	(x, y)	

T_d	E	$8C_3$	$3C_2$	$6S_4$	$6\sigma_d$		
A_1	1	1	1	1	1		$x^2+y^2+z^2$
A_2	1	1	1	−1	−1		
E	2	−1	2	0	0		$(2z^2-x^2-y^2, x^2-y^2)$
T_1	3	0	−1	1	−1	(R_x, R_y, R_z)	
T_2	3	0	−1	−1	1	(x, y, z)	(xy, xz, yz)

O_h	E	$8C_3$	$6C_2$	$6C_4$	$3C_2$ $(= C_4^2)$	i	$6S_4$	$8S_6$	$3\sigma_h$	$6\sigma_d$		
A_{1g}	1	1	1	1	1	1	1	1	1	1		$x^2 + y^2 + z^2$
A_{2g}	1	1	−1	−1	1	1	−1	1	1	−1		
E_g	2	−1	0	0	2	2	0	−1	2	0		$(2z^2 - x^2 - y^2, x^2 - y^2)$
T_{1g}	3	0	−1	1	−1	3	1	0	−1	−1	(R_x, R_y, R_z)	
T_{2g}	3	0	1	−1	−1	3	−1	0	−1	1		(xz, yz, xy)
A_{1u}	1	1	1	1	1	−1	−1	−1	−1	−1		
A_{2u}	1	1	−1	−1	1	−1	1	−1	−1	1		
E_u	2	−1	0	0	2	−2	0	1	−2	0		
T_{1u}	3	0	−1	1	−1	−3	−1	0	1	1	(x, y, z)	
T_{2u}	3	0	1	−1	−1	−3	1	0	1	−1		

$C_{\infty v}$	E	$2C_\infty^\phi$	$\ldots$	$\infty\sigma_v$		
$A_1 \equiv \Sigma^+$	1	1	$\ldots$	1	z	$x^2 + y^2, z^2$
$A_2 \equiv \Sigma^-$	1	1	$\ldots$	−1	R_z	
$E_1 \equiv \Pi$	2	$2\cos\phi$	$\ldots$	0	$(x, y)(R_x, R_y)$	(xz, yz)
$E_2 \equiv \Delta$	2	$2\cos 2\phi$	$\ldots$	0		$(x^2 - y^2, xy)$
$E_3 \equiv \Phi$	2	$2\cos 3\phi$	$\ldots$	0		
$\ldots$	$\ldots$	$\ldots$	$\ldots$	$\ldots$		

$D_{\infty h}$	E	$2C_\infty^\phi$	$\ldots$	$\infty\sigma_v$	i	$2S_\infty^\phi$	$\ldots$	∞C_2		
Σ_g^+	1	1	$\ldots$	1	1	1	$\ldots$	1		$x^2 + y^2, z^2$
Σ_g^-	1	1	$\ldots$	−1	1	1	$\ldots$	−1	R_z	
Π_g	2	$2\cos\phi$	$\ldots$	0	2	$-2\cos\phi$	$\ldots$	0	(R_x, R_y)	(xz, yz)
Δ_g	2	$2\cos 2\phi$	$\ldots$	0	2	$2\cos 2\phi$	$\ldots$	0		$(x^2 - y^2, xy)$
$\ldots$	$\ldots$	$\ldots$	$\ldots$	$\ldots$	$\ldots$	$\ldots$	$\ldots$	$\ldots$		
Σ_u^+	1	1	$\ldots$	1	−1	−1	$\ldots$	−1	z	
Σ_u^-	1	1	$\ldots$	−1	−1	−1	$\ldots$	1		
Π_u	2	$2\cos\phi$	$\ldots$	0	−2	$2\cos\phi$	$\ldots$	0	(x, y)	
Δ_u	2	$2\cos 2\phi$	$\ldots$	0	−2	$-2\cos 2\phi$	$\ldots$	0		
$\ldots$	$\ldots$	$\ldots$	$\ldots$	$\ldots$	$\ldots$	$\ldots$	$\ldots$	$\ldots$		

Appendix 4
The electromagnetic spectrum

The frequency of electromagnetic radiation is related to its wavelength by the equation:

$$\text{Wavelength}\,(\lambda) = \frac{\text{Speed of light}\,(c)}{\text{Frequency}\,(\nu)}$$

where $c = 3.0 \times 10^8\,\text{m}\,\text{s}^{-1}$.

$$\text{Wavenumber}\,(\bar{\nu}) = \frac{1}{\text{Wavelength}}$$

with units in cm^{-1} (pronounced 'reciprocal centimetre')

Energy (E) = Planck's constant (h) × Frequency (ν) where $h = 6.626 \times 10^{-34}\,\text{J}\,\text{s}$

(continued over the page)

The energy given in the last column is measured per mole of photons.

Frequency ν/Hz	Wavelength λ/m	Wavenumber $\bar{\nu}/\text{cm}^{-1}$	Type of radiation	Energy $E/\text{kJ mol}^{-1}$
10^{21}	10^{-13}	10^{11}		10^{9}
10^{20}	10^{-12}	10^{10}	γ-ray	10^{8}
10^{19}	10^{-11}	10^{9}		10^{7}
10^{18}	10^{-10}	10^{8}	X-ray	10^{6}
10^{17}	10^{-9}	10^{7}		10^{5}
10^{16}	10^{-8}	10^{6}	Vacuum ultraviolet	10^{4}
10^{15}	10^{-7}	10^{5}	Ultraviolet	10^{3}
			Visible	
10^{14}	10^{-6}	10^{4}	Near infrared	10^{2}
10^{13}	10^{-5}	10^{3}		10^{1}
10^{12}	10^{-4}	10^{2}	Far infrared	$10^{0} = 1$
10^{11}	10^{-3}	10^{1}		10^{-1}
10^{10}	10^{-2}	$10^{0} = 1$	Microwave	10^{-2}
10^{9}	10^{-1}	10^{-1}		10^{-3}
10^{8}	$10^{0} = 1$	10^{-2}		10^{-4}
10^{7}	10^{1}	10^{-3}		10^{-5}
10^{6}	10^{2}	10^{-4}		10^{-6}
10^{5}	10^{3}	10^{-5}	Radiowave	10^{-7}
10^{4}	10^{4}	10^{-6}		10^{-8}
10^{3}	10^{5}	10^{-7}		10^{-9}

Visible:
Violet ≈ 400 nm
Blue
Green
Yellow
Orange
Red ≈ 700 nm

Appendix 5
Naturally occurring isotopes and their abundances

Data from *WebElements* by Mark Winter. Further information on radioactive nuclides can be found using the Web link www.webelements.com

Element	Symbol	Atomic number, Z	Mass number of isotope (% abundance)
Actinium	Ac	89	artificial isotopes only; mass number range 224–229
Aluminium	Al	13	27(100)
Americium	Am	95	artificial isotopes only; mass number range 237–245
Antimony	Sb	51	121(57.3), 123(42.7)
Argon	Ar	18	36(0.34), 38(0.06), 40(99.6)
Arsenic	As	33	75(100)
Astatine	At	85	artificial isotopes only; mass number range 205–211
Barium	Ba	56	130(0.11), 132(0.10), 134(2.42), 135(6.59), 136(7.85), 137(11.23), 138(71.70)
Berkelium	Bk	97	artificial isotopes only; mass number range 243–250
Beryllium	Be	4	9(100)
Bismuth	Bi	83	209(100)
Boron	B	5	10(19.9), 11(80.1)
Bromine	Br	35	79(50.69), 81(49.31)
Cadmium	Cd	48	106(1.25), 108(0.89), 110(12.49), 111(12.80), 112(24.13), 113(12.22), 114(28.73), 116(7.49)
Caesium	Cs	55	133(100)
Calcium	Ca	20	40(96.94), 42(0.65), 43(0.13), 44(2.09), 48(0.19)
Californium	Cf	98	artificial isotopes only; mass number range 246–255
Carbon	C	6	12(98.9), 13(1.1)
Cerium	Ce	58	136(0.19), 138(0.25), 140(88.48), 142(11.08)
Chlorine	Cl	17	35(75.77), 37(24.23)
Chromium	Cr	24	50(4.345), 52(83.79), 53(9.50), 54(2.365)
Cobalt	Co	27	59(100)
Copper	Cu	29	63(69.2), 65(30.8)
Curium	Cm	96	artificial isotopes only; mass number range 240–250
Dysprosium	Dy	66	156(0.06), 158(0.10), 160(2.34), 161(18.9), 162(25.5), 163(24.9), 164(28.2)
Einsteinium	Es	99	artificial isotopes only; mass number range 249–256
Erbium	Er	68	162(0.14), 164(1.61), 166(33.6), 167(22.95), 168(26.8), 170(14.9)
Europium	Eu	63	151(47.8), 153(52.2)
Fermium	Fm	100	artificial isotopes only; mass number range 251–257
Fluorine	F	9	19(100)
Francium	Fr	87	artificial isotopes only; mass number range 210–227
Gadolinium	Gd	64	152(0.20), 154(2.18), 155(14.80), 156(20.47), 157(15.65), 158(24.84), 160(21.86)

Element	Symbol	Atomic number, Z	Mass number of isotope (% abundance)
Gallium	Ga	31	69(60.1), 71(39.9)
Germanium	Ge	32	70(20.5), 72(27.4), 73(7.8), 74(36.5), 76(7.8)
Gold	Au	79	197(100)
Hafnium	Hf	72	174(0.16), 176(5.20), 177(18.61), 178(27.30), 179(13.63), 180(35.10)
Helium	He	2	3(<0.001), 4(>99.999)
Holmium	Ho	67	165(100)
Hydrogen	H	1	1(99.985), 2(0.015)
Indium	In	49	113(4.3), 115(95.7)
Iodine	I	53	127(100)
Iridium	Ir	77	191(37.3), 193(62.7)
Iron	Fe	26	54(5.8), 56(91.7), 57(2.2), 58(0.3)
Krypton	Kr	36	78(0.35), 80(2.25), 82(11.6), 83(11.5), 84(57.0), 86(17.3)
Lanthanum	La	57	138(0.09), 139(99.91)
Lawrencium	Lr	103	artificial isotopes only; mass number range 253–262
Lead	Pb	82	204(1.4), 206(24.1), 207(22.1), 208(52.4)
Lithium	Li	3	6(7.5), 7(92.5)
Lutetium	Lu	71	175(97.41), 176(2.59)
Magnesium	Mg	12	24(78.99), 25(10.00), 26(11.01)
Manganese	Mn	25	55(100)
Mendelevium	Md	101	artificial isotopes only; mass number range 247–260
Mercury	Hg	80	196(0.14), 198(10.02), 199(16.84), 200(23.13), 201(13.22), 202(29.80), 204(6.85)
Molybdenum	Mo	42	92(14.84), 94(9.25), 95(15.92), 96(16.68), 97(9.55), 98(24.13), 100(9.63)
Neodymium	Nd	60	142(27.13), 143(12.18), 144(23.80), 145(8.30), 146(17.19), 148(5.76), 150(5.64)
Neon	Ne	10	20(90.48), 21(0.27), 22(9.25)
Neptunium	Np	93	artificial isotopes only; mass number range 234–240
Nickel	Ni	28	58(68.27), 60(26.10), 61(1.13), 62(3.59), 64(0.91)
Niobium	Nb	41	93(100)
Nitrogen	N	7	14(99.63), 15(0.37)
Nobelium	No	102	artificial isotopes only; mass number range 250–262
Osmium	Os	76	184(0.02), 186(1.58), 187(1.6), 188(13.3), 189(16.1), 190(26.4), 192(41.0)
Oxygen	O	8	16(99.76), 17(0.04), 18(0.20)
Palladium	Pd	46	102(1.02), 104(11.14), 105(22.33), 106(27.33), 108(26.46), 110(11.72)
Phosphorus	P	15	31(100)
Platinum	Pt	78	190(0.01), 192(0.79), 194(32.9), 195(33.8), 196(25.3), 198(7.2)
Plutonium	Pu	94	artificial isotopes only; mass number range 234–246
Polonium	Po	84	artificial isotopes only; mass number range 204–210
Potassium	K	19	39(93.26), 40(0.01), 41(6.73)
Praseodymium	Pr	59	141(100)
Promethium	Pm	61	artificial isotopes only; mass number range 141–151
Protactinium[†]	Pa	91	artificial isotopes only; mass number range 228–234
Radium	Ra	88	artificial isotopes only; mass number range 223–230

[†]See discussion in Section 27.5.

Element	Symbol	Atomic number, Z	Mass number of isotope (% abundance)
Radon	Rn	86	artificial isotopes only; mass number range 208–224
Rhenium	Re	75	185(37.40), 187(62.60)
Rhodium	Rh	45	103(100)
Rubidium	Rb	37	85(72.16), 87(27.84)
Ruthenium	Ru	44	96(5.52), 98(1.88), 99(12.7), 100(12.6), 101(17.0), 102(31.6), 104(18.7)
Samarium	Sm	62	144(3.1), 147(15.0), 148(11.3), 149(13.8), 150(7.4), 152(26.7), 154(22.7)
Scandium	Sc	21	45(100)
Selenium	Se	34	74(0.9), 76(9.2), 77(7.6), 78(23.6), 80(49.7), 82(9.0)
Silicon	Si	14	28(92.23), 29(4.67), 30(3.10)
Silver	Ag	47	107(51.84), 109(48.16)
Sodium	Na	11	23(100)
Strontium	Sr	38	84(0.56), 86(9.86), 87(7.00), 88(82.58)
Sulfur	S	16	32(95.02), 33(0.75), 34(4.21), 36(0.02)
Tantalum	Ta	73	180(0.01), 181(99.99)
Technetium	Tc	43	artificial isotopes only; mass number range 95–99
Tellurium	Te	52	120(0.09), 122(2.60), 123(0.91), 124(4.82), 125(7.14), 126(18.95), 128(31.69), 130(33.80)
Terbium	Tb	65	159(100)
Thallium	Tl	81	203(29.52), 205(70.48)
Thorium	Th	90	232(100)
Thulium	Tm	69	169(100)
Tin	Sn	50	112(0.97), 114(0.65), 115(0.36), 116(14.53), 117(7.68), 118(24.22), 119(8.58), 120(32.59), 122(4.63), 124(5.79)
Titanium	Ti	22	46(8.0), 47(7.3), 48(73.8), 49(5.5), 50(5.4)
Tungsten	W	74	180(0.13), 182(26.3), 183(14.3), 184(30.67), 186(28.6)
Uranium	U	92	234(0.005), 235(0.72), 236(99.275)
Vanadium	V	23	50(0.25), 51(99.75)
Xenon	Xe	54	124(0.10), 126(0.09), 128(1.91), 129(26.4), 130(4.1), 131(21.2), 132(26.9), 134(10.4), 136(8.9)
Ytterbium	Yb	70	168(0.13), 170(3.05), 171(14.3), 172(21.9), 173(16.12), 174(31.8), 176(12.7)
Yttrium	Y	39	89(100)
Zinc	Zn	30	64(48.6), 66(27.9), 67(4.1), 68(18.8), 70(0.6)
Zirconium	Zr	40	90(51.45), 91(11.22), 92(17.15), 94(17.38), 96(2.8)

Appendix 6
Van der Waals, metallic, covalent and ionic radii

Data are given for the *s*-, *p*- and first row *d*-block elements. The ionic radius varies with the charge and coordination number of the ion; a coordination number of 6 refers to octahedral coordination, and of 4 refers to tetrahedral unless otherwise specified. Data for the heavier *d*-block metals and the lanthanoids and actinoids are listed in Tables 22.1 and 27.1.

	Element	Van der Waals radius, r_v / pm	Metallic radius for 12-coordinate metal, r_{metal} / pm	Covalent radius, r_{cov} / pm	Ionic radius		
					Ionic radius, r_{ion} / pm	Charge on ion	Coordination number of the ion
Hydrogen	H	120[†]		37[‡]			
Group 1	Li		157		76	1+	6
	Na		191		102	1+	6
	K		235		138	1+	6
	Rb		250		149	1+	6
	Cs		272		170	1+	6
Group 2	Be		112		27	2+	4
	Mg		160		72	2+	6
	Ca		197		100	2+	6
	Sr		215		126	2+	8
	Ba		224		142	2+	8
Group 13	B	208		88			
	Al		143	130	54	3+	6
	Ga		153	122	62	3+	6
	In		167	150	80	3+	6
	Tl		171	155	89	3+	6
					159	1+	8
Group 14	C	185		77			
	Si	210		118			
	Ge			122	53	4+	6
	Sn		158	140	74	4+	6
	Pb		175	154	119	2+	6
					65	4+	4
					78	4+	6
Group 15	N	154		75	171	3−	6
	P	190		110			
	As	200		122			
	Sb	220		143			
	Bi	240	182	152	103	3+	6
					76	5+	6
Group 16	O	140		73	140	2−	6
	S	185		103	184	2−	6
	Se	200		117	198	2−	6
	Te	220		135	211	2−	6

[†] The value of 120 pm may be an overestimate; an analysis of intermolecular contacts in organic structures suggests a value of 110 pm. *See*: R.S. Rowland and R. Taylor (1996) *J. Phys. Chem.*, vol. 100, p. 7384.

[‡] Sometimes it is more appropriate to use a value of 30 pm in organic compounds.

	Element	Van der Waals radius, r_v / pm	Metallic radius for 12-coordinate metal, r_{metal} / pm	Covalent radius, r_{cov} / pm	Ionic radius		
					Ionic radius, r_{ion} / pm	Charge on ion	Coordination number of the ion
Group 17	F	135		71	133	1−	6
	Cl	180		99	181	1−	6
	Br	195		114	196	1−	6
	I	215		133	220	1−	6
Group 18	He	99					
	Ne	160					
	Ar	191					
	Kr	197					
	Xe	214					
First row *d*-block elements	Sc		164		75	3+	6
	Ti		147		86	2+	6
					67	3+	6
					61	4+	6
	V		135		79	2+	6
					64	3+	6
					58	4+	6
					53	4+	5
					54	5+	6
					46	5+	5
	Cr		129		73	2+	6 (low-spin)
					80	2+	6 (high-spin)
					62	3+	6
	Mn		137		67	2+	6 (low-spin)
					83	2+	6 (high-spin)
					58	3+	6 (low-spin)
					65	3+	6 (high-spin)
					39	4+	4
					53	4+	6
	Fe		126		61	2+	6 (low-spin)
					78	2+	6 (high-spin)
					55	3+	6 (low-spin)
					65	3+	6 (high-spin)
	Co		125		65	2+	6 (low-spin)
					75	2+	6 (high-spin)
					55	3+	6 (low-spin)
					61	3+	6 (high-spin)
	Ni		125		55	2+	4
					44	2+	4 (square planar)
					69	2+	6
					56	3+	6 (low-spin)
					60	3+	6 (high-spin)
	Cu		128		46	1+	2
					60	1+	4
					57	2+	4 (square planar)
					73	2+	6
	Zn		137		60	2+	4
					74	2+	6

Appendix 7
Pauling electronegativity values (χ^P) for selected elements of the periodic table

Values are dependent on oxidation state.

Group 1	Group 2		Group 13	Group 14	Group 15	Group 16	Group 17
H 2.2							
Li 1.0	Be 1.6		B 2.0	C 2.6	N 3.0	O 3.4	F 4.0
Na 0.9	Mg 1.3		Al(III) 1.6	Si 1.9	P 2.2	S 2.6	Cl 3.2
K 0.8	Ca 1.0		Ga(III) 1.8	Ge(IV) 2.0	As(III) 2.2	Se 2.6	Br 3.0
Rb 0.8	Sr 0.9	(*d*-block elements)	In(III) 1.8	Sn(II) 1.8 Sn(IV) 2.0	Sb 2.1	Te 2.1	I 2.7
Cs 0.8	Ba 0.9		Tl(I) 1.6 Tl(III) 2.0	Pb(II) 1.9 Pb(IV) 2.3	Bi 2.0	Po 2.0	At 2.2

Appendix 8

Ground state electronic configurations of the elements and ionization energies

Data are given for the first five ionizations.[†] $IE(n)$ in kJ mol^{-1} for the processes:

$IE(1)$ $\text{M(g)} \longrightarrow \text{M}^+\text{(g)}$

$IE(2)$ $\text{M}^+\text{(g)} \longrightarrow \text{M}^{2+}\text{(g)}$

$IE(3)$ $\text{M}^{2+}\text{(g)} \longrightarrow \text{M}^{3+}\text{(g)}$

$IE(4)$ $\text{M}^{3+}\text{(g)} \longrightarrow \text{M}^{4+}\text{(g)}$

$IE(5)$ $\text{M}^{4+}\text{(g)} \longrightarrow \text{M}^{5+}\text{(g)}$

Atomic number, Z	Element	Ground state electronic configuration	$IE(1)$	$IE(2)$	$IE(3)$	$IE(4)$	$IE(5)$
1	H	$1s^1$	1312				
2	He	$1s^2 = \text{[He]}$	2372	5250			
3	Li	$\text{[He]}2s^1$	520.2	7298	11820		
4	Be	$\text{[He]}2s^2$	899.5	1757	14850	21010	
5	B	$\text{[He]}2s^2 2p^1$	800.6	2427	3660	25030	32830
6	C	$\text{[He]}2s^2 2p^2$	1086	2353	4620	6223	37830
7	N	$\text{[He]}2s^2 2p^3$	1402	2856	4578	7475	9445
8	O	$\text{[He]}2s^2 2p^4$	1314	3388	5300	7469	10990
9	F	$\text{[He]}2s^2 2p^5$	1681	3375	6050	8408	11020
10	Ne	$\text{[He]}2s^2 2p^6 = \text{[Ne]}$	2081	3952	6122	9371	12180
11	Na	$\text{[Ne]}3s^1$	495.8	4562	6910	9543	13350
12	Mg	$\text{[Ne]}3s^2$	737.7	1451	7733	10540	13630
13	Al	$\text{[Ne]}3s^2 3p^1$	577.5	1817	2745	11580	14840
14	Si	$\text{[Ne]}3s^2 3p^2$	786.5	1577	3232	4356	16090
15	P	$\text{[Ne]}3s^2 3p^3$	1012	1907	2914	4964	6274
16	S	$\text{[Ne]}3s^2 3p^4$	999.6	2252	3357	4556	7004
17	Cl	$\text{[Ne]}3s^2 3p^5$	1251	2298	3822	5159	6540
18	Ar	$\text{[Ne]}3s^2 3p^6 = \text{[Ar]}$	1521	2666	3931	5771	7238
19	K	$\text{[Ar]}4s^1$	418.8	3052	4420	5877	7975
20	Ca	$\text{[Ar]}4s^2$	589.8	1145	4912	6491	8153
21	Sc	$\text{[Ar]}4s^2 3d^1$	633.1	1235	2389	7091	8843
22	Ti	$\text{[Ar]}4s^2 3d^2$	658.8	1310	2653	4175	9581
23	V	$\text{[Ar]}4s^2 3d^3$	650.9	1414	2828	4507	6299
24	Cr	$\text{[Ar]}4s^1 3d^5$	652.9	1591	2987	4743	6702

[†] Values are from several sources, but mostly from the *Handbook of Chemistry and Physics* (1993) 74th edn, CRC Press, Boca Raton, FL, and from the NIST Physics Laboratory, Physical Reference Data. The values in kJ mol^{-1} are quoted to four significant figures or less depending upon the accuracy of the original data in eV. A conversion factor of $1 \text{ eV} = 96.485 \text{ kJ mol}^{-1}$ has been applied.

Atomic number, Z	Element	Ground state electronic configuration	IE(1)	IE(2)	IE(3)	IE(4)	IE(5)
25	Mn	$[\text{Ar}]4s^2 3d^5$	717.3	1509	3248	4940	6990
26	Fe	$[\text{Ar}]4s^2 3d^6$	762.5	1562	2957	5290	7240
27	Co	$[\text{Ar}]4s^2 3d^7$	760.4	1648	3232	4950	7670
28	Ni	$[\text{Ar}]4s^2 3d^8$	737.1	1753	3395	5300	7339
29	Cu	$[\text{Ar}]4s^1 3d^{10}$	745.5	1958	3555	5536	7700
30	Zn	$[\text{Ar}]4s^2 3d^{10}$	906.4	1733	3833	5730	7970
31	Ga	$[\text{Ar}]4s^2 3d^{10} 4p^1$	578.8	1979	2963	6200	
32	Ge	$[\text{Ar}]4s^2 3d^{10} 4p^2$	762.2	1537	3302	4411	9020
33	As	$[\text{Ar}]4s^2 3d^{10} 4p^3$	947.0	1798	2735	4837	6043
34	Se	$[\text{Ar}]4s^2 3d^{10} 4p^4$	941.0	2045	2974	4144	6590
35	Br	$[\text{Ar}]4s^2 3d^{10} 4p^5$	1140	2100	3500	4560	5760
36	Kr	$[\text{Ar}]4s^2 3d^{10} 4p^6 = [\text{Kr}]$	1351	2350	3565	5070	6240
37	Rb	$[\text{Kr}]5s^1$	403.0	2633	3900	5080	6850
38	Sr	$[\text{Kr}]5s^2$	549.5	1064	4138	5500	6910
39	Y	$[\text{Kr}]5s^2 4d^1$	599.8	1181	1980	5847	7430
40	Zr	$[\text{Kr}]5s^2 4d^2$	640.1	1267	2218	3313	7752
41	Nb	$[\text{Kr}]5s^1 4d^4$	652.1	1382	2416	3700	4877
42	Mo	$[\text{Kr}]5s^1 4d^5$	684.3	1559	2618	4480	5257
43	Tc	$[\text{Kr}]5s^2 4d^5$	702	1472	2850		
44	Ru	$[\text{Kr}]5s^1 4d^7$	710.2	1617	2747		
45	Rh	$[\text{Kr}]5s^1 4d^8$	719.7	1744	2997		
46	Pd	$[\text{Kr}]5s^0 4d^{10}$	804.4	1875	3177		
47	Ag	$[\text{Kr}]5s^1 4d^{10}$	731.0	2073	3361		
48	Cd	$[\text{Kr}]5s^2 4d^{10}$	867.8	1631	3616		
49	In	$[\text{Kr}]5s^2 4d^{10} 5p^1$	558.3	1821	2704	5200	
50	Sn	$[\text{Kr}]5s^2 4d^{10} 5p^2$	708.6	1412	2943	3930	6974
51	Sb	$[\text{Kr}]5s^2 4d^{10} 5p^3$	830.6	1595	2440	4260	5400
52	Te	$[\text{Kr}]5s^2 4d^{10} 5p^4$	869.3	1790	2698	3610	5668
53	I	$[\text{Kr}]5s^2 4d^{10} 5p^5$	1008	1846	3200		
54	Xe	$[\text{Kr}]5s^2 4d^{10} 5p^6 = [\text{Xe}]$	1170	2046	3099		
55	Cs	$[\text{Xe}]6s^1$	375.7	2234	3400		
56	Ba	$[\text{Xe}]6s^2$	502.8	965.2	3619		
57	La	$[\text{Xe}]6s^2 5d^1$	538.1	1067	1850	4819	5940
58	Ce	$[\text{Xe}]4f^1 6s^2 5d^1$	534.4	1047	1949	3546	6325
59	Pr	$[\text{Xe}]4f^3 6s^2$	527.2	1018	2086	3761	5551
60	Nd	$[\text{Xe}]4f^4 6s^2$	533.1	1035	2130	3898	
61	Pm	$[\text{Xe}]4f^5 6s^2$	538.8	1052	2150	3970	
62	Sm	$[\text{Xe}]4f^6 6s^2$	544.5	1068	2260	3990	
63	Eu	$[\text{Xe}]4f^7 6s^2$	547.1	1085	2404	4120	
64	Gd	$[\text{Xe}]4f^7 6s^2 5d^1$	593.4	1167	1990	4245	
65	Tb	$[\text{Xe}]4f^9 6s^2$	565.8	1112	2114	3839	
66	Dy	$[\text{Xe}]4f^{10} 6s^2$	573.0	1126	2200	3990	
67	Ho	$[\text{Xe}]4f^{11} 6s^2$	581.0	1139	2204	4100	

Atomic number, Z	Element	Ground state electronic configuration	$IE(1)$	$IE(2)$	$IE(3)$	$IE(4)$	$IE(5)$
68	Er	$[Xe]4f^{12}6s^2$	589.3	1151	2194	4120	
69	Tm	$[Xe]4f^{13}6s^2$	596.7	1163	2285	4120	
70	Yb	$[Xe]4f^{14}6s^2$	603.4	1175	2417	4203	
71	Lu	$[Xe]4f^{14}6s^25d^1$	523.5	1340	2022	4366	
72	Hf	$[Xe]4f^{14}6s^25d^2$	658.5	1440	2250	3216	
73	Ta	$[Xe]4f^{14}6s^25d^3$	728.4	1500	2100		
74	W	$[Xe]4f^{14}6s^25d^4$	758.8	1700	2300		
75	Re	$[Xe]4f^{14}6s^25d^5$	755.8	1260	2510		
76	Os	$[Xe]4f^{14}6s^25d^6$	814.2	1600	2400		
77	Ir	$[Xe]4f^{14}6s^25d^7$	865.2	1680	2600		
78	Pt	$[Xe]4f^{14}6s^15d^9$	864.4	1791	2800		
79	Au	$[Xe]4f^{14}6s^15d^{10}$	890.1	1980	2900		
80	Hg	$[Xe]4f^{14}6s^25d^{10}$	1007	1810	3300		
81	Tl	$[Xe]4f^{14}6s^25d^{10}6p^1$	589.4	1971	2878	4900	
82	Pb	$[Xe]4f^{14}6s^25d^{10}6p^2$	715.6	1450	3081	4083	6640
83	Bi	$[Xe]4f^{14}6s^25d^{10}6p^3$	703.3	1610	2466	4370	5400
84	Po	$[Xe]4f^{14}6s^25d^{10}6p^4$	812.1	1800	2700		
85	At	$[Xe]4f^{14}6s^25d^{10}6p^5$	930	1600	2900		
86	Rn	$[Xe]4f^{14}6s^25d^{10}6p^6 = [Rn]$	1037				
87	Fr	$[Rn]7s^1$	393.0	2100	3100		
88	Ra	$[Rn]7s^2$	509.3	979.0	3300		
89	Ac	$[Rn]6d^17s^2$	499	1170	1900		
90	Th	$[Rn]6d^27s^2$	608.5	1110	1930	2780	
91	Pa	$[Rn]5f^27s^26d^1$	568	1130	1810		
92	U	$[Rn]5f^37s^26d^1$	597.6	1440	1840		
93	Np	$[Rn]5f^47s^26d^1$	604.5	1130	1880		
94	Pu	$[Rn]5f^67s^2$	581.4	1130	2100		
95	Am	$[Rn]5f^77s^2$	576.4	1160	2160		
96	Cm	$[Rn]5f^77s^26d^1$	578.0	1200	2050		
97	Bk	$[Rn]5f^97s^2$	598.0	1190	2150		
98	Cf	$[Rn]5f^{10}7s^2$	606.1	1210	2280		
99	Es	$[Rn]5f^{11}7s^2$	619	1220	2330		
100	Fm	$[Rn]5f^{12}7s^2$	627	1230	2350		
101	Md	$[Rn]5f^{13}7s^2$	635	1240	2450		
102	No	$[Rn]5f^{14}7s^2$	642	1250	2600		
103	Lr	$[Rn]5f^{14}7s^26d^1$	440 (?)				

Appendix 9
Electron affinities

Approximate enthalpy changes, $\Delta_{EA}H(298\ K)$, associated with the gain of one electron by a gaseous atom or anion. A negative enthalpy (ΔH), but a positive electron affinity (EA), corresponds to an exothermic process (see Section 1.10).
$\Delta_{EA}H(298\ K) \approx \Delta U(0\ K) = -EA$

	Process	$\approx \Delta_{EA}H\ /\ kJ\ mol^{-1}$
Hydrogen	$H(g) + e^- \longrightarrow H^-(g)$	-73
Group 1	$Li(g) + e^- \longrightarrow Li^-(g)$	-60
	$Na(g) + e^- \longrightarrow Na^-(g)$	-53
	$K(g) + e^- \longrightarrow K^-(g)$	-48
	$Rb(g) + e^- \longrightarrow Rb^-(g)$	-47
	$Cs(g) + e^- \longrightarrow Cs^-(g)$	-45
Group 15	$N(g) + e^- \longrightarrow N^-(g)$	≈ 0
	$P(g) + e^- \longrightarrow P^-(g)$	-72
	$As(g) + e^- \longrightarrow As^-(g)$	-78
	$Sb(g) + e^- \longrightarrow Sb^-(g)$	-103
	$Bi(g) + e^- \longrightarrow Bi^-(g)$	-91
Group 16	$O(g) + e^- \longrightarrow O^-(g)$	-141
	$O^-(g) + e^- \longrightarrow O^{2-}(g)$	$+798$
	$S(g) + e^- \longrightarrow S^-(g)$	-201
	$S^-(g) + e^- \longrightarrow S^{2-}(g)$	$+640$
	$Se(g) + e^- \longrightarrow Se^-(g)$	-195
	$Te(g) + e^- \longrightarrow Te^-(g)$	-190
Group 17	$F(g) + e^- \longrightarrow F^-(g)$	-328
	$Cl(g) + e^- \longrightarrow Cl^-(g)$	-349
	$Br(g) + e^- \longrightarrow Br^-(g)$	-325
	$I(g) + e^- \longrightarrow I^-(g)$	-295

Appendix 10
Standard enthalpies of atomization (Δ_aH°) of the elements at 298 K

Enthalpies are given in kJ mol^{-1} for the process:

$$\frac{1}{n}E_n(\text{standard state}) \longrightarrow E(g)$$

Elements (E) are arranged according to their position in the periodic table. The lanthanoids and actinoids are excluded. The noble gases are omitted because they are monatomic at 298 K.

1	2	3	4	5	6	7	8	9	10	11	12	13	14	15	16	17
H 218																
Li 161	Be 324											B 582	C 717	N 473	O 249	F 79
Na 108	Mg 146											Al 330	Si 456	P 315	S 277	Cl 121
K 90	Ca 178	Sc 378	Ti 470	V 514	Cr 397	Mn 283	Fe 418	Co 428	Ni 430	Cu 338	Zn 130	Ga 277	Ge 375	As 302	Se 227	Br 112
Rb 82	Sr 164	Y 423	Zr 609	Nb 721	Mo 658	Tc 677	Ru 651	Rh 556	Pd 377	Ag 285	Cd 112	In 243	Sn 302	Sb 264	Te 197	I 107
Cs 78	Ba 178	La 423	Hf 619	Ta 782	W 850	Re 774	Os 787	Ir 669	Pt 566	Au 368	Hg 61	Tl 182	Pb 195	Bi 210	Po ≈146	At 92

Appendix 11

Selected standard reduction potentials (298 K)

The concentration of each aqueous solution is $1 \, mol \, dm^{-3}$ and the pressure of a gaseous component is $1 \, bar$ ($10^5 \, Pa$). (Changing the standard pressure to $1 \, atm$ ($101 \, 300 \, Pa$) makes no difference to the values of E^o at this level of accuracy.) Each half-cell listed contains the specified solution species at a concentration of $1 \, mol \, dm^{-3}$; where the half-cell contains $[OH]^-$, the value of E^o refers to $[OH^-] = 1 \, mol \, dm^{-3}$, hence the notation $E^o_{[OH^-]=1}$ (see Box 8.1).

Reduction half-equation	E^o or $E^o_{[OH^-]=1}/V$
$Li^+(aq) + e^- \rightleftharpoons Li(s)$	-3.04
$Cs^+(aq) + e^- \rightleftharpoons Cs(s)$	-3.03
$Rb^+(aq) + e^- \rightleftharpoons Rb(s)$	-2.98
$K^+(aq) + e^- \rightleftharpoons K(s)$	-2.93
$Ca^{2+}(aq) + 2e^- \rightleftharpoons Ca(s)$	-2.87
$Na^+(aq) + e^- \rightleftharpoons Na(s)$	-2.71
$La^{3+}(aq) + 3e^- \rightleftharpoons La(s)$	-2.38
$Mg^{2+}(aq) + 2e^- \rightleftharpoons Mg(s)$	-2.37
$Y^{3+}(aq) + 3e^- \rightleftharpoons Y(s)$	-2.37
$Sc^{3+}(aq) + 3e^- \rightleftharpoons Sc(s)$	-2.03
$Al^{3+}(aq) + 3e^- \rightleftharpoons Al(s)$	-1.66
$[HPO_3]^{2-}(aq) + 2H_2O(l) + 2e^- \rightleftharpoons [H_2PO_2]^-(aq) + 3[OH]^-(aq)$	-1.65
$Ti^{2+}(aq) + 2e^- \rightleftharpoons Ti(s)$	-1.63
$Mn(OH)_2(s) + 2e^- \rightleftharpoons Mn(s) + 2[OH]^-(aq)$	-1.56
$Mn^{2+}(aq) + 2e^- \rightleftharpoons Mn(s)$	-1.19
$V^{2+}(aq) + 2e^- \rightleftharpoons V(s)$	-1.18
$Te(s) + 2e^- \rightleftharpoons Te^{2-}(aq)$	-1.14
$2[SO_3]^{2-}(aq) + 2H_2O(l) + 2e^- \rightleftharpoons 4[OH]^-(aq) + [S_2O_4]^{2-}(aq)$	-1.12
$[SO_4]^{2-}(aq) + H_2O(l) + 2e^- \rightleftharpoons [SO_3]^{2-}(aq) + 2[OH]^-(aq)$	-0.93
$Se(s) + 2e^- \rightleftharpoons Se^{2-}(aq)$	-0.92
$Cr^{2+}(aq) + 2e^- \rightleftharpoons Cr(s)$	-0.91
$2[NO_3]^-(aq) + 2H_2O(l) + 2e^- \rightleftharpoons N_2O_4(g) + 4[OH]^-(aq)$	-0.85
$2H_2O(l) + 2e^- \rightleftharpoons H_2(g) + 2[OH]^-(aq)$	-0.82
$Zn^{2+}(aq) + 2e^- \rightleftharpoons Zn(s)$	-0.76
$Cr^{3+}(aq) + 3e^- \rightleftharpoons Cr(s)$	-0.74
$S(s) + 2e^- \rightleftharpoons S^{2-}(aq)$	-0.48
$[NO_2]^-(aq) + H_2O(l) + e^- \rightleftharpoons NO(g) + 2[OH]^-(aq)$	-0.46
$Fe^{2+}(aq) + 2e^- \rightleftharpoons Fe(s)$	-0.44
$Cr^{3+}(aq) + e^- \rightleftharpoons Cr^{2+}(aq)$	-0.41
$Ti^{3+}(aq) + e^- \rightleftharpoons Ti^{2+}(aq)$	-0.37
$PbSO_4(s) + 2e^- \rightleftharpoons Pb(s) + [SO_4]^{2-}(aq)$	-0.36

Reduction half-equation	E° or $E^{\circ}_{[OH^-]=1}$ / V
$Tl^+(aq) + e^- \rightleftharpoons Tl(s)$	-0.34
$Co^{2+}(aq) + 2e^- \rightleftharpoons Co(s)$	-0.28
$H_3PO_4(aq) + 2H^+(aq) + 2e^- \rightleftharpoons H_3PO_3(aq) + H_2O(l)$	-0.28
$V^{3+}(aq) + e^- \rightleftharpoons V^{2+}(aq)$	-0.26
$Ni^{2+}(aq) + 2e^- \rightleftharpoons Ni(s)$	-0.25
$2[SO_4]^{2-}(aq) + 4H^+(aq) + 2e^- \rightleftharpoons [S_2O_6]^{2-}(aq) + 2H_2O(l)$	-0.22
$O_2(g) + 2H_2O(l) + 2e^- \rightleftharpoons H_2O_2(aq) + 2[OH]^-(aq)$	-0.15
$Sn^{2+}(aq) + 2e^- \rightleftharpoons Sn(s)$	-0.14
$Pb^{2+}(aq) + 2e^- \rightleftharpoons Pb(s)$	-0.13
$Fe^{3+}(aq) + 3e^- \rightleftharpoons Fe(s)$	-0.04
$2H^+(aq, 1\ mol\ dm^{-3}) + 2e^- \rightleftharpoons H_2(g, 1\ bar)$	0
$[NO_3]^-(aq) + H_2O(l) + 2e^- \rightleftharpoons [NO_2]^-(aq) + 2[OH]^-(aq)$	$+0.01$
$[S_4O_6]^{2-}(aq) + 2e^- \rightleftharpoons 2[S_2O_3]^{2-}(aq)$	$+0.08$
$[Ru(NH_3)_6]^{3+}(aq) + e^- \rightleftharpoons [Ru(NH_3)_6]^{2+}(aq)$	$+0.10$
$[Co(NH_3)_6]^{3+}(aq) + e^- \rightleftharpoons [Co(NH_3)_6]^{2+}(aq)$	$+0.11$
$S(s) + 2H^+(aq) + 2e^- \rightleftharpoons H_2S(aq)$	$+0.14$
$2[NO_2]^-(aq) + 3H_2O(l) + 4e^- \rightleftharpoons N_2O(g) + 6[OH]^-(aq)$	$+0.15$
$Cu^{2+}(aq) + e^- \rightleftharpoons Cu^+(aq)$	$+0.15$
$Sn^{4+}(aq) + 2e^- \rightleftharpoons Sn^{2+}(aq)$	$+0.15$
$[SO_4]^{2-}(aq) + 4H^+(aq) + 2e^- \rightleftharpoons H_2SO_3(aq) + H_2O(l)$	$+0.17$
$AgCl(s) + e^- \rightleftharpoons Ag(s) + Cl^-(aq)$	$+0.22$
$[Ru(OH_2)_6]^{3+}(aq) + e^- \rightleftharpoons [Ru(OH_2)_6]^{2+}(aq)$	$+0.25$
$[Co(bpy)_3]^{3+}(aq) + e^- \rightleftharpoons [Co(bpy)_3]^{2+}(aq)$	$+0.31$
$Cu^{2+}(aq) + 2e^- \rightleftharpoons Cu(s)$	$+0.34$
$[VO]^{2+}(aq) + 2H^+(aq) + e^- \rightleftharpoons V^{3+}(aq) + H_2O(l)$	$+0.34$
$[ClO_4]^-(aq) + H_2O(l) + 2e^- \rightleftharpoons [ClO_3]^-(aq) + 2[OH]^-(aq)$	$+0.36$
$[Fe(CN)_6]^{3-}(aq) + e^- \rightleftharpoons [Fe(CN)_6]^{4-}(aq)$	$+0.36$
$O_2(g) + 2H_2O(l) + 4e^- \rightleftharpoons 4[OH]^-(aq)$	$+0.40$
$Cu^+(aq) + e^- \rightleftharpoons Cu(s)$	$+0.52$
$I_2(aq) + 2e^- \rightleftharpoons 2I^-(aq)$	$+0.54$
$[S_2O_6]^{2-}(aq) + 4H^+(aq) + 2e^- \rightleftharpoons 2H_2SO_3(aq)$	$+0.56$
$H_3AsO_4(aq) + 2H^+(aq) + 2e^- \rightleftharpoons HAsO_2(aq) + 2H_2O(l)$	$+0.56$
$[MnO_4]^-(aq) + e^- \rightleftharpoons [MnO_4]^{2-}(aq)$	$+0.56$
$[MnO_4]^-(aq) + 2H_2O(aq) + 3e^- \rightleftharpoons MnO_2(s) + 4[OH]^-(aq)$	$+0.59$
$[MnO_4]^{2-}(aq) + 2H_2O(l) + 2e^- \rightleftharpoons MnO_2(s) + 4[OH]^-(aq)$	$+0.60$
$[BrO_3]^-(aq) + 3H_2O(l) + 6e^- \rightleftharpoons Br^-(aq) + 6[OH]^-(aq)$	$+0.61$
$O_2(g) + 2H^+(aq) + 2e^- \rightleftharpoons H_2O_2(aq)$	$+0.70$
$[BrO]^-(aq) + H_2O(l) + 2e^- \rightleftharpoons Br^-(aq) + 2[OH]^-(aq)$	$+0.76$
$Fe^{3+}(aq) + e^- \rightleftharpoons Fe^{2+}(aq)$	$+0.77$
$Ag^+(aq) + e^- \rightleftharpoons Ag(s)$	$+0.80$
$[ClO]^-(aq) + H_2O(l) + 2e^- \rightleftharpoons Cl^-(aq) + 2[OH]^-(aq)$	$+0.84$
$2HNO_2(aq) + 4H^+(aq) + 4e^- \rightleftharpoons H_2N_2O_2(aq) + 2H_2O(l)$	$+0.86$
$[HO_2]^-(aq) + H_2O(l) + 2e^- \rightleftharpoons 3[OH]^-(aq)$	$+0.88$

Reduction half-equation	$E^{\ominus}$ or $E^{\ominus}_{[OH^-]=1}$ / V
$[NO_3]^-(aq) + 3H^+(aq) + 2e^- \rightleftharpoons HNO_2(aq) + H_2O(l)$	+0.93
$Pd^{2+}(aq) + 2e^- \rightleftharpoons Pd(s)$	+0.95
$[NO_3]^-(aq) + 4H^+(aq) + 3e^- \rightleftharpoons NO(g) + 2H_2O(l)$	+0.96
$HNO_2(aq) + H^+(aq) + e^- \rightleftharpoons NO(g) + H_2O(l)$	+0.98
$[VO_2]^+(aq) + 2H^+(aq) + e^- \rightleftharpoons [VO]^{2+}(aq) + H_2O(l)$	+0.99
$[Fe(bpy)_3]^{3+}(aq) + e^- \rightleftharpoons [Fe(bpy)_3]^{2+}(aq)$	+1.03
$[IO_3]^-(aq) + 6H^+(aq) + 6e^- \rightleftharpoons I^-(aq) + 3H_2O(l)$	+1.09
$Br_2(aq) + 2e^- \rightleftharpoons 2Br^-(aq)$	+1.09
$[Fe(phen)_3]^{3+}(aq) + e^- \rightleftharpoons [Fe(phen)_3]^{2+}(aq)$	+1.12
$Pt^{2+}(aq) + 2e^- \rightleftharpoons Pt(s)$	+1.18
$[ClO_4]^-(aq) + 2H^+(aq) + 2e^- \rightleftharpoons [ClO_3]^-(aq) + H_2O(l)$	+1.19
$2[IO_3]^-(aq) + 12H^+(aq) + 10e^- \rightleftharpoons I_2(aq) + 6H_2O(l)$	+1.20
$O_2(g) + 4H^+(aq) + 4e^- \rightleftharpoons 2H_2O(l)$	+1.23
$MnO_2(s) + 4H^+(aq) + 2e^- \rightleftharpoons Mn^{2+}(aq) + 2H_2O(l)$	+1.23
$Tl^{3+}(aq) + 2e^- \rightleftharpoons Tl^+(aq)$	+1.25
$2HNO_2(aq) + 4H^+(aq) + 4e^- \rightleftharpoons N_2O(g) + 3H_2O(l)$	+1.30
$[Cr_2O_7]^{2-}(aq) + 14H^+(aq) + 6e^- \rightleftharpoons 2Cr^{3+}(aq) + 7H_2O(l)$	+1.33
$Cl_2(aq) + 2e^- \rightleftharpoons 2Cl^-(aq)$	+1.36
$2[ClO_4]^-(aq) + 16H^+(aq) + 14e^- \rightleftharpoons Cl_2(aq) + 8H_2O(l)$	+1.39
$[ClO_4]^-(aq) + 8H^+(aq) + 8e^- \rightleftharpoons Cl^-(aq) + 4H_2O(l)$	+1.39
$[BrO_3]^-(aq) + 6H^+(aq) + 6e^- \rightleftharpoons Br^-(aq) + 3H_2O(l)$	+1.42
$[ClO_3]^-(aq) + 6H^+(aq) + 6e^- \rightleftharpoons Cl^-(aq) + 3H_2O(l)$	+1.45
$2[ClO_3]^-(aq) + 12H^+(aq) + 10e^- \rightleftharpoons Cl_2(aq) + 6H_2O(l)$	+1.47
$2[BrO_3]^-(aq) + 12H^+(aq) + 10e^- \rightleftharpoons Br_2(aq) + 6H_2O(l)$	+1.48
$HOCl(aq) + H^+(aq) + 2e^- \rightleftharpoons Cl^-(aq) + H_2O(l)$	+1.48
$[MnO_4]^-(aq) + 8H^+(aq) + 5e^- \rightleftharpoons Mn^{2+}(aq) + 4H_2O(l)$	+1.51
$Mn^{3+}(aq) + e^- \rightleftharpoons Mn^{2+}(aq)$	+1.54
$2HOCl(aq) + 2H^+(aq) + 2e^- \rightleftharpoons Cl_2(aq) + 2H_2O(l)$	+1.61
$[MnO_4]^-(aq) + 4H^+(aq) + 3e^- \rightleftharpoons MnO_2(s) + 2H_2O(l)$	+1.69
$PbO_2(s) + 4H^+(aq) + [SO_4]^{2-}(aq) + 2e^- \rightleftharpoons PbSO_4(s) + 2H_2O(l)$	+1.69
$Ce^{4+}(aq) + e^- \rightleftharpoons Ce^{3+}(aq)$	+1.72
$[BrO_4]^-(aq) + 2H^+(aq) + 2e^- \rightleftharpoons [BrO_3]^-(aq) + H_2O(l)$	+1.76
$H_2O_2(aq) + 2H^+(aq) + 2e^- \rightleftharpoons 2H_2O(l)$	+1.78
$Co^{3+}(aq) + e^- \rightleftharpoons Co^{2+}(aq)$	+1.92
$[S_2O_8]^{2-}(aq) + 2e^- \rightleftharpoons 2[SO_4]^{2-}(aq)$	+2.01
$O_3(g) + 2H^+(aq) + 2e^- \rightleftharpoons O_2(g) + H_2O(l)$	+2.07
$XeO_3(aq) + 6H^+(aq) + 6e^- \rightleftharpoons Xe(g) + 3H_2O(l)$	+2.10
$[FeO_4]^{2-}(aq) + 8H^+(aq) + 3e^- \rightleftharpoons Fe^{3+}(aq) + 4H_2O(l)$	+2.20
$H_4XeO_6(aq) + 2H^+(aq) + 2e^- \rightleftharpoons XeO_3(aq) + 3H_2O(l)$	+2.42
$F_2(aq) + 2e^- \rightleftharpoons 2F^-(aq)$	+2.87

Appendix 12
Selected bond enthalpy terms

Bond	Bond enthalpy / kJ mol^{-1}	Bond	Bond enthalpy / kJ mol^{-1}	Bond	Bond enthalpy / kJ mol^{-1}
H–H	436	F–F	159	C–F	485
C–C	346	Cl–Cl	242	C–Cl	327
C=C	598	Br–Br	193	C–Br	285
C≡C	813	I–I	151	C–I	213
Si–Si	226	C–H	416	C–O	359
Ge–Ge	186	Si–H	326	C=O	806
Sn–Sn	152	Ge–H	289	C–N	285
N–N	159	Sn–H	251	C≡N	866
N=N	≈400	N–H	391	C–S	272
N≡N	945	P–H	322	Si–O	466
P–P	200	As–H	247	Si=O	642
P≡P	490	O–H	464	N–F	272
As–As	177	S–H	366	N–Cl	193
O–O	146	Se–H	276	N–O	201
O=O	498	F–H	570	P–F	490
S–S	266	Cl–H	432	P–Cl	319
S=S	425	Br–H	366	P–O	340
Se–Se	193	I–H	298	S–F	326

Answers to non-descriptive problems

Full methods of working for all problems are given in the accompanying *Solutions Manual*. Where *no* answer is given below, guidelines are given in the *Solutions Manual*.

CHAPTER 1

1.1 Each isotope: 24 e, 24 p; 26, 28, 29 and 30 n, respectively.

1.2 Only one isotope, e.g. P, Na, Be.

1.3 (a) $^{27}_{13}$Al, 13 p, 13 e, 14 n; (b) $^{79}_{35}$Br, 35 p, 35 e, 44 n; $^{81}_{35}$Br, 35 p, 35 e, 46 n; (c) $^{54}_{26}$Fe, 26 p, 26 e, 28 n; $^{56}_{26}$Fe, 26 p, 26 e, 30 n; $^{57}_{26}$Fe, 26 p, 26 e, 31 n; $^{58}_{26}$Fe, 26 p, 26 e, 32 n.

1.4 Assume ^{3}H can be ignored; % ^{1}H = 99.2, % ^{2}H = 0.8.

1.6 (a) 1.0×10^{-4} m, far infrared; (b) 3.0×10^{-10} m, X-ray; (c) 6.0×10^{-7} m, visible.

1.7 (a), (e) Lyman; (b), (d) Balmer; (c) Paschen.

1.8 266 kJ mol^{-1}

1.10 For $n = 2$, $r = 211.7$ pm; for $n = 3$, $r = 476.4$ pm.

1.11 (a) Energy increases; (b) size increases.

1.12 (a) $n = 6$, $l = 0$, $m_l = 0$; (b) $n = 4$, $l = 2$, $m_l = -2$; $n = 4$, $l = 2$, $m_l = -1$; $n = 4$, $l = 2$, $m_l = 0$; $n = 4$, $l = 2$, $m_l = 1$; $n = 4$, $l = 2$, $m_l = 2$.

1.13 (a) Same value of n; (b) same value of l; (c) different values of m_l; $n = 4$, $l = 1$, $m_l = -1$; $n = 4$, $l = 1$, $m_l = 0$; $n = 4$, $l = 1$, $m_l = 1$.

1.14 (a) 1; (b) 3; (c) 1; (d) 2; (e) 0; (f) 2.

1.16 (a) $n = 1$, $l = 0$, $m_l = 0$; (b) $n = 4$, $l = 0$, $m_l = 0$; (c) $n = 5$, $l = 0$, $m_l = 0$.

1.17 $n = 3$, $l = 1$, $m_l = -1$; $n = 3$, $l = 1$, $m_l = 0$; $n = 3$, $l = 1$, $m_l = 1$.

1.18 7; $4f$; $n = 4$, $l = 3$, $m_l = -3$; $n = 4$, $l = 3$, $m_l = -2$; $n = 4$, $l = 3$, $m_l = -1$; $n = 4$, $l = 3$, $m_l = 0$; $n = 4$, $l = 3$, $m_l = 1$; $n = 4$, $l = 3$, $m_l = 2$; $n = 4$, $l = 3$, $m_l = 3$.

1.19 (b); (e).

1.21 -146 kJ mol^{-1}; same energy.

1.22 $n = 1$, $E = -1312$; $n = 2$, $E = -328.0$; $n = 3$, $E = -145.8$; $n = 4$, $E = -82.00$; $n = 5$, $E = -52.50 \text{ kJ mol}^{-1}$; the larger is the value of n, the higher (less negative) the energy level; the energy levels get closer together as n increases.

1.23 Spin-paired designated by $m_s = \pm\frac{1}{2}$: $n = 5$, $l = 1$, $m_l = -1$, $m_s = \pm\frac{1}{2}$; $n = 5$, $l = 1$, $m_l = 0$, $m_s = \pm\frac{1}{2}$; $n = 5$, $l = 1$, $m_l = 1$, $m_s = \pm\frac{1}{2}$.

1.24 $1s < 2s < 3s < 3p < 3d < 4p < 6s < 6p$.

1.26 Core electrons written in [ ]: (a) $[1s^2 2s^2 2p^6]3s^1$; (b) $[1s^2]2s^2 2p^5$; (c) $[1s^2]2s^2 2p^3$; (d) $[1s^2 2s^2 2p^6 3s^2 3p^6]4s^2 3d^1$.

1.28 $1s^2 2s^2 2p^1$; $n = 1$, $l = 0$, $m_l = 0$; $m_s = \frac{1}{2}$; $n = 1$, $l = 0$, $m_l = 0$; $m_s = -\frac{1}{2}$; $n = 2$, $l = 0$, $m_l = 0$; $m_s = \frac{1}{2}$; $n = 2$, $l = 0$, $m_l = 0$; $m_s = -\frac{1}{2}$; $n = 2$, $l = 1$, $m_l = 0$ (or +1 or −1); $m_s = \frac{1}{2}$ or $-\frac{1}{2}$.

1.30 Energy level diagrams similar to Fig. 1.15 showing the configurations: (a) $2s^2 2p^5$; (b) $3s^2 3p^1$; (c) $3s^2$.

1.32 (a) $Sn^{3+}(g) \longrightarrow Sn^{4+}(g) + e^-$; endothermic; (b) $Al(g) \longrightarrow Al^{3+}(g) + 3e^-$.

1.33 Group 1.

1.36 (a) $+657 \text{ kJ mol}^{-1}$.

1.45 (b) $5.09 \times 10^{14} \text{ s}^{-1}$.

CHAPTER 2

2.2 (a) Single; (b) single; (c) double; (d) single.

2.5 (a) and (c)

2.6 (b) VB theory predicts all to be diamagnetic.

2.7 (a) Single; (b) single; (c) double; (d) triple; (e) single.

2.9 (a) $\frac{1}{2}$, 1; (b) yes (H_2 and $[He_2]^{2+}$ are isoelectronic).

2.10 (b) O_2, 2.0; $[O_2]^+$, 2.5; $[O_2]^-$, 1.5; $[O_2]^{2-}$, 1.0. (c) O_2, $[O_2]^+$ and $[O_2]^-$.

2.13 (a) $[NO_2]^-$; (b) $[O_2]^{2-}$; (c) $[BrF_6]^-$.

2.15 (a) Polar, $N^{\delta-}-H^{\delta+}$; (b) polar, $F^{\delta-}-Br^{\delta+}$; (c) slightly polar, $C^{\delta-}-H^{\delta+}$; (d) polar, $P^{\delta+}-Cl^{\delta-}$; (e) non-polar.

2.16 HF and $[OH]^-$; CO_2 and $[NO_2]^+$; NH_3 and $[H_3O]^+$; $SiCl_4$ and $[AlCl_4]^-$.

2.17 (a) Bent; (b) tetrahedral; (c) trigonal pyramidal; (d) trigonal bipyramidal; (e) trigonal pyramidal; (f) pentagonal bipyramidal; (g) linear; (h) bent; (i) trigonal planar.

2.19 (a) Bent, polar; (b) linear, non-polar; (c) bent, polar; (d) trigonal planar, non-polar; (e) trigonal bipyramidal, non-polar; (f) planar, polar; (g) planar, non-polar; (h) linear, polar.

2.20 (a) Trigonal planar; no isomers; (b) tetrahedral; no isomers; (c) trigonal bipyramidal; Me, axially or equatorially sited; (d) octahedral; *cis* or *trans*.

2.22 (a) *trans*; (b) NSF_3, no lone pair on S; (c) three lone pairs prefer to occupy equatorial sites in trigonal bipyramidal arrangement.

2.23 (a) Square-based pyramidal molecule; (b) $4s$ electron in K better shielded from nuclear charge; (c) BI_3, no lone pair.

2.24 (a) 2nd electron removed from positively charged ion; (b) *trans* isomer converted to *cis*; (c) degenerate HOMO $\pi_g^*(3p_x)^1 \pi_g^*(3p_y)^1$.

2.27 (a) $[PCl_4]^+$, tetrahedral, no stereoisomers; $[PCl_3F_3]^-$, octahedral, stereoisomers.

2.28 (a) 1; (b) 1; (c) 1; (d) 2; (e) 2.

2.30 (b) SO_2 bent; NH_3 trigonal pyramidal; N_2O linear; CH_4 tetrahedral; CO_2 linear; (c) O_2 (diradical); NO; (f) group 18 elements.

2.32 (b) Both polar.

CHAPTER 3

3.1 (a) Trigonal planar; non-polar; (b) bent; polar; (c) trigonal pyramidal; polar; (d) linear; non-polar; (e) tetrahedral; polar.

3.3 (a) C_8; (b) C_2; (c) C_5; (d) C_3.

3.4 Bent; E, C_2, σ_v and σ_v'.

3.5 C_2 axis bisecting the O—O bond.

3.6 Labels are C_3, C_2 ($\times 3$), σ_h and σ_v ($\times 3$).

3.7 (a) Lose C_3 axis, two C_2 axes, two σ_v planes; (b) lose C_2 axis, σ_v plane; (c) σ_h plane.

3.8 (a) NH_3, PBr_3, $[SO_4]^{2-}$; (b) SO_3; $AlCl_3$; $[NO_3]^-$.

3.9 $[ICl_4]^-$; XeF_4.

3.10 (a) 2 (see-saw); (b) 2 (bent); (c) 9 (octahedral); (d) 2 (see-saw); (e) 2 (bent); (f) 4 (trigonal planar).

3.11 (a) Ethane-like; (b) staggered; (c) yes, at the midpoint of the Si—Si bond; (d) eclipsed; (e) no.

3.12 (a) No; (b) no; (c) yes; (d) no; (e) no; (f) yes; (g) yes; (h) no.

3.14 C_{3v}.

3.15 Linear.

3.16 C_{4v}.

3.17 Structure is T-shaped.

3.19 (a) and (e) T_d; (b) and (d) C_{3v}; (c) C_{2v}.

3.20 (a) C_{2v}; (b) yes.

3.21 I_h.

3.22 (a) 3; (b) 9; (c) 4; (d) 3; (e) 6.

3.23 (a) 3; (b) 2; (c) 4; (d) 3; (e) 2; (f) 3.

3.26 T_d

3.28 (a) 18; (b) B_{1u}; B_{2u}; B_{3u}.

3.30 A_{1g}; B_{1g}; E_u (IR active).

3.33 (a) D_{3h}; A_1'; A_1'; A_2''; E'; (b) 3 Raman bands.

3.42 (e) 4.3×10^{15} molecules.

CHAPTER 4

4.3 250–400 nm UV; 400–750 nm VIS; 392 nm in UV but absorption broad tails into visible.

4.5 $x = 2$; $y = 4$.

4.6 $[bpyH]^+Cl^-$.

4.7 $x = 2$.

4.11 $[Ph_3PO + H]^+$; 154.0 arises from $[NOBA + H]^+$.

4.14 (a) Most intense peak set to 100%; (b) $[M + H]^+$, $[M + Na]^+$.

4.15 (a) $+2$; (b) $[M - Cl]^+$; (c) $[M - Cl + L + H]^+$ (ligand is L^-).

4.16 (a) Positive mode: $[Me_4Sb]^+$; negative mode $[Ph_2SbCl_4]^-$; (b) $[Me_4Sb]^+$, tetrahedral; $[Ph_2SbCl_4]^-$, octahedral, *trans* and *cis* isomers; mass spectrum gives no isomer information.

4.17 $[PtCl_2L_2]$; $[M+Na]^+$; $[M+H]^+$; $[M-Cl]^+$; $[L+H]^+$.

4.25 (a) 569 nm is in visible; appears purple; (b) 569 nm; (c) 0.96.

4.29 J_{PF} and $J_{PH} \gg J_{HH}$ (for directly attached pairs of nuclei).

4.30 2 ^{13}C environments; each ^{13}C couples to three equivalent ^{19}F; larger value of J_{CF} is due to ^{19}F directly attached to ^{13}C and smaller J_{CF} is long-range coupling.

4.31 Doublet for Ph_2PH with large J_{PH}; singlet for PPh_3.

4.32 (a) Coupling of ^{31}P to 9 equivalent ^{1}H; (b) doublet $J_{PH} = 2.7$ Hz.

4.33 (a) Coupling to two equivalent ^{1}H gives triplet; (b) only 4.7% of the terminal ^{1}H are attached to ^{29}Si; observe singlet with overlapping doublet ($J_{SiH} = 194$ Hz); relative intensities of three lines 2.35 : 95.3 : 2.35.

4.34 (a) Binomial quartet; coupling to three equivalent ^{1}H; (b) doublet of quartets; coupling to one ^{31}P (gives doublet) and to three equivalent ^{1}H (gives quartet).

4.35 (a) See-saw; (b) static structure at 175 K contains two equatorial and two axial F giving two triplets (J_{FF}); at 298 K, a fluxional process renders all ^{19}F equivalent.

4.36 Consistent for all except (b); VSEPR predicts PF_5 is trigonal bipyramidal with two F environments, ratio 2 : 3.

4.39 $SiCl_4$, $SiCl_3Br$, $SiCl_2Br_2$, $SiClBr_3$ and $SiBr_4$ present.

4.40 3 ^{31}P environments 2 : 1 : 2, with $J(^{31}P-^{31}P)$.

4.41 Doublet (satellites) superimposed on singlet.

4.42 One signal each for SeS_7, 1,2-Se_2S_6, 1,3-Se_2S_6; 2 for 1,2,3-Se_3S_5 and 1,2,3,4-Se_4S_4; 3 for 1,2,4- and 1,2,5-Se_3S_5.

4.43 Coupling to ^{11}B, $I = \frac{3}{2}$; 1 : 1 : 1 : 1 quartet.

4.44 Me group exchange on NMR timescale.

4.46 (a) Coupling to ^{31}P; (b) 33.8% of ^{31}P couples to ^{195}Pt ($I = \frac{1}{2}$).

4.47 (a) Triplet arises from CH_2 group; coupling to two equivalent ^{31}P; (b) 3.

4.54 (a) 4; (c) $A(^{63}Cu) = 140$ G; $A(^{65}Cu) = 150$ G; (d) $g = 2.07$.

4.55 (a) $g = 2.00$; (b) Fig. 4.40a; (c) from Fig. 4.40a, $A_1 = A_2 = 30$ G; from Fig. 4.40b, $A_1 = 50$ G, $A_2 = 10$ G.

4.57 (a) $g_1 = 4.25$; $g_2 = 3.43$.

4.65 (a) Complete enrichment gives a shift to 2130 cm^{-1}

CHAPTER 5

5.1 (c) $\psi_{sp\,hybrid} = c_1\psi_{2s} + c_2\psi_{2p_x}$ and $\psi_{sp\,hybrid} = c_3\psi_{2s} - c_4\psi_{2p_x}$; for $2s$, $c_1 = c_3$ and normalization means that for $2s$: $c_1{}^2 + c_3{}^2 = 1$; since $c_1 = c_3$, $c_1 = c_3 = 1/\sqrt{2}$.

5.2 (b) Start with three equations with nine coefficients; $c_1 = c_4 = c_7$ and normalization means $c_1{}^2 + c_4{}^2 + c_7{}^2 = 1$, giving $c_1 = c_4 = c_7 = 1/\sqrt{3}$. Other values of c_n determined likewise.

5.4 (a) Diagrams should show the combinations: $(s + p_x + d_{x^2 - y^2})$; $(s - p_x + d_{x^2 - y^2})$; $(s + p_y - d_{x^2 - y^2})$; $(s - p_y - d_{x^2 - y^2})$; (b) each is 25% s, 50% p, 25% d.

5.5 (a) sp^3; (b) sp^2d; (c) sp^3; (d) sp^3; (e) sp^3d; (f) sp^3d^2; (g) sp; (h) sp^2.

5.6 (a) sp^2; (b) sp^3.

5.7 (a) Trigonal bipyramidal.

5.8 (c) $[CO_3]^{2-}$ is isoelectronic and isostructural with $[NO_3]^-$; answer should resemble worked example 5.2.

5.9 (a) Linear; (b) sp; (c) σ-bond formation using C sp and O sp^2; leaves two orthogonal $2p$ orbitals on C; form a π-bond using a $2p$ orbital on each O; (d) 2; (e) see **5A**; yes.

(5A)

5.15 $[NH_4]^+$ is isoelectronic with CH_4; the description of bonding in $[NH_4]^+$ is essentially the same as that for CH_4.

5.16 (a) Ignoring lone pairs, see **5B**; no, all 2c-2e bonds; (b) from MO diagrams: bond order in $I_2 = 1$; bond order in $[I_3]^+ = 1$ (MO diagram similar to that for H_2O); bond order in $[I_3]^- = \frac{1}{2}$ (MO diagram similar to that for XeF_2).

(5B)

5.22 (a) sp^3; (b) T_d.

5.23 (a) One $2p$ per C; (b) a_{2u}, e_g, b_{2u}.

5.24 (b) D_{3h}.

5.25 sp^2; diagram (a), π-bonding, a_2''; diagram (b), non-bonding, one of e'' set; diagram (c) C—O σ^*, a_1'.

5.27 (a) $[H_3O]^+$, C_{3v}; C_2H_4, D_{2h}; CH_2Cl_2, C_{2v}; SO_3, D_{3h}; CBr_4, T_d; $[ICl_4]^-$, D_{4h}; HCN, $C_{\infty v}$; Br_2, $D_{\infty h}$; (b) staggered; (c) a_1'.

5.28 (a) O_h; D_{3h}.

5.29 (b) C_{3v}.

5.30 (a) 19.59% by wt; (f) coordinate bond.

CHAPTER 6

6.2 (a) 12; (b) 12; (c) 8; (d) 12 (same as ccp); (e) 6.

6.3 (a) Higher temp. form is the bcc lattice; polymorphism; (b) see text for $\beta \longrightarrow \alpha$-Sn transition.

6.4 (a) $\frac{1}{n}Co_n(s) \longrightarrow Co(g)$.

6.14 (b) $-662\,kJ\,mol^{-1}$.

6.15 $\Delta_{lattice}H^\circ(298\,K) = -2050\,kJ\,mol^{-1} \approx \Delta U(0\,K)$.

6.16 (a) $609\,kJ\,mol^{-1}$; (b) $657\,kJ\,mol^{-1}$.

6.18 (a) $-621.2\,kJ\,mol^{-1}$; (b) $-632.2\,kJ\,mol^{-1}$.

6.19 Exothermic: (a); (e).

6.21 (a) 4; (b) $1.79 \times 10^{-22}\,cm^3$; (c) $2.17\,g\,cm^{-3}$.

6.22 (b) $7.29 \times 10^{-23}\,cm^3$; $6.81\,g\,cm^{-3}$; 2%.

6.23 (a) Vacant Ca^{2+} and Cl^- sites must be in $1:2$ ratio; (b) see Fig. 6.28; (c) Ag^+ and Cd^{2+} similar size; replacement of Ag^+ by Cd^{2+} gives charge imbalance countered by an extra Ag^+ vacancy.

6.25 Creation of positive hole as electron hops from Ni^{2+} to Ni^{3+}; as more Li^+ incorporated, more Ni^{3+} sites created, and conductivity increases.

6.26 For small deviations from stoichiometry, $[Ni^{2+}]$ and $[O^{2-}]$ are nearly constant, so $K = [Ni^{3+}]^4[\square_+]^2/p(O_2)$. Since $[\square_+] = \frac{1}{2}[Ni^{3+}]$, $K \propto [Ni^{3+}]^6/p(O_2)$ with conductivity $\propto [Ni^{3+}]$ and hence $\propto p(O_2)^{1/6}$.

6.29 (a) Phase change, bcc to fcc.

6.30 See Fig. 21.5; $Re = 8 \times \frac{1}{8}$; $O = 12 \times \frac{1}{4}$.

6.32 Na, metal; CdI_2, layered structure; octahedral site, 6-coordinate; Ga-doped Si, extrinsic semiconductor; Na_2S, antifluorite structure; perovskite, double oxide; CaF_2, fluorite structure; GaAs, intrinsic semiconductor; wurtzite and zinc blende, polymorphs; SnO_2, cassiterite.

CHAPTER 7

7.1 (a) 0.18; (b) 3.24×10^{-7}.

7.2 Smallest pK_a refers to loss of first proton and so on.

7.4 (b) $pK_b(1) = 3.29$; $pK_b(2) = 6.44$.

7.9 (a) Basic; (b) amphoteric; (c) acidic; (d) acidic; (e) amphoteric; (f) acidic; (g) amphoteric; (h) amphoteric.

7.11 (a) $[Ag^+][Cl^-]$; (b) $[Ca^{2+}][CO_3^{2-}]$; (c) $[Ca^{2+}][F^-]^2$.

7.12 (a) $\sqrt{K_{sp}}$; (b) $\sqrt{K_{sp}}$; (c) $\sqrt[3]{\dfrac{K_{sp}}{4}}$.

7.13 $2.40 \times 10^{-4}\,g$.

7.15 (a) 5.37×10^{-13}; (b) 1.10×10^5.

7.16 (a) F^-; $[SO_4]^{2-}$; $[Fe(OH_2)_5(OH)]^{2+}$; NH_3; (b) H_2SO_4; $[PH_4]^+$; NH_3; HOBr; (c) $[VO(OH_2)]^{2+}$ (or VO^{2+}).

7.20 (a) $1.37 \times 10^{-5}\,g$ per $100\,g\,H_2O$; (b) $2.01 \times 10^{-11}\,g$ per $100\,g$ solution.

7.25 (a) $K_2 = \dfrac{[M(OH_2)_4L_2{}^{z+}]}{[M(OH_2)_5L^{z+}][L]}$; $K_4 = \dfrac{[M(OH_2)_2L_4{}^{z+}]}{[M(OH_2)_3L_3{}^{z+}][L]}$

(b) $\beta_2 = \dfrac{[M(OH_2)_4L_2{}^{z+}]}{[M(OH_2)_6{}^{z+}][L]^2}$; $\beta_4 = \dfrac{[M(OH_2)_2L_4{}^{z+}]}{[M(OH_2)_6{}^{z+}][L]^4}$

7.26 (b) -50; -46; $-34\,kJ\,mol^{-1}$.

7.27 (a) 3; (b) 3; (c) 3; (d) 4; (e) 6.

7.28 (a) Hard Co^{3+}; hardness: O, $N > P > As$-donor; (b) hard Zn^{2+} favours complex formation with hard F^-; (c) hard Cr^{3+} combined with relatively soft P-donor gives relatively weak $Cr-P$ bonds.

7.29 (a) Soft $Pd(II)$ favours soft donor atoms; chelate effect is factor for bidentate ligands; (b) $EDTA^{4-}$ is hexadentate with hard N and O-donors, forms five chelate rings in $[M(EDTA)]^{n-}$; hard donors favour M^{3+}.

7.30 (a) H_2O can act as acid or base; (c) $2.17 \times 10^{-3}\,g$.

7.31 (a) Li^+ smallest group 1 M^+ ion with highest charge density; (b) six chelate rings; (c) $Au^+(aq) + 2[CN]^-(aq) \rightleftharpoons [Au(CN)_2]^-$; $-222\,kJ\,mol^{-1}$.

7.33 (b) $K_{sp} = 10^{-41.5}$

7.34 (a) $K_a = 4.57 \times 10^{-8}$.

7.36 (e) $K_1 = 10^7$; $K_2 = 5.0 \times 10^5$.

CHAPTER 8

8.1 (a) Ca, $+2$; O, -2; (b) H, $+1$; O, -2; (c) H, $+1$; F, -1; (d) Fe, $+2$; Cl, -1; (e) Xe, $+6$; F, -1; (f) Os, $+8$; O, -2; (g) Na, $+1$; S, $+6$; O, -2; (h) P, $+5$; O, -2; (i) Pd, $+2$; Cl, -1; (j) Cl, $+7$; O, -2; (k) Cr, $+3$; H, $+1$; O, -2.

8.2   (a) Cr, +6 to +3; (b) K, 0 to +1; (c) Fe, +3 to 0; Al, 0 to +3; (d) Mn, +7 to +4.

8.3   All redox reactions *except* for (c), (e) and (h); for redox, red = reduced, ox = oxidized: (a) N, red; Mg, ox; (b) N, ox; O, red; (d) Sb, ox; F in F_2, red; (f) C, ox; O in O_2, red; (g) Mn, red; two Cl, ox.

8.4   Changes are: (a) N, $2 \times (-3)$; Mg, $3 \times (+2)$; (b) N, $2 \times (+2)$; O, $2 \times (-2)$; (d) Sb, $+2$; F, $2 \times (-1)$; (f) C, $2 \times (+2)$; O, $2 \times (-2)$; (g) Mn, -2; Cl, $2 \times (+1)$.

8.5   (a) $2Ag^+(aq) + Zn(s) \longrightarrow 2Ag(s) + Zn^{2+}(aq)$; $E^o_{cell} = 1.56\,V$; $\Delta G^o = -301\,kJ$ per mole of reaction; (b) $Cl_2(aq) + 2Br^-(aq) \longrightarrow 2Cl^-(aq) + Br_2(aq)$; $E^o_{cell} = 0.27\,V$; $\Delta G^o = -52.1\,kJ$ per mole of reaction; (c) $[Cr_2O_7]^{2-}(aq) + 14H^+(aq) + 6Fe^{2+}(aq) \longrightarrow 2Cr^{3+}(aq) + 7H_2O(l) + 6Fe^{3+}(aq)$; $E^o_{cell} = 0.56\,V$; $\Delta G^o = -324\,kJ$ per mole of reaction.

8.7   (a) $+1.48$; (b) $+1.34$; (c) $+1.20\,V$.

8.8   (a) $1.08\,V$; (b) $-208\,kJ\,mol^{-1}$; (c) kinetically stable; additives act as catalysts.

8.9 $0.34\,V$.

8.10   (a) $+0.74\,V$; (b) less easily (ΔG^o is less negative).

8.11 $-0.15\,V$.

8.13 $K \approx 10^{39}$.

8.14   (c).

8.15 $\Delta G^o(298\,K) = 41.5\,kJ\ mol^{-1}$; disproportionation of precipitated CuCl is thermodynamically unfavourable.

8.18   (a)
$$[VO_2]^+ \xrightarrow{+0.99} [VO]^{2+} \xrightarrow{+0.34} V^{3+} \xrightarrow{-0.26} V^{2+} \xrightarrow{-1.18} V$$
(b) No species disproportionates.

8.20   (a) $1.22\,V$.

8.22   (a) $\Delta_f G^o(K^+, aq) = -282.7\,kJ\,mol^{-1}$; $\Delta_f G^o(F^-, aq) = -276.9\,kJ\,mol^{-1}$; (b) $-21.8\,kJ\,mol^{-1}$; (c) $\Delta_{sol}G^o$ is significantly negative, and so the solubility of KF in water is relatively high.

8.23 6.00×10^{-29}.

8.24   (a) $[ClO_4]^-$; (b) Cl^-.

8.26   (a) $+1.84\,V$.

8.28   (a) $-0.78\,V$; (b) $0.06\,V$.

8.29   (b) $+0.95\,V$; $+1.54\,V$; (c) $+0.59\,V$; $-56.9\,kJ\,mol^{-1}$.

8.30   (a) $\beta([Fe(phen)_3]^{3+})/\beta([Fe(phen)_3]^{2+}) = 1.2 \times 10^{-6}$; (b) $[MnO_4]^{3-}$ is unstable with respect to disproportionation.

8.35   (b) $E^o_{anode} = -1.25\,V$; $E^o_{cathode} = +0.34\,V$; (c) $1.59\,V$; $-307\,kJ\,mol^{-1}$.

8.37   (a) $0.97\,V$; $-560\,kJ\,mol^{-1}$ for the reaction $3Ag_2S + 2Al \rightleftharpoons 6Ag + 3S^{2-} + 2Al^{3+}$; (b) $Al_2S_3 + 6H_2O \longrightarrow 2Al(OH)_3 + 3H_2S$.

CHAPTER 9

9.3   Polar: (a); (b); (c); (d); (e); (f); (h); (i); (j).

9.4   (a) $2KI + Zn(NH_2)_2$; (b) $K_2[Zn(NH_2)_4]$; (c) $GeH_4 + 2MgBr_2 + 4NH_3$; (d) $[NH_4]^+ + [CH_3CO_2]^-$; (e) Na_2O_2; NaO_2; (f) $K[HC\equiv C] + NH_3$; in aqu. sol., CH_3CO_2H only *partially* dissociates.

9.5   (a) $Zn + 2NaNH_2 + 2NH_3 \longrightarrow Na_2[Zn(NH_2)_4] + H_2$
$[Zn(NH_2)_4]^{2-} + 2[NH_4]^+ \longrightarrow Zn(NH_2)_2 + 4NH_3$
$Zn(NH_2)_2 + 2NH_4I \longrightarrow [Zn(NH_3)_4]I_2$.
(b) In water: $2K + 2H_2O \longrightarrow 2KOH + H_2$; in liquid NH_3, at low concentrations: form $K^+(NH_3) + e^-(NH_3)$;
on standing, $2NH_3 + 2e^- \longrightarrow 2[NH_2]^- + H_2$.

9.6   (a) H_2NNH_2; (b) Hg_3N_2; (c) O_2NNH_2; (d) $MeNH_2$; (e) $OC(NH_2)_2$; (f) $[Cr(NH_3)_6]Cl_3$.

9.7 $AlF_3 + NaF \longrightarrow Na[AlF_4]$ (soluble in liquid HF)
$Na[AlF_4] + BF_3 \longrightarrow AlF_3$ (precipitate) $+ Na[BF_4]$.

9.8   Species formed: (a) $[ClF_2]^+ + [HF_2]^-$; (b) $[MeOH_2]^+ + [HF_2]^-$; (c) $[Et_2OH]^+ + [HF_2]^-$; (d) $Cs^+ + [HF_2]^-$; (e) $Sr^{2+} + 2[HF_2]^-$; (f) $[H_2F]^+ + [ClO_4]^-$.

9.9   (a) $H_2S_2O_7 + H_2SO_4 \longrightarrow [H_3SO_4]^+ + [HS_2O_7]^-$; (b) relatively strong acid.

9.12   (a) $Ph_2C=CH_2 + HCl \rightleftharpoons [Ph_2CCH_3]^+ + Cl^-$; equilibrium then upset by: $Cl^- + BCl_3 \longrightarrow [BCl_4]^-$ with an increase in conductivity but further addition of BCl_3 has no effect.
(b) $A = K_2SnF_6$; $x = 4$.

9.15   (a) Terminal and bridge Al–Cl are 2c-2e bonds; localized bonding; (b) $[Al_2Cl_7]^- + AlCl_3 \rightleftharpoons [Al_3Cl_{10}]^-$.

9.20   (a) BF_3; SbF_5; (b) oxidizing agent and F^- acceptor; (c) $Na + N_2O_4 \longrightarrow NO(g) + NaNO_3$.

9.21 $[\mathbf{I}]^- = [Ga(NH_2)_4]^-$; $[\mathbf{II}]^- = [Ga(NH)_2]^-$.

9.22   (a) $2SbCl_3 \rightleftharpoons [SbCl_2]^+ + [SbCl_4]^-$; (b) $AgNO_3 + NOCl \longrightarrow AgCl + N_2O_4$; (c) $Cr(NH_2)_3$, $[Cr(NH_3)_6]^{3+}$, $[Cr(NH_2)_4]^-$.

9.24   (b) Trigonal bipyramidal and square-based pyramidal; little difference in energy.

CHAPTER 10

10.2 (b)

10.3 $1:1:1$ three-line signal.

10.4 Sample contains small amounts of CD_2HCN; $^1H-^2H$ spin–spin coupling gives $1:2:3:2:1$ signal; CDH_2CN and CH_3CN present in negligible amounts.

10.5 React $D_2O + AlCl_3$ to prepare DCl; then $Li[AlH_4] + DCl$; accurate measurement of M_r, or of density of water formed on combustion.

10.6 In dilute solutions, *tert*-BuOH $\approx$ monomeric; $3610 \, cm^{-1}$ due to $\nu(OH)$; in more concentrated solutions, hydrogen-bonded association weakens covalent O—H bond; band (broad) is shifted to lower frequency.

10.7 $MCl + HCl \rightleftharpoons M[HCl_2]$ equilibrium position is governed by relative lattice energies of MCl and $M[HCl_2]$.

10.10 (a) $KH + NH_3 \longrightarrow KNH_2 + H_2$; $KH + EtOH \longrightarrow KOEt + H_2$.

10.11 (a) $2H_2O \longrightarrow 2H_2 + O_2$;
(b) $2LiH \longrightarrow 2Li + H_2$;
(c) $CaH_2 + H_2O \longrightarrow Ca(OH)_2 + H_2$;
(d) $Mg + 2HNO_3 \longrightarrow Mg(NO_3)_2 + H_2$
(e) $2H_2 + O_2 \longrightarrow 2H_2O$;
(f) $CuO + H_2 \overset{\Delta}{\longrightarrow} Cu + H_2O$.

10.12 H_2O_2 is kinetically stable.

10.13 (b) Mg: 6-coordinate; octahedral; H: 3-coordinate; trigonal planar.

10.14 Ratio coordination numbers $Al:H$ $6:2$; stoichiometry $1:3$.

10.19 (b) $-3557 \, kJ \, mol^{-1}$

10.20 (b) Symmetrical $O\cdots H\cdots O$ in $[H_5O_2]^+$ unit; four H_2O hydrogen bonded (asymmetrical interactions likely) to H atoms of central $[H_5O_2]^+$; (c) symmetric stretch IR inactive for D_{3h} XY_3, but active for C_{3v} XY_3.

10.21 (a) $-401 \, kJ \, mol^{-1}$.

10.22 (b) $SiH_4 + LiAlCl_4$; $H_2 + K[PPh_2]$; $LiAlH_4 + 3LiCl$.

10.23 BeH_2, polymeric chain; $[PtH_4]^{2-}$, square planar; NaH, saline hydride; $[NiH_4]^{4-}$, M(0); $[PtH_6]^{2-}$, M(IV); $[TcH_9]^{2-}$, tricapped trigonal prismatic; $HfH_{2.1}$, non-stoichiometric; AlH_3, 3D lattice with octahedral metals.

10.24 (a) Hydrogen-bonded, wurtzite-like structure for both; (b) viscosity decreases as number of hydrogen bonds per molecule decreases; (c) stronger hydrogen bonding in dimer in vapour phase than in liquid lowers $\Delta_{vap}S$; (d) $pK_a(2)$ for maleic acid larger because hydrogen-bonded interaction hinders H^+ dissociation:

10.26 (a) 7.47%; (b) 5.60%.

10.28 (e) Electrodeposition (electrolysis and collection at cathode).

CHAPTER 11

11.1 (b) ns^1.

11.6 (a) $^{40}_{19}K \xrightarrow{\text{electron capture}} {}^{40}_{18}Ar$; (b) $0.57 \, dm^3$; assume decay is by the pathway in (a), but see eq. 11.2.

11.8 Gives LiF and NaI.

11.9 Halide exchange between $[PtCl_4]^{2-}$ and KBr or KI.

11.12 (a) N^{3-} wholly in unit, Li^+ per unit $= 6 \times \frac{1}{3} = 2$; (b) consider both layers 1 and 2 to obtain Li_3N.

11.14 Disproportionation.

11.15 (a) $[O_2]^-$; (b) $[O_2]^{2-}$; (c) $[O_3]^-$; (d) $[N_3]^-$; (e) N^{3-}; (f) Na^-.

11.17 (a) $[C\equiv N]^-$ isoelectronic with CO; bonding as in CO (Section 2.7); (b) as for KOH (Section 11.6).

11.19 (a) $NaH + H_2O \longrightarrow NaOH + H_2$;
(b) $KOH + CH_3CO_2H \longrightarrow [CH_3CO_2]K + H_2O$;
(c) $2NaN_3 \longrightarrow 2Na + 3N_2$;
(d) $K_2O_2 + 2H_2O \longrightarrow 2KOH + H_2O_2 \longrightarrow 2KOH + H_2O + \frac{1}{2}O_2$;
(e) $NaF + BF_3 \longrightarrow Na[BF_4]$;
(f) Cathode: $K^+ + e^- \longrightarrow K$; anode: $2Br^- \longrightarrow Br_2 + 2e^-$;
(g) Cathode: $2H_2O + 2e^- \longrightarrow 2[OH]^- + H_2$; anode: $2Cl^- \longrightarrow Cl_2 + 2e^-$.

11.22 (a) Li_2CO_3; (b) Na_2O; (c) $0.0588 \, mol \, dm^{-3}$.

11.23 (a) $K_2SO_4 + H_2O$; (b) $NaHSO_3$, or $Na_2SO_3 + H_2O$; (c) $K[C_2H_5O] + H_2O$; (d) $Na[(CH_3)_2HCO] + H_2$; (e) $NaHCO_3$, or $Na_2CO_3 + H_2O$; (f) HCO_2Na; (g) $Cs_2[C_2O_4] + 2H_2O$; (h) $NaBH_4 + NaCl$.

11.24 (a) $-18 \, kJ \, mol^{-1}$; (b) NaCl.

11.25 (a) $Li_3N + 3H_2O \longrightarrow 3LiOH + NH_3$; (b) $M = Li$; $A = Li_2O$; $B = H_2$.

11.26 (a) For gas-phase species, bond order $= 0$.

11.27 (b) Soluble: $NaNO_3$; $RbNO_3$, Cs_2CO_3, Na_2SO_4, $LiCl$.

11.28 Li_3N, direct combination of elements, layer structure; $NaOH$, neutralizes HNO_3, no gas evolved; Cs, reacts explosively with H_2O; Cs_7O, suboxide; Li_2CO_3, sparingly soluble; $NaBH_4$, reducing agent; Rb_2O, basic and antifluorite structure; Li, highest IE_1 of group 1 metals.

CHAPTER 12

12.2 $Ca(OH)_2 = 1.05 \times 10^{-2}\,mol\,dm^{-3}$; $Mg(OH)_2 = 1.12 \times 10^{-4}\,mol\,dm^{-3}$; relative solubilities $= 94 : 1$.

12.3 (a) $3Mg + N_2 \xrightarrow{\Delta} Mg_3N_2$;
(b) $Mg_3N_2 + 6H_2O \longrightarrow 2NH_3 + 3Mg(OH)_2$.

12.4 (a) Mg^{2+} replace Na^+ ions, and $[C\equiv C]^{2-}$ replace Cl^- in $NaCl$ lattice; $[C\equiv C]^{2-}$ is not spherical, so elongation along one axis; (b) free rotation of $[CN]^-$ in $NaCN$ means $[CN]^-$ ion is pseudo-spherical.

12.5 (a) $[NH_4]_2[BeF_4] \xrightarrow{\Delta} BeF_2 + 2NH_4F$
(b) $2NaCl + BeCl_2 \longrightarrow Na_2[BeCl_4]$
(c) $BeF_2 \xrightarrow{water} [Be(OH_2)_4]^{2+} + 2F^-$

12.6 (a) See **12A**; sp^2; (b) See **12B**.

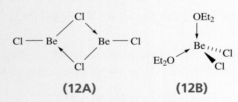

(12A)   (12B)

12.7 (a) See Fig. 6.22; (b) per unit cell, two Mg^{2+} and four F^- ions, giving $1:2$ $Mg^{2+}:F^-$ ratio.

12.9 (a) $CaCl_2$ forms a hydrate; $CaH_2 + H_2O \longrightarrow Ca(OH)_2 + H_2$.

12.10 (a) See discussion of disproportionation of CaF in Section 6.16; (b) dissolve each in dilute HCl, measure $\Delta_r H^\circ$, and apply Hess cycle.

12.11 (a) SrO_2 and H_2O_2, conjugate base and acid respectively; HCl and $SrCl_2$, conjugate acid and base respectively; (b) base + weak acid: $BaO_2 + 2H_2O \longrightarrow Ba(OH)_2 + H_2O_2$.

12.12 (a) $MO + H_2O \longrightarrow M(OH)_2$: Sr, $\Delta_r H^\circ = -81.5$; Ba, $\Delta_r H^\circ = -105.7\,kJ\,mol^{-1}$.

12.13 (a) Bubble CO_2 through limewater;

(b) $Ca(OH)_2(aq) + CO_2(g) \longrightarrow CaCO_3(s) + H_2O(l)$;
(c) white precipitate, 'milky' appearance.

12.16 A complex such as $[MgOMg]^{2+}$ or its hydrate formed in $MgCl_2(aq)$.

12.19 (b) Formation of $[Be(OH_2)_4]^{2+}$ thermodynamically favourable; (c) phase change hcp $\longrightarrow$ bcc.

12.20 (a) Antifluorite structure for Na_2S; (b) C^{2-}, N^- and O are isoelectronic; (c) formation of $[Be(OH)_4]^{2-}$; (d) high mp, stability at high temperatures.

12.21 (a) $Ca(OH)_2 + H_2$; (b) $2BeH_2 + LiCl + AlCl_3$; (c) $C_2H_2 + Ca(OH)_2$; (d) $BaSO_4 + H_2O_2$; (e) $2HF + Ca(HSO_4)_2$; (f) $MgO_2 + H_2O$; (g) $MgO + CO_2$; (h) $MgO + Mg_3N_2$.

12.22 (a) $M = Sr$; $A = [Sr(NH_3)_6]$; $B = Sr(NH_2)_2$; $C = H_2$; (b) $X = Ca$; $D = Ca(OH)_2$.

12.23 (a) $CaI_2(THF)_4$; $BaI_2(THF)_5$; $r(Ba^{2+}) > r(Ca^{2+})$; (b) sparingly soluble: $BaSO_4$, $MgCO_3$, $Mg(OH)_2$, CaF_2; soluble, no reaction: $BeCl_2$, $Mg(ClO_4)_2$, $BaCl_2$, $Ca(NO_3)_2$; react with water: $CaO \longrightarrow Ca(OH)_2$, $SrH_2 \longrightarrow Sr(OH)_2$.

CHAPTER 13

13.2

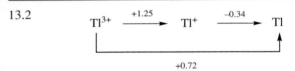

13.4 (a) $B_2O_3(s) + 3Mg(s) \xrightarrow{\Delta} 2B(s) + 3MgO(s)$
(b) Al_2O_3 is amphoteric, Fe_2O_3 is basic; only Al_2O_3 reacts, leaving solid Fe_2O_3:
$Al_2O_3(s) + 3H_2O(l) + 2NaOH(aq) \longrightarrow 2Na[Al(OH)_4](aq)$;
(c) $2Na[Al(OH)_4](aq) + CO_2(g) \longrightarrow Al_2O_3 \cdot 3H_2O(s) + Na_2CO_3(aq) + H_2O(l)$

13.5 (a) See Fig. 4.21; (b) $1:1:1:1$ multiplet; (c) doublet $[J(^{11}B-^{31}P)]$ of quartets $[J(^{11}B-^1H)]$; (d) singlet.

13.6 $\Delta_r H^\circ = -851.5\,kJ$ per mole of Fe_2O_3 (or Al_2O_3); enough energy released to melt the iron formed.

13.9 (a) $Me_3N\cdot BH_3$ forms; ^{11}B NMR spectrum of $THF\cdot BH_3$ and $Me_3N\cdot BH_3$ shows two $1:3:3:1$ quartets, at different chemical shifts; (b) no; no change in ^{11}B or ^{31}P NMR spectra; (c) yes; monitor solution by ^{11}B NMR spectroscopy; (d) formation of complex through one or two $P \longrightarrow B$ bonds; use ^{31}P or ^{11}B NMR spectroscopy.

13.11 Coupling to 12 equivalent ^{19}F; ratio $= 924 : 1$.

13.12 (a) $[B(CF_3)_4]^- + [H_3O]^+ \longrightarrow (F_3C)_3BCO + 3HF$.

13.13 (a) Attack by H_2O on larger Al (but not B) possible; (b) reaction steps are

(i) $B_2H_6 \underset{\text{fast}}{\rightleftharpoons} 2BH_3$, (ii) $BH_3 + H_2O \overset{\text{slow}}{\longrightarrow}$ products; (c) $B(OH)_3 + 2[HF_2]^- \longrightarrow [BF_4]^- + 2H_2O + [OH]^-$.

13.14 (a) $B(OEt)_3 + 3HCl$; (b) $EtOH \cdot BF_3$; (c) $B(NHPh)_3 + 3HCl$; (d) KBF_4 (ionic salt).

13.15 (a) $Na_3[AlF_6]$; (b) $CaTiO_3$; (c) rewrite $Na_3[AlF_6]$ as $Na_2[NaAlF_6] \equiv NaXF_3$; cryolite has perovskite structure with $\frac{2}{3}Na$ in Ca sites, and $Al + \frac{1}{3}Na$ in Ti sites.

13.16 (a) $[MBr_6]^{3-}$, octahedral; $[MCl_5]^{2-}$, trigonal bipyramidal; $[MBr_4]^-$, tetrahedral; (b) crystal packing effects; (c) $TlCl_3 + H_2N(CH_2)_5NH_2 + 2HCl$; $2TlCl_3 + 3CsCl$; (d) monomeric $GaCl_2$ would be paramagnetic; $Ga[GaCl_4]$ contains diamagnetic Ga^+ and $[GaCl_4]^-$ ions.

13.17 (a) $AlF_3 + 3F^- \longrightarrow [AlF_6]^{3-}$; on adding BF_3, formation of $[BF_4]^-$ causes displacement and precipitation of AlF_3. (b) Data indicate common species for $GaCl_2$ and $GaCl_3/HCl$; i.e. $[GaCl_4]^-$. (c) Solid TlI_3 is $Tl^+[I_3]^-$; hydrated Tl_2O_3 is insoluble, and oxidation of $Tl^+(aq)$ to solid Tl_2O_3 is much easier than to $Tl^{3+}(aq)$; I_2 is oxidant.

13.18 (a) At 298 K, terminal and bridging H involved in dynamic process; process persists at 203 K; (b) all ^{11}B nuclei equivalent; quintet due to coupling of ^{11}B nucleus to four equivalent ^1H nuclei (exchange of terminal and bridging H); (c) IR timescale $\neq$ NMR timescale.

13.19 C_{2h}.

13.22 (a) Cyclic $Et_3B_3O_3$; $(Me_2N)_2B-B(NMe_2)_2$; $K[SbF_6] + (C_2F_5)_3BF$.

13.24 Use localized 2c-2e bonds; coordinate $N \longrightarrow Al$ bonds.

13.25 Cyclic $[Cl_2GaP(H)Si^tBu_3]_2$.

13.26 B_5H_9, *nido*-cage, square-based pyramid with four bridging H; $[B_8H_8]^{2-}$, *closo*-dodecahedron; $C_2B_{10}H_{12}$, *closo*-icosahedron; *nido*-$[B_6H_9]^-$, pentagonal pyramid with three bridging H atoms; $C_2B_{10}H_{12}$ could have C atoms adjacent (1,2-isomer), or apart (1,7- and 1,12-isomers).

13.27 (a) Adding two electrons means parent deltahedron changes from $n = 6$ (for B_5H_9) to $n = 7$ (for B_5H_{11}); predict a change from *nido* to *arachno*. (b) Anion is dynamic in solution, all eight H equivalent and 'see' every B.

13.28 (a) 1-BrB_5H_8, isomerizing to 2-BrB_5H_8; (b) $B_4H_8(PF_3) + H_2$; (c) $K[1-BrB_5H_7] + H_2$; (d) $4B(OR)_3 + MeB(OR)_2 + 11H_2$ (the B$-$C bond is not hydrolysed).

13.30 (a) $Ga^+ + [I_3]^- \longrightarrow Ga^{3+} + 3I^-$;
$Ga^+ + Br_2 \longrightarrow Ga^{3+} + 2Br^-$;
$Ga^+ + 2[Fe(CN)_6]^{3-} \longrightarrow Ga^{3+} + 2[Fe(CN)_6]^{4-}$;
$Ga^+ + 2[Fe(bpy)_3]^{3+} \longrightarrow Ga^{3+} + 2[Fe(bpy)_3]^{2+}$;
(b) $[Tl(CN)_4]^-$, $Tl(CN)_3$.

13.31 (a) Al, $\approx 82\,000$ ppm; Mg, $\approx 24\,000$ ppm; (b) oxidation: H, $8 \times (-1$ to $0)$; reduction: Ga, $2 \times (+3$ to $0)$, $1 \times (+3$ to $+1)$.

13.32 (a)

(b) $\mathbf{A} = (Cl_2B)_3BCO$.

13.33 (b) N,N',N'',S,S',S''- and N,N',N'',N''',O,O',O''-donors.

13.34 (a) At 223 K, static structure, six BH_{term} and one μ_3-H over a B_3-face; capping H fluxional over B_6-cage at 297 K but no exchange with H_{term}; $J(BH_{term}) \gg J(BH_{cap})$; (b) $\mathbf{X} = [NH_4][GaF_4]$.

CHAPTER 14

14.4 $[C_{60}]^{n-}$: $(8 \times \frac{1}{8}) + (6 \times \frac{1}{2}) = 4$; K^+: $9 + (12 \times \frac{1}{4}) = 12$; $[C_{60}]^{n-} : K^+ = 1:3$.

14.6 (a) Mg_2C_3 and CaC_2 contain $[C=C=C]^{4-}$ and $[C\equiv C]^{2-}$ ions respectively; ThC_2 contains $[C_2]^{4-}$; TiC is an interstitial carbide; (b) $[NH_4]Br$ acts as an acid in liquid NH_3; (c) Si$-$H (or Si$-$D) is not broken in rate-determining step, and presumably $[OH]^-$ attacks Si.

14.8 (a) Linear; (b) linear; (c); trigonal pyramidal; (d) trigonal bipyramidal; (e) tetrahedral at Si; bent at O; (f) octahedral; (g) octahedral; (h) tetrahedral.

14.9 (a) $[Sn_9Tl]^{3-}$ possesses 11 pairs of electrons for cluster bonding; *closo* cage; (b) two isomers because Tl could occupy one of the two different sites.

14.11 (a) $GeCl_4 + 2H_2O \longrightarrow GeO_2 + 4HCl$; GeO_2 is dimorphic, rutile and quartz forms;

(b) $SiCl_4 + 4NaOH \longrightarrow Na_4SiO_4 + 4HCl$; discrete $[SiO_4]^{4-}$ ion not present, polymeric species;
(c) $CsF + GeF_2 \longrightarrow Cs[GeF_3]$; trigonal pyramidal $[GeF_3]^-$ ions;
(d) $2SiH_3Cl + H_2O \longrightarrow (H_3Si)_2O + 2HCl$;
(e) $2SiF_4 + 4H_2O \longrightarrow SiO_2 + 2[H_3O]^+ + [SiF_6]^{2-} + 2HF$; octahedral $[SiF_6]^{2-}$;
(f) $2[Bu_4P]Cl + SnCl_4 \longrightarrow [Bu_4P]_2[SiCl_6]$; octahedral $[SiCl_6]^{2-}$.

14.12 Splittings are due to $^{119}Sn-^{19}F$ couplings; each species is octahedral: (a) single F; environment; (b) *trans*- and *cis*-isomers both have two equivalent F sites; (c) A = *mer*-isomer with two F sites (1:2); B = *fac*-isomer with three equivalent F; (d) A = *trans*-Cl giving four equivalent F; B = *cis*-Cl, giving two F environments (2:2); (e) two F environments (1:4); (f) six equivalent F.

14.13 (a) $[Sn(OH)_6]^{2-} + H_2$; (b) $PbSO_4$; (c) Na_2CS_3; (d) $-SiH_2O-$ polymers; (e) $ClCH_2SiH_3$.

14.14 (a) Dissolve each in conc HF(aq), measure $\Delta_r H^o$, and apply Hess's law; (b) Si–Si and Si–H bond energies from $\Delta_c H^o$ for Si_2H_6 and SiH_4; apply Pauling relationship; (c) determine Pb(IV) by allowing it to oxidize I^- and titrating I_2 formed with thiosulfate (or heat with HCl, pass Cl_2 into KI(aq), and titrate I_2 formed against thiosulfate).

14.15 At 1000 K, CO is more thermodynamically stable than SnO_2; C reduces SnO_2 at 1000 K but not at 500 or 750 K.

14.16 (a) Fe^{2+} replaces Mg^{2+} with no structural change (r_{ion}, see Appendix 6); (b) see Fig. 14.22 to see that Al^{3+} can replace Si^{4+} with electrical neutrality conserved by Ca^{2+} replacing Na^+; (c) Al^{3+} can replace Si^{4+} in silica structure with interstitial Li^+ maintaining electrical neutrality.

14.17 $I = (CN)_2$; $II = CS_2$; $III = CO_2$; all $D_{\infty h}$.

14.18 KCN(aq) is very alkaline owing to hydrolysis; $[CN]^-$ competes unsuccessfully with $[OH]^-$ for Al^{3+}.

14.19 (a) $NH_3 + H_2CO_3$; then forms $CO_2 + H_2O$; (b) same as (a); (c) $NH_3 + H_2CO_2S$; then forms $OCS + H_2O$.

14.20 (a) Trigonal planar (D_{3h}); (c) IR active: A_2'', E'.

14.21 (a) T_d, (b) D_{3h}, (c) $D_{\infty h}$, (d) C_{2v}.

14.23 (b) Linear; (c) Sn_4F_4-ring, each Sn has a lone pair; localized Sn–F single bonds.

14.24 SiF_4, gas, tetrahedral molecules; Si, semiconductor; Cs_3C_{60}, superconducting at 40 K; SnO, amphoteric; $[Ge_9]^{4-}$, Zintl ion; GeF_2, carbene analogue; $[SiO_4]^{4-}$, Ca^{2+} salt is component of cement; PbO_2, acidic oxide; $Pb(NO_3)_2$, water-soluble salt not decomposed; SnF_4, sheet structure, octahedral Sn.

14.25 (b) $Pb(NO_3)_2$, $PbCl_2$, $Pb(O_2CCH_3)_2$; (c) $230\,cm^{-1}$, bending mode; $1917\,cm^{-1}$, $\nu(CN)$; $1060\,cm^{-1}$, $\nu(CCl)$.

14.26 (a) $NaCl + H_3GeOCH_3$; (b) $CaNCN + C$; (c) $Mg(OH)_2 + SiH_4$ + higher silanes; (d) $KF + Si$; (e) $[Ge(1,2-O_2C_6H_4)_3]^{2-}$; (f) $2SiH_3I + O_2$; (g) see eq. 14.13; (h) $Na_2[Sn(OH)_6]$.

14.27 (c) $[C_2O_4]^{2-}$, D_{2h}; $[C_2S_4]^{2-}$, D_{2d}.

14.28 (a) Formation of $[K(crypt-222)]^+$; (c) coupling to 12 ^{207}Pb (22.1%, $I = \frac{1}{2}$); 1720 Hz.

CHAPTER 15

15.1 (a) 0; (b) +5; (c) +3; (d) +4; (e) +2; (f) −3; (g) −1; (h) 0; (i) +5; (j) +3; (k) +5.

15.2 (a) +932; (b) −274; (c) −450 kJ per mole of reaction.

15.4 (a) $Ca_3P_2 + 6H_2O \longrightarrow 3Ca(OH)_2 + 2PH_3$;
(b) $NaOH + NH_4Cl \longrightarrow NaCl + NH_3 + H_2O$;
(c) $Mg(NO_3)_2 + 2NH_3 + 2H_2O \longrightarrow Mg(OH)_2(s) + 2NH_4NO_3$;
(d) $AsH_3 + 4I_2 + 4H_2O \longrightarrow H_3AsO_4 + 8HI$;
(e) $PH_3 + KNH_2 \xrightarrow{\text{liquid } NH_3} KPH_2 + NH_3$.

15.5 (a) HCl(aq) is fully ionized; solutions of NH_3 contain dissolved NH_3; (b) $[NH_4][NH_2CO_2]$ is salt of very weak acid.

15.7 $HNO_3(aq) + 6H^+(aq) + 6e^- \rightleftharpoons NH_2OH(aq) + 2H_2O(l)$
$[BrO_3]^-(aq) + 6H^+(aq) + 6e^- \rightleftharpoons Br^-(aq) + 3H_2O(l)$

$NH_2OH(aq) + [BrO_3]^-(aq) \longrightarrow HNO_3(aq) + Br^-(aq) + H_2O(l)$.

15.8 (a) $3NaNH_2 + NaNO_3 \longrightarrow NaN_3 + 3NaOH + NH_3$;
(b) Na with liquid NH_3;
(c) $2NaN_3 + Pb(NO_3)_2 \longrightarrow Pb(N_3)_2 + 2NaNO_3$.

15.9 (a) Species include $[CN_2]^{2-}$, $[NO_2]^+$, $[NCO]^-$; (b) bonding scheme similar to that for CO_2.

15.10 (b) Unit cell contains two complete As, and $[(4 \times \frac{1}{4}) + (8 \times \frac{1}{8})]$ Ni = 2 Ni; i.e. 1:1.

15.11 (a) Electron diffraction, or vibrational spectroscopy; (b) Raman (not IR) spectroscopy.

15.12 $F_5S^{\bullet}$, $Cl^{\bullet}$, $ON^{\bullet}$.

15.13 (b) Assume *spherical* $[PCl_4]^+$ and $[PCl_6]^-$ ions.

15.14 Each ion contains equivalent F centres: (a) doublet (coupling to ^{31}P); (b) 1:1:1:1:1:1 signal (coupling to ^{121}Sb) superimposed on a 1:1:1:1:1:1:1:1:1:1

signal (coupling to ^{123}Sb), relative abundances ^{121}Sb:^{123}Sb $\approx 1:1$.

15.15 Eq. 15.64 (Cu oxidized, N reduced); 15.73 (N oxidized, Co reduced); 15.111 (N in one HNO_2 oxidized, N in two HNO_2 reduced); 15.123 (Au oxidized, N in HNO_3 reduced).

15.16 Assume static structures: (a) *cis*-$[PF_4(CN)_2]^-$, triplet of triplets; *trans*-$[PF_4(CN)_2]^-$, quintet; (b) *mer*-$[PF_3(CN)_3]^-$, doublet of triplets; *fac*-$[PF_3(CN)_3]^-$, quartet.

15.17 Three isomers; two F environments (2 : 1) in each isomer.

15.18 (a) $[PCl_4]^+[SbCl_6]^-$; (b) $K^+[AsF_6]^-$; (c) $[NO]^+[SbF_6]^-$; (d) $[H_2F]^+[SbF_6]^-$; tendency for Sb–F–Sb bridge formation may give $[Sb_2F_{11}]^-$ or higher association.

15.19 (a) See Fig. 15.13b and **15.41**; $[Sb_2F_{11}]^-$, no lone pairs, 12 electrons in valence shell of each Sb(V) centre; $[Sb_2F_7]^-$, one lone pair and four bonding pairs per Sb(III) gives trigonal bipyramidal arrangements with equatorial lone pairs; (b) chains with octahedral Bi(III).

15.20 $[NO]^+$ is isoelectronic with CO and MO diagram is similar; NO has one more electron than $[NO]^+$ and this occupies a π^*-MO; frequency of vibration depends on force constant, which increases as bond strengthens.

15.21 **B** = N_2O.

15.23 (a) Triple-rutile lattice; (b) need three rutile-type unit cells to give unambiguous description of lattice; (c) O: 3-coordinate; Fe: 6-coordinate; Sb: 6-coordinate; (d) Fe: one central + eight corners = 2 Fe; Sb: two central + eight edge = 4 Sb; O: ten central + four face = 12 O; stoichiometry = $2:4:12 = 1:2:6$.

15.24 (a) $[P_3O_{10}]^{5-}$ gives two signals in ^{31}P NMR spectrum (rel. integrals 2:1); $[P_4O_{13}]^{6-}$ gives two signals of equal integral. (b) AsF_5 is isostructural with PF_5, **15.32**; two ^{19}F peaks of relative integrals 3:2 (eq:ax) will coalesce at a higher temperature if rapid exchange occurs. (c) Refer to Fig. 15.23a, replacing three Cl by NMe_2; three possible isomers giving one, two or three signals in the ^{1}H NMR spectrum; or use ^{31}P NMR.

15.25 (a) If 2Ti(III) $\longrightarrow$ 2Ti(IV), change in oxidation state for N is -1 to -3; product is NH_3. (b) If 2Ag(I) $\longrightarrow$ 2Ag(0), change in oxidation state for P is $+3$ to $+5$; product is $[PO_4]^{3-}$. (c) If

2I(0) $\longrightarrow$ 2I(-1) twice, change in oxidation state for P is $+1$ to $+3$ to $+5$, i.e. $H_3PO_2 \longrightarrow H_3PO_3 \longrightarrow H_3PO_4$.

15.26 (a) Tetrahedral; (b) planar; (c) trigonal pyramidal at N, bent at O; (d) tetrahedral; (e) trigonal bipyramidal with axial F atoms.

15.27 (a) $K^{15}NO_3$ + Al, NaOH(aq) $\longrightarrow$ $^{15}NH_3$; pass over Na; (b) oxidize $^{15}NH_3$ [see part (a)] with CuO or NaOCl; (c) $K^{15}NO_3$ + Hg, H_2SO_4 $\longrightarrow$ ^{15}NO; combine with Cl_2, $AlCl_3$.

15.28 (a) Reduce to $^{32}P_4$; treat with NaOH(aq); (b) $^{32}P_4$ [see part (a)] + limited Cl_2; hydrolyse the product; (c) $^{32}P_4$ [see part (a)] + excess S to $^{32}P_4S_{10}$; treat with Na_2S.

15.29 **D** = N_2.

15.30 Combination of Al + P is isoelectronic with Si + Si.

15.32 (a) $J(^{11}B–^{31}P)$; ^{31}P, $I = \frac{1}{2}$; ^{11}B, $I = \frac{3}{2}$.

15.33 (a) $[PI_4]^+[GaBr_4]^-$; (b) $[P(OH)Br_3]^+[AsF_6]^-$; (c) $2PbO + 4NO_2 + O_2$; (d) $K[PH_2] + H_2$; (e) $NH_3 + 3LiOH$; (f) $H_3AsO_3 + H_2SO_4$; (g) $BiOCl + 2HCl$; (h) $H_3PO_3 + 3HCl$.

15.34 (b) Bi behaves as a typical metal; (c) $[\textbf{X}]^- = [Fe(NO_3)_4]^-$.

15.35 (a) Doublet (939 Hz) of doublets (731 Hz) of quintets (817 Hz); (b) $[BiF_7]^{2-}$ as expected from VSEPR; (b) $[SbF_6]^{3-}$ must have stereochemically inactive lone pair.

15.36 (b) **A** = $AsOCl_3$; C_{3v} consistent with monomer; C_{2h} consistent with dimer (structure **15.37**).

15.37 (a) Decomposition to NO_2.

15.38 (b) -1332 kJ mol^{-1}.

15.39 (b) $PO(NMe_2)_3$; $6Me_2NH + POCl_3 \longrightarrow PO(NMe_2)_3 + 3Me_2NH_2Cl$.

15.40 PCl_3F_2, non-polar; PCl_2F_3, polar; $PClF_4$, polar.

15.41 (a) $OC(NH_2)_2 + 2HNO_2 \longrightarrow 2N_2 + CO_2 + 3H_2O$; (b) $[NO_2]^- + H_2NSO_3H \longrightarrow N_2 + [HSO_4]^- + H_2O$.

15.42 (a) $12P + 10KClO_3 \longrightarrow 3P_4O_{10} + 10KCl$; (b) $P_4S_3 + 5KClO_3 + \frac{1}{2}O_2 \longrightarrow P_4O_{10} + 3SO_2 + 5KCl$.

15.43 (b) **A** = $\{N=P(OCH_2CF_3)_2\}_n$; **B** = $\{N=P(NHMe)_2\}_n$; **C** = $\{N=P(NHCH_2CO_2Et)_2\}_n$.

CHAPTER 16

16.1 (b) ns^2np^4.

16.2 $^{209}_{83}$Bi(n,γ)$^{210}_{83}$Bi $\xrightarrow{\beta^-}$ $^{210}_{84}$Po.

16.3   Anode: $4[OH]^-(aq) \longrightarrow O_2(g) + 2H_2O(l) + 4e^-$; cathode: $2H^+(aq) + 2e^- \longrightarrow H_2(g)$.

16.4 $8E \longrightarrow 4E_2$: $\Delta_r H^o = -1992\,kJ\,mol^{-1}$ for $E = O$, and $-1708\,kJ\,mol^{-1}$ for $E = S$; $8E \longrightarrow E_8$: $\Delta_r H^o = -1168\,kJ\,mol^{-1}$ for $E = O$, and $-2128\,kJ\,mol^{-1}$ for $E = S$.

16.6   (a) $E^o_{cell} = 1.08\,V$, so $\Delta_r G^o$ is negative; (b) $59.9\,g\,dm^{-3}$.

16.7   (a) $2Ce^{4+} + H_2O_2 \longrightarrow 2Ce^{3+} + O_2 + 2H^+$; (b) $2I^- + H_2O_2 + 2H^+ \longrightarrow I_2 + 2H_2O$.

16.8   (a) $Mn(OH)_2 + H_2O_2 \longrightarrow MnO_2 + 2H_2O$; (b) MnO_2 will catalyse decomposition of H_2O_2: $2H_2O_2 \longrightarrow 2H_2O + O_2$.

16.9   Helical chains are chiral.

16.11   (a) Bent; (b) trigonal pyramidal; (c) bent; (d) see-saw; (e) octahedral; (f) bent at each S (two isomers).

16.12   (a) SF_4 is an F^- donor or acceptor; BF_3 is an F^- acceptor; CsF is source of F^-; (b) gives RCF_3.

16.16 $[TeF_7]^-$ is pentagonal bipyramidal; binomial octet in ^{125}Te NMR spectrum means it is fluxional on NMR timescale; ^{19}F NMR spectrum, singlet for F atoms attached to non-spin active Te; 0.9% ^{123}Te and 7.0% ^{125}Te couple to give two doublets, i.e. satellites.

16.17   (a) All isoelectronic and isostructural; (b) isoelectronic: CO_2, SiO_2 and $[NO_2]^+$; isostructural CO_2 and $[NO_2]^+$; isoelectronic SO_2 and TeO_2, but not isostructural; (c) all isoelectronic, but only SO_3 and $[PO_3]^-$ are isostructural; (d) all isoelectronic and isostructural.

16.18   (a) SO_3, trigonal planar; $[SO_3]^{2-}$, trigonal pyramidal.

16.20   (a) Reaction required is:
$[SO_4]^{2-} + 8H^+ + 8e^- \rightleftharpoons S^{2-} + 2H_2O$; this is assisted by very high $[H^+]$ and very low solubility of CuS. (b) Expected from VSEPR. (c) White precipitate is $Ag_2S_2O_3$, dissolves forming $[Ag(S_2O_3)_3]^{5-}$; disproportionation of $[S_2O_3]^{2-}$

$[S_2O_3]^{2-} + H_2O \longrightarrow S^{2-} + [SO_4]^{2-} + 2H^+$

brought about by removal of S^{2-} as insoluble Ag_2S.

16.21   (a) $[S_2O_4]^{2-} + 2Ag^+ + H_2O \longrightarrow [S_2O_5]^{2-} + 2Ag + 2H^+$; (b) $[S_2O_4]^{2-} + 3I_2 + 4H_2O \longrightarrow 2[SO_4]^{2-} + 6I^- + 8H^+$.

16.22 $SO_2(OH)(NH_2)$.

16.24 S_2O, **16.42**; $[S_2O_3]^{2-}$, **16.59**; NSF, **16.65**; NSF$_3$, **16.66**; $[NS_2]^+$, **16.73**; S_2N_2, **16.71**.

16.25   Planar; 6π-electron, $(4n+2)$ Hückel system.

16.26 S_∞, chiral polymer; $[S_2O_8]^{2-}$, strong oxidizing agent; $[S_2]^-$, blue, paramagnetic; S_2F_2, two monomeric isomers; Na_2O, antifluorite structure; $[S_2O_6]^{2-}$, contains weak S–S bond; PbS, black, insoluble solid; H_2O_2, disproportionates in presence of Mn^{2+}; HSO_3Cl, explosive with H_2O; $[S_2O_3]^{2-}$, strong reducing agent; H_2S, toxic gas; SeO_3, tetramer in solid.

16.27   (a) CuS ppt; forms soluble $Na_2[CuS_2]$; (b) $H_2O + SO_2 \longrightarrow H_2SO_3$; $SO_2 + H_2SO_3 + 2CsN_3 \longrightarrow Cs_2S_2O_5 + 2HN_3$.

16.28   (a) $[SF_3]^+[SbF_6]^-$; (b) HSO_3F; (c) $2NaCl + H_2S_4$; (d) $[HSO_4]^- + 2I^- + 2H^+$; (e) $NSF + Cs[AsF_6]$; (f) $H_2SO_5 + HCl$; (g) $SO_2 + [SO_4]^{2-}$.

16.29   (b) See Fig. 10.8 and discussion; (c) Al_2Se_3, SF_4, SeO_2; kinetically stable: SF_6.

16.30   (a) Planar; (b) $d(Se–Se) < 2r_{cov}$; suggests some π-character.

16.31   (b) Formation of three products, $TeF_{4-n}(CN)_n$ with $n = 3$, 2 and 0.

16.33   (a) Bent; (c) 3; (d) $2335\,cm^{-1}$.

16.34   (a) $CaCO_3$ (or $MgCO_3$) $+ H_2SO_4 \longrightarrow CaSO_4$ (or $MgSO_4$) $+ CO_2 + H_2O$.

CHAPTER 17

17.2   (a) $2X^- + Cl_2 \longrightarrow X_2 + 2Cl^-$ ($X = Br$ or I); (b) see scheme for the Downs process in Section 10.2; to prevent recombination of Na and Cl_2; (c) $F_2 + H_2 \longrightarrow 2HF$; explosive chain reaction.

17.3   Lone pair–lone pair repulsions between O and F weaken bond.

17.6   ClF, 170; BrF, 185; BrCl, 213; ICl, 232; IBr, 247 pm; agreement with Table 17.3 good where $[\chi^P(Y) - \chi^P(X)]$ is small.

17.7   (a) $2AgCl + 2ClF_3 \longrightarrow 2AgF_2 + Cl_2 + 2ClF$ (AgF$_2$, not AgF, because ClF_3 is a very strong oxidizing agent); (b) $2ClF + BF_3 \longrightarrow [Cl_2F]^+[BF_4]^-$; (c) $CsF + IF_5 \longrightarrow Cs^+[IF_6]^-$; (d) $SbF_5 + ClF_5 \longrightarrow [ClF_4]^+[SbF_6]^-$ or $2SbF_5 + ClF_5 \longrightarrow [ClF_4]^+[Sb_2F_{11}]^-$; (e) $Me_4NF + IF_7 \longrightarrow [Me_4N]^+[IF_8]^-$; (f) $K[BrF_4] \xrightarrow{\Delta} KF + BrF_3$

17.9   (a) Square planar; (b) bent; (c) see-saw; (d) pentagonal bipyramidal; (e) planar (see **17.8**); (f) octahedral; (g) square-based pyramidal.

17.10 (a) BrF_5: doublet and quintet (J_{FF}), rel. int. 4:1; $[IF_6]^+$: singlet; (b) BrF_5 likely to be fluxional, high-temperature limiting spectrum is singlet; $[IF_6]^+$: singlet at all temperatures.

17.12 368 nm (UV), charge transfer band; 515 nm (visible), $\sigma^* \longleftarrow \pi^*(I_2)$.

17.13 (a) Charge transfer complex with S---I–I interaction; (b) 1:1; Beer–Lambert Law, and Job's method; (c) transfer of charge weakens the I–I bond.

17.14 (a) See-saw; (b) see **17.28**, (c) bent, (d) square-based pyramid.

17.15 (a) In cold alkali:
$Cl_2 + 2NaOH \longrightarrow NaCl + NaOCl + H_2O$;
in hot alkali:
$3Cl_2 + 6NaOH \longrightarrow NaClO_3 + 5NaCl + 3H_2O$;
(b) $[IO_4]^- + 2I^- + H_2O \longrightarrow [IO_3]^- + I_2 + 2[OH]^-$;
$[IO_3]^- + 5I^- + 6H^+ \longrightarrow 3I_2 + 3H_2O$;
(c) $[IO_4]^-$.

17.17 (a) $[ClO_3]^- + 6Fe^{2+} + 6H^+ \longrightarrow Cl^- + 6Fe^{3+} + 3H_2O$;
(b) $[IO_3]^- + 3[SO_3]^{2-} \longrightarrow I^- + 3[SO_4]^{2-}$
(partial reduction also possible);
(c) $[IO_3]^- + 5Br^- + 6H^+ \longrightarrow 2Br_2 + IBr + 3H_2O$.

17.18 (a) Determine total chlorine by addition of excess of I^- and titration with thiosulfate; only HCl is a strong acid so concentration can be determined by pH measurement. $\Delta_r H^\circ$ found by measuring K at different temperatures.
(b) Neutralize solution of weighed amount of oxide with $NaHCO_3$ and titrate I_2 against thiosulfate; add excess dilute HCl and titrate again.
(c) Raman spectroscopy to find stretching frequency, that of $[Cl_2]^- < Cl_2$.

17.19 (a) HF vapour is polymeric, hydrogen bonds not broken on vaporization; those in H_2O are.
(b) Iodide complex with Ag^+ must be more stable than chloride complex.

17.20 (a) N–H$\cdots$F hydrogen bond formation; structure similar to that of ice. (b) For the product HX, HI has weakest bond.

17.21 (b) Cl^-; (c) $[ClO_4]^-$; (d) if H^+ is involved in the half-equation, E depends on $[H^+]$ (Nernst equation).

17.22 (a) $10CsF + I_2O_5 + 3IF_5 \longrightarrow 5Cs_2IOF_5$;
$5CsF + I_2O_5 + 3IF_5 \longrightarrow 5CsIOF_4$; not redox;
(b) $-150\,kJ\,mol^{-1}$.

17.23 (b) Interactions involving $\pi_g^*(2p_x)^1 \pi_g^*(2p_y)^1$ of O_2 and $\pi_g^*(3p_x)^2 \pi_g^*(3p_y)^1$ of $[Cl_2]^+$ give in-plane (σ-type) and out-of-plane (π-type) bonding interactions.

17.24 (b)

17.25 $HClO_4$, strong acid; CaF_2, prototype structure; I_2O_5, anhydride of HIO_3; ClO_2, radical; $[BrF_6]^+$, requires powerful fluorinating agent; $[IF_6]^-$, distorted octahedral; HOCl, weak acid; $C_6H_6 \cdot Br_2$, charge transfer complex; ClF_3, used to fluorinate uranium; RbCl, solid contains octahedral chloride; I_2Cl_6, halogen in square planar environment.

CHAPTER 18

18.2 He_2, $\sigma(1s)^2\sigma^*(1s)^2$; $[He_2]^+$, $\sigma(1s)^2\sigma^*(1s)^1$.

18.3 Linear XeF_2; square planar XeF_4; distorted octahedral XeF_6.

18.4 Eight bonding pairs and one lone pair; stereochemically inactive lone pair.

18.5 (a) From hydrolysis of XeF_2; $\Delta_f H^\circ$(HF, 298 K) is known. (b) Use thermochemical cycle relating $[XeF_2(s)]$, $[XeF_2(g)]$, $[Xe(g) + 2F(g)]$, $[Xe(g) + F_2(g)]$.

18.6 Consider $Xe + Cl_2 \longrightarrow XeCl_2$ versus F analogue; weaker Xe–Cl than Xe–F bond; stronger Cl–Cl than F–F bond.

18.7 From Born–Haber cycle assuming lattice energies of XeF and CsF $\approx$ equal.

18.8 $[XeO_6]^{4-}$, octahedral; $XeOF_2$, T-shaped; $XeOF_4$, square pyramidal (O apical); XeO_2F_2, see-saw; XeO_2F_4, octahedral; XeO_3F_2, trigonal bipyramidal (axial F).

18.9 (a) $CsF + XeF_4 \longrightarrow Cs[XeF_5]$;
(b) $SiO_2 + 2XeOF_4 \longrightarrow SiF_4 + 2XeO_2F_2$
or: $SiO_2 + XeOF_4 \longrightarrow SiF_4 + XeO_3$;
(c) $XeF_2 + SbF_5 \longrightarrow [XeF][SbF_6]$
or: $2XeF_2 + SbF_5 \longrightarrow [Xe_2F_3][SbF_6]$
or: $XeF_2 + 2SbF_5 \longrightarrow [XeF][Sb_2F_{11}]$;
(d) $2XeF_6 + 16[OH]^- \longrightarrow [XeO_6]^{4-} + Xe + O_2 + 8H_2O + 12F^-$;
(e) $2KrF_2 + 2H_2O \longrightarrow 2Kr + O_2 + 4HF$.

18.11 (a) Product is $[XeF]^+[RuF_6]^-$; band at $600\,cm^{-1}$ arises from ν(Xe–F); (b) $[XeF_5][RuF_6]$ with Xe---F–Ru interaction.

18.12 $\mathbf{A} = [F_2C=CXeCl]^+[BF_4]^-$.

18.14 (a) Doublet assigned to two F_{term}; triplet due to F_{bridge}; $J(F_{term}-F_{bridge})$.

18.15 (a) $[KrF][AuF_6] + Kr$; (b) $Rb[HXeO_4]$;
(c) $Xe + Cl_2 + [XeF][Sb_2F_{11}] + SbF_5$;
(d) $Kr(OTeF_5)_2 + BF_3$;
(e) $[C_6F_5Xe]^+[CF_3SO_3]^- + Me_3SiF$;
(f) $[C_6F_5Xe]^+ + C_6F_5IF_2$.

18.17 KrF_2 $D_{\infty h}$; symmetric stretch is IR inactive.

18.18 XeF_2, 3c-2e interaction, Xe–F bond order $= \frac{1}{2}$; $[XeF]^+$, σ-bonding MO, Xe–F bond order $= 1$.

CHAPTER 19

19.3 Trend in E^o values irregular across period; variation in ionization energies is not enough to account for variation in E^o.

19.6 (a) Ions generally too small; (b) charge distribution; (c) oxidizing power of O and F; (apply electroneutrality principle in b and c).

19.7 (a) +2; d^5; (b) +2; d^6; (c) +3; d^6; (d) +7; d^0; (e) +2; d^8; (f) +3; d^1; (g) +3; d^2; (h) +3; d^3.

19.8 (a) Linear; (b) trigonal planar; (c) tetrahedral; (d) trigonal bipyramidal *or* square-based pyramidal; (e) octahedral.

19.10 (a) Two, axial (2 C) and equatorial (3 C); (b) low-energy fluxional process; Berry pseudo-rotation.

19.12 Tripodal ligand; trigonal bipyramidal with central N of ligand and Cl in axial sites.

19.13 (a) Aqueous solutions of $BaCl_2$ and $[Co(NH_3)_5Br][SO_4]$ give $BaSO_4$ ppt; aqueous solutions of $AgNO_3$ and $[Co(NH_3)_5(SO_4)]Br$ give $AgBr$ ppt; only free ion can be precipitated; (b) needs quantitative precipitation of free Cl^- by $AgNO_3$; (c) Co(III) salts are ionization isomers; Cr(III) salts are hydration isomers; (d) *trans*- and *cis*-$[CrCl_2(OH_2)_4]$.

19.14 (a) $[Co(bpy)_2(CN)_2]^+[Fe(bpy)(CN)_4]^-$; $[Fe(bpy)_2(CN)_2]^+[Co(bpy)(CN)_4]^-$; $[Fe(bpy)_3]^{3+}[Co(CN)_6]^{3-}$; (b) *trans*- and *cis*-$[Co(bpy)_2(CN)_2]^+$, and *cis*-$[Co(bpy)_2(CN)_2]^+$ has optical isomers; similarly for $[Fe(bpy)_2(CN)_2]^+$; $[Fe(bpy)_3]^{3+}$ has optical isomers.

19.15 Ignoring conformations of the chelate rings: (a) four depending on orientations of the Me groups; (b) two.

19.16 8; Δ metal configuration with $(\delta\delta\delta)$, $(\delta\delta\lambda)$, $(\delta\lambda\lambda)$ or $(\lambda\lambda\lambda)$; similarly for Λ. All are related as diastereoisomers except those in which every chiral centre has changed configuration, e.g. Δ-$(\delta\delta\lambda)$ and Λ-$(\lambda\lambda\delta)$.

19.17 (a) Optical; (b) geometrical (*cis* and *trans*), and the *cis*-isomer has optical isomers; (c) geometrical (*trans* and *cis*) as square planar; (d) no isomers; *cis* arrangement; (e) geometrical (*trans* and *cis*); *cis* isomer has optical isomers.

19.18 (a) IR spectroscopy; (b) as for (a); ^{195}Pt is NMR active and ^{31}P NMR spectra of the *cis*- and *trans*-isomers show satellites with J_{PtP} *cis* > *trans*; (c) ^{31}P NMR spectroscopy, *fac*-isomer has one P environment, *mer*-isomer has two; Rh is spin-active, observe doublet for *fac* (J_{RhP}); for *mer*-isomer, observe doublet of triplets (J_{RhP} and $J_{PP'}$) and doublet of doublets (J_{RhP} and $J_{PP'}$) with relative integrals 1:2.

19.19 All octahedral; (a) *mer* and *fac*; (b) *cis* and *trans*, plus enantiomers for *cis*-isomer; (c) only *mer*-isomer.

19.20 (b) Enantiomers; (c) **A** = *mer*-$[CoL_3]$; **B** = *fac*-$[CoL_3]$.

19.21 *trans*-$[RuCl_2(dppb)(phen)]$ forms; it slowly converts to *cis*-$[RuCl_2(dppb)(phen)]$.

19.22 *cis*-$[PdBr_2(NH_3)_2]$ (square planar) has two IR active Pd–N stretching modes; in *trans*-$[PdBr_2(NH_3)_2]$, only the asymmetric mode is IR active.

19.23 (a) Bidentate coordination through O^-, either O_{term}/O_{term} or O_{term}/O_{middle}; coordination through 2 O-donors from one PO_4 unit is unlikely.

19.24 (a) $[Fe(bpy)_3]^{2+}$, $[Cr(ox)_3]^{3-}$, $[CrF_6]^{3-}$, $[Ni(en)_3]^{2+}$, $[Mn(ox)_2(OH_2)_2]^{2-}$, $[Zn(py)_4]^{2+}$, $[CoCl_2(en)_2]^+$; (b) ionic, unrealistic: Mn^{7+}, O^{2-}; charges of Mn^+ and $O^{\frac{1}{2}-}$ suggest bonding is largely covalent.

19.25 (a) Chiral: *cis*-$[CoCl_2(en)_2]^+$, $[Cr(ox)_3]^{3-}$, $[Ni(phen)_3]^{2+}$, *cis*-$[RuCl(py)(phen)_2]^+$; (b) $[Pt(SCN-S)_2(Ph_2PCH_2PPh_2)]$, singlet; $[Pt(SCN-N)_2(Ph_2PCH_2PPh_2)]$, singlet; $[Pt(SCN-S)(SCN-N)(Ph_2PCH_2PPh_2)]$, doublet, $J(^{31}P-^{31}P)$.

19.26 (a) N = chiral centre; (b) linear $[Ag(NH_3)_2]^+$; tetrahedral $[Zn(OH)_4]^{2-}$; (c) coordination isomerism.

19.27 (a) Tetrahedral; trigonal planar; monocapped trigonal prism; tricapped trigonal prism; square planar; linear; (b) cubic coordination for Cs^+ in CsCl; in complexes, more usual to find dodecahedral or square antiprismatic, less often hexagonal bipyramidal.

CHAPTER 20

20.2 Green is absorbed; appears purple.

20.3 (a) *N*-donors; bidentate; may be monodentate; (b) *N*-donors; bidentate; (c) *C*-donor; monodentate;

may bridge; (d) *N*-donor; monodentate; may bridge;
(e) *C*-donor; monodentate; (f) *N*-donors; bidentate;
(g) *O*-donors; bidentate; (h) *N*- or *S*-donor;
monodentate; (i) *P*-donor; monodentate.

20.4 $Br^- < F^- < [OH]^- < H_2O < NH_3 < [CN]^-$

20.5 (a) $[Cr(OH_2)_6]^{3+}$ (higher oxidation state);
(b) $[Cr(NH_3)_6]^{3+}$ (stronger field ligand);
(c) $[Fe(CN)_6]^{3-}$ (higher oxidation state);
(d) $[Ni(en)_3]^{2+}$ (stronger field ligand);
(e) $[ReF_6]^{2-}$ (third row metal);
(f) $[Rh(en)_3]^{3+}$ (second row metal).

20.6 (a) No possibility in d^8 case of promoting an electron
from a fully occupied t_{2g} orbital to an *empty* e_g
orbital; (c) magnetic data (μ_{eff}).

20.8 (a) Octahedral, low-spin d^5; (b) octahedral, low-spin
d^3; (c) octahedral, high-spin d^4; (d) octahedral, high-
spin d^5; (e) square planar, d^8; (f) tetrahedral, d^7;
(g) tetrahedral, d^8.

20.10 (b) $F^- < H_2O < NH_3 < en < [CN]^- < I^-$.

20.11 (a) In Co^{2+}, t_2 orbitals all singly occupied; in
tetrahedral Cu^{2+}, t_2 orbitals asymmetrically filled
and complex suffers Jahn–Teller distortion;
(b) Jahn–Teller effect in excited state $t_{2g}^3 e_g^3$
arising when electron is promoted from ground
state $t_{2g}^4 e_g^2$.

20.12 $^3P_0 < {}^3P_1 < {}^3P_2 < {}^1D_2 < {}^1S_0$.

20.14 d^{10} gives only ^1S; ground (and only) term is $^1S_0$; Zn^{2+}
or Cu^+.

20.16 $J = 2, 3, 4$; degeneracy is $2J + 1$; see Fig. 20.28.

20.17 (a) $^2T_{2g}$, ^2E_g; (b) does not split; becomes $^3T_{1g}$;
(c) $^3T_{1g}$, $^3T_{2g}$, $^3A_{2g}$.

20.18 (a) See table below; E and T_2; (b) see Table 20.7;
tetrahedral: A_2, T_2 and T_1; octahedral: A_{2g}, T_{2g}
and T_{1g}.

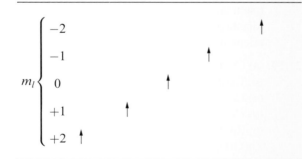

20.19 (a) $10\,000\ cm^{-1} = 1000\ nm$; $30\,000\ cm^{-1} = 333\ nm$;
(b) 400–700 nm; $25\,000$–$14\,285\ cm^{-1}$;
(c) $[Ni(OH_2)_6]^{2+}$: green; $[Ni(NH_3)_6]^{2+}$: purple;
(d) H_2O weaker field ligand than NH_3; relative
energies of transitions are estimated from Orgel
diagram: $[Ni(OH_2)_6]^{2+} < [Ni(NH_3)_6]^{2+}$; $E \propto$ wavenumber or $E \propto 1/$wavelength.

20.20 (a) Cr(III) is d^3, so three bands; (b) *trans*-
$[Co(en)_2F_2]^+$ has centre of symmetry, *cis* has not;
charge transfer (CT) from Cl^- to Co^{3+} probably
accounts for more intense colour of chlorido
complex; CT for F^- is most unlikely.

20.21 (a) Ru(III) is easier to reduce than Os(III); (b) bpy
easily accepts an electron, so electron transfer is
from M^{2+} to L.

20.22 $[Ti(OH_2)_6]^{3+}$ is d^1, Jahn–Teller effect in excited
state; $[Ti(OH_2)_6]^{2+}$ is d^2.

20.24 (a) $^4T_{2g} \longleftarrow {}^4T_{1g}(F)$; $^4T_{1g}(P) \longleftarrow {}^4T_{1g}(F)$; $^4A_{2g} \longleftarrow$
$^4T_{1g}(F)$;
(b) $7900\ cm^{-1}$; method applicable only to limiting
case where field strength is very weak.

20.26 $x = $ (a) 4; (b) 3; (c) 2; assume one can ignore
magnetic moment associated with orbital angular
momentum.

20.28 (a) $1.73\ \mu_B$; (b) take into account spin–orbit
coupling.

20.29 Octahedral Ni^{2+} (d^8) should have no orbital
contribution; tetrahedral Ni^{2+} will have an orbital
contribution, so $\mu_{eff} \neq \mu$(spin-only); all electrons
paired in square planar Ni^{2+}.

20.30 (a) Yes, octahedral d^3; (b) no, octahedral d^2; (c) no,
high-spin, octahedral d^6.

20.31 (a), (c) and (e) are diamagnetic.

20.33 Normal spinel would have tetrahedral Ni^{2+} with two
octahedral Mn^{3+}; inverse spinel would have
tetrahedral Mn^{3+}, octahedral Mn^{3+} and octahedral
Ni^{2+}; compare LFSE values:

LFSE tet. Ni^{2+} + oct. Mn^{3+} = $-15\,622\ cm^{-1}$

LFSE oct. Ni^{2+} + tet. Mn^{3+} = $-13\,933\ cm^{-1}$

Predict normal spinel; factor not accounted for is
Jahn–Teller effect for Mn^{3+}; predict normal spinel
by small margin, actual structure is inverse spinel.

20.34 (a) Difference in LFSE on going from octahedral
aqua ion to tetrahedral chlorido complex much less
for Co^{2+} (d^7) than Ni^{2+} (d^8); (b) indicates
$H_4[Fe(CN)_6]$ is weak acid in respect of fourth
dissociation constant; H^+ complexing of

$[Fe(CN)_6]^{4-}$ makes reduction easier; (c) LFSE plays a *minor* part; there is a loss of LFSE on reduction of Mn^{3+}, gain on reduction of Fe^{3+} and loss on reduction of Cr^{3+}; the decisive factor is large value of IE_3 for Mn.

20.35 (b) $[Fe(CN)_6]^{3-} > [Fe(CN)_6]^{4-} > [Fe(OH_2)_6]^{2+}$; (c) yes; $e^4 t_2^4$.

20.36 (a) $[CrI_6]^{4-}$, $[Mn(ox)_3]^{3-}$, both high-spin d^4; (b) $[NiBr_4]^{2-}$, d^8, tetrahedral; $[PdBr_4]^{2-}$, d^8, square planar.

20.38 (a) Both spin-allowed, but Laporte-forbidden transitions; non-centrosymmetric $[CoCl_4]^{2-}$ has larger ε_{max}; (b) hidden under higher energy charge transfer band; $17\,200\,cm^{-1}$, $^3T_{2g} \leftarrow {}^3T_{1g}(F)$; $25\,600\,cm^{-1}$, $^3T_{1g}(P) \leftarrow {}^3T_{1g}(F)$; (c) paramagnetic, tetrahedral; diamagnetic, square planar (probably *trans*).

20.39 (a) (i) Ti^{3+}, V^{4+}; (ii) e.g. Re^{6+}, W^{5+}, Tc^{6+}; temperature has greatest effect on ions in (ii); (b) F^-, σ- and π-donor; CO, σ-donor, π-acceptor; NH_3, σ-donor.

20.41 (a) $[Pc]^{2-}$ is tetradentate, macrocyclic ligand so forms 4 chelate rings when it binds Cu^{2+}; (b) blue; (f) to enhance solubility in H_2O.

CHAPTER 21

21.3 Ether is chelating ligand; $[BH_4]^-$ ligand may be mono-, bi- or tridentate; suggest three bidentate $[BH_4]^-$.

21.4 (a) Li_2TiO_3 must have NaCl structure, i.e. $[Li^+]_2Ti^{4+}[O^{2-}]_3$; Li^+, Ti^{4+} and Mg^{2+} are about the same size; electrical neutrality must be maintained; (b) E° for $Ti^{4+} + e^- \rightleftharpoons Ti^{3+}$ is $+0.1$ V at pH 0, so one might think that in alkali no reaction with Ti^{3+}; but TiO_2 extremely insoluble and H_2 evolution also upsets the equilibrium.

21.5 Yellow ammonium vanadate in acidic solution contains $[VO_2]^+$; reduction by SO_2 gives blue $[VO]^{2+}$; Zn reduction to purple V^{2+}.

21.6 $2VBr_3 \xrightarrow{\Delta} VBr_4 + VBr_2$; $2VBr_4 \xrightarrow{\Delta} 2VBr_3 + Br_2$; with removal of Br_2, VBr_2 is final product.

21.7 Compound is an alum containing $[V(OH_2)_6]^{3+}$, octahedral d^2; μ(spin-only) $= 2.83\,\mu_B$; three bands for d^2 ion.

21.8 $[Cr(\mathbf{21.78})]$; hexadentate N,N',N'',O,O',O''; *fac*-isomer.

21.9 Cr should be oxidized to Cr^{3+} but air should have no further action.

21.10 (a) Colorimetry (for $[MnO_4]^-$) or gas evolution (CO_2); (b) autocatalysis.

21.12 (a) Mössbauer spectrum; (b) show Fe^{3+}(aq) changes colour at high $[Cl^-]$ and also changes colour if Cl^- displaced by F^-; (c) treat ppt with acid to give MnO_2 and $[MnO_4]^-$, determine both with oxalic acid in strongly acidic solution.

21.13 (a) $2Fe + 3Cl_2 \longrightarrow 2FeCl_3$;
(b) $Fe + I_2 \longrightarrow FeI_2$;
(c) $2FeSO_4 + 2H_2SO_4 \longrightarrow Fe_2(SO_4)_3 + SO_2 + 2H_2O$;
(d) $[Fe(OH_2)_6]^{3+} + [SCN]^- \longrightarrow$
$[Fe(OH_2)_5(SCN\text{-}N)]^{2+} + H_2O$;
(e) $[Fe(OH_2)_6]^{3+} + 3[C_2O_4]^{2-} \longrightarrow [Fe(C_2O_4)_3]^{3-} + 6H_2O$; on standing, the Fe(III) oxidizes oxalate;
(f) $FeO + H_2SO_4 \longrightarrow FeSO_4 + H_2O$;
(g) $FeSO_4 + 2NaOH \longrightarrow$
$Fe(OH)_2(\text{precipitate}) + Na_2SO_4$.

21.14 (a) Compare lattice energy determined from Born–Haber cycle with that interpolated from values for MnF_2 and ZnF_2; (b) $K \approx 10^{35}$.

21.15 $Co^{II}Co^{III}{}_2O_4$: in *normal* spinel the Co^{3+} ions occupy octahedral sites, favoured for low-spin d^6 (LFSE).

21.16 (a) $[Co(en)_2Cl_2]^+$ is low-spin d^6 so diamagnetic; $[CoCl_4]^{2-}$ is d^7, tetrahedral, $e^4 t_2^3$, no orbital contribution expected; μ(spin-only) $= 3.87\,\mu_B$; here, spin–orbit coupling appears not to be important; (b) values $> \mu$(spin-only); due to spin–orbit coupling; μ_{eff} *inversely* related to ligand field strength.

21.17 (a) Green ppt is hydrated $Ni(CN)_2$; yellow solution contains $[Ni(CN)_4]^{2-}$, and red $[Ni(CN)_5]^{3-}$; (b) $K_2[Ni(CN)_4]$ reduced to give $K_4[Ni_2(CN)_6]$ (see **21.54**) or $K_4[Ni(CN)_4]$.

21.18 Gives octahedral *trans*-$[Ni(\mathbf{L})_2(OH_2)_2]$ (paramagnetic) then square planar $[Ni(\mathbf{L})_2]$ (diamagnetic); isomerism involves relative orientations of Ph groups in $\mathbf{L}$.

21.19 (a) $CuSO_4 + 2NaOH \longrightarrow Cu(OH)_2(s) + Na_2SO_4$;
(b) $CuO + Cu + 2HCl \longrightarrow 2CuCl + H_2O$;
(c) $Cu + 4HNO_3(\text{conc}) \longrightarrow Cu(NO_3)_2 + 2H_2O + 2NO_2$;
(d) $Cu(OH)_2 + 4NH_3 \longrightarrow [Cu(NH_3)_4]^{2+} + 2[OH]^-$;
(e) $ZnSO_4 + 2NaOH \longrightarrow Zn(OH)_2(s) + Na_2SO_4$;
$Zn(OH)_2(s) + 2NaOH \longrightarrow Na_2[Zn(OH)_4]$;
(f) $ZnS + 2HCl \longrightarrow H_2S + ZnCl_2$.

21.20 (b) $[Pd(Hdmg)_2]$ analogous to $[Ni(Hdmg)_2]$.

21.21 HCl can act in two ways: preferential complexing of Cu^{2+} by Cl^-, and diminution of reducing power of

SO$_2$ because of [H$^+$] in equilibrium:

$$[SO_4]^{2-} + 4H^+ + 2e^- \rightleftharpoons SO_2 + 2H_2O$$

Try effect of replacing HCl by (a) saturated LiCl or another very soluble chloride; (b) HClO$_4$ or another very strong acid which is not easily reduced.

21.22   (a) Square planar; (b) tetrahedral; (c) tetrahedral. Distinguish by magnetic data.

21.23   (a)   [MnO$_4$]$^-$;   (b)   [MnO$_4$]$^{2-}$;   (c)   [Cr$_2$O$_7$]$^{2-}$; (d) [VO]$^{2+}$; (e) [VO$_4$]$^{3-}$ (*ortho*), [VO$_3$]$^-$ (*meta*); (f) [Fe(CN)$_6$]$^{3-}$. Permanganate.

21.26   **X** = K$_3$[Fe(ox)$_3$]·3H$_2$O;   analysis   gives ox^{2-} : Fe = 3 : 1, hence 3K$^+$ needed, and 3H$_2$O to make 100%.

$$[Fe(ox)_3]^{3-} + 3[OH]^- \longrightarrow$$
$$\tfrac{1}{2}Fe_2O_3 \cdot H_2O + 3ox^{2-} + H_2O$$

$$2K_3[Fe(ox)_3] \longrightarrow 2Fe[ox] + 3K_2[ox] + 2CO_2$$

[Fe(ox)$_3$]$^{3-}$ is chiral but reaction with [OH]$^-$ suggests anion may be too labile to be resolved into enantiomers.

21.27   **A** = [Co(DMSO)$_6$][ClO$_4$]$_2$;
B = [Co(DMSO)$_6$][CoCl$_4$].

21.28   Cu^{2+} + H$_2$S $\longrightarrow$ CuS + 2H$^+$; very low solubility product of CuS allows its precipitation in acid solution. Reduction is:

$$[SO_4]^{2-} + 4H^+ + 2e^- \longrightarrow SO_2 + 2H_2O$$

but also: [SO$_4$]$^{2-}$ + 8H$^+$ + 8e$^-$ $\longrightarrow$ S^{2-} + 4H$_2$O with the very high [H$^+$] and insolubility of CuS combining to bring about the second reaction.

21.29   (a) 2BaFeO$_4$ + 3Zn $\longrightarrow$ Fe$_2$O$_3$ + ZnO + 2BaZnO$_2$; (c) Fe^{2+}(S$_2$)$^{2-}$, 1 : 1 ratio.

21.30   (a) High-spin Co^{3+}, $t_{2g}^4 e_g^2$; orbital contribution to μ_{eff} and for more than half-filled shell, $\mu_{eff} > \mu$(spin-only); (b) assume oxidation of [O$_2$]$^{2-}$ ligand; 1e-oxidation removes electron from $\pi_g^*(2p_x)^2 \pi_g^*(2p_y)^2$ level; bond order increases; (c) [Ni(acac)$_3$]$^-$; *cis*-[Co(en)$_2$Cl$_2$]$^+$.

21.31   (a) Lowest to highest energy: $^3T_{2g} \longleftarrow ^3A_{2g}$; $^3T_{1g}(F) \longleftarrow ^3A_{2g}$; $^4T_{1g}(P) \longleftarrow ^3A_{2g}$; (b) Jahn–Teller effect: CuF$_2$, d^9; [CuF$_6$]$^{2-}$ and [NiF$_6$]$^{3-}$, low-spin d^7; (c) [*trans*-VBr$_2$(OH$_2$)$_4$]Br·2H$_2$O; octahedral cation.

21.32   (a) V≡V ($\sigma^2\pi^4\delta^0$); (b) reducing agent; (c) decrease; electron added, giving $\sigma^2\pi^4\delta^1$.

21.33   (a) [NiL$_2$]$^{2+}$, d^8 versus [NiL$_2$]$^{3+}$, low-spin d^7, Jahn–Teller distorted; (b) low-spin d^6 diamagnetic; Fe(III) impurities, d^5, paramagnetic; (c) tautomers;

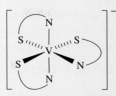

21.34   (a) Square-based pyramidal; oxido ligand in apical position; (b) [V(O)(OH$_2$)$_5$]$^{2+}$.

21.35   (a) [Cr(OH$_2$)$_6$]$^{3+}$; at low pH, H$^+$ present in solution prevents dissociation of H$^+$ from [Cr(OH$_2$)$_6$]$^{3+}$ (Le Chatelier).

21.36   (a) O,N,N',O'-donor; (b) enhances solubility in H$_2$O; (d) 600 nm is in visible region.

CHAPTER 22

22.3   (a)   Assume   CrCl$_2$ and   WCl$_2$ have   same structure;   calculate $\Delta_{lattice}H^o$(CrCl$_2$)   and estimate $\Delta_{lattice}H^o$(WCl$_2$)   using $\Delta U \propto 1/r$; $\Delta_{lattice}H^o$(WCl$_2$) $\approx$ −2450   to   −2500 kJ mol^{-1}; Born–Haber cycle gives $\Delta_f H^o$(WCl$_2$) $\approx$ +353 to +403 kJ mol^{-1}.

22.4   (a) Same 3D structure and same unit cell size but A_r Hf $\gg$ Zr; (b) Nb(IV) is d^1; NbF$_4$ has no Nb–Nb, but NbCl$_4$ and NbBr$_4$ contain pairs of Nb atoms.

22.5   (a) Cs[NbBr$_6$]; (b) K$_2$[TaF$_7$] or K$_3$[TaF$_8$] more likely   than   K[TaF$_6$]   under   conditions   given; (c) [Nb(bpy)F$_5$] is one possible product; (d) MF$_5$ (M = Nb, Ta), tetramer; NbBr$_5$, dimer; [NbBr$_6$]$^-$, octahedral;   [TaF$_7$]$^{2-}$,   monocapped   octahedron; [TaF$_8$]$^{3-}$,   square   antiprism;   [Nb(bpy)F$_5$], pentagonal bipyramid possible.

22.6   TaS$_2$, Ta(IV) and S^{2-}, 1 : 2 stoichiometry; FeS$_2$, Fe(II) and [S$_2$]$^{2-}$, 1 : 1 stoichiometry.

22.7   [Cl$_3$M(μ-Cl)$_3$MCl$_3$]$^{3-}$; no Cr−Cr bonding (Cr(III) is d^3); W≡W bond pairs up metal electrons.

22.8   (a) [Mo$_6$Cl$_8$]Cl$_2$Cl$_{4/2}$ = [Mo$_6$(μ_3-Cl)$_8$]$^{4+}$ with two extra terminal Cl *trans* to each other, and four equatorial Cl involved in bridging; [Mo$_6$Cl$_8$]Cl$_2$Cl$_{4/2}$ = [Mo$_6$Cl$_8$]Cl$_{2+2}$ = Mo$_6$Cl$_{12}$ = MoCl$_2$;   (b)   W = s^2d^4;   valence   electrons = 36 + 8 − 4 = 40; 16 used for eight M−Cl; 24 left for 12 W−W single bonds.

22.11   Re≡Re bond, eclipsed ligands; description as for Cr≡Cr.

22.12 $ReCl_4$ (**22.41**), Re—Re; Re_3Cl_9, Re=Re; $[Re_2Cl_8]^{2-}$, Re≡Re; $[Re_2Cl_9]^-$, Re—Re; $[Re_2Cl_4(\mu\text{-}Ph_2PCH_2CH_2PPh_2)_2]$, Re≡Re.

22.15 (b) *fac*- and *mer*-isomers; ^{31}P NMR spectroscopy is diagnostic; ^1H decoupled spectrum of *fac*-isomer, a singlet; for *mer*-isomer, triplet and doublet (J_{PP}). [Hydride signals in ^1H NMR spectra also diagnostic.]

22.16 (a) O_h; (b) one absorption; only T_{1u} mode is IR active.

22.17 IR spectroscopic data show H or D is present:

$$[RhBr_3(AsMePh_2)_3] \underset{Br_2}{\overset{H_3PO_2}{\rightleftharpoons}} [RhBr_2H(AsMePh_2)_3]$$

22.18 (a) β-$PdCl_2$ (**22.79**) related to $[Nb_6Cl_{12}]^{2+}$ but no M—M bonding in Pd_6 core.

22.19 (a) X-ray diffraction definitive; *cis*- and *trans*-$[PtCl_2(NH_3)_2]$ distinguished by dipole moments and IR spectroscopy; $[Pt(NH_3)_4][PtCl_4]$ is a 1:1 electrolyte; (b) $[(H_3N)_2Pt(\mu\text{-}Cl)_2Pt(NH_3)_2]Cl_2$ is 1:2 electrolyte; no $\nu(Pt-Cl)_{terminal}$ absorptions in IR spectrum.

22.20 (a) $K_2[PtI_4]$, square planar anion; (b) *cis*-$[PtCl_2(NH_3)_2]$, square planar; (c) $[PtCl_2(phen)]$, square planar and bidentate ligand, so *cis*; (d) $[PtCl(tpy)]Cl$, square planar, tridentate tpy; (e) $K_2[Pt(CN)_4]$, square planar anion, stacked in solid state.

22.21 For *trans*-$[PdCl_2(R_2P(CH_2)_nPR_2)]$ to form, $(CH_2)_n$ chain must be long; smaller chains give *cis*-monomer. Dimer with *trans*-arrangement:

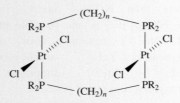

22.22 (a) Bulky $EtNH_2$ ligands prevent cation–anion stacking, so discrete ions; (b) complex $[Ag_2I]^+$ more stable than $[Ag_2Cl]^+$ (see Table 7.9); (c) equilibrium involved is: $Hg^{2+} + Hg \rightleftharpoons [Hg_2]^{2+}$ rather than: $Hg^{2+} + Hg \rightleftharpoons 2Hg^+$.

22.25 (a) Soft Hg(II)–soft *S*-donors; (b) 4-coordinate (assuming no solvent coordinated); d^{10} Hg(II) could be 4-, 5- or 6-coordinate; (c) triplet for α-CH_2 with ^{119}Hg satellites; quintet for β-CH_2.

22.27 (a) $[PCl_4^+]_3[ReCl_6^{2-}][ReCl_6^-]$; (b) $J(^{19}F-^{187}Os)$; ^{187}Os, 1.64%, $I = \frac{1}{2}$.

22.28 (a) $[NH_4]_3[HfF_7]$ is $[NH_4]_2[HfF_6] + NH_4F$; for 7-coordination, see Fig. 19.8; (b) $NbCl_5$ is Cl-bridged dimer with one ^{93}Nb environment; similarly for $NbBr_5$; halide exchange can introduce asymmetry and two ^{93}Nb environments.

22.29 (b) $[NH_4]_3[PMo_{12}O_{40}]$.

22.30 (a) Octahedral, low-spin d^6; square planar d^8; d^0; see worked example 22.2; (b) in ^{77}Se NMR spectra, $J(^{77}Se-^{103}Rh)$; in ^{13}C NMR spectra, singlets assigned to $[SeCN]^-$ ligands and doublet, $J(^{13}C_{CN}-^{103}Rh)$.

22.31 (a) $[NO]^-$; (b) 4 $[C_2O_4]^{2-}$ are each bidentate to each of two Mo centres, linking four Mo_2L_2 units into a 'square'; mass spectrometry to distinguish $[3+3]$ from $[4+4]$.

22.35 (a) $+3$ in $[RuCl_4(im)(DMSO)]^-$; $+3$ in $[RuCl_4(Ind)_2]$; $+2$ in $[RuCl_2(DMSO)_2(Biim)]$; $+3$ in $[RuCl_3(DMSO)(Biim)]$.

CHAPTER 23

23.1 (a) $MeBr + 2Li \xrightarrow{Et_2O} MeLi + LiBr$;
(b) $Na + (C_6H_5)_2 \xrightarrow{THF} Na^+[(C_6H_5)_2]^-$;
(c) $^nBuLi + H_2O \longrightarrow {}^nBuH + LiOH$;
(d) $Na + C_5H_6 \longrightarrow Na^+[C_5H_5]^-$; i.e. Na[Cp].

23.4 (a) $Mg + 2C_5H_6 \longrightarrow (\eta^5\text{-}C_5H_5)_2Mg$ (i.e. Cp_2Mg);
(b) $MgCl_2 + LiR \longrightarrow RMgCl + LiCl$ or $MgCl_2 + 2LiR \longrightarrow R_2Mg + 2LiCl$;
(c) $RBeCl \xrightarrow{LiAlH_4} RBeH$.

23.5 To make each Mg centre 4-coordinate, $n = 4$.

23.6 (a) Smaller K when steric demands of R smaller; dimer favoured.

23.7 (a) $Al_2Me_6 + 6H_2O \longrightarrow 2Al(OH)_3 + 6CH_4$
(b) $nAlR_3 + nR'NH_2 \longrightarrow (RAlNR')_n + 2nRH$ (e.g. $n = 2$);
(c) $Me_3SiCl + Na[C_5H_5] \longrightarrow Me_3Si(\eta^1\text{-}C_5H_5) + NaCl$;
(d) $2Me_2SiCl_2 + Li[AlH_4] \longrightarrow$
$$2Me_2SiH_2 + LiCl + AlCl_3.$$

23.9 Anthracene (**L**) and K give $K^+[(L)]^-$; radical anion acts as a reducing agent, Sn(IV) $\longrightarrow$ Sn(II) (regenerating anthracene); KBr is second product.

23.10 (a) Chain similar to **23.34**; octahedral; (b) chain; trigonal bipyramidal; (c) monomeric; tetrahedral; (d) monomeric; octahedral.

23.11 (a) Et_3SnOH or $(Et_3Sn)_2O$; (b) $(\eta^1\text{-}Cp)Et_3Sn$; (c) $(Et_3Sn)_2S$; (d) Et_3PhSn; (e) $Et_3SnSnEt_3$.

23.12 (a) Tilt angle of C_5-rings increases as the steric demands of R increase.

23.13 $\mathbf{A} = Br_2InCHBr_2 \cdot C_4H_8O_2$;
$\mathbf{B} = [Ph_4P]^+{}_3[HC(InBr_3)_3]^{3-}$.

23.15 (a) $Me_3Sb\cdot BH_3$; (b) Me_3SbO; (c) Me_3SbBr_2; (d) Me_3SbCl_2; $[Me_6Sb]^-$; (e) Me_4SbI; (f) Me_3SbBr_2; $Me_3Sb(OEt)_2$.

23.24 (a) Coparallel rings result in non-polar molecule; observed dipole moment implies rings are tilted; (b) $(\eta^5\text{-}C_5Me_5)_2Be$, all Me groups equivalent; $(\eta^5\text{-}C_5HMe_4)(\eta^1\text{-}C_5HMe_4)Be$ in solid; in solution, molecule fluxional with equivalent rings: two Me environments and equivalent CH protons.

23.25 $\mathbf{A} = [RP=PRMe]^+[CF_3SO_3]^-$ $(R = 2,4,6\text{-}^tBu_3C_6H_2)$; $\mathbf{B} = RMeP\text{-}PMeR$.

23.26 (a) $MeC(CH_2SbCl_2)_3 + 6Na \longrightarrow MeC(CH_2Sb)_3 + 6NaCl$; Sb–Sb bond formation.

23.27 (a) $[(\eta^5\text{-}C_5Me_5)Ge]^+[MCl_3]^-$ (M = Ge or Sn); (b) δ 121.2 (C_{ring}), 9.6 (CMe) ppm; (c) molecular ion $= [C_{10}H_{15}Ge]^+$, Ge has five isotopes; (d) trigonal pyramidal $[MCl_3]^-$.

23.28 (a) $\mathbf{X} = Et_3Bi$; $\mathbf{Y} = EtBiI_2$; chain has μ-I linking 5-coordinate Bi; (b) Ar_4Te, Ar_3TeCl and Ar_2TeCl_2 initially formed; disproportionation: $Ar_4Te + Ar_2TeCl_2 \longrightarrow Ar_4TeCl_2 + Ar_2Te$; then, $Ar_4TeCl_2 + 2LiAr \longrightarrow Ar_6Te + 2LiCl$;
(c)

23.29 Red RPhSn in equilibrium with green $RSnSnRPh_2$; change in temperature shifts equilibrium right or left.

23.31 $t_{1/2} = 2.1$ y TBT; 1.9 y DBT; 1.1 y MBT.

23.33 (a) $2RSH \longrightarrow RSSR + 2H^+ + 2e^-$;
(b) $Me_2AsO(OH) \longrightarrow$
$Me_2AsOH + H_2O + 2H^+ + 2e^-$.

CHAPTER 24

24.3 (a) $[V(CO)_6]^-$ and $Cr(CO)_6$ isoelectronic; greater negative charge leads to more back-donation; (b) 4-Me group does not affect cone angle but in 2-position, makes ligand more bulky; (c) $Me_3NO + CO \longrightarrow Me_3N + CO_2$; MeCN occupies vacant site but easily replaced by PPh_3; (d) free $HC\equiv CH$ is linear; back-donation from Os reduces C–C bond order, making C more sp^2-like.

24.4 (b) Shift consistent with metal–hydride; ^1H nucleus of bridging H couples to four equivalent ^{31}P nuclei (100%, $I = \frac{1}{2}$) to give binomial quintet.

24.5 Each $C_5Me_4SiMe_3$ contains three Me environments; δ/ ppm 0.53 (Me_{Si}), 1.41 (CH_2 in THF), 2.25 (Me_{ring}), 2.36 (Me_{ring}), 3.59 (CH_2 in THF), 4.29 (fluxional hydrides, coupling to 4 equivalent ^{89}Y).

24.7 Doublet (200 Hz) of doublets (17 Hz) with ^{195}Pt satellites.

24.8 (a) Significant population of π^*-MO causes C–C bond to lengthen; (b) replacement of THF ligand by PPh_3; (c) in $Fe(CO)_5$, 2025 and 2000 cm^{-1} due to ν_{CO}; PPh_3 poorer π-acceptor than CO.

24.11 $Fe(CO)_5 + Na_2[Fe(CO)_4] \longrightarrow CO + Na_2[Fe_2(CO)_8]$; $[(OC)_4Fe\text{-}Fe(CO)_4]^{2-}$ isoelectronic and isostructural with *solution* structure of $Co_2(CO)_8$.

24.14 $Os_7(CO)_{21}$: capped octahedral; $[Os_8(CO)_{22}]^{2-}$: bicapped octahedral.

24.15 Electron counts: (a) 86; (b) 60; (c) 72; (d) 64; (e) 48; (f) 48; (g) 86; (h) 48; (i) 60.

24.16 (a) $Os_5(CO)_{18}$ has 76 electrons; three edge-sharing triangles $= (3 \times 48) - (2 \times 34) = 76$; (b) Ir–Ir bond *between* clusters is 2c-2e; two 60-electron tetrahedra.

24.17 (a) $Fe(CO)_4(\eta^2\text{-}C_2H_4)$, trigonal bipyramidal, equatorial C_2H_4; (b) $Na[Re(CO)_5]$; anion trigonal bipyramidal; (c) $Mn(CO)_4(NO)$; trigonal bipyramidal (two isomers possible); (d) $HMn(CO)_5$; octahedral; (e) $Ni(CO)_3(PPh_3)$ or $Ni(CO)_2(PPh_3)_2$; tetrahedral.

24.18 For CO insertion, 25% product is $Mn(CO)_5Me$ (no ^{13}CO) and 75% is $Mn(^{13}CO)(CO)_4Me$ with ^{13}CO *cis* to Me.

24.20 $\mathbf{A} = (OC)_4Cr\{Ph_2P(CH_2)_4PPh_2\}$; $\mathbf{B} = (OC)_4Cr\{μ\text{-}Ph_2P(CH_2)_4PPh_2\}Cr(CO)_4$; each $LCr(CO)_4$ unit has C_{4v} symmetry, see Table 3.5.

24.25 (a) Deprotonation of imidazolium cation; (b) $Ru_3(CO)_{11}L$ where L = *N*-heterocyclic carbene.

24.27 (a) $[(\eta^5\text{-}Cp)_2Fe]^+[FeCl_4]^-$;
(b) $(\eta^5\text{-}Cp)Fe(\eta^5\text{-}C_5H_4C(O)Ph)$; $(\eta^5\text{-}C_5H_4C(O)Ph)_2Fe$;
(c) $[(\eta^5\text{-}Cp)Fe(\eta^6\text{-}C_6H_5Me)]^+[AlCl_4]^-$;
(d) $NaCl + (\eta^5\text{-}Cp)FeCo(CO)_6$.

24.28 ^1H NMR spectroscopy; η^5-Cp gives singlet, $\eta^5\text{-}C_5H_4C(O)Me$ gives singlet (Me) and two multiplets. Could also use ^{13}C NMR spectroscopy.

24.30 (a) 18-electron rule suggests L acts as 4-electron donor. (b) L becomes η^6 (see eq. 24.129, left side); (c) $[Ph_3C]^+$ abstracts H^-.

24.33 (a) 48 electrons; no; unsaturated 46-electron species with Os=Os bond.

24.34 (a) Wade: 7 electron pairs, predict octahedral Os_6-cage with interstitial B; total electrons available = 86, not consistent with the open cage observed; $H_3Os_6(CO)_{16}B$ is exception to both electron-counting rules; (b) σ-donation, π-back-donation in $[W(CO)_6]$; in $[Ir(CO)_6]^{3+}$, σ-donation dominates; $\bar{\nu}_{CO}$ for cation > neutral complex.

24.35 $A = Na[Ir(CO)_4]$, tetrahedral anion; $B^- = [Ir(CO)_3(SnPh_3)_2]^-$, trigonal bipyramid; *trans*-SbPh$_3$ likely on steric grounds.

24.38 (a) Square planar; (d) oxidation state should be +1 or +3, starting with +1 in *cic*-$[IrI_2(CO)_2]$.

CHAPTER 25

25.1 (a) First, formation of active catalytic species; step 1 = oxidative addition; step 2 = alkene insertion; step 3 = β-elimination; step 4 = elimination of HX; (b) no β-H present.

25.3 Grubbs' catalyst; see Figs. 25.4 and 25.5.

25.6 (a) PhMeC=CHPh; H_2C=C(CO$_2$H)(NHC(O)Me); (b) $\approx$8% *S* and 92% *R*.

25.7 (a) Base cycle on inner part of Fig. 25.11; (b) regioselectivity is *n*:*i* ratio; greater selectivity to linear aldehyde at lower temperature.

25.8

(I) highest yield, by alkene isomerization and then as in Fig. 25.11; (III) lowest yield (sterically hindered); (II) formed as secondary product with both (I) and (III).

25.9 (a) See Figs. 25.7 and 25.8.

25.10 (a) Similar IR absorptions indicate similar amounts of back-donation to CO ligands and so similar charge distribution in complexes; (b) complex needs to be water-soluble; Na^+ salt best choice.

25.12 (a) Active 16e-complex is NiL_3, so dissociation step is important; *K* depends on steric factors; (b) $NiL_3 \xrightarrow{HCN} Ni(H)(CN)L_3 \xrightarrow{-L} Ni(H)(CN)L_2 \xrightarrow{CH_2=CHCH=CH_2} Ni(\eta^3\text{-}C_4H_7)(CN)L_2$;

(c) transfer of CN to give either **25A** or **25B**; linear alkene is needed for the commercial process.

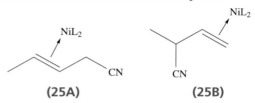

(25A)   **(25B)**

25.13 (a) 46 electron count, so unsaturated; (b) addition of alkene to an $Os(CO)_3$ vertex; transfer of one cluster H to give σ-bonded alkyl bound to Os at C(2); β-elimination gives alkene, *E*- or *Z*-isomer.

25.16 (a) Increases yield of SO_3; (b) reduces yield.

25.17 (b) V: strong chemisorption of N, nitride formation; Pt: high $\Delta G^{\ddagger}$ for N_2 adsorption; Os: rare and expensive compared with catalyst used (Fe_3O_4).

25.19 (a) Metallocene catalysts are homogeneous compared with the heterogeneous Ziegler–Natta catalysts.

25.23 (a) Asymmetric hydrogenations; ligand is chiral; (b) catalyst soluble in hexane, and catalyst recovery after phase separation.

25.24 (a) $4NH_3 + 6NO \longrightarrow 5N_2 + 6H_2O$; (b) see **25.3**;

$A =$

25.26 (a) No; each Rh is 18-electron centre.

25.27 (a) Acetic acid; (b) converts MeOH to MeI; (c) see eq. 10.14; (d) $A = [IrI_3(CO)_2H]^-$; (Ir ox. state +3); $B = H_2$; $C + D = CO_2 + 2HI$; Ir ox. states: +1 in $[Ir(CO)_2I_2]^-$; +3 in $[Ir(CO)_2I_4]^-$.

CHAPTER 26

26.4 Consider usual square planar rate law, eq. 26.12 with k_{obs} given by eq. 26.14; suggest pathways are:

$$[Rh(cod)(PPh_3)_2]^+ + py \xrightarrow{k_2} [Rh(cod)(PPh_3)(py)]^+ + PPh_3$$

competes with:

$$[Rh(cod)(PPh_3)_2]^+ + S \xrightarrow{k_1} [Rh(cod)(PPh_3)S]^+ + PPh_3$$
$$[Rh(cod)(PPh_3)S]^+ + py \xrightarrow{fast} [Rh(cod)(PPh_3)(py)]^+ + S$$

Plot of k_{obs} *vs* [py] is linear; gradient = k_2 = 322 dm^3 mol^{-1} s^{-1}; intercept = k_1 = 25 s^{-1}.

26.5 (b) *trans*-$[PtCl_2(PEt_3)_2]$ and Cl^-.

26.6 (a) As Fig. 26.4 with $L^1 = L^3 = L$, and $L^2 = X = Cl$; (b) rearrangement of 5-coordinate intermediate may be possible, giving *cis* + *trans*-$[PtL_2ClY]^+$.

26.7 See eqs. 26.14 and 26.12; line passes close to the origin, so k_1 must be very small; therefore, k_1 (solvent) pathway is not very important.

26.8 $\Delta H^{\ddagger} = +43 \text{ kJ mol}^{-1}$; $\Delta S^{\ddagger} = -84.1 \text{ J K}^{-1} \text{ mol}^{-1}$.

26.9 Positive $\Delta V^{\ddagger}$ suggests dissociative (*D* or I_d); the rate law suggests associative mechanism; apply Eigen–Wilkins mechanism to account for apparent second order kinetics.

26.10 (a) Step 1, only one product possible; *trans*-effect of $Cl^- > H_2O$, so specific isomer formation observed; (b) $[RhCl_5(OH_2)]^{2-}$ from *trans*-$[RhCl_4(OH_2)_2]^- + Cl^-$, or from $[RhCl_6]^{3-} + H_2O$; *cis*-$[RhCl_4(OH_2)_2]^-$ from $[RhCl_5(OH_2)]^{2-} + H_2O$ (*trans*-effect of Cl^-); *fac*-$[RhCl_3(OH_2)_3]$ from *cis*-$[RhCl_4(OH_2)_2]^- + H_2O$ (*trans*-effect of Cl^-).

26.11 All group 9, d^6; magnitude of Δ_{oct} increases down group.

26.13 Inversion at N; simple amines cannot be resolved.

26.14 These are $acac^-$-type ligands; common mechanism involving dissociation of one end of chelate and reformation of Co–O bond; this may exchange $C(O)CH_3$ and $C(O)CD_3$ groups.

26.15 $$-\frac{d[SCN^-]}{dt} = \left(k_1 + \frac{k_2 K_1}{[H^+]}\right)[Fe][SCN^-]$$
$$- \left(k_{-1} + \frac{k_{-2} K_2}{[H^+]}\right)[Fe(SCN)]$$
where $[Fe] = [Fe(OH_2)_6^{3+}]$ and $[Fe(SCN)] = [Fe(OH_2)_5(SCN)^{2+}]$.

26.16 First step involves breaking one Co–O bond in carbonato chelate ring; $H_2(^{18}O)$ fills vacant site; protonation of pendant carbonate-O atom.

26.18 Both sets of data are the same within experimental error; (a) $\Delta H^{\ddagger} = 128 \text{ kJ mol}^{-1}$; $\Delta S^{\ddagger} = 95 \text{ J K}^{-1} \text{ mol}^{-1}$; (b) data consistent with racemization by dissociative process.

26.19 *Dcb* mechanism; $[NH_2]^-$ in NH_3 is analogous to $[OH]^-$ in H_2O.

26.22 **I**: both low-spin, similar Ru–N bond lengths; **II**: $[Co(NH_3)_6]^{3+}$ is low-spin, becomes high-spin (and has longer Co–N) after reduction; $\Delta_r G^o$ helps reaction; **III**: see text discussion, Section 26.5; $\Delta G^o = 0$ for self-exchanges **I** and **III**.

26.24 (a) $[PtCl_3(NH_3)]^-$, *cis*-$[PtCl_2(NH_3)_2]$; (b) *cis*-$[Co(en)_2Cl(OH_2)]^{2+}$; (c) $[Fe(NO)(OH_2)_5]^{2+}$.

26.25 (b) Change from *trans,trans* to *cis,cis*-conformation.

26.26 ^{31}P NMR spectroscopy; electronic spectroscopy; take into account rate of reaction *vs* timescale of the method chosen.

26.27 (a) Dissociative pathway.

CHAPTER 27

27.2 $^2F_{5/2}$; $2.54 \mu_B$.

27.3 Consider cycle for: $3LnX_2 \longrightarrow 2LnX_3 + Ln$; for a given Ln, difference in lattice energy between $3LnX_2$ and $2LnX_3$ is the governing factor, and is least when X is largest.

27.4 Determine electrical conductivity.

27.5 (a) Near constancy originates in small variation in metal ion size which affects interactions with H_2O and $[EDTA]^{4-}$ similarly; hexadentate $[EDTA]^{4-}$ has four *O*-donors and so ΔH^o for replacement of H_2O is small. (b) Complex formation by anions: $Cl^- > [SO_4]^{2-} > [NO_3]^- > [ClO_4]^-$. (c) $BaCeO_3$ is a mixed oxide.

27.8 Hard Ln^{3+} suggests $[NCS]^-$ *N*-bonded; $[Ln(NCS)_6]^{3-}$, octahedral; 8-coordinate $[Ln(NCS)_7(OH_2)]^{4-}$ could be dodecahedral, square antiprismatic, cubic or distorted variants (hexagonal bipyramidal less likely); $[Ln(NCS)_7]^{4-}$ could be pentagonal bipyramidal, capped octahedral, or distorted variants.

27.9 (b) Sandwich complexes $[(\eta^8\text{-}C_8H_8)_2Sm]^-$ (K^+ salt) and $[(\eta^8\text{-}C_8H_8)_2Sm]^{2-}$.

27.10 (b) Zn amalgam should reduce Np(VI) to Np(III); O_2 at pH 0 should oxidize Np(III) to $[NpO_2]^+$ and some $[NpO_2]^{2+}$ (oxidation might be slow).

27.11 U(VI) $\longrightarrow$ U(IV) after aeration; UF_4 formed and then: $2UF_4 + O_2 \longrightarrow UF_6 + UO_2F_2$.

27.12 (a) UF_6; (b) $PaCl_5$, then $PaCl_4$; (c) UO_2; (d) $UCl_4 + UCl_6$; (e) $U(OC_6H_2\text{-}2,4,6\text{-}Me_3)_3$.

27.14 (a) **A**, ^{239}U; **B**, ^{239}Np; **C**, ^{239}Pu; (b) **D**, ^{241}Pu; **E**, ^{241}Am; **F**, ^{242}Cm.

27.15 (a) **A**, $^{253}_{99}Es$; (b) **B**, $^{244}_{94}Pu$; (c) **C**, $^{249}_{98}Cf$; (d) **D**, $^{248}_{96}Cm$; (e) **E**, $^{249}_{98}Cf$.

27.16 (a) All Th(IV) compounds: $Th^{4+}(I^-)_2(e^-)_2$, $Th^{4+}(I^-)_3(e^-)$ and ThI_4; (b) solid state salts contain linear UO_2 *unit* with other ligands in equatorial plane; (c) monomer only if R is very bulky, e.g. R = $2,6\text{-}^tBu_2C_6H_3$.

27.17 (a) $(\eta^5\text{-Cp})_3\text{ThRu(CO)}_2(\eta^5\text{-Cp})$; $(\eta^5\text{-Cp})_3\text{ThCHMeEt}$; $(\eta^5\text{-Cp})_3\text{ThCH}_2\text{Ph}$; (b) bulkier organic ligand hinders redistribution reaction; (c) to give $(\eta^5\text{-C}_5\text{Me}_5)(\eta^8\text{-C}_8\text{H}_8)\text{U(THF)}_x$ (in practice, $x = 1$).

27.18 (a) $\text{UCl}_4 + 4(\eta^3\text{-C}_3\text{H}_5)\text{MgCl}$ in Et_2O;

(b) $\text{U}(\eta^3\text{-C}_3\text{H}_5)_4 + \text{HCl} \longrightarrow$
$\text{U}(\eta^3\text{-C}_3\text{H}_5)_3\text{Cl} + \text{CH}_3\text{CH}{=}\text{CH}_2$.

27.22 (a) $\mathbf{A} = [fac\text{-}(\mathbf{27.19}\text{-}N,N',N'')\text{ScCl}_3]$; $\mathbf{B} = [fac\text{-}(\mathbf{27.19}\text{-}N,N',N'')\text{ScMe}_3]$; +3.

27.23 (a) For f^6, $S = 3$, $L = 3$, $J = 0$, $g = 1$; $\mu_{\text{eff}} = g\sqrt{J(J+1)} = 0$; (b) $\mathbf{A} = [(\text{THF})_2\text{ClO}_2\text{U}(\mu\text{-Cl})_2\text{UO}_2\text{Cl(THF)}_2]$; $\mathbf{B} = [\text{UO}_2(\text{THF})_2(\text{O-2,6-}^{\text{t}}\text{Bu}_2\text{C}_6\text{H}_3)_2]$; $\mathbf{C} = [\text{UO}_2\text{Cl}_2(\text{OPPh}_3)_2]$; all $trans\text{-UO}_2$ units.

27.24 (a) Let $\mathbf{27.20} = \text{HL}$; $\mathbf{A} = \text{LiL}$; $\mathbf{B} = \text{LTbBr}_2$.

27.25 (a) $n = 6$; (b) side-chain enhances solubility in fats/hydrocarbons; (c) L^{6-} is $N,N',N'',O,O',O'',O''',O''''$-donor + H_2O gives 9-coordinate Gd^{3+}; (d) one asymmetric C atom; (f) If $\text{M} = \text{Na}_3[\text{GdL(OH}_2)]$: m/z 981 $[\text{M} - \text{H}_2\text{O} + \text{Na}]^+$; 959 $[\text{M} - \text{H}_2\text{O} + \text{H}]^+$; 937 $[\text{M} - \text{H}_2\text{O} - \text{Na} + 2\text{H}]^+$; 915 $[\text{M} - \text{H}_2\text{O} - 2\text{Na} + 3\text{H}]^+$.

27.26 (a) Hard La^{3+} with hard O-donor; (b) deprotonates Hma; (c) $[\text{M(ma)}_3 + \text{Na}]^+$ ($\text{M} = \text{La}$ or Eu); (d) consistent with $[\text{Eu(ma)}_3]$; 6-coordinate Eu^{3+}; (e) only La^{3+} is diamagnetic.

CHAPTER 28

28.1 AgI is a solid Ag^+ ion conductor; passage of Ag^+ (not e^-) occurs through solid electrolyte.

28.3 V_6O_{13} reversibly intercalates Li^+;

$$x\text{Li}^+ + \text{V}_6\text{O}_{13} + x\,e^- \underset{\text{charge}}{\overset{\text{discharge}}{\rightleftharpoons}} \text{Li}_x\text{V}_6\text{O}_{13}.$$

28.5 (a) Fermi level is in conduction band of host; metallic-like electrical conductivity.

28.10 (a) See Fig. 28.17; (b) Mo $4d$ electrons localized in Mo_6-clusters; band structure; bands of Mo $4d$ character close to Fermi level.

28.14 (a) $\text{Li}_x\text{V}_2\text{O}_5 + \text{I}_2$; (b) CaWO_4; (c) Sr_2FeO_4 (or SrFeO_3).

28.15 (a) Bi_2O_3, V_2O_5, CaO; (b) Cu_2O, MoO_3, Y_2O_3; (c) Li_2O, In_2O_3; (d) RuO_2, Y_2O_3.

28.22 (a) See Fig. 11.4; $\text{Li}_3\text{As} < \text{Li}_3\text{P} < \text{Li}_3\text{N}$.

28.23 (b) H_2NNMe_2 is H atom donor to facilitate $^{\text{t}}\text{BuH}$ elimination; see Table 28.5 and discussion.

28.24 (a) F^- vacancies, giving holes for F^- migration; (b) for metal and semiconductor, see Figs. 6.10 and 6.11.

CHAPTER 29

29.3 Octahedral complex with three catecholate ligands; the Cr^{3+} complex (d^3) is kinetically inert, so solution studies practicable.

29.4 (a) Soft S-donors compatible with soft metal ion; (b) protein binding sites coordinate several metals in cluster units; (c) similar C_3N_2 heterocyclic rings present in each.

29.10 (a) Cu^{2+} blue; Cu^+, colourless; (b) changes in conformation of metal-binding pocket alters coordination environment and also reduction potential; (c) CO blocks O_2 binding site by coordinating to Fe^{2+}, but $[\text{CN}]^-$ favours Fe^{3+} and binds tightly to cytochrome haem.

29.12 (a) $[\text{Fe(SPh)}_4]^{2-}$ models $\text{Fe}\{\text{S(Cys)}\}_4$-site; for Fe^{2+} and Fe^{3+}, μ(spin-only) values are 4.90 and 5.92 μ_B; (b) spinach ferredoxin is a [2Fe–2S] system with an $\text{Fe}_2(\mu\text{-S})_2\{\text{S(Cys)}\}_4$ core; (c) **29.31** models half of FeMo cofactor; Mössbauer data consistent with delocalization of charge.

29.13 (a) Middle two; 4Fe(III) and 4Fe(II) states are not accessed; (b) 3Fe(III)·Fe(II) $\rightleftharpoons$ 2Fe(III)·2Fe(II).

29.22 (a) Metallothioneins; typically S-donor Cys.

29.23 (a) Imidazole rings mimic His residues; (b) tripodal ligand encourages formation of *tetrahedral* $[\text{Zn}(\mathbf{29.34})(\text{OH})]^+$.

29.25 (a) Haemoglobin contains four haem units; cooperativity leads to $K_4 \gg K_1$; (b) catalyses oxidation of H_2O to O_2 in green plants and algae; chlorophyll.

29.26 (a) Isoelectronic relationships: CH_2, NH and O; CO and $[\text{CN}]^-$; (b) m/z 1413 $[\text{M} + 12\text{H}]^{12+}$... 807 $[\text{M} + 21\text{H}]^{21+}$.

29.29 Chelating, multidentate ligand; complex has high K.

29.30 (a) Disulfide bonds; oxidative coupling of Cys residues; (b) all contain soft S-donors; Cd^{2+} is soft metal ion; (c) Hg^{2+}, Zn^{2+}.

Index

Note: (B) indicates text in a Box, (F) a Figure, (N) a footnote, (T) a Table, and (WE) a Worked Example. Alphabetization is in word-by-word order (e.g. 'alkyl ligands' is sorted before 'alkylaluminium derivatives', and '*d* orbitals' before 'Daniell cell'). Greek letters are listed at the beginning of the relevant alphabetical section (e.g. α/A, β/B, δ/D, η/E, γ/G, λ/L, μ/M, π/P, ρ/R, σ/S, etc.– see Appendix 1 (p. 1110) for details). Prefixed phrases (such as *d*-block/ *f*-block/*p*-block/*s*-block elements) are located at the beginning of the relevant alphabetical section (after any Greek letters).